COLD SPRING HARBOR SYMPOSIA ON QUANTITATIVE BIOLOGY

VOLUME LXXII

L2SINN

www.cshl-symposium.org

Institutions that have purchased the hardcover edition of this book are entitled to online access to the Symposium Web site. Please contact your institution's library to gain access to the Web site. The site contains the full text articles from the 2007 Symposium and the Symposia held in 1998–2006 as well as archive photographs and selected papers from the 63-year history of the annual Symposium.

If a token number is given above, you will need to activate the token to gain access. If you have previously registered and activated your online account for any prior volume without the use of a token, you do not need to register again.

If you do not have an account number or a token number or are experiencing access problems, please contact Kathy Cirone, CSHL Press Subscription Manager, at 1-800-843-4388, extension 4044 (Continental U.S. and Canada), 516-422-4100 (all other locations), cironek@cshl.edu, or subscriptionfeedback@symposium.org.

COLD SPRING HARBOR SYMPOSIA ON QUANTITATIVE BIOLOGY

VOLUME LXXII

Clocks and Rhythms

www.cshl-symposium.org

Meeting Organized by Bruce Stillman, David Stewart, and Terri Grodzicker
COLD SPRING HARBOR LABORATORY PRESS
2007

COLD SPRING HARBOR SYMPOSIA ON QUANTITATIVE BIOLOGY VOLUME LXXI

©2007 by Cold Spring Harbor Laboratory Press
International Standard Book Number 978-0-87969-822-5 (cloth)
International Standard Book Number 978-0-87969-823-2 (paper)
International Standard Serial Number 0091-7451
Library of Congress Catalog Card Number 34-8174

Printed in the United States of America

COLD SPRING HARBOR SYMPOSIA ON QUANTITATIVE BIOLOGY
Founded in 1933 by
REGINALD G. HARRIS
Director of the Biological Laboratory 1924 to 1936
Previous Symposia Volumes

I (1933) Surface Phenomena
II (1934) Aspects of Growth
III (1935) Photochemical Reactions
IV (1936) Excitation Phenomena
V (1937) Internal Secretions
VI (1938) Protein Chemistry
VII (1939) Biological Oxidations
VIII (1940) Permeability and the Nature of Cell Membranes
IX (1941) Genes and Chromosomes: Structure and Organization
X (1942) The Relation of Hormones to Development
XI (1946) Heredity and Variation in Microorganisms
XII (1947) Nucleic Acids and Nucleoproteins
XIII (1948) Biological Applications of Tracer Elements
XIV (1949) Amino Acids and Proteins
XV (1950) Origin and Evolution of Man
XVI (1951) Genes and Mutations
XVII (1952) The Neuron
XVIII (1953) Viruses
XIX (1954) The Mammalian Fetus: Physiological Aspects of Development
XX (1955) Population Genetics: The Nature and Causes of Genetic Variability in Population
XXI (1956) Genetic Mechanisms: Structure and Function
XXII (1957) Population Studies: Animal Ecology and Demography
XXIII (1958) Exchange of Genetic Material: Mechanism and Consequences
XXIV (1959) Genetics and Twentieth Century Darwinism
XXV (1960) Biological Clocks
XXVI (1961) Cellular Regulatory Mechanisms
XXVII (1962) Basic Mechanisms in Animal Virus Biology
XXVIII (1963) Synthesis and Structure of Macromolecules
XXIX (1964) Human Genetics
XXX (1965) Sensory Receptors
XXXI (1966) The Genetic Code
XXXII (1967) Antibodies
XXXIII (1968) Replication of DNA in Microorganisms
XXXIV (1969) The Mechanism of Protein Synthesis
XXXV (1970) Transcription of Genetic Material
XXXVI (1971) Structure and Function of Proteins at the Three-dimensional Level
XXXVII (1972) The Mechanism of Muscle Contraction
XXXVIII (1973) Chromosome Structure and Function
XXXIX (1974) Tumor Viruses
XL (1975) The Synapse
XLI (1976) Origins of Lymphocyte Diversity
XLII (1977) Chromatin
XLIII (1978) DNA: Replication and Recombination
XLIV (1979) Viral Oncogenes
XLV (1980) Movable Genetic Elements
XLVI (1981) Organization of the Cytoplasm
XLVII (1982) Structures of DNA
XLVIII (1983) Molecular Neurobiology
XLIX (1984) Recombination at the DNA Level
L (1985) Molecular Biology of Development
LI (1986) Molecular Biology of *Homo sapiens*
LII (1987) Evolution of Catalytic Function
LIII (1988) Molecular Biology of Signal Transduction
LIV (1989) Immunological Recognition
LV (1990) The Brain
LVI (1991) The Cell Cycle
LVII (1992) The Cell Surface
LVIII (1993) DNA and Chromosomes
LIX (1994) The Molecular Genetics of Cancer
LX (1995) Protein Kinesis: The Dynamics of Protein Trafficking and Stability
LXI (1996) Function & Dysfunction in the Nervous System
LXII (1997) Pattern Formation during Development
LXIII (1998) Mechanisms of Transcription
LXIV (1999) Signaling and Gene Expression in the Immune System
LXV (2000) Biological Responses to DNA Damage
LXVI (2001) The Ribosome
LXVII (2002) The Cardiovascular System
LXVIII (2003) The Genome of *Homo sapiens*
LXIX (2004) Epigenetics
LXX (2005) Molecular Approaches to Controlling Cancer
LXXI (2006) Regulatory RNAs

Front Cover (*Paperback*): Cronos, 2000 (oil on canvas) by Lescaux, Bob © Private Collection/The Bridgeman Art Library. Design by Joseph Sherman.

All Cold Spring Harbor Laboratory Press publications may be ordered directly from Cold Spring Harbor Laboratory Press, 500 Sunnyside Boulevard, Woodbury, NY 11797-2924. Phone: 1-800-843-4388 in Continental U.S. and Canada. All other locations: (516) 422-4100. FAX: (516) 422-4097. E-mail: cshpress@cshl.edu. For a complete catalog of all Cold Spring Harbor Laboratory Press publications, visit our World Wide Web Site http://www.cshlpress.com/

Web Site Access: Institutions that have purchased the hardcover edition of this book are entitled to online access to the companion Web site at www.cshl-symposium.org. For assistance with activation, please contact Kathy Cirone, CSHL Press Subscription Manager, at cironek@cshl.edu.

Symposium Participants

ABBRUZZESE, ELVIRA, Dept. of Clinical Psychology and Psychotherapy, University of Zürich, Zürich, Switzerland

ABRAHAM, DIYA, Dept. of Neurology, University of California, San Francisco

AČIMOVIČ, JURE, Dept. of Medicine, Institute of Biochemistry, University of Ljubljana, Ljubljana, Slovenia

ALBRECHT, URS, Dept. of Medicine, Div. of Biochemistry, University of Fribourg, Fribourg, Switzerland

ALLADA, RAVI, Dept. of Neurobiology and Physiology, Northwestern University, Evanston, Illinois

ALTIMUS, CARA, Dept. of Biology, Johns Hopkins University, Baltimore, Maryland

AN, SUNGWON, Dept. of Biology and Neuroscience, Washington University, Saint Louis, Missouri

ANTOCH, MARINA, Dept. of Cancer Biology, Cleveland Clinic Foundation, Cleveland, Ohio

ARTYMYSHYN, ROMAN, Dept. of Target and Discovery, Lundbeck Research, Paramus, New Jersey

BAGGS, JULIE, Dept. of Pharmacology, University of Pennsylvania, Philadelphia

BAKER, CHRISTOPHER, Dept. of Genetics, Dartmouth Medical School, Hanover, New Hampshire

BANERJEE, DIYA, Dept. of Biological Sciences, Virginia Polytechnic Institute and State University, Blacksburg, Virginia

BARRIGA-MONTOYA, CAROLINA, Dept. de Fisiología, Facultad de Medicina, Universidad Nacional Autónoma de México, México

BELL-PEDERSEN, DEBORAH, Dept. of Biology, Texas A&M University, College Station, Texas

BENITO, JULIANA, Dept. of Biology and Biochemistry, University of Houston, Houston, Texas

BERSOT, ROSS, Bay City Capital, San Francisco, California

BHATTACHARJEE, YUDHIJIT, Science, Science Magazine, American Association for the Advancement of Science, Washington, D.C.

BJARNASON, GEORG, Dept. of Medical Oncology, Toronto-Sunnybrook Regional Cancer Centre, Toronto, Ontario, Canada

BLANCO, MARINA, Dept. of Anthropology, University of Massachusetts, Amherst

BLAU, JUSTIN, Dept. of Biology, New York University, New York, New York

BOURGERON, THOMAS, Dept. of Human Genetics and Cognitive Functions, Institut Pasteur, Paris, France

BOYAULT, CYRIL, Dept. of Neurobiology, Harvard Medical School, Boston, Massachusetts

BROWN, STEVEN, Dept. of Pharmacology and Toxicology, University of Zürich, Zürich, Switzerland

BRUNNER, MICHAEL, Dept. of Biochemistry, University of Heidelberg, Heidelberg, Germany

BURSZTYN, MICHAEL, Dept. of Internal Medicine, Mount-Scopus Campus, Hadassah-Hebrew University Medical Center, Jerusalem, Israel

BUSINO, LUCA, Dept. of Pathology, New York University, New York, New York

CAO, GUAN, Dept. of Cellular and Molecular Physiology, Yale University School of Medicine, New Haven, Connecticut

CARNIOL, KAREN, Cell, Cell Press, Cambridge, Massachusetts

CARRUTHERS, JR., CARL, Dept. of Biochemistry and Biophysics, Texas A&M University, College Station, Texas

CASSONE, VINCENT, Dept. of Biology, Texas A&M University, College Station, Texas

CASTANON-CERVANTES, OSCAR, Dept. of Neuroscience, Morehouse School of Medicine, Atlanta, Georgia

CHANG, ALEXANDER, Dept. of Molecular Genetics, Southwestern Medical Center, University of Texas, Dallas

CHAPPELL, PATRICK, Dept. of Zoology, Oregon State University, Corvallis, Oregon

CHEN, CHEN-HUI, Dept. of Genetics, Dartmouth College, Hanover, New Hampshire

CHEN, INES, Nature Structural and Molecular Biology, Nature Publishing Group, New York, New York

CHEN, ZHENG (JAKE), Dept. of Biochemistry, Southwestern Medical Center, University of Texas, Dallas

CHÉVEZ, ESTRELLA, Dept. de Biologia Celular y Fisiología, Instituto de Investigaciones Biomédicas, Universidad Nacional Autónoma de México, México

CHOE, JOONHO, Dept. of Biological Sciences, Korea Advanced Institute of Science and Technology, Daejeon, South Korea

CHORY, JOANNE, Lab of Plant Biology, Howard Hughes Medical Institute, The Salk Institute for Biological Studies, La Jolla, California

COLLINS, BEN, Dept. of Biology, New York University, New York, New York

CONSTANCE, CARA, Dept. of Biology, College of the Holy Cross, Worcester, Massachusetts

COOPER, HOWARD, Dept. of Chronobiology, Stem Cell and Brain Research Institute, Institut National de la Santé et de la Recherche Médicale, University of Lyon, Bron, France

CORMIER, CATHERINE, Cold Spring Harbor Laboratory, Cold Spring Harbor, New York

CZEISLER, CHARLES, Dept. of Medicine, Div. of Sleep Medicine, Brigham & Women's Hospital, Harvard Medical School, Boston, Massachusetts

DANIELS, SUSAN, Dept. of Biochemistry and Molecular Biology, Health Science Center, University of Texas, Houston

DAVIDSON, ALEC, Neuroscience Institute, Morehouse School of Medicine, Atlanta, Georgia
DAVIS, FRED, Dept. of Biology, Northeastern University, Boston, Massachusetts
DEBRUYNE, JASON, Dept. of Neurobiology, Medical School, University of Massachusetts, Worcester
DECOURSEY, PATRICIA, Dept. of Biological Sciences, University of South Carolina, Columbia
DE HARO, LUCIANO, Dept. of Regulatory Biology, The Salk Institute for Biological Studies, La Jolla, California
DE PAULA, RENATO, Dept. of Biology, Texas A&M University, College Station, Texas
DESTICI, EUGIN, Dept. of Cell Biology and Genetics, Erasmus University Medical Centre, Rotterdam, The Netherlands
DITTY, JAYNA, Dept. of Biology, University of St. Thomas, St. Paul, Minnesota
DONG, GUOGANG, Dept. of Biology, Texas A&M University, College Station, Texas
DOWLATSHAD, HAMID, Dept. of Biology, Northeastern University, Boston, Massachusetts
DOYLE, SUSAN, Dept. of Biology, University of Virginia, Charlottesville
DUFFIELD, GILES, Dept. of Biological Sciences, University of Notre Dame, Notre Dame, Indiana
DUFFY, JEANNE, Dept. of Sleep Medicine, Brigham & Women's Hospital, Harvard Medical School, Boston, Massachusetts
DUNLAP, JAY, Dept. of Genetics, Dartmouth Medical School, Hanover, New Hampshire
ECKER, JEN, Dept. of Biology, Johns Hopkins University, Baltimore, Maryland
EDERY, ISAAC, Dept. of Molecular Biology and Biochemistry, Rutgers University, Piscataway, New Jersey
EDWARDS, KIERON, Dept. of Biological Sciences, University of Edinburgh, Edinburgh, Scotland, United Kingdom
ESPOSITO, JOSEPH, Dept. of Cancer Biology, Roswell Park Cancer Institute, Buffalo, New York
EVANS, RONALD, Howard Hughes Medical Institute, The Salk Institute for Biological Studies, La Jolla, California
FINK, MARTINA, Institute of Biochemistry, Faculty of Medicine, University of Ljubljana, Slovenia
FORGER, DANIEL, Dept. of Mathematics, University of Michigan, Ann Arbor
FUJII, SHINSUKE, Dept. of Molecular Genetics and Microbiology, Duke University Medical Center, Durham, North Carolina
GACHON, FRÉDÉRIC, Equipe Avenir, Institut National de la Santé et de la Recherche Médicale, Institute de Génétique Humaine, Centre National de la Recherche Scientifique, Montpellier, France
GALL, ANDREW, Dept. of Psychology, University of Iowa, Iowa City
GARY, SYDNEY, Banbury Center, Cold Spring Harbor Laboratory, Cold Spring Harbor, New York
GATFIELD, DAVID, Dept. of Molecular Biology, University of Geneva, Geneva, Switzerland
GERY, SIGAL, Division of Hematology and Oncology, Cedars-Sinai Medical Center, Los Angeles, California
GIBBS, JULIE, Faculty of Life Science, University of Manchester, Manchester, United Kingdom
GIEBULTOWICZ, JADWIGA, Dept. of Zoology, Oregon State University, Corvallis, Oregon
GOEL, NAMNI, Dept. of Psychiatry, School of Medicine, University of Pennsylvania, Philadelphia
GOLDEN, SUSAN, Dept. of Biology, Texas A&M University, College Station, Texas
GOOCH, VAN, Dept. of Science and Mathematics, University of Minnesota, Morris
GOOLEY, JOSHUA, Dept. of Sleep Medicine, Brigham & Women's Hospital, Harvard Medical School, Boston, Massachusetts
GOTTER, ANTHONY, Dept. of Sleep Disorders and Schizophrenia, Merck Research Laboratories, West Point, Pennsylvania
GREEN, CARLA, Dept. of Biology, University of Virginia, Charlottesville
GREENSPAN, RALPH, Dept. of Experimental Neurobiology, The Neurosciences Institute, San Diego, California
GRIMALDI, BENEDETTO, Dept. of Pharmacology, University of California, Irvine
GRODZICKER, TERRI, Cold Spring Harbor Laboratory Press, Woodbury, New York
GUARENTE, LEONARD, Dept. of Biology, Massachusetts Institute of Technology, Cambridge, Massachusetts
GUILDING, CLARE, Dept. of Neuroscience, University of Manchester, Manchester, United Kingdom
HALL, JEFFREY, Dept. of Biology, Brandeis University, Waltham, Massachusetts
HANNIBAL, JENS, Dept. of Clinical Biochemistry, Rigshospitalet, Copenhagen, Denmark
HARADA, TETSUO, Lab. of Environmental Physiology, Faculty of Education, Kochi University, Kochi, Japan
HARDIN, PAUL, Dept. of Biology, Texas A&M University, College Station, Texas
HARNISH, ERICA, Dept. of Central Nervous System Research, Sanofi-Aventis, Bridgewater, New Jersey
HARRISINGH, MARIE, Dept. of Molecular and Cellular Physiology, Yale University School of Medicine, New Haven, Connecticut
HARTLEY, PAUL, Dept. of Cardiovascular Sciences, Queen's Medical Research Institute, University of Edinburgh, Edinburgh, Scotland, United Kingdom
HASTINGS, J. WOODLAND, Dept. of Cellular and Molecular Biology, Harvard University, Cambridge, Massachusetts
HASTINGS, MICHAEL, Dept. of Neurobiology, Medical Research Council Laboratory for Molecular Biology, Cambridge, United Kingdom
HATTAR, SAMER, Dept. of Biology, Johns Hopkins University, Baltimore, Maryland
HELFRICH-FÖRSTER, Charlotte, Institute of Zoology, University of Regensburg, Regensburg, Germany
HIRAYAMA, JUN, Dept. of Pharmacology, University of California, Irvine
HOFSTETTER, JOHN, Dept. of Psychiatry, Veterans Administration, Indiana University, Indianapolis, Indiana
HOGENESCH, JOHN, Dept. of Pharmacology, School of Medicine, University of Pennsylvania, Philadelphia
HOLTMAN, CAROLYN, Dept. of Biology, Texas A&M University, College Station, Texas
HONG, CHRISTIAN, Dept. of Genetics, Dartmouth Medical School, Hanover, New Hampshire
HOOVEN, LOUISA, Dept. of Zoology, Oregon State University, Corvallis, Oregon

HORN, JACQUELINE, Dept. of Neuroscience, Queens Medical Research Institute, University of Edinburgh, Edinburgh, Scotland, United Kingdom
HRUSHESKY, WILLIAM, Dept. of Medical Oncology, W.J.B. Dorn Veteran's Administration Medical Center, Columbia, South Carolina
HUGHES, MICHAEL, Dept. of Pharmacology, University of Pennsylvania, Philadelphia
HULL, JOSEPH, Surrey Sleep Research Centre, School of Biomedical and Molecular Sciences, University of Surrey, Guildford, Surrey, United Kingdom
HUNDAHL, CHRISTIAN, Institute for Neuroscience and Phamacology, Panum Institute, Copenhagen, Denmark
IIGO, MASAYUKI, Dept. of Applied Biochemistry, Utsunomiya University, Utsunomiya, Japan
ISKRA-GOLEC, IRENA, Institute of Psychological Sciences, University of Leeds, Leeds, United Kingdom
IUVONE, P. MICHAEL, Dept. of Pharmacology, Emory University, Atlanta, Georgia
IZUMO, MARIKO, Dept. of Neurobiology and Physiology, Northwestern University, Evanston, Illinois
JACKSON, CHAD, Dept. of Pharmacology, Emory University, Atlanta, Georgia
JAKIMO, ALAN, Dept. of Law, Hofstra University, Hempstead, New York
JAKUBCAKOVA, VLADIMIRA, Institute for Genes and Behavior, Max-Planck-Institute for Biophysical Chemistry, Göttingen, Germany
JOHNSON, CARL, Dept. of Biological Sciences, Vanderbilt University, Nashville, Tennessee
JOHO, ROLF, Center for Basic Neuroscience, Southwestern Medical Center, University of Texas, Dallas
JONES, KENNETH, Dept. of Cellular Sciences, Lundbeck Research USA, Paramus, New Jersey
JUD, CORINNE, Div. of Biochemistry, University of Fribourg, Fribourg, Switzerland
KAASIK, KRISTA, Dept. of Neurology, University of California, San Francisco
KADENER, SEBASTIÁN, Dept. of Biology, Brandeis University, Waltham, Massachusetts
KAGEYAMA, RYOICHIRO, Institute for Virus Research, Kyoto University, Kyoto, Japan
KAY, STEVEN, Dept. of Cell Biology, The Scripps Research Institute, La Jolla, California
KEENE, JACK, Dept. of Molecular Genetics and Microbiology, Duke University, Durham, North Carolina
KENNEDY, WILLIAM, Merck, North Wales, Pennsylvania
KENYON, CYNTHIA, Dept. of Biochemistry and Biophysics, University of California, San Francisco
KIESSLING, SILKE, Institute for Genes and Behavior, Max-Plack-Institute for Biophysical Chemistry, Göttingen, Germany
KIM, TAE SUNG, Dept. of Plant Pathology, Cornell University, Ithaca, New York
KLEVECZ, ROBERT, Dept. of Biology, Dynamics Systems Group, Beckman Research Institute, City of Hope Medical Center, Duarte, California
KNUTTI, DARKO, Dept. of Neurobiology, Harvard Medical School, Boston, Massachusetts
KO, CAROLINE, Dept. of Neurobiology and Physiology, Northwestern University, Evanston, Illinois
KO, GLADYS, Dept. of Veterinary Integrative Biosciences, Texas A&M University, College Station, Texas
KOEFFLER, H. PHILLIP, School of Medicine, Cedars-Sinai Medical Center, University of California, Los Angeles
KOIKE, NOBUYA, Lab. of Chronogenomics, Mitsubishi Kagaku Institute of Life Sciences, Tokyo, Japan
KOKOLUS, WILLIAM, Dept. of Research and Administration, Fountain Diagnostic Test Component Company, Kenmore, New York
KON, NAOHIRO, Dept. of Biophysics and Biochemistry, Graduate School of Science, University of Tokyo, Tokyo, Japan
KONDO, MARI, Tokyo, Japan
KONDO, NORIAKI, Dept. of Chronogenomics, Mitsubishi Kagaku Institute of Life Sciences, Tokyo, Japan
KONDO, TAKAO, Div. of Biological Science, Graduate School of Science, Nagoya University, Nagoya, Japan
KONDRATOV, ROMAN, Dept. of Biological, Geological and Environmental Sciences, Cleveland State University, Cleveland, Ohio
KOWALSKA, ELZBIETA, Dept. of Pharmacology and Toxicology, University of Zürich, Zürich, Switzerland
KRAMER, ACHIM, Lab. of Chronobiology, Institute of Medical Immunology, Charité Universitätsmedezin, Humboldt University, Berlin, Germany
KUHLMAN, SANDRA, Cold Spring Harbor Laboratory, Cold Spring Harbor, New York
KURABAYASHI, NOBUHIRO, Dept. of Biophysics and Biochemistry, Graduate School of Science, University of Tokyo, Tokyo, Japan
LAKIN-THOMAS, PATRICIA, Dept. of Biology, York University, Toronto, Ontario, Canada
LAMBREGHTS, RANDY, Dept. of Genetics, Dartmouth Medical School, Hanover, New Hampshire
LAMIA, KATJA, Dept. of Neurobiology, Harvard Medical School, Boston, Massachusetts
LANDE-DINER, LAURA, Dept. of Cell Biochemistry and Human Genetics, Hebrew University Medical School, Jerusalem, Israel
LARRONDO, LUIS, Dept. of Genetics, Dartmouth Medical School, Hanover, New Hampshire
LATHAM, KRISTIN, Dept. of Zoology, Oregon State University, Corvallis, Oregon
LEE, CHENG CHI, Dept. of Biochemistry and Molecular Biology, Health Science Center, University of Texas, Houston
LEE, JONGBIN, Dept. of Biological Sciences, Korea Advanced Institute of Science and Technology, Daejeon, South Korea
LEE, KWANGWON, Dept. of Plant Pathology, Cornell University, Ithaca, New York
LÉVI, FRANCIS, Lab. of Biological Rhythms and Cancer, Hôpital Paul Brousse, Villejuif, France
LEWY, ALFRED, Dept. of Psychiatry, Sleep and Mood Disorders, Oregon Health and Science University, Portland, Oregon
LI, SANSHU, Dept. of Biology, York University, Toronto, Ontario, Canada
LI, WANHE, Cold Spring Harbor Laboratory, Cold Spring Harbor, New York
LIN, JIANDIE, Dept. of Cell and Developmental Biology, Life Sciences Institute, University of Michigan, Ann Arbor
LIU, YI, Dept. of Physiology, Southwestern Medical Center, University of Texas, Dallas

LiWang, Andy, Dept. of Biochemistry and Biophysics, Texas A&M University, College Station, Texas
Lockley, Steven, Dept. of Sleep Medicine, Brigham & Women's Hospital, Harvard Medical School, Boston, Massachusetts
Loros, Jennifer, Dept. of Biochemistry, Dartmouth Medical School, Hanover, New Hampshire
Loudon, Andrew, Faculty of Life Sciences, University of Manchester, Manchester, United Kingdom
Lyons, Lisa, Dept. of Biology and Biochemistry, University of Houston, Houston, Texas
Mackey, Shannon, Dept. of Biology, Texas A&M University, College Station, Texas
Mahoney, Carrie, Dept. of Biology, University of Massachusetts, Amherst
Malzahn, Erik, Biochemistry Center, University of Heidelberg, Heidelberg, Germany
Marcheva, Biliana, Dept. of Medicine, Northwestern University, Evanston, Illinois
Markiewicz, Ewa, School of Biological and Biomedical Sciences, University of Durham, Durham, United Kingdom
Markson, Joseph, Dept. of Molecular and Cellular Biology, Howard Hughes Medical Institute, Harvard University, Cambridge, Massachusetts
Matos, Maria, Dept. of Biology, Johns Hopkins University, Baltimore, Maryland
Matsumoto, Ken, Lab. of Chronogenomics, Mitsubishi Kagaku Institute of Life Sciences, Tokyo, Japan
Mayeda, Aimee, Dept. of Psychiatry, Veterans Administration, Indiana University, Indianapolis, Indiana
McClung, Colleen, Dept. of Psychiatry, Southwestern Medical Center, University of Texas, Dallas
McDonnell, Kevin, Cold Spring Harbor Laboratory, Cold Spring Harbor, New York
McEachron, Donald, School of Biomedical Engineering, Science, and Health Systems, Drexel University, Philadelphia, Pennsylvania
McKnight, Steven, Dept. of Biochemistry, Southwestern Medical Center, University of Texas, Dallas
McLoughlin, Sarah, Conway Institute, University College Dublin, Dublin, Ireland
McMahon, Douglas, Dept. of Biological Sciences, Vanderbilt University, Nashville, Tennessee
McMaster, Andrew, Centre for Molecular Medicine, University of Manchester, Manchester, United Kingdom
Mehra, Arun, Dept. of Genetics, Dartmouth Medical School, Hanover, New Hampshire
Menaker, Michael, Dept. of Biology, University of Virginia, Charlottesville
Menet, Jerome, Dept. of Biology, Howard Hughes Medical Institute, Brandeis University, Waltham, Massachusetts
Meng, Qing, Faculty of Life Sciences, University of Manchester, Manchester, United Kingdom
Meredith, Andrea, Dept. of Physiology, School of Medicine, University of Maryland, Baltimore
Merrow, Martha, Dept. of Chronobiology, University of Groningen, Haren, The Netherlands
Mihalcescu, Irina, Lab. de Spectrométrie Physique, Université Joseph Fourier, Grenoble, France
Millar, Andrew, Institute of Molecular Plant Sciences, University of Edinburgh, Edinburgh, Scotland, United Kingdom
Milne, Charlotte, Div. of Developmental Biology, National Institute for Medical Research, London, United Kingdom
Moulager, Mickael, Lab. Arago, Unité Mixte de Recherche, Centre National de la Recherche Scientifique, Université Paris, Banyuls sur Mer, France
Moynihan Ramsey, Kathryn, Dept. of Medicine, Northwestern University, Evanston, Illinois
Münch, Mirjam, Dept. of Medicine, Div. of Sleep Medicine, Brigham & Women's Hospital, Boston, Massachusetts
Muskus, Michael, School of Biological Sciences, University of Missouri, Kansas City
Naef, Felix, Dept. of Computational Systems Biology, Swiss Institute for Experimental Cancer Research, Ecole Polytechnique Fédérale de Lausanne, Lausanne, Switzerland
Neiss, Andrea, Biochemistry Center, University of Heidelberg, Heidelberg, Germany
Nikiforov, Anastasia, Dept. of Molecular and Cellular Biology, University of Massachusetts, Amherst
Nishiwaki, Taeko, Div. of Biological Science, Graduate School of Science, Nagoya University, Nagoya, Japan
Nitabach, Michael, Dept. of Cellular and Molecular Physiology, Yale University School of Medicine, New Haven, Connecticut
Nordheim, Alfred, Dept. of Molecular Biology, Eberhard-Karls University of Tübingen, Tübingen, Germany
Oates, Andrew, Dept. of Molecular Cell Biology and Genetics, Max-Planck-Institute for Cell Biology and Genetics, Dresden, Germany
Okamura, Hitoshi, Dept. of Systems Biology, Graduate School of Medicine, Kyoto University, Kyoto, Japan
Oster, Henrik, Wellcome Trust Centre for Human Genomics, University of Oxford, Oxford, United Kingdom
Oyama, Tokitaka, Div. of Biological Science, Graduate School of Science, Nagoya University, Nagoya, Japan
Padmanabhan, Kiran, Dept. of Neurobiology, Harvard Medical School, Boston, Massachusetts
Pagano, Michele, Dept. of Pathology, New York University Medical Center, New York, New York
Panda, Satchin, Dept. of Regulatory Biology, The Salk Institute, La Jolla, California
Park, Junghea, Dept. of Neurobiology and Physiology, Northwestern University, Evanston, Illinois
Park, Yong-Ju, Tropical Biosphere Research Center, University of the Ryukyus, Okinawa, Japan
Paterson, Janice, Centre for Cardiovascular Science, University of Edinburgh, Edinburgh, Scotland, United Kingdom
Paul, Matthew, Dept. of Neurology, Medical School, University of Massachusetts, Worcester
Paulose, Jiffin, Dept. of Biology, Texas A&M University, College Station, Texas
Pendergast, Julie, Dept. of Biological Sciences, Vanderbilt University, Nashville, Tennessee
Perry, Michael, Bay City Capital, San Francisco, California
Piggins, Hugh, Faculty of Life Sciences, University of Manchester, Manchester, United Kingdom

PLIKUS, MAKSIM, Dept. of Pathology, University of Southern California, Los Angeles, California
POLITOPOULOU, GALATIA, Dept. of Hematology and Oncology, Rhode Island Hospital, Brown University, Providence, Rhode Island
POURQUIÉ, OLIVIER, Stowers Institute for Medical Research, Kansas City, Missouri
PRICE, JEFFREY, School of Biological Sciences, University of Missouri, Kansas City
PTÁČEK, LOUIS, Dept. of Neurology, Howard Hughes Medical Institute, University of California, San Francisco
PULIVARTHY, SANDHYA RANI, Dept. of Regulatory Biology, The Salk Institute, San Diego, California
RAFF, MARTIN, Medical Research Council Laboratory for Molecular Cell Biology, University College London, London, United Kingdom
RAMJI, NAILA, Harvard University, Cambridge, Massachusetts
RAWASHDEH, OLIVER, Dept. of Biology and Biochemistry, University of Houston, Houston, Texas
REPPERT, STEVEN, Dept. of Neurobiology, Medical School, University of Massachusetts, Worcester
RIHEL, JASON, Dept. of Molecular and Cellular Biology, Harvard University, Cambridge, Massachusetts
RIPPERGER, JÜRGEN, Div. of Biochemistry, University of Fribourg, Fribourg, Switzerland
ROBLES, CHARO, Dept. of Neurobiology, Harvard Medical School, Boston, Massachusetts
ROENNEBERG, TILL, Dept. of Medical Psychology, University of Munich, Munich, Germany
ROGULJA, DRAGANA, Dept. of Genetics, Rockefeller University, New York, New York
ROSBASH, MICHAEL, Dept. of Biology, Howard Hughes Medical Institute, Brandeis University, Waltham, Massachusetts
ROSENWASSER, ALAN, Dept. of Psychology, University of Maine, Orono
ROSSETTI, STEFANO, Dept. of Cancer Genetics, Roswell Park Cancer Institute, Buffalo, New York
ROSSMANN, MARLIES, Cold Spring Harbor Laboratory, Cold Spring Harbor, New York
RUDIC, R. DANIEL, Dept. of Pharmacology and Toxicology, Medical College of Georgia, Augusta, Georgia
RUVKUN, GARY, Dept. of Molecular Biology, Massachusetts General Hospital, Harvard Medical School, Boston, Massachusetts
SAAFIR, TALIB, Dept. of Anatomy and Neurobiology, Morehouse School of Medicine, Atlanta, Georgia
SACCHI, NICOLETTA, Dept. of Cancer Biology, Roswell Park Cancer Institute, Buffalo, New York
SACHDEVA, UMA, Dept. of Cancer Biology, University of Pennsylvania, Philadelphia
SADACCA, AMANDA, Dept. of Neurobiology, Harvard Medical School, Boston, Massachusetts
SAEZ, LINO, Dept. of Genetics, Rockefeller University, New York, New York
SAHAR, SAURABH, Dept. of Pharmacology, University of California, Irvine
SAITHONG, TREENUT, Dept. of Biological Sciences, University of Edinburgh, Edinburgh, Scotland, United Kingdom
SANCAR, AZIZ, Dept. of Biochemistry and Biophysics, School of Medicine, University of North Carolina, Chapel Hill
SAPER, CLIFFORD, Dept. of Neurology, Beth Israel Deaconess Medical Center, Boston, Massachusetts
SASSONE-CORSI, PAOLO, Dept. of Pharmacology, University of California, Irvine
SCHIBLER, ULRICH, Dept. of Molecular Biology, University of Geneva, Geneva, Switzerland
SCHWARTZ, MICHAEL, Dept. of Biology, University of Washington, Seattle
SCHWARTZ, WILLIAM, Dept. of Neurology, Medical School, University of Massachusetts, Worcester
SCOTT, FIONA, Dept. of Neuroscience, University of Manchester, Manchester, United Kingdom
SEAY, DANIEL, Lab. of Genetics, Rockefeller University, New York, New York
SEHGAL, AMITA, Dept. of Neuroscience, Howard Hughes Medical Institute, University of Pennsylvania, Philadelphia
SHIEH, KUN-RUEY, Dept. of Neuroscience, Tzu Chi University, Hualien, Taiwan
SIDDIQUI, KHALID, Cold Spring Harbor Laboratory, Cold Spring Harbor, New York
SILVER, RAE, Dept. of Anatomy and Cell Biology, College of Physicians & Surgeons, Columbia University, New York, New York
SMITH, DANIEL, Dept. of Neuroscience, Lundbeck Research USA, Paramus, New Jersey
SMITH, RACHELLE, Dept. of Biology, University of Utah, Salt Lake City
SOMERS, DAVID, Dept. of Plant Cellular and Molecular Biology, Plant Biotechnology Center, Ohio State University, Columbus, Ohio
SPECTOR, DAVID, Cold Spring Harbor Laboratory, Cold Spring Harbor, New York
SPROUSE, JEFFREY, CNS Discovery Group, Global Research and Development, Pfizer, Groton, Connecticut
STANEWSKY, RALF, Dept. of Biological and Chemical Sciences, Queen Mary, University of London, London, United Kingdom
STAVROPOULOS, NICHOLAS, Dept. of Genetics, Rockefeller University, New York, New York
STEWART, DAVID, Meetings and Courses Programs, Cold Spring Harbor Laboratory, Cold Spring Harbor, New York
STILLMAN, BRUCE, President and CEO, Cold Spring Harbor Laboratory, Cold Spring Harbor, New York
STOLERU, DAN, Dept. of Biology, Howard Hughes Medical Institute, Brandeis University, Waltham, Massachusetts
STORCH, KAI-FLORIAN, Dept. of Neurobiology, Harvard Medical School, Boston, Massachusetts
STRATMANN, MARKUS, Dept. of Molecular Biology, University of Geneva, Geneva, Switzerland
TAFTI, MEHDI, Center for Integrative Genomics, University of Lausanne, Lausanne, Switzerland
TAKAHASHI, JOSEPH, Dept. of Neurobiology and Physiology, Howard Hughes Medical Institute, Northwestern University, Evanston, Illinois
TAKEDA, MAKIO, Div. of Molecular Science, Graduate School of Science and Technology, Kobe University, Kobe, Japan
TAKEMURA, AKIHIRO, Tropical Biosphere Research Center, University of the Ryukyus, Okinawa, Japan

TAMANINI, FILIPPO, Dept. of Genetics, Erasmus University, Rotterdam, The Netherlands

TERAUCHI, KAZUKI, Div. of Biological Science, Graduate School of Science, Nagoya University, Nagoya, Japan

THOMAS, ROBERT, Dept. of Medicine, Beth Israel Deaconess Medical Center, Boston, Massachusetts

THOMPSON, CAROL, Dept. of Neuroscience, Allen Institute for Brain Science, Seattle, Washington

TOTH, REKA, Dept. of Plant Developmental Biology, Max-Planck-Institute for Plant Breeding Research, Cologne, Germany

TU, BENJAMIN, Dept. of Biochemistry, Southwestern Medical Center, University of Texas, Dallas

UEDA, HIROKI, Center for Developmental Biology, RIKEN Kobe Institute, Kobe, Japan

VAN DER HORST, GIJSBERTUS, Dept. of Cell Biology and Genetics, Erasmus University, Rotterdam, The Netherlands

VAN DER LINDEN, ALEXANDER, Dept. of Biology, Brandeis University, Waltham, Massachusetts

VAN DER SCHALIE, ELLENA, Dept. of Biology, University of Virginia, Charlottesville

VIRSHUP, DAVID, Huntsman Cancer Institute, University of Utah, Salt Lake City

VITALINI, MICHAEL, Dept. of Biology, Texas A&M University, College Station, Texas

VON SCHANTZ, MALCOLM, Centre for Chronobiology, School of Biomedical and Molecular Sciences,University of Surrey, Guildford, Surrey, United Kingdom

WANG, CHAO-YUNG, Vascular Medicine Research Unit, Brigham & Women's Hospital, Boston, Massachusetts

WANG, HAN, Dept. of Zoology, Stephenson Research and Technology Center, University of Oklahoma, Norman

WANG, JING, Institute of Diabetes, Obesity, and Metabolism, University of Pennsylvania, Philadelphia

WANG, ZHENGHUI, Dept. of Molecular Cell Biology, Sunnybrook Health Sciences Centre, Toronto, Ontario, Canada

WEINERT, DIETMAR, Dept. of Zoology, Institute of Biology, Martin-Luther-Universität Halle-Wittenberg, Halle, Germany

WEITZ, CHARLES, Dept. of Neurobiology, Harvard Medical School, Boston, Massachusetts

WILLIAMS, STANLY, Dept. of Biology, University of Utah, Salt Lake City

WITKOWSKI, JAN, Banbury Center, Cold Spring Harbor Laboratory, Cold Spring Harbor, New York

WOOD, PATRICIA, Dept. of Oncology and Research, W.J.B. Dorn Veteran's Administration Medical Center, Columbia, South Carolina

WOTUS, CHERYL, Dept. of Biology, University of Washington, Seattle

WRESCHNIG, DANIEL, Dept. of Biology, Northeastern University, Brighton, Massachusetts

WU, JOSEPH, Dept. of Psychiatry, College of Medicine, University of California, Irvine

WU, YING, Dept. of Cellular and Molecular Physiology, Yale University School of Medicine, New Haven, Connecticut

XU, CHUNSU, Dept. of Neurobiology and Behavior, State University of New York, Stony Brook

XU, KANYAN, Dept. of Neuroscience, Howard Hughes Medical Institute, University of Pennsylvania, Philadelphia

YAMAZAKI, SHIN, Dept. of Biological Sciences, Vanderbilt University, Nashville, Tennessee

YANAGISAWA, MASASHI, Dept. of Molecular Genetics, Howard Hughes Medical Institute, Southwestern Medical Center, University of Texas, Dallas

YANG, XIAOMING, Dept. of Pathology and Microbiology, University of South Carolina, Columbia

YANG, XIAOYONG, Lab. of Gene Expression, The Salk Institute for Biological Studies, La Jolla, California

YELAMANCHILI, SOWMYA, Lab. of Regulatory Biology, The Salk Institute for Biological Studies, La Jolla, California

YEOM, MIJUNG, Dept. of Biological Sciences, Vanderbilt University, Nashville, Tennessee

YIN, LEI, Dept. of Endocrinology, School of Medicine, University of Pennsylvania, Philadelphia

YOKOYAMA, CHARLES, Neuron, Cell Press and Elsevier, Cambridge, Massachusetts

YOO, SEUNG-HEE, Dept. of Neurobiology and Physiology, Northwestern University, Evanston, Illinois

YOON, HOJUNG, Dept. of Health Informatics, University of Minnesota, Minneapolis

YOSHITANE, HIKARI, Dept. of Biophysics and Biochemistry, Graduate School of Science, University of Tokyo, Tokyo, Japan

YOUNG, MICHAEL, Dept. of Genetics, Howard Hughes Medical Institute, Rockefeller University, New York, New York

YU, WANGJIE, Dept. of Biology, Texas A&M University, College Station, Texas

ZATZ, MARTIN, Journal of Biological Rhythms, Society for Research on Biological Rhythms, Sage Publications, Bethesda, Maryland

ZHU, YONG, Dept. of Epidemiology and Public Health, Yale University, New Haven, Connecticut

ZORAN, MARK, Dept. of Biology, Center for Research on Biological Clocks, Texas A&M University, College Station, Texas

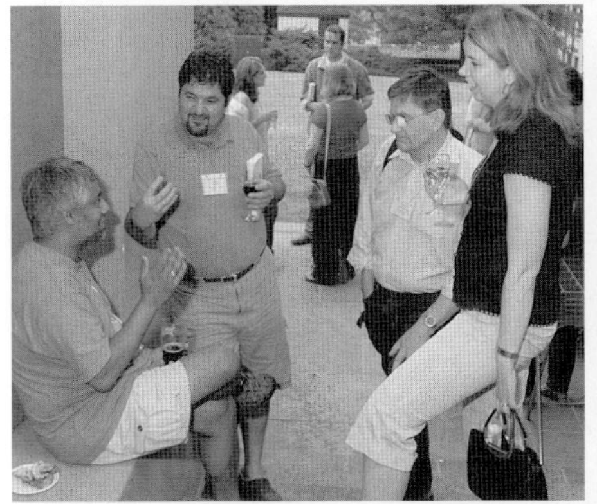

First row: C. Czeisler, Dorcus Cummings Lecture
Second row: S. Hattar, A. Mehra, R. Artymyshyn, C. McClung; M. Tafti, R. Greenspan; S. Brown, M. Merrow, T. Roenenberg
Third row: M. Young, R. Klevecz; F. Naef, T. Kondo

First row: H. Okamura, J. Takahashi; A. Millar, D. Bell-Pedersen; M. Zatz, M. Menaker
Second row: J. Keene, K. Carniol; M. Rosbash, D. Gatfield; S. Kuhlman, C. Saper
Third row: G. Cao, R. Stanewsky; M. Hughes, J. Hogenesch, J. Baggs
Fourth row: N. Kondo, P. DeCoursey; W. Schwartz, J. Loros, C. Weitz

First row: T. Roenneberg, P. Sassone-Corsi; G. Bjarnason, F. Lévi; S. Reppert, A. Sancar
Second row: E. Kowalska, J. Blau; J. Dunlap, M. Brunner
Third row: A. Millar, J.W. Hastings; S. McKnight; C. Green, M. Hastings, S. Golden
Fourth row: L. Ptáček, U. Schibler, S. Reppert; T. Oyama, D. Somers

First row: J. Pendergast, M. Yeon; J. Acimovic, A. Gall; C. Barriga-Montoya, E. Chévez
Second row: J. Loros, J. Paulose, J. Dunlop, V. Cassone, M. Zatz; S. Daniels, H. Piggins
Third row: C. Mahoney, A. Nikiforov; D. McMahon, A. Rosenwasser; C. Kenyon
Fourth row: M. Antoch; B. Stillman, M. Raff; R. Silver, R. Klevecz, P. Wood

First row: J.W. Hastings, P. DeCoursey, M. Menaker; Lunch at Blackford Hall
Second row: J. Hall; S. Gery, H.P. Koeffler; S. Yelamanchili, E. Harnish
Third row: J. Gooley, S. Lockley, J. Hull; M. Macaluso, H. Wang
Fourth row: P. Hardin, J. Benito, C. Helfrich-Förster; C. Weitz, C. Johnson

First row: Symposium Group Gathering
Second row: A. Davidson, C. Jackson; M. Rosbash; T. Saafir at the Poster Session
Third row: Symposium Interviews: S. Gary, M. Raff; J. Witkowski, J. Watson, A. Lewy

Foreword

The history of research into circadian rhythms can be traced back to the French astronomer Jean Jacques Ortous de Mairan, who conducted experiments with plants grown in the dark in the early 1700s. His observations started a slow march joined by luminaries such as Carolus Linnaeus and Charles Darwin interested in whether and how animal and plants measured time. The descriptive era of circadian biology continued into the latter half of the 20th century and was well summarized by the last Symposium held on the topic in 1960, in which the opening address by Erwin Bunning concluded that "thus far, however, such facts have not enabled us to draw far-reaching conclusions about the nature of the [biological] clock."

As with so many other disciplines in biology, the descriptive era has been revolutionized by the molecular era. Since the discovery and cloning of the first clock gene, *period*, more than 20 years ago, tremendous progress has been made about the nature of the clock and how it functions in a wide variety of different plants and animals. Research previously limited to describing "the hands of the clock" has been enormously successful in recent years in describing the inner anatomy and mechanism of the clock in individual cells and in the whole organism.

Many in the chronobiology community are now, more than ever, attempting to place the molecular and cellular details of the oscillatory machinery in the broader context of cellular physiology—for example, the cell cycle—or organismal behaviors—for example, sleep and circadian behavior, and how these may in turn be affected by environment and/or disease. It is noteworthy that Christoph Wilhelm Hufeland wrote in his 1796 treatise *The Art of Prolonging Life,* "The period of twenty-four hours formed by the regular revolution of our Earth...is apparent in all diseases...it is, as it were, the unit of our natural chronology."

Our decision to focus the 72nd Symposium on circadian and related rhythms reflects the tremendous advances that molecular and cellular approaches have yielded thus far, a decision taken in part thanks to an open letter from 18 prominent scientists in the field. The Symposium was arranged to cover a range of themes associated with biological clocks and rhythms, ranging in scale from the molecular to the whole organism, and addressed how aberrant function of the clock may have a causative role in diverse human diseases and conditions. The Symposium provided a unique synthesis of the exciting progress being made in the field of chronobiology not only for the Symposia attendees, but also for a wider global audience via interviews freely available on the World Wide Web and, we anticipate, for readers of these Proceedings.

In organizing this Symposium, we relied on the assistance of Steve Kay, Steve McKnight, and Ueli Schibler, in particular, for suggestions for speakers. We also wish to thank the first evening speakers Joe Takahashi, Louis Ptáček, Amita Sehgal, and Ron Evans for providing an overview of the areas to be covered. This year's Reginald Harris Lecture was delivered by Martin Raff on intracellular timers in oligodendrocyte precursor cells. We particularly wish to thank Michael Menaker—one of three participants along with Patricia DeCoursey and "Woody" Hastings, who attended both 1960 and 2007 Symposia—for delivering a masterful and eloquent summary of the current state of the field, and to Charles Czeisler who enlightened a mixed audience of scientists, lay friends, and neighbors with his Dorcas Cummings lecture on "Work Hours, Sleep, and Safety: Physician Heal Thyself."

This Symposium was attended by 316 scientists from more than 20 countries, and the program included an opening introductory workshop, 72 oral presentations, and 142 poster presentations. Essential funds to run this meeting were obtained from the National Institute of Neurological Disorders and Stroke, a branch of the National Institutes of Health. In addition, financial help from the corporate benefactors, sponsors, affiliates, and contributors of our meetings program is essential for these Symposia to remain a success and we are most grateful for their continued support.

We wish to thank Val Pakaluk and Mary Smith in the Meetings and Courses Program office for their efficient help in organizing the Symposium. Joan Ebert and Rena Steuer in the Cold Spring Harbor Laboratory Press, headed by John Inglis, ensured that this volume would be produced. We thank them for their dedication to producing high-quality publications.

Bruce Stillman
David Stewart
Terri Grodzicker
January 2007

Sponsors

This meeting was funded in part by the **National Cancer Institute,** a branch of the **National Institutes of Health.**

Contributions from the following companies provide core support for the Cold Spring Harbor meetings program.

Corporate Patron

Pfizer, Inc.

Corporate Benefactors

Amgen, Inc.
GlaxoSmithKline
Merck Research Laboratories
Novartis Institutes for BioMedical Research

Corporate Sponsors

Abbott Laboratories
Applied Biosystems
BioVentures, Inc.
Bristol-Myers Squibb Company
Diagnostic Products Corporation
Forest Laboratories, Inc.
GE Healthcare Bio-Sciences
Genentech, Inc.
Hoffmann-La Roche, Inc.
Johnson & Johnson Pharmaceutical Research & Development, LLC
Kyowa Hakko Kogyo Co., Ltd.
Merck Research Laboratories
New England BioLabs, Inc.
OSI Pharmaceuticals, Inc.
Pall Corporation
Sanofi-Aventis
Schering-Plough Research Institute

Plant Corporate Associates

ArborGen
Monsanto Company
Pioneer Hi-Bred International, Inc.

Corporate Affiliates

Abcam, Ltd.
Agilent Technologies
USB Corporation

Corporate Contributors

Cell Signaling Technology
Epicentre Biotechnologies
Hybrigenics, SA
Illumina
inGenious Targeting Laboratory, Inc.
IRx Therapeutics, Inc.
Millipore

Foundations

Albert B. Sabin Vaccine Institute, Inc.
Hudson-Alpha Institute for Biotechnology

Contents

Genetics of Rhythms

Entrainment and Peripheral Clocks

Systems Approaches to Biological Clocks

Models

Development, Proliferation, and Aging

Biological Rhythms Workshop I: Introduction to Chronobiology

S.J. KUHLMAN,* S.R. MACKEY,†‡ AND J.F. DUFFY§

*Cold Spring Harbor Laboratory, Cold Spring Harbor, New York 11724; †Department of Biology, The Center for Research on Biological Clocks, Texas A&M University, College Station, Texas 77843-3258; §Division of Sleep Medicine, Harvard Medical School and Brigham and Women's Hospital, Boston, Massachusetts 02115

In this chapter, we present a series of four articles derived from a Introductory Workshop on Biological Rhythms presented at the 72nd Annual Cold Spring Harbor Symposium on Quantitative Biology: Clocks and Rhythms. A diverse range of species, from cyanobacteria to humans, evolved endogenous biological clocks that allow for the anticipation of daily variations in light and temperature. The ability to anticipate environmental variation promotes optimal performance and survival. In the first article, Introduction to Chronobiology, we present a brief historical time line of how circadian concepts and terminology have emerged since the early observation of daily leaf movement in plants made by an astronomer in the 1700s. Workshop Part IA provides an overview of the molecular basis for rhythms generation in several key model organisms, Workshop Part IB focuses on how biology built a brain clock capable of coordinating the daily timing of essential brain and physiological processes, and Workshop Part IC gives key insight into how researchers study sleep and rhythms in humans.

INTRODUCTION

As a consequence of the Earth's rotation about its axis approximately every 24 hours, most organisms on this planet are subjected to predictable fluctuations of light and temperature. A diverse range of species, from cyanobacteria to humans, evolved endogenous biological clocks that allow for the anticipation of these daily variations. Thus, our internal physiology and function are fundamentally intertwined with this geophysical cycle. In fact, it was an astronomer, Jean-Jacques d'Ortous deMairan, rather than a biologist, who provided early insight into this evolutionary relationship between internal physiology and the geophysical cycle. deMairan (1729) made the observation that daily leaf movements in heliotrope plants continue in constant darkness. To emphasize the endogenous or self-sustained nature of biological clocks, Franz Halberg in 1959 coined the term *circadian* (Latin: *circa* = about; *dies* = day) to refer to daily rhythms that are truly endogenously generated, i.e., rhythms having a period of about 24 hours that continue to oscillate in the absence of any environmental input (Chandrashekaran 1998). Rhythm generation is now understood to be an intrinsic property of single cells, driven by an intracellular molecular oscillator based on transcriptional/posttranslational negative feedback loops. Under normal conditions, endogenous oscillations are synchronized to the environment, and it is generally thought that biological clocks provide an adaptive advantage by ensuring that an organism's internal biochemical and physiological processes, in addition to behavior, are optimally adapted to the local environment.

‡*Present address:* Department of Biology, St. Ambrose University, Davenport, Iowa 52803.

HISTORICAL TIME LINE

As early as 1729, it was documented that this daily rhythmic behavior was likely endogenously generated, and it was also near this time that Carl von Linne (1707–1778) constructed a "floral clock" noting the predictability of petal opening and closing times of various species of flowers (Chandrashekaran 1998). However, it was not until 200 years later that Erwin Bünning provided the first evidence for the genetic basis of circadian rhythms generation by demonstrating that period length is heritable in bean plants (Bünning 1935). Bünning (1936) also put forth an influential hypothesis that circadian oscillators can be used to measure seasonal changes in addition to measuring daily cycles and pointed out the adaptive significance of tracking seasonal changes. Thus, the field of circadian rhythms originated from keen observation of plants, and it was not until later that the first observations of endogenously driven rhythms in bacteria (Mitsui et al. 1986), single-cell eukaryotes (Sweeney and Hastings 1957), insects (Beling 1929), birds (Kramer 1952), rodents (Richter 1922), primates (Simpson and Gaibraith 1906), and humans (Aschoff and Wever 1962) were discovered.

A breakthrough in understanding the genetic basis for rhythms generation was made by Ronald Konopka and Seymour Benzer (1971) using a mutant screen in *Drosophila melanogaster*. Mutagenized flies were examined for the persistence of two circadian behaviors: pupal eclosion and locomotor activity. Flies displayed one of three categorical mutant phenotypes: a lengthening of circadian period, a shortening of period, or arrhythmia. All phenotypes were complemented by a single locus, now referred to as the *Period* gene. Shortly after this discovery in fruit flies, the *Frequency* gene was shown to be essential for rhythms in conidiation to persist in the filamentous fungus *Neurospora crassa* (Feldman and Hoyle 1973). These surprising results showed that single-gene mutations could disrupt a complex behavior and, together with the wonderful discovery of a heritable timing mutation in hamsters by Martin Ralph and Michael Menaker (1988), provided the rationale for pursuing a large-scale mutant screen in mice (Vitaterna et al. 1994). Subsequently, using a variety of strategies, dozens of clock genes have been discovered in both prokaryotic and eukaryotic systems, including cyanobacteria, fungi, plants, insects, and mammals. Even human rhythms are potently altered by clock gene mutations (Jones et al. 1999). The specific clock gene players in these diverse systems are described in more detail in Part IA. A striking common principle emerges from examining these diverse clock gene systems, i.e., all organisms seem to have evolved transcriptional/posttranslational feedback loops to ensure high-amplitude, near 24-hour, rhythms generation. Furthermore, the identification of specific clock genes has led to the development of real-time bioluminescent and fluorescent reporters that have the spatial resolution to track rhythmicity at the level of the functional unit of rhythms generation, i.e., the single cell (Hastings et al. 2005). Such resolution is essential to understand how individual oscillators are coupled within a population of rhythmic cells.

In the midst of this clock gene explosion, however, there is an important lesson to be learned from a pivotal experiment conducted using clock components of cyanobacteria. In 2005, Nakajima et al. (2005) reconstituted a circadian oscillator in a test tube using only cyanobacterial proteins and ATP. The lesson: It is possible to construct a near 24-hour oscillator in the absence of gene transcription.

It is clear from the work of many talented and dedicated scientists that most organisms inherit the innate ability to keep track of time on a 24-hour scale. How do organisms use their timing devices? What are the negative consequences if organisms are no longer able to effectively use their clocks, such as in disease or aging? We know from the work of Karl von Frisch and Beling that bees use their clocks to visit flowers at the appropriate time of day so that they can feed when the flower is open. Work from Kramer shows that birds use their biological clock during migration to help compensate for the changing position of the sun throughout the day. For an inspiring account of these classic studies, we recommend reading Chapter 1 in Moore-Ede et al. (1982). Additionally, DeCoursey et al. (2000) showed that chipmunks use an innate biological clock to properly time foraging to avoid predation. More recently, studies on monarch butterflies by Steve Reppert (2006) revealed that the circadian clock likely participates in initiating migration by tracking seasonal changes. Future studies in the field of circadian rhythms will continue to explore the wonderful and creative ways in which organisms use their biological clocks to coordinate internal function and navigate through the environment (for mammalian review, see Buijs and Kalsbeek 2001).

CIRCADIAN TERMINOLOGY AND GENERAL METHODS

Here, we define some commonly used terms in the field. "Black-box" experimental designs are frequently used to probe the mechanisms underlying clock function (Moore-Ede et al. 1982). For example, deMairan (1729) used a black-box approach in his heliotrope plant study. The plant was treated as a system whose internal components were unknown (i.e., black box), but whose function was studied by assaying the observable output of leaf movement in response to perturbations caused by environmental inputs (the light/dark cycle). To a large extent, the following terminology developed over the years to precisely report the results of black-box experiments.

Observable, or measurable, output rhythms of the circadian timing system, such as leaf movements, are defined as *overt outputs*. In the case of animals, wheel-running activity and levels of circulating hormones in the blood are two commonly assayed overt outputs. In addition, a system can be composed of multiple oscillators, in which the output signal of one oscillator influences the circadian properties of another oscillator. The output signal can then be referred to as a *coupling signal*. In practice, the distinction between an overt rhythm and coupling signal may depend on the experimenter's perspective.

A rhythm is considered to be circadian if the oscillation has a period of approximately 24 hours and continues in constant conditions, such as constant light (referred to as LL) or constant darkness (DD) (Fig. 1A). The inability of a rhythm to continue under constant conditions implies that the rhythm is driven by external time cues (or a *zeitgeber*, German for time-giver), rather than generated internally. It is important to keep in mind that when rhythmicity is measured at the tissue or population level, the loss of an overt rhythm due to experimental perturbation may not be the result of loss of rhythmicity per se. Rather, individual oscillations may continue, but they are not observed due to phase differences among oscillating units (Fig. 2).

The time needed for one circadian oscillation to occur under constant conditions is known as the *free-running period* (FRP). Under normal conditions, circadian rhythms are not free-running, rather, they are synchronized to the local environment. The process by which a rhythm synchronizes to an external cycle is referred to as *entrainment*. Interestingly, although circadian rhythms persist under constant conditions, the FRP can vary slightly in response to changing light intensities. Diurnal organisms display a slightly shorter FRP (i.e., faster clock) in higher light intensities than they do in low light; nocturnal organisms exhibit the inverse response with a longer FRP (i.e., slower clock) in high light than in dim light. This phenomenon, originally described by Jürgen Aschoff, is known as "Aschoff's Rule" (Aschoff and Wever 1962). FRP can also vary based on the light history that the organism experienced before being placed under constant conditions. The influence of light history on FRP is referred to as an *aftereffect*. Clearly, when assaying

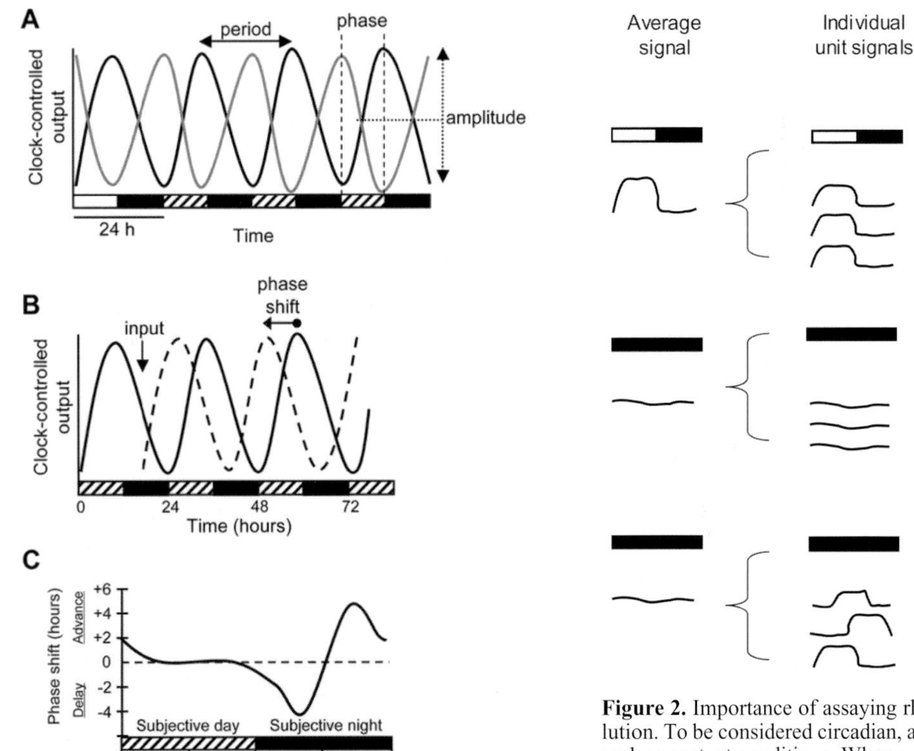

Figure 1. Properties of a circadian rhythm. (*A*) Behavioral circadian rhythms can be entrained by external stimuli, such as a light/dark cycle, and will persist with a near 24-hour period in the absence of environmental cues, such as constant darkness. Properties of the rhythm that are commonly measured are period, phase, and amplitude. Period is the duration of time to complete one cycle. It is typically measured from peak to peak, but it can be measured from any specific position on the curve. Phase is the relative position on the curve (e.g., the peak) in reference to a particular time, such as time placed in constant darkness. Amplitude is the measurement of the recorded output from the midline of the curve to either the peak or trough. (*B*) The phase of a circadian rhythm can be reset by the stimuli to which it entrains. In this case, exposure to a stimulus (input) rapidly lowers the level of the rhythmic variable (*dashed line*), which recovers to a rhythm with a shifted phase as compared to the curve that did not receive the input (*solid line*). (*X* axis) Time in the light/dark cycle is depicted by alternating white and black bars, and time under constant conditions is depicted by alternating hatched and black bars. Level of clock-controlled output is displayed on the *Y* axis. (*C*) Phase-response curves measure the magnitude and direction of phase-dependent responses to brief exposures of an external stimulus. The *X* axis represents the circadian time at which a light pulse is applied to an organism; the *Y* axis shows the change in phase of the circadian-controlled output in response to the light pulse. Positive shifts are indicated as advances, and negative phase shifts are delays. A brief light pulse given to an organism during the subjective day (the organism's own internal day) produces little to no phase response, light in the early subjective night (the organism's own internal early night) produces phase-delay shifts, and light in the late subjective night produce phase-advance shifts.

Figure 2. Importance of assaying rhythms at a single-cell resolution. To be considered circadian, a rhythm must be maintained under constant conditions. When examining rhythms at the tissue level in which the tissue is composed of multiple cells, it is important to realize that the absence of rhythmicity under constant conditions could be due to two possibilities: either the loss of rhythmicity (indicating the rhythm is not circadian) or desynchrony among oscillators. In the later case, individual rhythms may actually be circadian; however, the assay has insufficient resolution to demonstrate the presence of rhythmicity. To distinguish between the two possibilities, assays sensitive to the functional unit of rhythms generation must be used.

overt rhythms and calculating the FRP, the influence of these factors should be taken into consideration (see also Part IC, Definition of Sleep).

In addition to being endogenously generated, there are two other properties of circadian rhythms that are often the focus of experimental studies: (1) the ability of the rhythm to be reset, or phase-shifted, by transient exposure to time cues such as light and (2) temperature-compensation, defined below (Pittendrigh 1981; Dunlap et al. 2004).

Phase-shifting and the Phase-response Curve

The phase of a rhythm can be shifted (or reset) by transient exposure to certain environmental cues (Fig. 1B). Because FRPs are close to, but not exactly, 24 hours, the rhythm must be reset each day in order to avoid falling out of phase with the external environmental cycle. As discussed earlier, many organisms use the daily fluctuation of light and darkness as the resetting signal to entrain their rhythms to the 24-hour day. An early study by Hastings and Sweeney (1958) in the dinoflagellate *Gonyaulax polyhedra* demonstrated a property common among circadian systems: The biological clock is not equally sensitive to light at all times of day. As has been shown subsequently for many organisms that use light as an entraining signal, light exposure in the early subjective night produces a phase-delay shift (i.e., the rhythm is delayed to a later hour), whereas light exposure in the late subjective night produces a phase-advance shift (the rhythm is advanced to an earlier hour), and light during the subjective day produces very small or no phase shifts. Thus, the response to light is dependent on the organism's internal sense of time. The magnitude and direction of these responses can be plotted with respect to the organism's own circadian phase at which the light was presented. Such a plot is referred to as a *phase-response curve* (PRC) (Fig. 1C).

It is important to realize that there are cases in which it is possible that exposure to an external cue will simply transiently perturb the level of the rhythm being assayed, without leading to a persistent shift in the rhythm. This condition is referred to as *masking*. Mechanistically, masking can arise because sensory pathways may act on a specific behavior independent of the clock (for an example in mammals, see Part IB).

Temperature-compensation

For a biological clock to be reliable in a natural setting, its period should *not* change much despite changes in ambient temperature. Morning events should occur at essentially the same time, independent of weather conditions the night before. This property, termed temperature-compensation, stems from the observation that the value of the FRP changes very little during different temperatures within the organism's physiological range. The effect of changes in temperature on the rate of most biochemical reactions is measured by a Q_{10} value, which is defined as the ratio of the rate of a given process at one temperature to the rate at a temperature 10°C lower. The Q_{10} value of the period of a circadian rhythm remains near 1, as opposed to other known biochemical reactions that have Q_{10} values of 2 or 3.

It is important to note that temperature-compensation is not the same as temperature insensitivity. Temperature itself can be can be a strong zeitgeber in some organisms (Liu et al. 1998). Individual reactions within the clock are undoubtedly affected by changes in temperature, but the system as a whole is buffered such that the output of rhythmic behavior does not show large variations in its FRP.

Concept of Clock

What is a circadian clock and how has the concept of a "clock" facilitated research? As described in one account (Chandrashekaran 1998), the now widespread use of the term clock was in part inspired by Gustav Kramer's studies on time compensation of the sun compass in birds. Shortly thereafter in 1960, the Cold Spring Harbor Symposium of Quantitative Biology was boldly titled "Biological Clocks." Today, the term "clock" is used pervasively to emphasize the endogenous nature of rhythms generation, that circadian rhythms are innate rather than learned phenomena, and importantly, to imply that a primary function of circadian rhythms is to measure time (Pittendrigh 1961; Moore-Ede et al. 1982). It is this conceptual framework that prompted the design of shifted-schedule or jet-lag experiments and helps us understand the relationship between seasonal adaptation and circadian rhythms generation.

Conceptually, the components of a circadian clock can be broken down into three basic elements: an input pathway, a pacemaker, and an output pathway (Fig. 3). At the heart of a circadian clock is the pacemaker, a central oscillator or a network of coupled oscillators, that is entrainable and has the ability to synchronize downstream targets. The temporal information produced by the pacemaker is interpreted by the output pathways, which then regulate the timing of metabolic and behavioral processes. For the oscillator to maintain synchrony with the environment, input pathways must relay external timing cues to the pacemaker. It is important to realize that conceptually, one can discuss these three elements as distinct entities; however, in biological terms, one protein or physiological process can subserve multiple roles.

The 2007 Cold Spring Harbor 72nd Symposium on Clocks and Rhythms included presentations that covered

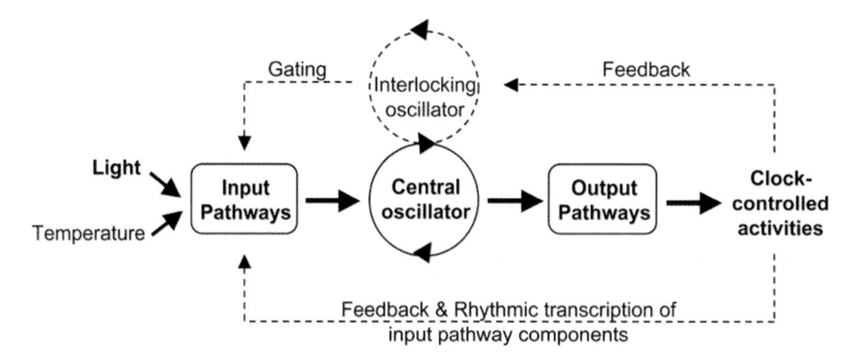

Figure 3. Representation of circadian clock divisions. A circadian clock can be depicted as having input pathways, a central oscillator (or pacemaker), and output pathways. The central oscillator produces the endogenous biological rhythm and can be synchronized with the environment via input pathways through cues such as light or temperature. Output pathways convey the clock's rhythms to downstream targets and drive overtly rhythmic activities. Some circadian systems consist of more elaborate pathways (shown as *dashed lines*) that include multiple, interlocking oscillators and positive or negative feedback from clock-controlled activities to oscillator and/or input components. (Modified, with permission, from Gardner et al. 2006 [© The Biochemical Society].)

an amazingly diverse range of organisms (from bacteria to humans) and technical approaches. Despite the diversity, there are fundamental concepts that bind the field together. In the following chapter, we use the three-division framework of input, rhythm generator/pacemaker, and output to introduce the reader to key circadian molecular players and physiological processes. The goal of this workshop is to concisely present the unifying concepts of the field in order to foster lively discussion and critical evaluation of the data and strategies used in the study of biological clocks and rhythms. A list of abbreviations used within this Workshop Review is provided below (Table 1).

ACKNOWLEDGMENTS

We thank Drs. Marty Zatz, Mirjam Münch, and Sean Cain for comments on earlier versions of the manuscript.

Table 1. List of Abbreviations

Abbreviation	Meaning	Abbreviation	Meaning
5HT	serotonin neurotransmitter	ipRGC	intrinsically photosensitive retinal ganglion cells
ARAS	ascending reticular activating system	IPSP	inhibitory postsynaptic potential
Arc	arcuate nucleus	JET	JETLAG
AVP	arginine vasopressin	KaiC~P	phosphorylated KaiC
BF	basal forebrain	LabA	low amplitude and bright
bHLH	basic helix-loop-helix	LC	locus coeruleus
BMAL1	brain and muscle ARNT-like protein 1	LD	light/dark cycle
C-box	Clock box	LdpA	light-dependent period
CAMK1	calcium/calmodulin-dependent protein kinase 1	LH	lateral hypothalamus
CBS	CCA1-binding site	LHY	LATE ELONGATED HYPOCOTYL
CCA1	CIRCADIAN CLOCK ASSOCIATED 1	LL	constant light
CikA	circadian input kinase	LUX	LUX ARRHYTHMO
CK(1, 2)	casein kinase (1, 2)	NE	norepinephrine neurotransmitter
CLK	CLOCK	NREM	nonrapid eye movement
CO	CONSTANS	PACAP	pituitary adenylate-cyclase-activating polypeptide
COP1	CONSTITUTIVELY PHOTOMORPHOGENIC 1	PAS	PER-ARNT-SIMS
CRE	cyclic-AMP response element	PDF	pigment-dispersing factor
CREB	CRE binding	PDP1ε	PAR DOMAIN PROTEIN 1ε
CRY	CRYPTOCHROME	PER	PERIOD
CYC	CYCLE	Pex	period extender
DBMIB	2,5-dibromo-3-methyl-6-isopropyl-*p*-benzoquinone	PHY	PHYTOCHROME
DBT	DOUBLE-TIME	PP1(2A)	protein phosphatase 1 (2A)
DD	constant darkness	PPT	pedunculopontine nuclei
DET	DE-ETIOLATED 1	PRD-4	Period-4
DIC	differential interference contrast	PRR	pseudo-response regulator
DMH	dorsal medial hypothalamus	REM	rapid eye movement
DMV	dorsal motor nucleus of the vagus	RHT	retinohypothalamic tract
DR	dorsal raphenuclei	RORE	ROR element
EE	evening element	RpaA	regulator of phycobiliosome-associated
EEG	electroencephalogram	SasA	*Synechococcus* adaptive sensor
ELF	EARLY FLOWERING	SCN	suprachiasmatic nucleus
FDR	fast-delayed rectifier	SGG	SHAGGY
FFT	fast Fourier transform	sPVZ	subparaventricular zone
FLO	FRQ-less oscillator	SPY	SPINDLY
FRH	FRQ-interacting RNA helicase	SWA	slow-wave activity
FRP	free-running period	SWS	slow-wave sleep
FRQ	FREQUENCY	TIC	TIME FOR COFFEE
FT	FLOWERING LOCUS T	TIM	TIMELESS
GABA	γ-aminobutyric acid	TOC1	TIMING OF CAB EXPRESSION 1
GFP	green fluorescent protein	TMN	tuberomammilary neurons
GHT	geniculohypothalamic tract	TTX	tetrodotoxin
GI	GIGANTEA	VIP	vasoactive polypeptide
GnRH	gonadotropin-releasing hormone	VLP	ventrolateral preoptic nucleus
GRP	gastrin-releasing peptide	VP-box	VRI/PDP1ε box
HAT	histone acetyltransferase	vPVN	paraventricular nucleus of the hypothalamus
hPVN	hypothalamic paraventricular neurons	VRI	VRILLE
Hz	hertz (cycles per second)	VVD	VIVID
IGL	intergeniculate leaflet	WC	WHITE COLLAR
		WCC	WHITE COLLAR COMPLEX
		ZTL	ZEITLUPE

REFERENCES

Aschoff J. and Wever R. 1962. Spontanperiodik des menschen bei ausschluss aller zeitgeber. *Naturwissenschaften* **49:** 337.

Beling I. 1929. Uber das zeitgedachtnis der bienen. *Z. Vgl. Physiol.* **9:** 259.

Buijs R.M. and Kalsbeek A. 2001. Hypothalamic integration of central and peripheral clocks. *Nat. Rev. Neurosci.* **2:** 521.

Bünning E. 1935. Zur kenntis der erblichen tagesperiodiztat bei den primarblattern von *Phaseolus multiflorus*. *Jahrb. Wiss. Bot.* **81:** 411.

———. 1936. Die endogene tagesrhythmik als grundlage der photoperiodischen reaktion. *Ber. Dtsch. Bot. Ges.* **54:** 590.

Chandrashekaran M.K. 1998. Biological rhythms research: A personal account. *J. Biosci.* **23:** 545.

DeCoursey P.J., Walker J.K., and Smith S.A. 2000. A circadian pacemaker in free-living chipmunks: Essential for survival? *J. Comp. Physiol. A* **186:** 169.

deMairan J. 1729. Observation botanique. *Histoire de L'Academie Royale des Sciences,* p. 35.

Dunlap J.C., Loros J.J., and DeCoursey P.J. 2004. *Chronobiology: Biological timekeeping*. Sinauer Associates, Sunderland, Massachusetts.

Feldman J.F. and Hoyle M.N. 1973. Isolation of circadian clock mutants of *Neurospora crassa*. *Genetics* **75:** 605.

Gardner M.J., Hubbard K.E., Hotta C.T., Dodd A.N., and Webb A.A. 2006. How plants tell the time. *Biochem. J.* **397:** 15.

Hastings J.W. and Sweeney B.A. 1958. A persistent diurnal rhythm of luminescence in *Gonyaulax polyedra*. *Biol. Bull.* **115:** 440.

Hastings M.H., Reddy A.B., McMahon D.G., and Maywood E.S. 2005. Analysis of circadian mechanisms in the suprachiasmatic nucleus by transgenesis and biolistic transfection. *Methods Enzymol.* **393:** 579.

Jones C.R., Campbell S.S., Zone S.E., Cooper F., DeSano A., Murphy P.J., Jones B., Czajkowski L., and Ptáĉek L.J. 1999. Familial advanced sleep-phase syndrome: A short-period circadian rhythm variant in humans. *Nat. Med.* **5:** 1062.

Konopka R.J. and Benzer S. 1971. Clock mutants of *Drosophila melanogaster*. *Proc. Natl. Acad. Sci.* **68:** 2112.

Kramer G. 1952. Experiments on bird orientation. *Naturwissenschaften* **94:** 265.

Liu Y., Merrow M., Loros J.J., and Dunlap J.C. 1998. How temperature changes reset a circadian oscillator. *Science* **281:** 825.

Mitsui A., Kumazawa S., Takahashi A., Ikemoto H., Cao S., and Arai T. 1986. Strategy by which nitrogen-fixing unicellular cyanobacteria grow photoautotrophically (letter). *Nature* **323**: 720.

Moore-Ede M.C., Sulzman F.M., and Fuller C.A. 1982. *The clocks that time us*. Harvard University Press, Cambridge, Massachusetts.

Nakajima M., Imai K., Ito H., Nishiwaki T., Murayama Y., Iwasaki H., Oyama T., and Kondo T. 2005. Reconstitution of circadian oscillation of cyanobacterial KaiC phosphorylation in vitro. *Science* **308:** 414.

Pittendrigh C.S. 1961. Circadian rhythms and the organization of living systems. *Cold Spring Harbor Symp. Quant. Biol.* **25:** 159.

———. 1981. Circadian systems: General perspective and entrainment. In *Handbook of behavioral neurobiology: Biological rhythms* (ed. J. Aschoff), pp. 57 and 95. Plenum Press, New York.

Ralph M.R. and Menaker M. 1988. A mutation of the circadian system in golden hamsters. *Science* **241:** 1225.

Reppert S.M. 2006. A colorful model of the circadian clock. *Cell* **124:** 233.

Richter C.P. 1922. A behavioristic study of the activity of the rat. *Comp. Psychol. Monogr.* **1:** 1.

Simpson S. and Gaibraith J.J. 1906. Observations on the normal temperature of the monkey and its diurnal variation, and on the effect of changes in the daily routine on this variation. *Trans. R. Soc. Edinb.* **45:** 65.

Sweeney B.M. and Hastings J.W. 1957. Characteristics of the diurnal rhythm of luminescence in *Gonyaulax polyedra*. *J. Cell. Comp. Physiol.* **49:** 115.

Vitaterna M.H., King D.P., Chang A.M., Kornhauser J.M., Lowrey P.L., McDonald J.D., Dove W.F., Pinto L.H., Turek F.W., and Takahashi J.S. 1994. Mutagenesis and mapping of a mouse gene, *Clock*, essential for circadian behavior. *Science* **264:** 719.

Biological Rhythms Workshop IA: Molecular Basis of Rhythms Generation

S.R. Mackey*

Department of Biology, The Center for Research on Biological Clocks, Texas A&M University, College Station, Texas 77843-3258

Current circadian models are based on genetic, biochemical, and structural data that, when combined, provide a comprehensive picture of the molecular basis for rhythms generation. These models describe three basic elements—input pathways, oscillator, and output pathways—to which each molecular component is assigned. The lines between these elements are often blurred because some proteins function in more than one element of the circadian system. The end result of these molecular oscillations is the same in each system (near 24-hour timing), yet the proteins involved, the interactions among those proteins, and the regulatory feedback loops differ. Here, the current models for the molecular basis for rhythms generation are described for the prokaryotic cyanobacterium *Synechococcus elongatus* as well as the eukaryotic systems *Neurospora crassa*, *Drosophila melanogaster*, *Arabidopsis thaliana*, and mammals (particularly rodents).

INTRODUCTION

Internally generated, near 24-hour rhythms of behavior are observed in numerous organisms ranging from prokaryotic cyanobacteria to mammals. Driving the overt 24-hour rhythmic behaviors are oscillations that occur at the molecular level. The molecular components can be assigned to one of three basic elements within the biological clock: input pathways that relay synchronizing environmental cues to the central oscillator, the oscillator that maintains internal time, or output pathways that transduce temporal information from the oscillator to clock-controlled processes. However, as more is discovered about the complexity of the various circadian systems, the distinctions among these elements become blurred because one protein may act in more than one element of the system. In this section, the similarities and differences of the molecular components among five different circadian systems—*Synechococcus elongatus*, *Neurospora crassa*, *Drosophila melanogaster, Arabidopsis thaliana,* and mammals (rodents)—are described and compared.

PROKARYOTIC CIRCADIAN CLOCK SYSTEMS

Until the 1980s, circadian clocks were believed to exist only in eukaryotes. Prokaryotes, with life spans much shorter than that of a full circadian cycle, were thought to have no need to maintain a 24-hour timing mechanism. However, investigation into the existence of two incompatible processes—oxygenic photosynthesis and oxygen-sensitive nitrogen fixation—that occur in some unicellular cyanobacteria led to the discovery that these processes are separated temporally by an endogenous timekeeping mechanism, with photosynthesis occurring during the day and nitrogen fixation at night (Mitsui et al. 1986). Further studies have shown that the circadian clock in cyanobacteria regulates the timing of amino acid uptake (Chen et al. 1991), global gene expression (Liu et al. 1995), chromosome condensation (Smith and Williams 2006), and cell division (Sweeney and Borgese 1989; Mori and Johnson 2000).

Synechococcus elongatus

Oscillator. The cyanobacterial central oscillator is composed of the KaiA, KaiB, and KaiC proteins. Accumulation of *kaiA* (expressed from its own promoter) and *kaiBC* (expressed as a dicistronic message from one promoter upstream of *kaiB*) mRNA occurs in a circadian manner with peak expression 12 hours after entering constant light conditions (Fig. 1) (Ishiura et al. 1998). Maximum levels of fluctuating KaiB and KaiC protein occurs 4–6 hours after peak mRNA abundance, whereas KaiA levels remain relatively constant throughout the circadian cycle (Xu et al. 2000). The *kai* genes and their protein products exhibit regulatory feedback (Fig. 1) (Ishiura et al. 1998), but this transcriptional/translational regulation appears to be a dispensable layer of reinforcement on a posttranslational clock in the cyanobacterium. Specific *cis* elements in the *kaiA* and *kaiBC* promoters are not required to preserve circadian regulation of gene expression (Xu et al. 2003; Ditty et al. 2005); instead, the cyanobacterial timing mechanism is controlled by posttranslational modifications to the KaiC protein.

An oscillation in KaiC phosphorylation occurs in the presence of only KaiA, KaiB, KaiC, and ATP in a temperature-compensated circadian rhythm in vitro (Nakajima et al. 2005). The beautiful simplicity of this test tube oscillator demonstrates the importance of the Kai-based posttranslational timing circuit and, along with promoter-replacement experiments in vivo, dismisses the requirement for the transcriptional/translational regulatory feedback loop in the cyanobacterial system.

*Present Address: Department of Biology, St. Ambrose University, Davenport, Iowa 52803

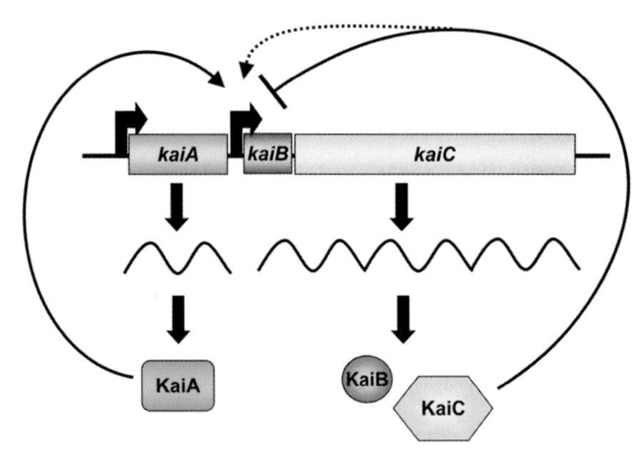

Figure 1. Regulation of the *kai* locus. Transcription of the *kaiA* and *kaiBC* genes is under the control of the circadian clock with peak expression levels that occur at subjective dusk under constant light conditions. One promoter (*bent arrow*) drives expression of the *kaiA* gene, and another drives transcription of both the *kaiB* and *kaiC* genes to create a *kaiBC* dicistron. Overexpression of KaiA causes an increase in the level of expression of *kaiBC*, whereas overexpression of KaiC inhibits its own expression; together, these data suggest that KaiA and KaiC are positive and negative regulators, respectively, of the *kaiBC* promoter. A basal level of KaiC is required for its own expression (*dotted line*), which implies an activating role for KaiC in maintaining *kaiBC* levels. However, this transcriptional regulation and any specific *cis* elements within the *kai* promoters are not required for proper clock function and are likely an additional layer of reinforcement to a posttranslational oscillator. (*Arrows* and *perpendicular lines*) Positive and negative regulation, respectively.

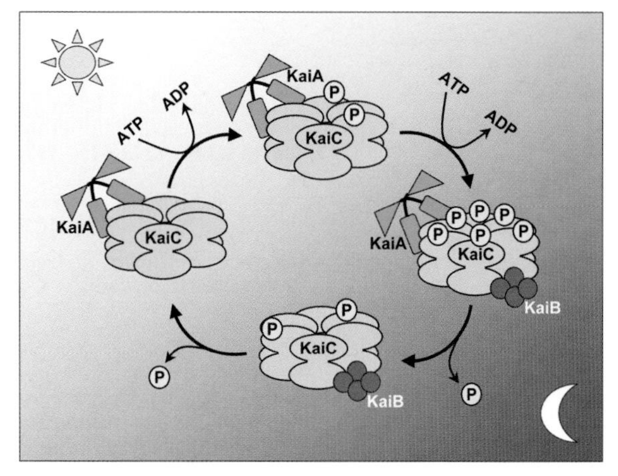

Figure 2. A posttranslational Kai oscillator. Daily rhythms in KaiC phosphorylation are aided by interactions with KaiA and KaiB. During the day, KaiA associates with KaiC to stimulate the autophosphorylation of KaiC. Near subjective dusk, KaiC is in a hyperphosphorylated state and is thought to undergo a conformational change that allows it to have a higher binding affinity for KaiB. Interaction with KaiB during the night results in the dephosphorylation of KaiC. Once in its hypophosphorylated state at subjective dawn, KaiC will again interact with KaiA and the cycle begins anew. (Reprinted, with permission, from Mackey and Golden 2007 [© Elsevier].)

KaiC forms an ATP-dependent homohexamer and autophosphorylates at adjacent serine and threonine residues (Pattanayek et al. 2004; Xu et al. 2004). The phosphorylation state of KaiC fluctuates in a circadian manner in vivo, yet this progressive phosphorylation of KaiC does not appear to lead to its rapid degradation (Iwasaki et al. 2002). During morning, KaiA dimers enhance the autophosphorylation of KaiC; KaiA and phosphorylated KaiC (KaiC~P) form a complex to maintain elevated levels of KaiC~P for the remainder of the day. At night, KaiB joins the KaiA/KaiC~P complex to negate the positive effect of KaiA on KaiC autophosphorylation and accelerate the dephosphorylation of KaiC~P to KaiC. The complex disassociates into its components before dawn and the cycle starts over for the next day (Fig. 2) (for review, see Golden and Canales 2003; Iwasaki and Kondo 2004; Williams 2006).

Input. No true photoreceptor has yet been identified in *S. elongatus* that is dedicated to transducing external light information to the central oscillator. Instead, the cyanobacterial clock system is sensitive to changes in cellular redox state, which reflects the flux of light driving photosynthesis (Fig. 3). The light-dependent period (LdpA) protein contains iron-sulfur clusters that allow LdpA to detect changes in the redox state of the cell, which are interpreted as changes in light intensity, to modulate the period of the internal oscillation (i.e., obey Aschoff's rule) (Katayama et al. 2003; Ivleva et al. 2005). Cells that lack *ldpA* maintain the shorter period associated with high light, regardless of actual light intensity. Resetting the phase of the rhythm in response to abrupt external stimuli, such as pulses of darkness or temperature, occur through the circadian input kinase (CikA) protein (Schmitz et al. 2000). Accumulation of CikA occurs during the night (Ivleva et al. 2006), and low levels of CikA protein are present at higher light intensities. Levels of CikA protein are locked at the low level that is associated with high light in strains deficient in *ldpA* (Ivleva et al. 2005). Degradation of CikA can be stimulated by the direct binding of a quinone analog 2,5-dibromo-3-methyl-6-isopropyl-*p*-benzoquinone (DBMIB) to the pseudo-receiver domain of the protein (Ivleva et al. 2006). DBMIB affects electron transport in the photosynthetic apparatus to produce an excess of electrons (reduction), which reflects the redox state of the cell. These results further implicate cellular metabolism and redox state as important factors for clock synchronization.

The period extender (Pex) protein binds to the *kaiA* promoter and likely represses expression of *kaiA* (Fig. 3) (Kutsuna et al. 1998; Takai et al. 2006a). Pex delays the internal oscillation when cells are subjected to a light/dark cycle in order to remain in-phase with that cycle (Takai et al. 2006a). Pex accumulates during the night and is undetectable in the light, which suggests that the repression of *kaiA* expression also has a role in delaying the phase of the oscillation in natural daily cycles.

Both CikA and LdpA have been shown to be part of the Kai protein complex (along with the output protein SasA, discussed below), and at least CikA influences the phosphorylation state of KaiC (Ivleva et al. 2005, 2006). The assembly and disassembly of this large heteromultimeric protein complex, termed the "periodosome," appear to be a driving force behind the cyanobacterial circadian system as a whole.

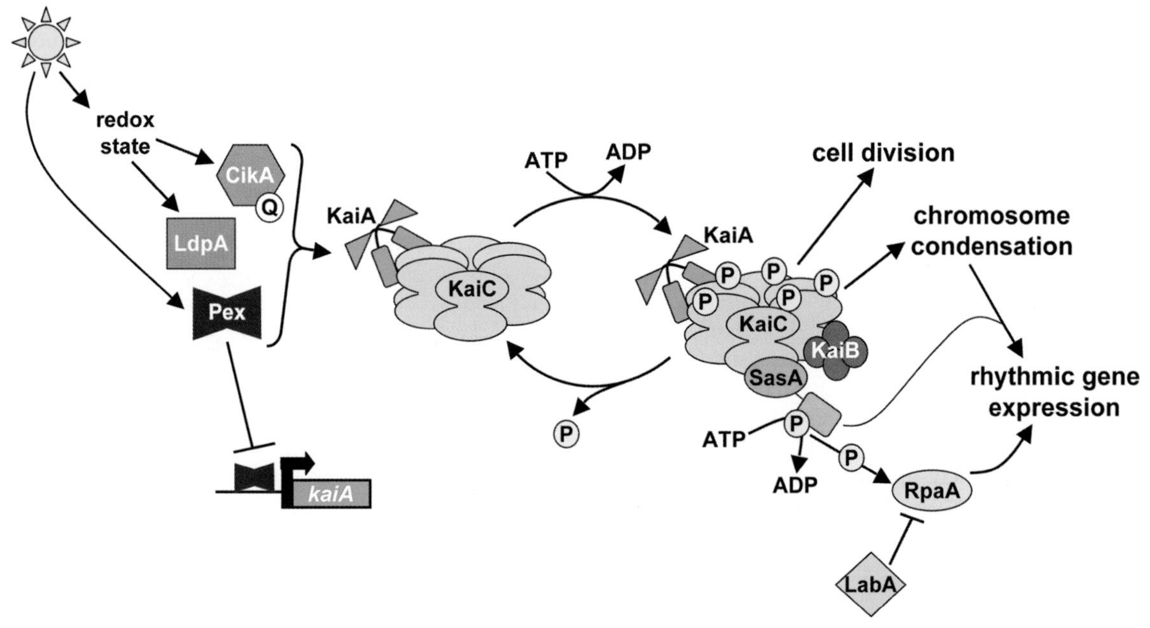

Figure 3. A clock model for the *S. elongatus* system. The Kai-based posttranslational oscillator is synchronized with the external light/dark cycle through input pathway components CikA, LdpA, and Pex. Changes in cellular redox state, which changes with light quantity, is sensed via iron-sulfur clusters of LdpA and direct binding of a quinone (Q) to CikA. Pex is predicted to repress the expression of *kaiA* through direct binding of the *kaiA* promoter. The oscillation of KaiC phosphorylation is achieved through the conflicting actions of KaiA, which stimulates KaiC autokinase activity, and KaiB, which decreases the positive effect of KaiA. Temporal information is transduced from KaiC to SasA. SasA transfers its phosphoryl group to RpaA, which is predicted to activate RpaA; LabA is thought to repress the activity of RpaA. Together, the *S. elongatus* clock controls many cellular activities, including cell division, chromosome condensation, and rhythmic gene expression. (*Arrows* and *perpendicular lines*) Positive and negative regulation, respectively. (Reprinted, with permission, from Mackey and Golden 2007 [© Elsevier].)

Output. The cyanobacterial circadian clock regulates global gene expression, such that every promoter tested drives expression of luciferase reporter genes in a 24-hour rhythm (for review, see Johnson 2004; Woelfle and Johnson 2006). This global control is likely the result of the Kai-dependent circadian rhythm in chromosome compaction that alters the availability of promoter regions to transcriptional machinery (Fig. 4) (Smith and Williams 2006). *Synechococcus* adaptive sensor (SasA) receives temporal information through direct interaction with KaiC in the periodosome complex, which stimulates the autophosphorylation of SasA (Fig. 3) (Smith and Williams 2006). Active, phosphorylated SasA transfers its phosphoryl group to the regulator of phycobiliosome-associated (RpaA) protein, which is predicted to act as a transcription factor to affect expression of downstream clock-controlled genes (Takai et al. 2006b). SasA is not required for rhythmic chromosome compaction, but it is necessary for information generated by the compaction to be translated into rhythmic promoter activity (Smith and Williams 2006).

The SasA/RpaA signal transduction pathway acts as the activation pathway; removal of either component lowers the overall levels of KaiC and decreases expression of bioluminescence reporters (Takai et al. 2006b). The low-amplitude and bright (LabA) protein is proposed to act in a repression pathway to feed temporal information

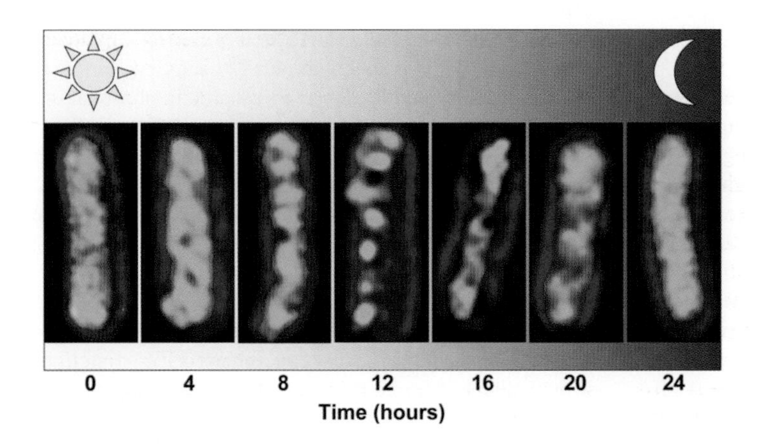

Figure 4. Chromosome compaction in *S. elongatus*. Deconvolved fluorescence microscopy images. (*Red*) Autofluorescence from *S. elongatus*; (*green*) DAPI-stained DNA of *S. elongatus* cells. Samples were taken at the time indicated on the *Y* axis during a 12 hours:12 hours light/dark cycle. The DNA begins in as decompacted chromosomes at 0 hour, arranges into distinct "nucleoid" regions in anticipation of the light-to-dark transition (12 hours), and returns to a diffuse state at 24 hours. (Figure courtesy of R.M. Smith and S.B. Williams, Department of Biology, University of Utah, Salt Lake City.)

through RpaA in an output pathway separate from that of SasA (Taniguchi et al. 2007). The connection between LabA and RpaA is likely indirect.

The cyanobacterial circadian clock persists in cells with doubling times much shorter (as fast as 6 hours) than a full circadian cycle (Kondo et al. 1997); yet, the population of cells remains in-phase with one another through unknown mechanisms (Mihalcescu et al. 2004). Cell division is gated by the endogenous circadian rhythm, such that there is a "forbidden" phase that stretches from late day to early-to-midnight (Mori et al. 1996), which is coincident with the formation of the periodosome and with the transition from a decondensed to condensed nucleoid.

S. elongatus does not fix nitrogen (Herrero et al. 2001), and without the need to separate photosynthesis and nitrogen fixation, the selective pressure to possess a robust circadian oscillator that regulates transcription is less obvious. Nonetheless, in a competitive environment, *S. elongatus* cells with a free-running period (FRP) that closely matches that of the external light/dark cycle possesses a growth advantage over strains whose FRP differs from that of the external cycle and over strains that lack a functional clock; however, the growth rates of cultures maintained individually are indistinguishable, even in different light/dark cycles (Ouyang et al. 1998).

EUKARYOTIC CIRCADIAN CLOCK SYSTEMS

Although the end result of maintaining an internal timing mechanism is the same for both prokaryotic and eukaryotic organisms, many differences exist between the two broad categories. Regulation of transcription by the clock is widespread, yet the global regulation of transcription from all promoters is unique to the cyanobacterium. Likewise, posttranslational mechanisms are key regulators in maintaining circadian timing in each system, yet the posttranslational oscillator of *S. elongatus* is unique and will likely remain unique given the importance of promoter elements in clock gene function in the eukaryotic systems.

One main difference is the means by which oscillators are reset in response to changes in the daily environment. Photoreceptors dedicated to receiving light information for the oscillator have not been identified in cyanobacteria; instead, the cells appear to interpret changes in cellular redox state as an indirect measurement of light intensity. The eukaryotic organisms for which the molecular mechanisms of the endogenous oscillator have been described possess photoreceptors that sense light to directly influence the steady-state level of core oscillator components to alter the phase angle of the rhythm.

The presence of a nucleus in eukaryotic organisms provides, perhaps not surprisingly, another critical difference in their internal timing mechanisms as compared to the prokaryotic system. Compartmentalization and translocation between the cytoplasm and nucleus are critical processes in the progression of the eukaryotic clock. The internal oscillators of eukaryotic systems are controlled by a common mechanism: transcriptional/posttranslational regulatory feedback loops (Fig. 5) (for review, see Bell-Pedersen et al. 2005). In these loops, genes that encode the positive effectors are transcribed and their protein products translocate to the nucleus to activate transcription of genes that encode the negative elements. As the concentration of the negative elements increases, they inhibit the activity of the positive effectors, which, in turn, decreases the level of expression of the negative elements. Progressive phosphorylation of the negative elements induces degradation, which leads to a decrease in their relative amounts. This decline relieves the inhibition of the positive effector proteins and the cycle can begin anew. An additional feedback loop exists in some oscillators that includes activating and inhibiting elements that compete for availability to the promoter region of (at least) one of the genes that encodes a positive effector protein. The combined, interlocking transcriptional/posttranslational feedback loops within the circadian oscillator require approximately 24 hours to complete one cycle.

Drosophila melanogaster

A working model of the molecular components of the *Drosophila* system is depicted in Figure 6 (for review, see Bell-Pedersen et al. 2005; Hardin 2006; Taghert and Shafer 2006).

Oscillator. Two interlocking feedback loops have been described in the *Drosophila* circadian oscillator. The first loop consists of the positive elements CLOCK (CLK) and CYCLE (CYC), which are transcription factors that contain basic-helix-loop-helix (bHLH) DNA-binding domains, PER-ARNT-SIMS (PAS) dimerization domains, and the negative elements PERIOD (PER) and TIMELESS (TIM). The *Clk* gene is rhythmically transcribed with peak expression occurring near dawn, whereas *cyc* mRNA levels remain constant. At midday, CLK/CYC heterodimers bind to E boxes in the promoter regions of the *per* and *tim* genes (as well as those of *vrille* [*vri*] and *PAR domain 1ε* [*Pdp1ε*], discussed below) to activate their transcription. The four protein products of these genes influence the activity of CLK protein and *Clk* expression levels to create the regulatory feedback loops necessary for generating 24-hour internal time.

Activation of *per* and *tim* expression by CLK/CYC leads to increased mRNA levels with peak accumulation occurring during the early evening. However, peak protein levels do not occur until late night due to the phosphorylation-induced degradation of PER protein by the kinase DOUBLE-TIME (DBT). DBT is a homolog of casein kinase 1ε (CK1ε) in mammals (discussed below). Dephosphorylation of PER occurs by protein phosphatase 2A (PP2A). These opposing activities alter PER stability, which is also dependent on its interactions with DBT and TIM. The PER/DBT complex binds TIM, which stabilizes PER by blocking DBT-dependent phosphorylation. Nuclear translocation of PER/DBT and TIM occurs independently and is promoted by casein kinase 2 (CK2) and Shaggy (SGG), which phosphorylate PER and TIM, respectively. After nuclear translocation, PER/DBT and PER/DBT/TIM complexes can inhibit CLK function through direct interaction and DBT-dependent phosphorylation of CLK, which helps to release the CLK/CYC het-

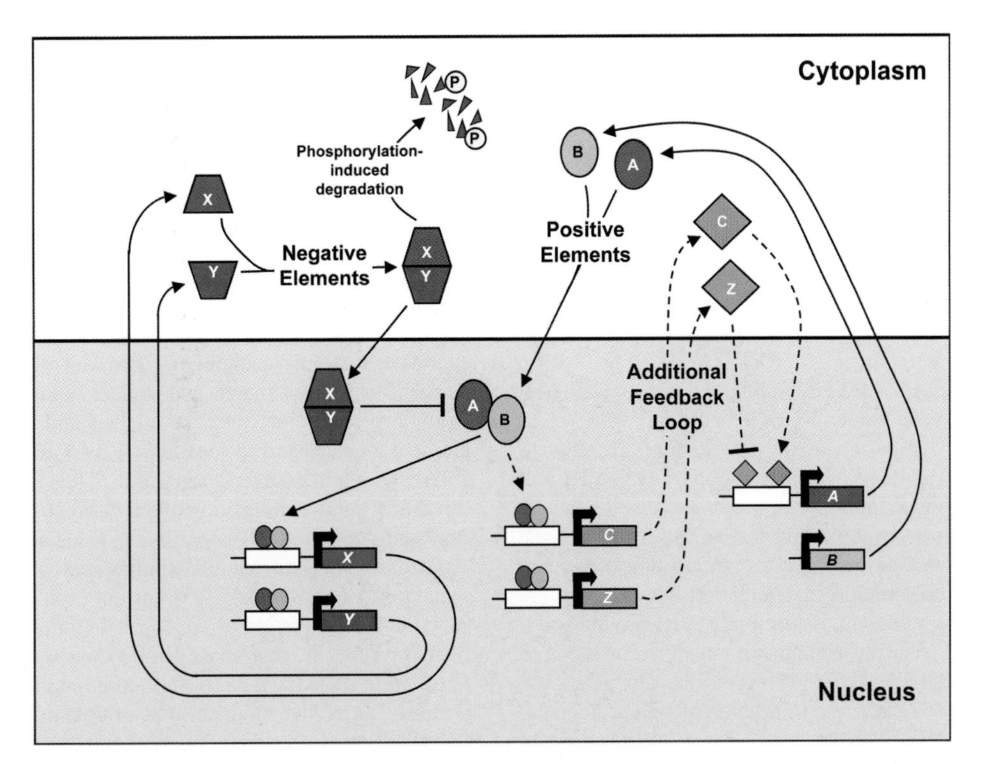

Figure 5. Transcription/translation regulatory feedback loops: A common theme in the generation of eukaryotic circadian rhythms. Many circadian systems are composed of positive and negative elements that are involved in feedback loops. Positive elements (*ovals*) interact and bind to upstream elements (*white rectangles*) in the promoter regions (*bent arrows*) of genes that encode negative elements to activate their expression. These negative components (*tetrahedrons*) interact to inhibit the activity of the positive elements, which ultimately leads to a decrease in expression of the negative elements. Progressive phosphorylation (*P*) of negative elements leads to their degradation and a relief of the inhibition of the positive effectors to allow the cycle to start again. Other components (*diamonds*) may be present that bind to the promoters of genes that encode the positive elements to comprise an additional feedback loop. (*Arrows* and *perpendicular lines*) Positive and negative regulation, respectively.

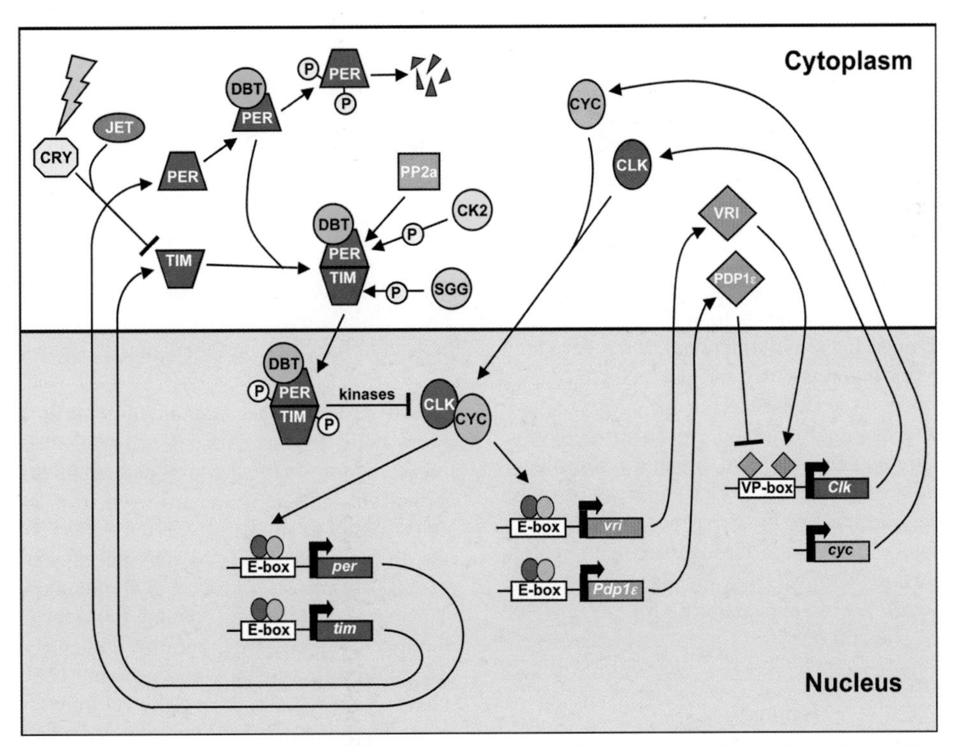

Figure 6. Model for the circadian clock of *D. melanogaster*. See Figure 5 for basic mechanism and text for details specific to the *Drosophila* oscillator. (*Vertical ovals*) Positive elements CLK and CYC; (*white rectangles*) E-box elements; (*bent arrows*) promoters; (*tetrahedrons*) negative elements PER and TIM; (*circles*) kinase proteins DBT, SGG, and CK2; (*square*) PP2a phosphatase; (*octagon*) CRY; (*horizontal oval*) JET; (*diamonds*) additional feedback components VRI and PDP1ε. (*P*) Phosphorylation. (*Arrows* and *perpendicular lines*) Positive and negative regulation, respectively.

erodimers from the E-box promoter sequences and reduce expression of *per* and *tim* (as well as that of *vri* and *Pdp1ε*). The progressive phosphorylation of PER by DBT ultimately results in the degradation of PER and the trough in its protein accumulation occurs in early morning. In contrast, there is a rhythm in the phosphorylation of CLK protein, but the overall protein level does not cycle. Coincident with the timing of PER degradation, TIM protein levels decrease due to light-induced degradation via the blue light photoreceptor CRYPTOCHROME (CRY). With the PER/DBT/TIM inhibition of CLK lifted and new CLK protein being made, the cycle begins anew.

Additional regulation of the *Clk* locus is thought to occur through a second feedback loop via the VRI and PDP1ε bZIP transcription factor proteins, as well as a predicted constitutive activator protein. In the nucleus, CLK/CYC heterodimers activate transcription of *vri* and *Pdp1ε* with peak mRNA accumulation occurring in the early night and midnight, respectively. VRI protein accumulates quickly in the cytoplasm, enters the nucleus to bind the VRI/PDP1ε box (VP box) in the *Clk* promoter, and inhibits *Clk* transcription. As levels of VRI protein dissipate and levels of PDP1ε protein rise, PDP1ε is thought to displace VRI from the VP box to enhance expression of *Clk* in the late night (which ultimately results in peak *Clk* mRNA levels at dawn).

Input. Entrainment of the *Drosophila* clock is mediated through (at least) the CRY-dependent degradation of TIM (Ceriani et al. 1999; Emery et al. 2000). In constant darkness, when TIM protein levels are rising, brief exposure to light results in a decrease in TIM protein to trough levels and an overall phase delay, whereas a light pulse during the time when TIM protein is declining causes a phase advance. Light exposure causes a conformational change in CRY that allows it to bind TIM. TIM is phosphorylated on at least one tyrosine residue and interacts with the JETLAG (JET) protein, which targets TIM for ubiquitination and degradation by the proteasome pathway (Koh et al. 2006).

Output. The regulatory feedback loops that comprise the molecular oscillator in flies is found in various tissues (e.g., head, wings, legs, and antennae), each of which appear to function autonomously and can be directly reset by light (for review, see Bell-Pedersen et al. 2005). These oscillators control pupal eclosion, courtship behavior, locomotor activity behavior, and olfactory responses. Although there does not appear to be a hierarchy of "master" and "slave" pacemakers, there may be communication among oscillators in the brain through the neuropeptide pigment-dispersing factor (PDF), which is secreted in the dorsal brain with a circadian pattern (for review, see Taghert and Shafer 2006).

Mammals

The current model of the molecular components of the mammalian clock system is portrayed in Figure 7 (for review, see Gachon et al. 2004; Ko and Takahashi 2006; Kuhlman and McMahon 2006).

Molecular oscillator. Like the *Drosophila* system, the mammalian clock is described as having two interlocking regulatory feedback loops. In the first loop, the positive effectors are CLOCK and BMAL1 (brain and muscle ARNT-like protein 1), which is the mammalian homolog of *Drosophila*'s CYC. CLOCK and BMAL1 are bHLH-PAS domain-containing transcription factors that form heterodimers in the cytoplasm, enter the nucleus, and bind to the E-box sequence in the promoters of the genes that encode the negative elements. BMAL1 contains both nuclear localization signals and nuclear export signals that allow it to shuttle between the nucleus and cytoplasm to promote the nuclear translocation of CLOCK (Kwon et al. 2006). In mammals, the negative limb of this feedback loop consists of some combination of the *Period* (in mice, *Per1, Per2*, and *Per3*) and *Cryptochrome* (*Cry1* and *Cry2*) genes. To date, the role of a mammalian protein that is similar in sequence to *Drosophila*'s TIM has not been firmly established in the generation of circadian rhythms. Instead, the CRY proteins in mammals have taken on the role of *Drosophila*'s TIM; CRYs form heterodimers with the PER proteins, enter the nucleus, and inhibit the activity of CLOCK/BMAL1 complexes. Without CLOCK/BMAL1 to activate transcription of the *Per* and *Cry* genes, levels of *Per* and *Cry* transcripts, and their respective protein products, decline. Degradation of PER and CRY is induced by progressive phosphorylation by casein kinase 1δ (CK1δ) and CK1ε, which leads to a release of the inhibition of CLOCK/BMAL1, and the cycle starts over. Interestingly, in contrast to *Drosophila*, the photoreceptor properties of CRY proteins in mammals have not yet been demonstrated.

A second feedback loop regulates the expression of the *Bmal1* gene. CLOCK/BMAL1 heterodimers enter the nucleus and bind to the E-box sequences in the promoter regions of genes that encode the retinoic-acid-related orphan nuclear receptors REV-ERBα and RORα, which compete for the ROR element (RORE) in the *Bmal1* promoter. The ROR family (α, β, and γ) of proteins activates *Bmal1* expression, and REV-ERBα and REV-ERBβ repress expression of *Bmal1*. This regulation of *Bmal1* maintains a robust rhythm of activity in vivo. Proteins with sequences similar to those of the ROR and REV-ERB families exist in *Drosophila*, but whether they participate in the clock system is not yet known.

SCN as master pacemaker. In contrast to the autonomous oscillators that exist in *Drosophila* cells and tissues, the oscillations that exist in tissues in the mammalian circadian clock system are arranged in a hierarchy. The suprachiasmatic nucleus (SCN) in the anterior hypothalamus of the brain is the "master" pacemaker. Removal of the SCN results in a loss of rhythmic outputs, including sleep/wake cycle, wheel-running behavior, hormone production, core body temperature, and accumulation of clock-controlled mRNA and protein. The master pacemaker in the SCN is reset daily by light signals that are transmitted through the retinohypothalamic tract and retinal ganglion cells to cause cyclic-AMP response element binding (CREB) protein-mediated induction of *Per1* transcription (discussed in Part IB). The SCN then transmits these environmental timing cues to the peripheral "slave"

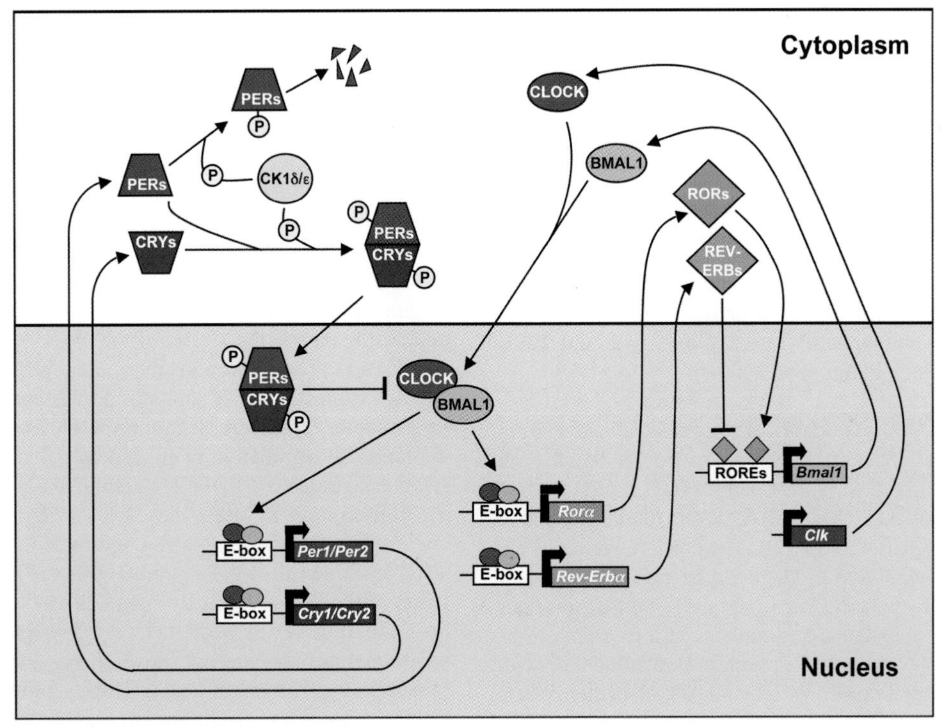

Figure 7. Model for the circadian clock in mammals. See Figure 5 for basic mechanism and text for details specific to the mammalian oscillator. (*Horizontal ovals*) Positive elements CLOCK and BMAL1; (*white rectangles*) E-box elements; (*bent arrows*) promoters; (*tetrahedrons*) negative elements PERs and CRYs; (*circle*) kinase proteins CK1δ and CK1δ; (*diamonds*) additional feedback components RORs and REV-ERBs. (*P*) Phosphorylation. (*Arrows* and *perpendicular lines*) Positive and negative regulation, respectively.

oscillators in mammalian tissues via polysynaptic and humoral pathways to coordinate and synchronize their oscillations with that of the master SCN oscillator (discussed in Part IB).

Output via histone modification. Transcriptional activation of targets of the CLOCK/BMAL1 heterodimer results from the histone acetyltransferase (HAT) activity of CLOCK itself (Doi et al. 2006). Histone acetylation promotes transcription through the modification of histones to allow for the "opening" of condensed chromatin, which provides access to the transcriptional machinery (Fig. 8). The HAT activity of CLOCK is necessary for the activation of transcription of clock genes *Per* and *Cry* and is thus

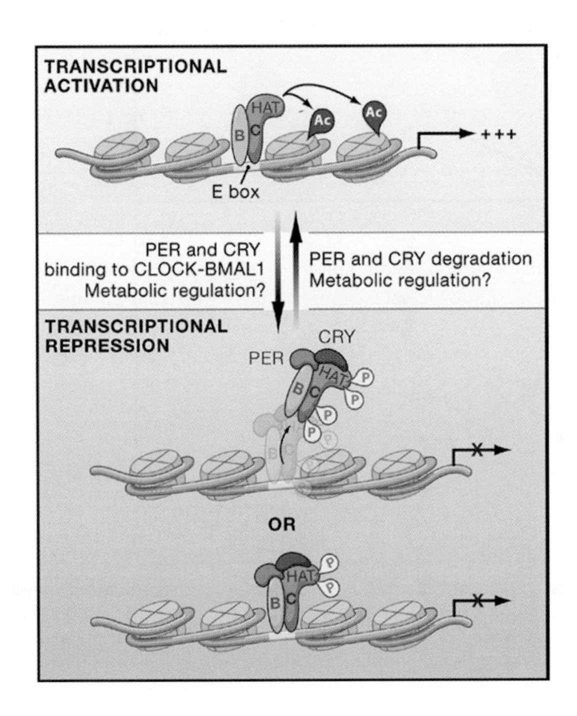

Figure 8. Model for the regulation of rhythms in CLOCK-BMAL1-dependent transcription. Transcriptional activation requires E-box binding by CLOCK(C)-BMAL1(B) heterodimers and CLOCK histone acetyltransferase (HAT)-dependent acetylation of histone H3 (Ac). Transcriptional repression is mediated by several events. Frist, PER and CRY bind to CLOCK-BMAL1. Formation of this complex can inhibit CLOCK HAT activity by promoting CLOCK phosphorylation (P) or inducing a conformation change in CLOCK-BMAL1. Binding of PER and CRY to CLOCK-BMAL1 may or may not cause the dissociation of the complex from E boxes. In either case, loss of CLOCK HAT events start with the degradation of PER and CRY. Phosphorylated CLOCK is then either dephosphorylated or degraded and resynthesized. CLOCK forms heterodimers with newly synthesized BMAL1, this complex binds to E boxes, and CLOCK HAT acetylates histones to activate transcription. The metabolic state of the cell may also modulate transcriptional activity by promoting either transcriptional activation or repression. (Figure and legend reprinted, with permission, from Hardin and Yu 2006 [© Elsevier].)

believed to be essential for the generation and maintenance of endogenous circadian timing in mammals. The sequence similarity among CLOCK homologs in other mammals and insects (including *Drosophila*) hints at a conserved mechanism for transcriptional regulation of core clock genes by posttranslational histone modifications.

Neurospora crassa

The current model for the internal timekeeping mechanism of *Neurospora crassa* is depicted in Figure 9 (for review, see Dunlap and Loros 2006; Liu and Bell-Pedersen 2006; Vitalini et al. 2006).

Oscillator. The oscillator of the *Neurospora* system consists of a single, well-characterized feedback loop composed of the positive elements WHITE COLLAR-1 (WC-1) and WHITE COLLAR-2 (WC-2) and the negative element FREQUENCY (FRQ). The positive elements are PAS-domain-containing, GATA-type transcription factors encoded by the *wc-1* and *wc-2* genes. Neither *wc-1* nor *wc-2* mRNA abundance accumulates in a rhythmic fashion, yet WC-1, but not WC-2, protein levels fluctuate with a near 24-hour period. WC-1 and WC-2 proteins heterodimerize in the cytoplasm to form the WHITE COLLAR COMPLEX (WCC). This complex translocates to the nucleus and binds directly to the Clock box (C box) of the *frq* promoter to activate *frq* expression. The level of *frq* mRNA increases and peaks midday; peak levels of FRQ protein occur 4–6 hours after peak mRNA levels. FRQ protein forms homodimers that complex with the FRQ-interacting RNA helicase (FRH). Stability of FRQ is determined by the conflicting actions of numerous kinases and phosphatases. The kinase proteins casein kinase I and II (CKI, CKII), calcium/calmodulin-dependent protein kinase 1 (CAMK1), and Period-4 (PRD-4) progressively phosphorylate FRQ, which facilitates the interaction of hyperphosphorylated FRQ with the ubiquitin ligase FWD-1 to target FRQ for degradation; the protein phosphatases 1 (PP1) and 2A (PP2A) dephosphorylate FRQ to allow for its accumulation. The FRQ/FRH complex enters the nucleus to inhibit the action of the WCC through phosphorylation. The hyperphosphorylated WCC is inactive and no longer capable of supporting transcription from the *frq* promoter. The decrease in *frq* mRNA level, along with the targeted degradation of FRQ protein and dephosphorylation of the WCC by PP2A, results in a reactivation of *frq* transcription to allow the cycle to begin again. It is unclear how these events are coordinated. FRQ protein also has a role (directly or indirectly) in the positive regulation of the WC proteins by stabilizing *wc-2* mRNA and up-regulating WC-1 protein levels through a posttranscriptional mechanism (not shown in Fig. 9); this positive feedback loop is thought to add stability and robustness to the *Neurospora* internal timing oscillation. FRQ, however, is not required for the rhythmic accumulation of WC-1 protein. The basis of the WC-1 cycle is unclear and may be related to a second oscillator.

At least one additional circadian oscillator exists in *Neurospora* that does not require the FRQ protein, i.e., a

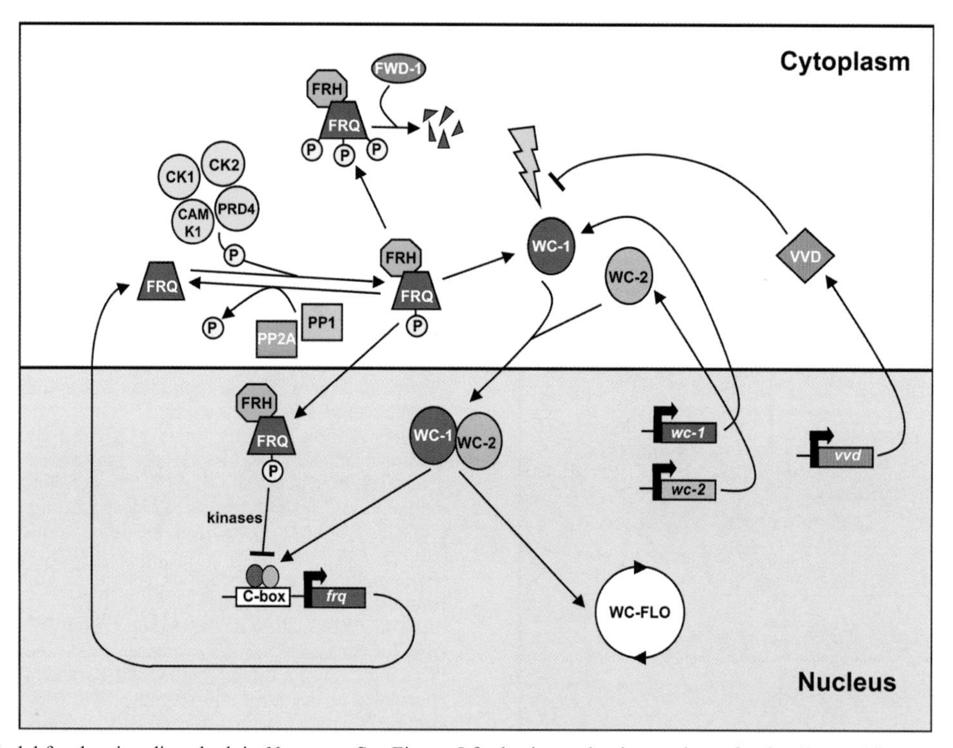

Figure 9. Model for the circadian clock in *N. crassa*. See Figure 5 for basic mechanism and text for details specific to the *Neurospora* oscillator. (*Vertical ovals*) Positive elements WC-1 and WC-2; (*white rectangles*) Clock-box elements; (*bent arrows*) promoters; (*tetrahedron*) negative element FRQ; (*circles*) kinase proteins CK1, CK2, CAMK1, and PRD4; (*squares*) PP1 and PP2a phosphatases; (*octagon*) FRH; (*horizontal oval*) FWD-1; (*diamond*) additional feedback component VVD. (*P*) Phosphorylation. (*Arrows* and *perpendicular lines*) Positive and negative regulation, respectively.

FRQ-less oscillator (FLO). The WC-1 and WC-2 proteins function in one FLO to form the WC-FLO (Fig. 9). This oscillator controls temperature-compensated molecular rhythms in mRNA accumulation of at least one clock-controlled gene, *ccg-16*, with a period near 24 hours under constant conditions in the absence of *frq*.

Input. Time-of-day information can be perceived by the *Neurospora* circadian system through light or temperature cues. WC-1 bound to the flavin FAD serves as a blue light photoreceptor that interacts with WC-2 to form a "light-response WC complex" that binds to light-response elements in the promoters of light-regulated genes, including those in the *frq* promoter, to activate their transcription. The abrupt change in the *frq* transcript is reflected in protein accumulation such that a light pulse during the time when the *frq* transcript is increasing will cause a phase advance (Fig. 10A), and the same stimulus given when the *frq* transcript is on the decline will result in a phase delay (Fig. 10B). The protein VIVID (VVD) helps to dampen the light-responsive increase in FRQ protein at dawn in natural light/dark cycles such that the clock can maintain its internal autoregulatory feedback loop during the day and reset its clock at dusk each day by influencing *frq* mRNA turnover.

Temperature also resets the phase of the *Neurospora* rhythm through posttranscriptional changes that affect FRQ protein level and protein form (long or short). Absolute levels of FRQ protein are higher at warmer temperatures than at cooler temperatures. In response to a temperature step-up, the current level of FRQ becomes the trough of the "new" rhythm; a temperature step-down results in the current level of FRQ becoming the "new" peak (Fig. 10C). Either condition resets the clock to a new phase as determined by the external temperature.

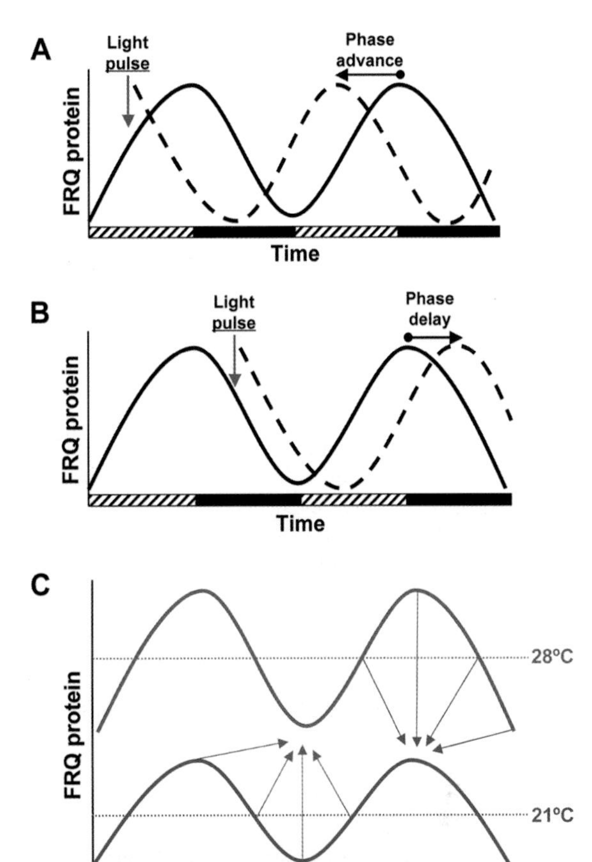

Figure 10. Light and temperature resetting of the *N. crassa* circadian oscillator. FRQ protein levels fluctuate in constant darkness with peak expression levels occurring near the transition between subjective day (*hatched bars* on *X* axis) and subjective night (*black bars* on *X* axis). Exposure to light causes an increase in the expression of the *frq* gene and subsequent FRQ protein, such that a light pulse during the time when FRQ protein is rising will cause a phase advance as in *A* and a light pulse during the time when FRQ protein is declining will cause a phase delay as shown in *B*. (*C*) FRQ protein levels are higher at warmer temperatures (*red curve*) than at cooler temperatures (*blue curve*). After a temperature step-up (from 21°C to 28°C), the current level of FRQ becomes the trough of the new rhythm (*red arrows*). After a temperature step-down (from 28°C to 21°C), the current level of FRQ protein becomes the trough of the resulting rhythm (*blue arrows*).

Output. The overt rhythm of asexual spore formation (conidiation) is the best-characterized output from the *Neurospora* circadian system. However, many other processes are under clock control including, but not limited to, lipid metabolism, development, CO_2 evolution, enzyme activity, and the rhythmic regulation of mRNA and protein accumulation. To date, more than 180 candidate clock-controlled genes have been identified (Vitalini et al. 2006). Recent work has implicated the clock in regulation of components of cellular mitogen-activated protein kinase (MAPk) signaling pathways that sense and respond to environmental stress. MAPk pathways are ubiquitous in eukaryotic systems and clock control over these universal regulatory systems would allow for numerous downstream genes to be affected in a circadian manner.

Arabidopsis thaliana

The interlocking molecular feedback loops that comprise the *Arabidopsis* circadian system are portrayed in Figure 11 (for review, see Mizuno and Nakamichi 2005; Gardner et al. 2006; McClung 2006).

Oscillator. The *Arabidopsis* circadian oscillator contains (at least) three interconnected feedback loops, each of which includes the MYB-like transcription factors LATE ELONGATED HYPOTCOTL (LHY) and CIRCADIAN CLOCK ASSOCIATED 1 (CCA1). In the first loop, the positive element TIMING OF CAB EXPRESSION 1 (TOC1; also known as PRR1) activates transcription of *LHY* and *CCA1*, whose protein products translocate to the nucleus to bind to the evening element (EE) of the *TOC1* promoter and directly inhibit its expression. As the level in *TOC1* mRNA, and subsequently TOC1 protein, declines, CCA1 and LHY levels also decrease, which relieves the inhibition on the *TOC1* promoter and the cycle can start again with peak TOC1 abundance occurring at subjective dawn. However, genetic evidence suggests that the TOC1/CCA1/LHY loop is

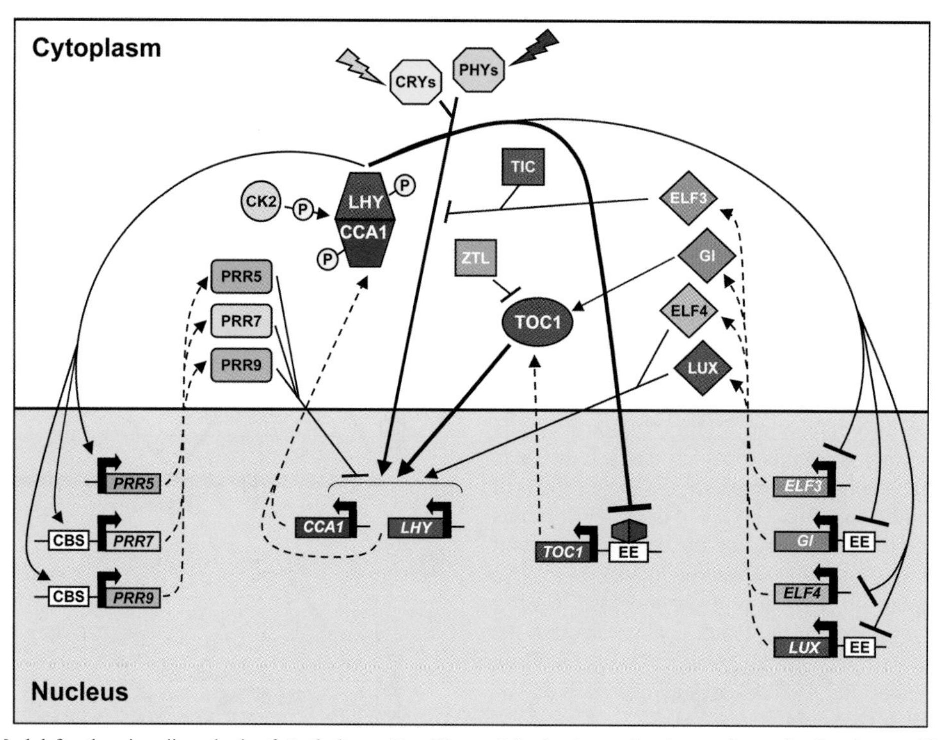

Figure 11. Model for the circadian clock of *A. thaliana*. See Figure 5 for basic mechanism and text for details specific to the three interlocking *Arabidopsis* feedback loops. Loop 1 is composed of positive element TOC1 (*horizontal oval*) and negative elements LHY and CCA1 (*tetrahedrons*) that bind to the evening element (EE; *white rectangle*) of the *TOC1* promoter. Loop 2 includes LHY and CCA1 as positive elements that activate transcription by binding to the CCA1-binding site (CBS; *white rectangles*) element of the *PRR5/7/9* genes, whose protein products negatively affect *CCA1* and *LHY* expression. Loop 3 contains LHY/CCA1 negatively inhibiting expression of a subset of genes that encodes the proteins ELF3, GI, ELF4, and LUX (*diamonds*), which feed back to affect the expression or activity of the LHY, CCA1, and TOC1 proteins. Additional components include CK2 kinase (*circle*), CRYs and PHYs (*octagons*), and inhibitory proteins TIC and ZTL (*squares*). (*P*) Phosphorylation. (*Arrows* and *perpendicular lines*) Positive and negative regulation, respectively.

incomplete and that the GIGANTEA (GI) protein is likely an additional component in this feedback loop that is involved in the activation of *TOC1*. The promoter of *GI* contains multiple EEs that would allow for inhibition by LHY/CCA1 to regulate its expression. Both LHY and CCA1 are phosphorylated by casein kinase II (CK2), which is necessary for clock function in vivo, but its role in the degradation of these proteins is not fully understood. LHY is targeted for degradation by the proteins DE-ETIOLATED 1 (DET1) and CONSTITUTIVELY PHOTOMORPHOGENIC 1 (COP1); TOC1 is targeted for degradation by a ZEITLUPE (ZTL)-dependent mechanism.

The second loop includes three pseudo-response regulators (*PRR5, PRR7*, and *PRR9*), whose respective proteins have distinct and nonoverlapping roles, but together are involved in the inhibition of *LHY* and *CCA1*, and activation of *TOC1* expression. CCA1, in turn, binds directly to the CCA1-binding site (CBS) in the promoters of *PRR7* and *PRR9* to activate their transcription.

The precise role of the PRR proteins in the clock is still unclear because they appear to be involved in light and/or temperature entrainment in addition to their role in maintaining internal timing. TOC1 (PRR1) likely has a critical role in regulating gene expression because it has been shown to interact with a number of bHLH transcription factors, including PHYTOCHROME INTERACTING FACTOR proteins (PIF3, PIF4) that are known to interact with PHYB and PIF3-LIKE proteins (PIL1, PIL2, PIL5, PIL6) that are involved in light signal transduction.

In a third loop, the MYB transcription factor LUX ARRYTHMO (LUX) activates expression of *CCA1* and *LHY*, which bind to the EE in the *LUX* promoter to inhibit its expression. Additionally, EARLY FLOWERING 4 (ELF4) is required for rhythmic expression of *CCA1* and *LHY*, but not *TOC1*, and transcription of *ELF4* is repressed by CCA1/LHY.

Input. Entrainment of the *Arabidopsis* clock occurs primarily through light. Red light responses occur through PHYTOCHROME proteins PHYA, PHYB, PHYD, PHYE, and likely PHYC. Each of the *PHY* mRNAs fluctuate in a circadian fashion, but only abundance of PHYA, PHYB, and PHYC proteins oscillate, which may account for the varying response to red light throughout the circadian cycle. The *PRR9* gene is rapidly and transiently induced by white or red light in a phytochrome-dependent manner; PRR9 interacts with TOC1 through their pseudo-receiver domains, which provides additional evidence for the involvement of the PRR family in light signal transduction pathways. Blue light is transduced through PHYA as well as the CRYPTOCHROME (CRY1 and CRY2)

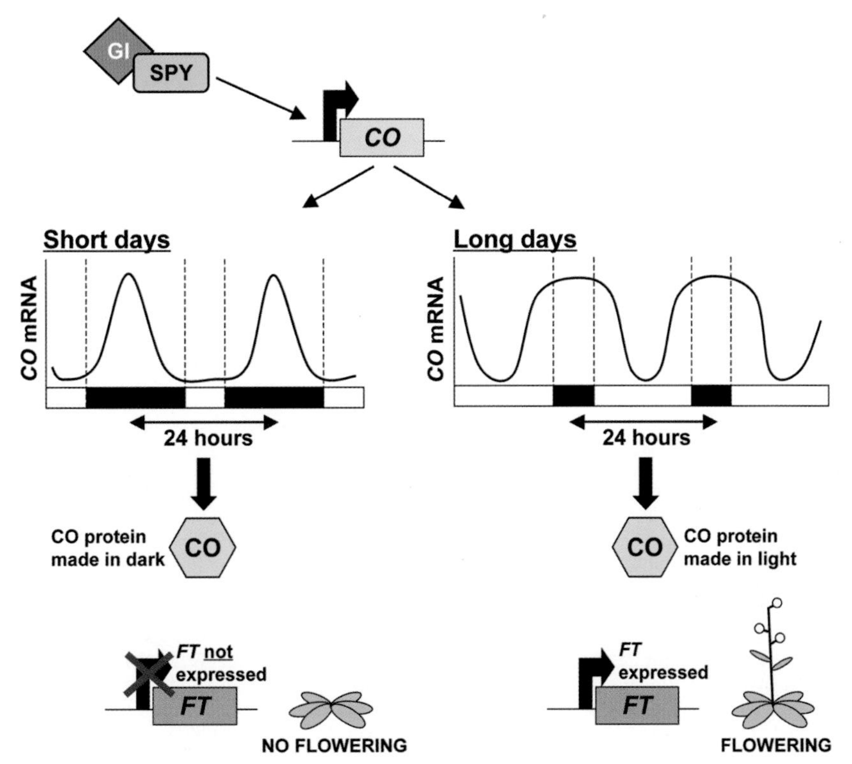

Figure 12. Circadian regulation of flowering in *A. thaliana*. During the short days of winter, the peak of *CO* mRNA and subsequent CO protein occurs during the night and does not promote the expression of the *FT* gene; no flowering occurs. During the long days of summer, the pattern of *CO* expression changes such that the peak of expression is more broad. This alteration results in high *CO* expression in the early and late day. CO protein, which is activated by light, is active during these long days and activates expression of *FT*, which leads to flowering. (Figure and legend modified, with permission, from Putterill et al. 2004 [© Wiley-Liss, a subsidiary of John Wiley & Sons].)

proteins. ELF3 and TIME FOR COFFEE (TIC) abrogate light input to the clock. Specifically, ELF3 gates light input at dusk to allow for increased sensitivity by the clock to light at dawn.

Output. Temporal information from the oscillator controls biochemical/physiological pathways that regulate photosynthesis, cotyledon and leaf movement, rate of hypocotyl elongation, stomatal movement, and rhythmic accumulation of mRNA and protein that are involved in these processes. Although molecular output pathways in most organisms are not clearly understood, circadian regulation of the flowering pathway in *Arabidopsis* has been defined (Fig. 12) (for review, see Putterill et al. 2004). GI and SPINDLY (SPY) interact to activate circadian-regulated expression of *CONSTANS* (*CO*). In short days, a narrow peak of *CO* occurs during the night; in long days, the peak of *CO* becomes more broad to allow the peak to extend within the light period at dawn and in late afternoon. Because CO protein is activated by light, this coincidence of peak expression during the light phase produces sufficient active CO to induce FLOWERING LOCUS T (FT), which interacts with the bZIP transcription factor FD to activate expression of genes necessary for the transition from vegetative growth to flowering.

The enhanced fitness of plants that maintain internal time has been demonstrated in a number of different experiments. *Arabidopsis* plants with an endogenous circadian period that closely matches that of a given light/dark cycle produce more chlorophyll and fix more carbon (i.e., possess a photosynthetic advantage) than mutant plant lines that generate a period that is shorter or longer than the environmental cycle (Dodd et al. 2005). Plant lines that exist in different geographic regions are subjected to different light/dark cycles. The natural variation in the circadian timing of the plants correlates with the day length at the respective latitudes (Michael et al. 2003). This correlation suggests the evolutionary importance of tightly regulating internal timing to match the surrounding environment. Additional evidence for the adaptive advantage of maintaining an endogenous timekeeper is seen in plants that overexpress the core clock component CCA1 or LHY; these plants no longer regulate transcription in an anticipatory manner of light-to-dark transitions (Green et al. 2002). Consequently, these plants flower under long-day conditions later than their wild-type counterparts and are less viable under short-day conditions.

ACKNOWLEDGMENTS

I thank D. Bell-Pedersen, V.M. Cassone, S.S. Golden, P.E. Hardin, C.R. McClung, M.W. Vitalini, and M. Zatz for careful reading of sections of this manuscript.

REFERENCES

Bell-Pedersen D., Cassone V.M., Earnest D.J., Golden S.S., Hardin P.E., Thomas T.L., and Zoran M.J. 2005. Circadian rhythms from multiple oscillators: Lessons from diverse organisms. *Nat. Rev. Genet.* **6:** 544.

Ceriani M.F., Darlington T.K., Staknis D., Mas P., Petti A.A., Weitz C.J., and Kay S.A. 1999. Light-dependent sequestration of TIMELESS by CRYPTOCHROME. *Science* **285:** 553.

Chen T.-H., Chen T.-L., Hung L.-M., and Huang T.-C. 1991. Circadian rhythm in amino acid uptake by *Synechococcus* RF-1. *Plant Physiol.* **97:** 55.

Ditty J.L., Canales S.R., Anderson B.E., Williams S.B., and Golden S.S. 2005. Stability of the *Synechococcus elongatus* PCC 7942 circadian clock under directed anti-phase expression of the *kai* genes. *Microbiology* **151:** 2605.

Dodd A.N., Salathia N., Hall A., Kevei E., Toth R., Nagy F., Hibberd J.M., Millar A.J., and Webb A.A. 2005. Plant circadian clocks increase photosynthesis, growth, survival, and competitive advantage. *Science* **309:** 630.

Doi M., Hirayama J., and Sassone-Corsi P. 2006. Circadian regulator CLOCK is a histone acetyltransferase. *Cell* **125:** 497.

Dunlap J.C. and Loros J.J. 2006. How fungi keep time: Circadian system in *Neurospora* and other fungi. *Curr. Opin. Microbiol.* **9:** 579.

Emery P., Stanewsky R., Hall J.C., and Rosbash M. 2000. A unique circadian-rhythm photoreceptor. *Nature* **404:** 456.

Gachon F., Nagoshi E., Brown S.A., Ripperger J., and Schibler U. 2004. The mammalian circadian timing system: From gene expression to physiology. *Chromosoma* **113:** 103.

Gardner M.J., Hubbard K.E., Hotta C.T., Dodd A.N., and Webb A.A. 2006. How plants tell the time. *Biochem. J.* **397:** 15.

Golden S.S. and Canales S.R. 2003. Cyanobacterial circadian rhythms - Timing is everything. *Nat. Rev. Microbiol.* **1:** 191.

Green R.M., Tingay S., Wang Z.Y., and Tobin E.M. 2002. Circadian rhythms confer a higher level of fitness to *Arabidopsis* plants. *Plant Physiol.* **129:** 576.

Hardin P.E. 2006. Essential and expendable features of the circadian timekeeping mechanism. *Curr. Opin. Neurobiol.* **16:** 686.

Hardin P.E. and Yu W. 2006. Circadian transcription: Passing the HAT to CLOCK. *Cell* **125:** 424.

Herrero A., Muro-Pastor A.M., and Flores E. 2001. Nitrogen control in cyanobacteria. *J. Bacteriol.* **183:** 411.

Ishiura M., Kutsuna S., Aoki S., Iwasaki H., Andersson C.R., Tanabe A., Golden S.S., Johnson C.H., and Kondo T. 1998. Expression of a gene cluster *kaiABC* as a circadian feedback process in cyanobacteria. *Science* **281:** 1519.

Ivleva N.B., Bramlett M.R., Lindahl P.A., and Golden S.S. 2005. LdpA: A component of the circadian clock senses redox state of the cell. *EMBO J.* **24:** 1202.

Ivleva N.B., Gao T., LiWang A.C., and Golden S.S. 2006. Quinone sensing by the circadian input kinase of the cyanobacterial circadian clock. *Proc. Natl. Acad. Sci.* **103:** 17468.

Iwasaki H. and Kondo T. 2004. Circadian timing mechanism in the prokaryotic clock system of cyanobacteria. *J. Biol. Rhythms* **19:** 436.

Iwasaki H., Nishiwaki T., Kitayama Y., Nakajima M., and Kondo T. 2002. KaiA-stimulated KaiC phosphorylation in circadian timing loops in cyanobacteria. *Proc. Natl. Acad. Sci.* **99:** 15788.

Johnson C.H. 2004. Global orchestration of gene expression by the biological clock of cyanobacteria. *Genome Biol.* **5:** 217.

Katayama M., Kondo T., Xiong J., and Golden S.S. 2003. *ldpA* encodes an iron-sulfur protein involved in light-dependent modulation of the circadian period in the cyanobacterium *Synechococcus elongatus* PCC 7942. *J. Bacteriol.* **185:** 1415.

Ko C.H. and Takahashi J.S. 2006. Molecular components of the mammalian circadian clock. *Hum. Mol. Genet.* (spec. no. 2) **15:** R271.

Koh K., Zheng X., and Sehgal A. 2006. JETLAG resets the *Drosophila* circadian clock by promoting light-induced degradation of TIMELESS. *Science* **312:** 1809.

Kondo T., Mori T., Lebedeva N.V., Aoki S., Ishiura M., and Golden S.S. 1997. Circadian rhythms in rapidly dividing cyanobacteria. *Science* **275:** 224.

Kuhlman S.J. and McMahon D.G. 2006. Encoding the ins and outs of circadian pacemaking. *J. Biol. Rhythms* **21:** 470.

Kutsuna S., Kondo T., Aoki S., and Ishiura M. 1998. A period-extender gene, *pex*, that extends the period of the circadian clock in the cyanobacterium *Synechococcus* sp. strain PCC 7942. *J. Bacteriol.* **180:** 2167.

Kwon I., Lee J., Chang S.H., Jung N.C., Lee B.J., Son G.H., Kim K., and Lee K.H. 2006. BMAL1 shuttling controls transactivation and degradation of the CLOCK/BMAL1 heterodimer. *Mol. Cell. Biol.* **26:** 7318.

Liu Y. and Bell-Pedersen D. 2006. Circadian rhythms in *Neurospora crassa* and other filamentous fungi. *Eukaryot. Cell* **5:** 1184.

Liu Y., Tsinoremas N.F., Johnson C.H., Lebedeva N.V., Golden S.S., Ishiura M., and Kondo T. 1995. Circadian orchestration of gene expression in cyanobacteria. *Genes Dev.* **9:** 1469.

Mackey S.R. and Golden S.S. 2007. Winding up the cyanobacterial circadian clock. *Trends Microbiol.* **15:** 381.

McClung C.R. 2006. Plant circadian rhythms. *Plant Cell* **18:** 792.

Michael T.P., Salome P.A., Yu H.J., Spencer T.R., Sharp E.L., McPeek M.A., Alonso J.M., Ecker J.R., and McClung C.R. 2003. Enhanced fitness conferred by naturally occurring variation in the circadian clock. *Science* **302:** 1049.

Mihalcescu I., Hsing W., and Leibler S. 2004. Resilient circadian oscillator revealed in individual cyanobacteria. *Nature* **430:** 81.

Mitsui A., Kumazawa S., Takahashi A., Ikemoto H., and Arai T. 1986. Strategy by which nitrogen-fixing unicellular cyanobacteria grow photoautotrophically. *Nature* **323:** 720.

Mizuno T. and Nakamichi N. 2005. *Pseudo*-response regulators (PRRs) or *true* oscillator components (TOCs). *Plant Cell Physiol.* **46:** 677.

Mori T. and Johnson C.H. 2000. Circadian control of cell division in unicellular organisms. *Prog. Cell Cycle Res.* **4:** 185.

Mori T., Binder B., and Johnson C.H. 1996. Circadian gating of cell division in cyanobacteria growing with average doubling times of less than 24 hours. *Proc. Natl. Acad. Sci.* **93:** 10183.

Nakajima M., Imai K., Ito H., Nishiwaki T., Murayama Y., Iwasaki H., Oyama T., and Kondo T. 2005. Reconstitution of circadian oscillation of cyanobacterial KaiC phosphorylation *in vitro*. *Science* **308:** 414.

Ouyang Y., Andersson C.R., Kondo T., Golden S.S., and Johnson C.H. 1998. Resonating circadian clocks enhance fitness in cyanobacteria. *Proc. Natl. Acad. Sci.* **95:** 8660.

Pattanayek R., Wang J., Mori T., Xu Y., Johnson C.H., and Egli M. 2004. Visualizing a circadian clock protein: Crystal structure of KaiC and functional insights. *Mol. Cell* **15:** 375.

Putterill J., Laurie R., and Macknight R. 2004. It's time to flower: The genetic control of flowering time. *Bioessays* **26:** 363.

Schmitz O., Katayama M., Williams S.B., Kondo T., and Golden S.S. 2000. CikA, a bacteriophytochrome that resets the cyanobacterial circadian clock. *Science* **289:** 765.

Smith R.M. and Williams S.B. 2006. Circadian rhythms in gene transcription imparted by chromosome compaction in the cyanobacterium *Synechococcus elongatus*. *Proc. Natl. Acad. Sci.* **103:** 8564.

Sweeney B.M. and Borgese M.B. 1989. A circadian rhythm in cell division in a prokaryote, the cyanobacterium *Synechococcus* WH7803. *J. Phycol.* **25:** 183.

Taghert P.H. and Shafer O.T. 2006. Mechanisms of clock output in the *Drosophila* circadian pacemaker system. *J. Biol. Rhythms* **21:** 445.

Takai N., Ikeuchi S., Manabe K., and Kutsuna S. 2006a. Expression of the circadian clock-related gene *pex* in cyanobacteria increases in darkness and is required to delay the clock. *J. Biol. Rhythms* **21:** 235.

Takai N., Nakajima M., Oyama T., Kito R., Sugita C., Sugita M., Kondo T., and Iwasaki H. 2006b. A KaiC-associating SasA-RpaA two-component regulatory system as a major circadian timing mediator in cyanobacteria. *Proc. Natl. Acad. Sci.* **103:** 12109.

Taniguchi Y., Katayama M., Ito R., Takai N., Kondo T., and Oyama T. 2007. *labA:* A novel gene required for negative feedback regulation of the cyanobacterial circadian clock protein KaiC. *Genes Dev.* **21:** 60.

Vitalini M.W., de Paula R.M., Park W.D., and Bell-Pedersen D. 2006. The rhythms of life: Circadian output pathways in *Neurospora*. *J. Biol. Rhythms* **21:** 432.

Williams S.B. 2006. A circadian timing mechanism in the cyanobacteria. *Adv. Microb. Physiol.* **52:** 229.

Woelfle M.A. and Johnson C.H. 2006. No promoter left behind: Global circadian gene expression in cyanobacteria. *J. Biol. Rhythms* **21:** 419.

Xu Y., Mori T., and Johnson C.H. 2000. Circadian clock-protein expression in cyanobacteria: Rhythms and phase setting. *EMBO J.* **19:** 3349.

———. 2003. Cyanobacterial circadian clockwork: Roles of KaiA, KaiB and the *kaiBC* promoter in regulating KaiC. *EMBO J.* **22:** 2117.

Xu Y., Mori T., Pattanayek R., Pattanayek S., Egli M., and Johnson C.H. 2004. Identification of key phosphorylation sites in the circadian clock protein KaiC by crystallographic and mutagenetic analyses. *Proc. Natl. Acad. Sci.* **101:** 13933.

Biological Rhythms Workshop IB: Neurophysiology of SCN Pacemaker Function

S.J. KUHLMAN
Cold Spring Harbor Laboratory, Cold Spring Harbor, New York 11724

Pacemakers are functional units capable of generating oscillations that synchronize downstream rhythms. In mammals, the suprachiasmatic nucleus (SCN) of the hypothalamus is a circadian pacemaker composed of individual neurons that intrinsically express a near 24-hour rhythm in gene expression. Rhythmic gene expression is tightly coupled to a rhythm in spontaneous firing rate via intrinsic daily regulation of potassium current. Recent progress in the field indicates that SCN pacemaking is a specialized property that emerges from intrinsic features of single cells, structural connectivity among cells, and activity dynamics within the SCN. The focus of this chapter is on how Nature built a functional pacemaker from many individual oscillators that is capable of coordinating the daily timing of essential brain and physiological processes.

INTRODUCTION

The mammalian SCN is a bilateral nucleus that contains approximately 20,000 densely packed neurons and is localized to the hypothalamus, just dorsal to the optic chiasm (Fig. 1). Individual SCN neurons are capable of generating self-sustained oscillations; however, it is the SCN ensemble as a whole that encodes luminance history and generates rhythmic output signals that serve to organize the timing of essential brain and physiological processes. The SCN is considered a pacemaker because it both generates self-sustained oscillations and synchronizes downstream targets to the environmental light/dark cycle. The SCN itself is entrained (synchronized) to the external light cycle via a light-input pathway that directly innervates the SCN. Pacemaking is a specialized property that emerges from intrinsic features of single cells, structural connectivity among cells, and activity dynamics within the SCN. The first section of this chapter focuses on how these three factors interact to produce a functional pacemaker. In the section following, evidence that the SCN is indeed a pacemaker capable of controlling downstream targets is reviewed, and in the final section, the potential physiological significance of circadian timing in animals is discussed.

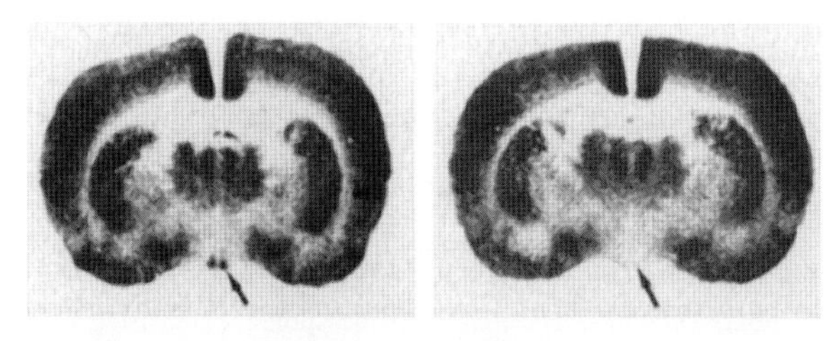

Figure 1. Day-night difference in ^{14}C-labeled deoxyglucose uptake in the SCN. Coronal sections of the rodent brain; *arrow* points to the bilateral SCN in the hypothalamus. (*Left*) Daytime glucose uptake is high, as indicated by the strong signal generated from an injection of radioactively labeled glucose into the rodent during the day. (*Right*) Nighttime glucose uptake is low in animals injected during the night. (Adapted, with permission, from Schwartz and Gainer 1977 [©AAAS].)

The SCN pacemaker is a model system in which one might say that the "black box" has been broken open; we now have the techniques to peer into the black box. A major advance since the 1960 Cold Spring Harbor "Biological Clocks" meeting is that it is now experimentally possible to directly probe mechanisms of pacemaker function.

BUILDING A PACEMAKER

Coupling the Intracellular Feedback Loop to Overt Output: The Role of Electrical Activity

A defining characteristic of mammalian SCN neurons is their ability to express a cell-autonomous rhythm in electrical activity. Electrical activity in neurons is produced by the opening of membrane-spanning ion channels, and the opening of these channels changes the voltage of the neuron. There is a voltage threshold (called the spike threshold) that, if reached, causes the neuron's membrane voltage to rapidly depolarize, and the cell is transiently flooded with calcium due to the opening of voltage-sensitive Ca^{2+} channels. The influx of calcium causes neurotransmitters or neuropeptides to be released from the presynaptic terminal of the neuron. The transient deflection in membrane voltage is referred to as a "spike" and is dependent on Na^{+} channels that are sensitive to the neurotoxin tetrodotoxin (TTX). SCN cells spontaneously generate spikes, and the frequency at which these spikes are generated varies with time of day. Spike rate (or firing rate) is high during the day and low at night for both nocturnal and diurnal mammals. Blocking expression of electrical activity with TTX in the SCN disrupts circadian behaviors such as locomotor activity (Schwartz et al. 1987); therefore, it is generally accepted that the rhythm in spike rate from the SCN serves to couple the transcription-translation feedback loop that occurs in the SCN to overt behavioral output.

In vivo recordings by Inouye and Kawamura (1979) first revealed the circadian rhythm in electrical activity from populations of SCN neurons. Subsequent work demonstrated that the rhythm in spike rate is self-sustained and does not require additional synaptic or humoral drive (Green and Gillette 1982; Groos and Hendriks 1982; Inouye and Kawamura 1982; Shibata et al. 1982). Because the generation of circadian rhythms are thought to be based on an intracellular transcription-translation feedback loop (discussed in Part IA), it is predicted that individual SCN cells should be competent oscillators, i.e., the rhythm in firing rate should be cell-autonomous.

In a landmark study, Welsh et al. (1995) provided strong evidence that, indeed, single cells are competent oscillators. Subsequent multielectrode array studies confirmed this finding (Liu et al. 1997; Herzog et al. 1998; Honma et al. 1998). The firing rates of individual SCN neurons dispersed at low density onto multielectrode array dishes can be continuously monitored (Fig. 2A). Dispersed neurons spike spontaneously, and many neurons express a rhythm in the frequency of spontaneous firing, with a period of about 24 hours. Importantly, neurons within the same culture oscillate at different phases and express different circadian periods (Fig. 2A). The most direct evidence for cell-autonomous rhythm generation comes from a nonvertebrate preparation. Completely isolated retinal basal neurons from the marine mollusk *Bulla gouldiana* express a circadian rhythm in membrane voltage and potassium conductance (Michel et al. 1993).

Our molecular understanding of rhythms generation (discussed in Part IA) provides valuable tools for testing hypotheses of clock function. Genetic mutation of a single clock gene can disrupt rhythmic expression of all clock genes. For example, mice that are heterozyogous for a point mutation in the *Clock* locus display a lengthening of the circadian period (Vitaterna et al. 1994). If the molecular feedback loop is cell-autonomous, the period of individual cells from the *Clock* mutant is predicted to also be lengthened as compared to that of wild-type cells (Fig. 2B). Indeed, recordings of SCN neurons dispersed onto multielectrode arrays demonstrated that the period of the rhythm in spontaneous firing rate is longer in mice heterozygous for the *Clock* mutation than that of wild type (Herzog et al. 1998; Nakamura et al. 2002). Similarly, a mutation in casein kinase 1ε (called *tau*) shortens the period of locomotor rhythmicity and also shortens the rhythm in spontaneous firing rate in single SCN neurons (Fig. 2B) (Liu et al. 1997). On the basis of this evidence, it is generally concluded that the intracellular molecular feedback loop imposes rhythmicity to downstream targets outside of the SCN through the expression of a rhythm in the frequency of electrical activity in individual SCN neurons.

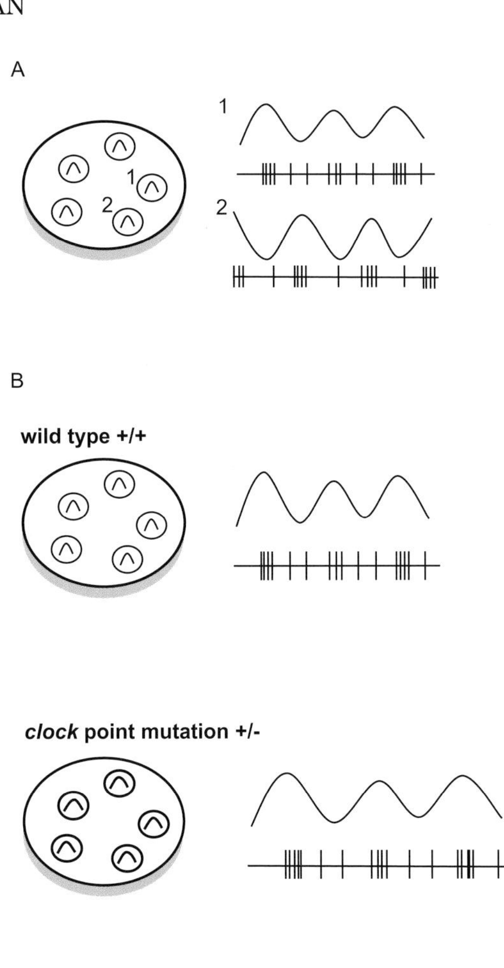

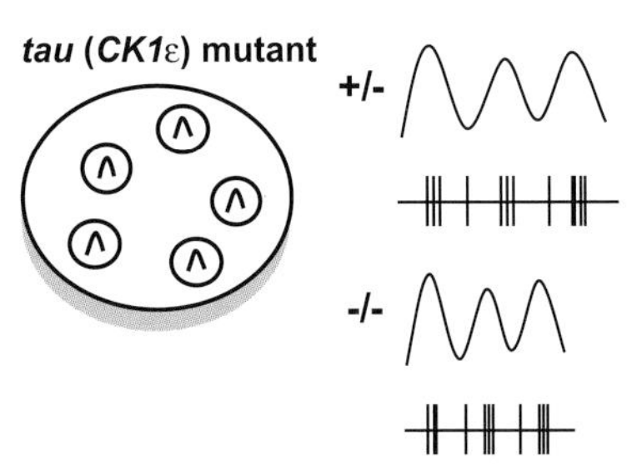

Figure 2. Evidence that the circadian rhythm in spontaneous firing rate of SCN neurons is cell-autonomous. (*A*) SCN neurons dispersed at low density in multielectrode dishes exhibit a rhythm in spontaneous firing rate. Importantly, the waveforms of individual neurons are not synchronized; thus, circadian rhythmicity does not require cell communication. Vertical lines represent the occurrence of a spike, and the amplitude of the waveform represents spike rate. (*B*) Mutations in core clock genes alter both the period of locomotor rhythmicity and the period of the rhythm in spontaneous firing rate. Spontaneous firing in wild-type control neurons oscillates with a period of approximately 23.5 hours (*top*), whereas spontaneous firing in neurons heterozygous for a point mutation in the *Clock* locus oscillates with a longer period, 25 hours (*middle*). The period of the spontaneous firing rate rhythm in neurons having only one functional copy of casein kinase 1ε (CK1ε) is short, 21 hours, and the firing rate rhythm is even shorter in the homozygous mutant, 19 hours.

Ionic Basis for the Rhythm in Spontaneous Firing Rate

In this section, we review the physiological basis of the rhythm in spontaneous firing. The key concept here is that the spontaneous rhythm in firing rate is a property intrinsic to individual cells. In subsequent sections, we discuss how these individual cells are wired together and how activity dynamics within the SCN contribute to the SCN's ability to measure time.

Many types of neurons are capable of spontaneously firing spikes in the absence of synaptic transmission (Hausser et al. 2004). The rate of spontaneous activity itself can be bidirectionally modulated, thus providing an additional level of regulation. Within a circuit, the source of rate modulation can be synaptic (Nelson et al. 2003; Smith and Otis 2003) or intrinsic in origin, as is the case for SCN neurons (Pennartz et al. 2002; Kuhlman and McMahon 2004).

To produce spontaneous activity, intrinsic currents must interact to depolarize the neuron's voltage to spike threshold, elicit a spike, and repolarize the membrane to a negative potential from which the next spike can be initiated. In other words, the neuron's voltage must fluctuate in a regenerative process. Different cell types use different strategies to generate this regenerative spontaneous activity. For example, in some neurons of the thalamus, the regenerative process results from dynamic recruitment and activation of hyperpolarization-activated channels and T-type Ca^{2+} channels (McCormick and Huguenard 1992). In other neurons of the cerebellum, persistent Na^{+} currents are essential for regenerative spontaneous firing (Raman et al. 2000; Taddese and Bean 2002; Do and Bean 2003).

SCN neurons are unique compared to other neuron types; in addition to firing spontaneously, the *rate* of spontaneous firing is modulated in a rhythmic manner. Therefore, there are two distinct aspects of spontaneous spike activity in SCN neurons that must be addressed: (1) the identification of processes that drive the daily rhythm in spike rate and (2) the identification of ion channels that contribute to the regenerative process that underlies spontaneous firing in general. Indeed, there are two dissociable mechanisms. One mechanism—the *subthreshold* regulation of resting membrane potential and basal potassium conductance—is likely the primary driver of the rhythm in spike rate. The second mechanism—regulation of *suprathreshold* spike-associated conductances mediated by channels such as voltage-dependent potassium channels—facilitates regenerative spontaneous firing and determines the shape of the spike waveform (Fig. 3).

Circadian rhythm in subthreshold basal potassium conductance. During the night phase, SCN neurons are hyperpolarized relative to neurons during the day phase (de Jeu et al. 1998; Schaap et al. 1999; Pennartz et al. 2002; Kuhlman and McMahon 2004). Daily hyperpolarization is accompanied by an increase in potassium ion conductance (i.e., more potassium channels are open during the night). Importantly, the rhythms in membrane potential and basal potassium conductance are maintained in constant dark conditions (Kuhlman and McMahon 2004).

How do these changes contribute to the rhythm in firing rate? An increase in potassium conductance during the night drives the resting membrane potential closer to

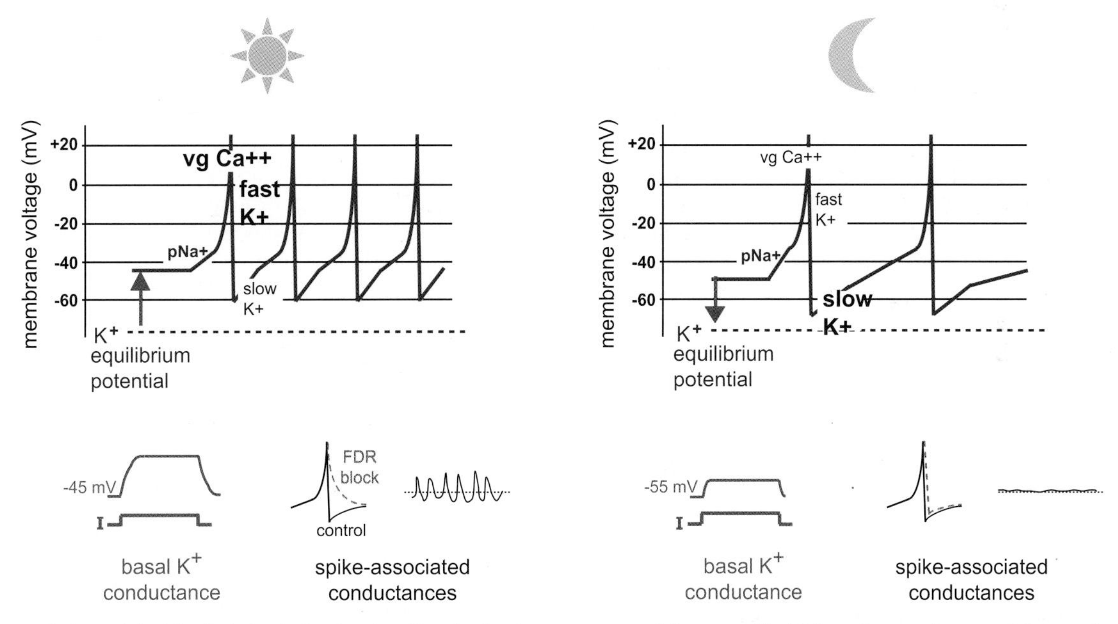

Figure 3. A model of the ionic basis for the circadian rhythm in spontaneous firing rate. (*Top*) The voltage trajectory of the membrane potential (*blue*) fluctuates in a regenerative manner due to the opening and closing of voltage-dependent spike-associated conductances (*black type*) during the day (*left*) and night (*right*). Persistent Na^{+} channels (pNa^{+}) continuously exert a depolarizing force toward spike threshold. Once spike threshold is reached, voltage-gated Na^{+} channels and Ca^{2+} (vg Ca^{2+}) channels open, causing the neuron to "spike." The membrane potential repolarizes due to the opening of fast-delayed rectifier K^{+} channels (fast K^{+}, FDR) and slower acting K^{+} channels (slow K^{+}), the voltage-gated Na^{+} and Ca^{+2} channels close before the next spike is initiated. Conductance through some channels is dependent on time of day: (*bold, large type*) high conductance; (*small type*) low conductance. (*Bottom*) Experimentally quantified daily changes in ionic conductances. There is a daily change in membrane potential (Δ10 mV), as predicted by the model. Additional evidence for daily rhythm in basal K^{+} conductance comes from the following observation: For the same current (I) injected into a cell, the voltage deflection is large during the day, a signature that basal K^{+} channels are closed (low conductance), whereas the voltage deflection is small during the night, a signature that basal K^{+} channels are opened (high conductance). The properties labeled in *red* are believed to underlie the rhythm in spike rate. Spike-associated conductances (indicated in *black*) enable the neuron to express a rhythm in spontaneous firing rate. Blockade of FDR conductances during the day broadens the spike (*dashed lines*), whereas blockade at night has no effect. Fast membrane potential oscillations mediated by L-type Ca^{2+} channels hover around spike threshold (*dotted line*) and are expressed only during the day. These transient Ca^{2+} channel openings may facilitate high firing rates or perhaps promote increased Ca^{2+} entry during the day.

the equilibrium potential for potassium ions (Fig. 3, red arrow), which is hyperpolarized relative to the resting membrane potential and pulls the cell's voltage further away from spike threshold. Thus, increased potassium conductance is expected to decrease spike rate. Additionally, low input resistance (high conductance) means that the voltage deflection in response to a given level of stimulation is small, compared to the same level of stimulation applied to a neuron with higher input resistance (Fig. 3). The daily depolarization in membrane voltage is thought to drive the rhythm in spike rate due to decreased potassium conductance. However, the specific potassium channels that underlie this daily change remain to be identified.

Spike-associated conductances. SCN neurons express a complement of ionic currents that generate spontaneous firing at rates up to 10–15 Hz during the day phase. As stated earlier, the voltage trajectory of the neuron must follow a regenerative process for spontaneous activity to be maintained. Many specific channels mediating this function have been identified in SCN neurons, and in some cases, it has been determined that the channels are modulated in a rhythmic manner.

One requirement for regenerative spiking is the presence of an intrinsic drive to bring the neuron to spike threshold. In SCN neurons, slowly inactivating, persistent sodium channels contribute to drive the cell to threshold (Pennartz et al. 1997; Jackson et al. 2004; Kononenko et al. 2004). It is unlikely that these currents are regulated in a circadian manner, although diurnal modulation by cAMP is possible (Kononenko and Dudek 2006). A second intrinsic drive to threshold may come from a depolarizing oscillation in Ca^{2+} conductance. Pennartz et al. (2002) identified the presence of TTX-resistant, high-frequency Ca^{2+} oscillations (2–8 Hz) that are mediated by L-type Ca^{2+} channels. These oscillations cause the membrane voltage to rapidly fluctuate around spike threshold and are regulated in a diurnal manner, present during the day phase only (Fig. 3). The daytime depolarized membrane voltage described in the earlier section also brings the cell closer to spike threshold, which, together with the high input resistance state, means that small changes in Ca^{2+} currents can have a large impact on the voltage trajectory. Therefore, the circadian rhythm in subthreshold basal potassium conductance and spike-associated mechanisms functionally complement one another. However, experimentally, the two mechanisms can be dissociated, demonstrating that they are distinct mechanistically. Suppression of fast Ca^{2+} oscillations by L-type channel blockade does not disrupt the rhythm in basal potassium conductance (Pennartz et al. 2002). The reciprocal experiment—testing whether the rhythm in L-type Ca^{2+} channel-mediated fast oscillations persists in the absence of the rhythm in basal K^+ conductance—has not been done because it requires the identification of the precise channels that underlie the rhythm in basal K^+ conductance. Nonetheless, the above experiment is sufficient to conclude that circadian regulation of subthreshold basal K^+ conductance and spike-associated conductances are distinct processes.

A second requirement for regenerative spiking is a sufficiently rapid repolarization to ensure brief interspike intervals. Rapidly activating voltage-gated potassium channels subserve this function and have been characterized in SCN neurons (Bouskila and Dudek 1995). Recently, a fast-delayed rectifier (FDR) potassium current in SCN neurons was identified that is regulated in a rhythmic manner (Itri et al. 2005). Blockade of this channel has a significant effect on spike shape specifically during the day (Fig. 3), and FDR channel function is required for the expression of rhythmic spontaneous spike activity.

Phase Shifts: Perturbation of the Transcription-Translation Feedback Loop

Before moving on to discuss structural connectivity among cells, we first review the physiology of resetting, because mechanisms of phase-shifting are believed to underlie entrainment and are therefore critical for pacemaker function. As briefly mentioned in Part IA, the transcription-translation feedback loop can be reset by light-mediated *trans*-activation of key clock genes such as *Period1* (*Per1*). Light is the primary entraining agent of the SCN pacemaker, and photic information is directly conveyed to the SCN by axons projecting from intrinsically photosensitive retinal ganglion cells (ipRGCs) (for review, see Berson 2003). Nocturnal light pulses initiate a signaling cascade that ultimately results in the stable shift (resetting) of the transcription-translation feedback loop via several calcium-dependent signaling pathways that lead to the phosphorylation of the cAMP response element–binding (CREB) protein, which binds to cAMP response elements (CRE) in the promoter region of responsive genes (e.g., *Per1*) to activate their expression (Morse and Sassone-Corsi 2002; Dziema et al. 2003). Thus, the light-input pathway regulates transcription of core clock genes via *cis*-response elements that are distinct from CLOCK/BMAL1-binding response elements (Fig. 4A).

Again, electrophysiological methods were used to verify specific predictions of the transcription-translation feedback loop model. Because light pulses are capable of activating core clock gene expression, it was predicted that the firing rate should also be reset in those cells that receive the phase-shifting light pulse. Indeed, it was demonstrated that the neurons in which *Per1* is specifically induced by light also have elevated firing rates (Kuhlman et al. 2003). In fact, as expected, a tight correlation between *Per1* activation and neuronal firing rate exists in entrained conditions (Fig. 4B). In both long days (14 hours light:10 hours dark [14:10 LD]) and balanced days (12:12 LD), firing rate correlates well with levels of *Per1* promoter activity on a cell-by-cell basis (Quintero et al. 2003; Kuhlman and McMahon 2004). What other predictions of the transcription-translation feedback model can be tested?

It is worth noting that resetting and rhythms generation are generally thought of as two distinct processes. However, as depicted in Figure 4A, the two processes ultimately converge at the level of gene transcription. Thus, agents that are experimentally classified as input/resetting agents may in fact have a potent role in rhythms generation, particularly if such agents are rhyth-

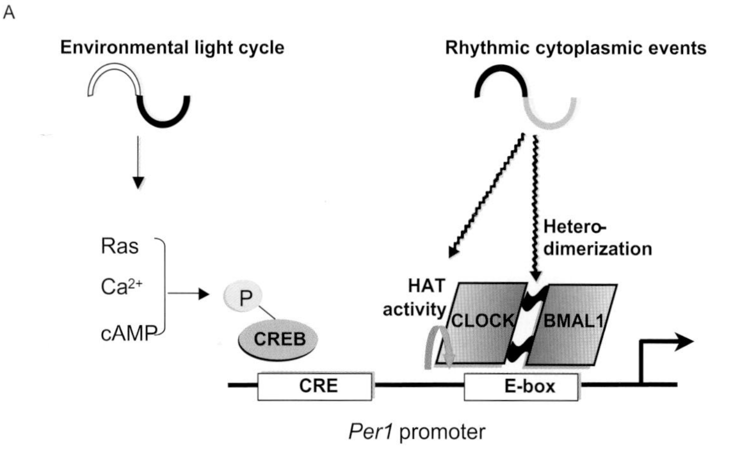

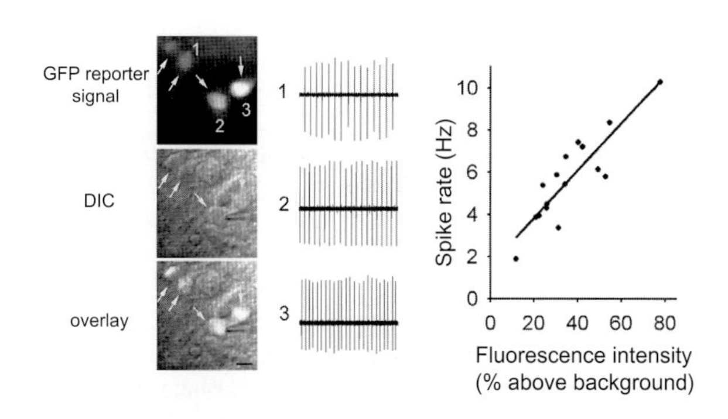

Figure 4. *Period1 trans*-activation is tightly coupled to spike output. (*A*) *Period1* gene transcription can be activated via the binding of phosphorylated CREB to cAMP response elements (CREs) or the binding of CLOCK-BMAL1 heterodimers to E-box response elements within the *Period1* promoter. The CLOCK protein itself possesses histone acetyltransferase (HAT) activity. High HAT activity relaxes chromatin structure, thus promoting increased transcription (see also Fig. 8 in Part IA). (*B*) Regardless of the route of activation, the level of *Period1 trans*-activation is correlated with firing rate on a cell-by-cell basis. The level of GFP (green fluorescent protein) reporter intensity (*top panel*) correlates with spike rate (*right column*): Cell 1, low intensity, fires at a rate of 1.5 Hz, whereas cell 3, high intensity, fires at a rate of 5 Hz. Shown here are neurons from an acutely prepared SCN slice during the day. Neurons are visualized for recording using differential interference contrast (DIC) microscopy; the recording pipette is patched onto cell 2 (*middle panel*). The digital intensity image is overlaid with the DIC image (*bottom panel*). Similar correlations are reported in acutely prepared SCN after a phase-shifting light pulse (not shown). Bar: 10 μm, trace 10 seconds.

mically expressed during normal intact conditions. Likewise, agents that are classified as output signals can also feed back via input pathways to promote high-amplitude rhythms generation (Nitabach et al. 2002; Yamaguchi et al. 2003; Aton et al. 2005; Lundkvist et al. 2005; Maywood et al. 2006). We present Figure 5 as a tool to help visualize this concept. The clock system can be potentially viewed as a network of coupled oscillators, each oscillation producing an output (or coupling signal) that then feeds into another oscillator to promote high-amplitude (and reliable) rhythms generation.

Such a view may help to conceptualize the following experimental findings: Coupling between oscillators within the SCN is dependent on VIP (vasoactive intestinal polypeptide) signaling (Cutler et al. 2003; Aton et al. 2005; Maywood et al. 2006) and electrical activity (Yamaguchi et al. 2003). In the absence of VIP signaling, the phase relationship among oscillators breaks down, and individual oscillators become desynchronized from one another. Furthermore, it appears that the number of competent oscillators decreases in the absence of VIP signaling, indicating that in some cell types, VIP signaling is required for rhythms generation. Thus, an emerging model in the field is that in order for intracellular rhyth-

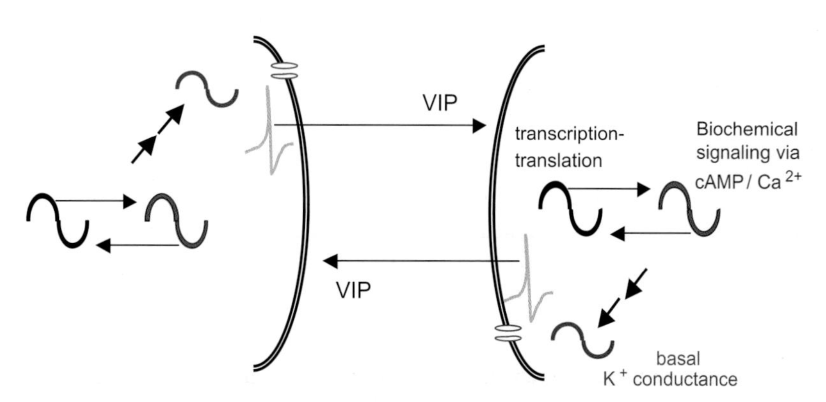

Figure 5. Oscillations are coupled. Output signals, such as electrical activity or signaling molecules (VIP in this example), serve as input signals to neighboring cell oscillators. Likewise, it is possible that within a cell, oscillations are coupled by signaling products of calcium-dependent G-protein-coupled pathways such as cAMP.

micity to remain self-sustained in the intact system, the transcription-translation feedback loop likely requires oscillations in biochemical signaling to reenforce high-amplitude rhythmicity.

Interestingly, an alternative E-box-binding transcription factor complex, NPAS2/BMAL1, can contribute to rhythms generation in SCN neurons under some conditions (Debruyne et al. 2007). In vitro, NPAS2/BMAL1 heterodimer transcriptional activation is sensitive to NADP(H) redox states (Rutter et al. 2001). Given that SCN cells exhibit a robust circadian rhythm in glucose uptake (see Fig. 1) (Schwartz and Gainer 1977), changes in metabolic activity could potentially regulate clockwork gene expression via redox-sensitive transcription factors (Rutter et al. 2002).

Structural Connectivity

Input. The photopigment melanopsin is expressed in about 3% of retinal ganglion cells and is responsible for converting the energy of photons to an intracellular chemical signaling cascade capable of depolarizing the ganglion cell to spike threshold. This small subset of retinal ganglion cells projects their axons to the SCN, forming what is called the retinal-hypothalamic tract (RHT, Fig. 6) (for review, see Moore 1996). RHT terminals release both pituitary adenylate-cyclase-activating polypeptide (PACAP) and glutamate. Stimulation of the RHT depolarizes SCN neurons; thus, PACAP and glutamate are considered to be excitatory in the SCN (for review, see Coogan and Piggins 2004). The rate of spike output from the light-sensitive ganglion cells is linearly related to luminance level. Thus, melanopsin$^+$ retinal ganglion cells represent an early stage of luminance coding and transduce information to the SCN regarding ambient lighting conditions that the animal encounters in the environment. RHT terminal density is highest in the ventral SCN. It is important to note that the RHT is not the sole afferent input carrying photic information to the SCN. There is input from the geniculohypothalamic tract (GHT) originating from the intergeniculate leaflet (IGL), the pretectum (Mikkelsen and Vrang 1994), and serotonergic projections originating from the midbrain median Raphe nucleus (Morin et al. 2006) that indirectly convey photic information in addition to internal brain-state information.

Connectivity within the SCN. To understand neuronal wiring within the SCN, one must appreciate that the SCN is composed of heterogeneous cell types (Fig. 6) for review, see Antle and Silver 2005). The SCN can be organized into two loosely define regions (Abrahamson and Moore 2001): (1) the vetrolateral or *core* region and (2) the dorsomedial or *shell* region. Neurons expressing VIP and gastrin-releasing peptide (GRP) are located primarily in the ventrolateral region and generally receive RHT input in addition to input from internal sources, whereas neurons expressing arginine vasopressin (AVP) are located in the dorsomedial region and receive input from internal brain nuclei, including limbic, hypothalamic, and brain stem areas. Most, if not all, SCN neurons express

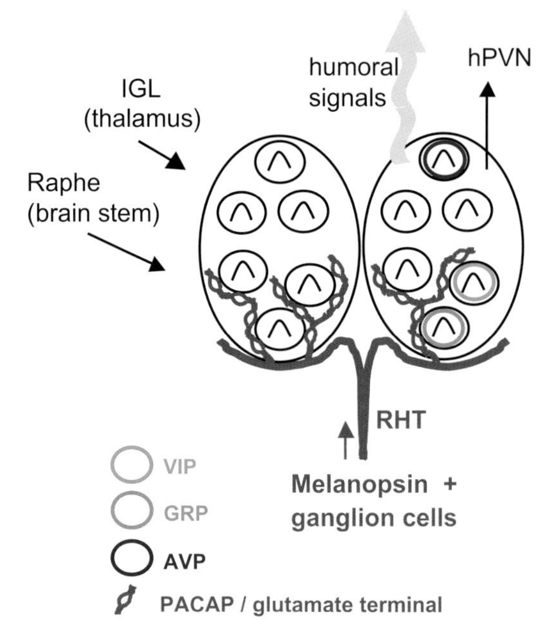

Figure 6. SCN input and output pathways. Light information reaches the SCN directly via the RHT, which is composed of melanopsin-positive ganglion cell axon fibers from the retina, and also indirectly via input from the Raphe nucleus and IGL. The SCN sends GABAergic output to the hPVN, among other targets (see Fig. 8), in addition to releasing diffusible humoral output signals such as TGF-α and prokineticin 2. Not all SCN neurons are directly retinorecipient, the RHT preferentially innervates VIP- and GRP-positive neurons. (VIP) Vasoactive intestinal polypeptide; (GRP) gastrin-releasing peptide; (AVP) arginine vasopressin; (PACAP) pituitary adenylate-cyclase-activating polypeptide; (RHT) retinal-hypothalamic tract; (hPVN) hypothalmic parventricular nucleus; (IGL) intergeniculate nucleus.

GABA (γ-amino-*n*-butyric acid) as their neurotransmitter (Moore and Speh 1993).

Communication between the core and shell subdivisions is mediated by axonal projections from the core to shell (Abrahamson and Moore 2001). The extent to which the shell communicates with the core is unknown. When core VIP or GRP neurons are activated, their terminals most likely release VIP or GRP and the inhibitory neurotransmitter GABA (for review, see Aton and Herzog 2005). Indirect evidence suggests that GABA release from the core to the shell is required for entrainment (Albus et al. 2005). However, direct demonstration using electrophysiological recording methods that SCN cells communicate with one another via GABAergic synapses is lacking.

Local communication among cells is potentially mediated by GABAergic transmission (Strecker et al. 1997) and gap junctions (Long et al. 2005). However, the specifics of how this communication gives rise to the patterns of activity dynamics observed in the SCN (described in the next section) are not well understood. For example, the relative proportion of GABAergic inputs that a single SCN neuron receives from within the SCN versus from outside the SCN is unknown. Furthermore, the functional relevance of gap-junction-mediated electrical coupling is unknown. To ultimately understand pacemaker function, future studies are needed to map out the functional wiring among distinct SCN cell types. Fortunately, the tools and techniques to do this are in existence.

Output. Here, we briefly highlight some outputs of the SCN to emphasize that as a population, SCN neurons are not functionally equivalent in terms of signaling to downstream targets (for a complete description of SCN projections to other brain regions, see Saper et al. 2005). For example, a subset of SCN neurons sends GABAergic axon projections to the hypothalamic paraventricular nucleus (hPVN) (Buijs and Kalsbeek 2001), whereas VIP neurons project to gonadotropin-releasing hormone neurons (GnRH) to regulate the timing of the luteinizing hormone (LH) surge (for review, see de la Iglesia and Schwartz 2006).

In addition to structural outputs, some SCN neurons release humoral factors, including transforming growth factor (TGF-α) and prokineticin 2 (Kramer et al. 2001; Cheng et al. 2002). Humoral release of signaling molecules represents an important mechanism for regulating downstream targets, as demonstrated by a classic study by Silver et al. (1996). Encapsulated SCN transplants maintain the capacity to restore behavioral rhythmicity in SCN-lesioned hosts (Silver et al. 1996). Thus, from these studies and others (Buijs et al. 2006), converging evidence indicates that SCN neurons regulate multiple downstream targets in the brain, raising the possibility that distinct cells types target and regulate distinct organs and downstream processes (Buijs and Kalsbeek 2001). It is interesting to speculate that such an organization allows for the sequential regulation of differentially timed internal physiological processes.

Dynamics: Constructing a Pacemaker from Many Oscillating Parts

In the previous sections, we discuss how to build individual oscillators using a molecular feedback loop and ion channels; we also discuss how these oscillators are connected to the rest of the brain. Now we discuss how to put the oscillators together to form a functional pacemaker. To address this issue requires that the waveforms of many oscillators must be simultaneously monitored within the SCN ensemble. Technically, this can has been accomplished using (1) mutlielectrode arrays to monitor firing rate of individual cells (Herzog et al. 1997; Nakamura et al. 2001), (2) multiunit electrophysiological recordings (Mrugala et al. 2000; Schaap et al. 2003; Vansteensel et al. 2003), and (3) transgenic reporter mice and rats harboring clock-gene-driven fluorescent (Kuhlman et al. 2000) or bioluminescent (Yamazaki et al. 2000; Yamaguchi et al. 2001; Wilsbacher et al. 2002; Yoo et al. 2004) reporters of circadian gene activity for real-time imaging of circadian gene expression.

On the basis of the diagram in Figure 4A, one may guess that because the *cis*-response elements for both entrainment (i.e., CRE) and rhythms (i.e., E box) generation coexist in single cells, entrainment is cell-autonomous. In other words, each oscillator is predicted to have the ability to detect and respond to light input and shift the transcription-translation feedback loop accordingly. However, not all SCN neurons receive direct input from the RHT. Therefore, not surprisingly, the response to light is diverse among cells, both in firing rate (Meijer et al. 1998; Kuhlman et al. 2003) and in *Per1* induction (Yan et al. 1999; Albus et al. 2005; Nakamura et al. 2005); the light pathway preferentially activates the retinorecipient region of the SCN. Furthermore, there is substantial evidence that the rhythms expressed by SCN tissue as a whole are a composite of single-cell rhythms that are narrower in waveform and distributed across the day (Fig. 7A). The time of peak electrical and peak transcriptional activity varies among cells, and individual cell waveforms are not synchronized to a single, average waveform (Quintero et al. 2003; Schaap et al. 2003; Yamaguchi et al. 2003; Inagaki et al. 2007). In fact, it has been directly observed that light input is capable of reorganizing the phase relationship among individual oscillators (Quintero et al. 2003; Inagaki et al. 2007). These observations suggest that mammalian entrainment is an emergent property and requires distinct coupling mechanisms among SCN neurons.

In addition to functioning as a circadian pacemaker, the SCN also functions as a seasonal clock capable of encoding day length (Sumova et al. 1995, 2003; Messager et al. 1999; Mrugala et al. 2000; Nüsslein-Hildesheim et al. 2000; Schaap et al. 2003). A major target of the SCN's seasonal output is the pineal gland. This gland releases melatonin during the night, and the duration of nighttime release reflects photoperiod length. One mechanism by which the SCN encodes day length appears to be via regulating phase relationships among individual SCN neurons (Inagaki et al. 2007; VanderLeest et al. 2007). In short days (8:16 LD), peak times of firing activity are relatively synchronized, restricted to a narrow temporal window. In contrast, in long days (16:8 LD), peak times are broadly distributed (Fig. 7B). Similarly, peak times of gene expression depend on photoperiod length (Inagaki et al. 2007).

To establish that the SCN is truly encoding day length by the phase relationship among oscillators, versus passively reflecting luminance history as an epiphenomenon, phase reorganization must be reflected in a downstream target, e.g., the behavioral rhythm in locomotor activity. A recent study in which single-cell waveforms of *Per1* activation were monitored from a cohort of animals that express three different behavioral states provides additional confirmation that phase reorganization is behaviorally relevant (Ohta et al. 2005). Ohta et al. (2005) used constant light to induce three different behavioral states in the cohort; individual animals exhibited either (1) long-period locomotor activity, (2) arrhythmic locomotor activity, or (3) "split" locomotor activity in which the animal displayed two distinct times of wheel-running behavior in 24 hours. The phase distribution of oscillating neurons did indeed correlate with the behavioral state of the animal (Fig. 7C). Importantly, the animals in this experiment were exposed to the same light stimulus, avoiding confounds of "masking," in which light directly reduces locomotor activity independent of daily rhythmicity.

The results of Ohta et al. (2005) also demonstrate the importance of assaying rhythmicity at the level of the functional unit of rhythms generation (the single cell) in

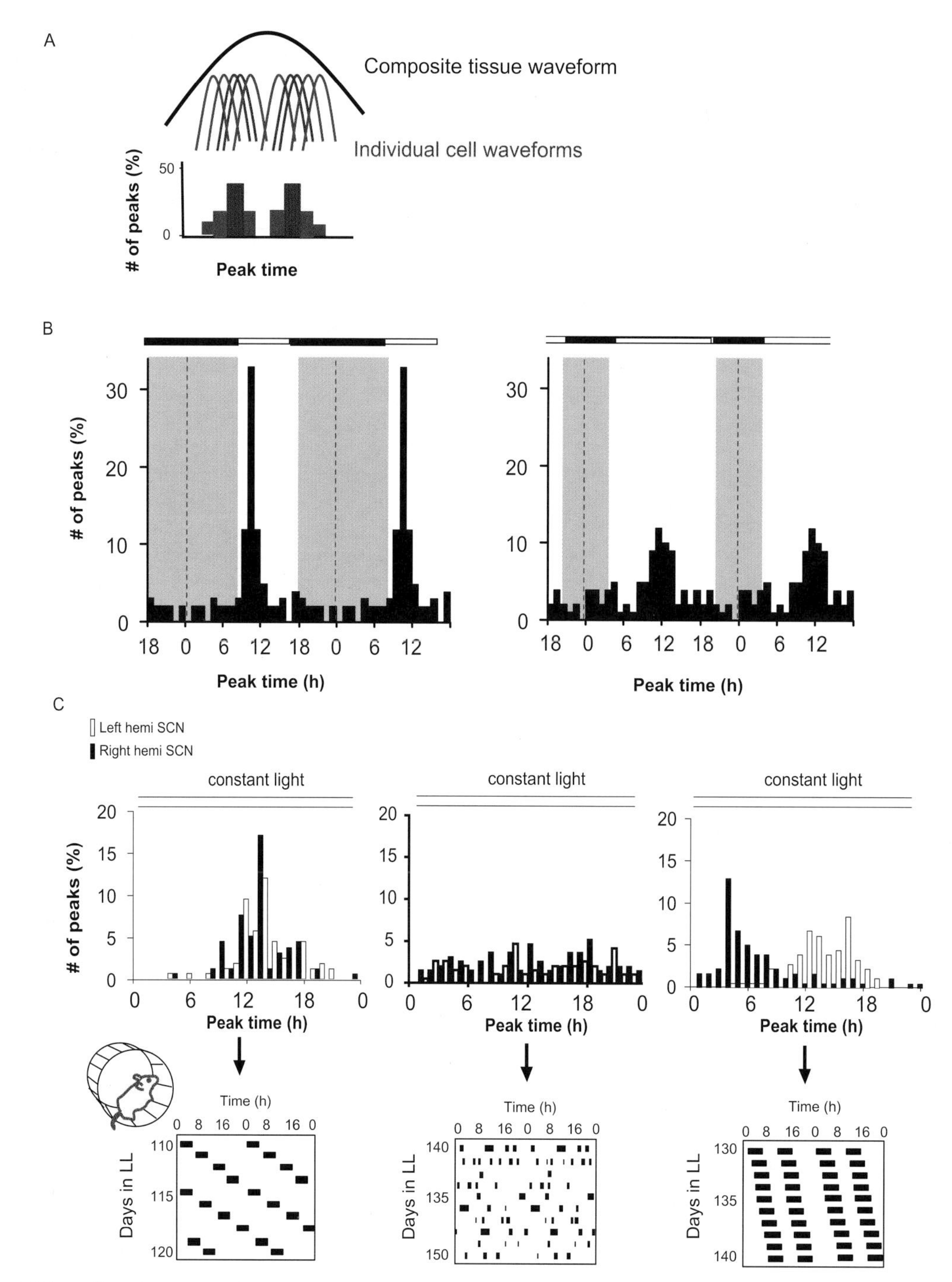

Figure 7. Phase diversity among SCN neurons encodes day length. (*A*) The composite waveform of SCN tissue, as measured by tissue-level in situ hybridizations for clock gene mRNAs or as measured by multiunit recording of the electrical activity of many neurons, is actually composed of individual cell waveforms that are narrower in width than the population waveform. The waveform distribution can be easily represented in a histogram plot of peak times (*bottom plot*). (*B*) The peak times of electrical activity waveforms of SCN neurons are narrowly distributed in short days. However, the distribution is experience-dependent and broadens when animals are housed in long days. Peak times, in hours (h), are plotted relative to midnight (*red vertical dashed line*). (Adapted, with permission, from VanderLeest et al. 2007 [© Elsevier].) (*C*) Variations in phase distribution are reflected in animal behavior. Constant light induces three distinct behavioral states: long period (*bottom, left*), arrhythmic (*bottom center*), or "split" (*bottom right*). The peak times of *Period1* reporter waveforms redistribute in a manner consistent with behavior. Peak times, in hours (h), are plotted relative to time in vitro. (Adapted, with permission, from Ohta et al. 2005 [©Nature Publishing Group].)

addition to measuring rhythmicity at the population level. Previous studies examining the effects of constant light (LL) on SCN tissue as a whole demonstrated that LL induces arrhythmicity in electrical activity and gene expression (Shibata et al. 1984; Mason 1991; Beaule et al. 2003; Sudo et al. 2003). On the basis of the results of Ohta et al. (2005), we can now interpret these earlier findings to mean that LL induces *desynchrony* among individual oscillators, rather than *arrhythmicity* of molecular oscillations (see also Fig. 2 in Kuhlman et al., this volume). Individual cells remain rhythmic under LL conditions.

Phase heterogeneity among individual oscillator cells is not limited to mammalian pacemakers. Studies from the laboratory of Michael Rosbash demonstrate that the phase relationship among individual oscillator cells contributes to encoding day length in *Drosophila* (Stoleru et al. 2005, 2007). Although the full physiological significance of phase diversity among pacemaker cells remains to be fully investigated, the evidence to date demonstrates that coupling among oscillator cells is a substrate for encoding day length.

HIERARCHICAL ORGANIZATION OF BODY OSCILLATORS

Many tissues and organs within the body display circadian rhythms in gene expression and in secretion of signaling molecules. Strikingly, when isolated in culture, many of these oscillations continue. Examples of such peripheral "clocks" are depicted in Figure 8. The SCN is considered a pacemaker capable of synchronizing peripheral oscillations due to the following evidence: (1) Classic transplant and lesion studies, such as the work of Ralph at al. (1990). In this 1990 study, SCN tissue was transplanted from casein kinase 1ε mutant hamsters (*tau*) into SCN-lesioned wild-type hosts, and the host animal's locomotor rhythm took on the donor's period (Ralph et al. 1990). (2) Circadian rhythms in corticosterone (Moore and Eichler 1972), locomotor and drinking behavior (Stephan and Zucker 1972), and body temperature (Refinetti et al. 1994) are lost in SCN-lesioned animals.

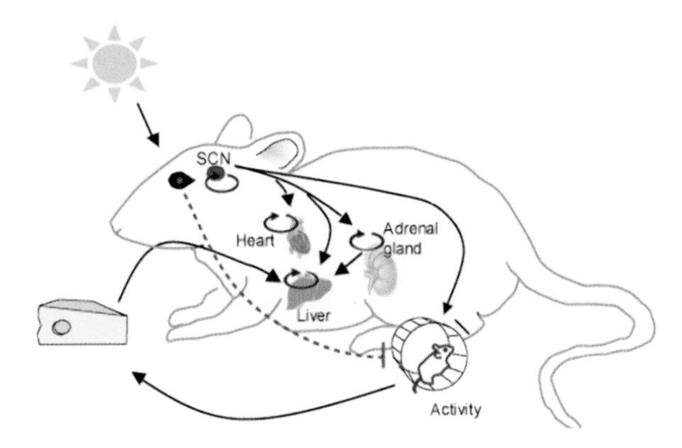

Figure 8. Hierarchical organization of body clocks. Many tissues throughout the body continue to oscillate for a few cycles when isolated in vitro; however, the oscillations eventually dampen. The SCN is generally considered to be the primary organizer of peripheral body oscillators, functioning to coordinate timing among the various tissue oscillators and to synchronize oscillations among the cell-autonomous oscillators within a given tissue. Importantly, there is a food-entrainable oscillator that is maintained in SCN-lesioned animals exposed to a feeding schedule in constant darkness. Feedback likely involves glucocorticoid release from the adrenal gland into the bloodstream that is then sensed in brain regions outside of the SCN (feedback not shown). In the intact system, the food-entrainable oscillator and the SCN oscillate in a coordinated manner due to SCN regulation of locomotor activity; the animal will only eat when it is awake, timed primarily by the SCN.

Cues from the SCN are hypothesized to coordinate the timing of peripheral clocks and to impose cellular synchrony within a given tissue. As a test of this hypothesis, peripheral tissues were isolated and cultured from SCN-lesioned animals. As expected, it was found that tissues oscillated out of phase with one another. However, within a given tissue, it was found that oscillations among individual cells were coherent, contrary to expectations (Yoo et al. 2004), suggesting the presence of organ-specific synchronizers (Stratmann and Schibler 2006). This later result could be an artifact of the dissection and culturing procedure. It has been demonstrated that changing culture medium is sufficient to reset desynchronized rhythms. Therefore, the presence of coherent oscillations in vitro does not necessarily mean that cells were coherent in vivo. To further explore the role of the SCN in regulating cellular synchrony within a tissue, Guo et al. (2006) examined *Per1:Bmal1* mRNA expression ratios in SCN-lesioned and control animals. *Per1* and *Bmal1* oscillate 180° out of phase from each other; therefore, if there is cellular synchrony within a tissue, one would expect to find a large peak in the *Per1:Bmal* ratio during the course of a cycle. In control animals, the *Per1:Bmal1* ratio attained a large value during the course of 24 hours, signifying coherence among cellular oscillators. In contrast, the *Per1:Bmal1* ratio did not attain a large value in SCN-lesioned animals, signifying desynchrony among cellular oscillators (Guo et al. 2006). Taken together, the evidence indicates that cues from the SCN coordinate the timing of peripheral clocks and impose cellular synchrony within a given peripheral tissue.

The SCN is not the only brain circuit capable of synchronizing internal physiology to environmental cycles. When rodents are housed in a light cycle and given restricted food access during the day so that the time of feeding is out of phase with the time they would normally be active and eat, the presence of a food-entrainable oscillator is revealed (Damiola et al. 2000; Stokkan et al. 2001). The food-entrainable circuit includes the following brain areas: the subparaventricular zone (Abrahamson and Moore 2001; Deurveilher and Semba 2005), dorsomedial nucleus of the hypothalamus, and paraventricular nucleus of the hypothalamus (see Fig. 8) (Gooley et al. 2006; Mieda et al. 2006). The food-entrainable oscillator regulates rhythmic expression of metabolic enzymes in the liver, thus "priming" internal physiology for food intake (Buijs et al. 2006). Under normal conditions, the SCN's ability to regulate locomotor activity ensures that the animal will eat during its waking phase; therefore, the SCN can be viewed as indirectly coordinating liver metabolism with other body oscillations via behavior. The rhythm in liver metabolism may itself reenforce robust rhythmicity via regulating redox state and NPAP2/BMAL1 transcriptional activity (Rutter et al. 2001, 2002).

Importantly, although there is a hierarchical organization, regulation of body clocks is not unidirectional. There are multiple feedback pathways to the brain, including sensitivity to blood-borne hormones by the arcuate nucleus that projects to the SCN (Fig. 9). Functionally, it has been demonstrated that sleep state modulates the firing rate of SCN neurons, perhaps via input from the Raphe nucleus (Deboer et al. 2003).

PHYSIOLOGICAL SIGNIFICANCE OF THE SCN PACEMAKER: TEMPORAL ORGANIZATION OF INTERNAL PHYSIOLOGY

The SCN serves to temporally organize a diverse range of physiological processes, including metabolism, growth, and cortical arousal. It is speculated that such coordination allows multicellular organisms to optimize performance during waking and to optimize tissue repair and memory consolidation during sleep. The SCN achieves this regulation via output pathways (Fig. 9A). Prominent output pathways include projections to the subparaventricular zone (sPVZ) located just dorsal to the SCN, the hypothalmic paraventricular nucleus (hPVN), the dorsomedial hypothalamus (DMH), and the arcuate nucleus (Arc) (for review, see Hungs and Mignot 2001; Saper et al. 2005).

The hPVN relays SCN signals to the autonomic nervous system via distinct sympathetic-parasympathetic pathways. Corticosterone release from the adrenal cortex located just above the kidney is regulated by the sympathetic pathway (Fig. 9A, highlighted light green). Temporal regulation of rate-limiting metabolic enzymes in the liver, including cytochrome P450s, heme biosynthesis, and mitochrondial function, is mediated in part via the parasympathetic pathway (Fig. 9A, highlighted dark green) (for review, see Hungs and Mignot 2001; Saper et al. 2005; Buijs et al. 2006). The hPVN also relays SCN signals to the pineal gland to produce rhythmic release of melatonin during the night (not shown in Fig. 9A).

In addition to regulating food uptake via a pathway to the hPVN as described in the previous section, the DMH

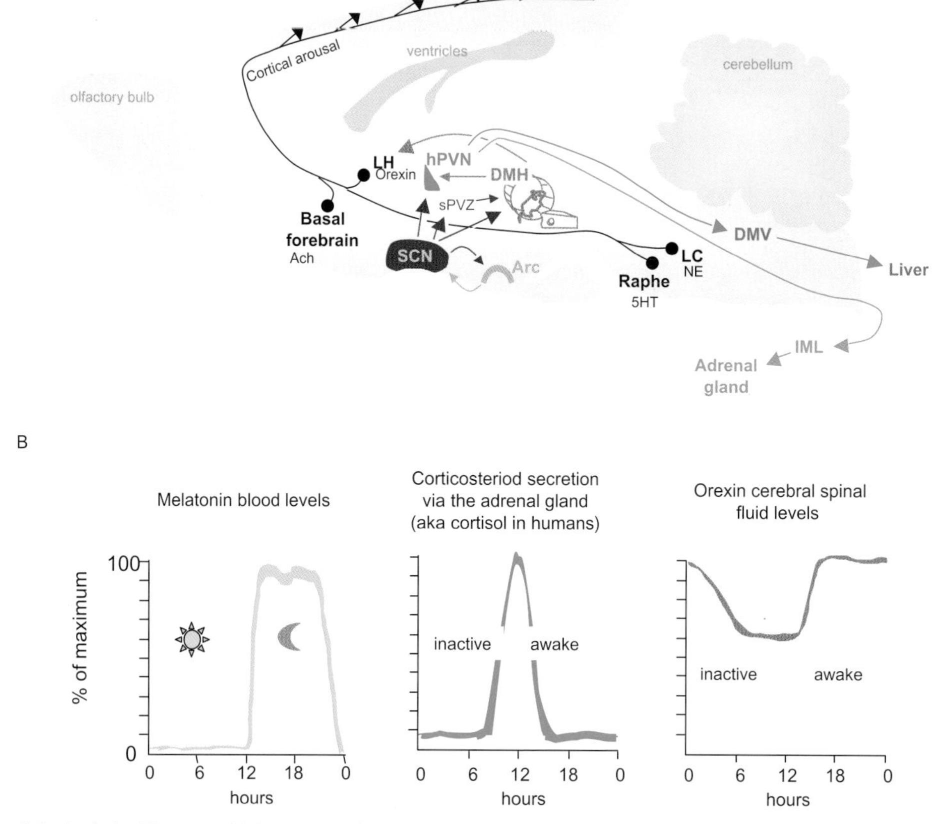

Figure 9. Physiological significance of SCN output signals: Coordination of body metabolism to optimize performance during waking. (*A*) The SCN (*red*) sends extensive projections to the paraventricular nucleus of the hypothalamus (*green*, hPVN), subparaventricular zone (*pink*, sPVZ), and dorsomedial hypothalamus (*brown*, DMH). The body is likely "primed" for food intake during the day via the SCN→sPVZ→DMH→hPVN circuit that ultimately regulates synthesis and secretion of corticosterone from the adrenal cortex and glucose production in the liver. Cortical arousal is regulated in part via the SCN→DMH→LH circuit. The SCN is reciprocally connected with the arcuate nucleus (Arc). The Arc contains receptors responsive to blood-borne hormones and therefore represents a key feedback pathway to the SCN. The SCN also receives input from other brain areas and thus functions as an integrator of both environmental and internal signals. (DMV) Dorsal motor nucleus of the vagus; (IML) intermediolateral column of spinal cord; (LC) locus coeruleus; (LH) lateral hypothalamus; (Raphe) Raphe nucleus; (Ach) acetylcholine neurotransmitter; (NE) norepinephrine neurotransmitter; (5HT) serotonin neurotransmitter. (*B*) Examples of various body rhythms controlled by the SCN. Melatonin is low during the day and high at night. Corticosterone and orexin levels are coupled to the behavioral state of the animal, high during wakefulness and low during sleep. Note that the timing and shape of the waveforms are different. Distinct cell types and heterogeneity of phase within the SCN may serve to regulate the diverse timing of these targets.

also sends projections to the lateral hypothalamus (LH) to regulate orexin secretion and cortical arousal (Fig. 9A, highlighted black). Traditionally, the ascending arousal system was thought to regulate cortical function via the neuromodulators: acetylcholine, serotonin, norepinephrin, in addition to dopamine and histamine. In 1999, a new regulatory pathway was added to the list. It was demonstrated that disruption of orexin signaling causes symptoms of narcolepsy (Chemelli et al. 1999; Lin et al. 1999). Orexins may represent a key link between SCN pacemaker function and arousal. Orexin neurons are active during waking (Bayer et al. 2004; Lee et al. 2005; Mileykovskiy et al. 2005), and there is a circadian rhythm in orexin cerebral spinal fluid levels (Fig. 9B) that is lost in SCN-lesioned animals (Deboer et al. 2004; Zhang et al. 2004).

The SCN temporally organizes processes that are associated with inactive/awake states via the orexin pathway and the corticosterone hypothalamic-adrenal axis. The SCN also temporally organizes processes that are specifically associated with the light cycle such as melatonin secretion (Fig. 9B). Thus, the SCN coordinates a number of differentially timed rhythmic events. Coordination of peripheral oscillations is achieved via distinct, identifiable output circuits that have received much research attention. Of course, to effectively coordinate differentially timed rhythms, the SCN must also receive feedback regarding the state of the various body oscillations. Feedback to the SCN is an area of research receiving more recent attention. It may well be that pathways classically described as serving "nonphotic resetting" have an important role in feedback, enabling the SCN to integrate peripheral signals, and thus promote proper organization of diverse body rhythms.

ACKNOWLEDGMENTS

We thank Drs. Marty Zatz, Doug McMahon, Mirjam Münch, and Sean Cain for comments on earlier versions of the manuscript.

REFERENCES

Abrahamson E.E. and Moore R.Y. 2001. Suprachiasmatic nucleus in the mouse: Retinal innervation, intrinsic organization and efferent projections. *Brain Res.* **916:** 172.

Albus H., Vansteensel M.J., Michel S., Block G.D., and Meijer J.H. 2005. A GABAergic mechanism is necessary for coupling dissociable ventral and dorsal regional oscillators within the circadian clock. *Curr. Biol.* **15:** 886.

Antle M.C. and Silver R. 2005. Orchestrating time: Arrangements of the brain circadian clock. *Trends Neurosci.* **28:** 145.

Aton S.J. and Herzog E.D. 2005. Come together, right...now: Synchronization of rhythms in a mammalian circadian clock. *Neuron* **48:** 531.

Aton S.J., Colwell C.S., Harmar A.J., Waschek J., and Herzog E.D. 2005. Vasoactive intestinal polypeptide mediates circadian rhythmicity and synchrony in mammalian clock neurons. *Nat. Neurosci.* **8:** 476.

Bayer L., Serafin M., Eggermann E., Saint-Mleux B., Machard D., Jones B.E., and Muhlethaler M. 2004. Exclusive postsynaptic action of hypocretin-orexin on sublayer 6b cortical neurons. *J. Neurosci.* **24:** 6760.

Beaule C., Houle L.M., and Amir S. 2003. Expression profiles of PER2 immunoreactivity within the shell and core regions of the rat suprachiasmatic nucleus: Lack of effect of photic entrainment and disruption by constant light. *J. Mol. Neurosci.* **21:** 133.

Berson D.M. 2003. Strange vision: Ganglion cells as circadian photoreceptors. *Trends Neurosci.* **26:** 314.

Bouskila Y. and Dudek F.E. 1995. A rapidly activating type of outward rectifier K^+ current and A-current in rat suprachiasmatic nucleus neurones. *J. Physiol.* **488:** 339.

Buijs R.M. and Kalsbeek A. 2001. Hypothalamic integration of central and peripheral clocks. *Nat. Rev. Neurosci.* **2:** 521.

Buijs R.M., Scheer F.A., Kreier F., Yi C., Bos N., Goncharuk V.D., and Kalsbeek A. 2006. Organization of circadian functions: Interaction with the body. *Prog. Brain Res.* **153:** 341.

Chemelli R.M., Willie J.T., Sinton C.M., Elmquist J.K., Scammell T., Lee C., Richardson J.A., Williams S.C., Xiong Y., Kisanuki Y., Fitch T.E., Nakazato M., Hammer R.E., Saper C.B., and Yanagisawa M. 1999. Narcolepsy in orexin knockout mice: Molecular genetics of sleep regulation. *Cell* **98:** 437.

Cheng M.Y., Bullock C.M., Li C., Lee A.G., Bermak J.C., Belluzzi J., Weaver D.R., Leslie F.M., and Zhou Q.Y. 2002. Prokineticin 2 transmits the behavioural circadian rhythm of the suprachiasmatic nucleus. *Nature* **417:** 405.

Coogan A.N. and Piggins H.D. 2004. MAP kinases in the mammalian circadian system: Key regulators of clock function. *J. Neurochem.* **90:** 769.

Cutler D.J., Haraura M., Reed H.E., Shen S., Sheward W.J., Morrison C.F., Marston H.M., Harmar A.J., and Piggins H.D. 2003. The mouse VPAC2 receptor confers suprachiasmatic nuclei cellular rhythmicity and responsiveness to vasoactive intestinal polypeptide in vitro. *Eur. J. Neurosci.* **17:** 197.

Damiola F., Le Minh N., Preitner N., Kornmann B., Fleury-Olela F., and Schibler U. 2000. Restricted feeding uncouples circadian oscillators in peripheral tissues from the central pacemaker in the suprachiasmatic nucleus. *Genes Dev.* **14:** 2950.

Deboer T., Vansteensel M.J., Detari L., and Meijer J.H. 2003. Sleep states alter activity of suprachiasmatic nucleus neurons. *Nat. Neurosci.* **6:** 1086.

Deboer T., Overeem S., Visser N.A., Duindam H., Frolich M., Lammers G.J., and Meijer J.H. 2004. Convergence of circadian and sleep regulatory mechanisms on hypocretin-1. *Neuroscience* **129:** 727.

Debruyne J.P., Weaver D.R., and S.M. Reppert. 2007. CLOCK and NPAS2 have overlapping roles in the suprachiasmatic circadian clock. *Nat. Neurosci.* **10:** 543.

de Jeu M., Hermes M., and Pennartz C. 1998. Circadian modulation of membrane properties in slices of rat suprachiasmatic nucleus. *Neuroreport* **9:** 3725.

de la Iglesia H.O. and Schwartz W.J. 2006. Minireview. Timely ovulation: Circadian regulation of the female hypothalamo-pituitary-gonadal axis. *Endocrinology* **147:** 1148.

Deurveilher S. and Semba K. 2005. Indirect projections from the suprachiasmatic nucleus to major arousal-promoting cell groups in rat: Implications for the circadian control of behavioural state. *Neuroscience* **130:** 165.

Do M.T. and Bean B.P. 2003. Subthreshold sodium currents and pacemaking of subthalamic neurons: Modulation by slow inactivation. *Neuron* **39:** 109.

Dziema H., Oatis B., Butcher G.Q., Yates R., Hoyt K.R., and Obrietan K. 2003. The ERK/MAP kinase pathway couples light to immediate-early gene expression in the suprachiasmatic nucleus. *Eur. J. Neurosci.* **17:** 1617.

Gooley J.J., Schomer A., and Saper C.B. 2006. The dorsomedial hypothalamic nucleus is critical for the expression of food-entrainable circadian rhythms. *Nat. Neurosci.* **9:** 398.

Green D.J. and Gillette R. 1982. Circadian rhythm of firing rate recorded from single cells in the rat suprachiasmatic brain slice. *Brain Res.* **245:** 198.

Groos G. and Hendriks J. 1982. Circadian rhythms in electrical discharge of rat suprachiasmatic neurones recorded in vitro. *Neurosci. Lett.* **34:** 283.

Guo H., Brewer J.M., Lehman M.N., and Bittman E.L. 2006. Suprachiasmatic regulation of circadian rhythms of gene expression in hamster peripheral organs: Effects of transplant-

ing the pacemaker. *J. Neurosci.* **26:** 6406.
Hausser M., Raman I.M., Otis T., Smith S.L., Nelson A., du Lac S., Loewenstein Y., Mahon S., Pennartz C., Cohen I., and Yarom Y. 2004. The beat goes on: Spontaneous firing in mammalian neuronal microcircuits. *J. Neurosci.* **24:** 9215.
Herzog E.D., Takahashi J.S., and Block G.D. 1998. *Clock* controls circadian period in isolated suprachiasmatic nucleus neurons. *Nat. Neurosci.* **1:** 708.
Herzog E.D., Geusz M.E., Khalsa S.B., Straume M., and Block G.D. 1997. Circadian rhythms in mouse suprachiasmatic nucleus explants on multimicroelectrode plates. *Brain Res.* **757:** 285.
Honma S., Shirakawa T., Katsuno Y., Namihira M., and Honma K. 1998. Circadian periods of single suprachiasmatic neurons in rats. *Neurosci. Lett.* **250:** 157.
Hungs M. and Mignot E. 2001. Hypocretin/orexin, sleep and narcolepsy. *Bioessays* **23:** 397.
Inagaki N., Honma S., Ono D., Tanahashi Y., and Honma K.-I. 2007. Separate oscillating cell groups in mouse suprachiasmatic nucleus couple photoperiodically to the onset and end of daily activity. *Proc. Natl. Acad. Sci.* **104:** 7664.
Inouye S.T. and Kawamura H. 1979. Persistence of circadian rhythmicity in a mammalian hypothalamic "island" containing the suprachiasmatic nucleus. *Proc. Natl. Acad. Sci.* **76:** 5962.
———. 1982. Characteristics of a circadian pacemaker in the suprachiasmatic nucleus. *J. Comp. Physiol.* **146:** 153.
Itri J.N., Michel S., Vansteensel M.J., Meijer J.H., and Colwell C.S. 2005. Fast delayed rectifier potassium current is required for circadian neural activity. *Nat. Neurosci.* **8:** 650.
Jackson A.C., Yao G.L., and Bean B.P. 2004. Mechanism of spontaneous firing in dorsomedial suprachiasmatic nucleus neurons. *J. Neurosci.* **24:** 7985.
Kononenko N.I. and Dudek F.E. 2006. Persistent calcium current in rat suprachiasmatic nucleus neurons. *Neuroscience* **138:** 377.
Kononenko N.I., Shao L.R., and Dudek F.E. 2004. Riluzole-sensitive slowly inactivating sodium current in rat suprachiasmatic nucleus neurons. *J. Neurophysiol.* **91:** 710.
Kramer A., Yang F.C., Snodgrass P., Li X., Scammell T.E., Davis F.C., and Weitz C.J. 2001. Regulation of daily locomotor activity and sleep by hypothalamic EGF receptor signaling. *Science* **294:** 2511.
Kuhlman S.J. and McMahon D.G. 2004. Rhythmic regulation of membrane potential and potassium current persists in SCN neurons in the absence of environmental input. *Eur. J. Neurosci.* **20:** 1113.
Kuhlman S.J., Quintero J.E., and McMahon D.G. 2000. GFP fluorescence reports Period 1 circadian gene regulation in the mammalian biological clock. *Neuroreport* **11:** 1479.
Kuhlman S.J., Silver R., Le Sauter J., Bult-Ito A., and McMahon D.G. 2003. Phase resetting light pulses induce Per1 and persistent spike activity in a subpopulation of biological clock neurons. *J. Neurosci.* **23:** 1441.
Lee M.G., Hassani O.K., and Jones B.E. 2005. Discharge of identified orexin/hypocretin neurons across the sleep-waking cycle. *J. Neurosci.* **25:** 6716.
Lin L., Faraco J., Li R., Kadotani H., Rogers W., Lin X., Qiu X., de Jong P.J., Nishino S., and Mignot E. 1999. The sleep disorder canine narcolepsy is caused by a mutation in the hypocretin (orexin) receptor 2 gene. *Cell* **98:** 365.
Liu C., Weaver D.R., Strogatz S.H., and Reppert S.M. 1997. Cellular construction of a circadian clock: Period determination in the suprachiasmatic nuclei. *Cell* **91:** 855.
Long M.A., Jutras M.J., Connors B.W., and Burwell R.D. 2005. Electrical synapses coordinate activity in the suprachiasmatic nucleus. *Nat. Neurosci.* **8:** 61.
Lundkvist G.B., Kwak Y., Davis E.K., Tei H., and Block G.D. 2005. A calcium flux is required for circadian rhythm generation in mammalian pacemaker neurons. *J. Neurosci.* **25:** 7682.
Mason R. 1991. The effects of continuous light exposure on Syrian hamster suprachiasmatic (SCN) neuronal discharge activity in vitro. *Neurosci. Lett.* **123:** 160.
Maywood E.S., Reddy A.B., Wong G.K., O'Neill J.S., O'Brien J.A., McMahon D.G., Harmar A.J., Okamura H., and Hastings M.H. 2006. Synchronization and maintenance of timekeeping in suprachiasmatic circadian clock cells by neuropeptidergic signaling. *Curr. Biol.* **16:** 599.
McCormick D.A. and Huguenard J.R. 1992. A model of the electrophysiological properties of thalamocortical relay neurons. *J. Neurophysiol.* **68:** 1384.
Meijer J.H., Watanabe K., Schaap J., Albus H., and Detari L. 1998. Light responsiveness of the suprachiasmatic nucleus: Long-term multiunit and single-unit recordings in freely moving rats. *J. Neurosci.* **18:** 9078.
Messager S., Ross A.W., Barrett P., and Morgan P.J. 1999. Decoding photoperiodic time through Per1 and ICER gene amplitude. *Proc. Natl. Acad. Sci.* **96:** 9938.
Michel S., Geusz M.E., Zaritsky J.J., and Block G.D. 1993. Circadian rhythm in membrane conductance expressed in isolated neurons. *Science* **259:** 239.
Mieda M., Williams S.C., Richardson J.A., Tanaka K., and Yanagisawa M. 2006. The dorsomedial hypothalamic nucleus as a putative food-entrainable circadian pacemaker. *Proc. Natl. Acad. Sci.* **103:** 12150.
Mikkelsen J.D. and Vrang N. 1994. A direct pretectosuprachiasmatic projection in the rat. *Neuroscience* **62:** 497.
Mileykovskiy B.Y., Kiyashchenko L.I., and Siegel J.M. 2005. Behavioral correlates of activity in identified hypocretin/orexin neurons. *Neuron* **46:** 787.
Moore R.Y. 1996. Entrainment pathways and the functional organization of the circadian system. *Prog. Brain Res.* **111:** 103.
Moore R.Y. and Eichler V.B. 1972. Loss of a circadian adrenal corticosterone rhythm following suprachiasmatic lesions in the rat. *Brain Res.* **42:** 201.
Moore R.Y. and Speh J.C. 1993. GABA is the principal neurotransmitter of the circadian system. *Neurosci. Lett.* **150:** 112.
Morin L.P., Shivers K.Y, Blanchard J.H., and Muscat L. 2006. Complex organization of mouse and rat suprachiasmatic nucleus. *Neuroscience* **137:** 1285.
Morse D. and Sassone-Corsi P. 2002. Time after time: Inputs to and outputs from the mammalian circadian oscillators. *Trends Neurosci.* **25:** 632.
Mrugala M., Zlomanczuk P., Jagota A., and Schwartz W.J. 2000. Rhythmic multiunit neural activity in slices of hamster suprachiasmatic nucleus reflect prior photoperiod. *Am. J. Physiol. Regul. Integr. Comp. Physiol.* **278:** R987.
Nakamura W., Honma S., Shirakawa T., and Honma K. 2001. Regional pacemakers composed of multiple oscillator neurons in the rat suprachiasmatic nucleus. *Eur. J. Neurosci.* **14:** 666.
———. 2002. Clock mutation lengthens the circadian period without damping rhythms in individual SCN neurons. *Nat. Neurosci.* **5:** 399.
Nakamura W., Yamazaki S., Takasu N.N., Mishima K., and Block G.D. 2005. Differential response of Period 1 expression within the suprachiasmatic nucleus. *J. Neurosci.* **25:** 5481.
Nelson A.B., Krispel C.M., Sekirnjak C., and du Lac S. 2003. Long-lasting increases in intrinsic excitability triggered by inhibition. *Neuron* **40:** 609.
Nitabach M.N., Blau J., and Holmes T.C. 2002. Electrical silencing of *Drosophila* pacemaker neurons stops the free-running circadian clock. *Cell* **109:** 485.
Nüsslein-Hildesheim B., O'Brien J.A., Ebling F.J., Maywood E.S., and Hastings M.H. 2000. The circadian cycle of mPER clock gene products in the suprachiasmatic nucleus of the Siberian hamster encodes both daily and seasonal time. *Eur. J. Neurosci.* **12:** 2856.
Ohta H., Yamazaki S., and McMahon D.G. 2005. Constant light desynchronizes mammalian clock neurons. *Nat. Neurosci.* **8:** 267.
Pennartz C.M., Bierlaagh M.A., and Geurtsen A.M. 1997. Cellular mechanisms underlying spontaneous firing in rat suprachiasmatic nucleus: Involvement of a slowly inactivating component of sodium current. *J. Neurophysiol.* **78:** 1811.
Pennartz C.M., de Jeu M.T., Bos N.P., Schaap J., and Geurtsen

A.M. 2002. Diurnal modulation of pacemaker potentials and calcium current in the mammalian circadian clock. *Nature* **416:** 286.

Quintero J.E., Kuhlman S.J., and McMahon D.G. 2003. The biological clock nucleus: A multiphasic oscillator network regulated by light. *J. Neurosci.* **23:** 8070.

Ralph M.R., Foster R.G., Davis F.C., and Menaker M. 1990. Transplanted suprachiasmatic nucleus determines circadian period. *Science* **247:** 975.

Raman I.M., Gustafson A.E., and Padgett D. 2000. Ionic currents and spontaneous firing in neurons isolated from the cerebellar nuclei. *J. Neurosci.* **20:** 9004.

Refinetti R., Kaufman C.M., and Menaker M. 1994. Complete suprachiasmatic lesions eliminate circadian rhythmicity of body temperature and locomotor activity in golden hamsters. *J. Comp. Physiol. A* **175:** 223.

Rutter J., Reick M., and McKnight S.L. 2002. Metabolism and the control of circadian rhythms. *Annu. Rev. Biochem.* **71:** 307.

Rutter J., Reick M., Wu L.C., and McKnight S.L. 2001. Regulation of clock and NPAS2 DNA binding by the redox state of NAD cofactors. *Science* **293:** 510.

Saper C.B., Scammell T.E., and Lu J. 2005. Hypothalamic regulation of sleep and circadian rhythms. *Nature* **437:** 1257.

Schaap J., Albus H., VanderLeest H.T., Eilers P.H., Detari L., and Meijer J.H. 2003. Heterogeneity of rhythmic suprachiasmatic nucleus neurons: Implications for circadian waveform and photoperiodic encoding. *Proc. Natl. Acad. Sci.* **100:** 15994.

Schaap J., Bos N.P., de Jeu M.T., Geurtsen A.M., Meijer J.H., and Pennartz C.M. 1999. Neurons of the rat suprachiasmatic nucleus show a circadian rhythm in membrane properties that is lost during prolonged whole-cell recording. *Brain Res.* **815:** 154.

Schwartz W.J. and Gainer H. 1977. Suprachiasmatic nucleus: Use of 14C-labeled deoxyglucose uptake as a functional marker. *Science* **197:** 1089.

Schwartz W.J., Gross R.A., and Morton M.T. 1987. The suprachiasmatic nuclei contain a tetrodotoxin-resistant circadian pacemaker. *Proc. Natl. Acad. Sci.* **84:** 1694.

Shibata S., Liou S., Ueki S., and Oomura Y. 1984. Influence of environmental light-dark cycle and enucleation on activity of suprachiasmatic neurons in slice preparations. *Brain Res.* **302:** 75.

Shibata S., Oomura Y., Kita H., and Hattori K. 1982. Circadian rhythmic changes of neuronal activity in the suprachiasmatic nucleus of the rat hypothalamic slice. *Brain Res.* **247:** 154.

Silver R., LeSauter J., Tresco P.A., and Lehman M.N. 1996. A diffusible coupling signal from the transplanted suprachiasmatic nucleus controlling circadian locomotor rhythms. *Nature* **382:** 810.

Smith S.L. and Otis T.S. 2003. Persistent changes in spontaneous firing of Purkinje neurons triggered by the nitric oxide signaling cascade. *J. Neurosci.* **23:** 367.

Stephan F.K. and Zucker I. 1972. Circadian rhythms in drinking behavior and locomotor activity of rats are eliminated by hypothalamic lesions. *Proc. Natl. Acad. Sci.* **69:** 1583.

Stokkan K.A., Yamazaki S., Tei H., Sakaki Y., and Menaker M. 2001. Entrainment of the circadian clock in the liver by feeding. *Science* **291:** 490.

Stoleru D., Peng Y., Nawathean P., and Rosbash M. 2005. A resetting signal between *Drosophila* pacemakers synchronizes morning and evening activity. *Nature* **438:** 238.

Stoleru D., Nawathean P., Fernandez M.P., Menet J.S., Ceriani M.F., and Rosbash M. 2007. The *Drosophila* circadian network is a seasonal timer. *Cell* **129:** 207.

Stratmann M. and Schibler U. 2006. Properties, entrainment, and physiological functions of mammalian peripheral oscillators. *J. Biol. Rhythms* **21:** 494.

Strecker G.J., Wuarin J.P., and Dudek F.E. 1997. $GABA_A$-mediated local synaptic pathways connect neurons in the rat suprachiasmatic nucleus. *J. Neurophysiol.* **78:** 2217.

Sudo M., Sasahara K., Moriya T., Akiyama M., Hamada T., and Shibata S. 2003. Constant light housing attenuates circadian rhythms of mPer2 mRNA and mPER2 protein expression in the suprachiasmatic nucleus of mice. *Neuroscience* **121:** 493.

Sumova A., Jac M., Sladek M., Sauman I., and Illnerova H. 2003. Clock gene daily profiles and their phase relationship in the rat suprachiasmatic nucleus are affected by photoperiod. *J. Biol. Rhythms* **18:** 134.

Sumova A., Travnickova Z., Peters R., Schwartz W.J., and Illnerova H. 1995. The rat suprachiasmatic nucleus is a clock for all seasons. *Proc. Natl. Acad. Sci.* **92:** 7754.

Taddese A. and Bean B.P. 2002. Subthreshold sodium current from rapidly inactivating sodium channels drives spontaneous firing of tuberomammillary neurons. *Neuron* **33:** 587.

VanderLeest H.T., Houben T., Michel S., Deboer T., Albus H., Vansteensel M.J., Block G.D., and Meijer J.H. 2007. Seasonal encoding by the circadian pacemaker of the SCN. *Curr. Biol.* **17:** 468.

Vansteensel M.J., Yamazaki S., Albus H., Deboer T., Block G.D., and Meijer J.H. 2003. Dissociation between circadian Per1 and neuronal and behavioral rhythms following a shifted environmental cycle. *Curr. Biol.* **13:** 1538.

Vitaterna M.H., King D.P., Chang A.M., Kornhauser J.M., Lowrey P.L., McDonald J.D., Dove W.F., Pinto L.H., Turek F.W., and Takahashi J.S. 1994. Mutagenesis and mapping of a mouse gene, *Clock*, essential for circadian behavior. *Science* **264:** 719.

Welsh D.K., Logothetis D.E., Meister M., and Reppert S.E. 1995. Individual neurons dissociated from rat suprachiasmatic nucleus express independently phased circadian firing rhythms. *Neuron* **14:** 697.

Wilsbacher L.D., Yamazaki S., Herzog E.D., Song E.J., Radcliffe L.A., Abe M., Block G., Spitznagel E., Menaker M., and Takahashi J.S. 2002. Photic and circadian expression of luciferase in *mPeriod1-luc* transgenic mice *in vivo*. *Proc. Natl. Acad. Sci.* **99:** 489.

Yamaguchi S., Isejima H., Matsuo T., Okura R., Yagita K., Kobayashi M., and Okamura H. 2003. Synchronization of cellular clocks in the suprachiasmatic nucleus. *Science* **302:** 1408.

Yamaguchi S., Kobayashi M., Mitsui S., Ishida Y., van der Horst G.T.J., Suzuki M., Shibata S., and Okamura H. 2001. View of a mouse clock gene ticking. *Nature* **409:** 684.

Yamazaki S., Numano R., Abe M., Hida A., Takahashi R., Ueda M., Block G.D., Sakaki Y., Menaker M., and Tei H. 2000. Resetting central and peripheral circadian oscillators in transgenic rats. *Science* **288:** 682.

Yan L., Takekida S., Shigeyoshi Y., and Okamura H. 1999. Per1 and Per2 gene expression in the rat suprachiasmatic nucleus: Circadian profile and the compartment-specific response to light. *Neuroscience* **94:** 141.

Yoo S.H., Yamazaki S., Lowrey P.L., Shimomura K., Ko C.H., Buhr E.D., Siepka S.M., Hong H.K., Oh W.J., Yoo O.J., Menaker M., and Takahashi J.S. 2004. PERIOD2::LUCIFERASE real-time reporting of circadian dynamics reveals persistent circadian oscillations in mouse peripheral tissues. *Proc. Natl. Acad. Sci.* **101:** 5339.

Zhang S., Zeitzer J.M., Yoshida Y., Wisor J.P., Nishino S., Edgar D.M., and Mignot E. 2004. Lesions of the suprachiasmatic nucleus eliminate the daily rhythm of hypocretin-1 release. *Sleep* **27:** 619.

Biological Rhythms Workshop IC: Sleep and Rhythms

M.Y. Münch, S.W. Cain, and J.F. Duffy

Division of Sleep Medicine, Department of Medicine, Brigham and Women's Hospital and Division of Sleep Medicine, Harvard Medical School, Boston, Massachusetts 02115

Rhythms of sleep and wakefulness (typically measured as rest/activity rhythms) are among the most prominent of biological rhythms and therefore were among the first to be recorded in early chronobiological studies. These rhythms can provide useful information about the central biological clock, although an appreciation of the problems associated with using rest/activity to infer central clock function is important in the design and interpretation of chronobiological experiments in both animals and humans. Here, we review the anatomical and neurophysiologic bases of sleep regulation in mammals as well as similarities and differences between the sleep of humans and that of other organisms. We outline how human sleep is measured, the role of the circadian system in models of human sleep regulation, and human circadian rhythm sleep disorders. Although the function of sleep is still not completely understood, sleep has a critical role for human health, and we have attempted to outline the role that the circadian timing system has in regulating human sleep and in contributing to sleep disorders.

HISTORICAL USE OF REST/ACTIVITY RHYTHMS IN CHRONOBIOLOGY

The rest/activity cycle is one of the most prominent rhythms in most animal species (Fig. 1). Because this behavioral rhythm is easy to observe and measure, the rest/activity cycle has been used for decades as an index of clock function. Rest/activity cycles are also often used as an index of sleep/wake function, although the agreement between how closely rest/activity matches actual sleep/wake will depend on the method used to measure rest/activity and the species in question. In this chapter, we outline how the rest/activity cycle has been used as a measure of clock function in animal and human circadian rhythm studies, how the sleep/wakefulness cycle is controlled by the circadian timing system, and how rest/activity and sleep/wake differ between typical rodent models used in circadian research and humans.

TIMING OF THE REST/ACTIVITY CYCLE

Rest/Activity in the 24-Hour Natural Day

As described in the Biological Rhythms Workshop Introduction, in nature, most organisms are exposed to a 24-hour cycle of light and dark, and this daily fluctuation of light and dark is the most powerful periodic environmental stimulus for entraining circadian rhythms. In most organisms, the periodic light/dark (LD) cycle normally synchronizes (or entrains) the near 24-hour endogenous circadian period to the exact 24-hour period of the environmental cycle (Pittendrigh 1961; Pittendrigh and Daan 1976a; Orth et al. 1979; Albers et al. 1984). Because the timing of the rest/activity cycle is determined (at least in part) by the circadian clock, the cycle of rest and activity (sleeping and wakefulness) is expressed as a 24-hour entrained rhythm in most organisms.

Although all organisms share the 24-hour LD cycle that entrains the circadian clock, and therefore the timing of the rest/activity cycle, the exact timing of rest/sleep and activity/wake varies between species. The most obvious difference in the timing of rest and activity is between nocturnal and diurnal organisms. Nocturnal animals have most of their activity during the biological night and spend most of the day resting, whereas diurnal animals are mostly active during the day and rest at night. The physiological mechanism of diurnality and nocturnality remains a mystery. Within nocturnal and diurnal groups, the exact timing of rest and activity can vary greatly among species and even among individuals within a species. This fact is only too apparent when one is awo-

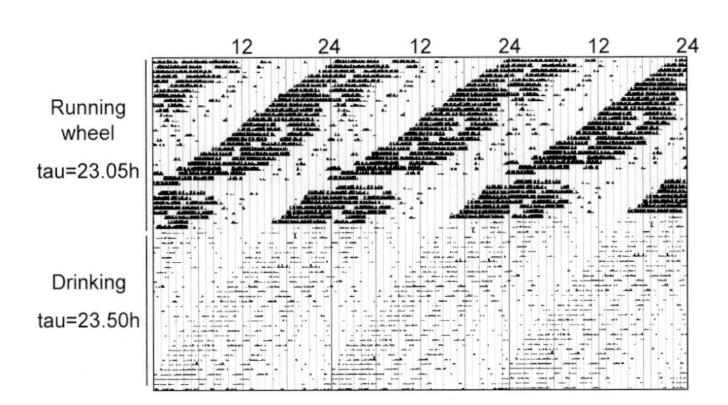

Figure 1. Triple-plotted activity/rest record from a C57B6 mouse studied in DD. As is typical in mice, activity onset is earlier each day than the previous day, indicating a period shorter than 24 hours. (*Upper half*) Activity recordings are from a running wheel; (*lower half*) animal's running wheel was locked and the activity is from drinking at the water spout. The period was shorter when the animal was able to run in the wheel (23.05 hours vs. 23.50 hours), indicating that running-wheel activity was affecting the observed circadian period.

ken by the chirping of early-rising diurnal birds or kept awake by noisy neighbors!

The timing of a circadian-clock-controlled physiological or behavioral rhythm with respect to the LD cycle is referred to as the "phase angle of entrainment." One such example of a phase angle is the time of activity onset relative to dawn in diurnal animals. In the natural environment, this phase angle of entrainment is determined by the endogenous period of the clock and its sensitivity to the phase-shifting effects of light. Given that both the endogenous period and light sensitivity are not static, nor is the day-to-day exposure to environmental light, phase angle can take on very different values among species, among individuals within a species, and even within a single individual.

Rest/Activity Independent of Light/Dark Cycles

Evaluation of rest/activity-sleep/wake timing in animals. It is a defining characteristic of endogenous circadian rhythms that they persist under constant conditions, without rhythmic input from external entraining agents such as the LD cycle. Initially, constant conditions were defined as constant environmental (external) conditions, hence long-term studies of animals in constant darkness (DD) or constant light (LL). In both DD and LL, circadian rhythms in rest and activity have been found to "free-run" with an endogenous period different from that of the 24-hour day. To place animals in a constant environment is still a widely used paradigm to quantify circadian phenotypes. The assumption underlying such experiments is that removal of periodic environmental cycles is sufficient to uncover the contribution of the circadian pacemaker to the observed rhythmicity. As we point out below, this assumption may not always be true.

Exposure to LL and DD does not have equivalent effects on the timing of rest and activity. The rest/activity cycle is different in DD and LL, and within LL conditions, the observed free-running period depends on light intensity ("Aschoff's rule"; Aschoff 1961). For nocturnal mammals, according to "Aschoff's rule," with increased light intensity, the period of the rest/activity cycle becomes longer and the length of time spent active decreases. For diurnal animals, the reverse tends to occur, with increasing light resulting in a shorter rest/activity cycle period and a longer active episode within each cycle. Furthermore, although the circadian period is usually stable upon entering DD, it has been found to take up to 60 days to reach a steady state while in LL (Pittendrigh and Daan 1976b).

Perhaps the most striking difference between the response to DD and LL is the effect on the integrity of the rest/activity rhythm. Although animals have been found to maintain stable rest/activity patterns throughout their life span in DD or very dim light (see, e.g., Richter 1968; Davis and Viswanathan 1998; Duffy et al. 1999a), bright LL is often associated with disruption of the rest/activity cycle, as outlined in Biological Rhythms Workshop IB (Pittendrigh and Daan 1976b; Honma and Hiroshige 1978; Shibata et al. 1984; Mason 1991; de la Iglesia et al. 2000; Jagota et al. 2000; Beaule et al. 2003; Sudo et al. 2003). This is not surprising, because DD represents an absence of light input to the clock, whereas with bright LL, there is constant exposure of the biological clock to a powerful entraining/resetting agent.

Given the disruptive effects of LL, when attempting to measure the endogenous rhythm in the rest/activity cycle of laboratory animals, the gold-standard method has been the measure of activity in DD. This method, however, is not without drawbacks. Activity itself has been demonstrated to be a powerful entraining agent. Phase shifts to acute presentation of a novel running wheel, eliciting intense locomotion, have been demonstrated in the hamster (Reebs and Mrosovsky 1989). Self-selected activity is also associated with an alteration of rest/activity timing. Increased locomotor activity has been shown to shorten the period of the rest/activity cycle in mice (Edgar et al. 1991) and rats (Yamada et al. 1988). An example of the effect of locomotor activity on the period of the rest/activity rhythm is presented in Figure 1, which shows the rest/activity rhythm of a mouse with access to a running wheel and with the running wheel locked. The period of the rest/activity rhythm was shorter when the animal was allowed access to the running wheel and lengthened when the wheel was locked and only drinking rhythms were measured (see also Cheng et al. 2004).

Feedback to the circadian system from the behavioral state of an animal is not limited to activity. Sleep has also been shown to affect the timing of the biological clock. Sleep deprivation has been shown to shift the phase of the rest/ activity cycle (Antle and Mistlberger 2000). Sleep can also modulate the firing rate of suprachiasmatic nucleus (SCN) neurons (Deboer et al. 2003). The timing of both sleep and activity feeds back onto the core biological clock and affects the timing of these behaviors. It is therefore important to consider the behavior of the experimental organism when measuring circadian activity patterns and to understand that controlling environmental conditions does not consequently create a constant internal environment.

Although the timing of the rest/activity cycle tends to be fairly stable in DD, light exposure history has been found to have an effect on the period of the cycle once the light is removed. The period-altering effect of light exposure was first reported by Pittendrigh in 1960 (Pittendrigh 1961). Entrainment to LD cycles of various lengths has been shown to affect the circadian period such that the endogenous timing approaches that of the environment. For example, in a classic experiment, Pittendrigh and Daan (1976b) entrained mice to LD cycles of 23 and 25 hours, with half of each cycle light and the other half darkness. They increased or decreased the LD cycle by 10 minutes each day for 18 days, resulting in a final LD cycle of 20 or 28 hours, respectively. Following this, the mice were released into DD to have their circadian period assessed. The period was longer in mice following entrainment to a 28-hour LD cycle than in mice entrained to a 20-hour LD cycle, and these period values were longer and shorter, respectively, than the estimates of period for this species after exposure to 24-hour LD cycles. Such history-dependent effects on period, called "aftereffects of entrainment," to longer and shorter than 24-hour day lengths have also been reported in hamsters (Reebs and Doucet 1997), albino rats (Stewart et al. 1990), and more recently humans (Scheer et al. 2007).

Aftereffects can also occur in response to different photoperiods. Using an LD ratio of 1:23 versus 18:6, four of five white-footed mice showed a shorter rest/activity cycle period after entrainment to the longer photoperiod (Pittendrigh and Daan 1976b), and the effects on period were observed to last for at least 300 cycles (nearly a year!) in some mice.

Aftereffects of light history on the timing of the rest/activity cycle have been demonstrated even after single exposures to light. Despite the wide array of species studied, findings are strikingly consistent: Aftereffects of light pulses on the timing of the rest/activity cycle depend on the magnitude and direction of the phase shift induced by the light. Delaying phase shifts produce a lengthening of period, and advancing light pulses produce a shorter period. The magnitude of the effect on period is positively correlated with the magnitude of phase shift produced (Pittendrigh 1961; DeCoursey 1964; Eskin 1971; Kramm 1971; Gerkema et al. 1993; Sharma and Daan 2002). The likely function of aftereffects is to stabilize entrainment when exposed to fluctuations of light input (Pittendrigh and Daan 1976b; DeCoursey 1989; Beersma et al. 1999), thus maintaining a fixed phase angle of entrainment to the environment.

Together, these various findings demonstrate that even when an organism is studied under constant conditions, the recent history and internal physiological state, as well as the organism's periodic behavior, can influence the central pacemaker, thus affecting current observations and the outcome of any experimental manipulation.

Evaluation of rest/activity timing in humans. Early attempts to measure the endogenous timing of the human rest/activity cycle were performed by Aschoff and Wever (1962) in an isolation unit free from time cues. Several variables (including rest/activity) were shown to have a daily rhythm of about 25.1 hours. Longer-term experiments conducted in isolation units and caves also found that human circadian rhythms had a period close to 25 hours (Wever 1979, 1989; Aschoff and Wever 1981).

In retrospect, we now understand that the period estimates derived from these experiments were systematically confounded by the influence of self-selected light exposure (for review, see Czeisler 1995). This is because these early human studies were performed with two major differences from animal studies. First, many of the human studies were not conducted under constant lighting conditions (LL or DD), but instead in LD cycles. Second, unlike studies in other organisms where the organism under study had no control over the lighting, in these human studies, the experimental subjects were given control over some or all of the lighting, and the lighting levels used were bright enough (as we now know) to cause effects on the circadian system (Boivin et al. 1996; Zeitzer et al. 2000). This led to several factors that confounded the period estimates, although these confounds were not realized at the time.

First, humans, unlike most other organisms, have monophasic sleep (see below). This means that sleep generally occurs in one long bout each day, with waking occupying the remainder of the day. Light exposure (the most powerful environmental signal to the internal clock) is therefore present at some circadian phases and absent at others when the lights are switched on upon waking and turned off for sleep. Even when lighting is kept on continuously, there is a change in light input to the central clock associated with eyelid closure upon sleep.

Second, under entrained conditions, the circadian rhythm of sleep propensity shows a paradoxical relationship to usual sleep/wake timing. Unlike what might be expected, there is a strong circadian drive for *wakefulness* just before usual bedtime in humans and a peak drive for sleep near the end of the usual sleep episode (see below for more detail). Humans who are studied in time-free environments and who are allowed to self-select their sleep/wake times routinely remain awake later than they do under entrained conditions (Czeisler et al. 1981; Zulley et al. 1981; Klerman et al. 1996). When they do so, the additional evening light exposure that they receive produces systematic phase-delay shifts due to the phase-dependent effects of light on the circadian system (Honma and Honma 1988; Czeisler et al. 1989; Minors et al. 1991; Khalsa et al. 2003). These systematic phase delays are further exacerbated by the fact that the sleep/dark episode is now extended into the usual morning time, when light exposure would normally counteract evening light exposure and produce a phase advance (for more detail, see Klerman et al. 1996).

The net result of these factors was that there was an overestimation of the period of the circadian system in those "free-running" studies in humans (Klerman et al. 1996). Development of a protocol (the "forced desynchrony" protocol [Czeisler et al. 1990] first proposed by Nathaniel Kleitman in the 1930s [Kleitman 1939]), in which these confounds are minimized, has subsequently shown that the period of the human circadian system is much closer to 24 hours (Middleton et al. 1996; Hiddinga et al. 1997; Waterhouse et al. 1998; Carskadon et al. 1999; Czeisler et al. 1999; Kelly et al. 1999; Wyatt et al. 1999) and far less variable among individuals than those initial free-running studies had suggested.

DEFINITION OF SLEEP

Sleep as an Altered Behavioral State

The alternating succession of sleep (or rest) and wakefulness is a ubiquitous, evolutionarily conserved behavior, and has been shown in invertebrates, vertebrates, and humans (Campbell and Tobler 1984; Tobler 2005). In invertebrates and lower vertebrates (fish and amphibians), sleep is assessed by quiet resting behavior, typical body position, reduced reactivity to external stimuli, and by the relatively rapid reversibility of these states. In birds, reptiles, and mammals, sleep and wakefulness are further ascertained in characteristically dynamic changes of electrical brain activity. More recently, it has also become possible to assess sleep- and wake-dependent differences in brain activity of very small invertebrates such as fruit flies (*Drosophila melanogaster*; Nitz et al. 2002). These characteristics allow the differentiation between sleep and quiet wakefulness and between sleep and other vegetative states such as torpor, hibernation, or coma.

In most animals, the entire brain enters the sleep state, but this does not occur in all animals. In aquatic mammals such as cetaceans (whales and dolphins), only one brain hemisphere sleeps at a time, and the other half remains awake (Mukhametov et al. 1977). By sleeping with only half of their brain at once, these aquatic mammals can periodically swim to the surface to breathe during sleep. Interestingly, in those animals that exhibit aquatic *and* terrestrial living across their life span (e.g., fur seals; subfamily *Arctocephalinae*), sleep patterns change from unihemispheric to symmetric hemispheric sleep and vice versa, depending on the medium in which the animal is currently living (Siegel 2005).

Monophasic versus Polyphasic Sleep

One of the major differences between rest/activity and sleep/wake cycles of typical rodent models of circadian rhythmicity and humans is in the distribution of sleep and wakefulness across the 24-hour day. Most animal species have polyphasic sleep, with sleep and wakefulness distributed in frequent bouts, often with a predominance during the dark (diurnal species) or light (nocturnal species) phase (Tobler et al. 1990a). For example, the nocturnal Syrian hamster (*Mesocricetus auratus*) spends about 78% of the light period and 51% of the dark period sleeping (Tobler and Jaggi 1987), and Sprague-Dawley rats (an outbred strain from *Rattus norvegicus*) sleep during about 69% of the light period and 27% of the dark period (Franken et al. 1991). The rabbit (family Leporidae) spends 55% of the light period and 40% of the dark period sleeping (Tobler et al. 1990b), whereas the diurnal Siberian chipmunk (*Tamias sibericus*) sleeps during only 27.5% of the light period but spends 74% of the dark period sleeping (Dijk and Daan 1989). Guinea pigs (*Cavia porcellus*) seem to have little phase preference, spending 30% of the dark phase and 35% of the light phase asleep (Tobler et al. 1993).

In contrast, adult humans and a few other primates (e.g., some species of New World monkeys) exhibit monophasic sleep, spending nearly all of the light phase awake and most of the dark phase (depending on LD cycle duration) asleep, resulting in only one (or two) major bouts of sleep and wakefulness each 24-hour cycle. A major difference between these primates and other animals is that they can maintain wakefulness for an extended duration without any special environmental influence or intervention, with their consolidated sleep likely resulting from their ability to consolidate wakefulness. Due to artificial illumination, modern humans control their LD cycle, typically spending about one third of each 24-hour day sleeping. Whether the distribution of sleep and wakefulness into one bout per day, and the relative duration of sleep (about one third) and wakefulness (about two thirds) within the daily cycle, reflects a biological process or is instead an artifact of access to artificial lighting is not entirely clear, but some laboratory studies provide hints. Wehr (1991, 1996, 2001) conducted a study in which they kept human subjects in bed in complete darkness for 14 hours each day for 4 weeks. After many nights of increased sleep duration (attributed to recovering from prior sleep deprivation), most of the subjects established a pattern in which they slept for about 8.4 hours per night, but this was divided into two major bouts, with several hours of wake in the middle of each 14-hour night. These findings suggest that although the relative sleep/wake ratio (1:2) seen in adults may reflect an average biological sleep need, the relative consolidation of adult human sleep into one major bout per day may be an artifact of photoperiod and/or how long we attempt to or are able to remain awake each day and that under the right conditions, human sleep may tend to be polyphasic.

In designing and interpreting studies of circadian rhythmicity, study conditions that result in sleep/wake and light/dark cycles being potentially confounded, and/or when rest/activity could have a feedback effect on the system under study, the effects of sleep itself and the species differences in sleep become important considerations.

Measurement of Human Sleep Architecture and Sleep EEG

Before technology allowed for the measurement of brain electrical activity (electroencephalogram, EEG), it was thought that the brain, like the rest of the body, was inactive during sleep. The EEG, first described in rabbits and monkeys by Caton in 1894, and later in humans (Berger 1929), reflects summated potentials of large cortical neuron groups, recorded from the surface of the scalp or from the cortex, in a voltage-time domain. Brain wave activity differs between sleep and wakefulness (Loomis et al. 1935, 1937), and during sleep itself, there are substantial differences in the neuronal firing pattern with ultradian changes across the sleep episode (Dement and Kleitman 1957).

A normal 8-hour sleep episode in humans consists of four to five sleep cycles, with each cycle lasting for about

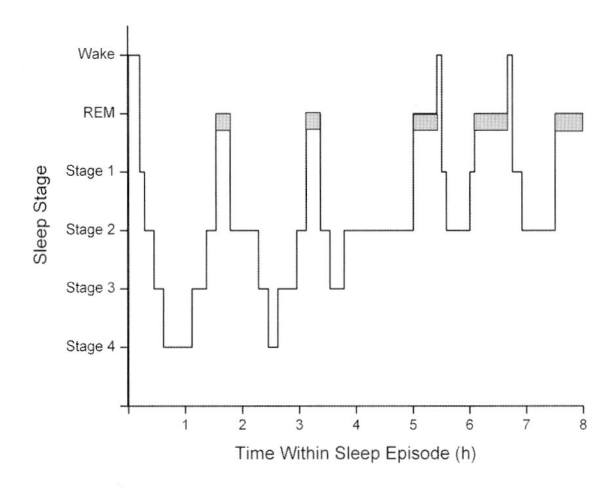

Figure 2. Schematic representation of the human sleep episode. The *x* axis indicates time within the sleep episode, and the *y* axis indicates the vigilance state. As is typical for a healthy, well-rested young adult, sleep onset occurs within 10–30 minutes of attempting to initiate sleep, and there is a quick procession through the lighter NREM stages to deep NREM sleep and then into REM sleep. As the night progresses, NREM and REM sleep cycles alternate about four to six times, with REM episodes getting progressively longer and NREM episodes containing fewer of the deeper stages toward the end of the night. There are a few brief awakenings, and spontaneous awakening occurs mostly from REM (as shown here).

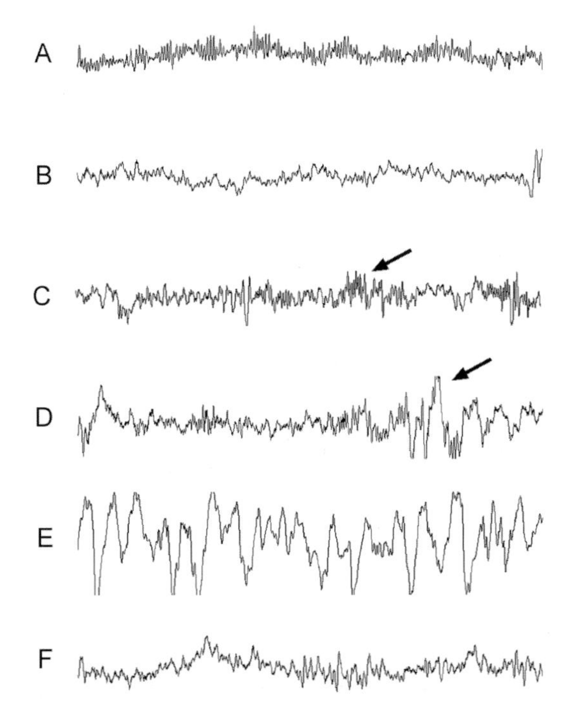

Figure 3. EEG waves in different vigilance states recorded from a 23-year-old man. (*A*) Alpha waves characteristic of quiet wakefulness with eyes closed; (*B*) Stage 1; (*C,D*) Stage 2. In panel *C*, the arrow indicates a sleep spindle, and in panel *D*, the arrow indicates a "K complex"; both of these EEG waves are characteristic of Stage 2. (*E*) Slow waves characteristic of Stages 3 and 4 sleep; (*F*) REM sleep. Note that as sleep progresses, the EEG slows down and the amplitude increases, reflecting synchronization of the underlying cortical neurons.

90–100 minutes (Fig. 2). These cycles are composed of two distinct states, rapid eye movement (REM) sleep and nonrapid eye movement (NREM) sleep. Visual scoring of the sleep EEG is defined according to standardized criteria (Rechtschaffen and Kales 1968). According to these somewhat arbitrary criteria, brain waves in the frequency range between 8 and 12 Hz (alpha waves) typically occur during quiet rest with eyes closed. At the transition from wakefulness to sleep, there is a gradual lowering in frequency and concomitant increase of amplitude in the EEG, and this general slowing of EEG frequency and increase of EEG amplitude continues as sleep initially progresses, reflecting a rising level of synchronization in cortical neurons. As these EEG changes progress, specific EEG frequency and amplitude criteria are used to subdivide sleep into NREM Stages 1 (the transition between wakefulness and sleep) through 4 (deep sleep) (Fig. 3).

Stages 3 and 4 together are commonly referred to as slow-wave sleep (SWS) and are defined as containing >20% delta waves (EEG waves with a frequency between 0.75 and 4.5 Hz and a peak-to-trough difference of at least 75 μV). There is clear evidence that delta waves are generated through synchronized rhythmic thalamocortical circuit activity, whereas very slow oscillations (<1 Hz) are excited within the neocortex (Steriade et al. 1993). Additional phasic EEG events, such as sleep spindles (12–14 Hz, see arrow in Fig. 3C) and K complexes (e.g., during Stage 2, see arrow in Fig. 3D) or vertex sharp transients (Stage 1) are also typical during NREM sleep.

REM sleep (also known as paradoxical sleep) was first described by Aserinsky and Kleitman (1953) and is mainly characterized by rapid eye movements (measured in the electrooculogram), a loss of muscle tone (measured in the electromyogram), and a low-voltage mixed EEG frequency (1–7 Hz) pattern (Fig. 4).

As the alternation of NREM and REM sleep continues across the night, the proportion of each sleep stage changes, with most SWS at the beginning of the night and longer REM sleep episodes and more Stage 2 sleep toward the end of the night. Visual scoring of the sleep polysomnogram (Rechtschaffen and Kales 1968) subdivides sleep into discrete stages (1–4, REM) and thus allows only limited quantification of the sleep EEG and the underlying physiological changes in the neuronal firing pattern of the cortex. There are more quantitative methods of analyzing the sleep EEG, such as fast Fourier transformation (FFT), which enables continuous analyses

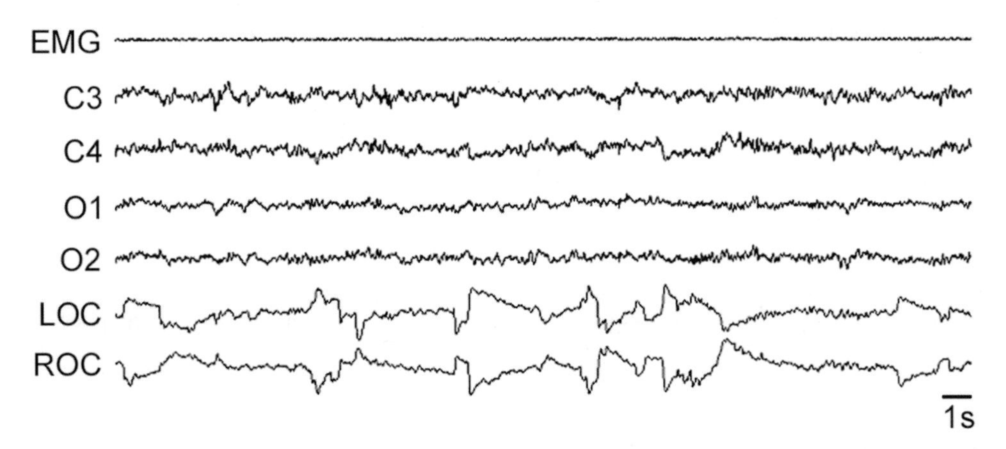

Figure 4. Thirty-second recording during REM sleep from a healthy 72-year-old woman. The upper recording line shows the electromyogram (EMG), the next four recording lines show the EEGs recorded from central (C3, C4) and occipital (O1, O2) sites, and the two lower recording lines show the electrooculograms (EOGs) from the left (LOC) and right (ROC) eyes. Characteristic of REM sleep, EMG activity is extremely low and there are several rapid eye movements within this 30-second epoch. The typical low-voltage mixed EEG frequency (1–7 Hz) pattern can also be seen here.

in the power/frequency domain (Dietsch 1932; Borbély et al. 1981).

It is now clear that during sleep the human brain exhibits regional differences in electrical activity (EEG power density in different frequency bands) (Werth et al. 1996; Finelli et al. 2001; Knoblauch et al. 2003), and there are age-dependent differences in this (Landolt and Borbély 2001; Münch et al. 2004). In addition, there is both EEG and neuroimaging evidence that during sleep, the brain shows use- and experience-dependent deactivation and reactivation in different areas (Braun et al. 1997; Hofle et al. 1997; Maquet et al. 1997), and such regional variation in activity during sleep may be associated with learning and memory-related processes in which sleep has a role (Kattler et al. 1994; Huber et al. 2004; Schmidt et al. 2006).

Neuroanatomic and Neurophysiologic Basis of Sleep

The ascending reticular activating system (ARAS) of the brain stem, first described by Moruzzi and Magoun (1949), is involved in maintaining cortical activation and behavioral arousal during wakefulness. The ARAS largely originates from a series of well-defined cell groups, including the laterodorsal tegmental (LDT) and pedunculopontine (PPT) nuclei, the locus coeruleus (LC), the dorsal and medial raphé nuclei (DR), the tuberomammilary neurons (TMN), and dopaminergic neurons in the periaequatorial grey matter (Saper et al. 2001). In animals, these nuclei of the ARAS in the brain stem receive various inputs from visceral, somatic, and sensory systems and project mainly via two different pathways to the cerebral cortex (for review, see Jones 2000). Excitatory neurons of the dorsal pathway in the ARAS project from the upper brain stem via relay and reticular neurons of the thalamus to the cerebral cortex. The ventral pathway projects directly to the cortex via the lateral hypothalamus (LH) and the basal forebrain (BF) (Jones 2000; Saper et al. 2005b). The firing rate of glutaminergic ascending reticular neurons and monoaminergic cell groups, as well as excitatory peptidergic neurons of the posterior hypothalamus and the LH (which synthesizes the wake-promoting hormone orexin/hypocretin), is crucial for maintaining cortical activation and behavioral arousal during wakefulness (Jones 2000; Saper et al. 2005b). The intralaminar and midline nuclei of the thalamus are also believed to have a role in cortical arousal (Berendse and Groenewegen 1991; Saper et al. 2005a).

During sleep, neurons in the ventrolateral preoptic nucleus (VLPO) in the anterior hypothalamus inhibit neuronal activation of the ARAS, BF, and cerebral cortex (Sherin et al. 1996; Jones 2000; Saper et al. 2005a). The VLPO receives afferents from each cell group of the monoaminergic system via the subparaventricular zone of the dorsomedial hypothalamus, thus mostly indirectly receiving input from the SCN. There are also mutual connections between the VLPO and orexin neurons (Chou et al. 2002). The VLPO consists of at least two subregions: a dense cell cluster projecting to the TMN and a diffuse cell region (eVLPO), which projects to the LC, DR (Lu et al. 2000, 2002), and probably also REM sleep-activating sites in the mesopontine tegmentum (Lu et al. 2006).

In the thalamus, progressive hyperpolarization of reticular neurons gates thalamocortical inputs to the cortex. When a certain level of hyperpolarization is achieved, those neurons change their firing pattern from single-spike mode to a rhythmic burst mode (Steriade et al. 1993). These oscillations are then relayed on other nuclei within the thalamus and lead to rhythmic inhibitory postsynaptic potentials in thalamocortical neurons. Consequently, the thalamocortical neurons fire rebound bursts of action potentials that are transferred to the cortical pyramidal cells, where they in turn induce excitatory postsynaptic potentials, thereby generating EEG sleep spindles. This results in an inverse firing pattern in reticular and thalamocortical neurons during these oscillations. Depending on the degree of hyperpolarization in the reticular cells, either sleep spindles or slow-wave oscillations are generated. Therefore, the progressive hyperpolarization of thalamocortical cells at sleep onset first leads to the generation of sleep spindle oscillations, which are replaced by slow-wave oscillations as deepening of sleep proceeds and a certain low-voltage range in thalamocortical neurons is reached. These slow-wave oscillations become synchronized in the cortex in large neuronal assemblies by a network of synchronizing factors (Steriade et al. 1993; Amzica and Steriade 1998; Steriade 2003).

The switch between REM and NREM sleep is thought to be modulated through interaction of monoaminergic with cholinergic neurons, although these mechanisms and their function are not fully understood. The monoaminergic neurons fire faster during wakefulness than during NREM sleep and show virtually no firing activity during REM sleep, whereas the cholinergic system is most active during wakefulness and REM sleep by preventing reticular cells in the thalamus from hyperpolarization (Aston-Jones and Bloom 1981).

Saper et al. (2001) have proposed the so-called "flip-flop" model of sleep and wakefulness (see Fig. 5). In this model, monoaminergic nuclei such as the TMN, LC, and DR promote wakefulness by direct excitatory effects on the cortex and by inhibition of sleep-promoting neurons of the VLPO. During sleep, the VLPO inhibits monoaminergic-mediated arousal regions through GABAergic and galaninergic projections (Saper et al. 2001). Thus, intermediate states between sleep and wakefulness are prevented by the reciprocal inhibition of VLPO neurons and monoaminergic cell groups that concomitantly disinhibit and reinforce their own firing rates. Orexin-containing neurons seem to have an important stabilizing role in the proposed flip-flop mechanism. In addition, it has also been suggested that there is a flip-flop switch for REM/NREM sleep, with mutually inhibitory neurons in the brain stem, also under the influence of orexin (Lu et al. 2006). Several neuronal pathways of the SCN to and from sleep and wake-regulatory neurons have been described in the last few years (Deboer et al. 2003; Saper et al. 2005b), and there is strong evidence that information from peripheral oscillators, such as feeding and temperature regulation and even emotional states, are integrated in the modulation of the endogenous circadian timekeeping system in animals and humans (for review, see Saper et al. 2005b).

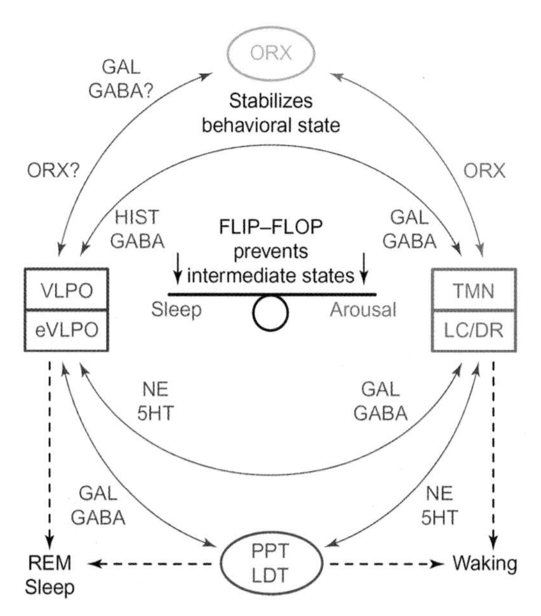

Figure 5. A flip-flop model for reciprocal interactions between sleep- and wake-promoting brain regions. (*Red*) Inhibitory pathways; (*green*) excitatory pathways. (DR) Dorsal raphé nucleus; (HIST) histamine; (LC) locus coeruleus; (LDT) laterodorsal tegmental nuclei; (PPT) pedunculopontine tegmental nuclei; (REM) rapid eye movement; (TMN) tuberomammillary nucleus; (VLPO) ventrolateral preoptic nucleus. A flip-flop switch promotes the system to be in one state or the other and prevents the system from being in an intermediate state for long durations. (Reprinted, with permission, from Saper et al. 2001 [© Elsevier].)

SLEEP REGULATION IN HUMANS

Interaction of Two Major Regulatory Processes: Borbély-Daan Model of Process C and Process S

Several models have been proposed to understand how consolidated sleep and waking can be achieved in humans. In the early 1980s, Borbély proposed a model with two interacting processes—a homeostatic Process S and a circadian Process C—that counterbalance each other to allow for a long consolidated bout of sleep each night and a long bout of wakefulness during each 24-hour day (Borbély 1982; Daan et al. 1984).

According to this two-process model, Process S (the level of sleep pressure) builds up during wakefulness with respect to the duration of time awake (Fig. 6, lower panel) and then that sleep pressure is dissipated during the course of the following sleep episode. Markers of this homeostatic process are the increase of EEG theta activity (4.5–8 Hz) during wakefulness (Cajochen et al. 1995, 1999b; Finelli et al. 2000) and the decrease of EEG slow-wave activity (0.75–4.5 Hz) during the following sleep episode (Werth et al. 1996; Cajochen et al. 1999a; Finelli et al. 2001). Process S has been shown to be operative in both animals and humans, and the predictions of the model, based on mathematical simulations, fit the experimental data (Borbély 1982; Daan et al. 1984; Achermann et al. 1993). The neurobiological correlates of the homeostatic process are not known, and although sleep homeostasis may be functionally linked to restoration, the underlying mechanisms remain to be elucidated (Benington and Heller 1995). There is mounting evidence that genetic factors (Retey et al. 2005; Viola et al. 2007) and processes related to synaptic plasticity (Tononi and Cirelli 2003, 2006) are also involved in homeostatic sleep regulation.

The circadian process (Process C) in the model describes the 24-hour periodicity of sleep propensity resulting from the neuronal and/or humoral influence of the biological clock, with strong sleep propensity at some and weaker or absent sleep propensity at other circadian phases. This waveform of sleep propensity in humans is associated with melatonin secretion from the pineal gland (Lavie 1997), and the circadian sleep propensity waveform is influenced by light via the retinohypothalamic tract to the SCN (see Biological Rhythms Workshop IB). As has been reported in many animal species (see Biological Rhythms Workshop IB), the biological night in humans seems to be extended when photoperiods are shorter (Wehr 1991, 1996, 2001).

Another phenomenon in humans is that of different chronotypes, whereby the relative sleep timing of some individuals is earlier (morning types) or later (evening types) than average (Kleitman 1939). This preference for different sleep times has been shown to be associated with differences in circadian period (Duffy et al. 2001), with individuals with shorter circadian periods showing a pref-

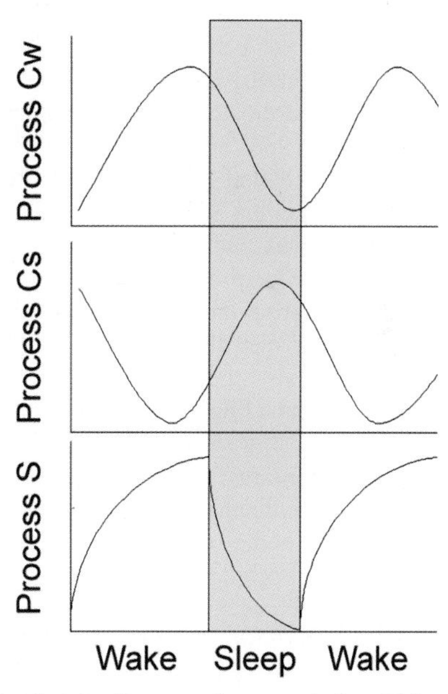

Figure 6. Model of human sleep regulation. This schematic incorporates the "two-process model" proposed by Borbély (Borbély 1982; Daan et al. 1984), with features of the "opponent process" model of Edgar (Edgar et al. 1993; Dijk and Edgar 1999) and additional features proposed by others (for review, see Mistlberger 2005). (*Lower panel*) Borbély's proposed homeostatic Process S builds up during wakefulness and dissipates during sleep (*shaded box*). (*Upper panel*) Wake-promoting function of Process C (from signals from the SCN) as proposed by Edgar peaks toward the end of the wake episode (Cw), acting to counteract the buildup of sleep pressure during waking. (*Middle panel*) Another proposed function of the circadian system, which is to actively promote sleep toward the end of the sleep episode to allow for a long consolidated bout of sleep (Cs).

erence for morningness and those with longer periods showing a preference for eveningness. Consistent with these period differences are reports that morning and evening types have differences in their phase angle of entrainment, such that they are not only sleeping at different clock times, but also sleeping at different biological times (Duffy et al. 1999b; Baehr et al. 2000). Morningness-eveningness has been shown to be associated with clock gene expression (Katzenberg et al. 1998; Archer et al. 2003; Carpen et al. 2005, 2006; Mishima et al. 2005), lending further support to the idea that chronotype preference can be the result of fundamental differences in the circadian system. More recently, human chronotypes have also been shown to differ in their homeostatic sleep regulation (Mongrain et al. 2006).

Regulation of Sleep by the Circadian Pacemaker

As described in detail in Biological Rhythms Workshop IB, the SCN represent the endogenous circadian pacemaker, and one role this group of neurons has is to synchronize the circadian oscillations of behavior. Among other factors, SCN lesions result in arrhythmicity of the rest/activity and sleep/wake cycles in animals. In fact, although SCN lesions cannot be done in humans, the anterior hypothalamus was identified as a locus of sleep/wake control as early as the 1920s, based on data from patients with tumors in that region of the brain (Fulton and Bailey 1929; von Economo 1930). Such patients continue to show patterns of sleep and wakefulness, although no longer on a consistent 24-hour basis and often no longer in consolidated bouts (Fulton and Bailey 1929; von Economo 1930; Schwartz et al. 1986; Cohen and Albers 1991; Bloch et al. 2005; Triarhou 2006), suggesting a role for the human SCN in maintaining 24-hour patterns of sleep and wakefulness.

The Borbély-Daan Two-process Model hypothesized a Process C emanating from the circadian pacemaker, and later researchers confirmed that it was indeed the SCN serving such a function by studying sleep/wake regulation in SCN-lesioned animals. Edgar (Edgar et al. 1993) studied the circadian control of sleep/wake function in SCN-lesioned squirrel monkeys (*Saimiri sciureus*, diurnal animals who exhibit monophasic sleep) and proposed an "Opponent Process" model of sleep/wake regulation to describe how the SCN promotes wakefulness during the subjective day (for review, see Dijk and Edgar 1999) to counterbalance the buildup of sleepiness. Others have since extended this by suggesting that the SCN not only promotes wakefulness during the subjective day, but also promotes sleep during the subjective night (for review, see Mistlberger 2005; Fuller et al. 2006).

A precise interaction between sleep/wake homeostatic (Process S) and circadian processes is required to maintain wakefulness throughout the biological day and sleep throughout the biological night. As homeostatic sleep pressure builds throughout wakefulness, the circadian pacemaker is thought to send an increasingly strong wake-promoting signal, allowing wakefulness to be maintained (Fig. 6, upper panel). Similarly during sleep, as sleep pressure dissipates across the night, the biological clock may send a strong sleep-promoting signal, allowing sleep to be maintained (Fig. 6, middle panel). Evidence supporting both processes (C and S) in these models came from studies using the forced desynchrony protocol.

In forced desynchrony studies, the timing of the sleep/wake cycle is uncoupled from the timing of the circadian system by scheduling human subjects to live on a rest/activity schedule that is much shorter (typically 20 hours) or much longer (typically 28 hours) than the near 24-hour period of the underlying circadian system. This protocol results in sleep and wake episodes distributed across all circadian phases, with similar scheduled durations of wakefulness preceding each sleep episode. Therefore, homeostatic (Process S) and circadian phase-dependent (Process C) effects on sleep can be separated and quantified.

In the first such study, Dijk and Czeisler (1994) found that the sleep/wake homeostat and the circadian pacemaker contributed about equally to sleep in young men. As had been suggested by earlier studies (Weitzman et al. 1974a,b; Carskadon and Dement 1975; Webb and Agnew 1975; Lavie 1986), the circadian sleep propensity rhythm was lowest near habitual bedtime under entrained conditions ("wake maintenance zone"), and circadian sleep propensity was greatest at phases corresponding to usual wake time under entrained conditions. Furthermore, they showed that it was only with a very precise relative timing between these two sleep regulatory processes that sleep could be maintained for a consolidated 8–9-hour sleep episode.

CIRCADIAN RHYTHM SLEEP DISORDERS

Circadian rhythm sleep disorders represent situations in which the relationship between the timing of sleep and wakefulness and the timing of the underlying circadian system is misaligned. This misalignment can be transient, such as in jet lag or during night shift work, or the misalignment can be chronic, as in advanced or delayed sleep phase disorders (ASP, DSP) or non-24-hour sleep/wake disorder. In jet lag and during night shift work, sleep and wake are attempted at nonoptimal biological times, resulting in difficulty falling asleep and/or difficulty in maintaining sleep. In contrast, ASP and DSP patients often do sleep at appropriate biological times, but those times are inappropriate socially, and when the patients try to change their sleep timing to a more acceptable social time, they have trouble remaining awake (ASP), falling asleep (DSP), or maintaining sleep. In most cases, patients with these disorders have a reduced sleep duration, resulting in decrements in performance, mood, and alertness during their wake episodes (for a recent review, see Lack and Wright 2007).

A similar situation affects many totally blind individuals. Due to a lack of light input to the biological clock, many blind people are unable to entrain to the 24-hour day, and thus experience non-24-hour sleep/wake disorder (Sack et al. 1992). The circadian system of these blind individuals typically cycles in and out of synchrony with the time of the external world, and thus at some times, they are able to sleep at night and maintain alertness throughout the day, whereas at other times, they have great difficulty sleeping at night and remaining awake during the day.

CHRONOPHARMACOLOGY

The biological clock surely serves to do more than properly time sleep and wakefulness. Coordination of temporal information throughout the whole organism is thought to be crucial for maintaining optimal health, with the central pacemaker located in the SCN serving to maintain synchrony among peripheral oscillators and with the external environment. This suggests that disruptions of this internal synchrony will not only affect the function of many organ systems, but also affect how the whole organism responds to challenges, and there is evidence of this from both animal and human studies. Experiments performed in the early 1980s demonstrated that the circadian time of administering a lethal drug dose to rats could influence mortality or long-term survival. Rats were administered *cis*-diamminedichlor-oplatinum (cisplatin) at different times of the day in an "LD_{50}" test, and depending on the time of drug administration, there were differences of threefold to eightfold 50% mortality rate and a nearly threefold difference in long-term survival (Hrushesky et al. 1982). In more recent studies, when such cytotoxic medications are given to human cancer patients, the drugs show a higher efficacy when administered at an ideal circadian time, allowing a reduced dose to be used, thereby reducing side effects (see, e.g., von Roemeling et al. 1990). The circadian timing of drug administration likely contributes to pharmacokinetic variability, which in the past has been attributed only to such factors as age, gender, ethnicity, or organ functioning (Undevia et al. 2005). In fact, many clinicians and scientists now suggest a custom-tailored therapy for an individual patient, including assessing circadian phase and taking into account the sleep/wake cycle (for review, see Lévi and Schibler 2007).

CONCLUSIONS

As we have outlined in this review, rhythms of sleep and wakefulness (typically measured as rest/activity rhythms) are among the most prominent of biological rhythms and therefore were among the first to be recorded in early chronobiological studies. These rhythms provide useful information about the central biological clock, but there are problems inherent in their use, and an understanding of those potential problems is important in the design and interpretation of chronobiological experiments. Furthermore, a more comprehensive understanding of sleep/wake-regulatory differences among species is important for the design and interpretation of chronobiological experiments and for extrapolating observations from animal studies to studies of human chronobiological disorders.

We have also described the anatomical and neurophysiological bases of sleep regulation in mammals and briefly outlined how sleep in humans is measured. The interaction of circadian with homeostatic sleep regulating processes to allow for consolidated sleep and wakefulness in humans has been described. Finally, we briefly summarized how circadian rhythm sleep disorders in humans occur. Although the function of sleep is still not completely understood, sleep has a critical role for human health, and we have attempted to outline the role of the circadian timing system in human sleep and sleep disorders, as well as similarities and differences between the sleep of humans and that of other organisms.

ACKNOWLEDGMENTS

We thank K. Huard for assistance in preparing the manuscript. M.Y.M. is supported by postdoctoral fellowships from the Novartis Foundation (Switzerland) and the L. & T. LaRoche Foundation (Switzerland); S.W.C. is supported by a postdoctoral fellowship from the Natural Sciences and Engineering Research Council of Canada; and J.F.D. is supported by National Institutes of Health grants AT002571, HL08978, AG06072, and AG09975.

REFERENCES

Achermann P., Dijk D.J., Brunner D.P., and Borbély A.A. 1993. A model of human sleep homeostasis based on EEG slow-wave activity: Quantitative comparison of data and simulations. *Brain Res. Bull.* **31:** 97.

Albers H.E., Lydic R., and Moore-Ede M.C. 1984. Role of the suprachiasmatic nuclei in the circadian timing system of the squirrel monkey. II. Light-dark cycle entrainment. *Brain Res.* **300:** 285.

Amzica F. and Steriade M. 1998. Electrophysiological correlates of sleep delta waves. *Electroencephalogr. Clin. Neurophysiol.* **107:** 69.

Antle M.C. and Mistlberger R.E. 2000. Circadian clock resetting by sleep deprivation without exercise in the Syrian hamster. *J. Neurosci.* **20:** 9326.

Archer S.N., Robilliard D.L., Skene D.J., Smits M., Williams A., Arendt J., and von Schantz M. 2003. A length polymorphism in the circadian clock gene *Per3* is linked to delayed sleep phase syndrome and extreme diurnal preference. *Sleep* **26:** 413.

Aschoff J. 1961. Exogenous and endogenous components in circadian rhythms. *Cold Spring Harbor Symp. Quant. Biol.* **25:** 11.

Aschoff J. and Wever R. 1962. Spontanperiodik des menschen bei ausschluss aller zeitgeber. *Naturwissenschaften* **49:** 337.

———. 1981. The circadian system of man. In *Biological rhythms: Handbook of behavioral neurobiology* (ed. J. Aschoff), p. 311. Plenum Press, New York.

Aserinsky E. and Kleitman N. 1953. Eye movements during sleep. *Fed. Proc.* **12:** 6.

Aston-Jones G. and Bloom F.E. 1981. Activity of norepinephrine-containing locus coeruleus neurons in behaving rats anticipates fluctuations in the sleep-waking cycle. *J. Neurosci.* **1:** 876.

Baehr E.K., Revelle W., and Eastman C.I. 2000. Individual differences in the phase and amplitude of the human circadian temperature rhythm: With an emphasis on morningness-eveningness. *J. Sleep Res.* **9:** 117.

Beaule C., Houle L.M., and Amir S. 2003. Expression profiles of PER2 immunoreactivity within the shell and core regions of the rat suprachiasmatic nucleus: Lack of effect of photic entrainment and disruption by constant light. *J. Mol. Neurosci.* **21:** 133.

Beersma D.G., Daan S., and Hut R.A. 1999. Accuracy of circadian entrainment under fluctuating light conditions: Contributions of phase and period responses. *J. Biol. Rhythms* **14:** 320.

Benington J.H. and Heller H.C. 1995. Restoration of brain energy metabolism as the function of sleep. *Prog. Neurobiol.* **45:** 347.

Berendse H.W. and Groenewegen H.J. 1991. Restricted cortical termination fields of the midline and intralaminar thalamic nuclei in the rat. *Neuroscience* **42:** 73.

Berger H. 1929. Über das elektroenzephalogramm beim menschen. *Arch. Psychiatrie Nervenkrankheiten* **87:** 527.

Bloch K.E., Brack T., and Wirz-Justice A. 2005. Transient short free running circadian rhythm in a case of aneurysm near the suprachiasmatic nuclei. *J. Neurol. Neurosurg. Psychiatry* **76:** 1178.

Boivin D.B., Duffy J.F., Kronauer R.E., and Czeisler C.A. 1996. Dose-response relationships for resetting of human circadian clock by light. *Nature* **379:** 540.

Borbély A.A. 1982. A two process model of sleep regulation. *Hum. Neurobiol.* **1:** 195.

Borbély A.A., Baumann F., Brandeis D., Strauch I., and Lehmann D. 1981. Sleep deprivation: Effect of sleep stages and EEG power density in man. *Electroencephalogr. Clin. Neurophysiol.* **51:** 483.

Braun A.R., Balkin T.J., Wesensten N.J., Carson R.E., Varga M., Baldwin P., Selbie S., Belenky G., and Herscovitch P. 1997. Regional cerebral blood flow throughout the sleep-wake cycle an $H_2{}^{15}O$ PET study. *Brain* **120:** 1173.

Cajochen C., Foy R., and Dijk D.J. 1999a. Frontal predominance of relative increase in sleep delta and theta EEG activity after sleep loss in humans. *Sleep Res. Online* **2:** 65.

Cajochen C., Brunner D.P., Krauchi K., Graw P., and Wirz-Justice A. 1995. Power density in theta/alpha frequencies of the waking EEG progressively increases during sustained wakefulness. *Sleep* **18:** 890.

Cajochen C., Khalsa S.B.S., Wyatt J.K., Czeisler C.A., and Dijk D.J. 1999b. EEG and ocular correlates of circadian melatonin phase and human performance decrements during sleep loss. *Am. J. Physiol.* **277:** R640.

Campbell S.S. and Tobler I. 1984. Animal sleep: A review of sleep duration across phylogeny. *Neurosci. Biobehav. Rev.* **8:** 269.

Carpen J.D., Archer S.N., Skene D.J., Smits M., and von Schantz M. 2005. A single-nucleotide polymorphism in the 5′-untranslated region of the *hPER2* gene is associated with diurnal preference. *J. Sleep Res.* **14:** 293.

Carpen J.D., von Schantz M., Smits M., Skene D.J., and Archer S.N. 2006. A silent polymorphism in the *PER1* gene associates with extreme diurnal preference in humans. *J. Hum. Genet.* **51:** 1122.

Carskadon M.A. and Dement W.C. 1975. Sleep studies on a 90-minute day. *Electroencephalogr. Clin. Neurophysiol.* **39:** 145.

Carskadon M.A., Labyak S.E., Acebo C., and Seifer R. 1999. Intrinsic circadian period of adolescent humans measured in conditions of forced desynchrony. *Neurosci. Lett.* **260:** 129.

Cheng H.Y., Obrietan K., Cain S.W., Lee B.Y., Agostino P.V., Joza N.A., Harrington M.E., Ralph M.R., and Penninger J.M. 2004. Dexras1 potentiates photic and suppresses nonphotic responses of the circadian clock. *Neuron* **43:** 715.

Chou T.C., Bjorkum A.A., Gaus S.E., Lu J., Scammell T.E., and Saper C.B. 2002. Afferents to the ventrolateral preoptic nucleus. *J. Neurosci.* **22:** 977.

Cohen R.A. and Albers H.E. 1991. Disruption of human circadian and cognitive regulation following a discrete hypothalamic lesion: A case study. *Neurology* **41:** 726.

Czeisler C.A. 1995. The effect of light on the human circadian pacemaker. *Ciba Found. Symp.* **183:** 254.

Czeisler C.A., Allan J.S., and Kronauer R.E. 1990. A method for assaying the effects of therapeutic agents on the period of the endogenous circadian pacemaker in man. In *Sleep and biological rhythms: Basic mechanisms and applications to psychiatry* (ed. J. Montplaisir and R. Godbout), p. 87. Oxford University Press, New York.

Czeisler C.A., Richardson G.S., Zimmerman J.C., Moore-Ede M.C., and Weitzman E.D. 1981. Entrainment of human circadian rhythms by light-dark cycles: A reassessment. *Photochem. Photobiol.* **34:** 239.

Czeisler C.A., Kronauer R.E., Allan J.S., Duffy J.F., Jewett M.E., Brown E.N., and Ronda J.M. 1989. Bright light induction of strong (type 0) resetting of the human circadian pacemaker. *Science* **244:** 1328.

Czeisler C.A., Duffy J.F., Shanahan T.L., Brown E.N., Mitchell J.F., Rimmer D.W., Ronda J.M., Silva E.J., Allan J.S., Emens J.S., Dijk D.J., and Kronauer R.E. 1999. Stability, precision, and near-24-hour period of the human circadian pacemaker. *Science* **284:** 2177.

Daan S., Beersma D.G.M., and Borbély A.A. 1984. Timing of human sleep: Recovery process gated by a circadian pacemaker. *Am. J. Physiol.* **246:** R161.

Davis F.C. and Viswanathan N. 1998. Stability of circadian timing with age in Syrian hamsters. *Am. J. Physiol.* **275:** R960.

Deboer T., Vansteensel M.J., Detari L., and Meijer J.H. 2003. Sleep states alter activity of suprachiasmatic nucleus neurons. *Nat. Neurosci.* **6:** 1086.

DeCoursey P.J. 1964. Function of a light response rhythm in hamsters. *J. Cell. Comp. Physiol.* **63:** 189.

———. 1989. Photoentrainment of circadian rhythms: An ecologist viewpoint. In *Circadian clocks and ecology* (ed. T. Hiroshige and K. Honma), p. 187. Hokkaido University Press, Sapporo, Japan.

de la Iglesia H., Meyer J., Carpino A., Jr., and Schwartz W.J. 2000. Antiphase oscillation of the left and right suprachiasmatic nuclei. *Science* **290:** 799.

Dement W.C. and Kleitman N. 1957. Cyclic variations in EEG during sleep and their relation to eye movements, body motility, and dreaming. *Electroencephalogr. Clin. Neurophysiol.* **9:** 673.

Dietsch G. 1932. Fourier analyse von elektroenkephalogrammen des menshen. *Pflüger's Arch. Ges. Physiol.* **230:** 106.

Dijk D.J. and Czeisler C.A. 1994. Paradoxical timing of the circadian rhythm of sleep propensity serves to consolidate sleep and wakefulness in humans. *Neurosci. Lett.* **166:** 63.

Dijk D.J. and Daan S. 1989. Sleep EEG spectral analysis in a diurnal rodent: *Eutamias sibiricus*. *J. Comp. Physiol. A* **165:** 205.

Dijk D.J. and Edgar D.M. 1999. Circadian and homeostatic control of wakefulness and sleep. In *Regulation of sleep and wakefulness* (ed. F.W. Turek and P.C. Zee), p. 111. Marcel Dekker, New York.

Duffy J.F., Rimmer D.W., and Czeisler C.A. 2001. Association of intrinsic circadian period with morningness-eveningness, usual wake time, and circadian phase. *Behav. Neurosci.* **115:** 895.

Duffy J.F., Viswanathan N., and Davis F.C. 1999a. Free-running circadian period does not shorten with age in female Syrian hamsters. *Neurosci. Lett.* **271:** 77.

Duffy J.F., Dijk D.J., Hall E.F., and Czeisler C.A. 1999b. Relationship of endogenous circadian melatonin and temperature rhythms to self-reported preference for morning or evening activity in young and older people. *J. Investig. Med.* **47:** 141.

Edgar D.M., Dement W.C., and Fuller C.A. 1993. Effect of SCN lesions on sleep in squirrel monkeys: Evidence for opponent processes in sleep-wake regulation. *J. Neurosci.* **13:** 1065.

Edgar D.M., Kilduff T.S., Martin C.E., and Dement W.C. 1991. Influence of running wheel activity on free-running sleep/wake and drinking circadian rhythms in mice. *Physiol. Behav.* **50:** 373.

Eskin A. 1971. Some properties of the system controlling the circadian activity rhythm of sparrows. In *Biochronometry* (ed. M. Menaker), p. 55. National Academy of Sciences, Washington, D.C.

Finelli L.A., Borbély A.A., and Achermann P. 2001. Functional topography of the human nonREM sleep electroencephalogram. *Eur. J. Neurosci.* **13:** 2282.

Finelli L.A., Baumann H., Borbély A.A., and Achermann P. 2000. Dual electroencephalogram markers of human sleep homeostasis: Correlation between theta activity in waking and slow-wave activity in sleep. *Neuroscience* **101:** 523.

Franken P., Dijk D.J., Tobler I., and Borbély A.A. 1991. Sleep deprivation in rats: Effects on EEG power spectra, vigilance states, and cortical temperature. *Am. J. Physiol.* **261:** R198.

Fuller P.M., Gooley J.J., and Saper C.B. 2006. Neurobiology of the sleep-wake cycle: Sleep architecture, circadian regulation

and regulatory feedback. *J. Biol. Rhythms* **21:** 482.
Fulton J.F. and Bailey P. 1929. Tumors in the region of the third ventricle: Their diagnosis and relation to pathological sleep. *J. Nerv. Ment. Dis.* **69:** 1.
Gerkema M.P., Daan S., Wilbrink M., Hop M.W., and van der Leest F. 1993. Phase control of ultradian feeding rhythms in the common vole (*Microtus arvalis*): The roles of light and the circadian system. *J. Biol. Rhythms* **8:** 151.
Hiddinga A.E., Beersma D.G.M., and van den Hoofdakker R.H. 1997. Endogenous and exogenous components in the circadian variation of core body temperature in humans. *J. Sleep Res.* **6:** 156.
Hofle N., Paus T., Reutens D., Fiset P., Gotman J., Evans A.C., and Jones B.E. 1997. Regional cerebral blood flow changes as a function of delta and spindle activity during slow wave sleep in humans. *J. Neurosci.* **17:** 4800.
Honma K. and Honma S. 1988. A human phase response curve for bright light pulses. *Jpn. J. Psychiatry Neurol.* **42:** 167.
Honma K.I. and Hiroshige T. 1978. Endogenous ultradian rhythms in rats exposed to prolonged continuous light. *Am. J. Physiol.* **235:** R250.
Hrushesky W.J., Lévi F.A., Halberg F., and Kennedy B.J. 1982. Circadian stage dependence of cis-diamminedichloroplatinum lethal toxicity in rats. *Cancer Res.* **42:** 945.
Huber R., Ghilardi M.F., Massimini M., and Tononi G. 2004. Local sleep and learning. *Nature* **430:** 78.
Jagota A., de la Iglesia H., and Schwartz W.J. 2000. Morning and evening circadian oscillations in the suprachiasmatic nucleus *in vitro*. *Nat. Neurosci.* **3:** 372.
Jones B.E. 2000. Basic mechanisms of sleep/wake states. In *Principles and practice of sleep medicine* (ed. M.H. Kryger et al.), p. 134. W.B. Saunders, Philadelphia, Pennsylvania.
Kattler H., Dijk D.J., and Borbély A.A. 1994. Effect of unilateral somatosensory stimulation prior to sleep on the sleep EEG in humans. *J. Sleep Res.* **3:** 159.
Katzenberg D., Young T., Finn L., Lin L., King D.P., Takahashi J.S., and Mignot E. 1998. A clock polymorphism associated with human diurnal preference. *Sleep* **21:** 569.
Kelly T.L., Neri D.F., Grill J.T., Ryman D., Hunt P.D., Dijk D.J., Shanahan T.L., and Czeisler C.A. 1999. Nonentrained circadian rhythms of melatonin in submariners scheduled to an 18-hour day. *J. Biol. Rhythms* **14:** 190.
Khalsa S.B.S., Jewett M.E., Cajochen C., and Czeisler C.A. 2003. A phase response curve to single bright light pulses in human subjects. *J. Physiol.* **549:** 945.
Kleitman N. 1939. *Sleep and wakefulness.* University of Chicago Press, Chicago, Illinois.
Klerman E.B., Dijk D.J., Kronauer R.E., and Czeisler C.A. 1996. Simulations of light effects on the human circadian pacemaker: Implications for assessment of intrinsic period. *Am. J. Physiol.* **270:** R271.
Knoblauch V., Martens W., Wirz-Justice A., Krauchi K., and Cajochen C. 2003. Regional differences in the circadian modulation of human sleep spindle characteristics. *Eur. J. Neurosci.* **18:** 155.
Kramm K. 1971. "Circadian activity in the antelope ground squirrel *Ammospermophilus leucurus*." Ph.D. thesis, University of California, Irvine.
Lack L.C. and Wright H.R. 2007. Chronobiology of sleep in humans. *Cell. Mol. Life Sci.* **64:** 1205.
Landolt H.-P. and Borbély A.A. 2001. Age-dependent changes in sleep EEG topography. *Clin. Neurophysiol.* **112:** 369.
Lavie P. 1986. Ultrashort sleep-waking schedule. III. "Gates" and "forbidden zones" for sleep. *Electroencephalogr. Clin. Neurophysiol.* **63:** 414.
———. 1997. Melatonin: Role in gating nocturnal rise in sleep propensity. *J. Biol. Rhythms* **12:** 657.
Lévi F. and Schibler U. 2007. Circadian rhythms: Mechanisms and therapeutic implications. *Annu. Rev. Pharmacol. Toxicol.* **47:** 593.
Loomis A.L., Harvey E.N., and Hobart G.A. III. 1935. Potential rhythms of the cerebral cortex during sleep. *Science* **81:** 597.
———. 1937. Cerebral states during sleep, as studied by human brain potentials. *J. Exp. Psychol.* **21:** 127.
Lu J., Greco M.A., Shiromani P., and Saper C.B. 2000. Effect of lesions of the ventrolateral preoptic nucleus on NREM and REM sleep. *J. Neurosci.* **20:** 3830.
Lu J., Sherman D., Devor M., and Saper C.B. 2006. A putative flip-flop switch for control of REM sleep. *Nature* **441:** 589.
Lu J., Bjorkum A.A., Xu M., Gaus S.E., Shiromani P.J., and Saper C.B. 2002. Selective activation of the extended ventrolateral preoptic nucleus during rapid eye movement sleep. *J. Neurosci.* **22:** 4568.
Maquet P., Degueldre C., Delfiore G., Aerts J., Pèters J.-M., Luxen A., and Franck G. 1997. Functional neuroanatomy of human slow wave sleep. *J. Neurosci.* **17:** 2807.
Mason R. 1991. The effects of continuous light exposure on Syrian hamster suprachiasmatic (SCN) neuronal discharge activity *in vitro*. *Neurosci. Lett.* **123:** 160.
Middleton B., Arendt J., and Stone B.M. 1996. Human circadian rhythms in constant dim light (8 lux) with knowledge of clock time. *J. Sleep Res.* **5:** 69.
Minors D.S., Waterhouse J.M., and Wirz-Justice A. 1991. A human phase-response curve to light. *Neurosci. Lett.* **133:** 36.
Mishima K., Tozawa T., Satoh K., Saitoh H., and Mishima Y. 2005. The 3111T/C polymorphism of hClock is associated with evening preference and delayed sleep timing in a Japanese population sample. *Am. J. Med. Genet. B Neuropsychiatr. Genet.* **133:** 101.
Mistlberger R.E. 2005. Circadian regulation of sleep in mammals: Role of the suprachiasmatic nucleus. *Brain Res. Brain Res. Rev.* **49:** 429.
Mongrain V., Carrier J., and Dumont M. 2006. Circadian and homeostatic sleep regulation in morningness-eveningness. *J. Sleep Res.* **15:** 162.
Moruzzi G. and Magoun H.W. 1949. Brain stem reticular formation and activation of the EEG. *Electroencephalogr. Clin. Neurophysiol.* **1:** 455.
Mukhametov L.M., Supin A.Y., and Polyakova I.G. 1977. Interhemispheric asymmetry of the electroencephalographic sleep patterns in dolphins. *Brain Res.* **134:** 581.
Münch M., Knoblauch V., Blatter K., Schroder C., Schnitzler C., Krauchi K., Wirz-Justice A., and Cajochen C. 2004. The frontal predominance in human EEG delta activity after sleep loss decreases with age. *Eur. J. Neurosci.* **20:** 1402.
Nitz D.A., van Swinderen B., Tononi G., and Greenspan R.J. 2002. Electrophysiological correlates of rest and activity in *Drosophila melanogaster*. *Curr. Biol.* **12:** 1934.
Orth D.N., Besser G.M., King P.H., and Nicholson W.E. 1979. Free-running circadian plasma cortisol rhythm in a blind human subject. *Clin. Endocrinol.* **10:** 603.
Pittendrigh C.S. 1961. Circadian rhythms and the circadian organization of living systems. *Cold Spring Harbor Symp. Quant. Biol.* **25:** 159.
Pittendrigh C.S. and Daan S. 1976a. A functional analysis of circadian pacemakers in nocturnal rodents. IV. Entrainment: Pacemaker as clock. *J. Comp. Physiol. A* **106:** 291.
———. 1976b. A functional analysis of circadian pacemakers in nocturnal rodents. I. The stability and lability of spontaneous frequency. *J. Comp. Physiol. A* **106:** 223.
Rechtschaffen A. and Kales A. 1968. *A manual of standardized terminology, techniques and scoring system for sleep stages of human subjects.* U.S. Government Printing Office, Washington, D.C.
Reebs S.G. and Doucet P. 1997. Relationship between circadian period and size of phase shifts in Syrian hamsters. *Physiol. Behav.* **61:** 661.
Reebs S.G. and Mrosovsky N. 1989. Effects of induced wheel running on the circadian activity rhythms of Syrian hamsters: Entrainment and phase response curve. *J. Biol. Rhythms* **4:** 39.
Retey J.V., Adam M., Honegger E., Khatami R., Luhmann U.F., Jung H.H., Berger W., and Landolt H.P. 2005. A functional genetic variation of adenosine deaminase affects the duration and intensity of deep sleep in humans. *Proc. Natl. Acad. Sci.* **102:** 15676.
Richter C.P. 1968. Inherent twenty-four hour and lunar clocks of a primate: The squirrel monkey. *Commun. Behav. Biol.* **1:** 305.
Sack R.L., Lewy A.J., Blood M.L., Keith L.D., and Nakagawa

H. 1992. Circadian rhythm abnormalities in totally blind people: Incidence and clinical significance. *J. Clin. Endocrinol. Metab.* **75:** 127.

Saper C.B., Chou T.C., and Scammell T. 2001. The sleep switch: Hypothalamic control of sleep and wakefulness. *Trends Neurosci.* **24:** 726.

Saper C.B., Scammell T.E., and Lu J. 2005a. Hypothalamic regulation of sleep and circadian rhythms. *Nature* **437:** 1257.

Saper C.B., Lu J., Chou T.C., and Gooley J. 2005b. The hypothalamic integrator for circadian rhythms. *Trends Neurosci.* **28:** 152.

Scheer F.A., Wright K.P., Jr., Kronauer R.E., and Czeisler C.A. 2007. Plasticity of the intrinsic period of the human circadian timing system. *PLoS ONE* **2:** e721.

Schmidt C., Peigneux P., Muto V., Schenkel M., Knoblauch V., Münch M., de Quervain D.J.-F., Wirz-Justice A., and Cajochen C. 2006. Encoding difficulty promotes postlearning changes in sleep spindle activity during napping. *J. Neurosci.* **26:** 8976.

Schwartz W.J., Busis N.A., and Hedley-Whyte E.T. 1986. A discrete lesion of ventral hypothalamus and optic chiasm that disturbed the daily temperature rhythm. *J. Neurol.* **233:** 1.

Sharma V.K. and Daan S. 2002. Circadian phase and period responses to light stimuli in two nocturnal rodents. *Chronobiol. Int.* **19:** 659.

Sherin J.E., Shiromani P.J., McCarley R.W., and Saper C.B. 1996. Activation of ventrolateral preoptic neurons during sleep. *Science* **271:** 216.

Shibata S., Liou S., Ueki S., and Oomura Y. 1984. Influence of environmental light-dark cycle and enucleation on activity of suprachiasmatic neurons in slice preparations. *Brain Res.* **302:** 75.

Siegel J.M. 2005. Clues to the functions of mammalian sleep. *Nature* **437:** 1264.

Steriade M. 2003. The corticothalamic system in sleep. *Front. Biosci.* **8:** d878.

Steriade M., McCormick D.A., and Sejnowski T.J. 1993. Thalamocortical oscillations in the sleeping and aroused brain. *Science* **262:** 679.

Stewart K.T., Rosenwasser A.M., Levine J.D., McEachron D.L., Volpicelli J.R., and Adler N.T. 1990. Circadian rhythmicity and behavioral depression. II. Effects of lighting schedules. *Physiol. Behav.* **48:** 157.

Sudo M., Sasahara K., Moriya T., Akiyama M., Hamada T., and Shibata S. 2003. Constant light housing attenuates circadian rhythms of mPer2 mRNA and mPER2 protein expression in the suprachiasmatic nucleus of mice. *Neuroscience* **121:** 493.

Tobler I. 2005. Phylogeny of sleep regulation. In *Principles and practice of sleep medicine* (ed. M.H. Kryger et al.), p. 77. Elsevier/Saunders, Philadelphia, Pennsylvania.

Tobler I. and Jaggi K. 1987. Sleep and EEG spectra in the Syrian hamster (*Mesocricetus auratus*) under baseline conditions and following sleep deprivation. *J. Comp. Physiol. A* **161:** 449.

Tobler I., Dijk D.J., and Borbély A.A. 1990a. Comparative aspects of sleep regulation in three species. In *Sleep '90* (ed. J. Horne), p. 349. Pontenagel Press, Bochum, Germany.

Tobler I., Franken P., and Jaggi K. 1993. Vigilance states, EEG spectra, and cortical temperature in the guinea pig. *Am. J. Physiol.* **264:** R1125.

Tobler I., Franken P., and Scherschlicht R. 1990b. Sleep and EEG spectra in the rabbit under baseline conditions and following sleep deprivation. *Physiol. Behav.* **48:** 121.

Tononi G. and Cirelli C. 2003. Sleep and synaptic homeostasis: A hypothesis. *Brain Res. Bull.* **62:** 143.

———. 2006. Sleep function and synaptic homeostasis. *Sleep Med. Rev.* **10:** 49.

Triarhou L.C. 2006. The percipient observations of Constantin von Economo on encephalitis lethargica and sleep disruption and their lasting impact on contemporary sleep research. *Brain Res. Bull.* **69:** 244.

Undevia S.D., Gomez-Abuin G., and Ratain M.J. 2005. Pharmacokinetic variability of anticancer agents. *Nat. Rev. Cancer* **5:** 447.

Viola A.U., Archer S.N., James L.M., Groeger J.A., Lo J.C.Y., Skene D.J., von Schantz M., and Dijk D.J. 2007. *PER3* polymorphism predicts sleep structure and waking performance. *Curr. Biol.* **17:** 1.

von Economo C. 1930. Sleep as a problem of localization. *J. Nerv. Ment. Dis.* **71:** 249.

von Roemeling R., Portuese E., Salzer M., Solis O., Bennett J., Sothern R., Hoffmann D., and Sedlacek H. 1990. Improved therapeutic index of cisplatin analogue: B-85-0040 by circadian timing. *Prog. Clin. Biol. Res.* **341A:** 11.

Waterhouse J., Minors D., Folkard S., Owens D., Atkinson G., MacDonald I., Reilly T., Sytnik N., and Tucker P. 1998. Light of domestic intensity produces phase shifts of the circadian oscillator in humans. *Neurosci. Lett.* **245:** 97.

Webb W. and Agnew H. 1975. Sleep efficiency for sleep-wake cycles of varied length. *Psychophysiology* **12:** 637.

Wehr T.A. 1991. The durations of human melatonin secretion and sleep respond to changes in daylength (photoperiod). *J. Clin. Endocrinol. Metab.* **73:** 1276.

———. 1996. A 'clock for all seasons' in the human brain. *Prog. Brain Res.* **111:** 321.

———. 2001. Photoperiodism in humans and other primates: Evidence and implications. *J. Biol. Rhythms* **16:** 348.

Weitzman E.D., Fukushima D., Nogeire C., Hellman L., Sassin J., Perlow M., and Gallagher T.F. 1974a. Studies on ultradian rhythmicity in human sleep and associated neuroendocrine rhythms. In *Chronobiology* (ed. L.E. Scheving and F. Halberg), p. 505. Igaku Shoin, Tokyo, Japan.

Weitzman E.D., Nogeire C., Perlow M., Fukushima D., Sassin J., McGregor P., Gallagher T.F., and Hellman L. 1974b. Effects of a prolonged 3-hour sleep-wake cycle on sleep stages, plasma cortisol, growth hormone, and body temperature in man. *J. Clin. Endocrinol. Metab.* **38:** 1018.

Werth E., Achermann P., and Borbély A.A. 1996. Brain topography of the human sleep EEG: Antero-posterior shifts of spectral power. *Neuroreport* **20:** 123.

Wever R.A. 1979. *The circadian system of man: Results of experiments under temporal isolation.* Springer-Verlag, New York.

———. 1989. Light effects on human circadian rhythms: A review of recent Andechs experiments. *J. Biol. Rhythms* **4:** 161.

Wyatt J.K., Ritz-De Cecco A., Czeisler C.A., and Dijk D.-J. 1999. Circadian temperature and melatonin rhythms, sleep, and neurobehavioral function in humans living on a 20-h day. *Am. J. Physiol.* **277:** R1152.

Yamada N., Shimoda K., Ohi K., Takahashi S., and Takahashi K. 1988. Free-access to a running wheel shortens the period of free-running rhythm in blinded rats. *Physiol. Behav.* **42:** 87.

Zeitzer J.M., Dijk D.-J., Kronauer R.E., Brown E.N., and Czeisler C.A. 2000. Sensitivity of the human circadian pacemaker to nocturnal light: Melatonin phase resetting and suppression. *J. Physiol.* **526:** 695.

Zulley J., Wever R., and Aschoff J. 1981. The dependence of onset and duration of sleep on the circadian rhythm of rectal temperature. *Pflüger's Arch.* **391:** 314.

A Cyanobacterial Circadian Clock Based on the Kai Oscillator

T. KONDO
Division of Biological Science, Graduate School of Science, Nagoya University and SORST/CREST, Japan Science and Technology Agency, Furo-cho, Chikusa-ku, Nagoya 464-8602, Japan

In the cyanobacterium *Synechococcus elongatus* PCC 7942, the products of three genes (*kaiA*, *kaiB*, and *kaiC*) have been identified as essential components of the circadian clock. Recently, we reconstituted the self-sustainable circadian oscillation of the KaiC phosphorylation state by incubating purified KaiC with KaiA, KaiB, and ATP. This in vitro oscillation persisted for at least three cycles and the period was compensated against temperature changes. Period lengths observed in vivo in various *kaiC* mutants were consistent with those measured using in vitro mixtures containing the respective mutant KaiC proteins. These results demonstrate that the oscillation of KaiC phosphorylation is the primary pacemaker of the cyanobacterial circadian clock and reveal a novel function of proteins as timing devices that govern cellular metabolism. We further analyzed four aspects of the KaiC phosphorylation cycle in vitro: the interactions among KaiA, KaiB, and KaiC; the functions of the two phosphorylation sites, the energetics that determine the circadian period, and the mechanisms that synchronize the components of the Kai oscillator. From these analyses, we have proposed a circadian program consisting of the three proteins that keeps biological time in a living cell.

INTRODUCTION

From bacteria and fungi to plants and animals, circadian clocks are ubiquitous endogenous biological timing mechanisms that adapt to daily alterations under environmental conditions. Cyanobacteria are the simplest organisms known to exhibit circadian rhythms. Thus, the cyanobacterium *Synechococcus elongatus* PCC 7942 (hereafter, *Synechococcus*) is an experimental model for the study of molecular mechanisms underlying these types of clocks (Kondo et al. 1993). Saturation mutagenesis revealed that three adjacent genes (*kaiA*, *kaiB*, and *kaiC*) and their protein products are essential for circadian rhythm generation. The levels of *kaiBC* mRNA and KaiC protein oscillate in a circadian manner, the latter of which is delayed by about 6 hours. Continuous overexpression of KaiC nullifies circadian rhythms through an amelioration of *kaiBC* expression, whereas a transient increase in KaiC sets the phase of the rhythm. These observations led to a model of the cyanobacterial transcription-translation-based oscillator (TTO) that explains prokaryotic circadian rhythm generation (Ishiura et al. 1998); in this model, KaiC and KaiA negatively and positively regulate *kaiBC* transcription, respectively.

This TTO model, however, did not explain how this loop obtains the characteristics unique to a circadian clock, i.e., circadian periodicity and stability against alterations in temperature and metabolic activity. A number of physiological studies have shown that these characteristics are essential for the oscillator to contribute to a fitness of organisms in their environment. Therefore, understanding the molecular basis of these features is an important goal in circadian biology. Our extensive screening for clock mutants in *Synechococcus* revealed that most of the period mutations were mapped to the *kai* gene cluster. In particular, some single-amino-acid substitutions in KaiC result in periods as short as 14 hours or as long as 60 hours (Fig. 1). To date, mutations in other genes of *Synechococcus* have only resulted in small changes in the circadian period, even if such mutation affects the amplitude of the rhythm severely. Therefore, KaiC is the primary determinant of the circadian period of cyanobacteria. One candidate for the molecular process that determines the circadian period is the phosphorylation of KaiC, which exhibits a robust circadian rhythm in its phosphorylation level in vivo (Iwasaki et al. 2002).

Although experimental evidence suggested that KaiC phosphorylation is an important contributor to circadian oscillation, it was not clear how this phosphorylation mediated circadian rhythm generation. A breakthrough was obtained when we examined the phosphorylation state of KaiC in prolonged darkness (Fig. 2a), which showed that the phosphorylation state of KaiC robustly oscillated in the cell with a 24-hour period, even under conditions in which neither transcription nor translation of *kaiBC* was permitted (Tomita et al. 2005). This finding raised doubts about the TTO model in cyanobacteria. The fact that this oscillation exhibited stability against temperature and metabolic activity changes suggested that the pacemaker of the cyanobacterial circadian system was not a transcription-translation feedback loop but was instead the KaiC phosphorylation cycle itself. We then attempted to reconstitute an oscillating cycle of KaiC phosphorylation in vitro. Incubating KaiC with KaiA, KaiB, and ATP produced a self-sustainable circadian oscillation of KaiC phosphorylation (Fig. 2b) (Nakajima et al. 2005). This in vitro oscillation of KaiC phosphorylation persisted for at least three cycles and the period exhibited temperature compensation. Furthermore, changes in the circadian rhythm period length observed in vivo in various *kaiC* mutant strains were consistent with those measured in vitro when the incubations were performed with the respective mutant KaiC proteins (Fig. 2c). Therefore, the oscillation of KaiC phosphorylation was identified as the primary pacemaker of the cyanobacterial circadian clock.

How do the three Kai proteins interact to produce a pre-

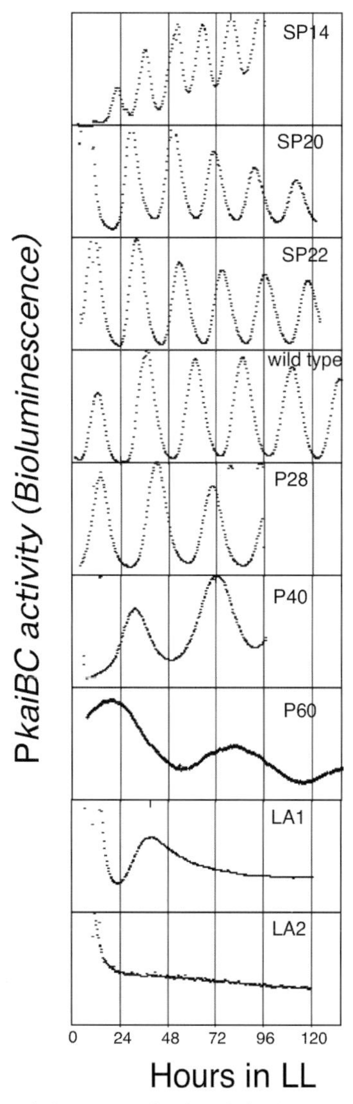

Figure 1. Period mutants of KaiC. Bioluminescence rhythm was examined using a bacterial luciferase reporter fused to the *kaiBC* promoter of period mutants of KaiC. All of the mutations were mapped to the *kaiC*-coding region and resulted in single-amino-acid substitutions. For the loci and details of the mutant phenotypes, see Ishiura et al. (1998).

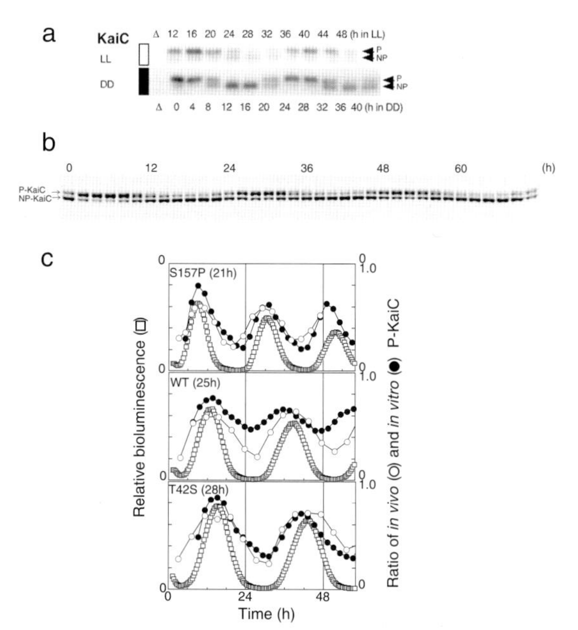

Figure 2. KaiC phosphorylation rhythms in vivo and in vitro. (*a*) Profiles of KaiC expression under constant light (LL) or constant dark (DD) conditions. The level of KaiC was estimated using the density of the signal on western blots of cell extracts taken at various time points under LL or DD conditions. The upper band in each lane represents phosphorylated KaiC, whereas the lower band is dephosphorylated KaiC (Tomita et al. 2005). (*b*) KaiC phosphorylation rhythm in vitro. The three Kai proteins were synthesized in *Escherichia coli*, purified by chromatography, and mixed with 1 mM ATP. Every 2 hours, aliquots were taken, subjected to SDS-PAGE, and detected with Coomassie brilliant blue (CBB). (*Upper band*) Phosphorylated KaiC; (*lower band*) dephosphorylated KaiC. (*c*) KaiC phosphorylation rhythms in KaiC mutants. In vivo bioluminescence profiles and the ratios of phosphorylated KaiC in vivo and in vitro are shown for wild-type KaiC, the S157P mutant, and the T42S mutant. The period estimated from the bioluminescence rhythms are also shown for each strain.

cise timing apparatus in the test tube? This chapter summarizes our recent studies about the mechanism underlying the KaiC phosphorylation cycle. Note also that this is a previously unrecognized phenomenon mediated by proteins, namely, proteins functioning as timing devices.

KAI PROTEIN COMPLEX DYNAMICS IN THE KAIC PHOSPHORYLATION CYCLE IN VITRO

Because our in vitro system is simple, we were able to quantitatively address the molecular dynamics of these three proteins. To elucidate how KaiC phosphorylation oscillation is regulated by interactions among the Kai proteins, we examined the associations of KaiA and KaiB with KaiC using immunoprecipitation analysis and gel-filtration chromatography (Kageyama et al. 2006). In this way, we confirmed that the sequences of Kai protein interactions and KaiC phosphorylation in vitro are similar to those observed in vivo (Kageyama et al. 2003; Kitayama et al. 2003).

Quantitative assessment of the biochemical data revealed the following dynamic protein associations: More than 95% of KaiA associated with KaiC throughout the cycle, whereas only 20–30% of the KaiC hexamers were associated with KaiA at any point in time. Because the ratio of KaiA dimers to KaiC hexamers in the mixture was 1:1, threefold to fourfold more KaiA dimers participated in the interaction with KaiC. Thus, the association between KaiA and KaiC should not be considered to be a simple one-to-one interaction. On the other hand, approximately 80% of the KaiC subunits were phosphorylated and dephosphorylated at the peaks and troughs of the cycle, respectively. If a stable association with KaiA is responsible for the phosphorylation and dephosphorylation of KaiC, this range of KaiC phosphorylation in the cycle did not agree with the results. To solve this difficulty, we propose that KaiA dimers associate with KaiC hexamers without forming stable complexes. Instead, KaiA dimers repeatedly associate with and dissociate from KaiC hexamers, as if jumping from one KaiC to another (Fig. 3). Through this repeated action, KaiA could gradu-

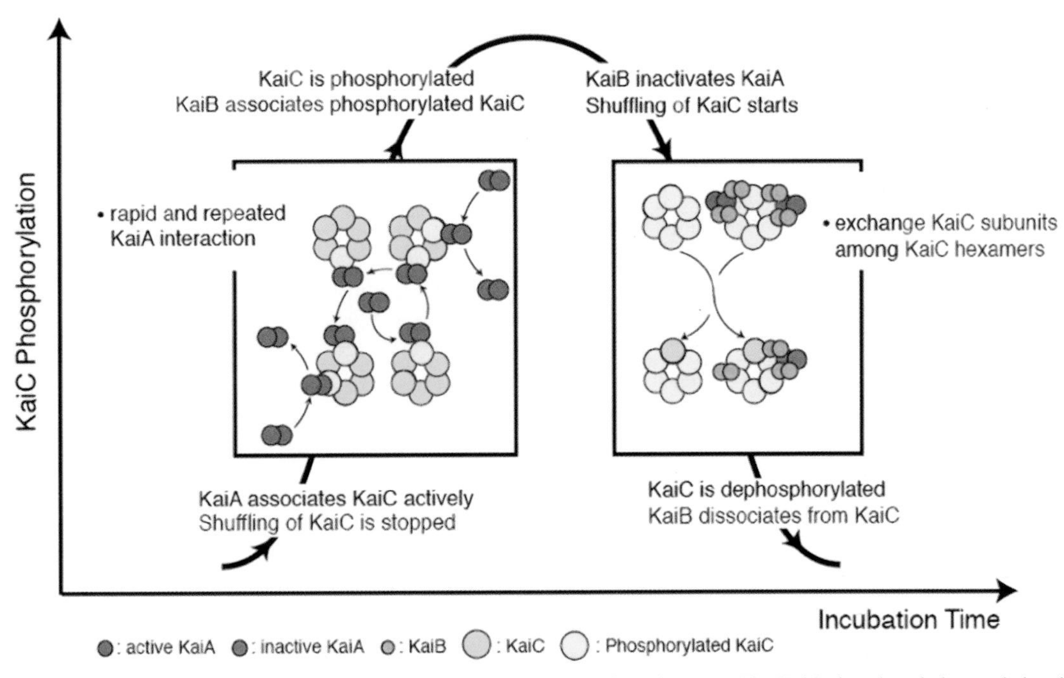

Figure 3. Associations between the three Kai proteins in an in-vitro-reconstituted system. The KaiC phosphorylation cycle is schematically illustrated on a plot of the incubation time versus KaiC phosphorylation. (*Left frame*) Principal protein dynamics for phosphorylation; (*right frame*) principal protein dynamics for dephosphorylation. See text for a more detailed explanation.

ally phosphorylate the KaiC subunits in the hexamers, resulting in 80% of the subunits being phosphorylated.

Increased phosphorylation of the KaiC hexamer could result in changes in the KaiC hexamer state, possibly through the addition of the charged phosphate groups. These changes could result in an increased affinity of KaiB for KaiC. In fact, KaiB tended to associate with the highly phosphorylated KaiC hexamers, forming the uniform complexes observed using gel-filtration chromatography. KaiC-bound KaiB traps and inactivates KaiA, promoting KaiC dephosphorylation. During the dephosphorylation phase, 60% or more of the KaiC hexamers remained free, whereas the remaining 40% associated with KaiA and KaiB. With the exception of the KaiA-KaiC complexes, phosphorylated KaiC hexamers in any complexes are likely to be dephosphorylated, as are the free KaiC hexamers. After decreasing the degree of KaiC hexamer phosphorylation, KaiB dissociates from the hexamer and liberates KaiA, which then participates in the next round of KaiC phosphorylation (Fig. 3).

In the presence of ATP, KaiC forms a hexamer (Hayashi et al. 2003). We examined the potential for KaiC monomer exchange between two KaiC hexamers using FLAG-tagged KaiC. We trapped KaiC hexamers containing KaiC-FLAG at various time points after the components of the oscillator were mixed. If KaiC monomers shuffled among other hexamers, a greater number of KaiC hexamers (up to twofold) would be pulled down using anti-FLAG antibodies. Actually, twice as much KaiC was precipitated following incubations for 4 hours or longer. This result clearly indicates that the KaiC monomers were shuffled between the hexamers in a manner that correlated with the temporal characteristics of the KaiC hexamers (Fig. 3). The shuffling of KaiC monomers may allow the dephosphorylation to be equalized among KaiC hexamers with different configurations. We discuss this issue in a later section of this chapter. Similar results of Kai protein interaction and KaiC monomer exchange were also reported using different methods (Pattanayek et al. 2006). The shuffling of KaiC monomers between different phosphorylation states was also predicted to be important for the stability of the oscillator (Emberly and Wingreen 2006).

A SEQUENTIAL PROGRAM OF KaiC PHOSPHORYLATION

Previously, we identified two phosphorylation sites in KaiC at serine 431 (S431) and threonine 432 (T432), both of which are required for rhythm generation in vivo (Nishiwaki et al. 2004; Xu et al. 2004). To individually examine the phosphorylation states of S431 and T432, we employed nanoflow-liquid chromatography coupled with electrospray ionization-mass spectrometry and improved SDS-PAGE (Fig. 4a) (Nishiwaki et al. 2007). We found that the KaiC phosphorylation cycle was a sequence of four steps as shown in Figure 4b. To clarify the mechanisms underlying the sequential reactions, we mutated KaiC at the phosphorylation sites, introducing an alanine or glutamic acid (E)/aspartic acid (D) residue to mimic the dephosphorylated or phosphorylated states, respectively. Using these mutants, we found that the phosphorylation state of each residue regulated the phosphorylation/dephosphorylation of the other residue. As illustrated in Figure 4b, the results suggested that the product of step 1 (T432 phosphorylation) allowed the subsequent reaction in step 2 (S431 phosphorylation) to proceed efficiently. Similarly, in steps 3 and 4, dephosphorylated T432 facilitated the dephosphorylation of S431, resulting in the

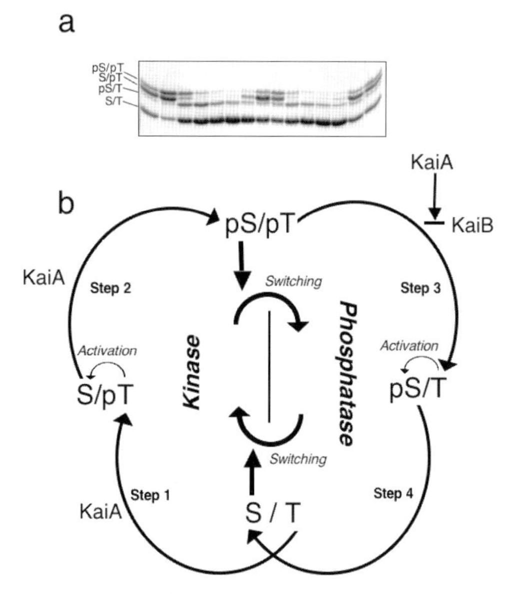

Figure 4. Phosphorylation profiles of the two phosphorylation sites of KaiC. (*a*) Aliquots of the reaction mixtures were collected at 4-hour intervals, mixed with [γ-^{32}P]ATP, and incubated for an additional 30 minutes. Samples were subjected to SDS-PAGE. After CBB staining, four phosphorylated states of KaiC were detected. Phosphate was incorporated mainly into the two upper bands. (*b*) Model of the KaiC phosphorylation cycle. The KaiC phosphorylation cycle consists of four sequential reactions. During steps 1 and 2, KaiC acts as an autokinase, whereas during steps 3 and 4, KaiC functions as an autophosphatase. Phosphorylation of S431 switches KaiC from an autokinase to an autophosphatase, whereas dephosphorylation of S431 to form the completely dephosphorylated form (S/T) switches KaiC from an autophosphatase to an autokinase. Phosphorylation of T432 promotes the phosphorylation of S431, whereas dephosphorylation of S431 promotes dephosphorylation of T432.

completely dephosphorylated S/T form. Thus, once KaiA and ATP promote the phosphorylation of T432, the four steps of the cycle are programmed to proceed within an individual KaiC molecule.

Importantly, between steps 2 and 3, the phosphorylated S431 switches KaiC activity from an autokinase to an autophosphatase. In the same way, switching from the autophosphatase activity to the autokinase activity takes place between steps 4 and 1 with the dephosphorylation of S431. Note that in either step, phosphorylation/dephosphorylation of T432 precedes the change in the S431 phosphorylation state. This mechanism facilitates the in vitro oscillation of KaiC phosphorylation. It is also important to note that the dual phosphorylation mechanism would contribute to the hysteresis in the switching, which may be caused by allosteric changes in the structure of KaiC (Fig. 4b).

We found that KaiA and KaiB had little effect on the dephosphorylation of KaiC-DT (S at 431 was replaced with D) or KaiC-SE (T at 432 was replaced with E), suggesting that the switching of KaiC from an autokinase to an autophosphatase is predominantly regulated by the phosphorylation state of KaiC (step 3). Furthermore, our previous results from a radioactive phosphate uptake assay showed that a small amount of phosphate was incorporated into KaiC even when KaiA was not included in the reaction (Iwasaki et al. 2002; Kitayama et al. 2003). We also confirmed that the ratio of phosphorylated serine and threonine was not altered by the presence of KaiA, whereas KaiA increased the overall level of phosphorylation at both residues (Iwasaki et al. 2002). Dephosphorylation of S431 (step 4) proceeded efficiently in the absence of KaiA and KaiB. Taken together, these observations indicate that KaiC potentially can progress through all of the phosphorylation cycle steps without KaiA and KaiB. In other words, dual phosphorylation sites of KaiC make KaiC function as a "flywheel" for the phsophorylation cycle.

THE ATPASE ACTIVITY OF KAIC SERVES AS THE BASIC TIMING MECHANISM OF THE CIRCADIAN CLOCK

We then examined the molecular basis that defines the circadian period length. Because the circadian period length and its stability over a range of temperatures are essential for an adaptive circadian clock (Dunlap et al. 2004), the mechanism that defines the period length is a critical feature of a circadian oscillator. In contrast, a self-sustained oscillating loop could be more generally found for a number of processes in living cells.

KaiC, which has two ATP-binding motifs (Ishiura et al. 1998), was expected to display a robust ATPase activity because the hexameric structure of KaiC resembles that of a number of DNA-associated ATPases (Leipe et al. 2000). However, so far, KaiC only shows slow kinase and phosphatase activities. By examining ATP consumption during the KaiC phosphorylation cycle, we found that KaiC hydrolyzes 15–20 ATP molecules per cycle (Terauchi et al. 2007). Although the ATPase activity of KaiC was extremely low, this activity was essential for the phosphorylation rhythm to persist. To examine the ATPase activity precisely, we monitored ADP production in the reaction mixture (Fig. 5a). We found that the ATPase activity of KaiC oscillated with a circadian period when KaiA and KaiB were present in the mixture. Interestingly, even in the absence of KaiA and KaiB, the ATPase activity remained constant at a level that corresponded to the average level observed under rhythmic conditions (14.5 molecules day^{-1} = 1.7 x 10^{-4} molecules s^{-1}). As was the case for the phosphorylation activity of KaiC (Iwasaki at al. 2002), KaiA stimulated the ATPase activity of KaiC and KaiB decreased the ATPase activity (Kitayama et al. 2003). We also found that a truncated variant of KaiC (KaiC-CI; residues 1–250 of KaiC), which lacked the phosphorylation sites, showed 70% of the basal activity, demonstrating that ATP hydrolysis was not entirely dependent on the kinase activity.

The thermal sensitivity of the activity was consistent with the thermal sensitivity of the cyanobacterial clock in vivo (Terauchi et al. 2007). As was done for the KaiC phosphorylation rhythm, we confirmed that the ATPase activity of KaiC in the presence of KaiA and KaiB was temperature-compensated in the range from 25°C to 35°C (Fig. 5b). Surprisingly, the ATPase activity in KaiC-only incubations showed a considerably stronger level of temperature compensation; i.e., the activity was fairly constant at all temperatures examined with a Q_{10} coefficient of approximately

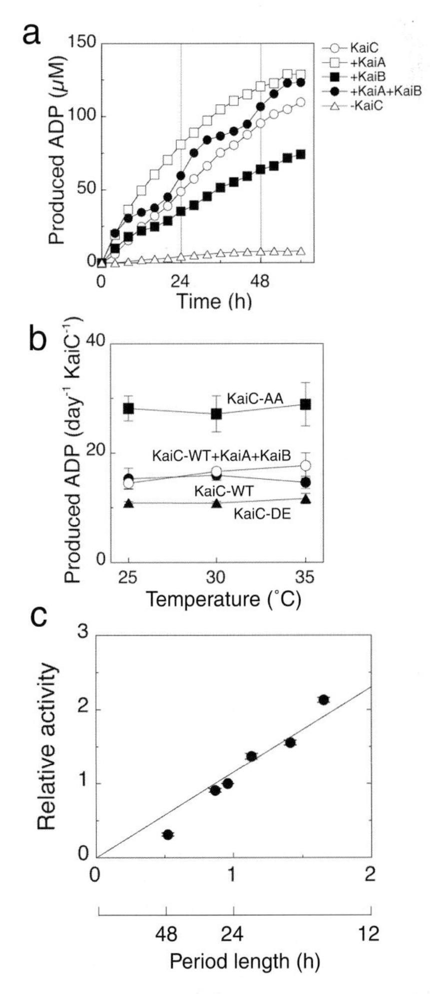

Figure 5. ATPase activity of KaiC. (*a*) KaiC-mediated ADP accumulation. KaiC was incubated under standard conditions with or without KaiA or KaiB at 30°C. The ADP concentration in the reaction mixture was measured using high-performance liquid chromatography. The ADP level in a mixture containing only KaiA and KaiB (*open triangles*) is shown as a negative control. (*b*) Temperature compensation of the ATPase activity of KaiC. KaiC protein was incubated with 1 mM ATP at 25°C, 30°C, or 35°C in the presence (*open circles*) or absence (*closed circles*) of KaiA and KaiB. KaiC-AA (*squares*) and KaiC-DE (*triangles*) were also examined under these conditions. (*c*) ATPase activity of KaiC and the circadian frequency. The ATPase activities of wild-type KaiC and five KaiC period mutant proteins (T42S, S157P, A251V, R393C, and F470Y) were measured in the absence of KaiA and KaiB. The ATPase activities of KaiC are plotted against the frequencies of the in vivo bioluminescence rhythms (reciprocal of the period length). The activity level observed for wild-type KaiC was set at 1.0.

1.0. This result indicated that the temperature compensation of the KaiC ATPase activity is an inherent property of the KaiC molecule. Furthermore, the ATPase activities of KaiC-AA (mimics dephosphorylated KaiC) and KaiC-DE (mimics phosphorylated KaiC) were also temperature-compensated, whereas levels of ATPase activity were altered by these mutations. This result indicates that the temperature compensation of the ATPase activity is not dependent on the KaiC phosphorylation state.

We selected three mutants from more than 900 *kaiC* mutants generated by polymerase chain reaction (PCR)-based mutagenesis as "pure" period mutants. Each of these mutants had a bioluminescence profile that was identical to that of the wild-type strain, except for the period length. All other rhythmic phenotypes were almost identical to those of the wild-type strain (K. Imai et al., unpubl.). Using these mutants, we found that the period lengths of the expression rhythms of *kaiBC* in vivo were consistent with those of the respective in vitro KaiC phosphorylation rhythms (Nakajima et al. 2005). We measured the ATPase activities of the short-period mutant proteins (S157P and F470Y) and the long-period mutant and plotted the activities of the KaiC variants against the frequencies of the in vivo oscillations (period^{-1}). The plot clearly indicated that the frequencies were directly proportional to the ATPase activities (Fig. 5c), i.e., the basal ATPase activity of KaiC dictates the circadian period length of cyanobacteria: The higher the activity, the faster the clock ticks. In particular, the linear correlation indicates that the circadian pacemaker depends directly on the energy provided by ATP hydrolysis. In other words, the same amount of energy (hydrolysis of 15 ATP molecules per KaiC monomer) is required for one period of the circadian cycle. This finding represents the first description of a simple biochemical reaction acting as a circadian timekeeper.

ATPases interact with various partners to convert the energy of ATP hydrolysis into mechanical forces (Ye et al. 2004). Like other members of the RecA superfamily, KaiC contains a RecA-type nucleotide-binding domain, which converts the chemical energy of ATP into mechanical energy that is harnessed to move the protein along macromolecules. To the best of our knowledge, the activity of KaiC (15 ATPs day^{-1}) is substantially lower than any other protein in this family; e.g., RuvB—a protein that is involved in DNA recombination—hydrolyzes 8 x 10^3 ATP day^{-1} even in the absence of substrate DNA (Marrione and Cox 1995). To explain the extraordinarily weak and temperature-compensated activity of KaiC, we hypothesize that the energy of ATP hydrolysis is not transferred to another molecule but is instead passed to KaiC itself, thereby lowering the observed activity (Fig. 6).

Similar to many RecA-like ATPases, KaiC forms a hexameric ring (Pattanayek et al. 2004); these RecA-like ATPases undergo conformational changes in response to ATP binding and hydrolysis (Wang 2004; Ye at al. 2004). Conformational changes in ATPases, such as p97 and ClpB, have been observed during their ATPase cycles (Rouiller et al. 2002; Lee et al. 2007). Thus, after the formation of the hexamer structure, hydrolysis of ATP in KaiC may result in an immediate conformational change in the hexamer that decreases the ATPase activity to an uncommonly low level. Moreover, this autoregulation of the ATPase activity could be the basis for the temperature independence of the activity, because as the activity increases with increasing temperatures, the degree to which the activity is inhibited also increases. Analysis of the submolecular mechanisms underlying the ATPase reaction in the KaiC hexamer should provide valuable information regarding the molecular basis of the circadian period. In addition, the similarities with the stability of an

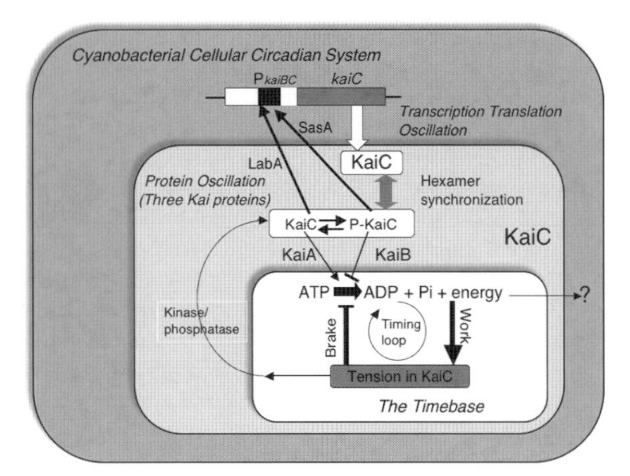

Figure 6. Cyanobacterial circadian system is defined by KaiC ATPase activity. A schematic diagram of the KaiC ATPase activity, the phosphorylation/dephosphorylation cycle, and the transcription-translation cycle. See text for a detailed explanation.

electric feedback amplifier should also be noted. When an amplifier forms a negative feedback loop that suppresses its native gain (open-loop gain) to 1/30 or lower of the native value, the closed gain of the feedback circuit should become constant and insensitive to a drift in the open-loop gain. Assuming that the open-loop gain is the native ATPase activity of KaiC, which is much higher than the observed level, the extremely weak but temperature-compensated properties of the KaiC ATPase activity can be explained.

How is the ATPase activity of KaiC phenotypically expressed as the circadian period of the phosphorylation cycle? It is important to note that the ATP hydrolysis and phosphorylation activities of KaiC may influence each other (Fig. 6). In fact, the ATPase activity was influenced by KaiA, KaiB, and the phosphorylation state of KaiC. Moreover, these three factors may regulate the ATPase activity both negatively and positively, because the associations between these proteins and KaiC and the phosphorylation of KaiC are rhythmic (Kageyama et al. 2006). On the other hand, the ATPase activity may influence the kinase activity of KaiC, possibly by inducing a conformational change in the structure of either monomeric or hexameric KaiC, as shown by the finding that both activities of KaiC oscillate with the same phase angle. Therefore, these mutual couplings between the ATPase and kinase activities of KaiC may generate a self-sustained oscillation through two processes that resonate with a circadian period (Fig. 6). Note also that such coupling would be based on the duplicated structure of KaiC, which combines two activities together into a single protein molecule.

These couplings appear to conform to the principles of an electric back-coupled oscillator, in which alternating energy transfer between two components is partially coupled with a feedback circuit containing two components that periodically resonate with a time constant defined by the interaction between the two components. For example, energy that alternately accumulates as capacitance and inductance generates a damping oscillation that serves as the time base of a self-sustained oscillation when it is coupled with feedback circuits. Thus, the ATPase activity of KaiC and the accumulation of stress in the structure of KaiC could function as the time base of the phosphorylation rhythm when it is coupled with the kinase/phosphatase activity of KaiC through interactions with KaiA and KaiB. The transition of the conformational states during the ATPase cycle may strain the KaiC structure, which would be followed by relaxation of the tension, as was suggested to occur in the GroEL allosteric transitions (Hyeon et al. 2006).

AUTONOMOUS SYNCHRONIZATION OF THE KAIC PHOSPHORYLATION RHYTHM

In the in vitro reconsitution experiments, mixing three proteins with ATP should reset the oscillation. However, under continuous temperature conditions, the oscillation is expected to be a damped oscillation, because, apparently, there are no mechanisms to sustain the oscillation under our constant temperature conditions. We examined the KaiC phosphorylation cycle during prolonged incubations. As shown in Figure 7a, the KaiC phosphorylation rhythm persisted for 10 days without damping as long as the system was replenished with ATP. This suggests that the individual oscillating KaiC units were precisely synchronized in the in vitro system. We then tested the potential synchronization of the KaiC phosphorylation rhythm by mixing together samples with six different phases and then observing KaiC phosphorylation in the mixture (Fig. 7b). If KaiC phosphorylation in each sample oscillated independently, the overall phosphorylation ratio within the mixture should have averaged out to a constant value. We, however, found that almost immediately after mixing the samples, the mixture exhibited a phosphorylation rhythm with an amplitude that was comparable to those observed in the original individual samples. This observation suggests that the phosphorylation of the individual KaiC units was rapidly synchronized (Ito et al. 2007).

We then prepared two KaiC proteins that were labeled with different fluorescent tags that allowed us to trace their respective phosphorylation states. We combined two samples of the fluorescent KaiC proteins with different oscillatory phases and observed the phosphorylation of the labeled KaiC proteins. After the samples were mixed, we found that the KaiC sample that was originally in the dephosphorylation phase was dephosphorylated as was observed for the original sample. In contrast, the KaiC sample originally in the phosphorylation phase either was dephosphorylated or was inhibited from proceeding with the phosphorylation reaction. After the phosphorylation ratios in the two KaiC samples equalized, they began to be phosphorylated synchronously.

We previously demonstrated that KaiC monomers shuffle between the hexamers (Kageyama et al. 2006). When KaiC hexamers in the same phase were combined, pull-down assays with wild-type and FLAG-tagged KaiC demonstrated that monomer shuffling occurred only during a limited 4-hour period early in the dephosphorylation phase. Shuffling among KaiC hexamers in the early dephosphorylation phase would allow the phosphorylation states to synchronize, leading to the robust oscillation

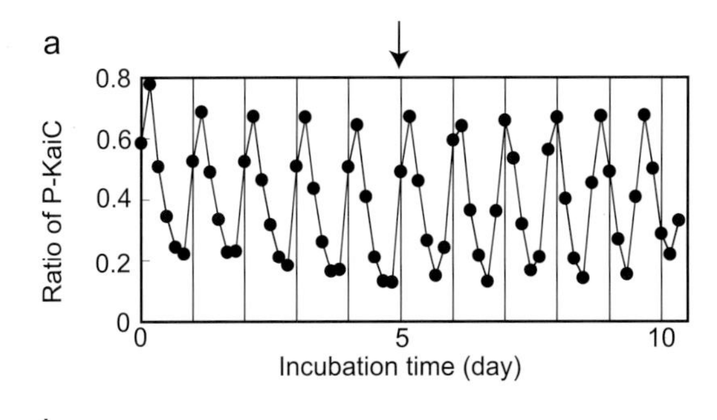

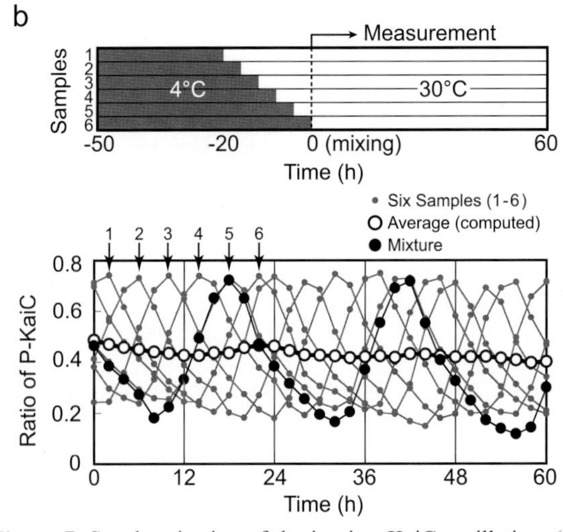

Figure 7. Synchronization of the in vitro KaiC oscillation. (*a*) Solution containing recombinant Kai proteins was incubated for 10 days under standard conditions. Five days after the incubation began, ATP was added to the solution to a final concentration of 1 mM (*arrow*). (*b*) After the standard sample was kept at 4°C for more than 30 hours, the temperature was raised to 30°C at various time points. At time 0, aliquots of six samples (samples 1–6) with different phases were combined. (*Gray dots*) Phosphorylation ratios of the six samples; (*black circles*) phosphorylation ratios of the mixture. (*Open circles*) Arithmetic mean of the phosphorylation ratios from the six samples at each time point. The *x* axis represents the period of time after mixing.

shown in Figure 7a. In short, the inevitable variance between the phosphorylation cycles of individual hexamers caused by thermal noise can be reset during each cycle. We also found that during the shuffling phase, KaiC monomers were able to shuffle between hexamers in different phases. This suggested that KaiC hexamers that contain a heterogeneous population of KaiC phosphorylation states are also likely to exchange monomers. Our model suggests that KaiC monomers within a hexamer in the phosphorylation phase switch to the dephosphorylation phase when monomers in the dephosphorylation phase are integrated into the hexamer. This process explains the results observed when KaiC proteins with different phosphorylation phases are mixed in vitro.

It is likely that the synchronization of the KaiC phosphorylation rhythm observed in vitro contributes to some extent to the robustness of the circadian rhythm in cyanobacterial cells. Because cells grown under constant light are able to divide more than twice a day (Kondo et al. 1997), each cell doubles the amount of Kai proteins due to de novo synthesis at least twice within one circadian cycle. Because newly synthesized KaiC protein is not phosphorylated, the KaiC phosphorylation rhythm in rapidly growing cells is constantly subjected to perturbations due to these new KaiC molecules. Although these perturbations would be expected to damp the KaiC oscillation, the amplitude and period of the phosphorylation cycle in rapidly dividing cells are similar to those in cells at a stationary phase or in cells under dark conditions. As observed in vitro, it is likely that newly synthesized KaiC proteins are subjected to an entraining process during the dephosphorylation phase, which results in a synchronization of the phosphorylation state and reaction direction at the beginning of next phosphorylation phase. Thus, the robustness of the circadian system in living cells may be principally achieved at the level of a chemical oscillator (Fig. 6).

Studies examining the circadian rhythms in single cells revealed that the circadian clock of individual cyanobacterial cells is extremely (Mihalcescu et al. 2004) robust when compared with other cellular oscillatory systems that are based on transcriptional networks. Oscillatory systems based on transcriptional regulatory networks are inevitably influenced by fluctuating concentrations of gene products due to the stochastic nature of transcriptional processes (Elowitz et al. 2002) and various external perturbations, such as cell division. The precision of mammalian circadian rhythms is thought to be achieved through cell–cell communication between oscillating cells. Although cyanobacterial cells do not communicate with each other with respect to circadian rhythm (Mihalcescu et al. 2004), communication (i.e., monomer shuffling) between the oscillating units (KaiC hexamers) renders the circadian oscillation of individual cells highly stable and precise. Such precision requires the participation of a relatively large number of oscillating units; indeed, an individual cyanobacterial cell contains approximately 10,000 KaiC proteins (Kitayama et al. 2003), a number that is comparable to the number of neurons in the mammalian suprachiasmatic nucleus.

THE KAIC-BASED CELLULAR CIRCADIAN SYSTEM

As shown above, the Kai-based chemical oscillator robustly functions in vitro, similar to a pendulum that is not connected to a downstream gear, allowing it to run without any external disturbances. On the other hand, the Kai-based oscillator in the cells is connected to a downstream process. Thus, the cellular circadian system involves transcription-translation processes; these processes supply the clock components that vary with the rate of cell proliferation. Even under such conditions, the circadian clock of the individual cell is extremely precise. To understand the intracellular dynamics of the KaiC-based circadian system, the following issues need to be examined: (1) the mechanism by which the Kai-based chemical oscillator drives circadian rhythms in genome-wide transcription in cyanobacteria; (2) how rhythms in Kai protein synthesis and degradation as well as those in intracellular metabolic conditions are coupled with the Kai-based oscillator; and (3) how the Kai-based oscillator is entrained to day/night alterations in the environment.

The linkage between KaiC and transcription is not mediated by a single pathway. One possibility is that KaiC activates gene expression via a two-component regulatory system, which includes the KaiC-binding histidine kinase SasA (Iwasaki et al. 2000). In *sasA*-inactivated strains, *kaiBC* expression is markedly reduced, and circadian transcription rhythms are severely attenuated. Although SasA is not required for a basic oscillation, it may function as a connection between the KaiC phosphorylation cycle and transcription. To elucidate this pathway, we recently identified the DNA-binding protein RpaA as a cognate response regulator of SasA (Takai et al. 2006). Circadian transcription was severely attenuated in *rpaA* mutant cells, and the phosphotransfer activity from SasA to RpaA was dependent on the circadian state of the coexisting Kai protein complexes in vitro. Thus, we have proposed a model in which the SasA-RpaA two-component system mediates time signals from the enzymatic oscillator to drive genome-wide transcription rhythms in cyanobacteria.

The mechanism and significance of negative feedback regulation of *kaiBC*, however, have not been fully elucidated. We recently reported that the novel gene *labA* is required for the negative feedback regulation by KaiC. Disruption of *labA* abolished the transcriptional repression caused by overexpression of KaiC and elevated the trough levels of circadian gene expression, resulting in a low-amplitude phenotype (Taniguchi et al. 2007). In contrast, overexpression of *labA* significantly lowered the levels of circadian gene expression. Furthermore, genetic analysis indicated that *labA* and *sasA* function in parallel pathways to regulate *kaiBC* expression. These results suggest that the temporal information derived from the KaiC-based oscillator diverges into a LabA-dependent negative regulatory pathway and a SasA-dependent positive regulatory pathway. It is likely that quantitative information from KaiC is transmitted to RpaA through LabA, whereas SasA mediates the state of the KaiC-based oscillator. It should be noted that disruption of either pathway did not affect the circadian period as much as it affected the transcriptional levels of the downstream targets. These observations imply that both types of regulations are necessary for robust cellular rhythms, whereas the circadian pacemaker is dominated by the biochemistry of KaiC.

Another important characteristic of a circadian clock is entrainment of the circadian oscillators. We observed a synchronization of the circadian phosphorylation of the KaiC hexamers, which resulted from monomer shuffling. This mechanism contributes to the robustness of the circadian oscillation in the cell by entraining de-novo-synthesized KaiC to the previously established oscillation. A more dynamical assessment in cells based on the rhythmic synthesis and degradation of Kai proteins, however, would allow a quantitative understanding of the cyanobacterial oscillator. On the basis of classical observation of circadian rhythm, we have two models for the entrainment of the circadian clock. An evaluation of the intracellular dynamics of Kai protein synthesis, degradation, and cellular localization would address the nonparametric model. In addition, note that temperature would be an effective factor to entrain the circadian oscillation of cyanobacteria. Mori et al. (2007) reported phase shifting due to temperature shifts, whereas we found that in vitro KaiC phosphorylation rhythms can be entrained to a temperature cycle (T. Yoshida et al., unpubl.). On the other hand, changes in intracellular factors, such as the ATP level, the ratio of KaiA to KaiC, pH, and the concentrations of other metabolites produced by photosynthesis, can be used to test the parametric entrainment model; the effect of these parameters on the ATPase activity of KaiC would be the basis of this entrainment model.

CONCLUSIONS

After identifying the KaiC phosphorylation rhythm in vitro, we have elucidated the dynamics and mechanism of this autonomous oscillating system with biochemical and functional approaches. We found that the interactions between the three proteins are coupled with the phosphorylation program that is installed in the two phosphorylation sites of KaiC like a molecular flywheel. These unique intermolecular and intramolecular mechanisms are critical for the Kai protein oscillator of the cyanobacterial circadian clock.

Moreover, we found that the circadian period length and its stability are derived from the ATPase activity of KaiC. Note that our findings imply that the circadian period length, which should be a product of natural selection, is based on ATP hydrolysis, rather than the delicate balance of the time delay of clock gene expression. This strategy of an oscillator as an accurate time base is similar to what is observed in many electronic or physical oscillators. In these systems, mutually interacting components that can transform energy into different forms (e.g., inductance and capacitance or tension and inertia) define the time base of a self-sustaining oscillator network.

Our results reviewed here provide a basis for studying the physical machinery of this unique protein-based timer. Associations between the three Kai proteins, phosphorylation of KaiC, and the KaiC ATPase activity may all be governed by intramolecular distortions of the KaiC structure. Fortunately, the atomic structures of the three Kai proteins have already been examined using X-ray crystallography and nuclear magnetic resonance (NMR) analyses (Golden et al. 2007). An understanding of KaiC at the atomic level, however, requires precise information regarding how the structure is modified with respect to KaiC functions. Thus, several snapshots of these configurations are crucial to allow the time profiles of the structural changes to be elucidated. It is important that these structural analyses be accompanied by an examination of their functional significance. Fortunately, more than 1000 KaiC mutants with previously characterized phenotypes are available for these types of analyses. It should be noted that the identification of KaiC as a time base is a novel function of proteins in general.

The dynamics of the Kai oscillator in the cell should also be examined for a complete understanding of the circadian system in these cells (see Fig. 6). We have shown that Kai oscillators are able to synchronize with each other due to monomer exchange between KaiC hexamers. The Kai oscillator, however, is subjected to various fluctuations in the rate of KaiC turnover. In a cell, KaiC turnover is subjected to negative feedback regulation, and the roles of KaiC in gene expression are not simply mediated by its phosphorylation status. Further studies of the feedback regulation as well as the integrated dynamics of the Kai oscillator and intracellular metabolism will elucidate mechanisms underlying the robustness of the cellular clock and its entrainment to day/night alterations.

ACKNOWLEDGMENTS

I thank the current and former lab members who contributed to our studies of the cyanobacterial circadian clock. We also thank *Synechococcus elongates*, which has developed a unique process to cope with the 24-hour periodicity observed on this planet. This research was supported in part a grant-in-aid from the Ministry of Education, Culture, Sports, Science and Technology of Japan (15GS0308 to T.K.).

REFERENCES

Dunlap J.C., Loros J.J., and DeCoursey P.J. 2004. *Chronobiology: Biological timekeeping*. Sinauer, Sunderland, Massachusetts.

Elowitz M.B., Levine A.J., Siggia E.D., and Swain P.S. 2002. Stochastic gene expression in a single cell. *Science* **297:** 1183.

Emberly E. and Wingreen S.N. 2006. Hourglass model for a protein-based circadian oscillator. *Phys. Rev. Lett.* **96:** 038303.

Golden S.S., Cassone V.M., and LiWang A. 2007. Shifting nanoscopic clock gears. *Nat. Struct. Mol. Biol.* **14:** 362.

Hayashi F., Suzuki H., Iwase R., Uzumaki T., Miyake A., Shen J.R., Imada K., Furukawa Y., Yonekura K., Namba K., and Ishiura M. 2003. ATP-induced hexameric ring structure of the cyanobacterial circadian clock protein KaiC. *Genes Cells* **8:** 287.

Hyeon C., Lorimer G.H., and Thirumalai D. 2006. Dynamics of allosteric transitions in GroEL *Proc. Natl. Acad. Sci.* **103:** 18939.

Ishiura M., Kutsuna S., Aoki S., Iwasaki H., Andersson C.R., Tanabe A., Golden S.S., Johnson C.H., and Kondo T. 1998. Expression of a gene cluster *kaiABC* as a circadian feedback process in cyanobacteria. *Science* **281:** 1519.

Ito H., Kageyama H., Mutsuda M., Nakajima M., Oyama T., and Kondo T. 2007. Origin of the resilience of the cyanobacterial circadian clock. *Nat. Struct. Mol. Biol.* (in press).

Iwasaki H., Nishiwaki T., Kitayama Y., Nakajima M., and Kondo T. 2002. KaiA-stimulated KaiC phosphorylation in circadian timing loops in cyanobacteria. *Proc. Natl. Acad. Sci.* **99:** 15788.

Iwasaki H., Williams S.B., Kitayama Y., Ishiura M., Golden S.S., and Kondo T. 2000. A KaiC-interacting sensory histidine kinase, SasA, necessary to sustain robust circadian oscillation in cyanobacteria. *Cell* **101:** 223.

Kageyama H., Kondo T., and Iwasaki H. 2003. Circadian formation of clock protein complexes by KaiA, KaiB, KaiC and SasA in cyanobacteria *J. Biol. Chem.* **278:** 2388.

Kageyama H., Nishiwaki T., Nakajima M., Iwasaki H., Oyama T., and Kondo T. 2006. Cyanobacterial circadian pacemaker: Kai protein complex dynamics in the KaiC phosphorylation cycle *in vitro*. *Mol. Cell* **23:** 161.

Kitayama Y., Iwasaki H., Nishiwaki T., and Kondo T. 2003. KaiB functions as an attenuator of KaiC phosphorylation in the cyanobacterial circadian clock system. *EMBO J.* **22:** 2127.

Kondo T., Mori T., Lebedeva N.V., Aoki S., Ishiura M., and Golden S.S. 1997. Circadian rhythms in rapidly dividing cyanobacteria. *Science* **275:** 224.

Kondo T., Strayer C.A., Kulkarni R.D., Taylor W., Ishiura M., Golden S.S., and Johnson C.H. 1993. Circadian rhythms in prokaryotes luciferase as a reporter of circadian gene expression in cyanobacteria. *Proc. Natl. Acad. Sci.* **90:** 5672.

Lee S., Choi J.-M., and Tsai F.T.F. 2007. Visualizing the ATPase cycle in a protein disaggregating machine: Structural basis for substrate binding by ClpB. *Mol. Cell* **25:** 261.

Leipe D.D., Aravind L., Grishin N.V., and Koonin E.V. 2000. The bacterial replicative helicase DnaB evolved from a RecA duplication *Genome Res.* **10:** 5.

Marrione P.E. and Cox M.M. 1995. RuvB protein-mediated ATP hydrolysis: Functional asymmetry in the RuvB hexamer *Biochemistry* **34:** 9809.

Mihalcescu I., Hsing W., and Leibler S. 2004. Resilient circadian oscillator revealed in individual cyanobacteria. *Nature* **430:** 81.

Mori T., Williams D.R., Byrne M.O., Qin X., Egli M., Mchaourab H.S., Stewart P.L., and Johnson C.H. 2007. Elucidating the ticking of an in vitro circadian clockwork. *PLoS Biol.* **5:** e93.

Nakajima M., Imai K., Ito H., Nishiwaki T., Murayama Y., Iwasaki H., Oyama T., and Kondo T. 2005. Reconstitution of circadian oscillation of cyanobacterial KaiC phosphorylation *in vitro*. *Science* **308:** 414.

Nishiwaki T., Satomi Y., Kitayama Y., Terauchi K., Kiyohara R., Takao T., and Kondo T. 2007. A sequential program of dual phosphorylation of KaiC as a basis for circadian rhythm in cyanobacteria. *EMBO J.* **26:** 4029.

Nishiwaki T., Satomi Y., Nakajima M., Lee C., Kiyohara R., Kageyama H., Kitayama Y., Temamoto M., Yamaguchi A., Hijikata A., Go M., Iwasaki H., Takao T., and Kondo T. 2004. Role of KaiC phosphorylation in the circadian clock system of *Synechococcus elongatus* PCC 7942. *Proc. Natl. Acad. Sci.* **101:** 13927.

Pattanayek R., Wang J., Mori T., Xu Y., Johnson C.H., and Egli M. 2004. Visualizing a circadian clock protein: Crystal structure of KaiC and functional insights. *Mol. Cell* **15:** 375.

Pattanayek R., Williams D.R., Pattanayek S., Xu Y., Mori T., Johnson C.H., Stewart P.L., and Egli M. 2006. Analysis of KaiA-KaiC protein interactions in the cyano-bacterial circadian clock using hybrid structural methods. *EMBO J.* **25:** 2017.

Rouiller I., DeLaBarre B., May A.P., Weis W.I., Brunger A.T., Milligan R.A., and Wilson-Kubalek E.M. 2002. Conformational changes of the multifunction p97 AAA ATPase during its ATPase cycle. *Nat. Struct. Biol.* **12:** 950.

Takai N., Nakajima M., Oyama T., Kito R., Sugita C., Sugita M., Kondo T., and Iwasaki H. 2006. A KaiC-associating SasA-RpaA two-component regulatory system as a major circadian timing mediator in cyanobacteria. *Proc. Natl. Acad. Sci.* **103:** 12109.

Taniguchi Y., Katayama M., Ito R., Takail N., Kondo T., and Oyama T. 2007. *labA:* A novel gene required for negative feedback regulation of the cyanobacterial circadian clock protein KaiC. *Genes Dev.* **21:** 60.

Terauchi K., Kitayama Y., Nishiwaki T., Miwa K., Murayama Y., Oyama T., and Kondo T. 2007. The ATPase activity of KaiC determines the basic timing for circadian clock of cyanobacteria. *Proc. Natl. Acad. Sci.* (in press).

Tomita J., Nakajima M., Kondo T., and Iwasaki H. 2005. No transcription-translation feedback in circadian rhythm of KaiC phosphorylation. *Science* **307:** 251.

Wang J. 2004. Nucleotide-dependent domain motions within rings of the RecA/AAA(+) superfamily. *J. Struct. Biol.* **148:** 259.

Xu Y., Mori T., Pattanayek R., Pattanayek S., Egli M., and Johnson C.H. 2004. Identification of key phosphorylation sites in the circadian clock protein KaiC by crystallographic and mutagenetic analyses. *Proc. Natl. Acad. Sci.* **101:** 13933.

Ye J., Osborne A.R., Groll M., and Rapoport T.A. 2004. RecA-like motor ATPases—Lessons from structures *Biochim. Biophys. Acta* **1659:** 1.

A Circadian Clock in *Neurospora:* How Genes and Proteins Cooperate to Produce a Sustained, Entrainable, and Compensated Biological Oscillator with a Period of about a Day

J.C. DUNLAP, J.J. LOROS,* H.V. COLOT, A. MEHRA, W.J. BELDEN, M. SHI, C.I. HONG, L.F. LARRONDO, C.L. BAKER, C.-H. CHEN, C. SCHWERDTFEGER, P.D. COLLOPY, J.J. GAMSBY, AND R. LAMBREGHTS

*Department of Genetics, *Department of Biochemistry, Dartmouth Medical School, Hanover, New Hampshire 03755*

Neurospora has proven to be a tractable model system for understanding the molecular bases of circadian rhythms in eukaryotes. At the core of the circadian oscillatory system is a negative feedback loop in which two transcription factors, WC-1 and WC-2, act together to drive expression of the *frq* gene. WC-2 enters the promoter region of *frq* coincident with increases in *frq* expression and then exits when the cycle of transcription is over, whereas WC-1 can always be found there. FRQ promotes the phosphorylation of the WCs, thereby decreasing their activity, and phosphorylation of FRQ then leads to its turnover, allowing the cycle to reinitiate. By understanding the action of light and temperature on *frq* and FRQ expression, the molecular basis of circadian entrainment to environmental light and temperature cues can be understood, and recently a specific role for casein kinase 2 has been found in the mechanism underlying circadian temperature-compensation. These data promise molecular explanations for all of the canonical circadian properties of this model system, providing biochemical answers and regulatory logic that may be extended to more complex eukaryotes including humans.

INTRODUCTION

We know a great deal about how circadian clocks work in the cells of higher organisms because ideas and models for circadian oscillators developed in model systems have proven to be eminently extensible to other related living organisms. Fungi are the group of organisms most closely evolutionarily related to both animals and higher plants. It has been a while since they all diverged; the organisms that eventually gave rise to both fungi and animals diverged from plants long before Gondwanaland split up (http://www.palaeos.com/) (Dunlap 1999), and some years later (between 1 and 1.5 billion) the fungi and animals diverged (Heckman et al. 2001). However, it seems that by then some decisions had been made about how to build a clock, and these were kept probably because they worked. Indeed, both the logic and many of the molecules underlying the circadian clock in *Neurospora* are also found in animals. For this reason, *Neurospora* has proven to be a useful and durable model for understanding circadian biology. In this chapter, we briefly review the emergence of *Neurospora* as a salient model for clocks, how the central transcription-translation feedback loop (TTFL) works, and current views and opinions regarding how the TTFL can account for the canonical properties that make a rhythm circadian.

DEFINING CHARACTERISTICS OF CIRCADIAN RHYTHMS

Beatrice Sweeney defined a true circadian rhythm in this way: "A circadian rhythm is an oscillation in a biochemical, physiological, or behavioral function which under conditions in nature has a period of exactly 24 hours, in phase with the environmental light and darkness, but which continues to oscillate with a period of approximately but usually not exactly 24 hours" (Sweeney 1976). Thus, these rhythms are phased by the environment but are not a simple response to it; the period also exhibits the temperature, nutritional, and pH compensation that distinguish circadian rhythms from cell-cycle-regulated phenomena and from other approximately 24-hour metabolic or developmental rhythms whose period lengths are often quite temperature- or medium-dependent. Overall, it is these few characteristics of a rhythm—persistence under constant conditions, having a period length of about a day, resettable by brief interruptions of this constant regimen and compensated so that the period length varies only a little under different conditions of ambient temperature or nutrition—that define a rhythm as being circadian and unite it with similar rhythms found in a large number of organisms. In turn, these are the characteristics that molecular models for the clock need to be able to explain: How do the known circadian clock genes and proteins act and interact to yield a molecular cycle with these characteristics?

A CIRCADIAN RHYTHM IN *NEUROSPORA*

Under just the right conditions, *Neurospora* (like all fungi examined) can be made to express any of a multitude of discernible rhythms in growth rate, morphology, and even gene expression (Ingold 1971; Bünning 1973). These rhythms comprise a rich biology as they are probably adaptive to the organism's success, but most of them are not characterized as true circadian rhythms because they lack one or more of the defining characteristics. However, Pittendrigh et al. (1959) found a rhythm in asexual spore production (conidiation) that met the circadian criteria. It is likely that *Neurospora* would never have risen to any prominence as a circadian model had not Malcolm Sargent and colleagues tamed the system. Their initial landmark success (Sargent et al. 1966) was the identification of a novel strain (*band*, recently shown to be an allele of *ras-1*; Belden et al. 2007b) that clarified circadian output. *band* was shown to be so useful that it was incorporated into virtually every strain used for rhythm studies in *Neurospora*; by making it trivial to follow the clock, *band* truly spawned a field. During the next few years, the basic characteristics of the rhythm with respect to light entrainment (Sargent and Briggs 1967) and medium composition (Sargent and Kaltenborn 1972) were worked out. At 25°C, the period length of the *Neurospora* circadian clock was about 22 hours, and it was temperature-compensated, varying from nearly 23 hours at 20°C to a bit under 21 hours at 30°C. Likewise, it is close to the same when cultures are grown with different amounts of glucose or at different pHs, and it lengthens by about 2 hours when grown, for instance, on histidine (Sargent and Kaltenborn 1972). Thus, the period of the cycle exhibits temperature, nutritional, and pH compensation. Environmental stimuli including light and temperature can reset the phase of the clock, which is by all criteria circadian in nature.

Sargent's pioneering work made the system reproducible and tractable and facilitated subsequent work by Feldman's lab on the genetics of the clock (Feldman et al. 1979). Although never prolific, Feldman's work has well stood the test of time in that over the next decade they genetically identified many of the core clock genes that have proven to be informative toward understanding the inner workings of the circadian TTFL, *frq*, *prd-1* through *4*, and *chr* (Feldman et al. 1979; Loros and Feldman 1986). The identification of these mutant strains allowed molecular analysis of the *Neurospora* circadian system to be initiated (shortly after that of *Drosophila*) in the mid-1980s; this resulted in the cloning of the clock gene *frequency (frq)* by 1986 (McClung et al. 1989) and the completion of targeted screens for clock-controlled genes (as well as the coining of this term) (Loros et al. 1989). Molecular manipulations of *frq* expression in the 1990s (Aronson et al. 1994a) proved *frq* to encode a central component of the core oscillator itself, rather than just a clock-controlled gene, and many additional clock mutants have since been characterized and cloned through analysis of the factors that regulate FRQ expression and abundance. Knowledge that the oscillation in *frq* expression was the equivalent of the operation of the circadian oscillator gave rise to studies on the effect of light on *frq* expression and the elucidation of the means through which fungal, and later animal, clocks are reset by light (Crosthwaite et al. 1995; Shigeyoshi et al. 1997). It was already known that two genes, *wc-1* and *wc-2*, were required for light sensing so it was logical to see if they were needed for *frq* induction by light. Surprisingly, analysis of *wc-1* and *wc-2* mutant strains proved them to be arrhythmic as well as truly blind (Crosthwaite et al. 1997; Collett et al. 2002) and led to the discovery of the first PAS-PAS protein heterodimer as a transcriptional activator in a circadian feedback loop (Crosthwaite et al. 1997). This type of complex was soon afterward also reported in mammals (see, e.g., Antoch et al. 1997) and flies (Allada et al. 1998; Darlington et al. 1998; Rutila et al. 1998) and is now a set piece in circadian feedback loops. A firm genetic and regulatory framework was in place by the mid- to late 1990s, a model in which a heterodimer of WC-1 and WC-2 drove expression of *frq* and FRQ in turn acted to reduce the activity of its WC activators. Since this time, steady progress has been made in understanding the molecular bases for sustainability of the rhythm, period length, resetting of the circadian system by light and temperature cues, and gating of input cues. Most recently, *Neurospora* is proving to be a valuable system for examining the role of coupled feedback loops in clocks and for defining global features of circadian output.

Neurospora now represents a salient model system for the analysis and molecular dissection of circadian oscillatory systems. Through examination of this cellular oscillator, and comparative physiology and molecular biology with analogous animals clocks, we can begin to discern threads of similarity in oscillator design principles as well as in molecular components. In some cases, as with the casein kinases CK1 and CK2, it seems likely that the cell simply used the most common workhorse kinases for modulating circadian proteins in all systems. However, in other cases where a variety of choices were possible—for instance in the layout of the feedback loop (as a heterodimeric activator whose activity is depressed by its product), the mode of protein–protein interaction in the activator (PAS:PAS), or even the E3 ubiquitin ligase used to regulate clock protein turnover—it seems likely that the strong similarities observed between fungi and animals may signal shared evolutionary origins for circadian systems and the oscillators that underlie them.

MOLECULAR BASES OF CIRCADIAN CLOCKS IN EUKARYOTIC MODELS: TRANSCRIPTION-TRANSLATION FEEDBACK LOOPS

The *Neurospora* circadian system includes a genuine clock in which a molecular negative feedback loop based on transcription and translation runs its course over about a day. Between the period in the 1980s when the first clock genes were being cloned and mid-1990s when the basic layout of the feedback loops became clear, circadian cycles were generally viewed in much the same way as we view the cell cycle, where a series of events happen one after the other over the course of a day. Such a cycle is quite distinct from the way we view the clock cycle now, as a single step loop where the long time constant can be described by the length of time it takes to turn on gene(s) (*frq*, or *per* and *tim* in flies, or the *per*s and *cry*s in mammals) plus the length of time it takes the protein products of these genes (FRQ, PER, or PERs and Crys) to turn off their activators (respectively, WC-1/WC-2, Cyc/Clk, or BMAL1/CLOCK). Figure 1 compares the basic layout of the circadian feedback loops in *Drosophila*, *Neurospora*, and mammals (the core feedback loops are shaded).

A brief aside is warranted here to consider whether the core TTFL described above is indeed the core of the clock

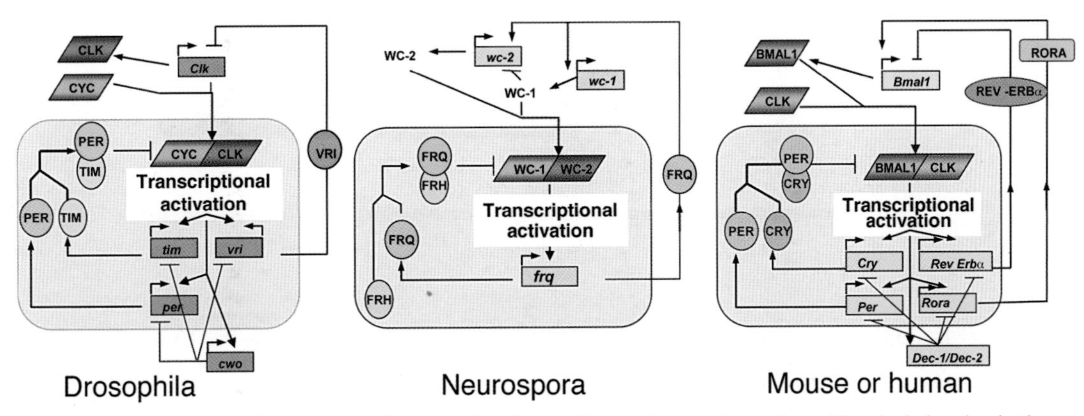

Figure 1. Similar regulatory relationships constitute the circadian oscillators in model organisms. The shaded region in the center of each loop covers the core transcription-translation feedback loop that underlies circadian rhythms. Regulatory relationships outside this core are helpful but not essential for circadian rhythmicity. (Adapted, with permission, from Bell-Pedersen et al. 2005 [© Nature Publishing Group].)

or just a reflection of output. This possibility is raised by the recent data from *Synechococcus* showing that a compensated circadian cycle of phosphorylation can be reconstituted in vitro from just the KaiA, KaiB, and KaiC proteins (Nakajima et al. 2005). Extensive work from the premolecular era in the late 1970s and early 1980s showed conclusively that daily timed protein synthesis was needed for operation of the clock, even in anucleate *Acetabularia*. In *Neurospora*, no level of constant *frq* expression has been found to be capable of supporting rhythmicity. Although carefully constructed scenarios might be mounted to question the obvious conclusion from these data (that this daily cycle is essential for the clock), in the real world, it seems undeniable that the TTFL does occur and that it drives whatever more fundamental loop(s) might exist. As explained above and below, the TTFL sets both period length and phase, and all existing molecular explanations for all the canonical circadian properties in all eukaryotes derive from it. Although considering alternative models is a useful academic exercise that may also shed light on the evolutionary origin of rhythmicity, TTFLs remain the best and only available option at this time for understanding how clocks work in cells in nature.

OVERVIEW OF THE *NEUROSPORA* CLOCK: HOW THE MOLECULAR ELEMENTS IN THE CIRCADIAN OSCILLATOR KEEP TIME

There are just a few components in the *Neurospora* cycle whose roles are well understood and several more whose actions are needed but are as yet poorly understood. The proteins WHITE COLLAR-1 (WC-1) and WHITE COLLAR-2 (WC-2) act together as a transcription factor (the WCC) to drive expression of the gene encoding another protein, FREQUENCY (FRQ); once made, FRQ acts to turn down the activity of WC-1 and WC-2. A series of kinases (CK1, Gorl et al. 2001; CK2, Yang et al. 2002; and PRD-4/checkpoint kinase 2, Pregueiro et al. 2006) and phosphatases (PP1 and PP2A, Heintzen and Liu 2007) regulate FRQ stability as well as WCC activity and stability. Phosphorylation of FRQ makes it attractive to FWD-1, the substrate-recruiting subunit of an SCF-type E3 ubiquitin ligase (He et al. 2003); FWD-1 is the ortholog of the Slimb protein that performs a similar function in the *Drosophila* clock (Ko et al. 2002). Ubiquitinated FRQ is degraded in the proteasome. Indeed, once the daily cycle of FRQ synthesis and turnover was known (Garceau et al. 1997), roles for proteins such as these were anticipated. An unexpected entrant to the list of proteins in necessary supporting roles is the FRQ-interacting RNA helicase FRH, a protein that copurifies with FRQ (Cheng et al. 2005) and is essential for the cell as well as for the clock. These various molecular components interact over time in the cell, providing a glimpse of the molecular basis of subjective time.

The *frq* Gene and Regulation of Its Expression

frq was the second clock gene identified (Feldman et al. 1979) and the second to be cloned (McClung et al. 1989). The synthesis and processing of the *frq* transcripts are themselves of interest because *frq* has one of the most complex expression patterns of any gene known in microbes.

By the late subjective night when most of the FRQ protein in the cell has become unstable and is being degraded, WC-1 and WC-2 act together at a specific sequence in the *frq* promoter (the Clock Box, Froehlich et al. 2003) to drive transcription (Fig. 2). The WC proteins also bind to a second site, the proximal light regulatory element (PLRE), to mediate the light induction of *frq* (Froehlich et al. 2002) that serves to entrain the clock as described below. As the names imply, the Clock Box is the *cis*-acting site that is essential for rhythmicity, whereas the PLRE mediates most of the light induction, although the Clock Box (distal LRE) contributes to this (Froehlich et al. 2002).

Chromatin immunoprecipitation (ChIP) studies (Belden et al. 2007a) are consistent with this view, but they have also turned up some surprising regulation. WC-1 has always been shown to exist with WC-2 in solution as the White Collar Complex, so the assumption when ChIP

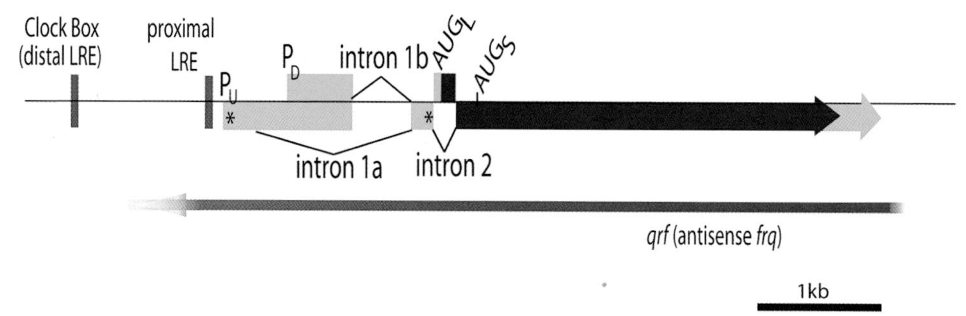

Figure 2. Complex regulation and splicing involved in the expression of *frq*. The Clock Box and PLRE are regulatory elements upstream of P_U and P_D, the upstream and downstream promoters, respectively. (*Green*) Parts of the primary transcript that are retained after splicing; (*asterisks*) upstream ORFs that influence the amount of FRQ made; (*purple*) FRQ ORF. Temperature-regulated splicing of intron 2 governs whether AUG_L or AUG_S is used to initiate FRQ. (*Red*) *qrf*, the antisense transcript (see text for details). (Adapted, with permission, from Dunlap 2006 [© ASBMB].)

experiments were undertaken was that WC-1 and WC-2 would be entering and leaving the Clock Box as a unit at clock phases associated with increasing and decreasing *frq* expression. Instead, WC-1 can always be found at both the PLRE and the Clock Box, and treatment with light results in only a minor increase in binding. In contrast to this, binding of WC-2 is highly regulated: In the dark, there is a high-amplitude cycle of WC-2 binding to the Clock Box, with binding beginning to rise in the mid-subjective night and peaking around subjective dawn, a timing consistent with the activation of *frq* expression. Similarly, short light treatments dramatically increase the binding of WC-2 to the PLRE and are associated with a decrease of histone H3 acetylation at the transcription start site. Not surprisingly, this regulated expression of *frq* is associated with remodeling of the chromatin in this area. Light elicits a very clear opening up of the chromatin at the transcription start site as assessed by micrococcal nuclease digestion, and there is an apparent but low-amplitude cycle of nucleosome remodeling at the Clock Box with, as might be expected, the peak of occupancy at the trough of *frq* expression and the low point at the peak of *frq* expression around subjective dawn.

To ascertain the means through which these changes were effected, all of the 19 homologs to yeast Swi2/Snf2-like ATP-dependent chromatin remodeling enzymes have been disrupted using new methods for high-efficiency gene replacements (Colot et al. 2006), and two enzymes were found to be important for operation of the clock. One is a homolog of the yeast *Fun30*, mouse *Etl1*, and human *SMARCAD* genes; Fun30p is known to remodel chromatin in vitro (C. Wu, pers. comm.), but its function(s) in the cell is still largely unknown. Because of its role in circadian regulation, this first gene was called *clockswitch-1* (*csw-1*). The second is the homolog of the mammalian *CHD2* and yeast *Chd1* genes (W.J. Belden, unpubl.). For the *csw-1* knockout, there is a rhythm of *frq* expression but this rhythm is not overtly circadian and degrades as time goes on. It is not clear, based on the banding rhythm, whether the strains are completely arrhythmic. The loss of amplitude is the result of a failure of WC-2 to exit the Clock Box, such that the levels of *frq* and FRQ expression never decline to baseline but remain somewhat elevated. There appears to be no defect in the strong increase in WC-2 binding to the PLRE as a result of light exposure. For the *chd-2* knockout, rhythms in *frq* expression are seen, although overall, *frq* is at higher levels and is seen at subjective dusk, a time when *frq* is not normally present. Taken together, these data provide the first picture of the events happening at the *frq* promoter. At this point, the simplest interpretation is that these daily CSW-1 and CHD-2-mediated events assist in the chromatin remodeling that is important for the correctly timed daily expression of *frq*, which is in turn essential for overt rhythmicity to appear.

As a result of all this regulation at the promoter, the *frq* transcripts begin to appear late in the subjective night. Sense *frq* RNAs (Fig. 2, in green) arise from two distinct start sites (P_U and P_D) (Froehlich et al. 2002, 2003; Colot et al. 2005). A first intron can be spliced using one of two 5′ donor sites (1a and 1b) with the same 3′ acceptor, and the second intron is alternatively spliced in a temperature-dependent manner, altogether giving rise to six major identifiable transcripts whose abundance reflects environmental conditions (Colot et al. 2005; Diernfellner et al. 2005). The alternatively spliced second intron contains the first AUG codon of the FRQ ORF (AUG_L), so if intron 2 is removed, FRQ begins from AUG_S. The result is that this temperature-influenced splicing determines the mix of short and long FRQ proteins (purple) (Garceau et al. 1997; Liu et al. 1997; Colot et al. 2005; Diernfellner et al. 2005).

The amount of each protein also varies with temperature (Liu et al. 1997) as a result of the action not only of splicing, but also of two upstream open reading frames (uORFs). The original unspliced transcript would contain a number of these (Garceau 1996) but most are removed with intron 1; however, even after splicing, one remains at the very 5′ end of all transcripts along with a second one just upstream of the 5′ splice junction of intron 2 (Colot et al. 2005). A function of uORFs in regulating FRQ translation was predicted by Liu et al. (1997) and later confirmed by reverse genetics (Diernfellner et al. 2005). Both long and short FRQ are needed for the best robust rhythmicity, but at temperatures of less than 22°C, more short FRQ is present and less overall FRQ is needed, whereas above 26°C, higher overall levels of predominantly large FRQ are used (Liu et al. 1997). Despite this regulation, distinct molecular activities and/or functions for the FRQ isoforms are not yet described; lFRQ and sFRQ appear to have indistinguishable stabilities (Liu et al. 1997) and both support a functional if not entirely wild-type clock if they are present in sufficient amounts. An earlier report (Diernfellner et al. 2005) suggesting that the temperature-modulated ratio of lFRQ and sFRQ had a role in determining temperature-compensation has not held up (Diernfellner et al. 2007; see below); instead, this complicated temperature regulation of forms and amounts seems to help in keeping the phase of the rhythm steady across a temperature range (Dunlap and Loros 2006). As described more fully below, compensation appears to derive from a balancing of synthesis and turnover of clock components, especially FRQ (Ruoff et al. 2005), and is strongly influenced by levels of certain kinases.

A long antisense *frq* transcript (*qrf,* for *frq* backward; red in Fig. 2) is also rhythmically expressed at low levels. This transcript shows peak expression at a phase opposite to that of *frq* and additionally is light induced, but the molecular basis of its activities is completely unknown. Interestingly, it does not appear to encode a protein but does appear to have a role in ensuring precise entrainment to light/dark cues (Kramer et al. 2003).

What Does FRQ Do and with What Other Proteins Does It Act?

Within a few hours of the appearance of its message in the late subjective night/early subjective morning (around point A in Fig. 3), FRQ is translated (Garceau et al. 1997). The proteins dimerize (Heintzen and Liu 2007), move to the nucleus (Luo et al. 1998), and interact with the RNA helicase FRH (Cheng et al. 2005). All of the FRQ in the cell appears to be complexed with FRH, although the reverse is not true; FRH appears to have many roles in the cell, and a substantial pool of FRH without FRQ exists. FRH is an essential protein, a member of the SKI2 subfamily of DEAD-box-containing RNA helicase proteins. It is highly similar to yeast Mtr4p (BLASTP ~ e-209), which is part of a yeast nuclear polyadenylation complex (LaCava et al. 2005) and a cofactor of the exosome, a complex of 3′-5′ exonucleases that have a role in 3′-end processing, RNA maturation, and quality control; Mtr4p is also involved in mRNA export. The only published mutant of *frh* is an engineered partial loss-of-function strain in which expression was reduced by coexpression of a hairpin RNA (Cheng et al. 2005), so it has not yet been possible to genetically separate its functions. Interaction studies using engineered domain deletions of FRQ suggest that FRH mediates FRQ-WCC interactions, but how this might play into a role with the exosome and/or poly(A) addition or 3′-end formation can only be guessed.

It is in the nucleus where FRQ fulfills its first and major role in the clock. Specific interactions occur between the FRQ/FRH complex and the WCC (before point B in Fig. 3) (Denault et al. 2001; Merrow et al. 2001; Cheng et al. 2005). The original observations of Crosthwaite et al. (1997) were that both WC-1 and WC-2 were required for expression of *frq*, and this gave rise to the model wherein FRQ would act to reduce their ability to activate *frq* transcription (Crosthwaite et al. 1997; Froehlich et al. 2003). This is now understood to happen through the ability of FRQ to induce or promote the phosphorylation of WC-1 and WC-2 (He and Liu 2005; Schafmeier et al. 2005; Heintzen and Liu 2007). The recent detailed analysis of the events at the *frq* promoter using the ChIP analysis cited above, showing that WC-1 is always bound to DNA,

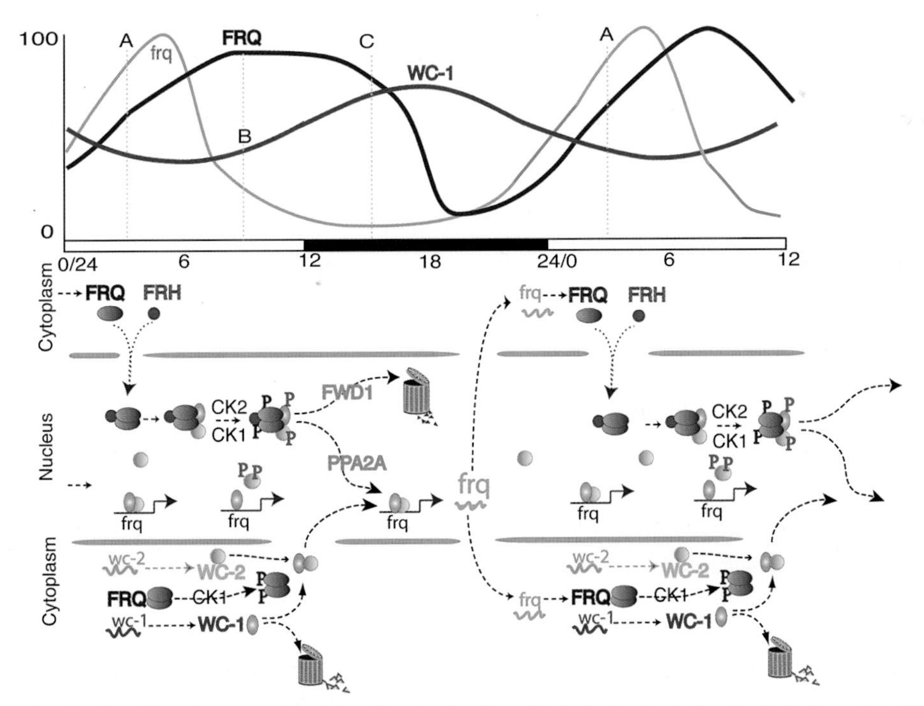

Figure 3. Molecular details within the *Neurospora* circadian mechanism. (*Top*) Cycles in the levels of clock-pertinent RNAs and proteins. (*Bottom*) Locations of RNAs and proteins important for the circadian oscillator are shown as a function of time, from left to right, beginning at subjective dawn (CT 0). The trash can indicates the proteasome. (Adapted, with permission, from Dunlap 2006 [© ASBMB].)

whereas WC-2 presence is profoundly rhythmic (Belden et al. 2007a), can be seen in this light as the direct result of this cyclic phosphorylation. Both CK1 and CK2 appear to be important for this, because mutants that reduce either their activities or amounts result in enhanced binding of the WCC to the *frq* promoter DNA that, moreover, cannot be reduced simply through overexpression of FRQ (Heintzen and Liu 2007). By mid-subjective day, the FRQ/FRH-promoted phosphorylation reduces WCC activity and WC-2 binding to their lowest levels (Lee et al. 2000; Froehlich et al. 2003; He and Liu 2005; Belden et al. 2007a). This results in dampened expression of *frq* RNA such that its level begins to fall. FRQ synthesis continues as long as *frq* mRNA is present, so the peak in FRQ levels is broad and occurs near the end of the subjective day and into the early night.

As soon as FRQ appears, it is also phosphorylated, giving rise to multiple phosphorylation isoforms and driving its eventual turnover (Garceau et al. 1997). Several kinases are known to be involved—CK1, CK2, CAMK1, and PRD-4 (when activated by DNA damage)—and they determine the stability of FRQ (Liu et al. 2000; Gorl et al. 2001; Yang et al. 2002; Pregueiro et al. 2006). Phosphorylation of FRQ has a central role in determining the length of the circadian cycle (Liu et al. 2000), because it is this phosphorylation, or series of progressive phosphorylations, that facilitates the interaction of FRQ with the SCF-type ubiquitin ligase FWD-1. This interaction, in turn, targets FRQ for turnover in the nucleosome (around point C in Fig. 3); FWD-1 is later recycled through the action of the COP9 signalosome (Heintzen and Liu 2007). FRQ phosphorylation may also influence FRQ/WCC interactions (Heintzen and Liu 2007). It is now clear that environmentally induced changes in phosphorylation can affect period and also reset the clock (see, e.g., Pregueiro et al. 2006) as described more fully below.

Although the dominant role of FRQ is its exclusively nuclear action in depressing its own synthesis, a third function of FRQ relies on its continued presence in the cytoplasm as well as on its phosphorylation. Late in the subjective day, at a point probably triggered by previous phosphorylations, FRQ becomes phosphorylated at serines 885 and 887, and this allows it to promote accumulation of WC-1 (Schafmeier et al. 2006). WC-1 is stabilized by its interaction with WC-2, so it may be that appropriately phosphorylated FRQ fosters the assembly of the WCC. In this way, the low-amplitude (and apparently dispensable) rhythm in WC-1 (Lee et al. 2000; Heintzen and Liu 2007) arises from a relatively invariant pool of spliced *wc*-1 mRNA. Additional mechanisms must contribute to this low-amplitude WC-1 rhythm, however, because it is still seen in *frq*-null strains (dePaula et al. 2006). Finally, FRQ also promotes the expression of *wc-2* mRNA (Heintzen and Liu 2007), although WC-2 levels are constitutively high in any case (Denault et al. 2001).

The end result of all of these actions of FRQ is the stable and robust cycle illustrated in Figures 1 and 3: Early action results in negative feedback and late action promotes the appearance of both WC-1 and WC-2 so that WCC is maintained at an elevated level but is moreover held inactive through the action of FRQ in promoting the phosphorylation of the WCC by CK1 and CK2 and perhaps in promoting turnover of WC-2 on the *frq* promoter. Eventually, the precipitous phosphorylation-mediated turnover of FRQ (point C, Fig. 3) releases the WCC; it can then be recycled to an active state through the action of protein phosphatases or replaced by newly synthesized and assembled WCC, and it can reinitiate the transcription of *frq* mRNA in the next cycle.

Knowledge of this feedback loop goes a long way toward explaining one of the central tenets of the circadian clock: its very long time constant. The long period of the circadian day is largely accounted for by the long time it takes for FRQ to be phosphorylated and turned over. Moreover, this process clearly requires the action of more than one kinase, suggesting the possibility that they must work in tandem or sequentially in such a way that they further delay each other's actions. Interestingly, the kinetics of FRQ synthesis would appear to have relatively little to do with the length of the circadian day. As an aside, this view of the circadian cycle as a FRQ cycle provides some conceptual problems for output. In the late 1980s, early 1990s view of the clock as a series of sequential actions, it was easy to see how different activities timed to different phases of the clock cycle could be regulated. Although it was not appreciated at the time, this is now understood as the way in which different cyclins act to produce cell-cycle-stage-specific activities. However, such phase determination is not possible with the clock as a single long feedback loop. It may be that ancillary and slave oscillators have been so widely used in circadian systems to circumvent just this problem and thereby to allow accurate control of activities at a variety of circadian phases.

MOLECULAR EXPLANATIONS FOR ENTRAINMENT

Entrainment describes the process by which the phase of the clock is adjusted in response to time-of-day-specific environmental cues so that the day phase of the circadian cycle coincides with the day phase of the external world and the night phase with the night. *Neurospora* was the first system in which mechanisms for entrainment by light (Crosthwaite et al. 1995) and by temperature (Liu et al. 1998) were described, and the descriptions for each fell directly out of the knowledge of how the circadian feedback loop operates. Likewise, once the resetting response was understood, the means by which the clock could impact resetting to provide circadian gating could be investigated, and this was also first done in *Neurospora* (Heintzen et al. 2001). Consistent with the fact that the overall logic behind the feedback loop in fungi is similar to that in mammals, aspects of the light response appear to be conserved in mammals (Shigeyoshi et al. 1997). At present, too little is known about temperature responses in mammalian cells in culture to understand whether the mechanism underlying temperature resetting in cellular rhythms is also conserved. For a description of resetting by light, the nature of the photoreceptor, the mechanism of entrainment, and a clue to the mechanism of gating are each considered separately.

In Addition to Its Role in the Clock, WC-1 is the Blue Light Photoreceptor

Most light responses described in *Neurospora* are specific to blue light and no red or far-red responses, such as those seen in other fungi and higher plants, have been seen. Blue-light-induced phenotypes are blocked in mutants of either *white collar-1* (*wc-1*) or *white collar-2* (*wc-2*) (Ballario and Macino 1997), suggesting that WC-1 and WC-2 are the principal if not the sole components for blue light responses in *Neurospora*. This is perhaps surprising based on genomic data that have revealed the presence of two phytochromes (Froehlich et al. 2005), one of which is circadianly regulated, as well as a strongly light-induced cryptochrome gene (Froehlich 2002); however, light-related phenotypes for these genes remain lacking.

Light induces mycelial carotenogenesis, resulting in the characteristic yellow-orange color of *Neurospora*. Mutations in genes affecting this light induction (*wc-1* and *wc-2*) result in white mycelia and were among the first mutations described in this organism. It was hypothesized early on that these genes encoded photoreceptors (Harding and Shropshire 1980), but it was not possible to make a distinction between photoreceptors and those proteins required to transduce the signal. Macino and colleagues (Ballario et al. 1996; Linden and Macino 1997) cloned the genes and they and other investigators described the protein interactions in vitro and in vivo (Ballario et al. 1998; Denault et al. 2001; Heintzen and Liu 2007). On the basis of the fact that the WC proteins strongly resembled transcriptional activators, Macino and colleagues proposed that *wc-1* and *wc-2* encoded the transcription factors mediating light-induced gene expression, and that WC-1 itself might be the photoreceptor.

Macino's guess was correct: WC-1 is the protein that perceives the light signal and initiates the photoresponse (Froehlich et al. 2002; He et al. 2002). The chromophore that actually absorbs the light is FAD, and it is bound into a specialized PAS domain called a LOV (for light, oxygen, and voltage sensing) domain. Studies of other LOV-domain-containing photoreceptors including those of fungi (Crosson and Moffat 2001; Zoltowski et al. 2007) suggest that absorption of blue light causes FAD to undergo a transient covalent interaction with WC-1 inducing a conformational change in the protein. Light causes the formation of multimers of the WC proteins and enhanced transcriptional activation of WCC-bound light-responsive promoters (Froehlich 2002). Froehlich et al. (2002) showed such a light-induced change in DNA binding of the WCC, using in-vitro-transcribed and -translated extracts primed only with *wc-1* and *wc-2* and dependent only on addition of FAD, indicating that these three components are sufficient for light-regulated DNA binding of these transcriptional activators.

How Light Resets the Clock

For entrainment to work, light must delay the clock into the previous day when seen early in the night and advance the clock into the next day when seen late at night. Thus, the molecular basis of resetting by light requires that the same photic cue have opposite effects on the timing mechanism depending on when in the cycle light is perceived (Crosthwaite et al. 1995). As noted above, *frq* mRNA peaks in the morning and this defines molecular morning (Aronson et al. 1994b; Crosthwaite et al. 1995). As a result, actions that increase *frq* to peak levels will tend to reset the clock to morning regardless of when they occur. Light acts rapidly through the WCC bound to the two light-response elements (LREs) within the *frq* promoter to induce *frq* (Crosthwaite et al. 1995, 1997; Froehlich et al. 2002). A plausible model for resetting (Fig. 4) works in this way: Through the subjective evening and early night, when *frq* is falling, light induction of *frq* rapidly sends the clock back toward the time of peak levels (morning) yielding a phase-delay, whereas in the late night when *frq* mRNA levels are rising to a midmorning peak, rapid induction of *frq* by light to its peak advances the clock to a point corresponding to morning (Crosthwaite et al. 1995). This remarkably simple mechanism provides a simple explanation for how resetting works: It is the same molecular mechanism—light induction of a clock component—and it is the dynamics of the feedback loop itself that causes this action to be interpreted as an advance or a delay. This is also in large part how mammalian clocks are reset by light (Shigeyoshi et al. 1997). On reflection, it is clear that the feedback loop could also be reset through light-triggered decay of a clock component that peaks at night; because *per* peak expression occurs at night, we correctly predicted (Crosthwaite et al. 1995) that the *Drosophila* clock would be reset in this way, as is the case (see, e.g., Hunter-Ensor et al. 1996).

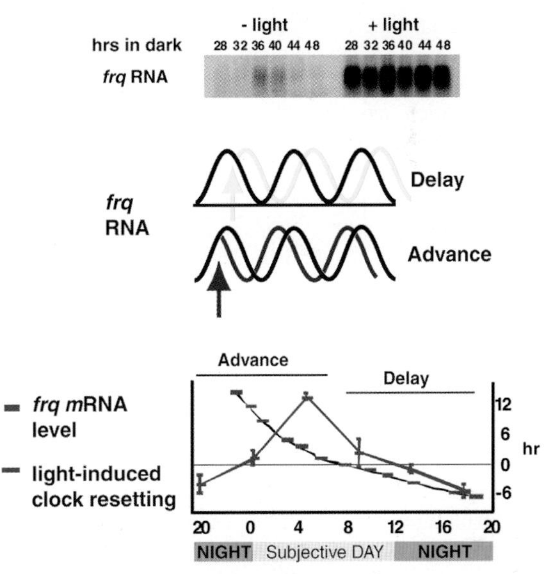

Figure 4. How light resets the clock. (*Top*) Cycle in *frq* RNA abundance and effect of brief exposure to light at any time. (*Middle*) *Black* represents the cycle of *frq* transcription; *colored lines* show the effect of transient light induction of *frq* on the steady-state rhythm. (*Bottom*) The cycle in *frq* transcript levels juxtaposed with response of the clock to light, showing how light seen while *frq* is rising leads to advances, and light seen while *frq* levels are falling leads to delays (see text for details). (Adapted, with permission, from Dunlap 1999 [© Elsevier].)

Although the model implies a wonderful simplicity in the responses to light, this may be misleading because the total light response is not just transcriptional. There is gating, as described below, but beyond this is the fact that FRQ accumulates and is phosphorylated in the light but is not significantly degraded until transfer to dark (Collett et al. 2002; Tan et al. 2004). Also in the dark, *frq* and FRQ synthesis reflects the feedback loop: If after transfer to darkness, there is not enough FRQ to complete the negative feedback loop, still more FRQ will be made. However, if enough FRQ is present, it will immediately begin to be phosphorylated and to progress toward degradation. Thus, depending on the duration of light treatment before darkness, the synthesis/decay kinetics of FRQ will vary substantially and the time separating *frq* transcription and the peak of FRQ can change by many hours. Additionally, *frq* is acutely induced by light and FRQ does not block this light induction (Crosthwaite et al. 1995). Furthermore, FRQ is phosphorylated in a complex manner as described above, some of it regulating turnover and some influencing activity or interactions with other proteins (see, e.g., Liu et al. 2000; Heintzen and Liu 2007), so depending on how long a strain has been in light, the FRQ present is qualitatively different. Finally, again as described, a light-inducible antisense *frq* message is expressed out-of-phase with the sense transcript; its loss results in much stronger phase-shifting in response to light, consistent with an important role in light responses by the clock (Kramer et al. 2003).

Gating of Light Responses

The circadian clock can modulate its own input; i.e., the response to a standard stimulus is actively modulated depending on time of day through use of an additional clock-associated feedback loop. This process is known as gating and is provided by the *vivid* (*vvd*) gene and VVD protein, which acts to modulate the WC photoreceptor complex. For any photoresponsive cell with just a single photoreceptor having typical reciprocity for duration and intensity (see, e.g., the *Neurospora* photoreceptor; Crosthwaite et al. 1995), after some minutes in moderate light, the response would be saturated and the cell would be unable to "see" any change in intensity; VVD confers this ability, a process known as photoadaptation (Schwerdtfeger and Linden 2003). Like WC-1, VVD is a member of the PAS protein superfamily (Heintzen et al. 2001) and also a photoreceptor that binds a flavin, in this case an FMN chromophore (Schwerdtfeger and Linden 2003; Zoltowski et al. 2007). When the flavin absorbs blue light, it undergoes a transient covalent addition to the VVD protein, resulting in structural changes that are propagated to the surface of the protein and end with the release of an amino-terminal helix (Zoltowski et al. 2007). This in some way causes VVD to change other proteins, likely kinases, that modulate the activity of the WCC and thereby regulate its activity, although this has not yet been verified biochemically. Significant expression of VVD is restricted to the first day in constant darkness and is controlled in part by the clock (Heintzen et al. 2001), thus accounting for the finding that light signals elicit reduced clock-resetting effects when delivered on the first day after a light-to-dark transfer (Dharmananda 1980). VVD is not required for circadian rhythmicity, but this immediate and transient repressor contributes to circadian entrainment by making dark-to-light transitions more discrete and is important for setting the phase of the overt rhythm. Specifically, VVD allows the dark-to-light transition to be less influential for purposes of phase determination so that the light-to-dark transition at the end of the day determines phase (Elvin et al. 2005), and it also has a major role in ensuring that the phase of the overt rhythm is more or less constant across a range of temperatures (Hunt et al. 2007).

TEMPERATURE EFFECTS ON THE CLOCK

There are three distinguishable aspects to the response of the circadian system to temperature. First, all clocks, including that of *Neurospora*, will entrain to temperature steps and cycles, with steps up in temperature resetting the clock in a manner similar to light pulses (see, e.g., Francis and Sargent 1979). Second, there are physiological temperature limits for operation of the clock that lie within the limits for growth. Third, the circadian period length is more or less the same when measured at different constant temperatures, a phenomenon known as "temperature-compensation." Compelling molecular models exist for the first two of these temperature effects as being mediated through the amount and perhaps the form of FRQ protein made, there being no independent temperature sensor (Nowrousian et al. 2003).

Temperature Resetting

Unlike the case with light, clock resetting by temperature steps is understood in terms of posttranscriptional regulation. Indeed, *frq* transcript levels are influenced little by temperature, but as described above (Liu et al. 1997), as the temperature increases, so does the level of FRQ. As a result, FRQ levels oscillate around a higher mean at higher temperatures; oscillations with peaks of FRQ in the late day to early evening and troughs in the late night continue, but the number of molecules of FRQ associated with a given "time of day" is different at different temperatures. For instance, Merrow et al. (1997) determined that at peak at 25°C there are about 30 molecules of FRQ in the nucleus, fewer than this number at 20°C and more at 30°C (Fig. 5). Because at the lowest point in the curve (dawn) at 28°C (the upper curves in Fig. 5) there are more molecules of FRQ per nucleus than at the highest point in the curve (dusk) at 21°C, a shift in temperature corresponds literally to a step to a different time and a shift in the state of the clock (Liu et al. 1998) even in the absence of any synthesis or turnover of clock components. Following the step, the dynamics of the feedback loop again take control so that the relative levels of *frq* and FRQ are adjusted in terms of the new temperature: If there is enough FRQ to shut off WCC, then no more is made and molecular time is around evening, but if more FRQ is needed to complete the feedback and shut off the WCC, then more is made and molecular time is

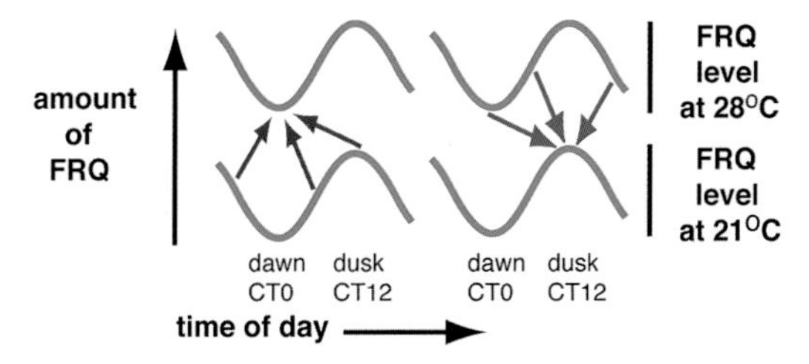

Figure 5. A possible mechanism for how temperature resets the clock. (*Green*) Cycles in FRQ levels at different temperatures. (*Red arrows*) Step up at any time leaves the clock with too little FRQ to close the feedback loop, so the clock is reset to the time corresponding to low FRQ, subjective dawn. (*Blue arrows*) Step down at any time leaves the clock with enough FRQ to close the feedback loop, so the clock is reset to the time corresponding to high FRQ, around subjective evening (see text for details). (Adapted, with permission, from Dunlap 1999 [© Elsevier].)

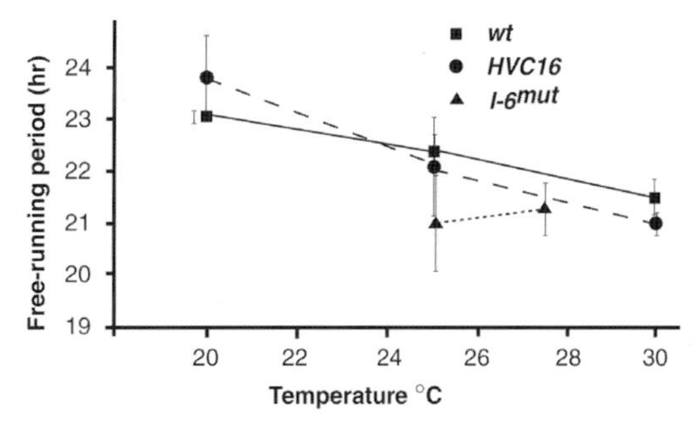

Figure 6. Strains expressing only long FRQ are not deficient in temperature-compensation. Period length is plotted as a function of temperature from a wild-type strain and from two independently engineered strains that express only long FRQ (*HVC16* from Colot et al. 2005; *l-6*^{mut} from Diernfellner et al. 2005). The error bars indicate +/– one standard deviation; $n = 6$. Both strains clearly show compensation. See text for details.

around dawn. As implicit in this discussion, because temperature acts directly on core components rather than being transduced by a photoreceptor, temperature can be a stronger resetting cue than light (Liu et al. 1998).

Physiological Limits for Rhythmicity

As noted above, the range of temperatures over which the clock keeps time reflects temperature influences on both the total amount of FRQ and the ratio of the two FRQ forms (Colot et al. 2005; Diernfellner et al. 2005). Temperature-regulated splicing can (at lower temperatures) remove the second intron in *frq* (see Fig. 2) so that translation initiates at codon 100 of the FRQ open reading frame (ORF). In addition, the absolute level of FRQ protein is controlled by ambient temperature (Liu et al. 1997), in part by the two uORFs in the 5′UTR (untranslated region) of *frq* (Fig. 2). Strains that can make only one FRQ form show a reduced temperature range permissive for rhythmicity, suggesting that the synthesis of two FRQ forms is a novel adaptive mechanism to extend the physiological temperature range over which the clock can function (Garceau et al. 1997; Liu et al. 1997). Additionally, the clock can function with either form of FRQ exclusively if there is enough of it (Colot et al. 2005; Diernfellner et al. 2005), and there is no difference in stability of the two forms (Liu et al. 1997). It should be noted, however, that the circadian rhythms in strains expressing exclusively either long or short FRQ are clearly temperature-compensated (Fig. 6), contrary to a prior suggestion (Diernfellner et al. 2005). In collaborative work (Diernfellner et al. 2007), we have examined the lFRQ-only and sFRQ-only strains reported by Diernfellner et al. (2005) and by us (Colot et al. 2005), as well as new strains, and we all agree that the basis of temperature-compensation does not lie in regulation of the relative levels at which the two isoforms of FRQ are made or in properties specific to one of the two isoforms. Instead of a role in compensation, the differential temperature-dependent synthesis of lFRQ and sFRQ may provide a means to fine-tune the period or sculpt the frq mRNA waveform in response to changes in ambient temperature (Diernfellner et al. 2007).

Temperature-Compensation

Temperature-compensation is the third way that clocks respond to temperature; it is the last of the canonical circadian properties for which a molecular explanation has not existed, and indeed it has been a difficult problem to study. Precedents show that many period mutants are associated with partial loss of temperature-compensation for (relatively) obvious reasons. For instance, in *Drosophila*, the dosage of PER influences period length (Smith and Konopka 1982), so mutations that influence the stability of the protein in a temperature-dependent manner will have compensation phenotypes. Likewise, mutations that create an unstable or an unusually stable clock component, either inherent to the protein or secondarily, can have a temperature-compensation phenotype. An example of a secondary effect is the PRD-4 protein in *Neurospora*. *prd-4* is the *Neurospora* ortholog of the gene encoding the DNA-damage-activated protein checkpoint kinase 2 that can phosphorylate and destabilize FRQ (Pregueiro et al. 2006). A mutation that causes PRD-4 to become active even in the absence of DNA damage results in constant elevated activity of this kinase, resulting in temperature-dependent destabilization of FRQ and a partial loss of temperature-compensation. It is not true, however, that every mutation affecting period necessarily affects compensation; for instance, the short-period mutants *frq*^{1} and *frq*^{2} are normal in this respect (Feldman et al. 1979).

Because of the likelihood that loss of compensation mutations would not be informative, we instead followed up two distinct mutations that resulted in *enhanced* compensation, *prd-3* and *chr*. Normally, the *Neurospora* clock shows a modest undercompensation phenotype whereby the period length gets slightly shorter as the temperature increases up to about 30°C. Beyond this point, compensation appears to be largely ineffective because the slope of the period versus temperature curve becomes much steeper. In *prd-3* and *chr,* this profile shows overcompensation or extended compensation (Feldman et al. 1979). Both genes were cloned by SNP (single-nucleotide polymorphism) mapping and, surprisingly, they encode separate subunits of the same holoenzyme, casein kinase 2

(CK2): *chr* harbors an R265C mutation in the β1-subunit tail region of CK2, which has a role in substrate recognition and is highly conserved among the fungi, and *prd-3* harbors a Y43H mutation in the phosphate anchor region associated with ATP binding near the catalytic core of the α subunit. This circumstantial evidence strongly suggests a role for CK2 in the temperature-compensation process.

Although a role in compensation would be novel, CK2 is, of course, an enzyme with a well-appreciated role in many circadian systems. The β subunit CKB3 in *Arabidopsis* associates with CCA1 and overexpression of CKB3 leads to a shorter period (Sugano et al. 1998). In *Neurospora*, FRQ has been shown to be phosphorylated in vitro by CK2 and a partial loss-of-function mutation lengthens period (Yang et al. 2002); finally, in *Drosophila*, two long-period mutants, *Andante* and *Timekeeper*, define CK2β and CK2α, respectively. Thus, CK2 is well positioned in the clock to have a role in compensation, and several approaches have been used to test this possibility. We completed gene disruptions for each subunit to establish definitive nulls and determined that spores lacking CK2α will not germinate, whereas loss of CK2β1 (Δ*ckb-1*) yields viable spores that fail to band rhythmically. This genetic background provides the perfect context in which to assess the role of CK2 dosage in compensation. A transgene was constructed in which the *quinic acid-2* (*qa-2*) promoter was used to drive expression of *ckb-1* at a level proportional to the amount of the inducer, quinic acid (QA), in the medium; this promoter has previously been extensively used to drive regulated expression in *Neurospora* (see, e.g., Aronson et al. 1994b; Merrow et al. 1997). In the context of Δ*ckb-1* at 25°C, this *qa-ckb-1* construct rescues rhythmic banding even with no added inducer due to a very low level of baseline expression. However, the period is quite long under these conditions (28–30 hours), and as the level of CK2β1 is increased, the period drops steadily to within the wild-type range (Fig. 7, left panel). More interesting is the effect of ambient temperature on this phenomenon. Across a physiological temperature range, strains expressing high levels of CK2β1 are slightly undercompensated in a manner typical for the wild-type clock. As CK2β1 dosage drops, compensation passes through a range of better than normal compensation and as dosage approaches the limit for rhythmic banding, compensation changes to overcompensated, fully mimicking the *ckb-1^chr^* overcompensation phenotype. Importantly, however, there are no mutant proteins in these experiments; only the dose of wild-type proteins is being controlled.

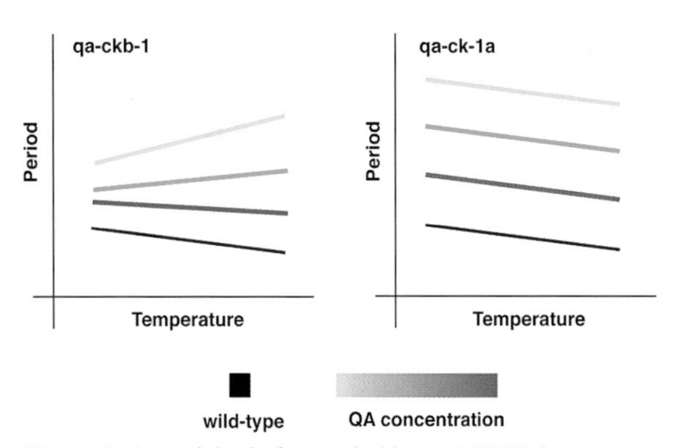

Figure 7. A special role for casein kinase 2 (CK2) in temperature-compensation. Strains bearing inducible copies of the gene encoding the β subunit of CK2 (*qa-ckb-1*) or CK1 (*qa-ck-1a*) were grown at different levels of inducer and at different temperatures. (*Left*) As levels of *ckb-1* decrease leading to less CK2 activity, the period length increases and compensation moves from the slight undercompensation typical of wild type to distinct overcompensation. (*Right*) As CK1 levels drop, the period gets longer, but in contrast to the behavior seen with CK2, the mode of compensation remains like that of wild type. These illustrations depict unpublished data from A. Mehra et al. (in prep.).

These data establish a clear role for CK2 in effecting compensation, but it may be that any mutation in a FRQ kinase will alter compensation. This was tested by performing similar experiments with CK1, the other principal kinase acting on FRQ. Because CK1 is essential, a knockin strategy was used to replace the normal *ck-1a* promoter with the *qa-2* promoter, thus placing CK1 expression under QA control. As seen in Figure 7 (right panel), reducing the dosage of CK1 has profound effects on period length, serving to increase the period to more than 30 hours. However, and importantly, the temperature-compensation profile was not changed; even the long-period clocks show the mild undercompensation seen for wild type. Similar experiments with consistent results were performed with protein phosphatase 1 (PP1). These data indicate that the role of kinases in compensation is special and that CK2 has such a role. In vitro studies on the activity of the CK2 enzyme have shown that even at low dosage, CK2 displays a normal temperature-activity profile; i.e., CK2 is more active at higher temperatures, suggesting that it should phosphorylate FRQ to a greater extent. However, and paradoxically, even though CK2 activity increases at higher temperatures, when examined at low CKB-1 levels, FRQ degradation is seen to actually decrease as temperature increases. Because the kinetics of FRQ turnover is a major determinant of period length, these data are consistent with the observation of increasing period as a function of temperature (i.e., overcompensation) at low CK2 levels; moreover, these results recapitulate the original *chr* phenotype of reduced FRQ turnover and longer period at higher temperatures.

These data are at once surprising and not easy to reconcile with a simple notion of compensation as a result of kinase action. A variety of data beginning with those of Liu et al. (2000) have shown and reconfirmed a central role for FRQ phosphorylation in determining its turnover and of FRQ turnover in determining period length. The simple interpretation of these data would be that higher temperature must yield more kinase activity (as we have confirmed), which ought to yield more rapid turnover and a shorter period. This is seen at high kinase activity; however, at low *ckb-1* levels, the reverse is seen: Turnover actually decreases and period length increases at higher temperatures despite greater activity of CK2. Although a complete mechanistic explanation for this is lacking, it remains true that this paradoxical behavior reflects exactly the behavior needed to bring about compensation, namely, a process that acts to decrease the pace of the clock at higher temperatures and that normally counters the drive of CK2 for turnover. The focus of these competing activities is FRQ, whose turnover kinetics dictates period length. One possible way to rationalize these results would be the existence of more than one "class" of CK2-influenced phosphorylation sites on FRQ, one class that simply leads to turnover (and longer periods at lower CK2 doses) and another "class" whose phosphorylation might be competed by an unknown additional activity. In this scenario, compensation is achieved by the balance of these activities, and at low doses of CK2, the competing activity wins out, leading to overcompensation. A prediction that follows is that mutation of one or more of the phosphorylation sites in the second class would lead to an extended or overcompensation phenotype like that seen in *ckb-1*^{chr}. We are now testing this hypothesis.

ACKNOWLEDGMENTS

This work was supported by grants from the National Institutes of Health to J.C.D. (GM34985 and GM068087) and to J.C.D. and J.J.L. (GM083336). L.F.L. is a Pew Latin American Fellow.

REFERENCES

Allada R., White N.E., So W.V., Hall J.C., and Rosbash M. 1998. A mutant *Drosophila* homolog of mammalian *CLOCK* disrupts circadian rhythms and transcription of *period* and *timeless*. *Cell* **93:** 805.

Antoch M., Soog E., Chang A., Vitaterna M., Zhao Y., Wilsbacher L., Sangoram A., King D., Pinto L., and Takahashi J. 1997. Functional identification of the mouse circadian *CLOCK* gene by transgenic BAC rescue. *Cell* **89:** 655.

Aronson B.D., Johnson K.A., and Dunlap J.C. 1994a. The circadian clock locus *frequency*: A single ORF defines period length and temperature compensation. *Proc. Natl. Acad. Sci.* **91:** 7683.

Aronson B., Johnson K., Loros J.J., and Dunlap J.C. 1994b. Negative feedback defining a circadian clock: Autoregulation in the clock gene *frequency*. *Science* **263:** 1578.

Ballario P. and Macino G. 1997. White collar proteins: PASsing the light signal in *Neurospora crassa*. *Trends Microbiol.* **5:** 458.

Ballario P., Talora C., Galli D., Linden H., and Macino G. 1998. Roles in dimerization and blue light photoresponse of the PAS and LOV domains of *Neurospora crassa* WHITE COLLAR proteins. *Mol. Microbiol.* **29:** 719.

Ballario P., Vittorioso P., Magrelli A., Talora C., Cabibbo A., and Macino G. 1996. *White collar-1*, a central regulator of blue-light responses in *Neurospora crassa*, is a zinc-finger protein. *EMBO J.* **15:** 1650.

Belden W.J., Loros J.J., and Dunlap J.C. 2007a. Execution of the circadian negative feedback loop in *Neurospora* requires the ATP-dependent chromatin-remodeling enzyme CLOCK-SWITCH. *Mol. Cell* **25:** 587.

Belden W.J., Larrondo L.F., Froehlich A.C., Shi M., Chen C.-H., Loros J.J., and Dunlap J.C. 2007b. The *band* mutation in *Neurospora crassa* is a dominant allele of *ras-1* implicating RAS-signaling in circadian output. *Genes Dev.* **21:** 1494.

Bell-Pedersen D., Cassone V.M., Earnest D.J., Golden S.S., Hardin P.E., Thomas T.L., and Zoran M.J. 2005. Circadian rhythms from multiple oscillators: Lessons from diverse organisms. *Nat. Rev. Genet.* **6:** 544.

Bünning E. 1973. *The physiological clock*. Springer-Verlag, New York.

Cheng P., He Q., He Q., Wang L., and Liu Y. 2005. Regulation of the *Neurospora* circadian clock by an RNA helicase. *Genes Dev.* **19:** 234.

Collett M.A., Garceau N., Dunlap J.C., and Loros J.J. 2002. Light and clock expression of the *Neurospora* clock gene frequency is differentially driven by but dependent on WHITE COLLAR-2. *Genetics* **160:** 149.

Colot H.V., Loros J.J., and Dunlap J.C. 2005. Temperature-modulated alternative splicing and promoter use in the circadian clock gene *frequency*. *Mol. Biol. Cell* **16:** 5563.

Colot H.V., Park G., Turner G.E., Ringelberg C., Crew C.M., Litvinkova L., Weiss R.L., Borkovich K.A., and Dunlap J.C. 2006. A high-throughput gene knockout procedure for *Neurospora* reveals functions for multiple transcription factors. *Proc. Natl. Acad. Sci.* **103:** 10352.

Crosson S. and Moffat K. 2001. Structure of a flavin-binding plant photoreceptor domain: Insights into light mediated signal transduction. *Proc. Natl. Acad. Sci.* **98:** 2995.

Crosthwaite S.C., Dunlap J.C., and Loros J.J. 1997. *Neurospora wc-1* and *wc-2:* Transcription, photoresponses, and the origins of circadian rhythmicity. *Science* **276:** 763.

Crosthwaite S.C., Loros J.J., and Dunlap J.C. 1995. Light-induced resetting of a circadian clock is mediated by a rapid increase in *frequency* transcript. *Cell* **81:** 1003.

Darlington T.K., Wager-Smith K., Ceriani M.F., Staknis D., Gekakis N., Steeves T., Weitz C.J., Takahashi J.S., and Kay S.A. 1998. Closing the circadian loop: CLOCK induced transcription of its own inhibitors, *per* and *tim*. *Science* **280:** 1599.

Denault D.L., Loros J.J., and Dunlap J.C. 2001. WC-2 mediates WC-1-FRQ interaction within the PAS protein-linked circadian feedback loop of *Neurospora crassa*. *EMBO J.* **20:** 109.

dePaula R.M., Lewis Z.A., Greene A.V., Seo K.S., Morgan L.W., Vitalini M.W., Bennett L., Gomer R.H., and Bell-Pedersen D. 2006. Two circadian timing circuits in *Neurospora crassa* share components and regulate distinct rhythmic processes. *J. Biol. Rhythms* **21:** 159.

Dharmananda S. 1980. "Studies of the circadian clock of *Neurospora crassa:* Light-induced phase shifting." Ph.D. thesis, University of California, Santa Cruz.

Diernfellner A.C., Schafmeier T., Merrow M.W., and Brunner M. 2005. Molecular mechanism of temperature sensing by the circadian clock of *Neurospora crassa*. *Genes Dev.* **19:** 1968.

Diernfellner A., Colot H.V., Dintsis O., Loros J.J., Dunlap J.C., and Brunner M. 2007. Long and short isoforms of *Neurospora* clock protein FRQ support temperature compensated circadian rhythms. *FEBS Lett.* (in press).

Dunlap J.C. 1999. Molecular bases for circadian clocks. *Cell* **96:** 271.

———. 2006. Proteins in the *Neurospora* circadian clockworks. *J. Biol. Chem.* **281:** 28489.

Dunlap J.C. and Loros J.J. 2006. How fungi keep time: Circadian system in *Neurospora* and other fungi. *Curr. Opin. Microbiol.* **9:** 579.

Elvin M., Loros J.J., Dunlap J.C., and Heintzen C. 2005. The PAS/LOV protein VIVID supports a rapidly dampened daytime oscillator that facilitates entrainment of the *Neurospora* circadian clock. *Genes Dev.* **19:** 2593.

Feldman J.F., Gardner G.F., and Dennison R.A. 1979. Genetic analysis of the circadian clock of *Neurospora*. In *Biological rhythms and their central mechanism* (ed. M. Suda), p. 57. Elsevier, Amsterdam.

Francis C. and Sargent M.L. 1979. Effects of temperature perturbations on circadian conidiation in *Neurospora*. *Plant Physiol.* **64:** 1000.

Froehlich A.C. 2002. "Light and circadian regulation in *Neurospora crassa*." Ph.D. thesis, Dartmouth College, Hanover, New Hampshire.

Froehlich A.C., Loros J.J., and Dunlap J.C. 2002. WHITE COLLAR-1, a circadian blue light photoreceptor, binding to the *frequency* promoter. *Science* **297:** 815.

———. 2003. Rhythmic binding of a WHITE COLLAR containing complex to the *frequency* promoter is inhibited by FREQUENCY. *Proc. Natl. Acad. Sci.* **100:** 5914.

Froehlich A.C., Noh B., Vierstra R.D., Loros J., and Dunlap J.C. 2005. Genetic and molecular analysis of phytochromes from the filamentous fungus *Neurospora crassa*. *Eukaryot. Cell* **4:** 2140.

Garceau N. 1996. "Molecular and genetic studies on the *frq* and *ccg-1* loci of *Neurospora*." Ph.D. thesis, Dartmouth College, Hanover, New Hampshire.

Garceau N., Liu Y., Loros J.J., and Dunlap J.C. 1997. Alternative initiation of translation and time-specific phosphorylation yield multiple forms of the essential clock protein FREQUENCY. *Cell* **89:** 469.

Gorl M., Merrow M., Huttner B., Johnson J., Roenneberg T., and Brunner M. 2001. A PEST-like element in FREQUENCY determines the length of the circadian period in *Neurospora crassa*. *EMBO J.* **20:** 7074.

Harding R.W. and Shropshire W.J. 1980. Photocontrol of carotenoid biosynthesis. *Annu. Rev. Plant Physiol.* **31:** 217.

He Q. and Liu Y. 2005. Molecular mechanism of light responses in *Neurospora:* From light-induced transcription to photoadaptation. *Genes Dev.* **19:** 2888.

He Q., Cheng P., Yang Y., He Q., Yu Q., and Liu Y. 2003. FWD-1 mediated degradation of FREQUENCY in *Neurospora* establishes a conserved mechanism for circadian clock regulation. *EMBO J.* **22:** 4421.

He Q., Cheng P., Yang Y., Wang L., Gardner K., and Liu Y. 2002. WHITE COLLAR-1, a DNA binding transcription factor and a light sensor. *Science* **297:** 840.

Heckman D.S., Geiser D.M., Eidell B.R., Stauffer R.L., Kardos N.L., and Hedges S.B. 2001. Molecular evidence for the early colonization of land by fungi and plants. *Science* **293:** 1129.

Heintzen C. and Liu Y. 2007. The *Neurospora crassa* circadian clock. *Adv. Genet.* **58:** 25.

Heintzen C., Loros J.J., and Dunlap J.C. 2001. VIVID, gating and the circadian clock: The PAS protein VVD defines a feedback loop that represses light input pathways and regulates clock resetting. *Cell* **104:** 453.

Hunt S.M., Elvin M., Crosthwaite S.K., and Heintzen C. 2007. The PAS/LOV protein VIVID controls temperature compensation of circadian clock phase and development in *Neurospora crassa*. *Genes Dev.* **21:** 1964.

Hunter-Ensor M., Ousley A., and Sehgal A. 1996. Regulation of the *Drosophila* protein TIMELESS suggests a mechanism for resetting the circadian clock by light. *Cell* **84:** 677.

Ingold C.T. 1971. *Fungal spores*. Clarendon Press, Oxford, United Kingdom.

Ko H.W., Jiang J., and Edery I. 2002. Role for Slimb in the degradation of *Drosophila* Period protein phosphorylated by Doubletime. *Nature* **420:** 673.

Kramer C., Loros J.J., Dunlap J.C., and Crosthwaite S.K. 2003. Role for antisense RNA in regulating circadian clock function in *Neurospora crassa*. *Nature* **421:** 948.

LaCava J., Houseley J., Saveanu C., Petfalski E., Thompson E., Jacquier A., and Tollervey D. 2005. RNA degradation by the exosome is promoted by a nuclear polyadenylation complex. *Cell* **121:** 713.

Lee K., Loros J.J., and Dunlap J.C. 2000. Interconnected feedback loops in the *Neurospora* circadian system. *Science* **289:** 107.

Linden H. and Macino G. 1997. White collar-2, a partner in blue-light signal transduction, controlling expression of light-regulated genes in *Neurospora crassa*. *EMBO J.* **16:** 98.

Liu Y., Loros J., and Dunlap J.C. 2000. Phosphorylation of the *Neurospora* clock protein FREQUENCY determines its degradation rate and strongly influences the period length of the circadian clock. *Proc. Natl. Acad. Sci.* **97:** 234.

Liu Y., Garceau N., Loros J.J., and Dunlap J.C. 1997. Thermally regulated translational control mediates an aspect of temperature compensation in the *Neurospora* circadian clock. *Cell* **89:** 477.

Liu Y., Merrow M., Loros J.J., and Dunlap J.C. 1998. How temperature changes reset a circadian oscillator. *Science* **281:** 825.

Loros J.J. and Feldman J.F. 1986. Loss of temperature compensation of circadian period length in the *frq-9* mutant of *Neurospora crassa*. *J. Biol. Rhythms* **1:** 187.

Loros J.J., Denome S.A., and Dunlap J.C. 1989. Molecular cloning of genes under the control of the circadian clock in *Neurospora*. *Science* **243:** 385.

Luo C., Loros J.J., and Dunlap J.C. 1998. Nuclear localization is required for function of the essential clock protein FREQUENCY. *EMBO J.* **17:** 1228.

McClung C.R., Fox B.A., and Dunlap J.C. 1989. The *Neurospora* clock gene *frequency* shares a sequence element with the *Drosophila* clock gene *period*. *Nature* **339:** 558.

Merrow M., Garceau N., and Dunlap J.C. 1997. Dissection of a circadian oscillation into discrete domains. *Proc. Natl. Acad. Sci.* **94:** 3877.

Merrow M., Franchi L., Dragovic Z., Gorl M., Johnson J., Brunner M., Macino G., and Roenneberg T. 2001. Circadian regulation of the light input pathway in *Neurospora crassa*. *EMBO J.* **20:** 307.

Nakajima M., Imai K., Ito H., Nishiwaki T., Murayama Y., Iwasaki H., Oyama T., and Kondo T. 2005. Reconstitution of circadian oscillation of cyanobacterial KaiC phosphorylation in vitro. *Science* **308:** 414.

Nowrousian M., Duffield G.E., Loros J.J., and Dunlap J.C. 2003. The *frequency* gene is required for temperature-dependent regulation of many clock-controlled genes in *Neurospora crassa*. *Genetics* **164:** 922.

Pittendrigh C.S., Bruce V.G., Rosenzweig N.S., and Rubin M.L. 1959. A biological clock in *Neurospora*. *Nature* **184:** 169.

Pregueiro A., Liu Q., Baker C., Dunlap J.C., and Loros J.J. 2006. The *Neurospora* checkpoint kinase 2: A regulatory link between the circadian and cell cycles. *Science* **313:** 644.

Ruoff P., Loros J.J., and Dunlap J.C. 2005. The relationship between FRQ-protein stability and temperature compensation in the *Neurospora* circadian clock. *Proc. Natl. Acad. Sci.* **102:** 17681.

Rutila J.E., Suri V., Le M., So W.V., Rosbash M., and Hall J.C. 1998. *CYCLE* is a second bHLH-PAS clock protein essential for circadian rhythmicity and transcription of *Drosophila period* and *timeless*. *Cell* **93:** 805.

Sargent M.L. and Briggs W.R. 1967. The effect of light on a circadian rhythm of conidiation in *Neurospora*. *Plant Physiol.* **42:** 1504.

Sargent M.L. and Kaltenborn S.H. 1972. Effects of medium composition and carbon dioxide on circadian conidiation in *Neurospora*. *Plant Physiol.* **50:** 171.

Sargent M.L., Briggs W.R., and Woodward D.O. 1966. The circadian nature of a rhythm expressed by an invertaseless strain of *Neurospora crassa*. *Plant Physiol.* **41:** 1343.

Schafmeier T., Kaldi K., Diernfellner A., Mohr C., and Brunner M. 2006. Phosphorylation-dependent maturation of *Neurospora* circadian clock protein from a nuclear repressor toward a cytoplasmic activator. *Genes Dev.* **20:** 297.

Schafmeier T., Haase A., Kaldi K., Scholz J., Fuchs M., and Brunner M. 2005. Transcriptional feedback of *Neurospora* circadian clock gene by phosphorylation-dependent inactivation of its transcription factor. *Cell* **122:** 235.

Schwerdtfeger C. and Linden H. 2003. VIVID is a flavoprotein and serves as a fungal blue light photoreceptor for photoadaptation. *EMBO J.* **22:** 4846.

Shigeyoshi Y., Taguchi K., Yamamoto S., Takeida S., Yan L., Tei H., Moriya S., Shibata S., Loros J.J., Dunlap J.C., and Okamura H. 1997. Light-induced resetting of a mammalian circadian clock is associated with rapid induction of the *mPer1* transcript. *Cell* **91:** 1043.

Smith R.F. and Konopka R.J. 1982. Effects of dosage alterations at the *per* locus on the period of the circadian clock of *Drosophila*. *Mol. Gen. Genet.* **185:** 30.

Sugano S., Andronis C., Green R., Wang Z., and Tobin E. 1998. Protein kinase CK2 interacts with and phosphorylates the *Arabidopsis* circadian clock-associated gene 1 protein. *Proc. Natl. Acad. Sci.* **95:** 11020.

Sweeney B.M. 1976. Circadian rhythms, definition and general characterization. In *The molecular basis of circadian rhythms* (report of the Dahlem Workshop) (ed. J.W. Hastings and H.-G. Schweiger), p. 77. Abakon-Verlagsgesellschaft, Berlin.

Tan Y., Dragovic Z., Merrow M., and Roenneberg T. 2004. Entrainment dissociates transcription and translation of a circadian clock gene in *Neurospora*. *Curr. Biol.* **14:** 433.

Yang Y., Cheng P., and Liu Y. 2002. Regulation of the *Neurospora* circadian clock by casein kinase II. *Genes Dev.* **16:** 994.

Zoltowski B.D., Schwerdtfeger C., Widom J., Loros J.J., Bilwes A.M., Dunlap J.C., and Crane B.R. 2007. Conformational switching in the fungal light sensor Vivid. *Science* **316:** 1054.

A PER/TIM/DBT Interval Timer for *Drosophila*'s Circadian Clock

L. SAEZ,* P. MEYER,† AND M.W. YOUNG*

**Laboratory of Genetics, The Rockefeller University, New York, New York 10021; †Department of Microbiology, College of Physician & Surgeons, Columbia University, New York, New York 10032*

Circadian rhythms in *Drosophila* are supported by a negative feedback loop, in which PERIOD (PER) and Timeless (TIM) shut down their own transcription as they translocate once a day from the cytoplasm of clock-containing cells to the nucleus. Period length is partially determined by an interval of cytoplasmic retention of the TIM and PER proteins. To study this process, we examined PER/TIM/Doubletime (DBT) physical interactions and nuclear translocation by imaging individual cultured *Drosophila* cells. Using live cell video microscopy and green fluorescent protein (GFP) tags, we observed dynamic patterns of stability and localization for DBT, PER, and TIM that resembled those previously found in vivo. These studies suggest that a cytoplasmic interval timer regulates nuclear translocation of these proteins. The cultured cell assay provides a potent system to study interactions among new and known genes involved in the generation of circadian behavior.

INTRODUCTION

That single genes could make a substantial contribution to circadian rhythms was first demonstrated by Konopka and Benzer (1971). Upon screening mutagenized *Drosophila* for changes in patterns of daily eclosion, they identified three mutations affecting a single X-chromosomal locus, *period* (*per*). One mutant, per^0, was arrhythmic, whereas *per-short* (per^S) and *per-long* (per^L) mutants were rhythmic with 19-hour and 29-hour periods, respectively. Locomotor activity was also affected, and each *per* mutation produced equivalent effects on eclosion and locomotor activity rhythms. More than a decade passed before the molecular identity of *period* was established (Bargiello and Young 1984; Reddy et al. 1984; Jackson et al. 1986; Citri et al. 1987). Yet, the primary sequence of the encoded protein provided little information about its function in circadian rhythm generation.

In further work, the spatial expression of *per* was found to be relatively widespread, localized to the nucleus and cytoplasm of several tissues (Siwicki et al. 1988; Saez and Young 1988), and was not limited to the head, which was known to be the site of *per*'s regulation of circadian locomotor behavior (Konopka et al. 1983). Most importantly, PER abundance in photoreceptor cells varied, with lowest levels observed in the middle of the day and highest levels in the middle of the night (Siwicki et al. 1988). This observation prompted an examination of the temporal expression of *per* mRNA. RNA abundance cycled with a circadian rhythm in fly heads, with specifically altered periods in the mutants (Hardin et al. 1990).

TIMELESS AND TEMPORAL DELAYS IN THE CIRCADIAN CLOCK

Because high levels of PER protein were correlated with times of low *per* expression, it was proposed that PER proteins negatively regulate *per* RNA accumulation (Hardin et al. 1990). A basis for oscillating feedback, and two new classes of molecular cycling (affecting protein stability and subcellular localization), emerged with the discovery of a second-chromosome clock gene called *timeless (tim)* (Sehgal et al. 1994; Vosshall et al. 1994). The *tim* null mutation (tim^0) eliminated *per* RNA and protein oscillations in addition to causing arrhythmic eclosion and locomotor activity (Sehgal et al. 1994). Like *per*, *tim* mRNA and protein were found to oscillate in a circadian fashion and these oscillations were abolished in per^0 and tim^0 mutants (Sehgal et al. 1995). Although the peak of *tim* mRNA accumulation was slightly earlier than that of *per*, PER and TIM proteins accumulated with the same phase, indicating a link between these two genes (Sehgal et al. 1995). Subsequently it was shown that the PER and TIM proteins physically interact in vitro, in yeast and *Drosophila* cells (Gekakis et al. 1995; Saez and Young 1996). Interdependent functions of *per* and *tim* were also indicated by the finding that PER and TIM coexpression was required for the nuclear accumulation of either protein, and that PER stability was dependent on TIM (Vosshall et al. 1994; Price et al. 1995; Hunter-Ensor et al. 1996; Myers et al. 1996). Thus, with the discovery of *tim* and the regulated nuclear accumulation of PER and TIM (see also Curtin et al. 1995), a model was proposed that incorporated the concept of a specific temporal delay to explain how negative feedback could generate oscillations in gene expression. This delay was originally thought to be determined by the slow kinetics of association of PER and TIM in the cytoplasm, such that once dimerized, PER/TIM complexes would enter the nucleus and negatively regulate their own expression (cf. Sehgal et al. 1995).

The isolation of several additional clock genes has helped to refine this mechanism (for review, see Young and Kay 2001; Hardin 2006). For example, some of the more recently identified genes regulate the phosphorylation or dephosphorylation of PER and TIM, affecting PER/TIM subcellular localization and stability (Harms et al. 2003; Bae and Edery 2006).

S2 CELLS AND NUCLEAR TRANSLOCATION OF PER AND TIM

To better understand the regulation of PER/TIM interactions and their nuclear translocation, we began a new phase of work employing cultured *Drosophila* cells (Saez and Young 1996). Subsequently, this cell line (S2; Schneider's line 2), has become an important tool for investigating the function of *Drosophila*'s circadian clock using transcription assays (Darlington et al. 1998), immunoprecipitation assays (Ceriani et al. 1999), degradation assays (Naidoo et al. 1999), and assays of phophorylation and dephosphorylation (cf. Ko et al. 2002; Sathyanarayanan et al. 2004).

S2 cells were originally derived from dissociated embryos 35 years ago (Schneider 1972) and do not have a functional circadian clock. They express *doubletime* (*dbt*), *casein kinase 2* (*CK2*), and *cycle* (*cyc*) (Nawathean et al. 2005; L. Saez, unpubl.) but lack expression of other known clock genes such as *Clock (Clk)* (Darlington et al. 1998), *per*, and *tim* (see Fig. 1) (Saez and Young 1996). Ectopic expression of *Clock* induces the expression of *tim*, and sometimes *per* in individual S2 cell lines (see Fig. 1) (McDonald and Rosbash 2001; Lim et al. 2007).

Initially, we established immunocytochemically in fixed S2 cells that PER and TIM must be coexpressed to achieve efficient nuclear localization of either protein. Nuclear accumulation of both proteins was found to be complete approximately 6 hours after their initial synthesis (Saez and Young 1996). Expression of PER alone led to its accumulation in the cytoplasm with a half-life of approximately 10 hours as determined by pulse-chase experiments (see Fig. 1). TIM shuttles between the nucleus and cytoplasm with most cells showing cytoplasmic rather than nuclear accumulation (Ashmore et al. 2003; Meyer et al. 2006). Amino acid sequences that mediate PER/TIM interaction and cytoplasmic localization of PER and TIM (CLD; cytoplasmic localization domains), were also identified (Saez and Young 1996). Although these early studies reproduced elements of the temporal delay seen in nuclear translocation in vivo, and the inter-dependence of PER and TIM to achieve the delay, they gave little information about the kinetics of the PER/TIM interaction and did not provide the necessary resolution to establish a mechanism for regulated nuclear entry.

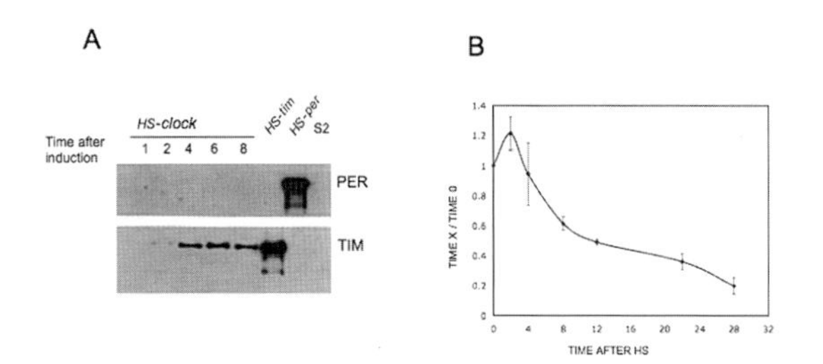

Figure 1. Response of endogenous clock genes in S2 cells to transient expression of Clock. (*A*) S2 cells were transfected with a plasmid containing a heat shock (*HS*) promoter fused to *Clock* cDNA; 48 hours after transfection, cells were heat-shocked for 30 minutes at 37°C, and an equal amount of the culture was collected at the times (hr) indicated. *HS-tim* and *HS-per* are controls from S2 cells expressing heat-shock-induced PER or TIM. Following protein extraction and SDS-PAGE, proteins were transferred to a membrane and probed with anti-PER (*top*) or anti-TIM antibodies (*bottom*). (*B*) Half-life of PER in S2 cells. S2 cells transfected with *HS-per* were incubated in Schneider M3 media (Schneider 1972) without methionine and cysteine for 1 hour. Cells were heat-shocked for 30 minutes at 37°C, allowed to recover for 30 minutes at 25°C, and resuspended in media containing [^{35}S]methionine-cysteine for 30 minutes. Cells were collected by centrifugation and resuspended in complete M3 media (Sigma-Aldrich), and equal amounts of the growing culture were taken every 2 hours thereafter. ^{35}S-labeled PER was immunoprecipitated from each sample and subjected to PAGE. The amount of ^{35}S-labeled PER recovered from each sample was expressed as the percentage remaining from the initial time point. The graph represents the average values obtained from four experiments.

PER AND TIM INTERACTION DOES NOT INDUCE NUCLEAR TRANSLOCATION

To individually trace PER and TIM movements in the cytoplasm and to assess the role of PER/TIM interaction in nuclear translocation, we developed a single-cell, live imaging assay in S2 cells (Meyer et al. 2006). We individually or jointly expressed in a single cell, using a heat shock promoter, PER and TIM tagged at their carboxyl termini with cyan fluorescent protein (CFP) and yellow fluorescent protein (YFP), respectively.

Initially, we followed PER-CFP for approximately 10 hours, when independently expressed in S2 cells. PER-CFP remained in the cytoplasm for the duration of the assay. Independent expression of TIM-YFP resulted in mainly cytoplasmic accumulation, with about 10% of the cells showing nuclear YFP (Meyer et al. 2006). These results are consistent with our previous observations using fixed S2 cells. We also confirmed that leptomycin-B abolishes nuclear export of TIM in S2 cells (Ashmore et al. 2003). However, our findings with PER differed from certain studies of subcellular localization that coexpressed *Clk* and *per* in S2 cells (Chang and Reppert 2003; Nawathean and Rosbash 2004). As previously discussed (Meyer and Young 2007), it seems likely that endogenous TIM that is produced in response to coexpressed *Clk* would influence PER nuclear accumulation in such studies.

We next measured the onset of nuclear accumulation of PER and TIM when coexpressed in the same cell. Figure 2 shows time-lapse images of PER-CFP and TIM-YFP that were collected over several hours following heat shock induction of both proteins. The images show that PER and TIM were evenly distributed in the cytoplasm for a period of about 4 hours, followed by a gradual shift of both proteins to discrete cytoplasmic foci surrounding the nucleus. The onset of nuclear accumulation of PER and TIM occurred abruptly (e.g., ~340 minutes after heat shock induction for TIM in this cell), with the phase of nuclear accumulation lasting less than 1 hour (Fig. 2) (Meyer et al. 2006). Surprisingly, the onset of nuclear translocation proved to be largely independent of PER and TIM levels in this system (Meyer et al. 2006; Meyer and Young 2007). Even in the presence of leptomycin-B, TIM, which otherwise would accumulate in the nucleus, remained for approximately 5.5 hours in the cytoplasm in PER-TIM-expressing cells, presumably due to a rapid association with PER that retains TIM in the cytoplasm. The treatment with leptomycin-B did not affect the onset

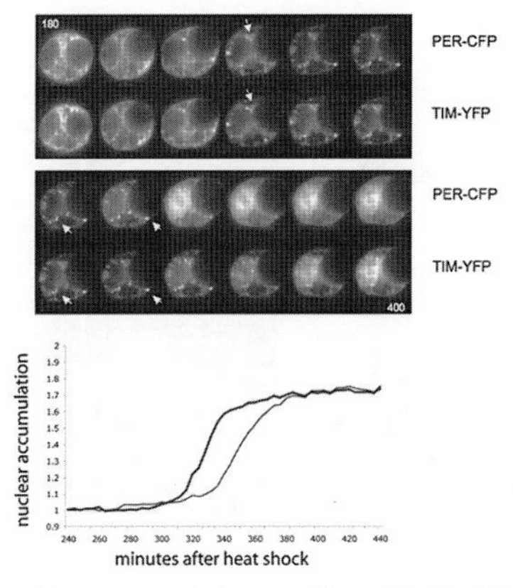

Figure 2. Nuclear accumulation profiles of PER-CFP and TIM-YFP in a single S2 cell. (*Top two panels*) Time-lapse images of a cell expressing PER-CFP and TIM-YFP. Starting at the top left, images were taken every 20 minutes, with the first image collected 180 minutes after heat shock. The last image (*bottom right*) was collected 400 minutes after heat shock. Arrows indicate the cytoplasmic foci whose formation preceded nuclear entry. The graph below these images shows the nuclear accumulation profiles (fluorescence measurements) of PER-CFP (*thick line*) and TIM-YFP (*thin line*). In this example, the onset of nuclear accumulation of PER was advanced compared to TIM by about 25 minutes. (Redrawn, with permission, from Meyer et al. 2006 [© AAAS].)

of nuclear entry in PER + TIM experiments (Meyer et al. 2006). These results indicate that nuclear translocation is not simply initiated by PER/TIM interaction, as corroborated by FRET analysis (see below).

PER AND TIM NUCLEAR ACCUMULATION PROFILES ARE DIFFERENT

Another intriguing observation came from a study of the individual patterns of nuclear accumulation of PER and TIM. A detailed analysis showed that nuclear accumulation profiles can be separated into three categories: In about 30% of the cells, PER and TIM have similar onsets and profiles of nuclear accumulation; in about 40% of the cells, PER and TIM had the same onsets of nuclear entry, but PER was transferred to nuclei more rapidly than TIM; in the remaining (~30%) cells, profiles of PER and TIM nuclear accumulation were similar, but the PER-CFP onset was advanced with respect to TIM (cf. Fig. 2). Thus, in approximately 70% of the cells analyzed, PER accumulated in the nucleus ahead of TIM (Meyer et al. 2006). This condition is similar to that previously observed in lateral neurons (pacemaker cells) of larval and adult brains, in which PER was detected in the nucleus before TIM (Shafer et al. 2002, 2004; Ashmore et al. 2003). These observations indicate that PER and TIM enter the nucleus independently.

FRET ANALYSIS OF PER AND TIM INTERACTION

Tagging PER and TIM with derivatives of GFP allowed us to develop a fluorescence resonant energy transfer (FRET) assay to measure PER/TIM interactions in a single cell. FRET is an indication of close proximity of two proteins, as intermolecular distances of less than 10 nm are needed for resonant energy transfer from one fluorophore to another (Forster 1948). In our S2 cell experiments, maximum levels of FRET were observed without a measurable delay following PER-CFP and TIM-YFP induction (see Fig. 3), suggesting rapid formation of PER/TIM complexes. Levels of FRET remained constant until the onset of PER and TIM nuclear translocation. A rapid decline in FRET was observed as the proteins moved to the nucleus. The interval over which FRET declined matched that of the independently measured nuclear translocation of PER and TIM in the same cell. Together with the divergent profiles of nuclear accumulation of PER and TIM, the loss of FRET suggests that following a prolonged physical association in the cytoplasm, PER and TIM dissociate and enter the nucleus independently.

PER^L MUTATION DELAYS NUCLEAR ENTRY WITHOUT AFFECTING THE PROFILE OF PER/TIM BINDING

As the molecular behaviors of PER and TIM in S2 cells retained many properties previously observed in vivo, we sought a further test of this correlation by examining the behavior of proteins produced by two clock mutants, *per*^{L} and *tim*^{UL}, that generate long-period behavioral rhythms by different mechanisms. The mutation in *per*^{L} is a single-amino-acid change (V242A) within the PAS (PER-ARNT-SM) domain (Baylies et al. 1987) that delays

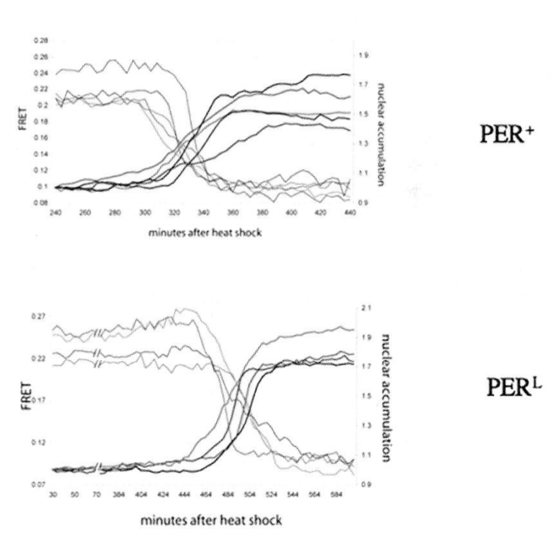

Figure 3. Profiles of FRET and nuclear translocation associated with PER$^+$/TIM and PER^L/TIM expression. (*Top and bottom panels*) Whole-cell FRET (*thin lines*) and nuclear accumulation (*thick lines*) profiles of PER-CFP and PER^L-CFP, respectively, in five (*top*) and four (*bottom*) S2 cells expressing TIM-YFP. The decline in FRET levels coincides with the time of nuclear entry at approximately 330 minutes for PER-CFP (PER$^+$) and about 500 minutes for PER^L-CFP (PER^L). The bottom panel also shows the level of whole-cell FRET beginning 30 minutes after induction. For each panel, *thick* and *thin lines* of the same color represent measurements of the same cell. (Redrawn, with permission, from Meyer et al. 2006 [© AAAS].

nuclear translocation of PER and TIM by 4 hours in pacemaker cells of the brain (Curtin et al. 1995), causing a 28-hour behavioral rhythm (Konopka and Benzer 1971). tim^{UL} is a single-amino-acid substitution that delays the turnover of PER and TIM in the nucleus, giving a 33-hour circadian period without affecting the timing of nuclear translocation (Rothenfluh et al. 2000).

The onset of nuclear entry for PER/TIM^{UL} was quite similar to the onset observed for PER/TIM, indicating that tim^{UL} does not affect nuclear entry in vivo or in S2 cells (Meyer et al. 2006). The per^{L} mutation, on the other hand, could faithfully reproduce in S2 cells the increased delay observed in vivo, shifting the onset of nuclear accumulation of PER and TIM to about 8.5 hours (Fig. 3) (Meyer et al. 2006). Thus, a single-amino-acid change in PER appears to affect the function of an interval timer that regulates cytoplasmic retention of PER and TIM (Meyer et al. 2006). FRET, in cells expressing the wild-type and the mutant proteins, showed comparable rates of emergence and decay, indicating that association and dissociation kinetics of the PER/TIM complex are not substantially affected in the mutants. Because dissociation of the PER/TIM complex appears to be required for nuclear entry, it is possible that PER^{L} alters a step needed to trigger dissociation from the PER^{L}/TIM complex, delaying nuclear entry of both proteins.

DBT, A NUCLEAR PROTEIN, IS RETAINED BY PER IN THE CYTOPLASM AND AFFECTS PER STABILITY

We next focused our attention on DBT, another clock protein that associates with PER and the PER/TIM complex (Kloss et al. 1998, 2001). *dbt* is expressed throughout the circadian cycle. Nevertheless, in lateral neurons and in photoreceptor cells, DBT is nuclear in *per* and *tim* null mutants and shows a circadian subcellular distribution, cycling between the cytoplasm and nucleus in a fashion that parallels the behavior of PER accumulation in wild-type fly heads (Kloss et al. 2001). In vivo, DBT physically interacts with PER but is only found associated with TIM in PER/TIM complexes (Kloss et al. 2001). DBT-dependent phosphorylation destabilizes PER, promoting PER's degradation by the ubiquitin-proteosome degradation pathway (Price et al. 1998; Ko et al. 2002). TIM promotes PER stability, apparently by suppressing PER's phosphorylation by DBT (Kloss et al. 1998; Price et al. 1998; Ko et al. 2002). Thus, DBT-dependent phosphorylation is believed to affect the timing of nuclear entry in part by regulating the stability of PER. In turn, PER influences the subcellular distribution of DBT through formation of DBT/PER and DBT/PER/TIM complexes (Kloss et al. 2001).

To study DBT in cell culture, we tagged it with YFP and induced DBT-YFP expression using a heat shock promoter. Consistent with observations made in vivo with per^{0} mutants, DBT-YFP in S2 cells localized exclusively to the nucleus when expressed alone (Fig. 4A) or in the presence of TIM-CFP (Fig. 4B, top). Coexpresion of DBT-YFP and PER-CFP immediately relocalized DBT from the nucleus to the cytoplasm, presumably due to a

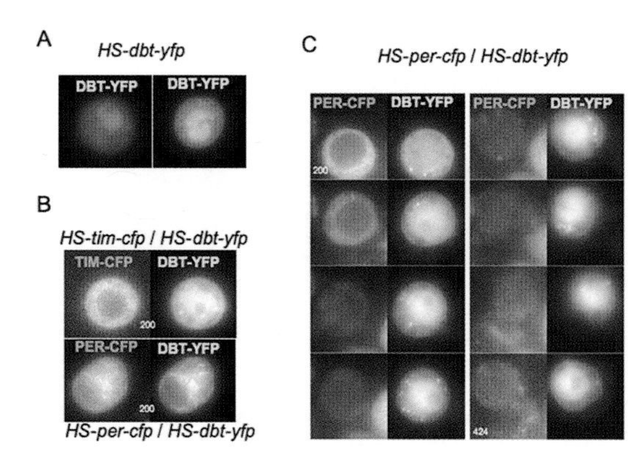

Figure 4. Doubletime and PER interact in S2 cells. (*A*) Subcellular localization of DBT in S2 cells. DBT-YFP in S2 cells accumulates in the nucleus immediately following induction. (*B*) DBT interacts with PER and not with TIM. Images show cells expressing TIM-CFP and DBT-YFP (*top*) and PER-CFP and DBT-YFP (*bottom*). Approximately 3 hours after induction, TIM-CFP remains in the cytoplasm and DBT localizes to the nucleus. Coexpression of PER and DBT localized both proteins to the cytoplasm. (*C*) Time-lapse images of a single S2 cell expressing PER-CFP and DBT-YFP. As previously observed in vivo, in the absence of TIM, DBT promotes PER degradation. Images shown were taken every 32 minutes.

physical association with cytoplasmic PER (see Fig. 4B, bottom). Because DBT affects the stability of PER, we collected time series data from these S2 cells over several hours (Fig. 4C). PER and DBT accumulated in the cytoplasm for approximately 180 minutes following their heat shock induction. Thereafter (~30 minutes), a progressive and rapid degradation of PER was observed with a relocalization of DBT to the nucleus. Because the degradation of PER did not occur immediately after its induction, and DBT was initially confined to the cytoplasm with PER, binding and a progressive phosphorylation of PER by DBT may be required for PER degradation. Endogenous DBT, which is found at low levels in S2 cells, did not have a detectable effect on the stability of PER in our experiments: PER, when expressed alone in S2 cells, was stable for at least 10 hours after induction, and *dbt* RNA interference (RNAi) did not alter the pattern of PER stability or the kinetics of nuclear entry in PER/TIM-expressing S2 cells (data not shown). Although progressive phosphorylation and degradation of PER were observed in our studies when PER was constitutively expressed and newly formed DBT was added through an inducible promoter, it was also shown by other investigators that coexpression of PER and DBT by a constitutive promoter can give constant low levels of PER that is highly phosphorylated (Ko et al. 2002; Ko and Edery 2005). These observations indicate that DBT affects PER stability and that patterns of DBT localization can be influenced by PER in S2 cells as previously observed in vivo.

PER AND DBT ENTER THE NUCLEUS IN *TIMELESS*-EXPRESSING S2 CELLS

To see whether TIM will protect PER from DBT-dependent degradation, we coexpressed PER-CFP and DBT-YFP in TIM-expressing cells. Figure 5 shows time-

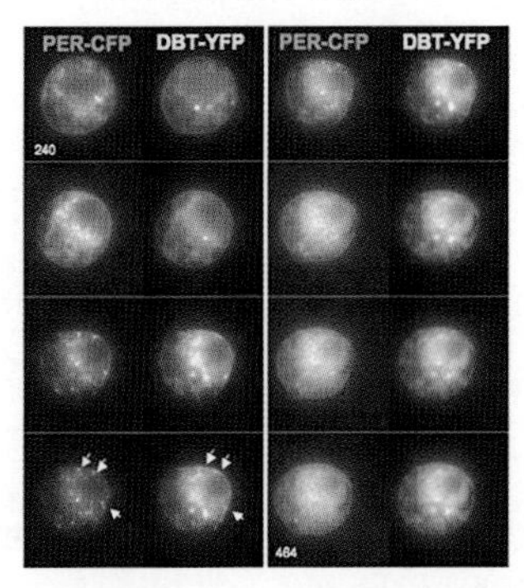

Figure 5. Coordinated nuclear accumulation of PER, TIM, and DBT. S2 cells expressing TIM from a heat shock promoter were cotransfected with *HS-per-cfp* and *HS-dbt-yfp*; 48 hours after transfection, cells were heat-shocked at 37°C for 30 minutes and imaged for 464 minutes. Time-lapse images of a single cell expressing PER-CFP and DBT-YFP show that both proteins enter the nucleus with an onset similar to that previously seen for PER and TIM. Arrowheads indicate the formation of cytoplasmic PER/DBT foci before nuclear accumulation.

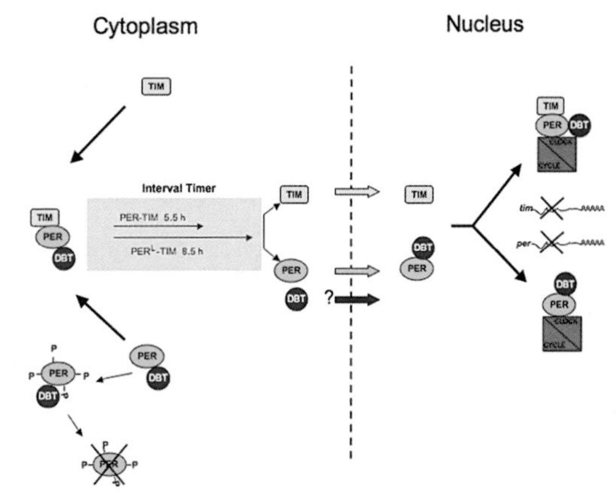

Figure 6. Model for regulated nuclear accumulation of PER, DBT, and TIM in S2 cells. Wild-type PER and TIM proteins associate with little or no delay and thereafter remain in the cytoplasm for about 5.5 hours. During this time, PER and TIM progressively associate in distinct foci, which are lost as PER and TIM dissociate and move to the nucleus. In the nucleus, PER, DBT, and possibly TIM, associate with Clock/Cycle complexes, inhibiting transcription of *Clock/cycle*-dependent genes including *per* and *tim* (see also Kim et al. 2007; Nawathean et al. 2007). The duration of the PER/TIM complex in the cytoplasm is affected by PER^L which delays the onset of nuclear accumulation by about 3 hours. A similar delay is observed in vivo (Curtin et al. 1995) and is thought to generate the abnormal (28 hours) period length of per^L flies. In addition to these interactions, DBT association leads to phosphorylation and destabilization of PER in the absence of TIM. Our studies have not determined whether DBT and PER move to the nucleus independently or as a complex.

lapse images of PER-CFP and DBT-YFP accumulation in these cells. Destabilization of PER by DBT was suppressed by TIM expression, because PER was observed to be stable in these cells for more than 8 hours, a period sufficient for complete DBT-dependent degradation of PER in the absence of TIM. As in the case of PER and TIM coexpression, DBT and PER progressively accumulated in the cytoplasm, forming discrete foci that presumably contain PER-CFP, DBT-YFP, and TIM (see small arrows in Fig. 5). The kinetics of nuclear accumulation of PER and DBT in these TIM-expressing cells has not been studied in detail but appears to be similar to that of PER and TIM, occurring at about 5–6 hours after induction. These results indicate that S2 cells recapitulate many of the in vivo behaviors of PER, TIM, and DBT. A summary of PER/TIM/DBT interactions and functions is indicated in the model in Figure 6. Further work will be required to determine patterns of PER degradation, whether PER and DBT independently enter nuclei, and profiles of FRET in cells coexpressing the three proteins.

THE INTERVAL TIMER

Although PER and TIM binding is not sufficient to trigger nuclear entry, physical association appears to be a prerequisite for their timed nuclear accumulation. The formation of the PER/TIM complex could, perhaps, allow subsequent posttranslational modifications or the binding of factors that determine the specific duration of its cytoplasmic phase.

The formation of cytoplasmic foci is especially intriguing as these are not observed when PER, or TIM, or DBT is expressed alone. The formation of these foci always precedes the onset of nuclear accumulation and they disappear as nuclear accumulation proceeds. The foci represent the final cytoplasmic phase of PER/TIM/DBT accumulation. Their contents may be further explored by a program of protein tagging that includes the remaining clock proteins, as well as proteins known to be involved in either cytosolic trafficking or transport to the nucleus.

CONCLUSIONS

We have shown that a simple nonoscillating cell culture system can be used to study the molecular features of a key step in the circadian clockworks. This single-cell assay not only recapitulates many of the in vivo behaviors of PER and TIM and their mutants, but it has also provided a revised model of the *Drosophila* circadian clock.

ACKNOWLEDGMENTS

We are grateful to A. North (Director, The Rockefeller University Bio-Imaging Resource Center) for interest, advice, and helpful discussion. This work was supported by the National Institutes of Health grant GM 54339 to M.W.Y.

REFERENCES

Ashmore L.J., Sathyanarayanan S., Silvestre D.W., Emerson M.M., Schotland P., and Sehgal A. 2003. Novel insights into the regulation of the timeless protein. *J. Neurosci.* **23:** 7810.

Bae K. and Edery I. 2006. Regulating a circadian clock's period, phase and amplitude by phosphorylation: Insights from *Drosophila*. *J. Biochem.* **140:** 609.

Bargiello T.A. and Young M.W. 1984. Molecular genetics of a biological clock in *Drosophila*. *Proc. Natl. Acad. Sci.* **81:** 2142.

Baylies M.K., Bargiello T.A., Jackson F.R., and Young M.W. 1987. Changes in abundance or structure of the per gene product can alter periodicity of the *Drosophila* clock. *Nature* **326:** 390.

Ceriani M.F., Darlington T.K., Staknis D., Mas P., Petti A.A., Weitz C.J., and Kay S.A. 1999. Light-dependent sequestration of TIMELESS by CRYPTOCHROME. *Science* **285:** 506.

Chang D.C. and Reppert S.M. 2003. A novel C-terminal domain of *Drosophila* PERIOD inhibits dCLOCK:CYCLE-mediated transcription. *Curr. Biol.* **13:** 758.

Citri Y., Colot H.V., Jacquier A.C., Yu Q., Hall J.C., Baltimore D., and Rosbash M.A. 1987. A family of unusually spliced biologically active transcripts encoded by a *Drosophila* clock gene. *Nature* **326:** 42.

Curtin K.D., Huang Z.J., and Rosbash M. 1995. Temporally regulated nuclear entry of the *Drosophila* period protein contributes to the circadian clock. *Neuron* **14:** 365.

Darlington T.K., Wager-Smith K., Ceriani M.F., Staknis D., Gekakis N., Steeves T.D., Weitz C.J., Takahashi J.S., and Kay S.A. 1998. Closing the circadian loop: CLOCK-induced transcription of its own inhibitors per and tim. *Science* **280:** 1599.

Forster T. 1948. Intermolecular energy migration and fluorescence. *Ann. Physics* **2:** 55.

Gekakis N., Saez L., Delahaye-Brown A.M., Myers M.P., Sehgal A., Young M.W., and Weitz C.J. 1995. Isolation of timeless by PER protein interaction: Defective interaction between timeless protein and long-period mutant PERL. *Science* **270:** 732.

Hardin P.E. 2006. Essential and expendable features of the circadian timekeeping mechanism. *Curr. Opin. Neurobiol.* **16:** 686.

Hardin P.E., Hall J.C., and Rosbash M. 1990. Feedback of the *Drosophila* period gene product on circadian cycling of its messenger RNA levels. *Nature* **343:** 536.

Harms E., Young M.W., and Saez L. 2003. CK1 and GSK3 in the *Drosophila* and mammalian circadian clock. *Novartis Found. Symp.* **253:** 267.

Hunter-Ensor M., Ousley A., and Sehgal A. 1996. Regulation of the *Drosophila* protein timeless suggests a mechanism for resetting the circadian clock by light. *Cell* **84:** 677.

Jackson F.R., Bargiello T.A., Yun S.H., and Young M.W. 1986. Product of per locus of *Drosophila* shares homology with proteoglycans. *Nature* **320:** 185.

Kim E.Y., Ko H.W., Yu W., Hardin P.E., and Edery I. 2007. A DOUBLETIME kinase binding domain on the *Drosophila* PERIOD protein is essential for its hyperphosphorylation, transcriptional repression, and circadian clock function. *Mol. Cell. Biol.* **27:** 5014.

Kloss B., Rothenfluh A., Young M.W., and Saez L. 2001. Phosphorylation of period is influenced by cycling physical associations of double-time, period, and timeless in the *Drosophila clock. Neuron* **30:** 699.

Kloss B., Price J.L., Saez L., Blau J., Rothenfluh A., Wesley C.S., and Young M.W. 1998. The *Drosophila* clock gene double-time encodes a protein closely related to human casein kinase Iepsilon. *Cell* **94:** 97.

Ko H.W. and Edery I. 2005. Analyzing the degradation of PERIOD protein by the ubiquitin-proteasome pathway in cultured *Drosophila* cells. *Methods Enzymol.* **393:** 394.

Ko H.W., Jiang J., and Edery I. 2002. Role for Slimb in the degradation of *Drosophila* Period protein phosphorylated by Doubletime. *Nature* **420:** 673.

Konopka R.J. and Benzer S. 1971. Clock mutants of *Drosophila melanogaster. Proc. Natl. Acad. Sci.* **68:** 2112.

Konopka R., Wells S., and Lee T. 1983. Mosaic analysis of a *Drosophila* clock mutant. *Mol. Gen. Genet.* **190:** 284.

Lim C., Lee J., Choi C., Doh E., and Choe J. 2007. Functional role of CREB-binding protein in the circadian clock system of *Drosophila. Mol. Cell. Biol.* **27:** 4876.

McDonald M.J. and Rosbash M. 2001. Microarray analysis and organization of circadian gene expression in *Drosophila. Cell* **107:** 567.

Meyer P. and Young M.W. 2007. The 2006 Pittendrigh/Aschoff Lecture: New roles for old proteins in the *Drosophila* circadian clock. *J. Biol. Rhythms* **22:** 283.

Meyer P., Saez L., and Young M.W. 2006. PER-TIM interactions in living *Drosophila* cells: An interval timer for the circadian clock. *Science* **311:** 226.

Myers M.P., Wager-Smith K., Rothenfluh-Hilfiker A., and Young M.W. 1996. Light-induced degradation of TIMELESS and entrainment of the *Drosophila* circadian clock. *Science* **271:** 1736.

Naidoo N., Song W., Hunter-Ensor M., and Sehgal A. 1999. A role for the proteasome in the light response of the timeless clock protein. *Science* **285:** 1737.

Nawathean P. and Rosbash M. 2004. The doubletime and CKII kinases collaborate to potentiate *Drosophila* PER transcriptional repressor activity. *Mol. Cell* **13:** 213.

Nawathean P., Menet J.S., and Rosbash M. 2005. Assaying the *Drosophila* negative feedback loop with RNA interference in S2 cells. *Methods Enzymol.* **393:** 610.

Nawathean P., Stoleru D., and Rosbash M. 2007. A small conserved domain of *Drosophila* PERIOD is important for circadian phosphorylation, nuclear localization, and transcriptional repressor activity. *Mol. Cell. Biol.* **27:** 5002.

Price J.L., Dembinska M.E., Young M.W., and Rosbash M. 1995. Suppression of PERIOD protein abundance and circadian cycling by the *Drosophila* clock mutation timeless. *EMBO J.* **14:** 4044.

Price J.L., Blau J., Rothenfluh A., Abodeely M., Kloss B., and Young M.W. 1998. double-time is a novel *Drosophila* clock gene that regulates PERIOD protein accumulation. *Cell* **94:** 83.

Reddy P., Zehring W. A., Wheeler D. A., Pirrotta V., Hadfield C., Hall J. C., and Rosbash M. 1984. Molecular analysis of the period locus in *Drosophila melanogaster* and identification of a transcript involved in biological rhythms. *Cell* **38:** 701.

Rothenfluh A., Saez L., and Young M. W. 2000. A TIMELESS-independent function for PERIOD proteins in the *Drosophila* clock. *Neuron* **26:** 505.

Saez L. and Young M.W. 1988. In situ localization of the per clock protein during development of *Drosophila melanogaster. Mol. Cell. Biol.* **8:** 5378.

———. 1996. Regulation of nuclear entry of the *Drosophila* clock proteins period and timeless. *Neuron* **17:** 911.

Sathyanarayanan S., Zheng X., Xiao R., and Sehgal A. 2004. Posttranslational regulation of *Drosophila* PERIOD protein by protein phosphatase 2A. *Cell* **116:** 603.

Schneider I. 1972. Cell lines derived from late embryo stages of *Drosophila melanogaster. J. Embryol. Exp. Morphol.* **27:** 353.

Sehgal A., Price J.L., Man B., and Young M.W. 1994. Loss of circadian behavioral rhythms and per RNA oscillations in the *Drosophila* mutant timeless. *Science* **263:** 1603.

Sehgal A., Rothenfluh-Hilfiker A., Hunter-Ensor M., Chen Y., Myers M.P., and Young M.W. 1995. Rhythmic expression of timeless: A basis for promoting circadian cycles in period gene autoregulation. *Science* **270:** 808.

Shafer O.T., Rosbash M., and Truman J.W. 2002. Sequential nuclear accumulation of the clock proteins period and timeless in the pacemaker neurons of *Drosophila melanogaster. J. Neurosci.* **22:** 5946.

Shafer O.T., Levine J.D., Truman J.W., and Hall J.C. 2004. Flies by night: Effects of changing day length on *Drosophila*'s circadian clock. *Curr. Biol.* **14:** 424.

Siwicki K.K., Eastman C., Petersen G., Rosbash M., and Hall J.C. 1988. Antibodies to the period gene product of *Drosophila* reveal diverse tissue distribution and rhythmic changes in the visual system. *Neuron* **1:** 141.

Vosshall L.B., Price J.L., Sehgal A., Saez L., and Young M.W. 1994. Block in nuclear localization of period protein by a second clock mutation, timeless. *Science* **263:** 1606.

Young M.W. and Kay S.A. 2001. Time zones: A comparative genetics of circadian clocks. *Nat. Rev. Genet.* **2:** 702.

Transcriptional Feedback and Definition of the Circadian Pacemaker in *Drosophila* and Animals

M. ROSBASH,* S. BRADLEY,† S. KADENER,† Y. LI,† W. LUO,* J.S. MENET,* E. NAGOSHI,† K. PALM,* R. SCHOER,† Y. SHANG,* AND C.-H.A. TANG†

**Biology Department, Howard Hughes Medical Institute, Brandeis University, Waltham, Massachusetts 02454;*
†Biology Department, Brandeis University, Waltham, Massachusetts 02454

The modern era of *Drosophila* circadian rhythms began with the landmark Benzer and Konopka paper and its definition of the *period* gene. The recombinant DNA revolution then led to the cloning and sequencing of this gene. This work did not result in a coherent view of circadian rhythm biochemistry, but experiments eventually gave rise to a transcription-centric view of circadian rhythm generation. Although these circadian transcription-translation feedback loops are still important, their contribution to core timekeeping is under challenge. Indeed, kinases and posttranslational regulation may be more important, based in part on recent in vitro work from cyanobacteria. In addition, kinase mutants or suspected kinase substrate mutants have unusually large period effects in *Drosophila*. This chapter discusses our recent experiments, which indicate that circadian transcription does indeed contribute to period determination in this system. We propose that cyanobacteria and animal clocks reflect two independent origins of circadian rhythms.

INTRODUCTION

Ron Konopka and Seymour Benzer are justifiably credited with initiating genetic studies of circadian rhythms in *Drosophila melanogaster* in 1971 (Konopka and Benzer 1971). They described three alleles of a single gene they called *period* (*per*), with fast clock properties (per^S), slow clock properties (per^L), and no apparent clock at all (per^0). These compelling phenotypes, coupled with the remarkable and fortuitous unfolding of the recombinant DNA revolution during that same decade, inspired a collaboration between my lab and Jeff Hall's to identify the *period* (*per*) gene and its function. Pranitha Reddy, a graduate student in my laboratory, led the effort to clone *per* DNA, and Will Zhering, a postdoc of Jeff's, led the effort to rescue the arrhythmic *per0* phenotype. This was the first rescue of a behavioral gene in any organism and arguably the first gene rescue of consequence that provided bona fide gene identification (Reddy et al. 1984; Zehring et al. 1984). An independent, parallel effort in Mike Young's laboratory at Rockefeller achieved these two goals at the same time (Bargiello and Young 1984; Bargiello et al. 1984).

In those early days of gene identification, the *period* gene product (PER) was a pioneer protein, meaning that its sequence did not provide strong clues about its function. This statement does not do justice to what is a more complicated tale, as it is more accurate to say that the few clues present in the sequence led us and the Young laboratory in a wrong direction, toward proteoglycans (Shin et al. 1985; Jackson et al. 1986; Reddy et al. 1986). However, a finding made principally by Steve Crews, then a postdoc in the Goodman lab at Stanford, raised a different possibility to me and my colleagues at Brandeis. Steve cloned and sequenced the transcription factor *single-minded* (*sim*), which turned out to have homology with PER (Crews et al. 1988). Although the region in question was small and had no known function or contribution to transcription (the domain was subsequently called PAS after the three founding members of the family: SIM, ARNT, and PER; Reyes et al. 1992), it suggested that PER might function to modulate gene expression at the transcriptional level, a very different role from that of proteoglycans.

We spent the next few years examining both hypotheses: proteoglycan function and the regulation of gene expression. The signature motif in PER that connected it with proteoglycans was a GT (glycine-threonine) repeat region. We discovered, however, that this motif was not well-conserved among *Drosophila* species, in contrast with other highly conserved regions of PER (Colot et al. 1988). More importantly, the motif was not necessary for PER function: Transgenic flies carrying a *period* gene with a deletion of this region were perfectly rhythmic (Yu et al. 1987).

Because these results made us doubt the proteoglycan hypothesis, a coherent gene expression picture began to emerge with the discovery that *period* mRNA levels undergo circadian oscillations under constant darkness (DD) as well as normal (LD) conditions. My postdoc Paul Hardin also showed that the phase and period of the *per* mRNA cycling are sensitive to the missense mutations in the *period*-encoded protein that advance, delay (speed up, slow down), or eliminate behavioral rhythms (Hardin et al. 1990). This *per* mRNA regulation was shown to be predominantly transcriptional (Hardin et al. 1992) and led to the proposal that PER inhibits its own gene expression and that this negative feedback loop is central to circadian timing. In a study designed to distinguish between the proteoglycan and the gene expression hypotheses, we collaborated with the Benzer laboratory to assay PER subcellular localization in fly brains by immunoelectron microscopy. The data indicated that PER was predominantly nuclear (Liu et al. 1992), a result consistent with a direct role for PER in the transcriptional regulation. Indeed, the PAS domain of PER was shown to be a protein–protein interaction motif (Huang et al. 1993), and we imagined that a PAS-containing transcription factor was directly contacted and inhibited by PER.

This idea, as well as the role of the transcriptional feedback loop in circadian rhythms, was strengthened by the identification and cloning of the PAS-domain-containing *Drosophila* transcription factors CLOCK and CYCLE (Allada et al. 1998; Darlington et al. 1998; Rutila et al. 1998). This aspect of the fly story was preceded by the landmark identification and cloning of the mammalian circadian gene CLOCK (Antoch et al. 1997; King et al. 1997). The connections between flies and mammals were also made in that same year by the discovery of mammalian PERs (Sun et al. 1997; Tei et al. 1997). Taken together, the data indicated that a similar feedback system exists in *Drosophila* and mammals, and this situation is still largely true today.

Transcriptional feedback is also a feature of other circadian systems, much more distantly related to flies and mammals. Four years after the Hardin et al. (1990) feedback paper, a *Neurospora* study described mRNA oscillations of the key circadian gene *frequency* (*frq*) influenced by feedback from the FRQ protein (Aronson et al. 1994). In the two other major genetic systems with bona fide circadian rhythms, plants and cyanobacteria, feedback regulation at the level of transcription is prominent, and key transcription factor mutants affect circadian period in both organisms (Dunlap 1999).

The *Drosophila* transcriptional feedback loop is much more complex than this early work suggested. For example, the landmark identification and characterization of the *timeless* (*tim*) gene, the second clock gene identified in the *Drosophila* system, showed that *tim* mRNA also undergoes circadian oscillations and that TIM is a heterodimeric partner of PER and participates in transcriptional feedback regulation (Gekakis et al. 1995; Myers et al. 1995; Sehgal et al. 1995; Zeng et al. 1996). In addition, many kinases participate in modifying PER and TIM and contribute to circadian timing (Kloss et al. 1998; Price et al. 1998; Lin et al. 2002), a phenomenon recognized in the first cycling western blots of PER (Edery et al. 1994). A similar possibility exists in mammals as well, because kinases clearly have a key role in this system (Lowrey et al. 2000; Gallego et al. 2006).

The importance of phosphorylation to the circadian regulation of transcription has increased in prominence because of some breathtaking in vitro experiments in the cyanobacterial system. Recombinant versions of the three clock proteins KaiA, KaiB, and KaiC were incubated in vitro and shown to undergo circadian oscillations. Evident was an approximately 24-hour cycle of protein–protein associations as well as KaiC phosphorylation and dephosphorylation, which are even temperature-compensated (Nakajima et al. 2005; Tomita et al. 2005). The autokinase and autophosphatase activities of KaiC are integral features of this posttranslational timing system (Nishiwaki et al. 2007; Rust et al. 2007; Terauchi et al. 2007). So a current view is that the massive transcriptional regulation that this protein cycle directs in vivo, including transcriptional oscillations of the three Kai mRNAs, has a predominantly output function (downstream from the pacemaker) and/or its effects on timekeeping are relatively subtle compared to the phosphorylation/dephosphorylation cycle intrinsic to the three Kai proteins.

These stunning experiments force a serious consideration of the possibility that transcriptional regulation may be essential only for circadian output in flies and mammals. In other words, the transcriptional feedback loop may serve primarily to drive the oscillation of the hundreds or thousands of output mRNAs that are under clock control. In this view, the circadian transcription of *per* and *tim* also reflects this output feature and may have no more than a minor influence on the central timekeeping machinery. Is this true? Is the posttranscriptional regulation of PER, TIM (or CRY), CLK, and CYC (or BMAL) the sole key to circadian timing?

Although a definitive answer is lacking at present, there are other indications that the answer for the *Drosophila* system may be yes. A key experiment was reported several years ago by Sehgal and colleagues. Behavioral rhythmicity was reported in a strain in which both PER and TIM are generated from constitutive promoters, i.e., without the possibility of transcriptional feedback on these two key circadian genes (Yang and Sehgal 2001). Although behavioral rhythmicity of this strain was poor, it was clearly present. Moreover, PER oscillations could still be detected by antibody staining within brain circadian neurons (Yang and Sehgal 2001). The inescapable conclusion would appear to be that circadian function can take place without transcriptional feedback on *per* and *tim*, which is almost certainly dependent on the normal *per* and *tim* promoters. This focuses attention on the posttranslational regulation of PER and TIM, the interaction and regulation of the multiple kinases and phosphatases that modify these two key clock proteins. This view is congruent not only with the cyanobacterial system, but also with the fact that the strongest mutants in the fly system, those with the greatest period effects, are in the kinases or are suspected phosphorylation substrate mutants in PER and TIM (Kloss et al. 1998; Price et al. 1998; Lin et al. 2002). Nonetheless, recent work from our lab and others indicates that the core transcriptional feedback loop is important not only for overt behavioral rhythmicity (output) in *Drosophila*, but also for proper period determination as well as circadian amplitude.

This view is based in part on a characterization of a CYC-VP16 fusion protein in the *Drosophila* system (S. Kadener et al., in prep.). VP16 is a potent and well-studied viral transcriptional activator. It imparts to the CLK-CYC-VP16 complex enhanced (>5X) transcriptional activity relative to CLK-CYC, which derives most if not all of its transcription activator activity from CLK. This increase is manifest in tissue culture (S2) cells as well as in flies expressing CYC-VP16. These strains also have increased levels of CLK-CYC direct target gene mRNAs as well as a short period, implicating circadian transcription in period determination (S. Kadener et al., in prep.). The results therefore indicate that the level of transcriptional activation on natural promoters in vivo is sensitive to the nature and number of activation regions. The robust behavioral and molecular rhythms of CYC-VP16 flies more generally indicate that CLK-CYC-VP16 circadian function, including the mechanism(s) that temporally activate or repress transcription of this hyperactive complex, must be similar to those that regulate the activity of the wild-type CLK-CYC complex. Because the VP16 activation domain almost certainly functions differently from the CLK poly(Q) region, this indicates that the recruitment of specific activator and/or repressor proteins is unlikely to have a prominent role in the circadian regulation of transcription. A more likely mechanism involves the cyclical inhibition of CLK-CYC DNA binding. Importantly, this notion is consistent with recent chromatin immunoprecipitation results from the mammalian system as well as the fly system (Brown et al. 2005; Yu et al. 2006).

This correlation between increased transcriptional activity of the CLK-CYC complex and period shortening fits with several other pieces of data from the *Drosophila* system. An increase in *per* gene dose leads to flies with short periods. There is a decrease of approximately 0.5 hour for each additional gene copy up to three to four copies, which have an approximately 22–23-hour period (Smith and Konopka 1982; Baylies et al. 1987; Hamblen-Coyle et al. 1992). In addition, a hemizygous deletion that includes *clock* lengthens circadian period by about 0.5 hour (Allada et al. 1998). Although this deletion removes more DNA than just *clock* (including the adjacent clock gene *pdp1*), recent evidence is consistent with the idea that the amount of *clock* mRNA (and level of primary gene transcription) affects circadian period: A transgenic copy of *clock* shortens circadian period of otherwise wild-type flies by about 0.5 hour (J.S. Menet, pers. comm.). These observations are qualitatively similar to those showing the increase in transcription and period shortening caused by expression of CYC-VP16 in flies.

For the reasons described above, the strong effect of CYC-VP16 on period length might be due to a CYC-VP16-mediated change in the timing or level of *per* transcription. To test this possibility, we assayed the period of CYC-VP16-expressing flies in the context of UAS-*per*, i.e., a *period* gene that can be driven by Gal4 but not by CLK-CYC or CLK-CYC-VP16. Importantly, Sehgal and coworkers have previously shown that UAS-PER can rescue the arrhythmic per^{01} genotype with the pan-neuronal *elav*-Gal4 driver (Yang and Sehgal 2001), and we verified this finding (data not shown). Importantly, the *elav*-Gal4 driver in combination with UAS-CYC-VP16 (and a wild-type *per* gene) also manifests the approximately 2-hour period shortening. However, these two transgenes in combination with the UAS-PER and per^{01} fail to appreciably shorten period. This indicates that a major contributor to CLK-CYC-VP16 period shortening is indeed an increase in the levels and/or timing of *per* transcription. We also note the broad distribution of periods in individual flies from genotypes containing the UAS-PER: per^{01} combination compared to the tighter distribution in genotypes containing a *per* promoter; this is an additional indication that the strain without a *per* promoter is highly abnormal and that cyclic *per* transcription contributes to proper period determination.

An additional indication that period shortening is not simply due to an increase in *per* mRNA levels (and PER levels) by the more potent CYC-VP16 molecule is that an increase in PER dose with a UAS-PER transgene slightly increases rather than decreases period (data not shown). This is consistent with literature showing that overexpression of a UAS-*per* transgene does not shorten period (Kaneko et al. 2000; Yang and Sehgal 2001; Nawathean et al. 2007). The simplest interpretation is therefore that the shorter period of CYC-VP16 flies is due to an altered timing of *per* (and *tim*?) transcription. A steeper increase may reflect the enhanced potency of CLK-CYC-VP16, and a steeper decrease may reflect a faster accumulation of active PER repressor. The results more generally suggest that circadian transcription contributes to core circadian function in *Drosophila*.

Yet how are the strains missing the *per* and *tim* promoters rhythmic (Yang and Sehgal 2001)? Because other work in our lab during the past few years has highlighted the importance of brain circuitry to behavioral rhythmicity (Peng et al. 2003; Stoleru et al. 2004, 2005, 2007), we suggest that individual neurons from this per^{01}, *elav*-Gal4; UAS-PER strain might be even more impaired than indicated by the behavioral rhythms of this strain, i.e., circadian brain circuitry might help to compensate for poor core circadian function within individual cells. This is analogous to the superior circadian performance of behavioral rhythmicity and the suprachiasmatic nucleus (SCN) from mutant mouse strains compared to individual tissue culture cells (MEFs) derived from the same mutant strains (Liu et al. 2007). To put this into perspective, we are not certain to what extent individual circadian neurons even from a wild-type strain will manifest free-running rhythms in dissociated culture; to our knowledge, there is no report of such an experiment. On the basis of the importance of the *per* promoter in our recent CYC-VP16 experiments (S. Kadener et al., in prep.), a prediction is that individual dissociated neurons from the *per* promoterless strain will be substantially more impaired than wild-type neurons.

The conclusion that the core *Drosophila* transcriptional feedback loop is important not only for overt rhythmicity (output), but also for circadian amplitude was previously suggested and based on a different mutation in the core clock gene *clock*, Clk^{ar} (Allada et al. 2003). The mutation in this transcription factor gives rise to weak CLK activity, and homozygous Clk^{ar} flies have a reduced transcriptional amplitude of oscillating direct CLK-target mRNAs; this strain is also arrhythmic, even in LD (Allada et al. 2003).

The wider conclusion of the CYC-VP16 studies, the importance of the core transcriptional loop to period determination as well as circadian amplitude, is reinforced by our recent identification and characterization of a new clock gene, *clockwork orange* (*cwo*) (Kadener et al. 2007). It encodes a transcriptional repressor that synergizes with PER and inhibits CLK-mediated activation. Consistent with this function, the mRNA profiles of CLK–direct target genes manifest lower-amplitude oscillations in mutant flies, due predominantly to higher trough values. Because rhythmicity fails to persist in DD and there is little or no effect on average mRNA levels in the *cwo*-deficient strain, transcriptional oscillation amplitude appears to be linked to rhythmicity. The *cwo* mutant flies are long period before they become arrhythmic, consistent with delayed repression indicated by the RNA profiles. These findings and others suggest that CWO synergizes with PER to help terminate CLK-CYC-mediated transcription of direct target genes in the late night, a function that contributes to period determination as well as to rhythmicity. In this context of transcription and rhythmicity, it is interesting to note that all five of the validated CLK-CYC–direct target genes (*per, tim, vri, pdp1, cwo*) are known or suspected transcription factors.

We note that *cwo* was identified and characterized independently in two other *Drosophila* studies (Lim et al. 2007; Matsumoto et al. 2007). Finally, the presence of two *cwo* orthologs in mammals (*Dec1* and *Dec2*) suggests a similar synergistic repression mechanism in this system, which would then be a collaboration between DEC and the CRY-PER repressor (Honma et al. 2002).

There is, however, an additional possibility for *cwo* function, as well as for the likely mammalian orthologs *Dec1* and *Dec2*. This is based on the many new candidate CLK-target genes in fly heads identified by the CLK-GR strategy (Fig. 1A,B). Intriguingly, a significant fraction of these genes are nonoscillating based on microarray studies of fly head RNA (Kadener et al. 2007). However, the S2 cell assays predict that most of them are bona fide CLK targets rather than a CLK-GR artifact (Fig. 1C,D); they may predominantly reflect *clk* function in noncircadian cell types. Indeed, a recent study reported that CLK expression is not restricted to circadian neurons in the fly brain (Houl et al. 2006).

This idea of noncircadian cell CLK function follows from the characterization of these new CLK–direct target genes (Kadener et al. 2007). For example, a clue to the difference between these new genes and canonical target genes is their response to the CWO repressor protein. In our hands, canonical clock genes respond to CWO addition but with only a modest reduction in CLK-CYC-mediated transcription (Kadener et al. 2007). This is exemplified by the *luc* reporter gene driven by the *tim* promoter (Fig. 2). (We suspect that the different results of Ueda and colleagues [Matsumoto et al. 2007], the apparent strong repression of CWO addition, is due to a different experimental protocol.) However, CWO works very effectively as a CLK-CYC inhibitor-competitor when it is added along with a small and limiting amount of PER (Fig. 2).

Our interpretation is that CWO is effective in collaborating with PER, because PER reduces the ability of CLK-CYC to bind to DNA (perhaps by promoting CLK phosphorylation; Yu et al. 2006). CWO can then compete effectively with the PER-weakened CLK-CYC complex for E-box access. In the absence of PER, CLK-CYC has a higher binding constant to canonical clock gene E boxes, so CWO does not compete well with CLK-CYC for access to canonical clock gene E boxes.

In contrast, the new direct target genes are better inhib-

A.

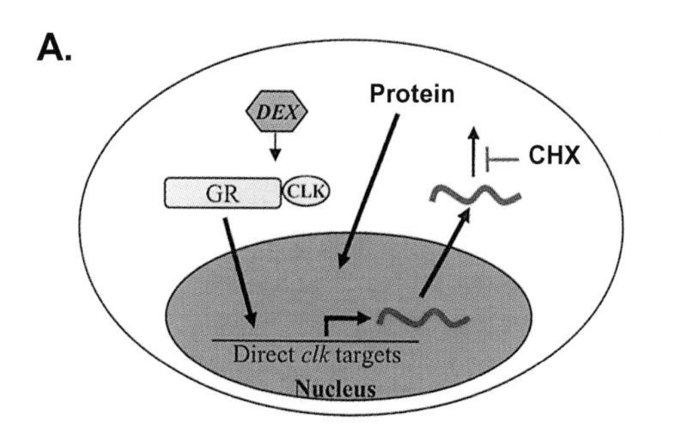

B.

Gene	TGT	Gene	TGT
tim	0.83	CG17100	0.25
CG15095	0.72	Picot	0.24
tim	0.65	CG3850	0.24
HLHm3	0.36	CG15096	0.23
per	0.36	CG6231	0.23
CG30035	0.33	CG18135	0.21
CG3348	0.33	GM02743	0.20
CG3764	0.32	EP2237	0.19
HLHmbeta	0.30	CG6767	0.19
vri	0.29	CG32170	0.18
CG4798	0.27	CG13624	0.17
CG10999	0.26	Mmp1	0.17
CG8008	0.26	CG13868	0.15
Pdp1	0.25	CG11086	0.13

C.

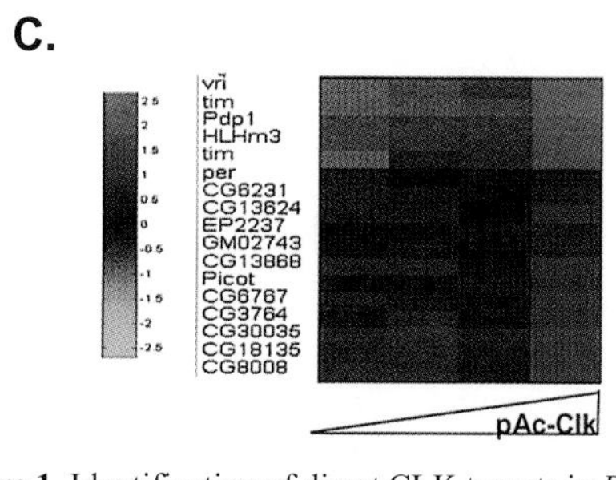

D.

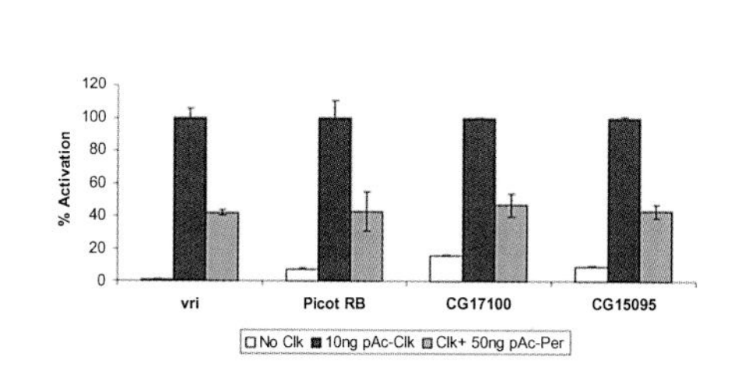

Figure 1. Identification of direct CLK targets in *Drosophila* S2 cells and fly heads. (*A*) Diagram illustrating the approach for the identification of direct CLK targets from *Drosophila* S2 cells and fly heads. (Dex) Dexamethasone; (GR) ligand-binding domain of the glucocorticoid receptor; (CHX) cycloheximide. (*B*) Top 28 direct CLK targets identified by the approach described in *A*. (TGT) Targetness; this index was obtained by averaging the relative stimulation by dexamethasone from S2 cells and fly heads. (*C*) CLK protein expression activates most of the direct CLK-GR targets. Transient transfections were performed with varying amounts of pAc-*Clk* plasmid in S2 cells (0, 10, 30, and 100 ng). After 48 hours, cells were harvested and total RNA isolated. Microarray analysis was performed using the *Drosophila* 2.0 genomic Affymetrix chips. (*D*) Effect of CLK and PER expression on *vri*-Luc, *picot*-Luc, *CG15095*-Luc, and *CG17100*-Luc reporters on S2 cells. pAc-*Clk* and pAc-*per* refers to CLK- and PER-expressing plasmids, respectively. In all cases, cotransfection with pCopia-Renilla luciferase was performed to normalize for cell number, transfection efficiency, and general transcription effects.

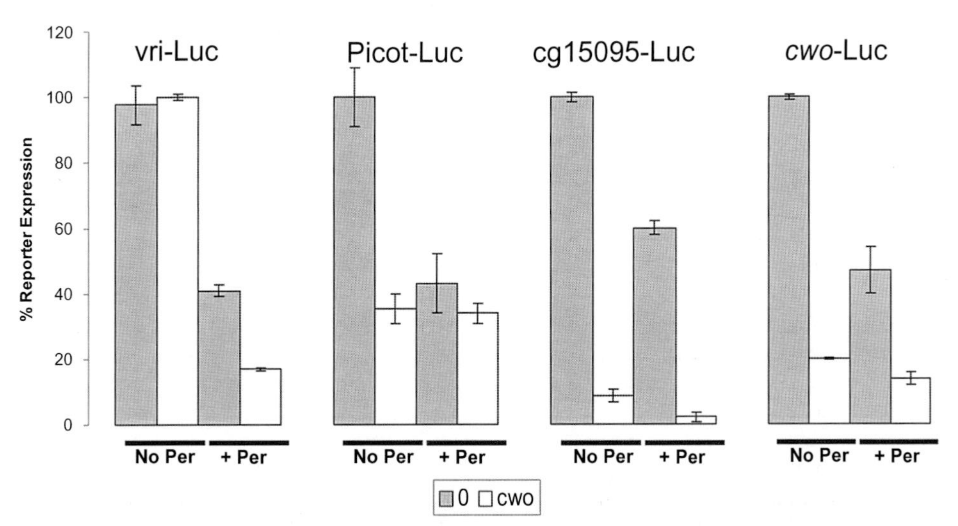

Figure 2. Effect of PER and CWO expression on CLK-mediated transcriptional activation. "No Per" indicates that 10 ng of pAc-*Clk* was cotransfected with 0 or 100 ng of pAc-*cwo* and the corresponding reporters. "+ Per" indicates that the transfection was performed as before, except that 50 ng of pAc-*per* was also cotransfected. A representative experiment is shown. (Duplicates for each condition were performed.)

ited by CWO alone under identical conditions (Fig. 2). Perhaps the E-box arrangements of these new CLK-target genes are different from those associated with canonical clock gene promoters. For example, CLK-CYC molecules may interact and therefore form very stable multimers on canonical clock gene E boxes (there appear to be closely spaced, multiple E boxes associated with the canonical clock gene promoters that are well-characterized), whereas they may interact less well with other E-box arrangements. At the risk of extending this speculation further, the argument suggests that CWO is of intermediate affinity, lower than that of CLK-CYC on optimally arranged E boxes, but higher than that of CLK-CYC on the other E-box arrangements. CWO would then serve to suppress transcription of the new direct target genes in clock cells.

To test this hypothesis, we manipulated the complex arrangement of E boxes and near-E boxes (*ter* boxes) within the *tim* promoter (Fig. 3). The deletion constructs with simpler arrangements of E boxes (fewer and/or fur-

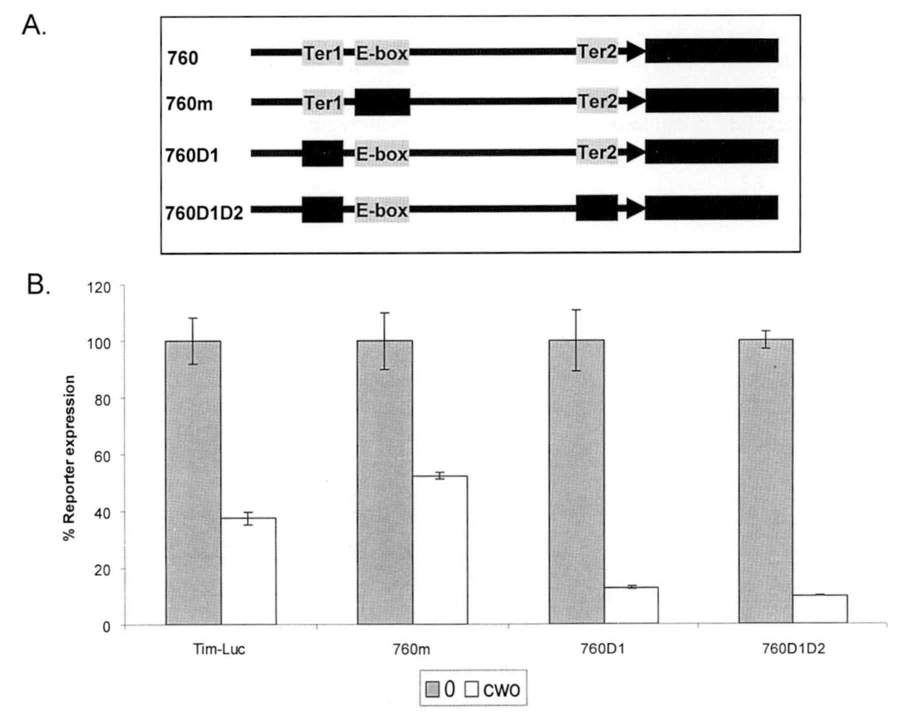

Figure 3. Different E-box arrangements alter the sensitivity of the *tim* promoter to *cwo* expression. (*A*) Schematic of the reporters used. (*B*) Effect of *cwo* expression (100 ng of pAc-*cwo*) on reporter expression. All experiments were performed in presence of 100 ng of pAc-*per*.

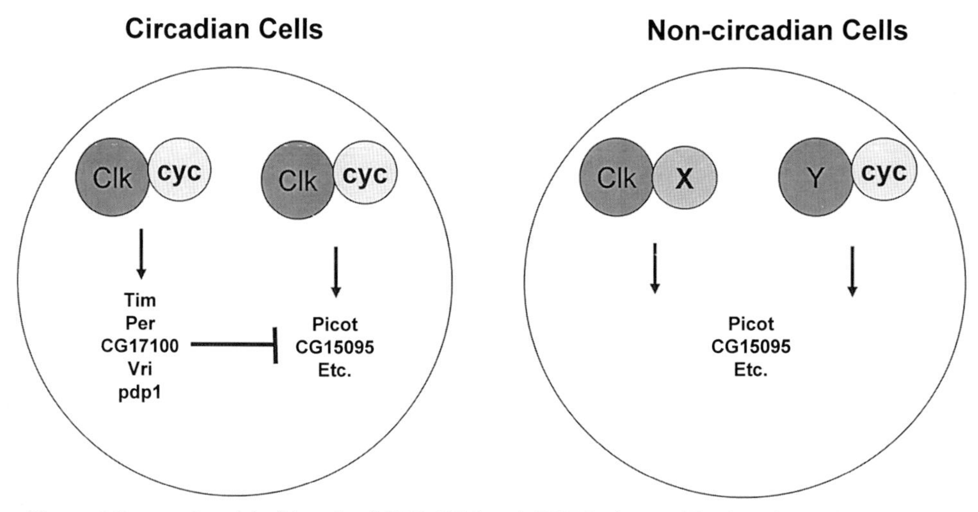

Figure 4. Proposed model of the role of CLK, CYC, and CWO in the specification of circadian cell types.

ther apart) are generally more sensitive to CWO. This suggests that it does indeed repress promoters that are not optimally designed to work with CLK-CYC. CWO therefore has a less restrictive E-box specificity and may therefore contribute to clock cell specification by preventing CLK from expressing inappropriate genes in bona fide clock cells (Fig. 4). We therefore suggest that the increased expression of the new direct target genes in the *cwo* mutant strain (Fig. 5) is due to significantly increased expression in circadian cells; the modest increase (Fig. 5) would then reflect unchanged levels in nonclock cells and dramatically increased clock cell levels. These are kept very low by CWO repression in wild-type strains.

These ideas fit with some of our previous work showing that ectopic CLK expression generates circadian clocks at additional locations within the brain (Zhao et al. 2003). The new insights into transcriptional circuitry presented here suggest that these additional locations are normal sites of CYC expression and that the induction of CLK in these cells creates clock neuron identity in part through the induction of CLK-CYC–direct target gene proteins including CWO and its repression of inappropriate transcription.

This role of CLK in clock cell specification is analogous to the role of PAX6 in eye specification and begs the question: How did a gene involved in intracellular circadian circuitry acquire a role in cell-type specification? We suggest that the answer is also similar to PAX6, which influences the transcription of rhodopsin and has a role in eye specification. Presumably, an ancient PAX6 molecule had a role in the transcription of one or more rhodopsin progenitors in single-cell organisms before multicellularity arose some 600 million years ago. Added later to this initial function was a more complex tissue specification role with the recruitment of other direct target genes as well as regulatory loops (Pichaud et al. 2001). We imag-

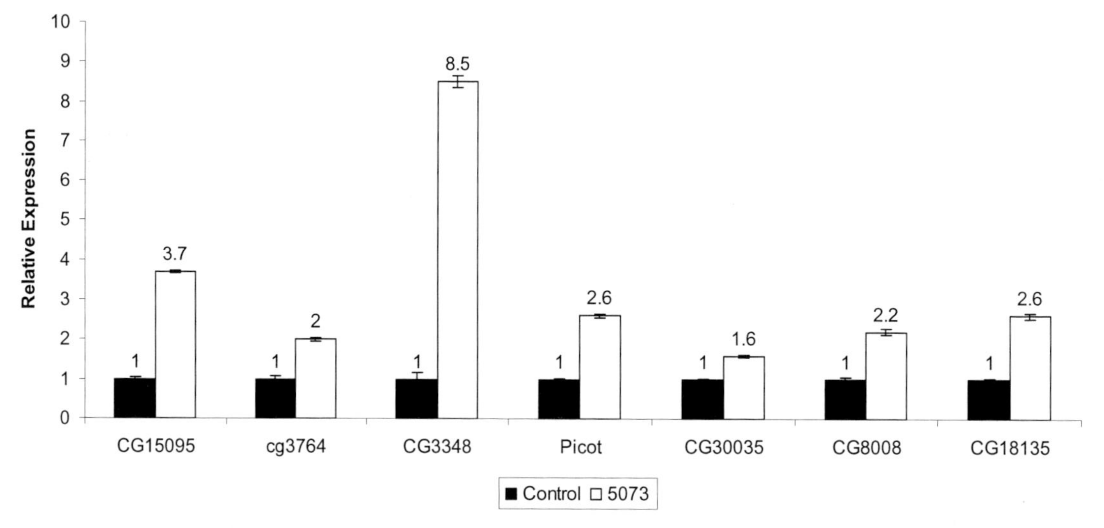

Figure 5. Noncircadian CLK–direct target levels are significantly higher in *cwo*-deficient flies. The expression value for each gene was obtained by averaging the microarray-based expression level across six time points. The values were then normalized to the ones obtained in control flies.

ine a similar early role in circadian rhythmicity for an ancient CLK-CYC heterodimer and its primary targets that we suspect were already present and functioning long ago in the clocks of our single-cell ancestors. More complex regulatory features, including those involving brain circuitry, were added more recently.

These evolutionary speculations suggest that circadian clocks have arisen at least twice in evolution: once in animal progenitors and once in cyanobacteria progenitors. (There are good arguments that circadian clocks have arisen more than twice, but it is best to keep things simple here.) Separate origins are supported by the total lack of sequence conservation between cyanobacterial and animal circadian clock proteins. It is also consistent with what appear to be different cellular properties of the two systems: Individual bacterial cells keep excellent circadian time, essentially indistinguishable from a bacterial culture (Mihalcescu et al. 2004), whereas individual eukaryotic cells (separated SCN cells, for example) show substantially more variation in period than an intact SCN or the organism (Welsh et al. 1995). Consistent with this view, animal clock properties appear to be more generally dependent on system or network properties (Liu et al. 2007). This difference in single-cell precision and intercellular communication echoes two additional differences between the systems: (1) rhythmic transcriptional oscillations are global in cyanobacteria, because most if not all genes are controlled by the same fundamental mechanism (Liu et al. 1995; Woelfle and Johnson 2006); in contrast, animal genes under circadian transcriptional regulation are regulated by different factors with only a minority apparently affected directly by the CLK-CYC heterodimer (McDonald and Rosbash 2001). (2) The KaiC kinase and phosphatase appear to be unique and of singular importance to timekeeping (McClung 2007; Terauchi et al. 2007). In contrast, animal rhythms have recruited several common enzymes involved in many other cellular processes (Kloss et al. 1998; Price et al. 1998; Lin et al. 2002). On the basis of the work summarized here as well, the circadian clockworks that govern intracellular animal rhythms may use a more equitable and complicated division of labor between gene expression and posttranslational regulatory mechanisms than the cyanobacterial clockworks. The intricacies of gene expression regulation may even explain why the strongest period mutations are in kinase genes, i.e., gene expression feedback loops may be more highly regulated (buffered) and therefore less easy to respond to single-gene mutations than mutations in kinases that affect clock protein turnover. We speculate that these differences between animal and cyanobacterial clocks reflect their independent evolutionary origins as well as the development of multicellularity.

ACKNOWLEDGMENTS

We thank the many past members of the Rosbash lab who made invaluable contributions to the development of the fly clock system as well as to the success of our laboratory more generally. These also include some outstanding administrators, especially Heather Felton and Lise-Anne Monaghan. We are also grateful for support from the National Institutes of Health and the Howard Hughes Medical Institute.

REFERENCES

Allada R., Kadener S., Nandakumar N., and Rosbash M. 2003. A recessive mutant of *Drosophila Clock* reveals a role in circadian rhythm amplitude. *EMBO J.* **22:** 3367.

Allada R., White N., So W., Hall J., and Rosbash M. 1998. A mutant *Drosophila* homolog of mammalian *Clock* disrupts circadian rhythms and transcription of *period* and *timeless*. *Cell* **93:** 791.

Antoch M.P., Song E.-J., Chang A.-M., Vitaterna M.H., Zhao Y., Wilsbacher L.D., Sangoram A.M., King D.P., Pinto L.H., and Takahashi J.S. 1997. Functional identification of the mouse circadian *Clock* gene by transgenic BAC rescue. *Cell* **89:** 655.

Aronson B.D., Johnson K.A., Loros J.J., and Dunlap J.C. 1994. Negative feedback defining a circadian clock: Autoregulation of the clock gene frequency. *Science* **263:** 1578.

Bargiello T.A. and Young M.W. 1984. Molecular genetics of a biological clock in *Drosophila*. *Proc. Natl. Acad. Sci.* **81:** 2142.

Bargiello T.A., Jackson F.R., and Young M.W. 1984. Restoration of circadian behavioural rhythms by gene transfer in *Drosophila*. *Nature* **312:** 752.

Baylies M.K., Bargiello T.A., Jackson F.R., and Young M.W. 1987. Changes in abundance or structure of the *per* gene product can alter periodicity of the *Drosophila* clock. *Nature* **326:** 390.

Brown S.A., Ripperger J., Kadener S., Fleury-Olela F., Vilbois F., Rosbash M., and Schibler U. 2005. PERIOD1-associated proteins modulate the negative limb of the mammalian circadian oscillator. *Science* **308:** 693.

Colot H.V., Hall J.C., and Rosbash M. 1988. Interspecific comparison of the *period* gene of *Drosophila* reveals large blocks of non-conserved coding DNA. *EMBO J.* **7:** 3929.

Crews S.T., Thomas J.B., and Goodman C.S. 1988. The *Drosophila single-minded* gene encodes a nuclear protein with sequence similarity to the *per* gene product. *Cell* **52:** 143.

Darlington T.K., Wager-Smith K., Ceriani M.F., Staknis D., Gekakis N., Steeves T.D., Weitz C.J., Takahashi J.S., and Kay S.A. 1998. Closing the circadian loop: CLOCK-induced transcription of its own inhibitors *per* and *tim*. *Science* **280:** 1599.

Dunlap J.C. 1999. Molecular bases for circadian clocks. *Cell* **96:** 271.

Edery I., Zwiebel L.J., Dembinska M.E., and Rosbash M. 1994. Temporal phosphorylation of the *Drosophila* period protein. *Proc. Natl. Acad. Sci.* **91:** 2260.

Gallego M., Eide E.J., Woolf M.F., Virshup D.M., and Forger D.B. 2006. An opposite role for tau in circadian rhythms revealed by mathematical modeling. *Proc. Natl. Acad. Sci.* **103:** 10618.

Gekakis N., Saez L., Delahaye-Brown A.-M., Myers M.P., Sehgal A., Young M.W., and Weitz C.J. 1995. Isolation of timeless by PER protein interaction: Defective interaction between timeless protein and long-period mutant PERL. *Science* **270:** 811.

Hamblen-Coyle M.J., Wheeler D.A., Rutila J.E., Rosbash M., and Hall J.C. 1992. Behavior of period-altered circadian rhythm mutants of *Drosophila* in light: Dark cycles (*Diptera: Drosophilidae*). *J. Insect Behav.* **5:** 417.

Hardin P.E., Hall J.C., and Rosbash M. 1990. Feedback of the *Drosophila* period gene product on circadian cycling of its messenger RNA levels. *Nature* **343:** 536.

———. 1992. Circadian oscillations in period gene mRNA levels are transcriptionally regulated. *Proc. Natl. Acad. Sci.* **89:** 11711.

Honma S., Kawamoto T., Takagi Y., Fujimoto K., Sato F., Noshiro M., Kato Y., and Honma K. 2002. *Dec1* and *Dec2* are regulators of the mammalian molecular clock. *Nature* **419:** 841.

Houl J.H., Yu W., Dudek S.M., and Hardin P.E. 2006. *Drosophila* CLOCK is constitutively expressed in circadian oscillator and non-oscillator cells. *J. Biol. Rhythms* **21:** 93.

Huang Z.J., Edery I., and Rosbash M. 1993. PAS is a dimerization domain common to *Drosophila* Period and several transcription factors. *Nature* **364:** 259.

Jackson F.R., Bargiello T.A., Yun S.-H., and Young M.W. 1986. Product of *per* locus of *Drosophila* shares homology with proteoglycans. *Nature* **320:** 185.

Kadener S., Stoleru D., McDonald M., Nawathean P., and Rosbash M. 2007. *Clockwork Orange* is a transcriptional repressor and a new *Drosophila* circadian pacemaker component. *Genes Dev.* **21:** 1675.

Kaneko M., Park J., Cheng Y., Hardin P., and Hall J. 2000. Disruption of synaptic transmission or clock-gene-product oscillations in circadian pacemaker cells of *Drosophila* cause abnormal behavioral rhythms. *J. Neurobiol.* **43:** 207.

King D.P., Zhao Y., Sangoram A.M., Wilsbacher L.D., Tanaka M., Antoch M.P., Steeves T.D., Vitaterna M.H., Kornhauser J.M., Lowrey P.L., Turek F.W., and Takahashi J.S. 1997. Positional cloning of the mouse circadian *clock* gene. *Cell* **89:** 641.

Kloss B., Price J.L., Saez L., Blau J., Rothenfluh-Hilfiker A., Wesley C.S., and Young M.W. 1998. The *Drosophila* clock gene *double-time* encodes a protein closely related to human casein kinase Iε. *Cell* **94:** 97.

Konopka R.J. and Benzer S. 1971. Clock mutants of *Drosophila melanogaster*. *Proc. Natl. Acad. Sci.* **68:** 2112.

Lim C., Chung B.Y., Pitman J.L., McGill J.J., Pradhan S., Lee J., Keegan K.P., Choe J., and Allada R. 2007. *clockwork orange* encodes a transcriptional repressor important for circadian-clock amplitude in *Drosophila*. *Curr. Biol.* **17:** 1082.

Lin J.M., Kilman V.L., Keegan K., Paddock B., Emery-Le M., Rosbash M., and Allada R. 2002. A role for casein kinase 2α in the *Drosophila* circadian clock. *Nature* **420:** 816.

Liu A.C., Welsh D.K., Ko C.H., Tran H.G., Zhang E.E., Priest A.A., Buhr E.D., Singer O., Meeker K., Verma I.M., Doyle F.J., III, Takahashi J.S., and Kay S.A. 2007. Intercellular coupling confers robustness against mutations in the SCN circadian clock network. *Cell* **129:** 605.

Liu X., Zwiebel L.J., Hinton D., Benzer S., Hall J.C., and Rosbash M. 1992. The *period* gene encodes a predominantly nuclear protein in adult *Drosophila*. *J. Neurosci.* **12:** 2735.

Liu Y., Tsinoremas N.F., Johnson C.H., Lebedeva N.V., Golden S.S., Ishiura M., and Kondo T. 1995. Circadian orchestration of gene expression in cyanobacteria. *Genes Dev.* **9:** 1469.

Lowrey P.L., Shimomura K., Antoch M.P., Yamazaki S., Zemenides P.D., Ralph M.R., Menaker M., and Takahashi J.S. 2000. Positional syntenic cloning and functional characterization of the mammalian circadian mutation *tau*. *Science* **288:** 483.

Matsumoto A., Ukai-Tadenuma M., Yamada R.G., Houl J., Uno K.D., Kasukawa T., Dauwalder B., Itoh T.Q., Takahashi K., Ueda R., Hardin P.E., Tanimura T., and Ueda H.R. 2007. A functional genomics strategy reveals *clockwork orange* as a transcriptional regulator in the *Drosophila* circadian clock. *Genes Dev.* **21:** 1687.

McClung C.R. 2007. A cyanobacterial circadian clock is based on the intrinsic ATPase activity of KaiC. *Proc. Natl. Acad. Sci.* **104:** 16727.

McDonald M.J. and Rosbash M. 2001. Microarray analysis and organization of circadian gene expression in *Drosophila*. *Cell* **107:** 567.

Mihalcescu I., Hsing W., and Leibler S. 2004. Resilient circadian oscillator revealed in individual cyanobacteria. *Nature* **430:** 81.

Myers M.P., Wager-Smith K., Wesley C.S., Young M.W., and Sehgal A. 1995. Positional cloning and sequence analysis of the *Drosophila* clock gene, *timeless*. *Science* **270:** 805.

Nakajima M., Imai K., Ito H., Nishiwaki T., Murayama Y., Iwasaki H., Oyama T., and Kondo T. 2005. Reconstitution of circadian oscillation of cyanobacterial KaiC phosphorylation in vitro. *Science* **308:** 414.

Nawathean P., Stoleru D., and Rosbash M. 2007. A small conserved domain of *Drosophila* PERIOD is important for circadian phosphorylation, nuclear localization, and transcriptional repressor activity. *Mol. Cell. Biol.* **27:** 5002.

Nishiwaki T., Satomi Y., Kitayama Y., Terauchi K., Kiyohara R., Takao T., and Kondo T. 2007. A sequential program of dual phosphorylation of KaiC as a basis for circadian rhythm in cyanobacteria. *EMBO J.* **26:** 4029.

Peng Y., Stoleru D., Levine J.D., Hall J.C., and Rosbash M. 2003. *Drosophila* free-running rhythms require intercellular communication. *PLoS Biol.* **1:** E13.

Pichaud F., Treisman J., and Desplan C. 2001. Reinventing a common strategy for patterning the eye. *Cell* **105:** 9.

Price J.L., Blau J., Rothenfluh-Hilfiker A., Abodeely M., Kloss B., and Young M.W. 1998. *double-time* is a novel *Drosophila* clock gene that regulates PERIOD protein accumulation. *Cell* **94:** 83.

Reddy P., Jacquier A.C., Abovich N., Petersen G., and Rosbash M. 1986. The *period* clock locus of *D. melanogaster* codes for a proteoglycan. *Cell* **46:** 53.

Reddy P., Zehring W.A., Wheeler D.A., Pirrotta V., Hadfield C., Hall J.C., and Rosbash M. 1984. Molecular analysis of the *period* locus in *Drosophila melanogaster* and identification of a transcript involved in biological rhythms. *Cell* **38:** 701.

Reyes H., Reisz-Porszasz S., and Hankinson O. 1992. Identification of the Ah receptor nuclear translocator protein (Arnt) as a component of the DNA binding form of the Ah receptor. *Science* **256:** 1193.

Rust M.J., Markson J.S., Lane W.S., Fisher D.S., and O'Shea E.K. 2007. Ordered phosphorylation governs oscillation of a three-protein circadian clock. *Science* **318:** 809.

Rutila J.E., Suri V., Le M., So W.V., Rosbash M., and Hall J.C. 1998. CYCLE is a second bHLH-PAS clock protein essential for circadian rhythmicity and transcription of *Drosophila period* and *timeless*. *Cell* **93:** 805.

Sehgal A., Rothenfluh-Hilfiker A., Hunter-Ensor M., Chen Y., Myers M., and Young M.W. 1995. Rhythmic expression of *timeless*: A basis for promoting circadian cycles in *period* gene autoregulation. *Science* **270:** 808.

Shin H.-S., Bargiello T.A., Clark B.T., Jackson F.R., and Young M.W. 1985. An unusual coding sequence from a *Drosophila* clock gene is conserved in vertebrates. *Nature* **317:** 445.

Smith R.F. and Konopka R.J. 1982. Effects of dosage alterations at the *per* locus on the period of the circadian clock of *Drosophila*. *Mol. Gen. Genet.* **185:** 30.

Stoleru D., Peng Y., Agosto J., and Rosbash M. 2004. Coupled oscillators control morning and evening locomotor behaviour of *Drosophila*. *Nature* **431:** 862.

Stoleru D., Peng Y., Nawathean P., and Rosbash M. 2005. A resetting signal between *Drosophila* pacemakers synchronizes morning and evening activity. *Nature* **438:** 238.

Stoleru D., Nawathean P., de la Paz Fernández M., Menet J.S., Ceriani M.F., and Rosbash M. 2007. The *Drosophila* circadian network is a seasonal timer. *Cell* **129:** 207.

Sun Z.S., Albrecht U., Zhuchenko O., Bailey J., Eichele G., and Lee C.C. 1997. *RIGUI,* a putative mammalian ortholog of the *Drosophila period* gene. *Cell* **90:** 1003.

Tei H., Okamura H., Shigeyoshi Y., Fukuhara C., Ozawa R., Hirose M., and Sakaki Y. 1997. Circadian oscillation of a mammalian homologue of the *Drosophila period* gene. *Nature* **389:** 512.

Terauchi K., Kitayama Y., Nishiwaki T., Miwa K., Murayama Y., Oyama T., and Kondo T. 2007. ATPase activity of KaiC determines the basic timing for circadian clock of cyanobacteria. *Proc. Natl. Acad. Sci.* **104:** 16377.

Tomita J., Nakajima M., Kondo T., and Iwasaki H. 2005. No transcription-translation feedback in circadian rhythm of KaiC phosphorylation. *Science* **307:** 251.

Welsh D.K., Logothetis D.E., Meister M., and Reppert S.M. 1995. Individual neurons dissociated from rat suprachiasmatic nucleus express independently phased circadian firing rhythms. *Neuron* **14:** 697.

Woelfle M.A. and Johnson C.H. 2006. No promoter left behind: Global circadian gene expression in cyanobacteria. *J. Biol. Rhythms* **21:** 419.

Yang Z. and Sehgal A. 2001. Role of molecular oscillations in generating behavioral rhythms in *Drosophila*. *Neuron* **29:** 453.

Yu Q., Colot H.V., Kyriacou C.P., Hall J.C., and Rosbash M. 1987. Behaviour modification by *in vitro* mutagenesis of a variable region within the *period* gene of *Drosophila*. *Nature* **326:** 765.

Yu W., Zheng H., Houl J.H., Dauwalder B., and Hardin P.E. 2006. PER-dependent rhythms in CLK phosphorylation and E-box binding regulate circadian transcription. *Genes Dev.* **20:** 723.

Zehring W.A., Wheeler D.A., Reddy P., Konopka R.J., Kyriacou C.P., Rosbash M., and Hall J.C. 1984. P-element transformation with *period* locus DNA restores rhythmicity to mutant, arrhythmic *Drosophila melanogaster*. *Cell* **39:** 369.

Zeng H., Qian Z., Myers M.P., and Rosbash M. 1996. A light-entrainment mechanism for the *Drosophila* circadian clock. *Nature* **380:** 129.

Zhao J., Kilman V.L., Keegan K.P., Peng Y., Emery P., Rosbash M., and Allada R. 2003. *Drosophila* clock can generate ectopic circadian clocks. *Cell* **113:** 755.

Genetic and Molecular Analysis of the Central and Peripheral Circadian Clockwork of Mice

E.S. Maywood,* J.S. O'Neill,* A.B. Reddy,* J.E. Chesham,* H.M. Prosser,†
C.P. Kyriacou,‡ S.I.H. Godinho,§ P.M. Nolan,§ and M.H. Hastings*

*Division of Neurobiology, MRC Laboratory of Molecular Biology, Cambridge CB2 0QH, United Kingdom; †The Wellcome Trust Sanger Institute, Wellcome Trust Genome Campus, Hinxton, Cambridge CB10 1SA, United Kingdom; ‡Department of Genetics, University of Leicester, Leicester, LE1 7RH, United Kingdom; §MRC Mammalian Genetics Unit, Harwell, Oxfordshire OX11 0RD, United Kingdom

A hierarchy of interacting, tissue-based clocks controls circadian physiology and behavior in mammals. Preeminent are the suprachiasmatic nuclei (SCN): central hypothalamic pacemakers synchronized to solar time via retinal afferents and in turn responsible for internal synchronization of other clocks present in major organ systems. The SCN and peripheral clocks share essentially the same cellular timing mechanism. This consists of autoregulatory transcriptional/posttranslational feedback loops in which the *Period* (*Per*) and *Cryptochrome* (*Cry*) "clock" genes are negatively regulated by their protein products. Here, we review recent studies directed at understanding the molecular and cellular bases to the mammalian clock. At the cellular level, we demonstrate the role of F-box protein Fbxl3 (characterized by the afterhours mutation) in directing the proteasomal degradation of Cry and thereby controlling negative feedback and circadian period of the molecular loops. Within SCN neural circuitry, we describe how neuropeptidergic signaling by VIP synchronizes and sustains the cellular clocks. At the hypothalamic level, signaling via a different SCN neuropeptide, prokineticin, is not required for pacemaking but is necessary for control of circadian behavior. Finally, we consider how metabolic pathways are coordinated in time, focusing on liver function and the role of glucocorticoid signals in driving the circadian transcriptome and proteome.

INTRODUCTION

The previous decade has witnessed major, indeed astonishing, advances in our understanding of the molecular genetic and cellular bases to circadian timing in mammals (Reppert and Weaver 2002; Lowrey and Takahashi 2004). The SCN of the hypothalamus were first revealed as the "body clock" more than 30 years ago (Weaver 1998). They remain preeminent as the pacemaker responsible for coordinating circadian physiology across the organism and synchronizing it to solar time by retinally mediated entrainment. But in addition to the SCN, we now appreciate that most major organ systems and diverse cell types contain local circadian clockworks and that SCN-dependent synchronization of these local clocks and their dependent transcriptomes orchestrates daily physiology (Akhtar et al. 2002; Panda et al. 2002). The SCN are therefore no longer thought of as coercing a passive periphery into rhythmic behavior; rather, they harness the intrinsic rhythmicity of organs, both sustaining their amplitude and setting their phase. This precise and elaborate temporal coordination of physiology underlies the marked daily prevalence of morbidity in cardiovascular, metabolic, and other diseases (Hastings et al. 2003). Moreover, circadian disturbance, be it environmental or genetic, is linked to malignancy and metabolic and mental illness. Understanding how the SCN operate as a pacemaker tissue and how their output coordinates circadian physiology are outstanding questions with considerable relevance to human health. They also provide a cardinal example of how regulated gene expression can control complex behavior—the Holy Grail of molecular neurobiology. The purpose of this chapter is to review recent studies from our laboratories that have addressed these issues. Our broad strategy has been to use real-time imaging of circadian gene expression to explore molecular timekeeping in genetically modified mice. We offer specific examples illustrating how the period of the central and peripheral clocks is determined by the rate of proteasomal degradation of circadian proteins, how neuropeptidergic signaling synchronizes and sustains cellular circadian pacemaking in the SCN, and how behavioral control by the SCN is mediated via neuropeptides not required for pacemaking but necessary for downstream signaling. Finally, we consider how SCN outputs coordinate metabolically relevant pathways, focusing on liver function. It is likely that the relevance of clocks to medicine will expand far beyond sleep disorders to encompass metabolic disturbances of many kinds. Understanding the cellular and molecular genetic bases of peripheral circadian physiology is therefore a strategic aim of the field.

SETTING CELLULAR CIRCADIAN PERIOD BY REGULATED PROTEASOMAL DEGRADATION

The cellular oscillator of the SCN is viewed as a series of interlocked transcriptional/posttranslational feedback loops (Reppert and Weaver 2002; Lowrey and Takahashi 2004). Expression of the genes encoding the negative regulators Cryptochrome (Cry) and Period (Per) is *trans*-activated at the start of the circadian day by complexes containing the positive factors, Clock and Bmal1, that act via E-box *cis*-regulatory elements. This is followed, with a delay, by progressive accumulation of Per and Cry proteins in the nucleus of SCN neurons (Kume et al. 1999; Field et al. 2000). By the beginning of circadian night, Per and Cry protein abundance is maximal, and negative regulation of the *Per* and *Cry* genes is evidenced by declining levels of the relevant mRNAs, which fall to a nadir by the end of circadian night. In the absence of further transcriptional activity, remaining Per and Cry proteins decline, again tracking the mRNA profile with a delay of about 4 hours. As Per and Cry protein levels reach a minimum around circadian dawn, negative regulation of their genes is released and the cycle starts anew. This core cycle is stabilized by accessory loops involving rhythmic expression of Rev-Erbα the clearance of which at the end of circadian night activates *Bmal1* (by derepression), augmenting the renewed surge in E-box-mediated drive.

The molecular details of these loops, including the nature of the interactions among the proteins, DNA regulatory sequences, and histone modification are described elsewhere in this volume, but a central question remains: What sets the period of the clock to 24 hours? Clearly, this is an emergent property conferred by various factors and processes. For example, a mutation that compromises the *trans*-activational function of Clock slows period to about 28 hours, whereas mutations of Per2 or casein kinase I (CKI) that alter Per2 stability accelerate period to about 20 hours. To identify further determinants of circadian period, and by implication components of the core molecular clockwork, a dominant *N*-ethyl-*N*-nitrosourea (ENU) mutagenesis screen for period mutations is being conducted in mice (Bacon et al. 2004). In G_1 progeny of ENU-mutagenized animals, a single founder animal was identified with a significantly lengthened period (~0.4 hours) of its rest-activity cycle. Inheritance testing and intercrossing revealed that the mutation, called "after hours" (*Afh*), was semidominant with a period of about 27 hours in homozygotes (Fig. 1A). To test the effect of the mutation on the core molecular loop, *Afh* mice were crossed into the Per2::Luciferase reporter line (Yoo et al. 2004), and bioluminescence was recorded in SCN organotypical slice cultures. The prolongation of period seen for the behavioral rhythm was also evident in the molecular rhythm in vitro, showing that the mutation affected the core SCN clockwork (Fig. 1B,C,D). Moreover, the mutation had a global effect on timekeeping, with significant lengthening of period across major organ systems including kidney, lung, and liver monitored in vitro (Fig. 1E). To assess the molecular basis of the *Afh* phenotype, in vivo expression of circadian proteins and mRNA was examined in SCN and peripheral tissues. Homozygotes exhibited a dramatic reduction in Per protein abundance and Per and Cry mRNA levels in SCN and liver (Fig. 2A,B). At the protein level, however, the mutation had a complex effect on Cry levels. In the SCN, the cycle of Cry immunoreactivity was not different from that of wild type, even though mRNA expression was markedly reduced. In the liver, peak levels of Cry (assessed by western blot) during circadian night were reduced but during circadian day levels were elevated.

The *Afh* mutation was mapped to a gene encoding an F-box protein with leucine-rich repeats, Fbxl3 (Godinho et

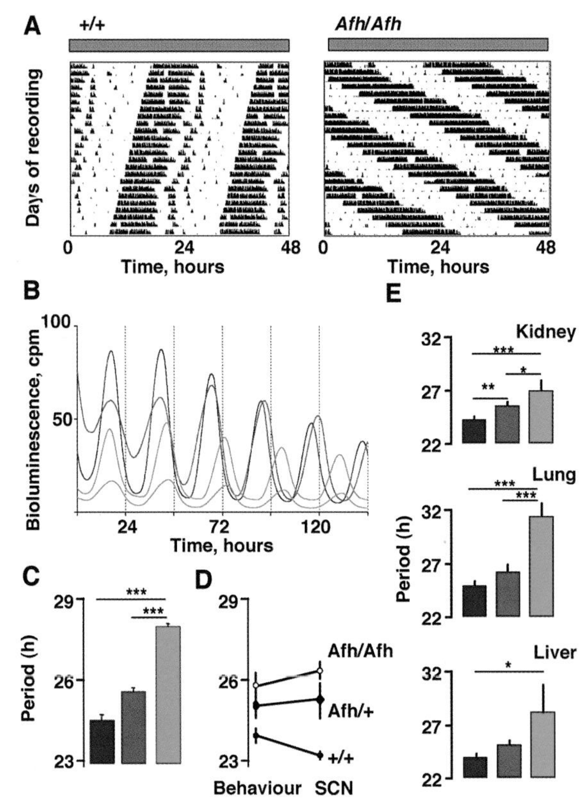

Figure 1. The afterhours long-period mutant. (*A*) Representative double-plotted actograms of wheel running from wild-type (*left*) and homozygous *Afh* mutant mice. (*B*) Representative bioluminescence recordings of PER2::LUC expression in organotypic SCN slice cultures from wild-type (*blue*), heterozygous (*red*) and homozygous (*green*) *Afh* mutants. (*C*) Mean (+S.E.M.) circadian period of bioluminescence rhythms from SCN of wild-type and *Afh* mutant neonatal SCN. (*D*) Comparison of circadian periods of rest-activity cycles in vivo, and SCN bioluminescence rhythms in vitro from adult wild-type and *Afh* mutant mice. (*E*) Mean (+S.E.M.) circadian periods of bioluminescence rhythms from tissue explants of wild-type and *Afh* mutant mice. (Redrawn from Godinho et al. 2007 [AAAS].)

al. 2007). These adapter proteins bind to substrates destined for proteasomal degradation, using the leucine-rich repeat (LRR) to specify substrate and their integral F box to bind to a Skp-Cullin ubiquitin ligase complex (Fig. 2C). These interactions allow assembly of a polyubiquitin chain onto the target, facilitating subsequent proteasomal degradation. The *Afh* mutation is a cysteine-to-serine substitution at residue 358 within a highly conserved region of the LRR which would therefore likely affect substrate affinity of Fbxl3. This issue was resolved with the biochemical identification of Cry1 and Cry2 as native substrates for Fbxl3 and loss of affinity for these proteins in the mutant (Busino et al. 2007). Moreover, in both cell line and native tissues, using either recombinant or native Cry, it was shown that the consequence of the $Fbxl3^{Afh}$ mutation was increased stability of Cry proteins, presumably because they were no longer targeted effectively to the proteasome (Fig. 2D).

These biochemical observations provide a ready explanation of the phenotype at both gene expression and behavioral levels. Failure to clear Cry at the end of circa-

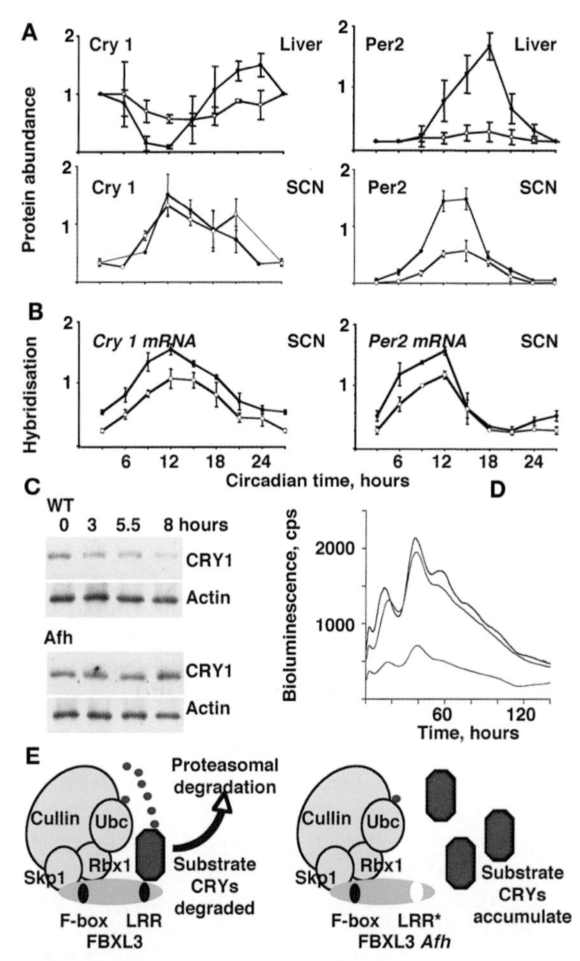

Figure 2. Altered dynamics of protein and mRNA expression within the clockwork of *Afh* mutant mice. (*A*) Abundance of Cry1 and Per2 proteins in liver and SCN (assessed by western blot and immunostaining, respectively) from wild-type (*closed symbols*) and *Afh* homozygous mutant (*open symbols*) mice (mean +S.E.M.). (*B*) Expression of Cry1 and Per2 mRNA in SCN (assessed by in situ hybridization) from wild-type (*closed symbols*) and *Afh* homozygous mutant (*open symbols*) mice (mean +S.E.M.). (*C*) Western blots of Cry1 levels in organotypic lung slice culture of wild-type and *Afh* homozygous mutant mice following treatment with cycloheximide. (*D*) Representative bioluminescence recordings from NIH-3T3 cells transfected with *mPer2::Lucferase* reporter and either empty vector (*black*), wild-type Fbxl3 (*blue*) or *Afh* mutant Fbxl3 (*red*). (*E*) Schematic representation of the role of Fbxl3 in ubiquitination of Cry and the impact of the *Afh* mutation thereon. (Redrawn from Godinho et al. 2007 [AAAS]; and M.H. Hastings et al., unpubl.)

dian night leads to a prolongation of the phase of negative feedback onto E-box-mediated gene expression. Consequently, mRNA levels of *Per* and *Cry* (and other clock-controlled genes) are suppressed for longer and/or more severely so that on the following cycle, peak levels of Per and Cry proteins are reduced. To test this presumed impact of the $Fbx3^{Afh}$ mutation on E-box-mediated gene expression, NIH-3T3 fibroblasts were transfected with either wild-type or mutant Fbxl3, and luciferase expression driven either by *Per2* or *Bmal1* promoter fragments was recorded (Ueda et al. 2002). Overexpressed $Fbxl3^{Afh}$ markedly repressed the *Per2* reporter (Fig. 2D), whereas wild-type Fbxl3 had no effect. Neither construct suppressed the *Bmal1* reporter, but the mutant extended period from 21.5 ±0.1 hours (empty vector) and 21.1 ±0.2 (wild-type Fbxl3) to 23.3 ±0.2 hours ($Fbxl3^{Afh}$ mutation) (mean ±S.E.M., $n = 4$). The more pronounced and extended suppression of gene expression in late circadian night arising from the failure to clear Cry proteins prolongs the circadian cycle such that the escape from negative feedback takes longer than in wild type. This selective prolongation of negative feedback was evident in the bioluminescence curves recorded from SCN slices, in which the nadir of Per2 expression was extended in homozygotes, and is sufficient to explain their longer behavioral period.

These studies, and the complementary work on the "overtime" mutation of Fbxl3 (Siepka et al. 2007), identify Fbxl3 as a regulator of Cry protein stability through its role in targeting Cry for proteasomal degradation, and hence Fbxl3 is a determinant of circadian period. The corresponding F-box adapter protein for Per is likely to be βTrCP, and inappropriate phosphorylation of Per proteins via mutations in Per2 or CKIε and CKIδ accelerates their degradation and thereby shortens circadian period (Gallego and Virshup 2007). The contribution of upstream kinases, including forms of CKI, to the licensing of Cry for ubiquitination remains to be determined. What is clear from studies in plants, fungi, fruit flies, and now mice is that regulated proteasomal degradation of circadian proteins mediated by F-box-containing ubiquitin ligases is a critical step in tuning circadian period to the solar cycle. As such, circadian clocks join the wider context of biological timing in which oscillatory processes, such as the cell division cycle, depend on very tight temporal control of the production, posttranslational modification, and ultimate clearance of proteins.

NEUROPEPTIDE SIGNALING AND THE SYNCHRONIZATION OF CIRCADIAN CLOCK CELLS IN THE SCN

The demonstration of free-running electrophysiological rhythms in SCN neurons dispersed in culture was instrumental in formulating the cell-autonomous model of circadian timing (Welsh et al. 1995). The natural condition of SCN neurons, however, is to be embedded within circuits, and thus a central question to understanding the SCN as a tissue-based clock is to ask how the individual oscillatory cells synchronize their molecular pacemakers and thereby broadcast a coherent circadian signal throughout the organism. Although early studies implicated GABAergic signaling in this coordination, more recently, the SCN neuropeptide vasoactive intestinal polypeptide (VIP) and its SCN target the $VPAC_2$ receptor (encoded by the *Vipr2* gene) have been shown to mediate cellular synchronization within the SCN (Harmar et al. 2002; Colwell et al. 2003). In mice lacking the *Vipr2* gene or the VIP peptide, behavioral rhythms appear normal under a regular lighting schedule but are severely perturbed on transfer to free-running conditions (Fig. 3A). Locomotor activity immediately jumps forward into putative circadian day, successive activity bouts are poorly coordinated, activity onset is irregular from day to day, and multiple but transient periodicities ranging beyond the circadian norm are evident. Loss of effective circadian control to behavior while mice are under the lighting schedule is evidenced by the acute induction of locomotor activity upon unscheduled exposure to darkness during the light phase. These behavioral perturbations are accompanied in vivo by suppressed and apparently arrhythmic circadian *Per* and *Cry* expression across the SCN, as assessed by in situ hybridization and immunostaining. The origin of the behavioral disturbances lies in disrupted molecular clockwork within the SCN, as revealed by bioluminescence recordings from *Vipr2* knockout SCN slices (Maywood et al. 2006). Not only is circadian coherence lost in the mutant tissue, but the overall level of *Per*-driven luciferase expression is also reduced by an order of magnitude (Fig. 3B). The reduction in mean expression level across the tissue indicates that the phenotype of the *Vipr2*$^{-/-}$ SCN is not merely a desynchronization of otherwise high-amplitude cellular oscillators; rather, the amplitude of the individual circadian clockworks is reduced. This is confirmed by microscopic imaging, using charge-coupled device (CCD) cameras to monitor circadian gene expression in individual SCN neurons across the slice (Fig. 3C). Whereas cells in the wild-type slice exhibit high amplitude, synchronized, and sustained circadian cycles, in the mutant slices, appreciably fewer cells are detectable by bioluminescence and of those that are, the level of circadian gene expression is appreciably reduced and the cycles are poorly defined and asynchronous across the slice. These effects of the *Vipr2*$^{-/-}$ mutation are also evident with electrical rhythms in both dispersed cultures and acute SCN slices, presumably a consequence of impaired molecular timekeeping (Aton et al. 2005; Brown et al. 2005). Furthermore, in VIP deficient tissues, electrophysiological synchrony can be briefly restored with VIP agonists (Aton et al. 2005, 2006).

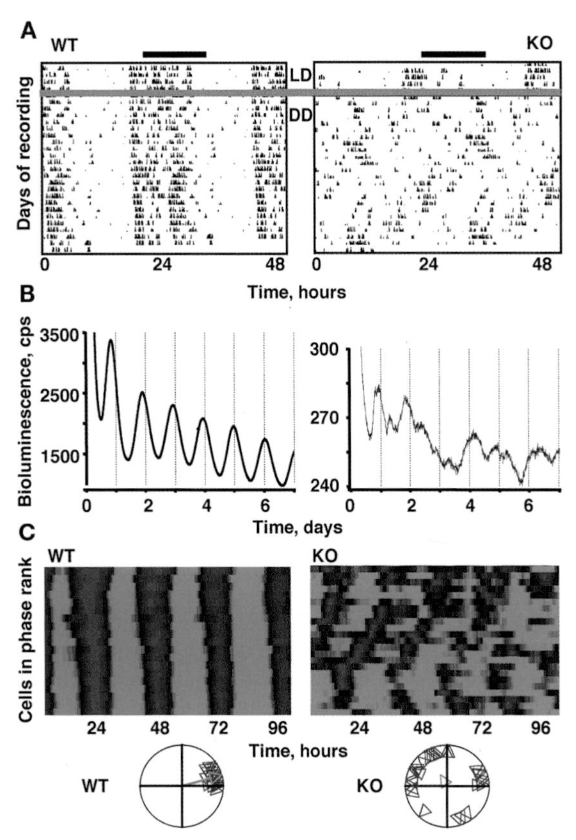

Figure 3. Disrupted circadian behavior and cellular circadian synchrony in SCN of mice lacking the $VPAC_2$ receptor for VIP. (*A*) Double-plotted representative actograms of wheel running in wild-type and homozygous *Vipr2* mutant (KO) mice under a lighting schedule (LD) and then free-running (DD). (*B*) Recordings of mPer1::luciferase activity in SCN slice cultures from animals depicted in *A*. (*C*) Raster plots of circadian luciferase expression in 25 representative cells from wild type and homozygous *Vipr2* mutant SCN slices. Accompanying Rayleigh plots illustrate cellular synchrony in wild type and its absence in mutant SCN. (Redrawn from Maywood et al. 2006 [Elsevier]; and M.H. Hastings et al., unpubl.)

It is therefore clear that VIP-mediated signals released by retinorecipient cells of the SCN core both synchronize and sustain the molecular clockwork of neurons right across the SCN. What is the cellular basis for this? $VPAC_2$ is positively coupled to adenylyl cyclase (AC) and in some tissues can thereby regulate Ca^{2+} signals (Hagen et al. 2006). Compromised SCN gene expression (and hence disrupted circadian behavior) likely arises, therefore, from a deficiency in intracellular cAMP and possibly Ca^{2+} signaling within the SCN. Consistent with this, direct activation of AC with forskolin acutely induces circadian gene expression in *Vipr2*$^{-/-}$ slices (Fig. 4A). Furthermore, acute depolarization of the mutant slice by the elevation of extracellular potassium levels, which would trigger influx of Ca^{2+}, or addition of the SCN neuropeptide gastrin-releasing peptide (GRP) transiently activates Per expression and synchronizes cellular rhythms (Maywood et al. 2006). In the absence of endogenous VIP signaling, however, this imposed synchrony is progressively lost as oscillator cells free-run and drift out of phase (Fig. 4B). These observations raise several interesting questions about the SCN circuitry and cellular timekeeping. First, what are the relationships between membrane depolarization, intracellular signaling via calcium and cAMP, and circadian gene expression? Given that the SCN cell is a neuron, is electrical activity, membrane depolarization, and Ca^{2+} signaling a necessary condition for the molecular clockwork to run (Lundkvist et al. 2005)? If so, what are the specific roles of particular neuropeptidergic inputs? Does VIP facilitate the cellular clock solely by activating AC or does a simultaneous or consequential change in membrane depolarization and Ca^{2+} signaling further enhance *Per* expression? Certainly, electrical silencing of the SCN by tetrodotoxin severely compromises circadian gene expression within individual cells and also leads to their desynchronization (Yamaguchi et al. 2003; Maywood et al. 2007). This raises a second issue: Are the amplitude of cellular circadian gene expression and interneuronal synchrony interdependent, as predicted by models of cellular timekeeping across the SCN (Bernard et al. 2007)? If so, VIP may be a critical neurochemical mediator of these relationships, representing both the output of one cellular

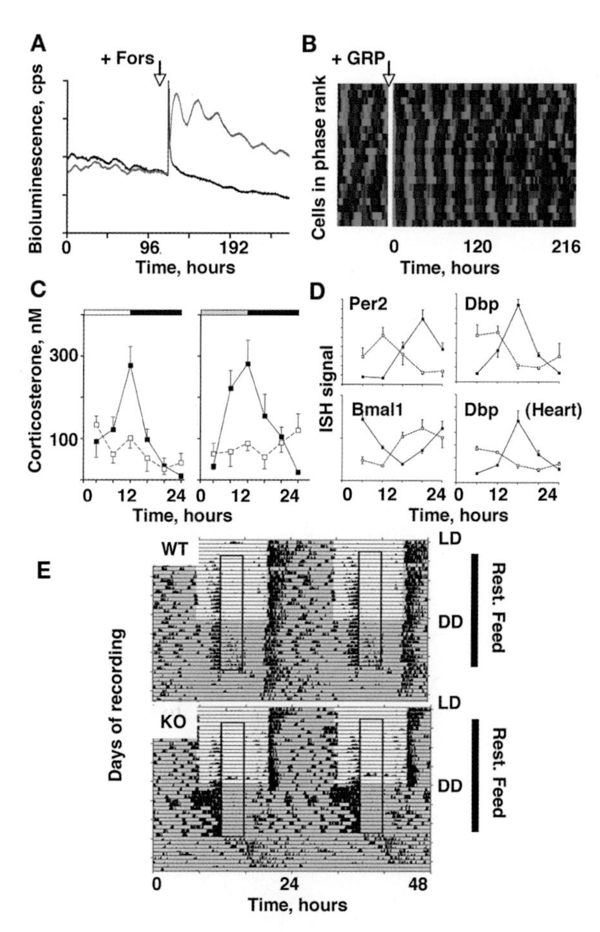

Figure 4. Circadian synchronization in SCN and periphery of $VPAC_2$ receptor null mice. (*A*) Representative bioluminescence recordings from *Vipr2*$^{-/-}$ SCN slices treated with either vehicle (*black line*) or forskolin (10 µM, *gray line*). (*B*) Raster plots of circadian luciferase expression in 25 representative cells from homozygous *Vipr2*$^{-/-}$ SCN slice, before and after treatment with 100 nM GRP. (*C*) Daily (*left*) and circadian (*right*) rhythms of corticosterone levels in wild type (mean +S.E.M., *closed symbols*) and their absence in *Vipr2*$^{-/-}$ mice (*open symbols*). (*D*) Circadian gene expression in liver and heart of wild-type (*closed symbols*, mean +S.E.M.) and their phase advance in *Vipr2*$^{-/-}$ mice (*open symbols*). (*E*) Double-plotted, representative wheel-running actograms of wild-type and *Vipr2*$^{-/-}$ mice, subjected to restricted feeding under both LD and DD. (Redrawn from Maywood et al. 2006 [Elsevier]; Sheward et al. 2007 [Society for Neuroscience]; and M.H. Hastings et al., unpubl.)

oscillator and the (synchronizing) input to its neighbors in the circuit. Hence, it is the ability of SCN neurons to communicate via VIP (and other?) signals that makes the entire nucleus such a robust and precise timekeeper as cellular synchrony and amplitude are reciprocally supportive. Indeed, the power of these circuit-coupling effects is revealed by their ability to compensate for clock gene mutations which at the level of isolated cells would compromise the clockwork (Liu et al. 2007). The consequence of obligate synchronization and interdependence, however, is that when interneuronal signals are compromised, the cellular clockwork of the SCN becomes far less robust than that of fibroblasts, which oscillate perfectly well in the absence of synchronizing cues (Welsh et al. 2004).

CIRCADIAN ENTRAINMENT IN *VIPR2*$^{-/-}$ MICE

In wild-type animals, light pulses relayed by the retina reset the SCN clockwork by acutely increasing levels of *Per* expression, initially in the retinorecipient core and then in the shell. This involves a glutamatergic signaling cascade via NMDA (*N*-methyl-D-aspartate) receptors and mitogen-activated protein kianse (MAPK), ultimately activating (by phosphorylation) the calcium/cyclic AMP response element (CRE)-binding protein (CREB) (Hastings et al. 2003). Both *Per1* and *Per2* carry CRE sequences through which pCREB can induce their expression. A similar cascade also activates the immediate-early gene c-*fos* in the retinorecipient SCN, but in all cases, these responses are gated so that light is only effective when delivered during circadian night. In the *Vipr2*$^{-/-}$ mouse, however, light pulses activate MAPK, CREB phosphorylation, *Per*, and c-*fos* expression in circadian day just as effectively as they do in circadian night (Hughes et al. 2004; Maywood et al. 2007). Loss of circadian timekeeping arising from the absence of VIP signaling therefore leaves open the gate for retinal activation of the core SCN neurons. Hence, VIP signals are necessary for circadian regulation of the core clockwork but not for its acute regulation by glutamatergic retinal inputs.

Compromised SCN timekeeping in $VPAC_2$ null mice also has marked consequences for downstream physiological rhythms. Female *Vipr2*$^{-/-}$ mice show irregular estrous cycles on a light/dark cycle, a disruption that is exacerbated by continuous darkness, and accompanied by phase advanced cycles of *Per*, *Cry*, and *Bmal* expression in the uterus (Dolatshad et al. 2006). Circadian rhythms of corticosterone secretion are also lost, the peak secretion at the start of circadian night being absent and levels remaining permanently low (Fig. 4C). This is not, however, a generic suppression of adrenal activation, because behavioral arousal by restricted feeding acutely increases corticosterone levels (Sheward et al. 2007).

The circadian cycle of corticosteroid secretion is an important internal synchronizer of circadian rhythms in the periphery, especially the liver (Balsalobre et al. 2000; Reddy et al. 2007; see below). In *Vipr2*$^{-/-}$ mice, clock gene expression remains rhythmic in the liver, in contrast to its absence from the SCN. The entire program, however, is phase-advanced by 4–6 hours relative to wild type, reflecting the combined effects of an advanced feeding schedule and the loss of rhythmic corticosterone, which usually opposes phase shifts of the liver clockwork by feeding cycles (Le Minh et al. 2001). The maintenance of rhythmical feeding in these "SCN-disabled" mice is an intriguing finding and is consistent with the view that feeding cycles can be regulated by other hypothalamic regions, especially the dorsomedial nucleus (Gooley et al. 2006). Indeed, there is clear evidence that *Vipr2*$^{-/-}$ mice retain a food-entrainable oscillator. Under a regime of restricted feeding, they exhibit very strong anticipatory behavior, an accompanying surge in circulating corticosterone levels, and rhythmic *Per* expression in the liver peaking with anticipated feeding time. Thus, food intake remains an effective synchronizing agent in *Vipr2*$^{-/-}$ mice, capable of controlling circadian behavior, corticosteroid secretion, and the clockwork of peripheral tissues, and although VIP signaling is critical for the function of the SCN pacemaker, it is not required for the food entrainable oscillator.

PROKINETICIN SIGNALING AND CIRCADIAN OUTPUT CONTROL BY THE SCN

Effective functioning of the extra-SCN oscillators in brain and peripheral tissues is dependent on their coordination by the SCN. Some of the physiological effects of the *Vipr2* null mutation may arise from impaired signaling by VIP-containing projections from SCN neurons to hypothalamic targets, for example, gonadotrophin hormone-releasing hormone neurons that control the estrous cycle. In addition, neuronal transplantation studies indicate that diffusible paracrine factors may signal circadian time to SCN targets, and transforming growth factor-α (TGF-α) (Kramer et al. 2005) and the neuropeptide cardiotrophin-like cytokine (Kraves and Weitz 2006) have recently been identified as likely mediators. Attention has also focused on Prokineticin 2 (Prok2), an 81-amino-acid secretory peptide first identified in gut but also strongly expressed in the SCN. Prok2 is a typical clock-regulated gene; its transcript levels in the SCN are highly circadian and driven by Clock/Bmal via E boxes. It is also regulated by light, presumably via a glutamatergic MAPK cascade. Pharmacological studies indicated a role for Prok2 in the suppression of daytime activity in nocturnal rodents (Cheng et al. 2002), but in both day-active and night-active species, its peak of expression in the SCN is in circadian daytime, indicating that it signals time rather than activity state (Lambert et al. 2005). Its receptor (Prokr2) is widely expressed in both the SCN and SCN target areas (Fig. 5A), including the medial thalamus, arcuate nucleus, and dorsomedial hypothalamus (Cheng et al. 2002).

To test the potential role of Prokr2 in the mediation of circadian outflow, we created a null mutation (Fig. 5A) (Prosser et al. 2007). Locomotor activity rhythms in null mice were severely disturbed even on a lighting cycle. The onset of activity around lights-off was poorly defined, and there was very little wheel running or general movement in the first part of the dark phase. In contrast, spontaneous activity at the end of the dark phase was comparable to that of wild-type mice. Upon release into continuous dim light, the residual activity remained rhythmic, evidence of an underlying circadian regulation, but it was poorly organized and activity onset lacked definition. This disturbance within circadian night was also evident in telemetric recordings of core body temperature (Fig. 5C). The typical sustained nocturnal elevation observed in wild-type mice was absent from mutants, the profile being bimodal with a brief elevation at the beginning of the night, a prolonged decline, and then finally a small increase at the end of circadian night coincident with the spontaneous surge in locomotor activity. Loss of Prokr2 signaling therefore altered the pattern of circadian behavior and physiology, an effect comparable to that seen with targeted deletion of the gene encoding the Prok2 ligand (Li et al. 2006).

Given the precedent of the $Vipr2^{-/-}$ mutant, the Prokr2 null phenotype may have arisen from disordered molecular timekeeping in the SCN. Bioluminescence recordings from SCN slices, however, confirmed that the molecular timekeeper is as robust in the null mutant as in the wild type (Fig. 5D). Although these SCN rhythms provide an explanation for the free-running behavioral and temperature rhythms, the coherence of the SCN pacemaker is at odds with the disturbed nocturnal patterning observed in vivo, i.e., altered behavioral waveform was not accompanied by changes in the SCN waveform. Prokr2 signaling is not therefore necessary for cellular pacemaking or synchrony in the SCN, despite its high level of expression therein. Rather, the data are consistent with a role for Prokr2 in mediating circadian control by a functional SCN over behavioral and/or physiological rhythms. More specifically, it appears that Prokr2-mediated signals are important for signaling specifically nocturnal events. It remains to be determined how this particular peptidergic signal is integrated into other SCN outputs, although one

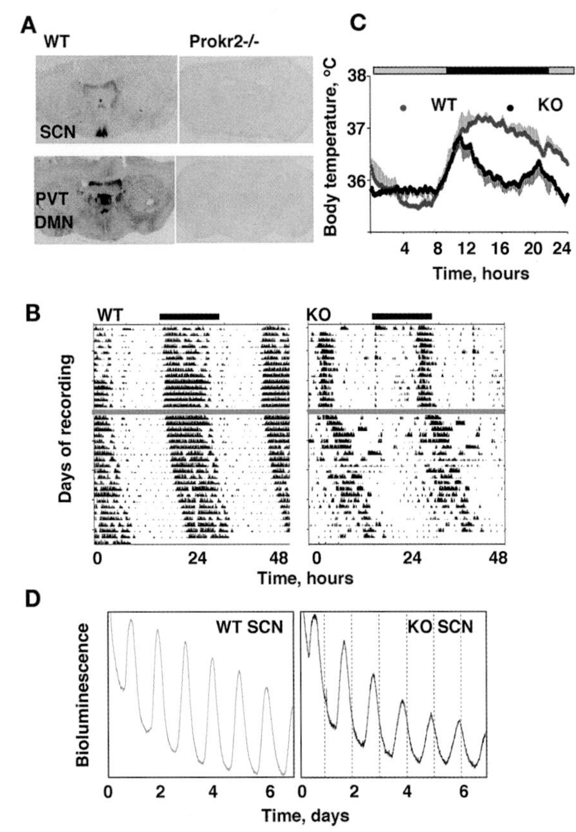

Figure 5. Compromised circadian behavior and thermoregulation, but a functional SCN pacemaker in mice lacking prokineticin receptor 2. (*A*) Autoradiographic distribution of Prok2 analog binding in mouse forebrain. (PVT) Paraventricular thalamus; (DMN) dorsomedial hypothalamus. (*B*) Double-plotted, representative wheel-running actograms of wild-type (*left*) and Prokr2 null (KO) mice entrained to LD (*above gray line*) and free-running under DD (*below gray line*). (*C*) Representative mean daily profile (+S.E.M.) of core body temperature of wild-type (*red*) and Prokr2 null (KO, *black*) mice. (*D*) Representative Per1::luciferase bioluminescence rhythms from wild-type (*left*) and Prokr2 null (*right*) SCN slices. (Redrawn from Prosser et al. 2007 [National Academy of Sciences].)

possibility is that a series of such signals is released across the circadian cycle, particular circadian times being identified by a phase-specific neurochemical "cocktail."

CIRCADIAN ORGANIZATION IN PERIPHERAL TISSUES

Understanding the molecular underpinnings of circadian physiology has been revolutionized by genome-based analytical techniques, not the least of which are "gene-chip" microarray methods by which the entire transcriptome of a tissue can be examined across circadian time. For any particular tissue, 5–10% of transcripts are under circadian regulation (Akhtar et al. 2002; Panda et al. 2002; Storch et al. 2002; Ueda et al. 2002), but with the exception of the canonical core clock genes and ubiquitous output genes such as albumin D-element-binding protein (*Dbp*), the circadian genes differ markedly between tissues as they tend to represent key regulators of the particular tissue's metabolic functions. Moreover, because circadian integration at the level of the organism essentially involves alternation between catabolic and anabolic phases, within any particular tissue, rhythmic genes fall into discrete, functionally coherent circadian clusters (Fig. 6A). Consequently, the transcriptomic profile of a tissue follows a precise temporal program, with particularly marked changes associated with anticipated dawn and dusk.

In the liver, for example, genes associated with glycolysis/gluconeogenesis, nitrogen metabolism, cytoskeleton, xenobiotic metabolism, vesicle trafficking, and solute carrier transporters are circadian, thereby enabling the organ to prepare for anticipated changes in ingestion, nutrient availability, and nutrient mobilization over the rest/activity cycle. Within a functional grouping, for example, various tubulin isoforms, there is clear coordination of their mRNA abundance, emphasizing the elegance and sophistication of circadian synchronization (Fig. 6B). Of particular interest is circadian regulation of cyclins and other cell division factors (Reddy et al. 2005). For example, *Wee-1* expression is circadian in many peripheral tissues, including both liver and kidney (Fig. 6C), contributing to circadian regulation of cell division and growth (Matsuo et al. 2003). This pattern is ultimately dependent on SCN signals because circadian coordination of *Wee-1* (and most other genes) is lost in the livers of SCN-ablated mice (Fig. 6C). The basis for this control lies within functional E boxes in the *Wee-1* gene, which are activated by Clock and Bmal1 and suppressed by Cry (Fig. 6D). *Wee-1* therefore illustrates the hierarchy of circadian regulation for clock-controlled genes in peripheral tissues: Within peripheral cells, the intracellular clockwork drives its expression, but for gene expression to be coherent at the tissue level, these clocks are regulated by SCN-derived cues.

Having established the complexity of circadian gene expression within peripheral tissues, the question arises as to how it is synchronized? The contribution of rhythmic glucocorticoids (GCs) was noted above (Balsalobre et al. 2000), and in circadian mutant mice, including *Vipr2* knockouts, loss of rhythmic corticosterone is associated with altered rhythms of gene expression in peripheral tis-

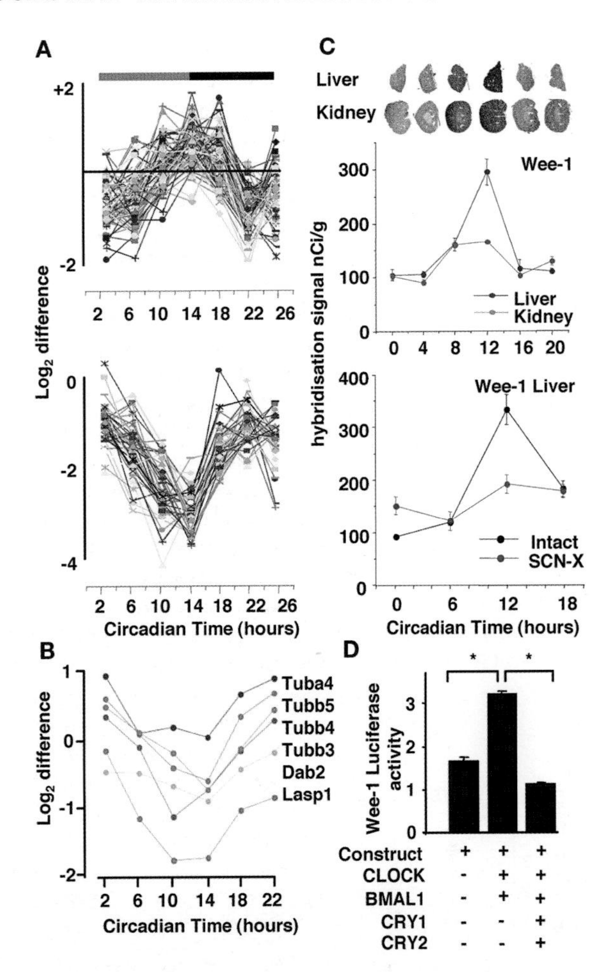

Figure 6. Circadian transcriptome and circadian metabolism. (*A*) Representative antiphasic clusters of circadian transcripts from livers of mice sampled across the circadian cycle. (*B*) Coordinated circadian expression of genes encoding cytoskeletal subunits. (*C*) *Wee-1* as a clock-controlled gene: Circadian expression in liver and kidney analyzed by in situ hybridization is lost in SCN-ablated mice (Mean +S.E.M.). (*D*) Luciferase reporter assays reveal Clock/Bmal-dependent transcription of *Wee-1* and its suppression by Cry. (Redrawn from Akhtar et al. 2002 [Elsevier]; and M. Hastings et al., unpubl.)

sues (Sheward et al. 2007). It is difficult, however, to test the role of any single factor in intact mice because its action cannot be isolated from those of other SCN-dependent synchronizing cues. We therefore rendered mice arrhythmic by ablation of the SCN, a procedure that would remove all synchronizing cues and cause desynchrony of cellular clocks in peripheral tissues. Mice then received an acute injection of the GC receptor agonist dexamethasone (DEX). Our prediction was that when applied against the arrhythmic background of SCN-ablated mice, an effective GC stimulus would resynchronize multiple cellular clocks in the tissue, driving them to a unique phase and thereby establishing a common temporal pattern in groups of animals sampled at different times following injection. By analyzing samples by DNA microarray, it would be possible to identify which elements of the liver circadian transcriptome are subject to acute regulation by GC. On the basis of our own and other studies, we identified a target population of 366 circadian transcripts that were rhythmic in tissues from intact mice and showed no circadian variation following SCN ablation. Of these, 57% responded to a single activation by DEX (Reddy et al. 2007), individual genes showing rapid or delayed activation, or suppression (Fig. 7A). This circadian resynchronization included induction of *Per1*, *Cry1*, and *Bmal1*, all of which carry GC response elements (GRE) and so are direct targets for circulating corticosterone. Similarly, two thirds of the responsive circadian genes also carry GREs, whereas a partially overlapping 38% carry E boxes. Consequently, activation of the circadian transcriptome by GC can occur by at least two overlapping routes: direct regulation of genes through GREs and indirect regulation via E boxes following GRE-mediated activation of the core clockwork. Further transcriptional cascades are also clock/GC-regulated, for example, those dependent on Dbp which does not carry GREs but is nevertheless GC responsive, likely an indirect effect mediated through its E-box sequences. To identify further circadian transcriptional cascades sensitive to GC, we screened our set of GC-sensitive circadian genes for transcription-factor-binding sites and identified a high frequency of targets (39% vs. 7% in noncircadian genes) of hepatocyte nuclear factor 4α (HNF4α) (Fig. 7B). HNF4α is a pivotal liver transcription factor that regulates numerous metabolic pathways, including ureagenesis, the production of serum proteins, and the activity of cytochrome P450 genes (Odom et al. 2004). It carries both GREs and E boxes, and correspondingly, its expression in the liver is circadian, lost on SCN ablation and restored by GC (Fig. 7C,D). Moreover, expression of a number of its downstream targets, including ornithine transcarbamylase (a central factor in nitrogen metabolism), is circadian (Fig. 7E). Rhythmicity of these genes is dependent on the SCN and is also lost in HNF4α null liver, highlighting the contribution of HNF4α to circadian coordination of the transcriptome.

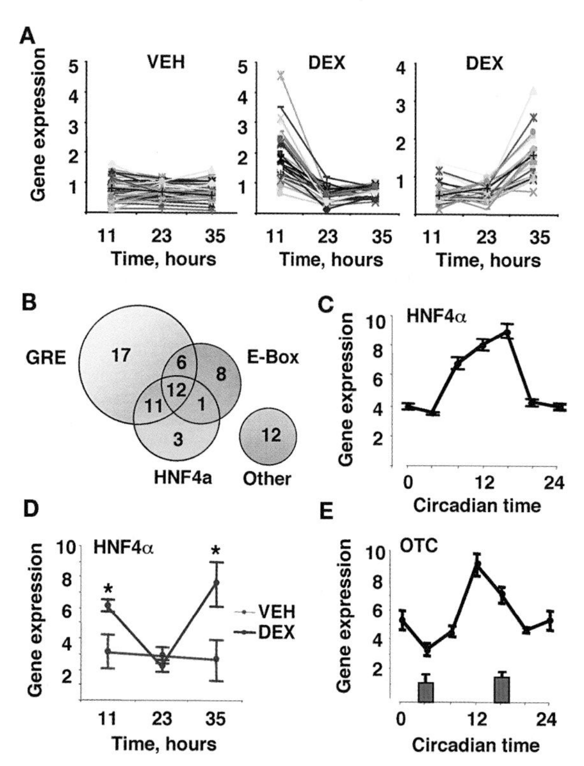

Figure 7. Coordination of the liver circadian transcriptome by glucocorticoid signaling. (*A*) Relative expression levels of circadian genes in livers of SCN-ablated mice are arrhythmic with vehicle treatment (*left*), and activated by DEX (mean +S.E.M.) either acutely (*middle*) or with a delay (*right*). (*B*) Frequency of regulatory elements in promoter and introns of circadian GC-regulated genes in liver. (*C*) Circadian expression of *HNF4α* mRNA in mouse liver (mean +S.E.M.). (*D*) *HNF4α* mRNA is arrhythmic in liver of SCN ablated mice and induced by DEX (mean +S.E.M., $*p < 0.05$ vs. vehicle). (*E*) Circadian expression of *HNF4α* target gene *OTC* in liver of wild-type mice and its loss in *HNF4α* null liver (*green bars*) (mean +S.E.M.). (Redrawn from Reddy et al. 2007 [AASLD].)

Having characterized these circadian transcriptional cascades, a critical question is their relationships to protein expression: Are they faithfully translated? By combining two-dimensional gel electrophoresis with mass spectroscopy, we have analyzed the circadian proteome of liver (Reddy et al. 2006). Although coverage is more limited than with a microarray, this approach revealed that 135 (21%) of 642 "spots" resolved in the gel were under circadian regulation. Of these, it was possible to identify 49 circadian proteins that were the products of 39 unique genes. These were principally enzymes and showed distinct phase clustering between circadian day and circadian night (Fig. 8A). Western blots confirmed circadian regulation for a number of proteins and also showed their rhythmic expression to be sensitive to circadian mutations (Fig. 8B). As with the transcriptomic analysis, a number of circadian factors were rate-limiting and/or vital components of key pathways, including carbohydrate and nitrogen metabolism (Fig. 8C). By comparison with published transcriptomic data and de novo assays of mRNA expression, it is evident that almost half (44%) of circadian proteins do not exhibit a circadian transcript, indicating that posttranscriptional and posttranslational mechanisms are in turn circadian and play a significant part in sculpting the circadian proteome (Reddy et al. 2006). Potential mediators of the former include the deadenylase nocturnin. Encoded by a clock-regulated gene, it is thought to control mRNA stability, and a series of clock-regulated metabolic genes are dys-regulated in nocturnin knockout mice (Green et al. 2007). MicroRNAs provide a further level of posttranscriptional circadian regulation, modifying mRNA expression by altering its stability and sequestration and/or directly impeding translation. MicroRNAs are active in the SCN (Cheng et al. 2007) and likely in peripheral clockworks as well. Evidence of circadian control over posttranslational modification is provided by peroxiredoxin 6. In the liver, this protective antioxidant factor is expressed rhythmically at the mRNA level and is present as a series of phosphoproteins that cycle in their abundance (Reddy et al. 2006). Remarkably, two isoforms cycle in antiphase to each other, with one in phase with the transcript and the other not. Thus, evidence is accumulating that a diversity of posttranscriptional regulatory mechanisms define the circadian proteome and thereby the unique daily metabolic programs of individual tissues.

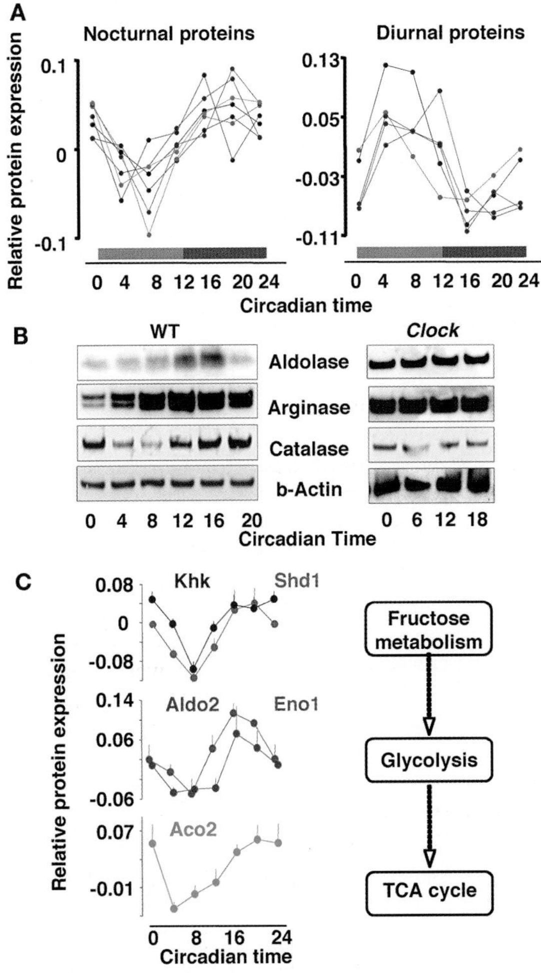

Figure 8. Circadian regulation of the liver proteome. (*A*) Representative plots of nocturnal and diurnal clusters of circadian proteins. (*B*) Validation by western blot of circadian expression of liver enzymes and loss of the rhythm in liver from *Clock* mutant mice. (*C*) Representative metabolically relevant circadian regulated enzymes from mouse liver (mean +S.E.M.). (Redrawn from Reddy et al. 2006 [Elsevier].)

CONCLUSION

Notwithstanding the recent progress described here and elsewhere in this volume, major questions remain regarding the molecular and cellular bases of circadian timing in mammals. First, although components of the feedback loops and the cellular factors that determine their behavior are being individually characterized, it is unclear how these components interact at a molecular level, forming multimeric complexes with emergent properties that specify and reflect circadian time. We have no knowledge of circadian protein structure and no biochemically validated quantitative model for the molecular circadian oscillator of mammals. Second, the relationship between the core molecular loops and neural activity in the SCN is unresolved, but one possible outcome is that far from being an autonomous clock cell, SCN neurons are more proficient pacemakers when operating within a circuit. Such necessary properties and circuit functions of the SCN wait to be determined. Third, we know that the SCN have anatomically and neurochemically diverse outputs, but we do not understand the nature of the temporal signals they convey: Are they graded and continuous, binary "stop" and "start," specific for a particular target, a property of individual SCN cells, or a population code, etc.? Finally, the major challenge will be to exploit newly acquired knowledge of the circadian control of vital cellular processes—cell division, nutrient metabolism, xenobiotic detoxification, etc.—to better manage and prevent major systemic illness.

ACKNOWLEDGMENTS

This work is supported by the Medical Research Council, the Biotechnology and Biological Sciences Research Council, The Wellcome Trust, and the FP6 EUCLOCK programme.

REFERENCES

Akhtar R.A., Reddy A.B., Maywood E.S., Clayton J.D., King V.M., Smith A.G., Gant T.W., Hastings M.H., and Kyriacou C.P. 2002. Circadian cycling of the mouse liver transcriptome, as revealed by cDNA microarray, is driven by the suprachiasmatic nucleus. *Curr. Biol.* **12:** 540.

Aton S.J., Huettner J.E., Straume M., and Herzog E.D. 2006. GABA and Gi/o differentially control circadian rhythms and synchrony in clock neurons. *Proc. Natl. Acad. Sci.* **103:** 19188.

Aton S.J., Colwell C.S., Harmar A.J., Waschek J., and Herzog E.D. 2005. Vasoactive intestinal polypeptide mediates circadian rhythmicity and synchrony in mammalian clock neurons. *Nat. Neurosci.* **8:** 476.

Bacon Y., Ooi A., Kerr S., Shaw-Andrews L., Winchester L., Breeds S., Tymoska-Lalanne Z., Clay J., Greenfield A.G., and Nolan P.M. 2004. Screening for novel ENU-induced rhythm, entrainment and activity mutants. *Genes Brain Behav.* **3:** 196.

Balsalobre A., Brown S.A., Marcacci L., Tronche F., Kellendonk C., Reichardt H.M., Schutz G., and Schibler U. 2000. Resetting of circadian time in peripheral tissues by glucocorticoid signaling. *Science* **289:** 2344.

Bernard S., Gonze D., Cajavec B., Herzel H., and Kramer A. 2007. Synchronization-induced rhythmicity of circadian oscillators in the suprachiasmatic nucleus. *PLoS Comput. Biol.* **3:** e68.

Brown T.M., Hughes A.T., and Piggins H.D. 2005. Gastrin-releasing peptide promotes suprachiasmatic nuclei cellular rhythmicity in the absence of vasoactive intestinal polypeptide-VPAC$_2$ receptor signaling. *J. Neurosci.* **25:** 11155.

Busino L., Bassermann F., Maiolica A., Lee C., Nolan P.M., Godinho S.I., Draetta G.F., and Pagano M. 2007. SCFFbxl3 controls the oscillation of the circadian clock by directing the degradation of cryptochrome proteins. *Science* **316:** 900.

Cheng H.Y., Papp J.W., Varlamova O., Dziema H., Russell B., Curfman J.P., Nakazawa T., Shimizu K., Okamura H., Impey S., and Obrietan K. 2007. microRNA modulation of circadian-clock period and entrainment. *Neuron* **54:** 813.

Cheng M.Y., Bullock C.M., Li C., Lee A.G., Bermak J.C., Belluzzi J., Weaver D.R., Leslie F.M., and Zhou Q.Y. 2002. Prokineticin 2 transmits the behavioural circadian rhythm of the suprachiasmatic nucleus. *Nature* **417:** 405.

Colwell C.S., Michel S., Itri J., Rodriguez W., Tam J., Lelievre V., Hu Z., Liu X., and Waschek J.A. 2003. Disrupted circadian rhythms in VIP- and PHI-deficient mice. *Am. J. Physiol. Regul. Integr. Comp. Physiol.* **285:** R939.

Dolatshad H., Campbell E.A., O'Hara L., Maywood E.S., Hastings M.H., and Johnson M.H. 2006. Developmental and reproductive performance in circadian mutant mice. *Hum. Reprod.* **21:** 68.

Field M.D., Maywood E.S., O'Brien J.A., Weaver D.R., Reppert S.M., and Hastings M.H. 2000. Analysis of clock proteins in

mouse SCN demonstrates phylogenetic divergence of the circadian clockwork and resetting mechanisms. *Neuron* **25:** 437.
Gallego M. and Virshup D.M. 2007. Post-translational modifications regulate the ticking of the circadian clock. *Nat. Rev. Mol. Cell Biol.* **8:** 139.
Godinho S.I., Maywood E.S., Shaw L., Tucci V., Barnard A.R., Busino L., Pagano M., Kendall R., Quwailid M.M., Romero M.R., O'neill J., Chesham J.E., Brooker D., Lalanne Z., Hastings M.H., and Nolan P.M. 2007. The after-hours mutant reveals a role for Fbxl3 in determining mammalian circadian period. *Science* **316:** 897.
Gooley J.J., Schomer A., and Saper C.B. 2006. The dorsomedial hypothalamic nucleus is critical for the expression of food-entrainable circadian rhythms. *Nat. Neurosci.* **9:** 398.
Green C.B., Douris N., Kojima S., Strayer C.A., Fogerty J., Lourim D., Keller S.R., and Besharse J.C. 2007. Loss of Nocturnin, a circadian deadenylase, confers resistance to hepatic steatosis and diet-induced obesity. *Proc. Natl. Acad. Sci.* **104:** 9888.
Hagen B.M., Bayguinov O., and Sanders K.M. 2006. VIP and PACAP regulate localized Ca^{2+} transients via cAMP-dependent mechanism. *Am. J. Physiol. Cell Physiol.* **291:** C375.
Harmar A.J., Marston H.M., Shen S., Spratt C., West K.M., Sheward W.J., Morrison C.F., Dorin J.R., Piggins H.D., Reubi J.C., Kelly J.S., Maywood E.S., and Hastings M.H. 2002. The VPAC(2) receptor is essential for circadian function in the mouse suprachiasmatic nuclei. *Cell* **109:** 497.
Hastings M.H., Reddy A.B., and Maywood E.S. 2003. A clockwork web: Circadian timing in brain and periphery, in health and disease. *Nat. Rev. Neurosci.* **4:** 649.
Hughes A.T., Fahey B., Cutler D.J., Coogan A.N., and Piggins H.D. 2004. Aberrant gating of photic input to the suprachiasmatic circadian pacemaker of mice lacking the $VPAC_2$ receptor. *J. Neurosci.* **24:** 3522.
Kramer A., Yang F.C., Kraves S., and Weitz C.J. 2005. A screen for secreted factors of the suprachiasmatic nucleus. *Methods Enzymol.* **393:** 645.
Kraves S. and Weitz C.J. 2006. A role for cardiotrophin-like cytokine in the circadian control of mammalian locomotor activity. *Nat. Neurosci.* **9:** 212.
Kume K., Zylka M.J., Sriram S., Shearman L.P., Weaver D.R., Jin X., Maywood E.S., Hastings M.H., and Reppert S.M. 1999. mCRY1 and mCRY2 are essential components of the negative limb of the circadian clock feedback loop. *Cell* **98:** 193.
Lambert C.M., Machida K.K., Smale L., Nunez A.A., and Weaver D.R. 2005. Analysis of the prokineticin 2 system in a diurnal rodent, the unstriped Nile grass rat (*Arvicanthis niloticus*). *J. Biol. Rhythms* **20:** 206.
Le Minh N., Damiola F., Tronche F., Schutz G., and Schibler U. 2001. Glucocorticoid hormones inhibit food-induced phase-shifting of peripheral circadian oscillators. *EMBO J.* **20:** 7128.
Li J.D., Hu W.P., Boehmer L., Cheng M.Y., Lee A.G., Jilek A., Siegel J.M., and Zhou Q.Y. 2006. Attenuated circadian rhythms in mice lacking the prokineticin 2 gene. *J. Neurosci.* **26:** 11615.
Liu A.C., Welsh D.K., Ko C.H., Tran H.G., Zhang E.E., Priest A.A., Buhr E.D., Singer O., Meeker K., Verma I.M., Doyle F.J., Takahashi J.S., and Kay S.A. 2007. Intercellular coupling confers robustness against mutations in the SCN circadian clock network. *Cell* **129:** 605.
Lowrey P.L. and Takahashi J.S. 2004. Mammalian circadian biology: Elucidating genome-wide levels of temporal organization. *Annu. Rev. Genomics Hum. Genet.* **5:** 407.
Lundkvist G.B., Kwak Y., Davis E.K., Tei H., and Block G.D. 2005. A calcium flux is required for circadian rhythm generation in mammalian pacemaker neurons. *J. Neurosci.* **25:** 7682.
Matsuo T., Yamaguchi S., Mitsui S., Emi A., Shimoda F., and Okamura H. 2003. Control mechanism of the circadian clock for timing of cell division in vivo. *Science* **302:** 255.
Maywood E.S., O'Neill J.S., Chesham J.E., and Hastings M.H. 2007. The circadian clockwork of the suprachiasmatic nuclei: Analysis of a cellular oscillator that drives endocrine rhythms. *Endocrinology*. doi:10.1210/en.2007-0660.
Maywood E.S., Reddy A.B., Wong G.K., O'Neill J.S., O'Brien J.A., McMahon D.G., Harmar A.J., Okamura H., and Hastings M.H. 2006. Synchronization and maintenance of timekeeping in suprachiasmatic circadian clock cells by neuropeptidergic signaling. *Curr. Biol.* **16:** 599.
Odom D.T., Zizlsperger N., Gordon D.B., Bell G.W., Rinaldi N.J., Murray H.L., Volkert T.L., Schreiber J., Rolfe P.A., Gifford D.K., Fraenkel E., Bell G.I., and Young R.A. 2004. Control of pancreas and liver gene expression by HNF transcription factors. *Science* **303:** 1378.
Panda S., Antoch M.P., Miller B.H., Su A.I., Schook A.B., Straume M., Schultz P.G., Kay S.A., Takahashi J.S., and Hogenesch J.B. 2002. Coordinated transcription of key pathways in the mouse by the circadian clock. *Cell* **109:** 307.
Prosser H.M., Bradley A., Chesham J.E., Ebling F.J., Hastings M.H., and Maywood E.S. 2007. Prokineticin receptor 2 (Prokr2) is essential for the regulation of circadian behavior by the suprachiasmatic nuclei. *Proc. Natl. Acad. Sci.* **104:** 648.
Reddy A.B., Wong G.K., O'Neill J., Maywood E.S., and Hastings M.H. 2005. Circadian clocks: Neural and peripheral pacemakers that impact upon the cell division cycle. *Mutat. Res.* **574:** 76.
Reddy A.B., Maywood E.S., Karp N.A., King V.M., Inoue Y., Gonzalez F.J., Lilley K.S., Kyriacou C.P., and Hastings M.H. 2007. Glucocorticoid signaling synchronizes the liver circadian transcriptome. *Hepatology* **45:** 1478.
Reddy A.B., Karp N.A., Maywood E.S., Sage E.A., Deery M., O'Neill J.S., Wong G.K., Chesham J., Odell M., Lilley K.S., Kyriacou C.P., and Hastings M.H. 2006. Circadian orchestration of the hepatic proteome. *Curr. Biol.* **16:** 1107.
Reppert S.M. and Weaver D.R. 2002. Coordination of circadian timing in mammals. *Nature* **418:** 935.
Sheward W.J., Maywood E.S., French K.L., Horn J.M., Hastings M.H., Seckl J.R., Holmes M.C., and Harmar A.J. 2007. Entrainment to feeding but not to light: Circadian phenotype of $VPAC_2$ receptor-null mice. *J. Neurosci.* **27:** 4351.
Siepka S.M., Yoo S.H., Park J., Song W., Kumar V., Hu Y., Lee C., and Takahashi J.S. 2007. Circadian mutant Overtime reveals F-box protein FBXL3 regulation of cryptochrome and period gene expression. *Cell* **129:** 857.
Storch K.F., Lipan O., Leykin I., Viswanathan N., Davis F.C., Wong W.H., and Weitz C.J. 2002. Extensive and divergent circadian gene expression in liver and heart. *Nature* **417:** 78.
Ueda H.R., Chen W., Adachi A., Wakamatsu H., Hayashi S., Takasugi T., Nagano M., Nakahama K., Suzuki Y., Sugano S., Iino M., Shigeyoshi Y., and Hashimoto S. 2002. A transcription factor response element for gene expression during circadian night. *Nature* **418:** 534.
Weaver D.R. 1998. The suprachiasmatic nucleus: A 25-year retrospective. *J. Biol. Rhythms* **13:** 100.
Welsh D.K., Yoo S.H., Liu A.C., Takahashi J.S., and Kay S.A. 2004. Bioluminescence imaging of individual fibroblasts reveals persistent, independently phased circadian rhythms of clock gene expression. *Curr. Biol.* **14:** 2289.
Welsh D.W., Logothetis D.E., Meister M., and Reppert S.M. 1995. Individual neurons dissociated from rat suprachiasmatic nucleus express independently-phased firing rhythms. *Neuron* **14:** 697.
Yamaguchi S., Isejima H., Matsuo T., Okura R., Yagita K., Kobayashi M., and Okamura H. 2003. Synchronization of cellular clocks in the suprachiasmatic nucleus. *Science* **302:** 1408.
Yoo S.H., Yamazaki S., Lowrey P.L., Shimomura K., Ko C.H., Buhr E.D., Siepka S.M., Hong H.K., Oh W.J., Yoo O.J., Menaker M., and Takahashi J.S. 2004. PERIOD2::LUCIFERASE real-time reporting of circadian dynamics reveals persistent circadian oscillations in mouse peripheral tissues. *Proc. Natl. Acad. Sci.* **101:** 5339.

The Multiple Facets of Per2

U. ALBRECHT, A. BORDON, I. SCHMUTZ, AND J. RIPPERGER
Department of Medicine, Division of Biochemistry, University of Fribourg, 1700 Fribourg, Switzerland

The *Period 2* (*Per2*) gene is an important component of the circadian system. It appears to be not only part of the core oscillator mechanism, but also part of the input and output pathways of the clock. Because of its involvement at multiple levels of the circadian system, *Per2* needs to meet a variety of different demands. We discuss how *Per2* might be able to fulfill multiple functions by reviewing known facts and combine this with speculations based on these facts. This might provide new views about *Per2* function and help to better understand diseases that are rooted in the circadian system.

INTRODUCTION

Life on earth has used the sun as a reference point to time biological processes over the 24 hours of a day. To optimize energy expenditure and uptake, a mechanism developed to predict the cyclic availability of light. Because the period length of one cycle of such a timing device is circadian (circa diem = about 1 day) and not exactly 24 hours, it was termed a circadian clock. Due to its slight imprecision, the clock must be reset periodically to serve as a reliable predictor of solar time. Continuous clock adaptation is also needed because the earth's orbit around the sun generates the seasons that manifest themselves in alterations of day length over the year. To adapt to these changes, the circadian clock is connected to mechanisms that allow it to stay in tune with nature. Sensory organs communicate environmental time information via signaling pathways to the clock, thereby synchronizing the internal circadian oscillators with the environment. Therefore, the circadian clock can be viewed as the link between the environment and the genetic and biochemical machinery of an organism.

At the core of the mammalian molecular clock are two transcriptional activators (*Clock* and *Bmal1*) and factors negatively acting on them (*Per* and *Cry*). The heterodimer of *Clock/Bmal1* binds to E-box elements present in the *Per* and *Cry* promoters leading to their expression. The proteins enter the nucleus and disrupt *Clock/Bmal1* action thereby suspending the transcription of their own genes and of the *Rev-erbα* gene. This gene codes for an orphan nuclear receptor that negatively regulates *Bmal1* gene expression (Wijnen and Young 2006).

Because *Per* genes are at the core of the circadian clock mechanism, it is of interest to understand the function of these genes and proteins at the molecular level. Alterations in *Per* gene function affect many biochemical processes such as the cell cycle and metabolism. These in turn are linked to physiological process such as aging, brain dysfunction, and development of cancer. Hence, understanding *Per* function will unravel solutions to influence the aging process and provide insights into the treatment of cancer, depression, and metabolic diseases. In this chapter, we focus on the *Per2* gene in particular, because its malfunction has widespread effects on the organism, but its molecular function is largely unknown.

GENERAL FEATURES OF *PER2*

The *Per2* gene has been identified searching the Genbank database using human *Per1* (Rigui) for comparison (Albrecht et al. 1997). An open reading frame encoded by a human cDNA designated as KIAA0347 (Nagase et al. 1997) had high sequence homology with *Per1* and was designated *Per2* (Albrecht et al. 1997; Shearman et al. 1997). Human *Per2* was mapped to chromosome 2q37.3 (Toh et al. 2001). This gene was also found on chromosome 2 in zebra fish, opossum, and the chimpanzee (2b), whereas it is localized on chromosome 1 in the mouse, on chromosome 9 in the rat, on chromosome 12 in the macaque, and on scaffold 55 in *Xenopus tropicalis*. Homologs of *Per2* have also been identified in birds, lizards, mole rats, horse, vole, and sheep, but their chromosomal localization has not been determined. Here, we only consider the human and mouse *Per2* genes, both containing 23 exons, coding for transcripts of 6220 bp or 5805 bp and proteins of 1255 amino acids or 1257 amino acids, respectively. It appears that the mammalian *Per* gene paralogs probably emerged through gene duplications from an ancestral gene, resulting in four paralogs of which *Per4* was lost (von Schantz et al. 2006). Interestingly, mouse *Per2* appears to be more similar to *Drosophila per* than to mouse *Per1* (Albrecht et al. 1997) or *Per3*.

REGULATION OF *PER2* GENE EXPRESSION

The *Per2* gene is expressed in a circadian manner in the suprachiasmatic nucleus (SCN), and its expression is delayed by 4 hours as compared to *Per1* (Albrecht et al. 1997). *Per2* gene expression is not restricted to the brain but is expressed in almost all tissues. Furthermore, *Per2* expression in the SCN of mice can be strongly induced by light in the early subjective night but not in the late subjective night in contrast to *Per1* (Fig. 1a,b) (Yan and Silver 2002). Interestingly, *Per2* is also induced by blue light in the early subjective night in oral mucosa of humans (Cajochen et al. 2006). Apart from light, *Per2* can also be regulated by nonphotic stimuli such as 5-HT1A/7 receptor agonists (Horikawa et al. 2000), neuropeptide Y (Fukuhara et al. 2001), vasoactive intestinal polypeptide (Nielsen et al. 2002), glucocorticoids (Segall et al. 2006), and ghrelin (Yannielli et al. 2007).

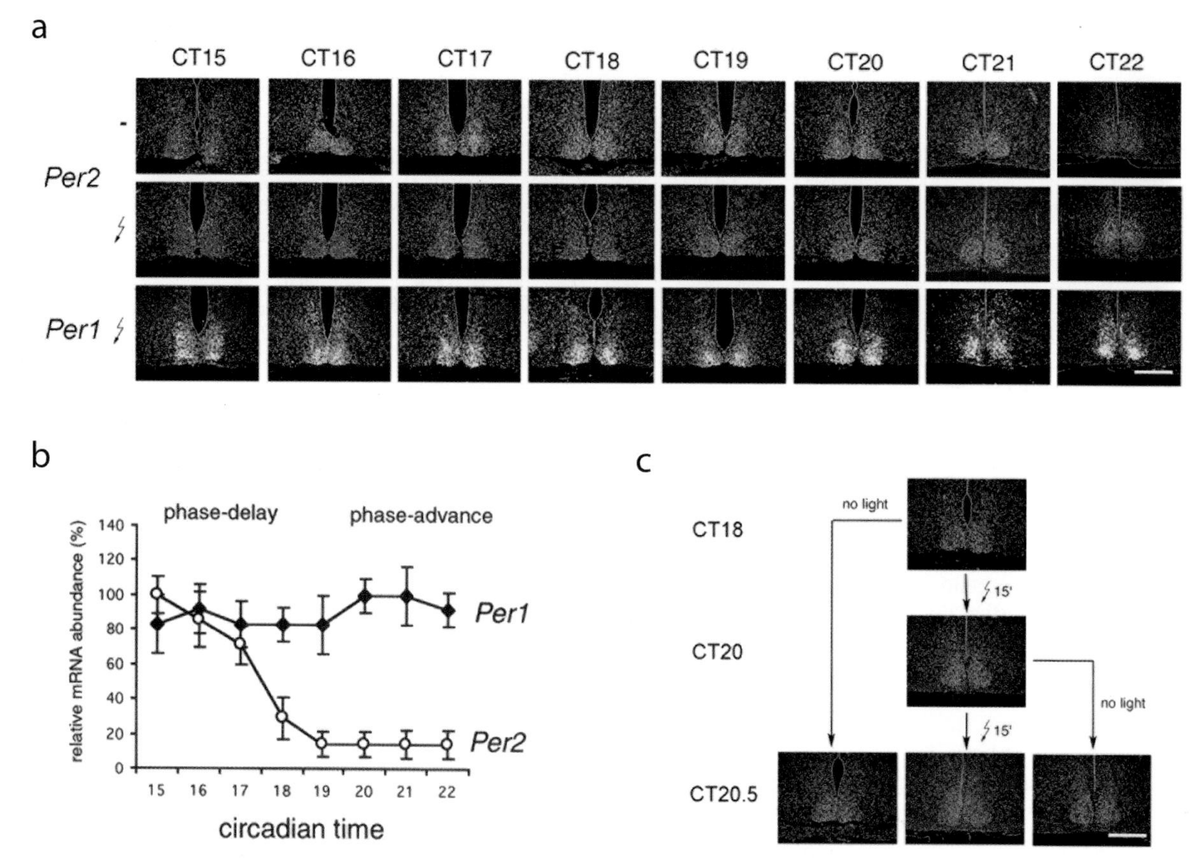

Figure 1. Light-induced levels of *Per2* correlate with the behavioral phase-delay domain. (*a*) Induction of *Per1* (*yellow*) and *Per2* (*red*) by a 15-minute light pulse applied at the circadian times (CTs) indicated (*blue* = cell nuclei stained with Hoechst dye). Endogenous *Per2* expression in control mice is low except for CT15, where residual expression is still detectable (*top row*). A light pulse induces this gene in a CT-dependent fashion (*center row*). Endogenous *Per1* expression is not expressed at any of the CTs examined, but it is induced by a light pulse in a CT-independent fashion (*bottom row*). Bar, 500 μm. (*b*) Graphical representation of the relationship between *Per1* and *Per2* induction and phase delay evoked by a light pulse. The abscissa plots the circadian time and the ordinate depicts the relative mRNA abundance with the maximal value for *Per1* set to 100%. (*Closed circles*) *Per1* expression levels in the suprachiasmatic nucleus (SCN); (*open circles*) *Per2* expression levels in the SCN. In the case of *Per2*, gene induction is correlated with the domain in the subjective night in which animals respond with a phase-delay to a light pulse. Each time point shows the mean ± S.E.M. of three different animals. *Per1* and *Per2* induction was assessed at each CT on adjacent sections of the same animal. (*c*) Effect of a double-light pulse at CT18 on *Per2* expression in the SCN. A single 15-minute light pulse applied at CT18 induced *Per2* weakly (see also *a*). However, a second 15-minute pulse evokes strong *Per2* expression. The control SCN that had not been exposed to light does not exhibit *Per2* expression. Bar, 500 μm.

These characteristics indicate an intricate manner of transcriptional regulation of the *Per2* gene. Toward a system-level understanding of the transcriptional circuitry regulating circadian clocks, Ueda et al. (2005) identified a number of clock-controlled elements in the *Per2* promoter. A noncanonical E-box enhancer (CATGTG, –497 in human and –163 in mouse, and CACGTT, –356 in human and –23 in mouse) drives circadian expression of *Per2* through *Clock/Bmal1*-mediated transcriptional activation (Fig. 2a) (Yoo et al. 2005; Akashi et al. 2006). Interestingly, this regulation appears to be tissue-specific, because absence of *Bmal1* in the liver leaves cyclic *Per2* expression unaffected, whereas absence of *Bmal1* in other tissues abolishes this expression (McDearmon et al. 2006). Tissue-specific regulation of *Per2* is underscored by the finding that its expression in the SCN and the liver is differentially affected by nutritional factors (Iwanaga et al. 2005).

Another important regulatory element appears to be a DBP/E4BP4-binding element (D-box). It influences *Per2* expression by a repressor-antiphasic-to-activator mechanism, which generates high-amplitude transcriptional activity (Ueda et al. 2005). The mouse *Per2* promoter contains two D-boxes, one at –151 (A site) and the other at +197 (B site) (Fig. 2a) with E4BP4 acting predominantly at the B site as a repressor (Ohno et al. 2007b). To start transcription, the tightly packed DNA in the promoter region of *Per2* must be opened. Rhythms in histone H3 acetylation and RNA polymerase II binding were observed to be synchronous with the *Per2* mRNA rhythms, indicating that histone acetylation might influence the expression of this gene (Etchegaray et al. 2003).

Acute induction of *Per2* seems to be mediated by a number of signaling pathways converging on the binding of phosphorylated cAMP-responsive element (CRE)-binding protein (CREB) on the CRE element present in the mouse *Per2* promoter at –1606 (Fig. 2a). However, this CRE element makes the *Per2* promoter much less responsive, in contrast to similar elements in the *Per1* promoter (Travnickova-Bendova et al. 2002). It appears

a

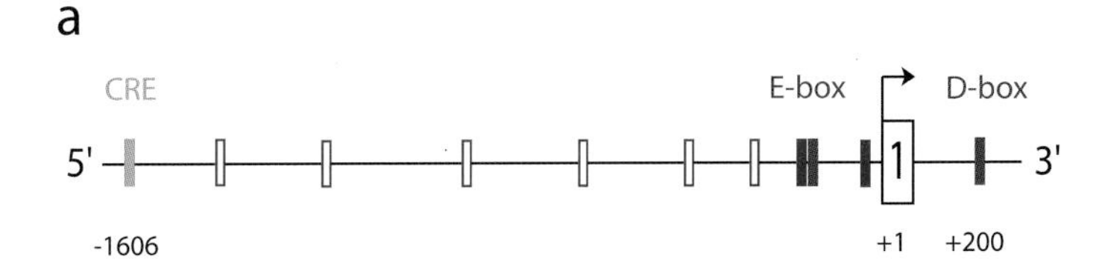

b

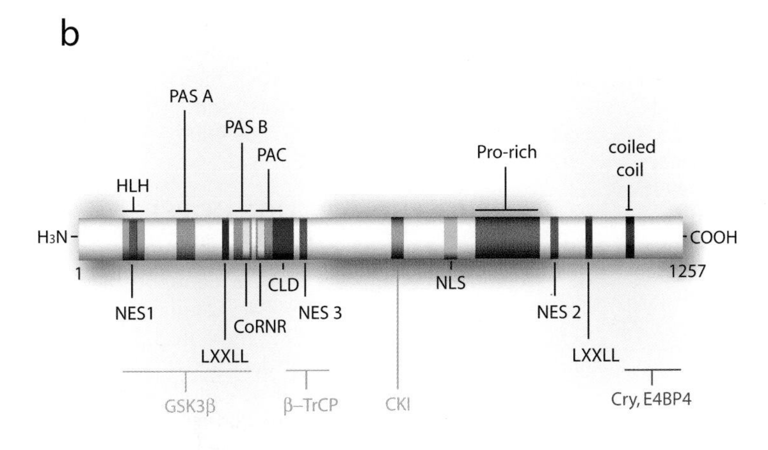

Figure 2. Murine *Per2* promoter and structural domains of mouse PER2 protein. (*a*) *Per2* promoter with CRE element (*green*), D boxes (*blue*), and E boxes (*red*). (*Filled red box*) E boxes sufficient for circadian expression of the gene; (*open black box*) exon 1; (*arrow*) transcription start site. Numbers below indicate the number of nucleotides upstream (negative values) or downstream (positive values) from the transcription start site (+1). (*b*) Structural domains and functional motifs in the PER2 protein. The PAS domain consists of PAS A, PAS B, and the PAC motifs (*gray*). (HLH) Helix-loop-helix motif (*green*); (NES1, 2, and 3) nuclear export sequence (*pink*); (L*XX*LL) motif found in coactivators to associate with nuclear receptors (*red*); (CoRNR) motif found in corepressors to associate with nuclear receptors (*yellow*); (CLD) cytoplasmic localization domain; (CKI) casein-kinase-binding site (*orange*); (NLS) nuclear localization sequence (*cyan*); (Pro-rich) proline-rich sequence (*brown*); (coiled-coil) dimerization domain (*purple*). (Yellow-shaded areas) Potential protein-binding domains; (*orange-shaded areas*) localization of phosphorylation sites.

that in the mouse, Per1 acts as an adapter to entrain the circadian clock to changing light/dark cycles by regulating PER2 protein (Masubuchi et al. 2005). This is underscored by our finding that *Per2* can be induced in the late subjective night after a priming light pulse (Fig. 1c). Per1 or another factor, which must be induced first, is probably needed for acute *Per2* gene expression.

Taken together, the *Per2* promoter responds to circadian clock factors (CLOCK/BMAL1), factors providing feedback from peripheral clocks (DBP/E4BP4), and to a lesser extent, factors immediately inducing expression (pCREB). There are many more potential regulatory elements in the *Per2* promoter; however, their functional importance has to be established (Kornmann et al. 2007).

STRUCTURAL DOMAINS AND FUNCTIONAL MOTIFS IN THE PER2 PROTEIN

The PER2 protein contains several structural domains illustrated in Figure 2b. The mouse *Per2* gene product contains an amino-terminal helix-loop-helix (HLH) motif that is probably not binding to DNA, because the preceding amino acids are not basic. A large region termed the PAS (PER, ARNT, Sim) domain that contains two imperfect repeats (PAS A and PAS B) followed by a PAS-like PAC motif can be found carboxy-terminal to the HLH motif. The PAS domains in animals are implicated in protein-protein interaction and dimerization (Huang et al. 1993; Gekakis et al. 1995; Yildiz et al. 2005). Glycogen synthase kinase-3β (Gsk3-β) has been reported to interact with the PAS A and PAS B domains of PER2 (Iitaka et al. 2005). However, PAS/PAC domains are speculated to serve as ligand-binding domains (Ponting and Aravind 1997) that could bind heme and serve as oxygen sensors. Indeed, heme has been identified as a prosthetic group of PER2 (Kaasik and Lee 2004), allowing the protein to sense redox status. Other evidence from bacteria suggests that PAS/PAC domains may act as light sensors resulting from FAD (flavin-adenine dinucleotide) binding (Hill et al. 1996; Bibikov et al. 1997). Although FAD binding has not been shown for PER2, it is known that *Per2* mutant mice, containing a PER2 protein lacking the carboxy-terminal half of the PAS B and the PAC domain (Zheng et al. 1999), show an altered behavioral response to light (Albrecht et al. 2001). Hence, it appears that the PAS/PAC domain in PER2 might enable this protein to sense cyclical variations in blue light and redox status.

Toward its carboxy-terminal end, the PER2 protein contains a coiled coil (Fig. 2b). This motif usually contains a repeated seven-amino-acid residue pattern forming a helix that can arrange with a helix of another protein to form a dimer that is held together by the burial of the hydrophobic surfaces in the helix dimer, hiding them from the water-filled environment. Interaction of the carboxy-terminal domain of PER2 with CRY (Miyazaki et al. 2001) and the leucine-zipper transcription factor E4BP4 has been reported (Ohno et al. 2007a). Hence, it appears that PER2 has the PAS domain plus a second structural feature that allows interaction with other proteins. Of note is that located between the PAS domain and the coiled-coil is a proline-rich sequence (Fig. 2b, brown). Proline residues make the amino acid chain less structured (no helices). Therefore, it appears that this proline-rich sequence divides the PER2 protein into two structural entities that also might fold back for binding of one protein. If both domains interact with different proteins, PER2 could serve as a scaffold bringing proteins in proximity to each other to exert a specific function. This is actually not a new concept looking at how hormone receptors and their coactivators and corepressors work (McKenna et al. 1999; Stallcup et al. 2003).

Interestingly, PER2 has some features found in coactivators and corepressors. For instance, *LXX*LL motifs (Fig. 2b, red)—where L is leucine and *X* can be any amino acid—are found in both potential protein–protein interaction domains of PER2. This motif is found in different members of coactivators interacting with nuclear receptors as, for example, the steroid hormone receptor coactivator-1 (SRC-1) (Onate et al. 1995). SRC-1 interacts with different nuclear receptors including the progesterone receptor, estrogen receptor, PPAR, or RXR, stimulating their transcriptional potential. Similar activities could be postulated for PER2 function; however, no interaction between PER2 and a nuclear receptor has been reported.

Analogous to the situation with coactivators, corepressors contain two motifs related to the L*XX*LL sequences (Hu and Lazar 1999; Nagy et al. 1999; Perissi et al. 1999). These motifs, which exhibit a consensus sequence of L*XX* I/H I *XXX* I/L, have also been termed CoRNR boxes. Such CoRNR boxes are present in the PAS/PAC domain of PER2, and it remains to be seen whether they give PER2 the potential to act as corepressor.

LOCALIZATION OF PER2 IN THE CELL

Transfection studies using COS-7 and NIH-3T3 cells revealed that exogenously expressed PER2 is localized in the cytoplasm and in the nucleus (Yagita et al. 2002). Deletion analysis in PER2 showed that this protein contains functional nuclear import and export signals (CLD, NLS, and NES1-3 in Fig. 2b) (Yagita et al. 2000, 2002). Coexpression with CRY1 or CRY2 proteins promotes nuclear entry of PER2 (Kume et al. 1999; Yagita et al. 2000, 2002; Miyazaki et al. 2001) but CRY1 is not required for nuclear import of PER2. However, the interaction between PER2 and CRY proteins appears to be important to prevent ubiquitylation of both proteins prolonging their presence in the cell (Yagita et al. 2002). Our own studies investigating the localization of PER2 in the cell using enhanced green fluorescent protein (EGFP)-tagged PER2 in nuclear escorting assays are largely in agreement with these previous studies (Fig. 3). We used human embryonic retinoblasts (HER) containing a partial adenovirus 5 (Ad5) genome. These cells are termed HER911 and have the advantage that they divide very slowly (about once in 40 hours; Fallaux et al. 1996) and therefore are very suitable to study localization of proteins over a longer period of time without cell division confounding the results. PER2-EGFP expressed in these cells is found in the nucleus, although some protein can be observed in the cytoplasm (Fig. 3). Other laboratories find PER2, when expressed alone, predominantly in the cytoplasm (Vielhaber et al. 2000). This discrepancy might be due to different cell lines used, indicating that modulation of the clock might be cell-type- and tissue-specific. Coexpression with CRY1 or CRY2 localizes PER2-EGFP entirely to the nucleus (Fig. 3, 2–4). Deletion of the CoRNR box in the PAS B/PAC domain leaves the protein largely in the cytoplasm, indicating that this motif has a role in nuclear escorting. Interestingly, CRY1 and CRY2 can transport this mutated form of PER2 to some extent to the nucleus (Fig. 3, 5–7). This indicates that either the CRY proteins interact at the CoRNR motif of PER2 for nuclear transport and/or that other proteins might influence nuclear translocation of PER2 through that motif. The deletion of the nuclear localization sequence (NLS) leaves the protein as expected in the cytoplasm. However, CRY1 and CRY2 can escort PER2 without NLS into the nucleus (Fig. 3, 9–11). Deletion of both the CoRNR motif and the NLS leaves the PER2 fusion protein largely in the cytoplasm, even when CRY proteins are coexpressed (Fig. 3, 12–14). These results suggest that the NLS in PER2 is not alone responsible for efficient nuclear localization, because deletion of the CoRNR motif in PER2 still containing the NLS leaves the protein in the cytoplasm. This indicates cooperativity between the NLS and CoRNR motifs with a possible involvement of CRY proteins. Similar results have been reported for PER3 (Yagita et al. 2000).

PHOSPHORYLATION OF PER2

Regulating the circadian clock through phosphorylation of its components appears to be a common theme in chronobiology (Edery et al. 1994; Tomita et al. 2005; Schafmeier et al. 2006). In PER2, more than 20 potential phosphorylation sites have been identified (Fig. 2b, orange shaded areas) (Schlosser et al. 2005; Vanselow et al. 2006) of which one site has received special attention in recent years. Patients with a specific form of familial advanced sleep phase syndrome have the mutation S662G in their PER2 protein, leading to a loss of binding of casein kinase 1ε/δ (CKIε/δ) and hypophosphorylation of PER2 (Toh et al. 2001). This leads to a shortened period length of the circadian clock. In vitro studies and mouse genetics revealed that alteration of this site (662 in human, 659 in mice) from S to G recapitulates the finding in humans. Furthermore, a change from S to D, mimicking constitutive phosphorylation charge and binding of CKIδ, increased phosphorylation of PER2 leading to a longer period length (Fig. 4) (Vanselow et al. 2006; Xu et al. 2007). This mutation also increased the amount of PER2 protein observed in cells and tissues, correlating phosphorylation with protein abundance and period length (Fig. 4) (Vanselow et al. 2006; Xu et al. 2007). PER2 protein abundance seems to be regulated at least in vitro by β-TrCP (β-transducin repeat-containing protein), an ubiquitin adapter protein, that binds CKIε phosphorylated PER2. This leads to polyubiquitination of PER2 followed by proteasome-mediated degradation (Eide et al. 2005).

In the future, it will be important to find the initiating kinase (red kinase in Fig. 4) that phosphorylates S662 (S659 in mice) and initiates the modulation of PER2 stability and variation of period length. A candidate could be Gsk-3β, a homolog of the *Drosophila* gene *shaggy* known to have a role in clock modulation (Martinek et al. 2001) and PER2 phosphorylation (Iitaka et al. 2005). However, it has not been determined where Gsk-3β phosphorylates PER2. The studies mentioned above all highlight the importance of the phosphorylation status of PER2 to modulate the clock. To balance the amount of phosphorylation, phosphatases seem to be critical. Protein phosphatase 1 (PP1) has been shown to remove phosphates on PER2, leading to its accelerated ubiquitin and proteasome-mediated degradation (Gallego et al. 2006). Although PER2 has no characteristic RVXF motif to which PP1 binds directly, regulators of PP1, such as I-1 and Darpp-32, whose expression varies during the circa-

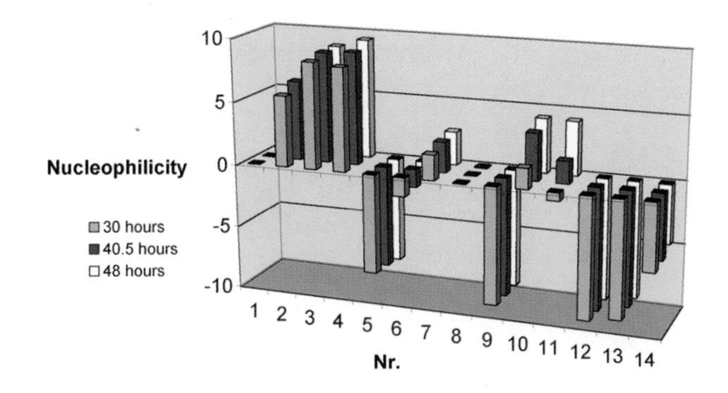

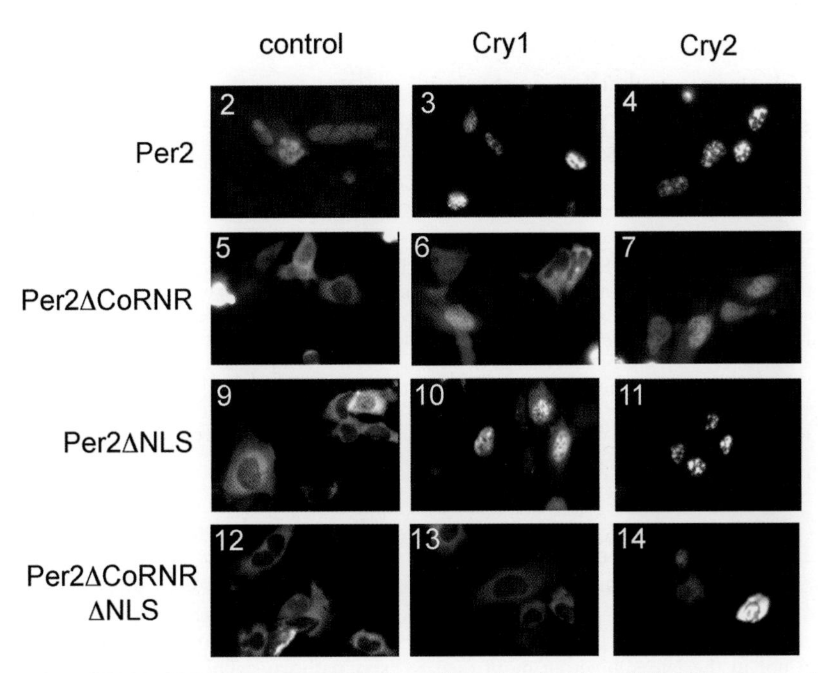

Figure 3. Nuclear escorting of PER2::GFP-fusion proteins. Human embryonic retinoblasts (HER911) were transfected with constructs coding for PER2::GFP or variants with deletions of the CoRNR and/or NLS motifs in PER2 (3 µg each). pcDNA3.1 vector containing no insert (control), CRY1, or CRY2 was cotransfected (3 µg each), and nucleophilicity of PER2::GFP-fusion proteins was assessed 30, 40.5, and 48 hours after transfection (*upper panel*). (*Lower panel*) Transfection scheme with typical illustrations 48 hours after transfection. (*Light yellow numbers*) Corresponding column in the summary graph in the upper panel. Columns 1 and 8 indicate the even distribution of the nonfused GFP protein in cells. Transfections were made with the linear polyethyleneimine method, and after fixation, the cells expressing GFP-tagged proteins were counted with the investigator not knowing the transfection scheme. The GFP-fusion proteins were assigned to five different categories according to their nuclear versus cytosolic fluorescence. N >> C, N > C, N = C, N < C, N << C with N standing for nuclear and C for cytosolic localization. For every time point, the experiment was performed three times.

dian cycle (Ueda et al. 2002), might be capable of targeting PP1 to PER2 in a time-dependent fashion.

A recent study reports that posttranscriptional regulation, such as phosphorylation, is sufficient to generate oscillations in a clock protein of cyanobacteria (Nakajima et al. 2005). Whether this could be the case in mammals was studied by inserting a single copy of exogenous *Per2* under the elongation factor-1α (EF-1α) promoter into NIH-3T3 fibroblasts, using a Flp-in system (Fujimoto et al. 2006). Interestingly, the exogenous PER2 protein oscillated without its coding mRNA cycling. Although the detailed mechanism regulating this process remains to be investigated, the bets are on phosphorylation and other posttranscriptional processes.

INVOLVEMENT OF PER2 IN BIOLOGICAL FUNCTIONS

Circadian Rhythms

The *Per2* gene has been discovered in a search for molecular components that make up the circadian clock. Subsequently, mutation analysis revealed that this gene has a prominent role in circadian clock function (Zheng et al. 1999, 2001; Bae et al. 2001) because lack of a functional PER2 protein resulted in gradual loss of circadian rhythmicity in wheel-running behavior as well as in cellular clock function (Brown et al. 2005). Interestingly, *Per2* affects the core clock mechanism by regulating *Bmal1* expression in a positive manner (Shearman et al.

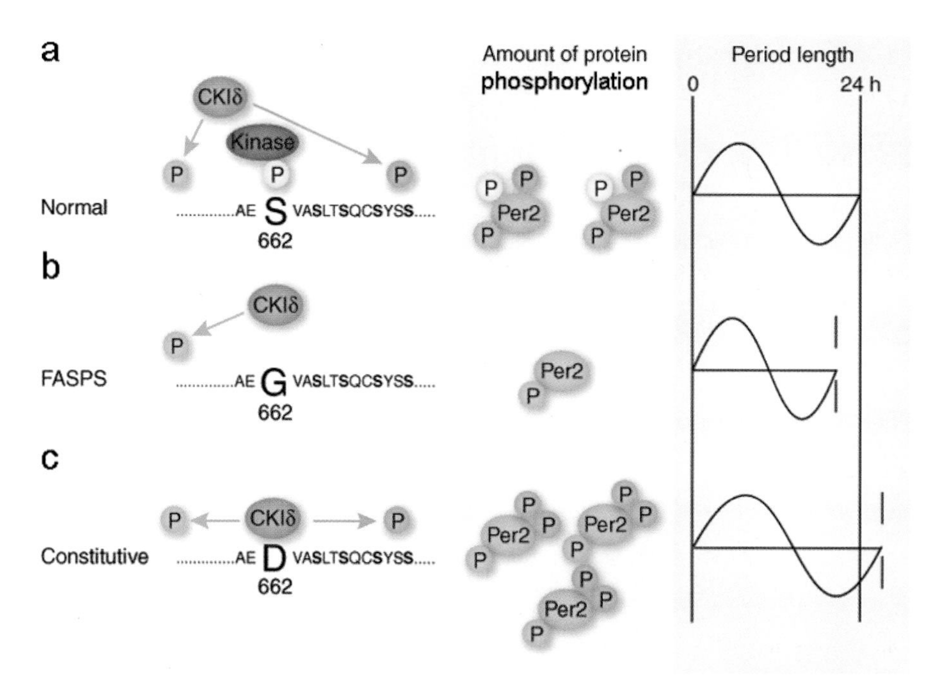

Figure 4. Differential phosphorylation of PER2 modulates period length of the circadian clock. (*a*) Amino acid sequence of normal human PER2 with significant serine (S) residues in bold. S662 is phosphorylated (*yellow*) by an unknown kinase (*red*), enabling casein kinase Iδ (CKIδ) to bind and phosphorylate adjacent serine residues (*blue path*). Independently of S662 phosphorylation, CKIδ phosphorylates other residues in PER2 (*green path*). This leads to intermediate levels of PER2 protein defining a period length of 24 hours. (*b*) Sequence of PER2 with the FASPS amino acid change S662G. CKIδ phosphorylates PER2 only through the *green path*. This leads to low levels of PER2 protein and a shortening of the clock period. (*c*) S662D mutation leads to constitutive binding of CKIδ and phosphorylation of adjacent S residues (*blue path*). Phosphorylation on other sites still occurs (*green path*). This leads to elevated levels of PER2 and a long-period length. (Reprinted, with permission, from Albrecht 2007 [©Nature Publishing Group].)

2000; Oster et al. 2002). Astonishingly, deletion of *Cry2* from *Per2* mutant mice reestablished circadian behavior (Oster et al. 2002). It remains to be seen how this is possible at the molecular level.

Because *Per2* is inducible by a light pulse in the early subjective night but not in the late subjective night (see Fig. 1a,b), it was speculated that it has a function in delaying clock phase. In agreement with that notion, *Per2* mutant mice show impaired clock resetting, especially in the delay domain (Albrecht et al. 2001; Spoelstra et al. 2004). These findings indicate a function of *Per2* in clock resetting. Consistent with such a function is the finding that a mutation in the human *Per2* gene, which abolishes normal phosphorylation of its protein, leads to familial advanced sleep phase syndrome (see above and Fig. 4b) (Toh et al. 2001). Mutation experiments in mice have confirmed that alteration of the PER2 phosphorylation pattern through mutation of the *Per2* gene leads to clock period alterations (Vanselow et al. 2006; Xu et al. 2007). A single-nucleotide polymorphism in the 5′-untranslated region of the human *Per2* gene has been associated with diurnal preference (Carpen et al. 2005), and a case study of a patient with recurrent hypersomnia revealed differences in *Per2* gene expression in remission and hypersomnia (Tomoda et al. 2003), further illustrating an involvement of this gene in clock adaptation and sleep regulation.

Cancer

Mice without functional PER2 are prone to develop cancer and display altered expression of genes involved in cell cycle regulation and tumor suppression such as cyclin D1, cyclin A, Myc, and Mdm2 (Fu et al. 2002). In particular, Myc is controlled directly by circadian regulators including *Per2*. Therefore, it appears that PER2 has a role in tumor suppression by regulating DNA-damage-responsive pathways. This notion is bolstered by the observation that expression of *Per1*, *Per2*, and *Per3* is deregulated in breast cancer tissue (Chen et al. 2005), although it is not clear whether this is the cause or the consequence of breast cancer. Improvement of tumor control in mice bearing Glasgow osteosarcoma was achieved by using seliciclib, a cyclin-dependent kinase inhibitor, which increased amplitudes of various clock genes including *Per2* (Iurisci et al. 2006). Furthermore, overexpression of *Per2* seems to induce cancer cell apoptosis (Hua et al. 2006), which will have to be confirmed by complementary studies. Interestingly, however, *Per2* was identified in a large-scale RNA interference (RNAi) screen in human cells to be a component of the p53 pathway (Berns et al. 2004), strongly supporting a role of *Per2* in tumor suppression. However, it is entirely unclear how PER2 is involved in that process mechanistically.

Immune System

Recent evidence suggests an involvement of the *Per2* gene in the immune system. Mice carrying a loss-of-function mutation of the *Per2* gene display a loss of interferon-γ (IFN-γ) mRNA cycling in the spleen (Arjona and Sarkar 2006). Furthermore, *Per2*-deficient mice are more resistant to lipopolysaccharide-induced endotoxic shock than wild-type mice. This is accompanied by decreased levels of proinflamatory cytokines such as IFN-γ and interleukin-1β in the serum. The impaired IFN-γ production seems to be attributable to defective natural killer cell function (Liu et al. 2006). This is an interesting observation in view of the involvement of *Per2* in cancer development described above, because IFN-γ activates macrophage and antitumor functions. How *Per2* impinges on the production of IFN-γ and function of natural killer cells remains to be discovered.

Cardiovascular System

Disturbed diurnal rhythm alters *Per2* gene expression and exacerbates cardiovascular disease in a mouse model of pressure overload cardiac hypertrophy. Upon resynchronization of the environmental light/dark cycle with the animal's internal cycle, cardiovascular parameters as well as *Per2* gene expression are normalized (Martino et al. 2007), suggesting a potential involvement of *Per2* in the cardiovascular system. In agreement with this notion is the finding that mice with a loss-of-function mutation in PER2 display altered vascular endothelial function due to a decreased production of nitric oxide and vasodilatory prostaglandins as well as an increased release of COX-1-derived vasoconstrictors in aortic rings (Viswambharan et al. 2007). Whether COX-1 is a direct target gene of PER2 and the clock mechanism or is regulated secondarily remains to be investigated. Interestingly, vascular endothelial growth factor (VEGF) appears to be regulated by clock components including PER2 as a negative regulator (Koyanagi et al. 2003). This finding is consistent with a role of PER2 as a tumor suppressor because it reduces expression of VEGF in hypoxic tumor cells, leading to reduction of vascularization of tumors.

Metabolism

That *Per2* is involved in metabolic control is suggested by a study investigating the hepatic proteome in normal and *Per2* mutant mice (Reddy et al. 2006). In this study, the authors find that many proteins that cycle in their expression over 24 hours lose this property if *Per2* is absent (e.g., aldolase, arginase, and catalase). The alteration of arginase in these mice is consistent with the finding that these animals have abnormal endothelial function due to an alteration in NO (nitric oxide) signaling (see above) (Viswambharan et al. 2007), because arginase is a critical enzyme in NO metabolism. Furthermore, *Per2* mutant mice show reduced endurance of muscles accompanied by increased levels of glycolytic enzymes in the anterior tibialis muscle, indicating a greater dependence on anaerobic metabolism under stress conditions (Bae et al. 2006). This is consistent with our own findings that *Per2* mutant mice show reduced muscle strength under stress conditions as revealed by the hanging wire test, in which the latency time for falling is much shorter for *Per2* mice compared to wild-type animals (Fig. 5).

A recent study highlights the function of PER2 in connecting the clock and metabolism by showing that PER2 not only is a part of the circadian clock influencing signaling and metabolic pathways, but also responds to systemic cues, thereby linking the clock and metabolism in a interdependent fashion (Kornmann et al. 2007). PER2 appears to be the cogwheel that connects the liver clock with the organism's physiology and biochemistry.

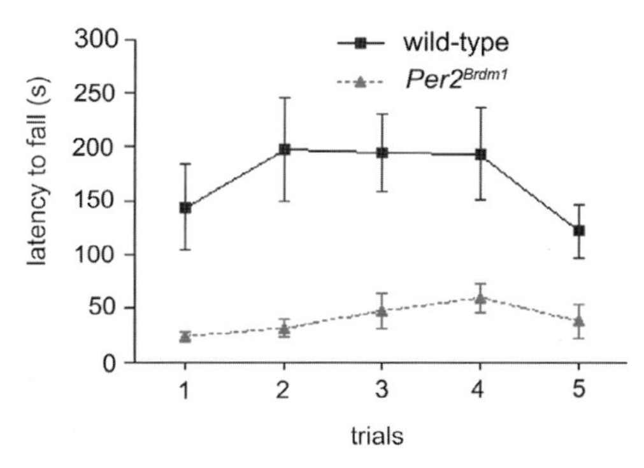

Figure 5. Muscle endurance in mice with and without functional PER2. Male mice (2–6 months old) were placed on a wire cage lid, which was turned upside down at a height of approximately 20 cm over a flat surface table. The time between the turning of the lid and the fall of the mouse was measured after the mice were familiar with this hanging-wire test. The latency to fall of wild-type (*closed squares*) and PER2 mutant (*gray triangles*) mice over five consecutive trials on 5 days was assessed. Two-way ANOVA with Bonferroni posthoc test revealed a significant difference for genotype with $p < 0.001$ but no significant influence of time. Data points are represented as mean ± S.E.M. with $n = 6$ for each genotype.

Nervous System

Apart from its function in the clock mechanism in the SCN, *Per2* appears to have additional functions in other brain areas. It is postulated that a food entrainable oscillator (FEO) resides in the brain, which is responsible for anticipatory activity in expectation of regularly scheduled meals. A mutation in *Per2* leads to loss of food anticipatory activity in mice, suggesting an important role of *Per2* in the FEO (Feillet et al. 2006). Because the mutation affects the whole organism, it remains to be shown that the observed loss-of-food anticipation in these mice is specific for *Per2* function in the brain and is not due to the role of *Per2* in the liver.

Interestingly, several studies have found that overeating shares neurobiological mechanisms with the addictive properties of drugs of abuse (for review, see Simerly 2006). Because PER2 not only affects food anticipatory behavior, but also modulates the effects of drugs of abuse (Abarca et al. 2002; Spanagel et al. 2005), we postulate that PER2 influences the neurobiological circuitry that is common to feeding signals and drugs. Because specific areas of the midbrain such as the arcuate nucleus and the ventral tegmental area are integrating these signals, a specific function of PER2 in these brain regions can be envisaged.

That PER2 has a central role in sleep, especially in familial advanced sleep phase syndrome, has been described above. Because sleep appears to involve remodeling of neuronal connectivity by strengthening or weakening synaptic connections, it is conceivable to suspect an involvement of PER2 in these processes. A recent study shows that PER2 is involved in gating light/dark information to vesicular glutamate transporter 1 (vGLUT1) content on synaptic vesicles (Yelamanchili et al. 2006). This leads to alterations in glutamate filling of these vesicles.

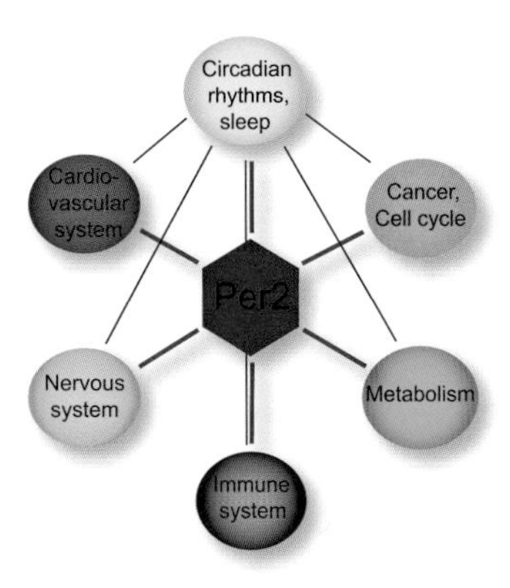

Figure 6. Schematic diagram illustrating various body functions influenced by PER2 (*red lines*) and other clock components (*black lines*).

Changing the amount of glutamate that can be released by neurons will impinge on synaptic strength. Hence, environmental signals such as light can have an impact on brain function. Therefore, one could assume that health problems in industrialized societies, where the natural day/night regime is largely ignored, are rooted in the misalignment between different clocks affecting normal brain function. As a consequence depression (Partonen et al. 2007), excessive alcohol consumption and overeating is observed.

CONCLUSIONS

In the past decade, many of the molecular components of the circadian clock have been identified in several organisms. The concept emerged that there is not only one clock ticking in the brain but virtually any tissue and every cell contains the cogwheels for a clock. How do these different clocks interact with each other to produce a coherent systemic rhythm useful for the entire organism? From the above description of the structural organization and function of PER2, it seems that this molecule unifies several characteristics necessary for a function to serve as an integrator (Fig. 6). Its promoter seems to be responsive to a plethora of transcriptional activators, the protein can be finely tuned in its stability and localization by posttranscriptional modifications, and it seems to interact with various proteins to mediate specific functions. Therefore, the next steps to understand the circadian system as a whole will be to study tissue-specific regulation of clock components and in particular PER2 function. Genetic tools to achieve this goal are available. At the protein level, resolving the protein structure will help to identify interacting partners, and potential pharmacological agents binding to PER2 might be discovered. This will help to find tissue-specific cooperation partners of this protein and provide explanations for the observed activator and repressor potential of PER2, elucidating the mysteries of this Janus-faced molecule.

ACKNOWLEDGMENTS

Support from the Swiss National Science Foundation, the Velux Foundation, and the European Union Project EUCLOCK is acknowledged.

REFERENCES

Abarca C., Albrecht U., and Spanagel R. 2002. Cocaine sensitization and reward are under the influence of circadian genes and rhythm. *Proc. Natl. Acad. Sci.* **99:** 9026.

Akashi M., Ichise T., Mamine T., and Takumi T. 2006. Molecular mechanism of cell-autonomous circadian gene expression of Period2, a crucial regulator of the mammalian circadian clock. *Mol. Biol. Cell* **17:** 555.

Albrecht U. 2007. Per2 has time on its side. *Nat. Chem. Biol.* **3:** 139.

Albrecht U., Sun Z.S., Eichele G., and Lee C.C. 1997. A differential response of two putative mammalian circadian regulators, mper1 and mper2, to light. *Cell* **91:** 1055.

Albrecht U., Zheng B., Larkin D., Sun Z.S., and Lee C.C. 2001. *mPer1* and *mPer2* are essential components for normal resetting of the circadian clock. *J. Biol. Rhythms* **16:** 100.

Arjona A. and Sarkar D.K. 2006. The circadian gene *mPer2* regulates the daily rhythm of IFN-gamma. *J. Interferon Cytokine Res.* **26:** 645.

Bae K., Jin X., Maywood E.S., Hastings M.H., Reppert S.M., and Weaver D. 2001. Differential functions of *mPer1*, *mPer2*, and *mPer3* in the SCN circadian clock. *Neuron* **30:** 525.

Bae K., Lee K., Seo Y., Lee H., Kim D., and Choi I. 2006. Differential effects of two period genes on the physiology and proteomic profiles of mouse anterior tibialis muscles. *Mol. Cells* **22:** 275.

Berns K., Hijmans E.M., Mullenders J., Brummelkamp T.R., Velds A., Heimerikx M., Kerkhoven R.M., Madiredjo M., Nijkamp W., Weigelt B., Agami R., Ge W., Cavet G., Linsley P.S., Beijersbergen R.L., and Bernards R. 2004. A large-scale RNAi screen in human cells identifies new components of the p53 pathway. *Nature* **428:** 431.

Bibikov S.I., Biran R., Rudd K.E., and Parkinson J.S. 1997. A signal transducer for aerotaxis in *Escherichia coli*. *J. Bacteriol.* **179:** 4075.

Brown S.A., Fleury-Olela F., Nagoshi E., Hauser C., Juge C., Meier C.A., Chicheportiche R., Dayer J.M., Albrecht U., and Schibler U. 2005. The period length of fibroblast circadian gene expression varies widely among human individuals. *PLoS Biol.* **3:** e338.

Cajochen C., Jud C., Munch M., Kobialka S., Wirz-Justice A., and Albrecht U. 2006. Evening exposure to blue light stimulates the expression of the clock gene PER2 in humans. *Eur. J. Neurosci.* **23:** 1082.

Carpen J.D., Archer S.N., Skene D.J., Smits M., and von Schantz M. 2005. A single-nucleotide polymorphism in the 5′-untranslated region of the hPER2 gene is associated with diurnal preference. *J. Sleep Res.* **14:** 293.

Chen S.T., Choo K.B., Hou M.F., Yeh K.T., Kuo S.J., and Chang J.G. 2005. Deregulated expression of the PER1, PER2 and PER3 genes in breast cancers. *Carcinogenesis* **26:** 1241.

Edery I., Zwiebel L.J., Dembinska M.E., and Rosbash M. 1994. Temporal phosphorylation of the *Drosophila* period protein. *Proc. Natl. Acad. Sci.* **91:** 2260.

Eide E.J., Woolf M.F., Kang H., Woolf P., Hurst W., Camacho F., Vielhaber E.L., Giovanni A., and Virshup D.M. 2005. Control of mammalian circadian rhythm by CKIε-regulated proteasome-mediated PER2 degradation. *Mol. Cell. Biol.* **25:** 2795.

Etchegaray J.P., Lee C., Wade P.A., and Reppert S.M. 2003. Rhythmic histone acetylation underlies transcription in the mammalian circadian clock. *Nature* **421:** 177.

Fallaux F.J., Kranenburg O., Cramer S.J., Houweling A., Van Ormondt H., Hoeben R.C., and Van Der Eb A.J. 1996. Characterization of 911: A new helper cell line for the titration and propagation of early region 1-deleted adenoviral vectors. *Hum. Gene Ther.* **7:** 215.

Feillet C.A., Ripperger J.A., Magnone M.C., Dulloo A., Albrecht U., and Challet E. 2006. Lack of food anticipation in Per2 mutant mice. *Curr. Biol.* **16:** 2016.

Fu L., Pelicano H., Liu J., Huang P., and Lee C.C. 2002. The circadian gene period 2 plays an important role in tumor suppression and DNA damage response in vivo. *Cell* **111:** 41.

Fujimoto Y., Yagita K., and Okamura H. 2006. Does mPER2 protein oscillate without its coding mRNA cycling?: Post-transcriptional regulation by cell clock. *Genes Cells* **11:** 525.

Fukuhara C., Brewer J.M., Dirden J.C., Bittman E.L., Tosini G., and Harrington M.E. 2001. Neuropeptide Y rapidly reduces Period 1 and Period 2 mRNA levels in the hamster suprachiasmatic nucleus. *Neurosci. Lett.* **314:** 119.

Gallego M., Kang H., and Virshup D.M. 2006. Protein phosphatase 1 regulates the stability of the circadian protein PER2. *Biochem. J.* **399:** 169.

Gekakis N., Saez L., Delahaye-Brown A.M., Myers M.P., Sehgal A., Young M.W., and Weitz C.J. 1995. Isolation of timeless by PER protein interaction: Defective interaction between timeless protein and long-period mutant PERL. *Science* **270:** 811.

Hill S., Austin S., Eydmann T., Jones T., and Dixon R. 1996. Azotobacter vinelandii NIFL is a flavoprotein that modulates transcriptional activation of nitrogen-fixation genes via a redox-sensitive switch. *Proc. Natl. Acad. Sci.* **93:** 2143.

Horikawa K., Yokota S., Fuji K., Akiyama M., Moriya T., Okamura H., and Shibata S. 2000. Nonphotic entrainment by 5-HT1A/7 receptor agonists accompanied by reduced Per1 and Per2 mRNA levels in the suprachiasmatic nuclei. *J. Neurosci.* **20:** 5867.

Hu X. and Lazar M.A. 1999. The CoRNR motif controls the recruitment of corepressors by nuclear hormone receptors. *Nature* **402:** 93.

Hua H., Wang Y., Wan C., Liu Y., Zhu B., Yang C., Wang X., Wang Z., Cornelissen-Guillaume G., and Halberg F. 2006. Circadian gene mPer2 overexpression induces cancer cell apoptosis. *Cancer Sci.* **97:** 589.

Huang Z.J., Edery I., and Rosbash M. 1993. PAS is a dimerization domain common to *Drosophila* period and several transcription factors. *Nature* **364:** 259.

Iitaka C., Miyazaki K., Akaike T., and Ishida N. 2005. A role for glycogen synthase kinase-3beta in the mammalian circadian clock. *J. Biol. Chem.* **280:** 29397.

Iurisci I., Filipski E., Reinhardt J., Bach S., Gianella-Borradori A., Iacobelli S., Meijer L., and Levi F. 2006. Improved tumor control through circadian clock induction by Seliciclib, a cyclin-dependent kinase inhibitor. *Cancer Res.* **66:** 10720.

Iwanaga H., Yano M., Miki H., Okada K., Azama T., Takiguchi S., Fujiwara Y., Yasuda T., Nakayama M., Kobayashi M., Oishi K., Ishida N., Nagai K., and Monden M. 2005. Per2 gene expressions in the suprachiasmatic nucleus and liver differentially respond to nutrition factors in rats. *JPEN J. Parenter. Enteral Nutr.* **29:** 157.

Kaasik K. and Lee C.C. 2004. Reciprocal regulation of haem biosynthesis and the circadian clock in mammals. *Nature* **430:** 467.

Kornmann B., Schaad O., Bujard H., Takahashi J.S., and Schibler U. 2007. System-driven and oscillator-dependent circadian transcription in mice with a conditionally active liver clock. *PLoS Biol.* **5:** e34.

Koyanagi S., Kuramoto Y., Nakagawa H., Aramaki H., Ohdo S., Soeda S., and Shimeno H. 2003. A molecular mechanism regulating circadian expression of vascular endothelial growth factor in tumor cells. *Cancer Res.* **63:** 7277.

Kume K., Zylka M.J., Sriram S., Shearman L.P., Weaver D.R., Jin X., Maywood E.S., Hastings M.H., and Reppert S.M. 1999. mCRY1 and mCRY2 are essential components of the negative limb of the circadian clock feedback loop. *Cell* **98:** 193.

Liu J., Mankani G., Shi X., Meyer M., Cunningham-Runddles S., Ma X., and Sun Z.S. 2006. The circadian clock Period 2 gene regulates gamma interferon production of NK cells in host response to lipopolysaccharide-induced endotoxic shock. *Infect. Immun.* **74:** 4750.

Martinek S., Inonog S., Manoukian A.S., and Young M.W. 2001. A role for the segment polarity gene shaggy/GSK-3 in the *Drosophila* circadian clock. *Cell* **105:** 769.

Martino T.A., Tata N., Belsham D.D., Chalmers J., Straume M., Lee P., Pribiag H., Khaper N., Liu P.P., Dawood F., Backx P.H., Ralph M.R., and Sole M.J. 2007. Disturbed diurnal rhythm alters gene expression and exacerbates cardiovascular disease with rescue by resynchronization. *Hypertension* **49:** 1104.

Masubuchi S., Kataoka N., Sassone-Corsi P., and Okamura H. 2005. Mouse Period1 (mPER1) acts as a circadian adaptor to entrain the oscillator to environmental light/dark cycles by regulating mPER2 protein. *J. Neurosci.* **25:** 4719.

McDearmon E.L., Patel K.N., Ko C.H., Walisser J.A., Schook A.C., Chong J.L., Wilsbacher L.D., Song E.J., Hong H.K., Bradfield C.A., and Takahashi J.S. 2006. Dissecting the functions of the mammalian clock protein BMAL1 by tissue-specific rescue in mice. *Science* **314:** 1304.

McKenna N.J., Lanz R.B., and O'Malley B.W. 1999. Nuclear receptor coregulators: Cellular and molecular biology. *Endocr. Rev.* **20:** 321.

Miyazaki K., Mesaki M., and Ishida N. 2001. Nuclear entry mechanism of rat PER2 (rPER2): Role of rPER2 in nuclear localization of CRY protein. *Mol. Cell. Biol.* **21:** 6651.

Nagase T., Ishikawa K., Nakajima D., Ohira M., Seki N., Miyajima N., Tanaka A., Kotani H., Nomura N., and Ohara O. 1997. Prediction of the coding sequences of unidentified human genes. VII. The complete sequences of 100 new cDNA clones from brain which can code for large proteins in vitro. *DNA Res.* **4:** 141.

Nagy L., Kao H.Y., Love J.D., Li C., Banayo E., Gooch J.T., Krishna V., Chatterjee K., Evans R.M., and Schwabe J.W. 1999. Mechanism of corepressor binding and release from nuclear hormone receptors. *Genes Dev.* **13:** 3209.

Nakajima M., Imai K., Ito H., Nishiwaki T., Murayama Y., Iwasaki H., Oyama T., and Kondo T. 2005. Reconstitution of circadian oscillation of cyanobacterial KaiC phosphorylation in vitro. *Science* **308:** 414.

Nielsen H.S., Hannibal J., and Fahrenkrug J. 2002. Vasoactive intestinal polypeptide induces per1 and per2 gene expression in the rat suprachiasmatic nucleus late at night. *Eur. J. Neurosci.* **15:** 570.

Ohno T., Onishi Y., and Ishida N. 2007a. The negative transcription factor E4BP4 is associated with circadian clock protein PERIOD2. *Biochem. Biophys. Res. Commun.* **354:** 1010.

———. 2007b. A novel E4BP4 element drives circadian expression of mPeriod2. *Nucleic Acids Res.* **35:** 648.

Onate S.A., Tsai S.Y., Tsai M.J., and O'Malley B.W. 1995. Sequence and characterization of a coactivator for the steroid hormone receptor superfamily. *Science* **270**: 1354.

Oster H., Yasui A., van der Horst G.T., and Albrecht U. 2002. Disruption of mCry2 restores circadian rhythmicity in mPer2 mutant mice. *Genes Dev.* **16:** 2633.

Partonen T., Treutlein J., Alpman A., Frank J., Johansson C., Depner M., Aron L., Rietschel M., Wellek S., Soronen P., Paunio T., Koch A., Chen P., Lathrop M., Adolfsson R., Persson M.L., Kasper S., Schalling M., Peltonen L., and Schumann G. 2007. Three circadian clock genes Per2, Arntl, and Npas2 contribute to winter depression. *Ann. Med.* **39:** 229.

Perissi V., Staszewski L.M., McInerney E.M., Kurokawa R., Krones A., Rose D.W., Lambert M.H., Milburn M.V., Glass C.K., and Rosenfeld M.G. 1999. Molecular determinants of nuclear receptor-corepressor interaction. *Genes Dev.* **13:** 3198.

Ponting C.P. and Aravind L. 1997. PAS: A multifunctional domain family comes to light. *Curr. Biol.* **7:** R674.

Reddy A.B., Karp N.A., Maywood E.S., Sage E.A., Deery M., O'Neill J.S., Wong G.K., Chesham J., Odell M., Lilley K.S., Kyriacou C.P., and Hastings M.H. 2006. Circadian orchestration of the hepatic proteome. *Curr. Biol.* **16:** 1107.

Schafmeier T., Kaldi K., Diernfellner A., Mohr C., and Brunner M. 2006. Phosphorylation-dependent maturation of *Neurospora* circadian clock protein from a nuclear repressor toward a cytoplasmic activator. *Genes Dev.* **20:** 297.

Schlosser A., Vanselow J.T., and Kramer A. 2005. Mapping of phosphorylation sites by a multi-protease approach with specific phosphopeptide enrichment and NanoLC-MS/MS analysis. *Anal. Chem.* **77:** 5243.

Segall L.A., Perrin J.S., Walker C.D., Stewart J., and Amir S. 2006. Glucocorticoid rhythms control the rhythm of expression of the clock protein, Period2, in oval nucleus of the bed nucleus of the stria terminalis and central nucleus of the amygdala in rats. *Neuroscience* **140:** 753.

Shearman L.P., Zylka M.J., Weaver D.R., Kolakowski L.F., and Reppert S.M. 1997. Two *period* homologs: Circadian expression and photic regulation in the suprachiasmatic nuclei. *Neuron* **19:** 1261.

Shearman L.P., Sriram S., Weaver D.R., Maywood E.S., Chaves I., Zheng B., Kume K., Lee C.C., van der Horst G.T., Hastings

M.H., and Reppert S.M. 2000. Interacting molecular loops in the mammalian circadian clock. *Science* **288:** 1013.

Simerly R. 2006. Feeding signals and drugs meet in the midbrain. *Nat. Med.* **12:** 1244.

Spanagel R., Pendyala G., Abarca C., Zghoul T., Sanchis-Segura C., Magnone M.C., Lascorz J., Depner M., Holzberg D., Soyka M., Schreiber S., Matsuda F., Lathrop M., Schumann G., and Albrecht U. 2005. The clock gene *Per2* influences the glutamatergic system and modulates alcohol consumption. *Nat. Med.* **11:** 35.

Spoelstra K., Albrecht U., van der Horst G.T., Brauer V., and Daan S. 2004. Phase responses to light pulses in mice lacking functional per or cry genes. *J. Biol. Rhythms* **19:** 518.

Stallcup M.R., Kim J.H., Teyssier C., Lee Y.H., Ma H., and Chen D. 2003. The roles of protein-protein interactions and protein methylation in transcriptional activation by nuclear receptors and their coactivators. *J. Steroid Biochem. Mol. Biol.* **85:** 139.

Toh K.L., Jones C.R., He Y., Eide E.J., Hinz W.A., Virshup D.M., Ptacek L.J., and Fu Y.-H. 2001. An hPer2 phosphorylation site mutation in familial advanced sleep phase syndrome. *Science* **291:** 1040.

Tomita J., Nakajima M., Kondo T., and Iwasaki H. 2005. No transcription-translation feedback in circadian rhythm of KaiC phosphorylation. *Science* **307:** 251.

Tomoda A., Joudoi T., Kawatani J., Ohmura T., Hamada A., Tonooka S., and Miike T. 2003. Case study: Differences in human Per2 gene expression, body temperature, cortisol, and melatonin parameters in remission and hypersomnia in a patient with recurrent hypersomnia. *Chronobiol. Int.* **20:** 893.

Travnickova-Bendova Z., Cermakian N., Reppert S.M., and Sassone-Corsi P. 2002. Bimodal regulation of *mPeriod* promoters by CREB-dependent signaling and CLOCK/BMAL1 activity. *Proc. Natl. Acad. Sci.* **99:** 7728.

Ueda H.R., Hayashi S., Chen W., Sano M., Machida M., Shigeyoshi Y., Iino M., and Hashimoto S. 2005. System-level identification of transcriptional circuits underlying mammalian circadian clocks. *Nat. Genet.* **37:** 187.

Ueda H.R., Chen W., Adachi A., Wakamatsu H., Hayashi S., Takasugi T., Nagano M., Nakahama K., Suzuki Y., Sugano S., Iino M., Shigeyoshi Y., and Hashimoto S. 2002. A transcription factor response element for gene expression during circadian night. *Nature* **418:** 534.

Vanselow K., Vanselow J.T., Westermark P.O., Reischl S., Maier B., Korte T., Herrmann A., Herzel H., Schlosser A., and Kramer A. 2006. Differential effects of PER2 phosphorylation: Molecular basis for the human familial advanced sleep phase syndrome (FASPS). *Genes Dev.* **20:** 2660.

Vielhaber E., Eide E., Rivers A., Gao Z.H., and Virshup D.M. 2000. Nuclear entry of the circadian regulator mPER1 is controlled by mammalian casein kinase I epsilon. *Mol. Cell. Biol.* **20:** 4888.

Viswambharan H., Carvas J.M., Antic V., Marecic A., Jud C., Zaugg C.E., Ming X.F., Montani J.P., Albrecht U., and Yang Z. 2007. Mutation of the circadian clock gene *Per2* alters vascular endothelial function. *Circulation* **115:** 2188.

von Schantz M., Jenkins A., and Archer S.N. 2006. Evolutionary history of the vertebrate period genes. *J. Mol. Evol.* **62:** 701.

Wijnen H. and Young M.W. 2006. Interplay of circadian clocks and metabolic rhythms. *Annu. Rev. Genet.* **40:** 409.

Xu Y., Toh K.L., Jones C.R., Shin J.Y., Fu Y.H., and Ptacek L.J. 2007. Modeling of a human circadian mutation yields insights into clock regulation by PER2. *Cell* **128:** 59.

Yagita K., Tamanini F., Yasuda M., Hoeijmakers J.H.J., van der Horst G.T.J., and Okamura H. 2002. Nucleocytoplasmic shuttling and mCRY-dependent inhibition of ubiquitylation of the mPER2 clock protein. *EMBO J.* **21:** 1301.

Yagita K., Yamaguchi S., Tamanini F., van der Horst G.T.J., Hoeijmakers J.H.J., Yasui A., Loros J.J., Dunlap J.C., and Okamura H. 2000. Dimerization and nuclear entry of mPER proteins in mammalian cells. *Genes Dev.* **14:** 1353.

Yan L. and Silver R. 2002. Differential induction and localization of mPer1 and mPer2 during advancing and delaying phase shifts. *Eur. J. Neurosci.* **16:** 1531.

Yannielli P.C., Molyneux P.C., Harrington M.E., and Golombek D.A. 2007. Ghrelin effects on the circadian system of mice. *J. Neurosci.* **27:** 2890.

Yelamanchili S.V., Pendyala G., Brunk I., Darna M., Albrecht U., and Ahnert-Hilger G. 2006. Differential sorting of the vesicular glutamate transporter 1 into a defined vesicular pool is regulated by light signaling involving the clock gene Period2. *J. Biol. Chem.* **281:** 15671.

Yildiz O., Doi M., Yujnovsky I., Cardone L., Berndt A., Hennig S., Schulze S., Urbanke C., Sassone-Corsi P., and Wolf E. 2005. Crystal structure and interactions of the PAS repeat region of the *Drosophila* clock protein PERIOD. *Mol. Cell* **17:** 69.

Yoo S.H., Ko C.H., Lowrey P.L., Buhr E.D., Song E.J., Chang S., Yoo O.J., Yamazaki S., Lee C., and Takahashi J.S. 2005. A noncanonical E-box enhancer drives mouse Period2 circadian oscillations in vivo. *Proc. Natl. Acad. Sci.* **102:** 2608.

Zheng B., Larkin D.W., Albrecht U., Sun Z.S., Sage M., Eichele G., Lee C.C., and Bradley A. 1999. The *mPer2* gene encodes a functional component of the mammalian circadian clock. *Nature* **400:** 169.

Zheng B., Albrecht U., Kaasik K., Sage M., Lu W., Vaishnav S., Li Q., Sun Z.S., Eichele G., Bradley A., and Lee C.C. 2001. Nonredundant roles of the mPer1 and mPer2 genes in the mammalian circadian clock. *Cell* **105:** 683.

Chromatin Remodeling and Circadian Control: Master Regulator CLOCK Is an Enzyme

B. GRIMALDI, Y. NAKAHATA, S. SAHAR, M. KALUZOVA, D. GAUTHIER, K. PHAM, N. PATEL, J. HIRAYAMA, AND P. SASSONE-CORSI

Department of Pharmacology, School of Medicine, University of California, Irvine, California 92697

The molecular machinery that governs circadian rhythmicity is based on clock gene products organized in regulatory feedback loops. Recently, we have shown that CLOCK, a master circadian regulator, has histone acetyltransferase activity essential for clock gene expression. The Lys-14 residue of histone H3 is a preferential target of CLOCK-mediated acetylation. As the role of chromatin remodeling in eukaryotic transcription is well recognized, this finding identified unforeseen links between histone acetylation and cellular physiology. Indeed, we have shown that the enzymatic function of CLOCK drives circadian control. We reasoned that CLOCK's acetyltransferase activity could also target nonhistone proteins, a feature displayed by other HATs. Indeed, CLOCK also acetylates a nonhistone substrate: its own partner, BMAL1. This protein undergoes rhythmic acetylation in the mouse liver, with a timing that parallels the down-regulation of circadian transcription of clock-controlled genes. BMAL1 is specifically acetylated on a unique, highly conserved Lys-537 residue. This acetylation facilitates recruitment of the repressor CRY1 to BMAL1, indicating that CLOCK may intervene in negative circadian regulation. Our findings reveal that the enzymatic interplay between two clock core components is crucial for the circadian machinery.

INTRODUCTION

The eukaryotic genome is organized in a nucleoprotein structure, chromatin, that enables a large variety of central processes such as regulation of gene expression, DNA repair, apoptosis, and cell division (Luger 2003; Felsenfeld and Groudine 2003). Changes in chromatin organization have a central role in ensuring that the storage, organization, and readout of the genetic information occurs in a proper spatial and temporal sequence during all biological processes. The basic repeating unit of chromatin, the nucleosome core particle, consists of 147 bp of DNA organized in approximately two superhelical turns of DNA wrapped around an octamer of core histones (two copies each of histone H2A, H2B, H3, and H4). Nucleosomes become organized in higher-order structures when interacting with H1 linker histones and more loosely with nonhistone chromatin-associated proteins (Luger 2003). A number of enzymatic processes lead to remodeling of the architecture of chromatin. These are dynamic changes, essential for the transition of chromatin from a condensed to a decondensed state, and vice versa, each state being permissive to specific cellular functions. These states somewhat correlate with the definition of "euchromatin" versus "heterochromatin," which most times identifies with "active" versus "inactive" states of gene expression, respectively (Strahl and Allis 2000; Felsenfeld and Groudine 2003).

Various cellular mechanisms operate that lead to modifications in chromatin structure: One type involves the active participation of ATP-dependent remodeling factors, such as Swi/Snf and NURF (Peterson and Workman 2000), and others implicate posttranslational covalent modifications of histones by specific enzymes. The amino-terminal tails of the core histones H2A, H2B, H3, and H4 are highly conserved and represent the protein domains where most of the covalent histone modifications occur. Since histone tails are thought to confer secondary and more flexible contacts with DNA, allowing for dynamic changes in the accessibility of the underlying genome, their modifications have been shown to contribute to chromatin remodeling and thereby to the control of a large array of nuclear processes (Cheung et al. 2000b; Felsenfeld and Groudine 2003). The amino-terminal domains of histones are subjected to a large variety of covalent modifications, such as acetylation (Grunstein 1997; Gregory et al. 2001; Roth et al. 2001), phosphorylation (Mahadevan et al. 1991; Sassone-Corsi et al. 1999), and methylation (Bannister et al. 2002), but also ADP-ribosylation (Huletsky et al. 1985) and ubiquitination (Sun and Allis 2002). In several cases, the combinatorial association of some specific modifications has been reported and coupled to unique nuclear functions (Cheung et al. 2000a; Lo et al. 2000), strongly suggesting that the coordinate action of multiple enzymes converges to promote physiological changes in chromatin organization (Lo et al. 2001; Merienne et al. 2001).

CHROMATIN REMODELING AND GENE EXPRESSION: AN ENZYMATIC AFFAIR

The four core histones have amino-terminal tails whose primary sequences are highly conserved from yeast to mammals. Remarkably, histone tails have a high density of residues that are prone to posttranslational modifications. For the most part, the enzymes that elicit these modifications are also conserved among species. Various position-specific modifications have been associated with distinct chromatin-based outputs (Strahl and Allis 2000; Felsenfeld and Groudine 2003). For example, modifications in the histone H3 amino-terminal tail have been coupled to transcriptional regulation (Lys-9/Lys-14 acetylation, Ser-10 phosphorylation), transcriptional silencing (Lys-9 methylation), histone deposition (Lys-9 acetylation), and chromosome condensation/segregation (Ser-10/Ser-28 phosphorylation). It has been proposed that various combinations of histone modifications may elicit differential regulation of the chromatin condensation state that corresponds to distinct biological responses (Fischle et al. 2003).

Histone phosphorylation is remarkable because it is directly linked to intracellular signaling pathways. Indeed, activation of a specific transduction system by an extracellular stimulus results in the stimulation of kinase cascades and phosphorylation at distinct sites on histones. Various serine-threonine kinases have been implicated for phosphorylating potential serine phosphoacceptor sites on histone H2A (Ser-1 and Ser-18), H2B (Ser14 and Ser-32), H3 (Ser-10 and Ser-28), and H4 (Ser-1). The distance between the serine residues in H2A, H2B, and H3 is conserved among all histones, with a constant spacing of 18 residues. The significance of this intriguing feature is unclear. Phosphorylation of Ser-10 in histone H3 has been studied in detail (Sassone-Corsi et al. 1999; Clayton et al. 2000) and found to be elicited by a number of kinases, all belonging to the AGC branch of cyclic nucleotide-regulated protein kinase, including RSK-2, MSK1, PKA, and IKKα (Nowak and Corces 2004).

Histone acetylation is believed to have a pivotal role in the modulation of chromatin structure associated with transcriptional activation (Grunstein 1997; Wade and Wolffe 1997; Kuo and Allis 1998; Struhl 1998; Workman and Kingston 1998). In support of this notion, a wide variety of nuclear proteins involved in transcriptional control have been demonstrated to possess intrinsic histone acetyltransferase (HAT) activity (Kouzarides 1999; Sterner and Berger 2000; Roth et al. 2001). In particular, a number of transcriptional coactivators, including GCN5 (Brownell et al. 1999), PCAF (Yang et al. 1996), CBP/p300 (Bannister and Kouzarides 1996; Ogryzko et al. 1996), SRC-1 (Spencer et al. 1997), and ACTR (Chen et al. 1997) are known to acetylate histones, thereby facilitating the *trans*-activation exerted by a number of DNA-binding transcription factors. Furthermore, HAT function has been ascribed also to TAFII250, a component of the TATA-box-binding TFIID complex of the basal transcription machinery (Mizzen et al. 1996), and to ATF-2, a sequence-specific DNA-binding transcription factor (Kawasaki et al. 2000). Thus, HATs constitute a family of proteins with remarkably diverse features.

Amino acid sequence analyses of HAT proteins reveal an important feature: HATs fall into distinct families that share relatively poor sequence similarity. For example, ACTR/SRC-1 is thought to constitute a unique class of HAT (Chen et al. 1997; Spencer et al. 1997), whereas p300/CBP displays only limited homology with the GCN5-related *N*-acetyltransferase superfamily (Martinez-Balbas et al. 1998). The MYST family of HATs is particularly interesting as these proteins show similarity with other acetyltransferases exclusively within the acetyl–coenzyme-A-binding motif (denominated "motif A") (Yamamoto and Horikoshi 1997). Accumulating evidence indicates that histone acetylation exerted by various classes of HATs contributes to plasticity in transcriptional control by increasing the dynamic changes in chromatin structure (Fischle et al. 2003).

Histone methylation has been linked mostly to gene silencing, although some examples of its role in transcriptional activation have been reported. Specifically, methylation of the Lys-4 of H3 has been found to be a positive mark for gene activation, possibly in combination to acetylation at Lys-14 (Berger 2007). Three forms of histone methylation exist, each one of them promoting a different regulatory function. Indeed, mono-, di-, and three-methylated lysines are obtained by the action of different types of histone methyltransferases (HMTs). Some of the enzymes devoted to the removal of the methyl groups, the demethyltransferases, have been implicated in neurological disorders (Shi 2007). Importantly it is the combined association of different posttranslational modifications that appears to bring unique biological functions, Thus, the identification of the enzymes involved in histone modifications, and their regulatory interplays, has far-reaching physiological consequences.

PERIPHERAL VERSUS CENTRAL CLOCKS: WHERE IS THE DIFFERENCE?

The finely controlled transcriptional regulation within the circadian system is absolutely remarkable. Indeed, more than 10% of all mammalian transcripts undergo circadian fluctuations in their expression levels (Akhtar et al. 2002; Duffield et al. 2002; Panda et al. 2002), underscoring that genome-wide mechanisms must operate in order to ensure such global transcriptional regulation. The highly specialized, temporally based regulation of gene expression that characterizes circadian oscillations elects the cellular clock as a prominent model for the study of dynamic regulations of chromatin remodeling (Crosio et al. 2000). Moreover, as circadian function is intimately coupled to physiological and metabolic control (Dunlap 1999; Rutter et al. 2001; Schibler and Naef 2005), clock-controlled chromatin reorganization is likely to reveal yet unexplored pathways linking histone modifications to cellular metabolism. In this context, the finding that peripheral tissues also contain independent clocks is of fundamental importance (Whitmore et al. 2000; Giebultowicz 2001; Stokkan et al. 2001; Schibler and Sassone-Corsi 2002). Peripheral clocks are not fully self-sustained and autonomous and, differently from the suprachiasmatic nucleus (SCN), require specific stimuli in order to sustain their circadian rhythms (Cermakian and Sassone-Corsi 2000; King and Takahashi 2000; Reppert and Weaver 2002). We favor a scenario where peripheral clocks are affected by physiological stimuli that may originate from the SCN and/or may be the result of SCN-mediated messages (Schibler and Sassone-Corsi 2002). This view has been substantiated by a number of studies that have identified growth factors, some steroids such as glucocorticoids, and retinoic acid to induce oscillations of clock genes and clock-controlled genes in cultured fibroblasts or peripheral tissues (Kramer et al. 2001, 2005). These observations underscore the presence of possible differences at the molecular level between the organization of the clock mechanism in the SCN and that in peripheral tissues. These may operate following some unique tissue-specific regulatory pathways since restricted access to food has an effect on peripheral rhythms without affecting the central pacemaker function of the SCN (Schibler and Sassone-Corsi 2002).

PLASTICITY IN CIRCADIAN REGULATION AND CHROMATIN REMODELING

How is the oscillatory expression of clock-controlled genes regulated, so that transcription-permissive chromatin states are dynamically established in a circadian time-specific manner? Interestingly, the activation of clock-controlled genes (CCGs) by CLOCK:BMAL1 has been shown to be coupled to circadian changes in histone acetylation at their promoters (Etchegaray et al. 2003; Curtis et al. 2004; Naruse et al. 2004; Nakahata et al. 2007). Specifically, histone H3 is acetylated in chromatin that encompasses the *Per1*, *Per2*, and *Cry1* promoters when these genes are actively transcribed. The molecular dissection of the CLOCK:BMAL1-mediated *trans*-activation mechanism is likely to provide significant information on how circadian regulation of histone acetylation is achieved. Several lines of genetic evidence also indicate that the carboxy-terminal glutamine-rich region of CLOCK exerts a central function in the circadian *trans*-activation of target genes in flies and mice (Antoch et al. 1997; King et al. 1997; Allada et al. 1998; Gekakis et al. 1998; Jin et al. 1999). In addition, H3 Ser-10 phosphorylation is involved in the transcriptional response to light in the SCN. Indeed, light-mediated signaling leads to nuclear responses that in turn influence the state of higher chromatin organization (Crosio et al. 2000). As the coupled modification of Ser-10 phosphorylation and Lys-14 acetylation on H3 has been shown to be a hallmark of transcriptional activation (Cheung et al. 2000a; Lo et al. 2000), it appears that the combined implication of HATs and other enzymatic pathways is crucial for circadian gene expression (Curtis et al. 2004; Ripperger and Schibler 2006). Finally, histone methylation may be important for circadian clock function as indicated by RNA interference (RNAi) experiments in cultured fibroblasts. The polycomb group protein EZH2 and methylation at H3 Lys-27 are found to be associated to *Per* gene promoters, extending the activity of the polycomb group proteins to the core clockwork mechanism of mammals (Etchegaray et al. 2006).

The finding that the central element of the core clock mechanism, the transcription factor CLOCK, has intrinsic HAT activity (Doi et al. 2006) has modified our view because this protein should not be simply considered a transcription factor but an enzyme (Fig. 1). This finding also further underscored the importance of the E box as a pivotal element in circadian control, because chromatin conformational changes may occur specifically and efficiently at predetermined locations identified by the presence of E boxes.

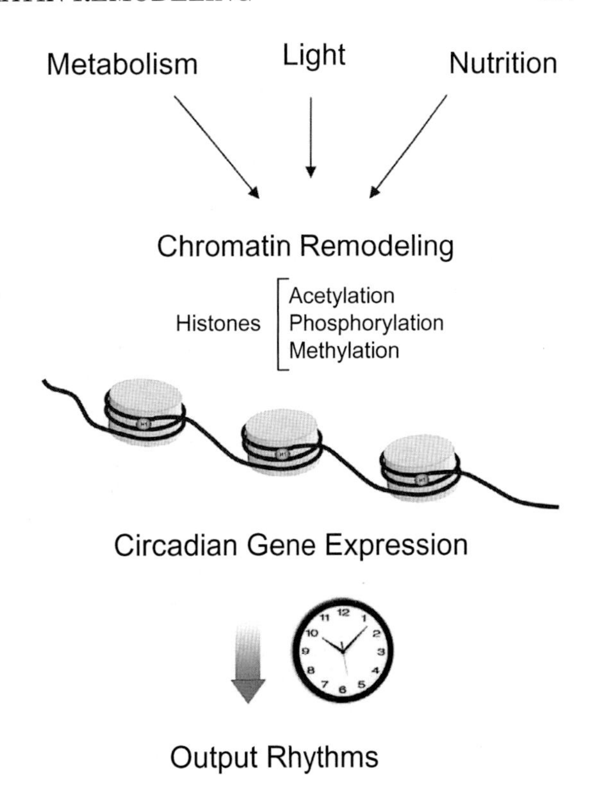

Figure 1. Schematic model of the role that chromatin remodeling has within the physiological pathways of circadian rhythmicity. Metabolic, nutritional, and environmental circadian cues are likely to modulate chromatin remodeling, likely via control of the HAT function of CLOCK by means of intracellular signaling. Furthermore, by regulating chromatin transitions, CLOCK exerts its function on a variety of output CCGs, thereby influencing cellular physiology and metabolism. Importantly, the expression of at least 10% of all mammalian transcripts oscillate in a circadian manner.

THE PROTEIN CLOCK IS AN ENZYME

The carboxy-terminal glutamine-rich region of CLOCK, a region implicated in *trans*-activation function (Allada et al. 1998; Gekakis et al. 1998), has some important features. Except for a polyglutamine stretch, no characteristic structural motifs were previously described in this region. A closer look revealed that the carboxy-terminal region of CLOCK displays a significant sequence homology with the carboxy-terminal domain of ACTR, a domain shown to have intrinsic HAT activity (Chen et al. 1997). In this region, at least six independent amino acid regions are found to share significant sequence similarity between the two proteins. Importantly, the amino acid residues common to CLOCK and ACTR are evolutionarily conserved in both proteins. It is also noteworthy that CLOCK and ACTR share a number of other structural features outside of the carboxy-terminal glutamine-rich region. These include the highly conserved basic helix-loop-helix (bHLH)-PAS domain at the amino termini, a NRID (nuclear receptor interaction domain), as well as serine-rich regions within the middle portion of both proteins. Although CLOCK is a significantly smaller protein as compared to ACTR, these common features result in a strikingly similar organizaton overall (Doi et al. 2006). Analysis of the primary sequence shows that BMAL1, the heterodimeric partner of CLOCK, which is also a bHLH-PAS domain-containing protein (Hirayama and Sassone-Corsi 2005), displays no similarity to the HAT region of ACTR.

An additional feature makes CLOCK quite unique as a HAT. Acetyl–coenzyme-A (CoA)-binding motifs are hallmarks of HAT proteins (Sterner and Berger 2000). Detailed sequence comparison between the acetyl–CoA-binding motifs of various HATs revealed that CLOCK contains a motif within the carboxy-terminal glutamine-rich region. This amino acid sequence stretch shares significant similarity to the so-called "motif A" in the HAT family denominated MYST (for its founding members MOZ, Ybf/2Sas3, Sas2, and Tip60). In particular, CLOCK shows high sequence similarity to yeast Esa1 and other MYST members, including yeast Sas3, fly MOF, and human Tip60. Importantly, residues that have been demonstrated to be involved in acetyl-CoA interaction by crystal structure analysis of the Esa1 protein (Yan et al. 2000) are shared with CLOCK. It is significant that these residues are all fully conserved in CLOCK proteins of various species. One intriguing feature of the motif A in CLOCK, when compared to MYST family members, is the insertion of five amino acids, also fully conserved among species. As compared to Esa1, the additional five amino acids would lengthen the loop comprised between β9 and α3, a region that is demonstrated to be exposed at the protein surface (Yan et al. 2000). Intriguingly, the putative acetyl–CoA-binding motif in *Drosophila* CLOCK lacks the five extra amino acids present in the vertebrate counterparts, which increases its similarity to Esa1 and other MYST family members. Finally, a number of acetyl–CoA-binding motifs from various *N*-acetyltransferases also show similarities. In conclusion, the structural features of CLOCK appear to identify it as a novel type of DNA-binding HAT. Indeed, the general organization is similar to a SRC/ACTR-type of HAT, but the acetyl–CoA-binding motif is more related to the MYST class. Thus, the unique combination of protein domains in CLOCK seems to form a somewhat hybrid HAT.

CIRCADIAN PHYSIOLOGY NEEDS A HAT

Our biochemical analyses have established that CLOCK is indeed a bona fide HAT with high specific enzymatic activity for H3 and H4, as it does not acetylate H2A and H2B. Site-directed mutagenesis indicated that H3 Lys14 is where most of the acetylation occurs (Doi et al. 2006). Most importantly, using an experimental system based on mouse embryonic fibroblast (MEF) cells derived from homozygous *Clock* mutant (*c/c*) mice (Vitaterna et al. 1994), we showed that the HAT function of CLOCK is essential for the circadian control of CCGs (Doi et al. 2006). As *Clock* is essential for circadian rhythm (Antoch et al. 1997; King et al. 1997), MEF *c/c* cells show no cyclic expression of clock genes (Pando et al. 2002). Importantly, ectopic expression of CLOCK is able to rescue the circadian expression of endogenous target genes. MEF *c/c* cells stably transfected with a CLOCK expression plasmid were subjected to a serum shock, a stimulus commonly used to trigger circadian gene transcription in a variety of cell lines (Balsalobre et al. 1998; Pando et al. 2002). Although the MEF *c/c* cells had no functional circadian clock, CLOCK ectopic expression restored circadian oscillation of the endogenous *Per1* gene and of the *Dbp* gene, an E-box-regulated circadian CCG output gene. On the contrary, ectopic expression of a HAT-deficient CLOCK failed to restore the circadian *trans*-activation of *mPer1* and *Dbp*, demonstrating that the HAT activity of CLOCK is necessary for circadian gene expression (Doi et al. 2006). These findings underscore the importance of chromatin remodeling in circadian regulation and reveal the molecular pathways by which such essential control is achieved (Nakahata et al. 2007).

BMAL1 IS ACETYLATED BY CLOCK

The finding that CLOCK is a HAT (Doi et al. 2006) suggested that the acetyltransferase enzymatic activity could also target other nonhistone proteins. This feature is displayed by other HATs (Glozak et al. 2005; Zhang and Dent 2005) and is demonstrated to have profound physiological significance. In a search for putative targets, we have found that CLOCK mediates the acetylation of its own heterodimerization partner, BMAL1 (Hirayama et al. 2007).

In a survey aimed at identifying proteins that may be acetylated rhythmically in vivo, we analyzed various clock proteins, such as BMAL1, CLOCK, and PER1 in the mouse liver at different zeitgeber times. Although, as expected, these proteins oscillate in abundance and phosphorylation levels (Lee et al. 2001; Matsuo et al. 2003), acetylation of BMAL1 displays a robust circadian oscillation with a peak at ZT15 (Hirayama et al. 2007). Significantly, the other clock proteins are not acetylated. Ongoing studies in our laboratory on a variety of nuclear proteins and transcription factors indicate that BMAL1 is one of the few substrates for CLOCK, underscoring the specificity of the assay. Importantly, CLOCK is directly responsible for BMAL1 acetylation in cultured mammalian cells (Hirayama et al. 2007).

From a mechanistic point of view, it is notable that heterodimerization between the two proteins seems to be required for BMAL1 acetylation. Indeed, an amino-terminally truncated CLOCK (CLOCKΔN) mutant that is unable to interact with BMAL1 because it lacks the PAS domains, although it still possesses HAT activity (Doi et al. 2006), did not induce BMAL1 acetylation. On the other hand, a CLOCK protein with three single-amino-acid mutations in the HAT domain (CLOCK-mutA) that we had previously shown to have reduced HAT activity (Doi et al. 2006) displayed a drastically reduced acetylation of BMAL1, demonstrating that the intrinsic HAT activity of CLOCK is required for BMAL1 acetylation.

HIGH SPECIFICITY: CLOCK ACETYLATES A SINGLE LYSINE IN BMAL1

In addition to acetylation, BMAL1 is posttranslationally modified by phosphorylation (Kondratov et al. 2003; Hirayama and Sassone-Corsi 2005) and SUMOylation

(Cardone et al. 2005). SUMOylation, as acetylation, occurs on lysine residues and thus we investigated whether an interplay converging on BMAL1 may exist between these two enzymatic pathways. We had previosuly identified four candidate lysines, K223, K229, K259, and K272, as subjected to SUMOylation in mouse BMAL1. Our analysis then demonstrated that K259 is the major in vivo SUMOylation site in BMAL1 (Cardone et al. 2005). We have found that none of these lysines are acetylated by CLOCK (Fig. 2), demonstrating that the target lysine residues for the two modifications are different. Thus, we embarked in the search for the lysines that are specifically acetylated by CLOCK. We first generated two carboxy-terminally truncated BMAL1 proteins (amino acids 1–282 and 1–469) (Fig. 2A). Both of these mutant proteins interact with CLOCK but are not acetylated, indicating that the target lysines for CLOCK-dependent acetylation must be located in the carboxyl terminus of mouse BMAL1 (amino acids 470–631; Fig. 2B). Somewhat conveniently, this region contains only four potential target lysines at positions 475, 494, 537, and 538, rendering our analysis quite accessible. Each lysine residue was mutated to arginine and tested for CLOCK-mediated acetylation. Strikingly, all mutant proteins are acetylated at levels comparable to wild-type BMAL1, with the exception of K537R (Fig. 2C). Specificity was confirmed also in an in vitro acetylation assay using bacterially purified GST-BMAL1 and a mutant GST-BMAL1-K537R in the presence of ^{3}H-labeled acetyl-CoA. Thus, K537 is the major acetylation site for CLOCK (Hirayama et al. 2007), and it is remarkable that this lysine is highly conserved among all vertebrate BMAL1s (Fig. 2A).

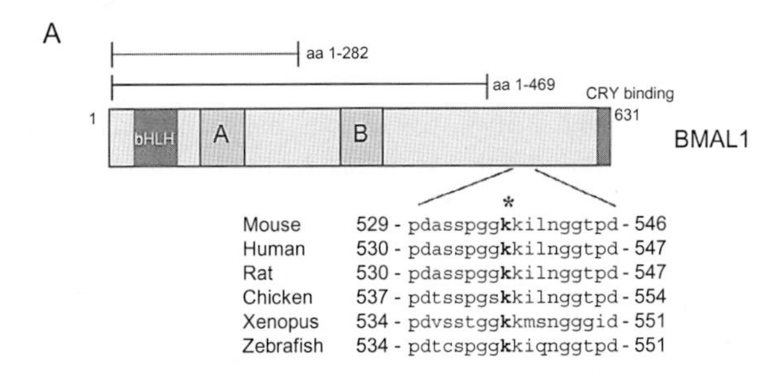

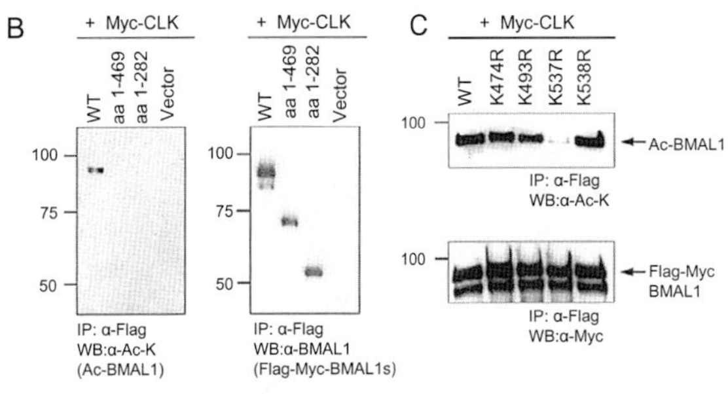

Figure 2. Highly specific acetylation of BMAL1. (*A*) Schematic representation of mBMAL1 protein showing the positions of bHLH, PAS A (A), PAS B (B), and CRY-binding site (CRY binding) (Hirayama and Sassone-Corsi 2005). Numbers indicate the amino acid residues in the mouse protein. The extent of two deletion mutants (amino acids 1–282 and 1–469) is shown as two gray bars on the top. The sequence alignment of BMAL1 with the target lysine for acetylation and its surrounding amino acids from various species is shown. (*Bold*) Target lysines. (*B,C*) Identification of the target lysine for acetylation. Cells were cotransfected with expression vectors for Myc-CLOCK and Flag-Myc-BMAL1 wild type or for each BMAL1 mutant described. Flag-Myc-BMAL1 proteins were immunoprecipitated, and acetylation was determined by western blotting using anti-pan-acetyl-Lys. Expression of precipitated BMAL1 was determined by western blotting using anti-BMAL1 (*B, right panel*) or anti-Myc (*C, bottom panel*) antibodies.

BMAL1 ACETYLATION IS ESSENTIAL FOR CIRCADIAN RHYTHMICITY

The unique acetylation profile and the remarkable specificity of CLOCK in targeting one single lysine in BMAL1 prompted us to establish the physiological relevance of BMAL1 acetylation. To investigate the requirement for circadian rhythmicity of BMAL1 acetylation, we performed rescue experiments using MEFs generated from *Bmal1*-null mice. Lack of BMAL1 expression in these MEFs results in a dysfunctional circadian clock and arrhythmic gene expression (Kondratov et al. 2003; Cardone et al. 2005). We used retroviruses expressing either wild-type BMAL1, an acetylation-deficient BMAL1 (K537R), or green fluorescent protein (GFP), and infected *Bmal1*$^{-/-}$ MEFs (Fig. 3). To synchronize the infected MEFs, we used dexamethasone (DEX), and circadian oscillation was monitored by a real-time bioluminescence assay based on an *mPer2* promoter-driven luciferase reporter vector (Nagoshi et al. 2004; Sato et al. 2006). Although wild-type BMAL1 rescued circadian *mPer2* expression, the BMAL1 (K537R) mutant was unable to do so. This experiment demonstrates that the acetylation of BMAL1 is essential for circadian gene regulation (Fig. 3).

Our interest then turned to the molecular mechanism by which BMAL1 acetylation influences circadian rhythmicity: What we found revealed a regulatory pathway highly significant for circadian physiology. Our analyses

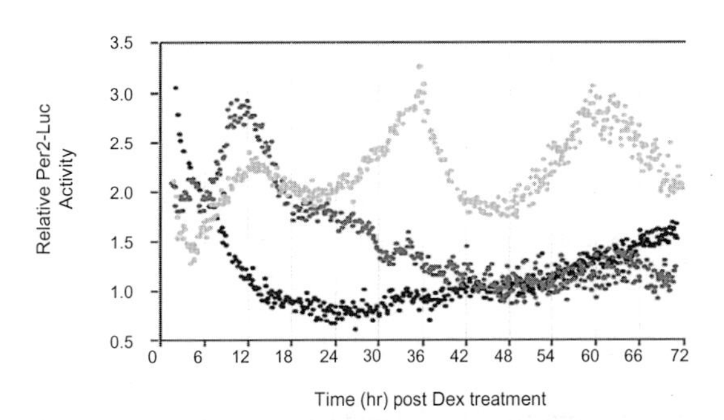

Figure 3. Acetylation of BMAL1 is essential to rescue the circadian rhythmicity in BMAL1-deficient cells. Retrovirus-infected *Bmal1*$^{-/-}$ MEFs were used to rescue circadian rhythmicity by expressing either a wild-type BMAL1 protein (*light gray dots*) or the BMAL1(K537R) mutant (*gray dots*) and compared with the lack of circadian oscillation in the noninfected *Bmal1*$^{-/-}$ MEFs (*black dots*). The *mPer2* promoter luciferase reporter plasmid was transfected into the retrovirus-infected cells, and *mPer2* promoter activity was monitored by a real-time bioluminescence assay. Expression levels were plotted as arbitrary units.

show that the BMAL1(K537R) mutant is recruited to the *mPer2* promoter with an efficiency equivalent to that of BMAL1. In addition, the K537R mutation has no effect on BMAL1 protein stability, capacity of association with CLOCK, and subcellular localization of the protein. Thus, we reasoned that BMAL1 acetylation could be involved in modulating CRY-mediated repression. This possibility would rationalize the results obtained in the rescue circadian experiments (Fig. 3) and is supported by the notion that impairment of CRY-mediated repression of the CLOCK:BMAL1 complex leads to loss of circadian rhythmicity (Sato et al. 2006). Indeed, the BMAL1(K537R) mutant showed a drastically reduced sensitivity to CRY1-mediated repression compared to wild-type BMAL1 (Fig. 4). In other terms, our findings indicate that acetylation of BMAL1 by CLOCK may be an essential regulatory switch as it facilitates CRY-dependent repression. On the basis of association studies, our working model hypothesizes that acetylation of BMAL1 at K537 induces a conformational switch that generates a better "docking site" for CRY.

CONCLUSIONS AND PERSPECTIVES

The notion that at least 10% of all cellular transcripts oscillate in a circadian manner underscores the importance of chromatin remodeling in the control of the circadian gene expression (Nakahata et al. 2007). The finding that CLOCK is a HAT pointed to unforeseen links between histone acetylation and cellular physiology. Indeed, because of these multiple implications, a number of exciting avenues of research are now evident. One of them relates to the search and finding of nonhistone targets for CLOCK. Our findings demonstrate that CLOCK exerts its enzymatic activity on another core component of the clock machinery in vivo in a circadian manner. As we demonstrated that BMAL1 enhances the intrinsic HAT activity of CLOCK (Doi et al. 2006), it is tempting to speculate that BMAL1 regulates its own acetylation by reciprocally controlling CLOCK enzymatic activity. Acetylation of proteins is recognized as an essential regulatory mechanism having both stimulatory and inhibitory effects on transcription (Workman and Kingston 1998; Sterner and Berger 2000). We have demonstrated that acetylation may operate at yet another level of control, because BMAL1 acetylation serves to increase the repressive function of another regulator, CRY. Our results also indicate that CLOCK enzymatic activity has a dual regulatory function. Indeed, we have shown that it contributes to the negative limb of the circadian feedback loop (Hirayama et al. 2007), whereas CLOCK-mediated acetylation of histones participates in the transcriptional stimulation of CCGs, acting within the positive limb of the loop (Doi et al. 2006). Thus, CLOCK enzymatic function contributes in multiple ways to the time-dependent regulation of transcription.

Future studies aimed at deciphering the structural determinants of HAT function, its regulation by BMAL1, and the possible intervention of additional regulators are on the way. Like other HATs, CLOCK is likely to associate with a number of nuclear proteins in a chromatin complex. The study of this complex will reveal additional control elements for circadian transcription, as well as putative histone deacetyltransferases (HDACs) involved

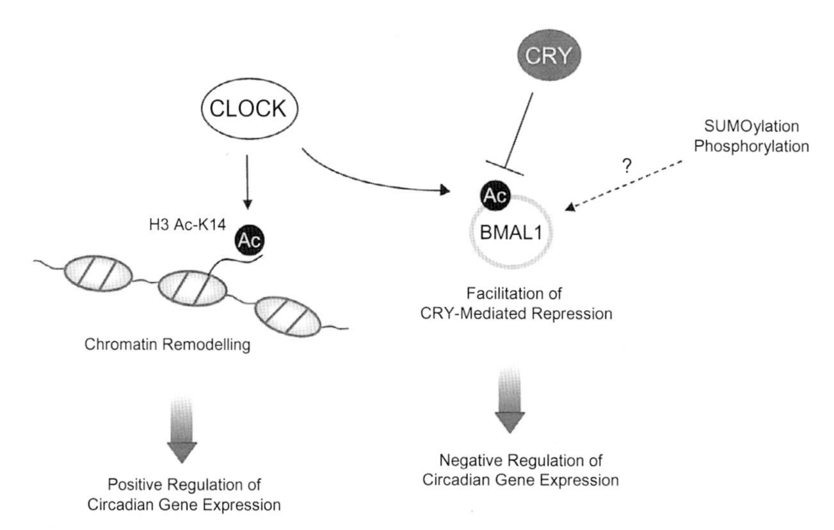

Figure 4. The enzymatic function of CLOCK governs the circadian machinery. Shown is a schematic model of CLOCK-mediated histone and nonhistone acetylation and its role within the physiological pathways of circadian rhythmicity. The HAT function of CLOCK regulates promoters of CCGs and clock genes (such as *Per1*) by inducing locally open organization of the chromatin. For example, acetylation on Lys-14 of histone H3 is thought to elicit chromatin remodeling by inducing a transcription-permissive state. Acetylation at Lys-14 could also be favored by additional concerted modifications of the histone tails. For example, phosphorylation at Ser-10, an event directly coupled to activation of intracellular signaling pathways, such as light-induced signals in SCN neurons (Crosio et al. 2000), induces a more efficient acetylation at Lys-14. We envisage a scenario where circadian control of chromatin remodeling by CLOCK may be influenced by the dynamic assembly of a multiprotein regulatory complex. In addition, CLOCK is responsible for BMAL1 acetylation at K537, an event that increases the repression potential by CRY of the CLOCK:BMAL1 complex. Thus, the acetyltransferase enzymatic activity of CLOCK has a dual function, by regulating the circadian machinery by targeting both histone and nonhistone proteins. Indeed, although acetylation of histone H3 results in the positive regulation of CCGs, acetylation of BMAL1 is involved in the negative repression by CRY.

in enzymatically counteracting the acetylation. Thus, many questions remain to be addressed. The exciting part is now to design experiments and gather the results that will elucidate the intricacies of how circadian control is linked to metabolism, cell cycle, and physiology. Uncovering the intimate links between chromatin remodeling and the circadian clock constitutes a conceptually novel challenge that will provide more excitement for future studies.

ACKNOWLEDGMENTS

Work in our laboratory is supported by the Cancer Research Coordinating Committee of the University of California and from the National Health Institute.

REFERENCES

Akhtar R.A., Reddy A.B., Maywood E.S., Clayton J.D., King V.M., Smith A.G., Gant T.W., Hastings M.H., and Kyriacou C.P. 2002. Circadian cycling of the mouse liver transcriptome, as revealed by cDNA microarray, is driven by the suprachiasmatic nucleus. *Curr. Biol.* **12:** 540.

Allada R., White N.E., So W.V., Hall J.C., and Rosbash M. 1998. A mutant *Drosophila* homolog of mammalian *Clock* disrupts circadian rhythms and transcription of *period* and *timeless*. *Cell* **93:** 791.

Antoch M.P., Song E.J., Chang A.M., Vitaterna M.H., Zhao Y., Wilsbacher L.D., Sangoram A.M., King D.P., Pinto L.H., and Takahashi J.S. 1997. Functional identification of the mouse circadian *Clock* gene by transgenic BAC rescue. *Cell* **89:** 655.

Balsalobre A., Damiola F., and Schibler U. 1998. A serum shock induces circadian gene expression in mammalian tissue culture cells. *Cell* **93:** 929.

Bannister A.J. and Kouzarides T. 1996. The CBP co-activator is a histone acetyltransferase. *Nature* **384:** 641.

Bannister A.J., Schneider R., and Kouzarides T. 2002. Histone methylation: Dynamic or static? *Cell* **109:** 801.

Berger S.L. 2007. The complex language of chromatin regulation during transcription. *Nature* **447:** 407

Brownell J.E., Mizzen C.A., and Allis C.D. 1999. An SDS-PAGE-based enzyme activity assay for the detection and identification of histone acetyltransferases. *Methods Mol. Biol.* **119:** 343.

Cardone L., Hirayama J., Giordano F., Tamaru T., Palvimo J.J., and Sassone-Corsi P. 2005. Circadian clock control by SUMOylation of BMAL1. *Science* **309:** 1390.

Cermakian N. and Sassone-Corsi P. 2000. Multilevel regulation of the circadian clock. *Nat. Rev. Mol. Cell Biol.* **1:** 59.

Chen H., Lin R.J., Schiltz R.L., Chakravarti D., Nash A., Nagy L., Privalsky M.L., Nakatani Y., and Evans R.M. 1997. Nuclear receptor coactivator ACTR is a novel histone acetyltransferase and forms a multimeric activation complex with P/CAF and CBP/p300. *Cell* **90:** 569.

Cheung P., Allis C.D., and Sassone-Corsi P. 2000a. Signaling to chromatin through histone modifications. *Cell* **103:** 263.

Cheung P., Tanner K.G., Cheung W.L., Sassone-Corsi P., Denu J.M., and Allis C.D. 2000b. Synergistic coupling of histone H3 phosphorylation and acetylation in response to epidermal growth factor stimulation. *Mol. Cell* **5:** 905.

Clayton A.L., Rose S., Barratt M.J., and Mahadevan L.C. 2000. Phosphoacetylation of histone H3 on c-*fos*- and c-*jun*-associated nucleosomes upon gene activation. *EMBO J.* **19:** 3714.

Crosio C., Cermakian N., Allis C.D., and Sassone-Corsi P. 2000. Light induces chromatin modification in cells of the mammalian circadian clock. *Nat. Neurosci.* **3:** 1241.

Curtis A.M., Seo S.B., Westgate E.J., Rudic R.D., Smyth E.M., Chakravarti D., FitzGerald G.A., and McNamara P. 2004. Histone acetyltransferase-dependent chromatin remodeling and the vascular clock. *J. Biol. Chem.* **279:** 7091.

Doi M., Hirayama J., and Sassone-Corsi P. 2006. Circadian regulator CLOCK is a histone acetyltransferase. *Cell* **125:** 497.

Duffield G.E., Best J.D., Meurers B.H., Bittner A., Loros J.J., and Dunlap J.C. 2002. Circadian programs of transcriptional activation, signaling, and protein turnover revealed by microarray analysis of mammalian cells. *Curr. Biol.* **12:** 551.

Dunlap J.C. 1999. Molecular bases for circadian clocks. *Cell* **96:** 271.

Etchegaray J.P., Lee C., Wade P.A., and Reppert S.M. 2003. Rhythmic histone acetylation underlies transcription in the mammalian circadian clock. *Nature* **421:** 177.

Etchegaray J.P., Yang X., DeBruyne J.P., Peters A.H., Weaver D.R., Jenuwein T., and Reppert S.M. 2006. The polycomb group protein EZH2 is required for mammalian circadian clock function. *J. Biol. Chem.* **281:** 21209.

Felsenfeld G. and Groudine M. 2003. Controlling the double helix. *Nature* **421:** 448.

Fischle W., Wang Y., and Allis C.D. 2003. Binary switches and modification cassettes in histone biology and beyond. *Nature* **425:** 475.

Gekakis N., Staknis D., Nguyen H.B., Davis F.C., Wilsbacher L.D., King D.P., Takahashi J.S., and Weitz C.J. 1998. Role of the CLOCK protein in the mammalian circadian mechanism. *Science* **280:** 1564.

Giebultowicz J.M. 2001. Peripheral clocks and their role in circadian timing: Insights from insects. *Philos. Trans. R. Soc. Lond. B Biol. Sci.* **356:** 1791.

Glozak M.A., Sengupta N., Zhang X., and Seto E. 2005. Acetylation and deacetylation of non-histone proteins. *Gene* **363:** 15.

Gregory P.D., Wagner K., and Horz W. 2001. Histone acetylation and chromatin remodeling. *Exp. Cell Res.* **265:** 195.

Grunstein M. 1997. Histone acetylation in chromatin structure and transcription. *Nature* **389:** 349.

Hirayama J. and Sassone-Corsi P. 2005. Structural and functional features of transcription factors controlling the circadian clock. *Curr. Opin. Genet. Dev.* **15:** 548.

Hirayama J., Sahar S., Grimaldi B., Tamaru T., Takamatsu K., Nakahata Y., and Sassone-Corsi P. 2007. CLOCK-mediated acetylation of BMAL1 controls circadian function. *Nature* (in press).

Huletsky A., Niedergang C., Frechette A., Aubin R., Gaudreau A., and Poirier G.G. 1985. Sequential ADP-ribosylation pattern of nucleosomal histones. ADP-ribosylation of nucleosomal histones. *Eur. J. Biochem.* **146:** 277.

Jin X., Shearman L.P., Weaver D.R., Zylka M.J., de Vries G.J., and Reppert S.M. 1999. A molecular mechanism regulating rhythmic output from the suprachiasmatic circadian clock. *Cell* **96:** 57.

Kawasaki H., Schiltz L., Chiu R., Itakura K., Taira K., Nakatani Y., and Yokoyama K.K. 2000. ATF-2 has intrinsic histone acetyltransferase activity which is modulated by phosphorylation. *Nature* **405:** 195.

King D.P. and Takahashi J.S. 2000. Molecular genetics of circadian rhythms in mammals. *Annu. Rev. Neurosci.* **23:** 713.

King D.P., Zhao Y., Sangoram A.M., Wilsbacher L.D., Tanaka M., Antoch M.P., Steeves T.D., Vitaterna M.H., Kornhauser J.M., Lowrey P.L., Turek F.W., and Takahashi J.S. 1997. Positional cloning of the mouse circadian clock gene. *Cell* **89:** 641.

Kondratov R.V., Chernov M.V., Kondratova A.A., Gorbacheva V.Y., Gudkov A.V., and Antoch M.P. 2003. BMAL1-dependent circadian oscillation of nuclear CLOCK: Posttranslational events induced by dimerization of transcriptional activators of the mammalian clock system. *Genes Dev.* **17:** 1921.

Kouzarides T. 1999. Histone acetylases and deacetylases in cell proliferation. *Curr. Opin. Genet. Dev.* **9:** 40.

Kramer A., Yang F.C., Karves S., and Weitz C.J. 2005. A screen for secreted factors of the suprachiasmatic nucleus. *Methods Enzymol.* **393:** 645.

Kramer A., Yang F.C., Snodgrass P., Li X., Scammell T.E., Davis F.C., and Weitz C.J. 2001. Regulation of daily locomotor activity and sleep by hypothalamic EGF receptor signaling. *Science* **294:** 2511.

Kuo M.H. and Allis C.D. 1998. Roles of histone acetyltransferases and deacetylases in gene regulation. *Bioessays* **20:** 615.

Lee C., Etchegaray J.P., Cagampang F.R., Loudon A.S., and Reppert S.M. 2001. Posttranslational mechanisms regulate the mammalian circadian clock. *Cell* **107:** 855.

Lo W.S., Duggan L., Emre N.C., Belotserkovskya R., Lane W.S., Shiekhattar R., and Berger S.L. 2001. Snf1—A histone kinase that works in concert with the histone acetyltransferase Gcn5 to regulate transcription. *Science* **293:** 1142.

Lo W.S., Trievel R.C., Rojas J.R., Duggan L., Hsu J.Y., Allis C.D., Marmorstein R., and Berger S.L. 2000. Phosphorylation of serine 10 in histone H3 is functionally linked in vitro and in vivo to Gcn5-mediated acetylation at lysine 14. *Mol. Cell* **5:** 917.

Luger K. 2003. Structure and dynamic behavior of nucleosomes. *Curr. Opin. Genet. Dev.* **13:** 127.

Mahadevan L.C., Willis A.C., and Barratt M.J. 1991. Rapid histone H3 phosphorylation in response to growth factors, phorbol esters, okadaic acid, and protein synthesis inhibitors. *Cell* **65:** 775.

Martinez-Balbas M.A., Bannister A.J., Martin K., Haus-Seuffert P., Meisterernst M., and Kouzarides T. 1998. The acetyltransferase activity of CBP stimulates transcription. *EMBO J.* **17:** 2886.

Matsuo T., Yamaguchi S., Mitsui S., Emi A., Shimoda F., and Okamura H. 2003. Control mechanism of the circadian clock for timing of cell division in vivo. *Science* **302:** 255.

Merienne K., Pannetier S., Harel-Bellan A., and Sassone-Corsi P. 2001. Mitogen-regulated RSK2-CBP interaction controls their kinase and acetylase activities. *Mol. Cell. Biol.* **21:** 7089.

Mizzen C.A., Yang X.J., Kokubo T., Brownell J.E., Bannister A.J., Owen-Hughes T., Workman J., Wang L., Berger S.L., Kouzarides T., Nakatani Y., and Allis C.D. 1996. The TAF(II)250 subunit of TFIID has histone acetyltransferase activity. *Cell* **87:** 1261.

Nagoshi E., Saini C., Bauer C., Laroche T., Naef F., and Schibler U. 2004. Circadian gene expression in individual fibroblasts: Cell-autonomous and self-sustained oscillators pass time to daughter cells. *Cell* **119:** 693.

Nakahata Y., Grimaldi B., Sahar S., Hirayama J., and Sassone-Corsi P. 2007. Signaling to the circadian clock: Plasticity by chromatin remodeling. *Curr. Opin. Cell Biol.* **19:** 230.

Naruse Y., Oh-hashi K., Iijima N., Naruse M., Yoshioka H., and Tanaka M. 2004. Circadian and light-induced transcription of clock gene *Per1* depends on histone acetylation and deacetylation. *Mol. Cell. Biol.* **24:** 6278.

Nowak S.J. and Corces V.G. 2004. Phosphorylation of histone H3: A balancing act between chromosome condensation and transcriptional activation. *Trends Genet.* **20:** 214.

Ogryzko V.V., Schiltz R.L., Russanova V., Howard B.H., and Nakatani Y. 1996. The transcriptional coactivators p300 and CBP are histone acetyltransferases. *Cell* **87:** 953.

Panda S., Antoch M.P., Miller B.H., Su A.I., Schook A.B., Straume M., Schultz P.G., Kay S.A., Takahashi J.S., and Hogenesch J.B. 2002. Coordinated transcription of key pathways in the mouse by the circadian clock. *Cell* **109:** 307.

Pando M.P., Morse D., Cermakian N., and Sassone-Corsi P. 2002. Phenotypic rescue of a peripheral clock genetic defect via SCN hierarchical dominance. *Cell* **110:** 107.

Peterson C.L. and Workman J.L. 2000. Promoter targeting and chromatin remodeling by the SWI/SNF complex. *Curr. Opin. Genet. Dev.* **10:** 187.

Reppert S.M. and Weaver D.R. 2002. Coordination of circadian timing in mammals. *Nature* **418:** 935.

Ripperger J.A. and Schibler U. 2006. Rhythmic CLOCK-BMAL1 binding to multiple E-box motifs drives circadian Dbp transcription and chromatin transitions. *Nat. Genet.* **38:** 369.

Roth S.Y., Denu J.M., and Allis C.D. 2001. Histone acetyltransferases. *Annu. Rev. Biochem.* **70:** 81.

Rutter J., Reick M., Wu L.C., and McKnight S.L. 2001. Regulation of clock and NPAS2 DNA binding by the redox state of NAD cofactors. *Science* **293:** 510.

Sassone-Corsi P., Mizzen C.A., Cheung P., Crosio C., Monaco L., Jacquot S., Hanauer A., and Allis C.D. 1999. Requirement of Rsk-2 for epidermal growth factor-activated phosphorylation of histone H3. *Science* **285:** 886.

Sato T.K., Yamada R.G., Ukai H., Baggs J.E., Miraglia L.J., Kobayashi T.J., Welsh D.K., Kay S.A., Ueda H.R., and Hogenesch J.B. 2006. Feedback repression is required for mammalian circadian clock function. *Nat. Genet.* **38:** 312.

Schibler U. and Naef F. 2005. Cellular oscillators: Rhythmic gene expression and metabolism. *Curr. Opin. Cell Biol.* **17:** 223.

Schibler U. and Sassone-Corsi P. 2002. A web of circadian pacemakers. *Cell* **111:** 919.

Shi Y. 2007. Histone lysine demethylases: Emerging roles in development, physiology and disease. *Nat. Rev. Genet.* **8:** 829.

Spencer T.E., Jenster G., Burcin M.M., Allis C.D., Zhou J., Mizzen C.A., McKenna N.J., Onate S.A., Tsai S.Y., Tsai M.J., and O'Malley B.W. 1997. Steroid receptor coactivator-1 is a histone acetyltransferase. *Nature* **389:** 194.

Sterner D.E. and Berger S.L. 2000. Acetylation of histones and transcription-related factors. *Microbiol. Mol. Biol. Rev.* **64:** 435.

Stokkan K.A., Yamazaki S., Tei H., Sakaki Y., and Menaker M. 2001. Entrainment of the circadian clock in the liver by feeding. *Science* **291:** 490.

Strahl B.D. and Allis C.D. 2000. The language of covalent histone modifications. *Nature* **403:** 41.

Struhl K. 1998. Histone acetylation and transcriptional regulatory mechanisms. *Genes Dev.* **12:** 599.

Sun Z.W. and Allis C.D. 2002. Ubiquitination of histone H2B regulates H3 methylation and gene silencing in yeast. *Nature* **418:** 104.

Vitaterna M.H., King D.P., Chang A.M., Kornhauser J.M., Lowrey P.L., McDonald J.D., Dove W.F., Pinto L.H., Turek F.W., and Takahashi J.S. 1994. Mutagenesis and mapping of a mouse gene, Clock, essential for circadian behavior. *Science* **264:** 719.

Wade P.A. and Wolffe A.P. 1997. Histone acetyltransferases in control. *Curr. Biol.* **7:** R82.

Whitmore D., Foulkes N.S., and Sassone-Corsi P. 2000. Light acts directly on organs and cells in culture to set the vertebrate circadian clock. *Nature* **404:** 87

Workman J.L. and Kingston R.E. 1998. Alteration of nucleosome structure as a mechanism of transcriptional regulation. *Annu. Rev. Biochem.* **67:** 545.

Yamamoto T. and Horikoshi M. 1997. Novel substrate specificity of the histone acetyltransferase activity of HIV-1-Tat interactive protein Tip60. *J. Biol. Chem.* **272:** 30595.

Yan Y., Barlev N.A., Haley R.H., Berger S.L., and Marmorstein R. 2000. Crystal structure of yeast Esa1 suggests a unified mechanism for catalysis and substrate binding by histone acetyltransferases. *Mol. Cell* **6:** 1195.

Yang X.J., Ogryzko V.V., Nishikawa J., Howard B.H., and Nakatani Y. 1996. A p300/CBP-associated factor that competes with the adenoviral oncoprotein E1A. *Nature* **382:** 319.

Zhang K. and Dent S.Y. 2005. Histone modifying enzymes and cancer: Going beyond histones. *J. Cell. Biochem.* **96:** 1137.

The Ancestral Circadian Clock of Monarch Butterflies: Role in Time-compensated Sun Compass Orientation

S.M. Reppert
Department of Neurobiology, University of Massachusetts Medical School, Worcester, Massachusetts 01605

The circadian clock has a vital role in monarch butterfly (*Danaus plexippus*) migration by providing the timing component of time-compensated sun compass orientation, which contributes to navigation to the overwintering grounds. The location of circadian clock cells in monarch brain has been identified in the dorsolateral protocerebrum (pars lateralis); these cells express PERIOD, TIMELESS, and a *Drosophila*-like cryptochrome designated CRY1. Monarch butterflies, like all other nondrosophilid insects examined so far, express a second *cry* gene (designated insect CRY2) that encodes a vertebrate-like CRY that is also expressed in pars lateralis. An ancestral circadian clock mechanism has been defined in monarchs, in which CRY1 functions as a blue light photoreceptor for photic entrainment, whereas CRY2 functions within the clockwork as the major transcriptional repressor of an intracellular negative transcriptional feedback loop. A CRY1-staining neural pathway has been identified that may connect the circadian (navigational) clock to polarized light input important for sun compass navigation, and a CRY2-positive neural pathway has been discovered that may communicate circadian information directly from the circadian clock to the central complex, the likely site of the sun compass. The monarch butterfly may thus use the CRY proteins as components of the circadian mechanism and also as output molecules that connect the clock to various aspects of the sun compass apparatus.

INTRODUCTION

During their spectacular fall migration, eastern North American monarch butterflies (*Danaus plexippus*) use a time-compensated sun compass to help them navigate to their overwintering sites in central Mexico (Reppert 2006). Because this navigational capability is genetically determined, I propose that the monarch butterfly can be used as a model to study the molecular and cellular basis of time-compensated sun compass orientation. The ultimate goal of such studies is to understand the molecular and anatomical mechanisms for clock-compass interactions that enable migrants to maintain a set flight bearing as the sun moves across the sky each day.

In the course of these studies, a novel circadian clock mechanism has been discovered in monarch butterflies that had not been described before in any other animal. In addition, some butterfly clock proteins delineate neural pathways that may connect the clock to the sun compass (or its inputs). But many questions remain about how the clock and compass interact to give rise to appropriately oriented flight behavior.

A TIME-COMPENSATED SUN COMPASS

A fascinating function of circadian clocks is their use in time-compensated sun compass orientation, a phenomenon that was first described decades ago by Karl von Frisch (1967) in foraging honeybees and by Gustav Kramer (1957) in migratory birds. The amazing navigational abilities of monarch butterflies are part of a genetic program initiated in the migrants, as those that make the trip south are at least two generations removed from the previous generation of migrants (Brower 1996). Time-compensated sun compass orientation is believed to be an essential component of navigation required for successful migration.

Time compensation is provided by a circadian clock that allows the butterflies to continually correct their flight direction relative to skylight parameters to maintain a fixed flight bearing in the south/southwesterly direction as the sun moves across the sky during the day (Fig. 1). The ability to successfully navigate requires that this underlying program is constantly recalibrated by environmental factors. For example, the circadian clock is standardized to local time by dawn and dusk, whereas the sun compass may be calibrated by geomagnetic forces and/or visualizing certain landmarks. Barometric pressure and the prevailing wind direction also have a marked influence on proper navigation and the progression of the migration (Reppert 2006).

That monarch butterflies use a time-compensated sun compass to help them navigate has been most convincingly shown using a flight simulator that allows study of flight trajectories from tethered butterflies during a sustained period of flight (Mouritsen and Frost 2002). The importance of the circadian clock in regulating the time-compensated component of flight orientation has been shown in two ways. First, clock-shift experiments, in which the timing of the daily light/dark cycle is either advanced or delayed, cause predictable alterations in the direction the butterflies fly (Perez et al. 1997; Mouritsen and Frost 2002; Froy et al. 2003). Second, constant light, which disrupts the molecular clock mechanism (see below), abolishes the time-compensated component of flight orientation (Fig. 1) (Froy et al. 2003).

What are the skylight cues sensed by the sun compass? Some studies suggest that monarch butterflies can use the skylight pattern of polarized light as a sun compass cue and that this information is utilized in a time-compensated manner (Hyatt 1993; Reppert et al. 2004). Moreover, polarized light relevant for proper orientation may be sensed through ultraviolet (UV) opsin-expressing photoreceptors in the dorsal rim area of the monarch eye, an

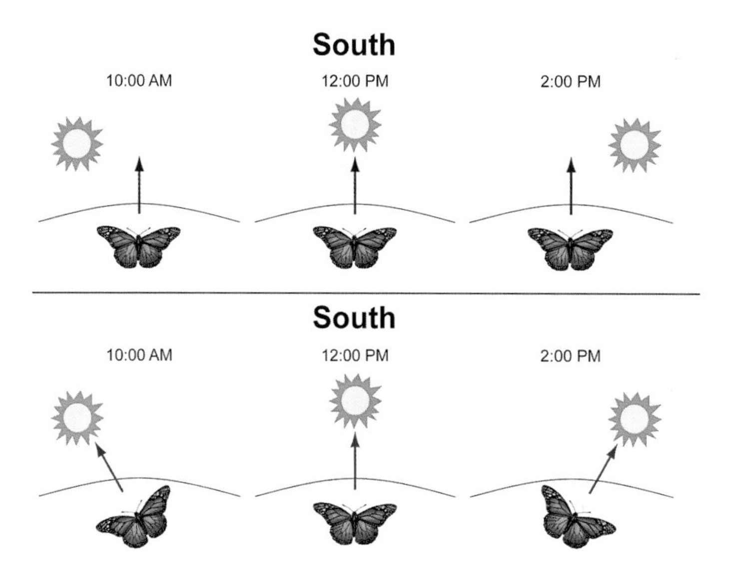

Figure 1. Lost without a clock. (*Upper panel*) Eastern North American monarch butterflies use a time-compensated sun compass to orient south during their fall migration. The circadian clock allows the butterflies to compensate for the movement of the sun. The migrants are thereby able to maintain a constant bearing in the southerly direction over the course of the day. (*Lower panel*) Monarch butterflies follow the sun without a functioning circadian clock. A broken circadian clock would disrupt migration south, and the butterflies would not be able to successfully travel to their overwintering grounds.

area in which photoreceptors are anatomically and physiologically specialized for polarized light detection (Sauman et al. 2005; Stalleicken et al. 2006).

Dorsal rim-sensed polarized light is not necessary for proper flight orientation in the flight simulator, as long as the sun can be seen (Stalleicken et al. 2005). This is consistent with the primary role of polarized light orientation occurring during cloudy days with some blue sky visible. In addition to polarized light, monarch butterflies likely use the sun itself and/or color gradients to orient (Reppert et al. 2004; Stalleicken et al. 2005). Recent electrophysiological studies of single neurons in locusts have shown that nonpolarized chromatic gradients are used coordinately with polarized light to provide more explicit skylight cues for navigation (Pfeiffer and Homberg 2007). These results may be salient for helping determine the relative or combined importance of various celestial cues for proper navigation in monarchs.

LOCATION OF CELLULAR CLOCKS IN BUTTERFLY BRAIN

To fully understand how a circadian clock is involved in time-compensated sun compass orientation in the butterfly, it is important to determine where the cellular clock actually resides in brain, to understand the molecular mechanism of the clock itself, and to delineate neural pathways that connect the clock to the compass. In *Drosophila* and mammals, the intracellular clock mechanism involves transcriptional feedback loops that drive persistent rhythms in mRNA and protein levels of key clock components (Reppert and Weaver 2002; Stanewsky 2003). The negative transcriptional feedback loop is essential for clockwork function and in *Drosophila* involves the transcription factors CLOCK (CLK) and CYCLE (CYC), which drive the expression of the *period* (*per*) and *timeless* (*tim*) genes. The resultant PER and TIM proteins heterodimerize and translocate back into the nucleus where PER inhibits CLK:CYC-mediated transcription. TIM appears to regulate PER protein stability and nuclear transport and is also necessary for photic responses that reset (entrain) the circadian clock. *Drosophila* cryptochrome (CRY), a flavoprotein, is colocalized with PER and TIM in clock cells and is a blue light photoreceptor involved in photic entrainment (Emery et al. 1998, 2000; Stanewsky et al. 1998).

Using a strategy that relied on the coexpression of PER, TIM, and a *Drosophila*-like CRY, designated CRY1, four cells in the dorsolateral region of monarch butterfly brain (the pars lateralis, PL) were identified as the putative location of a circadian clock (Fig. 2) (Sauman et al. 2005). Importantly, PER staining in the PL exhibits a robust 24-hour rhythm that is under circadian control and that is abolished by constant light. PER, TIM, and CRY1 are also colocalized in the central brain in large neurosecretory cells in the pars intercerebralis (PI), but the circadian control of PER levels is less apparent there (Fig. 2) (Sauman et al. 2005). Nonetheless, these PI cells may be part of a circadian network contributing to migratory behaviors.

THE DISCOVERY OF INSECT CRYPTOCHROME 2

In addition to CRY1, monarch butterflies also express a second *cry* gene, which encodes a light-insensitive protein, designated CRY2. CRY2 has potent repressive activity on CLK:CYC-mediated transcription, as the mouse CRY proteins do (Zhu et al. 2005). The finding of two functionally distinct *crys* in the butterfly, along with database searches, led to the recognition of the existence of *cry2* in every non-drosophilid insect so far examined (Yuan et al. 2007).

Drosophila expresses CRY1 only, whereas insects such as mosquitos and lepidopterans (butterflies and moths) express both CRY1 and CRY2. Surprisingly, the honeybee *Apis mellifera* and the beetle *Tribolium castaneum* genomes contain only CRY2 (as determined by BLAST searches of whole-genome sequences; Zhu et al. 2005; Rubin et al. 2006). This suggests that the core circadian oscillator has evolved throughout the insects such that at least three kinds of clocks exist: those containing both CRY1 and CRY2 (the ancestral state) as in monarch and mosquitoes; those containing only CRY1 as in *Drosophila*; and those containing CRY2 alone as in honeybee and beetle (Fig. 3) (Yuan et al. 2007).

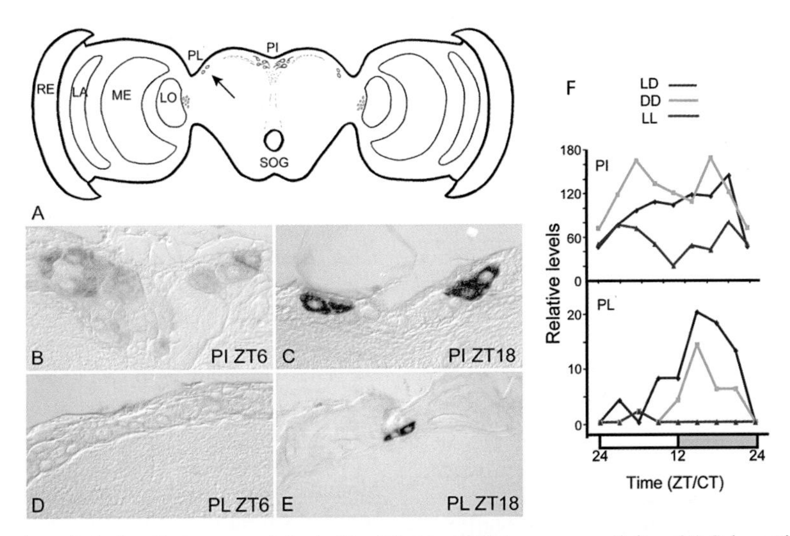

Figure 2. Cellular location of clock cells in monarch brain identified by PER immunoreactivity. (*A*) Schematic diagram of a frontal section illustrating the topography of PER-positive cells. (LA) Lamina; (ME) medulla; (LO) lobula; (PL) pars lateralis (*arrow*); (PI) pars intercerebralis; (SOG) suboesophageal ganglion; (RE) retina. (*B,C*) PER staining in a group of large neurosecretory cells in PI at zeitgeber time [ZT] 6 (B) and ZT 18 (*C*). (*D,E*) Daily osillation of PER immunoreactivity in two cells in PL at ZT 6 (*D*) and ZT 18 (*E*). (*F*) Semiquantitative assessment of PER immunostaining in PI or PL in a 12 hours light:12 hours dark (LD, *magenta*), the second day in constant darkness (DD, *green*), and the second day constant light (LL, *red*). Each value is the sum of intensity scores from three animals. (ZT) Zeitgeber time; (CT) circadian time. (Modified from Sauman et al. 2005.)

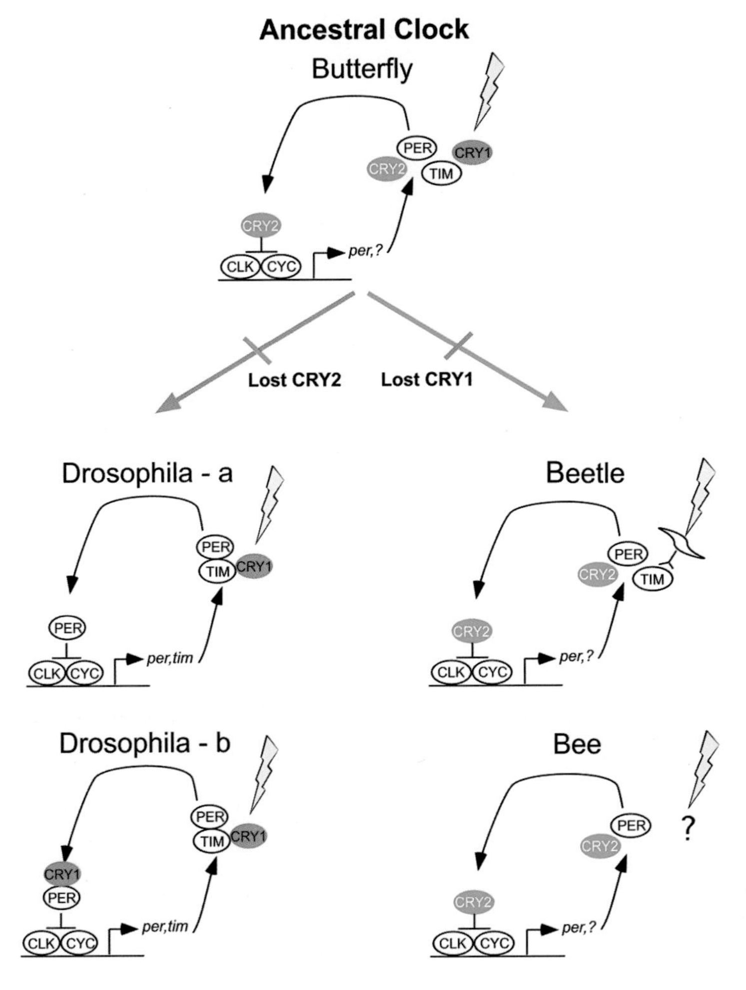

Figure 3. Insect clockwork models. Phylogenetic analyses show a least two rounds of gene duplication at the base of the metazoan radiation, as well as several losses, giving rise to the two *cry* gene families in insects (Yuan et al. 2007). With the existence of two functionally distinct CRYs in insects, three major types of clockwork models can be proposed: the ancestral clock (apparent in the monarch butterfly) in which both CRY1 (*orange oval*) and CRY2 (*green oval*) exist and function differentially within the clockwork; a derived clock (the *Drosophila* form, depicted below *orange arrow*) in which CRY2 has been lost and CRY1 only functions in the central brain clock as a circadian photoreceptor (*a*) (Emery et al. 1998) or in peripheral clocks as both a photoreceptor and central clock component (*b*) (Ivanchenko et al. 2001; Krishnan et al. 2001; Levine et al. 2002; Collins et al. 2006); and a derived clock (depicted below *green arrow*) in which CRY1 has been lost and only CRY2 exists and functions within the clockwork, as in beetles and bees. In beetles, CRY2 acts as a transcriptional repressor of the clockwork, and light input may be mediated through the degradation of TIM. In bees, which lack TIM (Rubin et al. 2006), CRY2 acts as a transcriptional repressor and novel light input pathways (?) are used to entrain the clock. (Modified from Yuan et al. 2007.)

All insect CRY2 proteins so far examined (including those of the bee and beetle) are potent repressors of CLK:CYC-mediated transcription in cell culture (Yuan et al. 2007). Importantly, the bee and beetle CRY2 proteins are not light-sensitive in culture, as assessed either by degradation of CRY2 or by derepression of inhibitory transcriptional activities. These results suggest that these species have novel light input pathways to their circadian clocks, perhaps opsin-based (see Spaethe and Briscoe 2005), as both lack CRY1 (Fig. 3) (Yuan et al. 2007).

THE ANCESTRAL CLOCK OF THE MONARCH BUTTERFLY

On the basis of the presence of two *cry* genes, the monarch butterfly utilizes an ancestral circadian clock, which has clockwork characteristics of both flies (through CRY1) and mice (through CRY2) (Fig. 4). In the butterfly, CRY1 functions primarily as a circadian photoreceptor (Song et al. 2007; Zhu et al. 2008b). CRY2, on the other hand, appears to function as a major transcriptional repressor of the core clock feedback loop in monarchs. In addition to being a potent repressor of CLK:CYC-mediated transcription in cell culture, monarch CRY2 is colocalized with the other clock proteins in the PL, and there it translocates to the nucleus at the appropriate time for transcriptional repression (Zhu et al. 2008b). Monarch PER does not inhibit CLK:CYC-mediate transcription, but it does stabilize CRY2 and may help translocate CRY2 to the nucleus (Zhu et al. 2008b). The ancestral circadian clockwork of monarch butterflies may be the prototype of a novel clock mechanism shared by those nondrosophilid invertebrates that express both *cry1* and *cry2*.

CLOCK-COMPASS NEURAL CONNECTIONS

The CRY-centric ancestral circadian clock defined in monarch butterflies may hold a key to understanding the regulation of time-compensated sun compass orientation. Indeed, a CRY1-staining neural pathway was found that could connect the circadian clock to polarized light input entering the brain (Fig. 5) (Sauman et al. 2005). The CRY1-positive fiber pathway ends in the posterior dorsolateral region of the medulla of the optic lobe, in the same location where the axons from dorsal rim photoreceptors terminate (Fig. 5). On the basis of studies in other insects, these photoreceptor axons would appear to communicate ultimately with the sun compass.

The site of the sun compass in insects now appears to be the central complex (Vitzthum et al. 2002; Heinze and Homberg 2007). The central complex is a midline structure consisting of the dorsally positioned protocerebral bridge and the more ventrally situated central body, which has upper and lower subdivisions. Recent studies in locusts and *Drosophila* have shown that the central complex is not only a control center for motor coordination, but also the actual site of the sun compass (for polarized skylight integration from both eyes and probably all skylight information) (Heinze and Homberg 2007), as well as being involved in visual pattern learning and recognition (Liu et al. 2006).

Our finding of a clock connection with the central complex in the monarch butterfly represents a major advance for beginning to understand its remarkable navigational capabilities. A dense arborization of CRY2 staining has been found in the central body of monarch butterflies, just

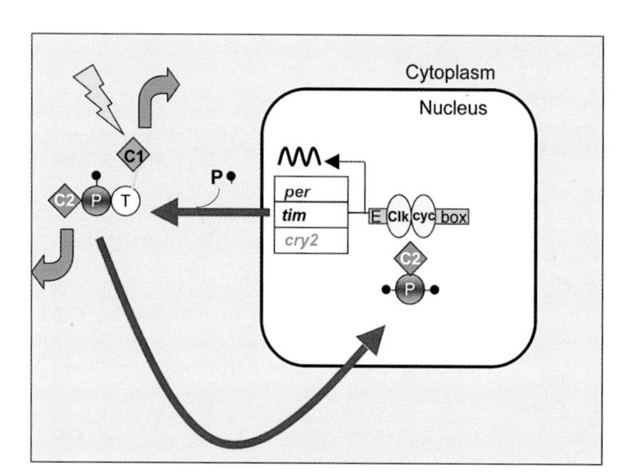

Figure 4. Proposed monarch butterfly circadian clock mechanism. The main gear of the clock mechanism in pars lateralis is an autoregulatory transcriptional feedback loop, in which CLK and CYC heterodimers drive the transcription of the *per*, *tim*, and *cry2* genes through E-box enhancer elements. TIM (T), PER (P), and CRY2 (C2) proteins form complexes in cytoplasm, and CRY2 is shuttled into the nucleus where it shuts down CLK:CYC-mediated transcription. PER is progressively phosphorylated and likely helps translocate CRY2 into nucleus. CRY1 (C1) is a circadian photoreceptor which, upon light exposure (*lightning bolt*) causes TIM degradation, allowing photic information to gain access to the central clock mechanism. (*Thick gray arrows*) Potential output functions for CRY1 and for CRY2. (Modified from Zhu et al. 2008b.)

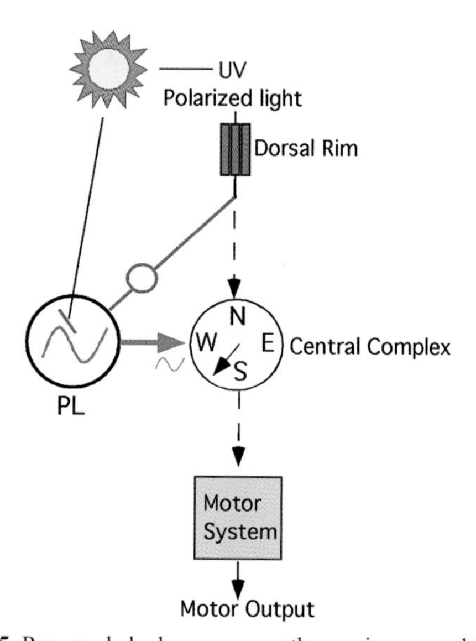

Figure 5. Proposed clock-compass pathways in monarch butterfly brain. A circadian clock in pars lateralis (PL) is entrained by light acting through CRY1 expressed in clock cells (*orange line*). A CRY1-positive fiber pathway (*orange*) connects the circadian clock to axons originating from polarized, UV-light-sensitive photoreceptors in the dorsal rim of the compound eye (Sauman et al. 2005). The circadian clock may also interact directly with the sun compass (in central complex) through a CRY2-positive fiber pathway (*green*) recently discovered. The sun compass ultimately controls motor output. (Modified from Zhu et al. 2008b.)

ventral to the protocerebral bridge (Zhu et al. 2008b). The CRY2 arbors in this area are under circadian control, with intense staining in the middle of the circadian night and little to no visible staining in middle of circadian day. Thus, the CRY2-positive neural pathway appears to be capable of providing circadian signals to the central complex (Fig. 5) (Zhu et al. 2008b). It is likely that the CRY2 fiber pathway to central complex originates from CRY2-positive cells in the PL, PI, or both, but this has not been definitively established. CRY2 may simply be marking the circadian pathway to the sun compass and/or the protein may be directly involved in rhythmic synaptic activity in that region.

MANY QUESTIONS REMAIN

As alluded to above, the central complex looms large as an area in need of focus for future studies of time-compensated sun compass orientation in monarchs. It will be important to define in more detail with confocal microscopy the anatomy of the central complex in monarch butterflies. Three-dimensional reconstruction can also be used to determine exactly where the CRY2 pathway to central complex arises. It will also be important to determine where the CRY2 pathway ends and whether it actually functionally communicates with the central complex.

Elegant electrophysiological studies in crickets and locust have begun to show how neurons in pertinent brain regions process skylight cues from both eyes relevant for proper orientation and subsequent navigational tasks. Studies in crickets have identified polarization-opponent interneurons in optic lobes, which process polarized light information from the orthogonally arranged photon-sensing microvilli in dorsal rim photoreceptors (Labhart and Meyer 2002). Studies in locusts have further shown that the columnar organization of cells in the protocerebral bridge of the central complex provides a topographical map of overhead polarized skylight patterns from both eyes (Vitzthum et al. 2002; Heinze and Homberg 2007). Moreover, interneurons in the anterior optic tubercle of the locust can process both polarization and color gradients for more precise skylight signals for navigation than could be obtained through polarized light signals alone (Pfeiffer and Homberg 2007). These studies provide a rich backdrop for potential electrophysiological correlates in monarch butterflies.

DNA microarray analysis could be used to determine genes important in central complex activity during different orientation states in tethered butterflies (Zhu et al. 2008a). Candidate genes could then be knocked down by RNAi to help determine function. The ultimate goal would be to selectively manipulate gene expression in the central complex (or other relevant brain regions) to understand the molecular logic behind the neuronal signaling patterns that have been identified and how they ultimately contribute to time-compensated flight orientation. Cracking the code of sun compass orientation will require a combination of molecular, anatomical, electrophysiological, and behavioral approaches in a single species. The monarch butterfly is ideally suited for such studies.

ACKNOWLEDGMENTS

I thank Haisun Zhu for help with Figure 1, Adriana Briscoe and David Weaver for comments on the manuscript, and members of my laboratory for helpful discussions.

REFERENCES

Brower L.P. 1996. Monarch butterfly orientation: Missing pieces of a magnificent puzzle. *J. Exp. Biol.* **199:** 93.

Collins B., Mazzoni E.O., Stanewsky R., and Blau J. 2006. *Drosophila* CRYPTOCHROME is a circadian transcriptional repressor. *Curr. Biol.* **16:** 441.

Emery P., So W.V., Kaneko M., Hall J.C., and Rosbash M. 1998. CRY, a *Drosophila* clock and light-regulated cryptochrome, is a major contributor to circadian rhythm resetting and photosensitivity. *Cell* **95:** 669.

Emery P., Stanewsky R., Helfrich-Förster C., Emery-Le M., Hall J.C., and Rosbash M. 2000. *Drosophila* CRY is a deep brain circadian photoreceptor. *Neuron* **26:** 493.

Froy O., Gotter A.L., Casselman A.L., and Reppert S.M. 2003. Illuminating the circadian clock in monarch butterfly migration. *Science* **300:** 1303.

Heinze S. and Homberg U. 2007. Maplike representation of celestial E-vector orientations in the brain of an insect. *Science* **315:** 995.

Hyatt M.B. 1993. "The use of polarization for migratory orientation by monarch butterflies." Ph.D. thesis, University of Pittsburgh, Pittsburgh, Pennsylvania.

Ivanchenko M., Stanewsky R., and Giebultowicz J.M. 2001. Circadian photoreception in *Drosophila:* Functions of cryptochrome in peripheral and central clocks. *J. Biol. Rhythms* **16:** 205-215.

Kramer G. 1957. Experiments on bird orientation and their interpretation. *Ibis* **99:** 196.

Krishnan B., Levine J.D., Lynch M.K., Dowse H.B., Funes P., Hall J.C., Hardin P.E., and Dryer S.E. 2001. A new role for cryptochrome in a *Drosophila* circadian oscillator. *Nature* **411:** 313.

Labhart T. and Meyer E.P. 2002. Neural mechanisms in insect navigation: Polarization compass and odometer. *Curr. Opin. Neurobiol.* **12:** 707.

Levine J.D., Funes P., Dowse H.B., and Hall J.C. 2002. Advanced analysis of a cryptochrome mutation's effects on the robustness and phase of molecular cycles in isolated peripheral tissues of *Drosophila. BMC Neurosci.* **3:** 5.

Liu G., Seiler H., Wen A., Zars T., Ito K., Wolf R., Heisenberg M., and Liu L. 2006. Distinct memory traces for two visual features in the *Drosophila* brain. *Nature* **439:** 551.

Mouritsen H. and Frost B.J. 2002. Virtual migration in tethered flying monarch butterflies reveals their orientation mechanisms. *Proc. Natl. Acad. Sci.* **99:** 10162.

Perez S.M, Taylor O.R., and Jander R. 1997. A sun compass in monarch butterflies. *Nature* **387:** 29.

Pfeiffer K. and Homberg U. 2007. Coding of azimuthal directions via time-compensated combination of celestial compass cues. *Curr. Biol.* **17:** 960.

Reppert S.M. 2006. A colorful model of the circadian clock. *Cell* **124:** 233.

Reppert S.M. and Weaver D.R. 2002. Coordination of circadian timing in mammals. *Nature* **418:** 935.

Reppert S.M., Zhu H., and White R.H. 2004. Polarized light helps monarch butterflies navigate. *Curr. Biol.* **14:** 155.

Rubin E.B., Shemesh Y., Cohen M., Elgavish S., Robertson H.M., and Bloch G. 2006. Molecular and phylogenetic analyses reveal mammalian-like clockwork in honey bee (*Apis mellifera*) and shed new light on the molecular evolution of the circadian clock. *Genome Res.* **16:** 1352.

Sauman I., Briscoe A.D., Zhu H., Shi D., Froy O., Stalleicken J., Yuan Q., Casselman A., and Reppert S.M. 2005. Connecting the navigational clock to sun compass input in monarch butterfly brain. *Neuron* **46:** 457.

Song S.H., Öztürk N., Denaro T.R., Arat N.O., Kao Y.T., Zhu H., Zhong D., Reppert S.M., and Sancar A. 2007. Formation and function of flavin anion radical in cryptochrome 1 blue-light photoreceptor of monarch butterfly. *J. Biol. Chem.* **282:** 17608.

Spaethe J. and Briscoe A.D. 2005. Molecular characterization and expression of the UV opsin in bumblebee: Three ommatidial subtypes in the retina and a new photoreceptor organ in lamina. *J. Exp. Biol.* **208:** 2347.

Stalleicken J., Labhart T., and Mouritsen H. 2006. Physiological characterization of the compound eye in monarch butterflies with focus on the dorsal rim area. *J. Comp. Physiol. A Neuroethol. Sens. Neural Behav. Physiol.* **192:** 321.

Stalleicken J., Mukhida M., Labhart T., Wehner R., Frost B., and Mouritsen H. 2005. Do monarch butterflies use polarized skylight for migratory orientation? *J. Exp. Biol.* **208:** 2399.

Stanewsky R. 2003. Genetic analysis of the circadian system in *Drosophila melanogaster* and mammals. *J. Neurobiol.* **54:** 111.

Stanewsky R., Kaneko M., Emery P., Beretta B., Wager-Smith K., Kay S.A., Rosbash M., and Hall J.C. 1998. The cryb mutation identifies cryptochrome as a circadian photoreceptor in *Drosophila*. *Cell* **95:** 681.

Vitzthum H., Muller M., and Homberg U. 2002. Neurons of the central complex of the locust *Schistocerca gregaria* are sensitive to polarized light. *J. Neurosci.* **22:** 1114.

von Frisch K. 1967. *The dance language and orientation of bees*. Belknap Press of Harvard University Press, Cambridge, Massachusetts.

Yuan Q., Metterville D., Briscoe A.D., and Reppert S.M. 2007. Insect cryptochromes: Gene duplication and loss define diverse ways to construct insect circadian clocks. *Mol. Biol. Evol.* **24:** 948.

Zhu H., Casselman A., and Reppert S.M. 2008a. Chasing migration genes: A brain expressed sequence tag resource for summer and migratory monarch butterflies (*Danaus plexippus*). *PLoS One* **3:** e1293.

Zhu H., Yuan Q., Briscoe A.D., Froy O., Casselman A., and Reppert S.M. 2005. The two CRYs of the butterfly. *Curr. Biol.* **15:** R953.

Zhu H., Sauman I., Yuan Q., Casselman A., Emery-Le M., Emery P., and Reppert S.M. 2008b. Cryptochromes define a novel circadian clock mechanism in monarch butterflies that may underlie sun compass navigation. *PLoS Biol.* **6:** e4.

Structure and Function of Animal Cryptochromes

N. ÖZTÜRK,* S.-H. SONG,* S. ÖZGÜR,* C.P. SELBY,* L. MORRISON,* C. PARTCH,[†]
D. ZHONG,[‡] AND A. SANCAR*

*Department of Biochemistry and Biophysics, University of North Carolina School of Medicine, Chapel Hill, North Carolina 27599; [†]Department of Biochemistry, University of Texas Southwestern Medical Center, Dallas, Texas 75390; [‡]Departments of Physics, Chemistry, and Biochemistry, Programs of Biophysics, Chemical Physics, and Biochemistry, Ohio State University, Columbus, Ohio 43210

Cryptochrome (CRY) is a photolyase-like flavoprotein with no DNA-repair activity but with known or presumed blue-light receptor function. Animal CRYs have DNA-binding and autokinase activities, and their flavin cofactor is reduced by photoinduced electron transfer. In *Drosophila*, CRY is a major circadian photoreceptor, and in mammals, the two CRY proteins are core components of the molecular clock and potential circadian photoreceptors. In mammals, CRYs participate in cell cycle regulation and the cellular response to DNA damage by controlling the expression of some cell cycle genes and by directly interacting with checkpoint proteins.

INTRODUCTION

CRYs are FAD-based blue light photoreceptors that control growth and development in plants and the circadian clock in animals and possibly in plants (Cashmore 2003; Lin and Shalitin 2003; Sancar 2003, 2004). CRYs have sequence and structural similarities to DNA photolyases (Fig. 1), which repair UV-induced DNA damage by a photoinduced cyclic electron transfer reaction (Sancar 2003; Kao et al. 2005). These two seemingly unrelated phenomena, circadian rhythm and DNA repair, may have had a common evolutionary origin (Pittendrigh 1993; Sancar 2000; Gehring and Rosbash 2003; Lowrey and Takahashi 2004). According to this "escape from light" hypothesis, in the distant past when more UV reached the surface of the earth, an aquatic organism used a blue-light photoreceptor (CRY) to restrict its S phase to the dark phase of the day (night) so as to minimize the harmful effects of DNA damage and to regulate the organism's vertical movement to and away from the surface of the water with daily (circadian) periodicity, thereby optimizing nutrient uptake and minimizing the extent of DNA damage. This same photoreceptor may have also been used to repair DNA damage (photolyase) that inevitably occurred under such conditions, especially in the early days of life on earth when more UV reached the surface because of the lack of the protective ozone layer. Blue light is best suited for both tasks since only blue light can penetrate to substantial depths in water. Consequently, this hypothesis suggests that the blue light photoreceptor carrying out these two functions diverged to give rise to the present-day photolyases and CRYs. It is conceivable that future research may uncover the "missing link" of the theory: a blue light photoreceptor with both circadian and DNA-repair functions.

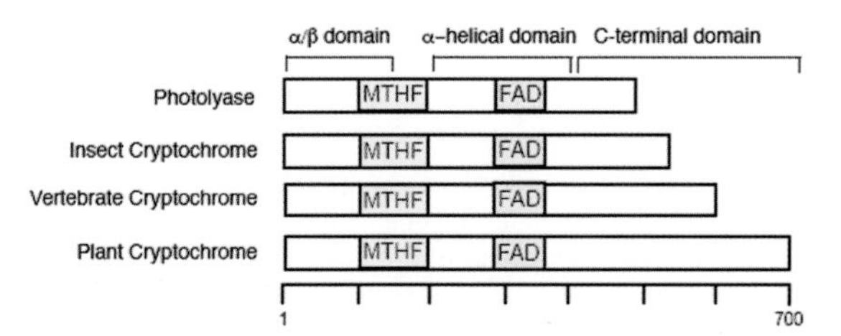

Figure 1. Schematic representation of the photolyase/CRY family proteins. The enzymes are 500–700 amino acids in length and have a modular structure with an amino-terminal α/β domain and a carboxy-terminal α-helical domain. In addition, most CRYs have carboxy-terminal extensions ranging in size from 40 to 250 amino acids. Representative examples of the major classes are shown. The insect CRY in this figure is Insect Type 1 CRY. The approximate binding sites of the two cofactors MTHF and FAD are indicated.

HISTORICAL PERSPECTIVE

Since the second half of the 19th century, plant biologists have known that blue light has a profound effect on growth, development, and phototropic movement of plants (Darwin 1881). However, efforts to identify the blue light photoreceptor initiating these responses were futile for a long period, and some plant biologists used the term "cryptochrome" as a generic name for this mysterious photoreceptor. The following explanation was given as a justification of the name: "The pigment system(s) responsible for many of the photoprocesses (as ascertained by action spectra) has been nicknamed 'cryptochrome' because of its importance in cryptogamic plants and its cryptic nature. This term, despised by many, will suffice us here just because it is shorter than other terms used, such as 'blue (UV) light photoreceptor,' and it will be a useful term until the pigments are identified" (Gressel 1977). Later work identified at least four classes of flavoproteins that mediate blue light responses in plants (Banerjee and Batschauer 2005): photolyase, the HY4 protein, phototropin, and the ZTL/ADO family. Of these, the photolyase was the first blue light photoreceptor to be identified as a flavoprotein in bacteria and many other species and has been extensively characterized (Sancar 2003). When the *Arabidopsis thaliana* HY4 gene, known to be required for inhibition of hypocotyl growth in response to blue light (Koornneef et al. 1980), was isolated and sequenced, it revealed high sequence homology with *Escherichia coli* photolyase, and hence, it was in retrospect, correctly speculated that HY4 encoded a blue light photoreceptor and not a signal transducer involved in blue light response (Ahmad and Cashmore 1993). Later, this protein was named cryptochrome 1 (Lin et al. 1995, 1998), and a second *Arabidopsis* protein that has high homology with CRY1 identified by genomics was named CRY2 (Lin and Shalitin 2003). The role of CRY in *Arabidopsis* growth (CRY1) and differentiation (CRY2) was well established by 1995 (see Guo et al. 1998). However, as late as 1997, it was thought that CRY had no role in circadian photoreception in plants (Millar and Kay 1997).

The first report implicating CRY in the circadian clock of any organism, plant or animal, came about from the study of DNA repair, in particular, the repair of UV damage in humans by photolyase. This issue had been controversial for nearly 25 years when Li et al. (1993) conducted an exhaustive study with a highly specific and sensitive assay, concluding that humans, like all placental mammals, lacked photolyase (Li et al. 1993). However, a 1995 release of a human expressed sequence tag (EST) list contained a "photolyase ortholog" entry (Adams et al. 1995). In light of this finding and the discovery of a photolyase in *Drosophila* and rattlesnake (Todo et al. 1993, 1996; Kim et al. 1996) that repairs the minor UV-induced lesion, the (6-4) photoproduct, in contrast to the classic photolyase that repairs cyclobutane pyrimidine dimers (CPDs), the earlier conclusion regarding the lack of photolyase in humans needed reevaluation. This was done by Hsu et al. (1996) who, in addition to the "photolyase ortholog" in public databases, discovered a second human photolyase gene. Human cells expressing both genes and recombinant proteins encoded by both genes were tested for CPD and (6-4) photolyase activities and were found to lack both. Moreover, the proteins encoded by these genes, like most photolyases (Johnson et al. 1988) and *Arabidopsis* CRY (Lin et al. 1995; Malhorta et al. 1995), contained FAD (flavin-adenine dinucleotide) and a pterin cofactor. Therefore, it was concluded that these proteins were not repair enzymes but that, like *Arabidopsis* CRYs, they performed non-repair-related blue light functions and were named human CRY1 and 2 (Hsu et al. 1996). In humans and most other animals, the two well-characterized, light-mediated reactions are vision and circadian entrainment. Because opsins are securely established as the vision pigment, it was suggested that the human CRYs might be circadian photoreceptors (Hsu et al. 1996; Zhao and Sancar 1997) and experiments were set up to test this prediction in the mouse.

The first experimental data linking CRY to the circadian clock was the finding that mouse CRY1 was highly expressed in the suprachiasmatic nucleus (SCN) (Fig. 2A) with an expression pattern that exhibited periodicity with a peak at ZT8 and nadir at ZT20 (Miyamoto and Sancar 1998, 1999). The functional proof that the mammalian CRY controlled the circadian clock came shortly afterward when $mCry2^{-/-}$ mice were generated and tested for circadian phenotypes (Fig. 2B). It was found that the mutant mice had a period about 1 hour longer than that of wild-type littermates and the CRY2 knockout exhibited greatly increased phase-shifts in response to light pulses (Thresher et al. 1998). These data firmly established CRY as a core clock protein regardless of its potential involvement in circadian photoreception. Nearly simultaneously, the strongest evidence to date for a circadian photoreceptive role of CRY in animals was obtained when it was discovered that a *Drosophila* mutant selected for reduced circadian photosensitivity contained a missense mutation in the *Drosophila* ortholog of CRY, DmCry (Emery et al. 1998; Stanewsky et al. 1998). In light of these developments, the *Arabidopsis* CRY mutants were tested for circadian photoreception, and it was found that under a particular lighting regimen, *Arabidopsis* CRYs also

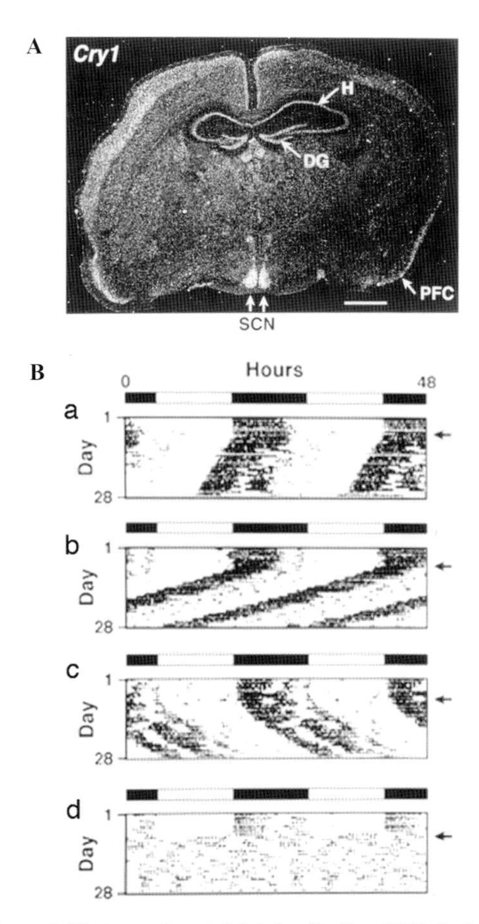

Figure 2. First experimental data implicating CRYs in the circadian clock. (*A*) Expression of mammalian CRY1 in the mouse SCN. Expression was measured by in situ hybridization. This sample was prepared at ZT6 when mammalian CRY1 expression is at its zenith; at ZT18 (or CT18) mammalian CRY1 is virtually undetectable. (SCN) Suprachiasmatic nucleus; (PFC) piriform cortex; (DG) dentate gyrus; (h) hippocampus. (*B*) Effect of CRY mutation on mouse circadian behavior. The locomotor activities of wild-type and mutant mice were recorded for 28 days. At the day indicated by arrows, the animals were switched from LD to DD conditions. (*a*) Wild-type: $\tau = 23.7$ hr; (*b*) $Cry1^{-/-}$: $\tau = 22.7$ hr; (*c*) $Cry2^{-/-}$: $\tau = 24.7$ hr; (*d*) $Cry1^{-/-}$;$Cry1^{-/-}$ (arrhythmic). (*A*, Reprinted, with permission, from Miyamoto and Sancar 1998; *B*, reprinted, with permission, from Vitaterna et al. 1999 [both © National Academy of Sciences].)

appeared to participate in circadian photoreception under a particular lighting regimen (Somers et al. 1998). Finally, mouse *Cry1*$^{-/-}$;*Cry2*$^{-/-}$ mutants were constructed and found to have lost circadian rhythm entirely (van der Horst et al. 1999; Vitaterna et al. 1999), thus expanding the conclusion based on the *Cry2*$^{-/-}$ mutant and consolidating the role of CRYs as core clock proteins in mammals (Fig. 2B). These results set the stage for the findings that mammalian CRYs interacted with all core clock proteins as revealed by yeast two-hybrid assay (Ceriani et al. 1999) and that mammalian CRYs acted as potent repressors of the CLOCK/BMAL1 transcriptional activator complex as revealed by reporter gene assays (Kume et al. 1999) and by the constitutive, elevated expression of mPer genes in *Cry1*$^{-/-}$;*Cry2*$^{-/-}$ mice (Vitaterna et al. 1999) and eventual development of transcriptional-translational feedback loop (TTFL) model for animal circadian clock (Gekakis et al. 1998; Young and Kay 2001; Reppert and Weaver 2002). As is apparent from the summary of the field given above, the identification of CRYs as circadian proteins followed a conventional and inductive scientific approach and was not, as suggested, "accidental" (Hunt and Sassone-Corsi 2007).

PHYLOGENY AND FUNCTIONAL CLASSIFICATION

To assess the evolution of the photolyase/CRY family, we performed an exhaustive search of annotated sequence databases, retrieving more than 250 sequences of photolyases and CRYs from all three kingdoms of life. Phylogenetic analysis by neighbor-joining and maximum parsimony methods grouped these sequences into eight major classes. A reduced tree with gene names is shown in Figure 3. Seven of these classes have been previously described and functionally characterized to varying degrees: class I and class II CPD photolyases, (6-4) photolyase, single-stranded DNA photolyase (previously called DASH CRYs), and plant, insect, and vertebrate CRYs. However, this analysis led to two unexpected findings: First, a group of novel independently segregating bacterial sequences was identified, comprising a new class (class III CPD photolyase), and second, vertebrate-like CRY sequences were discovered in nondrosophiloid insects. In vivo photoreactivation data on a class III CPD photolyase from *Caulobacter crescentus* indicate that this is indeed a CPD photolyase (see Partch 2006). Second, as first reported by Zhu et al. (2005) and Yuan et al. (2007), several nondrosophiloid insects such as mosquito, honeybee, and silkmoth possess a vertebrate-like CRY, suggesting that the origin of vertebrate CRY predates the last common ancestor shared by insects and vertebrates. The name of bilateral CRYs was proposed because they are present in at least two branches of Bilateria, animals defined by bilateral symmetry (Partch 2006). However, since the names of "Insect CRY1" and "Insect CRY2" have been used for the drosophiloid and vertebrate-type CRYs (Zhu et al. 2005; Yuan et al. 2007), we will follow that nomenclature. Current data indicate that Insect CRY1s are sensitive to photoinduced degradation in vivo, whereas Insect CRY2s are not, but instead, they function

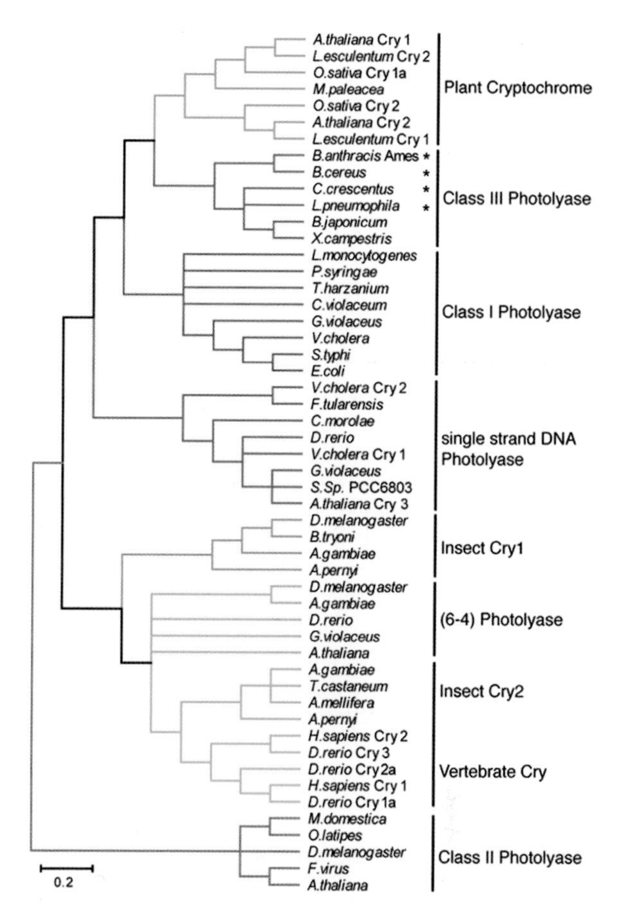

Figure 3. Evolutionary relationships in the photolyase/CRY family. Annotated photolyase and CRY sequences (>250 total sequences) from GenBank and Swiss-Prot databases were aligned using ClustalW. Alignments were manually verified, and an unrooted phylogenetic tree was generated using neighbor-joining methods (MEGA 4.1). Phylogenetic analysis of representative sequences are shown. Eight major classes are identified, including a novel group of photolyases (Class III, *purple*) that is a sister taxon to plant CRYs (*green*). The single-stranded DNA photolyase class was previously misclassified as CRY with Cry-DASH designation. However, recent work has shown that these are CPD photolyases specific for single-stranded DNA (Selby and Sancar 2006). Asterisks indicate organisms with sequenced genomes that possess this single photolyase gene and a literature report of photoreactivation. Bar represents residue substitutions per site.

as repressors of the Clock/Cycle complex. *Drosophila* possesses only Insect CRY1, the honeybee *Apis mellifera* has only Insect CRY2, and the monarch butterfly *Danaus plexippus* has both. It should also be noted that the assignment of the insect CRY1s as solely photoreceptors and the Insect CRY2/vertebrate CRY family exclusively as repressors is not universally accepted. There is credible evidence that DmCRY functions as a repressor (Collins et al. 2006). Similarly, there is considerable genetic evidence that vertebrate CRYs, in addition to the repressor activity, may function as photoreceptor/phototransducers (Thresher et al. 1998; Selby et al. 2000; Thompson et al. 2003; Tu et al. 2004). In addition to this phylogenetic classification, the photolyase/CRY family can be divided into functional classes. These classifications do not necessarily converge in all facets as evidenced by the fact that class I and class II CPD photolyases are more phylogenetically distant from each other than class I photolyases and plant CRYs, yet class I and class II photolyases perform exactly the same repair function.

STRUCTURES OF PHOTOLYASE AND CRYPTOCHROME

Photolyase/CRY family proteins are 50–80-kD proteins of 500–700 amino acids in length with two chromophores/cofactors (Fig. 4). One of the cofactors is always FAD, serving as the catalytic cofactor. The other cofactor serves as a photoantenna and is most commonly methenyltetrahydrofolate (MTHF) or, in rare instances, 8-hydroxy-5-deazaflavin (8-HDF) in organisms that synthesize this chromophore (Sancar 2003; Partch and Sancar 2005). Recently, it was reported that the *Thermus thermophilus* photolyase contains flavin mononucleotide (FMN) (Ueda et al. 2005) and photolyase from *Sulfolobus tokodaii* contains FAD (Fujihashi et al. 2007) as the second chromophore, suggesting that photolyase is capable of utilizing a variety of chromophores as photoantenna. Because of the high sequence and structural similarities between photolyase and CRY, it is generally assumed that CRYs have the same two cofactors as well. However, no CRY has been purified to date from its native source and those that have been purified as recombinant proteins contain FAD to varying levels and either trace amounts of MTHF or none at all (Lin et al. 1995; Malhotra et al. 1995; Özgür and Sancar 2003; Song et al. 2007). Hence, formal proof that CRYs contain MTHF, or any other secondary chromophore, is lacking.

CRYs diverge from photolyases in another significant aspect: Nearly all CRYs (but not CRY1 of *Sinapis alba*) possess carboxy-terminal domains beyond that of the photolyase homology region (PHR) ranging from 30 to 350 amino acids in length. Of significance, the sequences of these carboxy-terminal domains are not conserved from plants to animals (Partch and Sancar 2005). Biochemical and biophysical tests show that the carboxy-terminal domains of CRYs are highly unstructured when expressed alone (Partch et al. 2005; Kottke et al. 2006) but assume a rigid structure by interacting with the PHR domain (Lin and Shalitin 2003; Partch et al. 2005). Light-induced conformational change from order to disorder in AtCRY1 has been proposed to initiate the photosignaling reaction (Partch et al. 2005; Kottke et al. 2006; Yu et al. 2007).

Crystal structures of several photolyases are available and are quite similar (Huang et al. 2006; Fujihashi et al. 2007). In contrast, only the PHR domain of AtCRY1 has been crystallized (Brautigam et al. 2004). The structures of photolyases are characterized by two modular domains (Fig. 4): an amino-terminal α/β domain and a carboxy-terminal α-helical domain connected by a long interdomain loop. The catalytic FAD chromophore is bound within the α-helical domain in an unusual U-shaped conformation, with the isoalloxazine ring held in close proximity to the adenine ring, and the second chromophore is bound in a cleft located between the two domains close to the surface of the protein. Surface potential representation of the photolyase structure reveals a positively charged DNA-binding groove running the length of the molecule. A hole of approximately 10 Å in diameter, located in the middle of this groove, allows access of solvent and oxygen to the FAD molecule. Additionally, this hole is of the right dimensions and polarity to allow entry of a pyrimidine dimer to within van der Waals contact distance of the isoalloxazine ring of FAD. The structure of AtCRY1 PHR domain structure is very similar to that of photolyase in many aspects, including the substrate-binding cavity; however, AtCRY1 lacks the positively charged DNA-binding groove, and in fact, many of the amino acid residues lining this groove are negatively charged which may partly be responsible for the lack of DNA-repair activity by plant CRYs (Brautigam et al. 2004). The other significant aspect of the AtCRY1 PHR crystal structure is the lack of MTHF or any other second chromophore.

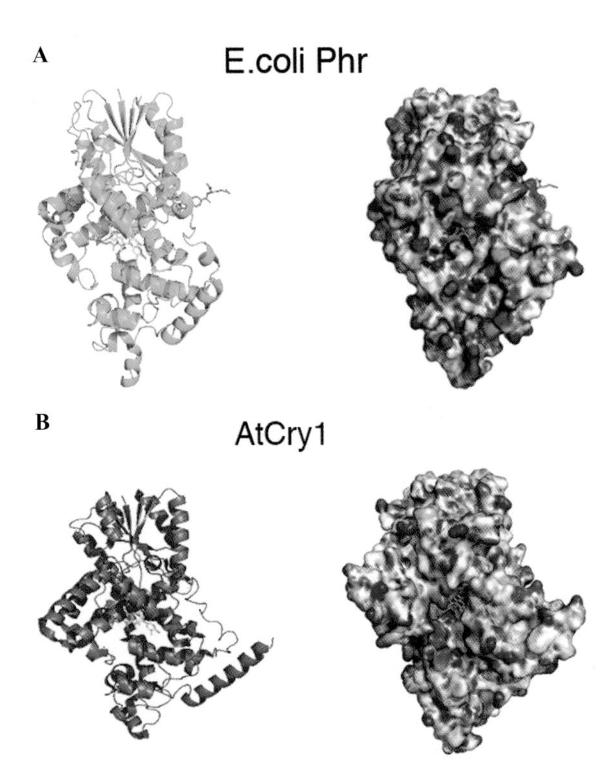

Figure 4. Crystal structures of the photolyase/CRY family. Both ribbon diagram and surface potential representations are shown. (*A*) *E. coli* photolyase; (*B*) *A. thaliana* CRY1-PHR domain. Note that the overall architectures are very similar including the hole leading to the FAD cofactor in the core of the α-helical domain. However, although photolyase possesses a positively charged DNA-binding groove running the length of the surface, this groove is mostly lined with negatively charged residues in AtCRY1. Note also that the crystal structure of AtCRY1 is lacking the MTHF cofactor.

REACTION MECHANISM OF PHOTOLYASE

Currently, three classes of flavin-based blue light photoreceptors are known, each utilizing flavin in different chemical forms or oxidation states (Banerjee and Batschauer 2005; Losi 2007): LOV proteins (FMN),

BLUF proteins (FAD), and CRY/photolyase ($FADH^-$). Blue light causes the formation of a blue-shifted FMN-cysteine C(4a)-thioladduct (λ_{max} = 340 nm) in LOV-domain-containing proteins such as phototropin and hydrogen-bond rearrangement of FAD and neighboring amino acids that causes a red-shift in FAD absorption (from between 365 and 445 to 371 and 460 nm) in BLUF-domain-containing proteins such as the *Euglena gracilius* photoactivated adenylyl cyclase (PAC). These could be potential models for the CRY photocycle. However, because of its evolutionary and structural relatedness to photolyase, CRY is more likely to have a photocycle similar to that of photolyase. For this reason, the photolyase photocycle is discussed below in some detail (Fig. 5).

Photolyase recognizes the 30° kink caused in DNA by a cyclobutane pyrimidine dimer (Pyr<>Pyr), and it flips out the dimer from within the duplex to the active site cavity of the enzyme to form a highly stable E•S complex. Light initiates catalysis: The MTHF (or 8-HDF) photoantenna absorbs a photon and transfers the energy to $FADH^-$ (the active form of flavin in photolyase) by Förster resonance energy transfer. The excited state flavin, $^1(FADH^-)^*$, transfers an electron to Pyr<>Pyr to generate a charge-separated radical pair (FADH°—Pyr<>Pyr$^{\circ -}$). The cyclobutane ring is split by cycloreversion and the flavin radical is restored to the catalytically competent $FADH^-$ form by back electron transfer following splitting of the cyclobutane ring. Significantly, at the end of the catalytic cycle, there is no change in the redox state of flavin or the substrate/product and hence the repair reaction is not a redox reaction. Following catalysis, the repaired dinucleotide no longer fits in the active-site pocket and is ejected back into the duplex, and the repaired DNA dissociates from the enzyme. It is thought that the (6-4) photolyase employs essentially the same mechanism as classical photolyase (Sancar 2003; Li et al. 2006).

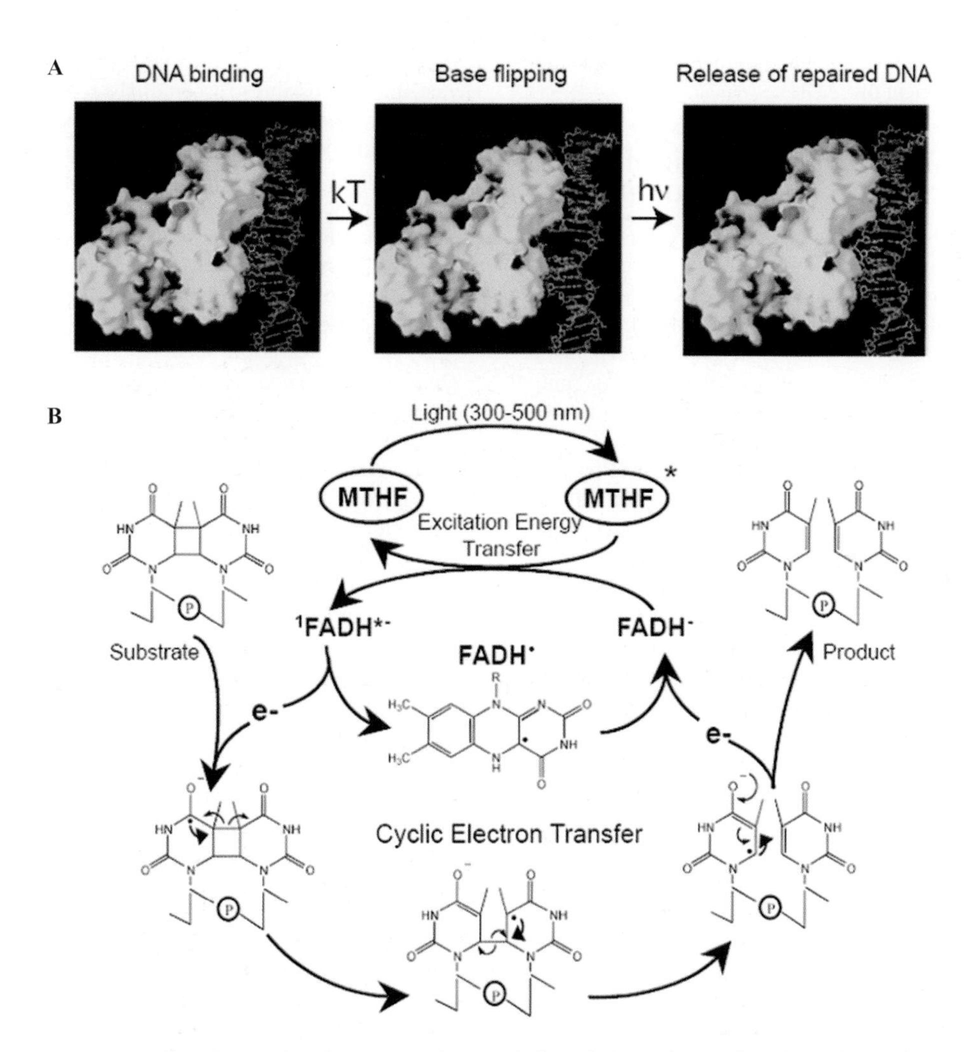

Figure 5. Reaction mechanism of DNA photolyase. (*A*) Substrate binding (dark reaction) and product release. The enzyme binds the DNA backbone around the cyclobutane pyrimidine dimer by random collision, forming a low-affinity complex. Then, it flips out the dimer into the active site cavity. The enzyme-substrate complex formation is a thermal (*k*T) reaction, independent of light. Following the light (*h*υ) reaction, the repaired dinucleotide is ejected from the active-site cavity and DNA dissociates from the enzyme. (*B*) Catalysis (light reaction). The photoantenna (MTHF) absorbs a photon and transfers the excitation energy to $FADH^-$ by FRET. The $^1(FADH^-)^*$ then transfers an electron to the cyclobutane dimer to generate a radical pair of flavin and pyrimidine dimer. The cyclobutane ring is split, and the electron returns to FADH° to regenerate catalytically active $FADH^-$. The reaction is a cyclic redox reaction with no net change in the redox status of either the enzyme or the substrate/product at the end of the reaction.

Finally, a comment should be made on the action spectra of photolyases. An action spectrum is a plot of the rate of a photochemical or photobiological reaction as a function of the wavelength eliciting the reaction (see Sancar 2000). Although the catalytic cofactor in photolyases is $FADH^-$ with λ_{max} ~360 nm and molar extinction coefficient ε ~6000 $M^{-1}cm^{-1}$, the action spectra of photolyases are dominated by the photoantennas that have much higher extinction coefficients and λ_{max} at longer wavelengths. Thus, in photolyases with MTHF cofactor (ε = 25,000 $M^{-1}cm^{-1}$) and λ_{max} = 380–415 nm (depending on the enzyme), the action spectra match that of MTHF absorption spectra. Similarly, in photolyases with 8-HDF as the second chromophore (ε = 40,000 $M^{-1}cm^{-1}$ and λ_{max} = 440 nm), the action spectrum maximum is at 440 nm. It should be noted that photolyases can carry out photorepair in the absence of the second chromophore because $FADH^-$ can directly absorb a photon and initiate catalysis. In this case, the action spectrum matches the absorption spectrum of $FADH^-$ (Payne and Sancar 1990) and efficiency of repair by an incident photon is lower than that of the holoenzyme because of the low extinction coefficient of $FADH^-$.

BIOCHEMICAL AND PHOTOCHEMICAL PROPERTIES OF ANIMAL CRYPTOCHROMES

Currently, it is known that Insect Type I CRYs function as circadian photoreceptors and that Insect Type II CRYs, like the evolutionarily related vertebrate CRYs, function as repressors of Clock/Cycle (CLOCK/BMAL in vertebrates) (Zhu et al. 2005; Yuan et al. 2007). Mechanistic details of both the photoreceptor and the repressor functions of CRYs are poorly understood; it is also unknown whether the two functions are mutually exclusive.

Below, some of the physical and biochemical properties of animal CRYs are discussed. Reference will be made to plant CRYs only when it is necessary to explain a particular property of animal CRYs.

Physical Properties

Spectroscopic properties. No native CRY has been purified to date due to their low abundance; instead, the CRYs have been expressed as recombinant proteins using bacterial, insect, or mammalian cells. CRYs purified in this manner contain little to no MTHF chromophore (Lin et al. 1995; Malhotra et al. 1995; Hsu et al. 1996; Song et al. 2007). With respect to flavin content, CRYs fall into two groups. The first group, which includes *Arabidopsis* CRYs and Insect CRY1s can be purified from bacterial and insect expression systems with essentially stoichiometric amounts of FAD (Lin et al. 1995; Malhotra et al. 1995; Bouly et al. 2003; Özgür and Sancar 2006; Song et al. 2007). In contrast, the second group, which includes vertebrate CRYs and Insect CRY2s, contains 1–2% FAD and trace amounts of MTHF when purified from bacterial (Hsu et al. 1996), insect (Özgür and Sancar 2006; Song et al. 2007), or mammalian (Özgür and Sancar 2003) expression systems. These findings have raised the legitimate question of whether these CRYs are in fact flavoproteins or whether they have only retained flavin-binding capacity as a functionally irrelevant evolutionary relic. To date, no photobiological activity has been associated with this group of CRYs; therefore, it is logical to conclude either that flavin binding by this group of CRYs is not relevant to their activities or that FAD has only a structural, and not catalytic, role. However, it must be noted that many photolyases, including the (6-4) photolyase of *Drosophila*, contain 1–5% FAD (Zhao and Sancar 1997) when expressed in a heterologous system, yet enzymatic activities of these enzymes are absolutely dependent on this cofactor. In light of all these considerations, we believe all CRYs contain FAD as a functional cofactor. Further work is needed to test this prediction.

A second question with respect to the chromophore/cofactor issue of CRYs is the redox status of FAD in the native CRY. Photolyases are known to utilize two-electron reduced deprotonated flavin ($FADH^-$) as the native cofactor (Sancar 2003; Selby and Sancar 2006); however, when purified under aerobic conditions, they may have the flavin in any of the three oxidation states, $FADH^-$, FADH° (blue neutral radical), and FAD_{ox} (Sancar 2003). Hence, it is not possible to ascertain the redox status of the cofactor in vivo by inspecting the redox status of the flavin cofactor of purified photolyase. So far, all of the plant and Insect Type I CRYs that have been purified contain flavin in the two-electron oxidized form, FAD_{ox} (Fig. 6). As argued for photolyases, this does not necessarily mean that the native pigment contains oxidized flavin. The resolution of this issue awaits the establishment of the CRY photocycle in an in vitro system.

Quarternary structure. All nonvisual photoreceptors that have been analyzed to date including phytochrome, phototropin, and BLUF proteins are homodimers (Christie 2007). It has been reported that AtCRY1 and AtCRY2 form homodimers and heterodimers that are essential for their in vivo function (Sang et al. 2005; Yu et al. 2007). In contrast, it was recently reported that *Drosophila* CRY is a monomer (Berndt et al. 2007), whereas coimmunoprecipitation experiments with recombinant mammalian CRYs suggest that they form homo- and heterodimers (Partch 2006). Further work with native mammalian CRYs is needed to resolve the issue of the quaternary structure of CRYs.

Biochemical Properties

Animal CRYs exhibit a number of macromolecular interactions and enzymatic activities that may be related to their light-independent signaling. The significance of some of these interactions are well-understood, whereas others require further investigation.

DNA binding. In contrast to the crystal structure of AtCRY1 (Brautigam et al. 2004), a computational model of human CRY2 revealed that the DNA-binding groove of photolyase is conserved in human CRYs (Özgür and Sancar 2003). Not surprisingly, human CRY2 bound weakly to double-stranded DNA and with higher affinity to single-stranded DNA nonspecifically (Fig. 7A), with a K_D ~5 x 10^{-9}, and with slightly higher affinity to a single-

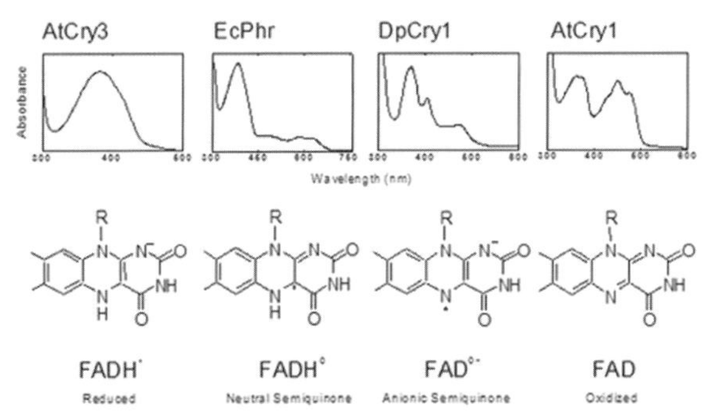

Figure 6. Spectroscopic and chemical properties of the FAD cofactor in representative members of the photolyase/CRY family. When the enzymes are purified, they may contain one or both chromophores, and the flavin cofactor may be in any of the three oxidation states (that do not necessarily represent the in vivo form of the cofactor) depending on the particular enzyme or the purification conditions. "AtCRY3" is the single-stranded DNA photolyase from *Arabidopsis*, which was misclassified as a CRY before discovery of its single-stranded DNA-specific photolyase activity. It contains stoichiometric MTHF and flavin in the two-electron-reduced and deprotonated ($FADH^-$) form. EcPhr is the *E. coli* photolyase. Even though this enzyme contains $FADH^-$ in its native state, the flavin is oxidized to the blue neutral radical (semiquinone) state when purified. The absorption spectrum shows the contribution of MTHF at 380 nm and of FADH° in the 450–700-nm range. DpCRY1 shows the absorption spectrum of the monarch butterfly CRY1 after light exposure. When the enzyme is purified in the dark, it exhibits a FAD_{ox} absorption spectrum. The enzyme is extremely sensitive to light, and upon light exposure, it is reduced to the flavin anion radical $FAD^{-\circ}$, which is thought to be the active form of this CRYI (Song et al. 2007). AtCRY1 shows the absorption spectrum of *Arabidopsis* CRY1 purified from an insect cell expression system, exhibiting the characteristic FAD_{ox} absorption spectrum. It is unclear whether this is the physiologically relevant form of the cofactor. Note that DpCRY1 and AtCRY1 contain only trace amounts of MTHF, which does not significantly contribute to the absorption spectrum.

stranded DNA oligomer containing a (6-4) photoproduct, but there was no photorepair of the lesion. Human CRY1 exhibited similar DNA-binding properties (Özgür and Sancar 2003). Currently, the significance of these findings is not known.

Protein–protein interactions. In the *Drosophila* circadian clock, CRY binds to Timeless (Tim) in a light-dependent manner (Ceriani et al. 1999) and promotes ubiquitylation of Tim by Jetlag SCF E3 ligase (Koh et al. 2006), and this leads to degradation. This reaction is important to phase setting. Mammalian CRYs interact with the integral clock proteins PER, CLOCK, and BMAL1 independently of light (Griffin et al. 1999) and repress the transcriptional activity of the CLOCK/ BMAL1 complex (Kume et al. 1999; Vitaterna et al. 1999) by an ill-defined mechanism. It should be noted that even though the binding of DmCRY to DmTim in a yeast two-hybrid assay is light-dependent and that of hCRY1 and hCRY2 to hPER and hCLOCK is light-independent, this should not be taken as evidence that mammalian CRYs have no light-dependent activity. As noted above, ectopically expressed mammalian CRYs contain essentially no flavin and therefore cannot be expected to be sensitive to light.

In addition to these key interactions that make up the negative arm of the core molecular clock, human CRYs interact with several other proteins that are involved in the clock and cell cycle checkpoints. Both human CRY1 and CRY2 bind to phosphoprotein phosphatase 5 (PP5) (Zhao and Sancar 1997) and inhibit PP5 activity toward autophosphorylated CKIε and in so doing inhibit phosphorylation and subsequent degradation of PER1 and PER2 (Partch et al. 2006). As a consequence, the CRY-PP5 interaction has a role in consolidating the core molecular clock. Similarly, FBXL3 E3 ubiquitin ligase binds to CRY1 and CRY2 and ubiquitylates them, leading to their timely degradation. In the absence of FBXL3, CRYs accumulate, leading to exaggerated inhibition of CLOCK/BMAL1 and abnormal circadian period and phase response (Busino et al. 2007; Godinho et al. 2007; Siepka et al. 2007). Finally, human CRYs interact with the Tim checkpoint/circadian protein and in so doing couple the circadian cycle to the cell cycle (Ünsal-Kaçmaz et al. 2005).

In *Arabidopsis*, both CRY1 and CRY2 interact directly with the E3 ubiquitin ligase COP1 in a light-independent manner through their carboxy-terminal domains (Wang et al 2001; Yang et al 2001). In the dark, the CRY-COP1 complex is active as an E3 enzyme, ubiquitylating transcription factors, such as HY5, and thus inhibiting transcription. When exposed to blue light, the CRY-COP1 complex does not dissociate, but apparently a light-induced conformational change in the carboxyl termini of CRY results in inhibition of COP1 activity and thus leads to transcriptional induction of blue-light-responsive

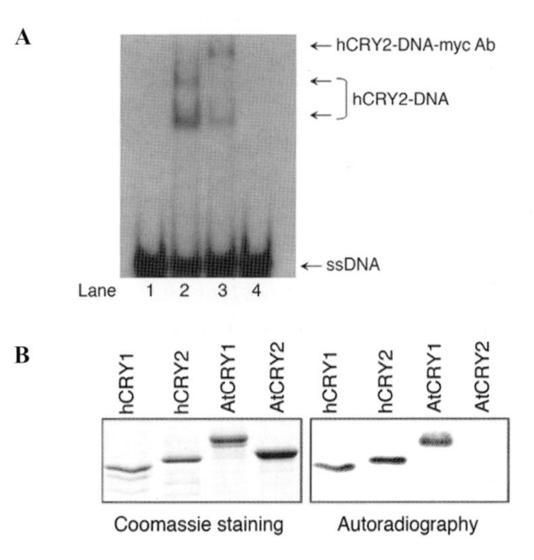

Figure 7. DNA binding and kinase activities of CRYs. (*A*) Electrophoretic mobility-shift assay showing binding of hCRY2 to single-stranded DNA. The DNA-protein complex is supershifted by antibodies to an epitope tag on human CRY2. The radiolabeled single-stranded DNA was incubated with *myc*-tagged human CRY2 (lanes *2* and *3*) or anti-*myc* antibodies (lanes *3* and *4*) before loading on the gel. (*B*) Kinase activities of *Arabidopsis* and human CRYs in vitro. Purified recombinant CRYs were incubated with [γ-^{32}P]ATP, separated on an SDS-PAGE gel, and then analyzed by Coomassie Blue staining (*left*) or autoradiography (*right*). (Reprinted, with permission, from Özgür and Sancar [2003] and Özgür and Sancar [2006] [both © American Chemical Society].)

genes in *Arabidopsis* that are involved in growth and differentiation (Yang et al 2000; Wang et al. 2001). It is thought that the CRY-COP1 interaction is unique to plant CRYs (Yang et al. 2001); however, more work is needed to find out whether or not mammalian CRYs bind COP1 and exert some control over protein degradation.

ATP binding and autokinase activity. Autophosphorylation is a common property of all photosensory pigments including phytochromes and phototropins (Christie 2007). However, it also appears that the autokinase activity is dispensable for both phytochrome (Matsushita et al. 2003) and phototropin (Kagawa et al. 2004) functions. It was found that both AtCRY1 and AtCRY2 are phosphorylated in vivo upon blue light exposure (Shalitin et al. 2002; 2003). Later, it was shown that purified AtCRY1 bound ATP stoichiometrically and that both AtCRY1 (Bouly et al. 2003; Shalitin et al. 2003) and human CRY1 (Bouly et al. 2003) were autophosphorylated in a FAD- and blue-light-dependent manner. In line with these biochemical findings, the crystal structure of the AtCRY1 PHR domain contained an ATP analog in the cavity leading to FAD where the pyrimidine dimer binds in photolyase (Brautigam et al. 2004). It is unclear, however, how this ATP could be used for autophosphorylation because there are no serine/threonine residues within reasonable distance of the binding site for chemical attack on ATP to achieve phosphorylation. However, the crystal structure is missing the carboxy-terminal domain, and hence, it is conceivable that a residue in the missing domain participates in the kinase activity.

Experiments with human CRYs cast some doubt about the role of the weak autokinase activity of CRYs in their function (Özgür and Sancar 2006). First, it was found that no correlation exists between the presence of FAD and the autokinase activity. The human CRY1 and CRY2 purified from baculovirus/insect cell system contained either no or only trace flavin, yet autophosphorylated to the same level as AtCRY1, which contained stoichiometric flavin (Fig. 7B). Second, purified AtCRY2 which is known to be phosphorylated in vivo in response to blue light was not phosphorylated in the dark or under blue light. Finally, the kinetics and extent of AtCRY1 autophosphorylation were essentially identical under blue light and in the dark under our conditions. The cause of these contradictory results is not known, and further work is needed to understand the significance of ATP binding to CRYs and the role, if any, of autokinase activity in CRY function.

The "Trp Triad" and Photoreduction

Excitation of flavin by blue light in the majority of flavoproteins leads to quenching of the excited state at a rate much faster than that of free flavin (see Sancar 2003). This is, in general, due to electron transfer from redox-active amino acids such as tryptophan, tyrosine, and histidine to the excited-singlet-state flavin that is a potent oxidant. This property of flavoproteins has been extensively used to study the physicochemical aspects of intraprotein electron transfer (Zhong 2007). Naturally, in enzymes that utilize ground-state flavin to catalyze redox reactions such as glucose oxidase, which carries out catalysis independently of light, the photoinduced reduction of flavin has no bearing on enzyme activity under physiological conditions. The significance of photoinduced electron transfer from aromatic amino acids to the flavin photoreceptor is more difficult to ascertain. Stringent criteria must be applied to determine whether photoreduction of the flavin is a side reaction with no biological relevance or a key step in the photocycle of these flavin-based photoreceptors. In *E. coli* photolyase, the "Trp triad," FADH°←Trp-382←Trp-359←Trp-306 is the predominant photoreduction pathway with Trp-306 being the ultimate electron donor (Fig. 8A) (Li et al. 1991; Park et al. 1995; Kavakli and Sancar 2004). Interestingly, the residues of this "Trp triad" are conserved in most DNA photolyases and CRYs (Kim et al. 1993; Lin and Shalitin 2003; Partch and Sancar 2005; Zeugner et al. 2005). It has been unequivocally shown that this pathway has no role in the photocycle of photolyase (Li et al. 1991; Kavaklı and Sancar 2004). In contrast, as will be discussed below, some studies have suggested that photoinduced electron

A

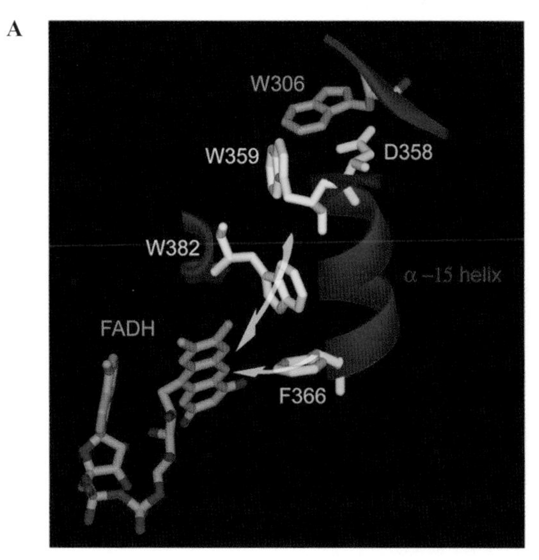

B

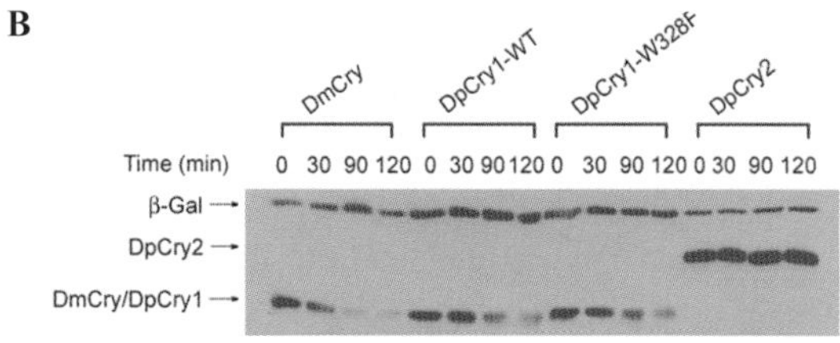

Figure 8. Trp triad and CRY function. (*A*) The "Trp triad" in *E. coli* photolyase. Excitation of FADH° by light leads to its photoreduction by electron transfer from Trp306 through either electron hopping (W382←W359←W306) or electron tunneling (F366←α15-helix←W306). Mutating Trp-306 to Phe-306 blocks both pathways. (*B*) Effect of blocking electron transfer to FAD_{ox} in Dp*Cry1* on its photoreception activity. Blocking the intraprotein electron transfer by the W328F mutation (the equivalent of *E. coli* Phr W306) has no effect on the photoreception activity of DpCRY1 as measured by its photoinduced degradation in S2 cells. (*A*, Reprinted, with permission, from Saxena et al. 2004 [(c) American Chemical Society]; *B*, reprinted, with permission, from Song et al. 2007 [© ASBMB].)

transfer is the key photophysical event in the CRY photocycle (Zeugner et al. 2005; Banerjee et al. 2007; Bouly et al. 2007).

FUNCTIONAL PROPERTIES OF ANIMAL CRYPTOCHROME

Cryptochrome Photocycle

The photocycle of a photosensory pigment is the sequence of physicochemical transitions that take place from the time of absorption of a photon to generation of a transient signaling state, followed by deactivation of the pigment to the ground state. Of the flavin-based photoreceptors, the photocycles of photolyase and phototropin are well-characterized. The former involves a photoinduced cyclic electron transfer (Sancar 2003) and the latter involves a photoinduced covalent bond formation between the protein and FAD, followed by a conformational change (Swartz et al. 2001; Harper et al. 2003). Therefore, it is tempting to consider the CRY photocycle in terms of one of these two general models. Because of the phylogenetic and structural relations, one is further tempted to think that the "photolyase model" is more likely to be applicable to CRYs. However, two observations indicate that the photolyase model cannot apply to CRY in the strict sense. First, photolyase binds its substrate in a light-independent reaction and then catalysis is initiated by light, whereas DmCRY binds to Tim only after excitation of CRY by light (Ceriani et al. 1999). Second, photolyase can undergo hundreds of photocycles without being degraded whereas all CRYs known to function as photoreceptors, with the exception of AtCRY1, undergo photoinduced degradation (Lin et al. 1998; Yuan et al. 2007).

In addition to the evidence against the "photolyase model" for the CRY photocycle, a number of reports appear to support a "phototropin-like model" involving photoinduced conformational change. Most of these studies have been done on *Arabidopsis* CRYs. First, it has been reported that AtCRY1 and AtCRY2 made in insect cells contain flavin in the FAD_{ox} state and that exposure of the insect cells expressing the CRYs to blue light reduces the FAD_{ox} to a FADH° neutral radical (Banerjee et al. 2007; Bouly et al. 2007). This was complemented by in vitro experiments that demonstrated that exposure of AtCRYs to light followed by a dark phase under aerobic conditions cause the following transitions:

$$FAD_{ox} \xrightarrow[e^-]{h\upsilon(\text{blue})} FADH^{\circ} \xrightarrow[e^-]{h\upsilon(\text{green})} FADH^- \xrightarrow[2e^-\ \searrow\ O_2]{dark} FAD_{ox}$$

This led to the conclusion that the CRY(FADH°) was the signaling state of *Arabidopsis* CRYs and that further exposure of the plant to green light where FADH° absorbs turned off the signal by reducing the radical to $FADH^-$. Conversely, dark incubation of CRY($FADH^-$) resulted in reoxidization back to the photoactive FAD_{ox} form. Second, it was reported that blue light activated the AtCRY1 kinase (Bouly et al. 2003) and that this activation was dependent on photoinduced electron transfer through the so-called "Trp triad." Mutations of one of the three tryptophans to phenylalanine blocked photoreduction and abolished light-induced kinase activity (Zeugner et al. 2005). Finally, it was shown by two independent methods that blue light causes significant conformational change in the carboxy-terminal domain of AtCRY1, which is critical for signaling in vivo (Partch et al. 2005; Kottke et al. 2006).

All of these observations are, however, subject to some serious caveats. First, any overproduced flavoprotein may be subject to photoreduction in the reducing milieu of insect cells, and hence, photoreduction of an overproduced protein need not necessarily be on the signaling pathway. Second, reports that blue light stimulates AtCRY1 kinase activity have not been reproducible in a recent exhaustive in vitro study (Özgür and Sancar 2006). Similarly, the report that blocking photoreduction by mutating the "Trp triad" abolishes the blue-light-induced kinase activity has a serious drawback: The mutant proteins had only 20% basal kinase (in the dark) activity compared to the wild-type enzyme (Zeugner et al. 2005), suggesting that the mutations must have caused partial misfolding of the proteins and making it difficult to interpret the lack of light stimulation, which again has not been reproducible. Third, this model predicts that the action spectrum for hypocotyl growth inhibition would be very similar to the absorption spectrum of FAD_{ox}. However, the actual action spectrum of hypocotyl growth inhibition in *Arabidopsis* is quite flat in the 390–480-nm range (Ahmad et al. 2002). Finally, DmCry missing the carboxy-terminal extension is capable of photoreception and phototransduction, suggesting that at least in some CRYs, the carboxy-terminal domain is not necessary for photosignaling (Busza et al. 2004).

More recently, it was reported that the photoreduction of the flavin in Insect CRY1 to anion radical (Denaro 2006; Berndt et al. 2007) and its reversal in the dark to FAD_{ox} constituted the photocycle of this class of CRYs (Berndt et al. 2007):

$$CRY1(FAD_{ox}) \underset{dark}{\overset{h\upsilon}{\rightleftharpoons}} CRY1(FAD^{-\circ})$$

However, the action spectrum of DmCRY does not match the absorption spectrum of FAD_{ox} (Van Vickle-Chavez and Van Gelder 2006), and the first experimental test of this model with the photosensitive monarch butterfly CRY, DpCRY1, did not support the model (Fig. 8B): Mutating the ultimate electron donor of the "Trp triad" blocked photoreduction of FAD_{ox} to $FAD^{\circ-}$ in vitro. However, the mutation had no effect in vivo on the photoreceptor activity of DpCry1 as revealed by blue-light-induced degradation of the CRY (Song et al. 2007). In light of this finding and the recently published action spectrum of DmCRY (a member of type 1 CRYs) that does not match the absorption spectrum of FAD_{ox} (VanVickle-Chavez and Van Gelder 2007), we conclude that most likely, FAD_{ox}—$h\upsilon$→$FAD^{-\circ}$ is not part of the insect CRY photocycle and that the active form of type I CRYs may actually contain $FAD^{-\circ}$.

Role of Cryptochrome in Magnetoreception

How organisms sense the earth's magnetic field to guide activities such as migration is not known. CRYs are currently considered candidate "magnetoreceptors" functioning as sensors via a postulated mechanism, "chemical magnetoreception," that has been gaining popularity (Ritz et al. 2000; Johnsen and Lohmann 2005). Theoretically, the weak magnetic field of the earth could influence the outcome of certain biochemical reactions by influencing the correlation of spin states of radical pair intermediates, such as those generated in photoactivated states of photolyases and possibly CRYs. This mechanism raises the possibility that CRYs have a substrate upon which they act and that the nature of the product or the rate of the photochemical reaction is influenced by the physical orientation of the CRY with respect to the magnetic field. For magnetic field lines to be sensed, this hypothesis requires that the CRYs be fixed in a uniform position within the architecture of the photoreceptive cells and tissue, for example, in unidirectional layers. In this way, all of the CRYs will be influenced uniformally by the magnetic field. Consequently, the rate of CRY signaling will vary as the position of the organism changes. Several aspects of migratory organisms are consistent with this mechanism. Light and eyes are required for certain migratory behaviors, and CRY has been found in photoreceptor tissues of several organisms. In the migratory bird garden warbler (*Sylvia borin*), CRY expression was localized in a subset of retinal ganglion cells associated with magnetic orientation behavior (Mouritsen et al. 2004). Other theories and potential magnetoreceptors exist; however, this is clearly a fertile field of research that will possibly include characterization of exciting new CRY functions as it develops.

Role of Cryptochrome in Cell Cycle Regulation

The circadian cycle and cell cycle are two global regulatory mechanisms that affect nearly all aspects of cellular physiology; therefore, it is to be expected that these two regulatory pathways might exhibit some overlap. In organisms ranging from *Chlamydomonas* (Nikaido and Johnson 2000) to zebra fish (Dekens et al. 2003) to humans (Bjarnason and Jordan 2000), the circadian rhythm affects the phasing of the cell cycle. Inversely, the cell cycle also gates the phase of the circadian cycle (Nagoshi et al. 2004) as a further indicator of the intimate connection between the two pathways. CRY affects the mammalian cell cycle checkpoints by two mechanisms. First, it interacts with the Timeless protein which, in turn, interacts with the damage sensor ATR and the signal transducer Chk1 kinase and in so doing, CRY directly participates in the DNA-damage checkpoint (Ünsal-Kaçmaz et al. 2005). Second, CRYs in their capacities as negative regulators of clock-controlled genes repress the expression of *Wee1* mitotic kinase, and as a consequence, in *Cry1*$^{-/-}$*;Cry2*$^{-/-}$ cells, Wee1 is elevated (Matsuo et al. 2003; Gauger and Sancar 2005). However, the effect of Wee1 elevation on cell cycle checkpoint may or may not be apparent depending on the cell type (Matsuo et al. 2003; Gauger and Sancar 2005); in fact, it was found that *Cry1*$^{-/-}$*;Cry2*$^{-/-}$ mice were not measurably different from the wild-type mice in terms of chronic effects of ionizing radiation (Gauger and Sancar 2005) and were more resistant to the acute genotoxic effects of cyclophosphamide (Gorbacheva et al. 2005; Kondratov et al. 2007). Finally, CRY regulates PP5 activity, which in turn regulates ATM activity (Partch et al. 2006; Yong et al. 2007), and through this pathway, CRY is expected to have a role in cellular response to ionizing radiation which is controlled to a large extent by ATM (Sancar et al. 2004). Clearly, the CRY–cell cycle checkpoint connection in particular and circadian cycle–cell cycle connection in general are fertile fields for further research and biomedical exploitation (Kondratov et al. 2007; Hunt and Sassone-Corsi 2007).

CONCLUSION

At present, the photochemical reaction initiating CRY signal transduction is not known. Two general models that might be referred to as the "photolyase model" (photoinduced cyclic electron transfer) and the "phototropin model" (photoinduced conformational change) have been proposed for the photocycle (Fig. 9). There are experimental data for and against each of these models, and further work is required to find out which of these models, if any, apply to the CRY photocycle. In addition to this

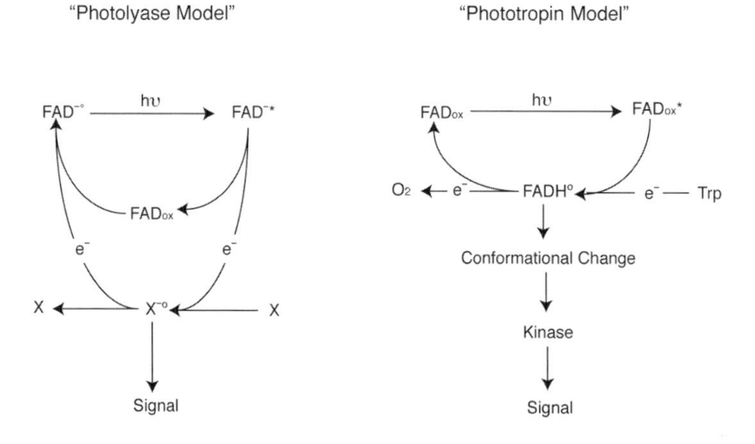

Figure 9. Two models for the CRY cycle. In the "Photolyase Model," photoinduced cyclic electron transfer from reduced or semireduced (shown) flavin to an unknown substrate generates a radical that initiates the signal; back electron transfer turns off the light signal (Song et al. 2007). In the "Phototropin Model," light-induced electron transfer from a Trp residue in the apoenzyme to FAD_{ox} generates a flavin radical concomitant with significant conformational change in the CRY that activates the CRY's autokinase activity and affects CRY's interactions with downstream proteins, such as COP1, modulating their activity to transmit signal (Bouly et al. 2007).

issue, the following questions remain to be addressed for a more comprehensive understanding of animal CRY structure and function: (1) Do Insect CRY1s act as transcriptional repressors? (2) Do vertebrate CRYs and Insect CRY2s function as photoreceptors? (3) Do vertebrate CRYs and Insect CRY2s contain FAD and if so what is the function of FAD? (4) Do CRYs from plant and animal sources have a second chromophore? (5) What is the role of CRY kinase activity in signaling?

ACKNOWLEDGMENTS

This work was supported by National Institutes of Health grant GM31082. This review is in large part based on papers by Özgür and Sancar (2003, 2006), Sancar (2004), Partch (2006), and Song et al. (2007).

REFERENCES

Adams M.D., Kerlavage A.R., Fleischmann R.D., Fuldner R.A., Bult C.J., Lee N.H., Kirkness E.F., Weinstock K.G., Gocayne J.D., and White O., et al. 1995. Initial assessment of human gene diversity and expression patterns based upon 83 million nucleotides of cDNA sequence. *Nature 377:* 3.

Ahmad M. and Cashmore A.R. 1993. HY4 gene of *A. thaliana* encodes a protein with characteristics of a blue-light photoreceptor. *Nature* **366:** 162.

Ahmad M., Grancher N., Heil M., Black R.C., Giovani B., Galland P., and Lardemer D. 2002. Action spectrum for cryptochrome-dependent hypocotyl growth inhibition in *Arabidopsis. Plant Physiol.* **129:** 774.

Banerjee R. and Batschauer A. 2005. Plant blue-light receptors. *Planta* **220:** 498.

Banerjee R., Schleicher E., Meier S., Munoz Viana R., Pokorny R., Ahmad M., Bittl R., and Batschauer A. 2007. The signaling state of *Arabidopsis* cryptochrome 2 contains flavin semiquinone. *J. Biol. Chem.* **282:** 14916.

Berndt A., Kottke T., Breitkreuz H., Dvorsky R., Hennig S., Alexander M., and Wolf E. 2007. A novel photoreaction mechanism for the circadian blue light photoreceptor *Drosophila* cryptochrome. *J. Biol. Chem.* **282:** 13011.

Bjarnason G.A. and Jordan R. 2000. Circadian variation of cell proliferation and cell cycle protein expression in man: Clinical implications. *Prog. Cell Cycle Res.* **4:** 193.

Bouly J.P., Giovani B., Djamei A., Mueller M., Zeugner A., Dudkin E.A., Batschauer A., and Ahmad M. 2003. Novel ATP-binding and autophosphorylation activity associated with *Arabidopsis* and human cryptochrome-1. *Eur. J. Biochem.* **270:** 2921.

Bouly J.P., Schleicher E., Dionisio-Sese M., Vandenbussche F., Van Der Straeten D., Bakrim N., Meier S., Batschauer A., Galland P., Bittl R., and Ahmad M. 2007. Cryptochrome blue light photoreceptors are activated through interconversion of flavin redox states. *J. Biol. Chem.* **282:** 9383.

Brautigam C.A., Smith B.S., Ma Z., Palnitkar M., Tomchick D.R., Machius M., and Deisenhofer J. 2004. Structure of the photolyase-like domain of cryptochrome 1 from *Arabidopsis thaliana. Proc. Natl. Acad. Sci.* **101:** 12142.

Busino L., Bassermann F., Maiolica A., Lee C., Nolan P.M., Godinho S.I., Draetta G.F., and Pagano M. 2007. SCF^{Fbxl3} controls the oscillation of the circadian clock by directing the degradation of cryptochrome proteins. *Science* **316:** 900.

Busza A., Emery-Le M., Rosbash M., and Emery P. 2004. Roles of the two *Drosophila* CRYPTOCHROME structural domains in circadian photoreception. *Science* **304:** 1503.

Cashmore A.R. 2003. Cryptochromes: Enabling plants and animals to determine circadian time. *Cell* **114:** 537.

Ceriani M.F., Darlington T.K., Staknis D., Mas P., Petti A.A., Weitz C.J., and Kay S.A. 1999. Light-dependent sequestration of TIMELESS by CRYPTOCHROME. *Science* **285:** 553.

Christie J.M. 2007. Phototropin blue-light receptors. *Annu. Rev. Plant Biol.* **58:** 21.

Collins B., Mazzoni E.O., Stanewsky R., and Blau J. 2006. *Drosophila* CRYPTOCHROME is a circadian transcriptional repressor. *Curr. Biol.* **16:** 441.

Darwin C. 1881. *The power of movement in plants.* Da Capo Press, New York.

Dekens M.P., Santoriello C., Vallone D., Grassi G., Whitmore D., and Foulkes N.S. 2003. Light regulates the cell cycle in zebrafish. *Curr. Biol.* **13:** 2051.

Denaro T.R. 2006. "Purification and characterization of the *Danaus plexippus* cryptochromes." M.S. thesis, University of North Carolina, Chapel Hill.

Emery P., So W.V., Kaneko M., Hall J.C., and Rosbash M. 1998. CRY, a *Drosophila* clock and light-regulated cryptochrome, is a major contributor to circadain rhythm resetting and photosensitivity. *Cell* **95:** 669.

Fujihashi M., Numoto N., Kobayashi Y., Mizushima A., Tsujimura M., Nakamura A., Kawarabayasi Y., and Miki K. 2007. Crystal structure of archaeal photolyase from *Sulfolobus tokodaii* with two FAD molecules: Implication of a novel light-harvesting cofactor. *J. Mol. Biol.* **365:** 903.

Gauger M.A. and Sancar A. 2005. Cryptochrome, circadian cycle, cell cycle checkpoints, and cancer. *Cancer Res.* **65:** 6828.

Gehring W. and Rosbash M. 2003. The coevolution of blue-light photoreception and circadian rhythms. *J. Mol. Evol.* **57:** S286.

Gekakis N., Staknis D., Nguyen H.B., Davis F.C., Wilsbacher L.D., King D.P., Takahashi J.S., and Weitz C.J. 1998. Role of CLOCK protein in the mammalian circadian mechanism. *Science* **280:** 1564.

Godinho S.I., Maywood E.S., Shaw L., Tucci V., Barnard A.R., Busino L., Pagano M., Kendall R., Quawailid M.M., Romero M.R., O'neill J., Chesham J.E., Brooker D., Lalanne Z., Hastings M.H., and Nolan P.M. 2007. The after-hours mutant reveals a role for Fbxl3 in determining mammalian circadian period. *Science* **316:** 897.

Gorbacheva V.Y., Kondratov R.V., Zhang R., Cherukuri S., Gudkov A.V., Takahashi J.S., and Antoch M.P. 2005. Circadian sensitivity to the chemotherapeutic agent cyclophosphamide depends on the funtional status of the CLOCK/BMAL1 transactivation complex. *Proc. Natl. Acad. Sci.* **102:** 3407.

Gressel J. 1977. Blue light photoreceptors. *Photochem. Photobiol.* **30:** 749.

Griffin E.A., Jr., Staknis D., and Weitz C.J. 1999. Light-independent role of CRY1 and CRY2 in the mammalian circadian clock. *Science* **286:** 768.

Guo H., Yang H., Mockler T.C., and Lin C. 1998. Regulation of flowering time by *Arabidopsis* photoreceptors. *Science* **279:** 1360.

Harper S.M., Neil L.C., and Gardner K.H. 2003. Structural basis of a phototropin light switch. *Science* **301:** 1541.

Hsu D.S., Zhao X., Zhao S., Kazantsev A., Wang R.P., Todo T., Wei Y.F., and Sancar A. 1996. Putative human blue-light photoreceptors hCRY1 and hCRY2 are flavoproteins. *Biochemistry* **35:** 13871.

Huang Y., Baxter R., Smith B.S., Partch C.L., Colbert C.L., and Deisenhofer J. 2006. Crystal structure of cryptochrome 3 from *Arabidopsis thaliana* and its implications for photolyase activity. *Proc. Natl. Acad. Sci.* **103:** 17701.

Hunt T. and Sassone-Corsi P. 2007. Riding tandem: Circadian clocks and the cell cycle. *Cell* **129:** 461.

Johnsen S. and Lohmann K.J. 2005. The physics and neurobiology of magnetoreception. *Nat. Rev. Neurosci.* **6:** 703.

Johnson J.L., Hamm-Alvarez S., Payne G., Sancar G.B., Rajagopalan K.V., and Sancar A. 1988. Identification of the second chromophore of *Escherichia coli* and yeast DNA photolyases as 5,10-methenyltetrahydrofolate. *Proc. Natl. Acad. Sci.* **85:** 2046.

Kagawa T., Kasahara M., Abe T., Yoshida S., and Wada M. 2004. Function analysis of phototropin2 using fern mutants deficient in blue light-induced chloroplast avoidance movement. *Plant Cell Physiol.* **45:** 416.

Kao Y.T., Saxena C., Wang L., Sancar A., and Zhong D. 2005. Direct observation of thymine dimer repair in DNA by photolyase. *Proc. Natl. Acad. Sci.* **102:** 16128.

Kavakli I.H. and Sancar A. 2004. Analysis of the role of intraprotein electron transfer in photoreactivation by DNA photolyase in vivo. *Biochemistry* **43:** 15103.

Kim S.T., Malhotra K., Taylor J.S., and Sancar A. 1996. Purification and partial characterization of (6-4) photoproduct DNA photolyase from *Xenopus laevis*. *Photochem. Photobiol.* **63:** 292.

Kim S.T., Sancar A., Essenmacher C., and Babcock G.T. 1993. Time-resolved EPR studies with DNA photolyase: Excited-state FADH° abstracts an electron from Trp-306 to generate $FADH^-$, the catalytically active form of the cofactor. *Proc. Natl. Acad. Sci.* **90:** 8023.

Koh K., Zheng X., and Sehgal A. 2006. JETLAG resets the *Drosophila* circadian clock by promoting light-induced degradation of TIMELESS. *Science* **312:** 1809.

Kondratov R.V., Gorbacheva V.Y., and Antoch M.P. 2007. The role of mammalian circadian proteins in normal physiology and genotoxic stress responses. *Curr. Top. Dev. Biol.* **78:** 173.

Koornneef M., Rolf E., and Spruit C.J.P. 1980. Genetic control of light-inhibited hypocotyl elongation in *Arabidopsis thaliana*. *Z. Pflanzenphysiol.* **100:** 147.

Kottke T., Batschauer A., Ahmad M., and Heberle J. 2006. Blue-light-induced changes in *Arabidopsis* cryptochrome 1 probed by FTIR difference spectroscopy. *Biochemistry* **45:** 2472.

Kume K., Zylka M.J., Sriram S., Shearman L.P., Weaver D.R., Jin X., Maywood E.S., Hastings M.H., and Reppert S.M. 1999. mCRY1 and mCRY2 are essential components of the negative limb of the circadian clock feedback loop. *Cell* **98:** 193.

Li J., Uchida T., Todo T., and Kitagawa T. 2006. Similarities and differences between cyclobutane pyrimidine dimer photolyase and (6-4) photolyase as revealed by resonance Raman spectroscopy: Electron transfer from the FAD cofactor to ultraviolet-damaged DNA. *J. Biol. Chem.* **281:** 25551.

Li Y.F., Heelis P.F., and Sancar A. 1991. Active site of DNA photolyase: Tryptophan-306 is the intrinsic hydrogen atom donor essential for flavin radical photoreduction and DNA repair in vitro. *Biochemistry* **30:** 6322.

Li Y.F., Kim S.T., and Sancar A. 1993. Evidence for lack of DNA photoreactivating enzyme in humans. *Proc. Natl. Acad. Sci.* **90:** 4389.

Lin C. and Shalitin D. 2003. Cryptochrome structure and signal transduction. *Annu. Rev. Plant Biol.* **54:** 469.

Lin C., Yang H., Guo H., Mockler T., Chen J., and Cashmore A.R. 1998. Enhancement of blue-light sensitivity of *Arabidopsis* seedlings by a blue light receptor cryptochrome 2. *Proc. Natl. Acad. Sci.* **95:** 2686.

Lin C., Robertson D.E., Ahmad M., Raibekas A.A., Jorns M.S., Dutton P.L., and Cashmore A.R. 1995. Association of flavin adenine dinucleotide with the *Arabidopsis* blue light receptor CRY1. *Science* **269:** 968.

Losi A. 2007. Flavin-based blue-light photosensors: A photophysics update. *Photochem. Photobiol.* (in press).

Lowrey P.L. and Takahashi J.S. 2004. Mammalian circadian biology: Elucidating genome-wide levels of temporal organization. *Annu. Rev. Genomics Hum. Genet.* **5:** 407.

Malhotra K., Kim S.T., Batschauer A., Dawut L., and Sancar A. 1995. Putative blue-light photoreceptors from *Arabidopsis thaliana* and *Sinapis alba* with a high degree of sequence homology to DNA photolyase contain the two photolyase cofactors but lack DNA repair activity. *Biochemistry* **34:** 6892.

Matsuo T.S., Yamaguchi S., Mitsui S., Emi A., Shimoda F., and Okamura H. 2003. Control mechanism of the circadian clock for timing of cell division in vivo. *Science* **302:** 255.

Matsushita T., Mochizuki N., and Nagatani A. 2003. Dimers of the N-terminal domain of phytochrome B are functional in the nucleus. *Nature* **424:** 571.

Millar A.J. and Kay S.A. 1997. The genetics of phototransduction and circadian rhythms in *Arabidopsis*. *Bioessays* **19:** 209.

Miyamoto Y. and Sancar A. 1998. Vitamin B2-based blue-light photoreceptors in the retinohypothalamic tract as the photoactive pigments for setting the circadian clock in mammals. *Proc. Natl. Acad. Sci.* **95:** 6097.

———. 1999. Circadian regulation of cryptochrome genes in the mouse. *Brain Res. Mol. Brain Res.* **71:** 238.

Mouritsen H., Janssen-Bienhold U., Liedvogel M., Feenders G., Stalleicken J., Dirks P., and Weiler R. 2004. Cryptochromes and neuronal-activity markers colocalize in the retina of migratory birds during magnetic orientation. *Proc. Natl. Acad. Sci.* **101:** 14294.

Nagoshi E.C., Saini C., Bauer C., Laroche T., Naef F., and Schibler U. 2004. Circadian gene expression in individual fibroblasts: Cell-autonomous and self-sustained oscillators pass time to daughter cells. *Cell* **119:** 693.

Nikaido S.S. and Johnson C.H. 2000. Daily and circadian variation in survival from ultraviolet radiation in *Chlamydomonas reinhardtii*. *Photochem. Photobiol.* **71:** 758.

Özgür S. and Sancar A. 2003. Purification and properties of human blue-light photoreceptor cryptochrome 2. *Biochemistry* **42:** 2926.

———. 2006. Analysis of autophosphorylating kinase activities of *Arabidopsis* and human cryptochromes. *Biochemistry* **45:** 13369.

Park H.W., Kim S.T., Sancar A., and Deisenhofer J. 1995. Crystal structure of DNA photolyase from *Escherichia coli*. *Science* **268:** 1866.

Partch C.L. 2006. "Signal transduction mechanisms of cryptochrome." Ph.D. thesis, University of North Carolina, Chapel Hill.

Partch C.L. and Sancar A. 2005. Photochemistry and photobiology of cryptochrome blue-light photopigments: The search for a photocycle. *Photochem. Photobiol.* **81:** 1291.

Partch C.L., Clarkson M.W., Özgür S., Lee A.L., and Sancar A. 2005. Role of structural plasticity in signal transduction by the cryptochrome blue-light photoreceptor. *Biochemistry* **44:** 3795.

Partch C.L., Shields K.F., Thompson C.L., Selby C.P., and Sancar A. 2006. Posttranslational regulation of the mammalian circadian clock by cryptochrome and protein phosphatase 5. *Proc. Natl. Acad. Sci.* **103:** 10467.

Payne G. and Sancar A. 1990. Absolute action spectrum of E-FADH2 and E-FADH2-MTHF forms of *Escherichia coli* DNA photolyase. *Biochemistry* **29:** 7715.

Pittendrigh C.S. 1993. Temporal organization: Reflections of a Darwinian clock-watcher. *Annu. Rev. Physiol.* **55:** 16.

Reppert S.M. and Weaver D.R. 2002. Coordination of circadian timing in mammals. *Nature* **418:** 935.

Ritz T., Adem S., and Schulten K. 2000. A model for photoreceptor-based magnetoreception in birds. *Biophys. J.* **78:** 707.

Sancar A. 2000. Cryptochrome: The second photoactive pigment in the eye and its role in circadian photoreception. *Annu. Rev. Biochem.* **69:** 31.

———. 2003. Structure and function of DNA photolyase and cryptochrome blue-light photoreceptors. *Chem. Rev.* **103:** 2203.

———. 2004. Regulation of the mammalian circadian clock by cryptochrome. *J. Biol. Chem.* **279:** 34079.

Sancar A., Lindsey-Boltz L.A., Ünsal-Kaçmaz K., and Linn S. 2004. Molecular mechanisms of mammalian DNA repair and the DNA damage checkpoints. *Annu. Rev. Biochem.* **73:** 39.

Sang Y., Li Q.H., Rubio V., Zhang Y.C., Mao J., Deng X.W., and Yang H.Q. 2005. N-terminal domain-mediated homodimerization is required for photoreceptor activity of *Arabidopsis* CRYPTOCHROME 1. *Plant Cell* **17:** 1569.

Saxena C., Sancar A., and Zhong D. 2004. Femtosecond dynamics of DNA photolyase: Energy transfer of antenna initiation and electron transfer of cofactor reduction. *J. Phys. Chem. B* **108:** 18026.

Selby C.P. and Sancar A. 2006. A cryptochrome/photolyase class of enzymes with single-stranded DNA-specific photolyase activity. *Proc. Natl. Acad. Sci.* **103:** 17696.

Selby C.P., Thompson C., Schmitz T.M., Van Gelder R.N., and Sancar A. 2000. Functional redundancy of cryptochromes and classical photoreceptors for nonvisual ocular photoreception in mice. *Proc. Natl. Acad. Sci.* **97:** 14697.

Shalitin D., Yu X., Maymon M., Mockler T., and Lin C. 2003. Blue light-dependent in vivo and in vitro phosphorylation of *Arabidopsis* cryptochrome 1. *Plant Cell* **15:** 2421.

Shalitin D., Yang H., Mockler T.C., Maymon M., Guo H., Whitelam G.C., and Lin C. 2002. Regulation of *Arabidopsis* cryptochrome 2 by blue-light-dependent phosphorylation. *Nature* **417:** 763.

Siepka S.M., Yoo S.H., Park J., Song W., Kumar V., Hu Y., Lee C., and Takahashi J.S. 2007. Circadian mutant overtime reveals F-box protein FBXL3 regulation of cryptochrome and period gene expression. *Cell* **129:** 1011.

Somers D.E., Devlin P.F., and Kay S.A. 1998. Phytochromes and cryptochromes in the entrainment of the *Arabidopsis* circadian clock. *Science* **282:** 1488.

Song S.H., Öztürk N., Denaro T.R., Arat N.O., Kao Y.T., Zhu H., Zhong D., Reppert S.M., and Sancar A. 2007. Formation and function of flavin anion radical in cryptochrome 1 blue-light photoreceptor of monarch butterfly. *J. Biol. Chem.* **282:** 13011.

Stanewsky R., Kaneko M., Emery P., Beretta B., Wager-Smith K., Kay S.A., Rosbash M., and Hall J. C. 1998. The cryb mutation identifies cryptochrome as a circadian photoreceptor in *Drosophila*. *Cell* **95:** 681.

Swartz T.E., Corchnoy S.B., Christie J.M., Lewis J.W., Szundi I., Briggs W.R., and Bogomolni R.A. 2001. The photocycle of a flavin-binding domain of the blue light photoreceptor phototropin. *J. Biol. Chem.* **276:** 36493.

Thompson C.L., Bowes Rickman C., Shaw S.J., Ebright J.N., Kelly U., Sancar A., and Rickman D.W. 2003. Expression of the blue-light receptor cryptochrome in the human retina. *Invest. Ophthalmol. Vis. Sci.* **44:** 4515.

Thresher R.J., Vitaterna M.H., Miyamoto Y., Kazantsev A., Hsu D.S., Petit C., Selby C.P., Dawut L., Smithies O., Takahashi J.S., and Sancar A. 1998. Role of mouse cryptochrome blue-light photoreceptor in circadian photoresponses. *Science* **282:** 1490.

Todo T., Ryo H., Yamamoto K., Toh H., Inui T., Ayaki H., Nomura T., and Ikenaga M. 1996. Similarity among the *Drosophila* (6-4)photolyase, a human photolyase homolog, and the DNA photolyase-blue-light photoreceptor family. *Science* **272:** 109.

Todo T., Takemori H., Ryo H., Ihara M., Matsunaga T., Nikaido O., Sato K., and Nomura T. 1993. A new photoreactivating enzyme that specifically repairs ultraviolet light-induced (6-4)photoproducts. *Nature* **361:** 371.

Tu D.C., Batten M.L., Palczewski K., and Van Gelder R.N. 2004. Non-visual photoreception in the chick iris. Science **306:** 129.

Ueda T., Kato A., Kuramitsu S., Terasawa H., and Shimada I. 2005. Identification and characterization of a second chromophore of DNA photolyase from *Thermus thermophilus* HB27. *J. Biol. Chem.* **280:** 36237.

Ünsal-Kaçmaz K., Mullen T.E., Kaufmann W.K., and Sancar A. 2005. Coupling of human circadian and cell cycles by the timeless protein. *Mol. Cell. Biol.* **25:** 3109.

van der Horst G.T., Muijtjens M., Kobayashi K., Takano R., Kanno S., Takao M., de Wit J., Verkerk A., Eker A.P., van Leenen D., Buijs R., Bootsma D., Hoeijmakers J.H., and Yasui A. 1999. Mammalian Cry1 and Cry2 are essential for maintenance of circadian rhythms. *Nature* **398:** 627.

VanVickle-Chavez S.J. and Van Gelder R.N. 2007. Action spectrum of *Drosophila* cryptochrome. *J. Biol. Chem.* **282:** 10561.

Vitaterna M.H., Selby C.P., Todo T., Niwa H., Thompson C., Fruechte E.M., Hitomi K., Thresher R.J., Ishikawa T., Miyazaki J., Takahashi J.S., and Sancar A. 1999. Differential regulation of mammalian period genes and circadian rhythmicity by cryptochromes 1 and 2. *Proc. Natl. Acad. Sci.* **96:** 12114.

Wang H., Ma L.G., Li J.M., Zhao H.Y., and Deng X.W. 2001. Direct interaction of *Arabidopsis* cryptochromes with COP1 in light control development. *Science* **294:** 154.

Yang H.Q., Tang R.H., and Cashmore A.R. 2001. The signaling mechanism of *Arabidopsis* CRY1 involves direct interaction with COP1. *Plant Cell* **13:** 2573.

Yang H.Q., Wu Y.J., Tang R.H., Liu D., Liu Y., and Cashmore A.R. 2000. The C termini of *Arabidopsis* cryptochromes mediate a constitutive light response. *Cell* **103:** 815.

Yong W., Bao S., Chen S., Li D., Sánchez E.R., and Shuo W. 2007. Mice lacking protein phosphatase 5 are defective in ataxia telangiectasia mutated (ATM)-mediated cell cycle arrest. *J. Biol. Chem.* **282:** 14690.

Young M.W. and Kay S.A. 2001. Time zones: A comparative genetics of circadian clocks. *Nat. Rev. Genet.* **2:** 702.

Yu X., Shalitin D., Liu X., Maymon M., Klejnot J., Yang H., Lopez J., Zhao X., Bendehakkalu K.T., and Lin C. 2007. Derepression of the NC80 motif is critical for the photoactivation of *Arabidopsis* CRY2. *Proc. Natl. Acad. Sci.* **104:** 7289.

Yuan Q., Metterville D., Briscoe A.D., and Reppert S.M. 2007. Insect cryptochromes: Gene duplication and loss define diverse ways to construct insect circadian clocks. *Mol. Biol. Evol.* **24:** 948.

Zeugner A., Byrdin M., Bouly J.P., Bakrim N., Giovani B., Brettel K., and

Ahmad M. 2005. Light-induced electron transfer in *Arabidopsis* cryptochrome-1 correlates with in vivo function. *J. Biol. Chem.* **280:** 19437.

Zhao S. and Sancar A. 1997. Human blue-light photoreceptor hCRY2 specifically interacts with protein serine/threonine phosphatase 5 and modulates its activity. *Photochem. Photobiol.* **66:** 727.

Zhu H., Yuan Q., Briscoe A.D., Froy O., Casselman A., and Reppert S.M. 2005. The two CRYs of the butterfly. *Curr. Biol.* **15:** R953.

Zhong D. 2007. Ultrafast catalytic processes in enzymes. *Curr. Opin. Chem. Biol.* **11:** 174.

Structure Function Analysis of Mammalian Cryptochromes

F. TAMANINI, I. CHAVES, M.I. BAJEK, AND G.T.J. VAN DER HORST
Department of Genetics, Erasmus University Medical Center, 3000 CA Rotterdam, The Netherlands

Members of the photolyase/cryptochrome family are flavoproteins that share an extraordinary conserved core structure (photolyase homology region, PHR), but the presence of a carboxy-terminal extension is limited to the cryptochromes. Photolyases are DNA-repair enzymes that remove UV-light-induced lesions. Cryptochromes of plants and *Drosophila* act as circadian photoreceptors, involved in light entrainment of the biological clock. Using knockout mouse models, mammalian cryptochromes (mCRY1 and mCRY2) were identified as essential components of the clock machinery. Within the mammalian transcription-translation feedback loop generating rhythmic gene expression, mCRYs potently inhibit the transcription activity of the CLOCK/BMAL1 heterodimer and protect mPER2 from 26S-protesome-mediated degradation. By analyzing a set of mutant mCRY1 proteins and photolyase/mCRY1 chimeric proteins, we found that the carboxyl terminus has a determinant role in mCRY1 function by harboring distinguished domains involved in nuclear import and interactions with other clock proteins. Moreover, the carboxyl terminus must cross-talk with the PHR to establish full transcription repression capacity in mCRY1. We propose that the presence of the carboxyl terminus in cryptochromes, which varies in sequence composition among mammalian, *Drosophila*, and plant CRYs, is critical for their different functions and possibly contributed to shape the different architecture and biochemistry of the clock machineries in these organisms.

INTRODUCTION: THE PHOTOLYASE/ CRYPTOCHROME PROTEIN FAMILY

Light/dark cycles, accompanied by the daily rise and fall in temperature, are well-known examples of environmental changes imposed by the rotation of the Earth around its axis. To anticipate these day/night cycles, most organisms have developed a self-sustained circadian clock that drives rhythmic changes in metabolism, physiology, and behavior. In mammals, the circadian system is composed of a master clock in the suprachiasmatic nuclei (SCN) of the hypothalamus in the brain and slave oscillators in peripheral tissues (Ralph et al. 1990; Reppert and Weaver 2002). Because the periodicity of the circadian pacemaker is not exactly 24 hours, it must be reset every day by environmental stimuli, among which light is the most predominant. Within the animal circadian system, cryptochromes hold a remarkable position because they have been shown to act either as integral components of transcription-translation feedback loops that make up the molecular oscillator or as light receptors required for photoentrainment of the circadian clock (Cashmore et al. 1999). Intriguingly, cryptochromes also show strong sequence and structural homology with photolyases, DNA-repair enzymes that remove UV-light-induced DNA lesions with the aid of visible light (Sancar 2003). Together, these proteins make up the photolyase/cryptochrome protein family of flavoproteins that share a highly homologous central core domain with two chromophores (photolyase homology region, PHR) and contain unique amino- and carboxy-terminal extensions, the latter likely providing the basis for totally distinct functions: (1) DNA repair (eukaryotic photolyases), (2) circadian photoentrainment (plant, zebra fish, and *Drosophila* cryptochromes), and (3) circadian clock gene expression (mammalian cryptochromes). Whereas the protein structure and reaction mechanism of photolyases have been largely resolved, relatively little such information is known about cryptochromes. Here, we summarize the current knowledge on the structure and function of cryptochromes, with emphasis on mammalian cryptochromes.

PHOTOLYASES AND DNA REPAIR

The light of our sun is essential for life on Earth. For example, visible light is mandatory for oxygen production by plants via photosynthesis. In mammals and other organisms, even the highly energetic UV component of sunlight is required for the synthesis of important substances such as vitamin D. In marked contrast to its beneficial effects, UV light causes DNA lesions that bring about cell death or, upon replication of a genome, cause somatic mutations (Friedberg 1996). Given the high levels of UV irradiation to which life forms in the Archeal world were exposed, it is not surprising that one of the first DNA-repair systems emerging during evolution, photoreactivation, focused on repair of UV-induced DNA damage.

Photoreactivation is a light-dependent enzymatic reaction in which photolyases directly revert cyclobutane pyrimidine dimers (CPDs) and pyrimidine 6-4 pyrimidone photoproduct (6-4PP) lesions into normal DNA bases. Photolyases are monomeric, single-chain proteins with a molecular mass of 50–65 kD, showing substrate specificity for either CPD or 6-4PP photolesions (Carell et al. 2001). The repair reaction starts with the light-independent binding of a single photolyase molecule to the DNA lesion, resulting in the formation of an enzyme-substrate complex. Next, the DNA dimer is split in a reaction requiring energy provided by (blue) light, after which the photolyase dissociates from the repaired DNA. Binding of photolyase to the DNA does not require the presence of light. Photolyases have been purified from a variety of species, and detailed information on the reaction mechanism is available (Tamada et al. 1997; Sancar 2003). Characteristic for photolyases is the presence of two different noncovalently bound chromophoric cofactors. The first chromophore, FAD, acts as a catalytic cofactor and is only biologically active in its fully reduced state. The latter is accomplished via a reaction called photoactivation, involving a unique feature of intraprotein electron transfer along a chain of three tryptophans over a distance of more than 13 Å, allowing the transfer of an electron to FAD (Aubert et al. 2000). The second chromophore, either a reduced folate or an 8-hydroxy-5-deazaflavin, serves as an auxillary light-harvesting antenna that transmits light energy to the FAD chromophore. Dimer splitting during photoreactivation occurs through electron donation. In the absence of the second chromophore, photolyase enzymes can still photosplit dimers, although the efficiency of the repair reaction is reduced. Comparison of photolyases revealed a strong conservation of amino acids involved in intraprotein electron transfer (during photoactivation) and FAD binding, as well as positively charged amino acids situated at the surface and potentially involved in DNA binding (Mees et al. 2004).

Although photolyases have been observed from bacteria and yeasts to plants and animals (including marsupials), the enzyme has not been identified in placental mammals and was apparently lost in evolution. Instead, mammals remove UV-induced CPDs and 6-4PPs (as well as other bulky strand-distorting adducts) by another evolutionary conserved repair mechanism, nucleotide excision repair (NER) (de Laat et al. 1999). Whereas 6-4PPs are rapidly repaired by NER, removal of CPDs is usually slow (for review, see Bohr et al. 1985). Despite their absence in placental mammals, we have shown that photolyases are functional in mice. Transgenic animals expressing the *Arabidopsis thaliana* (6-4PP)-photolyase and/or the *Potorous tridactylis* (rat-kangaroo) CPD-photolyase from the ubiquitous β-actin or keratinocyte-specific K14 promoter are viable and are capable of lesion-specifically repairing UV-induced 6-4PPs or CPDs in a light-dependent manner from the skin (Schul et al. 2002). We have shown that photoreactivation of CPDs in basal keratinocytes of UV-exposed animals dramatically reduced sunburn, mutation induction, and skin cancer formation (see Fig. 1). In marked contrast, photoreactivation of 6-4PPs hardly exerted any effect, which is likely due to the fact that NER repairs CPDs much more slowly than 6-4PPs (Garinis et al. 2005; Jans et al. 2005, 2006). Thus, these transgenic mice serve as tools to discriminate among the deleterious effects of each individual type of DNA damage and to understand their relative impact on skin cancer development, which is one of the most frequent causes of tumors in western society.

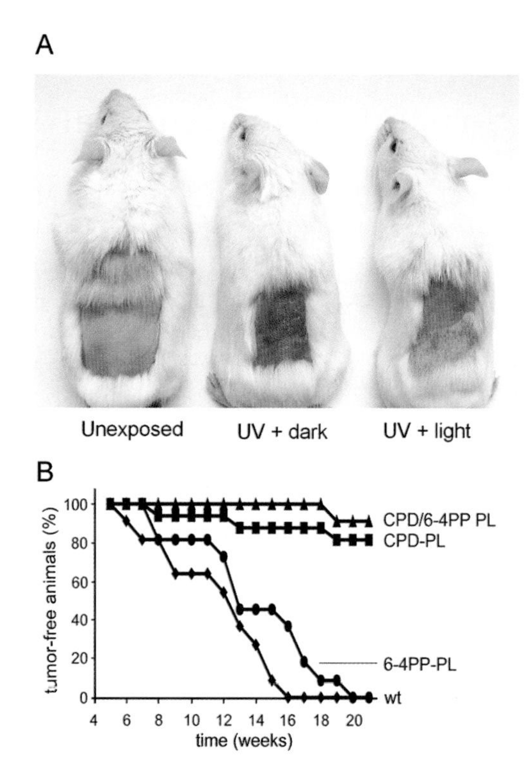

Figure 1. Effect of CPD photoreactivation on UV-induced erythema and carcinogenesis in the skin of β-act-CPD photolyase transgenic mice. (*A*) Appearance of the dorsal skin of non-UV-exposed (*left*), UV-exposed (*middle*), and UV-exposed/photoreactivated (*right*) animals, showing clear erythema when photoreactivation of UV-exposed animals is omitted. (*B*) Fraction of tumor-bearing wild-type and β-act-CPD photolyase (and β-act-6-4PP) transgenic mice in time after the first UV/photoreactivation treatment. (*A*, Reprinted, with permission, from Schul et al. 2002 [European Molecular Biology Organization]; *B*, reprinted, with permission, from Jans et al. 2005 [© Elsevier].)

CRYPTOCHROMES AND PHOTOTRANSDUCTION

Not only is light used by organisms as an energy source (e.g., photosynthesis and photoreactivation), but it also allows them to perceive and respond to their environment. Evidently, this requires the presence of photoreceptor proteins that transfer light information to the organism. Among such photoreceptor proteins are cryptochromes, initially discovered as plant blue light receptors (Cashmore 2003).

Arabidopsis cryptochrome 1 and 2 (AtCRY1 and AtCRY2) are nuclear proteins, resembling photolyases in having a central core domain (PHR) that binds two chromophores. Although lacking the nuclear and mitochondrial localization signal containing an amino-terminal extension that discriminates eukaryotic photolyases from their prokaryotic counterparts, they contain a unique carboxy-terminal extension with different length and amino acid composition. AtCRY proteins were first identified as blue light photoreceptors, involved in light-induced stomatal opening, anthocyanin production, flowering, and inhibition of hypocotyl elongations (Suarez-Lopez and Coupland 1998; Cashmore et al. 1999).

AtCRY proteins detect light signals via the PHR, resulting in activation of their carboxy-terminal domain (Sang et al. 2005). This allows the protein to directly interact with and inhibit the COP1 (constitutive photomorphogenic 1) protein, which serves as a critical component for activation of the photomorphogenic gene expression program in plants (Wang et al. 2001). Interestingly, overexpression of the carboxy-terminal tails of AtCRY1 and AtCRY2 is sufficient to cause constitutive CRY activity (e.g., repression of COP1) (Yang et al. 2000). Therefore, in plants, it seems that the PHR of AtCRYs is mainly a regulator of the carboxyl terminus, whereas the latter domain exerts major function in phototransduction. Furthermore, light activation of photoreceptor activity requires homodimerization of AtCRY proteins via amino-terminal domains in the PHR (Sang et al. 2005). Little is known about the mechanism of light activation of AtCRY, although (similar to photolyases) the activity of the protein depends on the redox status of the FAD chromophore (Bouly et al. 2007).

The occurrence of cryptochromes is not limited to plants. Emery et al. (1998) isolated a fly with a missense mutation at a highly conserved flavin-binding residue of *Drosophila* CRY (dCRY). Circadian experiments with this mutant strain (*cry*^{b}) revealed that when kept in constant darkness, flies no longer responded to brief light pulses that induce phase shifts (Stanewsky et al. 1998). In contrast, flies overexpressing a wild-type *cry* gene appeared to be hypersensitive to light-induced phase shifting. These data show that *Drosophila* CRY acts as a circadian photoreceptor for resetting the biological clock. Upon light activation, the dCRY protein directly interacts with clock proteins TIM and PER and causes proteasomal degradation of the protein complexes. In contrast, overexpression of a truncated dCRY protein, lacking the carboxyl terminus, causes constitutive CRY activity (Busza et al. 2004). Thus, in contrast to AtCRY, the carboxy-terminal domain of dCRY has a regulatory function, by preventing dCRY from being active in the dark.

CRYPTOCHROMES AND THE CIRCADIAN CORE CLOCK

In search of mammalian homologs of photolyases, we and other investigators have cloned two photolyase-like genes. Because the gene products do not possess photolyase activity and the PHRs lack the amino-terminal extension of eukaryotic photolyases and instead contain a carboxy-terminal extension also present in plant cryptochromes, the genes were designated *mCry1* and *mCry2* (van der Spek et al. 1996; Todo et al. 1997).

Studies with genetically modified mice in which the *mCry1* and/or *mCry2* genes were inactivated revealed that the two mouse cryptochromes are part of the molecular clockwork generating behavioral and molecular rhythms. Notably, single-knockout mice have opposing circadian phenotypes, as evident from the observation that *mCry1*$^{-/-}$ and *mCry2*$^{-/-}$ mice display short and long behavioral periodicity, respectively, as measured by voluntary wheel-running activity (van der Horst et al. 1999; Vitaterna et al. 1999). Remarkably, behavioral analysis of *mCry1/2* double-knockout mice indicated a complete loss of the circadian clock in these animals (van der Horst et al. 1999). Thus, mCRY1 and mCRY2 proteins not only have an antagonistic clock-adjusting function, but are also essential for maintenance of circadian rhythmicity.

Unexpectedly, given the anticipated photoreceptor function predicted on the basis of the homology with plant cryptochromes, *mCry1/2* double-knockout mice still show inhibition of wheel-running activity by light and induction of *mPer1* and *mPer2* expression upon exposure to light pulses known to reset the biological clock, suggesting that mCRY1 and mCRY2 are not required for photoentrainment or light masking of activity (Okamura et al. 1999).

How do CRY proteins fit in the circadian clockwork? Until the discovery of animal cryptochromes, the mammalian molecular oscillator was thought to be composed of *mPer1*, *mPer2*, and *mPer3* genes (homologs of the *Drosophila* period gene) and *clock* and *Bmal1* genes. In the transcription-translation feedback loop model, transcription of *mPer* genes is driven by a heterodimer of the CLOCK and BMAL proteins (positive loop) that binds to CACGTG E-box enhancer elements in the promoter of *Per* genes (Ripperger et al. 2000). Following translation of *mPer* mRNAs, mPER proteins were thought to inhibit CLOCK/BMAL1 and, accordingly, switch down transcription of their own genes (negative loop). After mPER protein levels are down again, transcription resumes and the next cycle can start. In line with their clock function, as suggested by animal experiments, the *mCry1* and *mCry2* genes are expressed in a circadian manner, whereas simultaneous inactivation of the *mCry* genes abolishes cycling of *mPer1* and *mPer2* expression (Okamura et al. 1999). A key observation in disclosing the function of cryptochrome genes was the finding that CRY proteins are much stronger inhibitors of CLOCK/BMAL1-driven transcription of E-box-containing reporter genes than mPER proteins (Kume et al. 1999). This observation placed CRY proteins unequivocally at the core of the circadian oscillator and pointed to them as being the most important factors in maintaining the negative feedback loop of the molecular oscillator, despite the fact that other proteins, such as DEC1 and DEC2, have also been proposed to contribute (Honma et al. 2002).

Interestingly, the function of mCRY1 and mCRY2 in the molecular oscillator is not exclusively confined to the level of transcription repression; the proteins also have a role in posttranslational control of other core clock proteins. Coimmunoprecipitation studies with transiently expressed clock proteins and yeast two-hybrid experiments have uncovered direct interactions between mCRY proteins and various other core oscillator components (i.e., mPER2, mPER1, CLOCK, and BMAL1) (Shearman et al. 2000; Yagita et al. 2002; Chaves et al. 2006). In addition, immunohistochemical and biochemical studies revealed synchronous circadian patterns of abundance, phosphorylation status, and nuclear localization of mCRY and mPER proteins (Lee et al. 2001). Particularly, the observation that *mPer2* mRNA levels are increased in *mCry1/2* double-knockout mice, whereas mPER2 protein could not be detected in the cytoplasm or the nucleus of SCN neurons (Shearman et al. 2000), suggested that mCRY proteins are necessary to stabilize mPER2. Indeed, physical interaction between mPER2 and mCRY proteins was shown to inhibit PER2 ubiquitylation and proteasomal degradation (and possibly changes in subcellular localization) of the mCRY-mPER complex (Yagita et al. 2002). Thus far, known protein–protein interactions of, as well as CLOCK/BMAL1-mediated transcription inhibition by, mCRY proteins have been shown to be independent from light (Griffin et al. 1999).

Unexpectedly, given the anticipated photoreceptor function predicted on the basis of homology with plant cryptochromes, *mCry1/2* double-knockout mice still show inhibition of wheel-running activity by light and induction of *mPer1* and *mPer2* expression upon exposure to light pulses known to reset the biological clock, suggesting that mCRY1 and mCRY2 are not required for photoentrainment or light masking of activity (Okamura et al. 1999). As such, mCRY proteins differ from dCRY proteins, which, in addition to functioning as a circadian photoreceptor, had been shown to be involved in core clock oscillations in peripheral tissues (Emery et al. 1998; Stanewsky et al. 1998).

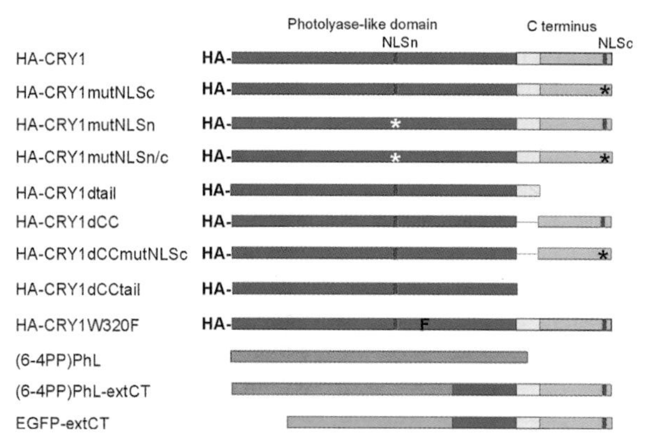

Figure 2. Schematic representation of (mutant) HA-CRY1 constructs, used for structure/function analysis of the mouse CRY1 protein. Mutations in the NLSc and NLSn are indicated by a star (*black* and *white*, respectively). (F) Mutation of tryptophan 320 to phenylalanine (W320F). (*Dark blue*) Photolyase-like PHR domain of mCRY1 (core domain); (*light blue*) carboxyl terminus (CT); (*yellow*) the coiled-coil domain.

STRUCTURE/FUNCTION ANALYSIS OF MAMMALIAN CRYPTOCHROMES

We have chosen to perform a detailed structure/function analysis of cryptochromes to (1) obtain mechanistic insight into the mode of action of CRY1 and CRY2 proteins in the circadian core oscillator, with special emphasis on the function of the carboxy-terminal extension, (2) determine whether photoreactivation and clock functions are mutually exclusive or whether it is possible to combine them in the same protein, and (3) determine how nature can use the same core sequence for completely different functions (e.g., photoreactivation by photolyases vs. clock/photoreceptor function of cryptochromes).

To understand the biochemical properties of mammalian cryptochromes, a panel of ten mutant mCRY1 expression constructs was generated, based on domains identified at the sequence level. We particularly paid attention to the most carboxy-terminal 100–120 amino acids of mCRY1, which most distinguishes the protein from mCRY2 as well as CRYs from other organisms (e.g., *Drosophila*) and which is lacking in photolyases. A schematic representation of the different mutant proteins, which include mutations in the nuclear localization signal (NLS) domains, deletion of the C-tail, deletion of the coiled coil, and mutation of tryptophan 320 to phenylalanine (in photolyase involved in intramolecular electron transport), is shown in Figure 2.

Using this panel of mutant constructs, we determined the domains involved in protein–protein interactions and subcellular localization of the protein and identified in the carboxyl terminus of mCRY1 the domain involved in association with mPER2 as well as with BMAL1 (Chaves et al. 2006; see Fig. 3). The interaction domain is represented by a coiled coil embedded in the carboxyl terminus, which is also present in *Xenopus* CRY (van der Schalie et al. 2007), but absent in *Drosophila* and plant CRYs. The carboxyl terminus was also shown to harbor a putative bipartite NLSc (Tamanini et al. 2005; Chaves et al. 2006). The PHR of mCRY1 contains a second NLS (NLSn), previously identified by Hirayama et al. (2003), which, however, is less potent than NLSc. Interestingly, combined deletion and/or mutagenesis of the NLS domains still allows the protein to reach the nucleus in an NLS-independent manner involving the coiled-coil domain. Previously, overexpressed mPER2 was shown to shuttle between the nucleus and cytoplasm through the combined action of NLS and nuclear export sequences (NES) (Vielhaber et al. 2001; Yagita et al. 2002), whereas coexpression with mCRY1 (or mCRY2) caused completed nuclear localization of mPER2 (Kume et al. 1999). Following our observation that mPER2 and BMAL1 can competitively bind to the coiled coil of mCRY1 and facilitate nuclear localization of mCRY1mutNLSc, we pro-

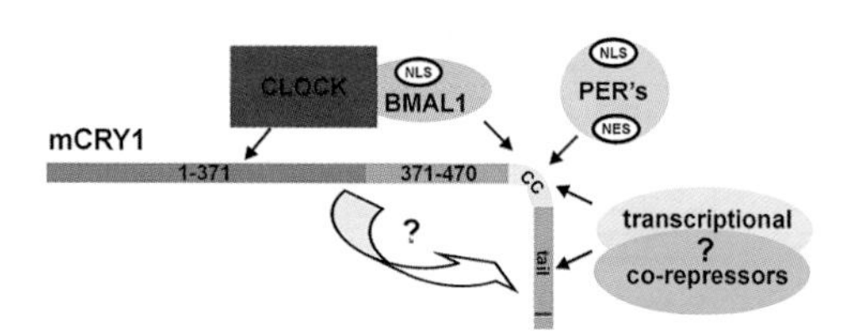

Figure 3. Model for the mechanism of action of mCRY1 within the mammalian core oscillator. The carboxyl terminus of mCRY1 is involved in association with mPER1 and mPER2 proteins and therefore regulates the stability and cellular localization of the latter proteins. The formation of the mCRY/mPER complex occurs possibly through coiled-coil (CC) interactions, because a predicted coiled coil is also present in the mCRY-binding region of mPER proteins (data not shown). To achieve full nuclear localization, this protein complex requires both the mCRY1-NLSc and mPER2-NLS to counteract the NES-mediated nuclear export of mPER2. This molecular mechanism may explain the synchrony in nuclear localization of these proteins observed in vivo. Moreover, the photolyase-like core domain of mCRY1 interacts with CLOCK, whereas the coiled coil is involved in association with BMAL1. Our data also suggest competition between mPER2 and BMAL1 for binding with mCRY1, which supports the concept that the periodicity of the oscillator depends on temporal abundance and strength of interaction between the partners, as well as on the transcription inhibitory capacity of each complex. The inhibitory action of mCRY1 is likely achieved through an intermolecular interaction, which may impose a structural change in the carboxyl terminus, thereby allowing this domain to recruit transcriptional corepressor complexes. (Reprinted, with permission, from Chaves et al. 2006 [© ASM].)

posed that the balance between the activities exerted by nuclear import and export signals harbored by CRY and PER2 determines the final subcellular localization of the complex at the steady-state level (Chaves et al. 2006).

Because the key function of mCRY proteins is to inhibit CLOCK/BMAL1-mediated transcription activation of E-box genes in the negative limb of the mammalian circadian core oscillator, we also analyzed the performance of the various mutant mCRY1 proteins in in vitro CLOCK/BMAL1 transcription assays. Although deletion of either the coiled coil or tail, or NLSc, did not affect the CLOCK/BMAL1 inhibitory capacity of mCRY1, complete removal of the carboxy-terminal extension (as in HA-CRY1dCCtail) rendered the protein inactive (Chaves et al. 2006). These data suggest that the "gain" of the unique carboxy-terminal extension during evolution has been critical in conferring core oscillator function to the PHR region. In line with this, although mutagenesis of conserved tryptophans (involved in intraprotein electron transport during photoactivation) abolishes photoreceptor function of *Drosophila* and butterfly CRY, it does not affect the CLOCK/BMAL1 transcription inhibitory capacity of mouse CRY1 (Zhu et al. 2005; Song et al. 2007).

Because the PHR of mCRY1 shares a high degree of homology with photolyase, it was therefore interesting to examine whether the carboxyl terminus of mCRY1 could convert a photolyase into a cryptochrome. Fusion of the carboxyl terminus of mCRY1 (amino acids 471–606) to *Arabidopsis thaliana* (6-4PP) photolyase did not result in a protein with CLOCK/BMAL1 inhibitory capacity. However, additional substitution of the last 100 amino acids of the photolyase PHR with that of the CRY1 PHR (amino acids 371–470) allowed the chimeric protein to inhibit CLOCK/BMAL1-driven transcription. When fused to enhanced green fluorescent protein (EGFP), the extended carboxyl terminus (amino acids 371–606) of mCRY1 was not able to inhibit CLOCK/BMAL1. This led us to suggest that the remainder of the photolyase/cryptochrome core domain is very important for the clock function of cryptochrome proteins, likely through a complex network of interactions and intrinsic (PHR-like) structural requirements for proper transcription inhibition (Chaves et al. 2006). Indeed, it was shown that the carboxyl termini of animal and plant CRYs are intrinsically unstructured and that a stable tertiary structure is achieved via intramolecular interaction with the PHR (Ozgur and Sancar 2003). In a comparable series of experiments, van der Schalie et al. (2007) showed that *Xenopus* CRY photolyase fusion proteins maintain CLOCK/BMAL1 inhibitory capacity, further underlining the need for proper structural folding.

FUTURE DIRECTIONS IN THE ANALYSIS OF MAMMALIAN CRYPTOCHROMES

Binding Partners of mCRY Proteins

Deletion analysis has shown that the putative coiled-coil domain in the carboxyl terminus of mCRY1 is required for interacting with mPER1/2, BMAL1 (Chaves et al. 2006), and TIMELESS (F. Tamanini and G.T. van der Horst, in prep.). The carboxyl terminus further contains domains involved in nuclear import (Chaves et al. 2006) and, at least for mCRY2, phosphorylation (Sanada et al. 2004; Harada et al. 2005). It is likely that in addition to the already identified proteins, other factors are interacting with the carboxyl terminus, such as kinases and phosphatases, including perhaps FXBL3, which is the E3 ubiquitin ligase recently shown to target mCRY1 and mCRY2 to the degradation through the 26S proteasome (Busino et al. 2007; Godinho et al. 2007; Siepka et al. 2007).

We are currently using amylose-bead-immobilized purified MBP-CRY1 and MBP-CRY2 carboxyl terminal domains to probe mouse tissue (e.g., liver) lysates to identify tail-binding proteins using MALDI-TOF, as well as more sophisticated methods (MALDI-TOF-TOF or ESI-Q-TOF) when necessary. We believe that the identification of protein partners will not only further elucidate the function of CRY proteins in the mammalian clock, but may also disclose new functions of these proteins.

Opposite Phenotypes of mCry1$^{-/-}$ and mCry2$^{-/-}$ Mice

Analysis of the phenotype of *mCry* mutant mice revealed that the *mCry1* gene drives long-period circadian clockworks, whereas the *mCry2* gene confers short periodicity to the circadian oscillator (Okamura et al. 1999; van der Horst et al. 1999). Although it is possible that the two genes/proteins could be expressed in different circadian time windows, CRY1 and CRY2 proteins reach the nuclear compartment in a synchronous manner (Field et al. 2000; Lee et al. 2001). We therefore favor the idea that the two proteins have opposing circadian effects because of sequence differences translating in different activities in and/or posttranslational regulation of the clock. Vanselow et al. (2006) launched the interesting hypothesis that short- and long-period phenotypes of *Cry1*$^{-/-}$ and *Cry2*$^{-/-}$ mutant mice could be the result of alterations in the phosphorylation patterns of PER proteins. The lack of CRY1 or CRY2 would differentially influence the phosphorylation state of PER proteins and thus could modulate the circadian period in opposite directions. In line with this idea, phosphorylation of mPER2 at different serines can have opposite effects on clock speed: Whereas phosphorylation of PER2 at the FASPS position and immediate downstream positions causes PER2 protein stabilization, phosphorylation by CKIε/δ at other sites leads to mPER2 protein degradation (Toh et al. 2001; Xu et al. 2007). In this scenario, the PER proteins are primarily responsible for proper timing of the mammalian circadian oscillator. To test this hypothesis, a dynamic and quantitative phosphorylation site mapping of PER proteins in *Cry1*$^{-/-}$ and *Cry2*$^{-/-}$ mouse cells or tissues would be needed.

Transcription Repression

Inhibition of the CLOCK-BMAL1 heterodimer is independent of light and is triggered at relatively low doses of CRY. Although the mechanism underlying CLOCK/BMAL1 repression is not known, some ideas have been proposed. One hypothesis is that mCRY proteins repress

CLOCK-BMAL1 by reducing CLOCK-BMAL1's affinity for the E box. Another idea is that CRYs inhibit CLOCK-BMAL1 by interacting with the heterodimer and then recruiting histone deacetylases or inhibiting histone acetylases (Etchegaray et al. 2003). Surely, an area of exploration would be the study of the impact of mCRYs on the posttranslational properties of CLOCK and BMAL1 (Dardente et al. 2007).

PHR Dimerization, a Key to Understanding the Evolution of the Mammalian Clock?

Recently, an illuminating report showed that homodimerization of the PHR of AtCRY1 was mandatory for subsequent light activation of its carboxyl terminus (CCT1) (Sang et al. 2005). The activated CCT1 would then bind and inhibit the E3 ubiquitin ligase COP1, thereby allowing the accumulation of a set of transcription factors (e.g., HY5) that initiate the photomorphogenic program (Wang et al. 2001; Yang et al. 2001). Thus far, all of our efforts to detect a functional homodimerization of mCRY proteins have failed (F. Tamanini and G.T. van der Horst, unpubl.). This lack of positive results is somewhat surprising because comparison of the PHR among members of the cryptochrome/photolyase family, as well as molecular modeling of the three-dimensional structures of the PHR, suggests an almost identical folding of the PHR in all cryptochromes. Moreover, interaction between the PHR of AtCRY1 and mCRY2 with their corresponding carboxyl termini has been shown to be an important event in establishing the stable tertiary structure of these proteins, suggesting a common biochemical mode of action for cryptochromes (Partch et al. 2005). Then why does mammalian mCRY1 not homodimerize as does its plant homolog? Given the difference in function, it is tempting to speculate that the necessity for homodimerization or heterodimerization is restricted to photoreceptor function. In mammals, mCRYs have a dominant role in the transcriptional-posttranslational feedback loops governing circadian rhythms in the SCN and peripheral tissues. With the exception of the SCN, which responds to light through neuronal signaling via the retinohypothalamic tract, the mammalian circadian oscillator does not respond to light. Therefore, one possibility is that the lack of homodimerization and heterodimerization of mCRY proteins underlines the absence of any photoreceptor function in this protein (although we cannot rule out light responsiveness of the mCRY PHR in the retina). If this interpretation is correct, the prediction would be that *Drosophila* CRY (Stanewsky 2002), which acts as both a circadian photoreceptor (like AtCRY) and a central clock component (like mCRYs), might have preserved the capacity to homodimerize as a regulatory mechanism for light signaling.

ACKNOWLEDGMENTS

We thank Dr. Andre Eker for stimulating discussion. This work was supported in part by grants from the Netherlands Organization for Scientific Research (ZonMW Vici 918.36.619 and NWO-CW 700.51.304), SenterNovem (BSIK03053), and the European Community (BrainTime QLG3-CT-2002-01829; EUCLOCK LSHG-CT-2006-018741) to G.T.J.vdH.

REFERENCES

Aubert C., Vos M.H., Mathis P., Eker A.P., and Brettel K. 2000. Intraprotein radical transfer during photoactivation of DNA photolyase. *Nature* **405:** 586.

Bohr V.A., Smith C.A., Okumoto D.S., and Hanawalt P.C. 1985. DNA repair in an active gene: Removal of pyrimidine dimers from the DHFR gene of CHO cells is much more efficient than in the genome overall. *Cell* **40:** 359.

Bouly J.P., Schleicher E., Dionisio-Sese M., Vandenbussche F., Van Der Straeten D., Bakrim N., Meier S., Batschauer A., Galland P., Bittl R., and Ahmad M. 2007. Cryptochrome blue light photoreceptors are activated through interconversion of flavin redox states. *J. Biol. Chem.* **282:** 9383.

Busino L., Bassermann F., Maiolica A., Lee C., Nolan P.M., Godinho S.I., Draetta G.F., and Pagano M. 2007. SCFFbxl3 controls the oscillation of the circadian clock by directing the degradation of cryptochrome proteins. *Science* **316:** 900.

Busza A., Emery-Le M., Rosbash M., and Emery P. 2004. Roles of the two *Drosophila* CRYPTOCHROME structural domains in circadian photoreception. *Science* **304:** 1503.

Carell T., Burgdorf L.T., Kundu L.M., and Cichon M. 2001. The mechanism of action of DNA photolyases. *Curr. Opin. Chem. Biol.* **5:** 491.

Cashmore A.R. 2003. Cryptochromes: Enabling plants and animals to determine circadian time. *Cell* **114:** 537.

Cashmore A.R., Jarillo J.A., Wu Y.J., and Liu D. 1999. Cryptochromes: Blue light receptors for plants and animals. *Science* **284:** 760.

Chaves I., Yagita K., Barnhoorn S., Okamura H., van der Horst G.T., and Tamanini F. 2006. Functional evolution of the photolyase/cryptochrome protein family: Importance of the C terminus of mammalian CRY1 for circadian core oscillator performance. *Mol. Cell. Biol.* **26:** 1743.

Dardente H., Fortier E.E., Martineau V., and Cermakian N. 2007. Cryptochromes impair phosphorylation of transcriptional activators in the clock: A general mechanism for circadian repression. *Biochem. J.* **402:** 525.

de Laat W.L., Jaspers N.G., and Hoeijmakers J.H. 1999. Molecular mechanism of nucleotide excision repair. *Genes Dev.* **13:** 768.

Emery P., So W.V., Kaneko M., Hall J.C., and Rosbash M. 1998. CRY, a *Drosophila* clock and light-regulated cryptochrome, is a major contributor to circadian rhythm resetting and photosensitivity. *Cell* **95:** 669.

Etchegaray J.P., Lee C., Wade P.A., and Reppert S.M. 2003. Rhythmic histone acetylation underlies transcription in the mammalian circadian clock. *Nature* **421:** 177.

Field M.D., Maywood E.S., O'Brien J.A., Weaver D.R., Reppert S.M., and Hastings M.H. 2000. Analysis of clock proteins in mouse SCN demonstrates phylogenetic divergence of the circadian clockwork and resetting mechanisms. *Neuron* **25:** 437.

Friedberg E.C. 1996. Relationships between DNA repair and transcription. *Annu. Rev. Biochem.* **65:** 15.

Garinis G.A., Mitchell J.R., Moorhouse M.J., Hanada K., de Waard H., Vandeputte D., Jans J., Brand K., Smid M., van der Spek P.J., Hoeijmakers J.H., Kanaar R., and van der Horst G.T. 2005. Transcriptome analysis reveals cyclobutane pyrimidine dimers as a major source of UV-induced DNA breaks. *EMBO J.* **24:** 3952.

Godinho S.I., Maywood E.S., Shaw L., Tucci V., Barnard A.R., Busino L., Pagano M., Kendall R., Quwailid M.M., Romero M.R., O'Neill J., Chesham J.E., Brooker D., Lalanne Z., Hastings M.H., and Nolan P.M. 2007. The after-hours mutant reveals a role for Fbxl3 in determining mammalian circadian period. *Science* **316:** 897.

Griffin E.A., Jr., Staknis D., and Weitz C.J. 1999. Light-independent role of CRY1 and CRY2 in the mammalian circadian clock. *Science* **286:** 768.

Harada Y., Sakai M., Kurabayashi N., Hirota T., and Fukada Y. 2005. Ser-557-phosphorylated mCRY2 is degraded upon synergistic phosphorylation by glycogen synthase kinase-3 β. *J. Biol. Chem.* **280:** 31714.

Hirayama J., Nakamura H., Ishikawa T., Kobayashi Y., and Todo T. 2003. Functional and structural analyses of cryptochrome. Vertebrate CRY regions responsible for interaction with the CLOCK:BMAL1 heterodimer and its nuclear localization. *J. Biol. Chem.* **278:** 35620.

Honma S., Kawamoto T., Takagi Y., Fujimoto K., Sato F., Noshiro M., Kato Y., and Honma K. 2002. Dec1 and Dec2 are regulators of the mammalian molecular clock. *Nature* **419:** 841.

Jans J., Schul W., Sert Y.G., Rijksen Y., Rebel H., Eker A.P., Nakajima S., van Steeg H., de Gruijl F.R., Yasui A., Hoeijmakers J.H., and van der Horst G.T. 2005. Powerful skin cancer protection by a CPD-photolyase transgene. *Curr. Biol.* **15:** 105.

Jans J., Garinis G.A., Schul W., van Oudenaren A., Moorhouse M., Smid M., Sert Y.G., van der Velde A., Rijksen Y., de Gruijl F.R., van der Spek P.J., Yasui A., Hoeijmakers J.H., Leenen P.J., and van der Horst G.T. 2006. Differential role of basal keratinocytes in UV-induced immunosuppression and skin cancer. *Mol. Cell. Biol.* **26:** 8515.

Kume K., Zylka M.J., Sriram S., Shearman L.P., Weaver D.R., Jin X., Maywood E.S., Hastings M.H., and Reppert S.M. 1999. mCRY1 and mCRY2 are essential components of the negative limb of the circadian clock feedback loop. *Cell* **98:** 193.

Lee C., Etchegaray J.P., Cagampang F.R., Loudon A.S., and Reppert S.M. 2001. Posttranslational mechanisms regulate the mammalian circadian clock. *Cell* **107:** 855.

Mees A., Klar T., Gnau P., Hennecke U., Eker A.P., Carell T., and Essen L.O. 2004. Crystal structure of a photolyase bound to a CPD-like DNA lesion after in situ repair. *Science* **306:** 1789.

Okamura H., Miyake S., Sumi Y., Yamaguchi S., Yasui A., Muijtjens M., Hoeijmakers J.H., and van der Horst G.T. 1999. Photic induction of mPer1 and mPer2 in cry-deficient mice lacking a biological clock. *Science* **286:** 2531.

Ozgur S. and Sancar A. 2003. Purification and properties of human blue-light photoreceptor cryptochrome 2. *Biochemistry* **42:** 2926.

Partch C.L., Clarkson M.W., Ozgur S., Lee A.L., and Sancar A. 2005. Role of structural plasticity in signal transduction by the cryptochrome blue-light photoreceptor. *Biochemistry* **44:** 3795.

Ralph M.R., Foster R.G., Davis F.C., and Menaker M. 1990. Transplanted suprachiasmatic nucleus determines circadian period. *Science* **247:** 975.

Reppert S.M. and Weaver D.R. 2002. Coordination of circadian timing in mammals. *Nature* **418:** 935.

Ripperger J.A., Shearman L.P., Reppert S.M., and Schibler U. 2000. CLOCK, an essential pacemaker component, controls expression of the circadian transcription factor DBP. *Genes Dev.* **14:** 679.

Sanada K., Harada Y., Sakai M., Todo T., and Fukada Y. 2004. Serine phosphorylation of mCRY1 and mCRY2 by mitogen-activated protein kinase. *Genes Cells* **9:** 697.

Sancar A. 2003. Structure and function of DNA photolyase and cryptochrome blue-light photoreceptors. *Chem. Rev.* **103:** 2203.

Sang Y., Li Q.H., Rubio V., Zhang Y.C., Mao J., Deng X.W., and Yang H.Q. 2005. N-terminal domain-mediated homodimerization is required for photoreceptor activity of *Arabidopsis* CRYPTOCHROME 1. *Plant Cell* **17:** 1569.

Schul W., Jans J., Rijksen Y.M., Klemann K.H., Eker A.P., de Wit J., Nikaido O., Nakajima S., Yasui A., Hoeijmakers J.H., and van der Horst G.T. 2002. Enhanced repair of cyclobutane pyrimidine dimers and improved UV resistance in photolyase transgenic mice. *EMBO J.* **21:** 4719.

Shearman L.P., Sriram S., Weaver D.R., Maywood E.S., Chaves I., Zheng B., Kume K., Lee C.C., van der Horst G.T., Hastings M.H., and Reppert S.M. 2000. Interacting molecular loops in the mammalian circadian clock. *Science* **288:** 1013.

Siepka S.M., Yoo S.H., Park J., Song W., Kumar V., Hu Y., Lee C., and Takahashi J.S. 2007. Circadian mutant *Overtime* reveals F-box protein FBXL3 regulation of cryptochrome and period gene expression. *Cell* **129:** 1011.

Song S.H., Oztürk N., Denaro T.R., Arat N.O., Kao Y.T., Zhu H., Zhong D., Reppert S.M., and Sancar A. 2007. Formation and function of flavin anion radical in cryptochrome 1 blue-light photoreceptor of monarch butterfly. *J. Biol. Chem.* **282:** 17608.

Stanewsky R. 2002. Clock mechanisms in *Drosophila*. *Cell Tissue Res.* **309:** 11.

Stanewsky R., Kaneko M., Emery P., Beretta B., Wager-Smith K., Kay S.A., Rosbash M., and Hall J.C. 1998. The *cry*^{b} mutation identifies cryptochrome as a circadian photoreceptor in *Drosophila*. *Cell* **95:** 681.

Suarez-Lopez P. and Coupland G. 1998. Plants see the blue light. *Science* **279:** 1323.

Tamada T., Kitadokoro K., Higuchi Y., Inaka K., Yasui A., de Ruiter P.E., Eker A.P., and Miki K. 1997. Crystal structure of DNA photolyase from *Anacystis nidulans*. *Nat. Struct. Biol.* **4:** 887.

Tamanini F., Yagita K., Okamura H., and van der Horst G.T. 2005. Nucleocytoplasmic shuttling of clock proteins. *Methods Enzymol.* **393:** 418.

Todo T., Tsuji H., Otoshi E., Hitomi K., Kim S.T., and Ikenaga M. 1997. Characterization of a human homolog of (6-4) photolyase. *Mutat. Res.* **384:** 195.

Toh K.L., Jones C.R., He Y., Eide E.J., Hinz W.A., Virshup D.M., Ptacek L.J., and Fu Y.H. 2001. An hPer2 phosphorylation site mutation in familial advanced sleep phase syndrome. *Science* **291:** 1040.

van der Horst G.T., Muijtjens M., Kobayashi K., Takano R., Kanno S., Takao M., de Wit J., Verkerk A., Eker A.P., van Leenen D., Buijs R., Bootsma D., Hoeijmakers J.H., and Yasui A. 1999. Mammalian Cry1 and Cry2 are essential for maintenance of circadian rhythms. *Nature* **398:** 627.

van der Schalie E.A., Conte F.E., Marz K.E., and Green C.B. 2007. Structure/function analysis of *Xenopus* cryptochromes 1 and 2 reveals differential nuclear localization mechanisms and functional domains important for interaction with and repression of CLOCK-BMAL1. *Mol. Cell. Biol.* **27:** 2120.

van der Spek P.J., Kobayashi K., Bootsma D., Takao M., Eker A.P., and Yasui A. 1996. Cloning, tissue expression, and mapping of a human photolyase homolog with similarity to plant blue-light receptors. *Genomics* **37:** 177.

Vanselow K., Vanselow J.T., Westermark P.O., Reischl S., Maier B., Korte T., Herrmann A., Herzel H., Schlosser A., and Kramer A. 2006. Differential effects of PER2 phosphorylation: Molecular basis for the human familial advanced sleep phase syndrome (FASPS). *Genes Dev.* **20:** 2660.

Vielhaber E.L., Duricka D., Ullman K.S., and Virshup D.M. 2001. Nuclear export of mammalian PERIOD proteins. *J. Biol. Chem.* **276:** 45921.

Vitaterna M.H., Selby C.P., Todo T., Niwa H., Thompson C., Fruechte E.M., Hitomi K., Thresher R.J., Ishikawa T., Miyazaki J., Takahashi J.S., and Sancar S. 1999. Differential regulation of mammalian period genes and circadian rhythmicity by cryptochromes 1 and 2. *Proc. Natl. Acad. Sci.* **96:** 12114.

Wang H., Ma L.G., Li J.M., Zhao H.Y., and Deng X.W. 2001. Direct interaction of *Arabidopsis* cryptochromes with COP1 in light control development. *Science* **294:** 154.

Xu Y., Toh K.L., Jones C.R., Shin J.Y., Fu Y.H., and Ptacek L.J. 2007. Modeling of a human circadian mutation yields insights into clock regulation by PER2. *Cell* **128:** 59.

Yagita K., Tamanini F., Yasuda M., Hoeijmakers J.H., van der Horst G.T., and Okamura H. 2002. Nucleocytoplasmic shuttling and mCRY-dependent inhibition of ubiquitylation of the mPER2 clock protein. *EMBO J.* **21:** 1301.

Yang H.Q., Tang R.H., and Cashmore A.R. 2001. The signaling mechanism of *Arabidopsis* CRY1 involves direct interaction with COP1. *Plant Cell* **13:** 2573.

Yang H.Q., Wu Y.J., Tang R.H., Liu D., Liu Y., and Cashmore A.R. 2000. The C termini of *Arabidopsis* cryptochromes mediate a constitutive light response. *Cell* **103:** 815.

Zhu H., Yuan Q., Briscoe A.D., Froy O., Casselman A., and Reppert S.M. 2005. The two CRYs of the butterfly. *Curr. Biol.* **15:** R953.

The *Gonyaulax* Clock at 50: Translational Control of Circadian Expression

J.W. HASTINGS
Department of Molecular and Cellular Biology; Harvard University, Cambridge, Massachusetts 02138

The unicellular circadian clock of *Gonyaulax polyedra* (now renamed *Lingulodinium polyedrum*) has provided important insights concerning circadian rhythmicity. Many, perhaps most, of its key systems are circadian-controlled, ranging from bioluminescence and photosynthesis to motility, cell division, and the synthesis of many proteins, favoring the "master clock" concept. But different rhythms may have different free-running periods and different phase angles under different T cycles, observations not easily accommodated in a single oscillator model. *Gonyaulax* has a feature significantly different from that of other known systems, namely, that clock control of protein synthesis occurs at the translational level. With one mRNA, this involves a protein binding to a 22-nucleotide region in the 3′-untranslated region (3′UTR), but no similar regions have been found in other mRNAs. Pulses of protein synthesis inhibitors cause phase shifts, whereas inhibitors of protein phosphorylation administered chronically cause period changes. In *Gonyaulax* and other systems, low temperature results in arrhythmicity. A return to a permissive temperature results in a reinitiation of the rhythm, with the phase established by the time of increase, similar to the effect of bright light. Evidence for cellular communication via substance(s) in the medium has been obtained.

INTRODUCTION

I am pleased to participate in this second Cold Spring Harbor Symposium on Biological Clocks and thank the organizers for the opportunity to say a few words on the *Gonyaulax* clock, 50 years after my first publications with the late Beatrice Sweeney (Hastings and Sweeney 1957; Sweeney and Hastings 1957). At the time of the first Symposium in 1960, it was rare to have meetings on a specialized topic; the Cold Spring Harbor series were pioneering in that respect and highly productive.

At Harvard, the long-standing custom of a 15-minute chapel service every morning at 8:45 is still with us. Classes cannot be scheduled during this time. The service includes a 5-minute talk given by a member of the University. Mine have always been secular; in one I commented on how an unexplained phenomenon may require a lengthy exposition, whereas when understood, it can often be described very briefly. I recalled that in the 1960s I took four lectures to describe and explain oxidative phosphorylation, but in the 1980s, subsequent to Mitchell's discoveries, it took only a part of one lecture.

The biological clock field is in many respects still in a pre-Mitchell stage, with models having complicated schemes and arrows going in many directions. Clock genes and proteins, along with many other discoveries, have contributed much to understanding circadian mechanisms, but there is still a long way to go. Studies of the unicellular dinoflagellate *Gonyaulax* have also contributed significantly to progress in understanding the circadian mechanism. These include results that suggested the mechanism of temperature compensation (Hastings and Sweeney 1957), basic features of phase shifting by light and the first PRC (Hastings and Sweeney 1958), the action spectrum for light phase shifting (Hastings et al. 1961), and phase shifting by drugs and drug PRCs, as well as drugs affecting period (for summary, see Hastings 2001).

I review here three aspects of the *Gonyaulax* clock that are not understood in terms of a generalized circadian mechanism: translational control of the synthesis of proteins, the reversible loss of rhythmicity at low temperature, and evidence for cellular communication of circadian phase.

TRANSLATIONAL CONTROL

During the past decade, luciferases have been widely used as reporters for tracking circadian changes in gene expression. *Gonyaulax* was effectively the pioneer in this regard; it is a bioluminescent organism whose light emission depends on its own unique luciferase, and the synthesis of this luciferase is circadian-controlled.

In lab cultures, rhythms of luminescence can persist in constant dim light for many cycles—20 or 30—with a period that may be greater or less than 24 hours, depending on conditions of light, temperature, drugs, or other factors. The cells also exhibit rhythms in many other processes, with different phase angles: cell division, photosynthesis, cell aggregation, and the synthesis of many proteins (Hastings and Sweeney 1959; Hastings 1961).

The bioluminescence comes from small (0.4 μm) organelles enveloped in the vacuolar membrane called scintillons where the two proteins involved are localized (Nicolas et al. 1987, 1991). These are luciferase (LCF) and the luciferin (substrate)-binding protein (LBP), which are unreactive at pH 8 (Morse et al. 1989a). A mechanically initiated action potential in the membrane allows protons to enter the scintillons and a rapid flash (100 msec) occurs.

As visualized in Figure 1, the scintillons are synthesized and destroyed each day (Fritz et al. 1990), as are LCF and LBP, as shown by western blots (Johnson et al. 1984; Morse et al. 1989b). New protein synthesis is usually associated with new transcription, but *Gonyaulax*

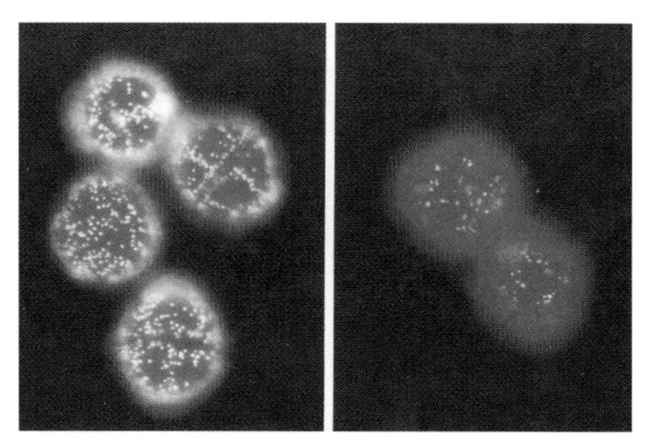

Figure 1. Fluorescence in vivo of *Gonyaulax* cells during night phase (*left*) and day phase (*right*) showing scintillons by their luciferin fluorescence (excitation wavelength, 405 nm) and their greater numbers at night. Background *red* is from the fluorescence of chlorophyll. (Photo by Carl H. Johnson.)

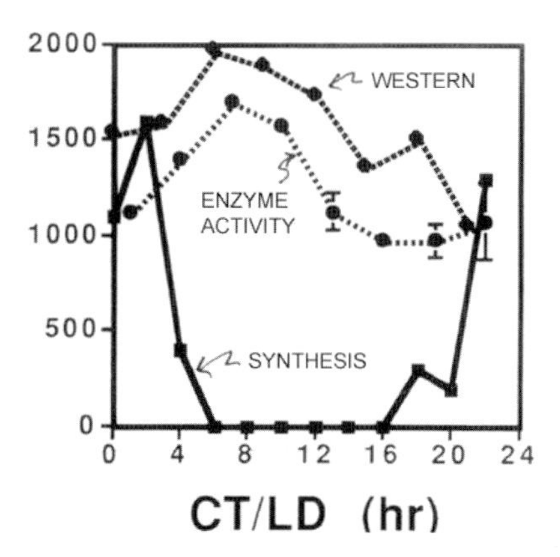

Figure 2. Synthesis rate, specific enzyme activity, and protein levels estimated by western analysis for GAPDH (synthesis rates from Marcovic et al. 1996) and westerns in relative units. Specific activities in nanomoles of NADPH formed per minute per milligram protein. (Modified, with permission, from Fagan et al. 1999 [© American Chemical Society].)

was a surprise; a daily approximately 5-hour bout of LBP synthesis occurred while the message for that protein remained constant (Morse et al. 1989b). Indeed, several *Gonyaulax* mRNAs have been found to have lifetimes of many hours (Rossini et al. 2003).

The luminescence-related proteins are not the only ones to be circadian-controlled at the translational level. Milos et al. (1990) showed on two-dimensional gels of extracts made at day and night phases that many proteins exhibited circadian changes in amount. However, equal amounts of protein were synthesized by in vitro translation from poly(A)$^+$ RNA extracted from cells at day and night phases, indicating that the message itself is present and fully capable at all times of the cycle. Morse and colleagues (Markovic et al. 1996) used pulse labeling to track the in vivo synthesis of a dozen or so of these circadian-controlled proteins and found that they fell into three acrophases, and later identified several of them.

One of them, GAPDH, like LBP, has constant message levels over the circadian cycle but illustrates nicely an additional point of interest in circadian biology (Fagan et al. 1999). Not all of the GAPDH is destroyed in a single cycle, so the rhythm of abundance is not as marked as it is with LBP or LCF (Fig. 2). Thus, synthesis might be strongly circadian, but if the protein has a lifetime of many days, its abundance would not exhibit a readily detectable rhythm.

Translational control is known to be an important regulatory mechanism in both bacteria and eukaryotes. It can be mediated by a variety of mechanisms, prominent among which are protein factors able to bind to either the 5′UTR or 3′UTR of an mRNA and act as either activators or repressors. In *Gonyaulax*, Mittag et al. (1994) made use of gel retardation and found proteins binding to a 22-nucleotide region in the 3′UTR of LBP containing seven UG repeats. They further reported that the binding activity cycles on a daily basis under constant conditions. A recent attempt to repeat these observations has not been successful (Lapointe and Morse 2008).

Further studies of translational control of circadian processes has been made by Mittag with her own group in Germany. She found that the 22-nucleotide sequence mRNA from *Gonyaulax* binds three proteins from extracts of the unicellular green alga *Chlamydmonas* in a circadian-dependent way (Mittag 1996). The most abundant, designated CHLAMY-1, was shown to be a heterodimer with three lysine homology motifs on one subunit and three RNA-recognition motifs on the other. A role in the circadian mechanism was implicated by the fact that the pattern of the photo-accumulation rhythm was altered in cells in which the individual subunits were either overexpressed or silenced (Iliev et al. 2006).

Evidence for translational control of circadian expression in a dinoflagellate has been obtained from microarray studies with a related luminous species, *Pyrocystis lunula*, about

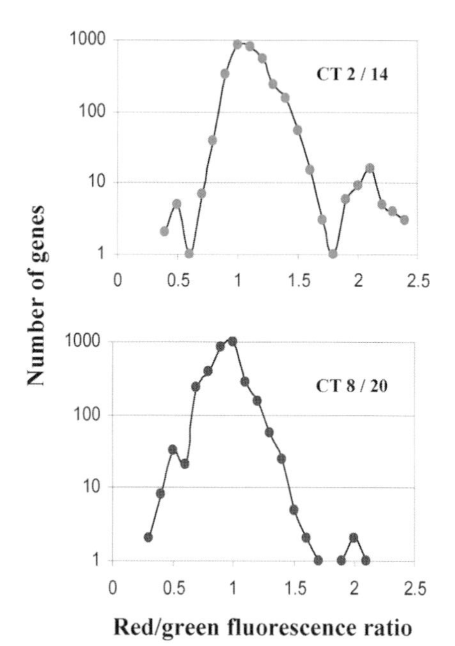

Figure 3. Expression ratios at circadian times 12 hours apart (*top*, CT2 and CT14; *bottom*, CT8 and CT20) for individual genes of *Pyrocystis lunula*, in bins of 0.1 ratio units. (Reprinted, with permission, from Okamoto and Hastings 2003 [© Wiley-Blackwell].)

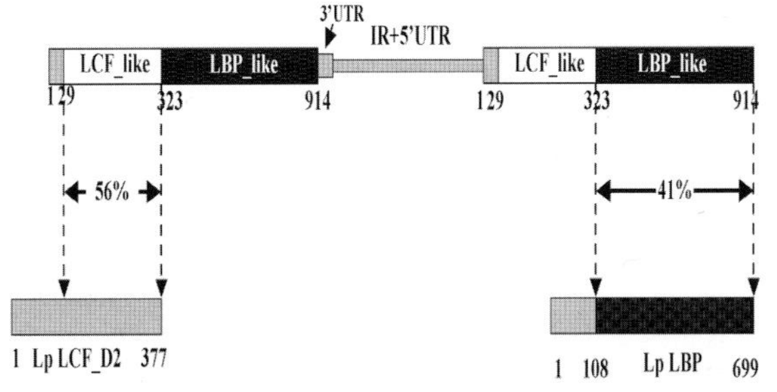

Figure 4. Like luciferase genes in all dinoflagellate species studied, those of *Noctiluca* occur in tandem (*top*), with different intergenic regions in different species (Liu and Hastings 2006). The *Noctiluca* gene contains both LCF-like and LBP-like regions in a single gene; the relationship to those in *Gonyaulax* is indicated at the bottom, The LCF-like domain shares a 56% sequence identity but is shorter at the amino terminus by 60-amino-acid residues than an individual domain of *Gonyaulax* LCF, where there are three such domains. The LBP-like region of 591-amino-acid residues has four repeat domains and is 41% identical to the comparable region of *Gonyaulax* LBP. (Reprinted, with permission, from Liu and Hastings 2007 [© National Academy of Sciences].)

3% of the approximately 2800 unique cDNAs screened were preferentially expressed by a factor of 2 or more at times of day 12 hours apart (Fig. 3), but none by a factor greater than 3 (Okamoto and Hastings 2003). About 50% of those sequenced could be identified with diverse known genes. Similar studies with *Gonyaulax* are in progress.

In the last decade, my lab has been especially concerned with the structures of the LBP and LCF genes and proteins, both of which have unusual repeat-domain components: LBP has four repeats and LCF three, and each of the latter is able to function independently as a luciferase (Li et al. 1997). A crystal structure has been obtained for one of the luciferase domains showing the putative substrate pocket and helices believed to be responsible for activity regulation by pH (Schultz et al. 2005). Luciferase genes are very similar in seven different photosynthetic species (Liu et al. 2004); in an eighth, *Noctiluca miliaris*, sequences are similar but arranged quite differently (Fig. 4) (Liu and Hastings 2007).

LOSS OF RHYTHMICITY AT A LOW TEMPERATURE

Gonyaulax luminescence in cells kept at temperatures below about 12°C exhibits no rhythm whatsoever, but if then returned at any time to a permissive temperature, the rhythm resumes with a typical pattern, with the phase determined by the time at which the temperature was increased (Njus et al. 1977). An explanation for this has not been obtained. It could be that a critical clock protein is cold-sensitive; for example, the subunits might dissociate at a low temperature.

Although there are many reports of loss of rhythmicity in the older literature (cited in Sweeney and Hastings 1961), I know of no recent studies of this kind. Do cyanobacterial rhythms exhibit this feature in vivo, and if so, is the in vitro rhythm of KaiC phosphorylation also reversibly lost at a low temperature? Such studies might provide important new insights into the circadian mechanism.

CELLULAR COMMUNICATION OF CIRCADIAN PHASE

It has often been considered that the circadian clock might be affected by humoral factors produced by the cells themselves. Some years ago, we performed experiments to investigate this possibility (Broda et al. 1985; Hastings et al. 1985). Two out-of-phase cultures were mixed and the luminescence rhythm was measured for about 10 days thereafter (Fig. 5). The maxima of the two came closer and closer over the 10 days, finally merging into a single peak. But in mixed cultures in which the medium was replaced every 4 days, the rhythms of the two remained distinct. Factors in the medium, either secreted or taken up, are implicated in this effect.

Even earlier than this, we had investigated with Neil Krieger the possibility that extracts of the suprachiasmatic nucleus could affect the *Gonyaulax* rhythm. The results were positive: A dramatic shortening of the period resulted. But cortical extracts gave the same effect, so it was judged not to be specifically related to the clock system.

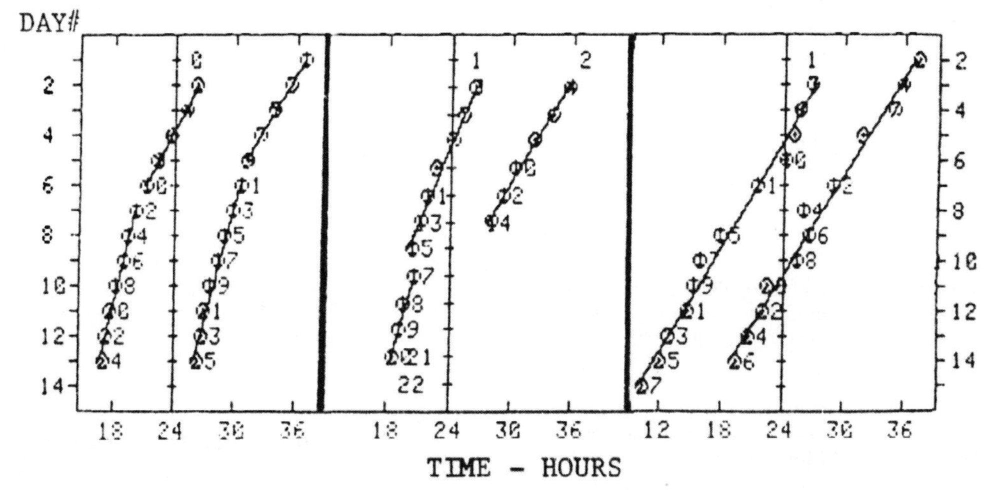

Figure 5. Phase plots of peaks of luminescence (glow) of two out-of-phase cultures measured separately in different vials but plotted on the same graph (*left*); cultures were mixed 50/50 and measured from a single vial (*center*) and mixed without medium and then replaced with fresh medium every 4 days. (Reprinted, with permission, from Hastings et al. 1985 [© Springer Science and Business Media].)

Roenneberg et al. (1988) pursued this matter and showed that muscle extracts had the same effect and finally discovered that it was due to the phosphagen creatine. He pursued the matter further and demonstrated an endogenous period-shortening substance, which he chemically identified as a cyclopropane carboxylic acid and named it gonyauline (Roenneberg et al. 1991). Synthetic gonyauline was prepared and shown to have similar activity.

Progress with the *Gonyaulax* system has been hampered for lack of a genetic system. However, it remains a valuable model system and promises to be of even greater value in the future after the structure and sequences of its genome are available. It is not widely appreciated that there is 40 times more DNA per *Gonyaulax* cell than in the human cell, so significant progress in determining sequence has been made only recently.

ACKNOWLEDGMENTS

I am deeply grateful to the many students and colleagues with whom I have been associated in this work. They have contributed many inspired ideas and important results and brought with them a sense of joy in the pursuit of knowledge, which has made the activity a great pleasure to me. Only some are cited in the text here, but my appreciation for the many others is similarly great. I dedicate this paper to the first of many students, Dr. Marlene Wiesner Karakashian, who started working with *Gonyaulax* in my laboratory at Northwestern University in 1955 and pursued scholarly research with zeal and integrity throughout her lifetime. She passed away on February 6, 2007.

REFERENCES

Broda H., Brugge D., Homma K., and Hastings J.W. 1985. Circadian communication between unicells? Effects on period by cell-conditioning of medium. *Cell Biophys.* **8:** 47.

Fagan T., Morse D., and Hastings J. W. 1999. Circadian synthesis of a nuclear-encoded chloroplast glyceraldehyde-3-phosphate dehydrogenase in the dinoflagellate *Gonyaulax polyedra* is translationally controlled. *Biochemistry* **38:** 7689.

Fritz L., Morse D., and Hastings J.W. 1990. The circadian bioluminescence rhythm of *Gonyaulax* is related to daily variations in the number of light emitting organelles. *J. Cell Sci.* **95:** 321.

Hastings J.W. 1961. Biochemical aspects of rhythms: Phase-shifting by chemicals. *Cold Spring Harbor Symp. Quant. Biol.* **25:** 131.

———. 2001. Cellular and molecular mechanisms of circadian regulation in the unicellular dinoflagellate *Gonyaulax polyedra*. In *Handbook of behavioral neurobiology,* vol. 12: *Circadian clocks* (ed. J. Takahashi et al.), p. 321. Plenum Press, New York.

Hastings J.W. and Sweeney B.M. 1957. On the mechanism of temperature independence in a biological clock. *Proc. Natl. Acad. Sci.* **43:** 804.

———. 1958. A persistent diurnal rhythm of luminescence in *Gonyaulax polyedra*. *Biol. Bull.* **115:** 440.

———. 1959. The *Gonyaulax* clock. In *Photoperiodism in plants and animals* (ed. R.B. Withrow), p. 567. A.A.A.S. Press, Washington, D.C.

Hastings J.W., Astrachan L., and Sweeney B.M. 1961. A persistent daily rhythm in photosynthesis. *J. Gen. Physiol.* **45:** 69.

Hastings J.W., Broda H., and Johnson C.H. 1985. Phase and period effects of physical and chemical factors. Do cells communicate? In *Temporal order* (ed. L. Rensing and N.I. Jaeger), p. 213. Springer-Verlag, Berlin.

Iliev D., Voytsekh O., Schmidt E.M., Fiedler M., Nykytenko A. and Mittag M. 2006. A heteromeric RNA-binding protein is involved in maintaining acrophase and period of the circadian clock. *Plant Physiol.* **142**: 797-806.

Johnson C.H., Roeber J., and Hastings J.W. 1984. Circadian changes in enzyme concentration account for rhythm of enzyme activity in *Gonyaulax*. *Science* **223:** 1428.

Lapointe M. and Morse D. 2008. Reassessing the role of a 3'UTR-binding translational inhibitor in regulation of the circadian bioluminescence rhythm in the dinoflagellate *Gonyaulax*. *Biol. Chem.* (in press).

Li L., Hong R., and Hastings J.W. 1997. Three functional luciferase domains in a single polypeptide chain. *Proc. Natl. Acad. Sci.* **94:** 8954.

Liu L. and Hastings J.W. 2006 Novel and rapidly diverging intergenic sequences between tandem repeats of the luciferase genes in seven dinoflagellate species. *J. Phycol.* **42:** 96.

———. 2007. Two different domains of the luciferase gene in the heterotrophic dinoflagellate *Noctiluca miliaris* occur as two separate genes in photosynthetic species. *Proc. Natl. Acad. Sci.* **104:** 696.

Liu L., Wilson T., and Hastings J.W. 2004. Molecular evolution of dinoflagellate luciferases, enzymes with three catalytic domains in a single polypeptide. *Proc. Natl. Acad. Sci.* **101:** 16555.

Markovic P., Roenneberg T., and Morse D. 1996. Phased protein synthesis at several circadian times does not change protein levels in *Gonyaulax*. *J. Biol. Rhythms* **11:** 57.

Milos P., Morse D., and Hastings J.W. 1990. Circadian control over synthesis of many *Gonyaulax* proteins is at the translational level. *Naturwissenschaften* **77:** 87.

Mittag M. 1996. Conserved circadian elements in phylogenetically diverse algae. *Proc. Natl. Acad. Sci.* **93:** 14401.

Mittag M., Lee D.-H., and Hastings, J.W. 1994. Circadian expression of the luciferin-binding protein correlates with the binding of a protein to its 3′ untranslated region. *Proc. Natl. Acad. Sci.* **91:** 5257.

Morse D., Pappenheimer A.M., and Hastings J.W. 1989a. Role of a luciferin binding protein in the circadian bioluminescent reaction of *Gonyaulax polyedra*. *J. Biol. Chem.* **264:** 11822.

Morse D., Milos P.M., Roux E., and Hastings J.W. 1989b. Circadian regulation of the synthesis of substrate binding protein in the *Gonyaulax* bioluminescent system involves translational control. *Proc. Natl. Acad. Sci.* **86:** 172.

Nicolas M.-T., Morse D., Bassot J.-M., and Hastings J.W. 1991. Colocalization of luciferin binding protein and luciferase to the scintillons of *Gonyaulax polyedra* revealed by immunolabeling after fast-freeze fixation. *Protoplasma* **160:** 159.

Nicolas M.-T., Nicolas G., Johnson C.H., Bassot J.-M., and Hastings J.W. 1987. Characterization of the bioluminescent organelles in *Gonyaulax polyedra* (dinoflagellates) after fast-freeze fixation and antiluciferase immunogold staining. *J. Cell Biol.* **105:** 723.

Njus D., McMurry L., and Hastings J.W. 1977. Conditionality of circadian rhythmicity: Synergistic effect of light and temperature. *J. Comp. Physiol.* **117:** 335.

Okamoto O.K. and Hastings J.W. 2003. Novel dinoflagellate clock-related genes identified through microarray analysis. *J. Phycol.* **39:** 519.

Roenneberg T., Nakamura H., and Hastings J.W. 1988. Creatine accelerates the circadian clock in the unicellular alga *Gonyaulax polyedra*. *Nature* **334:** 432.

Roenneberg T., Nakamura H., Cranmer L.D., Ryan K., Kishi Y., and Hastings J.W. 1991. Gonyauline: A novel endogenous substance shortening the period of the circadian clock of a unicellular alga. *Experientia* **47:** 103.

Rossini C., Taylor W.R., Fagan T.F., and Hastings J.W. 2003. Lifetimes of mRNAs for clock-regulated proteins in a dinoflagellate. *Chronobiol. Int.* **20:** 963.

Schultz W., Liu L., Cegielski M., and Hastings J.W. 2005. Crystal structure of a pH-regulated luciferase catalyzing the bioluminescent oxidation of open tetrapyrrole. *Proc. Natl. Acad. Sci.* **102:** 1378.

Sweeney B.M. and Hastings J.W. 1957. Characteristics of the diurnal rhythm of luminescence in *Gonyaulax polyedra*. *J. Cell. Comp. Physiol.* **49:** 115.

———. 1961. Effects of temperature upon diurnal rhythms. *Cold Spring Harbor Symp. Quant. Biol.* **25:** 87.

Posttranscriptional Regulation of Mammalian Circadian Clock Output

E. GARBARINO-PICO AND C.B. GREEN
Department of Biology, University of Virginia, Charlottesville, Virginia 22904

Circadian clocks are present in many different cell types/tissues and control many aspects of physiology. This broad control is exerted, at least in part, by the circadian regulation of many genes, resulting in rhythmic expression patterns of 5–10% of the mRNAs in a given tissue. Although transcriptional regulation is certainly involved in this process, it is becoming clear that posttranscriptional mechanisms also have important roles in producing the appropriate rhythmic expression profiles. In this chapter, we review the available data about posttranscriptional regulation of circadian gene expression and highlight the potential role of *Nocturnin* (*Noc*) in such processes. NOC is a deadenylase—a ribonuclease that specifically removes poly(A) tails from mRNAs—that is expressed widely in the mouse with high-amplitude rhythmicity. Deadenylation affects the stability and translational properties of mRNAs. Mice lacking the *Noc* gene have metabolic defects including a resistance to diet-induced obesity, decreased fat storage, changes in lipid-related gene expression profiles in the liver, and altered glucose and insulin sensitivities. These findings suggest that NOC has a pivotal role downstream from the circadian clockwork in the post-transcriptional regulation genes involved in the circadian control of metabolism.

INTRODUCTION

Daily oscillations in mRNA levels are a distinctive feature of circadian rhythms. Approximately 1–10% of the transcripts expressed in a particular cell/tissue are under circadian control, as observed in mammals, birds, insects, plants, fungus, and cell cultures (Duffield 2003; Sato et al. 2003; Lowrey and Takahashi 2004). In cyanobacteria, expression of the entire genome appears to be modulated by circadian clocks (Liu et al. 1995; Woelfle and Johnson 2006). The regulation of mRNA levels constitutes an intrinsic component of the molecular clock function and seems to be a widespread mechanism by which these oscillators control key enzymes, transcription factors, and regulators governing biosynthetic and metabolic pathways. As a consequence of these oscillations, several systemic signals, as well as many of the receptors required to decode these signals, are periodically produced for the orchestration of physiological and behavioral rhythms.

It is obvious that the abundance of a specific mRNA or protein depends on the balance between its synthesis and degradation. However, with regard to circadian oscillations of mRNA levels, there are two major prejudices that instinctively arise: Rhythmic transcription controls mRNA oscillations, and messenger content reflects the level and activity of the corresponding protein. These suppositions are true in many cases, but not always, and in general, other mechanisms contribute significantly to regulate both mRNA and protein levels. The levels of a particular mRNA can decrease even when its transcription increases if degradation rates increase more, and the abundance of a protein can change when its transcript level remains constant if its translation is silenced or enhanced. These apparent paradoxes are not just rhetorical speculations; there are many such examples that have been documented (see, e.g., Hastings 2001; Shu and Hong-Hui 2004; Reddy et al. 2006).

The control of gene expression is a complex process comprising several steps that are tightly regulated: (1) transcription, (2) mRNA processing (capping, splicing, polyadenylation, and quality control), (3) nuclear export, (4) sorting and transport (whereas most mRNAs are immediately translated, others are stored or translocated to specific cellular regions), (5) translation, and (6) mRNA degradation (Fig 1). Presently, we have evidence showing that some of these steps are regulated by clocks, but in theory, all of these events are possible points of circadian modulation.

Despite the considerations discussed above, it is important to point out that transcription is thought to be the primary source of many transcript oscillations (Harms et al. 2004). However, one should be careful before assuming it for a specific mRNA, and in general, transcriptional and posttranscriptional processes act coordinately in the regulation of expression programs. During recent years, it has become increasingly clear that mRNA decay and translational regulation have important consequences on gene expression and are tightly regulated (Wilusz et al. 2001; Gebauer and Hentze 2004; Wilusz and Wilusz 2004; Eulalio et al. 2007; Garneau et al. 2007; Mathews et al. 2007; Parker and Sheth 2007; Pillai et al. 2007). Microarray studies have shown that regulation of mRNA stability may account for as much as 50% of the variations in poly(A) mRNA levels and has a major role in regulating expression programs (for review, see Raghavan and Bohjanen 2004; Cheadle et al. 2005; Mata et al. 2005). Transcription has been extensively studied from a chronobiologic perspective, whereas posttranscriptional mechanisms have received much less attention.

Recently, we demonstrated that the rhythmic gene *Nocturnin* (*Noc*) encoded a deadenylase, an exoribonuclease specific for poly(A) tails of mRNAs (Green and Besharse 1996a; Baggs and Green 2003). To date, this is the only ribonuclease known to be under circadian control,

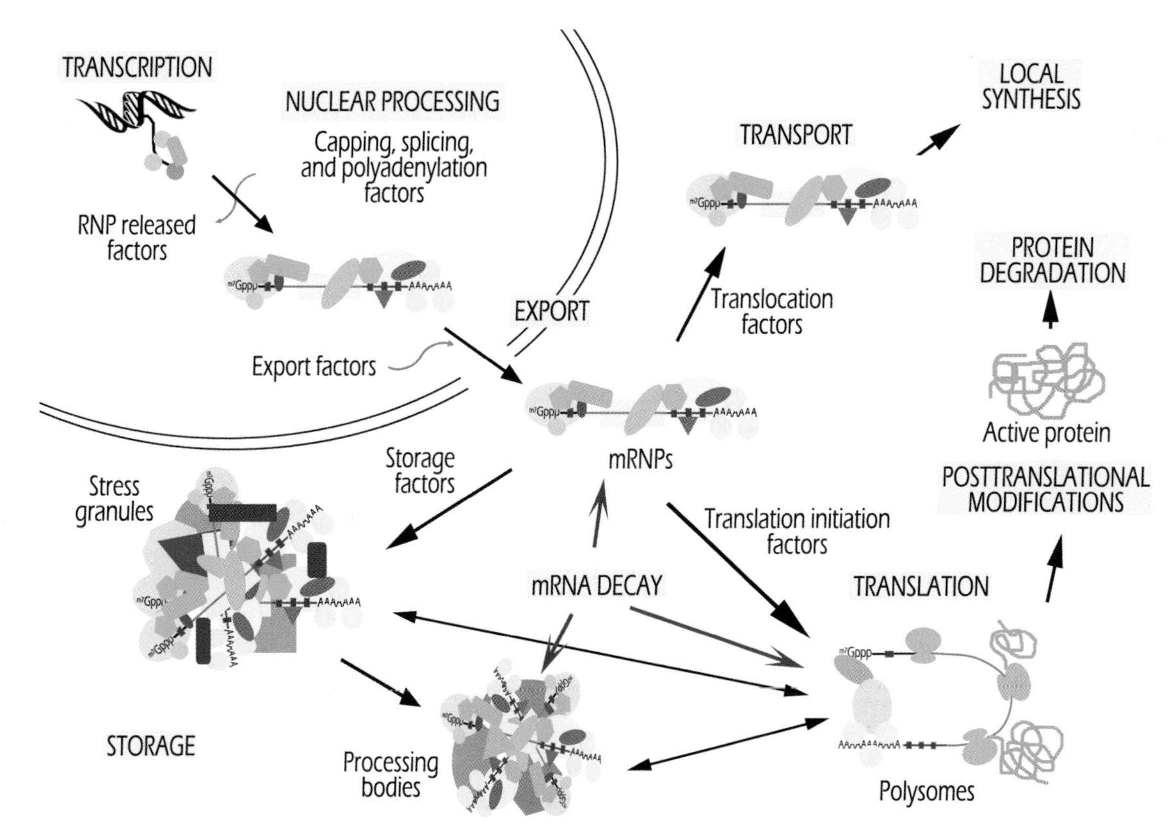

Figure 1. Circadian gene expression can be regulated at several levels. The control of gene expression is a complex process that includes several tightly regulated steps. Upon transcription, an mRNA interacts with RBPs, enzymes, scaffolding, and other regulatory proteins that together with noncoding RNAs determine its processing and finally its fate. The composition of these mRNPs changes concomitantly with the different mRNA-associated functions. Some of these mRNA-interacting factors have been shown to be circadianly regulated (see Table 1); however, several of the processes indicated in the boxes have not been studied from a chronobiology prospective.

providing a possible mechanism for rhythmic posttranscriptional control. In this chapter, we review what is currently known about the role of posttranscriptional regulation in shaping circadian gene expression patterns and discuss potential roles of *Noc* in these events. Posttranslational regulatory mechanisms are not discussed here.

EXAMPLES OF CIRCADIAN POSTTRANSCRIPTIONAL REGULATION

As stated above, posttranscriptional regulation includes numerous events (Fig. 1), many of which have not been approached from a chronobiologic perspective at all, and in general, we have a fragmented vision of how clocks may act at this level. Clock modulation of mRNA nuclear processing (including capping, splicing, polyadenylation, and messenger quality control), nuclear export, and sorting for storage or specific localization has not been examined. However, there are a number of reports that link posttranscriptional events to clock function. In one interesting example, the loss of function of Fragile-X mental retardation protein (FMRP), an RNA-binding protein (RBP), produces defects in clock output in *Drosophila* (Dockendorff et al. 2002). In addition, several examples demonstrate a role for regulation of mRNA decay and translation in circadian gene expression (summarized in Table 1). We review some of these examples in this section.

Arginine Vasopressin mRNA Poly(A) Length

Two decades ago, a daily regulation of vasopressin (*Avp*) mRNA poly(A) tail length in rat suprachiasmatic nuclei (SCN), but not in other brain areas, was reported (Robinson et al. 1988). The transcript levels of *Avp* are circadianly regulated specifically in SCN, showing the lowest levels at night (Uhl and Reppert 1986; Cagampang et al. 1994), when the mRNA species with short poly(A) tail length is observed (Robinson et al. 1988). Nuclear run-on experiments showed an approximately 30% diminution in transcription rate of *Avp* at night (Carter and Murphy 1992), which did not account for the total changes in mRNA levels. Because deadenylation (shortening of the poly(A) tail) was known to often trigger mRNA decay, these observations suggest that *Avp* mRNA decay rates are higher when the poly(A) tail is shorter, contributing to the rhythm of steady-state *Avp* transcript levels. On the other hand, *Avp* mRNA is translocated to dendrites for local synthesis (Mohr and Richter 2004), so the changes in poly(A) length may also influence this phenomenon, affecting the temporal production of this neuromodulator in synaptic regions.

Table 1. Factors Involved in Posttranscriptional Regulation of Circadian Expression

Factor	Organism	Motif (CDD#)[a]	Function	References
Enzymes				
NOCTURNIN (CCRN4L)	*Xenopus laevis*; *Mus musculus*	endonuclease/ exonuclease/ phosphatase (pfam03372)	deadenylase (poly(A)-specific ribonuclease); likely regulates the mRNA decay and/or translational properties of circadian-related transcripts; *mNoc*$^{-/-}$ mouse has metabolic alterations	Baggs and Green (2003); Garbarino-Pico et al. (2007); Green et al. (2007)
RNA-binding proteins				
AtGRP7	*Arabidopsis thaliana*	RNA recognition motif (RRM) (smart00360)	regulator of *Atgrp7* and *Atgrp8* mRNAs splicing; produces an unstable variant at specific times; involved in the regulation of abscisic acid and stress responses	Heintzen et al. (1997); Staiger et al. (2003); Cao et al. (2006)
CCTR	*Lingulodinium polyedrum* (formerly *Gonyaulax*)	n.d.	binds rhythmically to the 3′UTR (UG-repeat region) of *luciferin-binding protein* mRNA, repressing its translation and indirectly regulating bioluminescence rhythms	Mittag et al. (1994); Mittag (2003)
CHLAMY 1	*Chlamydomonas reinhardtii*	3x K homology RNA-binding domain; type I (KH-I) (cd00105); two conserved tryptophan domain (WW) (cd00201), and 3x RRM (smart00360)	binds rhythmically to 3′UTRs (UG-repeat region) of genes involved in N_2 and CO_2 metabolism; also implicated n the control of the phase angle and period of the circadian clock	Mittag (1996); Waltenberger et al. (2001); Zhao et al. (2004); Iliev et al. (2006)
hnRNP L	*Rattus norvegicus*	3x RRM (smart00360)	induces nocturnal *Aanat* mRNA degradation by binding *cis*-acting elements in its 3′UTR	Kim et al. (2005)
hnRNP Q	*Rattus norvegicus*	3x RRM (smart00360)	similar to hnRNP Q, but in addition binds another site in 5′UTR increasing *Aanat* translation rate	Kim et al. (2005, 2007)
hnRNP R	*Rattus norvegicus*	3x RRM (smart00360)	similar to hnRNP Q	Kim et al. (2005)
LARK	*Drosophila melanogaster*	2x RRM (smart00360); CCHC-type zinc finger (smart00343)	shows constant mRNA levels, but rhythmic protein accumulation; affects daily timing of adult eclosion but not clock properties or input mechanism	Newby and Jackson (1993, 1996); McNeil et al. (1998); Schroeder et al. (2003)
	Mus musculus	2x RRM (smart00360); CCHC-type zinc finger (pfam00098)	also shows constant mRNA levels, but rhythmic protein accumulation; binds 3′UTR of *mPer1* transcript enhancing its translation; its up- or down-regulation affects period in cell cultures	Kojima et al. (2007)
Regulatory RNAs				
Antisense *Frq* mRNAs	*Neurospora crassa*	n.a.	cycles in antiphase with sense *Frq* mRNA and is induced by light; affects circadian rhythms	Kramer et al. (2003)
miR-219	*Mus musculus*	n.a.	expression is circadianly regulated by CLOCK/BMAL1; its knockdown lengthens period	Cheng et al. (2007)
miR-132	*Mus musculus*	n.a.	it is induced by light, attenuating its entraining effects	Cheng et al. (2007)

[a]Conserved domain database, NCBI (Marchler-Bauer et al. 2007). n.d. indicates not determined; n.a. indicates not applicable.

Circadian Regulation of Splicing Variants

The *Arabidopsis* AtGRP7 is a circadian-expressed RBP that regulates the alternative splicing of its own message and, as consequence, its own expression rhythm (Staiger et al. 2003). When the protein accumulates, AtGRP7 binds a site in the 3′UTR (untranslated region) of its pre-mRNA that would induce a shift in splice site selection; the variant produced contains a premature stop codon and is very unstable, causing the reduction of the *Atgrp7* mRNA and protein levels. Additionally, AtGRP7 has a similar effect on the splicing of the *Atgrp8* transcript and may act on other gene products as well (Staiger et al. 2003).

The murine *Presenilin2* gene expresses different mRNA splice variants; two of them are rhythmic, whereas a third one is not (Belanger et al. 2006). It is not clear whether this is caused by circadian modulation of spliceosome machinery or of the splice variant stability. *Drosophila Period* (*Per*) represents another interesting example. It exhibits daily regulation in the splicing of an intron in the 3′UTR that shows seasonal changes (depending on photoperiod and temperature) affecting fly behavior (Collins et al. 2004; Majercak et al. 2004). Per is also regulated at additional posttranscriptional levels as discussed below. Alternative splicing of the *Frequency* (*Frq*) gene is also part of the adaptation of *Neurospora* molecular clock to temperature changes (Colot et al. 2005; Diernfellner et al. 2005).

Period (*Per*) mRNA

Frish et al. (1994) found that the promoter of *Drosophila Per* is not required for its circadian expression. This demonstration of a transcription-independent mRNA rhythm was particularly surprising because *Per* is part of the molecular clock machinery. The delivery of a promoterless-*dPer* transgene can actually restore locomotor activity oscillations in arrhythmic flies (Frisch et al. 1994). However, the amplitude of the mRNA content rhythm in the transgenic flies is significantly lower, suggesting that transcription does contribute to the normal oscillation. Indeed, the transcription rate of *dPer* shows a high-amplitude oscillation (Hardin et al. 1992; So and Rosbash 1997). In an elegant study, So and Rosbash (1997) showed that both transcriptional and posttranscriptional events regulate *dPe*r transcript rhythms. These authors compared the transcription rate temporal profile measured by nuclear run-on assays, with the mRNA content oscillation determined by RNase protection assays. This comparison revealed that the half-life of the *dPer* mRNA changes between twofold and fourfold over the course of the circadian cycle, demonstrating that the *dPer* mRNA decay rate is under circadian control. (The same study also showed that posttranscriptional events are involved in the generation of the mRNA rhythm of a clock-controlled gene called *Crg-1*.) *dPer* mRNA contains elements in its 5′ and 3′UTRs that affect PER protein cycling and fly behavior, probably through regulation of mRNA stability (Chen et al. 1998; Stanewsky et al. 2002). Additionally, it has been shown that translation is also under circadian regulation; two alternative splicing variants exhibit different translational properties, and cyclic *dPer* mRNA is not required for PER protein rhythmicity (Cheng et al. 1998; Cheng and Hardin 1998).

The mammalian homologs of *dPer* also have been shown to be regulated at the posttranscriptional level. The 3′UTR of human and murine *Per1* transcripts exhibit a high degree of homology, and in the mouse, this region has been shown to be involved in the regulation of *mPer1* expression (Wilsbacher et al. 2002; Kojima et al. 2003). mLARK, a circadian expressed RBP, binds an element in the *mPer1* 3′UTR resulting in enhanced translation (Kojima et al. 2007). Remarkably, both *dLark* and *mLark* show constant mRNA levels but rhythmic protein accumulation, denoting circadian posttranscriptional regulation of their expression (Newby and Jackson 1996; McNeil et al. 1998; Kojima et al. 2007).

The other two mammalian *Per* genes also have been shown to be regulated at the posttranscriptional level. *mPer2* expressed constitutively in NIH-3T3 cultures exhibits cyclic protein levels even while showing constant mRNA content, likely due to translational and/or proteolytic regulation (Yamamoto et al. 2005; Fujimoto et al. 2006; Nishii et al. 2006). The 3′UTR of *mPer3* mRNA also possesses a regulatory element(s) controlling its stability, but the *trans*-acting factor(s) responsible for this phenomenon has not yet been identified (Kwak et al. 2006).

Arylalkylamine *N*-acetyltransferase mRNA

The regulation of *Aanat* (arylalkylamine *N*-acetyltransferase) expression, a key enzyme in melatonin biosynthesis, constitutes another example of circadian control at multiple levels, including transcription (Baler et al. 1997; Foulkes et al. 1997), mRNA destabilization (Kim et al. 2005), and translation (Kim et al. 2007). Kim et al. (2005) found that the *Aanat* mRNA 3′UTR possesses destabilization elements. The circadianly expressed heterogeneous nuclear ribonucleoprotein (hnRNP) R, hnRNP Q, and hnRNP L induce the nocturnal *Aanat* mRNA degradation by binding those *cis*-acting elements (Kim et al. 2005). Furthermore, the 5′UTR of rat and sheep *Aanat* mRNAs contain an internal ribosome entry site (IRES) that confers rhythmic translation (Kim et al. 2007). IRESs allow ribosomes to bind transcripts and initiate translation in an alternative mechanism to the canonical cap-dependent scanning model (Jackson 2005; Elroy-Stein and Merrick 2007). hnRNP Q was also identified as one of the *trans*-acting factors responsible for the circadian activation of *Aanat* translation (Kim et al. 2007). Its expression levels, as wells as its binding to the *Aanat* mRNA 5′UTR, oscillates concomitantly with AANAT protein abundance.

Other Examples Involving mRNA Decay in Circadian Control of Gene Expression

In addition to the previous examples, several other mRNAs have been shown to be regulated posttranscriptionally; however, the mechanisms and *trans*-acting factors involved have not been identified. Studies done in *Arabidopsis thaliana* have provided several examples of mRNA rhythms controlled at transcript stability level: *Chlorophyll a/b binding 1* (Millar and Kay 1991), *Nitrate reductase 2* (Pilgrim et al. 1993) and *Ccr-like*, and *Senescence associated gene 1* (Lidder et al. 2005). Interestingly, some *Arabidopsis* clock-controlled genes (ccgs) which have mRNAs that are unstable contain a downstream (DST) element (apparently unique to plants) in their 3′UTR that confers the rapid decay property, but the factor acting on this element has not been characterized (Gutierrez et al. 2002; Lidder et al. 2005). In maize, two *Rubisco activase* genes expressing mRNAs with the same coding region but different 3′UTRs have been reported (Ayala-Ochoa et al. 2004). Both show daily changes in their levels but different degradation rates, presumably because distinct DST-like elements are present in their 3′UTRs. Rice *CatalaseA* (*CatA*) expression also was suggested to be regulated at the posttranscriptional level (likely via pre-mRNA stability), because a transgenic reporter gene driven by the promoter of *CatA*, showed different temporal patterns of mRNA levels when compared with the endogenous transcript (Iwamoto et al. 2000). However, it could not be ruled out that other gene regions outside the promoter affect transcription.

Circadian-regulated RBPs in Microalgae

In the dinoflagellate *Lingulodinium polyedrum* (formerly *Gonyaulax*) and the green algae *Chlamydomonas reinhardtii*, two related circadian-controlled RBPs were identified (for review, see Mittag 2003). In *L. polyedrum*, the circadian-controlled translational regulator (*Cctr*) protein product binds rhythmically to the 3′UTR of the luciferin-binding protein (*lbp*) mRNA regulating its translation and, consequently, the bioluminescence rhythm (Mittag et al. 1994). Using the *lbp* 3′UTR as bait, CHLAMY 1 was later identified in *C. reinhardtii* (Mittag 1996). Although this microalga does not express *lbp*, other mRNA targets were identified that contain a *cis*-acting element related to the one in *lbp*. The CHLAMY1 targets include enzymes and factors involved in nitrogen and carbon dioxide metabolism that are circadianly regulated (Waltenberger et al. 2001). CHLAMY 1 was also implicated in the control of the phase angle and period of the circadian clock (Iliev et al. 2006).

Circadian Regulation of Translation

In general, the examples of this kind of regulation are inferences from comparisons between mRNA and protein profiles, thus little mechanistic information is known. However, some RBPs involved in translational control have been identified (see above LARK, hnRNP Q, CCTR, and CHLAMY 1). Woodland Hastings's laboratory has been the pioneer in studying circadian posttranscriptional regulation in the dinoflagellate *L. polyedrum* (for review, see Hastings 2001). His laboratory showed that the accumulation of several proteins exhibit oscillations generated at the translational level and also was the first to identify a circadian-regulated RBP (CCTR, see above). In this organism, total protein synthesis in both cytoplasm and chloroplast shows circadian changes (Donner et al. 1985). Markovic et al. (1996) showed tenfold circadian changes in the synthesis rates of several proteins that exhibit constant levels, suggesting that this regulation is required for temporal organization of translation, rather than for generating a protein oscillation.

The *Neurospora* clock protein FRQ exists in two forms as a result of the use of alternative translation initiation sites (Garceau et al. 1997). And in *Arabidopsis*, the levels of the transcription factor LHY increase concurrently with the down-regulation of its mRNA during light induction, presumably acting to narrow the peak of this protein involved in the molecular clock (Kim et al. 2003).

Rhythmic Posttranscriptional Control by Noncoding RNAs

Recently, noncoding regulatory RNAs have been implicated in several functions related to mRNA metabolism, but there are only a few studies thus far involving them in circadian rhythms. In *Neurospora*, sense and antisense *Frq* mRNAs were identified that cycle in antiphase, and all are induced by light. Mutant strains not expressing these antisense transcripts show altered rhythms (Kramer et al. 2003). Natural antisense RNAs have also been found for mammalian *Rev-erbα* and *Antheraea pernyi* (silkmoth) *Per* (for discussion, see Crosthwaite 2004). A recent study reported for the first time the circadian regulation of a microRNA (miRNA); miR-219 rhythm is driven by CLOCK/BMAL1 and its knockdown lengthens circadian period (Cheng et al. 2007). Another miRNA, miR-132, responds to light affecting entrainment (Cheng et al. 2007).

Nocturnin

Finally, the recent determination that the rhythmic *Noc* gene encodes a deadenylase provides another mechanism for regulating circadian posttranscriptional output (Baggs and Green 2003). This is discussed in more detail below following a summary of the major mechanisms involved in mRNA turnover.

mRNA Decay Pathways

Many recent advances have been made in the understanding of mRNA-decay mechanisms and several excellent reviews on this topic have recently been published (Wilusz et al. 2001; Parker and Song 2004; Wilusz and Wilusz 2004; Garneau et al. 2007). In particular, the involvement of noncoding regulatory RNAs and the founding of novel cytoplasmic subdomains are revolutionizing the field and rapidly changing our view of mRNA degradation and translation control (Anderson and Kedersha 2006; Valencia-Sanchez et al. 2006; Eulalio et al. 2007; Mathews et al. 2007; Parker and Sheth 2007; Pillai et al. 2007).

Although a nascent transcript is still being synthesized, it is coated with regulatory factors including RBPs, enzymes, and noncoding regulatory RNAs. These complexes—ribonucleoparticles or ribonucleoproteins (mRNPs)—are very dynamic and change their composition with respect to the different functions associated with pre-mRNA processing, transcript translocation, translation, in some cases storage or specific localization, and finally, degradation (Fig. 1). Importantly, transcripts contain regulatory elements along their body including the 5′7-methylguanosine cap (m^7Gppp), the 5′UTR, the coding sequence, the 3′UTR, and the poly(A) tail. Generically, they are called *cis*-acting elements and constitute the sites for interacting with RBPs, the *trans*-acting factors. A picture of mRNA posttranscriptional processing is emerging that is analogous to the combinatorial model of transcription regulation: Whereas the activation of a gene depends on the arrangement of transcription factors bound to regulatory sites in its promoter, the fate of an mRNA depends on the combination of RBPs and noncoding regulatory RNAs associated with their *cis*-acting elements (Keene 2007). These factors recruit the enzymes and other regulatory and scaffold proteins that process the mRNA and control its fate.

The half-life of different mRNA species varies between minutes and days, some of them survive for only fractions of the cell cycle, whereas others last for many divisions. The stability of a particular transcript can change in different tissues or physiological conditions, in response to stimulation, across development, or along circadian time, as was exemplified in the previous section. In general, the levels of an mRNA correlate well with its translation rates; however, the number of examples where messages leave polysomes and are stored without being degraded is increasing. This constitutes a fast and economic mechanism for silencing the expression of a gene and such a message can later resume its translation independently of transcription and nuclear processing. For example, during the stress response, the translation machinery is co-opted to synthesize the factors involved in the cellular response, and the majority of the non-stress-related transcripts are stored in stress granules until their translation is again necessary (Anderson and Kedersha 2002).

In addition to participating in regulation of expression levels, mRNA decay is also involved in the antiviral defense and quality control of transcripts. These processes include double-stranded RNA responses or RNA interference, non-sense-mediated decay (NMD), nonstop decay (NSD), and nongo decay (NGD); these are not discussed here. The pathways responsible for "normal" mRNA degradation comprise deadenylation-dependent mRNA decay, deadenylation-independent mRNA decay, and endonuclease-mediated mRNA decay.

The cap and the poly(A) structures stabilize transcripts by their interaction with translation initiation factor 4E (eIF4E) and poly(A)-binding protein (PABP), respectively, protecting against exoribonuclease access and enhancing translation. The poly(A) tails of eukaryotic mRNAs usually contain 25–200 residues that interact with a number of regulators in addition to PABP (Mangus et al. 2003; Kuhn and Wahle 2004). It is currently believed that deadenylation-dependent mRNA decay is the pathway by which most transcripts are degraded in eukaryotic cells (Cao and Parker 2001). It initiates with the enzymatic shortening of the poly(A) tail, which is performed by deadenylases. The deadenylation step is very important because it is believed to constitute the rate-limiting step and the more-regulated step in the turnover of most transcripts. In addition, it is implicated in translational silencing. Several deadenylases have been identified (for review, see Parker and Song 2004); however, not much is known about deadenylase regulation and specificity. Some of these enzymes' activities are modulated by the presence or absence of the 5′ cap and/or the presence or absence of PABP in in vitro reactions. The spatial and temporal distribution of each deadenylase is another obvious aspect to consider and may impact the in vivo substrate specificity and the function of these enzymes. Recently, some RBPs and miRNAs have been implicated in the recruitment of deadenylases to specific targets.

After deadenylation, mRNA can be degraded in 3′→5′ direction by the exosome, a complex containing exonucleases and accessory proteins (Liu et al. 2006), followed by hydrolysis of the remaining m^7Gppp cap by the scavenger-decapping enzyme DCPS. Alternatively, the deadenylated messenger can be degraded in 5′→3′ direction being first decapped by DCP1 and DCP2 and then digested by the exoribonuclease XRN1.

Two other degradation pathways have been reported but appear to be much less common. Deadenylation-independent mRNA decay is triggered by decapping by DCP1 and DCP2, followed by degradation by XRN1. Endonucleolytic degradation bypasses the two major transcript stabilizers (cap and tail) by generating endonucleolytic products lacking the 5′ m^7Gppp or the 3′ poly(A), which are therefore susceptible to degradation by XRN1 and the exosome, respectively. The few riboendonucleases described thus far are associated with specific cellular compartments (i.e. endoplasmic reticulum, mitochondria, or polysomes) and are tightly regulated.

The discovery of new cytoplasmic subdomains where mRNPs accumulate, and a better understanding of the composition and function of previously described RNA granules, has modified our view of cytoplasmic mRNA metabolism and has provided a place for physical interaction between translation and mRNA decay factors (for review, see Anderson and Kedersha 2006; Kiebler and Bassell 2006; Eulalio et al. 2007; Parker and Sheth 2007). These RNA granules include processing bodies (PBs, also known as GW or DCP1 bodies), stress granules (SGs), neuronal granules, and maternal or germinal granules. They contain translationally silenced mRNAs, RBPs, translation and mRNA-degradation factors, scaffold proteins, and, in some cases, ribosomal subunits, miRNAs, and components of miRNA machinery. SGs and PBs are very dynamic structures that respond to physiological stimulation. It has been shown that transcripts can be redirected from polysomes to SGs and PBs or vice versa in response to different conditions, and as a consequence, the message is translated, stored, or degraded. PBs have been reported in yeast and mammalian cell cultures, but it is still not clear whether they have a prominent function in vivo. Importantly, in addition to their presence in PBs and SGs, the components of these structures are also dispersed in cytoplasm, and assembly of these foci is not a requisite for translational silencing, mRNA turnover, or storage.

There is considerable evidence for an interrelationship between translation and mRNA decay (for review, see Schwartz and Parker 2000). In general, when the translation of an mRNA is optimized, its turnover is decreased, and conversely, factors that induce transcript decay may diminish translation efficiency. Models have been proposed where the translation factors associated with the poly(A) tail and cap protect the transcript from deadenylases and decapping enzymes through competition or inhibition (Schwartz and Parker 2000). Another related observation is that PBs and SGs are disrupted by polysome stabilizers (i.e., cycloheximide) and their number and size increased by drugs that disassemble polysomes (i.e., puromycin). However, little is known about the mechanisms dictating when an mRNA switches from interacting with translation factors and ribosomes to being trapped by the decay machinery.

Discovery of *Nocturnin*

Noc was identified in a differential display screen performed in *Xenopus* retina searching for circadianly controlled mRNAs (Green and Besharse 1996a,b). It was named *nocturnin* by virtue of the high nocturnal increase that its transcript shows in the frog eye (Green and Besharse 1996a). The cDNA encoded a novel protein of 388 amino acids. Our laboratory initially studied this gene in *Xenopus* in an attempt to establish its function and significance in circadian rhythms (for review, see Baggs and Green 2006).

During development, *xNoc* mRNA is detected in different tissues and stages showing a complex expression pattern (K. Curran et al., unpubl.); in contrast, in adult frogs, it is predominantly expressed in retinal photoreceptor cells (Green and Besharse 1996a). This delimited localization is determined by a novel element identified in the *xNoc* promoter (photoreceptor-conserved element II; PCEII) that is active specifically in those cells (Liu and Green 2001). The circadian oscillation in *xNoc* mRNA levels is generated, at least in part, by concomitant variations in its transcription rates, as revealed by nuclear run-on analysis (Green and Besharse 1996a). The nocturnal increase of *xNoc* transcription is caused by the binding of phosphorylated CREB to a noncanonical CRE identified in the *xNoc* promoter (Liu and Green 2002). The xNOC protein accumulation also is circadian-regulated in *Xenopus* retina, consistently with the mRNA oscillation described (Baggs and Green 2003).

Originally, the only information we had regarding *xNoc* function was the spatial and temporal expression pattern described above, and its sequence. The carboxyl terminus of xNOC exhibits high similarity with the same region of the protein product of yeast Carbon catabolite repression 4 (*yCcr4*) (Green and Besharse 1996a). For this reason, *Noc* is also known as Carbon catabolite repression 4-like (*Ccrn4l*). yCCR4 is a subunit of the CCR4-NOT complex, a transcription factor involved in carbon metabolism regulation. To analyze whether xNOC was also a transcription factor, we determined the subcellular localization of both native and overexpressed tagged proteins. They were never found in the nucleus; xNOC was systematically located in cytoplasm (Baggs and Green 2003). Later, it was shown that yCCR4 also exhibits deadenylase activity (Chen et al. 2002; Tucker et al. 2002). After testing alternative hypotheses regarding xNOC biochemical function, Baggs and Green (2003) demonstrated that, like yCCR4, it is a Mg^{2+}-dependent poly(A)-specific ribonuclease (deadenylase). NOC belongs to the endonuclease/exonuclease/phosphatase family that includes DNase I, APE1, IP5P, CCR4, and CCR4-related proteins (Dlakic 2000; Dupressoir et al. 2001). Considering the xNOC temporal expression profile, the finding that xNOC is a deadenylase led to the formulation of the hypothesis that it regulates circadian gene expression at a posttranscriptional level by inducing the degradation or translational silencing of clock-related mRNAs (Baggs and Green 2003).

MOUSE NOCTURNIN IS CIRCADIANLY REGULATED AND BROADLY EXPRESSED

NOC is a highly conserved protein in animals, with homologs identified in mammals, birds, fishes, and insects (Dupressoir et al. 2001; Wang et al. 2001). *Caenorhabditis elegans* and *Arabidopsis thaliana* also have potentially homologous sequences, but it is not clear if they are true orthologs of NOC or rather of other yCCR4-related proteins.

The mouse *Noc* transcript contains an open reading frame (ORF) encoding a 429-amino-acid protein with a predicted molecular mass of 48 kD (Dupressoir et al. 1999; Wang et al. 2001). The message also possesses a conserved second ATG site that meets Kozak consensus requirements which might act as an alternative translation initiation site resulting in a protein of 365 amino acids and 41 kD (Wang et al. 2001). Western blots performed with two alternative antibodies generated against mNOC showed a unique band in mouse fibroblast (Garbarino-Pico et al. 2007) and tissue (E. Garbarino-Pico and C.B. Green, unpubl.) lysates. The electrophoretic mobility of the protein detected favors the possibility that the actual translation initiation site is the second one; however, this evidence is not conclusive and further experiments are required in order to resolve this issue.

The carboxy-terminal region of mNOC has a high degree of identity with the catalytic domain of yCCR4. In contrast, the amino-terminal regions of these proteins are dissimilar; mNOC lacks the two transcriptional activation domains and the leucine-rich repeat region of CCR4 (Dupressoir et al. 2001; Wang et al. 2001). The absence of these transcriptional activation domains and its cytoplasmic localization (J.E. Baggs et al., unpubl.) rule out the possibility that mNOC has a role as a transcription factor. As expected, mNOC (like xNOC) exhibits deadenylase activity (Garbarino-Pico et al. 2007). Molecular exclusion chromatography and immunoprecipation studies have shown that mNOC is included in complexes with other proteins (J.E. Baggs et al., unpubl.); it is likely that these partners have a role in modulating and targeting mNOC activity in vivo.

As in *Xenopus*, *mNoc* expression is also under circadian control showing high-amplitude rhythms in mRNA (Wang et al. 2001; Barbot et al. 2002) and protein (E. Garbarino-Pico and C.B. Green, unpubl.) levels. The wide spatial distribution of *Noc* in mouse contrasts with its localized expression in frog. The *mNoc* transcript has been detected in most mouse tissues, including brain, colon, heart, intestine, liver, lung, kidney, ovary, spleen, testis, thymus, and retina (Dupressoir et al. 1999; Wang et al. 2001; Barbot et al. 2002), as well as in fibroblast cultures (Garbarino-Pico et al. 2007). In the central nervous system, *Noc* mRNA was observed in the hypothalamic SCN and arcuate nucleus, olfactory bulb, piriform cortex, hippocampus, subiculum, cerebellum, and pineal gland (Wang et al. 2001). The broad expression in the mouse suggests that *mNoc* has a role in regulating transcripts involved in a wide range of functions.

mNoc IS AN IMMEDIATE-EARLY GENE

While investigating *mNoc* expression in NIH-3T3 cell cultures, we found that it is acutely induced by extracellular stimuli. We tested the effect of a serum shock, the phorbol ester TPA, forskolin, and dexamethasone because of their capabilities of entraining circadian clocks in fibroblast cultures (for review, see Nagoshi et al. 2005). *mNoc* mRNA levels show an approximately 30-fold increase after 2 hours of serum shock compared with serum-starved quiescent NIH-3T3 cells. This induction is also reflected in the protein levels which show an approximately ninefold induction 2.5 hours after the stimulation. Remarkably, the other four deadenylases reported in the mouse are not induced by this treatment. TPA also acutely stimulates *mNoc* transcript accumulation, but forskolin and dexamethasone do not (Garbarino-Pico et al. 2007).

The TPA *mNoc* acute induction is independent of protein synthesis and transient, with both the transcript and protein having a short half-life (Garbarino-Pico et al. 2007). These properties define an immediate-early gene (IEG): These genes primarily and directly respond to a stimulus or physiological change without requiring the synthesis of transcription factors or other regulators. Indeed, IEGs are the factors responsible for driving the cell response to new conditions through regulation of the expression of other genes, suggesting that *mNoc* is involved in such an adaptative response to stimuli. Because the change of expression programs involves the turning on of some genes and the silencing of others, it is natural that in addition to transcription factors, mRNA decay and silencing factors should also be involved in these responses. But again, as for circadian gene expression, the participation of transcription in response to extracellular stimulation has been extensively studied, whereas mRNA decay and translation considerably less so.

The acute response of *mNoc* to extracellular signals raises the question of whether its rhythmic expression is generated autonomously by the intracellular molecular clock or by a circadian systemic signal (note that these alternatives are not mutually exclusive). We still do not have a conclusive answer to this question, but several observations suggest that a daily induction by an extracellular factor, rather than by the intracellular clockwork, regulates *mNoc* circadian expression: (1) *mNoc* mRNA peaks at the same time in different tissues (Wang et al. 2001), whereas clock genes have different phases in different organs. (2) *mNoc* and *Dbp* (a CLOCK/BMAL1-driven gene) transcripts have different phase relationships in liver when comparing animals kept under a light/dark cycle versus constant darkness (Barbot et al. 2002). (3) Whereas two agents that entrain circadian clocks in cell cultures induce *mNoc* (serum and TPA), two others do not (forskolin and dexamethasone); in addition, the TPA induction does not require the synthesis of clock proteins (Garbarino-Pico et al. 2007). (4) Mice deprived of food for 1 day and maintained in constant light show a significantly lower nocturnal *mNoc* mRNA increase compared with controls (Barbot et al. 2002). (5) In transgenic mice in which the liver molecular clock function was specifically impaired, most liver clock-controlled genes lost their rhythmicity; however, *mNoc* is among the few mRNAs that still oscillate, presumably being driven by rhythmic systemic signals (Kornmann et al. 2007).

Many humoral factors, metabolites, and physiological signals show circadian oscillations in metazoans as a consequence of the feeding, activity, drinking, hormonal, and temperature rhythms (for discussion, see Schibler et al. 2003). It is not clear whether any of them induce *mNoc* expression. Interestingly, lowering blood pH in mice causes an acute induction of *Noc* expression in kidney proximal tubules (P.A. Preisig, pers. comm.). Both serum and TPA trigger many signaling cascades; thus, it is not possible to make a strong prediction about what induces *mNoc* in vivo. Sequence analysis of the promoter region of *mNoc* shows putative binding sites for SP1 and NF-κB (Dupressoir et al. 1999; M. Hurt and C.B. Green, unpubl.). There are also E-box sequences and other potential elements in the *mNoc* promoter region, but none of them have been functionally tested.

mNoc AND METABOLISM

To understand the role of *Noc* in vivo, mice lacking *Noc* (*Noc* KO) were generated by homologous recombination (Green et al. 2007). Homozygous *Noc* KO mice have normal circadian locomotor rhythms and normal expression of the core clock genes, suggesting that *Noc* is not part of the core circadian mechanism in mice. However, these mice have a striking metabolic phenotype, characterized by resistance to diet-induced obesity and resistance to hepatic steatosis. In addition, these animals have alterations in glucose and insulin sensitivity and in thermogenesis. As mentioned above, *Noc* is normally expressed with extremely high-amplitude rhythms in the liver, and examination of gene expression showed a number of changes in the profiles of key genes involved in lipid synthesis and/or utilization, several of which are themselves rhythmic. Many aspects of metabolism are known to be under the control of circadian clocks, and NOC presumably contributes to the clock's control of these processes through its impact on the posttranscriptional regulation of cycling mRNAs. Interestingly, a recent study by Reddy et al. (2006) found that almost the 50% of the rhythmic proteins in liver are translated from mRNAs with constitutive steady-state levels. This surprising result suggests that extensive circadian posttranscriptional control may be occurring in liver. The effect of NOC on liver metabolism may be through contributions to these types of processes.

CONCLUSIONS

There is an increasing body of evidence involving posttranscriptional control in the regulation of circadian clock output. However, we are still far away from knowing to what extent posttranscriptional events contribute to circadian gene expression. This is due in part because we usually measure steady-state level of mRNAs and proteins rather than transcription, mRNA decay, or translation rates. It has been shown that about 2–6% of *Arabidopsis* transcripts are under circadian control by using microarray technology (Harmer et al. 2000; Schaffer et al. 2001), surprisingly, when changes in transcriptional rates were measured by promoter-trap experiments, about 35% of the promoters were shown to be rhythmically controlled (Michael and McClung 2003). As described above, several circadian transcripts are already known to contain *cis*-acting elements interacting with circadian regulated RBPs, providing a mechanism to understand how clocks can control mRNA degradation or/and translation. Indeed, it may be that there is a specific *cis*-acting element(s) and a RBP(s) responsible for circadian mRNA degradation or translation, in a manner analogous to how CLOCK/BMAL1 acts on E boxes in the promoters of clock-controlled genes.

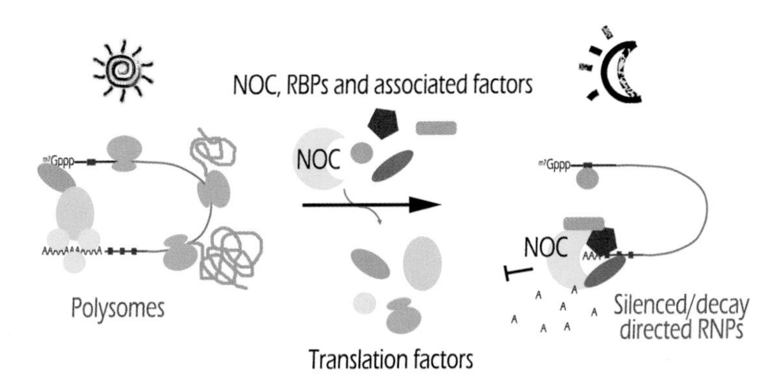

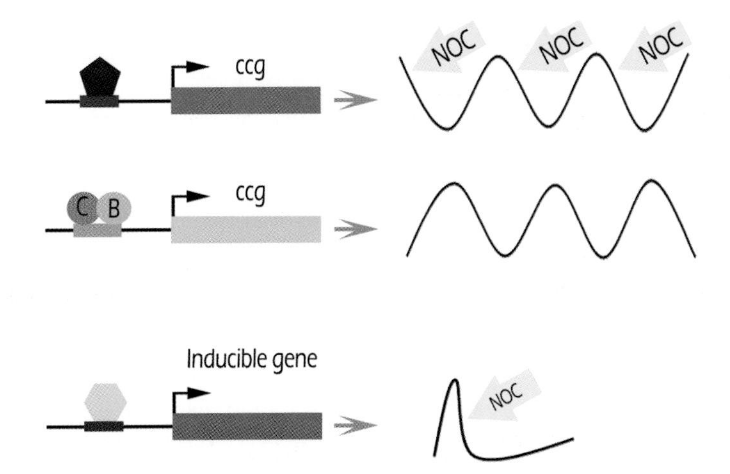

Figure 2. Working hypothesis for nocturnin's function. (*Top*) NOC is recruited by RBPs to specific mRNA targets during the night—when its levels increase—to remove their poly(A) tail and consequently induce their exit from active polysomes, possibly also triggering their degradation. (*Bottom*) NOC may participate in the generation of circadian expression rhythms by impeding translation of specific clock-controlled genes (ccgs) at night (*upper purple panel*). The temporal expression pattern obtained by this mechanism differs from that achieved by the CLOCK (C)/BMAL1(B)-mediated transcriptional activation which happens in the morning (*middle green panel*). Additionally, NOC may also be involved in the modulation of the response to acute stimulation (*lower red panel*).

The identification of NOC as a deadenylase provides one mechanism by which the circadian clock can posttranscriptionally regulate gene expression. NOC's expression patterns are unique among the known deadenylases, suggesting that it acts on specific target mRNAs that are important to the circadian system or in response to acute stimulation (Fig. 2). Because NOC does not have an obvious RNA-binding motif (outside of its substrate-binding catalytic pocket), we presume that it recognizes its targets via binding to RBPs. Deadenylation by NOC may result in destabilization or translational silencing of the message. In either case, NOC's activities would alter the protein composition of the cell. In the liver, this may help the clock shape the daily temporal profile in lipid storage or utilization.

Much still needs to be learned about the role of NOC in circadian posttranscriptional regulation and about the mechanism by which it functions. In particular, identification of NOC's mRNA targets in the different tissues in which it is expressed will be key. However, the identification of this rhythmic deadenylase provides an entrée into increased understanding of how the clock controls the protein expression patterns that drive the many physiological rhythms necessary for normal homeostasis.

ACKNOWLEDGMENTS

We thank Julie Baggs, Shuang Niu, Nicholas Douris, Patricia Preisig, Kris Curran, and Robert Alpern for sharing unpublished data.

REFERENCES

Anderson P. and Kedersha N. 2002. Stressful initiations. *J. Cell Sci.* **115:** 3227.

———. 2006. RNA granules. *J. Cell Biol.* **172:** 803.

Ayala-Ochoa A., Vargas-Suarez M., Loza-Tavera H., Leon P., Jimenez-Garcia L.F., and Sanchez-de-Jimenez E. 2004. In maize, two distinct ribulose 1,5-bisphosphate carboxylase/oxygenase activase transcripts have different day/night patterns of expression. *Biochimie* **86:** 439.

Baggs J.E. and Green C.B. 2003. Nocturnin, a deadenylase in *Xenopus laevis* retina: A mechanism for posttranscriptional control of circadian-related mRNA. *Curr. Biol.* **13:** 189.

———. 2006. Functional analysis of nocturnin: A circadian clock-regulated gene identified by differential display. *Methods Mol. Biol.* **317:** 243.

Baler R., Covington S., and Klein D.C. 1997. The rat arylalkylamine N-acetyltransferase gene promoter. cAMP activation via a cAMP-responsive element-CCAAT complex. *J. Biol. Chem.* **272:** 6979.

Barbot W., Wasowicz M., Dupressoir A., Versaux-Botteri C., and Heidmann T. 2002. A murine gene with circadian expression revealed by transposon insertion: Self-sustained rhythmicity in the liver and the photoreceptors. *Biochim. Biophys. Acta* **1576:** 81.

Belanger V., Picard N., and Cermakian N. 2006. The circadian regulation of Presenilin-2 gene expression. *Chronobiol. Int.* **23:** 747.

Cagampang F.R., Yang J., Nakayama Y., Fukuhara C., and Inouye S.T. 1994. Circadian variation of arginine-vasopressin messenger RNA in the rat suprachiasmatic nucleus. *Brain Res.* **24:** 179.

Cao D. and Parker R. 2001. Computational modeling of eukaryotic mRNA turnover. *RNA* **7:** 1192.

Cao S., Jiang L., Song S., Jing R., and Xu G. 2006. AtGRP7 is involved in the regulation of abscisic acid and stress responses in *Arabidopsis*. *Cell. Mol. Biol. Lett.* **11:** 526.

Carter D.A. and Murphy D. 1992. Nuclear mechanisms mediate rhythmic changes in vasopressin mRNA expression in the rat suprachiasmatic nucleus. *Brain Res.* **12:** 315.

Cheadle C., Fan J., Cho-Chung Y.S., Werner T., Ray J., Do L., Gorospe M., and Becker K.G. 2005. Stability regulation of mRNA and the control of gene expression. *Ann. N.Y. Acad. Sci.* **1058:** 196.

Chen J., Chiang Y.C., and Denis C.L. 2002. CCR4, a 3′-5′ poly(A) RNA and ssDNA exonuclease, is the catalytic component of the cytoplasmic deadenylase. *EMBO J.* **21:** 1414.

Chen Y., Hunter-Ensor M., Schotland P., and Sehgal A. 1998. Alterations of per RNA in noncoding regions affect periodicity of circadian behavioral rhythms. *J. Biol. Rhythms* **13:** 364.

Cheng H.Y., Papp J.W., Varlamova O., Dziema H., Russell B., Curfman J.P., Nakazawa T., Shimizu K., Okamura H., Impey S., and Obrietan K. 2007. microRNA modulation of circadian-clock period and entrainment. *Neuron* **54:** 813.

Cheng Y. and Hardin P.E. 1998. *Drosophila* photoreceptors contain an autonomous circadian oscillator that can function without period mRNA cycling. *J. Neurosci* **18:** 741.

Cheng Y., Gvakharia B., and Hardin P.E. 1998. Two alternatively spliced transcripts from the *Drosophila* period gene rescue rhythms having different molecular and behavioral characteristics. *Mol. Cell. Biol.* **18:** 6505.

Collins B.H., Rosato E., and Kyriacou C.P. 2004. Seasonal behavior in *Drosophila melanogaster* requires the photoreceptors, the circadian clock, and phospholipase C. *Proc. Natl. Acad. Sci.* **101:** 1945.

Colot H.V., Loros J.J., and Dunlap J.C. 2005. Temperature-modulated alternative splicing and promoter use in the circadian clock gene frequency. *Mol. Biol. Cell* **16:** 5563.

Crosthwaite S.K. 2004. Circadian clocks and natural antisense RNA. *FEBS Lett.* **567:** 49.

Diernfellner A.C., Schafmeier T., Merrow M.W., and Brunner M. 2005. Molecular mechanism of temperature sensing by the circadian clock of *Neurospora crassa*. *Genes Dev.* **19:** 1968.

Dlakic M. 2000. Functionally unrelated signalling proteins contain a fold similar to Mg^{2+}-dependent endonucleases. *Trends Biochem. Sci.* **25:** 272.

Dockendorff T.C., Su H.S., McBride S.M., Yang Z., Choi C.H., Siwicki K.K., Sehgal A., and Jongens T.A. 2002. *Drosophila* lacking dfmr1 activity show defects in circadian output and fail to maintain courtship interest. *Neuron* **34:** 973.

Donner B., Helmboldt-Caesar U., and Rensing L. 1985. Circadian rhythm of total protein synthesis in the cytoplasm and chloroplasts of *Gonyaulax polyedra*. *Chronobiol. Int.* **2:** 1.

Duffield G.E. 2003. DNA microarray analyses of circadian timing: The genomic basis of biological time. *J. Neuroendocrinol.* **15:** 991.

Dupressoir A., Barbot W., Loireau M.P., and Heidmann T. 1999. Characterization of a mammalian gene related to the yeast CCR4 general transcription factor and revealed by transposon insertion. *J. Biol. Chem.* **274:** 31068.

Dupressoir A., Morel A.P., Barbot W., Loireau M.P., Corbo L., and Heidmann T. 2001. Identification of four families of yCCR4- and Mg^{2+}-dependent endonuclease-related proteins in higher eukaryotes, and characterization of orthologs of yCCR4 with a conserved leucine-rich repeat essential for hCAF1/hPOP2 binding. *BMC Genomics* **2:** 9.

Elroy-Stein O. and Merrick W.C. 2007. Translation initiation via cellular internal ribosome entry sites. In *Translational control in biology and medicine* (ed. M.B. Mathews et al.), p. 155. Cold Spring Harbor Laboratory Press, Cold Spring Harbor, New York.

Eulalio A., Behm-Ansmant I., and Izaurralde E. 2007. P bodies: At the crossroads of post-transcriptional pathways. *Nat. Rev. Mol. Cell Biol.* **8:** 9.

Foulkes N.S., Borjigin J., Snyder S.H., and Sassone-Corsi P. 1997. Rhythmic transcription: The molecular basis of circadian melatonin synthesis. *Trends Neurosci.* **20:** 487.

Frisch B., Hardin P.E., Hamblen-Coyle M.J., Rosbash M., and Hall J.C. 1994. A promoterless period gene mediates behavioral rhythmicity and cyclical per expression in a restricted subset of the *Drosophila* nervous system. *Neuron* **12:** 555.

Fujimoto Y., Yagita K., and Okamura H. 2006. Does mPER2 protein oscillate without its coding mRNA cycling?: Post-transcriptional regulation by cell clock. *Genes Cells* **11:** 525.

Garbarino-Pico E., Niu S., Rollag M.D., Strayer C.A., Besharse J.C., and Green C.B. 2007. Immediate early response of the circadian polyA ribonuclease nocturnin to two extracellular stimuli. *RNA* **13:** 745.

Garceau N.Y., Liu Y., Loros J.J., and Dunlap J.C. 1997. Alternative initiation of translation and time-specific phosphorylation yield multiple forms of the essential clock protein FREQUENCY. *Cell* **89:** 469.

Garneau N.L., Wilusz J., and Wilusz C.J. 2007. The highways and byways of mRNA decay. *Nat. Rev. Mol. Cell Biol.* **8:** 113.

Gebauer F. and Hentze M.W. 2004. Molecular mechanisms of translational control. *Nat. Rev. Mol. Cell Biol.* **5:** 827.

Green C.B. and Besharse J.C. 1996a. Identification of a novel vertebrate circadian clock-regulated gene encoding the protein nocturnin. *Proc. Natl. Acad. Sci.* **93:** 14884.

———. 1996b. Use of a high stringency differential display screen for identification of retinal mRNAs that are regulated by a circadian clock. *Brain Res.* **37:** 157.

Green C.B., Douris N., Kojima S., Strayer C.A., Fogerty J., Lourim D., Keller S.R., and Besharse J.C. 2007. Loss of Nocturnin, a circadian deadenylase, confers resistance to hepatic steatosis and diet-induced obesity. *Proc. Natl. Acad. Sci.* **104:** 9888.

Gutierrez R.A., Ewing R.M., Cherry J.M., and Green P.J. 2002. Identification of unstable transcripts in *Arabidopsis* by cDNA microarray analysis: Rapid decay is associated with a group of touch- and specific clock-controlled genes. *Proc. Natl. Acad. Sci.* **99:** 11513.

Hardin P.E., Hall J.C., and Rosbash M. 1992. Circadian oscillations in period gene mRNA levels are transcriptionally regulated. *Proc. Natl. Acad. Sci.* **89:** 11711.

Harmer S.L., Hogenesch J.B., Straume M., Chang H.S., Han B., Zhu T., Wang X., Kreps J.A., and Kay S.A. 2000. Orchestrated transcription of key pathways in *Arabidopsis* by the circadian clock. *Science* **290:** 2110.

Harms E., Kivimae S., Young M.W., and Saez L. 2004. Posttranscriptional and posttranslational regulation of clock genes. *J. Biol. Rhythms* **19:** 361.

Hastings, J.W. 2001. Cellular and molecular mechanisms of circadian regulation in the unicellular dinoflagellate *Gonyaulax polyedra*. In *Handbook of behavioral neurobiology: Circadian clocks* (ed. J.S. Takahashi et al.), vol. 12, p. 321-334. Plenum Press, New York.

Heintzen C., Nater M., Apel K., and Staiger D. 1997. AtGRP7, a nuclear RNA-binding protein as a component of a circadian-regulated negative feedback loop in *Arabidopsis thaliana*. *Proc. Natl. Acad. Sci.* **94:** 8515.

Iliev D., Voytsekh O., Schmidt E.M., Fiedler M., Nykytenko A., and Mittag M. 2006. A heteromeric RNA-binding protein is involved in maintaining acrophase and period of the circadian clock. *Plant Physiol.* **142:** 797.

Iwamoto M., Higo H., and Higo K. 2000. Differential diurnal expression of rice catalase genes: The 5′-flanking region of CatA is not suficient for circadian control. *Plant Sci.* **151:** 39.

Jackson R.J. 2005. Alternative mechanisms of initiating translation of mammalian mRNAs. *Biochem. Soc. Trans.* **33:** 1231.

Keene J.D. 2007. RNA regulons: Coordination of post-transcriptional events. *Nat. Rev. Genet.* **8:** 533.

Kiebler M.A. and Bassell G.J. 2006. Neuronal RNA granules: Movers and makers. *Neuron* **51:** 685.

Kim J.Y., Song H.R., Taylor B.L., and Carre I.A. 2003. Light-regulated translation mediates gated induction of the *Arabidopsis* clock protein LHY. *EMBO J.* **22:** 935.

Kim T.D., Woo K.C., Cho S., Ha D.C., Jang S.K., and Kim K.T. 2007. Rhythmic control of AANAT translation by hnRNP Q in circadian melatonin production. *Genes Dev.* **21:** 797.

Kim T.D., Kim J.S., Kim J.H., Myung J., Chae H.D., Woo K.C., Jang S.K., Koh D.S., and Kim K.T. 2005. Rhythmic serotonin N-acetyltransferase mRNA degradation is essential for the maintenance of its circadian oscillation. *Mol. Cell. Biol.* **25:** 3232.

Kojima S., Hirose M., Tokunaga K., Sakaki Y., and Tei H. 2003. Structural and functional analysis of 3′ untranslated region of mouse Period1 mRNA. *Biochem. Biophys. Res. Commun.* **301:** 1.

Kojima S., Matsumoto K., Hirose M., Shimada M., Nagano M., Shigeyoshi Y., Hoshino S., Ui-Tei K., Saigo K., Green C.B., Sakaki Y., and Tei H. 2007. LARK activates posttranscriptional expression of an essential mammalian clock protein, PERIOD1. *Proc. Natl. Acad. Sci.* **104:** 1859.

Kornmann B., Schaad O., Bujard H., Takahashi J.S., and Schibler U. 2007. System-driven and oscillator-dependent circadian transcription in mice with a conditionally active liver clock. *PLoS Biol.* **5:** e34.

Kramer C., Loros J.J., Dunlap J.C., and Crosthwaite S.K. 2003. Role for antisense RNA in regulating circadian clock function in *Neurospora crassa*. *Nature* **421:** 948.

Kuhn U. and Wahle E. 2004. Structure and function of poly(A) binding proteins. *Biochim. Biophys. Acta* **1678:** 67.

Kwak E., Kim T.D., and Kim K.T. 2006. Essential role of 3′-untranslated region-mediated mRNA decay in circadian oscillations of mouse Period3 mRNA. *J. Biol. Chem.* **281:** 19100.

Lidder P., Gutierrez R.A., Salome P.A., McClung C.R., and Green P.J. 2005. Circadian control of messenger RNA stability. Association with a sequence-specific messenger RNA decay pathway. *Plant Physiol.* **138:** 2374.

Liu Q., Greimann J.C., and Lima C.D. 2006. Reconstitution, activities, and structure of the eukaryotic RNA exosome. *Cell* **127:** 1223.

Liu X. and Green C.B. 2001. A novel promoter element, photoreceptor conserved element II, directs photoreceptor-specific expression of nocturnin in *Xenopus laevis*. *J. Biol. Chem.* **276:** 15146.

———. 2002. Circadian regulation of nocturnin transcription by phosphorylated CREB in *Xenopus* retinal photoreceptor cells. *Mol. Cell. Biol.* **22:** 7501.

Liu Y., Tsinoremas N.F., Johnson C.H., Lebedeva N.V., Golden S.S., Ishiura M., and Kondo T. 1995. Circadian orchestration of gene expression in cyanobacteria. *Genes Dev.* **9:** 1469.

Lowrey P.L. and Takahashi J.S. 2004. Mammalian circadian biology: Elucidating genome-wide levels of temporal organization. *Annu. Rev. Genomics Hum. Genet.* **5:** 407.

Majercak J., Chen W.F., and Edery I. 2004. Splicing of the period gene 3′-terminal intron is regulated by light, circadian clock factors, and phospholipase C. *Mol. Cell. Biol.* **24:** 3359.

Mangus D.A., Evans M.C., and Jacobson A. 2003. Poly(A)-binding proteins: Multifunctional scaffolds for the post-transcriptional control of gene expression. *Genome Biol.* **4:** 223.

Marchler-Bauer A., Anderson J.B., Derbyshire M.K., DeWeese-Scott C., Gonzales N.R., Gwadz M., Hao L., He S., Hurwitz D.I., Jackson J.D., Ke Z., Krylov D., Lanczycki C.J., Liebert C.A., Liu C., Lu F., Lu S., Marchler G.H., Mullokandov M., Song J.S., Thanki N., Yamashita R.A., Yin J.J., Zhang D., and Bryant S.H. 2007. CDD: A conserved domain database for interactive domain family analysis. *Nucleic Acids Res.* (database issue). **35:** D237.

Markovic P., Roenneberg T., and Morse D. 1996. Phased protein synthesis at several circadian times does not change protein levels in *Gonyaulax*. *J. Biol. Rhythms* **11:** 57.

Mata J., Marguerat S., and Bahler J. 2005. Post-transcriptional control of gene expression: A genome-wide perspective. *Trends Biochem. Sci.* **30:** 506.

Mathews M.B., Sonenberg N., and Hershey J.W.B., Eds. 2007. *Translational control in biology and medicine*. Cold Spring Harbor Laboratory Press, Cold Spring Harbor, New York.

McNeil G.P., Zhang X., Genova G., and Jackson F.R. 1998. A molecular rhythm mediating circadian clock output in *Drosophila*. *Neuron* **20:** 297.

Michael T.P. and McClung C.R. 2003. Enhancer trapping reveals widespread circadian clock transcriptional control in *Arabidopsis*. *Plant Physiol.* **132:** 629.

Millar A.J. and Kay S.A. 1991. Circadian control of *cab* gene transcription and mRNA accumulation in *Arabidopsis*. *Plant Cell* **3:** 541.

Mittag M. 1996. Conserved circadian elements in phylogenetically diverse algae. *Proc. Natl. Acad. Sci.* **93:** 14401.

———. 2003. The function of circadian RNA-binding proteins and their *cis*-acting elements in microalgae. *Chronobiol. Int.* **20:** 529.

Mittag M., Lee D.H., and Hastings J.W. 1994. Circadian expression of the luciferin-binding protein correlates with the binding of a protein to the 3′ untranslated region of its mRNA. *Proc. Natl. Acad. Sci.* **91:** 5257.

Mohr E. and Richter D. 2004. Subcellular vasopressin mRNA trafficking and local translation in dendrites. *J. Neuroendocrinol.* **16:** 333.

Nagoshi E., Brown S.A., Dibner C., Kornmann B., and Schibler U. 2005. Circadian gene expression in cultured cells. *Methods Enzymol.* **393:** 543.

Newby L.M. and Jackson F.R. 1993. A new biological rhythm mutant of *Drosophila melanogaster* that identifies a gene with an essential embryonic function. *Genetics* **135:** 1077.

———. 1996. Regulation of a specific circadian clock output pathway by lark, a putative RNA-binding protein with repressor activity. *J. Neurobiol.* **31:** 117.

Nishii K., Yamanaka I., Yasuda M., Kiyohara Y.B., Kitayama Y., Kondo T., and Yagita K. 2006. Rhythmic post-transcriptional regulation of the circadian clock protein mPER2 in mammalian cells: A real-time analysis. *Neurosci. Lett.* **401:** 44.

Parker R. and Sheth U. 2007. P bodies and the control of mRNA translation and degradation. *Mol. Cell* **25:** 635.

Parker R. and Song H. 2004. The enzymes and control of eukaryotic mRNA turnover. *Nat. Struct. Mol. Biol.* **11:** 121.

Pilgrim M.L., Caspar T., Quail P.H., and McClung C.R. 1993. Circadian and light-regulated expression of nitrate reductase in *Arabidopsis*. *Plant Mol. Biol.* **23:** 349.

Pillai R.S., Bhattacharyya S.N., and Filipowicz W. 2007. Repression of protein synthesis by miRNAs: How many mechanisms? *Trends Cell Biol.* **17:** 118.

Raghavan A. and Bohjanen P.R. 2004. Microarray-based analyses of mRNA decay in the regulation of mammalian gene expression. *Brief. Funct. Genomics Proteomics* **3:** 112.

Reddy A.B., Karp N.A., Maywood E.S., Sage E.A., Deery M., O'Neill J.S., Wong G.K., Chesham J., Odell M., Lilley K.S., Kyriacou C.P., and Hastings M.H. 2006. Circadian orchestration of the hepatic proteome. *Curr. Biol.* **16:** 1107.

Robinson B.G., Frim D.M., Schwartz W.J., and Majzoub J.A. 1988. Vasopressin mRNA in the suprachiasmatic nuclei: Daily regulation of polyadenylate tail length. *Science* **241:** 342.

Sato T.K., Panda S., Kay S.A., and Hogenesch J.B. 2003. DNA arrays: Applications and implications for circadian biology. *J. Biol. Rhythms* **18:** 96.

Schaffer R., Landgraf J., Accerbi M., Simon V., Larson M., and Wisman E. 2001. Microarray analysis of diurnal and circadian-regulated genes in *Arabidopsis*. *Plant Cell* **13:** 113.

Schibler U., Ripperger J., and Brown S.A. 2003. Peripheral circadian oscillators in mammals: Time and food. *J. Biol. Rhythms* **18:** 250.

Schroeder A.J., Genova G.K., Roberts M.A., Kleyner Y., Suh J., and Jackson F.R. 2003. Cell-specific expression of the lark RNA-binding protein in *Drosophila* results in morphological and circadian behavioral phenotypes. *J. Neurogenet.* **17:** 139.

Schwartz D.C. and Parker R. 2000. Interaction of mRNA translation and mRNA degradation in *Saccharocmyces cerevisiae*. In *Translational control of gene expression* (ed. N. Sonenberg et al.), p. 807. Cold Spring Harbor Laboratory Press, Cold Spring Harbor, New York.

Shu Y. and Hong-Hui L. 2004. Transcription, translation, degradation, and circadian clock. *Biochem. Biophys. Res. Commun.* **321:** 1.

So W.V. and Rosbash M. 1997. Post-transcriptional regulation contributes to *Drosophila* clock gene mRNA cycling. *EMBO J.* **16:** 7146.

Staiger D., Zecca L., Wieczorek Kirk D.A., Apel K., and Eckstein L. 2003. The circadian clock regulated RNA-binding protein AtGRP7 autoregulates its expression by influencing alternative splicing of its own pre-mRNA. *Plant J.* **33:** 361.

Stanewsky R., Lynch K.S., Brandes C., and Hall J.C. 2002. Mapping of elements involved in regulating normal temporal period and timeless RNA expression patterns in *Drosophila melanogaster*. *J. Biol. Rhythms* **17:** 293.

Tucker M., Staples R.R., Valencia-Sanchez M.A., Muhlrad D., and Parker R. 2002. Ccr4p is the catalytic subunit of a Ccr4p/Pop2p/Notp mRNA deadenylase complex in *Saccharomyces cerevisiae*. *EMBO J.* **21:** 1427.

Uhl G.R. and Reppert S.M. 1986. Suprachiasmatic nucleus vasopressin messenger RNA: Circadian variation in normal and Brattleboro rats. *Science* **232:** 390.

Valencia-Sanchez M.A., Liu J., Hannon G.J., and Parker R. 2006. Control of translation and mRNA degradation by miRNAs and siRNAs. *Genes Dev.* **20:** 515.

Waltenberger H., Schneid C., Grosch J.O., Bareiss A., and Mittag M. 2001. Identification of target mRNAs for the clock-controlled RNA-binding protein Chlamy 1 from *Chlamydomonas reinhardtii*. *Mol. Genet. Genomics* **265:** 180.

Wang Y., Osterbur D.L., Megaw P.L., Tosini G., Fukuhara C., Green C.B., and Besharse J.C. 2001. Rhythmic expression of Nocturnin mRNA in multiple tissues of the mouse. *BMC Dev. Biol.* **1:** 9.

Wilsbacher L.D., Yamazaki S., Herzog E.D., Song E.J., Radcliffe L.A., Abe M., Block G., Spitznagel E., Menaker M., and Takahashi J.S. 2002. Photic and circadian expression of luciferase in mPeriod1-luc transgenic mice in vivo. *Proc. Natl. Acad. Sci.* **99:** 489.

Wilusz C.J. and Wilusz J. 2004. Bringing the role of mRNA decay in the control of gene expression into focus. *Trends Genet.* **20:** 491.

Wilusz C.J., Wormington M., and Peltz S.W. 2001. The cap-to-tail guide to mRNA turnover. *Nat. Rev. Mol. Cell Biol.* **2:** 237.

Woelfle M.A. and Johnson C.H. 2006. No promoter left behind: Global circadian gene expression in cyanobacteria. *J. Biol. Rhythms* **21:** 419.

Yamamoto Y., Yagita K., and Okamura H. 2005. Role of cyclic mPer2 expression in the mammalian cellular clock. *Mol. Cell. Biol.* **25:** 1912.

Zhao B., Schneid C., Iliev D., Schmidt E.M., Wagner V., Wollnik F., and Mittag M. 2004. The circadian RNA-binding protein CHLAMY 1 represents a novel type heteromer of RNA recognition motif and lysine homology domain-containing subunits. *Eukaryot. Cell* **3:** 815.

Biological Clocks and the Coordination Theory of RNA Operons and Regulons

J.D. Keene

Department of Molecular Genetics and Microbiology, Duke University Medical Center, Durham, North Carolina 27710

One of the regulatory models of circadian rhythms involves the oscillation of transcription and translation. Although transcription factors have been widely examined during circadian processes, posttranscriptional mechanisms are less well-studied. Several laboratories have used microarrays to detect changes in mRNA expression throughout the circadian cycle and have found that mRNAs encoding the RNA-binding proteins (RBPs) nocturnin and butyrate response factor (BRF1) undergo rhythmic changes. Nocturnin is a deadenylation enzyme that removes poly(A) from the 3′ ends of mRNAs, whereas BRF1 destabilizes mRNAs encoding early response gene (ERG) transcripts that contain AU-rich sequences in their 3′-untranslated regions (UTRs). Moroni and coworkers proposed that BRF1 functions as an oscillating posttranscriptional RNA operon (PTRO) that diurnally degrades ERG transcripts in peripheral organs (Keene and Tenenbaum 2002; Benjamin et al. 2006). The PTRO model posits that mRNAs can be members of one or more discrete functionally related subsets of mRNAs as determined by *cis* elements in mRNA and *trans*-acting RBPs or microRNAs that collectively recognize these *cis* elements (Keene 2007). This chapter describes the basis of posttranscriptional coordination by RNA operons and their potential for horizontal transfer among cells and discusses the potential for RBPs and microRNAs to participate in coordinating circadian rhythms and other biological clocks.

INTRODUCTION

In bacterial and eukaryotic cells, gene expression is regulated at both the transcriptional and translational levels. However, in bacteria, unlike that of eukaryotes, the transcription apparatus is directly coupled to the translation apparatus. Thus, bacterial transcripts begin translation even before transcription is completed because the ribosomes initiate translation on nascent transcripts (Beckwith 1996). Eukaryotic transcription and translation cannot be directly coupled because the nuclear membrane separates the chromosomes from the ribosomes requiring RNA processing and export between transcription and translation (Keene 2001; Maniatis and Reed 2002). Recent findings suggest that the steps between transcription and translation in eukaryotes are interconnected and that each step in the pathway is physically connected to the next (Fig. 1). Thus, as transcripts are transferred from the DNA template to the spliceosome, the splicing apparatus itself is coupled to the nuclear export machinery, and following export, mature transcripts are transported to cytoplasmic locations where their final fates are determined (for review, see Maniatis and Reed 2002). mRNAs contain sequence elements that help determine their fates both during and after export to the cytoplasm (Richter 1997; Keene 1999, 2001, 2003). Therefore, although the molecular interconnectivity of the steps between transcription and translation in eukaryotes provides indirect coupling of RNA-processing events, each mRNA and its bound proteins is assumed to function as a free agent. But is there coordination of groups of mRNAs in parallel along this pathway? The PTRO theory is based on evidence suggesting that the pathway of RNA processing is indeed coordinated and has implications for mechanisms of biological timing such as those involved in circadian rhythms and other biological clocks (Keene and Tenenbaum 2002; Keene 2007).

COUPLING AND COORDINATION OF GENE EXPRESSION FROM TRANSCRIPTION TO TRANSLATION

Control of gene expression at the level of transcription in mammalian cells appears to be regulated by transcription factors that bind specifically to multiple promoter elements (Orphanides and Reinberg 2002; Kosak and Groudine 2004). It is assumed that gene expression is coordinated by this mechanism, but there is not abundant experimental evidence to directly support this assumption. Models of transcriptional regulation state that multiple genes are activated as functional groups because they use common promoters, but the activation of transcription appears to occur in domains or neighborhoods where chromatin is modified to allow access to RNA polymerase and the transcription apparatus. For example, "ectopic expression" of genes that are physically located near one another in neighborhoods results from the decondensation of chromatin that makes genes accessible to activation (Rodriguez-Trelles et al. 2005; Aten and Kanaar 2006). Several studies have demonstrated that these ectopically expressed genes do not necessarily have related functions, and thus, the physical proximity of coexpressed genes does not correspond to the functional relatedness of those genes (Spellman and Rubin 2002; Lercher et al. 2004; Yanai et al. 2006). Thus, ectopically coexpressed genes are by definition temporally coordinated, but this has been interpreted to result as much from "leakage" of transcription as it does from mechanistic coregulation (Rodriguez-Trelles et al. 2005; Yanai et al.

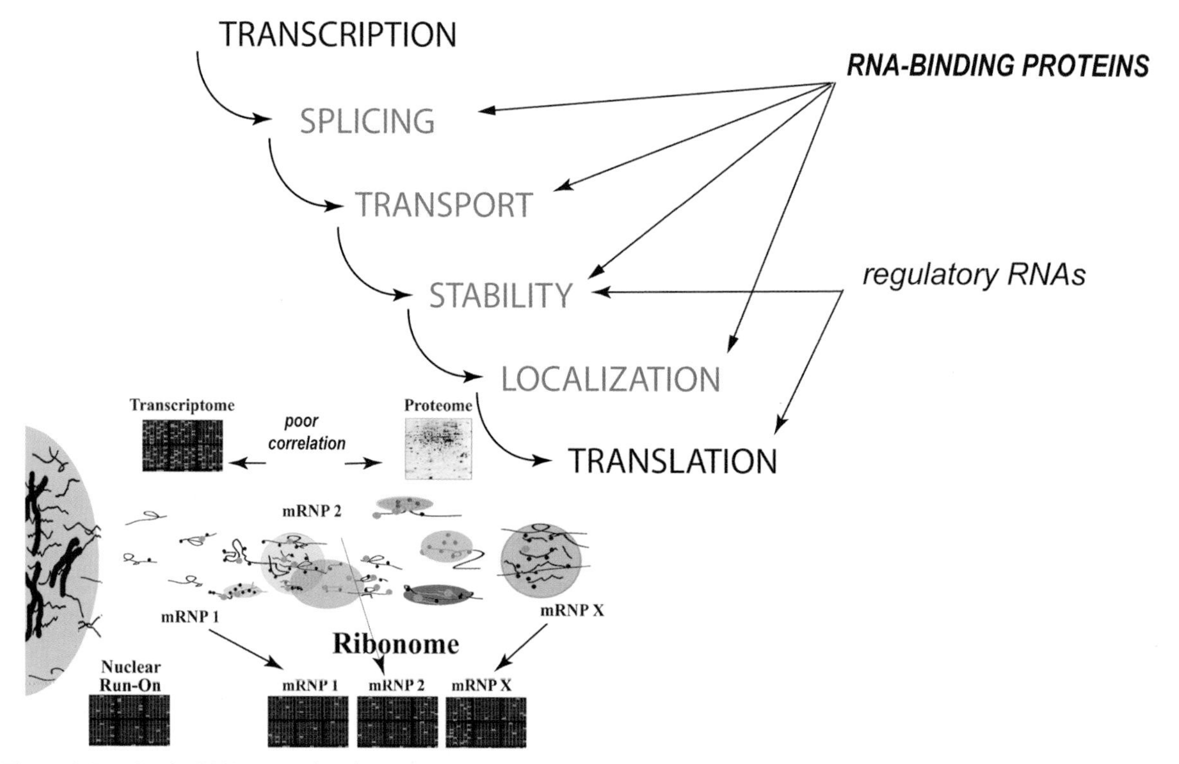

Figure 1. Levels of mRNA processing that are responsible for the interconnectivity of transcription to translation in eukaryotic cells that are regulated by RBPs and small regulatory RNAs such as microRNAs (Keene 2001, 2007). These processes are interconnected and coordinated by a dynamic RNP-driven process of regulation of multiple mRNAs in parallel. (*Inset*) Ribonome (the cellular ribonucleoprotein infrastructure) and the RIP-Chip approach of using microarray analysis of enriched RNPs and their associated mRNA subsets. A full analysis of the structural and functional relationships among mRNAs identified using RIP-Chip together with other informatics methods is termed ribonomic analysis (Tenenbaum et al. 2000, 2002, 2003).

2006). But the coordination of expression of functionally related groups of human genes is assumed to orchestrate complex biochemical pathways and developmental events such as "synexpression" (Niehrs and Pollet 1999). In fact, it is not known precisely how gene expression is coordinated in mammalian and other eukaryotic cells, but it is likely important for the proper performance of all cellular and developmental processes as well biological clocks.

REGULATION OF GENE EXPRESSION BY RNA-BINDING PROTEINS AND MICRORNAS

Although the assumption of a transcriptional basis for gene coordination in eukaryotes is widely assumed by most biologists, it has recently been discovered that posttranscriptional events as well can be coordinated (Keene 2007). Moreover, coordinated posttranscriptional regulation is important because these late events in the gene expression pathway dictate the final decisions of protein production regardless of how precisely transcription itself is coordinated (Fig. 1) (Keene 2001, 2007; Keene and Tenenbaum 2002). Indeed, evidence suggests that RBPs are multitargeted (Gao et al. 1994; Tenenbaum et al. 2000; for review, see Keene 2007) and such targeting has also been proposed more recently for microRNAs and possibly other small noncoding RNAs (Lewis et al. 2003, 2005; Bartel 2004).

MULTITARGETING OF MRNAS BY RBPS

One of the key requirements of coordinated gene expression is that the mRNA transcripts that express proteins are capable of being coregulated. For example, Gao et al. (1994) provided early evidence that RBPs may be targeted to multiple brain mRNAs in vitro. This idea of multitargeting was unique in that the mRNAs found to interact with ELAV/Hu proteins were believed to have related functions and potentially to respond together in neuronal or immune cells as ERG transcripts (Keene 1999; Brennan and Steitz 2001). These mRNAs were known for many years to contain a characteristic set of sequence elements that were AU-rich (ARE), and these genes encoded such proteins as c-Myc, c-Fos, interleukins, interferons, and many otherwise "dangerous" proteins that mammalian cells presumably need to keep under tight control (Shaw and Kamen 1986). A signature sequence of the ARE was the pentameric AUUUA or the larger but less-frequent seven-nucleotide consensus sequence UAUUUAU, but these elements were essentially U-rich and scattered with A and G residues (Levine et al. 1993; Gao et al. 1994). Most importantly, these regions of sequence similarity were never identical among these mRNAs but again represented a sequence of similar character. The lack of sequence identity made it exceedingly difficult to globally identify such elements among mRNAs using computational methods (Keene 2001). In essence, these early experiments were designed to allow the ELAV/Hu RBPs, rather than the computer, to inform us of their targeting specificity for multiple mRNAs. Thus, the demonstration by Gao et al. (1994) that the ELAV/HuB protein could bind to these A/G-U-rich sequences en masse led to the suggestion that functional consequences of such binding may coordinate functional outcomes by the ELAV/HuB RBPs in neural tissues. However, in vitro approaches do not address the key question of intracellular context of the protein and the RNAs with which it may be able to interact, and this rationale represents an ongoing experimental and technical effort in our laboratory and the field to develop methods that reveal these interactions in a functional cellular context.

The discovery and functional studies of the human ELAV/Hu RBP family followed from studies in *Drosophila* where Kalpana White and coworkers discovered the *elav* locus and demonstrated that the ELAV proteins were essential for neuronal differentiation in flies (Robinow et al. 1988). Complementary DNAs encoding four human neuronal ELAV/Hu proteins (HuA [HuR], HuB, HuC, and HuD) were subsequently cloned in the Furneaux and Keene laboratories (Szabo et al. 1991; Levine et al. 1993; King et al. 1994; Ma et al. 1996). As noted above, the RNA-binding specificity of ELAV/Hu family proteins was demonstrated for ARE-type sequences found in the proto-oncogene and cytokine ERG mRNAs (Levine et al. 1993; Gao et al. 1994). HuB protein was found to have a binding preference for short stretches of uridylates flanked by A or G, and sometimes even C residues (Levine et al. 1993; Gao et al. 1994). In addition, HuB was found to stabilize an endogenous ARE-containing target mRNA that encodes the glucose transporter 1 protein (GLUT1) in adipocytes and, at the same time, to increase its translation by 20–50-fold (Jain et al. 1997). Subsequently, all four of the Hu family proteins, including the ubiquitously expressed HuR protein (also called HuA and *ELAVl-1*), were found to bind similar ARE sequence elements and to stabilize reporter mRNAs containing AREs (for review, see Keene 1999; Brennan and Steitz 2001). In addition to the HuB activation of GLUT1 mRNA translation, HuB translational activation of neurofilament M (Antic et al. 1999), HuR translational activation of p53 (Mazan-Mamczarz et al. 2003), and HuR translational activation of the SIRT1 mRNA (Abdelmohsen et al. 2007) were subsequently demonstrated in the absence of effects on RNA stability. In many other cases, effects on RNA stability with subsequent effects on translation were observed (for review, see Brennan and Steitz 2001; Keene 2001).

MULTITARGETING OF mRNAs by microRNAs

MicroRNAs have also been suggested to act through a multitargeting mode (Lewis et al. 2003, 2005). The data in support of multitargeting by microRNAs were initially based on computational predictions using algorithms such as TargetScan, PicTar, and MirAnda, each of which has its own advantages and disadvantages. In general, however, all of these methods examine microRNAs and potential mRNA targets in large databases for sequence complementarity. The basis of the homology search is the conservation of microRNA-targeted mRNA sequences of approximately six to eight nucleotides that potentially bind to "seed" sequences (Lewis et al. 2003, 2005). This very small region of homology is restricted by searching databases for the conservation of the mRNA complementary sequence among species, thus improving the potential of the sequence match having biological significance. More recently, these methods were refined to consider other determinants of microRNA function on 3′UTR sequence elements and to provide a prediction model that quantitatively evaluates the context of the microRNA–mRNA interactions (Grimson et al. 2007). Subsequent attempts to directly demonstrate multitargeting have met with some degree of success, but issues regarding RNA structural constraints, "off-target effects," and cellular context remain to be understood. Thus, whether one is concerned with RBPs or microRNAs, it is essential to understand the in vivo intracellular context of the RNA–protein and RNA–RNA interactions. For this purpose, we devised methods to isolate ribonucleoprotein (RNP) complexes from cell extracts and to identify proteins and RNAs that associate with them (Fig. 2) (Keene et al. 2006).

Discrete subpopulations of mRNAs reside in RNP complexes and can be coordinately processed: spliced, transported, stabilized or degraded, localized, or translated into protein. All posttranscriptional steps involve the proper functioning of RBPs as members of these RNP complexes (see mRNP1, mRNP2, mRNPx in Fig. 1 inset). Our laboratory devised a method termed "RIP-Chip" or "RIP-on-Chip" (*R*NP-*I*mmuno*p*recipitation-micro*Chip*) and found that RBPs can interact with distinct subsets of the mRNAs that together encode functionally related proteins (Fig. 2) (Tenenbaum et al. 2000; for review, see Hieronymus and Silver 2004; Keene and

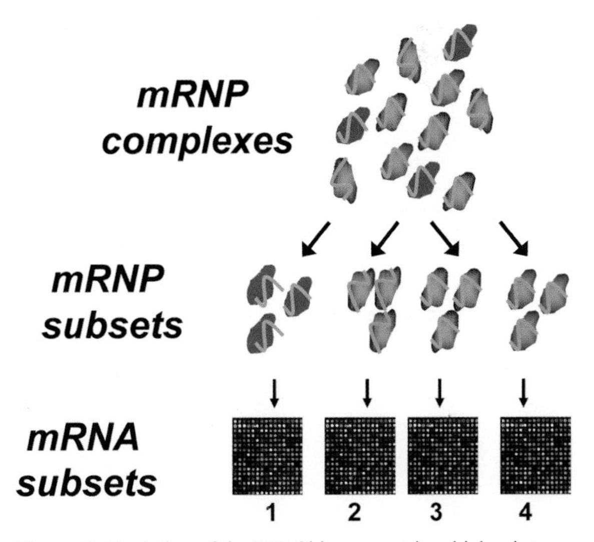

Figure 2. Depiction of the RIP-Chip concept in which mixtures of mRNPs are isolated from cell extracts and partitioned using antibodies or epitope tags that bind to proteins in the mRNP complexes. RNPs are released and the extracted RNAs are identified using microarrays or high-throughput sequencing. Small noncoding RNAs such as microRNAs that are associated with RNPs can be identified also using RIP-Chip (Keene et al. 2006).

Lager 2005; Keene et al. 2006). Data from many laboratories indicate that RBPs in yeast, flies, and mammals can function together to form gene expression networks regulating multiple functionally related mRNAs (for review, see Keene and Tenenbaum 2002; Hieronymus and Silver 2004; Keene and Lager 2005; Moore 2005; Keene 2007). Similar approaches in our laboratory also have indicated that specific subpopulations of microRNAs can be identified in association with the ELAV/HuR protein in human Jurkat cells (P.J. Lager and J.D. Keene, unpubl.).

RIBONOMIC ANALYSIS AND RIP-CHIP PROVIDE INSIGHTS INTO MULTITARGETING OF mRNAs

RIP-Chip is useful for reducing the complexity of transcriptomic data and at the same time allowing the simultaneous assessment of posttranscriptional regulation of multiple genes with the goal of better understanding the intracellular organization of mRNA transcripts (Figs. 1 and 2) (Tenenbaum et al. 2000, 2002). This has been called "ribonomics," because it allows one to determine the natural clustering of transcripts by RBPs en masse. The ribonomic approach is based on the use of RIP-Chip for purification of RNP complexes, followed by the identification of the mRNAs that are associated with the complexes on a microarray or by direct sequencing. The ribonomic analytical methods developed by Tenenbaum et al. (2000) consist of four major steps that are intended to derive relevant ribonomic information from RIP-Chip data. The association of proteins, mRNAs, and microRNAs with multiple mRNP complexes forms the "ribonome," and as described below, this kind of analysis provides functional information regarding the mRNA population of the cell (Fig. 3) (Brown et al. 2001; Eystathioy et al. 2002; Keene and Tenenbaum 2002; Hieronymus and Silver 2003).

The first step in the general ribonomic procedure is to isolate mRNPs using specific molecular tags, antibody epitopes, or other ligands that are associated with the mRNP complex using the RIP-Chip procedure (Fig. 2) (Tenenbaum et al. 2002, 2003; Keene et al. 2006). The conditions of the isolation can be optimized biochemically so that peripheral components of the mRNP complex are removed while leaving the RBP still associated with the mRNA targets in the RNP complex. If necessary, the RBP and its RNA targets can be cross-linked using a chemical such as formaldehyde or UV light, but this can introduce other complicating issues that are not addressed here for lack of space (see Penalva et al. 2004; Keene et al. 2006). In the second step of the general ribonomics procedure, the mRNA(s) in the isolated mRNP complex is identified. There are several methods available to identify the mRNAs including (1) the generation and sequencing of cDNA libraries from the isolated mRNAs, (2) the generation and sequencing of reverse transcription–polymerase chain reaction (RT-PCR) products, (3) specific probe RNase protection assays, or (4) microarray analysis. High-throughput sequencing procedures such as the "4-5-4" sequencing procedure that is a form of "mas-

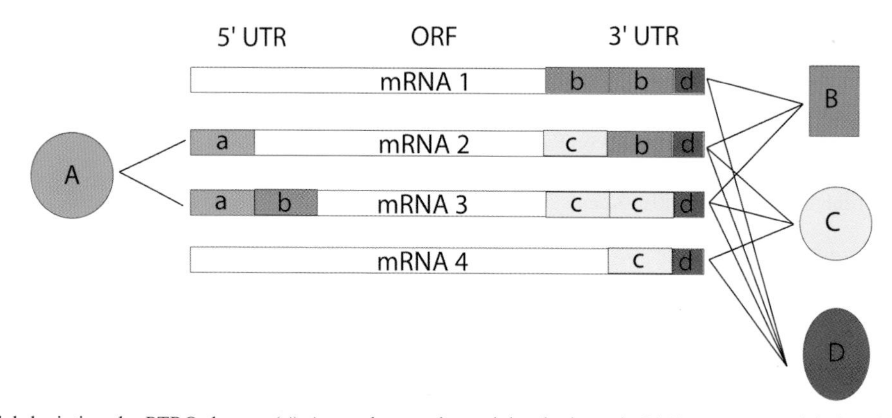

Figure 3. Model depicting the PTRO theory. (*A*) An analogy to bacterial polycistronic DNA operons and their collinear mRNAs is suggested, and it was proposed that in eukaryotes (*B*) monocistronic mRNAs are regulated by multiple RBPs and microRNAs to activate or repress distinct groups of mRNAs. The model accounts for multilevel regulation of functionally related groups of mRNAs in different combinations by *trans*-acting factors. This allows each mRNA to be a member of more than one RNA operon so that the encoded proteins can be produced coordinately in different combinations via distinct UTR codes (Keene and Tenenbaum 2002). (Reprinted, with permission, from Keene and Tenenbaum 2002 [© Elsevier].)

sively parallel sequential sequencing" (MPSS) or "single-molecule platform sequencing" (SMPS) can be used directly to provide complete and unbiased sequence evaluation of RIP-Chip-derived RNAs.

The third and fourth steps of the general ribonomics procedure involve determining the biological significance of the experimental outcome. These steps are more complex and require solving problems that can be technically challenging. If there are multiple targets observed in step two, the third step is used to determine which features that the RNAs found in the RNP subpopulation have in common. For example, is there a recognizable binding site for the RBP of interest (e.g., an ARE-type)? As described above, functionally similar mRNAs may have similar but not identical sequences in common for binding to RBPs, and these sequences are likely to be semiconserved among mRNAs in order for them to interact with the same *trans*-acting RBP. Tenenbaum et al. (2000) found that mRNAs identified by microarrays following immunoprecipitation of the ELAV/HuB protein contained ARE sequences of similar character that resemble known consensus RNA-binding sequences for ELAV/Hu proteins. This consensus was confirmed but further described within the context of a secondary structure by Lopez de Silanes et al. (2004) using RIP-Chip. Similar findings involving both primary and secondary mRNA structures were reported for the Fragile-X mental retardation protein by Brown et al. (2001) and for mRNAs bound to nuclear export proteins in yeast by Hieronymus and Silver (2003). Two of the most striking recent examples of subsets of mRNAs that associate with Pumilio RBPs are those that encode macromolecular structures such as the mitochondrion (Gerber et al. 2004) or the vacuolar proton-translocating V-type ATPase complex (Gerber et al. 2006), as well as multiple components that regulate anterioposterior patterning of *Drosophila* embryos (Gerber et al. 2006). Many other examples have been summarized by Keene and Lager (2005; Keene 2007). More recently, RIP-Chip was used with the histone mRNA stem-loop binding protein (SLBP) and demonstrated to interact with approximately 30 histone mRNAs with high specificity (Townley-Tilson et al. 2006). Five of the key histones involved in cell-cycle-specific coordination of chromatin reformation during S phase were temporarily associated in a manner that suggests a distinct PTRO (Fig. 4).

The fourth step is the most difficult of all because it is concerned with the functional relationships among the proteins encoded by the mRNAs that are identified. This is the most biologically important information to be obtained from the procedure and it may lead to a useful systems biological outcome. The best approach is to use various gene ontogeny (GO) databases such as Panther to search for

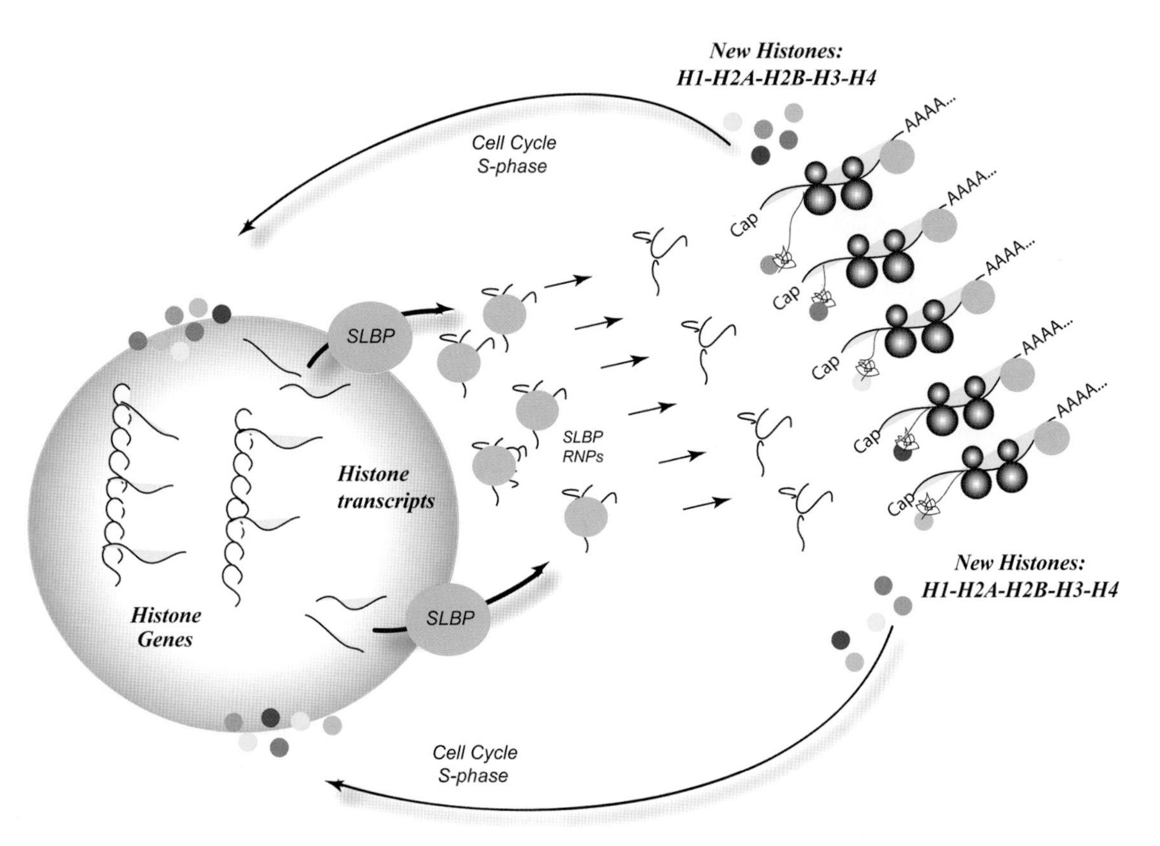

Figure 4. Depiction of the coordinated production of replication-dependent histone proteins H1-H2A-H2B and H4 encoded by a subset of mRNAs that are coregulated by the mammalian stem-loop-binding protein (SLBP) (Townley-Tilson et al. 2006). The histone genes are dispersed across the genome, but the SLBP binds to a 3′UTR element in the histone transcripts and coordinates the translation of the histone mRNAs. The mRNA subset that is bound to SLBP functions as a PTRO for temporal coordination of histones together with DNA synthesis and chromatin assembly during the cell cycle.

functional relationships and to then expand the analysis by linking multiple functional groupings together to visualize a potential RNA operon or higher-order regulon (Fig. 3) (Keene and Tenenbaum 2002; Keene 2007). The principle of a ribonomics analysis is that the association of a class of mRNAs with an RNP complex or an RBP serves as an indication that the mRNAs function together as a group. Cells may use this natural clustering of mRNAs to regulate the production of proteins that function together in a pathway or coordinate the synthesis of a biological machine such as a ribosome or spliceosome.

It should also be noted that other methods and experimental approaches have been used that have led to conclusions similar to those obtained using RIP-Chip (for review, see Keene and Lager 2005; Keene 2007). These include polysome gradient shifts, en masse microarray analysis of mRNA decay, and genetic methods in yeast and flies. Thus, in aggregate, the posttranscriptional PTRO model as depicted in Figure 3 and described below has been strongly supported by these various types of experimental data.

COORDINATED GENE EXPRESSION AND THE POSTTRANSCRIPTIONAL OPERON MODEL

The theory of the PTRO proposes that mammalian cells organize monocistronic mRNAs in functional groups much like polycistronic mRNAs cluster functionally related gene products in bacteria (Fig. 3) (Keene and Tenenbaum 2002). As noted above, in bacteria, the polycistronic DNA operons are highly efficient at coordinating the expression of functionally related proteins, but the mechanism is constrained by the inability of each protein to be expressed independently or as part of another functionally related subpopulation. Furthermore, these monocistronic mRNAs can be regulated independently of one another or in groups by *trans*-acting RBPs or microRNAs (represented by A, B, C, D) that bind *cis*-acting sequences (represented by a, b, c, d) in their 3′UTR and 5′UTR. This model provides regulatory independence of expression for proteins encoded by the monocistronic mRNAs to adopt new functions and to regulate the production of each functional version of the protein at distinct intracellular locations or under different temporal modes either alone or together with proteins of related function (Keene and Tenenbaum 2002). Thus, the independently functioning forms of each protein can be regulated posttranscriptionally via separate *cis*-acting elements in the mRNA. The power of this architecture for higher cells is that mRNA members of mRNA subpopulations can be reassembled in different combinations depending on the regulatory signals and growth state of the cell (Hieronymus and Silver 2004; Moore 2005; Keene 2007). Therefore, one can imagine how a modest number of human genes (e.g. <30,000) can be used in multiple combinations to organize complex functions required for the development of a multicellular organism. In sum, PTROs provide a mechanism to coordinate the final outcome of gene expression in eukaryotic cells (for review, see Hieronymus and Silver 2004; Wilusz and Wilusz 2004; Fan et al. 2005; Keene and Lager 2005; Moore 2005; Vemuri and Aristido 2005; Keene 2007).

POTENTIAL FOR HORIZONTAL TRANSFER OF PTROs

We have extended the PTRO theory based on recent evidence to include the potential for the horizontal transfer of PTROs (Keene 2007). This corollary to the PTRO theory proposes that exosomes and microvesicles contain PTROs, and the associated mRNAs and microRNAs are exchanged among these cells by exosome-mediated transfer (Fig. 5) (Valadi et al. 2007). More recently, Aliotta et al. (2007) demonstrated that RNAs contained in exosomes from radiation-treated lung cells could transfer a phenotype to bone marrow cells. In addition, conditioned media and secreted microvesicles derived from other cell types have been shown to alter the phenotypes of recipient cells. For example, embryonic stem cells were shown to produce RNA-containing microvesicles that could reprogram hematopoietic progenitor cells (Ratajczak et al. 2006). Similar claims of phenotypic transfer of mRNAs by microvesicles and exosomes from tumor cells to monocytes have been described recently (Baj-Krzyworzeka et al. 2006). This has exceedingly important implications for all biological systems. We theorize that specific mRNAs and microRNAs representing PTROs are involved in cell–cell communication during growth, differentiation, and environmental assault, including cellular repair and cell death, and potentially circadian rhythms. These processes take place within the cellular infrastructure of the ribonome (Figs. 1 and 5) (Keene 2001).

THE RIBONOME AND POSTTRANSCRIPTIONAL REGULATORY NETWORKS

The ribonome consists of thousands of RBPs and their associated mRNAs that encode those RBPs and other proteins. Each RBP-associated mRNA is capable of being regulated by other RBPs within the ribonome that can bind to it and determine its fate. Thus, each RBP's own mRNA can be translationally activated to replace that RBP only when the community of bound RBPs allows it (see reciprocal regulatory interactions between ribonome and proteome in Fig. 5). Thus, within the ribonome are regulators-of-regulators (i.e., RBPs acting on one another's mRNAs) to keep the environment in balance. Although some RBPs activate translation of mRNAs, other RBPs inhibit translation, thus providing safeguards to the network of overlapping feedback loops, feed-forward loops, and commands for self-replenishment of the ribonome to maintain a balanced yet dynamic entity. Although the ribonome is self-sustaining, it is also self-limiting given these regulatory loops, and therein resides its central functional role as a management site for cell growth and multicellular expansion (Fig. 5).

The concept of the ribonome as a central managerial entity in biological systems is consistent with both a closed and an open system because it allows its own infrastructure to be self-regulating while regulating multiple non-RBP mRNAs (Figs. 1 and 5). Thus, each of the RBPs in the system can regulate one another's mRNAs while also regulating multiple other mRNAs that do not

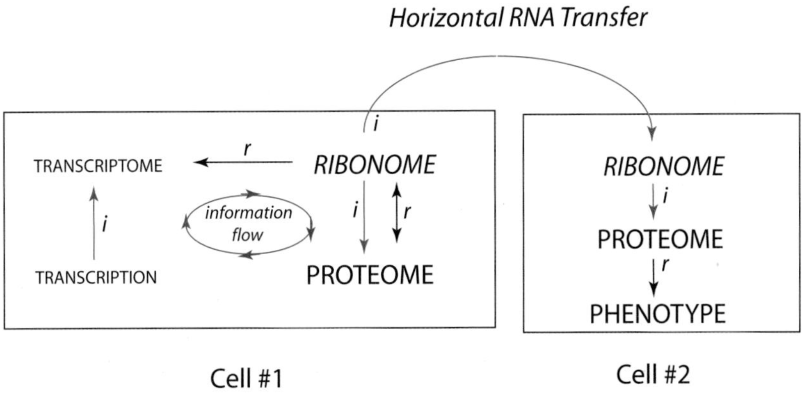

Figure 5. Depiction of a proposed mechanism of intercellular horizontal transfer of RNA. Cell #1 shows the interdependence of the ribonome and the proteome in both information transfer (*i*) and regulatory (*r*) modes. Information flows routinely from transcription to the proteome (*central circle*). Transcription feeds the transcriptome that is in turn shaped by *trans*-acting factors (e.g., RBPs and microRNAs) in the ribonome. The ribonome feeds information to the proteome via translation and thereby shapes its composition, but it also has a reciprocal regulatory relationship with it that sustains and limits RBP production within the ribonome (*double arrow*). Horizontal RNA transfer to the ribonome to Cell #2 (Valadi et al. 2007) via exosome-mediated exchange directly affects both its ribonome and its proteome through RNA delivery and translation, but thereby affecting the phenotype of the recipient cell (Baj-Krzyworzeka et al. 2006; Ratajczak et al. 2006; Aliotta et al. 2007.

necessarily encode RBPs. Thus, there is balanced control of multiple mRNAs that encode functionally related proteins, such as those that form macromolecular complexes and signaling pathways per the PTRO theory (Fig. 3) (Keene and Tenenbaum 2002). When a signal such as phosphorylation is directed at a specific RBP, it can affect the open system and at least temporarily override the regulatory effects of its neighboring RBPs in the ribonome (Intine et al. 2003 Benjamin et al. 2006; Abdelmohsen et al. 2007; Garbarino-Pico et al. 2007). Therefore, the "open" as well as the "closed" properties of the ribonome provide a dynamic quality of multilevel coordination that has not been appreciated by biologists in the past. Indeed, this theory posits that the ribonome is a central operating system of living organisms that communicates in several directions including feeding forward to the proteome, feeding back to the transcriptome, and feeding horizontally to other cells through exosomes or microvesicles (Fig. 5) (Keene 2007; Valadi et al. 2007). The cell-to-cell level of communication involves the horizontal transfer of genetic information as RNA. We propose that it is this power to transfer its self-sustaining, self-limiting regulatory information networks that gives the ribonome a central place in biology, making it vital to the origins and development of multicellular organisms.

POSTTRANSCRIPTIONAL REGULATION OF BIOLOGICAL CLOCKS

Biological clocks are logically among the most coordinated processes in living systems. From the mechanisms of circadian rhythms to the cell cycle, many studies have indicated examples of proteins and mRNAs that are posttranscriptionally regulated (Kim et al. 2002; Roenneberg and Merrow 2003; Majercak et al. 2004). In addition, some RBPs have been shown to undergo rhythmic changes during the 24-hour circadian period (Panda et al. 2002; Storch et al. 2002). For example, several laboratories have investigated changes in mRNA expression throughout the 24-hour circadian cycle in which the previously known proteins Period, Clock, Cry, and others were detected. Among the distilled data in these studies were cyclic changes in mRNAs that encode both transcription factors and RBPs. For example, transcription factors Zfp36 and Sox3 show circadian expression in mammalian organs, and the RNA regulators nocturnin (Ccr4 deadenylase) and butyrate response factor (BRF1) show cyclic circadian expression as well (Panda et al. 2002; Storch et al. 2002; Lidder et al. 2005). In addition, the RBP, LARK also cycles with circadian rhythms and was found to activate translation of the mRNA encoding the Period protein in the suprachiasmatic nucleus in the mouse brain (Kojima et al. 2007). Nocturnin, on the other hand, is a deadenylating enzyme that interacts with a subpopulation of mRNAs of yet unknown composition (Garbarino-Pico et al. 2007), whereas BRF1 specifically destabilizes mRNAs encoding immediate-early gene products under the control of a specific protein kinase (Stoecklin et al. 2002; Benjamin et al. 2006). With the demonstration that BRF1 cycles in a diurnal manner in peripheral organs, it is believed that it degrades immediate-early gene transcripts everyday. Moroni and coworkers (Benjamin et al. 2006) proposed that BRF1 functions as a circadian oscillating PTRO, thus offering an alternative or complementary mechanism of coordination in addition to transcription.

CONCLUSIONS

The proposed central role of the ribonome and the importance of RNA–protein and RNA–RNA interactions in regulating and, indeed, correlating outcomes of gene expression programs is based on evidence that such interactions are functionally related and contextually present in time and space in various cell types. How such interactions may affect biological clocks such as circadian rhythms in particular is unclear, but it is logical to predict that such coordinating principles like PTROs may have a role in oscillatory processes involving transcription and translation. However, the potential expansion of information transfer among surrounding cells in tissues and organs whether it involves PTROs per se presents a novel possibility for connections between biological clocks that depend on cycles in gene expression to work properly in both the central nervous system and the peripheral organs. Although evidence for such regulation has not been reported, it will be important to investigate this putative mode of information transfer as both a homeostatic and as an environmentally responsive system of intercellular communication and coordination.

REFERENCES

Abdelmohsen K., Pullmann R. Jr., Lal A., Kim H.H., Galban S., Yang X., Blethrow J.D., Walker M., Shubert J., Gillespie D.A., Furneaux H., and Gorospe M. 2007. Phosphorylation of HuR by Chk2 regulates SIRT1 expression. *Mol. Cell* **25:** 543.

Aliotta J.M., Sanchez-Guijo F.M., Dooner G.J., Johnson K.W., Dooner M.S., Greer K.A., Greer D., Pimentel J., Kolankiewicz L.M., Puente N., Faradyan S., Ferland P., Bearer E.L., Passero M.A., Adedi M., Colvin G.A., and Quesenberry P.J. 2007. Alteration of marrow cell gene expression, protein production and engraftment into lung by lung-derived microvesicles: A novel mechanism for phenotype modulation. *Stem Cells* (in press). **25:** 2245.

Antic D., Lu N., and Keene J.D. 1999. ELAV tumor antigen, Hel-N1, increases translation of neurofilament M mRNA and induces formation of neurites in human teratocarcinoma cells. *Genes Dev.* **13:** 449.

Aten J.A. and Kanaar R. 2006. Chromosomal organization: Mingling with the neighbors. *PLoS Biol.* **4:** e155.

Baj-Krzyworzeka M., Szatanek R., Weglarczyk K., Baran J., Urbanowicz B., Branski P., Ratajczak M.Z., and Zembala M. 2006. Tumour-derived microvesicles carry several surface determinants and mRNA of tumour cells and transfer some of these determinants to monocytes. *Cancer Immunol. Immunother.* **55:** 808.

Bartel D. P. 2004. MicroRNAs: Genomics, biogenesis, mechanism, and function. *Cell* **116:** 281.

Beckwith J. 1996. The operon: An historical account. In Escherichia coli *and* Salmonella *cellular and molecular biology* (ed. F.C. Neidhart), p. 1227. ASM Press, Washington, D.C.

Benjamin D., Schmidlin M., Min L., Gross B., and Moroni C. 2006. BRF1 protein turnover and mRNA decay activity are regulated by PKB at the same phosphorylation sites. *Mol. Cell. Biol.* **26:** 9497.

Brennan C.M. and Steitz J.A. 2001. HuR and mRNA stability. *Cell. Mol. Life Sci.* **58:** 266.

Brown V., Jin P., Ceman S., Darnell J. C., O'Donnell W. T., Tenenbaum S. A., Jin X., Feng Y., Wilkinson K. D., Keene J. D., Darnell R.B., and Warren S.T. 2001. Microarray identification of FMRP-associated brain mRNAs and altered mRNA translational profiles in fragile X syndrome. *Cell* **107:** 477.

Eystathioy T., Chan E.K.L., Griffith K., Tenenbaum S.T., Keene J.D., and Fritzler M.J. 2002. A phosphorylated cytoplasmic autoantigen, GW182, associates with a unique population of human mRNAs within novel cytoplasmic speckles. *Mol. Biol. Cell* **13:** 1338.

Fan J., Heller N.M., Gorospe M., Atasoy U., and Stellato C. 2005. The role of post-transcriptional regulation in chemokine gene expression in inflammation and allergy. *Eur. Respir. J.* **26:** 933.

Gao F., Carson C., Levine T.D., and Keene J.D. 1994. Selection of a subset of mRNAs from 3′UTR combinatorial libraries using neuronal RNA-binding protein, Hel-N1. *Proc. Natl. Acad. Sci.* **91:** 11207.

Garbarino-Pico E., Niu S., Rollag M.D., Strayer C.A., Besharse J.C., and Green C.B. 2007. Immediate early response of the circadian polyA ribonuclease nocturnin to two extracellular stimuli. *RNA* **13:** 745.

Gerber A.P., Herschlag D., and Brown P.O. 2004. Extensive association of functionally and cytotopically related mRNAs with Puf family RNA-binding proteins in yeast. *PLoS Biol.* **2:** e79.

Gerber A.P., Luschnig S., Krasnow MA, Brown P.O., and Herschlag D. 2006. Genome-wide identification of mRNAs associated with the translational regulator PUMILIO in *Drosophila melanogaster*. *Proc. Natl. Acad. Sci.* **103:** 4487.

Grimson A., Farh K.K.-H., Johnston W.K., Garrett-Engele P., Lim L.P., and Bartel D.P. 2007. MicroRNA targeting specificity in mammals: Determinants beyond seed pairing. *Cell* **27:** 91.

Hieronymus H. and Silver P.A. 2003. Genome-wide analysis of RNA-protein interactions illustrates specificity of the mRNA export machinery. *Nat. Genet.* **33:** 155.

———. 2004. A systems view of mRNP biology. *Genes Dev.* **18:** 2845.

Intine R.V., Tenenbaum S.A., Sakulich A.L., Keene J.D., and Maraia R.J. 2003. Differential phosphorylation and subcellular localization of La RNPs associated with precursor tRNAs and translation-related mRNAs. *Mol. Cell* **12:** 1301.

Jain R.G., Andrews L.G., McGowan K.M., Pekala P., and Keene J.D. 1997. Ectopic expression of Hel-N1, an RNA-binding protein, increases glucose transporter (GLUT1) expression in 3T3-L1 adipocytes. *Mol. Cell. Biol.* **17:** 954.

Keene J.D. 1999. Why is Hu where? Shuttling of early-response-gene messenger RNA subsets. *Proc. Natl. Acad. Sci.* **96:** 5.

———. 2001. Ribonucleoprotein infrastructure regulating the flow of genetic information between the genome and the proteome. *Proc. Natl. Acad. Sci.* **98:** 7018.

———. 2003. Organizing mRNA export. *Nat. Genet.* **33:** 111.

———. 2007. RNA regulons: Coordination of posttranscriptional events. *Nat. Rev. Genet.* **8:** 533.

Keene J.D. and Lager P.J. 2005. Post-transcriptional operons and regulons co-ordinating gene expression. *Chromosome Res.* **13:** 327.

Keene J.D. and Tenenbaum S.A. 2002 Eukaryotic mRNPs may represent posttranscriptional operons. *Mol. Cell* **9:** 1161.

Keene J.D., Komisarow J.M., and Friedersdorf M.B. 2006. RIP-Chip: The isolation and identification of mRNAs, microRNAs and protein components of ribonucleoprotein complexes from cell extracts. *Nat. Protoc.* **1:** 1.

Kim E.Y., Bae K., Ng F.S., Glossop N.R.J., Hardin P.E., and Edery I. 2002. *Drosophila* CLOCK protein is under posttranscriptional control and influences light-induced activity. *Neuron* **34:** 69.

King P.H., Levine T.D., Fremeau R.T., and Keene J.D. 1994. Mammalian homologs of *Drosophila* ELAV localized to a neuronal subset can bind *in vitro* to the 3′ UTR of mRNA encoding the Id transcriptional repressor. *J. Neurosci.* **14:** 1943.

Kojima S., Matsumoto K., Hirose M., Shimada M., Nagano M., Shigeyoshi Y., Hoshino S., Ui-Tei K., Saigo K., Green C.B., Sakaki Y., and Tei H. 2007. LARK activates posttranscriptional expression of an essential mammalian clock protein, PERIOD1. *Proc. Natl Acad. Sci.* **104:** 1859.

Kosak S.T. and Groudine M. 2004. Form follows function: The genomic organization of cellular differentiation. *Genes Dev.* **18:** 1371.

Lercher M.J., Chamary J.V., and Hurst L.D. 2004. Genomic regionality in rates of evolution is not explained by clustering of genes of comparable expression profile. *Genome Res.* **14:** 1002.

Levine T.D., Gao F., King P.H., Andrews L.G., and Keene J.D. 1993. Hel-N1: An autoimmune RNA-binding protein with specificity for 3′ uridylate-rich untranslated regions of growth factor mRNAs. *Mol. Cell. Biol.* **13:** 3494.

Lewis B.P., Burge C.B., and Bartel D.P. 2005. Conserved seed pairing, often flanked by adenosines, indicates that thousands of human genes are microRNA targets. *Cell* **120:** 15.

Lewis B.P., Shih I.H., Jones-Rhoades M.W., Bartel D.P., and Burge C.B. 2003. Prediction of mammalian micro RNA targets. *Cell* **115:** 787.

Lidder P., Gutierrez R.A., Salome P.A., McClung C.R., and Green P.J. 2005. Circadian control of messenger RNA stability. Association with a sequence-specific messenger RNA decay pathway. *Plant Physiol.* **138:** 2374.

Lopez de Silanes I., Zhan M., Lal A., Yang X., and Gorospe M. 2004. Identification of a target RNA motif for RNA-binding protein HuR. *Proc. Natl. Acad. Sci.* **101:** 2987.

Ma W.J., Chang S., Campbell C., Wright A., and Furneaux H. 1996. Cloning and characterization of HuR, a ubiquitously expressed Elav-like protein. *J. Biol. Chem.* **271:** 8144.

Maniatis T. and Reed R. 2002. An extensive network of coupling among gene expression machines. *Nature* **416:** 499.

Majercak J., Chen W.-F., and Edery I. 2004. Splicing of period 3′ terminal intron is regulated by light, circadian clock factors and phospholipase C. *Mol. Cell. Biol.* **24:** 3359.

Mazan-Mamczarz K., Galban S., Lopez de Silanes I., Martindale J.L., Atasoy U., Keene J.D., and Gorospe M. 2003. RNA-binding protein HuR enhances p53 translation after ultraviolet light irradiation. *Proc. Natl. Acad. Sci.* **100:** 8354.

Moore M.J. 2005. From birth to death: The complex lives of eukaryotic mRNAs. *Science* **309:** 1514.

Niehrs C. and Pollet N. 1999. Synexpression groups in eukaryotes. *Nature* **402:** 483.

Orphanides G. and Reinberg D. 2002. A unified theory of gene expression. *Cell* **108:** 439.

Panda S., Antoch M.P., Miller B.H., Su A.I., Schook A.B., Straume M., Schultz P.G., Kay S.A., Takahashi J.S., and Hogenesch J.B. 2002. Coordinated transcription of key pathways in the mouse by the circadian clock. *Cell* **109:** 307.

Penalva L.O., Tenenbaum S.A., and Keene J.D. 2004. Gene expression analysis of messenger RNP complexes. *Methods Mol. Biol.* **257:** 125.

Ratajczak J., Miekus K., Kucia M., Zhang J., Reca R., Dvorak P., and Ratajczak M.Z. 2006. Embryonic stem cell-derived microvesicles reprogram hematopoietic progenitors: Evidence for horizontal transfer of mRNA and protein delivery. *Leukemia* **20:** 847.

Richter J.D. 1997. *mRNA formation and function.* Academic Press, New York.

Robinow S., Campo A.R., Ya K.M., and White K. 1988. The elav gene product of *Drosophila*, required in neurons, has three RNP consensus motifs. *Science* **242:** 1570.

Rodriguez-Trelles F., Tarrio R., and Ayala F.J. 2005. Is ectopic expression caused by deregulatory mutations or due to gene-regulation leaks with evolutionary potential? *Bioessays* **27:** 592.

Roenneberg T. and Merrow M. 2003. The network of time: Understanding the molecular circadian system. *Curr. Biol.* **13:** R198.

Shaw G. and Kamen R. 1986. A conserved AU sequence from the 3′ untranslated region of GM-CSF mRNA mediates selective mRNA degradation. *Cell* **46:** 659.

Spellman P.T. and Rubin G.M. 2002. Evidence for large domains of similarly expressed genes in the *Drosophila* genome. *J. Biol.* **1:** 5.

Stoecklin G., Colombi M., Raineri I., Leuenberger S., Mallaun M., Schmidlin M., Gross B., Lu M., Kitamura T., and Moroni C. 2002. Functional cloning of BRF1, a regulator of ARE-dependent mRNA turnover. *EMBO J.* **21:** 4709.

Storch K.F., Lipan O., Leykin I., Viswanathan N., Davis F.C., Wong W.H., and Weitz C.J. 2002. Extensive and divergent circadian gene expression in liver and heart. *Nature* **417:** 78.

Szabo A., Dalmau J., Manley G., Rosenfeld M., Wong E., Henson J., Posner J.B., and Furneaux H.M. 1991. HuD, a paraneoplastic encephalomyelitis antigen, contains RNA-Binding domains and is homologous to Elav and Sex-lethal. *Cell* **67:** 325.

Tenenbaum S.A., Carson C.C., Atasoy U., and Keene J.D. 2003. Genome-wide regulatory analysis combining en masse nuclear run-ons (emRUNs) and ribonomic profiling. *Gene* **317:** 79.

Tenenbaum S.A., Carson C.C., Lager P.J. and Keene J.D. 2000. Identifying mRNA subsets in messenger ribonucleoprotein complexes by using cDNA arrays. *Proc. Natl. Acad. Sci.* **97:** 14085.

Tenenbaum S.A., Lager P.J., Carson C.C., and Keene J.D. 2002. Ribonomics: Identifying mRNA subsets in mRNP complexes using antibodies to RNA-binding proteins and genomic arrays. *Methods* **26:** 191.

Townley-Tilson W.H., Pendergrass S.A., Marzluff W.F., and Whitfield M.L. 2006. Genome-wide analysis of mRNAs bound to the histone stem-loop binding protein. *RNA* **12:** 1853.

Valadi H., Ekström K., Bossios A., Sjöstrand M., Lee J.J., and Lötvall J.O. 2007. Exosome-mediated transfer of mRNAs and microRNAs is a novel mechanism of genetic exchange between cells. *Nat. Cell Biol.* **9:** 654.

Vemuri G.N. and Aristidou A.A. 2005. Metabolic engineering in the -omics era: Elucidating and modulating regulatory networks. *Microbiol. Mol. Biol. Rev.* **69:** 197.

Wilusz C.J. and Wilusz J. 2004. Bringing the role of mRNA decay in the control of gene expression into focus. *Trends Genet.* **20:** 491.

Yanai I., Korbel J.O., Boue S., McWeeney S.K., Bork P., and Lercher M.J. 2006. Similar gene expression profiles do not imply similar tissue functions. *Trends Genet.* **22:** 132.

Role of Phosphorylation in the Mammalian Circadian Clock

K. VANSELOW AND A. KRAMER
Laboratory of Chronobiology, Charité Universitätsmedizin Berlin, 10115 Berlin, Germany

Circadian clocks regulate a wide variety of processes ranging from gene expression to behavior. At the molecular level, circadian rhythms are thought to be produced by a set of clock genes and proteins interconnected to form transcriptional-translational feedback loops. Rhythmic gene expression was formerly regarded as the major drive for rhythms in clock protein abundance, but recent findings underline the crucial importance of posttranslational mechanisms for both the generation and dynamics of circadian rhythms. In particular, the reversible phosphorylation of PER proteins—essential components within the negative feedback loop in *Drosophila* and mammals—seems to have a key role for the correct timing of nuclear repression. To understand how PER protein phosphorylation regulates the dynamics of the circadian oscillator, we have mapped endogenous phosphorylation sites in mPER2. Detailed investigation of the functional role of one particular phosphorylation site (Ser-659, which is mutated in the familial advanced sleep phase syndrome [FASPS]) led us propose a model of functionally different phosphorylation sites in PER2. This concept explains not only the FASPS phenotype, but also the effect of the *tau* mutation in hamster.

INTRODUCTION

Most light-sensitive organisms have evolved internal clocks that regulate daily rhythms in physiology, metabolism, and behavior. In a natural environment, these clocks are synchronized to external zeitgebers, such as the light/dark cycle or temperature cycles, to ensure a stable phase relationship between internal and external processes. Under constant conditions, oscillations are free-running with a period close to 24 hours, hence the term *circa-dian* clocks. Extensive studies in the last two decades have unraveled the molecular basis of circadian rhythmicity in a broad variety of organisms such as cyanobacteria, plants, *Neurospora, Drosophila*, and mammals. Conceptually, circadian rhythms are generated by delayed negative feedback loops within single cells. A set of so-called clock genes and clock proteins are interconnected to produce self-sustained circadian oscillations at the molecular level.

In mammals, a negative gene-regulatory feedback loop is at the heart of the circadian oscillator: CLOCK and BMAL1—bHLH (basic helix-loop-helix)–PAS (Period-Arnt-Single-minded)-containing transcription factors—heterodimerize and bind to E-box enhancer elements in the promoter region of the *Period* (*Per1, Per2*) and *Cryptochrome* (*Cry1*, *Cry2*) genes to activate their transcription. PER and CRY proteins, together with casein kinase 1ε/δ (CKIε/δ) and probably other proteins, form a large multimeric complex, the circadian feedback module (Brown et al. 2005a; Hofmann et al. 2006). It is believed that after some time, the complex enters the nucleus, where PER and CRY proteins inhibit their own synthesis by directly interacting with the CLOCK-BMAL1 heterodimer. At present, the exact composition of the module is unclear, although multiple additional proteins have been suggested to contribute to the periodic inhibition of CLOCK-BMAL1 activity. These include mammalian Timeless (Sangoram et al. 1998; Barnes et al. 2003), the bHLH transcription factors DEC1/2 (Honma et al. 2002), the PER1-associated proteins NONO and WDR5 (Brown et al. 2005a) as well as the CLOCK-interacting protein CIPC (Zhao et al. 2007).

Further feedback loops possibly contribute to the robustness of the system. In a positive feedback loop, the transcription of the *Bmal1* gene is regulated by rhythmic action of Rev-Erbα, the transcription of which is also controlled by CLOCK-BMAL1 via E-box elements (Preitner et al. 2002).

Critical for the generation of a self-sustained circadian oscillation with a period of about 24 hours is a time delay of several hours between the synthesis of PER and CRY proteins and their action as inhibitors of their own expression in the nucleus. Several mechanisms are discussed to participate in the generation of this delay. These mechanisms modulate abundance, localization, and activity of transcriptional inhibitors by regulating transcription, translation, and posttranslational modifications of important components of the circadian oscillator.

Up to now, many posttranslational mechanisms have been discovered to modulate circadian dynamics, although we are far from having a comprehensive view on the function of posttranslational modifications for the clock. Perhaps the most important posttranslational modification in eukaryotic cells is the phosphorylation of serine, threonine, and tyrosine residues. Such phosphorylations frequently act as molecular switches between active and inactive protein states either directly by regulating the activity (e.g., kinases) or indirectly by altering subcellular localization (e.g., transcription factors). In addition, phosphorylation often initiates protein degradation by the proteasome. Target proteins contain specific amino acid motifs, which, when phosphorylated, form a recognition sequence for the ubiquitin ligase complex.

Several important components of the mammalian circadian oscillator are known to be targets of posttranslational modifications. CLOCK, BMAL1, PER, and possibly also CRY proteins are phosphoproteins in vivo (Lee et al. 2001). Moreover, BMAL1 was shown to be sumoylated (Cardone et al. 2005) and, very recently, acetylated (Grimaldi et al., this volume). These modifications specifically alter the properties and functionality of the corresponding proteins by changing protein stability, subcellular localization, activity, or complex formation.

Hitherto, phosphorylation is the best-explored posttranslational mechanism within essentially all circadian model systems. Recent results from cyanobacteria even challenge the dogma of rhythmic transcriptional-translational feedback loops as the fundamental principle of rhythm generation: A purely posttranslational mechanism based on rhythmic phosphorylation and dephosphorylation of the hexameric KaiC protein is sufficient to generate and sustain circadian rhythms at least in vitro (Nakajima et al. 2005; Tomita et al. 2005). In flies and mammals, there is also evidence for the particular importance of posttranslational mechanisms: Constitutive expression of *dPer* in arrythmic *per*^{0} flies rescues arrhythmic behavior and leads to rhythmic PER protein levels (Yang and Sehgal 2001). Similarly, constitutive expression of mammalian *Per* genes in synchronized cultured cells also results in rhythmic PER protein abundance (Fujimoto et al. 2006; Nishii et al. 2006). Consequently, these data suggest that rhythmic PER protein abundance in vivo might be generated by a combination of rhythmic transcription of *Per* genes and posttranslational events. At least in the case of flies, the posttranslational mechanism seems to be sufficient for behavioral circadian rhythmicity.

This chapter focuses on the functional role of clock protein phosphorylation for the generation and the dynamics of circadian oscillations with a special emphasis on mammalian and *Drosophila* PER proteins.

Kinases and Phosphatases

Doubletime (DBT)—the *Drosophila* homolog of mammalian CKIε—was the first kinase identified to have an important functional role in the circadian system of *Drosophila*. Flies with mutations in the *dbt* gene have altered circadian periods at both the behavioral and molecular levels. In *dbt* mutant flies, the stability, phosphorylation pattern, and subcellular localization of dPER is modified (Kloss et al. 1998; Price et al. 1998). Subsequent studies showed an additional role for DBT in the phosphorylation of dCLOCK, thereby regulating its stability (Kim and Edery 2006; Yu et al. 2006). In addition, *Drosophila* casein kinase 2 (dCK2) was shown to phosphorylate dPER. This phosphorylation predominantly functions in the regulation of the subcellular localization of dPER by promoting its nuclear entry and accumulation, whereas its effects on dPER stability are rather weak (Lin et al. 2002, 2005; Akten et al. 2003). Moreover, Shaggy—the *Drosophila* homolog of mammalian glycogen synthase kinase 3β (GSK3β)—phosphorylates dTIM, the essential binding partner of dPER, thereby promoting the nuclear entry of the PER/TIM heterodimer (Martinek et al. 2001). The impact of Shaggy phosphorylation on TIM stability has not yet been elucidated, although it has been reported that phosphorylated TIM species are more prone to light-induced proteasomal degradation (Zeng et al. 1996).

The discovery of DBT's importance in the *Drosophila* circadian system initiated extensive studies to elucidate a possible conserved function of the DBT homologous kinases of the CKI family in the mammalian clockwork. Indeed, it has been demonstrated that CKIε and CKIδ are involved in the phosphorylation of the mammalian PER proteins (Keesler et al. 2000; Camacho et al. 2001; Akashi et al. 2002; Schlosser et al. 2005). Moreover, CKIε is able to phosphorylate BMAL1 and CRY proteins in vitro (Eide et al. 2002).

Recent studies showed that GSK3β also has a role in the mammalian circadian system, but interestingly, its targets do not seem to be conserved. Whereas Shaggy phosphorylates dTIM in *Drosophila*, substrates of mammalian GSK3β are CRY2 (Harada et al. 2005), Rev-erbα (Yin et al. 2006) and PER2 (Iitaka et al. 2005). Here, GSK3β-mediated phosphorylation regulates the stability of CRY2 and Rev-erbα as well as the subcellular localization of PER2.

It is still unclear whether additional kinases are involved in the generation of circadian rhythmicity in mammals. However, at least in the case of mPER2, it has been speculated that yet unknown kinases are involved in phosphorylation-dependent subcellular localization (Vanselow et al. 2006).

Phosphatases are the natural opponents of kinases: They ensure reversibility of phosphorylation-induced alterations in protein function. The role of phosphatases in the circadian clockwork is just emerging. So far, it has been demonstrated in *Drosophila* that PP2A dephosphorylates and thus stabilizes dPER (Sathyanarayanan et al. 2004). In mammals, PP1 was shown to dephosphorylate a mPER2 fragment in vitro. In addition, coexpression with a dominant-negative version of PP1 destabilizes this mPER2 fragment, indicating that dephosphorylation of PER proteins by PP1 may counteract CKIε/δ-induced proteasomal degradation (Gallego et al. 2006a). Furthermore, PP5 has been reported to regulate the kinase activity of CKIε by antagonizing its inhibitory autophosphorylation. Interestingly, the PP5-mediated activation of CKIε is inhibited by cryptochrome proteins (Partch et al. 2006), which may provide an explanation for the stabilizing effect of CRYs on PER proteins.

PER Phosphorylation

In the 1990s, a delayed negative feedback loop was proposed as the fundamental principle of circadian rhythm generation in *Drosophila*. This was primarily based on the striking 4–6-hour lag between the accumulation profiles of *Drosophila Per* mRNA and protein (Hardin et al. 1990), leading to speculations that a pure transcriptional mechanism is not sufficient to create this delay. When analyzing the circadian accumulation patterns of dPER and mammalian PERs by western blotting, a substantial electrophoretic mobility shift of the PER bands occurred, in addition to a high-amplitude variation in protein abundance. This mobility shift has been shown to be due to a gradual, circadian phase-dependent phosphorylation of the PER proteins (Edery et al. 1994; Lee et al. 2001). Interestingly, the timing and extent of the PER phosphorylation are modified in period-altering fly mutants (Edery et al. 1994), indicating that the phosphorylation status of the PER proteins reflects the progress in the circadian cycle.

Several properties of PER proteins turned out to be directly regulated by the degree of PER phosphorylation, such as stability, subcellular localization, and inhibitory activity (see below). Importantly, the phosphorylation of PER proteins in *Drosophila* and mammals is not mediated by a single kinase but is the result of a complex temporal and spatial interplay of several kinases and phosphatases.

Phosphorylation regulates stability. Already in one of the first studies on *Drosophila* PER protein, it has been noted that the dPER species with the highest degree of phosphorylation are present at times right before the drastic drop of protein abundance. This suggested a direct link between dPER phosphorylation and degradation (Edery et al. 1994). In *dbt*^{P} larvae (a *dbt* loss-of-function mutant), dPER is hypophosphorylated and accumulates to unusual high levels (Price et al. 1998), further indicating that phosphorylation destabilizes PER. Subsequently, it was shown that progressive phosphorylation of dPER triggers the binding of the F box/WD40-repeat protein Slimb. Slimb functions as a substrate-recognizing component of the ubiquitin ligase SCF complex and promotes the ubiquitination and subsequent proteasomal degradation of hyperphosphorylated PER (Grima et al. 2002; Ko et al. 2002). As *slimb* mutant flies are behaviorally arrhythmic, the time-specific degradation of PER proteins is an essential part of circadian rhythm generation and maintenance. It adds to the required time delay within the negative feedback loop by counteracting a premature cytoplasmic and nuclear accumulation of the inhibitory complex.

In a very similar manner, mammalian PER proteins are extensively phosphorylated by CKIε and CKIδ, thereby reducing the stability of the proteins by targeting them for proteasomal degradation (Keesler et al. 2000; Camacho et al. 2001; Akashi et al. 2002). PER proteins are recognized by the Slimb homologs β-TrCP1/2 as parts of the SCF ubiquitin ligase complex. The interactions between β-TrCP1/2 and their substrates occur via specific recognition sequences within the substrate. For both mPER2 and mPER1, a β-TrCP1/2-binding site has been proposed (Eide et al. 2005; Shirogane et al. 2005). Down-regulation of both endogenous β-TrCP1 and β-TrCP2 in synchronized fibroblasts results in a substantial lengthening of the circadian period. Furthermore, the expression of β-TrCP1/2 interaction-deficient PER2 variants in synchronized fibroblasts leads to a dramatic stabilization of PER2 protein as well as to a disruption of circadian rhythms (Reischl 2007).

In mammals, PER proteins are not the only clock components, whose stability is controlled by the proteasomal pathway. Very recently, the F-box protein Fbxl3 was shown to target CRY proteins for proteasomal degradation. Mice with mutations in Fbxl3 display a lengthened circadian locomotor activity rhythm, an increased stability of the CRY proteins, and a reduced expression of PER proteins (Busino et al. 2007; Godinho et al. 2007; Siepka et al. 2007). Given the relatively high CRY protein amounts as well as the rather small amplitude of CRY protein abundance rhythms, the molecular basis of Fbxl3's essential role in the circadian clock is still, at least in part, obscure. For example, it is unclear whether phosphorylation of CRY proteins is required for recognition by Fbxl3.

Phosphorylation regulates subcellular localization. Phosphorylation affects not only PER protein stability, but also its subcellular localization. For example, in *dbt* and *per* mutant flies, besides the accumulation profile, the nuclear import and nuclear clearance patterns of dPER are altered. Furthermore, Cyran et al. (2005) showed that phosphorylation by DBT retains PER in the cytoplasm, whereas hypophosphorylated dPER is able to enter the nucleus in *tim*^{01}; *dbt*^{P} double-mutant larvae and *tim*^{01}; *dbt*^{ar} double-mutant flies, respectively. Thus, proper dPER phosphorylation by DBT is crucial for correct timing of dPER nuclear entry and export. Besides DBT, the phosphorylation of CK2 has a clear impact on the subcellular localization of dPER. Mutations in the catalytic α or regulatory β subunit of CK2 lengthen the circadian period of flies and concomitantly decelerate the nuclear entry of the PER protein (Lin et al. 2002; Akten et al. 2003). A similar role for CK2 has also been established in the *Neurospora* and *Arabidopsis* clock (Sugano et al. 1999; Yang et al. 2002). Given this striking conservation in several phylogenetic kingdoms, an important function for CK2 has been proposed in the mammalian system, but not yet established.

In mammals, the influence of CKIε/δ phosphorylation on the subcellular localization of PER proteins is still not fully understood. Several independent studies have addressed this issue with ambiguous and sometimes contradictory results. In cultured cells, the subcellular distribution of mammalian PER proteins following coexpression with CKIε or CKIδ seems to depend on the respective PER paralog as well as on the cell line. Although a direct effect of CKIε/δ phosphorylation on subcellular localization can be detected, a consistent direction of this effect is not observed (Takano et al. 2000, 2004; Vielhaber et al. 2000; Akashi et al. 2002).

Phosphorylation regulates inhibitory activity. The activity of many eukaryotic transcription factors (e.g., NFAT, p53, HSF1, CREB, c-Jun, and Fos) is regulated by reversible phosphorylation (Holmberg et al. 2002). As PER proteins are believed to act as transcriptional repressors, it is conceivable that their inhibitory action on CLOCK:BMAL1 could be regulated by phosphorylation, as well. However, direct evidence for this is rare. Only one study demonstrates that phosphorylation of dPER by DBT and CK2 acts in a coordinate manner to potentiate PER repression activity. Importantly, this effect is not mediated by an alteration of the subcellular localization of the protein (Nawathean and Rosbash 2004). Similar results have not been reported in the mammalian system.

Circadian Phenotypes with Altered PER Phosphorylation

How important is the phosphorylation of PER proteins for the generation and dynamics of circadian rhythms? Probably very important, because several severe circadian phenotypes are due to mutations that alter the temporal and spatial phosphorylation patterns of PER proteins. Examples for known circadian phenotypes with altered PER phosphorylation in *Drosophila* and mammals are summarized in Table 1 and described in more detail below.

The first direct evidence of a genetic basis for circadian clocks came from the short- and long-period mutant flies (*per*^{S} and *per*^{L}) isolated by Konopka and Benzer (1971). Later, it was shown that in *per*^{S} mutants, the accumulation profile and nuclear clearance of dPER are advanced (Zerr et al. 1990; Edery et al. 1994), whereas in *per*^{L} flies, the nuclear entry of the PER protein is delayed (Curtin et al. 1995). Similar effects were observed in *dbt* mutants (e.g., *dbt*^{S} and *dbt*^{L}) (Kloss et al. 1998; Price et al. 1998).

In the mammalian circadian system, the most prominent examples of phosphorylation-based circadian phenotypes are the *tau* mutation in the hamster and the FASPS phenotype in humans. The *tau* mutation was discovered by Ralph and Menaker (1988) on the basis of a prominent shortening of the circadian locomotor activity rhythm—a 22-hour period in heterozygous animals and a 20-hour period in homozygous animals (Ralph and Menaker 1988). The mRNA profiles of *Per1* and *Per2* in the SCN (suprachiasmatic nucleus) of *tau* mutant hamsters show an earlier rise and decline and a reduction in the peak expression levels of both genes (Lowrey et al. 2000; Dey et al. 2005). Genetic analysis revealed that a single-nucleotide exchange in the coding region of the CKIε gene generates the *tau* mutant phenotype. This mutation

Table 1. Circadian Phenotypes with Altered PER Phosphorylation in *Drosophila* and Mammals

Locus/species	Mutant allele	Molecular lesion	Phenotype	Molecular impact	References
period gene, *Drosophila*	*per*^{S}	single-nucleotide exchange leads to Ser-589→Asn	short-period rhythms (19 hr) in eclosion and behavior	abundance and phosphorylation profiles of PER^S are advanced; premature nuclear clearance, but unaltered nuclear accumulation	Konopka and Benzer (1971); Yu et al. (1987); Zerr et al. (1990); Curtin et al. (1995)
period gene, *Drosophila*	*per*^{L}	single-nucleotide exchange leads to Val-243→Asp	long-period rhythms (29 hr) in eclosion and behavior	nuclear entry and clearance of PER^L is delayed	Konopka and Benzer (1971); Curtin et al. (1995)
doubletime gene, *Drosophila*	*dbt*^{S}	single-nucleotide exchange leads to Pro-47→Ser	short-period rhythms in eclosion (20 hr) and behavior (18 hr)	accelerated PER accumulation in the cytoplasm and disappearance from the nuclei; reduced kinase activity in vitro	Kloss et al. (1998); Price et al. (1998); Bao et al. (2001); Preuss et al. (2004)
doubletime gene, *Drosophila*	*dbt*^{L}	single-nucleotide exchange leads to Met-80→Ile	long-period rhythms (27 hr) in eclosion and behavior	accumulation pattern of PER is delayed; longer persistence in the declining phase of the circadian cycle; reduced kinase activity in vitro	Kloss et al. (1998); Price et al. (1998); Suri et al. (2000); Preuss et al. (2004)
doubletime gene, *Drosophila*	*dbt*^{P}	P-element insertion in intron 2 disrupts gene function	heterozygous: normal behavioral rhythms homozygous: lethal	hypophosphorylated PER accumulates to high levels in homozygous *dbt*^{P} larvae	Kloss et al. (1998); Price et al. (1998)
doubletime gene, *Drosophila*	*dbt*^{ar}	single-nucleotide exchange leads to His-126→Tyr	homozygous: arrhythmic behavior	hypo- and hyperphosphorylated PER accumulates to high levels; PER oscillation is stopped	Rothenfluh et al. (2000)
period 2 gene, human	*Per2S662G* (FASPS)	single-nucleotide exchange leads to Ser-662→Gly	FASPS, ~ 4-hr advance of circadian parameters, short endogenous period	hypophosphorylation of PER2 protein in vitro; reduced PER2 stability due to decreased nuclear retention	Toh et al. (2001); Vanselow et al. (2006)
casein kinase Iδ gene, human	*CKIδT44A* (FASPS)	single-nucleotide exchange leads to Thr-44→Ala	FASPS, ~ 4-hr advance of circadian parameters, short endogenous period	reduced kinase activity in vitro	Xu et al. (2005)
casein kinase Iε gene, hamster	*CKIεR178C* (*tau*)	single-nucleotide exchange leads to Arg-178→Cys	*tau* phenotype: short-period behavioral rhythms heterozygous: 22 hr heterozygous: 20 hr	reduced kinase activity in vitro, premature nuclear clearance of PER1 and PER2 in the SCN; increased phosphorylation of PER proteins in cells	Ralph and Menaker (1988); Lowrey et al. (2000); Dey et al. (2005); Gallego et al. (2006b)

results in an arginine-to-cysteine substitution at amino acid 178 of CKIε. In vitro, recombinantly produced *tau* kinase has a reduced activity, whereas the binding to mPER1 and mPER2 is not impaired (Lowrey et al. 2000). This findings suggest a possible reduction in the phosphorylation of the PER proteins by CKIε(*tau*) also in vivo. Unexpectedly, however, *tau* mutant hamsters exhibit robust circadian rhythms in both PER protein abundance and PER phosphorylation without substantial differences in the degree of hyperphosphorylation, although the overall amount of PER protein is reduced (Lee et al. 2001). Interestingly, an accelerated nuclear clearance of PER proteins in the SCN of *tau* mutant hamsters has been reported (Dey et al. 2005).

A second phosphorylation-based circadian phenotype is the FASPS phenotype in humans. The heredity transmission of FASPS follows an autosomal dominant pattern (Jones et al. 1999; Toh et al. 2001; Xu et al. 2005). Affected individuals are extreme morning larks with an approximately 4–5-hour advance of sleep onset and offset, temperature, and hormonal rhythms. The endogenous period of one affected individual was measured to be 23.3 hours (Jones et al. 1999), which is about 1 hour shorter than the average population (Czeisler et al. 1999). Up to now, genetic analyses have identified two single-nucleotide exchanges in human clock genes correlating with the FASPS phenotype. In one case, the human *Per2* gene is mutated, which leads to a serine-to-glycine exchange at residue 662 of hPER2 (Toh et al. 2001). This region of PER2 is highly conserved within all paralog PER proteins in mammals. Notably, the serine residue mutated in FASPS is the first residue in a cluster of serine and threonine residues forming a canonical CKIε/δ phosphorylation/recognition motif (pS/pT/pY$X_{(1-2)}$S/T; X = any amino acid) (Flotow et al. 1990). In vitro, the FASPS mutation causes hypophosphorylation of PER2 when phosphorylated with CKIε (Toh et al. 2001).

In a second form of FASPS, the human *CKIδ* gene is mutated. Here, a threonine-to-alanine substitution at amino acid 44 in the CKIδ protein has been identified. As in the case of *tau*, recombinantly produced CKIδ-T44A has reduced kinase activity in vitro. Interestingly, transgenic mice containing a human bacterial artificial chromosome (BAC) clone expressing the CKIδ-T44A mutant show a shortened circadian period, whereas the expression of hCKIδ-T44A in flies leads to a period lengthening (Xu et al. 2005), although the respective position is highly conserved between mammalian CKIδ and DBT. This suggests that reducing the activity of CKIδ can result in opposite period phenotypes depending on the cellular context and/or the substrates.

Until recently, the molecular mechanisms of both the *tau* and the FASPS phenotype were postulated as follows: In *tau*, the reduced kinase activity and in FASPS, the lack of a specific phosphorylatable residue in PER2 both result in hypophosphorylation of PER proteins. Because phosphorylation triggers degradation, it has been assumed that PER hypophosphorylation would lead to a stabilization of PER proteins and thus to a more rapid PER accumulation in the cytoplasm, followed by an advanced nuclear localization of PER proteins. An earlier repression of CLOCK-BMAL1 transcriptional activity then would lead to shorter circadian periods (Lowrey et al. 2000; Dey et al. 2005). However, we propose an alternative mechanism (see below). Although the available data strongly suggest a disturbance of PER phosphorylation as the molecular basis for *tau* and FASPS, direct evidence for altered phosphorylation patterns in vivo was still missing.

Mapping of Phosphorylation Sites in Clock Proteins

The examples described above underline the enormous importance of proper phosphorylation of PER proteins and possibly other clock components for the generation and dynamics of circadian oscillations. The circadian phase-dependent gradual decrease of PER's electrophoretic mobility indicates that several distinguishable PER phosphorylation species exist. In addition, the differential effects of some mutations on PER stability and subcellular localization argue for specific phosphorylation sites influencing these properties. To elucidate how phosphorylation at particular sites regulates the temporal and spatial properties of PER proteins, a detailed map of endogenous PER phosphorylation sites is needed. Knowing these sites would allow for the study of their specific function and may provide hints toward the identification of the kinases responsible for phosphorylation. Previous studies analyzed single phosphorylation sites in mPER1 with respect to their functional roles (Vielhaber et al. 2000; Takano et al. 2004). Until recently, however, no systematic effort to comprehensively map in vivo clock protein phosphorylation sites has been undertaken.

We have established a novel mass spectrometric technique that allows sensitive and comprehensive mapping of protein phosphorylation sites. We use this technique to analyze phosphorylation sites in clock proteins in a systematic manner. Immunoprecipitated proteins are subjected to in-gel digests with a set of four different proteases. Phosphopeptides are enriched on titanesphere columns and then analyzed by ultra-low-flow nanoLC-MS/MS (Schlosser et al. 2005).

The first clock protein that we analyzed was mouse PER2, because it is the best-characterized circadian phosphoprotein to date with a very pronounced temporal phosphorylation profile. For that purpose, we generated a HEK 293 cell line stably expressing carboxy-terminal V5-tagged mPER2 for efficient immunoprecipitation. The mass spectrometric approach mentioned above led to the identification of 21 (of 275 theoretically possible) endogenous phosphorylation sites in mPER2 (exclusively serine and threonine residues). Interestingly, we found the serine residue phosphorylated, which is mutated in FASPS (Vanselow et al. 2006). This is the first direct evidence that the FASPS site is indeed a target for phosphorylation in living cells. Generally, many of the detected phosphorylation sites are located close to the previously identified CKIε/δ-binding domain (amino acids 555–754) (Akashi et al. 2002) or surrounding the nuclear export sequence-2 (NES-2 amino acids 983–990) (Yagita et al. 2002). Further phosphorylation sites are concentrated at the carboxyl and amino termini, and no phosphorylation site was found in the PAS domain region.

To investigate the function of these 21 phosphorylation sites of mPER2, we established a reductionist model system, which is based on the expression of phosphosite-mutated mPER2 proteins in synchronized fibroblasts. Potential alterations in the oscillation properties are analyzed using luciferase-based circadian reporters. Fibroblast cell lines like Rat-1 and NIH-3T3 have been demonstrated to show self-sustained circadian oscillations with properties very similar to those of the SCN-intrinsic cellular oscillator (Balsalobre et al. 1998; Yagita et al. 2001; Brown et al. 2005b). In the first step, we generated an NIH-3T3 Flp-In host cell line harboring the Flp recombinase target (FRT) site (O'Gorman et al. 1991). This ensures the expression of different mPER2 variants from the same chromosomal location in the fibroblast genome. In a second step, expression cell lines of PER2 variants were generated by recombining mutant constructs into the FRT site of the NIH-3T3 genome. Thus, differences in the oscillation dynamics of the resulting expression cell lines or in the biochemical properties of the PER2 variants are directly caused by the PER2 protein mutations, and artifacts due to chromosomal positioning effects or variation of integration events were excluded.

We were particularly interested in investigating the functional role of the PER2 phosphorylation at the FASPS region, because this region has been implicated to be crucial for the development of FASPS. To this end, we generated different variants of mPER2 affecting the FASPS region (FASPS [S659G], mut-7 [S659G, S662A, S665A, S668A, S670A, S671A, TS672A], and S659D) by site-directed mutagenesis. These PER2 variants were then expressed in our NIH-3T3 Flp-In cell line. The circadian oscillations of the resulting cell lines were recorded continuously in a luminometer. Interestingly, the oscillation dynamics of PER2-FASPS expressing cells is very similar to the behavior of FASPS patients: an early phase of entrainment and a short free-running period (Fig. 1) (Vanselow et al. 2006).

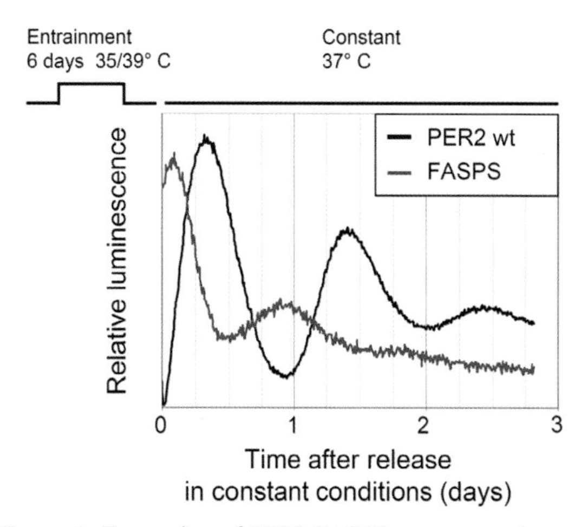

Figure 1. Expression of PER2-FASPS causes an advanced phase and a shorter period in oscillating NIH-3T3 fibroblasts after a several day temperature entrainment regime. Real-time circadian oscillations of luciferase reporter activity were recorded in a luminometer. (Reprinted, with permission, from Vanselow et al. 2006.)

Functional Different Phosphorylation Sites in mPER2: A Molecular Basis for FASPS and *tau*

We next wanted to elucidate the molecular processes that are altered in FASPS. In principle, there are several possibilities of how the phosphorylation at the FASPS position and presumably also the following serine and threonine residues of the CKIε/δ phosphorylation cluster might influence the oscillation dynamics. These include the regulation of PER2 stability, PER2 subcellular localization, and/or PER2 inhibition of CLOCK-BMAL1 activity.

We have analyzed all of these possibilities. Alterations in the ability of PER2-FASPS to inhibit *trans*-activation by CLOCK:BMAL1 were not observed. Surprisingly, however, we found that the stability of PER2-FASPS is not—as it might be expected—increased compared to PER2 wild-type protein but is substantially reduced. Although the protein half-life of PER2-wt is about 3 hours, the half-life of PER2-FASPS is reduced to about 1.5 hours. The destabilization is even more pronounced in the PER2-mut-7 variant, in which the whole CKIε/δ phosphorylation cluster is mutated, whereas the stability of the PER2-S659D mutant is not altered (Fig. 2) (Vanselow et al. 2006). Coexpression studies with CKIε revealed that the decreased stability of PER2-FASPS is due to its higher sensitivity toward CKIε-mediated proteasomal degradation. Notably, Takano and colleagues made similar observations when analyzing the corresponding sites in PER1. Coexpression of the PER1 mutant with CKIε resulted in a reduced stability and higher phosphorylation state (Takano et al. 2004; Takano and Nagai 2006). Thus, phosphorylation of PER proteins does not always trigger the degradation pathway but can have differential functions. Although phosphorylation in the FASPS region stabilizes PER proteins, phosphorylation at other residues promotes proteasomal degradation.

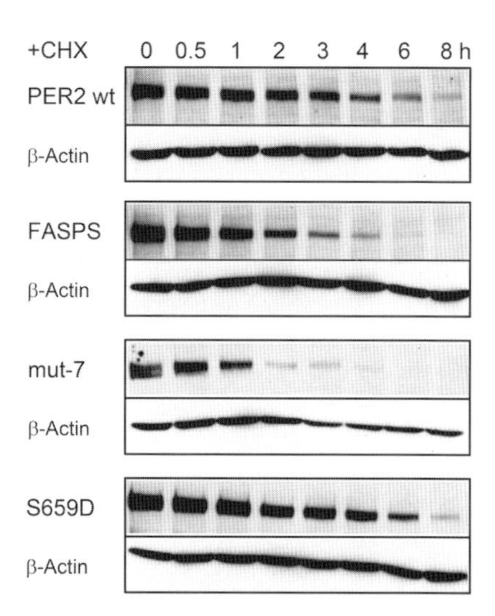

Figure 2. The FASPS mutation destabilizes the PER2 protein. NIH-3T3 cells stably expressing indicated PER2 variants were treated for up to 8 hours with the protein translation inhibitor cycloheximid (CHX). The PER2 protein amount was determined by SDS-PAGE/western blotting. (Reprinted, with permission, from Vanselow et al. 2006.)

Next, we wanted to answer the question of why phosphorylation in the FASPS region stabilizes PER2. To this end, we tested whether stabilization is an indirect consequence of an altered subcellular localization of PER2. Although we could not detect a different nuclear import rate of PER2-FASPS, we did observe an accelerated nuclear clearance of the PER2-FASPS protein (Fig. 3)

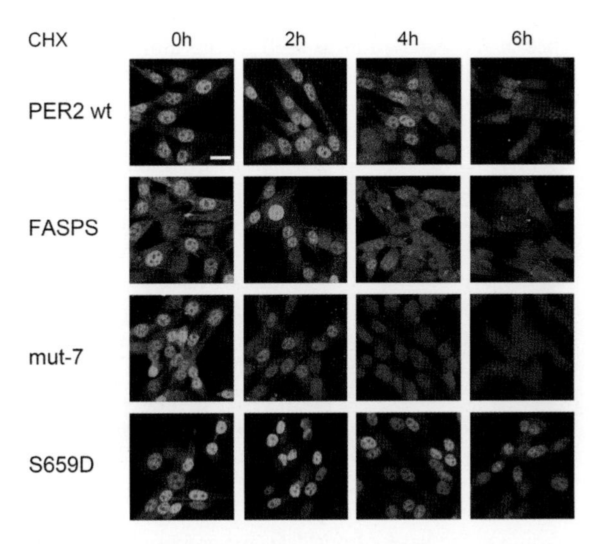

Figure 3. The FASPS mutation leads to a premature nuclear clearance. PER2 variant expressing NIH-3T3 cells were treated with the protein translation inhibitor cycloheximid (CHX) for the indicated times. Subcellular localization of PER2 proteins was visualized after immunostaining by confocal fluorescent microscopy. (Reprinted, with permission, from Vanselow et al. 2006.)

(Vanselow et al. 2006). This result is similar to observations for PER1 and PER2 in the SCN of *tau* mutant hamsters (Dey et al. 2005). Thus, phosphorylation at the FASPS region increases nuclear retention of PER2, thereby protecting it from cytoplasmic degradation by the proteasome. In the FASPS mutation, however, we propose that the mutant PER2 protein is prematurely exported, which leads to an earlier release of CLOCK-BMAL1 repression and earlier restart of a new cycle (Fig. 4) (Vanselow et al. 2006).

Xu et al. (2007) generated transgenic mice to study the molecular mechanisms underlying FASPS. The protein levels of PER2-FASPS (especially in the nuclear fraction) in the liver of those transgenic mice as well as in fibroblasts from skin biopsies of FASPS patients were found to be reduced as compared to controls. As in our study, this indicates a decreased stability of the PER2-FASPS protein. However, because the authors did not find a stabilizing effect of the proteasomal inhibitor MG132, they excluded the possibility of an altered stability of PER2-FASPS as the primary consequence of the mutation. Instead, they found a reduction of the *Per2* mRNA in FASPS and concluded that phosphorylation at the FASPS position has an impact on *Per2* transcription, a result we did not observe in our model system.

Functionally different phosphorylation sites in PER2 may also—at least qualitatively—explain the molecular mechanism for the *CKI*δ-based variant of FASPS as well as the *tau* phenotype in the hamster. In contrast to the *Per2*-based version of FASPS, *CKI*δ-based FASPS and *tau* could be explained by an altered ability of the kinases to phosphorylate PER2 (note, the reduced kinase activity

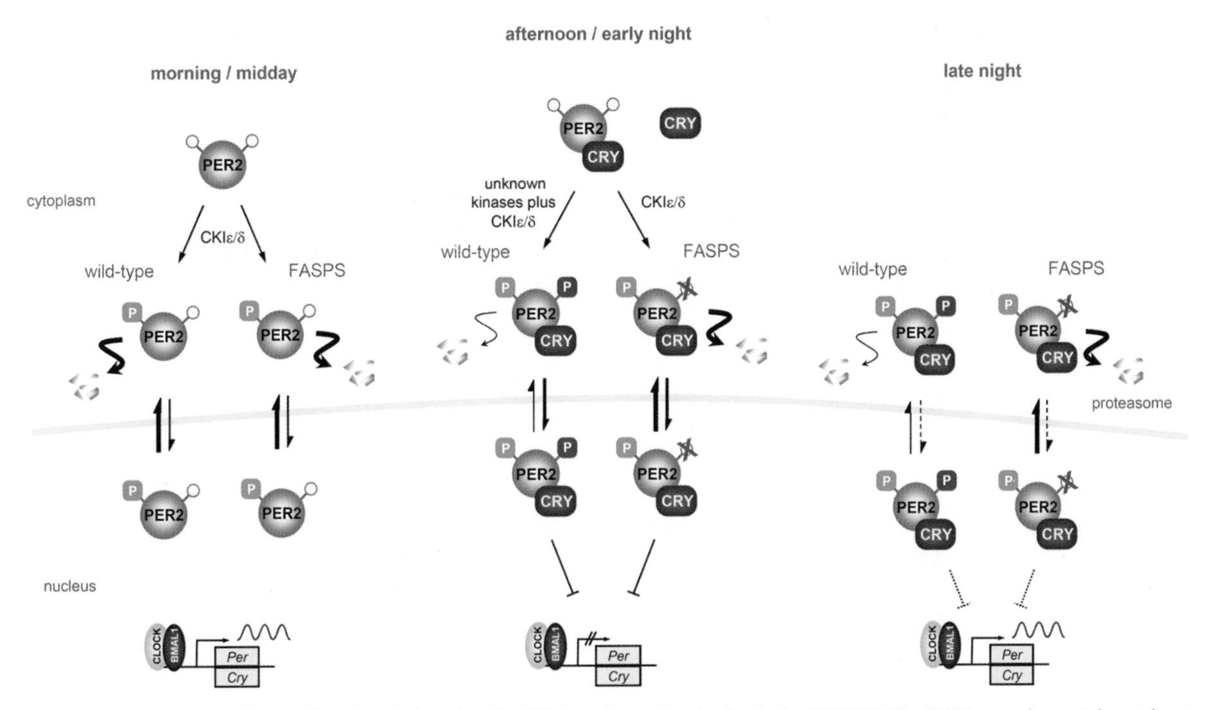

Figure 4. Functional different phosphorylation sites in PER2 as the molecular basis for FASPS. The PER2 protein contains at least two functionally different phosphorylation sites—one kind primarily mediating proteasomal degradation (*green*), and the other enhancing nuclear retention (*purple*). For mechanistic details, see the main text. (Reprinted, with permission, from Vanselow et al. 2006.)

of CKIδ-T44A and CKIε[*tau*] in vitro). Here, the key assumption is that the overall decrease in kinase activity does not lead to an evenly reduction of phosphorylation at all PER2 phosphorylation sites but to a specific decrease or inability to phosphorylate specific residues in the protein. According to our model, both CKIδ-T44A and CKIε(*tau*) are impaired in the phosphorylation of sites, which stabilize PER2 (maybe even at the FASPS region), whereas the phosphorylation at destabilizing degradation-promoting sites is unaffected or less affected. This may lead to a premature nuclear clearance of PER proteins and therefore to the observed early phase and short period of the FASPS individuals or *tau* hamsters, respectively.

To conceptualize our results, and to better understand their implications for circadian rhythm generation in a more quantitative manner, we constructed a mathematical model of PER2 phosphorylation in the circadian system (Vanselow et al. 2006). The model shows robust self-sustained oscillations with a period of 24 hours and delay of 6 hours between the *Per2* mRNA and PER2 protein rhythms. Furthermore, the simulation of an intervention with phosphorylation events that promote nuclear retention produces a short period and early phase as observed for FASPS patients. The model was also successfully applied to deduce testable predictions for the effects of specific system perturbations. For example, one biochemical prediction from the mathematical model is that the coexpression of PER2-wt with CKIε(*tau*) or CKIδ-T44A should destabilize the PER2-wt protein to a higher degree than the coexpression of PER2-wt with CKIε(wt). Concomitantly, the destabilizing effects of CKIε, CKIε(*tau*), and CKIδ-T44A on PER2-FASPS should be similar to each other. We tested these predictions for the *tau* kinase in cultured cells, and as predicted, the destabilizing effect of CKIε(*tau*) on PER2-wt was much more pronounced than that of the wild-type kinase. Furthermore, the destabilizing effect of CKIε(*tau*) on PER2-wt and PER2-FASPS was comparable.

In line with these results, Gallego et al. (2006b) found a higher destabilizing effect of the *tau* kinase on PER1 and PER2 in cell culture. The starting point of their experiments was mathematical modeling, which claimed an increased kinase activity (gain-of-function) for CKIε(*tau*) in vivo as the only possible explanation for the short-period phenotype. With respect to the *tau* mutation, our model and that of Gallego et al. (2006b) offer alternative but not mutually exclusive explanations. We favor a mechanism where a reduced kinase activity of mutant CKIε leads to an altered, less nuclear localization of PER proteins (as observed by Dey et al. 2005). Because phosphorylation-induced proteasomal degradation of PER proteins is primarily a cytosolic event (Vanselow et al. 2006), the overall phosphorylation of the PER proteins in *tau* hamsters may even increase (as observed by Gallego et al. 2006b). We propose that this increased phosphorylation is primarily due to a higher amount of PER substrate in the cytoplasm, rather than to an increased kinase activity. Nevertheless, the available data on the *tau* mutation fit both models. With respect to the *Per2*-based FASPS mutation, however, the cellular localization data (Vanselow et al. 2006) strongly favor the localization-dependent model that we postulate.

CONCLUSIONS

Experimental results from the last two decades underline the crucial importance of posttranslational mechanisms for both the generation and dynamics of circadian rhythms. Specifically, the reversible phosphorylation of the PER proteins in *Drosophila* and mammals seems to have a key role for the correct timing of nuclear repression. Animals with altered PER phosphorylation (by either a mutant PER protein or a mutant kinase) have striking circadian phenotypes at the behavioral as well as the molecular level. Biochemical analyses indicate that phosphorylation regulates not only the stability of PER proteins, but also their subcellular localization and inhibitory activity. Until recently, it was unknown which of the many potential phosphorylation sites of PER proteins are functionally relevant in vivo. Our aim is to perform a comprehensive phosphorylation function analysis of clock proteins. For that purpose, we developed an optimized mass spectrometric approach and were able to identify endogenous phosphorylation sites of mPER2. In a cell culture model system, we then studied the oscillation dynamics of cells expressing phosphosite-mutant PER proteins. This approach allowed us to analyze the molecular basis for the human FASPS. Remarkably, the phosphorylation site in PER2, which is mutated in FASPS, stabilizes the protein by means of nuclear retention and thus protects it from cytosolic degradation. On the basis of our data, we proposed a model of functionally different phosphorylation sites in PER2, which is also able to explain additional phosphorylation-based circadian phenotypes. We are currently investigating the role of other identified phosphorylation sites, especially with respect to potential destabilizing effects.

ACKNOWLEDGMENTS

We thank Ute Abraham for critical reading of the manuscript. Research in our lab is supported by the Deutsche Forschungsgemeinschaft (grants SFB 618, SFB 740) and the Sixth EU framework programme EUCLOCK.

REFERENCES

Akashi M., Tsuchiya Y., Yoshino T., and Nishida E. 2002. Control of intracellular dynamics of mammalian period proteins by casein kinase I epsilon (CKIepsilon) and CKIdelta in cultured cells. *Mol. Cell. Biol.* **22:** 1693.

Akten B., Jauch E., Genova G.K., Kim E.Y., Edery I., Raabe T., and Jackson F.R. 2003. A role for CK2 in the *Drosophila* circadian oscillator. *Nat. Neurosci.* **6:** 251.

Balsalobre A., Damiola F., and Schibler U. 1998. A serum shock induces circadian gene expression in mammalian tissue culture cells. *Cell* **93:** 929.

Bao S., Rihel J., Bjes E., Fan J.Y., and Price J.L. 2001. The *Drosophila* double-timeS mutation delays the nuclear accumulation of period protein and affects the feedback regulation of period mRNA. *J. Neurosci.* **21:** 7117.

Barnes J.W., Tischkau S.A., Barnes J.A., Mitchell J.W., Burgoon P.W., Hickok J.R., and Gillette M.U. 2003. Requirement of mammalian Timeless for circadian rhythmicity. *Science* **302:** 439.

Brown S.A., Ripperger J., Kadener S., Fleury-Olela F., Vilbois F., Rosbash M., and Schibler U. 2005a. PERIOD1-associated proteins modulate the negative limb of the mammalian circadian oscillator. *Science* **308:** 693.

Brown S.A., Fleury-Olela F., Nagoshi E., Hauser C., Juge C., Meier C.A., Chicheportiche R., Dayer J.M., Albrecht U., and Schibler U. 2005b. The period length of fibroblast circadian gene expression varies widely among human individuals. *PLoS Biol.* **3:** e338.

Busino L., Bassermann F., Maiolica A., Lee C., Nolan P.M., Godinho S.I., Draetta G.F., and Pagano M. 2007. SCFFbxl3 controls the oscillation of the circadian clock by directing the degradation of cryptochrome proteins. *Science* **316:** 900.

Camacho F., Cilio M., Guo Y., Virshup D.M., Patel K., Khorkova O., Styren S., Morse B., Yao Z., and Keesler G.A. 2001. Human casein kinase Idelta phosphorylation of human circadian clock proteins period 1 and 2. *FEBS Lett.* **489:** 159.

Cardone L., Hirayama J., Giordano F., Tamaru T., Palvimo J.J., and Sassone-Corsi P. 2005. Circadian clock control by SUMOylation of BMAL1. *Science* **309:** 1390.

Curtin K.D., Huang Z.J., and Rosbash M. 1995. Temporally regulated nuclear entry of the *Drosophila* period protein contributes to the circadian clock. *Neuron* **14:** 365.

Cyran S.A., Yiannoulos G., Buchsbaum A.M., Saez L., Young M.W., and Blau J. 2005. The double-time protein kinase regulates the subcellular localization of the *Drosophila* clock protein period. *J. Neurosci.* **25:** 5430.

Czeisler C.A., Duffy J.F., Shanahan T.L., Brown E.N., Mitchell J.F., Rimmer D.W., Ronda J.M., Silva E.J., Allan J.S., Emens J.S., Dijk D.J., and Kronauer R.E. 1999. Stability, precision, and near-24-hour period of the human circadian pacemaker. *Science* **284:** 2177.

Dey J., Carr A.J., Cagampang F.R., Semikhodskii A.S., Loudon A.S., Hastings M.H., and Maywood E.S. 2005. The tau mutation in the Syrian hamster differentially reprograms the circadian clock in the SCN and peripheral tissues. *J. Biol. Rhythms* **20:** 99.

Edery I., Zwiebel L.J., Dembinska M.E., and Rosbash M. 1994. Temporal phosphorylation of the *Drosophila* period protein. *Proc. Natl. Acad. Sci.* **91:** 2260.

Eide E.J., Vielhaber E.L., Hinz W.A., and Virshup D.M. 2002. The circadian regulatory proteins BMAL1 and cryptochromes are substrates of casein kinase Iepsilon. *J. Biol. Chem.* **277:** 17248.

Eide E.J., Woolf M.F., Kang H., Woolf P., Hurst W., Camacho F., Vielhaber E.L., Giovanni A., and Virshup D.M. 2005. Control of mammalian circadian rhythm by CKIepsilon-regulated proteasome-mediated PER2 degradation. *Mol. Cell Biol.* **25:** 2795.

Flotow H., Graves P.R., Wang A.Q., Fiol C.J., Roeske R.W., and Roach P.J. 1990. Phosphate groups as substrate determinants for casein kinase I action. *J. Biol. Chem.* **265:** 14264.

Fujimoto Y., Yagita K., and Okamura H. 2006. Does mPER2 protein oscillate without its coding mRNA cycling?: Post-transcriptional regulation by cell clock. *Genes Cells* **11:** 525.

Gallego M., Kang H., and Virshup D.M. 2006a. Protein phosphatase 1 regulates the stability of the circadian protein PER2. *Biochem. J.* **399:** 169.

Gallego M., Eide E.J., Woolf M.F., Virshup D.M., and Forger D.B. 2006b. An opposite role for tau in circadian rhythms revealed by mathematical modeling. *Proc. Natl. Acad. Sci.* **103:** 10618.

Godinho S.I., Maywood E.S., Shaw L., Tucci V., Barnard A.R., Busino L., Pagano M., Kendall R., Quwailid M.M., Romero M.R., O'Neill J., Chesham J.E., Brooker D., Lalanne Z., Hastings M.H., and Nolan P.M. 2007. The after-hours mutant reveals a role for Fbxl3 in determining mammalian circadian period. *Science* **316:** 897.

Grima B., Lamouroux A., Chelot E., Papin C., Limbourg-Bouchon B., and Rouyer F. 2002. The F-box protein slimb controls the levels of clock proteins period and timeless. *Nature* **420:** 178.

Harada Y., Sakai M., Kurabayashi N., Hirota T., and Fukada Y. 2005. Ser-557-phosphorylated mCRY2 is degraded upon synergistic phosphorylation by glycogen synthase kinase-3 beta. *J. Biol. Chem.* **280:** 31714.

Hardin P.E., Hall J.C., and Rosbash M. 1990. Feedback of the *Drosophila* period gene product on circadian cycling of its messenger RNA levels. *Nature* **343:** 536.

Hofmann K.P., Spahn C.M., Heinrich R., and Heinemann U. 2006. Building functional modules from molecular interactions. *Trends Biochem. Sci.* **31:** 497.

Holmberg C.I., Tran S.E., Eriksson J.E., and Sistonen L. 2002. Multisite phosphorylation provides sophisticated regulation of transcription factors. *Trends Biochem. Sci.* **27:** 619.

Honma S., Kawamoto T., Takagi Y., Fujimoto K., Sato F., Noshiro M., Kato Y., and Honma K. 2002. Dec1 and Dec2 are regulators of the mammalian molecular clock. *Nature* **419:** 841.

Iitaka C., Miyazaki K., Akaike T., and Ishida N. 2005. A role for glycogen synthase kinase-3beta in the mammalian circadian clock. *J. Biol. Chem.* **280:** 29397.

Jones C.R., Campbell S.S., Zone S.E., Cooper F., DeSano A., Murphy P.J., Jones B., Czajkowski L., and Ptacek L.J. 1999. Familial advanced sleep-phase syndrome: A short-period circadian rhythm variant in humans. *Nat. Med.* **5:** 1062.

Keesler G.A., Camacho F., Guo Y., Virshup D., Mondadori C., and Yao Z. 2000. Phosphorylation and destabilization of human period I clock protein by human casein kinase I epsilon. *Neuroreport* **11:** 951.

Kim E.Y. and Edery I. 2006. Balance between DBT/CKIepsilon kinase and protein phosphatase activities regulate phosphorylation and stability of *Drosophila* CLOCK protein. *Proc. Natl. Acad. Sci.* **103:** 6178.

Kloss B., Price J.L., Saez L., Blau J., Rothenfluh A., Wesley C.S., and Young M.W. 1998. The *Drosophila* clock gene double-time encodes a protein closely related to human casein kinase Iepsilon. *Cell* **94:** 97.

Ko H.W., Jiang J., and Edery I. 2002. Role for Slimb in the degradation of *Drosophila* Period protein phosphorylated by Doubletime. *Nature* **420:** 673.

Konopka R.J. and Benzer S. 1971. Clock mutants of *Drosophila melanogaster*. *Proc. Natl. Acad. Sci.* **68:** 2112.

Lee C., Etchegaray J.P., Cagampang F.R., Loudon A.S., and Reppert S.M. 2001. Posttranslational mechanisms regulate the mammalian circadian clock. *Cell* **107:** 855.

Lin J.M., Schroeder A., and Allada R. 2005. In vivo circadian function of casein kinase 2 phosphorylation sites in *Drosophila* PERIOD. *J. Neurosci.* **25:** 11175.

Lin J.M., Kilman V.L., Keegan K., Paddock B., Emery-Le M., Rosbash M., and Allada R. 2002. A role for casein kinase 2alpha in the *Drosophila* circadian clock. *Nature* **420:** 816.

Lowrey P.L., Shimomura K., Antoch M.P., Yamazaki S., Zemenides P.D., Ralph M.R., Menaker M., and Takahashi J.S. 2000. Positional syntenic cloning and functional characterization of the mammalian circadian mutation tau. *Science* **288:** 483.

Martinek S., Inonog S., Manoukian A.S., and Young M.W. 2001. A role for the segment polarity gene shaggy/GSK-3 in the *Drosophila* circadian clock. *Cell* **105:** 769.

Nakajima M., Imai K., Ito H., Nishiwaki T., Murayama Y., Iwasaki H., Oyama T., and Kondo T. 2005. Reconstitution of circadian oscillation of cyanobacterial KaiC phosphorylation in vitro. *Science* **308:** 414.

Nawathean P. and Rosbash M. 2004. The doubletime and CKII kinases collaborate to potentiate *Drosophila* PER transcriptional repressor activity. *Mol. Cell* **13:** 213.

Nishii K., Yamanaka I., Yasuda M., Kiyohara Y.B., Kitayama Y., Kondo T., and Yagita K. 2006. Rhythmic post-transcriptional regulation of the circadian clock protein mPER2 in mammalian cells: A real-time analysis. *Neurosci. Lett.* **401:** 44.

O'Gorman S., Fox D.T., and Wahl G.M. 1991. Recombinase-mediated gene activation and site-specific integration in mammalian cells. *Science* **251:** 1351.

Partch C.L., Shields K.F., Thompson C.L., Selby C.P., and Sancar A. 2006. Posttranslational regulation of the mammalian circadian clock by cryptochrome and protein phosphatase 5. *Proc. Natl. Acad. Sci.* **103:** 10467.

Preitner N., Damiola F., Lopez-Molina L., Zakany J., Duboule D., Albrecht U., and Schibler U. 2002. The orphan nuclear receptor REV-ERBalpha controls circadian transcription within the positive limb of the mammalian circadian oscillator. *Cell* **110:** 251.

Preuss F., Fan J.Y., Kalive M., Bao S., Schuenemann E., Bjes E.S., and Price J.L. 2004. *Drosophila* doubletime mutations which either shorten or lengthen the period of circadian rhythms decrease the protein kinase activity of casein kinase I. *Mol. Cell. Biol.* **24:** 886.

Price J.L., Blau J., Rothenfluh A., Abodeely M., Kloss B., and Young M.W. 1998. double-time is a novel *Drosophila* clock gene that regulates PERIOD protein accumulation. *Cell* **94:** 83.

Ralph M.R. and Menaker M. 1988. A mutation of the circadian system in golden hamsters. *Science* **241:** 1225.

Reischl S., Vanselow K., Westermark P.O., Thierfelder N., Maier B., Herzel H., and Kramer A. 2007. β-TrCP1-mediated degradation of PERIOD2 is essential for circadian dynamics. *J. Biol. Rhythms* **22:** 375.

Rothenfluh A., Abodeely M., and Young M.W. 2000. Short-period mutations of per affect a double-time-dependent step in the *Drosophila* circadian clock. *Curr. Biol.* **10:** 1399.

Sangoram A.M., Saez L., Antoch M.P., Gekakis N., Staknis D., Whiteley A., Fruechte E.M., Vitaterna M.H., Shimomura K., King D.P., Young M.W., Weitz C.J., and Takahashi J.S. 1998. Mammalian circadian autoregulatory loop: A timeless ortholog and mPer1 interact and negatively regulate CLOCK-BMAL1-induced transcription. *Neuron* **21:** 1101.

Sathyanarayanan S., Zheng X., Xiao R., and Sehgal A. 2004. Posttranslational regulation of *Drosophila* PERIOD protein by protein phosphatase 2A. *Cell* **116:** 603.

Schlosser A., Vanselow J.T., and Kramer A. 2005. Mapping of phosphorylation sites by a multi-protease approach with specific phosphopeptide enrichment and NanoLC-MS/MS analysis. *Anal. Chem.* **77:** 5243.

Shirogane T., Jin J., Ang X.L., and Harper J.W. 2005. SCFbeta-TRCP controls clock-dependent transcription via casein kinase 1-dependent degradation of the mammalian period-1 (Per1) protein. *J. Biol. Chem.* **280:** 26863.

Siepka S.M., Yoo S.H., Park J., Song W., Kumar V., Hu Y., Lee C., and Takahashi J.S. 2007. Circadian mutant Overtime reveals F-box protein FBXL3 regulation of cryptochrome and period gene expression. *Cell* **129:** 1011.

Sugano S., Andronis C., Ong M.S., Green R.M., and Tobin E.M. 1999. The protein kinase CK2 is involved in regulation of circadian rhythms in *Arabidopsis*. *Proc. Natl. Acad. Sci.* **96:** 12362.

Suri V., Hall J.C., and Rosbash M. 2000. Two novel doubletime mutants alter circadian properties and eliminate the delay between RNA and protein in *Drosophila*. *J. Neurosci.* **20:** 7547.

Takano A. and Nagai K. 2006. Serine 714 might be implicated in the regulation of the phosphorylation in other areas of mPer1 protein. *Biochem. Biophys. Res. Commun.* **346:** 95.

Takano A., Isojima Y., and Nagai K. 2004. Identification of mPer1 phosphorylation sites responsible for the nuclear entry. *J. Biol. Chem.* **279:** 32578.

Takano A., Shimizu K., Kani S., Buijs R.M., Okada M., and Nagai K. 2000. Cloning and characterization of rat casein kinase 1epsilon. *FEBS Lett.* **477:** 106.

Toh K.L., Jones C.R., He Y., Eide E.J., Hinz W.A., Virshup D.M., Ptacek L.J., and Fu Y.H. 2001. An hPer2 phosphorylation site mutation in familial advanced sleep phase syndrome. *Science* **291:** 1040.

Tomita J., Nakajima M., Kondo T., and Iwasaki H. 2005. No transcription-translation feedback in circadian rhythm of KaiC phosphorylation. *Science* **307:** 251.

Vanselow K., Vanselow J.T., Westermark P.O., Reischl S., Maier B., Korte T., Herrmann A., Herzel H., Schlosser A., and Kramer A. 2006. Differential effects of PER2 phosphorylation: Molecular basis for the human familial advanced sleep phase syndrome (FASPS). *Genes Dev.* **20:** 2660.

Vielhaber E., Eide E., Rivers A., Gao Z.H., and Virshup D.M. 2000. Nuclear entry of the circadian regulator mPER1 is controlled by mammalian casein kinase I epsilon. *Mol. Cell. Biol.* **20:** 4888.

Xu Y., Toh K.L., Jones C.R., Shin J.Y., Fu Y.H., and Ptacek L.J. 2007. Modeling of a human circadian mutation yields insights into clock regulation by PER2. *Cell* **128:** 59.

Xu Y., Padiath Q.S., Shapiro R.E., Jones C.R., Wu S.C., Saigoh N., Saigoh K., Ptacek L.J., and Fu Y.H. 2005. Functional consequences of a CKIdelta mutation causing familial advanced sleep phase syndrome. *Nature* **434:** 640.

Yagita K., Tamanini F., Der Horst G.T., and Okamura H. 2001. Molecular mechanisms of the biological clock in cultured fibroblasts. *Science* **292:** 278.

Yagita K., Tamanini F., Yasuda M., Hoeijmakers J.H., van der Horst G.T., and Okamura H. 2002. Nucleocytoplasmic shuttling and mCRY-dependent inhibition of ubiquitylation of the mPER2 clock protein. *EMBO J.* **21:** 1301.

Yang Y., Cheng P., and Liu Y. 2002. Regulation of the *Neurospora* circadian clock by casein kinase II. *Genes Dev.* **16:** 994.

Yang Z. and Sehgal A. 2001. Role of molecular oscillations in generating behavioral rhythms in *Drosophila*. *Neuron* **29:** 453.

Yin L., Wang J., Klein P.S., and Lazar M.A. 2006. Nuclear receptor Rev-erbalpha is a critical lithium-sensitive component of the circadian clock. *Science* **311:** 1002.

Yu Q., Jacquier A.C., Citri Y., Hamblen M., Hall J.C., and Rosbash M. 1987. Molecular mapping of point mutations in the period gene that stop or speed up biological clocks in *Drosophila melanogaster*. *Proc. Natl. Acad. Sci.* **84:** 784.

Yu W., Zheng H., Houl J.H., Dauwalder B., and Hardin P.E. 2006. PER-dependent rhythms in CLK phosphorylation and E-box binding regulate circadian transcription. *Genes Dev.* **20:** 723.

Zeng H., Qian Z., Myers M.P., and Rosbash M. 1996. A light-entrainment mechanism for the *Drosophila* circadian clock. *Nature* **380:** 129.

Zerr D.M., Hall J.C., Rosbash M., and Siwicki K.K. 1990. Circadian fluctuations of period protein immunoreactivity in the CNS and the visual system of *Drosophila*. *J. Neurosci.* **10:** 2749.

Zhao W.N., Malinin N., Yang F.C., Staknis D., Gekakis N., Maier B., Reischl S., Kramer A., and Weitz C.J. 2007. CIPC is a mammalian circadian clock protein without invertebrate homologues. *Nat. Cell Biol.* **9:** 268.

Posttranslational Regulation of *Neurospora* Circadian Clock by CK1a-dependent Phosphorylation

C. QUERFURTH, A. DIERNFELLNER, F. HEISE, L. LAUINGER, A. NEISS, Ö. TATAROGLU, M. BRUNNER, AND T. SCHAFMEIER

Heidelberg University Biochemistry Center, 69120 Heidelberg, Germany

Frequency (FRQ) and its transcriptional activator, the White Collar Complex (WCC), are essential components of interconnected feedback loops of the circadian clock of *Neurospora*. In a negative feedback loop, FRQ inhibits the WCC by recruiting casein kinase 1a (CK1a) and supporting its phosphorylation. In an interconnected positive loop, FRQ supports accumulation of high levels of WCC. Phosphorylation of clock proteins is crucial for the temporal and spatial coordination of these functions. We identified three isoforms of CK1a generated by alternative splicing that all interact with FRQ. Furthermore, we show that WC-2 is phosphorylated by CK1a in vitro and that WC-2 phosphorylation is inhibited in vivo by the CK1-specific inhibitor IC261. Finally, we demonstrate that CK1a activity regulates levels of WC-2.

INTRODUCTION

Circadian clocks are timekeeping devices that organize the physiology and behavior of most eukaryotic organisms in anticipation of environmental changes associated with the daily rhythm of earth rotation. On the molecular level, circadian clocks are cell-autonomous oscillators consisting of interconnected transcriptional and translational feedback loops. The oscillations are synchronized with the exogenous day by environmental cues ("zeitgebers") such as light and temperature. In the absence of zeitgebers, clock-driven oscillations persist with intriguingly precise periods, generating self-sustained subjective daily rhythms that usually deviate from 24 hours.

FRQ and the GATA-type zinc finger transcription factors White Collar-1 (WC-1) and WC-2 are core elements of interconnected feedback loops of the circadian clock of *Neurospora crassa* (Dunlap and Loros 2006; Liu and Bell-Pedersen 2006). WC-1 and WC-2 assemble and form the WCC that activates *frq* transcription. FRQ protein then inhibits the activity of WCC and thus feeds back on its own synthesis and regulates other genes controlled by WCC. Due to the feedback regulation, *frq* RNA and FRQ protein show robust circadian abundance rhythms. In an interconnected positive loop, FRQ supports accumulation of high levels of WCC, primarily on a posttranscriptional or posttranslational level (Lee et al. 2000; Cheng et al. 2001; Schafmeier et al. 2006). These apparently conflicting functions of FRQ are coordinated in a temporal and spatial fashion: Negative feedback is carried out by low levels of nuclear FRQ, whereas support of WCC levels requires progressive accumulation of high levels of FRQ in the cytosol (Schafmeier et al. 2006).

As a prerequisite for a robust and exactly timed circadian period, posttranscriptional and posttranslational regulation of the clock occurs on the level of mRNA processing (i.e., antisense RNA and alternative splicing), regulation of protein turnover and subcellular shuttling, as well as modulation of transcription factor activity (Diernfellner et al. 2005; Liu 2005; Brunner and Schafmeier 2006; Gallego and Virshup 2007). Phosphorylation is the most obvious modification of clock proteins (Schafmeier et al. 2005; Mizoguchi et al. 2006). Protein kinases and phosphatases that regulate circadian systems in a wide range of eukaryotic organisms have been described. Among them are CK1 and CK2, which are essential for a proper function of circadian systems in fungi, flies, and mammals (Kloss et al. 1998; Camacho et al. 2001; Blau 2003; Nawathean and Rosbash 2004; Eide et al. 2005; Knippschild et al. 2005; He et al. 2006; Gallego and Virshup 2007).

In the *Neurospora* genome, two genes coding for CK1 have been identified (Gorl et al. 2001). CK1a, a homolog of *Drosophila* Doubletime, was the first clock-regulating protein kinase identified (Gorl et al. 2001; He et al. 2006). CK1b, although important for normal growth and development, has no apparent clock-related function (Gorl et al. 2001; Yang et al. 2003). Several other kinases and phosphatases have been characterized that have distinct roles in the *Neurospora* circadian system. In particular, CK2, calcium/calmodulin-dependent kinase 1, checkpoint kinase 2, protein kinase C, and the protein phosphatases 1 and 2A functionally modulate clock proteins (Yang et al. 2001, 2002, 2004; Franchi et al. 2005; Pregueiro et al. 2006).

Here we summarize recent data concerning the role of CK1 in the circadian clock of *N. crassa*. Furthermore, we present evidence for a direct role of CK1 in WC-2 phosphorylation and discuss the existence of a clock-specific CK1 isoform.

RESULTS

Isoforms of CK1a in *N. crassa*

CK1s comprise a large family of evolutionary conserved eukaryotic kinases. They consist of a highly conserved amino-terminal catalytical domain of approximately 300 amino acid residues followed by a carboxy-terminal tail that is not conserved and variable in length. The carboxy-

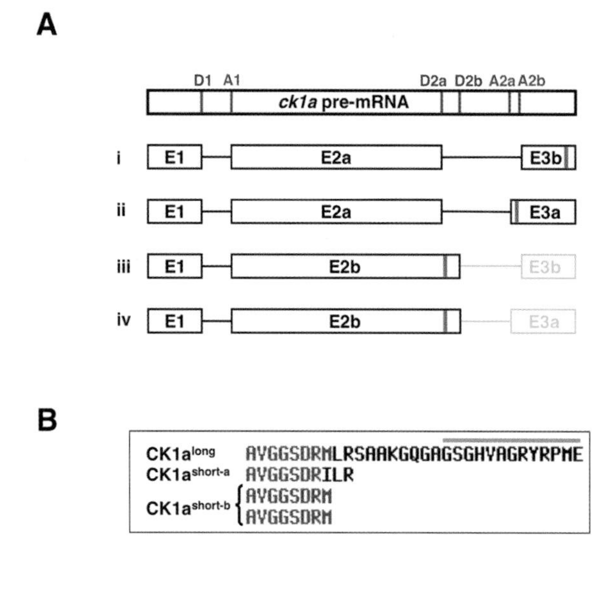

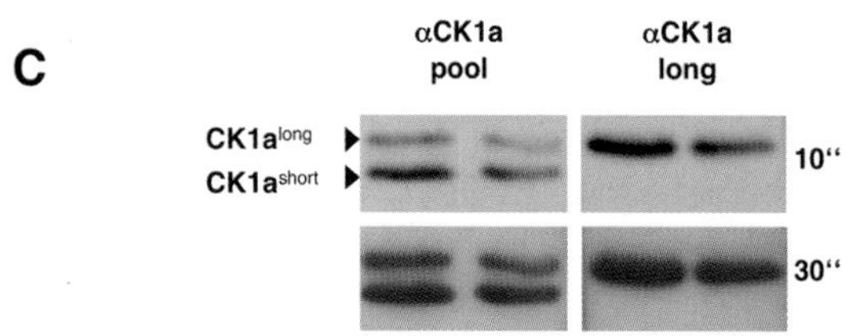

Figure 1. CK1a isoforms in *N. crassa*. (*A*) Schematic drawing of *ck1a* splice variants identified by cDNA sequencing (*i–iv*). In the unprocessed mRNA (*top*), splice donors (D) are shown as *blue lines* and acceptors (A) are shown as *red lines*. Each cDNA isoform contains three exons, referred to as E1, E2a/b, and E3a/b. (*Brown lines*) Stop codons used in the corresponding splice variants. For clarity, the 3′ region downstream from donor D2a is not to scale. (*B*) Carboxyl termini of corresponding gene products. Note that two isoforms (CK1a$^{short\text{-}b}$) are identical due to a stop codon upstream of D2b. (*Orange line*; isoform I) Epitope recognized by the CK1a-long antibody. (*Red*) Identical; (*blue*) similar amino acids. (*C*) Detection of CK1a by western blotting. (*Left panel*) The αCK1a-pool antiserum was raised against a fragment consisting of the 132 carboxy-terminal amino acid residues of isoform I and recognizes two bands in a *Neurospora* protein extract. Bands are labeled CK1a^{long} and CK1a^{short}, which contains CK1a$^{short\text{-}a}$ and CK1a$^{short\text{-}b}$. (*Right panel*) αCK1a-long antiserum specifically detects CK1a^{long}. Different exposures (10 sec and 30 sec) are shown.

terminal domains are thought to confer substrate specificity (Gross and Anderson 1998; Knippschild et al. 2005). Seven *ck1* genes are present in mammals (Knippschild et al. 2005), whereas the *Neurospora* genome contains only two genes, *ck1a* and *ck1b* (Gorl et al. 2001). Ck1a, which is most homologous to mammalian CK1ε/δ and *Drosophila* Doubletime (Dbt), is essential for viability and has, like its orthologs, a function in the circadian clock (Gorl et al. 2001). *ck1a* contains an intron in the 5′ portion of the open reading frame (ORF). Inspection of the genomic DNA additionally suggested the presence of two alternative splice donors, D2a and D2b, followed by two alternative acceptors, A2a and A2b. (Fig. 1A). Cloning and sequencing of *ck1a* cDNAs revealed that these splice sites are used, giving rise to four alternatively spliced mRNA species. D2a is located within the CK1 ORF immediately upstream of the stop codon in exon 2. The two mRNAs spliced at D2a encode a long (D2a/A2a) and medium size (D2a/A2b) isoform of CK1a, designated CK1a^{long} and CK1a$^{short\text{-}a}$ (Fig. 1B). Because D2b is located downstream from the stop codon in exon 2, the two corresponding mRNA isoforms (D2b/A2a and D2b/A2b) encode the same polypeptide, CK1a$^{short\text{-}b}$ (Fig. 1B). We generated an antibody against a fragment of CK1a comprising the carboxy-terminal 132-amino-acid residues. This antibody, αCK1a-pool, is predicted to recognize all isoforms of CK1a. The antibody detected two bands (Fig. 1C, left lanes). The upper band corresponds in size to CK1a^{long} and the lower band, to CK1a$^{short\text{-}a}$ and CK1a$^{short\text{-}b}$, which are not resolved by SDS-PAGE. Thus, CK1a$^{short\text{-}a}$ and CK1a$^{short\text{-}b}$ will henceforth be referred to as CK1a^{short}. Furthermore, we generated a peptide antibody against the extreme carboxyl terminus of CK1a^{long} (last 13 amino acids, epitope is marked in Fig. 1B). The antiserum recognized a single band of expected size (Fig. 1C, right lanes). Thus, we have generated useful tools for further analysis of CK1a.

Biochemical Properties of CK1a

Despite its simple structure, CK1 is regulated in multiple ways, including inhibition by autophosphorylation, subcellular distribution, and probably also dimerization (Longenecker et al. 1998). We wondered whether the *Neurospora* CK1a isoforms are differentially distributed and whether they interact with each other. Subcellular fractionation of light-grown mycelia revealed that the CK1a isoforms are distributed in the same manner between cytosol and nuclei. The subcellular localization was not affected in *frq*-deficient strains (Fig. 2A).

Because CK1 is supposed to form dimers, interaction of the different isoforms could be a possible reason for the equal distribution. To analyze whether CK1a forms heterodimers, we immunoprecipitated CK1a^{long} with an antibody specific to the carboxy-terminal tail and performed western analysis with αCK1a-pool. As shown in Figure 2B, CK1a^{long} was immunodepleted from the protein extract by affinity-purified αCK1a-long antiserum. In contrast, CK1a^{short} remained entirely in the supernatant. These data suggest that the CK1a^{long} does not form stable heterodimers with the medium-size and short isoforms.

To analyze whether CK1a forms dimers at all, we constructed a version of the *ck1a* gene that contains an amino-terminal double-FLAG tag. The splice sites giving rise to the CK1a isoforms were present in this recombinant gene. The tagged gene was expressed in a wild-type background, and immunoprecipitation was performed with αFLAG agarose. As shown in Figure 2C, the αFLAG agarose efficiently immunoprecipitated the FLAG-tagged isoforms of CK1a. In addition, endogenous CK1a isoforms were coprecipitated, indicating dimerization of CK1a. Together, the data suggest that CK1a has the potential to form homodimers, whereas heterodimers may not form or are less stable.

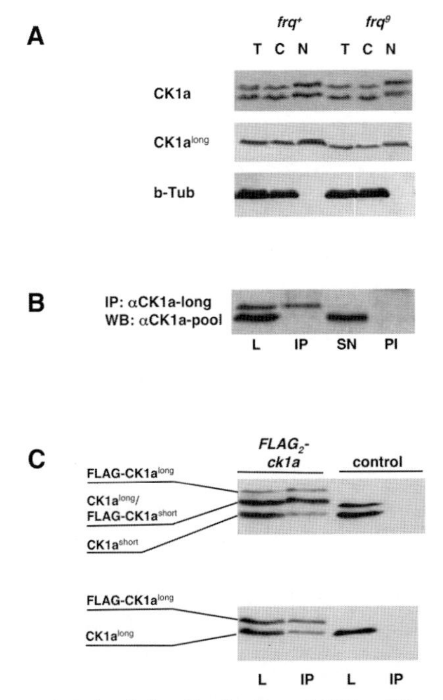

Figure 2. (*A*) Subcellular distribution of CK1a. (*Upper panel*) Western blot of subcellular fractions of *frq*$^{+}$ and *frq*^{9} mycelia decorated with αCK1a-pool antiserum. (*Middle panel*) Analysis of the same fractions with αCK1a-long antiserum. (*Lower panel*) Blots are redecorated with α-tubulin antiserum to control the fractionation. (T) Total; (C) cytosol; (N) nuclei. (*B*) Immunoprecipitation of total protein extracts with αCK1a-long antiserum. Western blots were decorated with αCK1a-pool antibody. Immunoprecipitation with preimmune serum was performed for control. (L) Load; (IP) immunoprecipitation; (SN) supernatant; (PI) preimmune serum. (*C*) Immunoprecipitaton with αFLAG agarose of extracts from a strain expressing FLAG-tagged CK1a isoforms in addition to endogenous CK1a. Western blots are immunodecorated with αCK1a-pool (*upper panel*) and αCK1a-long antiserum (*lower panel*).

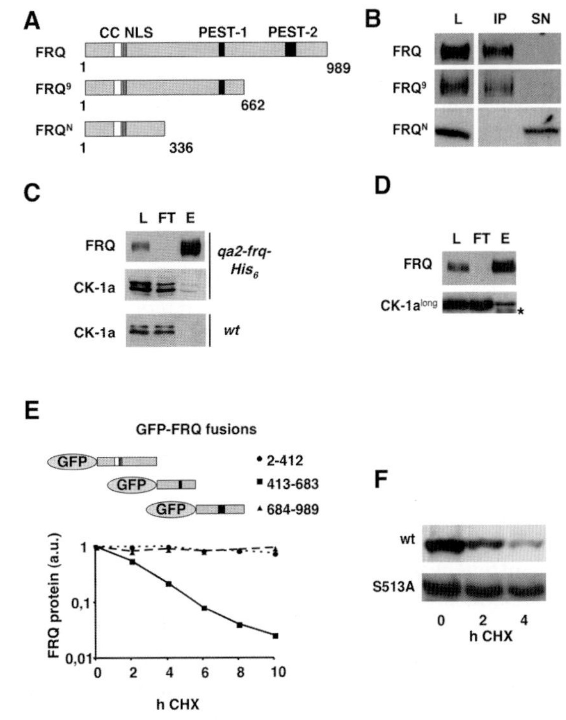

Figure 3. Interaction of CK1a with FRQ. (*A*) Scheme of FRQ constructs expressed for analysis of CK1a interaction. (CC) Coiled-coil domain; (NLS) nuclear localization sequence; (PEST) proline-, glutamate-, serine-, threonine-rich region. (*B*) Immunoprecipitation of total protein extracts with αCK1a-pool antiserum followed by western analysis. Immunodecoration was performed with monoclonal αFRQ antibody. (L) Load; (IP) immunoprecipitation; (SN) supernatant. (*C*) Ni-NTA purification of total protein extracts. (*Upper and middle panel*) Extract obtained from a strain expressing His_6-tagged FRQ; (*lower panel*) control experiment with wild-type extract. Western blots were decorated with αFRQ and αCK1a-pool antibody as indicated. (L) Load; (FT) flowthrough; (E) elution. (*D*) Same experiment as in *C*; western analysis with αCK1a-long antiserum. *Asterisk* indicates cross-reacting band enriched in the elution fraction. (*E*) The central domain of FRQ carries determinants for rapid protein turnover. FLAG-tagged GFP-fusion proteins with the amino-terminal, the middle, and the carboxy-terminal domain of FRQ are schematically outlined. FRQ-defficient strains expressing the fusion proteins were grown in constant light (LL) and transferred to medium containing 10 μg/ml cycloheximide (CHX), and samples were harvested after the indicated time points. Kinetics of degradation of GFP-FRQ–fusion proteins were determined by quantification of western blots. (*Circles*) Amino-terminal domain (2–412); (*squares*) central domain (413–683); (*triangles*) carboxy-terminal domain. (*F*) Exchange of S513 to alanine stabilizes the central portion of FRQ. Degradation kinetics of GFP-fusion proteins with the central portion of FRQ. (*Upper panel*) Wild type; (*lower panel*) S513 to alanine mutation. Western analysis and immunodecoration with αFLAG antibody are shown.

Interaction of CK1a with Clock Proteins

As reported previously, CK1a is in complex with FRQ, a central component of the circadian clock (Gorl et al. 2001). To identify the region of FRQ that interacts with CK1a, we analyzed strains expressing full-size FRQ, the carboxy-terminally truncated FRQ^{9} protein and an amino-terminal fragment of FRQ consisting of the first 366 amino acid residues (Fig. 3A). Extracts from light-grown mycelia were prepared and subjected to immunoprecipitation with αCK1a-pool antiserum (Fig. 3B). FRQ and FRQ^{9} copurified with αCK1a-pool antiserum, whereas the amino-terminal fragment of FRQ was not coprecipitated. These findings demonstrate that CK1a interacts with the central region of FRQ, in accordance with recent data showing that residues 488–496 of FRQ are crucial for CK1a binding (He et al. 2006).

To analyze whether the CK1a isoforms interact differently with FRQ, we used a strain that expresses a functional His-tagged version of FRQ under control of the *qa-2* promoter (Schafmeier et al. 2005). Extract prepared from light-grown mycelia was incubated with Ni-NTA resin. The His-tagged FRQ was depleted from the protein extract (Fig. 3C). A portion of CK1a specifically bound to the Ni-NTA resin when extract containing His-tagged FRQ was used (Fig. 3C,D). This confirms that CK1a is in a complex with FRQ. Both CK1a^{long} and CK1a^{short} were recovered at a ratio similar to that present in total extracts, suggesting that the isoforms interact equally well with

FRQ. The majority of CK1a did not bind to the Ni-NTA resin, indicating that it was not in complex with His-tagged FRQ. Because about 30% of FRQ coprecipitated when pull-down assays were performed with αCK1a-pool antiserum (Gorl et al. 2001), the data suggest that expression levels of CK1a are much higher than those of FRQ. CK1a has essential functions that are not related to the circadian clock.

In the course of a circadian day, FRQ is progressively phosphorylated and degraded with a half-time of 3–4 hours. The turnover kinetics is regulated by phosphorylation. In particular, putative phosphorylation sites in the central portion of FRQ have been reported to be crucial for FRQ stability. Thus, mutation of serine residue 513 (S513) to isoleucine (Liu et al. 2000) and deletion of the PEST-1 region (residues 540–566), which contains several putative phosphorylation sites (Gorl et al. 2001), led to a substantial stabilization of FRQ. We asked whether the central domain of FRQ carries the determinants that are sufficient to destabilize heterologous proteins and mediate their rapid turnover. Therefore, three fragments corresponding to the amino-terminal, carboxy-terminal, and the central portion of FRQ were fused to green fluorescent protein (GFP) and expressed in a *frq*-null background under control of the *frq* promoter. To assess the kinetics of turnover of these GFP-fusion proteins, cultures were incubated with cycloheximide (CHX) to block protein synthesis and samples were analyzed over a time course of 10 hours (Fig. 3E). The GFP fused with the central portion of FRQ was rapidly degraded with a half-life time of approximately 2 hours, whereas fusion proteins with the amino- and carboxy-terminal portion of FRQ were stable. When S513 was exchanged with alanine to abolish phosphorylation, the corresponding fusion protein with the central portion of FRQ was stable (Fig. 3F). In summary, the results demonstrate that the central portion of FRQ carries the determinants for rapid regulated protein turnover.

FRQ inhibits the activity of the WCC by mediating its phosphorylation (Schafmeier et al. 2005) and CK1a is, at least partially, responsible for the FRQ-dependent phosphorylation of WCC (He et al. 2006). We asked whether CK1a directly interacts with the WCC. WC-1 and WC-2 are GATA-type zinc finger proteins that bind efficiently to Ni-NTA resin. As shown in Figure 4A, both proteins were fully depleted from a *Neurospora* protein extract passed over a Ni-NTA column. However, CK1a did not bind to the affinity matrix, indicating that it is not in a stable complex with the WC proteins.

FRQ was expressed under control of the inducible *qa-2* promoter in a *frq*-null background. FRQ was efficiently induced over a time course of 8 hours (Fig. 4B). The phosphorylation status of WC-2 was analyzed by two-dimensional gel electrophoresis. WC-2, which was hypophosphorylated before induction of FRQ, was efficiently phosphorylated 4 hours after FRQ induction. This demonstrates that the WCC is phosphorylated in response to FRQ expression. CK1a is essential for the viability of *Neurospora* (Gorl et al. 2001). To access whether FRQ recruits CK1a to phosphorylate the WCC, cultures were treated with the CK1 inhibitor IC261 (Behrend et al. 2000).

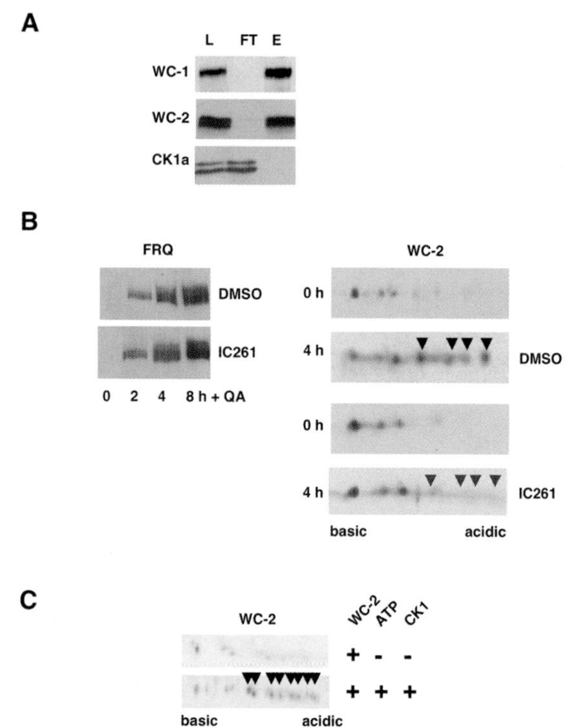

Figure 4. (*A*) CK1a is not associated with the WC proteins. The GATA-type zinc finger proteins WC-1 and WC-2 were affinity-purified with Ni-NTA agarose. Total extracts loaded on the Ni-NTA column (L), the flowthrough fraction (FT), and the bound proteins eluted with imidazole (E) were analyzed by western blotting with the indicated antisera. Copurification of CK1a with the WC proteins was not observed. (*B*) Inhibition of FRQ-dependent phosphorylation of WC-2 by the CK1 inhibitor IC261. Dark-grown frq^9 *qa-2frq* cultures were shifted to medium containing 0.3% quinic acid (QA) in the presence of 10 μM IC261 and DMSO (dimethylsulfoxide) for control. Total extracts were subjected to SDS-PAGE and immunoblotting with monoclonal αFRQ (*left*) or two-dimensional electrophoresis and analysis with αWC-2 (*right*). (*Black arrowheads*) Hyperphosphorylated WC-2 appearing in response to FRQ induction; (*red arrowheads*) expected positions of hyperphosphorylated WC-2 forms. (*C*) Phosphorylation of immunopurified WC-2 by recombinant CK1a in vitro. Hypophosphorylated WC-2 was immunoprecipitated from frq^{10} extracts and incubated with ATP and recombinant $CK1a^{long}$. Samples were analyzed by two-dimensional gel electrophoresis. Arrowheads indicate hyperphosphorylated forms of WC-2.

FRQ expression, as well as the overall phosphorylation pattern of FRQ, was not affected by IC261. This suggests that CK1a was not efficiently inhibited by IC261 and/or that other kinases are able to phosphorylate FRQ (Fig. 4B, left). However, IC261 interfered with FRQ-dependent phosphorylation of WC-2 (Fig. 4B, right). Hyperphosphorylated species of WC-2, which appeared 4 hours after FRQ induction in the control, were absent in the sample with IC261. This suggests that FRQ recruits CK1 and promotes WC-2 phosphorylation.

In an additional approach, we investigated whether purified recombinant CK1a is able to phosphorylate WC-2 in vitro. We purified hypophosphorylated WC-2 from a *frq*-null strain by immunoprecipitation. When the immunoprecipitate was incubated with recombinant CK1-His_6 and ATP, WC-2 was efficiently phosphorylated at multiple sites. Phosphorylation of WC-2 in vitro was more efficient than in vivo, suggesting that in a living cell, phosphorylation is regulated and/or antagonized by phosphatases (Fig. 4C). Taken together, our data present evidence for a direct phosphorylation of WC-2 by CK1a.

Effects of Constitutive Active and Dominant-negative Casein Kinase

For a further investigation of the function of CK1a in the *Neurospora* clock, we generated a constitutive active (CA) and a dominant-negative (DN) *Ck1a* allele, which can be expressed under control of the inducible *qa-2* promoter in a wild-type background. To obtain a CA allele, we inserted a stop codon (Q299^{TER}) to delete the regulatory carboxy-terminal domain. The DN form was generated by exchange of a conserved aspartic acid with asparagine (D131N) in accordance to a DN-CKI from *Xenopus* (Peters et al. 1999). To distinguish mutant and endogenous CK1a forms, we inserted an amino-terminal double-FLAG tag as shown in Figure 5A.

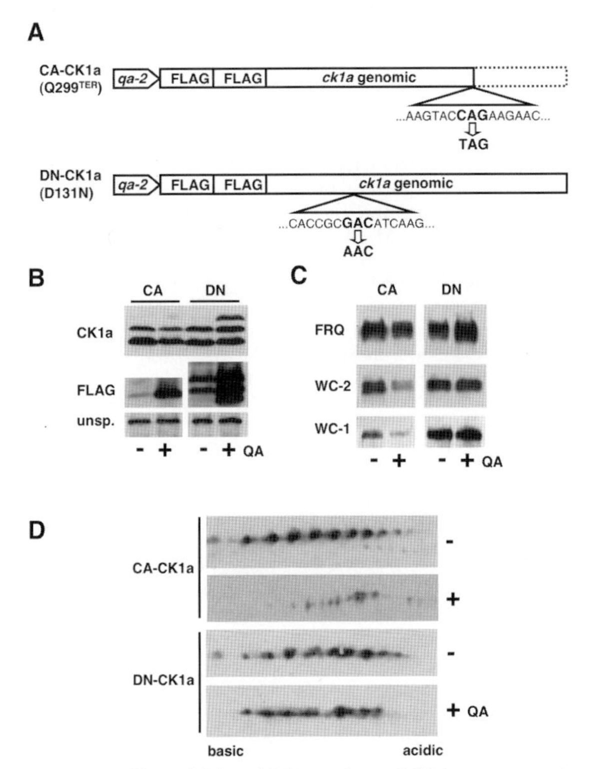

Figure 5. Effect of CA and DN versions of CK1a on expression levels and phosphorylation status of clock proteins. (*A*) Schematic drawing of FLAG-tagged *ck1a* constructs under control of the inducible *qa-2* promoter. The indicated point mutations were introduced to generate CA and DN CK1a. (*B*) Induction (6 hours) of CA-CK1a and DN-CK1a expression with quinic acid (QA). (*Upper panels*) Western blot of total protein extracts decorated with αCK1a-pool antiserum; (*middle panels*) same extracts were analyzed with monoclonal αFLAG antibody; (*lower panels*) unspecific band recognized by the αFLAG antibody confirms equal loading of the gel. (*C*) Expression of clock proteins in DN and CA mutants. Same extracts as in *B* were analyzed with αFRQ, αWC-2, or αWC-1 antibodies. (*D*) Two-dimensional analysis of extracts obtained from DN and CA mutants with αWC-2 antiserum.

After quinic acid (QA) induction, both mutant forms were detected in *Neurospora* protein extracts with an αFLAG antibody. Before induction, low levels were detected probably due to a weak leaky expression. After induction, expression levels increased (Fig. 5B). As expected, one form of the carboxy-terminally truncated CA-CK1a was detected and several splice isoforms of DN-CK1a were expressed. The DN-CK1a isoforms together with the endogenous CK1a were also detected by the αCK1a-pool antiserum, which was raised against the carboxy-terminal fragment of CK1a (132 amino acid residues). However, the short and medium splice isoforms of FLAG-tagged DN-CK1a comigrate with the long isoform of endogenous CK1a. Thus, only the long isoform of FLAG-tagged DN-CK1a can be clearly distinguished. Because the carboxyl terminus is lacking in the CA-CK1a, the protein cannot be detected with our αCK1a antisera.

Expression of CA-CK1a caused a slight reduction of FRQ and a pronounced reduction of WC-2 levels, suggesting a role of CK1a-dependent phosphorylation in the regulation of FRQ and WC-2 turnover (Fig. 5C). The DN form of the CK1a did not affect expression levels of FRQ and WC-2 (Fig. 5C).

To investigate the influence of CA-CK1a and DN-CK1a on the phosphorylation status of WC-2, we performed two-dimensional gel electrophoresis. Before induction of CA-CK1a and DN-CK1a, WC-2 was highly phosphorylated at multiple sites (Fig. 5D). When expression of CA-CK1a was induced by supplementing the growth medium with 0.3% QA for 6 hours, WC-2 protein levels decreased and mainly hyperphosphorylated species were detected. In particular, unphosphorylated and hypophosphorylated forms were completely absent (Fig. 5D, upper). Expression of DN-CK1a did not significantly affect expression and phosphorylation of WC-2 (Fig. 5D, lower), suggesting that it may not efficiently compete with the endogenous CK1a.

In summary, we show that expression of a CA-CK1a form results in hyperphosphorylation of WC-2 and a reduction of WC-2 protein levels. FRQ expression may also be affected by CA-CK1a, although to a much lesser extent.

CONCLUSIONS

We addressed the role of CK1a in the *Neurospora* circadian clock and identified four splice isoforms of *ck1a* mRNA. These mRNAs encode three polypeptides differing in length and primary structure of their extreme carboxyl termini. Carboxy-terminal variations were reported to determine substrate specificity of casein kinase isoforms (Knippschild et al. 2005). We show that the clock protein FRQ interacts with an affinity similar to that of the long and short isoforms of CK1a. Because CK1a has at least two known substrates among the clock proteins of *Neurospora*, i.e., FRQ and WCC, it is tempting to speculate whether the isoforms differ in specificity for phosphorylation of FRQ and WCC.

The CK1a-binding site in FRQ was recently mapped (He et al. 2006). Our data confirm that CK1a binds to the central portion of FRQ. Furthermore, our data show that the central domain of FRQ carries the determinants for rapid phosphorylation-dependent protein turnover. Thus, a fusion of the central domain to GFP is sufficient to promote rapid, regulated turnover. The central domain harbors, in addition to the CK1a interaction site, the PEST-1 domain and S513, both crucial elements for phosphorylation-dependent degradation (Liu et al. 2000; Gorl et al. 2001). The GFP-fusion protein with the central domain of FRQ was stabilized when S513 was exchanged to an alanyl residue. This demonstrates that the central portion of FRQ is a functional domain that regulates turnover via phosphorylation.

The phosphorylation of WCC depends on recruitment of CK1a by FRQ (He et al. 2006). This suggests that CK1a directly phosphorylates the WCC and implies a transient ternary complex of FRQ, WCC, and the kinase (Fig. 6).

A stable interaction of CK1a with the WCC was not detected. However, we provide evidence that WC-2 is a direct substrate of CK1a. Thus, WC-2 is phosphorylated by recombinant CK1a in vitro, demonstrating that it contains phosphorylation sites recognized by the kinase. In addition, IC261, a CK1-specific inhibitor, reduces phosphorylation of WC-2 in vivo. The apparent phosphorylation status of FRQ was not affected by IC261, suggesting that CK1a was still partially active and/or that other kinases substitute for CK1a.

Expression of a CA form of CK1a substantially reduced WC-2 expression. This observation suggests that both WCC activity (Schafmeier et al. 2005) and protein turnover are regulated via phosphorylation by CK1a. This is reminiscent to distinct CK1a phosphorylations of FRQ at PEST-1 and PEST-2 that regulate turnover and function, respectively (Schafmeier et al. 2006). Similarly, *Drosophila* PER and CLOCK proteins are also differentially regulated by phosphorylation, suggesting that similar functional aspects are regulated by phosphorylation of clock proteins (Kim et al. 2002; Kim and Edery 2006; Gallego and Virshup 2007). Whether the function and stability of clock proteins are regulated by distinct CK1a isoforms remains to be investigated.

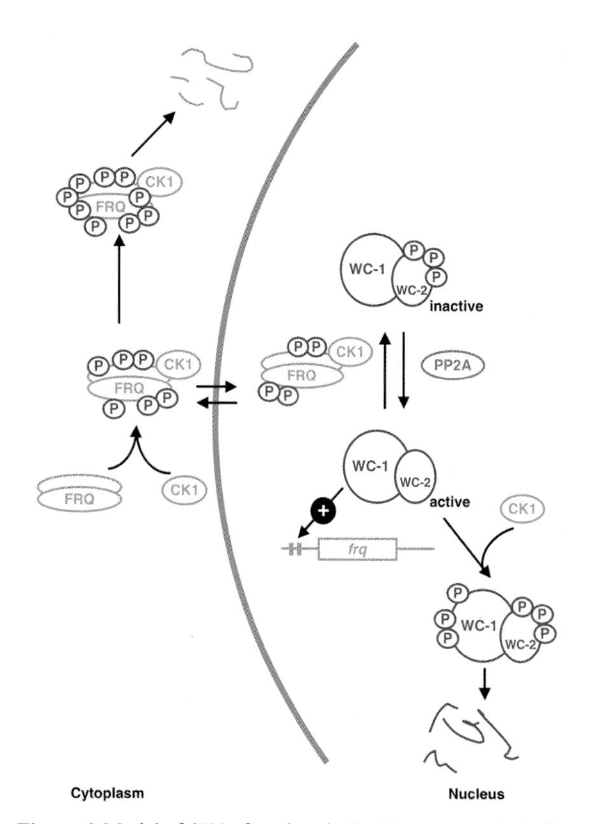

Figure 6. Model of CK1a functions in the *Neurospora* clock. For details, see text.

ACKNOWLEDGMENTS

We thank J. Scholz for excellent technical assistance. This work was supported by grants from Deutsche Forschungsgemeinschaft (BR 1375-1 and SFB 638) and by Fonds der Chemischen Industrie to M.B.

REFERENCES

Behrend L., Milne D.M., Stoter M., Deppert W., Campbell L.E., Meek D.W., and Knippschild U. 2000. IC261, a specific inhibitor of the protein kinases casein kinase 1-delta and -epsilon, triggers the mitotic checkpoint and induces p53-dependent postmitotic effects. *Oncogene* **19:** 5303.

Blau J. 2003. A new role for an old kinase: CK2 and the circadian clock. *Nat. Neurosci.* **6:** 208.

Brunner M. and Schafmeier T. 2006. Transcriptional and post-transcriptional regulation of the circadian clock of cyanobacteria and *Neurospora*. *Genes Dev.* **20:** 1061.

Camacho F., Cilio M., Guo Y., Virshup D.M., Patel K., Khorkova O., Styren S., Morse B., Yao Z., and Keesler G.A. 2001. Human casein kinase Idelta phosphorylation of human circadian clock proteins period 1 and 2. *FEBS Lett.* **489:** 159.

Cheng P., Yang Y., and Liu Y. 2001. Interlocked feedback loops contribute to the robustness of the *Neurospora* circadian clock. *Proc. Natl. Acad. Sci.* **98:** 7408.

Diernfellner A.C.R., Schafmeier T., Merrow M.W., and Brunner M. 2005. Molecular mechanism of temperature-sensing by the circadian clock of *Neurospora crassa*. *Genes Dev.* **19:** 1968.

Dunlap J.C. and Loros J.J. 2006. How fungi keep time: Circadian system in *Neurospora* and other fungi. *Curr. Opin. Microbiol.* **9:** 579.

Eide E.J., Kang H., Crapo S., Gallego M., and Virshup D.M. 2005. Casein kinase I in the mammalian circadian clock. *Methods Enzymol.* **393:** 408.

Franchi L., Fulci V., and Macino G. 2005. Protein kinase C modulates light responses in *Neurospora* by regulating the blue light photoreceptor WC-1. *Mol. Microbiol.* **56:** 334.

Gallego M. and Virshup D.M. 2007. Post-translational modifications regulate the ticking of the circadian clock. *Nat. Rev. Mol. Cell Biol.* **8:** 139.

Gorl M., Merrow M., Huttner B., Johnson J., Roenneberg T., and Brunner M. 2001. A PEST-like element in FREQUENCY determines the length of the circadian period in *Neurospora crassa*. *EMBO J.* **20:** 7074.

Gross S.D. and Anderson R.A. 1998. Casein kinase I: Spatial organization and positioning of a multifunctional protein kinase family. *Cell. Signal.* **10:** 699.

He Q., Cha J., He Q., Lee H.C., Yang Y., and Liu Y. 2006. CKI and CKII mediate the FREQUENCY-dependent phosphorylation of the WHITE COLLAR complex to close the *Neurospora* circadian negative feedback loop. *Genes Dev.* **20:** 2552.

Kim E.Y. and Edery I. 2006. Balance between DBT/CKIepsilon kinase and protein phosphatase activities regulate phosphorylation and stability of *Drosophila* CLOCK protein. *Proc. Natl.*

Acad. Sci. **103:** 6178.

Kim E.Y., Bae K., Ng F.S., Glossop N.R., Hardin P.E., and Edery I. 2002. *Drosophila* CLOCK protein is under posttranscriptional control and influences light-induced activity. *Neuron* **34:** 69.

Kloss B., Price J.L., Saez L., Blau J., Rothenfluh A., Wesley C.S., and Young M.W. 1998. The *Drosophila* clock gene double-time encodes a protein closely related to human casein kinase Iepsilon. *Cell* **94:** 97.

Knippschild U., Gocht A., Wolff S., Huber N., Lohler J., and Stoter M. 2005. The casein kinase 1 family: Participation in multiple cellular processes in eukaryotes. *Cell. Signal.* **17:** 675.

Lee K., Loros J.J., and Dunlap J.C. 2000. Interconnected feedback loops in the *Neurospora* circadian system. *Science* **289:** 107.

Liu Y. 2005. Analysis of posttranslational regulations in the *Neurospora* circadian clock. *Methods Enzymol.* **393:** 379.

Liu Y. and Bell-Pedersen D. 2006. Circadian rhythms in *Neurospora crassa* and other filamentous fungi. *Eukaryot. Cell* **5:** 1184.

Liu Y., Loros J., and Dunlap J.C. 2000. Phosphorylation of the *Neurospora* clock protein FREQUENCY determines its degradation rate and strongly influences the period length of the circadian clock. *Proc. Natl. Acad. Sci.* **97:** 234.

Longenecker K.L., Roach P.J., and Hurley T.D. 1998. Crystallographic studies of casein kinase I delta toward a structural understanding of auto-inhibition. *Acta Crystallogr. D Biol. Crystallogr.* **54:** 473.

Mizoguchi T., Putterill J., and Ohkoshi Y. 2006. Kinase and phosphatase: The cog and spring of the circadian clock. *Int. Rev. Cytol.* **250:** 47.

Nawathean P. and Rosbash M. 2004. The doubletime and CKII kinases collaborate to potentiate *Drosophila* PER transcriptional repressor activity. *Mol. Cell* **13:** 213.

Peters J.M., McKay R.M., McKay J.P., and Graff J.M. 1999. Casein kinase I transduces Wnt signals. *Nature* **401:** 345.

Pregueiro A.M., Liu Q., Baker C.L., Dunlap J.C., and Loros J.J. 2006. The *Neurospora* checkpoint kinase 2: A regulatory link between the circadian and cell cycles. *Science* **313:** 644.

Schafmeier T., Kaldi K., Diernfellner A., Mohr C., and Brunner M. 2006. Phosphorylation-dependent maturation of *Neurospora* circadian clock protein from a nuclear repressor toward a cytoplasmic activator. *Genes Dev.* **20:** 297.

Schafmeier T., Haase A., Kaldi K., Scholz J., Fuchs M., and Brunner M. 2005. Transcriptional feedback of *Neurospora* circadian clock gene by phosphorylation-dependent inactivation of its transcription factor. *Cell* **122:** 235.

Yang Y., Cheng P., and Liu Y. 2002. Regulation of the *Neurospora* circadian clock by casein kinase II. *Genes Dev.* **16:** 994.

Yang Y., Cheng P., Zhi G., and Liu Y. 2001. Identification of a calcium/calmodulin-dependent protein kinase that phosphorylates the *Neurospora* circadian clock protein FREQUENCY. *J. Biol. Chem.* **276:** 41064.

Yang Y., Cheng P., He Q., Wang L., and Liu Y. 2003. Phosphorylation of FREQUENCY protein by casein kinase II is necessary for the function of the *Neurospora* circadian clock. *Mol. Cell. Biol.* **23:** 6221.

Yang Y., He Q., Cheng P., Wrage P., Yarden O., and Liu Y. 2004. Distinct roles for PP1 and PP2A in the *Neurospora* circadian clock. *Genes Dev.* **18:** 255.

Posttranslational Control of the *Neurospora* Circadian Clock

J. Cha, G. Huang, J. Guo, and Y. Liu
Department of Physiology, University of Texas Southwestern Medical Center, Dallas, Texas 75390-9040

The eukaryotic circadian clocks are composed of autoregulatory circadian negative feedback loops that include both positive and negative elements. Investigations of the *Neurospora* circadian clock system have elucidated many of the basic mechanisms that underlie circadian rhythms, including negative feedback and light and temperature entrainment common to all eukaryotic clocks. The conservation of the posttranslational regulators in divergent circadian systems suggests that the processes mediating the modification and degradation of clock proteins may be the common foundation that allows the evolution of circadian clocks in eukaryotic systems. In this chapter, we summarize recent studies of the *Neurospora* circadian clock with emphasis on posttranslational regulation in the circadian negative feedback loop.

INTRODUCTION

Circadian clocks regulate a wide variety of physiological and molecular activities in almost all eukaryotic organisms and in certain prokaryotic organisms. *Neurospora* is an excellent experimental organism for studying circadian rhythms because of its relative simplicity and because the easily monitored circadian rhythm in asexual spore development (conidiation) has proven extremely useful for measuring the effects of mutations on clock function. Furthermore, systematic knockouts of a large number of genes are available from the *Neurospora* genome project (http://www.dartmouth.edu/~neurosporagenome/). In addition to the characterization of the clock, *Neurospora* has served to understand the light input pathway, temperature entrainment, and the output pathway of the clock (Liu 2003; Diernfellner et al. 2005; Liu and Bell-Pedersen 2006).

Like the circadian oscillator in *Drosophila* and mammals, the core circadian oscillator of *Neurospora* consists of an autoregulatory negative feedback loop with four core components: FREQUENCY (FRQ), FRH (an FRQ-interacting RNA helicase), WHITE COLLAR-1 (WC-1), and WC-2 (Dunlap 1999; Dunlap and Loros 2004; Liu and Bell-Pedersen 2006; Heintzen and Liu 2007). In this negative feedback loop, a complex of FRQ and FRH (called FFC) forms the negative limb of the loop, and the PER-ARNT-SIM (PAS) domain-containing transcription factors WC-1 and WC-2 form the WC complex (WCC) and acts as the positive element. The WCC binds to two *cis* elements in the promoter of *frq* to activate its transcription. On the other hand, FFC represses *frq* transcription by inhibiting the activity of WCC. After the progressive phosphorylation and degradation of FRQ, the reactivation of WCC leads to reactivation of *frq* transcription and the start of a new cycle. Thus, this negative feedback loop generates endogenous circadian oscillations of *frq* mRNA and FRQ protein that regulate rhythmicity close to 24 hours. In addition to its essential role in the circadian negative feedback loop, WC-1 is also the blue light photoreceptor responsible for circadian entrainment and all other known light responses, emphasizing the link between light input and the circadian oscillator (Froehlich et al. 2002; He et al. 2002; Cheng et al. 2003a; He and Liu 2005).

This chapter summarizes the progress that has been made in recent years in understanding the molecular basis of the *Neurospora* circadian negative feedback loop with the emphasis on roles of posttranslational modifications.

THE CIRCADIAN FEEDBACK LOOPS IN *NEUROSPORA*

Our current understanding of the regulation of the *Neurospora* circadian clock is depicted in Figure 1. In constant darkness around the subjective morning, WC-1 and WC-2 form a heterodimeric complex (D-WCC) that binds to the Clock box (C box) in the *frq* promoter, leading to the activation of *frq* transcription (Crosthwaite et al. 1997; Cheng et al. 2001a; Froehlich et al. 2003; He and Liu 2005). *frq* mRNA reaches its peak in the subjective day and FRQ protein amount peaks 4–6 hours later (Aronson et al. 1994; Garceau et al. 1997). After FRQ protein is synthesized, it dimerizes through its coiled-coil domain and forms a complex with FRH (Cheng et al. 2001b, 2005). In the nucleus, FFC inhibits D-WCC activity, resulting in a decrease in *frq* mRNA levels; *frq* mRNA level reaches a trough around the subjective early evening (Aronson et al. 1994; Merrow et al. 1997; Luo et al. 1998). As soon as FRQ is synthesized, it is progressively phosphorylated by several kinases and dephosphorylated by two phosphatases (Heintzen and Liu 2007). When FRQ becomes extensively phosphorylated, it interacts with FWD-1, an F box/WD-40 repeat-containing protein and the substrate-recruiting subunit of an SCF-type ubiquitin ligase complex, resulting in the ubiquitination and degradation of FRQ by the proteasome system (He et al. 2003). When FRQ levels drop below a certain threshold around subjective late night, D-WCC is no longer inhibited by FFC and *frq* transcription is reactivated to start a new cycle. As a result of this autoregulatory negative feedback loop, *frq* mRNA and FRQ protein levels oscillate with daily rhythms. These oscillations are critical for

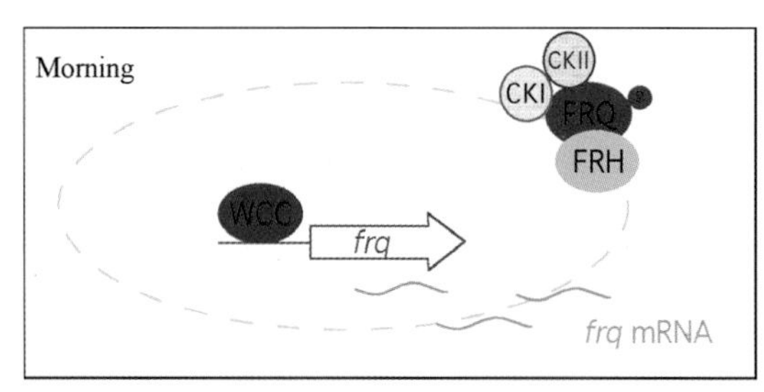

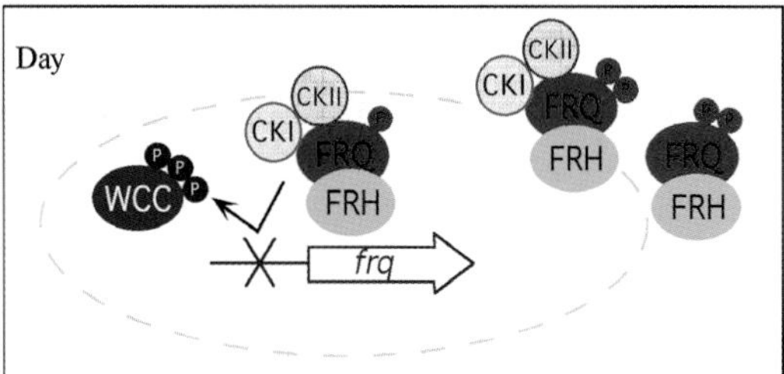

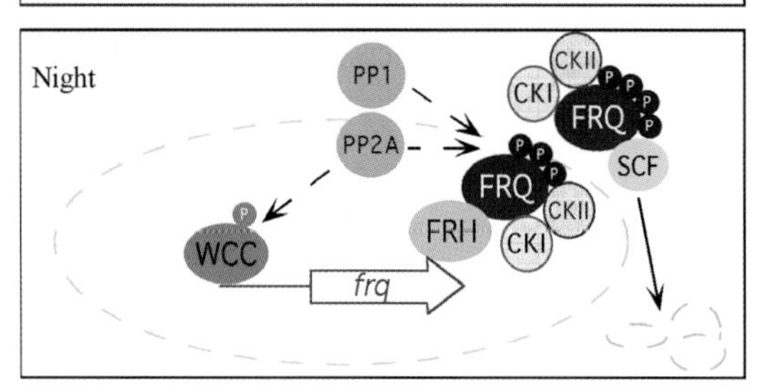

Figure 1. Current working model of *Neurospora* circadian clock. In the subjective morning, FRQ levels are low and WCs are hypophosphorylated. The hypophosphorylated WCC binds the C box to activate *frq* transcription. After FRQ protein is made, the homodimeric FRQ and FRH form FFC and the complex associates with CKI and CKII. In the subjective day, FFC interacts with WCC and recruits CKI and CKII to phosphorylate the WC proteins. As FRQ levels increase, WCC is extensively phosphorylated and therefore inactivated. This process results in the dissociation of WCC from the *frq* promoter and a decrease in *frq* transcription. At the same time, FRQ is progressively phosphorylated by CKI and CKII. The extensively phosphorylated FRQ is recognized by the SCF^{FWD-1} E3 ubiquitin ligase, leading to the ubiquitination and degradation of FRQ through the proteasome pathway. Through the night, FRQ levels decrease and the inactive hyperphosphorylated WCC is probably dephosphorylated by PP2A, which results in the reactivation of WCC and *frq* transcription in a new circadian cycle.

the normal circadian behavior of the organism (Garceau et al. 1997). The light and temperature entrainment of the *Neurospora* clock is caused by changes of FRQ expression levels after light and temperature treatments (Crosthwaite et al. 1995; Liu et al. 1998).

The central role of FRQ in the *Neurospora* circadian clock was highlighted by our recent finding that FRQ is a state variable of the *Neurospora* circadian clock (Huang et al. 2006). The induction of *frq* expression at a critical phase and with appropriate strength by light pulse, temperature step-up, or inducible expression alone triggers arrythmicity (also called singularity behavior) at the physiological and molecular levels. In addition, we showed that the amplitude of FRQ rhythm reflects the amplitude of the clock and determines the sensitivity of the clock to phase-resetting stimulus. Furthermore, our results suggest that the circadian singularity behavior is a two-step process: The critical treatment first drives the circadian negative feedback loop to a steady state, after which the cell populations become desynchronized.

Beside the repression of D-WCC activity in the dark, FRQ promotes the accumulation of WC-1 and WC-2, forming positive feedback loops that are interlocked with the negative loop (Lee et al. 2000; Cheng et al. 2001a, 2003b; Merrow et al. 2001). Such interlocked positive feedback loops are also shared by animal circadian systems (Glossop et al. 1999; Shearman et al. 2000). FRQ regulates WC-1 expression posttranscriptionally, whereas it promotes *wc-2* transcription. Our experiments in which *wc-1* or *wc-2* were overexpressed from an inducible promoter to different levels demonstrated that the positive feedback loops are important for maintaining the robustness and stability of the clock (Cheng et al. 2001a). It was recently shown that the phosphorylation of the PEST-2 region of FRQ is important for the accumulation of WC-1 but not WC-2 (Schafmeier et al. 2006). We showed recently that casein kinase 1a (CK-1a) is important for maintaining the steady-state levels of WC-1 and WC-2 (He et al. 2006). Together, these results suggest that the positive role of FRQ on WC levels is at least in part mediated by protein phosphorylation.

In addition to their essential role in the circadian negative feedback loop in the dark, WC-1 and WC-2 are required for all known light responses in *Neurospora*, including the entrainment of the circadian clock (Ballario et al. 1996; Crosthwaite et al. 1997; Cheng et al. 2003b; Liu 2003). WC-1 binds to the chromophore FAD through its photosensory LOV (light, oxygen, or voltage) domain, a specialized PAS domain, and functions as the blue light photoreceptor for light responses (He et al. 2002; Cheng et al. 2003b). Light triggers the formation of a large WC complex that binds to the promoters of light-inducible genes (Froehlich et al. 2002; He and Liu 2005), resulting in light-induced transcription and light responses.

ACTIVATION OF *frq* TRANSCRIPTION

Both WC-1 and WC-2 are PAS domain-containing transcription factors (WC-1 has three PAS domains, whereas WC-2 contains only one) with GATA-type zinc finger DNA-binding domains, and they primarily reside in the nucleus (Ballario et al. 1996; Linden and Macino 1997; Talora et al. 1999; Schwerdtfeger and Linden 2000). These proteins interact through the PASC domain of WC-1 and the PAS domain of WC-2 to form the heterodimeric D-WCC (Talora et al. 1999; Cheng et al. 2002, 2003b). WC-2 is required to maintain the steady-state level of WC-1, and WC-1 levels are very low in the *wc-2* null mutant and in mutants with disrupted WC-1/WC-2 interaction (Cheng et al. 2002). It is possible that WC-1 is unstable or cannot fold properly in the absence of WC-2. D-WCC binds to the C box in the *frq* promoter and activates *frq* transcription (Froehlich et al. 2003; He and Liu 2005). Its binding to the C box is rhythmic during a circadian cycle and the deletion of the C box abrogates rhythm of *frq* mRNA accumulation, indicating that D-WCC binding is essential for the function of the clock (Froehlich et al.

2003; He et al. 2006). *frq* mRNA and FRQ protein levels are extremely low in *wc* null mutants and the induction of WC expression in *wc* mutants from an ectopic locus leads to rapid induction of *frq* transcription (Crosthwaite et al. 1997; Cheng et al. 2001a). Together, these data indicate that D-WCC is the primary activator of *frq* transcription.

WCC reconstituted in vitro binds to the *frq* promoter in gel-shift assays without any additional components (Froehlich et al. 2003; He et al. 2005b). Mutational studies show that the zinc finger DNA-binding domains of both WC proteins are necessary to activate *frq* transcription in the dark (Crosthwaite et al. 1997; Cheng et al. 2002; Collett et al. 2002). However, the zinc finger domain of WC-1 is not required for its light-signaling function, and its LOV domain is not necessary for *frq* activation in the dark (He et al. 2002; Cheng et al. 2003b). Therefore, the light and dark functions of WC-1 can be molecularly separated.

INHIBITION OF WCC ACTIVITY BY FFC

Upon synthesis, FRQ self-associates through its amino-terminal coiled-coil domain, and this association is important for its function in the negative feedback loop (Cheng et al. 2001b). All FRQ proteins form complexes with FRH, an essential RNA helicase in *Neurospora* (Cheng et al. 2005). The formation of FFC is important for maintaining the steady-state level of FRQ; this is analogous to the stabilization of WC-1 by WC-2 (Cheng et al. 2002). FRH mediates the interaction between FRQ and WCC and interacts with WCC independently of FRQ. The down-regulation of FRH using RNA interference (RNAi) led to high levels of *frq* RNA, indicating that the negative feedback loop is abolished. These data indicate that FRH is an essential component of the negative feedback loop in the *Neurospora* circadian clock. The FRH homolog in yeast, Dob1p/Mtr4p, binds to RNA and functions as an essential cofactor for the exosome, an important regulator of RNA (rRNA and mRNA) metabolism (de la Cruz et al. 1998; Mitchell and Tollervey 2000; Hilleren and Parker 2003). Although it is not clear whether FRH functions as an RNA helicase in circadian regulation, it is likely that the daily fluctuation of the FFC may mediate circadian control of RNA processing and degradation in *Neurospora*.

FRQ, WC-1, and WC-2 are regulated by phosphorylation. After its synthesis, FRQ is immediately phosphorylated and becomes progressively more phosphorylated over time. It is finally degraded through the ubiquitin-proteasome pathway mediated by FWD-1 (Garceau et al. 1997; He et al. 2003, 2005a). Thus, in the dark, both the level and the phosphorylation status of FRQ are rhythmic. CK-1a, casein kinase II (CKII), and calcium/calmodulin-dependent protein kinase I (CAMK-1) are known to phosphorylate FRQ (Gorl et al. 2001; Yang et al. 2001, 2002, 2003; He et al. 2006).

Like FRQ, WC-1 and WC-2 are phosphorylated both in the dark and in the light to regulate WCC activity (He and Liu 2005; He et al. 2005b; Schafmeier et al. 2005, 2006). We previously identified five major in vivo WC-1 phosphorylation sites, located immediately downstream from the DNA-binding domain (He et al. 2005b). Mutation of these sites showed that these light-independent sites are critical for circadian clock function and that phosphorylation negatively regulates the D-WCC activity. The importance of WC phosphorylation in the circadian clock was later confirmed by the observation that the light-independent WC phosphorylation is FRQ-dependent (Schafmeier et al. 2005). In the *frq* null strain, both WC-1 and WC-2 are hypophosphorylated. In a wild-type strain, WC-2 exhibits a robust circadian rhythm of its phosphorylation profile. Importantly, the activation of *frq* transcription correlates with the hypophosphorylation of the WCs. Furthermore, we showed that the dephosphorylation of the *Neurospora* WCC significantly enhances its binding activity to the C box (He and Liu 2005). Together, these data suggest that FFC inhibits WCC activity by promoting its phosphorylation.

To understand the mechanism of how FFC inhibits WCC activity, we recently identified two kinases that can mediate the FRQ-dependent WC phosphorylation (He et al. 2006). CK-1a is the *Neurospora* homolog of the *Drosophila* DBT and it can phosphorylate FRQ in vitro (Gorl et al. 2001). More importantly, CK-1a was found to associate with FRQ (Gorl et al. 2001; Cheng et al. 2005), suggesting that it may phosphorylate FRQ in vivo. In addition, the association of FRQ and CK-1a raises the possibility that FRQ may recruit CK-1a to phosphorylate WCC. However, in vivo evidence for the involvement of CK-1a in the clock was previously not available because it is essential for cell survival in *Neurospora*. By mapping and mutating the FRQ–CK-1a interaction domain on FRQ, we showed that the FRQ–CK-1a interaction is essential for clock function (He et al. 2006). Importantly, both FRQ and WCs are hypophosphorylated in mutants with disrupted an FRQ–CK-1a interaction, indicating that CK-1a mediates the phosphorylation of FRQ and WCs. In addition, we found that WCC associates with CK-1a in an FRQ-dependent manner. To obtain direct genetic evidence for the role of CK-1a in the clock, we created a *ck-1a* knockin mutant, $ck\text{-}1a^{L}$, that carries a mutation equivalent to that of the *Drosophila* dbt^{L} mutation. In this mutant, FRQ and WCs are hypophosphorylated and the circadian rhythms exhibit long period (~32 hr) due to significant delay of FRQ progressive phosphorylation. These results indicate that CK-1a is a major kinase for both FRQ and WCs. The levels of WC-1 and WC-2 are low in the $ck\text{-}1a^{L}$ strain, indicating that CK-1a is important for the circadian-positive feedback loops. Despite the low WC levels in the $ck\text{-}1a^{L}$ strain, chromatin immunoprecipitation (ChIP) assays indicated that the hypophosphorylated WCC efficiently binds to the C box within the *frq* promoter and WCC cannot be efficiently inactivated by FRQ.

CKII is another kinase that we previously identified as a major kinase that phosphorylates FRQ in vivo (Yang et al. 2002, 2003). In addition to its role in promoting FRQ degradation as CK-1a, CKII is required for the repressor activity of FRQ, as indicated by high *frq* RNA levels in the CKII mutants despite their high FRQ protein levels. Like in the $ck\text{-}1a^{L}$ strain, we found that WC-1 and WC-2 are hypophosphorylated in WCC in the *cka* mutant strain in which the only catalytic subunit of CKII is disrupted

(He et al. 2006). In addition, WCC in this mutant constantly binds to *frq* promoter at higher levels despite high FRQ levels (He et al. 2006). Taken together, our results strongly suggest that FRQ closes the circadian negative feedback loop by mediating CK-1a and CKII phosphorylation of WCC. In this model, FRQ acts as the substrate-recruiting subunit of CK1 and CKII; the amount of FRQ determines the amount of kinases that can be recruited to phosphorylate WCC and thus the extent of WCC phosphorylation and its activity. A similar negative feedback process was also suggested in *Drosophila* (Kim and Edery 2006; Yu et al. 2006; Kim et al. 2007).

The *Neurospora* circadian negative feedback also involves phosphatase activities. We previously reported that the serine/threonine protein phosphatase 2A (PP2A) has an important role in the circadian negative feedback loop (Yang et al. 2004). The null mutation of one of the regulatory subunits of PP2A, *rgb-1*, leads to low *frq* mRNA and FRQ protein levels and a low-amplitude long-period rhythm. Such a conclusion was later confirmed by Schafmeier et al. (2005) showing that RGB-1 regulates the phosphorylation and activity of WC proteins. Thus, PP2A may regulate the circadian negative feedback process by dephosphorylating and reactivating WCC. Interestingly, the fly homolog RGB-1 is also an important clock component (Sathyanarayanan et al. 2004).

FRQ PHOSPHORYLATION AND DEGRADATION PATHWAY

The degradation and posttranslational modifications of FRQ have an essential role in period length determination and the function of the circadian negative feedback loop (Liu 2005; Liu and Bell-Pedersen 2006; Heintzen and Liu 2007). FRQ is progressively phosphorylated by several kinases, and its phosphorylation triggers its degradation through the ubiquitin-proteasome pathway (He et al. 2003). Mutation of FRQ phosphorylation sites was found to lengthen the period of the clock, suggesting that phosphorylation of FRQ promotes its turnover (Liu et al. 2000; Gorl et al. 2001; Yang et al. 2003; He et al. 2006). FRQ is phosphorylated by CK-1a, CKII, and CAMK-1 (Gorl et al. 2001; Yang et al. 2001, 2002, 2003; He et al. 2006). Like mutation of its *Drosophila* homolog, *ck-1a*^{L}, a hypomorphic mutation in *ck-1a* results in hypophosphorylation of FRQ, decreased rate of degradation, and long period of rhythm in the constant darkness (He et al. 2006). Disruption of the CKII catalytic subunit (*cka*) or regulatory subunit (*ckb-1*) in *Neurospora* also leads to hypophosphorylated FRQ and increased FRQ stability and results in abolishment of circadian rhythmicity or low-amplitude long-period rhythms, respectively (Yang et al. 2002, 2003). Interestingly, the phosphorylation events of FRQ appear to be independent of each other, suggesting that FRQ is phosphorylated by multiple kinases at multiple independent sites and each contributes to FRQ stability.

In addition to the kinases, protein phosphatases PP1 and PP2A also have important roles in regulating FRQ phosphorylation (Yang et al. 2004). Unlike PP2A, which appears to function mostly in the circadian negative feedback loop, PP1 regulates FRQ stability. A hypomorphic mutation of *ppp-1*, the catalytic subunit of PP1, leads to increased FRQ degradation rate, which results in rhythms of a short period and advanced phase. Thus, these phosphatases counteract kinases to regulate the phosphorylation status of FRQ. Taken together, these data suggest that the progressive phosphorylation of FRQ, regulated by multiple kinases and phosphatases at multiple independent sites, fine tunes the stability of FRQ and is a major determinant of the period length of the clock.

The phosphorylation-dependent degradation of FRQ is mediated by FWD-1, an F box/WD-40 repeat-containing protein and the *Neurospora* homolog of the *Drosophila* protein Slimb and mammalian β-TRCPs (Grima et al. 2002; He et al. 2003; Eide et al. 2005). FWD-1 physically interacts with phosphorylated forms of FRQ transiently and functions as the substrate recruiting subunit of the SCF-type ubiquitin (E3) ligase SCF^{FWD-1} to mediate FRQ ubiquitination (He et al. 2003, 2005a). In an *fwd-1* mutant strain, circadian rhythms are abolished and FRQ protein accumulates to high levels in its hyperphosphorylated forms. FWD-1 without its F box forms a stable complex with FRQ in vivo, suggesting that FWD-1 is a major component in the FRQ ubiquitination and degradation pathway.

The COP9 signalosome, a conserved multisubunit deneddylation complex in all eukaryotes, regulates the stability of the SCF^{FWD-1} complex in *Neurospora* and is thus critical to clock function (He et al. 2005a). The disruption of CSN leads to hyperneddylation of CULLIN-1 (a component of SCF complexes), which results in autoubiquitination and destruction of the SCF^{FWD-1} complex. This leads to low levels of SCF^{FWD-1} complex and impaired FRQ degradation and clock function.

CONSERVATION OF POSTTRANSLATIONAL REGULATION IN THE EUKARYOTIC CIRCADIAN SYSTEMS

The conservation of the posttranslational mechanisms is remarkable among different eukaryotic circadian systems from fungi to humans (Gallego and Virshup 2007; Heintzen and Liu 2007). Like FRQ and WCs in *Neurospora*, the animal and plant core clock components are also regulated by phosphorylation (Young and Kay 2001). FRQ and the animal PER proteins are all phosphorylated by CKI and CKII, dephosphorylated by the same phosphatases, and degraded by the ubiquitin/proteasome system using a conserved E3 ubiquitin ligase (Liu and Bell-Pedersen 2006). As in *Neurospora*, CKI is tightly associated with PER proteins in *Drosophila* and mammals (Kloss et al. 1998; Lee et al. 2001, 2004). Recently, the PER-DBT interaction domain was shown to be critical for PER phosphorylation, transcriptional repression, and circadian clock function (Kim et al. 2007). Together with the observations of the *Neurospora ck-1a*^{L} mutant strain, these results demonstrate the highly conserved role of FRQ and PER in recruitment of kinases. In mouse, the PER/CKI interaction appears to be critical for regulation of mammalian PER phosphorylation and the function of the clock (Lee et al. 2004). In humans, two types of familial advanced sleep phase syndromes (FASPS) are due to mutations of the human CKIδ or its phosphorylation sites on human PER2 (Toh et al. 2001; Xu et al. 2005). CKII in *Drosophila*, as in *Neurospora*, is required for the PER repressor function (Nawathean and Rosbash 2004), although how CKII functions to promote the repressor activity of PER is not clear. In *Arabidopsis*, where CKII was first implicated in clock function, CKII phosphorylates CCA1 and regulates its DNA-binding activity (Sugano et al. 1998).

The phosphorylation-dependent degradation of negative elements is also well conserved throughout the evolution. *Drosophila* PER proteins are phosphorylated by CKI and associate with Slimb (Grima et al. 2002; Ko et al. 2002; Eide et al. 2005). Another *Drosophila* locus, *jetlag*, contains an open reading frame (ORF) encoding an F-box protein with leucine-rich repeats and is important for light-induced degradation of TIM in the light entrainment pathway (Koh et al. 2006). Both biochemical and genetic approaches revealed that mammalian CRY and PER proteins are also regulated by an F-box protein, FBXL3 (Busino et al. 2007; Godinho et al. 2007; Siepka et al. 2007).

Like the WC proteins, the PAS-domain containing transcriptional factors that act as the positive components of the animal circadian negative feedback loops are also regulated by phosphorylation. The mammalian Bmal1 and CLK and the *Drosophila* CLK also exhibit robust rhythms in their phosphorylation profiles (Lee et al. 2001; Kim and Edery 2006; Yu et al. 2006; Kim et al. 2007). Similar to the case in *Neurospora*, the phosphorylation of CLK in *Drosophila* was recently found to be dependent on PER, DBT, and the PER-DBT inteaction. In addition, the CLK-CYC binding to the *per* E box correlates with the accumulation of hypophosphorylation of CLK and the hyperphosphorylation of CLK correlates with the peak of *per* mRNA repression. Furthermore, PP2A was also shown to be involved in the regulation of CLK phosphorylation. Together, these results strongly suggest that mechanisms of the circadian negative feedback process are highly conserved from *Neurospora* to animals.

CONCLUSION

Autonomous oscillations of clock components are generally mediated by posttranslational modifications, and overall, these modifications and the proteins involved are well conserved from fungi to humans. The identification of kinases and phosphatases of clock proteins and the studies of protein degradation pathway and the circadian negative feedback process have established the essential roles of protein phosphorylation in these processes and in clock functions. Furthermore, the identification of FRH as a core component of the circadian negative feedback loop raises the possibility of clock-controlled RNA metabolism. These recent advances in our understanding of circadian regulation in *Neurospora* have made its circadian oscillator one of the best understood among those in eukaryotes, and research in this organism has benefited parallel research in higher eukaryotes. The in vitro reconstitution of the cyanobacterial circadian clock indicated the fundamental importance of protein phosphorylation in clock regulation (Nakajima et al. 2005). Similar efforts in *Neurospora* and other eukaryotic organisms in the future should significantly enhance our current understanding of the mechanisms of eukaryotic circadian clocks.

ACKNOWLEDGMENTS

This work was supported by grants from National Institutes of Health (GM068496 and GM062591) and Welch Foundation to Y.L. Y.L. is the Louise W. Kahn Endowed Scholar in Biomedical Research at University of Texas Southwestern Medical Center.

REFERENCES

Aronson B.D., Johnson K.A., Loros J.J., and Dunlap J.C. 1994. Negative feedback defining a circadian clock: Autoregulation of the clock gene frequency. *Science* **263:** 1578.

Ballario P., Vittorioso P., Magrelli A., Talora C., Cabibbo A., and Macino G. 1996. White collar-1, a central regulator of blue light responses in *Neurospora,* is a zinc finger protein. *EMBO J.* **15:** 1650.

Busino L., Bassermann F., Maiolica A., Lee C., Nolan P.M., Godinho S.I., Draetta G.F., and Pagano M. 2007. SCFFbxl3 controls the oscillation of the circadian clock by directing the degradation of cryptochrome proteins. *Science* **316:** 900.

Cheng P., Yang Y., and Liu Y. 2001a. Interlocked feedback loops contribute to the robustness of the *Neurospora* circadian clock. *Proc. Natl. Acad. Sci.* **98:** 7408.

Cheng P., Yang Y., Gardner K.H., and Liu Y. 2002. PAS domain-mediated WC-1/WC-2 interaction is essential for maintaining the steady-state level of WC-1 and the function of both proteins in circadian clock and light responses of *Neurospora. Mol. Cell. Biol.* **22:** 517.

Cheng P., Yang Y., Heintzen C., and Liu Y. 2001b. Coiled-coil domain-mediated FRQ-FRQ interaction is essential for its circadian clock function in *Neurospora. EMBO J.* **20:** 101.

Cheng P., He Q., He Q., Wang L., and Liu Y. 2005. Regulation of the *Neurospora* circadian clock by an RNA helicase. *Genes Dev.* **19:** 234.

Cheng P., He Q., Yang Y., Wang L., and Liu Y. 2003a. Functional conservation of light, oxygen, or voltage domains in light sensing. *Proc. Natl. Acad. Sci.* **100:** 5938.

Cheng P., Yang Y., Wang L., He Q. and Liu Y. 2003b. WHITE COLLAR-1, a multifunctional *Neurospora* protein involved in the circadian feedback loops, light sensing, and transcription repression of wc-2. *J. Biol. Chem.* **278:** 3801.

Collett M.A., Garceau N., Dunlap J.C., and Loros J.J. 2002. Light and clock expression of the *Neurospora* clock gene frequency is differentially driven by but dependent on WHITE COLLAR-2. *Genetics* **160:** 149.

Crosthwaite S.K., Dunlap J.C., and Loros J.J. 1997. *Neurospora* wc-1 and wc-2: Transcription, photoresponses, and the origins of circadian rhythmicity. *Science* **276:** 763.

Crosthwaite S.K., Loros J.J., and Dunlap J.C. 1995. Light-induced resetting of a circadian clock is mediated by a rapid increase in frequency transcript. *Cell* **81:** 1003.

de la Cruz J., Kressler D., Tollervey D., and Linder P. 1998. Dob1p (Mtr4p) is a putative ATP-dependent RNA helicase required for the 3′ end formation of 5.8S rRNA in *Saccharomyces cerevisiae. EMBO J.* **17:** 1128.

Diernfellner A.C., Schafmeier T., Merrow M.W., and Brunner M. 2005. Molecular mechanism of temperature sensing by the circadian clock of *Neurospora crassa. Genes Dev.* **19:** 1968.

Dunlap J.C. 1999. Molecular bases for circadian clocks. *Cell* **96:** 271.

Dunlap J.C. and Loros J.J. 2004. The *Neurospora* circadian system. *J. Biol. Rhythms* **19:** 414.

Eide E.J., Woolf M.F., Kang H., Woolf P., Hurst W., Camacho F., Vielhaber E.L., Giovanni A., and Virshup D.M. 2005.

Control of mammalian circadian rhythm by CKIepsilon-regulated proteasome-mediated PER2 degradation. *Mol. Cell. Biol.* **25:** 2795.

Froehlich A.C., Loros J.J. and Dunlap J.C. 2003. Rhythmic binding of a WHITE COLLAR-containing complex to the frequency promoter is inhibited by FREQUENCY. *Proc. Natl. Acad. Sci.* **100:** 5914.

Froehlich A.C., Liu Y., Loros J.J., and Dunlap J.C. 2002. White Collar-1, a circadian blue light photoreceptor, binding to the frequency promoter. *Science* **297:** 815.

Gallego M. and Virshup D.M. 2007. Post-translational modifications regulate the ticking of the circadian clock. *Nat. Rev. Mol. Cell Biol.* **8:** 139.

Garceau N.Y., Liu Y., Loros J.J., and Dunlap J.C. 1997. Alternative initiation of translation and time-specific phosphorylation yield multiple forms of the essential clock protein FREQUENCY. *Cell* **89:** 469.

Glossop N.R., Lyons L.C., and Hardin P.E. 1999. Interlocked feedback loops within the *Drosophila* circadian oscillator. *Science* **286:** 766.

Godinho S.I., Maywood E.S., Shaw L., Tucci V., Barnard A.R., Busino L., Pagano M., Kendall R., Quwailid M.M., Romero M.R., O'Neill J., Chesham J.E., Brooker D., Lalanne Z., Hastings M.H., and Nolan P.M. 2007. The after-hours mutant reveals a role for Fbxl3 in determining mammalian circadian period. *Science* **316:** 897.

Gorl M., Merrow M., Huttner B., Johnson J., Roenneberg T., and Brunner M. 2001. A PEST-like element in FREQUENCY determines the length of the circadian period in *Neurospora crassa. EMBO J.* **20:** 7074.

Grima B., Lamouroux A., Chelot E., Papin C., Limbourg-Bouchon B., and Rouyer F. 2002. The F-box protein slimb controls the levels of clock proteins period and timeless. *Nature* **420:** 178.

He Q. and Liu Y. 2005. Molecular mechanism of light responses in *Neurospora:* From light-induced transcription to photoadaptation. *Genes Dev.* **19:** 2888.

He Q., Cheng P., He Q., and Liu Y. 2005a. The COP9 signalosome regulates the *Neurospora* circadian clock by controlling the stability of the SCFFWD-1 complex. *Genes Dev.* **19:** 1518.

He Q., Cha J., He Q., Lee H.C., Yang Y., and Liu Y. 2006. CKI and CKII mediate the FREQUENCY-dependent phosphorylation of the WHITE COLLAR complex to close the *Neurospora* circadian negative feedback loop. *Genes Dev.* **20:** 2552.

He Q., Cheng P., Yang Y., He Q., Yu H., and Liu Y. 2003. FWD1-mediated degradation of FREQUENCY in *Neurospora* establishes a conserved mechanism for circadian clock regulation. *EMBO J.* **22:** 4421.

He Q., Cheng P., Yang Y., Wang L., Gardner K.H., and Liu Y. 2002. White collar-1, a DNA binding transcription factor and a light sensor. *Science* **297:** 8403.

He Q., Shu H., Cheng P., Chen S., Wang L., and Liu Y. 2005b. Light-independent phosphorylation of WHITE COLLAR-1 regulates its function in the *Neurospora* circadian negative feedback loop. *J. Biol. Chem.* **280:** 17526.

Heintzen C. and Liu Y. 2007. The *Neurospora crassa* circadian clock. *Adv. Genet.* **58**: 25.

Hilleren P.J. and Parker R. 2003. Cytoplasmic degradation of splice-defective pre-mRNAs and intermediates. *Mol. Cell* **12:** 1453.

Huang G., Wang L., and Liu Y. 2006. Molecular mechanism of suppression of circadian rhythms by a critical stimulus. *EMBO J.* **25:** 5349.

Kim E.Y. and Edery I. 2006. Balance between DBT/CKIepsilon kinase and protein phosphatase activities regulate phosphorylation and stability of *Drosophila* CLOCK protein. *Proc. Natl. Acad. Sci.* **103:** 6178.

Kim E.Y., Ko H.W., Yu W., Hardin P.E., and Edery I. 2007. A DOUBLETIME kinase binding domain on the *Drosophila* PERIOD protein is essential for its hyperphosphorylation, transcriptional repression, and circadian clock function. *Mol. Cell. Biol.* **27:** 5014.

Kloss B., Price J.L., Saez L., Blau J., Rothenfluh A., Wesley C.S., and Young M.W. 1998. The *Drosophila* clock gene double-time encodes a protein closely related to human casein kinase Iepsilon. *Cell* **94:** 97.

Ko H.W., Jiang J., and Edery I. 2002. Role for Slimb in the degradation of *Drosophila* Period protein phosphorylated by Doubletime. *Nature* **420:** 673.

Koh K., Zheng X., and Sehgal A. 2006. JETLAG resets the *Drosophila* circadian clock by promoting light-induced degradation of TIMELESS. *Science* **312:** 1809.

Lee C., Weaver D.R., and Reppert S.M. 2004. Direct association between mouse PERIOD and CKIepsilon is critical for a functioning circadian clock. *Mol. Cell. Biol.* **24:** 584.

Lee C., Etchegaray J.P., Cagampang F.R., Loudon A.S., and Reppert S.M. 2001. Posttranslational mechanisms regulate the mammalian circadian clock. *Cell* **107:** 855.

Lee K., Loros J.J., and Dunlap J.C. 2000. Interconnected feedback loops in the *Neurospora* circadian system. *Science* **289:** 107.

Linden H. and Macino G. 1997. White collar 2, a partner in blue-light signal transduction, controlling expression of light-regulated genes in *Neurospora crassa. EMBO J.* **16:** 98.

Liu Y. 2003. Molecular mechanisms of entrainment in the *Neurospora* circadian clock. *J. Biol. Rhythms* **18:** 195.

———. 2005. Analysis of posttranslational regulations in the *Neurospora* circadian clock. *Methods Enzymol.* **393:** 379.

Liu Y. and Bell-Pedersen D. 2006. Circadian rhythms in *Neurospora crassa* and other filamentous fungi. *Eukaryot. Cell* **5:** 1184.

Liu Y., Loros J., and Dunlap J.C. 2000. Phosphorylation of the *Neurospora* clock protein FREQUENCY determines its degradation rate and strongly influences the period length of the circadian clock. *Proc. Natl. Acad. Sci.* **97:** 234.

Liu Y., Merrow M., Loros J.J., and Dunlap J.C. 1998. How temperature changes reset a circadian oscillator. *Science* **281:** 825.

Luo C., Loros J.J., and Dunlap J.C. 1998. Nuclear localization is required for function of the essential clock protein FRQ. *EMBO J.* **17:** 1228.

Merrow M.W., Garceau N.Y., and Dunlap J.C. 1997. Dissection of a circadian oscillation into discrete domains. *Proc. Natl. Acad. Sci.* **94:** 3877.

Merrow M., Franchi L., Dragovic Z., Gorl M., Johnson J., Brunner M., Macino G., and Roenneberg T. 2001. Circadian regulation of the light input pathway in *Neurospora crassa. EMBO J.* **20:** 307.

Mitchell P. and Tollervey D. 2000. Musing on the structural organization of the exosome complex. *Nat. Struct. Biol.* **7:** 843.

Nakajima M., Imai K., Ito H., Nishiwaki T., Murayama Y., Iwasaki H., Oyama T., and Kondo T. 2005. Reconstitution of circadian oscillation of cyanobacterial KaiC phosphorylation in vitro. *Science* **308:** 414.

Nawathean P. and Rosbash M. 2004. The doubletime and CKII kinases collaborate to potentiate *Drosophila* PER transcriptional repressor activity. *Mol. Cell* **13:** 213.

Sathyanarayanan S., Zheng X., Xiao R., and Sehgal A. 2004. Posttranslational regulation of *Drosophila* PERIOD protein by protein phosphatase 2A. *Cell* **116:** 603.

Schafmeier T., Kaldi K., Diernfellner A., Mohr C., and Brunner M. 2006. Phosphorylation-dependent maturation of *Neurospora* circadian clock protein from a nuclear repressor toward a cytoplasmic activator. *Genes Dev.* **20:** 297.

Schafmeier T., Haase A., Kaldi K., Scholz J., Fuchs M., and Brunner M. 2005. Transcriptional feedback of *Neurospora* circadian clock gene by phosphorylation-dependent inactivation of its transcription factor. *Cell* **122:** 235.

Schwerdtfeger C. and Linden H. 2000. Localization and light-dependent phosphorylation of white collar 1 and 2, the two central components of blue light signaling in *Neurospora crassa. Eur. J. Biochem.* **267:** 414.

Shearman L.P., Sriram S., Weaver D.R., Maywood E.S., Chaves I., Zheng B., Kume K., Lee C.C., van der Horst G.T., Hastings M.H., and Reppert S.M. 2000. Interacting molecular loops in the mammalian circadian clock. *Science* **288:** 1013.

Siepka S.M., Yoo S.H., Park J., Song W., Kumar V., Hu Y., Lee C., and Takahashi J.S. 2007. Circadian mutant Overtime reveals F-box protein FBXL3 regulation of cryptochrome and period gene expression. *Cell* **129:** 1011.

Sugano S., Andronis C., Green R.M., Wang Z.Y., and Tobin E.M. 1998. Protein kinase CK2 interacts with and phosphorylates the *Arabidopsis* circadian clock-associated 1 protein. *Proc. Natl. Acad. Sci.* **95:** 11020.

Talora C., Franchi L., Linden H., Ballario P., and Macino G. 1999. Role of a white collar-1-white collar-2 complex in blue-light signal transduction. *EMBO J.* **18:** 4961.

Toh K.L., Jones C.R., He Y., Eide E.J., Hinz W.A., Virshup D.M., Ptacek L.J. and Fu Y.H. 2001. An hPer2 phosphorylation site mutation in familial advanced sleep phase syndrome. *Science* **291:** 1040.

Xu Y., Padiath Q.S., Shapiro R.E., Jones C.R., Wu S.C., Saigoh N., Saigoh K., Ptacek L.J., and Fu Y.H. 2005. Functional consequences of a CKIdelta mutation causing familial advanced sleep phase syndrome. *Nature* **434:** 640.

Yang Y., Cheng P., and Liu Y. 2002. Regulation of the *Neurospora* circadian clock by casein kinase II. *Genes Dev.* **16:** 994.

Yang Y., Cheng P., Zhi G., and Liu Y. 2001. Identification of a calcium/calmodulin-dependent protein kinase that phosphorylates the *Neurospora* circadian clock protein FREQUENCY. *J. Biol. Chem.* **276:** 41064.

Yang Y., Cheng P., He Q., Wang L., and Liu Y. 2003. Phosphorylation of FREQUENCY protein by casein kinase II is necessary for the function of the *Neurospora* circadian clock. *Mol. Cell. Biol.* **23:** 6221.

Yang Y., He Q., Cheng P., Wrage P., Yarden O., and Liu Y. 2004. Distinct roles for PP1 and PP2A in the *Neurospora* circadian clock. *Genes Dev.* **18:** 255.

Young M.W. and Kay S.A. 2001. Time zones: A comparative genetics of circadian clocks. *Nat. Rev. Genet.* **2:** 702.

Yu W., Zheng H., Houl J.H., Dauwalder B., and Hardin P.E. 2006. PER-dependent rhythms in CLK phosphorylation and E-box binding regulate circadian transcription. *Genes Dev.* **20:** 723.

Posttranslational Photomodulation of Circadian Amplitude

D.E. Somers,* S. Fujiwara,* W.-Y. Kim,*† and S.-S. Suh*

*Department of Plant Cellular and Molecular Biology/Plant Biotechnology Center, Ohio State University Columbus, Ohio 43210, USA; †Environmental Biotechnology National Core Research Center, Gyeongsang National University, Jinju 660-701, South Korea

The transcription-translation feedback loops that form our current view of how the core mechanism of the clock operates is being challenged, as more and more posttranslational events are seen as essential to a full understanding of oscillator function. But in addition to phosphorylation, other processes may be involved. Here, a novel mechanism of posttranslational photomodulation of circadian amplitude is described that uniquely ties together light perception, protein stabilization, and proteolysis. In the process, the waveform of a core clock component is sharpened or "sculpted," resulting in appropriately high amplitude and proper phasing to obtain normal clock function.

INTRODUCTION

The earliest models that emerged in understanding eukaryotic clock function converged on a molecular view of the oscillator that involves a transcription-translation autoregulatory negative feedback loop (Young and Kay 2001; Bell-Pedersen et al. 2005). In this model, core transcription factors positively activate transcription of clock genes which are transcribed and translated into proteins in the cytoplasm. Certain of these cytoplasmic proteins undergo posttranslational modifications, generally phosphorylation, that facilitate their nuclear import. These factors then act to inhibit the positive activation of their own genes by interfering with the transcriptional activity of the dedicated core transcription factors. The cycle can begin again once the autoregulatory inhibitors are degraded and the transcriptional activators are free to bind.

This very basic scheme has been greatly elaborated on in both general and unique ways in each circadian system studied. It is now clear that more than one autoregulatory loop is present and that positive- and negative-acting branches are interlocked, providing greater stability to the system (Dunlap and Loros 2004; Hardin 2005, 2006; Dunlap 2006). Some of the core transcription factors are constitutively expressed, whereas others are under circadian control, with robust cycling of both mRNA and protein.

The focus here is the intersection between the light environment and the molecular components of the plant circadian system. Resetting of the clock occurs with each sunrise and, in some cases, sunset (Stoleru et al. 2007). How this occurs in each organism reflects the sensitivity of the circadian system to light. Here, two well-studied circadian systems with relatively direct light input pathways (*Drosophila* and *Neurospora*) are compared to new findings in the *Arabidopsis* system.

PHOTOENTRAINMENT IN *DROSOPHILA*

Classic assays of *Drosophila* rhythms used eclosion and then locomotor activity to follow clock function. Under constant light (LL), rhythmic activity of flies quickly diminishes, in striking contrast to the sustained robust cycles observed in extended darkness. Similarly, per-luciferase-based luminescence rhythms are robust in constant darkness (DD), and this assay was instrumental in the discovery of the first element in the entrainment pathway when the *cry*^{b} mutant was identified (Emery et al. 1998; Stanewsky et al. 1998). This mutation allows continued locomotor rhythmicity under LL, and mutant flies are recalcitrant to a phase-shifting light pulse. This finding identified cryptochrome, a protein already identified as a blue light photoreceptor in plants, as the first step in the light-signaling pathway to the clock in flies. As a loss-of-function mutation, *cry*^{b} indicates that light acts to disrupt clock, although pulses are effective, and essential, for photic entrainment (Dolezelova et al. 2007).

Subsequently, two-hybrid studies identified TIMELESS (TIM) as a CRY interaction partner. TIM had already been shown to be a core element of the fly clock, interacting with PERIOD (PER) to suppress transcription driven by the CLOCK/CYCLE heterodimer (Gekakis et al. 1995; Darlington et al. 1998). A light-dependent interaction between CRY and TIM in yeast, abrogated by the CRYB mutation, helped establish CRY as a bona fide circadian photoreceptor, with TIM as a direct interaction partner (Ceriani et al. 1999). In the fly, blue light absorption by CRY facilitates TIM interaction, leading to TIM phosphorylation and degradation via the proteasome (Naidoo et al. 1999; Hardin 2005). The F-box protein JETLAG binds TIM, possibly after release from CRY, and is the link to the proteasome-dependent degradation of TIM. Light activation of CRY appears to facilitate TIM release and phosphorylation, making it receptive to JETLAG interaction (Fang et al. 2007). In this way, light enters the fly clock system via CRY-dependent changes in TIM abundance. It is through the rapid light-dependent reduction of TIM that phase resetting and entrainment are effected, although other light-input pathways exist in other cells and tissues of the fly (Dolezelova et al. 2007).

PHOTOENTRAINMENT IN *NEUROSPORA*

In *Neurospora*, FREQUENCY (FRQ) and the WHITE COLLAR COMPLEX (WCC) are central elements of the circadian system (Liu 2003; Dunlap and Loros 2004; He and Liu 2005; Dunlap 2006). The WHITE COLLLAR-1 (WC-1) and WC-2 heterodimer (WCC) act to positively regulate transcription from the *frequency* promoter, as well as other light-regulated genes. WC-1 is the photoreceptive partner of the pair, containing an amino-terminal LOV domain that binds flavin adenine dinucleotide (FAD) that confers the blue-light-absorptive properties to the molecule (He et al. 2002). Upon blue light absorption, the WCC binds to the light-responsive elements (LRE) of the *frq* promoter, promoting transcription (Froehlich et al. 2002). At the same time, FRQ protein acts negatively to disrupt the complex, inhibiting its own transcription. But FRQ also promotes *wc-1* and *wc-2* transcription, positively acting to enhance *frq* activation (He and Liu 2005; Brunner and Schafmeier 2006). Hence, in this system, photocontrol of resetting and entrainment begins with transcriptional activation, via light-enhanced binding of the WCC to the *frq* promoter.

The WCC is required for *frq* transcription in the dark as well, but light absorption causes a transient increase in WCC-binding capacity, nicely accounting for an acute induction of FRQ and other light-regulated genes (He and Liu 2005). Following initial light absorption, WCC becomes progressively phosphorylated, which results in diminished affinity for LREs and a reduction in transcription from light-inducible promoters (He and Liu 2005; He et al. 2005). Hence, this direct light-enhanced DNA binding by a photoperceptive transcriptional complex (WCC) differs fundamentally from the light-enhanced degradation of protein, initiated by the CRY/TIM interaction in *Drosophila*.

PHOTOENTRAINMENT IN *ARABIDOPSIS*

The circadian system in plants involves numerous gene families, including Myb transcription factors (LHY and CCA1), pseudoresponse regulators that likely act as transcriptional cofactors (PRR1/TOC1, PRR3, PRR5, PRR7, PRR9), and two classes of classic red and blue photoreceptors (phytochromes and cryptochromes, respectively) (Mizuno and Nakamichi 2005; McClung 2006; Hotta et al. 2007). Mutations in both classes of photoreceptors lengthen free-running period, similar to the effects of reduced light intensity (Somers et al. 1998a; Devlin and Kay 2000). It is clear that these classic photoreceptors control light input to the plant clock, but no molecular mechanism has been found that directly links light-induced changes in either receptor class to interactions with known plant clock elements. Unlike the situation in *Drosophila*, no light-dependent interactor has been identified for either of the two plant cryptochromes (*cry1* and *cry2*) in the context of any blue light response in plants. Similarly, although the phytochromes are most likely light-activated kinases (Rockwell et al. 2006), a clock-linked phosphorylated substrate has not been identified.

Instead, a direct light-dependent connection between photoperception and a core plant clock protein has come from an unexpected source. ZEITLUPE (ZTL) is an F-box protein originally identified in a screen for long-period *Arabidopsis* mutants (Somers et al. 2000; Han et al. 2004). Carboxy-terminal to the F-box domain are six kelch repeats that fold into a β-propeller to form the substrate interaction domain. Unique to this class of F-box proteins is a LOV domain amino-terminal to the F box that is highly similar to the LOV domain of WC-1 and of another class of plant blue light receptors, the phototropins (Somers 2001, 2005; Crosson et al. 2003). The phototropins uniquely possess two LOV domains near the amino terminus, both of which bind flavin mononucleotide (FMN). Their blue light activation results in autophosphorylation that initiates a change in the level of the hormone auxin, resulting in differential growth toward the light (phototropism) (Briggs and Christie 2002).

Within the three-member ZTL family (ZTL, FKF1, and LKP2), only ZTL is unequivocally linked to the clock. LKP2 overexpression causes arrhythmicity, similar to ZTL overexpression, but *lkp2* mutants have no detectable circadian phenotype and the period of *lkp2 ztl* double mutants is similar to that of *ztl* mutants (Schultz et al. 2001; Somers et al. 2004 and unpubl.). *FKF1* transcription is clock-controlled, but at least one target is CDF1, which regulates the activity of CONSTANS, a key component of flowering time control (Imaizumi et al. 2005). *fkf1* mutants are late flowering but have no circadian defect (Nelson et al. 2000; Imaizumi et al. 2003).

In contrast, *ZTL* mRNA is constitutively expressed under all conditions. However, ZTL protein is rhythmic with a peak near ZT13, indicating some form of posttranscriptional circadian regulation (Somers et al. 2000; Kim et al. 2003). ZTL targets two members of the PRR family of pseudoresponse regulators for proteasome-dependent degradation: TOC1 (PRR1) and PRR5 (Mas et al. 2003a; Kiba et al. 2007; S. Fujiwara et al., unpubl.). Both are clock-controlled components of the plant circadian system, with a peak phase of expression near subjective dusk similar to the peak of ZTL abundance. Mutations in either TOC1 or PRR5 shorten period, and overexpression lengthens period (Somers et al. 1998b; Strayer et al. 2000; Makino et al. 2002; Sato et al. 2002; Mas et al. 2003b; Michael et al. 2003). This is similar to the long period of *ztl* mutations, consistent with the observed increase in TOC1 and PRR5 levels in *ztl* mutants (Mas et al. 2003a; Kiba et al. 2007). The phenotype of *ztl* mutants are fluence-rate-dependent, with a more severe long period at lower light intensities, relative to wild type (Somers et al. 2000, 2004). This suggests that ZTL has a stronger role at low fluence rates, and other photoreceptors feature more prominently in high light.

Further insights into how the ZTL LOV domain functions in the context of the full-length protein has come with the identification of GIGANTEA (GI) as a ZTL interaction partner. *GI* was first reported more than 45 years ago as a late-flowering mutant (Redei 1962). Subsequently, *gi* mutants were found to have circadian period defects, as well as aberrantly high starch accumulation and long hypocotyls (Table 1) (Eimert et al. 1995; Park et al. 1999; Huq et al. 2000; Tseng et al. 2004; Mizoguchi et al. 2005; Gould et al. 2006; Kim et al. 2007; Martin-Tryon et al. 2007; Oliverio et al. 2007). The large (1173 amino acids) GI protein has no recognizable domains that could suggest a mechanism for its action

Table 1. Effect of *gi* Alleles on Circadian Period and Reproductive and Morphological Development

	Period				Flowering time		
Allele	LL[a]	RR	BB	DD	LD	SD	Hypocotyl length
gi-1	short[1,2]	short[1]			late[6]	~WT[6]	long (LL, RR, BB, FR)[4,7]
gi-2	short/WT (leaf movement) [1,2] long (*cab::luciferase*)[1]				late[4,6]	late[4,6]	long (LL, RR, BB, FR)[4,7] or long (RR) and WT (DD, FR)[2]
gi-3	short (leaf movement)[2]			arrhythmic[5]	late[5,6]	late[6] or WT[5]	long (RR, SD)[4,5]
gi-4	short (leaf movement)[2]				late[6]	WT[6]	long (LL, RR, BB)[4]
gi-5	short (leaf movement)[2]				late[6]	late[6]	long (LL, RR, BB)[4]
gi-6	short (leaf movement)[2]				late[6]	~WT[6]	long (LL, RR, BB)[4]
gi-11	short (leaf movement)[3]				late[3,6]	late[3]	
gi-200	short (LL, R+B)[4]	short[4]	short[4]	short[4]	WT[4]	early[4]	long (LL, RR, BB)[4]
gi-201	~WT (LL), short (R+B)[4]	long[4]	short[4]		late[4]	WT[4]	long (LL, RR, BB)[4]
gi-596	long (R+B)[3]	long[3]			WT[3]		
gi-611	short (R+B)[3]	short[3]			WT[3]	early[3]	
GI-ox	short[5]			arrhythmic[5]	~WT[3,5]	early[5]	short (RR, SD)[5]

(LL) Constant white light; (RR) constant red light; (BB) constant blue light; (LD) long days (16 hours light/8 hours dark); (SD) short days (8 hours light/16 hours dark); (R+B) red plus blue light; (FR) continuous far red; (WT) wild type.
[a]References: [1]Park et al. 1999; [2]Tseng et al. 2004; [3]Gould et al. 2006; [4]Martin-Tryon et al. 2007; [5]Mizoguchi et al. 2005; [6]Fowler et al. 1999; [7]Huq et al. 2000.

(Fowler et al. 1999; Park et al. 1999). *gi* mutants are severely reduced in ZTL levels, but with no change in *ZTL* mRNA. Constitutive high expression of GI results in posttranscriptionally high ZTL accumulation, and the proteasome-dependent ZTL degradation rate is reduced in a GI overexpressor (Kim et al. 2003, 2007). Two-hybrid testing and in vivo coimmunoprecipitations show that ZTL and GI interact strongly through the ZTL LOV domain. Point mutations in the LOV domain alone severely abrogates in vivo interactions between GI and ZTL, whereas mutations in the F-box and kelch domains have little effect. Hence, it appears that GI functions as a posttranslational stabilizer of ZTL, through its interaction with the LOV domain (Kim et al. 2007).

Testing of the GI/ZTL interaction under different light conditions unequivocally demonstrates the photosensory function of the ZTL LOV domain. Quantitative coimmunoprecipitation assays show that blue light enhances the GI/ZTL interaction threefold compared to exposure to red light or darkness. Some point mutations in the LOV domain still allow blue-light-enhanced interactions but at a much reduced level. However, replacement of a key cysteine with alanine in the LOV region (C82A) completely eliminates blue-light-enhanced interactions, reducing the GI/ZTL interaction to uniformly low levels under all light conditions (Kim et al. 2007). This cysteine is conserved in all LOV domains and forms a thiocysteinyl adduct with FMN in the phototropin LOV moiety upon blue light absorbance (Salomon et al. 2000; Crosson and Moffat 2001). Thus, this residue is the critical defining element that confers photosensitivity to the LOV domain. Because, like the phototropins, the ZTL family binds FMN (Imaizumi et al. 2003), these results demonstrate that ZTL is a blue light circadian photoreceptor.

Because the blue-light-dependent effects on ZTL could also be mediated through the cryptochromes, we tested the levels of ZTL in the *cry1 cry2* background. Under all light conditions, we found no differences between wild type and the *cry* double mutant (Fig. 1). During a 12-hour blue light extension into the subjective dark period, ZTL levels were significantly higher than those observed in darkness or in red light in both the double mutant and wild type. Similar tests were conducted with the phytochrome mutants, and no effect on ZTL levels was observed (data not shown). These results indicate that all light-dependent effects on ZTL levels result entirely from ZTL photoabsorption properties.

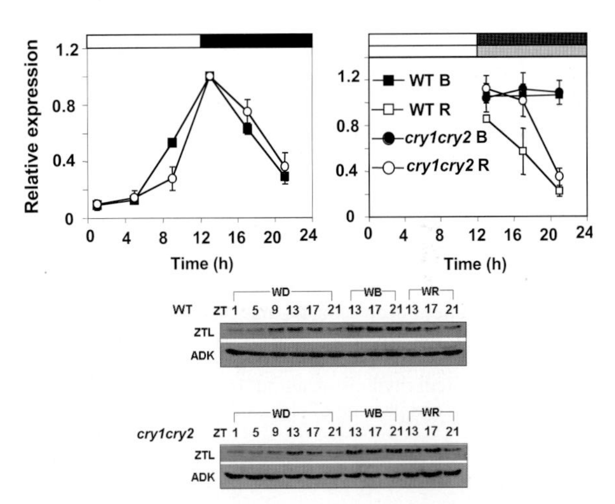

Figure 1. Protein extracts from wild type (Ler) and *cry1 cry2* mutants entrained in 12-hour light/12-hour dark cycles (WD) for 10 days and transferred to *dark* (D) (*left panel*), *blue* (B), or *red* (R) light (*right panel*) at ZT12 were immunoblotted and probed for ZTL abundance at the times indicated. Representative blots are shown below; adenosine kinase (ADK) was used as loading control. ZTL levels expressed relative to wild type at ZT13 (D). Means of three trials ±S.E.M.

POSTTRANSLATIONAL CONTROL OF CIRCADIAN CLOCK ELEMENTS WITHOUT PHOSPHORYLATION: A NOVEL MODE OF REGULATION

These recent data demonstrate that clock-controlled *GI* transcription, and consequent GI protein rhythms, confers a posttranslational rhythm on ZTL protein by virtue of their mutual cooperative stabilization (Fig. 2A,B) (Kim et al. 2007). Following on this, how does the blue-light-enhanced binding of GI and ZTL affect circadian cycling?

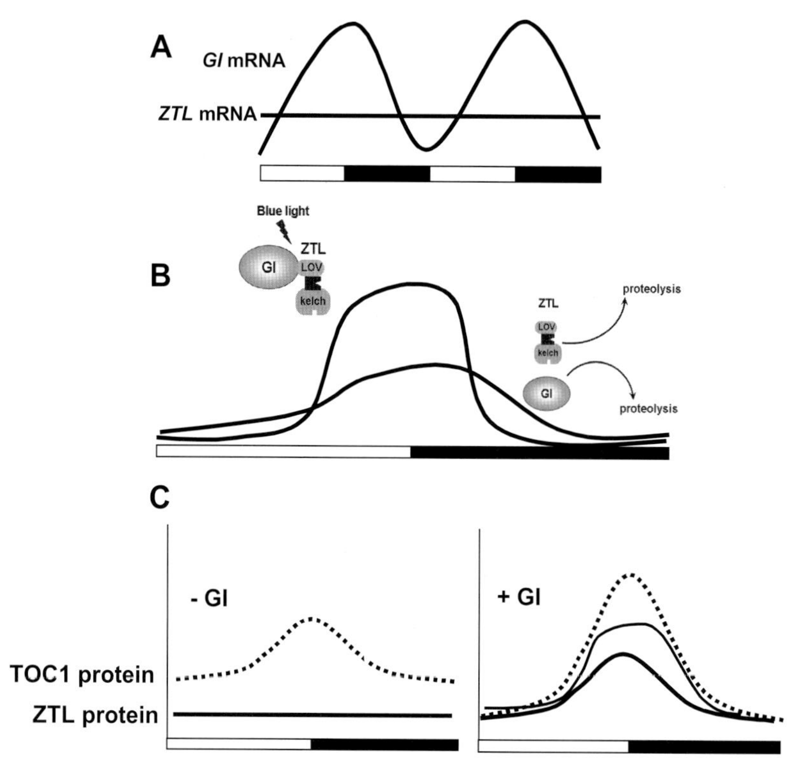

Figure 2. Posttranslational control of ZTL protein rhythms through blue-light-enhanced stabilization by GI. (*A*) Robust diurnal rhythms of GI message contrasts with constitutive expression of ZTL mRNA. (*B*) Increasing GI protein in the light (high amplitude) associates with ZTL, stabilizing both and allowing increasing ZTL and GI protein levels, following on the clock-controlled rise of GI mRNA. In darkness, decreasing GI message and reduced stabilization of the GI/ZTL interaction hasten clearance of both proteins from the system. (*C*) One consequence of ZTL oscillations is to sharpen the TOC1 protein profile (*dotted line*). In the absence of GI (–GI), ZTL levels are constitutively low. TOC1 protein oscillates with low amplitude, tracking its own message rhythms. In the wild type (+GI), ZTL protein rhythms, resulting from GI cycling (*thin line*), allow high-amplitude TOC1 protein cycles, resulting in a normal circadian period and phasing of clock-controlled outputs.

In contrast to the long-period *ztl* mutants (Kevei et al. 2006), *gi* mutants generally show a short free-running period, although some alleles cause long period for some outputs, and light quality is also a factor (Table 1) (Kim et al. 2007; Martin-Tryon et al. 2007). Therefore, GI likely affects other aspects of clock function in addition to its effects on ZTL. TOC1 protein in a *gi* null still cycles in light/dark but with much reduced amplitude (Kim et al. 2007). During the early part of the day (photoperiod), when TOC1 is normally low, levels are strongly increased compared to wild type. Later, during the dark phase (skotoperiod), TOC1 levels are somewhat lower in a GI-deficient background compared to wild type. This overall blunting of the TOC1 waveform points to the importance of ZTL cycling in maintaining the appropriate amplitude and sharpness of the TOC1 rhythm (Fig. 2C). In a *gi* null, ZTL levels are constitutively low throughout the light/dark cycle, different from a line completely lacking ZTL. If this protein is functional, this could partially explain why TOC1 levels are not pegged to maximal levels at all times, as seen in *ztl* null mutants (Mas et al. 2003a).

This aspect of the *gi* phenotype may also provide insight into the puzzle of why, in the wild type, ZTL and TOC1 proteins cycle in-phase. As the substrate of ZTL, TOC1 levels should be low at times of peak ZTL, and high when ZTL is minimal. Their synchronous phasing suggests that other factors may be involved. One attractive model to explain their similar peak levels depends on the light-mediated GI/ZTL complex dissociating in the dark. Although this will lead to subsequent ZTL degradation, until then, the newly freed ZTL will be available to bind and degrade TOC1. This scheme shows GI with the dual role of simultaneously stabilizing ZTL and protecting TOC1 from degradation through the binding of GI to ZTL during the photoperiod, only to release ZTL for action during the skotoperiod. Thus, the GI/ZTL interaction heightens the peak of ZTL through the stabilizing effect of GI and also heightens TOC1 peak expression through the sequestration of ZTL. Therefore, in *gi* mutants, peak TOC1 levels are lower during the skotoperiod than in wild type because although ZTL levels are low throughout the cycle, free ZTL may effectively be higher than normal soon after entrance into the skotoperiod because no GI is present to complex with ZTL. That there is any TOC1 cycling in *gi* lines at all likely arises from the tracking of *TOC1* mRNA rhythms.

In comparison to the fly and *Neurospora* light-input mechanisms, ZTL bears closer similarity to cryptochrome's role in *Drosophila*. In both systems, blue light facilitates a protein–protein interaction that leads to the degradation of a key component of the oscillator. In *Drosophila*, cryptochrome binds directly to that component (TIM) in a light-dependent way, resulting in increased TIM phosphorylation, which leads to better recognition of TIM by the F-box protein JETLAG. In *Arabidopsis*, light perception comes through the F-box protein itself (ZTL) but acts to facilitate binding of a stabilizing factor (GI) that leads to the appropriate phasing of peak ZTL. The ZTL/TOC1or ZTL/PRR5 interaction may require the phosphorylation of these two substrates, but there is no evidence that if this occurs, it is light-dependent. Similarly, there is no evidence that the GI/ZTL interaction is affected by phosphorylation of either protein.

Another key difference is that ZTL is most likely not the sole or even primary light-input pathway to entrainment and phase resetting in plants. Although there is good evidence that other ocular photoreceptors in flies can have a role in entraining locomotor activity, cryptochrome is clearly primary in its effect in controlling TIM levels acutely and in response to light (Dolezelova et al. 2007). In plants, as noted earlier, the *phy*s and *cry*s are essential in phase resetting, and other proteins, such as ELF3 and ELF4 are clearly involved in shaping the light responsiveness of the clock (Covington et al. 2001; Doyle et al. 2002; McWatters et al. 2007).

It is important to note that the effect of blue light absorption by ZTL is to stabilize the protein, not to initiate a light-dependent transduction chain. In this way, ZTL is not a photoreceptor in the "classic" sense of initiating a signaling cascade through a change in phosphorylation status of downstream components or through the activation of second messengers (e.g., rhodopsin). These signaling pathways act to amplify the original light input. The light-enhanced interaction of ZTL with GI is largely an act of self-preservation, and the partnering with GI is essential simply to maintain sufficiently high levels of ZTL, at the appropriate circadian phase, to diminish TOC1 and PRR5 levels.

PHOSPHORYLATION AND CLOCK FUNCTION

Phosphorylation in the *Drosophila* Clock

Protein maturation or stabilization controlled by a clock-regulated factor such as GI is a novel mechanism of conferring posttranslational regulation. More common is protein phosphorylation/dephosphorylation of clock components (Lee et al. 2001; Merrow et al. 2006; Gallego and Virshup 2007). In *Drosophila*, timely phosphorylation of TIM and PER regulates the timing of nuclear import as well as the susceptibility to degradation (Hardin 2005; Bae and Edery 2006). Casein kinase Iε (CKIε) associates with and phosphorylates PER in either the nucleus or cytoplasm when unbound to TIM, potentiating it for proteasome-dependent degradation. Mutations in CKIε were recovered as period mutants (*doubletime*), the first demonstration of the importance of phosphorylation in the *Drosophila* clock (Kloss et al. 1998; Price et al. 1998). This same kinase is found in the nucleus, acting on both PER and CLK, priming both for degradation (Bae and Edery 2006; Kim and Edery 2006). A likely collaborator of CKIε is casein kinase 2 (CK2), contributing to the progressive phosphorylation of PER in the cytoplasm and helping to regulate nuclear import (Nawathean and Rosbash 2004). Additionally, a third kinase, glycogen synthase kinase-3 (GSK-3), is essential in the phosphorylation and subsequent degradation of TIM by JETLAG, as well as the nuclear import of the TIM/PER complex (Naidoo et al. 1999; Martinek et al. 2001; Koh et al. 2006). Finally, in addition to these three kinases, protein phosphatase 2A (PP2A) and protein phosphatase 1 (PP1) act to remove phosphates from PER/CLK and TIM, respectively.

Phosphorylation in the *Neurospora* Clock

In *Neurospora*, phosphorylation has critical roles in a number of clock components. Progressive phosphorylation of FRQ by CKII, and likely other kinases (CAMK-1, CK1a), controls FRQ degradation through interaction with the F-box protein FWD-1 and is highly dependent on the phosphorylation status of certain key residues (He et al. 2003; Yang et al. 2003; Brunner and Schafmeier 2006). Two phosphatases, PP1 and PP2A dephosphorylate FRQ (Yang et al. 2004). PP1 positively regulates FRQ abundance, consistent with an antagonism toward the CKII activity that promotes FRQ degradation. Still unclear is how the opposing activities of these two protein classes work together to establish a particular state of FRQ phosphorylation during the circadian cycle and how critical this state is to FRQ function apart from dictating its degradation potential.

As noted earlier, WCC binding to the *frq* promoter is enhanced by blue light, but there is no evidence that this occurs through light-dependent phosphorylation. Indeed, hyperphosphorylation of WCC peaks 15–30 minutes after light exposure, but this is much later than the binding of WCC to LREs, which is detectable within 5 minutes (He and Liu 2005). Instead, light appears to potentiate WCC for subsequent phosphorylation which, paradoxically, reduces the affinity of WCC for LRE binding. Additionally, FRQ is required for this phosphorylation, which recruits or interacts with both CK1a and CKII at WCC (Schafmeier et al. 2005; He et al. 2006). Extensive FRQ-dependent phosphorylation (in complex with CK1a and CKII) of WCC results in inactivation and removal of WCC from the *frq* promoter. Increasing FRQ phosphorylation results in its degradation, and dephosphorylation of WCC, via PP2A, can reactivate the complex and *frq* transcription may begin again (He and Liu 2005; Schafmeier et al. 2005, 2006; He et al. 2006).

Phosphorylation in the *Arabidopsis* Clock

Although less well elaborated, phosphorylation does have a role in the plant circadian system. A series of papers concerned with CK2 and the *myb* transcription factor CCA1 have established a clear role for this kinase. Recombinant CK2 can phosphorylate CCA1 in vitro, and the DNA-binding activity of CCA1 from plant extracts requires CK2 phosphorylation, as demonstrated by CK2-specific inhibitors (Sugano et al. 1998). Overexpressors of a CK2 regulatory subunit (CKB3) have severely shortened periods (4 hours) with few other pleiotropic effects, further suggesting that CCA1 phosphorylation by CK2 is necessary for clock function (Sugano et al. 1999). Six key

CK2 phosphorylation sites in CCA1 were mutated (mCCA1) and overexpression of mCCA1 had no significant effect on period, in contrast to the arrhythmicity seen in wild-type CCA1-OX plants. Interestingly, CCA1 homodimerization was abolished by the mutations, which together with the previous in vitro DNA-binding results suggests that a phosphorylated CCA1 homodimer may be the functional product of this gene (Daniel et al. 2004). CCA1 and the related LHY both bind the TOC1 promoter to negatively regulate transcription (Alabadi et al. 2001), so CK2 activation of CCA1 may contribute to TOC1 repression.

WNK1, one of a nine-member *Arabidopsis* family of WNK serine/threonine kinases, related to mitogen-activated protein kinases (MAPK) was found to interact with PRR3 and PRR5 in a yeast two-hybrid assay (Murakami-Kojima et al. 2002; Nakamichi et al. 2002). WNK1 is able to phosphorylate PRR3 in vitro and is expressed rhythmically with the same phase as that of PRR3 and PRR5. This is strong circumstantial evidence that suggests a role for WNK1 in the plant circadian clock and possibly others of the family whose transcription is also rhythmic (Nakamichi et al. 2002). If true, this would mark a departure from other kinase families described earlier but put plants into close company with the mammalian circadian system, where a MAPK cascade has been implicated in resetting (Butcher et al. 2002; Weber et al. 2006).

CONCLUSION

It is becoming evident that the concept of the transcription-translation feedback loop needs further revision as the importance of posttranslational events such as phosphorylation becomes more apparent in eukaryotic clocks. Conservation of kinase and phosphatase classes across groups as diverse as flies and fungi—which find no common ground in transcriptional regulators—indicates just how fundamental this mode of regulation is. We are still uncovering many of the basic elements of the plant circadian clock, and phosphorylation will certainly have a key role in determining subcellular location and protein half-lives. However, entirely different regulatory mechanisms may emerge in the plant clock, as the ZTL/GI relationship has recently illustrated. Whether similar surprises will be found in other systems bears watching for.

ACKNOWLEDGMENTS

W.-Y.K. was supported in part by a grant from the MOST/KOSEF to the Environmental Biotechnology National Core Research Center (grant R15-2003-012-01001-0), Korea. S.F. was funded in part by Support for Long-term Visit from the Yamada Science Foundation. D.E.S. was supported by NSF IBN-0344377 and MCB-0544137.

REFERENCES

Alabadi D., Oyama T., Yanovsky M.J., Harmon F.G., Mas P., and Kay S.A. 2001. Reciprocal regulation between TOC1 and LHY/CCA1 within the *Arabidopsis* circadian clock. *Science* **293:** 880.

Bae K. and Edery I. 2006. Regulating a circadian clock's period, phase and amplitude by phosphorylation: Insights from *Drosophila*. *J. Biochem.* **140:** 609.

Bell-Pedersen D., Cassone V.M., Earnest D.J., Golden S.S., Hardin P.E., Thomas T.L., and Zoran M.J. 2005. Circadian rhythms from multiple oscillators: Lessons from diverse organisms. *Nat. Rev. Genet.* **6:** 544.

Briggs W.R. and Christie J.M. 2002. Phototropins 1 and 2: Versatile plant blue-light receptors. *Trends Plant Sci.* **7:** 204.

Brunner M. and Schafmeier T. 2006. Transcriptional and post-transcriptional regulation of the circadian clock of cyanobacteria and *Neurospora*. *Genes Dev.* **20:** 1061.

Butcher G.Q., Dziema H., Collamore M., Burgoon P.W., and Obrietan K. 2002. The p42/44 mitogen-activated protein kinase pathway couples photic input to circadian clock entrainment. *J. Biol. Chem.* **277:** 29519.

Ceriani M.F., Darlington T.K., Staknis D., Mas P., Petti A.A., Weitz C.J., and Kay S.A. 1999. Light-dependent sequestration of TIMELESS by CRYPTOCHROME. *Science* **285:** 553.

Covington M.F., Panda S., Liu X.L., Strayer C.A., Wagner D.R., and Kay S.A. 2001. Elf3 modulates resetting of the circadian clock in *Arabidopsis*. *Plant Cell* **13:** 1305.

Crosson S. and Moffat K. 2001. Structure of a flavin-binding plant photoreceptor domain: Insights into light-mediated signal transduction. *Proc. Natl. Acad. Sci.* **98:** 2995.

Crosson S., Rajagopal S., and Moffat K. 2003. The LOV domain family: Photoresponsive signaling modules coupled to diverse output domains. *Biochemistry* **42:** 2.

Daniel X., Sugano S., and Tobin E.M. 2004. CK2 phosphorylation of CCA1 is necessary for its circadian oscillator function in *Arabidopsis*. *Proc. Natl. Acad. Sci.* **101:** 3292.

Darlington T.K., Wager-Smith K., Ceriani M.F., Staknis D., Gekakis N., Steeves T.D.L., Weitz C.J., Takahashi J.S., and Kay S.A. 1998. Closing the circadian loop: CLOCK-induced transcription of its own inhibitors *per* and *tim*. *Science* **280:** 1599.

Devlin P.F. and Kay S.A. 2000. Cryptochromes are required for phytochrome signaling to the circadian clock but not for rhythmicity. *Plant Cell* **12:** 2499.

Dolezelova E., Dolezel D., and Hall J.C. 2007. Rhythm defects caused by newly engineered null mutations in *Drosophila*'s *cryptochrome* gene. *Genetics* **177:** 329.

Doyle M.R., Davis S.J., Bastow R.M., McWatters H.G., Kozma-Bognar L., Nagy F., Millar A.J., and Amasino R.M. 2002. The ELF4 gene controls circadian rhythms and flowering time in *Arabidopsis thaliana*. *Nature* **419:** 74.

Dunlap J.C. 2006. Proteins in the *Neurospora* circadian clockworks. *J. Biol. Chem.* **281:** 28489.

Dunlap J.C. and Loros J.J. 2004. The *Neurospora* circadian system. *J. Biol. Rhythms* **19:** 414.

Eimert K., Wang S.M., Lue W.I., and Chen J. 1995. Monogenic recessive mutations causing both late floral initiation and excess starch accumulation in *Arabidopsis*. *Plant Cell* **7:** 1703.

Emery P.T., So W.V., Kaneko M., Hall J.C., and Rosbash M. 1998. CRY, a *Drosophila* clock and light-regulated cryptochrome, is a major contributor to circadian rhythm resetting and photosensitivity. *Cell* **95:** 669.

Fang Y., Sathyanarayanan S., and Sehgal A. 2007. Post-translational regulation of the *Drosophila* circadian clock requires protein phosphatase 1 (PP1). *Genes Dev.* **21:** 1506.

Fowler S., Lee K., Onouchi H., Samach A., Richardson K., Morris B., Coupland G., and Putterill J. 1999. GIGANTEA: A circadian clock-controlled gene that regulates photoperiodic flowering in *Arabidopsis* and encodes a protein with several possible membrane-spanning domains. *EMBO J.* **18:** 4679.

Froehlich A.C., Liu Y., Loros J.J., and Dunlap J.C. 2002. White Collar-1, a circadian blue light photoreceptor, binding to the frequency promoter. *Science* **297:** 815.

Gallego M. and Virshup D.M. 2007. Post-translational modifications regulate the ticking of the circadian clock. *Nat. Rev. Mol. Cell Biol.* **8:** 139.

Gekakis N., Saez L., Delahaye-Brown A.M., Myers M.P., Sehgal A., Young M.W., and Weitz C.J. 1995. Isolation of

timeless by PER protein interaction: Defective interaction between *timeless* protein and long-period mutant PER^{L}. *Science* **270:** 811.
Gould P.D., Locke J.C., Larue C., Southern M.M., Davis S.J., Hanano S., Moyle R., Milich R., Putterill J., Millar A.J., and Hall A. 2006. The molecular basis of temperature compensation in the *Arabidopsis* circadian clock. *Plant Cell* **18:** 1177.
Han L., Mason M., Risseeuw E.P., Crosby W.L., and Somers D.E. 2004. Formation of an SCF complex is required for proper regulation of circadian timing. *Plant J.* **40:** 291.
Hardin P.E. 2005. The circadian timekeeping system of *Drosophila*. *Curr. Biol.* **15:** R714.
———. 2006. Essential and expendable features of the circadian timekeeping mechanism. *Curr. Opin. Neurobiol.* **16:** 686.
He Q. and Liu Y. 2005. Molecular mechanism of light responses in *Neurospora:* From light-induced transcription to photoadaptation. *Genes Dev.* **19:** 2888.
He Q., Cheng P., Yang Y., Yu H., and Liu Y. 2003. FWD1-mediated degradation of FREQUENCY in *Neurospora* establishes a conserved mechanism for circadian clock regulation. *EMBO J.* **22:** 4421.
He Q., Cha J., He Q., Lee H.C., Yang Y., and Liu Y. 2006. CKI and CKII mediate the FREQUENCY-dependent phosphorylation of the WHITE COLLAR complex to close the *Neurospora* circadian negative feedback loop. *Genes Dev.* **20:** 2552.
He Q., Cheng P., Yang Y., Wang L., Gardner K.H., and Liu Y. 2002. White Collar-1, a DNA binding transcription factor and a light sensor. *Science* **297:** 840.
He Q., Shu H., Cheng P., Chen S., Wang L., and Liu Y. 2005. Light-independent phosphorylation of WHITE COLLAR-1 regulates its function in the *Neurospora* circadian negative feedback loop. *J. Biol. Chem.* **280:** 17526.
Hotta C.T., Gardner M.J., Hubbard K.E., Baek S.J., Dalchau N., Suhita D., Dodd A.N., and Webb A.A. 2007. Modulation of environmental responses of plants by circadian clocks. *Plant Cell Environ.* **30:** 333.
Huq E., Tepperman J.M., and Quail P.H. 2000. GIGANTEA is a nuclear protein involved in phytochrome signaling in *Arabidopsis*. *Proc. Natl. Acad. Sci.* **97:** 9789.
Imaizumi T., Schultz T.F., Harmon F.G., Ho L.A., and Kay S.A. 2005. FKF1 F-box protein mediates cyclic degradation of a repressor of CONSTANS in *Arabidopsis*. *Science* **309:** 293.
Imaizumi T., Tran H.G., Swartz T.E., Briggs W.R., and Kay S.A. 2003. FKF1 is essential for photoperiodic-specific light signalling in *Arabidopsis*. *Nature* **426:** 302.
Kevei E., Gyula P., Hall A., Kozma-Bognar L., Kim W.Y., Eriksson M.E., Toth R., Hanano S., Feher B., Southern M.M., Bastow R.M., Viczian A., Hibberd V., Davis S.J., Somers D.E., Nagy F., and Millar A.J. 2006. Forward genetic analysis of the circadian clock separates the multiple functions of ZEITLUPE. *Plant Physiol.* **140:** 933.
Kiba T., Henriques R., Sakakibara H., and Chua N.H. 2007. Targeted degradation of PSEUDO-RESPONSE REGULATOR5 by a SCFZTL complex regulates clock function and photomorphogenesis in *Arabidopsis thaliana*. *Plant Cell* **19:** 2516.
Kim E.Y. and Edery I. 2006. Balance between DBT/CKIepsilon kinase and protein phosphatase activities regulate phosphorylation and stability of *Drosophila* CLOCK protein. *Proc. Natl. Acad. Sci.* **103:** 6178.
Kim W.Y., Geng R., and Somers D.E. 2003. Circadian phase-specific degradation of the F-box protein ZTL is mediated by the proteasome. *Proc. Natl. Acad. Sci.* **100:** 4933.
Kim W.Y., Fujiwara S., Suh S.S., Kim J., Kim Y., Han L., David K., Putterill J., Nam H.G., and Somers D.E. 2007. ZEITLUPE is a circadian photoreceptor stabilized by GIGANTEA in blue light. *Nature* **449:** 356.
Kloss B., Price J.L., Saez L., Blau J., Rothenfluh A., Wesley C.S., and Young M.W. 1998. The *Drosophila* clock gene double-time encodes a protein closely related to human casein kinase Iepsilon. *Cell* **94:** 97.
Koh K., Zheng X., and Sehgal A. 2006. JETLAG resets the *Drosophila* circadian clock by promoting light-induced degradation of TIMELESS. *Science* **312:** 1809.
Lee C., Etchegaray J.P., Cagampang F.R., Loudon A.S., and Reppert S.M. 2001. Posttranslational mechanisms regulate the mammalian circadian clock. *Cell* **107:** 855.
Liu Y. 2003. Molecular mechanisms of entrainment in the *Neurospora* circadian clock. *J. Biol. Rhythms.* **18:** 195.
Makino S., Matsushika A., Kojima M., Yamashino T., and Mizuno T. 2002. The APRR1/TOC1 quintet implicated in circadian rhythms of *Arabidopsis thaliana*. I. Characterization with APRR1-overexpressing plants. *Plant Cell Physiol.* **43:** 58.
Martin-Tryon E.L., Kreps J.A., and Harmer S.L. 2007. GIGANTEA acts in blue light signaling and has biochemically separable roles in circadian clock and flowering time regulation. *Plant Physiol.* **143:** 473.
Martinek S., Inonog S., Manoukian A.S., and Young M.W. 2001. A role for the segment polarity gene shaggy/GSK-3 in the *Drosophila* circadian clock. *Cell* **105:** 769.
Mas P., Kim W.Y., Somers D.E., and Kay S.A. 2003a. Targeted degradation of TOC1 by ZTL modulates circadian function in *Arabidopsis thaliana*. *Nature* **426:** 567.
Mas P., Alabadi D., Yanovsky M.J., Oyama T., and Kay S.A. 2003b. Dual role of TOC1 in the control of circadian and photomorphogenic responses in *Arabidopsis*. *Plant Cell* **15:** 223.
McClung C.R. 2006. Plant circadian rhythms. *Plant Cell* **18:** 792.
McWatters H.G., Kolmos E., Hall A., Doyle M.R., Amasino R.M., Gyula P., Nagy F., Millar A.J., and Davis S.J. 2007. ELF4 is required for oscillatory properties of the circadian clock. *Plant Physiol.* **144:** 391.
Merrow M., Mazzotta G., Chen Z., and Roenneberg T. 2006. The right place at the right time: Regulation of daily timing by phosphorylation. *Genes Dev.* **20:** 2629.
Michael T.P., Salome P.A., Yu H.J., Spencer T.R., Sharp E.L., McPeek M.A., Alonso J.M., Ecker J.R., and McClung C.R. 2003. Enhanced fitness conferred by naturally occurring variation in the circadian clock. *Science* **302:** 1049.
Mizoguchi T., Wright L., Fujiwara S., Cremer F., Lee K., Onouchi H., Mouradov A., Fowler S., Kamada H., Putterill J., and Coupland G. 2005. Distinct roles of GIGANTEA in promoting flowering and regulating circadian rhythms in *Arabidopsis*. *Plant Cell* **17:** 2255.
Mizuno T. and Nakamichi N. 2005. Pseudo-response regulators (PRRs) or true oscillator components (TOCs). *Plant Cell Physiol.* **46:** 677.
Murakami-Kojima M., Nakamichi N., Yamashino T., and Mizuno T. 2002. The APRR3 component of the clock-associated APRR1/TOC1 quintet is phosphorylated by a novel protein kinase belonging to the WNK family, the gene for which is also transcribed rhythmically in *Arabidopsis thaliana*. *Plant Cell Physiol.* **43:** 675.
Naidoo N., Song W., Hunter-Ensor M., and Sehgal A. 1999. A role for the proteasome in the light response of the timeless clock protein. *Science* **285:** 1737.
Nakamichi N., Murakami-Kojima M., Sato E., Kishi Y., Yamashino T., and Mizuno T. 2002. Compilation and characterization of a novel WNK family of protein kinases in *Arabiodpsis thaliana* with reference to circadian rhythms. *Biosci. Biotechnol. Biochem.* **66:** 2429.
Nawathean P. and Rosbash M. 2004. The doubletime and CKII kinases collaborate to potentiate *Drosophila* PER transcriptional repressor activity. *Mol. Cell* **13:** 213.
Nelson D.C., Lasswell J., Rogg L.E., Cohen M.A., and Bartel B. 2000. FKF1, a clock-controlled gene that regulates the transition to flowering in *Arabidopsis*. *Cell* **101:** 331.
Oliverio K.A., Crepy M., Martin-Tryon E.L., Milich R., Harmer S.L., Putterill J., Yanovsky M.J., and Casal J.J. 2007. GIGANTEA regulates phytochrome A-mediated photomorphogenesis independently of its role in the circadian clock. *Plant Physiol.* **144:** 495.
Park D., Somers D.E., Kim Y., Choy Y., Lim H., Soh M., Kim H., Kay S.A., and Nam H.G. 1999. Control of circadian rhythms and photoperiodic flowering by the *Arabidopsis* GIGANTEA gene. *Science* **285:** 1579.

Price J.L., Blau J., Rothenfluh A., Abodeely M., Kloss B., and Young M.W. 1998. double-time is a novel *Drosophila* clock gene that regulates PERIOD protein accumulation. *Cell* **94:** 83.
Redei G.P. 1962. Supervital mutants of *Arabidopsis. Genetics* **47:** 443.
Rockwell N.C., Su Y.S., and Lagarias J.C. 2006. Phytochrome structure and signaling mechanisms. *Annu. Rev. Plant Biol.* **57:** 837.
Salomon M., Christie J.M., Knieb E., Lempert U., and Briggs W.R. 2000. Photochemical and mutational analysis of the FMN-binding domains of the plant blue light receptor, phototropin. *Biochemistry* **39:** 9401.
Sato E., Nakamichi N., Yamashino T., and Mizuno T. 2002. Aberrant expression of the *Arabidopsis* circadian-regulated APRR5 gene belonging to the APRR1/TOC1 quintet results in early flowering and hypersensitiveness to light in early photomorphogenesis. *Plant Cell Physiol.* **43:** 1374.
Schafmeier T., Kaldi K., Diernfellner A., Mohr C., and Brunner M. 2006. Phosphorylation-dependent maturation of *Neurospora* circadian clock protein from a nuclear repressor toward a cytoplasmic activator. *Genes Dev.* **20:** 297.
Schafmeier T., Haase A., Kaldi K., Scholz J., Fuchs M., and Brunner M. 2005. Transcriptional feedback of *Neurospora* circadian clock gene by phosphorylation-dependent inactivation of its transcription factor. *Cell* **122:** 235.
Schultz T.F., Kiyosue T., Yanovsky M., Wada M., and Kay S.A. 2001. A role for LKP2 in the circadian clock of *Arabidopsis. Plant Cell* **13:** 2659.
Somers D.E. 2001. Clock-associated genes in *Arabidopsis:* A family affair. *Philos. Trans. R. Soc. Lond. B Biol. Sci.* **356:** 1745.
———. 2005. ZEITLUPE and the control of circadian timing. In *Light sensing in plants* (ed. M. Wada et al.), p. 347. Springer, Tokyo.
Somers D.E., Devlin P.F., and Kay S.A. 1998a. Phytochromes and cryptochromes in the entrainment of the *Arabidopsis* circadian clock. *Science* **282:** 1488.
Somers D.E., Kim W.Y., and Geng R. 2004. The F-box protein ZEITLUPE confers dosage-dependent control on the circadian clock, photomorphogenesis, and flowering time. *Plant Cell* **16:** 769.
Somers D.E., Schultz T.F., Milnamow M., and Kay S.A. 2000. *ZEITLUPE* encodes a novel clock-associated PAS protein from *Arabidopsis. Cell* **101:** 319.
Somers D.E., Webb A.A.R., Pearson M., and Kay S. 1998b. The short-period mutant, *toc1-1,* alters circadian clock regulation of multiple outputs throughout development in *Arabidopsis thaliana. Development* **125:** 485.
Stanewsky R., Kaneko M., Emery P., Beretta B., Wager-Smith K., Kay S.A., Rosbash M., and Hall J.C. 1998. The cryb mutation identifies cryptochrome as a circadian photoreceptor in *Drosophila. Cell* **95:** 681.
Stoleru D., Nawathean P., Fernandez M.L., Menet J.S., Ceriani M.F., and Rosbash M. 2007. The *Drosophila* circadian network is a seasonal timer. *Cell* **129:** 207.
Strayer C., Oyama T., Schultz T.F., Raman R., Somers D.E., Mas P., Panda S., Kreps J.A., and Kay S.A. 2000. Cloning of the *Arabidopsis* clock gene TOC1, an autoregulatory response regulator homolog. *Science* **289:** 768.
Sugano S., Andronis C., Green R.M., Wang Z.Y., and Tobin E.M. 1998. Protein kinase CK2 interacts with and phosphorylates the *Arabidopsis* circadian clock-associated 1 protein. *Proc. Natl. Acad. Sci.* **95:** 11020.
Sugano S., Andronis C., Ong M.S., Green R.M., and Tobin E.M. 1999. The protein kinase CK2 is involved in regulation of circadian rhythms in *Arabidopsis. Proc. Natl. Acad. Sci.* **96:** 12362.
Tseng T.S., Salome P.A., McClung C.R., and Olszewski N.E. 2004. SPINDLY and GIGANTEA interact and act in *Arabidopsis thaliana* pathways involved in light responses, flowering, and rhythms in cotyledon movements. *Plant Cell* **16:** 1550.
Weber F., Hung H.C., Maurer C., and Kay S.A. 2006. Second messenger and Ras/MAPK signalling pathways regulate CLOCK/CYCLE-dependent transcription. *J. Neurochem.* **98:** 248.
Yang Y., Cheng P., He Q., Wang L., and Liu Y. 2003. Phosphorylation of FREQUENCY protein by casein kinase II is necessary for the function of the *Neurospora* circadian clock. *Mol. Cell. Biol.* **23:** 6221.
Yang Y., He Q., Cheng P., Wrage P., Yarden O., and Liu Y. 2004. Distinct roles for PP1 and PP2A in the *Neurospora circadian* clock. *Genes Dev.* **18:** 255.
Young M.W. and Kay S.A. 2001. Time zones: A comparative genetics of circadian clocks. *Nat. Rev. Genet.* **2:** 702.

Circadian Output, Input, and Intracellular Oscillators: Insights into the Circadian Systems of Single Cells

J.J. Loros,*† J.C. Dunlap,† L.F. Larrondo,† M. Shi,† W.J. Belden,† V.D. Gooch,‡ C.-H. Chen,† C.L. Baker,† A. Mehra,† H.V. Colot,† C. Schwerdtfeger,† R. Lambreghts,† P.D. Collopy,† J.J. Gamsby,† and C.I. Hong†

*Department of Biochemistry, †Department of Genetics, Dartmouth Medical School, Hanover, New Hampshire 03755; ‡Division of Science and Mathematics, University of Minnesota, Morris, Minnesota 56267

Circadian output comprises the business end of circadian systems in terms of adaptive significance. Work on *Neurospora* pioneered the molecular analysis of circadian output mechanisms, and insights from this model system continue to illuminate the pathways through which clocks control metabolism and overt rhythms. In *Neurospora*, virtually every strain examined in the context of rhythms bears the *band* allele that helps to clarify the overt rhythm in asexual development. Recent cloning of *band* showed it to be an allele of *ras-1* and to affect a wide variety of signaling pathways yielding enhanced light responses and asexual development. These can be largely phenocopied by treatments that increase levels of intracellular reactive oxygen species. Although output is often unidirectional, analysis of the *prd-4* gene provided an alternative paradigm in which output feeds back to affect input. *prd-4* is an allele of checkpoint kinase-2 that bypasses the requirement for DNA damage to activate this kinase; FRQ is normally a substrate of activated Chk2, so in Chk2^{PRD-4}, FRQ is precociously phosphorylated and the clock cycles more quickly. Finally, recent adaptation of luciferase to fully function in *Neurospora* now allows the core FRQ/WCC feedback loop to be followed in real time under conditions where it no longer controls the overt rhythm in development. This ability can be used to describe the hierarchical relationships among FRQ-Less Oscillators (FLOs) and to see which are connected to the circadian system. The nitrate reductase oscillator appears to be connected, but the oscillator controlling the long-period rhythm elicited upon choline starvation appears completely disconnected from the circadian system; it can be seen to run with a very long noncompensated 60–120-hour period length under conditions where the circadian FRQ/WCC oscillator continues to cycle with a fully compensated circadian 22-hour period.

INTRODUCTION

The innate ability to gauge time of day juxtaposed with the perception of environmental light represents one of the most widespread and closely coupled forms of cellular, tissue, and organismal regulation across a broad range of taxonomic groups. Changes in light fluence and wavelength are detected with the harvesting of photons by chromophore-binding photoreceptive molecules allowing immediate detection of important environmental changes. Biological rhythms provide organisms with the ability to anticipate environmental changes arising from the Earth's rotation. The physiological and molecular mechanisms of daily biological rhythms, called circadian rhythms, have been the focus of study for more than a century, but the past quarter century has witnessed a wealth of information explaining the basis of these clocks, with microbial systems playing a key role in many regards. All circadian clocks are cellular in nature, a property first described in microbes (Hastings and Sweeney 1958). More recently, the filamentous fungus *Neurospora crassa* and the fruit fly *Drosophila melanogaster* have pioneered the description of feedback loops as the molecular basis of circadian rhythmicity.

Scientists first became interested in circadian rhythmicity due to the broad spectrum of biological processes that are controlled by the clock, resulting in time-of-day-specific activities and processes. Currently, a broad field of researchers has focused on studying these rhythms in many different organismal systems at many different levels. In the past two decades, much of the prominent work has focused on understanding molecular mechanisms involved in the core process that generates 24-hour rhythmicity, but understanding how the apparent daily time cues are perceived by the clock mechanism and then transduced into signals that regulate daily activities has more recently become of intense interest. The large and complex range of clock-regulated processes in multicellular organisms is well documented, although it is now understood that all of this complexity can be reduced to clock-regulated metabolism and physiology at the intracellular level (Loros et al. 2003), leading to a desire to understand the molecular underpinnings of intracellular regulation by circadian oscillators.

The well-understood filamentous fungus *N. crassa* has served as a basic model system to examine eukaryotic biology since the 1940s (Beadle and Tatum 1945), and it is also one of the earliest model systems to be shown to possess a bona fide circadian system (Pittendrigh and Bruce 1959). Close to two decades ago, we predicted that daily clock control of gene expression might be a major aspect of output (Loros et al. 1989). This has proved not only to be true for *Neurospora*, but also to be universally the case for all circadian systems (see, e.g., Liu et al. 1995; Harmer et al. 2000; Ceriani et al. 2002; Duffield et al. 2002; Panda et al. 2002). We performed the first systematic screen, using a novel technique at the time (subtractive hybridization), of two different mRNA populations, one isolated from morning and one from evening, to identify genes whose mRNAs were rhythmically abundant (Loros et al. 1989). These early genes were subsequently shown to be regulated at the level of transcriptional rate, suggesting that there would be *cis*-acting sequences in the promoter regions of these genes in addition to *trans*-acting factors that conferred circadian regulation (Loros and Dunlap 1991). Continuing studies using differential hybridization (Bell-Pedersen et al. 1996b), cDNA sequencing (Zhu et al. 2001), and cDNA and oligonucleotide microarrays (Correa et al. 2003; Nowrousian et al. 2003) have isolated several hundred more rhythmically regulated genes. Full-genome microarrays for the approximately 10,500 *Neurospora* genes have recently become available and are currently being used by more than one lab to examine clock regulation of gene expression at different points in the life cycle and under various growth conditions. In multicellular organisms, the identification of genes regulated by the clock was largely on a gene-by-gene basis until the advent of microarrays (Harmer et al. 2000; Duffield 2003) in addition to an innovative differential display protocol called ADDER used to isolate several cycling liver genes (Kornmann et al. 2001). These studies already have, and will continue to, dramatically improve our understanding of the global nature of circadian regulation on gene expression at the tissue and organ levels in plants and animals. Of great current interest is the mechanistic understanding of clock control of specific sets of genes at the level of individual cells, a more difficult problem to approach using in vivo samples. Adopting the same paradigms used in whole organisms, tissue cultures using individual cell types that display circadian properties (Balsalobre et al. 1998) have been recently leading the way (Duffield et al. 2002).

THE CIRCADIAN CLOCK AND THE CIRCADIAN SYSTEM

The haploid filamentous fungus *N. crassa* grows as incompletely septate, highly branched mycelia capable of fusing with separate but compatible strains, thereby commingling both cytoplasm and nuclei in a common cytoplasmic compartment. *Neurospora* is a superb organism for genetics and biochemistry with a well-defined genetic history, allowing the maintenance of lethal mutations, gene-dosage analysis, and complementation. The genome is about three times the size of yeast at 4.3×10^7 base pairs, encoding about 10,500 highly nonredundant genes expressed at different stages of the sexual and asexual life cycle. The clock in *Neurospora* is now monitored in a number of different ways. Historically and still of great utility is the monitoring of asexual conidiospore (conidia) development as the fungus grows on medium placed in hollow glass culture tubes called race tubes (Fig. 1) (Ryan et al. 1943). One consequence of growing in the enclosed environment of the race tube is that CO_2 levels become elevated, resulting in suppression of conidiation and masking of the periodic formation of conidia within the tube. A mutation called *band* (*bd*) (Sargent et al. 1966) overcomes this repression and has been used in *Neurospora* laboratory stocks in rhythm studies for the last 40 years. The *bd* locus has recently been cloned and found, as expected, not to alter function of the clock's central mechanism, and also not completely surprisingly, it was found to have global effects on the regulation of clock output (Belden et al. 2007). For molecular and biochemical analysis, *Neurospora* is grown in liquid culture (Nakashima 1981; Perlman et al. 1981; Loros et al. 1989; Loros and Dunlap 1991; Aronson et al. 1994b; Garceau et al. 1997). Mycelial disks in liquid culture (Shi et al. 2007) maintain endogenous rhythmicity and phase, even upon transfer to solid medium. The ability to use luciferase (Morgan et al. 2003) to follow either in vivo transcriptional or translational fusions of rhythmically expressed genes has recently become a valuable tool to monitor clock progress by the complete codon optimization of firefly luciferase (Gooch et al. 2007).

An Overview of the Circadian Oscillator

Genetic, molecular, and biochemical analyses of the *Neurospora* clock have led to truly extraordinary advances in our understanding of much of the organism's circadian properties, including the generation and sustainability of rhythmicity, phase resetting by light and temperature, and the means by which the clock controls metabolism and behavior. The following is an extremely brief overview of the molecular underpinnings: In *Neurospora*, as in all eukaryotes studied to date, a critical part is played by an autoregulatory, molecular feedback loop between the FREQUENCY protein(s) (FRQ); the White Collar transcription factors (WC-1 and WC-2) (Aronson et al. 1994a,b; Crosthwaite et al. 1997; Dunlap 1999; Lee et al. 2003); and at least one other associated protein, FRH (Cheng et al. 2005). Expression at the *frq* locus is activated by the heterodimeric White Collar Complex (WCC) (Crosthwaite et al. 1995; Ballario et al. 1996, 1998; Linden and Macino 1997; Linden et al. 1997; Talora et al. 1999; Collett et al. 2001, 2002; Denault et al. 2001; Cheng et al. 2002). Long and short forms of FRQ (Garceau et al. 1997), produced from temperature-regulated alternative splicing (Colot et al. 2005; Diernfellner et al. 2005), then feed back to block activation of *frq* expression (Aronson et al. 1994b; Garceau et al. 1997; Merrow et al. 1997; Froehlich et al. 2003), resulting in rhythmic waves of both *frq* mRNA and FRQ protein, highly regulated by both phosphorylation (Garceau et al. 1997; Liu et al. 2000; Schafmeier et al. 2006) and ubiquitination (He et al. 2003, 2005) over the course of the 24-hour day. The environment signals to the clock, most notably via light and temperature, result in phase resetting such that the organism is in synchrony with the external world (Crosthwaite et al. 1995; Liu et al. 1998; Kramer et al. 2003; Price-Lloyd et al. 2005; Hunt et al. 2007) (for more a detailed overview, see Dunlap et al., this volume). Finally, the FRQ/WCC feedback oscillator can signal temporal information to the cell via daily changes in transcript abundance of pertinent output genes (Loros et al. 1989). A major regulatory means to this end is through *cis*-acting sequences in promoters that confer daily changes of transcriptional rates (see, e.g., Loros and Dunlap 1991; Bell-Pedersen et al. 1996a).

Noncircadian Oscillators

Above, we alluded to the classic "input-oscillator-output" paradigm (Eskin 1979), but it is clear that organisms

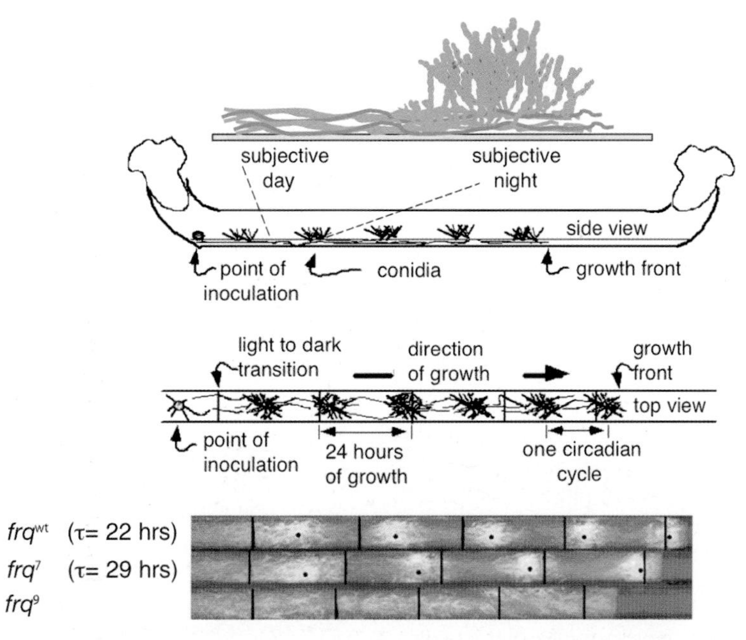

Figure 1. Analysis of *Neurospora* rhythms on race tubes. Glass tubes with growth medium are inoculated with fungal cultures on the left, grown for a day in the light, and transferred to darkness, which is interpreted as dusk (circadian time 12), synchronizing the circadian clock. As the culture grows, the clock controls development in the region of the growth front such that regions of undifferentiated surface hyphae alternate with regions of aerial hyphae over the course of the subjective day. The growth front leaves behind a region or "band" of aerial hyphae that will continue to differentiate into macroconidiospores, cell types biochemically distinct from the undifferentiated surface hyphae on either side. After several days of growth in constant darkness, the agar surface is covered with distinct conidial bands alternating with undifferentiated surface growth. The period length and phase of the clock are read from this pattern of growth, and mutants can be selected by screening for variations in the pattern. A long-period allele and loss-of-function allele of the *frequency* gene, frq^7 and frq^9, respectively, are shown. (Adapted, with permission, from Loros and Dunlap 2006 [© Springer-Verlag].)

contain a circadian system, not just an oscillator mechanism with linear inputs and outputs. Part of the system will doubtlessly include other oscillators (see, e.g., Dunlap and Loros 2004). As a field, regardless of organism, we are focusing more closely on individual cells, and as we do, examples of noncanonical rhythms are appearing more frequently. Oscillations can occur in any cellular pathway that has negative feedback regulation and a lag in at least one step, a common biological occurrence. Some of these oscillators will be unrelated and unconnected to the clock, and some may be part of the circadian system, either by having a necessary role in the clock oscillatory mechanism or by feeding into the clock, somewhere in the input or core oscillator, to alter properties of the clock. In addition, some will be "slave" oscillators (see below), controlled by the clock at some step within its own feedback to confer the properties of circadian period length and phase stability. Identification and elucidation of these alternative oscillators are major topics of interest to chronobiologists and are dealt with in depth below.

OUTPUT FROM THE CLOCK IS THE NATURE OF TEMPORAL INFORMATION

The utilitarian and therefore most important aspect of all clocks, often referred to as output, is the ability to invoke "time of dayness" onto the organism, such that it can predict daily changes in the environment to regulate its own changing metabolic needs over the course of the diurnal cycle. Most outputs may occur in a linear fashion and be distinct from the clock mechanism.

Neurospora has been developed as a paradigmatic system for understanding the physiological changes governed by the clock, including development of the hypothesis that changes in transcription would be a universal means by which the clock mechanism could regulate clock output. Testing this hypothesis led to the first genome-wide screens, using subtractive hybridization methodologies, for the daily changes in gene expression alluded to above and to coining the term "clock-controlled gene" (ccg) (Loros et al. 1989). The identity of hundreds of ccgs are currently known through a combination of differential hybridization, cDNA sequencing, cDNA and oligomer microarrays, and individual gene examination (Bell-Pedersen 2000; Correa et al. 2003; for review, see Loros et al. 2003; Nowrousian et al. 2003).

Isolation of the initial ccgs led to a first working model for defining output at the molecular level of gene expression. First, the endogenous rhythm of expression of a ccg would persist under constant conditions, reflecting control exerted by the clock, as opposed to external factors such as changes in light intensity. Second, a defining test was that the period length of the molecular rhythm would reflect the genotype of the strain. For example, in *Neurospora*, the first ccgs were shown to have changes in mRNA abundance with 22-hour period lengths in wild type and 29-hour period lengths when examined in the long-period strain *frq*[7]. The third initial criterion was that the inactivation of the ccg would have no effect on the clock, demonstrating that the gene represented a true molecular output in the circadian system as a whole. We now know that this third criterion is not universally true and although it may be true that most ccgs in any cell, tissue, or whole organism can be considered linear outputs, distinct from the mechanistic core of the oscillator mechanism, there exist important examples where an output can influence aspects of oscillator function, leading to coupled loops within the system.

With the early isolation of ccgs in several organisms, the question of ccg function emerged: What was the clock used for in different organisms and were there overlaps among organisms? It had been suggested that, as a general rule, clocks controlled "nonhousekeeping" processes that would be distinctly organism-dependent. This idea may have arisen due to the spectrum of output processes that had been studied before molecular analysis and that included such diverse biological functions as locomotor activity and behavior, leaf movement, asexual development, cardiovascular function, electrolyte balance, and timing of cell division. An observation surfacing from early clock genetics was that organisms were viable in the absence of operational clocks, leading to the thought that clock components, and possibly their regulatory outputs, would be found to be nonessential. We now know that virtually all aspects of biological function are controlled by the clock in *Neurospora* (Fig. 2) as well as in other systems examined (see, e.g., Liu et al. 1995; Harmer et al. 2000; Ceriani et al. 2002; Duffield et al. 2002; Panda et al. 2002; Wijnen and Young 2006). For example, the glycolytic gene encoding glyceraldehyde-3-phosphate dehydrogenase (GAPDH), the rate-limiting and first energy-harvesting enzyme in glycolysis, was identified early as a ccg in *Neurospora* (Bell-Pedersen et al. 1996b; Shinohara et al. 1998) and subsequently found in the dinoflagellate *Gonyaulax* (Fagan et al. 1999), chick retina (Bailey et al. 2004), mouse hepatocytes (Temme et al. 2000), and elsewhere (see, e.g., Iwasaki et al. 2004; Kamphuis et al. 2005).

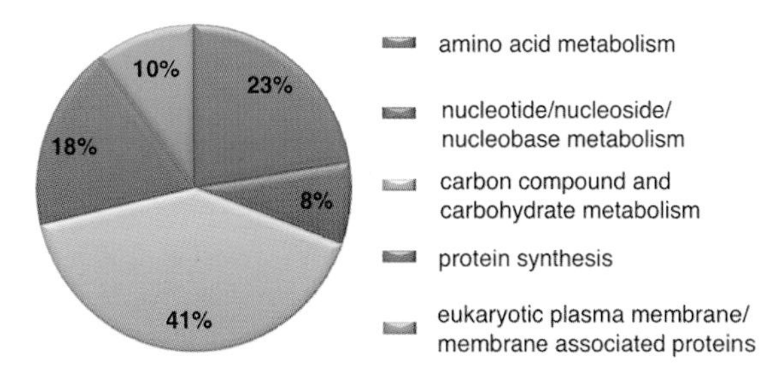

Figure 2. The clock regulates a broad variety of cellular functions in the *Neurospora* cell. RNAs were harvested over two circadian cycles and probed with oligomeric microarrays representing the approximately 10,500 genes from *N. crassa*. The results from two biologically independent sets of RNAs found 64% of the genes passed detection criteria with approximately 10% of those classified as clock-controlled genes (ccgs). About half of these displayed functional hits in the Functional Catalogue (http://mips.gsf.de/projects/funcat) (C.-H. Chen et al., unpubl.).

THE SPECTRUM OF CCGS WITHIN A SINGLE CELL TYPE MAY REFLECT GROWTH/EXTERNAL CONDITIONS

Certainly, GAPDH does not cycle in all cells nor under all conditions (Okamura et al. 1999; Kobayashi et al. 2004). An interesting feature of GAPDH, in addition to its role in glycolysis, is its ability to alter a cell's redox state through the reduction of NAD^+ to NADH. It may be that the clock is used to control GAPDH in specific cells or during specific oxidative/reduction conditions. In *Neurospora*, the spectrum of ccgs certainly reflects growth conditions. Early microarray studies identifying ccgs from high glucose (Nowrousian et al. 2003) versus lower glucose (Correa et al. 2003) cultures found largely nonoverlapping sets of cycling genes that ranged between 5% and 20%, respectively, of all genes on the array.

A lesson about the plasticity of circadian gene regulation within a single cell type and within the organism can be deduced from the identity of the *bd* gene, an allele commonly used in the background of laboratory stocks of *Neurospora* used for rhythms research. In the beginning years of *Neurospora* clock research, *bd* was identified as a spontaneous mutation. It was shown not to alter the canonical properties of the clock, but to significantly enhance the conidiation rhythm when strains were grown on race tubes (Sargent et al. 1966). Since this time, virtually all circadian research has been performed with *bd*-containing strains (for review, see Dunlap and Loros 2005; Liu and Bell-Pedersen 2006), although *bd* was known to display enhanced effects on both light and clock-regulated gene expression (Arpaia et al. 1993, 1995). A completed genome and single-nucleotide polymorphism (SNP) analysis fostered the identification of *bd* as a T791I point mutation in *ras-1* (Belden et al. 2007). RAS is a small, conserved, membrane-attached G protein, activated by exchange of GDP with GTP, which is modulated by guanine nucleotide exchange factors. It is a key player in numerous cellular signaling cascades, including those that link extracellular signals to gene expression (see, e.g., Mitin et al. 2005). The *ras-1*^{bd} hypermorph is somewhat more active in GDP/GTP exchange than the wild type (Belden et al. 2007). Not unexpectedly, genes involved in asexual development were found to be up-regulated in this altered function mutant. In addition, photo-induced expression of the *wc-1* gene is also increased in the *ras-1*^{bd} background. Importantly, in a *ras-1*^{bd} background, the C6 zinc cluster transcription factor *fluffy* (*fl*), which is both necessary and sufficient for asexual development (Bailey-Shrode and Ebbole 2004) and is itself rhythmically expressed (Correa and Bell-Pedersen 2002), is expressed at significantly enhanced levels both after light exposure and rhythmically in the dark. This enhanced expression is the cause of the increased visibility of rhythmic conidiation in *ras-1*^{bd}-containing strains (Fig. 3) (Belden et al. 2007).

While investigating the source of the *ras-1*^{bd} signal that results in changes in expression of light and clock-regulated genes, we found that altering the amount of reactive oxygen species (ROS) could also influence the banding of wild-type strains in race tubes. ROS are increasingly rec-

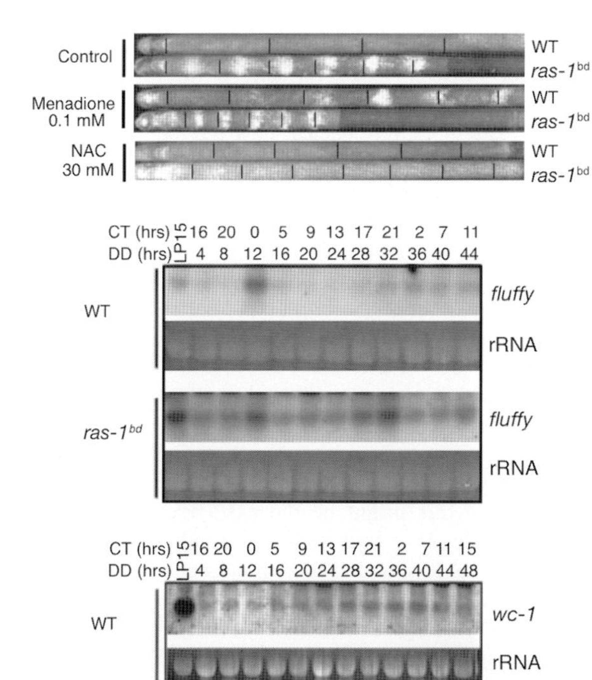

Figure 3. The expression of a transcription factor involved with conidiation is elevated in *ras-1*^{bd}. Race tubes are shown from the top down in groups of two: Wild type does not express rhythmic conidiation but *band* (= *ras-1*^{bd}) does; adding menadione, an oxidant, to wild type results in banding but does not affect *ras-1*^{bd}, other than slowing growth; adding the antioxidant *N*-acetyl cysteine (NAC) by itself eliminates banding in *ras-1*^{bd}, but has no effect on wild-type. Although reactive oxygen species (ROS) levels are not elevated in *ras-1*^{bd}, the mutant phenocopies changes in cellular ROS. Shown in the center and bottom panels are northern analysis of the transcription factors *fluffy*, required for conidiation in *Neurospora*, and *wc-1*, a core component in the *Neurospora* circadian system involved in clock regulation of conidiation. Both transcripts are up-regulated in the *ras-1*^{bd} strain. (Adapted, with permission, from Belden et al. 2007 [© Cold Spring Harbor Laboratory Press].)

ognized as important second messengers in cell signaling (D'Autréaux and Toledano 2007), and activated RAS is known to result in increased intracellular ROS levels (Irani et al. 1997). Conversely, changes in ROS may affect RAS signaling (see, e.g., Heo and Campbell 2006). The photoreceptor WC-1 and the clock component duplex WCC are thought to be sensitive to the redox states within the cell, as is carotenoid synthesis and development in *Neurospora* (Hansberg and Aguirre 1990). Additionally, increasing ROS levels may lead to increases in WCC transcriptional activity (Yoshida and Hasunuma 2004; Aguirre et al. 2005); all of these data suggest a credible connection among RAS signaling, cellular ROS levels, and clock-regulated gene expression. Coupled signaling between ROS and RAS in *Neurospora* has important implications for the study of circadian output regulation in all organisms. Significantly, conditions that alter metabolic state can result in different biological programs and may alter the ccg profile even in a single cell type. It is well established that the known complement of ccgs is different in different tissues from multicellular organisms. The *ras-1*^{bd} story indicates that within a specific cell or tissue type, changes in signaling or metabolic activity can alter the complement of genes regulated by the clock.

SOME OUTPUTS FEED BACK TO MODULATE CLOCK FUNCTION, RESULTING IN ADDITIONAL LOOPS

Most ccgs, when deleted from the system through gene inactivation, show no effect on clock function. However, some circadianly regulated outputs can change the workings of the clock. Two recent examples of different ways this can happen in *Neurospora* are detailed below. The first output feeds back to alter the way light information is received by the clock, and the second output feeds back to alter the biochemical nature of a clock component.

A Molecular Output Can Feed Back to Input

Research on *Neurospora* has had a leading role in understanding the molecular basis of how light resets the clock, a process called entrainment (Crosthwaite et al. 1995), and that includes gating (Heintzen et al. 2001). Much of the central mechanism of light entrainment in *Neurospora* is conserved in mammals (Shigeyoshi et al. 1997). Gating refers to the condition in which the clock regulates its own input such that it responds to an identical stimulus (e.g., a specific amount and wavelength of light) in nonidentical ways at different times of the circadian day. In gating, output and input become mechanistically merged. The WC-1 protein is the primary blue light photoreceptor for *Neurospora* (Froehlich et al. 2002; He et al. 2002). A covalently bound FAD chromophore in the WC-1 light-oxygen-voltage (LOV) domain absorbs photons, leading to rapid and strong activation of *frq* transcription by the WCC in response to light input and to clock resetting (Crosthwaite et al. 1997). The strength of this *frq*-induction signal is modulated by another LOV domain containing photoreceptive protein called VIVID (VVD) that gates the light information coming into the clock at different circadian times (Heintzen et al. 2001). The expression of *vvd* is clock-controlled, making *vvd* both a *ccg* and a clock input (Fig. 4). VVD is also responsible for the ability of *Neurospora* to sense changes in light intensity (Schwerdtfeger and Linden 2003; Schwerdtfeger et al. 2003), a process termed photoadaptation, and it additionally has a role in temperature compensation of phase (Heintzen and Liu 2007; Hunt et al. 2007). In response to light, a transient cysteine-flavin adduct forms in the LOV domain that promotes breaking of the hydrogen bonds holding an amino-terminal helix; this conformational change results in signaling (Zoltowski et al. 2007). In natural photoperiods of light and dark, the rhythmic output gene VVD effectively modulates the WCC transcriptional response to light on the *frq* promoter, permitting the clock to accurately keep time during

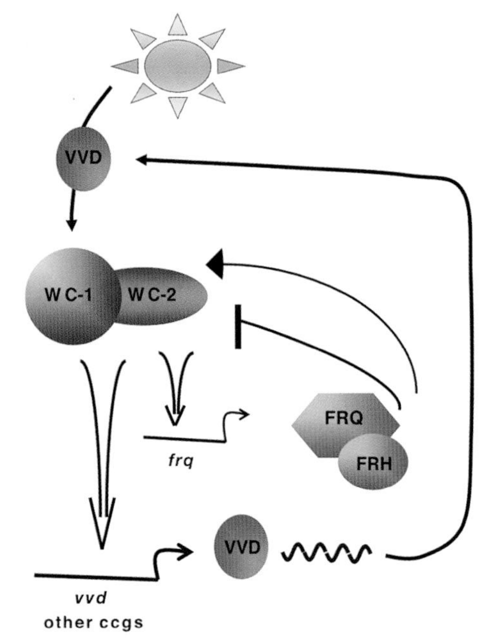

Figure 4. Coupled feedback loops within *Neurospora* may be part of the molecular mechanism of the clock or operate to connect output to input. Molecular clock components necessary for circadian rhythmicity, FRQ (Frequency), FRH (FRQ-interacting Helicase), WC-1 (White Collar-1), and WC-2 (White Collar-2), are shown in simplified coupled feedback relationships, acting through the *frq* promoter. The *vvd* gene, a ccg, is clock-regulated at the level of mRNA abundance. The blue light photoreceptor VVD modulates the light information coming into the clock at different times of day, resulting in circadian gating.

the day until the dusk reset (Elvin et al. 2005). Rhythmic control of outputs that feed back onto LOV-domain-containing inputs or oscillator components (Kim et al. 2007) may turn out to be a functionally conserved mechanism to modulate clock properties, as LOV-domain-containing photoreceptors are widely found.

A Molecular Output Can Directly Affect Clock Mechanisms

In the case of VVD, a gene deletion results in a largely normal clock, albeit with defects in phase control, where it is clear that the oscillator mechanism itself is not greatly perturbed. A recent and surprising example of a clock-regulated output with the ability to conditionally feed into the oscillator came through the cloning of a genetic mutation that altered period length and temperature compensation. The *prd-4* mutation was isolated in 1981 as a semidominant short-period-length mutation in *Neurospora* with a partial loss of temperature compensation (Gardner and Feldman 1981). Cloning and analysis of the gene responsible for this mutation in 2006 demonstrated *prd-4* to encode the *Neurospora* checkpoint kinase-2 (Chk2) bearing a S493L mutation in the catalytic domain (Pregueiro et al. 2006). Chk2 is a threonine/serine protein kinase required for cell cycle arrest in response to DNA damage (Matsuoka et al. 1998). A null mutation constructed by gene replacement (*prd-4*$^{K/O}$) displayed a completely wild-type clock, with no apparent defects in either compensation or period, although expression of *chk2* was found to be under clock control (Pregueiro et al. 2006). It was apparent that Chk2 was not required for the clock, yet the results made it entirely unclear how the S493L mutation could alter clock function so dramatically.

In many organisms, the clock is known to gate the cell cycle. This is known as the "GET effect," for *Gonyaulax*, *Euglena*, *Tetrahymena* (Ehret and Wille 1970), and has additionally been shown in several animal tissues, although never for *Neurospora*. It is clearly established in many species that DNA damage results in cell cycle arrest (see, e.g., Harrison and Haber 2006) and that mutagens, through DNA damage, can affect the cell cycle. Although it was unknown if DNA damage could reset the circadian clock, γ-irradiation had been found to induced several important clock genes (Fu et al. 2002; Lee 2005), suggesting that this might be the case. Exposure to the radiomimetic drug methylmethane sulfonate (MMS) results in double-stranded DNA breaks and was found to reset the clock in wild type but not in the *prd-4*$^{K/O}$ strain. When MMS was given at different times in the circadian cycle, the resulting phase-response curve (PRC) (Fig. 5) showed strong advance resetting during the subjective day in wild type but not in the *prd-4*$^{K/O}$ strain (Pregueiro et al. 2006).

What appears to be happening is this: Whenever Chk2 becomes activated by DNA damage, one of its normal substrates is FRQ. Phosphorylation of FRQ promotes its turnover. In Chk2$^{PRD\text{-}4}$, the mutation has resulted in a kinase with enhanced binding to FRQ, thereby partially bypassing the requirement for DNA damage to activate Chk2 as regards FRQ phosphorylation. As a result, in Chk2$^{PRD\text{-}4}$, FRQ is always precociously phosphorylated and the circadian period length is shortened. Chk2$^{PRD\text{-}4}$ is an example of a protein kinase not normally involved in operation of the clock but that regulates a clock protein in response to environmental damage to the cell.

ALTERNATIVE OSCILLATORS AND DISSECTION OF THE HIERARCHICAL ORGANIZATION OF THE CIRCADIAN SYSTEM

Oscillatory behavior is a major theme in living systems. Oscillations can naturally occur in any cellular pathway where there is negative feedback regulation with a lag. The vast majority of these oscillations do not meet the criteria for being a circadian rhythm. Much of the biochemistry of cellular metabolism can be described as feedback loops; many have short periods on the order of seconds and minutes, like feedbacks in the glycolytic oscillator, but many are also known with longer periods in the range of several hours. Often, such cycles are invisible because biochemical assays performed on a collection of cells, whether in a tissue or a dish, will report arrhythmicity if the cells are not in synchrony. Recall, for instance, that circadian rhythms in tissue culture were not found for years until the cells were appropriately synchronized (Balsalobre et al. 1998). As the clock field examines individual cells in more detail, such rhythms are appearing with greater frequency. One example of a long period but noncircadian, intracellular oscillation is the 5.5-hour sus-

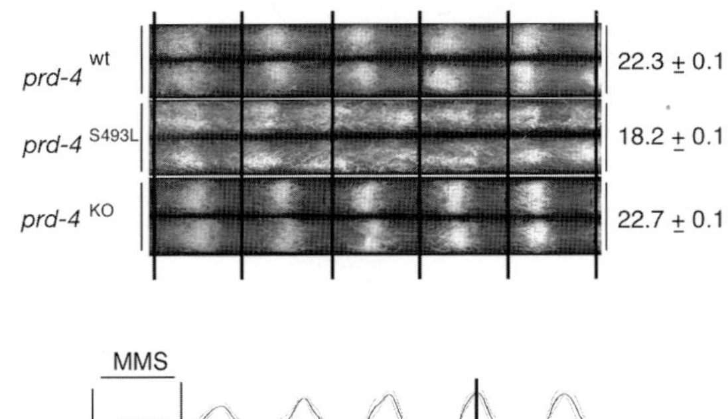

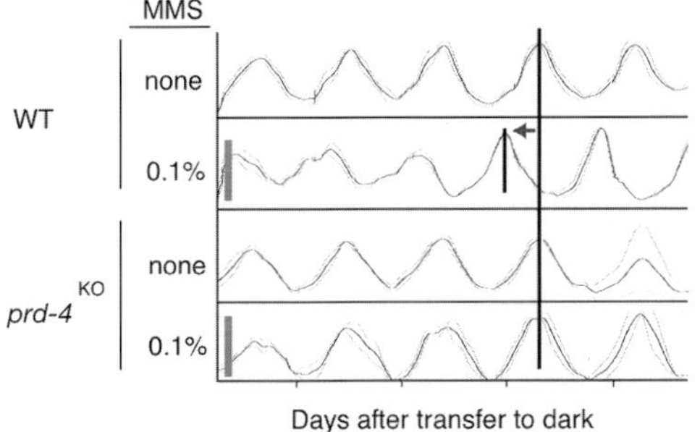

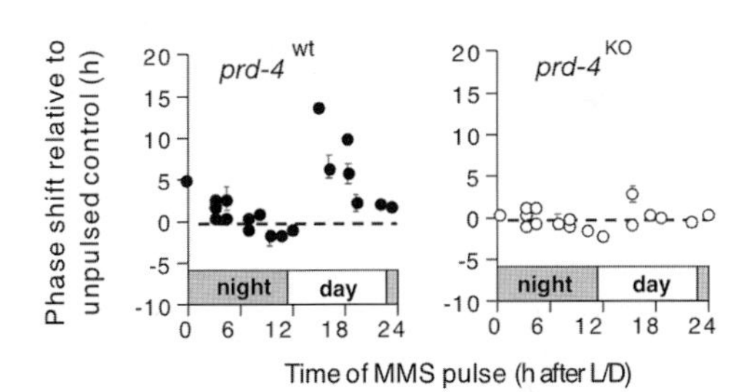

Figure 5. Clock resetting by a DNA-damaging agent requires Chk2(PRD-4). Race tubes (*top*) are shown in duplicate; wild type with an approximately 22-hour period length; the *chk2*$^{prd-4}$ mutant allele (with a serine-to-leucine change at amino acid position 493 of Chk2) isolated due to the short, approximately 18-hour-period length at 25°C; and a gene replacement knockout of the *chk2* gene showing a wild-type period length. (*Middle*) Traces from several race tubes illustrating that a 2-hour-long treatment with 0.1% methylmethane sulfonate (MMS), a radiomimetic DNA-damaging drug, phase-advances the clock in wild type but not in the *prd-4*^{ko}. A full phase-response curve (PRC) is shown (*bottom*), examining the effects of MMS treatment on circadian phase at different times over the subjective day. The wild-type strain at left is sensitive to and is reset by MMS at some times of day and not others. This response is lost in the *prd-4*^{ko} strain, indicating that Chk2(PRD-4) is required for phase resetting in response to MMS. (Adapted, with permission, from Pregueiro et al. 2006 [© AAAS].)

tained oscillation in nuclear levels of the transcription factor p53, and its negative regulator Mdm2 that is seen upon γ-irradiation (Lev Bar-Or et al. 2000; Geva-Zatorsky et al. 2006). Other examples of infradian rhythms include the redox rhythmicity in yeast that has been modeled as a possible evolutionary origin of circadian rhythms (Tu and McKnight 2006).

Colin Pittendrigh and Victor Bruce (1959) first hypothesized a type of oscillation that might come under control of the circadian pacemaker, calling it a slave oscillator. They noted that "any feedback loop in the organism is a potential slave oscillator and if the circadian pacemaker can make input to the loop, the slave will assume a circadian period and become a part of the temporal program that the pacemaker drives" (Pittendrigh 1981). Are there actually slave oscillators in *Neurospora* or other organisms? In many organisms, regulatory relationships among factors involved in nitrate assimilation can give rise to a feedback loop, such that nitrate reductase activity is rhythmic, first shown in *Gonyaulax* (Ramalho et al. 1995). Possibly, the best-described putative slave oscillator is the nitrate reductase (the NIT3 protein) rhythm in *Neurospora* (see, e.g., Lillo et al. 2001), the metabolic activity that regulates the conversion of assimilated nitrate to ammonia and then to glutamine in a negative feedback loop. NIT3 activity is rhythmic not only in wild-type strains, but also in *frq*-null and probably *wc-1*-null strains, and interestingly, even in constant light when the FRQ/WCC is not rhythmic (Fig. 6) (Christensen et al. 2004). Preliminary work examining nitrogen reductase activity at different temperatures gives some indication that the period length of the rhythm in a wild-type strain shows temperature compensation that may then be lost in the absence of FRQ/WCC. This would suggest that coupling to the clock allows the nitrogen reductase rhythm to display circadian characteristics of period length, compensation, and phase control in clock wild-type strains.

FRQ-LESS OSCILLATORS AND OTHER NONCIRCADIAN FEEDBACK LOOPS

Ancillary oscillators exist in organisms without functional circadian clocks and may display some circadian properties, but do they represent part of the working clock mechanism? Genetic, molecular, and biochemical approaches aimed at understanding FRQ/WCC regulatory loops have been highly successful in providing major insights into such canonical clock oscillator features as the production and maintenance of rhythmicity, light and temperature resetting, and phase control with respect to the physical diurnal cycle. Nevertheless, within any organism, there is a complex circadian system that is thought to encompass other feedbacks that result in oscillatory behavior. Some of these will be feedback loops with functions distinct from the clock mechanism. Some of these will be clock-regulated, and some will be examples of clock-regulated output that feeds back to some aspect of the clock, either directly to mechanism or to input as discussed above for Chk2^{PRD-4} and VVD. In *Neurospora*, oscillators that are unmasked when the FRQ/WCC feedback loop is eliminated have been referred to as FRQ-less oscillators or FLOs (Iwasaki and Dunlap 2000). To date, several FLOs have been identified, although their importance in terms of clock function has yet to be established. Among the fungi, a number of noncircadian rhythms in development have been described (for review, see Bünning 1973) in otherwise wild-type strains. An early example is the *clock* strain (Sussman et al. 1964, 1965), identified as producing a rhythm in growth when cultured in race tubes, but later shown to not be circadian (Feldman and Hoyle 1974).

More than two decades ago, a FLO rhythm was found in a clock-defective strain, *frq*^{9} (Loros 1984; Loros and Feldman 1986), that makes a truncated and nonfunctional

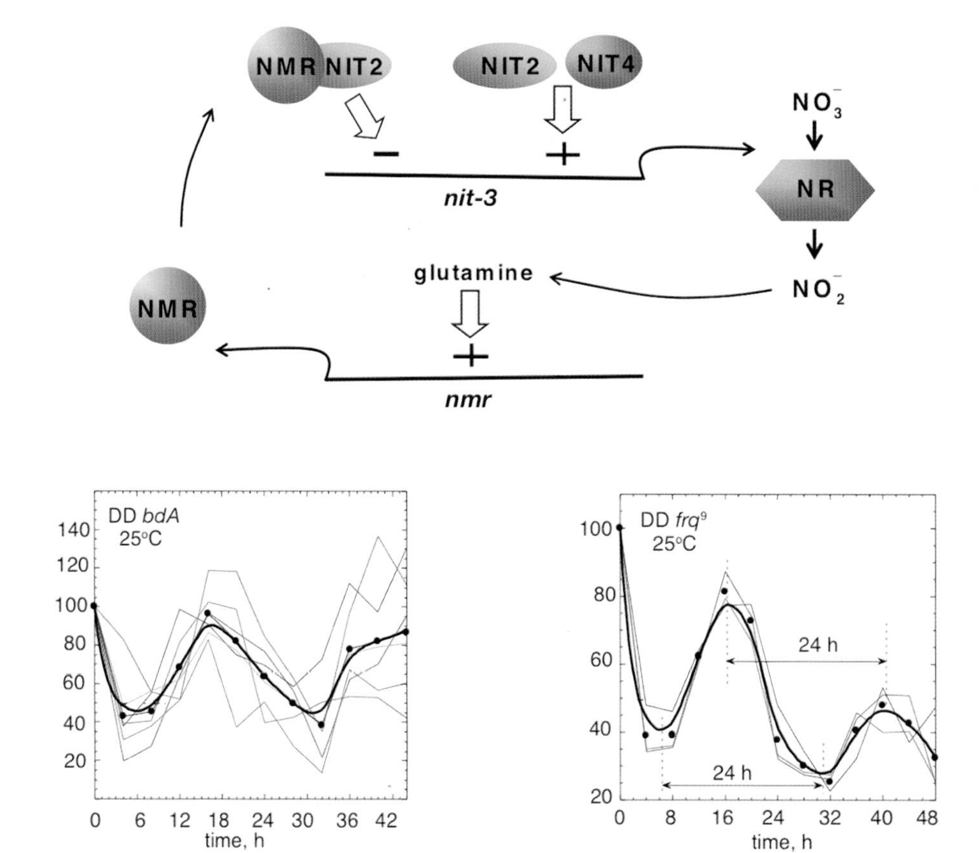

Figure 6. Nitrate reductase may be a slave oscillator. (*Top*) The effectors of nitrogen assimilation describe a negative feedback loop when nitrate is the sole carbon source. The positive regulators NIT2 and NIT4 are required for the expression of the *nit-3* gene encoding the nitrogen reductase (NR) enzyme. NR catalyzes the reduction of nitrate to nitrite, which is eventually converted to glutamine, thought to induce expression of the nitrogen metabolite regulator (NMR). NMR binds directly to NIT2, resulting in the inhibition of *nit-3* expression. NR activity displays endogenous oscillations with a period length of approximately 24 hours in constant darkness in both wild-type and loss-of-function *frq* alleles at 25°C (*bottom*). The loss-of-function *frq*^{9} strain may show changes in period length at other temperatures. (*Bottom* adapted, with permission, from Christensen et al. 2004 [© Sage Publications].)

version of the FRQ protein (Aronson et al. 1994a). Although the period length of the FLO-driven rhythm in conidial development can range from 12 to 34 hours, it can be manipulated to occur within the circadian range because the period is highly sensitive to changes in both temperature and nutrition. Several other FLOs have since been described (see, e.g., Mattern and Brody 1979; Lakin-Thomas 1998; Merrow et al. 1999; Correa et al. 2003; Granshaw et al. 2003; Christensen et al. 2004; dePaula et al. 2006; Lombardi et al. 2007; for review, see Dunlap et al. 2004; Vitalini et al. 2006), but none display the complete set of formal circadian properties nor have they been shown to affect the operation of the FRQ/WCC feedback loop. However, FLOs may exhibit some circadian properties (see, e.g., dePaula et al. 2006). The extent to which a particular FLO might be a manifestation of the circadian mechanism revealed by the removal of some clock components is an appealing idea, particularly because the appearance of noncircadian and ultradian rhythmic behavior has been found following genetic lesion of clock genes in other systems (see, e.g., Dowse et al. 1987; Hamblen-Coyle et al. 1989; Liu et al. 2007; Storch et al. 2007).

The *Drosophila* circadian system has been modeled to contain a master oscillator that drives or entrains slave oscillators directly involved with outputs (Pittendrigh and Bruce 1959), a model that may be pertinent to *Neurospora* (Iwasaki and Dunlap 2000; Merrow et al. 2001; Dunlap and Loros 2006) and other systems. Pittendrigh speculated that the use of several slave oscillators by the core clock would allow some aspects of the system to be open to evolutionary adjustment without altering other component parts and that the slaves, because they would normally be entrained by the master, need not have all circadian properties (Pittendrigh 1981). A reasonable interpretation of FLOs is that they represent a set of Pittendrighian slave oscillators coupled to the FRQ/WCC master oscillator; when the master oscillator is removed, the slaves run on their own in various noncircadian or partially circadian modes (Loros and Dunlap 2001; Dunlap and Loros 2004, 2005, 2006). A major and untested caveat to this model is that most FLO components have not been molecularly identified and therefore cannot be compromised or deleted to test for effects on the FRQ/WCC oscillator, so their involvement in the clock system is to date unanswered. Figure 7 summarizes the currently understood interrelationships among oscillators.

The most extensively studied FLO is the asexual

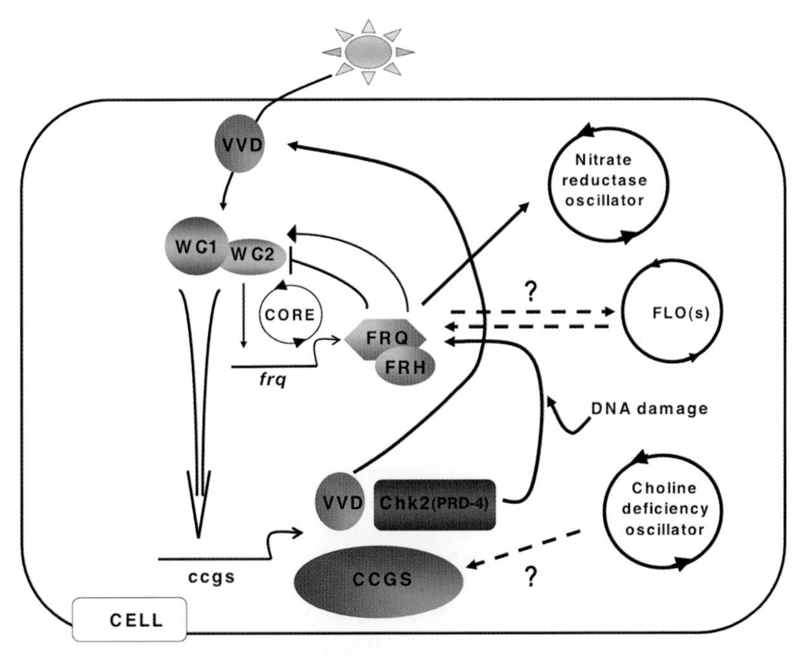

Figure 7. The circadian system is a collection of oscillators generating time information, integrating this with environmental information, whose function is to coordinate the life of the *Neurospora* cell. An expansion of Figure 4, this schematic is a representation of the relationships within the circadian system as a whole. Chk2 (checkpoint kinase-2) is shown as a ccg that can feed back, under conditions of DNA damage, to phosphorylate the clock component FRQ, thereby altering the phase of the clock. The nitrate reductase feedback loop is depicted as a classic slave oscillator, not required for clock function but under clock control. The choline deficiency oscillator is shown as an independent oscillatory system that may influence genes that are also clock-regulated. Finally, the FLO oscillator(s), where no molecular underpinnings are known, has unknown connections to the clock, but information in either direction may exist.

developmental rhythm that persists in the absence of FRQ (Loros and Feldman 1986; Loros et al. 1986; Aronson et al. 1994a). Under constant conditions, the rhythm only appears sporadically and under limited and specific conditions. The period can range from 12 to 35 hours depending on culture conditions, and phase is not maintained between individual cultures, reflected in the large standard deviation of the period length. The rhythm is not entrainable by light, but if it is the same FLO as that described by Merrow et al. (1999), it is entrained by temperature. It has been modeled as either a slave oscillator (a reflection of residual circuitry of the clock), not involved in the circadian system at all (Loros and Dunlap 2001; Dunlap and Loros 2004; see, e.g., Christensen et al. 2004), or as the underlying rhythm generator core to the circadian oscillator (Roenneberg and Merrow 1998; Merrow et al. 2001; Lakin-Thomas 2006a). Although a consortium of labs found the rhythm to be driven by temperature rather than entrained, as evidenced by peaks of development limited to the cold phase of a temperature cycle, and reported that the conidiation rhythm stopped as soon as the temperature cycle stopped (Pregueiro et al. 2005), others found the temperature cycle itself resulted in masking of a weaker, nonsustainable but entrainable component (Roenneberg et al. 2005; Lakin-Thomas 2006a). Despite all of the effort, the question of this FLO's potential role in any part of the circadian system remains unanswered. This will likely continue to be the case until a genetic or biochemical basis for the FLO is discovered, allowing it to be molecularly examined and disabled.

ANCILLARY OSCILLATORS CAN AFFECT THE SAME OUTPUTS AS THE CLOCK AND CAN MASK CIRCADIAN CONTROL

Work begun in the late 1970s described conidial banding rhythms in fungal strains with defects in lipid metabolism. One of these was called *cel*, a fatty acid chain elongation mutant that had defective circadian properties in period length and temperature compensation of the conidiation rhythm (see, e.g., Mattern and Brody 1979). Another mutant, *chol-1*, a morphological strain reparable by addition of choline, showed large changes in linear growth under limiting choline, manifesting as a rhythm in conidiation with extraordinarily long periods sometimes exceeding 100 hours (Lakin-Thomas 1996, 1998). Both the choline-starvation-induced and the *cel* rhythms, plus a recent mutation called *ult* (Lombardi et al. 2007), were shown to be largely independent of the FRQ/WCC oscillator. An interpretation has been that the FRQ/WCC feedback is dispensable in determinating conidiation period length and therefore that this feedback loop is not required for circadian rhythmicity (Lakin-Thomas 1998; Lakin-Thomas and Brody 2000), but instead provides input into the system (Lakin-Thomas 1998, 2006b). Another explanation is that perhaps lipid manipulations can change the coupling between the FRQ/WCC and FLO oscillators (Granshaw et al. 2003; Lombardi et al. 2007). An underlying assumption in this work is that the rhythm in conidiation is a true representation of the circadian oscillator, although as discussed above, not all developmental rhythms displayed by *Neurospora* are circadian nor are they necessarily even regulated by the circadian system.

To acquire a clearer picture of the FRQ/WCC oscillator with regard to these other oscillators, it is necessary to be able to follow both in the same culture over the same time frame. In this case, the activity of the FRQ/WCC feedback needed to be monitored at the same time that the long-period, choline-limited rhythm of conidiation was being expressed. This became possible with the development of a codon-optimized luciferase that has activity in *Neurospora* (Gooch et al. 2007). Using the *frq* promoter to drive luciferase expression allows the FRQ/WCC oscillator to be monitored under conditions when it no longer controls conidiation. Using luciferase to follow *frq* expression in race tubes of the *chol-1* strain under limiting choline showed the *chol-1,frqP-luc* strains replicating a long-period rhythm in conidiation of about 78 hours; this rhythm was *frq*-independent, as previously shown. A strain, *chol-1,* frq^7*,frqP-luc*, carrying the long-period frq^7 allele, showed a similar long-period rhythm of development on the race tube. However, when luciferase was monitored, the activity showed a clear 22-hour rhythm in the frq^+ strain and a long-period, 29-hour rhythm in the *chol*-1, frq^7, *frqP-luc* strain (Fig. 8). The choline-starvation-induced rhythm is known not to be temperature-compensated (Lakin-Thomas 1998), but the luciferase activity rhythm in the *chol-1, frqP-luc* strain demonstrated temperature compensation, equivalent to the temperature compensation of the FRQ/WCC oscillator (Shi et al. 2007). These experiments unequivocally demonstrated that the FRQ/WCC oscillator is functionally wild type under choline starvation in the *chol-1* strain, but it is no longer controlling the timing or expression of asexual development. These observations indicate that the choline-starvation-induced rhythm may reflect morphological cycles that mask or uncouple the circadian oscillator from development. It is clear, in the case of *chol-1*, that two distinctly different rhythms can coexist in the same culture, be of very different periodicities, and therefore not be controlled by the same oscillatory system (Fig. 7).

CONCLUSIONS

Circadian output comprises the functional end of circadian systems in terms of adaptive significance. Among fungi and animals, the regulatory logic of the core transcription-translation feedback loop is similar, and even in plants, where the core mechanism contains additional interlocked loops, a principal and primary form of output is the daily clock regulation of gene expression. For this reason, work on *Neurospora*, which pioneered the analysis of clock-controlled genes, continues to illuminate the pathways through which clocks can control metabolism and overt rhythms.

In addition to providing an overview of output in *Neurospora*, we have developed three recent stories. The first described the cloning of the *bd* gene and its identification as an allele of *ras-1*. In *Neurospora*, virtually every strain examined in the context of rhythms bears the *bd* allele that helps to clarify the overt rhythm in asexual development. *ras-1*^{bd} influences a wide variety of signaling pathways and results in both enhanced light responses and increased asexual development, altering the profile of ccgs in the cell and organism. Interestingly, both of these can be largely phenocopied by treatments that increase levels of intracellular reactive oxygen species.

A second story provides an example of an output from the clock that feeds back to affect circadian input. Cloning of *prd-4* showed it to be an allele of checkpoint kinase-2. Normally, Chk2 is quiescent until it is activated by

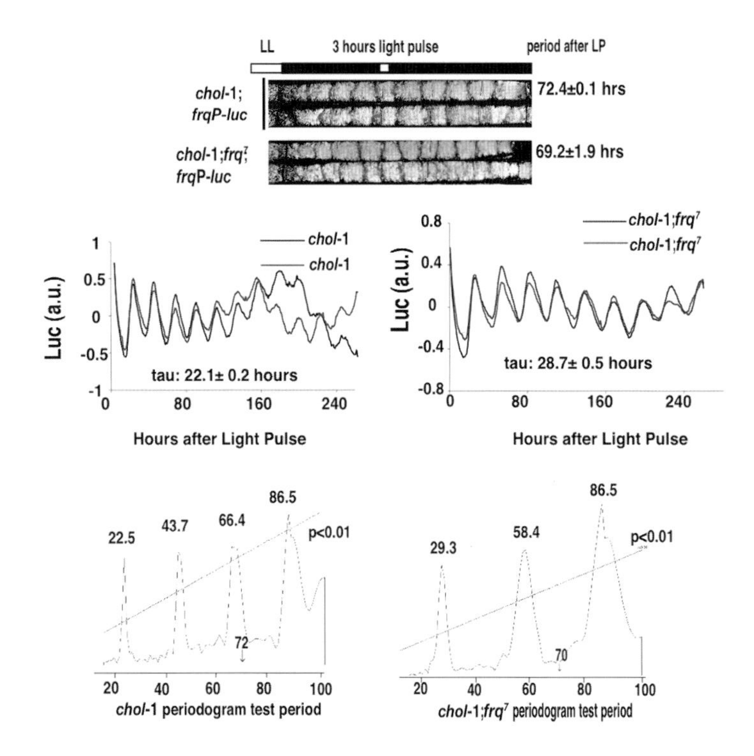

Figure 8. *frq* allele-dependent luciferase rhythms are circadian in long-period *chol-1* strains. Race tubes containing 12.5 μM luciferin and the *chol-1,*frq^+ or *cho-1,*frq^7 strains, both bearing a transcriptional fusion of the *frq* promoter to an optimized luciferase gene (*top*). Note the characteristically long period of development in the *chol-1* strains. Detrended signals in bioluminescence from each race tube express about 22- and 29-hour rhythms in the *chol-1,*frq^+ or *chol-1,*frq^7 strains, respectively (*middle*), running in the same cultures expressing the approximately 70-hour rhythms in development. The luciferase data were subjected to periodogram analysis (*bottom*), demonstrating strong 22.5-hour (wild-type) and 29.3-hour (frq^7) period components. The ability to follow both *chol-1* and *frq* rhythms has exposed independent oscillators in a common cytoplasm. (Reprinted, with permission, from Shi et al. 2007 [© National Academy of Sciences].)

ATM/ATR following DNA damage, and once activated, it phosphorylates a number of target genes, one of which is FRQ. The mutation in the Chk2$^{PRD\text{-}4}$ variant causes Chk2 to bypass the requirement for DNA damage to activate this kinase; as a result, FRQ is always precociously phosphorylated, resulting in more rapid turnover and a faster cycling clock with a short-period length. In the wild-type clock, Chk2 phosphorylates FRQ in response to DNA damage in a time-of-day-dependent manner, resulting in clock-regulated phase shifting.

Finally, with improved methods for the study of single cells and the synchronization of cells in culture, more and more noncircadian or circadian-coupled feedback loops are being described. Arguably one of the most closely examined circadian cells is *Neurospora,* in which at least nine distinct oscillators have been described in addition to the circadian FRQ/WCC loop. In clock wild-type cells, many of these are normally coupled to the clock and run with circadian period lengths, but when the clock is lesioned, for instance, through mutation of *frq* or *wc* genes, these FLOs can continue to cycle on their own. In doing so, some retain circadian characteristics, suggesting the possibility that they, like the FRQ/WCC loop, contribute to the core circadian oscillator. As long as the core feedback loop could be followed for long periods only through its control of the banding rhythm, it was difficult to study the FRQ/WCC loop simultaneously with other loops. However, the recent adaptation of luciferase to function at high levels in *Neurospora* now allows the core FRQ/WCC feedback loop to be followed in real time under conditions where it no longer controls the overt rhythm in development. This ability can be used to describe the hierarchical relationships among FLOs to see which are connected to and influence the circadian system. The nitrate reductase oscillator appears to be connected to the clock. In contrast, an interesting example arose from analysis of the oscillator controlling the long-period rhythm elicited upon choline starvation. This oscillator runs with a long 60–120-hour period length in conidia formation that is not compensated against changes in temperature or nutrition, however; the circadian FRQ/WCC oscillator continues to cycle with a fully compensated circadian 22-hour period. Over many days, the two oscillators cycle through all phase relationships with respect to one another, and periodogram analysis shows no influence of the choline deficiency oscillator (CDO) on the circadian cycle. These data strongly suggest that the CDO is an ancillary loop that can assume control of normally clock-controlled processes such as overt development and mask a wild-type circadian clock.

ACKNOWLEDGMENTS

This work was supported by grants from the National Institutes of Health to J.J.L. and J.C.D. (GM083336 and GM068087), J.C.D. (GM34985), and the Norris Cotton Cancer Center. L.F.L. is Pew Latin American fellow. We gratefully acknowledge the invaluable services of the Fungal Genetics Stock Center, University of Missouri, Kansas City.

REFERENCES

Aguirre J., Rios-Momberg M., Hewitt D., and Hansberg W. 2005. Reactive oxygen species and development in microbial eukaryotes. *Trends Microbiol.* **13:** 111.

Aronson B.D., Johnson K.A., and Dunlap J.C. 1994a. The circadian clock locus *frequency:* A single ORF defines period length and temperature compensation. *Proc. Natl. Acad. Sci.* **91:** 7683.

Aronson B., Johnson K., Loros J.J., and Dunlap J.C. 1994b. Negative feedback defining a circadian clock: Autoregulation in the clock gene *frequency*. *Science* **263:** 1578.

Arpaia G., Loros J.J., Dunlap J.C., Morelli G., and Macino G. 1993. The interplay of light and the circadian clock: Independent dual regulation of clock-controlled gene *ccg-2 (eas)*. *Plant Physiol.* **102:** 1299.

———. 1995. The circadian clock-controlled gene *ccg-1* is induced by light. *Mol. Gen. Genet.* **247:** 157.

Bailey M.J., Beremand P.D., Hammer R., Reidel E., Thomas T.L., and Cassone V.M. 2004. Transcriptional profiling of circadian patterns of mRNA expression in the chick retina. *J. Biol. Chem.* **279:** 52247.

Bailey-Shrode L. and Ebbole D.J. 2004. The *fluffy* gene of *Neurospora crassa* is necessary and sufficient to induce conidiophore development. *Genetics* **166:** 1741.

Ballario P., Talora C., Galli D., Linden H., and Macino G. 1998. Roles in dimerization and blue light photoresponse of the PAS and LOV domains of *Neurospora crassa* WHITE COLLAR proteins. *Mol. Microbiol.* **29:** 719.

Ballario P., Vittorioso P., Magrelli A., Talora C., Cabibbo A., and Macino G. 1996. *White collar-1,* a central regulator of blue-light responses in *Neurospora crassa*, is a zinc-finger protein. *EMBO J.* **15:** 1650.

Balsalobre A., Damiola F., and Schibler U. 1998. A serum shock induces circadian gene expression in mammalian culture cells. *Cell* **93:** 929.

Beadle G.W. and Tatum E.L. 1945. *Neurospora*. II. Methods of producing and detecting mutations concerned with nutritional requirements. *Am. J. Botan.* **32:** 678.

Belden W.J., Larrondo L.F., Froehlich A.C., Shi M., Chen C.-H., Loros J.J., and Dunlap J.C. 2007. The *band* mutation in *Neurospora crassa* is a dominant allele of *ras-1* implicating RAS-signaling in circadian output. *Genes Dev.* **21:** 1494.

Bell-Pedersen D. 2000. Understanding circadian rhythmicity in *Neurospora crassa:* From behavior to genes and back again. *Fungal Genet. Biol.* **29:** 1.

Bell-Pedersen D., Dunlap J.C., and Loros J.J. 1996a. Distinct *cis*-acting elements mediate clock, light, and developmental regulation of the *Neurospora crassa eas* (*ccg-2*) gene. *Mol. Cell. Biol.* **16:** 513.

Bell-Pedersen D., Shinohara M., Loros J., and Dunlap J.C. 1996b. Circadian clock-controlled genes isolated from *Neurospora crassa* are late night to early morning specific. *Proc. Natl. Acad. Sci.* **93:** 13096.

Bünning E. 1973. *The physiological clock*. Springer-Verlag, New York.

Ceriani M.F., Hogenesch J.B., Yanovsky M., Panda S., Straume M., and Kay S.A. 2002. Genome-wide expression analysis in *Drosophila* reveals genes controlling circadian behavior. *J. Neurosci.* **22:** 9305.

Cheng P., Yang Y., Gardner K.H., and Liu Y. 2002. PAS domain-mediated WC-1/WC-2 interaction is essential for maintaining the steady-state level of WC-1 and the function of both proteins in clock and light responses of *Neurospora*. *Mol. Cell. Biol.* **22:** 517.

Cheng P., He Q., He Q., Wang L., and Liu Y. 2005. Regulation of the *Neurospora* circadian clock by an RNA helicase. *Genes Dev.* **19:** 234.

Christensen M., Falkeid G., Hauge I., Loros J.J., Dunlap J.C., Lillo C., and Ruoff P. 2004. A *frq*-independent nitrate reductase rhythm in *Neurospora crassa*. *J. Biol. Rhythms* **19:** 280.

Collett M.A., Dunlap J.C., and Loros J.J. 2001. Circadian clock-specific roles for the light response protein WHITE COLLAR-2. *Mol. Cell. Biol.* **21:** 2619.

Collett M.A., Garceau N., Dunlap J.C., and Loros J.J. 2002. Light and clock expression of the *Neurospora* clock gene frequency is differentially driven by but dependent on WHITE COLLAR-2. *Genetics* **160:** 149.

Colot H.V., Loros J.J., and Dunlap J.C. 2005. Temperature-modulated alternative splicing and promoter use in the circadian clock gene *frequency*. *Mol. Biol. Cell* **16:** 5563.

Correa A. and Bell-Pedersen D. 2002. Distinct signaling pathways from the circadian clock participate in regulation of rhythmic conidiospore development in *Neurospora crassa*. *Eukaryot. Cell* **1:** 273.

Correa A., Lewis Z.A., Greene A.V., March I.J., Gomer R.H., and Bell-Pedersen D. 2003. Multiple oscillators regulate circadian gene expression in *Neurospora*. *Proc. Natl. Acad. Sci.* **100:** 13597.

Crosthwaite S.C., Dunlap J.C., and Loros J.J. 1997. *Neurospora wc-1* and *wc-2:* Transcription, photoresponses, and the origins of circadian rhythmicity. *Science* **276:** 763.

Crosthwaite S.C., Loros J.J., and Dunlap J.C. 1995. Light-induced resetting of a circadian clock is mediated by a rapid increase in *frequency* transcript. *Cell* **81:** 1003.

D'Autréaux B. and Toledano M. 2007. ROS as signalling molecules: Mechanisms that generate specificity in ROS homeostasis. *Nat. Rev. Mol. Cell Biol.* **8:** 813.

Denault D.L., Loros J.J., and Dunlap J.C. 2001. WC-2 mediates WC-1-FRQ interaction within the PAS protein-linked circadian feedback loop of *Neurospora crassa*. *EMBO J.* **20:** 109.

dePaula R.M., Lewis Z.A., Greene A.V., Seo K.S., Morgan L.W., Vitalini M.W., Bennett L., Gomer R.H., and Bell-Pedersen D. 2006. Two circadian timing circuits in *Neurospora crassa* share components and regulate distinct rhythmic processes. *J. Biol. Rhythms* **21:** 159.

Diernfellner A.C., Schafmeier T., Merrow M.W., and Brunner M. 2005. Molecular mechanism of temperature sensing by the circadian clock of *Neurospora crassa*. *Genes Dev.* **19:** 1968.

Dowse H.B., Hall J.C., and Ringo J.M. 1987. Circadian and ultradian rhythms in period mutants of *Drosophila melanogaster*. *Behav. Genet.* **17:** 19.

Duffield G.E. 2003. DNA microarray analyses of circadian timing: The genomic basis of biological time. *J. Neuroendocrinol.* **15:** 991.

Duffield G.E., Best J.D., Meurers B.H., Bittner A., Loros J.J., and Dunlap J.C. 2002. Circadian programs of transcriptional activation, signaling, and protein turnover revealed by microarray analysis of mammalian cells. *Curr. Biol.* **12:** 551.

Dunlap J.C. 1999. Molecular bases for circadian clocks. *Cell* **96:** 271.

Dunlap J.C. and Loros J.J. 2004. The *Neurospora* circadian system. *J. Biol. Rhythms* **19:** 414.

———. 2005. Analysis of circadian rhythms in *Neurospora:* Overview of assays and genetic and molecular biological manipulation. *Methods Enzymol.* **393:** 3.

———. 2006. How fungi keep time: Circadian system in *Neurospora* and other fungi. *Curr. Opin. Microbiol.* **9:** 579.

Dunlap J.C., Loros J.J., Denault D., Lee K., Froehlich A.F., Colot H., Shi M., and Pregueiro A. 2004. Genetics and molecular biology of circadian rhythms. In *Biochemistry and molecular biology: The Mycota III* (ed. R. Brambl and G.A. Marzluf), p. 209. Springer-Verlag, Berlin.

Ehret C.F. and Wille J.J. 1970. The photobiology of circadian rhythms in protozoa. In *Photobiology of microorganisms* (ed. P. Halldal), p. 369. Wiley, New York.

Elvin M., Loros J.J., Dunlap J.C., and Heintzen C. 2005. The PAS/LOV protein VIVID supports a rapidly dampened daytime oscillator that facilitates entrainment of the *Neurospora circadian* clock. *Genes Dev.* **19:** 2593.

Eskin A. 1979. Identification and physiology of circadian pacemakers. *Fed. Proc.* **38:** 2570.

Fagan T., Morse D., and Hastings J.W. 1999. Circadian synthesis of a nuclear-encoded chloroplast glyceraldehyde-3-phosphate dehydrogenase in the dinoflagellate *Gonyaulax polyedra* is translationally controlled. *Biochemistry* **38:** 7689.

Feldman J. and Hoyle M.N. 1974. A direct comparison between circadian and noncircadian rhythms in *Neurospora crassa*. *Plant Physiol.* **53:** 928.

Froehlich A.C., Loros J.J., and Dunlap J.C. 2002. White Collar-1, a circadian blue light photoreceptor, binding to the *frequency* promoter. *Science* **297:** 815.

———. 2003. Rhythmic binding of a WHITE COLLAR containing complex to the *frequency* promoter is inhibited by FREQUENCY. *Proc. Natl. Acad. Sci.* **100:** 5914.

Fu L., Pelicano H., Liu J., Huang P., and Lee C.-C. 2002. The circadian gene *Period2* plays an important role in tumor suppression and DNA damage response in vivo. *Cell* **111:** 41.

Garceau N., Liu Y., Loros J.J., and Dunlap J.C. 1997. Alternative initiation of translation and time-specific phosphorylation yield multiple forms of the essential clock protein FREQUENCY. *Cell* **89:** 469.

Gardner G.F. and Feldman J.F. 1981. Temperature compensation of circadian periodicity in clock mutants of *Neurospora crassa*. *Plant Physiol.* **68:** 1244.

Geva-Zatorsky N., Rosenfeld N., Itzkovitz S., Milo R., Sigal A., Dekel E., Yarnitzky T., Liron Y., Polak P., Lahav G., and Alon U. 2006. Oscillations and variability in the p53 system. *Mol. Syst. Biol.* **2:** 2006.0033.

Gooch V., Mehra A., Larrondo L., Fox J., Touroutoutoudis M., Loros J., and Dunlap J. 2008. Fully codon-optimized *luciferase* uncovers novel temperature characteristics of the *Neurospora* clock. *Eukaryot. Cell* **7:** 28.

Granshaw T., Tsukamoto M., and Brody S. 2003. Circadian rhythms in *Neurospora crassa:* Farnesol or geraniol allow expression of rhythmicity in the otherwise arrhythmic strains frq10, wc-1, and wc-2. *J. Biol. Rhythms* **18:** 287.

Hamblen-Coyle M., Konopka R.J., Zwiebel L.J., Colot H.V., Dowse H.B., Rosbash M., and Hall J.C. 1989. A new mutation at the *period* locus with some novel effects on circadian rhythms. *J. Neurogenet.* **5:** 229.

Hansberg W. and Aguirre J. 1990. Hyperoxidant states cause microbial cell differentiation by cell isolation from dioxygen. *J. Theor. Biol.* **142:** 201.

Harmer S.L., Hogenesch J.B., Straume M., Chang H.-S., Han B., Zhu T., Wang X., Kreps J.A., and Kay S.A. 2000. Orchestrated transcription of key pathways in *Arabidopsis* by the circadian clock. *Science* **290:** 2110.

Harrison J. and Haber J. 2006. Surviving the breakup: The DNA damage checkpoint. *Annu. Rev. Genet.* **40:** 209.

Hastings J.W. and Sweeney B.M. 1958. A persistent diurnal rhythm of luminescence in *Gonyaulax polyedra*. *Biol. Bull.* **115:** 440.

He Q., Cheng P., He Q., and Liu Y. 2005. The COP9 signalosome regulates the *Neurospora circadian* clock by controlling the stability of the SCFFWD-1 complex. *Genes Dev.* **19:** 1518.

He Q., Cheng P., Yang Y., He Q., Yu Q., and Liu Y. 2003. FWD-1 mediated degradation of FREQUENCY in *Neurospora* establishes a conserved mechanism for circadian clock regulation. *EMBO J.* **22:** 4421.

He Q., Cheng P., Yang Y., Wang L., Gardner K., and Liu Y. 2002. White Collar-1, a DNA binding transcription factor and a light sensor. *Science* **297:** 840.

Heintzen C. and Liu Y. 2007. The *Neurospora crassa* circadian clock. *Adv. Genet.* **58:** 25.

Heintzen C., Loros J.J., and Dunlap J.C. 2001. The PAS protein VIVID defines a clock-associated feedback loop that represses light input, modulates gating, and regulates clock resetting. *Cell* **104:** 453.

Heo J. and Campbell S.L. 2006. Ras regulation by reactive oxygen and nitrogen species. *Biochemistry* **45:** 2200.

Hunt S.M., Elvin M., Crosthwaite S.K., and Heintzen C. 2007. The PAS/LOV protein VIVID controls temperature compensation of circadian clock phase and development in *Neurospora crassa*. *Genes Dev.* **21:** 1964.

Irani K., Xia Y., Zweier J.L., Sollott S.J., Der C.J., Fearon E.R., Sundaresan M., Finkel T., and Goldschmidt-Clermont P.J. 1997. Mitogenic signaling mediated by oxidants in Ras-transformed fibroblasts. *Science* **275:** 1649.

Iwasaki H. and Dunlap J.C. 2000. Microbial circadian oscillatory systems in *Neurospora* and *Synechococcus:* Models for cellular clocks. *Curr. Opin. Microbiol.* **3:** 189.

Iwasaki T., Nakahama K., Nagano M., Fujioka A., Ohyanagi H., and Shigeyoshi Y. 2004. A partial hepatectomy results in altered expression of clock-related and cyclic glyceraldehyde 3-phosphate dehydrogenase (GAPDH) genes. *Life Sci.* **74:** 3093.

Kamphuis W., Cailotto C., Dijk F., Bergen A., and Buijs R.M. 2005. Circadian expression of clock genes and clock-controlled genes in the rat retina. *Biochem. Biophys. Res. Commun.* **330:** 18.

Kim W., Fujiwara S., Suh S., Kim J., Kim Y., Han L., David K., Putterill J., Nam H., and Somers D.E. 2007. ZEITLUPE is a circadian photoreceptor stabilized by GIGANTEA in blue light. *Nature* **449:** 356.

Kobayashi H., Oishi K., Hanai S., and Ishida N. 2004. Effect of feeding on peripheral circadian rhythms and behaviour in mammals. *Genes Cells* **9:** 857.

Kornmann B., Preitner N., Rifat D., Fleury-Olela F., and Schibler U. 2001. Analysis of circadian liver gene expression by ADDER, a highly sensitive method for the display of differentially expressed mRNAs. *Nucleic Acids Res.* **29:** E51.

Kramer C., Loros J.J., Dunlap J.C., and Crosthwaite S.K. 2003. Role for antisense RNA in regulating circadian clock function in *Neurospora crassa*. *Nature* **421:** 948.

Lakin-Thomas P. 1996. Effects of choline depletion on the circadian rhythm in *Neurospora crassa*. *Biol. Rhythm Res.* **27:** 12.

———. 1998. Choline depletion, *frq* mutations, and temperature compensation of the circadian rhythm in *Neurospora crassa*. *J. Biol. Rhythms* **13:** 268.

———. 2006a. Circadian clock genes *frequency* and *white collar-1* are not essential for entrainment to temperature cycles in *Neurospora crassa*. *Proc. Natl. Acad. Sci.* **103**: 4469.

———. 2006b. Transcriptional feedback oscillators: Maybe, maybe not. *J. Biol. Rhythms* **21:** 83.

Lakin-Thomas P.L. and Brody S. 2000. Circadian rhythms in *Neurospora crassa:* Lipid deficiencies restore robust rhythmicity to null *frequency* and *white-collar* mutants. *Proc. Natl. Acad. Sci.* **97:** 256.

Lee C.C. 2005. The circadian clock and tumor suppression by mammalian *period* genes. *Methods Enzymol.* **393:** 852.

Lee K., Dunlap J.C., and Loros J.J. 2003. Roles for WHITE COLLAR-1 in circadian and general photoperception in *Neurospora crassa*. *Genetics* **163:** 103.

Lev Bar-Or R., Maya R., Segel L., Alon U., Levine A., and Oren M. 2000. Generation of oscillations by the p53-Mdm2 feedback loop: A theoretical and experimental study. *Proc. Natl. Acad. Sci.* **97:** 11250.

Lillo C., Meyer C., and Ruoff P. 2001. The nitrate reductase circadian system. *Plant Physiol.* **125:** 1554.

Linden H. and Macino G. 1997. White collar-2, a partner in blue-light signal transduction, controlling expression of light-regulated genes in *Neurospora crassa*. *EMBO J.* **16:** 98.

Linden H., Ballario P., and Macino G. 1997. Blue light regulation in *Neurospora crassa*. *Fungal Genet. Biol.* **22:** 141.

Liu A., Welsh D., Ko C., Tran H., Zhang E., Priest A., Buhr E., Singer O., Meeker K., Verma I., Doyle F.J., III., Takahashi J., and Kay S. 2007. Intercellular coupling confers robustness against mutations in the SCN circadian clock network. *Cell* **129:** 605.

Liu Y. and Bell-Pedersen D. 2006. Circadian rhythms in *Neurospora crassa* and other filamentous fungi. *Eukaryot. Cell* **5:** 1184.

Liu Y., Loros J., and Dunlap J.C. 2000. Phosphorylation of the *Neurospora* clock protein FREQUENCY determines its degradation rate and strongly influences the period length of the circadian clock. *Proc. Natl. Acad. Sci.* **97:** 234.

Liu Y., Merrow M., Loros J.J., and Dunlap J.C. 1998. How temperature changes reset a circadian oscillator. *Science* **281:** 825.

Liu Y., Tsinoremas N., Johnson C., Lebdeva N., Golden S., Ishiura M., and Kondo T. 1995. Circadian orchestration of gene expression in cyanobacteria. *Genes Dev.* **9:** 1469.

Lombardi L., Schneider K., Tsukamoto M., and Brody S. 2007. Circadian rhythms in *Neurospora crassa:* Clock mutant effects in the absence of a *frq*-based oscillator. *Genetics* **175:** 1175.

Loros J.J. 1984. "Studies on *frq*-9, a recessive circadian clock mutant of *Neurospora crassa*." Ph.D. thesis, University of California, Santa Cruz.

Loros J.J. and Dunlap J.C. 1991. *Neurospora crassa* clock-controlled genes are regulated at the level of transcription. *Mol. Cell. Biol.* **11:** 558.

———. 2001. Genetic and molecular analysis of circadian rhythms in *Neurospora*. *Annu. Rev. Physiol.* **63:** 757.

———. 2006. Circadian rhythms, photobiology, and functional genomics in *Neurospora*. In *Fungal genomics: The Mycota* (ed. A.J.P. Brown), vol. 13, p.53. Springer-Verlag, Berlin.

Loros J.J. and Feldman J.F. 1986. Loss of temperature compensation of circadian period length in the *frq-9* mutant of *Neurospora crassa*. *J. Biol. Rhythms* **1:** 187.

Loros J.J., Denome S.A., and Dunlap J.C. 1989. Molecular cloning of genes under the control of the circadian clock in *Neurospora*. *Science* **243:** 385.

Loros J.J., Hastings J.W., and Schibler U. 2003. Adapting to life on a rotating world at the gene expression level. In *Chronobiology: Biological timekeeping* (ed. J.C. Dunlap et al.), p. 254. Sinauer Associates, Sunderland, Massachusetts.

Loros J.J., Richman A., and Feldman J.F. 1986. A recessive circadian clock mutant at the *frq* locus in *Neurospora crassa*. *Genetics* **114:** 1095.

Matsuoka S., Huang M., and Elledge S.J. 1998. Linkage of ATM to cell cycle regulation by the Chk2 protein kinase. *Science* **282:** 1893.

Mattern D. and Brody S. 1979. Circadian rhythms in *Neurospora crassa:* Effects of unsaturated fatty acids. *J. Bacteriol.* **139:** 977.

Merrow M., Brunner M., and Roenneberg T. 1999. Assignment of circadian function for the *Neurospora* clock gene *frequency*. *Nature* **399:** 584.

Merrow M., Garceau N., and Dunlap J.C. 1997. Dissection of a circadian oscillation into discrete domains. *Proc. Natl. Acad. Sci.* **94:** 3877.

Merrow M., Roenneberg T., Macino G., and Franchi L. 2001. A fungus among us: The *Neurospora crassa* circadian system. *Semin. Cell Dev. Biol.* **12:** 279.

Mitin N., Rossman K., and Der C. 2005. Signaling interplay in Ras superfamily function. *Curr. Biol.* **15:** R563.

Morgan L.W., Greene A.V., and Bell-Pedersen D. 2003. Circadian and light-induced expression of luciferase in *Neurospora crassa*. *Fungal Genet. Biol.* **38:** 327.

Nakashima H. 1981. A liquid culture system for the biochemical analysis of the circadian clock of *Neurospora*. *Plant Cell Physiol.* **22:** 231.

Nowrousian M., Duffield G.E., Loros J.J., and Dunlap J.C. 2003. The *frequency* gene is required for temperature-dependent regulation of many clock-controlled genes in *Neurospora crassa*. *Genetics* **164:** 922.

Okamura H., Miyake S., Sumi Y., Yamaguchi S., Yasui A., Muijtjens M., Hoeijmakers J.H., and van der Horst G.T. 1999. Photic induction of mPer1 and mPer2 in *cry*-deficient mice lacking a biological clock. *Science* **286:** 2531.

Panda S., Antoch M.P., Miller B., Su A., Schook A., Straume M., Schultz P., Kay S., Takahashi J., and Hogenesch J.B. 2002. Coordinated transcription of key pathways in the mouse by the circadian clock. *Cell* **109:** 307.

Perlman J., Nakashima H., and Feldman J. 1981. Assay and characteristics of circadian rhythmicity in liquid cultures of *Neurospora crassa*. *Plant Physiol.* **67:** 404.

Pittendrigh C.S. 1981. Circadian systems: Entrainment. In *Biological rhythms: Handbook of behavioral neurobiology* (ed. J. Aschoff), vol. 4, p. 95. Plenum, New York.

Pittendrigh C. and Bruce V. 1959. Daily rhythms as coupled oscillator systems and their relation to thermoperiodism and photoperiodism. In *Photoperiodism and related phenomena in plants and animals* (ed. R.B. Withrow), p. 475. AAAS, Washington D.C.

Pregueiro A., Liu Q., Baker C., Dunlap J.C., and Loros J.J. 2006. The *Neurospora* checkpoint kinase 2: A regulatory link between the circadian and cell cycles. *Science* **313:** 644.

Pregueiro A.M., Price-Lloyd N., Bell-Pedersen D., Heintzen C., Loros J.J., and Dunlap J.C. 2005. Assignment of an essential role for the *Neurospora frequency* gene in circadian entrainment to temperature cycles. *Proc. Natl. Acad. Sci.* **102:** 2210.

Price-Lloyd N., Elvin M., and Heintzen C. 2005. Synchronizing the *Neurospora crassa* circadian clock with the rhythmic environment. *Biochem. Soc. Trans.* **33:** 949.

Ramalho C.B., Hastings J.W., and Colepicolo P. 1995. Circadian oscillation of nitrate reductase activity in *Gonyaulax polyedra* is due to changes in cellular protein levels. *Plant Physiol.* **107:** 225.

Roenneberg T. and Merrow M. 1998. Molecular circadian oscillators: An alternative hypothesis. *J. Biol. Rhythms* **13:** 167.

Roenneberg T., Dragovic Z., and Merrow M. 2005. Demasking biological oscillators: Properties and principles of entrainment exemplified by the *Neurospora* circadian clock. *Proc. Natl. Acad. Sci.* **102:** 7742.

Ryan F.J., Beadle G.W., and Tatum E.L. 1943. The tube method for measuring the growth rate of *Neurospora. Am. J. Bot.* **30:** 784.

Sargent M.L., Briggs W.R., and Woodward D.O. 1966. The circadian nature of a rhythm expressed by an invertaseless strain of *Neurospora crassa. Plant Physiol.* **41:** 1343.

Schafmeier T., Kaldi K., Diernfellner A., Mohr C., and Brunner M. 2006. Phosphorylation-dependent maturation of *Neurospora* circadian clock protein from a nuclear repressor towards a cytoplasmic activator. *Genes Dev.* **20:** 297.

Schwerdtfeger C. and Linden H. 2003. VIVID is a flavoprotein and serves as a fungal blue light photoreceptor for photoadaptation. *EMBO J.* **22:** 4846.

Schwerdtfeger C., Loros J.J., Dunlap J.C., and Linden H. 2003. VIVID is a flavoprotein and serves as a fungal blue light photoreceptor for photoadaptation. *Fungal Genet. Newsl.* (suppl.) **50:** (Abstr. 236.)

Shi M., Larrondo L.F., Loros J.J., and Dunlap J.C. 2007. A developmental cycle masks output from the circadian oscillator under conditions of choline deficiency in *Neurospora. Proc. Natl. Acad. Sci.* **104:** 20102.

Shigeyoshi Y., Taguchi K., Yamamoto S., Takeida S., Yan L., Tei H., Moriya S., Shibata S., Loros J.J., Dunlap J.C., and Okamura H. 1997. Light-induced resetting of a mammalian circadian clock is associated with rapid induction of the *mPer1* transcript. *Cell* **91:** 1043.

Shinohara M., Loros J.J., and Dunlap J.C. 1998. Glyceraldehyde-3-phosphate dehydrogenase is regulated on a daily basis by the circadian clock. *J. Biol. Chem.* **273:** 446.

Storch K., Paz C., Signorovitch J., Raviola E., Pawlyk B., Li T., and Weitz C. 2007. Intrinsic circadian clock of the mammalian retina: Importance for retinal processing of visual information. *Cell Tissue Res.* **130:** 730.

Sussman A.S., Durkee T., and Lowrey R.J. 1965. A model for rhythmic and temperature-independent growth in the "clock" mutants of *Neurospora. Mycopathol. Mycol. Appl.* **25:** 381.

Sussman A.S., Lowrey R.J., and Durkee T. 1964. Morphology and genetics of a periodic colonial mutant of *Neurospora crassa. Am. J. Bot.* **51:** 243.

Talora C., Franchi L., Linden H., Ballario P., and Macino G. 1999. Role of a *white collar-1-white collar-2* complex in blue-light signal transduction. *EMBO J.* **18**: 4961.

Temme A., Ott T., Haberberger T., Traub O., and Willecke K. 2000. Acute-phase response and circadian expression of connexin26 are not altered in connexin32-deficient mouse liver. *Cell Tissue Res.* **300:** 111.

Tu B.P. and McKnight S.L. 2006. Metabolic cycles as an underlying basis of biological oscillations. *Nat. Rev. Mol. Cell Biol.* **7:** 696.

Vitalini M.W., de Paula R.M., Park W.D., and Bell-Pedersen D. 2006. The rhythms of life: Circadian output pathways in *Neurospora. J. Biol. Rhythms* **21:** 432.

Wijnen H. and Young M.W. 2006. Interplay of circadian clocks and metabolic rhythms. *Annu. Rev. Genet.* **40:** 409.

Yoshida Y. and Hasunuma K. 2004. Reactive oxygen species affect photomorphogenesis in *Neurospora crassa. J. Biol. Chem.* **279:** 6986.

Zhu H., Nowrousian M., Kupfer D., Colot H., Berrocal-Tito G., Lai H., Bell-Pedersen D., Roe B., Loros J.J., and Dunlap J.C. 2001. Analysis of expressed sequence tags from two starvation, time-of-day-specific libraries of *Neurospora crassa* reveals novel clock-controlled genes. *Genetics* **157:** 1057.

Zoltowski B.D., Schwerdtfeger C., Widom J., Loros J.J., Bilwes A.M., Dunlap J.C., and Crane B.R. 2007. Conformational switching in the fungal light sensor Vivid. *Science* **316:** 1054.

Principles and Problems Revolving Round Rhythm-related Genetic Variants

J.C. Hall, D.C. Chang, and E. Dolezelova
Department of Biology, Brandeis University, Waltham, Massachusetts 02454

Much of what is known about the regulation of circadian rhythms has stemmed from the induction, recognition, or manufacture of genetic variants. Such investigations have been especially salient in chronobiological analyses of *Drosophila*. Many starting points for elucidation of rhythmic processes operating in this insect entailed the isolation of mutants or the design of engineered gene modifications. Various features of the principles and practices associated with the genetic approach toward understanding clock functions, and chronobiologically related ones, are discussed from perspectives that are largely genetic as such, although intertwined with certain neurogenetic and molecular-genetic concerns when appropriate. Key themes in this treatment connect with the power and problems associated with multiply mutant forms of rhythm-related genes, with the opportunistic or problematical aspects of multigenic variants that are in play (sometimes surprisingly), and with a question as to how forceful chronogenetic inferences have been in terms of elucidating the mechanisms of circadian pacemaking.

What'dya say we bust up this joint?
Rodney Dangerfield (1980)

INTRODUCTION

Figuring out how circadian rhythms are regulated from within various organisms, ranging from microbes to mammals, has often been underpinned by genetic variants. These items have either been created or applied, or both. "Creation" is not always involved; because some rhythm variants, especially in *Drosophila*, are naturally occurring, as we shall see—over the course of a treatment that will discuss chrono*genetic* issues as such. Many such conceptual matters have arisen from genetic analysis of rhythms in *Drosophila*, which will figure most prominently in the piece, although analogous cases stemming from chronogenetic studies of mammals are also mentioned. Molecular mechanisms of circadian pacemaking, and the neural substrates of biological rhythms in animals, will not be objects of much focus, although these matters will come into play. Indeed, many molecular-chronobiological and chrononeurogenetic investigations of animals have been permitted by the generation or recognition of variants with abnormal circadian phenotypes.

THE POWER AND PROBLEMS ASSOCIATED WITH MULTIPLE ALLELES AT RHYTHM-RELATED GENETIC LOCI

The all-time classical rhythm mutants in *Drosophila melanogaster* are the three that were induced at the *period* (*per*) locus (Konopka and Benzer 1971). Each of them was independently isolated; thus, the separate mutations did not have to be allelic, but they turned out all to map to one X-chromosomal locus and failed to complement one another (see below).

The first such mutant found was per^{0} (now called per^{01}), based on aperiodic adult emergence (also known as eclosion). Later, per^{01} flies were shown also to exhibit arrhythmicity for adult behavior (Konopka and Benzer 1971). This mutant was isolated in 1968 (for the history, see Weiner 1999). The induction by chemical mutagenesis of per^{01} prompted extension of the study, which was published 3 years later by R.J. Konopka and S. Benzer, under the title "Clock mutants in *Drosophila*." However, an arrhythmic genetic variant need not have been altered in a clock gene, one whose functions subserve ongoing circadian pacemaking. Instead, per^{01} could have been brain damaged such that the neural substrates of the rhythmic attributes formed abnormally or not at all. In this regard, it is well known that physical destruction of a pacemaker structure in the mammalian brain leads to arrhythmic locomotion (see, e.g., Weaver 1998), which also can result from neuroanatomical injury caused by a mutation (see, e.g., Scheuch et al. 1982).

For *Drosophila*'s part, the isolation of two further *period* mutants was crucial, for they displayed altered circadian cycle durations, 5 hours shorter or longer than the normal approximately 24 hours (Konopka and Benzer 1971). These per^{S} and per^{L1} (née per^{L}) eclosion mutants, by their original definition, were revealed also to be adult locomotor variants. They "comapped" with each other and with per^{01} and were found to be allelic by virtue of noncomplementation. For example, per^{S}/per^{01} heterozygotes exhibited substantially shorter periods than do per^{S}/per^{+} flies. The latter display about 21.5-hour periodicity, signifying semidominance, and $per^{01}/+$ flies "run" slightly but consistently long, about 24.5 hours as opposed to about 24 hours (see, e.g., Smith and Konopka 1982). But *Drosophila*, whose only apparent functional *period* allele is per^{S}, exhibit 19–20-hour rhythms (Konopka and Benzer 1971). This assumes that the presence of per^{01}, including in a heterozygote involving per^{S}, provides no function (as was verified in subsequent genetically and molecularly based studies; see, e.g., Smith and Konopka 1981; Bargiello and Young 1984; Yu et al. 1987). The allelic interactions and gene-dosage effects implied by these findings are discussed later in terms of their potential forcefulness for inferring how the *period* gene acts as a potential contributor to circadian pacemaking.

If a gene can thus be altered so that such pacemaking is extant but "off," the idea is (and was) that this factor's normal function involves clock functioning as opposed to the formation of some piece of rhythm-related anatomy during development. It would follow that the *per*^{01} mutant is clockless at the functional level and likely to be normal for the pertinent morphology. Furthermore, because two separate circadian phenotypes come under the sway of the gene, one is encouraged to infer a "central" clock function as opposed to a process that operates on behalf of only a particular rhythmic attribute. In other words, if solely eclosion or adult behavior were mutationally altered, it could be supposed that the gene in question functions within a parochial "output pathway" originating at the core pacemaker; this pathway would end at the specific regulation of merely a particular rhythmic character.

Consider again the *per*^{S} and *per*^{L1} mutants. They were so dramatically altered for their cycle durations (in constant darkness or DD) that the implied *period* gene would not quit. But it is notable in this regard that one mutation at this locus causes only a mild circadian abnormality—*per*^{Clk}, a mutant that exhibits 22.5-hour periodicity (Dushay et al. 1990, 1992). What if *per*^{Clk}, by bad luck, had been the only mutation induced in the gene? The *per* story might have fallen by the wayside. This is what happened with regard to a little-known mutation on another *Drosophila* chromosome—*Toki*, which was induced by chemical mutagenesis and found to cause approximately 25-hour periodicities (Matsumoto et al. 1994). A certain degree of malaise set in, in the sense that an absence of substantial abnormality seemed to stimulate no further studies of the mutant or the gene that it defines. In addition, *Toki*'s genetic etiology was localized approximately to a region within the second chromosome of *D. melanogaster*. (All that could be said is that this mutation is not allelic to much better known rhythm mutations on chromosome *2*.) A second mutant allele of *Toki*, causing some sort of striking abnormality, might have held the gene in good stead.

We now turn to mouse chronogenetics, elements of which led to *Toki*-like scenarios. A handful of rhythm variants has been induced in this mammal (there are more and more such mutants as time goes by; see, e.g., Bacon et al. 2004; Siepka et al. 2007). One is the famed *Clock* (*Clk*) mutant, which causes severely abnormal rhythmicity when the mutation is homozygous (Vitaterna et al. 1994). Chemical induction of *Clk* prompted many studies of its mutational effects as well as analysis of the gene and its product's action (for review, see Lowrey and Takahashi 2004). In contrast, induction of the *Wheels* (*Whl*) mutant led to *Toki*-like melancholy, because this mouse mutant is only slightly abnormal for circadian period, and descendants of the original *Wheels* isolate were erratic for exhibition of such defects (Pickard et al. 1995). Furthermore, the mutation acts pleiotropically, owing to *Whl*-induced anatomical defects that seem to be unrelated to rhythm control (Pickard et al. 1995; Alavizadeh et al. 2001). An analogous murine case is provided by the *Wocko* rhythm mutant (Sollars et al. 1996), whose overall phenotypes (Crenshaw et al. 1991) were such that the investigators were "quits" in terms of digging deeply into the etiology of the less than exciting chronobiological abnormality.

This brings us back to *Drosophila* and to alterations of neural morphology, which can indeed be associated with rhythm mutants. The case in question involves the *disconnected* (*disco*) gene, which was recognized initially by a mutant exhibiting severely abnormal optic ganglia in the anterior central nervous system (CNS). *disco*^{1} (the first mutant of this type isolated) was one of several variants in *D. melanogaster* recognized by neuroanatomical screening per se; later, several of these mutants were shown to exhibit behavioral anomalies. Thus, taking the collection of brain-damaged variants and "screening through" them for rhythm defects disclosed *disco*^{1} as the only substantially abnormal type (Dushay et al. 1989; cf. Vosshall and Young 1995): Mutant cultures eclosed aperiodically and individual adults behaved arrhythmically. By this time in *Drosophila* chronogenetic history, the *per* gene had been cloned and was being assessed as to where it makes its products in fruit fly tissues (Liu et al. 1988; Saez and Young 1988; Siwicki et al. 1988). Among such locations were certain *laterally* located brain *neurons* (Siwicki et al. 1988), now called LNs (for review, see Hall 2005). Intriguingly, the apparent anatomical problems within a *disco*^{1} CNS extended a bit more deeply than visual system ganglia, because next to no *per*-expressing LNs were detectable in the adult brains of the mutant (Zerr et al. 1990; for many further data that speak to this point, see Helfrich-Förster 1998). So, it was inferred that functions operating within these CNS neurons underlie circadian phenotypes—plural, given the bidefective *disco*^{1} defects. This belied the notion that a mutant exhibiting more than one kind of rhythm deviation would necessarily define a circadian pacemaking factor.

What about the etiology of *disco*^{1}'s neurobiological and behavioral phenotypes (n = at least four abnormal ones)? Could it be that the combination of optic-ganglia and LN abnormalities, and their chronobiological consequences, does not have a simple underlying causation? It might be, for example, that *disco*^{1} itself leads to nonformation of the visual system elements (the most salient anatomical subnormality) but that something else in the "genetic background" causes the more subtle absence of LNs (or at least lack of *per*$^{+}$ expression in these approximately ten pairs of brain neurons). Another possibility is that *disco*^{1} does make both morphological aberrations manifest themselves but that this mutated allele is an odd duck that promotes these problems in some sort of epiphenomenological manner. What might this mean? Well, consider the *no-action-potential/temperature-sensitive* mutant: *Drosophila*, of either sex mind you, exhibit paralysis at high temperature, a condition that causes nerve conduction to fail (Wu et al. 1978). Later, it was shown that nap^{TS1} is mutated within a gene called *male-lethal* (*mle*); most mutations at the locus, by definition, cause only males to die during development (Kernan et al. 1991). The action potential mutation within the gene is apprehended to mediate some weird kind of "gain-of-function" defect that is quite apart from the workhorse role of mle^+.

In this context, it was warranted to test additional *disco* mutations to determine if they would cause dual defects—in visual system formation, on the one hand, and in terms of rhythm subnormalities, on the other hand. Comfortingly, it seemed, the $disco^2$ and $disco^3$ mutants, which were induced at a completely different institution and at a separate time compared with the origin of $disco^1$ (Steller et al. 1987), were similarly arrhythmic for eclosion and adult locomotion (Dushay et al. 1989). Discomfortingly, it turned out that the effects of independent "hits" within the *disconnected* gene were not being tested: $disco^2$ and $disco^3$ are molecularly identical to $disco^1$ (M. Freeman, pers. comm.), in that all three mutants are accounted for by the self-same nucleotide substitution (Heilig et al. 1991) within the gene's open reading frame (ORF).

This kind of quandary has surfaced for other genes as well, some of them rhythm-related: The seminal arrhythmia-inducing *period* mutation (Konopka and Benzer 1971; Baylies et al. 1987; Yu et al. 1987) is molecularly identical to per^{02} and per^{03} (Hamblen-Coyle et al. 1989), and per^{L1} has the same property vis-à-vis per^{L2} (Gailey et al. 1991). In *Neurospora crassa*, induction and isolation of the *frequency* (*frq*) mutants were crucial for kick-starting molecular elucidation of the circadian clock mechanism in this fungus (for review, see Dunlap et al. 2004). For a while, it seemed intriguing that period changes caused by *frq* mutations entailed "quantal" ways that such cycle durations could be altered: All three of the frq^2, frq^4, and frq^6 mutants exhibited 19-hour free-running periods, and two other such mutations—with separate allele numbers *7* and *8*, signifying that they were nominally isolated independently—each led to 29-hour periodicities. But the first three of these variants were identical to each other molecularly, and the second two had the same property in terms of their intra-ORF changes at the self-same site (Aronson et al. 1994). Incidentally, the notion that *Drosophila*'s *male-lethal* gene could commonly be mutated to cause nerve conduction failure collapsed when it was found that the original nap^{TS1} mutation at this locus was identical to the intragenic nucleotide substitution in nap^{TS} alleles *2* through *5* (Kernan et al. 1991). For *disco*'s part, the original allele number *1* was found to be identical not only to the aforementioned $disco^{2 \text{ and } 3}$, but also to mutation number *4* (cf. Steller et al. 1987; Heilig et al. 1991; M. Freeman, pers. comm.). To explain these annoying, or at least puzzling, cases, one could invoke intragenic "hot spots" that are susceptible to reinduction of mutations at the same sites in question. Alternatively, mutant screenings performed in part sloppily can lead to strain contaminants, whereby a supposedly new isolate in fact involves pulling aside an extant mutant already in the laboratory. Mercifully, with regard to interpreting the effects of $disco^1$ on *Drosophila* rhythms, a bona fide independent mutation was induced at the locus. This is $disco^{1656}$, for which an amino-acid-changing nucleotide substitution occurred at an ORF site different from that responsible for the $disco^1$ missense mutant (Heilig et al. 1991). And $disco^{1656}$ causes the same kind of locomotor arrhythmicity observed originally for $disco^{1=2=3}$ (Hardin et al. 1992).

This result led to some modest relief, in that the effects of these independently isolated *disconnected* mutations were mutually confirming: Two separate variants are better than one, at least in terms of discounting the possibility that one of them entails an occult mutation that would be difficult to elucidate in terms of connections between altered genotype and aberrant phenotype. The same situation is pointed to by the *Clk* mutation in mouse. It should not necessarily be apprehended as the most conventional type of mutant, many of which suffer from decrements or loss of gene product functions. Thus, the seminal mutation at this murine locus is an "antimorphic" variant (also known as a "dominant negative"). An allelic type with this property is inferred from the fact that a deletion of the *Clk* locus heterozygous with Clk^+ leads to no locomotor rhythm anomalies (King et al. 1997a), whereas the mutation over "+" by definition causes altered rhythmicity: The mutant strain was isolated among F_1 mice stemming from chemically mutagenized males (treated with a substance called ethylnitrosourea [ENU]) mated to genetically normal females (Vitaterna et al. 1994). This mutant type spawned a large number of phenogenetic investigations, which uncovered several abnormalities within and without the chronogenetic arena (see, e.g., Naylor et al. 2000; Low-Zeddies and Takahashi 2001; Turek et al. 2005; Oishi et al. 2006). All such studies were rooted in the effects of the one ENU-induced allele, which is accounted for by an intragenic change (call it "altered" or *alt*) that would lead to modified final products (CLK protein isoforms) (King et al. 1997b; cf. Antoch et al. 1997). At last, a second *Clk* mutation was created by knocking out the normal gene (DeBruyne et al. 2006); the rhythm phenotypes associated with homozygosity for this *Clk*-null allele (superscript "minus") are in a way "too normal." For example, locomotor periodicities of Clk^-/Clk^- mice are barely if at all different from those of wild-type individuals (DeBruyne et al. 2006), whereas Clk^{alt}/Clk^{alt} animals typically exhibit greater than 27-hour periods that often degrade into aperiodic locomotion in constant darkness (see, e.g., Vitaterna et al. 1994; King et al. 1997b). This situation is not a disaster, whereby the *Clk*-null's phenotypic properties would discount the significance of this gene's actions on behalf of murine chronobiology. But the dramatically different effects of the second mutant allele compared with the first one isolated have necessitated further inquiry (DeBruyne et al. 2007).

The opposite scenario is in force with regard to a molecular relative in the mouse of one of the two *timeless* (*tim*) genes in *Drosophila*. The mammalian "*mTim*" factor in question was shown to be involved in clockwork function, in conjunction with identifying the pertinent stretch of murine DNA in its normal form by similarity to what turned out to be *Drosophila*'s *tim2* gene (Benna et al.

2000), whose synonymous name is *timeout* (for review, see Gotter 2006). The initial mammalian experiments involved, in the main, transfection of *mTim* into cultured cells to interrogate effects on expression of other clock factors cointroduced into them (for review, see King and Takahashi 2000; Lowrey and Takahashi 2000). But these findings did not speak to the in vivo relevance of *mTim* functions. The latter were investigated by, as one might have guessed, knocking out the gene, which led, if you will, to *too much* of an abnormality: mice that die during development when the newly manufactured *mTim*$^-$ mutation was homozygous (Gotter et al. 2000). Furthermore, the viable *mTim*$^-$/+ type exhibited no locomotor rhythm abnormalities, i.e., no dosage effect of heterozygosity for an absence of the gene (Gotter et al. 2000). (Recall from above that a *per*-null variant does have a dosage effect that causes circadian period lengthening.) These murine findings led to disparagement of *mTim* as a pacemaking "player" (the title of the paper just cited is "A time-less function for mouse *timeless*"). And whether or not subsequent analysis of decrementing mTIM protein levels seemed to resurrect the rhythm-related significance of this gene product (Barnes et al. 2003), the fact that *mTim* encodes a developmentally essential factor should *not have undermined its potential meaning* from chronrobiological perspectives.

Here is the reason for this scolding: Several clock genes in *Drosophila* are *vital* ones, signified originally by induction of mutations at a locus called *doubletime* whose gene symbol is *dbt* (Price et al. 1998). The first-blush set of such variants allowed for viability, even when a *dbt* mutation was homozygous. By luck, a transposon (called a "*P* element") had inserted very near to this locus; this facilitated molecular identification of *dbt* as encoding a kinase subunit (Kloss et al. 1998). Additionally, homozygosity for *dbt*^{P} causes developmental lethality, and neither that mutated allele nor a deletion of the gene over *dbt*$^+$ leads to abnormal rhythmicity (Price et al. 1998). Other investigators who unwittingly induced mutations at the same locus (which they called *discs overgrown*) did so squarely with regard to their lethal effects (Zilian et al. 1999). Therefore, the actions of *doubletime* are pleiotropic, *not dedicated* to rhythm control. Stated another way, this gene is a *versatile* factor, which is exploited during development for one or more reasons quite different from the manner by which the enzymatic function gets reused much later in the life cycle to support proper circadian pacemaking.

By now, four rhythm-related kinase subunits are known in *Drosophila*, all pointed to initially by the effects of genetic variants involving the genes encoding the relevant polypeptides. The second of these genetic factors to emerge was *shaggy* (*sgg*, also know as *zw3*), which was reidentified in a transgene-based screen for "overexpression" effects on circadian locomotor periods (Martinek et al. 2001). The *sgg* gene had long been known to be able to mutate to lethality (see, e.g., Shannon et al. 1972), to cause interesting defects in developmental "pattern formation" (see, e.g., Simpson and Carteret 1989), and to encode a particular category of a kinase subunit (see, e.g., Peifer et al. 1994).

MUTATIONS AT CHRONOGENETIC LOCI MORE COMPLICATED THAN ONE WOULD IMAGINE

An analogous enzymatically based substory emerged, when an old *Drosophila* rhythm mutant eventually was discovered at the molecular level. This occurred many years after the *Andante* (*And*) variant had been chemically induced to make the circadian clock run "moderately slow." (However, and as is so for any chronovariant at the outset, there was no way to infer that *And* is a pacemaker mutant per se.) Cycle durations of *And* adults or of flies emerging from late-developing cultures are only at most 2 hours different from the norm (Konopka et al. 1991). This seems to have militated against immediate or even vigorous attempts to take this gene to the molecular level. (Recall *Toki* and *Wheels* in this regard.) Furthermore, *And* came with a wing anatomical abnormality whose appearance was like that of the classical *dusky* (*dy*) mutants (Konopka et al. 1991). Sure enough, the mutational etiology of the *And*- and *dy*-like defect mapped to one genetic locus on the X chromosome—to the narrowly defined region where *dy* mutations (per se) were long known to be located (Konopka et al. 1991). Did some dissatisfaction set in thereby, owing to pleiotropy of the *And* mutant strain? Perhaps, but it nevertheless seemed necessary to test one of the original *dy* mutants for abnormal rhythmicity. There was none (Konopka et al. 1991). However, a contemporary analysis of newly induced *dy* mutants (which involves easier screening compared with rhythm monitoring) turned up a handful of them; about half of these turned out to be *And*-like in terms of longer than normal circadian periods (Newby et al. 1991). It was as if the "*And-dy*" locus were a complex genetic factor, whereby some of its mutations would be bidefective and others would cause both kinds of phenotypic changes. Or, maybe one was unknowingly dealing with physical overlap of otherwise unrelated stretches of chromosomal DNA, so that only some mutations at such a locus would damage both encoded functions.

These complications were unraveled in conjunction with molecular identification of the pertinent transcription units. It turned out that they are right next door to each other on the X chromosome *and* that the seminal *And* mutant had been doubly hit: One nucleotide substitution is in the *And* gene per se (Akten et al. 2003) and the other is in the *dy* gene (DiBartolomeis et al. 2002). The latter encodes a function that need not concern us here (cf. Roch et al. 2003), and the former encodes a kinase subunit. Moreover, the enzyme regulatory factor specified by *And* is developmentally vital, because certain recently identified mutations at this "unitary" (noncomplex) locus are lethal (Jauch et al. 2002). So *Andante* is a pleiotopically acting gene after all but not because of the *dusky*-ness exhibited by the original mutant. The rhythm and the wing characters have nothing to do with each other biologically. Yet, the newer *And*-with-*dy* variants are puzzling in this regard (Newby et al. 1991). Could each of them be a double mutant too? They have not been assessed in this (molecular) respect. Provisionally, one would guess that these mutant types, isolated as wing abnormals, were errant reisolations of the original doubly mutated *And* variant. This would be a further case of the "no hot spots" supposition invoked above to explain the bugaboos of multiple *discos*, per^0s, and so forth.

In any case, it might be interesting to muse about why induction of *And* involved a double mutation at two extremely nearby chromosomal cites. Could a mutagenesis event involve stochastic accessibility of the chemical agent to a local region of the chromosome? Opening up the chromatin in this manner, within the gamete in question, could lead to nearby double hits more than Poisson would predict. This speculation cannot be dismissed out of hand because of the following: The third "clock kinase" identified molecularly in *Drosophila* involved a chemically induced mutation called *Timekeeper*, symbol *Tik* (Lin et al. 2002). It has dominant effects on behavioral rhythmicity, but *Tik* is a recessive-lethal mutation at a locus encoding a kinase catalytic subunit different from those specified by *dbt* or *sgg*. (TIK is a companion polypeptide to the AND polypeptide; for review, see Blau 2003.) The *Tik* mutant is accounted for by an *intragenic double mutation* that leads to separate amino acid substitutions (Lin et al. 2002).

This kind of mutational event has happened at least one more time in the *Drosophila* chronogenetic business: The most recently reported *timeless* mutant in this insect (at a locus different from the enigmatic *timeout* locus) also turned out to be doubly mutated within the ORF (Wulbeck et al. 2005). Another intriguing feature of this tim^{bl} mutant is that it exhibits longer than normal locomotor periods, but eclosion rhythmicity is like that of wild type. In contrast, the original tim^{01} mutant is aperiodic for eclosion, by definition of how it was isolated, and is similarly arrhythmic for adult behavior (Sehgal et al. 1994). In addition, subsequently induced *tim* mutants that were tested for both characters exhibit parallel abnormalities (Rothenfluh et al. 2000). What if tim^{bl} had been the only mutation induced in this gene? The unidefective phenotype might have caused it to be subconsciously disparaged, such that the gene might have been viewed as a "mere output factor," able to mutate so that only one particular kind of circadian rhythmicity is changed. As was introduced above, a bidefective mutant seems easier to view as defining a central pacemaker factor. Thus, what flows outward from the effects of a mutation within such a gene (such as *per*, *And*, and *dbt*) would be multiple circadian abnormalities (at least more than one of them). This supposition is of course belied by the case of *disco*, mutation of which leads to both eclosion and locomotor defects. Changes of this gene do not involve clock defects, but instead involve neuroanatomical mutants that suffer damage to the neural substrates of both rhythmic characters.

INPUTS TO THE *DROSOPHILA* PACEMAKER: MORE ABOUT DIFFERENTIALLY VARYING ALLELES AND CONSIDERATIONS OF FURTHER GENETIC COMPLEXITIES

I stretched the truth a little by suggesting that the above-mentioned tim^{bl} mutant is defective for only one rhythm-related character. This variant was noticed originally by virtue of a "phase change" readout by luciferase reporting of clock gene cycling. Live flies can be monitored automatically for this property when they harbor a transgene in which a fair amount of "5′" DNA from a clock gene (such as *per*) is fused to *luc* (Brandes et al. 1996), the latter being luciferase-encoding cDNA derived from a beetle (also known as firefly). After tim^{bl} was recognized by virtue of this molecular anomaly, it was found that the mutated form of TIM is relatively "unresponsive" to light. Such stimulation normally causes the concentration of this protein to crash, a hallmark of photically effected clock resetting in this organism (Dunlap et al. 2004). How "light gets to TIM" is crucially mediated, it seems, by the direct absorption of such stimulation by *Drosophila*'s version of cryptochrome (CRY), a substance that is activated mainly by blue light (see, e.g., Sancar 2000).

One way that this process began to be elucidated stemmed from the first mutation induced in the CRY-encoding gene ($n = 1$ in this insect, compared with the two or more harbored by various vertebrate species). Thus, the cry^b mutant allele was induced and then recognized because it eliminated *per-luc* cycling (Stanewsky et al. 1998). Second- and third-stage testing of the mutation's effects showed that *tim-luc* cycling was similarly flattened. In addition, the normal daily cycle of TIM protein abundance was eliminated as well, whereby the apparent level of the protein stayed high during the daytime (Stanewsky et al. 1998). In other words, light-induced TIM disappearance was disallowed in cry^b flies, as if a normal level of CRY is necessary to interact with TIM and cause it to be targeted for degradation. This supposition was contemporaneously supported by several additional kinds of experiments (whose details need not concern us, although for a review, see Hall [2003]).

The mutant-based side of this particular input substory involved an eventual array of ambiguities. First, note that cry^b is a missense mutant, which would not necessarily be a null variant. Whereas little or no CRY^b polypeptide was recognizable in the initial immunohistochemical tests (Stanewsky et al. 1998), subsequent tests have been detecting more and more leaking out of residual CRY levels that emanate from this mutated form of the gene (Busza et al. 2004). A subsequently induced *cry* mutant also was not demonstrable as a loss-of-function variant at the molecular level (Busza et al. 2004). The second and related point concerns the fact that the original cry^b mutant has been subjected to most of the requisite chrono*biological* testing. Here, too, light responsiveness of the mutated flies has suggested that they retain "some" CRY-mediated functions. These phenotypes involve, for example, the solid photic resettability of singly mutant cry^b adults. However, when this mutation was combined with "externally blinding" variants involving other fruit fly genes, the ability of doubly mutant individuals to resynchronize their locomotor rhythms to altered light/dark (LD) cycles was much more severely attenuated (see, e.g., Stanewsky et al. 1998; Veleri et al. 2007). Nevertheless, these CRY-depleted plus otherwise blind animals could be reentrained to an appreciable degree. One of the issues at hand, which now will be summarized all too briefly, is that there are "multiple input routes to the clock" in terms of the heads of such pathways being light-responsive. The separate routes are supported by actions of different kinds of "visual genes," on the one hand, and by different anatomical structures, on the other hand (for review, see Helfrich-Förster 2002). One of the latter such entities involves "deep-brain" photoreception (Emery et al. 2000b). It is mediated by CRY's presence within several of the aforementioned pacemaker neurons, those that contain clock gene products, the PDF neuropeptide, or both (for review, see Hall 2003).

To make a fly "circadian blind" therefore necessitates at least double damage to the photic input system, which is usually achieved genetically (although see also Ohata et al. 1998). But what if whatever genetic combination involved includes a residually functioning *cry* gene product? Interpretation of modest resettability is confounded by the non-null state of CRY functioning, juxtaposed with the possibility that some additional unknown input route is in play. In any case, it seemed warranted to expand the allelic series for *cry* by one more step. Therefore, the gene was "targeted" for knocking out, as is nicely achievable in *Drosophila* these days by transgene-based tactics (Bi and Rong 2003; Venken and Bellen 2005). True null variants (allelic designations *0* = zero) were thereby created (Dolezelova et al. 2007). But light responsiveness of the clock input system was found not to be thoroughgoingly blinded, even when a *cry*^{0} mutation was combined with another mutation that does (or should) render all peripherally located photoreceptive structures unresponsive (Dolezelova et al. 2007). It was concluded that, indeed, an "extra" input route functions within the system—one that could not be dependent on CRY as it cooperates with phototransduction processes believed to subserve light inputs to all of the relatively peripheral receptor cells known so far. (For the record, those entities are located in the compound eyes, the ocelli, an "extraocular" photoreceptor found at the periphery of the optic ganglia, and possibly some nonLN cells located near the dorsal extremity of the brain.)

Eclosion rhythmicity in *Drosophila* keeps being mentioned, even though this character has largely been replaced by locomotor monitoring as the workhorse bioassay of periodic phenomena operating at the whole-organismal level (for review, see Hall 2003; Price 2005). In this regard (sort of) studies performed long ago on late-developing cultures indicated that synchronization of them involves a relatively simple photic input. First, genetically normal *Drosophila* (of the *pseudoobscura* species) dietarily depleted of rhodopsin could be "entrained" or reentrained to eclose in an appropriately periodic manner, and there was no reduction in photic sensitivity (Zimmerman and Goldsmith 1971). In contrast, adult flies whose opsin-based photoreception is compromised exhibit subnormal sensitivity, even when they are CRY-enabled (see, e.g., Ohata et al. 1998; Stanewsky et al. 1998). The second point is that spectral sensitivity for light-induced eclosion entrainment was shown many years ago to exhibit a plateau "in the blue" (Frank and Zimmerman 1969; Klemm and Ninnemann 1976). A 21st century prediction, therefore, was that *cry* mutant cultures should not be synchronizable for periodic eclosion by exposure of late-developing animals to LD cycles or that populations of mutated flies striving to emerge rhythmically (with near-dawn peaks) in that photic condition could not exhibit such periodicity. Defective exhibition of the latter characteristic was reported as a result of one study: LD arrhythmicity for *cry*^{b} eclosion (Myers et al. 2003). But another study showed that this mutant type exhibits strongly periodic adult emergence, after being developmentally entrained via LD cycles (including through the metamorphic period) and then left to free-run in constant darkness (DD) during the times of actual eclosion events (Mealey-Ferarra et al. 2003). Even adding a blinding mutation to a *cry*^{b}-containing genotype (which disallows LD synchronization of clock neurons within the *larval* brain) left the doubly mutant animals in a rhythmically eclosing state (Mealey-Ferarra et al. 2003).

Could it have been that the differing experimental protocols in these conflicting studies were also "off" in some unknown manner with regard to the photic and other (local) conditions? If so, and if the non-null *cry*^{b} type hovers at a sharp margin of entrainability for eclosion rhythmicity, separate sets of results could be at variance. Thus, it was thought that analysis of *cry*^{0} for periodic eclosion might resolve the conundrum. The outcome was that metamorphosing CRY-less flies exposed to LD cycles emerged in solidly rhythmic ways, when the subsequent conditions remained (environmentally) periodic or were shifted to DD (Dolezelova et al. 2007). A new finding was that LD-entrained cultures of the *cry*^{0} mutants did not go arrhythmic when shifted to constant light (LL) during elcosion monitoring (Dolezelova et al. 2007). In contrast, wild-type *Drosophila* emerge aperiodically under this condition (see, e.g., Chandrashekaran and Loher 1969). Therefore, CRY was concluded to be utterly *dispensable* in terms of functioning on the input side of this process, which operates at the transition between late metamorphosis and early adult behavior. However, this blue-absorbing protein is somehow *involved*, or else its elimination would have left *cry*^{0} cultures-slash-flies in a responsive state insofar as the arrhythmia-inducing effects of constant light are concerned. Furthermore, the periodic eclosion of this null mutant type *in* LD was slightly aberrant: Adult-emergence peaks occurred slightly before dawn, whereas those defining rhythmic wild-type eclosion and that exhibited by *cry*^{b} flies occurred slightly after the dark/light (DL) transition (Dolezelova et al. 2007).

At all events, dedicated generation of null alleles for one of the key genes acting on the input side—and combining a given such *cry*^{0} mutation with those known to be null with respect to any of the companion visual genes—did not crack the cases in question. But it was arguably important to try, as opposed to endlessly milking the relatively dull tools provided by *cry* missense mutants.

There are additional genetically defined functions operating on behalf of photically mediated chrono-inputs in *Drosophila*. One of them is an "F-box" protein (cf. Siepka et al. 2007) encoded by the fly's *jetlag* gene, which specifies a factor that contributes to light-induced TIM degradation (Koh et al. 2006). Mutations at the *jet* locus were encountered somewhat serendipitously, at least in part because they were naturally occurring variants segregating in laboratory strains of *D. melanogaster*. A key effect of such a mutation was to render behaving flies largely unresponsive to the normally arrhythmia-inducing effects of LL. (By the way, anomalous retention of locomotor rhythmicity under this photic condition is also a property of *cry* mutants, as was first shown by Emery et al. [2000a].)

In a companion study to the *jetlag* study just cited, mapping the etiology of "LL rhythms" indicated, however, that the genetic difference underlying the phenotypic difference (vis-à-vis arrhythmicity of wild type in LL) was *not* located in the chromosomal interval that harbors *jet* (Peschel et al. 2006). The requisite meiotic recombinants involved the following crucially considered genotypes, any of which included a *jet* mutation: one allelic type at the *tim* locus versus an alternative allele. (The *tim* gene is located on the same chromosome as *jet*, nearby.) These *tim* "isoalleles" (Rosato et al. 1997) are defined by one form that produces two TIM proteins (both encoded by the so-called *LS-tim* allele), whereas the other one makes only one form of the polypeptide (the short one called S-TIM). Only when a *jet* mutation was combined with *LS-tim* did the abnormally rhythmic behavior in LL occur (Peschel et al. 2006). Meanwhile, it has been revealed that LS-TIM is relatively unresponsive to light, degradation-wise, at least in part because this form of the protein seems physically to interact with light-stimulated CRY in a relatively weak manner (Sandrelli et al. 2007).

The upshot of these studies is that one needs to mind the genetic background! A given singly mutant type might or might not lead to the expected defect, unless it is conjoined with a separate player, one that, by itself, might cause a minimally appreciable abnormality for the chronobiological character in question. (Indeed, in the case a hand, neither the *LS-tim* nor the *S-tim* isoallelic types retains anomalous rhythmicity in constant light.) The "background warning" should be especially in force if neither of the other variants is rarely met. Indeed, both of the *tim* isoalleles are commonly encountered, including and especially in natural populations, where they vary in a geographically systematic manner: The *LS-tim* form is found at relatively higher frequencies in northern latitudes (Tauber et al. 2007). This harks back to the "*per* clines," previously established to exist in natural populations analyzed north-to-south with regard to the intragenic length of a coding sequence that specifies a series of threonine-glycine amino acid repeats, located approximately in the middle of a given PER polypeptide (Costa et al. 1992; Sawyer et al. 2006).

MIND YOUR GENETIC BACKGROUND ALSO WHEN STUDYING OUTPUTS FROM THE CLOCK

The matter of pathways that connect central pacemaking to the distal regulation of revealed rhythmicity was introduced earlier. Increasing numbers of output-pathway endpoints are being recognized in *Drosophila* (for review, see Hall 2005). One to arrive on the scene in relatively recent years is the control of sleep versus wakefulness. This phenomenon goes deeper into a matter of rest versus activity, in part because of the regulatory factors contributing to sleeping and to recovery from sleep deprivation (for review, see Greenspan et al. 2001; Hendricks 2003). Nevertheless, sleep/wake cycles come under the sway of circadian clock control as well. It would follow that isolation of sleep mutants might define genes that function on *an* output pathway operating downstream from the clock; such mutants might thereby be "specifically" defective and would not exhibit, for example, abnormalities of eclosion rhythmicity or that which underlies varying sensitivity of an appendage to odor stimuli (see, e.g., Krishnan et al. 2001). So, a screen for newly induced sleep-defective mutants turned up one; mapping the etiology of the "mini-sleep" character suggested that what became mutated was the famed *Shaker* (*Sh*) gene, which encodes a particular category of potassium channel subunits (Cirelli et al. 2005). Indeed, the *mini-sleep* mutation involved a novel site change within the *Sh* locus. These investigators wondered whether only this particular amino acid substitution within a SH polypeptide would cause the sleep anomaly, imagining in turn that any "hyperexcitable" mutant involving this gene would not necessarily be so fidgety that it would not sleep as much as normal flies do during a given daily cycle. Thus, they tested some of the classical *Sh* mutants for sleeplessness, but most of these variants tested as normal. What, however, if these long-maintained mutant strains had accumulated *genetic modifiers* that act to ameliorate would-be effects of the main (*Sh*) mutation? This is a frequently invoked shibboleth in the biogenetic business. But the supposition had real meaning here, because outcrossing the old *Sh* mutants to genetically normal *Drosophila* and reextracting shakerness from among the segregants led to substrains that exhibited the mini-sleep character. Apparently, the kind of change within the gene that leads to generic hyperexcitability will also cause restlessness. A further question was whether the connection between being hyper and sleeping minimally is so nonspecific that mutations in other genes that encode different kinds of potassium channel subunits would cause the same abnormality. The answer seemed to be no, for neither *Hyperkinetic* (*Hk*) nor *seizure* mutants slept abnormally (Cirelli et al. 2005). Elements of this state of affairs were dubious when this study was reported, because the "non-*Shaker*" potassium channel variants were not outcrossed to ask whether the normal sleep phenotypes determined for some these "other" potassium channel mutants were influenced by factors lurking in the genetic backgrounds. Sure enough, a subsequent study undermined the hope for a *Shaker*-o-centric view of the connection between hyperexcitability and sleep anomalies, because *Hk* mutations were said to cause sleep-duration decrements, *and* the *Hyperkinetic* variants in question had been outcrossed to regularize their genetic backgrounds with the requisite control strains (Bushey et al. 2007).

An opposite kind of genetic modifier problem turned out to undermine the interpretability of another kind of output mutant in *Drosophila*. The question eventually asked was whether the *abnormality* is contributed to by factors other than the "main" mutations. Those gene changes had occurred long ago in a gene called *ebony* (*e*), because the flies exhibit dark body color. Various pigment mutants of this sort turned out also to be neurotransmitter variants (not surprising, because the biochemical pathways subserving melanin production overlap with those that manufacture amine-containing transmitter substances). Several different *e*-mutated alleles were found to make flies largely arrhythmic for adult locomotion, but all such mutants exhibited normally periodic eclosion (Newby and Jackson 1991). Meanwhile, the *e* gene was becoming ever better characterized as to how its product, β-alanyl synthase, regulates levels of dopamine and histamine (see, e.g., Borycz et al. 2002). It was as if one or both such substances were involved in intercellular communication along an anatomical pathway that links the proximal outputs of the clock to downstream entities that modulate walking behavior of the mature animal. An odd twist to this substory is that e^{+} seems to function on behalf of some kind of neurochemical processes operating in *Drosophila* glia (see, e.g., Richardt et al. 2002). This prompts mention of the fact that at least some of the clock genes in this insect are expressed in hosts of glial cells distributed throughout most CNS ganglia (Siwicki et al. 1988; Ewer et al. 1992; Helfrich-Förster 1995; Kaneko and Hall 2000). Why? Unknown—although it is worth mentioning further that the chronobiological meaning of glial cells harbored within a pacemaker structure in mammalian brains (Weaver 1998) was signified by at least one study (Prosser et al. 1994). In context of the current article being largely historical, it must be noted that the matter of "clock glia" potentially contributing to the regulation of actual *Drosophila* chronobiology (see especially Ewer et al. 1992) is one of many matters that are not properly a part of current concerns.

That aside, we plunge ahead to continue considering *ebony*-related neurochemistry. Thus, a companion set of *Drosophila* mutants to the *e* mutants are those altered in the *tan* (*t*) gene. These body color variants are subnormal for β-alanyl hydrolase; once again, dopamine and histamine levels are compromised compared with concentrations of these substances in wild-type flies (Borycz et al. 2002). We therefore surmised that testing *t* mutants for rhythmic characters was in order. Putative effects of the t^{1} and t^{5} mutations on eclosion and locomotor rhythmicty were assessed. Both mutant alleles supported normality for both characters (Fig. 1), unlike the reported effects of *e* mutations on the latter phenotype (alone). This outcome must be regarded as provisional, however: The *t* mutants (Fig. 1) will have to be outcrossed to wild type, followed by re-extraction of individuals with anomalous body color, and then retesting the t^{1}- and t^{5}-mutated substrains for both rhythm phenotypes.

Speaking of outcrossing and now focusing on the mutant type that created the clock-output scenario in question: First, we found that both e^{1} and e^{11} mutants exhibited normal eclosion profiles (Fig. 2A), as originally reported (Newby and Jackson 1991). However, the first of these mutant alleles allowed for normal locomotor rhythmicity (Fig. 2A,D), quite at variance with the earlier study. For e^{11}, the rest/activity cycles of adults were "weak" overall (Fig. 2C,D), allowing one to cling to the notion that the gene has a chronobiological role. However, outcrossing e^{1} and e^{11} to wild type (seven times) and reextracting the requisite darkly pigmented substrains showed that adults taken from them were "strong" and otherwise normal for locomotor rhythms (Fig. 2E,F).

In our hands, therefore, the connection between pigmentation, *these* neurotransmitters, and regulation of rhythmic behavior is solely a matter of one or more factors harbored in the genetic background of a particular *ebony* mutant. One soft feature of this analysis is that whatever these factors may be is unknown—as to whether few or many rhythm-undermining variants are harbored in the e^{11} strain, let alone where within the genome these putative output factors would be located.

A different tack is taken in this arena nowadays. The genetic tactic in question involves noticing that strains which vary among each other for "some" phenotypic character, followed by performing crosses between strains, which can result in specifiable genetic segregants and recombinants. The specificity at hand means that quantitative trait loci (QTL) can be chromosomally mapped: How many such factors are pointed to, and where is each located within genome (at least roughly)? To which features of the overall phenotype do they contribute (speaking to the fact that locomotor rhythmicity involves a variety of quantifiable attributes, such as the "free-running circadian period" and "phase angle of entrainment" metrics dealt with in the study cited shortly below)? What is the magnitude of a given QTL's contribution to the character difference observed by monitoring behavior of the starting strains? A few chronogenetic investigations of this sort have been

Figure 1. Adult emergence and locomotor rhythms of *tan* mutants. (*A*) Emergence of flies from metamorphosis, monitored in constant darkness (DD) as described by Mealey-Ferrara et al. (2003), led to the eclosion profiles in the left-hand two columns; high-frequency components were filtered in the second column (cf. Levine et al. 2002). These time courses were analyzed as plotted in the right-hand three columns via maximum entropy spectral analysis (MESA) and the two other formal treatments of the data plotted rightward (for principles and practices for these methods, see Levine et al. 2002). The two *tan* mutants tested are indicated by their superscripted allele designations. The +/+ genotype refers to a tan^{+} wild-type control, which was not isogenic with the mutants. (*B*) Locomotor activity of the tan^{1} mutant, monitored in DD as described by Dolezelova et al. (2007), for example. The second row gives an example of behavior performed by a female heterozygous for the X-chromosomal t^{1} mutation and a deletion (*Df*) of the genetic locus. The third row exemplifies behavior of a t^{1}/t^{+} heterozygote. The left-hand column displays double-plotted actograms (days 1–2 of locomotor movements on the top line of a given plot, days 2–3 on the second line, etc.) Formal analyses of these behavioral records (right-hand three columns) were again performed as described by Levine et al. (2002). (*C*) Locomotor activity of the tan^{5} mutant; labels and plot types are the same as those in *B*. (*D*) Relative robustness of locomotor rhythms exhibited by "average flies" expressing various *tan* alleles (the *n* indicates numbers of individual flies monitored). The homozygous and heterozygous such genotypes (see *B* and *C*) are displayed across the abscissa. The ordinate values are a measure of "rhythm strength," as described by Levine et al. (2002), including a cutoff line horizontally coursing across the plot; RS values above that line imply "significantly" periodic behavior.

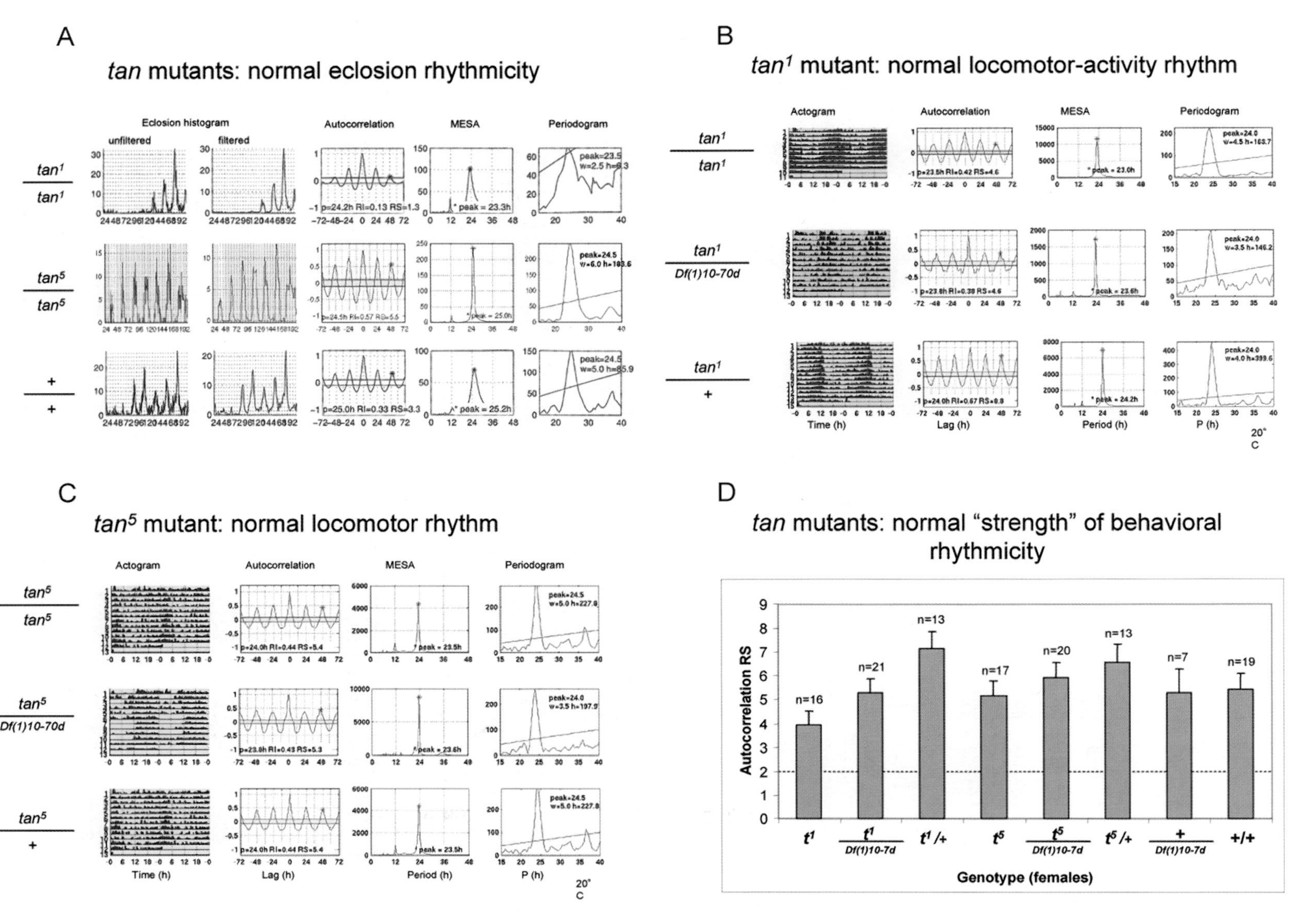

Figure 1. *(See facing page for legend.)*

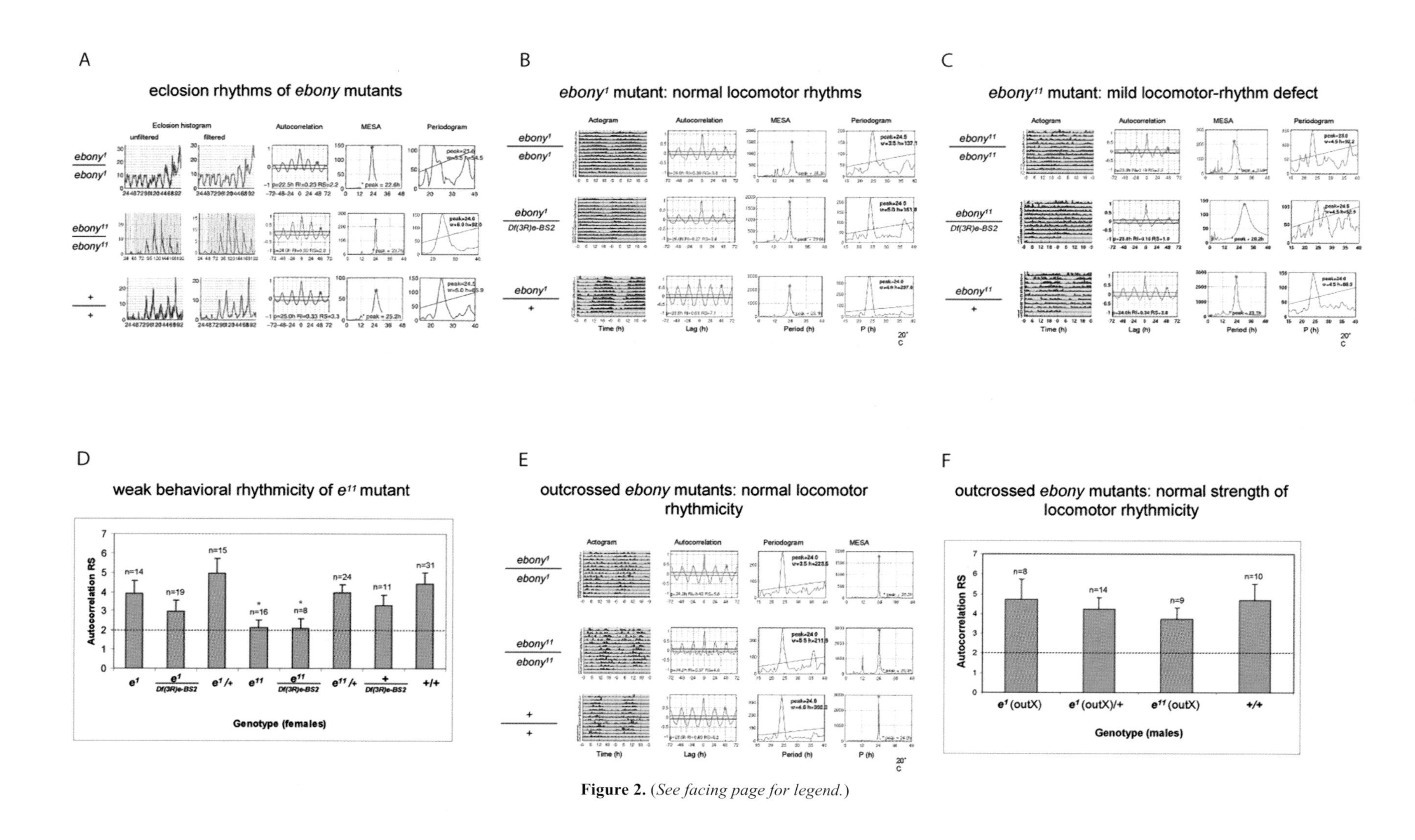

Figure 2. *(See facing page for legend.)*

performed, notably in mouse. A quite extensive one uncovered 14 QTLs, scattered among most of the rodent's 20 chromosomes (Shimomura et al. 2001). By this time, chromosomal sites that harbor several of the "main" clock genes in this mammal were known; additional such loci were pinned down as an accompaniment to this QTL analyses, i.e., the intrachromosomal sites corresponding to clock genes originally identified molecularly in their normal forms (Shimomura et al. 2001). Three of these genes that are among the most conspicuous chronogenetic factors in mammals are murine (*m*) relatives of *Drosophila*'s *period* gene (the former being known as *mPer1* through *3*, i.e., three separate genes). It could be regarded as disappointing that no QTL (in the study just cited) apparently corresponded to a previously known rhythm-related locus. Several of the latter have been subjected to gene knockouts, and such null mutations were frequently found to cause subnormalities or anomalies associated with murine locomotor rhythmicity (as nicely summarized by the list contained within Stanewsky 2003). For example, knocking out *mPer1* and *mPer2* and then combining the two nulls lead to severely subnormal rhythms of wheel-running behavior (for review, see Stanewsky 2003).

One idea underlying the QTL approach was that homing in on certain rhythm-related genetic loci would suggest that they entail milder genotypic changes within some of the "major" clock genes such as *mPer* genes. Such *gedanken* outcomes could independently confirm the chronobiological meaning of such genes, along with enhancing one's appreciation of what a given such factor is "about" in terms of how it contributes to a particular feature of circadian rhythmicity. Inasmuch as this kind of picture could not be painted by the results of the Shimomura et al. (2001) study, a different and just as good inference suggests itself: Tapping into naturally occurring variation can lead to identification of genetic factors *heretofore unknown* to be involved in the process. Bad luck could have militated against mutating these loci to isolate mutants exhibiting whatever rhythm irregularity was being screened for. That said, newly induced mutants that are saved *and* followed up tend to show rather substantial differences from the norm (prompting rereferrals to Vitaterna et al. 1994; Bacon et al. 2004; Siepka et al. 2007). It is therefore arguable that a QTL associated with modest variation for the character in question (Shimomura et al. 2001) might not get delved into deeply, unless more severely mutated allelic changes at the locus will be induced or otherwise encountered later on.

BACK TO THE BEGINNING: MULTIPLE GENOTYPIC VARIANTS INVOLVING *DROSOPHILA*'S *PERIOD* GENE AND THEIR IMPLICATIONS FOR FUNCTIONING OF ITS PRODUCT

Mentioning the mouse *Period* genes (shortly above) and their engineered mutations was done judiciously, for it brings us back to the seminal mutants of this kind and to the first clock factor that was identified in concrete form. In this light, the *period* mutants in *Drosophila* and the locus it defined were subjected to rather extensive phenogenetic analyses. The plural noun just invoked refers to more than extensive chronobiological testing of a given individual mutation's effects. In addition, many *per*-related genetic *combinations* were produced to ask whether the effects of such genotypes might lead to inferences about how the gene functions. For example, *per* mutations that allow for rhythmic behavior were made heterozygous for the normal allele; all such combinations indicated the aforementioned semidominance of these mutant alleles. Furthermore, the period alterations—observed for instance in per^S/+ or per^{L1}/+ heterozygotes—involved departures form the norm greater than those caused by per^{01}/+ (Konopka and Benzer 1971; Smith and Konopka 1982). It follows that the effects of the period-shortening and -lengthening alleles are worse than that of the "zero" mutation. That the latter might indeed be null was suggested, early on, by the fact that a deletion of the gene, over +, leads to the same modest period lengthening as observed for per^{01}/+ *Drosophila* (see, e.g., Smith and Konopka 1982).

This particular result points to "cytogenetic" studies of *period* that formed a part of this analytical process. A further feature of such manipulations was to assess the effect of a duplication of the chromosomal region con-

Figure 2. Adult emergence and locomotor activity rhythms of *ebony* mutants. (*A*) Periodic eclosion of two *e* mutants (distinguished by the different superscripts), compared with a wild-type control (+/+), the flies of which were not isogenic with the *e* mutants. Plotting and analytical tactics are the same as those in Fig. 1A. (*B*) Free-running locomotor activity rhythms of flies (monitored in DD), which were homozygous for e^1, heterozygous for that mutation and a deletion (*Df*) of the locus, or heterozygous for e^1 and the normal (+) allele. Plotting and analytical tactics are the same as those in Fig. 1B. (*C*) Locomotor rhythmicity influenced by the e^{11} mutant allele harbored in a long-inbred strain. Columns, rows, and genotypic labels are the same as those in *B* (cf. Fig. 1B,C). (*D*) Robustness of locomotor rhythmicity influenced by the e^{11} allele. The metric and the plotting conventions are the same as those in Fig. 1D. The asterisks designate significantly reduced rhythm strength caused by homozygosity for e^{11} or "uncoverage" of that mutation's effect by a deletion (*Df*). (*E*) Locomotor rhythmicity influenced by e^1 or e^{11} after flies carrying a given such mutation were outcrossed to a wild-type strain, followed by crossing the resulting heterozygotes ("over +") to each other and selection of homozygous mutant individuals among the offspring; those *e*/*e* flies were outcrossed again to +/+ to create round-2 worth of heterozygotes, further enriched for the wild-type genetic background. This crossing procedure was repeated such that seven rounds of *e*/+ heterozygosity were in force; after the last such round, mutant homozygotes were selected (as usual), and it was these e^1/ e^1 or e^{11}/e^{11} flies and their true-breeding homogygous descendants that were tested for locomotor rhythmicity. Examples of such behavior are displayed (along with the analytical plots) as in *B* and *C*. (*F*) Strong locomotor rhythms associated with flies carrying e^1 or e^{11} in, or derived from, the outcrossed (outX) strains. The one "derived" type is the e^1/+ heterozygote generated by crossing (outX) homozygotes to the wild-type strain used in the outcrossing (= "isogenization") procedure. The +/+ control (average rhythm strength) bar resulted from testing flies from that wild-type strain. The four ordinate values (cf. Fig. 1D and Levine et al. 2002) were statistically indistinguishable from one another.

taining *per*$^+$, whereby increased dosage of the normal form of this gene led to shorter than normal periodicity (Smith and Konopka 1982), a finding that prompted the notion that *per*^{S} might be a "hypermorphic" mutation, leading to an enhanced level of the encoded function. But this was belied by a combination of allele effect and cytogenetic analysis: If the *per*^{S}/+ genotype is making "lots more product," expression of the + allele in this heterozygote would contribute to the overall (heightened) concentration of whatever the gene encodes. If this supposition has meaning, a *per*^{S} overdeletion (*per*$^-$) heterozygote would produce slightly less product and result in a somewhat longer cycle duration compared with *per*^{S}/+. But the latter genotype leads to about 21.5-hour periodicity, whereas *per*^{S}/*per*$^-$ flies run at about 20 hours (as cited and interpreted by Coté and Brody 1986). Thus, a better way to view this subscenario is to assume that the effect of *per*^{S} is not to "add" to the effect of the + allele but instead to *interfere* with "normal function" (cf. Coté and Brody 1986). The hedge just quoted is that this might have to do with the product of the normal *per* gene or with factors specified by other rhythm-related genes, or both.

The upshot of all these phenogenetics was the following: (1) One job of the *period* gene product is to *interact* with other chronoactive factors (suggested by the dominant negativity of period-altering mutations, again as exemplified by the phenotype of *per*^{S}/+ flies), and (2) the level of *per*'s product is *rate-limiting* for period control (implied by the gene-dosage effects, which by the way do not obtain for some of the other clock genes in *Drosophila*, such as *timeless* or *doubletime*).

To what extent did any of these phenogenetic phenomena suggest what the *period* gene product actually is? Intriguingly, nothing whatsoever suggested itself in this regard. Any and all of the analytical outcomes just summarized could have allowed one to speculate that the presumed PER polypeptide possesses some sort of enzymatic function or that it could be inserted into cellular membranes, or it could be "anything."

Distinctly different kinds of phenogenetic stories have been told in biogenetic history (referring to genetically based investigations that have been aimed at understanding physiological, developmental, and neurobiological processes as opposed to hard-core genetic processes). Take the *lactose* operon in *Escherichia coli*. The early days (years, really) of studying it were rooted in recognition of many different kinds of mutations at this complex bacterial locus. Furthermore, different allelic types that mapped to a given subsite within the locus were recovered. Pitting a given mutant phenotype against the normal inducibility of *lac*-encoded protein products led, in time, to a magnificently conceived scheme as to how the regulatory factors comprising portions of the locus operate, i.e., what these "*cis*"- and "*trans*"-acting entities would consist of at the concrete level. In this regard, when examinations of *lac*-contained or -encoded regulatory items became molecular, elucidation of those factors was arguably anticlimactic, thanks to the broad and deep analyses of genotype with phenotype connections that had been previously performed (Jacob 1997). An analogous metazoan case is provided by a famous subset of the many developmental mutants known in *Drosophila*. Again, "pure" phenogenetics of the manner by which key mutations alter pattern formation of the developing animal indicated that the corresponding genes were "selector" factors (see Chapter 5 in Lawrence 1992). It was just a matter of time for these genes to be identified at the DNA level, and the chance that "clone and sequence" would *not* have revealed these sequences to encode gene-regulatory factors (transcription ones) was nil (Lawrence 1992).

In marked contrast, the *period* clock gene in *Drosophila* had to be cloned, absent anyone's wherewithal to deduce the nature and functionality of its product. Furthermore, sequencing the genes ORF led to no clues—or worse—as to what the encoded polypeptide is about (for an early review of what happened over the course of about 10 years postcloning, see Hall 1995). This initially featureless protein had to be elucidated for the circadian-pacemaking role it plays, largely by analyzing the manner by which *per* gene products are expressed. This means, for instance, that PER is found within the nuclei of neural cells, naturally (although in many other tissue types as well), and, from a temporal perspective, the levels of *per* mRNA and of PER protein were found to exhibit systematic daily oscillations (again see Hall 1995). These facts stimulated further inquiries that began to crack the case (as surveyed by Dunlap et al. 2004), a case so filled with cracks that perhaps one more would not hurt. The "one more" in question entails, in a nutshell, a scheme in which cyclically fluctuating PER feeds back to impinge on the transcribability of the first-stage gene product. Therefore, such mRNA cycling, as controlled in part by PER cycling, became viewed as a core component of the oscillator mechanism (Hall 1995; Dunlap et al. 2004).

The meaningfulness of *per* product cycling as it gets controlled transciptionally has been called into question, however. First, an empirical study was designed to "drive" *per* expression or that of a *timeless*-coding sequence, or both, in temporally constitutive ways. The relevant transgenic strains were set up to activate these clock gene product generations via a heterologous transcription factor whose own production was controlled by a gene-regulating stretch of DNA that is "always on" (Yang and Sehgal 2001). Therefore, *per* and *tim* mRNA levels would be temporally flat, at least in terms of how the primary gene products are transcribed. The transgene combinations were introduced into genetic backgrounds that harbored *per*- or *tim*-null mutations (or again, both, as the case may be). Even when the only DNA sequences that could generate functional PER and TIM were encoded by constitutively generated mRNAs, varying proportions of the adult flies exhibited locomotor rhythmicity (Yang and Sehgal 2001). Therefore, transcriptionally supported rhythmicities of these coding sequences were deemed not necessary for periodic biological readout.

These results have inspired some who hover round the clock molecular stories that have been told for various organisms to disclaim as follows: "Transcriptional feedback oscillators: Maybe, maybe not..." (Lakin-Thomas 2006). An appreciable portion of what this author argued (under the title just quoted) involved denigration of the standard view of rhythm regulation in *Drosophila*, which assumes (to quote this author further) that it is "'important' for function" for the final gene products in question to feed back in order that they can modulate cyclically varying generation of the molecules that encode them (that specify, in this case, PER and TIM proteins). Lakin-Thomas (2006) did give a nod to one feature of the Yang and Sehgal (2001) results, which was that "rhythms in the double constitutive flies [see Fig. 3] are somewhat weak." Nevertheless, this iconoclast went on to proclaim that "I take the glass half-full attitude: as many as half [sic] were rhythmic" (Lakin-Thomas 2006). With around "half" the animals in question performing rhythmically at the behavioral level (see Fig. 3) (referring to some of Yang and Sehgal's transgenic strains), such a result was surmised to mean: that's it! Any rhythmicity eked out of a situation in which transcriptional control of molecular cycling cannot be operating was viewed as undermining the original "case-cracking" discoveries of the transcriptional feedback oscillators (which, again, are laid out in their simplest form within the review by Hall 1995).

What are we to make of these sour notes? First, some further empiricism: Wondering how "full" is the "glass," we retested the triply transgenic strain designed to make *per*- and *tim*-coding sequences simultaneously transcribed in a constitutive manner. The "rescue" of per^{01}/tim^{01} arrhythmicity (whereby either such null mutation alone is sufficient to cause that property) was such that a mere 19% of the locomotor-monitored flies exhibited significantly periodic behavior (Fig. 3A). This less-than-half outcome is worse than what was reported from the results of several analogous locomotor tests (Yang and Sehgal 2001). Furthermore, about one third of the rhythmic "doubly constitutive" individuals that behaved rhythmically did so in an ambiguous manner, notably by exhibiting more than one periodic component in conjunction with their free-running cyclical behavior (see examples in Fig. 3B–E). Even worse, in a way, the unambiguously rhythmic cases involved cycle durations distributed over an approximately 6-hour range (Fig. 3F), even though the numerical boundaries were in the circadian ballpark. In marked contrast, genetically normal *Drosophila* yield locomotor periods that are tightly clustered near 24 hours (see, e.g., Hall 2003, 2005; Price 2005).

This calls into question whether a given abnormal rhythm-related genotype and its attendant phenotype are what we are all about investigatorily. In other words, are we not attempting to deduce from such genetic experiments what is necessary for *rhythm normality*? Should we not at least consider that certain genetically effected abnormalities are rather uninformative, especially when the outcome is a crude caricature of the normal process? Minimally, the rhythmic behavioral glass seems so far from "half-full" in the current case (Fig. 3) that one can surmise a prime role for transcriptional control of *molecular* cyclings if anything in hailing distance of normally periodic *biology* is to occur.

Furthermore, or more specifically, it appears that *all levels* at which clock gene products are regulated, such that they exhibit daily oscillations, are correlated with wild-type rhythmic phenotypes. The following are some experimental results that speak to this issue: (1) When a *per* transgene lacks its 5′ regulatory sequence, the *transcription rate* at which the encoded RNA is produced is temporally flat (So and Rosbash 1997), but (2) the *steady-state abundance* of *per* mRNA cycles nonetheless, as do the apparent levels of PER protein (Frisch et al. 1994, pointing to many subsequent studies that revealed PER cycling to be controlled in part posttranslationally, as shown seminally by Cheng and Hardin 1998), (3) the transgenic situation in question (Frisch et al. 1994) is enough to support routine rhythmicity of the adult flies, but their period and phase values are appreciably "off" the norms, and (4) additional kinds of *per*-manipulated transgenic types showed that control of RNA production at the transcriptional level alone is sufficient for this "level" of gene product to cycle rather robustly (referring to reporter DNA sequences that had been fused to *per* regulatory sequences), but the temporal dynamics were distinctly aberrant unless the pertinent heterologous DNA sequence (*luc*) was fused to well more than half of the *per*-coding sequence (Stanewsky et al. 1997). The latter set of results means that—fair enough—transcriptionally controlled cycling of clock gene expression does not tell the entire tale. But Points 1 and 2 indicate that the level of control, at the primary stage of gene expression, is far from unimportant.

All of this said, the issues remain up in the air. At least it is the case that mediating production of factors encoded by certain key clock genes—in a systematically varying manner, along with tweaking the temporal dynamics of these molecules after they are primarily generated—will remain objects of much further study. A nice recent example shows how these issues can and must be analyzed, in part from the perspective of transcriptional control of circadian oscillators (Kadener et al. 2007). This alludes to the fact that retesting whatever set of naysaying results that remain in the wind (Yang and Sehgal 2001) is insufficient (Fig. 3 notwithstanding).

At all events, it is time to wind things up because it appears as if the verbal wranglings with which the foregoing passages are replete has strayed rather far from the chronogenetic theme of the piece.

ACKNOWLEDGMENTS

We thank many colleagues with whom, since the late 1970s, several of the investigations mentioned have been performed, in particular Ronald J. Konopka, Charalambos P. Kyriacou, Melanie J. Hamblen, William A. Zehring, Kathleen K. Siwicki, John Ewer, Mitchell S. Dushay, Brigitte Frisch, Christian Brandeis, Maki Kaneko, Ralf Stanewsky, Charlotte Helfrich-Förster, and Joel D. Levine. Special thanks go to Michael Rosbash and Steve A. Kay and collaborating co-workers in their respective laboratories. Most of the studies performed by the authors and their associates have been supported by grants from the National Institutes of Health.

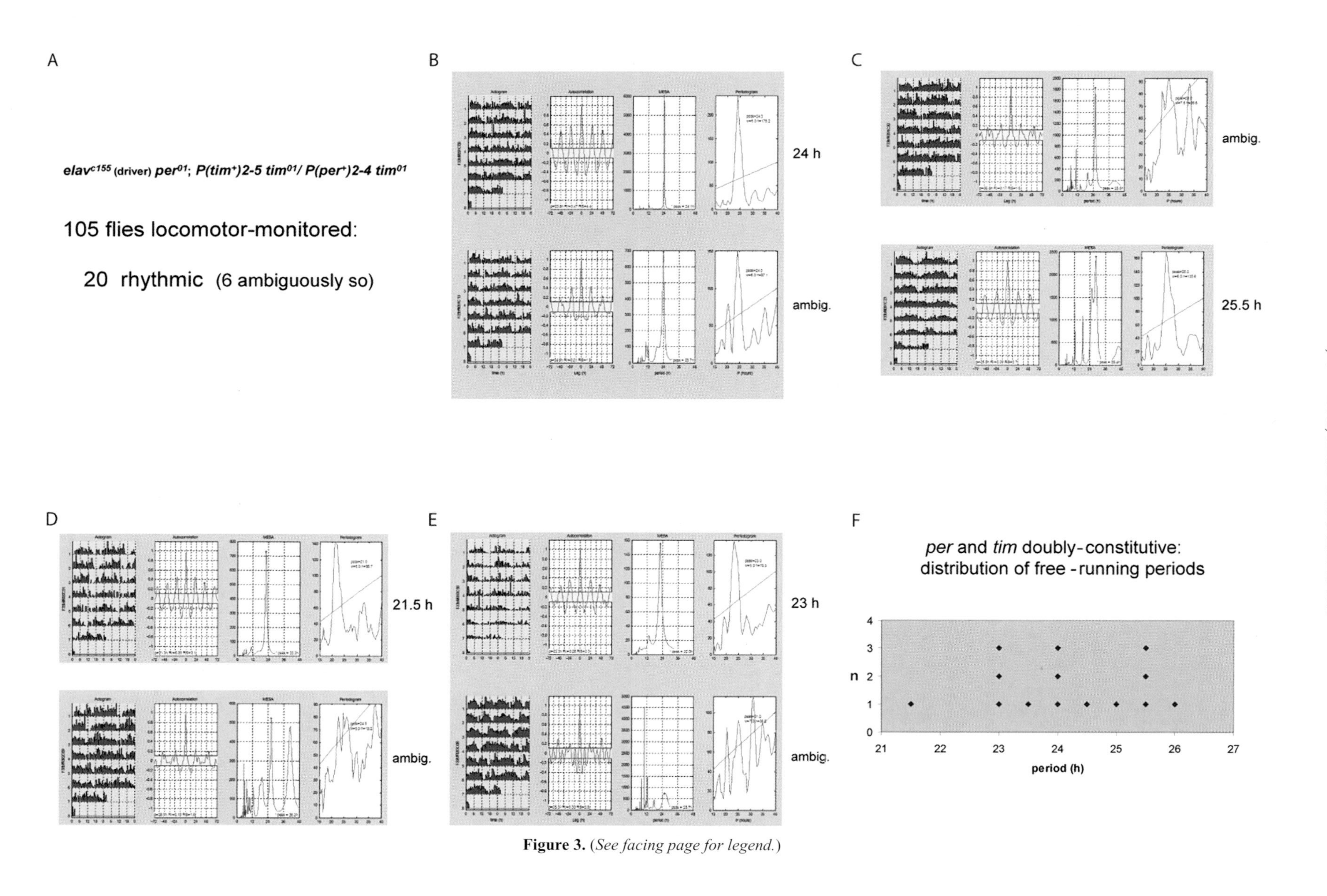

Figure 3. (*See facing page for legend.*)

REFERENCES

Akten B., Jauch E., Genova G.K., Kim E.Y., Edery I., Raabe T., and Jackson F.R. 2003. A role for CK2 in the *Drosophila* circadian oscillator. *Nat. Neurosci.* **6:** 251.

Alavizadeh A., Kiernan A.E., Nolan P., Lo C., Steel K.P., and Bucan M. 2001. The *Wheels* mutation in the mouse causes vascular, hindbrain, and inner ear defects. *Dev. Biol.* **234:** 244.

Antoch M.P., Song E.-J., Change A.-M., Vitaterna M.H., Zhao Y., Wilsbacher L.D., Sangoram A.M., King D.P., Pinto L.H., and Takahashi J.S. 1997. Functional identification of the mouse circadian *Clock* gene by transgenic BAC rescue. *Cell* **89:** 655.

Aronson B.D., Johnson K.A., and Dunlap J.C. 1994. Circadian clock locus frequency: Protein encoded by a single open reading frame defines period length and temperature compensation. *Proc. Natl. Acad. Sci.* **91:** 7683.

Bacon Y., Ooi A., Kerr S., Shaw-Andrews L., Winchester L., Breeds S., Tymoska-Lalanne Z., Clay J., Greenfield A.G., and Nolan P.M. 2004. Screening for novel ENU-induced rhythm, entrainment and activity mutants. *Genes Brain Behav.* **3:** 196.

Bargiello T.A. and Young M.W. 1984. Molecular genetics of a biological clock in *Drosophila. Proc. Natl. Acad. Sci.* **81:** 2142.

Barnes J.W., Tischkau S.A., Barnes J.A., Mitchell J.W., Burgoon P.W., Hickok J.R., and Gillette M.U. 2003. Requirement of mammalian Timeless for circadian rhythmicity (erratum in *Science* **302:** 1153). *Science* **302:** 439.

Baylies M.K., Bargiello T.A., and Young M.W. 1987. Changes in abundance and structure of the *per* gene product can alter periodicity of the *Drosophila* clock. *Nature* **326:** 390.

Benna C., Scannapieco P., Piccin A., Sandrelli F., Zordan M., Rosato E., Kyriacou C.P., Valle G., and Costa R. 2000. A second *timeless* gene in *Drosophila* shares greater sequence similarity with mammalian *tim. Curr. Biol.* **10:** R512.

Bi X. and Rong Y.S. 2003. Genome manipulation by homologous recombination in *Drosophila. Brief. Funct. Genomics Proteomics* **2:** 142.

Blau J. 2003. A new role for an old kinase: CK2 and the circadian clock. *Nat. Neurosci.* **6:** 208.

Borycz J., Borycz J.A., Loubani M., and Meinertzhagen I.A. 2002. *tan* and *ebony* genes regulate a novel pathway for transmitter metabolism at fly photoreceptor terminals. *J. Neurosci.* **22:** 10549.

Brandes C., Plautz J.D., Stanewsky R., Jamison C.F., Straume M., Wood K.V., Kay S.A., and Hall J.C. 1996. Novel features of Drosophila *period* transcription revealed by real-time luciferase reporting. *Neuron* **16:** 687.

Bushey D., Huber R., Tononi G., and Cirelli C. 2007. *Drosophila Hyperkinetic* mutants have reduced sleep and impaired memory. *J. Neurosci.* **27:** 5384.

Busza A., Emery-Le M., Rosbash M., and Emery P. 2004. Roles of the two *Drosophila* CRYPTOCHROME structural domains in circadian photoreception. *Science* **304:** 1503.

Chandrashekaran, M.K. and Loher W. 1969. The effect of light intensity on the circadian rhythm of eclosion in *Drosophila pseudoobscura. Z. Vgl. Physiol.* **62:** 337.

Cheng Y. and Hardin P.E. 1998. *Drosophila* photoreceptors contain an autonomous circadian oscillator that can function without *period* mRNA cycling. *J. Neurosci.* **18:** 741.

Cirelli C., Bushey D., Hill S., Huber R., Kreber R., Ganetzky B., and Tononi G. 2005. Reduced sleep in *Drosophila Shaker* mutants. *Nature* **434:** 1087.

Costa R., Peixoto A.A., Barbujani G., and Kyriacou C.P. 1992. A latitudinal cline in a *Drosophila* clock gene. *Proc. R. Soc. Lond. B Biol. Sci.* **250:** 43.

Coté G.G., and Brody S. 1986. Circadian rhythms in *Drosophila melanogaster:* Analysis of period as a function of gene dosage at the *per* (period) locus. *J. Theor. Biol.* **121:** 487.

Crenshaw E.B., III, Ryan A., Dillon S.R., Kalla K., and Rosenfeld M.G. 1991. *Wocko*, a neurological mutant generated in a transgenic mouse pedigree. *J. Neurosci.* **11:** 1524.

DeBruyne J.P, Weaver D.R., and Reppert S.M. 2007. CLOCK and NPAS2 have overlapping roles in the suprachiasmatic circadian clock. *Nat. Neurosci.* **10:** 543.

DeBruyne J.P., Noton E., Lambert C.M., Maywood E.S., Weaver D.R., and Reppert S.M. 2006. A clock shock: Mouse CLOCK is not required for circadian oscillator function. *Neuron* **50:** 465.

DiBartolomeis S.M., Akten B., Genova G., Roberts M.A., and Jackson F.R. 2002. Molecular analysis of the *Drosophila miniature-dusky* (*m-dy*) gene complex: *m-dy* mRNAs encode transmembrane proteins with similarity to *C. elegans* cuticulin. *Mol. Genet. Genomics* **267:** 564.

Dolezelova E., Dolezel D., and Hall J.C. 2007. Rhythm defects caused by newly engineered null mutations in Drosophila's *cryptochrome* gene. *Genetics* **177:** 329.

Dunlap J.C., Loros J.J., and Decoursey P.P. 2004. *Chronobiology: Biological timekeeping.* Sinauer Associates, Sunderland, Massachusetts.

Dushay M.S., Rosbash M., and Hall J.C. 1989. The *disconnected* visual system mutations in *Drosophila* drastically disrupt circadian rhythms. *J. Biol. Rhythms* **4:** 1.

———. 1992. Mapping the *Clock* mutation rhythm mutation to the *period* locus of *Drosophila melanogaster* by germline transformation. *J. Neurogenet.* **8:** 173.

Dushay M.S., Konopka R.J., Orr D., Greenacre M.L., Kyriacou C.P., Rosbash M., and Hall J.C. 1990. Phenotypic and genetic analysis of *Clock*, a new circadian rhythm mutant in *Drosophila melanogaster*. *Genetics* **125:** 557.

Emery P., Stanewsky R., Hall J.C., and Rosbash M. 2000a. *Drosophila* cryptochrome — A unique circadian-rhythm photoreceptor. *Nature* **404:** 456.

Emery P., Stanewsky R., Helfrich-Förster C., Emery-Le M., Hall J.C., and Rosbash M. 2000b. *Drosophila* CRY is a deep brain circadian photoreceptor. *Neuron* **26:** 493.

Ewer J., Frisch B., Hamblen-Coyle M.J., Rosbash M., and Hall

Figure 3. Behavior of *Drosophila* whose transcription of the *period* and *timeless* genes was engineered to be temporally constitutive. (*A*) Flies of the triply transgenic genotype indicated at the top were monitored for DD behavior (cf. Figs. 1 and 2). This transgenic type was created by Yang and Sehgal (2001) and supplied by A. Sehgal. The genetic background of that strain included a loss-of-function *period* mutation (*per*^{01}), but the *timeless* allele was genetically normal; thus, crosses were performed to generate a substrain in which a *tim*-null allele (*tim*^{01}) was homozygous on the second chromosome. This derivative strain also retained the "driver" transgene (containing a DNA sequence derived from the "pan-neuronally" expressed *elav* gene), along with the two "drivee" transgenes (containing, respectively, a *tim*$^{+}$-derived cDNA and a *per*$^{+}$-derived one). This multiply variant type of fly is designed to transcribe the TIM- and PER-protein-coding drivee transgenes in a manner that does not systematically vary across a given day (owing to *elav* being expressed in such a "constant" manner). However, immunoreactivity for TIM and PER do "cycle" to define about 24-hour periodicities (Yang and Sehgal 2001), owing to posttranslational effects on the levels of these polypeptides (see, e.g., Cheng and Hardin 1998). The proportion of rhythmic individuals was determined (after DD monitoring), according to the metrics used to create Figs. 1 and 2. (*B*–*E*) Examples of locomotor activity plots (as in Figs. 1 and 2), including cases involving a singly periodic component (as indicated by the circadian "h" values) and others that were ambiguous (ambig.), because more than one such component was analytically extracted (shown here via periodiograms, again cf. Figs. 1 and 2). With reference to *A*, 6 of the 20 "significantly rhythmic" individuals gave ambiguous outcomes. (*F*) Distribution of free-running cycle durations, resulting from analysis of the 14 behavioral records (cf. *A*) that yielded single-period values. The rather wide spread of these periods for the "*per*$^{+}$/*tim*$^{+}$-constitutive" transgenics (among the low proportion of such flies that behaved rhythmically) is distinctly different from normal circadian behavior: Nearly all wild-type flies locomotor-monitored in this manner are not only significantly periodic, but their free-running periods are nearly always in the range of 23.5 to 24.5 hours (see, e.g., Hall 2003, 2005).

J.C. 1992. Expression of the *period* clock gene within different cells types in the brain of *Drosophila* adults and mosaic analysis of these cells' influence on circadian behavioral rhythms. *J. Neurosci.* **12:** 3321.

Frank K.D. and Zimmerman W.F. 1969. Action spectra for phase shifts of a circadian rhythm in *Drosophila*. *Science* **163:** 688.

Frisch B., Hardin P.E., Hamblen-Coyle M.J., Rosbash M., and Hall J.C. 1994. A promoterless *period* gene mediates behavioral rhythmicity and cyclical per expression in a restricted subset of the *Drosophila* nervous system. *Neuron* **12:** 555.

Gailey D.A., Villella A., and Tully T. 1991. Reassessment of the effects of biological rhythm mutations on learning in *Drosophila melanogaster*. *J. Comp. Physiol. A* **169:** 685.

Gotter A.L. 2006. A Timeless debate: Resolving TIM's noncircadian roles with possible clock function. *Neuroreport* **17:** 1229.

Gotter A.L., Manganaro T., Weaver D.R., Kolakowski L.F., Jr., Possidente B., Sriram S., MacLaughlin D.T., and Reppert S.M. 2000. A time-less function for mouse *timeless*. *Nat. Neurosci.* **8:** 755.

Greenspan R.J., Tononi G., Cirelli C., and Shaw P.J. 2001. Sleep and the fruit fly. *Trends Neurosci.* **24:** 142.

Hall J.C. 1995. Tripping along the trail to the molecular mechanisms of biological clocks. *Trends Neurosci.* **18:** 230.

———. 2003. Genetics and molecular biology of rhythms in *Drosophila* and other insects. *Adv. Genet.* **48:** 1.

———. 2005. Systems approaches to biological rhythms in *Drosophila. Methods Enzymol.* **393:** 61.

Hamblen-Coyle M., Konopka R.J., Zwiebel L.J., Colot H.V., Dowse H.B., Rosbash M., and Hall J.C. 1989. A new mutation at the *period* locus of *Drosophila melanogaster* with some novel effects on circadian rhythms. *J. Neurogenet.* **5:** 229.

Hardin P.E., Hall J.C., and Rosbash M. 1992. Behavioral and molecular analyses suggest that circadian output is disrupted by *disconnected* mutants in *D. melanogaster*. *EMBO J.* **11:** 1.

Heilig J.S., Freeman M., Laverty T., Lee K.J., Campos A.R., Rubin G.M., and Steller H. 1991. Isolation and characterization of the *disconnected* gene of *Drosophila melanogaster*. *EMBO J.* **10:** 809.

Helfrich-Förster C. 1995. The period clock gene is expressed in CNS neurons which also produce a neuropeptide that reveals the projections of circadian pacemaker cells within the brain of *Drosophila melanogaster*. *Proc. Natl. Acad. Sci.* **92:** 612.

———. 1998. Robust circadian rhythmicity of *Drosophila melanogaster* requires the presence of lateral neurons: A brain-behavioral study of *disconnected* mutants. *J. Comp. Physiol. A* **182:** 435.

———. 2002. The circadian system of *Drosophila melanogaster* and its light input pathways. *Zoology* **105:** 297.

Hendricks J.C. 2003. Sleeping flies don't lie: The use of *Drosophila melanogaster* to study sleep and circadian rhythms. *J. Appl. Physiol.* **94:** 1660.

Jacob F. 1997. The operon after 25 years (translation). *C.R. Acad. Sci. III* **320:** 199.

Jauch E., Melzig J., Brkulj M., and Raabe T. 2002. In vivo functional analysis of *Drosophila* protein kinase casein kinase 2 (CK2) β-subunit. *Gene* **298:** 29.

Kadener S., Stoleru D., McDonald M., Nawathean P., and Rosbash M. 2007. *Clockwork orange* is a transcriptional repressor and a new *Drosophila* circadian pacemaker component. *Genes Dev.* **21:** 1675.

Kaneko M. and Hall J.C. 2000. Neuroanatomy of cells expressing clock genes in *Drosophila:* Transgenic manipulation of the *period* and *timeless* genes to mark the perikarya of circadian pacemaker neurons and their projections. *J. Comp. Neurol.* **422:** 66.

Kernan M.J., Kuroda M.I., Kreber R., Baker B.S., and Ganetzky B. 1991. *nap^{ts}*, a mutation affecting sodium channel activity in *Drosophila,* is an allele of *mle*, a regulator of X chromosome transcription. *Cell* **66:** 949

King D.P. and Takahashi J.S. 2000. Molecular genetics of circadian rhythms in mammals. *Ann. Rev. Neurosci.* **23:** 713.

King D.P., Vitaterna M.H., Chang A.-M., Dove W.F., Pinto L.H., Turek F.W., and Takahashi J.S. 1997a. The mouse *Clock* mutation behaves as an antimorph and maps within the W^{19H} deletion, distal of *Kit*. *Genetics* **146:** 1049.

King D.P., Zhao Y., Sangoram A.M., Wilsbacher L.D., Tanaka M., Antoch M.P., Steeves T.D.L., Vitaterna M.H., Kornhauser J.M., Lowrey P.L., Turek F.W., and Takahashi J.S. 1997b. Positional cloning of the mouse circadian *Clock* gene. *Cell* **89:** 641.

Klemm E. and Ninnemann H. 1976. Detailed action spectrum for the delay shift in pupae emergence of *Drosophila pseudoobscura*. *Photochem. Photobiol.* **24:** 369.

Kloss B., Price J.L., Saez L., Blau J., Rothenfluh A., Wesley C.S., and Young M.W. 1998. The *Drosophila* clock gene double-time encodes a protein closely related to human casein kinase Iepsilon. *Cell* **94:** 97.

Koh K., Zheing H., and Sehgal A. 2006. JETLAG resets the *Drosophila* circadian clock by promoting light-induced dedgradatin of TIMELESS. *Science* **312:** 1809.

Konopka R.J. and Benzer S. 1971. Clock mutants of *Drosophila melanogaster*. *Proc. Natl. Acad. Sci.* **68:** 2112.

Konopka R.J., Smith R.F., and Orr D. 1991. Characterization of *Andante,* a new *Drosophila* clock mutant, and its interactions with other clock mutants. *J. Neurogenet.* **7:** 103.

Krishnan B., Levine J.D., Lynch K.S., Dowse H.B., Funes P., Hall J.C., Hardin P.E., and Dryer S.E. 2001. A novel role for cryptochrome in a *Drosophila* circadian oscillator. *Nature* **411:** 313.

Lakin-Thomas P.L. 2006. Transcriptonal feedback oscillators: Maybe, maybe not... *J. Biol. Rhythms* **21:** 83.

Lawrence P.A. 1992. *The making of a fly*. Blackwell Scientific, Oxford, United Kingdom.

Levine J.D., Funes P., Dowse H.B., and Hall J.C. 2002. Signal analysis of behavioral and molecular cycles. *BMC Neurosci.* **3:** 1.

Lin J.M., Kilman V.L, Keegan K., Paddock B., Emery-Le M., Rosbash M., and Allada R. 2002. A role for casein kinase 2alpha in the *Drosophila* circadian clock. *Nature* **420:** 816.

Liu X., Lorenz L., Yu Q., Hall J.C., and Rosbash M. 1988. Spatial and temporal expression of the *period* gene in *Drosophila melanogaster*. *Genes Dev.* **2:** 228.

Low-Zeddies S.S. and Takahashi J.S. 2001. Circadian analysis of the *Clock* mutation in mice shows that complex cellular intergration determines circadian behavior. *Cell* **105:** 25.

Lowrey P.L. and Takahashi J.S. 2000. Genetics of the mammalian circadian system: Photic entrainment, circadian pacemaker mechanisms, and posttranslational regulation. *Annu. Rev. Genet.* **34:** 533.

———. 2004. Mammalian circadian biology: Elucidating genome-wide levels of temporal organization. *Annu. Rev. Genomics Hum. Genet.* **5:** 407.

Martinek S., Inonog S., Manoukian A.S., and Young M.W. 2001. A role for the segment polarity gene *shaggy*/GSK-3 in the *Drosophila* circadian clock. *Cell* **105:** 769.

Matsumoto A., Motoshige T., Murata T., Tomioka K., Tanimura T., and Chiba Y. 1994. Chronobiological analysis of a new clock mutant, *Toki*, in *Drosophila melanogaster*. *J. Neurogenet.* **9:** 141.

Mealey-Ferrara M.L., Montalvo A.G., and Hall J.C. 2003. Effects of combining a cryptochrome mutation with other visual-system variants on entrainment of locomotor and adult-emergence rhythms in *Drosophila*. *J. Neurogenet.* **17:** 171.

Myers E.M., Yu J., and Sehgal A. 2003. Circadian control of eclosion: Interaction between a central and peripheral clock in *Drosophila melanogaster*. *Curr. Biol.* **13:** 526.

Naylor E., Bergmann B.M., Krauski K., Zee P.C., Takahashi J.S., Vitaterna M.H., and Turek F.W. 2000. The circadian clock mutation alters sleep homeostasis in the mouse. *J. Neurosci.* **20:** 8138.

Newby L.M. and Jackson F.R. 1991. *Drosophila ebony* mutants have altered circadian activity rhythms but normal eclosion rhythms. *J. Neurogenet.* **7:** 85.

Newby L.M., White L., DiBartolomeis S.M., Walker B.J., Dowse H.B., Ringo J.M., Khuda N., and Jackson F.R. 1991.

Mutational analysis of the *Drosophila miniature-dusky* (*m-dy*) locus: Effects on cell size and circadian rhythms. *Genetics* **128:** 571.

Ohata K., Nishiyama H., and Tsukahara Y. 1998. Action spectrum of the circadian clock photoreceptor in *Drosophila melanogaster*. In *Biological clocks. Mechanisms and applications* (ed. Y. Touitou), p. 167. Elsevier, Amsterdam, The Netherlands.

Oishi K., Ohkura N., Wakabayashi M., Shirai H., Sato K., Matsuda J., Atsumi G., and Ishida N. 2006. CLOCK is involved in obesity-induced disordered fibrinolysis in *ob/ob* mice by regulating *PAI-1* gene expression. *J. Thromb. Haemost.* **4:** 1774.

Peifer M., Pai L.-M., and Casey M. 1994. Phosphorylation of the *Drosphila* adherens junction protein ARMADILLO: Roles for WINGLESS and ZESTE-WHITE 3 kinase. *Dev. Biol.* **166:** 543.

Peschel N., Veleri S., and Stanewsky R. 2006. *Veela* defines a molecular link between Cryptochrome and Timeless in the light-input pathway to *Drosophila*'s circadian clock. *Proc. Nat. Acad. Sci.* **103:** 17313.

Pickard G.E., Sollars P.J., Rincjick E.M., Nolan P.M., and Bucan M. 1995. Mutagenesis and behavioral screening for altered circadian activity identifies the mouse mutant, *Wheels*. *Brain Res.* **705:** 255.

Price J.L. 2005. Genetic screens for clock mutants in *Drosophila*. *Methods Enzymol.* **393:** 35.

Price J.L., Blau J., Rothenfluh A., Abodeely M., Kloss B., and Young M.W. 1998. *double-time* is a new *Drosophila* clock gene that regulates PERIOD protein accumulation. *Cell* **94:** 83.

Prosser R.A., Edgar D.M., Heller H.C., and Miller J.D. 1994. A possible glial role in the mammalian circadian clock. *Brain Res.* **643:** 296.

Richardt A., Rybak J., Stortkuhl K.F., Meinertzhagen I.A., and Hovemann B.T. 2002. Ebony protein in the *Drosophila* nervous system: Optic neuropile expression in glial cells. *J. Comp. Neurol.* **452:** 93.

Roch F., Alonso C.R., and Akam M. 2003. *Drosophila miniature* and *dusky* encode ZP proteins required for cytoskeletal reorganisation during wing morphogenesis *J. Cell Sci.* **116:** 1199.

Rosato E., Trevisan A., Sandrelli F., Zordan M., Kyriacou C.P., and Costa R. 1997. Conceptual translation of *timeless* reveals alternative initiating methionines in *Drosophila*. *Nucleic Acids Res.* **25:** 455.

Rothenfluh A., Abodeely M., Price J.L., and Young M.W. 2000. Isolation and analysis of six *timeless* alleles that cause short- or long-period circadian rhythms in *Drosophila*. *Genetics* **156:** 665.

Saez L. and Young M.W. 1988. In situ localization of the per clock protein during development of *Drosophila melanogaster*. *Mol. Cell. Biol.* **8:** 5378.

Sancar A. 2000. Cryptochrome: The second photoactive pigment in the eye and its role in circadian photoreception. *Annu. Rev. Biochem.* **69:** 31.

Sandrelli F., Tauber E., Pegoraro M., Mazzotta G., Cisotto P., Landskron J., Stanewsky R., Piccin P., Rosato E., Zordan M., Costa R., and Kyriacou C.P. 2007. A molecular basis for natural selection at the *timeless* locus in *Drosophila melanogaster*. *Science* **316:** 1898.

Sawyer L.A., Sandrelli F., Pasetto C., Peixoto A.A., Rosato E., Costa R., and Kyriacou C.P. 2006. The *period* gene Thr-Gly polymorphism in Australia and African *Drosophila melanogaster* population: Implications for selection. *Genetics* **174:** 465.

Scheuch G.C., Johnson W., Connor R.L., and Silver J. 1982. Investigation of circadian rhythms in a genetically anophthalmic mouse strain: Correlation of activity patterns with suprachiasmatic nuclei hypogenesis. *J. Comp. Physiol. A* **149:** 333.

Sehgal A., Price J.L., Man B., and Young M.W. 1994. Loss of circadian behavioral rhythms and *per* RNA oscillations in the *Drosophila* mutant *timeless*. *Science* **263:** 1603.

Shannon M.P., Kaufman T.C., Shen M.W., and Judd B.H. 1972. Lethality patterns and morphology of selected lethal and semi-lethal mutations in the *zeste-white* region of *Drosophila melanogaster*. *Genetics* **72:** 615.

Shimomura K., Low-Zeddies S.S., King D.P., Steeves T.D., Whiteley A., Kushla J., Zemenides P.D., Lin A., Vitaterna M.H., Churchill G.A., and Takahashi J.S. 2001. Genome-wide epistatic interaction analysis reveals complex genetic determinants of circadian behavior in mice. *Genome Res.* **11:** 959.

Siepka, S.M., Yoo S.-H., Park J., Song W., Kumar V., Hu Y., Lee C., and Takahashi J.S. 2007. Circadian mutant *overtime* Reveals F-box protein FBXL3 regulation of *Cryptochrome* and *Period* gene expression. *Cell* **129:** 1011.

Simpson P. and Carteret C. 1989. A study of *shaggy* reveals spatial domains of expression of *achaete-scute* alleles on the thorax of *Drosophila*. *Development* **106:** 57.

Siwicki K.K., Eastman C., Petersen G., Rosbash M., and Hall J.C. 1988. Antibodies to the *period* gene product of *Drosophila* reveal diverse tissue distribution and rhythm changes in the visual system. *Neuron* **1:** 141.

Smith R.F. and Konopka R.J. 1981. Circadian clock phenotypes of chromosome aberrations with a breakpoint at the *per* locus. *Mol. Gen. Genet.* **183:** 243.

———. 1982. Effects of dosage alterations at the *per* locus on the circadian clock of *Drosophila*. *Mol. Gen. Genet.* **185:** 30.

So W.V. and Rosbash M. 1997. Post-transcriptional regulation contributes to *Drosophila* clock gene mRNA cycling. *EMBO J.* **16:** 7146.

Sollars P.J., Ryan A., Oglivie M.D., and Pickard G.E. 1996. Altered circadian rhythmicity in the *Wocko* mouse, a hyperactive transgenic mutant. *Neuroreport* **7:** 1245.

Stanewsky R. 2003. Genetic analysis of the circadian system in *Drosophila melanogaster* and mammals. *J. Neurobiol.* **54:** 111.

Stanewsky R., Jamison C.F., Plautz J.D., Kay S.A., and Hall J.C. 1997. Multiple circadian-regulated elements contribute to cycling *period* gene expression in *Drosophila*. *EMBO J.* **16:** 5006.

Stanewsky R., Kaneko M., Emery P., Beretta B., Wager-Smith K., Kay S.A., Rosbash M., and Hall J.C. 1998. The cry^b mutation identifies cryptochrome as a circadian photoreceptor in *Drosophila*. *Cell* **95:** 681.

Steller H., Fischbach K.-F., and Rubin G.M. 1987. *disconnected:* A locus required for neuronal pathway formation in the visual system of *Drosophila*. *Cell* **50:** 1139.

Tauber E., Zordan M., Sandrelli F., Pergoraro M., Osterwalder N., Breda C., Daga A., Selmin A., Monger K., Benna C., Rosato E., Kyriacou C.P., and Costa R. 2007. Natural selection favors a newly derived *timeless* allele in *Drosophila melanogaster*. *Science* **316:** 1895.

Turek F.W., Joshu C., Kohsaka A., Lin E., Ivanova G., McDearmon E., Laposky A., Losee-Olson S., Easton A., Jensen D.R., Eckel R.H., Takahashi J.S., and Bass J. 2005. Obesity and metabolic syndrome in circadian *Clock* mutant mice. *Science* **308:** 1043.

Veleri S., Rieger D., Helfrich-Förster C., and Stanewsky R. 2007. Hofbaur-Buchner eyelet affects circadian photosensitivity and coordinates PER and TIM expression in *Drosophila* clock neurons. *J. Biol. Rhythms* **22:** 29.

Venken K.J.T. and Bellen H.J. 2005. Emerging technologies for gene manipulation in *Drosophila melanogaster* (erratum in *Nat. Rev. Genet.* **6:** 340). *Nat. Rev. Genet.* **6:** 167.

Vitaterna M.H., King D.P., Chang A.-M., Kornhauser J.M., Lowrey P.L., McDonald J.D., Dove W.F., Pinto L.H., Turek F.W., and Takahashi J.S. 1994. Mutagenesis and mapping of a mouse gene essential for circadian behavior. *Science* **264:** 719.

Vosshall L.B. and Young M.W. 1995. Circadian rhythms in *Drosophila* can be driven by *period* expression in a restricted group of central brain cells. *Neuron* **15:** 345.

Weaver D.R. 1998. The suprachiasmatic nucleus: A 25-year retrospective. *J. Biol. Rhythms* **13:** 100.

Weiner J. 1999. *Time, love, memory: A great biologist and his quest for the origins of behavior*. Alfred A. Knopf, New York.

Wu C.F., Ganetzky B., Jan L.Y., and Jan Y.N. 1978. A *Drosophila* mutant with a temperature-sensitive block in nerve conduction. *Proc. Natl. Acad. Sci.* **75:** 4047.

Wulbeck C., Szabo G., Shafer O.T., Helfrich-Förster C., and Stanewsky R. 2005. The novel *Drosophila* tim^{blind} mutation affects behavioral rhythms but not periodic eclosion. *Genetics* **169:** 751.

Yang Z. and Sehgal A. 2001. Role of molecular oscillations in generating behavioral rhythms in *Drosophila*. *Neuron* **29:** 453.

Yu Q., Jacquier A.C., Citri Y., Hamblen M., Hall J.C., and Rosbash M. 1987. Molecular mapping of point mutations in the *period* gene that stop or speed up biological clocks in *Drosophila melanogaster*. *Proc. Natl. Acad. Sci.* **84:** 784.

Zerr D.M., Hall J.C., Rosbash M., and Siwicki K.K. 1990. Circadian fluctuations of *period* protein immunoreactivity in the CNS and the visual system of *Drosophila*. *J. Neurosci.* **10:** 2749.

Zilian O., Frei E., Burke R., Brentrup D., Gutjahr T., Bryant P.J., and Noll M. 1999. *double-time* is identical to *discs overgrown*, which is required for cell survival, proliferation and growth arrest in *Drosophila* imaginal discs. *Development* **126:** 5409.

Zimmerman W.F. and Goldsmith T.H. 1971. Photosensitivity of the circadian rhythm and of visual receptors in carotenoid-depleted *Drosophila*. *Science* **171:** 1167.

Synchronization of the *Drosophila* Circadian Clock by Temperature Cycles

F.T. GLASER* AND R. STANEWSKY*†

*Institute of Zoology, University of Regensburg, 93040 Regensburg, Germany; †School of Biological and Chemical Sciences, Queen Mary College, University of London, London, E1 4NS, United Kingdom

The natural light/dark and temperature cycles are considered to be the most prominent factors that synchronize circadian clocks with the environment. Understanding the principles of temperature entrainment significantly lags behind our current knowledge of light entrainment in any organism subject to circadian research. Nevertheless, several effects of temperature on circadian clocks are well understood, and similarities as well as differences to the light-entrainment pathways start to emerge. This chapter provides an overview of the temperature effects on the *Drosophila* circadian clock with special emphasis on synchronization by *temperature cycles*. As in other organisms, such temperature cycles can serve as powerful time cues to synchronize the clock. Mutants that specifically interfere with aspects of temperature entrainment have been isolated and will likely help to reveal the underlying mechanisms. These mechanisms involve transcriptional and posttranscriptional regulation of clock genes. For synchronization of fly behavior by temperature cycles, the generation of a whole organism or systemic signal seems to be required, even though individual fly tissues can be synchronized under isolated culture conditions. If true, the requirement for such a signal would reveal a fundamental difference to the light-entrainment mechanism.

INTRODUCTION

Most organisms can synchronize their circadian clocks to light/dark or temperature cycles if the respective other zeitgeber is held constant. Given that in nature both light and temperature fluctuate in a daily manner, it is more than likely that under natural conditions, both cues—and likely many others (see, e.g., Levine et al. 2002b)—are used by the organisms to time their clocks. In most mammals, blood temperature changes in a daily fashion by about 1.5°C, and this low amplitude rhythm likely serves as a zeitgeber for peripheral clocks (Brown et al. 2002; Kornmann et al. 2007). Therefore, to fully understand "entrainment," it is necessary to study how these two major "physical" rhythms synchronize circadian clocks separately and in combination. So far, the main focus has been on light entrainment in both vertebrates and invertebrates (see, e.g., Hall 2003; Foster et al. 2007). An astonishing complexity of light-input pathways and of molecules assembling them has been revealed in both systems, consisting of classical visual photoreceptors as well as specialized circadian photopigments and light-responsive cells. The final degree of complexity is very likely still ahead of us, given that studies have so far been mainly conducted under strict 12-hour light:12-hour dark conditions, ignoring the changes of light quality and quantity during twilight. This is somewhat ironic, because light conditions during twilight represent a particularly potent zeitgeber (Roenneberg and Foster 1997). It therefore seems time to switch to more natural entrainment conditions in the lab (Bachleitner et al. 2007) or even for conducting studies "in the wild."

The latter would naturally also include temperature fluctuations. Isolated analysis of temperature entrainment in the lab has the small advantage of almost never being abrupt, because for technical reasons, incubators or experimental rooms need some time to adapt a new temperature set point. Although they most likely do not accurately reflect the heating or cooling times in nature (which can also vary dramatically), it seems nevertheless to be a "more natural" scenario compared with applying rectangular light/dark cycles. The downsides, and clearly a complicating issue, are the many effects temperature changes exert on organisms in general and on the circadian clock in particular. First, there is the heat shock response, which leads to major changes in the transcriptional and translational profile (Morimoto 1998, 2002). In flies, brief (30 minute) 37°C heat shocks stably phase-shift locomotor rhythmicity (Edery et al. 1994; Sidote et al. 1998; Kaushik et al. 2007), and it has been recently shown that the blue light photoreceptor Cryptochrome (Cry) is necessary for this response (Kaushik et al. 2007). Surprisingly, the same "resetting complex" of clock proteins is formed after both light and temperature pulses, but it is not likely that the same complex also forms during temperature entrainment, which operates normally in the absence of Cry (Stanewsky et al. 1998; Glaser and Stanewsky 2005; Kaushik et al. 2007). Moreover, it has been shown that prolonged (6–12 hours) heat pulses of 29°C can phase-shift the clock in wild-type and *cry*^{b} flies, further questioning whether Cry has a role in temperature entrainment under physiological conditions (Busza et al. 2007).

Second, circadian clocks are temperature-compensated, meaning that the free-running period of an organism is more or less constant at different ambient temperatures as long as they are within the physiological range of the organism (e.g., no heat shock). Although several interesting models have been developed, and clock genes have been shown "to be important" for temperature-compensation, the mechanisms are not understood on a molecular level (Huang et al. 1995; Leloup and Goldbeter 1997; Sawyer et al. 1997; Ruoff et al. 2005; Hunt et al.

2007). Cry has also been implicated in temperature compensation, because a mutant form of Cry, encoded by the *cry^b* mutation, largely restores the deficit of compensation associated with a mutation in the clock gene *period* (*per^L*) (Konopka et al. 1989; Kaushik et al. 2007). It is not clear whether this "rescue" is only a consequence of altered binding properties between the clock protein Timeless (Tim) and the mutated and perhaps structurally altered Cry^B and Per^L proteins. The recent observation that Cry loss-of-function mutations are perfectly temperature-compensated makes it rather unlikely that Cry is also relevant for temperature-compensation in wild-type flies (Dolezelova et al. 2007).

Third, in *Drosophila*, constant low or elevated temperatures (within the physiological range) result in a different behavioral activity pattern during light/dark (LD) cycles. The typical dusk activity occurs earlier at cold temperatures and later at warm temperatures. This effect is enhanced by short and long photoperiods, respectively (Majercak et al. 1999), which suggested that the combined response to temperature and photoperiod allows the fly to adjust its behavior to the different seasons of the year. More recent work has revealed that this adaptation is mediated via a splicing event in the last intron of the *per* gene, which ultimately results in an earlier accumulation of Per protein in short and cold days versus long and warm days (Collins et al. 2004; Majercak et al. 2004; also see below).

All of the temperature effects just described somehow alter or involve clock gene products. In addition, and as in the case of light, temperature changes also elicit direct or driven responses, which can influence (mask) behaviors that are normally clock-controlled. For example, in temperature cycles, fly locomotor activity shows rapid increases immediately after the temperature transitions even in flies that do not possess a functional clock (Fig. 1b) (see, e.g., Wheeler et al. 1993; Glaser and Stanewsky 2005), and transcription of many *Drosophila* genes can be driven by temperature cycles (Boothroyd et al. 2007; see below).

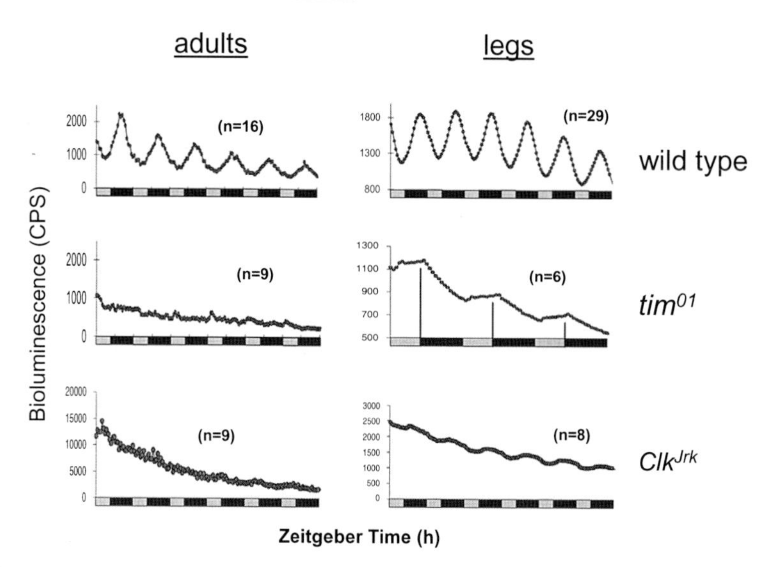

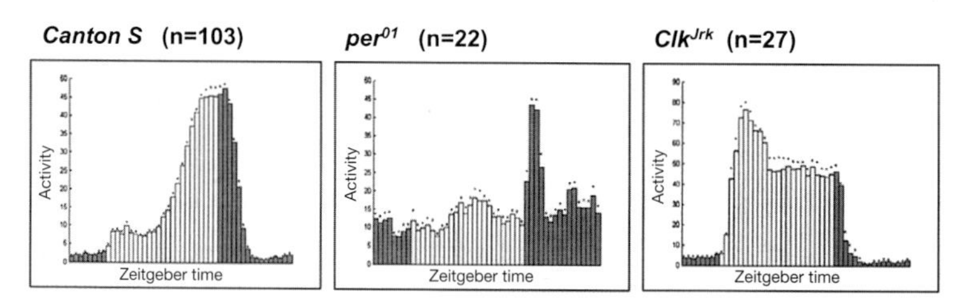

Figure 1. Temperature entrainment in wild-type and clock mutant flies in LL. (*a*) Bioluminescence rhythms recorded from whole flies or legs carrying *per-luc* transgenics encoding two thirds (*BG-luc*: wild type and *Clk^Jrk*) or all (*XLG-luc*: *tim^01*) of Per fused to luciferase. (*Gray bars*) Warm phase (25°C); (*black bars*) cold phase (18°C). (*b*) Daily locomotor activity averaged for 5–7 days. (*White bars*) Warm phase (25°C); (*gray bars*) cold phase (18°C). Each bar shows behavioral activity within a 30-minute interval. Dots above columns indicate S.E.M.

In light of the above-mentioned temperature effects on the circadian clock, especially the ability of the clock to actively compensate for different temperatures, it seems remarkable that daily temperature fluctuations can serve as a robust zeitgeber in many organisms (Dunlap et al. 2004). Even more astonishing, the amplitude of the temperature cycle necessary to elicit entrainment can be tiny, often only 1–3°C (see, e.g., Wheeler et al. 1993; Lahiri et al. 2005). In mammals, temperature cycles with a 1.5°C amplitude are able to induce rhythmic *per1* and *per2* expression in suprachiasmatic nucleus (SCN) glia cells (Prolo et al. 2005), and the daily body temperature cycles seem to be capable of synchronizing circadian transcripts in mouse liver and fibroblasts (Brown et al. 2002; Kornmann et al. 2007). We discuss here our current and very limited knowledge of temperature entrainment in *Drosophila* and where possible put this into context with other known effects of temperature on the circadian clock.

DISCUSSION

Location of Thermal Receptors/Tissue-autonomous Temperature Reception

It is known that thermal receptors for temperature preference distribution are located in the outer antennal segment of adult flies (Sayeed and Benzer 1996; Zars 2001). Flies lacking this part of the antenna are no longer able to identify their preferred temperature (25°C) and will randomly distribute in a temperature gradient. As such, the antennae were an initial candidate to also serve as a temperature reader for the circadian clock. We therefore analyzed temperature synchronization in flies, in which the outer segment of the antenna had been removed surgically or had been transformed into leg structures by the homeotic mutation *spineless^{aristapedia}* (Glaser and Stanewsky 2005). For this, we introduced a *period-luciferase* (*per-luc*) transgene into the genetic background of the antenna-deficient flies and determined if it would be possible to synchronize *per-luc* oscillations with temperature cycles. This was indeed possible, demonstrating that the receptors responsible for adult temperature preference behavior are not necessary for circadian synchronization. In addition, the temperature preference mutant *bizarre* was analyzed, whose antennae appear normal (Sayeed and Benzer 1996; Zars 2001) but which shows morphological brain abnormalities (D. Kretzschmar, pers. comm.). Again, *bizarre* flies showed normal synchronization of clock gene cycling to temperature cycles, indicating that brain structures important for propagating the temperature information from the antenna are not crucial for circadian temperature entrainment (Glaser and Stanewsky 2005).

Although these results imply that the antennal temperature receptors are not involved in clock synchronization, they do not rule out their involvement. For example, it can easily be imagined that the situation is similar to that of light entrainment, where both the image-forming photoreceptors in the compound eye *and* the dedicated circadian photoreceptor molecule Cry contribute (see, e.g., Stanewsky et al. 1998; Helfrich-Förster et al. 2001; Veleri et al. 2007). Clock synchronization by LD cycles is almost normal in flies that lack compound eyes (but contain functional Cry), or in flies either mutant or lacking the *cry* gene but having normal eyes (Stanewsky et al. 1998; Dolezelova et al. 2007). Similarly, for temperature entrainment, multiple receptor systems could synchronize the clock in parallel.

Temperature Reception Is Tissue Autonomous in Flies

To localize circadian temperature receptors in the fly, we analyzed the ability of various isolated tissues of *per-luc* flies to synchronize to temperature cycles. Surprisingly, all external tissues isolated so far showed robust *per-luc* oscillations neatly entrained to the cyclic temperature zeitgeber (Glaser and Stanewsky 2005). Therefore, the situation seems to be similar to that for light entrainment, where it has been shown that isolated tissues can be synchronized by LD cycles (see, e.g., Emery et al. 1997; Plautz et al. 1997; Ivanchenko et al. 2001; Levine et al. 2002a). Moreover, in zebra fish, it has been demonstrated that temperature entrainment occurs on a cell-autonomous level, suggesting that this may also be the case in flies (Lahiri et al. 2005).

In the case of light, Cry has been implicated as being the crucial factor for synchronization of the peripheral tissues, but it is clear that other photoreceptors also contribute, because isolated tissues synchronize to LD cycles in the absence of functional Cry (Ivanchenko et al. 2001; Levine et al. 2002a; Dolezelova et al. 2007). Given that Cry does have a role in synchronization of the clock, we also tested its potential involvement in temperature synchronization. Per and Tim proteins isolated from heads (and therefore mainly reflecting protein levels in the photoreceptor cells of the compound eye) of *cry^b* flies kept in temperature cycles and in constant dark (DD) show an approximately 50% reduction in abundance compared to those isolated from wild-type flies (Stanewsky et al. 1998). Nevertheless, both Per and Tim exhibit robust temperature-entrained oscillations, which were never observed under LD conditions in *cry^b* flies (Stanewsky et al. 1998). Similarly, *per-luc* oscillations can be synchronized in the face of *cry^b* in temperature cycles but not in LD cycles (Stanewsky et al. 1998; Glaser and Stanewsky 2005). These results clearly demonstrate that Cry is not crucial for temperature entrainment and confirm its role as a dedicated circadian photoreceptor. This is an important point to stress, because Cry has recently been implicated to function in temperature-compensation and in the response to heat pulses of 37°C (Kaushik et al. 2007; see below).

Clock Mutants and Temperature Entrainment

In *Neurospora crassa*, it has been a matter of controversial findings and discussions whether or not loss-of-function mutants of the central clock gene *frequency* (*frq*) can entrain to temperature cycles (Merrow et al. 1999; Pregueiro et al. 2005; Roenneberg et al. 2005). Another way to address this question is to create a similar situation in a different organism, where it might be easier to obtain a clear answer. In the case of *Drosophila*, this problem was tackled by K. Tomioka and his group in Japan, who studied temperature entrainment in mutants of the clock genes *per, tim, Clock* (*Clk*), and *cycle* (*cyc*). Overall, their experiments indicated a clear requirement of a functional circadian clock for proper temperature entrainment (Matsumoto et al. 1998; Yoshii et al. 2002). In one particular study (Yoshii et al. 2002), they exposed wild-type flies and the rhythm mutants per^{01}, tim^{01}, Clk^{Jrk}, and cyc^{01} (Hall 2003) to 25°C:30°C temperature cycles with varying periods (T) under constant light (LL) and DD conditions (T = 8 to T = 32 hours) with an equal length of warm and cold periods. Using this protocol, they demonstrated that the phase of the major behavioral activity peak in wild-type flies is dependent on the period of the temperature cycle, indicating proper entrainment as opposed to a simple reaction to the temperature changes. This dependency was only observable in LL; in DD, wild-type flies synchronized to T = 24 but not to any of the other T cycles (Yoshii et al. 2002). The findings suggest "better clock function" in LL compared to DD during temperature entrainment, which is quite surprising given that LL causes arrhythmicity at constant temperatures (Konopka et al. 1989; also see below).

Interestingly, in both per^{01} and tim^{01} flies, the phase of the activity peak depended on the T cycle in DD, suggesting that some clock functions remain intact in these mutants. This was in contrast to the Clk^{Jrk} and cyc^{01} mutants, which always showed peak activity about 2.5 hours after the onset of the warm phase (response to cold-to-warm transition) (Yoshii et al. 2002). A similar variation between the effects of mutations in the negative limb of the molecular feedback loop versus the positive limb was observed after exposing wild-type and mutant flies to temperature steps of 10°C during LL: A single temperature step from 20°C to 30°C induced a single behavioral activity bout approximately 9 hours after the step in wild-type flies (Yoshii et al. 2007). This activity bout was also observed in per^{01} and weakly in tim^{01} flies, but not in the Clk^{Jrk} and cyc^{01} mutants. In wild type, a temperature step from 30°C to 20°C induced a behavioral *rhythm*, with recurring activity bouts approximately every 21.5 hours. This rhythm could not be observed in any of the four clock mutants under study, indicating that it requires a functional clock. This was further demonstrated by shortening the period of this rhythm in per^{S} and its lengthening in per^{L} mutants, whereas the phase of the single activity peak observed after the 20°C → 30°C step up was not affected by the period-altering mutants in a meaningful manner.

Similarly, on a molecular level, these temperature steps induced alterations and oscillations in the clock gene mRNAs of the *per, tim, Clk, vrille,* and *Pdp1* genes in wild-type flies. These changes were abolished by the Clk^{Jrk} mutants (which cannot be synchronized by temperature cycles; see above), except that *per* RNA decreased or increased after the 20°C → 30°C and the 30°C → 20°C step, respectively, but stayed constant thereafter. Interestingly, in per^{01} flies, which show some behavioral adjustments to temperature cycles and steps (see above), the mRNAs of the five clock genes analyzed responded with either an increase or decrease (or both) after a temperature step down or step up, respectively. But in no case was a sustained oscillation of any clock gene mRNA maintained in a per^{01} mutant background, consistent with the idea that any remaining timing mechanism operating independent of *per* is of an hourglass type and not self-sustained (Yoshii et al. 2007).

In this regard, it is interesting to note that we also observed low-amplitude oscillations of *per-luc* oscillations in several clock mutant flies and their isolated tissues (Fig. 1a) (Glaser and Stanewsky 2005), but we were not able to demonstrate any behavioral synchronization in any of the mutants effecting the negative or positive limb of the clock (Fig. 1b) (Glaser and Stanewsky 2005). Moreover, in a microarray study of *Drosophila* genes expressed rhythmically in temperature cycles, Boothroyd et al. (2007) discovered that in tim^{01} flies, some of the crucial clock or clock-input genes (*per, tim, Clk,* and *cry*) are rhythmically expressed during a temperature cycle but not after release into constant conditions.

Considering all of the available data, it seems that flies are not able to synchronize properly to temperature cycles in the absence of any of the canonical clock gene products. It also seems clear that mutation of the negative limb of the fly clock (*per* and *tim*) allows for some features of synchronization to persist, but this was never comparable to the wild-type situation.

Mutants for Circadian Temperature Reception (*nocte, norpA*)

nocte. Genetic screens were very successful in both identifying genes constituting the core circadian clock and defining light-input factors (Hall 2003). It was therefore predictable that similar screens aimed at identifying components of the temperature-entrainment pathway would also succeed. Such screens and the resulting mutations are very desirable, given that it is not clear at present in any organism how temperature entrainment of the circadian clock is accomplished or which classes of genes and proteins may be involved (except for the possible involvement of heat shock genes in mammals) (Kornmann et al. 2007).

Indeed, by performing a random chemical mutagenesis screen for variants that abolish *per-luc* cycling induced by temperature cycles, we were able to isolate the temperature-entrainment mutant *nocte* (*no circadian temperature entrainment*) (Glaser and Stanewsky 2005). *nocte* mutant flies exhibit only weak (if any) *per-luc* oscillations in temperature cycles (both under LL and DD conditions) but robust and entrained oscillations in LD cycles. The mutant strain also exhibits normal behavioral rhythms in LD and DD and normal temperature-compensation, but it shows defects in synchronization to temperature cycles. The *nocte* mutant therefore represents a mirror image of the cry^{b} mutation, which shows defects in light-entrainment cycles but not in temperature cycles (Stanewsky et al. 1998).

Meiotic mapping experiments placed the *nocte* gene in the X-chromosomal interval 9A2-9D3. We eventually succeeded in identifying the mutated gene, which will be described elsewhere (F. Glaser et al., unpubl.). In the context of the current review, it seems sufficient to note that the sequence of the gene and the deduced protein was quite a surprise. As noted above, we had expected to get at least a hint of the factors and perhaps mechanisms involved in temperature entrainment, but this was not the case. The deduced protein encoded by the *nocte* gene is very large (~2300 amino acids) and contains no obvious homologies with any other proteins or domains! Nevertheless, we hope that by analyzing the *nocte* expression pattern and by spatially controlled manipulation of its function in certain parts of the fly, we will be able to gain insights into possible mechanisms involved in temperature entrainment. In a way, the situation is similar to that after cloning of *period*. There too, the sequence information did not lead to immediate insights or ideas about potential clock mechanisms. Instead, it sparked an army of researchers to study its function, eventually revealing its secrets, its partners, and the basic mechanism of the circadian clock (and not just that of flies). So, in a lot of ways, it was luck that *period* was the first circadian locus to be isolated after random mutagenesis. Or as the great Jeffrey Hall once asked: "*Would we all be here if the first fly rhythm mutant had turned out to be a kinase?*"

norpA. The only other gene known to have a role in temperature entrainment is *norpA* (*no receptor potential A*), which encodes the enzyme phospholipase C-β (PLC-β). The protein is famous for being a crucial component of the phototransduction cascade operating in the compound eye photoreceptors, and *norpA* loss-of-function mutants are physiologically blind (Pearn et al. 1996). Mutations in this gene were also applied to demonstrate the involvement of compound eyes in light entrainment of the circadian clock, both as single mutants and in combination with cry^b (Stanewsky et al. 1998; Emery et al. 2000; Helfrich-Förster et al. 2001; Mealey-Ferrara et al. 2003).

Surprisingly, *norpA* has also been shown to have a role in seasonal adaptation of the fly clock mediated by light and temperature changes. In shorter days and colder temperatures (early spring, late fall), flies are mainly active during the day, whereas in the summer (long days, warm nights), activity occurs in the early-to-late night (Majercak et al. 1999). Interestingly, this seasonal behavior is linked to the splicing of an intron in the 3′-untranslated region (UTR) of *per*. During short photoperiods and cold temperatures, splicing is enhanced, leading to an earlier rise of *per* mRNA and protein, which is likely responsible for the earlier behavioral activity peak under cold and short-day conditions (Majercak et al. 1999). In *norpA* mutants, splicing of the *per* intron occurs in the "cold" mode, even if the flies are kept under warm and long-day conditions (Collins et al. 2004; Majercak et al. 2004). Levels of the spliced *per* transcript are also high in DD, indicating that the role of *norpA* in the regulation of *per* splicing is independent of light. Moreover, light pulses cause a decrease of the spliced transcript, indicating that PLC-β function is not important for the phototransduction mechanism involved in this process (Collins et al. 2004; Majercak et al. 2004). Therefore, *norpA* seems to be important for the temperature sensing involved in regulation of splicing, i.e., flies are now unable to respond with reduced *per* splicing to warmer temperatures (in fact, they still respond, but much less compared to wild-type flies) (Collins et al. 2004; Majercak et al. 2004).

This rather temperature-specific effect of *norpA* prompted us to test if PLC-β is also involved in temperature entrainment. Indeed, neither *per-luc* expression nor locomotor activity synchronized to temperature cycles; in fact, the *norpA* mutant phenotypes were indistinguishable from those of *nocte* (Glaser and Stanewsky 2005). We observed these phenotypes in the context of two independently isolated *norpA* alleles, which makes it rather unlikely that the temperature defects were caused by a secondary hit on the *norpA* chromosome. This is noteworthy, because such events occur relatively frequently and have let to misinterpretations in several cases (see Hall et al., this volume). We then determined the level of splicing for *per*'s last intron in both *nocte* and *norpA* mutants during temperature-entrainment conditions in order to see if this seasonal timing mechanism also operates on a day-to-day basis. The ratio of spliced versus unspliced *per* RNA in flies exposed to LL and temperature cycles was indistinguishable between two control strains and the *nocte* mutant flies (Fig. 2) (Glaser and Stanewsky 2005). Identical splicing levels between a temperature entrainment mutant and wild-type flies suggest that splicing of *per*'s last intron is not involved in daily temperature entrainment. In contrast, *norpA* mutant flies showed the same increased level of the spliced *per* transcript during temperature entrainment as was described previously for other environmental conditions (Fig. 2) (Glaser and Stanewsky 2005; see above). This confirms the previous observation that *norpA* mutants "lock" *per* splicing in the "cold" mode. But interestingly, we did not observe higher levels of the spliced transcript associated with the colder period: Peak levels were observed at the beginning of the warm phase and at the end of the cold phase, whereas the trough occurred around the transition from the warm to the cold phase (Fig. 2) (Glaser and Stanewsky 2005).

So far, our results support the idea that both *norpA* and *nocte* abolish daily temperature entrainment of the fly clock. On the basis of their identical phenotypes, we speculate that both operate in the same pathway, which—based on the gene product encoded by *norpA*—likely involves a G-protein-coupled signal transduction cascade. Furthermore, analysis of *per* splicing in both mutants during temperature-entrainment conditions strongly suggests that this temperature-controlled splicing event is not involved in day-by-day temperature entrainment.

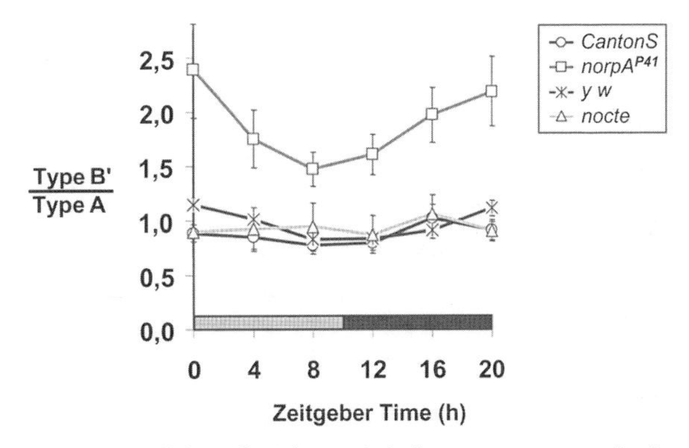

Figure 2. Splicing of *per* intron 8 during temperature cycles in LL: *Canton S* and *y w* are two different control stocks regularly used by many laboratories. Levels of the spliced (B′) and unspliced (A) *per* transcripts were determined by quantitative PCR, and their ratio was plotted as indicated. (Modified from the supplemental data of Glaser and Stanewsky 2005.)

Further analysis of *per-luc* temperature entrainment in *norpA* and *nocte* mutant-isolated tissues should reveal if theses mutants interfere with entrainment on a cellular/tissue level as opposed to the whole-organism level. This kind of analysis will also demonstrate if the two gene products are involved in temperature perception per se or rather in the processing of the various temperature signals perceived through the various tissues. No matter what the answer will be, both mutants specifically abolish temperature entrainment on a molecular (whole fly) and behavioral level, and their functional analysis will likely contribute significantly toward understanding how this zeitgeber synchronizes the circadian clock.

Role of Transcriptional and Posttranscriptional Mechanisms

Several experiments indicate that temperature primarily synchronizes molecular clock components by posttranscriptional mechanisms. For example, Yoshii et al. (2007) found that Per protein showed a different accumulation profile after temperature steps compared to its mRNA, whereas that of Tim protein closely followed its mRNA. Using various *per-luc* transgenics, we noticed that only those transgenes that did encode for at least a portion of the PER protein gave rise to robustly synchronized bioluminescence rhythms in adult flies kept in temperature cycles (Glaser and Stanewsky 2005). All transgenes containing only 5′UTRs from the *per* or *tim* genes showed no or only weakly synchronized luminescence rhythms (Fig. 3) (Glaser and Stanewsky 2005).

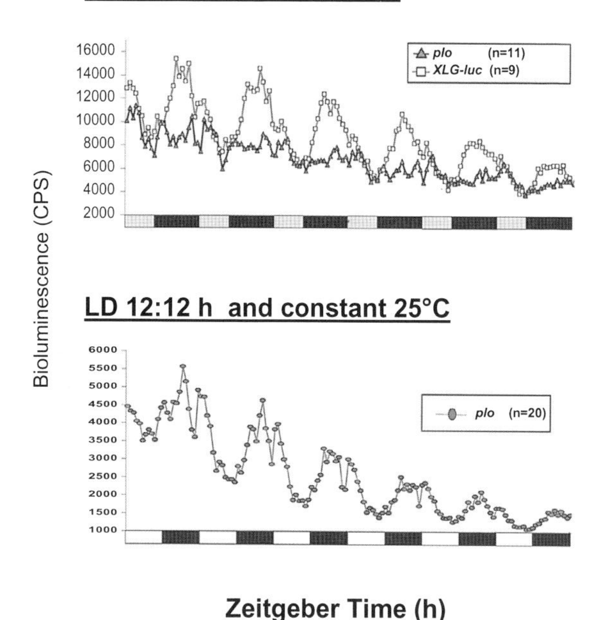

Figure 3. Transcriptional versus posttranscriptional control: Flies carrying transgenes containing only the *per* promoter fused to *luciferase* cDNA (*plo*) or additionally the entire Per-coding sequence (*XLG-luc*) were recorded under the conditions indicated above the luciferase recordings. For experimental details of such recordings, see Glaser and Stanewsky (2005).

This is in clear contrast to the results of the same transgenics analyzed under LD conditions and constant temperature, where all lines exhibited robust rhythmicity (Fig. 3) (Stanewsky et al. 2002). We do not have a reporter line to monitor luminescence from a Tim-Luc protein fusion, but because the *tim-luc* promoter line does not synchronize its expression to temperature cycles, whereas the Tim protein does (Stanewsky et al. 1998; Glaser and Stanewsky 2005; Yoshii et al. 2005), it seems clear that Tim oscillations are also primarily synchronized by posttranscriptional mechanisms.

Despite these findings pointing to a prominent role of posttranscriptional mechanisms in temperature entrainment, a recent study revealed substantial rhythmic mRNA regulation using whole-genome microarrays (Boothroyd et al. 2007). These authors discovered that the majority of the transcripts that oscillated in temperature cycles and DD conditions where temperature-driven. For example, in the background of the tim^{01} mutation, which abolishes rhythmic expression of clock or clock-regulated genes, 939 transcripts were rhythmically expressed (i.e., temperature-driven). Under LD and constant temperature conditions, only 72 transcripts turned out to be light-driven in the same tim^{01} genetic background. To determine the number of circadian transcripts that can be synchronized by thermocycles, the authors also analyzed transcript rhythmicity in wild-type flies after transition from temperature cycles to constant temperatures. Surprisingly, the set of remaining rhythmic transcripts ($n = 143$) did overlap substantially with those that remained rhythmic after entrainment to LD cycles ($n = 172$). Further testing by northern blot analysis revealed that six randomly picked transcripts of the above transcripts, which initially appeared to show entrainment only to thermocycles, also synchronized to LD cycles. In contrast, three transcripts that appeared to synchronize only to LD cycles after microarray analyses were confirmed to be light-specific by northern blot.

In an apparent contradiction to our results (e.g., Fig. 3), Boothroyd et al. (2007) demonstrated temperature-entrained mRNA rhythms of all clock gene transcripts that also oscillate in LD cycles (*per, tim, Clk, vri, cry,* and *Pdp1*). The discrepancy could be explained by the difference in the experimental design. We analyzed expression in whole animals, whereas Boothroyd et al. (2007) analyzed mRNAs extracted from fly heads. As discussed above for the temperature-entrainment mutants *nocte* and *norpA*, it is possible that temperature entrainment requires the generation of a temporally "organized" signal throughout the whole body, which may rely on posttranscriptional mechanisms. In isolated tissues, on the other hand, transcriptional rhythms may be visible and synchronized by temperature cycles. In fact, we also observed *per* and *tim* mRNA rhythmicity in heads of wild-type flies kept in LL and temperature cycles (Fig. 4). It will be interesting to see whether the same result applies for other isolated body parts, and if mRNA rhythms also persist in isolated body parts of the two temperature-entrainment mutants (see above).

Interestingly, clock gene mRNAs are also driven into rhythmicity by temperature cycles in a clock mutant background, but the induced rhythms are out-of-phase with

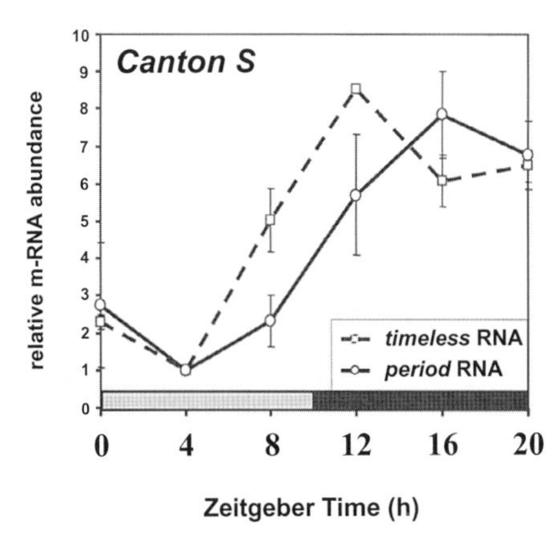

Figure 4. Whole-head mRNA expression levels during temperature cycles in LL: *per* and *tim* mRNA levels in heads of wild-type *Canton-S* flies were determined by quantitative PCR using a light cycler. (*Gray bars*) Warm (25°C); (*black bars*) cold (18°C) periods. Experiments were repeated three times and error bars indicate S.E.M.

those observed in flies containing a functional clock (Boothroyd et al. 2007). Therefore, the circadian clock somehow has to suppress the temperature-induced mRNA oscillations or their potential consequences, again pointing toward the importance of posttranscriptional mechanisms.

Similar to the temperature-regulated *per* RNA splicing event, two different *tim* transcripts were discovered (Boothroyd et al. 2007). During and after LD entrainment, the previously undetected tim^{cold} transcript is more abundant at 18°C, whereas the other (the canonical *tim* transcript) is more abundant at 25°C. During temperature entrainment, tim^{cold} appears to cycle in-phase with the *per* transcript, but the classical *tim* transcript exhibits a phase-delay (Boothroyd et al. 2007). Again, our findings seem to be different because we did not observe a phase delay of the *tim* message compared to *per*, but instead, we did observe a mild phase-advance (Fig. 4). Our analysis was performed with real-time polymerase chain reaction (PCR) using primers common to both *tim* transcripts, but this still cannot easily explain the different results. Perhaps it makes a difference that the *tim* phase-delay was observed during temperature cycles in DD (Boothroyd et al. 2007), whereas the similar temporal *tim* and *per* RNA profiles were observed in LL (Fig. 4).

More importantly, the tim^{cold} transcript was found to retain the last *tim* intron. In contrast to the alternative splicing event involving the last *per* intron that only affects the accumulation of *per* RNA, the tim^{cold} transcript is predicted to encode a truncated Tim protein. In fact, such a truncated protein was observed on western blots during the light portion of an LD cycle at a constant temperature of 18°C, but not at 25°C (Boothroyd et al. 2007). Further work will be necessary to fully understand the function of alternative *tim* splicing and the two Tim proteins in temperature entrainment.

Transcriptional rhythms of potential core clock genes could also be induced by temperature cycles in zebra fish cells and larvae (Lahiri et al. 2005). Similar to behavioral rhythms in *Drosophila*, RNA rhythms in fish were synchronized even if the amplitude of the thermocycle was as low as 2°C (Wheeler et al. 1993; Lahiri et al. 2005). Moreover, it was shown that the abundance and phosphorylation status of zebra fish Clk1 is different at constant low (20°C) versus high (30°C) temperatures during LD cycles (Lahiri et al. 2005). Although this does not directly address the question of whether these differences at the protein level contribute to temperature entrainment, it does suggest that posttranscriptional regulation also contributes to thermal entrainment in vertebrates.

Temperature Entrainment in LL versus DD

In LL and constant temperatures, molecular and behavioral rhythmicity breaks down (Konopka et al. 1989; Marrus et al. 1996), presumably because of the light-induced and Cry-mediated degradation of Tim (see, e.g., Stanewsky et al. 1998; Ceriani et al. 1999; Busza et al. 2004). Nevertheless, robust behavioral synchronization and oscillations of *per* and *tim* gene products are observed in LL and temperature cycles (Glaser and Stanewsky 2005; Matsumoto et al. 1998; Miyasako et al. 2007; Yoshii et al. 2002, 2005). Moreover, it has been repeatedly observed that temperature cycles are a stronger zeitgeber when they are applied in LL compared to DD. Under both environmental conditions, light input into the clock should be clearly separated from temperature entrainment, which makes it difficult to comprehend the enhancing effect of LL. The first hint that this effect indeed exists came from the observation that flies carrying the period-altering mutations per^S and per^L synchronized their behavior nicely to temperature cycles in LL, whereas they free-ran in DD (Matsumoto et al. 1998). Later, the same research team found that wild-type flies synchronized better to temperature cycles in LL compared to DD (see above) (Yoshii et al. 2002). Finally, we noted that although *per-luc* luminescence rhythms can be temperature-entrained in DD and LL, rhythmicity is consistently more robust in LL compared to DD (Glaser and Stanewsky 2005 and unpubl.).

It therefore seems that factors which are present in DD and absent in LL somehow interfere with, or dampen, temperature entrainment. A good candidate for such a factor is Cry, which is degraded by light and accumulates in DD (Emery et al. 1998). Indeed, bioluminescence analysis of two different *per-luc* transgenics in the cry^b background compared to wild-type mutant background suggests that Cry indeed interferes with temperature entrainment: In LL and temperature cycles, *BG-luc* transgenics, in which the *per* promoter and two thirds of the Per-coding region are fused to *luciferase*, seem to exhibit a higher-amplitude cycling when they carry the cry^b mutant (Glaser and Stanewsky 2005).

A more prominent effect of this mutant was observed when its consequences on the expression of a promoterless *per-luc* construct were tested. The *8.0-luc* transgenics have previously been shown to express this construct predominantly in subgroups of the "dorsal neurons" (DNs) (Veleri et al. 2003). When crossed to a cry^b mutant background, they showed a drastic increase in amplitude of luciferase activity in LL and temperature cycles (Fig. 5). This

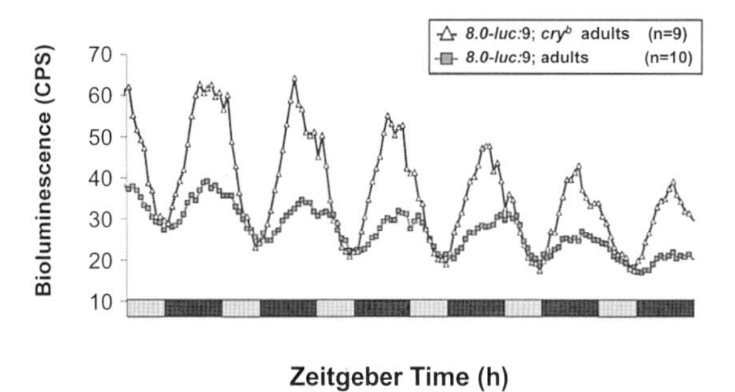

Figure 5. Cry interferes with temperature entrainment: Bioluminescence rhythms of *8.0-luc* transgenic flies were recorded in a normal and cry^b mutant background during temperature cycles in LL. (*Gray bars*) Warm (25°C); (*black bars*) cold (18°C) periods. For details regarding the *8.0-luc* reporter construct, see Veleri et al. (2003) and Wülbeck et al. (2005).

strongly indicates that Cry interferes with this process. To confirm Cry's suppressing role in this process, experiments also need to be conducted in DD and temperature cycles. We would expect that the amplitude of *per-luc* oscillations increases in the face of cry^b compared to the wild-type situation. The observed amplitude increase by cry^b in LL—a condition that reduces Cry levels substantially—could be explained by a further reduction of Cry levels compared to those present in cry^b flies during DD.

In this regard, it is also interesting to note that a recent study found that a subset of the pacemaker neurons in the central brain of the fly seems to inhibit rapid behavioral synchronization to temperature cycles (Busza et al. 2007). Perhaps LL creates a situation in the fly brain similar to removal of the PDF-expressing lateral neurons does, because in both cases, synchronization to temperature cycles is enhanced (Busza et al. 2007; see above).

In any case, the current data suggest that Cry has a negative impact on temperature entrainment, which would also explain the behavioral results that implicate an enhancement of zeitgeber strength in LL (see above). A corollary of such a scenario would be that the Tim protein—normally degraded by light—can accumulate because of a depletion of Cry. Perhaps the dual role of Cry contributes to the order of zeitgeber strength in wild-type flies, where light is generally considered to be superior over temperature.

Neural Substrates

Several sets of clock neurons in the fly brain are responsible for mediating rhythmic locomotor behavior under LD and DD conditions (see, e.g., Veleri et al. 2003; Grima et al. 2004; Stoleru et al. 2004, 2005), and multiple light-input pathways synchronize clock protein expression in these neurons, either directly (see, e.g., Klarsfeld et al. 2004; Veleri et al. 2007) or through communication between neurons (see, e.g., Stoleru et al. 2005). Thus, for thermocycles to synchronize behavior, the temperature signals must somehow regulate clock gene expression in the clock neurons. Indeed, temperature cycles (in LL) induce Per cycling in all known clock neuronal groups, with a temporal profile very similar to that observed after LD entrainment (Yoshii et al. 2005). Robust Per rhythms were also found in a group of "lateral posterior neurons" (LPNs), in which Per is difficult to detect during LD entrainment (Kaneko and Hall 2000; Helfrich-Förster 2005; Yoshii et al. 2005) and which therefore may have a more prominent role in temperature entrainment. Interestingly, Pdf^{01} mutant flies, or flies in which the *Pdf*-expressing cells have been ablated, are still able to synchronize to temperature cycles behaviorally (Yoshii et al. 2005; Busza et al. 2007). This suggests that PDF-negative clock neurons (including the LPN) may be more important. To test this, *disco* mutant flies, which lack the PDF neurons and the dorsal lateral neurons (LNd), but not the dorsal neurons (DN) and the LPNs (Zerr et al. 1990; Kaneko and Hall 2000; Yoshii et al. 2005), were analyzed in temperature cycles. They were found to exhibit only weakly synchronized behavioral rhythms and also exhibited synchronized PER expression in LPN and DN cells (Yoshii et al. 2005), suggesting that these neurons mediate at least some aspects of temperature-entrained behavior. These initial observations were confirmed by a recent study, in which Tim expression was analyzed in flies that were exposed to a combined LD and a 6-hour advanced temperature cycle (25°C:20°C) (Miyasako et al. 2007). Tim expression in the LPNs and DNs was mainly synchronized to the temperature cycle, whereas it followed the LD schedule in the two groups of LN cells.

Independent evidence for the importance of the LPNs and the dorsally located LNd and DN cells stems from experiments where the ventrally located PDF-expressing neurons were ablated without compromising temperature entrainment (Busza et al. 2007). Interestingly, weak temperature entrainment was observed even after eliminating the PDF cells and the LNd, strongly suggesting that the LPN and DN neurons are important mediators of temperature entrainment. As discussed above, the PDF neurons most likely serve to modulate the temperature response of the other neurons in order to prevent the system from being hypersensitive to temperature changes (Busza et al. 2007).

Preliminary results obtained with the *nocte* and *norpA* mutants indicate that Per expression in all clock neurons can still be synchronized by temperature cycles, although subtle differences exist (F. Glaser and R. Stanewsky, unpubl.). It remains to be seen if these differences are responsible for the lack of behavioral synchronization observed in these mutants (Glaser and Stanewsky 2005), or if the mutants affect a different step of the temperature-entrainment mechanism (see above). Although it was originally reported that isolated brains "in culture" are able to synchronize *per-luc* oscillations to temperature cycles (Glaser and Stanewsky 2005), further experiments suggest that this is not the case. Instead, it seems that Per expression in isolated brains simply responds to temperature changes (H. Sehadova et al., unpubl.). If true, this would further imply that synchronization of clock gene expression in the behaviorally important clock neurons is synchronized *indirectly* by temperature signals reaching the brain from the periphery. If true, this would demonstrate a fundamental difference between the light- and temperature-entrainment pathways.

CONCLUSIONS

Determining how temperature cycles synchronize the circadian clock remains a problematic issue for any organism, including—and as discussed here in detail—*Drosophila*. This is due in part to the direct effects different temperatures exert on organisms, which includes changes in the amounts of clock molecules. A further complication arises from the fact that in flies, light is the major zeitgeber, and temperature cycles therefore probably have only a supportive role in entrainment overall. It is also clear that both entrainment pathways must be integrated at some level, and our findings suggest that Cry may be involved in this process (Fig. 5). Interestingly, Cry seems also to be involved in temperature-compensation and in the responses of circadian clock molecules to heat shock pulses (Kaushik et al. 2007), indicating that this flavoprotein is involved in both light and temperature regulation of the circadian clock.

Further analysis of the existing mutations that interfere with temperature entrainment will be necessary to gain deeper insight into the underlying mechanisms. Because the multiple effects of temperature on the circadian clock share some components, this will in parallel help to elucidate the mechanisms underlying temperature-compensation and seasonal adaptation (although as discussed above, one mechanism regulating the latter process seems not to be important for daily entrainment).

Finally, from our own work a picture seems to emerge in which the generation of an integrated whole-organism signal is necessary to obtain synchronized clock output (i.e., behavioral rhythms). In other words, although the clocks of individual cells and tissues can be synchronized by thermo cycles, this does not guarantee synchronization of the whole organism. It will be important to resolve whether the currently available temperature mutants interfere with the generation of such a signal and how they do it. Moreover, if these mutants do not interfere with circadian temperature reception as such, further mutant screens will be necessary to understand the reception mechanism.

ACKNOWLEDGMENTS

We thank members of the lab for critical reading of the manuscript. This work was supported by the Deutsche Forschungsgemeinschaft grants Sta 421/3-3 and Sta 421/6-6 given to R.S. Our work is supported by EUCLOCK, an Integrated Project (FP6) funded by the European Commission.

REFERENCES

Bachleitner W., Kempinger L., Wülbeck C., Rieger D., and Helfrich-Förster C. 2007. Moonlight shifts the endogenous clock of *Drosophila melanogaster*. *Proc. Natl. Acad. Sci.* **104:** 3538.

Boothroyd C.E., Wijnen H., Naef F., Saez L., and Young M.W. 2007. Integration of light and temperature in the regulation of circadian gene expression in *Drosophila*. *PLoS Genet.* **3:** e54.

Brown S.A., Zumbrunn G., Fleury-Olela F., Preitner N., and Schibler U. 2002. Rhythms of mammalian body temperature can sustain peripheral circadian clocks. *Curr. Biol.* **12:** 1574.

Busza A., Murad A., and Emery P. 2007. Interactions between circadian neurons control temperature synchronization of *Drosophila* behavior. *J. Neurosci.* **27:** 10722.

Busza A., Emery-Le M., Rosbash M., and Emery P. 2004. Roles of the two *Drosophila* CRYPTOCHROME structural domains in circadian photoreception. *Science* **304:** 1503.

Ceriani M.F., Darlington T.K., Staknis D., Mas P., Petti A.A., Weitz C.J., and Kay S.A. 1999. Light-dependent sequestration of TIMELESS by CRYPTOCHROME. *Science* **285:** 553.

Collins B.H., Rosato E., and Kyriacou C.P. 2004. Seasonal behavior in *Drosophila melanogaster* requires the photoreceptors, the circadian clock, and phospholipase C. *Proc. Natl. Acad. Sci.* **101:** 1945.

Dolezelova E., Dolezel D., and Hall J.C. 2007. Rhythm defects caused by newly engineered null mutations in *Drosophila*'s *cryptochrome* gene. *Genetics* **177:** 329.

Dunlap J.C., Loros J.J., and DeCoursey P.J. 2004. *Chronobiology: Biological timekeeping*. Sinauer Associates, Sunderland, Massachusetts.

Edery I., Rutila J.E., and Rosbash M. 1994. Phase shifting of the circadian clock by induction of the *Drosophila* Period protein. *Science* **263:** 237.

Emery I.F., Noveral J.M., Jamison C.F., and Siwicki K.K. 1997. Rhythms of *Drosophila period* gene expression in culture. *Proc. Natl. Acad. Sci.* **94:** 4092.

Emery P., So W.V., Kaneko M., Hall J.C., and Rosbash M. 1998. CRY, a *Drosophila* clock and light-regulated cryptochrome, is a major contributor to circadian rhythm resetting and photosensitivity. *Cell* **95:** 669.

Emery P., Stanewsky R., Helfrich-Förster C., Emery-Le M., Hall J.C., and Rosbash M. 2000. *Drosophila* CRY is a deep brain circadian photoreceptor. *Neuron* **26:** 493.

Foster R.G., Hankins M.W., and Peirson S.N. 2007. Light, photoreceptors, and circadian clocks. *Methods Mol. Biol.* **362:** 3.

Glaser F.T. and Stanewsky R. 2005. Temperature synchronization of the *Drosophila* circadian clock. *Curr. Biol.* **15:** 1352.

Grima B., Chelot E., Xia R., and Rouyer F. 2004. Morning and evening peaks of activity rely on different clock neurons of the *Drosophila* brain. *Nature* **431:** 869.

Hall J.C. 2003. Genetics and molecular biology of rhythms in *Drosophila* and other insects. *Adv. Genet.* **48:** 1.

Helfrich-Förster C. 2005. Neurobiology of the fruit fly's circadian clock. *Genes Brain Behav.* **4:** 65.

Helfrich-Förster C., Winter C., Hofbauer A., Hall J.C., and Stanewsky R. 2001. The circadian clock of fruit flies is blind after elimination of all known photoreceptors. *Neuron* **30:** 249.

Huang Z.J., Curtin K.D., and Rosbash M. 1995. PER protein interactions and temperature compensation of a circadian clock in *Drosophila*. *Science* **267:** 1169.

Hunt S.M., Elvin M., Crosthwaite S.K., and Heintzen C. 2007. The PAS/LOV protein VIVID controls temperature compensation of circadian clock phase and development in *Neurospora crassa*. *Genes Dev.* **21:** 1964.

Ivanchenko M., Stanewsky R., and Giebultowicz J.M. 2001. Circadian photoreception in *Drosophila:* Functions of *cryptochrome* in peripheral and central clocks. *J. Biol. Rhythms* **16:** 205.

Kaneko M. and Hall J.C. 2000. Neuroanatomy of cells expressing clock genes in *Drosophila:* Transgenic manipulation of the *period* and *timeless* genes to mark the perikarya of circadian pacemaker neurons and their projections. *J. Comp. Neurol.* **422:** 66.

Kaushik R., Nawathean P., Busza A., Murad A., Emery P., and Rosbash M. 2007. PER-TIM interactions with the photoreceptor *cryptochrome* mediate circadian temperature responses in *Drosophila*. *PLoS Biol.* **5:** e146.

Klarsfeld A., Malpel S., Michard-Vanhee C., Picot M., Chelot E., and Rouyer F. 2004. Novel features of *cryptochrome*-mediated photoreception in the brain circadian clock of *Drosophila*. *J. Neurosci.* **24:** 1468.

Konopka R.J., Pittendrigh C., and Orr D. 1989. Reciprocal behaviour associated with altered homeostasis and photosen-

sitivity of *Drosophila* clock mutants. *J. Neurogenet.* **6:** 1.
Kornmann B., Schaad O., Bujard H., Takahashi J.S., and Schibler U. 2007. System-driven and oscillator-dependent circadian transcription in mice with a conditionally active liver clock. *PLoS Biol.* **5:** e34.
Lahiri K., Vallone D., Gondi S.B., Santoriello C., Dickmeis T., and Foulkes N.S. 2005. Temperature regulates transcription in the zebrafish circadian clock. *PLoS Biol.* **3:** e351.
Leloup J.C. and Goldbeter A. 1997. Temperature compensation of circadian rhythms: Control of the period in a model for circadian oscillations of the PER protein in *Drosophila*. *Chronobiol. Int.* **14:** 511.
Levine J.D., Funes P., Dowse H.B., and Hall J.C. 2002a. Advanced analysis of a *cryptochrome* mutation's effects on the robustness and phase of molecular cycles in isolated peripheral tissues of *Drosophila*. *BMC Neurosci.* **3:** 5.
———. 2002b. Resetting the circadian clock by social experience in *Drosophila melanogaster*. *Science* **298:** 2010.
Majercak J., Chen W.F., and Edery I. 2004. Splicing of the *period* gene 3′-terminal intron is regulated by light, circadian clock factors, and phospholipase C. *Mol. Cell. Biol.* **24:** 3359.
Majercak J., Sidote D., Hardin P.E., and Edery I. 1999. How a circadian clock adapts to seasonal decreases in temperature and day length. *Neuron* **24:** 219.
Marrus S.B., Zeng H., and Rosbash M. 1996. Effect of constant light and circadian entrainment of per^S flies: Evidence for light-mediated delay of the negative feedback loop in *Drosophila*. *EMBO J.* **15:** 6877.
Matsumoto A., Matsumoto N., Harui Y., Sakamoto M., and Tomioka K. 1998. Light and temperature cooperate to regulate the circadian locomotor rhythm of wild type and period mutants of *Drosophila melanogaster*. *J. Insect Physiol.* **44:** 587.
Mealey-Ferrara M.L., Montalvo A.G., and Hall J.C. 2003. Effects of combining a *cryptochrome* mutation with other visual-system variants on entrainment of locomotor and adult-emergence rhythms in *Drosophila*. *J. Neurogenet.* **17:** 171.
Merrow M., Brunner M., and Roenneberg T. 1999. Assignment of circadian function for the *Neurospora* clock gene *frequency*. *Nature* **399:** 584.
Miyasako Y., Umezaki Y., and Tomioka K. 2007. Separate sets of cerebral clock neurons are responsible for light and temperature entrainment of *Drosophila* circadian locomotor rhythms. *J. Biol. Rhythms* **22:** 115.
Morimoto R.I. 1998. Regulation of the heat shock transcriptional response: Cross talk between a family of heat shock factors, molecular chaperones, and negative regulators. *Genes Dev.* **12:** 3788.
———. 2002. Dynamic remodeling of transcription complexes by molecular chaperones. *Cell* **110:** 281.
Pearn M.T., Randall L.L., Shortridge R.D., Burg M.G., and Pak W.L. 1996. Molecular, biochemical, and electrophysiological characterization of *Drosophila norpA* mutants. *J. Biol. Chem.* **271:** 4937.
Plautz J.D., Kaneko M., Hall J.C., and Kay S.A. 1997. Independent photoreceptive circadian clocks throughout *Drosophila*. *Science* **278:** 1632.
Pregueiro A.M., Price-Lloyd N., Bell-Pedersen D., Heintzen C., Loros J.J., and Dunlap J.C. 2005. Assignment of an essential role for the *Neurospora frequency* gene in circadian entrainment to temperature cycles. *Proc. Natl. Acad. Sci.* **102:** 2210.
Prolo L.M., Takahashi J.S., and Herzog E.D. 2005. Circadian rhythm generation and entrainment in astrocytes. *J. Neurosci.* **25:** 404.
Roenneberg T. and Foster R.G. 1997. Twilight times: Light and the circadian system. *Photochem. Photobiol.* **66:** 549.
Roenneberg T., Dragovic Z., and Merrow M. 2005. Demasking biological oscillators: Properties and principles of entrainment exemplified by the *Neurospora* circadian clock. *Proc. Natl. Acad. Sci.* **102:** 7742.
Ruoff P., Loros J.J., and Dunlap J.C. 2005. The relationship between FRQ-protein stability and temperature compensation in the *Neurospora* circadian clock. *Proc. Natl. Acad. Sci.* . **102:** 17681.
Sawyer L.A., Hennessy J.M., Peixoto A.A., Rosato E., Parkinson H., Costa R., and Kyriacou C.P. 1997. Natural variation in a *Drosophila* clock gene and temperature compensation. *Science* **278:** 2117.
Sayeed O. and Benzer S. 1996. Behavioral genetics of thermosensation and hygrosensation in *Drosophila*. *Proc. Natl. Acad. Sci.* **93:** 6079.
Sidote D., Majercak J., Parikh V., and Edery I. 1998. Differential effects of light and heat on the *Drosophila* circadian clock proteins PER and TIM. *Mol. Cell. Biol.* **18:** 2004.
Stanewsky R., Lynch K.S., Brandes C., and Hall J.C. 2002. Mapping of elements involved in regulating normal temporal *period* and *timeless* RNA expression patterns in *Drosophila melanogaster*. *J. Biol. Rhythms* **17:** 293.
Stanewsky R., Kaneko M., Emery P., Beretta B., Wager-Smith K., Kay S.A., Rosbash M., and Hall J.C. 1998. The cry^b mutation identifies *cryptochrome* as a circadian photoreceptor in *Drosophila*. *Cell* **95:** 681.
Stoleru D., Peng Y., Agosto J., and Rosbash M. 2004. Coupled oscillators control morning and evening locomotor behaviour of *Drosophila*. *Nature* **431:** 862.
Stoleru D., Peng Y., Nawathean P., and Rosbash M. 2005. A resetting signal between *Drosophila* pacemakers synchronizes morning and evening activity. *Nature* **438:** 238.
Veleri S., Rieger D., Helfrich-Förster C., and Stanewsky R. 2007. Hofbauer-Buchner eyelet affects circadian photosensitivity and coordinates TIM and PER expression in *Drosophila* clock neurons. *J. Biol. Rhythms* **22:** 29.
Veleri S., Brandes C., Helfrich-Förster C., Hall J.C., and Stanewsky R. 2003. A self-sustaining, light-entrainable circadian oscillator in the *Drosophila* brain. *Curr. Biol.* **13:** 1758.
Wheeler D.A., Hamblen-Coyle M.J., Dushay M.S., and Hall J.C. 1993. Behavior in light-dark cycles of *Drosophila* mutants that are arrhythmic, blind, or both. *J. Biol. Rhythms* **8:** 67.
Wülbeck C., Szabo G., Shafer O.T., Helfrich-Förster C., and Stanewsky R. 2005. The novel *Drosophila* tim^{blind} mutation affects behavioral rhythms but not periodic eclosion. *Genetics* **169:** 751.
Yoshii T., Fujii K., and Tomioka K. 2007. Induction of *Drosophila* behavioral and molecular circadian rhythms by temperature steps in constant light. *J. Biol. Rhythms* **22:** 103.
Yoshii T., Sakamoto M., and Tomioka K. 2002. A temperature-dependent timing mechanism is involved in the circadian system that drives locomotor rhythms in the fruit fly *Drosophila melanogaster*. *Zoolog. Sci.* **19:** 841.
Yoshii T., Heshiki Y., Ibuki-Ishibashi T., Matsumoto A., Tanimura T., and Tomioka K. 2005. Temperature cycles drive *Drosophila* circadian oscillation in constant light that otherwise induces behavioural arrhythmicity. *Eur. J. Neurosci.* **22:** 1176.
Zars T. 2001. Two thermosensors in *Drosophila* have different behavioral functions. *J. Comp. Physiol. A* **187:** 235
Zerr D.M., Hall J.C., Rosbash M., and Siwicki K.K. 1990. Circadian fluctuations of Period protein immunoreactivity in the CNS and the visual system of *Drosophila*. *J Neurosci* **10:** 2749.

What Is There Left to Learn about the *Drosophila* Clock?

J. Blau, F. Blanchard, B. Collins, D. Dahdal, A. Knowles, D. Mizrak, and M. Ruben
Department of Biology, New York University, New York, New York 10003

Circadian rhythms offer probably the best understanding of how genes control behavior, and much of this understanding has come from studies in *Drosophila*. More recently, genetic manipulation of clock neurons in *Drosophila* has helped identify how daily patterns of activity are programed by different clock neuron groups. Here, we review some of the more recent findings on the fly molecular clock and ask what more the fly model can offer to circadian biologists.

INTRODUCTION

A forward genetic screen conducted by Ron Konopka while a Ph.D. student in Seymour Benzer's lab more than 35 years ago identified three different mutations that dramatically affected circadian (~24 hours) rhythms in pupal eclosion (Konopka and Benzer 1971). All three mutations mapped to a single gene that was named *period* (*per*) because the mutations either affected the length of the behavioral period or completely abolished rhythmicity altogether. At the time, the isolation of these mutants was a surprise to many because it was not clear that mutation of a single gene could have such a profound effect on complex animal behavior. Indeed, this is not always the case: Mice in which only one of the three mouse *Per* genes are eliminated are initially rhythmic, although mutation of *mPer1* and *mPer2* together mimics the arrhythmicity of *Drosophila per* null mutants (for review, see Stanewsky 2003).

Work during the next three decades led to the cloning of *Drosophila per* and an understanding of PER protein's integral role in an intracellular molecular clock. The importance of the *Drosophila* work cannot be underestimated: During the sequencing of human chromosome 17, one cDNA was identified that had sequence homology with fly *per* (Sun et al. 1997). This of course suggested a role in the human clock, but without flies, the human *Per* genes could still be in search of a function. Instead, a polymorphism in *hPer2* has been associated with a human sleep disorder (Toh et al. 2001), and mice lacking *Per2* show increased tumor rates, suggesting a hierarchical relationship between the circadian clock and the cell cycle (Fu et al. 2002). The multiple functions associated with mammalian *Per* genes make sense given the presence of molecular clock genes in many tissues. In turn, this is of great interest to basic scientists and clinicians alike and has implications for the daily timing of therapeutic interventions, discussed by Hunt and Sassone-Corsi (2007).

In contrast to mammals, the *Drosophila per* gene and its associated intracellular clocks are largely confined to peripheral sensory neurons and a small subset of central brain neurons. Thus, circadian studies in *Drosophila* have a neurobiological perspective as fly clocks mainly regulate rhythms, sensory perception, and behavior. We start with a discussion of the already detailed understanding of the molecular clock in *Drosophila*. We then discuss how recent findings about clock neuron coordination of daily rhythms in locomotor activity are changing the way we view the fly circadian clock.

MOLECULAR CLOCKS: TRANSCRIPTIONAL REGULATION IN THE FIRST LOOP

Our current view of the *Drosophila* molecular clock has two interconnected transcriptional feedback loops (Fig. 1). The first loop, involving *per*, is a simple negative feedback loop (for review, see Blau 2001). Two tran-

A

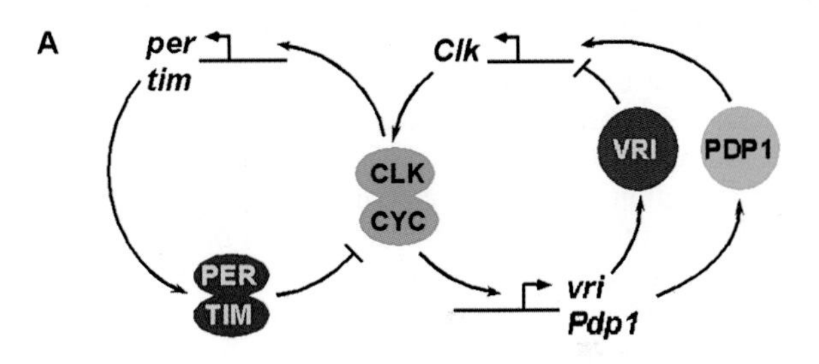

B

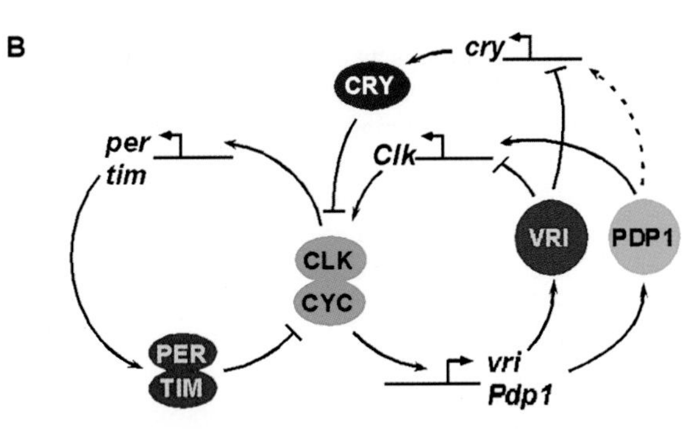

Figure 1. Model of the molecular clock feedback loops in *Drosophila*. The clock is composed of two interconnected transcriptional-translational feedback loops. (*A*) In s-LN$_{V}$s, *cry* is not required for rhythms and is omitted from the model, although it acts as a cell-autonomous photoreceptor in these cells. (*B*) In peripheral clocks, CRY has an important clock role and acts as a transcriptional repressor of CLK/CYC activity in conjunction with PER. Because *cry* expression is regulated by VRI and because VRI- and PDP1-binding sites are very similar, *cry* is probably also regulated by PDP1 (*dashed line*).

scriptional activators, Clock (CLK) and Cycle (CYC) heterodimerize and bind to specific E-box sequences in the *per* and *timeless* (*tim*) promoters. PER and TIM proteins interact in the cytoplasm and then enter the nucleus. Once inside the nucleus, PER inhibits CLK/CYC activity by removing them from DNA (Yu et al. 2006); thus, PER protein inhibits further expression of *per* and *tim*. PER is progressively phosphorylated and then degraded, allowing CLK/CYC to reactivate *per* and *tim* expression and start the next cycle of the clock. Delays that slow down PER accumulation and PER/TIM nuclear entry help make each cycle of a conceptually simple feedback loop last for 24 hours and presumably help the clock oscillate as opposed to come to equilibrium.

Originally, PER and TIM were thought to enter the nucleus together because their subcellular localization showed essentially identical timing in flies, first appearing in the cytoplasm and then translocating into the nucleus. Furthermore, TIM is largely cytoplasmic in *per* null mutant flies, and a stable PER-fusion protein is cytoplasmic in *tim* null mutants. However, this simple model has been challenged recently.

The first upset was when two studies showed that PER could be detected in pacemaker neuron nuclei before TIM (Shafer et al. 2002, 2004). The viability of the simple model was partly rescued when Ashmore et al. (2003) showed that although TIM could shuttle in and out of the nucleus, even in a *per* null mutant, it seems to have a shorter dwelling time in the nucleus. Unfortunately, for the simple model, we showed that PER is constitutively nuclear in the absence of its tightly associated protein kinase, Doubletime (DBT), even in a *tim* null mutant (Cyran et al. 2005). The final nail in the simple PER/TIM translocation model was experiments in *Drosophila* S2 cells in culture transfected with *per* and *tim* expression plasmids that found that although PER and TIM associate for a considerable time in the nucleus, they often dissociate just before nuclear entry (Meyer et al. 2006).

TIM, however, is certainly required for PER's cytoplasmic accumulation (which necessarily precedes PER nuclear entry) and *tim* is an essential clock gene because *tim* null mutant flies are molecularly and behaviorally arrhythmic. Many different *tim* alleles carrying single point mutations profoundly affect period length (ranging from 20 to 33 hours) and hence the timing of the molecular clock. The longest of these *tim* mutants (tim^{UL}; Rothenfluh et al. 2000) appears to increase the association of PER and TIM in the nucleus. One interpretation of this finding is that PER and TIM normally interact in the nucleus, and this is magnified in tim^{UL} mutants. Thus, the dissociation of PER and TIM observed in S2 cells before nuclear entry may only be a transient event or could even represent a peculiarity of this system. Ultimately, it will be necessary for the fine-resolution imaging of PER and TIM described in S2 cells to be applied to fly clock neurons to answer these mechanistic details.

Importantly, the delay in nuclear entry of PER and TIM observed in clock cells seems to be an intrinsic feature of these proteins because it occurs even in S2 cells in vitro (Meyer et al. 2006). Furthermore, the per^{Long} mutation originally identified by Konopka and Benzer (1971) lengthens the time taken for nuclear entry in S2 cells (Meyer et al. 2006), accurately reflecting what happens in per^{L} pacemaker neurons in vivo (Siwicki et al. 1988).

MOLECULAR CLOCKS: POSTTRANSCRIPTIONAL REGULATION IN THE FIRST LOOP

A substantial amount of posttranscriptional regulation exists in this first clock loop and presumably contributes to delay in nuclear entry of PER and TIM. The DBT and casein kinase II (CK2) protein kinases probably regulate PER stability and determine the timing of its nuclear entry (Price et al. 1998; Suri et al. 2000; Lin et al. 2002; Akten et al. 2003). DBT forms a stable interaction with PER and translocates to the nucleus along with PER. Inside the nucleus, PER is progressively phosphorylated, presumably by DBT.

Inside the nucleus, maximal PER phosphorylation takes several hours and is probably so slow because of a balance between DBT phosphorylating PER and protein phosphatase 2A (PP2A) dephosphorylating PER. There is excellent genetic evidence supporting this balancing model, hidden in the supplementary data of a paper by Sathyanarayanan et al. (2004). These authors found that overexpression in clock cells of *Widerborst* (*Wdb*), a nuclear PP2A regulatory subunit, lengthens the period by about 0.8 hours. Flies heterozygous for a dominant *dbt* allele (dbt^{g}) have a period almost 4 hours longer than wild type (27.6 hours). But overexpression of *wdb* in a dbt^{g} background adds another 3 hours to the period (now 30.3 hours), a synergistic interaction. The well-described biochemical functions of DBT and PP2a and the numerical output of locomotor assays allowed Sathyanarayanan et al. (2004) to make a clear biochemical prediction on the basis of this genetic interaction: The synergy arises from increasing PP2A activity via increased Wdb levels at the same time as decreasing DBT kinase activity via dbt^{g}. Overall, this supports the idea that PER stability and hence period length is a balance between PER phosphorylation via DBT and dephosphorylation via PP2A.

A second substrate for DBT and PP2A is the CLK transcription factor itself, whose activity is decreased when phosphorylated (Kim and Edery 2006; Yu et al. 2006). This adds extra significance to the stable interaction of DBT and PER: One function of PER may be to bring DBT into close proximity with CLK, leading to CLK phosphorylation and inactivation, presumably an integral part of the repression mechanism.

Shaggy (Sgg) is a third protein kinase that regulates clock speed. Overexpression of *sgg* in clock cells speeds up the clock, whereas reduced *sgg* levels slow down the clock (Martinek et al. 2001). Although *sgg* clearly regulates period length, it is unknown if *sgg* is required for rhythmicity because *sgg* is essential for fly development and animals without *sgg* function do not survive until adulthood. However, *sgg* is of great interest to the circadian field for at least two reasons: (1) extracellular signals regulate Sgg activity, suggesting that Sgg could link external stimuli to the intracellular clock (supported by the work of Yuan et al. 2005), and (2) Sgg is the major target of Lithium, one of the most effective agents against bipolar disorder, suggesting that bipolar disorder could even be a disorder of the circadian system, at least in part.

Martinek et al. (2001) proposed that increased Sgg levels speed up the clock by directly promoting TIM phosphorylation, leading to earlier nuclear entry of both PER and TIM. However, in a recent paper, Stoleru et al. (2007) revealed that the dramatic period-shortening associated with increased *sgg* expression requires the circadian photoreceptor Cryptochrome (CRY): *sgg* overexpression in a cry^b mutant background gives only a 1-hour period-shortening compared to 3.5 hours in a cry^+ background. Why should a photoreceptor have such a large effect on the period length of flies in constant darkness?

MOLECULAR CLOCKS: WHEN IS A PHOTORECEPTOR MORE THAN JUST A PHOTORECEPTOR?

Before the observations of Stoleru et al. (2007), we had asked the same question for two reasons: (1) one of the major differences between the *Drosophila* and mammalian clocks is the function of CRY: a photoreceptor in flies, but a transcriptional repressor in mammals and (2) if CRY functions solely as a photoreceptor, then it was unclear why the molecular clocks in the majority of clock-containing cells in the fly should become arrhythmic in *cry* mutants. One could imagine that light is required to start the clock. However, this is not the case, at least for the brain clocks driving behavior, because flies kept in constant darkness for their entire development are rhythmic as adults (Sehgal et al. 1992).

Although the usual criteria for a gene to be considered a core clock component are that mutations affect adult locomotor behavior in constant darkness, it was already striking that *cry* mutants had lost circadian rhythms in antennal sensitivity (Krishnan et al. 2001). In addition, because these rhythms are driven by clocks within the antenna, this suggested that *cry* is required for the peripheral antennal clocks to function (Tanoue et al. 2004) and, by analogy to the mammalian clock, could involve CRY functioning as a transcriptional repressor. We tested this idea by examining in which state the clock had stopped in the eye peripheral clocks and found that it was similar to the clocks in *per* mutants: Four direct CLK/CYC targets were constitutively derepressed. To balance this, we performed a gain-of-function experiment and found that overexpression of CRY together with PER stopped the clock with constitutively low levels of CLK/CYC activity, although overexpression of either CRY alone or, perhaps surprisingly, PER alone had little effect on molecular oscillations (Collins et al. 2006).

Taken together, this indicates that CRY can be considered a clock component at least in some clock cells. Perhaps understandably, our bias in the field is to focus on mutations that alter adult locomotor rhythms and these usually affect the molecular clock in the major pacemaker neurons: the small ventral lateral neurons (s-LN_Vs). The behavior of *cry* mutants in constant darkness (DD) is very similar to wild-type fly behavior (Stanewsky et al. 1998), suggesting that molecular rhythms in the s-LN_Vs are largely unaffected by mutating *cry*. Indeed, the paper by Stanewsky et al. (1998) was one piece of evidence that pointed to the s-LN_Vs as the major pacemaker neurons because their clocks were still rhythmic in *cry* mutants. However, some other clocks in brain neurons are stopped in *cry* mutants, suggesting that *cry* functions as a transcriptional repressor in a subset of central brain clock cells that excludes the s-LN_Vs. This fits with the observation that the effects of Sgg on rhythms in constant light (LL) are stronger in flies in which Sgg is overexpressed in all clock neurons except LN_Vs (Stoleru et al. 2007). However, it is not yet clear how Sgg functions via CRY to regulate the clock.

MOLECULAR CLOCKS: A SECOND CLOCK LOOP

cry RNA levels peak around dawn, perhaps to make the cells in which CRY is produced most sensitive to light as the sun rises. *Clk* RNA levels also oscillate in-phase with *cry*, and these rhythms are antiphase to those of *per* and *tim*, whose RNA levels peak close to dusk. How are these antiphase rhythms generated?

Using a now Stone Age technique called differential display, we identified a rhythmically expressed transcriptional repressor—*vrille* (*vri*)—which is a direct CLK/CYC target like *per* and *tim* (Blau and Young 1999). Transgenic overexpression of *vri* in clock cells either lengthened the period or prevented behavioral rhythms, depending on whether a weak or strong *UAS-vri* transgene was used. At the molecular level, the strong *UAS-vri* transgene stopped the clock in a state similar to that of *Clk* mutants.

Furthermore, in wild-type flies, VRI protein levels peaked as *Clk* RNA levels were at their lowest, suggesting that VRI protein could feed back to repress *Clk* expression. Experiments in flies and in vitro supported this model, thereby identifying one component of a second feedback loop in the clock, which Glossop et al. (1999) had previously predicted. Glossop et al. (2003) also came to similar conclusions about the regulation of *Clk* by VRI, and they also showed that *cry* was a second direct target of VRI.

If VRI is the *Clk* repressor, then what activates *Clk* expression? In 2003, we published a paper identifying *Pdp1* as a transcriptional activator for *Clk* (Cyran et al. 2003). We showed that in vitro PDP1 and VRI compete for binding to at least one site in the *Clk* promoter. In vivo, we found that VRI and PDP1 protein levels are rhythmic, but PDP1 levels peak 3–6 hours after VRI. Thus, we proposed that in the late day and early evening, VRI is present and represses *Clk* transcription, whereas later at night as VRI levels fall, PDP1 activates *Clk* expression (see Fig. 1). Furthermore, although *Pdp1* null mutants develop abnormally slowly and do not survive until adulthood, the clock in their larval LN_Vs could be analyzed. We found that the clock stops in the LN_Vs of *Pdp1* null mutant larvae, with constitutively low levels of *tim* RNA and PER protein, again similar to both *vri* overexpression and *Clk* mutants and indicating that *Pdp1* is an essential clock component.

Genetic interactions supported opposite roles for VRI and PDP1 in the clock. The period length in flies heterozygous for either a *vri* or *Pdp1* null mutation is altered slightly from wild type: *vri* heterozygous flies have about 0.7-hour shorter periods than wild type, whereas *Pdp1* heterozygotes have about 0.5-hour longer periods (Blau and Young 1999; Cyran et al. 2003). More dramatic changes were seen when combining alleles. The weakest of the *UAS-vri* overexpression transgenes gives 25-hour rhythms, but removing one copy of *Pdp1* in addition leads to 27-hour rhythms, although with considerable variation, with some flies having as long a period as 28.5 hours (Cyran et al. 2003). Although constitutive expression of *vri* via the Gal4-UAS system is obviously artificial, the synergistic interaction between increased *vri* and decreased *Pdp1* strongly supported the idea that VRI and PDP1 have opposing roles in the clock, in a manner analagous to that described earlier for DBT and PP2a for PER phosphorylation (see above).

There could be several functions for this second clock loop. One possible role is to add robustness to the molecular clock. This is supported by studies of mice lacking *Rev-erbα*, which has an analogous role to VRI in the second mammalian clock loop. *Rev-erbα* knockout mice have slightly shorter periods than wild-type mice overall, but the period is much more variable from mouse to mouse compared to wild-type controls (Preitner et al. 2002). A second role of the second clock loop could be to express genes important for outputs of the molecular clock with phase rhythms opposite those controlled by CLK/CYC (discussed below).

MOLECULAR CLOCKS: A CHALLENGE TO THE STATUS OF PDP1 IN THE SECOND LOOP

The importance of PDP1 in the second clock loop was recently challenged by Benito et al. (2007), who made *UAS-Pdp1ε* and *UAS-Pdp1ε* RNA interference (RNAi) transgenes to respectively overexpress or knock down PDP1ε levels specifically in clock neurons, allowing the analysis of adult behavior in flies. These authors found that either increasing or reducing PDP1ε levels only in the *Pdf*-expressing sLN$_V$s made flies largely arrhythmic. However, one surprising finding was that the molecular clock in the s-LN$_V$s keeps running in DD.

These results led Benito et al. (2007) to suggest that whereas *Pdp1* is important for circadian behavior, its major function is in the regulation of outputs from the clock, rather than in regulating the core clock because this self-sustains even with dramatically altered *Pdp1* levels. What could account for the different interpretations resulting from analyzing a *Pdp1* null mutant that prevents normal development (Cyran et al. 2003) and *Pdp1* RNAi in clock neurons (Benito et al. 2007)? There are at least two possibilities:

1. Because RNAi often produces knockdowns rather than complete null phenotypes, the manipulations of Benito et al. (2007) may have removed most but not all PDP1. Because this still produces behavioral arrhythmicity, it may be that PDP1-regulated behavioral output genes are indeed more sensitive to PDP1 reductions than the molecular clock, because multiple transcriptional and posttranslational controls support molecular clock rhythms (discussed earlier). A little PDP1 goes a long way within the clock by this argument, as long as there is enough to give some *Clk* expression to allow the clock to run. In support of the idea that a *Pdp1* null mutant is stronger than *Pdp1* RNAi, we noticed that *Pdf* RNA is absent from s-LN$_V$s in both *Clk* and *Pdp1* mutants (Cyran et al. 2003), whereas Benito et al. (2007) found that *Pdp1* RNAi did not dramatically affect the level of PDF peptide immunostaining, which was used as a control to quantify molecular clock oscillations.
2. There may be other defects associated with development in *Pdp1* null mutants that affect s-LN$_V$ function. This idea was raised by Benito et al. (2007) who suggested that our inability to detect molecular clock oscillations in *Pdp1* null mutant LN$_V$s might be due to defective development. However, in the first submission of our manuscript, we had included a figure showing that in light/dark cycles (LD), it is possible to detect TIM correctly localized to the nucleus at night in LN$_V$s, even without *Pdp1* (included here as Fig. 2). This indicates that LN$_V$s are present and even function in LD; however, this is a light-driven rhythm because *tim* RNA oscillations stop on the first day in DD (Cyran et al. 2003). Thus, we concluded that *Pdp1* is required for the molecular clock in DD, although the strong effects of light somehow bypass the requirement for *Pdp1*. This is reminiscent of the light-driven molecular clocks that stop in DD in a strong *dbt* hypomorph (Price et al. 1998) and in electrically silenced neurons (Nitabach et al. 2002). The molecular basis for any of these light-driven rhythms remains unclear. The effect of light on the clock can also be seen in the peripheral clocks in the eye, which show strong rhythms in LD but damp rapidly in DD.

In summary, it is clear that *Pdp1* is important for circadian rhythms, and further research on this topic will aim to explain current discrepancies. One important point to note is that in flies in which *Pdp1ε* RNAi is driven by *tim-gal4*, the rhythmic flies have approximately 2-hour longer periods than wild type (Benito et al. 2007), again consistent with a central clock regulatory role for *Pdp1*. However, the idea that *Pdp1* is very important in regulating behavioral outputs from s-LN$_V$s (Benito et al. 2007) is certainly interesting because PDP1's maximal activity at the end of the night coincides with the time at which s-LN$_V$s drive the onset of morning locomotor activity (see below). Thus, transcriptional regulation of clock outputs by VRI and PDP1 may be key to understanding how a molecular clock is linked to behavioral rhythms.

NEUROBIOLOGY AND THE *DROSOPHILA* CIRCADIAN CLOCK

The majority of clocks in flies are found in neurons. In contrast, many different mammalian tissues possess clocks leading to daily physiological rhythms in addition to behavioral rhythms. Thus, studies of *Drosophila* circadian rhythms have a largely neurobiological focus, and it

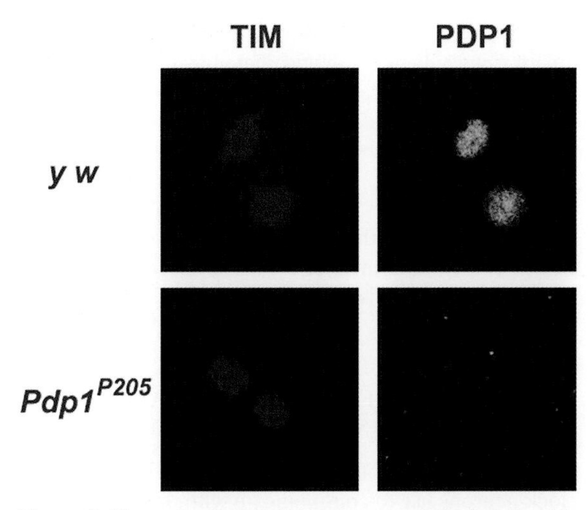

Figure 2. The molecular clock is at least partially functional in light/dark cycles in *Pdp1* null mutants. Control (*y w*) and $Pdp1^{P205}$ null mutant larvae were entrained in LD cycles, dissected at ZT22 and stained for TIM (*red*) and PDP1 (*green*) proteins. Larval LN_Vs were identified by their characteristic clustering of four cells in the center of the brain. Images were taken by confocal microscopy. LN_Vs in both genotypes show TIM in the nucleus, although PDP1 is missing from $Pdp1^{P205}$ mutants. Note that TIM cannot be detected in $Pdp1^{P205}$ mutant LN_Vs on the first day in constant darkness (Cyran et al. 2003).

is in this arena that *Drosophila* is likely to make key contributions to the circadian field in the future.

Some of the fly molecular clocks are found in sensory neurons, for example, in the photoreceptors that make up the adult eye. However, these clocks do not maintain rhythmicity under constant conditions. To our knowledge, it has not been established whether they simply lose synchrony from one another and so the overall population loses rhythmicity or whether the individual clocks run down and stop. Thus, although the clocks in the photoreceptor cells run in light/dark, they cannot really be considered bona fide circadian clocks because at best they drift out of phase under constant conditions, and the hallmark of a circadian clock is to keep accurate 24-hour rhythms in DD. Nevertheless, the ease of obtaining large amounts of tissue from these photoreceptor cells made possible biochemical studies of the fly clock.

In contrast, flies can keep precise rhythms of locomotor activity under constant conditions for several weeks, and the molecular clocks in the central brain neurons continue to show strong rhythms. This has often been taken to mean that the brain neurons driving this behavior have very precise cell-autonomous circadian clocks. However, the recent identification of circadian neural networks in the brain (Grima et al. 2004; Stoleru et al. 2004, 2005) means that we may have to reconsider this idea. Indeed, mammalian pacemaker neurons in the suprachiasmatic nucleus (SCN) show strong self-sustaining rhythms in electrical activity even when dissociated in culture (Welsh et al. 1995). However, these electrical rhythms are quite variable in period when comparing cells isolated from one SCN. This contrasts with the very similar behavioral periods found when comparing different animals and suggested that the precision of circadian behavior in mammals comes partly from the coupling of pacemaker neurons.

Is there any evidence for equivalent coupling in *Drosophila*? There have not yet been any reports of rhythms in isolated fly clock neurons to test their ability to sustain accurate rhythms autonomously. However, a number of in vivo experiments implicate strong effects of one clock neuron group on another. Indeed, it is even possible that the difference between the damping clocks in the eye in DD and the sustained central brain clocks lies in these non-cell-autonomous effects on the clock. As described below, signals from other clock neurons either may reinforce the molecular clock loops to keep them running accurately or may even be required for the clock to keep ticking.

NEUROBIOLOGY: ARE THERE CELL-AUTONOMOUS CIRCADIAN CLOCKS IN *DROSOPHILA*?

One example of cell nonautonomy involves the neuropeptide pigment dispersing factor (PDF), which is produced in the LN_Vs and is likely released from s-LN_V termini in a rhythmic manner to signal circadian time of day information to downstream neurons. In *Pdf* null mutants, or in the absence of the LN_Vs themselves, flies do not anticipate dawn and instead show a startle response to the abrupt lights-on transition (Renn et al. 1999). Hence, in LD cycles, the LN_Vs signal via PDF to drive increased locomotor activity in anticipation of light at dawn, and the PDF cells are also known as morning (M) cells.

At the other end of the day, wild-type flies also anticipate lights-off at dusk by increasing their activity, and this behavior requires a second set of clock neurons that probably includes the dorsal lateral neurons (LN_Ds), dorsal neuron group 1 (DN_1s), and one PDF-ve s-LN_V—collectively known as evening or E cells (Grima et al. 2004; Stoleru et al. 2004). Anticipation of dusk indicates a circadian response, and this was confirmed by the requirement for *per* expression in these E cells (Grima et al. 2004; Stoleru et al. 2004).

Interestingly, the peak of activity at dusk is advanced by approximately 1 hour *Pdf* in null mutants (Renn et al. 1999). This is accompanied by an advanced phase of the molecular clock in the LN_D subgroup of E cells in *Pdf* null mutants (Lin et al. 2004). Because PDF is not produced in LN_Ds, this is a non-cell-autonomous effect of PDF on their clock. However, the initial reports localizing the PDF receptor (PDFR) found that most LN_Ds do not synthesize PDFR (Hyun et al. 2005; Mertens et al. 2005), so how the PDF signal is transmitted to LN_Ds remains to be identified.

Nonautonomy of the clocks in E cells in general was further supported in experiments by Stoleru et al. (2005) in which only the LN_V clock was accelerated by overexpression of *sgg*. When the phase of the clock in all brain clock neurons was measured via *tim* RNA, these authors found that the clocks in the LN_Ds, DN_1s, and DN_3s were also advanced, coming into phase with the s-LN_V clock. Thus, the LN_D, DN_1, and DN_3 clocks were altered in a non-cell-autonomous manner.

In the opposite experiment, Stoleru et al. (2005) tested the effect of accelerating the phase of the clock in non-*Pdf*-expressing cells again via overexpression of *sgg* in all clock neurons except s-LN_Vs. Here, they found that the molecular clock in the LN_Ds, DN_1s, and DN_3s was largely unaffected by this manipulation, presumably because the dominant s-LN_V clock was still running with a 24-hour period. Because *sgg* expression in all clock neurons speeds up the phase of all neuronal clocks, the pace of the clocks in the LN_Ds, DN_1s, and DN_3s is capable of being increased at least when the s-LN_V clock is accelerated (Stoleru et al. 2005).

Finally, when Stoleru et al. (2005) overexpressed *sgg* only in the DN_2 clock neurons, they accelerated both the DN_2 clock *and* the large-LN_V (l-LN_V) clock. Although this had no effect on the period of locomotor behavior, it again demonstrates a non-cell-autonomous clock effect in this case with the DN_2s determining the timing of the l-LN_V clock.

Taking all of this together, one could make the case that the s-LN_Vs and perhaps the DN_2s are the only cells that possess cell-autonomous clocks in *Drosophila* and that they determine clock time for the rest of the brain clock neurons. However, a careful study by Lin et al. (2004) showed that the s-LN_V clocks drift out of synchrony from one another in *Pdf* null mutants in DD. Because PDF is a secreted molecule, this provides evidence for noncell autonomy of even the s-LN_V clock. Potentially, PDF could synchronize the s-LN_Vs via an autocrine signal back to the cell from which it was released, via cell–cell communication between neighboring s-LN_Vs or perhaps most likely via the clock neural circuit.

So if the s-LN_V clock is not completely cell-autonomous either, is there any evidence for non-PDF clock neurons communicating with s-LN_Vs? Yes. In their original classification of M and E cells, Grima et al. (2004) and Stoleru et al. (2004) found that rescue of *per* only to E cells was sufficient to drive both morning and evening peaks of activity, even though the M cells did not have a functional clock. In contrast, ablation of M cells leads to complete loss of morning activity (Renn et al. 1999). The simplest interpretation of these experiments is that E cells communicate to M cells to drive the release of PDF at the correct time for morning activity, and anatomical studies support the idea of communication between E and M cells (Stoleru et al. 2004; Shafer et al. 2006). However, a functional clock only in E cells cannot support long-term rhythmicity in DD, underscoring the importance of M cells in DD (Grima et al. 2004; Stoleru et al. 2004). So although the PDF+ve M cells seem to have the most cell-autonomous of all of the clocks in the brain, there is evidence for communication between E and M cells (see Fig. 3).

NEUROPEPTIDES AS SIGNALS IN BRAIN CLOCK NEURONS

It will be important to identify not only which signals clock neurons use to communicate with one another, but also how these signals are transduced to the molecular clock to alter phase. So far, we know of only three neurotransmitters among all the different clock neuron groups: PDF in the LN_Vs, IPNamide in a subgroup of DN_1 cells (DN_{1a}; Shafer et al. 2006), and neuropeptide F (NPF) in a subset of LN_Ds, although only in male flies (Lee et al. 2006). Of these, only PDF has so far been assigned a circadian role (Renn et al. 1999), but there is also evidence that LN_Vs use an alternative unidentified signal(s) to mediate larval light avoidance (Mazzoni et al. 2005) and to regulate adult cocaine sensitivity (Tsai et al. 2004).

Intriguingly, all of the identified signals are neuropeptides, which are also important in signaling within the SCN. Mice lacking a key neuropeptide receptor in the SCN (the $VPAC_2$ receptor) are arrhythmic (Harmar et al. 2002) and individual cells in the SCN of these mutants either have lost rhythms completely or are desynchronized from one another (Aton et al. 2005; Maywood et al. 2006). Thus, communication among clock neurons via neuropeptides seems to be conserved across species. The demonstration that coupling among clock neurons in the SCN can override genetic defects in their individual clocks (Liu et al. 2007) supports the importance of understanding clock neuron communication.

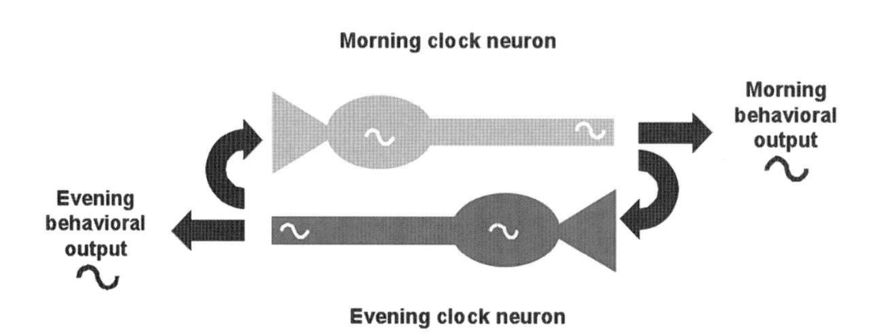

Figure 3. A model for cell–cell interactions that promote precision of behavioral rhythms and morning and evening activity. A highly simplified view of the clock neural circuit involving two neurons. Each cell has a functional clock that helps drive rhythmic neuronal signals. The morning (*light gray*) neuron promotes morning activity, and its output signals also modify the phase of the evening (*dark gray*) neuron. Similarly, signals from the evening neuron regulate both evening activity and input to the morning neuron. The behavioral output signals and the input signals to the other neuron could either be the same or they could be different.

CONCLUSIONS

Molecular genetic analysis has made circadian rhythms the best understood behavior at the molecular level. New approaches that more accurately treat the clock cell as a neuron should also make circadian rhythms the best understood behavior at the neuronal and circuit levels. Signaling between clock neurons seems to augment the molecular loops and keep their precision in the absence of environmental signals. We believe that the ability to manipulate gene expression in precisely defined subsets of clock neurons in *Drosophila* and to measure changes at molecular and behavioral levels means that the fly still has a major role in the circadian field.

ACKNOWLEDGMENTS

We thank Esteban Mazzoni for help with confocal microscopy in Figure 2.

REFERENCES

Akten B., Jauch E., Genova G.K., Kim E.Y., Edery I., Raabe T., and Jackson F.R. 2003. A role for CK2 in the *Drosophila* circadian oscillator. *Nat. Neurosci.* **6:** 251.

Ashmore L.J., Sathyanarayanan S., Silvestre D.W., Emerson M.M., Schotland P., and Sehgal A. 2003. Novel insights into the regulation of the Timeless protein. *J. Neurosci.* **23:** 7810.

Aton S.J., Colwell C.S., Harmar A.J., Waschek J., and Herzog E.D. 2005. Vasoactive intestinal polypeptide mediates circadian rhythmicity and synchrony in mammalian clock neurons. *Nat. Neurosci.* **8:** 476.

Benito J., Zheng H., and Hardin P.E. 2007. *PDP1epsilon* functions downstream of the circadian oscillator to mediate behavioral rhythms. *J. Neurosci.* **27:** 2539.

Blau J. 2001. The *Drosophila* circadian clock: What we know and what we don't know. *Semin. Cell Dev. Biol.* **12:** 287.

Blau J. and Young M.W. 1999. Cycling *vrille* expression is required for a functional *Drosophila* clock. *Cell* **99:** 661.

Collins B., Mazzoni E., Stanewsky R., and Blau J. 2006. *Drosophila* CRYPTOCHROME is a circadian transcriptional repressor. *Curr. Biol.* **16:** 441.

Cyran S.A., Yiannoulos G., Buchsbaum A.M., Saez L., Young M.W., and Blau J. 2005. The Double-Time protein kinase regulates the subcellular localization of the *Drosophila* clock protein Period. *J. Neurosci.* **25:** 5430.

Cyran S.A., Buchsbaum A.M., Reddy K.L., Lin M.-C., Glossop N.R., Hardin P.E., Young M.W., Storti R.V., and Blau J. 2003. *vrille*, *Pdp1*, and *dClock* form a second feedback loop in the *Drosophila* circadian clock. *Cell* **112:** 329.

Fu L., Pelicano H., Liu J., Huang P., and Lee C. 2002. The circadian gene *Period2* plays an important role in tumor suppression and DNA damage response in vivo. *Cell* **111:** 41.

Glossop N.R., Lyons L.C., and Hardin P.E. 1999. Interlocked feedback loops within the *Drosophila* circadian oscillator. *Science* **286:** 766.

Glossop N.R., Houl J.H., Zheng H., Ng F.S., Dudek S.M., and Hardin P.E. 2003. VRILLE feeds back to control circadian transcription of *Clock* in the *Drosophila* circadian oscillator. *Neuron* **37:** 249.

Grima B., Chelot E., Xia R., and Rouyer F. 2004. Morning and evening peaks of activity rely on different clock neurons of the *Drosophila* brain. *Nature* **431:** 869.

Harmar A.J., Marston H.M., Shen S., Spratt C., West K.M., Sheward W.J., Morrison C.F., Dorin J.R., Piggins H.D., Reubi J.C., Kelly J.S., Maywood E.S., and Hastings M.H. 2002. The VPAC(2) receptor is essential for circadian function in the mouse suprachiasmatic nuclei. *Cell* **109:** 497.

Hunt T. and Sassone-Corsi P. 2007. Riding tandem: Circadian clocks and the cell cycle. *Cell* **129:** 461.

Hyun S., Lee Y., Hong S.T., Bang S., Paik D., Kang J., Shin J., Lee J., Jeon K., Hwang S., Bae E., and Kim J. 2005. *Drosophila* GPCR Han is a receptor for the circadian clock neuropeptide PDF. *Neuron* **48:** 267.

Kim E.Y. and Edery I. 2006. Balance between DBT/CKIepsilon kinase and protein phosphatase activities regulate phosphorylation and stability of *Drosophila* CLOCK protein. *Proc. Natl. Acad. Sci.* **103:** 6178.

Konopka R.J. and Benzer S. 1971. Clock mutants of *Drosophila melanogaster*. *Proc. Natl. Acad. Sci.* **68:** 2112.

Krishnan B., Levine J.D., Sisson K., Dowse H.B., Funes P., Hall J.C., Hardin P.E., and Dryer S.E. 2001. A new role for Cryptochrome in a *Drosophila* circadian oscillator. *Nature* **411:** 313.

Lee G., Bahn J.H., and Park J.H. 2006. Sex- and clock-controlled expression of the neuropeptide F gene in *Drosophila*. *Proc. Natl. Acad. Sci.* **103:** 12580.

Lin J.M., Kilman V.L., Keegan K., Paddock B., Emery-Le M., Rosbash M., and Allada R. 2002. A role for casein kinase 2α in the *Drosophila* circadian clock. *Nature* **420:** 816.

Lin Y., Stormo G.D., and Taghert P.H. 2004. The neuropeptide pigment-dispersing factor coordinates pacemaker interactions in the *Drosophila* circadian system. *J. Neurosci.* **24:** 7951.

Liu A.C., Welsh D.K., Ko C.H., Tran H.G., Zhang E.E., Priest A.A., Buhr E.D., Singer O., Meeker K., Verma I.M., Doyle F.J., III, Takahashi J.S., and Kay S.A. 2007. Intercellular coupling confers robustness against mutations in the SCN circadian clock network. *Cell* **129:** 605.

Martinek S., Inonog S., Manoukian A.S., and Young M.W. 2001. A role for the segment polarity pene *shaggy*/GSK-3 in the *Drosophila* circadian clock. *Cell* **105:** 769.

Maywood E.S., Reddy A.B., Wong G.K., O'Neill J.S., O'Brien J.A., McMahon D.G., Harmar A.J., Okamura H., and Hastings M.H. 2006. Synchronization and maintenance of timekeeping in suprachiasmatic circadian clock cells by neuropeptidergic signaling. *Curr. Biol.* **16:** 599.

Mazzoni E.O., Desplan C., and Blau J. 2005. Circadian pacemaker neurons transmit and modulate visual information to control a rapid behavioral response. *Neuron* **45:** 293.

Mertens I., Vandingenen A., Johnson E.C., Shafer O.T., Li W., Trigg J.S., De Loof A., Schoofs L., and Taghert P.H. 2005. PDF receptor signaling in *Drosophila* contributes to both circadian and geotactic behaviors. *Neuron* **48:** 213.

Meyer P., Saez L., and Young M.W. 2006. PER-TIM interactions in living *Drosophila* cells: An interval timer for the circadian clock. *Science* **311:** 226.

Nitabach M.N., Blau J., and Holmes T.C. 2002. Electrical silencing of *Drosophila* pacemaker neurons stops the free-running circadian clock. *Cell* **109:** 485.

Preitner N., Damiola F., Lopez-Molina L., Zakany J., Duboule D., Albrecht U., and Schibler U. 2002. The orphan nuclear receptor REV-ERBα controls circadian transcription within the positive limb of the mammalian circadian oscillator. *Cell* **110:** 251.

Price J.L., Blau J., Rothenfluh A., Abodeely M., Kloss B., and Young M.W. 1998. *double-time* is a novel *Drosophila* clock gene that regulates PERIOD protein accumulation. *Cell* **94:** 83.

Renn S.C., Park J.H., Rosbash M., Hall J.C., and Taghert P.H. 1999. A *pdf* neuropeptide gene mutation and ablation of PDF neurons each cause severe abnormalities of behavioral circadian rhythms in *Drosophila*. *Cell* **99:** 791.

Rothenfluh A., Young M.W., and Saez L. 2000. A TIMELESS-independent function for PERIOD proteins in the *Drosophila* clock. *Neuron* **26:** 505.

Sathyanarayanan S., Zheng X., Xiao R., and Sehgal A. 2004. Posttranslational regulation of *Drosophila* PERIOD protein by protein phosphatase 2A. *Cell* **116:** 603.

Sehgal A., Price J., and Young M.W. 1992. Ontogeny of a biological clock in *Drosophila melanogaster*. *Proc. Natl. Acad. Sci.* **89:** 1423.

Shafer O.T., Rosbash M., and Truman J.W. 2002. Sequential nuclear accumulation of the clock proteins Period and Timeless in the pacemaker neurons of *Drosophila*

melanogaster. *J. Neurosci.* **22:** 5946.
Shafer O.T., Helfrich-Förster C., Renn S.C., and Taghert P.H. 2006. Reevaluation of *Drosophila melanogaster's* neuronal circadian pacemakers reveals new neuronal classes. *J. Comp. Neurol.* **498:** 180.
Shafer O.T., Levine J.D., Truman J.W., and Hall J.C. 2004. Flies by night: Effects of changing day length on *Drosophila's* circadian clock. *Curr. Biol.* **14:** 424.
Siwicki K.K., Eastman C., Petersen G., Rosbash M., and Hall J.C. 1988. Antibodies to the *period* gene product of *Drosophila* reveal diverse tissue distribution and rhythmic changes in the visual system. *Neuron* **1:** 141.
Stanewsk R. 2003. Genetic analysis of the circadian system in *Drosophila melanogaster* and mammals. *J. Neurobiol.* **54:** 111.
Stanewsky R., Kaneko M., Emery P., Beretta B., Wager-Smith K., Kay S.A., Rosbash M., and Hall J.C. 1998. The *cry*^{b} mutation identifies Cryptochrome as an essential circadian photoreceptor in *Drosophila*. *Cell* **95:** 681.
Stoleru D., Peng Y., Agosto J., and Rosbash M. 2004. Coupled oscillators control morning and evening locomotor behaviour of *Drosophila*. *Nature* **431:** 862.
Stoleru D., Peng Y., Nawathean P., and Rosbash M. 2005. A resetting signal between *Drosophila* pacemakers synchronizes morning and evening activity. *Nature* **438:** 238.
Stoleru D., Nawathean P., Fernandez de la Paz M., Menet J.S., Ceriani M.F., and Rosbash M. 2007. The *Drosophila* circadian network is a seasonal timer. *Cell* **129:** 207.
Sun Z.S., Albrecht U., Zhuchenko O., Bailey J., Eichele G., and Lee C.C. 1997. RIGUI, a putative mammalian ortholog of the *Drosophila period* gene. *Cell* **90:** 1003.
Suri V., Hall J.C., and Rosbash M. 2000. Two novel *doubletime* mutants alter circadian properties and eliminate the delay between RNA and protein in *Drosophila*. *J. Neurosci.* **20;** 7547.
Tanoue S., Krishnan P., Krishnan B., Dryer S.E., and Hardin P.E. 2004. Circadian clocks in antennal neurons are necessary and sufficient for olfaction rhythms in *Drosophila*. *Curr. Biol.* **14:** 638.
Toh K.L., Jones C., He Y., Eide E.J., Hinz W.A., Virshup D.M., Ptáček L.J., and Fu Y.H. 2001. An hPer2 phosphorylation site mutation in familial advanced sleep phase syndrome. *Science* **291:** 1040.
Tsai L., Bainton R.J., Blau J., and Heberlein U. 2004. *Lmo* mutants reveal a novel role for circadian pacemaker neurons in cocaine-induced behaviors. *PLoS Biol.* **2:** 2122.
Welsh D.K., Logothetis D.E., Meister M., and Reppert S.M. 1995. Individual neurons dissociated from rat suprachiasmatic nucleus express independently phased circadian firing rhythms. *Neuron* **14:** 697.
Yu W., Zheng H., Houl J.H., Dauwalder B., and Hardin P.E. 2006. PER-dependent rhythms in CLK phosphorylation and E-box binding regulate circadian transcription. *Genes Dev.* **20:** 723.
Yuan Q., Lin F., Zheng X., and Sehgal A. 2005. Serotonin modulates circadian entrainment in *Drosophila*. *Neuron* **47:** 115.

Genetics and Neurobiology of Circadian Clocks in Mammals

S.M. Siepka,*†‡ S.-H. Yoo,*†‡ J. Park,*† C. Lee,§ and J.S. Takahashi*†¶§

*Howard Hughes Medical Institute, †Center for Functional Genomics, ¶Department of Neurobiology and Physiology, Northwestern University, Evanston, Illinois 60208; §Department of Biological Sciences, College of Medicine Florida State University, Tallahassee, Florida 32306

In animals, circadian behavior can be analyzed as an integrated system, beginning with genes and leading ultimately to behavioral outputs. In the last decade, the molecular mechanism of circadian clocks has been unraveled primarily by the use of phenotype-driven (forward) genetic analysis in a number of model systems. Circadian oscillations are generated by a set of genes forming a transcriptional autoregulatory feedback loop. In mammals, there is a "core" set of circadian genes that form the primary negative feedback loop of the clock mechanism (*Clock/Npas2, Bmal1, Per1, Per2, Cry1, Cry2,* and *CK1ε*). A further dozen candidate genes have been identified and have additional roles in the circadian gene network such as the feedback loop involving *Rev-erbα*. Despite this remarkable progress, it is clear that a significant number of genes that strongly influence and regulate circadian rhythms in mammals remain to be discovered and identified. As part of a large-scale *N*-ethyl-*N*-nitrosourea mutagenesis screen using a wide range of nervous system and behavioral phenotypes, we have identified a number of new circadian mutants in mice. Here, we describe a new short-period circadian mutant, *part-time* (*prtm*), which is caused by a loss-of-function mutation in the *Cryptochrome1* (*Cry1*) gene. We also describe a long-period circadian mutant named *Overtime* (*Ovtm*). Positional cloning and genetic complementation reveal that *Ovtm* is encoded by the F-box protein FBXL3, a component of the SKP1–CUL1–F-box protein (SCF) E3 ubiquitin ligase complex. The *Ovtm* mutation causes an isoleucine to threonine (I364T) substitution leading to a loss of function in FBXL3 that interacts specifically with the CRYPTOCHROME (CRY) proteins. In *Ovtm* mice, expression of the PERIOD proteins PER1 and PER2 is reduced; however, the CRY proteins CRY1 and CRY2 are unchanged. The loss of FBXL3 function leads to a stabilization of the CRY proteins, which in turn leads to a global transcriptional repression of the *Per* and *Cry* genes. Thus, *Fbxl3*^{Ovtm} defines a molecular link between CRY turnover and CLOCK/BMAL1-dependent circadian transcription to modulate circadian period.

INTRODUCTION

The mechanism of circadian oscillators in mammals is generated by a cell-autonomous autoregulatory transcription-translation feedback loop (Reppert and Weaver 2002; Lowrey and Takahashi 2004; Ko and Takahashi 2006). In the primary negative feedback loop, the basic helix-loop-helix (bHLH)-PAS transcription factors, CLOCK (and its paralog NPAS2) and BMAL1 (ARNTL), dimerize and activate transcription of the *Period* (*Per1, Per2*) and *Cryptochrome* (*Cry1, Cry2*) genes (Antoch et al. 1997; King et al. 1997; Gekakis et al. 1998; Kume et al. 1999; Bunger et al. 2000; DeBruyne et al. 2007). As the PER proteins accumulate, they form complexes with the CRY proteins, translocate into the nucleus, and interact with the CLOCK/BMAL1 complex to inhibit their own transcription (Lee et al. 2001). This leads to a fall in the inhibitory complex through turnover, and the cycle starts again with a new round of CLOCK/BMAL1-activated transcription. Additional pathways in the circadian gene network such as the second negative feedback loop (involving *Rev-erbα*) in the positive limb of the oscillator are thought to add robustness to the circadian mechanism (Preitner et al. 2002; Sato et al. 2004). Finally, posttranslational modifications have critical roles in regulating the turnover, cellular localization, and activity of circadian clock proteins (Lowrey et al. 2000; Eide et al. 2005; Gallego and Virshup 2007).

Despite this progress, it is clear that a significant number of genes that strongly influence and regulate circadian rhythms in mammals remain to be discovered and identified (Shimomura et al. 2001; Takahashi 2004). Forward genetic screens have been one of the most effective tools for circadian gene discovery (Takahashi et al. 1994; Vitaterna et al. 1994; Takahashi 2004), and we have used this approach to screen the mouse genome for circadian rhythm mutants generated in the Neurogenomics Project in the Center for Functional Genomics at Northwestern University (Vitaterna et al. 2006).

MUTAGENESIS, SCREENING, AND IDENTIFICATION OF THE *PRTM* and *OVTM* GENES

In an *N*-ethyl-*N*-nitrosourea (ENU) recessive screen using the BTBR T$^+$ tf/J (BTBR/J) inbred mouse strain (Siepka and Takahashi 2005), we identified two mutants with short (21.4 hours) and long (25.8 hours) circadian periods in constant darkness (DD) (Fig. 1). These two mutants were named *part-time (prtm)* and *Overtime (Ovtm)* (Siepka et al. 2007), respectively. Genetic mapping of *prtm* places this mutant on chromosome 10 in the region of *Cry1* (Fig. 2A). Complementation tests of *prtm* with a *Cry1* null allele show that *prtm* is a new allele of *Cry1* (Fig. 2B). Sequencing of the *Cry1* gene in *prtm* mutants reveals a T to C mutation in the second position of the splice donor site of exon 2 causing readthrough and premature termination in intron 2 (Fig. 2C). Crosses of *prtm* with *Cry2* null mutants to produce double homozygous mutants show that these mice are arrhythmic, similar

‡These authors contributed equally to this work.

to that seen in *Cry1/Cry2* double-mutant mice (Vitaterna et al. 1999). Thus, *prtm* is a loss-of-function allele of *Cry1* and serves as a validation of the genetic screen.

The second mutant, *Ovtm*, maps to a 1.7-cM interval on chromosome 14 (Fig. 3) (Siepka et al. 2007). This region corresponds to a 4-Mb interval and contains 18 open reading frames, none of which corresponds to previously known circadian clock genes (Fig. 4A) (Lowrey and Takahashi 2004). We sequenced all annotated exons for the 18 candidate genes in the *Ovtm* interval and found only a single nonsynonymous point mutation within the coding region of *Fbxl3*. There is a single-base transition from A to G in exon 5 of *Fbxl3* in *Ovtm* mice as compared to wild-type BTBR/J mice. This mutation cosegregated perfectly with the long-period phenotype of *Ovtm*/*Ovtm* mice. The point mutation converts amino acid residue 364 from isoleucine to threonine in FBXL3 (Fig. 4A,B). This isoleucine residue is highly conserved in FBXL3 from vertebrates and in the mouse paralog FBXL21 (Fig. 4B). FBXL3 is a member of the F-box protein family with leucine-rich repeats (LRR) which is defined by its founding member, SKP2 (S-phase kinase-associated protein-2) (FBXL1) (Jin et al. 2004). SKP2 is the F-box protein moiety in the SKP1–CUL1–F-box-protein $(SCF)^{SKP2}$ E3 ubiquitin ligase complex that mediates the recognition and ubiquitination of the CDK2 inhibitor, $p27^{Kip1}$, to target it for proteasomal degradation (Cardozo and Pagano 2004). FBXL3 has 11 LRRs and can align with SKP2, which has 10 LLRs based on its protein structure; however, the carboxyl termini of SKP2 and FBXL3 are not conserved likely due to the recognition of different substrates (Hao et al. 2005). The *Ovtm* I364T mutation occurs in the carboxyl terminus of FBXL3 between LRR10 and LRR11 where the alignment with SKP2 becomes divergent (Fig. 4B). Because the LRR domains of F-box proteins are involved with substrate recognition with the SCF complex, we hypothesized that the I364T mutation could alter the interaction of FBXL3 with its substrates.

Because we isolated only one mutant allele of *Fbxl3*, and either a second independent allele, rescue, or functional evidence is required for proof in positional cloning (Takahashi et al. 1994), we used genetic complementation tests to confirm that *Ovtm* was allelic with *Fbxl3*. The crosses show that *Ovtm* and an *Fbxl3* gene trap (GT) fail to complement each other, thus providing independent and definitive evidence that *Ovtm* is an allele of *Fbxl3* (Fig. 4C). Interestingly, the period length of GT/*Ovtm* mice is indistinguishable from *Ovtm* homozygotes, suggesting that the *Ovtm* mutant allele is likely a hypomorphic or loss-of-function allele.

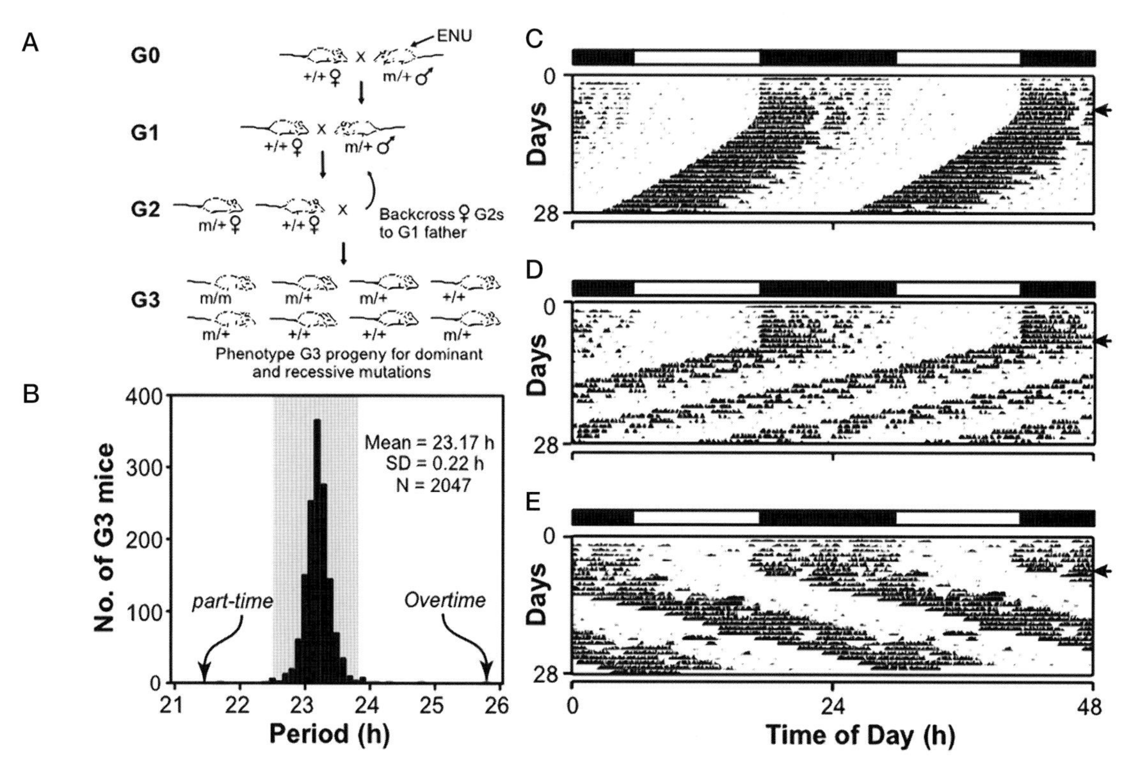

Figure 1. ENU mutagenesis screen. (*A*) Mutant mouse production. Male BTBR/J mice were treated with the chemical mutagen *N*-ethyl-*N*-nitrosourea (ENU) (1 x 250 mg/kg) and G_1 offspring were used to breed three-generation pedigrees to make ENU-induced mutations homozygous as described previously (Siepka and Takahashi 2005). (*B*) Histogram distribution of free-running period values for 3198 mice screened. (*Gray shaded area*) ±3 standard deviations (S.D.) from the mean. The original *part-time* and *Overtime* mutants are indicated by arrows. (*C*) Representative actogram of a wild-type BTBR/J mouse. The actogram is double-plotted where 48 hours of activity are represented on each horizontal line. The mice were kept on a LD12:12 cycle (*bar above*) for the first 7 days and then released into DD for 21 days (*arrowhead on the right*). (*D*) Actogram of the original *prtm* G3 mouse. The animal (a *prtm* homozygote) had a free-running period of 21.4 hours. (*E*) Actogram of the original *Ovtm* G3 mouse. The animal (an *Ovtm* homozygote) had a free-running period of 25.83 hours.

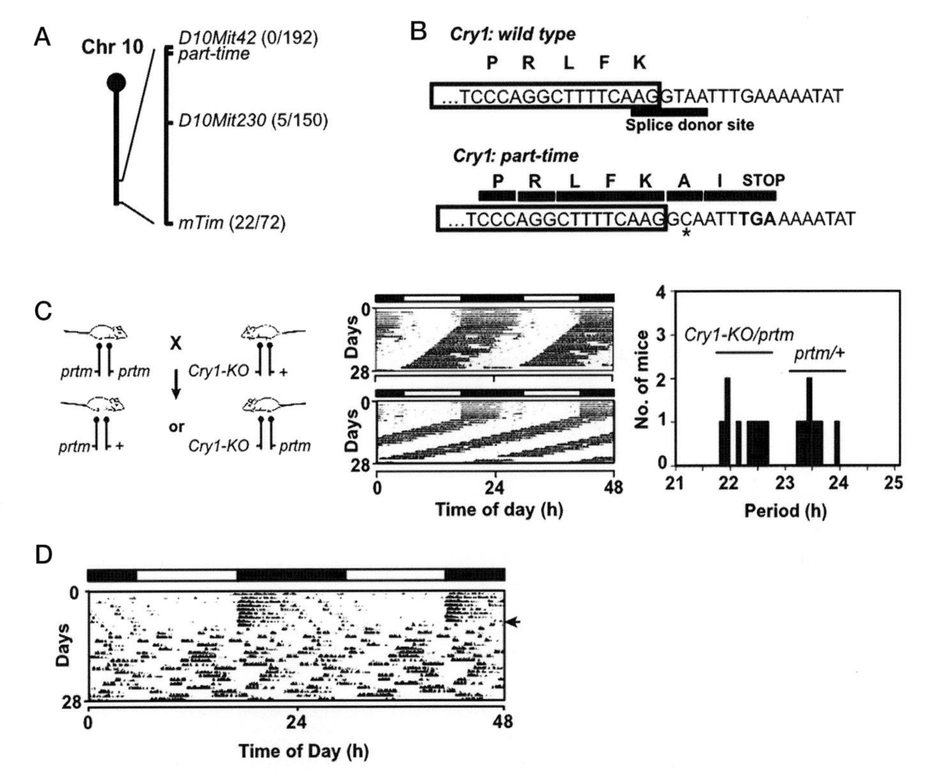

Figure 2. Genetic mapping and cloning of the *part-time* mutant. (*A*) Genetic mapping of *prtm* to chromosome 10. (*B*) The *prtm* mutation occurs within the splice donor site at the 3′ end of exon 2 of *Cry1*. (*C*) Complementation test for *prtm* and *Cry1*. (*Left panel*) Mating scheme for the complementation test. *prtm* mice were crossed to heterozygous *Cry1* knockout (*mCry1-KO*) mice. Circadian behavior was recorded for 15 progeny. The actograms are representative of *prtm*/+ (*top*) and *Cry-1KO/prtm* (*bottom*) mice. Period histogram distribution: (*left*) *Cry1-KO/prtm* mice (mean 22.15 hours; S.D. = 0.31; $n = 8$); (*right*) *prtm*/+ mice (mean = 23.4 hours; S.D. = 0.23; $n = 7$). Student's *t*-test (unequal variances) shows a significant difference between the two populations (DF = 13; $T = -9.12$; $p = 5.2 \times 10^{-7}$). (*D*) Representative actogram of a *prtm/prtm*, $Cry2^{-/-}$ double-mutant mouse showing an arrhythmic phenotype similar to that seen with $Cry1^{-/-}$, $Cry2^{-/-}$ double-mutant mice.

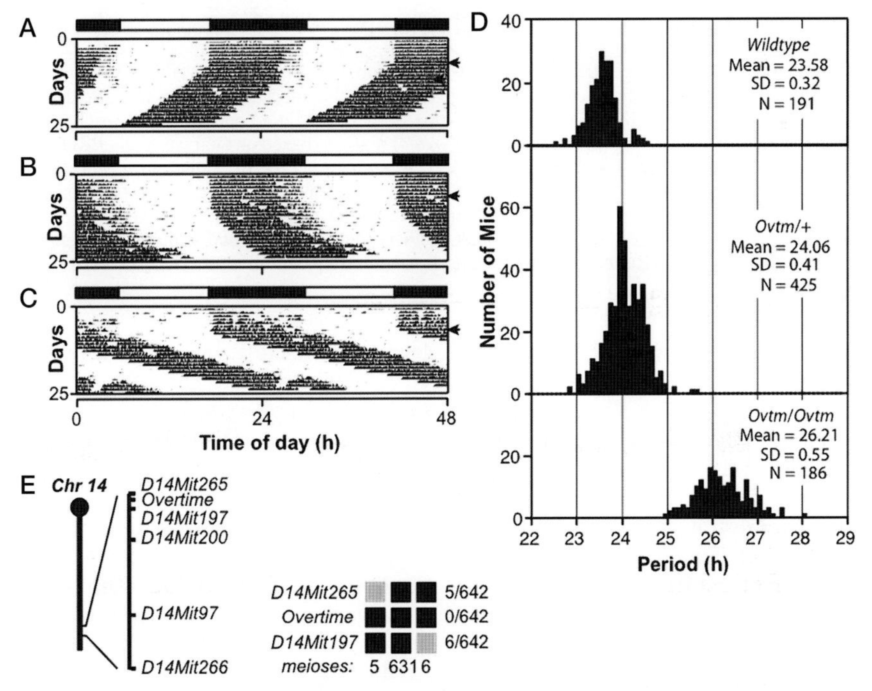

Figure 3. Semidominant phenotype of *Overtime* and genetic mapping. (*A*) Representative actograms of wild-type, (*B*) *Ovtm*/+, and (*C*) *Ovtm/Ovtm* [BTBR/J x C57BL/6J]F_2 mice. The actograms are plotted as described in Figure 1. (*D*) Period distribution of F_2 intercross progeny. The three panels from top to bottom represent wild-type, *Ovtm*/+, and *Ovtm/Ovtm* mice, respectively. (*E*) *Ovtm* maps between D14Mit265 and D14Mit197 on chromosome 14. Haplotypes of the 321 *Ovtm/Ovtm* F2 intercross progeny (642 meioses) are on the right. (*Black boxes*) BTBR/J alleles; (*gray boxes*) heterozygous alleles (BTBR/J and C57BL/6J). The number of recombinants per total meioses is indicated to the right of the haplotype map. (Reprinted, with permission, from Siepka et al. 2007 [©Elsevier].)

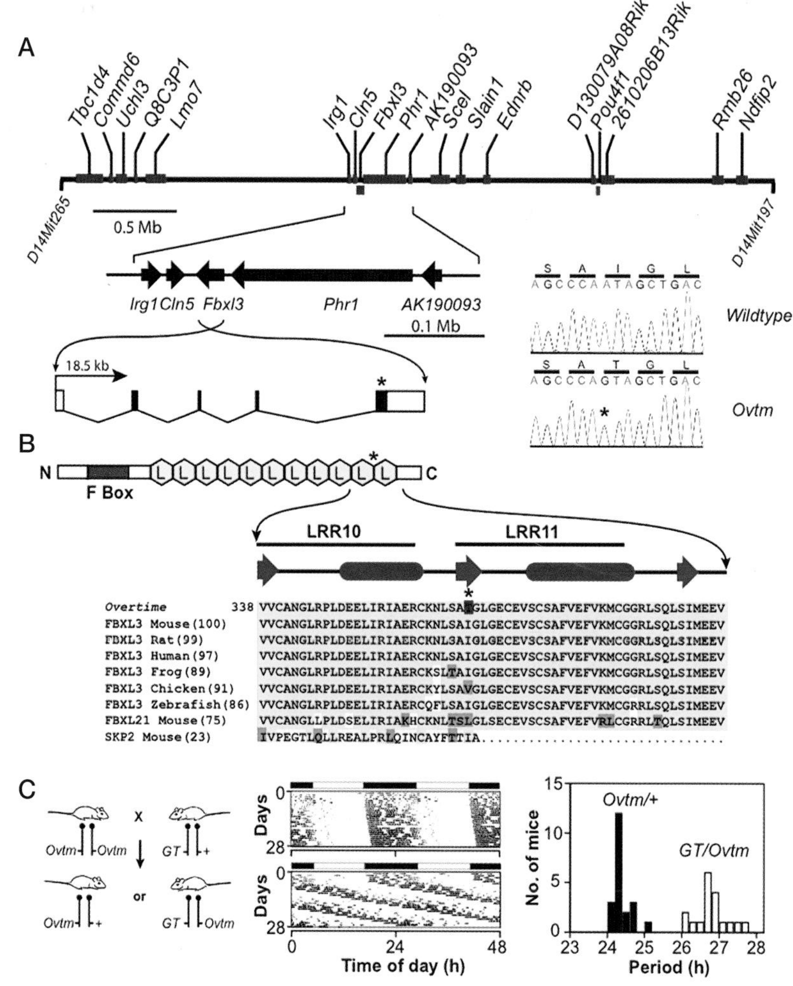

Figure 4. Positional cloning of *Overtime* and identification of *Fbxl3* mutation. (*A*) Physical map of the *Ovtm* interval. *Ovtm* maps to a 4-Mb region of chromosome 14. (*Red blocks*) Eighteen candidate genes within the interval. *Asterisks* indicate the location of the *Ovtm* mutation. (*B*) FBXL3 is an F-box protein. FBXL3 contains one F-box domain (*red box*) and 11 leucine-rich regions (LRR) (*yellow hexagons*) (Jin et al. 2004). β-strand (*blue arrow*) and α-helical (*red ovals*) regions (based on analysis using PROF in the PredictProtein server; Rost et al. 2004) of LRR10 and LRR11 are indicated above the sequence. Protein alignment: (*Yellow*) amino acid identity; (*green*) conservative substitutions; (*white*) nonconservative substitutions; (*red*) I364T *Ovtm* mutation. (*C*) Complementation test for *Ovtm* and *Fbxl3*. (*Left panel*) Mating scheme for the complementation assay. *Ovtm* mice were crossed to heterozygous *Fbxl3* gene-trap (GT) mice. Actograms are representative of *Ovtm*/+ (*top*) and GT/*Ovtm* (*bottom*) mice. Period histogram distribution: (*Black bars*) *Ovtm*/+ mice; (*white bars*) GT/*Ovtm* mice. (Reprinted, with permission, from Siepka et al. 2007 [©Elsevier].)

EFFECTS OF THE OVERTIME MUTATION ON CIRCADIAN CLOCK GENE EXPRESSION

Because FBXL3 is likely a component of an SCF E3 ubiquitin ligase complex, we examined the in vivo expression patterns of circadian clock proteins in mouse tissues to explore whether *Ovtm* might alter their abundance by affecting degradation. Figure 5A shows expression patterns of the clock proteins, CRY1, CRY2, PER1, PER2, CLOCK, and BMAL1, in liver and cerebellum. In wild-type mice, there were low-amplitude rhythms of CRY1 and CRY2 and high-amplitude rhythms of PER1 and PER2 as reported previously (Lee et al. 2001). In *Ovtm* liver tissue, CRY1 and CRY2 protein patterns were not significantly altered; however, PER1 and PER2 levels were significantly reduced (Fig. 4B). In the cerebellum, the effects of *Ovtm* were more striking. Although CRY1 levels were not different, CRY2 levels were significantly elevated in *Ovtm* mice, consistent with the hypothesis that CRY degradation is impaired. In addition, there were very clear reductions in the levels of PER1 and PER2. The reduction of PER1 and PER2 levels in *Ovtm* mice is unexpected and counterintuitive. We would have expected to see an increase rather than a decrease in protein abundance if the PER proteins were targets of FBXL3 because the *Ovtm* mutation is a loss-of-function mutation. This suggests that it is unlikely that the PER proteins are targets of FBXL3 and that the reduction in PER levels could occur as a consequence of the negative feedback on CLOCK/BMAL1-dependent transcription.

To explore the reasons for the reduction in PER1 and PER2 protein abundance, we profiled the in vivo circadian mRNA expression patterns for *Cry1, Cry2, Per1, Per2*, and *Dbp* in the liver and cerebellum of mice main-

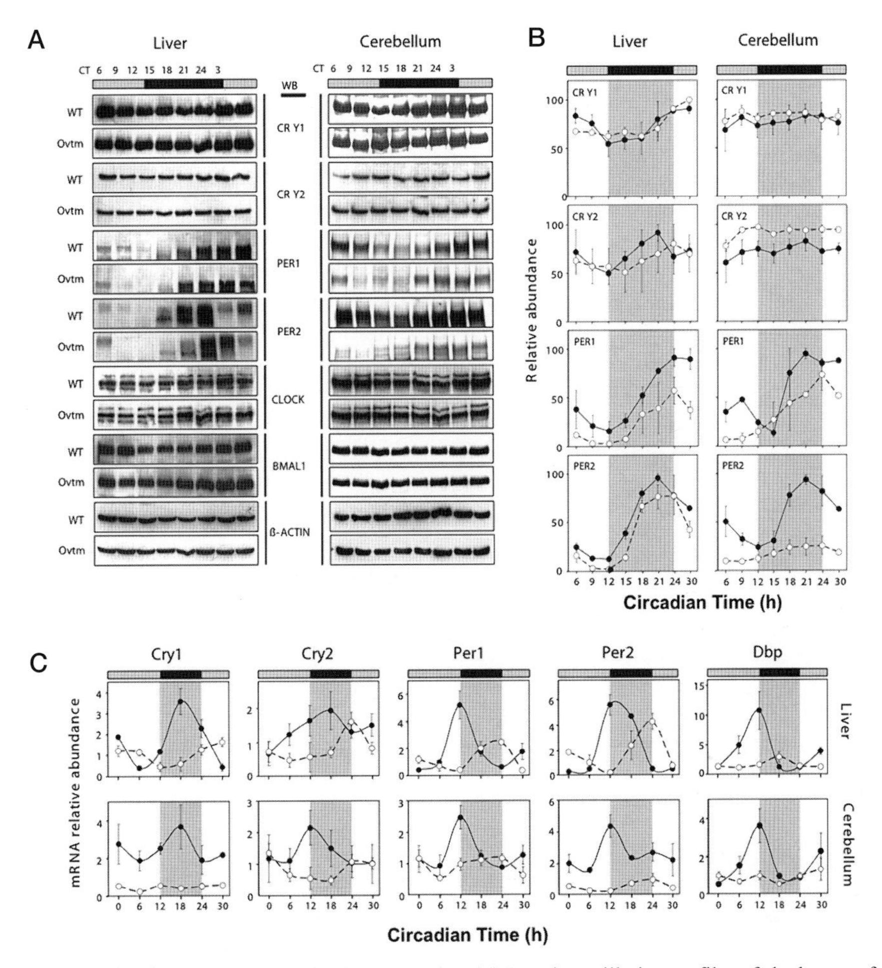

Figure 5. Altered circadian clock gene expression in *Overtime* mice. (*A*) Protein oscillation profiles of clock genes from liver and cerebellum. Wild-type and *Ovtm* mutant tissues were collected at indicated circadian times. Western blotting was performed on total protein extracts with indicated antibodies. (*B*) Quantification of proteins from liver and cerebellum. (*Closed circles*) Normalized values from wild type; (*open circles*) normalized values from *Ovtm* mice. (*C*) Real-time RT-PCR (reverse transcriptase–polymerase chain reaction) analysis for clock gene expression in wild-type and *Ovtm* mice. (*Closed circles*) Wild type; (*open circles*) values from *Ovtm* mice. All cycling genes show a significant reduction of mRNA level from *Ovtm* mice compared to wild-type mice in liver and cerebellum except for *Per2* in liver. (Reprinted, with permission, from Siepka et al. 2007 [©Elsevier].)

tained in DD. As shown in Figure 5C, the *Ovtm* mutation caused significant reductions in the mRNA abundance of all of these cycling transcripts, with the strongest effects being seen with *Cry1* and *Per2* in the cerebellum. At the mRNA level, both a delay in the peak time and a reduction in abundance can be seen. Importantly, although CRY1 and CRY2 protein levels were not lower in *Ovtm* mice, the corresponding mRNA levels for *Cry1* and *Cry2* are significantly reduced in both tissues. In addition, mRNA levels for the cycling CLOCK target gene, *Dbp* (Ripperger and Schibler 2006), were very strongly reduced in *Ovtm* mice. Thus, the mRNA profiling experiments point to an interesting and unexpected consequence of the *Ovtm* mutation: a reduction in steady-state mRNA expression of *Cry1*, *Cry2*, *Per1*, *Per2*, and *Dbp*, which are all transcriptional targets of the CLOCK/BMAL1 complex (Gekakis et al. 1998; Kume et al. 1999; Yoo et al. 2005; Ripperger and Schibler 2006).

Comparison of the effects of *Ovtm* on protein versus mRNA abundance suggests that there are two different effects on the expression of the CRY and PER proteins. The PER protein levels appear to be reduced as a consequence of reduced transcript levels. In contrast, the CRY protein levels are *not* reduced even in the face of reduced transcript levels. This suggests that potential reductions in CRY protein levels caused by reduced *Cry* transcript levels could be compensated by a reduction in protein degradation.

INTERACTION OF OVTM WITH CIRCADIAN CLOCK PROTEINS

The Pagano laboratory has found that FBXL3 targets CRY proteins for ubiquitination and degradation (Busino et al. 2007). To confirm these results and determine whether the *Ovtm* mutation affects interactions with CRY, we examined the interaction of FBXL3 or OVTM with circadian clock proteins by immunoprecipitation assays. Both FBXL3 and OVTM interacted strongly with native CRY1 and CRY2 proteins. Very weak or no interaction of FBXL3 was seen with PER1 and PER2, especially in comparison with that seen between the PERs and βTrCP1, an F-box protein known to interact with the PERs (Fig. 6A) (Eide et al. 2005; Shirogane et al. 2005). In all experiments, there was a discernibly stronger interaction of the CRY proteins with FBXL3 relative to OVTM, but the difference was subtle.

We also used tagged proteins in coimmunoprecipitation assays in 293A cells which are easily transfected and express relatively low levels of clock proteins. Both FBXL3 and OVTM interacted strongly with CRY1 and CRY2 (Fig. 6B). To explore the weak interaction of FBXL3 with PER proteins, tagged *Per* constructs were also tested. All three PER proteins showed interactions with FBXL3 and OVTM, however, the strongest interactions were seen with PER2. Because PER2 interacts very strongly with CRY1 (Griffin et al. 1999; Kume et al. 1999; Lee et al. 2001), it is likely that the interactions seen here with FBXL3 may be indirect via CRY1.

EFFECTS OF OVTM ON CRY DEGRADATION

To determine whether OVTM is less efficient than FBXL3 in inducing the degradation of CRY1, we compared the effects of FBXL3 and OVTM on the stability of CRY1 following cycloheximide treatment to prevent de novo protein synthesis in transfected cells (Fig. 6C). In 293A cells, FBXL3 expression leads to the degradation of CRY1, and this degradation is blocked by the 26S proteasomal inhibitor MG132 (Fig. 6C). OVTM is less effective than FBXL3 in causing CRY1 degradation under these conditions. Interestingly, the turnover of FBXL3 is also

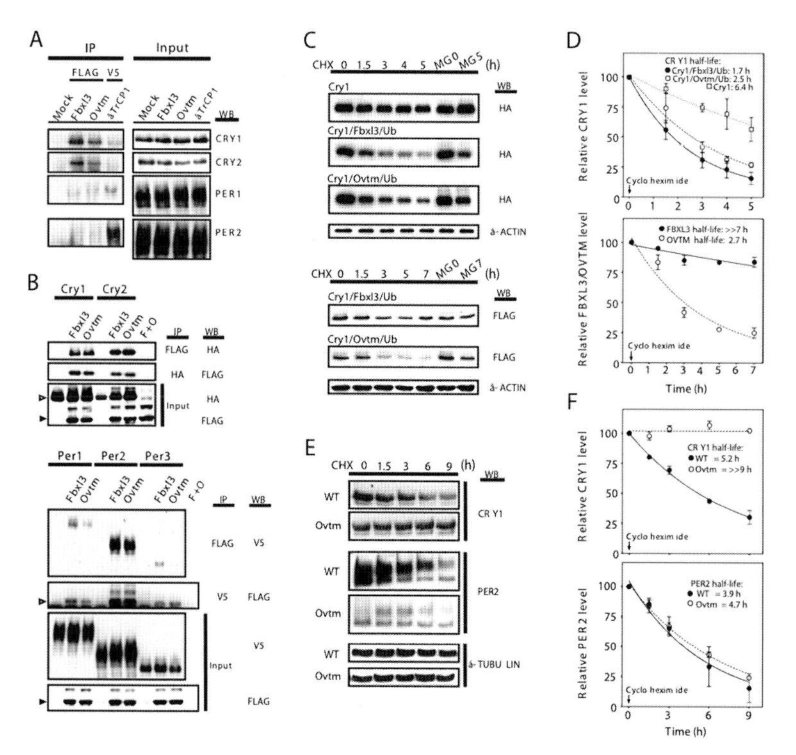

Figure 6. Interaction of FBXL3 and OVTM with circadian clock proteins. (*A*) NIH-3T3 cells were transfected with FLAG-*Fbxl3*, FLAG-*Ovtm*, and V5-β*TrCP1*. Immunoprecipitation was performed with anti-Flag or anti-V5 antibody. Native immunoprecipitated proteins were further analyzed by western blotting with anti-PER or anti-CRY antibodies. (*B*) Confirmation of interaction between FBXL3 with CRY and PER2. (*Top panel*) 293A cells were transfected with *Cry*-HA and FLAG-*Fbxl3*, FLAG-*Ovtm*. F+O is cotransfection of FLAG-*Fbxl3*, FLAG-*Ovtm*. (*Open arrowheads*) CRY; (*closed arrowheads*) FBXL3 or OVTM. (*Lower panel*) *Per*-V5 were cotransfected with FLAG-*Fbxl3*, FLAG-*Ovtm*. (*Open arrowheads*) Immunoprecipitated FBXL3 or OVTM; (*closed arrowheads*) FBXL3 or OVTM from input. (*C*) Effects of *Fbxl3* and *Ovtm* on CRY1 and FBXL3 protein stability. *Cry1*-HA was cotransfected with FLAG-*Fbxl3* or FLAG-*Ovtm* with HA-Ubiquitin (Ub). Abundance of CRY1, FBXL3, and OVTM was measured by western blotting. (*Upper panel*) Inhibition of CRY degradation by MG132 treatment plus CHX treatment; (*lower panel*) *Ovtm* mutation causes accelerated degradation of FBXL3 through the proteasomal degradation pathway. β-actin was used as a loading control. (*D*) Quantitation of the effects of *Fbxl3* and *Ovtm* on CRY1 and FBXL3 protein stability. (*Upper panel*) Effects of FBXL3 on CRY1 stability in 293A cells; (*open squares*) *Cry1* only; (*closed circles*) *Cry1* cotransfected with *Fbxl3*; (*open circles*) *Cry1* cotransfected with *Ovtm* mutant. The half-life of CRY1 is reduced by either FBXL3 (half-life: 1.7 hours) or OVTM (half-life: 2.5 hours). OVTM coexpression was less effective in CRY1 degradation compared to FBXL3. (*Lower panel*) Effect of the *Ovtm* mutation on the half-life of FBXL3; (*closed circles*) FBXL3; (*open circles*) OVTM. In the presence of CRY1, the *Ovtm* mutation caused a reduction in the protein stability of OVTM. (*E*) Stability of endogenous CRY1 and PER2 in *Ovtm* ear fibroblast cells. Protein extracts from harvested cells were analyzed with western blotting using anti-CRY1 and anti-PER2 antibody. α-tubulin was used as a loading control. (*F*) Half-life measurements of endogenous CRY1 and PER2 proteins in wild-type and *Ovtm* fibroblasts. (*Upper panel*) Time course of CRY1 levels following addition of CHX; (*closed circles*) wild-type fibroblasts; (*open circles*) *Ovtm* cells. CRY1 degradation is much more rapid in wild-type fibroblasts (half-life: 5.2 hours) than in *Ovtm* fibroblasts (>>9 hours). (*Lower panel*) Half-life measurements of PER2 proteins in wild-type and *Ovtm* fibroblasts; (*closed circles*) wild-type fibroblasts; (*open circles*) *Ovtm* fibroblasts. (Reprinted, with permission, from Siepka et al. 2007 [©Elsevier].)

affected by the OVTM mutation (Fig. 6C, bottom). FBXL3 is relatively stable with a half-life of greater than 7 hours, whereas OVTM has a much shorter half-life of 2.7 hours (Fig. 6D). Therefore, there are two effects of the OVTM mutation: a reduction in proteasome-mediated CRY1 degradation and a decreased stability of the OVTM protein itself, both of which could contribute to a loss-of-function phenotype.

To determine whether these changes in CRY1 stability seen in transfected cells are physiological, we used fibroblasts prepared from either wild-type or *Ovtm* mice and determined the half-lives of native CRY1 and PER2 proteins in these cells. As shown in Figure 6E,F, the half-life of CRY1 in *Ovtm* fibroblasts is extremely long (>>9 hours) as compared to wild-type cells (half-life = 5.2 hours). The overall levels of PER2 in *Ovtm* fibroblasts were very low, similar to that seen in the cerebellum. When the half-life of PER2 was determined, however, there was no detectable difference in the half-life of PER2 in wild-type and *Ovtm* cells (Fig. 6F). Thus, these experiments in fibroblasts from wild-type and *Ovtm* mice show that native CRY1, but not PER2, turnover is specifically altered by the *Ovtm* mutation.

CONCLUSIONS

We have shown that the ENU-induced *Overtime* mutant is caused by an I364T mutation in the mouse FBXL3 protein (Siepka et al. 2007), a member of the F-box protein with leucine-rich repeats family (Jin et al. 2004). The OVTM protein is less efficient than FBXL3 in degrading CRY1, thus providing genetic evidence that FBXL3 appears likely to be a primary F-box protein within an SCF E3 ubiquitin ligase complex (Cardozo and Pagano 2004) that targets the CRY proteins for degradation in the proteasome (Fig. 7). The I364T OVTM mutation lengthens circadian periodicity approximately 2.5 hours in mice. We propose that the phenotypic effects of the *Ovtm* mutation occur primarily through two mechanisms: (1) loss of FBXL3 function leading to stability of CRY1 protein and (2) repression of CLOCK/BMAL1-dependent transcriptional activation. These two processes lead to a striking reduction in the expression of the PER proteins which is caused by a reduction in transcription of the *Per* genes. In contrast, the levels of CRY are not reduced by the *Ovtm* mutation despite lower rates of *Cry* transcription. Because the transcription of the *Cry1* gene is strongly attenuated by the *Ovtm* mutation, it is surprising that CRY1 protein levels are not lower. The low rate of CRY1 protein degradation in *Ovtm* tissues must offset the lower synthesis of CRY1 so that the steady-state abundance of CRY1 protein is similar in *Ovtm* and wild-type mice. Importantly, however, because the turnover rate of CRY1 is reduced, the clearance of CRY1 will be prolonged even if the initial steady-state abundance levels are comparable. This would then lead to a prolongation of the CRY-dependent repression phase of the circadian cycle (Godinho et al. 2007). If such a prolongation extended CRY repression for 2–3 hours, the period-lengthening phenotype seen in *Ovtm* mice would follow as a consequence.

These results highlight the significance of an SCF^{FBXL3} E3 ubiquitin ligase complex in regulating the stability and kinetics of CRY degradation. The specificity of the FBXL3 interaction with the CRY proteins is striking and

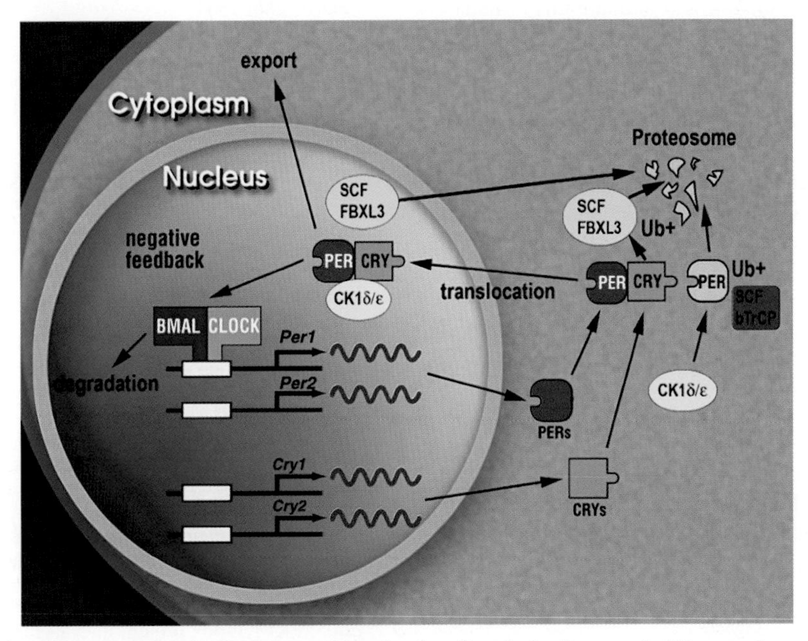

Figure 7. Model of the primary negative feedback loop within the circadian clock of mammals. Diagram shows the primary negative autoregulatory feedback loop that constitutes the circadian oscillator mechanism in mammals. The BMAL1-CLOCK heterodimeric complex activates transcription of the *Per1, Per2, Cry1*, and *Cry2* genes. The PER and CRY proteins accumulate in the cytoplasm and interact with each other and then translocate into the nucleus with CK1δ/ε where the complex interacts with BMAL1-CLOCK to inhibit the transcription of the *Per* and *Cry* genes. The turnover of the PER and CRY proteins are selectively regulated by interaction with the F-box proteins, β-TrCP and FBXL3, respectively, to target them for ubiquitination and subsequent degradation by the proteasome.

suggests that additional F-box proteins may regulate other circadian clock components. The first example of an F-box protein having a role in circadian rhythmicity was the *Arabidopsis* gene *ZEITLUPE* (*ZTL*) that encodes an F-box protein with an amino-terminal LOV domain and carboxy-terminal kelch repeats (Somers et al. 2000). *ZTL* targets the *Arabidopsis* clock protein, TOC1, for degradation by the proteasome and is thought to regulate circadian period by controlling TOC1 stability (Mas et al. 2003). In addition, the F-box protein, FKF1, mediates the cyclic degradation of CDF1, a repressor of the photoperiodic gene *CONSTANS* (Imaizumi et al. 2005), and the F-box protein, AFR, is a positive regulator of phytochrome-A-mediated light signaling in *Arabidopsis* (Harmon and Kay 2003). FBXL3 is the second example of a mammalian F-box protein regulating the circadian clock proteins. The first example is β*TrCP*, which has been shown to interact directly with the PER proteins (Eide et al. 2005; Shirogane et al. 2005). Evidence for β*TrCP* in the circadian pathway first emerged from *Drosophila* in which it was shown that *Slimb*, the ortholog of β*TrCP*, regulated circadian expression of PER and TIM (Grima et al. 2002; Ko et al. 2002). Interestingly, in *Neurospora*, the ortholog of β*TrCP*, FWD1, regulates the degradation of the clock protein, FREQUENCY (He et al. 2003). More recently, JETLAG, the *Drosophila* ortholog of Fbxl15, has been shown to have a critical role in light-induced TIM degradation by the proteasome (Koh et al. 2006). In *Drosophila*, two different SCF complexes appear to control TIM levels: a circadian pathway involving *Slimb* and a light-dependent pathway involving JET (Koh et al. 2006). It will be interesting see whether similar types of mechanisms are conserved in mammals. Because of the differences in the roles of the PER and CRY proteins in *Drosophila* and in mammals (Allada et al. 2001; Young and Kay 2001), where CRY is primarily a circadian repressor (not a photoreceptor), FBXL3 appears to function in a circadian SCF complex-mediated pathway. Unlike *Drosophila* PER, the PER1 and PER2 proteins in mammals are transcriptionally induced by light in the suprachiasmatic nucleus (SCN). It will be interesting to see whether β*TrCP* functions in a circadian or in a light-dependent SCF pathway for PER degradation (analogous to the TIM protein in *Drosophila*).

Using a forward genetics approach in mice, we have identified FBXL3 as a new molecular component of the negative feedback loop that generates circadian rhythmicity. Ironically, in the same screen, we found a second mutation in *Cry1*, the substrate for FBXL3. These experiments highlight the important role of CRY1 in the regulation of circadian period in which loss of function leads to shortened periodicity, whereas its overexpression by virtue of stabilization leads to lengthened periodicity.

ACKNOWLEDGMENTS

This chapter reports previously unpublished work describing the identification and candidate gene cloning of the *part-time* mutant and includes a shortened version of the identification and positional cloning of the *Overtime* mutant originally published by Siepka et al. (2007) in *Cell* (©Elsevier). This research was supported by National Institutes of Health grants U01 MH61915, P50 MH074924, and R01 MH078024 to J.S.T. and R01 NS053616 to C.L. J.S.T. is an investigator in the Howard Hughes Medical Institute.

REFERENCES

Allada R., Emery P., Takahashi J.S., and Rosbash M. 2001. Stopping time: The genetics of fly and mouse circadian clocks. *Annu. Rev. Neurosci.* **24:** 1091.

Antoch M.P., Song E.J., Chang A.M., Vitaterna M.H., Zhao Y., Wilsbacher L.D., Sangoram A.M., King D.P., Pinto L.H., and Takahashi J.S. 1997. Functional identification of the mouse circadian *Clock* gene by transgenic BAC rescue. *Cell* **89:** 655.

Bunger M.K., Wilsbacher L.D., Moran S.M., Clendenin C., Radcliffe L.A., Hogenesch J.B., Simon M.C., Takahashi J.S., and Bradfield C.A. 2000. Mop3 is an essential component of the master circadian pacemaker in mammals. *Cell* **103:** 1009.

Busino L., Bassermann F., Maiolica A., Lee C., Nolan P.M., Godinho S.I., Draetta G.F., and Pagano M. 2007. SCFFbxl3 controls the oscillation of the circadian clock by directing the degradation of cryptochrome proteins. *Science* **316:** 900.

Cardozo T. and Pagano M. 2004. The SCF ubiquitin ligase: Insights into a molecular machine. *Nat. Rev. Mol. Cell Biol.* **5:** 739.

DeBruyne J.P., Weaver D.R., and Reppert S.M. 2007. CLOCK and NPAS2 have overlapping roles in the suprachiasmatic circadian clock. *Nat. Neurosci.* **10:** 543.

Eide E.J., Woolf M.F., Kang H., Woolf P., Hurst W., Camacho F., Vielhaber E.L., Giovanni A., and Virshup D.M. 2005. Control of mammalian circadian rhythm by CKIepsilon-regulated proteasome-mediated PER2 degradation. *Mol. Cell. Biol.* **25:** 2795.

Gallego M. and Virshup D.M. 2007. Post-translational modifications regulate the ticking of the circadian clock. *Nat. Rev. Mol. Cell Biol.* **8:** 139.

Gekakis N., Staknis D., Nguyen H.B., Davis F.C., Wilsbacher L.D., King D.P., Takahashi J.S., and Weitz C.J. 1998. Role of the CLOCK protein in the mammalian circadian mechanism. *Science* **280:** 1564.

Godinho S.I., Maywood E.S., Shaw L., Tucci V., Barnard A.R., Busino L., Pagano M., Kendall R., Quwailid M.M., Romero M.R., O'Neill J., Chesham J.E., Brooker D., Lalanne Z., Hastings M.H., and Nolan P.M. 2007. The after-hours mutant reveals a role for Fbxl3 in determining mammalian circadian period. *Science* **316:** 897.

Griffin E.A., Jr., Staknis D., and Weitz C.J. 1999. Light-independent role of CRY1 and CRY2 in the mammalian circadian clock. *Science* **286:** 768.

Grima B., Lamouroux A., Chelot E., Papin C., Limbourg-Bouchon B., and Rouyer F. 2002. The F-box protein slimb controls the levels of clock proteins Period and Timeless. *Nature* **420:** 178.

Hao B., Zheng N., Schulman B.A., Wu G., Miller J.J., Pagano M., and Pavletich N.P. 2005. Structural basis of the Cks1-dependent recognition of p27(Kip1) by the SCF(Skp2) ubiquitin ligase. *Mol. Cell* **20:** 9.

Harmon F.G. and Kay S.A. 2003. The F box protein AFR is a positive regulator of phytochrome A-mediated light signaling. *Curr. Biol.* **13:** 2091.

He Q., Cheng P., Yang Y., Yu H., and Liu Y. 2003. FWD1-mediated degradation of FREQUENCY in *Neurospora* establishes a conserved mechanism for circadian clock regulation. *EMBO J.* **22:** 4421.

Imaizumi T., Schultz T.F., Harmon F.G., Ho L.A., and Kay S.A. 2005. FKF1 F-box protein mediates cyclic degradation of a repressor of CONSTANS in *Arabidopsis*. *Science* **309:** 293.

Jin J., Cardozo T., Lovering R.C., Elledge S.J., Pagano M., and Harper J.W. 2004. Systematic analysis and nomenclature of mammalian F-box proteins. *Genes Dev.* **18:** 2573.

King D.P., Zhao Y., Sangoram A.M., Wilsbacher L.D., Tanaka M., Antoch M.P., Steeves T.D., Vitaterna M.H., Kornhauser

J.M., Lowrey P.L., Turek F.W., and Takahashi J.S. 1997. Positional cloning of the mouse circadian *Clock* gene. *Cell* **89:** 641.

Ko C.H. and Takahashi J.S. 2006. Molecular components of the mammalian circadian clock. *Hum. Mol. Genet.* (suppl. 2) **15:** R271.

Ko H.W., Jiang J., and Edery I. 2002. Role for Slimb in the degradation of *Drosophila* Period protein phosphorylated by Doubletime. *Nature* **420:** 673.

Koh K., Zheng X., and Sehgal A. 2006. JETLAG resets the *Drosophila* circadian clock by promoting light-induced degradation of TIMELESS. *Science* **312:** 1809.

Kume K., Zylka M.J., Sriram S., Shearman L.P., Weaver D.R., Jin X., Maywood E.S., Hastings M.H., and Reppert S.M. 1999. mCRY1 and mCRY2 are essential components of the negative limb of the circadian clock feedback loop. *Cell* **98:** 193.

Lee C., Etchegaray J.P., Cagampang F.R., Loudon A.S., and Reppert S.M. 2001. Posttranslational mechanisms regulate the mammalian circadian clock. *Cell* **107:** 855.

Lowrey P.L. and Takahashi J.S. 2004. Mammalian circadian biology: Elucidating genome-wide levels of temporal organization. *Annu. Rev. Genomics Hum. Genet.* **5:** 407.

Lowrey P.L., Shimomura K., Antoch M.P., Yamazaki S., Zemenides P.D., Ralph M.R., Menaker M., and Takahashi J.S. 2000. Positional syntenic cloning and functional characterization of the mammalian circadian mutation tau. *Science* **288:** 483.

Mas P., Kim W.Y., Somers D.E., and Kay S.A. 2003. Targeted degradation of TOC1 by ZTL modulates circadian function in *Arabidopsis thaliana*. *Nature* **426:** 567.

Preitner N., Damiola F., Lopez-Molina L., Zakany J., Duboule D., Albrecht U., and Schibler U. 2002. The orphan nuclear receptor REV-ERBalpha controls circadian transcription within the positive limb of the mammalian circadian oscillator. *Cell* **110:** 251.

Reppert S.M. and Weaver D.R. 2002. Coordination of circadian timing in mammals. *Nature* **418:** 935.

Ripperger J.A. and Schibler U. 2006. Rhythmic CLOCK-BMAL1 binding to multiple E-box motifs drives circadian Dbp transcription and chromatin transitions. *Nat. Genet.* **38:** 369.

Rost B., Yachdav G., and Liu J. 2004. The PredictProtein server. *Nucleic Acids Res.* **32:** W321.

Sato T.K., Panda S., Miraglia L.J., Reyes T.M., Rudic R.D., McNamara P., Naik K.A., FitzGerald G.A., Kay S.A., and Hogenesch J.B. 2004. A functional genomics strategy reveals Rora as a component of the mammalian circadian clock. *Neuron* **43:** 527.

Shimomura K., Low-Zeddies S.S., King D.P., Steeves T.D., Whiteley A., Kushla J., Zemenides P.D., Lin A., Vitaterna M.H., Churchill G.A., and Takahashi J.S. 2001. Genome-wide epistatic interaction analysis reveals complex genetic determinants of circadian behavior in mice. *Genome Res.* **11:** 959.

Shirogane T., Jin J., Ang X.L. and Harper J.W. 2005. SCFbeta-TRCP controls clock-dependent transcription via casein kinase 1-dependent degradation of the mammalian period-1 (Per1) protein. *J. Biol. Chem.* **280:** 26863.

Siepka S.M. and Takahashi J.S. 2005. Forward genetic screens to identify circadian rhythm mutants in mice. *Methods Enzymol.* **393:** 219.

Siepka S.M., Yoo S.H., Park J., Song W., Kumar V., Hu Y., Lee C., and Takahashi J.S. 2007. Circadian mutant *Overtime* reveals F-box protein FBXL3 regulation of cryptochrome and period gene expression. *Cell* **129:** 1011.

Somers D.E., Schultz T.F., Milnamow M., and Kay S.A. 2000. ZEITLUPE encodes a novel clock-associated PAS protein from *Arabidopsis*. *Cell* **101:** 319.

Takahashi J.S. 2004. Finding new clock components: Past and future. *J. Biol. Rhythms* **19:** 339.

Takahashi J.S., Pinto L.H., and Vitaterna M.H. 1994. Forward and reverse genetic approaches to behavior in the mouse. *Science* **264:** 1724.

Vitaterna M.H., Pinto L.H., and Takahashi J.S. 2006. Large-scale mutagenesis and phenotypic screens for the nervous system and behavior in mice. *Trends Neurosci.* **29:** 233.

Vitaterna M.H., King D.P., Chang A.M., Kornhauser J.M., Lowrey P.L., McDonald J.D., Dove W.F., Pinto L.H., Turek F.W., and Takahashi J.S. 1994. Mutagenesis and mapping of a mouse gene, Clock, essential for circadian behavior. *Science* **264:** 719.

Vitaterna M.H., Selby C.P., Todo T., Niwa H., Thompson C., Fruechte E.M., Hitomi K., Thresher R.J., Ishikawa T., Miyazaki J., Takahashi J.S., and Sancar A. 1999. Differential regulation of mammalian period genes and circadian rhythmicity by cryptochromes 1 and 2. *Proc. Natl. Acad. Sci.* **96:** 12114.

Yoo S.H., Ko C.H., Lowrey P.L., Buhr E.D., Song E.J., Chang S., Yoo O.J., Yamazaki S., Lee C., and Takahashi J.S. 2005. A noncanonical E-box enhancer drives mouse Period2 circadian oscillations in vivo. *Proc. Natl. Acad. Sci.* **102:** 2608.

Young M.W. and Kay S.A. 2001. Time zones: A comparative genetics of circadian clocks. *Nat. Rev. Genet.* **2:** 702.

The Biology of the Circadian Ck1ε *tau* Mutation in Mice and Syrian Hamsters: A Tale of Two Species

A.S.I. Loudon,* Q.J. Meng,* E.S. Maywood,† D.A. Bechtold,* R.P. Boot-Handford,* and M.H. Hastings†

*Faculty of Life Sciences, University of Manchester, Manchester M13 9PT; †Division of Neurobiology, MRC Laboratory of Molecular Biology, Cambridge, CB2 OQH

The *tau* mutation in the Syrian hamster resides in the enzyme casein kinase 1 ε (CK1ε), resulting in a dramatic acceleration of wheel-running activity cycles to about 20 hours. *tau* also impacts growth, energy, metabolism, feeding behavior, and circadian mechanisms underpinning seasonal timing, causing accelerated reproductive and neuroendocrine responses to photoperiodic changes. Modeling and experimental studies suggest that *tau* acts as a gain of function on specific residues of PER, consistent with hamster studies showing accelerated degradation of PER in the suprachiasmatic nucleus in the early circadian night. We have created null and *tau* mutants of *Ck1ε* in mice. Circadian period lengthens in CK1$\varepsilon^{-/-}$, whereas CK1$\varepsilon^{tau/tau}$ shortens circadian period of behavior in vivo in a manner nearly identical to that of the Syrian hamster. CK1$\varepsilon^{tau/tau}$ also accelerates molecular oscillations in peripheral tissues, demonstrating its global circadian role. CK1ε^{tau} acts by promoting degradation of both nuclear and cytoplasmic PERIOD, but not CRYPTOCHROME, proteins. Our studies reveal that *tau* acts as a gain-of-function mutation, to accelerate degradation of PERIOD proteins. *tau* has consistent effects in both hamsters and mice on the circadian organization of behavior and metabolism, highlighting the global impact of this mutation on mammalian clockwork in brain and periphery.

INTRODUCTION

The past decade has witnessed spectacular advances in our understanding of the genetic basis of circadian timing in mammals. The circadian pacemakers within the suprachiasmatic nucleus (SCN) of the hypothalamus provide a crucial function in conducting a circadian repertoire throughout the body, acting on peripheral oscillators in all major body organs and tissues, with profound effects on general systemic physiology (Reppert and Weaver 2002; Hastings et al. 2003; Lowrey and Takahashi 2004; Saper et al. 2005). In both brain and periphery, the molecular clockwork operates as a series of interlocked autoregulatory feedback loops in which CLOCK:BMAL1 heterodimers bind to E-box DNA sequences contained within genes encoding transcriptional repressors PER and CRY. Following their accumulation in the cytoplasm, PER:CRY complexes translocate to the nucleus after a delay of several hours and repress the activity of constitutively bound CLOCK:BMAL1 complexes. These inhibitory complexes are then degraded following a further delay, and the consequent derepression of CLOCK:BMAL1 activity initiates the next circadian cycle of Per and Cry transcription.

We have now come to recognize the all-pervasive nature of circadian timing in biology, and the next challenge is to unravel the mechanisms and pathways involved in mediating the effects of the clocks on physiology and metabolism. The general prevalence of many diseases is also marked by a strong circadian component in morbidity, and circadian mutants are often characterized by unexpected side effects in a wide range of pathologies, including malignancy and metabolic and cardiovascular defects. Mutations in various elements of this core molecular clock have been described, many of which result in either arrhythmia or alteration in period of rest/activity and sleep cycles. Genetically mediated desynchronization of central pacemaker function is also strongly implicated as the primary causal mechanism of familial advanced sleep phase syndrome (FASPS), and unraveling the complexities of how circadian mutations act on both central and peripheral pacemakers will reveal important new insight into both the normal regulation of sleep and the pathologies associated with sleep disruption.

As in many areas of biology, much of our recent understanding has come from the use of genetically modified mice, allowing studies of incredible sophistication in a species that has an "amenable" genome. Many of the chapters in this volume are devoted to a detailed exposition of the molecular regulation of circadian timing and the impact of clocks on both normal physiology and disease processes. Here, we describe the biology of the first bona fide circadian mutation ever discovered in a mammal, the *tau* mutation of the Syrian hamster. We explore the impact that this timing mutation has on both the daily and seasonal biology of this species and then describe the same mutation in genetically modified mice. Here, we are able to use the power of mouse genetics to demonstrate some unexpected features of the *tau* gene on central timing processes in the SCN, which reveal underpinning mechanisms controlling activity and sleep-wake cycles, with clear implications for our own species.

THE DISCOVERY OF THE *TAU* MUTATION IN THE SYRIAN HAMSTER

The *tau* mutation was discovered in 1988 by Michael Menaker and colleagues in an individual animal delivered as part of a routine laboratory shipment of Syrian hamsters (*Mesocricetus auratus*; Ralph and Menaker 1988). Subsequent breeding studies revealed that *tau* acts as a semidominant mutation, shortening the wheel-running period to approximately 22 hours in the heterozygote and to 20 hours in the homozygote. This spectacular and robust phenotype remains the defining feature of this iconic circadian mutant. Further studies using reciprocal SCN transplants between *tau* and wild-type animals provided definitive evidence that the SCN is the dominant pacemaker determining circadian wheel-running period (Ralph et al. 1990). These were extended in subsequent experiments that used microencapsulation of transplanted tissue to demonstrate that a diffusible signal of SCN origin is still capable of establishing circadian behavioral output (Silver et al. 1996). Studies of retinal melatonin rhythms also revealed an acceleration in hormone rhythms in tissue derived from *tau* mutants and provided crucial early evidence for the existence of additional SCN-independent oscillators (Tosini and Menaker 1996).

Following a heroic effort by Joe Takahashi and colleagues, the gene was mapped and identified as a mutation in the enzyme casein kinase 1ε (CK1ε) (Lowrey et al. 2000). The team employed a comparative syntenic mapping approach, using genetically directed representational difference analysis to identify polymorphic markers that are tightly linked to a monogenic trait. This technique offered the advantage that prior knowledge of chromosomal location of the trait is unnecessary, an essential prerequisite in the unmapped mouse genome of that time. CK1ε was identified as the candidate; in mice, the gene is located on chromosome 15 and in humans, on chromosome 22. The mutation was identified as a single-base-pair C-to-T substitutional mutation, resulting in an arginine-to-cysteine transition in a highly conserved region at position 178 of CK1ε. The seven mammalian members of the CK1 family are a unique group of serine/threonine enzymes with closely conserved identity in the catalytic domain. The mutation itself occurs within the phosphate recognition domain formed by Arg-178, Gly-215, and Lys-229. The *Drosophila* ortholog of Ck1ε is the *double-time* (*dbt*) gene (Kloss et al. 1998), which has been identified in a mutagenesis screen for aberrant circadian phenotypes. Missense mutations in *dbt* are known to confer either long or short periods on circadian behavior in flies (Price et al. 1998). In mammals, the nearest relatives of *dbt* are two closely related members of the CK1 family: CK1ε and CK1δ. These studies thus placed CK1 as a key component in circadian timing in mammals and flies. Before exploring in more detail how CK1ε may accelerate circadian timing, we need first to review the impact that this mutation has upon physiology and, in particular, the endocrine and seasonal timing system of the Syrian hamster.

THE IMPACT OF *TAU* ON NEUROENDOCRINE AND SEASONAL TIMING IN HAMSTERS

Circadian rhythmicity is an important feature of the endocrine axis of mammals. Endocrine products of the pineal, pituitary, and adrenal glands all show marked circadian cycles in synthesis and secretion, and this allows the endocrine system to impose a functionally significant temporal order on distal target tissues (Hastings et al. 2007; Maywood et al. 2007) . The SCN is known to have a central role in synchronizing circadian events throughout the endocrine system, and lesions of the SCN have long been known to disrupt rhythmic hormone secretion (Moore and Eichler 1972). The accelerated circadian clock of the *tau* mutant Syrian hamster, together with its larger body size allowing serial blood sampling, has provided an excellent opportunity to characterize the impact of this circadian mutation on hormone secretion. Furthermore, because Syrian hamsters are seasonally breeding mammals, the *tau* mutation offers an ideal test bed to define the role of circadian timing in the control of the seasonal photoperiodic response.

Ultradian and Daily Endocrine Rhythms

The luteinizing hormone (LH) is under tight circadian control, and in female mammals, the time of LH surge is strongly gated by the circadian clock, with the consequence that SCN lesions severely disrupt LH secretion and ovulation (Chappell 2005; de la Iglesia and Schwartz 2006). In ovariectomized estrogen-implanted female hamsters, the gated LH surge is of normal duration and occurs daily at the same relative circadian time (CT6–8). By serial sampling during two circadian cycles in *tau* mutant animals, it was possible to show that the LH surge interval is accelerated such that it occurs every 20 hours in homozygote *tau* mutants (Lucas et al. 1999). Moreover, the surge itself is abnormal, with a significant twofold to threefold reduction in amplitude (Fig. 1f). *tau* also impacts high-frequency oscillators regulating episodic hormone release. For instance, in ovariectomized female hamsters, the ultradian oscillators regulating both episodic LH and cortisol secretion (~25–35-minute oscillators) are significantly decelerated in the *tau* mutant (Loudon et al. 1994), a marked contrast with the acceleration of circadian cycles. It is possible that altered ultradian rhythmicity may reflect action of *tau* with hypothalamic GnRH neurons, because studies of GnRH cell lines reveal significant effects of clock gene mutations on endogenous episodic secretion (Chappell 2005). Our studies of hamsters reveal that although the overall circadian timing of the estrogen-induced LH surge appears to be accelerated in a manner equivalent to that of wheel-running behavior, both ultradian episodic LH and cortisol secretion and the LH surge amplitudes are highly abnormal in *tau* mutants. Together, these studies suggest that genes involved in circadian timing may have a key role in timing of endocrine release and secretion not only at the level of timed daily hormones, but also on noncircadian ultradian endocrine oscillator systems. It is still unclear whether abnormal neuroendocrine responses arise as a result of general pleiotropic effects of "clock" mutations or whether they reflect hitherto undefined actions of circadian timing on the endocrine axis in general.

In mammals, pineal melatonin secretion is strongly regulated by the circadian clock. Melatonin provides the brain with an internal hormonal representation of day length and is known to be the key signal involved in timing the mam-

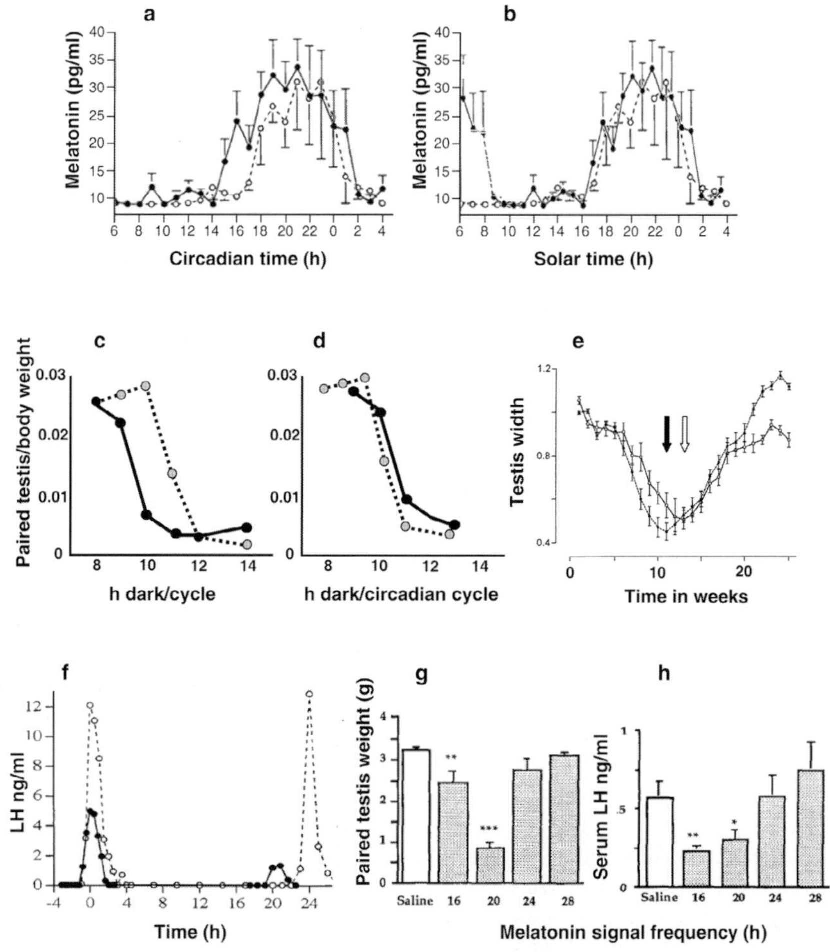

Figure 1. Seasonal and endocrine responses of *tau* mutant hamsters. (*a*) Melatonin rhythms of *tau* (*solid line*) and wild-type (*dotted line*) hamsters, collected every circadian hour, replotted in (*b*) against a solar hour time base. Data are replotted in *c* against a circadian time base. (*d*) Seasonal changes in testis size in *tau* and wild-type hamsters maintained in constant darkness (DD) for 25 weeks. (*e*) Paired testis weight in proportion to body size for *tau* mutant (*solid line*) and wild-type (*dotted line*) hamsters maintained on 20-hour (*tau*) or 24-hour (wild type) LD cycles. *Arrows* indicate the time of maximal testicular regression for *tau* (*solid*) and wild type (*open arrow*). (*f*) Representative gated circadian LH surges from *tau* mutant (*solid line*) and wild-type (*dotted line*) ovariectomized estrogen-implanted animals sampled during two circadian cycles. (*g,h*) Testicular and LH responses of *tau* mutant hamsters to a range of long-duration (10 hours) melatonin signals administered at differing frequencies. (*a,b,f,* Redrawn, with permission, from Lucas et al. 1999; *c,e,* redrawn, with permission, from Stirland et al. 1996b; *d,* redrawn, with permission, from Loudon et al. 1998; *g,h,* redrawn, with permission, from Stirland et al. 1996a.)

malian neuroendocrine response to seasonal photoperiodic change: The longer nights of winter are encoded as a longer duration of the nocturnal melatonin signal (Malpaux et al. 2001). The pinealocyte itself is regulated by multisynaptic sympathetic outflow from the SCN/PVN (paraventricular nucleus), resulting in *trans*-synaptic release of noradrenaline and consequent regulation of *N*-acetyltransferase, the rate-limiting step in nocturnal melatonin production (Deguchi and Axelrod 1972; Axelrod 1974). In *tau* hamsters, the circadian onset of melatonin secretion occurs approximately 3 hours earlier (relative to activity onset) than in wild types, but curiously has a normally phased offset relative to activity onset (Lucas et al. 1999). The consequence is that despite exhibiting an accelerated circadian cycle of melatonin production (i.e., once every 20 hours), the net duration of nocturnal secretion per cycle is not significantly different in real time for each genotype. This is illustrated in Figure 1a,b where the data are plotted on both a circadian and a solar time basis. Using pineal explants, we have shown that this advance in secretion onset is not due to an altered time course of response to noradrenaline, suggesting that the timing of sympathetic activation itself and perhaps phasing of SCN-mediated output may be advanced at this specific phase of the circadian cycle (Lucas et al. 1999). This has more general implications for the manner in which the *tau* mutation accelerates the circadian axis, which we explore in detail below in the context of the mouse model of *tau*.

Seasonal and Photoperiodic Time Measurement

Syrian hamsters are seasonal breeders, responsive to photoperiod and melatonin. Because the melatonin signal is generated by the circadian axis, this allows us to ask how the *tau* mutation might impact on photoperiodic time measurement and responsiveness to melatonin. Using testicular regression as an index of photoresponsiveness, adult male *tau* mutants maintained on 24-hour light/dark (LD) cycles fail to exhibit a short-day response and undergo testicular involution, even when exposed to very short photoschedules. Instead, they exhibit a long-day-like default response (Shimomura et al. 1997). This occurs because the intrinsic 20-hour period of *taus* prevents their stable entrainment to the prevailing 24-hour LD cycle and thus prevents the generation of a melatonin profile representative of short days. However, when adult *tau* males are maintained in LD cycles on a 20-hour time base, they do entrain effectively, and when exposed to varying durations of darkness, melatonin-dependent testicular regression can be triggered by photoschedules of 10 hours light/10 hours dark. This contrasts with the threshold of 12 hours of darkness per cycle needed to trigger short-day responses in wild types and suggests that critical day-length responses are proportional to the circadian phenotype, rather than absolute (Fig. 1c,d) (Stirland et al. 1996b). One explanation of this difference is that the 10-hour night is represented by a longer melatonin profile in *taus* than it is in wild types due to the earlier onset of melatonin secretion (~1 hour, rather than 4 hours after lights off). Hence, 10 hours of darkness in *taus* may result in the generation of the same melatonin duration as 12 hours of darkness in wild-type hamsters.

taus thus measure the passage of seasonal time using a 20-hour time base. One consequence of this is that in 20-hour lighting regimes or constant darkness (DD), melatonin signals are generated once every 20 hours (rather than every 24 hours), and correspondingly, the rate of short-day-induced physiological change is accelerated. For instance, when *tau* and wild-type hamsters are maintained in DD for 25 weeks, testicular regression occurs, followed by the onset of spontaneous refractoriness and recrudescence. *tau* mutants, however, exhibit a significantly greater rate of testicular regression, directly in proportion to the acceleration of the circadian clock (i.e., by ~20%), with the net result that they reach the nadir of the cycle 2 weeks earlier than do wild-type animals. This corresponds approximately to the time required to experience the same number of melatonin signals as lead to regression in wild type (Fig. 1e) (Loudon et al. 1998). Collectively, these data are therefore compatible with the hypothesis that acceleration in the generation of circadian melatonin signal frequency in *taus* leads to accelerated reproductive and neuroendocrine responses. Complementary results have been obtained in pinealectomized wild-type hamsters exposed to inhibitory long-duration melatonin signals. When these signals are administered every 20 hours, they lead to a more rapid onset of testicular regression (Maywood et al. 1990). Hence, these particular differences between *tau* and wild-type hamsters in photoperiodic responses to day-length and programmed melatonin signals can be explained by the impact of the mutation on circadian entrainment of the pineal melatonin profile and its photoperiodic modulation, rather than any effect of the mutation on downstream responses to melatonin.

To test whether the *tau* mutation has altered the definition of what constitutes a long-duration (short-night) melatonin signal or how the mutation may impair interpretation of melatonin signals of differing frequencies, we exposed pinealectomized *tau* and wild-type hamsters to programmed melatonin infusions every 20 hours of 8-hour duration (which are known to be inhibitory in this species) or signals of 20% shorter duration (6.7 hour), reduced by the same proportion by which the circadian period is shortened (Stirland et al. 1995), measuring testicular regression, serum LH, and prolactin as an end point. Both genotypes exhibited a short-day-like response, but in *taus*, the magnitude of response to the short-duration 6.7-hour signal was greater, with significantly reduced testis size and lowered LH levels. This implies that the mutation has altered the time base against which an inhibitory melatonin signal is measured. We have extended these studies to ask whether *tau* has similarly altered the time base against which melatonin signal frequencies are measured. Using the pinealectomized melatonin-infusion paradigm, we used repeated (inhibitory) long-duration (10 hours) melatonin signals administered at a range of differing signal frequencies, from 16 to 28 hours during a 6-week period (Stirland et al. 1996a). Whereas signal frequencies of 20 hours induced both significant testicular regression and suppression of LH, 16-, 24-, or 28-hour signals failed to do so. In contrast, in wild-type animals, both 20- and 24-hour infusions are effective (Fig. 1g,h). These experiments have clearly revealed that modulo-24-hour melatonin signals are not "decoded" as equivalent to short days in *tau* mutants, with the clear implication that the underlying ultradian interval timer that responds to the melatonin signal is itself running at an accelerated rate (see Fig. 6).

These experiments might usefully be considered in light of recent work revealing the operation of melatonin-induced circadian clock gene rhythms at a melatonin neuroendocrine target site, the pituitary *pars tuberalis* (PT). Here, the seasonal phasing of individual components of the circadian clockwork is regulated by changes in the duration of the melatonin signal (Lincoln et al. 2002, 2003). Specifically, *Cry1* is strongly induced by melatonin, and onset of expression thus coincides with the early night (Hazlerigg et al. 2004), whereas the decline in melatonin at dawn leads to onset of expression of *Per1* (Messager et al. 1999; Lincoln et al. 2002). As a result, the melatonin-regulated clock gene rhythm in the PT target site operates as an internal coincidence timer, the interval between Cry and Per peaks being inversely proportional to day length (Lincoln et al. 2002). This has been hypothesized to lead to seasonal changes in PER/CRY-mediated repression/derepression of downstream E-box-regulated genes and hence seasonal changes in a neuroendocrine output (Lincoln et al. 2003). There is no convincing evidence that the PT operates as an autonomous circadian oscillator, rather that the molecular components of the circadian clock have been used to provide a transcriptional pathway to decode the melatonin signal. Such decoding can occur at a range of melatonin signal frequencies operating well beyond the normal range of entrainment of the SCN, because wild-type hamsters are able to read melatonin signals at 20-hour frequencies (Maywood et al. 1990). However, the alteration in response of *taus* to melatonin signal frequency strongly implies that changes in the kinetics of the local molecular components of the circadian clock at a melatonin target site may underpin failure to respond to modulo-24-hour melatonin signals. To gain insight into how the *tau* mutation accelerates the molecular clockwork in the Syrian hamster, we have investigated the dynamics of *Per1* and *Per2* genes and their products during the circadian cycle.

HOW DOES THE *TAU* MUTATION ACCELERATE CIRCADIAN TIMING?

PERIOD Protein Turnover in the *tau* Mutant Hamster

The original paper describing the hamster mutation (Lowrey et al. 2000) proposed that *tau* acts as a loss of function, resulting in hypophosphorylation of target proteins. This proposition is, however, difficult to reconcile with current models of clock function, in which phosphorylation is a necessary prerequisite for targeted proteosomal degradation. One clear prediction from the biology of CK1ε is that the *tau* mutation may impact on the dynamics of clock protein turnover. To test this, we tracked both mRNA and protein for Per1 and Per2 across the circadian cycle in the SCN (Fig. 2) (Dey et al. 2005). Our data revealed that in wild-type hamsters, the rhythmic expression of PER protein exhibited a characteristic phase lag of 2–4 hours behind the mRNA cycle, with accumulation of nuclear PER proteins at the end of the subjective day and into the night. This is consistent with earlier publications in mice (Field et al. 2000). In *tau* mutant hamsters, there was no effect on the relative phasing of *per* mRNA rhythms nor on the rate or phasing of accumulation of PER proteins (counter to the predictions of accelerated accumulation based on the fly models of *dbt* action) (Price et al. 1998). In contrast, there was an accelerated loss of immunoreactive PER, which occurs several hours earlier in the night. As a consequence of this accelerated degradation, mRNA and protein levels declined simultaneously, in contrast to wild types, which exhibit a pronounced lag in protein loss. A more detailed study taking place during the early night revealed that much of the loss of PER proteins occurred during the first 2.5 to 3 hours of the early night following the mRNA and protein peaks (Dey et al. 2005). These data are therefore consistent with a model in which CK1ε^{tau} leads to destabilization and degradation of nuclear PER proteins during the early subjective night, resulting in an overall acceleration of circadian period by compression of early-night phases. Such a model explains the earlier onset of melatonin secretion in *tau* mutant hamsters.

Models for Action of *Ck1*ε in Circadian Timing

In addition to periodic transcription, posttranslational modifications of clock proteins are now recognized as central components involved in the operation of molecular clockworks (Lee et al. 2001; Gallego and Virshup 2007; Mignot and Takahashi 2007). In particular, reversible phosphorylation provides a potential mechanism for the regulated formation of protein complexes, their nuclear entry, and their ultimate degradation via ubiquitination pathways, each step of which introduces delays into the feedback loop. Current interest now focuses on the role that protein phosphorylation may have in tuning the circadian oscillator to a period of 24 hours. Recent mathematical modeling studies by Forger, Virshup, and colleagues have proposed an alternative model of action to that of Lowrey et al. (2000), namely, that CK1ε^{tau} acts as a gain of function on target PER1 and PER2 proteins (Gallego et al. 2006). These authors modeled the possible actions of all known mutations of CK1 in mammals and flies to identify potential components that may contribute to period shortening. All models involving hypophosphorylation lead to lengthened, rather than shortened, period and thus

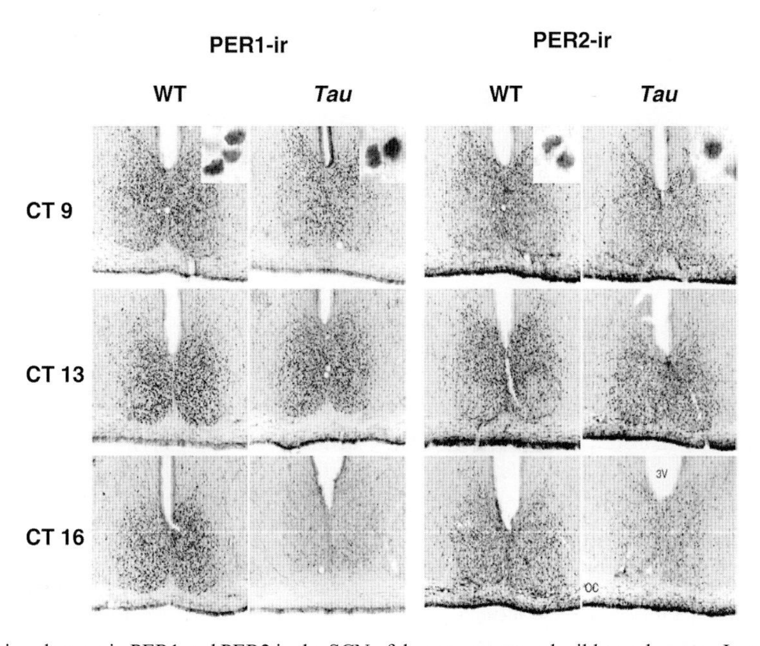

Figure 2. Immunoreactive changes in PER1 and PER2 in the SCN of the *tau* mutant and wild-type hamster. In *tau* mutants, PER proteins rapidly decline in the early nocturnal phase. (*Insets*) Nuclear localization for PER. (Redrawn, with permission, from Dey et al. 2005.)

did not comply with the known biology. However, modeling of *tau* as a gain of function on target proteins leads to the prediction that period would shorten and that target proteins would decay at an accelerated rate. (This topic is extensively reviewed elsewhere in this volume.) The Forger/Virshup gain-of-function model was then tested experimentally in cell lines, and this revealed that CK1ε^{tau} does act as a highly specific gain-of-function mutation, which increases in vivo phosphorylation of circadian PER1 and PER2 proteins (Gallego et al. 2006). A further prediction from this model was that specific targeting of PER would lead to increased degradation, altered stability, and accelerated protein turnover, with the main effect operating on nuclear localized PER proteins (Gallego and Virshup 2007). This model is entirely consistent with our earlier data on the *tau* mutant hamster showing the accelerated clearance of PER at a specific phase of the cycle in *tau* mutants (Dey et al. 2005).

The *tau* Mutant Mouse Model

To test predictions for action of CK1 on core clockwork and to make use of the full power of mouse genetics, we have recently developed a mouse model for the hamster *tau* mutation. We adapted the *loxP-Cre* strategy to produce mice carrying the *tau* allele within the *Ck1ε* gene and then used these mice subsequently to use *cre*-mediated in vivo disruption of exon 4, which encodes the catalytic domain of CK1ε (Lowrey et al. 2000). This generated a frameshift resulting in a *Ck1ε* null allele (Meng et al. 2008). Wheel-running studies revealed that CK1ε^{tau} mice exhibit significant shortening of circadian period in a dose-dependent manner (see Fig. 3). This is equivalent to a reduction in circadian period of approximately 1.80 hours per copy of CK1ε^{tau}, and, remarkably, this is virtually identical to the behavioral phenotype of this mutation in Syrian hamsters (Ralph and Menaker 1988). In contrast, knockout mice exhibited a small but significant period lengthening compared to their wild-type counterparts (~18 min/cycle) (Fig. 3a). Importantly, *tau* hemizygote and heterozygote (CK1ε$^{tau/+}$) mice exhibited similar periods (Fig. 3a,b), which reveals that a single wild-type copy of *Ck1ε* provides little protective effect in the face of a *tau* allele. These data are therefore consistent with the hypothesis that *tau* acts as a gain-of-function mutation. Furthermore, the relatively mild circadian phenotype of the knockout (slight lengthening of period, but normal intensity of activity) implies that the wild-type copy of *Ck1ε* may make relatively little contribution to normal circadian period.

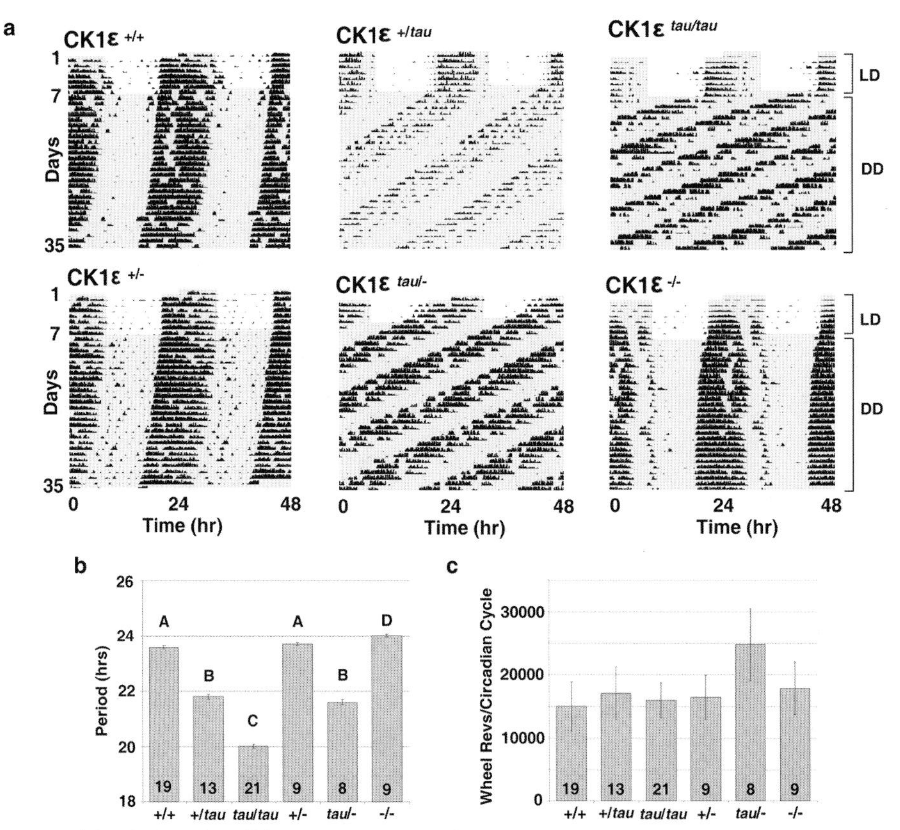

Figure 3. Locomotor activity rhythms for wild-type, *tau* mutant, and knockout mice. (*a*) Representative wheel-running activity records (actograms) for different genotypes are shown in double-plotted format. Each horizontal line represents 48 hours, with the second day plotted to the right and one cycle below the first cycle. Animals were run in 24-hour cycles of 12-hour light:12-hour dark (LD) and then in DD. The timing of the LD cycles is indicated by the alternating white and gray areas of the actogram. (*b*) Periodogram estimates of period for each genotype as mean ± S.E.M.; the number of animals is indicated within each bar. Bars with different letters show a significant difference; bars with the same letter are not significantly different. (*c*) Activity levels (mean ± S.E.M.) during the first 4 weeks of DD for each genotype as determined by total wheel-running revolutions per circadian cycle. (Redrawn, with permission, from Meng et al. 2008.)

We have extended these studies by investigating the electrophysiological properties of the SCN oscillator using acutely prepared SCN slices (Meng et al. 2008). In each genotype (CK1$\varepsilon^{+/+}$, CK1$\varepsilon^{tau/+}$, CK1$\varepsilon^{tau/tau}$, CK1$\varepsilon^{-/-}$), there was a close concordance with its period of multiunit and single-unit recordings in each slice (data not shown), indicating that the effects of CK1ε^{tau} on behavioral period likely arise from corresponding changes in circadian patterns of electrical activity of SCN neurons (Brown et al. 2005). Thus, the creation of the CK1ε^{tau} mutation in mice and subsequent gene deletion results in a novel allelic series of circadian periods that range from 20.0 (CK1$\varepsilon^{tau/tau}$) to 24.0 (CK1$\varepsilon^{-/-}$) hours and are reflected in both altered patterns of behavior and altered physiology of the SCN.

The acceleration of circadian period of activity/rest cycles as well as SCN firing rates is likely to be reflected in the altered dynamics within the molecular feedback loops that underpin circadian timekeeping within the SCN. To test this, we used a PER2::luciferase (PER2::LUC) protein fusion reporter mouse model (Yoo et al. 2004) that allows circadian monitoring of luciferase activity, accurately reporting the underlying PER2 molecular oscillator. This reporter was crossed with CK1ε^{tau} mice, and SCN slices were recorded using bioluminescence recording (Meng et al. 2008). These studies revealed oscillations that closely matched the earlier recorded behavioral rhythms for each genotype (Fig. 4a,b): 24.4 ± 0.15, 21.9 ± 0.08, and 20.2 ± 0.17 hours in CK1$\varepsilon^{+/+}$, CK1$\varepsilon^{tau/+}$, and CK1$\varepsilon^{tau/tau}$ slices, respectively. The overall shortening of period of the SCN is reflected at the single-cell level. For instance, when PER2 expression of individual SCN neurons is tracked using a CCD (charged-coupled device) camera, the resulting circadian period of single neurons also closely matches the period of the whole slice (Fig. 4c). From this, it is clear that CK1ε^{tau} has no discernible effect on the intercellular synchrony nor on the regional distribution of circadian gene expression across the SCN and is consistent with a model whereby CK1ε^{tau} accelerates all SCN oscillators in an equivalent manner.

Action in Peripheral Tissues

Little is known of the penetrance of many circadian mutants to peripheral tissues, and in the case of CK1ε^{tau}, the expression pattern is not known in peripheral tissues. We have tested whether endogenous CK1ε has a global role in setting clock speed in tissues outside the nervous system. Circadian PER2::LUC activity recording of organotypic slices of pituitary, lung, and kidney and primary lung fibroblast cultures reveals a significant shortening of period by CK1ε^{tau} (Fig. 4b,d). The magnitude of the effect is, however, highly variable across tissues, revealing only partial penetrance of the mutation (Meng et al. 2008). Although wild-type periods of peripheral tissues are close to 24 hours, in CK1ε^{tau} mice, periods are generally longer than those in the SCN by up to 2 hours, and as a result, clock speed is not accelerated to the same extent in these peripheral tissues. This suggests that mutations such as CK1ε^{tau} may induce a considerable degree of internal desynchrony in the peripheral circadian repertoire, because phased output signals driven by the SCN oscillator may act on distal oscillators of relatively longer peri-

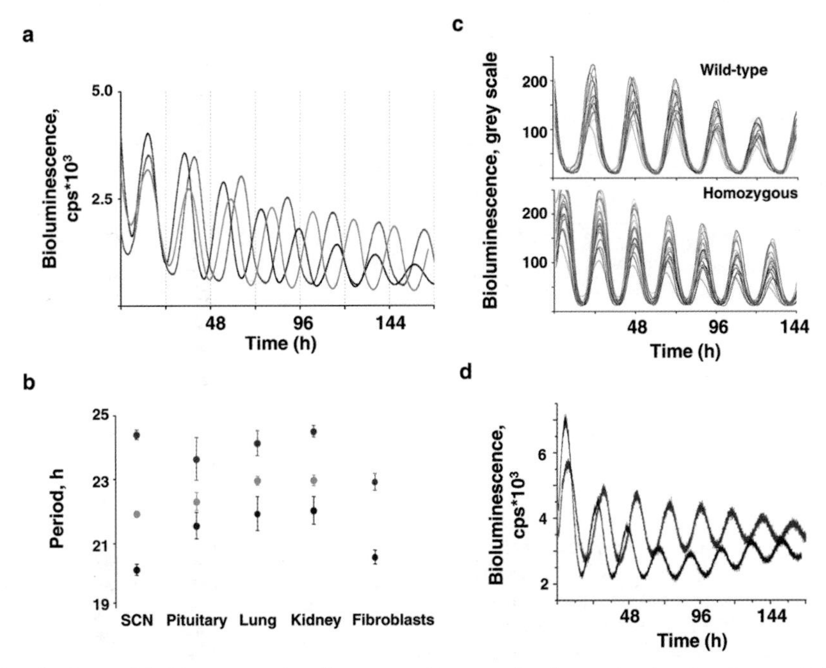

Figure 4. CK1ε^{tau} in mice has a global impact on circadian pacemaking, accelerating molecular circadian oscillators in SCN, peripheral tissues, and primary fibroblasts. (*a*) Representative PER2::LUC bioluminescence oscillations in wild-type (red), heterozygote (green), and homozygote (blue) organotypic SCN slices. (*b*) Periodicity of tissues from wild-type (*red*), heterozygote (*green*), and homozygote (*blue*) *tau* PER2::LUC mice (mean ± S.E.M.; wild type, $n = 4$–6; heterozygote, $n = 10$–11; homozygote, $n = 5$–6). (*c*) CCD recordings of SCN PER2::LUC expression at the single-cell level for wild-type and homozygote SCN slices (20 per slice; $n = 3$ slices for each genotype). (*d*) Representative traces of PER2::LUC expression in primary fibroblast cultures of wild-type (*red*) and *tau* mutant (*blue*) mice. (Redrawn, with permission, from Meng et al. 2008.)

ods. One clear prediction is that peripheral clock gene expression may exhibit a different phase relationship to the central SCN oscillator. This issue is unexplored in mice, but in *tau* hamsters, we observed significant phase differences of Per1 and Bmal1 mRNA expression compared to wild type in the heart (Dey et al. 2005).

Which Circadian Clock Proteins Does *tau* Target?

Relatively little is known of the biochemical basis of CK1ε action or of the proteins targeted. To address this, we have explored the action of CK1ε on PER and CRY proteins in both cell lines and primary tissues. Initially, using COS-7 cells, we used real-time fluorescence video microscopy to track PER2::YFP degradation following blockade of de novo protein synthesis by treatment with cycloheximide (CHX). Here, a clear prediction of the gain-of-function model is that coexpression with CK1ε^{tau} kinase should accelerate PER2 degradation relative to the wild type. The data are consistent with this model (Fig. 5a), with significantly greater degradation in the face of the CK1ε^{tau} kinase. It appears that these effects are specific to PER, because CRY1::CFP degradation is not affected by any variants of CK1ε (Fig. 5b). This indicates that CK1ε^{tau} is selective for PER2. To explore whether degradation is dependent on nuclear export, we also used leptomycin B (LMB) to retain PER2::YFP in the nuclear compartment. Such treatments do not change the overall pattern of accelerated PER2 degradation by CK1ε^{tau} (Fig. 5c) and confirm that CK1ε^{tau} acts within the nucleus. Similar dynamics appear to operate on PER1 proteins. For instance, using NIH-3T3 cells, we have retained PER1 in the nucleus or cytoplasm and examined the action of CK1ε^{tau} on degradation rates (Meng et al. 2008). Nuclear-trapped PER1 was achieved by either mutation of a nuclear export signal (mtNES) or insertion of an additional copy of the nuclear localization signal (+NLS), whereas retention of PER1 in the cytoplasmic compartment was induced by use of a mutant form (PER1$_{1-823}$), which lacked the native nuclear localization signal, resulting in cytoplasmic retention. In all cases, CK1ε^{tau} caused accelerated degradation not only of wild-type PER1 as previously reported (Gallego et al. 2006), but also of both nuclear-trapped PER1 and cytoplasmic PER1. Together, these cell-line data clearly reveal that CK1ε^{tau} targets both PER1 and PER2 but not CRY1 proteins, and this selective targeting occurs regardless of nuclear or cytoplasmic localization.

Studies of Endogenous Protein Degradation

We have extended these studies to cells and tissue derived from mice to test whether studies of recombinant protein stability accurately reflect endogenous events. Western blots of native PER1 and CRY1 expression in wild-type and *tau* mutant fibroblast cultures following CHX treatment reveal that the overall protein expression of PER1 is lower in *tau* fibroblasts (Meng et al. 2008). Furthermore, hyperphosphorylated forms of PER1 (upper bands) display a significantly shorter half-life in *tau* mutant cells. These data are consistent with the concept that hyperphosphorylated PER1 may be preferentially tar-

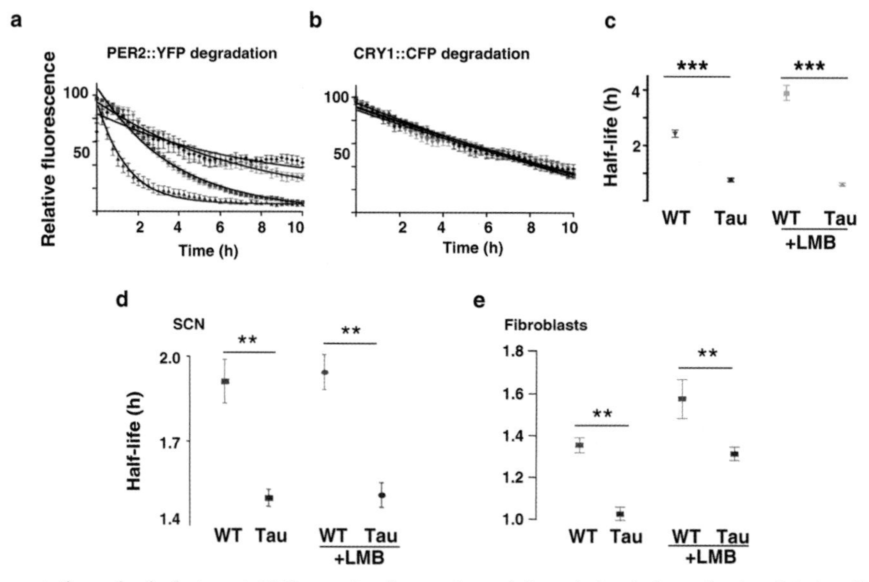

Figure 5. The *tau* mutation selectively targets PER proteins for accelerated degradation independently of their cellular localization. (*a,b*) Decay of PER2::YFP or CRY1::CFP fluorescence in COS-7 cells cotransfected with wild-type (*red*), *tau* (*blue*), kinase-dead (*green*) CK1ε, or empty plasmid (*black*). Cells were treated with CHX (20 μg/ml) 0.5 hour before recording. (*c*) Half-lives for PER2::YFP in presence of wild-type or *tau* CK1ε. PER2::YFP-transfected COS-7 cells were treated with vehicle or leptomycin B (LMB, 10 ng/ml) 1 hour before CHX (20 μg/ml). (*Red*) CK1ε wild type + vehicle; (*yellow*) wild type + LMB; (*blue*) *tau* + vehicle; (*green*) *tau* + LMB (mean ±95% confidence limits, $n = 14–33$, two to four independent experiments, ***$p < 0.001$). (*d,e*) Half-lives for PER2::LUC bioluminescence from SCN slices and lung fibroblasts exposed to CHX in the presence or absence of LMB. Data are normalized to the peak level of expression (time 0 after CHX) and to the minimum after 8 hours and plotted as mean ±S.E.M. CK1ε^{tau} significantly accelerated loss of PER2::LUC signal regardless of LMB treatment. **$p < 0.01$ by *t*-test. (Redrawn, with permission, from Meng et al. 2008.)

geted to the proteasome for degradation (Gallego and Virshup 2007). In contrast, the overall expression level and CHX-stimulated decline of CRY1 was unaffected by CK1ε^{tau}. To obtain a more dynamic view of PER degradation, bioluminescence of PER2::LUC was recorded from SCN and lung fibroblasts treated with CHX at the peak of PER2 expression. This revealed that CK1ε^{tau} significantly shortened the half-life of PER2::LUC expression for both SCN and fibroblasts. Moreover, CK1ε^{tau} was still effective at accelerating PER2::LUC degradation in the presence of LMB (Fig. 5d,e), consistent with the recombinant protein studies that showed a nuclear action for the mutation. Together, these data provide a convincing demonstration that CK1ε^{tau} facilitates the accelerated degradation of endogenous PER (but not CRY) proteins in both the SCN and primary cells and that blockade of nuclear export does not attenuate the *tau* phenotype.

HOW DOES AN ACCELERATED CLOCK ACT ON CIRCADIAN OUTPUTS?

Our earlier studies on the *tau* mutant hamster demonstrated that CK1ε^{tau} has altered the dynamics of PER protein turnover, specifically during the early-night phase of the circadian cycle. Other events (i.e., the rise of PER proteins and increase in mRNA expression) are normally phased and occur at the same solar time in both *tau* and wild-type animals. This early decay of the negative regulator in the SCN (PER proteins) therefore allows for early onset of nocturnal events such as melatonin secretion. Altered dynamics of PER turnover may also offer an explanation for changes in the time base of response to repeated melatonin signals, although in the absence of a detailed understanding of how melatonin regulates these proteins at target sites, such considerations must remain speculative. One prediction from studies of PER turnover is that in the CK1ε^{tau} mouse, we might expect to find similar changes in the circadian dynamics of PER. We have tested this in the SCN, using the PER2::LUC reporter and aligning waveforms of SCN slices normalized to the peak or nadir of luciferase activity to identify relative changes in the rates of expression of the underlying protein. When aligned at peak expression, there is a significantly accelerated decline in signal in CK1ε^{tau} slices, resulting in an advance at this phase of the cycle of about 3.4 hours, compared to wild-type slices (Fig. 6) (Meng et al. 2008). Thus, much of the acceleration of period in the CK1$\varepsilon^{tau/tau}$ mouse can be attributed to a selective early truncation of molecular events during this nocturnal phase. These data are therefore entirely consistent with our earlier studies of the *tau* hamster (Dey et al. 2005).

Impact on Metabolism and Activity

It appears that acceleration of molecular clockwork at a specific phase may be associated with altered circadian dynamics of behavior and physiology. For instance, in the CK1ε^{tau} mouse, the overall level of activity as reflected in total wheel-running revolutions is, somewhat remarkably, not reduced by the mutation (see Fig. 3c). This is also

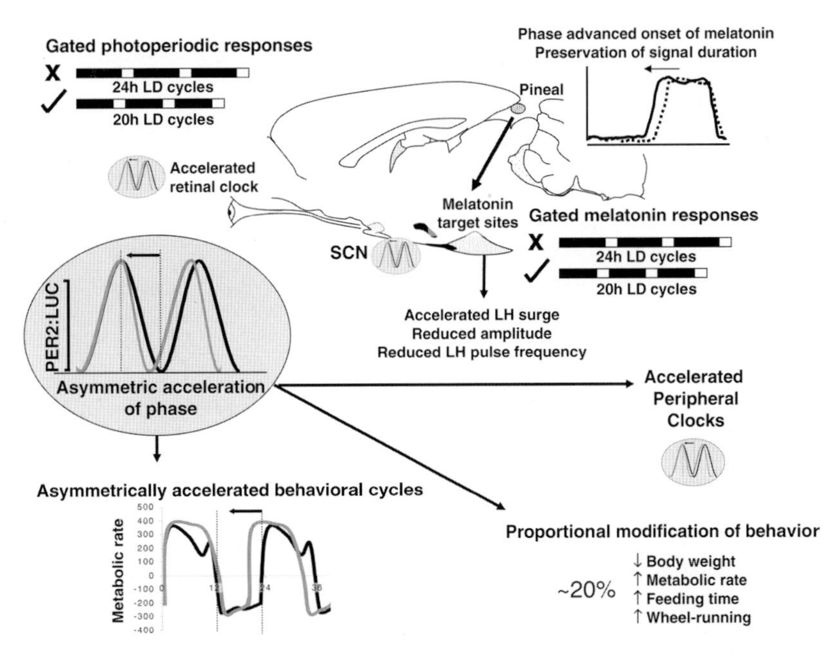

Figure 6. Summary of the impact of CK1ε^{tau} on circadian timing in Syrian hamsters and mice. In hamsters, the *tau* mutation alters the gated response to photoperiod and melatonin signals and also impacts neuroendocrine outputs (i.e., LH, cortisol). In mice, *tau* accelerates behavioral cycles in a manner identical to that of hamsters. In both species, this is associated with accelerated degradation of SCN PER and electrical firing rates. In both species, PER rhythms are asymmetric, with an accelerated decline in protein in the early nocturnal phase. This asymmetry results in selective compression of the diurnal phase of the cycle, such that nocturnal events (wheel-running intensity, melatonin rhythms in hamsters, and metabolic cycles in mice) are relatively unimpaired. In both species, metabolic rate is increased and body weight is reduced. Peripheral oscillators are not accelerated to the same extent as that in the SCN, leading (in hamsters) to abnormal phasing of peripheral clockwork. (PER2:LUC and metabolic rate are redrawn, with permission, from Meng et al. 2008.)

reflected in an altered pattern of circadian energy expenditure. In CK1ε^{tau} mice, a compression of circadian rhythms of oxygen consumption (VO_2) closely matches periods determined earlier by wheel running (Meng et al. 2008). However, the pattern of oxygen consumption is highly asymmetric, such that the shortening of circadian period is almost entirely associated with a compression of the inactive (subjective day) phase (Fig. 6). The net consequence of these altered metabolic dynamics is that overall metabolic rate is increased in CK1ε^{tau} mice by approximately 20%.

An identical phenotype has been reported for the *tau* mutant hamster. Specifically, resting metabolic rate is elevated (relative to body weight) by about 20% in homozygote *tau* hamsters (Oklejewicz et al. 1997), the same extent to which circadian period is accelerated, such that hamsters with only one copy of the *tau* allele exhibit a resting metabolic rate midway between the homozygote and wild-type animals. As a consequence, when metabolic rate is calculated per circadian cycle, no statistical differences are observed between genotypes. Oklejewicz et al. (2001) extended their studies of energy balance in *tau* hamsters by detailing feeding behavior and meal patterning. *tau* hamsters exhibited similar meal sizes, whereas total feeding time was elevated in the mutant animals. As for metabolic rate, meal frequency and total time feeding were similar between genotypes when calculated relative to the circadian cycle. Although these findings imply that both energy intake (feeding) and expenditure (metabolic rate) are proportionally increased in *tau* animals, altered dynamics of growth and body size are also observed in hamsters. This is not a trivial consequence of the lighting schedules in which the animals are reared, because *tau* hamsters grow significantly more slowly and achieve lower adult weights even when maintained, with wild-type animals as controls, in DD from birth (Lucas et al. 2000). Although we have not verified the growth rates of mutant mice during a full life cycle, adult CK1ε^{tau} mice are also typically smaller than their wild-type counterparts.

A further nonintuitive but nonetheless intriguing feature of our data is that the duration of molecular night is relatively compressed due to the accelerated clearance of PER2 after its peak expression at CT12, but the duration of nocturnal behavioral and physiological processes (CBT, melatonin secretion) is apparently normal, as measured in solar time. Consequently, because the offsets are not governed by the molecular cycle, they occur several hours later than the compressed molecular cycle would predict: nocturnal processes spilling over into the start of the molecular day. Hence, the behavioral day (defined by activity offset and onset) is shorter in the mutant, even though the underlying determining change is shortening of the molecular night. These strongly asymmetric effects of CK1ε^{tau} on molecular and behavioral timing likely underlie the pronounced disturbances of circadian timekeeping (dampened amplitude, altered phasing) widely observed in the peripheral tissues of *tau* mutant hamsters (Dey et al. 2005) and may contribute to poor growth and general morbidity associated with the mutation in hamsters (Lucas et al. 2000). The variable responses of peripheral tissues to the *tau* mutation are intriguing and may suggest tissue-specific differences in the relative expression of CK1ε or CK1δ, or cellular differences in the manner in which PER proteins are trafficked and degraded.

Importantly, these novel phenotypes indicate that mutations in humans that alter circadian period may have complex and unpredictable effects on overall circadian organization and metabolism. This can arise either because the mutations have direct biochemical actions on metabolism or because the genetic dislocation of global circadian programs compromises the optimal functioning of particular tissues and metabolic processes, even if the mutant gene is not involved in such processes directly. Consequently, conditions such as FASPS, first recognized as sleep disorders (Toh et al. 2001; Xu et al. 2005; Vanselow et al. 2006), may in fact comprise multifactorial metabolic and physiological disturbances, echoing those recently associated with long-term rotational shift work (Knutsson 2003; Sookoian et al. 2007). This may also have implications for the development of novel pharmacological approaches to the regulation of circadian timing and phasing of sleep, because any new drugs that target these primary sleep-related complaints may also need to be tested on a wide range of circadian systems.

ACKNOWLEDGMENTS

We thank the many current and past colleagues who have worked with us over the years, including Michael Menaker, Anne Stirland, Rob Lucas, Andrei Semikhodskii, Hugh Piggins and Tim Brown, Joita Dey, Martin Sladek, Jake Lebiecki, Joe Takahashi and Seung-Hee Yoo, David Virshup and Monica Gallego, Larisa Logonova, Julia Grosse, and Jo Chesham. The work was supported by research grants from the BBSRC, MRC, and EUCLOCK.

REFERENCES

Axelrod J. 1974. The pineal gland: A neurochemical transducer. *Science* **184:** 1341.

Brown T.M., Hughes A.T., and Piggins H.D. 2005. Gastrin-releasing peptide promotes suprachiasmatic nuclei cellular rhythmicity in the absence of vasoactive intestinal polypeptide-VPAC2 receptor signaling. *J. Neurosci.* **25:** 11155.

Chappell P.E. 2005. Clocks and the black box: Circadian influences on gonadotropin-releasing hormone secretion. *J. Neuroendocrinol.* **17:** 119.

de la Iglesia H.O. and Schwartz W.J. 2006. Minireview: Timely ovulation: Circadian regulation of the female hypothalamo-pituitary-gonadal axis. *Endocrinology* **147:** 1148.

Deguchi T. and Axelrod J. 1972. Control of circadian change of serotonin *N*-acetyltransferase activity in the pineal organ by the β-adrenergic receptor. *Proc. Natl. Acad. Sci.* **69:** 2547.

Dey J., Carr A.J., Cagampang F.R., Semikhodskii A.S., Loudon A.S., Hastings M.H., and Maywood E.S. 2005. The *tau* mutation in the Syrian hamster differentially reprograms the circadian clock in the SCN and peripheral tissues. *J. Biol. Rhythms* **20:** 99.

Field M.D., Maywood E.S., O'Brien J.A., Weaver D.R., Reppert S.M., and Hastings M.H. 2000. Analysis of clock proteins in mouse SCN demonstrates phylogenetic divergence of the circadian clockwork and resetting mechanisms. *Neuron* **25:** 437.

Gallego M. and Virshup D.M. 2007. Post-translational modifications regulate the ticking of the circadian clock. *Nat. Rev. Mol. Cell Biol.* **8:** 139.

Gallego M., Eide E.J., Woolf M.F., Virshup D.M., and Forger D.B. 2006. An opposite role for *tau* in circadian rhythms revealed by mathematical modeling. *Proc. Natl. Acad. Sci.* **103:** 10618.

Hastings M., O'Neill J.S., and Maywood E.S. 2007. Circadian clocks: Regulators of endocrine and metabolic rhythms. *J. Endocrinol.* **195:** 187.

Hastings M.H., Reddy A.B., Garabette M., King V.M., Chahad-Ehlers S., O'Brien J., and Maywood E.S. 2003. Expression of clock gene products in the suprachiasmatic nucleus in relation to circadian behavior. *Novartis Found. Symp.* **253:** 203.

Hazlerigg D.G., Andersson H., Johnston J.D., and Lincoln G. 2004. Molecular characterization of the long-day response in the Soay sheep, a seasonal mammal. *Curr. Biol.* **14:** 334.

Kloss B., Price J.L., Saez L., Blau J., Rothenfluh A., Wesley C.S., and Young M.W. 1998. The *Drosophila* clock gene *double-time* encodes a protein closely related to human casein kinase Iε. *Cell* **94:** 97.

Knutsson A. 2003. Health disorders of shift workers. *Occup. Med.* **53:** 103.

Lee C., Etchegaray J.P., Cagampang F.R., Loudon A.S., and Reppert S.M. 2001. Posttranslational mechanisms regulate the mammalian circadian clock. *Cell* **107:** 855.

Lincoln G.A., Andersson H., and Loudon A. 2003. Clock genes in calendar cells as the basis of annual timekeeping in mammals—A unifying hypothesis. *J. Endocrinol.* **179:** 1.

Lincoln G., Messager S., Andersson H., and Hazlerigg D. 2002. Temporal expression of seven clock genes in the suprachiasmatic nucleus and the pars tuberalis of the sheep: Evidence for an internal coincidence timer. *Proc. Natl. Acad. Sci.* **99:** 13890.

Loudon A.S., Ihara N., and Menaker M. 1998. Effects of a circadian mutation on seasonality in Syrian hamsters (*Mesocricetus auratus*). *Proc. Biol. Sci.* **265:** 517.

Loudon A.S., Wayne N.L., Krieg R., Iranmanesh A., Veldhuis J.D., and Menaker M. 1994. Ultradian endocrine rhythms are altered by a circadian mutation in the Syrian hamster. *Endocrinology* **135:** 712.

Lowrey P.L. and Takahashi J.S. 2004. Mammalian circadian biology: Elucidating genome-wide levels of temporal organization. *Annu. Rev. Genomics Hum. Genet.* **5:** 407.

Lowrey P.L., Shimomura K., Antoch M.P., Yamazaki S., Zemenides P.D., Ralph M.R., Menaker M., and Takahashi J.S. 2000. Positional syntenic cloning and functional characterization of the mammalian circadian mutation *tau*. *Science* **288:** 483.

Lucas R.J., Stirland J.A., Mohammad Y.N., and Loudon A.S. 2000. Postnatal growth rate and gonadal development in circadian *tau* mutant hamsters reared in constant dim red light. *J. Reprod. Fertil.* **118:** 327.

Lucas R.J., Stirland J.A., Darrow J.M., Menaker M., and Loudon A.S. 1999. Free running circadian rhythms of melatonin, luteinizing hormone, and cortisol in Syrian hamsters bearing the circadian *tau* mutation. *Endocrinology* **140:** 758.

Malpaux B., Migaud M., Tricoire H., and Chemineau P. 2001. Biology of mammalian photoperiodism and the critical role of the pineal gland and melatonin. *J. Biol. Rhythms* **16:** 336.

Maywood E.S., O'Neill J.S., Chesham J.E., and Hastings M.H. 2007. Minireview: The circadian clockwork of the suprachiasmatic nuclei—Analysis of a cellular oscillator that drives endocrine rhythms. *Endocrinology* **148:** 5624.

Maywood E.S., Buttery R.C., Vance G.H., Herbert J., and Hastings M.H. 1990. Gonadal responses of the male Syrian hamster to programmed infusions of melatonin are sensitive to signal duration and frequency but not to signal phase nor to lesions of the suprachiasmatic nuclei. *Biol. Reprod.* **43:** 174.

Meng Q.J., Logunova L., Maywood E.S., Gallego M., Lebiecki J., Brown T.M., Sládek M., Semikhodskii A.G., Glossop N.R.J., Piggins H.D., Chesham J.E., Bechtold D.A., Yoo S.H., Takahashi J.S., Virshup D.M., Boot-Handford R.P., Hastings M.H., and Loudon A.S.I. 2008. Setting clock speed in mammals: The CK1ε *tau* mutation in mice accelerates the circadian pacemaker by selectively destabilizing PERIOD proteins. *Neuron* (in press).

Messager S., Ross A.W., Barrett P., and Morgan P.J. 1999. Decoding photoperiodic time through Per1 and ICER gene amplitude. *Proc. Natl. Acad. Sci.* **96:** 9938.

Mignot E. and Takahashi J.S. 2007. A circadian sleep disorder reveals a complex clock. *Cell* **128:** 22.

Moore R.Y. and Eichler V.B. 1972. Loss of a circadian adrenal corticosterone rhythm following suprachiasmatic lesions in the rat. *Brain Res.* **42:** 201.

Oklejewicz M., Overkamp G.J., Stirland J.A., and Daan S. 2001. Temporal organization of feeding in Syrian hamsters with a genetically altered circadian period. *Chronobiol. Int.* **18:** 657.

Oklejewicz M., Hut R.A., Daan S., Loudon A.S., and Stirland A.J. 1997. Metabolic rate changes proportionally to circadian frequency in *tau* mutant Syrian hamsters. *J. Biol. Rhythms.* **12:** 413.

Price J.L., Blau J., Rothenfluh A., Abodeely M., Kloss B., and Young M.W. 1998. *double-time* is a novel *Drosophila* clock gene that regulates PERIOD protein accumulation. *Cell* **94:** 83.

Ralph M.R. and Menaker M. 1988. A mutation of the circadian system in golden hamsters. *Science* **241:** 1225.

Ralph M.R., Foster R.G., Davis F.C., and Menaker M. 1990. Transplanted suprachiasmatic nucleus determines circadian period. *Science* **247:** 975.

Reppert S.M. and Weaver D.R. 2002. Coordination of circadian timing in mammals. *Nature* **418:** 935.

Saper C.B., Scammell T.E., and Lu J. 2005. Hypothalamic regulation of sleep and circadian rhythms. *Nature* **437:** 1257.

Shimomura K., Nelson D.E., Ihara N.L., and Menaker M. 1997. Photoperiodic time measurement in *tau* mutant hamsters. *J. Biol. Rhythms* **12:** 423.

Silver R., LeSauter J., Tresco P.A., and Lehman M.N. 1996. A diffusible coupling signal from the transplanted suprachiasmatic nucleus controlling circadian locomotor rhythms. *Nature* **382:** 810.

Sookoian S., Gemma C., Fernandez Gianotti T., Burgueno A., Alvarez A., Gonzalez C.D., and Pirola C.J. 2007. Effects of rotating shift work on biomarkers of metabolic syndrome and inflammation. *J. Intern. Med.* **261:** 285.

Stirland J.A., Mohammad Y.N., and Loudon A.S. 1996a. A mutation of the circadian timing system (*tau* gene) in the seasonally breeding Syrian hamster alters the reproductive response to photoperiod change. *Proc. Biol. Sci.* **263:** 345.

Stirland J.A., Hastings M.H., Loudon A.S., and Maywood E.S. 1996b. The *tau* mutation in the Syrian hamster alters the photoperiodic responsiveness of the gonadal axis to melatonin signal frequency. *Endocrinology* **137:** 2183.

Stirland J.A., Grosse J., Loudon A.S., Hastings M.H., and Maywood E.S. 1995. Gonadal responses of the male *tau* mutant Syrian hamster to short-day-like programmed infusions of melatonin. *Biol. Reprod.* **53:** 361.

Toh K.L., Jones C.R., He Y., Eide E.J., Hinz W.A., Virshup D.M., Ptáček L.J., and Fu Y.H. 2001. An hPer2 phosphorylation site mutation in familial advanced sleep phase syndrome. *Science* **291:** 1040.

Tosini G. and Menaker M. 1996. Circadian rhythms in cultured mammalian retina. *Science* **272:** 419.

Vanselow K., Vanselow J.T., Westermark P.O., Reischl S., Maier B., Korte T., Herrmann A., Herzel H., Schlosser A., and Kramer A. 2006. Differential effects of PER2 phosphorylation: Molecular basis for the human familial advanced sleep phase syndrome (FASPS). *Genes Dev.* **20:** 2660.

Xu Y., Padiath Q.S., Shapiro R.E., Jones C.R., Wu S.C., Saigoh N., Saigoh K., Ptáček L.J., and Fu Y.H. 2005. Functional consequences of a CKIΔ mutation causing familial advanced sleep phase syndrome. *Nature* **434:** 640.

Yoo S.H., Yamazaki S., Lowrey P.L., Shimomura K., Ko C.H., Buhr E.D., Siepka S.M., Hong H.K., Oh W.J., Yoo O.J., Menaker M., and Takahashi J.S. 2004. PERIOD2::LUCIFERASE real-time reporting of circadian dynamics reveals persistent circadian oscillations in mouse peripheral tissues. *Proc. Natl. Acad. Sci.* **101:** 5339.

Novel Insights from Genetic and Molecular Characterization of the Human Clock

L.J. Ptáček,*† C.R. Jones,‡ and Y.-H. Fu*

*Department of Neurology, University of California, San Francisco, California 94158-2324;
†Howard Hughes Medical Institute, University of California, San Francisco, California 94158-2324;
‡Department of Neurology, University of Utah, Salt Lake City, Utah 84132-2450

Biological rhythms govern the ebb and flow of life on planet Earth. Animals have an internal timekeeping mechanism that precisely regulates 24-hour rhythms of body function and behavior and synchronizes them to the day/night cycle. Circadian pacemakers trigger behavioral and physiological processes that dictate our daily rhythms. Despite the importance of the circadian clock to all aspects of our physiology and behavior, the opportunity to probe the human circadian clock only recently became possible with the recognition of Mendelian circadian variants in people (familial advanced sleep phase syndrome, FASPS). We have now cloned several genes and identified mutations causing FASPS. Study of these genes and the proteins they encode and engineering of the human mutations into mouse models are allowing study of this fascinating phenotype and yielding novel insights into circadian regulation in humans. Ultimately, such work will allow us to understand the similarities *and* differences between the human clock and those of model organisms. In addition, recent studies have also linked disruption of the circadian clock with numerous ailments, including cancer, cardiovascular diseases, asthma, and learning disorders. Thus, studying the molecular mechanism of human circadian rhythmicity will have an enormous impact on our understanding of human health and disease. It should also lead to new strategies for pharmacological manipulation of the human clock to improve the treatment of jet lag, various clock-related sleep and psychiatric disorders, and other human diseases.

HUMAN BEHAVIORAL GENETICS

Studies of genetic behaviors in humans have been hampered by the complexity of behavioral phenotypes, in addition to locus and allelic heterogeneity that is certain to exist. The identification of a Mendelian circadian variant of humans (FASPS) was an exciting advance in the field, as the existence of such families suggested that cloning of a human behavioral gene was, in fact, a reasonable undertaking (Jones et al. 1999). The variants found in an extant human population have been consistent with life on our planet. In contrast, forward genetic screens that have been performed in other organisms have focused on mutants of large effect. Many of these would likely be selected against in evolution. Thus, in human FASPS, the a priori expectation is that period variants in humans would be smaller in amplitude than those identified in genetic screens.

In model organisms, most mutation screens have focused on measuring the circadian period and identifying alleles that lead to very short or very long τ. Period measurements are used because they are relatively easy to perform and easily quantitated in organisms such as *Drosophila* and mice. However, measuring periods in humans is very challenging. Consequently, it would be very difficult to get periods on many family members from the same pedigree as well as from many other pedigrees. Thus, in the new field of human circadian genetics, the focus has been on phenotypes affecting phase (phase advance: "morning larks" or phase delay: "night owls"). Phase alterations can result from period alterations but, theoretically, could also be caused by altered phase angle of entrainment or coupling of clock to behavioral outputs. Theoretically, phase advance/delay may result from mutations that affect period *and/or* phase angle of entrainment.

FASPS was identified because of a very dramatic advance of the sleep phase that was recognized initially in a single individual. In contrast to the phase advance seen in the ASPS of aging, FASPS typically has its onset in childhood or young adult life and persists into old age, sometimes apparently worsening with aging (Jones et al. 1999). The phase was advanced 4–6 hours in FASPS subjects compared to controls, but sleep quality and quantity were normal. Subsequently, we have identified a large number of families segregating dominant FASPS alleles.

Identification of FASPS in a large cohort of FASPS families thus provided the opportunity to search for genes and variants that cause period alterations as well as phase angle of entrainment variants in the general population.

LESSONS FROM GENETIC AND MOLECULAR DISSECTION OF CLOCKS IN THE MODEL ORGANISMS

Extensive work in a number of model organisms has led to a model of the circadian molecular clock consisting of interlocked negative and positive transcriptional/translational feedback loops that oscillate with a period of approximately 24 hours. Not surprisingly, there are shared elements from the clocks of diverse organisms, from *Neurospora* to *Drosophila* and rodents. However, it is also not surprising that the molecular components that have been identified in mammals have elucidated a somewhat more complex clock mechanism, presumably with more precise regulation of more complex organisms. It is also interesting that there are differences between the clocks of invertebrates and vertebrates, and it is very likely that differences exist even between humans and rodents. Most rodents have circadian periods of less than 24 hours, whereas human periods tend to be slightly longer than 24 hours. The molecular basis of phase differences is not known; mice are nocturnal, whereas humans are diurnal organisms. Therefore, the ability to begin to probe the genetic basis of human clock function was an exciting and important advance, as it allows one to begin to understand not only the similarities, but also the *differences* that exist between clocks in humans and other organisms.

Advanced Sleep Phase Syndrome and Aging

The advanced sleep phase syndrome of aging (ASPS of aging) is defined as a tendency to awake and go to sleep earlier than desired as one gets older. It has recently been demonstrated that this tendency begins at about the age of 20 but tends to become of greater amplitude as individuals age (Roenneberg et al. 2004). Many people tend to go to sleep and awaken earlier as they age but are not troubled by this sleep phase advance. Although it is likely that this results from the same phenomenon causing ASPS of aging, clinicians do not refer to this as ASPS of aging because it is not troublesome to many subjects.

Familial Advanced Sleep Phase Syndrome

FASPS was first recognized in a large Utah pedigree, segregating an allele that led to a lifelong tendency to wake up and sleep at very early times (biological tendency to awaken spontaneously between 1:00 a.m. and 5:30 a.m. and to fall asleep spontaneously between 5:30 p.m. and 8:30 p.m.). To identify the underlying biological preference, it is necessary to dissect away the psychosocial and familial-cultural factors that impinge on the innate tendency. Furthermore, environmental factors such as artificial light, alcohol, and caffeine also modify expression of the preferred biological sleep/wake times.

The trait was recognized as early as 8 years of age in one subject, but frequently, the subjects to do not meet formal criteria until their mid twenties. This is likely because the delayed sleep phase syndrome (DSPS) of adolescence can in many cases mask the strong phase advance of FASPS. The trait segregates as a highly penetrant autosomal dominant trait (Jones et al. 1999). Affected individuals from this first pedigree were studied and had normal sleep quality and quantity. Melatonin and temperature rhythms, as well as sleep and wake times, were advanced by 4–6 hours in affected subjects relative to controls. The proband from the family underwent a free-running period measurement and was shown to have a period nearly 1 hour shorter than age- and gender-matched controls (Jones et al. 1999).

Analysis of subjects and families included the Horne-Ostberg questionnaire, although it is now clear that the Munich Chronotype questionnaire (Zavada et al. 2005) may be better for assessing phase in individuals from FASPS families. In addition, subjects completed a general sleep-screening questionnaire to rule out sleep disorders such as obstructive sleep apnea, narcolepsy, or restless leg syndrome, which might lead to excessive daytime sleepiness and early sleep times. The Beck depression inventory and neuropsychology interview were important to assess whether subjects had major depression, which could lead to early morning awakening. In contrast to the early morning awakening common in patients with depression, FASPS individuals wake up feeling energetic and refreshed. Finally, all subjects undergo a 1-hour structured interview that attempts to disentangle the many psychosocial, familial-cultural, and self-imposed decisions that humans tend to overlay on their underlying biologic sleep and wake tendencies.

POSITIONAL CLONING OF THE FIRST HUMAN CIRCADIAN RHYTHM GENE

Period 2

Genetic study of the Utah family described by Jones et al. led to the mapping of the FASPS allele to chromosome 2q (Toh et al. 2001). At the time of the mapping, very little genomic sequence was available for chromosome 2q, and thus, physical mapping of the 3–4-Mb region of chromosome 2q was necessary and led to identification of more than 40 cDNAs within the critical region. Sequencing of these candidate cDNAs identified one to be a homolog of the *Drosophila period* gene. Sequencing of this gene (*hPER2*) revealed a serine-to-glycine mutation at position 662 that segregated with affected members in the family (Toh et al. 2001). There was one small branch containing three "affected" individuals that did not carry the mutation. On the basis of in vitro functional studies and, ultimately, animal modeling of the human mutation, it was shown that these individuals represented FASPS phenocopies. Although these individuals had a strong morning lark tendency, they were the least advanced of any of the individuals who had been classified as affected in this family. These phenocopies underscore the challenges of behavioral genetics where strong morning lark phenotypes in individuals were not caused by the FASPS mutation segregating in this family. In addition, they emphasize the power of large Mendelian families for mapping and identifying genes despite the challenges of behavioral genetics.

This serine-to-glycine mutation led to hypophosphorylation of the PER2 polypeptide by casein kinase I (CKI). Coimmunoprecipitation experiments demonstrated that the region harboring this mutation was a point where CKI interacts with PER2. Sequence alignment of mammalian period proteins also showed that the mutant serine residue was the first of five that were separated from one another by two amino acids. This motif is consistent with a recognition site for CKI. Finally, replacement of the serine at position 662 with acidic residues was shown to reconstitute the gel-shift due to phosphorylation of downstream residues when the PER2 polypeptide was treated with recombinant-purified CKI (Toh et al. 2001).

Casein Kinase Iδ

With the increasing density of the human genome sequence across the genome and a growing collection of well-characterized FASPS subjects, it then became possible to sequence candidate gene exons to search for mutations in other FASPS families. The second mutation found was in a moderate-sized family with a strong FASPS allele in five individuals. An additional sixth individual was a 14-year-old boy at the time of his assessment, and although he was a strong morning lark for a teenager, he did not meet the strict clinical criteria of FASPS (Jones et al. 1999; Toh et al. 2001). All six of these individuals carried a missense mutation in the casein kinase Iδ (CKIδ) gene in a highly conserved region of the protein. The mutation occurred six amino acids carboxyl terminal to the catalytic lysine (K38) of CKIδ. This residue was completely conserved from the *Drosophila* double-time protein through human CKI. This mutation led to hypophosphorylation of phosvitin and α-casein, as well as mammalian PER2 proteins (Xu et al. 2005). Generation of bacterial artificial chromosome (BAC) transgenic mice carrying the T44A mutation revealed a shortening of period of approximately 0.4 hours (Xu et al. 2005).

Phosphorylation of the region downstream from S662 by CKI thus represents a convergence of two different genes (*CKI* and *Per2*) where mutations cause the same FASPS phenotype. In the case of PER2, an inability to phosphorylate S662 (because it has mutated to a glycine) leads to hypophosphorylation, despite the presence of wild-type CKI. In the case of the τ mutant hamster and the human FASPS family with the CKIδ mutation, hypophosphorylation results from a mutant kinase (CK1ε in the case of the τ mutant hamster and CK1δ in a human FASPS family).

IN VITRO BIOCHEMICAL EXPERIMENTS TO ASSESS PHOSPHORYLATION IN THE REGION DOWNSTREAM FROM PER2 S662

Oligopeptides with the sequence including S662 and downstream residues were synthesized with or without a phosphate covalently linked to the serine corresponding to position 662 in PER2. Qualitative assays measuring incorporation of ^{32}P revealed that there was robust phosphorylation of the peptide with the covalently linked phosphate but not for the peptide without. Subsequently, a quantitative assay was used to demonstrate that approximately 4 moles of phosphate were added per mole of PER2 substrate with the covalently linked phosphate (Xu et al. 2007). Phosphopeptide-mapping experiments showed that the tyrosine and threonine residues in the oligopeptides were not phosphorylated, consistent with the prediction that serial serine phosphorylations occurred at every third amino acid downstream from position 662. This experiment also suggests that CKI does not phosphorylate S662 but, rather, that a priming event is required due to phosphorylation by another kinase before CK1 can phosphorylate the four downstream serines (Xu et al. 2007).

PER2 MICE CARRYING THE HUMAN FASPS PER2 MUTATION RECAPITULATES AN FASPS PHENOTYPE

Transgenesis was accomplished using a human BAC that contained the entire *Per2* gene with more than 50 kb of genomic sequence upstream of the ATG start site. The advantage of this strategy is that the genomic gene is large and, with intronic and intragenic genomic sequences, is typically buffered from the position effects that are so common with transgenesis using cDNAs. Furthermore, the cis-regulatory elements are frequently included in the upstream sequence, leading to expression patterns that are faithful to the endogenous genes.

A wild-type *hPER2* transgenic was generated as a control for increased PER2 dosage and for the use of the human gene. Homologous recombination in bacteria was then used to generate the S662G mutation. Finally, a third transgene was generated encoding an aspartate at position 662, to mimic a constitutively phosphorylated S662. Multiple lines of each of these transgenes were generated, and the period of wild-type transgenics carrying one to three copies of the *hPer2* transgene had a completely normal period. In contrast, the S662G transgenic animals had a period that was approximately 2 hours shorter than controls, whereas the S662D transgenic animals had a period that was approximately one-half-hour longer than wild-type. When studied in a light/dark 12:12 paradigm, the S662G mice were phase-advanced and began running on the wheel 4–6 hours before the lights were turned off and decreasing activity hours before the lights came back on.

Further study of the cycling of various mRNAs was accomplished in liver and fibroblast extracts from these animals and showed that both the mouse and human *Per2* mRNA was phase-advanced and decreased in levels in the S662G mice and phase-delayed and of higher amplitude in S662D mice when compared to wild-type controls (Xu et al. 2007). Contrary to the expectation, the S662G mutation did not appear to manifest its effects through decreased protein stability but through decreased transcription of the *Per2* gene. The mutant *hPer2* protein had a dominant effect on decreased transcription of both *mPer2* and *hPer2*.

CK1δ DOSAGE EXPERIMENTS OF S662G AND S662D TRANSGENIC PER2 MICE

Heterozygosity for a null mutation in CKIδ does not have an effect on period (Xu et al. 2005) (homozygous knockout is lethal). Transgenesis using a wild-type CKIδ BAC transgene with one to three copies per genome also does not have an effect on period. However, the CKIδ transgene, when crossed onto PER2 S662G transgenic mice led to a further shortening of period, whereas crossing a null allele for CKIδ onto the same transgenic animals led to a less short period than the *Per2* mutant alone. In addition, crossing one null CK1δ allele onto the S662D *Per2* transgenic mouse led to further prolongation of the period. Taken together, these data suggest a model whereby CKI is exerting opposite effects through phosphorylation of different sites. Phosphorylations downstream from residue 662 lead to increased transcription and increased PER2 protein, whereas phosphorylation at another site (presumably in PER2, although the possibility remains that this could be through an indirect effect) leads to increased PER2 degradation and lower PER2 levels. Changing CK1δ dosage therefore has little effect when wild-type PER2 is present. However, with a mutation at position 662, CKIδ is unable to activate transcription and therefore PER2 levels are lower and the normal process of degradation continues, leading to decreased PER2 levels at all time points in the circadian day.

FUTURE DIRECTIONS AND PERSPECTIVES

Following on many discoveries using model organisms, tremendous insights into the workings of circadian clocks have been realized. During the last decade, these types of genetic experiments have now been extended into humans with recognition and characterization of a Mendelian circadian phenotype in humans and cloning of genes causing the FASPS phenotype. It is interesting that some elements of the circadian clock are conserved across very large evolutionary times, but certainly, there is evidence that differences also occur between the clocks of different species. Studying genes and mutations identified in humans with strong circadian phenotypes is focusing attention on interesting regions of these proteins and allowing biological experiments that will lead to greater insights not only into the workings of the circadian clock mechanism, but also its relationship to other metabolic and biochemical systems. Because the circadian clock regulates many biological processes, including (but not limited to) the immune system, cardiopulmonary systems, and cell cycle regulation, it is likely that knowledge of the workings of the human clock will come to bear on our understanding not only of health, but also of the relationship of the circadian clock and disease. Generation of human circadian mutations in other organisms and comparing phenotypes across different species are certain to provide interesting information regarding similarities and differences that the human clock shares with different organisms.

APPLICATION OF KNOWLEDGE TO THE SPECTRUM OF CHRONOTYPES IN THE GENERAL HUMAN POPULATION

With the identification of genes known to function in the circadian system and with the recognition of nonsynonymous genetic variants that exist in some of these genes, some groups have set out to look for associations of certain clock gene alleles with extreme chronotypes (morning larks or night owls). To date, a number of positive associations have been recognized (Katzenberg et al. 1998; Ebisawa et al. 2001; Robilliard et al. 2002; Archer et al. 2003; Takano et al. 2004; Carpen et al. 2005, 2006; Mishima et al. 2005). Although in each case, the actual effect of one allele and the strength of the effect in the association of that allele with either morningness or eveningness have been statistically significant, they also tend to be relatively small effects. Thus, it is likely that the broad spectrum of sleep time preferences in the population, separate from the Mendelian FASPS and FDSPS families, results from the combination of different alleles in a number of different circadian genes that add up to manifest in different tendencies in different individuals.

It is also likely that there are genetic contributions not only to circadian period, but to total sleep requirement, tendency to jet lag when crossing time zones, plasticity of the homeostatic mechanism in responding to sleep deprivation, and the phase angle of entrainment. Understanding these will require either identification of mutant genes from model organisms that can be tested in the human population or identification of Mendelian phenotypes of these types that will allow the kind of genetic approaches outlined here for FASPS.

We have now recognized a growing number of autosomal dominant families with FDSPS. Given that DSPS is more common in the general population, it is not yet clear how difficult it will be to identify genetic variants contributing to FDSPS, but efforts are under way to identify such mutations.

REFERENCES

Archer S.N., Robilliard D.L., Skene D.J., Smits M., Williams A., Arendt J., and von Schantz M. 2003. A length polymorphism in the circadian clock gene Per3 is linked to delayed sleep phase syndrome and extreme diurnal preference. *Sleep* 26: 413.

Carpen J.D., Archer S.N., Skene D.J., Smits M., and von Schantz M. 2005. A single-nucleotide polymorphism in the 5′-untranslated region of the hPER2 gene is associated with diurnal preference. *J. Sleep Res.* **14:** 293.

Carpen J.D., von Schantz M., Smits M., Skene D.J., and Archer S.N. 2006. A silent polymorphism in the PER1 gene associates with extreme diurnal preference in humans. *J. Hum. Genet.* **51:** 1122.

Ebisawa T., Uchiyama M., Kajimura N., Mishima K., Kamei Y., Katoh M., Watanabe T., Sekimoto M., Shibui K., Kim K., Kudo Y., Ozeki Y., Sugishita M., Toyoshima R., Inoue Y., Yamada N., Nagase T., Ozaki N., Ohara O., Ishida N., Okawa M., Takahashi K., and Yamauchi T. 2001. Association of structural polymorphisms in the human period3 gene with delayed sleep phase syndrome. *EMBO Rep.* **2:** 342.

Jones C.R., Campbell S.S., Zone S.E., Cooper F., DeSano A., Murphy P.J., Jones B., Czajkowski L., and Ptáček L.J. 1999. Familial advanced sleep-phase syndrome: A short-period circadian rhythm variant in humans (see comments). *Nat. Med.* **5:** 1062.

Katzenberg D., Young T., Finn L., Lin L., King D.P., Takahashi J.S., and Mignot E. 1998. A CLOCK polymorphism associated with human diurnal preference. *Sleep* **21:** 569.

Mishima K., Tozawa T., Satoh K., Saitoh H., and Mishima Y. 2005. The 3111T/C polymorphism of hClock is associated with evening preference and delayed sleep timing in a Japanese population sample. *Am. J. Med. Genet. B Neuropsychiatr. Genet.* **133:** 101.

Robilliard D.L., Archer S.N., Arendt J., Lockley S.W., Hack L.M., English J., Leger D., Smits M.G., Williams A., Skene D.J., and Von Schantz M. 2002. The 3111 Clock gene polymorphism is not associated with sleep and circadian rhythmicity in phenotypically characterized human subjects. *J. Sleep Res.* **11:** 305.

Roenneberg T., Kuehnle T., Pramstaller P.P., Ricken J., Havel M., Guth A., and Merrow M. 2004. A marker for the end of adolescence. *Curr. Biol.* **14:** R1038.

Takano A., Uchiyama M., Kajimura N., Mishima K., Inoue Y., Kamei Y., Kitajima T., Shibui K., Katoh M., Watanabe T., Hashimotodani Y., Nakajima T., Ozeki Y., Hori T., Yamada N., Toyoshima R., Ozaki N., Okawa M., Nagai K., Takahashi K., Isojima Y., Yamauchi T., and Ebisawa T. 2004. A missense variation in human casein kinase I epsilon gene that induces functional alteration and shows an inverse association with circadian rhythm sleep disorders. *Neuropsychopharmacology* **29:** 1901.

Toh K.L., Jones C.R., He Y., Eide E.J., Hinz W.A., Virshup D.M., Ptáček L.J., and Fu Y.H. 2001. An hPer2 phosphorylation site mutation in familial advanced sleep phase syndrome. *Science*

291: 1040.
Xu Y., Toh K.L., Jones C.R., Shin J.Y., Fu Y.H., and Ptáček L.J. 2007. Modeling of a human circadian mutation yields insights into clock regulation by PER2. *Cell* **128:** 59.
Xu Y., Padiath Q.S., Shapiro R.E., Jones C.R., Wu S.C., Saigoh N., Saigoh K., Ptáček L.J., and Fu Y.H. 2005. Functional consequences of a CKIdelta mutation causing familial advanced sleep phase syndrome. *Nature* **434:** 640.
Zavada A., Gordijn M.C., Beersma D.G., Daan S., and Roenneberg T. 2005. Comparison of the Munich Chronotype Questionnaire with the Horne-Ostberg's Morningness-Eveningness score. *Chronobiol. Int.* **22:** 267.

Circadian Entrainment of *Neurospora crassa*

M. Merrow* and T. Roenneberg[†]
*The Biological Center, University of Groningen, 9750AA Haren, The Netherlands
[†]The Institute for Medical Psychology, University of Munich, 80366 Munich, Germany

The circadian clock evolved under entraining conditions, yet most circadian experiments and much circadian theory are built around free-running rhythms. The interpretation of entrainment experiments is certainly more complex than that of free-running rhythms due to the relationship between exogenous and endogenous cycles. Here, we systematically describe entrainment in the simplest of the traditional eukaryotic model systems in circadian research, *Neurospora crassa*. This fungus forms a mass of spores (bands of conidia) each day. Over a wide range of photoperiods, these bands begin to appear at midnight, suggesting integration of neither dawn nor dusk signals alone. However, when symmetrical light/dark cycles (T cycles, each with 50% light) are applied, dusk determines the time of conidiation with a uniform, period-dependent delay in phase. This "forced" synchronization appears to be specific for the zeitgeber light because similar experiments, but using temperature, result in systematic entrainment, with bands appearing relatively later in shorter cycles and earlier in longer cycles. We find that the molecular mechanism of entrainment primarily concerns posttranscriptional regulation. Finally, we have used *Neurospora* to investigate acute effects of zeitgeber stimuli known as "masking."

INTRODUCTION

One of the key properties of circadian clocks specified by Pittendrigh (1961) 47 years ago at Cold Spring Harbor was that of a self-sustained, free-running, approximately 24-hour rhythm. This is a surprising characteristic on which to focus, considering that circadian systems evolved under cycles of light and darkness, as well as those of temperature, wind, humidity, and many other environmental factors. Although the Earth's rotation frequency has changed over the course of evolution, circadian rhythms hardly ever had the possibility to free-run. Thus, the experimentally generated free-run is a consequence of how circadian systems *must* be put together to function as the adaptive timers that they are. Nevertheless, Pittendrigh's list of properties overwhelmingly concerned circadian rhythms under constant conditions, dealing explicitly with entrainment in only one item. We propose that an understanding of the clock under entrained conditions may reveal mechanisms and genes relevant to clock functions in the real world beyond those discoverable under constant conditions.

Circadian entrainment describes a biological rhythm that is synchronized with the external physical environment (Johnson et al. 2003; Roenneberg et al. 2003). The most dominant zeitgeber appears to be light, not surprising, as it is the most dependable representation of time of day (compared with temperature or humidity, for example). The coordination of the two oscillators (the Earth's rotation and the circadian clock) obeys well-established rules, long known from the example of two physical oscillators entraining each other (like two pendulum clocks on the same wall, which can influence each other) (Huygens 1673). In the case of the circadian system, the physical cycle is obviously inflexible, so the biological oscillation must find the appropriate, stable phase relationship within the external cycle. A longer circadian rhythm will settle later in the cycle than a shorter one (Pittendrigh and Daan 1976; Duffy et al. 2001). A very strong zeitgeber may synchronize by driving or forcing the clock, rather than by entraining it. In this case, the phase will remain fixed relative to the zeitgeber despite changing cycle structure (e.g., length or photoperiod). Thus, comparison of cycles with different structures is one of the tricks to reveal circadian entrainment (Bruce 1961). This can involve changing either the length of the cycle (T) or the length or strength of, for example, the photoperiod within the cycle (often at $T = 24$ hours). Cycles can consist of full zeitgeber periods (e.g., a continuous signal throughout the photoperiod and scotoperiod) or skeletons, whereby a relatively short pulse of a signal is applied to simulate both on and off transitions.

Besides light, there are a multitude of other signals in the environment that can influence entrainment or that, under certain conditions, even can override entrainment by light. Combined light and temperature cycles can show synchronization to either zeitgeber depending on their respective amplitudes (Liu et al. 1998; Roenneberg and Merrow 2001), but without titration experiments, a single observation cannot distinguish entrainment versus masking by either zeitgeber. Similar to the situation of light and temperature in *Neurospora*, rhythmic gene expression in the mouse liver is indirectly controlled by light, via the suprachiasmatic nucleus (SCN), when food is administered ad libitum, but when food is restricted antiphase to normal activity, gene expression in the liver shifts accordingly (Damiola et al. 2000; Stokkan et al. 2001). Presumably, feeding, or rather circulating nutrients, can override the SCN signals for the clock in liver cells. Understanding entrainment on these levels—using multiple zeitgebers that show varied temporal structure—will be relevant to designing healthy work schedules in our 24/7 society, as well as understanding response characteristics to entrainment in the different seasons.

ENTRAINMENT PROPERTIES OF *NEUROSPORA*

For several reasons, we have chosen the fungal genetic model system, *Neurospora crassa*, to systematically investigate entrainment properties: It is haploid, thus untangling genetic principles is simpler compared to diploid organisms. *Neurospora* lends itself to high-throughput experiments and is inexpensive to grow and easy to handle (it can "sit" on the lab bench for weeks without apparent injury); both temperature and light have been used as zeitgebers, enabling comparison of entrainment characteristics with multiple zeitgebers. The power of this system is supported by several other contributions to this volume (Jay Dunlap #95 NYR; Jennifer Loros #10 NYR; Querfurth et al.; de Paula et al.; Cha et al.; all this volume).

The circadian clock in *Neurospora* has been elucidated largely from assays concerning asexual spore (conidia) formation as it grows along the surface of agar media. Mutant screens have revealed a number of *Neurospora* clock genes (genes in which a mutation changes fundamental clock properties). The first to be identified was *frequency* (*frq*) (McClung et al. 1989). FRQ is part of a genetic network that was the prototype for demonstrating the principle of the negative feedback loop (Hardin et al. 1990; Aronson et al. 1994), the molecular mechanism that is thought to be fundamental to the molecular "clockwork." Other components of this loop include *white collar-1* and *-2* (*wc-1, wc-2*), which encode a blue light photoreceptor and its essential cofactor, respectively. The WHITE COLLAR complex regulates transcription of *frq*. Considering their dominant effects on clock function, FRQ and the WHITE COLLAR complex are generally thought to be the core clock components, forming the "oscillator" that receives inputs and regulates outputs. Traditionally, the circadian system is separated into input–oscillator–output. In *Neurospora*, input and oscillator appear to be fused into one set of components. This principle was already suggested by Menaker, Takhashi, and Eskin (Menaker et al. 1978).

Mutations in *frq* alter the free-running period of spore formation in *Neurospora* (Fig. 1a) (Feldman and Hoyle 1976). The correlation of free-running period under constant conditions and the entrained phase under zeitgeber conditions (Pittendrigh and Daan 1976; Duffy et al. 2001) can be used to investigate fundamental principles of entrainment: How does a single-base-pair change (in the entire genome) lead to altered daily timing? As expected, frq^1, a short-period allele of the *frequency* locus, shows an advanced entrained phase in all entrainment protocols performed, regardless of zeitgeber modality (e.g., light or temperature, see Fig. 1) (Merrow et al. 1999; Tan et al. 2004), whereas a long-period allele, frq^7, always shows a delayed phase. In *Neurospora*, we have looked at T cycles that use equal lengths in "on" and "off" times within the zeitgeber (e.g., 50% light and darkness [photoperiod and scotoperiod] or warm and cold [thermoperiod and cryoperiod]; Fig. 1b,c) as well as cycles of the same length (T = 24 hours) with changing length in photoperiods (Fig. 1d). In all cases, phase and period correlate.

A closer inspection of entrainment in light/dark cycles reveals that phase appears driven and not entrained in symmetrical T cycles of different lengths (Fig. 1c) (Merrow et al. 1999): For all *Neurospora* strains tested, conidiation occurs after a constant strain-specific delay following dusk, much like a reset that occurs upon release from constant light into constant darkness. For such a driven response, a self-sustained rhythm would not be necessary! Interestingly, the phase moves away from the theoretical driven line in cycles about 24 hours at least for the wild-type strain. In contrast to T cycles, in 24-hour cycles with different photoperiods, conidiation starts at about midnight for the wild-type strain and proportionally earlier or later for the short- and long-period mutants, respectively (Fig. 1d) (Tan et al. 2004). This series of experiments shows that conidiation in *Neurospora* is not simply driven by light.

The information obtained in T cycles versus those recorded in photoperiods does not appear to be consistent; we still do not fully understand the driven phase angles of *Neurospora* in T cycles with light as a zeitgeber. The midnight lock of the entrained phase demonstrates that this simple system takes information either from dawn *and* dusk or from the entire structure of the light/dark cycle, rather than simply responding to either dawn or dusk. Dawn and dusk signals *could* be transduced by different photoreceptors, from different kinetic properties of the same photoreceptor signaling pathway, or from divergent consequences of the same light signal on distinct oscillators. So far, two blue light photoreceptors have been shown to be involved in circadian rhythms in this system (WC-1 and another flavin-binding protein, VIVID; Froehlich et al. 2002; He et al. 2002; Schwerdtfeger and Linden 2003), and more photoreceptors are indicated by sequence in the *Neurospora* genome and by functional analyses (Bieszke et al. 1999a,b; Heintzen et al. 2001; Froehlich et al. 2002, 2005; He et al. 2002; Galagan et al. 2003).

In contrast to T cycles with light, temperature cycles yield systematic entrainment (Fig. 1b), with all strains showing a later phase in a short cycle and an earlier one in a long cycle (Merrow et al. 1999). Furthermore, the expected phase relationship between *frq* strains with short and long free-running periods is evident.

ENTRAINMENT OF THE CLOCK AT THE MOLECULAR LEVEL

In view of the different strategies that the *Neurospora* clock uses for entrainment in temperature (T cycles) and light (photoperiod cycles), we looked for differences in the regulation of clock components at the molecular level. We followed the level of *frq* RNA, as its regulation by light was one of the first mechanisms proposed for how light phase-shifts the clock (Crosthwaite et al. 1995). In temperature T cycles, the phase of expression of *frq* RNA correlates with the phase of conidia formation (Fig. 2) (Merrow et al. 1999). In entraining light cycles (T = 24 hours), however, neither accumulation nor disappearance of *frq* RNA correlates with entrained phase (Fig. 3a) (Tan

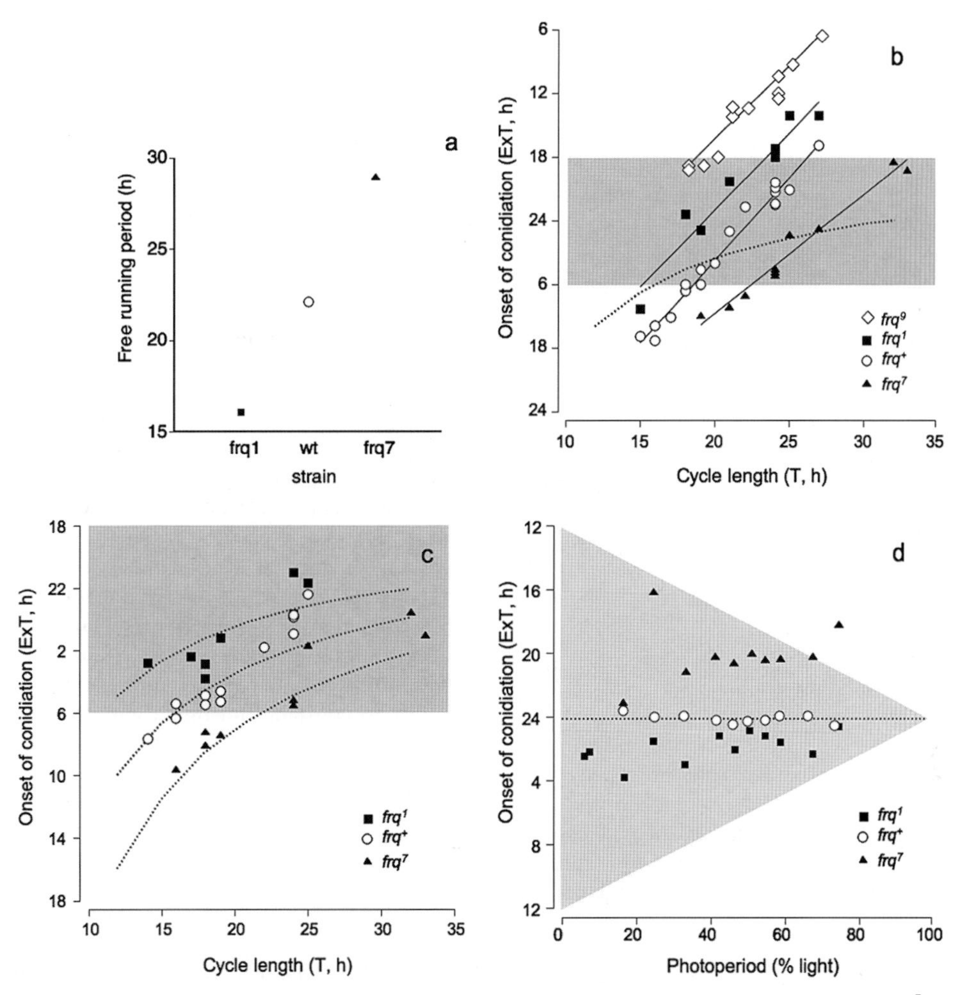

Figure 1. Characteristics of *N. crassa* in circadian protocols. In all panels: (*open circles*) *frq*$^+$; (*closed triangles*) *frq*^{7}; (*closed squares*) *frq*^{1}. (*a*) The free-running period in constant darkness is 22 hours for the *band* (*bd*) strain (*frq*$^+$), 16 hours for *bd frq*^{1}, and 29 hours for *bd frq*^{7}. (*b*) Circadian entrainment of *Neurospora* in T cycles using temperature as a zeitgeber. The onset of conidiation is plotted, in hours, according to external time (*ExT0* = midnight) (Daan et al. 2002). (*Open circles*) *frq*$^+$; (*closed triangles*) *frq*^{7}; (*closed squares*) *frq*^{1}; (*open diamonds*) *frq*^{9}. (*Dotted line*) Theoretically derived plot for a nonentrained but driven output (see *c*). (*Gray area*) Cold phases of the cycles. (*c*) T cycles using light as a zeitgeber. The onset of conidiation is plotted, in hours, according to external time. (*Dotted lines*) Theoretically derived plots for a nonentrained but driven output, using 5.5 hours following lights off for *frq*^{1}, 8 hours following lights off for *frq*$^+$, and 11 hours after lights off for *frq*^{7}. (*d*) Light cycles of $T = 24$ hours with varying photoperiods. The onset of conidiation is plotted, in hours, according to external time. (*Gray area*) Darkness; (*open areas*) light period; (*dotted line*) midnight.

et al. 2004). *frq* RNA is rapidly induced whenever the light comes on, and it decreases when lights are turned off. Although the light responsiveness of the *frq* transcript level was confirmed in these results (and extended to its immediate degradation in darkness), *frq* appears to have lost its role as a clock component in entrainment by light/dark cycles. Our results address the difficulties in extending results from light-pulse/release experiments to entrainment with light/dark cycles. We also investigated other light-regulated RNAs that show different kinetics relative to *frq* (*wc-1*, *vvd*, *al-1*, and *con-10*), some of which are considered clock genes or clock-regulated output genes. None of these genes showed transcriptional kinetics that could be associated with the respective entrained phases of conidiation.

In contrast to the results found in light cycles concerning transcriptional regulation, the kinetics of FRQ protein did correlate with the entrained phase. The daily initiation of conidiation always coincided with the point of half-maximal decline of FRQ levels (Fig. 3b). These results suggest that posttranscriptional regulation is more important for entrainment by full zeitgeber cycles than transcriptional control.

On one hand, this is no surprise in light of the elaboration of the importance of posttranslational chemistry such as phosphorylation (Kloss et al. 1998; Liu et al. 2000; Yang et al. 2002; Nakajima et al. 2005; Schaffmeier et al. 2005). On the other hand, in all systems, a transcriptional feedback loop is absolutely essential for a normal clock function. What are we missing? For instance, there are at least two *frq* transcripts, and it could be that only one of them is light-induced and is simply masking a functional

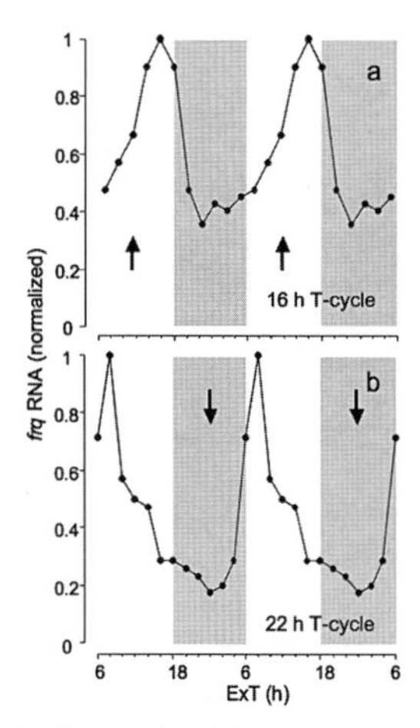

Figure 2. *frq* RNA is entrained in temperature cycles. The *frq* RNA levels are double-plotted after normalization within each time series according to ribosomal RNA levels, followed by adjustment of the maximum value to 1.0. (*a*) The 16-hour T cycle; (*b*) the 22-hour T cycle. (*Gray area*) Cool phase (22°C); (*open area*) warm phase (27°C). The arrows show where the conidial band is developing relative to the RNA profiles.

and entrained expression of another less-abundant transcript, the kinetics of which could be associated with entrainment. However, as we understand the system now, the (predominant) light-induced *frq* transcript(s) can clearly be ruled out as a state variable of the *Neurospora* clock under entrained conditions, as much as it appears crucial for self-sustained rhythmicity of conidiation in constant darkness (Aronson et al. 1994).

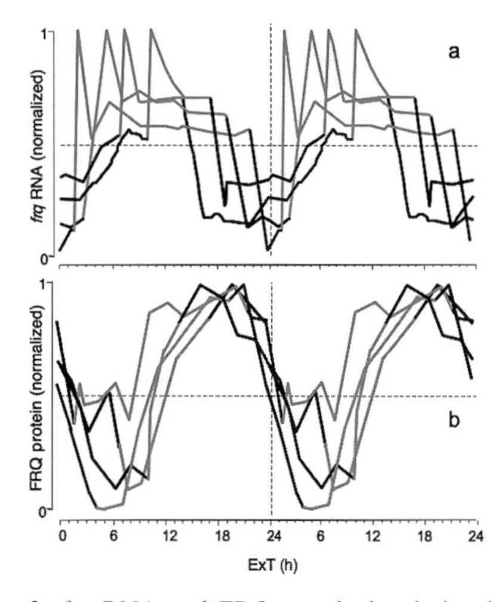

Figure 3. *frq* RNA and FRQ protein levels in photoperiod cycles, with $T = 24$ hours. The highest value in each individual profile is normalized to 1.0, and the rest of the values are adjusted accordingly. The time series are double-plotted, with *dotted lines* indicating midnight (*ExT* = 0/24 on the *x* axis) or half of the peak value of normalized expression (0.5 on the *y* axis). (*a*) *frq* RNA levels reflect the photoperiod. (*Thick gray line*) Data points corresponding to incubation in the light phase; (*thin black lines*) data points corresponding to the dark phase incubation. *ExT0* = midnight (Daan et al. 2002). (*b*) FRQ protein levels reflect conidiation in entrainment with light. The lines are drawn according to *a*.

REVEALING A MULTIOSCILLATOR SYSTEM WITH ENTRAINMENT OF "CLOCK-NULL" STRAINS

Most clock genes have earned their status based on their loss of free-running circadian rhythm, implying a central and critical function in this behavior. However, modeling reveals how an input component upstream of an oscillator can drastically change circadian properties, leading even to arrhythmicity (Roenneberg and Merrow 1998). The fact that many circadian rhythms are arrhythmic in constant light demonstrates the strong impact of inputs on the entire clock system. So, any interpretation of arrhythmicity resulting from a genetic mutation should be extended to the input components of the clock as well as to components that may couple the oscillator to the output pathway (Roenneberg and Merrow 2001).

There are several ways to test whether a clock component is on an input pathway versus being central to the oscillator mechanism. One of them compares different constant conditions (e.g., constant light of different intensities or constant darkness) to see if rhythmicity is restored in one of them. When *mPeriod2* and *mClock* mouse mutants that are arrhythmic in constant darkness are kept in constant dim light, they show free-running activity rhythms. They also respond to changing fluence rate (Aschoff's rule; Aschoff 1979), by systematically changing their circadian period (Spoelstra et al. 2002; Steinlechner et al. 2002). In *Neurospora*, clock-null mutants can show rhythms when incubated on media containing different nutrient concentrations (Dragovic et al. 2002; Granshaw et al. 2003).

An alternative approach to discern input from oscillator is to look for features of circadian entrainment in clock-null mutants that are otherwise arrhythmic under constant conditions. In light cycles, the *frq* null mutant strains cannot be synchronized, but temperature cycles yield rhythmic conidial bands that are nearly antiphase to those found in wild-type strains (Merrow et al. 1999). In a series of T cycles, the phase of conidiation in these mutants advances with increasing cycle length, reflecting entrainment of a circadian oscillator (or of a multioscillator network). This indicates that free-running rhythmicity in constant darkness may utilize different components than entrainment in temperature cycles. The *frq* gene is a key clock component for the former but apparently dispensable for the latter and for free-running rhythms in constant light (de Paula et al. 2006).

The more we probe circadian systems under different conditions, the more it becomes apparent that we are dealing with a network of feedback loops and oscillators (Roenneberg and Merrow 2003). This is supported by the findings that certain combinations of mutations in independent clock genes can result in the restoration of clock properties as was shown in *Drosophila*, when the cry^b mutation was crossed into a per^0 fly, and in mice, when *mCry2mPer2* animals were generated (Oster et al. 2002; Collins et al. 2005). In *Neurospora*, temperature-compensated molecular oscillations were revealed in *frq* knockout strains (Correa et al. 2003). Collectively—rhythms and evidence of entrainment without key clock genes—

should make us reevaluate the role of the transcriptional feedback loop in both entrained and free-running circadian systems. Clearly, in all model organisms studied to date, it is critical for proper timing, but evidence is accumulating to support the idea that there are clock components which are difficult to identify using the traditional genetic tools and the protocols that have brought us so far (Lakin-Thomas 2006).

MASKING IS PART OF ENTRAINMENT

Stable entrainment is typically conditional. It depends on the length of the zeitgeber cycle being within the range of entrainment (Aschoff and Wever 1962) and on sufficient zeitgeber strength. If a zeitgeber is too weak, it will not stably entrain and if it is too strong, it will simply drive the system. Furthermore, zeitgebers can have a number of additional effects on the system that do not strictly concern timing. The term "masking" describes the regulation of a circadian behavior by a zeitgeber in a manner that is not consistent with circadian entrainment (Mrosovsky 1999). A common example is the effect of light on nocturnal rodents, which—at least in the laboratory—remain inactive until dark (Mrosovsky et al. 2001). Release from light/dark cycles to constant darkness often reveals that the clock would have initiated activity somewhat earlier and that it was masked by light. This is presumably an adaptive function, an override of the circadian system.

We have characterized masking in *Neurospora* in several ways. First, we tested for frequency demultiplication: When an entraining cycle length is approximately half of the free-running period, a robust circadian clock entrains to every other cycle, rather than to the short cycle that lies well outside its range of entrainment (Bruce 1961). In 12-hour temperature cycles, one band of conidia appears every 24 hours when the *Neurospora* wild-type strain is used, whereas two conidial bands per 24 hours are produced by the *frq* null strain (Merrow et al. 1999). This is reminiscent of the driven conidiation of the wild-type strain in T cycles when light is used as a zeitgeber. With closer inspection, however, the two bands produced by the *frq* null strain in 24 hours appear to be asymmetrical, suggesting that one is a masking response and the other circadian (Roenneberg et al. 2005). Dose-response titration of the zeitgeber confirms this explanation, showing increasing masking with higher mean temperatures and a higher-amplitude temperature cycle, in both wild-type and mutant strains.

A second exploration of masking made use of air pulses and their induction of conidia (perhaps by not allowing CO_2 to accumulate in the race tube or, alternatively, due to the physical stimulation of air flow over hyphae) (Roenneberg et al. 2005). Under otherwise constant conditions, air pulses once per 18 hours yield a conidial band, whereas circadian conidiation continues with its endogenous period of 22 hours (in wild type). When the masking conidiation and the circadian conidiation coincide, the amplitude of the signal increases. The *frq* null mutant strain shows a similar pattern except that the masking response is larger relative to the clock mutant's endogenous conidiation pattern; the free-running period in this case is shorter than the wild-type strain, and once the endogenously and exogenously stimulated conidiation come together, they do not immediately separate. These observations also show how the circadian clock affects the amplitude of the masking response.

Finally, in a multilab repetition of temperature cycles in *Neurospora*, a data set was generated that, at first analysis, did not show circadian entrainment of *frq* null strain (Pregueiro et al. 2005). Through comparison with the original data set, the waveform was interrogated to extract masked elements of the synchronization (Roenneberg et al. 2005). Retrospectively, the zeitgeber conditions (temperature transitions, relative humidity) were different enough to result in the differences in conidiation patterns through masking.

These experiments suggest that entrainment will always show some degree of masking. In understanding entrainment, on the one hand, we have to understand how to demask entrainment data (to uncover the pure circadian components of entrainment). On the other hand, we have to appreciate more the role of masking itself in contributing to entrainment in the real world.

DISCUSSION

It is remarkable how extensively one can describe entrainment of circadian clocks by rules that are based on the synchronization of simple mechanical oscillators. This was already known when the pioneers of the field met at Cold Spring Harbor in 1960: Woody Hastings had published the first light-pulse PRC 2 years before (Hastings and Sweeney 1958). Forty-seven years later, the apparent power of this simple explanatory approach is even more remarkable, given that the overwhelming theme of the 2007 Cold Spring Harbor Meeting on Clocks and Rhythms was the complexity of circadian systems. In addition, compexity derives from the temporal structure of the entraining environment in nature, which includes multiple zeitgebers whose precise qualitites and quantities vary from day to day. Furthermore, the impact of environmental signals on the circadian system is modulated by the clock itself (zeitnehmers; Roenneberg and Merrow 1998). Thus, it is surprising that PRCs can be classified by a relatively small number of forms as an indication of how entrainment works.

Here, we have described our first attempts to apply the holistic approaches of systems biology on understanding entrainment. Using a relatively simple and inexpensive system, we are generating a "surface" of circadian entrainment by light (all combinations of endogenous period, zeitgeber period, photoperiod, and zeitgeber strength). We are currently generating the same circadian surface for temperature and plan to construct surfaces with different combinations of zeitgeber modalities. Eventually, the effects of different nutrients on surface characteristics can be investigated. We expect this approach will identify hitherto unknown properties of circadian entrainment. Although some of these may be special for *Neurospora* spore formation, others will be common to all clocks. Protocols developed in this manner will also help to identify new clock genes that may not be able to be discovered under standard circadian protocols, such as 12:12 light/dark cycles or constant conditions.

These protocols, so far, operate at the level of the organism (entrainment of conidiation) or of the gene (entrainment at the molecular level). In addition, there are intermediate levels within systems. Clocks in cells, tissues, and organs are entrained by endogenous zeitgeber signals that may not strictly correlate with (represent) exogenous zeitgeber signals. The result is a set of internal phase angles of which we know little but which will be critical to the functioning of the system as a whole. Understanding entrainment at all of these levels will eventually help us to treat the pathologies of entrainment in humans associated with shift work, jet lag, or weak zeitgeber exposure in modern city life.

Our results have already changed how we view entrainment by light in *Neurospora*. Although former experiments suggested that release from light resets the clock, we find—when looking at the entire surface—that the clock integrates more features of the zeitgeber structure. Nevertheless, the simple mechanical oscillator rules apparently often still apply. The strong correlation between free-running period and phase of entrainment (chronotype) is systematically present, regardless of (period) mutant and zeitgeber conditions. Yet, there are apparently exceptions, as demonstrated in *Neurospora* by the *vivid* mutant. Although they show a wild-type-like free-running period, their phase is delayed, either on release from light to constant darkness (Heintzen et al. 2001; Shrode et al. 2001) or under entrained conditions (C. Boesl et al., unpubl.). This is probably not an isolated case but a rather common case regarding the principles of entrainment. The identification of genes that influence both phase of entrainment *and* period versus those that influence either period *or* phase alone will be an excellent phenotyping tool for understanding the molecular mechanisms of the clock.

Our systems approach to entrainment has so far concentrated on the physiological outputs of the system. In the future, this must be extended to the molecular level with high-throughput methods (i.e., reporter genes or sensitive physicochemical methods such as LC/MS). In addition, we have to find ways to assess "small effects" that are a critical aspect of how complex traits, such as circadian clocks, are regulated. Entrainment of circadian clocks—with genetic control mechanisms that extend from big to small effects—is an excellent systems biology question. A post hoc analysis of microarray data showed that essentially all transcripts in mammalian cells, as was previously shown for *Synechococcus* (Liu et al. 1995), are oscillating with a circadian period (Ptitsyn et al. 2007). The oscillating cellular machinery will likely include many zeitnehmer feedbacks with small (sometimes very small) effects. To identify these is a challenge that will benefit from experiments with constant *and* entraining conditions. We have begun to attack this question by creating surfaces involving all parameters influencing entrainment and by measuring physiological outputs. Now, this has to integrate transcriptome, proteome, and phosphorylome data in the context of various entraining conditions.

ACKNOWLEDGMENTS

Our work is supported by the Dutch Science Foundation (the NWO), The Hersenen Stichting, the German Science Foundation (the DFG), EUCLOCK, a Sixth Framework Programme of the European Union, ClockWORK, a Daimler-Benz-Stiftung network, the Dr.-Meyer-Struckmann-Stiftung, and the Rosalind Franklin Fellowship Program of the University of Groningen.

REFERENCES

Aronson B.D., Johnson K.A., Loros J.J., and Dunlap J.C. 1994. Negative feedback defining a circadian clock: Autoregulation of the clock gene *frequency*. *Science* **263:** 1578.

Aschoff J. 1979. Circadian rhythms: Influences of internal and external factors on the period measured under constant conditions. *Z. Tierpsychol.* **49:** 225.

Aschoff J. and Wever R. 1962. Über Phasenbeziehungen zwischen biologischer Tagesperiodik und Zeitgeberperiodik. *Z. Vgl. Physiol.* **46:** 115.

Bieszke J.A., Spudich E.N., Scott K.L., Borkovich K.A., and Spudich J.L. 1999a. A eukaryotic protein, NOP-1, binds retinal to form an archaeal rhodopsin-like photochemically reactive pigment. *Biochemistry* **38:** 14138.

Bieszke J.A., Braun E.L., Bean L.E., Kang S., Natvig D.O., and Borkovich K.A. 1999b. The *nop-1* gene of *Neurospora crassa* encodes a seven transmembrane helix retinal-binding protein homologous to archaeal rhodopsins. *Proc. Natl. Acad. Sci.* **96:** 8034.

Bruce V. 1961. Environmental entrainment of circadian rhythms. *Cold Spring Harbor Symp. Quant. Biol.* **25:** 29.

Collins B.H., Dissel S., Gaten E., Rosato E., and Kyriacou C.P. 2005. Disruption of Cryptochrome partially restores circadian rhythmicity to the arrhythmic period mutant of *Drosophila. Proc. Natl. Acad. Sci.* **102:** 19021.

Correa A., Lewis Z.A., Greene A.V., March I.J., Gomer R.H., and Bell-Pedersen D. 2003. Multiple oscillators regulate circadian gene expression in *Neurospora*. *Proc. Natl. Acad. Sci.* **100:** 13597.

Crosthwaite S.K., Loros J.J., and Dunlap J.C. 1995. Light-induced resetting of a circadian clock is mediated by a rapid increase in *frequency* transcript. *Cell* **81:** 1003.

Daan S., Merrow M., and Roenneberg T. 2002. External time–internal time. *J. Biol. Rhythms* **17:** 107.

Damiola F., Minh N.L., Preitner N., Kornmann B., Fleury-Olela F., and Schibler U. 2000. Restricted feeding uncouples circadian oscillators in peripheral tissues from the central pacemaker in the suprachiasmatic nucleus. *Genes Dev.* **14:** 2950.

de Paula R.M., Lewis Z.A., Greene A.V., Seo K.S., Morgan L.W., Vitalini M.W., Bennett L., Gomer R.H., and Bell-Pedersen D. 2006. Two circadian timing circuits in *Neurospora crassa* cells share components and regulate distinct rhythmic processes. *J. Biol. Rhythms* **21:** 159.

Dragovic Z., Tan Y., Görl M., Roenneberg T., and Merrow M. 2002. Light reception and circadian behavior in 'blind' and 'clock-less' mutants of *Neurospora crassa*. *EMBO J.* **21:** 3643.

Duffy J.F., Rimmer D.W., and Czeisler C.A. 2001. Association of intrinsic circadian period with morningness-eveningness, usual wake time, and circadian phase. *Behav. Neurosci.* **115:** 895.

Feldman J.F. and Hoyle M.N. 1976. Complementation analysis of linked circadian clock mutants of *Neurospora crassa*. *Genetics* **82:** 9.

Froehlich A.C., Liu Y., Loros J.J., and Dunlap J.C. 2002. WHITE COLLAR-1, a circadian blue light photoreceptor, binding to the *frequency* promoter. *Science* **297:** 815.

Froehlich A.C., Noh B., Vierstra R.D., Loros J., and Dunlap J.C. 2005. Genetic and molecular analysis of phytochromes from the filamentous fungus *Neurospora crassa*. *Eukaryot. Cell* **4:** 2140.

Galagan J.E., Calvo S.E., Borkovich K.A., Selker E.U., Read N.D., Jaffe D., FitzHugh W., Ma L.J., Smirnov S., Purcell S., Rehman B., Elkins T., Engels R., Wang S., Nielsen C.B., Butler J., Endrizzi M., Qui D., Ianakiev P., Bell-Pedersen D., Nelson M.A., Werner-Washburne M., Selitrennikoff C.P., Kinsey J.A., and Braun E.L., et al. 2003. The genome sequence of the filamentous fungus *Neurospora crassa*. *Nature* **422:** 859.

Granshaw T., Tsukamoto M., and Brody S. 2003. Circadian rhythms in *Neurospora crassa:* Farnesol or geraniol allow expression of rhythmicity in otherwise arrhythmic strains frq10, wc-1, and wc-2. *J. Biol. Rhythms* **18:** 287.

Hardin P.E., Hall J.C., and Rosbash M. 1990. Feedback of the *Drosophila period* gene product on circadian cycling of its messenger RNA levels. *Nature* **343:** 536.

Hastings J.W. and Sweeney B.M. 1958. A persistent diurnal rhythm of luminescence in *Gonyaulax polyedra*. *Biol. Bull.* **115:** 440.

He Q., Cheng P., Yang Y., Wang L., Gardner K.H., and Liu Y. 2002. WHITE COLLAR-1, a DNA binding transcription factor and a light sensor. *Science* **297:** 840.

Heintzen C., Loros J.J., and Dunlap J.C. 2001. The PAS protein VIVID defines a clock-associated feedback loop that represses light input, modulates gating, and regulates clock resetting. *Cell* **104:** 453.

Huygens C. 1673. *Horologium oscillatorium*. Apud F. Muguet, Paris.

Johnson C.H., Elliott J.A., and Foster R. 2003. Entrainment of circadian programs. *Chronobiol. Int.* **20:** 741.

Kloss B., Price J.L., Saez L., Blau J., Rothenfluh A., Wesley C.S., and Young M.W. 1998. The *Drosophila* clock gene *double-time* encodes a protein closely related to human casein kinase epsilon. *Cell* **94:** 97.

Lakin-Thomas P.L. 2006. Transcriptional feedback oscillators: Maybe, maybe not... *J. Biol. Rhythms* **21:** 83.

Liu Y., Loros J., and Dunlap J.C. 2000. Phosphorylation of the *Neurospora* clock protein FREQUENCY determines its degradation rate and strongly influences the period length of the circadian clock. *Proc. Natl. Acad. Sci.* **97:** 234.

Liu Y., Merrow M., Loros J.L., and Dunlap J.C. 1998. How temperature changes reset a circadian oscillator. *Science* **281:** 825.

Liu Y., Tsinoremas N.F., Johnson C.H., Lebedeva N.V., Golden S.S., Ishiura M., and Kondo T. 1995. Circadian orchestration of gene expression in cyanobacteria. *Genes Dev.* **9:** 1469.

McClung C.R., Fox B.A., and Dunlap J.C. 1989. The *Neurospora* clock gene *frequency* shares a sequence element with the *Drosophila* clock gene *period*. *Nature* **339:** 558.

Menaker M., Takahashi J.S., and Eskin A. 1978. The physiology of circadian pacemakers. *Annu. Rev. Physiol.* **40:** 501.

Merrow M., Brunner M., and Roenneberg T. 1999. Assignment of circadian function for the *Neurospora* clock gene *frequency*. *Nature* **399:** 584.

Mrosovsky N. 1999. Masking: History, definitions, and measurement. *Chronobiol. Int.* **16:** 415.

Mrosovsky N., Lucas R., and Foster R. 2001. Persistence of masking responses to light in mice lacking rods and cones. *J. Biol. Rhythms* **16:** 585.

Nakajima M., Imai K., Ito H., Nishiwaki T., Murayama Y., Iwasaki H., Oyama T., and Kondo T. 2005. Reconstitution of circadian oscillation of cyanobacterial KaiC phosphorylation in vitro. *Science* **308:** 414.

Oster H., Yasui A., van der Horst G.T.J., and Albrecht U. 2002. Disruption of *mCry2* restores circadian rhythmicity in *mPer2* mutant mice. *Genes Dev.* **16:** 2633.

Pittendrigh C.S. 1961. Circadian rhythms and the circadian organization of living systems. *Cold Spring Harbor Symp. Quant. Biol.* **25:** 159.

Pittendrigh C.S. and Daan S. 1976. A functional analysis of circadian pacemakers in nocturnal rodents. IV. Entrainment: Pacemaker as clock. *J. Comp. Physiol. A* **106:** 291.

Pregueiro A., Price-Lloyd N., Bell-Pedersen D., Heintzen C., Loros J.J., and Dunlap J.C. 2005. Assignment of an essential role for the *Neurospora* frequency gene in circadian entrainment to temperature cycles. *Proc. Natl. Acad. Sci.* **102:** 2210.

Ptitsyn A.A., Zvonic S., and Gimble J.M. 2007. Digital signal processing reveals circadian baseline oscillation in majority of mammalian genes. *PLoS Comput. Biol.* **3:** e120.

Roenneberg T. and Merrow M. 1998. Molecular circadian oscillators: An alternative hypothesis. *J. Biol. Rhythms* **13:** 167.

———. 2001. The role of feedbacks in circadian systems. In *Zeitgebers, entrainment and masking of the circadian system* (ed. K. Honma et al.), p. 113. Hokkaido University Press, Sapporo, Japan.

———. 2003. The network of time: Understanding the molecular circadian system. *Curr. Biol.* **13:** R198.

Roenneberg T., Daan S., and Merrow M. 2003. The art of entrainment. *J. Biol. Rhythms* **18:** 183.

Roenneberg T., Dragovic Z., and Merrow M. 2005. Demasking biological oscillators: Properties and principles of entrainment exemplified by the *Neurospora* circadian clock. *Proc. Natl. Acad. Sci.* **102:** 7742.

Schaffmeier T., Haase A., Kaldi K., Scholz J., Fuchs M., and Brunner M. 2005. Transcriptional feedback of *Neurospora* circadian clock gene by phosphorylation-dependent inactivation of its transcription factor. *Cell* **122:** 235.

Schwerdtfeger C. and Linden H. 2003. VIVID is a flavoprotein and serves as a fungal blue light photoreceptor for photoadaptation. *EMBO J.* **22:** 4846.

Shrode L.B., Lewis Z.A., White L.D., Bell-Pedersen D., and Ebbole D.J. 2001. *vvd* is required for light adaptation of conidiation-specific genes of *Neurospora crassa*, but not circadian conidiation. *Fungal Genet. Biol.* **32:** 169.

Spoelstra K., Okeljewicz M., and Daan S. 2002. Restoration of self-sustained rhythmicity by the mutant clock allele in mice in constant illumination. *J. Biol. Rhythms* **17:** 520.

Steinlechner S., Jacobmeier B., Scherbarth F., Dernbach H., Kruse F., and Albrecht U. 2002. Robust circadian rhythmicity of *Per1* and *Per2* mutant mice in constant light and dynamics of *Per1* and *Per2* gene expression under long and short photoperiods. *J. Biol. Rhythms* **17:** 202.

Stokkan K.A., Yamazaki S., Tei H., Sakaki Y., and Menaker M. 2001. Entrainment of the circadian clock in the liver by feeding. *Science* **291:** 490.

Tan Y., Dragovic Z., Roenneberg T., and Merrow M. 2004. Entrainment of the circadian clock: Translational and post-translational control as key elements. *Curr. Biol.* **14:** 433.

Yang Y., Cheng P., and Liu Y. 2002. Regulation of the *Neurospora* circadian clock by casein kinase II. *Genes Dev.* **16:** 994.

Constant Darkness Is a Mammalian Biological Signal

C.C. LEE
Department of Biochemistry and Molecular Biology, University of Texas Health Science Center, Houston, Texas 77030

Environmental light is a potent modulator of mammalian circadian rhythm and expression of clock genes. Constant darkness (DD) is regarded as a "free-running" circadian state. In nature, hibernating mammals encounter constant darkness (DD) seasonally. Circadian expression of enzymes involved in fat catabolism, procolipase (CLP) and pancreatic-lipase-related protein 2 (PLRP2), were identified in many peripheral organs of mice during DD but not during regular light/dark (LD) cycles. Circulating 5′-adenosine monophosphate (5′-AMP) was associated with DD-activated gene expression. Synthetic 5′-AMP, when injected into LD mice, activated procolipase expression in their peripheral organs and the animals become severely hypothermic, both key features of hibernating mammals. These findings identified a circadian-regulated metabolic cycle in mammals that may be associated with hypometabolic behaviors such as hibernation and torpor.

INTRODUCTION

Hypometabolism of cells and organs as a way to reduce ischemic damage holds enormous clinical applications (Kabon et al. 2003). Natural hypometabolism can be observed in animals that are able to undertake torpor, hibernation, or estivation. These behaviors, which reflect the different degrees of hypometabolism in diverse animal species, are thought to have very similar physiological controls (Heldmaier et al. 2004). However, little is known about the biochemical and cellular mechanisms underlying these behaviors in mammals.

In mammals, association between daily torpor and the body temperature rhythms implicates the circadian clock in the temporal control of this behavior (Perret and Aujard 2001; Heller and Ruby 2004). The observation that the ablation of the suprachiasmatic nucleus (SCN), the central endogenous clock synchronizer, abolished the torpor rhythm further implicates the circadian clock in such behavior (Ruby et al. 2002). Diverse mammalian species, from bears to primates such as the Malagasy lemurs, are known to undertake severe hypometabolic states such as torpor and hibernation (Heldmaier et al. 2004). These observations suggest the underlying biochemical processes for hypometabolism could be preserved in many mammalian species including humans.

It is widely recognized that mammals seasonally encounter prolonged DD during hibernation (Ruby et al. 1996; Heldmaier et al. 2004). Although regarded as a "free-running" state, the possibility that DD is a biological signal in mammals has never been excluded. Hence, a project was initiated to probe into the possibility of identifying genes activated in one aspect of the environments encountered during hibernation such as DD.

PERIPHERAL ORGAN GENES ACTIVATED BY DD ENVIRONMENT

Using a mouse gene library, a microarray comparison of gene expression was undertaken using liver mRNA from an animal kept in 48-hour DD (12:12-hour dark:dark) and an animal kept in a regular LD 12:12-hour cycle. (Zhang et al. 2006). This microarray comparison identified a gene encoding procolipase that was highly expressed in the liver from the animal kept in DD. Previous studies have shown that expression of the procolipase gene is highly specific to the pancreas and gastrointestinal tract, consistent with its role in the catabolism of dietary fat into fatty acids (Lowe 1977). However, in hibernating ground squirrels, expression of procolipase mRNA was activated in other peripheral organs, suggesting that our findings could be linked to such hypometabolic behavior (Squire et al. 2003). Northern blot analysis confirmed that the procolipase gene was activated in a circadian manner in peripheral organs when mice were kept in DD but not during LD cycles (Fig. 1a). Time course studies of procolipase expression over 48 hours in various tissues demonstrated a highly synchronized circadian control (Fig. 1b). Colipase is necessary for the enzymatic activity of pan-

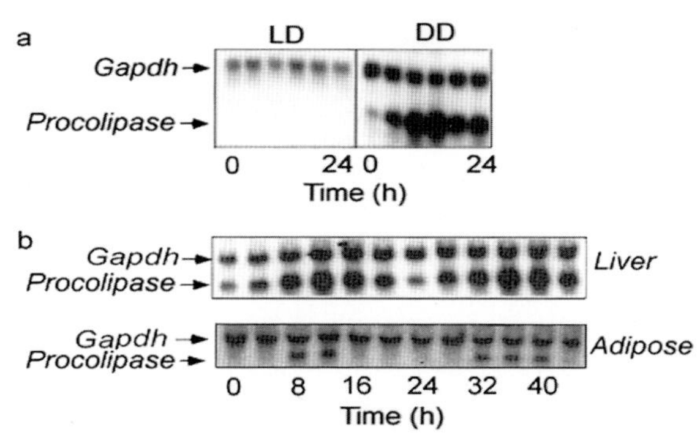

Figure 1. Activation of procolipase expression by DD. (*a*) Northern blot analysis showed the liver mRNA species of *Gapdh*- and procolipase-obtained mice kept in12:12-hour LD or after 2 days in DD cycles or constant darkness environment. (*b*) A time course of procolipase expression in liver and adipose tissue during constant darkness. A housekeeping gene *Gapdh* was used as an internal control for relative mRNA levels. The respective genes were identified by radiolabeled cDNA probe for *Gapdh* and *procoliapse*, respectively.

creatic lipases (Squire et al. 2003). Expression analysis of the gene encoding PLRP2 demonstrated that it was coordinately expressed with procolipase (Zhang et al. 2006). These observations indicated that DD-activated gene regulation is part of a complex biological program and is not unique to procolipase. Exposing mice kept in DD to white light resulted in the shut down of the procolipase expression in the various organs (Zhang et al. 2006). Together, these findings implicated a circulatory factor in the activation of procolipase expression during DD.

IDENTIFICATION OF 5′-AMP AS A MEDIATOR OF PROCOLIPASE EXPRESSION

On the basis of the above observations, the endogenous regulator was rationalized to have the following properties: (1) It is a circulatory molecule that must display circadian profile in its activity or levels and (2) it could act either as an activator or as a repressor. If it is an activator, then its injection into mice kept in the LD cycle will result in the induction of procolipase expression in the major organs. However, as a repressor, its injection into DD mice will abolish its procolipase expression in the peripheral organs.

Through a series of careful experiments, our search for this circulatory molecule was narrowed to the soluble nonpolypeptide aqueous fraction of blood extracts. Fractionation of the blood extracts by reverse-phase high-performance liquid chromatography (HPLC) revealed four highly reproducible peaks, of which peak 2 had a robust circadian pattern (Fig. 2). This peak also showed a strong diurnal profile with lower amplitude in LD mice (Zhang et al. 2006). Its characterization revealed a strong spectral absorption at 260 nm, suggesting that this molecule was a nucleoside or nucleotide. Comparison of its fractionation distance on HPLC to chemical nucleotide standards demonstrated that it matches exactly to synthetic 5′-AMP (Fig. 3a). Identification of this molecule as 5′-AMP was confirmed by enzymatic analysis with snake venom nucleotidase, which dephosphorylated 5′-AMP into adenosine (Fig. 3b,c).

To demonstrate that 5′-AMP was the endogenous molecule associated with DD activation of procolipase expression, synthetic 5′-AMP was injected into LD mice. By reverse transcriptase–polymerase chain reaction (RT-PCR), procolipase expression was detected in all tissues sampled with the exception of the brain from LD mice given 5′-AMP (Fig. 4a). In contrast, saline-injected LD animals display expression of procolipase only in the pancreas and the stomach, consistent with previous observations (Lowe 1977; Squire et al. 2003). These findings demonstrate that the elevated circulating level of 5′-AMP was indeed associated with procolipase expression.

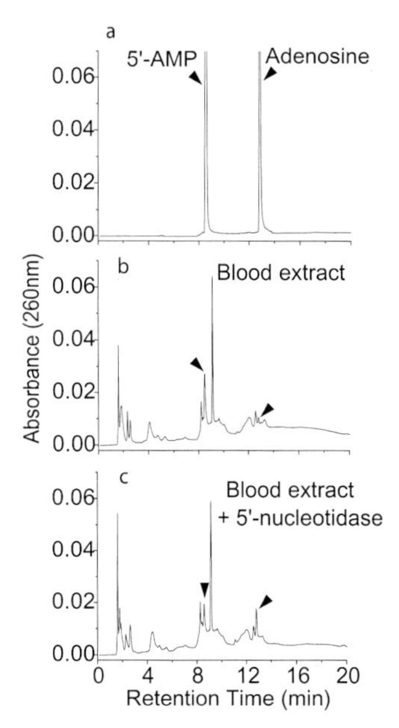

Figure 3. Identification of 5′-AMP as the circulatory factor. (*a*) Separation of chemical standards of 5′-AMP and adenosine under similar HPLC conditions. (*b*) Fractionation of blood extract; (*c*) fractionation of the same blood extract after treatment by snake vernom nucleotidase. Note the reciprocal change in peak sizes of 5′-AMP with adenosine.

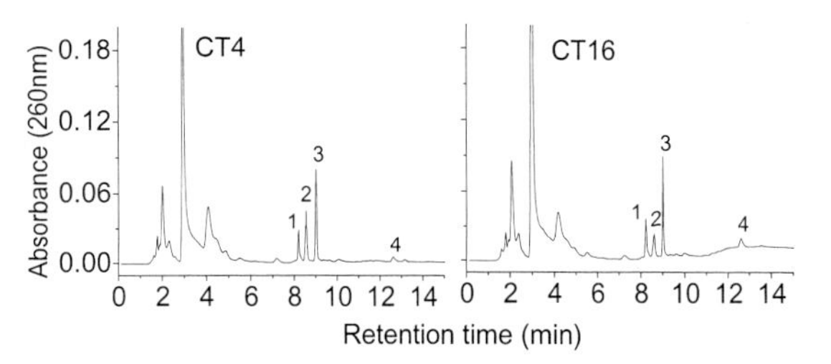

Figure 2. Identification of a circadian circulatory factor in blood. Extract was prepared from blood obtained from mice kept in 48-hour DD at CT4 and CT16. The aqueous phase was fractionated by HPLC, revealing four highly reproducible peaks (1, 2, 3, and 4). Note the robust circadian profile of peak 2.

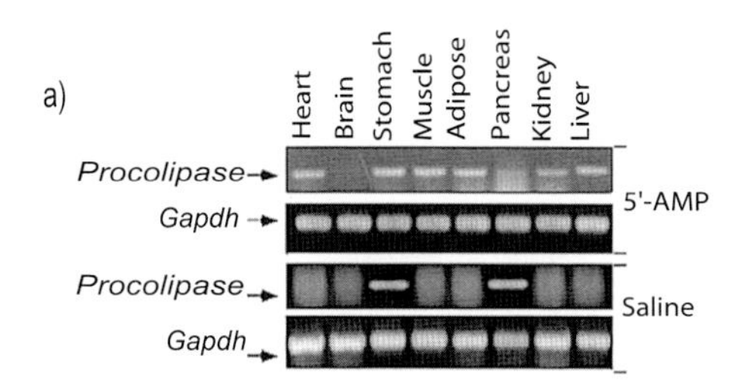

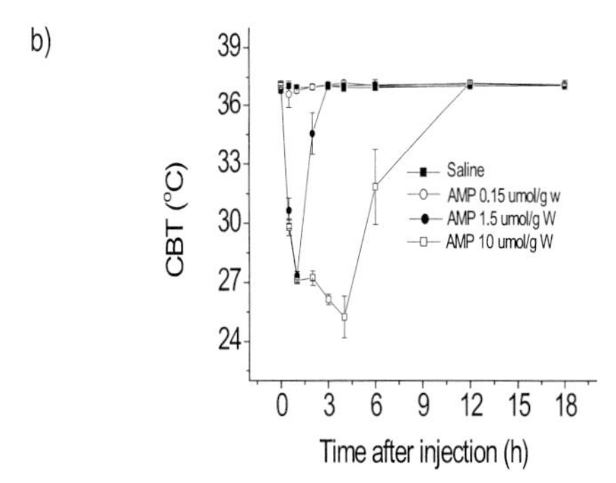

Figure 4. Induction of procolipase expression and hypothermia by synthetic 5′-AMP. (*a*) RT-PCR detection of procolipase and *Gapdh* expression in mice kept in 12:12-hour LD cycles given saline or 5′-AMP. (*b*) Hypothermic response of mice injected with various dosages of 5′-AMP.

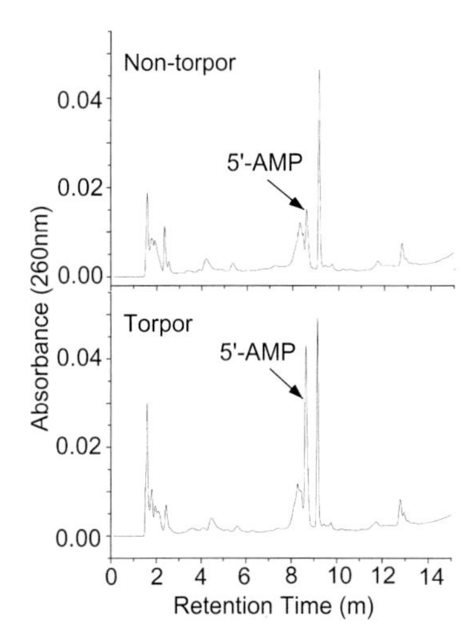

Figure 5. Rise in blood 5′-AMP levels during fasting-induced torpor. (*a*) HPLC fractionation of blood extract obtained from DD mice that were fed and not in torpor. (*b*) HPLC fractionation of blood extract obtained from a DD mouse that was in torpor induced by fasting.

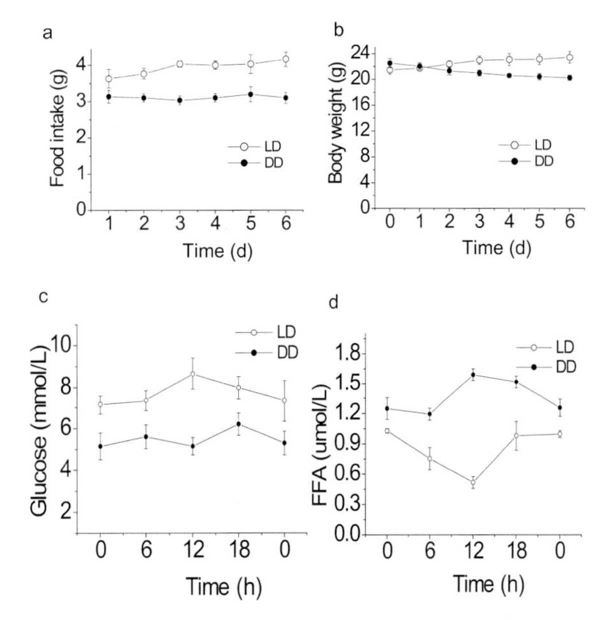

Figure 6. Effects of DD on metabolic status. (*a*) Daily food intake of mice in LD and DD cycles; (*b*) body weight of mice during LD and DD cycles; (*c*) temporal rhythm of blood glucose during LD and DD cycles; (*d*) temporal blood fatty acids level during LD and DD cycles. Note that all measurements were performed after DD mice had been in DD for 48 hours.

The induction of procolipase by 5′-AMP has a prolonged time course, suggesting that its role in gene activation was likely indirect in vivo. Unexpectedly, mice that were given 5′-AMP were also severely hypothermic, with core body temperature as low as 25°C when kept in ambient room temperature (AET) of about 24°C (Fig. 4b). This hypothermic state was transient, and core body temperature was restored several hours later with no apparent shortcomings. These observations raised the possibility that 5′-AMP may have a similar role in natural torpor behavior.

This possibility was addressed by fasting mice kept in DD to induce natural torpor. HPLC analysis revealed that mice in torpor induced by fasting have a highly elevated circulatory 5′-AMP level. In contrast, mice in a similar environment that did not enter torpor and were fed have much lower circulatory 5′-AMP (Fig. 5). Whereas the kinetics of natural torpor was slower than that induced by synthetic 5′-AMP, this can be accounted for by the slower build up of natural 5′-AMP generated by fasting and its impact on thermal regulatory inhibition. Together, these observations revealed that an increase in natural 5′-AMP was associated with fasting-induced torpor.

CONSTANT DARKNESS AND ENDOGENOUS METABOLIC RHYTHM

Cessation of food intake and the generation of endogenous energy from fat are some of the physiological hallmarks of an animal in a deep hypometabolic state such as hibernation. The targeted activation of procolipase by DD must be physiological because procolipase mRNA encodes two peptides that are important for these hypometabolic events. The amino-terminal sequence of procolipase is a pentapeptide (VPDPR) that is posttranslationally cleaved from the colipase enzyme. This VPDPR peptide, named enterostatin, has been shown to act as a satiety inhibitor (Erlanson-Albertsson and Larsson 1988). A predicted outcome of the activated procolipase expression during the DD cycle would be an increase in VPDPR peptide production, which in turn should decrease satiety. Our studies revealed that DD mice consumed less food than LD mice during a 7-day study period (Fig. 6a). These observations were consistent with previous observations of rats displaying lower satiety in DD compared to those maintained in LD cycles (Stoynev and Ikonomov 1983). Consistent with the decrease in solid food intake was the observation that the body weight of the DD animals was moderately lower than that of mice kept in LD cycles over the same corresponding period (Fig. 6b). The broad activation of procolipase expression during DD cycles suggested that fat catabolism must be activated in these peripheral organs. A predicted outcome of such an increase in fat catabolism would be elevated serum free fatty acids. Indeed, DD mice have higher levels of free fatty acids in their serum compared to LD animals (Fig. 6d). These findings are consistent with previous observations that large mammals kept in DD have higher serum-free fatty acids than those maintained in LD environment (Alila-Johansson et al. 2004). The observed increase in the use of fatty acids as fuel is complemented by a decline in blood glucose levels of DD animals compared to LD mice (Fig. 6c). This difference in preference of fatty acids over glucose as fuel in DD mice is reminiscent of animals in hibernation.

Membrane-anchored and circadian-regulated ecto-5′nucleotidase controls the circulatory 5′-AMP levels and mediates its intracellular action (von Mayersbach and Klaushofer 1979; Uchiyama 1983; Thompson et al. 1989). The extracellular ecto-5′nucleotidase dephospho-

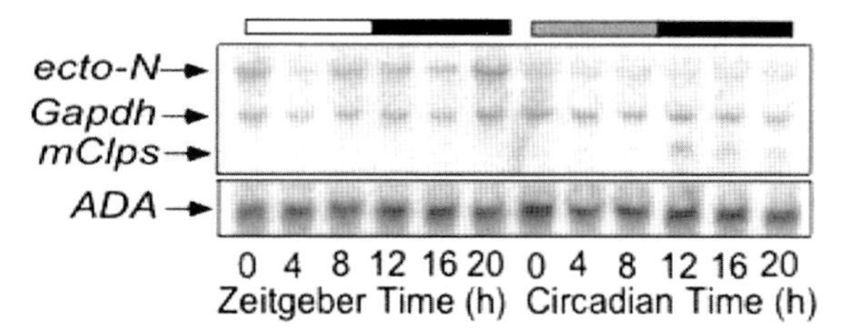

Figure 7. Inhibition of ecto-5′nucleotidase mRNA expression by DD. Shown here is northern blot analysis for expression of ecto-5′nucleotidase (*ecto-N*), *Gapdh*, procolipase (*mClps*), and adenosine deaminase (*ADA*) in mice liver mRNA obtained every 4 hours from animals kept in LD or DD cycles. The mRNA expression levels were identified by radiolabeled cDNA probe for the respective genes.

rylates 5′-AMP to adenosine, which is taken rapidly into the cell by nucleoside transporters (Thorn and Jarvis 1996). Intracellular adenosine is primarily phosphorylated to 5′-AMP by adenosine kinase because its K_m for adenosine is one or two orders of magnitude lower than that of adenosine deaminase (*ADA*) (Schnebli et al. 1967; Arch and Newsholme 1978). Northern blot analysis demonstrates that the expression of the ecto-5′nucleotidase mRNA in LD mice liver was regulated in a circadian manner and was significantly dampened in DD animals (Fig. 7). In contrast, *ADA* mRNA expression was unaffected by changes in lighting regime. The repression of ecto-5′ nucleotidase expression by DD is opposite to the activation observed for procolipase and its enzymatic partners. These observations demonstrate that DD regulation of gene expression was not narrowly restricted to those of procolipase and its enzymatic partners but is coordinated to involve regulators of circulatory adenylate levels.

DISCUSSION

Although ATP is the cellular energy currency, 5′-AMP occupies the unique biochemical position that determines salvage or catabolism of cellular adenine nucleotides (Lehninger 1977). The adenylate biochemical equilibrium, ATP + 5′-AMP ↔ 2ADP, which is regulated by the enzyme adenylate kinase, controls cellular energy charge. In addition, 5′-AMP can act as either a positive or a negative allosteric regulator of enzymes that controls glucose homeostasis such as fructose 1,6 phosphatase (*FDP*) and phosphofructose kinase (*PFK*), respectively. *FDP* is a rate-limiting enzyme for gluconeogenesis and it converts fructose 1,6 phosphates to fructose 6-phosphate. *FDP* binding of 5′-AMP will inhibit its enzymatic activity, thereby limiting gluconeogenesis. On the reverse direction, *PFK* is a rate-limiting enzyme for glycolysis. *PFK* converts fructose 6-phosphate into fructose 1,6 phosphate utilizing an ATP molecule. Unlike *FDP*, the activity of *PFK* is enhanced by 5′-AMP thereby increasing the rate of glycolysis. For both enzymes, ATP has the opposite allosteric effects to 5′-AMP because these adenylate nucleotides bind competitively to the same regulatory motif (Kemp and Gunasekera 2002). Therefore, a change in the ATP to 5′-AMP ratio will regulate the cellular energy demand and is directly linked to biochemical processes that regulate glucose production and its utilization. In turn, gluconeogenesis performed primarily in the liver is largely dependent on ATP and NADH generated by fat catabolism. Our observations indicate that the activation of procolipase expression is linked to a change in amplitude of blood glucose levels (Zhang et al. 2006). The current observations raise the question of why the animal maintains such a complex cellular mechanism to maintain circadian rhythm for circulating 5′-AMP. Could it be that the temporal changes in circulating 5′-AMP levels was reflecting the state of the energy demands in vivo? That is, 5′-AMP is a necessary pivotal metabolic signal for the organism to maintain metabolic homeostasis.

In conclusion, our studies demonstrated that induction of procolipase expression by a DD cycle via 5′-AMP is part of a complex biochemical and physiological process that accompanies a change in environmental conditions.

ACKNOWLEDGMENTS

This work is supported in part by a National Institutes of Health Director Pioneer Award to the author. The author is indebted to Drs. J. Zhang and K. Kassik for their contributions to this chapter.

REFERENCES

Alila-Johansson A., Eriksson L., Soveri T., and Laakso M.L. 2004. Daily and annual variations of free fatty acid, glycerol and leptin plasma concentrations in goats (*Capra hircus*) under different photoperiods. *Comp. Biochem. Physiol. A Mol. Integr. Physiol.* **138:** 119.

Arch J.R. and Newsholme E.A. 1978. Activities and some properties of 5′-nucleotidase, adenosine kinase and adenosine deaminase in tissues from vertebrates and invertebrates in relation to the control of the concentration and the physiological role of adenosine. *Biochem. J.* **174:** 965.

Erlanson-Albertsson C. and Larsson A. 1988. The activation peptide of pancreatic procolipase decreases food intake in rats. *Regul. Pept.* **22:** 325.

Heldmaier G., Ortmann S., and Elvert R. 2004. Natural hypometabolism during hibernation and daily torpor in mammals. *Respir Physiol. Neurobiol.* **141:** 317.

Heller H.C and Ruby N.F. 2004. Sleep and circadian rhythms in mammalian torpor. *Annu. Rev. Physiol.* **66:** 275.

Kabon B., Bacher A., and Spiss C.K. 2003. Therapeutic hypothermia. *Best Pract. Res. Clin. Anaesthesiol.* **17:** 551.

Kemp R.G. and Gunasekera D. 2002. Evolution of the allosteric ligand sites of mammalian phosphofructo-1-kinase. *Biochemistry* **41:** 9426.

Lehninger A.L. 1977. *Biochemistry: The molecular basis of cell structure and function,* 2nd edition, p 623. Worth, New York.

Lowe M.E. 1977. Molecular mechanisms of rat and human pancreatic triglyceride lipases. *J. Nutr.* **127:** 549.

Perret M. and Aujard F. 2001. Daily hypothermia and torpor in a tropical primate: Synchronization by 24-h light-dark cycle. *Am. J. Physiol. Regul. Integr. Comp. Physiol.* **281:** R1925.

Ruby N.F., Dark J., Heller H.C., and Zucker I. 1996. Ablation of suprachiasmatic nucleus alters timing of hibernation in ground squirrels. *Proc. Natl. Acad. Sci.* **93:** 9864.

Ruby N.F., Dark J., Burns D.E., Heller H.C., and Zucker I. 2002. The suprachiasmatic nucleus is essential for circadian body temperature rhythms in hibernating ground squirrels. *J. Neurosci.* **22:** 357.

Schnebli H.P., Hill D.L., and Bennett L.L., Jr. 1967. Purification and properties of adenosine kinase from human tumor cells of type H. Ep. No. 2. *J. Biol. Chem.* **242:** 1997.

Squire T.L., Lowe M.E., Bauer V.W., and Andrews M.T. 2003. Pancreatic triacylglycerol lipase in a hibernating mammal. II. Cold-adapted function and differential expression. *Physiol. Genomics* **16:** 131.

Stoynev A.G. and Ikonomov O.C. 1983. Effect of constant light and darkness on the circadian rhythms in rats. I. Food and water intake, urine output and electrolyte excretion. *Acta Physiol. Pharmacol. Bulg.* **9:** 58.

Thompson L.F., Ruedi J.M., Glass A., Low M.G., and Lucas A.H. 1989. Antibodies to 5′-nucleotidase (CD73), a glycosyl-phosphatidylinositol-anchored protein, cause human peripheral blood T cells to proliferate. *J. Immunol.* **143:** 1815.

Thorn J.A. and Jarvis S.M. 1996. Adenosine transporters. *Gen. Pharmacol.* **27:** 613.

Uchiyama Y. 1983. A histochemical study of variations in the localization of 5′-nucleotidase activity in the acinar cell of the rat exocrine pancreas over the twenty-four hour period. *Cell Tissue Res.* **230:** 411.

von Mayersbach H. and Klaushofer K. 1979. Circadian variations of 5′-nucleotidase activity in rat liver. *Cell. Mol. Biol. Incl. Cyto Enzymol.* **24:** 73.

Zhang J., Kaasik K., Blackburn M.R., and Lee C.C. 2006. Constant darkness is a circadian metabolic signal in mammals. *Nature* **439:** 340.

Entrainment of the Human Circadian Clock

T. Roenneberg* and M. Merrow[†]
*Centre for Chronobiology, Institute for Medical Psychology, University of Munich, 80336 Munich, Germany;
[†]Department of Chronobiology, University of Groningen, 9750AA Haren, The Netherlands

Humans are an excellent model system for studying entrainment of the circadian clock in the real world. Unlike the situation in laboratory experiments, entrainment under natural conditions is achieved by different external signals as well as by internal signals generated by multiple feedbacks within the system (e.g., behavior-dependent light and temperature changes, melatonin levels, or regular nutrient intake). Signals that by themselves would not be sufficient zeitgebers may contribute to entrainment in conjunction with other self-sufficient zeitgeber signals (e.g., light). The investigation of these complex zeitgeber interactions seems to be problematic in most model systems and strengthens the human system for circadian research.

Here, we review our endeavors measuring human entrainment in real life, predominantly with the help of the Munich ChronoType Questionnaire (MCTQ). The large number of participants in our current MCTQ database allows accurate quantification of the human phase of entrainment (chronotype) and how it depends on age or sex. We also present new data showing how chronotype depends on natural light exposure. The results indicate the importance of zeitgeber strength on human entrainment and help in understanding the differences in chronotype, e.g., between urban and rural regions.

INTRODUCTION

Entrainment is the most common and most important state for circadian systems. Although clock research has acquired a host of fundamental knowledge about free-running rhythms as well as their entrainment under relatively artificial laboratory conditions, we know fairly little about how clocks are synchronized to their cyclic environment in the real world. Unlike in most laboratory experiments, the onset and offset of light are gradual and are not constant in intensity or in spectral composition throughout the natural day. In addition, alternations of day and night are always accompanied by temperature changes (with a certain lag), whereas light entrainment is recorded in constant temperatures in the laboratory. Natural entrainment is probably a complex interaction of multiple zeitgebers, among them light, temperature, and nutrition.

There is no doubt that light is the most potent zeitgeber for the biological clock, but other environmental signals also entrain, especially in poikilotherms. It has even been hypothesized that temperature is a stronger zeitgeber than light for the *Neurospora* clock (Liu et al. 1998; see also Doyle and Menaker, this volume), but the relative power of these two signals very much depends on their respective strength (Roenneberg and Merrow 2001). It is noteworthy, however, that temperature cycles entrain the circadian rhythm in this fungus even in constant light, which normally renders the *Neurospora* clock arrhythmic (Roenneberg and Merrow 2001). Nutrients can also entrain some circadian systems; nitrate, for example, resets the circadian clock in the unicellular alga *Gonyaulax polyedra* (Roenneberg and Rehman 1996).

The ecology of *Gonyaulax* gives an indication of how multiple zeitgebers may interact in the real world. Light, temperature, and nutrients have all been shown to be zeitgebers for the *Gonyaulax* clock in laboratory experiments. Both quantity and quality of light can act independently as zeitgebers in this alga because as in many plants (Roenneberg and Merrow 2000) and animals (see Doyle et al. 2006; Doyle and Menaker, this volume), light reaches the *Gonyaulax* clock via more than one light receptor (Roenneberg and Hastings 1988; Deng and Roenneberg 1997). In the ocean, *Gonyaulax* migrates daily over large distances from nutrient-poor surface waters to nutrient-rich depths (Roenneberg et al. 1989). Over the course of their natural day and during their daily migration, the algae encounter significant changes in all of the three zeitgebers which potentially all contribute to entraining its circadian system. Due to the lack of experimental possibilities, however, it is unlikely that we will ever be able to understand how these multiple inputs act as a composite zeitgeber signal for *Gonyaulax* in the real world. Similar limitations are true for practically all circadian model systems, except perhaps for one—the human circadian clock.

Circadian research has predominantly investigated self-sufficient zeitgebers, i.e., those that each and by themselves can entrain the clock under laboratory conditions. However, this might not be enough for understanding entrainment in the real world. It is "standard operating procedure" to carefully separate the acute (masking) responses elicited by the organism's environment as well as by its own behavior from the endogenous changes controlled by the circadian system. Yet, masking might be part of entrainment in the real world (Mrosovsky 1999). Although both temperature cycles and periodic feeding can coordinate daily behavior in mammals, this synchronization has often been interpreted as masking, at least for the activity rhythm driven by the SCN (suprachiasmatic nucleus) (Honma et al. 1983). On the other hand, both temperature (Brown et al. 2002) and nutrient (Stokkan et al. 2001; Schibler et al. 2003) cycles can entrain the molecular rhythms in mammalian cells and tissues. Thus, both endogenous temperature changes and oscillating nutrient levels in the blood may actually be part of the entrainment process.

Nonphotic zeitgeber effects in humans are often studied in blind people. There are several types of blindness: (1) lack of visual perception, (2) lack of residual light perception (conscious or unconscious), and (3) lack of physiological light responses (such as suppression of melatonin). Activity-rest rhythms can still be entrained in the first two types, whereas individuals tend to run free in the last group despite being submitted to strong social time cues. This suggests that the efficacy of nonphotic zeitgebers in humans depends on functional light perception possibly by creating light/dark cycles via behavior (Czeisler et al. 1986; Honma et al. 2003). However, some individuals in the third group still entrain to the 24-hour day (Sack et al. 1992; Czeisler et al. 1995; Lockley et al. 1997). Their free-running periods might be very close to 24 hours, thereby allowing adjustments by nonphotic time cues that would be too weak to entrain longer or shorter rhythms (Klerman et al. 1998; Mistlberger and Skene 2005). The closer a free-running period is to 24 hours, the more readily weak periodic signals, such as activity-dependent temperature changes and/or scheduled food intake, might contribute to entrainment. It follows that "entrained" blind people within the third group would start to run free if they were exposed to schedules longer or shorter than 24 hours and, conversely, that free-running blind people would start to entrain if exposed to schedules closer to their free-running period.

In any case, examples of blind individuals indicate that entrainment will surely involve multiple zeitgebers, some of which may not be self-sufficient and thus, their contribution to entrainment will depend on the presence of other zeitgebers. In view of the many open questions concerning the details of entrainment in the real world, it seems that—if the circadian clock was compared to a car—we certainly know a lot about the engine but little about the car and how it drives it on real roads, in real traffic.

HUMAN PHASE OF ENTRAINMENT (CHRONOTYPE)

The formalisms behind entrainment are surprisingly simple. The circadian system in all its complexity appears to behave very much like a simple mechanical oscillator. Depending on its phase, the circadian clock responds differently to a zeitgeber stimulus (Roenneberg et al. 2003a, 2005a), either by advancing, delaying, or not responding at all.

As with other genetic traits, circadian properties depend on specific genotypes. Different variants of "clock" genes (Young and Kay 2001; Roenneberg and Merrow 2003) are associated, for example, with the period length of the circadian rhythm under constant conditions. In a given population, free-running periods are distributed around a species-specific mean that has been shown in rodents (Pittendrigh and Daan 1976) and humans (e.g., see Wever 1979; Klerman 2001; Dijk and Lockley 2002). Genetic variation also contributes to the interindividual differences of the circadian clock under entrained conditions (Katzenberg et al. 1998; Ebisawa et al. 2001; Toh et al. 2001; Carpen et al. 2006; Hamet and Tremblay 2006; Viola et al. 2007). Individuals adopt a specific temporal relationship to the zeitgeber (e.g., the time difference between dawn and wake-up, core body temperature minimum, or melatonin onset). This relationship between external and internal time is called *phase of entrainment*, and when individuals differ in this trait, they are referred to as different chronotypes (Roenneberg et al. 2003b).

We have developed an instrument, the Munich ChronoType Questionnaire (MCTQ) to assess individual phase of entrainment with simple questions (Roenneberg et al. 2003b, 2004, 2005b, 2007a). The MCTQ asks about sleep and activity times, such as when do you go to bed, how long do you need to fall asleep, when do you wake up. The same set of questions is asked separately for work and for free days. The MCTQ has been validated with highly significant correlations by more than 700 sleep logs, by actimetry, and by correlations to biochemical rhythms such as melatonin and cortisol (T. Roenneberg et al., in prep.). In addition, the phase of entrainment determined by the MCTQ correlates well (Zavada et al. 2005) with the score produced by the Morningness-Eveningness Questionnaire (MEQ) (Horne and Östberg 1976).

To define chronotype quantitatively, a single phase marker has to be extracted from the different times queried in the MCTQ. We initially used mid-sleep on free days (MSF, the half-way point between sleep onset and sleep end) as a definition of chronotype (Roenneberg et al. 2003b). The distribution of MSF within our database (currently more than 60,000 individuals, mainly central Europeans) is almost normal with a slight overrepresentation of later chronotypes (Fig. 1). Although human chronotypes cluster around a mean phase of entrainment, the differences between extreme types span over half of the day.

Average sleep timing and sleep duration are essentially independent traits, i.e., the distribution of sleep duration is similar for early types and late types or vice versa the distribution of chronotypes is similar for short and for long sleepers (Roenneberg et al. 2007b). However, when sleep duration is analyzed separately for work and for free days, striking differences become apparent (Fig. 2). This result suggests that both duration and timing of sleep on free

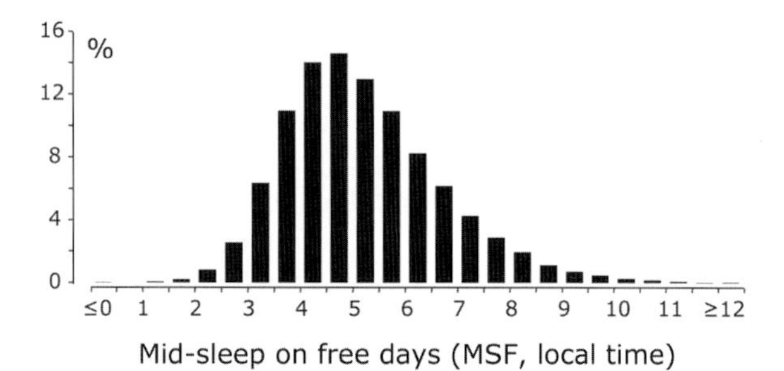

Figure 1. Distributions of chronotypes judged from mid-sleep on free days (MSF). The MCTQ database currently comprises more than 60,000 individuals, mainly from Germany, Switzerland, Austria, and The Netherlands.

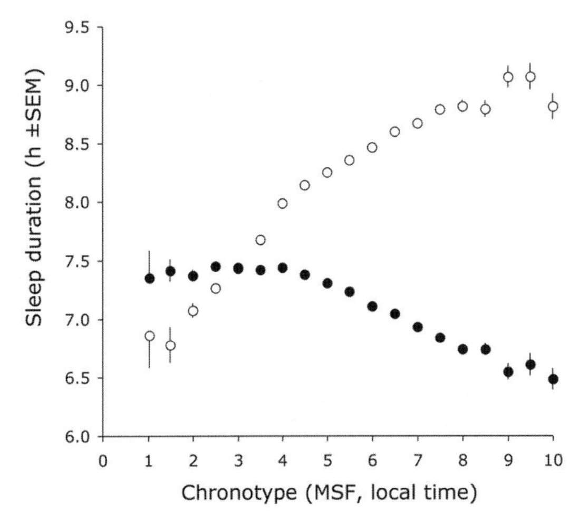

Figure 2. Relationship between chronotype (mid-sleep on free days, MSF) and sleep duration analyzed separately for work and free days (*closed circles* and *open circles*, respectively). Early chronotypes are sleep-deprived on free days, whereas late chronotypes sleep less than their weekly average on workdays. People with an MSF of 3 a.m. (e.g., those who sleep from 11 p.m. to 7 a.m., or midnight to 6 a.m.) are the only chronotypes who show no difference in sleep duration between work and free days. Vertical bars represent the standard error of the mean (S.E.M.) in each category (to avoid overlap, they are in some cases only drawn in one direction); most errors are smaller than the data points.

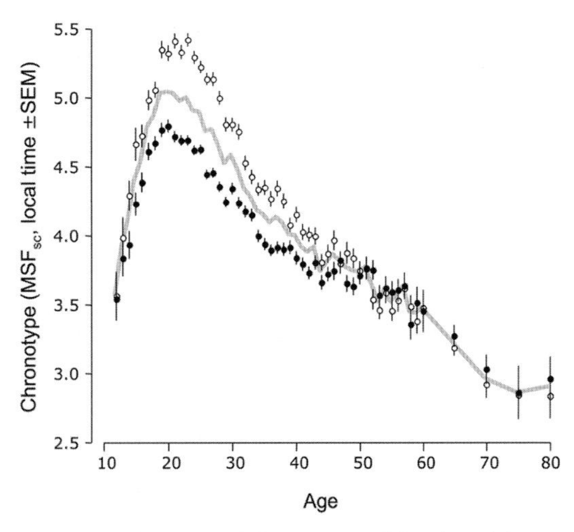

Figure 3. Chronotype (MSF, corrected for sleep debt accumulated over the workweek, MSF$_{sc}$; see text) depends on age. These changes are highly systematic and are different for males and females: (*closed circles*) females; (*open circles*) males; (*gray line*) averages for the entire population. The first data points represent the averages for subjects aged 12 or younger. Between ages 12 and 60, data were averaged for each year of age, whereas those showing the mean chronotype for subjects above 60 years of age are averaged over groups of 5 years. Vertical lines represent standard error of the mean (S.E.M.).

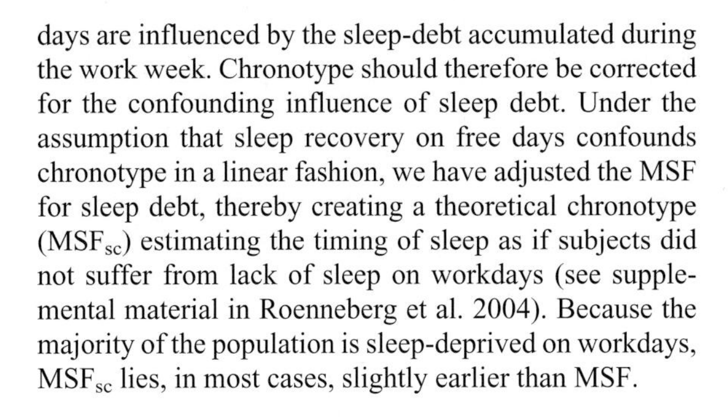

days are influenced by the sleep-debt accumulated during the work week. Chronotype should therefore be corrected for the confounding influence of sleep debt. Under the assumption that sleep recovery on free days confounds chronotype in a linear fashion, we have adjusted the MSF for sleep debt, thereby creating a theoretical chronotype (MSF$_{sc}$) estimating the timing of sleep as if subjects did not suffer from lack of sleep on workdays (see supplemental material in Roenneberg et al. 2004). Because the majority of the population is sleep-deprived on workdays, MSF$_{sc}$ lies, in most cases, slightly earlier than MSF.

CHRONOTYPE, SEX, AND AGE

Chronotype depends not only on genetic (Toh et al. 2001; Vink et al. 2001; Archer et al. 2003) and environmental factors (Roenneberg et al. 2003b), but also on age (Carskadon et al. 1999; Dijk et al. 2000; Duffy and Czeisler 2002; Park et al. 2002; Roenneberg et al. 2003). The large MCTQ database accumulated with our survey allows examination of this age dependency as an epidemiological phenomenon (Roenneberg et al. 2004). Within each age group, the shape and width of the chronotype distribution (MSF$_{sc}$) are similar to that of the general population. Across different age groups, however, their respective means vary systematically (Fig. 3). Children are generally earlier chronotypes, progressively delaying during development, reaching a maximum in "lateness" at about the age of 20, and then becoming earlier again with increasing age.

The general phenomenon that females tend to mature earlier than males in many developmental parameters is also apparent for the ontogeny of chronotype (Fig. 3). Women reach their maximum in lateness at about 19.5 years of age, whereas men continue to delay their sleep until at about the age of 21 (Roenneberg et al. 2004). As a consequence, men are, on average, later chronotypes than women for most of adulthood (see also, Adan and Natale 2002). This sex difference disappears at approximately age 50, which coincides with the average age of concluded menopause (Hollander et al. 2001; Greer et al. 2003). People over 60 years of age are on average even earlier chronotypes than children (are today).

NATURAL DAYLIGHT IS THE PREDOMINANT ZEITGEBER FOR THE HUMAN CLOCK

In our MCTQ survey, addresses and postal codes are provided by the large majority of participants. With the help of this information, we are able to address the question: What zeitgebers entrain the human clock in real life?

Within a given time zone, people live according to a common social time, whereas dawn and dusk progress continuously from east to west. This creates discrepancies between, for example, the time the sun reaches the zenith and noon (by social time). Because we predominantly arrange our lives according to the social clock, we are largely unaware of these discrepancies—but is the human circadian clock similarly oblivious of sun time? If humans were entrained by social time, average sleep-wake behavior should be the same across a time zone from its eastern to its western border. If it was, however, entrained by sun time, a gradual change should be detectable or at least some systematic deviation from the time zone constancy.

A total of 21,600 responses were selected from the

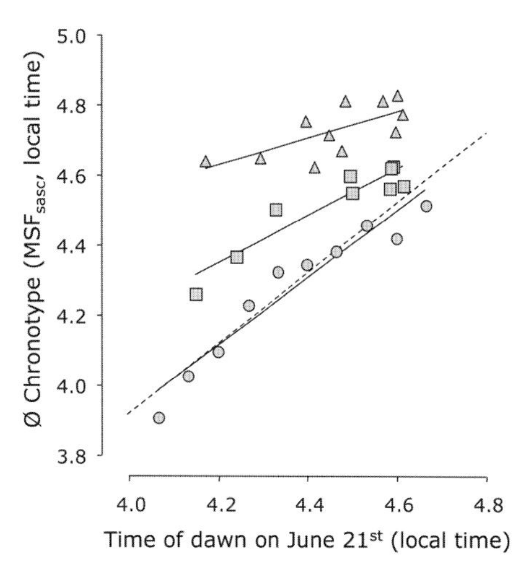

Figure 4. Chronotype, averaged for each of the 11 longitudes from western to eastern Germany (*circles*), highly correlates with the respective time of dawn; for reference, dawn times on the longest day were chosen; the *dashed line* represents the 1:1 relationship. Chronotype is given as the MSF corrected for sleep debt, age, and sex (MSF_{sasc}; see Roenneberg et al. 2007a). The data (*circles*) represent people living in areas with no more than 300,000 inhabitants. Although the average chronotype in larger cities—(*squares*) up to 500,000; (*triangles*) above 500,000—is later and less tightly coupled to sun time, the correlations are still significant.

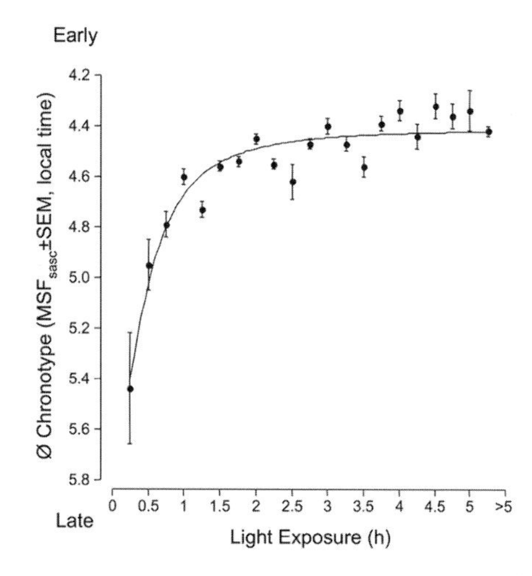

Figure 5. Average chronotype (MSF_{sasc}) depends on the daily light exposure, i.e., the average amount of time subjects spend outside without a roof above their head ($r = 0.96$; $p << 0.0001$; total $N = 41{,}232$). Error bars indicate standard error of the mean (S.E.M.). Note that unlike Figs. 3 and 4, the ordinate showing chronotype is plotted in reverse order from early at the top to late at the bottom.

MCTQ database (at that time comprising slightly more than 40,000 entries) which all contained a German postal code *and* the correct name of the corresponding location to unambiguously allow geographical mapping (Roenneberg et al. 2007a). The comparison between the chronotype distributions for each longitude clearly showed that the human circadian clock is predominantly entrained by sun time rather than by social time. In villages and towns with no more than 300,000 inhabitants (82% of the German population), the average chronotype moves proportionally with the progression of dawn (Fig. 4).

In the 20 larger cities of Germany, the correlation with sun time is weaker (although still significant) and, on average, chronotype is later. This observation could be explained if inhabitants of large cities were exposed to a weaker zeitgeber (less natural light during the day and more artificial light during the night). Weaker zeitgebers predictably lead to a later chronotype (Roenneberg et al. 2003a, 2004) in all individuals whose period—under free-running conditions—is longer than 24 hours (which is the case in the majority of humans).

To obtain some estimate of the zeitgeber strength to which an individual is exposed, the MCTQ also contains a question about how much time participants spend outside without a roof above their head. Although this question can only be answered as a coarse estimate, one can presume that it correlates with reality, especially when based on the high numbers present in our database. Figure 5 shows that chronotype correlates with the amount of time an individual spends outdoors. Chronotype progressively advances by more than an hour when people spend up to 2 hours outside per day; 42% of the population in our database fall into this category, stressing the fact that industrialization means living inside. Beyond 2 hours of natural light exposure, chronotype changes very little. This is important because the sleep debt accumulated by late types over the workweek (see Fig. 2) would be greatly reduced if those individuals could fall asleep an hour earlier by spending more time outside.

Figures 4 and 5 show that sun time is the dominant zeitgeber for entraining the human clock. Sunlight exerts its effects through timing (i.e., creates a longitudinal gradient) and also via zeitgeber strength: People living in larger towns depend less on daylight timing and are, on average,

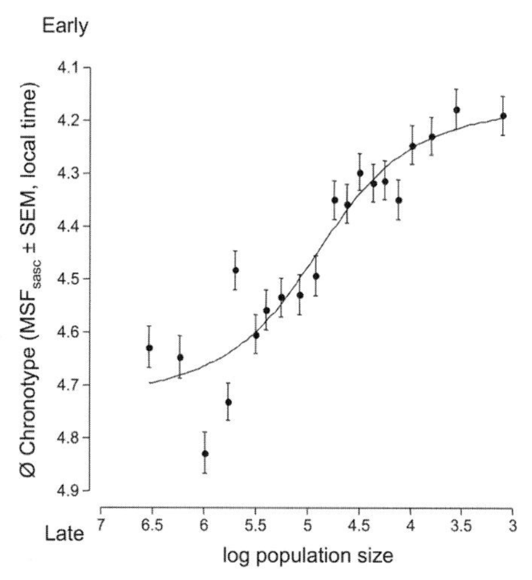

Figure 6. Chronotype depends on population size. The database, sorted by population size, was divided into equal bins of 1080 individuals each, for which average population size and average chronotype (MSF_{sasc}) were calculated. The abscissa is plotted on a logarithmic scale and in reverse order with large cities on the left and villages on the right, assuming that this represents the amount of light inhabitants are exposed to ($r = 0.94$; $p << 0.0001$; total $N = 21{,}600$). Vertical bars represent standard error of the mean (S.E.M.).

later chronotypes. Figure 4 separated the German population into three categories according to the population of their place of residence. A more continuous dependency of chronotype on population size is shown in Figure 6. According to these results, the chronotype of city dwellers is, on average, more than half an hour later than that of people living in the countryside.

THE CIRCADIAN CLOCK: A HYPOTHESIS ABOUT TEMPORAL ENVIRONMENT

Our findings emphasize that individual circadian time rather than social external time should be considered in scientific studies, in school and work schedules, or in medical considerations. Individual time reflects, in turn, the strong dependence of the circadian clock on light, reminding us that this biological system concerns perception in a classical sense. One could argue that its ability to continue without external input (free-run) in addition to its power to anticipate recurring events within the external daily structure are features of an active rather than a perceptive machinery. However, perception is rarely just a passive response; our visual perception of the world, for example, is predominantly based on updating endogenous hypotheses. A good example of hypothesis-driven perception is disorientation upon waking in an unfamiliar setting (e.g., a hotel room). We are disoriented despite the fact that we very well know the room in which we went to bed as well as the room that our brain obviously has provided as the hypothesis to be updated (Roenneberg 1997). After what seems to be a long time, the perceptual deadlock is resolved when the brain provides a new hypothesis leading to an instantaneous recognition of where we are.

This example shows how the brain provides an intrinsic hypothesis about the environment's spatial structures, and the circadian system provides an intrinsic hypothesis of a temporal structure of our environment—the "day." In both cases, hypotheses are continually updated by external and internal information. Not only do these complex interactions result in what we call entrainment, but they can also produce after-effects. Recent, regular time structures of the environment are incorporated into the current temporal hypothesis and thereby shape the responses to new temporal information. The long time it takes to alter an existing hypothesis indicates the robustness of the intrinsic components contributing to perception; this holds for the perception of both spatial and temporal structures. Many factors are integrated to form the "circadian hypothesis," which is then continually updated by the current environmental information.

In view of the complexity and the adaptiveness of such a system, the concept of a stable intrinsic period (Czeisler et al. 1999; Wyatt et al. 1999) seems to be an oxymoron. That we can measure a self-sustained rhythm in temporal isolation may be remarkable to us, but free-running rhythms merely reflect how clocks evolved to optimally fulfill their function as a temporal perception apparatus under natural, entrained conditions. Although free-running period and phase of entrainment are often strongly correlated, natural selection could only act upon the latter.

Most of us consider the relationship between free-running period and entrained phase as causal: Entrained phase *depends* on free-running period. This statement is simply wrong. Phase is merely associated with but does not depend on period! The way individuals behave in everyday life depends on their personalities (this is the analogy to the natural, entrained state). The way individuals behave under the influence of alcohol also depends on their personalities (this is the analogy to the unnatural state of free-running rhythms). But the way individuals behave in everyday life does not depend on the way they behave under the influence of alcohol.

Beyond the erroneous conclusion about a causal period-phase dependency, their strong correlation is a one-way logic that only holds under special conditions: Only if this relationship is investigated in a homogeneous genetic background, a manipulation of components contributing to the trait "free-running period" will also penetrate in the trait "phase of entrainment" (chronotype; see also Merrow and Roenneberg, this volume). Yet, many other external and internal factors, besides free-running period, contribute to chronotype: sensitivity to the zeitgeber stimulus, properties of the transduction pathway, coupling between oscillators and to outputs within the system, and many more. Thus, a generalization of this correlation is built on soft grounds; in genetically heterogeneous populations or even in experiments investigating quantitative trait loci (QTL), one would not necessarily expect this correlation to hold (Michael et al. 2003).

If there was an "intrinsic period," one might also presume an "intrinsic phase." We have shown here the adaptive qualities of human chronotype (for similar results in a simple fungus, see Merrow and Roenneberg, this volume). In humans, phase depends at least on age, sex, and light environment. The latter is influenced by many different factors, such as longitude, latitude, time of year, place of residence, and even profession (office vs. outdoors). To investigate the genetic basis of the human circadian clock, chronotype must be corrected for all of these factors because they all do not work at the genetic level. Because the free-running period reflects (not *is*) part of the mechanism that determines (and adaptively changes) chronotype, its values should be confined to a well-defined and relatively narrow range. However, the term "intrinsic period" only makes sense if it is associated with a distinct value. What are the conditions that produce this value? Forced desynchrony has been used to determine an individual's value of "intrinsic period" (Czeisler et al. 1999). As argued in the Introduction, multiple external factors contribute to entrainment even then if they cannot act as self-sufficient zeitgebers. In addition, entrainment depends on internal factors, both concerning prior history (which influence the current hypothesis) and current internal states. In a forced desynchrony protocol, many of these change and thereby influence the length of the period that has broken away from the imposed schedule. Behavior-dependent light exposure (even if the light changes are too weak to entrain by themselves), activity-dependent temperature changes and regular food intake—to name only few—all influence what is interpreted as "intrinsic period."

ACKNOWLEDGMENTS

Our work is supported by the 6th European Framework Programme EUCLOCK (018741) and the Daimler-Benz-Stiftung project CLOCKWORK.

REFERENCES

Adan A. and Natale V. 2002. Gender differences in morningness-eveningness preference. *Chronobiol. Int.* **19:** 709.

Archer S.N., Robilliard D.L., Skene D.J., Smits M., Williams A., Arendt J., and von Schantz M. 2003. A length polymorphism in the circadian clock gene Per3 is linked to delayed sleep phase syndrome and extreme diurnal preference. *Sleep* **26:** 413.

Brown S.A., Zumbrunn G., Fleury-Olela F., Preitner N., and Schibler U. 2002. Rhythms of mammalian body temperature can sustain peripheral circadian clocks. *Curr. Biol.* **12:** 1574.

Carpen J.D., von Schantz M., Smits M., Skene D.J., and Archer S.N. 2006. A silent polymorphism in the PER1 gene associates with extreme diurnal preference in humans. *J. Hum. Genet.* **51:** 1122.

Carskadon M.A., Labyak S.E., Acebo C., and Seifer R. 1999. Intrinsic circadian period of adolescent humans measured in conditions of forced desynchrony. *Neurosci. Lett.* **260:** 129.

Czeisler C.A., Shanahan T.L., Kerman E.B., Martens H., Brotman D.J., Emens J.S., Klein T., and Rizzo J.F. 1995. Suppression of melatonin secretion in some blind patients by exposure to bright light. *N. Engl. J. Med.* **332:** 6.

Czeisler C.A., Allan J.S., Strogatz S.H., Ronda J.M., Sanchez R., Rios C.D., Freitag W.O., Richardson G.S., and Kronauer R.E. 1986. Bright light resets the human circadian pacemaker independent of the timing of the sleep-wake cycle. *Science* **233:** 667.

Czeisler C.A., Duffy J.F., Shanahan T.L., Brown E.N., Mitchel J.F., Rimmer D.W., Ronda J.M., Silva E.J., Allan J.S., Emens J.S., Dijk D.-J., and Kronauer R.E. 1999. Stability, precision, and near-24-hour period of the human circadian pacemaker. *Science* **284:** 2177.

Deng T.-S. and Roenneberg T. 1997. Photobiology of the *Gonyaulax* circadian system. II. Allopurinol inhibits blue light effects. *Planta* **202:** 502.

Dijk D.-J. and Lockley S.W. 2002. Integration of human sleep-wake regulation and circadian rhythmicity. *J. Appl. Physiol.* **92:** 852.

Dijk D.-J., Duffy J.F., and Czeisler C.A. 2000. Contribution of circadian physiology and sleep homeostasis to age-related changes in human sleep. *Chronobiol. Int.* **17:** 285.

Doyle S.E., Castrucci A.M., McCall M., Provencio I., and Menaker M. 2006. Nonvisual light responses in the Rpe65 knockout mouse: Rod loss restores sensitivity to the melanopsin system. *Proc. Natl. Acad. Sci.* **103:** 10432.

Duffy J.F. and Czeisler C.A. 2002. Age-related change in the relationship between circadian period, circadian phase, and diurnal preference in humans. *Neurosci. Lett.* **318:** 117.

Ebisawa T., Uchiyama M., Kajimura N., Mishima K., Kamei Y., Katoh M., Watanabe T., Sekimoto M., Shibui K., Kim K., Kudo Y., Ozeki Y., Sugishita M., Toyoshima R., Inoue Y., Yamada N., Nagase T., Ozaki N., Ohara O., Ishida N., Okawa M., Takahashi K., and Yamauchi T. 2001. Association of structural polymorphisms in the human *period3* gene with delayed sleep phase syndrome. *EMBO Rep.* **2:** 342.

Greer W., Sandridge A.L., and Chehabeddine R.S. 2003. The frequency distribution of age at natural menopause among Saudi Arabian women. *Maturitas* **46:** 263.

Hamet P. and Tremblay J. 2006. Genetics of the sleep-wake cycle and its disorders. *Metabolism* (suppl. 2) **55:** S7.

Hollander L.E., Freeman E.W., Sammel M.D., Berlin J.A., Grisso J.A., and Battistini M. 2001. Sleep quality, estradiol levels, and behavioral factors in late reproductive age women. *Obstet. Gynecol.* **98:** 391.

Honma K., von Goetz C., and Aschoff J. 1983. Effects of restricted daily feeding on free running circadian rhythms in rats. *Physiol. Behav.* **30:** 905.

Honma K., Hashimoto S., Nakao M., and Honma S. 2003. Period and phase adjustments of human circadian rhythms in the real world. *J. Biol. Rhythms* **18:** 261.

Horne J.A. and Östberg O. 1976. A self-assessment questionnaire to determine morningness-eveningness in human circadian rhythms. *Int. J. Chronobiol.* **4:** 97.

Katzenberg D., Young T., Finn L., Lin L., King D.P., Takahashi J.S., and Mignot E. 1998. A CLOCK polymorphism associated with human diurnal preference. *Sleep* **21:** 569.

Klerman E.B. 2001. Non-photic effects on the circadian system: Results from experiments in blind and sighted individuals. In *Zeitgebers, entrainment and masking of the circadian system* (ed. K. Honma et al.), p. 155. Hokkaido University Press, Sapporo, Japan.

Klerman E.B., Rimmer D.W., Dijk D.-J., Kronauer R.E., Rizzo J.F.I., and Czeisler C.A. 1998. Nonphotic entrainment of the human circadian pacemaker. *Am. J. Physiol.* **274:** R991.

Liu Y., Merrow M., Loros J.L., and Dunlap J.C. 1998. How temperature changes reset a circadian oscillator. *Science* **281:** 825.

Lockley S.W., Skene D.J., Tabandeh H., Bird A.C., Defrance R., and Arendt J. 1997. Relationship between napping and melatonin in the blind. *J. Biol. Rhythms* **12:** 16.

Michael T.P., Salome P.A., Yu H.J., Spencer T.R., Sharp E.L., McPeek M.A., Alonso J.M., Ecker J.R., and McClung C.R. 2003. Enhanced fitness conferred by naturally occurring variation in the circadian clock. *Science* **302:** 1049.

Mistlberger R.E. and Skene D.J. 2005. Nonphotic entrainment in humans? *J. Biol. Rhythms* **20:** 339.

Mrosovsky N. 1999. Masking: History, definitions, and measurement. *Chronobiol. Int.* **16:** 415.

Park Y.M., Matsumoto K., Seo Y.J., Kang M.J., and Nagashima H. 2002. Changes of sleep or waking habits by age and sex in Japanese. *Percept. Mot. Skills* **94:** 1199.

Pittendrigh C.S. and Daan S. 1976. A functional analysis of circadian pacemakers in nocturnal rodents. V. Pacemaker structure: A clock for all seasons. *J. Comp. Physiol. A* **106:** 333.

Roenneberg T. 1997. Zeiträume, innere Uhren und Zeitgeber. *du* **10:** 01.00.

Roenneberg T. and Hastings J.W. 1988. Two photoreceptors influence the circadian clock of a unicellular alga. *Naturwissenschaften* **75:** 206.

Roenneberg T. and Merrow M. 2000. Circadian clocks: Omnes viae Romam ducunt. *Curr. Biol.* **10:** R742.

———. 2001. The role of feedbacks in circadian systems. In *Zeitgebers, entrainment and masking of the circadian system* (ed. K. Honma et al.), p. 113. Hokkaido University Press, Sapporo, Japan.

———. 2003. The network of time: Understanding the molecular circadian system. *Curr. Biol.* **13:** R198.

Roenneberg T. and Rehman J. 1996. Nitrate, a nonphotic signal for the circadian system. *FASEB J.* **10:** 1443.

Roenneberg T., Colfax G.N., and Hastings J.W. 1989. A circadian rhythm of population behavior in *Gonyaulax polyedra*. *J. Biol. Rhythms* **4:** 201.

Roenneberg T., Daan S., and Merrow M. 2003a. The art of entrainment. *J. Biol. Rhythms* **18:** 183.

Roenneberg T., Dragovic Z., and Merrow M. 2005a. Demasking biological oscillators: Properties and principles of entrainment exemplified by the *Neurospora* circadian clock. *Proc. Natl. Acad. Sci.* **102:** 7742.

Roenneberg T., Kumar C.J., and Merrow M. 2007a. The human circadian clock entrains to sun time. *Curr. Biol.* **17:** R44.

Roenneberg T., Wirz-Justice A., and Merrow M. 2003b. Life between clocks: Daily temporal patterns of human chronotypes. *J. Biol. Rhythms* **18:** 80.

Roenneberg T., Tan Y., Dragovic Z., Ricken J., Kuehnle T. and Merrow M. 2005b. Chronoecology from fungi to humans. In *Biological rhythms* (ed. K. Honma et al.), p. 73. Hokkaido University Press, Sapporo, Japan.

Roenneberg T., Kuehnle T., Juda M., Kantermann T., Allebrandt K., Gordijn M., and Merrow M. 2007b. Epidemiology of the human circadian clock. *Sleep Med. Rev.* Epub ahead of print.

Roenneberg T., Kuehnle T., Pramstaller P.P., Ricken J., Havel M., Guth A., and Merrow M. 2004. A marker for the end of adolescence. *Curr. Biol.* **14:** R1038.

Sack R.L., Lewy A.J., Blood M.L., Keith L.D., and Nakagawa H. 1992. Circadian rhythm abnormalities in totally blind people: Incidence and clinical significance. *J. Clin. Endocrinol. Metab.* **75:** 127.

Schibler U., Ripperger J., and Brown S.A. 2003. Peripheral circadian oscillators in mammals: Time and food. *J. Biol. Rhythms* **18:** 250.

Stokkan K.A., Yamazaki S., Tei H., Sakaki Y., and Menaker M. 2001. Entrainment of the circadian clock in the liver by feeding. *Science* **291:** 490.

Toh K.L., Jones C.R., He Y., Eide E.J., Hinz W.A., Virshup D.M., Ptacek L.J., and Fu Y.H. 2001. An *hPer2* phosphorylation site mutation in familial advanced sleep phase syndrome. *Science* **291:** 1040.

Vink J.M., Groot A.S., Kerkho G.A., and Boomsma D.I. 2001. Genetic analysis of morningness and eveningness. *Chronobiol. Int.* **18:** 809.

Viola A.U., Archer S.N., James L.M., Groeger J.A., Lo J.C., Skene D.J., von Schantz M. and Dijk D.-J. 2007. PER3 polymorphism predicts sleep structure and waking performance. *Curr. Biol.* **17:** 613.

Wever R. 1979. *The Circadian system of man*. Springer, Berlin, Germany.

Wyatt J.K., Ritz-de Cecco A., Czeisler C.A., and Dijk D.-J. 1999. Circadian temperature and melatonin rhythms, sleep, and neurobiological function in humans living on a 20-h day. *Am. J. Physiol.* **277:** R1152.

Young M.W. and Kay S.A. 2001. Time zones: A comparative genetics of circadian clocks. *Nat. Rev. Genet.* **2:** 702.

Zavada A., Gordijn M.C.M., Beersma D.G.M., Daan S., and Roenneberg T. 2005. Comparison of the Munich Chronotype Questionnaire with the Horne-Östberg's Morningness-Eveningness score. *Chronobiol. Int.* **22:** 267.

Peripheral Clocks: Keeping Up with the Master Clock

E. KOWALSKA AND S.A. BROWN
University of Zurich, Institute for Pharmacology and Toxicology, 8057 Zurich, Switzerland

Circadian clocks influence most aspects of physiology and behavior, so perhaps it is not surprising that circadian oscillators exist in nearly all mammalian cells. These cells remain synchronized to the outside world in hierarchical fashion, with a "master clock" tissue in the suprachiasmatic nucleus of the hypothalamus receiving light input from the retina and then conveying timing information to "slave" clocks in peripheral tissues. Recent research has highlighted both the similarities and differences between central and peripheral clocks and provided new insight into their communication. Above all, however, this parallelism of clockwork has provided a unique opportunity to study at the cellular level a regulatory mechanism that affects complex behaviors.

INTRODUCTION

In most organisms, circadian rhythms have a key role in the regulation of numerous aspects of physiology and behavior. The circadian clock can be found in organisms ranging from cyanobacteria to complex vertebrates such as mouse and zebra fish. Although the evolutionary advantage of maintaining such a molecular clock is still controversial, it is clear that it confers a fitness advantage for some simple organisms under selective pressure. For example, cyanobacteria that possess an oscillator with a period length tuned to their environment easily outgrow those that do not (Woelfle et al. 2004), and the endogenous period length of wild fruit flies has been shown to change with latitude (Costa et al. 1992). In more complex organisms, mutations in clock genes can lead to cancer and infertility (Fu et al. 2002; Miller et al. 2004) and are correlated with various depressive and sleep disorders (Cermakian and Boivin 2003). It is unclear, however, whether these phenotypes are related directly to the clock or to other functions of clock genes.

At least in metazoan organisms, the circadian clock was for a long time believed to be a complex neuronal phenomenon. A central clock tissue—the suprachiasmatic nucleus (SCN) of the brain hypothalamus in mammals, the pineal gland in birds and reptiles, and the lateral neurons of *Drosophila*—was believed to synchronize circadian processes throughout the body via presumably electrical cues to other brain regions. The first evidence that these cues might be primarily hormonal in nature came from pioneering work by Silver et al. (1996), who showed that an implanted SCN encased in porous plastic material could rescue the circadian rhythms of an SCN-lesioned animal. Other experiments revealed that the basis of this clock is actually cell-autonomous and non-electrical (Welsh et al. 1995).

Soon afterward in 1997, S.A. Kay's group showed that in *D. melanogaster*, explanted parts of the body possess independent photoreceptive circadian clocks (Plautz et al. 1997). Cell-autonomous circadian clocks were operative throughout the body. Even serum-shocked immortalized rat fibroblasts, isolated over 35 years previously, were observed to have circadian expression of clock genes (Balsalobre et al. 1998). Subsequent experiments with transgenic Per1::luciferase rats showed that these clocks in fact exist in most tissues of the mammalian body (Yamazaki et al. 2000).

Many further investigations by numerous laboratories have demonstrated that the molecular principle of circadian clocks in metazoans is probably based on interlocking negative transcriptional feedback loops within the cell (Bell-Pedersen et al. 2005). In simpler organisms, each cell-autonomous clock is individually light-sensitive and is therefore independently entrained by the environment. In mammals, however, this synchronization happens in strictly hierarchical fashion to ensure that clocks throughout the whole organism remain properly synchronized. First, an external timing cue (principally light) sets the phase in the central pacemaker, the SCN. This bilateral nucleus contains several thousand independently cycling but locally coupled neurons. Subsequently, the SCN projects its rhythms onto cell-autonomous clocks of similar mechanism in peripheral tissues. The result is synchronous circadian transcription in peripheral tissues with a constant phase delay compared to the SCN (Fig. 1).

COMMUNICATION BETWEEN CENTRAL AND PERIPHERAL OSCILLATORS

Despite increasing knowledge of the mechanism of the circadian clock and its entrainment, the way in which it controls circadian physiology and gene expression is far from clear. The basic signaling between the core oscillator and peripheral clocks probably involves a mixture of direct hormonal cues such as glucocorticoids and indirect cues such as cyclic body temperature and food metabolites (Damiola et al. 2000; Le Minh et al. 2001; Stokkan et al. 2001; Brown et al. 2002). Although each of these cues can phase-shift peripheral oscillators without affecting the central clock in the SCN, the elimination of the circadian pattern in any one of these signals does not result in the loss of peripheral circadian gene expression. Hence,

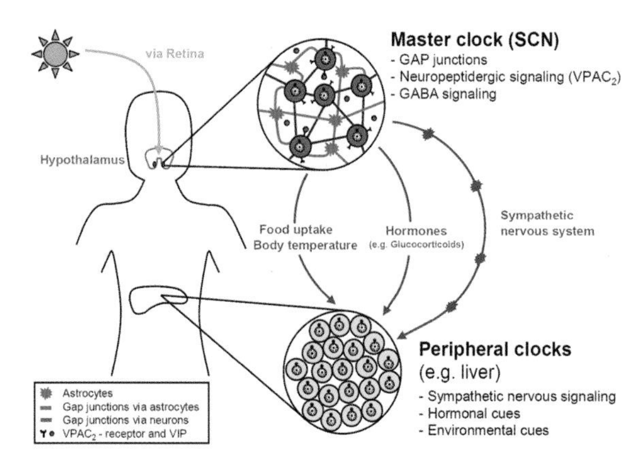

Figure 1. A model for clock hierarchy in mammals. (*Red*) Connections between different body clocks and their influence on other oscillators. In the hypothalamus (*blue*), the master clock sitting in the SCN is composed of pacemaker neurons (*purple*) which are interconnected and synchronized by neuropeptidergic signals and/or gap junctions via astrocytes (*green*) or neurons (*purple*). Light perceived via the retina and downstream signaling through the retinohypothalamic tract (*yellow*), is responsible for adjusting clock phase in the pacemaker neurons. In peripheral clocks (*light blue*), as an example the liver clockwork, entrainment is mainly dependent on SCN downstream signaling via the sympathetic nervous system, hormones, and environmental cues (e.g., glucocorticoids).

each of these signals is either redundant or unimportant to circadian synchrony in vivo.

Recent research has even challenged the established hierarchy between the core oscillator and peripheral clocks. For example, the expression of the clock gene *Per1* could be directly induced in the adrenal gland via light in an SCN-dependent mechanism, suggesting the existence of a "shortcut" directly from light to some peripheral clocks. An intact sympathetic nervous system was essential to this process (Ishida et al. 2005). Tissue-specific clock disruptions have confirmed the existence of such direct circuits. Genetic disruption of circadian rhythms in liver results in the abolition of circadian transcription of some liver genes, but not of others, including the clock gene *Per2* (Kornmann et al. 2007). Similarly, the section of the vagus nerve resulted in elimination of oscillations both in *Per2* expression and in acetylcholine receptor protein levels in the respiratory tract (Bando et al. 2007).

The current working model for circadian clocks is thus a multifaceted one in which the SCN communicates with peripheral oscillators via several pathways. These peripheral oscillators can in turn directly control circadian genes either via transcription factor cascades or via the same *cis*-acting elements that control clock genes in general. Finally, some further peripheral circadian gene expression and physiology appears to be controlled not by peripheral clocks, but directly by the SCN via nervous stimuli.

SIMILARITIES AND DIFFERENCES BETWEEN CENTRAL AND PERIPHERAL OSCILLATORS

Considerable speculation has centered on the fundamental nature of clock architecture in SCN neurons and in other tissues. The same basic oscillator components exist in both central and peripheral oscillators, and both are capable of robust cell-autonomous oscillations. Most genetic mutations that affect central oscillator function have similar qualitative effects upon peripheral oscillators (Yagita et al. 2001; Pando et al. 2002). Nevertheless, these effects are often exaggerated in peripheral oscillators, pointing to possible differences. For example, deletion of the *Per1* gene results in a shortening of the circadian period of behavior by 1 hour, but the period of circadian gene expression in isolated *Per1*$^{-/-}$ fibroblasts is 4 hours shorter (Brown et al. 2005b).

One possible reason for this difference could arise at the level of expression of clock components themselves. For example, it has recently been shown that deletion of the important circadian transcriptional activator CLOCK in mice does not abolish circadian behavioral rhythmicity (Debruyne et al. 2006). These authors speculate that in the SCN, the function of CLOCK can be substituted by the NPAS2 protein (Debruyne et al. 2007). Because NPAS2 shows a tissue-specific expression pattern, one might suppose that explanted peripheral tissues that do not express NPAS2 would be severely attenuated even though the SCN was not.

Another obvious difference between SCN and peripheral oscillators is that whereas explanted SCN oscillators appear to possess the ability to continue oscillations indefinitely, oscillators in explanted peripheral tissues dampen rapidly (Yamazaki et al. 2000; Yoo et al. 2004). In principal, this experimental observation could arise either through attenuation of clock oscillations in each cell or via gradually increasing desynchrony among clocks in adjacent cells due to differences in cell-autonomous endogenous period length. Fluorescent or bioluminescent imaging of fibroblast cells in culture firmly supports the latter hypothesis: Individual fibroblasts show long-duration circadian oscillations (each of slightly differing period) but fail to synchronize to one another without external stimuli (Nagoshi et al. 2004; Welsh et al. 2004). Although fibroblasts in culture clearly lose synchrony, the same question is less clear in vivo. Confirming the in vitro experiment above, SCN-lesioned hamsters show constant, intermediate levels of clock genes in peripheral organs, an observation that implies cellular desynchrony within each organ (Guo et al. 2006). In contrast, SCN-ablated mice after several days display large phase differences in individual tissues of an animal and among different animals, suggesting the opposite (Yoo et al. 2004).

This discrepancy aside, the clearly superior synchrony among SCN neurons compared to peripheral cells and tissues is likely the result of better intercellular coupling, rather than greater clock precision. Dissociated SCN neurons, like fibroblasts, demonstrate significant heterogeneity in period length and phase (Welsh et al. 1995, 2004). In intact SCN tissue, three clearly defined intercellular coupling methods exist: gap junctions, peptidergic signaling using the VIP neuropeptide and the VPAC2 receptor, and GABA signaling. Elimination of either of these first two pathways results in significant circadian impairments in vivo (Liu and Reppert 2000; Harmar 2003; Long et al. 2005; Maywood et al. 2006).

Current views divide the SCN into at least two functional suboscillators: the dorsal SCN and the ventral SCN. Interrupting the connection between them results in loss of synchrony in the dorsal part of the SCN but leaves the ventral part perfectly synchronized (Yamaguchi et al. 2003). It is thought that the ventral SCN receives timing information from the retinohypothalamic tract and subsequently communicates this information to the dorsal SCN neurons. Such a bipartite organization might further stabilize SCN oscillation.

Altogether, experimental evidence and mathematical modeling suggest that intercellular coupling could explain the resistance of the SCN—and by inference circadian behavior—to mutations that more severely attenuate peripheral oscillators of similar molecular makeup (Bernard et al. 2007; Liu et al. 2007).

PERIPHERAL OSCILLATORS AS PROBES OF CIRCADIAN CLOCK FUNCTION

Differences in both clock gene expression and intercellular coupling likely exist between peripheral and central oscillators. Nevertheless, self-autonomous peripheral clocks could provide an important model system for the elucidation of many aspects of clock function that are more difficult or impossible to study in the central SCN oscillator itself, especially in human beings. In principle, peripheral clocks provide two advantages over the study of the whole organism or of the central clock in the SCN: accessibility to experimental manipulation and availability in homogeneous large quantities. Multiple laboratories have exploited these aspects for both biochemical and genetic studies into the mammalian circadian oscillator.

For example, by labeling the clock protein PER1 with peptide epitopes and then expressing it in fibroblasts, our laboratory was able to purify a PER1-containing protein complex that contained cryptochromes (proteins previously identified as important to clock function), as well as two other novel proteins, WDR5 and NONO. Fibroblasts were then used as easy model systems in which to study the function of these two proteins. RNA interference (RNAi)-based knockdown of NONO protein levels demonstrated NONO to be essential to circadian rhythms in these cells, and knockdown of WDR5 demonstrated that this protein was necessary for histone methylation at circadian clock loci (Brown et al. 2005a).

Of course, the observation that WDR5 and NONO are important to clock function in fibroblasts does not permit an immediate generalization to the whole organism. The final "acid test" of validity remains the analysis of the whole organism. Usually, this test is performed via a mouse knockout model. Such a knockout is generated by homologous recombination in embryonic stem (ES) cells, which are then injected into a mouse blastocyst to create a chimera—a time-consuming and costly process. Because these ES cells are pluripotent, their differentiation into other cell types that exhibit circadian rhythmicity permits the rapid screening of generated cells for circadian phenotypes, at least if the targeted gene has a phenotype at the heterozygous or hemizygous level. Coupled with "gene-trap" approaches to generate nonfunctional alleles, such an ES cell differentiation approach could be used as a rapid screen for new X-linked clock genes (E. Kowalska and S.A. Brown, unpubl.).

Fibroblast oscillators have also been used as functional tools to identify the underlying mechanism of human mutations that cause circadian disorders. For example, familial advanced sleep phase syndrome (FASPS) has been mapped in one family to a point mutation in the *Per2* locus (Toh et al. 2001). By expressing the mutant allele in fibroblasts, Vanselow et al. (2006) were recently able to characterize the nature of this defect at a molecular level, as well as to recapitulate the advanced phase of the behavioral phenotype of this mutation by measuring the transcriptional phase of FASPS fibroblast cells under entrained conditions.

OUTLOOK: POTENTIAL USES OF PERIPHERAL CLOCKS TO CHARACTERIZE HUMAN DISORDERS

Such an application of peripheral cells to verify or study human phenotypes could potentially impact patient care and diagnosis in a clinical setting. Although the human circadian oscillator has been characterized extensively at a behavioral level, the difficulty and cost of maintaining subjects under controlled conditions to effect these measurements prevent their widespread use. Easily available peripheral tissues (blood, skin, hair) could provide a useful proxy. Primary cells from these tissues can be infected with lentiviral or adenoviral reporter vectors that permit bioluminescent readout of circadian gene expression, thereby enabling the investigator to monitor different properties of the molecular clock and characterize its function (Brown et al. 2005b).

For such studies to be possible, it is important to establish the relationship between circadian properties measured in peripheral tissues such as fibroblasts and those measured via human behavior. Although initial studies have shown excellent correlations between behavior in mice and the molecular properties of fibroblasts, further studies in human beings are necessary to validate these conclusions. Fibroblast period length per se is influenced by culture conditions such as temperature and the concentration of serum in their growth medium. Nevertheless, cells displaying short- and long-period lengths seem to retain their relative values under all conditions (Fig. 2). Thus, although comparisons of values from different laboratories may prove problematic, the assay as a whole shows great promise.

Specifically, peripheral oscillators as a model system might permit screening of patients with sleep disorders to determine which are due to molecular defects in the circadian clock. When peripheral cell cultures are kept under constant growth conditions, an estimate of free-running period length can be obtained. By placing them in entrained conditions—e.g., 24-hour temperature cycles—one can then look at entrained phase. It will be interesting to see how both of these properties correspond to behavior in human subjects. Finally, by using these properties as quantitative traits in human pedigrees or populations, genetic linkage or association studies should be possible,

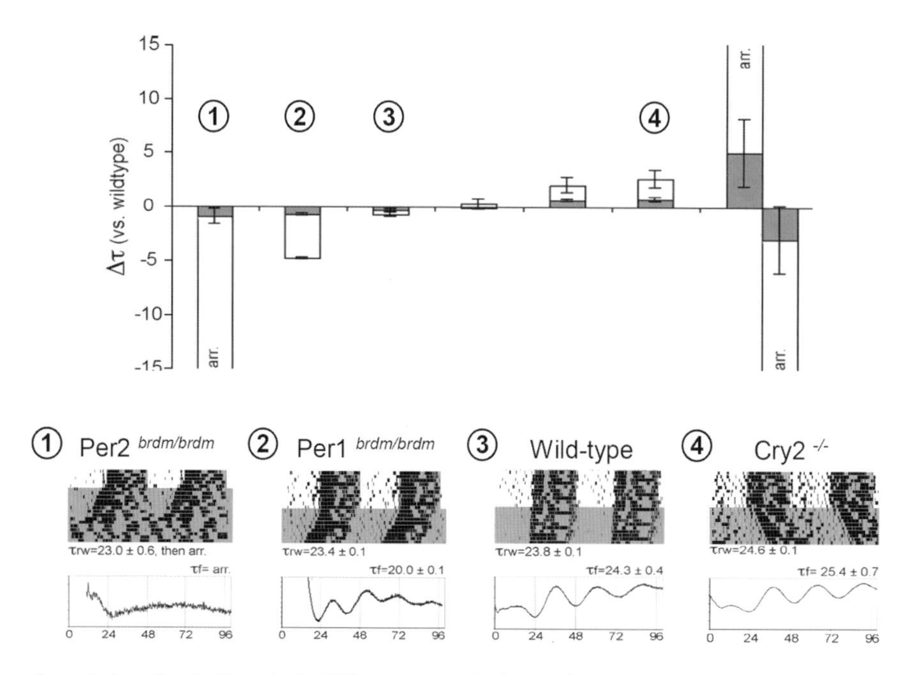

Figure 2. Genotypic variation of period lengths in different mutant backgrounds. The variation of circadian period length measured from fibroblasts (*white bars*) is compared to behavioral period length measured via running-wheel activity data (*light gray bars*), expressed as difference in hours from the 24-hour light cycle. (*Top panel*) Genotypes shown from left to right: $Per2^{brdm/brdm}$, $Per1^{brdm/brdm}$, *wild type*, $Cry2^{+/-}$, $Cry2^{-/-}$;$Per1^{brdm/brdm}$, $Cry2^{-/-}$, $Per2^{brdm/brdm}$;$Cry2^{-/-}$. (*Lower panel*) Representative running-wheel actograms from individual animals. (Adapted from Brown et al. 2005b [PLoS Biol.].)

enabling the discovery of modifier loci for human chronotype.

Recent studies all highlight the extent to which circadian clocks impact not only behavior, but also cellular processes such as cell division and metabolism. Peripheral oscillators could also provide an excellent model system in which to study these phenomena—for example, the involvement of the circadian clock in DNA-damage checkpoint control, whose disregulation leads to cancer (Collis and Boulton 2007). Further investigations will undoubtedly ascertain not only the potential, but also the limits of this exciting model system.

ACKNOWLEDGMENTS

The authors would like to thank L. Cuninkova and A. Dumas for their helpful commentary regarding this manuscript. Work herein from the authors' laboratory was in part supported by grants from the Swiss National Science Foundation, EUClock, and the Desiree and Neils Yde Foundation.

REFERENCES

Balsalobre A., Damiola F., and Schibler U. 1998. A serum shock induces circadian gene expression in mammalian tissue culture cells. *Cell* **93:** 929.

Bando H., Nishio T., van der Horst G.T., Masubuchi S., Hisa, Y., and Okamura H. 2007. Vagal regulation of respiratory clocks in mice. *J. Neurosci.* **27:** 4359.

Bell-Pedersen D., Cassone V.M., Earnest D.J., Golden S.S., Hardin P.E., Thomas T.L., and Zoran M.J. 2005. Circadian rhythms from multiple oscillators: Lessons from diverse organisms. *Nat. Rev. Genet.* **6:** 544.

Bernard S., Gonze D., Cajavec B., Herzel H., and Kramer A. 2007. Synchronization-induced rhythmicity of circadian oscillators in the suprachiasmatic nucleus. *PLoS Comput. Biol.* **3:** e68.

Brown S.A., Zumbrunn G., Fleury-Olela F., Preitner N., and Schibler U. 2002. Rhythms of mammalian body temperature can sustain peripheral circadian clocks. *Curr. Biol.* **12:** 1574.

Brown S.A., Ripperger J., Kadener S., Fleury-Olela F., Vilbois F., Rosbash M., and Schibler U. 2005a. PERIOD1-associated proteins modulate the negative limb of the mammalian circadian oscillator. *Science* **308:** 693.

Brown S.A., Fleury-Olela F., Nagoshi E., Hauser C., Juge C., Meier C.A., Chicheportiche R., Dayer J.M., Albrecht U., and Schibler U. 2005b. The period length of fibroblast circadian gene expression varies widely among human individuals. *PLoS Biol.* **3:** e338.

Cermakian N. and Boivin D.B. 2003. A molecular perspective of human circadian rhythm disorders. *Brain Res. Rev.* **42:** 204.

Collis S.J. and Boulton S.J. 2007. Emerging links between the biological clock and the DNA damage response. *Chromosoma* **116:** 331.

Costa R., Peixoto A.A., Barbujani G., and Kyriacou C.P. 1992. A latitudinal cline in a *Drosophila* clock gene. *Proc. Biol. Sci.* **250:** 43.

Damiola F., Le Minh N., Preitner N., Kornmann B., Fleury-Olela F., and Schibler U. 2000. Restricted feeding uncouples circadian oscillators in peripheral tissues from the central pacemaker in the suprachiasmatic nucleus. *Genes Dev.* **14:** 2950.

Debruyne J.P., Weaver D.R., and Reppert S.M. 2007. CLOCK and NPAS2 have overlapping roles in the suprachiasmatic circadian clock. *Nat. Neurosci.* **10:** 543.

Debruyne J.P., Noton E., Lambert C.M., Maywood E.S., Weaver D.R., and Reppert S.M. 2006. A clock shock: Mouse CLOCK is not required for circadian oscillator function. *Neuron* **50:** 465.

Fu L., Pelicano H., Liu J., Huang P., and Lee C. 2002. The circadian gene Period2 plays an important role in tumor suppression and DNA damage response in vivo. *Cell* **111:** 41.

Guo H., Brewer J.M., Lehman M.N., and Bittman E.L. 2006.

Suprachiasmatic regulation of circadian rhythms of gene expression in hamster peripheral organs: Effects of transplanting the pacemaker. *J. Neurosci.* **26:** 6406.
Harmar A.J. 2003. An essential role for peptidergic signalling in the control of circadian rhythms in the suprachiasmatic nuclei. *J. Neuroendocrinol.* **15:** 335.
Ishida A., Mutoh T., Ueyama T., Bando H., Masubuchi S., Nakahara D., Tsujimoto G., and Okamura H. 2005. Light activates the adrenal gland: Timing of gene expression and glucocorticoid release. *Cell Metab.* **2:** 297.
Kornmann B., Schaad O., Bujard H., Takahashi J.S., and Schibler U. 2007. System-driven and oscillator-dependent circadian transcription in mice with a conditionally active liver clock. *PLoS Biol.* **5:** e34.
Le Minh N., Damiola F., Tronche F., Schutz G., and Schibler U. 2001. Glucocorticoid hormones inhibit food-induced phase-shifting of peripheral circadian oscillators. *EMBO J.* **20:** 7128.
Liu A.C., Welsh D.K., Ko C.H., Tran H.G., Zhang E.E., Priest A.A., Buhr E.D., Singer O., Meeker K., Verma I.M., Doyle F.J., III, Takahashi J.S., and Kay S.A. 2007. Intercellular coupling confers robustness against mutations in the SCN circadian clock network. *Cell* **129:** 605.
Liu C. and Reppert S.M. 2000. GABA synchronizes clock cells within the suprachiasmatic circadian clock. *Neuron* **25:** 123.
Long M.A., Jutras M.J., Connors B.W., and Burwell R.D. 2005. Electrical synapses coordinate activity in the suprachiasmatic nucleus. *Nat. Neurosci.* **8:** 61.
Maywood E.S., Reddy A.B., Wong G.K., O'Neill J.S., O'Brien J.A., McMahon D.G., Harmar A.J., Okamura H., and Hastings M.H. 2006. Synchronization and maintenance of timekeeping in suprachiasmatic circadian clock cells by neuropeptidergic signaling. *Curr. Biol.* **16:** 599.
Miller B.H., Olson S.L., Turek F.W., Levine J.E., Horton T.H., and Takahashi J.S. 2004. Circadian clock mutation disrupts estrous cyclicity and maintenance of pregnancy. *Curr. Biol.* **14:** 1367.
Nagoshi E., Saini C., Bauer C., Laroche T., Naef F., and Schibler U. 2004. Circadian gene expression in individual fibroblasts: Cell-autonomous and self-sustained oscillators pass time to daughter cells. *Cell* **119:** 693.
Pando M.P., Morse D., Cermakian N., and Sassone-Corsi P. 2002. Phenotypic rescue of a peripheral clock genetic defect via SCN hierarchical dominance. *Cell* **110:** 107.
Plautz J.D., Kaneko M., Hall J.C., and Kay S.A. 1997. Independent photoreceptive circadian clocks throughout *Drosophila*. *Science* **278:** 1632.
Silver R., LeSauter J., Tresco P.A., and Lehman M.N. 1996. A diffusible coupling signal from the transplanted suprachiasmatic nucleus controlling circadian locomotor rhythms. *Nature* **382:** 810.
Stokkan K.A., Yamazaki S., Tei H., Sakaki Y., and Menaker M. 2001. Entrainment of the circadian clock in the liver by feeding. *Science* **291:** 490.
Toh K.L., Jones C.R., He Y., Eide E.J., Hinz W.A., Virshup D.M., Ptacek L.J., and Fu Y.H. 2001. An hPer2 phosphorylation site mutation in familial advanced sleep phase syndrome. *Science* **291:** 1040.
Vanselow K., Vanselow J.T., Westermark P.O., Reischl S., Maier B., Korte T., Herrmann A., Herzel H., Schlosser A., and Kramer A. 2006. Differential effects of PER2 phosphorylation: Molecular basis for the human familial advanced sleep phase syndrome (FASPS). *Genes Dev.* **20:** 2660.
Welsh D.K., Logothetis D.E., Meister M., and Reppert S.M. 1995. Individual neurons dissociated from rat suprachiasmatic nucleus express independently phased circadian firing rhythms. *Neuron* **14:** 697.
Welsh D.K., Yoo S.H., Liu A.C., Takahashi J.S., and Kay S.A. 2004. Bioluminescence imaging of individual fibroblasts reveals persistent, independently phased circadian rhythms of clock gene expression. *Curr. Biol.* **14:** 2289.
Woelfle M.A., Ouyang Y., Phanvijhitsiri K., and Johnson C.H. 2004. The adaptive value of circadian clocks: An experimental assessment in cyanobacteria. *Curr. Biol.* **14:** 1481.
Yagita K., Tamanini F., van Der Horst G.T., and Okamura H. 2001. Molecular mechanisms of the biological clock in cultured fibroblasts. *Science* **292:** 278.
Yamaguchi S., Isejima H., Matsuo T., Okura R., Yagita K., Kobayashi M., and Okamura H. 2003. Synchronization of cellular clocks in the suprachiasmatic nucleus. *Science* **302:**1408
Yamazaki S., Numano R., Abe M., Hida A., Takahashi R., Ueda M., Block G.D., Sakaki Y., Menaker M., and Tei H. 2000. Resetting central and peripheral circadian oscillators in transgenic rats. *Science* **288:** 682.
Yoo S.H., Yamazaki S., Lowrey P.L., Shimomura K., Ko C.H., Buhr E.D., Siepka S.M., Hong H.K., Oh W.J., Yoo O.J., Menaker M., and Takahashi J.S. 2004. PERIOD2: LUCIFERASE real-time reporting of circadian dynamics reveals persistent circadian oscillations in mouse peripheral tissues. *Proc. Natl. Acad. Sci.* **101:** 5339.

Physiological Importance of a Circadian Clock Outside the Suprachiasmatic Nucleus

K.-F. Storch,*† C. Paz,*† J. Signorovitch,‡, E. Raviola,† B. Pawlyk,§
T. Li,§ and C.J. Weitz†

†Department of Neurobiology, Harvard Medical School, Boston, Massachusetts 02115;
‡Department of Biostatistics, Harvard School of Public Health, Boston, Massachusetts 02115;
§Massachusetts Eye & Ear Infirmary, Berman-Gund Laboratory, Department of Ophthalmology,
Harvard Medical School, Boston Massachusetts 02114

Circadian clocks are widely distributed in mammalian tissues, but little is known about the physiological functions of clocks outside the suprachiasmatic nucleus of the brain. The retina has an intrinsic circadian clock, but its importance for vision is unknown. Here, we show that mice lacking *Bmal1*, a gene required for clock function, had abnormal retinal transcriptional responses to light and defective inner retinal electrical responses to light, but normal photoreceptor responses to light and retinas that appeared structurally normal as observed by light and electron microscopy. We generated mice with a retina-specific genetic deletion of *Bmal1*, and they had defects of retinal visual physiology essentially identical to those of mice lacking *Bmal1* in all tissues and lacked a circadian rhythm of inner retinal electrical responses to light. Our findings indicate that the intrinsic circadian clock of the retina regulates retinal visual processing in vivo.

INTRODUCTION

Circadian clocks are endogenous oscillators that drive daily rhythms of physiology and behavior. In mammals, the circadian clock mechanism is built upon a molecular feedback loop in which the CLOCK-BMAL1 transcription factor drives expression of its PER and CRY inhibitors (Gekakis et al. 1998; Ko and Takahashi 2006). The clock generates circadian rhythms cell-autonomously, and it is thought to generate rhythms of physiology in large part by driving rhythms of transcription of output genes (Akhtar et al. 2002; Duffield et al. 2002; Panda et al. 2002; Storch et al. 2002). The adaptive significance of circadian clocks likely lies in their ability to allow anticipatory responses to predictable daily variations in the environment (Ouyang et al. 1998).

It has long been known that the circadian clock regulating behavior in mammals is located in the suprachiasmatic nucleus (SCN) of the brain (Ko and Takahashi 2006). Recently, it has become clear that circadian clocks are distributed in mammalian tissues, present at sites such as the retina (Tosini and Menaker 1996), multiple brain regions (Abe et al. 2002), and in many peripheral tissues (Balsalobre et al. 1998; Damiola et al. 2000; Yamazaki et al. 2000). It is thought that clocks outside the SCN have physiological functions, but to date, few studies have addressed this question (Durgan et al. 2006; McDearmon et al. 2006). In *Drosophila,* there is compelling evidence that a clock in the antenna drives rhythms of olfactory sensitivity (Tanoue et al. 2004).

Retinas from a wide range of vertebrates, including amphibians (Besharse and Iuvone 1983), birds (Pierce et al. 1993), and mammals (Tosini and Menaker 1996), contain a circadian clock. In *Xenopus* retina, photoreceptors are the circadian clock cells (Cahill and Besharse 1993; Hayasaka et al. 2002), whereas in the mammalian retina, circadian clock cells are found in the inner retina (Witkovsky et al. 2003; Gustincich et al. 2004; Ruan et al. 2006; Storch et al. 2007).

Fundamental retinal processes are under circadian control, including photoreceptor disc shedding (LaVail and Ward 1978), release of melatonin and dopamine (Doyle et al. 2002), and retinal electrical responses to light, manifested as a circadian rhythm of one or more components of the electroretinogram (ERG) (Manglapus et al. 1998; Barnard et al. 2006). At present, little is known about the biochemical pathways under circadian control in the mammalian retina or the molecular mechanisms that modulate retinal physiological responses to light. A circadian rhythm of melatonin release is driven autonomously from the retina (Tosini and Menaker 1996), but it is not yet known to what extent circadian rhythms of retinal electrical activity in response to light reflect the action of a local retinal clock or the action of a remote clock, such as the SCN. Evidence from nonmammalian vertebrates suggests that circadian rhythms of retinal electrical responses to light are driven at least in part from the brain (Miranda-Anaya et al. 2002) or pineal (McGoogan and Cassone 1999).

EXPERIMENTAL PROCEDURES

Mice and tissue collection. For 3-day microarray experiments, 108 adult male CBA/CaJ mice (The Jackson Laboratory) were entrained to a 12:12-hour light/dark (LD) cycle for 3 weeks. During the dark phase, 54 mice were transferred to DD and the remainder were kept in

*These authors contributed equally.

Cold Spring Harbor Symposia on Quantitative Biology. Volume LXXII. Cold Spring Harbor Laboratory Press 978-08796-822-5 (cloth); -823-2 (paper). (This paper is adapted, with permission, from Storch et al. 2007. *Cell* **130:** 730 [© Elsevier].)

LD. At 4-hour intervals, three mice from each group were euthanized (CO_2), and eyes were collected, frozen in liquid nitrogen, and stored (–80°C). Total RNA was purified separately from the eyes of each mouse (Trizol, Invitrogen, and RNAeasy, QIAGEN), and equal aliquots of RNA from the three pairs of eyes collected at a single time point were pooled. Studies were performed in accordance with the protocol approved by the Harvard Medical School Standing Committee on Animals.

Microroarray analysis. Samples were hybridized to Affymetrix mouse 430.2 arrays representing the complete mouse genome (one array per time point). Fluorescence images were normalized to median brightness (Li and Wong 2001), and model-based expression values were computed using a DNA-Chip Analyzer (Li and Wong 2003). In each 3-day, 18-array data set, probe sets that were not classified as "expressed" (http://www. affymetrix.com/support/technical/technotes/statistical_reference_guide.pdf) in at least three arrays were excluded, as were probe sets that had three or more missing expression values. The square-root transformation was applied to all expression values to make the error distribution more symmetric while limiting amplification of low-level noise. (For a detailed description of procedures for identifying rhythmic expression, see Storch et al. 2007.) To assess nonrhythmic transcripts, expression values determined by the DNA-Chip Analyzer (Li and Wong 2003) for the six arrays corresponding to the six time points from wild-type mice in LD were used to select a "constitutively expressed" set of genes. Genes with low mean expression value were excluded (<200 units; range: 0–10,900 units) and the remaining genes with a near-constant level of expression across the six arrays were selected (standard deviation of expression <5%, yielding 3,047 genes). Microarray data have been deposited in the ArrayExpress database (Accession number E-TABM-285).

Light and electron microscopy of retinas. For light microscopy, eyecups of three mice of each genotype were fixed (2% formaldehyde in 0.15 M Sørensen phosphate buffer at pH 7.4), cryoprotected (20% sucrose in phosphate-buffered saline [PBS]), and frozen in monochlorodifluomethane. Sections parallel to and including the horizontal meridian were obtained and stained with 300 nM 4′,6-diamidino-2-phenylindole (DAPI) in PBS. Fluorescence was detected with a Zeiss LSM 510 Meta confocal system and Zeiss Axioplan-2 microscope. Because photoreceptor activity was unaltered in *Bmal1*$^{-/-}$ mice, the thickness of the deep retinal layers was normalized to that of the outer nuclear layer (ONL) in each retina in order to avoid errors due to obliquity of sections. For electron microscopy, two mice of each genotype were perfused with 2% formaldehyde, 2.5% glutaraldehyde in 0.15 M Sørenson phosphate buffer at pH 7.4, followed by 1% OsO_4 and 1.5% potassium ferrocyanide, and stained en bloc with 1% uranyl acetate. Thick sections of the eyecup were stained with toluidine blue and examined with a Nikon Eclipse E600 microscope. Thin sections were stained with uranyl and lead, and micrographs were obtained with a JEOL x100 electron microscope.

Quantitative RT-PCR. Total RNA from whole eyes or dissected eye fractions (retinal, lens, and the rest of the eye) was transcribed into cDNA using random hexamers and Superscript reverse transcriptase (Invitrogen). cDNA derived from 25 ng of total RNA was PCR-amplified in a PTC-200 thermocycler with a Chromo4 module (MJ Research) using SYBR green (IQ SYBR Green Supermix, Bio-Rad) according to the manufacturer's instructions. Templates were amplified with the annealing temperature initially lowered from 70°C to 60°C in two-cycle increments followed by 30 cycles of the following three steps: 20 seconds at 94°C, 20 seconds at 60°C, and 30 seconds at 72°C. For quantification, the threshold cycle number difference was calculated based on the sample with the lowest expression, and the values obtained were 2^n-transformed. Input RNA concentration or the expression level of hypoxanthine guanine phosphoribosyl transferase-1 was used to normalize expression values.

Electroretinography. ERG recordings were performed as described by Lu et al. (2001). Briefly, mice were dark-adapted overnight before recording. Mice were anesthetized and their pupils dilated, and responses were recorded with a chloride silver wire loop placed on the cornea. ERG responses were elicited with 10-?s flashes of white light (1.37 x 10^5 cd/m^2) presented in a Ganzfeld dome at 1-minute intervals in darkness. Mice were then adapted to a background illumination (12 ft.-lamberts, 10 minutes) and responses were elicited by 1-Hz flashes of white light (1.37 x 10^5 cd/m^2). Circadian components of the light-adapted ERG were obtained from mice in constant light as described by Barnard et al. (2006), except the intensity of light was 300 Lux.

SCN lesions. Adult C57BL/6J mice were anesthetized with ketamine/xylazine (45 mg/kg/5 mg/kg), and an electrode (RNE-300X, Rhodes Medical Instruments) was lowered stereotaxically through the skull at the midsagittal sinus (anteroposterior –0.45 mm from Bregma) with the final tip position at 6.0 mm below the skull. Lesions were generated with constant current (2 mA, 10 seconds; D.C. Constant Lesion Maker, Grass Instruments). One week after surgery, locomotor activity in DD was monitored, and only mice showing arrhythmic behavior for more than 4 weeks were selected for ERG studies. Only mice with histologically verified complete SCN lesions were included in the analysis (data not shown).

Generation of conditional Bmal1 mice. A cassette containing neomycin acetyltransferase (*neo*) flanked by two *FRT* sites and a single *loxP* site (EcoRI–BamHI fragment from vector PL452; Liu et al. 2003) was inserted into a bacterial artificial chromosome (BAC) encompassing the *Bmal1* locus (RPCI23-331K23, BACPAC resources, Children's Hospital Oakland Research Institute) by homologous recombination in bacteria (Liu et al. 2003). The insertion site was immediately upstream of the PacI site in the intron 3′ of the exon encoding the bHLH domain (exon 8). A 10.8-kb PmlI-fragment containing the Neo cassette with flanking genomic regions was subcloned into the HpaI–SmaI-digested targeting vector pKOII (Bardeesy et al. 2002), which contained a diphtheria toxin expression cassette. An oligonucleotide comprising a single *loxP* site preceded by a BamHI restriction site was cloned into the SanDI site of the intron preceding exon 8, resulting in the final targeting vector. Electroporation of embryonic stem (ES) cells, selection of neomycin-resistant clones, and injection of ES cells into blastocysts were performed using standard methods. The Neo cassette was removed by mating chimeras to mice carrying a Flpe knockin at the Rosa26 locus (Farley et al. 2000). Genotyping was performed using multiplex PCR and primers L1, ACTGGAAGTAACTTTATCAAACTG; L2, AATCCGCCTGCCTACTGCCTCC (reverse primer); and R4, GGGTGGAGTATGATATGACC. For assessing the efficiency of *loxP* recombination, retinal genomic DNA was digested with AvrII, and Southern blot hybridization was performed using a 350-bp probe specific for sequences immediately following the 3′ flanking region used for targeting.

In situ hybridization, histology, and western blots. For in situ hybridization, dissected retinas were fixed (cold 4% formaldehyde in PBS, 10 minutes), embedded (Tissue-Tek), cut frozen (20 ?m), and processed as described by Kraves and Weitz (2006). Sequences of riboprobes are available upon request. β-galactosidase activity in 20-?m brain or retina sections was detected with 5-bromo-4-chloro-3-indolyl β-D-galactopyranoside (X-gal). Western analysis was performed with rabbit anti-BMAL1 antiserum (1:500).

RESULTS

Daily Rhythms of Retinal Gene Expression

To gain a view of molecular regulation in the mammalian retina by a circadian clock, light, or both, we performed whole-genome microarray studies in mice to identify genes with an approximately 24-hour rhythm of expression during a 3-day interval in constant darkness (DD) or in a 12:12-hour LD cycle. Because of the role of melatonin in retinal function (Doyle et al. 2002), we used CBA/CaJ mice, a strain that makes melatonin (Goto et al. 1989) and does not have retinal degeneration. To optimize the efficiency of tissue collection, we used whole eyes for RNA extraction.

To identify rhythmic variations in gene expression, we computed the best-fit function that models the expression of a gene across the 3-day, 18-time-point microarray profile as as approximately 24-hour rhythmic pattern; we did not assume a simple waveform. For some genes, the best-fitting rhythmic function makes a good fit, for most, a poor fit. Next, we randomly permuted the order of the 18 time points for each gene 50,000 times. With 18 time points, random permutation of the time series will degrade the fit of a truly rhythmic profile but will have little or no effect on the fit of noisy or flat profiles (Storch et al. 2007). This procedure allows all or any subset of the 45,101 probe sets on the array to be ranked quantitatively for rhythmicity, and one of its key advantages is that a statistical threshold for rhythmicity can be set according to any desired false-discovery rate, defined as the percentage of genes ranking above the threshold that can be accounted for by noise.

In DD, we identified 277 genes with a circadian rhythm of expression at a moderate threshold corresponding to a 15% false-discovery rate, (Fig. 1A, top, DD). Phases of peak expression around the clock were represented about equally (Fig. 1A, top, DD), and the data set included genes with known circadian regulation in other tissues, including clock components. Overall, genes expressed in the retina made a major contribution to the data set (Storch et al. 2007), and the genes represent a wide range of functions, including synaptic transmission, photoreceptor signaling, intercellular communication, and regulation of the cytoskeleton and chromatin (Storch et al.

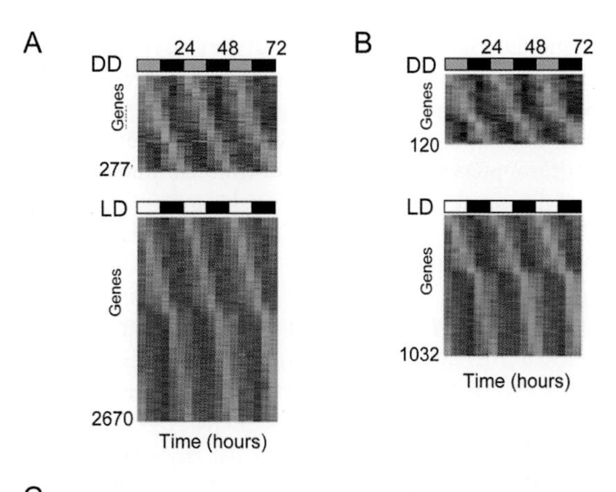

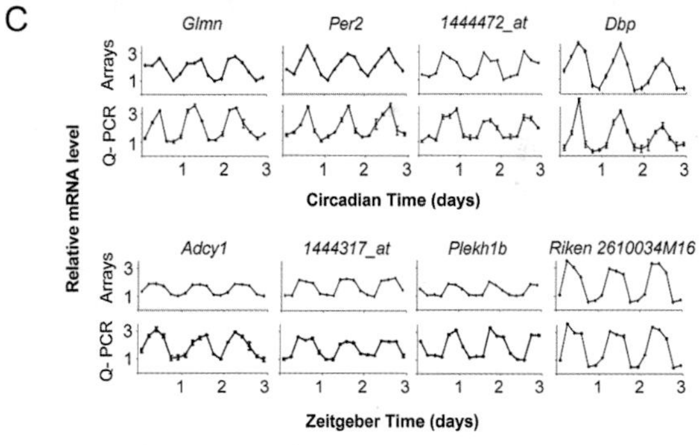

Figure 1. Genes with 24-hour rhythms of expression in the mouse eye. (*A*,*B*) Three-day expression profiles in which each column represents a time point and each row represents a gene, with genes ordered by phase of peak expression. Light shades represent expression values above the mean for a gene; dark shades below the mean. Total time in constant darkness (DD) or in a light/dark cycle (LD) is indicated at top, and the bars represent subjective day and night in DD or light and dark conditions in LD. The number of genes is indicated at the lower left. (*A*) Threshold: 15% false-discovery rate. (*B*) Threshold: 5% false-discovery rate. (*C*) Validation of microarray profiles by Q-PCR. (*Top*) DD; (*bottom*) LD. Shown are comparisons of individual profiles from microarrays (arrays) and Q-PCR from the same RNA samples. For Q-PCR, the mean ($N = 3$) and S.E.M. are shown (some error bars cannot be seen at this scale). Relative expression levels are plotted in arbitrary linear units. (*Glmn*) Glomulin; (*Per2*) Period 2; (*Dbp*) D-site albumin promoter-binding protein; (*Adcy1*) adenylate cyclase 1; (*Plekh1b*) pleckstrin-homology 1b.

2007). For some genes, expression was limited to photoreceptors, likely reflecting non-cell-autonomous regulation that depends on clock cells in the inner retina or elsewhere. The rhythmic data set included secreted factors expressed exclusively in the inner retina, candidates for circadian signals from inner retinal clock cells to photoreceptors (Storch et al. 2007).

In LD, at the same statistical threshold, we identified 2670 genes with rhythmic expression (Fig. 1A, bottom, LD). Included were 80% of the genes with rhythmic expression in DD and many genes with known expression in the retina. In addition to the approximately ninefold greater number of genes showing rhythmic expression in LD than in DD, the distribution of phases in LD differed markedly from that in DD, with somewhat more than half of the genes showing a peak of expression during the night and the rest showing about equal phase distribution (Fig. 1A, bottom, LD). Both of these differences between the two conditions were observed at more stringent thresholds (Fig. 1B), indicating that they do not arise from chance inclusion of noise. These results suggest that LD cycles drive expression of a large number of genes in the retina and, in particular, a large cluster of genes with a nighttime peak of expression. Additional analyses indicated that any loss of circadian synchrony among mice or cells during the 3-day period in DD did not contribute substantially to the differences between the DD and LD data sets (Storch et al. 2007).

For validation, we selected 26 genes with rhythmic expression from DD and 28 from LD, as well as genes from each classified as nonrhythmic, for assessment by quantitative reverse transcriptase PCR (Q-PCR). In all cases but one, Q-PCR confirmed the rhythmic profiles (for examples, see Fig. 1C). Microarray analysis thus accurately identified genes with rhythmic expression.

Importance of *Bmal1* for Retinal Gene Expression Rhythms in a Light/Dark Cycle

Given that the retina is a dedicated photosensory organ, the approximately ninefold excess of genes exhibiting rhythmic expression in LD compared to DD could simply result from regulation of many genes by light, independently of circadian clock function. To test this expectation, we compared temporal profiles of ocular gene expression in LD in wild-type mice and *Bmal1*$^{-/-}$ (*Mop3*$^{-/-}$) (Bunger et al. 2000) littermates, which lack an essential component of the clock in all tissues and consequently are expected to lack all clock function. Genes regulated purely by LD cycles would be expected to retain full rhythms of expression in *Bmal1*$^{-/-}$ mice. In contrast, genes regulated by LD cycles in a manner that depends on clock function would be expected to exhibit altered regulation in LD, even if the genes are primarily driven by light and do not exhibit detectable rhythmic expression in DD. In initial Q-PCR studies of eyes from *Bmal1*$^{-/-}$ mice in DD, we observed loss of rhythms of expression of *Per1*, *Per2*, *Dbp*, and *Rev-erbα*, indicating that *Bmal1* is required for circadian rhythms in the eye, as expected (Storch et al. 2007).

It is important to note that the null mutation of *Bmal1* can affect clock-regulated processes in two ways: by eliminating the circadian feedback loop and thus abolishing circadian rhythms (Bunger et al. 2000) and by reducing expression of output genes driven directly by the CLOCK-BMAL1 transcription factor (Panda and Hogenesch 2004). Therefore, any phenotypes observed in *Bmal1*$^{-/-}$ mice could result from a loss of rhythmicity or from a reduction in one or more clock-driven transcriptional outputs, as well as by hypothetical noncircadian functions of *Bmal1*. For the purpose of discussion of *Bmal1* mutants here and below, we define "circadian clock function" in this broad sense, including both intrinsic rhythmicity and transcriptional outputs of the clock mechanism.

First, to check for any confounding retinal developmental defects in *Bmal1*$^{-/-}$ mice, we studied retinas from adult *Bmal1*$^{-/-}$ mice and wild-type littermates by light and electron microscopy. Retinas from the two genotypes were indistinguishable in general architecture, cellular organization and density, complement of organelles, and distribution and structure of synaptic specializations of the principal retinal cell types (Fig. 2). Although we cannot exclude the possibility of subtle quantitative differences, these results make it highly unlikely that a developmental or structural defect underlies any abnormalities of retinal gene expression or function in *Bmal1*$^{-/-}$ mice.

We next performed microarray analysis to compare ocular gene expression profiles of wild-type mice and *Bmal1*$^{-/-}$ littermates (on a C57BL/6 background) in LD at 4-hour intervals during one daily cycle. To interpret these 1-day data sets, we used information from the 3-day experiment shown in Figure 1 (CBA/CaJ mice). For each of the genes showing rhythmic expression in LD in the 3-day data set, we examined the 1-day profile in wild-type or *Bmal1*$^{-/-}$ littermates in LD. A 1-day expression profile was classified as rhythmic if it showed a phase and waveform like that in the 3-day experiment, defined as a significantly closer match to the mean of its corresponding 3-day profile than to independent Gaussian noise (i.e., it behaved statistically like a "fourth day" of the rhythm).

In wild-type C57BL/6 mice (littermates of *Bmal1*$^{-/-}$ mice), many genes showed rhythmic regulation matching that in the 3-day experiment (Fig. 3A; the threshold, for a match was set at a stringent 5% false-discovery rate). In contrast, at the same threshold, very few genes qualified as rhythmic in LD in *Bmal1*$^{-/-}$ littermates. A comparison of 1-day expression profiles classified as rhythmic in wild-type mice with profiles of the same genes from the mutants revealed substantial disorganization in the mutants, although weak features similar to the wild-type pattern were apparent (Fig. 3A). This result held true across different statistical thresholds (data not shown). Inspection of many individual gene profiles and validation of 20 by Q-PCR indicated that about 60% showed flat or noisy profiles, about 30% showed apparent regulation by light but with reduced amplitude and altered waveform, and about 10% showed fully preserved rhythms (Fig. 3B). In contrast, the *Bmal1* mutation had no detectable effect on the expression of more than 3000 genes that are expressed constitutively in the eye in LD (Fig. 3C).

Bmal1 thus has a significant role in the light-dependent regulation of genes in the eye, most of which were not

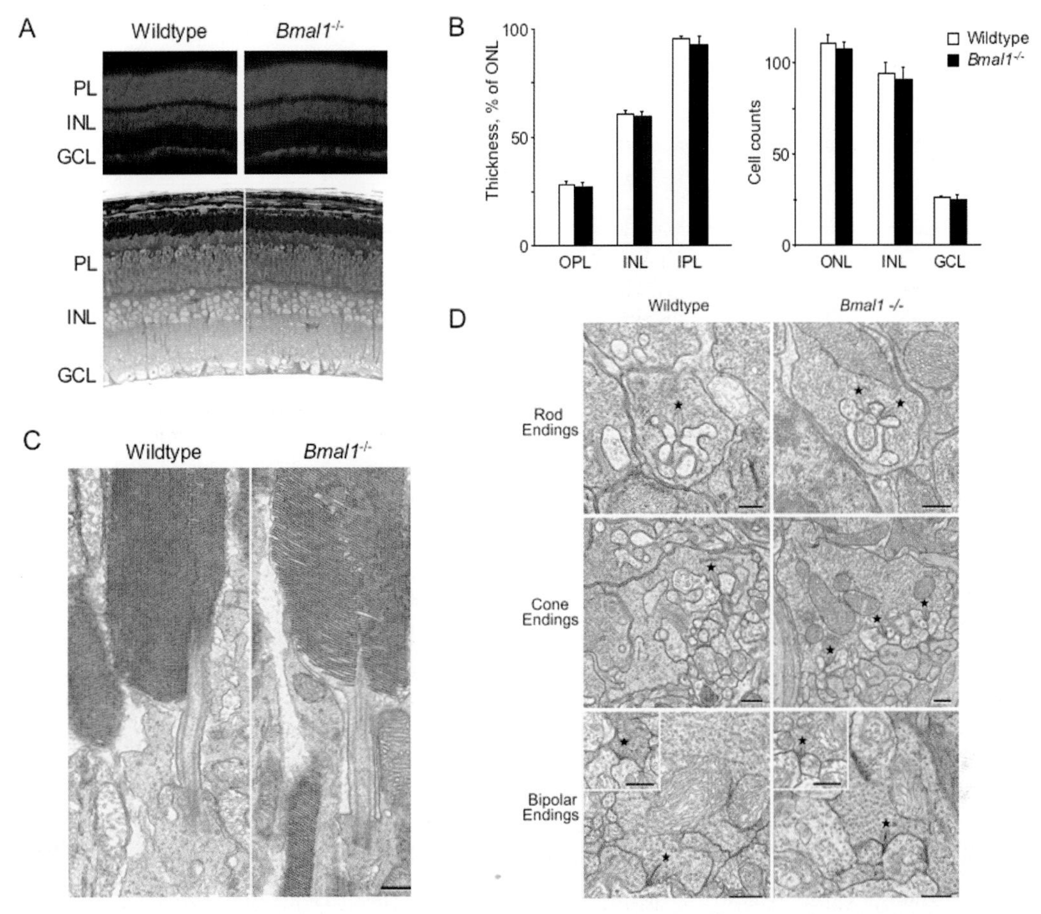

Figure 2. Normal retinal architecture, cellular organization, and ultrastructure in *Bmal1*$^{-/-}$ mutant mice. (*A, Top*) Fluorescence images of retina sections from adult littermate wild-type or *Bmal1*$^{-/-}$ mutant mice stained with ethidium bromide to show cell nuclei. Organization of nuclear layers was indistinguishable in the two genotypes. (*Bottom*) Thick plastic sections of retinas from adult littermate wild-type or *Bmal1*$^{-/-}$ mice stained with toluidine blue to show cell morphology and structure. No difference between genotypes was observed. (PL) Photoreceptor layer; (INL) inner nuclear layer; (GCL) ganglion cell layer. (*B*) Quantitative comparison of thickness (*left*) and cell densities (*right*) (mean and S.E.M.; $N = 3$) of retinal layers (at midperiphery) of adult wild-type and *Bmal1*$^{-/-}$ littermate mice (see Experimental Procedures). Results for central retina were similar (data not shown). There were no significant differences between genotypes. (OPL) Outer plexiform layer; (IPL) inner plexiform layer. Cell counts: for ONL, INL per 5000-?m^2 area; GCL per 200-?m segment of retina. (*C*) Representative electron micrographs of rod photoreceptors from adult littermate wild-type or *Bmal1*$^{-/-}$ mice. No difference between genotypes was observed in the fine structure of outer and inner segments. (*D*) Representative electron micrographs from adult littermate wild-type or *Bmal1*$^{-/-}$ mice. No differences between genotypes were evident in the morphology of synaptic endings of rods, cones, rod bipolars, or cone bipolars (*insets*) or in the complement of synaptic vesicles or structure of ribbon synapses (*asterisks*). (*C,D*) Bars, 500 nm.

detectably rhythmic in DD. These results raise the possibility that a circadian clock broadly regulates transcriptional responses to light in the retina. Among genes with robust rhythms in both DD and LD were multiple histones and at least one histone deacetylase, genes with known actions in chromatin remodeling (Fig. 3D). Thus, it seems plausible that loss of circadian regulation of chromatin might underlie the aberrant transcriptional responses to LD cycles in *Bmal1*$^{-/-}$ mice.

Importance of *Bmal1* for Retinal Electrical Activity in Response to Light

Given the importance of *Bmal1* for retinal gene expression in LD, it seemed plausible that retinal visual physiology might be abnormal in *Bmal1*$^{-/-}$ mice. We therefore performed ERG studies (Lu et al. 2001) to compare retinal electrical activity in response to light in wild-type and *Bmal1*$^{-/-}$ littermate mice (C57BL/6). ERG responses were recorded during midday hours, when the mouse light-adapted ERG b-wave amplitude is at or near the maximum in its daily rhythm (Barnard et al. 2006). Mice of the two genotypes were studied in alternating order to minimize possible effects of circadian time. Under dark-adapted conditions (rod pathway ERG), the amplitude of the ERG a-wave, the summed electrical response of photoreceptors, showed no significant difference between genotypes (Fig. 4A,B). However, the b-wave, the postreceptor electrical response, was distinctly abnormal in *Bmal1*$^{-/-}$ mice, reduced in amplitude by 30% ($P < 0.03$; two-tailed *t*-test; $N = 6$ for each genotype). The selective reduction of b-wave amplitude is underscored by the highly significant

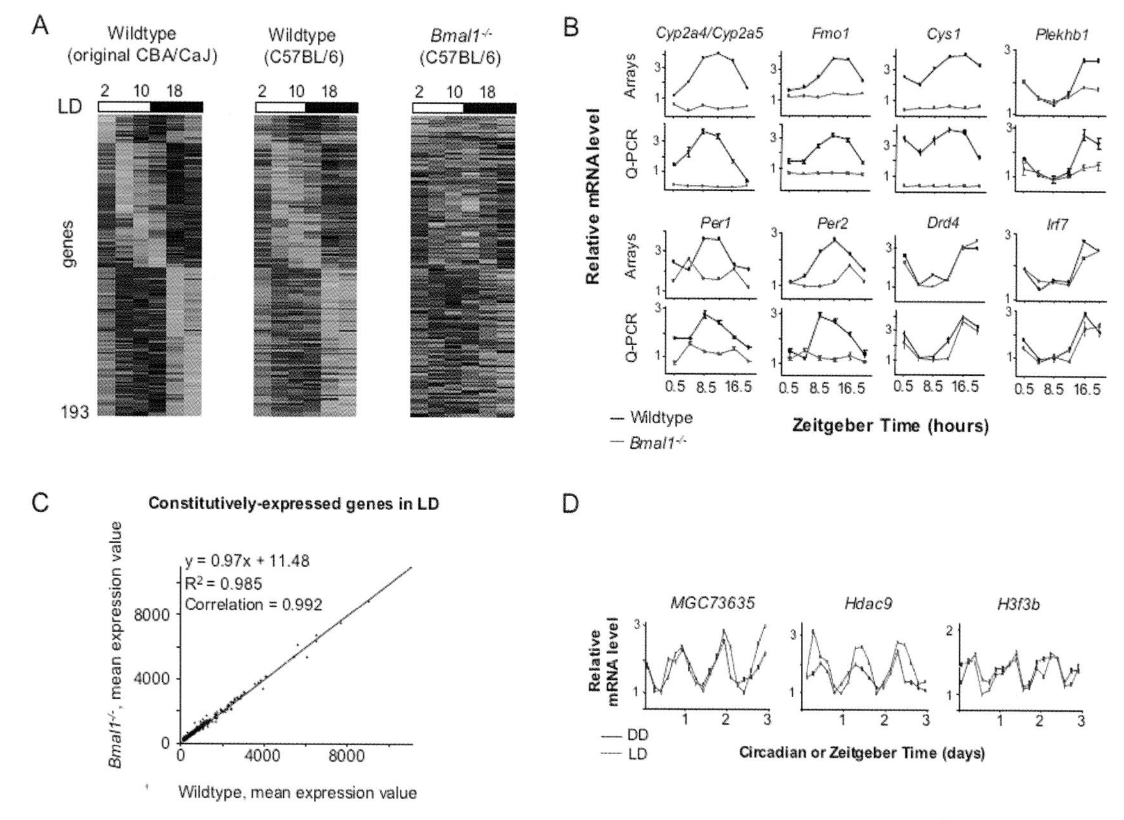

Figure 3. Deficient rhythmic gene expression in LD in the eyes of *Bmal1*$^{-/-}$ mice. (*A*) One-day temporal profiles comparing ocular gene expression in LD for wild-type CBA/CaJ mice (the mean of 3 days; Fig. 1) and wild-type and littermate *Bmal1*$^{-/-}$ mice (C57BL/6); labels are the same as those in Fig. 1. The same genes are shown in all three profiles; genes are ordered by the phase of peak expression in wild-type CBA/CaJ mice (*left*), and the brightness scale for expression is the same in all profiles. (*B*) Validation of microarray profiles by Q-PCR. Shown are examples of genes in *Bmal1*$^{-/-}$ mice with flat, altered, or normal rhythmic expression in comparison with wild-type littermates (C57BL/6). (*Cyp2a4/5*) Cytochrome P450-2a4/5; (*Fmo1*) flavin-containing monooxygenase-1; (*Cys1*) cystin 1; (*Plekh1b*) plekstrin-homology 1b; (*Per1*) Period 1; (*Drd4*) D4 dopamine receptor; (*Irf7*) interferon regulatory factor 7. (*C*) No effect of *Bmal1* deletion on mean expression of 3047 constitutively expressed genes in the eye in LD (see Experimental Procedures). Genes are plotted in order of increasing mean expression from the wild-type data set. (*D*) Daily rhythms of ocular expression of chromatin remodeling genes detected by microarray (Fig. 1). (*MGC73635*) Similar to histone 2a; (*Hdac9*) histone deacetylase 9; (*H3f3b*) H3 histone, family 3B. For *B–D*, relative expression values are plotted in arbitrary linear units.

decrease of the ratio of b-wave to a-wave amplitude in *Bmal1*$^{-/-}$ mice ($P < 0.0005$) (Fig. 4B). Under light-adapted conditions (cone pathway ERG), the reduction of b-wave amplitude in *Bmal1*$^{-/-}$ mice was even greater, 60% ($P < 0.015$) (Fig. 4A,B). Thus, in the mutants, the daytime ERG b-wave response is diminished such that it resembles the lower nighttime amplitude of wild-type mice (Barnard et al. 2006; see below). These results indicate that *Bmal1* is important for inner retinal processing of visual stimuli but not for rod photoreceptor electrical responses.

To determine whether the SCN clock contributes to retinal performance, we made SCN lesions in wild-type mice and compared ERG responses in the behaviorally arrhythmic SCN-lesioned mice and intact controls (C57BL/6). No significant differences were found between groups for a- or b-wave amplitudes under either dark- or light-adapted conditions (Fig. 4C; representative of $N = 5$ for each group). Thus, the SCN circadian clock is not required for normal daytime retinal electrical activity in response to light. Given the ERG abnormality of *Bmal1*$^{-/-}$ mice (Fig. 4A,B), any circadian clock important for retinal physiological function must therefore be outside the SCN.

Retina-specific Deletion of *Bmal1* and Loss of Retinal Circadian Clock Function

To examine the role of *Bmal1* and circadian clock function specifically within the retina, we generated mice with a conditional allele of *Bmal1*. Upon the action of Cre recombinase, the exon encoding the BMAL1 basic helix-loop-helix (bHLH) domain is deleted (Fig. 5A,B), resulting in a mutation nearly identical to the original null mutation in *Bmal1*$^{-/-}$ mice (Bunger et al. 2000). In validation studies, mice homozygous for the conditional allele had circadian behavioral rhythms indistinguishable from that of wild-type littermates, whereas after crossing in a ubiquitously acting Cre transgene (Schwenk et al. 1995), the mice showed a complete loss of circadian behavioral rhythms and emergence of ultradian behavior (Fig. 5C), copying the *Bmal1*$^{-/-}$ phenotype (Bunger et al. 2000). Thus, the *Bmal1* conditional allele has wild-type activity unless acted upon by Cre recombinase, after which it acts as a *Bmal1* null mutation.

To generate a retina-specific *Bmal1* mutation, we obtained a mouse line carrying a CHX10-Cre transgene,

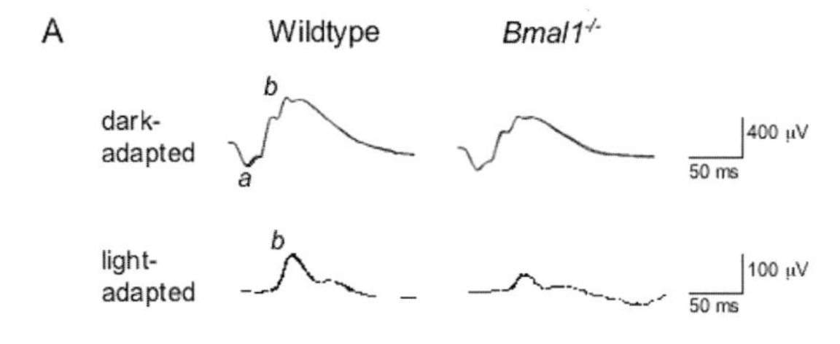

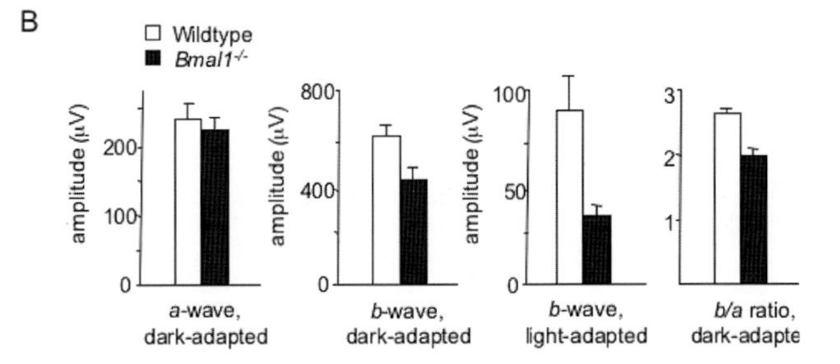

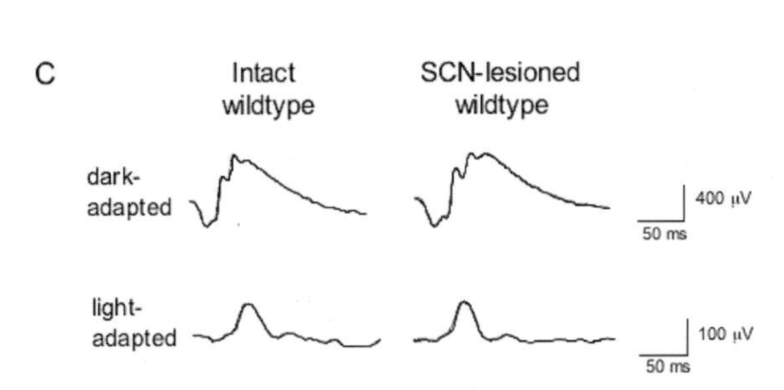

Figure 4. Defective retinal electrical activity in response to light in *Bmal1*$^{-/-}$ mice but not SCN-lesioned wild-type mice. (*A*) Electroretinogram (ERG) traces in response to a flash of light for adult wild-type and littermate *Bmal1*$^{-/-}$ mice under dark- and light-adapted conditions; a and b waves are labeled on wild-type traces. (*B*) Quantification of ERG responses in wild-type and *Bmal1*$^{-/-}$ littermates. Shown are mean and S.E.N.; $N = 6$ for each genotype. (*C*) Representative ERG responses of intact wild-type and SCN-lesioned wild-type mice (each $N = 5$). No significant differences between groups were found for a- or b-wave amplitudes. Mice were C57BL/6, and ERGs were performed between zeitgeber time 4 and 9; wild-type and mutant (or intact and SCN-lesioned) mice were studied in alternating order.

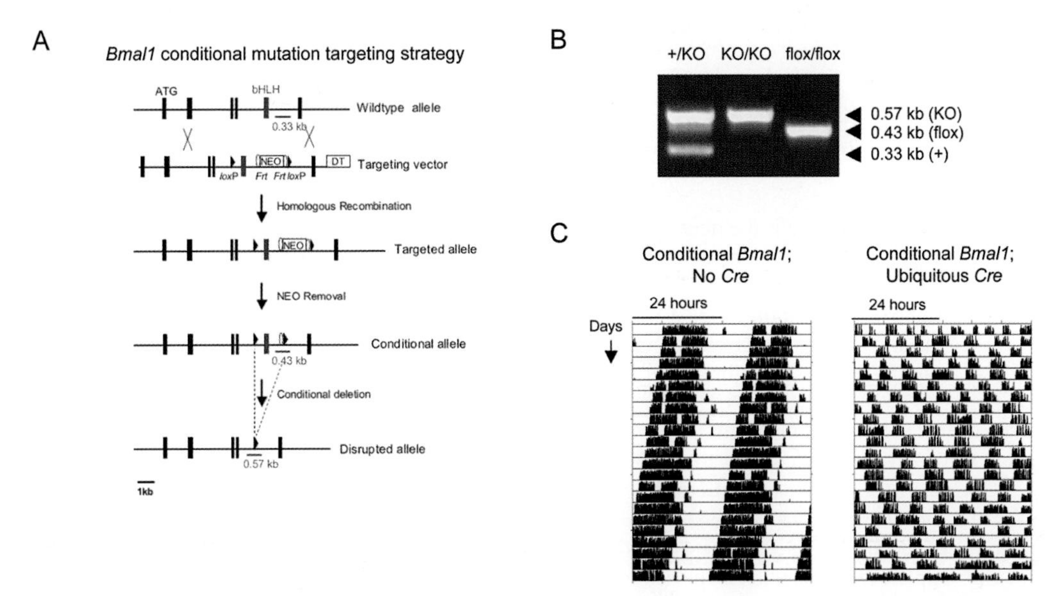

Figure 5. Conditional *Bmal1* allele. (*A*) Targeting strategy and conditional disruption. (*Closed boxes*) Exons; (ATG) translation start site; (bHLH) exon encoding basic helix-loop-helix domain; (NEO) neomycin resistance marker; (DT) diphtheria toxin cassette; (*triangles*) *loxP* sites; (*ovals*) Frt sites. Bars with kilobase (kb) markers are sites and sizes of PCR products diagnostic of genotypes. (*B*) PCR products amplified from mouse tail DNA demonstrating the indicated genotypes. (+) Wild-type allele; (KO) disrupted allele (from ubiquitously acting Cre); (flox) conditional allele. (*C*) Functional validation of *Bmal1* condition allele: conditional loss of circadian rhythms of locomotor activity. Shown are representative double-plotted records of wheel-running activity in DD of mice homozygous for the conditional *Bmal1* allele with or without a ubiquitously acting Cre. Tick mark heights correspond to the number of running-wheel revolutions in a 6-minute bin.

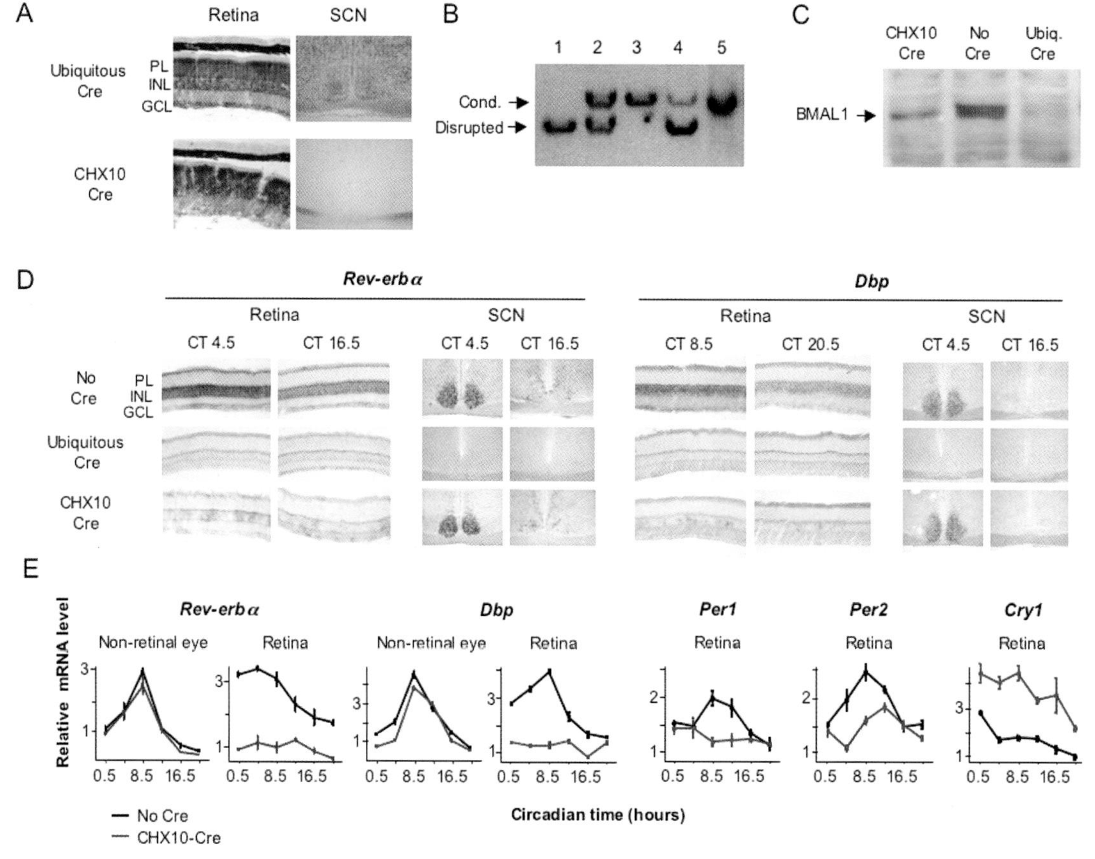

Figure 6. Retina-specific loss of *Bmal1* function. (*A*) Retina-specific Cre activity from CHX10-Cre transgene. Sections of retina and SCN from indicator mice showing blue precipitate reporting Cre recombinase activity. For CHX10-Cre, blue stain at bottom of the brain is from retinal ganglion cell projections. (PL) Photoreceptor layer; (INL) inner nuclear layer; (GCL) ganglion cell layer. (*B*) Retina-specific disruption of *Bmal1* conditional allele. Genomic Southern showing fragments diagnostic of the conditional or disrupted *Bmal1* alleles. (*1–4*) Retina DNA; (*1*) homozygous *Bmal1* conditional, ubiquitous Cre; (*2*) heterozygous for disrupted allele; (*3*) homozygous *Bmal1* conditional, no Cre; (*4*) homozygous *Bmal1* conditional, CHX10-Cre; (*5*) hypothalamus DNA from same mouse as in *4*. (*C*) Loss of BMAL1 protein from the retina. Anti-BMAL1 western blot of protein extracts from retinas of homozygous conditional *Bmal1* mice with the indicated Cre. (*D*) Loss of *Bmal1* function in retina but not SCN. In situ hybridization to retina and SCN sections showing expression of *Bmal1*-dependent genes *Rev-erbα* and *Dbp* in homozygous conditional *Bmal1* mice carrying the indicated Cre transgene. Circadian times (CT) correspond to the peak (*left*) or trough (*right*) of transcript rhythms. (*E*) Loss of molecular rhythms in retina. Temporal expression profiles (Q-PCR) of the indicated genes in mice homozygous for the conditional *Bmal1* allele with or without CHX10-Cre. Relative expression levels are plotted in arbitrary linear units.

previously shown to act throughout the neural retina but to have little or no activity in the brain or other tissues (Rowan and Cepko 2004). Crossing CHX10-Cre to an indicator line (Soriano 1999) demonstrated Cre activity throughout the retina and little or no activity in nonretinal ocular tissues or the SCN and other brain regions, as expected (Fig. 6A).

We next generated mice homozygous for the *Bmal1* conditional allele that carry a single copy of the CHX10-Cre transgene. As expected, these mice showed a retina-specific disruption of the *Bmal1* conditional allele (Fig. 6B). The residual nonrecombined allele (Fig. 6B, lane 4) is likely the result of trace incomplete recombination in the neural retina and to DNA from retinal vasculature and contaminating retinal pigment epithelium, in which CHX10-Cre is not expected to act. In addition, the mice showed the expected substantial loss of BMAL1 protein from the retina (Fig. 6C), with residual protein also likely derived from trace incomplete recombination in the neural retina and from retinal vascular and pigment epithelial cells. Interestingly, the residual BMAL1 appears to comprise only the faster-migrating form of the protein (Fig. 6C), suggesting the possibility that BMAL1 has different posttranslational modifications in neural and nonneural retinal cells.

To examine *Bmal1* function, we monitored transcripts for *Rev-erbα* and *Dbp*, clock-regulated genes that depend on CLOCK-BMAL1 activity for expression (Ripperger et al. 2000). In conditional *Bmal1* mice without Cre, in situ hybridization demonstrated the expected circadian rhythm of expression of *Rev-erbα* and *Dbp* in both the retina and the SCN (Fig. 6D). With ubiquitously acting Cre, expression of *Rev-erbα* and *Dbp* was lost in both the retina and the SCN (Fig. 6D). With CHX10-Cre, *Rev-erbα* and *Dbp* showed loss of expression in the retina, except for variable small patches showing weak expression, whereas the SCN showed normal circadian rhythms of expression for both genes (Fig. 6D). As expected, circadian rhythms of behavior were normal (data not shown). The patchy residual retinal expression of *Rev-erbα* and *Dbp* is consistent with the residual nonrecombined allele in the retina (Fig. 6B) and the trace retinal mosaicism previously reported for CHX10-Cre (Rowan and Cepko 2004).

In conditional *Bmal1* mice without Cre, Q-PCR demonstrated a circadian profile of *Rev-erbα* and *Dbp* expression in both the retina and nonretinal ocular tissues, as expected, whereas with CHX10-Cre, only nonretinal ocular tissues showed a circadian rhythm (Fig. 6E). In addition, retinal rhythmic profiles of *Per1* and *Per2* expression were defective (Fig. 6E), and *Cry1*, often weakly rhythmic or noisy in non-SCN sites (Storch et al. 2002), was abnormally high (Fig. 6E), as expected in the absence of *Bmal1* (Kondratov et al. 2006). Together, these results indicate that conditional *Bmal1* mice have wild-type *Bmal1* function and robust retinal circadian rhythms, whereas in the presence of CHX10-Cre, the mice show a retina-specific loss of *Bmal1* function that does not support demonstrable retinal circadian rhythms. It is possible, however, that some cells within patches of residual retinal *Bmal1* function retain rhythms.

Importance of Retinal *Bmal1* for Circadian Rhythm of Retinal Responses to Light

To test the importance of retinal *Bmal1* for retinal physiology in vivo, we first compared daytime ERG responses of homozygous conditional *Bmal1* mice (controls) and littermate homozygous conditional *Bmal1* mice with a single copy of CHX10-Cre (*Ret-Bmal1*$^{-/-}$; all mice were C57BL/6 x 129 hybrids). Under dark-adapted conditions, the amplitude of the a-wave showed no significant difference between genotypes (Fig. 7A,B). In contrast, the amplitude of the b-wave was reduced by 27% in *Ret-Bmal1*$^{-/-}$ mice (Fig. 7A,B) ($P < 0.012$; two-tailed *t*-test; $N = 7$ for each genotype), producing a highly significant reduction in the ratio of b-wave to a-wave amplitude ($P < 0.0004$) (Fig. 7B). Under light-adapted conditions, the reduction in b-wave amplitude in *Ret-Bmal1*$^{-/-}$ mice was even greater, 44% ($P < 0.018$) (Fig. 7A,B). Despite the difference in genetic background, the ERG phenotype of *Ret-Bmal1*$^{-/-}$ mice was essentially identical to that of *Bmal1*$^{-/-}$ mice (Fig. 4A,B), indicating that retinal *Bmal1* function fully accounts for the role of *Bmal1* in retinal electrical responses to light.

Recent work demonstrates that robust circadian regulation of the mouse light-adapted ERG b-wave amplitude is revealed under constant light (LL) conditions (Barnard et al. 2006). To determine the importance of retinal *Bmal1* function for this rhythm, we compared light-adapted ERG responses of conditional *Bmal1* controls and *Ret-Bmal1*$^{-/-}$ littermates in LL at times of high (CT6) or low (CT18) b-wave amplitude (it is not yet known if these circadian times represent the true peak and trough of the rhythm). Controls showed the expected circadian rhythm of b-wave responses (Fig. 7C,D), with significantly higher amplitudes at CT6 than CT18 (Fig. 7D). In contrast, *Ret-Bmal1*$^{-/-}$ mice had no detectable circadian rhythm of b-wave amplitude, with responses at both circadian times apparently fixed at a low amplitude, somewhat lower than the amplitude at CT18 in controls (Fig. 7C,D). In addi-

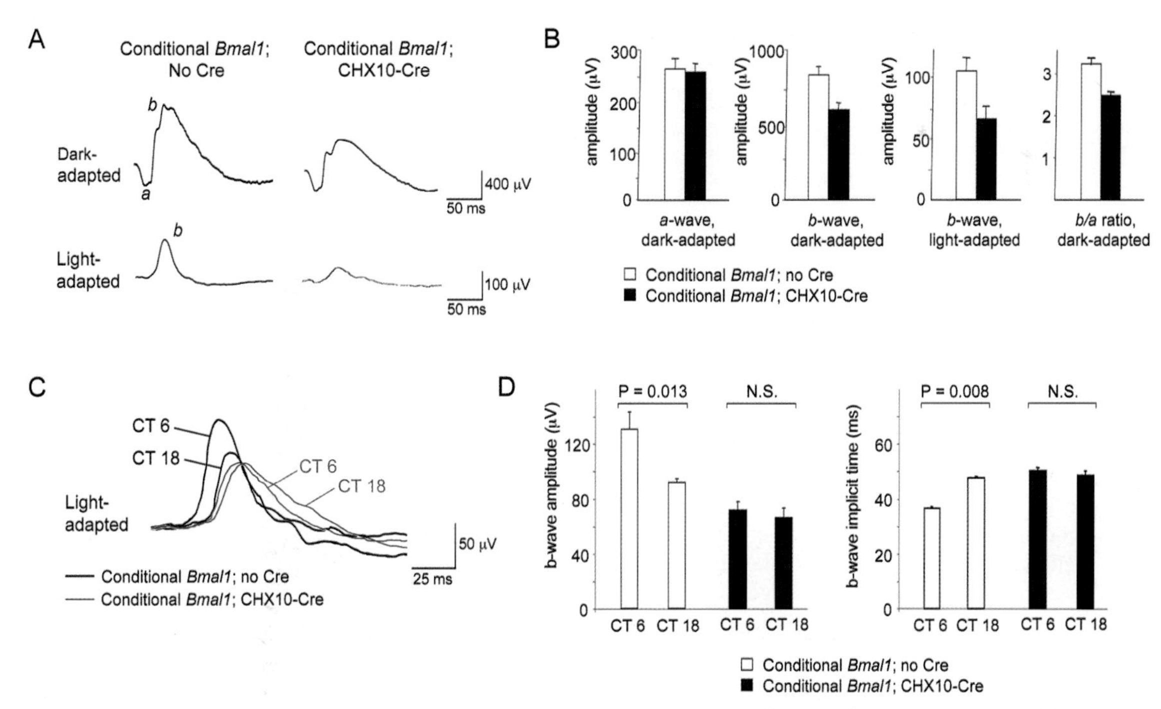

Figure 7. Loss of circadian rhythm of inner retinal electrical activity in response to light in mice with a retina-specific deletion of *Bmal1*. (*A*) Examples of daytime ERG traces in response to a flash of light for conditional *Bmal1* littermates with or without the CHX10-Cre. (*B*) Quantification of ERG responses (mean and S.E.M.; $N = 7$ for each genotype). (*C*) Examples of ERG traces in response to a flash of light for conditional *Bmal1* littermates with or without CHX10-Cre at two circadian times (CT) in constant light (LL). (*D*) Quantification of ERG b-wave amplitudes and implicit times (mean and S.E.M.; $N = 8$ for controls and $N = 7$ for *ret-Bmal1*). *P* values, *t*-tests. (N.S.) Not significant.

tion, control mice exhibited a circadian rhythm of b-wave implicit time (the interval between the flash of light and the peak of the b-wave), a rhythm that has also been shown to be robust in DD (Barnard et al. 2006). In *Ret-Bmal1*$^{-/-}$ mice, this rhythm was undetectable, and b-wave implicit times appeared to be fixed at a value similar to CT18 in controls, much like b-wave amplitude (Fig. 7C,D). These results indicate that *Bmal1* function in the retina is required for circadian rhythms of inner retinal processing of visual stimuli. More specifically, retinal *Bmal1* function is required to promote inner retinal visual processing during the subjective day, strongly suggesting that the rhythm in retinal electrophysiological function is driven by an intrinsic retinal circadian clock.

DISCUSSION

Our results indicate that in the mammalian eye, hundreds of genes, many of which are expressed in the retina, are controlled by a circadian clock. We found that thousands more are regulated rhythmically in LD cycles, including a large cluster of more than 1000 genes with a nighttime peak of expression. The rhythmically expressed genes are associated with a wide range of functions, and the findings provide a rich context in which to investigate molecular processes underlying the impact of circadian clock function and light on retinal physiology and metabolism. Although less pronounced than that reported here, an excess of transcripts with daily cycles in LD compared to DD has been observed in chick retina (Bailey et al. 2004) and in *Drosophila* head (Ceriani et al. 2002; Lin et al. 2002; Wijnen et al. 2006), a structure predominantly composed of the eyes.

In *Bmal1*$^{-/-}$ mice, a large fraction of genes that normally show rhythmic ocular expression exclusively in LD exhibited nonrhythmic expression or reduced amplitude. In contrast, constitutively expressed genes showed unaltered expression in *Bmal1*$^{-/-}$ mice. These results indicate that *Bmal1* has an important role in light-dependent but not global gene regulation in the retina. Although we cannot exclude a possible noncircadian role of *Bmal1* in transcriptional responses to light, several lines of evidence suggest that defective circadian regulation likely underlies the phenotype. Circadian control of light-dependent gene induction in the mammalian retina has been documented (Masana et al. 1996; Humphries and Carter 2004), regulation of chromatin by histone H3 has been implicated in light-dependent gene induction in the SCN (Crosio et al. 2000), and our finding that multiple histones (including histone H3 family members) and other chromatin remodeling genes exhibit circadian expression in the eye suggests that chromatin in the retina is under clock control. Together, these observations suggest that regulators of chromatin state function in the retina as clock outputs in the control of transcriptional responses to light and that in the absence of *Bmal1* one or more of these rhythmic regulators is deficient.

Our results indicate that *Bmal1* function within the retina is required for circadian rhythms of inner retinal visual processing, specifically for the circadian rhythms of the light-adapted ERG b-wave amplitude and b-wave implicit time. In the absence of retinal *Bmal1*, both rhythms appear to be fixed at or near the low values typical for subjective night, indicating that retinal *Bmal1* function is required for the facilitation of inner retinal visual processing during subjective day. This phenotype is strikingly similar to the unexpected ERG phenotype recently described for mice lacking melanopsin (Barnard et al. 2006), the photopigment of photoreceptive ganglion cells mediating circadian entrainment to LD cycles (Hattar et al. 2003). The essentially identical loss of circadian rhythms of retinal processing in two independent retina-specific mutations affecting the circadian system makes it very likely that circadian clock function within the retina directly drives rhythms of retinal visual physiology, and it provides support for the idea that melanopsin activity is required to synchronize a population of retinal clock cells controlling inner retinal electrical responses to light (Barnard et al. 2006).

A mild ERG b-wave amplitude defect has been reported for mice lacking the D4 dopamine receptor (Nir et al. 2002). Given the importance of dopamine in regulating inner retinal network properties (Gustincich et al. 1997) and evidence that retinal dopaminergic amacrine cells are circadian clock cells (Witkovsky et al. 2003; Gustincich et al. 2004; Ruan et al. 2006; Storch et al. 2007), it is possible that the loss of clock function within dopaminergic amacrine cells contributes to the ERG phenotype of *Ret-Bmal1*$^{-/-}$ mice.

Circadian clocks in mammals are widely distributed, but except for the SCN clock known to regulate behavior, their physiological functions in vivo are largely mysterious. The studies described here indicate that an intrinsic retinal circadian clock regulates retinal visual processing in vivo and that it does so autonomously, with no detectable contribution from the SCN or other clocks. Together with observations emerging from other tissue-specific manipulations of clock function (Durgan et al. 2006; McDearmon et al. 2006), our work provides evidence that circadian clocks outside the SCN contribute important physiological functions in vivo. Thus, over evolutionary time, different cell types have likely recruited the circadian clock mechanism inherited from a single-celled ancestor for control of specialized tissue-specific processes.

ACKNOWLEDGMENTS

We thank Susan Dymecki and Connie Cepko for data on CHX10-Cre; Christopher Bradfield for *Bmal1*$^{-/-}$ (*Mop3*$^{-/-}$) mice; Ueli Schibler for anti-BMAL1 antiserum; the Gene Manipulation Core of the Developmental Disabilities Research Center of Children's Hospital, Boston for expert service, and Ming Liu and Xiao Ling Long for expert technical assistance. This work was supported by grants from the National Institutes of Health to C.J.W. (NS055831), E.R. (EY001344), and T.L. (EY10309).

REFERENCES

Abe M., Herzog E.D., Yamazaki S., Straume M., Tei H., Sakaki Y., Menaker M., and Block G.D. 2002. Circadian rhythms in isolated brain regions. *J. Neurosci.* **22:** 350.

Akhtar R.A., Reddy A.B., Maywood E.S., Clayton J.D., King

V.M., Smith A.G., Gant T.W., Hastings M.H., and Kyriacou C.P. 2002. Circadian cycling of the mouse liver transcriptome, as revealed by cDNA microarray, is driven by the suprachiasmatic nucleus. *Curr. Biol.* **12:** 540.

Bailey M.J., Beremand P.D., Hammer R., Reidel E., Thomas T.L., and Cassone V.M. 2004. Transcriptional profiling of circadian patterns of mRNA expression in the chick retina. *J. Biol. Chem.* **279:** 52247.

Barnard A.R., Hattar S., Hankins M.W., and Lucas R.J. 2006. Melanopsin regulates visual processing in the mouse retina. *Curr. Biol.* **16:** 389.

Balsalobre A., Damiola F., and Schibler U. 1998. A serum shock induces circadian gene expression in cultured Rat-1 fibroblasts. *Cell* **93:** 929.

Bardeesy N., Sinha M., Hezel A.F., Signoretti S., Hathaway N.A., Sharpless N.E., Loda M., Carrasco D.R., and DePinho R.A. 2002. Loss of the Lkb1 tumour suppressor provokes intestinal polyposis but resistance to transformation. *Nature* **419:** 162.

Besharse J.C. and Iuvone P.M. 1983. Circadian clock in *Xenopus* eye controlling retinal serotonin N-acetyltransferase. *Nature* **305:** 133.

Bunger M.K., Wilsbacher L.D., Moran S.M., Clendenin C., Radcliffe L.A., Hogenesch J.B., Simon M.C., Takahashi J.S., and Bradfield C.A. 2000. Mop3 is an essential component of the master circadian pacemaker in mammals. *Cell* **103:** 1009.

Cahill G.M. and Besharse J.C. 1993. Circadian clock functions localized in *Xenopus* retinal photoreceptors. *Neuron* **10:** 573.

Ceriani M.F., Hogenesch J.B., Yanovsky M., Panda S., Straume M., and Kay S.A. 2002. Genome-wide expression analysis in *Drosophila* reveals genes controlling circadian behavior. *J. Neurosci.* **22:** 9305.

Crosio C., Cermakian N., Allis C.D., and Sassone-Corsi P. 2000. Light induces chromatin modification in cells of the mammalian circadian clock. *Nat. Neurosci.* **3:** 1241.

Damiola F., Le Minh N., Preitner N., Kornmann B., Fleury-Olela F., and Schibler U. 2000. Restricted feeding uncouples circadian oscillators in peripheral tissues from the central pacemaker in the suprachiasmatic nucleus. *Genes Dev.* **14:** 2950.

Doyle S.E., Grace M.S., McIvor W., and Menaker M. 2002. Circadian rhythms of dopamine in mouse retina: The role of melatonin. *Vis. Neurosci.* **19:** 593.

Duffield G.E., Best J.D., Meurers B.H., Bittner A., Loros J.J., and Dunlap J.C. 2002. Circadian programs of transcriptional activation, signaling, and protein turnover revealed by microarray analysis of mammalian cells. *Curr. Biol.* 12: 551.

Durgan D.J., Trexler N.A., Egbejimi O., McElfresh T.A., Suk H.Y., Petterson L.E., Shaw C.A., Hardin P.E., Bray M.S., Chandler M.P., Chow C.W., and Young M.E. 2006. The circadian clock within the cardiomyocyte is essential for responsiveness of the heart to fatty acids. *J. Biol. Chem.* **281:** 24254.

Farley F.W., Soriano P., Steffen L.S., and Dymecki S.M. 2000. Widespread recombinase expression using FLPeR (flipper) mice. *Genesis* **28:** 106.

Gekakis N., Staknis D., Nguyen H.B., Davis F.C., Wilsbacher L.D., King D.P., Takahashi J.S., and Weitz C.J. 1998. Role of the CLOCK protein in the mammalian circadian mechanism. *Science* **280:** 1564.

Goto M., Oshima I., Tomita T., and Ebihara S. 1989. Melatonin content of the pineal gland in different mouse strains. *J. Pineal Res.* **7:** 195.

Gustincich S., Feigenspan A., Wu D.K., Koopman L.J., and Raviola E. 1997. Control of dopamine release in the retina: A transgenic approach to neural networks. *Neuron* **18:** 723.

Gustincich S., Contini M., Gariboldi M., Puopolo M., Kadota K., Bono H., LeMieux J., Walsh P., Carninci P., Hayashizaki Y., Okazaki Y., and Raviola E. 2004. Gene discovery in genetically labeled single dopaminergic neurons of the retina. *Proc. Natl. Acad. Sci.* **101:** 5069.

Hattar S., Lucas R.J., Mrosovsky N., Thompson S., Douglas R.H., Hankins M.W., Lem J., Biel M., Hofmann F., Foster R.G., and Yau K.W. 2003. Melanopsin and rod-cone photoreceptive systems account for all major accessory visual functions in mice. *Nature* **424:** 76.

Hayasaka N., LaRue S.I., and Green C.B. 2002. In vivo disruption of *Xenopus* CLOCK in the retinal photoreceptor cells abolishes circadian melatonin rhythmicity without affecting its production levels. *J. Neurosci.* **22:** 1600.

Humphries A. and Carter D.A. 2004. Circadian dependency of nocturnal immediate-early protein induction in rat retina. *Biochem. Biophys. Res. Commun.* **320:** 551.

Ko C.H. and Takahashi J.S. 2006. Molecular components of the mammalian circadian clock. *Hum. Mol. Genet.* **15:** R271.

Kondratov R.V., Shamanna R.K., Kondratova A.A., Gorbacheva V.Y., and Antoch M.P. 2006. Dual role of the CLOCK/BMAL1 circadian complex in transcriptional regulation. *FASEB J.* **20:** 530.

Kraves S. and Weitz C.J. 2006. A role for cardiotrophin-like cytokine in the circadian control of mammalian locomotor activity. *Nat. Neurosci.* **9:** 212.

LaVail M.M. and Ward P.A. 1978. Studies on the hormonal control of circadian outer segment disc shedding in the rat retina. *Invest. Ophthalmol. Vis. Sci.* **17:** 1189.

Li C. and Wong W.H. 2001. Model-based analysis of oligonucleotide arrays: Model validation, design issues and standard error application. *Genome Biol.* **2:** research0032.1.

———. 2003. DNA-Chip Analyzer (dChip). In *The analysis of gene expression data: Methods and software* (ed. G. Parmigiani et al.), p. 120. Springer, New York.

Lin Y., Han M., Shimada B., Wang L., Gibler T.M., Amarakone A., Awad T.A., Stormo G.D., Van Gelder R.N., and Taghert P.H. 2002. Influence of the period-dependent circadian clock on diurnal, circadian, and aperiodic gene expression in *Drosophila melanogaster*. *Proc. Natl. Acad. Sci.* **99:** 9562.

Liu P., Jenkins N.A., and Copeland N.G. 2003. A highly efficient recombineering-based method for generating conditional knockout mutations. *Genome Res.* **13:** 476.

Lu C., Peng Y.W., Shang J., Pawlyk B.S., Yu F., and Li T. 2001. The mammalian retinal degeneration B2 gene is not required for photoreceptor function and survival. *Neuroscience* **107:** 35.

Manglapus M.K., Uchiyama H., Buelow N.F., and Barlow R.B. 1998. Circadian rhythms of rod-cone dominance in the Japanese quail retina. *J. Neurosci.* **18:** 4775.

Masana M.I., Benloucif S., and Dubocovich M.L. 1996. Light-induced c-fos mRNA expression in the suprachiasmatic nucleus and the retina of C3H/HeN mice. *Brain Res. Mol. Brain. Res.* **42:** 93.

McDearmon E.L., Patel K.N., Ko C.H., Walisser J.A., Schook A.C., Chong, J.L., Wilsbacher L.D., Song E.J., Hong H.K., Bradfield C.A., and Takahashi J.S. 2006. Dissecting the functions of the mammalian clock protein BMAL1 by tissue-specific rescue in mice. *Science* **314:** 1304.

McGoogan J.M. and Cassone V.M. 1999. Circadian regulation of chick electroretinogram: Effects of pinealectomy and exogenous melatonin. *Am. J. Physiol.* **277:** R1418.

Miranda-Anaya M., Bartell P.A., and Menaker M. 2002. Circadian rhythm of iguana electroretinogram: The role of dopamine and melatonin. *J. Biol. Rhythms* **17:** 526.

Nir I., Harrison J.M., Haque R., Low M.J., Grandy D.K., Rubinstein M., and Iuvone P.M. 2002. Dysfunctional light-evoked regulation of cAMP in photoreceptors and abnormal retinal adaptation in mice lacking dopamine D4 receptors. *J. Neurosci.* **22:** 2063.

Ouyang Y., Andersson C.R., Kondo T., Golden S.S., and Johnson C.H. 1998. Resonating circadian clocks enhance fitness in cyanobacteria. *Proc. Natl. Acad. Sci.* **95:** 8660.

Panda S. and Hogenesch J.B. 2004. It's all in the timing: Many clocks, many outputs. *J. Biol. Rhythms* **19:** 374.

Panda S., Antoch M.P., Miller B.H., Su A.I., Schook A.B., Straume M., Schultz P.G., Kay S.A., Takahashi J.S., and Hogenesch J.B. 2002. Coordinated transcription of key pathways in the mouse by the circadian clock. *Cell* **109:** 307.

Pierce M.E., Sheshberadaran H., Zhang Z., Fox L.E., Applebury M.L., and Takahashi J.S. 1993. Circadian regulation of iodopsin gene expression in embryonic photoreceptors in retinal cell culture. *Neuron* **10:** 579.

Ripperger J.A., Shearman L.P., Reppert S.M., and Schibler U. 2000. CLOCK, an essential pacemaker component, controls expression of the circadian transcription factor DBP. *Genes Dev.* **14:** 679.

Rowan S. and Cepko C.L. 2004. Genetic analysis of the homeodomain transcription factor Chx10 in the retina using a novel multifunctional BAC transgenic mouse reporter. *Dev. Biol.* **271:** 388.

Ruan G.X., Zhang D.Q., Zhou T., Yamazaki S., and McMahon D.G. 2006. Circadian organization of the mammalian retina. *Proc. Natl. Acad. Sci.* **103:** 9703.

Schwenk F., Baron U., and Rajewsky K. 1995. A cre-transgenic mouse strain for the ubiquitous deletion of loxP-flanked gene segments including deletion in germ cells. *Nucleic Acids Res.* **23:** 5080.

Soriano P. 1999. Generalized lacZ expression with the ROSA26 Cre reporter strain. *Nat. Genet.* **21:** 70.

Storch K.F., Lipan O., Leykin I., Viswanathan N., Davis F.C., Wong W.H., and Weitz C.J. 2002. Extensive and divergent circadian gene expression in liver and heart. *Nature* **417:** 78.

Storch K.F., Paz C., Signorovitch J., Raviola E., Pawlyk B., Li T., and Weitz C.J. 2007. Intrinsic circadian clock of the mammalian retina: Importance for retinal processing of visual information. *Cell* **130:** 730.

Storey J.D., Taylor J.E., and Siegmund D. 2004. Strong control, conservative point estimation and simultaneous conservative consistency of false discovery rates: A unified approach. *J. R. Stat. Soc. Ser. B* **66:** 187.

Tanoue S., Krishnan P., Krishnan B., Dryer S.E., and Hardin P.E. 2004. Circadian clocks in antennal neurons are necessary and sufficient for olfaction rhythms in *Drosophila*. *Curr. Biol.* **14:** 638.

Tosini G. and Menaker M. 1996. Circadian rhythms in cultured mammalian retina. *Science* **272:** 419.

Wijnen H., Naef F., Boothroyd C., Claridge-Chang A., and Young M.W. 2006. Control of daily transcript oscillations in *Drosophila* by light and circadian clock. *PLoS Genet.* **2:** e39.

Witkovsky P., Veisenberger E., LeSauter J., Yan L., Johnson M., Zhang D.Q., McMahon D., and Silver R. 2003. Cellular location and circadian rhythm of expression of the biological clock gene Period 1 in the mouse retina. *J. Neurosci.* **23:** 7670.

Yamazaki S., Numano R., Abe M., Hida A., Takahashi R., Ueda M., Block G.D., Sakaki Y., Menaker M., and Tei H. 2000. Resetting central and peripheral circadian oscillators in transgenic rats. *Science* **288:** 682.

Regulation of Circadian Gene Expression in Liver by Systemic Signals and Hepatocyte Oscillators

B. KORNMANN,*† O. SCHAAD,‡ H. REINKE,* C. SAINI,* AND U. SCHIBLER*

*Departments of Molecular Biology and NCCR Frontiers in Genetics, Sciences III, University of Geneva 30, CH-1211 Geneva-4, Switzerland; †Howard Hughes Medical Institute/University of California, San Francisco, Genentech Hall N316-600, San Francisco, California 94158-2517; ‡Departments of Biochemistry and NCCR Frontiers in Genetics, Sciences II, University of Geneva 30, CH-1211 Geneva-4, Switzerland

The mammalian circadian timing system has a hierarchical structure, in that a master pacemaker located in the suprachiasmatic nuclei (SCN) coordinates slave oscillators present in virtually all body cells. In both the SCN and peripheral organs, the rhythm-generating oscillators are self-sustained and cell-autonomous, and it is likely that the molecular makeup of master and slave oscillators is nearly identical. However, due to variations in period length, the phase coherence between peripheral oscillators in intact animals must be established by daily signals emanating directly or indirectly from the SCN master clock. The synchronization of individual cellular clocks in peripheral organs is probably accomplished by immediate-early genes that interpret the cyclic systemic signals and convey this phase information to core clock components. This model predicts that circadian gene expression in peripheral organs can be influenced either by systemic signals emanating from the SCN master clock, local oscillators, or both. We developed a transgenic mouse strain in which hepatocyte clocks are only operative when the tetracycline analog doxycycline is added to the food or drinking water. The genome-wide mapping of genes whose cyclic expression in liver does not depend on functional hepatocyte oscillators unveiled putative signaling pathways that may participate in the phase entrainment of peripheral clocks.

INTRODUCTION

In mammals, virtually all aspects of physiology and behavior are influenced by the circadian clock (for review, see Gachon et al. 2004). According to knowledge gained during the past 10 years on the subject, the term "clock" is actually somewhat misleading, because nearly every body cell contains its own circadian oscillator. Thus, in organ explants or serum-shocked fibroblasts kept in tissue-culture medium, the expression of clock genes oscillates with a period length close to 24 hours (Balsalobre et al. 1998; Yamazaki et al. 2000; Nagoshi et al. 2004; Welsh et al. 2004; Yoo et al. 2004). The molecular makeup of the clockwork circuitry is probably shared by SCN neurons and cultured fibroblasts. In both, circadian oscillators function in a self-sustained and cell-autonomous manner, the phase relationship between the expression cycles of different clock genes is practically identical, and clock gene mutations elicit related phenotypes (Yagita et al. 2001; Brown et al. 2005). There is an important difference, however, when these cellular oscillators are studied in the tissue context. Although the clocks of SCN neurons are tightly coupled, those of peripheral cells do not appear to communicate with each other to any significant extent. As a consequence, neurons of organotypic SCN slice cultures maintain phase coherence, whereas oscillators of cultured peripheral cells or tissue explants rapidly desynchronize (Liu et al. 2007). Likewise, the cellular oscillators of peripheral tissues are quickly put out-of-phase in SCN-lesioned animals, suggesting that they must be phase-entrained daily by signals emanating from the SCN (Yoo et al. 2004; Guo et al. 2006). As outlined below, the SCN master pacemaker, which itself is synchronized by the photoperiod, appears to set the phase in peripheral oscillators by employing a multitude of different timing cues (zeitgebers). Elegant parabiosis studies by Bittman and coworkers suggest that at least for liver and kidney cells, some of these zeitgebers are probably blood-borne signals, such as hormones and metabolites (Guo et al. 2005). Thus, in a parabiosed pair composed of an intact and an SCN-lesioned hamster, circadian liver and kidney gene expression is synchronized in both parabiosed partners. Work by the same group also suggests that the SCN-phase entrains peripheral timekeepers on the single cell rather than the organ level, at least in liver (Guo et al. 2006). In this study, the authors experimentally scrutinized the following strong prediction: If phase coherence was maintained within the liver of SCN-lesioned animals, the livers of different SCN-lesioned animals should display vastly different ratios of mRNAs issued from antiphasically expressed genes (e.g., *Per1* and *Bmal1*). Yet, the mean values of *Per1* mRNA to *Bmal1* mRNA ratios were found to be very similar for all examined animals, and the standard deviations of these mean values were quite small. This result is only compatible with a scenario in which individual hepatocytes within an SCN-lesioned animal do not exhibit significant phase preference.

The coexistence of cyclic systemic cues and local oscillators implies that circadian transcription in peripheral tissues could be regulated by two different—albeit not mutually exclusive—mechanisms. One class of rhythmically expressed genes, dubbed systemically driven genes (Kornmann et al. 2007), may be under the direct control of systemic signals, such as diurnal hormones. For example, glucocorticoid-responsive genes would be expected to be cyclically transcribed, because the plasma levels of the glucocorticoid receptor ligands cortisol or corticosterone oscillate with a robust daily amplitude (Lejeune-Lenain et al. 1987 and references therein). The transcription of another class of genes, perhaps best described as cell-autonomously clock-controlled genes, may be driven by core components of local oscillators, such as BMAL1, CLOCK, CRYs, and PERs, or transcription factors such as the PAR basic leucine zipper (bZIP) proteins DBP, HLF, and TEF, whose circadian accumulation and/or activity depends on these factors. As argued later in this chapter, the discrimination between systemically driven and clock-controlled genes is not just a matter of semantics. The first class of genes may actually contain immediate-early genes that participate in the synchronization of peripheral clocks, and the identification of such genes may therefore provide important insight into the signaling cascades involved in this process.

FEEDING/FASTING CYCLES ARE DOMINANT ZEITGEBERS FOR CLOCKS IN LIVER AND OTHER PERIPHERAL ORGANS

Genome-wide transcriptome-mapping studies revealed that the fraction of diurnally accumulating liver transcripts amounts to 2–10%, depending on the stringency of algorithms used for the analyses of microarray hybridization data (Akhtar et al. 2002; Duffield et al. 2002; Panda et al. 2002; Storch et al. 2002; Walker and Hogenesch 2005). Many of these genes encode enzymes involved in the processing and detoxification of food components, suggesting that the temporal coordination of metabolism is a major task of hepatocyte clocks. If so, it would make sense to tune the phase of circadian liver clocks to the metabolic cycles in order to optimize the temporal coordination of catabolic and anabolic processes. Damiola et al. (2000) and Stokkan et al. (2001) therefore examined whether daily feeding/fasting cycles might serve as zeitgebers for the oscillators operative in liver and, perhaps, other peripheral tissues. To this end, food was offered only during a restricted time window outside the normal (nocturnal) activity phase, and the phase of circadian gene expression was determined. The results showed that the inversion of feeding cycles from nighttime feeding to daytime feeding provoked a complete inversion in the liver and some other tissues, such as pancreas, kidney, and heart. In contrast, feeding cycles had no significant effect on the circadian phase of the SCN master clock and a somewhat less dramatic effect on the phase of lung oscillators. In both studies, the liver was found to respond more quickly to the imposed feeding regimens than other tissues examined.

On the basis of the observations described above, we suspect that under normal circumstances, the SCN master timekeeper synchronizes clocks in peripheral tissues primarily through behavioral rhythms. In the case of liver and some other metabolically active tissues, rest/activity cycles may drive feeding rhythms, and the absorption and processing of nutrients may elicit a number of strong timing cues for peripheral oscillators. Clearly, however, the SCN may also act on peripheral cell types through more direct zeitgebers, such as body temperature rhythms (Brown et al. 2002) and cyclically secreted hormones (Balsalobre et al. 2000a,b; Le Minh et al. 2001). The phase entrainment of the liver clock by behavioral and more direct SCN outputs under normal conditions is schematically illustrated in Figure 1A (top panel). Imposed feeding rhythms can completely uncouple peripheral oscillators from the SCN master pacemaker, and the timing signals provoked by feeding cycles must therefore be dominant over more directly controlled systemic zeitgebers. This scheme makes some clear predictions that can be experimentally tested: (1) The kinetics of phase uncoupling after imposing an anticyclic feeding regimen should be slower than the phase recoupling after termination of the feeding regimen, because the feeding-related zeitgebers are only antagonistic with the "direct timing signals" during the forced-feeding schedule. Damiola et al. (2000) have examined and validated this prediction. (2) If a "direct signaling pathway" is eliminated (Fig. 1A, lower left panel), the kinetics of feeding-induced phase inversion should be faster than in the presence of all "direct signaling routes" (green curve vs. black curve in Fig. 1B), because the competitiveness of feeding-related versus direct zeitgebers is now enhanced. Again, this scenario has been examined and validated for glucocorticoids, a strong phase-shifting agent whose cyclic production and secretion are controlled by the SCN via the hypothalamus-pituitary-adrenal (HPA) axis (Le Minh et al. 2001). (3) Obliteration of a food-related zeitgeber pathway should slow down the feeding-induced phase inversion (Fig. 1A, lower right panel and red curve in Fig. 1B), because the impaired indirect timing mechanisms now suffer a fiercer competition from antagonistic timing cues directly controlled by the SCN. No food-related timing signals have yet been identified in an unambiguous manner, and this prediction could hence not yet be scrutinized. Obviously, a more detailed experimental evaluation of the model proposed in Figure 1 requires profound knowledge on the molecular makeup of the signaling pathways that participate in both behavioral (indirect) synchronization routes and directly SCN-driven phase entrainment mechanisms. Below, we propose an experimental system that hopefully will contribute to the deciphering of the various input pathways implicated in the synchronization of peripheral oscillators.

Feeding rhythms consist of an absorptive phase, during which food is ingested and used to build energy stores such as glycogen granules and fat droplets, and a postabsorptive phase, during which the energy stores are used. The metabolism is obviously quite different during these two phases, and the question thus arises of whether the most efficient molecular zeitgebers for peripheral clocks are produced during the absorptive or the postabsorptive phase. At least for rodents, the latter may be more important than the former for the synchronization of hepatocyte clocks. This conclusion is based on the temporal analysis of hepatic gene expression in mice and voles subjected to ultradian and circadian feeding regimens, respectively (van der Veen et al. 2006). Mice display a strongly circadian behavior, and as nocturnal animals, they consume about 80% of their food during the dark phase when fed ad libitum. When mice received only 75% of the normal calorie intake as portions offered

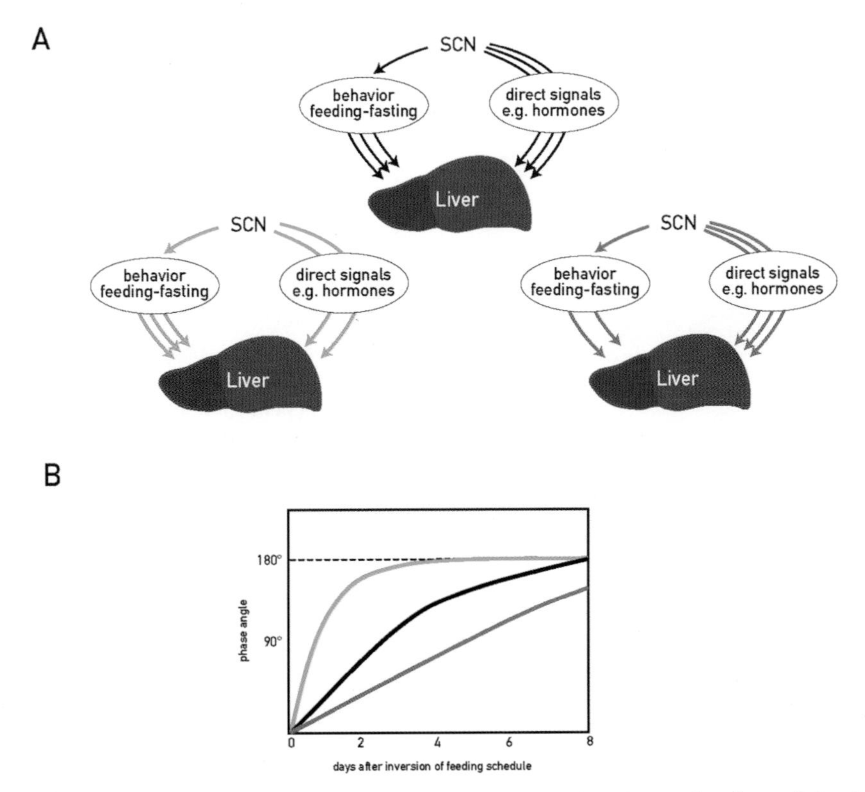

Figure 1. Direct and indirect phase entrainment routes for hepatocyte oscillators. The SCN synchronizes cellular circadian oscillators in liver (and other peripheral tissues) indirectly through circadian behavior (feeding rhythms) and more directly via controlling cyclic hormone secretion or via establishing body temperature oscillations. The scheme presented in *A* arbitrarily assumes that both routes employ three signaling pathways (*top scheme*). When an inverted feeding regimen is imposed on such animals (rats or mice), the kinetics of phase inversion follows the hypothetical black curve in *B*. Feeding inversion does not affect the phase of the SCN and thus the phase of direct zeitgeber cues. Therefore, after inverting the phase of the feeding/fasting cycles, the direct and indirect synchronization signals are in conflict. The phase inversion kinetics in liver depends on the competition between direct and indirect zeitgeber signals. If one (or more) of the direct signaling pathways are inactivated, the food-regimen-induced phase inversion is expected to be accelerated (*left scheme in A*, *green curve in B*). Conversely, if a feeding-dependent pathway is impaired, the kinetics of food-regimen-induced phase inversion is expected to be slowed down (*right scheme in A*, *red curve in B*).

at 150-minute intervals (ultradian feeding regimen), they eagerly ingested each meal immediately after it was provided by the feeding machine. This ultradian feeding regimen had little influence (a small phase advance) on the amplitude and magnitude of clock gene expression in the liver, suggesting that more direct SCN-controlled timing cues synchronize hepatocyte clocks when mice are feeding throughout the day (van der Veen et al. 2006). Common voles (*Microtus arvalis*) normally forage and feed at 150-minute intervals and hence are considered to be ultradian animals (Gerkema et al. 1990; Gerkema and van der Leest 1991). Under these conditions, circadian clock gene expression was found to be flat at intermediate levels in liver and kidney, suggesting that the cells of peripheral organs are not synchronized when voles feed at their normal ultradian schedule (van der Veen et al. 2006). However, when 8-hour daily fasting periods were imposed upon these animals, their livers and kidneys displayed robustly circadian clock gene expression. The timing signals provoked by the fasting periods in laboratory rodents remain to be identified, but it is tempting to speculate that fibroblast growth factor 21 (FGF-21) is one of them (see below).

MANY CHEMICAL AND PHYSICAL SIGNALS CAN SYNCHRONIZE CIRCADIAN OSCILLATORS IN CULTURED FIBROBLASTS

Cultured fibroblasts offer a simple experimental system to demonstrate the efficacy of body temperature rhythms as circadian timing cues. Originally, we thought that such low-amplitude temperature oscillations would sustain but not establish the synchronization of fibroblast oscillators (Brown et al. 2002). However, by reexamining the impact of body temperature rhythms on the phase entrainment with more sensitive techniques during extended time periods, we realized that a substantial fraction of fibroblasts can be synchronized by such temperature oscillations (C. Saini and U. Schibler, unpubl.).

Although fibroblast clocks are synchronized when implanted under the skin in the neck region of mice, it turned out to be difficult to use them as an experimental system for the characterization of chemical signals. Obviously, serum contains a myriad of growth factors, cytokines, metabolites, and hormones, and as shown in Figure 2A, experiments performed in several laboratories have indicated that many signaling pathways can syn-

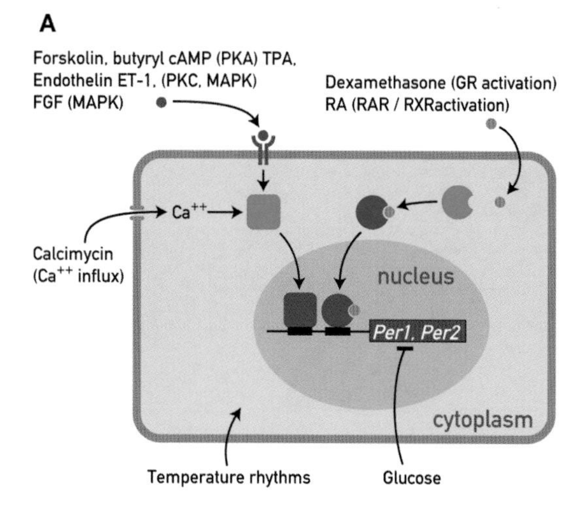

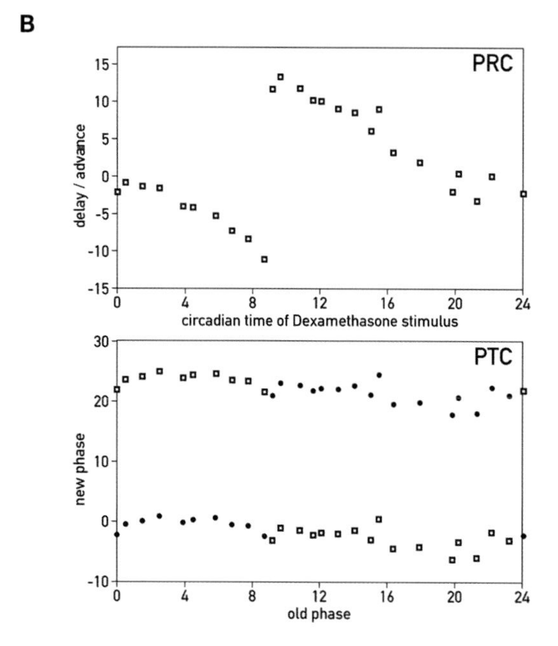

Figure 2. Many signaling pathways can synchronize the circadian oscillators of cultured fibroblasts. (*A*) The synchronization of cultured fibroblasts can be accomplished by a short treatment of cells with chemicals activating a variety of known signaling pathways, metabolites, and artificially imposed body temperature rhythms. All of these entrainment pathways probably involve the up- or down-regulation of *Per1* and *Per2*, and some of them act through the activation of various protein kinases (given in parentheses) and transcription factors (e.g., CREB, *blue square*). Ligand-bound nuclear receptors, such as the glucocorticoid receptor, may directly bind to hormone-responsive DNA response elements within the promoter and enhancer regions of *Per* genes. (*B*) Phase-response curve (PRC, *upper panel*) and phase transition curve (PTC, *lower panel*) of dexamethasone-treated NIH-3T3 cells expressing luciferase from a *Bmal1* promoter. Dexamethasone pulses were applied to parallel cultures approximately every hour. The resulting phase shifts in the daily luminescence cycles were plotted against circadian time. The data presented in the PRC are represented as a phase-transition curve (PTC), in which the new phase is plotted against the old phase. *Black dots* and *open squares* can be considered as phase-shifts measured on 2 consecutive days. For simplicity, all points are plotted on both days. Note that all new phases are nearly identical irrespective of when cells were treated with dexamethasone. Hence, the slope of the PTC is near zero, and the corresponding phase-response curve is called type-zero PRC. (*B*, Reprinted, with permission, from Nagoshi et al. 2004 [© Elsevier].)

chronize circadian oscillators in cultured fibroblasts (Gachon et al. 2004). Moreover, fibroblast clocks exhibit a so-called type-zero phase-response curve (PRC) (Nagoshi et al. 2004). When they are synchronized using a serum shock and then phase-shifted by a short pulse of dexamethasone at different times during the day, the slope of the phase transition curve (PTC), in which the new phase is plotted against the old phase, is close to zero (see PRC and PTC in Fig. 2B). This exemplifies the high sensitivity of fibroblast oscillators to phase-shifting cues and explains why they can readily be synchronized. However, these phase-shifting properties also render it difficult to measure the effect of, say, a blood-borne zeitgeber signal in a dose-dependent manner.

Given these difficulties, we have decided to design experimental strategies aimed at the identification of physiological zeitgebers in live animals. One such approach is presented in the next section.

A MOUSE WITH A CONDITIONALLY ACTIVE LIVER CLOCK

In principle, circadian gene expression in peripheral tissues can be regulated by systemic cues and/or local oscillators. Systemic cues (e.g., hormones) are interpreted by their sensors (i.e., hormone receptors), and this leads to the activation of immediate-early genes. If the blood-borne systemic signaling component oscillates during the day and if its average concentration is equal or lower than the dissociation equilibrium constant (K_D) determining its interaction with its receptor, the transcription rates of the signal-dependent immediate-early gene should also oscillate. Moreover, if an immediate-early (or early) gene controlled by this pathway was itself a cock component (such as PER1 or PER2), its system-controlled cyclic expression might influence the phase of the cycling signaling component. Therefore, the separation of system-driven and oscillator-dependent immediate-early (or early) genes would possibly provide insight into the molecular signaling pathways involved in the phase entrainment of peripheral clocks.

We thus attempted to design a transgenic mouse model that would allow us to discriminate between system- and oscillator-driven genes in hepatocytes. For this purpose, we established a mouse in which local hepatocyte oscillators can be switched off and on at will by providing or not the tetracycline analog doxycycline (Dox) in the food or drinking water (Kornmann et al. 2007). In this system, a rat REV-ERBα (rREV)-encoding cDNA expression vector was brought under the control of a liver-specifically expressed Tet activator that can only bind to its operators within the rREV cDNA transgene promoter in the absence of Dox (Tet-off system). Under these conditions, overproduced rREV occupies its RORE cognate elements within the *Bmal1* promoter and thereby represses *Bmal1* transcription. As BMAL1 is indispensable for circadian oscillator function, the hepatocyte oscillators are not operational in the absence of Dox. If Dox is added to the food (or drinking water), the Tet activator changes conformation and loses its affinity for the Tet operators within the rREV transgene promoter. Hence, rREV no longer accumulates, *Bmal1* transcription is up-regulated, and hepatocyte oscillators become functional (Fig. 3A). An *mPer2::luc* reporter allele, engineered by Takahashi and colleagues by inserting a fire fly luciferase open reading frame into the resident *mPer2* locus by homologous recombination (Yoo et al. 2004), was crossed into the Tet-activator/rREV double transgenic mice. This allowed us to record circadian luminescence cycles in tissue explants in real time. As shown by the data presented in Figure 3B (right panels), liver slices from Tet-activator/rREV/*mPer2::luc* triple transgenic mice displayed circadian luminescence cycles only when Dox was added to the culture medium. As expected, Dox treatment had little effect on circadian *mPer2::luc* expression in control mice without the Tet-activator and rREV transgenes (Fig. 3B, left panels). Importantly, the phases of *mPer2-luc*-dependent luminescence cycles were nearly identical in *mPer2::luc* control mice and Tet-activator/rREV/*mPer2::luc* mice when these animals received Dox for 48 hours before they were sacrificed (Fig. 3B, bottom panels). However, Dox-activated liver explants from untreated Tet-activator/rREV/*mPer2::luc* mice produced luminescence cycles with a delayed phase (Fig. 3B, middle panels). Presumably, the decay of transgene-encoded rREV repressor and the accumulation of BMAL1 to appropriate levels lasted several hours. All in all, these experiments provided proof of principle for the Dox dependence of circadian oscillator function in Tet-activator/rREV/ *mPer2::luc* mice, and, by inference, in Tet-activator/ rREV mice.

SYSTEMICALLY AND OSCILLATOR-DRIVEN GENES

Encouraged by these results, we performed genome-wide transcriptome-profiling experiments in mice that did or did not receive Dox as a food accompaniment. To this end, untreated and Dox-treated mice were sacrificed every 4 hours during 2 entire days, and the liver RNAs from these animals were hybridized to 430 2.0 Affymetrix oligonucleotide arrays representing the majority of protein-coding genes. Circadian transcripts were identified by applying data analysis procedures that use relatively stringent algorithms (see legend to Fig. 4). The circadian transcripts identified in the Dox-treated mice are expected to be products of both system-dependent and oscillator-dependent genes. The class of genes encoding these mRNAs contained core clock genes (e.g., *mPer1, mPer2, Bmal1, mCry1, Dec1, Dec2, Ror*γ, *Rev-erb*α, and *Rev-erb*β) and clock-controlled genes (e.g., *Dbp, E4bp4, Tef,* and *Hlf*). In addition and in keeping with related studies, many genes encoding enzymes involved in the metabolism of proteins, carbohydrates, lipids, cholesterol, bile acids, and xenobiotics were found to be cyclically expressed. Given the high stringency of the algorithms we used to extract circadian genes from the Affymetrix microarray data, our analysis revealed somewhat fewer genes (350 genes represented by 470 feature sets) than other studies (Panda et al. 2002; Storch et al. 2002).

The genes identified in mice not receiving Dox as a food supplement are expected to be issued from system-driven, oscillator-independent genes (Fig. 3C). This inter-

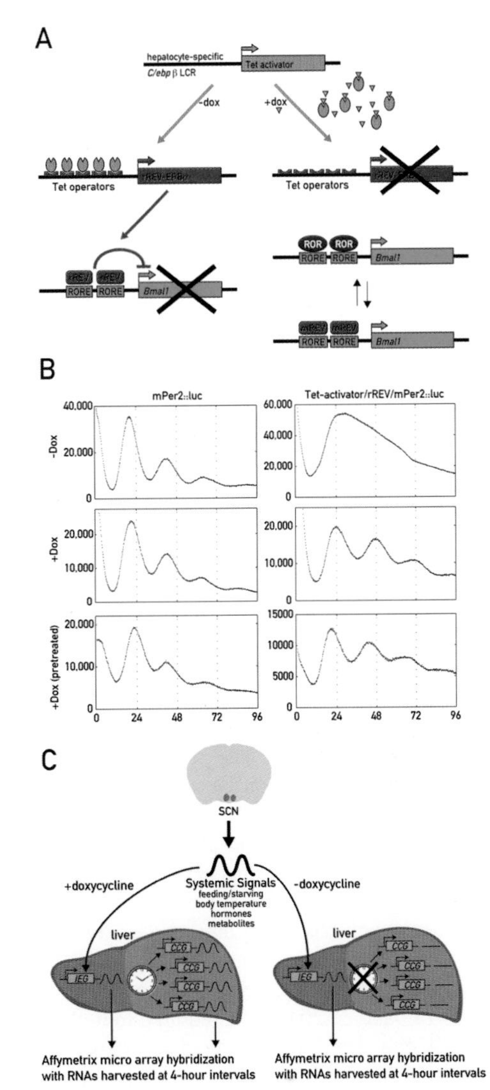

Figure 3. Systemic and oscillator-driven gene expression in transgenic mice with conditionally active liver clocks. (*A*) Hepatocyte oscillators can be arrested by the overexpression of REV-ERBα (rREV for rat REV-ERBα). Hepatocyte-specific, Dox-dependent expression of HA-REV-ERBα was achieved by placing a 5′-HA-tagged REV-ERBα cDNA transgene under the control of seven tetracycline responsive elements (TREs). In the liver of mice expressing the Tet-responsive *trans*-activator from the hepatocyte-specific *C/ebp*β*-LAP* locus control region (LCR), rREV-ERBα transcription is constitutively repressed in the absence of the tetracycline analog dox (Tet-off system). The constitutively overexpressed rREV repressor now effectively competes throughout the day with ROR activators for the occupancy at RORE elements. This leads to an attenuation of circadian oscillator function, because *Bmal1* is required for circadian rhythm generation. In the presence of Dox, the rREV-ERBα is silent, and circadian *Bmal1* transcription is regulated like in wild-type mice through alternating binding of ROR activators and endogenous murine REV-ERBα/REV-ERBβ repressors (mREV) to RORE elements. (*B*) Liver slices from *mPer2::luc* (*left*) and Tet-activator/rREV/*mPer2::luc* (*right*) were cultured in luciferin-containing medium in the absence (–Dox) or presence of 10 ng/μl doxycycline (+Dox). Luminescence was recorded using photomultiplier tubes. –Dox and +Dox samples are from the same animal, +Dox (pretreated) samples are from mice that have received two intraperitoneal injections of Dox 48 and 24 hours before being sacrificed. (*C*) A genome-wide search for liver clock-driven and systemically regulated gene (which comprise "immediate-early genes," IEG). Note that only systemically driven genes continue to be expressed in a diurnal fashion in the absence of Dox.

pretation should, however, be qualified by some notes of caution. In our Tet-activator/rREV mice, the Tet-activator transgene is expressed specifically in hepatocytes. Although these make up the majority of the liver mass, about 10–20% of the liver RNA is probably derived from Kupffer cells, bile duct cells, endothelial cells, and liver fibroblasts. In these cells, *Bmal1* accumulation is supposedly normal, even in the absence of Dox. Furthermore, even hepatocytes of untreated animals may still accumulate low levels of BMAL1, and these might be sufficient for the residual transcription of those CLOCK-BMAL1 target genes that contain very high-affinity E-box-binding motifs. To limit the number of such "false positives" in the identification of systemically driven transcripts, we selected mRNAs whose average expression levels were similar in untreated and Dox-treated animals (ratio –Dox/+Dox between 0.7 and 1.5). These mRNAs constituted the majority (60%) of liver transcripts whose expression remained rhythmic in Dox-free mice. The 61 genes encoding such transcripts were represented by 79 feature sets and are displayed in Table 1 and in the phase map of Figure 4B (right panels). As expected, virtually all of these transcripts also accumulate in a rhythmic fashion in the livers of Dox-treated mice. Interestingly, the systemically driven genes can be grouped into functional categories (Table 1; Fig. 4B,C). The genes of two categories fall into defined phase clusters. Genes involved in cholesterol synthesis are maximally expressed between ZT17 and ZT20, and genes encoding chaperones (heat shock proteins) are maximally expressed between ZT18 and ZT00.

As nocturnal animals, mice are active between ZT12 and ZT00 and hence consume most of their food during this time period. Increased hepatic cholesterol synthesis between ZT17 and ZT20 may be required to replenish the cholesterol converted to bile acids. Indeed, *Cyp7a1* mRNA, encoding the rate-limiting enzyme of bile acid synthesis, is maximally expressed at ZT10, shortly before the cholesterol synthesis is expected to reach maximal rates (G. Le Martelot and U. Schibler, unpubl.). Conceivably, the conversion of cholesterol to bile acids reduces membrane cholesterol levels in the endoplasmic reticulum, and this triggers the activation of SREBP, the transcription factor regulating most of the cholesterol-related genes listed in Table 1 as a consequence of cholesterol depletion. In keeping with these conjectures, the proteolytic activation of SREBP has been found to follow a robustly diurnal pattern (Brewer et al. 2005; G. Le Martelot and U. Schibler, unpubl.).

Heat shock proteins (HSPs) participate in the response to proteotoxic stress, caused by elevated temperature and/or chemical insults (Burel et al. 1992; Voellmy 1996). As chaperones, HSPs bind to and refold denatured cellular proteins or target ubiquitin ligases to damaged proteins, so that these can be recognized and degraded by the proteasome (Hirsch et al. 2006). Heat shock transcription factor 1 (HSF1) is a major sensor of elevated temperature and noxious chemicals, such as reactive oxygen species (ROS) (Liu et al. 1996; Voellmy and Boellmann 2007). It normally is sequestered in an inert cytoplasmic protein complex, consisting of HSP90 and small cochaperones. Upon exposing cells to a heat stress or treating them with certain chemicals, denatured proteins compete with HSF1 for binding to HSP90. As a consequence, HSF1 gets released from chaperones, becomes phosphorylated by various protein kinases, migrates to the nucleus, and binds as a homotrimer to heat shock elements (HSEs) within the promoters of multiple HSPs. In a second step, the *trans*-activation capacity of HSF1 becomes further stimulated through phosphorylation by calcium/calmodulin-dependent kinase 2β (CAMK2β), which leads to the transcription of HSF1 target genes (Holmberg et al. 2001). Of note, *Camk2*β mRNA was also found among the systemically driven transcripts, and the phase of *Camk2*β expression is compatible with its function in HSF1 activation (see Table 1).

PUTATIVE CANDIDATE SIGNALING PATHWAYS INVOLVED IN THE PHASE ENTRAINMENT OF PERIPHERAL CLOCKS

As pointed out earlier in this chapter, some of the systemically driven genes may participate in the synchronization of peripheral clocks. Of note, the hepatic expression of *mPer2*, a bona fide component of the molecular circadian oscillator, follows a robust diurnal rhythm in the absence of Dox in the animal but not in tissue explants (Kornmann et al. 2007). We hence surmise that systemic cues drive its rhythmic transcription in the absence of naturally occurring BMAL1 levels. Although we have not yet unambiguously identified the molecular mechanisms responsible for the systemic regulation of *mPer2* expression, based on several observations, we suspect that HSF1 may participate in this endeavor: (1) The cyclic accumulation of *mPer2* mRNA, like *Hsp* mRNAs, reach maximal levels when body temperature is highest, (2) in liver and lung explants of *mPer2::luc* mice, mPer2-LUC expression is transiently increased by a heat shock, (3) the circadian expression of clock gene expression in cultured fibroblasts can be synchronized by body temperature rhythms, and (4) during this process, the expression of *mPer2* is synchronized more rapidly than that of other clock genes (C. Saini and U. Schibler, unpubl.). We wish to reiterate, however, that temperature cycles may not be the only cues leading to a rhythmic activation of HSF1. For example, both feeding and redox potential have been reported to stimulate the activity of HSF1 (Ahn and Thiele 2003; Katsuki et al. 2004). These findings are of particular interest, given the dominant role of feeding/fasting cycles in the synchronization of oscillators in liver and other peripheral tissues.

Fibroblast growth factor 21 (*Fgf-21*) is another systemically driven gene that aroused our interest. The expression of FGF-21 has recently been shown to be controlled by peroxisome proliferator-activated receptor α (PPARα) through fasting (Badman et al. 2007; Inagaki et al. 2007; Lundasen et al. 2007). Indeed, both the expression and the activity of PPARα are subject to diurnal regulation (Waring 1970; Lemberger et al. 1996; Canaple et al. 2006; F. Gachon and U. Schibler, unpubl.). Free fatty acids are natural activators of PPARα, and we observed that the hepatic concentrations of both *Ppar*α mRNA and several

Table 1. List of System-driven Diurnally Expressed Genes

Affy gene name	Full gene name	Putative gene function	Ratio −Dox/+Dox
Hspa4l	Heat shock 70-kD protein-4-like (HSP110 gene family)	chaperone, osmotolerance	1.5
Atf3	Activating transcription factor 3	stress-induced by cytokines, genotoxic agents, amino acid starvation	1.5
Afp	α-fetoprotein	transport of lipophilic compounds	1.5
Actg1	Actin, γ, cytoplasmic 1	cytoskeleton	1.5
Hspca	HSP1, α, HSP90α	chaperone, regulator of NRs and other TFs	1.4
Dnclc1	Dynein, cytoplasmic, light chain 1	dynein microtubule motor	1.4
Tuba4	Tubulin, α 4A	tubulin cytoskeleton, microtubules	1.4
Hspa4	Heat shock 70-kD protein 4	chaperone	1.3
Klhl24	Kelch-like 24	?	1.3
Tuba4	Tubulin, α 4A	tubulin cytoskeleton, microtubules	1.3
Afp	α-fetoprotein	transport of lipophilic compounds	1.3
Hspca	HSP1, α, HSP90α	chaperone, regulator of NRs and other TFs	1.3
Ccrn4l	CCR4 carbon catabolite repression 4-like, nocturnin	mRNA deadenylase?	1.3
1431214_at			1.3
Ldb1	LIM domain-binding 1	cofactor of TFs (e.g., Nkx5), development	1.2
1423672_at			1.2
Trpm4	Transient receptor potential cation channel, subfamily M, member 4	cellular sensors of Ca^{++}, temperature	1.2
Hddc3	HD-domain-containing 3		1.2
Tubb6	Tubulin, β6	tubulin cytoskeleton, microtubules	1.2
Heca	Headcase homolog (*Drosophila*)	basic protein, *Drosophila* imaginal disk cell proliferation	1.1
Cyp51	Sterol 14 α-demethylase (cholesterol synthesis)	cholesterol synthesis	1.1
Hspa1a	Heat shock 70-kD protein 1A	chaperone	1.1
Hspa8	Heat shock 70-kD protein 8	chaperone	1.1
Fus	Fusion involved in t(12;16) in malignant liposarcoma (TLS)	RNA/DNA-binding protein	1.1
Hspca	HSP1, α, HSP90α	chaperone, regulator of NRs and other TFs	1.1
Slc25a15	Slc25a15 solute carrier family 25	mitochondrial carrier ornithine transporter	1.1
Hspa1a	Heat shock 70-kD protein 1A	chaperone	1.1
Hsp110	HSP110	chaperone, apoptosis regulator	1.0
Fbxl20	F-box and leucine-rich repeat protein 20	ubiquitination, substrate receptor	1.0
Chordc1	Cysteine and histidine-rich domain (CHORD)-containing, zinc-binding protein 1	chromatin regulator? histone methyl transferase? associated with HSP90	1.0
Hsp110	HSP110	chaperone, apoptosis regulator	1.0
Fus	Fusion involved in t(12;16) in malignant liposarcoma (TLS)	RNA/DNA-binding protein	1.0
Ris2	Retroviral integration site 2 (Human homolog = CDT1)	DNA replication licensing factor, complex with geminine	1.0
Hspa8	Heat shock 70-kD protein 8	chaperone	1.0
Cyp2b10	Cytochrome p450, family 2, subfamily b, polypeptide 10	xenobiotic detoxification, CAR target gene	1.0
Cyp51	Sterol 14 α-demethylase (cholesterol synthesis)	cholesterol synthesis	1.0
Per2	Period 2	circadian core clock component	1.0
Rnf6	RNF6 ring finger protein	ubiquitin ligase regulates LIMK1→actin polymerase	1.0
Fgf21	FGF-21 fibroblast growth factor 21	MAPK activator	1.0
1453303_at			1.0
Hspa1a	Heat shock 70-kD protein 1A	chaperone	1.0
Per2	Period 2	circadian core clock component	1.0

(*Continued on following page.*)

Table 1. (*Continued*)

Affy gene name	Full gene name	Putative gene function	Ratio −Dox/+Dox
Alas1	Aminolevulinate, δ-, synthase 1	heme synthesis	0.9
Rnf125	Ring finger protein 125	ubiquitin ligase, inhibits RIG-1 signaling, dsRNA sensor	0.9
Erbb3	v-*erb*-b2 erythroblastic leukemia viral oncogene homolog 3	EGF receptor family, HGF receptor, PI3K induction	0.9
Rbl2	Retinoblastoma-like 2 (=p130)	cell cycle regulator, cancer	0.9
Tgoln1	(=Ttgn1) *trans*-Golgi network protein	protein trafficking, EGFR endocytosis?	0.9
Slc25a25	Slc25a15 solute carrier family 25	mitochondrial carrier ornithine transporter	0.9
1431213_a_at			0.9
Hmgcs1	3-hydroxy-3-methylglutaryl–coenzyme A synthase 1	cholesterol synthesis	0.9
Errfi1	ERBB receptor feedback inhibitor 1 (=Mig6 ortholog)	inhibitor of EGFR and MAPK signaling	0.9
Sqle	SQLE squalene epoxidase	cholesterol metabolism	0.9
Stip1	stress-induced-phosphoprotein 1	HSP70/HSP90-organizing protein	0.9
Efna1	Ephrin-A1	ephrin signaling, ligand of EphA2 receptor	0.9
Cirbp	Cold-inducible RNA-binding protein	cold stress regulator?	0.9
Fbxo21	F-box-only protein 21	ubiquitination, substrate receptor	0.9
Tcp11l2	Tcp11l2 t-complex 11 (mouse)-like 2	?	0.9
Errfi1	ERBB receptor feedback inhibitor 1 (=Mig6 ortholog)	inhibitor of EGFR and MAPK signaling	0.9
Ddx46	DEAD (Asp-Glu-Ala-Asp) box polypeptide 46	RNA helicase	0.9
Per2	Period 2	circadian core clock component	0.9
Calcoco1	Calcium binding and coiled-coil domain 1	transcriptional coactivator	0.9
March7	Membrane-associated ring finger (C3HC4) 7	membrane-bound ubiquitin ligase, endos., Endos. transport	0.9
Sc4mol	Sterol-C4-methyl oxidase-like	cholesterol synthesis	0.8
1455892_x_at			0.8
Ctgf	Connective tissue growth factor	extracellular-matrix-bound signaling factor, activated by SRF, repressed by soluble actin	0.8
1452418_at			0.8
1435084_at			0.8
Hmgcs1	3-hydroxy-3-methylglutaryl–coenzyme A synthase 1	cholesterol synthesis	0.8
Ass1	Argininosuccinate synthetase 1	arginine synthesis, urea cycle, induced by amino acid starvation	0.8
Idi1	Isopentenyl-diphosphate δ isomerase	cholesterol synthesis	0.8
Fbxo21	F-box-only protein 21	ubiquitination, substrate receptor	0.8
Hmgcs1	3-hydroxy-3-methylglutaryl–coenzyme A synthase 1	cholesterol synthesis	0.8
Alas1	Aminolevulinate, δ-, synthase 1	heme synthesis	0.8
Fdps	Farnesyl diphosphate synthetase	isoprenoid metabol, FGF signaling?	0.8
1428531_at			0.7
Enpp3	Ectonucleotide pyrophosphatase/phosphodiesterase 3	hydrolysis of extracellular nucleotides	0.7
Idi1	Isopentenyl-diphosphate δ isomerase	cholesterol synthesis	0.7
Alas1	Aminolevulinate, δ-, synthase 1	heme synthesis	0.7
Lss	Lanosterol synthase (2,3-oxidosqualene-lanosterol cyclase)	cholesterol synthesis	0.7
Camk2b	Calcium/calmodulin-dependent protein kinase (CaM kinase) II β	activation of HSF1, inhibition of CREB	0.7

Liver transcripts whose accumulation continues to be diurnal in the absence of functional hepatocyte oscillators are listed according to the ratio of average hybridization signals obtained in the absence and presence of doxycycline. The putative functions of these genes are indicated in column 3.

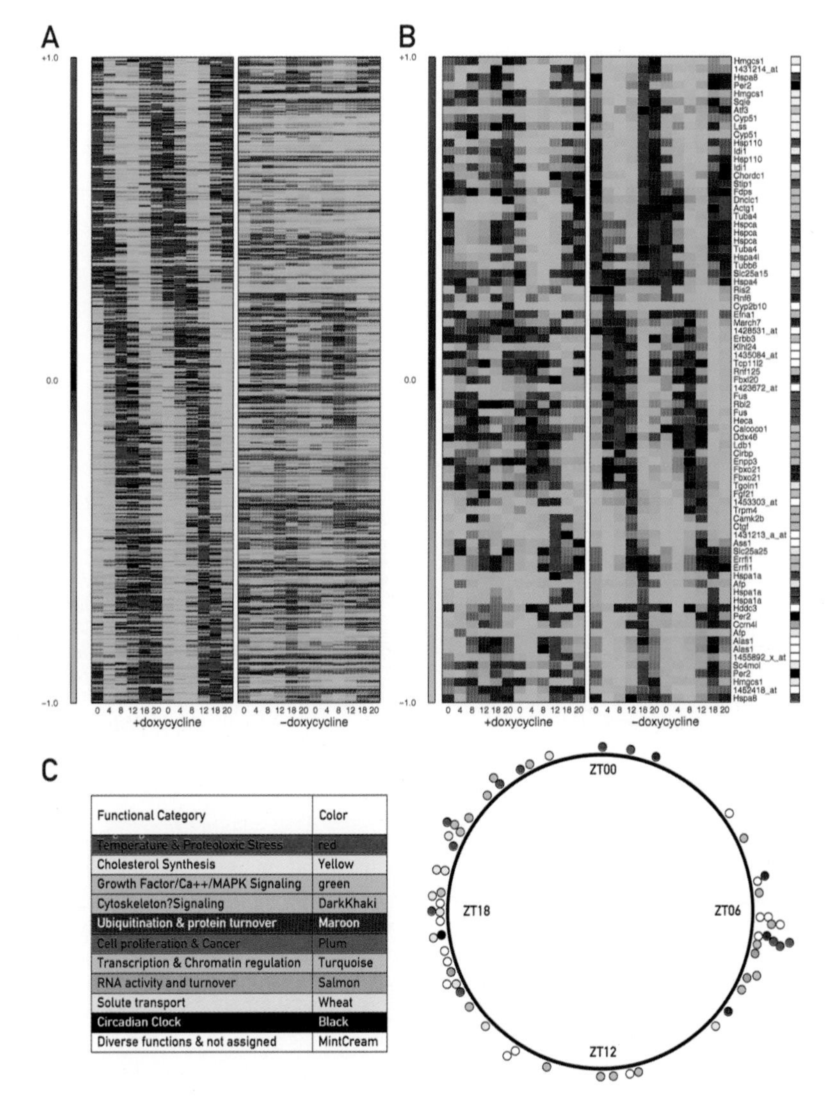

Functional Category	Color
Temperature & Proteotoxic Stress	red
Cholesterol Synthesis	Yellow
Growth Factor/Ca++/MAPK Signaling	green
Cytoskeleton?Signaling	DarkKhaki
Ubiquitination & protein turnover	Maroon
Cell proliferation & Cancer	Plum
Transcription & Chromatin regulation	Turquoise
RNA activity and turnover	Salmon
Solute transport	Wheat
Circadian Clock	Black
Diverse functions & not assigned	MintCream

Figure 4. Phase maps of cyclically expressed transcripts in the presence and absence of Dox. Phase maps of genes selected from temporal Affymetrix data sets using the algorithms described by Kornmann et al. (2007). Briefly, the data sets composed of genome-wide hybridization signals obtained for liver RNA prepared from Tet-activator/rREV mice sacrificed at 4-hour intervals during 2 days were screened for circadian transcripts using a p-value of 0.05 for Fourier analysis and an amplitude of equal or greater than two. (*A*) Phase map of cyclic transcripts selected according to these criteria from the data sets obtained for Dox-treated animals. Note that most (but not all) of the transcripts selected in this way lose robust circadian accumulation in the absence of Dox. (*B*) Phase map of cyclic transcripts selected according to these criteria from the data sets obtained for untreated animals (–Dox). Note that most of the transcripts selected in this way also show a daily accumulation cycle in the presence of Dox, as expected. The color code bar on the right hand side of the panel indicates the putative functions of the proteins encoded by these systemically regulated genes. (*C*) Color code for functional gene categories for the transcripts displayed in *B* and circular phase map of these transcripts.

free fatty acids culminate toward the end of the postabsorptive phase (ZT08 to ZT12). Fibroblast growth factors signal through tyrosin receptor kinases that subsequently can activate various protein kinases, including mitogen-activated protein kinases (MAPKs) (Eswarakumar et al. 2005; Kurosu et al. 2007). MAPK phosphorylates serine and threonine residues in many protein substrates, including the cAMP-responsive element-binding protein (CREB). Both *mPer1* and *mPer2* are CREB target genes, and the activation of CREB has been shown to have an important role in the synchronization of circadian oscillators in SCN neurons (Morse and Sassone-Corsi 2002; Dziema et al. 2003; Hastings and Herzog 2004) and cultured fibroblasts (Balsalobre et al. 2000a; Yagita and Okamura 2000). In contrast to other fibroblast growth factors, FGF-21 has a weak heparin-binding domain and is thus not trapped in the extracellular matrix (Kurosu et al. 2007). Hence, it may act as a paracrine factor in the liver and as a hormone in distant tissues, such as muscle, fat, and kidney. As mentioned above, liver clocks get food-entrained more rapidly than clocks in other peripheral tissues, and the liver-specific expression of FGF-21 may provide a rational explanation for this finding.

Figure 5 schematically illustrates the speculations proposed above on phase entrainment signaling pathways operative in peripheral tissues. Conceivably, PER1 and

A

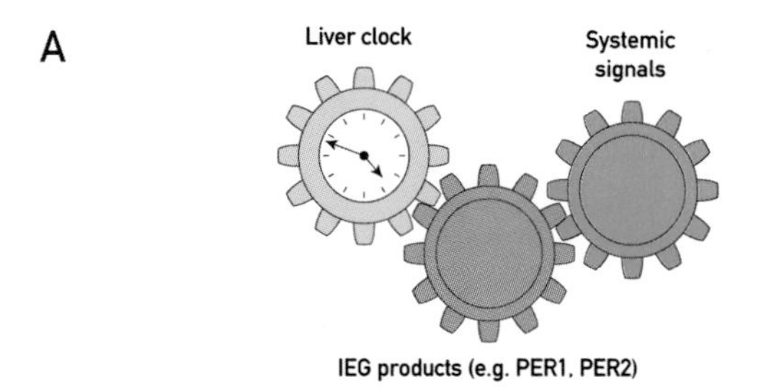

B

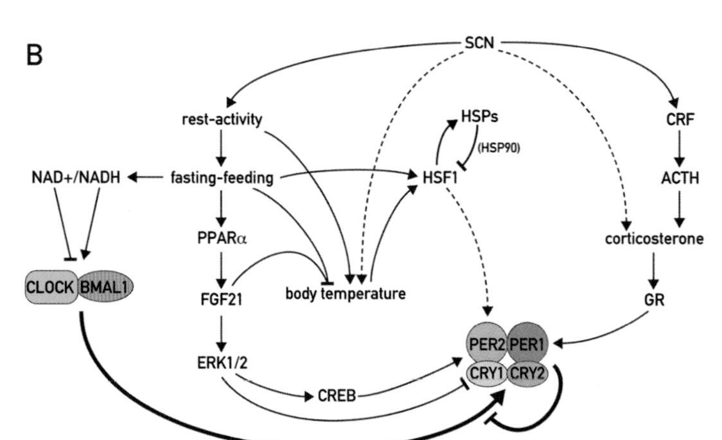

Figure 5. Interaction between system- and oscillator-driven genes. (*A*) Cogwheel diagram illustrating how systemically driven immediate-early genes, such as those encoding mPER1 and mPER2, could synchronize local circadian oscillators. Such components can serve as both core clock components and sensors of systemic zeitgeber cues controlled by the SCN. (*B*) Hypothetical scheme of molecular input pathways for peripheral circadian oscillators (see text for explanation).

PER2 serve as both cogwheels of the clockwork circuitry and immediate-early genes in its synchronization (Fig. 5A). Many molecular signaling pathways may contribute to this synchronization, and as hypothesized in Figure 5B, they are likely to be intertwined. For example, the feeding-dependent cyclic expression of FGF-21, in conjunction with fasting, may decrease body temperature and thereby down-regulate HSF1 activity during the second half of the postabsorptive phase. Furthermore, MAPK may not only increase PER1 and PER2 expression, but also inhibit the repressive activity of cryptochrome proteins on CLOCK-BMAL1-mediated transcription. In addition to these molecular signaling pathways depending on feeding behavior, glucocorticoids (and perhaps other hormones) are likely to serve as more direct messengers of the SCN in its endeavor to keep the phase of peripheral oscillators tuned (Balsalobre et al. 2000b; Le Minh et al. 2001).

CONCLUSIONS AND PERSPECTIVES

Overt daily cycles in behavior and physiology depend on the synchronization of countless cellular oscillators in the SCN and peripheral organs. The deciphering of molecular signaling pathways involved in the establishment of phase coherence in multicellular organisms such as mammals is thus a major challenge in circadian rhythm research. Although many signaling cascades can synchronize circadian clocks in cultured cells, the molecular routes by which the SCN phase entrains peripheral clocks remain subject to speculation. Their unambiguous identification requires a combination of state-of-the-art biochemical and genetic approaches. We have presented here our initial attempts toward the detection of immediate-early genes that participate in the interpretation of systemic signals in liver. In conjunction with studies recently published by us and other investigators (see above), these experiments hint toward a role of body temperature rhythms, FGF-21 signaling, and glucocorticoids in the daily phase resetting of peripheral clocks. Undoubtedly, additional pathways will soon join the list. This redundancy poses a major problem in the dissection of entrainment pathways, because the elimination of only one signaling pathway may only have no or only minor consequences on the steady-state phase. As illustrated in Figure 1A and as demonstrated by the ablation of glucocorticoid receptor signaling in liver, the kinetics of feeding-induced phase resetting is a much more sensitive readout in genetic loss-of-function studies. The recording of phase-resetting kinetics at a high temporal resolution requires novel experimental approaches, in which the phase of circadian gene expression can be monitored in real time and in live animals. Our laboratory is therefore investing a major effort toward the development of whole-animal imaging technologies that allow the measurement of circadian fluorescent protein expression in organs of freely moving mice.

ACKNOWLEDGMENTS

We thank Nicolas Roggli for preparing the artwork. Research in our laboratory was supported by the Swiss National Science Foundation (through an individual research grant to U.S. and the National Center of Competence in Research program Frontiers in Genetics), the State of Geneva, the Louis Jeantet Foundation of Medicine, the Bonizzi-Theler Stiftung, and the Sixth

European Framework Project EUCLOCK. H.R. is the recipient of a Human Frontiers Science Program (HFSP) long-term postdoctoral fellowship.

REFERENCES

Ahn S.G. and Thiele D.J. 2003. Redox regulation of mammalian heat shock factor 1 is essential for Hsp gene activation and protection from stress. *Genes Dev.* **17:** 516.

Akhtar R.A., Reddy A.B., Maywood E.S., Clayton J.D., King V.M., Smith A.G., Gant T.W., Hastings M.H., and Kyriacou C.P. 2002. Circadian cycling of the mouse liver transcriptome, as revealed by cDNA microarray, is driven by the suprachiasmatic nucleus. *Curr. Biol.* **12:** 540.

Badman M.K., Pissios P., Kennedy A.R., Koukos G., Flier J.S., and Maratos-Flier E. 2007. Hepatic fibroblast growth factor 21 is regulated by PPARalpha and is a key mediator of hepatic lipid metabolism in ketotic states. *Cell Metab.* **5:** 426.

Balsalobre A., Damiola F., and Schibler U. 1998. A serum shock induces circadian gene expression in mammalian tissue culture cells. *Cell* **93:** 929.

Balsalobre A., Marcacci L., and Schibler U. 2000a. Multiple signaling pathways elicit circadian gene expression in cultured Rat-1 fibroblasts. *Curr. Biol.* **10:** 1291.

Balsalobre A., Brown S.A., Marcacci L., Tronche F., Kellendonk C., Reichardt H.M., Schutz G., and Schibler U. 2000b. Resetting of circadian time in peripheral tissues by glucocorticoid signaling. *Science* **289:** 2344.

Brewer M., Lange D., Baler R., and Anzulovich A. 2005. SREBP-1 as a transcriptional integrator of circadian and nutritional cues in the liver. *J. Biol. Rhythms* **20:** 195.

Brown S.A., Zumbrunn G., Fleury-Olela F., Preitner N., and Schibler U. 2002. Rhythms of mammalian body temperature can sustain peripheral circadian clocks. *Curr. Biol.* **12:** 1574.

Brown S.A., Fleury-Olela F., Nagoshi E., Hauser C., Juge C., Meier C.A., Chicheportiche R., Dayer J.M., Albrecht U., and Schibler U. 2005. The period length of fibroblast circadian gene expression varies widely among human individuals. *PLoS Biol.* **3:** e338.

Burel C., Mezger V., Pinto M., Rallu M., Trigon S., and Morange M. 1992. Mammalian heat shock protein families. Expression and functions. *Experientia* **48:** 629.

Canaple L., Rambaud J., Dkhissi-Benyahya O., Rayet B., Tan N.S., Michalik L., Delaunay F., Wahli W., and Laudet V. 2006. Reciprocal regulation of brain and muscle Arnt-like protein 1 and peroxisome proliferator-activated receptor alpha defines a novel positive feedback loop in the rodent liver circadian clock. *Mol. Endocrinol.* **20:** 1715.

Damiola F., Le Minh N., Preitner N., Kornmann B., Fleury-Olela F., and Schibler U. 2000. Restricted feeding uncouples circadian oscillators in peripheral tissues from the central pacemaker in the suprachiasmatic nucleus. *Genes Dev.* **14:** 2950.

Duffield G.E., Best J.D., Meurers B.H., Bittner A., Loros J.J., and Dunlap J.C. 2002. Circadian programs of transcriptional activation, signaling, and protein turnover revealed by microarray analysis of mammalian cells. *Curr. Biol.* **12:** 551.

Dziema H., Oatis B., Butcher G.Q., Yates R., Hoyt K.R., and Obrietan K. 2003. The ERK/MAP kinase pathway couples light to immediate-early gene expression in the suprachiasmatic nucleus. *Eur. J. Neurosci.* **17:** 1617.

Eswarakumar V.P., Lax I., and Schlessinger J. 2005. Cellular signaling by fibroblast growth factor receptors. *Cytokine Growth Factor Rev.* **16:** 139.

Gachon F., Nagoshi E., Brown S.A., Ripperger J., and Schibler U. 2004. The mammalian circadian timing system: From gene expression to physiology. *Chromosoma* **113:** 103.

Gerkema M.P. and van der Leest F. 1991. Ongoing ultradian activity rhythms in the common vole, *Microtus arvalis,* during deprivations of food, water and rest. *J. Comp. Physiol.* **168:** 591.

Gerkema M.P., Groos G.A., and Daan S. 1990. Differential elimination of circadian and ultradian rhythmicity by hypothalamic lesions in the common vole, *Microtus arvalis. J. Biol. Rhythms* **5:** 81.

Guo H., Brewer J.M., Lehman M.N., and Bittman E.L. 2006. Suprachiasmatic regulation of circadian rhythms of gene expression in hamster peripheral organs: Effects of transplanting the pacemaker. *J. Neurosci.* **26:** 6406.

Guo H., Brewer J.M., Champhekar A., Harris R.B., and Bittman E.L. 2005. Differential control of peripheral circadian rhythms by suprachiasmatic-dependent neural signals. *Proc. Natl. Acad. Sci.* **102:** 3111.

Hastings M.H. and Herzog E.D. 2004. Clock genes, oscillators, and cellular networks in the suprachiasmatic nuclei. *J. Biol. Rhythms* **19:** 400.

Hirsch C., Gauss R., and Sommer T. 2006. Coping with stress: Cellular relaxation techniques. *Trends Cell Biol.* **16:** 657.

Holmberg C.I., Hietakangas V., Mikhailov A., Rantanen J.O., Kallio M., Meinander A., Hellman J., Morrice N., MacKintosh C., Morimoto R.I., Eriksson J.E., and Sistonen L. 2001. Phosphorylation of serine 230 promotes inducible transcriptional activity of heat shock factor 1. *EMBO J.* **20:** 3800.

Inagaki T., Dutchak P., Zhao G., Ding X., Gautron L., Parameswara V., Li Y., Goetz R., Mohammadi M., Esser V., Elmquist J.K., Gerard R.D., Burgess S.C., Hammer R.E., Mangelsdorf D.J., and Kliewer S.A. 2007. Endocrine regulation of the fasting response by PPARalpha-mediated induction of fibroblast growth factor 21. *Cell Metab.* **5:** 415.

Katsuki K., Fujimoto M., Zhang X.Y., Izu H., Takaki E., Tanizawa Y., Inouye S., and Nakai A. 2004. Feeding induces expression of heat shock proteins that reduce oxidative stress. *FEBS Lett.* **571:** 187.

Kornmann B., Schaad O., Bujard H., Takahashi J.S., and Schibler U. 2007. System-driven and oscillator-dependent circadian transcription in mice with a conditionally active liver clock. *PLoS Biol.* **5:** e34.

Kurosu H., Choi M., Ogawa Y., Dickson A.S., Goetz R., Eliseenkova A.V., Mohammadi M., Rosenblatt K.P., Kliewer S.A., and Kuro O.M. 2007. Tissue-specific expression of beta KLOTHO and fibroblast growth factor receptor isoforms determines metabolic activity of FGF19 and FGF21. *J. Biol. Chem.* **282:** 26687.

Le Minh N., Damiola F., Tronche F., Schutz G., and Schibler U. 2001. Glucocorticoid hormones inhibit food-induced phase-shifting of peripheral circadian oscillators. *EMBO J.* **20:** 7128.

Lejeune-Lenain C., Van Cauter E., Desir D., Beyloos M., and Franckson J.R. 1987. Control of circadian and episodic variations of adrenal androgens secretion in man. *J. Endocrinol. Invest.* **10:** 267.

Lemberger T., Saladin R., Vazquez M., Assimacopoulos F., Staels B., Desvergne B., Wahli W., and Auwerx J. 1996. Expression of the peroxisome proliferator-activated receptor alpha gene is stimulated by stress and follows a diurnal rhythm. *J. Biol. Chem.* **271:** 1764.

Liu A.C., Welsh D.K., Ko C.H., Tran H.G., Zhang E.E., Priest A.A., Buhr E.D., Singer O., Meeker K., Verma I.M., Doyle F.J., III, Takahashi J.S., and Kay S.A. 2007. Intercellular coupling confers robustness against mutations in the SCN circadian clock network. *Cell* **129:** 605.

Liu A.Y., Lee Y.K., Manalo D., and Huang L.E. 1996. Attenuated heat shock transcriptional response in aging: Molecular mechanism and implication in the biology of aging. *EXS* **77:** 393.

Lundasen T., Hunt M.C., Nilsson L.M., Sanyal S., Angelin B., Alexson S.E., and Rudling M. 2007. PPARalpha is a key regulator of hepatic FGF21. *Biochem. Biophys. Res. Commun.* **360:** 437.

Morse D. and Sassone-Corsi P. 2002. Time after time: Inputs to and outputs from the mammalian circadian oscillators. *Trends Neurosci.* **25:** 632.

Nagoshi E., Saini C., Bauer C., Laroche T., Naef F., and Schibler U. 2004. Circadian gene expression in individual fibroblasts: Cell-autonomous and self-sustained oscillators pass time to daughter cells. *Cell* **119:** 693.

Panda S., Antoch M.P., Miller B.H., Su A.I., Schook A.B.,

Straume M., Schultz P.G., Kay S.A., Takahashi J.S., and Hogenesch J.B. 2002. Coordinated transcription of key pathways in the mouse by the circadian clock. *Cell* **109:** 307.
Stokkan K.A., Yamazaki S., Tei H., Sakaki Y., and Menaker M. 2001. Entrainment of the circadian clock in the liver by feeding. *Science* **291:** 490.
Storch K.F., Lipan O., Leykin I., Viswanathan N., Davis F.C., Wong W.H., and Weitz C.J. 2002. Extensive and divergent circadian gene expression in liver and heart. *Nature* **417:** 78.
van der Veen D.R., Minh N.L., Gos P., Arneric M., Gerkema M.P., and Schibler U. 2006. Impact of behavior on central and peripheral circadian clocks in the common vole *Microtus arvalis,* a mammal with ultradian rhythms. *Proc. Natl. Acad. Sci.* **103:** 3393.
Voellmy R. 1996. Sensing stress and responding to stress. *EXS* **77:** 121.
Voellmy R. and Boellmann F. 2007. Chaperone regulation of the heat shock protein response. *Adv. Exp. Med. Biol.* **594:** 89.
Walker J.R. and Hogenesch J.B. 2005. RNA profiling in circadian biology. *Methods Enzymol.* **393:** 366.
Waring M. 1970. Variation of the supercoils in closed circular DNA by binding of antibiotics and drugs: Evidence for molecular models involving intercalation. *J. Mol. Biol.* **54:** 247.
Welsh D.K., Yoo S.H., Liu A.C., Takahashi J.S., and Kay S.A. 2004. Bioluminescence imaging of individual fibroblasts reveals persistent, independently phased circadian rhythms of clock gene expression. *Curr. Biol.* **14:** 2289.
Yagita K. and Okamura H. 2000. Forskolin induces circadian gene expression of rPer1, rPer2 and dbp in mammalian rat-1 fibroblasts. *FEBS Lett.* **465:** 79.
Yagita K., Tamanini F., van Der Horst G.T., and Okamura H. 2001. Molecular mechanisms of the biological clock in cultured fibroblasts. *Science* **292:** 278.
Yamazaki S., Numano R., Abe M., Hida A., Takahashi R., Ueda M., Block G.D., Sakaki Y., Menaker M., and Tei H. 2000. Resetting central and peripheral circadian oscillators in transgenic rats. *Science* **288:** 682.
Yoo S.H., Yamazaki S., Lowrey P.L., Shimomura K., Ko C.H., Buhr E.D., Siepka S.M., Hong H.K., Oh W.J., Yoo O.J., Menaker M., and Takahashi J.S. 2004. PERIOD2::LUCIFERASE real-time reporting of circadian dynamics reveals persistent circadian oscillations in mouse peripheral tissues. *Proc. Natl. Acad. Sci.* **101:** 5339.

Integrating the Circadian Oscillator into the Life of the Cyanobacterial Cell

S.S. Golden

Center for Research on Biological Clocks, Department of Biology, Texas A&M University, College Station, Texas 77843-3258

In two decades, the study of circadian rhythms in cyanobacteria has gone from observations of phenomena in intractable species to the development of a model organism for mechanistic study, atomic-resolution structures of components, and reconstitution of a circadian biochemical oscillation in vitro. With sophisticated biochemical, biophysical, genetic, and genomic tools in place, the circadian clock of the unicellular cyanobacterium *Synechococcus elongatus* is poised to be the first for which a systems-level understanding can be achieved.

INTRODUCTION

Initial observations of what are now accepted as clock-controlled processes in cyanobacteria were reported in the mid-1980s by scientists working with natural isolates of diverse species (Ditty et al. 2003). Diurnal rhythms in nitrogen fixation, photosynthesis, amino acid uptake, and cell division were not immediately interpreted as the consequence of an endogenous biological clock. At least two groups recognized that these rhythms were reminiscent of eukaryotic circadian cycles (Sweeney and Borgese 1989; Huang et al. 1990), and T.-C. Huang's group performed experiments to demonstrate that rhythms of amino acid uptake in *Synechococcus* (now *Cyanothece*) RF-1 persist under constant conditions, reset the timing of peak activity in response to an environmental signal, and are temperature-compensated (Huang et al. 1990). With these three criteria fulfilled, a cyanobacterial rhythm was demonstrated to share the hallmarks of eukaryotic circadian rhythms (Dunlap et al. 2004).

EMERGENCE OF *SYNECHOCOCCUS ELONGATUS* PCC 7942 AS A PROKARYOTIC MODEL ORGANISM FOR CIRCADIAN RESEARCH

None of the strains in which rhythmic behavior was first identified is amenable to genetic manipulation, which left the investigation of the cyanobacterial circadian mechanism at a standstill until a genetic model system could be developed. The cyanobacterium *S. elongatus* had not been reported to exhibit any rhythmic behaviors, but *S. elongatus* PCC 7942 was then and remains among the best-developed cyanobacterial genetic systems, and our laboratory was at the forefront of developing molecular tools and protocols for the species (Golden et al. 1987; Golden 1988). M.R. Schaefer in our laboratory (M.R. Schaefer, unpubl.) had noted rhythmic reporter expression in *S. elongatus* during growth under continuous culture conditions over several days. Technician C.A. Strayer had already created marine bacterial luciferase gene fusions as a tool for studying light-regulated photosynthesis genes, based on work in *Streptomyces* by A. Schauer (Schauer et al. 1988). Moreover, we were aware of the reports of circadian rhythmicity in *Synechococcus* (*Cyanothece*) RF-1 through interactions with our colleague V. Cassone. Circadian biologists C.H. Johnson (Vanderbilt University) and T. Kondo (then at the National Institute for Basic Biology in Okazaki, Japan) were in search of a tractable microbial circadian system, and they contacted our lab with the premise that perhaps all cyanobacteria have a clock and that reporter genes might turn a genetically tractable strain into a circadian model organism, as had been done by the S.A. Kay laboratory for the plant *Arabidopsis thaliana* (Millar et al. 1992). A cyanobacterial genetic model was born in 1992, when expression of luciferase, driven by the promoter for the photosynthesis gene *psbAI*, proved to be robustly rhythmic (Kondo et al. 1993).

During the next few years, an international team established baseline data and sophisticated manipulation and analysis tools for circadian rhythms in *S. elongatus*: T. Kondo and colleague M. Ishiura in Okazaki were joined by students S. Aoki and S. Kutsuna; C.H. Johnson at Vanderbilt by student Y. Liu; and our lab at Texas A&M by postdoctoral scientists N.F. Tsinoremas and C.R. Andersson and student N.V. Lebedeva. Together, these researchers quickly showed that luciferase accurately reports gene expression in *S. elongatus* (Liu et al. 1995a), that essentially all loci in the organism are expressed rhythmically (Liu et al. 1995b), and that circadian rhythms of gene expression continue even when cells are dividing faster than once per day (Kondo et al. 1997). A critical advance was the design by T. Kondo of an instrument and software that could acquire bioluminescence data from thousands of colonies over the course of several days, allowing the detection of mutants defective in specific circadian properties (Kondo et al. 1994). This custom instrument, affectionately referred to in the Golden and Johnson labs as the "Kondotron" (and more modestly as "the turntable" in Japan) was used extensively in the Kondo-Ishiura laboratory. Large-scale screening capabilities were extended to the Johnson and Golden labs using, respectively, Kondotron technology and the Packard TopCount luminometer that had been developed for circadian use by the S.A. Kay laboratory (Strayer et al. 2000).

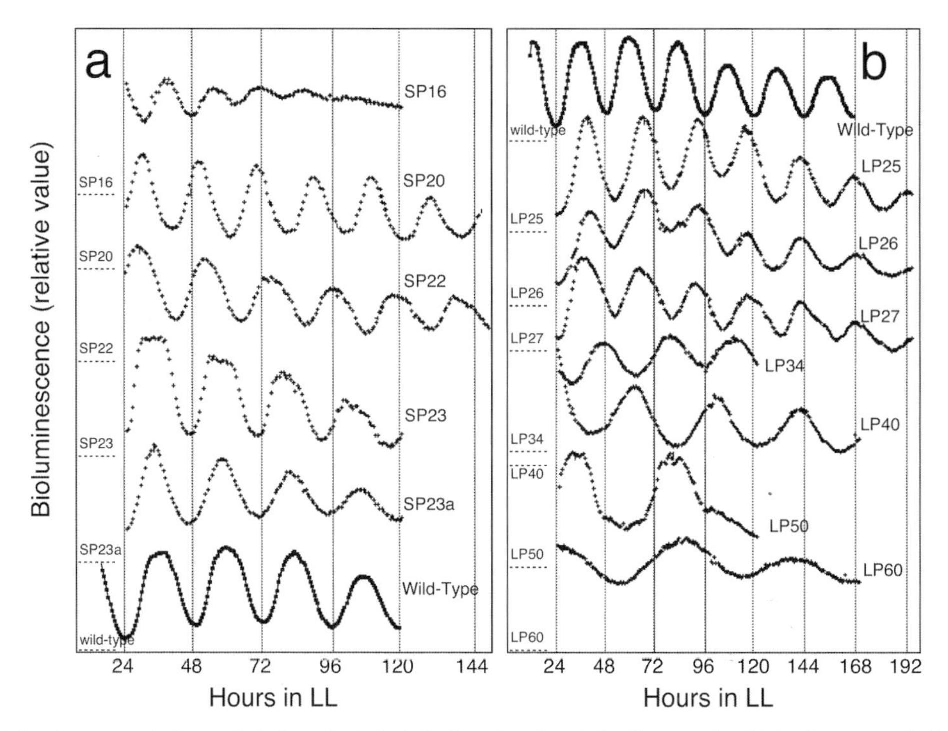

Figure 1. Bioluminescence rhythms of clock mutants that display altered periods. Free-running bioluminescence rhythms of period mutants together with that of wild type under standard assay conditions (36 min white light at 46 μE $m^{-2}s^{-1}$ followed by 9 min of darkness, 30°C) are shown. (*a*) Short-period mutants and wild type; (*b*) long-period mutants and wild type. For all experiments, colonies were grown in LL, given a 12-hour dark pulse, then returned to LL for the measurement of rhythmicity depicted here. Vertical scales for bioluminescence intensity are approximately equivalent. The baseline (dark) levels for the traces are shown (*dashed lines*) on the left vertical axes that are labeled with the name of the mutant. (Figure and legend reprinted, with permission, from Kondo et al. 1994 [AAAS].)

The first mutants to emerge from large-scale mutagenesis procedures turned out to be the most central (Kondo et al. 1994); however, they were not the first whose loci were identified (Tsinoremas et al. 1996; Kutsuna et al. 1998), owing to the greater ease of retrieving physically tagged loci than chemically induced mutant alleles. Kondotron screening yielded a plethora of mutants with altered circadian periods or arrhythmic reporter gene expression after chemical mutagenesis (Fig. 1) (Kondo et al. 1994). Strikingly, all mutant phenotypes were complemented or otherwise altered by the introduction of a single segment of *S. elongatus* DNA from a library (Ishiura et al. 1998). The overlapping segment among several positive library clones carried three open reading frames that did not match any characterized genes in sequence databases and bore no resemblance to cloned clock genes from eukaryotic organisms. They were named *kaiA*, *kaiB*, and *kaiC* for the Japanese kanji Kai, which implies a cycle. Subsequent work in the Kondo lab would show conclusively that their products comprise a circadian oscillator (Nakajima et al. 2005).

THE KAI OSCILLATOR OF *S. ELONGATUS*

The *kaiABC* locus is arranged as three adjacent genes, with a promoter preceding *kaiA* and *kaiB* (Ishiura et al. 1998). The *kaiB* and *kaiC* genes are expressed as a dicistronic operon. Because overexpression of *kaiA* stimulates activity of the *kaiBC* promoter and overexpression of *kaiC* represses it, parallels were drawn to the transcription-translation feedback loops of eukaryotic circadian clock models (Ishiura et al. 1998). From the beginning, however, some of the data did not fit this scheme. For example, the influence of *kaiC* on *kaiBC* expression is not solely negative; KaiC is necessary for full expression of the *kaiBC* promoter. Subsequent work would show that the KaiABC oscillator uses a fundamentally posttranslational mechanism of action (Xu et al. 2003; Nakahira et al. 2004; Ditty et al. 2005; Tomita et al. 2005).

The *kaiA*, *kaiB*, and *kaiC* nucleotide sequences yielded little predictive information regarding their functions (Ishiura et al. 1998). Their only homologs at the time of their initial description in 1997 were unannotated open reading frames in the first entirely sequenced cyanobacterial genome, that of *Synechocystis* PCC 6803 (http://bacteria.kazusa.or.jp/cyanobase/). Among dozens of finished cyanobacterial genome sequences available today (http://www.genomesonline.org/gold.cgi), the *kaiB* and *kaiC* genes are present and adjacently arranged in all but one, *Gloeobacter violaceus* PCC 7421, which is also the only known strain to lack thylakoid membranes to house the photosynthetic machinery (http://bacteria.kazusa.or. jp/cyanobase/). Potential homologs of *kaiB* and *kaiC* are also present in some green and purple photosynthetic bacteria and some Archaea, but *kaiA* appears to be exclusively cyanobacterial (Dvornyk et al. 2003). Some cyanobacteria, notably, the abundant *Prochlorococcus* species of the open ocean, also lack *kaiA*.

Only KaiC had some predictable activities. Its Walker box-P-loop motifs suggested nucleotide binding (Ishiura et al. 1998), the similarity of its amino- and carboxy-terminal halves suggested an iteration in the structure, and general resemblance to the RecA/DNAB superfamily (Leipe et al. 2000; Vakonakis and LiWang 2004) hinted at a multimeric ring structure. Indeed, KaiC binds ATP, autophosphorylates on serine and threonine residues, and forms hexameric rings (Mori et al. 2002; Hayashi et al. 2003; Nishiwaki et al. 2004; Pattanayek et al. 2004; Vakonakis et al. 2004b). The featureless sequences of KaiA and KaiB required that their functions be inferred indirectly by their influence on KaiC phosphorylation.

Structural analysis revealed that KaiA has two independently folded domains, of which the carboxy-terminal portion (C-KaiA) is dimerized and shares with the full-length KaiA protein the activity of stimulating KaiC phosphorylation (Williams et al. 2002; Uzumaki et al. 2004; Vakonakis et al. 2004b). The lack of bioinformatics data for this domain reflected genuine novelty, as C-KaiA established a new self-contained, X-class, four-helix bundle fold (Vakonakis et al. 2004b). The amino-terminal domain (N-KaiA), not involved in direct interaction with KaiC and not universally present among diverse cyanobacterial species, shares conservation in structure, although not in primary sequence, with signal transduction receiver domains (Williams et al. 2002). The biochemical function of N-KaiA is different from that of bona fide receiver domains as it lacks the active-site residues needed for phosphoryl transfer, which is the established activity of receivers. More likely, associations at the interface of amino- and carboxy-terminal domains modify interaction of the carboxy-terminal dimer in its interaction with KaiC. Such a mechanism provides an entry point via N-KaiA for external stimuli to impinge on the Kai oscillator (Fig. 2a).

The interaction of KaiA with KaiC occurs specifically with a peptide at the extreme carboxyl terminus of KaiC (Vakonakis and LiWang 2004). The recent structural determination of KaiC reveals that part of this peptide is tucked into the main hexameric ring as a loop, with an extension protruding from the doughnut (Pattanayek et al. 2004). The straightened conformation of this KaiC segment in complex with C-KaiA (Vakonakis and LiWang 2004), the receiver-fold identity of N-KaiA, and the observation that N-KaiA affects KaiC phosphorylation (Williams et al. 2002) (apparently without direct interactions) all come together in the following model, developed in collaboration with A. LiWang, that connects the input pathways of the clock with the oscillator in *S. elongatus*. We propose that an input pathway impinges on N-KaiA, altering the interface between the KaiA amino- and carboxy-terminal domains and thereby modifying the dimeric structure of C-KaiA (Fig. 2a). This conformational change affects the affinity of C-KaiA for the carboxy-terminal extension on KaiC. In this model, the function of KaiA is to grab the protruding peptide from KaiC and pull the loop out of the folded domain, thereby stimulating the autokinase activity of KaiC (Fig. 2b). In this way, oscillator activity could be hastened or slowed by the input pathways and their influence over KaiA-KaiC-binding activity.

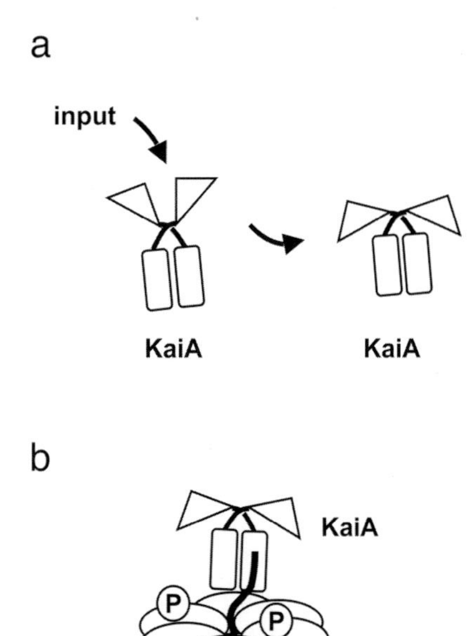

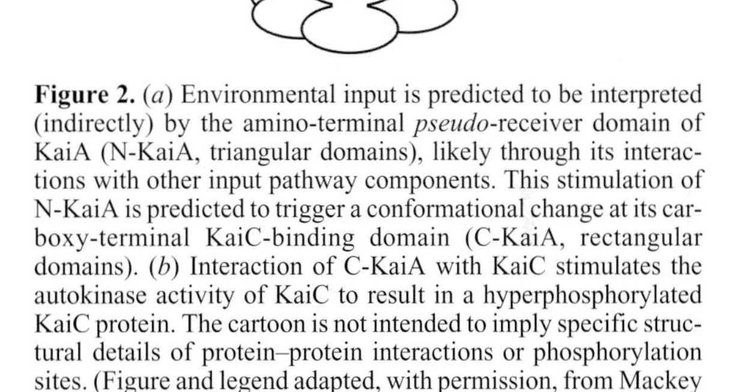

Figure 2. (*a*) Environmental input is predicted to be interpreted (indirectly) by the amino-terminal *pseudo*-receiver domain of KaiA (N-KaiA, triangular domains), likely through its interactions with other input pathway components. This stimulation of N-KaiA is predicted to trigger a conformational change at its carboxy-terminal KaiC-binding domain (C-KaiA, rectangular domains). (*b*) Interaction of C-KaiA with KaiC stimulates the autokinase activity of KaiC to result in a hyperphosphorylated KaiC protein. The cartoon is not intended to imply specific structural details of protein–protein interactions or phosphorylation sites. (Figure and legend adapted, with permission, from Mackey and Golden 2007 [© Elsevier].)

KaiB remains the least understood of the Kai oscillator proteins. It is multimeric in structure, having been reported by different groups to be a dimer (Garces et al. 2004) or a tetramer (Hitomi et al. 2005; Iwase et al. 2005); this lack of agreement may reflect a difference among species, genuine alternative quaternary structures of the protein, or simply differences in interpretation of structural data. The known function of KaiB is to block KaiA-induced autophosphorylation of hyperphosphorylated but not hypophosphorylated KaiC (Williams et al. 2002; Kitayama et al. 2003; Xu et al. 2003; Mori et al. 2007). The combined effects of KaiA and KaiB on KaiC phosphorylation, and the significance of these activities for circadian oscillation, were shown most dramatically by the T. Kondo lab, who reconstituted in vitro a temperature-compensated oscillation of KaiC phosphorylation from a simple mixture of the Kai proteins and ATP (Nakajima et al. 2005). The cycling appearance of a slower-migrating phosphorylated KaiC band recurs with a cycle time of 22–24 hours. C.H. Johnson and colleagues have visualized protein complexes over a time course of this in vitro reaction and confirmed that KaiB comes into play after KaiA is already associated with KaiC (Mori et al. 2007). This finding was predicted by the fact that KaiB does not seem to stimulate dephosphorylation of KaiC in the absence of KaiA (Williams et al. 2002; Kitayama et al. 2003; Xu et al. 2003).

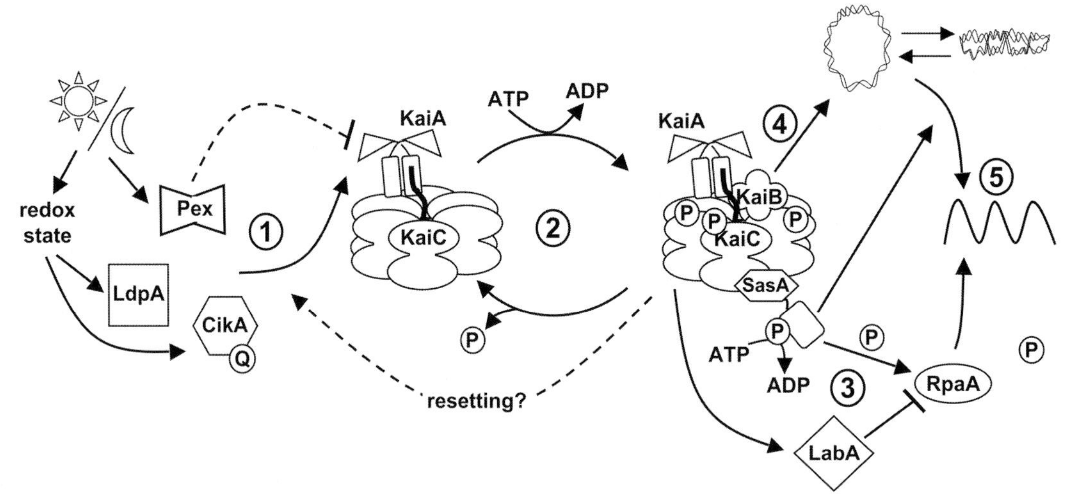

Figure 3. Schematic representation of the internal timing mechanism in *S. elongatus*. (*1*) The Kai-based oscillator is synchronized with the external environment through input pathways. Daily cues, such as fluctuations in light, are interpreted through the input pathway components Pex (described in Kutsuna et al. 1998), LdpA, and CikA. LdpA and CikA likely sense changes in light quantity via modifications in metabolism that effect changes in the cellular redox state. LdpA is a redox-active Fe-S protein, and CikA directly binds a quinone molecule (Q), which alters CikA abundance in the cell. Pex protein represses expression from, the promoter of *kaiA* (Kutsuna et al. 2007). The combined information from these proteins is predicted to feed into and slow down the chemical oscillation of the Kai proteins because the absence of any one of the input pathway components results in a shorter circadian rhythm, reflective of a faster clock. (*2*) The oscillation in KaiC phosphorylation results from the opposing actions of KaiA, which stimulates KaiC autokinase activity, and KaiB, which abolishes the positive effect of KaiA. Increasing evidence supports the hypothesis that the phosphorylation state of KaiC affects the ability of the clock to reset properly. (*3*) The SasA protein interacts with KaiC through its amino terminus (hexagonal domain), which is similar in sequence, but not structure, to KaiB. Temporal information is transduced from the Kai oscillator to SasA through the stimulation of SasA autophosphorylation by KaiC. RpaA is activated by SasA via transfer of a phosphoryl group. LabA acts as a mediator of negative regulation from KaiC and is predicted to repress RpaA function (described in Takai et al. 2006). (*4*) Additionally, the Kai oscillator regulates the compaction rhythm of the cyanobacterial nucleoid (double-strand loop), which would control accessibility of transcriptional machinery to promoter regions. (*5*) SasA is not required for rhythmic chromosome compaction, but it is necessary (as is RpaA) for overt rhythmicity of gene expression. Arrows and perpendicular lines indicate positive and negative regulation, respectively. The cartoon is not intended to imply specific structural details of protein–protein interactions or phosphorylation sites. (Figure and legend adapted, with permission, from Mackey and Golden 2007 [© Elsevier].)

CONNECTING THE OSCILLATOR WITH THE OUTSIDE WORLD

No data have, as yet, confirmed that input pathways converge on N-KaiA. However, some input pathway components have been identified (Fig. 3), and they copurify from cellular extracts with KaiA (Ivleva et al. 2005, 2006). The first is LdpA, for light-dependent period, which is an iron-sulfur cluster-containing protein (Katayama et al. 2003). Consistent with its cofactors, LdpA undergoes reduction-oxidation transitions (Ivleva et al. 2005). The protein is named for the phenotype of a null mutant, which has a fixed circadian period regardless of ambient light intensity. Wild-type *S. elongatus* follows Aschoff's Rule, such that the period is shorter under high-light conditions and longer under lower light, with a range of about 24–25 hours. The *ldpA* mutant is locked at the short end of the range (Katayama et al. 2003). Overexpression of LdpA lengthens circadian period by about 1 hour (Ivleva et al. 2005).

Another factor that is needed to transmit environmental cues to the circadian clock is CikA (circadian input kinase), a histidine protein kinase whose activity is required for resetting the phase of circadian rhythms (Schmitz et al. 2000). A *cikA* null mutant is "blind" to a 5-hour dark pulse administered during free-running conditions in constant light; in contrast, the wild type will reset the timing of the next peak by up to 8 hours, depending on the circadian time at which the dark pulse is administered. When either LdpA or CikA is expressed with a 6-histidine affinity tag and retrieved from cyanobacterial cells, both KaiA and KaiC copurify with them (Ivleva et al. 2005, 2006). Likewise, His-tagged KaiC retrieves CikA (Ivleva et al. 2006). These data provide biochemical evidence that the input and oscillator divisions of the clock are physically associated in *S. elongatus*.

CikA includes a GAF, a histidine protein kinase (HPK), and a receiver-like domain (Mutsuda et al. 2003). Although GAF is a chromophore-binding domain in phytochrome-type photoreceptors, it does not seem to serve that function in CikA: The cysteine residue necessary for bilin attachment is absent, and CikA purified from *S. elongatus* does not carry a covalently bound cofactor. The HPK domain conforms in both sequence and biochemical activity with well-characterized members of this family from diverse bacteria. The receiver-like domain is recognizable in sequence as a member of that class, but it lacks the crucial aspartate residue needed to perform receiver phosphoryltransfer activity. Thus, like N-KaiA, the carboxy-terminal domain of CikA is a *pseudo*-receiver (PsR) (Mutsuda et al. 2003; Gao et al. 2007).

The PsR domain of CikA serves several important functions in the protein (Ivleva et al. 2006; Zhang et al. 2006). It represses kinase activity, likely by associating with the HPK domain as would a bona fide receiver and thereby occluding the active-site histidine (Zhang et al. 2006; Gao et al. 2007). Removal of PsR markedly increases autophosphorylation activity of the HPK domain (Zhang et al. 2006). Fusion of a fluorescent protein, ZsGreen, to the amino terminus of CikA produces a fully functional protein that localizes to the pole of the cell. Removal of the PsR domain blocks localization and left the fluorescent fusion distributed throughout the cell. These results are consistent with genetic evidence that the PsR domain is used to interact with one or more partners and that those components are related to the clock. Whereas a *cikA* null mutant still exhibits circadian rhythms, albeit with a reduced amplitude and shortened circadian period, a strain in which *cikA* is overexpressed is arrhythmic (Mutsuda et al. 2003; Zhang et al. 2006). This cessation of rhythmicity is dependent on the presence of the PsR domain in the overexpressed protein, suggesting that this domain titrates out components that are essential for the circadian cycle (Zhang et al. 2006). Most recently, the PsR domain was shown to bind a quinone cofactor, implicating CikA as a direct redox sensor (Ivleva et al. 2006).

Despite several forward genetic screens aimed at identification of clock-related photoreceptors and the mutation of candidate genes of this class, there are no data to suggest that light information is transmitted directly to the clock in *S. elongatus*. The known input components implicate a cellular redox state, which varies as a function of photosynthetic electron transport, as a proxy for light in setting the clock (Ivleva et al. 2005, 2006). The extensive photosynthetic apparatus, so central to this obligate photoautotroph, may fulfill the role of photoreceptor for the circadian system.

Work is in progress to identify the links between the known input components LdpA and CikA and the circadian oscillator. Candidate pathway constituents are under investigation to determine whether there is a specific flow of environmental information from these factors to the amino-terminal domain of KaiA as proposed in the model (S.R. Mackey and S.S. Golden, unpubl.). We are developing fusions of oscillator proteins to the yellow fluorescent protein (YFP) in order to determine cellular localization of oscillator proteins during the circadian cycle (G. Dong and S.S. Golden, unpubl.).

CONNECTING THE OSCILLATOR WITH THE FUNCTIONS IT CONTROLS

In *S. elongatus*, the best-documented clock-controlled process is the global expression of genes throughout the genome (Liu et al. 1995b). Two cellular mechanisms are thought to contribute to rhythmic gene expression (Fig. 3). A signal transduction pathway composed of an HPK protein called SasA (*Synechococcus* adaptive sensor) and its probable response regulator partner RpaA (regulator of phycobilisome-associated) provide a direct route for information to flow from the oscillator to gene expression (Iwasaki et al. 2000; Takai et al. 2006). SasA, whose presumptive sensor domain resembles KaiB at the primary sequence level but not in structure (Iwasaki et al. 2000; Garces et al. 2004; Vakonakis et al. 2004a; Hitomi et al. 2005), interacts physically with KaiC (Iwasaki et al. 2000). This interaction greatly stimulates the autokinase activity of SasA (Iwasaki et al. 2000) and the phosphorylation of RpaA (Takai et al. 2006). The latter protein carries motifs characteristic of DNA-binding proteins. However, a specific binding site for RpaA has not been determined. Mutants defective for either SasA or RpaA have very attenuated rhythmicity of gene expression, although oscillation of expression from the *kaiBC* promoter, with a shortened period and reduced amplitude, is evident (Iwasaki et al. 2000; Takai et al. 2006). SasA does not affect the phosphorylation of KaiC in vitro, indicating that the flow of information is from the oscillator to SasA (Smith and Williams 2006).

Visualization of the cyanobacterial nucleoid during the circadian cycle has revealed that the chromosome compacts and decompacts rhythmically (Smith and Williams 2006). Such a comprehensive change in genome topology could explain the universal rhythmicity of promoter activity in this organism and the residual rhythmicity observed in *sasA* and *rpaA* mutants. The chromosome compaction cycle is dependent on the *kai* genes but does not require SasA (Smith and Williams 2006). Thus, the clock appears to include output pathways from the oscillator to gene expression using both direct transcription modulation, via SasA and RpaA, and indirectly through wholesale remodeling of the chromosome. The targets of the former pathway, and the mediators of the latter, have yet to be determined.

DISCOVERY OF NEW CONNECTIONS BETWEEN THE CLOCK AND CELLULAR ACTIVITIES

Our understanding of the circadian oscillator and the greater clock in *S. elongatus* has come a long way in a relatively short period of time, but we still do not know what the oscillator really does, other than perform rhythmic phosphorylation of KaiC. Although many large-scale mutant hunts have been performed in *S. elongatus* and many genes related to the clock have been identified, more participants are predicted to exist in forging connection of the circadian oscillator to the cell it serves. The diminutive size of the *S. elongatus* genome renders feasible a comprehensive examination of each locus for its potential effect on circadian timing. A functional genomics project is under way in which we have mutagenized segments of the genome with transposons to achieve saturation of mutant alleles (Holtman et al. 2005). Individual genes that carry insertion mutations are introduced into bioluminescent reporter strains where they recombine with native chromosomal alleles; each mutant is then screened for alterations in circadian gene expression. Thus far, 95% of loci have been mutagenized, and mutants detected for 700 different loci, approximately one quarter of the genome, have been screened for circadian defects (Holtman et al., unpubl.). From the screened pool, 71 genes not previously associated with the clock have been identified as affecting circadian gene expression; most of these cause long or short circadian periods when mutated or change the phase of peak reporter gene expression. None that cause complete arrhythmia have been identified, other than the known *kaiA*, *kaiB*, and *kaiC* genes. The newly identified genes fall into various categories, including many loci of unknown function, some of which are conserved among diverse cyanobacteria and some that are apparently unique to *S. elongatus*. Some loci of particular interest are predicted to encode redox-active proteins, supporting an emerging picture of redox, rather than light itself, as the chief environment input currency in the cell.

Many of the mutants that show circadian phenotypes carry insertion alleles that cannot be segregated entirely from their wild-type counterparts, indicating that the mutated locus is essential for viability (Holtman et al. 2005). In these situations, the phenotype is observed in merodiploid strains that carry both alleles and presumably have decreased expression of the target gene. Because *S. elongatus* carries multiple copies of its chromosome, a strain with a mixed population of mutant and wild-type chromosomes may carry a range of ratios of alleles, which is influenced by selective pressure for an antibiotic resistance gene on the transposon.

The first clock-affecting gene found in the genomics screen reinforces a common theme found in eukaryotic circadian systems: Proteolysis is an important factor for maintaining normal circadian rhythms in vivo. The *clpP2* and *clpX* genes form an operon, in which the upstream gene encodes the ClpP2 protease and the downstream gene encodes the ClpX chaperone that delivers substrates to the enzyme. Neither can be inactivated completely in *S. elongatus*, but merodiploids affected in either locus have long-period phenotypes (Holtman et al. 2005). Conditional expression of an antisense RNA that is complementary to and spans the intergenic region of the dicistronic RNA phenocopies the (merodiploid) mutant phenotype. Because bacteria do not possess the dicer pathway that is responsible for gene silencing, the gene suppression is temporary and presumably results from direct interference with transcription or translation, or stimulates a pathway that nonspecifically recognizes double-stranded RNA. The ability to conditionally suppress genes that are essential for viability provides a valuable tool for assessing the contributions to circadian timing of genes that tie the clock to critical cellular processes.

Although we do not yet know how the circadian clock orchestrates the life of the cell, we do know that its activity is important for the organism. The wealth of circadian period mutants generated in the early 1990s provided the means to test the extent to which a circadian clock confers an adaptive advantage. Mutant and wild-type strains that have different free-running periods grow equally well when grown in different T cycles (11 hr light:11 hr dark; 15 hr light:15 hr dark). However, big differences in survival emerge when these strains are mixed in equal numbers and challenged to compete in days of different lengths (Ouyang et al. 1998). Under the pressure of competition, the strain whose free-running period is closest to that of the T cycle always emerges as the predominant strain. Together, these findings suggest that cyanobacteria actively influence the growth of others in the population and that resonance of circadian period with the prevailing daily cycle confers a fitness advantage.

HOW REPRESENTATIVE IS *S. ELONGATUS*?

S. elongatus is one representative among a handful of cyanobacterial strains that are amenable to genetic manipulation (Koksharova and Wolk 2002). Of these, it has the smallest genome, the most limited metabolic range, and no developmental programs. It is a freshwater mesophile that has been isolated from environments in Texas and California (PCC 6301 and PCC 7942, Pasteur Culture Collection, http://www.pasteur.fr/recherche/banques/PCC/index.html). The cyanobacterial clade encompasses very diverse species, including multicellular organisms that differentiate specialized cell types, species that help to form desert crusts, denizens of mats in alkaline hot springs, abundant phytoplankton in the open ocean, and strains that thrive in Antarctica.

The age of genomics has particularly favored the cyanobacterial clade, with more than two dozen genomes from diverse species available. Sequenced genomes vary in size from 1.7 to 12 Mbp (http://www.genomesonline.org/gold.cgi), and genome size correlates roughly with the range of metabolic and developmental capabilities of a particular species (Rippka et al. 1979; Herdman et al. 2001). The diversity of lifestyles and genetic makeup suggests that cyanobacteria will show variations in circadian machinery. For example, the presence of multiple copies of *kaiB* and *kaiC* in some species, which does not correlate with genome size, suggests the action of multiple oscillators or, at minimum, a more complicated oscillatory cycle than has been proposed in *S. elongatus*. Some input pathway features are certain to vary among species. KaiA in *Nostoc punctiforme* and *Anabaena* sp. PCC 7120 lacks the amino-terminal domain of the *S. elongatus* protein (Williams et al. 2002), and the oceanic *Prochlorococcus* species do not encode a KaiA protein at all (Dvornyk et al. 2003). The CikA kinase, which is essential for circadian resetting in *S. elongatus*, has no clear homologs in other cyanobacteria. Variations among species in the mechanism for the transfer of environmental light information to the clock are not surprising, because their light-harvesting antenna structures and pigments for photosynthesis are also diverse (Glazer 1989; Ting et al. 2002). Thus, although the findings in *S. elongatus* have been enlightening and instructive, it should not be assumed that all cyanobacterial circadian clocks will operate like the *S. elongatus* clock. Variations in circadian mechanism at least as notable as those that distinguish the clocks of fruit flies and mice should be expected.

Potential homologs of the KaiB and KaiC proteins of the *S. elongatus* circadian oscillator are encoded in the genomes of some other bacteria and some Archaea (Dvornyk et al. 2003). A question still unanswered is whether these proteins serve circadian functions in their hosts or perhaps perform ancestral, nonclock activities. Moreover, many clades in the bacterial domain do not have *kai* homologs, but they might have undiscovered, nonhomologous clocks, just as plants and fungi have circadian oscillators all their own. Only concerted research efforts in other prokaryotic models will reveal the extent of circadian biological timing in the Tree of Life.

CONCLUSIONS

The choice of *S. elongatus* as the cyanobacterial circadian model was, in retrospect, quite fortuitous. Robust and high-amplitude rhythms of bioluminescence were observed from the first luciferase reporter gene that was tested (*psbAI* promoter), and these clear rhythms facilitated circadian screening. Another popular genetically tractable cyanobacterium, *Synechocystis* sp. strain PCC 6803, was found to exhibit lower amplitude rhythms in similar assays (Aoki et al. 1995, 2002). Moreover, now that the oscillator genes have been identified, the presence of additional copies of *kaiB* and *kaiC* in the *Synechocystis* genome (Dvornyk et al. 2003; Chen 2007) suggests that redundancy might have complicated forward genetics had that organism been chosen initially.

The efforts of many talented students and senior scientists have contributed to the richly developed clock model system that we enjoy today with *S. elongatus*. Among these, the creative technological innovations of T. Kondo, which drove the development of phenotypic screening (Kondo et al. 1994) and the in vitro oscillator system, merit special notice (Nakajima et al. 2005). Clock analysis in this organism is now at a very sophisticated stage, with structural detail (Garces et al. 2004; Pattanayek et al. 2004; Ye et al. 2004; Hitomi et al. 2005), mathematical models (Mori et al. 2007), global genomics tools (Holtman et al. 2005), and cellular imaging probes (Zhang et al. 2006). Fifteen years of progress in describing the parts that comprise the oscillator have led us to the threshold of understanding how the cell harnesses this ticking mechanism to build a clock that tunes the life of a cyanobacterium to thrive in the world it inhabits.

ACKNOWLEDGMENTS

I am deeply indebted to C.H. Johnson and T. Kondo for the invitation to join the world of chronobiologists and for their longtime collaboration and friendship. I also gratefully acknowledge the extensive role of M. Ishiura, who contributed richly to the progress described here. They and their students have shared hopes, ideas, and reagents with my group for over a decade. My colleague A. LiWang is the very personification of an effective and enthusiastic collaborator, and the work in my lab owes an ever-increasing debt to him and his students as our ideas and expertise continue to mesh. The members of the Center for Research on Biological Clocks at Texas A&M provide guidance, inspiration, and collegiality on a daily basis. My own students have given tirelessly and enthusiastically of themselves in investigating the clock while retaining “a feeling for the organism” (Keller 1983). I thank them all, with special nods to C.R. Andersson, N.F. Tsinoremas, S.B. Williams, and S.R. (Mini-Me) Mackey. Work in my lab is supported by grants from the National Institutes of Health (P01 NS039546 and R01 GM062419) and the Department of Energy (DE-FG02-04ER15558).

REFERENCES

Aoki S., Kondo T., and Ishiura M. 1995. Circadian expression of the *dnaK* gene in the cyanobacterium *Synechocystis* sp. strain PCC 6803. *J. Bacteriol.* **177:** 5606.

———. 2002. A promoter-trap vector for clock-controlled genes in the cyanobacterium *Synechocystis* sp. PCC 6803. *J. Microbiol. Methods* **49:** 265.

Chen Y. 2007. “Functional genomics of the unicellular cyanobacterium *Synechococcus elongatus* PCC 7942.” Ph.D. thesis, Texas A & M University, College Station, Texas.

Ditty J.L., Williams S.B., and Golden S.S. 2003. A cyanobacterial circadian timing mechanism. *Annu. Rev. Genet.* **37:** 513.

Ditty J.L., Canales S.R., Anderson B.E., Williams S.B., and Golden S.S. 2005. Stability of the *Synechococcus elongatus* PCC 7942 circadian clock under directed anti-phase expression of the *kai* genes. *Microbiology* **151:** 2605.

Dunlap J.C., Loros J.J., and DeCoursey P.J., eds. 2004. *Chronobiology: Biological timekeeping*. Sinauer Associates, Sunderland, Massachusetts.

Dvornyk V., Vinogradova O., and Nevo E. 2003. Origin and evolution of circadian clock genes in prokaryotes. *Proc. Natl. Acad. Sci.* **100:** 2495.

Gao T., Zhang X., Ivleva N.B., Golden S.S., and LiWang A. 2007. NMR structure of the *pseudo*-receiver domain of CikA. *Protein Sci.* **16:** 465.

Garces R.G., Wu N., Gillon W., and Pai E.F. 2004. *Anabaena* circadian clock proteins KaiA and KaiB reveal a potential common binding site to their partner KaiC. *EMBO J.* **23:** 1688.

Glazer A.N. 1989. Light guides. Directional energy transfer in a photosynthetic antenna. *J. Biol. Chem.* **264:** 1.

Golden S.S. 1988. Mutagenesis of cyanobacteria by classical and gene-transfer-based methods. *Methods Enzymol.* **167:** 714.

Golden S.S., Brusslan J., and Haselkorn R. 1987. Genetic engineering of the cyanobacterial chromosome. *Methods Enzymol.* **153:** 215.

Hayashi F., Suzuki H., Iwase R., Uzumaki T., Miyake A., Shen J.R., Imada K., Furukawa Y., Yonekura K., Namba K., and Ishiura M. 2003. ATP-induced hexameric ring structure of the cyanobacterial circadian clock protein KaiC. *Genes Cells* **8:** 287.

Herdman M., Castenholz R.W., Iteman I., Waterbury J.B., and Rippka R. 2001. Subsection I: (Formerly **Chroococcales** Wettstein 1924, emend. Rippka, Deruelles, Waterbury, Herdman and Stanier 1979). In *Bergey's manual of systematic bacteriology* (ed. D.R. Boone et al.), p. 721. Springer-Verlag, New York.

Hitomi K., Oyama T., Han S., Arvai A.S., and Getzoff E.D. 2005. Tetrameric architecture of the circadian clock protein KaiB. A novel interface for intermolecular interactions and its impact on the circadian rhythm. *J. Biol. Chem.* **280:** 19127.

Holtman C.K., Chen Y., Sandoval P., Gonzales A., Nalty M.S., Thomas T.L., Youderian P., and Golden S.S. 2005. High-throughput functional analysis of the *Synechococcus elongatus* PCC 7942 genome. *DNA Res.* **12:** 103.

Huang T.-C., Tu J., Chow T.J., and Chen T.-H. 1990. Circadian rhythm of the prokaryote *Synechococcus* sp. RF-1. *Plant Physiol.* **92:** 531.

Ishiura M., Kutsuna S., Aoki S., Iwasaki H., Andersson C.R., Tanabe A., Golden S.S., Johnson C.H., and Kondo T. 1998. Expression of a gene cluster *kaiABC* as a circadian feedback process in cyanobacteria. *Science* **281:** 1519.

Ivleva N.B., Bramlett M.R., Lindahl P.A., and Golden S.S. 2005. LdpA: A component of the circadian clock senses redox state of the cell. *EMBO J.* **24:** 1202.

Ivleva N.B., Gao T., LiWang A.C., and Golden S.S. 2006. Quinone sensing by the circadian input kinase of the cyanobacterial circadian clock. *Proc. Natl. Acad. Sci.* **103:** 17468.

Iwasaki H., Williams S.B., Kitayama, Y., Ishiura M., Golden S.S., and Kondo T. 2000. A KaiC-interacting sensory histidine kinase, SasA, necessary to sustain robust circadian oscillation in cyanobacteria. *Cell* **101:** 223.

Iwase R., Imada K., Hayashi F., Uzumaki T., Morishita M., Onai K., Furukawa Y., Namba K., and Ishiura M. 2005. Functionally important substructures of circadian clock protein KaiB in a unique tetramer complex. *J. Biol. Chem.* **280:** 43141.

Katayama M., Kondo T., Xiong J., and Golden S.S. 2003. *ldpA* encodes an iron-sulfur protein involved in light-dependent modulation of the circadian period in the cyanobacterium *Synechococcus elongatus* PCC 7942. *J. Bacteriol.* **185:** 1415.

Keller E.F. 1983. *A feeling for the organism: The life and work of Barbara McClintock.* Adonis Press, Hillsdale, New York.

Kitayama Y., Iwasaki H., Nishiwaki T., and Kondo T. 2003. KaiB functions as an attenuator of KaiC phosphorylation in the cyanobacterial circadian clock system. *EMBO J.* **22:** 2127.

Koksharova O. and Wolk C. 2002. Genetic tools for cyanobacteria. *Appl. Microbiol. Biotechnol.* **58:** 123.

Kondo T., Mori T., Lebedeva N.V., Aoki S., Ishiura M., and Golden S.S. 1997. Circadian rhythms in rapidly dividing cyanobacteria. *Science* **275:** 224.

Kondo T., Tsinoremas N.F., Golden S.S., Johnson C.H., Kutsuna S., and Ishiura M. 1994. Circadian clock mutants of cyanobacteria. *Science* **266:** 1233.

Kondo T., Strayer C.A., Kulkarni R.D., Taylor W., Ishiura M., Golden S.S., and Johnson C.H. 1993. Circadian rhythms in prokaryotes: Luciferase as a reporter of circadian gene expression in cyanobacteria. *Proc. Natl. Acad. Sci.* **90:** 5672.

Kutsuna S., Kondo T., Aoki S., and Ishiura M. 1998. A period-extender gene, *pex*, that extends the period of the circadian clock in the cyanobacterium *Synechococcus* sp. strain PCC 7942. *J. Bacteriol.* **180:** 2167.

Kutsyna S., Kondo T., Ikegami H., Uzumaki T., Katayama M., and Ishiura M. 2007. The circadian clock-related gene *pex* regulates a negative *cis* element in the *kaiA* promoter region. *J. Bacteriol.* **189:** 7690.

Leipe D.D., Aravind L., Grishin N.V., and Koonin E.V. 2000. The bacterial replicative helicase DnaB evolved from a RecA duplication. *Genome Res.* **10:** 5.

Liu Y., Golden S.S., Kondo T., Ishiura M., and Johnson C.H. 1995a. Bacterial luciferase as a reporter of circadian gene expression in cyanobacteria. *J. Bacteriol.* **177:** 2080.

Liu Y., Tsinoremas N.F., Johnson C.H., Lebedeva N.V., Golden S.S., Ishiura M., and Kondo T. 1995b. Circadian orchestration of gene expression in cyanobacteria. *Genes Dev.* **9:** 1469.

Mackey S.R. and Golden S.S. 2007. Winding up the cyanobacterial circadian clock. *Trends Microbiol.* **15:** 381.

Millar A.J., Short S.R., Chua N.H., and Kay S.A. 1992. A novel circadian phenotype based on firefly luciferase expression in transgenic plants. *Plant Cell* **4:** 1075.

Mori T., Saveliev S.V., Xu Y., Stafford W.F., Cox M.M., Inman R.B., and Johnson C.H. 2002. Circadian clock protein KaiC forms ATP-dependent hexameric rings and binds DNA. *Proc. Natl. Acad. Sci.* **99:** 17203.

Mori T., Williams D.R., Byrne M.O., Qin X., Egli M., McHaourab H.S., Stewart P.L., and Johnson C.H. 2007. Elucidating the ticking of an *in vitro* circadian clockwork. *PLoS Biol.* **5:** e93.

Mutsuda M., Michel K.P., Zhang X., Montgomery B.L., and Golden S.S. 2003. Biochemical properties of CikA, an unusual phytochrome-like histidine protein kinase that resets the circadian clock in *Synechococcus elongatus* PCC 7942. *J. Biol. Chem.* **278:** 19102.

Nakahira Y., Katayama M., Miyashita H., Kutsuna S., Iwasaki H., Oyama T., and Kondo T. 2004. Global gene repression by KaiC as a master process of prokaryotic circadian system. *Proc. Natl. Acad. Sci.* **101:** 881.

Nakajima M., Imai K., Ito H., Nishiwaki T., Murayama Y., Iwasaki H., Oyama T., and Kondo T. 2005. Reconstitution of circadian oscillation of cyanobacterial KaiC phosphorylation *in vitro*. *Science* **308:** 414.

Nishiwaki T., Satomi Y., Nakajima M., Lee C., Kiyohara R., Kageyama H., Kitayama Y., Temamoto M., Yamaguchi A., Hijikata A., Go M., Iwasaki H., Takao T., and Kondo T. 2004. Role of KaiC phosphorylation in the circadian clock system of *Synechococcus elongatus* PCC 7942. *Proc. Natl. Acad. Sci.* **101:** 13927.

Ouyang Y., Andersson C.R., Kondo T., Golden S.S., and Johnson C.H. 1998. Resonating circadian clocks enhance fitness in cyanobacteria. *Proc. Natl. Acad. Sci.* **95**: 8660.

Pattanayek R., Wang J., Mori T., Xu Y., Johnson C.H., and Egli M. 2004. Visualizing a circadian clock protein: Crystal structure of KaiC and functional insights. *Mol. Cell* **15:** 375.

Rippka R., Deruelles J., Waterbury J.B., Herdman M., and Stanier R.Y. 1979. Generic assignments, strain histories and properties of pure cultures of cyanobacteria. *J. Gen. Microbiol.* **111:** 1.

Schauer A., Ranes M., Santamaria R., Guijarro J., Lawlor E., Mendez C., Chater K., and Losick R. 1988. Visualizing gene expression in time and space in the filamentous bacterium *Streptomyces coelicolor*. *Science* **240:** 768.

Schmitz O., Katayama M., Williams S.B., Kondo T., and Golden S.S. 2000. CikA, a bacteriophytochrome that resets the cyanobacterial circadian clock. *Science* **289:** 765.

Smith R.M. and Williams S.B. 2006. Circadian rhythms in gene transcription imparted by chromosome compaction in the cyanobacterium *Synechococcus elongatus*. *Proc. Natl. Acad. Sci.* **103:** 8564

Strayer C., Oyama T., Schultz T.F., Raman R., Somers D.E., Mas P., Panda S., Kreps J.A., and Kay S.A. 2000. Cloning of the *Arabidopsis* clock gene TOC1, an autoregulatory response regulator homolog. *Science* **289:** 768.

Sweeney B.M. and Borgese M.B. 1989. A circadian rhythm in cell division in a prokaryote, the cyanobacterium *Synechococcus* WH7803. *J. Phycol.* **25:** 183.

Takai N., Nakajima M., Oyama T., Kito R., Sugita C., Sugita M., Kondo T., and Iwasaki H. 2006. A KaiC-associating SasA-RpaA two-component regulatory system as a major circadian timing mediator in cyanobacteria. *Proc. Natl. Acad. Sci.* **103:** 12109.

Ting C.S., Rocap G., King J., and Chisholm S.W. 2002. Cyanobacterial photosynthesis in the oceans: The origins and significance of divergent light-harvesting strategies. *Trends Microbiol.* **10:** 134.

Tomita J., Nakajima M., Kondo T., and Iwasaki H. 2005. No transcription-translation feedback in circadian rhythm of KaiC phosphorylation. *Science* **307:** 251.

Tsinoremas N.F., Ishiura M., Kondo T., Andersson C.R., Tanaka K., Takahashi H., Johnson C.H., and Golden S.S. 1996. A sigma factor that modifies the circadian expression of a subset of genes in cyanobacteria. *EMBO J.* **15:** 2488.

Uzumaki T., Fujita M., Nakatsu T., Hayashi F., Shibata H., Itoh N., Kato H., and Ishiura M. 2004. Crystal structure of the C-terminal clock-oscillator domain of the cyanobacterial KaiA protein. *Nat. Struct. Mol. Biol.* **11:** 623.

Vakonakis I. and LiWang A.C. 2004. Structure of the C-terminal domain of the clock protein KaiA in complex with a KaiC-derived peptide: Implications for KaiC regulation. *Proc. Natl. Acad. Sci.* **101:** 10925.

Vakonakis I., Klewer D.A., Williams S.B., Golden S.S., and LiWang A.C. 2004a. Structure of the N-terminal domain of the circadian clock-associated histidine kinase SasA. *J. Mol. Biol.* **342:** 9.

Vakonakis I., Sun J., Wu T., Holzenburg A., Golden S.S., and LiWang A.C. 2004b. NMR structure of the KaiC-interacting C-terminal domain of KaiA, a circadian clock protein: Implications for KaiA-KaiC interaction. *Proc. Natl. Acad. Sci.* **101:** 1479.

Williams S.B., Vakonakis I., Golden S.S., and LiWang A.C. 2002. Structure and function from the circadian clock protein KaiA of *Synechococcus elongatus:* A potential clock input mechanism. *Proc. Natl. Acad. Sci.* **99:** 15357.

Xu Y., Mori T., and Johnson C.H. 2003. Cyanobacterial circadian clockwork: Roles of KaiA, KaiB and the kaiBC promoter in regulating KaiC. *EMBO J.* **22:** 2117.

Ye S., Vakonakis I., Ioerger T.R., LiWang A.C., and Sacchettini J.C. 2004. Crystal structure of circadian clock protein KaiA from *Synechococcus elongatus*. *J. Biol. Chem.* **279:** 20511.

Zhang X., Dong G., and Golden S.S. 2006. The *pseudo*-receiver domain of CikA regulates the cyanobacterial circadian input pathway. *Mol. Microbiol.* **60:** 658.

The Yeast Metabolic Cycle: Insights into the Life of a Eukaryotic Cell

B.P. Tu and S.L. McKnight

Department of Biochemistry, University of Texas Southwestern Medical Center, Dallas, Texas 75390-9038

The budding yeast *Saccharomyces cerevisiae* undergoes robust oscillations in oxygen consumption during continuous growth under nutrient-limited conditions. Comprehensive microarray studies reveal that more than half of the yeast genome is expressed periodically as a function of these respiratory oscillations, thereby specifying an extensively orchestrated program responsible for regulating numerous cellular outputs. Here, we summarize the logic of the yeast metabolic cycle (YMC) and highlight additional cellular processes that are predicted to be compartmentalized in time. Certain principles of temporal orchestration as seen during the YMC might be conserved across other biological cycles.

INTRODUCTION

The budding yeast *S. cerevisiae* has been known for decades to be capable of exhibiting various modes of oscillatory behavior, both in cell-free extracts and during continuous growth (Chance et al. 1964; von Meyenburg 1969; Satroutdinov et al. 1992; Richard 2003; Klevecz et al. 2004). We recently studied a continuous culture system that revealed a robust approximately 4–5-hour YMC that occurs under nutrient-limited growth conditions (Tu et al. 2005). When prototrophic yeast cells are grown in culture to a high density, starved for a short period, and then continuously fed low concentrations of glucose, the population of cells becomes highly synchronized and undergoes robust oscillations of oxygen consumption (Fig. 1). During these oscillations, cells proceed through phases where they rapidly consume oxygen, followed by longer phases where they consume much less oxygen (Fig. 1). The period length of these metabolic cycles under such growth conditions is typically about 4–5 hours, but it can vary depending on the rate of glucose addition (Tu et al. 2005). Such cycles persist as long as continuous concentrations of glucose are supplied to the cells.

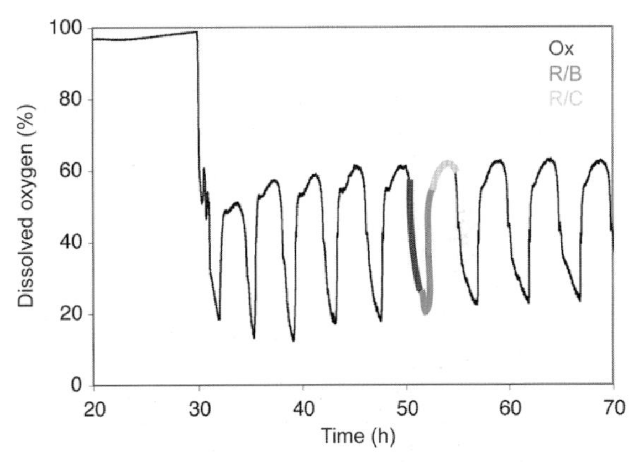

Figure 1. The yeast metabolic cycle. During continuous culture growth, budding yeast exhibit robust metabolic cycles as measured by oxygen consumption. These cycles are composed of three major phases: OX (oxidative), RB (reductive, building), and RC (reductive charging). Shown are approximately 4–5-hour metabolic cycles exhibited by the wild-type diploid strain CEN.PK upon continuous feeding of glucose.

Remarkably, more than half of yeast genes (~57%) are expressed periodically as a precise function of the YMC (Tu et al. 2005). Genes that encode proteins with a common function often display similar temporal expression patterns, and different classes of genes are up-regulated at entirely different temporal windows of the YMC. Moreover, genes that encode proteins having functions that are associated with energy and metabolism tend to be expressed with exceptionally precise and robust periodicity, suggesting that these cycles are intrinsically metabolic (Tu et al. 2005).

Analysis of the temporal gene expression profiles revealed three superclusters of gene expression, thereby defining three primary phases of the YMC: OX (oxidative, respiratory), RB (reductive, building), and RC (reductive, charging) (Fig. 1) (Tu et al. 2005). Different categories of genes peak during each phase, and cells successively pass through each of the three phases of the YMC in every cycle. The temporal gene expression data predict that the YMC controls precisely when certain cellular and metabolic events occur (e.g., respiration, mitochondria biogenesis, ribosome biogenesis, cell division, fatty acid oxidation, glycolysis, and autophagy), indicating that key cellular processes in a simple eukaryotic cell are compartmentalized in time (Tu et al. 2005). Such temporal compartmentalization might provide a means for the cell to perform a multitude of metabolic processes in a more coordinated and efficient fashion and help minimize futile reactions, especially under nutrient-limited growth environments. Periodic gene expression patterns prototypical of both the circadian cycle and the YMC signify that organisms have developed a number of sophisticated strategies to take advantage of the dimension of time.

THREE METABOLIC PHASES IN THE LIFE OF A YEAST CELL

Inspection of the genes that are maximally expressed in each phase of the YMC has provided insight into the cellular outputs and events that occur during particular temporal windows of the YMC. The OX phase is relatively brief compared to the reductive RB and RC phases and is characterized by a period of intense respiration (Fig. 1). Many ribosomal genes and amino acid biosynthetic genes are induced in the OX phase, suggesting that this phase is dedicated to establishing the protein synthesis machinery for growth and the preparation of cell division (Tu et al. 2005). Both ribosome and amino acid biosynthesis, which are highly energetically demanding, might be coordinated with the OX phase in order to benefit from copious amounts of ATP that are produced from respiration. Many genes encoding components of the nucleolus, small nuclear RNAs (snRNAs), and transfer RNA (tRNA) synthetases are also up-regulated during OX phase, consistent with the concept that it represents a temporal window devoted to growth. Interestingly, many genes involved in various aspects of RNA processing and degradation are also highly up-regulated during the OX phase (Tu et al. 2005). Furthermore, many mRNA transcripts are at their lowest levels toward the end of OX phase. These observations suggest that many transcripts are actively being degraded during OX phase, perhaps in response to oxidative damage or in preparation for cell division.

The RB (reductive, building) phase follows the OX phase and is characterized by a significant decrease in the rate of oxygen consumption (Fig. 1). During the midst of RB phase, oxygen consumption almost entirely ceases, which is marked by a sudden spike in dissolved oxygen levels. During this phase, the overwhelming majority of genes encoding mitochondrial proteins are up-regulated (Tu et al. 2005). The collective up-regulation of these mitochondrial gene products suggests that cells might be rebuilding their mitochondria during this reductive temporal window, perhaps in response to the previous period of intense respiration in the OX phase. Furthermore, many genes that have functions associated with cell division are also up-regulated during RB phase (Tu et al. 2005). These gene products include histones, spindle pole body components, and gene products known to be important for the initiation of cell divison. As predicted by the microarray data, the initiation of cell division during the YMC is strictly gated to this reductive RB phase, as normal wild-type cells are never observed to divide during OX phase (Tu et al. 2005). Approximately 40–50% of cells enter the cell cycle in each metabolic cycle (Tu et al. 2005).

Interestingly, cell cycle mutants that are forced to partake in cell division during the OX phase of the YMC were found to display higher spontaneous mutation rates (Chen et al. 2007). These observations suggest that restriction of DNA replication and cell division to the reductive phases serves to minimize oxidative damage to DNA, thereby reducing the rate of spontaneous mutation (Tu et al. 2005; Chen et al. 2007). Restriction of DNA replication to a reductive metabolic environment was also found to occur during short-period, 40-minute yeast oscillations (Klevecz et al. 2004). The gating of the cell cycle by the YMC might be formally analagous to gating of the cell cycle by the circadian cycle, first observed in cyanobacteria and later in mammals (Mori et al. 1996; Kondo et al. 1997; Matsuo et al. 2003; Nagoshi et al. 2004). These findings provide substantial evidence that there has been selective pressure to confine DNA replication to optimal temporal windows. Overall, the RB phase is associated with the induction of many mitochondrial and cell cycle genes during a period of lower oxygen consumption.

The RC (reductive, charging) phase is the longest of the three phases and immediately follows RB phase (Fig. 1). During this temporal window, many gene products that specify functions associated with nonrespiratory modes of metabolism are up-regulated (Tu et al. 2005), including peroxisomal proteins, enzymes involved in the storage and breakdown of carbohydrates, and enzymes necessary for ethanol utilization. The predicted outcome of these reactions is the production of acetyl-CoA units, which is the substrate for respiration, a hallmark of the impending OX phase. These observations indicate that during RC phase, yeast cells become dependent on these additional metabolic strategies for energy production and preparation for OX phase.

Many genes associated with starvation and stress-associated responses are also highly up-regulated during RC phase (Tu et al. 2005), including components of the vacuole, proteasome, and ubiquitination machinery and genes required for autophagy. In addition, many heat shock protein (HSP) genes are also highly induced. These observations predict the increased and regulated turnover of proteins and organelles during RC phase. Thus, this RC temporal window appears to be dedicated to the rebuilding and recharging of the cell, perhaps in preparation for the subsequent OX phase and a new round of growth. As such, vacuole-mediated catabolism and autophagy might in fact occur normally during the life of a yeast cell under nutrient-poor growth conditions.

PREDICTION AND ANNOTATION OF GENE FUNCTION BASED ON YMC EXPRESSION PROFILES

The temporal expression pattern of a gene during the YMC can often provide insights into the biological function of the encoded protein. For example, the open reading frame *YHR075c* (*PPE1* or *MRPS2*) was previously identified to encode a mitochondrial ribosomal protein of the small subunit. Upon alignment of the expression patterns of all known genes encoding components of the mitochondrial ribosomes, all of them displayed highly similar temporal expression patterns except for *PPE1*, which peaks at a time distinct from that of the others (Tu et al. 2005). Subsequent studies have shown that *PPE1* instead encodes a protein with carboxyl methyl esterase activity and might have been misannotated as a mitochondrial ribosomal protein (Wu et al. 2000).

As a second example, the temporal expression profiles of periodic heat shock protein genes (*HSP*) reveals that two of them (*HSP10*, *HSP60*) are expressed maximally in the RB phase, whereas the majority of others peak during the RC phase (Tu et al. 2005). Hsp10p and Hsp60p are localized to the mitochondria which is entirely consistent with their expression peaks in the RB phase, which is when other mitochondria gene products tend to be induced. Thus, a simple phase assignment based on expression during the YMC can be useful in predicting the function of uncharacterized genes.

Inspection of the temporal expression profiles of the YMC reveals that genes having a common function occasionally exhibit slightly distinct moments of expression, most notably cell cycle genes. Using a deconvolution-based algorithm, the precise peaks of expression of cell cycle transcriptionally regulated genes were timed to a resolution of about 2 minutes (Rowicka et al. 2007). These data have presented a landscape of the transcriptional events that occur during the yeast cell cycle at a remarkably fine level of detail, and they reveal that many genes are transcribed just when their protein product is needed (a phenomenon termed "just-in-time synthesis"). Future application of such algorithms will provide additional insights into the precision with which cells have temporally optimized their transcriptional output.

BEYOND TRANSCRIPTION

The periodic transcription of genes during the YMC will undoubtedly direct many periodic outputs in the cell. In addition to respiration, cell division, and mitochondrial homeostasis, numerous other fundamental processes are predicted to be temporally regulated by the YMC. As an example, electron microscopic analysis of cells isolated during different temporal windows of the YMC have demonstrated that the vacuole exhibits dynamic changes in morphology during metabolic oscillation (Tu et al. 2005). During the late RC and Ox phases, the vacuole becomes more prominent and markers of autophagy become visible. These observations suggest that autophagy and vacuole-mediated catabolism might be normal aspects of the life cycle of a cell under nutrient-poor growth conditions.

The many oscillating gene expression patterns predict that the YMC should differentially control metabolic state (Tu and McKnight 2006). Certain metabolites might exhibit periodic fluctuation and, in turn, have a reciprocal role in regulating the YMC. To determine whether cyclic changes in metabolic state might occur during the YMC, both liquid chromatography–tandem mass spectrometry (LC-MS/MS) and two-dimensional gas chromatography/time-of-flight mass spectrometry (GCxGC-TOFMS) metabolite profiling methods were used to monitor the intracellular concentrations of common metabolites at different time points of the cycles (Tu et al. 2007). The results of these surveys show that many metabolites, including amino acids, nucleotides, and carbohydrates, oscillate in abundance with a periodicity precisely matching that of the YMC (Tu et al. 2007). From analysis of these metabolite profiling data, the logic of metabolite oscillation largely matches that predicted by the YMC transcript data set. Moreover, the metabolite profiling extends this logic in ways not inherently obvious by inspection of gene expression profiles. These results imply that many cellular processes will be intimately coupled to these cyclic changes in a metabolic or redox state and thereby executed more optimally and efficiently.

SIMILARITIES BETWEEN THE YMC AND CIRCADIAN CYCLE

The circadian cycle temporally orchestrates many aspects of organismal metabolism and physiology (Rutter et al. 2002; Lowrey and Takahashi 2004). Genome-wide expression studies reveal that the circadian clock drives the periodic expression of many genes known to control metabolic state, including those that encode the rate-limiting enzymes of numerous metabolic processes (Harmer et al. 2000; Claridge-Chang et al. 2001; McDonald and Rosbash 2001; Panda et al. 2002). In mice, the expression of mitochondrial oxidative phosphorylation genes in the suprachiasmatic nucleus (SCN) oscillates in a circadian fashion (Panda et al. 2002), which indicates that respiration might be periodically up-regulated in a manner similar to the periodic bursts of respiration that are a hallmark of the YMC. The uptake of 2-deoxyglucose is known to be robustly periodic as a function of the day/night cycle in the SCN (Schwartz and Gainer 1977). Such data indicate that fundamental metabolic processes in SCN neurons fluctuate as a function of the circadian cycle.

Less well understood is the cyclic relationship between metabolism and circadian rhythm, whereby metabolism can, in turn, directly feed back to entrain rhythm. Restricted feeding can reset the phase of circadian gene expression in the liver, which demonstrates that food is a potent zeitgeber for the peripheral oscillators (Damiola et al. 2000; Stokkan et al. 2001). Moreover, several studies have suggested that small-molecule metabolites can directly modulate the activity of core clock components (Rutter et al. 2001; Dioum et al. 2002). However, the precise biochemical mechanisms by which food entrainment occurs remain poorly understood.

These numerous connections between metabolism and the circadian cycle parallel the cardinal properties of the YMC (Tu and McKnight 2006). In both cycles, the periodic expression of genes directs a variety of cellular and metabolic outputs. Both cycles can, in turn, be modulated by feeding conditions and metabolism. A preliminary analysis revealed that among the circadian periodic genes of *Drosophila melanogaster* that encode proteins with clear homologs in yeast, most of these yeast genes were also highly periodic during the YMC (Table 1). Such correlative observations might be indicative of either evolutionary conservation of periodic gene expression or a convergent solution to the same biological challenge.

Much like the YMC, the circadian cycle thus has a key role in optimizing cellular and metabolic output via temporal compartmentalization. In turn, metabolic state will likely alternate during the course of each approximately

Table 1. Circadian Periodic Genes in *Drosophila* and Their Yeast Homologs

Drosophila	Yeast	Function	Periodicity score (σ)
Fdxh	*YAH1*	iron-sulfur protein homologous to human adrenodoxin	4.52
HSP26	*HSP26*	heat shock protein 26	4.34
cat	*CTA1*	catalase A	4.34
INO-1	*INO1*	L-myo-inositol-1-phosphate synthase	4.10
CG9748	*DBP1*	putative ATP-dependent RNA helicase, DEAD box protein	4.02
CG7828	*ULA1*	required for activation of RUB1 (neddylation)	3.95
Septin1	*CDC11*	component of septin ring required for cytokinesis	3.59
Nsf2	*SEC18*	involved in protein transport between ER and Golgi	3.12
CG6145	*YEL041W*	strong similarity to Utr1p	3.11
GC8468	*MCH5*	similarity to human X-linked PEST-containing transporter	3.04
Cyp4d21	*ERG11*	cytochrome P450 lanosterol 14a-demethylase	3.04
CG3021	*SLM3*	tRNA-specific 2-thiouridylase	2.99
Pdh	*YMR226C*	similarity to ketoreductases	2.75
CG4963	*MRS4*	mitochondrial iron transporter	2.52
CG7288	*SAD1*	zinc finger protein involved in pre-mRNA splicing	0.93
moira	*SWI3*	transcription factor	0.54
zw	*ZWF1*	glucose-6-phosphate dehydrogenase	–0.57

Of the approximately 158 genes that are expressed periodically as a function of the circadian cycle in the fly head (Claridge-Chang et al. 2001), 17 have clear yeast homologs. Of these, 14 of 17 are highly periodically expressed (>2 σ) as a function of the YMC. (We thank A. Kudlicki and M. Rowicka for assistance in the preparation of this table.)

24-hour period and feed back to modulate the activity of the circadian machinery. These considerations form an explicit prediction that the circadian cycle is, most fundamentally, a metabolic cycle (Rutter et al. 2002; Tu and McKnight 2006).

OSCILLATIONS IN COMMON LABORATORY STRAINS OF YEAST?

The prototrophic, genetically tractable CEN.PK strain, first adopted by the European yeast community (van Dijken et al. 2000), is our strain of choice for the study of these robust, approximately 4–5-hour metabolic cycles. Do common laboratory strains, such as the S288C and W303 strain backgrounds, also exhibit metabolic cycles? These popular yeast strains have been domesticated and contain mutations in multiple metabolic genes that serve as selectable nutritional markers. Although these auxotrophic markers have greatly facilitated the development of genetic screens and tools, they might compromise the output of cellular and metabolic pathways in ways not inherently obvious and elicit compensatory responses that are not typical of wild strains of yeast. For example, a laboratory strain that cannot synthesize adenine, uracil, or methionine will be absolutely dependent on supplementation of these metabolites for growth, which will undoubtedly alter flux through various biosynthetic pathways and hence the regulation of particular metabolic processes. Thus far, even using the parental S288C prototrophic strain, we have only observed low-amplitude oscillations of a period of about 80 minutes (B. Tu, unpubl.), which seem quite different from the robust approximately 4–5-hour cycles exhibited by the prototrophic CEN.PK strain. Thus, we speculate that many common laboratory strains have been compromised in their ability to oscillate, specifically due to domestication and the presence of auxotrophic nutritional markers.

LOG-PHASE, NUTRIENT-RICH GROWTH VS. CONTINUOUS, NUTRIENT-POOR CHEMOSTAT GROWTH

How do the chemostat growth conditions used to observe the YMC compare to traditional log-phase growth conditions? Although the use of log-phase growth conditions is convenient and often informative, they might not always be representative of the conditions encountered by yeast in the wild. Yeast typically grow in colonies, where the cell density is very high and the distance between cell neighbors is very small. In the wild, yeast colonies will likely be exposed to environments where nutrients are poor, unlike the nutrient-rich conditions of log phase. Thus, the dense population of cells in the chemostat that undergo the YMC might approximate a colony exposed to a nutrient-poor environment. Under these challenging conditions, yeast cells deploy a more extensive assortment of regulatory strategies that are not typical of laboratory strains grown at log phase. Under nutrient-rich growth conditions, yeast tend not to perform mitochondrial respiration. Likewise, many classes of genes, including those encoding products required for respiration and fatty acid oxidation, are minimally expressed in log phase (Ghaemmaghami et al. 2003). In contrast, almost all genes required for mitochondrial and peroxisomal function are not only expressed, but also coordinately activated during precise temporal windows of the YMC (Tu et al. 2005). Thus, the growth conditions (i.e., high cell density and nutrient-limited) used to observe the YMC are perhaps more representative of what yeast cells encounter in the wild. Moreover, the use of chemostats for yeast growth enables the maintenance of constant pH, temperature, aeration, and nutrient levels, thereby ensuring continuous, homogeneous growth conditions that are not easily achievable by use of batch cultures grown under high-glucose conditions. In doing so, the use of chemostats has enabled the observation of aspects of the life of a yeast cell that would otherwise be either obscure or not even utilized. Indeed, as the famous physicist Leo Szilard, inventor of the chemostat, once predicted: "a study of this slow-growth phase by means of the chemostat promises to yield information of some value on metabolism, regulatory processes, adaptations, and mutations of microorganisms." (Novick and Szilard 1950)

CONCLUSIONS

In summary, the yeast metabolic cycle is composed of an intricate transcriptional program that specifies a variety of periodic cellular outputs during the life of a yeast cell. In doing so, many fundamental cellular and metabolic processes are precisely compartmentalized in time, enabling cells to optimize metabolic output and improve fitness under nutrient-poor growth conditions. As a consequence of temporal compartmentalization, cyclic changes in metabolic state occur during the YMC, which might have an essential reciprocal role in the establishment and maintenance of metabolic oscillation. The numerous similarities between the YMC and circadian cycle suggest that certain principles of temporal orchestration might be conserved across other biological cycles and further indicate that the circadian cycle can be viewed as fundamentally a metabolic cycle. It is hopeful that future studies of the yeast metabolic cycle will continue to contribute to our understanding of the mechanisms of temporal compartmentalization and the basis of the circadian and other biological cycles.

ACKNOWLEDGMENTS

The authors thank funding support from a National Institutes of Health Director's Pioneer Award (S.L.M.), unrestricted funds from an anonymous donor (S.L.M.), a Helen Hay Whitney Foundation postdoctoral fellowship (B.P.T.), a Sara & Frank McKnight Foundation Fellowship (B.P.T.), and a Burroughs Wellcome Fund Career Award in Biomedical Sciences (B.P.T.).

REFERENCES

Chance B., Estabrook R.W., and Ghosh A. 1964. Damped sinusoidal oscillations of cytoplasmic reduced pyridine nucleotide in yeast cells. *Proc. Natl. Acad. Sci.* **51:** 1244.

Chen Z., Odstrcil E.A., Tu B.P., and McKnight S.L. 2007. Restriction of DNA replication to the reductive phase of the metabolic cycle protects genome integrity. *Science* **316:** 1916.

Claridge-Chang A., Wijnen H., Naef F., Boothroyd C., Rajewsky N., and Young M.W. 2001. Circadian regulation of gene expression systems in the *Drosophila* head. *Neuron* **32:** 657.

Damiola F., Le Minh N., Preitner N., Kornmann B., Fleury-Olela F., and Schibler U. 2000. Restricted feeding uncouples circadian oscillators in peripheral tissues from the central pacemaker in the suprachiasmatic nucleus. *Genes Dev.* **14:** 2950.

Dioum E.M., Rutter J., Tuckerman J.R., Gonzalez G., Gilles-Gonzalez M.A., and McKnight S.L. 2002. NPAS2: A gas-responsive transcription factor. *Science* **298:** 2385.

Ghaemmaghami S., Huh W.K., Bower K., Howson R.W., Belle A., Dephoure N., O'Shea E.K., and Weissman J.S. 2003. Global analysis of protein expression in yeast. *Nature* **425:** 737.

Harmer S.L., Hogenesch J.B., Straume M., Chang H.S., Han B., Zhu T., Wang X., Kreps J.A., and Kay S.A. 2000. Orchestrated transcription of key pathways in *Arabidopsis* by the circadian clock. *Science* **290:** 2110.

Klevecz R.R., Bolen J., Forrest G., and Murray D.B. 2004. A genomewide oscillation in transcription gates DNA replication and cell cycle. *Proc. Natl. Acad. Sci.* **101:** 1200.

Kondo T., Mori T., Lebedeva N.V., Aoki S., Ishiura M., and Golden S.S. 1997. Circadian rhythms in rapidly dividing cyanobacteria. *Science* **275:** 224.

Lowrey P.L. and Takahashi J.S. 2004. Mammalian circadian biology: Elucidating genome-wide levels of temporal organization. *Annu. Rev. Genomics Hum. Genet.* **5:** 407.

Matsuo T., Yamaguchi S., Mitsui S., Emi A., Shimôda F., and Okamura H. 2003. Control mechanism of the circadian clock for timing of cell division in vivo. *Science* **302:** 255.

McDonald M.J. and Rosbash M. 2001. Microarray analysis and organization of circadian gene expression in *Drosophila*. *Cell* **107:** 567.

Mori T., Binder B., and Johnson C.H. 1996. Circadian gating of cell division in cyanobacteria growing with average doubling times of less than 24 hours. *Proc. Natl. Acad. Sci.* **93:** 10183.

Nagoshi E., Saini C., Bauer C., Laroche T., Naef F., and Schibler U. 2004. Circadian gene expression in individual fibroblasts: Cell-autonomous and self-sustained oscillators pass time to daughter cells. *Cell* **119:** 693.

Novick A. and Szilard L. 1950. Description of the chemostat. *Science* **112:** 715.

Panda S., Antoch M.P., Miller B.H., Su A.I., Schook A.B., Straume M., Schultz P.G., Kay S.A., Takahashi J.S., and Hogenesch J.B. 2002. Coordinated transcription of key pathways in the mouse by the circadian clock. *Cell* **109:** 307.

Richard P. 2003. The rhythm of yeast. *FEMS Microbiol. Rev.* **27:** 547.

Rowicka M., Kudlicki A., Tu B.P., and Otwinowski Z. 2007. High-resolution timing of cell cycle gene expression. *Proc. Natl. Acad. Sci.* September 7.

Rutter J., Reick M., and McKnight S.L. 2002. Metabolism and the control of circadian rhythms. *Annu. Rev. Biochem.* **71:** 307.

Rutter J., Reick M., Wu L.C., and McKnight S.L. 2001. Regulation of clock and NPAS2 DNA binding by the redox state of NAD cofactors. *Science* **293:** 510.

Satroutdinov A.D., Kuriyama H., and Kobayashi H. 1992. Oscillatory metabolism of *Saccharomyces cerevisiae* in continuous culture. *FEMS Microbiol. Lett.* **77:** 261.

Schwartz W.J. and Gainer H. 1977. Suprachiasmatic nucleus: Use of 14C-labeled deoxyglucose uptake as a functional marker. *Science* **197:** 1089.

Stokkan K.A., Yamazaki S., Tei H., Sakaki Y., and Menaker M. 2001. Entrainment of the circadian clock in the liver by feeding. *Science* **291:** 490.

Tu B.P. and McKnight S.L. 2006. Metabolic cycles as an underlying basis of biological oscillations. *Nat. Rev. Mol. Cell Biol.* **7:** 696.

Tu B.P., Kudlicki A., Rowicka M., and McKnight S.L. 2005. Logic of the yeast metabolic cycle: Temporal compartmentalization of cellular processes. *Science* **310:** 1152.

Tu B.P., Mohler R.E., Liu J.C., Dombek K.M., Young E.T., Synovec R.E., and McKnight S.L. 2007. Cyclic changes in metabolic state during the life of a yeast cells. *Proc. Natl. Acad. Sci.* October 16.

van Dijken J.P., Bauer J., Brambilla L., Duboc P., Francois J.M., Gancedo C., Giuseppin M.L., Heijnen J.J., Hoare M., Lange H.C., Madden E.A., Niederberger P., Nielsen J., Parrou J.L., Petit T., Porro D., Reuss M., van Riel N., Rizzi M., Steensma H.Y., Verrips C.T., Vindelov J., and Pronk J.T. 2000. An interlaboratory comparison of physiological and genetic properties of four *Saccharomyces cerevisiae* strains. *Enzyme Microb. Technol.* **26:** 706.

von Meyenburg H.K. 1969. Energetics of the budding cycle of *Saccharomyces cerevisiae* during glucose limited aerobic growth. *Arch. Microbiol.* **66:** 289.

Wu J., Tolstykh T., Lee J., Boyd K., Stock J.B., and Broach J.R. 2000. Carboxyl methylation of the phosphoprotein phosphatase 2A catalytic subunit promotes its functional association with regulatory subunits in vivo. *EMBO J.* **19:** 5672.

Complexity of the *Neurospora crassa* Circadian Clock System: Multiple Loops and Oscillators

R.M. de Paula,* M.W. Vitalini,* R.H. Gomer,† and D. Bell-Pedersen*
*Center for Research on Biological Clocks and Department of Biology, Department of Biology, Texas A&M University, College Station, Texas 77843; †Department of Biochemistry and Cell Biology, Rice University, Houston, Texas 77005

Organisms from bacteria to humans use a circadian clock to control daily biochemical, physiological, and behavioral rhythms. We review evidence from *Neurospora crassa* that suggests that the circadian clock is organized as a network of genes and proteins that form coupled evening- and morning-specific oscillatory loops that can function automously, respond differently to environmental inputs, and regulate phase-specific outputs. There is also evidence for coupled morning and evening oscillator loops in plants, insects, and mammals, suggesting conservation of clock organization. From a systems perspective, fungi provide a powerful model organism for investigating oscillator complexity, communication between oscillators, and addressing reasons why the system has evolved to be so complex.

INTRODUCTION

A diverse group of organisms, ranging from bacteria to humans, display daily rhythms in gene expression, protein levels, physiology, and behavior that are controlled by endogenous circadian clocks. These clocks are composed of limit-cycle oscillators based on molecular feedback loops, input pathways that transduce external information to the oscillators, and output pathways that allow the oscillators to temporally organize cellular and behavioral processes to specific times of the day.

Considerable effort has gone into understanding the molecular, biochemical, and physical properties of the circadian clock. In eukaryotic organisms, the molecular oscillators contain multiple interlocked feedback loops consisting of both positive and negative elements (Glossop et al. 1999; Lee et al. 2000; Shearman et al. 2000; Cheng et al. 2001a; Glossop and Hardin 2002; Preitner et al. 2002; Cyran et al. 2003; Locke et al. 2006). Several lines of evidence point toward the existence of multiple distinct oscillators in cells and/or tissues of organisms contributing to circadian timing. First, there exist free-running rhythms with different period lengths in the same organism (Morse et al. 1994; Sai and Johnson 1999; Cambras et al. 2007), and there is residual rhythmicity in strains that are defective in known oscillator components (Loros et al. 1986; Stanewsky et al. 1998; Emery et al. 2000; Collins et al. 2005). Second, some tissue-specific oscillators are constructed differently from the core oscillators located in the brain (Stanewsky et al. 1998; Ivanchenko et al. 2001; Krishnan et al. 2001; Hardin et al. 2003; Collins et al. 2005). Thus, oscillator complexity may arise both intracellularly and intercellularly in organisms with differentiated tissues.

There are at least three possible reasons why organisms would benefit from having multiple distinct oscillators or interlocking clock loops. First, the interlocked loops might have evolved to allow the circadian system to be more robust—in other words, immune or resistant to stochastic noise or to small perturbations of the parameters. This is supported by the observation that alteration of the levels of components of the loops results in less precise clocks (Cheng et al. 2001a; Preitner et al. 2002; Locke et al. 2006). Second, multiple coupled loops could add to the degree of flexibility of the network by providing a mechanism for the environment to entrain the different loops with different input signals, so that, for instance, there could be one clock that is reset by light input and a different clock that is reset by a temperature shift (Rand et al. 2006). Third, increased flexibility of the system could also be achieved using multiple distinct oscillators that respond differently to environmental input and regulate different phase-specific outputs, so that, for example, morning-specific outputs would be controlled by an oscillator with one set of light and temperature inputs and evening-specific outputs would be controlled by a different oscillator with a different set of light and temperature input parameters.

The circadian clock system of *Neurospora crassa* has been studied extensively and is one of the best-understood circadian models (Loros and Dunlap 2001; Liu and Bell-Pedersen 2006; Vitalini et al. 2006; Heintzen and Liu 2007). Armed with a detailed description of core *N. crassa* FRQ/WCC feedback loop, we are now in a unique position to extend our current view of the oscillator to accommodate a growing body of evidence showing residual oscillations in the absence of the core FRQ/WCC oscillator components. We suggest that similar to plants and animals, the FRQ/WCC oscillator contains multiple coupled feedback loops that are important for precision and flexibility of the system.

THE INTERLOCKED FRQ/WCC FEEDBACK LOOPS

N. crassa displays an easily observable 22-hour rhythm in asexual spore development (conidiation) when cultures are grown in constant darkness (DD) (Pittendrigh et al.

1959), as well as rhythms in gene expression (Loros et al. 1989), metabolism (Shinohara et al. 1998), pheromone production (Bobrowicz et al. 2002), stress response (Shinohara et al. 2002), and other processes (for review, see Vitalini et al. 2006).

In *N. crassa*, the FRQ/WCC oscillator is considered to be the core circadian oscillator necessary for generating many of the observed circadian rhythms, including the developmental rhythm. The FRQ/WCC oscillator is known to be composed of an autoregulatory, transcriptional/translational feedback loop involving the *frequency* (*frq*) and *white collar* (*wc-1*, *wc-2*) genes and their protein products (Loros and Dunlap 2001; Liu and Bell-Pedersen 2006). The WC-1 blue light photoreceptor (Froehlich et al. 2002; He et al. 2002) forms a complex with WC-2 (white collar complex; WCC) that binds the *frq* promoter and directly activates transcription of the *frq* gene (Froehlich et al. 2003). Levels of FRQ protein then slowly increase. FRQ dimerizes (Cheng et al. 2001b) and forms a complex with FRH (a FRQ-interacting RNA helicase) (Cheng et al. 2005) that binds to, and promotes the phosphorylation of, the WCC by several kinases, including casein kinase I (CKI) and CKII (Schafmeier et al. 2005; He et al. 2006). Once hyperphosphorylated, activity of the WCC is inhibited such that it is unable to activate transcription of *frq*. This inhibition results in reduced *frq* transcript levels and a decrease in FRQ protein production, forming the negative part of the feedback loop. Subsequent phosphorylation-induced decay of FRQ, in conjunction with dephosphorylation of the WCC by protein phosphatase 2A (Schafmeier et al. 2005), releases the inhibitory affect on the WCC and leads to reactivation of *frq* transcription, allowing the cycle to start again the next day (Liu et al. 2000).

FRQ also acts in a positive feedback loop, maintaining levels of *wc-2* mRNA and WC-1 and WC-2 proteins (Lee et al. 2000; Cheng et al. 2001a, 2002). To investigate the importance of the positive feedback loop on clock function, *wc-1* or *wc-2* mRNA levels were artificially induced in cells (Cheng et al. 2001a). As the concentration of WC-1 or WC-2 increased, the amplitudes of the developmental rhythm and FRQ protein cycles increased. This suggested that the positive loop confers stability and robustness to the FRQ/WCC oscillator (Lee et al. 2000; Cheng et al. 2001a; Schafmeier et al. 2006).

The conflicting roles of FRQ in positive and negative feedback are now beginning to be understood in terms of spatial location and posttranslational modification (Schafmeier et al. 2006). Negative feedback through WCC is accomplished by nuclear FRQ that is hypophosphorylated and thus begins early after FRQ is expressed. Progressive phosphorylation of FRQ appears to trigger a switch of FRQ from a nuclear repressor to a cytoplasmic activator of WC-1 protein accumulation (Schafmeier et al. 2006). This regulation may be through FRQ-mediated phophorylation of cytosolic WC-1 or WC-2, which could enhance their assembly and stability. However, the mechanism by which FRQ regulates the levels of *wc-2* mRNA is currently not known.

Although not well understood, WC-1 and WC-2 also regulate each other to form an additional loop in the FRQ/WCC oscillator. WC-2 stabilizes WC-1 protein by forming the WCC, and WC-1 negatively regulates transcription of *wc-2* (Cheng et al. 2002, 2003).

A NEW LOOP IN THE FRQ-WCC OSCILLATOR NETWORK?

Several labs have demonstrated residual oscillations in *N. crassa* cells in the absence of FRQ or WC-1 and WC-2 (Loros et al. 1986; Merrow et al. 1999; Ramsdale and Lakin-Thomas 2000; Correa et al. 2003; Granshaw et al. 2003; Christensen et al. 2004; He et al. 2005; de Paula et al. 2006). These data suggested the existence of FRQ-independent oscillators. The term FRQ-less oscillator (FLO) has been coined to collectively describe the putative oscillators responsible for such rhythms (Iwasaki and Dunlap 2000).

To begin to investigate the FLOs, microarrays were used to identify genes that cycle with a circadian period in the absence of FRQ (Correa et al. 2003). Three evening-specific clock-controlled genes (ccgs) that displayed a daily rhythm in mRNA accumulation in FRQ-null strains were identified. One of these, W06H2 (now called *ccg-16*), was confirmed by northern assays to accumulate rhythmic mRNA in the absence of FRQ and under conditions in which the conidiation rhythm is abolished, such as during growth in constant light (LL) or when *frq* is constitutively overexpressed from an inducible promoter (de Paula et al. 2006). In wild-type strains, the oscillator that is responsible for generating the *ccg-16* rhythm responds to both temperature and light cues for synchronization; however, in the absence of FRQ, the oscillator responds better to temperature cues. Surprisingly, the *ccg-16* mRNA rhythm requires functional WC-1 and WC-2, suggesting that the oscillator responsible for *ccg-16* rhythms is coupled either directly or indirectly to the FRQ/WCC oscillator. To distinguish the *ccg-16* FLO from other potential *N. crassa* oscillators, it has been called the WC-dependent FLO (WC-FLO).

Importantly, the *ccg-16* mRNA rhythm is temperature-compensated in the FRQ-null strain (Q_{10} of 0.9), suggesting that FRQ is not absolutely required for temperature compensation. However, this conclusion needs to be taken with a certain level of caution because it was based on RNA blots using 4-hour time points to measure *ccg-16* mRNA rhythms at three different temperatures. Sampling of more time points using a *ccg-16*:luciferase reporter (a generous gift from J. Dunlap and J. Loros) in both FRQ^+ and FRQ-null strains is currently being conducted to confirm these results.

Together, these data established that the oscillator driving *ccg-16* mRNA rhythms is not inhibited by high or low FRQ levels. In addition, *ccg-16* mRNA rhythms can be observed in both LL and DD, suggesting that the WC-FLO is a circadian oscillator that is not inhibited by high or low light levels. This is unlike the FRQ/WCC oscillator, which is functional only in the first 12–16 hours of LL following synchronization in light/dark cycles (Elvin et al. 2005).

Because WC-1 and WC-2 are required for rhythmic *ccg-16* mRNA levels in LL, we hypothesized that WC-1 protein levels might also be rhythmic under these same conditions (WC-2 is not normally rhythmically expressed [Denault et al. 2001]). Indeed, a robust rhythm in WC-1 levels was also observed in wild-type strains grown in LL and in ΔFRQ strains grown in DD when the cultures were synchronized by a temperature transition (de Paula et al. 2006). Furthermore, two forms of cycling WC-1 were observed in wild-type strains, a large (~135 kD) and a small (~127 kD) form. The large form typically displayed a more robust rhythm than the small form, and only the large form was observed in ΔFRQ strains. Interestingly, the amplitude of the WC-1 rhythm was qualitatively higher in ΔFRQ strains, as compared the FRQ^+ strains. The small form of WC-1 is likely due to alternative initiation from an in-frame methionine codon located within the WC-1 open reading frame at codon 88 (Cheng et al. 2003; Kaldi et al. 2006). The roles of the different WC-1 forms in the circadian clock are currently under investigation.

These results demonstrated that although FRQ is involved in maintaining the overall levels of WC-1 protein in the cell (Lee et al. 2000; Cheng et al. 2001a) and in generating the small form of WC-1, FRQ is not essential for rhythmic WC-1 levels. We speculate that components of the WC-FLO control WC-1 rhythms (see below). However, WC-1 rhythms were previously shown to have a long-period oscillation in the 29-hour period frq^7 mutant strain (Lee et al. 2000; Cheng et al. 2001a), suggesting that when the FRQ/WCC oscillator is functional, it can influence the period of the WC-1 rhythm. This may be through coupling of the FRQ/WCC oscillator to the WC-FLO.

The function of the CCG-16 protein is unknown. The FRQ/WCC oscillator functions normally in a ΔCCG-16 strain. In addition, WC-1 protein accumulates rhythmically in DD in a ΔFRQ ΔCCG-16 strain, suggesting that CCG-16 is not a component of the WC-FLO that generates the WC-1 rhythm. Thus, *ccg-16* is likely an output of the WC-FLO.

Although the WC-FLO can function autonomously, the requirement of WC-1 and WC-2 for both the FRQ-based oscillator and the WC-FLO suggests that the two oscillators are coupled either directly or indirectly by the WC proteins. This situation is strikingly analogous to the two coupled oscillators proposed to explain the different responses of rhythmic pupal eclosion in *Drosophila* to light and temperature treatments and later used to explain splitting behavior in rodents in LL (Pittendrigh and Bruce 1959; Pittendrigh 1961; Pittendrigh and Daan 1976). In this model, a central morning oscillator (M) is light-entrainable and is coupled to a temperature-entrainable evening oscillator (E). While both oscillators are autonomous and can be directly entrained, the E oscillator requires the M oscillator for complete circadian properties and for proper phasing. Temperature directly entrains the E oscillator, and the E oscillator feeds back on the M oscillator to bring the system to equilibrium. In *N. crassa*, the M oscillator would be equivalent to the light-entrainable FRQ/WCC oscillator that controls morning-specific ccgs and development, and the E oscillator would represent the WC-FLO that controls evening-specific ccgs, such as *ccg-16* (Fig. 1). The WC-FLO would thus derive phase information from two sources: directly from environmental cues and indirectly through the light and temperature-responsive FRQ/WCC oscillator.

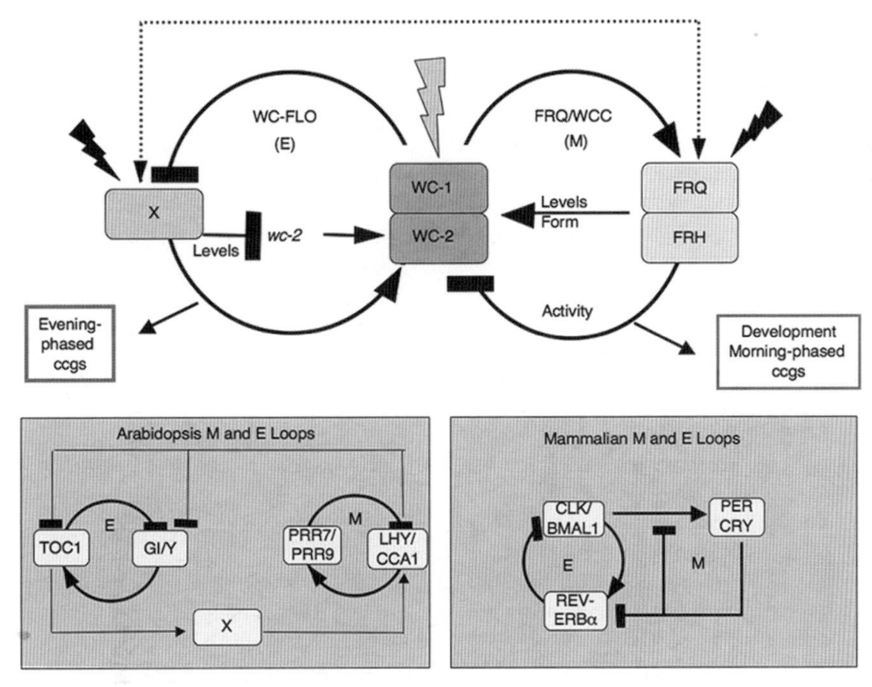

Figure 1. Speculative working model for the coupled E and M multiloop oscillator of *N. crassa.* For details of the model, see text. (*Light-gray jagged arrow*) Light input into the oscillator mechanism; (*black jagged arrows*) temperature inputs to both the FRQ/WCC oscillator and the WC-FLO; (*black arrows*) positive regulation; (*bars*) negative regulation. The *dotted line* between FRQ and X may explain residual developmental rhythms in the absence of FRQ or WC proteins. (*Bottom left*) Model of the three-loop *Arabidopsis* clock that contains M and E oscillators (Locke et al. 2006); (*bottom right*) intertwined E and M loop model of the mammalian clock (Leloup and Goldbeter 2004).

THE PREDICTED EVENING AND MORNING MULTILOOP ARCHITECTURE OF PLANTS AND ANIMALS SUGGESTS A NEW WORKING MODEL FOR THE *N. CRASSA* CLOCK

A three-loop clock, comprising a system of intracellularly coupled M and E oscillators, was recently predicted for the *Arabidopsis* clock based on experimental evidence and mathematical modeling (Locke et al. 2006). This model accounts for the residual rhythmicity that can be observed in mutations of the core clock component TOC1 (Fig. 1). The model predicts the existance of two autonomous short-period oscillators, plus a third loop connecting the two. An M oscillator contains PRR7/9 and LHY/CCA1, and an E oscillator contains TOC1, GI, and an unknown component Y. The two oscillators are predicted to be coupled together in a loop that involves negative regulation of TOC1 by LHY/CCA and positive regulation of LHY/CCA1 from TOC1 through an unknown component X.

Similar to plants, there is also evidence in *Drosophila* (Stoleru et al. 2004) and mammals (Jagota et al. 2000; Daan et al. 2001) for separate control of M and E process. But unlike plants, the M and E oscillators of flies and mammals are present in different cells and are thus coupled together via cell–cell signaling. It is not known if M and E are two genetically distinct oscillators or two identical oscillators that are set to different phases using different input pathways.

Intertwined intracellular loops, similar to that proposed in plants, may also exist in mammals. Modeling of mammalian molecular clockwork revealed the existence of a new source of oscillations arising from negative regulation of *Bmal1* expression by *Rev-Erbα* (Fig. 1) (Leloup and Goldbeter 2004). In the absence of PER, oscillations of *Bmal1* mRNA persist. Furthermore, in the absence of the negative *Bmal1/Rev-Erbα* loop, BMAL1-dependent oscillations of PER and CRY also persist. The two oscillator loops are coupled through CLOCK-BMAL1. The first oscillator involves an indirect negative feedback exerted through the inhibitory binding of PER-CRY to CLOCK-BMAL1 complexes, which activates expression of *per* and *cry* mRNA. The second oscillator involves negative feedback exerted by CLOCK-BMAL1, via REV-ERBα, on the expression of *Bmal1*. Circadian oscillations that persist in *Rev-Erbα* knockout mice (Preitner et al. 2002), along with observations of restored rhythmicity by an extended light pulse in *mPer1/mPer2*-deficient mice (Bae and Weaver 2007), provide experimental support for the multiloop model.

On the basis of similarities of the FRQ/WCC oscillator and the WC-FLO to the coupled M and E oscillators of plants, flies, and mammals, we speculate that the WC-FLO constitutes a temperature-responsive evening loop of the FRQ/WCC oscillator (Fig. 1). This working model proposes that the WCC represses an unknown component X, which in turn activates the WCC. We predict that this activation is in part due to new synthesis of the large form of WC-1 protein by rhythmically active X protein because large WC-1 accumulates rhythmically in the absence of FRQ. The model predicts that the levels or activity of X would be low when the WCC is active and high before the peak in WC-1 levels. X would also be predicted to peak antiphase to FRQ, providing an opportunity to regulate outputs at the opposite phase of the cycle (Fig. 1). When FRQ is overexpressed in cells, *ccg-16* mRNA and WC-1 rhythms are maintained (de Paula et al. 2006), and the levels of WCC increase. But most of the WCC would be phosphorylated and inactive for positive regulation of *frq* (He et al. 2006; Schafmeier et al. 2006). However, phosphorylation of the WC proteins might facilitate repression of X, allowing the WC-FLO cycle to continue. For this reason, we favor the WCC inhibiting X rather than WCC activating X, and X inhibiting the WCC to form the WC-FLO feedback loop. The levels of WC proteins are lower in FRQ-null strains; thus, FRQ is thought to positively regulate WC protein levels posttrancriptionally. In addition, the levels of *wc-2* mRNA are lower in FRQ-null strains, suggesting FRQ is somehow involved in transcriptional regulation of *wc-2*. Because it is hard to imagine how a single protein might carry out both of these activities directly, we wildly speculate that the WC-FLO would have one of these roles, such as in transcriptional repression of *wc-2*. When FRQ is absent, the levels of WCC would decrease. This would result in an increase in the activity of X, which would in turn result in lower levels of *wc-2* transcription. Finally, we predict that X would be responsive to a temperature input, allowing the WC-FLO to be directly responsive to temperature. Obviously, to test this model, we need to identify X and experiments are currently in progress to do this.

Coupling of the FRQ/WCC oscillator and WC-FLO through the WCC proteins would likely add additional stability to the system while allowing each oscillator the flexibility to control phase-dependent outputs. In the absence of external light and temperature cycles, the integrity of the circadian system would depend on this coupling to maintain identical frequencies and the appropriate phasing of the oscillators. To begin to test if both the FRQ/WCC oscillator and the WC-FLO are required for stable *ccg-16* mRNA rhythms, the phase, amplitude, and period of the *ccg-16* rhythms in wild type versus a ΔFRQ strain grown in DD or LL were compared (R. de Paula et al., unpubl.). As predicted, in the ΔFRQ strain, the amplitude of the *ccg-16* mRNA rhythms showed a significantly increased variation between experiments as compared to wild-type strains grown in DD (Fig. 2). These data support the notion that when the circadian oscillator network is intact, the precision of the *ccg-16* rhythms is increased.

EVIDENCE FOR MORE FLOS IN *N. CRASSA*

In *N. crassa* cultures grown with nitrate as the sole nitrogen source, daily rhythms in nitrate reductase activity are observed in DD or LL in wild-type strains, as well as in strains that lack FRQ or WC-1 (Christensen et al. 2004). These data indicated that a nitrate reductase oscillator (NRO), which can run in the absence of a functional FRQ/WCC oscillator and is separate from the WC-FLO, generates rhythms in nitrate reductase activity. However, in the absence of WC-1, the nitrate reductase activity

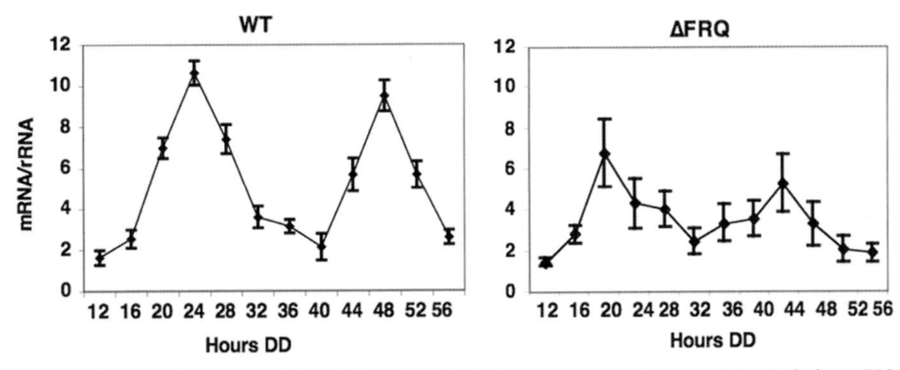

Figure 2. FRQ/WCC oscillator maintains precision of the WC-FLO that regulates *ccg-16* rhythmicity. Relative mRNA levels for *ccg-16* from wild-type and ΔFRQ cultures synchronized by a shift from 30°C to 25°C and grown in DD were determined by densitometry of northern blots and plotted versus time (hours in DD). Values are mean ± S.E.M. from six separate experiments. For each strain, at each timepoint, the data were normalized to the average value at that timepoint. The standard deviation of the normalized data for wild type was 0.427, and the standard deviation for ΔFRQ was 0.554. An F-test on the normalized data showed that the variance of the *ccg-16* levels in ΔFRQ was significantly higher than the variance in wild type with $p = 0.022$.

rhythms are of low amplitude and more variable, suggesting some connection to the FRQ/WCC;WC-FLO oscillator. It is not yet known if the NRO is temperature-compensated or entrained by light or temperature.

Other evidence for multiple oscillators in *N. crassa* cells comes from observations of residual developmental rhythms in strains that lack FRQ, WC-1, or WC-2 in DD (Loros et al. 1986; Aronson et al. 1994; Lakin-Thomas 1998, 2006; Merrow et al. 1999; Lakin-Thomas and Brody 2000; He et al. 2005; Pregueiro et al. 2005). However, the rhythms are defective in one or more of the canonical circadian oscillator properties—the rhythms are not stable, do not persist, are not temperature-compensated, and are not entrained by LD cycles. These data suggest that an intact FRQ/WCC oscillator is required for complete circadian properties of the developmental rhythm. Although somewhat controversial, the developmental rhythms appear to be entrained by temperature cycles (Merrow et al. 1999; Pregueiro et al. 2005; Roenneberg et al. 2005; Lakin-Thomas 2006). Thus, another possibility is that the residual developmental rhythms arise from parts of the FRQ/WCC ocillator or WC-FLO oscillator that remain intact and respond to temperature in the mutant strains. This would be best explained if FRQ and X are connected to each other in another loop, similar to the three-loop clock of *Arabidopsis* (Fig. 1).

To date, no progress has been made in identifying the molecular constituents of the FLO(s) responsible for the residual developmental rhythmicty. This lack of progress can primarily be attributed to the inherent variability of the FLO rhythms, making it difficult to screen for mutants that disrupt the FLO oscillations. However, a recent breakthrough may change our ability to characterize the molecular components responsible for the residual oscillations. The period of the FLO conidiation rhythm can be stabilized, through an unknown mechanism, by the addition of farnesol or gerianiol, two intermediates of the sterol synthesis pathway, to the growth medium (Granshaw et al. 2003). Similar to the variable rhythm on unsupplemented medium, this stabilized rhythm still lacks some of the canonical properties of circadian oscillators, including temperature compensation and light entrainment, but the rhythm can be reset by temperature pulses.

The ability to observe a consistent developmental rhythm in the absence of a functional FRQ/WCC oscillator now paves the way for identifying FLO components. Toward this end, the influence of other known clock-affecting mutations, in combination with FRQ or WC-2 deletion mutations, on the stabilized FLO developmental rhythm is being assayed (Lombardi et al. 2007). The ability to consistently observe a rhythm in the absence of the FRQ/WCC oscillator will also allow for the application of genetic screens for mutations that alter the FLO.

CONCLUSIONS

Enormous progress has been made in understanding the molecular mechanisms of the core circadian oscillators that were originally identified in genetic screens for mutations that alter a specific rhythmic behavior. We are now in a unique position to refine our view of the clock and incorporate information demonstrating residual rhythmicity in the absence of core clock components into working models, as carried out here for the *N. crassa* clock system. We suspect that these loops were not targeted in original mutant screens because they regulate different output pathways from those assayed, such as development rhythms in fungi (Feldman and Hoyle 1973) and eclosion and locomotion rhythms in flies (Konopka and Benzer 1971).

Experimental data and modeling support an organization of the core clock as a network of coupled oscillators, which can function when some of its parts are missing to regulate some aspects of rhythmicity. On their own, the individual loops appear to be less robust; this may be in part due to the components becoming more sensitive to cellular noise. Evidence is also accumulating to suggest the existence of genetically distinct circadian oscillators that can function as core oscillators to regulate specific outputs. In microbial organisms, the complexity exists within the cell, however; in muticellular organisms, complexity can occur both between and within cells.

We clearly still have a lot more work to do, the most important of which is to identify components of the new loops. The genetically tractable *N. crassa* clock system will continue to be an important model for identifying additional components and for developing a better understanding of the roles of a multi-oscillator clock.

ACKNOWLEDGMENTS

This research was supported in part by the National Science Foundation under a grant (PHY05-51164) to the Kavli Institute for Theoretical Physics, and National Institutes of Health grants (P01 NS39546 and R01 GM58529) to D.B.-P.

REFERENCES

Aronson B.D., Johnson K.A., and Dunlap J.C. 1994. Circadian clock locus *frequency:* Protein encoded by a single open reading frame defines period length and temperature compensation. *Proc. Natl. Acad. Sci.* **91:** 7683.

Bae K. and Weaver D.R. 2007. Transient, light-induced rhythmicity in mPer-deficient mice. *J. Biol. Rhythms* **22:** 85.

Bobrowicz P., Pawlak R., Correa A., Bell-Pedersen D., and Ebbole D.J. 2002. The *Neurospora crassa* pheromone precursor genes are regulated by the mating type locus and the circadian clock. *Mol. Microbiol.* **45:** 795.

Cambras T., Weller J.R., Angles-Pujoras M., Lee M.L., Christopher A., Diez-Noguera A., Krueger J.M., and de la Iglesia H.O. 2007. Circadian desynchronization of core body temperature and sleep stages in the rat. *Proc. Natl. Acad. Sci.* **104:** 7634.

Cheng P., Yang Y., and Liu Y. 2001a. Interlocked feedback loops contribute to the robustness of the *Neurospora* circadian clock. *Proc. Natl. Acad. Sci.* **98:** 7048.

Cheng P., Yang Y., Gardner K.H., and Liu Y. 2002. PAS domain-mediated WC-1/WC-2 interaction is essential for maintaining the steady-state level of WC-1 and the function of both proteins in circadian clock and light responses of *Neurospora*. *Mol. Cell. Biol.* **22:** 517.

Cheng P., Yang Y., Heintzen C., and Liu Y. 2001b. Coiled-coil domain-mediated FRQ-FRQ interaction is essential for its circadian clock function in *Neurospora*. *EMBO J.* **20:** 101.

Cheng P., He Q., He Q., Wang L., and Liu Y. 2005. Regulation of the *Neurospora* circadian clock by an RNA helicase. *Genes Dev.* **19:** 234.

Cheng P., Yang Y., Wang L., He Q., and Liu Y. 2003. WHITE COLLAR-1, a multifunctional *Neurospora* protein involved in the circadian feedback loops, light sensing, and transcription repression of *wc-2*. *J. Biol. Chem.* **278:** 3801.

Christensen M.K., Falkeid G., Loros J.J., Dunlap J.C., Lillo C., and Ruoff P. 2004. A nitrate-induced frq-less oscillator in *Neurospora crassa*. *J. Biol. Rhythms.* **19:** 280.

Collins B.H., Dissel S., Gaten E., Rosato E., and Kyriacou C.P. 2005. Disruption of cryptochrome partially restores circadian rhythmicity to the arrhythmic period mutant of *Drosophila*. *Proc. Natl. Acad. Sci.* **102:** 19021.

Correa A., Lewis Z.A., Greene A.V., March I.J., Gomer R.H., and Bell-Pedersen D. 2003. Multiple oscillators regulate circadian gene expression in *Neurospora*. *Proc. Natl. Acad. Sci.* **100:** 13597.

Cyran S.A., Buchsbaum A.M., Reddy K.L., Lin M.C., Glossop N.R., Hardin P.E., Young M.W., Storti R.V., and Blau J. 2003. vrille, Pdp1, and dClock form a second feedback loop in the *Drosophila* circadian clock. *Cell* **112:** 329.

Daan S., Albrecht U., van der Horst G.T., Illnerova H., Roenneberg T., Wehr T.A., and Schwartz W.J. 2001. Assembling a clock for all seasons: Are there M and E oscillators in the genes? *J. Biol. Rhythms* **16:** 105.

de Paula R.M., Lewis Z.A., Greene A.V., Seo K.S., Morgan L.W., Vitalini M.W., Bennett L., Gomer R.H., and Bell-Pedersen D. 2006. Two circadian timing circuits in *Neurospora crassa* cells share components and regulate distinct rhythmic processes. *J. Biol. Rhythms* **21:** 159.

Denault D.L., Loros J.J., and Dunlap J.C. 2001. WC-2 mediates WC-1-FRQ interaction within the PAS protein-linked circadian feedback loop of *Neurospora*. *EMBO J.* **20:** 109.

Elvin M., Loros J.J., Dunlap J.C., and Heintzen C. 2005. The PAS/LOV protein VIVID supports a rapidly dampened daytime oscillator that facilitates entrainment of the *Neurospora* circadian clock. *Genes Dev.* **19:** 2593.

Emery P., Stanewsky R., Hall J.C., and Rosbash M. 2000. A unique circadian-rhythm photoreceptor. *Nature* **404:** 456.

Feldman J.F. and Hoyle M.N. 1973. Isolation of circadian clock mutants of *Neurospora crassa*. *Genetics* **75:** 605.

Froehlich A.C., Loros J.J., and Dunlap J.C. 2003. Rhythmic binding of a WHITE COLLAR-containing complex to the frequency promoter is inhibited by FREQUENCY. *Proc. Natl. Acad. Sci.* **100:** 5914.

Froehlich A.C., Liu Y., Loros J.J., and Dunlap J.C. 2002. White Collar-1, a circadian blue light photoreceptor, binding to the *frequency* promoter. *Science* **297:** 815.

Glossop N.R. and Hardin P.E. 2002. Central and peripheral circadian oscillator mechanisms in flies and mammals. *J. Cell Sci.* **115:** 3369.

Glossop N.R., Lyons L.C., and Hardin P.E. 1999. Interlocked feedback loops within the *Drosophila* circadian oscillator. *Science* **286:** 766.

Granshaw T., Tsukamoto M., and Brody S. 2003. Circadian rhythms in *Neurospora crassa:* Farnesol or geraniol allow expression of rhythmicity in the otherwise arrhythmic strains *frq*^{10}, *wc-1*, and *wc-2*. *J. Biol. Rhythms* **18:** 287.

Hardin P.E., Krishnan B., Houl J.H., Zheng H., Ng F.S., Dryer S.E., and Glossop N.R. 2003. Central and peripheral circadian oscillators in *Drosophila*. *Novartis Found. Symp.* **253:** 140.

He Q., Cheng P., He Q., and Liu Y. 2005. The COP9 signalosome regulates the *Neurospora* circadian clock by controlling the stability of the SCFFWD-1 complex. *Genes Dev.* **19:** 1518.

He Q., Cha J., He Q., Lee H.C., Yang Y., and Liu Y. 2006. CKI and CKII mediate the FREQUENCY-dependent phosphorylation of the WHITE COLLAR complex to close the *Neurospora* circadian negative feedback loop. *Genes Dev.* **20:** 2552.

He Q., Cheng P., Yang Y., Wang L., Gardner K.H., and Liu Y. 2002. White collar-1, a DNA binding transcription factor and a light sensor. *Science* **297:** 840.

Heintzen C. and Liu Y. 2007. The *Neurospora crassa* circadian clock. *Adv. Genet.* **58:** 25.

Ivanchenko M., Stanewsky R., and Giebultowicz J.M. 2001. Circadian photoreception in *Drosophila:* Functions of cryptochrome in peripheral and central clocks. *J. Biol. Rhythms* **16:** 205.

Iwasaki H. and Dunlap J.C. 2000. Microbial circadian oscillatory systems in *Neurospora* and *Synechococcus:* Models for cellular clocks. *Curr. Opin. Microbiol.* **3:** 189.

Jagota A., de la Iglesia H.O., and Schwartz W.J. 2000. Morning and evening circadian oscillations in the suprachiasmatic nucleus *in vitro*. *Nat. Neurosci.* **3:** 372.

Kaldi K., Gonzalez B.H., and Brunner M. 2006. Transcriptional regulation of the *Neurospora* circadian clock gene *wc-1* affects the phase of circadian output. *EMBO Rep.* **7:** 199.

Konopka R.J. and Benzer S. 1971. Clock mutants of *Drosophila melanogaster*. *Proc. Natl. Acad. Sci.* **68:** 2112.

Krishnan B., Levine J.D., Lynch M.K., Dowse H.B., Funes P., Hall J.C., Hardin P.E., and Dryer S.E. 2001. A new role for cryptochrome in a *Drosophila* circadian oscillator. *Nature* **411:** 313.

Lakin-Thomas P.L. 1998. Choline depletion, *frq* mutations, and temperature compensation of the circadian rhythm in *Neurospora crassa*. *J. Biol. Rhythms* **13:** 268.

———. 2006. Circadian clock genes *frequency* and *white collar-1* are not essential for entrainment to temperature cycles in *Neurospora crassa*. *Proc. Natl. Acad. Sci.* **103:** 4469.

Lakin-Thomas P.L. and Brody S. 2000. Circadian rhythms in *Neurospora crassa:* Lipid deficiencies restore robust rhythmicity to null *frequency* and *white-collar* mutants. *Proc. Natl. Acad. Sci.* **97:** 256.

Lee K., Loros J.J., and Dunlap J.C. 2000. Interconnected feedback loops in the *Neurospora* circadian system. *Science* **289:** 107.

Leloup J.C. and Goldbeter A. 2004. Modeling the mammalian circadian clock: Sensitivity analysis and multiplicity of oscillatory mechanisms. *J. Theor. Biol.* **230:** 541.

Liu Y. and Bell-Pedersen D. 2006. Circadian rhythms in *Neurospora crassa* and other filamentous fungi. *Eukaryot. Cell* **5:** 1184.

Liu Y., Loros J., and Dunlap J.C. 2000. Phosphorylation of the *Neurospora* clock protein FREQUENCY determines its degradation rate and strongly influences the period length of the circadian clock. *Proc. Natl. Acad. Sci.* **97:** 234.

Locke J.C., Kozma-Bognar L., Gould P.D., Feher B., Kevei E., Nagy F., Turner M.S., Hall A., and Millar A.J. 2006. Experimental validation of a predicted feedback loop in the multi-oscillator clock of *Arabidopsis thaliana*. *Mol. Syst. Biol.* **2:** 59.

Lombardi L., Schneider K., Tsukamoto M., and Brody S. 2007. Circadian rhythms in *Neurospora crassa:* Clock mutant effects in the absence of a *frq*-based oscillator. *Genetics* **175:** 1175.

Loros J.J. and Dunlap J.C. 2001. Genetic and molecular analysis of circadian rhythms in *Neurospora*. *Annu. Rev. Physiol.* **63:** 757.

Loros J.J., Denome S.A., and Dunlap J.C. 1989. Molecular cloning of genes under control of the circadian clock in *Neurospora*. *Science* **243:** 385.

Loros J.J., Richman A., and Feldman J.F. 1986. A recessive circadian clock mutation at the *frq* locus of *Neurospora crassa*. *Genetics* **114:** 1095.

Merrow M., Brunner M., and Roenneberg T. 1999. Assignment of circadian function for the *Neurospora* clock gene *frequency*. *Nature* **399:** 584.

Morse D., Hastings J.W., and Roenneberg T. 1994. Different phase responses of the two circadian oscillators in *Gonyaulax*. *J. Biol. Rhythms* **9:** 263.

Pittendrigh C.S. 1961. Circadian rhythms and the circadian organization of living things. *Cold Spring Harbor Symp. Quant. Biol.* **25:** 159.

Pittendrigh C.S. and Bruce V.G. 1959. Daily rhythms as coupled oscillator systems and their relation to thermoperiodism and photoperiodism. In *Photoperiodism and related phenomena in plants and animals* (ed. Withrow), p. 475. AAAS, Washington, D.C.

Pittendrigh C. and Daan S. 1976. A functional analysis of circadian pacemakers in nocturnal rodents. V. Pacemaker structure: A clock for all seasons. *J. Comp. Physiol.* **106:** 333.

Pittendrigh C.S., Bruce B.G., Rosensweig N.S., and Rubin M.L. 1959. Growth patterns in *Neurospora*. *Nature* **184:** 169.

Pregueiro A.M., Price-Lloyd N., Bell-Pedersen D., Heintzen C., Loros J.J., and Dunlap J.C. 2005. Assignment of an essential role for the *Neurospora frequency* gene in circadian entrainment to temperature cycles. *Proc. Natl. Acad. Sci.* **102:** 2210.

Preitner N., Damiola F., Lopez-Molina L., Zakany J., Duboule D., Albrecht U., and Schibler U. 2002. The orphan nuclear receptor REV-ERBalpha controls circadian transcription within the positive limb of the mammalian circadian oscillator. *Cell* **110:** 251.

Ramsdale M. and Lakin-Thomas P.L. 2000. sn-1,2-diacylglycerol levels in the fungus *Neurospora crassa* display circadian rhythmicity. *J. Biol. Chem.* **275:** 27541.

Rand D.A., Shulgin B.V., Salazar J.D., and Millar A.J. 2006. Uncovering the design principles of circadian clocks: Mathematical analysis of flexibility and evolutionary goals. *J. Theor. Biol.* **238:** 616.

Roenneberg T., Dragovic Z., and Merrow M. 2005. Demasking biological oscillators: Properties and principles of entrainment exemplified by the *Neurospora* circadian clock. *Proc. Natl. Acad. Sci.* **102:** 7742.

Sai J. and Johnson C.H. 1999. Different circadian oscillators control Ca(2+) fluxes and lhcb gene expression. *Proc. Natl. Acad. Sci.* **96:** 11659.

Schafmeier T., Kaldi K., Diernfellner A., Mohr C., and Brunner M. 2006. Phosphorylation-dependent maturation of *Neurospora* circadian clock protein from a nuclear repressor toward a cytoplasmic activator. *Genes Dev.* **20:** 297.

Schafmeier T., Haase A., Kaldi K., Scholz J., Fuchs M., and Brunner M. 2005. Transcriptional feedback of *Neurospora* circadian clock gene by phosphorylation-dependent inactivation of its transcription factor. *Cell* **122:** 235.

Shearman L.P., Sriram S., Weaver D.R., Maywood E.S., Chaves I., Zheng B., Kume K., Lee C.C., van der Horst G.T., Hastings M.H., and Reppert S.M. 2000. Interacting molecular loops in the mammalian circadian clock. *Science* **288:** 1013.

Shinohara M.L., Loros J.J., and Dunlap J.C. 1998. Glyceraldehyde-3-phosphate dehydrogenase is regulated on a daily basis by the circadian clock. *J. Biol. Chem.* **273:** 446.

Shinohara M.L., Correa A., Bell-Pedersen D., Dunlap J.C., and Loros J.J. 2002. *Neurospora clock-controlled gene 9* (*ccg-9*) encodes trehalose synthase: Circadian regulation of stress responses and development. *Eukaryot. Cell* **1:** 33.

Stanewsky R., Kaneko M., Emery P., Beretta B., Wager-Smith K., Kay S.A., Rosbash M., and Hall J.C. 1998. The *cry*^{b} mutation identifies cryptochrome as a circadian photoreceptor in *Drosophila*. *Cell* **95:** 681.

Stoleru D., Peng Y., Agosto J., and Rosbash M. 2004. Coupled oscillators control morning and evening locomotor behaviour of *Drosophila*. *Nature* **431:** 862.

Vitalini M.W., de Paula R.M., Park W.D., and Bell-Pedersen D. 2006. The rhythms of life: Circadian output pathways in *Neurospora*. *J. Biol. Rhythms* **21:** 432.

The Diurnal Project: Diurnal and Circadian Expression Profiling, Model-based Pattern Matching, and Promoter Analysis

T.C. MOCKLER,* T.P. MICHAEL,† H.D. PRIEST,* R. SHEN,* C.M. SULLIVAN,* S.A. GIVAN,* C. MCENTEE,† S.A. KAY,‡ AND J. CHORY†§

**Department of Botany and Plant Pathology and Center for Genome Research and Biocomputing, Oregon State University, Corvallis, Oregon 97331; †Plant Biology Laboratory, The Salk Institute for Biological Studies, La Jolla, California 92037; ‡Department of Cell and Developmental Biology, University of California at San Diego, La Jolla, California 92093-0116; §Howard Hughes Medical Institute, The Salk Institute for Biological Studies, La Jolla, California 92037*

The DIURNAL project (http://diurnal.cgrb.oregonstate.edu/) provides a graphical interface for mining and viewing diurnal and circadian microarray data for *Arabidopsis thaliana*, poplar, and rice. The database is searchable and provides access to several user-friendly Web-based data-mining tools with easy-to-understand output. The associated tools include HAYSTACK (http://haystack.cgrb.oregonstate.edu/) and ELEMENT (http://element.cgrb.oregonstate.edu/). HAYSTACK is a model-based pattern-matching algorithm for identifying genes that are coexpressed and potentially coregulated. HAYSTACK can be used to analyze virtually any large-scale microarray data set and provides an alternative method for clustering microarray data from any experimental system by grouping together genes whose expression patterns match the same or similar user-defined patterns. ELEMENT is a Web-based program for identifying potential *cis*-regulatory elements in the promoters of coregulated genes in *Arabidopsis*, poplar, and rice. Together, DIURNAL, HAYSTACK, and ELEMENT can be used to facilitate cross-species comparisons among the plant species supported and to accelerate functional genomics efforts in the laboratory.

INTRODUCTION

Various databases containing gene expression data and Web-based tools for analyzing microarray data have emerged as valuable resources for many aspects of plant research. These resources include AtGenExpress (Schmid et al. 2005; http://www.weigelworld.org/resources/microarray/AtGenExpress/), ArrayExpress (Brazma et al. 2003; http://www.ebi.ac.uk/arrayexpress/), Botany Array Resource (BAR; Toufighi et al. 2005; http://www.bar.utoronto.ca/), GENEVESTIGATOR (Zimmermann et al. 2004; https://www.genevestigator.ethz.ch/), GEO (Barrett et al. 2007; http://www.ncbi.nlm.nih.gov/geo/), NASC Arrays (Craigon et al. 2004; http://affymetrix.arabidopsis.info/), PlexDB/Barleybase (Shen et al. 2005; http://www.plexdb.org/), TAIR (Garcia-Hernandez et al. 2002; Rhee et al. 2003; http://www.arabidopsis.org/), and VirtualPlant (http://virtualplant.org). Here, we introduce the DIURNAL project, consisting of a database containing diurnal and circadian expression data for approximately 22,800 *Arabidopsis* genes collected during 11 experiments, a Web interface for accessing this database, and complementary Web-based tools for analyzing microarray data and promoter sequences.

DIURNAL and its associated tools differ from other resources in several ways. DIURNAL is focused on plant-diurnal- and circadian-clock-regulated gene expression and currently supports *Arabidopsis*, rice, and poplar, three plant species with high-quality annotated genome sequences (*Arabidopsis* Genome Initiative 2000; International Rice Genome Sequencing Project 2005; Tuskan et al. 2006). A novel tool called HAYSTACK allows users to mine large microarray expression data sets by searching for specific user-defined patterns of expression. HAYSTACK is designed to find rare occurrences of very specific patterns in large data sets and provides an alternative method for clustering microarray data by grouping genes whose expression patterns match the same or similar HAYSTACK patterns. In addition to clustering genes based on the user-supplied model patterns, HAYSTACK can be used to identify genes that exhibit a pattern of expression similar to that of a particular gene of interest. Although we have used HAYSTACK in the DIURNAL project to identify diurnal and circadian regulated genes, this tool can be used to compare any large-scale data set representing at least three samples (e.g., treatments, genotypes, and time points) against a set of user-supplied model patterns. The resulting lists of coregulated genes in the HAYSTACK output can then be used with a third tool, ELEMENT, which is an enumerative promoter analysis program that analyzes the upstream regions of *Arabidopsis*, poplar, or rice genes to discover overrepresented elements that may represent novel transcription-factor-binding sites. Together, the DIURNAL suite of tools allows researchers to go from expression data to *cis*-regulatory elements and place this into the context of diurnal and circadian control, an important layer of regulation in *Arabidopsis*, with 90% of genes exhibiting diurnal or circadian regulation under at least one growth condition (T. Michael et al., in prep.).

THE DIURNAL DATABASE

Organisms experience daily environmental changes in light (photocycles) and temperature (thermocycles) that vary by season and latitude. Consequently, organisms have evolved an endogenous circadian clock with a period of about 24 hours, which ensures that internal biological processes are appropriately synchronized with the daily changes in the environment (Michael et al. 2003; Woelfle et al. 2004; Dodd et al. 2005). Transcript abundance can be found peaking at almost every hour during the day/night cycle, and this regulation forms the foundation for time-of day-specific biological activities.

The *Arabidopsis* microarray data available in the DIURNAL database was collected using the Affymetrix ATH1 GeneChip microarray platform, which represents about 22,800 genes as annotated by TAIR (ftp://ftp.arabidopsis.org/home/tair/Microarrays/Affymetrix). To identify cycling genes, we used these 336 model patterns with HAYSTACK to interrogate Affymetrix ATH1 GeneChip data sets for 11 time courses with 12 time points each. These data are presented and analyzed in a manuscript that has been submitted for publication (T. Michael et al., in prep.). Plant material and growth conditions for the *Arabidopsis* time courses are described at http://diurnal.cgrb.oregonstate.edu/diurnal_details.html.

Interface Design, Features, and Navigation

Layouts, fonts, color schemes, and navigation were designed to make the interface user-friendly. The main page of DIURNAL contains a link to an "About Diurnal" page that summarizes the project. This page has a menu bar with links to the other tools of the project: ELEMENT and HAYSTACK. An additional tool, ORTHOMAP, provides predicted orthologs, homologs, or simple best BLAST matches for *Arabidopsis*, rice, and poplar and links to the respective external annotation resources at TAIR, TIGR, or JGI. This tool allows a user to obtain this information for a gene or group of genes of interest, without having to browse different databases or conduct sequence comparisons. The menu bar and links box appear on every page to enable efficient browsing between tools. In addition, the tools are connected in such a way that, for example, a list of genes identified by a query of DIURNAL or a list of predicted orthologs identified by a query of ORTHOMAP can be sent to ELEMENT for promoter analysis. The databases underlying DIURNAL, ELEMENT, and ORTHOMAP are based on MySQL, and the Web interfaces have been implemented using the Apache HTML delivery engine (http://www.apache.org) Perl and Mason (http://www.masonhq.com).

DIURNAL: A Diurnal/Circadian Gene Expression Data-mining Tool

DIURNAL is useful for a researcher who is interested in the diurnal or circadian expression profiles of a particular set of genes, either for diurnal/circadian experiments or for planning other types of experiments. The interface allows users to input a list of genes and select among the experiments in the database. The data resulting from the query is displayed graphically and available for download as a text file. In addition, we enable direct exporting of gene lists resulting from queries into ELEMENT for promoter analysis.

The entry page of DIURNAL includes a text box for entering locus identifiers and check boxes for selecting experimental conditions/array data sets. This page represents the "Basic Search," which has options for displaying the best-fitting HAYSTACK model and the expression profiles for multiple genes on a single graph, and for normalizing the expression levels among multiple genes being graphed. The normalization option makes it easier to compare the phases and waveforms of expression for sets of genes with large differences in expression levels. The basic search returns a graph or graphs of the diurnal or circadian expression profiles for the gene(s) submitted by the user (Fig. 1A). Below the graph is a table that summarizes the condition, phase, best-fitting model, and correlation value representing the quality of the match between the model and the experimental data. Above each graph are links to the raw text data and to the annotation page for the particular gene. By clicking on the graph, the user is presented with a larger version of the graph that can be downloaded in PNG image file format.

The left side of all pages within DIURNAL contains a menu box with a link to "Advanced Search." On the Advanced Search page, a user can query the database to return all genes matching a set of user-defined criteria. The user can define the phase, model, and correlation cutoff values and use a pulldown menu to select an experimental condition to query. Results are represented as a table listing genes matching the query criteria (Fig. 1B). Each row of the table represents a single gene and the fields of the table contain the Affymetrix probe-set identifier, the locus identifier, correlation value, phase of expression based on the best-fitting HAYSTACK model, and the name of the best-fitting model. The rows are linked to a detailed graphical display of the time course expression profile for the particular gene. As in the basic search, the graph image and raw data are downloadable, and a link is provided to the respective annotation database. An advanced search allows the user to cluster cycling probe sets/genes by their phase or waveform (best model) and thus can generate gene lists suitable for input to ELEMENT or another promoter analysis algorithm. To facilitate such analysis, we have provided an option to send the gene list resulting from a DIURNAL advanced search directly to ELEMENT for prediction of potential *cis*-regulatory elements.

When a user enters a list of genes and chooses the desired options, the microarray data are selected for display by looking up the appropriate Affymetrix probe-set identifier in a table within our database. There is not always a unique relationship between the Affymetrix probe-set identifier and a locus identifier, so more than one row of the table may correspond to the same gene, but representing data from different probe sets. The data generated by the query is displayed on the appropriate results pages, depending on the options chosen and whether a basic or advanced search was performed.

Another output option of DIURNAL is a tab-delimited

A

Diurnal Results Page

Welcome to the Diurnal Results page.
You can view the source data for any graph by clicking the header.
Both the graphs and the source data are downloadable, by right clicking on either the link or the graph, and selecing 'save as'.
Also, you may click on any chart to enlarge it.

Gene: AT1G01060 **[Graph source data - AT1G01060 Information at TAIR]**

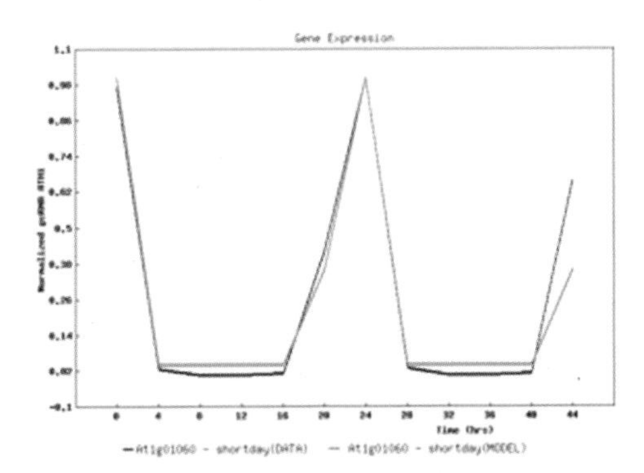

Condition	Phase	Model	Correlation
shortday	23	ct23-spike	0.971688

B

Affy ID	Locus ID	Correlation	Phase	Best Model
267199_at	AT2G30990	0.962899	13	ct13-spike
264482_at	AT1G77210	0.962727	13	ct13-spike
250777_at	AT5G05440	0.958629	13	ct13-spike
256433_at	AT3G10985	0.951229	13	ct13-spike
262870_at	AT1G64710	0.945601	13	ct13-spike
266825_at	AT2G22890	0.944285	13	ct13-spike
263461_at	AT2G31800	0.943772	13	ct13-spike
261668_at	AT1G18500	0.942345	13	ct13-spike
252924_at	AT4G39070	0.940817	13	ct13-spike
258404_at	AT3G17465	0.940213	13	ct13-spike
267523_at	AT2G30600	0.939136	13	ct13-spike
259813_at	AT1G49860	0.937439	13	ct13-spike
258531_at	AT3G06720	0.935411	13	ct13-spike
257206_at	AT3G16530	0.934895	13	ct13-spike

Figure 1. DIURNAL. (*A*) Example of the output of the DIURNAL Basic Search showing a graph of a gene expression profile for AT1G01060 under the short-day condition with a summary of model match statistics. (*B*) Example of the HTML table output from the DIURNAL Advanced Search showing genes and model match statistics fitting the search criteria. Clicking on a row of the table displays a graph for the gene and its best-fitting HAYSTACK model.

plain text file that contains the expression profile data for queried probe sets/genes. The data presented in this file are the same data as those used to generate the graphs, with each row of data representing a single gene or probe set. The data columns list the probe-set identifier, locus identifier, best model, the experimental condition/data set, and correlation, whether the row is data (DATA) or fitted model (MODEL), followed by the series of expression values for the time course.

DIURNAL may be used in several different ways by a biologist. First, the tool can be used to characterize the diurnal or circadian temporal expression pattern of a gene of interest, especially by querying the diurnal conditions that approximate standard laboratory conditions such as short days or long days. Second, it can be used to identify genes coexpressed and potentially coregulated with a gene of interest. This type of analysis can suggest potential protein interactions as well because it has been demonstrated that gene expression among interacting proteins can coevolve (Fraser et al. 2004). In vivo experiments altering gene expression or using the two-hybrid system can further elucidate potential interactions among the genes identified. Finally, DIURNAL provides a method of clustering coexpressed genes whose promoters may be analyzed with either ELEMENT or other algorithms.

Because the DIURNAL interface can accept lists of multiple genes for comparison, it can also be used to predict functional redundancies among members of a gene family. For example, a user could submit a list of gene identifiers for a family of genes and use the DIURNAL output to classify them according to their temporal expression patterns. Example of such analyses are shown in Figure 2. Members of the C2C2-YABBY transcription factor family function to specify abaxial cell fate in *Arabidopsis* (Siegfried et al. 1999). A query of DIURNAL using the gene identifiers for the six members of the C2C2-YABBY TF family reveals diurnal expression profiles consistent with functional redundancy among some family members (Fig. 2A). Five out of six family members are represented on the Affymetrix ATH1 microarrays, and three out of these five cycle under the long-day condition. YAB1 (At2g45190) and YAB3 (At4g00180) share similar expression profiles, with peak expression occurring just before dawn, at phases 23 and 22, respectively. The similar phasing of expression of these two family members could indicate functional redundancy at this time of day. In contrast, YABBY5 (At2g26580) cycles with a phase of 16 hours after dawn, whereas CRC (AT1G69180) and YAB4 (AT1G23420) do not cycle. Another example involves an 11-gene family of trehalose-6-phosphate synthases involved in biosynthesis of trehalose, a sugar whose accumulation is implicated in drought stress tolerance (Leyman et al. 2001; Karim et al. 2007). All 11 family members are represented on the ATH1 microarrays. In short days, 7 out of 11 cycle and 6 of these are phased to 18–19 hours after dawn (Fig. 2B). In thermocycles (LLHC), 8 out of 11 cycle and 4 of these are phased to 8 hours after dawn (not shown in figure). These examples illustrate how diurnal expression profiles can help clarify potential functional redundancies among subgroups within a gene family.

HAYSTACK: A MODEL-BASED PATTERN-MATCHING ALGORITHM

HAYSTACK (http://haystack.cgrb.oregonstate.edu/) is a program for identifying genes whose expression levels behave similarly across all samples in a microarray data set. Multiple approaches have been developed for organizing and inferring patterns emerging from microarray data (for review, see Belacel et al. 2006). Conventional microarray-clustering approaches are based on identifying distinct or separable groups of genes based on a distance metric (Hierarchical, K-Means, Self-organizing Maps, Support Vector Machines), or principal component analysis. In contrast, our approach is hypothesis-driven and depends on predefined models to identify statistically similar groups of coexpressed genes. By identifying rare occurrences of specific biologically relevant expression patterns in the experimental data, we are able to dramatically reduce the search space for subsequent analyses, including promoter analysis to identify important *cis*-regulatory elements.

HAYSTACK uses a pattern-matching algorithm to identify genes whose expression patterns fit a user-defined model. HAYSTACK is designed to find rare occurrences of very specific patterns in a large data set and provides an alternative method for clustering microarray data, by grouping genes whose expression patterns match the same or similar HAYSTACK patterns. The algorithm is based on determining the Pearson correlation coefficient between gene expression profiles and user-defined models. HAYSTACK determines the correlation of an experimental data series with each supplied model pattern and applies a series of statistical tests and ad hoc filters to identify genes of interest and their corresponding best-fitting model. We have used HAYSTACK to compare microarray time course data against a collection of diurnal/circadian models to identify cycling genes. The Web version of HAYSTACK (http://haystack.cgrb.oregonstate.edu/) can be used to compare a large-scale data set against a set of user-supplied model patterns to search for biologically relevant patterns in the data.

We are interested in time-of-day specific and circadian transcriptional networks; therefore, we have focused on time course data to highlight the simplicity and power of HAYSTACK. We developed multiple cycling patterns based on diurnal and circadian time courses available in the literature: asymmetric, rigid, spike, cosine, sine, and/or box-like patterns (Harmer et al. 2000; Smith et al. 2004; Blasing et al. 2005; Edwards et al. 2006). To capture both cycling and phase information in the time course data, HAYSTACK patterns were used to mine data from 11 *Arabidopsis* time courses. The resulting analyses are available through the DIURNAL interface.

To use HAYSTACK, a user starts by uploading a file containing the model patterns (which can be easily constructed using a text editor or Microsoft Excel) and a file containing the microarray data of interest, arranged as a data series in the same format as the model patterns. The user then selects optional statistical criteria and ad hoc filters, and HAYSTACK calculates the correlation coefficient between the expression values across the microarray

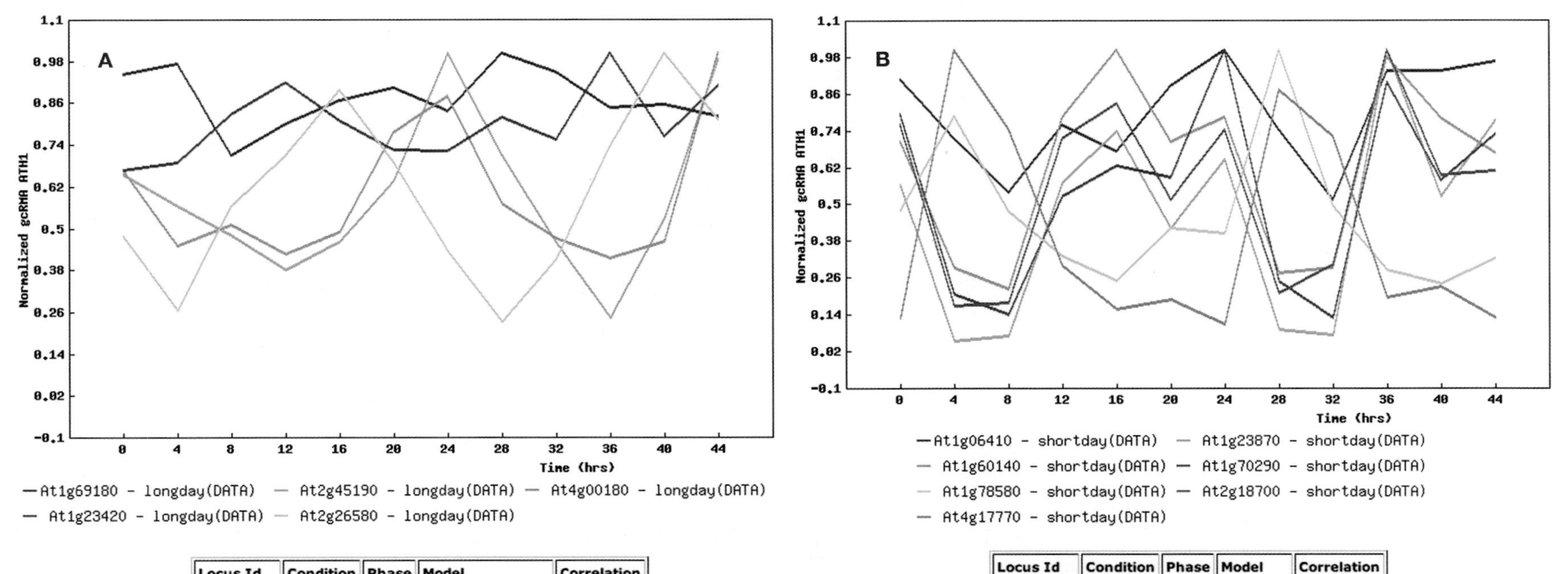

Locus Id	Condition	Phase	Model	Correlation
AT1G69180	longday	3	ct03-spike	0.667899
AT2G45190	longday	23	cos_per_24_ph_23	0.847533
AT4G00180	longday	22	ct22-spike	0.901396
AT1G23420	longday	12	ct12-spike	0.736117
AT2G26580	longday	16	ct16-rigid	0.976838

Locus Id	Condition	Phase	Model	Correlation
AT1G06410	shortday	19	ct19-box2	0.838949
AT1G23870	shortday	18	ct18-box2	0.893498
AT1G60140	shortday	18	ct18-box2	0.932179
AT1G70290	shortday	18	ct18-box2	0.885764
AT1G78580	shortday	5	ct05-spike	0.928022
AT2G18700	shortday	18	ct18-box2	0.891818
AT4G17770	shortday	6	ct06-spike	0.967791

Figure 2. DIURNAL. (*A*) Graph and summary table of diurnal expression profiles of the C2C2-YABBY transcription factor family in long days. (*B*) Graph and summary table of diurnal expression profiles of the trehalose-6-phosphate synthase family in short days. For clarity, expression profiles of genes that do not cycle in this condition have been omitted.

data set for that gene and each of the user-supplied models. HAYSTACK is not limited to any particular microarray platform, genome annotation, or even expression values. It can be used to compare any large-scale data set representing at least three samples (e.g., treatments, genotypes, and time points) against the user-supplied model patterns. A minimum of three data points are necessary for the program to perform valid correlation calculations between the model and experimental data series. The program returns to the user (via e-mail) a link to a results file containing those genes satisfying the user-specified criteria (Fig. 3A). The results from HAYSTACK may be viewed in additional ways. First, the results can be passed directly to a plotting program on the HAYSTACK Web site for easy visual inspection of the data series and corresponding best-match model (Fig. 3B). Alternatively, the results are available in a text file format that may be downloaded and viewed separately.

HAYSTACK can be applied intuitively in several different ways by a biologist. For example, to generate leads for further characterization in the laboratory, a researcher may want to identify all genes in a microarray data set that exhibit an expression profile similar to a particular pattern. Alternatively, a researcher may want to identify genes coexpressed with a particular gene of interest. In this case, the supplied model pattern could be the expression profile of the gene of interest. A third use for HAYSTACK is to seed promoter analysis using ELEMENT or other promoter element discovery programs such as Promomer (Toufighi et al. 2005), SIFT (Hudson and Quail 2003), or MotifSampler (Thijs et al. 2001, 2002). An interesting potential use of HAYSTACK

A

Haystack Results Page

sort by: (•) Data ID ( ) Best Model ( ) Correlation ( ) T-statistic ( ) T-prob Sort

datatype	data-id	best model	correlation	t-statistic	t-prob	point 1	point 2	point 3	point 4	point 5	point 6	point 7	point 8	point 9	point 10	point 11	point 12
DATA	AT1G01060	ct01-spike	0.986737	19.2223	1.5807e-09	2243.22	481.2	7.19888	4.78697	10.6717	150.279	1904.97	483.612	8.46222	6.00153	10.8894	135.684
MODEL	AT1G01060	ct01-spike	0.986737	19.2223	1.5807e-09	2009.52	676.143	9.45715	9.45715	9.45715	9.45715	2009.52	676.143	9.45715	9.45715	9.45715	9.45715
DATA	AT1G01120	ct01-spike	0.954332	10.1016	7.2436e-07	1187.93	871.941	417.74	431.762	539.167	601.483	1160.63	594.5	407.266	359.107	438.074	460.516
MODEL	AT1G01120	ct01-spike	0.954332	10.1016	7.2436e-07	1184.43	702.783	461.962	461.962	461.962	461.962	1184.43	702.783	461.962	461.962	461.962	461.962
DATA	AT1G01470	ct11-rigid	0.93741	8.51268	3.4032e-06	666.668	1055.24	1257.8	1504.93	1035.58	1024.44	824.042	1050.68	1474.04	1497.35	1053.51	806.671
MODEL	AT1G01470	ct11-rigid	0.93741	8.51268	3.4032e-06	738.417	1031.08	1323.74	1470.07	1177.41	884.749	738.417	1031.08	1323.74	1470.07	1177.41	884.749
DATA	AT1G01500	ct09-box1	0.961736	11.1004	3.029e-07	16.9362	52.0911	75.0569	60.3221	20.4559	19.406	13.906	44.4155	79.249	68.6977	17.2291	17.2611
MODEL	AT1G01500	ct09-box1	0.961736	11.1004	3.029e-07	13.1714	54.0426	67.6663	67.6663	26.7951	13.1714	13.1714	54.0426	67.6663	67.6663	26.7951	13.1714
DATA	AT1G01520	ct04-spike	0.99274	26.101	7.8432e-11	11.7446	67.3882	5.606	7.32917	9.423	6.27926	11.1671	60.218	7.2421	4.61058	7.53232	5.77256
MODEL	AT1G01520	ct04-spike	0.99274	26.101	7.8432e-11	7.67067	63.8031	7.67067	7.67067	7.67067	7.67067	7.67067	63.8031	7.67067	7.67067	7.67067	7.67067
DATA	AT1G01540	ct22-spike	0.971989	13.078	6.479e-08	507.897	305.716	264.305	263.683	312.525	500.089	527.356	318.02	234.453	277.001	309.147	481.343
MODEL	AT1G01540	ct22-spike	0.971989	13.078	6.479e-08	504.171	285.606	285.606	285.606	285.606	504.171	504.171	285.606	285.606	285.606	285.606	504.171
DATA	AT1G01710	ct05-spike	0.908336	6.86785	2.1811e-05	43.5761	88.3527	59.0485	51.1082	44.6455	32.5425	37.1494	70.3272	60.5696	48.1761	44.1561	35.6909
MODEL	AT1G01710	ct05-spike	0.908336	6.86785	2.1811e-05	42.85	80.7787	55.4929	42.85	42.85	42.85	42.85	80.7787	55.4929	42.85	42.85	42.85
DATA	AT1G01820	ct15-box2	0.920648	7.45746	1.0832e-05	156.438	99.71	233.061	258.122	192.254	267.124	171.151	114.502	253.253	247.011	241.602	233.995
MODEL	AT1G01820	ct15-box2	0.920648	7.45746	1.0832e-05	190.339	98.258	236.379	236.379	236.379	236.379	190.339	98.258	236.379	236.379	236.379	236.379
DATA	AT1G02300	ct09-spike	0.920982	7.47529	1.0611e-05	248.915	305.551	1214.18	919.968	565.218	538.209	313.017	287.087	1146.41	815.452	518.659	379.696
MODEL	AT1G02300	ct09-spike	0.920982	7.47529	1.0611e-05	423.351	423.351	1237.91	694.87	423.351	423.351	423.351	423.351	1237.91	694.87	423.351	423.351
DATA	AT1G02475	ct12-rigid	0.922183	7.54019	9.8496e-06	344.432	446.926	523.909	536.711	502.167	446.676	302.202	485.569	486.607	634.236	542.159	455.529
MODEL	AT1G02475	ct12-rigid	0.922183	7.54019	9.8496e-06	353.338	434.842	516.345	597.849	516.345	434.842	353.338	434.842	516.345	597.849	516.345	434.842
DATA	AT1G02640	ct22-spike	0.965483	11.7218	1.8206e-07	182.664	52.0703	47.2754	40.1909	40.6823	171.633	178.068	48.5434	37.7738	44.8032	33.9005	252.54
MODEL	AT1G02640	ct22-spike	0.965483	11.7218	1.8206e-07	196.226	43.155	43.155	43.155	43.155	196.226	196.226	43.155	43.155	43.155	43.155	196.226
DATA	AT1G02750	ct22-spike	0.933269	8.2167	4.6551e-06	133.721	55.3044	59.269	71.6809	75.0109	116.825	101.988	63.6852	72.8554	66.2262	62.4566	105.998
MODEL	AT1G02750	ct22-spike	0.933269	8.2167	4.6551e-06	114.633	65.8111	65.8111	65.8111	65.8111	114.633	114.633	65.8111	65.8111	65.8111	65.8111	114.633
DATA	AT1G02816	ct04-spike	0.983746	17.3244	4.3474e-09	146.694	330.463	128.255	143.202	133.157	135.967	168.847	317.425	114.509	126.817	134.733	133.351
MODEL	AT1G02816	ct04-spike	0.983746	17.3244	4.3474e-09	136.553	323.944	136.553	136.553	136.553	136.553	136.553	323.944	136.553	136.553	136.553	136.553
DATA	AT1G02820	ct05-spike	0.98225	16.5594	6.7344e-09	35.6421	1009.01	113.502	9.42576	15.5821	21.9633	34.8405	1025.22	262.281	10.6041	19.3711	17.4676

B

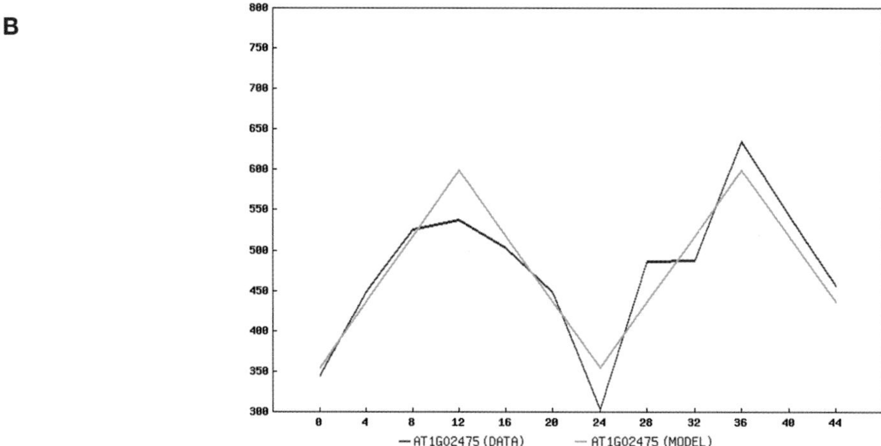

Figure 3. HAYSTACK. (*A*) Example of an HTML table of HAYSTACK output showing genes, their best matching models, and summary statistics. (*B*) Example of a graph of gene expression profile and best-fitting HAYSTACK model, derived by clicking on the row of data for a gene shown on the HTML table in *A*.

involves using it to search nonmicroarray data such as the significance statistics for potential *cis* elements generated by enumerative promoter-searching tools such as ELEMENT. We used this approach to search the serialized *Z*-scores for overrepresented words identified in our diurnal/circadian studies and thus identified co-occurring elements that form the basis of diurnal/circadian transcriptional network modules (see Fig. 5 below) (T. Michael et al., in prep.).

Other Web-based bioinformatics tools such as the "Expression Angler" or "Sample Angler" at BAR (Toufighi et al. 2005) can identify genes that respond similarly across samples (i.e., genes whose expression profiles are highly correlated), whereas HAYSTACK provides a model-based method for clustering microarray data—genes that are returned with best matches to the same model are potentially coregulated. Some qualifications must be made regarding the results from HAYSTACK, and some of the user options are intended to overcome these shortcomings. For example, spurious model/data matches involving data series in which all of the points are below the background level (noise) are possible. To address this possibility, we implemented a "Background Cutoff," which is the minimum acceptable value for the highest value in the data series. Another possible basis for spurious matches involves situations in which a data series is highly correlated with a model, but the difference between the maximum and minimum values in the data series is insignificant. This problem arises because the Pearson correlation is amplitude-independent. We address this issue by providing a "Fold Cutoff" option that allows a user to determine the minimum acceptable fold difference (i.e., max/min) between the maximum and minimum values in the data series. These options provide flexibility for users to decide what parameters make sense in the context of their experiments.

ELEMENT: A TOOL FOR IDENTIFYING POTENTIAL *CIS*-REGULATORY ELEMENTS IN PLANTS

The regulation of gene expression in eukaryotes is largely mediated by transcription factors (TFs) that bind within regulatory regions (promoters) upstream of the coding sequence. Transcription factors recognize specific DNA motifs, bind, and in turn interact with each other and the basal transcriptional machinery to regulate the expression of adjacent genes. With the recent availability of high-quality sequenced and annotated genomes, large public microarray databases, and easy access to microarray technology for individual laboratories, there is a need for bioinformatics tools to predict components of transcriptional networks, including transcription-factor-binding sites. Groups of coexpressed genes identified using microarrays may be coregulated and thus can form the foundation for analyses of promoter sequences to identify important *cis*-regulatory elements.

A number of algorithms have been developed to identify known and putative regulatory elements in the promoter sequences of coregulated genes (for review, see Rombauts et al. 2003; Tompa et al. 2005). The fundamental assumption underlying all of these computational approaches is that coregulated genes should contain similar regulatory motifs in their promoters, and these motifs should be significantly overrepresented in the set of coregulated promoters. There are two general computational approaches for identifying potential *cis*-regulatory elements. One approach is an enumerative method, and the other is an alignment method. The alignment methods are exemplified by programs that use a Gibbs sampling method (Thijs et al. 2002).

The enumerative methods estimate the probability of occurrence of short DNA sequences, or "words," by comparing the count in a set of coregulated sequences to an expected count based on random sampling or statistical modeling of a background distribution (van Helden et al. 1998; Hudson and Quail 2003; Kreps et al. 2003; Marino-Ramirez et al. 2004; Nemhauser et al. 2004; Koussevitzky et al. 2007). Therefore, each algorithm requires some background model to calculate an expected frequency for each word. The composition of the sequences underlying the background model is critical because the various features (e.g., exons, introns, and intergenic regions) within a genome exhibit different oligomer compositions. Both enumerative and alignment approaches have been applied to promoter analysis in plants, and the putative coregulated sequences chosen for analysis were typically derived from hierarchical clustering or other analyses of microarray data (Harmer et al. 2000; Chen et al. 2002; Hudson and Quail 2003; Hulzink et al. 2003; Nemhauser et al. 2004; Koussevitzky et al. 2007).

The goal of the ELEMENT program is to provide a user-friendly Web-based tool that uses the enumerative method to identify statistically overrepresented 3–8 mer DNA words in a group of coexpressed genes in *Arabidopsis*, rice, or poplar. An earlier version of ELEMENT (Nemhauser et al. 2004) calculated *Z*-scores for each DNA word based on a comparison of the number of occurrences of that word in the upstream sequences of the coexpressed genes against a background distribution derived by random sampling of the upstream sequences of all genes represented on the microarray. The current Web-based version of ELEMENT has been improved considerably. It supports *Arabidopsis*, poplar, and rice and allows a user to choose various promoter lengths for analysis and to apply a false discovery rate (FDR) filter as desired (Benjamini and Hochberg 1995; Storey and Tibshirani 2003). There are currently a few other Web-based enumerative promoter analysis tools for *Arabidopsis*, such as the BAR Promomer (Toufighi et al. 2005), SIFT (Hudson and Quail 2003), and TAIR's motiffinder (http://www.arabidopsis.org/tools/bulk/motiffinder/index.jsp). Besides ELEMENT, we are aware of no other Web-based bioinformatics tools for analysis of rice and poplar promoters.

The ELEMENT platform consists of databases of putative *Arabidopsis*, rice, and poplar regulatory DNAs, word statistics for all 3–8 mer DNA words occurring in these promoter sequences, software implemented in Perl to analyze promoters and apply statistical screening criteria, a series of accessory scripts to summarize the results of these analyses, and a Web interface implemented in Mason and HTML. ELEMENT uses a database of precalculated statistics for 3–8 mer words in the promoters of all genes represented on the *Arabidopsis*, rice, and poplar Affymetrix arrays to estimate the *Z*-score, a measure of the distance in standard deviations of a sample from the mean, for each word. The observed frequency of a particular 3–8 mer word in a group of promoters is compared with the expected frequency of that word derived from the promoters of randomly sampled genes represented on the Affymetrix microarrays for the species, thus providing a background model based on sequences of an appropriate composition for the species and promoter length under consideration. For example, the background model statistics for *Oryza sativa* ssp. *Japonica* are derived from the frequencies of all 3–8 mer words in the upstream sequences of 34,967 non-transposable-element-related *Japonica* rice genes represented on the Affymetrix rice microarrays.

ELEMENT takes as an input a list of standard locus identifiers for the respective species. We have also provided direct links from DIURNAL and ORTHOMAP so that lists of coexpressed genes or groups of predicted orthologs/homologs may be sent directly to ELEMENT for analysis. The user selects the species and promoter length to be analyzed and adjusts the false-discovery rate and minimum number of occurrences for overrepresented elements as desired. ELEMENT uses, as its reference sets, the 500 bp, 1 kbp, 2 kbp, and 3 kbp upstream of annotated gene models of *Arabidopsis*, rice, and poplar. Because ELEMENT relies on genome annotations, and the transcription start sites are not always well annotated, some gene models lack annotated 5′-untranslated regions (5′UTRs), and thus, in some cases the 3′ end of the predicted promoter region is defined as the beginning of the open reading frame.

ELEMENT returns to the user (via e-mail) a link to results files containing (1) a table of overrepresented elements and their corresponding statistics, with any user-selected filtering criteria applied (Fig. 4A), and (2) a visual alignment of the significant elements aligned to the promoter sequences (Fig. 4B). The latter analysis enables a user to easily identify clusters of nested or partially overlapping overrepresented words or groups of significant words and their relative positions within the promoters. A user may also decide to apply no filtering, in which case, the program returns the statistics for all 3–8 mers.

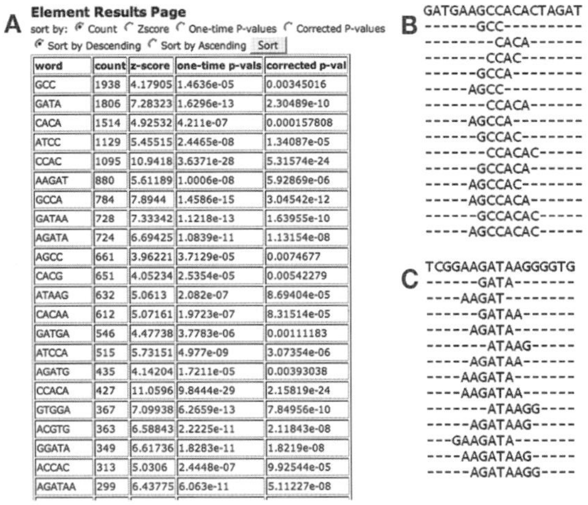

A

Element Results Page

sort by: Count Zscore One-time P-values Corrected P-values

Sort by Descending Sort by Ascending Sort

word	count	z-score	one-time p-vals	corrected p-val
GCC	1938	4.17905	1.4636e-05	0.00345016
GATA	1806	7.28323	1.6296e-13	2.30489e-10
CACA	1514	4.92532	4.211e-07	0.000157808
ATCC	1129	5.45515	2.4465e-08	1.34087e-05
CCAC	1095	10.9418	3.6371e-28	5.31574e-24
AAGAT	880	5.61189	1.0006e-08	5.92869e-06
GCCA	784	7.8944	1.4586e-15	3.04542e-12
GATAA	728	7.33342	1.1218e-13	1.63955e-10
AGATA	724	6.69425	1.0839e-11	1.13154e-08
AGCC	661	3.96221	3.7129e-05	0.0074677
CACG	651	4.05234	2.5354e-05	0.00542279
ATAAG	632	5.0613	2.082e-07	8.69404e-05
CACAA	612	5.07161	1.9723e-07	8.31514e-05
GATGA	546	4.47738	3.7783e-06	0.00111183
ATCCA	515	5.73151	4.977e-09	3.07354e-06
AGATG	435	4.14204	1.7211e-05	0.00393038
CCACA	427	11.0596	9.8444e-29	2.15819e-24
GTGGA	367	7.09938	6.2659e-13	7.84956e-10
ACGTG	363	6.58843	2.2225e-11	2.11843e-08
GGATA	349	6.61736	1.8283e-11	1.8219e-08
ACCAC	313	5.0306	2.4448e-07	9.92544e-05
AGATAA	299	6.43775	6.063e-11	5.11227e-08

B

```
GATGAAGCCACACTAGAT
-------GCC---------
---------CACA------
--------CCAC-------
-------GCCA--------
------AGCC---------
--------CCACA------
------AGCCA--------
-------GCCAC-------
--------CCACAC-----
-------GCCACA------
------AGCCAC-------
------AGCCACA------
-------GCCACAC-----
------AGCCACAC-----
```

C

```
TCGGAAGATAAGGGGTG
-------GATA-------
----AAGAT--------
-------GATAA------
------AGATA-------
--------ATAAG-----
------AGATAA------
----AAGATA-------
----AAGATAA------
--------ATAAGG----
------AGATAAG-----
---GAAGATA-------
----AAGATAAG-----
------AGATAAGG----
```

Figure 4. ELEMENT. (*A*) Example of an HTML table of ELEMENT output showing statistically overrepresented *cis* elements in a set of 347 promoters of *Arabidopsis* genes up-regulated in *gun1* and *gun5* mutants (Koussevitzky et al. 2007). (*B*) Close-up of ELEMENT visual output showing alignment of a cluster of CCAC-containing overrepresented words in the promoter of AT2G35960. (*C*) Close-up of ELEMENT visual output showing alignment of a cluster of GATA-containing overrepresented words in the promoter of AT1G22630.

Results from ELEMENT may be viewed in several ways. First, the results can be viewed on the ELEMENT Web site in an HTML table format (Fig. 4A) that can be sorted online by word, number of occurrences, or significance statistics (*Z*-score, *p*-value, or corrected *p*-value). Alternatively, the results are available as a text file that may be downloaded for import into a spreadsheet or other program. The results tables also flag overrepresented elements that match known *cis* elements in the PLACE (Higo et al. 1998) or PlantCARE (Rombauts et al. 1999) databases. Therefore, in addition to predicting novel promoter elements, ELEMENT also finds known *cis* elements.

To demonstrate the utility of ELEMENT, we have used it to identify both known and novel regulatory motifs that function in the diurnal/circadian regulation of *Arabidopsis* gene expression. We analyzed 11 diurnal and circadian time courses in the reference plant *Arabidopsis*. By using HAYSTACK, we were able to identify putative coexpressed/coregulated genes for each phase of the day. The list of genes in each phase served as the input for ELEMENT, which identified overrepresented 3–8 mer DNA words in 500-bp upstream regions.

Our ELEMENT analysis revealed multiple variants of the Morning Element (ME) (Fig. 5A) (Harmer and Kay 2005), Evening Element (EE), and GATA (Fig. 5B) (Schindler and Cashmore 1990; Anderson and Kay 1995; Harmer et al. 2000), and G-box (Fig. 5C) (Giuliano et al. 1988; Michael and McClung 2002, 2003; Hudson and Quail 2003), all promoter elements previously characterized as being involved in light or circadian regulation of gene expression. ELEMENT also predicted previously unknown *cis*-regulatory elements not identified by other methods (T. Michael et al., in prep.). We have validated a group of these predicted elements using in vivo luciferase imaging, specifying their activity to a phase of day during which no diurnal- or circadian-associated *cis* elements had previously been identified. These predictions of novel diurnal/circadian-associated elements were robust enough to be well-conserved in rice and poplar. Therefore, using ELEMENT, we were able to predict both known and novel *cis*-regulatory elements and define specific aspects of their activity that were previously unknown.

CONCLUSIONS AND FUTURE DIRECTIONS

In conclusion, DIURNAL, HAYSTACK, and ELEMENT are a multifunctional, user-friendly, and complementary collection of Web-based tools. Our Affymetrix microarray data for plant diurnal and circadian time courses are accessible through the DIURNAL interface, which provides a powerful means to query and visualize the data. Two additional tools, HAYSTACK and ELEMENT, provide useful methods for querying microarray data. HAYSTACK analysis constitutes an alternative

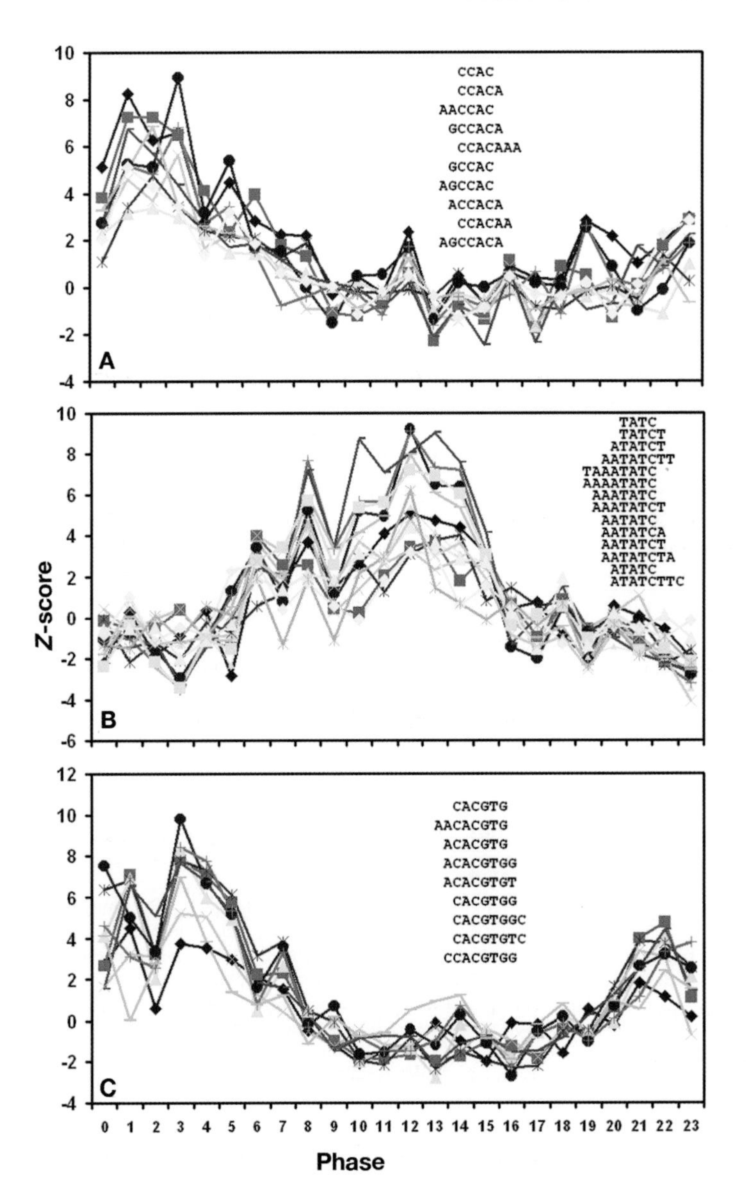

Figure 5. Examples of *Z*-score profiles and overrepresented word clusters for known light- and circadian-associated motifs under diurnal conditions. (*A*) Cluster of overrepresented words and their *Z*-score profiles under the light/dark (12:12) condition (LDHH_SM). The clustered words correspond to Morning Element (ME) variants. (*B*) A cluster of overrepresented words and their *Z*-score profiles under the long-day condition. The clustered words represent Evening Element (EE) and GATA variants. (*C*) A cluster of overrepresented words and their *Z*-score profiles under the short-day condition. The clustered words represent G-Box variants.

method for clustering microarray data by grouping together genes whose expression patterns match the same or similar user-defined HAYSTACK patterns. HAYSTACK can be used to compare virtually any large-scale data set against a set of user-supplied model patterns. ELEMENT identifies predicted *cis*-regulatory elements within the promoters of groups of coexpressed genes from *Arabidopsis*, rice, or poplar and can be used to facilitate cross-species comparisons of promoter architecture.

Data from our poplar and rice microarray time courses will be added to the DIURNAL database over the next year. We plan to add diurnal and circadian time course data for additional species when such data become available. We also plan to expand ELEMENT to support promoter analysis for additional plant species with sequenced genomes or genomes that will be sequenced in the near future, including sorghum, medicago, maize, *Arabidopsis lyrata*, papaya, and *Brachypodium distachyon*. We hope that these resources will help to accelerate functional genomics efforts directed at the mechanisms underlying diurnal and circadian biology in plants. For example, in *Arabidopsis*, the ability to order knockout or knockdown T-DNA insertion lines directly from stock centers makes it possible to rapidly perform in vivo tests of in silico predictions derived from HAYSTACK analysis of microarray data. Promoter element predictions from ELEMENT can in turn be tested experimentally to identify functional *cis*-regulatory elements.

ACKOWLEDGMENTS

This work was supported by a National Science Foundation Plant Genome grant (DBI 0605240) (T.C.M, J.C., and S.A.K), the Howard Hughes Medical Institute, and National Institutes of Health (NIH) grants GM56006 and GM67837 (S.A.K.) and GM52413 and GM62932 (J.C.). T.P.M and T.C.M were supported by Ruth L. Kirschstein NIH Postdoctoral Fellowships.

REFERENCES

Anderson S.L. and Kay S.A. 1995. Functional dissection of circadian clock- and phytochrome-regulated transcription of the *Arabidopsis CAB2* gene. *Proc. Natl. Acad. Sci.* **92:** 1500.

Arabidopsis Genome Initiative. 2000. Analysis of the genome sequence of the flowering plant *Arabidopsis thaliana. Nature* **408:** 796.

Barrett T., Troup D.B., Wilhite S.E., Ledoux P., Rudnev D., Evangelista C., Kim I.F., Soboleva A., Tomashevsky M., and Edgar R. 2007. NCBI GEO: Mining tens of millions of expression profiles: Database and tools update. *Nucleic Acids Res.* (database issue) **35:** D760.

Belacel N., Wang Q., and Cuperlovic-Culf M. 2006. Clustering methods for microarray gene expression data. *OMICS* **10:** 507.

Benjamini Y. and Hochberg Y. 1995. Controlling the false discovery rate: A practical and powerful approach to multiple testing. *J. R. Stat. Soc. B* **57:** 289.

Blasing O., Gibon Y., Gunther M., Hohne M., Morcuende R., Osuna D., Thimm O., Usadel B., Scheible W., and Stitt M. 2005. Sugars and circadian regulation make major contributions to the global regulation of diurnal gene expression in *Arabidopsis. Plant Cell* **17:** 3257.

Brazma A., Parkinson H., Sarkans U., Shojatalab M., Vilo J., Abeygunawardena N., Holloway E., Kapushesky M., Kemmeren P., Lara G.G., Oezcimen A., Rocca-Serra P., and Sansone S.A. 2003. ArrayExpress: A public repository for microarray gene expression data at the EBI. *Nucleic Acids Res.* **31:** 68.

Chen W., Provart N.J., Glazebrook J., Katagiri F., Chang H.S., Eulgem T., Mauch F., Luan S., Zou G., Whitham S.A., Budworth P.R., Tao Y., Xie Z., Chen X., Lam S., Kreps J.A., Harper J.F., Si-Ammour A., Mauch-Mani B., Heinlein M., Kobayashi K., Hohn T., Dangl J.L., Wang X., and Zhu T. 2002. Expression profile matrix of *Arabidopsis* transcription factor genes suggests their putative functions in response to environmental stresses. *Plant Cell* **14:** 559.

Craigon D.J., James N., Okyere J., Higgins J., Jotham J., and May S. 2004. NASCArrays: A repository for microarray data generated by NASC's transcriptomics service. *Nucleic Acids Res.* (database issue) **32:** D575.

Dodd A.N., Salathia N., Hall A., Kevei E., Toth R., Nagy F., Hibberd J.M., Millar A.J., and Webb A.A. 2005. Plant circadian clocks increase photosynthesis, growth, survival, and competitive advantage. *Science* **309:** 630.

Edwards K.D., Anderson P.E., Hall A., Salathia N.S., Locke J.C., Lynn J.R., Straume M., Smith J.Q., and Millar A.J. 2006. *FLOWERING LOCUS C* mediates natural variation in the high-temperature response of the *Arabidopsis* circadian clock. *Plant Cell* **18:** 639.

Fraser H.B., Hirsh A.E., Wall D.P., and Eisen M.B. 2004. Coevolution of gene expression among interacting proteins. *Proc. Natl. Acad. Sci.* **101:** 9033.

Garcia-Hernandez M., Berardini T.Z., Chen G., Crist D., Doyle A., Huala E., Knee E., Lambrecht M., Miller N., Mueller L.A., Mundodi S., Reiser L., Rhee S.Y., Scholl R., Tacklind J., Weems D.C., Wu Y., Xu I., Yoo D., Yoon J., and Zhang P. 2002. TAIR: A resource for integrated *Arabidopsis* data. *Funct. Integr. Genomics* **2:** 239.

Giuliano G., Pichersky E., Malik V.S., Timko M.P., Scolnik P.A., and Cashmore A.R. 1988. An evolutionarily conserved protein binding sequence upstream of a plant light-regulated gene. *Proc. Natl. Acad. Sci.* **85:** 7089.

Harmer S. and Kay S. 2005. Positive and negative factors confer phase-specific circadian regulation of transcription in *Arabidopsis. Plant Cell* **17:** 1926.

Harmer S.L., Hogenesch J.B., Straume M., Chang H.-S., Han B., Zhu T., Wang X., Kreps J.A., and Kay S.A. 2000. Orchestrated transcription of key pathways in *Arabidopsis* by the circadian clock. *Science* **290:** 2110.

Higo K., Ugawa Y., Iwamoto M., and Higo H. 1998. PLACE: A database of plant *cis*-acting regulatory DNA elements. *Nucleic Acids Res.* **26:** 358.

Hudson M. and Quail P. 2003. Identification of promoter motifs involved in the network of phytochrome A-regulated gene expression by combined analysis of genomic sequence and microarray data. *Plant Physiol.* **133:** 1605.

Hulzink R.J., Weerdesteyn H., Croes A.F., Gerats T., van Herpen M.M., and van Helden J. 2003. In silico identification of putative regulatory sequence elements in the 5′-untranslated region of genes that are expressed during male gametogenesis. *Plant Physiol.* **132:** 75.

International Rice Genome Sequencing Project. 2005. The map-based sequence of the rice genome. *Nature* **436:** 793.

Karim S., Aronsson H., Ericson H., Pirhonen M., Leyman B., Welin B., Mantyla E., Palva E.T., Van Dijck P., and Holmstrom K.O. 2007. Improved drought tolerance without undesired side effects in transgenic plants producing trehalose. *Plant Mol. Biol.* **64:** 371.

Koussevitzky S., Nott A., Mockler T.C., Hong F., Sachetto-Martins G., Surpin M., Lim J., Mittler R., and Chory J. 2007. Signals from chloroplasts converge to regulate nuclear gene expression. *Science* **316:** 715.

Kreps J., Budworth P., Goff S., and Wang R. 2003. Identification of putative plant cold responsive regulatory elements by gene expression profiling and a pattern enumeration algorithm. *Plant Biotechnol. J.* **1:** 345.

Leyman B., Van Dijck P., and Thevelein J.M. 2001. An unexpected plethora of trehalose biosynthesis genes in *Arabidopsis thaliana. Trends Plant Sci.* **6:** 510.

Marino-Ramirez L., Spouge J.L., Kanga G.C., and Landsman D. 2004. Statistical analysis of over-represented words in human promoter sequences. *Nucleic Acids Res.* **32:** 949.

Michael T.P. and McClung C.R. 2002. Phase-specific circadian clock regulatory elements in *Arabidopsis. Plant Physiol.* **130:** 627.

———. 2003. Enhancer trapping reveals widespread circadian clock transcriptional control in *Arabidopsis thaliana. Plant Physiol.* **132:** 629.

Michael T.P., Salome P.A., Yu H.J., Spencer T.R., Sharp E.L., McPeek M.A., Alonso J.M., Ecker J.R., and McClung C.R. 2003. Enhanced fitness fonferred by naturally occurring variation in the circadian clock. *Science* **302:** 1049.

Nemhauser J.L., Mockler T.C., and Chory J. 2004. Interdependency of brassinosteroid and auxin signaling in *Arabidopsis. PLoS Biol.* **2:** e258.

Rhee S.Y., Beavis W., Berardini T.Z., Chen G., Dixon D., Doyle A., Garcia-Hernandez M., Huala E., Lander G., Montoya M., Miller N., Mueller L.A., Mundodi S., Reiser L., Tacklind J., Weems D.C., Wu Y., Xu I., Yoo D., Yoon J., and Zhang P. 2003. The *Arabidopsis* Information Resource (TAIR): A model organism database providing a centralized, curated gateway to *Arabidopsis* biology, research materials and community. *Nucleic Acids Res.* **31:** 224.

Rombauts S., Dehais P., Van Montagu M., and Rouzé P. 1999. PlantCARE, a plant *cis*-acting regulatory element database. *Nucleic Acids Res.* **27:** 295.

Rombauts S., Florquin K., Lescot M., Marchal K., Rouzé P., and van de Peer Y. 2003. Computational approaches to identify promoters and *cis*-regulatory elements in plant genomes. *Plant Physiol.* **132:** 1162.

Schindler U. and Cashmore A.R. 1990. Photoregulated gene expression may involve ubiquitous DNA binding proteins. *EMBO J.* **9:** 3415.

Schmid M., Davison T., Henz S., Pape U., Demar M., Vingron M., Scholkopf B., Weigel D., and Lohmann J. 2005. A gene expression map of *Arabidopsis thaliana* development. *Nat. Genet.* **37:** 501.

Shen L., Gong J., Caldo R.A., Nettleton D., Cook D., Wise R.P., and Dickerson J.A. 2005. BarleyBase: An expression profiling database for plant genomics. *Nucleic Acids Res.* (database issue) **33:** D614.

Siegfried K.R., Eshed Y., Baum S.F., Otsuga D., Drews G.N., and Bowman J.L. 1999. Members of the YABBY gene family specify abaxial cell fate in *Arabidopsis. Development* **126:** 4117.

Smith S., Fulton D., Chia T., Thorneycroft D., Chapple A., Dunstan H., Hylton C., Zeeman S., and Smith A. 2004.

Diurnal changes in the transcriptome encoding enzymes of starch metabolism provide evidence for both transcriptional and posttranscriptional regulation of starch metabolism in *Arabidopsis* leaves. *Plant Physiol.* **136:** 2687.

Storey J.D. and Tibshirani R. 2003. Statistical significance for genomewide studies. *Proc. Natl. Acad. Sci.* **100:** 9440.

Thijs G., Lescot M., Marchal K., Rombauts S., De Moor B., Rouzé P., and Moreau Y. 2001. A higher-order background model improves the detection of promoter regulatory elements by Gibbs sampling. *Bioinformatics* **17:** 1113.

Thijs G., Marchal K., Lescot M., Rombauts S., De Moor B., Rouzé P., and Moreau Y. 2002. A Gibbs sampling method to detect overrepresented motifs in the upstream regions of coexpressed genes. *J. Comput. Biol.* **9:** 447.

Tompa M., Li N., Bailey T.L., Church G.M., De Moor B., Eskin E., Favorov A.V., Frith M.C., Fu Y., Kent W.J., Makeev V.J., Mironov A.A., Noble W.S., Pavesi G., Pesole G., Regnier M., Simonis N., Sinha S., Thijs G., van Helden J., Vandenbogaert M., Weng Z., Workman C., Ye C., and Zhu Z. 2005. Assessing computational tools for the discovery of transcription factor binding sites. *Nat. Biotechnol.* **23:** 137.

Toufighi K., Brady S.M., Austin R., Ly E., and Provart N.J. 2005. The Botany Array Resource: e-Northerns, Expression Angling, and promoter analyses. *Plant J.* **43:** 153.

Tuskan G.A., DiFazio S., Jansson S., Bohlmann J., Grigoriev I., Hellsten U., Putnam N., Ralph S., Rombauts S., Salamov A., Schein J., Sterck L., Aerts A., Bhalerao R.R., Bhalerao R.P., Blaudez D., Boerjan W., Brun A., Brunner A., Busov V., Campbell M., Carlson J., Chalot M., Chapman J., and Chen G.L., et al. 2006. The genome of black cottonwood, *Populus trichocarpa* (Torr. & Gray). *Science* **313:** 1596.

van Helden J., Andre B., and Collado-Vides J. 1998. Extracting regulatory sites from the upstream region of yeast genes by computational analysis of oligonucleotide frequencies. *J. Mol. Biol.* **281:** 827.

Woelfle M., Ouyang Y., Phanvijhitsiri K., and Johnson C. 2004. The adaptive value of circadian clocks: An experimental assessment in cyanobacteria. *Curr. Biol.* **14:** 1481.

Zimmermann P., Hirsch-Hoffmann M., Hennig L., and Gruissem W. 2004. GENEVESTIGATOR. *Arabidopsis* microarray database and analysis toolbox. *Plant Physiol.* **136:** 2621.

Systems Biology of Mammalian Circadian Clocks

H.R. UEDA
*Laboratory for Systems Biology and Functional Genomics Unit,
Center for Developmental Biology, Riken, Kobe, Hyogo 650-0047, Japan*

Systems Biology is a natural extension of molecular biology and can be defined as biology after identification of key gene(s). Systems-biological research is hence seen as a multistage process, beginning with the comprehensive identification and quantitative analysis of individual system components and their networked interactions and leading to the ability to control existing systems toward the desired state and design new ones based on an understanding of structure and underlying dynamical principles. In this chapter, we take mammalian circadian clocks as a model system and describe systems-biological approaches, including the identification of clock-controlled genes, clock-controlled *cis* elements, and clock transcriptional circuits driven by functional genomics; the parameter change of clock components followed by quantitative measurement; and the dynamic and quantitative perturbation of the clock and its application to one of the fundamental but yet-unsolved questions: singularity behavior of clocks. As perspective for systems-biological investigations, we also introduce the system-level dynamical questions related to the core of clocks, including delay, nonlinearity, temperature-compensation and synchronization of mammalian circadian oscillator(s), and the system-level information problems related to clocks in the environment, including the internal representation of light change through perfect adaptation and internal representation of day length through photoperiodism in mammals.

INTRODUCTION

Systems Biology as "Biology after Identification"

Recent large-scale efforts in genome sequencing, expression profiling, and functional screening have produced an embarrassment of riches for life science researchers, and biological data can now be accessed in quantities that are orders of magnitude greater than were available even a few years ago. The growing need for interpretation of data sets, as well as the accelerating demand for their integration to a higher-level understanding of life, has set the stage for the advent of systems biology (Kitano 2002a,b), in which biological processes and phenomena are approached as complex and dynamic systems. Systems biology is a natural extension of molecular biology and can be defined as "biology after identification of key gene(s)." We see systems-biological research as a multistage process, beginning with the comprehensive identification and quantitative analysis of individual system components and their networked interactions and leading to the ability to control existing systems toward the desired state and design new ones based on an understanding of structure and underlying dynamical principles (Fig. 1).

Mammalian Circadian Clock as a Model System

We have taken the mammalian circadian clock as an initial model system that exhibits system-level dynamical and structural properties in order to develop research strategies and technologies for studies of complex and dynamic biological systems. The mammalian circadian clock consists of complexly integrated feedback and feedforward loops (Reppert and Weaver 2002) and also exhibits well-defined dynamical properties (Dunlap et al. 2004), including (1) endogenous oscillations of an approximately 24-hour period, (2) entrainment to external environmental changes (temperature and light cycle), (3) temperature-compensation over a wide range of temperature, and (4) synchronization of multiple cellular clocks despite the inevitable molecular noise. All of these dynamical properties would be difficult to elucidate without utilizing such system-level approaches. In addition to its advantage as a basic model system for systems-biological research, the function of the circadian clock is intimately involved in the control of metabolic and physiological processes (Panda et al. 2002; Reppert and Weaver 2002), and its dysregulation is associated with the onset and development of numerous human diseases, including sleep disorders, depression, and dementia. An improved understanding at the systems level promises to provide biomedical and clinical investigators with a powerful new arsenal for attacking these conditions.

Development of Systems-Biological Approaches and Their Application to Clocks

Attempts to elucidate the design principles of complex and dynamic biological systems such as the mammalian circadian clock may require (1) identification of whole-network structure through comprehensive (genome-wide) screening (*system identification*), (2) prediction and validation to derive the design principle through the accurate measurement of network behaviors (*system analysis*), (3) repair and control of network state toward the desired state through the precise perturbation of its components (*system control*) and ultimately, (4) reconstruction and design of new systems based on the design principles derived from the identified structure and observed dynamics of the original network (*system design*). To develop these systems-biological approaches, we have focused mainly on the development and evaluation of

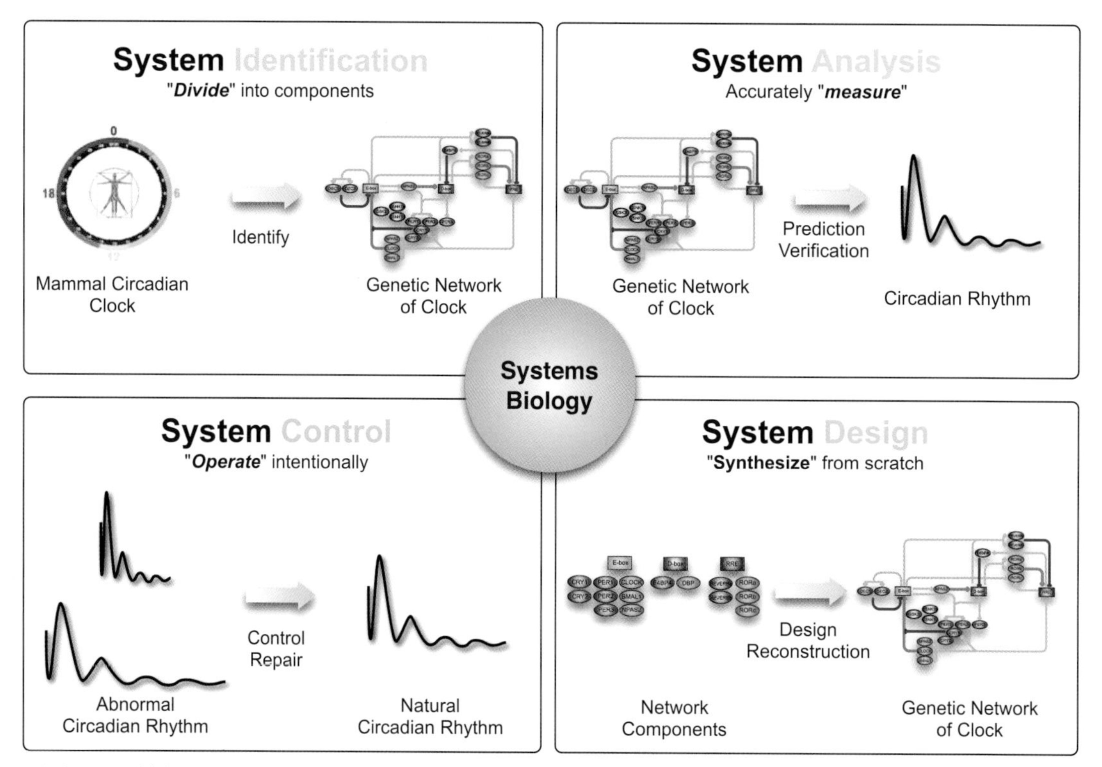

Figure 1. Systems biology. Systems-biological research starts with comprehensive identification (*upper left panel*). In this step, individual system components and their networked interactions are comprehensively identified. In the second step, to derive the design principle of a target system, the behavior of the system is predicted and validated through an accurate measurement with perturbations (*upper right panel*). An understanding of the design principle of the system is essential to derive the method of controlling the system toward the desired state (*lower left panel*). Finally, the level of understanding is confirmed by reconstruction of the system (*lower right panel*).

technologies and strategies. For each of these processes, we have been able to report several strategies and technologies as well as their application to specific questions in mammalian circadian clocks. This has included the identification of clock-controlled genes, clock-controlled *cis* elements, and clock transcriptional circuits driven by functional genomics (Ueda et al. 2002c, 2005) in *system identification*, parameter change of clock components by quantitative measurement (Sato et al. 2006) in *system analysis*, dynamic and quantitative perturbation of clock and its application to one of the fundamental but yet-unsolved questions, and singularity behavior of clocks (Ukai et al. 2007) in *system control*. As for the effort of *system design*, we recently initiated our current project on in cellulo and in vitro reconstruction of mammalian circadian circuits. In the following sections, we describe in detail these strategies and/or technologies developed for each stage of systems-biological research, in addition to their application to the specific question of mammalian circadian clocks.

SYSTEMS-BIOLOGICAL APPROACHES

Identification of Clocks

Following the completion of genome projects for species such as mouse and human, genome-wide resources such as small interfering RNA (siRNA) or cDNA libraries have undergone considerable expansion. Development of high-throughput technologies has also assisted in the efficient use of these resources. These genome-wide resources and technologies and genome-associated information currently allow us to comprehensively identify system components of interest (system identification).

Circadian clocks of multicellular organisms consist of complex integrated regulatory loops with positive or negative regulators known as clock genes (Wuarin and Schibler 1990; King et al. 1997; Zylka et al. 1998; van der Horst et al. 1999; Vitaterna et al. 1999; Bunger et al. 2000; Lowrey et al. 2000; Bae et al. 2001; Mitsui et al. 2001; Reick et al. 2001; Zheng et al. 2001; Honma et al. 2002; Preitner et al. 2002; Sato et al. 2004). The transcriptional regulation network of these genes forms a circadian clock oscillator, which is known to control output genes and to affect physiological and metabolic processes (Panda et al. 2002; Reppert and Weaver 2002). Although some transcriptional regulations of identified clock genes have been the subject of previous studies, a system-level understanding of circadian clocks remains to be elucidated. In this section, we provide results of our system identification of mammalian circadian clocks (Ueda et al. 2002c, 2005).

Identification of the mammalian clock circuit. The mammalian circadian master clock is primarily located in the suprachiasmatic nucleus (SCN) (Reppert and Weaver 2002). Transcript analyses have indicated that circadian clocks are not restricted to the SCN but are found in several tissues (Yamazaki et al. 2000) including liver and cultured fibroblast cells such as Rat-1 (Balsalobre et al. 1998; Yagita et al. 2001) or NIH-3T3 (Akashi and Nishida 2000) cells. The mechanisms underlying circadian rhythms are also known to be conserved across species (Dunlap et al. 2004). At the basic core of the clock lies a transcriptional/translational feedback loop (Gekakis et al. 1998; Kume et al. 1999; Shearman et al. 2000), whose primary components are known as "clock genes" (Wuarin and Schibler 1990; King et al. 1997; Zylka et al. 1998; van der Horst et al. 1999; Vitaterna et al. 1999; Bunger et al. 2000; Lowrey et al. 2000; Bae et al. 2001; Mitsui et al. 2001; Reick et al. 2001; Zheng et al. 2001; Honma et al. 2002; Preitner et al. 2002; Sato et al. 2004). For example, in the mouse system, the transcription factors CLOCK and BMAL1 proteins dimerize and directly and indirectly activate transcription of the *Per* and *Cry* genes through E-box elements (5′-CACGTG-3′) (Gekakis et al. 1998; Kume et al. 1999). The PER and CRY proteins accumulate in the cytosol and are then translocated following phosphorylation into the nucleus where they inhibit the activity of CLOCK and BMAL1 (Reppert and Weaver 2002). The turnover of the inhibitory PER and CRY proteins leads to a new cycle of activation by CLOCK and BMAL1 via E-box elements. Despite the reporting of many transcriptional regulations of each gene, however, an overview of the circadian clock core network remains to be put forward.

Complicated networks cannot be elucidated without access to both (1) comprehensive identification of network circuits and (2) accurate measurement of system dynamics. In a previous attempt to comprehensively identify the circadian clock core network, we first quantitatively and comprehensively measured genome-wide gene expression using a high-density oligonucleotide probe array (Lipshutz et al. 1999) and identified genes showing circadian oscillation with characteristic expression patterns through biostatistics (Fig. 2). The second step involved comprehensively determining the transcription start sites (TSSs) (Suzuki et al. 2001) and conserved noncoding regions to construct a genome-wide promoter/ enhancer database. Using these data, we predicted that there was a relationship between expression patterns of identified genes and DNA regulatory elements on their promoter/enhancer regions. We found that clock-controlled elements (CCEs), E boxes (5′-CACGTG-3′; Hogenesch et al. 1998), E′ boxes (5′-CACGTT-3′; Ueda et al. 2005; Yoo et al. 2005), D boxes (5′ -TTATG[C/T]AA-3′; Falvey et al. 1996), or Rev-response elements (RREs) (5′-[A/T]A[A/T]NT [A/G]GGTCA-3′; Harding and Lazar 1993) are distributed throughout the oscillatory genes.

To determine the role of these elements in the circadian clock, we utilized a cell culture system with which we can monitor circadian rhythms in transcriptional dynamics using a destabilized luciferase (d*Luc*) reporter driven by clock-controlled promoters (Fig. 3A). In this cell culture system—named "in cellulo cycling assay"—we transiently transfected reporter constructs into cultured Rat-1 cells and stimulated them with dexamethasone and measured their bioluminescences. Dexamethasone was administrated to induce macroscopic circadian oscillations in the cultured cells. Through the genome-wide searching described above, we found CCEs on 16 clock/clock-controlled gene promoter/enhancers. Then, using the in cellulo cycling assay system, we were able to reveal that functionally and evolutionary conserved E/E′ boxes are located on noncoding regions of nine genes (*Per1*, *Per2*, *Cry1*, *Dbp*, *Rorγ*, *RevErbAα*/*Nr1d1*, *RevErbAβ*/*Nr1d2*, *Dec1*/*Bhlhb2*, and *Dec2*/*Bhlhb3*), D boxes on those of seven genes (*Per1*, *Per2*, *Per3*, *RevErbAα*, *RevErbAβ*, *Rorα*, and *Rorβ*), and RREs on those of six genes (*Bmal1*/*Arntl*, *Clock*, *Npas2*, *Cry1*, *E4bp4*/*Nfil3*, and *Rorγ*). On the basis of this functional and conserved transcriptional regulatory mechanism, we succeeded in drawing transcriptional circuits underlying mammalian circadian rhythms (Fig. 3B) (Ueda et al. 2005).

Our analysis further suggested that regulation of E/E′ boxes is the topological vulnerability point in mammalian circadian clocks. We functionally verified this concept using in cellulo cycling assay system (Fig. 3C). Overexpression of repressors of E/E′ box regulation (CRY1; Kume et al. 1999), RRE regulation (REVERBAα; Preitner et al. 2002; Ueda et al. 2002c), or D-box regulation (E4BP4; Mitsui et al. 2001) affected circadian rhythmicity in *Per2* or *Bmal1* promoter activity. The effects were different, however, between each repressor, and the severest effect was observed when the E/E′ box was attacked. Such different modes of effects cannot be explained by mere quantitative differences in the strength of these three repressors, indicating that there is some qualitative difference between E/E′ box, D-box, and RRE regulation in circadian rhythmicity (Ueda et al. 2005).

Analysis of Clocks

To derive the design principles of a system of interest, it is important to validate the behavior of a predicted system through an accurate measurement with several types of perturbations (system analysis). In our efforts to identify the circadian clock system, we have succeeded in drawing the circadian transcriptional circuits and revealed the topological importance of the morning element, E box. In this section, we describe our effort to validate the hypothesis that transcriptional feedback repression through the E box is required for mammalian clock function. In this study, we collaborated with Dr. John B. Hogenesch's and Dr. Steve A. Kay's groups to *change the molecular parameter* for feedback repression by functional genomics and then *tested the cellular phenotype* caused by this parameter change by our in cellulo cycling assay system. Through this study, we have demonstrated the necessity of transcriptional repression for circadian clock function (Sato et al. 2006).

Negative feedback as the heart of the transcriptional circuit in mammalian clocks. Circadian clocks have been proposed as consisting of autoregulatory loops in which transcriptional feedback and regulated protein turnover are used to maintain a 24-hour periodicity (Dunlap 1999;

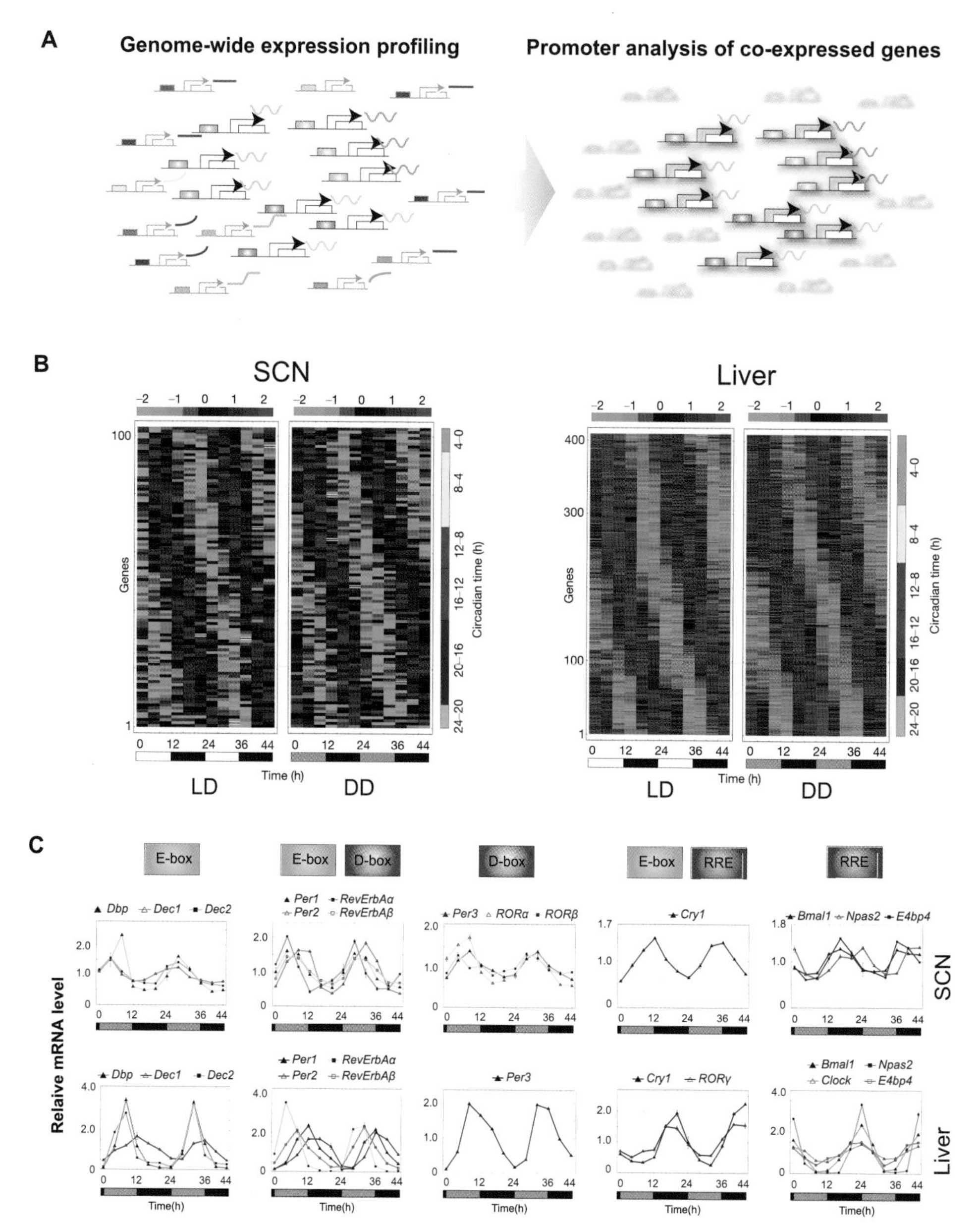

Figure 2. (*A*) Strategy for identification of clock-controlled elements (CCEs). Gene expression information was obtained by performing comprehensive expression profiling (*left panel*). Through statistical analysis, genes with a special characteristic pattern of expression (circadian oscillation) were selected (*right panel*, oscillatory genes are indicated by their color). DNA regulatory elements for specific issue (i.e., expression timing) were predicted by combining expression pattern information and transcriptional regulatory element information from promoter regions. (*B*) Genome-wide expression profiles in mouse central (SCN, *left panel*) and peripheral (liver, *right panel*) clocks. Total RNA were extracted every 4 hours during light/dark cycles (LD) or constant darkness (DD) over 2 days and used to determine genome-wide gene expression profiles with an Affymetrix mouse high-density oligonucleotide probe array. Data were normalized so that the average signal intensity and standard deviation over 12-point time courses were 0.0 and 1.0, respectively. Columns represent time points, and rows represent genes that were organized by peak time. Colors in descending order from *red* to *black* to *green* represent the normalized data. From the obtained data, we identified a set of genes rhythmically expressed under both LD and DD. We classified 101 genes in the SCN and 393 genes in the liver as "significantly rhythmic under both LD and DD." (*C*) Temporal expression profiles of transcription factors in the SCN (*upper panel*) and liver (*lower panel*) under DD conditions. Relative mRNA levels under DD conditions of the indicated genes were measured with quantitative polymerase chain reaction (Q-PCR) assay, in which *GAPDH* expression was used as an internal control. Data were normalized so that the average copy number (Q-PCR) over a 12-point time course was 1.0. Circadian expression of transcription factors having functional and evolutionary conserved E boxes (*Dbp, Dec1*, and *Dec2*), both E boxes/E′ boxes and D boxes (*Per1, Per2, RevErbA*α, and *RevErbA*β), D boxes (*Per3*, *Ror*α, and *Ror*β), RREs (*Bmal1*, *Clock*, *Npas2*, and *E4bp4*), both E boxes/E′ boxes and RREs (*Cry1* and *Ror*γ) on their noncoding regions. *Clock* and *Ror*γ were constitutively expressed in the SCN. (Modified from Ueda et al. 2002c, 2005.)

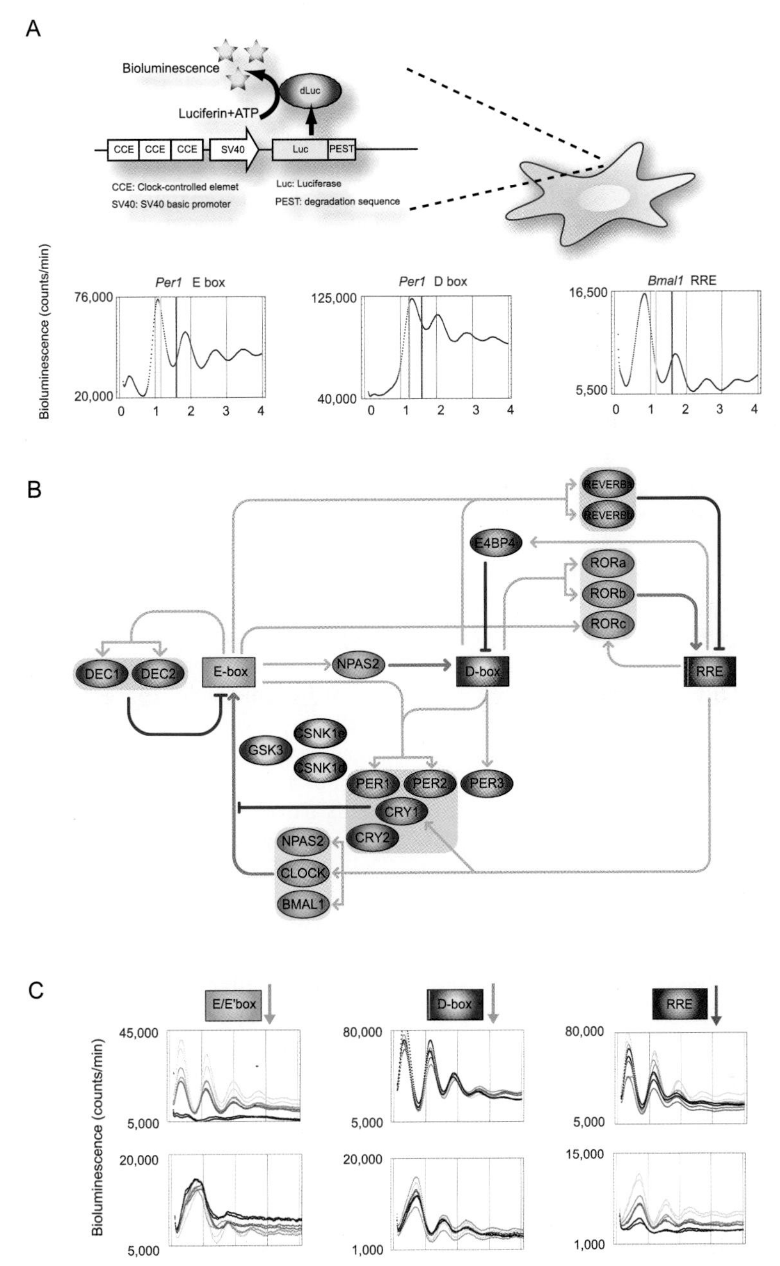

Figure 3. (*A*) Schematic overview of the experiment. Cultured mammalian cells (Rat-1) were transfected with d*Luc* under the regulation of CCE and SV40 basic promoter. The circadian change of bioluminescence was monitored by a PMT detector over several days (*upper panel*). Representative circadian rhythms of bioluminescence from wild-type CCE fused to the SV40 basic promoter driving a d*Luc* reporter. The circadian bioluminescence phase from the *Per2* promoter and that of the *Bmal1* promoter are marked by *yellow* and *purple* lines, respectively (*bottom panels*). (*B*) Schematic representation of the transcriptional network of the mammalian circadian clock. (*Ellipsoids*) Genes; (*rectangles*) CCEs. Transcriptional-translational activation and repression are depicted as *gray*, *green*, and *red* lines, respectively. (*C*) Effect of repression on each CCEs. The E/E′ boxes, D box, and RRE are repressed by overproduction of CRY1, E4BP4, and REVERBAα, respectively. The consequences of those repressions were monitored by *Per2*-d*Luc* (*upper panel*) and *Bmal1*-d*Luc* (*lower panel*). (d*Luc*) Destabilized luciferase; (CCE) clock-controlled elements; (PMT) photomultiplier tube. (Modified from Ueda et al. 2005.)

Young and Kay 2001; Reppert and Weaver 2002). The universal necessity for transcriptional feedback repression, however, arose as a question mainly from recent studies about the cyanobacterial circadian rhythms in which repression was shown not to be necessary (Nakajima et al. 2005; Tomita et al. 2005). Circadian feedback repression in mammals is believed to be mediated by CRYPTOCHROME (CRY1, CRY2) (Kume et al. 1999; van der Horst et al. 1999; Vitaterna et al. 1999) and PERIOD (PER1 and PER2) (Tei et al. 1997; Zheng et al. 1999, 2001) proteins. CRY and PER proteins are hypothesized to autoregulate their own expression by repressing the heterodimeric complex of the basic helix-loop-helix (bHLH) PERARNT-SIM (PAS) domain transcriptional activators CLOCK and BMAL1, which bind to E-box elements in the CRY (Etchegaray et al. 2003) and PER (Gekakis et al. 1998; Ueda et al. 2005; Yoo et al. 2005) promoters. However, direct evidence for the requirement of CRY-mediated repression of CLOCK/BMAL1 transcriptional activity in the maintenance of circadian clock function has yet to be presented. Here, in a fruitful collaboration with the J.B. Hogenesch and the S.A. Kay groups, we have successfully shown that feedback repression is actually required for mammalian circadian clock function (Sato et al. 2006).

To determine the requirement of feedback repression in circadian clock function, we sought to identify the CLOCK alleles that were insensitive to CRY1 repression but maintained normal transcriptional activity. Our collaborative partners first generated a library of approximately 6000 random point mutations of human alleles for both CLOCK and BMAL1 and then screened clones individually in cell-based reporter assays with wild-type *Bmal1* cDNA and a *Per1* promoter-*luciferase* (*Per1-Luc*) construct (Gekakis et al. 1998) in the presence of cotransfected *Cry1*. Of the CLOCK and BMAL1 clones screened, several reproducibly maintained threefold or greater reporter activity in the presence of CRY1 compared with wild-type alleles. Notably, these clones demonstrated transcriptional activities similar to those of wild type in the absence of cotransfected *Cry1*, suggesting that these mutations do not cause overt alterations in the heterodimerization, nuclear localization, DNA-binding, and *trans*-activation properties of the mutant CLOCK/BMAL1 complex.

The prevailing transcriptional feedback model (Fig. 4A) predicts that impairment of CRY-mediated repression should have marked effects on circadian expression of the *Per* genes. This notion is supported by in vivo observations that expression of *Per1* and *Per2* is constitutively elevated in *Cry1/Cry2* double-knockout mice (Okamura et al. 1999; Vitaterna et al. 1999). To determine whether these mutations in CLOCK and BMAL1 cause phenotypic changes in circadian gene expression, we performed an in cellulo cycling assay (Ueda et al. 2002b, 2005) (see also above, Identification of Clocks). Mouse NIH-3T3 cells were transfected with plasmids harboring destabilized luciferase (d*Luc*) driven by *Per2* or SV40 basic promoters along with the BMAL1 and CLOCK mutant allele and then monitored using the in cellulo cycling assay (Fig. 4B, left). Cotransfection of wild-type CLOCK and BMAL1 did not substantially alter rhythmicity compared with empty vector transfection, as their period lengths were 21.4 hours ±0.4. In contrast, when compared with wild-type CLOCK/BMAL1, transfection of either of CLOCK or BMAL1 mutant alleles resulted in substantial impairment of circadian rhythmicity after one or two cycles of oscillations. Notably, cotransfection of the CRY-insensitive mutant CLOCK and BMAL1 together resulted in the loss of circadian *Per2* promoter activity. We were therefore able to demonstrate that the transcriptional repression of CLOCK/BMAL1 by CRY was required for circadian E-box activity.

In addition to *Per* and *Cry*, the rhythmic expression of *Bmal1* mRNA is also under circadian clock regulation (Shearman et al. 2000). The *Bmal1* promoter used in this study, however, does not have E-box sites but instead contains RRE (Preitner et al. 2002; Ueda et al. 2002c), whose activities are reciprocally controlled by the rhythmically expressed transcriptional repressor REVERBAα (Preitner et al. 2002) and activator RORα (Sato et al. 2004). As an additional test for circadian clock function, we examined the effects of mutant CLOCK and BMAL1 on rhythmic RRE activity by in cellulo cycling assays with a *Bmal1*-d*Luc* reporter. Similar to the results with the *Per2*-d*Luc* reporter, transfection of single CLOCK or BMAL1 mutants resulted in the decreased amplitude of cycling of *Bmal1*-d*Luc* activity compared with wild-type CLOCK/ BMAL1 transfection. Moreover, this decrease in cycling amplitude was further exacerbated upon cotransfection of the double-mutant heterodimer (Fig. 4B, right). These results indicate that transcriptional repression of CLOCK/BMAL1 by CRY is also required for circadian BMAL1 expression through RRE, which is dependent on the transcriptional, translational, and post-translational actions of endogenous cellular factors.

Arrhythmic *Per2* expression seen from a population of cells expressing the double-mutant CLOCK/BMAL1 complex may be due to the disruption of oscillator function or a lack of synchrony between individual rhythmic cells. To address these possibilities, we measured quantitative imaging of *Per2*-d*Luc* reporter activity from individual NIH-3T3 cells by using an approach similar to that used in analyzing *Bmal1* reporter rhythms from single cells (Welsh et al. 2004). As with the whole-well assays, the median reporter activity for the population of imaged individual cells coexpressing wild-type CLOCK/BMAL1 oscillated rhythmically (Fig. 4C, left). In contrast, the population of individual *Clock/Bmal1* mutant cells (Fig. 4C, right) was visibly arrhythmic. Individual reporter activities from single wild-type cells were rhythmic, as expected, whereas individual *Clock/Bmal1* double-mutant cells showed arrhythmic reporter activities. These differences in activity patterns were evaluated by two independent statistical methods that score the circadian rhythmicity of experimental time-course data.

This is the first study that shows functional necessity for feedback repression in the mammalian circadian clock. Proof of this necessity is important, as recent studies of the cyanobacterial clock found that circadian oscillations in protein phosphorylation can be maintained in the absence of transcriptional feedback repression in vivo (Tomita et al. 2005) and with purified proteins in vitro (Nakajima et al. 2005). We therefore sought to formally test the requirement of CRY-mediated transcriptional feedback repression in mammalian circadian clock function by developing and implementing a new unbiased cellular genetics approach that uses robust mutagenesis techniques and mammalian cell-based screening. Our data presented here provide direct evidence that CRY-mediated feedback repression of the CLOCK/BMAL1 complex is required for mammalian clock function. Although it is likely that the mammalian clock is governed by a combination of both transcriptional and non-

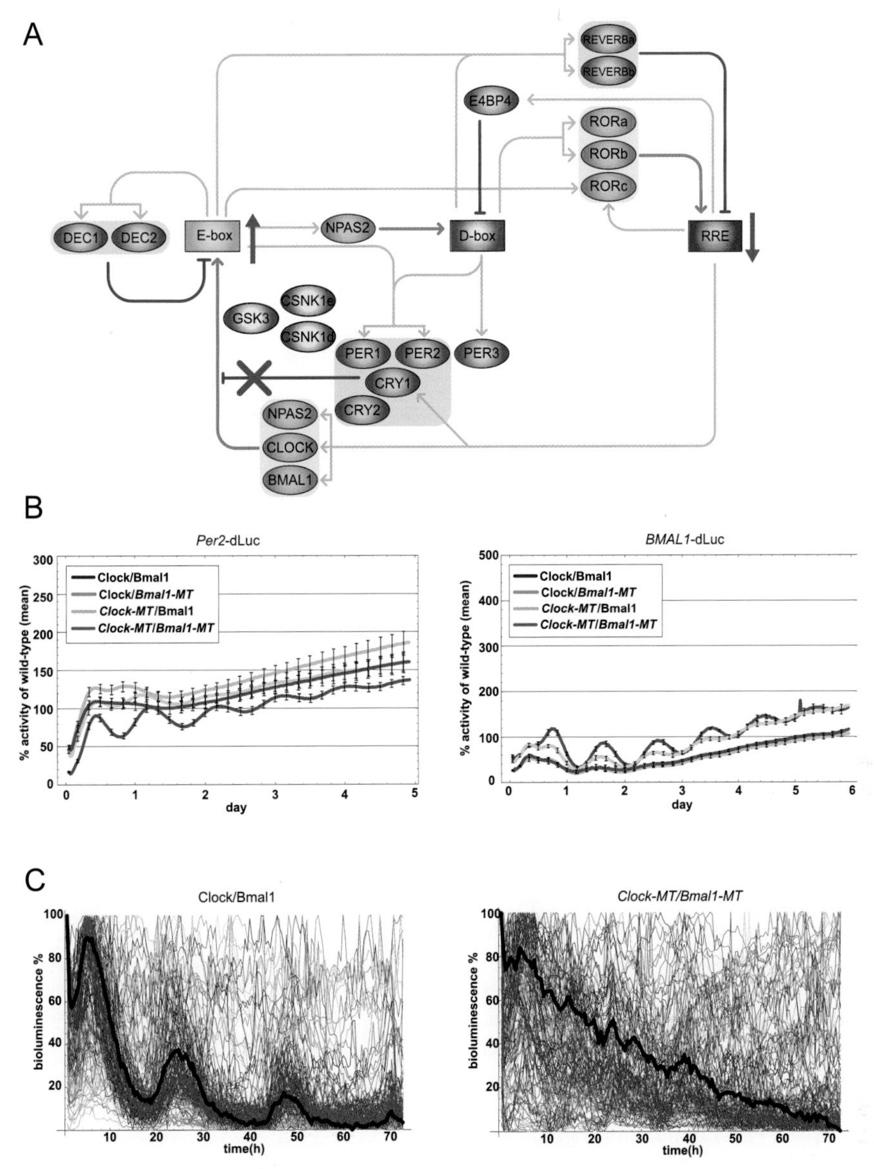

Figure 4. (*A*) Schematic representation of the transcriptional network of the mammalian circadian clock. (*Ellipsoids*) Genes; (*rectangles*) CCE. Transcriptional-translational expression, activation and repression are depicted as *gray*, *green*, and *red* lines, respectively. (*B*) Coexpression of CLOCK/BMAL1 mutant heterodimers that are insensitive to CRY repression ablates circadian E-box and RRE activities in NIH-3T3 cells. Plasmids expressing Flag-tagged *Clock* and *Bmal1* alleles were transiently cotransfected with the *Per2-dLuc* reporter plasmid into NIH-3T3 cells. (*Left panel*) *Per2* promoter activities in NIH-3T3 cells transfected with single or double CRY1-insensitive *Clock* and *Bmal1* mutants were monitored over 5 days. (*Right panel*) *Bmal1* promoter activities in NIH-3T3 cells transfected with single or double CRY-insensitive mutants of *Clock* and *Bmal1* were monitored over 6 days. All reporter activities were normalized such that the median wild-type luciferase activity over the time course was 100%. (*C*) Coexpression of CLOCK/BMAL1 mutant heterodimers impairs circadian rhythmicity in individual cells. *Per2-Luc* reporter activities from individual NIH-3T3 cells (*n* = 133) transfected with Flag-tagged wild-type CLOCK/BMAL1 or double-mutant *Clock-MT/Bmal1-MT* were monitored over 3 days. Reporter activities from each wild-type or double-mutant cells were normalized such that the maximum bioluminescence value was set to 100% for each panel. The mean reporter activity for all analyzed single cells at each time point is indicated by a black line. (Modified from Sato et al. 2006.)

transcriptional feedback mechanisms, any residual circadian properties that remain upon uncoupling of transcriptional feedback are insufficient to maintain circadian transcriptional output and molecular clock function. Finally, we predict that the application of cellular genetics technology will have a significant impact on mammalian biology as similar approaches have had on prokaryotic and yeast biology.

Control of Clocks

System control aims to regulate the target system toward the desired state through the precise perturbation of its components. To achieve this, it is necessary to develop an assay system that can be controlled with dynamic and quantitative perturbation.

The circadian clock is known to be entrainable by

external cues such as light. The light information is transmitted to the circadian clock through sensing mechanisms containing photoreceptors, and as a result of light pulse, the clock system shows a drastic change of its dynamics. In this section, we show the success of control of the oscillating clock system in individual cultured cells via artificial light-sensing mechanisms. We also applied this photoperturbation system to one of the long-standing and unsolved biological phenomena known as the singularity behavior of circadian clocks (Ukai et al. 2007).

Singularity behaviors of circadian clocks. Circadian clocks exhibit various dynamic properties, making them difficult to elucidate without quantitative perturbation and precise measurement of their dynamics. One of the most fundamental but yet-unsolved dynamic properties of circadian clocks is singularity behavior, in which robust circadian rhythmicity can be abolished after a certain critical stimulus, such as light and temperature pulses applied at the appropriate timing and strength. Since the first report of singularity behavior in *Drosophila pseudoobscura* by Arthur T. Winfree (1970), circadian clock singularities have been experimentally observed in various organisms including unicells such as *Gonyaulax* (Taylor et al. 1982), *Euglena* (Malinowski et al. 1985), *Chlamydomonas* (Johnson and Kondo 1992), fungi (Huang et al. 2006), insects (Winfree 1980), plants (Engelmann et al. 1978; Covington et al. 2001), and mammals (Jewett et al. 1991; Honma and Honma 1999), suggesting that this behavior is a shared property of an extremely broad range of circadian clocks (Dunlap et al. 2004).

Although singularity behavior has been widely observed, little is known about the underlying mechanisms. Since such behaviors were experimentally observed at the multicell level (i.e., the collective behavior of unicells or the physiological or locomoter activity of a multicellular organisms), two alternative single-cell-level mechanisms have been proposed to explain their collapse to singularity: (1) arrhythmicity of individual clocks (Fig. 5A) or (2) desynchronization of individual rhythmically oscillating clocks (Fig. 5B) (Winfree 1975, 1980). In the former mechanism, individual clocks become arrhythmic, i.e., the amplitude of the individual cells is substantially attenuated by the application of the critical light pulse. In contrast, in the latter mechanism of desynchronization, the phases of individual clocks are diversified by the critical light pulse. Although both mechanisms can explain substantial suppression of the multi-cell-level amplitude of circadian rhythm, there is a fundamental difference in dynamical properties between the two in that the oscillations of individual cells are impaired in the former, whereas individual cells maintain their oscillations in the latter. Importantly, although many researchers have observed multi-cell-level singularity behavior in various organisms, it remains elusive whether arrhythmicity or desynchronization of individual clocks underlies the singularity behavior of circadian clocks.

Determination of the underlying mechanism for singularity behaviors of circadian clocks may require adjustable perturbation, because the ability of a critical stimulus to drive circadian clocks into singularity depends on its timing and strength. Various stimuli, such as reagents and temperature, have been reported to directly reset mammalian cellular clocks (Balsalobre et al. 1998, 2000a,b; Akashi and Nishida 2000; Yagita and Okamura 2000; Yagita et al. 2001; Brown et al. 2002; Hirota et al. 2002; Tsuchiya et al. 2003, 2005). Unfortunately, it is difficult, but not impossible, for these factors to achieve the requisite flexibility in timing and strength. In contrast to perturbations achieved by the use of reagents or temperature change, photoperturbation provides an ideal range of adjustability in timing and strength. Although most mammalian cells cannot sense light, recent studies have shown that mammalian cells (Neuro-2a; HEK293) become photoresponsive following the introduction of an exogenous G-protein-coupled photoreceptor, melanopsin (Melyan et al. 2005; Qiu et al. 2005). It was reported that photostimulation of melanopsin triggers a release of intracellular calcium mediated through the G_q-protein signaling pathway, and, importantly, there are several reports that mammalian cellular clocks can be reset by the G_q-protein signaling pathway involving a release of intracellular calcium (Balsalobre et al. 2000a; Tsuchiya et al. 2005). These

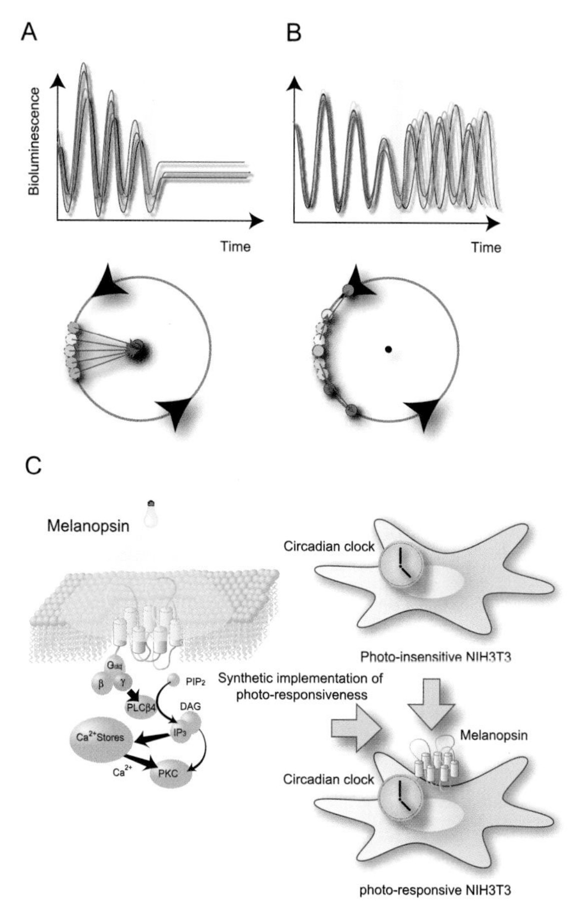

Figure 5. (*A–B*) Schematic diagrams of two alternative single-cell-level mechanisms for multi-cell-level singularity behavior. Arrhythmicity (*A*) and desynchronization (*B*) of individual cellular clocks are indicated. (*C*) Synthetic implementation of photoresponsiveness within mammalian clock cells. Schematic representation of melanopsin-dependent photoresponsive NIH-3T3 cells and the known Gq signaling pathway. (Modified from Ukai et al. 2007.)

results suggest that melanopsin-dependent photoperturbation may enable the adjustable and quantitative perturbation of mammalian cellular clocks by changing intracellular calcium level.

To experimentally reveal the underlying mechanism of singularity behavior of mammalian cells, we synthetically implemented photoresponsive mammalian cells by exogenously introducing a G_q-protein-coupled photoreceptor—melanopsin in our recent study (Fig. 5C), see also our recent study (Ukai et al. 2007). We then devised a high-throughput monitoring system with a light-exposure unit to continuously and quantitatively monitor the effect of photoperturbation on the state of cellular clocks. Using this system, we revealed that a critical light pulse drives cellular clocks into a singularity behavior in which robust circadian rhythmicity can be abolished after a certain stimulus. Theoretical analysis and subsequent single-cell-level observation consistently predicted and directly proved that the desynchronization of individual cellular clocks underlies this singularity behavior. We also constructed a theoretical framework to explain why singularity behaviors have been experimentally observed in various organisms and proposed desynchronization as a plausible mechanism for the observable singularity of circadian clocks. Importantly, these in cellulo and in silico findings are further supported by our in vivo observations that desynchronization actually underlies the multi-cell-level amplitude decrease in the rat SCN induced by the critical light pulses that can predispose organisms to transient amplitude decrease in their locomotor activity.

Historically speaking, to elucidate the underlying mechanism of singularity behaviors, Arthur T. Winfree and other investigaters conducted a two-pulse experiment and revealed the "unclocklike" behavior of circadian clock, in which the critical light pulse inducing the singularity behavior seemed to decrease the amplitude of circadian clock without affecting its frequency, apparently arguing against both the simple limit-cycle model and its resulting prediction that the arrhythmicity would underlie the singularity behavior. To explain this "unclocklike" behavior of circadian clock, he proposed the "clockshop" hypothesis, in which an organism-level circadian clock consists of multiple circadian oscillators with substantial fluctuations, and predicted that the desynchronization of individual circadian oscillators would underlie the singularity behavior (multi-cell-level amplitude decrease). He was unable to directly test this prediction, however, as there was no way to observe single-cell-level circadian rhythmicity at that time. In our recent study (Ukai et al. 2007), his prediction on the desynchronization was directly proved at least in the mammalian circadian system more than 30 years after he originally proposed it in his review (Winfree 1975).

Design of Clocks

The next step is system design—reconstruction and design of new systems based on the design principles that have been revealed through the efforts of a combination strategy of system identification, system analysis, and system control. In this stage, we can validate the sufficiency of the hypothesis derived from the identified structure or observed dynamics. To test the sufficiency of the design principles derived from the identified transcriptional circuits underlying mammalian circadian clocks (Fig. 3B), we are trying to extend our in cellulo cycling assay system (Fig. 3A) to the "physical simulator," with which we will be able to implement artificial transcriptional circuits of interest. We are planning to use this in cellulo system to prove the sufficiency of the components we predict in the natural circadian phase-controlling mechanism.

Alternatively, we can also take a radical and fundamental approach for system design of mammalian circadian clocks. In the next section, we introduce, as a radical approach, our current project on the in vitro reconstruction of the mammalian circadian clock from scratch.

Reconstitution of clocks. The molecular mechanisms underlying circadian clocks in many organisms have been studied for many years (Dunlap 1999; Young and Kay 2001; Reppert and Weaver 2002). Most of the molecular-level observations in previous studies suggest that every clock system has the translational-transcriptional negative feedback loop as a central oscillator, in which positive regulators such as BMAL1/CLOCK in mammals directly or indirectly activate the transcription of its negative regulators such as *Pers*/*Crys*, via a *cis*-acting DNA-element-like E box (Gekakis et al. 1998; Kume et al. 1999; Shearman et al. 2000).

Although comprehensive studies about the molecular mechanism of clocks have been reported, the basic mechanism for the fundamental nature of circadian clocks is still unclear. Specifically, the mechanism for autonomous oscillation, the most fundamental mechanism of clocks, remains to be fully understood.

Recently, these fundamental questions in eukaryotic clocks have become evident from studies concerning the circadian clock of cyanobacteria, known as the simplest organism possessing a circadian clock. Although ubiquitous molecular behaviors concerning the circadian clock, such as negative feedback regulation of clock genes, circadian oscillation of accumulation of mRNA and clock proteins, and phosphorylation of clock proteins, are also observed in cyanobacteria, the robust circadian oscillation of the phosphorylation state of KaiC, a circadian clock protein, was reconstituted by mixing only three cyanobacterial clock proteins and ATP in a test tube (Fig. 6) (Nakajima et al. 2005). The circadian oscillation of KaiC phosphorylation was therefore proved to be the central oscillator of the cyanobacterial circadian clock. A number of elegant studies about the cyanobacterial circadian clock evoked the importance of using a biochemical approach of the circadian clock for fundamental questions, such as autonomous oscillation and temperature-compensation.

Clock-related genes and proteins have been almost completely identified in some model organisms such as mammals, *Drosophila*, and *Neurospora*, as well as in cyanobacteria (Dunlap 1999; Young and Kay 2001; Reppert and Weaver 2002; Dunlap et al. 2004). In addition to the identification of clock-related components, we have clarified the dynamic nature of circadian clocks.

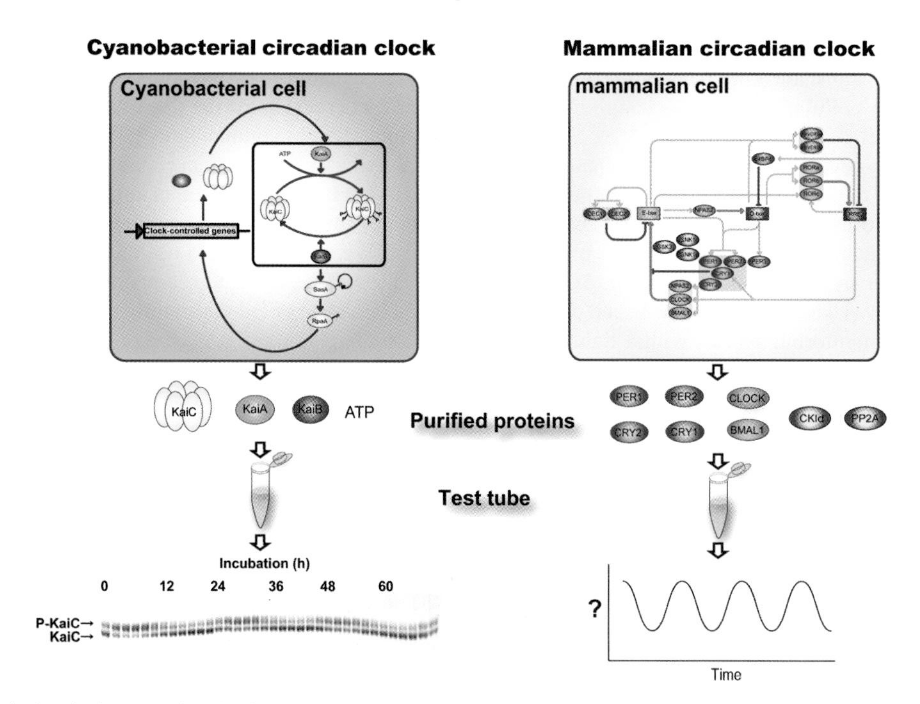

Figure 6. Biochemical approach to the fundamental questions of the circadian clock. KaiC, one of the three cyanobacterial clock proteins (KaiA, KaiB, and KaiC), has both autokinase and autophosphatase activity, and its phosphorylation state shows circadian rhythm in vivo. When three Kai proteins and ATP are mixed in a test tube, temperature-compensated KaiC phosphorylation rhythm is reconstituted (*left*). Because the biochemical properties of the mammalian clock proteins are limited, it is thought that the fundamental nature of the mammalian circadian clock is also dependent on these clock proteins (*right*). (Modified from Nakajima et al. 2005; panel on the periodic KaiC phosphorylation is courtesy of Dr. M. Nakajima.)

These circumstances in circadian clock studies led us to reevaluate the negative feedback model of the circadian clock by biochemical methods. We have succeeded in the purification of some clock proteins through the expression system of *Escherichia coli*. Furthermore, we are now trying to characterize some basic reactions in the mammalian circadian clock, such as the enzymatic properties of casein kinase Iε (CKIε) (Lowrey et al. 2000) and some protein-protein interactions. Although the mammalian circadian clock system is more complicated than that of cyanobacteria, we are now trying to reconstitute the mammalian circadian clock to demonstrate the essential mechanism of autonomous circadian oscillation and the other fundamental questions.

PERSPECTIVES: SYSTEM-LEVEL QUESTIONS

In the previous sections, we introduced a series of systems-biological approaches as well as their application to the specific questions such as singularity behaviors of mammalian circadian clocks. This has provided us with the groundwork to take a further step forward. It is high time to fully integrate these approaches to realize a system-level understanding of the mammalian circadian clock. The success of these approaches will therefore be measured by the importance and number of system-level questions that are solved. In the following sections, we list several system-level questions in mammalian circadian clocks (Fig. 7). We first describe the dynamical problems related to the *core of clocks*, including the *delay* in feedback repression; *high-amplitude* oscillations generated by transcriptional response of *nonlinearity*; *temperature-compensation* in elementary processes of the circadian oscillation; and *synchronization of clocks* against inevitable fluctuations of phase and period in multiple circadian oscillators. We then introduce information problems related to *clocks in the environment*, including the *internal representation of light change*, especially through a mechanism known as *perfect adaptation* and the *internal representation of day length* by *photoperiodism*.

Core of Clocks

In the mammalian circadian clocks, it is assumed that the delay in transcriptional feedback repression has a pivotal role in generating circadian oscillations based on the basic control theory. However, the underlying molecular mechanism to generate the delay still remains elusive. We first describe this dynamical question on delay. Nonlinearity in transcriptional feedback repression also has an important role in generating high-amplitude circadian oscillation itself as well as the high-amplitude output oscillations. Thus, as the second dynamical question, we pick up this nonlinearity problem. In addition to these dynamical properties, the core of mammalian circadian clocks also exhibit well-defined dynamical properties, including temperature-compensation and synchronization of multiple cellular clocks. We describe these two dynamical questions in the subsequent sections.

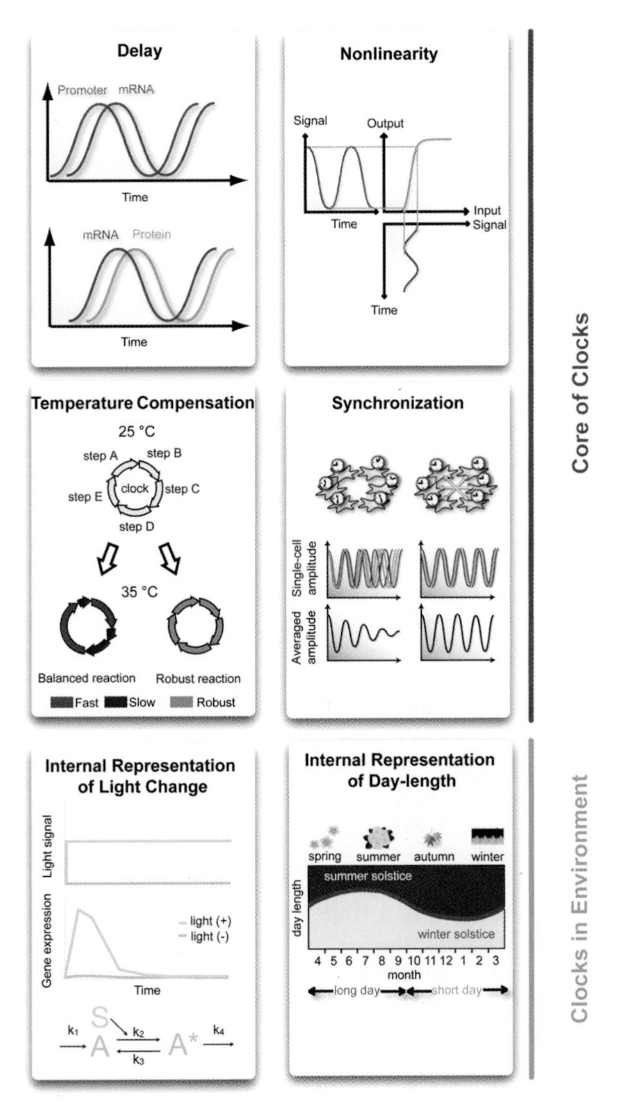

Figure 7. System-level questions in mammalian circadian clocks. Questions related to core of clocks: Delay in feedback repression seems to have an important role in the generation of oscillations. Although the nonlinearity involved in generating a high amplitude of transcription has been implicitly supposed in mammalian circadian clock, its mechanism so far remains unsolved. Although temperature-compensation is one of the most mysterious dynamical properties of clocks, its mechanism so far remains unsolved. The synchronization mechanism to generate synchrony of the multiple cellular clocks has been proposed theoretically but is not yet proven. Questions related to clocks in environment: Central clock tissue can internally represent a change of environmental light signal probably through perfect adaptation. Several mechanisms for perfect adaptation have been proposed theoretically but are not yet proven in mammalian circadian clocks. Organisms can also internally represent day length through photoperiodism. Although a gating mechanism is expected, its mechanism so far remains unsolved.

Delay. System identification and system analysis of mammalian circadian clocks revealed that transcriptional feedback repression mediated through the E box has an important role in generating transcriptional circadian oscillation. Interestingly, this transcriptional feedback repression accompanies delay in elementary processes, which is consistent with the basic control theory predicting that the negative feedback loop with a certain delay can generate the oscillation with twofold period of delay. In mammalian circadian clocks, delay has been observed at least in two processes (Fig. 7, top right panel). One is the delay between mRNA amounts and protein amounts. For example, expression levels of *Per1* and *Per2* mRNA are several hours earlier than those of PER1 and PER2 proteins (Lee et al. 2001), respectively. According to the basic control theory, this delay between mRNA and protein may be one of the determinants of duration of circadian oscillation, although there has so far been no definitive proof. The other process is the delay between promoter activity and mRNA amount. For example, promoter activity of *Cry1* is several hours earlier than the expression level of *Cry1* mRNA (Etchegaray et al. 2003; Ueda et al. 2005). Because the *Cry1* gene is one of the strongest repressors for E-box-mediated transcription, this delay also seems to be one of the determinants of duration of circadian oscillation.

A delayed process, where an input signal is just shifted to an output signal by a certain period without change of its shape, is completely different from a merely slow process, where an input signal is usually transformed to a dulled output signal. Thus, the generation of delay requires certain molecular mechanisms. However, molecular mechanisms to generate delays and hence oscillations still remain elusive in mammalian circadian clocks.

Nonlinearity. Nonlinearity is another dynamical property closely associated with delay because a slow process with certain nonlinearity can restore the original sharp signal from a dulled signal through amplification (Fig. 7, top right panel). Sigmoidal response (Fig. 7, top right panel), one of the nonlinear responses, is important for this amplification. Sigmoidal response is also important to generate high-amplitude circadian oscillations in clock output where a sharp input signal is usually dulled through a transfer process. Such sigmoidal responses can amplify the dulled signal to restore the original sharp signal and hence generate high-amplitude outputs.

Several molecular mechanisms can be expected to achieve a sigmoidal response. One class of molecular mechanisms is called as *cooperativity* (Monod et al. 1963), where multiple subunits work together to accomplish the entire process. In this cooperativity process, a Hill coefficient, which describes the strength of nonlinearity in a sigmoidal response, is usually smaller than the number of subunits, and hence usually approximately 2 (Hill coefficient 1 indicates a linear response). On the other hand, the other class of molecular mechanisms called *ultrasensitivity* (Goldbeter and Koshland 1981) has a potential to generate strong nonlinearity, with a Hill coefficient of more than 2. This switch-like behavior, where the output signal increases abruptly when an input signal exceeds a certain threshold, was originally proposed theoretically in antagonistic enzymatic reactions, where an enzyme competes with another enzyme acting on the same molecule (Goldbeter and Koshland 1981). However, molecular mechanisms to generate such a sigmoidal response and hence high-amplitude oscillations still remain elusive in mammalian circadian clocks.

Temperature-compensation. A robust 24-hour period against environmental changes, such as temperature and nutrition, is an important element for the circadian clock. The robustness of the period against temperature change is known as *temperature-compensation*, which is one of the intriguing aspects of the circadian clock. Typical biochemical reactions, such as enzymatic reactions, show temperature dependence, which is represented by a Q10 value of about 2. In contrast, the period of circadian rhythm is independent of or compensated against temperature change (Q10 value of the period is about 1.). The importance of temperature-compensation in poikilotherm can be easily understood. Although this aspect of nature in the mammalian circadian clock has been controversial, it has been confirmed in cultured mammalian cells (Izumo et al. 2003; Tsuchiya et al. 2003). Temperature-compensation has proved to be one of the general elements of circadian clocks from cyanobacteria to humans. Despite our increasing knowledge of the molecular mechanism of circadian clocks, however, it is difficult to explain how circadian clocks sustain such a constant period against temperature change.

Theoretical studies (Ruoff et al. 1997; Kurosawa and Iwasa 2005) have proposed a *balanced reaction model* (Fig. 7, middle left panel) for explaining this nature, in which increasing kinetic parameters of some reactions in the negative feedback loop lead to a shortening of the period and others lead to prolonging the period. As a result, the period of the circadian rhythm is unchanged against temperature change by a balance between positive and negative effects of reactions on the period. These theoretical explanations seem to be plausible, but the balance of effects on the period length between basic reactions can be easily broken by perturbations, such as inhibitors and point mutations on clock proteins. Furthermore, these theoretical models seem inconsistent with the fact that many circadian clock mutants show a diverse period but sustain temperature-compensation. How do circadian clocks acquire the robustness against temperature change?

There are some implications for understanding the mechanism in studies of the cyanobacterial circadian clock. The circadian rhythm of KaiC phosphorylation can be reconstituted in a test tube, demonstrating that the KaiC phosphorylation rhythm is the central oscillator of the cyanobacterial circadian clock (Nakajima et al. 2005). The reaction rate of KaiC phosphorylation in a test tube is unaffected by temperature change despite the fact that it is a biochemical reaction. These observations indicate that the robustness of circadian oscillation in cyanobacterial cells depends on the biochemical properties of three clock proteins. Although the mechanism of the KaiC phosphorylation cycle remains unclear, these findings suggest a *robust reaction model* (Fig. 7, middle left panel) for temperature-compensation of circadian clocks, at least in cyanobacteria.

In summary, there exist two possible models of temperature-compensations of the circadian clock. In the balanced reaction model, increasing kinetic parameters of some basic reactions lead to shortening the period, and those of other basic reactions lead to prolonging the period. These effects are cancelled out so that the period is sustained constantly. In the balanced reaction model, temperature-compensation is expected as a emergent network property of multiple clock components. In the robust reaction model, such as the cyanobacterial circadian clock, temperature-compensation of the circadian clock is caused by reactions, of which kinetic parameters are independent of temperature change. In the robust reaction model, temperature-compensation is expected as a molecular property of clock components. According to this model, the system-level understanding of temperature-compensation extensively depends on the investigation of biochemical property of clock components.

Synchronization of clocks. The core of mammalian circadian clocks located in the SCN consists of multiple autonomous single-cell oscillators. The individual cellular oscillators in the SCN produce coherent circadian rhythm despite the inevitable internal noise. However, the synchronization mechanism to couple individual cellular oscillators is largely unknown. As a possible model of the intercellular synchronization of circadian cellular clocks, we previously presented a multicellular stochastic model with an intercellular synchronization factor that emulates the synchronization of individual cellular oscillators (Ueda et al. 2002a). In this model, owing to a synchronizing factor, the neighboring cells correct each other's differences to stay in the proper rhythm, and thus, coherent oscillation of individual cellular clock is retained even in the presence of internal noise (Fig. 7, middle right panel). On the other hand, in the model without such a synchronizing factor, coherent oscillation of individual cellular clocks is abolished despite the identical initial conditions. The averaged amplitude is reduced due to the internal noise and the absence of external time cues such as light and temperature (Fig. 7, middle right panel).

In our model, a synchronization factor is supposed to be secreted from the individual cellular oscillators and then affect the circadian state of neighboring cells. The model predicts that if the synchronization factor is secreted during the subjective day, then the synchronization factor will induce the light-type phase-response in the neighboring cells (Ueda et al. 2002a). On the other hand, it is predicted that if the synchronization factor is secreted during the subjective night, then it will induce the dark-type phase-response in the neighboring cells.

Interestingly, recent studies reported a candidate synchronization factor, vasoactive intestinal polypeptide (VIP), which seems to possess the predicted dynamical properties described above. VIP is a peptide-type neurotransmitter that is secreted during the subjective day in the SCN and is known to induce a light-type phase-response in the circadian clock in a cultured SCN slice (Reed et al. 2001). Moreover, the VIP receptor VPAC2 is also expressed in the SCN. Actually, it has been reported that loss of VIP or VPAC2 disrupted synchrony between each oscillator neuron in the SCN (Aton et al. 2005; Maywood et al. 2006). Seen in this light, VIP has an important role in provoking the synchronous oscillation of cells at the SCN via VPAC2 receptor signaling. If our theory is correct, we may be able to reconstruct the synchronization of individual cellular oscillators in cultured cells such as Rat-1 and NIH-3T3, which exhibits no detectable synchronization among cells, by exogenously expressing VIP and the VPAC2 receptor under the control of a clock-controlled promoter at the appropriate timing in cultured cells. The theory predicts that VIP expressed during the subjective day will induce the efficient coupling among individual cellular oscillators, but VIP expressed at other times such as subjective night will not. The experimental verification of these predictions may lead to design principles in the synchronization of mammalian circadian clocks.

Clocks in Environment

Mammalian circadian clocks are surrounded by an environment and thereby receive external stimuli such as light and temperature to extract environmental information on the state of the earth. In the following sections, we discuss how the living organism accurately extracts and internally represents change of light signals and the day length.

Internal representation of light change: Perfect adaptation. Living organisms sense changes of external environment and entrain their internal circadian clocks by utilizing environmental change as a time cue. The most well-known time cue is light, and dawn and dusk light mark the advance and delay of the circadian clock, respectively. These rapid illumination changes that occur at dawn and dusk act on the resetting of circadian rhythm in the SCN, which is known as *nonparametric entrainment* (Johnson et al. 2003). How does the central clock tissue internally represent the change of the light signals in order to achieve nonparametric entrainment?

One of the biologically plausible mechanisms to sense environmental change is known as *perfect adaptation* (Hao et al. 2007), where the system can sense not the absolute value but the relative change of external signals. Three possible mathematical models have been proposed for (near) perfect adaptation: (1) feed-forward loop, (2) feedback loop, and (3) activation-dependent inactivation. The feed-forward loop model has activation and inactivation pathways to receptor A from signal S. Signal S both activates and inactivates receptor A. Suppose that there are fast activation and slow inactivation of receptor A, then the active form of receptor A* would be perfectly adapted, i.e., A* would rapidly respond to the step-up of signal S but quickly go back to its original level. It is also noteworthy that both activation and inactivation of receptor A by signal S must have a similar dependency on signal S in order to achieve the perfect adaptation in the feed-forward loop model. The second model is the feedback loop model where signal S activates receptors, and then the active form receptor A* inhibits the receptor activation or synthesis. We note that the feedback loop model only achieves near perfect adaptation, where A* rapidly responds to the step-up of signal S and goes back near to its original level (but this is not perfect). The third and most robust model is the active-dependent inactivation model, where signal S activates the receptor. Then, only activated receptor A* is degraded or inactivated (Fig. 7, bottom left panel).

To solve this information problem on the internal representation of light change, it will be a limiting process to identify the "perfect adaptation" gene, which only responds to the change of the light. When the step function of light is applied, this type of genes will be rapidly induced once but quickly go back to its original expression level (Fig. 7, bottom left panel). Following the identification of such genes, precise measurement and quantitative perturbation of its expression dynamics will be critical for the system-level understanding of its design principle.

Internal representation of day length: Photoperiodism. Living organisms also measure the day length (i.e., the duration of daily light time) and hence sense the season of external environment because day length is longer in summer and shorter in winter. This alternation of day length induces a seasonal physiological and metabolic change called photoperiodism (Dunlap et al. 2004). Living organisms are believed to demonstrate photoperiodism through circadian clock function because only certain periods in a day (photoinducible phase) are sensitive to light signals and have a critical role in measurement of day length. However, it is still largely unknown how the specific brain region can internally represent the day length through interplay between the circadian clock and light (or dark) signals, especially in mammals (Fig. 7, bottom right panel).

In photoperiodism research on animals, birds such as the Japanese quail are often used because of their dynamic and rapid response to day length change. In birds, it is known that light exposure in a certain phase (photoinducible phase) can induce a photoperiodic response. Japanese quail have an approximately 4-hour photoinducible phase that starts about 12–16 hours after the beginning of the light period. When the day length becomes longer (season changes to breeding period), female Japanese quail begin to lay eggs and males' testicles start to mature. During the breeding period, their testis enlarge to more than 100 times that of the nonbreeding period. At the hormonal regulation level, triiodothyronine (T3), an active form of enzyme converted from deiodination of thyroxine (T4) prohormone via Dio2 activity, promotes secretion of gonadotropin-releasing hormone (GnRH) and hence testicular maturation.

Scientific efforts to reveal the induction mechanism of photoperiodic reaction in Japanese quail have achieved great success in recent years. To summarize, the median eminence (ME) in the hypothalamus was found to be the responsible area for photoperiodism formation. In the ME, light stimulation induces *Dio2* at the photoinducible phase. On the other hand, *Dio2* is not induced in the nonphotoinducible phase. As clock gene oscillation has been reported in these areas, the circadian clock is also thought to exist in these regions. As there is also a time-dependent gating system (i.e., photoinducible phase), the relationship between *Dio2* gene regulation and the circadian clock system has already been the subject of discussion (Yoshimura et al. 2003). In mammals, the pineal body, which is known as a melatonin-synthesizing organ, is important in photoperiodic reaction. Because melatonin synthesis occurs during the night and is repressed by light, melatonin level represents the dark length information. Interestingly, expression of *Dio2* in ME, which highly expresses a melatonin receptor, is also under the influence of melatonin in mammals. Thus, the mammals and birds share a photoperiodic mechanism.

To solve this information problem on the internal representation of day length and season, it will be a limiting process to identify the "photoperiodic" gene, which is rapidly induced (or repressed) in response to the longer day length, and to regulate *Dio2* and other downstream genes. After the identification of such a gene, precise measurement and quantitative perturbation of its gating expression dynamics will be critical for the system-level understanding of its design principle.

CONCLUSION

Following the identification of the key clock genes, there was increasing demand for higher-order understanding of design principles in mammalian circadian clocks. In this chapter, we described several approaches, beginning with comprehensive identification (system identification) and quantitative analysis (system analysis) of individual clock components and their networked interactions, leading to the ability to control existing systems toward the desired state (system control) and the design of new ones based on an understanding of structure and underlying dynamical principles (system design). We also listed several dynamical and information problems including delay, nonlinearity, temperature-compensation, and synchronization of mammalian circadian oscillator(s), as well as the internal representation of light change through perfect adaptation and internal representation of day length through photoperiodism in mammals. We strongly believe that it is now high time to fully integrate the systems-biological approaches for the solution of the system-level questions.

ACKNOWLEDGMENTS

We thank our collaborators. *System identification of clocks*: collaboration with Drs. W. Chen, A. Adachi, H. Wakamatsu, S. Hayashi, T. Takasugi, M. Nagano, K. Nakahama, Y. Suzuki, S. Sugano, M. Iino, Y. Shigeyoshi, S. Hashimoto, M. Sano, and M. Machida. *System analysis of clocks*: collaboration with Drs. T.K. Sato, R.G. Yamada, H. Ukai, J.E. Baggs, L.J. Miraglia, T.J. Kobayashi, D.K. Welsh, S.A. Kay, and J.B. Hogenesch. *System control of clocks*: collaboration with Drs. H. Ukai, T.J. Kobayashi, M. Nagano, K. Masumoto, M. Sujino, T. Kondo, K. Yagita, and Y. Shigeyoshi. *System design of clocks*: collaboration with Drs. M. Nakajima, M. Ishida, M. Ukai-Tadenuma, and T. Kasukawa. *Molecular-timetable method*: collaborative work with Drs. W. Chen, Y. Minami, S. Honma, K. Honma, M. Iino, and S. Hashimoto. *Theoretical study on synchronization of clocks*: collaboration with K. Hirose and M. Iino. We also thank Drs. I. Nikaido, H. Ukai, R.G. Yamada, T.J. Kobayashi, M. Nakajima, and Y. Minami as well as A. Wada for creation of the figures and critical reading of manuscript. This research was supported by intramural grant-in-aid from the Center for Developmental Biology (CDB), Director's Fund from CDB, President's Fund from Riken, grant-in-aid for Genome-network Project and Scientific Research on Priority Areas "Systems Genomics" from MEXT, Japan, and grant-in-aid for NEDO project from METI, Japan.

REFERENCES

Akashi M. and Nishida E. 2000. Involvement of the MAP kinase cascade in resetting of the mammalian circadian clock. *Genes Dev.* **14:** 645.

Aton S.J., Colwell C.S., Harmar A.J., Waschek J., and Herzog E.D. 2005. Vasoactive intestinal polypeptide mediates circadian rhythmicity and synchrony in mammalian clock neurons. *Nat. Neurosci.* **8:** 476.

Bae K., Jin X., Maywood E.S., Hastings M.H., Reppert S.M., and Weaver D.R. 2001. Differential functions of mPer1, mPer2, and mPer3 in the SCN circadian clock. *Neuron* **30:** 525.

Balsalobre A., Damiola F., and Schibler U. 1998. A serum shock induces circadian gene expression in mammalian tissue culture cells. *Cell* **93:** 929.

Balsalobre A., Marcacci L., and Schibler U. 2000a. Multiple signaling pathways elicit circadian gene expression in cultured Rat-1 fibroblasts. *Curr. Biol.* **10:** 1291.

Balsalobre A., Brown S.A., Marcacci L., Tronche F., Kellendonk C., Reichardt H.M., Schutz G., and Schibler U.

2000b. Resetting of circadian time in peripheral tissues by glucocorticoid signaling. *Science* **289:** 2344.

Brown S.A., Zumbrunn G., Fleury-Olela F., Preitner N., and Schibler U. 2002. Rhythms of mammalian body temperature can sustain peripheral circadian clocks. *Curr. Biol.* **12:** 1574.

Bunger M.K., Wilsbacher L.D., Moran S.M., Clendenin C., Radcliffe L.A., Hogenesch J.B., Simon M.C., Takahashi J.S., and Bradfield C.A. 2000. Mop3 is an essential component of the master circadian pacemaker in mammals. *Cell* **103:** 1009.

Covington M.F., Panda S., Liu X.L., Strayer C.A., Wagner D.R., and Kay S.A. 2001. ELF3 modulates resetting of the circadian clock in *Arabidopsis*. *Plant Cell* **13:** 1305.

Dunlap J.C. 1999. Molecular bases for circadian clocks. *Cell* **96:** 271.

Dunlap J.C., Loros J.J., and DeCoursey P.J., eds. 2004. *Chronobiology: Biological timekeeping*. Sinauer Associates, Sunderland, Massachusetts.

Engelmann W., Johnsson A., Karlsson H.G., Kobler R., and Schimmel M.-L. 1978. Attenuation of the petal movement rhythm in kalanchoe with light pulses. *Physiol. Plant.* **43:** 68.

Etchegaray J.P., Lee C., Wade P.A., and Reppert S.M. 2003. Rhythmic histone acetylation underlies transcription in the mammalian circadian clock. *Nature* **421:** 177.

Falvey E., Marcacci L., and Schibler U. 1996. DNA-binding specificity of PAR and C/EBP leucine zipper proteins: A single amino acid substitution in the C/EBP DNA-binding domain confers PAR-like specificity to C/EBP. *Biol. Chem.* **377:** 797.

Gekakis N., Staknis D., Nguyen H.B., Davis F.C., Wilsbacher L.D., King D.P., Takahashi J.S., and Weitz C.J. 1998. Role of the CLOCK protein in the mammalian circadian mechanism. *Science* **280:** 1564.

Goldbeter A. and Koshland D.E., Jr. 1981. An amplified sensitivity arising from covalent modification in biological systems. *Proc. Natl. Acad. Sci.* **78:** 6840.

Hao N., Behar M., Elston T.C., and Dohlman H.G. 2007. Systems biology analysis of G protein and MAP kinase signaling in yeast. *Oncogene* **26:** 3254.

Harding H.P. and Lazar M.A. 1993. The orphan receptor Rev-ErbA alpha activates transcription via a novel response element. *Mol. Cell. Biol.* **13:** 3113.

Hirota T., Okano T., Kokame K., Shirotani-Ikejima H., Miyata T., and Fukada Y. 2002. Glucose down-regulates Per1 and Per2 mRNA levels and induces circadian gene expression in cultured Rat-1 fibroblasts. *J. Biol. Chem.* **277:** 44244.

Hogenesch J.B., Gu Y.Z., Jain S., and Bradfield C.A. 1998. The basic-helix-loop-helix-PAS orphan MOP3 forms transcriptionally active complexes with circadian and hypoxia factors. *Proc. Natl. Acad. Sci.* **95:** 5474.

Honma S. and Honma K. 1999. Light-induced uncoupling of multioscillatory circadian system in a diurnal rodent, Asian chipmunk. *Am. J. Physiol.* **276:** R1390.

Honma S., Kawamoto T., Takagi Y., Fujimoto K., Sato F., Noshiro M., Kato Y., and Honma K. 2002. Dec1 and Dec2 are regulators of the mammalian molecular clock. *Nature* **419:** 841.

Huang G., Wang L., and Liu Y. 2006. Molecular mechanism of suppression of circadian rhythms by a critical stimulus. *EMBO J.* **25:** 5349.

Izumo M., Johnson C.H., and Yamazaki S. 2003. Circadian gene expression in mammalian fibroblasts revealed by real-time luminescence reporting: Temperature compensation and damping. *Proc. Natl. Acad. Sci.* **100:** 16089.

Jewett M.E., Kronauer R.E., and Czeisler C.A. 1991. Light-induced suppression of endogenous circadian amplitude in humans. *Nature* **350:** 59.

Johnson C.H. and Kondo T. 1992. Light pulses induce "singular" behavior and shorten the period of the circadian phototaxis rhythm in the CW15 strain of *Chlamydomonas*. *J. Biol. Rhythms* **7:** 313.

Johnson C.H., Elliott J.A., and Foster R. 2003. Entrainment of circadian programs. *Chronobiol. Int.* **20:** 741.

King D.P., Zhao Y., Sangoram A.M., Wilsbacher L.D., Tanaka M., Antoch M.P., Steeves T.D., Vitaterna M.H., Kornhauser J.M., Lowrey P.L., Turek F.W., and Takahashi J.S. 1997. Positional cloning of the mouse circadian clock gene. *Cell* **89:** 641.

Kitano, H. 2002a. Computational systems biology. *Nature* **420:** 206.

———. 2002b. Systems biology: A brief overview. *Science* **295:** 1662.

Kume K., Zylka M.J., Sriram S., Shearman L.P., Weaver D.R., Jin X., Maywood E.S., Hastings M.H., and Reppert S.M. 1999. mCRY1 and mCRY2 are essential components of the negative limb of the circadian clock feedback loop. *Cell* **98:** 193.

Kurosawa G. and Iwasa Y. 2005. Temperature compensation in circadian clock models. *J. Theor. Biol.* **233:** 453.

Lee C., Etchegaray J.P., Cagampang F.R., Loudon A.S., and Reppert S.M. 2001. Posttranslational mechanisms regulate the mammalian circadian clock. *Cell* **107:** 855.

Lipshutz R.J., Fodor S.P., Gingeras T.R., and Lockhart D.J. 1999. High density synthetic oligonucleotide arrays. *Nat. Genet.* (suppl. 1) **21:** 20.

Lowrey P.L., Shimomura K., Antoch M.P., Yamazaki S., Zemenides P.D., Ralph M.R., Menaker M., and Takahashi J.S. 2000. Positional syntenic cloning and functional characterization of the mammalian circadian mutation tau. *Science* **288:** 483.

Malinowski J.R., Laval-Martin D.L., and Edmunds L.N., Jr. 1985. Circadian oscillators, cell cycles, and singularities: Light perturbations of the free-running rhythm of cell division in *Euglena*. *J. Comp. Physiol. B* **155:** 257.

Maywood E.S., Reddy A.B., Wong G.K., O'Neill J.S., O'Brien J.A., McMahon D.G., Harmar A.J., Okamura H., and Hastings M.H. 2006. Synchronization and maintenance of timekeeping in suprachiasmatic circadian clock cells by neuropeptidergic signaling. *Curr. Biol.* **16:** 599.

Melyan Z., Tarttelin E.E., Bellingham J., Lucas R.J., and Hankins M.W. 2005. Addition of human melanopsin renders mammalian cells photoresponsive. *Nature* **433:** 741.

Mitsui S., Yamaguchi S., Matsuo T., Ishida Y., and Okamura H. 2001. Antagonistic role of E4BP4 and PAR proteins in the circadian oscillatory mechanism. *Genes Dev.* **15:** 995.

Monod J., Changeux J.P., and Jacob F. 1963. Allosteric proteins and cellular control systems. *J. Mol. Biol.* **6:** 306.

Nakajima M., Imai K., Ito H., Nishiwaki T., Murayama Y., Iwasaki H., Oyama T., and Kondo T. 2005. Reconstitution of circadian oscillation of cyanobacterial KaiC phosphorylation in vitro. *Science* **308:** 414.

Okamura H., Miyake S., Sumi Y., Yamaguchi S., Yasui A., Muijtjens M., Hoeijmakers J.H., and van der Horst G.T. 1999. Photic induction of mPer1 and mPer2 in cry-deficient mice lacking a biological clock. *Science* **286:** 2531.

Panda S., Hogenesch J.B., and Kay S.A. 2002. Circadian rhythms from flies to human. *Nature* **417:** 329.

Preitner N., Damiola F., Lopez-Molina L., Zakany J., Duboule D., Albrecht U., and Schibler U. 2002. The orphan nuclear receptor REV-ERBalpha controls circadian transcription within the positive limb of the mammalian circadian oscillator. *Cell* **110:** 251.

Qiu X., Kumbalasiri T., Carlson S.M., Wong K.Y., Krishna V., Provencio I., and Berson D.M. 2005. Induction of photosensitivity by heterologous expression of melanopsin. *Nature* **433:** 745.

Reed H.E., Meyer-Spasche A., Cutler D.J., Coen C.W., and Piggins H.D. 2001. Vasoactive intestinal polypeptide (VIP) phase-shifts the rat suprachiasmatic nucleus clock in vitro. *Eur. J. Neurosci.* **13:** 839.

Reick M., Garcia J.A., Dudley C., and McKnight S.L. 2001. NPAS2: An analog of clock operative in the mammalian forebrain. *Science* **293:** 506.

Reppert S.M. and Weaver D.R. 2002. Coordination of circadian timing in mammals. *Nature* **418:** 935.

Ruoff P., Rensing L., Kommedal R., and Mohsenzadeh S. 1997. Modeling temperature compensation in chemical and biological oscillators. *Chronobiol. Int.* **14:** 499.

Sato T.K., Panda S., Miraglia L.J., Reyes T.M., Rudic R.D.,

McNamara P., Naik K.A., FitzGerald G.A., Kay S.A., and Hogenesch J.B. 2004. A functional genomics strategy reveals Rora as a component of the mammalian circadian clock. *Neuron* **43:** 527.

Sato T.K., Yamada R.G., Ukai H., Baggs J.E., Miraglia L.J., Kobayashi T.J., Welsh D.K., Kay S.A., Ueda H.R., and Hogenesch J.B. 2006. Feedback repression is required for mammalian circadian clock function. *Nat. Genet.* **38:** 312.

Shearman L.P., Sriram S., Weaver D.R., Maywood E.S., Chaves I., Zheng B., Kume K., Lee C.C., van der Horst G.T., Hastings M.H., and Reppert S.M. 2000. Interacting molecular loops in the mammalian circadian clock. *Science* **288:** 1013.

Suzuki Y., Taira H., Tsunoda T., Mizushima-Sugano J., Sese J., Hata H., Ota T., Isogai T., Tanaka T., Morishita S., Okubo K., Sakaki Y., Nakamura Y., Suyama A., and Sugano S. 2001. Diverse transcriptional initiation revealed by fine, large-scale mapping of mRNA start sites. *EMBO Rep.* **2:** 388.

Taylor W., Krasnow R., Dunlap J.C., Broda H., Hastings J.W. 1982. Critical pulses of anisomycin drive the circadian oscillator in *Gonyaulax* towards its singularity. *J. Comp. Physiol.* **148:** 11.

Tei H., Okamura H., Shigeyoshi Y., Fukuhara C., Ozawa R., Hirose M., and Sakaki Y. 1997. Circadian oscillation of a mammalian homologue of the *Drosophila* period gene. *Nature* **389:** 512.

Tomita J., Nakajima M., Kondo T., and Iwasaki H. 2005. No transcription-translation feedback in circadian rhythm of KaiC phosphorylation. *Science* **307:** 251.

Tsuchiya Y., Akashi M., and Nishida E. 2003. Temperature compensation and temperature resetting of circadian rhythms in mammalian cultured fibroblasts. *Genes Cells* **8:** 713.

Tsuchiya Y., Minami I., Kadotani H., and Nishida E. 2005. Resetting of peripheral circadian clock by prostaglandin E2. *EMBO Rep.* **6:** 256.

Ueda H.R., Hirose K., and Iino M. 2002a. Intercellular coupling mechanism for synchronized and noise-resistant circadian oscillators. *J. Theor. Biol.* **216:** 501.

Ueda H.R., Matsumoto A., Kawamura M., Iino M., Tanimura T., and Hashimoto S. 2002b. Genome-wide transcriptional orchestration of circadian rhythms in *Drosophila*. *J. Biol. Chem.* **19**: 14048.

Ueda H.R., Chen W., Minami Y., Honma S., Honma K., Iino M., and Hashimoto S. 2004. Molecular-timetable methods for detection of body time and rhythm disorders from single-time-point genome-wide expression profiles. *Proc. Natl. Acad. Sci.* **101:** 11227.

Ueda H.R., Hayashi S., Chen W., Sano M., Machida M., Shigeyoshi Y., Iino M., and Hashimoto S. 2005. System-level identification of transcriptional circuits underlying mammalian circadian clocks. *Nat. Genet.* **37:** 187.

Ueda H.R., Chen W., Adachi A., Wakamatsu H., Hayashi S., Takasugi T., Nagano M., Nakahama K., Suzuki Y., Sugano S., Iino M., Shigeyoshi Y., and Hashimoto S. 2002c. A transcription factor response element for gene expression during circadian night. *Nature* **418:** 534.

Ukai H., Kobayashi T.J., Nagano M., Masumoto K., Sujino M., Kondo T., Yagita Y., Shigeyoshi Y., and Ueda H.R. 2007. Melanopsin-dependent photo-perturbation reveals desynchronization underlying the singularity of mammalian circadian clocks. *Nat. Cell Biol.* **(in press)**. **[AU: Update?]**

van der Horst G.T., Muijtjens M., Kobayashi K., Takano R., Kanno S., Takao M., de Wit J., Verkerk A., Eker A.P., van Leenen D., Buijs R., Bootsma D., Hoeijmakers J.H., and Yasui A. 1999. Mammalian Cry1 and Cry2 are essential for maintenance of circadian rhythms. *Nature* **398:** 627.

Vitaterna M.H., Selby C.P., Todo T., Niwa H., Thompson C., Fruechte E.M., Hitomi K., Thresher R.J., Ishikawa T., Miyazaki J., Takahashi J.S., and Sancar A. 1999. Differential regulation of mammalian period genes and circadian rhythmicity by cryptochromes 1 and 2. *Proc. Natl. Acad. Sci.* **96:** 12114.

Welsh D.K., Yoo S.H., Liu A.C., Takahashi J.S., and Kay S.A. 2004. Bioluminescence imaging of individual fibroblasts reveals persistent, independently phased circadian rhythms of clock gene expression. *Curr. Biol.* **14:** 2289.

Winfree A.T. 1970. Integrated view of resetting a circadian clock. *J. Theor. Biol.* **28:** 327.

———. 1975. Unclocklike behaviour of biological clocks. *Nature* **253:** 315.

———. 1980. *The geometry of biological time*. Springer-Verlag, New York.

Wuarin J. and Schibler U. 1990. Expression of the liver-enriched transcriptional activator protein DBP follows a stringent circadian rhythm. *Cell* **63:** 1257.

Yagita K. and Okamura H. 2000. Forskolin induces circadian gene expression of rPer1, rPer2 and dbp in mammalian rat-1 fibroblasts. *FEBS Lett.* **465:** 79.

Yagita K., Tamanini F., van der Horst G.T., and Okamura H. 2001. Molecular mechanisms of the biological clock in cultured fibroblasts. *Science* **292:** 278.

Yamazaki S., Numano R., Abe M., Hida A., Takahashi R., Ueda M., Block G.D., Sakaki Y., Menaker M., and Tei H. 2000. Resetting central and peripheral circadian oscillators in transgenic rats. *Science* **288:** 682.

Yoo S.H., Ko C.H., Lowrey P.L., Buhr E.D., Song E.J., Chang S., Yoo O.J., Yamazaki S., Lee C., and Takahashi J.S. 2005. A noncanonical E-box enhancer drives mouse Period2 circadian oscillations in vivo. *Proc. Natl. Acad. Sci.* **102:** 2608.

Yoshimura T., Yasuo S., Watanabe M., Iigo M., Yamamura T., Hirunagi K., and Ebihara S. 2003. Light-induced hormone conversion of T4 to T3 regulates photoperiodic response of gonads in birds. *Nature* **426:** 178.

Young M.W. and Kay S.A. 2001. Time zones: A comparative genetics of circadian clocks. *Nat. Rev. Genet.* **2:** 702.

Zheng B., Larkin D.W., Albrecht U., Sun Z.S., Sage M., Eichele G., Lee C.C., and Bradley A. 1999. The mPer2 gene encodes a functional component of the mammalian circadian clock. *Nature* **400:** 169.

Zheng B., Albrecht U., Kaasik K., Sage M., Lu W., Vaishnav S., Li Q., Sun Z.S., Eichele G., Bradley A., and Lee C.C. 2001. Nonredundant roles of the mPer1 and mPer2 genes in the mammalian circadian clock. *Cell* **105:** 683.

Zylka M.J., Shearman L.P., Weaver D.R., and Reppert S.M. 1998. Three period homologs in mammals: Differential light responses in the suprachiasmatic circadian clock and oscillating transcripts outside of brain. *Neuron* **20:** 1103.

High-resolution Time Course Analysis of Gene Expression from Pituitary

M. HUGHES,*† L. DEHARO,‡† S.R. PULIVARTHY‡, J. GU,* K. HAYES,*
S. PANDA,‡ AND J.B. HOGENESCH*

*Department of Pharmacology, Institute for Translational Medicine and Therapeutics, University of Pennsylvania School of Medicine, Philadelphia, Pennsylvania 19104;
‡Regulatory Biology, Salk Institute for Biological Studies, La Jolla, California 92186

In both the suprachiasmatic nucleus (SCN) and peripheral tissues, the circadian oscillator drives rhythmic transcription of downstream target genes. Recently, a number of studies have used DNA microarrays to systematically identify oscillating transcripts in plants, fruit flies, rats, and mice. These studies have identified several dozen to many hundred rhythmically expressed genes by sampling tissues every 4 hours for 1, 2, or more days. To extend this work, we have performed DNA microarray analysis on RNA derived from the mouse pituitary sampled every hour for 2 days. COSOPT and Fisher's G-test were used at a false-discovery rate of less than 5% to identify more than 250 genes in the pituitary that oscillate with a 24-hour period length. We found that increasing the frequency of sampling across the circadian day dramatically increased the statistical power of both COSOPT and Fisher's G-test, resulting in considerably more high-confidence identifications of rhythmic transcripts than previously described. Finally, to extend the utility of these data sets, a Web-based resource has been constructed (at http://wasabi.itmat.upenn.edu/circa/mouse) that is freely available to the research community.

INTRODUCTION

The central circadian oscillator in mammals is located in a small cluster of neurons in the hypothalamus called the suprachiasmatic nucleus (SCN) (Stratmann and Schibler 2006). These neurons have an endogenous circadian clock that oscillates with a period of approximately 24 hours (Ko and Takahashi 2006). In addition, these neurons receive information from retinal ganglion cells, permitting them to entrain the clock to environmental light cues (Cermakian and Sassone-Corsi 2002). Peripheral tissues have endogenous circadian clocks as well and may be entrained by a number of different stimuli (for review, see Stratmann and Schibler 2006). However, these peripheral oscillators are subordinate to the central circadian clock in the SCN, which is responsible for integrating environmental input and orchestrating the biological rhythms of the entire organism (Schibler and Sassone-Corsi 2002).

Consequently, many complex physiologies in an organism show regular oscillations during the course of a single day. For example, rhythms of sleep and arousal are controlled by the circadian clock, as well as rhythms in food consumption, blood pressure, body temperature, and metabolism (for review, see Hastings et al. 2003; Curtis and Fitzgerald 2006). These rhythms have direct and indirect implications for human health. The metabolism and efficacy of drugs depend heavily on the time of day they are administered (Lis et al. 2003; Antoch et al. 2005; Halberg et al. 2006) . Moreover, disruptions of the circadian clock have been shown to increase susceptibility to cancer, heart disease, and metabolic disorders while directly causing serious sleep disorders and influencing mental illness (Klerman 2005; Curtis and Fitzgerald 2006; Halberg et al. 2006; Levi and Schibler 2007).

Despite considerable effort, the link between the molecular oscillations of the core circadian clock and rhythms of organismal physiology remains poorly understood (Hastings et al. 2003). As a first step toward bridging this gap, many groups, including ours, have used DNA microarrays to systematically identify genes that oscillate during the course of a signal day (for examples of this work, see Harmer et al. 2000; Claridge-Chang et al. 2001; McDonald and Rosbash 2001; Akhtar et al. 2002; Ceriani et al. 2002; Duffield et al. 2002; Lin et al. 2002; Panda et al. 2002; Storch et al. 2002; Ueda et al. 2002; Kornmann et al. 2007). Typically, these studies have analyzed RNA samples isolated every 4 hours over the course of 2 or more days and used curve-fitting algorithms, Fourier analysis, or autocorrelation tests to identify rhythmic transcripts (Table 1). Using these methodologies, several hundred genes have been found under circadian regulations in mouse tissues and rat fibroblasts, as well as in plants (*Arabidopsis*) and fruit flies (*Drosophila*) (Table 1). These data and analyses

Table 1. Previous Circadian Microarray Studies Have Predominately Used 4-hour Time Resolution

Study resolution	Year	Tissue	Time (hour)
Harmer et al.	2000	*Arabidopsis*	4
Claridge-Chang et al.	2001	fly heads	4
McDonald and Rosbash	2001	fly heads	4
Ceriani et al.	2002	fly heads and bodies	4
Lin et al.	2002	fly heads	4
Akhtar et al.	2002	mouse liver	4
Storch et al.	2002	mouse liver and heart	4
Panda et al.	2002	mouse SCN and liver	4
Ueda et al.	2002	mouse SCN and liver	4
Duffield et al.	2002	Rat-1 fibroblasts	4
Kornmann et al.	2007	mouse liver	4

†These authors contributed equally to this work.

represented an important first step in elucidating the link between molecular oscillations of the core circadian clock and downstream physiology.

Several controversies, however, have emerged as a consequence of this work (Etter and Ramaswami 2002; Duffield 2003; Hayes et al. 2005). For example, most transcriptional profiling studies were able to identify many of the core components of the circadian clock, although these studies missed a number of bona fide oscillating genes and clock components (Harmer et al. 2000; Ceriani et al. 2002; Panda et al. 2002). This suggests that the conventional design of these experiments may have been insufficiently powerful to identify every significant circadian gene, as well as illustrated a traditional weakness of chip-based analysis: sensitivity. Similarly, there was surprisingly small overlap between the rhythmically expressed genes identified in different studies (Etter and Ramaswami 2002). In part, this discrepancy can be explained by the use of different experimental designs, strains of organisms, array platforms, and condensation algorithms, as well as the analytic methods used to detect cycling genes (Walker and Hogenesch 2005). These examples illustrate the high false-positive and high false-negative rates in these original studies, and taken as a whole, they suggest that additional work is necessary to precisely define the subset of the transcriptome regulated by the circadian clock.

TRANSCRIPTIONAL PROFILING AT HIGH TEMPORAL RESOLUTION

To address this issue, a second-generation microarray analysis was performed to identify genes whose transcripts are under circadian regulation. Wild-type mice were entrained to a 12 hours light:12 hours dark (12:12 L:D) schedule before release into constant darkness conditions. Tissue samples from the pituitary of four mice were collected every hour for 2 full days, and RNA samples were analyzed using Affymetrix microarrays. To identify rhythmically expressed genes, these data were subsequently analyzed using COSOPT, a method that relies on curve fitting and permutation analysis, and Fisher's G-test, based on the Fourier transform. Following the initial analyses, the false-discovery rate (FDR) was calculated for all transcripts using both methods in R, a high-level programming language used for statistical computing.

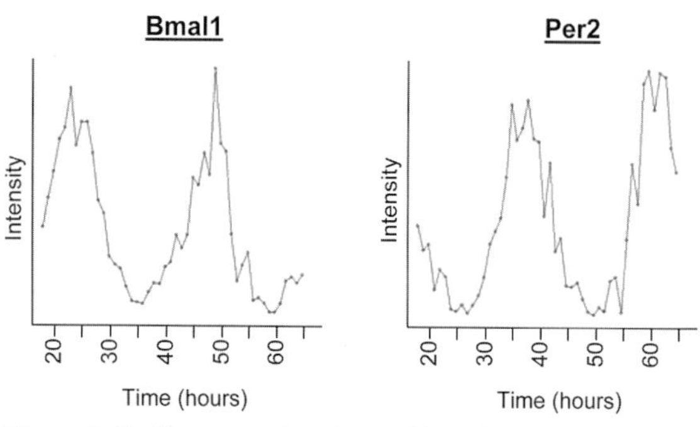

Figure 1. Cycling transcripts detected by microarray analysis in the pituitary. Tissue samples from the pituitary were collected from wild-type mice every hour for 2 days. RNA was purified from these samples and profiled on Affymetrix microarrays. *Bmal1* (*left*) and *Per2* (*right*) show robust circadian oscillations through the entire time course.

Table 2. Cycling transcripts in Pituitary Identified at Different False-discovery Rates

FDR	COSOPT	Fisher's G-test	Both
< 0.05	334	1152	274
< 0.01	131	316	120

Using this analysis, we found considerable evidence for robust circadian rhythms (Fig. 1). These included confidently detecting the circadian oscillations of known core clock transcripts. For example, two components of the core oscillator, Bmal1 and Clock (Ko and Takahashi 2006), as well as many other components of the circadian clock, including *Rev-erbα*, *Rorα*, *Per2*, *Cry1*, and *Cry2* were among the genes with the lowest p values by both statistical tests (Fig. 1 and data not shown). Importantly, the phase relationship among these components in both the liver and pituitary was in excellent agreement with previous expectations and was resolved to a preciseness that was not possible in previous studies.

Prompted by earlier work on meta-analysis of circadian gene expression data from DNA arrays, we sought to identify a cohort of oscillating transcripts using two distinct statistical methods: COSOPT and Fisher's G-test. Each identified several hundred rhythmic transcripts in pituitary samples at highly restrictive FDRs (Table 2). Importantly, there was considerable overlap between the transcripts identified by both tests, increasing the confidence that these genes are bona fide outputs of the circadian clock. All told, 274 transcripts were identified as rhythmically expressed by both COSOPT and Fisher's G-test at an FDR of 5%; when the FDR is reduced to 1%, 120 genes were found to be rhythmic by both algorithms (Table 2). As expected, the vast majority of these genes show period lengths of approximately 24 hours (Fig. 2).

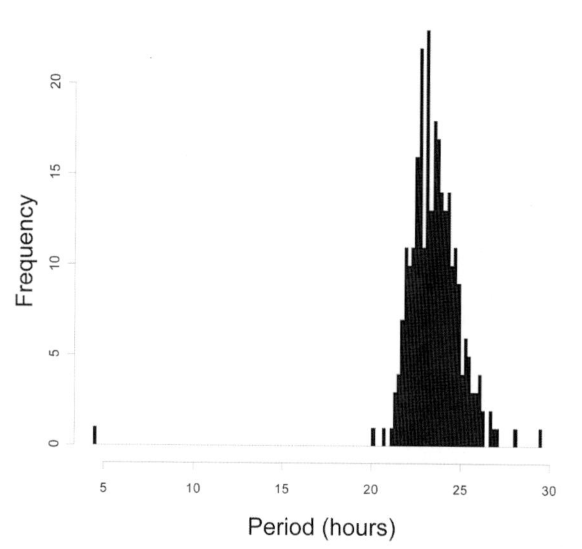

Figure 2. COSOPT and Fisher's G-test predominantly detect rhythmic transcripts with periods of approximately 24 hours. In the pituitary, several hundred rhythmic transcripts were detected at a false-discovery rate of <0.05.

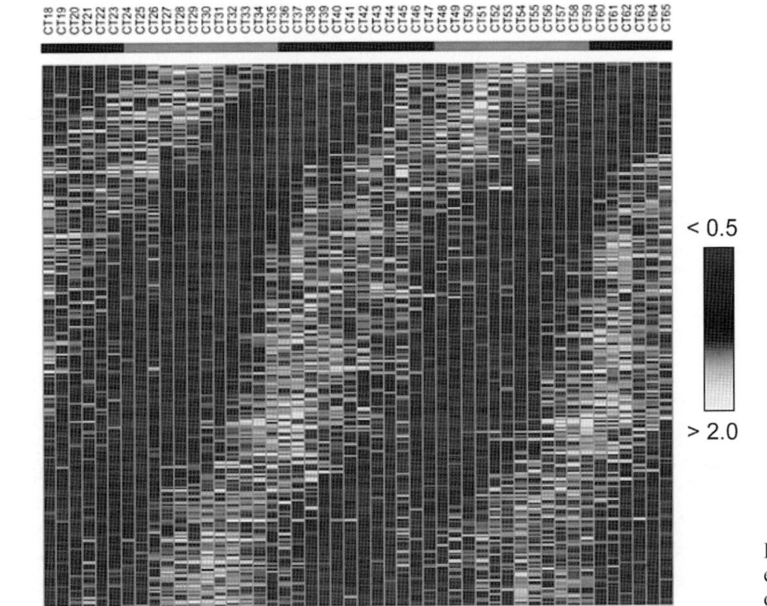

Figure 3. Phases of circadian transcripts span the entire circadian day. Rhythmic transcripts detected by COSOPT and Fisher's G-test at a false-discovery rate of <0.05 were plotted on heat maps.

Taken as a whole, these data suggest that more than 100 genes in the mouse pituitary oscillate in a circadian manner. Interestingly, the phases of these circadian genes show an approximately even distribution throughout a single day (Fig. 3), indicating that the output of the clock is operating nearly equally at every hour in these tissues.

STATISTICAL ANALYSIS

The design of the original studies by our group and others was somewhat arbitrarily derived from northern and western blot analysis studies predominantly aimed at the study of one or a few genes. The inherent weakness in this approach for DNA array analysis is the multiple testing problem; instead of one or a few genes being analyzed, tens of thousands were analyzed, which greatly amplified the likelihood of both false-positive and false-negative errors (type I and type II errors). The availability of the high-resolution time course experiment enabled empirical statistical analyses to determine the optimal time frequency of sampling. By analyzing expression data from 48 time points across 2 complete days, the present study successfully identified several hundred rhythmically expressed genes in the mouse pituitary (Table 2). These data indicate that the power of statistical tests to determine rhythmicity is dramatically improved through modest increases in sampling frequency. One of the key questions was the number of time points in a 48-hour time course necessary to robustly detect circadian transcription. In an era of tight research funding and relatively expensive DNA arrays, the best balance between sensitivity, selectivity, and cost would provide a tremendous benefit in the detection of cycling genes.

To evaluate this issue, we performed a number of simulations to assess the success of COSOPT and Fisher's G-test in identifying rhythmic transcripts at different sampling densities. To this end, we randomly selected time points from our data set at 1-, 2-, 3-, and 4-hour intervals and employed COSOPT and Fisher's G-test to identify oscillating genes. We discovered that there is a considerable statistical advantage in increasing the frequency of sampling (Fig. 4). At a wide range of FDR (0.01 to 0.4), both COSOPT and Fisher's G-test show approximately logarithmic improvements in detecting rhythmic transcripts as sampling density increases (Fig. 4). At low sampling resolutions (less than or equal to every 3 hours), Fisher's G-test performs considerable better than COSOPT, particularly at relatively high FDR (compare graphs in Fig. 4). Interestingly, both algorithms perform similar efficacies at high sampling resolutions, which is consistent with our expectation that sufficiently high-quality data sets render their downstream analysis insensitive to the particular algorithm used. These data indicate that the detection of transcripts under circadian control using conventional algorithms with a low rate of false discovery necessitates a sampling density of at least once every 2 hours over 2 full days and best balances cost and data quality.

BIOINFORMATICS RESOURCES

Several software resources were developed in the course this work and are publicly available to facilitate circadian gene expression analysis. COSOPT requires an input data step that is transcript-centric, meaning that the number of input files rises linearly with the number of genes being analyzed. To streamline this process, we developed a Perl script that automatically converts microarray data (in *.txt or *.csv format) into the files necessary for COSOPT. In addition, we have written an R script that utilizes the GeneTS package, calculates the average periodogram of the data, and finds period length of transcription.

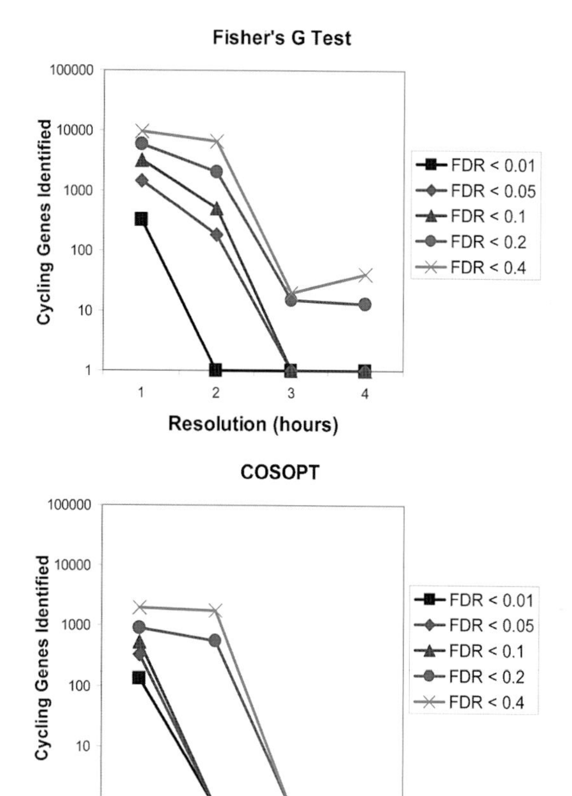

Figure 4. Increasing the frequency of sampling improves the performance of Fisher's G-test and COSOPT algorithms. Using high-resolution circadian time course, simulations were performed to determine how many cycling transcripts could be detected at any given false-discovery rate (FDR). Both Fisher's G-test (*top*) and COSOPT (*bottom*) showed significant improvements in the number of cycling transcripts detected as the frequency of sampling is increased from 4 hours to 1 hour.

The identification of novel circadian genes in the pituitary will enable further studies on the molecular mechanism of peripheral oscillators as well as additional work on the output of the circadian clock. To facilitate the distribution of this data set, we have created a Web-based interface available at http://wasabi.itmat.upenn.edu/circa/mouse (Fig. 5). This interface, affectionately known as "Wasabi," is written in Ruby on Rails, implemented on a Linux server, and permits the user to search for the transcriptional profile of any gene in either the liver or pituitary (Fig. 5, top). At the same time, q values (representing the false-discovery rate) and p values (representing the probability that a transcript was identified as rhythmic by chance alone) from COSOPT and Fisher's G-test, as well as phase and period length, can be use to filter queries (Fig. 5, top).

Figure 5 shows an example of a query submitted to Wasabi. The user is searching for the transcriptional profile of *Per2* in the liver, filtered by COSOPT q value (<0.05) and period length (>22 hours). The resulting profile is shown in Figure 5 (bottom left); the transcriptional profile of *Per2* is plotted as intensity (expression level) versus time. Additionally, information on the statistical analysis of this transcript, as well as external Web links to information on *Per2*, and its genomic locus are available within this interface (Fig. 5, bottom right). By streamlining the storage and retrieval of gene expression data, we anticipate Wasabi will become an important resource for the circadian community, analogous to SymAtlas, a Web-based resource describing the tissue-specific expression of most mammalian genes (Su et al. 2004).

ANALYSIS OF CYCLING TRANSCRIPTS

As previously discussed, a considerable gap exists in our understanding between the molecular mechanisms of

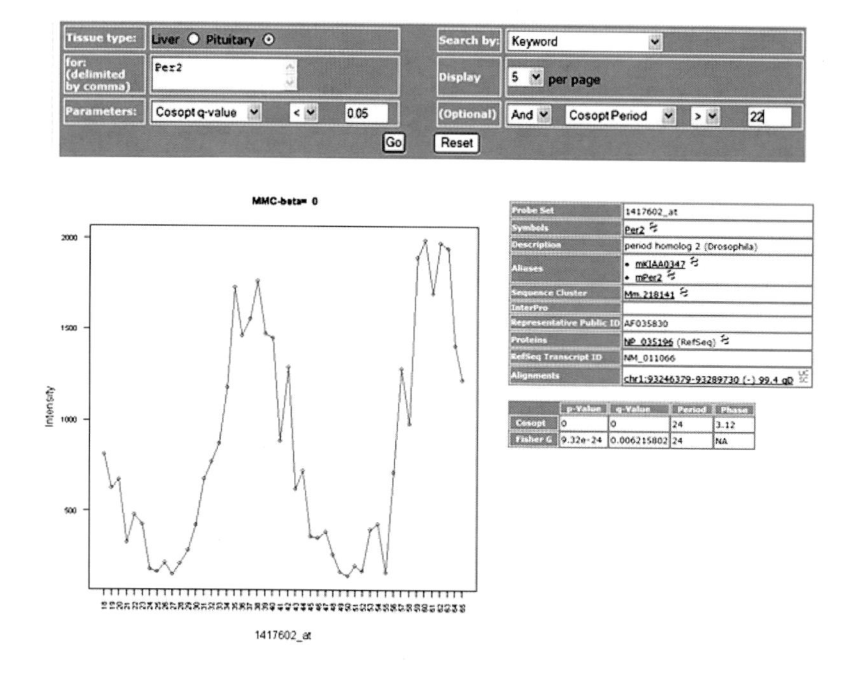

Figure 5. "Wasabi" is a Web-based resource available to the circadian community. The high-resolution microarray data set is available to the public at http://wasabi.itmat.upenn.edu/circa/mouse. This interface permits the user to search for any transcript or probe set using filters based on p and q values as well as period and phase (*top*). Data are displayed as a plot of intensity versus time (*bottom left*), and information on the gene as well as COSOPT and Fisher's G-test statistics is supplied (*bottom right*).

the circadian clock and its role in regulating rhythmic physiology (Etter and Ramaswami 2002; Duffield 2003; Hastings et al. 2003). High-throughput gene expression analysis enabled by DNA arrays and informatics tools can help to bridge this gap. In its simplest form, this analysis can consist of identifying a rate-limiting enzyme in a biochemical pathway as a target of the clock, inferring that the pathway is clock-regulated, and testing the hypothesis. Even more power can be generated by analyzing groups (rather than single) of transcripts for their coherent action in cellular pathways. For example, we can use informatics technologies to infer the physiological function of subsets of rhythmically expressed genes. We have used Ingenuity pathway analysis (IPA) to identify networks of interrelated circadian genes in the pituitary (Fig. 6). As expected, the highest confidence network identified (14 genes, $p < 10^{-8}$) focused on the canonical circadian clock (Fig. 6, top). In this network, BMAL1 / ARNTL and Clock form a pair of central nodes that link together familiar components of the clock, including *Per*, *Cry*, and *Rev-erb* genes. The similarity of this network to conventional models of clock mechanics suggests that IPA is a useful tool for identifying functionally related genes.

Interestingly, another significant network identified by IPA ($p < 10^{-4}$) seems to pivot around genes involved in aldehyde metabolism (Fig. 6, bottom). Of the 17 aldehyde dehydrogenase genes in mice, 5 show significant rhythmicity in the pituitary (Fig. 6, bottom). Previous work has established the role of the hypothalamic-pituitary-adrenal (HPA) axis in modulating alcohol sensitivity (Gianoulakis 1998). Moreover, considerable efforts have been made to document circadian variation in the organismal response to alcohol (Wasielewski and Holloway 2001). However, to the best of our knowledge, no study has addressed the role of ALDH (aldehyde dehydrogenase) genes in the pituitary let alone their role in modulating circadian rhythms of alcohol metabolism. Moreover, several members of the ALDH family have been shown to have important roles in the metabolism of GABA (γ-amino-*n*-butyric acid) (Vasiliou et al. 2004), an inhibitory neurotransmitter known to have a role in susceptibility to alcoholism (Morrow et al. 2006). On the basis of this ingenuity network analysis, we suggest that the circadian regulation of ALDH genes in the pituitary may be an important component of the organismal response to alcohol. Thus, the combined use of both informatics tools and the Wasabi database may prove to be a valuable strategy for generating testable hypotheses in subsequent studies of circadian clock output and its relationship to clock-regulated physiology.

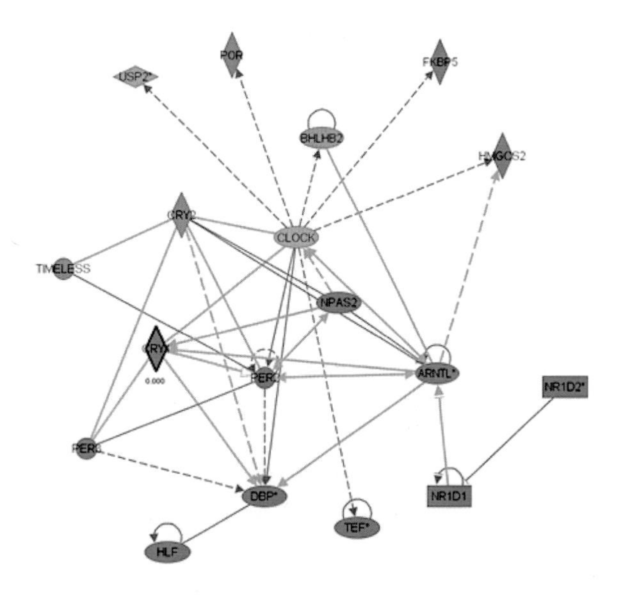

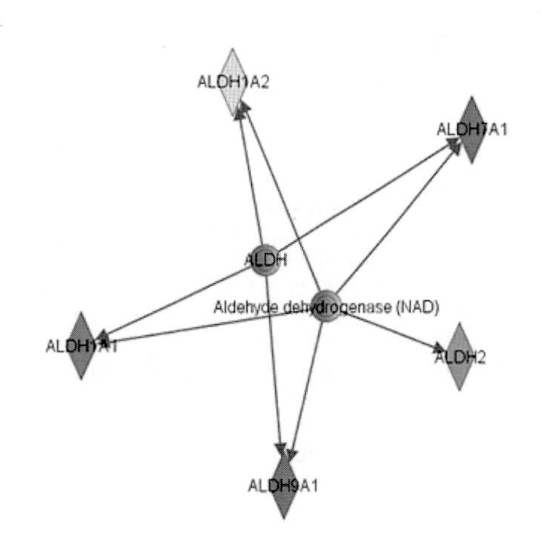

Figure 6. Network analysis of circadian transcripts suggests a putative role in cell cycle regulation. Ingenuity pathway analysis (IPA) was used to identify genetic and biochemical interactions between cycling transcripts in the pituitary. Genes involved in aldehyde metabolism cell cycle regulation were among the most highly represented group identified by IPA. Labeled polygons represent gene products and lines represent functional interactions.

CONCLUSIONS

Recently, more than 70 studies have used microarray analysis to identify genes whose transcription is under circadian control. To extend on this work, we have analyzed RNA samples isolated from the pituitary of wild-type mice every hour over the course of 2 complete days. By increasing the frequency of sampling to once every hour, more than 250 genes in the pituitary were identified as circadianly regulated. Simulations based on this data set indicate that the power of conventional algorithms for detecting rhythmic transcripts is dramatically improved by increasing the sampling density of transcriptional profiling studies. Consequently, we recommend that future studies sample tissues at least once every 2 hours for 48 hours as the best balance in managing costs and robustly identifying circadian genes. Software tools to analyze circadian data sets and the results of this transcriptional profiling study have been made available to the circadian community in a fully searchable Web resource: Wasabi.

ACKNOWLEDGMENTS

We thank members of the Hogenesch and Panda laboratories for helpful comments through the execution of these experiments and the preparation of the manuscript. The microarray facilities at Scripps Research Institute and the University of Pennsylvania were indispensable for the collection and analysis of these data. J.B.H. is supported by the National Institute of Neurological Disease and Stroke (1R01NS054794), and the National Institute of Mental Health (P50 MH074924-01, awarded to Joseph S. Takahashi, Northwestern University). S.P. is supported by the Pew Scholars Program in Biomedical Science and the Whitehall Foundation.

REFERENCES

Akhtar R.A., Reddy A.B., Maywood E.S., Clayton J.D., King V.M., Smith A.G., Gant T.W., Hastings M.H., and Kyriacou C.P. 2002. Circadian cycling of the mouse liver transcriptome, as revealed by cDNA microarray, is driven by the suprachiasmatic nucleus. *Curr. Biol.* **12:** 540.

Antoch M.P., Kondratov R.V., and Takahashi J.S. 2005. Circadian clock genes as modulators of sensitivity to genotoxic stress. *Cell Cycle* **4:** 901.

Ceriani M.F., Hogenesch J.B., Yanovsky M., Panda S., Straume M., and Kay S.A. 2002. Genome-wide expression analysis in *Drosophila* reveals genes controlling circadian behavior. *J. Neurosci.* **22:** 9305.

Cermakian N. and Sassone-Corsi P. 2002. Environmental stimulus perception and control of circadian clocks. *Curr. Opin. Neurobiol.* **12:** 359.

Claridge-Chang A., Wijnen H., Naef F., Boothroyd C., Rajewsky N., and Young M.W. 2001. Circadian regulation of gene expression systems in the *Drosophila* head. *Neuron* **32:** 657.

Curtis A.M. and Fitzgerald G.A. 2006. Central and peripheral clocks in cardiovascular and metabolic function. *Ann. Med.* **38:** 552.

Duffield G.E. 2003. DNA microarray analyses of circadian timing: The genomic basis of biological time. *J. Neuroendocrinol.* **15:** 991.

Duffield G.E., Best J.D., Meurers B.H., Bittner A., Loros J.J., and Dunlap J.C. 2002. Circadian programs of transcriptional activation, signaling, and protein turnover revealed by microarray analysis of mammalian cells. *Curr. Biol.* **12:** 551.

Etter P.D. and Ramaswami M. 2002. The ups and downs of daily life: Profiling circadian gene expression in *Drosophila*. *Bioessays* **24:** 494.

Gianoulakis C. 1998. Alcohol-seeking behavior: The roles of the hypothalamic-pituitary-adrenal axis and the endogenous opioid system. *Alcohol Health Res. World* **22:** 202.

Halberg F., Cornelissen G., Ulmer W., Blank M., Hrushesky W., Wood P., Singh R.K., and Wang Z. 2006. Cancer chronomics III. Chronomics for cancer, aging, melatonin and experimental therapeutics researchers. *J. Exp. Ther. Oncol.* **6:** 73.

Harmer S.L., Hogenesch J.B., Straume M., Chang H.S., Han B., Zhu T., Wang X., Kreps J.A., and Kay S.A. 2000. Orchestrated transcription of key pathways in *Arabidopsis* by the circadian clock. *Science* **290:** 2110.

Hastings M.H., Reddy A.B., and Maywood E.S. 2003. A clockwork web: Circadian timing in brain and periphery, in health and disease. *Nat. Rev.* **4:** 649.

Hayes K.R., Baggs J.E., and Hogenesch J.B. 2005. Circadian clocks are seeing the systems biology light. *Genome Biol.* **6:** 219.

Klerman E.B. 2005. Clinical aspects of human circadian rhythms. *J. Biol. Rhythms* **20:** 375.

Ko C.H. and Takahashi J.S. 2006. Molecular components of the mammalian circadian clock. *Hum. Mol. Genet.* (spec. no. 2) **15:** R271.

Kornmann B., Schaad O., Bujard H., Takahashi J.S., and Schibler U. 2007. System-driven and oscillator-dependent circadian transcription in mice with a conditionally active liver clock. *PLoS Biol.* **5:** e34.

Levi F. and Schibler U. 2007. Circadian rhythms: Mechanisms and therapeutic implications. *Annu. Rev. Pharmacol. Toxicol.* **47:** 593.

Lin Y., Han M., Shimada B., Wang L., Gibler T.M., Amarakone A., Awad T.A., Stormo G.D., Van Gelder R.N., and Taghert P.H. 2002. Influence of the period-dependent circadian clock on diurnal, circadian, and aperiodic gene expression in *Drosophila melanogaster*. *Proc. Natl. Acad. Sci.* **99:** 9562.

Lis C.G., Grutsch J.F., Wood P., You M., Rich I., and Hrushesky W.J. 2003. Circadian timing in cancer treatment: The biological foundation for an integrative approach. *Integr. Cancer Ther.* **2:** 105.

McDonald M.J. and Rosbash M. 2001. Microarray analysis and organization of circadian gene expression in *Drosophila*. *Cell* **107:** 567.

Morrow A.L., Porcu P., Boyd K.N., and Grant K.A. 2006. Hypothalamic-pituitary-adrenal axis modulation of GABAergic neuroactive steroids influences ethanol sensitivity and drinking behavior. *Dialogues Clin. Neurosci.* **8:** 463.

Panda S., Antoch M.P., Miller B.H., Su A.I., Schook A.B., Straume M., Schultz P.G., Kay S.A., Takahashi J.S., and Hogenesch J.B. 2002. Coordinated transcription of key pathways in the mouse by the circadian clock. *Cell* **109:** 307.

Schibler U. and Sassone-Corsi P. 2002. A web of circadian pacemakers. *Cell* **111:** 919.

Storch K.F., Lipan O., Leykin I., Viswanathan N., Davis F.C., Wong W.H., and Weitz C.J. 2002. Extensive and divergent circadian gene expression in liver and heart. *Nature* **417:** 78.

Stratmann M. and Schibler U. 2006. Properties, entrainment, and physiological functions of mammalian peripheral oscillators. *J. Biol. Rhythms* **21:** 494.

Su A.I., Wiltshire T., Batalov S., Lapp H., Ching K.A., Block D., Zhang J., Soden R., Hayakawa M., Kreiman G., Cooke M.P., Walker J.R., and Hogenesch J.B. 2004. A gene atlas of the mouse and human protein-encoding transcriptomes. *Proc. Natl. Acad. Sci.* **101:** 6062.

Ueda H.R., Matsumoto A., Kawamura M., Iino M., Tanimura T., and Hashimoto S. 2002. Genome-wide transcriptional orchestration of circadian rhythms in *Drosophila*. *J. Biol. Chem.* **277:** 14048.

Vasiliou V., Pappa A., and Estey T. 2004. Role of human aldehyde dehydrogenases in endobiotic and xenobiotic metabolism. *Drug Metab. Rev.* **36:** 279.

Walker J.R. and Hogenesch J.B. 2005. RNA profiling in circadian biology. *Methods Enzymol.* **393:** 366.

Wasielewski J.A. and Holloway F.A. 2001. Alcohol's interactions with circadian rhythms. A focus on body temperature. *Alcohol Res. Health* **25:** 94.

Nuclear Receptors, Metabolism, and the Circadian Clock

X. YANG,* K.A. LAMIA,* R.M. EVANS
Gene Expression Laboratory, Howard Hughes Medical Institute, The Salk Institute for Biological Studies, La Jolla, California 92037

As ligand-dependent transcription factors, the nuclear receptor superfamily governs a remarkable array of rhythmic physiologic processes such as metabolism and reproduction. To provide a "molecular blueprint" for nuclear receptor function in circadian biology, we established a diurnal expression profile of all mouse nuclear receptors in critical metabolic tissues. Our finding of broad expression and tissue-specific oscillation of nuclear receptors along with their key target genes suggests that diurnal nuclear receptor expression may contribute to established rhythms in metabolic physiology and that nuclear receptors may be involved in coupling peripheral circadian clocks to divergent metabolic outputs. Conversely, nuclear receptors may serve peripheral clock input pathways, integrating signals from the light-sensing central clock in the suprachiasmatic nucleus and other environmental cues, such as nutrients and xenobiotics. Interplay between the core circadian clock and nuclear receptors may define a large-scale signaling network that links biological timing to metabolic physiology.

INTRODUCTION

The prototypes of the nuclear receptor (NR) superfamily were identified as the steroid receptors that mediate gene transcription in response to steroid hormone signaling (Evans 1988). To date, a total of 48 human NR genes have been identified, including classic endocrine receptors for steroid hormones, thyroid hormones, and Vitamin A and D derivatives, and a large number of orphan receptors whose ligands and physiological functions were initially unknown (Giguere et al. 1988; Mangelsdorf et al. 1995; Giguere 1999). The past decade has witnessed stunning advances in orphan receptor research, largely owing to the identification of dietary lipids and metabolites as the ligands for a number of orphan receptors and establishing these adopted orphan receptors as lipid sensors that activate transcriptional programs for metabolic homeostasis (Chawla et al. 2001). For approximately half of the NR superfamily members, no ligands have yet been identified. However, their diverse roles in development, reproduction, and general metabolism have begun to be uncovered.

Many aspects of mammalian physiology are robustly rhythmic. Among these rhythmic phenomena, reproductive physiology, glucose and lipid homeostasis, and toxin clearance broadly depend on hormones and metabolites that serve as ligands for NRs. For example, periodic variations in estrogen and progesterone levels drive female menstrual cycles, whereas a morning surge in cortisol boosts energy production. A large array of metabolites, such as glucose, free fatty acids, and cholesterol and bile acids, also exhibit daily fluctuation, which is thought to be controlled by the intrinsic circadian timing system and in turn may regulate metabolic rhythms through adopted orphan receptors (Back et al. 1969; Yang et al. 2006).

The mammalian circadian timing system comprises a central pacemaker in the suprachiasmatic nucleus (SCN) of the hypothalamus and numerous peripheral tissue oscillators. The central clock is directly entrained by light from the retina via the retinohypothalamic tract, whereas the peripheral oscillators can be synchronized either by neuronal and hormonal signals from the central clock or by other environmental cues such as daily feeding/fasting or activity/rest cycles (Kohsaka and Bass 2007; Levi and Schibler 2007). This hierarchy of circadian clocks coordinates daily cycles of physiology and behavior, allowing animals to adapt to predictable changes in the environment, likely promoting fitness, health, and longevity of the organism.

In this chapter, we discuss molecular and functional links between nuclear receptors and circadian clocks, with emphasis on implications for metabolic physiology (Fig. 1).

NRs WITHIN THE CORE CLOCKWORK

Circadian clocks are self-sustained, cell-autonomous molecular oscillators. The current view of the clockwork is two interlocked transcriptional/posttranslational feedback loops comprising a battery of transcriptional activators and repressors (Hardin 2006; Ko and Takahashi 2006). A heterodimeric complex of BMAL1 and either CLOCK or NPAS2 activate the transcription of *Period* genes (*Per1, Per2,* and *Per3*) and *Cryptochrome* genes (*Cry1 and Cry2*) by recognizing E-box *cis*-regulatory elements in their promoters. Upon accumulating to a critical concentration, PER and CRY move to the nucleus and inhibit the transcription of their own genes by blocking BMAL1-CLOCK/NPAS2 activity. *Clock* expression is generally constant, whereas rhythmic transcription of *Bmal1* is driven by a second feedback loop, involving the orphan nuclear receptors RORα and REV-ERBα (Preitner et al. 2002; Sato et al. 2004).

The ROR (α, β, γ) and REV-ERB (α and β) proteins represent closely related families of NRs that recognize similar *cis* response elements (ROREs) on target genes (Forman et al. 1994). RORs act as constitutive transcriptional activators, whereas REV-ERBs are constitutive

* These authors contributed equally to this work.

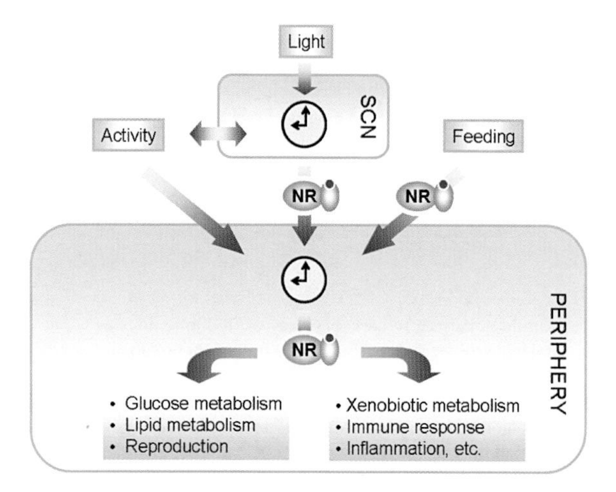

Figure 1. Potential roles of nuclear receptors in the circadian timing system. The central clock in the suprachiasmatic nucleus (SCN) is directly entrained by light to the solar time. The clocks in peripheral tissues can be synchronized either by neuronal and hormonal signals from the central clock or by other environmental cues such as daily feeding and activity cycles. NRs may function in multiple layers of this circadian clock hierarchy. Some NRs such as RORα and REV-ERBα are components of the core clockwork. By sensing fat-soluble hormones, vitamins, and dietary lipids in the circulation, NRs may transmit circadian signals from the central clock and/or the feeding cycle into the peripheral clocks. NRs may also act as circadian output factors to drive rhythmic physiologic processes including metabolism, immune response, and reproduction.

repressors. As a result, RORα and REV-ERBβ regulate *Bmal1* transcription in an opposing manner by competitively binding to ROREs in its promoter (Preitner et al. 2002; Sato et al. 2004; Akashi and Takumi 2005). Alternate action of these two NRs over each light/dark cycle leads to rhythmic expression of *Bmal1*. Closing the feedback loop, BMAL1-CLOCK directly regulates the transcription of *Rev-erbα* via E-box elements in its promoter (Preitner et al. 2002). The promoter of the *Rev-erbα* gene also contains a functional RORE through which it is repressed by itself and induced by RORα (Delerive et al. 2002; Raspe et al. 2002). This additional layer of regulation may ensure antiphase expression of *RORα* and *Rev-erbα*, thus enhancing the precision, robustness, or sustainability of the clock.

The functions of closely related REV-ERBβ, RORβ, and RORγ isoforms are not well understood. We found that expression of *Rev-erbβ* mRNA is similar to that of *Rev-erbα* in both tissue distribution and temporal profile (Yang et al. 2006). Although REV-ERBα has been more intensively studied with regard to circadian rhythms and the mechanisms by which it can repress *Bmal1* transcription have been described in some molecular detail (Yin and Lazar 2005), it has been reported that REV-ERBβ is capable of repressing *Bmal1* transcription with a strength similar to that of REV-ERBα (Guillaumond et al. 2005). Thus, REV-ERBβ may contribute to *Bmal1* expression in a manner similar and partially redundant to REVERBα. Mice with a null mutation of the *Rev-erbα* locus were found to have a slightly shorter free-running period in constant darkness (DD) and increased phase shifts in response to light pulses compared to wild-type controls (Preitner et al. 2002). Perhaps mice lacking both *Rev-erbα* and *Rev-erbβ* function would display a stronger disruption of circadian function. Further studies on *Rev-erbα*- and *Rev-erbβ*-deficient mouse models would illuminate the separate and distinct roles of the REV-ERB family members in the functioning of the circadian oscillator.

In contrast to the uniform diurnal expression patterns of *Rev-erbα* and *Rev-erbβ* across tissues, *Rorα*, β, and γ isoforms exhibit diverse temporal and spatial expression patterns (Akashi and Takumi 2005; Guillaumond et al. 2005; Bookout et al. 2006; Yang et al. 2006). Although widely expressed, *Rorα* transcripts robustly cycle in the SCN but are seemingly arrhythmic or display low-amplitude diurnal rhythms of expression in the four peripheral tissues that we examined (Ueda et al. 2002; Yang et al. 2006). *Staggerer* mice, which lack functional RORα, have a decreased period of free-running locomotor activity in DD and slightly reduced *Bmal1* expression in the SCN (Sato et al. 2004; Akashi and Takumi 2005). The expression of other core clock genes is unaffected. This modest phenotype may be due to partial compensation by RORβ, which is also highly expressed in the SCN with a circadian rhythm of expression similar to that of the *Period* gene transcripts (Sumi et al. 2002). Genetic disruption of *Rorβ* increases the free-running period in mice (Andre et al. 1998; Masana et al. 2007). *Rorγ* mRNA is highly expressed in the periphery and cycles in selective tissues (Yang et al. 2006). Mice lacking functional RORγ have not been described.

The nuclear receptor coactivator PGC1α, which was originally cloned as a cofactor of the PPAR family of NRs and has been shown to bind and regulate the activity of multiple NRs (Puigserver et al. 1998; Knutti et al. 2000), was recently shown to coactivate RORα and RORγ on the *Bmal1* promoter RORE (Liu et al. 2007). PGC1α$^{-/-}$ mice seem to have a slightly increased free-running period in DD, although it is unclear whether this result is statistically significant (Liu et al. 2007). In contrast, RORα mutant *Staggerer* mice have a slightly decreased free-running period of locomotor activity, suggesting that a mild effect of PGC1α loss on free-running period would not be due primarily to loss of coactivation of RORα but may result from the loss of coactivation of RORβ, RORγ, or other NRs. PGC1β$^{-/-}$ mice exhibit decreased nocturnal locomotor activity under light/dark conditions, which may be due to a disturbance of the circadian system (Sonoda et al. 2006).

An earlier report suggested a role for the nuclear receptor Ear2 (also known as COUP-TFIII) in the maintenance of circadian rhythmicity because *Ear2*$^{-/-}$ mice were found to have increased error in the timing of locomotor activity onsets and their behavioral rhythms degraded in response to lower intensities of constant light (LL) compared to wild-type controls (Warnecke et al. 2005). However, the same animals exhibited distorted architecture of several neuronal structures, raising the concern that their disrupted circadian behaviors may be due to developmental defects rather than to disruption of the circadian clock in adult animals. Conditional deletion of Ear2/COUP-TFIII in adult neurons would resolve this question. In our study, transcripts encoding all three members of the COUP-TF family of NRs were expressed arrhythmically in all tissues examined (Yang et al. 2006).

Transcripts encoding the highly related nuclear receptors NGFI-B, NOR1 and NURR1 (also known as NR4A1, NR4A2, and NR4A3) displayed striking high-amplitude diurnal patterns of expression in most tissues examined in our study. The *Ngfi-b* transcript was also reported to be strongly induced by light in the hamster SCN (Morris et al. 1998), which led to the hypothesis that NGFI-B may have a role in either light entrainment or maintenance of circadian rhythmicity. However, *Ngfi-b*$^{-/-}$ mice did not exhibit any disruption of circadian locomotor behavior under constant conditions and displayed normal behavioral phase shifting in response to a variety of light stimuli (Kilduff et al. 1998). Because there are three highly homologous receptors in the NR4A family and they have similar diurnal rhythms of expression, they are likely redundant for some functions, possibly including circadian clock entrainment. The function of the ubiquitous, strongly diurnal expression of the NR4A receptors remains unexplained.

NRs MAY MEDIATE ENTRAINMENT OF PERIPHERAL CLOCKS BY METABOLIC SIGNALS

Elegant experiments within the last decade established that fasting and feeding patterns are the primary determinant of the timing of peripheral circadian clocks (Damiola et al. 2000; Stokkan et al. 2001). However, the mechanisms by which feeding time sets peripheral clocks remain obscure and may include both neural and humoral components. Circulating factors that are involved in the entrainment of peripheral clocks by feeding time would be expected to display robust diurnal rhythms in the circulation that are altered by changing the time of restricted feeding. Many NR ligands meet these criteria, including glucocorticoids, triiodothyronine (T3), thyroxine (T4), retinoic acid, and dietary lipids. One or many of such molecules may determine the timing of all peripheral clocks; alternatively, unique factors or combinations of factors, possibly including one or more NR ligands, may entrain various peripheral organ clocks.

Glucocorticoids

Circulating glucocorticoids display strong diurnal rhythms, and glucocorticoids have been shown to induce the expression of circadian transcripts in cultured cells (Balsalobre et al. 2000a,b), which made these steroid hormones attractive candidates for peripheral entrainment cues. However, recent evidence argues against their role as a major time cue for peripheral clocks (Le Minh et al. 2001). Similar detailed analysis of the potential for other NR ligands to entrain peripheral circadian clocks has not yet been done and would be required to make conclusions about their physiological roles in peripheral clock entrainment.

Thyroid Hormones

Thyroid stimulating hormone (TSH), T3, and T4 are robustly rhythmic in the circulation in rodents, and the circadian rhythm of TSH has been shown to be driven by circadian clock function in human subjects (Allan and Czeisler 1994). Furthermore, the phases of T3 and T4 rhythms are completely reversed by reversed-phase restricted feeding (Ahima et al. 1998). Finally, thyroidectomy alters the diurnal expression of clock genes outside of the SCN (Amir and Robinson 2006).

Retinoic Acid

Retinoic acid is the ligand for heterodimeric NR complexes containing RAR and RXR. RAR and RXR associate with CLOCK and NPAS2 (also known as MOP4) in a hormone-dependent fashion (McNamara et al. 2001). The same study also found that retinoic acid could inhibit CLOCK-mediated transcription both in cultured muscle cells and in cardiovascular organs from intact animals. Moderate phase shifts of peripheral clocks in the heart and aorta were observed after injection of retinoic acid. An independent screen tested 299 peptides and bioactive lipids for the ability to mediate entrainment of circadian rhythms in cultured fibroblasts expressing *Per2*-luciferase. Of the 12 targets active in their screen, 3 were all *trans*-retinoic acid, 9-*cis* retinoic acid, and 13-*cis* retinoic acid (Nakahata et al. 2006). Moreover, there is evidence that retinoic acid signaling might be involved in light sensing and the central clock resetting (Thompson et al. 2004; Fu et al. 2005).

PPAR Ligands

The intestinal synthesis and degradation of oleylethanolamide (OEA) occur diurnally and are regulated by food intake (Fu et al. 2003, 2007). OEA is an endogenous ligand for PPARα and inhibits appetite via PPARα activation (Fu et al. 2003), presumably at hypothalamic sites. The timing and regulation of OEA synthesis and degradation are consistent with a potential role in the entrainment of peripheral circadian clocks. PPARα regulates the transcription of both *Bmal1* and *Rev-erbα* via PPRE *cis*-regulatory elements in their promoters (Canaple et al. 2006), suggesting a potential mechanism of contribution to peripheral clock entrainment. In addition, the endogenous PPARγ ligand 15-deoxy-Delta12,14-prostaglandin J2 was another of the 12 candidates found to entrain circadian rhythms in cultured fibroblasts by real-time monitoring of *Per2*-luciferase rhythms, further emphasizing the potential of dietary responsive NRs to influence the clock (Nakahata et al. 2006).

Non-NR Ligands

It is also possible that NRs participate in the entrainment of peripheral clocks in response to nutrient-derived signals that are not direct NR ligands such as glucose and insulin. Degradation of REV-ERBα protein is blocked by GSK3-mediated phosphorylation (Yin et al. 2006). Because

GSK3 activity is inhibited in response to acute bouts of feeding by insulin-stimulated AKT-mediated phosphorylation, degradation of REV-ERBα in response to feeding may contribute to the entrainment of peripheral clocks.

NRs AS CIRCADIAN EFFECTORS OF METABOLISM

Our survey of the diurnal expression profile of all 49 mouse nuclear receptors reveals wide expression and tissue-specific oscillation of NRs in a variety of metabolic tissues (Yang et al. 2006). From this analysis, we are not able to determine which cycling NR transcripts are driven directly by peripheral circadian clocks and which are responsive to physiological rhythms or to secondary clock-driven transcription factors. Nor did we measure NR protein levels and are thus unable to say which NRs are rhythmic at the protein level. However, there is evidence that several NR transcripts are directly regulated by CLOCK/BMAL1, and a few NR protein rhythms have been described that closely follow the rhythmic expression of the corresponding transcripts. The dynamic and coordinated changes in NR expression along with expression of their key target genes suggest that NRs may contribute to the regulation of divergent metabolic readouts by peripheral circadian clocks.

The NR transcripts most likely to be directly regulated by peripheral circadian clocks are those with peak expression at ZT 4 or ZT 8 in peripheral metabolic tissues, when CLOCK/BMAL1 heterodimers are most active in those tissues. The rhythmic expression of NR transcripts that oscillate in a tissue-specific manner may be driven by CLOCK-BMAL1 but may require additional cofactors that are expressed in a tissue-specific manner. We discuss a few examples of NRs that appear most likely to be directly controlled by peripheral circadian clocks and that would be expected to have a large impact on metabolic physiology, thus potentially linking peripheral circadian clocks to physiological rhythms.

PPARs

The roles of the PPAR family of transcription factors in various aspects of mammalian metabolic physiology are well established (Lee et al. 2003; Evans et al. 2004). Although the amplitudes of their diurnal expression patterns in some of the tissues that we examined are low, the rhythmicity of at least PPARα is likely to be physiologically relevant. Both the PPARα transcript and protein oscillate in mouse liver with a time of peak expression consistent with regulation by local circadian clocks (Lemberger et al. 1996). *Pparα* transcription can be directly regulated by CLOCK and BMAL1 in vitro and in the liver and intestine in vivo (Inoue et al. 2005; Oishi et al. 2005; Canaple et al. 2006). As discussed above, the endogenous PPARα ligand OEA is synthesized diurnally in the gut epithelium, probably in response to nutrient availability. Perhaps the rhythmic expression of PPARα in the liver synergizes with OEA production to increase the amplitude of PPARα activity on its target promoters. This mechanism may also have a role in the diurnal regulation of appetite, as OEA activation of PPARα in the central nervous system (CNS) is a satiety signal. We did not measure the transcription of NRs in the CNS, so it is not clear whether a similar amplification of the signal is likely there.

CAR

The so-called constitutive androstane receptor (CAR) would be an interesting candidate for circadian regulation as it is a potent modulator of xenobiotic metabolism (Qatanani and Moore 2005). A recent elegant study showed that transcription of *CAR* is regulated by the PARbZIP family of transcription factors, including DBP, HLF, and TEF, which in turn are regulated by CLOCK:BMAL1-dependent transcription and indeed are among the transcripts with the highest amplitude of circadian transcription in many organs (Gachon et al. 2006). In control animals, *CAR* expression peaks in the early night (ZT 12), whereas compound null mutations of *Dbp*, *Hlf*, and *Tef* result in loss of diurnal expression of *CAR* and its target genes, including many cytochrome family enzymes involved in the clearance of exogenous compounds by the liver. Physiologically, $Dbp^{-/-}$; $Tef^{-/-}$; $Hlf^{-/-}$ animals have increased liver weight and reduced tolerance for anesthetic and chemotherapeutic agents. The combined diurnal expression of CAR with peak expression during the night phase when mice are actively ingesting food and diurnal sensitivity to exogenous agents that are both lost upon loss of the PARbZIP transcription factors suggests that these proteins have a critical role in optimizing circadian timing of toxin clearance to the phase of food ingestion.

SHP

We found that the *Shp* (small heterodimeric partner) transcript is robustly rhythmic in the liver but is not detectably expressed in adipose tissues or skeletal muscle (Yang et al. 2006). The SHP promoter contains E boxes and can be directly activated by CLOCK and BMAL1 (Oiwa et al. 2007). Furthermore, the ability of CLOCK and BMAL1 to drive transcription from the *Shp* promoter was found to be increased more than fivefold by coexpression of the nuclear receptor LRH-1 (Oiwa et al. 2007), which we found to be highly expressed in the liver compared to the other tissues examined, probably accounting for the tissue-specific oscillation of *Shp* transcription that we observed.

SHP is an orphan nuclear receptor that dimerizes with other NRs, including LXRs, FXR, and PXR, and represses their activities. Multiple NR partners of SHP have been shown to regulate the transcription of *Cyp7a1* and *Cyp8b2*, among other transcripts involved in metabolizing dietary lipids and exogenous toxins (Schoonjans and Auwerx 2002). The loss of SHP function alone is sufficient to significantly increase the expression of *Cyp7a1* and *Cyp8b2* in the liver. Furthermore, when $Shp^{-/-}$ mice are fed a diet high in cholesterol and/or cholic acid, their expression of detoxifying enzymes is strikingly higher than in control animals under similar conditions and they avoid the hepatic steatosis suffered by control animals (Wang et al. 2003). Taken together, these results imply that the liver-specific rhythmic expression of *Shp* is probably driven by local circadian clocks and is expected to drive rhythmic repression of multiple NRs, thus contributing to daily rhythms in clearance of excess dietary lipids and other toxins.

COMPLEXITY AND SPECIFICITY OF CIRCADIAN REGULATION BY NRs

In metabolic pathways, certain circadian responsive target genes can be subject to direct regulation by multiple NRs and other factors. For example, evidence suggests that the *Cyp7a1* gene is regulated by at least six NRs: LXRα, LRH-1, FXRα, PXR, RXRα, and SHP (Chawla et al. 2000; Schoonjans and Auwerx 2002). Conversely, individual NRs participate in gene regulation in diverse cellular pathways. For example, PPARγ in adipocytes regulates genes involved in lipogenesis and lipid storage, glucose uptake, energy expenditure, and adipokine production (Lee et al. 2003; Evans et al. 2004). Furthermore, cross-talk within the NR superfamily has been widely documented. In the liver, the *Shp* gene can be regulated by FXRα, LRH-1, and ERRγ (Lu et al. 2000; Sanyal et al. 2002), whereas the *Rev-erbα* gene can be regulated by RORα, PPARα, PPARγ, and LXR (Gervois et al. 1999; Raspe et al. 2002; Fontaine et al. 2003). These NRs also interact functionally with other classes of transcriptional regulators such as SREBP-1c (Repa et al. 2000; Chen et al. 2004). Taken together, we propose that a large pool of NRs in any given peripheral tissue comprises an interlaced transcriptional network that coordinates multiple metabolic pathways in response to circadian and other cues.

In this complex regulatory network, how does the oscillation of an array of NRs give rise to specific rhythmic outputs? Several mechanisms could be involved. First, as suggested by our results and others (Panda et al. 2002), NRs appear to preferentially target rate-limiting genes in a metabolic pathway for circadian regulation. Second, for a battery of NRs that target a single gene, cycling of a minimal number of nuclear receptors seems to be sufficient to boost the oscillation of the target gene. This notion is supported by our observation that, among six nuclear receptors that are known to regulate *Cyp7a1*, only SHP is rhythmically expressed and appears to be the primary contributor to *Cyp7a1* cycling. Third, for NRs that act as heterodimers, cyclic expression of one subunit may be sufficient for periodic changes in the heterodimer activity. Because RXRs serve as the partners for many other NRs (Mangelsdorf and Evans 1995), any dramatic changes in their levels would be detrimental. Indeed, we found that all RXR subtypes are continually expressed at fairly constant levels. In the absence of the cyclical induction of an RXR ligand, this places the critical regulation on the partner and/or its partner ligand; here, the PPARs may be one example.

MULTIPLE LOOPS BETWEEN NRs AND THE CORE CLOCK

In the above sections, we have depicted a scheme in which various NRs may be involved in circadian input and output pathways, as well as in the core clock mechanism. Feedback regulation is implicated at many points in circadian signal transduction. Not only does the core clock involve double interlocked feedback loops (Glossop et al. 1999; Shearman et al. 2000), but this type of feedback may also regulate communication between the core clockwork and input/output pathways. For example, PPARα has been proposed to serve the input pathway by promoting *Bmal1* and *Rev-erbα* expression (Gervois et al. 1999; Canaple et al. 2006). Conversely, BMAL1 and CLOCK directly regulate *Pparα* transcription (Inoue et al. 2005; Oishi et al. 2005; Canaple et al. 2006). Xenobiotic metabolism is an important output of the circadian clock and xenobiotic compounds may, in turn, act as NR ligands to alter core clock gene expression (Claudel et al. 2007). We postulate that NRs and the core clock components are integrated into multiple feedback loops, which constitute a large-scale signaling network.

The signaling network connecting NRs and circadian clocks may serve a number of functions. First, as the double loops in the core clock are likely vulnerable to stochastic perturbation, additional loops involving NRs may improve the precision and robustness of peripheral clocks' 24-hour oscillations. In addition, this NR-clock network may sense a broad range of external cues, such as light, diet, and stress, by hormonal mechanisms, and integrate multiple input signals. In addition, NRs in this network regulate a variety of outputs that may link the core clock to diverse molecular genetic programs, thus orchestrating metabolism and physiology over the light/dark cycle. Finally, nearly half of the NR superfamily, including members of the ROR, REV-ERB, COUPTF, NGFI-B, RAR, RXR, TR, GR, PPAR, and ERR subfamilies, are ubiquitously expressed in all tested tissues (Bookout et al. 2006); other NRs are expressed either constitutively or diurnally in a tissue-specific manner. By regulating tissue-specific transcriptional programs, those NRs may create local versions of the circadian network.

CONCLUDING REMARKS

During the last decade, remarkable progress has been made in uncovering the molecular basis of the core circadian clockwork. However, we are just beginning to understand its biological inputs and outputs at a level ranging from gene expression to physiology and metabolism. In this chapter, we discussed evidence suggesting that the NR superfamily might constitute critical signaling cascades involved in coupling the circadian clock to divergent physiological outputs as well as in the clock entrainment by various zeitgebers. We have proposed that a conserved large-scale circadian network may emerge from feedback regulation between the NR and core clock genes. However, the existence and architecture of such NR-clock signaling network have yet to be determined. Although rhythmic expression is prevalent in the NR superfamily, it remains unknown which NRs (if any) are involved in peripheral clock resetting, which are solely responsive to the ticking output of the clock and which can manifest a diurnal rhythm as a result of other oscillatory cues.

Efficacy of NR signaling depends on the availability of receptors as well as their ligands. Hormonal ligands are usually produced in specific tissues such as the thyroid and adrenal glands and are delivered to target tissues via the circulation, although some active ligands can be generated in local tissues. Important questions yet to be answered include how the circadian clock affects endocrine and locally produced autocrine ligands and how ligand cycling is correlated with receptor cycling in regulation of rhythmic physiological processes. Furthermore, it remains elusive whether feeding/fasting cycles give rise to cyclic accumulation of dietary lipids and metabolites in the body, which would serve as ligands to stimulate the activity of adopted orphan receptors in a diurnal manner, thereby inducing cyclic expression of their target genes.

Understanding the mechanisms by which cell-autonomous clocks are in sync at the tissue, organ, and system levels is a major challenge in the field of circadian biology. A plausible model is that, by sensing light, the central clock dictates rhythmic activity of the endocrine system which in turn entrains whole-body physiology to the light/dark cycle. Conditions of restricted feeding may overcome such SCN-driven entrainment by directly altering some endocrine functions. Nuclear hormone receptors represent a family of candidates that may be involved in the synchronization mechanisms of individual organs. Overall, the NR signaling system and the circadian timing system appear to be integrated at many levels (see Fig. 1), and it will be interesting to further elucidate their connections as we move toward understanding circadian inputs and outputs in greater detail.

ACKNOWLEDGMENTS

This work was supported by Atlas Grant U19DK62434-01 (R.M.E.) and a NRSA postdoctoral fellowship (X.Y.) from the National Institutes of Health, a Merck fellowship from the Life Sciences Research Foundation (K.A.L.), and a short-term fellowship from the European Molecular Biology Organization (X.Y.). R.M.E. is an Investigator of the Howard Hughes Medical Institute and the March of Dimes Chair in Molecular and Developmental Biology.

REFERENCES

Ahima R.S., Prabakaran D., and Flier J.S. 1998. Postnatal leptin surge and regulation of circadian rhythm of leptin by feeding. Implications for energy homeostasis and neuroendocrine function. *J. Clin. Invest.* **101:** 1020.

Akashi M. and Takumi T. 2005. The orphan nuclear receptor RORalpha regulates circadian transcription of the mammalian core-clock Bmal1. *Nat. Struct. Mol. Biol.* **12:** 441.

Allan J.S. and Czeisler C.A. 1994. Persistence of the circadian thyrotropin rhythm under constant conditions and after light-induced shifts of circadian phase. *J. Clin. Endocrinol. Metab.* **79:** 508.

Amir S. and Robinson B. 2006. Thyroidectomy alters the daily pattern of expression of the clock protein, PER2, in the oval nucleus of the bed nucleus of the stria terminalis and central nucleus of the amygdala in rats. *Neurosci. Lett.* **407:** 254.

Andre E., Conquet F., Steinmayr M., Stratton S.C., Porciatti V., and Becker-Andre M. 1998. Disruption of retinoid-related orphan receptor beta changes circadian behavior, causes retinal degeneration and leads to vacillans phenotype in mice. *EMBO J.* **17:** 3867.

Back P., Hamprecht B., and Lynen F. 1969. Regulation of cholesterol biosynthesis in rat liver: Diurnal changes of activity and influence of bile acids. *Arch. Biochem. Biophys.* **133:** 11.

Balsalobre A., Marcacci L., and Schibler U. 2000a. Multiple signaling pathways elicit circadian gene expression in cultured Rat-1 fibroblasts. *Curr. Biol.* **10:** 1291.

Balsalobre A., Brown S.A., Marcacci L., Tronche F., Kellendonk C., Reichardt H.M., Schutz G., and Schibler U. 2000b. Resetting of circadian time in peripheral tissues by glucocorticoid signaling. *Science* **289:** 2344.

Bookout A.L., Jeong Y., Downes M., Yu R.T., Evans R.M., and Mangelsdorf D.J. 2006. Anatomical profiling of nuclear receptor expression reveals a hierarchical transcriptional network. *Cell*, **126:** 789.

Canaple L., Rambaud J., Dkhissi-Benyahya O., Rayet B., Tan N.S., Michalik L., Delaunay F., Wahli W., and Laudet V. 2006. Reciprocal regulation of brain and muscle Arnt-like protein 1 and peroxisome proliferator-activated receptor alpha defines a novel positive feedback loop in the rodent liver circadian clock. *Mol. Endocrinol.* **20:** 1715.

Chawla A., Saez E., and Evans R.M. 2000. "Don't know much bile-ology." *Cell* **103:** 1.

Chawla A., Repa J.J., Evans R.M., and Mangelsdorf D.J. 2001. Nuclear receptors and lipid physiology: opening the X-files. *Science* **294:** 1866.

Chen G., Liang G., Ou J., Goldstein J.L., and Brown M.S. 2004. Central role for liver X receptor in insulin-mediated activation of Srebp-1c transcription and stimulation of fatty acid synthesis in liver. *Proc. Natl. Acad. Sci.* **101:** 11245.

Claudel T., Cretenet G., Saumet A., and Gachon F. 2007. Crosstalk between xenobiotics metabolism and circadian clock. *FEBS Lett.* **581:** 3626.

Damiola F., Le Minh N., Preitner N., Kornmann B., Fleury-Olela F., and Schibler U. 2000. Restricted feeding uncouples circadian oscillators in peripheral tissues from the central pacemaker in the suprachiasmatic nucleus. *Genes Dev.* **14:** 2950.

Delerive P., Chin W.W., and Suen C.S. 2002. Identification of Reverb(alpha) as a novel ROR(alpha) target gene. *J. Biol. Chem.* **277:** 35013.

Evans R.M. 1988. The steroid and thyroid hormone receptor superfamily. *Science* **240:** 889.

Evans R.M., Barish G.D., and Wang Y.X. 2004. PPARs and the complex journey to obesity. *Nat. Med.* **10:** 355.

Fontaine C., Dubois G., Duguay Y., Helledie T., Vu-Dac N., Gervois P., Soncin F., Mandrup S., Fruchart J.C., Fruchart-Najib J., and Staels B. 2003. The orphan nuclear receptor Rev-Erbalpha is a peroxisome proliferator-activated receptor (PPAR) gamma target gene and promotes PPARgamma-induced adipocyte differentiation. *J. Biol. Chem.* **278:** 37672.

Forman B.M., Chen J., Blumberg B., Kliewer S.A., Henshaw R., Ong E.S., and Evans R.M. 1994. Cross-talk among ROR alpha 1 and the Rev-erb family of orphan nuclear receptors. *Mol. Endocrinol.* **8:** 1253.

Fu J., Astarita G., Gaetani S., Kim J., Cravatt B.F., Mackie K., and Piomelli D. 2007. Food intake regulates oleoylethanolamide formation and degradation in the proximal small intestine. *J. Biol. Chem.* **282:** 1518.

Fu J., Gaetani S., Oveisi F., Lo Verme J., Serrano A., Rodriguez De Fonseca F., Rosengarth A., Luecke H., Di Giacomo B., Tarzia G., and Piomelli D. 2003. Oleylethanolamide regulates feeding and body weight through activation of the nuclear receptor PPAR-alpha. *Nature* **425:** 90.

Fu Y., Zhong H., Wang M.H., Luo D.G., Liao H.W., Maeda H., Hattar S., Frishman L.J., and Yau K.W. 2005. Intrinsically photosensitive retinal ganglion cells detect light with a vitamin A-based photopigment, melanopsin. *Proc. Natl. Acad. Sci.* **102:** 10339.

Gachon F., Olela F.F., Schaad O., Descombes P., and Schibler U. 2006. The circadian PAR-domain basic leucine zipper transcription factors DBP, TEF, and HLF modulate basal and inducible xenobiotic detoxification. *Cell Metab.* **4:** 25.

Gervois P., Chopin-Delannoy S., Fadel A., Dubois G., Kosykh V., Fruchart J.C., Najib J., Laudet V., and Staels B. 1999. Fibrates increase human REV-ERBalpha expression in liver via a novel peroxisome proliferator-activated receptor response element. *Mol. Endocrinol.* **13:** 400.

Giguere V. 1999. Orphan nuclear receptors: From gene to function. *Endocr. Rev.* **20:** 689.

Giguere V., Yang N., Segui P., and Evans R.M. 1988. Identification of a new class of steroid hormone receptors. *Nature* **331:** 91.

Glossop N.R., Lyons L.C., and Hardin P.E. 1999. Interlocked feedback loops within the *Drosophila* circadian oscillator. *Science* **286:** 766.

Guillaumond F., Dardente H., Giguere V., and Cermakian N. 2005. Differential control of Bmal1 circadian transcription by REV-ERB and ROR nuclear receptors. *J. Biol. Rhythms* **20:** 391.

Hardin P.E. 2006. Essential and expendable features of the circadian timekeeping mechanism. *Curr. Opin. Neurobiol.* **16:** 686.

Inoue I., Shinoda Y., Ikeda M., Hayashi K., Kanazawa K., Nomura M., Matsunaga T., Xu H., Kawai S., Awata T., Komoda T., and Katayama S. 2005. CLOCK/BMAL1 is involved in lipid metabolism via transactivation of the peroxisome proliferator-activated receptor (PPAR) response element. *J. Atheroscler. Thromb.* **12:** 169.

Kilduff T.S., Vugrinic C., Lee S.L., Milbrandt J.D., Mikkelsen J.D., O'Hara B.F., and Heller H.C. 1998. Characterization of the circadian system of NGFI-A and NGFI-A/NGFI-B deficient mice. *J. Biol. Rhythms* **13:** 347.

Knutti D., Kaul A., and Kralli A. 2000. A tissue-specific coactivator of steroid receptors, identified in a functional genetic screen. *Mol. Cell. Biol.* **20:** 2411.

Ko C.H. and Takahashi J.S. 2006. Molecular components of the mammalian circadian clock. *Hum. Mol. Genet.* (spec. no. 2) **15:** R271.

Kohsaka A. and Bass J. 2007. A sense of time: How molecular clocks organize metabolism. *Trends Endocrinol. Metab.* **18:** 4.

Lee C.H., Olson P., and Evans R.M. 2003. Minireview: Lipid metabolism, metabolic diseases, and peroxisome proliferator-activated receptors. *Endocrinology* **144:** 2201.

Lemberger T., Saladin R., Vazquez M., Assimacopoulos F., Staels B., Desvergne B., Wahli W., and Auwerx J. 1996. Expression of the peroxisome proliferator-activated receptor alpha gene is stimulated by stress and follows a diurnal rhythm. *J. Biol. Chem.* **271:** 1764.

Le Minh N., Damiola F., Tronche F., Schutz G., and Schibler U. 2001. Glucocorticoid hormones inhibit food-induced phase-shifting of peripheral circadian oscillators. *EMBO J.* **20:** 7128.

Levi F. and Schibler U. 2007. Circadian rhythms: Mechanisms and therapeutic implications. *Annu. Rev. Pharmacol. Toxicol.* **47:** 593.

Liu C., Li S., Liu T., Borjigin J., and Lin J.D. 2007. Transcriptional coactivator PGC-1alpha integrates the mammalian clock and energy metabolism. *Nature* **447:** 477.

Lu T.T., Makishima M., Repa J.J., Schoonjans K., Kerr T.A., Auwerx J., and Mangelsdorf D.J. 2000. Molecular basis for feedback regulation of bile acid synthesis by nuclear receptors. *Mol. Cell* **6:** 507.

Mangelsdorf D.J. and Evans R.M. 1995. The RXR heterodimers and orphan receptors. *Cell* **83:** 841.

Mangelsdorf D.J., Thummel C., Beato M., Herrlich P., Schutz G., Umesono K., Blumberg B., Kastner P., Mark M., Chambon P., and Evans R.M. 1995. The nuclear receptor superfamily: The second decade. *Cell* **83:** 835.

Masana M.I., Sumaya I.C., Becker-Andre M., and Dubocovich M.L. 2007. Behavioral characterization and modulation of circadian rhythms by light and melatonin in C3H/HeN mice homozygous for the RORbeta knockout. *Am. J. Physiol. Regul. Integr. Comp. Physiol.* **292:** R2357.

McNamara P., Seo S.P., Rudic R.D., Sehgal A., Chakravarti D., and FitzGerald G.A. 2001. Regulation of CLOCK and MOP4 by nuclear hormone receptors in the vasculature: A humoral mechanism to reset a peripheral clock. *Cell* **105:** 877.

Morris M.E., Viswanathan N., Kuhlman S., Davis F.C., and Weitz C.J. 1998. A screen for genes induced in the suprachiasmatic nucleus by light. *Science* **279:** 1544.

Nakahata Y., Akashi M., Trcka D., Yasuda A., and Takumi T. 2006. The in vitro real-time oscillation monitoring system identifies potential entrainment factors for circadian clocks. *BMC Mol. Biol.* **7:** 5.

Oishi K., Shirai H., and Ishida N. 2005. CLOCK is involved in the circadian transactivation of peroxisome-proliferator-activated receptor alpha (PPARalpha) in mice. *Biochem. J.* **386:** 575.

Oiwa A., Kakizawa T., Miyamoto T., Yamashita K., Jiang W., Takeda T., Suzuki S., and Hashizume K. 2007. Synergistic regulation of the mouse orphan nuclear receptor SHP gene promoter by CLOCK-BMAL1 and LRH-1. *Biochem. Biophys. Res. Commun.* **353:** 895.

Panda S., Antoch M.P., Miller B.H., Su A.I., Schook A.B., Straume M., Schultz P.G., Kay S.A., Takahashi J.S., and Hogenesch J.B. 2002. Coordinated transcription of key pathways in the mouse by the circadian clock. *Cell* **109:** 307.

Preitner N., Damiola F., Lopez-Molina L., Zakany J., Duboule D., Albrecht U., and Schibler U. 2002. The orphan nuclear receptor REV-ERBalpha controls circadian transcription within the positive limb of the mammalian circadian oscillator. *Cell* **110:** 251.

Puigserver P., Wu Z., Park C.W., Graves R., Wright M., and Spiegelman B.M. 1998. A cold-inducible coactivator of nuclear receptors linked to adaptive thermogenesis. *Cell* **92:** 829.

Qatanani M. and Moore D.D. 2005. CAR, the continuously advancing receptor, in drug metabolism and disease. *Curr. Drug Metab.* **6:** 329.

Raspe E., Mautino G., Duval C., Fontaine C., Duez H., Barbier O., Monte D., Fruchart J., Fruchart J.C., and Staels B. 2002. Transcriptional regulation of human Rev-erbalpha gene expression by the orphan nuclear receptor retinoic acid-related orphan receptor alpha. *J. Biol. Chem.* **277:** 49275.

Repa J.J., Liang G., Ou J., Bashmakov Y., Lobaccaro J.M., Shimomura I., Shan B., Brown M.S., Goldstein J.L., and Mangelsdorf D.J. 2000. Regulation of mouse sterol regulatory element-binding protein-1c gene (SREBP-1c) by oxysterol receptors, LXRalpha and LXRbeta. *Genes Dev.* **14:** 2819.

Sanyal S., Kim J.Y., Kim H.J., Takeda J., Lee Y.K., Moore D.D., and Choi H.S. 2002. Differential regulation of the orphan nuclear receptor small heterodimer partner (SHP) gene promoter by orphan nuclear receptor ERR isoforms. *J. Biol. Chem.* **277:** 1739.

Sato T.K., Panda S., Miraglia L.J., Reyes T.M., Rudic R.D., McNamara P., Naik K.A., FitzGerald G.A., Kay S.A., and Hogenesch J.B. 2004. A functional genomics strategy reveals Rora as a component of the mammalian circadian clock. *Neuron* **43:** 527.

Schoonjans K. and Auwerx J. 2002. A sharper image of SHP. *Nat. Med.* **8:** 789.

Shearman L.P., Sriram S., Weaver D.R., Maywood E.S., Chaves I., Zheng B., Kume K., Lee C.C., van der Horst G.T., Hastings M.H., and Reppert S.M. 2000. Interacting molecular loops in the mammalian circadian clock. *Science* **288:** 1013.

Sonoda J. Mehl I.R., Chong L.-W., Nofsinger R.N., and Evans R.M. 2006. PGC1b controls mitochondrial metabolism to modulate circadian activity, adaptive thermogenesis, and hepatic steatosis. *Proc. Natl. Acad. Sci.* **104:** 5223.

Stokkan K.A., Yamazaki S., Tei H., Sakaki Y., and Menaker M. 2001. Entrainment of the circadian clock in the liver by feeding. *Science* **291:** 490.

Sumi Y., Yagita K., Yamaguchi S., Ishida Y., Kuroda Y., and Okamura H. 2002. Rhythmic expression of ROR beta mRNA in the mice suprachiasmatic nucleus. *Neurosci. Lett.* **320:** 13.

Thompson C.L., Selby C.P., Van Gelder R.N., Blaner W.S., Lee J., Quadro L., Lai K., Gottesman M.E., and Sancar A. 2004. Effect of vitamin A depletion on nonvisual phototransduction pathways in cryptochromeless mice. *J. Biol. Rhythms* **19:** 504.

Ueda H.R., Chen W., Adachi A., Wakamatsu H., Hayashi S., Takasugi T., Nagano M., Nakahama K., Suzuki Y., Sugano S., Iino M., Shigeyoshi Y., and Hashimoto S. 2002. A transcription factor response element for gene expression during circadian night. *Nature* **418:** 534.

Wang L., Han Y., Kim C.S., Lee Y.K., and Moore D.D. 2003. Resistance of SHP-null mice to bile acid-induced liver damage. *J. Biol. Chem.* **278:** 44475.

Warnecke M., Oster H., Revelli J.P., Alvarez-Bolado G., and Eichele G. 2005. Abnormal development of the locus coeruleus in Ear2(Nr2f6)-deficient mice impairs the functionality of the forebrain clock and affects nociception. *Genes Dev.* **19:** 614.

Yang X. Downes M., Yu R.T., Bookout A.L., He W., Straume M., Mangelsdorf D.J., and Evans R.M. 2006. Nuclear receptor expression links the circadian clock to metabolism. *Cell* **126:** 801.

Yin L. and Lazar M.A. 2005. The orphan nuclear receptor Rev-erbalpha recruits the N-CoR/histone deacetylase 3 corepressor to regulate the circadian Bmal1 gene. *Mol. Endocrinol.* **19:** 1452.

Yin L., Wang J., Klein P.S., and Lazar M.A. 2006. Nuclear receptor Rev-erbalpha is a critical lithium-sensitive component of the circadian clock. *Science* **311:** 1002.

Bacterial Circadian Programs

C.H. JOHNSON
Department of Biological Sciences, Vanderbilt University, Nashville, Tennessee 37235

Twenty years ago, it was widely believed that prokaryotes were too "simple" to have evolved circadian programs. Since that time, however, the cyanobacterial circadian system has progressed from a curiosity to a major model system for analyzing clock phenomena. In addition to globally regulating gene expression, cyanobacteria are one of the only systems in which the adaptive fitness of a circadian system has been rigorously evaluated. Moreover, cyanobacteria are the only clock system in which all essential proteins of the core oscillator have been crystallized and structurally determined, namely, the KaiA, KaiB, and KaiC proteins. A biochemical oscillator can be reconstituted in vitro with these three purified Kai proteins and displays the key properties of temperature-compensated rhythmicity. This result spectacularly demonstrates that a strictly posttranslational clock is sufficient to elaborate circadian phenomena and that a transcription-translation feedback loop is not obligatory. The conjunction of structural information on essential clock proteins with a defined system that reconstitutes circadian oscillations in vitro leads to a turning point whereby biophysical and biochemical approaches bring analyses of circadian clockwork to an unprecedented level of molecular detail.

INTRODUCTION

At the original Cold Spring Harbor Symposium on Biological Clocks (47 years ago), there was already an appreciation that microbial organisms were capable of circadian rhythmicity. That volume featured papers on circadian rhythms in the microbes *Euglena* (Bruce 1961), *Gonyaulax* (Hastings 1961; Sweeney 1961; Sweeney and Hastings 1961), and *Paramecium* (Ehret 1961). Several of those papers have been influential for many years. For example, the paper by J.W. Hastings on biochemical aspects has been widely cited as evidence for the relative insensitivity of circadian oscillators to drugs/chemicals (Hastings 1961). The paper by B.M. Sweeney on single *Gonyaulax* cells provided suggestive evidence that circadian rhythms might be a cellular phenomenon and not an emergent property of populations of cells (Sweeney 1961). And the paper by Sweeney and Hastings on temperature compensation is still cited regularly for its suggestion that temperature compensation might be mechanistically accomplished by coupled biochemical reactions (Sweeney and Hastings 1961). During the past 47 years, salient discoveries in circadian rhythmicity have used microbial organisms (Edmunds 1988; Johnson and Kondo 2001). The two eukaryotic microbes that are most commonly studied today for circadian investigations are the fungus *Neurospora crassa* and the green alga *Chlamydomonas reinhardtii*, whose circadian properties were first reported in 1959 for *Neurospora* (Pittendrigh et al. 1959) and in 1970 for *Chlamydomonas* (Bruce 1970). Undoubtedly, the reason that those two microbes have been selected for continued study is that both classical and molecular genetic approaches are possible.

Nevertheless, with the increasing ability in mammalian systems to apply techniques that were heretofore only possible in microbes, coupled with the increasing pressure from funding agencies to do "translational research" rather than basic research, the interest in nonmammalian models for the study of circadian rhythms has waned somewhat. This attitude might be particularly applied to the case of bacterial model systems—the subject of this paper—because the key clock genes that are involved in prokaryotic circadian rhythms (at least, in cyanobacteria; Ishiura et al. 1998) have no homologs in the genomes of mammals, insects, or fungi, nor do the clock genes in eukaryotes have obvious homologs in bacteria. Despite that perception, however, the bacterial circadian system has in the interval of 15 years blossomed from a curiosity to one of the most important model systems—one that has enabled insights and approaches that were technically impossible elsewhere. The goal of this chapter is to briefly describe the insights resulting from the study of bacterial (especially cyanobacterial) clocks, to demonstrate what those studies have uniquely told us, and to speculate upon the future of bacterial clock studies and what they might tell us about eukaryotic clocks.

THE DISCOVERY OF CIRCADIAN CLOCKS IN BACTERIA

Before 1986, it was generally thought that circadian rhythms were exclusively a eukaryotic phenomenon (Ehret 1961; Johnson et al. 1996). A few reports that suggested that bacteria might have circadian clocks were not persuasive because the described rhythms were noisy and temperature compensation had not been demonstrated (Halberg and Conner 1961; Sturtevant 1973). That prokaryotic cells were too "simple" to express circadian behavior became a dogma (Johnson et al. 1996), despite the fact that there were few publications that directly tested the proposition (one exception was Taylor 1979). Some models relied on this dogma as an a priori assumption upon which eukaryotic intracellular organelles were conceived to be a necessary prerequisite for a circadian mechanism (Ehret and Trucco 1967; Schweiger and Schweiger 1977; Kippert 1986).

During 1985–1986, three papers were published that began to depose the "no clocks in proks" dogma (Stal and Krumbein 1985; Grobbelaar et al. 1986; Mitsui et al. 1986). These various groups discovered that cyanobacteria which fix nitrogen (*Oscillatoria* sp. strain 23; *Synechococcus* spp. Miami BG 43511 and 43522; *Synechococcus* sp. RF-1) display daily rhythms of nitrogen fixation in LD (light/dark) and in LL (light/light). In particular, the group of Huang and coworkers was apparently the first to clearly recognize that cyanobacteria were exhibiting circadian rhythms and, in a series of publications beginning in 1986, demonstrated the salient characteristics of circadian rhythms in the unicellular freshwater cyanobacterium, *Synechococcus* sp. RF-1 (Grobbelaar et al. 1986; Huang and Grobbelaar 1995).

A crucial characteristic that needed to be demonstrated for cyanobacteria before they could be accepted into the circadian fold was that of temperature compensation. This criterion was first satisfied by studies of two species of *Synechococcus:* the marine *Synechococcus* WH7803 and the freshwater *Synechococcus* RF-1. Sweeney and Borgese (1989) showed that WH7803 displays temperature-compensated rhythms of cell division, whereas Chen et al. (1991) found temperature-compensated rhythms of amino acid uptake in the freshwater cyanobacterium isolated from rice fields, *Synechococcus* RF-1. These findings along with those of others unseated the eukaryotic-centric dogma of chronobiology (Johnson et al. 1996). Despite the fact that this dogma is a relic of the past, however, cyanobacteria remain the only prokaryotic system for which circadian organization is proven at this time (see below).

Our laboratory and those of our collaborators—Drs. Takao Kondo, Susan Golden, and Masahiro Ishiura—extended the studies on circadian programming to non-nitrogen-fixing cyanobacteria that were genetically malleable (Johnson et al. 1996). Our first studies used a strain of *Synechococcus elongatus* PCC 7942 transformed with a luminescence reporter construct, which is the fusion of the *psbAI* promoter with a bacterial luciferase cassette (P_{psbAI}::*luxAB;* Kondo et al. 1993). This strain of cyanobacteria is notable because of the ease with which exogenous DNA can be transformed and homologously recombined into the chromosome (Andersson et al. 2000). Because the circadian clock turns P_{psbAI} on and off rhythmically, this first reporter strain of *S. elongatus* glowed rhythmically. This choice of strain and promoter was fortuitous. Subsequent experiments using other strains/species of cyanobacteria have found rhythms (Aoki et al. 1997), but the reporters in those strains are not bright. And even in *S. elongatus*, other promoters do not show such a robust rhythm of luminescence (Liu et al. 1995). The combination of the P_{psbAI}::*luxAB* reporter and the *S. elongatus* strain remains one of the most robustly rhythmic combinations in cyanobacteria, even after 15 years of intensive research. Dr. Kondo was able to exploit the bright and robust luminescence rhythm by designing clever apparatuses to monitor the rhythms. For liquid cultures, Dr. Kondo enlisted and modified the automated photomultiplier apparatus that was originally designed for endogenously bioluminescent algae (Kondo et al. 1993). For monitoring colonies, Drs. Kondo and Ishiura discovered that the rhythms of single colonies could be tracked with a charge-coupled device (CCD) camera (Kondo and Ishiura 1994). Dr. Kondo used those observations to design an innovative turntable/CCD camera apparatus for mutant screening of single colonies, enabling the simultaneous screening of up to 12,000 colonies in a single experiment (Kondo et al. 1994).

The coincidence of good fortune, clever ideas, and hard work has transformed the *S. elongatus* system into one of the best-characterized circadian clock systems, even though it is a relative newcomer to molecular clock analyses; only 20 years ago, no one believed that prokaryotes were capable of circadian rhythmicity (Johnson et al. 1996). Therefore, the *S. elongatus* system has rapidly caught up with and in several areas has surpassed the eukaryotic systems, as described below.

CIRCADIAN ORCHESTRATION OF GLOBAL GENE EXPRESSION

With bacteria finally accepted as "members of the circadian club," an early question of interest was how pervasive was clock control of gene expression in the *S. elongatus* system. Was *psbAI* an isolated example of a gene that was regulated by the circadian mechanism or were more genes under its control in *S. elongatus*? In eukaryotes, the number of genes regulated by the circadian clock ranges between 5% and 15%, but eukaryotic genetic organization and regulation are very different from that of prokaryotes. For example, with the exception of a few genes on plasmids, all the genes of *S. elongatus* are arrayed upon a single circular chromosome. Might they be regulated coordinately?

This question was answered by a sensitive and comprehensive promoter-trap experiment (Liu et al. 1995). A promoterless luciferase gene set was randomly inserted throughout the genome; whenever the luciferase gene set inserted into a genomic locus that was correctly positioned and oriented to a promoter, luciferase was expressed and the colonies glowed. The luminescent patterns of more than 800 independent colonies were analyzed. We were astonished to observe that *all* of the glowing colonies displayed circadian rhythms with the same period (Liu et al. 1995). The pattern of the rhythmic expression differed among the promoters, both in terms of phasing and in waveform (Liu et al. 1996). The promoter activity for a ribosomal RNA gene (*rrnA*) was likewise rhythmic (Liu et al. 1995), as well as heterologous promoters, such as an *Escherichia coli* promoter (P_{conII}) that is transcribed rhythmically when inserted into the cyanobacterial chromosome (Katayama et al. 1999). Apparently, the cyanobacterial clock globally controls gene expression, in this case, the activity of all promoters. This global expression is regulated by one of the key cyanobacterial clock proteins (KaiC; see below) (Nakahira et al. 2004). It is possible that these patterns are mediated by daily changes in the topology of the single chromosome which modulate the activity of all promoters in the chromosome (Mori and Johnson 2001a; Smith and Williams 2006; Woelfle and Johnson 2006). If so, the cyanobacterial chromosome might be envisioned as an oscillating nucleoid, or "oscilloid" that regulates all promoters by torsion-sensitive transcription (Johnson 2004; Woelfle and Johnson 2006; M.A. Woelfle et al. 2007).

TWO TIMING CIRCUITS IN THE SAME CELL? CELL DIVISION VERSUS CIRCADIAN OSCILLATORS

There is a long history of study of the interplay between the cell division cycle and the circadian system in microbes (Edmunds 1988; Mori and Johnson 2000); this is a topic that vertebrate biologists appear to have recently "rediscovered" (Hunt and Sassone-Corsi 2007). In microbes, the bottom line is that the circadian clockwork is independent of cell division timing (otherwise, how could the circadian clock keep accurate track of environmental time?) but that the circadian clock gates the timing of cell division by having a circadian checkpoint in the cell division cycle. This circadian checkpoint assures that cell division will occur at an optimal environmental phase.

The independence of the circadian system from cell division cycling is clear in *S. elongatus*. These bacteria express robust circadian rhythms of gene expression when dividing with doubling times much faster than once per 24 hours (at least down to doubling times of 6–12 hours; Mori et al. 1996: Kondo et al. 1997). They also express excellent rhythms when a key cell division cycle protein, FtsZ, is overexpressed: Division is inhibited but growth continues to produce long "spaghetti-like" cells (Fig. 1) (Mori and Johnson 2001b). Therefore, the circadian clock appears to ignore the status of the cell division cycle in cyanobacteria. However, the timing of cell division is *not* independent of the circadian clock; even when the cells are dividing rapidly, the circadian clock rhythmically slows the rate of cell division every night. Even in constant light, the rate of cell division is slowed in the early subjective night as though there is a circadian checkpoint that forbids division in the early night (Mori et al. 1996). Apparently, the circadian clockwork is well-buffered and stable against significant changes of the intracellular milieu so that it can accomplish its raison d'etre—keeping track of environmental phase and regulating cellular events to occur at the optimal time. Consequently, the circadian clockwork in *S. elongatus* gates cell division and gene expression, but its timing circuit is independent of the cell division cycle.

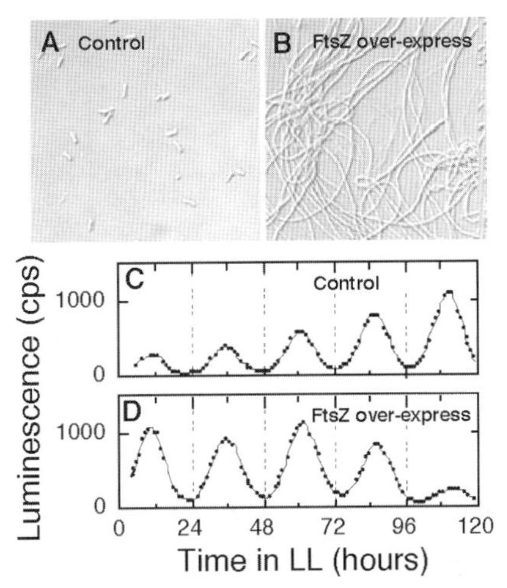

Figure 1. Circadian rhythms of gene expression continue in dividing versus nondividing cells of *S. elongatus*. Cell division of growing cyanobacteria is stopped by overexpression of the cell septum protein FtsZ. (*A,B*) Morphology of control cells (*A*) and cells in which FtsZ is overexpressed (*B*). (*C,D*) Luminescence rhythms in control cells (*C*) and cells in which division has been halted by FtsZ overexpression (*D*). Luminescence rhythms are monitored from reporters of luciferase-promoter fusions (Kondo et al. 1993). The reporter in *C* and *D* is the *kaiBC* promoter driving bacterial luciferase expression. (Modified, with permission, from Mori and Johnson 2001b [© American Society for Microbiology].)

ADAPTIVE SIGNIFICANCE OF CIRCADIAN TIMING IN CYANOBACTERIA

The first studies of cyanobacterial clocks used diazotrophic strains that fix nitrogen at night and photosynthesize in the day (Grobbelaar et al. 1986; Mitsui et al. 1986). Because photosynthesis generates oxygen that inhibits the nitrogen-fixing enzyme nitrogenase, this day/night "temporal separation" seemed to be a clear case of the adaptiveness of circadian programming. This hypothesis would predict that cyanobacterial growth in constant light would be slower than in a light/dark cycle because nitrogen fixation would be inhibited under these conditions, and therefore the growing cells might rapidly become starved for metabolically available nitrogen. The problem is that cyanobacteria grow perfectly well in constant light; in fact, they grow faster in constant light than in light/dark cycles, presumably because of the extra energy they derive from the additional photosynthesis. This result is inconsistent with the "temporal separation" hypothesis. This is an example of "adaptive storytelling" (Johnson 2005). A more rigorous test was needed.

We therefore tested the adaptive significance of circadian programs in the *S. elongatus* by competing different strains against each other in different laboratory environments (Ouyang et al. 1998; Woelfle et al. 2004). For asexual microbes such as *S. elongatus*, differential growth of one strain under competition with other strains is a good measure of reproductive fitness. In pure culture, the strains grew at about the same rate in constant light and in light/dark cycles, so there did not appear to be a significant advantage or disadvantage in having different circadian periods when the strains were grown individually. The fitness test was to mix different strains together and to grow them in competition to determine if the composition of the population changes as a function of time. The laboratory environments were different kinds of light/dark (LD) cycles or constant light. The cultures were diluted at intervals to allow growth to continue.

In one series of experiments, wild-type cells were competed against an apparently arhythmic strain (CLAb). As shown in Figure 2B, the arhythmic strain was rapidly defeated by wild type in LD 12:12, but under competition

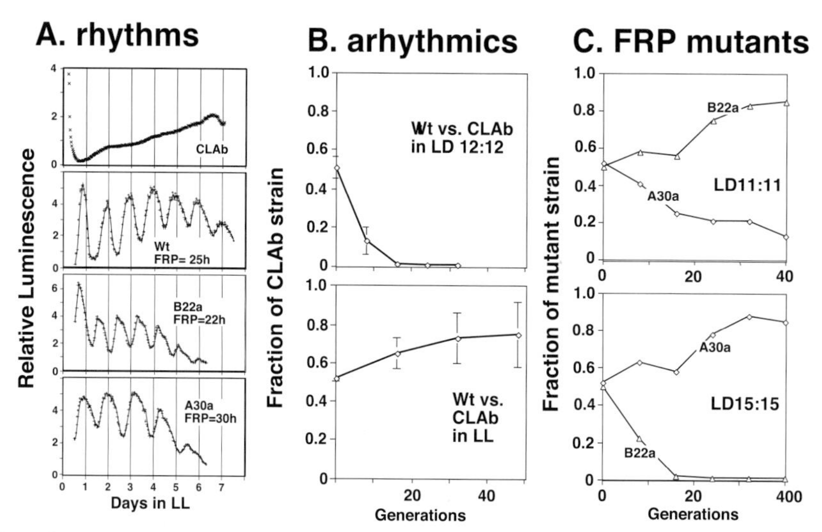

Figure 2. Competition of clock-modified strains with wild type in mixed cultures of *S. elongatus*. (*A*) Circadian phenotypes of luminescence emission from wild type, the clock-disrupted strain CLAb, and period mutants (mutation in *kaiB* [B22a, period ~22 hours] and *kaiA* [A30a, period ~30 hours]). (*B*) Competition between wild type and CLAb in LD 12:12 (*upper*) and LL (*lower*) plotted as the fraction of mutant in the mixed culture versus the estimated number of generations (mean ±S.D.). (*C*) Kinetics of competition between wild type and the period mutant strains in mixed cultures exposed to LD 11:11 (*upper*) or LD 15:15 (*lower*). Data are plotted as the fraction of the mutant strain in the mixed culture versus the estimated number of generations. (FRP) Free-running period; (LD) light/dark cycle; (LL) constant light. (Modified, with permission, from Woelfle et al. 2004 [© Elsevier].)

in constant light, the arhythmic strain grew slightly better than wild type (Woelfle et al. 2004). Therefore, the clock system does not appear to confer an intrinsic value for cyanobacteria under constant conditions. In another series of experiments, period mutants were used to answer the question: Does having a period that is similar to the period of the environmental cycle enhance fitness? The circadian phenotypes of the strains shown in Figure 2A had free-running periods of about 22 hours (B22a) and 30 hours (A30a) (wild type has a period of about 25 hours under these conditions). When each of the strains was mixed with another strain and grown together in competition, a pattern emerged that depended on the frequency of the light/dark cycle and the circadian period. When grown on a 22-hour cycle (LD 11:11), the 22-hour-period mutant could overtake wild type in the mixed cultures (Fig. 2C). On the other hand, in a 30-hour cycle (LD 15:15), the 30-hour-period mutant could defeat wild type (Fig. 2C). On a "normal" 24-hour cycle (LD 12:12), the wild-type strain could outcompete either mutant (Ouyang et al. 1998; Woelfle et al. 2004). Note that over many cycles, each of these light/dark conditions have equal amounts of light and dark (which is important because photosynthetic cyanobacteria derive their energy from light); it is only the frequency of light versus dark that differs among the light/dark cycles.

These data clearly show that the strain whose period most closely matched that of the light/dark cycle eliminated the competitor. Because the mutant strains could defeat the wild-type strain in light/dark cycles in which the periods are similar to their endogenous periods, the differential effects that were observed are likely to result from the differences in the circadian clock. Our results show that an intact clock system whose free-running period is consonant with the environment significantly enhances fitness of cyanobacteria in rhythmic environments; however, this same clock system provides no adaptive advantage in constant environments and may even be slightly detrimental to this organism. We still do not know the mechanistic basis for these competitive effects in cyanobacteria, especially because pure cultures of each strain seem to have no decrement in growth. Does this mean that the basis of the competition results is in the interaction between strains in mixed cultures? Despite the fact that we do not know the mechanism by which one strain outcompetes another, however, our competition approach inspired similar studies with the plant *Arabidopsis*, where the competitive advantage under conditions of clock/environment consonance was attributed to differences in photosynthetic efficiency (Dodd et al. 2005).

STRUCTURAL BIOLOGY OF CIRCADIAN CLOCK PROTEINS

A mutation/complementation analysis identified the three key clock genes in *S. elongatus*: *kaiA*, *kaiB*, and *kaiC*, which are clustered together on the chromosome (Ishiura et al. 1998). As described below, the proteins encoded by these genes are necessary and sufficient for circadian rhythmicity (at least, for in vitro rhythmicity). The years 2004–2005 marked a dramatic turning point in the circadian clock field, with the nearly simultaneous reporting of the three-dimensinal structures of the KaiA, KaiB, and KaiC proteins—the first clock proteins from any system to have their full-length structures defined. Cyanobacteria remain the only clock system in which all essential proteins of the core oscillator have been crystallized and structurally determined. This event is important because structural information enables analyses at a truly molecular level; now, we can interpret the effect of mutations on the structure and function of the clock proteins. We can also make precise predictions about the ways in which clock proteins interact and influence each other's activity.

The structure of cyanobacterial KaiA was studied by several groups, and solved by solution nuclear magnetic resonance (NMR) (Vakonakis et al. 2004) and X-ray crystallography (Garces et al. 2004; Ye et al. 2004). The 2.0-Å crystal structure of KaiA from *S. elongatus* reveals that the protein is composed of two independently folded domains connected by a linker (Fig. 3) (Ye et al. 2004). The amino-terminal pseudoreceiver domain has a fold similar to that of bacterial response regulators (despite a lack of sequence similarity), whereas the carboxyl-terminal four-helix bundle domain is novel and forms the interface of the twofold-related homodimer. The crystal structure of cyanobacterial KaiB revealed an α-β meander motif (Garces et al. 2004; Hitomi et al. 2005). The fold shows close resemblance to the α-β motif of thioredoxin. Although the folds of KaiA versus KaiB are clearly different, their size and some surface features of the physiologically relevant dimers are very similar. Notably, the functionally important residues Arg-69 of KaiA and Arg-23 of KaiB align well in space. The apparent structural similarities suggest that KaiA and KaiB may compete for a potential common binding site on KaiC (Garces et al. 2004; Hitomi et al. 2005), and these similarities may explain KaiB's ability to antagonize the effects of KaiA upon KaiC phosphorylation.

My laboratory collaborated with our structural colleagues Drs. Martin Egli, Rekha Pattanayek, and Jimin Wang to determine the crystal structure of the core protein of the circadian clock system from *S. elongatus*, the KaiC homohexamer, at 2.8-Å resolution (Pattanayek et al. 2004). The structure resembles a double doughnut with a central pore that is partially sealed at one end (Fig. 3). This crystal structure revealed ATP-binding sites, intersubunit organization, a scaffold for Kai protein complex formation, the location of critical KaiC mutations, and evolutionary relationships to other proteins such as DnaB and RecA. The KaiCI and KaiCII domains of each subunit adopt similar conformations, but their ATP-binding pockets exhibit significant differences that are likely of functional importance. Previously identified mutations that affect rhythmicity and length of period map to the subunit interface and KaiA-binding regions. The most important binding site of KaiA to KaiC appears to be on the surface of the CII terminus and to 22-amino-acid "tentacles" that extend from CII (Vakonakis and LiWang 2004; Pattanayek et al. 2006). In addition to facilitating our understanding of KaiA-KaiC interaction, the KaiC structure also sheds light on the mechanism of rhythmic phosphorylation of KaiC by identifying T432, S431, and T426 in KaiCII as sites at which KaiC is phosphorylated (Xu et al. 2004); an independent mass spectrometry approach also identified T432 and S431 but not T426 (Nishiwaki et al. 2004). The important role of these residues is confirmed by the loss of rhythmicity in T432A, S431A, and T426A single mutants (Nishiwaki et al. 2004; Xu et al. 2004).

THE CYANOBACTERIAL CLOCKWORK: A TRANSCRIPTIONAL-TRANSLATIONAL FEEDBACK LOOP? OR NOT?

At the time that the central cyanobacterial clock genes *kaiABC* were reported (Ishiura et al. 1998), it appeared that the cyanobacterial clockwork was a transcriptional-translational feedback loop (TTFL) oscillator as had been pro-

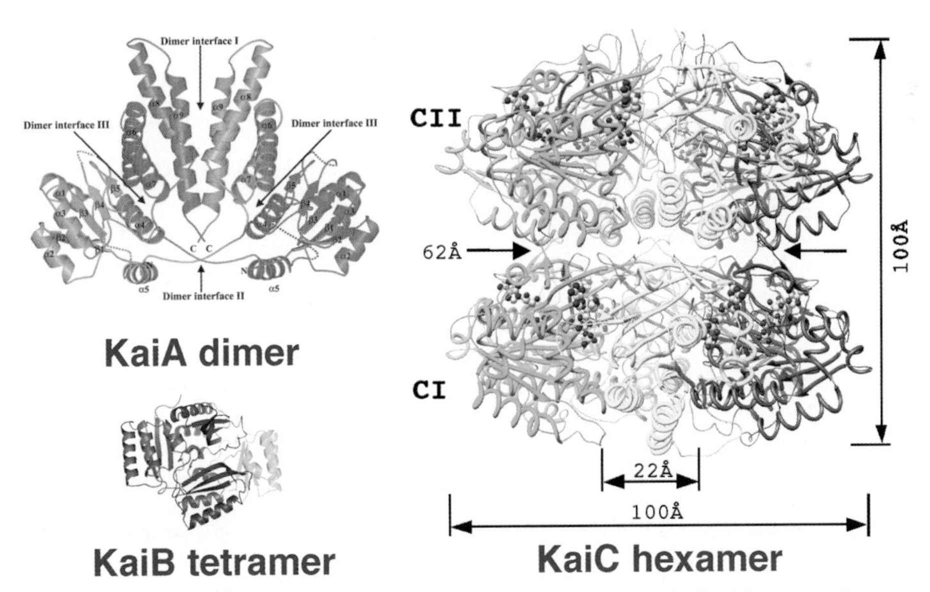

Figure 3. Ribbon representations of the three essential cyanobacterial clock proteins determined from crystal structures. (*Upper left*) Dimer of KaiA from *S. elongatus*. Different colors denote the two different KaiA molecules in the dimer. (*Lower left*) Tetramer of the KaiB from *Synechocystis* sp. PCC6803; different colors indicate the four KaiB molecules in the tetramer. (*Right*) Hexamer of KaiC from *S. elongatus* viewed from the side. Individual subunits are colored differently; the slimmer waist region of the particle demarcates the CI and CII domains. ATP molecules are drawn in a ball-and-stick mode with atoms colored *gray*, *red*, *blue*, and *yellow* for carbon, oxygen, nitrogen, and phosphorus, respectively. (*Upper left,* modified, with permission, from Ye et al. 2004 [© American Society for Biochemistry and Molecular Biology]; *lower left,* modified, with permission, from Hitomi et al. 2005 [© American Society for Biochemistry and Molecular Biology]; *right,* modified, with permission, from Pattanayek et al. 2004 [© Elsevier].)

posed for the circadian mechanism in various eukaryotic organisms (Dunlap 1999). This inference was based on the fact that the same kind of evidence used to support a TTFL oscillator in eukaryotes also held true for the cyanobacterial clock, namely: (1) rhythms of abundance for mRNAs and proteins encoded by "clock genes," (2) negative feedback of clock proteins on their gene's transcription, and (3) phase setting by overexpression of clock genes, and others. However, recent experiments have led to a major reevaluation of the roles of the TTFL versus a specifically posttranslational oscillator in the cyanobacterial system. The first indication that the cyanobacterial TTFL model was flawed was the demonstration that the promoters driving *kaiABC* gene expression could be replaced with nonspecific heterologous promoters without disturbing the circadian rhythm (Xu et al. 2003; Nakahira et al. 2004). If the TTFL model were correct in its original formulation, the Kai proteins would be expected to feedback-inhibit their own transcription through direct or indirect interaction with their specific promoters. By that line of reasoning, if the *kaiABC* promoters were replaced with nonspecific heterologous promoters, the TTFL model would predict that the central feedback loop would be interrupted. But this did not happen in the cyanobacterial case (Xu et al. 2003; Nakahira et al. 2004).

Studies of cyanobacterial cells in dark/dark more directly attacked the TTFL model. Because *S. elongatus* is an obligate photoautotroph, transcription and translation are shut down in darkness, but Tomita et al. (2005) discovered that the rhythm of KaiC phosphorylation continues robustly in dark/dark. They also discovered that pharmacological inhibition of transcription or translation similarly had little effect upon the circadian rhythm of KaiC phosphorylation. Therefore, neither transcription nor translation was needed to accomplish the circadian rhythm of posttranslational modification of KaiC. The coup de grâce for the usual formulation of the TTFL model came with the stunning demonstration that the three Kai proteins could be combined in a test tube and the circadian rhythm of KaiC phosphorylation merrily churned along! (Nakajima et al. 2005) This in vitro rhythm was even temperature-compensated, so the temperature-compensation mechanism was also encoded in the characteristics of the three Kai proteins and the nature of their interaction. Clearly, transcriptional and translational feedback is not necessary to build a circadian clockwork. Does this mean that transcriptional and translational feedback has no role in the larger circadian system in cyanobacteria? Not necessarily. It is still possible—even likely—that the rhythms of Kai protein abundance can feed into the posttranslational oscillator to promote robustness, and the posttranslational oscillator regulates the timing of transcription-translation to occur in the optimal phase so as to promote that robustness. If so, it may be that the posttranslational oscillator is embedded in a TTFL; the posttranslational oscillator is the process that most directly determines the dynamics of the circadian system, but the TTFL can affect the system by providing a secondary feedback loop. The elucidation of the relative contributions of the posttranslational KaiABC oscillator versus the TTFL circuit is therefore of key importance for future studies.

VISUALIZING THE TICKING OF THE CYANOBACTERIAL IN VITRO CLOCKWORK

The availability of an in vitro system for analyzing the molecular nature of a circadian clockwork permits biophysical, biochemical, and structural analyses that were previously impossible. Given that the rubicon of three-dimensional structural knowledge had been crossed for the Kai proteins, we embraced these techniques and collaborated with Drs. Phoebe Stewart and Dewight Williams to use electron microscopy (EM) to analyze and quantify the time-dependent interaction among the three proteins (KaiA, KaiB, and KaiC) to elucidate the timing of the formation of complexes among the Kai proteins (Mori et al. 2007). We could visualize in the EM that KaiC existed in any of all possible combinations at the different phases of the in vitro oscillation, namely, free KaiC hexamers, KaiA•KaiC complexes, KaiB•KaiC complexes, and KaiA•KaiB•KaiC complexes, but the proportions of these complexes vary in a phase-dependent manner: Free KaiC hexamers predominate at all phases, about 10% of KaiC hexamers appear as KaiA•KaiC complexes at all phases, and KaiB•KaiC and KaiA•KaiB•KaiC complexes are clearly rhythmic with a peak in the KaiC dephosphorylation phase. These results were confirmed by two-dimensional native gel electrophoresis (Mori et al. 2007) and are consistent with the results obtained by the Kondo lab using other techniques (Kageyama et al. 2006). The Kondo lab had reported that KaiC hexamers appeared to be capable of exchanging monomers among the hexamers; therefore, we used the biophysical technique of fluorescence resonance energy transfer (FRET) to confirm that monomer exchange among KaiC hexamers occurs. Finally, we applied the first perturbation analyses of the in vitro oscillator by using temperature pulses to reset the phase of the KaiABC oscillator, thereby testing the resetting characteristics of this unique circadian oscillator (Mori et al. 2007). These structural, biochemical, and biophysical approaches have provided insights to a circadian clockwork at a level of molecular detail that was heretofore unthinkable (Kageyama et al. 2006, Mori et al. 2007).

MODELING THE IN VITRO OSCILLATOR

I have always been an advocate of the potential benefits of modeling toward opening our eyes to features of circadian rhythmicity that are not otherwise obvious. For example, I firmly believe that limit-cycle modeling of phase resetting has been enlightening (Johnson et al. 2003). Nonetheless, my laboratory has not contributed any mathematical model of circadian mechanisms because I felt that our knowledge of the "gears and cogs" of circadian oscillators was too rudimentary to provide the basis for a useful model of the circadian mechanism. Until now, that is! When the Kondo lab provided a circadian clockwork that depended only on three purified proteins in a test tube (Nakajima et al. 2005), I decided that the time had come for a predictive mathematical model that would be testable by biochemical, biophysical, and structural experiments. Therefore, the structural and biophysical data from our laboratory (Mori et al. 2007) and the biochemical data from the Kondo lab (Kageyama et al. 2006) were used by our physicist collaborator Dr. Mark Byrne to derive a dynamic model for the in vitro oscillator that accurately reproduced the rhythms of KaiABC complexes and of KaiC phosphorylation.

Figure 4 depicts this model, which stochastically simulates the kinetics of KaiC hexamers and the degree of phosphorylation of each monomer in every hexamer. Starting from a hypophosphorylated state of KaiC (state α), rapid association and disassociation of KaiA facilitate phosphorylation until the KaiC hexamer is hyperphosphorylated (state β). KaiB is assumed to bind with hyperphosphorylated KaiC; KaiB association with KaiC induces a simultaneous conformational change to a new state (KaiC*, state χ). The KaiC* hexamer (state χ) dephosphorylates to a relatively hypophosphorylated status (state δ) and relaxes to the original conformation (state α). Simultaneous with the phosphorylation cycle of a hexamer is the possibility of the exchange of monomers between any two hexamers in any of the states. The rates of this exchange may differ depending on the conformational state of the KaiC hexamers, their degree of phosphorylation, and their association with KaiA or KaiB (Fig. 4). We assume that KaiA stochastically binds and unbinds rapidly from KaiC hexamers and that KaiB associates/disassociates with the KaiC hexamer when the total degree of phosphorylation of the hexamer exceeds a threshold that places KaiC in state β (for a complete explanation of the model, see Mori et al. 2007).

Within a space of 2 months in early 2007, there were four publications (including ours) that independently derived a similar core KaiC model (Clodong et al. 2007; Mori et al. 2007; van Zon et al. 2007; Yoda et al. 2007). This does not mean that this core model is correct, but it does mean that the parameter space has been thoroughly explored and found to provide robust oscillations. An important difference among the models is the mechanism invoked to explain the excellent maintenance of high-amplitude phosphorylation rhythms in vitro. This phenomenon is significant because it requires that the phosphorylation status of each individual KaiC hexamer be synchronized with that of other hexamers in the molecular population. If the hexamers are not synchronized, then the KaiC hexamers will independently free-run and will rapidly get out of phase with each other. In this case, each hexamer may be rhythmic, but the population of hexamers will become asynchronous. Therefore, a mechanism of maintaining synchrony is necessary to explain the sustained in vitro rhythms. In our model (Mori et al. 2007) and that of Yoda and coworkers (2007), phase-dependent exchange of KaiC monomers provides a potential mechanism for synchronizing the number of phosphates on different hexamers of a given conformation. Therefore, oscillation in the phosphorylation of the population of hexamers proceeds in a regulated fashion. The models of Clodong et al. (2007) and van Zon et al. (2007) explain KaiC population synchrony by proposing differential

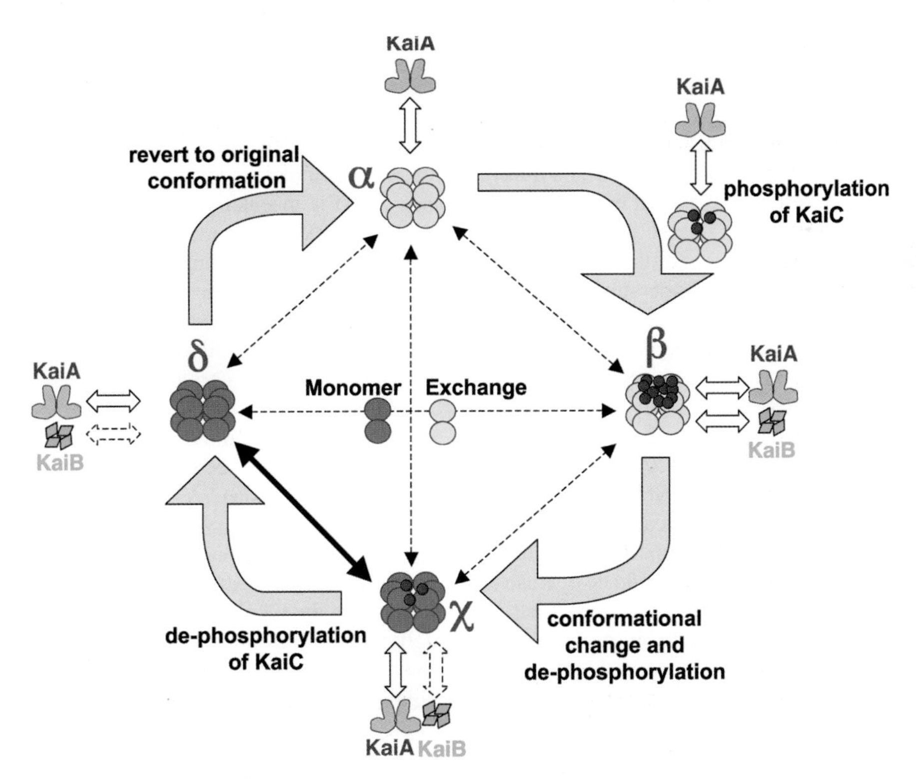

Figure 4. A model for the KaiABC in vitro oscillator. The diagram represents a mathematical model for the phosphorylation cycle of a KaiC hexamer and its association with KaiA and KaiB. A KaiC monomer is shown as a double circle that can form a hexamer. KaiC hexamers can associate/dissociate with KaiA and/or KaiB. KaiC hexamers are depicted in two conformational states: a default KaiC status (*light blue*) and an altered KaiC* state that has undergone a conformational change (*darker blue*). (*Red dots*) Phosphates attached to phosphorylation sites on KaiC. KaiC hexamers can exchange monomers between hexamers in any of the states, depicted by the double-headed arrows in the central region of the figure. The rate of monomer exchange is maximum during the KaiC dephosphorylation phase (*solid double-headed arrow*). (Modified, with permission, from Mori et al. 2007 [PLoS Biology].)

interaction between KaiA and KaiC that depends on the phosphorylation status of each individual KaiC hexamer. We favor the monomer exchange explanation because it relies upon an experimentally observed phenomenon, whereas the KaiA-KaiC explanation is hypothetical, but these two models of KaiC synchronization are not mutually exclusive and both may be operating.

THE NEXT CHAPTER IN THE BOOK OF BACTERIAL CLOCK STUDIES

The future is always difficult to predict, but my opinion is that further study of cyanobacterial clock systems will continue to contribute in two major areas: *mechanism* and *evolution/ecology*. In the realm of *evolution*, many unanswered questions remain. The experiments using competition between different strains of *S. elongatus* clearly demonstrate the adaptive significance of having a circadian system (Ouyang et al. 1998; Woelfle et al. 2004), but the mechanistic basis of this strong fitness advantage remains murky. Clarification of the mechanism of this selection could identify the key selective forces involved in the early evolution of circadian clocks, especially because cyanobacteria are thought to be one of the earliest life forms to have evolved on earth (Johnson and Kyriacou 2005). The *kai* genes that control circadian rhythmicity in *S. elongatus* are widely distributed among cyanobacterial species (Lorne et al. 2000). Indeed, *kaiB* and *kaiC* homologs are found widely among other eubacteria and even in the archaea (*kaiA* is less common and is therefore thought to be of more recent evolutionary origin; Dvornyk and Knudsen 2005). Do circadian clocks based on a Kai-type oscillator exist in other types of bacteria? It would be surprising if circadian oscillators do not exist in other bacteria. Cyanobacteria are absolutely dependent on the light/dark cycle for their energy (photosynthesis) and therefore it makes sense that a circadian system was a crucial adaptation; nevertheless, the daily cycle of light (including UV) and temperature must be of critical importance to other types of bacteria and therefore the occurence of *kai* homologs among other prokaryotes is probably a clue to the existence of bacterial clocks. In the case of *Rhodobacter*, there is already evidence for a circadian-like rhythm, but temperature compensation has not yet been demonstrated (Min et al. 2005).

Is the *kai*-based oscillator a prokaryotic phenomenon without evolutionary links to the circadian systems in eukaryotes? When we first started our research on the cyanobacterial clock, we hoped that it would be a "Rosetta Stone" for plant clocks based on the assumption that an ancestral clock from a cyanobacterium might have been transferred or conserved during the endosymbiotic events of plant evolution (Johnson 1994). Unfortunately, that hope was frustrated—we have not found *kai* homologs in chloroplast or nuclear genomes of plants (there does not even appear to be conservation of clock genes between eukaryotic algae such as *Chlamydomonas* and eukaryotic plants! [Mittag et al. 2005]). Nevertheless, it remains possible that there is an evolutionary link between prokaryotic and eukaryotic clocks but that tremendous divergence in the sequences of the relevant clock genes has obscured these links so that the links are undetectable by comparisons within the current repertoire of known prokaryotic and eukaryotic genomes. Perhaps fundamental aspects of clock biochemistry have been conserved between prokaryotes and eukaryotes? It is an intriguing thought.

In the *ecological* dimension, not only did cyanobacteria transform the atmosphere of Earth from a reducing to an oxidizing environment, but cyanobacteria continue to be a major (probably predominant) contributor to global photosynthesis. Undoubtedly, knowing more about how cyanobacteria accomplish and regulate photosynthesis is important, especially in the face of the impending global warming threat.

In the realm of clock *mechanism*, the convergence of structural, biophysical, and biochemical studies enabled by the combination of an in vitro system and structural information on key clock proteins will certainly give us the clearest picture in the near future of how "the blind watchmaker" (aka evolution) constructed a circadian clockwork. These insights will be at the most basic molecular level, including visualizing circadian conformational changes of single molecules. In addition to the elucidation of the in vitro oscillator, we need to define the role of a TTFL in cyanobacteria. It is likely that the rhythms of KaiB and KaiC protein abundance (Xu et al. 2003) can feed into the posttranslational oscillator to promote robustness, and the elucidation of the relative contributions of the posttranslational KaiABC oscillator *versus* the TTFL circuit is of key importance for future studies. The implications of these connections extend beyond the cyanobacterial system. If it is true that the bacterial posttranslational oscillator is embedded within a TTFL and is the underlying determinant of the properties of this circadian system, this organization could also be true for eukaryotic systems. One consequence of the cyanobacterial studies might therefore be to spark a major reevaluation of the evidence supporting a TTFL in eukaryotes—perhaps a totally protein-driven clockwork underlies circadian rhythmicity in eukaryotes (including humans) as well.

Finally, the question of *mechanism* within the larger clock system of cyanobacteria implicates inputs and outputs. For *input*, light/dark signals (especially pulses of darkness) and temperature changes can entrain the cyanobacterial clock system. Recent evidence suggests that environment-stimulated alterations of the intracellular redox status may be involved in these input pathways (Ivleva et al. 2006). A particular fascination of the cyanobacterial clock system is how the *output* rhythm of global gene expression is regulated. As mentioned above, these expression patterns may be mediated by daily changes in the topology of the chromosome such that the activity of all promoters are rhythmically modulated by torsion-sensitive transcription (Mori and Johnson 2001a; Johnson 2004; Smith and Williams 2006; Woelfle and Johnson 2006; Woelfle et al. 2007). In contrast to the cyanobacterial clock, eukaryotic clock control of gene expression is usually described in terms of regulation by the rhythmic activity of specific transcription factors. However, recent evidence indicates that trans- criptional factors that regulate the mammalian clock have an essential histone acetyltransferase activity (Etchegaray et al. 2003; Doi et al. 2006), suggesting that regulation of chromatin structure is an important mechanism for clock control of transcription in eukaryotes. Thus, regulation of chromosome structure may be a conserved feature of circadian clocks in general and not a curiosity of the cyanobacterial circadian system.

CONCLUSIONS

The studies on the cyanobacterial clock system have been pioneering circadian research and continue to be so. First, we now know that eukaryotic organization is not a requirement for the evolution of a circadian system. Cyanobacterial studies provided the first rigorous evidence for the adaptive fitness of circadian programs. We know from cyanobacterial investigations that a strictly posttranslational network of protein interactions is sufficient to confer circadian periodicity and temperature compensation. However, this posttranslational clockwork operates within a larger TTFL network, and the elucidation of the interplay between the posttranslational oscillator and the TTFL may enlighten us regarding eukaryotic clock mechanisms. Cyanobacterial clock proteins are the only ones that have been successfully crystallized and are therefore the only circadian timekeeping proteins to expose their structural secrets. The combination of an in vitro system and structural information has led to a true watershed in the analyses of circadian mechanism, foretelling biophysical and biochemical approaches that will bring circadian analyses of a circadian clockwork to an unprecedented level of molecular detail. Finally, scrutinizing the intriguing orchestration of gene expression by the cyanobacterial clock is likely to lead to new insights into basic mechanisms of gene expression. Are cyanobacterial circadian clocks a mere curiosity? No longer!

ACKNOWLEDGMENTS

Many scientists contributed to the story described here. Dr. Tsung-Hsien Chen inspired me to study cyanobacterial clocks, which Dr. Takao Kondo and I initiated during Dr. Kondo's sabbatical in my laboratory with the help of Dr. Susan Golden, and later Dr. Masahiro Ishiura. The labs of these four scientists (Kondo, Golden, Ishiura, and Johnson) are largely responsible for establishing the *S. elongatus* system. The cyanobacterial work in my laboratory was performed by Drs. Yao Xu, Tetsuya Mori, Yi Liu, Yan Ouyang, Mark Woelfle, Vladimir Podust, and Ximing Qin with the help of our collaborators Drs. Takao Kondo, Susan Golden, Masahiro Ishiura, Martin Egli, Phoebe Stewart, Hassane Mchaourab, David Piston, Michael Cox, Ross Inman, Rekha Pattanayek, Dewight Williams, and Mark Byrne. Dr. David McCauley has cheerfully consulted on statistical analyses and Dr. Terry Page has been a reliable "sounding board" for discussing circadiana over the years. Finally, I thank my mentors for being scientific muses and role models: J.W. Hastings, Michael Menaker, and Colin Pittendrigh (all of whom were participants at the 1960 Cold Spring Harbor Symposium).

REFERENCES

Andersson C.A., Tsinoremas N.F., Shelton J., Lebedeva N.V., Yarrow J., Min H., and Golden, S.S. 2000. Application of bioluminescence to the study of circadian rhythms in cyanobacteria. *Methods Enzymol.* **305:** 527.

Aoki S., Kondo T., Wada H., and Ishiura M. 1997. Circadian rhythm of the cyanobacterium *Synechocystis* sp. PCC 6803 in the dark. *J. Bacteriol.* **179:** 5751.

Bruce V.G. 1961. Environmental entrainment of circadian rhythms. *Cold Spring Harbor Symp. Quant. Biol.* **25:** 29.

———. 1970. The biological clock in *Chlamydomonas reinhardtii*. *J. Protozool.* **17:** 328.

Chen T.-H., Chen T.-L., Hung L.-M., and Huang T.-C. 1991. Circadian rhythm in amino acid uptake by *Synechococcus* RF-1. *Plant Physiol.* **97:** 55.

Clodong S., Duhring U., Kronk L., Wilde A., Axmann I., Herzel H., and Kollmann M. 2007. Functioning and robustness of a bacterial circadian clock. *Mol. Syst. Biol.* **3:** 90.

Dodd A.N., Salathia N., Hall A., Kevei E., Toth R., Nagy F., Hibberd J.M., Millar A.J., and Webb A.A. 2005. Plant circadian clocks increase photosynthesis, growth, survival, and competitive advantage. *Science* **309:** 630.

Doi M., Hirayama J., and Sassone-Corsi P. 2006. Circadian regulator CLOCK is a histone acetyltransferase. *Cell* **125:** 497.

Dunlap J.C. 1999. Molecular bases for circadian clocks. *Cell* **96:** 271.

Dvornyk V. and Knudsen B. 2005. Functional divergence of the circadian clock proteins in prokaryotes. *Genetica* **124:** 247.

Edmunds L.N. 1988. *Cellular and molecular bases of biological clocks*. Springer-Verlag, New York.

Ehret C.F. 1961. Action spectra and nucleic acid metabolism in circadian rhythms at the cellular level. *Cold Spring Harbor Symp. Quant. Biol.* **25:** 149.

Ehret C.F. and Trucco E. 1967. Molecular models for the circadian clock. I. The chronon concept. *J. Theor. Biol.* **15:** 240.

Etchegaray J.P., Lee C., Wade P.A., and Reppert S.M. 2003. Rhythmic histone acetylation underlies transcription in the mammalian circadian clock. *Nature* **421:** 177.

Garces R.G., Wu N., Gillon W., and Pai E.F. 2004. *Anabaena* circadian clock proteins KaiA and KaiB reveal potential common binding site to their partner KaiC. *EMBO J.* **23:** 1688.

Grobbelaar N., Huang T.-C., Lin H.Y., and Chow T.J. 1986. Dinitrogen-fixing endogenous rhythm in *Synechococcus* RF-1. *FEMS Microbiol. Lett.* **37:** 173.

Halberg F. and Conner R.L. 1961. Circadian organization and microbiology: Variance spectra and a periodogram on behavior of *Escherichia coli* growing in fluid culture. *Proc. Minn. Acad. Sci.* **29:** 227.

Hastings J.W. 1961. Biochemical aspects of rhythms: Phase shifting by chemicals. *Cold Spring Harbor Symp. Quant. Biol.* **25:** 131.

Hitomi K., Oyama T., Han S., Arvai A.S., and Getzoff E.D. 2005. Tetrameric architecture of the circadian clock protein KaiB: A novel interface for intermolecular interactions and its impact on the circadian rhythm. *J. Biol. Chem.* **280:** 19127.

Huang T.-C. and Grobbelaar N. 1995. The circadian clock in the prokaryote *Synechococcus* RF-1. *Microbiology* **141:** 535.

Hunt T. and Sassone-Corsi P. 2007. Riding tandem: Circadian clocks and the cell cycle. *Cell* **129:** 461.

Ishiura M., Kutsuna S., Aoki S., Iwasaki H., Andersson C.R., Tanabe A., Golden S.S., Johnson C.H., and Kondo T. 1998. Expression of a gene cluster *kaiABC* as a circadian feedback process in cyanobacteria. *Science* **281:** 1519.

Ivleva N.B., Gao T., LiWang A.C., and Golden S.S. 2006. Quinone sensing by the circadian input kinase of the cyanobacterial circadian clock. *Proc. Natl. Acad. Sci.* **103:** 17468.

Johnson C.H. 1994. Illuminating the clock: Circadian photobiology. *Semin. Cell Biol.* **5:** 355.

———. 2004. Global orchestration of gene expression by the biological clock of cyanobacteria. *Genome Biol.* **5:** 217.

———. 2005. Testing the adaptive value of circadian systems. *Methods Enzymol.* **393:** 818.

Johnson C.H. and Kondo T. 2001. Circadian rhythms in unicellular organisms. In *Handbook of behavioral neurobiology*, vol. 13: *Developmental psychobiology* (ed. J.S. Takahashi et al.), Chap. 3, p. 61. Plenum Press, New York.

Johnson C.H. and Kyriacou C.P. 2005. Clock evolution and adaptation: Whence and whither? In *Endogenous plant rhythms* (ed. A.J.W. Hall and H. McWatters), chap. 10, p. 237. Blackwell Publishing, Oxford, United Kingdom.

Johnson C.H., Elliott J.A., and Foster R.G. 2003. Entrainment of circadian programs. *Chronobiol. Int.* **20:** 741.

Johnson C.H., Golden S.S., Ishiura M., and Kondo T. 1996. Circadian clocks in prokaryotes. *Mol. Microbiol.* **21:** 5.

Kageyama H., Nishiwaki T., Nakajima M., Iwasaki H., Oyama T., and Kondo T. 2006. Cyanobacterial circadian pacemaker: Kai protein complex dynamics in the KaiC phosphorylation cycle *in vitro*. *Mol. Cell* **23:** 161.

Katayama M., Tsinoremas N.F., Kondo T., and Golden S.S. 1999. *cpmA*, a gene involved in an output pathway of the cyanobacterial circadian system. *J. Bacteriol.* **181:** 3516.

Kippert F. 1986. Endocytobiotic coordination, intracellular calcium signalling, and the origin of endogenous rhythms. *Ann. N.Y. Acad. Sci.* **503:** 476.

Kondo T. and Ishiura M. 1994. Circadian rhythms of cyanobacteria: Monitoring the biological clocks of individual colonies by bioluminescence. *J. Bacteriol.* **176:** 1881-1885.

Kondo T., Mori T., Lebedeva N.V., Aoki S., Ishiura M., and Golden S.S. 1997. Circadian rhythms in rapidly dividing cyanobacteria. *Science* **275:** 224.

Kondo T., Tsinoremas N.F., Golden S.S., Johnson C.H., Kutsuna S., and Ishiura M. 1994. Circadian clock mutants of cyanobacteria. *Science* **266:** 1233.

Kondo T., Strayer C.A., Kulkarni R.D., Taylor W., Ishiura M., Golden S.S., and Johnson C.H. 1993. Circadian rhythms in prokaryotes: Luciferase as a reporter of circadian gene expression in cyanobacteria. *Proc. Natl. Acad. Sci.* **90:** 5672.

Liu Y., Tsinoremas N.F., Golden S.S., Kondo T., and Johnson C.H. 1996. Circadian expression of genes involved in the purine biosynthetic pathway of the cyanobacterium *Synechococcus* sp. strain PCC 7942. *Mol. Microbiol.* **20:** 1071.

Liu Y., Tsinoremas N.F., Johnson C.H., Lebedeva N.V., Golden S.S., Ishiura M., and Kondo T. 1995. Circadian orchestration of gene expression in cyanobacteria. *Genes Dev.* **9:** 1469.

Lorne J., Scheffer J., Lee A., Painter M., and Miao V.P. 2000. Genes controlling circadian rhythm are widely distributed in cyanobacteria. *FEMS Microbiol. Lett.* **189:** 129.

Min H., Guo H., and Xiong J. 2005. Rhythmic gene expression in a purple photosynthetic bacterium, *Rhodobacter sphaeroides*. *FEBS Lett.* **579:** 808.

Mitsui A., Kumazawa S., Takahashi A., Ikemoto H., and Arai T. 1986. Strategy by which nitrogen-fixing unicellular cyanobacteria grow photoautotrophically. *Nature* **323:** 720.

Mittag M., Kiaulehn S., and Johnson C.H. 2005. The circadian clock in *Chlamydomonas reinhardtii:* What is it for? What is it similar to? *Plant Physiol.* **137:** 399.

Mori T. and Johnson C.H. 2000. Circadian control of cell division in unicellular organisms. *Prog. Cell Cycle Res.* **4:** 185.

———. 2001a. Circadian programming in cyanobacteria. *Semin. Cell Dev. Biol.* **12:** 271.

———. 2001b. Independence of circadian timing from cell division in cyanobacteria. *J. Bacteriol.* **183:** 2439.

Mori T., Binder B., and Johnson C.H. 1996. Circadian gating of cell division in cyanobacteria growing with average doubling times of less than 24 hours. *Proc. Natl. Acad. Sci.* **93:** 10183.

Mori T., Williams D.R., Byrne M.O., Qin X., Mchaourab H.S., Egli M., Stewart P.L., and Johnson C.H. 2007. Elucidating the ticking of an *in vitro* circadian clockwork. *PLoS Biol.* **5:** e93.

Nakahira Y., Katayama M., Miyashita H, Kutsuna S., Iwasaki H., Oyama T., and Kondo T. 2004. Global gene repression by KaiC as a master process of prokaryotic circadian system. *Proc. Natl. Acad. Sci.* **101:** 881.

Nakajima M., Imai K., Ito H., Nishiwaki T., Murayama Y., Iwasaki H., Oyama T., and Kondo T. 2005. Reconstitution of circadian oscillation of cyanobacterial KaiC phosphorylation *in vitro*. *Science* **308:** 414.

Nishiwaki T., Satomi Y., Nakajima M., Lee C., Kiyohara R., Kageyama H., Kitayama Y., Temamoto M., Yamaguchi A., Hijikata A., Go M., Iwasaki H., Takao T., and Kondo T. 2004. Role of KaiC phosphorylation in the circadian clock system of *Synechococcus elongatus* PCC 7942. *Proc. Natl. Acad. Sci.* **101:** 13927.

Ouyang Y., Andersson C.R., Kondo T., Golden S.S., and Johnson C.H. 1998. Resonating circadian clocks enhance fitness in cyanobacteria. *Proc. Natl. Acad. Sci.* **95:** 8660.

Pattanayek R., Wang J., Mori T., Xu Y., Johnson C.H., and Egli M. 2004. Visualizing a circadian clock protein: Crystal structure of KaiC and functional insights. *Mol. Cell* **15:** 375.

Pattanayek R., Williams D.R., Pattanayek S., Xu Y., Mori T., Johnson C.H., Stewart P.L., and Egli M. 2006. Analysis of KaiA-KaiC protein interactions in the cyano-bacterial circadian clock using hybrid structural methods. *EMBO J.* **25:** 2017.

Pittendrigh C.S., Bruce V.G., Rosensweig N.S., and Rubin M.L. 1959. Growth patterns in *Neurospora:* A biological clock in *Neurospora*. *Nature* **184:** 169.

Schweiger H.-G. and Schweiger M. 1977. Circadian rhythms in unicellular organisms: An endeavor to explain the molecular mechanism. *Int. Rev. Cytol.* **51:** 315.

Smith R.M. and Williams S.B. 2006. Circadian rhythms in gene transcription imparted by chromosome compaction in the cyanobacterium *Synechococcus elongatus*. *Proc. Natl. Acad. Sci.* **103:** 8564.

Stal L.J. and Krumbein W.E. 1985. Nitrogenase activity in the non-heterocystous cyanobacterium *Oscillatoria* sp. grown under alternating light-dark cycles. *Arch. Microbiol.* **143:** 67.

Sturtevant R.P. 1973. Circadian variability in *Klebsiella* demonstrated by cosinor analysis. *Int. J. Chronobiol.* **1:** 141.

Sweeney B.M. 1961. The photosynthetic rhythm in single cells of *Gonyaulax polyedra*. *Cold Spring Harbor Symp. Quant. Biol.* **25:** 145.

Sweeney B.M. and Borgese M.B. 1989. A circadian rhythm in cell division in a prokaryote, the cyanobacterium *Synechococcus* WH7803. *J. Phycol.* **25:** 183.

Sweeney B.M. and Hastings J.W. 1961. Effects of temperature upon diurnal rhythms. *Cold Spring Harbor Symp. Quant. Biol.* **25:** 87.

Taylor W.R. 1979. "Studies on the bioluminescent glow rhythm of *Gonyaulax polyedra*." Ph.D. thesis, University of Michigan, Ann Arbor.

Tomita J., Nakajima M., Kondo T., and Iwasaki H. 2005. Circadian rhythm of KaiC phosphorylation without transcription-translation feedback. *Science* **307:** 251.

Vakonakis I. and LiWang A.C. 2004. Structure of the C-terminal domain of the clock protein KaiA in complex with a KaiC-derived peptide: Implications for KaiC regulation. *Proc. Natl. Acad. Sci.* **101:** 10925.

Vakonakis I., Sun J., Wu T., Holzenburg A., Golden S.S., and LiWang A.C. 2004. NMR structure of the KaiC-interacting C-terminal domain of KaiA, a circadian clock protein: Implications for the KaiA-KaiC Interaction. *Proc. Natl. Acad. Sci.* **101:** 1479.

van Zon J.S., Lubensky D.K., Altena P.R., and ten Wolde P.R. 2007. An allosteric model of circadian KaiC phosphorylation. *Proc. Natl. Acad. Sci.* **104:** 7420.

Woelfle M.A. and Johnson C.H. 2006. No promoter left behind: Global orchestration of circadian gene expression in cyanobacteria. *J. Biol. Rhythms* **21:** 419.

Woelfle M.A., Ouyang Y., Phanvijhitsiri K., and Johnson C.H. 2004. The adaptive value of circadian clocks: An experimental assessment in cyanobacteria. *Curr. Biol.* **14:** 1481.

Woelfle M.A., Xu Y., Qin X., and Johnson C.H. 2007. Circadian rhythms of superhelical status of DNA in cyanobacteria. *Proc. Natl. Acad. Sci.* (in press).

Xu Y., Mori T., and Johnson C.H. 2003. Cyanobacterial circadian clockwork: Roles of KaiA, KaiB, and the *kaiBC* promoter in regulating KaiC. *EMBO J.* **22:** 2117.

Xu Y., Mori T., Pattanayek R., Pattanayek S., Egli M., and Johnson C.H. 2004. Identification of key phosphorylation sites in the circadian clock protein KaiC by crystallographic and mutagenetic analyses. *Proc. Natl. Acad. Sci.* **101:** 13933.

Ye S., Vakonakis I., Ioerger T.R., LiWang A.C., and Sacchettini J.C. 2004. Crystal structure of circadian clock protein KaiA from *Synechococcus elongatus*. *J. Biol. Chem.* **279:** 20511.

Yoda M., Eguchi K., Terada T.P., and Sasai M. 2007. Monomer-shuffling and allosteric transition in KaiC circadian oscillation. *PLoS ONE* **2:** e408.

Stochastic Phase Oscillators and Circadian Bioluminescence Recordings

J. ROUGEMONT* AND F. NAEF*†

*Swiss Institute of Bioinformatics, CH-1015 Lausanne, Switzerland; †Swiss Institute of Experimental Cancer Research, Ecole Polytechnique Fédérale de Lausanne, CH-1015 Lausanne, Switzerland

Cultured circadian oscillators from peripheral tissues were recently shown to be both cell-autonomous and self-sustained. Therefore, the dominant cause for amplitude reduction observed in bioluminescence recordings of cultured fibroblasts is desynchronization, rather than the damping of individual oscillators. Here, we review a generic model for quantifying luminescence signals from biochemical oscillators, based on noisy-phase oscillators. Our model incorporates three essential features of circadian clocks: the stability of the limit cycle, fluctuations, and intercellular coupling. The model is then used to analyze bioluminescence recordings from immortalized and primary fibroblasts. Fits to population recordings allow simultaneous estimation of the stability of the limit cycle (or equivalently, the stiffness of individual frequencies), the period dispersion, and the interaction strength between cells. Consistent with other work, coupling is found to be weak and insufficient to synchronize cells. Interestingly, we find that frequency fluctuations remain correlated for longer periods than one clock cycle, which is confirmed from individual cell recordings. We discuss briefly how to link the generic model with more microscopic models, which suggests mechanisms by which circadian oscillators resist fluctuations and maintain accurate timing in the periphery.

INTRODUCTION

The reciprocal interactions among environment, genes, and behavior that lead living organisms to evolve circadian clocks are inevitably very complex. Therefore, the attempt to use mathematics to model these phenomena may seem like a daunting enterprise. Yet, circadian biology is one of the earliest fields that has attracted modelers and continues to do so at an ever-increasing rate. One of the reasons why theoreticians have been attracted to the problem is that oscillatory trajectories can be abstracted by a few informative variables, e.g., phase and amplitude. While providing a coarser-grained description, a number of generally applicable concepts can then be explored, e.g., the measurement of phase-response curves to probe the structure of the underlying limit cycles. In this context, the theoretical prediction and subsequent measurement of phase singularities in hatching flies (Winfree 2001) provide a fascinating demonstration of the power of such approaches.

Although these earlier studies were guided by geometric or topological principles, the recent tendency has turned toward more-detailed molecularly inspired models. The latter describe the experimentally mapped clock genes and their mutual interactions in the language of chemical kinetics. These models are then solved or simulated using ordinary differential equations or stochastic versions thereof based on the master equation (Gillespie 1977).

Among the most recent models, Goldbeter (1995) proposed a rate equation model to describe the negative feedback of the Period protein in *Drosophila melanogaster*, including mRNA transcription, translation, and phosphorylation of the protein, as well as nuclear translocation. Subsequent models for circadian clocks in other species can be split somewhat artificially in two groups: (1) models proposing simplified low-dimensional toy models to focus on the role of common network motifs such as negative feedback loops (Barkai and Leibler 2000; Vilar et al. 2002; Gonze and Goldbeter 2006) and (2) more explicit and higher-dimensional models that implement the detailed biochemical processes known to date (Forger and Peskin 2003; Leloup and Goldbeter 2003; Locke et al. 2005). The latter usually contain a large number of unknown rate constants, some of which can be calibrated by imposing experimentally derived constraints such as period length or peak phases under various conditions such as mutant data. The second category of models allows us to make specific predictions (e.g., the influence of the hypophosphorylation of the Per protein in the family advanced sleep phase syndrome [FASPS]; Toh et al. 2001), but such a detailed approach produces a large number of apparently discordant models, each with many unknown parameters, to describe various specific circadian clocks (see discussion in Rand et al. 2006). The first approach, however, focuses on generic properties of cycling systems and seeks to develop an intuitive understanding of the fundamental structures required to explain experimentally observed phenomena such as the apparent loss of rhythmicity in tissue explants. Whereas the low-dimensional models can be fit precisely to experimental data, simulations can be used to relate the explicit biochemical variables of the detailed models to the effective summary variables on the simplified models, thus offering a rather complete view of the problem.

To be ultimately useful to the biologist and not serve merely as toys for theorists, the models should lead to the formulation of testable hypotheses. In Winfree's days, such a prediction was, for example, that there must exist a phase singularity if strong perturbation (type 0) phase-resetting is observed and the oscillator has limited cycle properties. He then went on to design an experiment to measure eclosion times in *Drosophila* that explicitly showed the phase singularity. Nowadays, predictive modeling studies based on rate equation models that are also tested are still rare.

Synthetic biology seems a good playground for such efforts; compare, e.g., the repressilator system in *Escherichia coli* (Elowitz and Leibler 2000). Recent studies indicated that modeling the circadian clock can guide identification of new clock components, as in the plant *Arabidopsis thaliana* (Locke et al. 2005). An often-used methodology to derive predictions is the so-called sensitivity analysis, where model properties are tracked while changing parameters. Typical properties are the period length (Leloup and Goldbeter 2004; Bagheri et al. 2007) or the oscillator stability (Rougemont and Naef 2007).

CLOCK HIERARCHIES, COUPLING, AND ENTRAINMENT

In higher organisms, timing is organized in a hierarchical manner, with cellular rhythms contributing to rhythmicity of whole organs, and organs being coordinated by a central clock that integrates external inputs such as light or temperature cues. In mammals, the central pacemaker resides in the suprachiasmatic nucleus (SCN) and consists of about 15,000 neurons. This hierarchical mode of transmitting timing information requires mechanisms that can reset or drive the oscillator phases, which is also called entrainment. In addition, cellular oscillators must be accurate enough, a property that can be achieved either by autonomous genetic circuits with sufficient stability properties or alternatively by intercellular coupling as in the SCN. Coupling is known to be effective in the SCN in mammals (Welsh et al. 2004), but its role in peripheral tissues is probably minor (Yoo et al. 2004; Guo et al. 2006; Rougemont and Naef 2007). In essence, stability and coupling are necessary to override noise that is ubiquitous at the cellular level; it is thus the relative balance of stability, coupling, and noise that will determine to what extent specific circadian output functions can be scheduled properly.

BIOLUMINESCENCE

Long before the availability of modern luciferase reporters, the circadian bioluminescent glow in the unicellular *Gonyaulax polyhydra* had been exploited to study the underlying properties of the oscillators. In particular, population measurements were modeled mathematically to estimate a high precision of about 1% in the individual periods (Njus et al. 1981). Importantly, this study emphasized the richness of population signals and proposed a model to extract information such as period dispersion from such signals.

Fluorescence microscopy and bioluminescence recordings have recently allowed the observation of self-sustained oscillators with single-cell resolution. This showed that oscillations are common in individual cells; in fact, they occur by default (Nagoshi et al. 2004; Welsh et al. 2004). Expectedly, the individual oscillators were also noisy, such that their period is not exactly constant but changes from one cycle to the next (Carr and Whitmore 2005). In the absence of intercellular interactions, this leads to desynchronization of the cells, i.e., the phases in a population drift apart until they cancel each other and the population appears globally arrhythmic.

COUPLING

Several studies explored the possibility of intercellular coupling in peripheral organs and in the SCN. In SCN neurons, coupling depends on synaptic transmission (Liu and Reppert 2000; Yamaguchi et al. 2003; Ohta et al. 2005; Maywood et al. 2006). In the periphery, Guo et al. (2006) have shown that circadian clocks in the liver of SCN-lesioned hamsters lose their phase coherence, whereas Yoo et al. (2004) observed residual oscillations of Per2 expression in various tissue explants for as long as 20 days. It is, however, difficult to determine if these oscillations reflect some intercellular coupling or whether the samples experience partial resetting after the explantation. Liu et al. (2007) have shown that population signals in lung and fibroblast cultures decay faster than in the SCN, for which actual intercellular coupling has been demonstrated, suggesting that there is no or negligible coupling in peripheral tissues.

Because of their high signal-to-noise values, bioluminescence recordings are suited for direct comparison with mathematical models (Izumo et al. 2003; Nagoshi et al. 2004). It is the purpose of this chapter to review a class of models known as phase models and their applications to the analysis of bioluminescence recordings. The focus of this chapter is on the theory and applications of noisy-phase models in which the individual oscillator frequencies fluctuate in time while cells are coupled, as in the Kuramoto (1984) model. Analysis of bioluminescence recordings from populations of mouse fibroblasts is considered as an example.

MATHEMATICS OF PHASE MODELS

Phase oscillators provide an abstraction of a limit-cycle oscillator where one does not worry about the underlying mechanisms generating the oscillations, i.e., the details in the biochemical interactions. Phase oscillators use a parameterization of limit-cycle oscillations in terms of phase and amplitude variables. So in a sense they are much simpler objects than the trajectories for the concentrations of activity levels of genes and proteins. Thus, they provide a coarse-grained description that can be most useful to study generic effects such as fluctuations, oscillator entrainment (forcing) and resetting, and oscillator coupling in cell populations. Phase models have also been popular to discuss synchronization properties in a population of oscillators. (Winfree 1967; Kuramoto 1984; Strogatz 2000; Mihalcescu et al. 2004; Rougemont and Naef 2006). From the Kuramoto study, it was known that gradually increasing the coupling would eventually overcome fluctuations and lead to a macroscopically synchronized state, which is intuitively quite plausible. Synchronyzation here does not mean that all oscillators point exactly in the same phase but that there is a preferred phase orientation or nonuniform phase distribution in the population. The uniformly distributed state is referred to as the incoherent, or desynchronized, state.

The Model

The fact that circadian clocks trigger oscillating biochemical species is taken as a given. There therefore

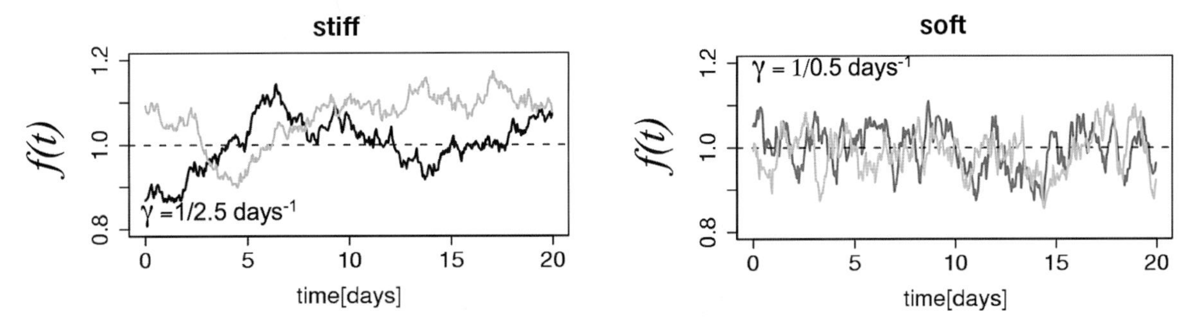

Figure 1. Stiff and soft stochastic frequency trajectories generated according to model in Equation 1. The variance σ (frequency dispersion) is the same for both cases, whereas the mean $\mu = 1$.

exists at least two variables that draw a cyclic shape when plotted against each other during one period. This cycle is reparametrized using a phase (φ) and a frequency (f) (variations in amplitude are not included in this model). Compared with Kuramoto's original model (Kuramoto 1984) that used static (fixed) frequencies, we generalize to frequencies that can also drift in time. This choice was motivated by experiments in zebra fish cells (Carr and Whitmore 2005) that illustrated how the frequency in individual cells drifts in time. A simple model to represent this possibility is to consider instantaneous frequencies which follow a Gaussian random process that is correlated in time. In other words, the frequencies cannot change too abruptly, and this is controlled by an inverse time constant γ.

If γ is small, the time constant is large and the oscillator is not very stable; on the contrary, very stable oscillators are characterized by short timescales or large γ. Such a frequency dynamics can be conveniently modeled as an Ornstein–Uhlenbeck process f(t) with the following three parameters:

$$\mathbf{E}[f(t)] = \mu, \quad \mathbf{E}[(f(t) - \mu)(f(s) - \mu)] = \sigma^2 e^{-\gamma|t-s|} \quad (1)$$

where **E**[] denotes the expectation. The mean $\mu = 1/24$ hours sets the circadian periodicity, the rate γ is the decay of fluctuations, and the variance σ is the amplitude of the fluctuations. Typical trajectories are shown in Figure 1. In independent oscillators, the phase is simply tied to the frequency as:

$$\frac{d}{dt}\varphi_i = f_i(t)$$

The phase dynamics is then also a Gaussian process with mean μt and time-dependent variance:

$$\sigma_\varphi^2(t) = \mathbf{E}[(\varphi(t) - \mu t)(\varphi(t) - \mu t)]$$
$$= \frac{2\sigma^2}{\gamma^2}(\gamma t + e^{-\gamma t} - 1)$$

In these models, the frequencies are coupled to a noise source with variance σ that approximates all noise sources in the cellular environment. This is different from the noise simulated in Gillespie-like algorithms, which only take into account state-space-dependent fluctuations due to small number of molecules.

Statistical Properties of Ensembles of Oscillators

We consider here N individual cells, each described by its own phase and frequency. The phases are initially taken to synchronized by imposing that $\varphi(t = 0) = 0$. The population signal represented as the average cosine of the phases is an important quantity that can be linked to the data (compare below). When the number of cells N is large and *the phases are independent*, this average can be related to the time-dependent variance in the phases as follows:

$$Z_N(t) = \frac{1}{N}\sum_{i=1}^{N}\cos(2\pi\varphi_i(t)) \overset{N\to\infty}{=} \cos(2\pi\mu t)e^{-\frac{1}{2}(2\pi\sigma_\varphi(t))^2}$$

This leads to an oscillation at the mean frequency, with an envelope function represented by an exponential of the variance of the phases. This envelope decays to zero if the variance in the phases keeps increasing, which is the case for nonsynchronized oscillators. Note that if the actual limit cycle is somewhat noncircular, the cosine could be replaced by a more general periodic function. This would in many cases only have a weak effect on the shape of the population average as the higher harmonics have a much larger decay rate in the population average. The population average has two limiting regimes:

$$e^{-\frac{1}{2}(2\pi\sigma_\varphi(t))^2} \approx \begin{cases} e^{-(2\pi\sigma)^2 t^2/2} & \text{for } t \to 0 \\ e^{-(2\pi\sigma/\gamma)^2(\gamma t - 1)} & \text{for } t \to \infty \end{cases} \quad (2)$$

At short times, the population behavior resembles that of static frequencies (indicated by the quadratic time dependence in the exponent), whereas in the long-time limit, we approach a phase diffusion (linear time dependence in the exponent). The dephasing dynamics connecting these two asymptotic formulae provides a way to estimate the relevant parameters from experimental data. In other words, the early decay resembles a bell shape and reflects the frequency dispersion, whereas the longer times reflect phase diffusion.

Interacting Phase Oscillators

We next introduce coupling among the oscillators. Here, we follow the Kuramoto model and use a simple sine of the phase difference between coupled cells. This introduces an additional parameter K to measure coupling strength:

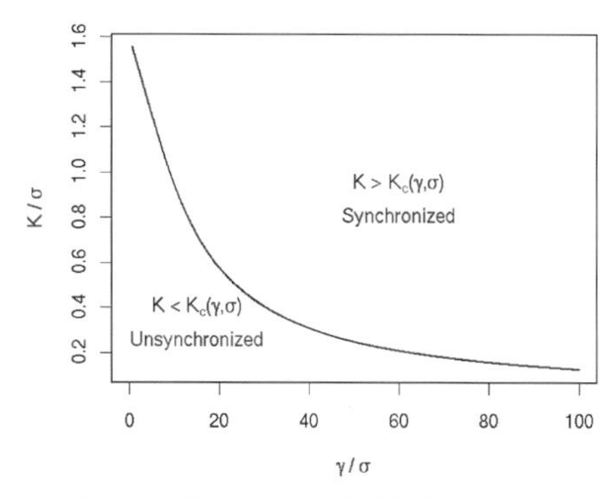

Figure 2. Synchronization thresholds in the coupled phase model. The critical coupling above which the population synchronizes is shown as a function of the damping rate γ. Coupling and damping are expressed in units of the frequency dispersion σ. (Adapted, with permission, from Rougemont and Naef 2006 [© American Physical Society].)

$$\begin{cases} \dfrac{d\varphi_i(t)}{d(t)} = f_i(t) + \dfrac{K}{|N_i|} \displaystyle\sum_{j \in N_i} \sin(2\pi(\varphi_j(t) - \varphi_i(t))) \\ \dfrac{df_i(t)}{dt} = -\gamma(f_i(t) - \mu) + \sqrt{(2\gamma)}\sigma\eta_i(t) \end{cases} \quad (3)$$

Note that the equation for f is only a convenient alternative description of the Ornstein–Uhlenbeck process introduced above. In this model, cell number i is coupled uniformly to a neighborhood N_i, and we considered the case of all-to-all coupling for simplicity. The noise source has been reparametrized such that $\eta_i(t)$ has unit variance. As in the Kuramoto model, we find a phase transition from a desynchronized state to a collectively synchronized state when the coupling is increased above a threshold K_c (Fig. 2) (Rougemont and Naef 2006). Here, synchronization corresponds to the statement that the amplitude of the variable $Z_N(t)$ defined above does not tend to zero at large times even for an infinite number of oscillators.

ANALYZING REAL BIOLUMINESCENCE CURVES

Recent recordings in immortalized (Nagoshi et al. 2004) and primary (Welsh et al. 2004) mouse fibroblasts showed that single cells from peripheral tissues generate cell-autonomous rhythms that could be resynchronized by a serum shock. These single-cell assays were backed up by comparing long (19 days) bioluminescence population recordings with a phase model for static and uncoupled oscillators (Nagoshi et al. 2004). This analysis showed that dephasing, rather than amplitude death, was the dominant cause for amplitude reduction during the 19-day recording. However, the static model underestimated the frequency dispersion. This is expected because the observed frequencies are drifting: Given equal dispersions, drifting frequencies have the property of keeping phase coherence longer, because each oscillator stays out of tune for a shorter average time (see Eq. 2 above).

Envelope Analysis Using Phase Models

We analyzed a 19-day bioluminescence recording from NIH-3T3 fibroblasts stably transfected with a luceriferase reporter inserted in the Bmal1 locus (cf. Fig. 3B in Nagoshi et al. 2004) using the above phase model. The data are reproduced here in Figure 3A. The detrended signal was fit to the prediction for the population average from Equation 3 (Fig. 3B). All parameters could be estimated reliably, and the error bars indicated in Figure 3C show that the model does not overfit the data. The frequency dispersion was found to be 0.1 days^{-1}, which is close to values measured in single cells (Nagoshi et al. 2004). Furthermore, the estimated parameter equals 0.64 ± 0.17, reflecting a frequency damping time of 1.56 days and implying that frequency disturbances take longer than a period length to decay. The model could also estimate intercell coupling which indicated that coupling in cell culture might be positive. However, the values estimated were clearly below the synchronization threshold (Fig. 3C). This analysis is detailed in Rougemont and Naef (2007), where it is also substantiated with the analysis of a second bioluminescence curve from Welsh et al. (2004) with similar conclusions.

Recordings in large populations produce signals with a very high signal-to-noise ratio by averaging out noncircadian variations over thousands of individual cells (see Fig. 3) (Welsh et al. 2004). The resulting clean signal has a typical decaying envelope that is a characteristic signature of the underlying dephasing dynamics as shown above. An alternative way of removing random fluctuations can be obtained by averaging the signal produced by a single cell over a large number of periods using the autocorrelation function. We described (Rougemont and Naef 2007) how the autocorrelation and the population average contain similar information and can both be used to fit informative parameters. In practice, however, it is more common to have relatively short recordings (up to 20 periods) of large populations (up to 5000 cells), which is more suitable to the population analysis than the autocorrelation calculations. Furthermore, only supercritical coupling strengths are visible in the autocorrelation; subcritically coupled populations are indistinguishable from uncoupled populations in this type of analysis (Fig. 4).

Coculture Experiment

Nagoshi et al. (2004) designed a coculture experiment to study the question of intercellular coupling directly. Short-period (20 hours) mutant feeder cells were cocultured with wild-type reporter cells, with a 20-fold relative excess of nonluminescent feeders. If there were significant coupling, one would predict that the reporter line would slow down its period. Bioluminescence recordings showed no significance phase-shifts over approximately 3 days. To show the potential effect of intercellular coupling in such an experiment, we plotted the expected phase-shift over time for several values of the coupling constant K. This is computed as follows:

$$\begin{aligned} \delta\varphi(t) &= \varphi_{\text{reporter}}(t) - \frac{t}{24} \\ &= K \int_0^t \sin(2\pi(\varphi_{\text{feeder}}(s) - \varphi_{\text{reporter}}(s)))ds \\ &= K \int_0^t \sin(2\pi(s(1/20 - 1/24) - \delta\varphi(s)))ds \end{aligned}$$

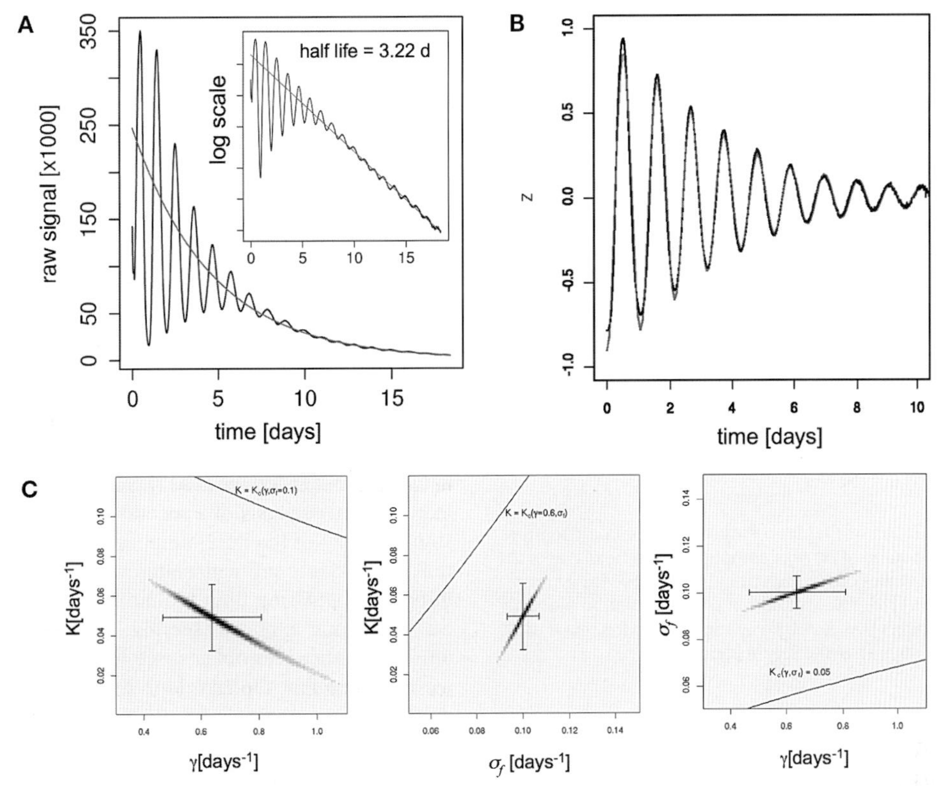

Figure 3. Analysis of population bioluminescence data. (*A*) Raw bioluminescence counts for a recording of 3 weeks. (*Inset*) Exponential trend with a half-life of 3.2 days. (*B*) Detrended signal (*black*) and the model best fit (*gray*). (*C*) Parameter estimates. (*Gray shades*) Posterior probabilities; (*black line*) lines of critical coupling. (Reprinted, with permission, from Rougemont and Naef 2007 [Nature Publishing Group].)

First note that the phase-shift does not increase monotonously due to the sine. In the subcritical region, the coupling induces only a slow dephasing that can remain very low for several days (Fig. 5). In particular, a value of $K = 0.05$ days^{-1} as estimated from population recordings would only shift phases forward by 2.4 hours after 3 days. Note that even a critical coupling of $K = 0.1183$ would still require about 10 days to reverse the phase relative to a free-running wild-type cell.

CORRESPONDENCE BETWEEN PHASE AND RATE EQUATION MODELS

Our analysis has shown that the stability of the frequencies (γ) leads to observable effects both in single cells, where it determines the frequency dispersion, and in population recordings of initially synchronized oscillators, where it shapes the long-term decay characteristics of the envelope. One can now ask how is stability determined by the network of coupled chemical reactions that generates the oscillations. The canonical way to discuss the stability of periodic orbits in dynamical systems is through the computation of Floquet multipliers (FMs), which determine the rate at which small perturbations decay back to the limit cycle. These multipliers can be computed from the system of ordinary differential equations (rate models) using computational methods (Doedel et al. 2001). We have explored the Floquet stability in a detailed model of the mammalian clock (Leloup and

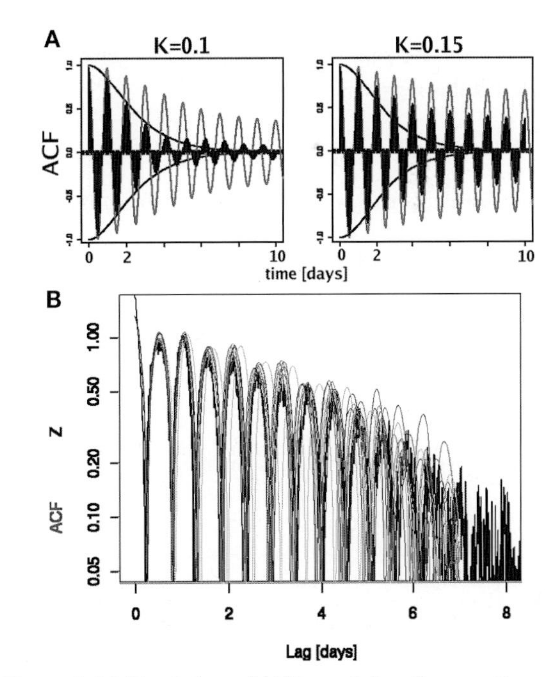

Figure 4. (*A*) Simulations of 1000 coupled oscillators with coupling strength below and above the synchronization threshold at $K = 0.118$. (*Black bars*) Autocorrelation function of one the oscillators; (*black line*) population average; (*gray line*) exact expression for an infinite population of uncoupled oscillators. (*B*) The population average (*thick black bars*) is superimposed on the autocorrelations of ten individual cells (*gray lines*) computed from measurements by Welsh et al. (2004).

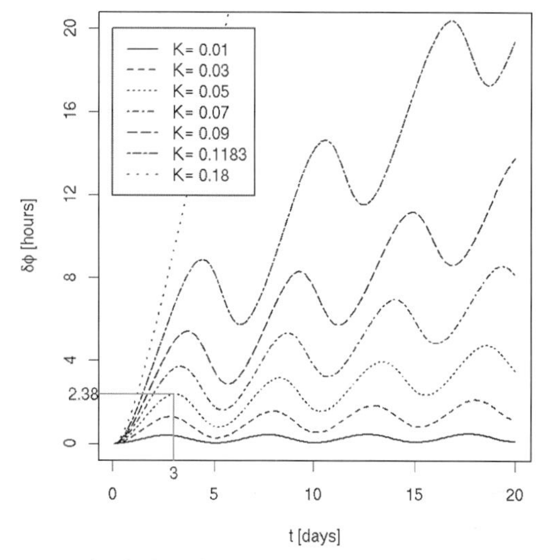

Figure 5. Simulation of the dephasing dynamics in a coculture experiment. The value of $\delta\phi$ was computed numerically over 20 days for several values of K. The critical value of K for synchronization is $K = 0.1183$. Our estimation from population bioluminescence recordings was $K = 0.05$, which leads to a phase-shift of 2.38 hours 3 days after the serum shock.

Goldbeter 2004) by varying single parameters in a window around their nominal values. This computation showed that most parameters have little effects on the stability as measured by FMs, whereas the transcription and phosphorylation rates of the *Period* gene (only one merged *Period* gene is included in the model) are among the parameters with the strongest influence (Rougemont and Naef 2007). Such analyses being relevant in the near proximity of the limit cycle, it remains to be investigated to what extent cellular noise sources are small enough such that this regime is indeed probed.

DISCUSSION

We presented how modeling of noisy-phase oscillators can be fruitful for circadian biology. In the models developed, we introduced the fewest number of parameters necessary to capture generic and experimentally observed phenomena such as drifting frequencies or oscillator coupling. The practical advantage is that the limited parameter set can be determined from currently available data. The downside is that the parameters are not necessarily easy to interpret in molecular terms. For example, explicit coupling mechanisms are not yet known in molecular detail, and we therefore used an effective interaction of phases in terms of a sine function. The molecular basis of stability and its interplay with cellular noise are perhaps a more readily accessible problem. First, we are beginning to obtain relatively detailed kinetic models on the circadian networks generating oscillations, and second, we also have the theoretical methodologies for sensitivity and stability analyses of differential equation models. We also know how to account for cellular fluctuations in these models through the use of stochastic simulation algorithms.

Our analysis of cultured cells showed that intercellular coupling is certainly too weak for synchronizing populations. Our best fits do, however, suggest a small residual coupling. Theoretically, the presence of a positive subcritical coupling may facilitate resynchronization of tissues by systemic cues in the event that the latter would be confined to relatively short-time windows. Although the systems are difficult to compare, a recent report in cyanobacteria used an ingenious mixing of two initially out-of-phase populations to directly measure coupling strength and found an upper bound for the coupling that is about 30 times smaller than our values (Amdaoud et al. 2007). A similar approach could be applied to mammalian cells.

A recent work (Liu et al. 2007) specifically addressed the role of coupling in the SCN compared to peripheral organs such as liver and lung. These authors showed that intercellular coupling is active (and required) in the SCN by showing that SCN explants from Cry2 mutant mice maintain a synchronized rhythmic expression of Per2, although dissociated neurons rapidly drifted out of phase. In peripheral organs, however, tissue explants displayed a similar decay of the Per2 bioluminescence at the population level than confluent cultures of dissociated fibroblasts. By comparing their bioluminescence recordings of SCN and lung explants (see Fig. 1) (Liu et al. 2007) to those of cultured fibroblasts (see Fig. 2) (Liu et al. 2007), we observed that the envelope of the population signal does not decay to zero in the SNC, whereas the lung explant and the cell culture have a visually similar decay rate, which in the light of our phase models is fully consistent with a high (supercritical) coupling in the SCN and a low (subcritical) coupling in the peripheral organs.

Most importantly, several interesting model systems exhibit molecular oscillations, e.g., the respiratory cycle in yeast or the segmentation clock in vertebrates, for which powerful molecular reporter techniques are being developed. It is thus desirable to have at hand a theoretical framework that allows one not only to extract the most pertinent information from the data, but also to bridge the gap between the often unknown or too complex microscopic description and a phenomenological lower-dimensional description in terms of few highly informative parameters. We believe that phase models do qualify for this purpose and forecast that they will continue to provide useful insights for circadian biologists.

ACKNOWLEDGMENTS

The authors acknowledge stimulating discussions with Benoit Kornmann and Ueli Schibler. This work was supported by a Swiss National Science Foundation grant to F.N.

REFERENCES

Amdaoud M., Vallade M., Weiss-Schaber C., and Mihalcescu I. 2007. Cyanobacterial clock, a stable phase oscillator with negligible intercellular coupling. *Proc. Natl. Acad. Sci.* **104:** 7051.

Bagheri N., Stelling J., and Doyle F.J., III. 2007. Quantitative performance metrics for robustness in circadian rhythms. *Bioinformatics* **23:** 358.

Barkai N. and Leibler S. 2000. Circadian clocks limited by noise. *Nature* **403:** 267.

Carr A.J. and Whitmore D. 2005. Imaging of single light-responsive clock cells reveals fluctuating free-running periods. *Nat.*

Cell Biol. **7:** 319.
Doedel E.J., Paffenroth R.C., Chapneys A.R., Fairgrieve T.F., Kutzetsov Y.A., Oldman B.E., Sandstede B., and Wang X.J. 2001. *AUTO2000: Continuation and bifurcation software for ordinary differential equations*. Technical report, California Institute of Technology, Pasadena, California.
Elowitz M.B. and Leibler S. 2000. A synthetic oscillatory network of transcriptional regulators. *Nature* **403:** 335.
Forger D.B. and Peskin C.S. 2003. A detailed predictive model of the mammalian circadian clock. *Proc. Natl. Acad. Sci.* **100:** 14806.
Gillespie D.T. 1977. Exact stochastic simulation of coupled chemical reactions. *J. Phys. Chem.* **81:** 2340.
Goldbeter A. 1995. A model for circadian oscillations in the *Drosophila* period protein (PER). *Proc. Biol. Sci.* **261:** 319.
Gonze D. and Goldbeter A. 2006. Circadian rhythms and molecular noise. *Chaos* **16:** 026110.
Guo H., Brewer J.M., Lehman M.N., and Bittman E.L. 2006. Suprachiasmatic regulation of circadian rhythms of gene expression in hamster peripheral organs: Effects of transplanting the pacemaker. *J. Neurosci.* **26:** 6406.
Izumo M., Johnson C.H., and Yamazaki S. 2003. Circadian gene expression in mammalian fibroblasts revealed by real-time luminescence reporting: Temperature compensation and damping. *Proc. Natl. Acad. Sci.* **100:** 16089.
Kuramoto Y. 1984. *Chemical oscillations, waves, and turbulence*. Springer-Verlag, Berlin.
Leloup J.C. and Goldbeter A. 2003. Toward a detailed computational model for the mammalian circadian clock. *Proc. Natl. Acad. Sci.* **100:** 7051.
———. 2004. Modeling the mammalian circadian clock: Sensitivity analysis and multiplicity of oscillatory mechanisms. *J. Theor. Biol.* **230:** 541.
Liu A.C., Welsh D.K., Ko C.H., Tran H.G., Zhang E.E., Priest A.A., Buhr E.D., Singer O., Meeker K., Verma I.M., Doyle F.J., III, Takahashi J.S., and Kay S.A. 2007. Intercellular coupling confers robustness against mutations in the SCN circadian clock network. *Cell* **129:** 605.
Liu C. and Reppert S.M. 2000. GABA synchronizes clock cells within the suprachiasmatic circadian clock. *Neuron* **25:** 123.
Locke J.C., Millar A.J., and Turner M.S. 2005. Modelling genetic networks with noisy and varied experimental data: The circadian clock in *Arabidopsis thaliana*. *J. Theor. Biol.* **234:** 383.
Maywood E.S., Reddy A.B., Wong G.K., O'Neill J.S., O'Brien J.A., McMahon D.G., Harmar A.J., Okamura H., and Hastings M.H. 2006. Synchronization and maintenance of timekeeping in suprachiasmatic circadian clock cells by neuropeptidergic signaling. *Curr. Biol.* **16:** 599.
Mihalcescu I., Hsing W., and Leibler S. 2004. Resilient circadian oscillator revealed in individual cyanobacteria. *Nature* **430:** 81.
Nagoshi E., Saini C., Bauer C., Laroche T., Naef F., and Schibler U. 2004. Circadian gene expression in individual fibroblasts: Cell-autonomous and self-sustained oscillators pass time to daughter cells. *Cell* **119:** 693.
Njus D., Gooch V.D., and Hastings J.W. 1981. Precision of the *Gonyaulax* circadian clock. *Cell Biophys.* **3:** 223.
Ohta H., Yamazaki S., and McMahon D.G. 2005. Constant light desynchronizes mammalian clock neurons. *Nat. Neurosci.* **8:** 267.
Rand D.A., Shulgin B.V., Salazar J.D., and Millar A.J. 2006. Uncovering the design principles of circadian clocks: Mathematical analysis of flexibility and evolutionary goals. *J. Theor. Biol.* **238:** 616.
Rougemont J. and Naef F. 2006. Collective synchronization in populations of globally coupled phase oscillators with drifting frequencies. *Phys. Rev. E Stat. Nonlin. Soft Matter Phys.* **73:** 011104.
———. 2007. Dynamical signatures of cellular fluctuations and oscillator stability in peripheral circadian clocks. *Mol. Syst. Biol.* **3:** 93.
Strogatz S.H. 2000. *Nonlinear dynamics and Chaos: With applications to physics, biology, chemistry and engineering*. Perseus Books, Cambridge, Massachusetts.
Toh K.L., Jones C.R., He Y., Eide E.J., Hinz W.A., Virshup D.M., Ptacek L.J., and Fu Y.H. 2001. An hPer2 phosphorylation site mutation in familial advanced sleep phase syndrome. *Science* **291:** 1040.
Vilar J.M., Kueh H.Y., Barkai N., and Leibler S. 2002. Mechanisms of noise-resistance in genetic oscillators. *Proc. Natl. Acad. Sci.* **99:** 5988.
Welsh D.K., Yoo S.H., Liu A.C., Takahashi J.S., and Kay S.A. 2004. Bioluminescence imaging of individual fibroblasts reveals persistent, independently phased circadian rhythms of clock gene expression. *Curr. Biol.* **14:** 2289.
Winfree A.T. 1967. Biological rhythms and the behavior of populations of coupled oscillators. *J. Theor. Biol.* **16:** 15.
———. 2001. *The Geometry of biological time,* 2nd edition. Springer, New York.
Yamaguchi S., Isejima H., Matsuo T., Okura R., Yagita K., Kobayashi M., and Okamura H. 2003. Synchronization of cellular clocks in the suprachiasmatic nucleus. *Science* **302:** 1408.
Yoo S.H., Yamazaki S., Lowrey P.L., Shimomura K., Ko C.H., Buhr E.D., Siepka S.M., Hong H.K., Oh W.J., Yoo O.J., Menaker M., and Takahashi J.S. 2004. PERIOD2::LUCIFERASE real-time reporting of circadian dynamics reveals persistent circadian oscillations in mouse peripheral tissues. *Proc. Natl. Acad. Sci.* **101:** 5339.

Reversible Protein Phosphorylation Regulates Circadian Rhythms

D.M. VIRSHUP,*†‡ E.J. EIDE,†§ D.B. FORGER,¶ M. GALLEGO,†
AND E. VIELHABER HARNISH†**

*Division of Hematology / Oncology, Department of Pediatrics, University of Utah, Salt Lake City, Utah 84112;
†Department of Oncological Sciences and the Huntsman Cancer Institute, University of Utah, Salt Lake City, Utah 84112; ¶Department of Mathematics, University of Michigan, Ann Arbor, Michigan 48109

Protein phosphorylation regulates the period of the circadian clock within mammalian cells. Circadian rhythms are an approximately 24-hour cycle that regulates key biological processes. Daily fluctuations of wakefulness, stress hormones, lipid metabolism, immune function, and the cell division cycle are controlled by the molecular clocks that function throughout our bodies. Mutations in regulatory components of the clock can shorten or lengthen the timing of the rhythms and have significant physiological consequences. The clock is formed by a negative feedback loop of transcription, translation, and inhibition of transcription. The precision of clock timing is controlled by protein kinases and phosphatases. Casein kinase Iε is a protein kinase that regulates the circadian clock by periodic phosphorylation of the proteins PER1 and PER2, controlling their stability and localization. The role of phosphorylation in regulating PER function in the clock has been explored in detail. Quantitative modeling has proven to be very useful in making important predictions about how changes in phosphorylation alter the clock's behavior. Quantitative data from biological studies can be used to refine the quantitative model and make additional testable predictions. A detailed understanding of how reversible protein phosphorylation regulates circadian rhythms and a detailed quantitative model that makes clear, testable, and accurate predictions about the clock and how we may manipulate it can have important benefits for human health. Pharmacological manipulation of rhythms could mitigate stress from jet lag, shift work, and perhaps even seasonal affective disorder.

INTRODUCTION

Circadian rhythms govern key physiologic processes, ranging from sleep-wake cycles, glucose, lipid and drug metabolism, heart rate, stress and growth hormones, and immunity, as well as basic cellular processes such as DNA repair and the timing of the cell division cycle (Chang and Reppert 2001; Fu et al. 2002; Reppert and Weaver 2002; Matsuo et al. 2003; Lowrey and Takahashi 2004; Antoch et al. 2005; Turek et al. 2005). The disruption of circadian rhythms causes significant physiologic stress, is frequently experienced in jet lag and night shift work, and has been linked to bipolar disorder (Mansour et al. 2005). Circadian changes in physiology can also be exploited to more efficiently treat disease. For example, because there is a fourfold increase in cholesterol biosynthesis in the liver at night, the cholesterol-lowering 3-hydroxy-3-methylglutaryl–coenzyme A (HMG-CoA) reductase inhibitors (statins such as lovastatin) are most effective when taken in the evening (e.g., see Wallace et al. 2003). Similarly, two studies found that children with leukemia who take oral maintenance chemotherapy (6-mercaptopurine and methotrexate) in the evening have a clear survival advantage over children who take these medications in the morning (Rivard et al. 1993; Schmiegelow et al. 1997). Mice with mutant circadian rhythms are at increased risk of obesity, metabolic syndrome, and cancer (Fu et al. 2002; Antoch et al. 2005; Turek et al. 2005). Thus, circadian regulation of physiology has many important consequences for health. A detailed quantitative model that makes clear, testable, and accurate predictions about the clock and how we may manipulate it can have important benefits for human health. Model predictions can guide drug development. Pharmacological manipulation of rhythms could mitigate stress from jet lag, shift work, and perhaps even seasonal affective disorder. The best drug targets in the clock are enzymes, and the enzymes that have the greatest effect on the clock are kinases, specifically casein kinase I (CKI).

Reversible protein phosphorylation is a key regulator of circadian rhythms. Mutations in *PERIOD*, *TIMELESS*, *CLOCK*, *CYCLE*, and the casein kinase Iε (CKIε) ortholog *DOUBLETIME* all alter circadian rhythm in *Drosophila*. Orthologs of the same genes regulate mammalian circadian rhythms. Similarly, the first mammalian circadian rhythms mutant identified, the *tau* hamster, also has a missense mutation in CKIε (Ralph and Menaker 1988; Lowrey et al. 2000). This *tau* mutation, R178C, decreases the V_{max} of CKIε eightfold. Mutations both in the closely related and apparently redundant CKIδ and in a CKI phosphorylation site in PER2, have been identified in human families with familial advanced sleep phase syndrome (FASPS) (Jones et al. 1999; Toh et al. 2001; Xu et al. 2005). However, we are only beginning to understand how these kinase mutations alter circadian period. One confusing point has been that mutations in *Drosophila* CKI (*Doubletime, Dbt*) that either cause *long*

Present addresses: ‡Program in Cancer and Stem Cell Biology, Duke-NUS Graduate Medical School Singapore, 2 Jalan Bukit Merah, Singapore 169547; §Fred Hutchison Cancer Research Center, Seattle, Washington 98109; **Sanofi-Aventis, Bridgewater, New Jersey 08807.

or *short* periods all *decrease* CKI activity in vitro (Suri et al. 2000; Eide and Virshup 2001; Preuss et al. 2004). Likewise, mutants with decreased mammalian CKIε and CKIδ activity (in vitro) may either lengthen (Eide et al. 2005) or shorten circadian periods (Lowrey et al. 2000; Xu et al. 2005). It has been difficult to reconcile how different mutations and inhibitors that apparently decrease CKI activity have opposite effects on circadian periods by previous models of the circadian clock. Because CKI represents one of the best drug targets in the circadian regulatory system, understanding if its inhibition speeds up or delays the clock is of central importance. In a collaboration between mathematical modeling and experimental biology, we have tested a detailed mathematical model of the clock and experimentally verified a key and unexpected prediction: The short-period mutations of CKI are in fact gain-of-function, not loss-of-function, mutations. We found that CKIε^{tau} accelerates the degradation of the PER proteins (Gallego et al. 2006b).

THE CORE CLOCK IS A NEGATIVE FEEDBACK LOOP REGULATED BY PROTEIN PHOSPHORYLATION

The heart of the mammalian clock is a transcription-translation negative feedback loop with a delay between the signal and the negative feedback (Fig. 1) (Lowrey and Takahashi 2004; Schibler and Naef 2005). The oscillations this establishes are then reinforced by additional positive feedback loops. The negative feedback is provided by PERIOD (PER1 and PER2) and CRYPTOCHROME (CRY1 and CRY2) proteins inhibiting the heterodimeric transcription factors CLOCK and BMAL1. CLOCK and BMAL1 drive expression both of circadian output genes and, importantly, their own negative regulators PER1, PER2, CRY1, and CRY2. The abundance of PER1 and PER2 is in turn controlled by protein phosphorylation by

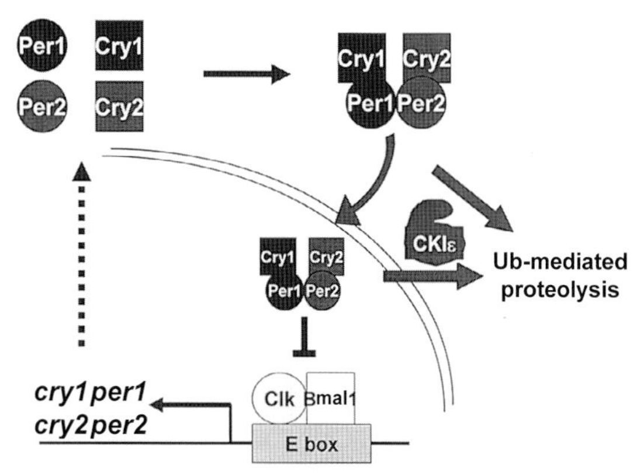

Figure 1. Circadian Rhythms are controlled by a phosphorylation-regulated negative feedback loop. Early in the circadian cycle, PER, CKI, and CRY proteins multimerize in the cytoplasm and then translocate to the nucleus to repress the CLK:BMAL1 transcription factor. Potential functional effects of CKIδ and CKIε (denoted CKIε for simplicity) include (1) degradation of PER early in the accumulation phase, delaying repression; (2) regulating PER nuclear entry of the inhibitor complex, or (3) promoting degradation of PER, thereby terminating repression. The stabilizing Rev-Erbα loop is not shown here.

CKIε and CKIδ. In cell-based assays, we have recently reported that CKIε and CKIδ accelerate the proteolysis of PER1 and PER2 by creating a binding site for the ubiquitin ligase β-TrCP (β-transducin repeat-containing protein) (Eide et al. 2005). This is likely to be a key element speeding up the clock, because both CKI inhibitors and proteasome inhibitors slow the clock. However, our result in cell-based systems is in direct conflict with the animal-based result that the *tau* mutation, shown to decrease CKIε activity in vitro, speeds up the clock. Building on a detailed and predictive quantitative model of circadian rhythms, we have resolved this conflict, providing new insights into basic clock properties (Gallego et al. 2006b).

PHOSPHATASES IN THE CLOCK

Regulated phosphorylation and degradation of PER is important in controlling the timing of the mammalian clock. Because protein phosphorylation is reversible, protein phosphatases may regulate circadian rhythms as well (Comolli et al. 1996, 2003; Sathyanarayanan et al. 2004; Yang et al. 2004). Protein phosphatase 1 (PP1) and protein phosphatase 2A (PP2A) have recently been implicated in core clock control. Mutations in PP1 and PP2A alter period in lower organisms, and specific PP2A regulators have been shown to regulate PER dynamics and period in *Drosophila* (Sathyanarayanan et al. 2004; Yang et al. 2004; Schafmeier et al. 2005). Our data show that phosphatases have a role in mammalian circadian rhythm regulation as well and strongly implicate PP1 (Gallego et al. 2006a). Oscillations in PP1 regulators are a potential route to regulating PP1 activity and PER degradation. A robust quantitative model of circadian rhythm will need to incorporate both kinase and phosphatase control of key regulatory steps in the clock.

A PRELIMINARY QUANTITATIVE MODEL OF THE MAMMALIAN CIRCADIAN CLOCK

How much of the mechanism of the mammalian circadian clock is known? Are the genes, proteins, and molecular interactions, which have been identified as key components in the mammalian circadian clock, sufficient to explain the robust 24-hour timing seen in mammalian cells? To answer these questions, we can convert all of the known genes, proteins, and molecular interactions into a quantitative model and see if this model reproduces the known properties of circadian timing. An initial attempt at modeling the mammalian circadian clock was conducted by Forger and Peskin (2003). Unlike other modeling studies that aim solely at parsimony or arriving at a model which is a good candidate for mathematical analysis, the goal of this work was to represent the available (at the time of the model's development) experimental data on the mammalian circadian clock as accurately as possible.

This model considers the dynamics of the clock proteins PER1, PER2, CRY1, CRY2, Rev-Erbα, CKIε, and CKIδ (although these last two kinases are treated as the same). CLK-BMAL1 is assumed to be constitutively bound to the promoters of the relevant genes, based on published experimental data (Kume et al. 1999;

Yamaguchi et al. 2000; Etchegaray et al. 2003). Rev-Erbα is assumed to form a dimer that can bind to the CRY1 promoter (Preitner et al. 2002). When one or more molecules of the PER/CRY complex (or Rev-Erbα) are bound to a promoter, transcription ceases (Etchegaray et al. 2003). The only posttranslational modification that the model currently includes is phosphorylation of PER1 and PER2. When these proteins are phosphorylated, they are targeted for ubiquitin-mediated degradation (Eide et al. 2005; Shirogane et al. 2005). The model assumes that phosphorylation allows the PER proteins to enter the nuclear of the cell (Vielhaber et al. 2000; Akashi et al. 2002). We also assumed that a alternative phosphorylation event blocked the nuclear import of PER1 (but not PER2) to match experimental data (Vielhaber et al. 2000; Yagita et al. 2002).

With these biological assumptions, a quantitative model was developed based on mass action (Forger and Peskin 2003). Variables are the concentrations of the various states of mRNAs and protein complexes and the probability that binding sites on a promoter are bound by transcription factors. This leads to a set of differential equations with multiple rate constants (Forger and Peskin 2003). These rate constants were estimated by choosing those that gave the best predictions of the time courses of the concentrations of clock proteins using a direct search method.

This model has made many important predictions of the effects of circadian clock mutations (Forger and Paydarfar 2004; Forger and Peskin 2004; Locke et al. 2005; Indic and Brown 2006; also see below). The model can also be simulated with the Gillespie Method, which accounts for the stochastic interaction of individual molecules in a cell (Forger and Peskin 2005). Using this method, predictions can be made about the accuracy of timing in the wild-type mammalian circadian clock and in the presence of mutations of clock genes. Although this model will need to be regularly updated reflecting both recent data and the results from experiments in this proposal, it presents a powerful tool for understanding clock function and interpreting experimental data.

EXPERIMENTAL ADVANCES

Our initial studies demonstrated that phosphorylation controlled the nuclear localization of mPER1, and we mapped phosphorylation-dependent nuclear localization sequences and nuclear export signals (Vielhaber et al. 2000). Novel nuclear export sequences were discovered in PER1 and PER2 that enable the PER proteins to be exported from the nucleus as well (Vielhaber et al. 2001). Because CKIε is in a multiprotein complex with multiple circadian proteins, we asked if other circadian regulators were targets of CKIε. We found that CRY1 and CRY2 are phosphorylated by CKIε but only when they bind to a PER/CKIε complex. BMAL1 is also an in vivo CKIε substrate. Knock down of CKIε by RNA interference (RNAi), expression of a dominant-negative CKIε (K38A), or use of a CKI inhibitor lead to decreased phosphorylation of BMAL1 in cells and decreased transcriptional activation activity of BMAL1:CLOCK (Eide et al. 2002; Gallego et al. 2006b).

PER2 PROTEIN STABILITY IS CONTROLLED BY PROTEIN PHOSPHORYLATION AND UBIQUITIN-MEDIATED PROTEASOMAL DEGRADATION

We and other investigators found that although PER1 nuclear localization is regulated by CKIε, PER2 localization is not altered by CKIε or CKIδ expression (Vielhaber et al. 2000; Akashi et al. 2002; Takano et al. 2004). This suggests that phosphorylation of PER2 has distinct regulatory effects. We next reported that CKI-regulated phosphorylation of PER2 changes its stability (Eide et al. 2005; Gallego et al. 2006b). The quantitative model indicates that control of PER2 abundance is likely to be a major control point in proper clock timing (Gallego et al. 2006b).

As we investigated how protein phosphorylation is a critical regulator of PER2 stability, we found that both kinases *and* protein phosphatases are important in this control (Fig. 2) (Gallego and Virshup 2007). Treatment of cells expressing full-length PER2 with a cell-permeable serine/threonine phosphatase inhibitor (either okadaic acid or calyculin A) causes rapid degradation of PER2 (Eide et al. 2005). Degradation requires the activity of CKI, because pretreatment of cells with the CKI inhibitor IC261 blocks PER2 degradation. PER2 degradation is ubiquitin-mediated and proteasome-dependent, because pretreatment with the proteasome inhibitor MG-132 blocks mPER2 degradation and causes the accumulation of polyubiquitinated PER2. An important cellular ubiquitin E3 ligase, the stem cell factor (SCF), recognizes phosphorylated proteins before their degradation. The substrate-recognition protein in this complex is the adapter protein β-TrCP (see model in Fig. 2). We showed that a dominant-negative form of β-TrCP (able to interact with phosphorylated substrates but lacking the "F box" and unable to interact with the remainder of the SCF) is able to bind to phosphorylated PER2 and block both its ubiquitination and proteolysis (Eide et al. 2005). This result was subsequently confirmed by others (Shirogane et al. 2005). We have identified the phosphatase that regulates PER2 phosphorylation stability as PP1 (Gallego et al. 2006a).

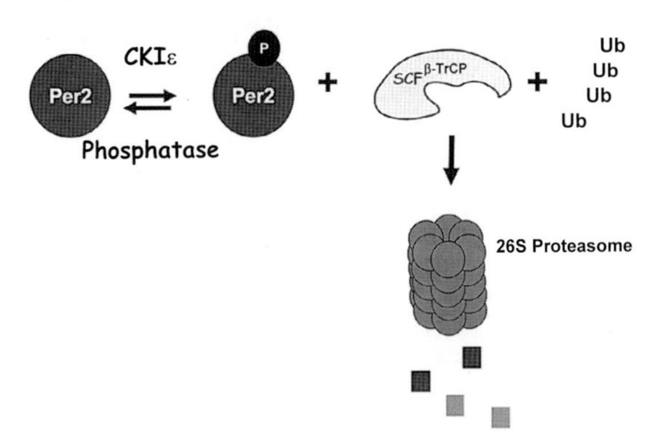

Figure 2. Phosphorylation controls the abundance of PER2. Phosphorylation of a specific site on PER2 is regulated by the balance of CKI and phosphatase activity. $SCF^{\beta\text{-TrCP}}$ binds to phosphorylated PER2 and drives ubiquitinylation and subsequent proteasome-dependent proteolysis. Destruction of PER2 will relieve inhibition of the CLK:BMAL1 complex shown in Fig. 1.

Studying CKIε-regulated PER2 stability, we found that β-TrCP interacts with PER2 via the sequence ^{477}SSGYGS^{482}, similar to the consensus β-TrCP recognition motif DSGϕXS, where ϕ is hydrophobic (Wu et al. 2003). Mutation to ^{477}**A**S**A**YGS^{482} (mPER2 [S477A/G479A]) abolished almost all detectable CKIε-dependent β-TrCP binding to PER2 and inhibited PER2 degradation (Eide et al. 2005). The simplest model from these data is that CKIε phosphorylation of PER2 at Ser-478 and Ser-482 creates a β-TrCP-binding site leading to ubiquitination and degradation of PER2.

A CELL-BASED ASSAY FOR CIRCADIAN RHYTHMS

A key step in the analysis of circadian rhythms is to test if biochemical phenomenon dissected in vitro and in cells actually affects clock function. Until recently, this has relied heavily on analysis of rhythms in whole animals. However, the observation that individual cells throughout the body have an intrinsic clock and that these clocks can be synchronized in cultured cells (Balsalobre et al. 1998; Yamazaki et al. 2000) has enabled a new generation of assays examining circadian rhythms. The initial method described by the Schibler lab assessed circadian gene expression by ribonuclease protection assay from cells harvested every 3 hours for 4 days (Balsalobre et al. 1998). We postulated that a similar cell-culture-based system with a real-time luciferase reporter would be a more robust and less labor-intensive method to measure circadian rhythms. Using 6.7 kb of the *mPER1* promoter driving luciferase (Yamazaki et al. 2000), stable Rat-1a cell transfectants were developed that display serum-inducible and rhythmic luciferase activity; these cells are denoted Rat-1 (*Per1::luc*). To follow circadian rhythms in real time, cells are incubated in medium supplemented with luciferin (described in Eide et al. 2005). The primary light-output data are collected in a 24-well luminometer and smoothed by subtracting the 24-hour running mean. Rat-1 (*Per1::luc*) cells show sustained and consistent circadian oscillations of luciferase abundance (Fig. 3). The period, phase, and amplitude are then quantitatively determined by Fourier transform of the data. The end result is a useful system that allows us to directly test both model predictions and molecular hypotheses (Eide et al. 2005).

THE PROTEASOME REGULATES MAMMALIAN CIRCADIAN RHYTHM

We tested two hypotheses about the effect of proteasome inhibition on clock activity. If this accelerated accumulation of PER2 early in the cycle, it might speed up the clock, whereas if it delayed degradation late in the cycle, it might prolong repression of CLOCK/BMAL1 and slow the clock. We found that proteasome inhibition caused a significant lengthening of period. This result suggests, as predicted by the quantitative model, that degradation of PER proteins is a key event *shortening* circadian period; therefore, inhibition of PER degradation lengthens period (see model in Fig. 1). However, a number of circadian regulators including both PER proteins are degraded by the proteasome and could be altered by proteasome inhibition. It remains critical to specifically test whether specific inhibition of PER2 degradation lengthens the period. The mutations in the F-box protein Fbxl3 indicate that impaired CRY protein degradation also lengthens period (Busino et al. 2007; Godinho et al. 2007; Siepka et al. 2007; Virshup and Forger 2007).

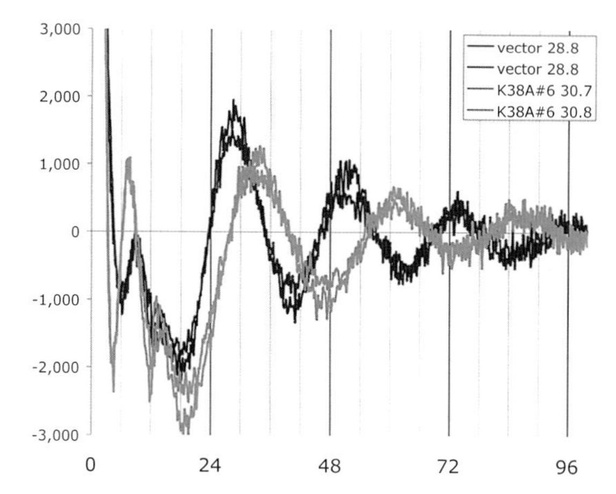

Figure 3. Genetic manipulation of circadian rhythm in cultured cells. Dominant-negative CKIε slows down the clock. Rat-1 (*per1::luc*) cells were transduced with retrovirus expressing CKIε (K38A) (*light gray*) or an empty vector (*black*). Independent isolates gave similar results. The *DbtS* P47S mutation engineered into CKIε shortens period. Two independent isolates (*light gray*) both showed period shortening.

INHIBITION OF CKIε UNEXPECTEDLY LENGTHENS CIRCADIAN PERIOD

To further test the hypothesis that CKIε phosphorylation of PER targets it for proteasomal degradation, the effect of the CKIδ/ε inhibitor IC261 was assessed. Using our newly developed cell-based system, we found the surprising result that inhibition of CKI both delayed PER degradation and *lengthened* the circadian period in cells (Eide et al. 2005). Similar results have now been reported for another CKI inhibitor in intact animals (Badura et al. 2007).

The *tau* mutant hamster has been instrumental in developing models of the circadian clock. First reported in 1988, the *tau* heterozygote has a period of 22 hours, whereas the homozygous mutant has a period of 20 hours (Ralph and Menaker 1988). The cause of the short period is a point mutation in hamster CKIε that *decreases* in vitro kinase activity (Lowrey et al. 2000). Subsequently, we contributed to the description of a human family with a short circadian period with a serine→glycine mutation in PER2 in the CKIε-binding domain (Jones et al. 1999; Toh et al. 2001). These two mutations strongly suggest that phosphorylation of PER2 by CKI is important for proper period length. Importantly, they predicted that decreased CKIε phosphorylation of PER2 should *shorten* mammalian period. However, our experiments described here indicate that this is not the case. Unlike the results expected from the *tau* hamster with low CKIε activity, *inhibition of CKI in cells lengthens period.*

One concern is that the effects of proteasome and

kinase inhibitors might be nonspecific. We therefore inhibited CKIε activity in cells by expression of a dominant-negative CKIε. We used a retrovirus driving expression of wild-type CKIε or dominant-negative CKIε (K38A) in our Rat-1 reporter cells. As Figure 3 shows, overexpression of dominant-negative CKIε lengthens the period. Ribonuclease protection assays showed that this effect was not limited to the *per1* promoter but affected the timing of expression of several circadian output genes (data not shown). This is consistent with the results of our inhibitor assay and indicates that the *critical effect of CKIε is to shorten the period* (hence, inhibition lengthens period). On the basis of model predictions, we postulated that shortening of the period is due to the ability of CKI to accelerate degradation of PER proteins.

Do all mutants of CKI that shorten period increase CKI activity in vivo? As a first step to testing this, we made Rat-1 (*Per1::luc*) cell lines expressing CKIε with the *Drosophila doubletime short* (*DbtS*) (P47S) mutation. The mutant proline 47 residue is highly conserved in the CKI family and the *DbtS* allele has the strongest effect of any period-shortening mutant (Kloss et al. 1998; Price et al. 1998). Like the K38A mutation, the *DbtS* mutation *decreases* CKIε kinase activity in vitro (Eide and Virshup 2001; Preuss et al. 2004). Importantly, although expression of the truly kinase-inactive CKIε (K38A) lengthens period, CKIε (*DbtS*) expression shortens period, similar to the effect it has in *Drosophila* (Fig. 3). Importantly, the different effects of distinct CKIε alleles on period in cell lines mirror the results in mutant animals. These data indicate that the Rat-1 (*Per1::luc*) cells provide an experimentally tractable model system to study the biochemical basis for the ability of CKIε to alter rhythm.

MATHEMATICAL MODELING PREDICTS AN OPPOSITE PHENOTYPE FOR *TAU*

Our experimental data obtained in Rat-1 (*Per1::luc*) cells treated with kinase and proteasome inhibitors or dominant-negative CKIε agree with the predictions of the quantitative model of the circadian molecular clock (Forger and Peskin 2003), arguing that the CKIε^{tau} mutant must be a gain of function, not a loss of function. Recall, however, that CKIε^{tau} has an eightfold decrease in activity in vitro (Lowrey et al. 2000; Preuss et al. 2004). Previously, several potential conceptual models had attempted to explain how a decrease in function of CKIδ/ε might cause the fast period phenotype of the *tau* mutation. However, when we decreased the rate of phosphorylation of PER1 or PER2 in the Forger–Peskin model, we consistently found that it predicted a *longer*, not a shorter, period: A 50% decrease in the rate of the primary phosphorylation of PER1 and PER2 lengthened the period of the model by 0.13 and 2.43 hours, respectively. Remarkably, in both cases, the Forger–Peskin model predicts that changes in CKI activity on PER2 stability have a far greater impact than changes in localization or in phosphorylation of PER1. None of the previously proposed explanations of *tau* are consistent with the quantitative model's predictions. In the detailed and otherwise accurate mathematical model of the clock, only increased kinase activity resulting in increased PER2 degradation was able to produce significant short-period phenotypes.

AN OPPOSITE ROLE FOR *TAU*

If the *tau* mutation in fact increased, rather than decreased, the phosphorylation and degradation of the PER proteins, a shorter period would be expected and this conflict would be resolved. It would also show that the model was both correct and useful. We therefore tested this model prediction—that CKIε^{tau} increased PER2 degradation—directly in cell-based assays (Fig. 4, right). The half-life of PER2 was measured in the presence of wild-type CKIε, CKIε^{tau}, or CKIε (K38A) (a kinase-inactive mutant). Although expression of wild-type CKIε modestly shortened PER2 half-life, CKIε^{tau} expression *markedly decreased PER2 half-life*. Similar results were seen for PER1 (Gallego et al. 2006b). CKIε (K38A) had no significant effect on PER abundance and half-life,

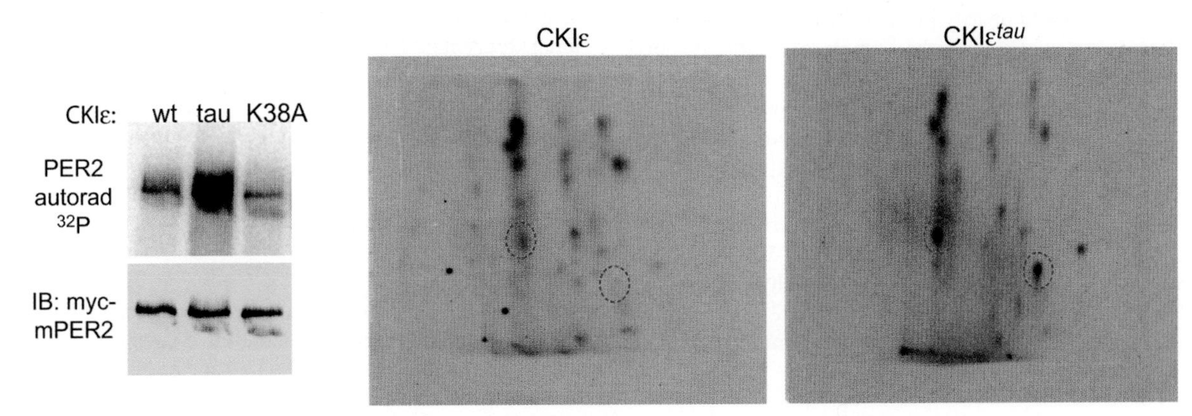

Figure 4. *tau* is a gain-of-function mutation in vivo that stimulates the site-specific phosphorylation of PER2. (*Left*) CKIε^{tau} expression increases the in vivo phosphorylation of PER2. PER2 was coexpressed with the indicated form of CKIε and cells were metabolically labeled with ^{32}P orthophosphate for 4 hours in the presence of MG132. *myc*-tagged PER2 was then immunoprecipitated, followed by autoradiography and immunoblotting. (*Right*) CKIε^{tau} increases phosphorylation of two specific peptides. The metabolically labeled PER2 was analyzed by trypsin/chymotrypsin digestion and two-dimensional phosphopeptide mapping. There may be two motifs whose phosphorylation is increased by CKIε^{tau}, or the protease digestion may cleave a single motif into half, creating two phosphopeptides with increased labeling.

notably distinct from the effect of CKIε^{tau}. These results indicate that in cells, the *tau* mutation is a *gain of function* that leads to the more rapid degradation of PER1 and PER2. This CKIε^{tau}-accelerated degradation still requires the β-TrCP/proteasome pathway, because it is inhibited by MG132, and a dominant-negative form of β-TrCP.

Because CKIε^{tau} causes the rapid degradation of PER2, it should be a gain-of-function mutation. This conflicts with reports that CKIε^{tau} has decreased kinase activity in vitro (Lowrey et al. 2000). Our data show that the *tau* mutation accelerates the degradation of PER1 and PER2 via the ubiquitin/proteasome pathway. We then tested the prediction of the Forger–Peskin model that CKIε^{tau} phosphorylated PER more efficiently than wild-type kinase in cells, rather than in vitro. Wild-type CKIε expression in cultured cells caused a small increase in both PER1 and PER2 phosphorylation, whereas the catalytically inactive CKIε (K38A) had little effect. Notably, in vivo, CKIε^{tau} caused markedly increased phosphorylation of PER proteins compared with wild-type kinase (Fig. 4, left). *This is consistent with CKIε^{tau} being a gain-of-function mutation in vivo (although not in vitro) specific to the PER proteins.*

Importantly, we found that CKIε^{tau} showed decreased activity on most substrates. We considered two potential mechanisms for the specific increased in vivo phosphorylation of PER2. CKIε^{tau} may selectively phosphorylate a site in PER that specifically regulates its degradation or CKIε^{tau} may globally increase phosphorylation of PER in vivo despite its decreased activity in vitro. To differentiate between these possibilities, in vivo metabolically ^{32}P-labeled PER2 (Fig. 4, right) and PER1 (not shown) were examined by two-dimensional phosphopeptide mapping (Firulli et al. 2004). Expression of wild-type CKIε did not lead to any apparent changes in the pattern of PER phosphorylation, indicating that endogenous CKIε is not rate-limiting for PER phosphorylation within the cells. Coexpression of CKIε^{tau} caused a marked increase in phosphorylation of two PER2 peptides. This result has been reproduced in four independent experiments. This suggests that two specific peptides within PER2 are phosphorylated much more efficiently by CKIε^{tau} than by wild-type CKIε.

We have preliminarily identified the specific phosphorylation sites controlling the stability of PER2. We first identified a CKI phosphorylation-dependent β-TrCP-binding site in PER2 (amino acids 477–482) (Eide et al. 2005). Enhanced CKIε^{tau} phosphorylation in this motif may promote PER2 degradation. To test this, we compared the effect of CKIε^{tau} expression on wild-type and S477A G479A PER2 that cannot bind β-TrCP (Eide et al. 2005). Mutation of these sites blocked CKIε^{tau}-induced degradation. This suggests the specific hypothesis that CKIε^{tau} has increased activity on the PER2 β-TrCP-binding motif, thereby creating a β-TrCP-binding site and accelerating PER2 degradation.

PROTEIN PHOSPHATASES REGULATE mPER2 DEGRADATION AND CIRCADIAN RHYTHM

The ability of cell-permeable phosphatase inhibitors okadaic acid and calyculin A to cause PER2 degradation suggested that cellular protein phosphatases can also regulate PER2 stability and hence circadian rhythms (Eide et al. 2005). Okadaic acid and calyculin A inhibit both PP1 and PP2A phosphatases. As an initial test to discriminate between PP1 and PP2A, we examined the stability of in-vitro-translated ^{35}S-labeled mPER2 in cell extracts in the presence of phosphatase inhibitors.

We found that PER2 was phosphorylated but not degraded over 3 hours in cell extracts, although addition of the specific PP1 inhibitor, inhibitor-2, caused a rapid degradation of the protein. This suggested that PER2 is a specific substrate of PP1, and PP1 can regulate the phosphorylation and degradation of PER2. PER2 can be dephosphorylated by PP1 immunoprecipitated from mouse brain, liver, and Rat-1a cells (Gallego et al. 2006a). To examine if PER2 and PP1 interact, PP1 and PER2 were coexpressed in HEK 293 cells. Immunoprecipitation of PP1 brought down PER2 and immunoprecipitation of PER2 brought down PP1. Thus, PP1 can interact with and dephosphorylate PER2.

If PP1 controls PER2 stability, then we postulate that a dominant-negative PP1 would enhance PER2 phosphorylation and accelerate its degradation. We generated a dominant-negative PP1 (D95N) and coexpressed it with PER2. Expression of dominant-negative but not wild-type green fluorescent protein (GFP)-tagged PP1 caused a marked decrease in PER2 abundance. This decrease can be blocked by dominant-negative β-TrCP and proteasome inhibitors, consistent with the hypothesis that PP1 can regulate phosphorylation and hence stability of PER2 (Gallego et al. 2006a).

Both PP1 and PP2A are regulated by tightly associated targeting/regulatory proteins (Gallego and Virshup 2005). Sehgal and coworkers (Sathyanarayanan et al. 2004) reported that in *Drosophila*, PP2A targeting proteins could regulate dPER stability during circadian rhythms, and they presented genetic evidence that PP2A was a clock regulator. We therefore examined synchronized Rat-1 cells, as well as mouse livers, for evidence of oscillations in both PP1 and PP2A targeting/regulatory proteins. We have detected significant circadian oscillation of PP1-binding proteins in both Rat-1 fibroblasts (Fig. 5, center) and mouse liver (Fig. 5, left). Immunoblots examining PP2A regulators have shown no circadian oscillation of any of these proteins (Fig. 5, right). The data, taken together, are consistent with the hypothesis that PP1 and an oscillating PP1-targeting protein regulate the phosphorylation and stability of PER2 in the mammalian circadian clock.

CONCLUSIONS

Reversible phosphorylation and regulated degradation of key circadian regulators have an important role in determining circadian timing. The degradation of PERIOD, and now CRYPTOCHROMES, in mammalian cells is clearly important, because perturbations in those rates have highly significant effects on the length of a cycle. The evidence to date indicate that both CKIδ and CKIε are important in the clock, although it is not clear if they have overlapping, identical, or distinct roles. Regulated destruction of PER protein is determined by a

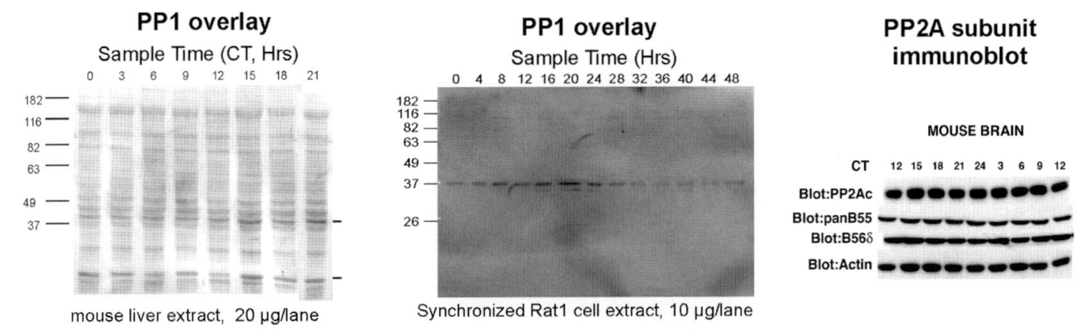

Figure 5. PP1 regulators vary during the circadian cycle. (*Left* and *center panels*) PP1 overlay detects PP1-binding proteins. Livers (*left*) and brains (*far right*) from mice kept in 12 hours light:12 hours dark. (*Middle*) Rat-1 cells synchronized with forskolin pulse were lysed at the indicated times. All samples were separated by SDS-PAGE and transferred to PVDF membranes. (*Left and center*) Membranes were incubated with the digoxygenin-labeled PP1 catalytic subunit (Beullens et al. 1999), and PP1-binding proteins were detected with antidigoxygenin antibodies (the overlay was performed in collaboration with the S. Shenolikar lab at Duke University). (*Right panel*) No evidence for circadian oscillations of PP2A regulators. Brain and liver (not shown) extracts were probed with antibodies to the PP2A regulatory subunits PP2Ac, B56δ, and a pan-B55 (PR55), as well as B56β, δ, and PR72 (not shown). No circadian variation was detected for any PP2A subunit in either brain or liver.

balance of CKI activity and protein phosphatase activity. There is a role for PP2A in regulating protein stability of PER in *Drosophila*, and we found a role for mammalian cells. The evidence from a number of laboratories that the rhythmic expression of PER and perhaps CRY is not essential to maintaining a rhythm supports a somewhat heretical hypothesis that rhythmic destruction of the regulators is the most numerically important regulator of the core clock.

ACKNOWLEDGMENTS

These studies were funded in part by R01GM060387 and the Huntsman Cancer Foundation.

REFERENCES

Akashi M., Tsuchiya Y., Yoshino T., and Nishida E. 2002. Control of intracellular dynamics of mammalian period proteins by casein kinase I epsilon (CKIepsilon) and CKIdelta in cultured cells. *Mol. Cell. Biol.* **22:** 1693.

Antoch M.P., Kondratov R.V., and Takahashi J.S. 2005. Circadian clock genes as modulators of sensitivity to genotoxic stress. *Cell Cycle* **4:** 901.

Badura L., Swanson T., Adamowicz W., Adams J., Cianfrogna J., Fisher K., Holland J., Kleiman R., Nelson F., Reynolds L., St. Germain K., Schaeffer E., Tate B., and Sprouse J. 2007. An inhibitor of casein kinase Iε induces phase delays in circadian rhythms under free-running and entrained conditions. *J. Pharmacol. Exp. Ther.* **322:** 730.

Balsalobre A., Damiola F., and Schibler U. 1998. A serum shock induces circadian gene expression in mammalian tissue culture cells. *Cell* **93:** 929.

Beullens M., Van Eynde A., Vulsteke V., Connor J., Shenolikar S., Stalmans W., and Bollen M. 1999. Molecular determinants of nuclear protein phosphatase-1 regulation by NIPP-1. *J. Biol. Chem.* **274:** 14053.

Busino L., Bassermann F., Maiolica A., Lee C., Nolan P.M., Godinho S.I., Draetta G.F., and Pagano M. 2007. SCFFbxl3 controls the oscillation of the circadian clock by directing the degradation of cryptochrome proteins. *Science* **316:** 900.

Chang D.C. and Reppert S.M. 2001. The circadian clocks of mice and men. *Neuron* **29:** 555.

Comolli J.C., Fagan T., and Hastings J.W. 2003. A type-1 phosphoprotein phosphatase from a dinoflagellate as a possible component of the circadian mechanism. *J. Biol. Rhythms* **18:** 367.

Comolli J., Taylor W., Rehman J., and Hastings J.W. 1996. Inhibitors of serine/threonine phosphoprotein phosphatases alter circadian properties in *Gonyaulax polyedra. Plant Physiol.* **111:** 285.

Eide E.J. and Virshup D.M. 2001. Casein kinase I: Another cog in the circadian clockworks. *Chronobiol. Int.* **18:** 389.

Eide E.J., Vielhaber E.L., Hinz W.A., and Virshup D.M. 2002. The circadian regulatory proteins BMAL1 and cryptochromes are substrates of casein kinase Iε (CKIε). *J. Biol. Chem.* **277:** 17248.

Eide E., Woolf M., Kang H., Woolf P., Camacho F., Vielhaber E., Giovanni A., and Virshup D. 2005. Control of mammalian circadian rhythm by CKIε-regulated proteasome-mediated PER2 degradation. *Mol. Cell. Biol.* **25:** 2795.

Etchegaray J.P., Lee C., Wade P.A., and Reppert S.M. 2003. Rhythmic histone acetylation underlies transcription in the mammalian circadian clock. *Nature* **421:** 177.

Firulli B.A., Virshup D.M., and Firulli A.B. 2004. Phosphopeptide mapping of proteins ectopically expressed in tissue culture cell lines. *Biol. Proced. Online* **6:** 16.

Forger D.B. and Paydarfar D. 2004. Starting, stopping, and resetting biological oscillators: In search of optimum perturbations. *J. Theor. Biol.* **230:** 521.

Forger D.B. and Peskin C.S. 2003. A detailed predictive model of the mammalian circadian clock. *Proc. Natl. Acad. Sci.* **100:** 14806.

———. 2004. Model based conjectures on mammalian clock controversies. *J. Theor. Biol.* **230:** 533.

———. 2005. Stochastic simulation of the mammalian circadian clock. *Proc. Natl. Acad. Sci.* **102:** 321.

Fu L., Pelicano H., Liu J., Huang P., and Lee C. 2002. The circadian gene Period2 plays an important role in tumor suppression and DNA damage response in vivo. *Cell* **111:** 41.

Gallego M. and Virshup D.M. 2005. Protein serine/threonine phosphatases: Life, death, and sleeping. *Curr. Opin. Cell Biol.* **17:** 197.

———. 2007. Post-translational modifications regulate the ticking of the circadian clock. *Nat. Rev. Mol. Cell Biol.* **8:** 139.

Gallego M., Kang H., and Virshup D.M. 2006a. Protein phosphatase 1 regulates the stability of the circadian protein PER2. *Biochem. J.* **399:** 169.

Gallego M., Eide E., Woolf M., Virshup D., and Forger D. 2006b. An opposite role for tau in circadian rhythms revealed by mathematical modeling. *Proc. Natl. Acad. Sci.* **103:** 10618.

Godinho S.I., Maywood E.S., Shaw L., Tucci V., Barnard A.R., Busino L., Pagano M., Kendall R., Quwailid M.M., Romero M.R., O'Neill J., Chesham J.E., Brooker D., Lalanne Z., Hastings M.H., and Nolan P.M. 2007. The after-hours mutant reveals a role for Fbxl3 in determining mammalian circadian period. *Science* **316:** 897.

Indic P. and Brown E.N. 2006. Characterizing the amplitude dynamics of the human core-temperature circadian rhythm using a stochastic-dynamic model. *J. Theor. Biol.* **239:** 499.
Jones C.R., Campbell S.S., Zone S.E., Cooper F., DeSano A., Murphy P.J., Jones B., Czajkowski L., and Ptacek L.J. 1999. Familial advanced sleep-phase syndrome: A short-period circadian rhythm variant in humans. *Nat. Med.* **5:** 1062.
Kloss B., Price J.L., Saez L., Blau J., Rothenfluh A., Wesley C.S., and Young M.W. 1998. The *Drosophila* clock gene double-time encodes a protein closely related to human casein kinase I epsilon. *Cell* **94:** 97.
Kume K., Zylka M.J., Sriram S., Shearman L.P., Weaver D.R., Jin X., Maywood E.S., Hastings M.H., and Reppert S.M. 1999. mCRY1 and mCRY2 are essential components of the negative limb of the circadian clock feedback loop. *Cell* **98:** 193.
Locke J.C., Millar A.J., and Turner M.S. 2005. Modelling genetic networks with noisy and varied experimental data: the circadian clock in *Arabidopsis thaliana. J. Theor. Biol.* **234:** 383.
Lowrey P.L. and Takahashi J.S. 2004. Mammalian circadian biology: Elucidating genome-wide levels of temporal organization. *Annu. Rev. Genomics Hum. Genet.* **5:** 407.
Lowrey P.L., Shimomura K., Antoch M.P., Yamazaki S., Zemenides P.D., Ralph M.R., Menaker M., and Takahashi J.S. 2000. Positional syntenic cloning and functional characterization of the mammalian circadian mutation tau. *Science* **288:** 483.
Mansour H.A., Monk T.H., and Nimgaonkar V.L. 2005. Circadian genes and bipolar disorder. *Ann. Med.* **37:** 196.
Matsuo T., Yamaguchi S., Mitsui S., Emi A., Shimoda F., and Okamura H. 2003. Control mechanism of the circadian clock for timing of cell division in vivo. *Science* **302:** 255.
Preitner N., Damiola F., Lopez-Molina L., Zakany J., Duboule D., Albrecht U., and Schibler U. 2002. The orphan nuclear receptor REV-ERBalpha controls circadian transcription within the positive limb of the mammalian circadian oscillator. *Cell* **110:** 251.
Preuss F., Fan J.Y., Kalive M., Bao S., Schuenemann E., Bjes E.S., and Price J.L. 2004. *Drosophila* doubletime mutations which either shorten or lengthen the period of circadian rhythms decrease the protein kinase activity of casein kinase I. *Mol. Cell. Biol.* **24:** 886.
Price J.L., Blau J., Rothenfluh A., Abodeely M., Kloss B., and Young M.W. 1998. *double-time* is a novel *Drosophila* clock gene that regulates PERIOD protein accumulation. *Cell* **94:** 83.
Ralph M.R. and Menaker M. 1988. A mutation of the circadian system in golden hamsters. *Science* **241:** 1225.
Reppert S.M. and Weaver D.R. 2002. Coordination of circadian timing in mammals. *Nature* **418:** 935.
Rivard G.E., Infante-Rivard C., Dresse M.F., Leclerc J.M., and Champagne J. 1993. Circadian time-dependent response of childhood lymphoblastic leukemia to chemotherapy: A long-term follow-up study of survival. *Chronobiol. Int.* **10:** 201.
Sathyanarayanan S., Zheng X., Xiao R., and Sehgal A. 2004. Posttranslational regulation of *Drosophila* PERIOD protein by protein phosphatase 2A. *Cell* **116:** 603.
Schafmeier T., Haase A., Kaldi K., Scholz J., Fuchs M., and Brunner M. 2005. Transcriptional feedback of *Neurospora* circadian clock gene by phosphorylation-dependent inactivation of its transcription factor. *Cell* **122:** 235.
Schibler U. and Naef F. 2005. Cellular oscillators: Rhythmic gene expression and metabolism. *Curr. Opin. Cell Biol.* **17:** 223.
Schmiegelow K., Glomstein A., Kristinsson J., Salmi T., Schroder H., and Bjork O. 1997. Impact of morning versus evening schedule for oral methotrexate and 6-mercaptopurine on relapse risk for children with acute lymphoblastic leukemia (Nordic Society for Pediatric Hematology and Oncology [NOPHO]). *J. Pediatr. Hematol. Oncol.* **19:** 102.
Shirogane T., Jin J., Ang X.L., and Harper J.W. 2005. SCFbeta-TRCP controls clock-dependent transcription via casein kinase 1-dependent degradation of the mammalian period-1 (Per1) protein. *J. Biol. Chem.* **280:** 26863.
Siepka S.M., Yoo S.H., Park J., Song W., Kumar V., Hu Y., Lee C., and Takahashi J.S. 2007. Circadian mutant Overtime reveals F-box protein FBXL3 regulation of cryptochrome and Period gene expression. *Cell* **129:** 1011.
Suri V., Hall J.C., and Rosbash M. 2000. Two novel doubletime mutants alter circadian properties and eliminate the delay between RNA and protein in *Drosophila. J. Neurosci.* **20:** 7547.
Takano A., Isojima Y., and Nagai K. 2004. Identification of mPer1 phosphorylation sites responsible for the nuclear entry. *J. Biol. Chem.* **279:** 32578.
Toh K.L., Jones C.R., He Y., Eide E.J., Hinz W.A., Virshup D.M., Ptacek L.J., and Fu Y.H. 2001. An hPer2 phosphorylation site mutation in familial advanced sleep-phase syndrome. *Science* **291:** 1040.
Turek F.W., Joshu C., Kohsaka A., Lin E., Ivanova G., McDearmon E., Laposky A., Losee-Olson S., Easton A., Jensen D.R., Eckel R.H., Takahashi J.S., and Bass J. 2005. Obesity and metabolic syndrome in circadian Clock mutant mice. *Science* **308:** 1043.
Vielhaber E.L., Duricka D., Ullman K.S., and Virshup D.M. 2001. Nuclear export of mammalian PERIOD proteins. *J. Biol. Chem.* **276:** 45921.
Vielhaber E., Eide E., Rivers A., Gao Z.-H., and Virshup D.M. 2000. Nuclear entry of the circadian regulator mPER1 is controlled by casein kinase I ε. *Mol. Cell. Biol.* **20:** 4888.
Virshup D.M. and Forger D.B. 2007. After hours keeps clock researchers CRYing Overtime. *Cell* **129:** 857.
Wallace A., Chinn D., and Rubin G. 2003. Taking simvastatin in the morning compared with in the evening: Randomised controlled trial. *Br. Med. J. (BMJ)* **327:** 788.
Wu G., Xu G., Schulman B.A., Jeffrey P.D., Harper J.W., and Pavletich N.P. 2003. Structure of a beta-TrCP1-Skp1-beta-catenin complex: Destruction motif binding and lysine specificity of the SCF(beta-TrCP1) ubiquitin ligase. *Mol. Cell* **11:** 1445.
Xu Y., Padiath Q.S., Shapiro R.E., Jones C.R., Wu S.C., Saigoh N., Saigoh K., Ptacek L.J., and Fu Y.H. 2005. Functional consequences of a CKIdelta mutation causing familial advanced sleep phase syndrome. *Nature* **434:** 640.
Yagita K., Tamanini F., Yasuda M., Hoeijmakers J.H., van der Horst G.T., and Okamura H. 2002. Nucleocytoplasmic shuttling and mCRY-dependent inhibition of ubiquitylation of the mPER2 clock protein. *EMBO J.* **21:** 1301.
Yamaguchi S., Mitsui S., Miyake S., Yan L., Onishi H., Yagita K., Suzuki M., Shibata S., Kobayashi M., and Okamura H. 2000. The 5′ upstream region of mPer1 gene contains two promoters and is responsible for circadian oscillation. *Curr. Biol.* **10:** 873.
Yamazaki S., Numano R., Abe M., Hida A., Takahashi R., Ueda M., Block G.D., Sakaki Y., Menaker M., and Tei H. 2000. Resetting central and peripheral circadian oscillators in transgenic rats. *Science* **288:** 682.
Yang Y., He Q., Cheng P., Wrage P., Yarden O., and Liu Y. 2004. Distinct roles for PP1 and PP2A in the *Neurospora* circadian clock. *Genes Dev.* **18:** 255.

Evolution of the Clock from Yeast to Man by Period-Doubling Folds in the Cellular Oscillator

R.R. KLEVECZ AND C.M. LI

Dynamic Systems Group, Department of Biology, Beckman Research Institute of The City of Hope Medical Center, Duarte, California 91010

Analysis of genome-wide oscillations in transcription reveals that the cell is an oscillator and an attractor and that the maintenance of a stable phenotype requires that maximums in expression in clusters of transcripts must be poised at antipodal phases around the steady state—this is the dynamic architecture of phenotype. Plots of the path through concentration phase space taken by all of the transcripts of *Saccharomyces cerevisiae* yield a simple three-dimensional surface. How this surface might change as period lengthens or as a cell differentiates is at the center of current work. We have shown that changes in gene expression in response to mutation or perturbation by drugs occur through a folding or unfolding of the surface described by this circle of transcripts and we suggest that the path from this 40-minute oscillation to the cell cycle and circadian rhythms takes place through a series of period-two or period-three bifurcations. These foldings in the surface of the putative attractor result in an increasingly dense set of nested trajectories in the concentrations of message and protein. Evolutionary advantage might accrue to an organism that could change period by changes in just one or a few genes as day length increased from 4 hours in the prebiotic Earth, through 8 hours during the expansion of photoautotrophs, to the present 24 hours.

INTRODUCTION

Recently, we described the first example of period-doubling behavior in the genome-wide transcriptional oscillation seen in yeast cultures growing at high cell densities and exhibiting continuous gated synchrony (Li and Klevecz 2006). We showed that this behavior was closely modeled by a simple modification of the Rossler attractor (Rossler 1976). From the dynamic systems perspective, we can take this rigorous model for describing the global behavior of the timekeeping oscillator and begin to build a gene-by-gene or regulon-by-regulon dynamic systems network. The model also immediately offers an accessible graphical representation of the genome-scale changes that can lead from a high-frequency timekeeper to the cell cycle and circadian rhythms. Cell-to-cell signaling in continuous cultures of the budding yeast *S. cerevisiae* leads to mutual entrainment or synchronization that is manifested as an oscillation in redox state and a genome-wide oscillation in transcription and metabolism (Klevecz et al. 2004; Tu et al. 2005; Murray et al. 2007). In turn, this transcriptional redox attractor cycle (TRAC) times, or gates, DNA replication and other cell cycle events. In this regard, the cell cycle is a developmental process timed by the TRAC. DNA replication is restricted to the reductive phase of the cycle and is initiated as levels of hydrogen sulfide rise. This is seen as an evolutionarily important mechanism for preventing oxidative damage to DNA during replication (Klevecz et al. 2004). Both exposure to drugs known to alter circadian rhythms and deletions of known clock genes in *S. cerevisiae* yield an increase in the period of the TRAC that follows a bifurcation path from 40 minutes to 3–4 hours. This finding resonates with an earlier observation of a 3–4-hour oscillation that gated cell cycle events in higher organisms (Klevecz 1976). We suggest that the genome-wide oscillation discovered in yeast is a primordial oscillator and that the differentiation pathway to multiple phenotypes is through a process characterized by period doubling and period-three bifurcations.

Several independent lines of evidence argue that the time sense in biological systems is not dissectible from the dynamic architecture of a cellular phenotype. Experimental evidence from expression arrays, metabalome analyses, and limited green fluorescent protein (GFP)-tagged proteins pushes us to conclude that "everything" oscillates. For those few constituents that cannot be shown to oscillate, we will point to dynamic systems theory, which says that as more things oscillate in a coupled system, the likelihood that everything oscillates becomes a certainty (Hess and Boiteux 1971). Finally, there is a strong evolutionary argument for expecting that the clock started out as a high-frequency oscillator. Circadian rhythms are always about 1 day, but the length of 1 day was not always 24 hours (Klevecz 1984).

WHEN THE DAY WAS EIGHT HOURS LONG

All heavenly bodies, from the most massive type-O and -B stars to the smallest asteroids, form and rotate with periods that are at the limit of their rotational instability, which ranges, depending on composition, from 80 minutes to 4 hours. The Earth was no exception. Early in the Earth's history, an impact with a massive planetoid created the Moon and resulted in the transfer of angular momentum from the Earth's rotation to the orbit of the Moon. The Earth slowed its rotation and the separation between the Earth and the Moon increased. Geophysical calculations put the period of the Earth's rotation at 8 hours 3.4 billion years ago (Turcotte et al. 1977), a time when photoautotrophs were plentiful and mixed colonies of *Cyanobacteria* were forming stromatolites. From this point onward, geophysical calculations are not needed to determine the period of the Earth's rotation because day length was recorded in daily growth rings that resulted from the swimming and settling of the algal colonies that comprise stromatolites (Panella 1972). Because tidal and annual rhythms are visible in cross-sections of stromatolites, it is possible to calculate that there were slightly more than 1000 days per year 3 billion years ago, yielding a day length of 8–9 hours. Because of the nearness of the Moon and the lack of a significant atmosphere, the day was punctuated by boiling and condensing tides and intense ultraviolet radiation. If ever there was a need for the progenitors of *Synechoccocus* to coordinate with the external environment, this was it. In such a world, with shallow seas, sensing and coordinating ones chemistry with the tides might well have been as important as circadian coordination. The evidence for changes in day length becomes quite plentiful beginning 600–800 million years ago, as metazoans expanded. Devonian and Ordovician corals and bivalves recorded the later recession of the Moon and the slowing of the Earth's rotation in daily periodicities in growth patterns. Wells (1963) was the first to realize that the history of the Earth's rotational period and the evolution of the Earth-Moon system could be charted by counting the number of daily increments per annual series added to skeletons of living and fossil corals. He found that whereas living corals add 365 daily increments per year, Devonian fossils added 400, yielding a day length at that time of 21.9 hours. Measurements from the Cambrian, 600 million years ago, give a day length of about 20.6 hours (Wells 1963).

On the basis of paleontological observations of growth ring differences: the existence of ultradian, often quantal, rhythmic expression in unicellular eukaryotes, simple metazoans, and mammalian cells in culture; and the clustering of period mutants in *Drosophila*, *Chlamydomonas*, and *Neurospora*, it is plausible to think that modern circadian rhythms have evolved from their initial primordial 4- or 8-hour period to their present length through a path similar to the TRAC oscillations seen in yeast. Let us first discuss the evidence for the existence of a 3–4 hour periodicity underlying circadian rhythms in modern higher organisms (Fig. 1) and then treat the molecular and dynamic details of period changes from 40 minutes to 4 hours as displayed in the yeast gated synchrony system.

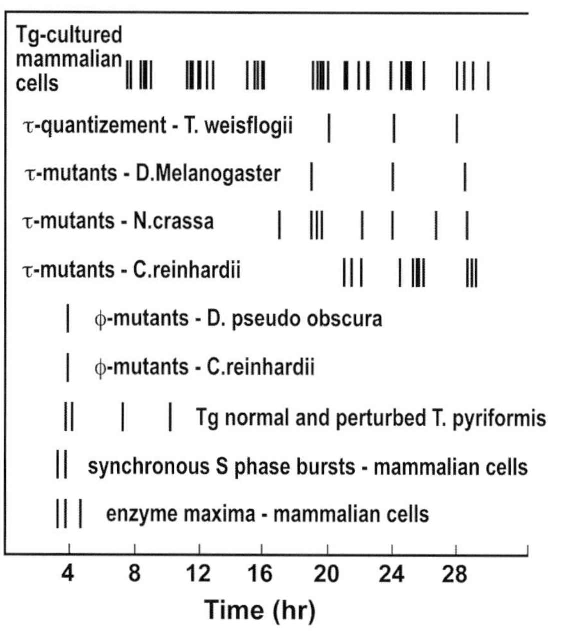

Figure 1. Expression of a fundamental oscillator underlying cell cycle and circadian rhythms. Each line represents a peak of occurrence. Starting from the top of the figure, (Line *1*) Quantized variation in generation times (Tg) of cultured mammalian cell lines. Generation times were determined from the published data on cells that were synchronized by mitotic selection or from time-lapse analyses of intermitotic times of random cultures. The list is not exhaustive, but it represents a sampling of papers published since 1961 in which the stated generation time could be directly confirmed in the data. Wherever possible, modal generation times were obtained, and reports stating only population doubling time were excluded (Klevecz 1976). (Line *2*) Polymodal distribution of generation times in the marine diatom *Thalassiosira weisflogii* growing in the circadian mode (Chisholm and Costello 1980). (Line *3*) Long- and short-period mutants of *Drosophila melanogaster* isolated by Konopka and Benzer (1971). (Line *4*) τ mutants of *Neurospora crassa* studied by Feldman and coworkers (Feldman and Hoyle 1973; Feldman et al. 1979; Feldman and Dunlap 1982). (Line *5*) Phototaxic τ mutant of *Chlamydomonas reinhardii* picked and isolated by Bruce (1972) from cultures treated with nitrosoguanidine. (Line *6*) Phase angle (Φ) early-eclosion mutants selected by Pittendrigh (1967) by continuous selection through 50 generations for early emerging *Drosophila pseudoobscura*. (Line *7*) Φ mutants of *C. reinhardii* isolated by Bruce (1972) from cultures treated and selected for period changes. (Line *8*) Oscillatory variations in generation times of *Tetrahymena pyriformis* perturbed by continuous incubation in low levels of actinomycin D. Normal generation time in these cultures is 4–4.5 hours (Jauker and Cleffmann 1970). (Line *9*) Interval between synchronous bursts in DNA synthesis in S phase of mammalian cells (Klevecz 1969; Klevecz and Kapp 1973; Collins 1978; Kapp et al. 1979; Holmquist et al. 1982) scored from a maximum slope of thymidine incorporation rate between peaks. (Line *10*) Intervals between peaks in maximum enzyme activity or levels in the cell cycle of synchronous hamster cells in culture (Klevecz et al. 1984).

QUANTIZED GENERATION TIMES, PERIOD MUTANT CLUSTERS, AND THE FOLDING OF THE TRAC

Before the development of methods for examining genome-wide expression, clues to the dynamic structure of the cell could only be assessed by measurement of small subsets of cellular constituents from synchronous systems (Klevecz and Ruddle 1968) or by intentional perturbation and measurement of the phase-response curves of events such as DNA replication and cell division to the perturbation (Klevecz et al. 1984). These early studies provided a sketch of the dynamic architecture of phenotype and offered a rigorous alternative to simple branched sequential models based on mutational analysis of the cell cycle (Hartwell et al. 1974). Quantized generation times, together with perturbation analyses, formed the experimental foundation of efforts to synthesize a model of the cell cycle in which such disparate concepts as checkpoints and limit cycles or complex attractors were fused (Shymko et al. 1984). The basic idea was that checkpoints represent subthreshold oscillations in an attractor that underlies the cell cycle (Klevecz et al. 1984). The oscillator that gave rise to gated cell divisions in mammalian cells was shown to be phase-responsive and temperature-compensated (Klevecz et al. 1978; Klevecz and King 1982). The quantized generation time model was extended to other cell types and to gating of circadian rhythm-based cell division in plants, dinoflagellates, and a variety of mammalian cells in culture. One prediction of the attractor models was that all cell cycle events would be gated by the attractor, and this period would be an integral submultiple of the cell cycle or circadian rhythm it timed.

Quantized generation times were the first direct evidence of a cellular clock, but the more recent findings that the continuous culture system in yeast appears to be timed by a similar oscillator that can be tuned or driven to "fold" (i.e., undergo a series of period-two or period-three bifurcations) and that cell cycle events in *S. cerevisiae* appear to be gated by this transcriptional cycle suggest that a similar phenomenon, although on a different timescale, is operating in all systems from yeast to mammalian cells. This realization has opened a new and experimentally more accessible path to investigations of synchronous gating and the role of oscillations in generating and maintaining a stable phenotype.

FROM MOLECULAR CONSTRAINTS TO A PRIMORDIAL CLOCK: THE PATH FROM 40 MINUTES TO 4 HOURS

The time required for a prokaryotic replication fork to go from origin to terminus is 40 minutes. This too seems to be constant in a large variety of prokaryotic cells and is thought to be the minimum length of time required to accurately replicate a chromosome containing the minimum amount of information necessary to create a cell (Donachie 1968; Bipatnath et al. 1998; Nordstrom and Dasgupta 2006). Primitive prokaryotes must have faced the challenge of coordinating this molecular constraint with the external environment. We think that the yeast-gated synchrony system, with its ability to alter period length in response to mutations, drug treatments, or media differences, offers a glimpse into this primordial past.

The case for the origin of eukaryotic cells by fusion of symbionts has been made in a general way by Margulis (1993). More specifically, Searcy et al. (1981) suggested that the earliest eukaryote consisted of an Archaeal host capable of producing H_2S from environmental sulphate and a proteobacterial H_2S oxidizing endosymbiont engulfed by phagocytosis (Searcy et al. 1981; Searcy 2003). One could suppose that this is expressed today in the time sharing and phase separation seen in the shift between the oxidative and reductive phases in *S. cerevisiae*. This approximately 40-minute cycle has been seen in essentially every simple unicellular eukaryote examined. The stability and reproducibility of the oscillation have made it possible to pursue the question of whether there is a genome-wide oscillation in transcription and, furthermore, to explore the extent to which models of genetic regulatory circuits, based on large arrays of coupled chaotic attractors, are predictive of the dynamics of phenotypic organization (Klevecz et al. 1992; Bolen et al. 1993).

Microarray analysis from a yeast continuous synchrony culture system (using our standard wild-type strain, IFO0233) shows a genome-wide oscillation in transcription (Fig. 2A). Maximums in transcript levels occur at three nearly equally spaced intervals in this approximately 40-minute cycle of respiration and reduction (Fig. 2B). Two temporal clusters (4679 of 5329) are maximally expressed during the reductive phase of the cycle, whereas a third cluster (650) is maximally expressed during the respiratory phase (Fig. 2C). Transcription is organized functionally into redox-state superclusters with genes known to be important in respiration or reduction being synthesized in opposite phases of the cycle. As we demonstrate, there is a phase lag between the synthesis of the transcript and its translation into protein, such that each phase set of transcripts includes transcripts for genes that might be expected to function maximally in the next phase of the cycle (R.R. Klevecz, unpubl.). Most recently, analysis by fast Fourier transforms, singular value decomposition, and self-organizing maps of data obtained using Affymetrix expression arrays on samples taken every 4 minutes through 3 or 4 cycles from the gated synchrony system found evidence for a genome-wide oscillation and a 40-minute period in more than 95% of genes scored as present in all samples (Li and Klevecz 2006).

Many of these results have been repeated for a CEN.PK strain that has a 4–5-hour respiratory cycle at lower glucose levels (Tu et al. 2005). Moreover, Murray et al. (2007), using the gated synchrony system with a 40-minute cycle, showed that metabolites are almost universally oscillatory. Currently, the TRAC is thought to be synchronized by a mechanism involving respiratory inhibition by the release of small bursts of H_2S and carbon flux tuning mediated by acetaldehyde (Murray et al. 2003). Although these two compounds seem to be sufficient to explain the synchronization of the respiratory/reductive cycle, approximately equal numbers of cells are gated into S phase in each turn of the TRAC (Figs. 2D,E) and the length of the cell cycle is increased over its minimum, suggesting that other signaling pathways, including those that involve quorum-sensing molecules, may have a role in coordinating the respiratory/transcriptional cycle with the cell cycle to achieve sustained oscillations in the population. The role of cell-to-cell signaling and the distinction between quorum sensing and quorum conflicts in regulating growth rate has been analyzed dynamically in a series of computer simulations from this laboratory (Klevecz et al. 1992).

GATING OF CELLS FROM G_1 INTO S PHASE BY THE TRAC

Although the respiratory oscillation has been investigated in a number of laboratories, its relationship to the cell cycle and to the gating of cells into S phase and mitosis was largely unexplored. On the basis of knowledge that the population doubling time in continuous cultures was greater than 8 hours, it was assumed that the respiratory oscillation was entirely uncoupled from the cell cycle. If the approximately 40-minute cycle is the fundamental timekeeping oscillator, then gene expression for the majority of the genes of yeast would show oscillations with an approximately 40-minute period, or an integral

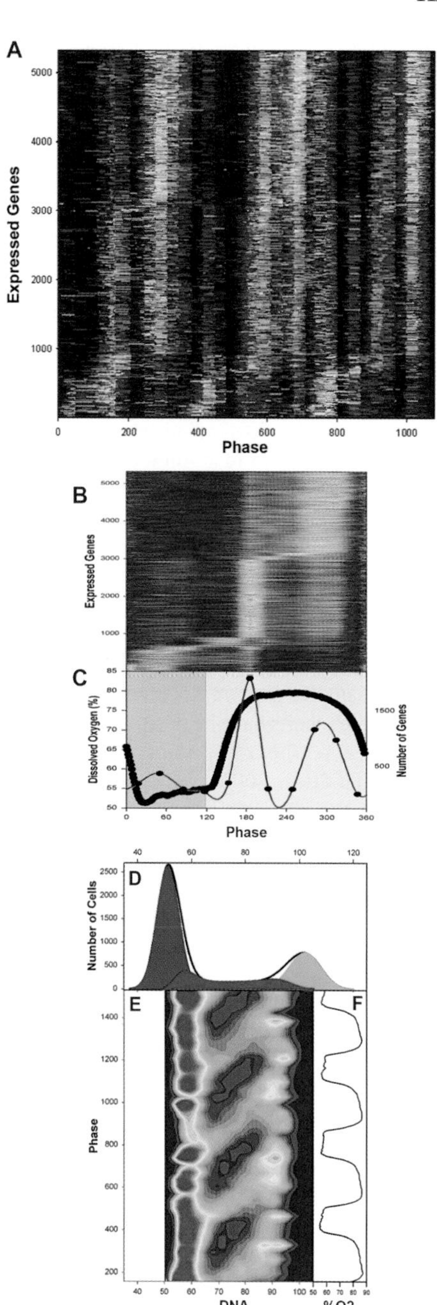

Figure 2. Genome-wide oscillations gate S phase. (*A*) Color contour (intensity) maps of the expression levels of the 5329 expressed genes are shown for 32 RNA samples through three cycles of the dissolved oxygen (DO) oscillation. To get a measure of the experiment to experiment reproducibility, the first ten samples were taken from a different experiment from those in samples 11–32. (*Orange*) High levels of expression; (*blue*) low levels of expression. Genes were scored as present based on the Affymetrix default settings and were included in the analysis if they were scored as present in each of the three cycles for at least 1 of the 32 samples. Values shown here were scaled by dividing the average expression level for each gene into each of the time-series samples for that gene. Transcripts were ordered according to their phase of maximum expression in the average of three replicates. Samples are identified according to their phase in the cycle (0–360°/cycle). (*B*) Consensus expression levels from three cycles of the TRAC oscillation. Sample phases are shown in reference to the DO curve (*thick black line, C*). Color scale: (*orange-red*) >1.6; (*dark blue*) <0.8. (*C*) Summary of the results for the time of maximum expression (*red line*) for the transcripts of *A*. The reductive phase is taken to be the period of minimal oxygen consumption (maximum DO, *yellow* background) in the interval between the minimum DO levels, and the respiratory phase is shown against a *blue* background. (*D*) Low cytometric analysis of S-phase gating in the reductive phase of the cycle. (*E*) An example of one-dimensional frequency histogram of DNA content. DNA content of the population for 39 samples taken at 4-minute intervals through four cycles of the TRAC have been stacked with respect to the time in the DO oscillation at which they were sampled as indicated in *F*. In the resulting two-dimensional color map, with *red* indicating more cells and *blue* representing fewer cells, the track of cells through S phase appears as a series of *green* bands moving left to right and upward on the diagonal. (*Red*) 350 cells; (*light green*) 200 cells; (*dark blue*) 0. (Reprinted, with permission, from Klevecz et al. 2004 [©Elsevier].)

multiple of that period. Moreover, if transcription is regulated coordinately with the oscillation, then cell cycle events might be similarly timed. In Figure 3 (top), a sketch of the relationship between the TRAC and the cell cycle is shown for a TRAC oscillation with a 40-minute period. As the period of the TRAC increases, the number of TRAC cycles per cell cycle would decrease, because the average generation time in the population as a whole is fixed by dilution rate. With decreasing numbers of TRAC cycles per cell cycle, the fraction of cells gated into S phase should increase. As Figure 3 (bottom) shows, this was found to be the case for IFO0233 wild-type cells with a 40-minute oscillation and also for CEN.PK cells growing with about 2; 2.5; and about 4-hour TRAC oscillations. The value in which 70% of cells were gated into S phase was taken from an IFO0233 culture that was synchronized by transiently lowering the pH of the medium to 2.

YEAST AS A STOCHASTIC TISSUE

The details of the cellular dynamics that lead to the emergence of redox and TRAC oscillations and the gating of cells into cell cycle stages are still not completely known. What seems clear is that at the cell densities required for emergence of the oscillation—between 2 x 10^8 and 8 x 10^8 cells/ml—the cultures are in effect tissues. The distance between cells is less than one cell diameter, creating the potential for the constant exchange of materi-

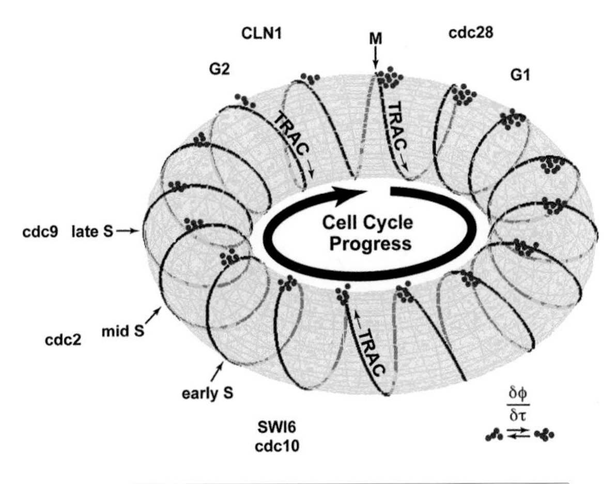

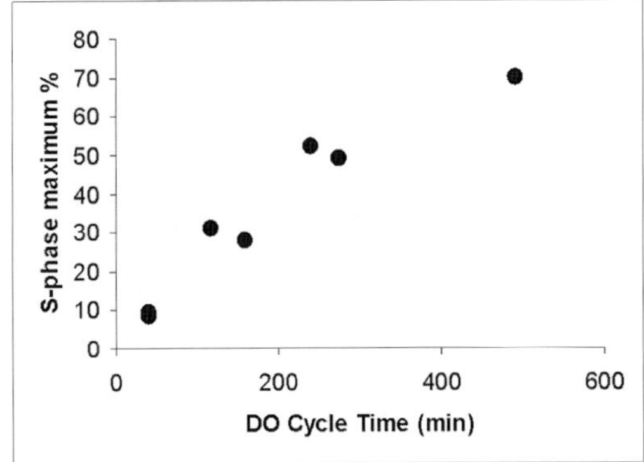

Figure 3. TRAC cycle and S-phase fraction. (*Top*) Sketch of the relationship between the TRAC and the cell cycle. *Red clusters* indicate that greater numbers of cells (*red dots*) signal to the population from G_1 phase compared with S and G_2. As the length of the DO oscillation increases, it is expected that the number of turns of the TRAC around the torus would decrease, thereby increasing the fraction of cells entering S phase in any turn of the TRAC. In the limit, when there are just two turns of the TRAC per cell cycle, the S-phase fraction should be maximal. Defining the relationship of the track to the cell cycle should give the information necessary to design culture conditions in which continuous cell cycle synchrony is maintained. (*Bottom*) Flow cytometric determined maximum S phase as a function of the length of the DO cycle.

als directly, as well as through diffusion. Collisions between cells are random and in some sense global rather than local, as in a mammalian tissue. Moreover, because of the balance between new cells appearing by division and the removal of cells by dilution, there is always a disproportionate fraction of newly divided cells: the exponential growth distribution. How the signaling of a cell that is not yet ready to replicate or divide affects a cell that would otherwise be ready to divide is central to understanding how cells with adequate nutrients are prevented from replicating and dividing with the minimal generation times. At lower cell densities, cell cycle times are in the range normally found for wild-type yeast, 2–4 hours. Under conditions that promote synchronization of their respiratory/reductive cycles, their generation times are 8 hours long. It is logical to first assume that the cultures are limited by some essential nutrient, even though the cultures are continuously receiving fresh medium with 2% glucose. Direct measurement of glucose levels in the cultures yields levels of glucose in the range of 50—250 μM. When the nutrient supply is shut off, cultures continue to oscillate for 6–12 hours. Continuous sampling at 4-minute intervals throughout the cycle shows a low-amplitude oscillation in glucose levels, with the level increasing slightly when the cells enter the reductive phase of the cycle. It seems likely that oscillations in batch cultures occur more frequently than is recognized. Routine detection of the oscillations relies on close monitoring of oxygen consumption. In standard medium, with 2% (110 mM) glucose, the Crabtree effect initially inhibits respiration, and most laboratories do not measure glucose levels, oxygen consumption, hydrogen sulfide production, and carbon dioxide production as we do. Most significantly, shaker cultures, which are by far the most common means of growing *S. cerevisiae*, are poorly oxygenated with dissolved oxygen (DO) levels of less than 2%.

Kinetically, the yeast stochastic tissue and a mammalian tissue, such as the epithelial cells of the gastrointestinal tract, are similar if on different timescales. In the gastrointestinal tract of mammals, the cell cycle time of a single cell is in the range of 3–8 days, even though a fraction of the cells in that tissue (typically 10–15%) divide each day, at the same time of day (Potten and Bullock 1983; Potten and Loeffler 1990). In the yeast gated synchrony system, where the TRAC is 40 minutes long, 8–10% of the cells divide in each turn of the cycle, even though the average cell cycle time of these cells is about 8 hours. When explanted to culture, mammalian cells exhibit an ability to grow with generation times much shorter than 8 days, typically 24 hours. Similarly, yeast cells that are diluted and refed with the conditioned medium divide with a 2-hour generation time. Under conditions favoring gated synchrony, cells that might otherwise initiate DNA synthesis or cell division are prevented from doing so by signals received from other cells that are in earlier phases of the cell cycle. The more mature cells are thus retarded in their progress by cell-to-cell communication and by virtue of their being embedded in a network of cells that is on average younger and therefore unready to replicate or divide. We call this a quorum conflict.

PERIOD DOUBLING AND THE EFFECT OF PHENELZINE ON GENOME-WIDE OSCILLATIONS

As discussed in detail in our recent publication (Li and Klevecz 2006), perturbation of the gated synchrony system in yeast with phenelzine, an antidepressant drug used in the treatment of affective disorders in humans, leads to a rapid lengthening in the period of the genome-wide transcriptional oscillation. Lithium and phenelzine are among the oldest of the mood stabilizing or antidepressant psychoactive drugs used in the treatment of bipolar and other affective disorders in humans. One effect of these drugs is a slight lengthening of the circadian rhythm in both normal volunteers and patients under controlled settings. In humans, as well as experimental systems as disparate as rodents and the plant *Kalanchoe*, treatment with these agents increases the period of the circadian clock by an amount just under 1 hour. Treatment of *S. cerevisiae* with phenelzine causes a similar 30–40-minute increase, such that the period of the oscillation in the respiratory/reductive cycle is nearly doubled from 40 to about 70 minutes. Close examination of the benchmark oscillation in DO in the yeast gated synchrony system indicates that the reductive phase is exactly doubled, whereas the length of the respiratory phase is unchanged by drug treatment. The doubling of the length of the reductive phase in DO is anticipated by a doubling in the number of maxima in expression of reductive phase transcripts. This doubling of the reductive phase follows a path that can be closely modeled by a modification of the simple system of equations represented by the Rossler attractor (Rossler 1976).

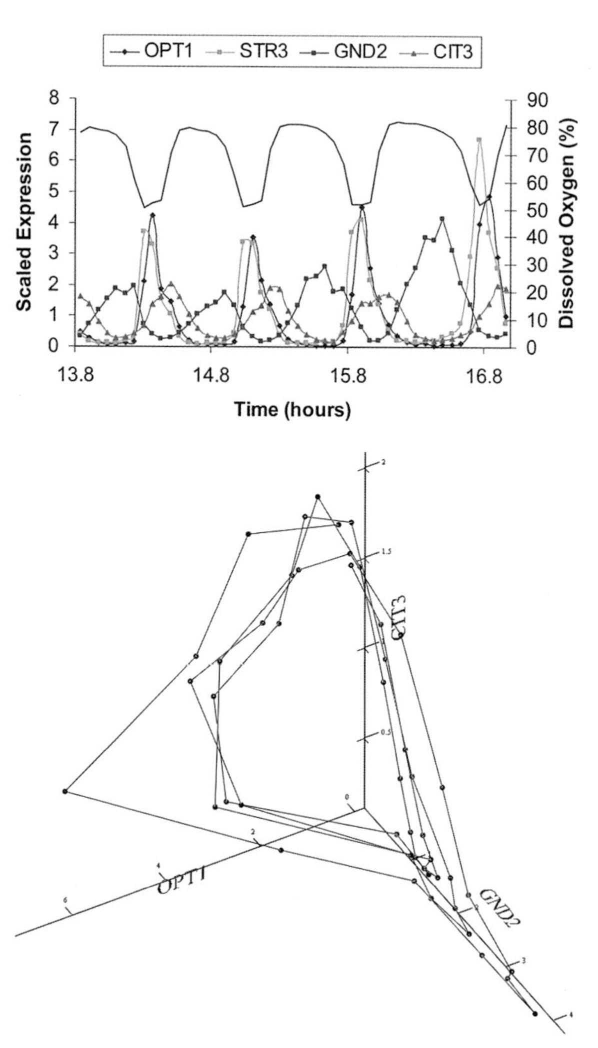

Figure 4. Super cluster swarms around the steady state. The expression patterns of four transcripts are shown. (*Top*) Two of the transcripts, OPT1 and STR3, are respiratory-phase transcripts that were so classified because their maximum expression occurs during the phase of high oxygen utilization. OPT1 and STR3 are shown to indicate the stability of the phase angle between the two transcripts. CIT3 is maximally expressed during the early reductive phase and GND2 is maximally expressed in the late reductive phase. Solid line indicates DO. (*Bottom*) The expression patterns of three of the transcripts are plotted together. If we take average of all of the transcripts whose maxima occur in early reductive, late reductive, or respiratory phase and plot these three sets of average values together, a plot very similar to that for the three genes emerges, as shown by Li and Klevecz (2006).

DESCRIBING AND RECONSTRUCTING AN ATTRACTOR

An attractor can be represented by a mathematical formalism. In the most familiar instances, these are a set of ordinary differential equations that represent the path taken through concentration phase space by the variables of the system over time (Li and Klevecz 2006). Rather than plotting the change in concentration or level of each of the variables versus time separately in a series of graphs, it is more informative to plot them relative to one another. Time then becomes implicit in the points along the trajectory taken by the system as a whole. In systems with large numbers of variables, it would seem that the resulting attractor would not lend itself to a simple graphical representation. But one of the most interesting and important attributes of complex systems is that they can often be accurately represented by attractors with very low numbers of dimensions. In this view, genes with similar patterns of expression are thought of as being in the same basin of attraction and can be visualized as circling around the steady state as swarms in a periodic path. These systems have a solid mathematical basis that has been well understood for some time. The rules of behavior are very specific and limited, and therefore, the predictions can be readily tested in well-controlled biological systems. Here, we have chosen to show just four of the transcripts from our most recent study (Fig. 4). These results can be compared with those of Figure 2D in Li and Klevecz (2006), where the averages of all transcripts from early and late reductive phase and respiratory phase are shown. This preferential lengthening of the reductive phase is also seen in strains such as CEN.PK, where the nominal period length is greater than that of IFO0233. The changes in expression, and the folding path from the approximately 40-minute oscillation to the 4-hour oscillation, are of greatest interest in the effort to map the surface of the attractor as it changes during the course of differentiation.

CLUSTERING OF PERIODS IN DIFFERING STRAINS AND MUTANTS OF THE TRAC

The use of deletion mutants, drugs, and media manipulation has yielded a spectrum of oscillatory modes and periods in a number of *S. cerevisiae* strains, with DO periods ranging from 20 minutes to 16 hours. However, without an analysis of protein levels or mRNA expression, it is difficult to say how the increase in period lengths seen in the DO curves would be manifested in the TRAC. Here, we show two examples of period doubling. The first is the doubling in the reductive phase alone of IFO0233 cultures treated with phenelzine (Fig. 5B) in comparison with untreated IFO0233 (Fig. 5A). The second example involves the deletion of *YBR239C* from the

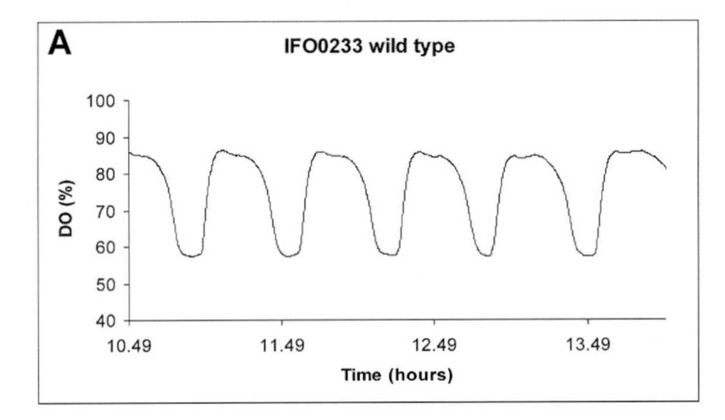

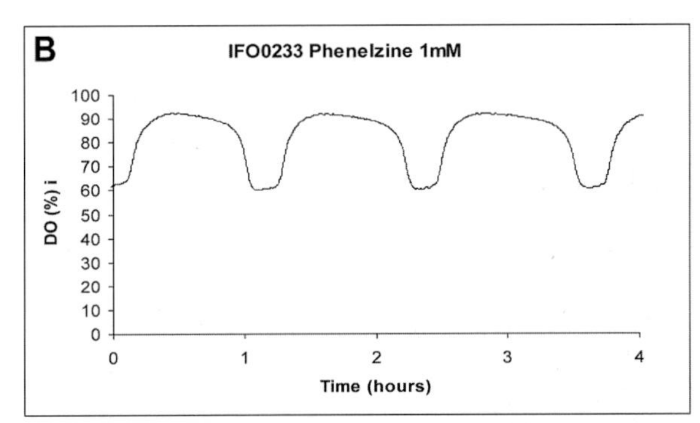

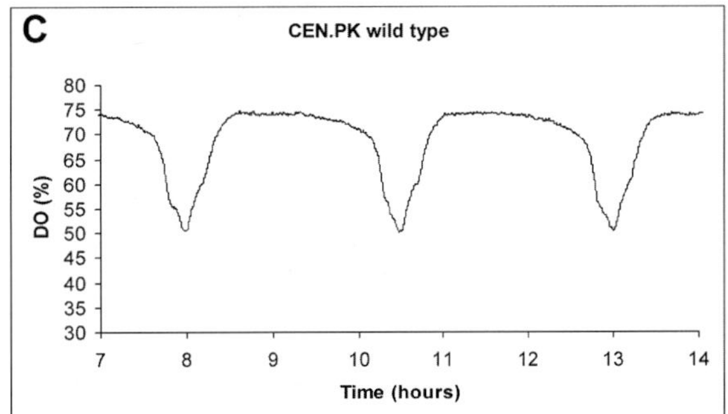

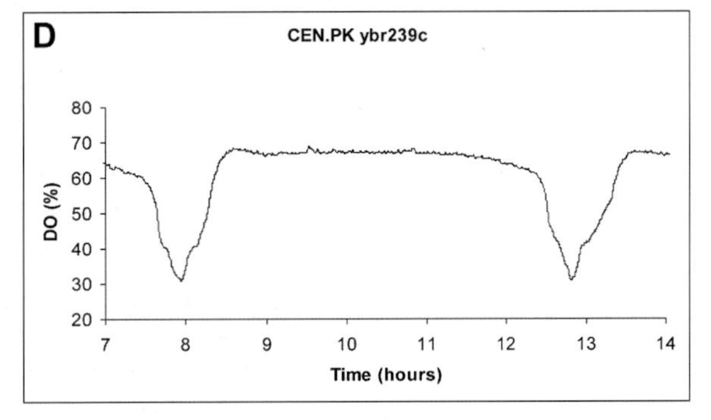

Figure 5. Reductive phase period doubling. The DO oscillations for the IFO0233 diploid wild type (*A*), the phenelzine-treated IFO0233 (*B*), the CEN.PK haploid (*C*), and the CEN.PK haploid Δ*ybr239c* deletion (*D*) are shown. The deletion of the *YBR239C* gene from the haploid increases the period of the oscillation from 154 minutes to 291 minutes. As in the phenelzine experiments, the change in period seems to take place principally in the reductive phase.

CEN.PK strain. In the haploid wild-type CEN.PK strain, the period of the DO oscillation is 154 minutes, whereas in the *YBR239C* deletion, it is approximately doubled to 291 minutes (Fig. 5C,D). It must be emphasized that the oscillations in both wild type and mutants are somewhat pleiodynamic and can show periods other than those shown here.

The path from the transcriptional cycle to the cell cycle and circadian rhythms appears to involve period-two and period-three bifurcations. With the help of a series of simple simulations (see Figure 6 in Li and Klevecz 2006), it is possible to show how an increase in the number of doublings in the attractor surface increases the time required for a system to complete one cycle, and, at the same time, the pattern of any trajectory within the surface becomes quite complex. This should have some general appeal because it explains how a system can have a high-amplitude oscillator, as might be expressed in a circadian length rhythm, while not obviously expressing high-amplitude oscillations in the controlling high-frequency oscillations. Nevertheless, the model predicts that close time series sampling in such a circadian system should uncover the high-frequency components embedded in the approximately 24-hour cycle. Moreover, these high-frequency components serve to increase the precision of the clock to permit precise timing of events at differing phases. We are hopeful that additional high-frequency analyses of the sort recently described by van der Veen et al. (2006) will be applied to high-frequency sampling of transcript and protein levels in circadian systems to test the validity of this model.

PERIOD DOUBLING: A PATH TO EVOLUTIONARY CHANGE REQUIRING MINIMAL DELETERIOUS MUTATIONS

Period doubling is a path for phenotypic change that requires minimal mutational change. A single-gene mutation can tug an entire pathway into a new basin of attraction, thereby altering phenotype and giving the system an opportunity to experiment without the cost of multiple mutations. Stable changes in expression can occur simply by moving the entire multigene complex into a new basin of attraction. One prediction is that genes which alter coupling strength and cell-to-cell or gene-to-gene signaling should show a greater substitution rate than those that are deeply embedded in the attractor basin. Put another way, to change the function of a gene that is an integral part of one of the core oscillations is likely to be difficult. A strength of this model is that it permits many genes to be expressed at very low levels without the consequent loss of temporal organization that would result if the organization of phenotype was stochastic.

CONSERVATION AND EVOLUTION OF PERIOD

The idea that certain functional domains are conserved through great evolutionary distances is the core of our understanding of modern molecular evolution. The mechanism that facilitated this conservation was gene duplication, as first suggested by Ohno (1970; Kellis et al. 2004). Moreover, that the rate of evolution of a protein is related to the number of its interactions suggests that evolutionary changes are constrained by the need to function as a system and may occur largely by coevolution.

What has only rarely been considered is the idea that a dynamic property of the system, in this case the period of the oscillation, might impose similar or greater constraints on the evolution of a protein and the network in which it is embedded. In such a case, one member of a regulon of genes that evolved to generate a precise period of biologically important duration cannot change without causing the function of the system of proteins to fail. It is the dynamic analog of a multiunit complex; changes in one gene might only be achieved by a compensatory change in another to give rise to a cycle of similar period to the original. In a phenotype in which "everything oscillates," it seems likely that to a greater or lesser extent, every gene product interacts with some or many other gene products. In such a globally coupled system, the constraints on period change might be expected to be extreme. On the basis of our recent work, a strong argument can be made for this idea. We suggest that one of the constraints on the evolution of the circadian period from its primordial approximately 8-hour period to the present is through a folding of the attractor surface to achieve an increase in period without requiring multiple changes in the regulatory and kinetic properties of a large number of genes or regulons.

By this hypothesis, genes that are part of the core regulatory complex would evolve most slowly, whereas those that function to alter period, such as genes involved in signaling and coupling pathways and their networks, would evolve more rapidly. A change in a single or a small number of genes involved in the signaling or diffusive coupling of genes might "tune" the core oscillatory elements to a new basin of attraction where functionality would endure, if at a slightly suboptimal level, giving the system as a whole the opportunity to evolve more slowly. For example, one might expect that genes containing the PAS (PER-ARNT-SM) domain, which is known to alter period in both circadian rhythms and the yeast transcriptional cycle, would be less conserved than those central to the core oscillator.

CONCLUSION: EVOLUTION OF CIRCADIAN RHYTHMS FROM A HIGH-FREQUENCY OSCILLATOR

We know that circadian rhythms cannot always have been 24 hours because the period of the Earth's rotation has slowed since the time when life first emerged. Tidal rhythms at that early time would have been about 4 hours, close to the period of the quantal oscillator described in mammalian cells and also equal to the period differences seen in mutants of the circadian clock. If a primitive, short-period oscillator with the capacity to entrain to sunlight emerged early, and if the ensemble of reactions specifying phase and period were fixed and integrated into the chemistry of the organism, as "frozen accidents" are fixed in primordial DNA sequences, some vestige of this primitive clock might still be expressed as a high-frequency rhythm in mammalian cells and yeast. If the path from high-frequency rhythms to the modern circadian clock follows the period-two or period-three bifurcation path predicted by the coupling of simple reaction-diffusion systems, and observed experimentally in yeast, then we might be close to a general theory of evolution of the circadian clock.

ACKNOWLEDGMENTS

This work was supported in part by grant GM81757-01. The authors thank Keely Walker for her help with the manuscript editing and Ian Marcus for his help with data analysis and computer simulations.

REFERENCES

Bipatnath M., Dennis P.P., and Bremer H. 1998. Initiation and velocity of chromosome replication in *Escherichia coli* B/r and K-12. *J. Bacteriol.* **180:** 265.

Bolen J.L., Duran O., and Klevecz R.R. 1993. Amplification and damping of deterministic noise in coupled cellular arrays. *Physica D* **67:** 245.

Bruce V.G. 1972. Mutants of the biological clock in *Chlamydomonas reinhardi*. *Genetics* **70:** 537.

Chisholm S.W. and Costello J.C. 1980. Influence of environmental factors and population composition on the timing of cell division in *Thalassiosira fluviatilis* (Bacillariophyceae) grown on light/dark cycles. *J. Phycol.* **16:** 375.

Collins J.M. 1978. Rates of DNA synthesis during the S-phase of HeLa cells. *J. Biol. Chem.* **253:** 8570.

Donachie W.D. 1968. Relationship between cell size and time of initiation of DNA replication. *Nature* **219:** 1077.

Feldman J.F. and Dunlap J.C. 1982. *Neurospora crassa:* A unique system for studying circadian clocks. *Photochem. Photobiol. Rev.* **7:** 319.

Feldman J.F. and Hoyle M.N. 1973. Isolation of circadian clock mutants of *Neurospora crassa*. *Genetics* **75:** 605.

Feldman J.F., Gardner G., and Denison R. 1979. Genetic analysis of the circadian clock of *Neurospora*. In *Biological rhythms and their central mechanism:* Naito Foundation Symposium, Tokyo, Japan, 1978 (ed. M. Suda et al.), p. 58. Elsevier, North Holland, Amsterdam.

Hartwell L.H., Culotti J., Pringle J.R., and Reid B.J. 1974. Genetic control of the cell division cycle in yeast. *Science* **183:** 46.

Hess B. and Boiteux A. 1971. Oscillatory phenomena in biochemistry. *Annu. Rev. Biochem.* **40:** 237.

Holmquist G., Gray M., Porter T., and Jordan J. 1982. Characterization of Giemsa dark- and light-band DNA. *Cell* **31:** 121.

Jauker F. and Cleffmann G. 1970. Oscillation of individual generation times in cells lines of *Tetrahymena pyriformis*. *Exp. Cell Res.* **62:** 477.

Kapp L.N., Millis A.J., and Pious D.A. 1979. Variation in S phase in synchronous human cell lines. *In Vitro* **15:** 669.

Kellis M., Birren B.W., and Lander E.S. 2004. Proof and evolutionary analysis of ancient genome duplication in the yeast *Saccharomyces cerevisiae*. *Nature* **428:** 617.

Klevecz R.R. 1969. Temporal order in mammalian cells. I. The periodic synthesis of lactate dehydrogenase in the cell cycle. *J. Cell Biol.* **43:** 207.

———. 1976. Quantized generation time in mammalian cells as an expression of the cellular clock. *Proc. Natl. Acad. Sci.* **73:** 4012.

———. 1984. Cellular oscillators as vestiges of a primitive circadian clock. In *Cell cycle clocks* (ed. L. Edmunds), p. 47. Marcel Dekker, New York.

Klevecz R.R. and Kapp L.N. 1973. Intermittent DNA synthesis and periodic expression of enzyme activity in the cell cycle of WI-38. *J. Cell Biol.* **58:** 564.

Klevecz R.R. and King G.A. 1982. Temperature compensation in the mammalian cell cycle. *Exp. Cell Res.* **140:** 307.

Klevecz R.R. and Ruddle F.H. 1968. Cyclic changes in enzyme activity in synchronized mammalian cell cultures. *Science* **159:** 634.

Klevecz R.R., Bolen J.L., Duran O. 1992. Self-organization in biological systems. *Int. J. Bifurcation Chaos* **2:** 941.

Klevecz R.R., Kauffman S.A., and Shymko R.M. 1984. Cellular clocks and oscillators. *Int. Rev. Cytol.* **86:** 97.

Klevecz R.R., Kros J., and Gross S.D. 1978. Phase response versus positive and negative division delay in animal cells. *Exp. Cell Res.* **116:** 285.

Klevecz R.R., Bolen J., Forrest G., and Murray D.B. 2004. A genomewide oscillation in transcription gates DNA replication and cell cycle. *Proc. Natl. Acad. Sci.* **101:** 1200.

Konopka R.J. and Benzer S. 1971. Clock mutants of *Drosophila melanogaster*. *Proc. Natl. Acad. Sci.* **68:** 2112.

Li C.M. and Klevecz R.R. 2006. A rapid genome-scale response of the transcriptional oscillator to perturbation reveals a period-doubling path to phenotypic change. *Proc. Natl. Acad. Sci.* **103:** 16254.

Margulis L. 1993. Origins of species: Acquired genomes and individuality. *Biosystems* **31:** 121.

Murray D.B., Beckmann M., and Kitano H. 2007. Regulation of yeast oscillatory dynamics. *Proc. Natl. Acad. Sci.* **104:** 2241.

Murray D.B., Klevecz R.R., and Lloyd D. 2003. Generation and maintenance of synchrony in *Saccharomyces cerevisiae* continuous culture. *Exp. Cell Res.* **287:** 10.

Nordstrom K. and Dasgupta S. 2006. Copy-number control of the *Escherichia coli* chromosome: A plasmidologist's view. *EMBO Rep.* **7:** 484.

Ohno S. 1970. *Evolution by gene duplication*. Springer, New York.

Panella G. 1972. Paleontological evidence on the earth's rotational history since early precambrian. *Astrophys. Space Sci.* **16:** 212.

Pittendrigh C.S. 1967. Circadian systems. I. The driving oscillation and its assay in *Drosophila pseudoobscura*. *Proc. Natl. Acad. Sci.* **58:** 1762.

Potten C.S. and Bullock J.C. 1983. Cell kinetic studies in the epidermis of the mouse. I. Changes in labeling index with time after tritiated thymidine administration. *Experientia* **39:** 1125.

Potten C.S. and Loeffler M. 1990. Stem cells: Attributes, cycles, spirals, pitfalls and uncertainties. Lessons for and from the crypt. *Development* **110:** 1001.

Rossler O.E. 1976. An equation for continuous chaos. *Phys. Lett.* **35A:** 397.

Searcy D.G. 2003. Metabolic integration during the evolutionary origin of mitochondria. *Cell Res.* **13:** 229.

Searcy D.G., Stein D.B., and Searcy K.B. 1981. A mycoplasma-like archaebacterium possibly related to the nucleus and cytoplasms of eukaryotic cells. *Ann. N.Y. Acad. Sci.* **361:** 312.

Shymko R.M., Klevecz R.R., and Kauffman S.A. 1984. The cell cycle as an oscillatory system. In *Cell cycle clocks* (ed. L.N. Edmunds, Jr.), p. 273. Marcel Dekker, New York.

Tu B.P., Kudlicki A., Rowicka M., and McKnight S.L. 2005. Logic of the yeast metabolic cycle: Temporal compartmentalization of cellular processes. *Science* **310:** 1152.

Turcotte D.L., Cisne J.L., and Nordmann J.C. 1977. On the evolution of the lunar orbit. *Icarus* **30:** 254.

van der Veen D.R., Minh N.L., Gos P., Arneric M., Gerkema M.P., and Schibler U. 2006. Impact of behavior on central and peripheral circadian clocks in the common vole *Microtus arvalis,* a mammal with ultradian rhythms. *Proc. Natl. Acad. Sci.* **103:** 3393.

Wells J.W. 1963. Coral growth and geochronometry. *Nature* **197:** 948.

Intracellular Developmental Timers

M. Raff

MRC Laboratory for Molecular Cell Biology and the Biology Department, University College London, London WC1E 6BT, United Kingdom

One of the most poorly understood aspects of animal development is how the timing of developmental events is controlled. In most vertebrate cell lineages, for example, precursor cells divide a limited number of times before they stop and terminally differentiate, but it is not known what controls when the cells stop dividing and differentiate. There is increasing evidence, however, that intracellular timers play an important part. Such cell-intrinsic timers are examples of intracellular developmental programs that change precursor cells over time. My colleagues and I have studied such intracellular timers and programs in rodent oligodendrocyte precursor cells (OPCs), as reviewed here.

INTRODUCTION

Our interest in developmental timing began with a surprising experimental result obtained by Erika Abney, an immunologist who joined our group in the late 1970s. We had earlier defined a set of cell-type-specific markers that allowed us to distinguish the major types of glial cells found in suspensions and cultures of rat central nervous system (CNS) cells: astrocytes, which are heterogeneous and have many functions; oligodendrocytes, which myelinate neuronal axons; and ependymal cells, which line the fluid-filled ventricles of the brain (Raff et al. 1979). Analyzing cell suspensions prepared from developing rat brains from embryonic day eleven (E11) through to birth at approximately E21, she determined when the differentiated cells of each type first appeared. She found that the first astrocytes appeared at E15–E16, the first ependymal cells at E17–E18, and the first oligodendrocytes at around the time of birth (E21–E22). Each cell type always first appeared in very small numbers and rapidly increased over the following days. Remarkably, when she isolated cells from E10 brain and cultured them in 10% fetal calf serum (FCS), the times of first appearance of these three cell types were the same as if the cells had been left in the developing brain; moreover, when cultures were prepared from E13 brain, all three cell types first appeared 3 days earlier, just as in vivo (Abney et al. 1981). These results are as surprising today as they were when we published them more than 25 years ago, because the axial cues and morphogen gradients that play such an important part in controlling cell specification in early animal development are presumably missing in these cultures. Moreover, it seems unlikely that the sequential cell–cell interactions that are thought to help time events in development could occur normally in such cultures.

Whatever the nature of the timing mechanisms involved, the finding that they apparently could operate normally in dissociated-cell culture over a 10-day period encouraged us to study them. The complexity of the brain cell cultures, however, was daunting, and so we turned to the developing optic nerve, which is much simpler, and we focused on the timing of oligodendrocyte development.

AN INTRACELLULAR TIMER IN OLIGODENDROCYTE PRECURSOR CELLS

The optic nerve contains no neurons, although it contains the axons of retinal ganglion neurons. The main cell types in the nerve are astrocytes and oligodendrocytes, with smaller numbers of macrophages (microglia), blood vessel cells, and glial precursor cells. Whereas the astrocytes in the optic nerve develop from the neuroepithelial cells of the optic stalk (the primordium of the nerve), the oligodendrocytes develop from precursor cells that migrate into the developing nerve from the brain, beginning before birth (Small et al. 1987). The oligodendrocyte precursor cells (OPCs) divide a limited number of times before they stop and terminally differentiate into postmitotic oligodendrocytes. As in the brain, the first oligodendrocytes appear in the rat optic nerve around the day of birth, and their numbers then progressively increase over the next 6 weeks (Miller et al. 1985). Our goal was to understand why the OPCs stop dividing and differentiate when they do. This is still not known for any type of mammalian precursor cell.

As was the case in embryonic brain cell cultures, we could reconstitute the normal timing of oligodendrocyte development in dissociated cell cultures of embryonic rat optic nerve containing 10% FCS (Raff et al. 1985). Mark Noble (then at the Institute of Neurology in London and now at Rochester Medical School) had shown that astrocytes secrete signal molecules that are required to stimulate OPCs to proliferate in culture and to prevent the premature differentiation of the OPCs into oligodencrocytes (Noble and Murray 1984). Both he and our Biology Department colleague Bill Richardson independently showed that the relevant signal molecule is platelet-derived growth factor-α (PDGF-α) (Noble et al. 1988; Richardson et al. 1988). Our three laboratories collaborated to show that PDGF can bypass the need for FCS and astrocytes and allow normal timing of oligodendrocyte development in sparse embryonic optic nerve cell cultures (Raff et al. 1988).

Sally Temple (then a Ph.D. student and now at Albany Medical School) showed that OPCs have a cell-intrinsic mechanism that helps determine when they stop dividing and differentiate (Temple and Raff 1986). She placed single OPCs from postnatal day 7 (P7) rat optic nerve onto monolayers of astrocytes in individual microwells and found that the isolated single cells divide a variable number of times before they differentiate, with a maximum of eight divisions. Crucially, she showed that all the progeny of an individual OPC stop dividing and differentiate at about the same time, even if the two daughter cells of an OPC are placed on astrocyte monolayers in separate microwells. These findings established that an intrinsic counting or timing mechanism is built into each OPC, and we spent much of the next 15 years trying to determine how the mechanism operates. This seemed a worthwhile effort, as it seemed likely that a similar mechanism probably operates in many types of precursor cells.

Ben Barres (then a postdoc and now at Stanford Medical School) advanced both the study and understanding of the cell-intrinsic mechanism in two important ways. First, he developed a sequential immunopanning method to purify OPCs to homogeneity from neonatal rat optic nerves (Barres et al. 1992). Second, he showed that in serum-free clonal-density cultures of purified OPCs, the intrinsic mechanism depends on thyroid hormone (TH), as well as on PDGF (Barres et al. 1994). Without PDGF, the cells prematurely stop dividing and differentiate. In the presence of PDGF but without TH, most OPCs fail to stop dividing and differentiate; if TH is added after the time when most OPCs would have differentiated had the hormone been present all along, the cells quickly stop dividing or differentiate. Ben's findings suggested that OPCs are able to count divisions or measure time in the absence of TH but that TH is required for the cells to withdraw from the cell cycle and differentiate when the intrinsic mechanism indicates it is time. Therefore, the counter or timer seems to consist of at least two functional components: (1) a counting or timing component, which depends on PDGF but not on TH and counts cell divisions or measures elapsed time, and (2) an effector component, which is regulated by TH and stops cell division and initiates differentiation when the counting or timing component indicates that it is time (Barres et al. 1994). Bögler and Noble reached a similar conclusion using a combination of PDGF and basic fibroblast growth factor-2 (FGF-2) to keep the OPCs dividing beyond their normal limit (Bögler and Noble 1994).

Why should TH regulate the intrinsic counting or timing mechanism in OPCs? Adult animals use hormones to help coordinate the behavior of their cells throughout the body, and so it is not surprising that developing animals use hormones such as TH to help coordinate the timing of development in their various organs and tissues. TH, for example, has been shown to coordinate the onset of myelination by oligodendrocytes and Schwann cells in the central and peripheral parts, respectively, of the developing auditory nerve (Knipper et al. 1998).

Fen-Biao Gao (then a postdoc and now at the Gladstone Institute of Neurological Disease in San Francisco) used the Barres method to purify OPCs from embryonic rat optic nerves to show that in serum- and extract-free clonal-density cultures containing PDGF and TH, the OPCs stop dividing and differentiate on the same schedule as they do in vivo (Gao et al. 1998). He also showed that, on average, OPCs purified from E18 optic nerve proliferate longer under these conditions than do OPCs purified from P7 nerve (Gao and Raff 1997), suggesting that the reason OPCs from P7 optic nerve go through a variable number of divisions before they differentiate (Temple and Raff 1986) is because they are at various stages of maturation. Most important, he showed that OPCs cultured at 33°C divide more slowly but stop dividing and differentiate earlier, after fewer divisions, than when they are cultured at 37°C (Gao and Raff 1997). This finding provided evidence that the cell-intrinsic mechanism does not depend on counting cell divisions but instead measures time in some other way, leading us to refer to the mechanism as an intracellular timer. But what is the other way, and why does the timer run faster at the lower temperature?

SOME PROTEIN COMPONENTS OF THE INTRACELLULAR TIMER

The first clue to the molecular nature of the intrinsic timer came from Béatrice Durand (then a postdoc and now at the Pasteur Institute in Paris). She showed that the amount of the cyclin-dependent protein kinase (Cdk) inhibitor $p27^{Kip1}$ (p27) progressively increases in the nucleus of purified OPCs as they proliferate in the presence of PDGF and the absence of TH (Durand et al. 1997). The amount of p27 protein reaches a plateau at the time when most of the cells would have stopped dividing if TH had been present; without TH, however, the cells continue to proliferate, despite the high levels of p27. Fen-Biao Gao and Béa found that the level of p27 protein rises faster at 33°C than at 37°C, suggesting that this may be one reason why the timer runs faster at the lower temperature (Gao et al. 1997). Béa then collaborated with Jim Roberts in Seattle, the head of one of the three laboratories that knocked the $p27^{Kip1}$ gene out in mice (Fero et al. 1996; Kiyokawa et al. 1996; Nakayama et al. 1996). She showed that in cultures containing PDGF and TH, p27-deficient OPCs divide for a day or two longer than wild-type OPCs before they stop dividing and differentiate, even though the cell cycle times are indistinguishable in the mutant and wild-type cells (Durand et al. 1998). This finding suggested that p27 is one component of the timer. Moreover, Béa found that the p27-deficient cells are defective in both the timing and effector components of the timer, suggesting that p27 plays a part in both components. Jim Apperly, a Ph.D. student, showed that overexpression of p27 accelerates the timer, providing further support for a role of p27 in the timing process (Tokumoto et al. 2002). Yasu Tokumoto (then a postdoc and now at the Research Institute of Cell Engineering in Japan) showed that p27 mRNA levels remain constant as the protein increases in proliferating OPCs (Tokumoto et al. 2002), suggesting that the increase in p27 protein over time depends on post-transcriptional mechanisms, which remain to be identified.

Mice without p27 are about 30% larger than wild-type mice and have increased cell numbers in all of their organs, as a result of increased cell proliferation rather than decreased cell death (Fero et al. 1996; Kiyokawa et al. 1996; Nakayama et al. 1996; Tokumoto et al. 2002). It therefore seems likely that p27 has a similar role in timing cell cycle withdrawal and differentiation in many mammalian cell lineages. A homolog of p27 is present in *Drosophila* and *Caenorhabditis elegans*, and if the gene encoding it is inactivated by mutation in either organism, cells go through an extra division or two in several cell lineages (de Nooij et al. 1996; Lane et al. 1996; Hong et al. 1998). Thus, Cdk inhibitors are probably involved in stopping the cell cycle at the appropriate time during development in all animals.

The p27 protein, however, is only a minor component of the timer that operates in OPCs, as the timer still works in p27-deficient OPCs; it just works inaccurately. The phenotype of mice deficient in another Cdk inhibitor, $p18^{Ink4c}$ (p18), is very similar to that of p27-deficient mice (Franklin et al. 1998). Yasu Tokumoto found that p18 protein is expressed in OPCs and increases much like p27 as OPCs proliferate in culture, suggesting that it is also part of the timer; like p27, its increase is controlled posttranscriptionally (Tokumoto et al. 2002). Dugas, Ibrahim, and Barres recently showed that $p57^{Kip2}$ is also an important component of the timer, but in this case, $p57^{Kip2}$ mRNA and protein increase in parallel as OPCs proliferate, suggesting that the increase in p57 protein in OPCs is controlled transcriptionally (Dugas et al. 2007).

Toru Kondo (then a postdoc and now in the Riken Center for Developmental Biology in Kobe) showed that the inhibitor of differentiation protein 4 (Id4) is also a component of the timer, although it works in a manner opposite to that of p18, p27, and p57 (Kondo and Raff 2000a). Id proteins inhibit basic helix-loop-helix gene regulatory proteins that are required for the differentiation of many types of precursor cells; in this way, they inhibit differentiation and promote proliferation of the precursors. Toru found that Id4 protein decreases as purified OPCs proliferate in the presence of PDGF and the absence of TH and that Id4 mRNA and protein decrease in parallel, suggesting that a transcriptional mechanism is probably responsible for the progressive decrease in Id4 protein. He also showed that overexpression of Id4 prolongs OPC proliferation and inhibits differentiation in the presence of PDGF and TH. Toru then collaborated with Fred Sablitzky at Nottingham University to show that neural stem cells isolated from the brains of Id4-deficient mouse embryos produce oligodendrocytes prematurely in culture (Marin-Husstege et al. 2006). Taken together, these findings suggest that Id4 is another component of the timer and that its progressive decrease helps control when OPCs stop dividing and differentiate.

The OPC timer, like other intracellular timers, is still poorly understood. It is clear, however, that it is complex and depends on the progressive increase of some intracellular proteins such as p18, p27, and p57 and the progressive decrease of others such as Id4. Both transcriptional and posttranscriptional controls have roles, but how these controls operate remains to be determined.

AN INTRINSIC MATURATION PROGRAM IN OPCs

The intracellular timer that helps control when OPCs withdraw from the cell cycle and differentiate seems to be only one part of a much more complex cell-intrinsic timing mechanism that changes many of the properties of OPCs over time.

Charles ffrench-Constant (then a medically trained Ph.D. student and now at Edinburgh University) found that there are small numbers of OPCs in cell suspensions prepared from adult rat optic nerves (ffrench-Constant and Raff 1986), suggesting that OPCs are present in the optic nerve throughout postnatal life. Mark Noble and his colleagues independently found these adult OPCs and characterized them in more detail (Wolswijk and Noble 1989). Julia Burne (a technician turned Ph.D. student) collaborated with Barbara Fulton in the Anatomy Department at University College London to visualize OPCs in the rat optic nerve at different times in development. They found that the cells become progressively more complex in morphology with age (Fulton et al. 1992), but it is not clear whether these changes reflect changes in the cells' environment with age, an intrinsic maturation program operating within the cells themselves, or both.

Fen-Biao Gao provided strong evidence that perinatal OPCs, at least, have an intrinsic developmental program that changes many aspects of the cell over time (Gao and Raff 1997). He first compared the properties of purified E18 OPCs to those of purified P7 OPCs (which are 10 days older) using time-lapse video recording of individual clones in culture. He found that in serum-free cultures containing PDGF but not TH, the embryonic OPCs have a simpler morphology than the P7 OPCs and divide and migrate faster; moreover, in the presence of PDGF and TH, the embryonic cells divide more times before differentiating than do the P7 cells. Remarkably, when he cultured purified E18 OPCs in PDGF without TH for 10 days (so that they were now the same age as freshly isolated P7 OPCs), he found that the embryonic cells had acquired all the properties of the P7 cells, indicating that developing OPCs have an intrinsic maturation program.

Dean Tang (then a postdoc and now at M.D. Anderson Cancer Center) and Yasu Tokumoto showed that purified P7 rat OPCs can proliferate in serum-free culture for more than 1 year in PDGF without TH. After many months in culture, Dean found that the OPCs start to express the glycolipid galactocerebroside (Tang et al. 2000), which we originally believed was expressed in the CNS only by oligodendrocytes and myelin (Raff et al. 1978). The galactocerebroside-expressing OPCs continue to proliferate and do not express other oligodendrocyte or myelin markers. The expression of galactocerebroside is unlikely to be a culture artifact, as Ben Barres and his colleagues showed earlier that OPCs in the rat optic nerve start to express galactocerebroside after months in vivo (Shi et al. 1998). Is it possible that the intrinsic maturation program in OPCs continues to change the cells for months? This would be remarkable and would raise the question of how such an extended program could work.

It is important to emphasize that the developmental programs and timers described here in OPCs are not set in stone. As already mentioned, for example, most OPCs ignore the timer that helps determine when they stop dividing and differentiate if TH is omitted from the culture medium (Barres et al. 1994). Moreover, Toru Kondo found that OPCs are not irreversibly committed to becoming oligodendrocytes or even glial cells: Transient exposure of purified OPCs in culture to bone morphogenetic protein 4 (BMP4), followed by FGF-2, rapidly reprograms the cells to become more like multipotential CNS neural stem cells, which can now produce both neurons and glia (Kondo and Raff 2000b). Remarkably, a 2-day treatment with BMP4 is enough to induce the transcription of a variety of genes that are normally expressed in neural stem cells but not in untreated OPCs (Kondo and Raff 2004).

A PUTATIVE TIMER PROTEIN IN SILKWORM

One of the most interesting intracellular timers ever reported seems to have received little if any attention, which is surprising, given its uniqueness and potential implications. It concerns the timing of diapause in the silkworm *Bombyx mori*, which Professor Hidenori Kai of Tottori University in Japan has been studying for more than 30 years. I have had nothing to do with this work, but I discuss it here because so few cell and developmental biologists seem to know about it.

In response to particular environmental conditions, many developing insects enter diapause, a period of developmental arrest (Denlinger 2002). In summer temperatures and long day lengths, for example, the developing eggs of *B. mori* are programmed by a hormone to enter diapause several days after the eggs are laid. Subsequent exposure to cold (~5°C) triggers the eggs to complete diapause, a process that takes 2 weeks if the eggs are cooled 2 days after they are laid and maintained at the low temperature for the entire period. Kai originally reported that an enzyme activity, assayed as an esterase, turns on after 2 weeks at 5°C and then turns off again within 24 hours (Kai and Nishi 1976). Remarkably, he later found that the same thing happens when he cooled an extract of the eggs in a test tube for 2 weeks (Kai et al. 1987). Even more remarkably, he subsequently showed that a purified preparation of the enzyme (now assayed as an ATPase) behaves similarly: The enzyme becomes active after 2 weeks at 5°C and then rapidly inactivates again (Kai et al. 1995). Kai calls the protein TIME (time-interval measuring enzyme), as it seems to measure the duration of the cold-induced diapause completion process (called diapause development). After completion, if the eggs are warmed, embryo development restarts, leading to hatching after about 12 days.

Two other properties of TIME are equally fascinating. First, the enzyme cannot be reactivated after it has turned on in the cold and turned off again, unless it is treated with a denaturant (6 M guanidine-HCl), in which case it responds to cooling in the same way as it does in freshly isolated extracts of diapause eggs (Kai et al. 1995). Second, TIME is physically associated in the developing egg with a small peptide called PIN (peptidyl inhibitory needle), which rapidly dissociates from TIME at 5°C, allowing the enzyme to start measuring time (Ti et al. 2005). The time period that PIN remains associated with TIME determines how long diapause development takes after cooling (Ti et al. 2006). It seems that at warm (summer) temperatures, PIN restructures TIME into a timekeeping conformation and does so progressively over days.

The amino acid sequences of TIME and PIN have been determined (Isobe et al. 2006). TIME contains 156 amino acids and PIN 38 amino acids, and both contain copper ions. TIME does not contain an obvious ATPase domain, but it does contain a functional Cu,Zn superoxide dismutase domain, the activity of which does not vary during diapause development. How TIME measures time remains a mystery, but it presumably involves a series of conformational changes that occur over many days at 5°C.

CONCLUSIONS

It is unlikely that circadian oscillators play an important part in timing events in animal development. Nonetheless, the intracellular programs that change developing animal cells over time and help control the timing of developmental events should be of some interest to biologists who study timing mechanisms in cells. They surely deserve more attention than they have received so far.

ACKNOWLEDGMENTS

I am extremely grateful to the postdocs, students, and associates mentioned, as well as those lab members not mentioned because their work fell outside the subject of this review; it was a privilege having them as friends and colleagues. I am also grateful to the Medical Research Council for providing my salary and research support for 31 years.

REFERENCES

Abney E.R., Bartlett P.P., and Raff M.C. 1981. Astrocytes, ependymal cells, and oligodendrocytes develop on schedule in dissociated cell cultures of embryonic rat brain. *Dev. Biol.* **83:** 301.

Barres B.A., Lazar M.A., and Raff M.C. 1994. A novel role for thyroid hormone, glucocorticoids and retinoic acid in timing oligodendrocyte development. *Development* **120:** 1097.

Barres B.A., Hart I.K., Coles H.S., Burne J.F., Voyvodic J.T., Richardson W.D., and Raff M.C. 1992. Cell death and control of cell survival in the oligodendrocyte lineage. *Cell* **70:** 31.

Bögler O. and Noble M. 1994. Measurement of time in oligodendrocyte-type-2 astrocyte (O-2A) progenitors is a cellular process distinct from differentiation or division. *Dev. Biol.* **162:** 525.

de Nooij J.C., Letendre M.A., and Hariharan I.K. 1996. A cyclin-dependent kinase inhibitor, Dacapo, is necessary for timely exit from the cell cycle during *Drosophila* embryogenesis. *Cell* **87:** 1237.

Denlinger D.L. 2002. Regulation of diapause. *Annu. Rev. Entomol.* **47:** 93.

Dugas J.C., Ibrahim A., and Barres B.A. 2007. A crucial role for p57(Kip2) in the intracellular timer that controls oligodendrocyte differentiation. *J. Neurosci.* **27:** 6185.

Durand B., Gao F.B., and Raff M. 1997. Accumulation of the cyclin-dependent kinase inhibitor p27/Kip1 and the timing of oligodendrocyte differentiation. *EMBO J.* **16:** 306.

Durand B., Fero M.L., Roberts J.M., and Raff M.C. 1998. p27Kip1 alters the response of cells to mitogen and is part of a cell-intrinsic timer that arrests the cell cycle and initiates differentiation. *Curr. Biol.* **8:** 431.

Fero M.L., Rivkin M., Tasch M., Porter P., Carow C.E., Firpo E., Polyak K., Tsai L.H., Broudy V., Perlmutter R.M., Kaushansky K., and Roberts J.M. 1996. A syndrome of multiorgan hyperplasia with features of gigantism, tumorigenesis, and female sterility in p27(Kip1)-deficient mice. *Cell* **85:** 733.

ffrench-Constant C. and Raff M.C. 1986. Proliferating bipotential glial progenitor cells in adult rat optic nerve. *Nature* **319:** 499.

Franklin D.S., Godfrey V.L., Lee H., Kovalev G.I., Schoonhoven R., Chen-Kiang S., Su L., and Xiong Y. 1998. CDK inhibitors p18(INK4c) and p27(Kip1) mediate two separate pathways to collaboratively suppress pituitary tumorigenesis. *Genes Dev.* **12:** 2899.

Fulton B.P., Burne J.F., and Raff M.C. 1992. Visualization of O-2A progenitor cells in developing and adult rat optic nerve by quisqualate-stimulated cobalt uptake. *J. Neurosci.* **12:** 4816.

Gao F.B. and Raff M. 1997. Cell size control and a cell-intrinsic maturation program in proliferating oligodendrocyte precursor cells. *J. Cell Biol.* **138:** 1367.

Gao F.B., Apperly J., and Raff M. 1998. Cell-intrinsic timers and thyroid hormone regulate the probability of cell-cycle withdrawal and differentiation of oligodendrocyte precursor cells. *Dev. Biol.* **197:** 54.

Gao F.B., Durand B., and Raff M. 1997. Oligodendrocyte precursor cells count time but not cell divisions before differentiation. *Curr. Biol.* **7:** 152.

Hong Y., Roy R., and Ambros V. 1998. Developmental regulation of a cyclin-dependent kinase inhibitor controls postembryonic cell cycle progression in *Caenorhabditis elegans. Development* **125:** 3585.

Isobe M., Kai H., Kurahashi T., Suwan S., Pitchayawasin-Thapphasaraphong S., Franz T., Tani N., Higashi K., and Nishida H. 2006. The molecular mechanism of the termination of insect diapause. 1. A timer protein, TIME-EA4, in the diapause eggs of the silkworm *Bombyx mori* is a metallo-glycoprotein. *Chembiochem.* **7:** 1590.

Kai H. and Nishi K. 1976. Diapause development in *Bombyx* eggs in relation to "esterase A" activity. *J. Insect Physiol.* **22:** 1315.

Kai H., Kawai T., and Kawai Y. 1987. A time-interval activation of esterase A4 by cold: Relation to the termination of embryonic diapause in the silkworm, *Bombyx mori. Insect Biochem.* **17:** 367.

Kai H., Kotani Y., Miao Y., and Azuma M. 1995. Time interval measuring enzyme for resumption of embryonic development in the silkworm, *Bombyx mori. J. Insect Physiol.* **41:** 905.

Kiyokawa H., Kineman R.D., Manova-Todorova K.O., Soares V.C., Hoffman E.S., Ono M., Khanam D., Hayday A.C., Frohman L.A., and Koff A. 1996. Enhanced growth of mice lacking the cyclin-dependent kinase inhibitor function of p27(Kip1). *Cell* **85:** 721.

Knipper M., Bandtlow C., Gestwa L., Kopschall I., Rohbock K., Wiechers B., Zenner H.P., and Zimmermann U. 1998. Thyroid hormone affects Schwann cell and oligodendrocyte gene expression at the glial transition zone of the VIIIth nerve prior to cochlea function. *Development* **125:** 3709.

Kondo T. and Raff M. 2000a. The Id4 HLH protein and the timing of oligodendrocyte differentiation. *EMBO J.* **19:** 1998.

———. 2000b. Oligodendrocyte precursor cells reprogrammed to become multipotential CNS stem cells. *Science* **289:** 1754.

———. 2004. Chromatin remodeling and histone modification in the conversion of oligodendrocyte precursors to neural stem cells. *Genes Dev.* **18:** 2963.

Lane M.E., Sauer K., Wallace K., Jan Y.N., Lehner C.F., and Vaessin H. 1996. Dacapo, a cyclin-dependent kinase inhibitor, stops cell proliferation during *Drosophila* development. *Cell* **87:** 1225.

Marin-Husstege M., He Y., Li J., Kondo T., Sablitzky F., and Casaccia-Bonnefil P. 2006. Multiple roles of Id4 in developmental myelination: Predicted outcomes and unexpected findings. *Glia* **54:** 285.

Miller R.H., David S., Patel R., Abney E.R., and Raff M.C. 1985. A quantitative immunohistochemical study of macroglial cell development in the rat optic nerve: In vivo evidence for two distinct astrocyte lineages. *Dev. Biol.* **111:** 35.

Nakayama K., Ishida N., Shirane M., Inomata A., Inoue T., Shishido N., Horii I., and Loh D.Y. 1996. Mice lacking p27(Kip1) display increased body size, multiple organ hyperplasia, retinal dysplasia, and pituitary tumors. *Cell* **85:** 707.

Noble M. and Murray K. 1984. Purified astrocytes promote the in vitro division of a bipotential glial progenitor cell. *EMBO J.* **3:** 2243.

Noble M., Murray K., Stroobant P., Waterfield M.D., and Riddle P. 1988. Platelet-derived growth factor promotes division and motility and inhibits premature differentiation of the oligodendrocyte/type-2 astrocyte progenitor cell. *Nature* **333:** 560.

Raff M.C., Abney E.R., and Fok-Seang J. 1985. Reconstitution of a developmental clock in vitro: A critical role for astrocytes in the timing of oligodendrocyte differentiation. *Cell* **42:** 61.

Raff M.C., Lillien L.E., Richardson W.D., Burne J.F., and Noble M.D. 1988. Platelet-derived growth factor from astrocytes drives the clock that times oligodendrocyte development in culture. *Nature* **333:** 562.

Raff M.C., Fields K.L., Hakomori S.I., Mirsky R., Pruss R.M., and Winter J. 1979. Cell-type-specific markers for distinguishing and studying neurons and the major classes of glial cells in culture. *Brain Res.* **174:** 283.

Raff M.C., Mirsky R., Fields K.L., Lisak R.P., Dorfman S.H., Silberberg D.H., Gregson N.A., Leibowitz S., and Kennedy M.C. 1978. Galactocerebroside is a specific cell-surface antigenic marker for oligodendrocytes in culture. *Nature* **274:** 813.

Richardson W.D., Pringle N., Mosley M.J., Westermark B., and Dubois-Dalcq M. 1988. A role for platelet-derived growth factor in normal gliogenesis in the central nervous system. *Cell* **53:** 309.

Shi J., Marinovich A., and Barres B.A. 1998. Purification and characterization of adult oligodendrocyte precursor cells from the rat optic nerve. *J. Neurosci.* **18:** 4627.

Small R.K., Riddle P., and Noble M. 1987. Evidence for migration of oligodendrocyte–type-2 astrocyte progenitor cells into the developing rat optic nerve. *Nature* **328:** 155.

Tang D.G., Tokumoto Y.M., and Raff M.C. 2000. Long-term culture of purified postnatal oligodendrocyte precursor cells. Evidence for an intrinsic maturation program that plays out over months. *J. Cell Biol.* **148:** 971.

Temple S. and Raff M.C. 1986. Clonal analysis of oligodendrocyte development in culture: Evidence for a developmental clock that counts cell divisions. *Cell* **44:** 773.

Ti X., Tani N., Isobe M., and Kai H. 2006. Time-measurement-regulating peptide PIN may alter a timer conformation of Time Interval Measuring Enzyme (TIME). *J. Insect Physiol.* **52:** 461.

Ti X., Tuzuki N., Tani N., Isobe M., and Kai H. 2005. The peptide PIN changes the timing of transitory burst activation of timer-ATPase TIME in accordance with diapause development in eggs of the silkworm, *Bombyx mori. J. Insect Physiol.* **51:** 1025.

Tokumoto Y.M., Apperly J.A., Gao F.B., and Raff M.C. 2002. Posttranscriptional regulation of p18 and p27 Cdk inhibitor proteins and the timing of oligodendrocyte differentiation. *Dev. Biol.* **245:** 224.

Wolswijk G. and Noble M. 1989. Identification of an adult-specific glial progenitor cell. *Development* **105:** 387.

Transcriptional Feedback Loop Regulation, Function, and Ontogeny in *Drosophila*

J. BENITO, H. ZHENG, F.S. NG, AND P.E. HARDIN
Center for Research on Biological Clocks, Department of Biology, Texas A&M University, College Station, Texas 77843-3258

The *Drosophila* circadian oscillator is composed of interlocked *period/timeless* (*per/tim*) and *Clock* (*Clk*) transcriptional feedback loops. These feedback loops drive rhythmic transcription having peaks at dawn and dusk during the daily cycle and function in the brain and a variety of peripheral tissues. To understand how the circadian oscillator keeps time and controls metabolic, physiological, and behavioral rhythms, we must determine how these feedback loops regulate rhythmic transcription, determine the relative importance of the *per/tim* and *Clk* feedback loops with regard to circadian oscillator function, and determine how these feedback loops come to be expressed in only certain tissues. Substantial insight into each of these issues has been gained from experiments performed in our lab and others and is summarized here.

INTRODUCTION

As in other organisms, the circadian clock in *Drosophila* controls daily rhythms in physiology, metabolism, and behavior through a self-sustaining oscillator that is synchronized to environmental cycles in light and temperature but keeps time even in the absence of environmental cues. The *Drosophila* circadian oscillator is composed of interlocked *period/timeless* (*per/tim*) and *Clock* (*Clk*) feedback loops that regulate rhythmic transcription in different phases of the circadian cycle (for review, see Hardin 2006; Yu and Hardin 2006). The transcriptional rhythms imparted by these feedback loops not only maintain circadian oscillator function, but also activate clock "output genes" that control physiological, metabolic, and behavioral processes in various tissues (Claridge-Chang et al. 2001; McDonald and Rosbash 2001; Ceriani et al. 2002; Y. Lin et al. 2002; Ueda et al. 2002; Wijnen et al. 2006). Because transcriptional regulation is critically important for circadian clock function in all organisms (Young and Kay 2001; Bell-Pedersen et al. 2005), a major focus within the circadian rhythms field has been to define the mechanisms through which transcriptional feedback loops mediate rhythmic transcription.

One of the best-characterized transcriptional feedback loops is the *per/tim* feedback loop in *Drosophila* (Fig. 1). This feedback loop is initiated when CYCLE (CYC) forms a heterodimer with CLK and binds E-box regulatory sequences to activate per and *tim* transcription around midday or zeitgeber time 6 (ZT6) (note: Zeitgeber time refers to time during a light-dark cycle where ZT0 is lights-on and ZT12 is lights-off [Darlington et al. 1998].) Although *per* and *tim* mRNAs accumulate to high levels around dusk (ZT12) (Hardin et al. 1990; Sehgal et al. 1995), PER and TIM proteins do not feed back to inhibit CLK-CYC transcription until midnight (ZT18) because of a phosphorylation-dependent delay in their accumulation and nuclear localization (Kloss et al. 1998, 2001; Price et al. 1998; Martinek et al. 2001; J.M. Lin et al. 2002; Akten et al. 2003). Degradation of PER and TIM in the early morning (ZT4) releases transcriptional inhibition to enable the next round of CLK-CYC transcriptional activation (Hunter-Ensor et al. 1996; Myers et al. 1996; Zeng et al. 1996; Naidoo et al. 1999; Grima et al. 2002; Ko et al. 2002). The *per/tim* feedback loop is necessary for, and intrinsically linked to, the *Clk* feedback loop, which drives rhythms in *Clk* mRNA expression that peak around dawn (ZT0) (Glossop et al. 1999). The *Clk* feedback loop is initiated when CLK-CYC binds E boxes to activate *vrille* (*vri*) and *PAR domain protein* 1ε (*Pdp1ε*) transcription (Cyran et al. 2003; Glossop et al. 2003). VRI protein accumulates to high levels in concert with *vri* mRNA during the early night (ZT15) and binds V/P boxes to inhibit *Clk* transcription (Cyran et al. 2003; Glossop et al. 2003). PDP1ε protein does not accumulate to high levels until late evening (ZT21), when it is thought to compete with decreasing levels of VRI for V/P-box binding to activate *Clk* transcription (Cyran et al. 2003). Although the *per/tim* and *Clk* feedback loops control rhythmic transcription within (and downstream from) the circadian oscillator, their relative contributions to oscillator function and overt rhythmicity are less well understood. Accumulating data from our lab and others suggest that the *Clk* feedback loop is not necessary for circadian oscillator function. Moreover, recent data from our lab argue that PDP1ε is a critical output factor, rather than a major *Clk* transcriptional activator, and suggests that key factors controlling rhythmic transcription within the *Clk* loop are yet to be identified.

Circadian feedback loop oscillators function in several clusters of brain neurons and numerous peripheral tissues in *Drosophila* (Hall 2003; Helfrich-Forster 2005). Because CLK-CYC heterodimers initiate both *per/tim* and *Clk* feedback loop function, factors that activate *Clk* and *cyc* would then be primary determinants of oscillator cell fate. However, the situation is more complicated because CLK is also expressed in cells that do not harbor circadian oscillators (i.e., nonoscillator cells) (Houl et al.

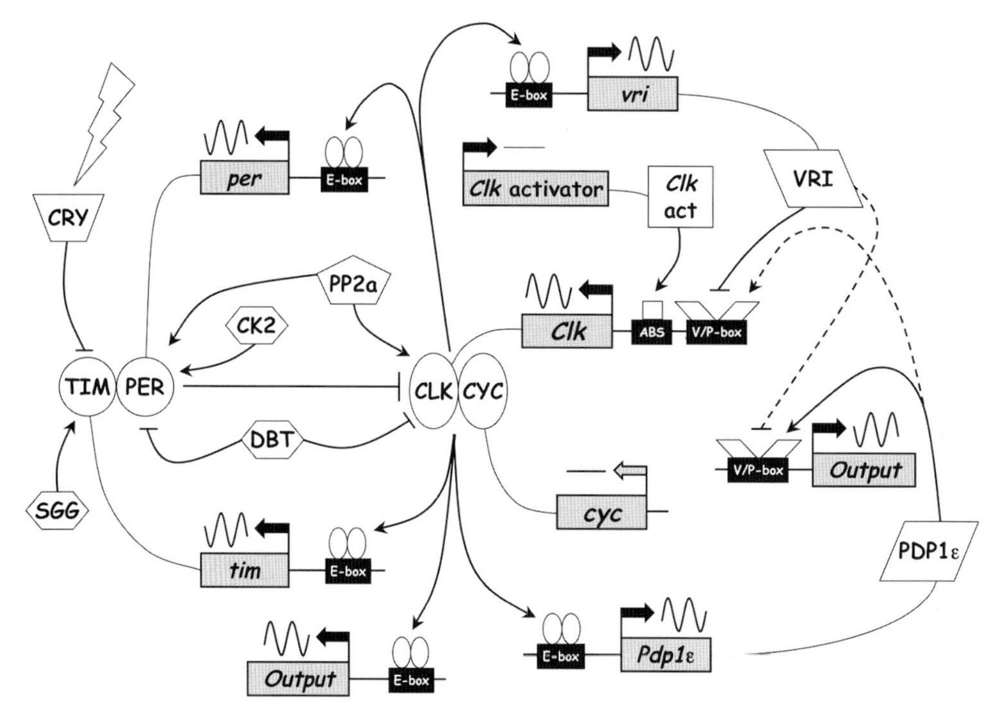

Figure 1. Model of the *Drosophila* circadian oscillator. The *per/tim* (*left half*) and *Clk* (*right half*) feedback loops are shown. Genes are depicted as *gray rectangles*, where the name of the gene is shown in italics. E box, VRI/PDP1ε box (V/P box), and *Clk* activator-binding site (ABS) transcriptional regulatory sequences are shown in *black boxes* upstream of the genes they control. (*Thick arrow*) Start of transcription. The pattern of transcription is shown above the gene: (*sinusoidal lines*) rhythmic transcription; (*straight line*) constant transcription. (*Thin lines*) Translation of proteins; (*white shapes*) proteins; (*thick lines with arrows*) positive regulation; (*thick lines with a bar at the end*) negative regulation. (*Dashed line with an arrow*) Weak positive regulation; (*dashed line with a bar*) proposed negative regulation. The lightening bolt represents blue light activation of CRY, which promotes TIM degradation. Gene symbols are as described in the text.

2006); thus, not all *Clk* activators initiate oscillator function. The spatial expression of CYC has not yet been determined, but it may also be expressed in nonoscillator cells. Circadian oscillators in various cells and tissues begin to function at different times during development. Oscillators in three clusters of brain neurons begin to function during the first larval instar (Sehgal et al. 1992), whereas oscillators in other groups of brain neurons and in peripheral tissues do not start to function until the late pupal/early adult stage (Liu et al. 1988; Kaneko et al. 1997). How these various cells and tissues are fated to contain circadian oscillators is not known, but determining when CLK and CYC expression is initiated during development will provide insight into when these cell-fate decisions are taking place. Moreover, such analysis may suggest candidate *Clk* activators that not only establish oscillator cell fate, but maintain *Clk* transcriptional activation, and thus *per/tim* and *Clk* feedback loop function, in adults.

THE SIGNIFICANCE OF RHYTHMIC TRANSCRIPTION AND *PER/TIM* AND *CLK* FEEDBACK LOOP FUNCTION

Since the initial discovery that molecular feedback loops control rhythmic transcription of core circadian clock components in *Drosophila* (Hardin et al. 1990, 1992), considerable effort has been directed toward defining the mechanisms that regulate rhythmic transcription. Consequently, we have a detailed, although not comprehensive, understanding of the factors and interactions that control rhythmic transcription. In contrast, relatively little attention has been paid to determining whether rhythmic transcription is necessary to maintain feedback loop function. Although some studies argue that rhythmic transcription is not necessary for circadian oscillator function, there is a growing consensus that rhythmic transcription is indeed required.

The key regulatory events governing circadian transcription within the *per/tim* loop are transcriptional activation by CLK-CYC heterodimers and transcriptional repression by PER-containing complexes (Fig. 1). Transcriptional activation and repression are temporally separated: Activation occurs from midday (ZT6) through mid-evening (ZT18) and repression begins around mid-evening and is relieved around midday (for review, see Yu and Hardin 2006). This separation of transcriptional activation and repression is primarily regulated by phosphorylation of CLK, PER, and TIM. CLK is rhythmically phosphorylated in phase with CLK-CYC transcriptional activity: Hypophosphorylated CLK predominates when CLK-CYC target genes are activated and hyperphosphorylated CLK predominates when CLK-CYC target genes are repressed (Kim and Edery 2006; Yu et al. 2006). CLK hyperphosphorylation is PER-dependent and occurs in concert with the accumulation of phosphorylated PER in CLK-CYC complexes in the nucleus (Lee et al. 1998; Kim and Edery 2006; Yu et al. 2006). DOUBLE-TIME (DBT) kinase-dependent phosphorylation promotes PER degradation (Price et al. 1998), whereas TIM stabilizes PER by blocking DBT phosphorylation (Price et al. 1995; Ko et al. 2002). Phosphorylation of PER by casein kinase 2 (CK2) and TIM by SHAGGY (SGG) kinase promotes PER and TIM nuclear localization, respectively (Martinek et al. 2001; J.M. Lin et al. 2002; Akten et al. 2003; Nawathean and Rosbash 2004). TIM is degraded around dawn, which facilitates DBT-dependent degradation of hyperphosphorylated PER and CLK and transcriptional activation by accumulating levels of hypophosphorylated CLK (Hunter-Ensor et al. 1996; Myers et al. 1996; Zeng et al. 1996; Ko et al. 2002; Kim and Edery 2006; Yu et al. 2006).

Because posttranslational regulation of PER and TIM mediates CLK phosphorylation and transcriptional activity, is rhythmic transcription of *per* and *tim* (and by extension *per* and *tim* mRNA cycling) necessary for circadian oscillator function? Transgenic flies expressing constitutive levels of either *per* or *tim* mRNAs had little effect on molecular and behavioral rhythms (Cheng and Hardin 1998; Yang and Sehgal 2001). Likewise, about half of the transgenic fly lines that express constant levels of both *per* and *tim* mRNAs retained molecular and behavioral rhythmicity, suggesting that mRNA rhythms are not necessary for circadian oscillator function (Yang and Sehgal 2001). However, the lack of rhythmicity in half the transgenic lines and severe decrements in molecular and behavioral rhythmicity in the others suggest that mRNA cycling is important, if not essential, for circadian oscillator function. The transgenic lines used to constitutively express *per* and *tim* mRNAs (Yang and Sehgal 2001) were recently retested and found to display less than 20% rhythmicity that was generally weak and in a broad range of circadian periodicities, indicating that *per* and *tim* mRNA rhythms are indeed required for circadian oscillator function in *Drosophila* (Hall et al., this volume).

In the *Clk* feedback loop, rhythms in the VRI/PDP1ε ratio are thought to control rhythmic transcription of *Clk* (Cyran et al. 2003). However, the resulting rhythm in *Clk* mRNA abundance does not give rise to cycling levels of CLK protein, although PER-dependent CLK phosphorylation cycles with a peak near dawn (Kim and Edery 2006; Yu et al. 2006). This dawn peak in CLK phosphorylation persists even after the phase of *Clk* mRNA cycling is reversed (Kim et al. 2002), indicating that *Clk* mRNA levels are not important for *per/tim* feedback loop function. In contrast, mutants that disrupt *per/tim* feedback loop function also disrupt the *Clk* feedback loop (Glossop et al. 1999), indicating that the *per/tim* loop is of primary importance. If the *Clk* feedback loop is not necessary for *per/tim* feedback loop function, then what is the purpose of the *Clk* feedback loop? One possibility is that the *Clk* loop functions to enhance the amplitude and robustness of the circadian oscillator. This possibility is supported by experiments in which disruption of the analogous feedback loop in mammals, the *Bmal1* loop, reduces the amplitude of molecular rhythms and the consolidation of locomotor activity rhythms (Sato et al. 2004). Another potential function for the *Clk* loop is to drive outputs that require transcriptional activation around dawn. Although no output genes have been identified that are directly under *Clk* feedback loop control (i.e., activated by PDP1ε and/or repressed by VRI), microarray screens have identified a large number of potential *Clk* feedback loop target genes whose mRNAs peak at dawn (Claridge-Chang et al. 2001; McDonald and Rosbash 2001; Ceriani et al. 2002; Y. Lin et al. 2002; Ueda et al. 2002; Wijnen et al. 2006).

REGULATION OF RHYTHMIC TRANSCRIPTION WITHIN THE *CLK* FEEDBACK LOOP

The *Clk* feedback loop was first proposed to explain the opposite phase cycling of *Clk* mRNA with respect to *per* and *tim* mRNAs (Glossop et al. 1999). *Clk* transcription is repressed when CLK-CYC activity is high from midday to midnight in wild-type flies and in per^{01} or tim^{01} mutants, whereas *Clk* transcription is activated when CLK-CYC activity is low from midnight to midday in wild-type flies and in Clk^{Jrk} or cyc^{01} mutants. The high levels of *Clk* mRNA in Clk^{Jrk} and cyc^{01} flies predict that *Clk* is activated by a factor that is not dependent on circadian oscillator function. In contrast, the low levels of *Clk* mRNA in per^{01} and tim^{01} mutants suggest that CLK-CYC activates a repressor.

One CLK-CYC-activated transcript identified in a screen for rhythmically expressed mRNAs encodes the bZIP (basic leucine zipper) repressor VRI (Blau and Young 1999). Although circadian phenotypes cannot be measured in *vri* null mutants due to developmental lethality (George and Terracol 1997), VRI overexpression represses *Clk* and CLK-CYC-dependent transcripts such as *per* and *tim* (Blau and Young 1999; Cyran et al. 2003; Glossop et al. 2003). VRI accumulates in phase with its mRNA from midday to early evening (Cyran et al. 2003; Glossop et al. 2003), consistent with the timing of *Clk* repression and the dependence of *Clk* repression on CLK-CYC activity (Fig. 1). Another rhythmically expressed transcript that is dependent on CLK-CYC, *Pdp1ε*, encodes a bZIP activator with high similarity to VRI (Claridge-Chang et al. 2001; McDonald and Rosbash 2001; Ceriani et al. 2002; Y. Lin et al. 2002; Ueda et al. 2002; Cyran et al. 2003; Wijnen et al. 2006). *Pdp1ε* mRNA and protein accumulate to high levels several hours later than *vri* mRNA and protein, suggesting that accumulating PDP1ε could activate *Clk* transcription late at night as VRI levels fall (Fig. 1). Consistent with this possibility, PDP1ε binds V/P-box elements in vitro, PDP1ε competes with VRI to activate *Clk* transcription in cell culture, and little or no *Clk* mRNA is detected in *Pdp1ε* null mutant flies (Cyran et al. 2003). Activation of *Clk* by PDP1ε was unexpected because *Clk* mRNA levels are high in Clk^{Jrk} and cyc^{01} flies even there is little or no PDP1ε. Although it is possible that small amounts of PDP1ε could predominate over low levels of VRI to activate *Clk* in Clk^{Jrk} or cyc^{01} mutants, this would contrast with the relationship between these proteins in per^{01} and tim^{01} flies, where high levels of VRI predominate over

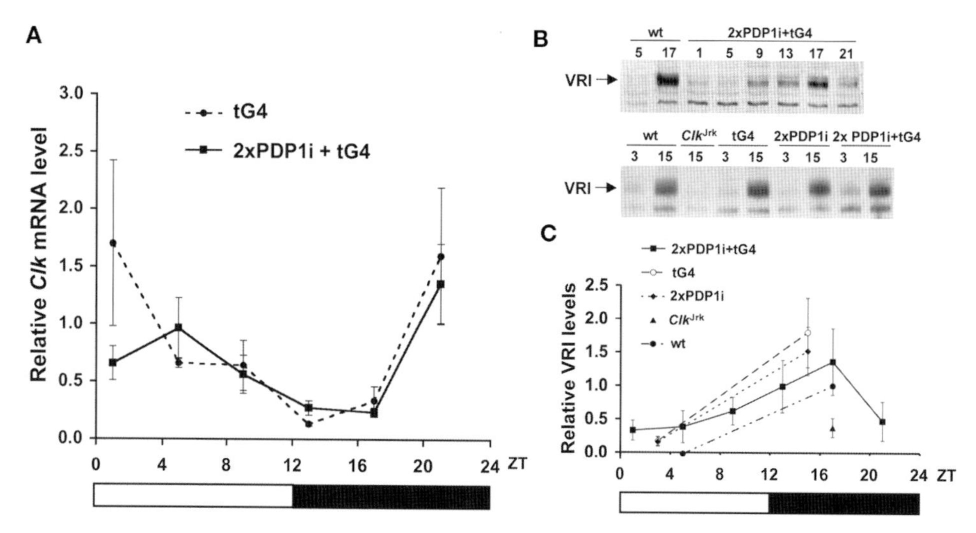

Figure 2. Constant low PDP1ε, levels do not disrupt *Clk* mRNA or VRI cycling. (*A*) qPCR of *Clk* mRNA from the heads of *w*;+/+;*tim*Gal4/+ (*tim*Gal4) and *w*;UAS-PDP1i/+;UAS-PDP1i/*tim*Gal4 (2xPDP1i+tG4) flies collected at the indicated times under LD conditions (ZT). Relative *Clk* mRNA levels were quantified as described in Benito et al. (2007). The *black* and *white bars* represent times when lights were on or off, respectively. (*B*) Western blots of head extracts from 2xPDP1i+tG4 and wild-type (wt), Clk^{Jrk}, *tim*Gal4 (tG4), and *w*;UAS-PDP1i/+;UAS-PDP1i/+ (2xPDP1i) control flies collected at the indicated times under LD conditions and probed with VRI antibody. (*C*) Quantification of VRI levels from the western blots in *B* and at least two independent repeats. VRI levels were quantified as described in Benito et al. (2007), and normalized to wild type at ZT17. (Reprinted, with permission, from Benito et al. 2007 [© The Society for Neuroscience].)

high levels of PDP1ε to repress *Clk* (Cyran et al. 2003). *Pdp1* produces multiple RNA and protein isoforms that are expressed in a variety of tissues including the central nervous system (CNS) (Reddy et al. 2000). Deletion of *Pdp1* leads to severe defects in growth, mitosis, and endoreplication that cause lethality during larval stages (Reddy et al. 2006). It is possible that loss of all PDP1 isoforms in the *Pdp1* deletion mutant alters clock neuron identity or development in larvae, and thus indirectly reduces *Clk* mRNA levels.

If *Clk* mRNA cycling is driven by competition between PDP1ε and VRI for V/P-box binding, then altering the PDP1ε/VRI ratio by reducing or increasing PDP1ε expression specifically in circadian clock cells should disrupt *Clk* mRNA cycling by producing a constant trough or peak levels of *Clk* mRNA, respectively. When PDP1ε was reduced to its normal trough level or below via RNA interference (RNAi), *Clk* mRNA continued to cycle similarly to the wild-type controls with the exception of a lower *Clk* mRNA level at ZT1 (Fig. 2). VRI levels continued to cycle in flies expressing low levels of PDP1ε, consistent with its role as a *Clk* repressor (Fig. 2). Similar results were seen when PDP1ε levels were increased to ≥5-fold above its normal peak: *Clk* mRNA and VRI protein levels continued to cycle with amplitudes comparable to wild-type controls (Fig. 3). These results demonstrate that PDP1ε levels, and thus the PDP1ε/VRI ratio, are not critical for *Clk* mRNA cycling or circadian oscillator function (Benito et al. 2007). Moreover, the wild-type levels of *Clk* mRNA in *Pdp1* RNAi and *Pdp1ε* overexpression strains suggest that PDP1ε is not the major *Clk* activator, although a minor role in *Clk* activation cannot be discounted (Benito et al. 2007).

FUNCTION OF PDP1ε WITHIN THE CIRCADIAN CLOCK

Because circadian oscillator function is not disrupted in flies with constant low or high levels of PDP1ε, perhaps PDP1ε functions to control clock output. To test this possibility, locomotor activity rhythms were measured in *Pdp1* RNAi, *Pdp1ε* overexpression, and control strains (Benito et al. 2007). Flies expressing low or high levels of PDP1ε show drastic reductions in behavioral rhythmicity compared to control strains (Table 1). Despite this loss in behavioral rhythmicity, oscillator function persists in locomotor activity pacemaker cells or lateral neurons (LNs), which indicates that PDP1ε controls clock output (Benito et al. 2007). Because PDP1ε is a rhythmically expressed bZIP transcriptional factor, it is likely to mediate behavioral rhythms by rhythmically activating effector genes. These PDP1ε-dependent rhythmic transcripts may represent some of those identified in previous microarray studies that peak in abundance around dawn (Claridge-Chang et al. 2001; McDonald and Rosbash 2001; Ceriani et al. 2002; Y. Lin et al. 2002; Ueda et al. 2002; Wijnen et al. 2006). These data suggest a revised model for clock function in *Drosophila* in which a constitutive activator and VRI mediate *Clk* mRNA cycling within the *Clk* feedback loop (Fig. 1). PDP1ε may have a minor role in regulating *Clk* transcription, but it is proposed to be a major regulator of output transcription required for (at least) behavioral rhythmicity (Benito et al. 2007). VRI may also function to control output transcription along with PDP1ε, but it is worth noting that VRI's role within the clock is based on gain-of-function experiments. Eliminating VRI function within oscillator cells

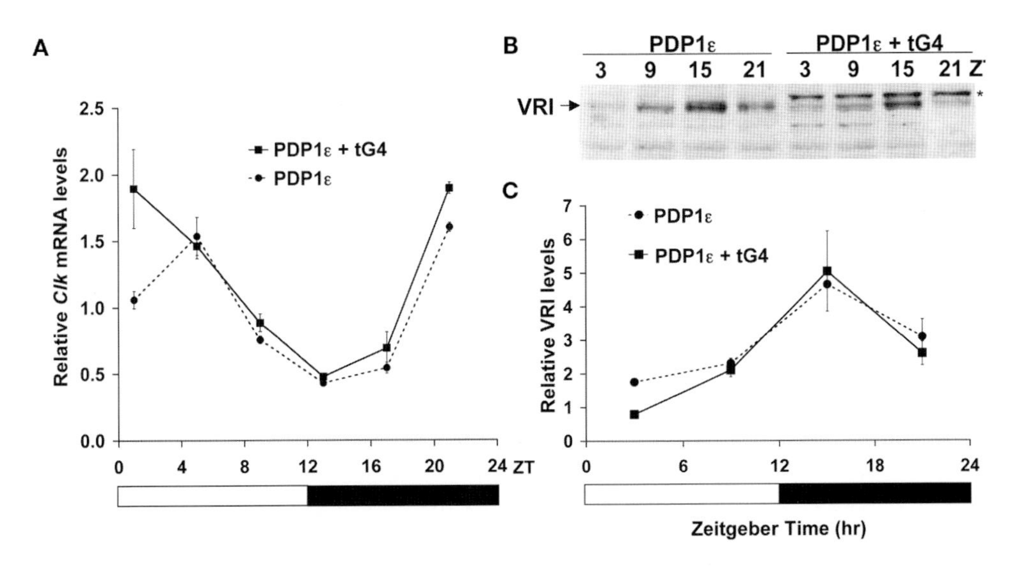

Figure 3. Overexpression of PDP1ε, in oscillator cells does not disrupt *Clk* mRNA or VRI cycling. (*A*) qPCR of *Clk* mRNA from the heads of *w*;UAS-PDP1ε/+;UAS-PDP1ε/+ (UPDP1ε) and *w*; UAS-PDP1ε/+;UAS-PDP1ε/*tim*Gal4 (UPDP1ε+tG4) flies were collected at the indicated times under LD conditions (ZT). Relative *Clk* mRNA levels were quantified as described by Benito et al. (2007). The *black* and *white bars* represent times when lights were on or off, respectively. (*D*) Western blot of head extracts from UPDP1ε and UPDP1ε+tG4 flies collected at the indicated times under LD conditions and probed with VRI antibody. The asterisk denotes PDP1ε band that did not wash off completely from the blot in *A*. (*E*) Quantification of VRI levels in the western blot in *B* and at least two independent repeats. VRI levels were quantified as described by Benito et al. (2007). (Reprinted, with permission, from Benito et al. 2007 [©The Society for Neuroscience].)

will more definitively pinpoint VRI's role within the clock.

The mammalian homologs of PDP1ε and VRI likely have an analogous role in regulating clock output. PDP1ε is related to three rhythmically expressed PAR domain-containing bZIP transcriptional activators in mammals: albumin gene site D-binding protein (DBP), thyroid embryonic factor (TEF), and hepatocyte leukemia factor (HLF) (Wuarin and Schibler 1990; Falvey et al. 1995; Fonjallaz et al. 1996). DBP knockout mice show lower locomotor activity levels and a shorter circadian period than wild-type animals, yet robust circadian oscillator function persists (Lopez-Molina et al. 1997). These results imply that DBP functions primarily to control clock output, consistent with PDP1ε function in flies. DBP binds to the same regulatory element as adenovirus E4 promoter ATF site-binding protein (E4BP4), a rhythmically expressed bZIP repressor related to VRI (Mitsui et al. 2001). Because E4BP4 is expressed in the opposite phase than DBP (Mitsui et al. 2001), sequential binding of E4BP4 and DBP could control rhythmic target gene transcription.

Table 1. Reducing and Increasing PDP1ε Levels Disrupts Locomotor Activity Rhythms

Genotype	Percent rhythmic (N)	Period (h) ± S.E.M.	Power ± S.E.M.	Activity ± S.E.M.
w;+/+;+/+	75.0 (172)	23.8 ± 0.03	78.9 ± 3.7	17.8 ± 0.6
w;+/+;*tim*-Gal4/+	80.5 (41)	24.6 ± 0.07	76.7 ± 5.4	16.7 ± 0.9
w;+/+;*pdf*-Gal4/+	57.0 (35)	24.2 ± 0.13	68.7 ± 7.7	14.0 ± 0.8
w;PDP1iA/+;PDP1iB/+	77.8 (38	23.4 ± 0.08	72.8 ± 5.3	28.3 ± 2.1
w;PDP1iA/+;PDP1iB/*tim*-Gal4	17.2 (29)	25.3 ± 0.23	35.3 ± 1.7	13.5 ± 1.0
w;PDP1iA/+;PDP1iB/*pdf*-Gal4	0 (27)	n.a.	n.a.	12.1 ± 0.7
w;+/+;+/+	96.0 (25)	23.5 ± 0.15	96.1 ± 6.4	19.9 ± 2.3
w;+/+;*tim*-Gal4/+	100 (31)	24.4 ± 0.21	78.4 ± 5.2	21.6 ± 2.0
w;+/+;*pdf*-Gal4/+	95.5 (22)	24.3 ± 0.09	76.4 ± 9.5	13.5 ± 2.3
w;PDP1εA/PDP1εB;+/+	73.9 (29)	24.5 ± 0.16	78.1 ± 11.7	27.1 ± 3.9
w;PDP1εA/+;PDP1εB/*tim*-Gal4	12.5 (31)	25.2 ± 0.12	24.5 ± 5.9	15.4 ± 1.8
w;PDP1εA/+;PDP1εB/*pdf*-Gal4	10.3 (29)	24.1 ± 0.02	27.2 ± 2.3	19.1 ± 1.5

Period and power were calculated for all rhythmic animals as described in Benito et al. (2007). Activity was calculated for all flies that survived to the end of the experiment, as described in Benito et al. (2007). n.a. indicates not applicable. (Reprinted, with permission, from Benito et al. 2007 [©The Society for Neuroscience].)

REGULATION OF *CLK* SPATIAL EXPRESSION

Circadian oscillator cells from *Drosophila* can be identified by the rhythmic expression of *per* and *tim* mRNA and protein (Hardin 1994; Plautz et al. 1997). In adults, oscillator cells have been identified in the brain, i.e., four clusters of lateral neurons (dorsal lateral neurons [LN_Ds]; small ventral lateral neurons [sLN_Vs]; large ventral lateral neurons [lLN_Vs]; and a lateral posterior neuron [LPN]) and three clusters of dorsal neurons (dorsal neuron 1 [DN1]; dorsal neuron 2 [DN2]; and dorsal neuron 3 [DN3]), and in peripheral tissues (e.g., the gut, fat body, Malpighian tubules, the antenna, compound eye, proboscis, wing, and leg) (Hall 2003; Helfrich-Forster 2005; Shafer et al. 2006). Although a subset of oscillator neurons in the brain control morning and evening activity and oscillator neurons in the antenna control olfactory responses (Krishnan et al. 1999; Grima et al. 2004; Stoleru et al. 2004), outputs from other adult brain and peripheral oscillator cells have not been characterized. Circadian control of tissue-specific processes in *Drosophila* and other animals is thought to enable coordinate temporal control of behavior, physiology, and metabolism. Despite the importance of oscillator function in different tissues, relatively little is known about how this tissue-specific pattern of oscillator expression is specified.

Because CLK-CYC initiates circadian oscillator function by activating *per*, *tim*, *vri*, and *Pdp1ε* expression (Fig. 1), coexpression of CLK and CYC necessarily determines which tissues contain circadian oscillators. Such coexpression could arise by (1) expression of both CLK and CYC specifically in oscillator cells, (2) broad expression of CLK and oscillator-specific expression of CYC, (3) broad expression of CYC and oscillator-cell-specific expression of CLK, or (4) broad expression of CLK and CYC that overlaps specifically in oscillator cells. Localization of CLK has provided insight into which of these scenarios for CLK and CYC coexpression is correct. In adults, CLK is expressed in all oscillator cells as expected, but it is also expressed in nonoscillator cells (Houl et al. 2006). This result was somewhat surprising because previous work had shown that ectopic expression of CLK in certain brain neurons could generate circadian oscillators (Zhao et al. 2003). Perhaps CLK is able to induce oscillator function only in cells that already express CYC and the kinases and phosphatases that mediate feedback repression by PER and TIM. CRY expression in ectopic oscillator cells, and thus the ability of these oscillators to entrain to light, may come from CLK-CYC-dependent induction of the candidate *cry* activator PDP1ε (Cyran et al. 2003). In any case, expression of CLK in nonoscillator cells indicates that oscillator cells are determined by either oscillator-cell-specific expression of CYC or the overlapping expression of CLK and CYC specifically in oscillator cells. These possibilities can be distinguished when reagents necessary to localize CYC are available.

Circadian oscillators in the brain start to be expressed at earlier times during development than those in peripheral tissues. Brain oscillator cells can be detected as early as the first larval instar and include precursors to the LNVs, DN2s, and a subset of DN1s (Sehgal et al. 1992; Kaneko et al. 1997; Helfrich-Forster 2005). PER expression in photoreceptors and the antenna is not detected until late during pupal development, indicating that functional oscillators are not expressed in these peripheral tissues until just before flies eclose (Liu et al. 1988). CLK expression precedes oscillator function in the larval brain and can be seen as early as embryonic stage 11, which indicates that brain oscillators are determined during mid embryogenesis (F.S. Ng et al., unpubl.).

Factors that activate *Clk* and/or *cyc* in oscillator cells have not yet been identified. Given the diverse array of brain cells and peripheral tissues that contain circadian oscillators, it is likely that multiple tissue-specific factors activate *Clk* and *cyc* in oscillator cells. *Clk* is activated in locomotor activity pacemaker neurons during embryogenesis, well before oscillator function is initiated in L1 larvae, and more than 1 week before these neurons control locomotor activity rhythms. Pacemaker neurons are unique in that they appear to be dedicated to controlling activity rhythms, yet are not required for locomotor activity per se. In contrast, peripheral oscillators are thought to control outputs inherent to a given tissue, where the tissue is necessary for the process being modulated. For instance, *Drosophila* olfactory sensory neurons (OSNs) control rhythms in electrophysiological responses to odors and are necessary for odor-dependent electrophysiological responses per se (Tanoue et al. 2004; Vosshall and Stocker 2007). *Clk* is presumably activated just before *per* and *tim* when OSNs and other clock-containing peripheral tissues are differentiating in late pupae (Liu et al. 1988). The activation of *Clk* during peripheral tissue differentiation implies that oscillator function is a core property of these tissues. When these activators are identified, it will be important to determine whether they also activate *cyc* and whether they maintain *Clk* activation in adults. Because circadian oscillators in the hypothalamic suprachiasmatic nucleus and peripheral tissues of mammals are analogous to the locomotor activity pacemaker neurons and peripheral oscillator tissues in flies, respectively, similar mechanisms may be responsible for specifying clock cell identity in mammals.

CONCLUSIONS

The core circadian timekeeping mechanism operates through interlocked *per/tim* and *Clk* feedback loops that regulate rhythmic transcription peaking around dusk and dawn, respectively. Although it is well established that phosphorylation-dependent rhythms in PER and TIM abundance are necessary for circadian oscillator function, mounting evidence indicates that rhythms in *per* and *tim* transcription are also required for circadian oscillator function. Presumably high levels of *per* and *tim* mRNAs at inappropriate times alters the overall levels, cycling phase, or activity of PER and TIM, thereby disrupting oscillator function. In contrast, reversing rhythms in *Clk* mRNA abundance disrupts neither circadian oscillator function nor CLK phosphorylation rhythms, indicating that *Clk* mRNA rhythms are dispensable. Taken together with the PER dependence of rhythms in CLK phosphorylation, these results argue that the *per/tim* feedback loop has a dominant role in maintaining circadian oscillator function.

Despite its subservient role within the *Drosophila* circadian oscillator, the *Clk* feedback loop regulates rhythmic transcription in a phase opposite to that of the *per/tim* feedback loop. *Clk* feedback-loop-dependent transcription of genes such as *Clk* was initially thought to be mediated by competition between VRI and PDP1ε for V/P boxes: VRI binds V/P boxes to repress *Clk* transcription at dusk because the VRI/PDP1ε ratio is high, and PDP1ε binds V/P boxes to activate *Clk* transcription at dawn because the VRI/PDP1ε ratio is low. However, we recently found that altering the VRI/PDP1ε ratio by reducing or increasing PDP1ε levels had little affect on *Clk* transcription or VRI cycling. Taken together with the constant high levels of *Clk* mRNA in mutants that express little if any VRI or PDP1ε (i.e., Clk^{Jrk} and cyc^{01}), this result suggests a model in which VRI periodically represses the constant activation of *Clk* transcription. Although PDP1ε does not greatly affect *Clk* transcription, PDP1ε is required for rhythms in locomotor activity. We hypothesize that PDP1ε has a major role in activating genes that control rhythmic outputs, perhaps in conjunction with VRI. Such a role is consistent with that proposed for DBP and E4BP4, the mammalian homologs of PDP1ε and VRI, respectively.

Circadian oscillators are expressed in several clusters of brain neurons and a variety of peripheral tissues. This spatial distribution is controlled by factors that activate *Clk* and *cyc* because CLK-CYC initiates circadian oscillator function. However, *Clk* expression is not restricted to oscillator cells, which suggests that CYC is either specifically expressed only in oscillator cells or that CYC is broadly expressed and specifically overlaps CLK expression in oscillator cells. Moreover, expression of CLK in nonoscillator cells indicates that CLK contributes to processes other than circadian oscillator function. Nevertheless, ectopic expression of *Clk* in certain brain cells can generate circadian oscillator function, presumably because these cells already express CYC and kinases/phosphatases normally found in oscillator cells. The developmental expression of CLK suggests that locomotor activity pacemaker cells are determined during mid embryogenesis. CLK appears to be activated in peripheral oscillator tissues around the time they differentiate in late pupae. Insight into oscillator cell determination and maintenance will come with the identification and characterization of *Clk* and *cyc* activators.

ACKNOWLEDGMENTS

This work was supported by National Institutes of Health grants NS051280 and NS052854 to P.E.H.

REFERENCES

Akten B., Jauch E., Genova G.K., Kim E.Y., Edery I., Raabe T., and Jackson F.R. 2003. A role for CK2 in the *Drosophila* circadian oscillator. *Nat. Neurosci.* **6:** 251.

Bell-Pedersen D., Cassone V.M., Earnest D.J., Golden S.S., Hardin P.E., Thomas T.L., and Zoran M.J. 2005. Circadian rhythms from multiple oscillators: Lessons from diverse organisms. *Nat. Rev. Genet.* **6:** 544.

Benito J., Zheng H., and Hardin P.E. 2007. PDP1epsilon functions downstream of the circadian oscillator to mediate behavioral rhythms. *J. Neurosci.* **27:** 2539.

Blau J. and Young M.W. 1999. Cycling *vrille* expression is required for a functional *Drosophila* clock. *Cell* **99:** 661.

Ceriani M.F., Hogenesch J.B., Yanovsky M., Panda S., Straume M., and Kay S.A. 2002. Genome-wide expression analysis in *Drosophila* reveals genes controlling circadian behavior. *J. Neurosci.* **22:** 9305.

Cheng Y. and Hardin P.E. 1998. *Drosophila* photoreceptors contain an autonomous circadian oscillator that can function without period mRNA cycling. *J. Neurosci.* **18:** 741.

Claridge-Chang A., Wijnen H., Naef F., Boothroyd C., Rajewsky N., and Young M.W. 2001. Circadian regulation of gene expression systems in the *Drosophila* head. *Neuron* **32:** 657.

Cyran S.A., Buchsbaum A.M., Reddy K.L., Lin M.C., Glossop N.R., Hardin P.E., Young M.W., Storti R.V., and Blau J. 2003. *vrille*, *Pdp1*, and *dClock* form a second feedback loop in the *Drosophila* circadian clock. *Cell* **112:** 329.

Darlington T.K., Wager-Smith K., Ceriani M.F., Staknis D., Gekakis N., Steeves T.D., Weitz C.J., Takahashi J.S., and Kay S.A. 1998. Closing the circadian loop: CLOCK-induced transcription of its own inhibitors *per* and *tim*. *Science* **280:** 1599.

Falvey E., Fleury-Olela F., and Schibler U. 1995. The rat hepatic leukemia factor (HLF) gene encodes two transcriptional activators with distinct circadian rhythms, tissue distributions and target preferences. *EMBO J.* **14:** 4307.

Fonjallaz P., Ossipow V., Wanner G., and Schibler U. 1996. The two PAR leucine zipper proteins, TEF and DBP, display similar circadian and tissue-specific expression, but have different target promoter preferences. *EMBO J.* **15:** 351.

George H. and Terracol R. 1997. The *vrille* gene of *Drosophila* is a maternal enhancer of *decapentaplegic* and encodes a new member of the bZIP family of transcription factors. *Genetics* **146:** 1345.

Glossop N.R., Lyons L.C., and Hardin P.E. 1999. Interlocked feedback loops within the *Drosophila* circadian oscillator. *Science* **286:** 766.

Glossop N.R., Houl J.H., Zheng H., Ng F.S., Dudek S.M., and Hardin P.E. 2003. VRILLE feeds back to control circadian transcription of *Clock* in the *Drosophila* circadian oscillator. *Neuron* **37:** 249.

Grima B., Chelot E., Xia R., and Rouyer F. 2004. Morning and evening peaks of activity rely on different clock neurons of the *Drosophila* brain. *Nature* **431:** 869.

Grima B., Lamouroux A., Chelot E., Papin C., Limbourg-Bouchon B., and Rouyer F. 2002. The F-box protein slimb controls the levels of clock proteins period and timeless. *Nature* **420:** 178.

Hall J.C. 2003. Genetics and molecular biology of rhythms in *Drosophila* and other insects. *Adv. Genet.* **48:** 1.

Hardin P.E. 1994. Analysis of *period* mRNA cycling in *Drosophila* head and body tissues indicates that body oscillators behave differently from head oscillators. *Mol. Cell. Biol.* **14:** 7211.

———. 2006. Essential and expendable features of the circadian timekeeping mechanism. *Curr. Opin. Neurobiol.* **16:** 686.

Hardin P.E., Hall J.C., and Rosbash M. 1990. Feedback of the *Drosophila period* gene product on circadian cycling of its messenger RNA levels. *Nature* **343:** 536.

———. 1992. Circadian oscillations in *period* gene mRNA levels are transcriptionally regulated. *Proc. Natl. Acad. Sci.* **89:** 11711.

Helfrich-Forster C. 2005. Neurobiology of the fruit fly's circadian clock. *Genes Brain Behav.* **4:** 65-76.

Houl J.H., Yu W., Dudek S.M., and Hardin P.E. 2006. *Drosophila* CLOCK is constitutively expressed in circadian oscillator and non-oscillator cells. *J. Biol. Rhythms* **21:** 93.

Hunter-Ensor M., Ousley A., and Sehgal A. 1996. Regulation of the *Drosophila* protein *timeless* suggests a mechanism for resetting the circadian clock by light. *Cell* **84:** 677.

Kaneko M., Helfrich-Forster C., and Hall J.C. 1997. Spatial and temporal expression of the *period* and *timeless* genes in the developing nervous system of *Drosophila:* Newly identified pacemaker candidates and novel features of clock gene product cycling. *J. Neurosci.* **17:** 6745.

Kim E.Y. and Edery I. 2006. Balance between DBT/CKIepsilon kinase and protein phosphatase activities regulate phosphorylation and stability of *Drosophila* CLOCK protein. *Proc. Natl. Acad. Sci.* **103:** 6178.

Kim E.Y., Bae K., Ng F.S., Glossop N.R., Hardin P.E., and Edery I. 2002. *Drosophila* CLOCK protein is under posttranscriptional control and influences light-induced activity. *Neuron* **34:** 69.

Kloss B., Rothenfluh A., Young M.W., and Saez L. 2001. Phosphorylation of *period* is influenced by cycling physical associations of *double-time*, *period*, and *timeless* in the *Drosophila* clock. *Neuron* **30:** 699.

Kloss B., Price J.L., Saez L., Blau J., Rothenfluh A., Wesley C.S., and Young M.W. 1998. The *Drosophila* clock gene *double-time* encodes a protein closely related to human casein kinase Iepsilon. *Cell* **94:** 97.

Ko H.W., Jiang J., and Edery I. 2002. Role for Slimb in the degradation of *Drosophila* Period protein phosphorylated by Doubletime. *Nature* **420:** 673.

Krishnan B., Dryer S.E., and Hardin P.E. 1999. Circadian rhythms in olfactory responses of *Drosophila melanogaster*. *Nature* **400:** 375.

Lee C., Bae K., and Edery I. 1998. The *Drosophila* CLOCK protein undergoes daily rhythms in abundance, phosphorylation, and interactions with the PER-TIM complex. *Neuron* **21:** 857.

Lin J.M., Kilman V.L., Keegan K., Paddock B., Emery-Le M., Rosbash M., and Allada R. 2002. A role for casein kinase 2alpha in the *Drosophila* circadian clock. *Nature* **420:** 816.

Lin Y., Han M., Shimada B., Wang L., Gibler T.M., Amarakone A., Awad T.A., Stormo G.D., Van Gelder R.N., and Taghert P.H. 2002. Influence of the *period*-dependent circadian clock on diurnal, circadian, and aperiodic gene expression in *Drosophila melanogaster*. *Proc. Natl. Acad. Sci.* **99:** 9562.

Liu X., Lorenz L., Yu Q.N., Hall J.C., and Rosbash M. 1988. Spatial and temporal expression of the *period* gene in *Drosophila melanogaster*. *Genes Dev.* **2:** 228.

Lopez-Molina L., Conquet F., Dubois-Dauphin M., and Schibler U. 1997. The DBP gene is expressed according to a circadian rhythm in the suprachiasmatic nucleus and influences circadian behavior. *EMBO J.* **16:** 6762.

Martinek S., Inonog S., Manoukian A.S., and Young M.W. 2001. A role for the segment polarity gene *shaggy*/GSK-3 in the *Drosophila* circadian clock. *Cell* **105:** 769.

McDonald M.J. and Rosbash M. 2001. Microarray analysis and organization of circadian gene expression in *Drosophila*. *Cell* **107:** 567.

Mitsui S., Yamaguchi S., Matsuo T., Ishida Y., and Okamura H. 2001. Antagonistic role of E4BP4 and PAR proteins in the circadian oscillatory mechanism. *Genes Dev.* **15:** 995.

Myers M.P., Wager-Smith K., Rothenfluh-Hilfiker A., and Young M.W. 1996. Light-induced degradation of TIMELESS and entrainment of the *Drosophila* circadian clock. *Science* **271:** 1736.

Naidoo N., Song W., Hunter-Ensor M., and Sehgal A. 1999. A role for the proteasome in the light response of the *timeless* clock protein. *Science* **285:** 1737.

Nawathean P. and Rosbash M. 2004. The doubletime and CKII kinases collaborate to potentiate *Drosophila* PER transcriptional repressor activity. *Mol. Cell* **13:** 213.

Plautz J.D., Kaneko M., Hall J.C., and Kay S.A. 1997. Independent photoreceptive circadian clocks throughout *Drosophila*. *Science* **278:** 1632.

Price J.L., Dembinska M.E., Young M.W., and Rosbash M. 1995. Suppression of PERIOD protein abundance and circadian cycling by the *Drosophila* clock mutation *timeless*. *EMBO J.* **14:** 4044.

Price J.L., Blau J., Rothenfluh A., Abodeely M., Kloss B., and Young M.W. 1998. *double-time* is a novel *Drosophila* clock gene that regulates PERIOD protein accumulation. *Cell* **94:** 83.

Reddy K.L., Rovani M.K., Wohlwill A., Katzen A., and Storti R.V. 2006. The *Drosophila* Par domain protein I gene, Pdp1, is a regulator of larval growth, mitosis and endoreplication. *Dev. Biol.* **289:** 100.

Reddy K.L., Wohlwill A., Dzitoeva S., Lin M.H., Holbrook S., and Storti R.V. 2000. The *Drosophila PAR domain protein 1* (*Pdp1*) gene encodes multiple differentially expressed mRNAs and proteins through the use of multiple enhancers and promoters. *Dev. Biol.* **224:** 401.

Sato T.K., Panda S., Miraglia L.J., Reyes T.M., Rudic R.D., McNamara P., Naik K.A., FitzGerald G.A., Kay S.A., and Hogenesch J.B. 2004. A functional genomics strategy reveals Rora as a component of the mammalian circadian clock. *Neuron* **43:** 527.

Sehgal A., Price J., and Young M.W. 1992. Ontogeny of a biological clock in *Drosophila melanogaster*. *Proc. Natl. Acad. Sci.* **89:** 1423.

Sehgal A., Rothenfluh-Hilfiker A., Hunter-Ensor M., Chen Y., Myers M.P., and Young M.W. 1995. Rhythmic expression of *timeless:* A basis for promoting circadian cycles in *period* gene autoregulation. *Science* **270:** 808.

Shafer O.T., Helfrich-Forster C., Renn S.C., and Taghert P.H. 2006. Reevaluation of *Drosophila melanogaster's* neuronal circadian pacemakers reveals new neuronal classes. *J. Comp. Neurol.* **498:** 180.

Stoleru D., Peng Y., Agosto J., and Rosbash M. 2004. Coupled oscillators control morning and evening locomotor behaviour of *Drosophila*. *Nature* **431:** 862.

Tanoue S., Krishnan P., Krishnan B., Dryer S.E., and Hardin P.E. 2004. Circadian clocks in antennal neurons are necessary and sufficient for olfaction rhythms in *Drosophila*. *Curr. Biol.* **14:** 638.

Ueda H.R., Matsumoto A., Kawamura M., Iino M., Tanimura T., and Hashimoto S. 2002. Genome-wide transcriptional orchestration of circadian rhythms in *Drosophila*. *J. Biol. Chem.* **277:** 14048.

Vosshall L.B. and Stocker R.F. 2007. Molecular architecture of smell and taste in *Drosophila*. *Annu. Rev. Neurosci.* **30:** 505.

Wijnen H., Naef F., Boothroyd C., Claridge-Chang A., and Young M.W. 2006. Control of daily transcript oscillations in *Drosophila* by light and the circadian clock. *PLoS Genet.* **2:** e39.

Wuarin J. and Schibler U. 1990. Expression of the liver-enriched transcriptional activator protein DBP follows a stringent circadian rhythm. *Cell* **63:** 1257.

Yang Z. and Sehgal A. 2001. Role of molecular oscillations in generating behavioral rhythms in *Drosophila*. *Neuron* **29:** 453.

Young M.W. and Kay S.A. 2001. Time zones: A comparative genetics of circadian clocks. *Nat. Rev. Genet.* **2:** 702.

Yu W. and Hardin P.E. 2006. Circadian oscillators of *Drosophila* and mammals. *J. Cell Sci.* **119:** 4793.

Yu W., Zheng H., Houl J.H., Dauwalder B., and Hardin P.E. 2006. PER-dependent rhythms in CLK phosphorylation and E-box binding regulate circadian transcription. *Genes Dev.* **20:** 723.

Zeng H., Qian Z., Myers M.P., and Rosbash M. 1996. A light-entrainment mechanism for the *Drosophila* circadian clock. *Nature* **380:** 129.

Zhao J., Kilman V.L., Keegan K.P., Peng Y., Emery P., Rosbash M., and Allada R. 2003. *Drosophila Clock* can generate ectopic circadian clocks. *Cell* **113:** 755.

Building the Spine: The Vertebrate Segmentation Clock

O. POURQUIÉ
Howard Hughes Medical Institute, Stowers Institute for Medical Research, Kansas City, Missouri 64110

One of the most striking characteristics of many animal and plant species is their organization in a series of periodically repeated anatomical modules. In animals, this particular patterning strategy of the body axis is termed segmentation, and it is observed in both vertebrates and invertebrates. Vertebrate segmentation has been associated with a molecular oscillator—the segmentation clock—whose existence had been predicted on theoretical grounds in the clock and wave-front model. The segmentation clock is proposed to generate pulses of signaling used for the positioning of segmental boundaries. Whereas several models have proposed that simple negative autoregulatory circuits involving the transcription repressors of the hairy and enhancer of split family constitute the clock pacemaker, recent microarray studies in mouse have identified a large network of oscillating signaling genes belonging to the Notch, Wnt, and FGF (fibroblast growth factor) pathways. Thus, significant progress has been made, but the molecular nature of the clockwork underlying the oscillator remains poorly understood. Few examples of oscillators exist in developmental biology, and the segmentation clock provides a unique model of periodic regulation in patterning.

INTRODUCTION

Segmentation has been an intense subject of study since the 19th century when it was first taken by Geoffroy Saint Hilaire as an argument to support the concept of the unity of the body plan among animals. In vertebrates, this particular mode of organization is especially conspicuous at the level of the serial arrangement of vertebrae in the spine. This segmental or metameric pattern is established during embryogenesis through the somitogenesis process (Pourquié 2001). Somites are epithelial blocks of mesoderm, containing the precursors of the vertebrae and the skeletal muscles that are formed rhythmically from the anterior tip of the presomitic mesoderm (PSM) on both sides of the embryonic axis. The rhythm of somite production is characteristic of the species and ranges from 30 minutes in zebra fish embryos to several hours in mammals. Somites form in a sequential anterior-to-posterior progression until a defined number, also characteristic of the species, is reached. This number ranges between approximately 30 in some fish and up to several hundred in snakes (Richardson et al. 1998).

THE CLOCK AND WAVE-FRONT MODEL

The 1970s was a particularly fertile period for theoretical biology and witnessed the birth of important conceptual models in developmental biology. Among the important models postulated at the time was the "clock and wave-front" model proposed by Jonathan Cooke and Chris Zeeman (1976). This model was inspired by the mathematical theory of Catastrophes originally developed by Rene Thom. It proposed that PSM cells forming a somite underwent an abrupt change in cellular properties that could be formalized by a particular type of mathematical catastrophe. Such a catastrophe can be simply represented as a bistable transition between two steady states, allowing an abrupt switch from one particular state to another (Goldbeter et al. 2007). To make the catastrophe periodic, Cooke and Zeeman had to postulate an oscillator controlling the response of PSM cells to the mechanism triggering the catastrophe. Following this pioneering work, a number of subsequent models were proposed, all differing in their modalities but, for many, also relying upon the conversion of a temporal oscillation into a spatial periodic pattern (for review, see Dale and Pourquié 2000; Kulesa et al. 2007). Predicting the behavior of such oscillatory systems is often beyond intuition and thus requires mathematical modeling (Pourquié and Goldbeter 2003).

THE SEGMENTATION CLOCK OSCILLATOR

The first evidence for the existence of an oscillatory process coupled to somitogenesis was provided by the illustration of the rhythmic expression of the *c-hairy1* mRNA in the chick embryo PSM (Palmeirim et al. 1997). During the formation of each somite, the PSM is swiped by a dynamic wave of *c-hairy1* mRNA expression. This wave of transcription activation does not, however, require any long-range signaling among PSM cells and proceeds even if the tissue is physically interrupted, suggesting that it is largely cell-autonomous. These transcriptional oscillations of *c-hairy1* occurring with the same periodicity as the somitogenesis process were proposed to reflect the existence of a molecular oscillator (termed the segmentation clock) acting in PSM cells as originally proposed in the clock and wave-front model. Experimental demonstration of the existence of the oscillator provided a striking example of the predictive value of such theoretical models. Subsequently, several other genes exhibiting such a cyclic behavior were identified in fish, frog, and mouse, suggesting that the oscillator is conserved in vertebrates (Holley et al. 2000; Jiang et al. 2000;

Jouve et al. 2000; Bessho et al. 2001b; Li et al. 2003; Dale et al. 2006). These genes are now referred to as cyclic genes, and the vast majority belong to the Notch, Wnt, and FGF signaling pathways.

FISH OSCILLATOR, TIME DELAY, AND NEGATIVE FEEDBACKS

I first discuss the fish oscillator and its modeling as it is believed to be simpler than the amniote oscillator. In fish, all of the cyclic genes identified thus far belong to the Notch pathway and comprise the Notch downstream targets *Her1* and *Her7*, which are homologous to the chick *hairy1*, as well as to the Notch ligand *DeltaC* (Holley et al. 2000; Jiang et al. 2000; Henry et al. 2002; Oates and Ho 2002). Large genetic screens carried out in zebra fish have identified a few mutants in which somitogenesis is disrupted (van Eeden et al. 1996). Most of these mutants show alterations in genes that belong to the Notch pathway (Rida et al. 2004; Holley 2007). These observations have led to the proposal of a central role for Notch signaling in the control of the oscillations. On the basis of the results of gain-of-function and loss-of-function experiments involving these Notch-related genes in the zebra fish embryo, Lewis (2003) proposed a simple oscillator model. This model is based on a central role of the transcriptional repressors Her1 and Her7, which can establish a simple negative feedback loop by negatively regulating their own promoter. By postulating the existence of a time delay between the production of the transcription factor and its binding on its own promoter, Lewis showed that oscillations exhibiting a period consistent with that of zebra fish somite formation could be obtained. Thus, in fish, the segmentation clock pacemaker is considered to essentially rely on this simple negative autoregulatory circuit (Lewis and Ozbudak 2007). So far in fish, only genes of the Notch pathway have been shown to oscillate, and the overlap with the mammalian cyclic genes is limited, suggesting that the mechanism underlying the fish oscillator could be quite different from the amniote oscillator discussed below.

THE AMNIOTE OSCILLATOR: AN OSCILLATING SIGNALING NETWORK

In amniotes such as chick or mouse, the situation appears to be somewhat different from that described in fish. Whereas a majority of the cyclic genes initially identified also belong to the Notch pathway (Palmeirim et al. 1997; Forsberg et al. 1998; McGrew et al. 1998; Aulehla and Johnson 1999; Jouve et al. 2000; Bessho et al. 2003), different but overlapping repertoires of genes appear to be used between fish and amniotes. For instance, no oscillations of the gene *lunatic fringe* are observed in zebra fish (Prince et al. 2001), whereas oscillations of the *hairy* homologs are detected in both species (Palmeirim et al. 1997; Holley et al. 2000; Jouve et al. 2000; Bessho et al. 2001a; Henry et al. 2002; Oates and Ho 2002). However, unlike that in fish, several other pathways appear to exhibit oscillations in the mouse PSM. For instance, *Axin2*, a key negative feedback inhibitor acting in the Wnt pathway, was shown to exhibit oscillations in the mouse PSM (Aulehla et al. 2003). More recently, a microarray screen carried out in mouse led to a more complex picture of the mouse oscillator (Dequeant et al. 2006). Analysis of a microarray time series encompassing one period of the segmentation clock using Fourier analysis led to the identification of a large group of periodic genes expressed in the mouse PSM (Dequeant et al. 2006). The first 40 cyclic genes identified by this strategy fall into two clusters oscillating in opposite phase. One cluster contains known genes of the Notch pathway, such as *Hes1*, but also genes of the Notch pathway that had not been previously associated with the oscillator, such as *Nrarp*. In addition, this cluster also contains known downstream targets of the FGF pathway, such as *Sprouty2*, suggesting that the FGF pathway oscillates in phase with Notch signaling in the posterior PSM. *Axin2* was found to oscillate in the cluster in opposite phase to the Notch/FGF cluster. The *Axin2* cluster contains several other genes of the Wnt pathway including the inhibitors *Dkk1* and *Dact1* and the Wnt targets *c-myc*, *SP5*, or *Tnfrsf19* (Dequeant et al. 2006 and references therein). Remarkably, the majority of the cyclic genes of the Wnt and Notch/FGF clusters are feedback inhibitors or targets of these pathways (Dequeant et al. 2006 and references therein). Thus, this microarray analysis increased by an order of magnitude the number of cyclic genes identified in the mouse but did not really challenge the biological coherence of the system. However, it suggests that rather than a set of simple negative feedback loops acting as the pacemaker of the system, the amniote segmentation clock relies on a complex network of signaling feedback inhibitors. This analysis further suggests that the role of the segmentation clock is to deliver coordinated pulses of Notch, FGF, and Wnt signaling that are in turn used for the appropriate patterning of the segments. How this periodic signaling is translated into a coordinated striped gene activation that defines the segmental domain and its boundaries is now becoming better understood (Morimoto et al. 2005; Saga 2007).

COUPLING THE OSCILLATORS: SYNCHRONIZATION OF THE OSCILLATIONS IN THE PSM

A remarkable property of the segmentation clock is the synchrony with which the oscillations occur in nearby cells, resulting in the smooth dynamic pattern of the transcription wave observed in the PSM. Experimental evidence now supports a role for cell–cell communication (Horikawa et al. 2006) and more specifically for the Notch signaling pathway in this synchronization process as first postulated in zebra fish (Jiang et al. 2000). In fish, a handful of segmentation mutants have been identified in the large genetic screens carried out during the 1990s (van Eeden et al. 1996). These mutants, which are almost all in Notch pathway components, show remarkably similar phenotypes with a variable number of anterior somites that form, followed by a disruption of the segmentation process (Holley 2007). Additionally, expression of the cyclic genes *Her1* and *DeltaC* in these mutants shows a typical salt-and-pepper pattern instead of the dynamic profile seen in wild-type embryos (Jiang et al. 2000). This led to the proposal that the role of Notch in PSM cells is to synchronize the oscillations, which are initiated by a different mechanism (Jiang et al. 2000). Hence, in the embryos mutant for the Notch pathway, oscillations are set initially and the first somites segment normally. However, because of the lack of Notch-dependent coupling, oscillations progressively drift out of synchrony, resulting in the progressive failure of segmentation observed in the mutants. The salt-and-pepper expression of the cyclic genes would reflect uncoordinated oscillations in PSM cells. Although a definitive demonstration of this hypothesis remains to be provided, recent transplantation experiments in zebra fish demonstrate that implanting cells from an embryo injected with morpholino against the *Her1* and *Her7* genes, which result in Notch ligand overepression, can desynchronize the waves of cyclic gene expression, resulting in the shifting of somitic boundaries on the injected side (Horikawa et al. 2006; Ishimatsu et al. 2007). These experiments strongly argue in favor of the role of Notch signaling in the synchronization of oscillations among nearby cells. Mathematical modeling of the Notch extracellular loop connected to the fish *Her1/Her7* oscillator further supports the proposed role of this pathway in coupling oscillations in nearby cells (Lewis 2003). Accordingly, dissociating the PSM cells—a procedure expected to impair Notch signaling, which requires cell–cell interactions—from chick embryos rapidly results in a loss of synchronized oscillations (Maroto et al. 2005). Similarly, dissociation of mouse PSM cells and analysis of their oscillations using a real-time luciferase-based reporter fused to the *Hes1* promoter show that dissociated cells exhibit very chaotic oscillatory patterns (Masamizu et al. 2006). Together, these data argue in favor of the role of the Notch pathway in synchronization of the oscillations among neighboring cells, thus resulting in the smooth oscillatory waves observed in the vertebrate PSM.

THE WAVE FRONT: TRANSLATING THE CLOCK PULSE INTO A SPATIAL PERIODIC ARRAY OF SEGMENTS

The wave front or determination front was proposed to be defined by a gradient of FGF and Wnt signaling that regresses in concert with axis elongation (Dubrulle et al. 2001; Sawada et al. 2001; Aulehla et al. 2003). At a particular threshold of FGF and Wnt signaling (the determination front), cells of the PSM become competent to respond to the segmentation clock (Dubrulle and Pourquié 2004; Delfini et al. 2005). Upon receiving the clock signal, competent PSM cells simultaneously activate expression of genes such as *Mesp2* in a striped domain, hence defining the future segment (Saga 2007). The role of genes of the Mesp2 transcription factor family, that respond to the periodic clock signal, has been shown to be critical for defining the future segment (Morimoto et al. 2005). Mesp2 controls a complex morphogenetic program resulting in boundary formation and specification of anterior and posterior somite compartments, ultimately leading to formation of the morphological segmental units (Saga and Takeda 2001; Saga 2007).

The process of segmentation in the embryo was shown to be tightly coordinated with that of axis elongation through the formation of the FGF signaling gradient in the posterior PSM (Dubrulle and Pourquié 2004). Transcription of the *fgf8* mRNA is restricted to the precursors of the PSM in the tail bud, and it ceases when their descendents enter the posterior PSM. Thus, as the axis elongates, cells become located progressively more anteriorly in the PSM, and their content in *fgf8* mRNA progressively decays. This results in the establishment of a dynamic gradient of *fgf8* mRNA which is converted into a graded ligand distribution and FGF activity (Sawada et al. 2001; Dubrulle and Pourquié 2004; Delfini et al. 2005). A similar mechanism is assumed to be responsible for establishing the Wnt gradient (Aulehla and Herrmann 2004). The determination front is constantly displaced posteriorly as a result of the progressive decay of the FGF/Wnt mRNA and proteins in PSM cells, and the speed of this displacement defines the speed of somitogenesis progression along the anteroposterior axis. Concomitantly, new cells produced by the tail bud and strongly expressing FGF and Wnt ligands are constantly added to the posterior PSM, accompanying the posterior elongation movement of the tail bud. The FGF and Wnt pathways that define gradients in the PSM also oscillate in this tissue (Aulehla et al. 2003; Dale et al. 2006; Dequeant et al. 2006), therefore arguing that the clock and the wave front are tightly coupled. The gradients also likely have an important role in directly controlling the timing of the arrest of the clock.

WHEN THE OSCILLATOR FAILS: CONGENITAL SCOLIOSIS

In humans, severe disruptions of the segmentation pattern of the vertebrae are generally called congenital scoliosis (Erol et al. 2004). Congenital scoliosis is a rare deformity of the spine, occurring in 1–2 per 10,000 births. It can negatively impact health, with progressive deformity resulting in cardiopulmonary compromise or neurological deficits. In addition, congenital scoliosis is often associated with anomalies of other organ systems, most commonly the renal, neural, and cardiac systems (Turnpenny et al. 2007). Clinical management of these patients is problematic primarily because it is difficult to predict the long-term behavior or associated anomalies of the other organ systems. Most forms of congenital scoliosis are thought to be sporadic in nature, but in fact, little information on familial incidence is available. Thus far, traditional linkage analysis in families with individuals affected by congenital scoliosis has led to the identification of three genes, all associated with the segmentation clock. Mutation of the genes *Dll3*, *Mesp2*, and *lunatic fringe* was shown to lead to familial forms of spondylocostal dysostosis—a particular form of congenital scoliosis (Bulman et al. 2000; Whittock et al. 2004; Sparrow et al. 2006). The fact that all the genes associated with familial congenital scoliosis thus far are linked to the segmentation clock suggests that these anomalies result from defects in the somitogenesis process (Turnpenny et al.

2007). This suggests that the segmentation clock oscillator also acts in human embryos to control segmentation. Studies of the somitogenesis process in mouse embryos point to a number of genes whose mutation results in phenotypes resembling human congenital scoliosis, thus providing interesting candidate genes that could carry mutations in the patients. Therefore, deciphering the segmentation clock mechanism in model organisms will certainly help improve our knowledge of these diseases, which in turn could lead to a better clinical management of these patients.

CONCLUSIONS

Studies of the segmentation clock oscillator in vertebrates have begun to shed light on the complex mechanism involved in generating the characteristic periodic pattern of the vertebrate body axis. Different molecular circuitries appear to have been established by amniotes and lower vertebrates. It remains possible that the oscillatory network identified thus far, which involves interlocked inhibitory feedback loops of the Notch, FGF, and Wnt pathways, merely reflects an output of a pacemaker that would be conserved in these different species. How general is the role of oscillators in generating periodic structures in metazoans remains to be explored. Other examples of spatial periodic structures generated by an oscillator include the well-characterized circadian pattern of sporulation in *Neurospora crassa*, which results in stripe formation (Dunlap and Loros 2004). The role of such oscillators in generating spatial periodicity in animal and vegetal species appears thus far to be limited to systems with polarized growth. No such clock has been identified in the fly embryo in which axial growth occurs after segmentation (Davis and Patel 1999). In contrast, other arthropods that exhibit a progressive mode of axis extension show a segmentation mode that appears to be related to that of vertebrate and hence, could involve a molecular oscillator. Therefore, the use of a traveling oscillator might reflect a widely used strategy of metazoans for generating periodic patterns.

ACKNOWLEDGMENTS

Work in the author's laboratory is supported by a National Institutes of Health grant (R02HD043158) and by Stowers Institute for Medical Research. O.P. is a Howard Hughes Medical Institute Investigator.

REFERENCES

Aulehla, A. and Herrmann B.G. 2004. Segmentation in vertebrates: Clock and gradient finally joined. *Genes Dev.* **18:** 2060.

Aulehla A. and Johnson R.L. 1999. Dynamic expression of lunatic fringe suggests a link between notch signaling and an autonomous cellular oscillator driving somite segmentation. *Dev. Biol.* **207:** 49.

Aulehla A., Wehrle C., Brand-Saberi B., Kemler R., Gossler A., Kanzler B., and Herrmann B.G. 2003. Wnt3a plays a major role in the segmentation clock controlling somitogenesis. *Dev. Cell* **4:** 395.

Bessho Y., Hirata H., Masamizu Y., and Kageyama R. 2003. Periodic repression by the bHLH factor Hes7 is an essential mechanism for the somite segmentation clock. *Genes Dev.* **17:** 1451.

Bessho Y., Miyoshi G., Sakata R., and Kageyama R. 2001a. *Hes7:* A bHLH-type repressor gene regulated by Notch and expressed in the presomitic mesoderm. *Genes Cells* **6:** 175.

Bessho Y., Sakata R., Komatsu S., Shiota K., Yamada S., and Kageyama R. 2001b. Dynamic expression and essential functions of Hes7 in somite segmentation. *Genes Dev.* **15:** 2642.

Bulman M.P., Kusumi K., Frayling T.M., McKeown C., Garrett C., Lander E.S., Krumlauf R., Hattersley A.T., Ellard S., and Turnpenny P.D. 2000. Mutations in the human delta homologue, DLL3, cause axial skeletal defects in spondylocostal dysostosis. *Nat. Genet.* **24:** 438.

Cooke J. and Zeeman E.C. 1976. A clock and wavefront model for control of the number of repeated structures during animal morphogenesis. *J. Theor. Biol.* **58:** 455.

Dale J.K. and Pourquié O. 2000. A clock-work somite. *Bioessays* **22:** 72.

Dale J.K., Malapert P., Chal J., Vilhais-Neto G., Maroto M., Johnson T., Jayasinghe S., Trainor P., Herrmann B., and Pourquié O. 2006. Oscillations of the snail genes in the presomitic mesoderm coordinate segmental patterning and morphogenesis in vertebrate somitogenesis. *Dev. Cell* **10:** 355.

Davis G.K. and Patel N.H. 1999. The origin and evolution of segmentation. *Trends Cell Biol.* **9:** M68.

Delfini M.C., Dubrulle J., Malapert P., Chal J., and Pourquié O. 2005. Control of the segmentation process by graded MAPK/ERK activation in the chick embryo. *Proc. Natl. Acad. Sci.* **102:** 11343.

Dequeant M.L., Glynn E., Gaudenz K., Wahl M., Chen J., Mushegian A., and Pourquié O. 2006. A complex oscillating network of signaling genes underlies the mouse segmentation clock. *Science* **314:** 1595.

Dubrulle J. and Pourquié O. 2004. Coupling segmentation to axis formation. *Development* **131:** 5783.

Dubrulle J., McGrew M.J., and Pourquié O. 2001. FGF signaling controls somite boundary position and regulates segmentation clock control of spatiotemporal Hox gene activation. *Cell* **106:** 219.

Dunlap J.C. and Loros J.J. 2004. The *Neurospora* circadian system. *J. Biol. Rhythms* **19:** 414.

Erol B., Tracy M.R., Dormans J.P., Zackai E.H., Maisenbacher M.K., O'Brien M.L., Turnpenny P.D., and Kusumi K. 2004. Congenital scoliosis and vertebral malformations: Characterization of segmental defects for genetic analysis. *J. Pediatr. Orthop.* **24:** 674.

Forsberg H., Crozet F., and Brown N.A. 1998. Waves of mouse Lunatic fringe expression, in four-hour cycles at two-hour intervals, precede somite boundary formation. *Curr. Biol.* **8:** 1027.

Goldbeter A., Gonze D., and Pourquié O. 2007. Sharp developmental thresholds defined through bistability by antagonistic gradients of retinoic acid and FGF signaling. *Dev. Dyn.* **236:** 1495.

Henry C.A., Urban M.K., Dill K.K., Merlie J.P., Page M.F., Kimmel C.B., and Amacher S.L. 2002. Two linked hairy/Enhancer of split-related zebrafish genes, her1 and her7, function together to refine alternating somite boundaries. *Development* **129:** 3693.

Holley S.A. 2007. The genetics and embryology of zebrafish metamerism. *Dev. Dyn.* **236:** 1422.

Holley S.A., Geisler R., and Nüsslein-Volhard C. 2000. Control of her1 expression during zebrafish somitogenesis by a delta-dependent oscillator and an independent wave-front activity. *Genes Dev.* **14:** 1678.

Horikawa K., Ishimatsu K., Yoshimoto E., Kondo S., and Takeda H. 2006. Noise-resistant and synchronized oscillation of the segmentation clock. *Nature* **441:** 719.

Ishimatsu K., Horikawa K., and Takeda H. 2007. Coupling cellular oscillators: A mechanism that maintains synchrony against developmental noise in the segmentation clock. *Dev. Dyn.* **236:** 1416.

Jiang Y.J., Aerne B.L., Smithers L., Haddon C., Ish-Horowicz D., and Lewis J. 2000. Notch signalling and the synchronization of the somite segmentation clock. *Nature* **408:** 475.

Jouve C., Palmeirim I., Henrique D., Beckers J., Gossler A., Ish-Horowicz D., and Pourquié O. 2000. Notch signalling is required for cyclic expression of the hairy-like gene HES1 in the presomitic mesoderm. *Development* **127:** 1421.

Kulesa P.M., Schnell S., Rudloff S., Baker R.E., and Maini P.K. 2007. From segment to somite: Segmentation to epithelialization analyzed within quantitative frameworks. *Dev. Dyn.* **236:** 1392.

Lewis J. 2003. Autoinhibition with transcriptional delay: A simple mechanism for the zebrafish somitogenesis oscillator. *Curr. Biol.* **13:** 1398.

Lewis J. and Ozbudak E.M. 2007. Deciphering the somite segmentation clock: Beyond mutants and morphants. *Dev. Dyn.* **236:** 1410.

Li Y., Fenger U., Niehrs C., and Pollet N. 2003. Cyclic expression of esr9 gene in *Xenopus* presomitic mesoderm. *Differentiation* **71:** 83.

Maroto M., Dale J.K., Dequeant M.L., Petit A.C., and Pourquié O. 2005. Synchronised cycling gene oscillations in presomitic mesoderm cells require cell-cell contact. *Int. J. Dev. Biol.* **49:** 309.

Masamizu Y., Ohtsuka T., Takashima Y., Nagahara H., Takenaka Y., Yoshikawa K., Okamura H., and Kageyama R. 2006. Real-time imaging of the somite segmentation clock: Revelation of unstable oscillators in the individual presomitic mesoderm cells. *Proc. Natl. Acad. Sci.* **103:** 1313.

McGrew M.J., Dale J.K., Fraboulet S., and Pourquié O. 1998. The lunatic fringe gene is a target of the molecular clock linked to somite segmentation in avian embryos. *Curr. Biol.* **8:** 979.

Morimoto M., Takahashi Y., Endo M., and Saga Y. 2005. The Mesp2 transcription factor establishes segmental borders by suppressing Notch activity. *Nature* **435:** 354.

Oates A.C. and Ho R.K. 2002. Hairy/E(spl)-related (Her) genes are central components of the segmentation oscillator and display redundancy with the Delta/Notch signaling pathway in the formation of anterior segmental boundaries in the zebrafish. *Development* **129:** 2929.

Palmeirim I., Henrique D., Ish-Horowicz D., and Pourquié O. 1997. Avian hairy gene expression identifies a molecular clock linked to vertebrate segmentation and somitogenesiss. *Cell* **91:** 639.

Pourquié O. 2001. Vertebrate somitogenesis. *Annu. Rev. Cell Dev. Biol.* **17:** 311.

Pourquié O. and Goldbeter A. 2003. Segmentation clock: Insights from computational models. *Curr. Biol.* **13:** R632.

Prince V.E., Holley S.A., Bally-Cuif L., Prabhakaran B., Oates A.C., Ho R.K., and Vogt T.F. 2001. Zebrafish lunatic fringe demarcates segmental boundaries. *Mech. Dev.* **105:** 175.

Richardson M.K., Allen S.P., Wright G.M., Raynaud A., and Hanken J. 1998. Somite number and vertebrate evolution. *Development* **125:** 151.

Rida P.C., Le Minh N., and Jiang Y.J. 2004. A Notch feeling of somite segmentation and beyond. *Dev. Biol.* **265:** 2.

Saga Y. 2007. Segmental border is defined by the key transcription factor Mesp2, by means of the suppression of notch activity. *Dev. Dyn.* **236:** 1450.

Saga Y. and Takeda H. 2001. The making of the somite: Molecular events in vertebrate segmentation. *Nat. Rev. Genet.* **2:** 835.

Sawada A., Shinya M., Jiang Y.J., Kawakami A., Kuroiwa A., and Takeda H. 2001. Fgf/MAPK signalling is a crucial positional cue in somite boundary formation. *Development* **128:** 4873.

Sparrow D.B., Chapman G., Wouters M.A., Whittock N.V., Ellard S., Fatkin D., Turnpenny P.D., Kusumi K., Sillence D., and Dunwoodie S.L. 2006. Mutation of the LUNATIC FRINGE gene in humans causes spondylocostal dysostosis with a severe vertebral phenotype. *Am. J. Hum. Genet.* **78:** 28.

Turnpenny P.D., Alman B., Cornier A.S., Giampietro P.F., Offiah A., Tassy O., Pourquié O., Kusumi K., and Dunwoodie S. 2007. Abnormal vertebral segmentation and the notch signaling pathway in man. *Dev. Dyn.* **236:** 1456.

van Eeden F.J., Granato M., Schach U., Brand M., Furutani-Seiki M., Haffter P., Hammerschmidt M., Heisenberg C.P., Jiang Y.J., Kane D.A., Kelsh R.N., Mullins M.C., Odenthal J., Warga R.M., Allende M.L., Weinberg E.S., and Nüsslein-Volhard C. 1996. Mutations affecting somite formation and patterning in the zebrafish, *Danio rerio*. *Development* **123:** 153.

Whittock N.V., Sparrow D.B., Wouters M.A., Sillence D., Ellard S., Dunwoodie S.L., and Turnpenny P.D. 2004. Mutated MESP2 causes spondylocostal dysostosis in humans. *Am. J. Hum. Genet.* **74:** 1249.

Ultradian Oscillators in Somite Segmentation and Other Biological Events

R. KAGEYAMA,*† S. YOSHIURA,*†‡ Y. MASAMIZU,*†§ AND Y. NIWA*†§

*Institute for Virus Research, Kyoto University, Kyoto 606-8507, Japan; †Japan Science and Technology Agency, Core Research for Evolutional Science and Technology, Kyoto 606-8507, Japan; ‡Kyoto University Graduate School of Biostudies, Kyoto 606-8502, Japan; §Kyoto University Graduate School of Medicine, Kyoto 606-8501, Japan

Somite formation occurs every 2 hours in mouse embryos by periodic segmentation of the anterior ends of the presomitic mesoderm, and this process is controlled by a biological clock called the segmentation clock. During this process, the basic helix-loop-helix gene *Hes7* is cyclically expressed, and each cycle leads to generation of a bilateral pair of somites. Both sustained expression and loss of expression of *Hes7* result in severe somite fusion, indicating that *Hes7* constitutes an essential component of the segmentation clock. Interestingly, expression of the related gene *Hes1* also oscillates with a periodicity of about 2 hours in many cell types. Both sustained *Hes1* expression and loss of Hes1 activity lead to retardation of the G_1 phase of the cell cycle, suggesting that *Hes1* oscillation with an ultradian rhythm is required for efficient cell proliferation. Both *Hes1* and *Hes7* oscillations are regulated by negative feedback and rapid degradation of their gene products. Strikingly, expression of other factors such as Stat-Socs and Smad signaling molecules also display ultradian rhythms. All of these data suggest that ultradian oscillations are more general responses than were previously thought and that oscillatory and sustained gene expression results in different biological outcomes.

INTRODUCTION

It is well known that many day and night activities have a circadian rhythm, which has a periodicity of about 24 hours. Some biological activities, however, are known to have rhythms with shorter periodicities called ultradian rhythms. For example, secretion of hormones and changes of plasma glucose levels are pulsatile with periodicities of a few hours (Tannenbaum and Martin 1976; Simon et al. 1987). Another example is somite formation, which occurs every 2 hours in the mouse and every several hours in humans. Recently, characterization of the molecular mechanism of periodic somite formation has been greatly advanced.

Somites are transient metameric structures located at either side of the neural tube, and they later give rise to the vertebral column, ribs, skeletal muscles, and subcutaneous tissues. Somites are formed by segmentation of the anterior ends of the presomitic mesoderm (PSM) (Fig. 1a). In mouse embryos, a bilateral pair of somites are formed every 2 hours, suggesting that this process is controlled by a biological clock with a 2-hour periodicity, and it has been called the segmentation clock (Bessho and Kageyama 2003; Pourquié 2003; Aulehla and Herrmann 2004; Gridley 2006). During this process, the basic helix-loop-helix (bHLH) genes *Hes1* and *Hes7* display oscillatory expression in PSM cells, leading to generation of a pair of somites after each cycle (Bessho et al. 2001a). In particular, *Hes7* is an essential component of the segmentation clock. In the absence of *Hes7*, somites become severely fused (Bessho et al. 2001b). Interestingly, *Hes1* also shows oscillatory expression in many cell types, in addition to PSM cells, suggesting that oscillation is not unique to PSM cells and that *Hes1* oscillation may regulate the timing of many biological events (Jouve et al. 2000; Hirata et al. 2002). It has been shown that other genes are also cyclically expressed by many cell types, raising the possibility that oscillatory expression is more general than was previously thought. In this chapter, we

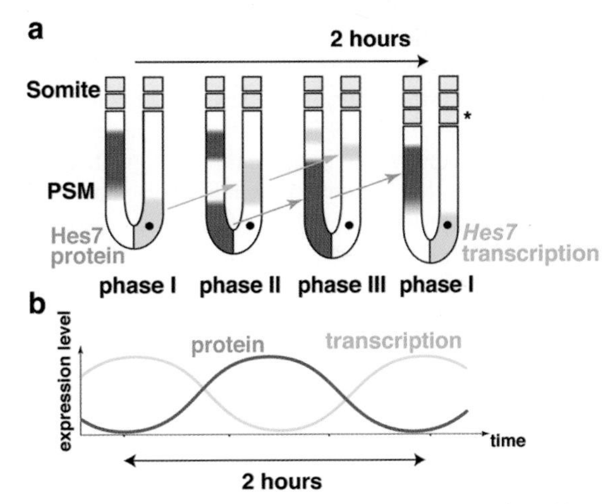

Figure 1. Hes7 oscillation in the somite segmentation. (*a*) In mouse embryos, a bilateral pair of somites are formed every 2 hours by segmentation of the anterior ends of the PSM (*asterisk*). Expression of the *Hes7* intron (corresponding to the region where the *Hes7* gene is transcribed, *green*) is initiated in the posterior PSM (phase I), and this expression region is then propagated into the anterior PSM (phase II), finally reaching S-1 (phase III). By the time S-1 becomes S0, *Hes7* intron expression disappears, and the next expression occurs again in the posterior PSM (i.e., it returns to phase I). The *Hes7* transcription region (*green*) and the Hes7-protein-expressing region (*blue*) are mutually exclusive in all three phases, suggesting that *Hes7* transcription occurs only when Hes7 protein is not expressed. S0 is the next somitic region formed, whereas S-1 is the region that becomes a somite following S0. Anterior is top. (*b*) Dynamic change in *Hes7* expression is the result of oscillatory expression in individual PSM cells (indicated by a dot in *a*). (Adapted, with permission, from Kageyama et al. 2007a [© John Wiley & Sons].)

discuss the mechanism and the significance of Hes1/Hes7 oscillations and other ultradian oscillators in biological events.

THE ROLES AND THE MECHANISM OF HES7 OSCILLATION IN THE SEGMENTATION CLOCK

Hes7 expression dynamically changes in PSM and has been classified as occuring in three phases (Fig. 1a). Expression of the *Hes7* intron (corresponding to the region where *Hes7* gene is transcribed) is initiated in the posterior PSM (phase I), and this region is then propagated into the anterior PSM (phase II), finally reaching S-1 (phase III) (Fig. 1a, green). By the time S-1 becomes S0 (the next somitic region formed), *Hes7* intron expression disappears, whereas the next expression of *Hes7* intron occurs again in the posterior PSM (i.e., it returns to phase I). This dynamic change in gene expression is the result of oscillatory expression of *Hes7* in individual PSM cells (Fig. 1b). Cells located laterally display oscillation in-phase, but this is delayed in anterior cells compared to posterior cells. As a result, the expression domain moves from the posterior to the anterior region.

Interestingly, the *Hes7* transcription region (Fig. 1a, green) and the Hes7 protein-expressing region (Fig. 1a, blue) are mutually exclusive in all three phases, suggesting that *Hes7* transcription occurs only when Hes7 protein is not expressed (Bessho et al. 2001a, 2003). In agreement with this notion, the Hes7 protein can repress *Hes7* transcription by directly binding to its own promoter (Bessho et al. 2003). Furthermore, in the absence of functional Hes7 protein, the *Hes7* gene is constitutively transcribed in the PSM, whereas *Hes7* transcription is constitutively repressed when Hes7 protein is stabilized (Bessho et al. 2003). These data together suggest that *Hes7* oscillation is regulated by negative autoregulation as follows (Fig. 2). In the PSM, Notch signaling induces expression of Hes7, which represses its own expression by binding to its own promoter. When the promoter is repressed, both *Hes7* mRNA and protein disappear rapidly because they are extremely unstable, allowing the next round of expression.

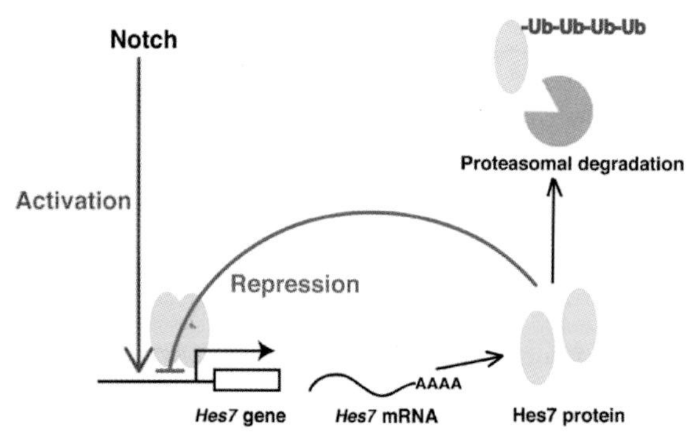

Figure 2. Hes7 oscillation regulated by negative autoregulation. Notch signaling induces expression of Hes7, which represses its own expression by binding to its own promoter. When the promoter is repressed, both Hes7 mRNA and protein disappear rapidly because they are extremely unstable, allowing the next round of expression. In this way, *Hes7* autonomously starts oscillatory expression.

Hes1 displays an expression pattern similar to that of *Hes7* in the PSM. Hes1, a transcriptional repressor, can repress its own expression by directly binding to the promoter (Sasai et al. 1992; Takebayashi et al. 1994), and *Hes1* oscillation is also regulated by this negative autoregulation (Hirata et al. 2002), although the expression is maintained in S0 and in the somitic regions, unlike *Hes7*.

It has been shown that expression of *Lunatic fringe* (*Lfng*), a glycosyltransferase gene that inhibits Notch activity (Brückner et al. 2000; Moloney et al. 2000), oscillates in phase with *Hes7* (Bessho et al. 2001b). In *Lfng*-null mice, *Hes7* expression still oscillates, although the expression is somewhat affected (Niwa et al. 2007), suggesting that *Hes7* oscillation does not depend on *Lfng* oscillation. In contrast, in *Hes7*-null mice, *Lfng* is constitutively expressed throughout the PSM (Bessho et al. 2001b). Conversely, stabilized Hes7 constitutively represses *Lfng* expression (Bessho et al. 2003). These data indicate that *Lfng* oscillation depends on periodic repression by Hes7.

Both *Hes7* and *Lfng* have an essential role in somite segmentation. In *Hes7*-null and *Lfng*-null mice, somites severely fuse (Evrard et al. 1998; Zhang and Gridley 1998; Bessho et al. 2001b). Interestingly, transgenic mice that constitutively express *Hes7* or *Lfng* throughout the PSM also display severe somite fusion (Serth et al. 2003; Hirata et al. 2004). Thus, both sustained expression and loss of expression of *Hes7* or *Lfng* disturb somite segmentation, suggesting that oscillatory expression of both genes is required for this process.

MATHEMATICAL SIMULATION OF HES7 OSCILLATION

Hes1/Hes7 oscillations can be mathematically simulated, based on the negative autoregulation mechanism described above (Jensen et al. 2003; Lewis 2003; Monk 2003; Hirata et al. 2004). The simulation for Hes7 oscillation consists of two differential equations (Hirata et al. 2004).

$$\frac{dp(t)}{dt} = am(t - T_p) - bp(t)$$

$$\frac{dm(t)}{dt} = f(p(t - T_m)) - cm(t)$$

$$f(p) = \frac{k}{1 + (\frac{p}{p_{crit}})^2}$$

where $p(t)$ and $m(t)$ are the quantities of functional Hes7 protein and mRNA, respectively, per cell at time t, and $f(p)$ is the rate of initiation of transcription, which is negatively regulated by a dimer of the protein p present at the time of initiation. a is the rate constant for translation, whereas b and c are the degradation rate constants for protein and mRNA, respectively.

According to this simulation, the instability of Hes7 protein is very important for sustained oscillation. The

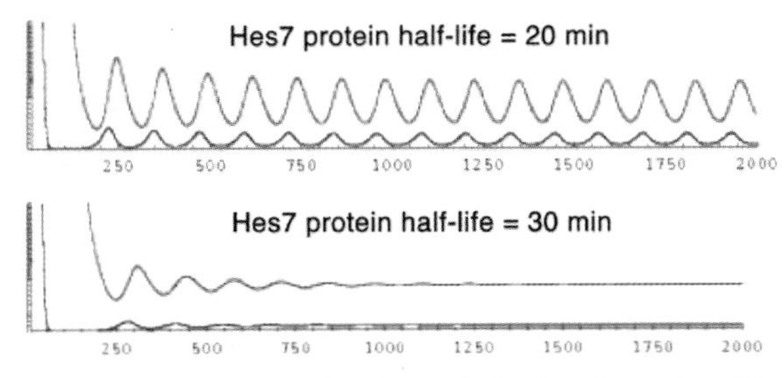

Figure 3. Stabilization of Hes7 protein leads to damped oscillation. The half-life of Hes7 protein is about 20 minutes, and this value is important for sustained oscillations. If the half-life becomes 30 minutes, Hes7 oscillations become damped after three to four cycles. (Adapted, with permission, from Hirata et al. 2004 [Nature Publishing Group].)

half-life of Hes7 protein is about 20 minutes, but if it reaches 30 minutes instead of 20 minutes, the simulation predicts that Hes7 oscillation becomes damped after several cycles (Fig. 3). However, this prediction is counterintuitive, which suggests that even if the half-life is 30 minutes, Hes7 oscillations could continue in a sustained manner, although the period would be a little longer than 2 hours. The prediction of the mathematical simulation was experimentally evaluated by making mutant mice that express stabilized Hes7 protein. Hes7 protein with a K14R mutation (the 14th amino acid residue [lysine] of Hes7 protein is mutated to arginine) has a normal repressor activity, but its half-life is about 30 minutes. A K14R point mutation was introduced into the *Hes7* locus of ES cells, and from these ES cells, the mutant mice that express Hes7 with the K14R mutation were generated. Interestingly, although three or four pairs of somites are segmented, the subsequent somites fuse severely in these mutant mice (Hirata et al. 2004). Furthermore, Hes7 oscillation becomes damped after three to four cycles (Hirata et al. 2004). These data agree well with the prediction of the mathematical simulation and indicate that intuition alone is insufficient and that mathematical modeling is required to understand the dynamics of Hes oscillations.

REAL-TIME MONITORING OF *HES* OSCILLATIONS

To further examine Hes oscillation patterns, we established a real-time imaging system for Hes1/Hes7 expression, where the Hes1/Hes7 promoter-driven ubiquitinated luciferase is used as a reporter (Masamizu et al. 2006; and our unpublished data). Because the normal luciferase protein has a half-life of several hours, it is too stable to monitor 2-hour cycle oscillations. Such a reporter protein would be accumulated after several cycles of oscillations. In contrast, the ubiquitinated luciferase has a half-life of about 10 minutes and can be used for monitoring oscillations with a 2-hour periodicity. The system with the ubiquitinated luciferase reporter successfully enabled observation of Hes1/Hes7 oscillations in real time in PSM, offering a powerful tool to understand the dynamics of Hes1/Hes7 oscillations. On the basis of real-time imaging, a spatiotemporal profile was reproduced (Fig. 4a) (Kageyama et al. 2007a). This profile suggests that each individual PSM cell seems to experience five cycles of Hes1 oscillations before entering the S0 region (the next somite region) about embryonic day 10.5. Interestingly, the period of the first three cycles is about 2 hours, whereas that of the fourth and fifth cycles is longer than 2 hours (Fig. 4b). Cells experiencing the first three cycles are located in the posterior PSM, whereas those experiencing the last two cycles are located in the anterior PSM, indicating that the period of Hes1 oscillations is different between the posterior and the anterior PSM. The different periods between the posterior and the anterior PSM could be regulated by Fgf signaling (Niwa et al. 2007), which forms a posterior-to-anterior gradient in the PSM, although the precise mechanism remains to be determined.

Hes1 expression oscillates in a stable manner in PSM cells but not in dissociated PSM cells. In these cells, the periods of Hes1 oscillations are variable from cycle to cycle, indicating that Hes1 oscillation in individual PSM cells is unstable (Masamizu et al. 2006). Thus, negative autoregulation can create only an unstable oscillator, suggesting that cell–cell communication and/or extracellular signals are important for stable Hes1 oscillations. We observed the same features in Hes7 oscillations (our unpublished data). It is likely that Notch signaling is involved in cell–cell communication (Jiang et al. 2000; Horikawa et al. 2006), although further analysis is required to clarify this issue.

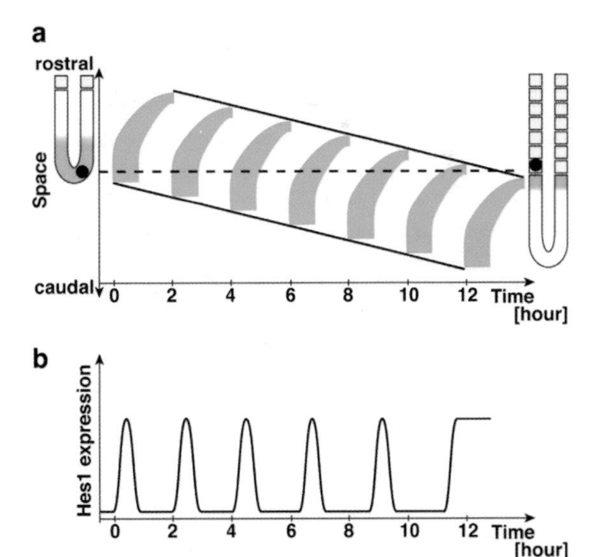

Figure 4. Spatiotemporal profile of Hes1 oscillation in PSM. (*a*) A spatiotemporal profile of Hes1 oscillation in PSM was reproduced, based on the real-time imaging analysis (Masamizu et al. 2006; Kageyama et al. 2007a). (*b*) Hes1 expression in individual PSM cells. Each PSM cell seems to experience five cycles of *Hes1* oscillation after getting out of the tail bud and before reaching the S0 region. The periodicity of the first three cycles is about 2 hours, whereas that of the last two cycles is longer, suggesting that *Hes1* oscillation is differentially regulated between the anterior and the posterior PSM. (Adapted, with permission, from Kageyama et al. 2007a [© John Wiley & Sons].)

Hes1 OSCILLATION IN NON-PSM CELLS

Interestingly, oscillatory expression is not unique to PSM cells. Hes1 is widely expressed by many cell types such as fibroblasts and neuroblasts (Kageyama et al. 2007b). After serum stimulation or Notch activation, Hes1 expression oscillates with a periodicity of about 2 hours in many cell types, suggesting that Hes1 may regulate many biological processes as a biological clock with a 2-hour cycle (Hirata et al. 2002). Both sustained Hes1 expression and knock down of Hes1 activity reduce proliferation of fibroblasts, suggesting that oscillatory Hes1 expression is required for efficient proliferation of fibroblasts (Yoshiura et al. 2007). Furthermore, persistent Hes1 expression also inhibits proliferation and differentiation of neuroblasts (Baek et al. 2006). It seems that Hes1 promotes certain steps but inhibits other steps of the cell cycle, although it remains to be determined at which steps of the cell cycle Hes1 regulation actually occurs.

Northern and western blot analyses indicate that in non-PSM cells, Hes1 oscillations seem to be damped after three to six cycles (Hirata et al. 2002). Real-time monitoring at the single-cell level revealed that Hes1 oscillation continues in each individual cell even after 2 days, but because the period of each cycle is variable from cycle to cycle and from cell to cell, Hes1 oscillation among cells soon goes out of synchrony (Masamizu et al. 2006). Thus, damping of Hes1 oscillation in the population is not due to damped oscillation in all individual cells, but it is due to desynchronization between cycling cells. Interestingly, this unstable period of Hes1 oscillation of fibroblasts is very similar to that of dissociated PSM cells.

OTHER ULTRADIAN OSCILLATORS

In addition to Hes1/Hes7, many other factors have been shown to display oscillatory expression in PSM cells (Dequéant et al. 2006). Similarly, several other factors have been shown to display oscillatory expression in non-PSM cells. NF-κB is trapped in the cytoplasm by IκB, but in the presence of inducing signals such as tumor necrosis factor-α (TNF-α,) IκB is degraded, and NF-κB is released and transferred into the nucleus, where it induces expression of many genes. One of them is IκB, which again traps NF-κB. However, if the inducing signals are still present, IκB is degraded and NF-κB is translocated to the nucleus again. As a result, the localization of NF-κB oscillates between the nucleus and cytoplasm (N-C oscillation) (Hoffmann et al. 2002; Nelson et al. 2004). It has been shown that p53 expression also oscillates due to the negative feedback of Mdm2 (Bar-Or et al. 2000; Lahav et al. 2004). These observations suggest the possibility that other molecules forming negative feedback loops can also display oscillatory responses.

To identify new ultradian oscillators, RNA was prepared from fibroblasts every 30 minutes after serum stimulation and subjected to microarray analyses, which revealed oscillatory responses of Socs3 and Smad6 (Yoshiura et al. 2007). Expression of both genes oscillates with a period of about 2 hours, although the peaks appear at different times. Socs3 is the downstream target of the Jak-Stat signaling (Starr et al. 1997; Levy and Darnell 2002; Yu and Jove 2004). Jak2 activates Stat3 by phosphorylation, and phosphorylated Stat3 (p-Stat3) forms a dimer, enters the nucleus, and induces expression of many genes (Fig. 5). One of them is Socs3, which antagonizes Jak-dependent activation of Stat3, thus forming a negative feedback loop. Because Socs3 expression oscillates, Jak-dependent activation of Stat3 should be periodically inhibited. In agreement with this notion, formation of p-Stat3 is also found to be oscillatory. Thus, the Jak-Stat3-Socs3 negative feedback loop generates p-Stat3/Socs3 oscillations (Fig. 5) (Yoshiura et al. 2007). Importantly, inhibition of p-Stat3/Socs3 oscillations abolishes Hes1 oscillation, suggesting that Hes1 oscillation is not solely regulated by negative feedback. p-Stat3 seems to periodically destabilize Hes1 protein, and this periodic destabilization may be important for Hes1 oscillation (Yoshiura et al. 2007). Conversely, it was previously shown that Hes1 promotes Jak-dependent activation of Stat3 by physical interaction (Kamakura et al. 2004). Thus, it is likely that Hes1 oscillation and p-Stat3/Socs3 oscillations are coupled with and depend on each other.

Smad6 is the downstream target of Smad1 signaling (Imamura et al. 1997; Massagué and Wotton 2000; ten Dijke et al. 2000). Signaling molecules such as bone morphogenetic protein (BMP) activate Smad1/5/8 by phosphorylation. Phosphorylated Smad1/5/8 (p-Smad1/5/8) up-regulates expression of downstream genes including Smad6, which then inhibits p-Smad1/5/8 formation, constituting a negative feedback loop (Fig. 6). Because Smad6 expression oscillates, Smad1/5/8 activation should be periodically inhibited. In agreement with this notion, formation of p-Smad1 is also found to be oscillatory. Thus, the Smad1-Smad6 negative feedback loop generates p-Smad1/Smad6 oscillations (Fig. 6) (Yoshiura et al. 2007). However, unlike p-Stat3/Socs3 oscillations, p-Smad1/Smad6 oscillations are not involved in Hes1 oscillation, suggesting that the p-Smad1/Smad6 oscillations function independently of Hes1 oscillation. In

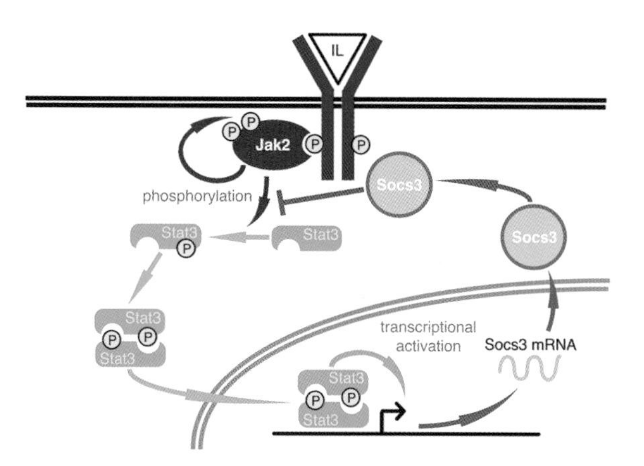

Figure 5. p-Stat3/Socs3 oscillations. Jak2 activates Stat3 by phosphorylation. Phosphorylated Stat3 (p-Stat3) enters the nucleus and induces expression of downstream genes. One of them is Socs3, which antagonizes formation of p-Stat3. This negative feedback loop generates oscillatory expression of p-Stat3 and Socs3.

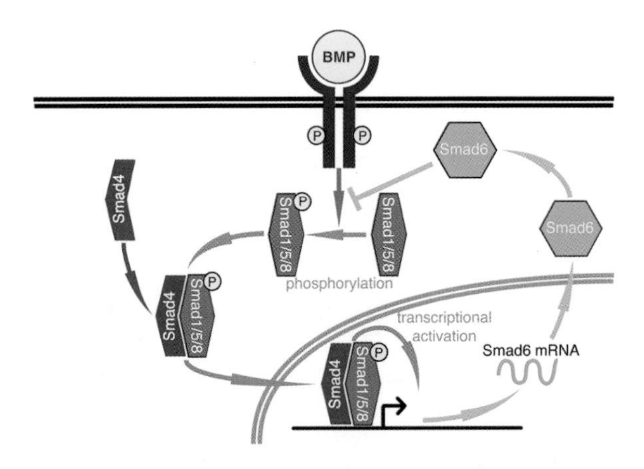

Figure 6. p-Smad1/Smad6 oscillations. Signaling molecules such as BMP activate Smad1/5/8 by phosphorylation. Phosphorylated Smad1/5/8 (p-Smad1/5/8) up-regulates expression of downstream genes including Smad6, which then inhibits p-Smad1/5/8 formation, constituting a negative feedback loop. Because Smad6 expression oscillates, Smad1/5/8 activation should be periodically inhibited. In agreement with this notion, formation of p-Smad1 is found to be oscillatory. The Smad1/5/8-Smad6 negative feedback loop generates p-Smad1/5/8 and Smad6 oscillations.

agreement with this notion, Hes1 and Socs3 oscillate in phase, whereas Hes1 and Smad6 do not.

POSSIBLE SIGNIFICANCE OF ULTRADIAN OSCILLATIONS

Although Hes1 oscillation is required for efficient cell proliferation, the expression modes of the downstream target genes are not well-characterized. In our microarray analyses, we did not identify oscillatory downstream target genes for Stat, Smad, and Hes oscillators. Expression of some downstream genes could oscillate (Fig. 7a), but synchrony of oscillation among cells might easily be lost. In this case, real-time monitoring of gene expression at the single-cell level is required. Hes1 oscillations would produce different conditions in different cells, leading to different responses. For example, when differentiation is induced, cells in a low Hes1 phase, but not those in a high Hes1 phase, would respond to the induction. Thus, Hes1 oscillations could make responsive and unresponsive cells to the same signals.

If the downstream target gene products are relatively stable, they do not respond in an oscillatory manner. For example, expression of such downstream factors can be maintained within a certain range by periodic down-regulation by Hes1 oscillation (Fig. 7b). If products of downstream target genes are even more stable, transcriptional induction by activators and periodic repression by Hes1 oscillation could lead to up-regulation of downstream target factors in a stepwise manner (Fig. 7c). In this case, if the expression level reaches a certain value, the next event could happen. Thus, the information about the number of cycles can be converted into the timing of the next event. Further analysis of downstream target genes will be required to reveal more precise significance and roles of ultradian oscillators.

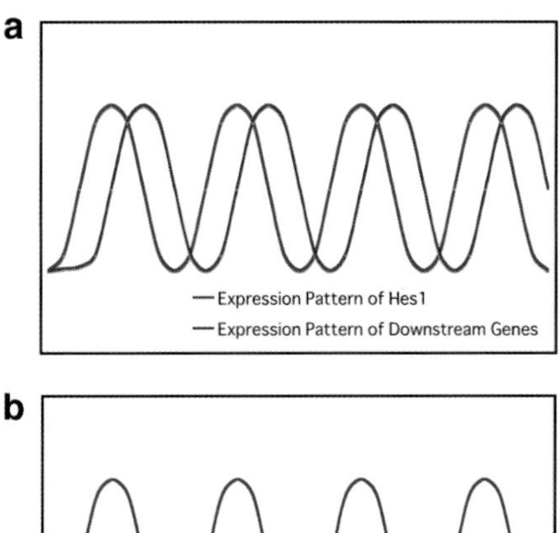

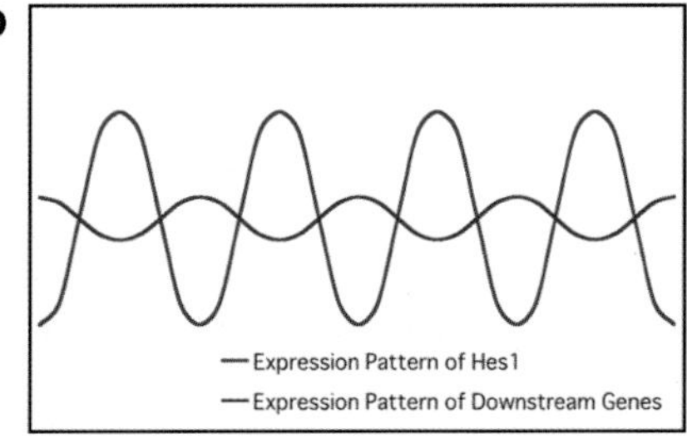

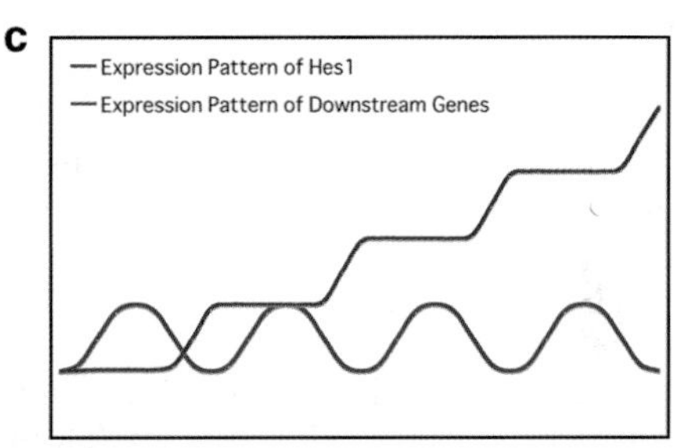

Figure 7. Downstream events of ultradian oscillators. Downstream factors for Hes1/Hes7 oscillations could display at least three different expression patterns. (*a*) If products of downstream genes are unstable, their expression could oscillate like Hes1 and Hes7. (*b*) If downstream factors are relatively stable. their expression can be maintained within a certain range by periodic down-regulation by Hes1/Hes7 oscillations. (*c*) If downstream factors are very stable, transcriptional induction by activators and periodic repression by Hes1/Hes7 oscillations could lead to up-regulation of downstream target factors in a stepwise manner. If the expression level reaches a certain value, the next event could happen. In this case, the information about the number of cycles can be converted into the timing of the next event.

CONCLUSIONS

It has been shown that many genes display expression with ultradian rhythms. In the case of Hes1, although the expression seems to be stationary on northern and western analyses without serum stimulation, it is actually oscillatory at the single-cell level. Oscillatory expression of Hes1 is required for efficient cell proliferation, as sustained expression retards the G_1 phase of the cell cycle. Similarly, oscillatory expression of Hes7 is required for somite segmentation, as sustained expression leads to severe somite fusion. Thus, oscillatory and sustained gene expression result in different biological outcomes. Ultradian oscillations are regulated by negative feedback, but it seems that the negative feedback loop alone generates only unstable oscillators. Cell–cell communication or other extracellular signals are required for stable oscillators. Reconstitution of such oscillators in non-PSM cells would be required to understand the basic mechanism of stable oscillations.

Other molecules forming negative feedback loops are also shown to display ultradian rhythms, suggesting that ultradian oscillations are more general responses than were previously thought. Further characterization of ultradian oscillators and their downstream targets will give us more insight into the understanding of how cells measure time in many biological processes.

ACKNOWLEDGMENTS

This work was supported by the grants-in-aid from the Genome Network Project and the Ministry of Education, Culture, Sports, Science and Technology of Japan. S.Y., Y.M., and Y.N. were supported by the 21st Century COE Program of the Ministry of Education, Culture, Sports, Science and Technology of Japan and by research fellowships of the Japan Society for the Promotion of Science for Young Scientists.

REFERENCES

Aulehla A. and Herrmann B.G. 2004. Segmentation in vertebrates: Clock and gradient finally joined. *Genes Dev.* **18:** 2060.

Baek J.H., Hatakeyama J., Sakamoto S., Ohtsuka T., and Kageyama R. 2006. Persistent and high levels of Hes1 expression regulate boundary formation in the developing central nervous system. *Development* **133:** 2467.

Bar-Or R.L., Maya R., Segel L.A., Alon U., Levine A.J., and Oren M. 2000. Generation of oscillations by the p53-Mdm2 feedback loop: A theoretical and experimental study. *Proc. Natl. Acad. Sci.* **97:** 11250.

Bessho Y. and Kageyama R. 2003. Oscillations, clocks and segmentation. *Curr. Opin. Genet. Dev.* **13:** 379.

Bessho Y., Hirata H., Masamizu Y., and Kageyama R. 2003. Periodic repression by the bHLH factor Hes7 is an essential mechanism for the somite segmentation clock. *Genes Dev.* **17:** 1451.

Bessho Y., Miyoshi G., Sakata R., and Kageyama R. 2001a. *Hes7:* A bHLH-type repressor gene regulated by Notch and expressed in the presomitic mesoderm. *Genes Cells* **6:** 175.

Bessho Y., Sakata R., Komatsu S., Shiota K., Yamada S., and Kageyama R. 2001b. Dynamic expression and essential functions of *Hes7* in somite segmentation. *Genes Dev.* **15:** 2642.

Brückner K., Perez L., Clausen H., and Cohen S. 2000. Glycosyltransferase activity of Fringe modulates Notch-Delta interactions. *Nature* **406:** 411.

Dequéant M.-L. Glynn E., Gaudenz K., Wahl M., Chen J., Mushegian A., and Pourquié O. 2006. A complex oscillating network of signaling genes underlies the mouse segmentation clock. *Science* **314:** 1595.

Evrard Y.A., Lun Y., Aulehla A., Gan L., and Johnson R.L. 1998. *lunatic fringe* is an essential mediator of somite segmentation and patterning. *Nature* **394:** 377.

Gridley T. 2006. The long and short of it: Somite formation in mice. *Dev. Dyn.* **235:** 2330.

Hirata H., Bessho Y., Kokubu H., Masamizu Y., Yamada S., Lewis J., and Kageyama R. 2004. Instability of Hes7 protein is crucial for the somite segmentation clock. *Nat. Genet.* **36:** 750.

Hirata H., Yoshiura S., Ohtsuka T., Bessho Y., Harada T., Yoshikawa K., and Kageyama R. 2002. Oscillatory expression of the bHLH factor Hes1 regulated by a negative feedback loop. *Science* **298:** 840.

Hoffmann A., Levchenko A., Scott M.L., and Baltimore D. 2002. The IκB-NF-κB signaling module: Temporal control and selective gene activation. *Science* **298:** 1241.

Horikawa K., Ishimatsu K., Yoshimoto E., Kondo S., and Takeda H. 2006. Noise-resistant and synchronized oscillation of the segmentation clock. *Nature* **441:** 719.

Imamura T., Takase M., Nishihara A., Oeda E., Hanai J., Kawabata M., and Miyazono K. 1997. Smad6 inhibits signalling by the TGF-β superfamily. *Nature* **389:** 622.

Jensen M.H., Sneppen K., and Tiana G. 2003. Sustained oscillations and time delays in gene expression of protein Hes1. *FEBS Lett.* **541:** 176.

Jiang Y.J., Aerne B.L., Smithers L., Haddon C., Ish-Horowicz D., and Lewis J. 2000. Notch signalling and the synchronization of the somite segmentation clock. *Nature* **408:** 475.

Jouve C., Palmeirim I., Henrique D., Beckers J., Gossler A., Ish-Horowicz D., and Pourquié O. 2000. Notch signalling is required for cyclic expression of the hairy-like gene *HES1* in the presomitic mesoderm. *Development* **127:** 1421.

Kamakura S., Oishi K., Yoshimatsu T., Nakafuku M., Masuyama N., and Gotoh Y. 2004. Hes binding to STAT3 mediates crosstalk between Notch and JAK-STAT signaling. *Nat. Cell Biol.* **6:** 547.

Kageyama R., Masamizu Y., and Niwa Y. 2007a. Oscillator mechanism of Notch pathway in the segmentation clock. *Dev. Dyn.* **236:** 1403.

Kageyama R., Ohtsuka T., and Kobayashi T. 2007b. The *Hes* gene family: Repressors and oscillators that orchestrate embryogenesis. *Development* **134:** 1243.

Lahav G., Rosenfeld N., Sigal A., Geva-Zatorsky N., Levine A.J., Elowitz M.B., and Alon U. 2004. Dynamics of the p53-Mdm2 feedback loop in individual cells. *Nat. Genet.* **36:** 147.

Levy D.E. and Darnell J.E., Jr. 2002. Stats: Transcriptional control and biological impact. *Nat. Rev. Mol. Cell Biol.* **3:** 651.

Lewis J. 2003. Autoinhibition with transcriptional delay: A simple mechanism for the zebrafish somitogenesis oscillator. *Curr. Biol.* **13:** 1398.

Masamizu Y., Ohtsuka T., Takashima Y., Nagahara H., Takenaka Y., Yoshikawa K., Okamura H., and Kageyama R. 2006. Real-time imaging of the somite segmentation clock: Revelation of unstable oscillators in the individual presomitic mesoderm cells. *Proc. Natl. Acad. Sci.* **103:** 1313.

Massagué J. and Wotton D. 2000. Transcriptional control by the TGF-β/Smad signaling system. *EMBO J.* **19:** 1745.

Moloney D.J., Panin V.M., Johnston S.H., Chen J., Shao L., Wilson R., Wang Y., Stanley P., Irvine K.D., Haltiwanger R.S., and Vogt T.F. 2000. Fringe is a glycosyltransferase that modifies Notch. *Nature* **406:** 369.

Monk N.A. 2003. Oscillatory expression of Hes1, p53, and NF-κB driven by transcriptional time delays. *Curr. Biol.* **13:** 1409.

Nelson D.E., Ihekwaba A.E., Elliott M., Johnson J.R., Gibney C.A., Foreman B.E., Nelson G., See V., Horton C.A., Spiller D.G., Edwards S.W., McDowell H.P., Unitt J.F., Sullivan E., Grimley R., Benson N., Broomhead D., Kell D.B., and White M.R. 2004. Oscillations in NF-κB signaling control the dynamics of gene expression. *Science* **306:** 704.

Niwa Y., Masamizu Y., Liu T., Nakayama R., Deng C.-X., and Kageyama R. 2007. The initiation and propagation of Hes7 oscillation are cooperatively regulated by Fgf and Notch signaling in the somite segmentation clock. *Dev. Cell* **13:** 298.

Pourquié O. 2003. The segmentation clock: Converting embryonic time into spatial pattern. *Science* **301:** 328.

Sasai Y., Kageyama R., Tagawa Y., Shigemoto R., and Nakanishi S. 1992. Two mammalian helix-loop-helix factors structurally related to *Drosophila hairy* and *Enhancer of split*. *Genes Dev.* **6:** 2620.

Serth K., Schuster-Gossler K., Cordes R., and Gossler A. 2003. Transcriptional oscillation of Lunatic fringe is essential for somitogenesis. *Genes Dev.* **17:** 912.

Simon C., Follenius M., and Brandenberger G. 1987. Postprandial oscillations of plasma glucose, insulin and C-peptide in man. *Diabetologia* **30:** 769.

Starr R., Willson T.A., Viney E.M., Murray L.J., Rayner J.R., Jenkins B.J., Gonda T.J., Alexander W.S., Metcalf D., Nicola N.A., and Hilton D.J. 1997. A family of cytokine-inducible inhibitors of signalling. *Nature* **387:** 917.

Takebayashi K., Sasai Y., Sakai Y., Watanabe T., Nakanishi S., and Kageyama R. 1994. Structure, chromosomal locus, and promoter analysis of the gene encoding the mouse helix-loop-helix factor HES-1: Negative autoregulation through the multiple N box elements. *J. Biol. Chem.* **269:** 5150.

Tannenbaum G.S. and Martin J.B. 1976. Evidence for an endogenous ultradian rhythm governing growth hormone secretion in the rat. *Endocrinology* **98:** 562.

ten Dijke P., Miyazono K., and Heldin C.-H. 2000. Signaling inputs converge on nuclear effectors in TGF-β signaling. *Trends Biosci.* **25:** 64.

Yoshiura S., Ohtsuka T., Takenaka Y., Nagahara H., Yoshikawa K., and Kageyama R. 2007. Ultradian oscillations of Stat, Smad and Hes1 expression in response to serum. *Proc. Natl. Acad. Sci.* **104:** 11292.

Yu H. and Jove R. 2004. The STATS of cancer: New molecular targets come of age. *Nat. Rev. Cancer* **4:** 97.

Zhang N. and Gridley T. 1998. Defects in somite formation in lunatic fringe-deficient mice. *Nature* **394:** 374.

The Role of Circadian Regulation in Cancer

S. Gery and H.P. Koeffler
Cedars-Sinai Medical Center, Division of Hematology/Oncology, University of California School of Medicine, Los Angeles, California 90048

Proper circadian regulation is essential for the well being of the organism, and disruption of circadian rhythms is associated with pathological conditions including cancer. In mammals, the core clock genes, *Per1* and *Per2*, are key regulators of circadian rhythms both in the central clock in the hypothalamous and in peripheral tissues. Recent findings revealed molecular links between *Per* genes and cellular components that control fundamental cellular processes such as cell division and DNA damage. New data also shed light on mechanisms by which circadian oscillators operate in peripheral organs to influence tissue-dependent metabolic and hormonal pathways. Circadian cycles are linked to basic cellular functions, as well as to tissue-specific processes through the control of gene expression and protein interactions. By controlling global networks such as chromatin remolding and protein families, which themselves regulate a broad range of cellular functions, circadian regulation impinges upon almost all major physiological pathways. These molecular insights illustrate how disregulation of circadian rhythms might influence the susceptibility to cancer development and provide further support for the emerging role of circadian genes in tumor suppression.

INTRODUCTION

Most living organisms exhibit behavioral and physiological circadian rhythms, allowing them to adapt to the daily cycle of light and dark. In mammals, key physiological processes, including sleep-wake cycles, glucose and lipid metabolism, hormone secretion, blood pressure, DNA-damage response, immunity, and cell cycle, are influenced by circadian rhythms. Furthermore, perturbations of these rhythms both in humans (as occurs in night shift workers) and animal experimental models have been associated with diverse pathogenic conditions, such as sleep disorders, depression, diabetes, obesity, and malignant transformation.

Circadian rhythms are driven by a master clock located in the hypothalamic suprachiasmatic nucleus (SCN) that synchronizes numerous subsidiary oscillators in peripheral tissues. The circadian clockwork in both the SCN and the peripheral cells is regulated by finely tuned transcription-translation feedback loops and posttranslational modifications that are maintained by a core set of clock genes (Shearman et al. 2000; Reppert and Weaver 2002; Schibler and Sassone-Corsi 2002; Ko and Takahashi 2006). The positive feedback loop involves two transcription factors, Clock and Bmal1 which dimerize and bind to E-boxes in the promoters of a large number of target genes. These include their own negative regulators Period (Per1, 2, and 3) and cryptochrome (Cry1 and Cry2). During the course of the day, Per and Cry proteins accumulate and multimerize in the cytoplasm. They then translocate to the nucleus in a phosphorylation-dependent manner where they repress the Clock-Bmal1 complex, thus forming the major negative circadian feedback loop. For a new cycle to begin, the inhibitory proteins must be removed. Per and Cry are phosphorylated and degraded, releasing the repression of the Clock-Bmal1 transcriptional activity. An additional feedback loop involves the orphan nuclear receptors Rev-erbα and Rora that, respectively, activate and repress Bmal1 expression. This adds stability and robustness to the clock, reinforcing the oscillations.

The same molecular mechanisms that regulate core clock components in the SCN, i.e., transcription-translation feedback loops and posttranslational modifications, are used by the clock to regulate various genes in peripheral tissues, such as throughout the body, fundamental cellular pathways like the cell cycle and metabolic cycles are under circadian regulation (Lowrey and Takahashi 2004). Expression profiling studies demonstrated that in any given tissue, up to 10% of transcripts are clock-controlled genes exhibiting circadian oscillation (Panda et al. 2002; Storch et al. 2002). The Clock-Bmal1 complex directly controls the expression of some of these genes, whereas others are indirectly regulated via circadian expression of relevant transcription factors. Significantly, some of these genes are key players in cell division, proliferation, and survival. With the realization of the importance of circadian regulation to the overall health of the organism, it becomes apparent why disruption of circadian rhythms may increase the risk for cancer. Recent findings are now beginning to shed light on the molecular mechanisms underlying the circadian-cancer connection.

THE CELL CYCLE

At a first glance, circadian rhythms, regulated by the earth's movement around the sun, and the cell division occurring within the confined boundaries of cells may not seem like obvious partners. However, a closer look reveals that these two systems have some intriguing similarities. Both the core circadian network and the cell cycle machinery are found in most cells, even in established cell lines. Both are intracellular systems that rely on phases of transcription-translation and protein modification-degradation. Another interesting finding is that like the 24-hour periodicity of biological clocks, mammalian cell division is approximately 1 day, and even in culture, most eukaryotic cells undergo division within 24 hours or so. Circadian gene expression in fibroblasts continues during cell division, and the circadian oscillator gates cytokinesis to defined time windows; in turn, mitosis can phase-shift the circadian cycle (Nagoshi et al. 2004; Reddy et al. 2005; Hunt and Sassone-Corsi 2007). Another instructive observation is that among the hundreds of proteins exhibiting circadian expression patterns, several are key regulators of the cell cycle and apoptosis. For example, c-Myc, a well-known cell cycle transcription factor, and Wee-1, a cell cycle checkpoint kinase, contain E-boxes in their promoters and are under direct transcriptional regulation by the Clock-Bmal1 complex (Fu et al. 2002; Matsuo et al. 2003). The levels of the apoptotic proteins BCL2 and BAX rhythmically oscillate in murine bone marrow (Granda et al. 2005).

Studies in mice with mutations in core clock genes clearly demonstrate the critical role for proper circadian regulation in cell cycle progression in peripheral tissues. After partial hepatectomy, livers from cry-deficient mice regenerate slower than those from normal mice. In this model, the circadian clock was shown to control the G_2/M transition by regulating the expression of wee1, which negatively regulates the Cdc2–cyclin B1 complex, the major cyclin complex governing G_2/M (Matsuo et al. 2003). *Per2* mutant mice have increased susceptibility to both spontaneous and radiation-induced tumor development. These mutant mice show aberrant temporal expression of cell cycle genes such as *p53*, c-*myc cyclin A*, and *mdm2*, leading to deregulated cell division and reduced levels of apoptosis (Fu et al. 2002). In clock mutant mice, several cell cycle inhibitory genes are up-regulated, e.g., *p21*$^{(WAF1)}$, *p27*$^{(Kip1)}$, *chk1*, *chk2*, and *atr1*, whereas proproliferative genes such as *jak2*, *ERα*, *akt1*, *cdk2*, cyclins *D3* and *E1*, and the transforming growth factor-β (TGF-β) and epiderminal growth factor (EGF) receptors are down-regulated. Fibroblasts derived from clock mutant embryos exhibit reduced DNA synthesis and cell proliferation compared with normal fibroblasts (Miller et al. 2007). We and other investigators have shown that ectopic expression of *Per1* and *Per2* in human cancer cell lines results in deregulated expression of cell cycle genes such as c-*myc*, *p21*, *wee1*, *cyclin B1*, and *cdc2*, associated with cell cycle arrest and apoptosis (Gery et al. 2005, 2006, 2007a,b; ; Hua et al. 2006). These data establish a strong link between the circadian clock and the cell cycle. Furthermore, they may explain how disruption of the clock may shift the cellular balance between proproliferative versus antiproliferative genes, increasing the risk for caner initiation and progression.

DNA DAMAGE

For early life forms, strong UV irradiation during the day must have been a constant source of DNA damage; hence, restricting the S phase of the cell cycle to nighttime may have had an adaptive value (Woelfle et al. 2004; Reddy et al. 2005). In current multicellular organisms as humans and other mammals, DNA damage leading to genomic instability is a major force driving cancer. To cope with this, cells have developed a network of checkpoint pathways allowing them to initiate either cell cycle arrest and DNA repair or apoptosis. Recent findings show that the circadian system impinges on some of the key players within these networks. Moreover, the relationship between the circadian clock and the DNA-damage response appears to persevere across species, as it is found in diverse organisms from plants to mammals. An early interesting clue on how clock genes may influence the cellular DNA-damage response came from the observation that the expression of *Per1*, *Per2*, *Clock*, *Cry1*, and *Bmal1* is induced in the liver of mice that undergo exposure to γ-radiation. This induction is absent in *Per2* mutant mice, suggesting a role of *Per2* in cellular pathways that govern the response to stress. The *Per2*-deficient mice are more sensitive to γ-radiation, showing premature hair graying, hair loss, and high frequency of development of lymphoma, attributed, at least partially, to decreased apoptosis in damaged cells. These findings raised the possibility that by regulating the DNA-damage response, *Per2* functions in tumor suppression (Fu et al. 2002; Lee 2006). Our studies focused on *Per1* and have shown that *Per1* has an important role in coupling the circadian cycle to the DNA-damage response and to growth control. In human cancer cells, γ-radiation induces *Per1* expression and nuclear localization. Overexpression of *Per1* sensitizes these cells to DNA-damage-induced apoptosis, whereas its inhibition has the opposite effect. This phenotype is associated with altered expression of c-Myc and p21 (Gery et al. 2006). Deregulated expression of c-*myc* was suggested to be a key factor leading to tumor development in *Per2* mutant mice. In addition, malignant growth in mice with disruption of circadian coordination has been suggested to result from abnormal c-*myc* expression (Filipski et al. 2006).

An even more intimate relationship between the circadian clock and the DNA-damage response was recently revealed by several groups who demonstrated that core clock proteins directly bind to major checkpoint proteins. The Atm kinase and its downstream effector Chk2 are activated by DNA double-strand breaks and phosphorylate target proteins that initiate either DNA repair and cell cycle arrest or apoptosis (Shiloh 2003). We have shown that in human colon cancer cells, endogenous Per1 interacts with endogenous Atm and Chk2. Furthermore, ectopic expression of Per1 is sufficient to induce phosphorylation of Chk2 in the absence of DNA damage, whereas silencing of Per1 expression impairs Chk2 phosphorylation, suggesting that Per1 has a role in the DNA-damage response by interacting with and activating the Atm checkpoint pathway (Gery et al. 2006). Parallel to our findings in mammalian cells, a recent study in the fungus *Neurospora* demonstrated that the kinase prd-4, an ortholog of mammalian Chk2, interacts with a central clock factor in *Neurospora*, FRQ, and this interaction induces FRQ phosphorylation. prd-4 is itself a clock-controlled gene, suggesting a regulatory loop coupling the circadian clock and the response to DNA damage. Even more so, it was demonstrated that the DNA-damaging agent methylmethane sulfonate (MMS) can act as strong clock-resetting signal in a manner that depends on the time of day and that *prd-4* is essential to resetting the clock (Pregueiro et al. 2006).

Anther clock gene that has been shown to interact directly with checkpoint proteins is *Timeless*. This protein is essential for circadian rhythms in *Drosophila* and is required for the maintenance of robust circadian rhythms in mice. Human *Timeless* sustains the DNA replication fork movement in the absence of damage, whereas its partner *Tipin* slows DNA replication in UV-damaged cells. *Timeless* was found to be in a complex that includes the Atm-related kinase Atr and Atrip, a substrate of ATR. In addition, down-regulation of *Timeless* in human cells compromises the S-phase checkpoint, demonstrating that a clock gene coordinates cell cycle progression thought the S-phase checkpoint in response to DNA damage (Unsal-Kacmaz et al. 2005, 2007; Yoshizawa-Sugata and Masai 2007). hClk2 is the human homolog of the *Caenorhabditis elegans* biological clock protein, CLK-2. A recent study demonstrated that hClk2 associates with a number of checkpoint proteins including Atr and Chk1. hClk2 promotes activation of the S-phase checkpoint and downstream DNA repair by preventing unscheduled Chk1 degradation by the proteasome (Collis et al. 2007). Together, these findings provide a molecular mechanism demonstrating how the circadian clock functions in the peripheral tissues to control the cell cycle, apoptosis, and malignant growth. An emerging theme from these studies is that both transcriptional regulation and direct protein-protein interactions couple the clock and the DNA-damage response in an evolutionary conserved manner. This suggests that these ancient systems may have evolved parallel to each other, helping cells make critical cell cycle and apoptotic decisions and, in doing so, contributing to the long-term survival of the organism.

CHROMATIN

Transcriptional activation and repression usually involves chromatin modifications such as acetylation and methylation. Chromatin changes are therefore an important normal cellular function, yet alterations in chromatin patterns occur under pathogenic conditions, including cancer (Jones and Baylin 2007). Human tumors show a massive overall loss of DNA methylation, whereas certain promoters acquire specific hypermethylation. Patterns of histone acetylation also display changes during cancer development. One consequence of these alterations is the silencing of tumor suppressor genes, owing to aberrant methylation of CpG islands and deacetylation of histones in their promoter regions. Epigenetic interventions, particularly those targeting histone deacetylase (HDAC), are among the most promising therapies for cancer, and HDAC inhibitors are already being used in the clinic. Moreover, because epigenetic changes occur early in tumorgenesis and are associated with distinctive cancer types, they could represent targets for chemoprevention and early diagnosis.

Finely tuned transcriptional regulation is at the heart of the circadian clock, and with approximately 10% of mammalian transcripts showing daily oscillations, it is clear that genome-wide architecture is required for proper maintenance of the clock transcriptional machinery. It was therefore hypothesized that chromatin remodeling may be a key element in the clock transcriptional regulation. Several recent exciting discoveries demonstrate that this indeed is the case. The finding that changes in histone modifications are an important step in light-induced gene expression in neurons of the hypothalamic SCN provided early evidence that dynamic chromatin remodeling occurs within the clock system (Crosio et al. 2000). Later studies showed that the cyclic binding and release of the CLOCK-BMAL1 activator complex to E boxes correlates with rhythmic changes in acetylation and methylation of surrounding DNA of target genes. Specifically, histone H3 is acetylated in chromatin that encompasses the promoters of *Per1*, *Per2*, *Cry1*, and the clock output gene, *Dbp*, when these genes are actively transcribed (Etchegaray et al. 2003; Curtis et al. 2004; Naruse et al. 2004; Ripperger and Schibler 2006). As histone acetylation promotes transcription, enzymes that acetylate histones, histone acetyltransferases (HATs), act as transcriptional coactivators by interactions with sequence-specific DNA-binding transcription factors. HAT activity, is in fact, required for Clock-Bmal1-dependent *trans*-activation, and as it turns out, Clock itself displays HAT activity with specificity to histones H3 and H4 (Doi et al. 2006).

Abnormal expression of circadian genes has been reported in human tumors including breast, pancreas, and endometrial cancers, as well as in hematological malignancies (Chen et al. 2005; Gery et al. 2005, 2006, 2007b; Yeh et al. 2005; Pogue-Geile et al. 2006; M.Y. Yang et al. 2006). In a number of studies, clock gene deregulation was associated with promoter hypermethylation. In a study of 55 Taiwanese women, disturbances in the expression of Per1 and Per2 was found in most (>95%) breast cancer samples, in comparison with nearby non-cancerous cells. The *Per* gene deregulation was not caused by genetic mutations but was probably due to methylation of the Per1 or Per2 promoters (Chen et al. 2005). Expression levels of *Per1* were significantly decreased in endometrial carcinoma, partly due to DNA methylation of Per1 promoter (Yeh et al. 2005). The expression levels of Per1, Per2, Per3, Cry1, Cry2, and Bmal1 were significantly impaired in both chronic phase and blast crisis of chronic myeloid leukemia (CML) samples compared with those in normal samples. Although no mutations were found within the coding region of Per3, the CpG islands in its promoter were methylated in all the CML samples. Likewise, the CpG islands of Per2 were also methylated in 40% of cases (M.Y. Yang et al. 2006). Using microarray analysis, we identified Per1 as a candidate tumor suppressor, epigenetically silenced in non-small-cell lung cancer (NSCLC). Per1 levels were low in a large panel of NSCLC patient samples and in NSCLC cell lines compared to normal lung tissue. The down-regulation of Per1 expression was associated with hypermethylation of the Per1 promoter. Furthermore, we found that aberrant acetylation of Per1 promoter is also a potential mechanism for silencing Per1 in cancer. In cancer cell lines from various tissues including lung, colon, and breast, Per1 expression could be induced by the histone deacetylase inhibitor, SAHA. In addition, SAHA increased acetylated histone H3 binding to the Per1 promoter in NSCLC cells (Gery et al. 2007b). The HAT function of Clock, essential for normal transcriptional regulation of clock genes like Per1, may be compromised by disruption of circadian rhythms. This in turn could lead to aberrant expression of clock genes, leaving cells more prone to the development of cancer.

Mutations in clock genes have been described in humans disorders (such as familial advanced sleep phase syndrome); thus far, they have not been associated with cancer. However, clock genes are down-regulated in many human tumors, and this has been attributed at least partly to aberrant methylation and acetylation of their promoters. Rhythmic chromatin changes in the promoters of clock genes are a normal part of the circadian transcriptional control. Variation in this process could disrupt not only the expression of core clock factors, but also the web of genes and cellular pathways that are under circadian control. Chromatin remodeling by the circadian system may have a role in the global epigenetic events occurring during carcinogenesis.

HORMONES

The secretion of many endocrine and metabolic hormones is controlled by the day/night cycle, and reciprocally circulating levels of hormones influence circadian rhythms. The pineal hormone melatonin, the main hormone relaying the circadian rhythm to the peripheral organs, is another clock component tying the circadian system to cancer. Melatonin appears to suppress tumor growth in animals and has been implicated in the pathogenesis of various human cancers, in particular, breast, colon, lung, melanoma, and leukemia. This protein is secreted by the pineal gland in the brain only in darkness.

Exposure to light at night, even briefly, decreases its production, and it has been suggested that lower levels of melatonin in the serum may contribute to tumor development (Ravindra et al. 2006). Interestingly, epidemiological studies have found a positive association between breast cancer incidence and overnight shift work (Hansen 2001). An increased worldwide incidence of breast cancer is hypothesized to be related to exposure to artificial light at night and suppression of melatonin secretion (Stevens 2005). Similarly, melatonin was reported to be lower in men with prostate cancer compared with healthy men (Bartsch et al. 1985). Although the molecular mechanism underlying these clock-cancer connections remain to be determined, data have suggested that reduced melatonin levels may increase hormone-related cancer risk by affecting the production or function of the gonadal steroid hormones estrogen and androgen, as well as their receptors.

The estrous cycle is an endocrine clock orchestrated by circulating levels of the ovarian steroids, estrogen and progesterone. As with other homeostatic processes, circadian regulation has a prominent role in this endocrine cycle. In the mammary gland, one of the target tissues of ovarian hormones, estrogen is essential for normal development and physiological functions, but it is also a potent mitogen (Yager and Davidson 2006). Thus, circadian regulation may be implicated not only in estrogen homeostasis, but also in hormone-related tumorgenesis in breast epithelia cells. The biological actions of estrogens are meditated mainly by the nuclear receptors, estrogen receptor α and β (ERα and ERβ), which bind to estrogen response elements in the regulatory regions of target genes and associate with basal transcription factors and tissue-specific cofactors to alter gene expression. Extensive studies have identified numerous proteins and processes involved in the ER signal transduction pathway (McDonnell and Norris 2002). Given that transcriptional regulation is a key feature of both the circadian machinery and the ER pathway and the proposed connections between the clock and breast cancer, we hypothesized that the ER signaling is yet another front on which core clock factors may exert their effects. Recently, we found that Per2 is a novel estrogen-inducible ERα cofactor. Per2 interacts with ERα, enhances receptor degradation, and suppresses estrogen-mediated transcription of ER target genes in breast cancer cells. In turn, Per2 itself is estrogen-inducible in these cells. Our results suggest a feedback mechanism to attenuate estrogen response that couples the circadian clock to the estrogen pathway. Overexpression of Per2 in breast cancer cell lines leads to significant growth inhibition, proliferation arrest, and apoptosis associated with deregulated expression of c-Myc, Cyclin D1, and p21, indicating that Per2 may function as a tumor suppressor in breast tissue (Gery et al. 2007a). Interestingly, Per2 was also suggested to have a role in normal mammary cell differentiation (Metz et al. 2006). Studies in rats have shown that estrogen differentially regulates the expression of *per1* and *per2* in specific brain regions as well as in some peripheral tissues (i.e., liver, kidney, and uterus) (Nakamura et al. 2005; Perrin et al. 2006). A circadian clock was reported to function in rat ovarian cells, further implicating circadian regulation in local endocrine tissues, as well as its role in the hypothalamo-pituitary-ovarian axis (Fahrenkrug et al. 2006). These results illustrate how circadian oscillators, although ubiquitously expressed, regulate gene expression and function in a cell-type-dependent manner, ensuring that within each tissue, relevant pathways are appropriately controlled (e.g., gonadal steroid hormones in reproductive tissues and glucose, and lipid metabolism in metabolic tissues).

The ER-Per2 interaction is not the first case where core clock factors have been shown to interact directly with members of the nuclear receptor family in peripheral tissues. Clock was found to interact in a hormone-dependent manner with some nuclear receptors, including RARα and RXRα. These interactions negatively regulate circadian-mediated transcriptional activation in vascular cells (McNamara et al. 2001). Another example of cross-talk between the circadian transcriptional system and nuclear receptors was recently reported in metabolically active tissues. A large number of nuclear receptors (28 of the 49 mouse nuclear receptors) were found to display tissue-specific oscillation in liver, skeletal muscle, and adipose tissues, providing an explanation for the well-recognized role of the circadian clock in energy balance and general metabolism. The nuclear receptors could provide a global network of proteins under circadian regulation through which environmental signals can be integrated to regulate endocrine and metabolic physiology (X. Yang et al. 2006). This network of proteins could also provide a route by which circadian control influences susceptibility to cancer development.

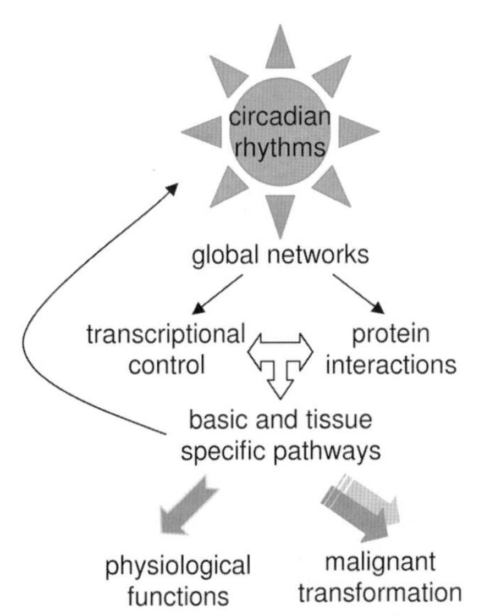

Figure 1. The circadian clock controls normal physiology and is associated with malignant transformation. Circadian clocks are found in most cells where they influence pathways critical to cell growth and survival, as well as tissue-specific functions through the control of gene expression and protein interactions. Genome-wide mechanisms such as remodeling of chromatin and links between clock components and superprotein families (like checkpoint proteins and nuclear receptors) may act in concert with systemic circadian regulation to communicate signals to numerous physiological processes. Disruption of this extensive system may contribute to tumorigenesis and cancer progression in more than one way.

CONCLUSION

The circadian system integrates environmental and endogenous signals; subsequently, by means of transcriptional regulation and protein interactions, circadian oscillators relay this information to fundamental as well as tissue-specific cellular pathways to synchronize the physiology and behavior of the organism (Fig. 1). A convergence of data demonstrate the pervasive role circadian regulation has on almost every aspect of normal cellular physiology, as well as many aspects of malignant transformation. As we learn more about the molecular workings of this complex system, the implications for disruption of circadian rhythms, often associated with modern lifestyle, become apparent. The extent to which circadian regulation contributes to carcinogenesis and the ways by which core clock factors modify gene expression and protein function require further exploration. Even so, far-reaching therapeutic strategies include the use of clock factors as new targets for pharmaceutical compounds and for chronotherapy as a way to optimize the efficacy of current therapies.

REFERENCES

Bartsch C., Bartsch H., Fluchter S.H., Attanasio A., and Gupta D. 1985. Evidence for modulation of melatonin secretion in men with benign and malignant tumors of the prostate: Relationship with the pituitary hormones. *J. Pineal Res.* **2:** 121.

Chen S.T., Choo K.B., Hou M.F., Yeh K.T., Kuo S.J., and Chang J.G. 2005. Deregulated expression of the PER1, PER2 and PER3 genes in breast cancers. *Carcinogenesis* **26:** 1241.

Collis S.J., Barber L.J., Clark A.J., Martin J.S., Ward J.D., and Boulton S.J. 2007. HCLK2 is essential for the mammalian S-phase checkpoint and impacts on Chk1 stability. *Nat. Cell Biol.* **9:** 391.

Crosio C., Cermakian N., Allis C.D., and Sassone-Corsi P. 2000. Light induces chromatin modification in cells of the mammalian circadian clock. *Nat. Neurosci.* **3:** 1241.

Curtis A.M., Seo S.B., Westgate E.J., Rudic R.D., Smyth E.M., Chakravarti D., FitzGerald G.A., and McNamara P. 2004. Histone acetyltransferase-dependent chromatin remodeling and the vascular clock. *J. Biol. Chem.* **279:** 7091.

Doi M., Hirayama J., and Sassone-Corsi P. 2006. Circadian regulator CLOCK is a histone acetyltransferase. *Cell* **125:** 497.

Etchegaray J.P., Lee C., Wade P.A., and Reppert S.M. 2003. Rhythmic histone acetylation underlies transcription in the mammalian circadian clock. *Nature* **421:** 177.

Fahrenkrug J., Georg B., Hannibal J., Hindersson P., and Gras S. 2006. Diurnal rhythmicity of the clock genes Per1 and Per2 in the rat ovary. *Endocrinology* **147:** 3769.

Filipski E., Li X.M., and Levi F. 2006. Disruption of circadian coordination and malignant growth. *Cancer Causes Control* **17:** 509.

Fu L., Pelicano H., Liu J., Huang P., and Lee C. 2002. The circadian gene Period2 plays an important role in tumor suppression and DNA damage response in vivo. *Cell* **111:** 41.

Gery S., Virk R., Chumakov K., Yu A., and Koeffler H.P. 2007a. The clock gene Per2 links the circadian system to the estrogen receptor. *Oncogene*. 2007. Jun 18; [Epub ahead of print].

Gery S., Gombart A.F., Yi W.S., Koeffler C., Hofmann W.K., and Koeffler H.P. 2005. Transcription profiling of C/EBP targets identifies Per2 as a gene implicated in myeloid leukemia. *Blood* **106:** 2827.

Gery S., Komatsu N., Baldjyan L., Yu A., Koo D., and Koeffler H.P. 2006. The circadian gene per1 plays an important role in cell growth and DNA damage control in human cancer cells. *Mol. Cell* **22:** 375.

Gery S., Komatsu N., Kawamata N., Miller C.W., Desmond J., Virk R.K., Marchevsky A., McKenna R., Taguchi H., and Koeffler H.P. 2007b. Epigenetic silencing of the candidate tumor suppressor gene Per1 in non-small cell lung cancer. *Clin. Cancer Res.* **13:** 1399.

Granda T.G., Liu X.H., Smaaland R., Cermakian N., Filipski E., Sassone-Corsi P., and Levi F. 2005. Circadian regulation of cell cycle and apoptosis proteins in mouse bone marrow and tumor. *FASEB J.* **19:** 304.

Hansen J. 2001. Light at night, shiftwork, and breast cancer risk. *J. Natl. Cancer Inst.* **93:** 1513.

Hua H., Wang Y., Wan C., Liu Y., Zhu B., Yang C., Wang X., Wang Z., Cornelissen-Guillaume G., and Halberg F. 2006. Circadian gene mPer2 overexpression induces cancer cell apoptosis. *Cancer Sci.* **97:** 589.

Hunt T. and Sassone-Corsi P. 2007. Riding tandem: Circadian clocks and the cell cycle. *Cell* **129:** 461.

Jones P.A. and Baylin S.B. 2007. The epigenomics of cancer. *Cell* **128:** 683.

Ko C.H. and Takahashi J.S. 2006. Molecular components of the mammalian circadian clock. *Hum. Mol. Genet.* **15:** R271.

Lee C.C. 2006. Tumor suppression by the mammalian Period genes. *Cancer Causes Control* **17:** 525.

Lowrey P.L. and Takahashi J.S. 2004. Mammalian circadian biology: Elucidating genome-wide levels of temporal organization. *Annu. Rev. Genomics Hum. Genet.* **5:** 407.

Matsuo T., Yamaguchi S., Mitsui S., Emi A., Shimoda F., and Okamura H. 2003. Control mechanism of the circadian clock for timing of cell division in vivo. *Science* **302:** 255.

McDonnell D.P. and Norris J.D. 2002. Connections and regulation of the human estrogen receptor. *Science* **296:** 1642.

McNamara P., Seo S.P., Rudic R.D., Sehgal A., Chakravarti D., and FitzGerald G.A. 2001. Regulation of CLOCK and MOP4 by nuclear hormone receptors in the vasculature: A humoral mechanism to reset a peripheral clock. *Cell* **105:** 877.

Metz R.P., Qu X., Laffin B., Earnest D., and Porter W.W. 2006.

Circadian clock and cell cycle gene expression in mouse mammary epithelial cells and in the developing mouse mammary gland. *Dev. Dyn.* **235:** 263.

Miller B.H., McDearmon E.L., Panda S., Hayes K.R., Zhang J., Andrews J.L., Antoch M.P., Walker J.R., Esser K.A., Hogenesch J.B., and Takahashi J.S. 2007. Circadian and CLOCK-controlled regulation of the mouse transcriptome and cell proliferation. *Proc. Natl. Acad. Sci.* **104:** 3342.

Nagoshi E., Saini C., Bauer C., Laroche T., Naef F., and Schibler U. 2004. Circadian gene expression in individual fibroblasts: Cell-autonomous and self-sustained oscillators pass time to daughter cells. *Cell* **119:** 693.

Nakamura T.J., Moriya T., Inoue S., Shimazoe T., Watanabe S., Ebihara S., and Shinohara K. 2005. Estrogen differentially regulates expression of Per1 and Per2 genes between central and peripheral clocks and between reproductive and nonreproductive tissues in female rats. *J. Neurosci. Res.* **82:** 622.

Naruse Y., Oh-hashi K., Iijima N., Naruse M., Yoshioka H., and Tanaka M. 2004. Circadian and light-induced transcription of clock gene Per1 depends on histone acetylation and deacetylation. *Mol. Cell. Biol.* **24:** 6278.

Panda S., Antoch M.P., Miller B.H., Su A.I., Schook A.B., Straume M., Schultz P.G., Kay S.A., Takahashi J.S., and Hogenesch J.B. 2002. Coordinated transcription of key pathways in the mouse by the circadian clock. *Cell* **109:** 307.

Perrin J.S., Segall L.A., Harbour V.L., Woodside B., and Amir S. 2006. The expression of the clock protein PER2 in the limbic forebrain is modulated by the estrous cycle. *Proc. Natl. Acad. Sci.* **103:** 5591.

Pogue-Geile K.L., Lyons-Weiler J., and Whitcomb D.C. 2006. Molecular overlap of fly circadian rhythms and human pancreatic cancer. *Cancer Lett.* **243:** 55.

Pregueiro A.M., Liu Q., Baker C.L., Dunlap J.C., and Loros J.J. 2006. The *Neurospora* checkpoint kinase 2: A regulatory link between the circadian and cell cycles. *Science* **4:** 313.

Ravindra T., Lakshmi N.K., and Ahuja Y.R. 2006. Melatonin in pathogenesis and therapy of cancer. *Indian J. Med. Sci.* **60:** 523.

Reddy A.B., Wong G.K., O'Neill J., Maywood E.S., and Hastings M.H. 2005. Circadian clocks: Neural and peripheral pacemakers that impact upon the cell division cycle. *Mutat. Res.* **574:** 76.

Reppert S.M. and Weaver D.R. 2002. Coordination of circadian timing in mammals. *Nature* **418:** 935.

Ripperger J.A. and Schibler U. 2006. Rhythmic CLOCK-BMAL1 binding to multiple E-box motifs drives circadian Dbp transcription and chromatin transitions. *Nat. Genet.* **38:** 369.

Schibler U. and Sassone-Corsi P. 2002. A web of circadian pacemakers. *Cell* **111:** 919.

Shearman L.P., Sriram S., Weaver D.R., Maywood E.S., Chaves I., Zheng B., Kume K., Lee C.C., van der Horst G.T., Hastings M.H., and Reppert S.M. 2000. Interacting molecular loops in the mammalian circadian clock. *Science* **288:** 1013.

Shiloh Y. 2003. ATM and related protein kinases: Safeguarding genome integrity. *Nat. Rev. Cancer* **3:** 155.

Stevens R.G. 2005. Circadian disruption and breast cancer: From melatonin to clock genes. *Epidemiology* **16:** 254.

Storch K.F., Lipan O., Leykin I., Viswanathan N., Davis F.C., Wong W.H., and Weitz C.J. 2002. Extensive and divergent circadian gene expression in liver and heart. *Nature* **417:** 78.

Unsal-Kacmaz K., Mullen T.E., Kaufmann W.K., and Sancar A. 2005. Coupling of human circadian and cell cycles by the timeless protein. *Mol. Cell. Biol.* **25:** 3109.

Unsal-Kacmaz K., Chastain P.D., Qu P.P., Minoo P., Cordeiro-Stone M., Sancar A., and Kaufmann W.K. 2007. The human Tim/Tipin complex coordinates an Intra-S checkpoint response to UV that slows replication fork displacement. *Mol. Cell. Biol.* **27:** 3131.

Woelfle M.A., Ouyang Y., Phanvijhitsiri K., and Johnson C.H. 2004. The adaptive value of circadian clocks: An experimental assessment in cyanobacteria. *Curr. Biol.* **14:** 1481.

Yager J.D. and Davidson N.E. 2006. Estrogen carcinogenesis in breast cancer. *N. Engl. J. Med.* **354:** 270.

Yang M.Y., Chang J.G., Lin P.M., Tang K.P., Chen Y.H., Lin H.Y., Liu T.C., Hsiao H.H., Liu Y.C., and Lin S.F. 2006. Downregulation of circadian clock genes in chronic myeloid leukemia: Alternative methylation pattern of hPER3. *Cancer Sci.* **97:** 1298.

Yang X., Downes M., Yu R.T., Bookout A.L., He W., Straume M., Mangelsdorf D.J., and Evans R.M. 2006. Nuclear receptor expression links the circadian clock to metabolism. *Cell* **126:** 801.

Yeh K.T., Yang M.Y., Liu T.C., Chen J.C., Chan W.L., Lin S.F., and Chang J.G. 2005. Abnormal expression of period 1 (PER1) in endometrial carcinoma. *J. Pathol.* **206:** 111.

Yoshizawa-Sugata N. and Masai H. 2007. Human Tim/Timeless-interacting protein, Tipin, is required for efficient progression of S phase and DNA replication checkpoint. *J. Biol. Chem.* **282:** 2729.

Cross-talks between Circadian Timing System and Cell Division Cycle Determine Cancer Biology and Therapeutics

F. LÉVI,*†‡ E. FILIPSKI,*† I. IURISCI,*† X.M. LI,*† AND P. INNOMINATO*†‡
*INSERM, U776 Rythmes biologiques et cancers, Hôpital Paul Brousse, Villejuif, F-94807, France; †Université Paris-Sud, UMR-S0776, Orsay, F-91405, France; ‡Assistance Publique-Hôpitaux de Paris, Chronotherapy Unit, Department of Oncology, Hôpital Paul Brousse, Villejuif, F-94807, France

The circadian clock orchestrates cellular functions over 24 hours, including cell divisions, a process that results from the cell cycle. The circadian clock and cell cycle interact at the level of genes, proteins, and biochemical signals. The disruption or the reinforcement of the host circadian timing system, respectively, accelerates or slows down cancer growth through modifications of host and tumor circadian clocks. Thus, cancer cells not only display mutations of cell cycle genes but also exhibit severe defects in clock gene expression levels or 24-hour patterns, which can in turn favor abnormal proliferation. Most of the experimental research actively ongoing in this field has been driven by the original demonstration that cancer patients with poor circadian rhythms had poor quality of life and poor survival outcome independently of known prognostic factors. Further basic research on the gender dependencies in circadian properties is now warranted, because a large clinical trial has revealed that gender can largely affect the survival outcome of cancer patients on chronotherapeutic delivery. Mathematical models further show that the therapeutic index of chemotherapeutic drugs can be optimized through distinct delivery profiles, depending on the initial host/tumor status and variability in circadian entrainment and/or cell cycle length. Clinical trials and systems-biology approaches in cancer chronotherapeutics raise novel issues to be addressed experimentally in the field of biological clocks. The challenge ahead is to therapeutically harness the circadian timing system to concurrently improve quality of life and down-regulate malignant growth.

INTRODUCTION

Dividing cells undergo a sequence of molecular and biochemical events that gate and monitor the traverse of the cell division cycle through four successive phases called Gap 1 (G_1), S (for DNA synthesis), G_2, and M (for mitosis). A complex gene and protein machinery regulates, gates, and times the transitions from one phase of this cycle to the next (Sanchez and Dynlacht 2005; Santamaría et al. 2007; Sclafani and Holzen 2007). Interconnected with the cell cycle are also the DNA repair/apoptosis/necrosis systems that limit genomic instability and prevent genetic mutations to accumulate and eventually result in malignant transformation (Shiloh 2003; Golstein and Kroemer 2006; Bartek and Lucas 2007). Indeed, the deregulation of the cell division cycle represents a main feature of malignant cells that can stem from cell cycle gene mutations as well as from hypoxia or altered available energy (Hanahan and Weinberg 2000; Keith and Simon 2007). Cell cycle phases also convey important therapeutic information regarding the cytotoxic potential of anticancer drugs (De Vita et al. 2007). For instance, S-phase cells are usually most susceptible to antimetabolites such as 5-fluorouracil, whereas antimitotic agents such as vinorelbine or taxanes exert greatest cytotoxicity on M-phase cells. Conversely, no cell cycle phase specificity seems to characterize the susceptibility for alkylating agents such as platinum complexes. Therefore, the timing of treatment delivery relative to cell cycle stages or events has guided the development of several chemotherapy protocols.

On another hand, 24-hour changes in cell divisions have long been known in healthy mammalian tissues from rodents or humans (Scheving 1959). The consequences of these rhythms for the determination of timing treatment delivery have also led to the development of cancer chronotherapeutics in order to minimize damage to healthy cells and to optimize malignant cell kill (Lévi 2001; Mormont and Lévi 2003).

CIRCADIAN GATING OF CELL DIVISION

Rhythmic DNA synthesis and mitotic activity have thus been demonstrated in most components of the hematopoietic and immune system, in all the segments of the gastro-intestinal tract, liver, skin, and cornea of rodents (Burns et al. 1976; Lakatua et al. 1983; Scheving et al. 1992). These rhythms have also been uncovered in human bone marrow, skin, and oral and rectal mucosae (Smaaland et al. 1991, 2002; Bjarnason et al. 2001; Bjarnason and Jordan 2002). The above 24-hour changes were first documented in laboratory rodents on light/dark (LD) synchronization and in humans on normal diurnal routine. Numerous experimental studies have shown that cell division rhythms were endogenous because they persisted in constant darkness (DD). Photoperiodic entrainment was further demonstrated through inverting the 24-hour pattern within 7–21 days of inversion of the LD 12:12 regimen. The robustness of the DNA synthesis and mitoses rhythms were further confirmed through their persistence despite ablation of adrenals, medulla, or pituitary (Scheving et al. 1992). Nevertheless, the 24-hour changes in cell divisions displayed altered rhythm characteristics in these animals. This observation suggested that these organs were not involved in the generation of the cell division rhythms, but rather in the circadian coordination and phase setting of dividing cells. Even the suprachiasmatic nuclei (SCN) do not seem to be indispensable for rhythmic divisions to occur in synchrony in mouse intestine or corneal epithelium (Scheving et al. 1983), as well as in mouse bone marrow (Filipski et al. 2004a). In a study involving 52 mice with histologically proven SCN ablation and 34 sham-operated controls kept under LD 12:12 synchronization, the rest-activity cycle was suppressed in all of the mice with SCN ablation whereas the plasma corticosterone rhythm persisted yet with a nonsinusoidal pattern, a damped amplitude, and a phase-advance by a few hours (Filipski et al. 2004a). The bone marrow proliferation rhythms appeared to be least affected by SCN ablation. Thus, the count in bone-marrow-nucleated cells as well as the proportions of cells in G_1, S, or G_2/M displayed highly statistically significant rhythms, with similar amplitudes and phases in mice with SCN ablation or sham operation (Fig. 1). In addition, bone marrow hematopoiesis was accelerated in the mice with ablated SCN, as shown by a nearly 25% increase in the 24-hour mean of bone-marrow-nucleated cell count and a doubling in the 24-hour mean of circulating neutrophil count (Filipski et al. 2004a). These findings indicated the ability of peripheral circadian oscillators, and in particular those that regulate cellular proliferation, to remain synchronized in the absence of the established central SCN pacemaker. In a separate study, mouse bone marrow cells were sampled and processed for liquid culture for 96 hours. Samples from the bone marrow liquid culture were obtained every 3–4 hours for up to 4 days and exposed to granulocyte macrophage–colony-stimulating factor (GM-CSF) on agarose culture, a method used to determine the number of bone marrow GM progenitors. The study revealed proliferative circadian rhythms in cultured bone marrow cells that persisted for 3–4 days (Bourin et al. 2002). Interestingly, the circadian maximum recurred daily near CT3, which is the time of peak ex vivo proliferative response of mouse bone marrow cells to GM-CSF exposure (Perpoint et al. 1995). The ability of peripheral oscillators in proliferating tissues to maintain circadian rhythms in the absence of a central pacemaker has been further demonstrated for gene expression in cultured dividing fibroblasts (Nagoshi et al. 2004) and in some tumors, yet with rapid dampening (Balsalobre et al. 1998; Delaunay and Laudet 2002). Usually, the circadian rhythms in nondividing cultured cells tend to fade away unless a stimulation (glucocorticoid or serum shock) or an environmental 24-hour cycle (temperature and light) is introduced after a few days in culture, suggesting the need for a regular resetting of free-running peripheral oscillators in order for their coordination to be maintained (Balsalobre et al. 1998; Balsalobre 2000; Brown et al. 2002). This indeed may also be the case for cultured proliferating tissues. However, the above results support the assumption that LD and/or other periodic signals can be conveyed to proliferative tissues, among other peripheral oscillators, via structures other than the SCN, and at least partly synchronize the molecular or cellular rhythms.

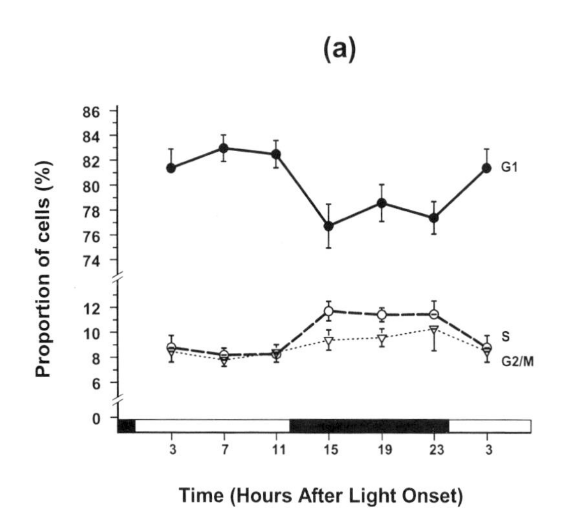

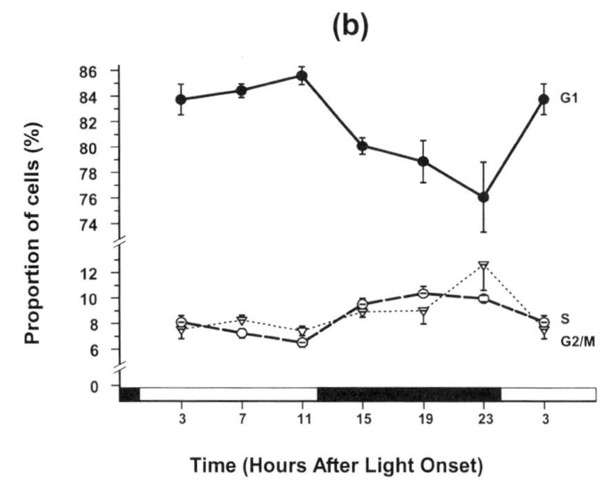

Figure 1. Circadian regulation of bone marrow proliferation rhythms in ♂ B6D2F1 mice kept in LD 12:12. Twenty-four hour changes in the proportions of whole bone marrow cells in G_1, S, or G_2-M phases of the cell cycle (mean ±S.E.M.), with light onset as time reference. (*a*) Circadian pattern in healthy mice; (*b*) persistent rhythms with similar patterns despite SCN ablation 4 weeks earlier. (Modified from Filipski et al. 2004a.)

DOWN-REGULATION OF TUMOR GROWTH BY THE CIRCADIAN TIMING SYSTEM

The very first results supporting the concept that the circadian timing system could down-regulate cancer progression stem from clinical investigations, where 24-hour rhythms are determined in patients and their patterns are correlated with clinical endpoints. These results have called for further confirmatory clinical investigations and experimental demonstrations of the relevance of the circadian timing system for controlling tumor progression.

CLINICAL STUDIES

The rest-activity rhythm is a well-established physiological output of the circadian timing system that can be easily and noninvasively recorded for several days with a wrist-worn monitor. A 3-day time series in rest-activity provide a window on the circadian timing system and can discriminate cancer patients with near-normal rhythms

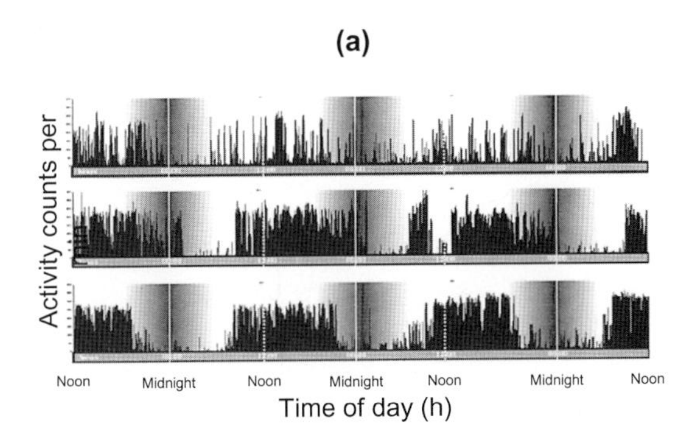

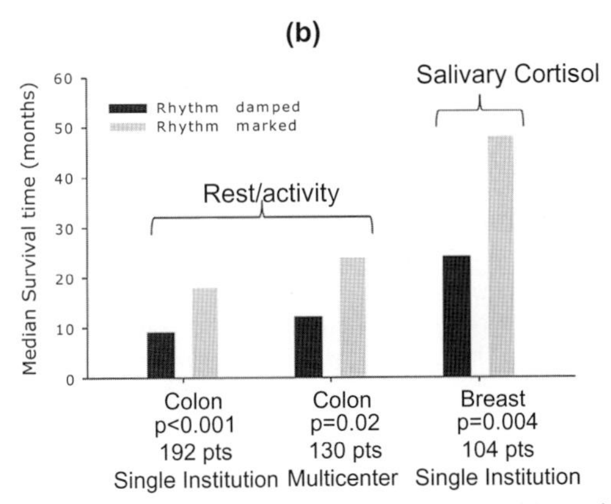

Figure 2. Relevance of circadian physiology for the survival outcome of cancer patients. Circadian physiology can be estimated with rest-activity monitoring for 3 days or more and/or by diurnal patterns in plasma or salivary cortisol concentration. Relations are shown between maintained or disrupted circadian physiology and survival outcome in patients with metastatic colorectal cancer or metastatic breast cancer. (*a*) Different rest-activity circadian patterns in three patients with metastatic colorectal cancer; (*b*) prediction of better survival with marked rhythm in rest-activity or salivary cortisol in cancer patients. (Adapted from Mormont et al. 2000; Sephton et al. 2000; P. Innominato et al., in prep.)

from those with clearly abnormal 24-hour patterns (Fig. 2a). Two rhythm parameters, autocorrelation coefficient r24 and dichotomy index I < O, respectively, estimate the regularity of the pattern over 24 hours and the relative amount of activity In-bed versus Out-of-bed (Mormont et al. 2000). In a single institution study, rest-activity rhythm as assessed with both parameters was found to be an independent predictor of survival for 192 patients with metastatic colorectal cancer, 62% of whom had received prior chemotherapy before circadian physiology assessment (Mormont et al. 2000). In a second multicenter international prospective study, rest-activity rhythm was confirmed as an independent predictor of survival in 130 patients with metastatic colorectal cancer who had not received any previous chemotherapy (P. Innominato et al., in prep.). In a third single institution study, salivary cortisol rhythm was estimated through repeated daily autosampling by 104 patients with metastatic breast cancer (Sephton et al. 2000). Large interpatient differences in daily patterns of serum or salivary cortisol were confirmed, with nearly flat patterns in some patients and clearly rhythmic patterns in others (Touitou et al. 1995). The salivary cortisol rhythm also independently predicted for survival in these patients with metastatic breast cancer (Sephton et al. 2000). Taken together, these three studies reveal that the risk of an earlier death from metastatic colorectal or breast cancer is significantly lower and the median survival is nearly twice as high in the patients with near-normal circadian physiology outputs as compared to those with damped or abnormal rhythms (Fig. 2b).

Experimental Demonstrations

In a first series of studies, mice kept in LD 12:12 were submitted to SCN ablation or sham operation and then received bilateral subcutaneous implants of a fragment of Glasgow osteosarcoma (GOS) or pancreatic adenocarcinoma (P03) and were monitored for tumor growth and survival. Following inoculation of healthy recipient mice, the doubling time is rapid for GOS (2–3 days) and slower for P03 (4–5 days). In the current study, the tumor grew significantly faster in the mice with bilateral SCN destruction (a posteriori documented) as compared to those that were sham-operated. These differences further translated into highly statistically significant differences in survival (Fig. 3a) (Filipski et al. 2002). In a second series of experiments, mice were submitted to chronic jet lag (CJL). The CJL regimen consisted in 8-hour advances of light onset of LD 12:12 every 2 days. It was selected as being the most disturbing one for the rest-activity rhythm. GOS was inoculated in mice exposed to CJL for 10 days (a condition that was maintained thereafter) or in mice kept in LD 12:12. Mice on CJL had complete suppression or severe dampening of their rest-activity and body temperature rhythms, whereas the serum corticosterone rhythm became bimodal (Filipski et al. 2004b). The tumor grew significantly faster in the mice on CJL as compared with those on LD 12:12, resulting in statistically significant differences in survival (Fig. 3b) (Filipski et al. 2004b). In a further experiment, the hypothesis that feeding synchronization could counterbalance the deleterious effect of CJL on malignant growth was tested. Mice to receive GOS were either kept in LD 12:12 or exposed to CJL alone or exposed to CJL with food availability regularly alternating between availability for 12 hours and no food for 12 hours. Feeding synchronization moderately slowed down malignant growth as compared to CJL (Filipski et al. 2005). In additional experiments, food availability was limited to 4 hours ("meal timing") for several weeks before GOS or P03 inoculation and thereafter. This procedure allowed the mice on meal timing to regain weight following initial loss, so that their body weights were similar to those of mice on LD 12:12 upon tumor inoculation. Meal timing, a procedure known to entrain peripheral clocks, proved to be an effective method to slow down tumor growth (Wu et al. 2004).

The results then emphasize that both an anatomical structure such as the SCN and lifestyle-related factors such as the LD exposure cycle and feeding schedules can impact on tumor growth rate. This effect can be mediated through host circadian physiology, central circadian coordination, and/or molecular clocks in healthy and/or malignant tissues.

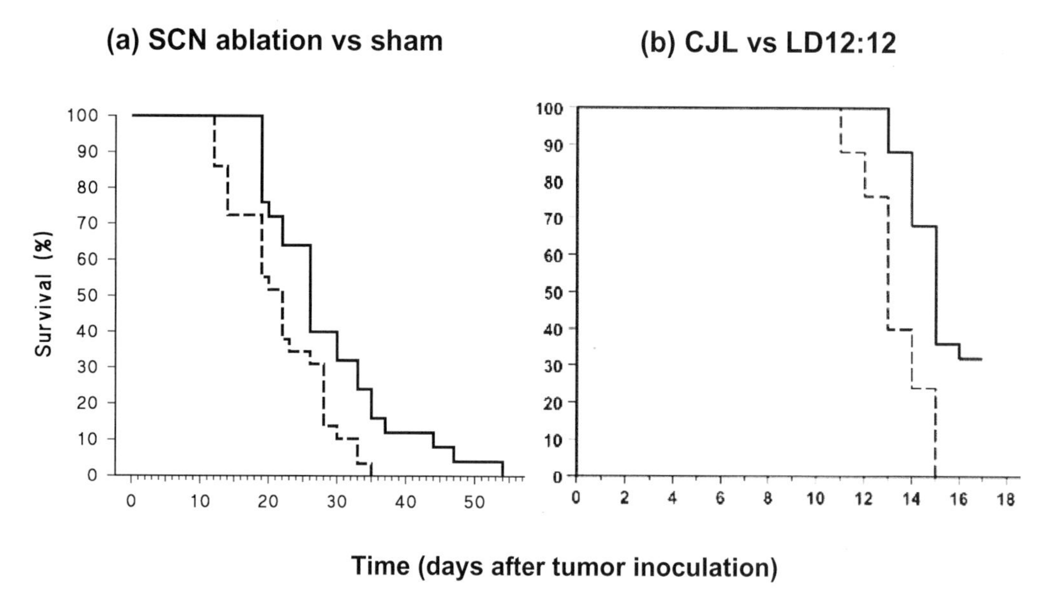

Figure 3. Relevance of the circadian timing system for experimental tumor progression and survival in ♂ B6D2F1 mice. (*a*) Survival curves of mice inoculated with GOS or pancreatic adenocarcinoma P03 4 weeks after ablation of SCN (*dashed line*) or sham-operated (*solid line*) and kept in LD 12:12 (log rank test adjusted for tumor type, $p = 0.006$) (Modified from Filipski et al. 2002.). (*b*) Survival curves of mice inoculated with GOS 10 days after beginning exposure to chronic jet lag (*dahed line*) or maintained on LD 12:12 (*solid line*) (log rank test, $p < 0.001$). (Modified from Filipski et al. 2004b.)

INTERACTIONS BETWEEN MOLECULAR CLOCK AND CELLULAR PROLIFERATION

At least three molecular mechanisms link molecular circadian clock with the cycling of cell division. The molecular clock controls *Wee1* transcription through an E-box-mediated mechanism. *Wee1* negatively controls the activity of CDK1/cyclin B1 that regulates the G_2/M transition (Matsuo et al. 2003). In addition, the BMAL1: CLOCK heterodimers activate *Per2* and *Rev-erbα* transcription and repress c-Myc transcription through E-box-mediated reactions in the c-*myc* gene P1 promoter (Fu and Lee 2003). *Per2* can also suppress c-*myc* expression indirectly by stimulating *Bmal1* transcription.

Both *Per1* and *Per2* as well as possibly other clock genes also control DNA repair through interactions with ATM and *mdm2* and maintain genome stability (Hunt and Sassone-Corsi 2007). However, circadian clock gene expression in bone marrow display damped or ablated patterns for *Bmal1* and persistent *Per1* and *Per2* rhythms (Granda et al. 2005; Tsinkalovsky et al. 2006).

CJL not only profoundly alters circadian physiology as previously mentioned, but also severely disrupts the molecular clock in the SCN and in liver. Thus, the rhythm in PER1 protein expression was suppressed in the SCN of mice on CJL, and so were the 24-hour rhythms in the mRNA expression of clock genes *Rev-erbα*, *Cry1*, and *Bmal1* in liver. However, *Per2* mRNA retains a circadian rhythm yet with a nearly 10-hour phase-advance as compared with mice on LD 12:12 (Filipski et al. 2005). Interestingly, CJL also resulted in the repression of *p53* and the derepression of c-*myc* mRNA expression in liver. Furthermore, the expression pattern of c-*myc* became prominently rhythmic in the liver of mice on CJL (Fig. 4a) (Filipski et al. 2005). *p53* is a tumor suppressor gene whose activation can induce cell cycle arrest, DNA repair or apoptosis, or senescence, depending on the differential activation of *p53* target genes (Vogelstein et al. 2000; Liu and Chen 2006). Recent data further reveal that the tumor suppression activities of *p53* could result from the control this transcription factor exerts on energy metabolism, including glycolysis and mitochondrial respiration (Bensaad and Vousden 2007). Conversely, c-*myc* is an oncogene that has been linked with many human cancers. This transcription factor in the basic helix-loop-helix zipper family regulates up to 15% of all cellular genes, promotes entry into the cell cycle, and transition from G_1 to S. c-*myc* also impinges on global chromatin structure both directly and indirectly via both the regulation of histone acetyltransferase (HAT) GCN5, which in turn globally acetylates chromosomal histones, and histone methylation (Knoepfler 2007).

The role of the circadian clock in mediating the effect of CJL on p53 and c-Myc transcription patterns is further supported by very similar findings in the liver of mice with a constitutive *Per2* mutation (Fig. 4b) (Fu and Lee 2003) . Taken together, the results suggest that genetic or functional disruption of the molecular circadian clock results in genomic instability and accelerated cellular proliferation, two conditions that favor carcinogenesis (Hanahan and Weinberg 2000). Indeed, exposure of mice with a constitutive *Per2* mutation to γ-radiation resulted in shortened survival and increased incidence of tumors as compared with normal mice with a similar genetic background (Fu et al. 2002). No such effect was found in mice with the *Cry1* and *Cry2* double mutation (Gauger and Sancar 2005). However, the in vivo part of this study was performed in double-mutant mice kept in LD 12:12, an environmental condition that dampens, yet does not disrupt, 24-hour physiology, whereas DD exposure does

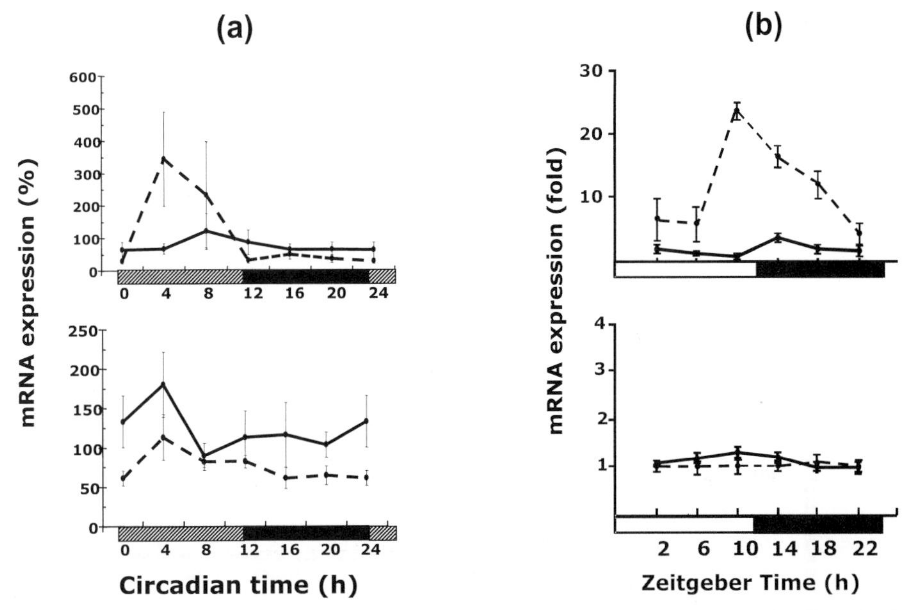

Figure 4. Downstream effects of molecular clock disruption on the mRNA expression patterns of c-Myc (*upper panels*) and p53 (*lower panels*) in mouse liver. (*a*) Chronic jet lag (*dashed lines*) versus LD 12:12 synchronization (*solid lines*). (Modified from Filipski et al. 2005.) (*b*) Constitutive *Per2* mutation (*dashed lines*) versus wild type (*solid lines*). (Modified from Fu et al. 2002.)

(Nagashima et al. 2005). Cultured fibroblasts from the *Cry* double mutants also did not display an altered response to radiation exposure (Gauger and Sancar 2005). However, recent data show that *Crys* do not seem to be required for normal circadian clock function in mouse fibroblasts (Fan et al. 2007). Thus, the hypothesis that core circadian clock disruption favors malignant growth irrespective of its mechanisms remains plausible.

The hypothetical molecular circadian clock (Lévi and Schibler 2007) can be estimated to be functional through the relative phase relations among three core clock genes whose transcription is regulated by one another: *Rev-erbα* down-regulates *Bmal1*, *Bmal1* up-regulates both *Rev-erbα* and *Per2*, and *Per2* down-regulates both *Rev-erbα* and its own transcription. This simplified model is built through the adjustment of a 24-hour cosine function to the mRNA expression data of each of the three genes in tissues sampled from mice kept in DD for 24–60 hours. Figure 5 depicts the circadian clock of GOS under various experimental conditions, in comparison with that in healthy liver. The acrophases (maxima in fitted 24-hour cosine function) of the mRNA rhythms occur at $CT5^{20}$ for *Rev-erbα*, $CT14^{20}$ for *Per2*, and $CT23^{40}$ for *Bmal1* in healthy mouse liver (Fig. 5a). In GOS, the clock gene transcription rhythms are ablated and not statistically significant (Fig. 5b).

Exposure of GOS-bearing mice to CJL results in further apparent alteration of the simplified tumor clock (Fig. 5c). However, feeding synchronization (FS 12:12) induces a statistically significant rhythm in the transcription of all three clock genes despite CJL. Yet, the mRNA rhythms of these three clock genes occur in coincidence near CT6, which suggests that the molecular clock remains partly abnormal (Fig. 5d). This is also the case for the liver of these tumor-bearing mice, which display near-normal rhythms in mRNA expression of *Rev-erbα* and *Per2*, yet without any significant *Bmal1* rhythm (Filipski et al. 2005).

Some drugs, such as the cyclin-dependent kinase inhibitor (CDKI) seliciclib, arrest cycling cells in both G_1-S and G_2-M stages by preventing assembly of CDK and cyclins through competing with ATP for the binding to the catalytic site of CDKs. Seliciclib was used as a pharmacologic tool to study the cross-talk between the circadian clock and cell division cycle in GOS-bearing mice. Seliciclib treatment did produce rhythmic clock gene expression patterns in advanced malignant tumors with otherwise disrupted or uncoordinated molecular clocks (Iurisci et al. 2006). More specifically, seliciclib induced a near-normal molecular clock in the tumors of mice dosed at zeitgeber time 3 (ZT3), i.e., in the early light span (Fig. 5e). No such effect was found in the tumors of mice receiving seliciclib at ZT19, i.e., in the second half of the dark span (Fig. 5f). The pharmacologic induction of the tumor clock is associated with the best antitumor effect, with both tumor clock induction and antitumor effects being greatest following treatment at ZT3 (Iurisci et al. 2006).

Circadian clock induction improves control of G_2/M gating through an enhancement of *Wee1* transcription, a gene that is unidirectionally controlled by CLOCK:BMAL1. In the case of seliciclib, the induction of the molecular clock involves inhibition of CK1δ/ε, a key determinant for circadian clock function. Thus, the CK1δ/ε mutation or alteration produces disorders of rest-activity or sleep-wakefulness rhythm both in rodents and in humans (Hastings et al. 2003; Xu et al. 2005). The inhibition of CK1δ/ε by seliciclib results in increased *Bmal1* transcription, an effect that likely results from decreased PER2 degradation and nuclear translocation.

Because highly coordinated sequential transcription is a major mechanism of circadian rhythms, programmed

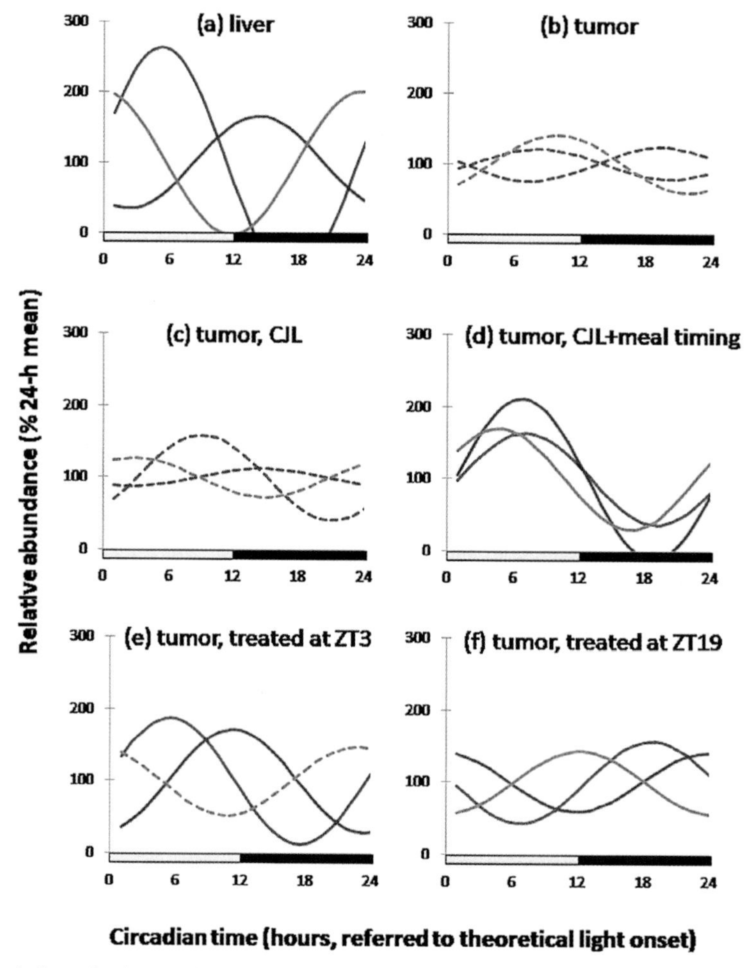

Figure 5. Molecular clock disruption in experimental tumor (GOS) at an advanced stage of growth. Effects of light, food, and drug on the circadian patterns in mRNA expression of *Rev-erbα* (*blue line*), *Per2* (*red line*), and *Bmal1* (*green line*). All studies in ♂ B6D2F1 mice. (*Solid line*) Statistically significant cosine fit for a 24-hour period; (*dashed line*) best-fitting yet nonsignificant 24-hour cosine function. (Modified from Filipski et al. [2005] and Iurisci et al. [2006].) (*a*) Liver clock, taken as a reference of a healthy functional peripheral clock in mice synchronized with LD 12:12. (*b*) Tumor clock (GOS) in mice synchronized with LD 12:12. (*c*) Tumor clock in mice on chronic jet lag for 10 days before and during the course of malignant growth. Chronic jet lag consisted of 8-hour advances of light onset of LD 12:12 every 2 days. (*d*) Tumor clock in mice on chronic jet lag (same as above) and meal timing, consisting of fixed daily food availability for a 12-hour span. (*e*) Tumor clock in mice synchronized with LD 12:12 following treatment with the cell cycle inhibitor seliciclib dosed at ZT3, i.e., the time that achieved best antitumor efficacy. (*f*) Tumor clock in mice synchronized with LD 12:12 following treatment with the cell cycle inhibitor seliciclib at ZT19, i.e., the time that achieved poorest antitumor efficacy.

food availability or daily seliciclib acts as a strong resetter of tumor cells that have lost synchrony in functional clocks, through transient inhibition of CK1δ/ε or other pathways within permissive time windows.

Recent studies in our group reveal that clock gene transcription is indeed rhythmic at an early stage of GOS growth, whereas the rhythms in *Rev-erbα*, *Per2*, and *Bmal1* mRNAs become ablated at a late stage of tumor growth. Concurrently, the circadian control of cell cycle phase distribution disappears and the rate of apoptotic cells decreases markedly (X.M. Li et al., unpubl.). Consistent with these data is the high and nonrhythmic expression of antiapoptotic protein BCL-2 in another advanced mouse tumor (mammary adenocarcinoma MA13/C) with presumably an altered molecular clock (Granda et al. 2005). Conversely, antiapoptotic BCL-2 and proapoptotic BAX, respectively, vary threefold and fivefold over 24 hours in mouse bone marrow. The circadian peak occurs at early light for BCL-2 and near mid dark for BAX, revealing the fine tuning of apoptosis circadian control in rapidly dividing healthy tissues (Granda et al. 2005).

The molecular clock seems to follow similar dynamics in other mouse tumor models. In early-stage mouse sarcoma180, clock gene mRNAs were rhythmic, with peaks occurring near the LD transition for *Per1* and *Per2* and at late dark/early light for *Bmal1* (Koyanagi et al. 2003). In a mouse breast cancer model at a more advanced stage of growth (early to late), *Per1* and *Per2* mRNAs lacked any circadian periodicity, whereas *Bmal1* mRNA remained rhythmic, yet with a 13-fold reduction in circadian amplitude (You et al. 2005). An elegant study performed in rats with diethylnitrosamine-induced hepatocarcinoma, a very slow growing tumor, further shows that the entrainment properties of *Per1* in cancerous liver tissue differ from those of the healthy liver tissue it originates from (Davidson et al. 2006).

Taken together, the results from these and other studies indicate that clock genes are expressed in most malignant tumors that are currently used as experimental cancer models. The tumor molecular clock appears to switch from near normal to disrupted along the course of tumor growth. Such clock disruption is associated with altered transcription patterns of cell cycle genes so that proliferation and genomic instability are favored, apoptosis is down-regulated, and malignant progression is accelerated.

The increased proliferation and decreased apoptosis that are associated with clock disruption in tumors can be reverted through overexpressing *Per1* or *Per2* in malignant cells (Gery et al. 2006; Hua et al. 2006, 2007). Thus, overexpression of *Per1*-sensitized human cancer cells to low-dose radiation that produced DNA-damage-induced apoptosis, whereas inhibition of *Per1* cells blunted apoptosis in these tumor cells (Gery et al. 2006). Radiation triggered an p53-independent increase in c-Myc expression and a decrease in p21 expression in the tumor cells that were overexpressing *Per1*, but not in those with blunted *Per1*. The mechanisms of *Per1* in the elicitation of the apoptotic response to radiation involve interactions with Chk2 and activation of the ATM checkpoint pathway (Gery et al. 2006). Timeless, another clock protein, also has an important role in the ATR pathway, another arm of the checkpoint network (Unsal-Kaçmaz et al. 2005). In addition, *Per1* overexpression decreases the levels of Wee1, cyclin B1, and CDK1 (Gery et al. 2006). The CDK1-cyclin B1 complex gates the G_2/M transition and it is negatively regulated by Wee1 (Matsuo et al. 2003). Similarly, overexpression of *Per2* reduced cellular proliferation and increased apoptosis in mouse Lewis lung carcinoma (LLC) and mammary carcinoma (EMT6) cell lines (Hua et al. 2006). *Per2* overexpression down-regulated c-Myc, Bcl-X(L), and Bcl-2 and up-regulated p53 and Bax, thus promoting apoptosis in both cancer cell lines. However, no such effect was found for NIH-3T3 cells (Hua et al. 2006). Furthermore, intratumoral *Per2* gene delivery significantly slowed down the growth of LLC transplanted in mice through inhibition of proliferating cell nuclear antigen (PCNA) expression and apoptosis induction (Hua et al. 2007).

A HYPOTHETICAL MODEL OF THE INTERACTIONS BETWEEN CIRCADIAN TIMING SYSTEM AND TUMOR PROLIFERATION

A synthetic view of the interactions between the circadian timing system and the cell division cycle is depicted in Figure 6, with a major emphasis on the cell cycle in cancer cells. The panel on the left part of the figure illustrates the fact that circadian physiology represents the most obvious rhythms that reflect circadian timing system function. The rhythms in rest-activity, core body temperature, feeding behavior, and hormones are generated or controlled by the SCN in the hypothalamus through diffusible signals that include epidermal growth factor (EGF), transforming growth factor-α (TGF-α), prokineticin-2, and cardiotrophin-like cytokine (Kramer et al. 2001; Cheng et al. 2002; Kraves and Weitz 2006).

The SCN can control the molecular clocks in peripheral tissues through sympathetic and parasympathetic pathways (Kalsbeek et al. 2006). These clocks are also redundantly regulated and coordinated through circadian physiology (Balsalobre et al. 2000; Brown et al. 2002; Kornmann et al. 2007). In turn, constitutive mutations in the molecular clock can alter both SCN functions and circadian physiology (Hastings et al. 2003). Molecular clocks in healthy cells control the cell division cycle through *Clock:Bmal1* regulation of several key cell cycle genes (Fu and Lee 2003; Matsuo et al. 2003). Circadian physiology can also and redundantly regulate cell cycle traverse (Dickmeis et al. 2007).

As indicated in the lower part of Figure 6, cancer cells are characterized by a deregulated cell cycle through numerous mutations that critically affect the transition

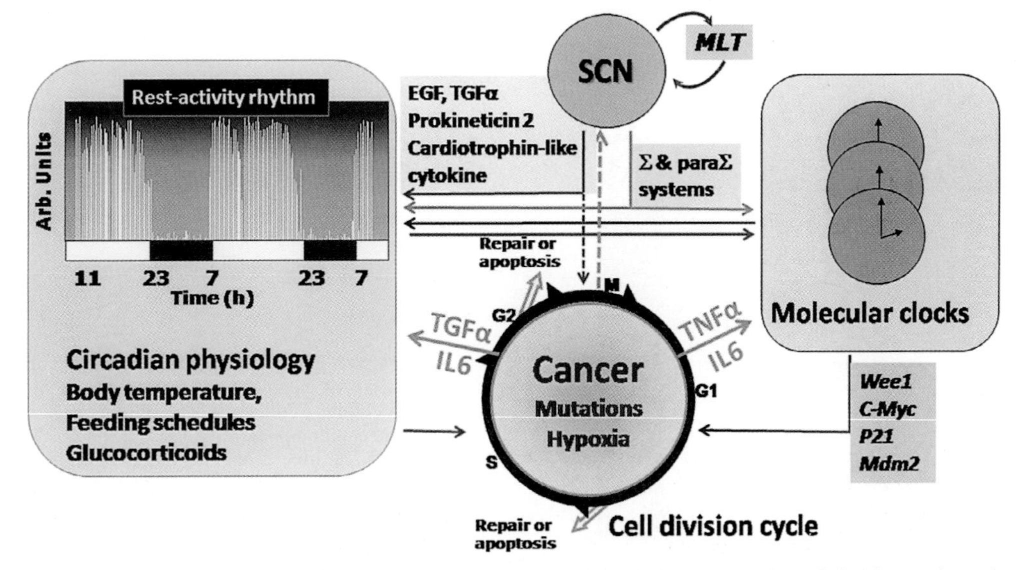

Figure 6. Schematic representation of the interactions among the circadian timing system, the cell division cycle, and cancer.

from G_1 to S in most tumors (Hanahan and Weinberg 2000). Hypoxia is also a classical feature that modifies both cellular proliferation (Keith and Simon 2007) and the circadian clock (Chilov et al. 2001; Tu and McKnight 2006). Furthermore, rhythmic metabolism can also drive rhythmic cell divisions (Chen et al. 2007).

Both gene mutations and metabolism alterations in cancer cells could thus account for defective circadian clock control of malignant cell proliferation. Additionally, malignant tumors can directly or indirectly produce cytokines, such as TGF-α, interleukin-6 (IL-6), or tumor necrosis factor-α (TNF-α) (Aggarwal et al. 2006; Rajput et al. 2007). These cytokines can interfere at various levels of the circadian timing system and result in circadian disruption or alteration (Kramer et al. 2001; Motzkus et al. 2002; Kraves and Weitz 2006; Cavadini et al. 2007). Indeed, severe blunting of the rest-activity rhythm was strongly related to increased serum levels of TGF-α, IL-6, and TNF-α in patients with metastatic colorectal cancer. Significant correlations were found between serum levels of TGF-α and IL-6, circadian patterns in wrist activity, and serum cortisol and tumor-related symptoms in 80 patients with metastatic colorectal cancer. These data support clinically relevant links between tumor cytokines and the circadian timing system (Rich et al. 2005; Rich 2007).

IMPLICATIONS FOR CANCER CHRONOTHERAPEUTICS

The circadian timing system controls cellular proliferation as well as drug metabolism over 24 hours through molecular clocks in each cell, circadian physiology, and the SCN, the hypothalamic pacemaker that coordinates circadian rhythms (Lévi and Schibler 2007). As a result, both the toxicity and efficacy of more than 30 anticancer agents vary by more than 50% as a function of dosing time in experimental models (Mormont and Lévi 2003). The circadian timing system also down-regulates malignant growth in several experimental models and in several clinical situations, as previously discussed.

Programmable-in-time infusion pumps and rhythmic physiology monitoring devices have made possible the application of chronotherapeutics to more than 2000 cancer patients without hospitalization. This treatment method consists in the chronomodulated administration of anticancer agents with appropriate selection of peak times of drug delivery within the 24-hour timescale. This strategy first revealed the antitumor efficacy of oxaliplatin, which is now a main drug used against colorectal cancer. In this disease, international clinical trials have shown a fivefold improvement in patient tolerability and near doubling of antitumor activity through chronomodulated administration, in comparison to constant-rate delivery of the same drug combination, 5-fluorouracil-leucovorin and oxaliplatin (Lévi et al. 1997; Lévi 2001). Recent clinical trials have further shown the relevance of the peak time of the chronomodulated delivery of these cancer medications along the 24 hours for achieving best tolerability: The incidence of severe adverse events varied up to fivefold as a function of the choice of when during the 24 hours the peak dose of the medications was timed. Gender was an important determinant of drug schedule tolerability in this trial (Lévi et al. 2007).

The role of gender on cancer chronotherapeutics is consistent with recent results from a large randomized clinical trial involving 554 patients with metastatic colorectal cancer, where gender also predicted survival outcome on chronotherapy but not on conventional drug delivery (Giacchetti et al. 2006). In this trial, the survival of men with colorectal cancer was significantly improved with the chronotherapy schedule that was applied, whereas that of women was significantly reduced as compared with conventional delivery. The "gender x schedule" interaction remained highly statistically significant following multivariate analyses, suggesting that gender could have an essential role in the determination of optimal chronotherapeutic schedule (Fig. 7) (Giacchetti et al. 2006). A recent study has further revealed that gender largely determines the circadian transcriptome in human oral mucosa, a finding that further supports the hypothe-

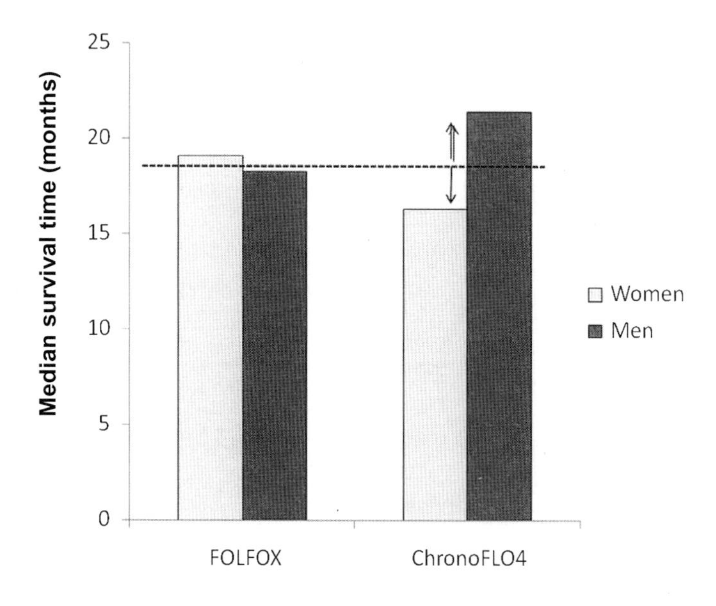

Figure 7. Relevance of gender for the survival outcome of patients with metastatic colorectal cancer receiving conventional delivery (FOLFOX) or chronomodulated infusion (ChronoFLO4). Median survival time of women and men on either delivery schedule in a randomized international trial in 554 patients (gender x schedule interaction test, $p < 0.0001$). (*Dotted line*) Median survival time for all patients on FOLFOX. (Modified from Giacchetti et al. 2006.)

sis of gender dependencies in optimal chronotherapeutics (Bjarnason et al. 2007).

Ongoing translational studies and technology developments are exploring new methods for tailoring cancer chronotherapeutics to the main rhythmic characteristics of the individual patient, an approach needed in view of the dynamic cross-talks between the circadian timing system and the cell cycle along the course of cancer processes. Thus, impaired expressions of the clock genes *Per1*, *Per2*, or *Per3* have been reported in human cancers originating from breast, lung, endometrium, pancreas, or colon, as well as in human myeloid leukemia (Chen et al. 2005; Gery et al. 2005, 2006, 2007a,b; Yeh et al. 2005; Shih et al. 2006; Krugluger et al. 2007). Promoter methylation defects may account for such impaired clock gene expression (Yeh et al. 2005). In human colorectal cancers, a close association has further been shown among decreased *Per1* expression, rapid cellular proliferation, and decreased expression of dihydropyrimidine dehydrogenase, the rate-limiting enzyme of 5-fluorouracil catabolism. In this study, an interaction with gender was also found (Krugluger et al. 2007). The yet limited human cancer data thus corroborate rather well the results from studies in mouse tumors regarding the occurrence of defective circadian clocks in malignant cells and its implications for tumor proliferation and susceptibility to therapeutic agents.

Recent approaches in theoretical models help identify and take into account the main factors that can impinge upon the success of cancer chronotherapeutics. Modeling based on experimental chronotherapeutic data is now confirming the role of an infusional circadian schedule of 5-fluorouracil and oxaliplatin upon toxicity and efficacy. The cell cycle automaton model reveals the key roles of variability in circadian entrainment and cell cycle length upon therapeutic activity of different sinusoidal delivery or constant-rate infusion schedules. This theoretical study pinpoints the need for novel biological assessments in the individual patient and his or her tumor in order to take full advantage of cancer chronotherapeutics (Altinok et al. 2007). A pharmacokinetic-pharmacodynamic model of chemotherapy with circadian periodic dimension investigates the effects of temporal drug-delivery dynamics on a population of tumor cells and its tolerance by a population of fast-renewing healthy cells. The model parameters are based on the experimental chronotherapeutics of mouse GOS with oxaliplatin, whose main toxicity target is the jejunal mucosa. Here also, the model shows the advantage of a periodic time-scheduled regimen, compared to the conventional continuous constant infusion of the same daily dose, when the biological time of peak infusion is correctly chosen. Furthermore, mathematical optimization methods of drug infusion flow, choosing tumor population minimization as the objective function and healthy tissue preservation as a constraint, reveal nonintuitive optimal dynamic schedules of drug delivery that depend on both host/tumor status and therapeutic objectives (Clairambault 2007).

CONCLUSIONS

The circadian timing system controls healthy cells and malignant proliferation. The genetic, epigenetic, and lifestyle factors that impinge on the several components of this system can interfere with the cross-talks between the circadian clock and the cell cycle. This can contribute to the large variability in outcome of patients harboring a similar cancer type. Therefore, targeting therapeutic delivery to the dynamics of the cross-talk among the circadian clock, the cell division cycle, and pharmacology pathways represents a new multidisciplinary challenge to concurrently improve quality of life and survival through personalized cancer chronotherapeutics.

ACKNOWLEDGMENTS

The research on circadian clocks in cancer is supported by the Association pour la Recherche sur le Temps Biologique et la Chronothérapie (ARTBC International) and the Association pour la Recherche sur le Cancer (ARC), hospital Paul Brousse, Villejuif (France), as well as the European Union through the Network of Excellence BIOSIM (Biosimulation: a new tool for drug development; contract LSHBCT- 2004-005137), and the Scientific Targeted Research Project TEMPO (Temporal Genomics for tailored chronotherapeutics; contract LSHG-ct-2006-037543).

REFERENCES

Aggarwal B.B., Shishodia S., Sandur S.K. Pandey M.K., and Sethi G. 2006. Inflammation and cancer: How hot is the link? *Biochem. Pharmacol.* **72:** 1605.

Altinok A., Lévi F., and Goldbeter A. 2007. A cell cycle automaton model for probing circadian patterns of anticancer drug delivery. *Adv. Drug Delivery Rev.* **59:** 1036.

Balsalobre A. 2000. Multiple signaling pathways elicit circadian gene expression in cultured Rat-1 fibroblasts. *Curr. Biol.* **10:** 1291.

Balsalobre A., Damiola F., and Schibler U. 1998. A serum shock induces circadian gene expression in mammalian tissue culture cells. *Cell* **93:** 929.

Balsalobre A., Brown S.A., Marcacci L., Tronche F., Kellendonk C., Reichardt H.M., Schutz G., and Schibler U. 2000. Resetting of circadian time in peripheral tissues by glucocorticoid signaling. *Science* **289:** 2344.

Bartek J. and Lukas J. 2007. DNA damage checkpoints: From initiation to recovery or adaptation. *Curr. Opin. Cell Biol.* **19:** 238.

Bensaad K. and Vousden K.H. 2007. P53: New roles in metabolism. *Trends Cell Biol.* **17:** 286.

Bjarnason G. and Jordan R. 2002. Rhythms in human gastrointestinal mucosa and skin. *Chronobiol. Int.* **19:** 129.

Bjarnason G., Seth A., Wang Z., Blanas N., Straume M., and Martino T. 2007. Diurnal rhythms in gene expression in human oral mucosa: Implications for gender differences in toxicity, response and survival and optimal timing of targeted therapy. 2007. In *Proceedings of the 43rd Annual Meeting ASCO* **25:** 98s. (Abstr. 2507.)

Bjarnason G., Jordan R., Wood P., Li Q., Lincoln D., Sothern R., Hrushesky W., and Ben-David Y. 2001. Circadian expression of clock genes in human oral mucosa and skin: Association with specific cell-cycle phases. *Am. J. Pathol.* **58:** 1793.

Bourin P., Ledain A., Beau J., Mille D., and Levi F. 2002. In-vitro circadian rhythm of murine bone marrow progenitor production. *Chronobiol. Int.* **19:** 57.

Brown S., Zumbrunn G., Fleury-Olela F., Preitner N., and Schibler U. 2002. Rhythms of mammalian body temperature can sustain peripheral circadian clocks. *Curr. Biol.* **12:** 1574.

Burns E.R., Scheving L.E., Pauly J.E., and Tsai T. 1976. Effect of altered lighting regimens, time-limited feeding, and presence of Ehrlich ascites carcinoma on the circadian rhythm in DNA synthesis of mouse spleen. *Cancer Res.* **36:** 1538.

Cavadini G., Petrzilka S., Kohler P., Jud C., Tobler I., Birchler T., and Fontana A. 2007. TNF-α suppresses the expression f clock genes by interfering with E-box-mediated transcription. *Proc. Natl. Acad. Sci.* **104:** 12843.

Chen S.T., Choo K.B., Hou M.F., Yeh K.T., Kuo S.J.. and Chang J.G. 2005. Deregulated expression of the PER1; PER2 and PER3 genes in breast cancers. *Carcinogenesis* **26:** 1241.

Chen Z., Odstrcil E.A., Tu B.P., and McKnight S.L. 2007. Restriction of DNA replication to the reductive phase of the metabolic cycle protects genome integrity. *Science* **316:** 1916.

Cheng M.Y., Bullock C.M., Li C., Lee A.G., Bermak J.C., Belluzzi J., Weaver D.R., Leslie F.M., and Zhou Q.Y. 2002. Prokineticin 2 transmits the behavioural circadian rhythm of the suprachiasmatic nucleus. *Nature* **417:** 405.

Chilov D., Hofer T., Bauer C., Wenger R.H., and Gassmann M. 2001. Hypoxia affects expression of circadian genes PER1 and CLOCK in mouse brain. *FASEB J.* **15:** 2613.

Clairambault J. 2007. Modelling oxaliplatin drug delivery to circadian rhythms in drug metabolism and host tolerance. *Adv. Drug Delivery Rev.* (in press).

Davidson A.J., Straume M., Block G.D., and Menaker M. 2006. Daily timed meals dissociate circadian rhythms in hepatoma and healthy host liver. *Int. J. Cancer* **118:** 1623.

Delaunay F. and Laudet V. 2002. Circadian clock and microarrays: Mammalian genome gets rhythm. *Trends Genet.* **18:** 595.

De Vita V., Hellman S., and Rosenberg S., Eds. 2007. *Cancer: Principles and practice of oncology*, 7th edition. Lippincott, Philadelphia, Pennsylvania.

Dickmeis T., Lahiri K., Nica G., Vallone D., Santoriello C., Neumann C.J., Hammerschmidt M., and Foulkes N.S. 2007. Glucocorticoids play a key role in circadian cell cycle rhythms. *PLOS Biol.* **5:** 179.

Fan Y., Hida A., Anderson D.A., Izumo M., and Johnson C.H. 2007. Cycling of CHRYPTOCHROME proteins is not necessary for circadian-clock function in mammalian fibroblasts. *Curr. Biol.* **17:** 1091.

Filipski E., Innominato P.F., Wu M.W., Li X.M., Iacobelli S., Xian L.J., and Lévi F. 2005. Effects of light and food schedules on liver and tumor molecular clocks. *J. Natl. Cancer Inst.* **97:** 507.

Filipski E., Delaunay F., King V.M., Wu M.W., Claustrat B., Gréchez-Cassiau A., Guettier C., Hastings M.H., and Lévi F. 2004a. Effects of chronic jet lag on malignant growth in mice. *Cancer Res.* **64:** 7879.

Filipski E., King V.M., Etienne M.C., Li X.M., Claustrat B., Granda T.G., Milano G., Hastings M.H., and Lévi F. 2004b. Persistent twenty-four hour changes in liver and bone marrow despite suprachiasmatic nuclei ablation in mice. *Am. J. Physiol. Regul. Integr. Comp. Physiol.* **287:** R844.

Filipski E., King V.M., Li X.M., Granda T., Mormont M.C., Liu X.H., Claustrat B., Hastings M., and Lévi F. 2002. Host circadian clock as a control point in tumor progression. *J. Natl. Cancer Inst.* **94:** 690.

Fu L. and Lee C.C. 2003. The circadian clock: pacemaker and tumour suppressor. *Nat. Rev Cancer* **3:** 350.

Fu L., Pelicano H., Liu J., Huang P., and Lee C.C. 2002. The circadian gene *Period2* plays an important role in tumor suppression and DNA damage response in vivo. *Cell* **111:** 41.

Gauger M.A. and Sancar A. 2005. Cryptochrome, circadian cycle, cell cycle checkpoints, and cancer. *Cancer Res.* **65:** 6828.

Gery S., Virk R.K., Chumakov K., Yu A., and Koeffler H.P. 2007a. The clock gene Per2 links the circadian system to the estrogen receptor. *Oncogene* (in press).

Gery S., Gombart A.F., Yi W.S., Koeffler C., Hofmann W.K., and Koeffler H.P. 2005. Transcription profiling of C/EBP targets identifies Per2 as a gene implicated in myeloid leukemia. *Blood* **106:** 2827.

Gery S., Komatsu N., Baldjyan L., Yu A., Koo D., and Koeffler H.P. 2006. The circadian gene per1 plays an important role in cell growth and DNA damage control in human cancer cells. *Mol. Cell* **22:** 375.

Gery S., Komatsu N., Kawamata N., Miller C.W., Desmond J., Virk R.K., Marchevsky A., Mckenna R., Taguchi H., and Koeffler H.P. 2007b. Epigenetic silencing of the candidate tumor suppressor gene Per1 in non-small cell lung cancer. *Clin. Cancer Res.* **13:** 1399.

Giacchetti S., Bjarnason G., Garufi C., Genet D., Iacobelli S., Tampellini M., Smaaland R., Focan C., Coudert B., Humblet Y., Canon J.L., Adenis A., Lo Re G., Carvalho C., Schueller J., Anciaux N., Lentz M.A., Baron B., Gorlia T., and Lévi F. (European Organisation for Research and Treatment of Cancer Chronotherapy Group). 2006. Phase III trial of 4-day chronomodulated vs 2-day conventional delivery of 5-fluorouracil, leucovorin and oxaliplatin as first line chemotherapy of metastatic colorectal cancer: The EORTC. *J. Clin. Oncol.* **24:** 3562.

Golstein P. and Kroemer G. 2006. Cell death by necrosis: Towards a molecular definition. *Trends Biochem. Sci.* **32:** 37.

Granda T.G., Liu X.H., Smaaland R., Cermakian N., Filipski E., Sassone-Corsi P., and Lévi F. 2005. Circadian regulation of cell cycle and apoptosis proteins in mouse bone marrow and tumour. *FASEB J.* **19:** 304.

Hanahan D. and Weinberg R.A. 2000. The hallmarks of cancer: A review. *Cell* **100:** 57.

Hastings M., Reddy A.B., and Maywood E.S. 2003. A clockwork web: Circadian timing in brain and periphery in health and disease. *Nat. Rev. Neurosci.* **4:** 649.

Hua H., Wang Y., Wan C., Liu Y., Zhu B., Wang X., Wang Z., and Ding J.M. 2007. Inhibition of tumorigenesis by intratumoral delivery of the circadian gene mPer2 in C57BL/6 mice. *Cancer Gene Ther.* **14:** 815.

Hua H., Wang Y., Wan C., Liu Y., Zhu B., Yang C., Wang X., Wang Z., Cornelissen-Guillaume G., and Halberg F. 2006. Circadian gene *mPer2* overexpression induces cancer cell apoptosis. *Cancer Sci.* **97:** 589.

Hunt T. and Sassone-Corsi P. 2007. Riding tandem: Circadian clocks and the cell cycle. *Cell* **129:** 461.

Iurisci I., Filipski E., Reinhardt J., Bach S., Gianella-Borradori A., Iacobelli S., Meijer L., and Lévi F. 2006. Improved tumor control through circadian clock induction by seliciclib, a cyclin-dependent kinase inhibitor. *Cancer Res.* **66:** 10720.

Kalsbeek A., Palm I.F., La Fleur S.E., Scheer F.A., Perreau-Lenz S., Ruiter M., Kreier F., Cailotto C., and Buijs R.M. 2006. SCN outputs and the hypothalamic balance of life. *J. Biol. Rhythms* **21:** 458.

Keith B. and Simon C. 2007. Hypoxia-inducible factors, stem cells and cancer. *Cell* **129:** 465.

Knoepfler P.S. 2007. Myc goes global: New tricks for an old oncogene. *Cancer Res.* **67:** 5061.

Kornmann B., Schaad O., Bujard H., Takahashi J.S., and Schibler U. 2007. System-driven and oscillator-dependent circadian transcription in mice with a conditionnlly active liver clock. *PLoS Biol.* **5:** 179.

Koyanagi S., Kuramoto Y., Nakagawa H., Aramaki H., Ohdo S., Soeda S., and Shimeno H.A. 2003. Molecular mechanism regulating circadian expression of vascular endothelial growth factor in tumor cells. *Cancer Res.* **63:** 7277.

Kramer A., Yang F.C., Snodgrass P., Li X., Scammell T.E., Davis F.C., and Weitz C.J. 2001. Regulation of daily locomotor activity and sleep by hypothalamic EGF receptor signaling. *Science* **294:** 2511.

Kraves S. and Weitz C.J. 2006. A role for cardiotrophin-like cytokine in the circadian control of mammalian locomotor activity. *Nat. Neurosci.* **9:** 212.

Krugluger W., Brandstaetter A., Kallay E., Schueller J., Krexner E., Kriwanek S., Bonner E., and Cross H.S. 2007. Regulation of genes of the circadian clock in human colon cancer: Reduced Period-1 and dihydropyrimidine dehydrogenase transcription correlates in high-grade tumors. *Cancer Res.* **67:** 7917.

Lakatua D.J., White M., Sackett-Lundeen L.L., and Haus E. 1983. Change in phase relations of circadian rhythms in cell proliferation induced by time-limited feeding in BALB/c X DBA/2F1 mice bearing a transplantable Harding-Passey tumor. *Cancer Res.* **43:** 4068.

Lévi F. 2001. Circadian chronotherapy for human cancers. *Lancet Oncol.* **2:** 307.

Lévi F. and Schibler U. 2007. Circadian rhythms: Mechanisms and therapeutic implications. *Annu. Rev. Pharmacol. Toxicol.* **47:** 593.

Lévi F., Zidani R., and Misset J.L. (International Organization for Cancer Chronotherapy). 1997. Randomized multicentre trial of chronotherapy with oxaliplatin, fluorouracil, and folinic acid in metastatic colorectal cancer. *Lancet* **350:** 681.

Lévi F., Focan C., Karaboué A., de la Valette V., Focan-Henrard D., Baron B., Kreutz M., and Giacchetti S. 2007. Implications of circadian clocks for the rhythmic delivery of cancer therapeutics. *Adv Drug Delivery Rev.* **59:** 1015.

Liu G. and Chen X. 2006. Regulation of the p53 transcriptional activity. *J. Cell. Biochem.* **97:** 448.

Matsuo T., Yamaguchi S., Mitsui S., Emi A., Shimoda F., and Okamura H. 2003. Control mechanism of the circadian clock for timing of cell division in vivo. *Science* **302:** 255.

Mormont M.C. and Lévi F. 2003. Cancer chronotherapy: Principles, applications, and perspectives. *Cancer* **97:** 155.

Mormont M.C., Waterhouse J., Bleuzen P., Giacchetti S., Jami A., Bogdan A., Lellouch J., Misset J.L., Touitou Y., and Lévi F. 2000. Marked 24-h rest-activity rhythms are associated with better quality of life, better response and longer survival in patients with metastatic colorectal cancer and good performance status. *Clin. Cancer Res.* **6:** 3038.

Motzkus D., Albrecht U., and Maronde E. 2002. The human Per 1 gene is inducible by interleukin-6. *J. Mol. Neurosci.* **18:** 105.

Nagashima K., Matsue K., Konishi M., Iidaka C., Miyazaki K., Ishida N., and Kanosue K. 2005. The involvement of Cry1 and Cry2 genes in the regulation of the circadian body temperature rhythm in mice. *Am. J. Physiol. Regul. Integr. Comp. Physiol.* **288:** R329.

Nagoshi E., Saini C., Bauer C., Laroche T., Naef F., and Schibler U. 2004. Circadian gene expression in individual fibroblasts: Cell-autonomous and self-sustained oscillators pass time to daughter cells. *Cell* **119:** 693.

Perpoint B., Le Bousse-Kerdiles C., Clay D., Smadja-Joffe F., Deprés-Brummer P., Laporte-Simitsidis S., Jasmin C., and Lévi F. 1995. In vitro chronopharmacology of recombinant mouse IL-3, mouse GM-CSF and human G-CSF on murine myeloid progenitor cells. *Exp. Hematol.* **23:** 362.

Rajput A., Koterba A.P., Kreisberg J.I., Foster J.M., Willson J.K.V., and Brattain M.G. 2007. A novel mechanism of resistance to epidermal growth factor receptor antagonism in vivo. *Cancer Res.* **67:** 665.

Rich T. 2007. Symptom clusters in cancer patients and their relation to EGFR ligand modulation of the circadian axis. *J. Support. Oncol.* **5:** 167.

Rich T., Innominato P.F., Boerner J., Mormont M.C., Iacobelli S., Baron B., Jasmin C., and Lévi F. 2005. Elevated serum cytokines correlated with altered behaviour, serum cortisol rhythm, and dampened 24 hour rest-activity patterns in patients with metastatic colorectal cancer. *Clin. Cancer Res.* **11:** 1757.

Sanchez I. and Dynlacht B.D. 2005. New insights into cyclins, CDKs, and cell cycle control: A review. *Semin. Cell Dev. Biol.* **16:** 311.

Santamaría D., Barrière C., Cerqueira A., Hunt S., Tardy C., Newton K., Cáceres J.F., Dubus P., Malumbres M., and Barbacid M. 2007. Cdk1 is sufficient to drive the mammalian cell cycle. *Nature* **448:** 811.

Scheving L.E. 1959. Mitotic activity in human epidermis. *Anat. Rec.* **135:** 7.

Scheving L.E., Tsai T.-H., Scheving L.A., Feuers R.J., and Kanabrocki E.L. 1992. Normal and abnormal cell proliferation in mice as it related to cancer. In *Biological rhythms in clinical and laboratory medicine* (ed. Y. Touitou and E. Haus), p. 566. Springer-Verlag, Berlin.

Scheving L.E., Tsai T.-H., Powell E.W., Pasley J.N., Halberg F., and Dunn J. 1983. Bilateral lesions of suprachiasmatic nuclei affect circadian rhythms in [^{3}H]-thymidine incorporation into deoxyribonucleic acid in mouse intestinal tract, mitotic index of corneal epithelium, and serum corticosterone. *Anat. Rec.* **205:** 239.

Sclafani R.A. and Holzen T.M. 2007. Cell cycle regulation of DNA replication. *Annu. Rev. Genet.* (in press).

Sephton S.E., Sapolsky R.M., Kraemer H.C., and Spiegel D. 2000. Diurnal cortisol rhythm as a predictor of breast cancer survival. *J. Natl. Cancer Inst.* **92:** 994.

Shih M.C., Yeh K.T., Tang R.P., Chen J.C. and Chang J.G. 2006. Promoter methylation in circadian genes of endometrial cancers detected by methylation specific PCR. *Mol. Carcinog.* **45**: 732.

Shiloh Y. 2003. ATM and related protein kinases: Safeguarding genome integrity. *Nat. Rev. Cancer* **3:** 155.

Smaaland R., Sothern R.B., Laerum O.D., and Abrahamsen J.F. 2002. Rhythms in human bone marrow and blood cells. *Chronobiol. Int.* **19:** 101.

Smaaland R., Laerum O.D., Lote K., Sletvold O., Sothern R.B., and Bjerknes R. 1991. DNA synthesis in human bonne marrow is circadian stage dependent. *Blood* **77:** 2603.

Touitou Y., Lévi F., Bogdan A., Benavides M., Bailleul F., and Misset J.L. 1995. Rhythm alteration in patients with metastatic breast cancer and poor prognostic factors. *J. Cancer Res. Clin. Oncol.* **121:** 181.

Tsinkalovsky O., Filipski E., Rosenlund B., Sothern R.B., Eiken H.G., Wu M.W., Claustrat B., Bayer J., Lévi F., and Laerum O.D. 2006. Circadian expression of clock genes in purified hematopoietic stem cells is developmentally regulated in mouse bone marrow. *Exp. Hematol.* **34:** 1248.

Tu B. and McKnight S.L. 2006. Metabolic cycles as an underlying basis of biological oscillations. *Nat. Rev. Mol. Cell Biol.* **7:** 696.

Unsal-Kaçmaz K., Mullen T.E., Kaufmann W.K., and Sancar A. 2005. Coupling of human circadian and cell cycles by the timeless protein. *Mol. Cell. Biol.* **25:** 3109.

Vogelstein B., Lane D., and Levine A.J. 2000. Surfing the p53 network. *Nature* **408:** 307.

Wu M.W., Li X.M., Xian L.J., and Lévi F. 2004. Effects of meal timing on tumor progression in mice. *Life Sci.* **75:** 1181.

Xu Y., Padiath Q.S., Shapiro R.E., Jones C.R., Wu S.C., Saigoh N., Saigoh K., Ptáček L.J., and Fu Y.H. 2005. Functional consequences of a CKIdelta mutation causing familial advanced sleep phase syndrome. *Nature* **434:** 640.

Yeh K.T., Yang M.Y., Liu T.C., Chen J.C., Chan W.L., Lin S.F., and Chang J.G. 2005. Abnormal expression of period 1 (PER1) in endometrial carcinoma. *J. Pathol.* **206:** 111.

You S., Wood P.A., Xiong Y., Kobayashi M., Du-Quiton J., and Hrushesky W.J.M. 2005. Daily coordination of cancer growth and circadian clock gene expression. *Breast Cancer Res. Treat.* **91:** 47.

The Clock Proteins, Aging, and Tumorigenesis

R.V. KONDRATOV* AND M.P. ANTOCH†

*BGES Department, Cleveland State University, Cleveland, Ohio 44115; †Department of Molecular and Cellular Biology, Roswell Park Cancer Institute, Buffalo, New York 14263

Many aspects of mammalian physiology and behavior are driven by an intrinsic timekeeping system that has an important role in synchronizing various biological processes within an organism and coordinating them with the environment. It is believed that deregulation of this coordination may cause the development of various pathologies. However, recent studies using mice deficient in individual components of the circadian system clearly demonstrated more complex interaction of the circadian system with various biological processes. The growing amount of evidence suggests that in addition to their roles in the core clock mechanism, some of the components of the molecular oscillator are involved in modulation of such diverse physiological processes as response to genotoxic stress, regulation of the cell cycle, aging, and carcinogenesis. These new data provide a mechanistic link between deregulation of the circadian system and/or some of its core components and the development of various pathologies, suggesting novel strategies for the disease treatment and prevention.

INTRODUCTION

Circadian Rhythms and Human Disease

The circadian clock is a universal intrinsic timekeeping system, which regulates many vital physiological and biochemical processes. In mammals, it is organized as a hierarchical network of molecular clocks that are operative in all tissues, with the master clock residing in the hypothalamic suprachiasmatic nucleus (SCN). The master clock is synchronized with the environment by daily changes in the light/dark cycles and transmits information regarding its phase to multiple tissue-specific clocks (Schibler and Sassone-Corsi 2002). Precise coordination of numerous tissue-specific clocks with each other and with the environment, as well as their proper functioning at the cellular level, is recognized now as important factors of an organism's well being. In humans, the malfunctioning of the circadian clock apparatus has been linked to various pathological syndromes, including sleep and mood disorders, jet lag, tumorigenesis (for review, see Kondratov et al. 2007; Sahar and Sassone-Corsi 2007), and, more recently, to hypertension and type-2 diabetes (Woon et al. 2007). Detailed phenotypic analysis of mice with targeted disruption or mutations in core clock genes further advanced our current view on the role of circadian proteins in pathophysiology. Thus, the growing amount of evidence suggests that core clock proteins may have distinct noncircadian functions important for maintaining tissue homeostasis under normal and stress conditions, and their malfunction may contribute to disease development (Ko and Takahashi 2006; Kondratov et al. 2007). Here, we review recent data from our and other laboratories that address complex interaction of mammalian circadian system with such fundamental biological processes as aging and tumorigenesis.

CLOCK/BMAL1 Transcriptional Complex as the Major Regulator of Circadian Output and a Modulator of the Response to Genotoxic Stress

The current view of the molecular circadian machinery underlying daily oscillations in physiology, metabolism, and behavior describes the CLOCK/BMAL1 transcriptional complex as the major regulator of both core clock function and circadian output. CLOCK and BMAL1 are two basic helix-loop-helix (bHLH)-PAS domain transcription factors that in a form of heterodimer activate expression of target genes through the E-box enhancer elements in their promoter regions. Among the genes induced by CLOCK/BMAL1 are those encoding their own repressors, PERIODs (PERs) and CRYPTOCHROMEs (CRYs), the regulators of *Bmal1* transcription (REV-ERB? and ROR?) as well as multiple clock-controlled genes (CCGs). The expression of CCGs is achieved either directly or indirectly through the activity of the E box containing noncircadian transcription factors (Gachon et al. 2006).

Among other vital physiological processes, the circadian system has been implicated in modulation of an organism's response to various genotoxic treatments. The initial observations that an organism's response to chemotherapeutic drugs and radiation varies depending on time of administration justified the use of chronotherapy in cancer patients (Levi 2001). Several successful clinical trials have conclusively demonstrated that the therapeutic index of any given treatment (the ratio of tumor response to treatment to the amount of damage caused to normal tissues) could be significantly improved by proper timing (Hrushesky 1985; Giacchetti 2002; Kobayashi et al. 2002). Recent studies performed on mice with targeted disruption of individual clock genes provide the mechanistic background for these initial observations.

The key role of major circadian proteins in genotoxic stress response was demonstrated by testing the sensitivity of wild-type, *Clock* mutant, *Bmal1*$^{-/-}$ knockout, and *Cry1*$^{-/-}$*Cry2*$^{-/-}$ double-knockout animals to toxicity induced by the chemotherapeutic drug cyclophosphamide. Wild-type mice display a robust daily rhythm in sensitivity to the drug. Importantly, morbidity and mortality associated with treatment were at their highest levels when cyclophosphamide was administered at the times of day corresponding to minimal functional activity of the CLOCK/BMAL1 complex and the lowest at the peak times of its activity. Consistently, animals with mutations or targeted disruption of the *Clock* or *Bmal1* genes that are characterized with constant low levels of CLOCK/BMAL1 transcriptional activity (*Clock/Clock* and *Bmal1*$^{-/-}$) showed high levels of drug sensitivity at all times tested. Moreover, animals with constant high levels of CLOCK/BMAL1 functional activity due to the lack of circadian repressors (*Cryptochrome* double-knockout animals) were extremely resistant to the treatment (Gorbacheva et al. 2005). These data suggest that drug sensitivity is affected by the functional status of major circadian *trans*-activation complex, which translates into different gene expression patterns of its direct and indirect targets. Indeed, as shown recently, many drug metabolizing and detoxification enzymes as well as a number of key regulators of cell cycle progression and genotoxic stress responses display circadian patterns of expression at either the mRNA or protein level (Fu et al. 2002; Panda et al. 2002; Storch et al. 2002; Gachon et al. 2006). Moreover, some of them, such as the *Wee1* kinase and c-*Myc* transcription factor, have been identified as direct transcriptional targets of the circadian CLOCK/BMAL1 *trans*-activation complex (Fu et al. 2002; Matsuo et al. 2003). As expected, the expression pattern of many of these genes is altered in tissues of circadian mutant mice or in cells with modulated expression of circadian proteins (Gauger and Sancar 2005; Miller et al. 2007). It has been proposed that the abnormal response to genotoxic stress due to the changes in the abundance and/or activity of the key regulators of cell cycle and apoptosis underlies a cancer-prone phenotype reported for mice with a mutation in the *Per2* gene (Fu et al. 2002).

The differential sensitivity of various circadian mutant mice to drug-induced toxicity provided the first example for contrasting physiological phenotypes displayed by behaviorally arrhythmic animals and suggested a more complex interconnection between the circadian system and physiology. Detailed analysis of animals and cells deficient in individual core clock components further confirmed the complexity and the multilevel character of circadian modulation of stress response.

It was thus shown that a deficiency in BMAL1 (positive component of the circadian transcriptional feedback loop) sometimes results in the up-regulation of CLOCK/BMAL1 transcriptional targets, suggesting that CLOCK/BMAL1 may also work as a transcriptional repressor (Kondratov et al. 2006b). Importantly, while serving as a repressor, CLOCK/BMAL1 can interfere with the ability of other transcription factors to activate their targets, thus attenuating a DNA-damage response. Hence, the circadian clock may act as a master switch that under normal conditions permits or restricts activation of various signaling pathways. Consistently, the deficiency of individual clock components will result in deregulation of inducibility of stress response pathways, which may cause the development of various pathologies, including premature aging and tumorigenesis (see below).

In addition to transcription-based regulation, core circadian proteins PERIOD1 and TIMELESS interact with components of the cell cycle checkpoint system such as ATM/Chk2 and ATR/Chk1, respectively, and are necessary for activation of Chk1 and Chk2 by DNA damage (for review, see Kondratov and Antoch 2007). Importantly, radiation-induced ATM activation followed by ATM-dependent phosphorylation of CHK2 is impaired in cells with specific suppression of PER1 with small interfering RNA (siRNA). These data suggest that PER1 may act either as a cofactor for the activation of ATM or as an adapter protein for recruiting ATM substrates. In addition, modulation of PER1 expression alters both the steady-state expression levels and IR-dependent induction of several cell cycle checkpoint-related genes (Gery et al. 2006).

Analysis of the expression pattern of clock genes in tumor cells also revealed significant differences. Thus, expression of *Per1* is down-regulated in colon cancer (Krugluger et al. 2007), breast cancers (Chen et al. 2005) and endometrial carcinoma (Yeh et al. 2005) partly due to gene inactivation by promoter methylation. It has been suggested that down-regulation of *Per1* may benefit the survival of cancer cells, thus promoting carcinogenesis (Chen et al. 2005). Consistent with this, expression of exogenous PER1 dramatically reduces the growth of several tested cancer cell lines, presumably by increasing the rate of apoptosis and inducing cell cycle arrest (Gery et al. 2006). Finally, as was shown recently, intratumoral delivery of another *Period* gene, *Per2*, inhibits tumorigenesis in C57BL/6 mice (Hua et al. 2007).

Taken together, these data establish a strong molecular link between the clock proteins, on the one hand, and response to DNA-damaging agents, cell cycle regulation, and carcinogenesis, on the other hand.

Premature Aging in BMAL1-deficient Mice

Detailed analysis of mice deficient in various components of the molecular clock revealed a variety of pathological changes, which sometimes are specific for each mutation, and the list of these pathologies is constantly growing. One of the most striking phenotypes has been described in BMAL1-deficient animals, which display various metabolic defects (Rudic et al. 2004), severe joint ankylosis (Bunger et al. 2005), and premature aging (Kondratov et al. 2006a). BMAL1 (brain and muscle ARNT-like protein 1, also known as MOP3, ARNT3, or ARNTL1) belongs to the family of bHLH-PAS domain transcription factors. In complex with its partner, CLOCK, it regulates expression of target genes. Targeted disruption of BMAL1 results in an immediate loss of behavioral rhythmicity, suggesting that BMAL1 is a core component of the mammalian circadian clock (Bunger et al. 2000).

In addition to its well-established role in circadian mechanisms, a growing amount of recent data suggests that BMAL1 is also directly involved in regulation of a variety of physiological processes. This evidence came from the detailed phenotypic characterization of *Bmal1*$^{-/-}$ mice performed in several laboratories, implying BMAL1 in the regulation of adipogenesis (Shimba et al. 2005), glucose and fat metabolism (Rudic et al. 2004), blood pressure, and heart rate (Curtis et al. 2007). In addition, both male and female mice are infertile (Kennaway 2005) and develop idiopathic calcification and ossification of joints later in age (Bunger et al. 2005). The molecular mechanisms leading to the development of these pathologies are still unclear; however, it is believed that observed deregulation in the endocrine system and increased sensitivity to DNA-damaging agents may contribute to overall pathogenesis (Gorbacheva et al. 2005).

Recently, we have demonstrated that *Bmal1*$^{-/-}$ mice have several phenotypic changes, which may be summarized as segmental progeria (Kondratov et al. 2006a). Being born with an expected ratio and indistinguishable from their wild-type littermates, at the age of 16–18 weeks, BMAL1-deficient animals start showing signs of growth retardation. As a result, at the age of 35–40 weeks, the knockout animals morphologically resemble aged mice and die when they are 37.0 ± 12.1 weeks old (Kondratov et al. 2006a).

The progeria-like phenotype in *Bmal1*$^{-/-}$ mice is accompanied with many known features of normal and premature aging in animals. The most prominent among them are sarcopenia (age-related reduction of a muscle mass due to a decrease in the number of muscle fibers and their diameter), osteoporosis (reduction of a bone mass), kidney and spleen shrinkage, significant reduction of visceral and subcutaneous fat, and changes in blood cell composition (increased amount of neutrophiles and monocytes and decreased amount of lymphocytes). In addition, *Bmal1*$^{-/-}$ mice develop various forms of cataracts and cornea inflammation and lose elasticity of skin and hair growth ability early in age (Kondratov et al. 2006a).

The molecular mechanisms underlying the early aging phenotype of *Bmal1*$^{-/-}$ mice are still unclear. Most likely, it results from the superposition of deficiencies in multiple pathways that are normally controlled by BMAL1, including the regulation of glucose and fat metabolism and homeostasis (Curtis et al. 2004; Shimba et al. 2005) and the modulation of the genotoxic stress response (Gorbacheva et al. 2005), as well as direct involvement of BMAL1 in the regulation of reactive oxygen and reactive nitrogen species (ROS/RNS) (Kondratov et al. 2006a). Age-related accumulation of oxidative damage in an organism has been long recognized as one of the major causes of aging and age-associated degenerative diseases (Balaban et al. 2005). Consistent with this view, the level of free radicals in several tissues of BMAL1-deficient animals appeared to be significantly higher than in the tissues of the age-matched wild-type littermates. Importantly, such age-dependent accumulation of ROS/RNS occurs in those tissues that demonstrate severe reduction in size (i.e., spleen and kidney). Moreover, suppression of BMAL1 levels by siRNA in a cell culture results in an increase of ROS concentration, suggesting a direct involvement of this core circadian protein in ROS homeostasis (Kondratov et al. 2006a).

Noteworthy, BMAL1 is highly conserved between species. Thus, the sequences of the zebra fish and human homologs of the protein are 84% identical, whereas mouse and human proteins are 99% identical. Consistent with this, BMAL1 has a similar functional role in molecular circadian oscillator in human, mouse, and fly (Bell-Pedersen et al. 2005). A recently reported 15% reduction in life span in flies with a mutation in the CYCLE protein (*Drosophila* homolog of the mammalian BMAL1) (Hendricks et al. 2003) suggests that the role of BMAL1 in aging is conserved through evolution as well.

Low Doses of Radiation Provoke Premature Aging in *Clock* Mutant Mice

Although animals with a mutation in the BMAL1 transcriptional partner CLOCK (*Clock* mutant mice) share some of the phenotypic characteristics of *Bmal1*-deficient animals (i.e., high sensitivity to drug-induced genotoxic stress) (Gorbacheva et al. 2005), they normally do not demonstrate any signs of premature aging. On the contrary, some of the phenotypes are completely opposite to those displayed by *Bmal1* knockout animals. Unlike *Bmal1*-deficient mice that show age-dependent loss of muscle and adipose tissue, *Clock* mutant animals develop obesity relatively early in their life. It is caused by various metabolic deregulations including hyperleptinemia, hyperlipidemia, hyperglycemia, and hypoinsulinemia (Turek et al. 2005).

Another circadian mutant that showed some of the features of the early aging are *Per2*$^{m/m}$ mice (Fu et al. 2002; Lee 2006), although there are some important differences. First, in *Per2*$^{m/m}$ mice, this phenotype is developed only after exposure of animals to a low dose of irradiation. Second, it is not that severe compared to BMAL1 deficiency and is manifested mainly by early hair graying. Finally, it is accompanied by a higher incidence of tumor formation (Fu et al. 2002). The latter finding supports previous epidemiological studies linking the disruption of the circadian system (due to abnormal working schedules and/or exposure to frequent changes in the light/dark regime) to tumorigenesis (Antoch et al. 2005; Levi and Schibler 2007). Because previous studies of *Clock* mutant mice did not identify either a higher incidence of spontaneous tumor development or premature aging, in order to explore the potential role of the CLOCK protein in these processes, we exposed wild-type and *Clock/Clock* mice to a low dose of ionizing radiation and monitored them for any changes in their gross appearance for 80 weeks following the treatment.

As we reported recently, both wild-type and *Clock/Clock* mice showed no differences in their acute responses to 4 Gy of total body irradiation. However, starting at about 20 weeks after the treatment, *Clock* mutant mice begin to show signs of higher morbidity manifested by progressive weight loss, kyphosis, development of cataracts and eye inflammation, early hair graying, and alopecia. As a result, the survival rate of *Clock/Clock* mice is significantly reduced compared to similarly treated wild-type mice, and during their life span, they developed a phenotype reminiscent to premature aging in BMAL1-deficient animals (Fig. 1). Detailed necropsy performed on animals of both genotypes suggested that the reduction in body weight in *Clock/Clock* mice resulted from shrinkage of several major organs (liver, spleen, and kidney) as well as from the loss of the adipose tissue. However, no differences in the incidence of tumors were detected in animals of both genotypes (M. Antoch et al., in prep.). Together, these data conclusively demonstrate that although under normal conditions *Clock/Clock* mice do not display signs of premature aging, exposure to genotoxic treatment such as ionizing radiation accelerates the aging program without affecting carcinogenesis (M. Antoch et al., in prep.).

A

B

Figure 1. Premature aging in *Bmal*$^{-/-}$ and *Clock/Clock* mice. (*A*) Gross appearance of 1-year-old wild-type (WT) and BMAL1-deficient mice (Kondratov et al. 2006a). (*B*) Gross appearance of wild-type (WT) and *Clock/Clock* mutant mice 80 weeks after exposure to 4 Gy of total body irradiation (M. Antoch et al., in prep.).

This observation is supported by gene expression experiments demonstrating that the *Clock* mutation results in down- regulation of several proproliferative genes and up-regulation of genes associated with growth arrest and apoptosis (Miller et al. 2007; M. Antoch et al., in prep.) predicting less cancer-prone phenotype. Consistent with this, "clock-less" mice lacking both *Cryptochrome* genes are indistinguishable from their wild-type littermates in their response to genotoxic treatments (Gauger and Sancar 2005; Gorbacheva et al. 2005). They do not show signs of premature aging or a higher rate of tumorigenesis even after exposure to ionizing radiation (Gauger and Sancar 2005).

The unique progeria-like phenotype described in BMAL1-deficient animals, radiation-induced aging in *Clock/Clock* mice, or cancer-prone phenotype in *Per2*$^{m/m}$ mutant mice raises an important question on the causal links of these phenotypes with the disruption of the circadian function. The growing list of diverse and sometimes even contrasting phenotypes displayed by behaviorally arrhythmic circadian mutant mice suggests that in addition to their involvement in the molecular clock mechanism, clock proteins may have important roles in maintaining normal tissue homeostasis, and these roles may be distinct from their circadian function (for review, see Kondratov et al. 2007). This becomes particular obvious in comparing the phenotypes of *Bmal1*$^{-/-}$ and *Clock/Clock* mice.

Indeed, both CLOCK and BMAL1 are required for the formation of an active transcriptional complex that has a central role in circadian-regulated processes. Consistent with this, both *Bmal1*$^{-/-}$ and *Clock/Clock* mice demonstrate disruption of circadian rhythmicity in behavior and gene expression (Vitaterna et al. 1994; Bunger et al. 2000), as well as some similar changes in metabolic pathways (Rudic et al. 2004). However, in addition to this, they develop a variety of unique phenotypic characteristics. Thus, *Clock* mutants are obesity-prone, whereas the BMAL1-deficient animals show a "lean" phenotype and defects in adipogenesis (Shimba et al. 2005). In addition, only *Bmal*$^{-/-}$ mice normally display premature aging (Kondratov et al. 2006a). These differences may have several explanations. Thus, *Clock* mutant mice still express functional protein with reduced *trans*-activation properties and intact transcriptional repressor function (Gekakis et al. 1998; Kondratov et al. 2006b), which makes this model different from the conventional knockout. Because mice with a complete lack of the CLOCK protein have been generated only recently (DeBruyne et al. 2006) and are still not fully characterized, a detailed phenotypic comparison of *Clock*$^{-/-}$ and *Bmal1*$^{-/-}$ mice is required to fully address the role of the CLOCK/BMAL1 transcriptional complex and its individual components in maintaining normal tissue homeostasis. In addition, both CLOCK and BMAL1 may interact with unique partners forming the complexes, whose functional activity is independent from the CLOCK/BMAL1 dimer. It has been shown that NPAS2, the closest homolog of CLOCK, which has different tissue specificity of expression, can partially compensate CLOCK deficiency (DeBruyne et al. 2007).

Accelerated aging normally observed in BMAL1-deficient mice and provoked by a low dose of radiation in *Per2* and *Clock/Clock* mutants supports a recently proposed model describing the complex role that the entire circadian system and its core components has under normal and various stress conditions (Kondratov et al. 2007). According to this model, in addition to pathological changes resulting from systemically deregulated circadian function (Fig. 2A), the deficiency of a particular circadian protein may result in the development of a unique set of pathological changes (Fig. 2B). At the same time, because the expression and activities of all components of molecular clock machinery are interrelated, the deficiency of a particular circadian protein will affect the activity of others as well. Therefore, in addition to primary pathology, these animals will have an increased risk for development of secondary pathologies caused by a deficiency of other circadian proteins. In addition, mutations of circadian proteins may affect their interaction with the environment, in which case pathological changes that are not displayed normally will be provoked by environmental stress (Fig. 2C). Accelerated aging of *Clock/Clock* mice after exposure to ionizing radiation provides an excellent example for such scenario.

CONCLUDING REMARKS

The results obtained in our and other laboratories during the past several years clearly demonstrate the importance of the circadian system in mammalian and human physiology. The spectra of pathological changes described in mice with a deficiency of individual components of the molecular clock only partially overlap. Most of the observed pathologies are unique for each individual

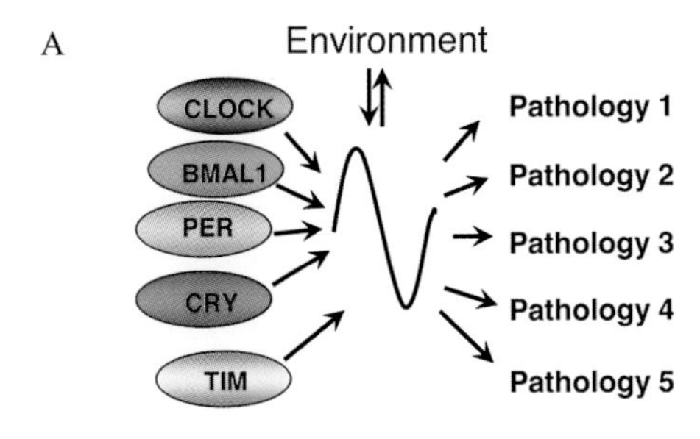

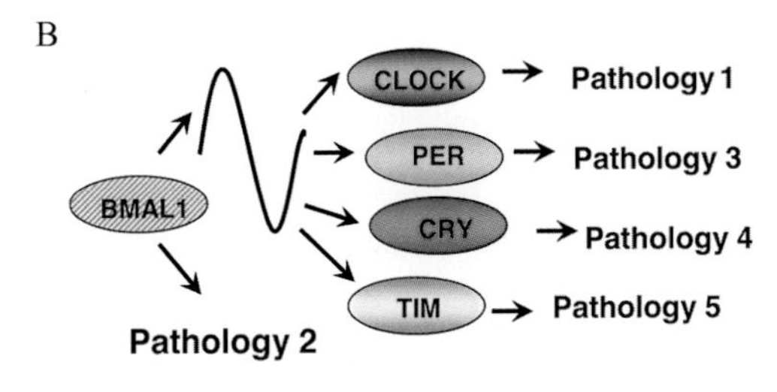

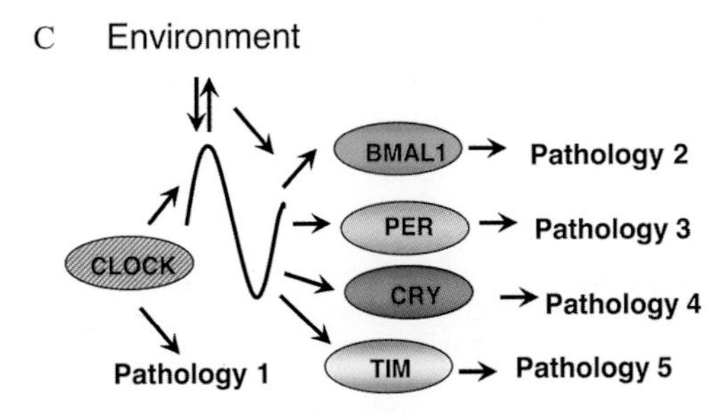

Figure 2. Role of the circadian system, circadian proteins, and organism/environment interactions in development of pathologies. (*A*) Deficiency in any circadian protein results in deregulation of the circadian system and desynchronization of multiple tissue-specific clocks with each other and with the environment, thus causing various pathological changes. (*B*) Individual circadian proteins have unique roles in maintaining normal tissue homeostasis that may be relatively independent from their roles in the molecular circadian oscillator. Impaired activity of a particular circadian protein (in this example, BMAL1) will result in a specific primary pathology (Pathology 2). At the same time, since BMAL1 is one of the core components of molecular clock, its deficiency will also affect entire circadian function, leading to a variety of secondary pathologies. (*C*) Improper interaction of an organism with the environment may affect the expression and/or activity of circadian proteins and induce specific pathologies without directly affecting clock function. (Reprinted, with permission from Kondratov et al. 2007 [© Elsevier].)

mutant, suggesting that at least some of them develop not as a result of circadian deregulation (because, in this case, they will be identical in all circadian mutants), but as a result of a deficiency in specific noncircadian function of an individual circadian protein.

Most importantly, these results provide molecular mechanisms explaining the increased risk of various diseases in shift workers and frequent time zone travelers. Traditionally, it was proposed that abnormal working schedules result in desynchronization of the tissue-specific molecular clocks within an organism and within an environment, which interferes with an organ's functions through a yet unknown mechanism. However, recent data suggest that exposure to frequent changes in the light/dark cycle may affect the expression of a particular circadian gene, which will cause the changes in the expression and/or activity of other molecular clock components. For example, the functional activity of BMAL1 depends on its expression (regulated by PERs, CRYs, and REV-ERB? proteins) and posttranslational modifications (regulated by CLOCK and CRY proteins). At the same time, BMAL1 is implicated in the control of ROS levels; therefore, any disturbance of the circadian system may have an effect on the activity of BMAL1, which, in turn, through ROS or other mechanisms, will change tissue homeostasis and increase risk of disease.

Taken together, the emerging data of the involvement of circadian proteins in the regulation of many processes in mammalian organisms demonstrate the dichotomy in biological consequences of disruption of the circadian clock with respect to many diseases, including aging and cancer. All of these findings make the circadian proteins perspective targets for the development of novel therapeutic approaches.

ACKNOWLEDGMENTS

We thank former members of the laboratory, Victoria Gorbacheva and Olena Vykhovanets, for 6 years of productive work that ensured the success of our research program. This work is supported by National Cancer Institute grant CA102522 and NIGMS grant GM075226 (to M.P.A).

REFERENCES

Antoch M.P., Kondratov R.V., and Takahashi J.S. 2005. Circadian clock genes as modulators of sensitivity to genotoxic stress. *Cell Cycle* **4:** 901.

Balaban R.S., Nemoto S., and Finkel T. 2005. Mitochondria, oxidants, and aging. *Cell* **120:** 483.

Bell-Pedersen D., Cassone V.M., Earnest D.J., Golden S.S., Hardin P.E., Thomas T.L., and Zoran M.J. 2005. Circadian rhythms from multiple oscillators: Lessons from diverse organisms. *Nat. Rev. Genet.* **6:** 544.

Bunger M.K., Walisser J.A., Sullivan R., Manley P.A., Moran S.M., Kalscheur V.L., Colman R.J., and Bradfield C.A. 2005. Progressive arthropathy in mice with a targeted disruption of the Mop3/Bmal-1 locus. *Genesis* **41:** 122.

Bunger M.K., Wilsbacher L.D., Moran S.M., Clendenin C., Radcliffe L.A., Hogenesch J.B., Simon M.C., Takahashi J.S., and Bradfield C.A. 2000. Mop3 is an essential component of the master circadian pacemaker in mammals. *Cell* **103:** 1009.

Chen S.T., Choo K.B., Hou M.F., Yeh K.T., Kuo S.J., and Chang J.G. 2005. Deregulated expression of the PER1, PER2 and PER3 genes in breast cancers. *Carcinogenesis* **26:** 1241.

Curtis A.M., Cheng Y., Kapoor S., Reilly D., Price T.S., and Fitzgerald G.A. 2007. Circadian variation of blood pressure and the vascular response to asynchronous stress. *Proc. Natl. Acad. Sci.* **104:** 3450.

Curtis A.M., Seo S.B., Westgate E.J., Rudic R.D., Smyth E.M., Chakravarti D., FitzGerald G.A., and McNamara P. 2004. Histone acetyltransferase-dependent chromatin remodeling and the vascular clock. *J. Biol. Chem.* **279:** 7091.

DeBruyne J.P., Weaver D.R., and Reppert S.M. 2007. CLOCK and NPAS2 have overlapping roles in the suprachiasmatic circadian clock. *Nat. Neurosci.* **10:** 543.

DeBruyne J.P., Noton E., Lambert C.M., Maywood E.S.,

Weaver D.R., and Reppert S.M. 2006. A clock shock: Mouse CLOCK is not required for circadian oscillator function. *Neuron* **50:** 465.

Fu L., Pelicano H., Liu J., Huang P., and Lee C. 2002. The circadian gene Period2 plays an important role in tumor suppression and DNA damage response in vivo. *Cell* **111:** 41.

Gachon F., Olela F.F., Schaad O., Descombes P., and Schibler U. 2006. The circadian PAR-domain basic leucine zipper transcription factors DBP, TEF, and HLF modulate basal and inducible xenobiotic detoxification. *Cell Metab.* **4:** 25.

Gauger M.A. and Sancar A. 2005. Cryptochrome, circadian cycle, cell cycle checkpoints, and cancer. *Cancer Res.* **65:** 6828.

Gekakis N., Staknis D., Nguyen H.B., Davis F.C., Wilsbacher L.D., King D.P., Takahashi J.S., and Weitz C.J. 1998. Role of the CLOCK protein in the mammalian circadian mechanism. *Science* **280:** 1564.

Gery S., Komatsu N., Baldjyan L., Yu A., Koo D., and Koeffler H.P. 2006. The circadian gene per1 plays an important role in cell growth and DNA damage control in human cancer cells. *Mol. Cell* **22:** 375.

Giacchetti S. 2002. Chronotherapy of colorectal cancer. *Chronobiol. Int.* **19:** 207.

Gorbacheva V.Y., Kondratov R.V., Zhang R., Cherukuri S., Gudkov A.V., Takahashi J.S., and Antoch M.P. 2005. Circadian sensitivity to the chemotherapeutic agent cyclophosphamide depends on the functional status of the CLOCK/BMAL1 transactivation complex. *Proc. Natl. Acad. Sci.* **102:** 3407.

Hendricks J.C., Lu S., Kume K., Yin J.C., Yang Z., and Sehgal A. 2003. Gender dimorphism in the role of cycle (BMAL1) in rest, rest regulation, and longevity in *Drosophila melanogaster*. *J. Biol. Rhythms* **18:** 12.

Hrushesky W.J. 1985. Circadian timing of cancer chemotherapy. *Science* 228: 73.

Hua H., Wang Y., Wan C., Liu Y., Zhu B., Wang X., Wang Z., and Ding J.M. 2007. Inhibition of tumorigenesis by intratumoral delivery of the circadian gene mPer2 in C57BL/6 mice. *Cancer Gene Ther.* **14:** 815.

Kennaway D.J. 2005. The role of circadian rhythmicity in reproduction. *Hum. Reprod. Update* 11: 91.

Ko C.H. and Takahashi J.S. 2006. Molecular components of the mammalian circadian clock. *Hum. Mol. Genet.* (spec. no. 2) **15:** R271.

Kobayashi M., Wood P.A., and Hrushesky W.J. 2002. Circadian chemotherapy for gynecological and genitourinary cancers. *Chronobiol. Int.* **19:** 237.

Kondratov R.V. and Antoch M.P. 2007. Circadian proteins in the regulation of cell cycle and genotoxic stress responses. *Trends Cell Biol.* **17:** 311.

Kondratov R.V., Gorbacheva V.Y., and Antoch M.P. 2007. The role of mammalian circadian proteins in normal physiology and genotoxic stress responses. *Curr. Top. Dev. Biol.* **78:** 173.

Kondratov R.V., Kondratova A.A., Gorbacheva V.Y., Vykhovanets O.V., and Antoch M.P. 2006a. Early aging and age-related pathologies in mice deficient in BMAL1, the core component of the circadian clock. *Genes Dev.* **20:** 1868.

Kondratov R.V., Shamanna R.K., Kondratova A.A., Gorbacheva V.Y., and Antoch M.P. 2006b. Dual role of the CLOCK/BMAL1 circadian complex in transcriptional regulation. *FASEB J.* **20:** 530.

Krugluger W., Brandstaetter A., Kallay E., Schueller J., Krexner E., Kriwanek S., Bonner E., and Cross H.S. 2007. Regulation of genes of the circadian clock in human colon cancer: Reduced period-1 and dihydropyrimidine dehydrogenase transcription correlates in high-grade tumors. *Cancer Res.* **67:** 7917.

Lee C.C. 2006. Tumor suppression by the mammalian *Period* genes. *Cancer Causes Control* **17:** 525.

Levi F. 2001. Circadian chronotherapy for human cancers. *Lancet Oncol.* **2:** 307.

Levi F. and Schibler U. 2007. Circadian rhythms: Mechanisms and therapeutic implications. *Annu. Rev. Pharmacol. Toxicol.* **47:** 593.

Matsuo T., Yamaguchi S., Mitsui S., Emi A., Shimoda F., and Okamura H. 2003. Control mechanism of the circadian clock for timing of cell division in vivo. *Science* **302:** 255.

Miller B.H., McDearmon E.L., Panda S., Hayes K.R., Zhang J., Andrews J.L., Antoch M.P., Walker J.R., Esser K.A., Hogenesch J.B., and Takahashi J.S. 2007. Circadian and CLOCK-controlled regulation of the mouse transcriptome and cell proliferation *Proc. Natl. Acad. Sci.* **104:** 3342.

Panda S., Antoch M.P., Miller B.H., Su A.I., Schook A.B., Straume M., Schultz P.G., Kay S.A., Takahashi J.S., and Hogenesch J.B. 2002. Coordinated transcription of key pathways in the mouse by circadian clock. *Cell* **109:** 307.

Rudic R.D., McNamara P., Curtis A.M., Boston R.C., Panda S., Hogenesch J.B., and Fitzgerald G.A. 2004. BMAL1 and CLOCK, two essential components of the circadian clock, are involved in glucose homeostasis. *PLoS Biol.* **2:** e377.

Sahar S. and Sassone-Corsi P. 2007. Circadian clock and breast cancer: A molecular link. *Cell Cycle* **6:** 1329.

Schibler U. and Sassone-Corsi P. 2002. A web of circadian pacemakers. *Cell* **111:** 919.

Shimba S., Ishii N., Ohta Y., Ohno T., Watabe Y., Hayashi M., Wada T., Aoyagi T., and Tezuka M. 2005. Brain and muscle Arnt-like protein-1 (BMAL1), a component of the molecular clock, regulates adipogenesis. *Proc. Natl. Acad. Sci.* **102:** 12071.

Storch K.F., Lipan O., Leykin I., Viswanathan N., Davis F.C., Wong W.H., and Weitz C.J. 2002. Extensive and divergent circadian gene expression in liver and heart. *Nature* **417:** 78.

Turek F.W., Joshu C., Kohsaka A., Lin E., Ivanova G., McDearmon E., Laposky A., Losee-Olson S., Easton A., Jensen D.R., Eckel R.H., Takahashi J.S., and Bass J. 2005. Obesity and metabolic syndrome in circadian Clock mutant mice. *Science* **308:** 1043.

Vitaterna M.H., King D.P., Chang A.M., Kornhauser J.M., Lowrey P.L., McDonald J.D., Dove W.F., Pinto L.H., Turek F.W., and Takahashi J.S. 1994. Mutagenesis and mapping of a mouse gene, Clock, essential for circadian behavior. *Science* **264:** 719.

Woon P.Y., Kaisaki P.J., Braganca J., Bihoreau M.T., Levy J.C., Farrall M., and Gauguier D. 2007. Aryl hydrocarbon receptor nuclear translocator-like (BMAL1) is associated with susceptibility to hypertension and type 2 diabetes. *Proc. Natl. Acad. Sci.* **104:** 14412.

Yeh K.T., Yang M.Y., Liu T.C., Chen J.C., Chan W.L., Lin S.F., and Chang J.G. 2005. Abnormal expression of period 1 (PER1) in endometrial carcinoma. *J. Pathol.* **206:** 111.

Sirtuins in Aging and Disease

L. GUARENTE
Department of Biology, Massachusetts Institute of Technology, Cambridge, Massachusetts 02139

Sirtuin genes function as anti-aging genes in yeast, *Caenorhabditis elegans*, and *Drosophila*. The NAD requirement for sirtuin function indicates a link between aging and metabolism, and a boost in sirtuin activity may in part explain how calorie restriction extends life span. In mammals, one of the substrates of the SIR2 ortholog, SIRT1, is a regulator of mitochondrial biogenesis, PGC-1α. Indeed, the putative SIRT1 activator resveratrol has been shown to stimulate mitochondrial biogenesis and deliver health benefits in treated mice. I explore here how mitochondrial biogenesis may have beneficial effects on aging and, perhaps, diseases of aging. In particular, I speculate that SIRT1-mediated mitochondrial biogenesis may reduce the production of reactive oxygen species, a possible cause of aging, and offer two possible mechanisms for this effect. An understanding of how calorie restriction works may lead to novel drugs to combat diseases of aging.

INTRODUCTION

Studies on aging in model systems have revealed a class of proteins termed sirtuins (homologs of yeast Sir2) that counteract aging (Blander and Guarente 2004). The founding member, yeast SIR2, extends the life span of mother cells by suppressing the formation of toxic rDNA circles (Sinclair and Guarente 1997) and by other mechanisms including management of oxidatively damaged molecules (Aguilaniu et al. 2003). The activity of SIR2 is increased by moderate calorie restriction (CR) and the *SIR2* gene is essential for the observed extension in life span by CR in many yeast strains (Lin et al. 2000, 2002; Anderson et al. 2003). In some yeast strains, Sir2 paralogs HST1 and HST2 also have a role in life extension by CR (Lamming et al. 2005).

The generality of sirtuins as anti-aging genes was shown by the life-extending effects of overexpression of SIR2 orthologs in *C. elegans* (Tissenbaum and Guarente 2001) and *Drosophila* (Rogina and Helfand 2004). Why are sirtuins pervasive regulators of aging in nature? An important clue was provided by the discovery that SIR2 and the human ortholog SIRT1 possess a unique biochemical activity—NAD-dependent protein deacetylase (Imai et al. 2000; Landry et al. 2000). This suggested that sirtuins are agents that regulate life span in accord with metabolism, which is in turn dictated by diet. The enzymatic activity of sirtuins is what actually prompted an examination of their possible role in CR. It now seems that that this role is quite general, because mammalian sirtuins, SIRT1 and SIRT4, have been clearly linked to several physiological processes regulated by diet (see below). This chapter discusses sirtuins and CR largely from an angle not usually sufficiently considered—the role of sirtuins in mitigating aging and diseases of aging by regulating mitochondrial function. Although many studies have linked SIRT1 to metabolic and stress-resistance pathways, the role of this sirtuin in regulating mitochondrial biogenesis may be central to the salutary effects of CR.

METABOLIC AND STRESS-SENSING PATHWAYS REGULATED BY SIRT1

Although the yeast Sir2p was shown to function by deacetylating histones, a large number of SIRT1 protein substrates have been identified that are nonhistone nuclear proteins. These tend to be transcription factors and cofactors that control a number of important physiological pathways in mammals. Figure 1 shows a partial listing of these substrates and groups them into two categories. The first group consists of regulators, the deacetylation of which by SIRT1 would confer resistance to cellular stress. For example, two of the early-defined substrates of SIRT1 are p53 (Luo et al. 2001; Vaziri et al. 2001) and FOXO proteins (Brunet et al. 2004; Motta et al. 2004), which respond to stress and mediate DNA repair and apoptosis. Most interestingly, the deacetylation of both p53 and FOXO proteins reduces apoptosis, whereas FOXO deacetylation also boosts expression of targets involved in DNA repair. It is very clear that fostering cellular repair and reducing cell loss would be consistent with an anti-aging effect on the organism, particularly in tissues comprising nondividing cells. Indeed, in neurons and cardiac myocytes, SIRT1 was shown to protect cells against stress-induced damage (Araki et al. 2004; J. Chen et al. 2005; Qin et al. 2006; Alcendon et al. 2007; Kim et al. 2007). The down-regulation of p53 is somewhat vexing from the perspective of cancer because CR is known to suppress cancer. Thus, one is compelled to consider the possibility that other SIRT1-driven anti-cancer mechanisms may exist to override any partial reduction in p53 activity.

The second category of proteins targeted by SIRT1 (Fig. 1) is involved in metabolism. In this group are regulators of adipocyte biology, including PPARγ and PGC-1α. PPARγ is a nuclear receptor that drives adipogenesis in white adipose tissue. By down-regulating PPARγ by docking with negative cofactors (Picard et al. 2004), SIRT1 is predicted to reduce adipogenesis. PGC-1α is a transcriptional coactivator that promotes mitochondrial biogenesis in the muscle and gluconeogenesis in the liver

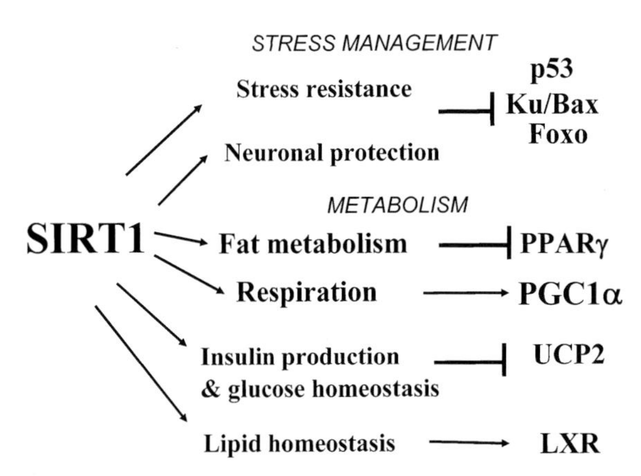

Figure 1. Targets of SIRT1 in mammalian cells. Targets known to be regulated and/or deacetylated by SIRT1 are grouped into two functional categories: stress resistance and metabolism. The p53, FOXO, and Ku (Cohen et al. 2004a) proteins are deacetylated by SIRT1 to reduce the tendency of cells to undergo apoptosis when exposed to damage. The deacetylation of FOXO proteins also boosts cellular capacity to repair the damage. Neuroprotection by SIRT1 overexpression and/or resveratrol treatment has been observed in vivo in mice and also in cultured neurons. On the metabolic side, SIRT1 represses PPARγ and deacetylates PGC-1α and LXRα and β to effect aspects of fat metabolism, respiration, and lipid homeostasis, as indicated. Regulation of insulin secretion in β cells occurs by repression of UCP2 and regulates glucose homeostasis. SIRT1 has also been shown to confer stress resistance to these cells by deacetylating FOXO1 (Kitamura et al. 2005).

(Lin et al. 2005). By deacetylating PGC-1α, SIRT1 has been shown to up-regulate its activity, thus potentiating increases in PGC-1α-dependent processes in cells (Rodgers et al. 2005; Gerhart-Hines et al. 2007). Note that most of these metabolic effects described above would sensitize the animal to insulin action and thus promote efficient metabolism of glucose.

In the β cells of the pancreas, SIRT1 has been shown to be a positive regulator of insulin secretion by repressing the mitochondrial uncoupling protein UCP2 (Moynihan et al. 2005; Bordone et al. 2006). Again, this effect would place SIRT1 on the side of efficient glucose clearance from blood. Another target of SIRT1 may be NF-κB, the deacetylation of which was reported to repress its activity (Yeung et al. 2004). Perhaps in this case, one functional role is to reduce the proinflammatory activity of macrophages driven by NF-κB, which again would be consistent with life extension. However, a complication of this latter mechanism may be a down-regulation of the innate immune response, and it should be fascinating to see more generally how sirtuins influence the immune system.

A CONSERVED CR PATHWAY THAT INCREASES MITOCHONDRIA

CR studies in yeast have been plagued by confusion sown by the use of different degrees of glucose deprivation. In the method employed by the labs of Guarente, Sinclair, and Lin, glucose is reduced from the canonical 2% to 0.5%, a regimen I will call moderate CR (Lin et al. 2000, 2002; Anderson et al. 2003; Easlon et al. 2007). The method used by the Kaeberlein/Kennedy labs used 0.05% glucose, which I will call severe CR (Kaeberlein et al. 2004, 2005b). Although the longevity engendered by moderate CR generally was shown to require SIR2, that of severe CR was not. A recent paper from Easlon et al. (2007) clarified the distinction between these protocols. Employing the yeast strain used by Kennedy, these authors showed in a single controlled experiment that the longevity engendered by moderate CR required sirtuin genes, whereas that induced by severe CR did not (Fig. 2). In the latter case, life extension was actually repressed by sirtuins, suggesting that severe CR may invoke novel pathways that are actually interfered with by sirtuins.

The question arises whether moderate or severe CR in yeast is a better model of CR in higher organisms. Another important distinction between the two yeast mechanisms is that moderate CR induced a boost in mitochondrial respiration, which was required for life span extension; deletion of CYT1 encoding cytochrome c_1 abolished the SIR2-dependent extension in life span in moderate CR (Lin et al. 2002). In contrast, severe CR did not require mitochondria at all (Kaeberlein et al. 2005a). Furthermore, triggering mitochondrial biogenesis by enforced expression of the transcription factor HAP4 gave rise to a long life span in 2% glucose that could not be further extended by moderate CR (Lin et al. 2002). This experiment shows that the boost in respiration is sufficient to extend the life span. More specifically, life span was also extended sim-

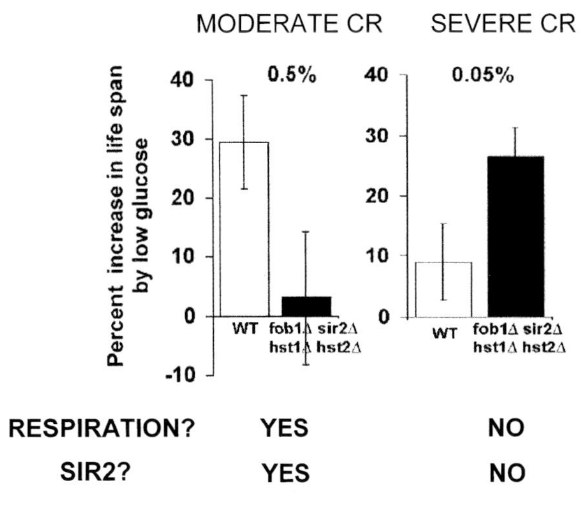

Figure 2. A requirement of *SIR2* genes for CR in yeast. This figure (Easlon et al. 2007) shows the effects of moderate CR (0.5% glucose) and severe CR (0.05% glucose) on life span of yeast mother cells. In the wild-type strain BKY4742 (Kaeberlein et al. 2005a) both moderate and severe CR extend life span, although moderate CR extends more (*open bars*). (*Closed bars*) Strain in which three sirtuin genes (*SIR2*, *HST1*, and *HST2*) have been deleted as well as the *FOB1* gene (to prevent rDNA recombination). In this strain, moderate CR no longer extends the life span, but severe CR works even better. This study shows that sirtuins are essential for the extension of life span by moderate CR, but they are not necessary and even antagonistic for extension of life span by severe CR. Furthermore, the figure indicates that moderate CR is known to require respiration for extension of the life span, whereas severe CR does not. (Reprinted, with permission, from Easlon et al. 2007 [©American Society for Biochemistry and Molecular Biology].)

ply by overexpressing the NADH dehydrogenases, NDE1 or NDI1, which donate electrons from NADH to the electron transport chain (Lin et al. 2002).

Three recent studies suggest that the induction of respiration by CR may be general (Fig. 3). First, CR mice (60–70% ad libitum feeding) showed an increase in mitochondria biogenesis and a higher rate of respiration than ad libitum–fed controls (Nisoli et al. 2005). Interestingly, these increases seemed to work via induction of endothelial nitric oxide synthase (eNOS) to increase the levels of nitric oxide, which the authors showed was an inducer of SIRT1 gene expression. Second, humans on a CR diet for 6 months showed increases in expression of mitochondrial proteins in muscle, as well as SIRT1 itself (Civitarese et al. 2007). Third, *C. elegans* on a regimen of dietary restriction increased their respiration, and blocking this increase with respiration inhibitors prevented the extension in life span (Bishop and Guarente 2007). In this last example, the boost in respiration throughout the animal was triggered by the action of the *skn-1* gene transcription factor acting in just two neurons in the head of the worm. The effects on respiration and life span are evidently triggered by an endocrine-based mechanism emanating from these two cells.

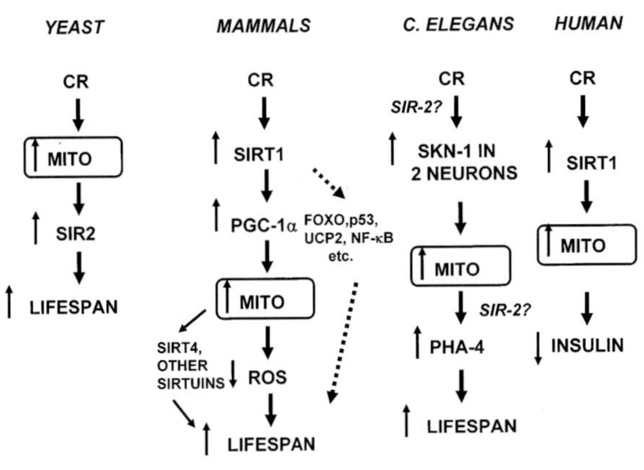

Figure 3. CR pathways across species. Shown is what is known about CR pathways in the four indicated species. In yeast, CR increases mitochondrial respiration, which activates SIR2 and extends the life span. In mammals, CR also activates SIRT1 in certain tissues (via increasing eNOS activity), which leads to deacetylation and activation of PGC-1α, increased mitochondrial biogenesis, and increased life span. We suggest that a key mechanism of this mammalian pathway may be a reduction in ROS by the increased mitochondrial biogenesis, as discussed in the text and Fig. 4. Effects of SIRT1 on other targets and effects of other sirtuins may also be important, as indicated. In *C. elegans*, CR up-regulates the SKN-1 transcription factor in the two ASI neurons in the head, which leads to a systemic increase in respiration accompanied by activation of the forkhead protein PHA-4 and an increase in life span. Possible positions where any of the four sirtuin genes could have roles in this pathway are indicated. In humans, 6-month CR was shown to lead to activation of expression of SIRT1 and mitochondrial proteins in muscle, as determined by analysis of punch biopsies. Physiological changes observed in rodents, such as reduction in blood insulin, are also observed in these people. For all four organisms, the increase in mitochondria is boxed to emphasize its position in the pathway relative to SIR2 orthologs; i.e., in yeast, SIR2 lies downstream from mitochondria, and in mammals, the ortholog SIRT1 lies upstream.

Upon inspection, there appears to be a major difference in the relationship between *sir2* genes and mitochondria in CR pathways of different organisms (Fig. 3). In yeast, *SIR2* lies downstream from mitochondria. This makes sense in light of the finding that the increase in respiration increases the NAD/NADH ratio, which is one of the mechanisms that can activate the SIR2 enzyme to extend life span (Lin et al. 2004). Another mechanism of SIR2 activation during CR is an increase in the expression of PNC1, an NAD salvage synthesis enzyme, which consumes the known SIR2 inhibitor, nicotinamide (Anderson et al. 2003). In mice, however, SIRT1 lies *upstream* of mitochondria. SIRT1 protein levels in muscle are induced by CR (Cohen et al. 2004b). As discussed above, one of the important substrates for SIRT1 is the transcriptional coactivator, PGC-1α, which is an inducer of mitochondrial biogenesis (Rodgers et al. 2005). By deacetylating PGC-1α and thereby increasing its activity, SIRT1 promotes the biogenesis of mitochondria, a topic expanded upon below. Although SIRT1 was shown to be required for at least one output of CR in mice (D. Chen et al. 2005), functional roles of sirtuins in CR in *C. elegans* and humans have not been shown as of this time.

The increase in SIRT1 activity in CR is also expected to affect other targets in mammalian cells. For example, deacetylation of SIRT1 described above like FOXO proteins and p53 may promote stress resistance during CR. However, the increase in PGC-1α and the resulting mitochondrial biogenesis may be the major driving force for the health benefits of CR and is the subject explored below.

The fact that SIRT1 functions upstream of mitochondrial biogenesis in mice does not preclude the possibility that sirtuins may also function downstream from (i.e., in response to) the increase. For example, the mitochondrial ADP-ribosyl-transferase SIRT4 is known to be down-regulated by CR, and it has been suggested that changes in the NAD/NADH ratio concomitant with changes in respiration rates may mediate this effect (Haigis et al. 2006). The down-regulation of SIRT4 during CR leads to an increase in the activity of its substrate for ADP-ribosylation, the crucial mitochondrial enzyme, glutamate dehydrogenase, which facilitates the entry of glutamine and glutamate into central metabolism. The use of amino acids as carbon and energy sources may be particularly important during CR. This logic may also apply to the other mitochondrial sirtuins, SIRT3 and SIRT5. A final point is that even SIRT1 itself may also function downstream from the mitochondrial respiration change, although this has not yet been shown.

HOW MIGHT MITOCHONDRIAL BIOGENESIS PROTECT AGAINST DAMAGE?

As mentioned above, it is possible that the function of sirtuins or other proteins can be regulated by respiration, and these downstream effects may mediate the benefits of CR. Another intriguing possibility is that mitochondrial biogenesis is salutary per se. Why might this be? One simple possibility is that mitochondria are subject to damage over time by toxic by-products of respiration, reactive oxygen species (ROS). ROS have long been proposed as a causative agent in aging because they damage proteins, DNA, and lipids in cells, and their effects are cumulative (Harmon 1956). Having more mitochondria might buffer over the progressive loss of these ATP-generating organelles over time and mitigate deleterious effects on diseases of aging (Wallace 2005).

Another more subtle possibility is that mitochondrial biogenesis reduces the *production* of ROS. Figure 4 shows possible mechanisms how SIRT1/PGC-1α-driven biogenesis of mitochondria may reduce ROS. A common misconception is that an increase in respiration would necessarily trigger an increase in production of ROS. In fact, ROS are produced when electrons are stalled on complexes I and III of the electron transport chain and combine with local oxygen to produce ROS (Barros et al. 2004; Barja 2007).

By this reckoning, there are at least two possible ways that increased mitochondria biogenesis may reduce ROS. First, electrons can stall when the charge gradient across the mitochondrial membrane generated by proton extrusion becomes too steep, for example, under conditions of excess energy and high levels of NADH (Fig. 4). It simply becomes more difficult for the electron transport chain to pump more protons out of mitochondria against this hyperpolarized gradient. An expanded mitochondrial surface would reduce the steepness of the charge gradient by spreading the charges over a larger area and, in this way, may prevent electron stalling and ROS production.

A second possible mechanism (Fig. 4) is that an increase in respiration might reduce local oxygen levels and thus reduce the generation of ROS by complexes I and III (Barja 2007). In this regard, the binding of oxygen to these complexes is likely to be much less avid than the bona fide O_2 using components of the chain, complex IV (cytochrome *c* oxidase), and complex V (ATPase), so a mild reduction in oxygen levels may selectively disfavor ROS production but not impact on respiration itself.

It remains to be seen how valid these ideas are. In the first place, the extent to which ROS themselves contribute to aging and diseases of aging is not yet clear. Second, we do not know for certain that mitochondrial biogenesis would be sufficient to reduce ROS production. What we do know is that CR reduces the production of ROS (Weindruch and Walford 1988) and, as described above, increases mitochondrial production. Investigating whether an increase in mitochondrial biogenesis per se is what causes reduced ROS during CR is a worthy goal for future research.

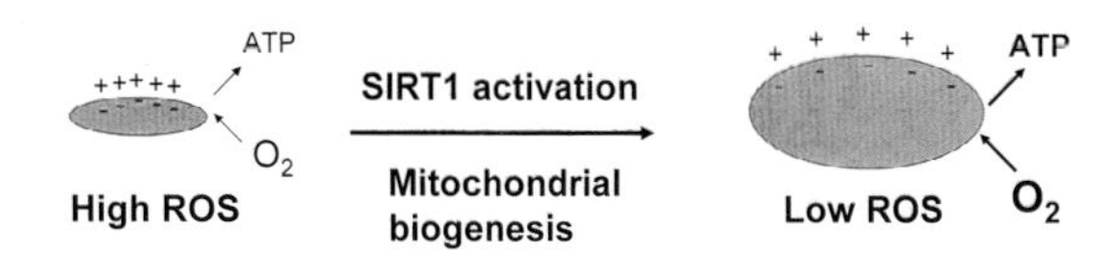

Small mito area **Steep proton gradient** **ET stalls at complex I & III** **$e^- + O_2$ = ROS**	Model 1	**Expanded Mito area** **Relaxed gradient** **ET flows** **Low ROS**
Low respiration **High local O_2** **More substrate for ROS at complex I and III**	Model 2	**High respiration** **Low local O_2** **Less substrate for ROS at complex I and III**

Figure 4. Relationship between mitochondrial biogenesis and ROS production. It is known that CR increases mitochondrial biogenesis and reduces the production of ROS. This figure suggests two possible mechanisms for a causal relationship between mitochondrial biogenesis and reduced ROS. In model 1, the increase in biogenesis is proposed to spread the charge gradient generated by proton pumping across the mitochondrial membrane over a larger surface area, thereby reducing the steepness of the charge gradient across the membrane. This reduction will ameliorate the stalling of electrons in the electron transport chain. This stalling is known to cause the production of ROS by complexes I and III of that chain because stalled electrons can leech off, combine with O_2, and form ROS. In model 2, the increase in mitochondrial biogenesis is proposed to increase respiration, thereby lowering local O_2 levels available for the production of ROS by complexes I and III. Because the components of the chain that use O_2 as a true substrate (complexes IV and V) should have a much higher affinity for O_2 than complexes I and III, a partial reduction in O_2 levels may reduce ROS production without compromising electron transport and respiration itself.

COULD SIRTUIN-DRIVEN MECHANISMS MITIGATE EFFECTS OF DIETARY EXCESS?

The metabolic and mitochondrial effects discussed above are very relevant to an enlarging health problem in the developed world—metabolic syndrome induced by overeating and a sedentary lifestyle (for review, see Guarente 2006). This syndrome is characterized by glucose intolerance and obesity and is associated with cardiovascular disease and a predisposition to diabetes. If SIRT1 mediates some of the effects of CR, could manipulating its activity pharmacologically provide health benefits broadly and particularly for people with metabolic syndrome?

First, one must worry that salutary effects of CR observed in rodents might not translate to primates. In fact, current primate studies show that many of the favorable metabolic changes induced by this diet in rodents also occur in monkeys, for example, high glucose tolerance (Mattison et al. 2007). Indeed, patients in the 6-month study of humans on a CR diet mentioned above showed improvements in glucose and fat homeostasis (Civitarese et al. 2007).

Several years ago, a high-throughput screen for SIRT1 activators identified several polyphenols, a class of stress-induced natural products made by plants (Howitz et al. 2003). One of these, resveratrol, is found in red wine. Although there has been some controversy about whether and how resveratrol activates SIRT1 in vitro, there have been numerous examples of biological activities of this compound that are SIR2-dependent in vivo. These include life span extension in yeast (Howitz et al. 2003), *C. elegans* (Viswanathan et al. 2005), and *Drosophila* (Wood et al. 2004) and neuroprotection (J. Chen et al. 2005) and fat reduction (Picard et al. 2004) in higher organisms.

Interestingly, two studies showed that dosing mice with resveratrol counteracted some of the adverse consequences of a high fat/high calorie diet, including shortened life span (Baur et al. 2006; Lagouge et al. 2006). One of these studies went on to show that the mice fed resveratrol underwent a considerable expansion in mitochondrial number in muscle (Lagouge et al. 2006). In the C2C12 myocyte cell line, mitochondrial biogenesis was shown to be under the control of transfected PGC-1α and SIRT1. Thus, this may be an example in which the pathway of SIRT1 activation detailed above, i.e., mitochondrial biogenesis, has a clear health benefit in mammals.

If indeed polyphenols or other compounds can be proven to be at least partial CR mimetics, one might expect some of the health benefits associated with this diet to accrue. In mice, CR protects against many of the diseases of aging, including cancer and neurodegenerative diseases (Ingram et al. 1987; Zhu et al. 1999; Wang et al. 2005). If SIRT1 or other sirtuins mediate the effects of CR, it seems likely that small molecules that can bind to them and modulate their activity similar to CR would provide benefit. In the extreme, one can imagine a magic bullet that could forestall or treat any disease for which aging is a significant risk factor, i.e., all of the major diseases, excluding infectious disease.

CONCLUSIONS

In summary, we are on the verge of understanding CR at a detailed molecular level, which may allow the development of powerful new drugs to provide some of the health benefits of this dietary regimen. Would such a class of new drugs mean that any responsibility to maintain a healthy life style would be obviated? The answer seems likely to be no. Any CR mimetic would likely move an individual some distance along the health spectrum toward the excellent health observed under CR conditions. For example, someone already in good health would obtain even better health. However, someone with metabolic syndrome or worse, diabetes, would likely be moved the same distance up the health spectrum to a place better than before, but will end up no where near as healthy as someone who began healthy. Thus, these drugs will not create *carte blanche* for irresponsible living. So live clean and hang on!

ACKNOWLEDGMENTS

I thank all of the scientists who have contributed to this work over the years. My lab is supported by the National Institutes of Health and The Glenn Foundation.

REFERENCES

Aguilaniu H., Gustafsson L., Rigoulet M., and Nystrom T. 2003. Asymmetric inheritance of oxidatively damaged proteins during cytokinesis. *Science* **299:** 1751.

Alcendon R.R., Gao S., Zhai P., Zablocki D., Holle E., Yu X., Tian B., Wagner T., Vatner S.F., and Sadoshima J. 2007. Sirt1 regulates aging and resistance to oxidative stress in the heart. *Circ. Res.* **100:** 1512.

Anderson R.M., Bitterman K.J., Wood J.G., Medvedik O., and Sinclair D.A. 2003. Nicotinamide and PNC1 govern lifespan extension by calorie restriction in *Saccharomyces cerevisiae*. *Nature* **423:** 181.

Araki T., Sasaki Y., and Milbrandt J. 2004. Increased nuclear NAD biosynthesis and SIRT1 activation prevent axonal degeneration. *Science* **305:** 1010.

Barja G. 2007. Mitochondrial oxygen consumption and reactive oxygen species production are independently modulated: Implications for aging studies. *Rejuvenation Res.* **10:** 215.

Barros M.H., Bandy B., Tahara E.B., and Kowaltowski A.J. 2004. Higher respiratory activity decreases mitochondrial reactive oxygen release and increases life span in *Saccharomyces cerevisiae*. *J. Biol. Chem.* **279:** 49883.

Baur J.A., Pearson K.J., Price N.L., Jamieson H.A., Lerin C., Kalra A., Prabhu V.V., Allard J.S., Lopez-Lluch G., Lewis K., Pistell P.J., Poosala S., Becker K.G., Boss O., Gwinn D., Wang M., Ramaswamy S., Fishbein K.W., Spencer R.G., Lakatta E.G., Le Couteur D., Shaw R.J., Navas P., Puigserver P., Ingram D.K., de Cabo R., and Sinclair D.A. 2006. Resveratrol improves health and survival of mice on a high-calorie diet. *Nature* **444:** 337.

Bishop N.A. and Guarente L. 2007. Two neurons mediate diet-restriction-induced longevity in *C. elegans*. *Nature* **447:** 545.

Blander G. and Guarente L. 2004. The Sir2 family of protein deacetylases (review). *Annu. Rev. Biochem.* **73:** 417.

Bordone L., Motta M.C., Picard F., Robinson A., Jhala U.S., Apfeld J., McDonagh T., Lemieux M., McBurney M., Szilvasi A., Easlon E.J., Lin S.J., and Guarente L. 2006. Sirt1 regulates insulin secretion by repressing UCP2 in pancreatic beta cells. *PLoS Biol.* **4:** e31.

Brunet A., Sweeney L.B., Sturgill J.F., Chua K.F., Greer P.L., Lin Y., Tran H., Ross S.E., Mostoslavsky R., Cohen H.Y., Hu L.S., Cheng H.L., Jedrychowski M.P., Gygi S.P., Sinclair D.A., Alt F.W., and Greenberg M.E. 2004. Stress-dependent regulation of FOXO transcription factors by the SIRT1 deacetylase. *Science* **303:** 2011.

Chen J., Zhou Y., Mueller-Steiner S., Chen L.F., Kwon H., Yi S., Mucke L., and Gan L. 2005. SIRT1 protects against microglia-dependent amyloid-beta toxicity through inhibiting NF-kappaB signaling. *J. Biol. Chem.* **280:** 40364.

Chen D., Steele A.D., Lindquist S., and Guarente L. 2005. Increase in activity during calorie restriction requires Sirt1. *Science* **310:** 1641.

Civitarese A.E., Carling S., Heilbronn L.K., Hulver M.H., Ukropcova B., Deutsch W.A., Smith S.R., and Ravussin E. 2007. Calorie restriction increases muscle mitochondrial biogenesis in healthy humans. *PLoS Med.* **4:** e76.

Cohen H.Y., Lavu S., Bitterman K.J., Hekking B., Imahiverobo T.A., Miller C., Frye R., Ploegh H., Kessler B.M, and Sinclair D.A. 2004a. Acetylation of the C terminus of Ku70 by CBP and PCAF controls Bax-mediated apoptosis. *Mol. Cell* **13:** 627.

Cohen H.Y., Miller C., Bitterman K.J., Wall N.R., Hekking B., Kessler B., Howitz K.T., Gorospe M., de Cabo R., and Sinclair D.A. 2004b. Calorie restriction promotes mammalian cell survival by inducing the SIRT1 deacetylase. *Science* **305:** 390.

Easlon E., Tsang F., Dilova I., Wang C., Lu S.P., Skinner C., and Lin S.J. 2007. The dihydrolipoamide acetyltransferase is a novel metabolic longevity factor and is required for calorie restriction-mediated life span extension. *J. Biol. Chem.* **282:** 6161.

Gerhart-Hines Z., Rodgers J.T., Bare O., Lerin C., Kim S.H., Mostoslavsky R., Alt F.W., Wu Z., and Puigserver P. 2007. Metabolic control of muscle mitochondrial function and fatty acid oxidation through SIRT1/PGC-1alpha. *EMBO J.* **26:** 1913.

Guarente L. 2006. Sirtuins as potential targets for metabolic syndrome. *Nature* **444:** 868.

Harman D. 1956. Aging: A theory based on free radical and radiation chemistry. *J. Gerontol.* **11:** 298.

Haigis M.C., Mostoslavsky R., Haigis K.M., Fahie K., Christodoulou D.C., Murphy A.J., Valenzuela D.M., Yancopoulos G.D., Karow M., Blander G., Wolberger C., Prolla T.A., Weindruch R., Alt F.W., and Guarente L. 2006. SIRT4 inhibits glutamate dehydrogenase and opposes the effects of calorie restriction in pancreatic beta cells. *Cell* **126:** 941.

Howitz K.T., Bitterman K.J., Cohen H.Y., Lamming D.W., Lavu S., Wood J.G., Zipkin R.E., Chung P., Kisielewski A., Zhang L.L., Scherer B., and Sinclair D.A. 2003. Small molecule activators of sirtuins extend *Saccharomyces cerevisiae* lifespan. *Nature* **425:** 191.

Imai S., Armstrong C.M., Kaeberlein M., and Guarente L. 2000. Transcriptional silencing and longevity protein Sir2 is an NAD-dependent histone deacetylase. *Nature* **403:** 795.

Ingram D.K., Weindruch R., Spangler E.L., Freeman J.R., and Walford R.L. 1987. Dietary restriction benefits learning and motor performance of aged mice.
J. Gerontol. **42:** 78.

Kaeberlein M., Kirkland K.T., Fields S., and Kennedy B.K. 2004. Sir2-independent life span extension by calorie restriction in yeast. *PLoS Biol.* **2:** E296.

Kaeberlein M., Hu D., Kerr E.O., Tsuchiya M., Westman E.A., Dang N., Fields S., and Kennedy B.K. 2005a. Increased life span due to calorie restriction in respiratory-deficient yeast. *PLoS Genet.* **1:** e69.

Kaeberlein M., Powers R.W.,III, Steffen K.K., Westman E.A., Hu D., Dang N., Kerr E.O., Kirkland K.T., Fields S., and Kennedy B.K. 2005b. Regulation of yeast replicative life span by TOR and Sch9 in response to nutrients. *Science* **310:** 1193.

Kim D., Nguyen M.D., Dobbin M.M., Fischer A., Sananbenesi F., Rodgers J.T., Delalle I., Baur J.A., Sui G., Armour S.M., Puigserver P., Sinclair D.A., and Tsai L.H. 2007. SIRT1 deacetylase protects against neurodegeneration in models for Alzheimer's disease and amyotrophic lateral sclerosis. *EMBO J.* **26:** 3169.

Kitamura Y.I., Kitamura T., Kruse J.P., Raum J.C., Stein R., Gu W., and Accili D. 2005. FoxO1 protects against pancreatic beta cell failure through NeuroD and MafA induction. *Cell Metab.* **2:** 153.

Lagouge M., Argmann C., Gerhart-Hines Z., Meziane H., Lerin C., Daussin F., Messadeq N., Milne J., Lambert P., Elliott P., Geny B., Laakso M., Puigserver P., and Auwerx J. 2006. Resveratrol improves mitochondrial function and protects against metabolic disease by activating SIRT1 and PGC-1alpha. *Cell* **127:** 1109.

Lamming D.W., Latorre-Esteves M., Medvedik O., Wong S.N., Tsang F.A., Wang C., Lin S.J., and Sinclair D.A. 2005. HST2 mediates SIR2-independent life-span extension by calorie restriction. *Science* **309:** 1861.

Landry J., Sutton A., Tafrov S.T., Heller R.C., Stebbins J., Pillus L., and Sternglanz R. 2000. The silencing protein SIR2 and its homologs are NAD-dependent protein deacetylases. *Proc. Natl. Acad. Sci.* **97:** 5807.

Lin S.J., Defossez P.A., and Guarente L. 2000. Requirement of NAD and Sir2 for life-span extension by calorie restriction in *Saccharomyces cerevisiae*. *Science* **289:** 2126.

Lin S.J., Ford E., Haigis M., Liszt G., and Guarente L. 2004. Calorie restriction extends yeast lifespan by lowering the level of NADH. *Genes Dev.* **18:** 12.

Lin S.J., Kaeberlein M., Andalis A.A., Sturtz L.A., Defossez P.A., Culotta V.C., Fink G.R., and Guarente L. 2002. Calorie restriction extends *Saccharomyces cerevisiae* lifespan by increasing respiration. *Nature* **418:** 344.

Lin J., Handschin C., and Spiegelman B.M. 2005. Metabolic control through the PGC-1 family of transcription coactivators (review). *Cell Metab.* **1:** 361.

Luo J., Nikolaev A.Y., Imai S., Chen D., Shiloh A., Guarente L., and Gu W. 2001. Negative control of p53 by Sir2a promotes cell survival under stress. *Cell* **107:** 137.

Mattison J.A., Roth G.S., Lane M.A., and Ingram D.K. 2007. Dietary restriction in aging nonhuman primates. *Interdiscip. Top. Gerontol.* **35:** 137.

Motta M.C., Divecha N., Lemieux M., Kamel C., Chen D., Gu W., Bultsma Y., McBurney M., and Guarente L. 2004. Mammalian SIRT1 represses forkhead transcription factors. *Cell* **116:** 551.

Moynihan K.A., Grimm A.A., Plueger M.M., Bernal-Mizrachi E., Ford E., Cras-Meneur C., Permutt M.A., and Imai S. 2005. Increased dosage of mammalian Sir2 in pancreatic beta cells enhances glucose-stimulated insulin secretion in mice. *Cell Metab.* **2:** 105.

Nisoli E., Tonello C., Cardile A., Cozzi V., Bracale R., Tedesco L., Falcone S., Valerio A., Cantoni O., Clementi E., Moncada S., and Carruba M.O. 2005. Calorie restriction promotes mitochondrial biogenesis by inducing the expression of eNOS. *Science* **310:** 314.

Picard F., Kurtev M., Chung N., Topark-Ngarm A., Senawong T., Machado De Oliveira R., Leid M., McBurney M.W., and Guarente L. 2004. Sirt1 promotes fat mobilization in white adipocytes by repressing PPAR-gamma. *Nature* **429:** 771.

Qin W., Yang T., Ho L., Zhao Z., Wang J., Chen L., Zhao W., Thiyagarajan M., MacGrogan D., Rodgers J.T., Puigserver P., Sadoshima J., Deng H., Pedrini S., Gandy S., Sauve A.A., and Pasinetti GM. 2006. Neuronal SIRT1 activation as a novel mechanism underlying the prevention of Alzheimer disease amyloid neuropathology by calorie restriction. *J. Biol. Chem.* **281:** 21745.

Rogina B. and Helfand S.L. 2004. Sir2 mediates longevity in the fly through a pathway related to calorie restriction. *Proc. Natl. Acad. Sci.* **101:** 15998.

Rodgers J.T., Lerin C., Haas W., Gygi S.P., Spiegelman B.M., and Puigserver P. 2005. Nutrient control of glucose homeostasis through a complex of PGC-1alpha and SIRT1. *Nature* **434:** 113.

Sinclair D.A. and Guarente L. 1997. Extrachromosomal rDNA circles: A cause of aging in yeast. *Cell* **91:** 1033.

Tissenbaum H.A. and Guarente L. 2001. Increased dosage of a sir-2 gene extends lifespan in *Caenorhabditis elegans*. *Nature* **410:** 227.

Vaziri H., Dessain S.K., Ng Eaton E., Imai S.I., Frye R.A., Pandita T.K., Guarente L., and Weinberg R.A. 2001. hSIR2(SIRT1) functions as an NAD-dependent p53 deacetylase. *Cell* **107:** 149.

Viswanathan M., Kim S.K., Berdichevsky A., and Guarente L. 2005. A role for SIR-2.1 regulation of ER stress response genes in determining *C. elegans* life span. *Dev. Cell* **9:** 605.

Wallace D.C. 2005. A mitochondrial paradigm of metabolic and degenerative diseases, aging, and cancer: A dawn for evolutionary medicine (review). *Annu. Rev. Genet.* **39:** 359.

Wang J., Ho L., Qin W., Rocher A.B., Seror I., Humala N., Maniar K., Dolios G., Wang R., Hof P.R., and Pasinetti G.M. 2005. Caloric restriction attenuates beta-amyloid neuropathology in a mouses model of Alzheimer's disease. *FASEB J.* **19:** 659.

Weindruch R. and Walford R. L. 1988. *The retardation of aging and disease by dietary restriction*. C.C. Thomas, Springfield, Illinois.

Wood J.G., Rogina B., Lavu S., Howitz K., Helfand S.L., Tatar M., and Sinclair D. 2004. Sirtuin activators mimic caloric restriction and delay ageing in metazoans. *Nature* **430:** 686.

Yeung F., Hoberg J.E., Ramsey C.S., Keller M.D., Jones D.R., Frye R.A., and Mayo M.W. 2004. Modulation of NF-kappaB-dependent transcription and cell survival by the SIRT1 deacetylase. *EMBO J.* **23:** 2369.

Zhu H., Guo Q., and Mattson M.P. 1999. Dietary restriction protects hippocampal neurons against the death-promoting action of a presenilin-1 mutation. *Brain Res.* **842:** 224.

Identification of *Caenorhabditis elegans* Genes Regulating Longevity Using Enhanced RNAi-sensitive Strains

A.V. SAMUELSON,*† R.R. KLIMCZAK,‡ D.B. THOMPSON,* C.E. CARR,*§ AND G. RUVKUN,*†

**Department of Molecular Biology, Massachusetts General Hospital, Boston, Massachusetts 02114; †Department of Genetics, Harvard Medical School, Boston, Massachusetts 02115; ‡Department of Molecular and Cellular Biology, University of California, Berkeley, California 94122; §Department of Earth, Atmospheric and Planetary Sciences, Massachusetts Institute of Technology, Cambridge, Massachusetts 02114*

A systematic genome-wide RNA interference screen was performed in the *Caenorhabditis elegans lin-15b;eri-1* strain, which has an enhanced response to double-stranded RNA including the nervous system, to identify life-span regulatory factors. In total, 16,757 genes were examined, revealing 115 gene inactivations that extended life span. A more stringent longitudinal analysis revealed 18 gene inactivations that induced the greatest increase in life span (10–90%), all of which extended life span when inactivated either in *eri-1* alone or in a second strain with an enhanced response to double-stranded RNA, *eri-3*. Most reduced the rate of aging, implying that animals aged more slowly. As was the case in previous studies, genes critical for metabolism caused the greatest extension of longevity. Extension of life span occurs through disparate mechanisms as increased resistance to thermal stress, oxidative damage, and decreased age pigment accumulation analysis of the 18 stronger positives failed to demonstrate a correlation between enhanced stress resistance and decreased lysosomal function. Consistently, *aps-3* and *lys-10*, two genes annotated to have lysosomal functions, extended life span when inactivated without enhancing stress resistance. The results of this study reinforce the importance of metabolism, mitochondrial and lysosomal functions, genomic stability, and stress resistance on animal life-span determination.

INTRODUCTION

From the nematode with an adult life span of approximately 20 days to the bowhead whale with an estimated life span of 200 years (Finch and Austad 2001), eukaryotic evolution has generated a large variety of organisms with a wide life-span range. Even more phylogenetically related animals demonstrate large differences within their life span: Laboratory rodents typically live for 2–3 years, whereas naked mole rats can live for more than 26 years (Buffenstein 2005). Genetic analysis of the cardinal organisms—worms, flies, and mice—has identified factors and pathways that impact organismal longevity.

New studies indicate that genetic factors have a large role in attaining exceptionally old age in humans. Perls et al. (2002) studied 2000 relatives who lived to 100 years. The study found that brothers of centenarians were 17 times more likely than a control group to reach the age of 100, whereas sisters of centenarians were eight times more likely to reach this age (Perls et al. 2002). These results indicate the presence of a "handful of genes that could be playing really substantial roles in the ability to get to very old age" (Perls et al. 2002).

The biochemical functions of genes associated with increased life span in *C. elegans* have been implicated to function in such processes as metabolism and free radical production, processes that are potentially significant to aging models within vertebrates (Finch and Ruvkun 2001).

An insulin-like signaling pathway regulates longevity and metabolism in *C. elegans* (Kimura et al. 1997). *daf-2* encodes the worm ortholog of the insulin/IGF-1 (insulin-like growth factor-1) receptor gene (Kimura et al. 1997). *C. elegans* with loss-of-function mutations in *daf-2* live up to three times longer than wild type (Dorman et al. 1995). High insulin signaling activates the DAF-2 receptor, which in turn activates an AGE-1 phosphoinositol-3 kinase (PI3K) to further activate the downstream kinases PDK-1, AKT-1 (PKB), and AKT-2 (PKB) (Dorman et al. 1995). The AKT-1 and AKT-2 kinases phosphorylate the DAF-16 transcription factor, preventing nuclear localization, subsequently resulting in a change in transcription of DAF-16 target genes. *daf-16* encodes two proteins with forkhead DNA-binding domains, the human orthologs of which are the FoxO factors (Ogg et al. 1997). Under conditions that favor reproductive growth rather than dauer arrest, DAF-16 is excluded from the nucleus, does not activate the genes involved in dauer arrest and long life span, or does not repress the genes responsible for inhibiting reproductive growth and short life span (Finch and Ruvkun 2001). Under nonoptimal growth conditions that favor dauer arrest, this kinase cascade is inactive and thus DAF-16 is nuclear and active. Some of the known regulated targets of DAF-16 include genes that express key free-radical detoxifying enzymes, for instance, the *sod-3* manganese superoxide dismutase (mnSOD), which converts toxic superoxide radicals and peroxides to less reactive products (Larsen 1993), in accordance with free-radical aging theory. Furthermore, DAF-16 together with HSF-1 activates the expression of specific genes encoding small heat shock proteins (sHSPs) (Hsu et al. 2003). These sHSPs act as molecular chaperones that inhibit protein aggregation and delay the onset of polyglutamine-expansion protein aggregation. Interestingly, enhanced expression of *shsp* was also found to increase longevity (Hsu et al. 2003).

Enhanced longevity of *daf-2* mutants is rescued (i.e., reduced to normal wild-type levels) by exclusive neuronal expression of *daf-2* but not by restoration of *daf-2* path-

  978-087969823-2

way activity to the muscles or intestines individually (Wolkow et al. 2000). Importantly, these results suggest that the nervous system is the critical tissue where *daf-2* insulin-like signaling regulates aging. Neurons themselves perhaps may be the most vulnerable to free-radical damage during aging. In one study in *Drosphila*, it was found that overexpression of copper/zinc superoxide dismutase within motor neurons contributed to life-span extension by 48% (Parkes et al. 1998). Laser ablation studies of distinct neurons further underscore the neuronal role of longevity regulation because certain sensory neurons were found to inhibit longevity, whereas others promoted longevity, most likely by influencing insulin/IGF-1 signaling (Alcedo and Kenyon 2004).

RNA interference (RNAi) has provided an efficient technology to discover genes that mediate life span. One of the critical issues concerning the use of RNAi is its cell specificity in gene silencing. Although the use of RNAi ensures a much less encumbered process of mutational "knockdown" gene analysis, delivery of RNAi itself is problematic. Previous RNAi wild-type studies have been inherently restricted by their limited function at gene inactivation in neurons. Performing an RNAi screen in a strain with enhanced neuronal response to double-stranded RNA (dsRNA) allows the detection of additional factors regulating longevity that perhaps were previously obscured in wild-type studies. A strain that exhibits increased RNAi sensitivity thus permits a more inclusive RNAi screen, including its neuronal constituents.

Two classical genetic lesions that induce enhanced RNAi phenotypes, *eri-1* and *lin15b*, have been used for RNAi screens in the nervous system (Kennedy et al. 2004; Sieburth et al. 2005; Wang et al. 2005). The DexDh exonuclease *eri-1* is an endogenous inhibitor of RNAi, whereas *lin-15b* is implicated in chromatin remodeling and also enhances the response to dsRNAs. *eri-3* acts in a complex with *eri-1* and like mutations in *eri-1*; *eri-3* mutations enhance the response to dsRNAs that are not effective in wild type (Duchaine et al. 2006).

MATERIALS AND METHODS

Strains. The *lin-15b(n744);eri-1(mg366)* mutant strain of *C. elegans*, with an enhanced RNAi phenotype, was used in the genome-wide functional genomic screen for increased longevity. Additional analysis was conducted using *eri-3(mg408), eri-1(mg366);daf-16(mgDf47),* and *eri-1(mg366).*

RNAi life-span screen. A large-scale systematic RNAi screen was performed with 16,757 *C. elegans* genes. dsRNA corresponding to each gene was introduced to the worm via an *Escherichia coli* strain expressing the dsRNAs from a cloning vector containing a selective ampicillin-resistant marker. Each RNAi colony was grown overnight on shakers in Luria broth with 50 µg/ml ampicillin at 37°C in 96-well plates and then seeded onto 24-well RNAi agar plates containing 5 mM isopropylthiogalactoside to induce the expression of the T7 RNA polymerase, which binds to promoter elements flanking the *C. elegans* DNA segment to in turn produce a dsRNA corresponding to that *C. elegans* DNA segment. The identities of the RNAi clones in Table 1 were confirmed by DNA sequencing.

To obtain synchronized colonies of animals, worms were grown on nematode growth medium plates (NMP) until they became gravid adults, upon which their eggs were extracted with treatment with a sodium hypochlorite/hydroxide solution. The eggs were then grown overnight at room temperature. Approximately 25 synchronized larval stage-1 (L1) worms were then added to each well. The worms were allowed to develop to larval stage 4 (L4) at 15°C and then 5-fluorodeoxyuridine (FUDR) solution was added to prevent further nematode reproduction by abating gene synthesis, upon which the plates were transferred to a 25°C incubator. Two controls were also grown alongside these colonies composed of worms fed *daf-2* RNAi and empty vector (positive and negative controls, respectively). Once all of the animals fed vector had died (~12 days after FUDR was added), the plates were then scored for life span. Worms were scored as dead when they fail to thrash about when immersed in water or when they failed to respond to gentle prodding by a platinum wire. Clones that exhibited at least a 10% survival rate were marked as positive.

Retest of positives from RNAi life-span screen. The positives obtained from the primary RNAi life-span screen were each retested in triplicate for increased survival at day-12 adulthood to ensure the removal of false positives within the preliminary data. Age-synchronized L1 animals were grown on each RNAi clone at 15°C until L4, at which point animals were moved to 25°C and FUDR was added to prevent reproduction (adulthood day 0). Additionally, positive and negative RNAi controls, *daf-2* and empty vector, respectively, were included on every 24-well plate. The previous screen's protocol was identically followed in screening these clones. The three retests consisted of a complete set of positives that were screened in a staggered fashion to ensure that there was minimal environmental variability within each single set of positives.

Life-span longitudinal assay. Instead of following mortality in one population of animals over time, enough multiple replica sets were created so that one set could be scored, on average, every other day. Similar to the primary screen, each RNAi colony was grown overnight in Luria broth with 50 µg/ml ampicillin and then seeded onto 24-well RNAi agar plates containing 5 mM isopropylthiogalactoside to induce dsRNA expression overnight at room temperature. About 25 synchronized L1s were added to each well and allowed to develop to L4 at 15°C; FUDR solution was then added to the RNAi plates and the plates were moved to 20°C (day = 0). For each 24-well plate, one well contained bacteria expressing dsRNA to an empty vector and another to *daf-2*. At each time point, one replica in the set was measured by both flooding the wells with water and gentle prodding with a platinum wire; animals that failed to move under either treatment were scored as dead. At each subsequent time point, survival in a different replica set was measured. The survival proportion at each time point for every gene inactivation was obtained from three independent experiments.

Table 1. 18 Gene Inactivations Extending Life Span in *C. elegans*

Code	Strain: gene	Age?	*lin-15b(n744);eri-1(mg366)* rel. life. +/– S.D.	*lin-15b(n744);eri-1(mg366)* MRDT	*eri-3(mg408)* rel. life. +/– S.D.	*eri-1(mg366)* rel. life. +/– S.D.	*daf-16(mgDf47); eri-1(mg366)* rel. life. +/– S.D.	Function
*	vector		1.00 +/– 0.02	1.07	1.00 +/– 0.05	1.00 +/– 0.03	1.00 +/– 0.04	control
#	*daf-2*	*	2.57 +/– 0.07	2.23	2.12 +/– 0.11	2.68 +/– 0.46	1.13 +/– 0.03	insulin/IGF-1 receptor
A	R12E2.2		1.19 +/– 0.02	1.26	1.20 +/– 0.05	1.54 +/– 0.05	1.06 +/– 0.05	membrane protein galactose-binding-like [KOG1396]
B	F36A2.7		1.59 +/– 0.03	2.32	1.85 +/– 0.08	1.63 +/– 0.08	1.31 +/– 0.04	unknown [LSE0817]
C	*cco-1*	*	1.82 +/– 0.04	2.93	1.78 +/– 0.07	1.80 +/– 0.06	1.33 +/– 0.05	cytochrome oxidase subunit Vb
D	F59C6.5	*	1.26 +/– 0.03	1.37	1.52 +/– 0.07	1.70 +/– 0.05	1.31 +/– 0.05	NADH-ubiquinone oxidoreductase
E	W09C5.8	*	1.66 +/– 0.04	2.07	1.63 +/– 0.07	1.97 +/– 0.12	1.33 +/– 0.05	cytochrome oxidase IV (complex IV)
F	*aps-3*		1.27 +/– 0.05	1.19	1.02 +/– 0.05	1.41 +/– 0.09	1.04 +/– 0.04	AP-3 subunit, intracellular membrane trafficking
G	F11G11.13		1.13 +/– 0.03	2.72	1.10 +/– 0.05	1.23 +/– 0.07	1.04 +/– 0.03	creatine kinases [KOG3581]
H	F22D3.5		1.09 +/– 0.02	1.63	1.08 +/– 0.05	1.11 +/– 0.07	1.09 +/– 0.05	unknown
I	B0491.5		1.35 +/– 0.03	1.55	1.54 +/– 0.06	1.64 +/– 0.05	1.19 +/– 0.04	similarity to mitochondrial NADH-ubiquinone oxidoreductase
L	F29C4.7		1.12 +/– 0.03	2.02	1.09 +/– 0.04	1.09 +/– 0.10	1.09 +/– 0.04	large RNA-binding protein, orthologous to human RBM15
M	*pat-6*	*	1.02 +/– 0.03	4.49	1.35 +/– 0.07	1.67 +/– 0.08	1.17 +/– 0.04	actinin-type actin-binding domain required for muscle assembly/function
N	T21D12.9		1.22 +/– 0.03	2.16	1.11 +/– 0.08	1.35 +/– 0.10	1.03 +/– 0.04	membrane glycoprotein LIG-1 [KOG4194]; Bestrophin [KOG3547]
O	W03G1.8		1.17 +/– 0.05	2.39	1.06 +/– 0.05	1.14 +/– 0.08	1.01 +/– 0.04	unknown
P	C06E7.1		1.49 +/– 0.03	2.53	1.07 +/– 0.05	1.47 +/– 0.07	0.98 +/– 0.04	*S*-adenosylmethionine synthetase
Q	*lys-10*		1.07 +/– 0.03	2.51	1.22 +/– 0.05	1.18 +/– 0.09	1.03 +/– 0.04	similarity to *Entemeba histolytica* lysozymes
R	K08E7.1		1.31 +/– 0.04	6.14	1.36 +/– 0.06	1.15 +/– 0.09	1.05 +/– 0.04	similarity to xxidation resistance protein 1
S	*srt-56*		1.17 +/– 0.07	0.69	0.94 +/– 0.05	1.27 +/– 0.10	1.09 +/– 0.04	7-transmembrane receptor
T	Y39B6A.3		1.46 +/– 0.07	1.32	1.61 +/– 0.06	1.34 +/– 0.06	1.15 +/– 0.04	Fe-S cluster biosynthesis protein ISA1 (HesB-like domain)
			Median life span (days adulthood 20°C) +/– S.D.					
*	vector		13.9 +/– 0.2		18.5 +/– 0.7	14.0 +/– 0.3	12.2 +/– 0.3	
#	*daf-2*	*	35.5 +/– 0.8		38.9 +/– 1.4	37.3 +/– 6.4	13.9 +/– 0.2	
A	R12E2.2		16.5 +/– 0.2		22.2 +/– 0.6	21.5 +/– 0.5	13.0 +/– 0.5	
B	F36A2.7		22.0 +/– 0.3		34.0 +/– 0.7	22.8 +/– 1.0	16.1 +/– 0.2	
C	*cco-1*	*	25.2 +/– 0.4		32.9 +/– 0.3	25.2 +/– 0.7	16.4 +/– 0.5	
D	F59C6.5	*	17.5 +/– 0.3		28.1 +/– 0.7	23.8 +/– 0.5	16.1 +/– 0.5	
E	W09C5.8	*	23.0 +/– 0.5		30.1 +/– 0.6	27.4 +/– 1.7	16.3 +/– 0.4	
F	*aps-3*		17.6 +/– 0.7		18.8 +/– 0.5	19.7 +/– 1.1	12.8 +/– 0.4	
G	F11G11.13		15.6 +/– 0.4		20.3 +/– 0.5	17.2 +/– 0.9	12.7 +/– 0.2	
H	F22D3.5		15.1 +/– 0.2		19.9 +/– 0.6	15.5 +/– 0.9	13.3 +/– 0.5	
I	B0491.5		18.7 +/– 0.3		28.3 +/– 0.3	23.0 +/– 0.5	14.6 +/– 0.3	
L	F29C4.7		15.6 +/– 0.4		20.0 +/– 0.4	15.2 +/– 1.3	13.3 +/– 0.3	
M	*pat-6*	*	14.1 +/– 0.3		24.9 +/– 0.9	23.3 +/– 0.9	14.3 +/– 0.2	
N	T21D12.9		16.9 +/– 0.4		20.4 +/– 1.4	18.8 +/– 1.4	12.6 +/– 0.3	
O	W03G1.8		16.2 +/– 0.7		19.5 +/– 0.6	15.9 +/– 1.1	12.4 +/– 0.3	
P	C06E7.1		20.6 +/– 0.3		19.8 +/– 0.6	20.5 +/– 0.9	12.0 +/– 0.4	
Q	*lys-10*		14.8 +/– 0.4		22.5 +/– 0.4	16.5 +/– 1.2	12.6 +/– 0.4	
R	K08E7.1		18.2 +/– 0.6		25.1 +/– 0.6	16.1 +/– 1.3	12.9 +/– 0.3	
S	*srt-56*		16.2 +/– 0.9		17.4 +/– 0.6	17.8 +/– 1.4	13.3 +/– 0.4	
T	Y39B6A.3		20.2 +/– 0.9		29.6 +/– 0.3	18.6 +/– 0.8	14.1 +/– 0.4	

Asterisk indicates previously published to extend *C. elegans* life span when inactivated.

Gompertz and logit fitting. The Gompertz model for age-specific mortality $M = M_0e^{Ga}$, with initial mortality M_0, aging coefficient G, and age a, can be shown to give the proportion dead $p = 1 - \exp[(M_0/G)(1 - e^{Ga})]$. In the logit model for proportion dead, $p = e^{b+ca}/(1 + e^{b+ca})$, the parameters b and c are analogous to, but different from, $\log(M_0)$ and G in the Gompertz model. Reflecting this limited analogy, we refer to b and c as initial mortality and the aging coefficient, respectively.

When analyzing data for a particular clone, sample proportions from different trials at the same age were pooled using population weighting. We performed population-weighted nonlinear least-squares Gompertz model fits using a Trust-Region Reflective Newton algorithm implemented by the MATLAB *fit()* function (The MathWorks, Inc., Natick, Massachusetts). We performed population-weighted logit fits using the MATLAB *glmfit()* function. We estimated the LD_{50}, or age at which $p = 0.5$, given by $\log[1 + \log(2)(G/M_0)]/G$ in the Gompertz model or $-b/a$ in the logit model.

Age pigment accumulation. To access whether a gene inactivation accelerated the accumulation of autofluorescent age pigment, each RNAi colony was grown overnight in Luria broth with 50 µg ml ampicillin and then seeded onto 6-well RNAi agar plates containing 5 mM isopropylthiogalactoside to induce dsRNA expression overnight at room temperature. Between 50 and 75 synchronized *eri-3(mg408)* L1s were added to each well and allowed to develop to L4 at 15°C; FUDR solution was then added to the RNAi plates and the plates were moved to 20°C (day = 0). The presence of intestinal fluorescence as a result of age pigment was assayed using a Discovery V12 Zeiss stereomicroscope with an incident wavelength of 420 nm. Images of representative animals were taken at 37x magnification. Quantification of age pigment levels at each time point was obtained using Axiovision Rel v4.5 software (Zeiss) of the intestinal fluorescence in ten animals between the pharynx and the vulva opening. Background worm fluorescence was subtracted by measuring fluorescence within the head, and the average fluorescence at adulthood days 3, 8, 10, and 13 was determined.

Heat shock and paraquot survival. Synchronized L1 *eri-3(mg408)* animals were grown on 24-well RNAi agar plates containing the appropriate dsRNA expressing bacteria, including a *daf-2* and empty vector control. Animals were incubated at 15°C until worms developed into L4-stage animals. Subsequently, FUDR was added and the worms were then incubated at 20°C. Animals were tested either for increased thermotolerance or for resistance to oxidative damage at day-5 adulthood. To measure heat shock survival, animals were shifted to 35°C for a period of 24 hours, after which they were shifted back to 20°C for 12 hours and then scored for viability. To measure increased resistance to oxidative damage, animals were collected into 1.5-ml Eppendorf tubes, washed twice in S-basal, and then treated with 50 µl of 300 mM paraquot for 6 hours at room temperature with vigorous shaking. After treatment, animals were washed twice in S-basal and returned to 24-well plates with fresh bacteria expressing the appropriate dsRNA; they were allowed to recover for 12 hours at 20°C, at which point viability was scored. Each experiment was conducted in triplicate.

RESULTS

Previous functional genomic feeding-based RNAi screens for extension of life span in *C. elegans* have identified approximately 170 gene inactivations (Lee et al. 2003; Hamilton et al. 2005; Hansen et al. 2005; Curran and Ruvkun 2007). However, overlap among the aging genes identified in these studies is minimal, but many of these gene inactivations cause barely detectable extensions of life span, suggesting that a very large number of gene inactivations can approach these borderline cutoffs. The weakness of the life extension of these gene inactivations could be due to poor efficacy of RNAi to inactivate a gene to extend life span or because these gene inactivations cause marginally interpretable increases in life span. Neuronal insulin signaling is key in the regulation of life span, and neurons are generally refractory to RNAi, suggesting that perhaps some of the gene inactivations that induce marginal increases in life span of wild type may cause more profound increases in life span in a strain with enhanced neuronal RNAi (Wolkow et al. 2000; Kennedy et al. 2004). Mutations have been identified in *C. elegans* that enhance the efficacy of both neuronal and global RNAi (*eri* mutation). A strain carrying a combination of two *eri* mutants, *lin-15b(n744)* and *eri-1(mg366)*, is most sensitive to RNAi and most enhanced for neuronal RNAi. In fact, this strain has been used in RNAi screens for synaptic components (Sieburth et al. 2005).

Strains harboring the *lin-15b* mutation, however, are short-lived (Fig. 1A, compare open squares to open circles and data not shown), implying either inherent sickness and/or accelerated aging. Demographic (Gompertz) analysis reveals that *lin-15b(n744);eri-1(mg366)* animals have almost a twofold exponential increase in the relative rate of aging and almost a fivefold increase in relative initial mortality rate (xIMR) (Fig. 1B,E compare open square to open circle), suggesting both an inherent increased mortality and accelerated aging. Interestingly, decreased *daf-2* function (RNAi) extends the relative life span of *lin-15b(n744);eri-1(mg366)* and wild-type animals to a similar extent, and they have a similar rate of aging (Fig. 1C,D,E, compare closed squares to closed circles). However, decreased insulin signaling decreased the xIMR sixfold in wild-type animals but there was no change in the xIMR of *lin-15b(n744);eri-1 (mg366)* animals (Fig. 1B). These results imply that decreased insulin signaling rescues the increased rate of aging of the *lin-15b* mutation but not the age-independent causes of increased mortality. Consistently, *lin-15b(n744); eri-1(mg366)* animals have accelerated age pigment accumulation (Fig. 2, compare A and B), which is suppressed by decreased insulin signaling (Fig. 2C). Collectively, these results suggest that the *lin-15b* mutation accelerates aging, which is suppressed by decreased insulin signaling.

Given that *daf-2* functions in neurons to influence life span, it is paradoxical that *daf-2* RNAi extends relative life

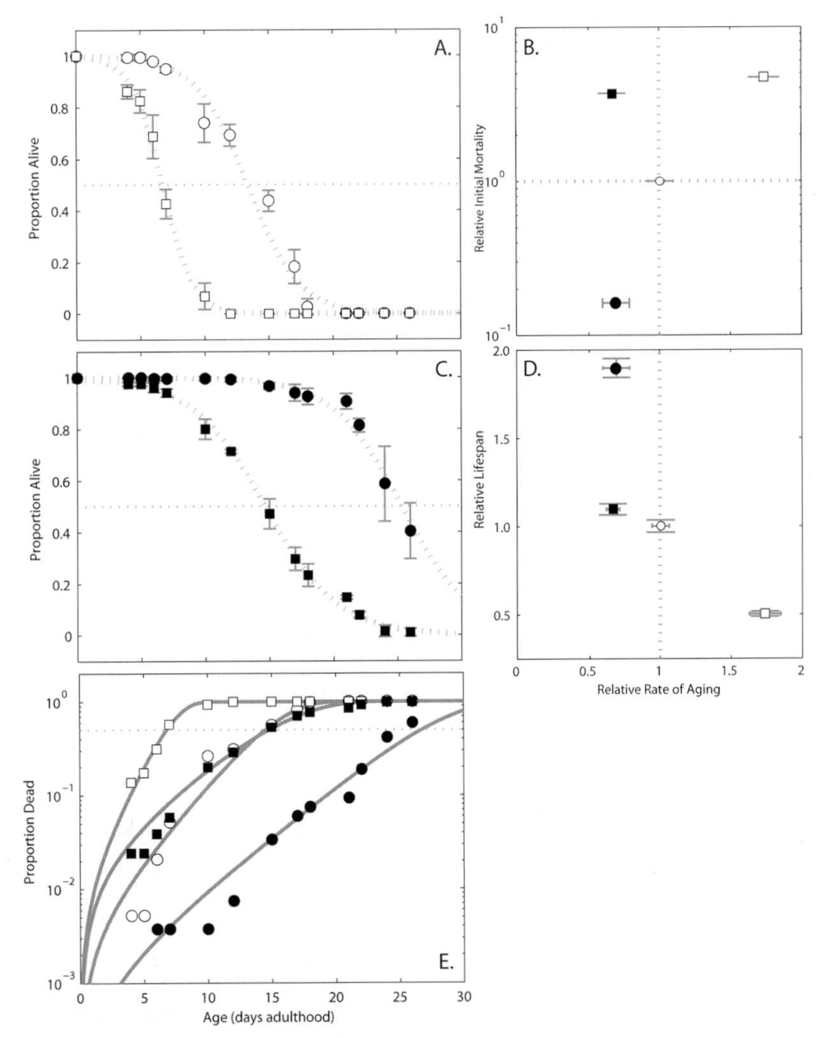

Figure 1. (*A*) Survival of *lin-15b(n744);eri-1(mg366)* (*squares*) and wild-type (N2) animals (*circles*) on empty vector control RNAi at 25°C. (*B*) Relative rate of aging versus relative initial mortality rate (compared to wild type) at 25°C. (*C*) Survival of *lin-15b(n744);eri-1(mg366)* (*squares*) and wild-type (N2) animals (*circles*) on *daf-2* RNAi at 25°C. (*D*) Relative rate of aging versus relative median life span (compared to wild type) at 25°C. (*E*) Gompertz fit to *A* and *C*.

span to a similar extent in *lin-15b(n744);eri-1(mg366)* and wild-type animals (Fig. 1D). One possible explanation is that decreased insulin signaling actually induces an increased sensitivity to RNAi (Wang et al. 2005); thus, RNAi to *daf-2* may make neurons of wild-type animals susceptible to *daf-2* RNAi in a feed-forward loop.

To identify other gene inactivations that can increase life span in an enhanced RNAi mutant background, a feeding-based RNAi library was used to systematically inactivate 16,757 annotated *C. elegans* genes in *lin-15b(n744);eri-1(mg366)* animals. Mortality was measured at the time when a population of animals on control

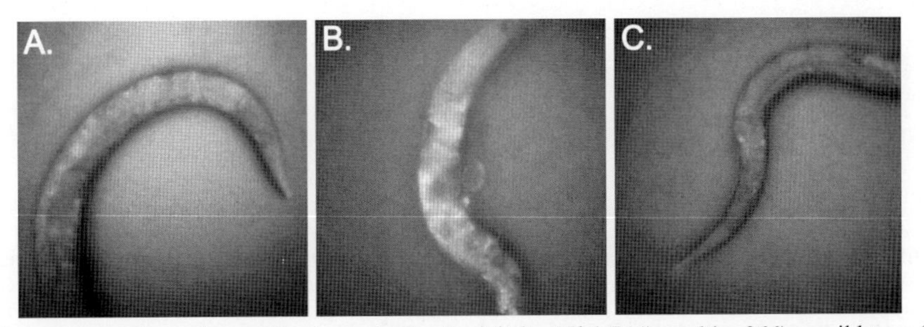

Figure 2. Accumulation of the fluorochrome lipofuscin in day-5 adult *lin-15b(n744);eri-1(mg366)* or wild-type (N2) nematodes as viewed under a fluorescent microscope at wavelength 420 nm. (*A*) Wild type (N2) treated with empty vector control RNAi; (*B*) *lin-15b(n744);eri-1(mg366)* treated with control; (*C*) *lin-15b(n744);eri-1(mg366)* treated with *daf-2* RNAi.

RNAi approached 100%. From the initial screen, 752 gene inactivations induced increased survival, albeit, in many cases, a just barely detectable increase in longevity. These gene inactivations were retested in triplicate for increased survival to reveal 115 reproducible gene inactivations that induce extended survival at a point when mortality of control animals approaches 100%. The 115 gene inactivations were tested in a longitudinal life-span analysis using replica sets at 20°C; this approach (Samuelson et al. 2007) has several advantages including noise reduction through independence of observations, decreased stress through elimination of repeated handling, and an overall extension of life span that expands absolute differences in life span between strains (at 20°C vs. 25°C).

Eighteen gene inactivations extending life span of *lin-15b(n744);eri-1(mg366)* animals were characterized in more detail, which included additional longitudinal life-span analysis in *lin-15b(n744);eri-1(mg366)* animals and demographic analysis (Fig. 3A,B). We tested whether a gene inactivation could extend life span in the absence of the *lin-15b* mutation (i.e., in a strain carrying only the *eri-1[mg366]* mutation) (Fig. 3C) or an unrelated *eri* strain (*eri-3[mg408]*) (Fig. 3D) and how the life-span extension observed in one strain compares to the extension in another. The additional analysis revealed that the majority of gene inactivations extended life span to a similar extent in *lin-15b(n744);eri-1(mg366), eri-1(mg366),* and *eri-3(mg408)* animals, implying that life-span extension was not strain-specific (Fig. 3E,F). Furthermore, the majority of gene inactivations reduced the relative rate of aging of *lin-15b(n744);eri-1(mg366)* animals (Fig. 3A, summarized in Table 1). Interestingly, inactivation of *pat-*

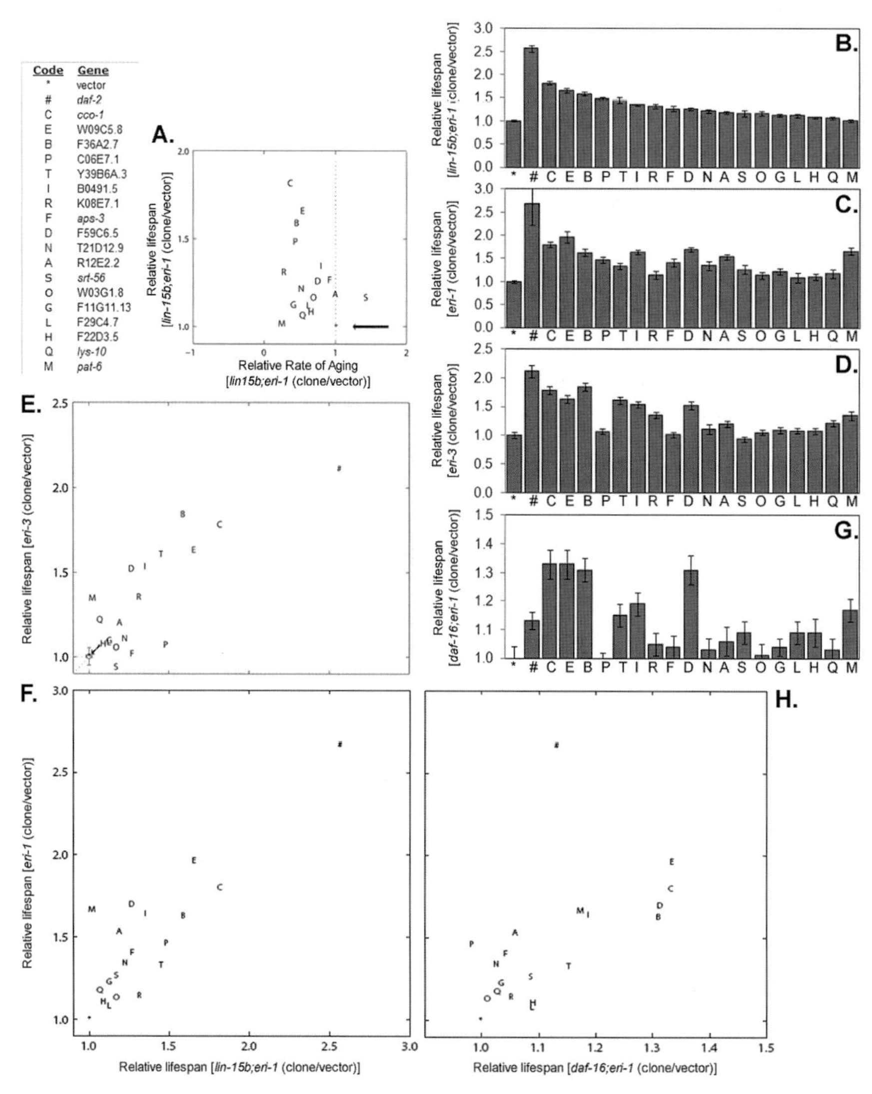

Figure 3. (*A*) Relative changes of life span versus relative rate of aging in *lin-15b(n744);eri-1(mg366)* animals after gene inactivation. (*B,C,D*) Relative life span of *lin-15b(n744);eri-1(mg366), eri-1(mg366),* and *eri-3(mg408)* animals, respectively; (*E*) relative life span of *eri-3(mg408)* versus *lin-15b(n744);eri-1(mg366)*; (*F*) relative life span of *eri-1(mg366)* versus *lin-15b(n744);eri-1(mg366)*; (*G*) relative life span of *daf-16(mgDf47);eri-1(mg366)*; (*H*) relative life span of *daf-16(mgDf47);eri-1(mg366)* versus *eri-1(mg366)*. Symbols representing each distinct gene inactivation are listed and relative change is compared to an empty vector control; for life span, it is the relative change in the median life span based on a logit curve best fit to the mortality data.

6 extended life span in *eri-1(mg366)* and *eri-3(mg408)* to a much greater extent than in *lin-15b(n744);eri-1(mg366)* animals, implying that the negative impact on life span caused by the *lin-15b* mutation is downstream from *pat-6* (compare clone M in Fig. 3E,F). Loss of *pat-6* function has previously been shown to extend life span (Hansen et al. 2005) and encodes a protein with an actinin-type actin-binding domain required for proper muscle assembly and function (Williams and Waterston 1994). The fact that RNAi clones failed to extend life span to a greater extent in *lin-15b(n744);eri-1(mg366)*, which is most sensitive to RNAi, compared to *eri-1(mg366)* alone and *eri-3(mg408)* suggests that decreased gene function in life-span extension, in either spatial or absolute terms, was not limiting. In contrast, because decreased *pat-6* function in *eri-1 (mg366)* animals extended life span to a greater extent than in *lin-15b(n744);eri-1(mg366)*, by focusing on gene inactivations with the greatest extension of life span, we may have failed to identify novel aging genes (i.e., due to the accelerated aging caused by the *lin-15b* mutation).

The reduced expression of metabolism genes was the largest annotated class of genes identified and reaffirms the link between energy production and longevity. By reducing cellular respiration, RNAi inactivation of mitochondrial genes induces significant life-span extension in *C. elegans*. Inactivation of W09C5.8, encoding cytochrome *c* oxidase IV, caused a 39% extension of mean life span relative to vector. Cytochrome *c* oxidase, the terminal enzyme in the respiratory pathway, is located in the inner membrane of the mitochondria. This transmembrane protein is the terminal electron acceptor in the electron transfer chain and further catalyzes the reduction of dioxygen to water, translocating protons across the mitochondrial membrane. The resulting electrochemical gradient is used by ATP synthase to synthesize ATP.

Inactivation of *cco-1* and F59C6.5, encoding cytochrome *c* oxidase subunit Vb and NADH ubiquinone oxidoreductase, respectively, also induced weaker extension of life span. The latter protein complex catalyzes the oxidation of NADH and the reduction of ubiquinone and further establishes a proton gradient across the mitochondrial membrane for ATP synthesis. The link between energy production and longevity is further demonstrated in the inactivation of F11G11.13 encoding an ATP:Guanido phosphotransferase, an enzyme that reversibly catalyzes the transfer of phosphate between ATP and various phosphogens, similar to human creatine kinase. In humans, this protein has a central role in energy transduction in tissues with large energy demands, including the heart and the brain.

The gene Y39B6A.3, encoding an HesB protein, is an enzyme that has been proposed to function in the assembly of iron-sulphur clusters in conjunction with the protein ferredoxin (Hochstrasser 2000). Iron-sulfur clusters are cofactors of numerous proteins with important functions in metabolism, electron transport, and regulation of gene expression. Biosynthesis of iron-sulfur complexes occurs inside the mitochondria. Interestingly, in *Saccharomyces cerevisiae,* deletion of *isa1* causes loss of mitochondrial DNA and respiratory deficiency (Jensen and Culotta 2000; Muhlenhoff et al. 2002), suggesting that loss of Y39B6A.3 may extend life span in *C. elegans* by decreasing mitochondrial reactive oxygen species (ROS) production.

Whether the extension of life span after gene inactivation was dependent on *daf-16*—the ortholog of the FoxO transcription factor and major output of decreased insulin-like signaling—was tested by measuring mortality in aging *daf-16(mgDf47);eri-1(mg366)* animals (Fig. 3G) and comparing the extension of relative life span to *eri-1 (mg366)* animals (Fig. 3H). Most gene inactivations, like *daf-2(RNAi),* required *daf-16* to extend life span. Consistent with previous findings, gene inactivations impairing mitochondrial function extended life span largely independently of *daf-16* function.

The accumulation of the fluorochrome age pigment within postmitotic cells is a recognized characteristic of aging, occurring at a rate directly proportional to age in adults. As an intralysosomal, polymeric substance, lipofuscin is composed of cross-linked protein residues formed from iron-catalyzed oxidative processes (Brunk and Terman 2002). Undegradable and unremovable by exocytosis, lipofuscin accumulates in postmitotic cells over time. In cell culture models, oxidative stress promotes lipofuscin accumulation, and thus assaying for the presence of this fluorochrome allows for not only the measurement of age over time, but also the determination of the animal's inherent resistance to oxidative stress (Brunk and Terman 2002). In previous studies, it was found that long-lived *daf-2* mutants are highly tolerant to multiple stress treatments, including hydrogen peroxide and heat shock (Larsen 1993), which collectively suggest that stress resistance and accumulation of age pigment may be linked. Generally, it is thought that this heightened stress resistance is a key component of extended life span in worms.

We tested whether age pigment accumulation, stress resistance, and aging were coregulated. To our surprise, with the exception of *daf-2(RNAi)*, decreased age pigment accumulation, increased thermotolerance, or resistance to oxidative damage did not correlate (Fig. 4). For instance, inactivation of *cco-1* or W09C5.8 resulted in decreased age pigment accumulation on a par with *daf-2(RNAi)* (Fig. 4A, compare clones C and E with #, respectively). However, in contrast to decreased *daf-2* function, inactivation of either *cco-1* or W09C5.8 failed to increase survival after heat shock (Fig. 4B) or paraquot treatment (Fig. 4C). Similarly, inactivation of Y39B6A.3, B0491.5, or K08E7.1 increased heat shock survival (Fig. 4B, clones T, I, and R) but resulted in little or no change in age pigment accumulation (Fig. 4A) or resistance to paraquot treatment (Fig. 4C). Similarly, a large increased resistance to oxidative damage failed to result in a large enhancement of resistance to heat shock or decrease in age pigment accumulation. Collectively, these findings imply that decreased lysosomal function (via age pigment accumulation), reactive oxygen species, and accumulation of misfolded proteins function through disparate, yet partially overlapping, mechanisms to promote aging.

DISCUSSION

A total of 115 reproducible positives comprising approximately 0.5% of the genome had at least 10%

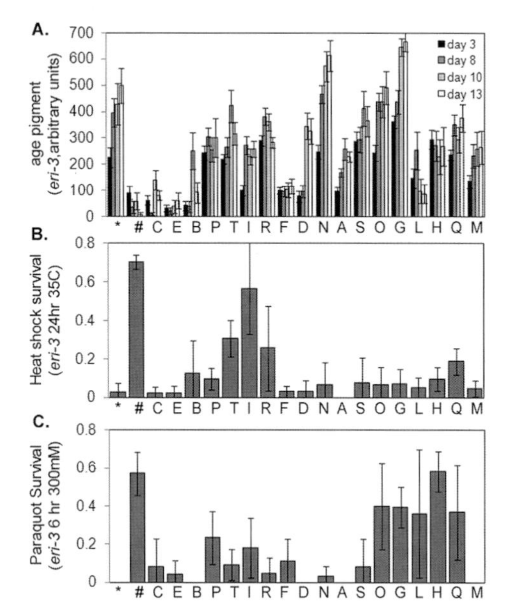

Figure 4. (*A*) Age pigment accumulation in *eri-3(mg408)* animals for each gene inactivation at adulthood day 3, 8, 10, and 13 (20°C), respectively. Error bars represent standard error. (*B*) Survival of *eri-3(mg408)* animals (day-5 adulthood) after 24-hour treatment at 35°C. Error bars represent the standard deviation of three independent experiments. (*C*) Survival of *eri-3(mg408)* animals (day-5 adulthood) after 6-hour treatment with 300 mM paraquot. Error bars represent the standard deviation of three independent experiments. Symbols representing each gene inactivation are the same as those used in Figure 3.

extended life span. Eighteen stronger candidates were examined more closely through a longitudinal life-span analysis, temporal accumulation of age pigment, heat shock, and paraquot survival assays. Collectively, genes associated with metabolic and mitochondrial functions were the most represented among positive candidates from the genomic screen. RNAi clones that target factors upstream of ATP synthase in the electron-transport chain would be expected to abrogate the formation of a proton gradient across the mitochondrial membrane, thus stifling free-radical formation. Posttran-scriptional silencing of other genes associated with energy production resulted in life-span extension as well, further underscoring this link between longevity and free-radical reduction in accordance with free-radical aging theory. Similar to observations seen in calorie-restricted environments, nematodes with compromised metabolic activity exhibit decreased ROS production, thus minimizing the damaging cellular effects of these highly reactive species.

Relative to wild type, which exhibits a mean adult life span of approximately 16 days at 25°C under vector control, *lin-15b(n744);eri-1(mg366)* displays a 42% reduction in mean adult life span at 9.3 days at 25°C for unknown reasons. Because the criteria for demarcating life-span extension was similar between our study and previous studies using wild-type *C. elegans* (i.e., a 10% survival ratio when all vector control had died), one would thus expect only relatively stronger promoters of longevity to surface within the *lin-15b(n744);eri-1(mg366)* screen. The window of opportunity to observe life-span extension in this strain is considerably reduced when compared to wild type.

The results from this study reinforce the importance of metabolism on animal life-span determination. The study of the aging process itself reveals that life-span determination is not an arbitrary occurrence within organisms but a deliberate process regulated by distinct factors and pathways that can be modified and manipulated. Ultimately, findings from this study may provide new insights into the mechanisms and components of life-span regulation concerning not only nematodes, but also higher organisms such as humans.

ACKNOWLEDGMENTS

Some strains were provided by the *Caenorhabditis* Genetics Center, which is funded by the National Institutes of Health (NIH) National Center for Research Resources. We thank members of the Ruvkun laboratory for scientific discussions and support. This research was funded by NIH grant 5-R01-AG16636.

REFERENCES

Alcedo J. and Kenyon C. 2004. Regulation of *C. elegans* longevity by specific gustatory and olfactory neurons. *Neuron* **41:** 45.

Brunk U.T. and Terman A. 2002. Lipofuscin: Mechanisms of age-related accumulation and influence on cell function. *Free Radic. Biol. Med.* **33:** 611.

Buffenstein R. 2005. The naked mole-rat: A new long-living model for human aging research. *J. Gerontol. A Biol. Sci. Med. Sci.* **60:** 1369.

Curran S.P. and Ruvkun G. 2007. Lifespan regulation by evolutionarily conserved genes essential for viability. *PLoS Genet.* **3:** e56.

Dorman J.B., Albinder B., Shroyer T., and Kenyon C. 1995. The age-1 and daf-2 genes function in a common pathway to control the lifespan of *Caenorhabditis elegans*. *Genetics* **141:** 1399.

Duchaine T.F., Wohlschlegel J.A., Kennedy S., Bei Y., Conte D., Jr., Pang K., Brownell D.R., Harding S., Mitani S., Ruvkun G., Yates J.R., III, and Mello C.C. 2006. Functional proteomics reveals the biochemical niche of *C. elegans* DCR-1 in multiple small-RNA-mediated pathways. *Cell* **124:** 343.

Finch C.E. and Austad S.N. 2001. History and prospects: Symposium on organisms with slow aging. *Exp. Gerontol.* **36:** 593.

Finch C.E. and Ruvkun G. 2001. The genetics of aging. *Annu. Rev. Genomics Hum. Genet.* **2:** 435.

Hamilton B., Dong Y., Shindo M., Liu W., Odell I., Ruvkun G., and Lee S.S. 2005. A systematic RNAi screen for longevity genes in *C. elegans*. *Genes Dev.* **19:** 1544.

Hansen M., Hsu A.L., Dillin A., and Kenyon C. 2005. New genes tied to endocrine, metabolic, and dietary regulation of lifespan from a *Caenorhabditis elegans* genomic RNAi screen. *PLoS Genet.* **1:** 119.

Hochstrasser M. 2000. Evolution and function of ubiquitin-like protein-conjugation systems. *Nat. Cell Biol.* **2:** E153.

Hsu A.L., Murphy C.T., and Kenyon C. 2003. Regulation of aging and age-related disease by DAF-16 and heat-shock factor. *Science* **300:** 1142.

Jensen L.T. and Culotta V.C. 2000. Role of *Saccharomyces cerevisiae* ISA1 and ISA2 in iron homeostasis. *Mol. Cell. Biol.* **20:** 3918.

Kennedy S., Wang D., and Ruvkun G. 2004. A conserved siRNA-degrading RNase negatively regulates RNA interference in *C. elegans*. *Nature* **427:** 645.

Kimura K.D., Tissenbaum H.A., Liu Y., and Ruvkun G. 1997. *daf-2,* an insulin receptor-like gene that regulates longevity

and diapause in *Caenorhabditis elegans*. *Science* **277:** 942.
Larsen P.L. 1993. Aging and resistance to oxidative damage in *Caenorhabditis elegans*. *Proc. Natl. Acad. Sci.* **90:** 8905.
Lee S.S., Lee R.Y., Fraser A.G., Kamath R.S., Ahringer J., and Ruvkun G. 2003. A systematic RNAi screen identifies a critical role for mitochondria in *C. elegans* longevity. *Nat. Genet.* **33:** 40.
Muhlenhoff U., Richhardt N., Gerber J., and Lill R. 2002. Characterization of iron-sulfur protein assembly in isolated mitochondria. A requirement for ATP, NADH, and reduced iron. *J. Biol. Chem.* **277:** 29810.
Ogg S., Paradis S., Gottlieb S., Patterson G.I., Lee L., Tissenbaum H.A., and Ruvkun G. 1997. The Fork head transcription factor DAF-16 transduces insulin-like metabolic and longevity signals in *C. elegans*. *Nature* **389:** 994.
Parkes T.L., Elia A.J., Dickinson D., Hilliker A.J., Phillips J.P., and Boulianne G.L. 1998. Extension of *Drosophila* lifespan by overexpression of human SOD1 in motorneurons. *Nat. Genet.* **19:** 171.
Perls T., Kunkel L.M., and Puca A.A. 2002. The genetics of exceptional human longevity. *J. Mol. Neurosci.* **19:** 233.
Samuelson A.V., Carr C.E., and Ruvkun G. 2007. Gene activities that mediate increased life span of *C. elegans* insulin-like signaling mutants. *Genes Dev.* **21:** 2976.
Sieburth D., Ch'ng Q., Dybbs M., Tavazoie M., Kennedy S., Wang D., Dupuy D., Rual J.F., Hill D.E., Vidal M., Ruvkun G., and Kaplan J.M. 2005. Systematic analysis of genes required for synapse structure and function. *Nature* **436:** 510.
Wang D., Kennedy S., Conte D., Jr., Kim J.K., Gabel H.W., Kamath R.S., Mello C.C., and Ruvkun G. 2005. Somatic misexpression of germline P granules and enhanced RNA interference in retinoblastoma pathway mutants. *Nature* **436:** 593.
Williams B.D. and Waterston R.H. 1994. Genes critical for muscle development and function in *Caenorhabditis elegans* identified through lethal mutations. *J. Cell Biol.* **124:** 475.
Wolkow C.A., Kimura K.D., Lee M.S., and Ruvkun G. 2000. Regulation of *C. elegans* life-span by insulinlike signaling in the nervous system. *Science* **290:** 147.

Circadian Photoreception in Vertebrates

S. DOYLE AND M. MENAKER
Department of Biology, University of Virginia, Charlottesville, Virginia 22936

To be adaptively useful, internal circadian clocks must be entrained (synchronized) to daily rhythms in the external world. The entraining process adjusts the period of the internal clock to 24 hours and its phase to a value that determines the organism's temporal niche (e.g., diurnal and nocturnal). For most vertebrates, the dominant environmental synchronizer is light. All vertebrates employ specialized photoreceptor cells to perceive synchronizing light signals, but mammals and nonmammalian vertebrates do this differently. Mammals concentrate circadian photoreceptors in the retina, employing rods, cones, and a subset of retinal ganglion cells that are directly photosensitive and contain an unusual photopigment (melanopsin). Nonmammalian vertebrates use photoreceptors located deep in the brain and in the pineal gland as well as others in the retina. Such photoreceptor extravagance is difficult to explain. It seems likely that the different photoreceptor classes in this elaborate sensory system may have specialized roles in entrainment. There is some evidence that this is in fact the case. Furthermore, this nonvisual "circadian" photoreceptive system also controls acute behavioral responses to light (masking), pupillary constriction, and photoperiodic regulation of reproductive state. We review some of the early work on birds and describe new findings that indicate specific roles for retinal rods, cones, and photosensitive retinal ganglion cells in mammals.

INTRODUCTION

In the Discussion following Tony Lees' paper (Lees 1961) from the 25th Cold Spring Harbor Symposium held in 1960, Donald Kennedy made a characteristically prescient remark. Lees' paper was in part about photoperiodism in aphids. Using fine-gauge metal tubes as light guides applied to various parts of the bodies of these tiny insects, Lees had shown that the photoreceptors involved were not in the eyes but were most probably in the brain itself. Kennedy, who himself had worked on "primitive" photoreceptors in invertebrates, said: "These very beautiful results, I think, call attention to the important business of specifying the light reception involved in photoperiodic control—or light entrainment of rhythms... . One might, in the light of such evidence, be cautious about assuming that photoperiodic control is mediated by 'obvious' photoreceptor structures in any given case." In the intervening 47 years, the study of photoreception by "nonobvious" structures in vertebrates has become a cottage industry.

The mechanisms by means of which the vertebrate eye and associated brain areas form visual images have long been the focus of persistent and highly successful research efforts. The very success in this understanding of image formation has led, until recently, to a lack of appreciation of other mechanisms for light perception possessed even by vertebrates with elaborate image-forming eyes. Evidence for the existence of such mechanisms has been accumulating slowly for some time. As early as 1911, von Frisch showed pigment color changes in the skin of minnows that persisted after removal of the eyes and pineal gland (von Frisch 1911). Scharrer (1928) subsequently demonstrated that blinded fish could be trained to associate feeding with a light stimulus. In 1935, Benoit and colleagues published the first in a series of papers in which they clearly showed that ducks whose eyes had been surgically removed could still perceive day length: Their gonads grew dramatically when they were kept in long (but not short) days. Although these observations were surprising and implied the existence of unidentified photoreceptors in unknown locations, they were not vigorously pursued, in part, because they were isolated cases with no common functional thread to connect them. One candidate structure, the pineal/parietal complex in amphibians and reptiles, had long been assumed to be photoreceptive (Leydig 1890), but interest in studying it was somewhat dampened by inability to identify a clear function despite many attempts.

In the mid-1960s, my students and I discovered serendipitously that house sparrows, whose eyes had been removed for other reasons, could still synchronize their activity rhythms to light cycles. Subsequent work by us and other investigators has established that such "extraretinal" entrainment is a property of the circadian systems of all nonmammalian vertebrates and depends on photoreceptors in several brain areas. In mammals, there are no "extraretinal" photoreceptors; however, circadian entrainment is accomplished in part by a population of nonvisual photosensitive retinal ganglion cells. Circadian entrainment thus finally provided the functional connection among several diverse nonvisual photoreceptive systems that was necessary to stimulated intense research. As work has progressed on these systems, other functions have been uncovered. In the future, it will be of particular interest to explore, especially in mammals, how the visual and nonvisual photoreceptive systems, which operate under very different physiological constraints, interact with each other in the regulation of these several functions. We have briefly reviewed here some of the early work in birds and related it to ongoing experiments on nonvisual functions of the mammalian retina.

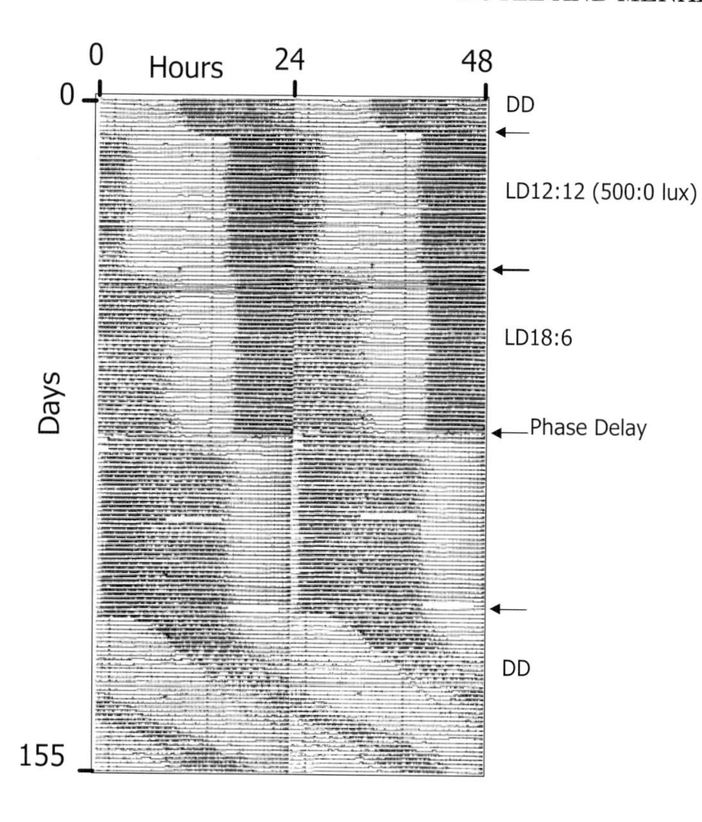

Figure 1. Perch-hopping activity of a blinded sparrow, maintained as indicated to the right of the record in constant darkness (DD), a 12-hour light:12-hour dark cycle (LD12:12), and an 18-hour light:6-hour dark cycle (LD18:6). On the day marked "phase shift," the LD16:8 cycle was delayed by 9.5 hours. The record is double plotted so that the righthand side of the record is displaced upward 1 day and each horizontal line equals 48 hours. (Reprinted from Menaker 1968a [National Academy of Sciences].)

HOUSE SPARROWS POSSESS EXTRARETINAL PHOTORECEPTORS COUPLED TO THE CIRCADIAN CLOCK CONTROLLING LOCOMOTOR ACTIVITY

We measured circadian rhythms of locomotor activity in house sparrows by continuously recording perch-hopping behavior (Menaker 1968a). Figure 1 shows a representative record from a bilaterally enucleated sparrow. For the first several days of the record, the sparrow was maintained in constant darkness and displayed a locomotor activity rhythm with a free-running period slightly longer than 24 hours, as is common in sighted sparrows. On day 7 of the record, a 12-hour light:12-hour dark cycle of 500 lux from white fluorescent light bulbs was imposed. The blinded sparrow readily entrained to this light:dark cycle and also responded normally to subsequent manipulations of the light:dark cycle (i.e., increasing the photoperiod to light:dark 16:8 on day 40 and delaying the light:dark cycle by 9.5 hours on day 79). When returned to constant darkness on day 117, the bird began to free-run once again. These experiments strongly suggested that house sparrows possessed extraretinal photoreceptor(s) mediating circadian entrainment to light. Further experiments ruled out the possibility that the birds were responding to another rhythmic feature in their environment, such as heat, noise, or electrical fields (Menaker 1968a). We also found the extraretinal photoreceptor to be surprisingly sensitive. Approximately one half of blinded sparrows entrained to a 12-hour light:12-hour dark cycle in which the light source was a small electroluminescent panel producing green light of 0.1 lux (Fig. 2) (Menaker 1968a).

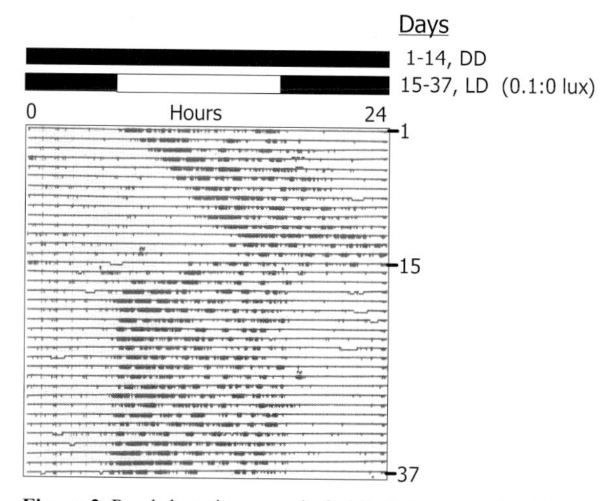

Figure 2. Perch-hopping record of a blinded sparrow. During the first 14 days of the record, the bird was maintained in constant darkness (DD) and free-ran with a period of approximately 25 hours. On day 15, the bird was exposed to a 12-hour light:12-hour dark cycle (LD), with 0.1 lux green light during the light phase. After 2–3 days of transients, the sparrow entrained to this regime. (Reprinted from Menaker 1968a [National Academy of Sciences].)

THE EXTRARETINAL PHOTORECEPTOR OF SPARROWS IS LOCATED WITHIN THE BRAIN

Early work by Benoit (1935) suggested that light directed to the hypothalamus affected photoperiodic testis growth in blinded ducks. These data and the relative transparency of the avian skull suggested to us that the extraretinal photoreceptor might be localized within the brain. To test this hypothesis, we conducted an experiment in which six enucleated sparrows were exposed to a 12-hour light:12-hour dark cycle of 0.02 lux, just below the threshold for entrainment (Fig. 3) (Menaker 1968b). We then plucked feathers from an approximately half-inch-diameter area on the birds' back. This treatment had no effect on activity rhythms, and the birds continued to free-run through the dim light:dark cycle. We next plucked feathers from a similarly sized area on the birds' heads and found that after this manipulation, the activity rhythms of all of the sparrows rapidly entrained to the light:dark cycle. About 4 weeks later, the feathers regrew and the bird began to drift out of entrainment. When these new feathers were plucked, the bird entrained again. After India ink was injected under the skin above the skull to decrease the effective light intensity reaching underlying tissues, the birds free-ran once more. Finally, when the ink deposit was scraped away, the birds reentrained to the light:dark cycle. Pinealectomy had no effect on the entrainment of blinded sparrows if subjected to the above manipulations (Menaker 1968b). These experiments demonstrate that manipulation of the light intensity reaching the brain of blinded sparrows affects the ability of the animals to entrain to light cycles and indicate that the extraretinal photoreceptor(s) are located within the brain. The discovery of extraretinal brain photoreceptors raised the question of whether the eyes, when present, are involved at all in photoperiodism, entrainment, and other circadian responses to light such as changes in free-running period in response to constant light and the effect of constant light in producing arrhythmicity. Interestingly, the role of the eyes and the extraretinal photoreceptors was found to vary for each of these responses.

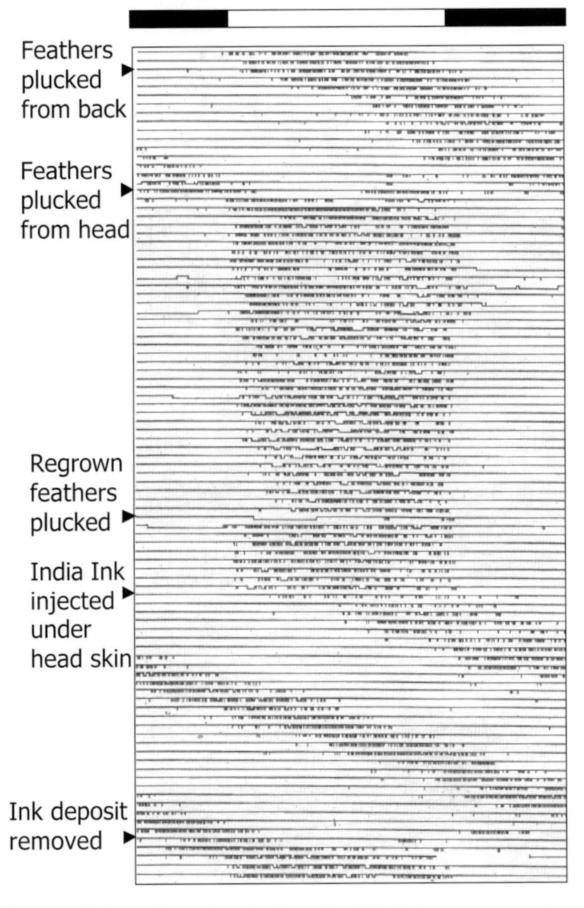

Figure 3. Perch-hopping rhythm of a blinded sparrow, maintained in a 12-hour light:12-hour dark cycle (0.02 lux green light during the light phase, as indicated at the top of the figure) for the duration of the record and, on the days indicated by the arrowheads, subjected to the manipulations described at the left of the record. (Reprinted from Menaker 1968b [National Academy of Sciences].)

The Eyes Contribute to Circadian Entrainment

As described above, approximately 50% of blinded sparrows entrained to a light:dark cycles of an intensity of 0.1 lux; however, all sighted birds entrain to light of this intensity. Therefore, the eyes, although not necessary for circadian entrainment, do appear to contribute light information for entrainment when they are present. The suspected contribution of both the eyes and extraretinal photoreceptors to circadian entrainment was conclusively demonstrated by McMillan et al. (1975c), who exposed intact birds to light:dark cycles of very dim intensity (0.03 lux). Under these conditions, 25% of the birds free-ran, suggesting that this intensity is close to threshold for entrainment in intact birds. McMillan then either plucked head feathers from free-running birds or injected carbon black between the skin and skull of entrained birds to either increase or decrease, respectively, the amount of light reaching brain photoreceptors. All plucked birds subsequently entrained and nearly half of the carbon-black-injected birds began to free-run. When blinded, birds that had remained entrained after carbon black injection began to free-run. The fact that photosensitivity for entrainment was greater in intact birds than in blinded or injected birds demonstrated that both the extraretinal photoreceptors and the eyes contribute to entrainment in intact birds.

The Eyes Contribute to Free-running Period Changes in Constant Light

In many animals, constant light has dramatic effects on the circadian clock, altering period length, and above certain thresholds, producing arrhythmicity. In sparrows, the free-running period (τ) shortens as light intensity is increased and lengthens as it is reduced. Like the situation in intact birds, blinded sparrows exhibit shortening of the free-running period of activity in response to increasing intensities of constant light (Fig. 4) (Menaker 1968a). Treatments altering the level of light input from the eyes or the extraretinal photoreceptors in intact and blinded birds affect the free-running period in the expected directions, showing that both the eyes and the extraretinal photoreceptors mediate the effects of constant light on the

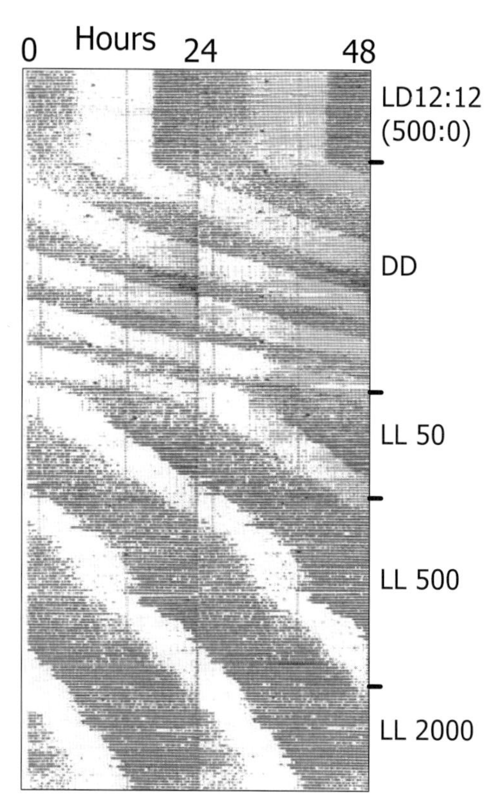

Figure 4. Perch-hopping record of a blinded sparrow, maintained in LD12:12 (500 lux), DD, and then exposed to constant light (LL) of three intensities as indicated at the right of the record. Note the large change in free-running period and the increase in total activity time upon transfer to from DD to LL, and smaller period changes as the intensity of LL is increased. In LL of 2000 lux, feathers were plucked from the bird's head, raising the effective intensity to 20,000 lux (feathers attenuate light intensity by a factor of 10). Note that the bird did not become arrhythmic. (Reprinted from Menaker 1968a [National Academy of Sciences].)

free-running period. Blinding and hooding to decrease the amount of light reaching the brain both lengthen τ, whereas head feather plucking or removal of the hood shortens τ (McMillan et al. 1975b).

The Eyes Are Necessary for LL-induced Arrhythmicity

In intact house sparrows, the intensity of constant light required to produce arrhythmicity is surprisingly low and lies between 10 and 100 lux (Menaker 1977). In constant dim light (1.0 lux), rhythmic intact birds will become arrhythmic when the light intensity reaching the brain photoreceptors is increased by plucking head feathers. Rhythmicity can then be restored with carbon black injection (McMillan et al. 1975a). However, blind birds will not become arrhythmic in constant light, even at effective intensities of 20,000 lux (Fig. 4). These results indicate that even though extraretinal photoreceptors contribute to the production of LL-induced arrhythmicity, the eyes are required for this effect of light on the clock.

The Eyes Are Not Involved in Photoperiodic Photoreception

In the wild, male sparrows undergo a dramatic annual cycle of gonadal growth. During spring, when the photoperiod increases to 14 hours or more of light per day, testis weight increases by approximately 500-fold (Fig. 5A) (Menaker 1971). We examined photoperiodic photoreception in the sparrow with respect to the role of the eyes. Both intact and blinded birds showed identical reproductive responses in terms of testis weight, histology, and rates of gonadal growth when exposed to long photoperiods (Menaker and Keatts 1968; Underwood and Menaker 1970). Furthermore, the threshold for the photoperiodic response was more than two orders of magnitude higher than the threshold for vision (Menaker 1977), suggesting that the eyes had little or no role in photoperiodic responses to light. A key experiment, shown in Figure 5B, was conducted in which sparrows with intact eyes were exposed to 16-hour light:8-hour dark cycles of intensity close to threshold for the photoperiodic reproductive response (Menaker et al. 1970; Menaker 1971). Head feathers were plucked from one group of birds to increase the amount of light reaching the brain, whereas India ink was injected beneath the head skin of another group of birds to decrease the light intensity reaching the brain. The testes of the plucked birds increased in size; however, despite the fact that their eyes were exposed to long days, those of the India-ink-injected group showed no increase compared to a control group of sparrows with small testes that had been maintained on short days. These results demonstrated that the eyes are not involved in photoperiodic photoreception.

MAMMALIAN PHOTOENTRAINMENT

In contrast to nonmammalian vertebrates, in which extraretinal photoreceptors are widespread, in mammals, circadian entrainment and other nonvisual light responses are mediated exclusively by the eyes (Nelson and Zucker 1981). Although it has been clear for some time that mammalian nonvisual photoreception was based in the retina, the photoreceptors involved have only recently been identified. In the early 1990s, Foster and colleagues examined circadian photoresponses in *rd/rd* mice with severe degeneration of the outer retina. Surprisingly, these mice, lacking all rods and most cones, showed circadian phase-shifting responses to light that were indistinguishable in sensitivity from those of wild-type control mice (Foster et al. 1991). Similar results were later obtained using mice with more complete lesions of the outer retina, *rdta;cl* (Freedman et al. 1999). The insensitivity of circadian photoreception to loss of rod and cone function implicated a novel nonrod, noncone photoreceptor. These photoreceptors were recently identified as a small subset of intrinsically photosensitive retinal ganglion cells (ipRGCs), containing the photopigment melanopsin (OPN4; Provencio et al. 1998, 2000; Berson et al. 2002; Hattar et al. 2002). "Triple-knockout" mice lacking rods,

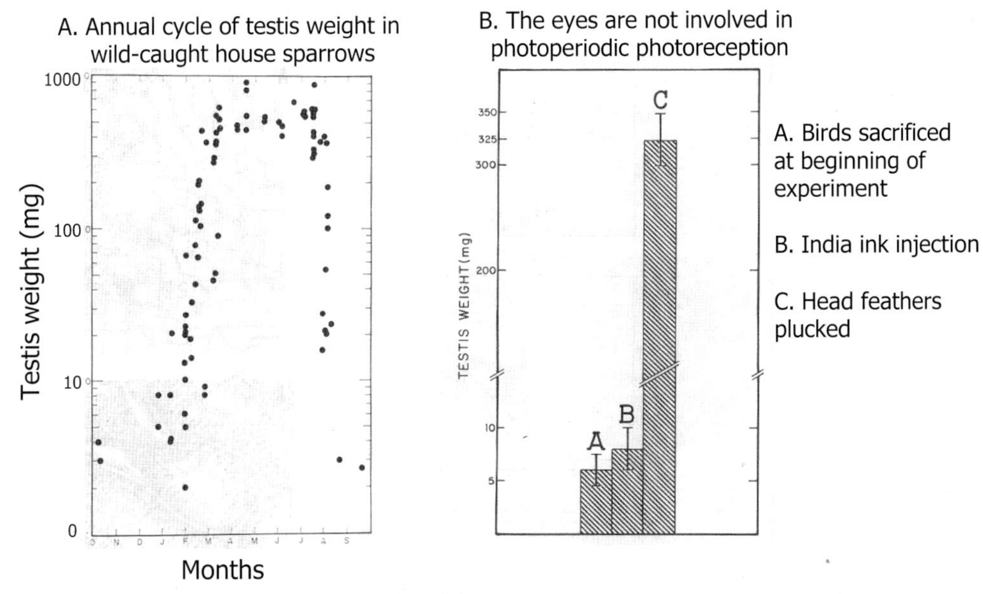

Figure 5. (*A*) Testis weight measured from individual wild-caught sparrows over the course of 1 year. Each point represents combined weight of both testes from a single bird. (*B*) Testis weight from intact sparrows exposed to a long-day photoperiod and subjected to manipulations (indicated at the right of the graph) to either decrease or increase the amount of light reaching the brain. (Reprinted from Menaker 1971 [National Academy of Sciences].)

cones, and melanopsin lose all circadian responses to light (Hattar et al. 2003; Panda et al. 2003). In melanopsin knockout mice, ipRGCs lose their photosensitivity (Hattar et al. 2002), yet circadian photoresponses, although attenuated, are not completely lost (Panda et al. 2002; Ruby et al. 2002). This suggests that rods, cones, and melanopsin-based photoreceptors all participate in circadian photoreception and raises questions regarding the specific roles of these photoreceptors and their possible interactions.

INTERACTION BETWEEN RODS AND ipRGCs

We have recently examined the effect of loss of RPE65, a key enzyme for recycling the 11-*cis*-retinaldehyde chromophore of rods and cones, on circadian photoreception in mice. *Rpe65*$^{-/-}$ mice lose cone function and retain only a small amount of rod function (Seeliger et al. 2001; Wenzel et al. 2007). In contrast to the intact circadian photoresponses reported in mouse models with more complete rod and cone dysfunction, we found that circadian phase-shifting responses in *Rpe65*$^{-/-}$ mice were greatly attenuated (Fig. 6) (Doyle et al. 2006). These results indicated that melanopsin cells as well as rods and cones had been affected by RPE65 loss and initially suggested that melanopsin might use RPE65 to recycle its chromophore. However, we subsequently generated *Rpe65*$^{-/-}$ mice carrying the *rdta* transgene (McCall et al. 1996), which produces rapid and complete ablation of rod photoreceptors. Photoreception in these mice is therefore limited to ipRGCs. Surprisingly, circadian entrainment and phase-shifting responses were restored to wild-type levels in *Rpe65*$^{-/-}$*;rdta* animals (Fig. 7). This restoration of circadian photosensitivity demonstrated that melanopsin can use a chromophore regeneration pathway independent of that used by visual photoreceptors. It also revealed an intriguing interaction between the rod and melanopsin systems, suggesting that in the *Rpe65*$^{-/-}$ retina, rods were somehow inhibiting ipRGC function.

ROD PATHWAYS AND CONTROL OF TEMPORAL NICHE

We eliminated melanopsin from *Rpe65*$^{-/-}$ retinas by generating double-knockout *Rpe65*$^{-/-}$*;Opn4*$^{-/-}$ mice. When maintained in a 12-hour light:12-hour dark cycle, we found that 80% of these mice displayed a very striking diurnal activity phenotype (Fig. 8); the remaining 20% free-ran through the light:dark cycle. Furthermore, after release into constant darkness, the activity onset of "diurnal" mice free-ran from the time of lights-on, indicating that their circadian activity rhythms were not simply masked but were entrained to the diurnal phase. The loss of the ability of *Rpe65*$^{-/-}$*;Opn4*$^{-/-}$ mice to maintain normal photoentrainment indicated that melanopsin provides the majority of the photic input to the circadian system of *Rpe65*$^{-/-}$ mice. However, the very striking diurnal phenotype also suggests that retinal input may have a greater role in determining temporal niche than previously believed. Because loss of RPE65 has been shown to eliminate cone function (Seeliger et al. 2001; Znoiko et al. 2005; Wenzel et al. 2007), rod pathways alone must medi-

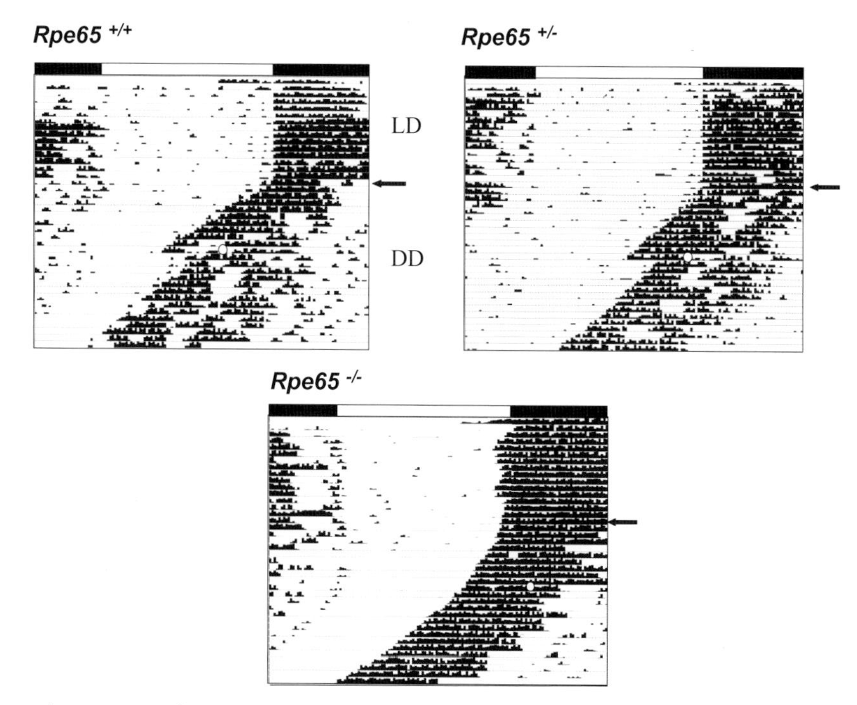

Figure 6. Representative actograms from an Rpe65 wild-type, heterozygote, and homozygous knockout. Mice were maintained in a 12-hour light:12-hour dark cycle (LD), indicated by black and white bars above each record and, on the day indicated by the arrow, were released into constant darkness (DD). After 10–12 days in DD, the animals were given a 15-minute light pulse (515 nm, 0.1 μW/cm^2) at circadian time (CT) 16 (*open circles*). (Reprinted from Doyle et al. 2006 [©National Academy of Sciences].)

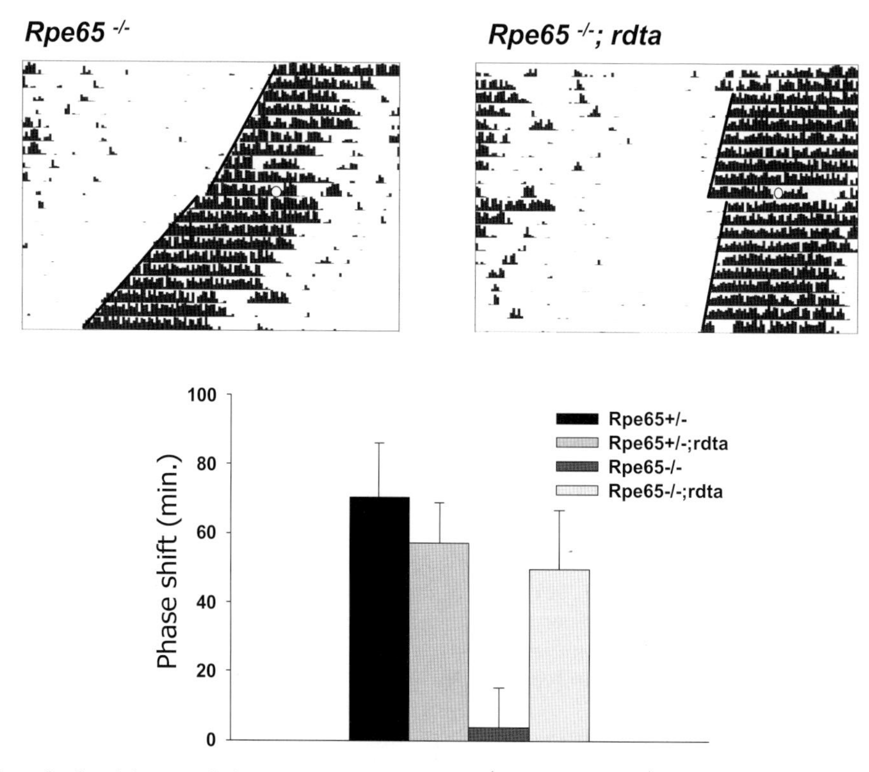

Figure 7. Running-wheel activity records from a representative *Rpe65$^{-/-}$* (*A*) and Rpe65$^{-/-}$*;rdta* (*B*) mouse, given a 15-minute light pulse (515 nm, 0.1 μW/cm^2) at CT 16 after free-running in constant darkness. Best-fit lines are drawn through activity onsets on the days before and after the light pulse (*open circles*). (*C*) Histogram showing mean phase shift (±S.E.M.). (Reprinted from Doyle et al. 2006 [© National Academy of Sciences].)

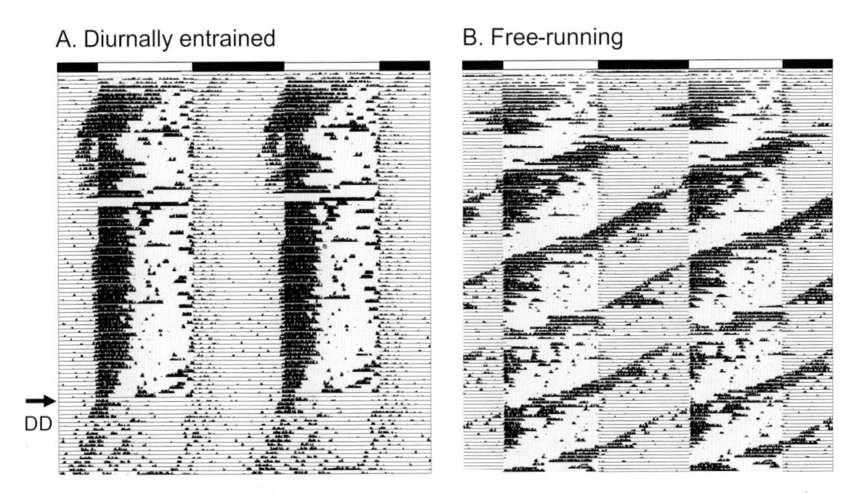

Figure 8. Double-plotted wheel-running records of a diurnally entrained (*A*) and a free-running (*B*) *Rpe65*$^{-/-}$;*Opn4*$^{-/-}$ mouse maintained in a 12-hour light:12-hour dark cycle. The light:dark cycle is indicated by *black* and *white bars* at the top of each record. The arrow in *A* indicates the day of release into constant darkness. (Reprinted from Doyle et al. 2006 [©National Academy of Sciences].)

ate the temporal niche switching of *Rpe65*$^{-/-}$;*Opn4*$^{-/-}$ mice. We are currently testing the hypothesis that dim light, below the sensitivity range of cones and melanopsin, can produce a similar phenotype in wild-type mice.

CONCLUSIONS

Synchronization of circadian clocks with the external environment is essential for their adaptive function. Photoreception provides a critical link between the environment and circadian clocks. However, circadian photoreceptors must accomplish a photoreceptive task, irradiance detection, which is qualitatively different from the fine spatial and temporal resolution carried out by the photoreceptors involved in image formation. This distinction is emphasized by the universality of the separation of visual and nonvisual photoreceptors. Specialized photoreceptors that convey information about environmental light conditions to the circadian system occur widely in invertebrates and in all five vertebrate classes.

A key question in our investigation of the circadian and photoperiodic responses to light in the house sparrow has been: What are the relative contributions of the eyes and the extraretinal photoreceptors? We have approached this question by comparing the similarities and differences between blinded and intact birds. Photoperiodic stimulation of testicular growth is mediated exclusively by extraretinal photoreceptors; the eyes do not have any role in this response. However, the eyes do participate in three circadian responses to light: entrainment, period change in LL, and arrhythmicity in LL. This last response cannot be produced in the absence of the eyes and is the only one of the above four "nonvisual" light responses that requires their input. Thus, four different responses to light involve three different patterns of interactions between the retinal and extraretinal inputs.

Experiments examining the roles of retinal and extraretinal photoreceptors in circadian photoreception have also been undertaken in lizards and have yielded results that are not only different from those of sparrows, but also vary among lizard species (Underwood and Menaker 1976). These interactions are likely to be quite different in different species and will most certainly depend on the specific photic niche occupied by the animal. The recent discovery of melanopsin has added a new and interesting level of complexity to the study of these interactions. Although it has a central role in mammalian circadian photoreception, its photoreceptive role in nonmammalian vertebrates has not been investigated. In nonmammalian vertebrates, melanopsin expression has been reported in subsets of horizontal, amacrine, and/or ganglion cells, depending on the species (Provencio et al. 1998; Bellingham et al. 2002; Drivenes et al. 2003; Jenkins et al. 2003; Chaurasia et al. 2005; Tomonari et al. 2007). Although the function(s) of mammalian melanopsin cells and their interactions with retinal rods and cones are an area of active research, it is virtually unexplored in nonmammalian vertebrates.

The existence of brain photoreceptors regulating photoperiodism and circadian entrainment has been described in many invertebrate groups and in all classes of vertebrates, with the exception of mammals (Shand and Foster 1999). In vertebrates, neurons immunoreactive for various opsins have been localized to the walls of brain ventricles in the hypothalamus and septum (Vigh et al. 2002). On the basis of this anatomical evidence, these cells have been assumed to be deep-brain photoreceptors. However, there has only been one report linking a specific population of putative deep-brain photoreceptors, in the hypothalamus of the ruin lizard *Podarcis sicula*, to circadian function (Pasqualetti et al. 2003). With the exception of this one case, we do not even know whether the opsins of these putative photoreceptor cells form functional photopigments, much less what type of phototransduction cascade might be involved. The neural and/or humoral inputs and outputs of these cells are also completely unexplored.

Mammals lack extraretinal photoreception, and the photoreceptors subserving visual and nonvisual functions coexist within the eye. The respective roles of the three photoreceptor classes that mediate nonvisual light responses in mammals—rods, cones, and melanopsin-containing ipRGCs—have been addressed to some extent by comparing the phenotypes of mice with degeneration or dysfunction of rods and/or cones to those of melanopsin knockout mice. As in nonmammalian vertebrates, there is redundancy between different photoreceptor types, and the role of the different photoreceptor classes appears to vary depending on the nonvisual response. For example, mice lacking both rod- and cone-based photoreception are able to phase shift and lengthen the free-running period in constant light with virtually unattenuated sensitivity when compared to wild-type mice (Foster et al. 1991; Argamaso et al. 1995; Freedman et al. 1999; Mrosovsky 2003; Panda et al. 2003). In melanopsin knockout mice, these responses are not eliminated but are substantially impaired (Panda et al. 2002; Ruby et al. 2002), suggesting that the primary photoreceptive input for circadian light responses comes from melanopsin. On the other hand, another nonvisual light response, the pupillary light reflex, appears to rely heavily on rod/cone input (Lucas et al. 2001). A caveat to the interpretation of these data is the question of whether developmental compensation occurs as a result of rod, cone, or melanopsin loss. This issue could be addressed through the use of conditional knockouts.

In addition to their intrinsic photoresponses, ipRGCs are activated by light signals originating in the rods and cones (Dacey et al. 2005; Perez-Leon et al. 2006; Wong et al. 2007). Rods and cones are thus capable of interacting with ipRGCs or providing separate input for nonvisual responses through a small number of conventional retinal ganglion cells that project to nonvisual centers in the brain (Gooley et al. 2003; Morin et al. 2003; Hattar et al. 2006). The pathways within the retina through which light information from rods and cones reach ipRGCs have just begun to be characterized. Recent studies indicate that rod and cone signals travel through both ON and OFF bipolar and amacrine pathways (Belenky et al. 2003; Viney et al. 2007; Wong et al. 2007). The effects of rod and cone input on nonvisual light responses are therefore likely to be complex and may differ in diurnal mammals with cone-dominated retinas. Although future studies using rod- and cone-specific knockouts will be necessary for a complete understanding of the contribution of rods versus cones to nonvisual physiology, recent behavioral studies suggest that manipulation of rod and cone pathways can have significant effects on entrainment. Abnormal phase angles of entrainment have been reported in $TR\beta_2^{-/-}$ mice lacking mid-wavelength cones (Dkhissi-Benyahya et al. 2007), and in $Rpe65^{-/-}$ mice (Doyle et al. 2006). The dramatic reversal of entrainment phase in $Rpe65^{-/-};Opn4^{-/-}$ mice underscores the importance of fully understanding these pathways.

The multiple extraretinal photoreceptors of nonmammalian vertebrates are often contained within tissues that are themselves circadian oscillators, e.g., the pineal gland of lamprey, birds, and lizards and the parietal eye of lizards (Zimmerman and Menaker 1979; Underwood 1990; Morita et al. 1992; Tosini and Menaker 1998). In zebra fish, peripheral oscillators in many organs and tissues are directly entrainable by light (Whitmore et al. 2000), a situation similar to that of *Drosophila melanogaster*, in which clock tissues all over the body use the flavin-based photoreceptor crypotochrome (Plautz et al. 1997; Emery et al. 2000). However, mammals lack extraretinal photoreceptors, with the possible exception of the neonatal pineal (Torres and Lytle 1989; Tosini et al. 2000). One attractive explanation for this difference in organization is the "nocturnal bottleneck" hypothesis, according to which the selective pressure for the shift from broadly distributed circadian photoreceptors to photoreceptors concentrated in the retina was the result of the enforced nocturnality of primitive mammals (Menaker et al. 1997). This hypothesis stems from the broader idea that the photic niche has a predominant role in shaping the circadian systems of vertebrates. The multiple extraretinal photoreceptors of nonmammalian vertebrates might therefore be necessary because they are each tuned to respond to different aspects of the animal's largely diurnal photic environment and/or to provide information directly to their closely associated oscillators. Indeed, "it may well be that the connection between light and the circadian system is a much more intimate one than is implied by the statement that light is the dominant natural entraining agent. In that case it might be possible to follow light into the heart of the mechanism" (Menaker et al. 1978). The strategy of following light input to the circadian system has been, and will in all probability continue to be, a particularly fruitful approach to understanding circadian organization (Gaston and Menaker 1968; Moore and Lenn 1972; Froehlich et al. 2002).

ACKNOWLEDGMENTS

The experimental work described in this paper has been supported over many years by grants from the National Science Foundation and the National Institutes of Health. The most recent work on mammalian retina has been supported by NIH grant MH56647.

REFERENCES

Argamaso S.M., Froehlich A.C., McCall M.A., Nevo E., Provencio I., and Foster R.G. 1995. Photopigments and circadian systems of vertebrates. *Biophys. Chem.* **56:** 3.

Belenky M.A., Smeraski C.A., Provencio I., Sollars P.J., and Pickard G.E. 2003. Melanopsin retinal ganglion cells receive bipolar and amacrine cell synapses. *J. Comp. Neurol.* **460:** 380.

Bellingham J., Whitmore D., Philp A.R., Wells D.J., and Foster R.G. 2002. Zebrafish melanopsin: Isolation, tissue localisation and phylogenetic position. *Brain Res. Mol. Brain Res.* **107:** 128.

Benoit J. 1935. Stimulation par la lumiere artificielle du developpement testiculaire chez des canards aveugles par enucleation des globes oculaire. *C.R. Soc. Biol.* **120:** 136.

Berson D.M., Dunn F.A., and Takao M. 2002. Phototransduction by retinal ganglion cells that set the circadian clock. *Science* **295:** 1070.

Chaurasia S.S., Rollag M.D., Jiang G., Hayes W.P., Haque R., Natesan A., Zatz M., Tosini G., Liu C., Korf H.W., Iuvone

P.M., and Provencio I. 2005. Molecular cloning, localization and circadian expression of chicken melanopsin (Opn4): Differential regulation of expression in pineal and retinal cell types. *J. Neurochem.* **92:** 158.

Dacey D.M., Liao H.W., Peterson B.B., Robinson F.R., Smith V.C., Pokorny J., Yau K.W., and Gamlin P.D. 2005. Melanopsin-expressing ganglion cells in primate retina signal colour and irradiance and project to the LGN. *Nature* **433:** 749.

Dkhissi-Benyahya O., Gronfier C., De Vanssay W., Flamant F., and Cooper H.M. 2007. Modeling the role of mid-wavelength cones in circadian responses to light. *Neuron* **53:** 677.

Doyle S.E., Castrucci A.M., McCall M., Provencio I., and Menaker M. 2006. Nonvisual light responses in the rpe65 knockout mouse: Rod loss restores sensitivity to the melanopsin system. *Proc. Natl. Acad. Sci.* **103:** 10432.

Drivenes O., Soviknes A.M., Ebbesson L.O., Fjose A., Seo H.C., and Helvik J.V. 2003. Isolation and characterization of two teleost melanopsin genes and their differential expression within the inner retina and brain. *J. Comp. Neurol.* **456:** 84.

Emery P., Stanewsky R., Helfrich-Forster C., Emery-Le M., Hall J.C., and Rosbash M. 2000. *Drosophila* cry is a deep brain circadian photoreceptor. *Neuron* **26:** 493.

Foster R.G., Provencio I., Hudson D., Fiske S., De Grip W., and Menaker M. 1991. Circadian photoreception in the retinally degenerate mouse (rd/rd). *J. Comp. Physiol. A* **169:** 39.

Freedman M.S., Lucas R.J., Soni B., von Schantz M., Munoz M., David-Gray Z., and Foster R. 1999. Regulation of mammalian circadian behavior by non-rod, non-cone, ocular photoreceptors. *Science* **284:** 502.

Froehlich A.C., Liu Y., Loros J.J., and Dunlap J.C. 2002. White collar-1, a circadian blue light photoreceptor, binding to the frequency promoter. *Science* **297:** 815.

Gaston S. and Menaker M. 1968. Pineal function: The biological clock in the sparrow? *Science* **160:** 1125.

Gooley J.J., Lu J., Fischer D., and Saper C.B. 2003. A broad role for melanopsin in nonvisual photoreception. *J. Neurosci.* **23:** 7093.

Hattar S., Liao H.W., Takao M., Berson D.M., and Yau K.W. 2002. Melanopsin-containing retinal ganglion cells: Architecture, projections, and intrinsic photosensitivity. *Science* **295:** 1065.

Hattar S., Kumar M., Park A., Tong P., Tung J., Yau K.W., and Berson D.M. 2006. Central projections of melanopsin-expressing retinal ganglion cells in the mouse. *J. Comp. Neurol.* **497:** 326.

Hattar S., Lucas R.J., Mrosovsky N., Thompson S., Douglas R.H., Hankins M.W., Lem J., Biel M., Hofmann F., Foster R.G., and Yau K.W. 2003. Melanopsin and rod-cone photoreceptive systems account for all major accessory visual functions in mice. *Nature* **424:** 76.

Jenkins A., Munoz M., Tarttelin E.E., Bellingham J., Foster R.G., and Hankins M.W. 2003. VA opsin, melanopsin, and an inherent light response within retinal interneurons. *Curr. Biol.* **13:** 1269.

Lees A.D. 1961. Some aspects of animal photoperiodism. *Cold Spring Harbor Symp. Quant. Biol.* **25:** 261.

Leydig F. 1890. Das parietalorgan der amphibien und reptilien. *Abh. Senckenb. Ges.* **16:** 441.

Lucas R.J., Douglas R.H., and Foster R.G. 2001. Characterization of an ocular photopigment capable of driving pupillary constriction in mice. *Nat. Neurosci.* **4:** 621.

McCall M.A., Gregg R.G., Merriman K., Goto Y., Peachey N.S., and Stanford L.R. 1996. Morphological and physiological consequences of the selective elimination of rod photoreceptors in transgenic mice. *Exp. Eye Res.* **63:** 35.

McMillan J.P., Elliot J.A., and Menaker M. 1975a. On the role of eyes and brain photoreceptors in the sparrow: Arrhythmicity in constant light. *J. Comp. Physiol.* **102:** 263.

———. 1975b. On the role of eyes and brain photoreceptors in the sparrow: Aschoff's rule. *J. Comp. Physiol.* **102:** 257.

McMillan J.P., Keatts H.C., and Menaker M. 1975c. On the role of eyes and brain photoreceptors in the sparrow: Entrainment to light cycles. *J. Comp. Physiol.* **102:** 251.

Menaker M. 1968a. Extraretinal light perception in the sparrow. I. Entrainment of the biological clock. *Proc. Natl. Acad. Sci.* **59:** 414.

———. 1968b. Light perception by extra-retinal receptors in the brain of the sparrow. In *Proceedings of the 76th Annual Convention of the American Psychological Association,* p. 299. American Psychological Association, Arlington, Virginia.

———. 1971. Synchronization with the photic environment via extraretinal receptors in the avian brain. In *Biochronometry: Proceedings of a Symposium* (ed. M. Menaker), p. 314. National Academy of Sciences, Washington, D.C.

———. 1977. Extraretinal photoreception. In *The science of photobiology* (ed. K.C. Smith), p. 227. Plenum, New York.

Menaker M. and Keatts H. 1968. Extraretinal light perception in the sparrow. II. Photoperiodic stimulation of testis growth. *Proc. Natl. Acad. Sci.* **60:** 146.

Menaker M., Moreira L.F., and Tosini G. 1997. Evolution of circadian organization in vertebrates. *Braz. J. Med. Biol. Res.* **30:** 305.

Menaker M., Takahashi J.S., and Eskin A. 1978. The physiology of circadian pacemakers. *Annu. Rev. Physiol.* **40:** 501.

Menaker M., Roberts R., Elliott J., and Underwood H. 1970. Extraretinal light perception in the sparrow. 3. The eyes do not participate in photoperiodic photoreception. *Proc. Natl. Acad. Sci.* **67:** 320.

Moore R.Y. and Lenn N.J. 1972. A retinohypothalamic projection in the rat. *J. Comp. Neurol.* **146:** 1.

Morin L.P., Blanchard J.H., and Provencio I. 2003. Retinal ganglion cell projections to the hamster suprachiasmatic nucleus, intergeniculate leaflet, and visual midbrain: Bifurcation and melanopsin immunoreactivity. *J. Comp. Neurol.* **465:** 401.

Morita Y., Tabata M., Uchida K., and Samejima M. 1992. Pineal-dependent locomotor activity of lamprey, *Lampetra japonica*, measured in relation to LD cycle and circadian rhythmicity. *J. Comp. Physiol. A* **171:** 555.

Mrosovsky N. 2003. Aschoff's rule in retinally degenerate mice. *J. Comp. Physiol. A* **189:** 75.

Nelson R.J. and Zucker I. 1981. Absence of extraocular photoreception in diurnal and nocturnal rodents exposed to direct sunlight. *Comp. Biochem. Physiol.* **69A:** 145.

Panda S., Sato T.K., Castrucci A.M., Rollag M.D., DeGrip W.J., Hogenesch J.B., Provencio I., and Kay S.A. 2002. Melanopsin (Opn4) requirement for normal light-induced circadian phase shifting. *Science* **298:** 2213.

Panda S., Provencio I., Tu D.C., Pires S.S., Rollag M.D., Castrucci A.M., Pletcher M.T., Sato T.K., Wiltshire T., Andahazy M., Kay S.A., Van Gelder R.N., and Hogenesch J.B. 2003. Melanopsin is required for non-image-forming photic responses in blind mice. *Science* **301:** 525.

Pasqualetti M., Bertolucci C., Ori M., Innocenti A., Magnone M.C., De Grip W.J., Nardi I., and Foa A. 2003. Identification of circadian brain photoreceptors mediating photic entrainment of behavioural rhythms in lizards. *Eur. J. Neurosci.* **18:** 364.

Perez-Leon J.A., Warren E.J., Allen C.N., Robinson D.W., and Lane Brown R. 2006. Synaptic inputs to retinal ganglion cells that set the circadian clock. *Eur. J. Neurosci.* **24:** 1117.

Plautz J.D., Kaneko M., Hall J.C., and Kay S.A. 1997. Independent photoreceptive circadian clocks throughout *Drosophila*. *Science* **278:** 1632.

Provencio I., Jiang G., De Grip W.J., Hayes W.P., and Rollag M.D. 1998. Melanopsin: An opsin in melanophores, brain, and eye. *Proc. Natl. Acad. Sci.* **95:** 340.

Provencio I., Rodriguez I.R., Jiang G., Hayes W.P., Moreira E.F., and Rollag M.D. 2000. A novel human opsin in the inner retina. *J. Neurosci.* **20:** 600.

Ruby N.F., Brennan T.J., Xie X., Cao V., Franken P., Heller H.C., and O'Hara B.F. 2002. Role of melanopsin in circadian responses to light. *Science* **298:** 2211.

Scharrer E. 1928. Die lichtempfindlichkeit blinder elritzen i. Untersuchungen uber das zwischenhirn der fische. *Z. Vgl. Physiol.* **7:** 1.

Seeliger M.W., Grimm C., Stahlberg F., Friedburg C., Jaissle G., Zrenner E., Guo H., Reme C.E., Humphries P., Hofmann F., Biel M., Fariss R.N., Redmond T.M., and Wenzel A. 2001.

New views on Rpe65 deficiency: The rod system is the source of vision in a mouse model of Leber congenital amaurosis. *Nat. Genet.* **29:** 70.

Shand J. and Foster R.G. 1999. The extraretinal photoreceptors of non-mammalian vertebrates. In *Adaptive mechanisms in the ecology of vision* (ed. S.N. Archer et al.), p. 197. Kluwer Academic, Dordrecht, The Netherlands.

Tomonari S., Takagi A., Noji S., and Ohuchi H. 2007. Expression pattern of the melanopsin-like (cOpn4m) and VA opsin-like genes in the developing chicken retina and neural tissues. *Gene Expr. Patterns* (in press). **AU: Update?**

Torres G. and Lytle L.D. 1989. Extraretinal mechanisms mediate light-induced changes in neonatal rat pineal gland N-acetyltransferase activity. *J. Pineal Res.* **7:** 211.

Tosini G. and Menaker M. 1998. Multioscillatory circadian organization in a vertebrate, *Iguana iguana*. *J. Neurosci.* **18:** 1105.

Tosini G., Doyle S., Geusz M., and Menaker M. 2000. Induction of photosensitivity in neonatal rat pineal gland. *Proc. Natl. Acad. Sci.* **97:** 11540.

Underwood H. 1990. The pineal and melatonin: Regulators of circadian function in lower vertebrates. *Experientia* **46:** 120.

Underwood H. and Menaker M. 1970. Photoperiodically significant photoreception in sparrows: Is the retina involved? *Science* **167:** 298.

———. 1976. Extraretinal photoreception in lizards. *Photophysiology* **23:** 227.

Vigh B., Manzano M.J., Zadori A., Frank C.L., Lukats A., Rohlich P., Szel A., and David C. 2002. Nonvisual photoreceptors of the deep brain, pineal organs and retina. *Histol. Histopathol.* **17:** 555.

Viney T.J., Balint K., Hillier D., Siegert S., Boldogkoi Z., Enquist L.W., Meister M., Cepko C.L., and Roska B. 2007. Local retinal circuits of melanopsin-containing ganglion cells identified by transsynaptic viral tracing. *Curr. Biol.* **17:** 981.

von Frisch K. 1911. Beitrage zur physiologie der pigmentzellen inder fischhaut. *Pflugers Arch.* **138:** 319.

Wenzel A., von Lintig J., Oberhauser V., Tanimoto N., Grimm C., and Seeliger M.W. 2007. Rpe65 is essential for the function of cone photoreceptors in nrl-deficient mice. *Invest. Ophthalmol. Vis. Sci.* **48:** 534.

Whitmore D., Foulkes N.S., and Sassone-Corsi P. 2000. Light acts directly on organs and cells in culture to set the vertebrate circadian clock. *Nature* **404:** 87.

Wong K.Y., Dunn F.A., Graham D.M., and Berson D.M. 2007. Synaptic influences on rat ganglion-cell photoreceptors. *J. Physiol.* **582:** 279.

Zimmerman N.H. and Menaker M. 1979. The pineal gland: A pacemaker within the circadian system of the house sparrow. *Proc. Natl. Acad. Sci.* **76:** 999.

Znoiko S.L., Rohrer B., Lu K., Lohr H.R., Crouch R.K., and Ma J.X. 2005. Downregulation of cone-specific gene expression and degeneration of cone photoreceptors in the *rpe65*$^{-/-}$ mouse at early ages. *Invest. Ophthalmol. Vis. Sci.* **46:** 1473.

Multiple Photoreceptors Contribute to Nonimage-forming Visual Functions Predominantly through Melanopsin-containing Retinal Ganglion Cells

A.D. GÜLER,* C.M. ALTIMUS,* J.L. ECKER,* AND S. HATTAR*†

*Department of Biology, Johns Hopkins University, Baltimore, Maryland 21218; †Department of Neuroscience, Johns Hopkins University–School of Medicine, Baltimore, Maryland 21205

In the absence of functional rod and cone photoreceptors, mammals retain the ability to detect light for a variety of physiological functions such as circadian photoentrainment and pupillary light reflex. This is attributed to a third class of photoreceptors, the intrinsically photosensitive retinal ganglion cells that express the photopigment melanopsin. Even though in the absence of rods and cones, mammals retain the ability to detect light for various nonimage-forming visual functions, rods and cones can compensate for the absence of the melanopsin protein in nonvisual light-dependent physiological behaviors. Several studies have addressed the relative contribution of each photoreceptor type to nonimage-forming visual functions; however, a comprehensive model for these interactions is far from complete. Under conditions where melanopsin-containing retinal ganglion cells were genetically ablated, image formation is maintained, whereas circadian photoentrainment and pupillary light reflex are severely impaired. The findings indicate that multiple photoreceptors contribute to nonimage-forming visual functions through signaling via melanopsin-containing retinal ganglion cells. Future studies will aim to determine more quantitatively the relative contributions of each retinal photoreceptor in signaling light for nonimage-forming visual functions.

INTRODUCTION

Daily and seasonal changes in the light/dark cycles of the earth due to the tilted rotation around its axis provide all organisms with temporal information that allows them to better adapt to their environment. Leaf senescence, which is preceded by the remarkable autumn foliage colors such as that seen at the Cold Spring Harbor Laboratory, results from the shortened day length during fall to allow plants to cope with pressures of the winter months (Thomas and Stoddart 1980). The ability of organisms to anticipate changes to the solar day with their internal biological clock permits them to coordinate their physiology for optimal performance throughout the day/night cycle. For the internal clock to be relevant, a consistent phase association with the solar day must be established. The synchrony between the light/dark cycle and circadian rhythms has been dubbed circadian photoentrainment, which requires the oscillator to receive input from photoreceptive molecules. This light reception can be an integral part of the oscillator cell or accomplished by a specialized light receptive organ(s). For instance, in mammals, the eye receives light not only for image formation, but also for irradiance-dependent functions including circadian photoentrainment.

For nearly a century, the retina, particularly the classical photoreceptors (rods and cones), has been extensively studied using a variety of mutant or lesioned animals to understand phototransduction and image formation. In vertebrates, rods and cones are the photoreceptors for image formation under dim- and bright-light conditions, respectively. Although studies on nonmammalian vertebrates during the 1960s revealed that extraocular organs, such as the pineal gland, were photoreceptive for circadian photoentrainment (Menaker 1968; Menaker et al. 1997), research on mammals blinded by eye removal demonstrated that all photoreception for photoentrainment and vision occurs exclusively in the mammalian retina (Nelson and Zucker 1981). Later, mice with outer retinal degeneration that lacked functional rods and cones were shown to photoentrain but not form visual images (Freedman et al. 1999; Lucas et al. 1999). Studies on blind humans also demonstrated that photoentrainment without the ability to form conscious vision is possible (Czeisler et al. 1995; Klerman et al. 2002). The ability of an animal to photoentrain in the absence of intact classical photoreceptors puzzled researchers and pointed to the presence of a distinct photoreceptor located beyond the outer retina that is responsible for photoentrainment.

The recent identification of the photopigment melanopsin (Opn4) in a unique subpopulation of the retinal ganglion cells resolved this long-standing conundrum (Provencio et al. 1998, 2000). Electrophysiological recordings from melanopsin-expressing ganglion cells (also known as intrinsically photosensitive retinal ganglion cells [ipRGCs]), and cells expressing melanopsin protein ectopically, clearly demonstrated that melanopsin is capable of responding to light and that it functions as a true photopigment (Rollag et al. 2000; Berson et al. 2002; Newman et al. 2003; Melyan et al. 2005; Panda et al. 2005; Qiu et al. 2005). The ipRGC response to light is less sensitive and more "sluggish" than the classical photoreceptors rods and cones (Berson et al. 2002). The ipRGCs tile the entire retina and project to nonimage-forming visual centers, including the suprachiasmatic nucleus (SCN) (Hattar et al. 2002), making them the premier candidate for the specialized cells of irradiance detection and signaling. The ipRGCs located in the inner retina slowly

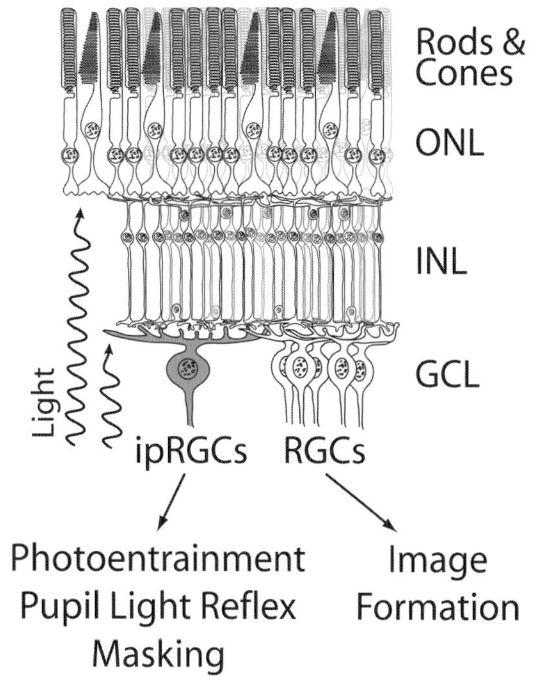

Figure 1. Model describing how rod/cone signaling might contribute to nonimage- and image-forming functions. Three classes of photoreceptors are illustrated: rods, cones, and the ipRGCs. (GCL) Ganglion cell layer; (INL) inner nuclear layer; (ONL) outer nuclear layer.

depolarize and fire action potentials during extended light stimulation, which allows them to signal directly (monosynaptically) to the SCN. This is in contrast to the rapidly activating and desensitizing responses observed in classical photoreceptors of the outer retina, which require a multisynaptic cascade to signal light information to the brain (Fig. 1). The initial physiological and morphological characterization of the ipRGCs and their relationship to the classical photoreceptors has been reviewed in detail elsewhere (Berson 2003).

SEVERAL PHOTORECEPTOR TYPES CONTRIBUTE TO NONIMAGE-FORMING FUNCTIONS

Brain centers associated with photoentrainment are predominantly innervated by retinal ganglion cells expressing melanopsin (Hattar et al. 2002, 2006). Therefore, researchers anticipated that the elimination of melanopsin protein would lead to an animal that has severe photoentrainment defects. Surprisingly, melanopsin knockout mice, generated in three independent laboratories, were able to photoentrain to a 24-hour light/dark cycle (Hattar et al. 2002; Panda et al. 2002; Ruby et al. 2002). This suggests that either rods/cones can compensate for the absence of the melanopsin protein or the presence of a fourth photopigment, such as cryptochromes (Selby et al. 2000), is required for photoentrainment. This issue was resolved with animals that had the G-protein α-subunit (Gnat1) of rods, cyclic nucleotide-gated channel (Cnga3) of cones, and melanopsin photopigment (Opn4) of ipRGCs genetically eliminated (*gnat1*$^{-/-}$; *cnga3*$^{-/-}$; *opn4*$^{-/-}$), blocking all three known photoreceptor cascades without major retinal degeneration. The triple mutant animals are unable to form images, but they also do not respond to light for all tested nonimage-forming functions including pupillary light reflex (PLR), circadian photoentrainment, and masking (Hattar et al. 2003). At the same time, Panda et al. (2003) demonstrated that animals that lack melanopsin protein and have outer retinal degeneration (*opn4*$^{-/-}$; *rd/rd*) do not photoentrain or detect light for other nonimage-forming functions. The ability of animals that lack either the melanopsin protein or outer retinal photoreceptive functions to photoentrain indicates that each photoreceptive system can compensate for the lack of the other for circadian photoentrainment. However, it is not clear from these models if and how the two photoreceptive systems interact.

Because photoreceptors respond optimally at a given wavelength of light and have distinct response profiles, researchers have reasoned that if a photoreceptor underlies a light-dependent behavior, the response profile of the behavior should match that of the photoreceptor. Before the discovery of the melanopsin photopigment, Takahashi et al. (1984) elegantly demonstrated that the maximal spectral sensitivity for phase-shifting response (an index for an animal's ability to signal light to the circadian pacemaker) in hamsters is near 500 nm. Although this is the optimal photosensitivity of rods, these authors noted not only that the threshold for this response is 6 log units above the threshold for rod light detection, but that the photoentrainment system is also capable of integrating light information for extended periods of time. Because this is atypical of classical photoreceptors, the authors implicated a novel photoreceptive system mediating the entrainment of mammalian circadian rhythms. Since then, researchers have shown that melanopsin is a part of this novel photoreceptive system (Hattar et al. 2002; Panda et al. 2002; Ruby et al. 2002). Although animals that lack the melanopsin protein showed attenuated phase-shifting responses, the maximal sensitivity of melanopsin cells is 480 nm (Yoshimura and Ebihara 1996; Hattar et al. 2003), which does not match the maximal spectral sensitivity of phase shifts in hamsters. This indicates that multiple photoreceptors contribute to phase shifting the circadian oscillator. Determining the relative contribution of each photoreceptor to photoentrainment will allow the creation of better models to understand how light interacts with the circadian pacemaker.

CURRENT MODELS ARE NOT SUFFICIENT TO EXPLAIN THE RELATIVE CONTRIBUTION OF EACH PHOTORECEPTOR TO CIRCADIAN PHOTOENTRAINMENT

Although the animals that lack the rod, cone, and melanopsin photoreceptive functions (*gnat1*$^{-/-}$; *cnga3*$^{-/-}$; *opn4*$^{-/-}$) (Hattar et al. 2003) or that lack melanopsin and have outer retinal degeneration (*rd/rd; opn4*$^{-/-}$) (Panda et al. 2003) demonstrated that the three photoreceptor types are sufficient for the proper operation of nonimage functions, there are still unexplained observations that are worth mentioning. In mice that lack the rod, cone, and melanopsin photoreceptive functions, a variable transient 5% response to light was reported in their pupil light reflex, indicating that these animals are not completely blind (Hattar et al. 2003). A possible source of this signal is transient membrane hyperpolarization of the receptor potential of rods and cones caused by charge movements associated with conformational changes of the visual pigments after photoisomerization (Cone 1967; Murakami and Pak 1970). Alternatively, a redundant inefficient signaling cascade in one of the mutant photoreceptor cell types might account for this response. For instance, the Cngb3 subunit is known to be expressed in cone cells and might be able to substitute for the eliminated Cnga3 and form a homomeric form of the cone cyclic nucleotide channel, hence leading to minuscule light responses. The other animal model, which utilized the melanopsin knockout in conjunction with the *rd/rd* mutation, did not show the residual transient light responses in PLR (Panda et al. 2003). This could indicate that the photoisomerization-dependent hyperpolarization in the classical photoreceptors is the cause for the residual 5% transient response. The *rd/rd* animals have a mutation in the rod phosphodiesterase (Pde6b) (Bowes et al. 1990), which causes the initial ablation of rods leading to a secondary yet incomplete degeneration of cones giving rise to opsin-positive perikarya (Foster et al. 1991). These functional aberrant cone photoreceptors are capable of surviving in the *rd/rd* animals for many months (Garcia-Fernandez et al. 1995). Provencio and Foster (1995) demonstrated that *rd/rd* animals can be phase-shifted by monochromatic light at 515 nm as well as 375 nm (around the maximal spectral sensitivity of MW-cones and S-cones, respectively) implicating that both S- and MW-cones participate in photoentrainment. Moreover, Dkhissi-Benyahya et al. (2007) recently demonstrated that the MW-cones are important in signaling light information for circadian photoentrainment. In light of these findings, it is surprising that the *rd/rd; opn4*$^{-/-}$ animals do not show any light responses in non-image-forming functions, and further investigation is warranted to settle the discrepancy between these animal models with respect to cone function. It is crucial to understand how it is possible that the remaining cone cells are incapable of signaling any light information in the *rd/rd; opn4*$^{-/-}$ animals. It is possible that *rd/rd; opn4*$^{-/-}$ animals have a retina that is rewired and incapable of signaling properly for nonimage-forming functions due to degeneration and the absence of the melanopsin protein. Alternatively, as demonstrated by Yoshimura and Ebihara (1996), the maximum light response in *rd/rd* animals is 480 nm, which matches the spectral sensitivity of melanopsin and implies that cone photoreceptors in *rd/rd* animals are not involved in phase shifting the circadian oscillator.

The role of cone cells in signaling light information for circadian photoentrainment becomes even more complicated when considering mice containing only functional cone photoreceptors. Mrosovsky and Hattar (2005) achieved this by knocking-out essential components of rod (*gnat1*$^{-/-}$) and melanopsin (*opn4*$^{-/-}$) phototransduction machinery, leaving the cone system intact. These animals, despite having a nonfunctional rod-signaling cascade, do not have rod degeneration. Therefore, cone cells do not suffer secondary degeneration as in *rd/rd* mutants. Albeit the sample size in these experiments was low, some of these animals demonstrated attenuated or possibly a complete lack of circadian photoentrainment. Furthermore, another subset of animals with only functional cones (*gnat1*$^{-/-}$*; opn4*$^{-/-}$) showed a preference for activity in the light portion of the cycle, i.e., became diurnal. These results indicate that cone input independent of rods is not sufficient to compensate for circadian photoentrainment in the absence of melanopsin protein. Dkhissi-Benyahya et al. (2007) studied mice mutated in the thyroid hormone receptor (TR$\beta^{-/-}$), which leads to selective elimination of MW cones. Animals without the MW cones have defects in detecting light for circadian responses, especially at longer wavelength and shorter duration light pulses. Taken together, the observations from the Mrosovsky and Dkhissi-Benyahya studies beg the question: How could cone photoreceptors have a major role in circadian phase-shifting yet fail to photoentrain the animals? To resolve this issue, phase-shifting studies in animals with only functional cones should be undertaken. In addition, because several mouse lines that have specific elimination of cone function (*cnga3*$^{-/-}$ or *cl* animals) are available, the results obtained with TR$\beta^{-/-}$ mice should be confirmed with the same light parameters that the Dkhissi-Benyahya study utilized. The contribution of cone input to circadian phase shifting is also shown in human studies. Lockley et al. (2003) showed that a 555-nm light that preferentially activates cone photoreceptors in humans is twofold less efficient in phase shifting the oscillator than a 460-nm light that preferentially activates the melanopsin photoreceptors. In conclusion, despite the fact that the role and relative contribution of cone photoreceptors in circadian photoentrainment are still not fully resolved, animals that lack rod function and melanopsin protein have weaker circadian photoentrainment than animals lacking melanopsin protein only. Therefore, do rod photoreceptors contribute to circadian photoentrainment?

One way to determine the role of rod photoreceptors in circadian photoentrainment is to use animals that have only functional rod cells intact (*cnga3*$^{-/-}$*; opn4*$^{-/-}$). However, the only available data in the literature about rod function comes from animals lacking the Rpe65 protein (Redmond et al. 1998). Rpe65 is the retinoid isomerase that is required for regeneration of chromophore for classical photoreceptors but is not essential for melanopsin function (Jin et al. 2005; Moiseyev et al. 2005; Redmond et al. 2005). Animals that lack this isomerase have rods as the only remaining functional classical photoreceptor, albeit with reduced photosensitivity (Fan et al. 2005). Doyle et al. (2006) demonstrated that animals that are null for melanopsin protein in addition to the Rpe65 gene (*opn4*$^{-/-}$*; rpe65*$^{-/-}$), despite having no cone function and reduced rod sensitivity, can photoentrain; moreover, some of these animals showed a preference for day activity and become diurnal. Their ability to photoentrain indicates that rods may be important for signaling light for nonimage-forming functions. This is surprising given that the threshold for the circadian light response is well above the threshold for rod light signaling.

So far, the current research in this area does not conclusively clarify the role of each photoreceptor to circadian light functions. These reports demonstrate that all three photoreceptor cell types work together in order to signal for irradiance-dependent functions. However, how can we determine the specific contribution of each photoreceptor for nonimage-forming functions? To answer this question, future studies should utilize multiple animal models that preserve photoreceptive systems individually and models that manipulate specific aspects of the retinal circuitry without affecting photoreception.

AN AMAZING FEAT: DECODING LIGHT INFORMATION FOR BOTH CONTRAST AND IRRADIANCE DETECTION

The intricate architecture of the mammalian retina provides precise pathways for light signaling for both image- and nonimage-forming functions. Although both functions require detection of light, they are distinct in the nature of light information attained. Although image formation involves the ability of the retina to differentiate between two juxtaposing points in the visual field, the nonimage-forming functions require the knowledge of the absolute environmental luminosity.

The light information that is detected by rods and cones is signaled via the inner nuclear layer of the retina to RGCs, the only output neurons of the retina (see Fig. 1). Two classes of RGCs, ON and OFF, either depolarize and activate by a light or a dark signal from the outer retina, respectively. Because the ipRGCs receive very few OFF-light "signals" (Brown and Silva 2004; Dacey et al. 2005; Wong et al. 2007), primarily ON-light responses will be discussed.

To form images, the retina absorbs light from the environment and deciphers this overwhelming amount of information into differences of contrast between objects and signals this information through retinotopic maps to the visual cortex. For this visual function, the total intensity of the light is not utilized as much as the relative contrasts, which allows us to be able to read not only outside under bright sunlight (50,000 lux), but also inside a dimly lit room (100 lux). The adaptation of cones to a wide range of light intensities (up to 9 log units) allows this remarkable ability of reading under such vastly different light conditions (Knox and Solessio 2006). Although very advantageous for reading a book, light adaptation with such vigor might cause uncertainty about the intensity of light from the perspective of the tissues receiving this information for photoentrainment, such as the SCN. In contrast to cone photoreceptors, rod photoreceptors can decode differences between even a few photons but only at low light intensity (scotopic conditions). Rods bleach at medium (mesopic) to high (photopic) light intensities and hence cannot detect further increases in light irradiances (Burns and Baylor 2001; Stockman and Sharpe 2006). Although the cone pathway under mesopic and photopic conditions is adapted and might not supply adequate irradiance information, the continuous bleached signal of rods could be used to indicate the presence or absence of light in the environment. Therefore, the signals from rods and cones could be decoded to measure irradiances even in the absence of melanopsin protein.

MELANOPSIN-CONTAINING RGCs ARE REQUIRED FOR NONIMAGE-FORMING FUNCTIONS

The role of melanopsin-expressing ganglion cells as photoreceptors has been well-studied, but their role as RGCs, responsible for mediating rod/cone-dependent signaling to the brain, has not been established independent of their photosensitivity. The ability of melanopsin knockout animals to photoentrain and constrict their pupil to light has been attributed to the light detected by rods and cones; however, the route used to signal to nonimage-forming centers by the classical photoreceptors for these functions has not been elucidated. Morphological studies indicate that although the majority of the projections to SCN are ipRGCs, some of these projections are of classical RGC origin (Gooley et al. 2001; Morin et al. 2003; Sollars et al. 2003). Furthermore, the olivary pretectal nucleus is highly innervated by the nonmelanopsin cells (Hattar et al. 2006). What is the role of the contrast-detecting classical RGCs in nonimage-forming functions? Do melanopsin expressing RGCs integrate all light input from the three types of photoreceptors and relay this information to nonimage-forming centers as decoders of irradiance?

To appreciate all of the functions carried out by ipRGCs, the elimination of not only the melanopsin photopigment, but the whole cell is necessary. Because only RGCs that do not express melanopsin survive, this ablation also indirectly determines the relative contribution of the classical RGCs to nonimage-forming functions. We accomplished the specific ablation of ipRGCs by expressing the attenuated form of diphtheria toxin subunit A (aDTA), a translational inhibitor, at the melanopsin locus (A.D. Güler et al., in prep.). As expected, animals carrying a single copy of aDTA ($opn4^{aDTA/+}$) demonstrate nearly complete loss of melanopsin-expressing cells with approximately 17% of the cells surviving at 6 months of age. As expected, the projections from these cells to their targets (i.e., SCN) are reduced in the aDTA animals.

Bilateral ocular injections of the cholera toxin B subunit revealed that the projections by classical RGCs remained intact in both nonimage- and image-forming centers, leaving only few fibers projecting to the SCN. When the ipRGCs were ablated to a greater extent by the generation of aDTA homozygotes ($opn4^{aDTA/aDTA}$), we observed a further elimination of RGC projections to the SCN, leaving few fibers at 1 year of age. This indicates that there are minimal classical RGC innervations to the SCN. These morphological observations indicate that light signaling for nonimage functions may be studied independent of image formation. In fact, we showed that the aDTA animals are capable of forming images quite normally using several classical vision tests.

THE SIGNAL FOR PLR CONVERGES AT THE LEVEL OF MELANOPSIN-EXPRESSING RGCs

Light-dependent pupil constriction regulates the amount of light that enters the eye based on overall illumination. When light enters the retina, light intensity information is sent to the olivary pretectal nucleus via RGCs, whose firing rate to the Edinger Westphal nuclei is modulated by irradiance. The oculomotor nerve originating from Edinger Westphal nuclei synapses to the ciliary ganglion, which controls pupil size (Zhang et al. 1996). Because irradiance detectors within the retina control the PLR, it was thought that the melanopsin-containing RGCs would be involved; however, Lucas et al. (2001) established that rods and cones are the main input for pupil constriction at low light intensities, whereas rods and cones are not able to execute full pupil constriction at higher light intensities in the absence of melanopsin protein. Therefore, the ability to fully constrict the pupil at higher light intensities depends on the intrinsic photosensitivity of the melanopsin-containing RGCs (Lucas et al. 2003). By utilizing the DTA mouse, we were able to determine whether rod- and cone-dependent pupil constriction acts through melanopsin-expressing cells or through the classical RGCs. In mice expressing aDTA from the melanopsin locus, pupil constriction at low light intensities was significantly weakened compared to that in wild-type mice. This suggested that the rod and cone signal necessary for pupil constriction is routed through the melanopsin cells even at low light intensities. In animals with nearly complete ablation of ipRGCs, the ability to constrict the pupil is attenuated to a greater extent at high light intensities than in animals that lack the melanopsin photoreceptor. These results indicate that melanopsin-containing RGCs are necessary for PLR at high and low light intensities by acting both as the primary light detectors and as conduits for rod/cone light information to the olivary pretectal nucleus.

IN THE ABSENCE OF MELANOPSIN-CONTAINING RGCs, CIRCADIAN PHOTOENTRAINMENT IS ABOLISHED

To fully assess the ability of ipRGC-ablated animals to photoentrain, we exposed them to a "jet-lag" paradigm (three 2-week 12:12 light/dark cycles advanced and delayed by 6 hours consecutively). Although wild-type animals had a close phase relationship with each light/dark cycle, the animals expressing aDTA segregated into two distinct phenotypic groups, neither of which photoentrain. The first group, group A, showed no light responses and completely free-ran similar to genetically engineered mice that lack all three photoreceptor types or bilaterally enucleated animals. Although the second group of animals, group B, was unable to photoentrain, they exhibited weak light responses with unstable and large phase angles to the light/dark cycles. The differences in light responses between these two animal groups may be attributed to the variability among mutants in the number of remaining melanopsin-expressing cells. Because phase-shifting experiments are a more quantitative indication of an animal's ability for circadian light response, we tested whether either group A or B mutant animals can be phase-shifted by a light stimulus that elicits approximately 2-hour phase delays at circadian time (CT) 16 in wild-type mice. Interestingly, neither mutant group showed any phase delays to this light stimulus. This reveals that the photoentrainment capability of animals that lack ipRGCs is severely diminished. The rod/cone-dependent light signal for photoentrainment observed in wild-type and melanopsin knockout animals is routed through the ipRGCs. Unlike rod/cone or melanopsin photoreceptive functions, signaling of light input to the SCN via the ipRGCs is required for photoentrainment.

THE CIRCADIAN OSCILLATOR IS NOT COMPROMISED IN aDTA ANIMALS

In constant darkness, wild-type mice are able to maintain a regular, approximately 23.6-hour rhythm in locomotor activity reflecting their endogenous clock. Animals lacking an endogenous circadian oscillator will not maintain a regular rhythm under constant conditions and will be arrhythmic. Because both groups A and B of the aDTA-expressing mice have intact circadian rhythms in constant darkness, we concluded that the endogenous clock in these animals is not affected by the attenuated light input to their SCN.

Aschoff's rule states that generally in nocturnal species, *tau* is positively correlated with light intensity in constant light, i.e., as the light intensity is increased, the period length of the rhythm will lengthen. Diurnal animals respond oppositely and will shorten their period as light intensity increases in constant light (Aschoff 1961; Daan 2000). As expected, wild-type mice in our experiments lengthen their free-running periods from 23.3 hours in constant dark to 25.5 hours in constant light. Although the group B mice do not show a significant difference in their period length between constant dark and light conditions, group A mice shorten their periods from 23.8 hours in constant dark to 23.4 hours in constant light. This trend toward a shortened period in constant light in animals that have only conventional RGCs remaining is characteristic of diurnal animals. Interestingly, mice with only functional cone photoreceptors ($gnat^{-/-}$; $opn4^{-/-}$) and $rpe65^{-/-}$; $opn4^{-/-}$ show diurnal tendencies (Mrosovsky and Hattar 2005; Doyle et al. 2006). In light of these findings, it would be interesting to determine whether these animal models with diurnal tendencies also have shortened periods under constant light conditions.

MASKING DEMONSTRATES THAT THE MELANOPSIN PROTEIN MAINTAINS PROLONGED RESPONSES TO LIGHT

Organisms are able to coordinate their activity to the light/dark cycles by either synchronizing their internal clock so that they are able to anticipate the light cycle (i.e., photoentrainment) or responding directly to light changes (masking). It is thought that these two responses work together to control daily cycles. Photoentrainment is mediated by light input from the retina to the SCN; however, the light pathway and brain regions mediating masking are unknown. The most-prominent defect observed in mice lacking melanopsin protein is their masking responses to light. To measure the ability of the melanopsin protein to detect light for prolonged periods in vivo, Mrosovsky and Hattar (2003) presented 3-hour light pulses that inhibit locomotor activity, 2 hours after the onset of activity in the dark. In animals that lack melanopsin protein, the initial rod/cone-mediated inhibitory effect on locomotor activity is not maintained compared to wild-type animals that sustain this masking response for the duration of the light pulse (Mrosovsky and Hattar 2003). This confirms that the melanopsin protein is an irradiance detector that can measure photons for a prolonged period of time.

Using a 7-hour light/dark cycle (3.5 hour:3.5 hour light/dark cycle) to which mice cannot readily entrain, melanopsin knockout mice confine approximately 75% of their activity to the dark portion compared to the wild-type mice at a light intensity that confines 98% of their activity to the dark phase (Mrosovsky and Hattar 2003). In this paradigm, animals that are unable to mask should distribute their activity equally between the light and the dark phases and have approximately 50% of their activity in the dark. At a light intensity where the wild-type animals confined 84% of their activity to the dark portion of the ultradian cycle, the $opn4^{aDTA/aDTA}$ mice that free-ran through all light paradigms (group A) confined 62% of their activity to the dark portion of the light/dark cycle. The mutant mice with weak light responses (group B) had slightly higher masking ability with 67% of their activity confined to the dark phase of the light/dark cycle. Because there exists a possibility that group A animals contain fewer melanopsin-expressing cells than group B animals, the positive correlation between the strength of this system and the animals' ability to mask suggests that melanopsin cells are responsible for masking responses. Because the melanopsin-containing RGCs have defined brain targets, this finding narrows the potential pathways responsible for masking.

CONCLUSIONS

The selective ablation of ipRGCs reveals a clear distinction between the function of the melanopsin protein and the melanopsin-containing RGCs. The ipRGCs are the main conduit for light information to nonimage-forming functions combining light responses from all three types of photoreceptors. Rods/cones contribute to nonimage functions mainly at low light intensities, whereas the melanopsin protein is important at high light intensities and especially for prolonged measurement of light. A delicate balance between all three photoreceptor systems coordinates physiologically important functions to light detection.

ACKNOWLEDGMENTS

We sincerely thank Drs. Marnie Halpern and Rejji Kuruvilla for valuable comments on the manuscript.

REFERENCES

Aschoff J. 1961. Exogenous and endogenous components in circadian rhythms. *Cold Spring Harbor Symp. Quant. Biol.* **25:** 11.

Berson D.M. 2003. Strange vision: Ganglion cells as circadian photoreceptors. *Trends Neurosci.* **26:** 314.

Berson D.M., Dunn F.A., and Takao M. 2002. Phototransduction by retinal ganglion cells that set the circadian clock. *Science* **295:** 1070.

Bowes C., Li T., Danciger M., Baxter L.C., Applebury M.L., and Farber D.B. 1990. Retinal degeneration in the rd mouse is caused by a defect in the beta subunit of rod cGMP-phosphodiesterase. *Nature* **347:** 677.

Brown R. and Silva A.J. 2004. Molecular and cellular cognition; the unraveling of memory retrieval. *Cell* **117:** 3.

Burns M.E. and Baylor D.A. 2001. Activation, deactivation, and adaptation in vertebrate photoreceptor cells. *Annu. Rev. Neurosci.* **24:** 779.

Cone R.A. 1967. Early receptor potential: Photoreversible charge displacement in rhodopsin. *Science* **155:** 1128.

Czeisler C.A., Shanahan T.L., Klerman E.B., Martens H., Brotman D.J., Emens J.S., Klein T., and Rizzo J.F., III. 1995. Suppression of melatonin secretion in some blind patients by exposure to bright light. *N. Engl. J. Med.* **332:** 6.

Daan S. 2000. The Colin S. Pittendrigh Lecture. Colin Pittendrigh, Jurgen Aschoff, and the natural entrainment of circadian systems. *J. Biol. Rhythms* **15:** 195.

Dacey D.M., Liao H.W., Peterson B.B., Robinson F.R., Smith V.C., Pokorny J., Yau K.W., and Gamlin P.D. 2005. Melanopsin-expressing ganglion cells in primate retina signal colour and irradiance and project to the LGN. *Nature* **433:** 749.

Dkhissi-Benyahya O., Gronfier C., De Vanssay W., Flamant F., and Cooper H.M. 2007. Modeling the role of mid-wavelength cones in circadian responses to light. *Neuron* **53:** 677.

Doyle S.E., Castrucci A.M., McCall M., Provencio I., and Menaker M. 2006. Nonvisual light responses in the Rpe65 knockout mouse: Rod loss restores sensitivity to the melanopsin system. *Proc. Natl. Acad. Sci.* **103:** 10432.

Fan J., Woodruff M.L., Cilluffo M.C., Crouch R.K., and Fain G.L. 2005. Opsin activation of transduction in the rods of dark-reared Rpe65 knockout mice. *J. Physiol.* **568:** 83.

Foster R.G., Provencio I., Hudson D., Fiske S., De Grip W., and Menaker M. 1991. Circadian photoreception in the retinally degenerate mouse (rd/rd). *J. Comp. Physiol. A* **169:** 39.

Freedman M.S., Lucas R.J., Soni B., von Schantz M., Munoz M., David-Gray Z., and Foster R. 1999. Regulation of mammalian circadian behavior by non-rod, non-cone, ocular photoreceptors. *Science* **284:** 502.

Garcia-Fernandez J.M., Jimenez A.J., and Foster R.G. 1995. The persistence of cone photoreceptors within the dorsal retina of aged retinally degenerate mice (rd/rd): Implications for circadian organization. *Neurosci. Lett.* **187:** 33.

Gooley J.J., Lu J., Chou T.C., Scammell T.E., and Saper C.B. 2001. Melanopsin in cells of origin of the retinohypothalamic tract. *Nat. Neurosci.* **4:** 1165.

Hattar S., Liao H.W., Takao M., Berson D.M., and Yau K.W. 2002. Melanopsin-containing retinal ganglion cells: Architecture, projections, and intrinsic photosensitivity. *Science* **295:** 1065.

Hattar S., Kumar M., Park A., Tong P., Tung J., Yau K.W., and Berson D.M. 2006. Central projections of melanopsin-expressing retinal ganglion cells in the mouse. *J. Comp. Neurol.* **497:** 326.

Hattar S., Lucas R.J., Mrosovsky N., Thompson S., Douglas R.H., Hankins M.W., Lem J., Biel M., Hofmann F., Foster R.G., and Yau K.W. 2003. Melanopsin and rod-cone photoreceptive systems account for all major accessory visual functions in mice. *Nature* **424:** 75.

Jin M., Li S., Moghrabi W.N., Sun H., and Travis G.H. 2005. Rpe65 is the retinoid isomerase in bovine retinal pigment epithelium. *Cell* **122:** 449.

Klerman E.B., Shanahan T.L., Brotman D.J., Rimmer D.W., Emens J.S., Rizzo J.F., III, and Czeisler C.A. 2002. Photic resetting of the human circadian pacemaker in the absence of conscious vision. *J. Biol. Rhythms* **17:** 548.

Knox B.E. and Solessio E. 2006. Shedding light on cones. *J. Gen. Physiol.* **127:** 355.

Lockley S.W., Brainard G.C., and Czeisler C.A. 2003. High sensitivity of the human circadian melatonin rhythm to resetting by short wavelength light. *J. Clin. Endocrinol. Metab.* **88:** 4502.

Lucas R.J., Douglas R.H., and Foster R.G. 2001. Characterization of an ocular photopigment capable of driving pupillary constriction in mice. *Nat. Neurosci.* **4:** 621.

Lucas R.J., Freedman M.S., Munoz M., Garcia-Fernandez J.M., and Foster R.G. 1999. Regulation of the mammalian pineal by non-rod, non-cone, ocular photoreceptors. *Science* **284:** 505.

Lucas R.J., Hattar S., Takao M., Berson D.M., Foster R.G., and Yau K.W. 2003. Diminished pupillary light reflex at high irradiances in melanopsin-knockout mice. *Science* **299:** 245.

Melyan Z., Tarttelin E.E., Bellingham J., Lucas R.J., and Hankins M.W. 2005. Addition of human melanopsin renders mammalian cells photoresponsive. *Nature* **433:** 741.

Menaker M. 1968. Extraretinal light perception in the sparrow. I. Entrainment of the biological clock. *Proc. Natl. Acad. Sci.* **59:** 414.

Menaker M., Moreira L.F., and Tosini G. 1997. Evolution of circadian organization in vertebrates. *Braz. J. Med. Biol. Res.* **30:** 305.

Moiseyev G., Chen Y., Takahashi Y., Wu B.X., and Ma J.X. 2005. RPE65 is the isomerohydrolase in the retinoid visual cycle. *Proc. Natl. Acad. Sci.* **102:** 12413.

Morin L.P., Blanchard J.H., and Provencio I. 2003. Retinal ganglion cell projections to the hamster suprachiasmatic nucleus, intergeniculate leaflet, and visual midbrain: Bifurcation and melanopsin immunoreactivity. *J. Comp. Neurol.* **465:** 401.

Mrosovsky N. and Hattar S. 2003. Impaired masking responses to light in melanopsin-knockout mice. *Chronobiol. Int.* **20:** 989.

———. 2005. Diurnal mice (*Mus musculus*) and other examples of temporal niche switching. *J. Comp. Physiol. A Neuroethol. Sens. Neural Behav. Physiol.* **191:** 1011.

Murakami M. and Pak W.L. 1970. Intracellularly recorded early receptor potential of the vertebrate photoreceptors. *Vision Res.* **10:** 965.

Nelson R.J. and Zucker I. 1981. Photoperiodic control of reproduction in olfactory-bulbectomized rats. *Neuroendocrinology* **32:** 266.

Newman L.A., Walker M.T., Brown R.L., Cronin T.W., and Robinson P.R. 2003. Melanopsin forms a functional short-wavelength photopigment. *Biochemistry* **42:** 12734.

Panda S., Nayak S.K., Campo B., Walker J.R., Hogenesch J.B., and Jegla T. 2005. Illumination of the melanopsin signaling pathway. *Science* **307:** 600.

Panda S., Sato T.K., Castrucci A.M., Rollag M.D., DeGrip W.J., Hogenesch J.B., Provencio I., and Kay S.A. 2002. Melanopsin (Opn4) requirement for normal light-induced circadian phase shifting. *Science* **298:** 2213.

Panda S., Provencio I., Tu D.C., Pires S.S., Rollag M.D., Castrucci A.M., Pletcher M.T., Sato T.K., Wiltshire T., Andahazy M., Kay S.A., Van Gelder R.N., and Hogenesch J.B. 2003. Melanopsin is required for non-image-forming photic responses in blind mice. *Science* **301:** 525.

Provencio I. and Foster R.G. 1995. Circadian rhythms in mice can be regulated by photoreceptors with cone-like characteristics. *Brain Res.* **694:** 183.

Provencio I., Jiang G., De Grip W.J., Hayes W.P., and Rollag M.D. 1998. Melanopsin: An opsin in melanophores, brain, and eye. *Proc. Natl. Acad. Sci.* **95:** 340.

Provencio I., Rodriguez I.R., Jiang G., Hayes W.P., Moreira E.F., and Rollag M.D. 2000. A novel human opsin in the inner retina. *J. Neurosci.* **20:** 600.

Qiu X., Kumbalasiri T., Carlson S.M., Wong K.Y., Krishna V., Provencio I., and Berson D.M. 2005. Induction of photosensitivity by heterologous expression of melanopsin. *Nature* **433:** 745.

Redmond T.M., Poliakov E., Yu S., Tsai J.Y., Lu Z., and Gentleman S. 2005. Mutation of key residues of RPE65 abolishes its enzymatic role as isomerohydrolase in the visual cycle. *Proc. Natl. Acad. Sci.* **102:** 13658.

Redmond T.M., Yu S., Lee E., Bok D., Hamasaki D., Chen N., Goletz P., Ma J.X., Crouch R.K., and Pfeifer K. 1998. Rpe65 is necessary for production of 11-cis-vitamin A in the retinal visual cycle. *Nat. Genet.* **20:** 344.

Rollag M.D., Provencio I., Sugden D., and Green C.B. 2000. Cultured amphibian melanophores: A model system to study melanopsin photobiology. *Methods Enzymol.* **316:** 291.

Ruby N.F., Brennan T.J., Xie X., Cao V., Franken P., Heller H.C., and O'Hara B.F. 2002. Role of melanopsin in circadian responses to light. *Science* **298:** 2211.

Selby C.P., Thompson C., Schmitz T.M., Van Gelder R.N., and Sancar A. 2000. Functional redundancy of cryptochromes and classical photoreceptors for nonvisual ocular photoreception in mice. *Proc. Natl. Acad. Sci.* **97:** 14697.

Sollars P.J., Smeraski C.A., Kaufman J.D., Ogilvie M.D., Provencio I., and Pickard G.E. 2003. Melanopsin and non-melanopsin expressing retinal ganglion cells innervate the hypothalamic suprachiasmatic nucleus. *Vis. Neurosci.* **20:** 601.

Stockman A. and Sharpe L.T. 2006. Into the twilight zone: The complexities of mesopic vision and luminous efficiency. *Ophthalmic Physiol. Opt.* **26:** 225.

Takahashi J.S., DeCoursey P.J., Bauman L., and Menaker M. 1984. Spectral sensitivity of a novel photoreceptive system mediating entrainment of mammalian circadian rhythms. *Nature* **308:** 186.

Thomas H. and Stoddart J.L. 1980. Leaf senescence. *Annu. Rev. Plant Physiol.* **31:** 83.

Wong K.Y., Dunn F.A., Graham D.M., and Berson D.M. 2007. Synaptic influences on rat ganglion-cell photoreceptors. *J. Physiol.* **582:** 279.

Yoshimura T. and Ebihara S. 1996. Spectral sensitivity of photoreceptors mediating phase-shifts of circadian rhythms in retinally degenerate CBA/J (rd/rd) and normal CBA/N (+/+) mice. *J. Comp. Physiol. A* **178:** 797.

Zhang H., Clarke R.J., and Gamlin P.D. 1996. Behavior of luminance neurons in the pretectal olivary nucleus during the pupillary near response. *Exp. Brain Res.* **112:** 158.

The Lateral and Dorsal Neurons of *Drosophila melanogaster*: New Insights about Their Morphology and Function

C. HELFRICH-FÖRSTER,* T. YOSHII,* C. WÜLBECK,* E. GRIESHABER,* D. RIEGER,* W. BACHLEITNER,* P. CUSUMANO,† AND F. ROUYER†

**University of Regensburg, Institute of Zoology, 93040 Regensburg, Germany;*
†Institut de Neurobiologie Alfred Fessard, CNRS UPR2216, 91198 Gif-sur-Yvette, France

This chapter summarizes our present knowledge about the master clock of the fruit fly at the neuronal level. The clock is organized in distinct groups of interconnected pacemaker neurons with different functions. All of these neurons appear to communicate with one another in order to produce the species-specific activity rhythm, which is organized in morning (M) and evening (E) activity bouts. These two activity components are differentially influenced by distinct groups of pacemaker neurons reminiscent of the Pittendrigh–Daan dual oscillator model. In the original work (Grima et al. 2004; Stoleru et al. 2004), the ventrolateral (LN_v) and dorsolateral (LN_d) plus some dorsal groups (DN) of clock neurons have been defined as M and E cells, respectively. We further specify that the clock neurons belong to the M and E oscillators and define a more complex picture of the *Drosophila* brain clock.

INTRODUCTION

During the past years, the fruit fly *Drosophila melanogaster* has been very helpful in understanding the neuronal organization and function of the circadian clock in the brain. As is the situation in the mammalian clock, the fly clock consists of a network of morphologically different neurons that have putatively different functions and outputs (Hall 2005; Helfrich-Förster 2006; Taghert and Shafer 2006). Fortunately, the *Drosophila* clock neurons are few in number (~150), and they can be easily genetically manipulated. Nevertheless, the activity rhythm of *Drosophila* is similarly as complex as that of mammals. It consists of two main activity bouts, one in the morning (M) and the other in the evening (E), that are timed in a sophisticated manner to dawn and dusk. Most significantly, two groups of neurons in the lateral brain—the ventral and dorsal lateral neurons—have been implicated in controlling M and E activity, respectively (Grima et al. 2004; Stoleru et al. 2004). The lateral neurons are regarded as the main pacemaker cells, because they were shown to be necessary and sufficient to drive behavioral rhythmicity in the absence of external time cues (Ewer et al. 1992; Frisch et al. 1994).

This chapter reviews our current knowledge about the morphology and function of these main clock neurons and other players. It includes morphological details that were not reported in previous papers (Shafer et al. 2006; Helfrich-Förster et al. 2007), and it shows that the functional division in the M and E oscillators is more complex than initially thought.

CLOCK NEURONS OF *DROSOPHILA MELANOGASTER*

The overall anatomical organization of *Drosophila's* pacemaker system was the topic of two recent papers and one recent review (Shafer et al. 2006; Taghert and Shafer 2006; Helfrich-Förster et al. 2007) and is only briefly summarized here (Fig. 1). According to their location in the brain, the clock neurons are traditionally divided into six groups: three dorsal ones (DN_{1-3}) in the superior brain and three lateral ones (LN_d, l-LN_v, and s-LN_v) in the anterior brain; but there is an additional group of lateral neurons in the posterior brain called the LPN (Kaneko et al. 1997; Kaneko and Hall 2000; Yoshii et al. 2005; Shafer et al. 2006). The projections of the LPN are still unknown, but they seem to overlap completely with those of the other neurons (O. Shafer and C. Helfrich-Förster, unpubl.).

The clock neurons have two main projection targets: the accessory medulla that appears to be the principal pacemaker center of insects and the dorsal brain that houses the hormonal center (pars intercerebralis and lateralis) of insects and also has connections to most brain areas (Homberg et al. 2003; Shiga 2003; for discussion, see Jaramillo et al. 2004; Lear et al. 2005a). The clock neurons form a well-defined fiber network in these brain areas, putatively allowing considerable cross-talk between them (Fig. 1).

Recent studies indicate that the classical division of the clock neurons is insufficient, because several of these groups are rather heterogeneous (Lee et al. 2006; Shafer et al. 2006; Helfrich-Förster et al. 2007). A further subdivision is particularly appropriate for s-LN_v, LN_d, DN_1, and DN_3. Of the s-LN_v, four cells express the neuropeptide pigment-dispersing factor (PDF) and one is PDF-negative (Kaneko et al. 1997). The projection of the PDF-negative fifth s-LN_v appears to be similar to that of the PDF-positive s-LN_v, but its neurotransmitter still remains elusive (Helfrich-Förster et al. 2007).

The six LN_d cells consist of three neurons that are strongly cryptochrome (CRY)-positive, whereas the remaining three cells appear to express less or no CRY (Klarsfeld et al. 2004; Picot et al. 2007; T. Yoshii, unpubl.). The CRY-positive LN_d cells are also revealed by the *Mai179-gal4* enhancer-trap line (see Fig. 4) (Siegmund and Korge 2001; Grima et al. 2004; Picot et al.

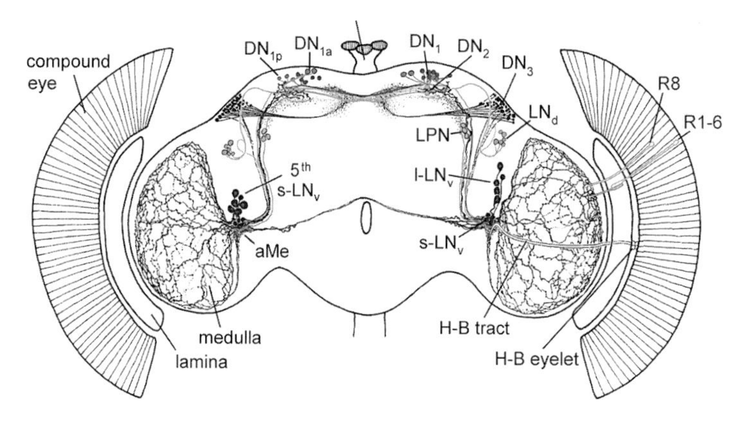

Figure 1. Arborization patterns of the lateral neurons (LN_ds, l-LN_vs, and s-LN_vs) and dorsal neurons (DN_1, DN_2, DN_3). Their fibers overlap in the dorsal brain and the accessory medulla (aMe). (Reprinted, with permission, from Helfrich-Förster et al. 2007 [© Wiley].)

2007), but the nature of the gene hosting the *Mai179* insertion is still unknown. Furthermore, it has been reported that in male flies, three LN_d cells express the neuropeptide F (NPF), a homolog of the mammalian neuropeptide Y (NPY) (Lee et al. 2006). It is unclear whether the three NPF-positive cells are identical to the CRY- and *Mai*-positive LN_d cells. The LN_d cells seem not only heterogeneous in their neurohistochemistry, but also heterogeneous in their arborizations (Helfrich-Förster et al. 2007). The three CRY- and Mai-positive cells appear to project ipsilaterally and contralaterally, whereas the axons of the remaining LN_d cells run only to the contralateral side of the brain (T. Yoshii et al., unpubl.).

DN_1 encompasses a large and obviously rather heterogeneous group of clock neurons. They consist of approximately 17 neurons that are spread in the dorsal brain. Two of the DN_1 are conspicuous because (1) they are already present in larvae, (2) they are the only DN_1 cells that lack the transcription factor *glass* and are present in gl^{60J} mutants, and (3) they are clearly located in an anterior position in the adult brain and thus separated from the others (Klarsfeld et al. 2004; Shafer et al. 2006). These cells were accordingly named $DN_{1anterior}$ (DN_{1a}). These two neurons express the neuropeptide IPN-amide (Shafer et al. 2006). Furthermore, they show anti-CRY immunoreactivity (Klarsfeld et al. 2004) and are strongly labeled by the *cry-gal4* line (#13) (Schafer et al. 2006; Helfrich-Förster et al. 2007). Similarly, CRY-positive cells are two further DN_1 cells that lie among the other DN_1 cells in the posterior brain and were named DN_{1p}. The DN_{1a} and the two DN_{1p} cells appear to show the same arborization pattern; they project ventrally to the ipsilateral accessory medulla (Fig. 1) (Shafer et al. 2006; Helfrich-Förster et al. 2007). Of the remaining DN_1 cells, some may run also to the accessory medulla, others may cross to the contralateral brain in a dorsal commissure, and others even may do both.

With about 40 neurons, the DN_3 cluster covers the largest number of clock neurons. Again, these neurons are heterogeneous in size and projection pattern (Shafer et al. 2006). Two DN_3 neurons with larger soma are labeled by the *cry-gal4* line (#13), and these neurons project ventrally to the ipsilateral accessory medulla, whereby their neurites overlap with those of the CRY-positive LN_d (Helfrich-Förster et al. 2007).

FUNCTIONAL DISSECTION OF THE FLY'S PACEMAKER SYSTEM

The Lateral Neurons as Master Clocks

As mentioned above, the current view is that the LN clusters (specifically, the s-LN_vs and LN_ds[1]) are *Drosophila's* main clock neurons, whereas the DN groups seem to be less important. This conclusion is based on experiments that restricted PER expression to certain subsets of clock neurons: Flies that expressed PER in the LN clusters but not in the DN clusters showed rhythmic locomotor activity under constant dark (DD) conditions (Frisch et al. 1994), whereas PER expression in brain cells other than the LN clusters was not sufficient to produce robust activity rhythms (Ewer et al. 1992). Furthermore, PER expression restricted to the PDF-expressing LN_v cells is sufficient to drive robust rhythms in DD (Fig. 2, top) (Grima et al. 2004), whereas rhythmic LN_d cells are sufficient to drive robust rhythmicity in LL (Fig. 2, bottom) (Picot at al. 2007). Conversely, the *disconnected* (*disco*) mutation that causes degeneration of many optic lobe neurons including the LN clusters during development, but that leaves the DN clusters intact, resulted in arrhythmic behavior under DD conditions (Fig. 3) (Steller et al. 1987; Dushay et al. 1989). Similarly, transgenic flies, in which the LN_v cells were ablated specifically by activation of cell death or apoptosis genes, became arrhythmic under DD conditions (Renn et al. 1999; Blanchardon et al. 2001). The same happened after electrically silencing the LN_v cells (Nitabach et al. 2002). Altogether, this points to the LN clusters as main pacemakers. But there is one major caveat. Nobody has so far shown that ablating all DN and leaving the LN intact will preserve behavioral rhythmicity. It is only known that elimination of the *glass*-positive DN_1 (in gl^{60J} mutants) does not lead to arrhythmicity (Helfrich-Förster et al. 2001; Klarsfeld et al 2004). Although PER expression in the DN is neither required nor sufficient for sustained

[1]The role of the l-LN_v cells is still unknown. These cells do not show apparent oscillations in clock proteins under DD conditions and are therefore not regarded as master clock neurons (Stanewsky et al. 1998; Yang and Sehgal 2001; Shafer et al. 2002; Lin et al. 2004)

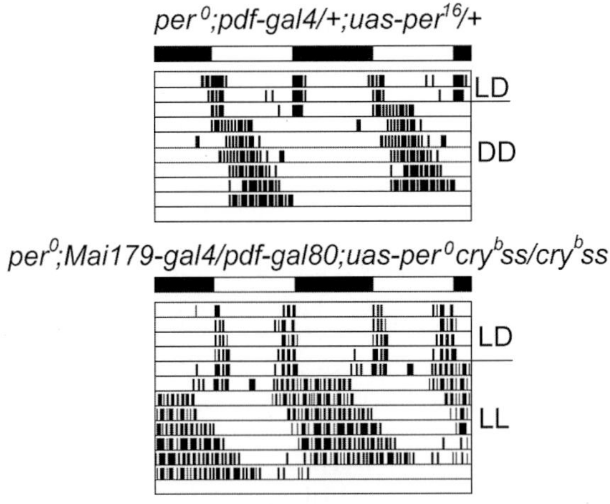

Figure 2. Averaged group actograms ($n = 16$) of flies, in which PER was restricted to the PDF-positive s-LN_v cells (only the M oscillator is functional; *top*) or to the fifth s-LN_v, the CRY-positive LN_d, and perhaps the DN_{1a} cells (predominantly the E oscillator is functional; *bottom*). (*Top*) The flies with a functional clock only in the M oscillator show only the M peak plus a lights-off peak under LD, and the free-run under subsequent DD conditions started from the M peak. (*Bottom*) The flies with a functional clock in the E oscillator also carry the cry^b mutation. They remain rhythmic under LL conditions, and activity clearly starts from the E-activity peak.

behavioral rhythms in DD, the function of DN_{1a}, DN_2, and DN_3 in the brain clock remains unknown.

The Dorsal Neurons as Poorly Characterized Players in the *Drosophila* Circadian System

Several studies have shown that the DN cells are involved in the control of behavioral rhythmicity. Most *disco* flies missing the LN clusters showed rather organized activity patterns under LD conditions with clearly recognizable M and E peaks (Fig. 2) (Hardin et al. 1992; Helfrich-Förster 1998). Some of them even showed residual rhythmicity for the first few days under constant conditions, indicating that they still possess a dampened circadian clock (Fig. 2). Indeed, PER levels were found to vary cyclically in the DN and in the glial cells of *disco* mutants (Zerr et al. 1990), and this cycling persisted even in the absence of the LNs for at least 2 days under constant conditions (Blanchardon et al. 2001). With the help of a luciferase-based *per* reporter system, a stable *per* cycling in the DN_2 and DN_3 clusters was revealed for many days under DD conditions (Veleri et al. 2003). This cycling continued even in absence of PER in the LN cells and was just self-sustained. The same may apply to the DN_1. Most recent studies showed that under certain circumstances, a subset of the DN_1 cells is clearly cycling and that these cells may even control rhythmic behavior under constant light (LL) conditions (Murad et al. 2007; Stoleru et al. 2007; see below for further discussion).

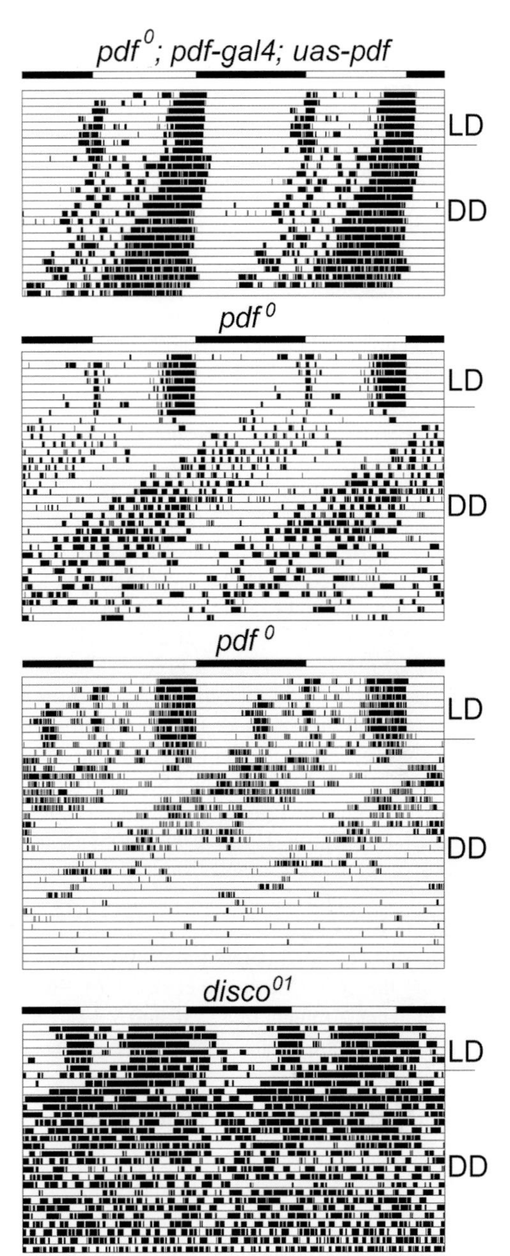

Figure 3. Activity patterns of one phenotypically wild-type fly (pdf^{01} mutant, in which PDF expression was rescued in the PDF neurons), two pdf^{01} mutants, and one $disco^1$ mutant. The first pdf^{01} mutant appears to lack the morning activity under LD conditions (except for a lights-on peak) and free-runs with short period under DD conditions before rhythmicity slowly vanishes. Most importantly, free-run started from the middle of the night. The second pdf^{01} fly showed activity in the second half in the night under LD conditions and the free-running activity in DD started indeed from this nocturnal activity. Thus, the midnight activity might relate to a strongly advanced M peak. The *disco* mutant, which lacked all LN_v and most probably also LN_d cells still showed M and E peaks under LD conditions, but became quickly arrhythmic after transfer into DD (for details, see Helfrich-Förster 1998).

Mysterious LPN with Unidentified Function

LPNs were originally described as cells that expressed the clock protein Timeless (TIM) but not PER (Kaneko and Hall 2000), and more recently, LPNs were confirmed as bona fide clock neurons expressing PER as well and specifically implicated in temperature entrainment (Yoshii et al. 2005; Shafer et al. 2006; Miyasako et al. 2007).

The Neuropeptide PDF as Internal Synchronizer

In sum, there is evidence that all clock neurons work together in controlling behavioral rhythmicity. They form a sophisticated interactive network, in which each neuronal cluster might have its specific role. This does not exclude a certain hierarchy in the network. The LN clusters appear to have a dominant role in the network, and this role becomes most evident in the absence of external time cues (e.g., under DD conditions). Most interestingly, the neuropeptide PDF, which is produced in most of the LN_v cells, has a prominent role in coordinating the oscillations of all clock neurons (Peng et al. 2003; Klarsfeld et al. 2004; Lin et al. 2004; for review, see Taghert and Shafer 2006). *pdf*^{01} mutants entrain to LD cycles showing rather normal activity patterns, except for an advanced E peak and a barely visible if not absent M peak (Fig. 3) (Renn et al. 1999; however, see below). When transferred to DD, these flies maintain a rhythm with a short period for at least 1 week under DD (Fig. 3). Then, the oscillations in the clock neurons become asynchronous and the flies become arrhythmic (Peng et al. 2003, Lin et al. 2004). These results indicate that the main role of PDF is to coordinate and synchronize the oscillations of the other clock neurons in the absence of external time cues. A similar role of PDF was also suggested for larger insects (Petri and Stengl 2001; Singaravel et al. 2003; Schneider and Stengl 2005).

PDF has no homolog in mammals, but most interestingly, the PDF receptor (Hyun et al. 2005; Lear et al. 2005b; Mertens et al. 2005) shows similarity to the mammalian VPAC2 receptor that binds the vasoactive intestinal polypeptide (VIP). VIP and VPAC2 are essential to produce a coherent oscillating output of the mammalian pacemaker center, the SCN (Maywood et al. 2006). Similar to the situation in *pdf*^{01} mutant flies, VIP and VPAC2 mutant mice become arrhythmic under constant conditions.

THE DUAL OSCILLATOR MODEL

Diurnal animals are not just active when light turns on and inactive when light turns off, but they entrain their clock in a sophisticated manner to the environmental conditions and they are well capable of adapting to seasonal changes in day length and temperature (Majercak et al. 1999; Rieger et al. 2003). Most diurnal species show a broader activity band under long summer days as compared to short winter days. If M and E activity components are present, the M component couples to dawn and the E oscillator couples to dusk, and a pronounced midday trough becomes visible in long summer days. This means that M and E activity components change their phase relationship to adapt to seasonal changes in day length. These adjustments can be best explained by assuming two oscillators with different properties: one accelerating and the other decelerating with increasing light (Pittendrigh and Daan 1976). In this two-oscillator model, the M oscillator would lock to dawn and the E oscillator would lock to dusk. The anatomical substrates of M and E oscillators remained unknown until Grima et al. (2004) and Stoleru et al. (2004) found that the LN_v and the LN_d cells are good candidates for M and E oscillators, respectively. Stoleru et al. (2004) genetically eliminated either the PDF-positive LN_v or the LN_d cells (plus the fifth s-LN_v and some DN cells) and found in the first case that the M peak disappeared and in the second case that the E peak vanished. Grima et al. (2004) limited *per* expression to the PDF-positive LN_vs and found that the flies showed a normal M peak but lacked the E peak (Fig. 2, top), whereas adding PER expression to some LN_d cells induced an evening peak. Both studies suggest that PDF-positive LN_v and the LN_d cells (plus at least the fifth s-LN_v) actually are the M and E oscillators proposed by Pittendrigh and Daan (1976). It was shown recently that the M and E cells are inhibited and activated, respectively, by light at the output level, suggesting a switch between M and E oscillators under LD conditions (Picot et al. 2007). Finally, Stoleru et al. (2007) proposed a clever model showing how M and E oscillators adapt to seasonal changes (the E cells being DN_1 neurons in this study). In this model, the M cells are "dark oscillators" that dominate under short photoperiods, whereas the E cells are "light oscillators" that dominate under long photoperiods. Our findings are principally in concert with the M/E oscillator model, but we have indications that the separation in M and E cells is far too simple. In the following, we describe our current view based on published and still unpublished findings.

The Nature of the M Cells

There are indications that the PDF-positive s-LN_v are not the only cells that generate the M peak under LD conditions but that some DN cells contribute. Stoleru et al. (2004) found that flies without a functional clock in the PDF-positive s-LN_v cells still showed M and E activity bouts. The same happened after restricting *per* expression to the DN cells and a few LN_d using a promoter-less *per* construct (Veleri et al. 2003). These data suggest that the DN cells contribute not only to the E peak, but also to the M peak. This is in agreement with the findings for *disco* mutants that lack all LNs but still show some bimodality under LD conditions (Fig. 3). Thus, the DNs alone can mediate aspects of normal activity under LD conditions, and a combination of DN and LN_d cells is sufficient to cause almost wild-type-like activity patterns under LD (Veleri et al. 2003; Stoleru et al. 2004). This applies also to a certain degree if PDF is absent. A closer look at the *pdf*^{01} mutants that should lack the M oscillator (Fig. 3) shows that these flies do not exhibit a clear M peak; however, several of them show elevated activity in the second half of the night as if they had an extremely advanced M peak. In agreement with such an advanced phase, the free-running activity of all *pdf*^{01} flies starts from this point when the flies are released into constant conditions (Fig. 3). Thus, a weakly coordinated M peak with advanced phase could be present in these "M-oscillator-less" flies. This advanced M peak could stem from the DN cells that contribute to the M oscillator.

To complicate the story, there are also indications that some of the LN_d cells may contribute to the M oscillator. Flies are very light-sensitive, and they phase-advance and phase-delay their M and E peaks not only under long days, but also when exposed to dim light during the night (Fig. 4A) (Bachleitner et al. 2007). This phase-advance and phase-delay was not only observed for behavior, but also for the molecular clocks in the LN. As expected, the s-LN_v cells phase-advanced and thus behaved as an M oscillator. But unexpectedly, the majority of LN_d cells also advanced (Fig. 4A). Only the fifth s-LN_v cell behaved as an E oscillator and phase-delayed. Thus, the LN_d cells as an entity did not behave as a bona fide E oscillator under these conditions. However, it is still possible that some LN_d cells acted as an E oscillator and phase-delayed. The peak in PER/TIM immunoreactivity in the LNd cells did considerably broaden when the flies were exposed to the nocturnal dim-light conditions. This is consistent with the idea that the LN_d cells are a heterogeneous cell group (see above) consisting of M and E oscillators.

To test this idea, we recorded flies that express PER only in the CRY- and Mai-positive LN_d cells and in the fifth s-LN_v cell (*per*^{0}*;Mai179-gal4/pdf-gal80;uas-per*^{16}/+ flies) under the LD-LM (light moonlight) paradigm. The CRY- and Mai179-positive LN_d cells are those that appear to act as E oscillators under LD and LL conditions (Fig. 2, bottom) (Grima et al. 2004). Indeed, these flies showed a normal E peak and no clearly visible M peak under LD conditions, although some activity before lights on was present (Fig. 4B). However, when the flies were exposed to nocturnal moonlight, a prominent advanced M peak became visible and after transfer into constant moonlight, free-running activity started from the M and E peak (Fig. 4B). This behavior is hard to reconcile with the existence of only the E oscillator. Either some of the three LN_d cells must behave as an M oscillator or additional clock neurons must express PER and have this role. Picot et al. (2007) found weak cycling PER expression in the DN_{1a} cells of a few brains under LD conditions, raising the possibility that

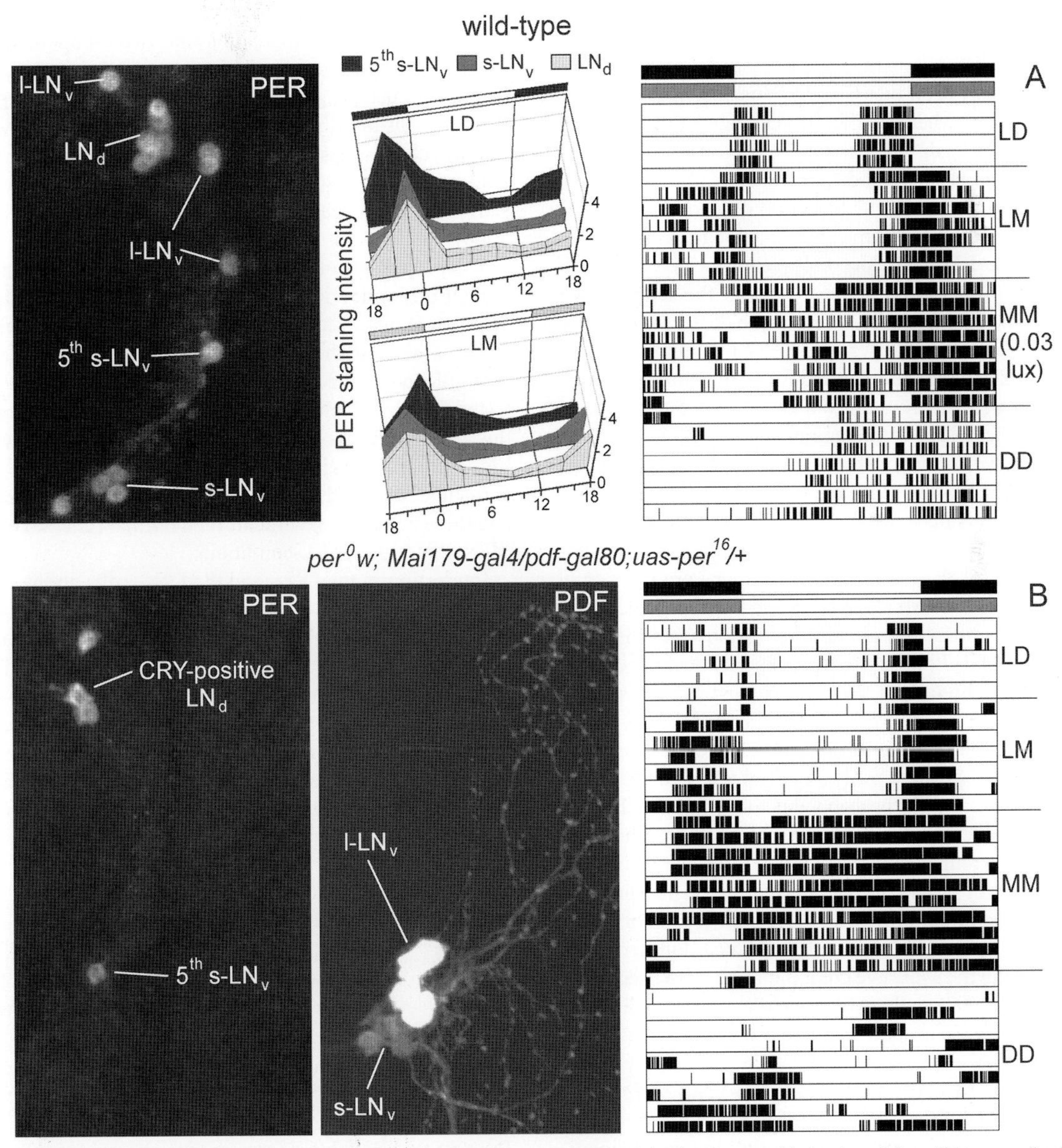

Figure 4. Under nocturnal dim light, flies shift M and E peaks into the night. (*A*) Histology and behavior of the wild type under light-dark (LD), light-moonlight (LM), constant moonlight (MM), and constant darkness (DD) (for details, see Bachleitner et al. 2007). (*B*) Flies with PER restricted to the CRY- and Mai179-positive LN_d and to the fifth s-LN_v still show M and E peaks under LM conditions and rhythmicity under MM but not under DD.

the DN_{1a} cells contribute to the M peak in LM. Future studies will have to determine whether these cells start to cycle robustly and with advanced phase under LM conditions or whether some of the LN_d cells behave as an M oscillator under LM conditions. The comparison between LD and LM conditions indicates that the early M activity peak is promoted by light, which was suggested to be a feature of the E cells (Murad et al. 2007; Stoleru et al. 2007; Picot et al. 2007).

There is a further study that is in favor of our idea that the LN_d cells are a mixture of M and E cells (Rieger et al. 2006). The M and E oscillator model predicts that the M oscillator will free-run with a short period and the E oscillator will free-run with a long period when animals are placed in constant light. Indeed, internal desynchronization into two free-running components (one with short period and the other with long period) could be observed in wild-type flies under dim-light conditions (Rieger et al. 2006). In *cry*^{b} mutants in which the immediate degradation of TIM by light is prevented, this internal desynchronization could even be observed at higher light intensities and it occurred simultaneously in all flies (Fig. 5) (Yoshii et al. 2004). This allowed testing whether PER/TIM cycling in the clock neurons is similarly desynchronized. Indeed it was. In agreement with the M and E model, the PDF-positive s-LN_v cells appeared to control the short-period component, whereas the fifth s-LN_v appeared to control the long-period component. Of the LN_d cells, one cell clearly behaved as an E oscillator showing the same phase as the fifth s-LN_v, whereas the others could not be unequivocally attributed to M or E oscillators (Rieger et al. 2006). In retrospect, we looked again at the stainings and found that three of the six LN_d cells were always stained together with the s-LN_v and three to four LN_d cells (including the one with larger soma that could be unequivocally assessed from the beginning) were stained together with the fifth s-LN_v (Fig. 5). Thus, three LN_d cells may act as an M oscillator and three LN_d cells may act as as E oscillator. In the future, we will determine which of these different LN_d groups are CRY- and Mai-positive and which express the neuropeptide NPF. Nonetheless, there are already indications that the three CRY- and Mai179-positive LN_d cells may act as E oscillators in LL. These cells showed a robust cycling in *cry*^{b} mutants under LL conditions that was in-phase with the fifth s-LN_v cell (Picot et al. 2007). In this experiment, the M cells (PDF-positive s-LN_v) had only a slightly earlier phase than the E cells, probably due to the fact the flies did not show evident internal desynchronization into short and long free-running components at that time (the recordings lasted only for 6 days and the cycling was determined on days 2 to 3; in the previous study, both oscillators were 180° out of phase not before days 4 to 5). Further-more, flies carrying the *white* eye color mutation were used in the Picot et al. (2007) study, whereas red-eyed flies, which are much more prone to splitting, were used in the Rieger et al. (2006) study.

When the M cells (PDF-positive s-LN_v) were made arrhythmic by expressing CRY in an otherwise *cry*^{b} background, the flies showed a long period under LL conditions (Murad et al. 2007; M. Picot et al., in prep.). The same situation occurs in a wild-type background after down-regulating CRY in all clock cells except for s-LN_v. These results are consistent with the previous findings where the E cells free-run with a long period (Rieger et al. 2006). However, the opposite prediction that flies showing oscillations only in the PDF-positive s-LN_v cells (with all other clock cells made arrhythmic by the above-mentioned methods) should show a short period under LL conditions was not fulfilled. Unexpectedly, these flies were behaviorally arrhythmic (Stoleru et al. 2007; Picot et al. 2007). The ultimate explanation of this result is that the PDF-positive s-LN_v cells are not capable of driving rhythmicity under LL conditions.

But how do these results fit with the simultaneous occurrence of short and long periods in *cry*^{b} mutants under LL conditions? One cannot exclude that the behavior of the multioscillator clock in LL is more complex than predicted by the E/M model (MO accelerates and EO slows down). One possibility is that the short-period component has nothing to do with the M oscillator, but stems from the E oscillator that splits into two components. In favor of this hypothesis is the finding that the short-period component usually appears to start from the evening peak (Fig. 5). But then, why do the M cells (PDF-positive s-LN_v) cycle in-phase with the short-period component? And why does a second short-period component often start from the M peak (Helfrich-Förster 2006; Rieger et al. 2006)?

We propose another hypothesis. There is one fundamental difference in the different experimental designs. Whereas Rieger et al. (2006) tested flies in which all clock neurons carried the *cry*^{b} phenotype and thus were all rhythmic in LL, Picot et al. (2007) and Stoleru et al. (2007) tested flies in which only the PDF-positive s-LN_v carried this phenotype. As outlined earlier, the PDF-positive s-LN_v cells are most likely not the only M cells. Some LN_d and some DN_1 cells seem also to belong to the M cells, and these cells may be crucial for a normal function of the M oscillator under LL conditions.

Interestingly, the DN_{1a} and the DN_{1p} cells are likely to be among the cells that are responsible for rhythmic behavior under LL conditions after overexpression of PER[2] (Murad et al. 2007). Our flies with limited PER expression in the three Mai-positive LN_d, the fifth s-LN_v (and possibly the DN_{1a} under certain conditions), were similarly rhythmic under constant moonlight but became arrhythmic after a transfer into DD (Fig. 4B). This raises the possibility that the DN_{1a} cells are important for rhythmicity in low light, in addition to being candidates for the

[2]This is not stated in the paper, but can be concluded from the strains the authors have used. They report that strains which overexpress PER in all clock cells (under control of the *tim-gal4* driver) are rhythmic under LL conditions. Flies that overexpress PER in all clock cells except the PDF-positive LN_v (*tim-gal4;pdf-gal80;uas-per*) are still rhythmic under LL conditions, indicating that the PDF-positive cells are not responsible for this rhythmicity. When PER is overexpressed in all clock cells except for those that express CRY (*tim-gal4;cry-gal80;uas-per*) the flies are arrhythmic in LL as are wild-type flies. This means that cells that are *cry-gal80*-positive, but *pdf*-negative, are responsible for LL rhythmicity. These are the LN_d, two DN_3, and the DN_{1a} and two DN_{1p}. Because Murad et al. (2007) found molecular oscillations under LL only in some DN_1, the LN_d, and DN_3 can be excluded as source for LL rhythmicity. Thus, the DN_{1a} and DN_{1p} are the best candidates.

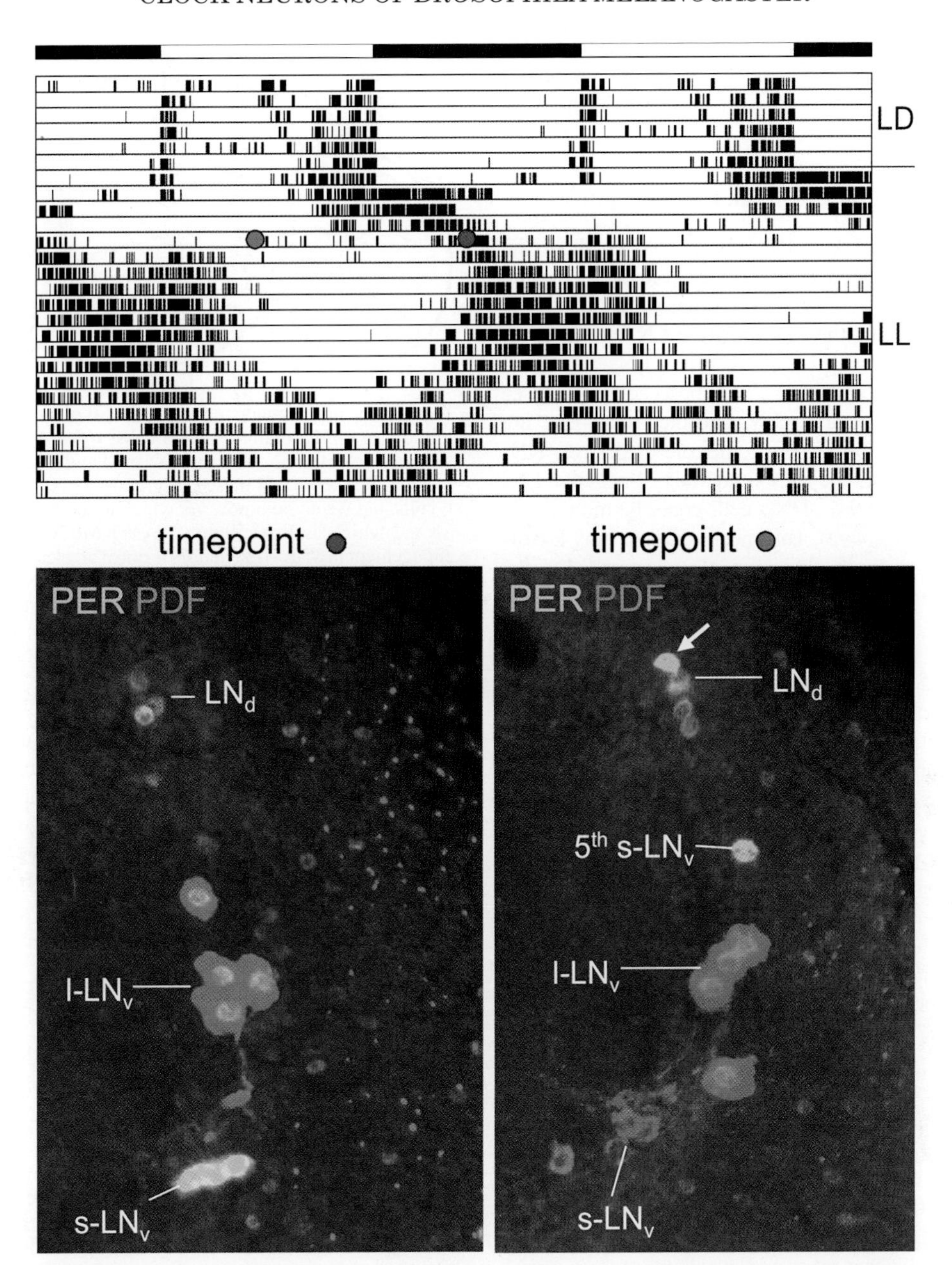

Figure 5. Internal desynchronization of a *cry^b* mutant in constant light (LL of 500 μW/cm^2) and the corresponding desynchronization in the lateral neurons on days 4–5 in LL. Flies were immunostained at the two indicated time points, and representative examples for PER and PDF labeling of the lateral neurons are shown below. Because PER staining is at its trough when activity is high, the cells that are strongly stained at the activity peak of the long component (*red point*) control the short-period component and those that are stained at the activity peak of the short compartment (*blue point*) control the long period (for details, see Rieger et al. 2006). Accordingly, the PDF-positive s-LN$_v$ and three LN$_d$ cells appear to govern the short-period component, whereas, the fifth s-LN$_v$ cell, a strongly stained LN$_d$ cell (*arrow*) and two to three other LN$_d$ cells appear to govern the long-period component.

M oscillator (see above). Most significantly, the DN$_{1a}$ cells have arborizations in close vicinity to the s-LN$_v$ terminals in the dorsal brain and to the dendrites of the s-LN$_v$ in the accessory medulla (see Fig. 1) (Shafer et al. 2006). Thus, the DN$_{1a}$ may mutually interact with the s-LN$_v$. It is quite likely that the normal function of the M oscillator depends on the DN$_{1a}$ and DN$_{1p}$ under LL conditions, whereas it does depend on the s-LN$_v$ under DD conditions. Both cell types are anatomically well suited to get light input from the compound eyes and/or the H-B eyelet via the accessory medulla. This light input might interfere with the interaction between the two cell types and might trigger the switch in dominance between both, as it has been proposed between the morning PDF-positive and evening PDF-negative LN (Picot et al. 2007). These ideas about a switch in pacemaker dominance between LL and DD are in line with those of Stoleru et al. (2007) that some clock neurons are the dominant pacemaker cells under short days and others, under long days. According to our results, it is, however, more likely that the dominant pacemakers under LL are not exclusively E cells but that some M cells (e.g., the DN_{1a}) are included in the "LL clock." In contrast, some E cells may contribute to rhythmicity under DD conditions.

The Nature of the E Cells

The E cells are less well defined than the M cells. In the Stoleru et al. (2004) study, E cells were ablated by using *cry-gal4;pdf-gal80;uas-hid* flies. In these flies, all CRY-positive cells except the PDF-positive LN_v are ablated. Thus, such flies lack all LN_d,[3] the fifth s-LN_v, the DN_{1a}, DN_{1p}, and two DN_3 cells. Among these cells, the three CRY- and Mai179-positive LN_d and the fifth s-LN_v are bona fide E oscillators, whereas the role of the two DN_3 cells is unknown, and the CRY- and Mai179-negative LN_d cells as well as the DN_{1a}, DN_{1p} cells could be more likely M oscillators (see above). This means that not only E cells were ablated in these flies, but also certain M cells. Furthermore, most probably not all E cells were removed, because other DN_1 (not the DN_{1a}, DN_{1p}) and some DN_2 and DN_3 cells seem to act as E oscillators. In later work, the E cells were manipulated using a *tim-gal4;pdf-gal80* driver (Stoleru et al. 2005, 2007). This driver indeed hits all E cells, but additionally also some M cells. How can we reinterpret the published findings taking this knowledge into account? First of all, one would expect that the still present E cells in *cry-gal4;pdf-gal80;uas-hid* flies would influence the activity rhythm. This is indeed true and has been discussed previously (Helfrich-Förster 2006). These flies still showed signs of bimodality in their activity pattern under LD. Besides the normal M peak, a strongly delayed E peak is visible. This peak is somehow comparable with the advanced M peak visible in flies without functioning PDF-positive s-LN_v (*pdf^{01}* mutants; see Fig. 3). Under DD conditions, the E peak is apparent as shoulder following the putative M peak (see Fig. 2d in Stoleru et al. 2004). This remnant E peak was previously discussed as consequence of the dominant PDF-positive s-LN_v cells that also control aspects of the E peak (see also Rieger et al. 2006; Taghert and Shafer 2006; Miyasako et al. 2007). After our present state of knowledge, we see this remnant E peak as output from the DN cells that act as an E oscillator.

CONCLUSIONS

We have depicted a rather complex picture in which several groups of clock neurons interact to produce *Drosophila's* typical M- and E-activity bouts. In this picture, the M and E cells are not confined to certain regional clusters of clock neurons, but instead are distributed among several different clusters. Figure 6 summarizes our current view. Certainly, this picture is still partly hypothetical. It relies on several preliminary results that must be confirmed in future studies. Nevertheless, we are approaching an adaptable layered system with emerging properties at multiple levels of organization. Without any doubt, working on the *Drosophila's* clock remains exciting.

[3]The used *cry-gal4* (#13) is expressed in all six LN_d. Therefore, all LN_d cells are absent in these flies, including the three cells that show no CRY immunoreactivity and are Mai179-negative.

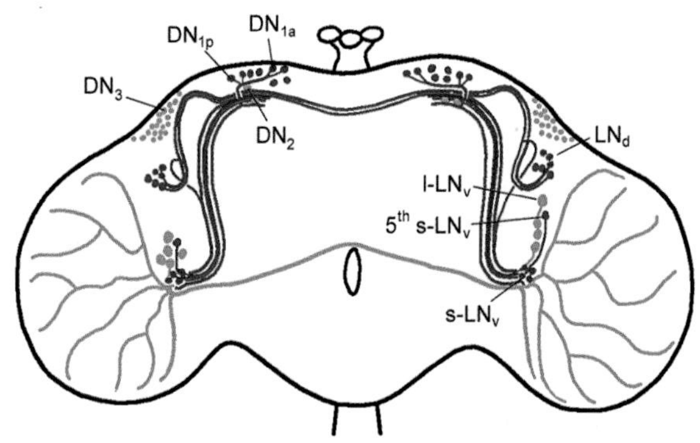

Figure 6. Semischematic representation of putative M and E cells in the *Drosophila* brain. (*Red*) M cells; (*blue*) E cells. Note that the division of the LN_d and the DN_1 in M and E cells is still hypothetical. We know that these cell groups are composed of M and E cells, but we do not know yet which individual cell behaves as M and which as E oscillator. (*Gray*) All cells for which the engagement in M or E oscillators is not at all clear. For better clarity, the LPN as well as the arborizations of the DN_2 and DN_3 cells are omitted. Only some of the DN_1 (~17 cells) are shown.

ACKNOWLEDGMENTS

We are thankful to Alois Hofbauer and André Klarsfeld for discussions and comments on the manuscript. The reported studies of the German group were supported by the Deutsche Forschungsgemeinschaft; that of the French group by Ministere de la Recherche and Agence Nationale de la Recherche. F.R. is supported by INSERM. Both groups were funded by the EU sixth framework program EUCLOCK.

REFERENCES

Bachleitner W., Kempinger L., Wülbeck C., Rieger D., and Helfrich-Förster C. 2007. Moonlight shifts the endogenous clock of *Drosophila melanogaster*. *Proc. Natl. Acad. Sci.* **104:** 3538.

Blanchardon E., Grima B., Klarsfeld A., Chelot E., Hardin P.E., Preat T., and Rouyer F. 2001. Defining the role of *Drosophila* lateral neurons in the control of circadian rhythms in motor activity and eclosion by targeted genetic ablation and PERIOD protein overexpression. *Eur. J. Neurosci.* **13:** 871.

Dushay M.S., Rosbash M., and Hall J.C. 1989. The *disconnected* visual system mutations in *Drosophila melanogaster* drastically disrupt circadian rhythms. *J. Biol. Rhythms* **4:** 1.

Ewer J., Frisch B., Hamblen-Coyle M.J., Rosbash M., and Hall J.C. 1992. Expression of the period clock gene within different cell types in the brain of *Drosophila* adults and mosaic analysis of these cells' influence on circadian behavioral rhythms. *J. Neurosci.* **12:** 3321.

Frisch B., Hardin P.E., Hamblen-Coyle M.J., Rosbash M., and Hall J.C. 1994. A promoterless period gene mediates behavioral rhythmicity and cyclical per expression in a restricted subset of the *Drosophila* nervous system. *Neuron* **12:** 555.

Grima B., Chelot E., Xia R., and Rouyer F. 2004. Morning and evening peaks of activity rely on different clock neurons of the *Drosophila* brain. *Nature* **431:** 869.

Hall J.C. 2005. Systems approaches to biological rhythms in *Drosophila*. *Methods Enzymol.* **393:** 61.

Hardin P.E., Hall J.C., and Rosbash M. 1992. Behavioral and molecular analyses suggest that circadian output is disrupted by *disconnected* mutants in *D. melanogaster*. *EMBO J.* **11:** 1.

Helfrich-Förster C. 1998. Robust circadian rhythmicity of *Drosophila melanogaster* requires the presence of lateral neurons: A brain-behavioral study of disconnected mutants. *J. Comp. Physiol. A* **182:** 435.

———. 2006. The neural basis of *Drosophila's* circadian clock.

Sleep Biol. Rhythms **4:** 224.

Helfrich-Förster C., Winter C., Hofbauer A., Hall J.C., and Stanewsky R. 2001. The circadian clock of fruit flies is blind after elimination of all known photoreceptors. *Neuron* **30:** 249.

Helfrich-Förster C., Shafer O.T., Wulbeck C., Grieshaber E., Rieger D., and Taghert P. 2007. Development and morphology of the clock-gene-expressing lateral neurons of *Drosophila melanogaster*. *J. Comp. Neurol.* **500:** 47.

Homberg U., Reischig T., and Stengl M. 2003. Neural organization of the circadian system of the cockroach *Leucophaea maderae*. *Chronobiol. Int.* **20:** 577.

Hyun S., Lee Y., Hong S.T., Bang S., Paik D., Kang J., Shin J., Lee J., Jeon K., Hwang S., Bae E., and Kim J. 2005. *Drosophila* GPCR *Han* is a receptor for the circadian clock neuropeptide PDF. *Neuron* **48:** 267.

Jaramillo A.M., Zheng X., Zhou Y., Amado D.A., Sheldon A., Sehgal A., and Levitan I.B. 2004. Pattern of distribution and cycling of SLOB, *Slowpoke* channel binding protein, in *Drosophila*. *BMC Neurosci.* **5:** 3.

Kaneko M. and Hall J.C. 2000. Neuroanatomy of cells expressing clock genes in *Drosophila:* Transgenic manipulation of the *period* and *timeless* genes to mark the perikarya of circadian pacemaker neurons and their projections. *J. Comp. Neurol.* **422:** 66.

Kaneko M., Helfrich-Förster C., and Hall J.C. 1997. Spatial and temporal expression of the *period* and *timeless* genes in the developing nervous system of *Drosophila:* Newly identified pacemaker candidates and novel features of clock gene product cycling. *J. Neurosci.* **17:** 6745.

Klarsfeld A., Malpel S., Michard-Vanhee C., Picot M., Chelot E., and Rouyer F. 2004. Novel features of cryptochrome-mediated photoreception in the brain circadian clock of *Drosophila*. *J. Neurosci.* **24:** 1468.

Lear B.C., Lin J.M., Keath J.R., McGill J.J., Raman I.M., and Allada R. 2005a. The ion channel narrow abdomen is critical for neural output of the *Drosophila* circadian pacemaker. *Neuron* **48:** 965.

Lear B.C., Merrill C.E., Lin J.M., Schroeder A., Zhang L., and Allada R. 2005b. A G protein-coupled receptor, groom-of-PDF, is required for PDF neuron action in circadian behavior. *Neuron* **48:** 221.

Lee G., Bahn J.H., and Park J. H. 2006. Sex- and clock-controlled expression of the neuropeptide F gene in *Drosophila*. *Proc. Natl. Acad. Sci.* **103:** 12580.

Lin Y., Stormo G.D., and Taghert P.H. 2004. The neuropeptide pigment-dispersing factor coordinates pacemaker interactions in the *Drosophila* circadian system. *J. Neurosci.* **24:** 7951.

Majercak J., Sidote D., Hardin P.E. and Edery I. 1999. How a circadian clock adapts to seasonal decreases in temperature and day length. *Neuron* **24:** 219.

Maywood E.S., Reddy A.B., Wong G.K., O'Neill J.S., O'Brien J.A., McMahon D.G., Harmar A.J., Okamura H., and Hastings M.H. 2006. Synchronization and maintenance of timekeeping in suprachiasmatic circadian clock cells by neuropeptidergic signaling. *Curr. Biol.* **16:** 599.

Mertens I., Vandingenen A., Johnson E.C., Shafer O.T., Li W., Trigg J.S., De Loof A., Schoofs L., and Taghert P.H. 2005. PDF receptor signaling in *Drosophila* contributes to both circadian and geotactic behaviors. *Neuron* **48:** 213.

Miyasako Y., Umezaki Y., and Tomioka K. 2007. Separate sets of cerebral clock neurons are responsible for light and temperature entrainment of *Drosophila* circadian locomotor rhythms. *J. Biol. Rhythms* **22:** 115.

Murad A., Emery-Le M., and Emery P. 2007. A subset of dorsal neurons modulates circadian behavior and light responses in *Drosophila*. *Neuron* **53:** 689.

Nitabach M.N., Blau J., and Holmes T.C. 2002. Electrical silencing of *Drosophila* pacemaker neurons stops the free-running circadian clock. *Cell* **109:** 485.

Peng Y., Stoleru D., Levine J.D., Hall J.C., and Rosbash M. 2003. *Drosophila* free-running rhythms require intercellular communication. *PLoS Biol.* **1:** E13.

Petri B. and Stengl M. 2001. Phase response curves of a molecular model oscillator: Implications for mutual coupling of paired oscillators. *J. Biol. Rhythms* **16:** 125.

Picot M., Cusumano P., Klarsfeld A., Ueda R., and Rouyer F. 2007. Light activates output from evening neurons and inhibits output from morning neurons in the *Drosophila* circadian clock. *PLoS Biol.* **5:** e315 (in press).

Pittendrigh C.S. and Daan S. 1976. A functional analysis of circadian pacemakers in nocturnal rodents. V. Pacemaker structure: A clock for all seasons. *J. Comp. Physiol. A* **106:** 333.

Renn S.C., Park J.H., Rosbash M., Hall J.C., and Taghert P.H. 1999. A *pdf* neuropeptide gene mutation and ablation of PDF neurons each cause severe abnormalities of behavioral circadian rhythms in *Drosophila*. *Cell* **99:** 791.

Rieger D., Stanewsky R., and Helfrich-Förster C. 2003. Cryptochrome, compound eyes, Hofbauer-Buchner eyelets, and ocelli play different roles in the entrainment and masking pathway of the locomotor activity rhythm in the fruit fly *Drosophila melanogaster*. *J. Biol. Rhythms* **18:** 377.

Rieger D., Shafer O.T., Tomioka K., and Helfrich-Förster C. 2006. Functional analysis of circadian pacemaker neurons in *Drosophila melanogaster*. *J. Neurosci.* **26:** 2531.

Schneider N.L. and Stengl M. 2005. Pigment-dispersing factor and GABA synchronize cells of the isolated circadian clock of the cockroach *Leucophaea maderae*. *J. Neurosci.* **25:** 5138.

Shafer O.T., Rosbash M., and Truman J.W. 2002. Sequential nuclear accumulation of the clock proteins period and timeless in the pacemaker neurons of *Drosophila melanogaster*. *J. Neurosci.* **22:** 594.

Shafer O.T., Helfrich-Förster C., Renn S.C., and Taghert P.H. 2006. Reevaluation of *Drosophila melanogaster's* neuronal circadian pacemakers reveals new neuronal classes. *J. Comp. Neurol.* **498:** 180.

Shiga S. 2003. Anatomy and functions of brain neurosecretory cells in *Diptera*. *Microsc. Res. Tech.* **62:** 114.

Siegmund T. and Korge G. 2001. Innervation of the ring gland of *Drosophila melanogaster*. *J. Comp. Neurol.* **431:** 481.

Singaravel M., Fujisawa Y., Hisada M., Saifullah A.S., and Tomioka K. 2003. Phase shifts of the circadian locomotor rhythm induced by pigment-dispersing factor in the cricket *Gryllus bimaculatus*. *Zool. Sci.* **20:** 1347.

Stanewsky R., Kaneko M., Emery P., Beretta B., Wagner-Smith K., Kay S.A., Rosbash M., and Hall J.C. 1998. The *cry^b* mutation identifies cryptochrome as a circadian photoreceptor in *Drosophila*. *Cell* **95:** 681.

Steller H., Fischbach K.F., and Rubin G.M. 1987. *Disconnected:* A locus required for neuronal pathway formation in the visual system of *Drosophila*. *Cell* **50:** 1139.

Stoleru D., Peng Y., Agosto J., and Rosbash M. 2004. Coupled oscillators control morning and evening locomotor behaviour of *Drosophila*. *Nature* **431:** 862.

Stoleru D., Peng Y., Nawathean P., and Rosbash M. 2005. A resetting signal between *Drosophila* pacemakers synchronizes morning and evening activity. *Nature* **438:** 238.

Stoleru D., Nawathean P., Fernandez Mde L., Menet J.S., Ceriani M.F., and Rosbash M. 2007. The *Drosophila* circadian network is a seasonal timer. *Cell* **129:** 207.

Taghert P.H. and Shafer O.T. 2006. Mechanisms of clock output in the *Drosophila* circadian pacemaker system. *J. Biol. Rhythms* **21:** 445.

Veleri S., Brandes C., Helfrich-Förster C., Hall J.C., and Stanewsky R. 2003. A self-sustaining, light-entrainable circadian oscillator in the *Drosophila* brain. *Curr. Biol.* **13:** 1758.

Yang Z. and Sehgal A. 2001. Role of molecular oscillations in generating behavioral rhythms in *Drosophila*. *Neuron* **29:** 453.

Yoshii T., Funada Y., Ibuki-Ishibashi T., Matsumoto A., Tanimura T., and Tomioka K. 2004. *Drosophila cry^b* mutation reveals two circadian clocks that drive locomotor rhythm and have different responsiveness to light. *J. Insect Physiol.* **50:** 479.

Yoshii T., Heshiki Y., Ibuki-Ishibashi T., Matsumoto A., Tanimura T., and Tomioka K. 2005. Temperature cycles drive *Drosophila* circadian oscillation in constant light that otherwise induces behavioural arrhythmicity. *Eur. J. Neurosci.* **22:** 1176.

Zerr D.M., Hall J.C., Rosbash M., and Siwicki K.K. 1990. Circadian fluctuations of *period* protein immunoreactivity in the CNS and the visual system of *Drosophila*. *J. Neurosci.* **10:** 2749.

Exploring Spatiotemporal Organization of SCN Circuits

L. YAN,* I. KARATSOREOS,* J. LESAUTER,† D.K. WELSH,§¶** S. KAY,§
D. FOLEY,‡ AND R. SILVER,*†§§

*Departments of Psychology, *Columbia University and †Barnard College, New York, New York 10027; §Division of Biological Sciences and ¶Department of Psychiatry, University of California, San Diego, La Jolla, California 92093; **Veterans Affairs San Diego Healthcare System, San Diego, California 92161; ‡Department of Economics, New School for Social Research, New York, New York 10003 and External Faculty, Santa Fe Institute, Sante Fe, New Mexico 87501; §§Department of Anatomy and Cell Biology, College of Physicians & Surgeons, Columbia University, New York, New York 10032*

Suprachiasmatic nucleus (SCN) neuroanatomy has been a subject of intense interest since the discovery of the SCN's function as a brain clock and subsequent studies revealing substantial heterogeneity of its component neurons. Understanding the network organization of the SCN has become increasingly relevant in the context of studies showing that its functional circuitry, evident in the spatial and temporal expression of clock genes, can be reorganized by inputs from the internal and external environment. Although multiple mechanisms have been proposed for coupling among SCN neurons, relatively little is known of the precise pattern of SCN circuitry. To explore SCN networks, we examine responses of the SCN to various photic conditions, using in vivo and in vitro studies with associated mathematical modeling to study spatiotemporal changes in SCN activity. We find an orderly and reproducible spatiotemporal pattern of oscillatory gene expression in the SCN, which requires the presence of the ventrolateral core region. Without the SCN core region, behavioral rhythmicity is abolished in vivo, whereas low-amplitude rhythmicity can be detected in SCN slices in vitro, but with loss of normal topographic organization. These studies reveal SCN circuit properties required to signal daily time.

INTRODUCTION

The Problem

Studies of the suprachiasmatic nucleus (SCN) and its function as the "master circadian pacemaker" offer an opportunity to examine multiple temporal and spatial scales so as to explore how complex responses emerge from cellular elements. Breakthroughs in molecular biology and in computer-based storage and analysis of enormous amounts of data have made it possible to measure brain structure, gene expression, and electrical activity simultaneously at many locations with high spatial and temporal resolution. The SCN is an unusual model system in the mammalian brain in that its function is well delineated and can be studied at cellular, tissue, and whole-organism levels. Finally, joint use of experimental and modeling approaches has led to new experimental studies that help to explore and explain seemingly unintuitive results.

The SCN and Its Function

The SCN is an endogenous master circadian pacemaker, whose phase is reset by photic cues via the retinohypothalamic tract (RHT) and which controls daily rhythms of behavior and physiology (Klein et al. 1991). Ablation of the SCN produces a loss of circadian rhythmicity at the whole-animal level (Moore and Eichler 1972; Stephan and Zucker 1972), with no recovery of function irrespective of the age at which the lesion is made (Mosko and Moore 1979). The SCN oscillates in the absence of input from the rest of the brain both in vivo (Inouye and Kawamura 1979) and in vitro (Green and Gillette 1982; Groos and Hendriks 1982; Shibata et al. 1982). Within the SCN, circadian oscillations can be measured in gene expression, electrical activity, metabolic activity, and neuropeptide release (Reppert and Weaver 2001). Furthermore, fetal SCN grafts can restore behavioral rhythms in SCN-lesioned animals (Lehman et al. 1987; Ralph et al. 1990). The SCN produces both synaptic and diffusible output signals (Hakim et al. 1991; Silver et al. 1996). Neural efferents appear to be necessary for SCN control of neuroendocrine responses but not for activity-dependent rhythms such as drinking, gnawing, and temperature regulation (Lehman et al. 1987; Hakim et al. 1991; Meyer-Bernstein et al. 1999). Finally, individual SCN neurons display circadian rhythmicity upon dispersion (Welsh et al. 1995). These experiments suggest the concept of the SCN as a brain clock comprising a feed-forward circuit of similar cellular elements that together produce a coherent circadian rhythm output that signals time of day information for the rest of the body (Fig. 1A).

The Tissue Is the Issue

The initial view of the SCN assigned to it nearly magical properties. Here, within the mammalian brain was a hypothalamic nucleus that was necessary and sufficient for a single function, and upon damage (even neonatally), this function could not be assumed by any other tissue. The early view of the SCN as a black box was a very useful heuristic in drawing attention to inputs and outputs of the master clock (see Fig. 1 in Pittendrigh and Bruce 1959; Eskin 1979). Since that time, numerous advances in the analysis of SCN organization have contributed to understanding the component cellular elements and neural circuits inside this black box (summarized in Fig. 1B–C). The investigation of SCN responses to naturally occurring and experimentally induced challenges has made it clear that robust function of the brain clock is dependent on its intricate neural circuitry (Fig. 1D–F).

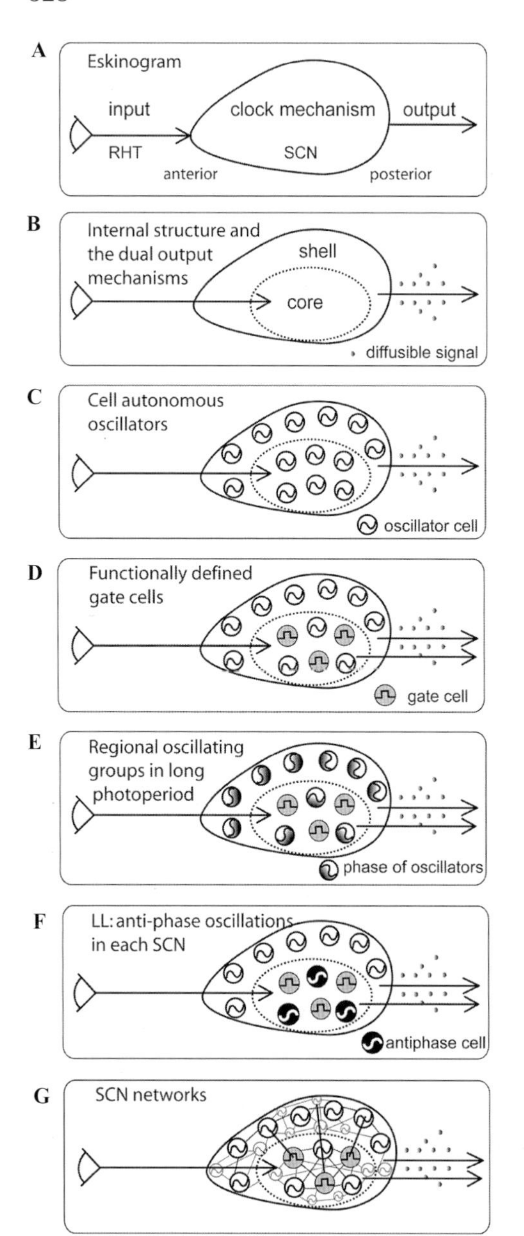

Figure 1. Building the brain clock: A historical perspective. (*A*) An early depiction of the circadian system involved a feed-forward mechanism with three components, namely, input, clock mechanism, and output (Pittendrigh 1961; Eskin 1979). This simple heuristic captures several key concepts of circadian timekeeping. (*B*) Delineation of the distinct core and shell compartments of the SCN began the effort of identifying components of the clock in the black box (Moore 1973; van den Pol and Tsujimoto 1985). Discovery that a diffusible signal was sufficient to maintain coherent activity rhythms (Silver et al. 1996) led to the search for such a signal. (*C*) In dispersed cell culture, individual SCN neurons show circadian rhythms in their firing rate with different phases, suggesting that circadian oscillation is a property of individual SCN neurons, rather than an emergent network property (Welsh et al. 1995). (*D*) The SCN is composed of functionally distinct cells. In addition to oscillator cells, some cells in the "core" region lack detectable oscillation of clock gene expression and electrical activity (Hamada et al. 2001; Jobst and Allen 2002; Karatsoreos et al. 2004). Cells of the core and shell region each project to SCN target sites (Abrahamson and Moore 2001; Kriegsfeld et al. 2004a). (*E*) Photoperiod induces both temporal and spatial changes in the circadian phase of oscillators within the SCN. In short photoperiods, the oscillators are mostly in-phase; in long photoperiods, the phases of the oscillators change along the rostrocaudal axis (Johnson 2005; Inagaki 2007). (*F*) New networks of SCN oscillators are seen under unusual environmental conditions. In the behaviorally split animal, the left and right SCN are in antiphase, and within each SCN, the core and shell of each SCN are in antiphase (Tavakoli-Nezhad and Schwartz 2005; Yan et al. 2005). (*G*) SCN network organization may resemble a small-world network with mostly local connections between nodes and with a few long-distance connections. On the other hand, it may consist of a locally connected network in which signals spread slowly from one layer of cells to the next.

We know relatively little of SCN circuits, although some general principles of network organization may be useful in conceptualizing available data (Fig. 1G).

Wiring economy or optimization has been suggested as an important general principle for brain organization because larger volumes raise metabolic cost, delay signal transduction, and attenuate signals (Ramón y Cajal et al. 1999). According to this principle, among the various functionally equivalent arrangements of neurons, the one having minimum wiring cost is most evolutionarily fit and therefore is most likely to be selected (Chklovskii and Koulakov 2004). The SCN is small, with small (8–10-μm diameter) densely packed neurons. Even though brain weight and total neuron number differ more than a 1000-fold between human and mouse (human vs. mouse, 1350 g vs. 450 mg; 85 billion vs. 75 million neurons, respectively; Williams and Herrup 1988), the number of neurons in the human SCN is only a fewfold greater (human vs. mouse, 45,000 vs. 10,500; Hofman and Swaab 1989; Abrahamson and Moore 2001). The slight change in the number of SCN neurons during evolution suggests computational performance of the SCN has reached an optimal level at a small size. (This observation also suggests that the SCN does not scale with brain size.) How does the SCN solve the conflict between wiring economy and the need to coordinate long-term, i.e., circadian, rhythmicity among neurons? If the SCN has properties of a small-world network (Watts and Strogatz 1998), one would expect to find local connections between nodes with a few long-distance connections (Fig. 1G). Restated, it would have a small number of connections per node but also a small total distance between any two nodes. Signals in a small-world network tend to propagate very rapidly through the whole network. The advantage of the small-world arrangement for a neural network is not only for wiring economy, but also to increase clustering, and some aspects of SCN circuits are consistent with these properties. For other aspects, particularly the function of the SCN in representing the time of day and season of the year, the evidence of spatiotemporal organization reviewed in this chapter is more consistent with a locally connected network in which signals spread rather slowly from one layer of connected nodes to the next. Such a locally connected network exploits the inherent periodicity of the oscillator cells so as to establish the timescale of the circadian timekeeping. (A "cellular automaton" approach to modeling a locally connected network oscillator is presented below.) In view of regional heterogeneity of the SCN, it is possible that both types of networks occur within the nucleus.

It is our thesis that "the tissue is the issue," i.e., SCN function can best be understood by delineating circuit properties of the SCN that render it a brain clock. Our objective is to explore how the individual cells that comprise the SCN produce its rhythmic output signal(s). The view that rhythmicity of the SCN is a product of individual, coupled oscillators has set the basis for physiological and modeling studies of the clock, as well as its inputs and outputs. We explore how the cells of the bilateral SCN are organized in space and time, and how unique clock functions are achieved by this organization of SCN cellular elements. The cells that constitute SCN circuits include diverse elements: cell-autonomous oscillators, slave oscillators, and cells that do not appear to oscillate at all (Lee et al. 2003; Hastings and Herzog 2004; Antle and Silver 2005). Thus, within the SCN, heterogeneous, independently phased cellular oscillators occur in a spatially and temporally ordered pattern, and this underlies the timekeeping function of the nucleus. Our goal is to review SCN neuroanatomy briefly, to describe changes in its organization following exposure to various photic stimuli, and to deconstruct its neural circuitry using in vivo and in vitro studies and associated modeling analyses.

SECTION THEMES

Brief Anatomy of SCN Focusing on Core and Shell Organization

Numerous features of SCN organization are consistent among mammals. Each side of this bilaterally symmetrical nucleus of rodents contains approximately 10,000 neurons, with characteristic topography of peptide content, afferent/efferent connections, and clock gene expression. From the time of the earliest studies, it was clear that the SCN is composed of two fundamentally different regions based on cell size and morphology, afferent and efferent connections, and neuropeptide phenotype (Moore 1983). Largely based on the rat SCN, these two regions were designated ventrolateral and dorsomedial. As comparative data emerged, the limited generality of these two "geographic" descriptors came into better focus. Because the topography of SCN organization differs among species, it became more useful to think of the SCN as composed of "core" and "shell" regions, based on physiological and functional criteria (Miller et al. 1996; Antle and Silver 2005). This brought the advantage of conceptually differentiating two regions but lacked precise localization. Because the anatomical loci suggested by the terms core and shell do not delineate SCN regions very precisely, these designations can lead to confusion (for discussion, see Morin and Allen 2006). Nevertheless, anatomical specializations are generally associated with functional specializations, and the core-shell distinction usefully sets the stage for exploring intra-SCN organization (Abrahamson and Moore 2001). As long it is clear which part of the nucleus is being described, and which criteria are being used, the core and shell terminology helps to analyze SCN organization and to conceptualize and communicate the results of such analyses.

Distribution of peptidergic phenotypes within the SCN. The study of core-shell functional specializations has been hampered by the fact that there is significant diversity among species in the peptidergic phenotypes of SCN cells (Morin and Allen 2006). The SCN shell contains neurons expressing arginine vasopressin (AVP) in most species, including mouse, rat, hamster, lemur and humans, although not in mink (Larsen and Mikkelsen 1993) or mole rat (Rosen et al. 2007). The SCN core is rich in vasoactive intestinal polypeptide (VIP) and gastrin-releasing peptide (GRP) in mouse, rat, and hamster (Morin et al. 1992; Moore 1996; Abrahamson and Moore 2001). Other peptides in the SCN core, however, are more variable among species (Card and Moore 1984; Hartwich et al. 1994; Silver et al. 1999; Abrahamson and Moore 2001), likely reflecting functional species specializations. For example, the hamster core contains calbindin (CalB) and substance P (SP), whereas the mouse core contains calretinin and neurotensin.

There is a broad association between peptidergic phenotype and clock function. The SCN shell region is delineated by AVP neurons and by rhythmic *Per* mRNA and AVP expression (Schwartz et al. 1983; Jin et al. 1999). The core region, bearing VIP and GRP cells, lacks detectable rhythmicity or shows very low-amplitude changes in clock gene expression (Fig. 2) (Antle and Silver 2005). In contrast to AVP, neither VIP nor GRP is rhythmically expressed in the SCN under constant conditions, although in light/dark (LD) cycles, GRP is higher in the day and VIP is higher at night (Inouye and Shibata 1994).

Inputs to SCN. The SCN core and shell have distinct afferent inputs indicative of fundamental differences in function. The core receives direct input from the retina through the retinohypothalamic tract (RHT), secondary visual input from the intergeniculate leaflet (IGL), and the lateral geniculate complex through the geniculohypothalamic tract (GHT), as well as input from the pretectal nuclei and from the midbrain raphe nucleus (Moore 1973; Moga and Moore 1997; Abrahamson and Moore 2001). The shell receives input from the basal forebrain, the cerebral cortex, the hippocampus, brainstem cholinergic nuclei, medullary noradrenergic areas, and several hypothalamic nuclei.

In some instances, the peptidergic phenotype of cells receiving afferent inputs is known. RHT fibers synapse onto VIP cells and GRP cells (rat; Ibata et al. 1989;

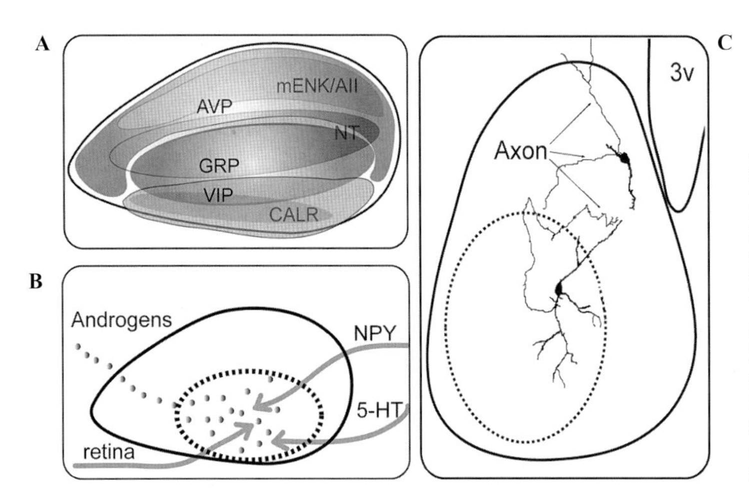

Figure 2. Anatomical organization of the SCN. (*A*) Distribution of the neuropeptides in the mouse SCN reconstructed from data in Abrahamson and Moore (2001). (*B*) Schematic summarizing the neuronal (*gray line*) and the hormonal (androgen, *gray dots*) input to the core (*dashed black line*) and shell SCN. The core region receives both hormonal and neural (retinal input, 5-HT from raphe and NPY from IGL) inputs. The shell region receives GAL and TH afferent fibers. (*C*) Biotin-filled GRP cell in the SCN core (Drouyer et al. 2005) and AVP cell in the shell (adapted from Pennartz et al. 1998). The GRP cell has an axon and dendrite extending into the shell.

Tanaka et al. 1997) or on calbindin and GRP cells (hamster; Bryant et al. 2000). Although some retinal fibers may be seen throughout the SCN, the greatest density is in the core region. Functional analysis of *Per* gene or c-Fos expression reveals that light-induced activation of SCN cells occurs first within the SCN core and later in the rest of the SCN, indicating a multistep propagation of a signal through the SCN (Fig. 1B and 4) (Yan and Silver 2002, 2004; Karatsoreos et al. 2004). The core is also a site of hormone action. In mouse, androgen receptors (AR) are concentrated in GRP cells of the core SCN, among other cell types, and SCN AR expression is modulated by circulating androgen (Karatsoreos et al. 2007).

Intra-SCN connections: Core and shell specializations. There is very little work done on intra-SCN connections of specific neuronal phenotypes, as the effort is hampered by the small size of the nucleus and of its neurons. Although the phenotype of particular cellular elements is unclear, it is known from electron microscopy studies that the SCN contains large numbers of local circuit axons or local collaterals of long projection axons (van den Pol 1991). Some of these axons appear to be devoted to communication among cells of the SCN, whereas others are axon collaterals of neurons that project outside the SCN and also terminate within the SCN. Indeed, about 80% of synaptic endings within the SCN survive transection of long efferent connections, indicating that SCN axons form extensive axonal circuits within the SCN. The results of the three studies available on known peptidergic phenotypes are consistent with the foregoing. Morphologically, AVP neurons have few axon collaterals, most of which remain inside the boundaries of the SCN but occasionally project to SCN target areas. Their dendritic branches are compact and bear numerous varicosities (Fig. 2C) (Pennartz et al. 1998). In studies using similar methods to study GRP cells in mouse, we find that biocytin-filled GRP cells have axon projections and dendritic arbors in the shell, and GRP dendrites make appositions with AVP neurons and their processes (Fig. 2C) (Drouyer et al. 2005). Furthermore, these intra-SCN circuits are directional. *Trans*-synaptic labeling of neuronal circuits indicates dense projections from core to shell but little reciprocal innervation (Leak et al. 1999), and this directionality is also seen in biotin-filled core cells (Jobst et al. 2004; Drouyer et al. 2005).

Gap junctions. In addition to synaptic connections, gap junctions also contribute to intra-SCN circuits. In the rat SCN, cells are coupled (Jiang et al. 1997) either within dorsomedial or within ventrolateral SCN, but cells in one region are not coupled to cells in the other region (Colwell 2000). Data from our lab are consistent with these findings: The majority (67%) of GRP cells in mouse SCN are dye-coupled to other cells in the core (Drouyer et al. 2005). In connexin-36 (gap junction protein) knockout mice, behavioral circadian rhythmicity is damped and the onset of activity at the transition to darkness is delayed, suggesting an important functional role for electrical coupling among SCN neurons (Long et al. 2005).

SCN outputs. SCN efferents have been amply reviewed, and we focus here on the question of core/shell differences/similarities. SCN outputs are densest to adjacent hypothalamic nuclei, especially to the subparaventricular zone (SPVZ), the preoptic area (POA), and the dorsomedial hypothalamus (DMH) (Watts et al. 1987; Leak and Moore 2001; Kriegsfeld et al. 2004a), with limited projections to the forebrain, thalamus, and periaqueductal grey. It has been proposed that the former nuclei serve as relays for SCN projections to cell groups throughout the brain (Moore 1996; Deurveilher and Semba 2005). On the basis of track tracing studies, it appears that core and shell SCN neurons project to all the same targets, although the relative densities of these projections differ (Leak and Moore 2001; Kriegsfeld et al. 2004a). For example, it seems that the medial SPVZ, DMH, and POA receive most of their input from the shell, whereas the lateral SPVZ receives most of its input from the core. This differential pattern of efferent connectivity permits distinct signals from core and shell to reach target sites in the brain.

Diffusible signals represent another SCN output. Polymer-encapsulated SCN fetal tissue grafts, producing diffusible output signals, are sufficient to restore behavioral but not endocrine rhythms (Fig. 1B) (Lehman et al.

1987; Silver et al. 1996; LeSauter and Silver 1998; Meyer-Bernstein et al. 1999). In addition, SCN2.2 cell lines induce rhythms of metabolic activity and clock gene expression in cocultured fibroblasts via a diffusible signal (Allen et al. 2001). A few candidate diffusible factors have been proposed including prokineticin 2 (Cheng et al. 2002; Prosser et al. 2007), transforming growth factor-α (TGF-α) (Kramer et al. 2001), and cardiotrophin-like cytokine (Kraves and Weitz 2006). It is not clear whether these diffusible signals emanate from core or shell, whether they are produced by particular subtypes of SCN cells, and whether additional diffusible factors remain to be identified.

Function of SCN Subregions

Distinct functions of core and shell. Use of immediate-early genes and clock genes as markers indicates that there are two fundamentally different responses of SCN neurons and that these are well aligned with "core" and "shell" compartments. In constant darkness (DD), rhythmic oscillation of *Per1* and *Per2* expression occurs in the shell (mouse, Shigeyoshi et al. 1997; rat, Yan et al. 1999; hamster, Hamada et al. 2001), but depending on the methods used, such oscillation is very low or not detectable in the core (hamster, Hamada et al. 2001; mouse, Karatsoreos et al. 2004). On the other hand, light-induced *Per1* and *Per2* expression is largely limited to the core SCN. Similarly, endogenous circadian rhythms of c-FOS and FRA-2 expression occur in the shell, but not in the core, whereas light-induced expression of c-FOS, FRA-2, and FOS-B occurs in the core, but not in the shell (Schwartz et al. 2000). These regional differences may be difficult to detect in some preparations as the VIP and GRP cells that delineate the core have very dense efferent projections that cover the entire SCN and extend beyond its borders (Fig. 3).

This core/shell difference is also seen in electrical activity rhythms. Both rat AVP cells (Schaap et al. 1999) and neurons lying in the AVP-rich region of mouse SCN (delineated by Per1-GFP) (Kuhlman and McMahon 2004) have a higher firing rate during the day than at night. In contrast, calbindin cells of the hamster core SCN (Jobst and Allen 2002) or GRP cells in a transgenic green fluorescent protein (GFP)-reporter mouse (Cloues et al. 2001) show no day/night rhythm in firing rate. In summary, at least some core cells are not detectably rhythmic in gene expression or electrical activity but are light responsive during the night. The light responsiveness of these cells is said to be "gated" (opened or closed) by the circadian clock because it occurs at night but not during the day. Although the gating mechanism is not known, the

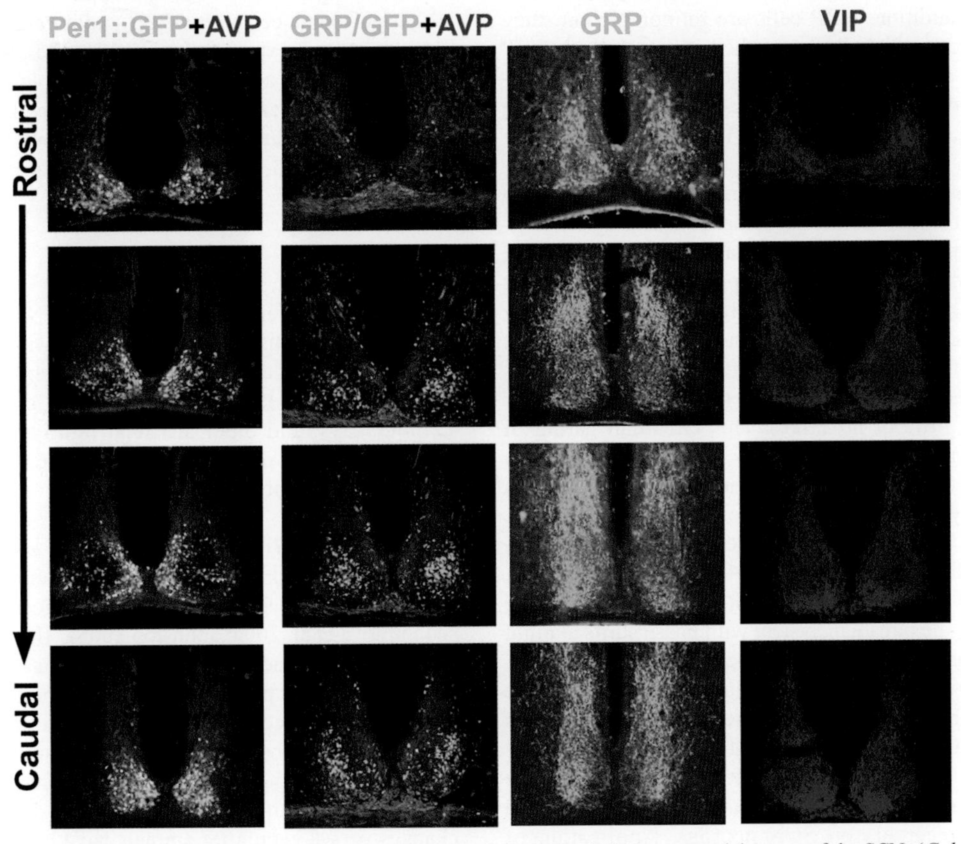

Figure 3. Photomicrographs showing immunostaining of SCN sections through the rostrocaudal extent of the SCN. (*Column 1*) Both Per1-GFP (*green*) and AVP (*red*) are colocalized in the shell region; (*column 2*) GRP cell bodies can be visualized in the GRP-GFP transgenic mouse. GRP-GFP (*green*) cells occur in the core SCN, as can be seen when the shell region is delineated by AVP (*red*); (*columns 3 and 4*) GRP and VIP fibers project extensively throughout the SCN, obscuring visualization of GRP somata. (Data in column 2 are taken, with permission, from Karatsoreos et al. 2004 [Society for Neuroscience].)

discovery of these cells sets the stage for exploring the functional importance of cellular diversity within the SCN (Figs. 1D and 3).

Role of core in SCN oscillation. Lesions of the SCN core region eliminate circadian rhythms of locomotor activity, drinking, gnawing, body temperature, and hormone secretion, even when substantial SCN tissue survives ablation (LeSauter and Silver 1999; Kriegsfeld et al. 2004b). The spared SCN tissue contains AVP cells, which are known cellular oscillators, capable of self-sustained oscillation in SCN slices (Noguchi et al. 2004) and even when cultured at low density in vitro (Murakami et al. 1991; Watanabe et al. 1993). In organotypic cultured SCN slices, surgical separation of the ventral and dorsal SCN region causes the neurons in the dorsal shell SCN to desynchronize, whereas the ventral core neurons retain coherent rhythmicity (Yamaguchi et al. 2003). These results suggest that cells of the SCN shell are not sufficient to maintain synchrony among the oscillator neurons and that core-to-shell communication is a requirement for rhythmicity of the tissue as a whole.

Several other lines of evidence support the suggestion that core cells are involved in synchronization and entrainment of SCN oscillators. Intercellular signaling through VIP and its receptor VPAC2 is critical in maintaining the synchrony of SCN cellular oscillators and circadian function within the SCN (Aton et al. 2005; Maywood et al. 2006). In addition, GRP cells are retinorecipient, they communicate with cells in the shell region through GRP receptors, and they function to keep SCN neurons synchronized during entrainment (McArthur et al. 2000; Aida et al. 2002; Karatsoreos et al. 2004; Antle et al. 2005).

Gating of light in the core. One of the fundamental features of circadian oscillation is that resetting by environmental signals depends on the phase at which the stimuli are applied. Thus, a light pulse applied in the early night (or early subjective night) produces phase-delays, whereas the same stimulus presented during the late night (or late subjective night) produces phase-advances. A light pulse applied during the day (or subjective day) typically has little or no effect. This circadian gating of responsiveness (or sensitivity) to photic stimulation is a central feature of all current models of entrainment, yet the underlying mechanisms are not known.

Several lines of evidence suggest that diurnal and circadian gating occur at the level of SCN neurons. Electrical stimulation of the RHT causes light-like changes in the circadian system with phase shifts at night but not during the day in both behavior (de Vries et al. 1994) and electrical activity in vitro (Shibata and Moore 1993). Similarly, light-induced clock gene expression occurs at night or subjective night but not during subjective day (Shigeyoshi et al. 1997). It appears that part of the gating mechanism involves *N*-methyl-D-aspartic acid (NMDA) receptors on SCN neurons. Application of NMDA causes light-like phase-shifts of the circadian rhythm in neuronal activity in the SCN in vitro (Ding et al. 1994; Shibata et al. 1994). The NMDA component of the evoked excitatory response is larger at night than during the day (Pennartz et al. 2001), and the circadian system gates the magnitude of NMDA-induced currents and Ca^{2+} transients (Colwell 2001). A hindrance in understanding these gated responses lies in our ignorance of which SCN neurons are involved.

Calbindin-containing cells of the hamster SCN receive direct RHT input (Bryant et al. 2000) and may participate in the gating of photic input. There is a circadian rhythm of subcellular localization of calbindin; whereas the protein is detected at all times in the cytoplasm, it is low or absent in the nucleus during the night. Under normal circumstances, light-induced behavioral phase-shifts and *Period* (*Per*) gene expression in the SCN occur only during the subjective night. Surprisingly, after administration of calbindin antisense oligodeoxynucleotides, both behavioral phase-shifts and light-induced *Per* are blocked during the subjective night and enhanced during the subjective day (Hamada et al. 2003). This evidence suggests that in hamsters, calbindin-containing cells of the SCN may participate in photic gating.

The core is important in gating light responsiveness, but information from photic signals must reach the shell if phase-shifts are to occur. After an acute light exposure, induction of the clock genes *Per1* and *Per2* initially occurs in the ventrolateral/core SCN and then spreads into the dorsomedial/shell region, with the exact pattern depending on the timing of light exposure (Yan and Silver 2002). During the delay zone of the phase-response curve, a light pulse induces both *mPer1* and *mPer2* in the core, but only *mPer2* expression spreads to the shell. In contrast, during the advance zone of the phase-response curve, when light causes advances of behavioral rhythms, a light pulse induces *mPer1* in the core which then spreads to the shell, but there is no significant induction of *mPer2* (Fig. 4).

SCN Plasticity

Whereas the foregoing describes the SCN response to a light pulse, patterns of gene expression within the SCN are also altered by exposure to various other lighting conditions, including long versus short days, constant light, 12-hour days, and other artificial lighting conditions. These phenomena point to the nature of the SCN circuitry underlying plasticity.

Spatiotemporal changes in various photoperiods. Regulation of clock gene expression and neuronal activity in the SCN is dependent on the photoperiod (for review, see Sumova et al. 2004; Johnston 2005). The durations of elevated clock gene expression and neuronal activity are expanded in long photoperiods but compressed in short photoperiods. Electrophysiological data show that in freely moving mice, rhythms in SCN multiunit activity (MUA) (high during subjective day) become compressed in short days and more distributed in long days, and these changes persist even after animals have been transferred into constant darkness (VanderLeest et al. 2007). Corresponding changes are observed in vitro in SCN slices taken from animals under these conditions (Mrugala et al. 2000; Inagaki et al. 2007; VanderLeest et al. 2007).

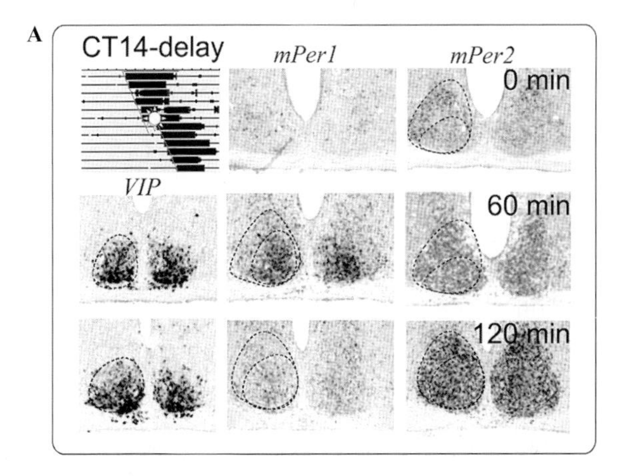

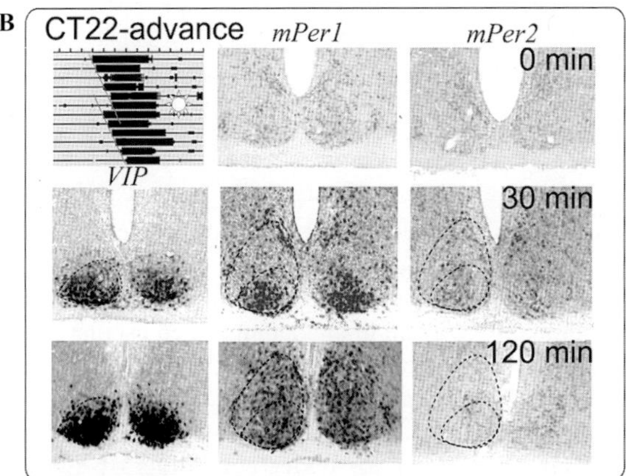

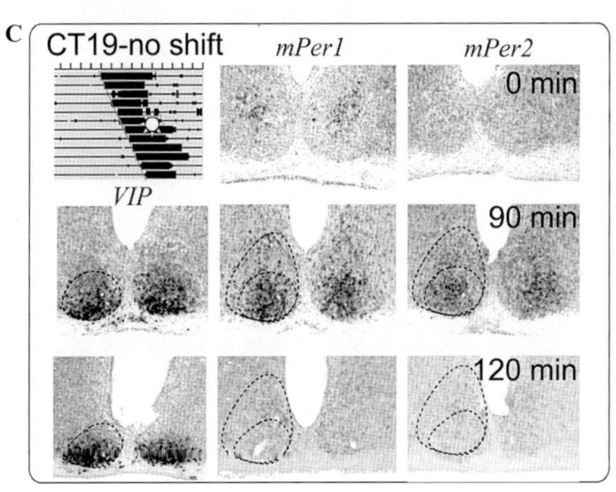

Figure 4. The photic response is gated in SCN. (*A*) After a light pulse at CT14, *mPer1* is induced only in the core SCN, whereas *mPer2* is initially induced in the core and then spreads throughout the SCN. This is correlated with a behavioral phase delay. (*B*) After a light pulse at CT22, *mPer1* is initially induced in the core and then spreads throughout the SCN. There is no *mPer2* induction. This is correlated with a behavioral phase advance. (*C*) After a light pulse at CT19, both *mPer1* and *mPer2* are transiently induced in the core SCN, without spreading into the shell. This produces no phase-shift. (Modified, with permission, from Yan and Silver 2002 [Wiley-Blackwell].)

A model of two separate, but mutually coupled, circadian oscillators has been proposed to explain photoperiodic responses of behavioral rhythms in nocturnal rodents: an evening oscillator, which drives the activity onset and entrains to dusk, and a morning oscillator, which drives the end of activity and entrains to dawn (Pittendrigh and Daan 1976). Measurements of clock gene expression indicate that circadian oscillation in the posterior SCN is phase-locked to the end of activity, whereas oscillations of some cells in the anterior SCN are phase-locked to the onset of activity (Hazlerigg et al. 2005; Johnston 2005; Inagaki et al. 2007).

Photoperiod also produces changes in the SCN core, acting on calbindin expression in the core of hamster SCN. Calbindin expression is negatively correlated with the day length: The number of calbindin-immunopositive neurons and calbindin mRNA levels are markedly increased in hamsters exposed to short photoperiods (light:dark, 6:18). Thus, calbindin neurons may be involved in the encoding of seasonal information by the SCN (Menet et al. 2003).

Spatiotemporal changes in response to constant light. In hamsters, constant light causes "splitting" of circadian rhythms, such that a single daily bout of activity separates into two components, 12 hours apart (Pittendrigh and Daan 1976), with antiphase circadian oscillations of gene expression in the left and right SCN (de la Iglesia et al. 2000). Given the functional heterogeneity of the SCN, in which ventrolateral but not dorsomedial neurons are retinorecipient, we studied how the two compartments respond to the constant lighting conditions that produce splitting. In the rostral and mid level of the hamster SCN, a region containing AVP cells, levels of FOS and PER1 expression are high in one unilateral SCN and low in the other, as previously reported (de la Iglesia et al. 2000). In addition, we found a novel network of oscillators in the core SCN that was not recognized by previous studies of split animals. In a core subregion of the caudal SCN, FOS and PER1 expression are high at phases when expression levels in rostral and mid SCN (on the same side) are low (Fig. 5) (Yan et al. 2005); i.e., the core and the shell regions within each SCN oscillate in antiphase. This suggests that the two antiphase oscillators postulated to underlie splitting (Daan and Berde 1978) may be the product of a new pattern of coherent rhythmicity in a newly emergent SCN network.

Spatiotemporal changes in response to bimodal (7:5:7:5) light/dark cycles. Desynchronized oscillations in different subregions of the SCN have also been reported under other artificial lighting conditions. Bimodal (LDLD) lighting cycles, in which there are two light periods and two dark periods in a 24-hour cycle, induce and entrain split circadian activity rhythms in rodents (Gorman and Elliott 2003). Under such a lighting cycle (LDLD 7:5:7:5), the *mPer1* level in the core SCN is higher than in the shell during one of the two light periods (morning photophase, M) and lower in the other light period (evening photophase; E) (Watanabe et al. 2007). In contrast, the *mBmal1* level in the core SCN is lower than in the shell in the M phase and higher in the E phase. In yet another variation on this theme, when housed in an 11:11-hour LD cycle, the ventral and dorsal regions of the rat SCN show separate circadian oscillations in the expression of clock genes, with different periods, that

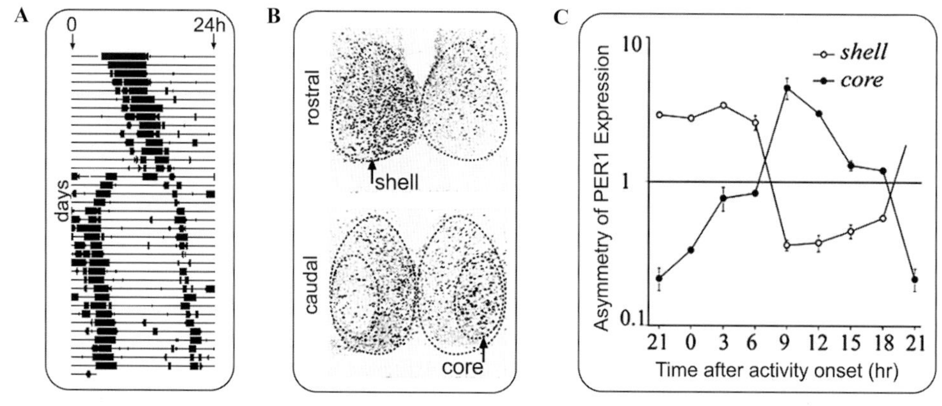

Figure 5. Constant light (LL) produces behavioral "splitting" (*A*) in which the animal's single daily bout of activity separates into two components, and the two components free-run until they become stably coupled 12 hours apart. LL also induces antiphase oscillation within each side of the SCN (*B*). PER1 is not uniform on either side of the SCN, but it is high in the left shell and low in the left core. In contrast, PER1 is low in the right shell and high in the right core. The asymmetry of the bilateral core and shell SCN was quantified in split animals killed every 3 hours (*C*). The *y* axis is in log scale. Error bars indicate S.E. (Modified, with permission, from Yan et al. 2005 [Society for Neuroscience].)

reflect two desynchronized components of locomotor activity (de la Iglesia et al. 2004).

Gates and Oscillators in the SCN

The discovery that the SCN is composed of functionally distinct cells, differentiated by their patterns of clock gene expression, forces a paradigm shift in understanding the cellular basis of plasticity and coherent rhythmicity of the master circadian pacemaker. The observation that SCN rhythmicity is not uniform or fixed, but is plastic in the face of environmental challenges, begs the question of its circuit organization. We now consider how SCN core cells maintain phase synchronization in the oscillator tissue. Importantly, in the absence of some mechanism of synchronization, oscillator cells with different periods will drift out of phase with each other, resulting in a loss of circadian rhythm for the SCN as a whole, despite the continuing oscillation of individual cells. Initial work on this problem centered on the idea that the cells could maintain synchronization through direct cell-to-cell coupling (Liu et al. 1997; Kunz and Achermann 2003; Bernard et al. 2007) or global coupling through a diffusible secreted factor (Garcia-Ojalvo et al. 2004; Gonze et al. 2005). However, the fact that the animal does not sustain circadian rhythms when the SCN core is ablated (LeSauter and Silver 1999; Kriegsfeld et al. 2004b) suggests that such direct coupling among the oscillator cells, if it has a role in synchronization, is not sufficient by itself to maintain synchronization.

Antle et al. (2003, 2007) proposed an alternative model for the maintenance of synchronization in SCN tissue, which gives a central role to the core cells acting as a "gate." The idea is that the core cells under certain conditions send a phase-resetting signal to the shell oscillators. This signal retards the phase of shell oscillators with advanced phase and advances the phase of shell oscillators with retarded phases, resulting in a reduction in the phase dispersion of the shell oscillators. The resetting signal in this model is triggered either by light or by a feedback signal (direct or indirect) from the circadian system. Computer simulations demonstrate the effectiveness of this gating mechanism in maintaining shell oscillator synchronization and show that the gating model is capable of explaining both entrainment of the SCN to daily light variations (when the gate is triggered by light) and free-running circadian behavior under conditions of constant darkness (when the gate is triggered by feedback from the shell). These simulations also produce phase-response curves to simulated light pulses that are similar to those observed in animals. The model makes the testable prediction that in the absence of the core, the shell oscillators will drift out of phase, much as is seen in peripheral tissues.

Spatiotemporal Oscillations in the SCN

Phase dispersion of SCN oscillators. It is not the case that in vivo, the SCN as a whole has a single phase of clock gene expression (Fig. 6). Instead, each cellular SCN oscillator has a characteristic phase, and the individual oscillators are temporally and spatially organized within the nucleus. During a circadian cycle in vivo, the clock genes *Per1* and *Per2* are first expressed in cells lying in the dorsomedial, periventricular part of the SCN (near the third ventricle) and then spread out to the central region and finally to the ventral region (Yan and Okamura 2002; Quintero et al. 2003; Yamaguchi et al. 2003). Although the majority of the SCN cellular oscillators (~50–60%) have similar phases, others do not (Quintero et al. 2003; Yamaguchi et al. 2003). Application of the protein synthesis inhibitor cycloheximide (CHX) stops the molecular clock and sets the SCN neurons to the same phase. However, once the CHX is removed, the cells gradually reestablish their original phase relationships (Yamaguchi et al. 2003). This suggests that the phase relationships among the cellular oscillators are an intrinsic property of SCN circuitry.

To examine more precisely the spatiotemporal dynam-

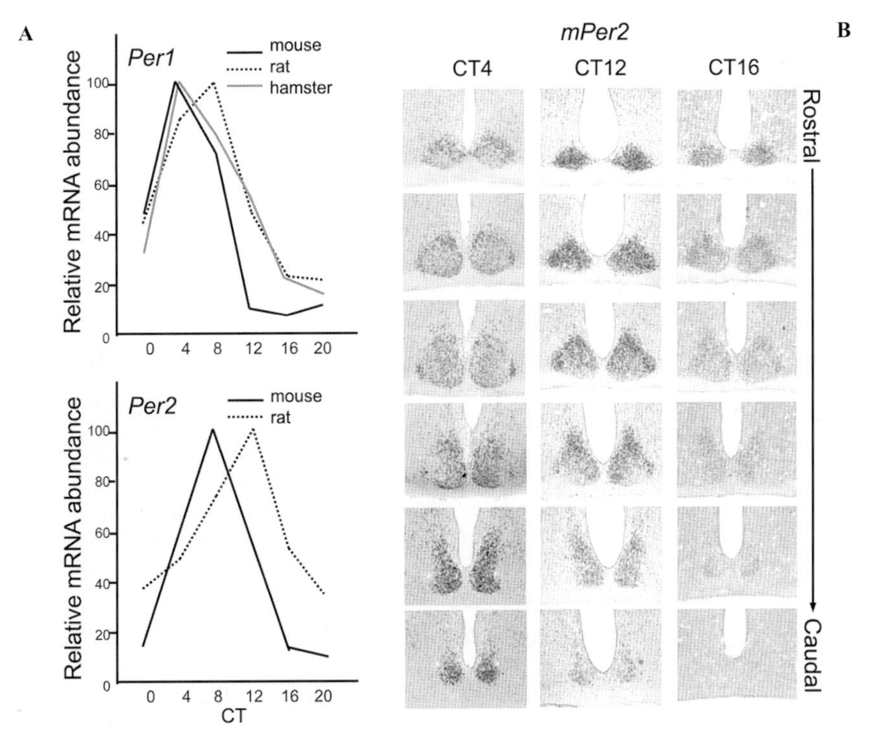

Figure 6. SCN organization: Temporal and spatial changes in clock gene expression. (*A*) Temporal profile of clock genes *Per1* and *Per2* in the SCN (mouse, Takumi et al. 1998; rat, Yan et al. 1999; hamster, Yamamoto et al. 2001). (*B*) Spatial changes of *mPer2* expression in mouse SCN. Representative photomicrographs showing digoxigenin in situ hybridization signals of *mPer2* at CT4, CT12, and CT16 from the rostral to caudal end of the SCN. At CT4, *mPer2* expression level is moderate in the rostral half and strong in the caudal half of the SCN. At CT12, *mPer2* expression is strong in the rostral half and moderate in the caudal half. At CT16, *mPer2* is moderate in the rostral half but very low in the caudal half of the SCN. This indicates that the SCN oscillators are spatially and temporally organized.

ics of SCN oscillations, we performed bioluminescence imaging of coronal SCN slices cultured from PER2::LUC mice, in which the endogenous *Per2* gene is replaced by a fusion of *Per2* with the firefly *luciferase* reporter (Yoo et al. 2004; Liu et al. 2007). This permitted real-time longitudinal monitoring of rhythmic expression of the PER2::LUC fusion protein, with high spatial and temporal resolution (~1 µm, 30 minutes), in various subregions of the SCN. Our most rostral or caudal slices lacked an obvious core in the bioluminescence images ($n = 2$). In these "core-less" slices, PER2::LUC expression was rhythmic but with low amplitude, and the relative phasing of expression among SCN regions was not consistent over time.

In the coreless slice shown in Figure 7A, bioluminescence is initially elevated in the dorsal, central, and dorsomedial SCN compared to other regions. Bioluminescence then expands to the whole SCN slice before regressing back to the dorsal, central, and dorsomedial regions. In the other core-less slice (not shown), at the trough time, there is more bioluminescence in the center of the SCN than elsewhere; this increases and expands to the whole SCN slice and then regresses back to the center. These data are in accordance with the *Per1-luc* data of Yamaguchi et al. (2003), who found that dorsal cells were no longer synchronous when isolated from the ventral core SCN by a knife cut.

In our other SCN slices, which contained both core and shell ($n = 3$), we observed strong PER2::LUC rhythms in all subregions, with relative phases that remained consistent over several circadian cycles and across all slices (Fig. 7B). Previous studies have also found differential phasing of cells in SCN slices. Quintero et al. (2003), using a *Per1-GFP* reporter in acute slices, found that lateral cells tended to peak earlier than medial cells. Yamaguchi et al. (2003), using a *Per1-luc* reporter in cultured slices, found a "dorsomedial-to-ventral spread" of activation. Careful examination of our high-resolution data, using a PER2::LUC reporter in cultured slices, revealed a more complex pattern. In each circadian cycle, PER2::LUC bioluminescence appears first in a small dorsomedial region (dark blue) near the third ventricle and then in lateral and ventromedial SCN (light blue, red). These three regions of early activation delineate the most peripheral aspects of the SCN shell, the "outer shell." This is followed by a modest but widespread activation throughout the SCN (with greater intensity in the shell than in the core) and, finally, a stronger activation of the central "inner shell," or "cap" (purple), just dorsal to the core. Overall, the weakest activation is in the core (yellow). While activation is peaking in the cap (purple), other regions are gradually declining in brightness. In contrast to this complex pattern of activation, the deactivation phase is less complex and appears as a contraction toward the central cap, which is the last area showing bioluminescence. However, we also observed a striking relationship between the time of peak activation of these

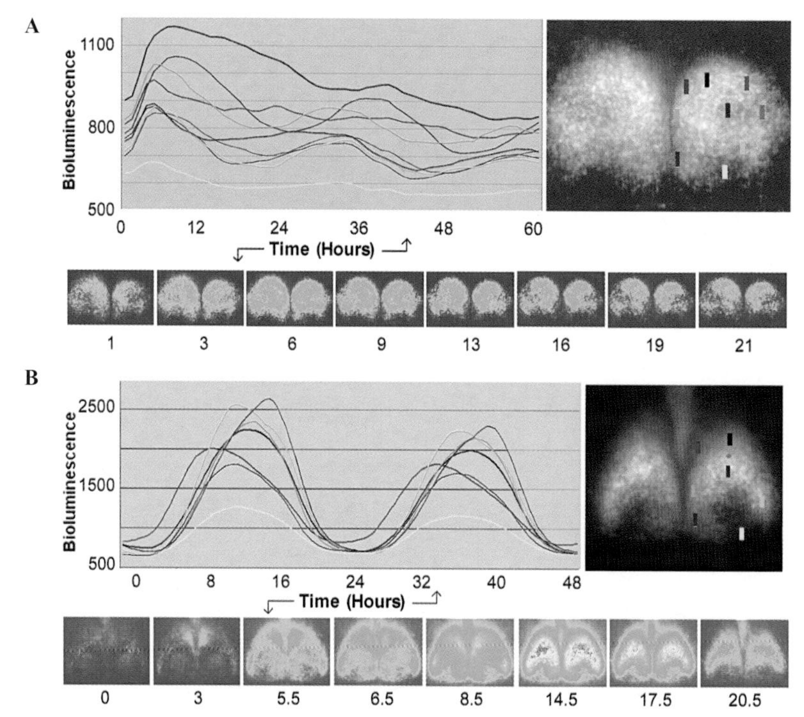

Figure 7. Spatiotemporal patterns of PER2::LUC bioluminescence in cultured SCN slices of wild-type mice. Monochrome images on the right show bioluminescence for SCN slices integrated over 24 hours. Time courses of bioluminescence are shown in *pseudocolored* images as well as in line plots for small SCN regions shown in matching *colored rectangles* (each 15 x 50 μm). (*Top panels, A*) Slice in the rostral SCN lacking a core; (*bottom panels, B*) slice containing both core and shell. The delineation of core and shell is based on the shape of SCN and optic tract, as well as minimal PER2::LUC expression in the core.

regions and their rate of deactivation: PER2::LUC expression declines most gradually in the early-peaking outer shell and most steeply in the late-peaking cap. In summary, there appears to be a complex wave of PER2::LUC activation in a series of steps emanating from the most dorsomedial SCN, spreading ventrally in the outer shell, and then centrally to the inner shell (cap); deactivation is simpler and entails a contraction towards the central cap.

Longitudinal in vitro studies using luminescent reporters thus nicely complement our extensive cross-sectional in vivo studies. The convergence of these studies indicates that despite its relatively weak circadian oscillations, the core region is important for maintaining an organized pattern of synchronized oscillations in SCN tissue, which has complex spatiotemporal dynamics.

Space and Time in SCN Circuits: Tides and Waves

The emergent property of coherent oscillation in SCN tissue is not reducible to its cellular elements alone. Mathematical models of coupling in multioscillator systems provide the basis for understanding coherent oscillation of SCN cellular oscillators (Kuramoto 1975) but lack a spatial dimension. The experimental evidence points to complex highly reproducible spatiotemporal patterns of SCN activity on a circadian basis. The gate oscillator model treats this problem at a high level of aggregation, distinguishing only the two main functional SCN compartments (Antle et al. 2003, 2007). The phase-resetting function used to represent the interaction of the gate and the oscillators summarizes what is a complex system of cell-level interactions. In particular, the phase-resetting function of the gating model cannot represent the temporal changes in individual oscillator states with respect to their connections in space, i.e., the spatiotemporal dimension of the SCN between core and shell and within the shell. Anatomically detailed studies of the SCN make it obvious that there is an orderly spatial pattern of activity over the circadian cycle. Depending on which data one examines, it appears that the spatial change can be described as either a wave (Figs. 6 and 7) (Yan and Okamura 2002) or a tide (Figs. 7 and 8) (Hamada et al. 2004).

Waves and tides are significantly different. A wave begins with the activation of one region of oscillator cells, which continue their cycles as the adjacent region activates, continuing progressively to the most distal region. In a wave, cells in the first activated region are the first to deactivate, and all regions remain active for the same proportion of the cycle. In contrast, in a tide, the population of cells in the first activated region are the last to deactivate, so that this area is active for a longer proportion of the cycle than are other areas. Waves and tides imply different underlying mechanisms of interoscillator interaction and require different mathematical models. The distinction between waves and tides is important because it reveals the spatiotemporal organization of the SCN oscillation, shows how its functional connectivity arises from the SCN's circuitry, and imposes constraints on the analysis of SCN data derived from slices and single cells.

One way to visualize and model the spatial changes in

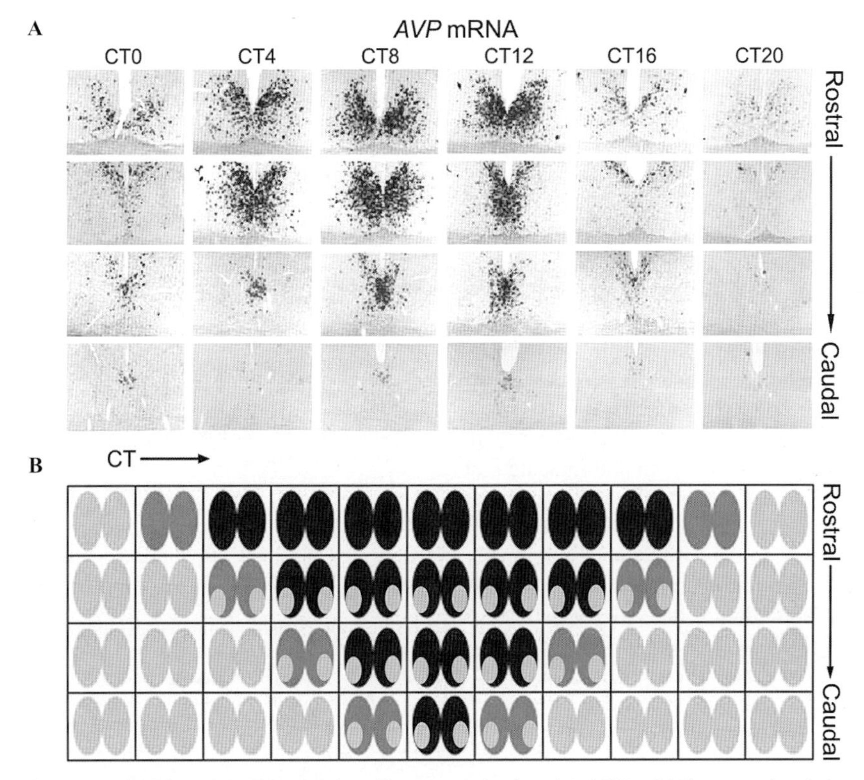

Figure 8. In vivo evidence and tidal model of the SCN oscillator organization. (*A*) AVP mRNA expression in hamster SCN detected by digoxigenin in situ hybridization. (*B*) Model showing expanding tide of a signal that lasts longer in the dorsal aspect than in the ventral aspect of the SCN. (*A*, Modified, with permission, from Hamada et al. 2004 [Wiley-Blackwell].)

the SCN is in the expression of AVP mRNA. The data shown in Figure 8A indicate that expression starts in one region of the SCN tissue, which activates neighboring cells, and these in turn activate their neighbors. But the region of initial activity does not turn off, as one would expect if those cells continued to advance their phase in a regular fashion, but remains active while the tide of activity spreads to the whole SCN. In fact, it is the last activated regions of the SCN that are the first to turn off, so that the region of activity retreats (like a tide) back to the initially active region, which is the last to turn off. A similar tide of expression occurs in PER::LUC expression in the SCN slice and is most obvious in the regression of bioluminescence to the "cap" (Fig. 7B). A testable implication of this pattern is that the population of cells in the initiating region remain longer in the expressing phase of their cycle than do the last cells to be activated. These observations suggest a simple, conceptual model of the circadian oscillation of the SCN, based on the tidal rather than the wave approach. The gate (triggered either by light or by a feedback signal) sends a (excitatory or inhibitory) signal to the oscillator cells. In the case of an excitatory signal, the gate triggers the first group of oscillator cells, which stimulates their expression. The first group of oscillator cells stimulates the next to express AVP mRNA, and so on, continuing out to the furthest SCN region. But the expression of mRNA in the second group of cells appears to slow down the oscillation of the first, so that the first continues to express AVP mRNA until the second group stops expressing AVP mRNA. The last group of cells have no neighbors to activate, nor to slow down their expression, so they turn off first, allowing the next-to-last group of cells to turn off. In this way, the AVP mRNA expression recedes back to the start point. If the effect of a downstream activated group of cells is to slow, but not stop, oscillation in the upstream cells, over a long "day," the initial group of cells might actually start to turn off before the ebbing tide reaches them.

Although there is no direct evidence for this type of inhibition by downstream cells, an interaction of this kind could explain the observed tidal pattern of activation of the shell. Our preliminary investigation shows that it is possible to create simple "cellular automata" that oscillate over space in a tidal-like pattern similar to the SCN (Fig. 8B). These models require that the cells be oscillators, but the connectivity of the system distorts the shapes of their oscillations. The point is that the tidal mechanism needs to know whether the upstream cells are starting up" or "closing down" in order to function. Because the model represents connections as a directed graph, it is extremely flexible and scalable. It is possible to represent tissue with many connections in complex three-dimensional configurations with this approach. One prediction from this conceptual model is that bioluminescence imaging of a sagittal section that retains the connectivity of the tissue in the rostrocaudal axis will reveal a tidal pattern of cell activation, rather than wavelike patterns. Another testable prediction is that this tide might be the product of increased dispersion of phase in cells of one region compared to another; alternatively, it could be the result of some cells having a longer duration of gene expression than others. The focus of future research will be the calibration of this model to experimentally observed spatiotemporal changes in the SCN.

OVERVIEW AND CONCLUSIONS

On the basis of anatomical, functional, and modeling studies, we propose that SCN circuit organization is key to understanding oscillator synchrony and that the connectivity among the functionally different neurons accounts for the coherent circadian expression pattern of the nucleus and its role as a master pacemaker. The analyses of spatial and temporal changes in SCN point to an important role for the core SCN region and indicate that the daily cycle of SCN activity involves a signal that spreads and then retracts in a wave or tidal pattern, such that the the "on" time of various oscillators is regionally organized. We further point to an important role of cells in the core area for maintaining the orderly pattern of activity and hypothesize that changes in the phase relationships among oscillator networks underlie the ability of the SCN to measure time and to respond in a robust manner to its internal and external environment.

ACKNOWLEDGMENTS

This work was supported by National Institutes of Health grants NS37919 and MH075045 to R.S., K08 MH067657 to D.K.W., and NSF DB1 320988 to Barnard College. We are grateful to Tina Tong and David Vernon for assistance in preparation of the data analysis and manuscript.

REFERENCES

Abrahamson E.E. and Moore R.Y. 2001. Suprachiasmatic nucleus in the mouse: Retinal innervation, intrinsic organization and efferent projections. *Brain Res.* **916:** 172.

Aida R., Moriya T., Araki M., Akiyama M., Wada K., Wada E., and Shibata S. 2002. Gastrin-releasing peptide mediates photic entrainable signals to dorsal subsets of suprachiasmatic nucleus via induction of Period gene in mice. *Mol. Pharmacol.* **61:** 26.

Allen G., Rappe J., Earnest D.J., and Cassone V.M. 2001. Oscillating on borrowed time: Diffusible signals from immortalized suprachiasmatic nucleus cells regulate circadian rhythmicity in cultured fibroblasts. *J. Neurosci.* **21:** 7937.

Antle M.C. and Silver R. 2005. Orchestrating time: Arrangements of the brain circadian clock. *Trends Neurosci.* **28:** 145.

Antle M.C., Kriegsfeld L.J., and Silver R. 2005. Signaling within the master clock of the brain: Localized activation of mitogen activated protein kinase by gastrin-releasing peptide. *J. Neurosci.* **25:** 2447.

Antle M.C., Foley D.K., Foley N.C., and Silver R. 2003. Gates and oscillators: A network model of the brain clock. *J. Biol. Rhythms* **18:** 339.

Antle M.C., Foley N.C., Foley D.K., and Silver R. 2007. Gates and oscillators. II. Zeitgebers and the network model of the brain clock. *J. Biol. Rhythms* **22:** 14.

Aton S.J., Colwell C.S., Harmar A.J., Waschek J., and Herzog E.D. 2005. Vasoactive intestinal polypeptide mediates circadian rhythmicity and synchrony in mammalian clock neurons. *Nat. Neurosci.* **8:** 476.

Bernard S., Gonze D., Cajavec B., Herzel H., and Kramer A. 2007. Synchronization-induced rhythmicity of circadian oscillators in the suprachiasmatic nucleus. *PLoS Comput. Biol.* **3:** e68.

Bryant D.N., LeSauter J., Silver R., and Romero M.T. 2000. Retinal innervation of calbindin-D28K cells in the hamster suprachiasmatic nucleus: Ultrastructural characterization. *J. Biol. Rhythms* **15:** 103.

Card J.P. and Moore R.Y. 1984. The suprachiasmatic nucleus of the golden hamster: Immunohistochemical analysis of cell and fiber distribution. *Neuroscience* **13:** 415.

Cheng M.Y., Bullock C.M., Li C., Lee A.G., Bermak J.C., Belluzzi J., Weaver D.R., Leslie F.M., and Zhou Q.Y. 2002. Prokineticin 2 transmits the behavioural circadian rhythm of the suprachiasmatic nucleus. *Nature* **417:** 405.

Chklovskii D.B. and Koulakov A.A. 2004. Maps in the brain: What can we learn from them? *Annu. Rev. Neurosci.* **27:** 369.

Cloues R.K., Silver R., and Sather W.A. 2001. Electrophysiological characterization of neurons expressing Calbindin-D28k in the mouse suprachiasmatic nucleus. Program No. 180.3. *Abstract Viewer/Itinerary Planner*. San Diego: *Society for Neuroscience*, 2001 (online).

Colwell C.S. 2000. Rhythmic coupling among cells in the suprachiasmatic nucleus. *J. Neurobiol.* **43:** 379.

———. 2001. NMDA-evoked calcium transients and currents in the suprachiasmatic nucleus: Gating by the circadian system. *Eur. J. Neurosci.* **13:** 1420.

Daan S. and Berde C. 1978. Two coupled oscillators: Simulations of the circadian pacemaker in mammalian activity rhythms. *J. Theor. Biol.* **70:** 297.

de la Iglesia H.O., Cambras T., Schwartz W.J., and Diez-Noguera A. 2004. Forced desynchronization of dual circadian oscillators within the rat suprachiasmatic nucleus. *Curr. Biol.* **14:** 796.

de la Iglesia H.O., Meyer J., Carpino A., Jr., and Schwartz W.J. 2000. Antiphase oscillation of the left and right suprachiasmatic nuclei. *Science* **290:** 799.

de Vries M.J., Treep J.A., de Pauw E.S.D., and Meijer J.H. 1994. The effects of electrical stimulation of the optic nerves and anterior optic chiasm on the circadian activity rhythm of the Syrian hamster: Involvement of excitatory amino acids. *Brain Res.* **642:** 206.

Deurveilher S. and Semba K. 2005. Indirect projections from the suprachiasmatic nucleus to major arousal-promoting cell groups in rat: Implications for the circadian control of behavioural state. *Neuroscience* **130:** 165.

Ding J.M., Chen D., Weber E.T., Faiman L.E., Rea M.A., and Gillette M.U. 1994. Resetting the biological clock: Mediation of nocturnal circadian shifts by glutamate and NO. *Science* **266:** 1713.

Drouyer E., LeSauter J., and Silver R. 2005. Phenotype matters: Anatomical characteristics of gastrin-releasing peptide cells of the suprachiasmatic nucleus. Program No. 312.9. *Abstract Viewer/Itinerary Planner*. Washington, D.C.: *Society for Neuroscience*, 2005 (online).

Eskin A. 1979. Identification and physiology of circadian pacemakers. Introduction. *Fed. Proc.* **38:** 2570.

Garcia-Ojalvo J., Elowitz M.B., and Strogatz S.H. 2004. Modeling a synthetic multicellular clock: Repressilators coupled by quorum sensing. *Proc. Natl. Acad. Sci.* **101:** 10955.

Gonze D., Bernard S., Waltermann C., Kramer A., and Herzel H. 2005. Spontaneous synchronization of coupled circadian oscillators. *Biophys. J.* **89:** 120.

Gorman M.R. and Elliott J.A. 2003. Entrainment of 2 subjective nights by daily light:dark:light:dark cycles in 3 rodent species. *J. Biol. Rhythms* **18:** 502.

Green D.J. and Gillette R. 1982. Circadian rhythm of firing rate recorded from single cells in the rat suprachiasmatic brain slice. *Brain Res.* **245:** 198.

Groos G. and Hendriks J. 1982. Circadian rhythms in electrical discharge of rat suprachiasmatic neurones recorded in vitro. *Neurosci. Lett.* **34:** 283.

Hakim H., DeBernardo A.P., and Silver R. 1991. Circadian

locomotor rhythms, but not photoperiodic responses, survive surgical isolation of the SCN in hamsters. *J. Biol. Rhythms* **6:** 97.

Hamada T., Antle M.C., and Silver R. 2004. Temporal and spatial expression patterns of canonical clock genes and clock-controlled genes in the suprachiasmatic nucleus. *Eur. J. Neurosci.* **19:** 1741.

Hamada T., LeSauter J., Venuti J.M., and Silver R. 2001. Expression of Period genes: Rhythmic and nonrhythmic compartments of the suprachiasmatic nucleus pacemaker. *J. Neurosci.* **21:** 7742.

Hamada T., LeSauter J., Lokshin M., Romero M.T., Yan L., Venuti J.M., and Silver R. 2003. Calbindin influences response to photic input in suprachiasmatic nucleus. *J. Neurosci.* **23:** 8820.

Hartwich M., Kalsbeek A., Pevet P., and Nurnberger F. 1994. Effects of illumination and enucleation on substance-P-immunoreactive structures in subcortical visual centers of golden hamster and Wistar rat. *Cell Tissue Res.* **277:** 351.

Hastings M.H. and Herzog E. D. 2004. Clock genes, oscillators, and cellular networks in the suprachiasmatic nuclei. *J. Biol. Rhythms* **19:** 400.

Hazlerigg D.G., Ebling F.J., and Johnston J.D. 2005. Photoperiod differentially regulates gene expression rhythms in the rostral and caudal SCN. *Curr. Biol.* **15:** R449.

Hofman M.A. and Swaab D.F. 1989. The sexually dimorphic nucleus of the preoptic area in the human brain: A comparative morphometric study. *J. Anat.* **164:** 55.

Ibata Y., Takahashi Y., Okamura H., Kawakami F., Terubayashi H., Kubo T., and Yanaihara N. 1989. Vasoactive intestinal peptide (VIP)-like immunoreactive neurons located in the rat suprachiasmatic nucleus receive a direct retinal projection. *Neurosci. Lett.* **97:** 1.

Inagaki N., Honma S., Ono D., Tanahashi Y., and Honma K. 2007. Separate oscillating cell groups in mouse suprachiasmatic nucleus couple photoperiodically to the onset and end of daily activity. *Proc. Natl. Acad. Sci.* **104:** 7664.

Inouye S.T. and Kawamura H. 1979. Persistence of circadian rhythmicity in a hypothalamic 'island' containing the suprachiasmatic nucleus. *Proc. Natl. Acad. Sci.* **76:** 5962.

Inouye S.T. and Shibata S. 1994. Neurochemical organization of circadian rhythm in the suprachiasmatic nucleus. *Neurosci. Res.* **20:** 109.

Jiang Z.G., Yang Y.Q., and Allen C.N. 1997. Tracer and electrical coupling of rat suprachiasmatic nucleus neurons. *Neuroscience* **77:** 1059.

Jin X., Shearman L.P., Weaver D.R., Zylka M.J., de Vries G.J., and Reppert S.M. 1999. A molecular mechanism regulating rhythmic output from the suprachiasmatic nucleus. *Cell* **96:** 57.

Jobst E.E. and Allen C.N. 2002. Calbindin neurons in the hamster suprachiasmatic nucleus do not exhibit a circadian variation in spontaneous firing rate. *Eur. J. Neurosci.* **16:** 2469.

Jobst E.E., Robinson D.W., and Allen C.N. 2004. Potential pathways for intercellular communication within the calbindin subnucleus of the hamster suprachiasmatic nucleus. *Neuroscience* **123:** 87.

Johnston J.D. 2005. Measuring seasonal time within the circadian system: Regulation of the suprachiasmatic nuclei by photoperiod. *J. Neuroendocrinol.* **17:** 459.

Karatsoreos I.N., Wang A., Sasanian J., and Silver R. 2007. A role for androgens in regulating circadian behavior and the suprachiasmatic nucleus. *Endocrinology* **148:** 5487.

Karatsoreos I.N., Yan L., LeSauter J., and Silver R. 2004. Phenotype matters: Identification of light-responsive cells in the mouse suprachiasmatic nucleus. *J. Neurosci.* **24:** 68.

Klein D.C., Moore R.Y., and Reppert S.M., Eds. 1991. *Suprachiasmatic nucleus. The mind's clock.* Oxford University Press, New York.

Kramer A., Yang F.C., Snodgrass P., Li X., Scammell T.E., Davis F.C., and Weitz C. J. 2001. Regulation of daily locomotor activity and sleep by hypothalamic EGF receptor signaling. *Science* **294:** 2511.

Kraves S. and Weitz C.J. 2006. A role for cardiotrophin-like cytokine in the circadian control of mammalian locomotor activity. *Nat. Neurosci.* **9:** 212.

Kriegsfeld L.J., LeSauter J., and Silver R. 2004a. Targeted microlesions reveal novel organization of the hamster suprachiasmatic nucleus. *J. Neurosci.* **24:** 2449.

Kriegsfeld L.J., Leak R.K., Yackulic C.B., LeSauter J., and Silver R. 2004b. Organization of suprachiasmatic nucleus projections in Syrian hamsters (*Mesocricetus auratus*): An anterograde and retrograde analysis. *J. Comp. Neurol.* **468:** 361.

Kuhlman S.J. and McMahon D.G. 2004. Rhythmic regulation of membrane potential and potassium current persists in SCN neurons in the absence of environmental input. *Eur. J. Neurosci.* **20:** 1113.

Kunz H. and Achermann P. 2003. Simulation of circadian rhythm generation in the suprachiasmatic nucleus with locally coupled self-sustained oscillators. *J. Theor. Biol.* **224:** 63.

Kuramoto A. 1975. Self-entrainment of a population of coupled nonlinear oscillators. *Lect. Notes Phys.* **39:** 420.

Larsen P.J. and Mikkelsen J.D. 1993. The suprachiasmatic nucleus of the mink (*Mustela vison*): Apparent absence of vasopressin-immunoreactive neurons. *Cell Tissue Res.* **273:** 239.

Leak R.K. and Moore R.Y. 2001. Topographic organization of suprachiasmatic nucleus projection neurons. *J. Comp. Neurol.* **433:** 312.

Leak R.K., Card J.P., and Moore R.Y. 1999. Suprachiasmatic pacemaker organization analyzed by viral transynaptic transport. *Brain Res.* **819:** 23.

Lee H.S., Nelms J.L., Nguyen M., Silver R., and Lehman M.N. 2003. The eye is necessary for a circadian rhythm in the suprachiasmatic nucleus. *Nat. Neurosci.* **6:** 111.

Lehman M.N., Silver R., Gladstone W.R., Kahn R.M., Gibson M., and Bittman E.L. 1987. Circadian rhythmicity restored by neural transplant. Immunocytochemical characterization of the graft and its integration with the host brain. *J. Neurosci.* **7:** 1626.

LeSauter J. and Silver R. 1998. Output signals of the SCN. *Chronobiol. Int.* **15:** 535.

———. 1999. Localization of a suprachiasmatic nucleus subregion regulating locomotor rhythmicity. *J. Neurosci.* **19:** 5574.

Liu A.C., Welsh D.K., Ko C.H., Tran H.G., Zhang E.E., Priest A.A., Buhr E.D., Singer O., Meeker K., Verma I.M., Doyle F.J., III, Takahashi J.S., and Kay S.A. 2007. Intercellular coupling confers robustness against mutations in the SCN circadian clock network. *Cell* **129:** 605.

Liu C., Weaver D.R., Strogatz S.H., and Reppert S.M. 1997. Cellular construction of a circadian clock: Period determination in the suprachiasmatic nucleus. *Cell* **91:** 855.

Long M.A., Jutras M.J., Connors B.W., and Burwell R.D. 2005. Electrical synapses coordinate activity in the suprachiasmatic nucleus. *Nat. Neurosci.* **8:** 61.

Maywood E.S., Reddy A.B., Wong G.K., O'Neill J.S., O'Brien J.A., McMahon D.G., Harmar A.J., Okamura H., and Hastings M.H. 2006. Synchronization and maintenance of timekeeping in suprachiasmatic circadian clock cells by neuropeptidergic signaling. *Curr. Biol.* **16:** 599.

McArthur A.J., Coogan A.N., Ajpru S., Sugden D., Biello S.M., and Piggins H.D. 2000. Gastrin-releasing peptide phase-shifts suprachiasmatic nuclei neuronal rhythms in vitro. *J. Neurosci.* **20:** 5496.

Menet J., Vuillez P., and Pevet P. 2003. Calbindin expression in the hamster suprachiasmatic nucleus depends on day-length. *Neuroscience* **122:** 591.

Meyer-Bernstein E.L., Jetton A.E., Matsumoto S.I., Markuns J.F., Lehman M.N., and Bittman E.L. 1999. Effects of suprachiasmatic transplants on circadian rhythms of neuroendocrine function in golden hamsters. *Endocrinology* **140:** 207.

Miller J.D., Morin L.P., Schwartz W.J., and Moore R.Y. 1996. New insights into the mammalian circadian clock. *Sleep* **19:** 641.

Moga M.M. and Moore R.Y. 1997. Organization of neural inputs to the suprachiasmatic nucleus in the rat. *J. Comp. Neurol.* **389:** 508.

Moore R.Y. 1973. Retinohypothalamic projection in mammals:

A comparative study. *Brain Res.* **49:** 403.

———. 1983. Organization and function of a central nervous system circadian oscillator: The suprachiasmatic hypothalamic nucleus. *Fed. Proc.* **42:** 2783.

———. 1996. Entrainment pathways and the functional organization of the circadian system. *Prog. Brain Res.* **111:** 103.

Moore R.Y. and Eichler V.B. 1972. Loss of circadian adrenal corticosterone rhythm following suprachiasmatic nucleus lesion in the rat. *J. Comp. Neurol.* **146:** 1.

Morin L.P. and Allen C.N. 2006. The circadian visual system, 2005. *Brain Res. Rev.* **51:** 1.

Morin L.P., Blanchard J., and Moore R.Y. 1992. Intergeniculate leaflet and suprachiasmatic nucleus organization and connections in the golden hamster. *Vis. Neurosci.* **8:** 219.

Mosko S.S. and Moore R.Y. 1979. Neonatal suprachiasmatic nucleus lesions: Effects on the development of circadian rhythms in the rat. *Brain Res.* **164:** 17.

Mrugala M., Zlomanczuk P., Jagota A., and Schwartz W.J. 2000. Rhythmic multiunit neural activity in slices of hamster suprachiasmatic nucleus reflect prior photoperiod. *Am. J. Physiol. Regul. Integr. Comp. Physiol.* **278:** R987.

Murakami N., Takamura M., Takahashi K., Utunomiya K., Kuroda H., and Etoh T. 1991. Long-term cultured neurons from rat suprachiasmatic nucleus retain the capacity for circadian oscillation of vasopressin release. *Brain Res.* **545:** 347.

Noguchi T., Watanabe K., Ogura A., and Yamaoka S. 2004. The clock in the dorsal suprachiasmatic nucleus runs faster than that in the ventral. *Eur. J. Neurosci.* **20:** 3199.

Pennartz C.M., Hamstra R., and Geurtsen A.M. 2001. Enhanced NMDA receptor activity in retinal inputs to the rat suprachiasmatic nucleus during the subjective night. *J. Physiol.* **532:** 181.

Pennartz C.M., Bos N.P., Jeu M.T., Geurtsen A.M., Mirmiran M., Sluiter A.A., and Buijs R.M. 1998. Membrane properties and morphology of vasopressin neurons in slices of rat suprachiasmatic nucleus. *J. Neurophysiol.* **80:** 2710.

Pittendrigh C.S. 1961. Circadian rhythms and the circadian organization of living things. *Cold Spring Harbor Symp. Quant. Biol.* **25:** 159.

Pittendrigh C.S. and Bruce V.G. 1959. Daily rhythms as coupled oscillator systems and their relation to thermoperiodism and photoperiodism. In *Photoperiodism and related phenomena in plants and animals* (ed. R.B. Withrow), p. 475. AAAS, Washington, D.C.

Pittendrigh C.S. and Daan S. 1976. A functional analysis of circadian pacemakers in nocturnal rodents. *J. Comp. Neurol.* **106:** 355.

Prosser H.M., Bradley A., Chesham J.E., Ebling F.J., Hastings M.H., and Maywood E.S. 2007. Prokineticin receptor 2 (Prokr2) is essential for the regulation of circadian behavior by the suprachiasmatic nuclei. *Proc. Natl. Acad. Sci.* **104:** 648.

Quintero J.E., Kuhlman S.J., and McMahon D.G. 2003. The biological clock nucleus: A multiphasic oscillator network regulated by light. *J. Neurosci.* **23:** 8070.

Ralph M.R., Foster R.G., Davis F.C., and Menaker M. 1990. Transplanted suprachiasmatic nucleus determines circadian period. *Science* **247:** 975.

Ramón y Cajal S., Pasik P., and Pasik T. 1999. *Texture of the nervous system of man and the vertebrates*. Springer, New York.

Reppert S.M. and Weaver D.R. 2001. Molecular analysis of mammalian circadian rhythms. *Annu. Rev. Physiol.* **63:** 647.

Rosen G.J., de Vries G.J., Goldman S.L., Goldman B.D., and Forger N.G. 2007. Distribution of vasopressin in the brain of the eusocial naked mole-rat. *J. Comp. Neurol.* **500:** 1093.

Schaap J., Bos N.P., de Jeu M.T., Geurtsen A.M., Meijer J.H., and Pennartz C.M. 1999. Neurons of the rat suprachiasmatic nucleus show a circadian rhythm in membrane properties that is lost during prolonged whole-cell recording. *Brain Res.* **815:** 154.

Schwartz W.J., Coleman R.J., and Reppert S.M. 1983. A daily vasopressin rhythm in rat cerebrospinal fluid. *Brain Res.* **263:** 105.

Schwartz W.J., Carpino A., Jr., de la Iglesia H.O., Baler R., Klein D.C., Nakabeppu Y., and Aronin N. 2000. Differential regulation of fos family genes in the ventrolateral and dorsomedial subdivisions of the rat suprachiasmatic nucleus. *Neuroscience* **98:** 535.

Shibata S. and Moore R.Y. 1993. Neuropeptide Y and optic chiasm stimulation affect suprachiasmatic nucleus circadian function in vitro. *Brain Res.* **615:** 95.

Shibata S., Oomura Y., Kita H., and Hattori K. 1982. Circadian rhythmic changes in neuronal activity in the suprachiasmatic nucleus of the rat hypothalamic slice. *Brain Res.* **247:** 154.

Shibata S., Watanabe A., Hamada T., Ono M., and Watanabe S. 1994. N-methyl-D-aspartate induces phase shifts in circadian rhythm of neuronal activity of rat SCN in vitro. *Am. J. Physiol.* **267:** R360.

Shigeyoshi Y., Taguchi K., Yamamoto S., Takekida S., Yan L., Tei H., Moriya T., Shibata S., Loros J.J., Dunlap J.C., and Okamura H. 1997. Light-induced resetting of a mammalian circadian clock is associated with rapid induction of the mPer1 transcript. *Cell* **91:** 1043.

Silver R., LeSauter J., Tresco P.A., and Lehman M.N. 1996. A diffusible coupling signal from the transplanted suprachiasmatic nucleus controlling circadian locomotor rhythms. *Nature* **382:** 810.

Silver R., Sookhoo A.I., LeSauter J., Stevens P., Jansen H.T., and Lehman M.N. 1999. Multiple regulatory elements result in regional specificity in circadian rhythms of neuropeptide expression in mouse SCN. *Neuroreport* **10:** 3165.

Stephan F.K. and Zucker I. 1972. Circadian rhythms in drinking behavior and locomotor activity are eliminated by suprachiasmatic lesions. *Proc. Natl. Acad. Sci.* **54:** 1521.

Sumova A., Bendova Z., Sladek M., Kovacikova Z., and Illnerova H. 2004. Seasonal molecular timekeeping within the rat circadian clock. *Physiol. Res.* (suppl. 1) **53:** S167.

Takumi T., Taguchi K., Miyake S., Sakakida Y., Takashima N., Matsubara C., Maebayashi Y., Okumura K., Takekida S., Yamamoto S., Yagita K., Yan L., Young M.L., and Okamura H. 1998. A light independent oscillatory gene *mPer3* in mouse SCN and OVLT. *EMBO J.* **17:** 4753.

Tanaka M., Hayashi S., Tamada Y., Ikeda T., Hisa Y., Takamatsu T., and Ibata Y. 1997. Direct retinal projections to GRP neurons in the suprachiasmatic nucleus of the rat. *Neuroreport* **8:** 2187.

Tavakoli-Nezhad M. and Schwartz W.J. 2005. c-Fos expression in the brains of behaviorally "split" hamsters in constant light: Calling attention to a dorsolateral region of the suprachiasmatic nucleus and the medial division of the lateral habenula. *J. Biol. Rhythms* **20:** 419.

van den Pol A.N. 1991. Glutamate and aspartate immunoreactivity in hypothalamic presynaptic axons. *J. Neurosci.* **11:** 2087.

van den Pol A.N. and Tsujimoto K.L. 1985. Neurotransmitters of the hypothalamic suprachiasmatic nucleus: Immunocytochemical analysis of 25 neuronal antigens. *Neuroscience* **15:** 1049.

VanderLeest H.T., Houben T., Michel S., Deboer T., Albus H., Vansteensel M.J., Block G.D., and Meijer J.H. 2007. Seasonal encoding by the circadian pacemaker of the SCN. *Curr. Biol.* **17:** 468.

Watanabe K., Koibuchi N., Ohtake H., and Yamaoka S. 1993. Circadian rhythms of vasopressin release in primary cultures of rat suprachiasmatic nucleus. *Brain Res.* **624:** 115.

Watanabe T., Naito E., Nakao N., Tei H., Yoshimura T., and Ebihara S. 2007. Bimodal clock gene expression in mouse suprachiasmatic nucleus and peripheral tissues under a 7-hour light and 5-hour dark schedule. *J. Biol. Rhythms* **22:** 58.

Watts A.G., Swanson L.W., and Sanchez-Watts G. 1987. Efferent projections of the suprachiasmatic nucleus. I. Studies using anterograde transport of *Phaseolus vulgaris* leucoagglutinin in the rat. *J. Comp. Neurol.* **258:** 204.

Watts D.J. and Strogatz S.H. 1998. Collective dynamics of 'small world' networks. *Nature* **393:** 440.

Welsh D.K., Logothetis D.E., Meister M., and Reppert S.M. 1995. Individual neurons dissociated from rat suprachiasmatic nucleus express independently phased circadian firing rhythms. *Neuron* **14:** 697.

Williams R.W. and Herrup K. 1988. The control of neuron number. *Annu. Rev. Neurosci.* **11:** 423.

Yamaguchi S., Isejima H., Matsuo T., Okura R., Yagita K., Kobayashi M., and Okamura H. 2003. Synchronization of cellular clocks in the suprachiasmatic nucleus. *Science* **302:** 1408.

Yamamoto S., Shigeyoshi Y., Ishida Y., Fukuyama T., Yamaguchi S., Yagita K., Moriya T., Shibata S., Takashima N., and Okamura H. 2001. Expression of the Per1 gene in the hamster: Brain atlas and circadian characteristics in the suprachiasmatic nucleus. *J. Comp. Neurol.* **430:** 518.

Yan L. and Okamura H. 2002. Gradients in the circadian expression of Per1 and Per2 genes in the rat suprachiasmatic nucleus. *Eur. J. Neurosci.* **15:** 1153.

Yan L. and Silver R. 2002. Differential induction and localization of mPer1 and mPer2 during advancing and delaying phase shifts. *Eur. J. Neurosci.* **16:** 1531.

———. 2004. Resetting the brain clock: Time course and localization of mPER1 and mPER2 protein expression in suprachiasmatic nuclei during phase shifts. *Eur. J. Neurosci.* **19:** 1105.

Yan L., Takekida S., Shigeyoshi Y., and Okamura H. 1999. Per1 and Per2 gene expression in the rat suprachiasmatic nucleus: Circadian profile and the compartment-specific response to light. *Neuroscience* **94:** 141.

Yan L., Foley N.C., Bobula J.M., Kriegsfeld L.J., and Silver R. 2005. Two antiphase oscillations occur in each suprachiasmatic nucleus of behaviorally split hamsters. *J. Neurosci.* **25:** 9017.

Yoo S.H., Yamazaki S., Lowrey P.L., Shimomura K., Ko C.H., Buhr E.D., Siepka S.M., Hong H.K., Oh W.J., Yoo O.J., Menaker M., and Takahashi J.S. 2004. PERIOD2::LUCIFERASE real-time reporting of circadian dynamics reveals persistent circadian oscillations in mouse peripheral tissues. *Proc. Natl. Acad. Sci.* **101:** 5339.

Inducible Clocks: Living in an Unpredictable World

C.B. SAPER AND P.M. FULLER
Department of Neurology, Division of Sleep Medicine, and Program in Neuroscience, Beth Israel Deaconess Medical Center, Harvard Medical School, Boston, Massachusetts 02215

All mammals have daily cycles of behavior (e.g., wake-sleep and feeding), and physiology (e.g., hormone secretion and body temperature). These cycles are typically entrained to the external light/dark cycle, but they can be altered dramatically under conditions of restricted food availability, changes in ambient temperature, or the presence of external stimuli such as predators. During the past 30 years, one of the best studied of these responses has been the entrainment of circadian rhythms to food availability. Experiments in rats and other rodents have provided evidence for a food-entrainable oscillator (FEO) in the mammalian circadian timing system (CTS). Until recently, however, very little was understood about the locus subserving the FEO or the functional interrelationship between the FEO and the master CTS pacemaker, the suprachiasmatic nucleus (SCN). We discuss here new data on the location of the FEO and suggest that it may involve an oscillator mechanism that is "induced" by starvation and refeeding.

The first reported evidence for an FEO in the mammalian CTS can be traced to the seminal observation of Richter (1922) that rats fed one meal a day increased activity before mealtime. Work during the past 40 years has extended Richter's initial observation by demonstrating that animals with restricted access to food exhibit marked and coordinated increases not only in locomotor activity, but in a host of other physiological and endocrinological variables, including body temperature, wakefulness, duodenal disaccharases, plasma ACTH/corticosteroids, ketones, and free fatty acids in anticipation of feeding time, i.e., preprandial elevations (Bolles and Stokes 1965; Krieger 1974; Nelson et al. 1975; Escobar et al. 1998; Stephan 2001; Gooley et al. 2006). The ability of limited food access to modulate potently the behavior and physiology of animals is perhaps not surprising when one considers the adaptive advantage of being able to predict the timing of food availability, i.e., optimal foraging. Some progress has been made in recent years in an effort to delineate (1) the locus of FEO, (2) afferent pathways mediating food entrainment, and (3) the molecular underpinnings of the FEO, but major gaps remain in our knowledge. We review here the earlier literature, in which investigators attempted to identify a site for the FEO mainly with lesions and which pinpoint the dorsomedial nucleus of the hypothalamus (DMH) as having a critical role. We then discuss the recent evidence that the DMH has a central role not only in a restricted feeding schedule (RFS), but also in light-entrained circadian rhythms, and thus is well placed to override the SCN signal, as its downstream effector site. We then review recent developments that have demonstrated that the DMH does not have appreciable cycling of clock genes when animals are fed ad libitum, but that the canonical clock genes are induced in the DMH, and continue to cycle, under conditions of an RFS.

Both light- and food-entrainable circadian rhythms share properties characteristic of oscillators, including limits of entrainment (*T* ~23–27 hours), free-running rhythms when isolated from environmental time cues, transients following phase-shifts, and phase-control, i.e., phase stability following removal of a periodic forcing signal (Bolles and Stokes 1965; Stephan et al. 1979; Boulos et al. 1980; Stephan 1986, 2001; Mistlberger 1994). Despite sharing similar properties, however, food- and light-entrainable circadian rhythms are not thought to be subserved by a common neural oscillator. For example, neurons of the SCN of the hypothalamus act as a light-entrainable oscillator, receiving synchronizing light input from the external light-dark cycle via the retinohypothalamic tract (RHT) and conveying, through both neural and humoral outputs, temporal information to other organ systems (for review, see Van Esseveldt et al. 2000). Lesions of the SCN disrupt virtually all neurobehavioral and physiological rhythms, a notable exception being the feeding rhythm (Moore and Eichler 1972; Krieger et al. 1977; Stephan et al. 1979; Stephan 1981; Moore-Ede et al. 1982; Edgar et al. 1993). Separate food- and light-based oscillators are also indicated by that have demonstrated that the rhythms of SCN clock gene expression and electrical activity remain entrained to the light-dark cycle during restricted feeding (Kalsbeek et al. 1998; Damiola et al. 2000; Hara et al. 2001; Stokkan et al. 2001; Wakamatsu et al. 2001; Schibler et al. 2003). Thus, available data support the concept that two separate oscillators, a light-entrainable oscillator in the SCN and a food-entrainable oscillator with an unknown neuroanatomical basis, subserve the CTS.

THE LOCUS OF THE FEO

An intensive experimental search for the neural/peripheral basis of the FEO began nearly 20 years ago. Remarkably, lesions of hypothalamic nuclei (anterior, suprachiasmatic, preoptic, periventricular, subparaventricular, paraventricular, lateral), extrahypothalamic locations (neocortex, accumbens, hippocampus, amygdala, area postrema), and the hypothalamic-pituitary-adrenal

(HPA) axis (median eminence, hypophysis, adrenal gland) have failed to block entrainment to temporally restricted windows of food availability (Krieger et al. 1977; Krieger 1980; Inouye 1982; Mistlberger and Rechtschaffen 1984; Davidson and Stephan 1999; Davidson et al. 2000; Mistlberger and Mumby 1992; Mistlberger and Rusak 1998; Stephan 2001; Mistlberger et al. 2003). In contrast, lesions including the ventromedial nucleus of the hypothalamus (VMH), the parabrachial nucleus (PBN), and most recently, the DMH, have been reported to produce long-lasting deficits in food entrainment (Inouye 1982; Mistlberger and Rechtschaffen 1984; Davidson et al. 2000; Gooley et al. 2006; Landry et al. 2006). We focus primarily on a more detailed analysis of the different lesion experiments that have claimed to find impairment of food entrainment.

In the first of the lesion experiments yielding "positive" results, Inouye (1982) found that lesions of the VMH impaired food entrainment. However, later work by Mistlberger and Rechtschaffen (1984) showed that rats sustaining large electrolytic lesions of the VMH recovered the ability to entrain by 14 21 weeks postlesion. More recently, Landry et al. (2006) placed large lesions in the DMH (which also involved the VMH to varying degrees) and found that the animals continued to exhibit anticipatory activity (in the form of food bin approaches, see below) under restricted feeding conditions, suggesting that like the VMH, the DMH is *not* the site of an FEO. These findings by Landry et al. must, however, be compared with the recent study by Gooley et al. (2006), who performed cell-specific lesions of the DMH and correlated the percentage of cell loss in the DMH and VMH with the amplitude of the anticipatory increases in wakefulness, body temperature, and locomotor activity. Whereas all of the measured response variables, i.e., waking, body temperature, and activity, increased sharply 3–4 hours before mealtime in nonlesioned animals (and persisted for two circadian cycles under fasted conditions), cell-specific lesions of the DMH, but not VMH, blocked these anticipatory increases. Moreover, by including incomplete and partial lesions in their analysis, Gooley et al. (2006) were able to demonstrate that the loss of entrainment to the food stimulus was directly proportional to the loss of DMH neurons. In the same experiment, VMH lesion controls were done and the correlations analysis revealed little if any dependent effect of food entrainment on surviving VMH neurons.

The disparate findings between the two investigator groups highlight the importance of incorporating into the analysis the effects of lesions on neighboring cell groups and their connections. Studies using electrolytic lesions are further confounded by tissue distortion, which makes it very difficult to do cell counts on the remaining neurons, and of course, the degree of damage to pathways connecting them to other structures cannot be assessed. Hence, the use of rigorous cell counting and multivariate statistics for analyzing cell-specific lesions adds a level of accuracy that cannot be achieved with electrolytic lesions. For example, as Landry et al. (2006) suggest, the small effects of lesions of the paraventricular hypothalamus or lateral hypothalamus on entrainment of food bin approaches, but not general locomotor activity, may be explained by inadvertent damage to adjacent structures, including the DMH.

Another difference in these studies that requires careful analysis is the method of assessing circadian output. In the Gooley et al. (2006) study, this was done by measuring body temperature and general locomotor activity, both recorded by implanted telemetric devices. These outputs are unrelated, in general, to the animal's food acquisition behavior and hence reasonable measures of overall circadian rhythms in a wide range of biological variables. In the Landry et al. (2006) study, the circadian behavior of the animals was measured in two ways: by nose pokes into the food bin and by breaks in a beam across the exit from the animal's tunnel-like housing at one end of the cage, between the housing and the food dish. The number of nose pokes during the light period was so small that in the end, this measure was not used (i.e., the animals determined whether there was food in the bin without performing nose pokes). Hence, the observation depended on the number of times the animal moved out of its tunnel into the cage. Importantly, then, movement out of the tunnel becomes part of the food approach behavior (animals cannot get to the food dish to inspect it without breaking the light beam). Animals under restricted feeding conditions are naturally quite hungry following a 16-hour fast, and it is to be expected that they will sleep less during that light cycle and that when they wake up, they will look to the food dish to see if there is food. This is not necessarily food entrainment of circadian rhythms, as the animal has a cue (its hunger and light if the feeding conditions are done under a light-dark cycle) and should have intact memory of where and when it last found food. In this paradigm, to avoid the confound of the animal simply looking for food when it is hungry, where and when it remembers the food to have been, it would be more accurate to measure an unrelated circadian output to determine the effects of the manipulation on circadian rhythms.

A counterargument might be that the animals continued to show anticipatory food approach behavior during a 72-hour fast following food restriction, suggesting an intact oscillator system. However, the degree of this anticipation was modest compared to the time the food was actually presented, and the animals would also still have intact peripheral biological clocks, such as ghrelin-releasing cells in the stomach, which might entrain to the new feeding cycle. Thus, again, use of a food-related output should be avoided when measuring the response of circadian rhythms to a food-restriction stimulus.

Similarly, for a small mammal to eat, it must first find food. Hence, foraging, to a rat, is the first component of feeding behavior. Other behaviors related to foraging, such as running (e.g., in a running wheel), are intrinsically related to food seeking. Although the original description of food-anticipatory activity in rats by Richter (1922) was measured with wheel running, we now know that circadian influences pervade all aspects of physiology and behavior. Hence, we would argue that food bin approaches, wheel running, and any other behavior that might be related to food seeking in rats should be avoided as a confounded measure in future work on food entrainment of circadian rhythms.

A second set of experiments that should be reviewed are those by Davidson et al. (2000), who studied the effects of both cell-specific ibotenic acid lesions and electrolytic lesions of the PBN on food entrainment. They reported that there was loss of both food bin approaches and an increase in body temperature in anticipation of restricted feeding. However, the body temperature effects were only measured in animals with electrolytic lesions. A much smaller study, using ibotenic acid lesions by Gooley et al. (2006), was unable to replicate the effects of parabrachial lesions on entrainment of either body temperature or general activity levels during an RFS. It is possible that the latter lesions may have been too small (although some of them were quite extensive), but in the absence of careful analysis of which parabrachial subnuclei, and which surrounding structures were involved in each animal, it is difficult to compare the results. In addition, it is possible, for the same reasons as the Landry et al. (2006) study reviewed above, that the use of food bin approaches as a measure of circadian rhythm of behavior may represent a confound in the Davidson et al. (2000) ibotenic acid lesion experiments. The electrolytic lesion experiments, in contrast, may involve body temperature rhythms by damaging an ascending or descending pathway (see last section).

To summarize, the current data based on analysis of cell-specific lesions, in which the actual locus of the cell injury can be assessed, provide strong support for the DMH having a critical role in entraining a wide range of basic behaviors to food restriction. DMH lesions by Gooley et al. (2006) demonstrated that the loss of a *specific* structure completely eliminates preprandial rises in all (measured) physiological and behavioral variables. At a minimum, these data firmly establish a critical role for the DMH in the expression of multiple SCN-generated rhythms in behavior and physiology. Taken together and coupled with the fact that the DMH receives input from systems that monitor food intake, energy balance, and food availability (Thompson and Swanson 1998; Elmquist et al. 1999), this makes the DMH, from a neuroanatomical perspective, uniquely poised to act as a hypothalamic "integrator" for circadian rhythms (Fig. 1) (Saper et al. 2005). The evidence for the role of the PBN, and other possible sources such as the nucleus of the solitary tract, requires further assessment. However, such experiments should use measures of circadian physiology or behavior that avoid contamination by food seeking.

INTERACTION OF THE FEO AND SCN: THE CIRCADIAN INTEGRATOR HYPOTHESIS

If the FEO can override the SCN in shaping many circadian rhythms, it must be able to interact with the SCN output pathway. For this reason, it is important to review recent advances that have elucidated how SCN signals regulate many circadian rhythms.

The outputs from the SCN to the cell groups in the brain that are concerned with regulating wake-sleep, feeding, and thermoregulation (e.g., the ventrolateral preoptic

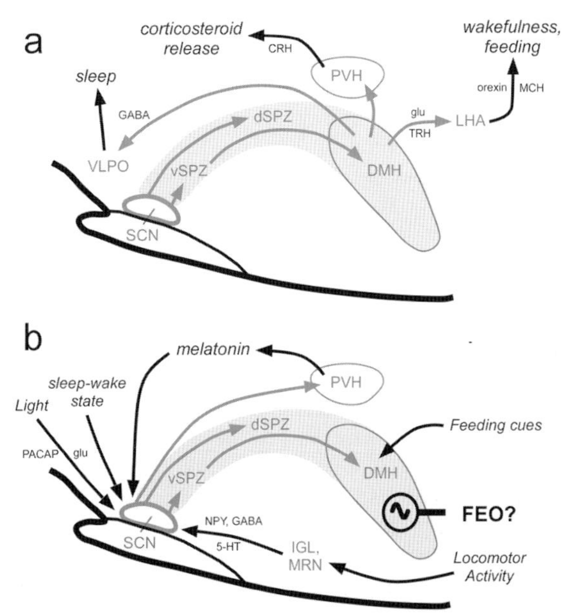

Figure 1. Hypothalamic circadian integrator model. (*a*) The master circadian clock in the SCN sends an indirect projection to the DMH via the SPZ that is critical for the circadian rhythm of sleep-wake, feeding, locomotion, and corticosteroid secretion. (*b*) This multistage regulation of circadian behavior in the hypothalamus allows for the integration of multiple time cues from the environment to shape daily patterns of physiology and behavior. Current evidence suggests the presence of a food-entrainable oscillator (FEO) in the DMH. (5-HT) 5-Hydroxytryptamine (serotonin); (CRH) corticotrophin-releasing hormone; (GABA) γ-amino butyric acid; (glu) glutamate; (DMH) dorsomedial hypothalamic nucleus; (dSPZ) dorsal subparaventricular zone; (IGL) intergeniculate leaftlet; (LHA) lateral hypothalamic area; (MRN) median raphe nucleus; (MCH) melanin-concentrating hormone; (NPY) neuropeptide Y; (PACAP) pituitary adenylate cyclase-activating polypeptide; (PVH) paraventricular hypothalamic nucleus; (SCN) suprachiasmatic nucleus; (TRH) thyrotropin-releasing hormone; (VLPO) ventrolateral preoptic nucleus; (vSPZ) ventral subparaventricular zone. (Adapted, with permission, from Fuller et al. 2006 [© SAGE Publications].)

nucleus, the lateral hypothalamic orexin neurons, or the medial preoptic area) are surprisingly meager, suggesting that the clock input to these vital functions must be relayed by some intermediate structure(s). The major output pathway from the SCN leaves the nucleus along its dorsal border and arches back through the medial hypothalamus, passing just beneath the paraventricular nucleus and terminating in the DMH (Watts et al. 1987). There are numerous synaptic boutons all along this pathway, suggesting that it may play an important part as a relay.

The region above the SCN and below the paraventricular nucleus that receives these outputs was named by Watts and Swanson (1987) the subparaventricular zone (SPZ). Lu et al. (2002) placed cell-specific cytotoxic lesions along this pathway to determine their effects on circadian rhythms. They found that lesions of the ventral SPZ (just above the SCN) caused loss of circadian rhythms of wake-sleep and locomotor activity but had minimal impact on body temperature rhythms. Conversely, lesions of the dorsal SPZ (just below the paraventricular nucleus) impaired circadian rhythms of body temperature but not wake-sleep or locomotion. In each case, the loss of circadian rhythms showed a strong and statistically significant correlation with cell loss in the corresponding area. This double dissociation means that neurons in the ventral SPZ must be critical for relaying SCN output necessary for circadian regulation of locomotor activity and wake-sleep cycles, whereas neurons in the dorsal SPZ must relay information necessary for body temperature rhythms.

Because wake-sleep, locomotor, and body temperature rhythms are usually viewed as intertwined (sleep occurs during the falling phase of body temperature, and locomotor activity generates body heat), this result was quite surprising. However, examining the data carefully, this discrepancy can be explained by the emergence of robust ultradian rhythms (with a 3–4-hr period) in animals with impairment of circadian rhythms. In other words, sleep still occurred predominantly in the falling phase of ultradian temperature variations in animals with ventral SPZ lesions, but this was now distributed across the circadian day, instead of being consolidated in the light period.

The SPZ, however, like the SCN, has relatively little output to the hypothalamic cell groups that regulate wake-sleep, feeding, or thermoregulation. By placing injections of an anterograde tracer into the ventral SPZ, Chou et al. (2003) were able to show that it projects heavily to the DMH. Injections of retrograde tracer into the DMH showed that the ventral SPZ contains about two to three times as many retrogradely labeled neurons as does the SCN itself. Thus, the ventral SPZ sits above the SCN and acts as an amplifier, reinforcing the SCN output to the DMH.

Chou et al. (2003) assessed the role of the DMH in shaping circadian rhythms by placing cell-specific cytotoxic lesions in the DMH. They found that the lesions caused the loss of circadian rhythms of sleep-wake, locomotion, feeding, and corticosteroid secretion. Interestingly, the levels of wake time, locomotor activity, and corticosteroid secretion were reduced overall. The levels of locomotor activity and corticosteroid secretion stayed at all times at the lowest level of the usual circadian cycle, suggesting that the role of the DMH is mainly to augment the arousal, locomotor, and corticosteroid secretion drives during the on period. The level of secretion of melatonin, in contrast, remained tightly fixed to the light-dark cycle and did not change at all. This observation correlates with the observation that melatonin secretion is driven by direct inputs from the SCN to the paraventricular nucleus, which drives sympathetic input to the pineal gland, whereas the SCN has minimal projections to the output areas controlling wake-sleep, thermoregulation, or corticosteroid secretion.

The DMH, in contrast, was found to have abundant projections to the ventrolateral preoptic nucleus, the lateral hypothalamus, the paraventricular nucleus, and the medial preoptic area (Thompson et al. 1996; Chou et al. 2003). The projection to the ventrolateral preoptic nucleus was predominantly GABAergic and that to the lateral hypothalamus predominantly contained glutamate or thyrotropin-releasing hormone (both excitatory) (Chou et al. 2003). Thus, the DMH is well positioned to be the final common output site for regulating many circadian rhythms, receiving the bulk of the outflow of the SCN and SPZ and supplying abundant input to the target areas necessary to control circadian rhythms.

AFFERENT AND EFFERENT PATHWAYS OF THE FEO

Beyond the anatomical locus of the FEO, at present, very little is understood regarding the afferent and efferent pathways mediating food entrainment. For the former, both neuronal and humoral inputs have been evaluated, with no candidate pathway yet identified. For example, food anticipatory responses are still intact following subdiaphragmatic vagotomy (to remove vagal inputs) and capsaicin-induced visceral deafferentation (to eliminate nonvagal visceral afferents) (Moreira and Krieger 1982; Comperatore and Stephan 1990; Davidson and Stephan 1998). Additional studies have demonstrated that olfactory (Coleman and Hay 1990; Davidson et al. 2001) and gustatory cues (Stephan and Davidson 1998) are not necessary for entrainment to restricted feeding. Collectively, these observations suggest humoral signals are communicating gut information (presumably to reflect feed intake), although as yet, no candidate factor has been identified. Both leptin and ghrelin, which signal energy status, and the major hypophyseal hormones, e.g., GH and TSH, have been ruled out as the sole signal (Davidson and Stephan 1999; Mistlberger and Marchantt 1999; Gooley et al. 2006). One of the few functional experiments to find loss of food entrainment involved use of nonnutrative feeding (Mistlberger and Rusak 1987). Thus, animals fed a nonnutrative (but palatable) diet, even with artificial sweeteners such as saccharin, fail to entrain to the stimulus (Mistlberger and Rusak 1987; Stephan and Davidson 1998). Hence, some feature of the nutritive value of the food must be signaled to the brain, but the mechanism of action remains unknown (and may involve multiple pathways). Finally, it is also possible that the FEO may be driven by multiple pathways and that the elimination of any one of them may be insufficient to prevent food entrainment. This problem will be easier to approach, however, once the site of the FEO is established.

Because the locus of the FEO is unknown, there is a paucity of data available regarding the efferent pathways by which phase information is communicated to peripheral tissues, although there is some, albeit conflicting, evidence that the orexin/hypocretin system of the lateral hypothalamus may have a role (Akiyama et al. 2004; Mieda et al. 2004). Glucocorticoid signaling has also been suggested to have a role in the entrainment of peripheral tissue oscillators to restricted feeding, but the glucocorticoid receptor is dispensable for entrainment of liver clock genes, and repeated daily injections of corticosterone (at a time approximating the hormone peak in food-restricted animals) also do not phase-shift liver clock gene expression (Balsalobre et al. 2000; Le Minh et al. 2001). It also remains possible that "food signals" function independently (at the level of the individual tissues, perhaps) of the FEO to entrain peripheral tissues. As an example, feeding may influence gene expression more directly in peripheral organs. Because the redox state of the cell, i.e., metabolic state, is thought to strongly influence the ability of CLOCK/BMAL1 and NPAS2/BMAL1 (see below) to bind DNA E-box elements, it is possible that high/low fuel, e.g., glucose titers, influences cell circadian gene expression directly (Rutter et al. 2001). However, because the SCN is phase-refractory to RFS, the phenomenon of redox state control of gene expression must be exclusive to other tissue oscillators (Damiola et al. 2000).

THE MOLECULAR FEO

The presence and circadian expression of clock genes has been documented in many central nervous system (CNS) nuclei and extra-CNS tissues (Yamazaki et al. 2000; Abe et al. 2002; Yamamoto et al. 2004). For example, the presence and circadian expression of mBmal1, mNpas2, mRev-erbα, mDbp, mRev-erbβ, mPer3, mPer1, and mPer2 (given in the temporal order of the rhythm peak) has been demonstrated in mouse heart, lung, liver, stomach, spleen, and kidney (Yamamoto et al. 2004). Unlike the SCN, however, these peripheral oscillators are not self-sustaining, because without periodic forcing (by unknown mechanisms) by the SCN, peripheral tissue oscillators show evidence of damping of circadian gene expression (Yamazaki et al. 2000). Nevertheless, the observation of similar constituent genetic elements and the phase angle of mRNA accumulation in peripheral tissues to those of the SCN strongly suggests conservation of function for the canonical clock genes in generating circadian rhythmicity in extra-SCN tissues (Yamazaki et al. 2000; Yamamoto et al. 2004). As previously indicated, and consistent with the neurobehavioral and physiological observations from SCN-lesioned animals, i.e., survival of food-entrainable rhythms, it has been demonstrated that RFS does not alter the regulation (i.e., phasing) of clock genes in the SCN (Damiola et al. 2000). However, placing mice on an RFS does alter clock gene expression (e.g., *Per1* and *Per2*) in the liver of both intact and SCN-lesioned mice (Damiola et al. 2000; Hara et al. 2001). It has been similarly demonstrated in transgenic (*Period1*-luciferase) rats that during restricted feeding, the SCN remains phase-locked to the light-dark cycle, whereas the liver and other organs synchronize clock gene expression to the RFS (Stokkan et al. 2001). Thus, the RFS uncouples circadian oscillations in peripheral tissues from the SCN. Despite these important observations, it has yet to be determined if the RFS-induced phase-shift of peripheral clock gene rhythms requires input from an endogenous clock (FEO) or is based on food metabolites/signal(s).

Operating under the assumption of functional clock gene conservation, several investigators have used circadian clock mutants to determine if the canonical circadian clock genes govern FEO rhythm generation. As an example, it was reported recently that the circadian rhythm of food-anticipatory activity is altered in cryptochrome-deficient ($mCry1^{-/-}mCry2^{-/-}$) mice (Iijima et al. 2005). Here, it was demonstrated that mCry-deficient mice are slower to establish a stable phase relationship with the RFS and exhibit more variable phase-control (as assessed during two circadian cycles of food deprivation subsequent to entrainment to the RFS). A close examination of the data reveals, however, that the mCry proteins are ultimately not necessary for the RFS-induced anticipatory activity, and it can thus be concluded that the mCry proteins are not necessary for entrainment of the FEO. Another important study evaluated the ability of the homozygous *Clock* mutant (Clk/Clk) mouse to entrain to an RFS (Pitts et al. 2003). Here, the authors demonstrated that Clk/Clk mutants not only entrained to the RFS, but also had a somewhat enhanced response to RFS. These authors thus proposed that (1) Clk/Clk mice may have a more effective FEO and (2) the *Clock* gene is dispensable for entraining the FEO. Although these findings suggested that the FEO was not a CLOCK-based oscillator (and hence unlike the SCN), it remained possible that the core components of the FEO and SCN are, with the exception of clock, identical. This is because another heterodimeric binding partner for BMAL1 (to drive the positive arm of transcription) is expressed in extra-SCN areas. This "other" binding partner, neuronal PAS domain protein 2 (NPAS2) is a paralog of clock (Garcia et al. 2000; Reick et al. 2001). Recently, it was reported that mice lacking NPAS2 do demonstrate significant deficits in entraining to an RFS, suggesting a possible role for NPAS2 in the molecular entrainment of the FEO (Dudley et al. 2003). Importantly, however, NPAS2 (–/–) mice do eventually exhibit entrainment to an RFS. However, these animals still have an intact *Clock* gene, which may be able to compensate for NPAS2 loss by interacting with BMAL1. This observation nevertheless highlights the need to evaluate NPAS2 expression in the DMH, as it may be a key component of the molecular algorithm for circadian rhythm generation in the FEO. Finally, a very recent study has reported that *Per2* mutant mice do not show food anticipation (Feillet et al. 2006). Ultimately, understanding which clock genes are expressed in the DMH during restricted feeding will be useful in determining how feeding enables this entrainment.

Although previous studies of clock gene mutant mice, with the exception of *Per2* mutants, have failed to identify a single mutation that could eliminate FEO, this may be because all of the clock genes examined so far have active paralogs that may substitute in a clock cycle. We therefore have been evaluating the ability of BMAL1/Mop3 null clock mutants to entrain to an RFS. BMAL1 has no functional paralog, and BMAL1 (–/–) mice have not, to date, been tested for deficits in food entrainment. The BMAL1 (–/–) mutant mouse is completely arrhythmic under ad lib feeding and demonstrates significant decrements in entrainment to light/dark cycles, demonstrating the absolute requirement for BMAL1 in generating and maintaining SCN-driven circadian rhythms (Bunger et al. 2000). Recent pilot data (P.M. Fuller and C.B. Saper, unpubl. 2006) suggest that BMAL1 (–/–) mutants do not entrain to an RFS (i.e., no preprandial elevations in body temperature or generalized locomotor activity), and thus, it is likely that a BMAL1-based clock is necessary for FEO function.

IS THE DMH THE FOOD ENTRAINABLE OSCILLATOR?

The foregoing data indicate that the DMH is necessary for the SCN output to entrain circadian rhythms of wake-sleep, locomotor activity, feeding, body temperature, and corticosteroid secretion. The first section implies that the DMH is also necessary for animals to adapt to an RFS. Although these data do support a role for the DMH in food entrainment of circadian rhythms, these studies do not definitively establish the DMH as the FEO of the CTS. For example, it is possible that an upstream oscillator provides inputs to the DMH that drives food-entrainable rhythms, including DMH c-Fos expression. Conversely, it is possible that the DMH may function as a requisite relay in the afferent limb of the food entrainment pathway, providing input to another brain region (i.e., located downstream) functioning as the FEO. Thus, to firmly establish the DMH as an FEO of the CTS, we must determine if the DMH fulfills the molecular requirements of an oscillator. For example, we predict that the molecular mechanisms of the DMH (as a *possible* FEO) are functionally conserved and mirror that of the SCN transcriptional-translational clock algorithm, with the possible exception that it may be the coinduction of NPAS2 (whose expression has never been systematically evaluated in the hypothalamus) and BMAL1, not CLOCK and BMAL1, which drives *Per* and *Cry* gene transcription.

Interestingly, recent work by Mieda et al. (2006) has provided evidence of restricted-feeding-induced clock gene expression in the DMH, providing support for the concept that the DMH may contain a functional oscillator system. They reported that although clock gene expression is common in the cortex and parts of the brainstem, it is not prevalent in the hypothalamus other than in the SCN under conditions of ad libitum feeding. However, after starving rats for 20 hours and then refeeding them, they reported the induction of *Per1* and *Per2* genes in the DMH and nucleus of the solitary tract, which was not present under ad libitum feeding conditions. Interestingly, these genes reached peak expression during the window of food availability, and the neurons in the DMH continued to express *Per1* rhythmically for two circadian cycles during a fast following an RFS, demonstrating phase-control by the RFS. *Per1* expression in the nucleus of the solitary tract was present only on the days the animals were fed, indicating the lack of an underlying oscillator system and suggesting induction based on vagal gastrointestinal signals related to feeding. Despite these compelling observations, demonstrating that the DMH is the FEO would require two additional pieces of information: Under an RFS, does the DMH express a suite of clock genes that form a self-sustained molecular oscillator and is the cycling of these clock genes in the DMH necessary to maintain food entrainment? We are currently testing this hypothesis by profiling clock gene expression in the DMH under ad lib and RFS conditions. Also interesting is the observation that *Per1* and *Per2* were mainly expressed in a small subnucleus, called the pars compacta of the DMH. The pars compacta cells are small and closely packed, highly reminiscent of the SCN itself. Little is known of its connections, but experiments by Thompson et al. (1996 and pers. comm.) suggest that it mainly has local connections with the adjacent output zones of the DMH, as defined above. The hypothesis would further suggest that the outputs from the compact part of the DMH must impinge upon the same output paths in the remainder of the DMH as are used by SCN inputs. In addition, these inputs must be able to override the SCN clock input. It is unlikely that they feed back on the SCN in any major way, as the SCN continues to cycle on the light entrainment schedule for many weeks during food entrainment (as long as the diet is not also hypocaloric).

FUTURE EXPERIMENTS

The availability of genetic approaches may soon allow us to determine if the clock in the DMH is both necessary and sufficient for expression of food entrainment. The experiments of Mieda et al. (2006) suggest that the food entrainment of circadian rhythms depends on the same suite of clock genes as drive the 24-hour rhythms in the SCN and other tissues. This can be tested by using animals with mutations of specific clock genes, which should have parallel effects on the SCN light-entrainable clock as well as the putative food-entrainable clock in the DMH.

In addition, as at least two clock gene mutants appear to be incapable of entrainment to restricted feeding, it may be possible to perform a rescue experiment. Current models for this approach include conditional knockins, in which the gene of interest contains a transcriptional blocker, surrounded by *loxP* sites. These animals would have a functional deletion of the clock gene, and this could be restored by a genetic recombination event caused by Cre recombinase. Cre recombinase could in turn be introduced into the DMH either by crossing with an animal line in which Cre was expressed under a regional promoter in the DMH or by injection with a viral vector containing the gene for Cre recombinase. Such approaches should soon allow us to test the hypothesis that the DMH clock is both necessary and sufficient for food entrainment.

REFERENCES

Abe M., Herzog E.D., Yamazaki S., Straume M., Tei H., Sakaki Y., Menaker M., and Block G.D. 2002. Circadian rhythms in isolated brain regions. *J. Neurosci.* **22:** 350.

Akiyama M., Yuasa T., Hayasaka N., Horikawa K., Sakurai T., and Shibata S. 2004. Reduced food anticipatory activity in genetically orexin (hypocretin) neuron-ablated mice. *Eur. J. Neuosci.* **20:** 3054.

Balsalobre A., Brown S.A., Marcacci L., Tronche F., Kellendonk C., Reichardt H.M., Schultz G., and Schibler U. 2000. Resetting of circadian time in peripheral tissues by glucocorticoid signaling. *Science* **289:** 2344.

Bolles R.C. and Stokes L.W. 1965. Rat's anticipation of diurnal and a-diurnal feeding. *J. Comp. Physiol. Psychol.* **60:** 290.

Boulos Z., Rosenwasser A.M., and Terman M. 1980. Feeding schedules and the circadian organization of behavior in the rat. *Behav. Brain Res.* **1:** 39.

Bunger M.K., Wilsbacher L.D., Moran S.M., Clendenin C., Radcliffe L.A., Hogenesch J.B., Simon M.C., Takahashi J.S., and Bradfield C.A. 2000. Mop3 is an essential component of the master circadian pacemaker in mammals. *Cell* **103:** 1009.

Chou T.C., Scammell T.E., Gooley J.J., Gaus S.E., Saper C.B., and Lu J. 2003. Critical role of dorsomedial hypothalamic nucleus in a wide range of behavioral circadian rhythms. *J. Neurosci.* **23:** 10691.

Coleman G.J. and Hay M. 1990 Anticipatory wheel-running in behaviorally anosmic rats. *Physiol. Behav.* **47:** 1145.

Comperatore C.A. and Stephan F.K. 1990. Effects of vagotomy on entrainment of activity rhythms to food access. *Physiol. Behav.* **47:** 671.

Damiola F., Minh L.N., Preitner N., Kornmann B., Fleury-Olela F., and Schibler U. 2000. Restricted feeding uncouples circadian oscillators in peripheral tissues from the central pacemaker in the suprachiasmatic nucleus. *Genes Dev.* **14:** 2950.

Davidson A.J. and Stephan F.K. 1998. Circadian food anticipation persists in capsaicin deafferented rats. *J. Biol. Rhythms* **13:** 422.

———. 1999. Feeding-entrained circadian rhythms in hypophysectomized rats with suprachiasmatic nucleus lesions. *Am. J. Physiol.* **277:** R1376.

Davidson A.J., Cappendijk S.L., and Stephan F.K. 2000. Feeding-entrained circadian rhythms are attenuated by lesions of the parabrachial region in rats. *Am. J. Physiol.* **278:** R1296.

Davidson A.J., Aragona B.J., Werner R.M., Schroeder E., Smith J.C., and Stephan F.K. 2001. Food anticipatory activity persists after olfactory bulb ablation in the rat. *Physiol. Behav.* **72:** 231.

Dudley C.A., Erbel-Sieler C., Estill S.J., Reick M., Franken P., Pitts S., and McKnight S.L. 2003. Altered patterns of sleep and behavioral adaptability in NPAS2-deficient mice. *Science* **301:** 379.

Edgar D.M., Dement W.C., and Fuller C.A. 1993. Effect of SCN lesions on sleep in squirrel monkeys: Evidence for opponent processes in sleep-wake regulation. *J. Neurosci.* **3:** 1065.

Elmquist J.K., Elias C.F., and Saper C.B. 1999. From lesions to leptin: Hypothalamic control of food intake and body weight. *Neuron* **22:** 221.

Escobar C., Diaz-Munoz M., Encinas F., and Aguilar-Roblero R. 1998. Persistence of metabolic rhythmicity during fasting and its entrainment by restricted feeding schedules in rats. *Am. J. Physiol.* **274:** R1309.

Feillet C.A., Ripperger J.A., Magnone M.C., Dulloo A., Albrecht U., and Challet E. 2006. Lack of food anticipation in Per2 mutant mice. *Curr. Biol.* **16:** 2016.

Fuller P.M., Gooley J.J., and Saper C.B. 2006. Neurobiology of the sleep-wake cycle: Sleep architecture, circadian regulation, and regulatory feedback. *J. Biol. Rhythms* **21:** 482.

Garcia J.A., Zhang D., Estill S.J., Michnoff C., Rutter J., Reick M., Scott K., Diaz-Arrastia R., and McKnight S.L. 2000. Impaired cued and contextual memory in NPAS2-deficient mice. *Science* **288:** 2226.

Hara R., Wan K., Wakamatsu H., Aida R., Moriya T., Akiyama M., and Shibata S. 2001. Restricted feeding entrains liver clock without participation of the suprachiasmatic nucleus. *Genes Cells* **6:** 269.

Gooley J.J., Schomer A., and Saper C.B. 2006. The dorsomedial hypothalamic nucleus is critical for the expression of food-entrainable circadian rhythms. *Nat. Neurosci.* **3:** 398.

Iijima M., Yamaguchi S., van der Horst GT, Bonnefont X., Okamura H., and Shibata S. 2005. Altered food-anticipatory activity rhythm in cryptochrome-deficient mice. *Neurosci. Res.* **52:** 166.

Inouye S.T. 1982. Ventromedial hypothalamic lesions eliminate anticipatory activities of restricted daily feeding schedules in the rat. *Brain Res.* **250:** 183.

Kalsbeek A., Van Heerikhuize J.J., Wortel J., and Buijs R.M. 1998. Restricted daytime feeding modifies suprachiasmatic nucleus vasopressin release in rats. *J. Biol. Rhythms* **13:** 18.

Krieger D.T. 1974. Food and water restriction shifts corticosterone, temperature, activity and brain amine periodicity. *Endocrinology* **95:** 1195.

———. 1980. Ventromedial hypothalamic lesions abolish food-shifted circadian adrenal and temperature rhythmciity. *Endocrinology* **106:** 649.

Krieger D.T., Hauser H., and Krey L.C. 1977. Suprachiasmatic nuclear lesions do not abolish food-shifted circadian adrenal and temperature rhythmicity. *Science* **197:** 398.

Landry G.J., Simon M.M., Webb I.C., and Mistlberger R.E. 2006. Persistence of a behavioral food-anticipatory circadian rhythm following dorsomedial hypothalamic ablation in rats. *Am. J. Physiol. Integr. Comp. Physiol.* **290:** R1527.

Le Minh N., Damiola F., Tronche F., Schütz G., and Schibler U. 2001. Glucocorticoid hormones inhibit food-induced phase-shifting of peripheral circadian oscillators. *EMBO J.* **20:** 7128.

Lu J., Bjorkum A.A., Xu M., Gaus S.E., Shiromani P.J., and Saper C.B. 2002. Selective activation of the extended ventrolateral preoptic nucleus during rapid eye movement sleep. *J. Neurosci.* **22:** 4568.

Mieda M., Williams S.C., Richardson J.A., Tanaka K., and Yanagisawa M. 2006. The dorsomedial hypothalamic nucleus as a putative food-entrainable circadian pacemaker. *Proc. Natl. Acad. Sci.* **103:** 12150.

Mieda M., Williams S.C., Sinton C.M., Richardson J.A., Sakurai T., and Yanagisawa M. 2004. Orexin neurons function in an efferent pathway of a food-entrainable circadian oscillator in eliciting food-anticipatory activity and wakefulness. *J. Neurosci.* **24:** 10493.

Mistlberger R.E. 1994. Circadian food-anticipatory activity: Formal models and physiological mechanisms. *Neurosci. Biobehav. Rev.* **18:** 171.

Mistlberger R.E. and Marchantt E.G. 1999. Enhanced food-anticipatory circadian rhythms in the genetically obese Zucker rat. *Physiol. Behav.* 66: 329.

Mistlberger R.E. and Mumby D.G. 1992. The limbic system and food-anticipatory circadian rhythms in the rat: Ablation and dopamine blocking studies. *Behav. Brain Res.* **47:** 159.

Mistlberger R.E. and Rechtschaffen A. 1984. Recovery of anticipatory activity to restricted feeding in rats with ventromedial hypothalamic lesions. *Physiol. Behav.* **33:** 227.

Mistlberger R.E. and Rusak B. 1987. Palatable daily meals entrain anticipatory activity rhythms in free-feeding rats: Dependence on meal size and nutrient content. *Physiol. Behav.* **41:** 219.

———. 1998. Food-anticipatory circadian rhythms in rats with paraventricular and lateral hypothalamic ablations. *J. Biol. Rhythms* **3:** 277.

Mistlberger R.E., Antle M.C., Kilduff T.S., and Jones M. 2003. Food- and light-entrained circadian rhythms in rats with hypocretin-2-saporin ablations of the lateral hypothalamus. *Brain Res.* **980:** 161.

Moore R.Y. and Eichler V.B. 1972. Loss of circadian adrenal corticosterone rhythm following suprachiasmatic lesions in the rat. *Brain Res.* **42:** 201.

Moore-Ede M., Sulzman F.M., and Fuller C.A. 1982. *The clocks that time us: Physiology of the circadian timing system in mammals.* Harvard University Press, Cambridge, Massachusetts.

Moreira A.C. and Krieger D.T. 1982. The effects of subdiaphragmatic vagotomy on circadian corticosterone rhythmicity in rats with continuous or restricted food access. *Physiol. Behav.* **28:** 789.

Nelson W., Scheving L., and Halberg F. 1975. Circadian rhythms in mice fed a single daily meal at different stage of lighting regimen. *J. Nutr.* **105:** 171.

Pitts S., Perone E., and Silver R. 2003. Food-entrained circadian rhythms are sustained in arrhythmic Clk/Clk mutant mice. *Am. J. Physiol. Regul. Integr. Comp. Physiol.* **285:** R57.

Reick M., Garcia J.A., Dudley C., and McKnight S.L. 2001. NPAS2: An analog of clock operative in the mammalian forebrain. *Science* **293:** 506.

Richter C.P. 1922. A behavioristic study of the activity of the rat. *Comp. Psychol. Monogr.* **1:** 1.

Rutter J., Reick M., Wu L.C., and McKnight S.L. 2001. Regulation of clock and NPAS2 DNA binding by the redox state of NAD cofactors. *Science* **293:** 510.

Saper C.B., Lu J., Chou T.C., and Gooley J. 2005. The hypothalamic integrator for circadian rhythms. *Trends Neurosci.* **28:** 152.

Schibler U., Ripperger J., and Brown S.A. 2003 Peripheral circadian oscillators in mammals: Time and food. *J. Biol. Rhythms* **18:** 250.

Stephan F.K. 1981. Limits of entrainment to periodic feeding in rats with suprachiasmatic lesions. *J. Comp. Physiol. A* **143:** 401.

———. 1986. Coupling between feeding- and light-entrainable circadian pacemakers in the rat. *Physiol. Behav.* **38:** 537.

———. 2001. Food-entrainable oscillators in mammals. In *Handbook of behavioral neurobiology: Circadian clocks* (ed. J.S. Takahashi et al.), vol. 12, p. 223. Kluwer Academic/Plenum, New York, New York.

Stephan F.K. and Davidson A.J. 1998. Glucose, but not fat, phase shifts the feeding-entrained circadian clock. *Physiol. Behav.* **65:** 277.

Stephan F.K., Swann J.M., and Sisk C.L. 1979. Anticipation of 24-hr feeding schedules in rats with lesions of the suprachiasmatic nucleus. *Behav. Neural Biol.* **25:** 346.

Stokkan K.A., Yamazaki S., Tei H., Sakaki Y. and Menaker M. 2001. Entrainment of the circadian clock in the liver by feeding. *Science* **291:** 490.

Thompson R.H. and Swanson L.W. 1998. Organization of inputs to the dorsomedial nucleus of the hypothalamus: A reexamination with Fluorogold and PHAL in the rat. *Brain Res. Brain Res. Rev.* **27:** 89.

Thompson R.H., Canteras N.S., and Swanson L.W. 1996. Organization of projections from the dorsomedial nucleus of the hypothalamus: A PHA-L study in the rat. *J. Comp. Neurol.* **376:** 143.

Van Esseveldt K.E., Lehman M.A., and Boer G.J. 2000. The suprachiasmatic nucleus and the circadian time-keeping system revisited. *Brain Res. Brain Res. Rev.* **33:** 34.

Wakamatsu H., Yoshinobu Y., Aida R., Moriya T., Akiyama M., and Shibata S. 2001. Restricted-feeding-induced anticipatory activity rhythm is associated with a phase-shift of the expression of mPer1 and mPer2 mRNA in the cerebral cortex and hippocampus but not in the suprachiasmatic nucleus of mice. *Eur. J. Neurosci.* **13:** 1190.

Watts A.G. and Swanson L.W. 1987. Efferent projections of the suprachiasmatic nucleus. II. Studies using retrograde transport of fluorescent dyes and simultaneous peptide immunohistochemistry in the rat. *J. Comp. Neurol.* **258:** 230.

Watts A.G., Swanson L.W., and Sanchez-Watts G. 1987. Efferent projections of the suprachiasmatic nucleus. I. Studies using anterograde transport of *Phaseolus vulgaris* leucoagglutinin in the rat. *J. Comp. Neurol.* **258:** 204.

Yamamoto T., Nakahata Y., Soma H., Akashi M., Mamine T., and Takumi T. 2004. Transcriptional oscillation of canonical clock genes in mouse peripheral tissues. *BMC Mol. Biol.* **5:** 18.

Yamazaki S., Numano R., Abe M., Hida A., Takahashi R., Ueda M., Block G.D., Sakaki Y., Menaker M., and Tei H. 2000. Resetting central and peripheral circadian oscillators in transgenic rats. *Science* **288:** 682.

Suprachiasmatic Nucleus Clock Time in the Mammalian Circadian System

H. Okamura
Department of Systems Biology, Graduate School of Pharmaceutical Sciences, Kyoto University, Kyoto 606-8501, Japan, and Division of Molecular Brain Science, Graduate School of Medicine, Kobe 650-0017, Japan

The integration of time from gene to system levels is an exciting feature of circadian biology. In mammals, clock cells in the suprachiasmatic nucleus (SCN) generate time by an autoregulatory transcription-(post)translational feedback loop. Clock activity in the SCN neurons is expressed as activity-dependent electric signals, which are coupled to those of other SCN neurons. The SCN spreads the time signals in a form of synchronized nerve impulses to central parasympathetic nuclei (e.g., dorsal motor nucleus of the vagus) and central sympathetic nuclei (e.g., intermediolateral cell column of the spinal cord). The vagal nerve innervates gastrointestinal and respiratory organs. Sympathetic signals to the adrenal gland are converted to hormonal (glucocorticoid) signals. Glucocorticoids released into the bloodstream bind to glucocorticoid receptors of peripheral organs, activate the mammalian *Per1* gene in systemic cells, and reset the time of body clocks. Thus, the SCN-evoked time generated by specific genes localized to the SCN is converted to neuronal and hormonal signals and synchronizes the clocks in the whole body.

INTRODUCTION

The discovery of mammalian clock genes and the core clock machinery in most organs has raised the possibility that virtually 60 trillion cells of the human body have oscillating ability using the same transcription-based core loop of clock genes (Balsalobe et al. 1998; Yagita et al. 2001; Schibler and Sassone-Corsi 2002). However, clock oscillation in the body is abolished after the lesion of the hypothalamic suprachiasmatic nucleus (SCN), although the ablation of any other region cannot abolish the rhythm (Moore and Eichler 1972; Stephan and Zuker 1972). Thus, only this specialized brain locus can work as a system clock. The special ability of the SCN for rhythm generation is also supported by transplantation or gene-rescue studies (Ralph et al. 1990; Sujino et al. 2003; McDearmon et al. 2006). Here, we describe the molecular mechanism that determines how time is generated in SCN cells and is tuned and transmitted to the whole body.

CELLULAR MOLECULAR OSCILLATOR

The SCN is composed of thousands of clock oscillating cells (Klein et al. 1991; Ibata et al. 1997; Yamaguchi et al. 2003). In each clock cell, rhythm is first generated by the intracellular molecular oscillator. This core oscillator is composed of an autoregulatory transcription-(post)translation-based feedback loop involving a set of clock genes, which was first proposed in the *Drosophila* oscillating gene *period* (Hardin et al. 1990). Following this basic concept, a number of clock genes have been identified in mammals since the discovery of mammalian *Per* and *Clock* genes in 1997. It is now speculated that positive factors CLOCK and BMAL1 form heterodimers that bind E-box elements to activate *Per1* and *Per2* transcription; then, negative factors PER1/PER2 dimerize with CRY1/CRY2 feedback in a delayed fashion to inhibit CLOCK/BMAL1 activity (Fig. 1) (Reppert and Weaver 2002; Okamura 2004). As a general rule of transcription machinery, it was recently confirmed that circadian transcription accompanies chromatin remodeling (Etchegaray et al. 2003; Doi et al. 2006).

Furthermore, there is growing evidence that clock proteins are regulated dynamically in both spatial (nuclear and cytoplasmic) and temporal (production and degradation) dimensions. The main clock oscillatory protein PER2 shuttles between the nucleus and the cytoplasm by using their own nuclear localization signals (NLS) and nuclear export sequences (NES) (Yagita et al. 2002). Shuttling of BMAL1 is also reported, and it has been demonstrated that the nuclear translocation of CLOCK is accompanied by BMAL1 (Kondratov et al. 2003; Kwon et al. 2006). During spatial transportation, clock proteins receive posttranscriptional modifications and targeted destruction (Gallego and Virshup 2007). Is transcriptional regulation or posttranslational regulation more important for the generation of cyclicity? By applying the Tet-Off and Flp-In system to a fibroblast cell line, we demonstrated that the intracellular endogenous clock system has the ability to modify the *Per2* gene posttranscriptionally to make PER2 proteins oscillate without its coding mRNA cycling (Yamamoto et al. 2005; Fujimoto et al. 2006). However, to obtain robustness and continuity, rhythmic regulation seems to be prerequisite at both the transcriptional and posttranslational regulation levels.

CELL CLOCK TO CELL FUNCTION

The cell clock coordinates the timing of the expression of a variety of genes with specific cellular functions. These so-called clock-controlled genes (*ccg*) are rhythmically regulated by the core clock loop with their E-box, D-

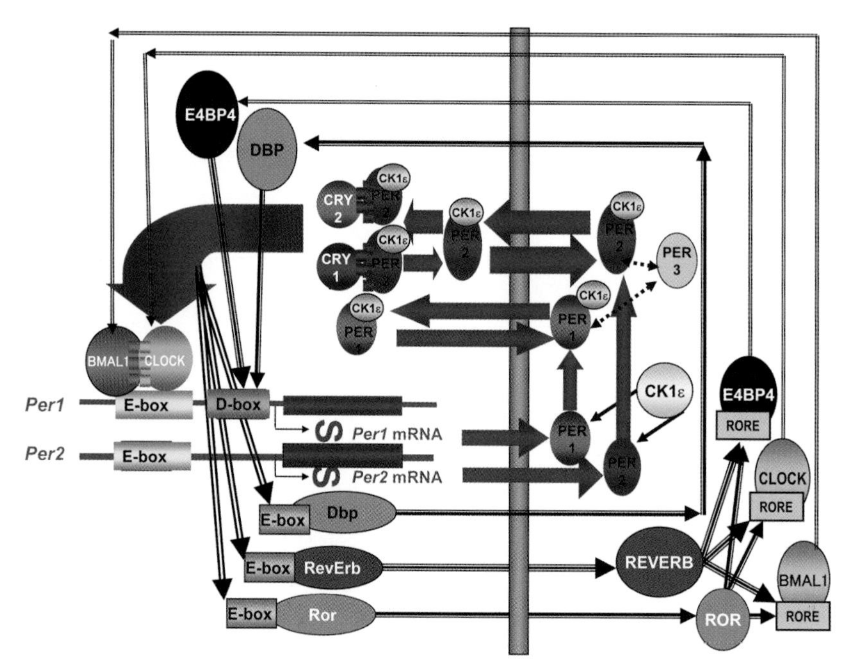

Figure 1. Core oscillatory loop of the mammalian circadian clock genes. The BMAL1/CLOCK heterodimer binds to the E box of the *Per1* and *Per2* genes to accelerate their transcription. PER2 protein is produced in the cytoplasm and then phospholylated by casein kinase 1ε (CK1ε). PER2 protein shuttles between the nucleus and cytoplasm via the CRM1/Exportin1 nuclear export system and is stabilized by the binding of CRY1 or CRY2. BMAL1/CLOCK also binds to the E box in clock-controlled genes (*ccg*) and accelerates their transcription. The core negative autoregulatory feedback loop provided by *Per1* and *Per2* runs on, accompanying the move of an accessory DBP/E4BP4, ROR, and REVERB protein loop.

box, and RORE regions. E box and D box are related to genes expressed during the day (Mitsui et al. 2001), and RORE is related to genes expressed at night. Vasopressin, albumin, cholesterol 7α hydroxylase, and cytochrome P450 are examples, and gene array studies have reported that about 5–10% of expressed genes show circadian rhythm (Panda et al. 2002; Storch et al. 2002). Among these *ccg* genes, it is interesting to note that there are bZIP (basic leucine zipper) transcription factors (*Dbp, E4bp4*) and nuclear receptors *Ror* (*Rorα*, *Rorβ*, *Rorγ*) and *Rev-Erb* (*RevErbα*, *RevErbβ*). These specialized genes form the subloops of core clock gene oscillation, which will contribute to the robustness and stability of the rhythm of core clock oscillation (Fig. 1).

Why is circadian time information needed for cellular function? The proteins are key players in cellular function, and for single functional proteins (5–30 nm in diameter), cell space is huge (10–50 μm in diameter). Thus, for proteins to perform effective cellular functions, cellular reactions should be organized in time and space. The intracellular clock oscillating loop may be useful in controlling cellular events for proper and adequate time organization. One example is the cell cycle. There is substantial evidence that circadian rhythms affect the timing of cell divisions in vivo (Bjarnason and Jordan 2000). Because some of these mitotic rhythms were shown to persist in constant darkness, it was concluded that they might be under the control of an endogenous clock. Matsuo et al. (2003) demonstrated that the circadian clock controls the G_2-to-S transition via the regulation of WEE1. Circadian clock signals may also control the timing of many basic cellular metabolisms yet to be elucidated.

SYNCHRONY OF SCN CLOCK CELLS

Per1-luc study of the SCN slice culture at microscopic resolution enables analysis of clock gene transcription at the cellular level (Yamaguchi et al. 2003). These SCN cells showed robust transcription rhythms with a period length of approximately 24 hours, with several hundreds of cells expressing *Per1* genes synchronously. Moreover, the individual oscillatory cells are arranged topographically: The phase leader with a shorter period length is located in the dorsomedial periventricular part of the SCN. A protein synthesis inhibitor (cycloheximide) sets all of the cell clocks to the same phase and, following withdrawal, intrinsic interactions among cell clocks reestablish the stable program of gene expression across the assemblage. Tetrodotoxin, which blocks action potentials, not only desynchronizes the cell population, but also suppresses the level of clock gene expression, demonstrating that neuronal network properties dependent on action potentials have a dominant role both in establishing cellular synchrony and in maintaining spontaneous oscillations across the SCN. Thus, the cell-rhythm oscillation generated by the core clock oscillatory loop is coupled and amplified by the ordered cell–cell communications in the SCN.

Does this transcriptional rhythm really occur in living, free-moving animals? An optical fiber insertion just on the SCN of living animals enables us to monitor the *Per1-luc* transcription activity in living animals (Yamaguchi et al. 2001). Using this system, we were able to continuously record light emission from the SCN of *Per1-luc* transgenic mice in vivo under constant dark conditions. The

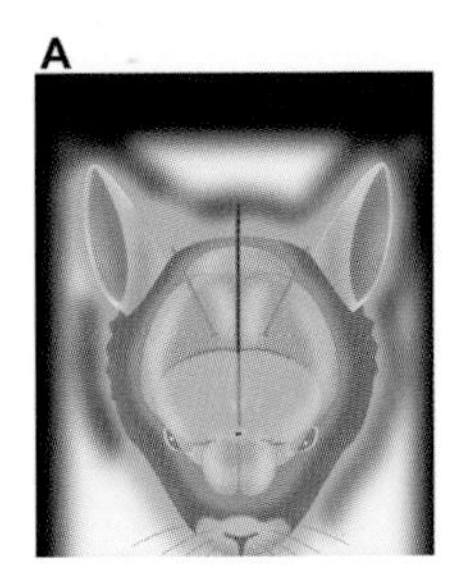

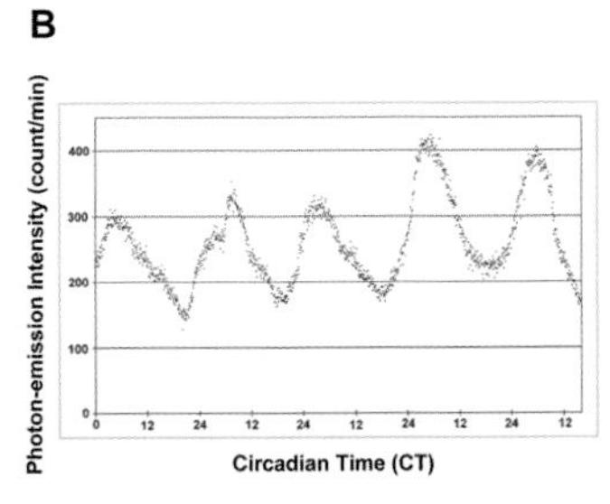

Figure 2. Monitoring gene expression of the SCN in living freely moving animals. (*A*) An illustration showing the experiment monitoring gene expression of SCN in living mice brain. A polymer optical fiber (500 µm in diameter; 0.5 numerical aperture) is inserted just above the SCN of a *mPer1-luc* transgenic mouse. The other end of the optical fiber (not shown) is connected to a photomultiplier tube operating as part of a photon-counting apparatus. The luciferase substrate luciferin dissolved in artificial cerebrospinal fluid is continuously infused through a lateral ventricle at a rate of 15 µl/hour. Luminescence was recorded under constant dark conditions. (*B*) Circadian fluctuation of luminescence in the SCN. One dot represents an average of the value of 5 minutes. Hatched and closed bars at the bottom of the figure represent subjective day and subjective night, respectively. This model system enables real-time gene expression to be monitored in the intact brain under physiological conditions. (*A*, drawn by Seiichi Takekida; *B*, reprinted, with permission, from Yamaguchi et al. 2001 [© Nature Publishing Group].)

luminescence showed a clear circadian fluctuation (Fig. 2), with a 1.5- to 2.5-fold amplitude and peaks and troughs at circadian time (CT) 4–6 and CT15–20, respectively (where CT0 is subjective dawn and CT12 is subjective dusk). Considering the expression profiles of *Per1* mRNA in the SCN (Shigeyoshi et al. 1997), the phase lag between luciferase mRNA and luminescence was only 0–2 hours. An important finding of this study is that the net expression of gene activity fluctuates in living animals, although with a substantial difference of circadian phase at each cell level (Yamaguchi et al. 2003). Moreover, luciferase peaks at subjective morning and troughs at subjective night are very similar to electric activity rhythm (Inouye and Kawamura 1979). Thus, the net clock gene expression reflects net electrical activity, which will be widely conducted to other brain areas.

SCN OUTPUTS AND REGULATION OF PERIPHERAL ORGANS

Do SCN signals really drive circadian clocks in the peripheral organs? Sujino et al. (2003) reported that SCN grafts from wild-type mice into mice that were "doubly arrhythmic," due to being both SCN-lesioned and *Cry1/Cry2* double-knockout mice, restored circadian locomotor activities. Grafts should include SCN tissue, because grafts from the cerebral cortex cannot restore rhythms to SCN-ablated animals (Sujino et al. 2003). Similarly, McDearmon et al. (2006) reported that the behavioral rhythm of arrhythmic BMAL1 knockout mice was normalized by brain-specific rescue but not by muscle-specific rescue. These studies strongly suggest that peripheral clocks are unnecessary for the generation and expression of behavioral rhythms and that the SCN is the generator of brain rhythms. The dominance of the SCN clock over peripheral oscillators was also demonstrated by Pando et al. (2002) using grafts from animals with genetically altered period lengths.

By what route are time signals conveyed from the SCN to peripheral tissues? More than a quarter of a century ago, autonomic changes of circadian rhythms were shown in sleep/wake cycles, as well as in cardiovascular, respiratory, and gastrointestinal functions, although the mechanisms of these changes are unknown. Most outputs of SCN neurons terminate to the subparaventricular zone (Klein et al. 1991). From there, signals are transmitted to central sympathetic (e.g., intermediolateral cell column) and parasympathetic (e.g., dorsal motor nucleus of the vagus) nuclei (Fig. 3) (Buijs and Kalsbeek 2001).

Parasympathetic vagal innervation is dominant in gastrointestinal and respiratory tracts. Bando et al. (2007) reported the existence of a clock-gene-mediated oscillating system in the respiratory gland whose rhythm was abolished after ablation of the SCN. To address the role of the vagal nerve on these respiratory clocks, these authors performed a unilateral vagotomy and found that circadian expression of mucin and PER2 protein levels was abolished in the (operated) ipsilateral side of the submucosal glands, although clear rhythm was detected in the (intact) contralateral side (Fig. 4). Because transcripts of a mucin-encoding gene did not show circadian rhythm, its regulation may not be at the level of production. Because muscarinic acethylcholine receptor genes *Chm2*, *Chm3*, and *Chm4* are under the control of the circadian clock, mucin will be rhythmically released through these receptors. Thus, in airway organs, the vagal nerve is the key mediator conveying SCN circadian signals to the airway glands.

Sympathetic regulation of clock genes is initially found in the liver. Terazono et al. (2003) clearly demonstrated that chemical sympathectomy damped clock gene oscillation and that adrenergic drug treatment resets the rhythm. The above finding suggests that both sympathetic and parasympathetic regulation convey SCN rhythm to visceral organs.

CONVERSION OF NERVE IMPULSE TO HORMONE SURGE IN THE ADRENAL GLAND

In the analysis of the sympathetic effect of clock genes on visceral organs, we devised the macroimaging of *Per1-luc* mice with a two-dimensional photon-counting camera, which enabled the detection of light-induced gene expression at each organ at one glance. Ishida et al. (2005) have revealed that light preferentially increases the transcription of *Per1* in the adrenal gland without any change in other visceral organs such as liver and kidney (Fig. 5).

Adrenal gene expression is not limited to the *Per1* gene; DNA microarray analysis has demonstrated 156 up-regulated genes and 39 down-regulated genes. Because adrenal gene expression was abolished after SCN lesioning or by the transaction of the adrenal sympathetic nerve, light signals were conveyed to the adrenal gland via the SCN-sympathetic nerve route. More interestingly, this gene expression accompanies the surge of plasma and

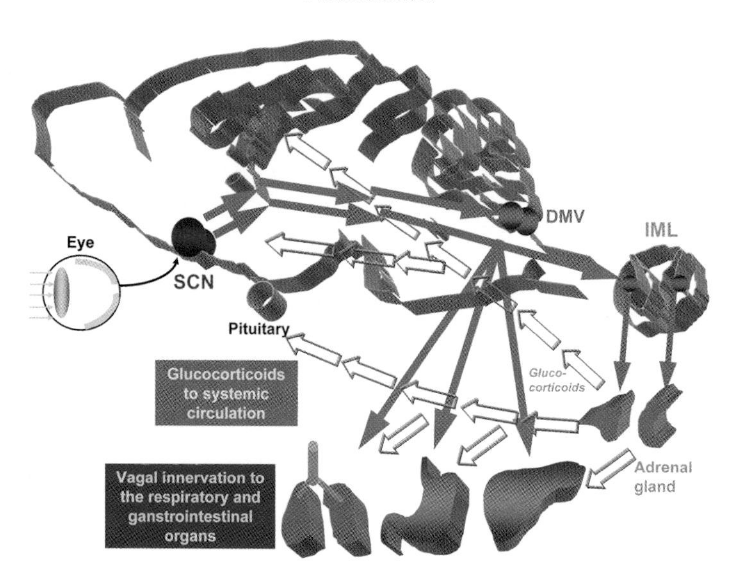

Figure 3. SCN-autonomic outputs. SCN signals are finally transmitted to the central sympathetic intermediolateral (IML) cell column of the spinal cord or to the central parasympathetic dorsal motor nucleus of the vagus (DMV). Signals to the adrenal gland are converted to glucocorticoid signals and spread to whole body by the blood vessels.

brain corticosterone levels after 60–90 minutes without activation of the hypothalamus-pituitary-adrenal (HPA) axis. This activation was completely different from the usual stress reaction: Corticosterone as well as ACTH levels were immediately increased (i.e., 2–5 minutes) to their maximal levels. The larger time lag of the response of corticosterone (60 minutes) also suggests a different mechanism in this process.

On which target does corticosterone act? The released corticosterone may bind to the glucocorticoid receptor (Balsalobre et al. 2000) and activate the *Per1* gene in systemic cells (Koyanagi et al. 2006), which will reset the time of the peripheral clock. Thus, the light-evoked corticosterone may have a role in transmitting environmental light signals to internal organs. In this light-induced situation, the SCN-sympathetic-adrenal system will also be used for the generation of circadian plasma glucocorticoid levels peaking at dawn. Thus, the adrenal gland acts as a transducer of SCN clock time from nerve impulse to hormone surge in the circadian timekeeping system.

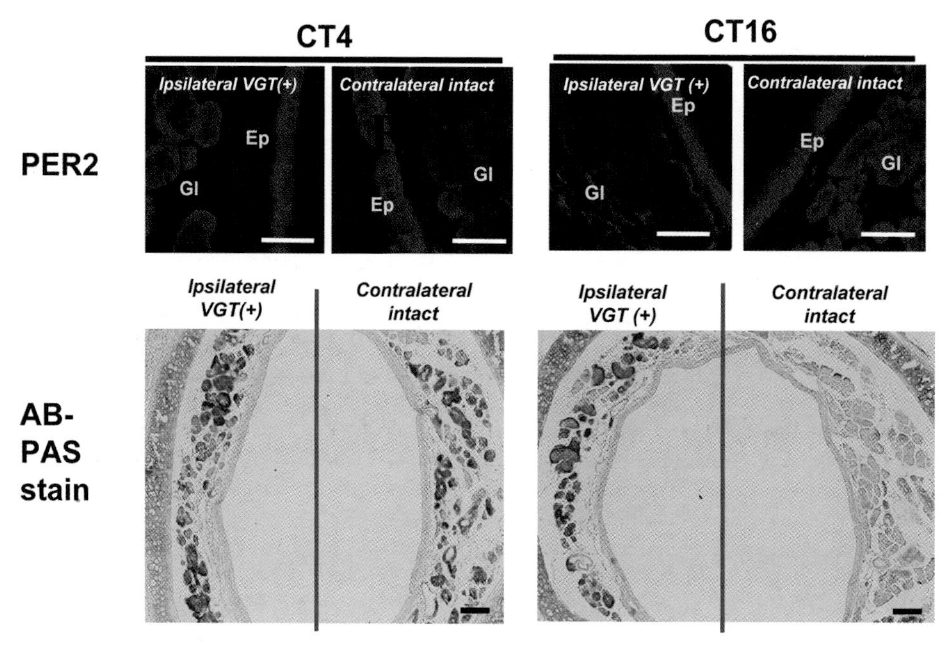

Figure 4. Vagotomy-induced changes in mucin levels and clock gene expression. (*Top*) Immunocytochemical detection of the PER2 protein at CT16 in the ipsilateral vagotomized (VGT[+]) and contralateral intact side of the trachea. Note the rhythmic expression of PER2 in the submucosal gland at the intact contralateral side. (*Bottom*) Mucin levels (Alcian blue/PAS staining) at CT4 and CT16 in the trachea of C57BL/6 mice. Bar, 100 µm. (Reprinted, with permission, from Bando et al. 2007 [© Society for Neuroscience].)

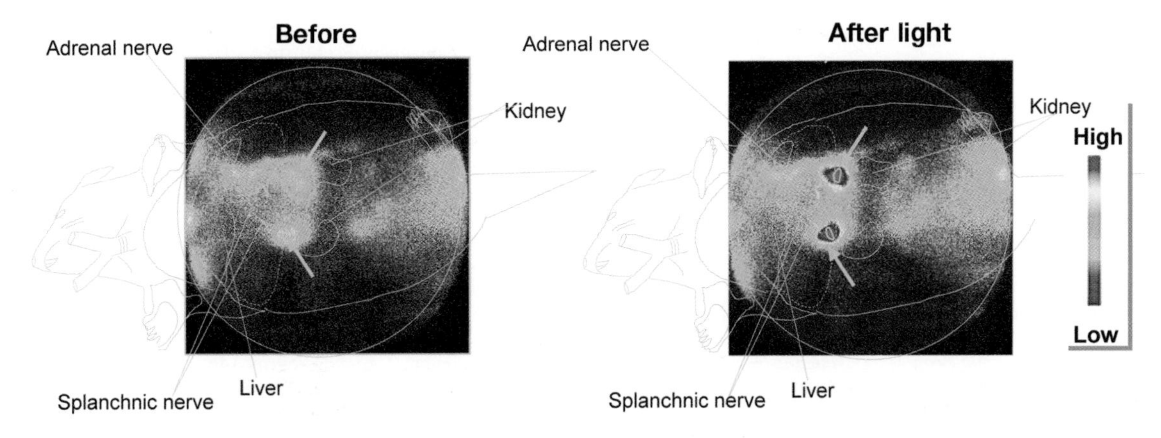

Figure 5. Light-induced *Per1-luc* luminescence in the adrenal gland. Luciferase bioluminescence in the abdomen of *Per1* promoter-luciferase (*Per1-luc*) transgenic mice before and after 60-minute light exposure. Note the high level of expression of luciferase luminescence in the adrenal gland (*arrows*) after light exposure. (Reprinted, with permission, from Ishida et al. 2005 [© Elsevier].)

SCN-GENE PROJECT FOR CIRCADIAN STUDY

As shown above, the mammalian circadian system is designed to integrate SCN clock time to body clock time. However, there has been little effort to isolate clock genes themselves from this nucleus. The only trial has been the DNA microarray analysis of SCN picked up by the brain microdissection technique applied to sliced brain (Panda et al. 2002). However, this technique has serious methodological problems because SCN is a very tiny nucleus, and contamination of adjacent brain structures (preoptic area, anterior hypothalamic area, retrochiasmatic area, and optic chiasma) in the tissue is inevitable.

We are studying the gene involved in the ubiquitin-proteasome system. We are interested in this system because we found that PER2 and CRY proteins were easily ubiquitinated and CRY-PER heterodimerization stabilized PER (Yagita et al. 2002) in the analysis of the fibroblast cell line having the core component of the biological clock (Yagita et al. 2001). Recently, F-box E3 ubiquitin ligase FBXL3 was isolated as the specific ligase bound to CRY1 and CRY2 proteins (Busino et al. 2007; Godinho et al. 2007; Siepka et al. 2007). Among nine genes (*Ubce5, UbcM4, Ube2v, Ube2d2, UchL1, UchL3, Ubp41, Ufd1L*, β-*TrCP*) having rhythmicity in a previously published DNA microarray study (Panda et al. 2002), we found that most of these enzymes were only faintly expressed in the SCN by in situ hybridization except for ubiquitin-conjugating enzyme 5 (Ubce5) and ubiquitin carboxy-terminal hydrolase L1 (UchL1), a dominant deubiquitinating salvaging enzyme (Dong et al. 2005). Thus, the microarray technique presently available does not strictly represent the gene profile of the SCN itself.

To overcome this problem, we began a project to isolate clock-associated genes from the SCN by histological screening (the SCN-Gene Project). DNA microarray (Affimetrix: GeneChip Mouse Genome 430 2.0 Array) analysis identified 12,379 genes. To determine whether these genes were really expressed in the SCN, we screened them by using in situ hybridization sampled subjectively during day and night. Genes that were highly and specially expressed in the SCN were selected and targeted in the next step, and the circadian phenotype of the mice was characterized.

Despite previous studies focusing on rhythmic gene expression, our project does not exclude constantly expressed genes, because we believe that constancy does not exclude involvement from rhythm generation. The feasibility of this project is supported by the finding that all known rhythm mutants have some defects in the SCN. In other words, genes expressed in the SCN are possibly involved in rhythm generation. Because the expressed gene number is limited, the SCN-Gene Project is expected to isolate a new clock gene.

CONCLUSIONS

The SCN is a small nucleus, occurring in pairs, located just above the optic chasma. It is astonishing that only about 10,000 of the oscillating cells in the SCN convey standard time to other brain regions and/or whole bodies. Gene expression and cell function in the SCN are specialized to generate circadian time. It is astonishing that *Per1, Per2, Cry1, Cry2, VIP*, and others that have profound effects on circadian rhythms are strongly and specifically expressed in this nucleus. We began a project to isolate new clock genes expressed in the SCN by histological analysis using the in situ hybridization technique. The SCN will be a link to analyze the integration mechanism of "time" in a vertical arrangement, providing a bridge between single genes and the living organism as a whole.

ACKNOWLEDGMENTS

This work was supported in part by The Scientific Grant of the 21st Century COE Program, and The Special Promotion Funds from Ministry of Education, Culture, Sports, Science, and Technology of Japan.

REFERENCES

Balsalobre A., Damiola F., and Schibler U. 1998. A serum shock induces circadian gene expression in mammalian tissue culture cells. *Cell* **93:** 929.

Balsalobre A., Brown S.S.A., Marcacci L., Tronche F., Kellendonk C., Reichardt H.M., Schulz G., and Schibler U. 2000. Resetting of circadian time in peripheral tissues by glucocorticoid signalling. *Science* **289:** 2344.

Bando H., Nishio T., van der Horst G.T.J., Masubuchi S., Hisa Y., and Okamura H. 2007. Vagal regulation of airway clocks in mice. *J. Neurosci.* **27:** 4359.

Bjarnason G.A. and Jordan R. 2000. Circadian variation of cell proliferation and cell cycle protein expression in man: Clinical implications. *Prog. Cell Cycle Res.* **4:** 193.

Buijs R.M. and Kalsbeek A. 2001. Hypothalamic integration of central and peripheral clocks. *Nat. Rev. Neurosci.* **2:** 521.

Busino L., Bassermann F., Maiolica A., Lee C., Nolan P.M., Godinho S.I., Draetta G.F., and Pagano M. 2007. SCFFbxl3 controls the oscillation of the circadian clock by directing the degradation of cryptochrome proteins. *Science* **316:** 900.

Doi M., Hirayama J., and Sassone-Corsi P. 2006. Circadian regulator CLOCK is a histone acetyltransferase. *Cell* **125:** 497.

Dong X., Yagita Y., Zhang J., and Okamura H. 2005. Expression of ubiquitin-related enzymes in the suprachiasmatic nucleus with special reference to ubiquitin carboxy-terminal hydrolase UchL1. *Biomed. Res.* **26:** 43.

Etchegaray J.P., Lee C., Wade P.A., and Reppert S.M. 2003. Rhythmic histone acetylation underlies transcription in the mammalian circadian clock. *Nature* **421:** 177.

Fujimoto Y., Yagita K., and Okamura H. 2006. Does mPER2 protein oscillate without its coding mRNA cycling?: Post-transcriptional regulation by cell clock. *Genes Cells* **11:** 525.

Gallego M. and Virshup D.M. 2007. Post-translational modifications regulate the ticking of the circadian clock. *Nat. Rev. Mol. Cell Biol.* **8:** 139.

Godinho S.I., Maywood E.S., Shaw L., Tucci V., Barnard A.R., Busino L., Pagano M., Kendall R., Quwailid M.M., Romero M.R., O'neill J., Chesham J.E., Brooker D., Lalanne Z., Hastings M.H., and Nolan P.M. 2007. The after-hours mutant reveals a role for Fbxl3 in determining mammalian circadian period. *Science* **316:** 897.

Hardin P.E., Hall J.C., and Rosbash M. 1990. Feedback of the *Drosophila* period gene product on circadian cycling of its messenger RNA levels. *Nature* **343:** 536.

Ibata Y., Tanaka M., Tamada Y., Hayashi S., Kawakami F., Takamatsu T., and Okamura H. 1997. The suprachiasmatic nucleus: A circadian oscillator. *Neuroscientist* **3:** 215.

Inouye S.-I.T. and Kawamura H. 1979. Persistence of circadian rhythmicity in a hypothalamic 'island' containing the suprachiasmatic nucleus. *Proc. Natl. Acad. Sci.* **76:** 5962.

Ishida A., Mutoh T., Ueyama T., Bando H., Masubuchi S., Nakahara D., Tsujimoto G., and Okamura H. 2005. Light activates the adrenal gland: Timing of gene expression and glucocorticoid release. *Cell Metab.* **2:** 297.

Klein D.C., Moore R.Y., and Reppert S.M. 1991. *Suprachiasmatic nucleus: The mind's clock.* Oxford University Press, New York.

Kondratov R.V., Chernov M.V., Kondratova A.A., Gorbacheva V.Y., Gudkov A.V., and Antoch M.P. 2003. BMAL1-dependent circadian oscillation of nuclear CLOCK: Posttranslational events induced by dimerization of transcriptional activators of the mammalian clock system. *Genes Dev.* **17:** 1921.

Koyanagi S., Okazawa S., Kuramoto Y., Ushijima K., Shimeno H., Soeda S., Okamura H., and Ohdo S. 2006. Chronic treatment with prednisolone represses the circadian oscillation of clock gene expression in mouse peripheral tissues. *Mol. Endocrinol.* **20:** 573.

Kwon I., Lee J., Chang S.H., Jung N.C., Lee B.J., Son G.H., Kim K., and Lee K.H. 2006. BMAL1 shuttling controls transactivation and degradation of the CLOCK/BMAL1 heterodimer. *Mol. Cell. Biol.* **26:** 7318.

Matsuo T., Yamaguchi S., Mitsui S., Emi A., Shimoda F., and Okamura H. 2003. Control mechanism of the circadian clock for timing of cell division. *Science* **302:** 255.

McDearmon E.L., Patel K.N., Ko C.H., Walisser J.A., Schook A.C., Chong J.L., Wilsbacher L.D., Song E.J., Hong H.K., Bradfield C.A., and Takahashi J.S. 2006. Dissecting the functions of the mammalian clock protein BMAL1 by tissue-specific rescue in mice. *Science* **314:** 1304.

Mitsui S., Yamaguchi S., Matsuo T., Ishida Y., and Okamura H. 2001. Antagonistic role of E4BP4 and PAR proteins in the circadian oscillatory mechanism. *Genes Dev.* **15:** 995.

Moore R.Y. and Eichler V.B. 1972. Loss of circadian adrenal corticosterone rhythm following suprachiasmatic nucleus lesion in the rat. *J. Comp. Neurol.* **146:** 1.

Okamura H. 2004. Clock genes and cell clocks: Roles, actions and mysteries. *J. Biol. Rhythms* **19:** 388.

Panda S., Antoch M.P., Miller B.H., Su A.I., Schook A.B., Straume M., Schultz P.G., Kay S.A., Takahashi J.S., and Hogenesch J.B. 2002. Coordinated transcription of key pathways in the mouse by the circadian clock. *Cell* **109:** 307.

Pando M.P., Morse D., Cermakian N., and Sassone-Corsi P. 2002. Phenotypic rescue of a peripheral clock genetic defect via SCN hierarchical dominance. *Cell* **110:** 107.

Ralph M.R., Foster R.G., Davis F.C., and Menaker M. 1990. Transplanted suprachiasmatic nucleus determines circadian period. *Science* **247:** 975.

Reppert S.M. and Weaver D. 2002. Coordination of circadian timing in mammals. *Nature* **418:** 935.

Schibler U. and Sassone-Corsi P. 2002. A web of circadian pacemakers. *Cell* **111:** 919.

Shigeyoshi Y., Taguchi K., Yamamoto S., Takekida S., Yan L., Tei H., Moriya T., Shibata S., Loros J.J., Dunlap J.C., and Okamura H. 1997. Light-induced resetting of a mammalian circadian clock is associated with rapid induction of the *mPer1* transcript. *Cell* **91:** 1043.

Siepka S.M., Yoo S.H., Park J., Song W., Kumar V., Hu Y., Lee C., and Takahashi J.S. 2007. Circadian mutant Overtime reveals F-box protein FBXL3 regulation of cryptochrome and period gene expression. *Cell* **129:** 1011.

Stephan F.K. and Zucker I. 1972. Circadian rhythms in drinking behavior and locomotor activity are eliminated by suprachiasmatic lesions. *Proc. Natl. Acad. Sci.* **54:** 1521.

Storch K.F., Lipan O., Leykin I., Viswanathan N., Davis F.C., Wong W.H., and Weitz C.J. 2002. Extensive and divergent circadian gene expression in liver and heart. *Nature* **417:** 78.

Sujino M., Matsumoto K., Yamaguchi S., van der Horst G., Okamura H., and Inouye S.I.T. 2003. Suprachiasmatic nucleus grafts restore circadian behavioral rhythms of genetically arrhythmic mice. *Curr. Biol.* **13:** 664.

Terazono H., Mutoh T., Yamaguchi S., Kobayashi M., Akiyama M., Udo R., Ohdo S., Okamura H., and Shibata S. 2003. Adrenergic regulation of clock gene expression in the mouse liver. *Proc. Natl. Acad. Sci.* **100:** 6795.

Yagita K., Tamanini F., van der Horst G., and Okamura H. 2001. Molecular mechanisms of the biological clock in cultured fibroblasts. *Science* **292:** 278.

Yagita K., Tamanini F., Yasuda M., Hoeijimakers J.H.J., van der Horst G.T.J., and Okamura H. 2002. Nucleocytoplasmic shuttling and mCRY dependent inhibition of ubiquitination of the mPER2 clock protein. *EMBO J.* **21:** 1301.

Yamaguchi S., Isejima H., Matsuo T., Okura R., Yagita K., Kobayashi M., and Okamura H. 2003. Synchronization of cellular clocks in the suprachiasmatic nucleus. *Science* **302:** 1408.

Yamaguchi S., Kobayashi M., Mitsui S., Ishida Y., van der Horst G.T.J., Suzuki M., Shibata S., and Okamura H. 2001. Real time monitoring of clock gene expression in the living mouse. *Nature* **409:** 684.

Yamamoto Y., Yagita Y., and Okamura H. 2005. Role of cyclic mPer2 expression in mammalian cellular clock. *Mol. Cell. Biol.* **25:** 1912.

Molecular Analysis of Sleep:Wake Cycles in *Drosophila*

A. SEHGAL,* W. JOINER,* A. CROCKER,* K. KOH,* S. SATHYANARAYANAN,[†]
Y. FANG,* M. WU,* J.A. WILLIAMS,[‡] AND X. ZHENG*

**Howard Hughes Medical Institute, Department of Neuroscience, University of Pennsylvania School of Medicine, Philadelphia, Pennsylvania 19104; [†]Molecular Oncology, Merck Research Laboratories, Boston, Massachusetts 02115; [‡]CABM, University of Medicine and Dentistry of New Jersey, Piscataway, New Jersey 08854*

Sleep is controlled by two major regulatory systems: a circadian system that drives it with a 24-hour periodicity and a homeostatic system that ensures that adequate amounts of sleep are obtained. We are using the fruit fly *Drosophila melanogaster* to understand both types of regulation. With respect to circadian control, we have identified molecular mechanisms that are critical for the generation of a clock. Our recent efforts have focused on the analysis of posttranslational mechanisms, specifically the action of different phosphatases that control the phosphorylation and thereby the stability and/or nuclear localization of circadian clock proteins period (PER) and timeless (TIM). Resetting the clock in response to light is also mediated through posttranslational events that target TIM for degradation by the proteasome pathway; a recently identified ubiquitin ligase, jet lag (JET), is required for this response. Our understanding of the homeostatic control of sleep is in its early stages. We have found that mushroom bodies, which are a site of synaptic plasticity in the fly brain, are important for the regulation of sleep. In addition, through analysis of genes expressed under different behavioral states, we have identified some that are up-regulated during sleep deprivation. Thus, the *Drosophila* model allows the use of cellular and molecular approaches that should ultimately lead to a better understanding of sleep biology.

INTRODUCTION

To adapt to the cyclic environment in which they live, organisms have evolved endogenous timekeeping mechanisms that synchronize their behavior and physiology with the environmental 24-hour cycle. Of the many behaviors that occur with a circadian (or ~24 hour) periodicity, the best known is perhaps the sleep/wake cycle. Sleep-like states are displayed in many different species, but the regulation and function of such a state are poorly understood. Although the internal circadian clock controls the 24-hour rhythmicity of sleep, it is not required for the manifestation of sleep. The latter is controlled by a homeostatic system that drives the need to sleep based on prior wakefulness. Simply put, the circadian system is largely responsible for controlling the timing of sleep, whereas the homeostatic system controls the amount of sleep. We are using a simple model system, the fruit fly, *Drosophila melanogaster*, to understand the molecular basis of both these systems.

CIRCADIAN RHYTHMS IN *DROSOPHILA*

Over the years, *Drosophila* has proved to be an extremely powerful model for the study of circadian rhythms (Hardin 2005). Currently, the assay of choice for measuring rhythms in *Drosophila* quantitates locomotor activity over time using an automated beam-break system. Flies are diurnal, and so most activity occurs during the day, with extended periods of rest (sleep) at night. The robust, approximately 24-hour rhythm observed in these behavioral assays provides an effective measurement for studying the genetic basis of the clock underlying this pattern of activity.

Molecular Basis of the Clock

In all organisms examined, the basic clock mechanism consists of one or more feedback loops in which cycling clock proteins rhythmically control the expression of their own mRNAs, thereby maintaining cycles of gene expression (Sehgal 2004). In *Drosophila*, the major cycling components of the principal loop are the period and timeless proteins (PER and TIM) (see Fig. 1) (Hardin 2005). The *per* and *tim* genes are actively transcribed by the transcriptional factors, Clock (CLK) and cycle (CYC), during the day, which leads to peak levels of the respective mRNAs in the early evening. At that point, the PER and TIM proteins start

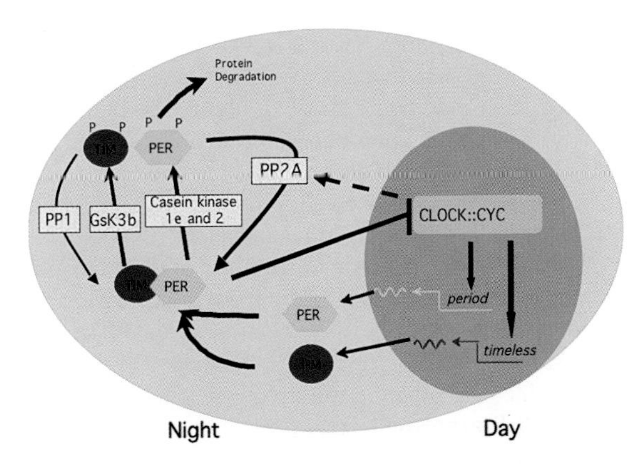

Figure 1. Model for the *Drosophila* circadian clock. The model depicts the classical transcription-translation feedback loop for *per* and *tim* and a posttranslational loop through which rhythmic phosphorylation of PER and TIM may be maintained. For the sake of simplicity, the transcription loop that regulates expression of *Clk* is not shown.

 978-087969823-2

to accumulate and, at about the middle of the night, translocate into the nucleus to inhibit the activity of CLK and CYC. TIM is degraded in the late night/early morning, and PER a few hours later, which relieves the repression on CLK/CYC and leads to the start of a new cycle (Hardin 2005).

It is clear that in order to generate a 24-hour rhythm from such a loop, multiple regulatory steps have to be built in. One such regulatory mechanism is phosphorylation; both PER and TIM are hyperphosphorylated at specific times of day and some of the kinases responsible for this phosphorylation have been identified (Fig. 1). PER is phosphorylated by casein kinase 1ε (CK1ε) and by casein kinase 2 and TIM is phosphorylated by glycogen synthase kinase 3β (GSK3β) (Kloss et al. 1998; Martinek et al. 2001; Lin et al. 2002; Akten et al. 2003). The different phosphorylation events regulate critical aspects of the feedback loop such as stability and nuclear expression of PER/TIM, indicating the importance of posttranslational control in the clock. In fact, transcriptional rhythms of *per* and *tim* may be dispensable to some extent. A few years ago, we generated flies in which levels of *per* and *tim* mRNAs were held constant and found that the two proteins continued to cycle and drive behavioral rhythms in many of these flies (Yang and Sehgal 2001). This does not imply that rhythmic transcription does not serve a purpose. It (1) probably strengthens the amplitude and promotes the penetrance of the rhythm (a smaller percentage of the flies with noncycling *per* and *tim* mRNAs showed behavioral rhythms), (2) confers temporal precision on to the rhythm (although the average period of these flies was ~24 hours, there was some variability from fly to fly), and (3) may regulate the differential response of the clock to light at different times of day (flies expressing noncycling *per* and *tim* mRNAs showed aberrant responses to pulses of light delivered at different times during the night) (Yang and Sehgal 2001).

Given that it is possible, however, to generate at least a partially functional clock without rhythmic expression of *per* and *tim* mRNAs, the question arises as to how this is achieved. The cycling of the proteins under these conditions is most likely driven by posttranslational events such as phosphorylation. However, expression of the regulatory kinases mentioned above is not known to cycle. This led us to investigate a role of phosphatases in controlling the rhythmic phosphorylation states of PER and TIM. We found that both proteins, but predominantly PER, are dephosphorylated by protein phosphatase 2A (PP2A) (Fig. 1) (Sathyanarayanan et al. 2004). This dephosphorylation increases the stability of PER and also promotes its nuclear localization. Overexpression of wild-type or dominant-negative versions of the PP2A catalytic subunit in wild-type flies results in loss of rhythms accompanied by either constitutively elevated or greatly reduced PER expression, respectively. In addition, circadian periodicity is altered by changes in the levels of specific regulatory subunits of PP2A. Thus, overexpression of the *widerborst* (*wdb*) subunit produces a long period, and loss or gain of function of the *twins* (*tws*) subunit also results in circadian phenotypes (Sathyanarayanan et al. 2004; and data not shown). Note that we previously reported that overexpression of *tws* shortens circadian period, but we have since found that only the overexpression of the PP2A catalytic subunit produces a short period. Overexpression of *tws* results in a long period as does overexpression of the other regulatory subunit, *wdb*. Importantly, mRNA levels of both regulatory subunits are expressed rhythmically, and the *tws* mRNA cycles with a high amplitude. This cycling is eliminated in cyc^0 mutants, indicating that cycling of the phosphatase subunits constitutes another loop in the central clock, and one which controls the cycling of PER and TIM at a posttranslational level. Note that because the cycling of *tws* occurs at the level of the mRNA, this loop may also require rhythmic transcription, although not of *per* and *tim*. Yet another transcription-based loop, which controls the expression of *Clk*, was described previously (Glossop et al. 1999, 2003; Cyran et al. 2003).

We have recently found that protein phosphatase 1 (PP1) is also important for clock function. PP1 dephosphorylates PER and TIM and increases their stability (Fig. 1) (Fang et al. 2007). In this case, the primary target appears to be TIM, which also affects PP1 activity on PER. Thus, in cultured cells, in the absence of TIM, PER is destabilized by mild inhibition of PP1, presumably due to its increased phosphorylation. The presence of TIM stabilizes PER and renders it more resistant to inhibition of PP1. The effect is specific for PP1 because TIM does not affect the dephosphorylation of PER by PP2A (Fang et al. 2007). The stabilizing effect of TIM most likely involves a change in the phosphorylation state of PER: It may down-regulate kinase activity on PER or up-regulate phosphatase activity. A similar situation occurs in flies where PER depends on TIM for stability, but the mechanisms involved are not known (Price et al. 1995).

Entrainment of the Clock to Light

Entrainment of the *Drosophila* clock to light is achieved by light-induced degradation of TIM (Hunter-Ensor et al. 1996; Myers et al. 1996; Zeng et al. 1996). This effect of light fits well with the model which postulates that levels of PER and TIM constitute time-of-day signals; it follows then that anything that changes the time of the clock would do so by changing the levels of these molecules. Light is transmitted to TIM by the circadian photoreceptor, cryptochrome (CRY) (Fig. 2), which is

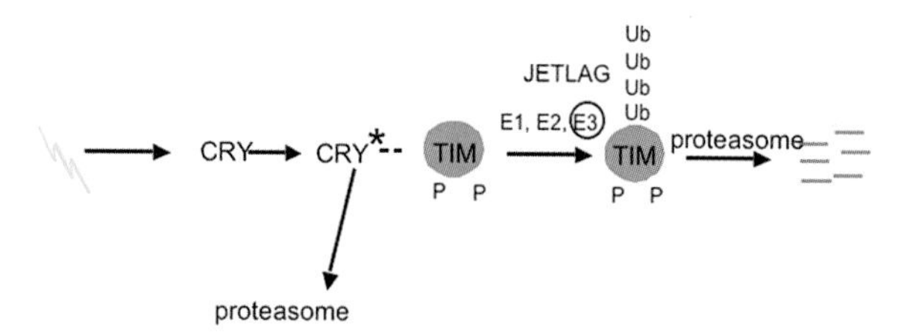

Figure 2. Schematic of the *Drosophila* light response. In response to light, the circadian photoreceptor, CRY, transmits a signal to TIM that results in the degradation of TIM through the ubiquitin-proteasome pathway and the action of a specific E3 ligase, JET. CRY is itself also degraded by light, but with a slower time course than TIM. The visual system (not shown here) can also entrain the clock to light:dark cycles, but the response to pulses of light and to constant light appears to require this pathway.

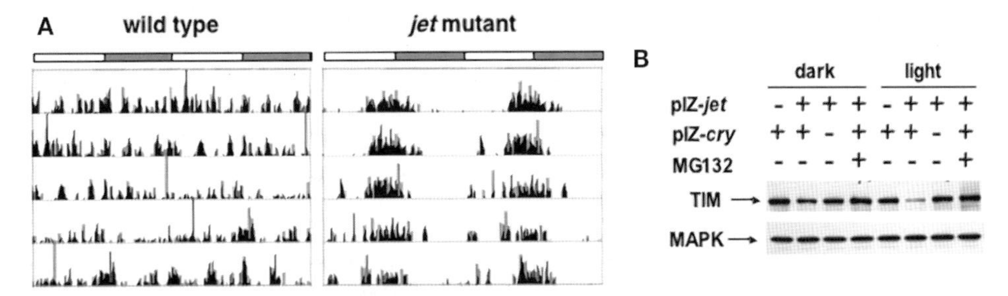

Figure 3. JET is required for the TIM response to light. (*A*) Activity records of *jet* flies in constant light. Wild-type flies (see representative example on the left) lose circadian rhythms in constant darkness, but *jet* mutants (representative example on right) retain rhythms under such conditions. (*B*) A cell culture assay for the circadian light response. TIM was transfected into *Drosophila* cultured cells along with other genes as indicated. When CRY and JET are coexpressed, TIM is degraded in response to light.

coexpressed with clock proteins in the central brain (Stanewsky et al. 1998; Emery et al. 2000). CRY is itself degraded in response to light, which may serve to terminate the light response (Lin et al. 2001).

Both CRY and TIM are degraded by the proteasome (Fig. 2) (Naidoo et al. 1999; Lin et al. 2001), but until recently, other molecular components of this response were not known. We recently identified one such component through the analysis of a mutation found in one of our fly stocks (Koh et al. 2006a). This particular stock showed reduced sensitivity to light such that it was rhythmic in the presence of constant light (wild-type flies are arrhythmic under these conditions) (Fig. 3A), and it took longer to adjust to a change in the light:dark schedule. This latter phenotype, which is analogous to extended jet lag, led us to term this mutant *jet lag* (*jet*). In addition to behavioral deficits, *jet* mutants were also aberrant in the molecular response to light, i.e., degradation of TIM by light was reduced in these flies. Cloning of the affected gene revealed that it encodes an E3 ligase, a molecule that typically targets specific substrates to the proteasome. In fact, JET turned out to be the E3 ubiquitin ligase that mediates the response of TIM to light. Using JET, we were able to reconstitute the TIM response to light in cultured cells. We transfected *Drosophila* S2 cells with TIM along with CRY and/or JET and found that upon exposure to light, TIM was degraded only if both CRY and JET were present (Fig. 3B) (Koh et al. 2006a). Mutant forms of JET, corresponding to the proteins in the *jet* mutant flies, were deficient in the cell culture assays mentioned above.

In the course of cloning *jet*, we found that multiple stocks in the laboratory carried a mutation in this gene. Mutations in *jet* may have been selected for because they facilitate adaptation to laboratory conditions, i.e., they allow flies to retain rhythms despite the irregular light conditions usually found in laboratories.

Other Extrinsic Factors That Affect the Clock

In addition to light, the clock can entrain to many other environmental factors such as temperature and social cues (Levine et al. 2002; Kaushik et al. 2007). In addition, metabolic activity may entrain the clock. Thus, mammals can be entrained to a restricted feeding paradigm (Damiola et al. 2000; Stokkan et al. 2001). Although this has not been demonstrated for *Drosophila*, our recent findings suggest that the *Drosophila* clock may be sensitive to metabolic activity. More specifically, we have found that an increase in oxidative stress compromises clock function. These findings were made with flies mutant for the metabolic gene *foxo* (Zheng et al. 2007). The FOXO protein is best known for its regulation by the insulin signaling pathway; it is a transcription factor that is excluded from the nucleus in response to insulin action (Neufeld 2003). When in the nucleus, FOXO controls the transcription of many genes, including those that have antioxidant activity (Kops et al. 2002). Thus, *foxo* mutants have increased oxidative stress.

We found that the cycling of clock genes in peripheral clocks (clocks in nonbrain tissues) is dampened in *foxo* mutants (Zheng et al. 2007). Rest:activity rhythms are normal, indicating that central clock function is intact (Fig. 4). However, in response to low concentrations of paraquat (which further increases oxidative stress), the amplitude of the molecular oscillation in the central clock cells is also reduced, and rest:activity rhythms are rapidly abolished (Fig. 4). In contrast, wild-type flies retain rhythms in the presence of paraquat for up to 3 weeks (Koh et al. 2006b). These effects of FOXO occur in the fat

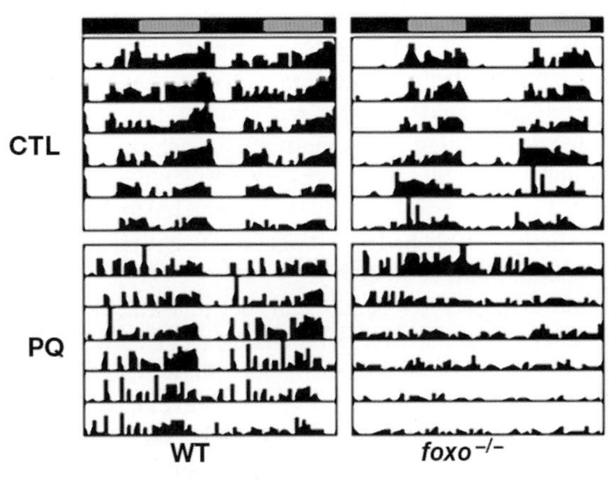

Figure 4. *foxo* mutants lose behavioral rhythms in the presence of paraquat. Activity rhythms were monitored for wild-type flies and *foxo* mutants in the presence or absence of 1 mM paraquat. Although wild-type flies are rhythmic under both conditions, *foxo* mutants lose rhythms when paraquat is added.

body, which is the fly equivalent of the liver and adipose tissue. Given that rest:activity rhythms are driven by the central clock in the brain, these data demonstrate that a peripheral metabolic tissue can affect clock function in the brain. Because aging is typically associated with an increase in oxidative stress, and we have found that circadian rhythms also break down with age (Koh et al. 2006b), we also examined the effects of age on *foxo* mutants. As one might expect, they show a premature breakdown of rhythms (Zheng et al. 2007).

SLEEP IN *DROSOPHILA*

The basic mechanisms, as well as most of the molecules, underlying circadian rhythms in *Drosophila* are conserved in mammals and have even been implicated in human circadian disorders (Toh et al. 2001; Xu et al. 2005). The success of the fly system in elucidating the mechanisms of circadian biology prompted us to ask whether the fly could also be a model for sleep. As noted above, the timing of sleep is controlled by the circadian system, but the need for sleep, which in turn determines the amount of sleep, is controlled by a homeostatic system. The nature of this homeostatic control is poorly understood, as is the function(s) served by sleep. There is the expectation, however, that an understanding of the molecular mechanisms underlying sleep and sleep homeostasis may reveal the function of sleep (Hendricks et al. 2000a), hence, the recent efforts to study sleep in organisms that lend themselves to genetic analysis.

To develop a *Drosophila* model for sleep, we sought to determine if criteria that have been proposed for sleep over the years are met by the rest phase in flies. Because electrophysiology experiments are difficult to conduct in the fly, we focused our efforts on behavioral characteristics of sleep. As described earlier, our circadian measurements had indicated the presence of well-consolidated daily rest periods. Thus, we knew that fly rest was a reversible state of behavioral quiescence regulated by the circadian clock. We also found that, like sleep, fly rest consists of long periods of behavioral immobility during which the arousal threshold (i.e., the minimum stimulus needed to invoke a response) is increased. More importantly, fly rest is controlled by homeostatic mechanisms which ensure that adequate amounts of rest are obtained. Thus, if flies are deprived of rest at night, they will make up for it by resting in the morning, a time at which they are normally active. These studies, and similar ones conducted simultaneously by another group, led to the fly rest state being established as a model for sleep (Hendricks et al. 2000b; Shaw et al. 2000).

Interestingly, fly sleep can be pharmacologically manipulated by the same neurochemicals that affect mammalian sleep (Hendricks et al. 2000b; Shaw et al. 2000). In addition, since the original model was developed, electrophysiological correlates of the fly sleep state have been described (Nitz et al. 2002). We recently discovered another aspect of fly rest that is shared with mammalian sleep—they both get fragmented with age. As flies get older, they show increased daytime sleep, decreased nighttime sleep, and decreased duration of sleep bouts (Koh et al. 2006b). Thus, in general, they have trouble maintaining a consolidated sleep state. Similar problems are known to occur in elderly humans, but the mechanisms are not well-understood (Pandi-Perumal et al. 2002). Our fly studies suggest that although circadian regulation may be disrupted with age, this is probably not the sole cause of the sleep disturbances in old flies. Young tim^{01} mutants have more fragmented sleep than their wild-type counterparts, but their sleep patterns also get worse with age (Koh et al. 2006b). Because tim^{01} flies have no clock to begin with, their age-induced defects must occur elsewhere, most likely in the homeostatic system.

The fly model for sleep allows us to use genetic approaches to ask fundamental questions about sleep. Perhaps the most important questions in basic sleep biology concern the cellular location of the sleep homeostat, and the molecules that comprise the homeostat. Work in our laboratory is directed toward addressing both of these questions.

Mapping Sleep-regulating Loci in the Fly Brain

Shortly after developing a fly model for sleep, we began to test candidate molecules for sleep-promoting or sleep-inhibiting effects. The first pathway we studied in this context was the cAMP/PKA (protein kinase A)/CREB (cAMP-response element-binding protein) pathway because we were interested in a possible connection between sleep and the consolidation of memory, and this pathway is implicated in learning and memory in all organisms studied (Mayford and Kandel 1999). We found that mutants of this pathway did indeed have effects on sleep, such that up-regulation of this pathway resulted in reduced sleep, whereas down-regulation was associated with increased sleep (Hendricks et al. 2001). The most dramatic phenotype was observed in flies that expressed a constitutively active form of protein kinase A (PKA). These flies had greatly reduced sleep.

To identify regions of the fly brain that are important for the regulation of sleep, we made use of the constitutively active PKA molecule (Joiner et al. 2006). We first expressed it under control of an inducible panneural promoter and verified that induction of PKA in adult flies is sufficient to reduce sleep. We then drove its expression with promoters that are expressed in different parts of the fly brain and assayed the effects on sleep. We found that the most dramatic sleep phenotypes were obtained when PKA was expressed in the mushroom bodies (MBs), which are a site of synaptic plasticity in the fly brain. The MBs are made up of fiber tracts that are organized into different lobes. Based on the region of the MB targeted, PKA expression either increased or decreased sleep (Fig. 5). We concluded that MBs contain two types of sleep-regulating neurons: those that promote sleep when PKA is increased and those that inhibit sleep under such conditions (Joiner et al. 2006).

To ascertain whether PKA acts in adult MB neurons to regulate sleep, we also expressed it under control of a drug-inducible MB promoter. This driver is expressed in the same pattern as the sleep-inhibiting promoter and, as expected, its induction in adult flies carrying the upstream

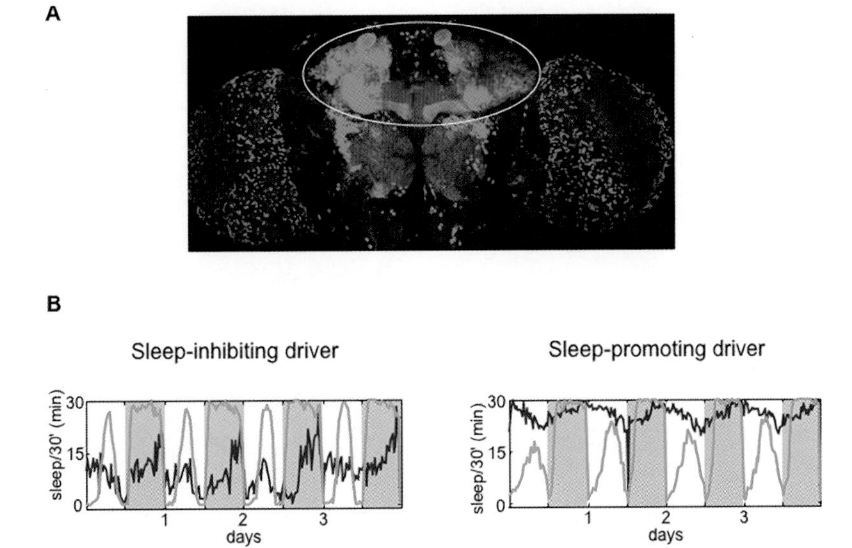

Figure 5. Mushroom bodies (MBs) are a sleep-regulating structure in the fly brain. (*A*) Location of the MBs in the fly brain. The image depicts GFP expression driven by a panneuronal driver (*elav*-Gene Switch). The MBs are encircled in white. The structures immediately below them are the antennal lobes. The punctate staining on the left and right corresponds to the optic lobes. As noted in the *text*, expression of PKA* with this driver inhibits sleep in a manner similar to that shown in the left profile in *B*. (*B*) Expression of a constitutively active PKA molecule (PKA*) is some areas of the MBs inhibits sleep (see sleep profile on the left), and in other areas it promotes sleep (profile on right). (*Blue trace*) Flies expressing PKA*; (*green trace*) controls.

activation sequence (UAS)-PKA transgenic resulted in reduced sleep. Analysis of the sleep architecture in flies expressing PKA under control of the inducible MB promoter revealed specific defects in the homeostatic mechanisms. Despite the reduction in sleep, the number of sleep bouts did not increase to compensate, suggesting a deficit in the expression of the homeostat (Joiner et al. 2006). However, the homeostatic accrual of a sleep-promoting signal was intact in these flies, based on the compensatory sleep observed when the drug was removed (Joiner et al. 2006).

We also used the MB promoter to probe the mechanisms perturbed by PKA action. To determine if PKA affects neuronal excitability, we directly manipulated electrical activity using the inducible promoter. To decrease firing, we employed a hyperpolarizing potassium channel, and to increase firing, we used a depolarizing sodium channel. Inducible expression of the former led to increased sleep, whereas the latter resulted in decreased sleep, mimicking the PKA phenotype. Thus, PKA most likely increases neuronal output in the sleep-inhibiting neurons of the MBs (Joiner et al. 2006).

Two other lines of evidence supported a role for the MBs in regulating sleep. (1) Ablation of the MBs decreased sleep (Joiner et al. 2006). It should be noted, however, that the decrease in sleep was smaller than that seen with other manipulations of the MBs, supporting the idea that MBs contain neurons with opposing effects on sleep. (2) We found that the effects of a serotonin receptor on sleep are mediated in MBs (Yuan et al. 2006). We proposed a model for how MBs regulate sleep taking into account these opposing influences and the effect of MB ablation. Our model postulates that the activity of sleep-promoting and sleep-inhibiting neurons in the MBs is

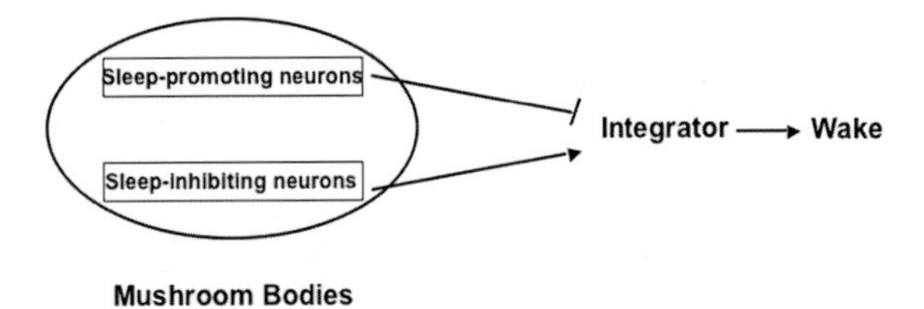

Figure 6. Model for control of sleep by the MBs. As noted in the text, we found that the MBs contain sleep-promoting and sleep-inhibiting neurons. Activity of these is most likely integrated to produce the overt behavioral state. In the default state (in the absence MBs), there is increased wakefulness. We propose that activity of the sleep-inhibiting neurons further increases wakefulness. The sleep-promoting neurons most likely exert an inhibitory influence on the default state and promote sleep. (Adapted from Joiner et al. 2006.)

integrated to produce the overt behavioral state (Fig. 6). The default mode of the hypothetical integrator promotes wakefulness; thus, ablation of the MBs results in reduced sleep. The sleep-inhibiting neurons increase the activity of the integrator to further promote the wake state. The sleep-promoting neurons inhibit the integrator to reduce wakefulness and promote sleep.

Molecules That Regulate Sleep

To identify molecules that regulate sleep, we are taking the following multiple approaches.

1. Testing candidate molecules. Candidates are picked either because they have been implicated in sleep by work in other model systems or because of their link with some hypothesized function for sleep. As noted

above, the cAMP pathway was tested as part of this approach.
2. Identifying targets of pharmacological reagents that affect sleep.
3. Conducting random mutagenesis screens for mutations that affect sleep and then cloning the genes mutated.
4. Looking for genes expressed differentially as a function of behavioral state.

It is believed that sleep-promoting factors build up when an organism has been awake for a long period of time. This belief is based on experiments done many years ago in which cerebrospinal fluid extracted from sleep-deprived goats was shown to promote sleep in rested (i.e., nonsleep-deprived) controls (Pappenheimer et al. 1967). The nature of the sleep-promoting molecules is not known, although attempts were made to purify such molecules (Pappenheimer et al. 1975). We are interested in identifying homeostatic factors that promote sleep and, to this end, looked for changes in gene expression during sleep deprivation and sleep rebound (compensatory sleep following deprivation) in flies (Williams et al. 2007). It is possible that regulation of the critical sleep-promoting molecules does not occur at the transcriptional level, but earlier studies indicated that gene expression profiles do change with altered sleep regulation, although it is not yet known if they are the cause or the effect of the altered sleep (Cirelli and Tononi 1999).

To identify changes in gene expression associated with sleep deprivation or sleep rebound, we conducted microarray analysis of the entire *Drosophila* genome (Williams et al. 2007). This resulted in the identification of a number of RNAs that are up-regulated or down-regulated during these behavioral states. The class of molecules that stood out among these was that of the immune response genes. *Drosophila* have an innate immune system that includes many components shared with the mammalian immune/inflammatory system (Brennan and Anderson 2004). Of note is the protein NF-κb that mediates many immune/inflammatory responses. We found that sleep deprivation of *Drosophila* results in the up-regulation of *Relish*, one of the *Drosophila* NF-κb genes, and of many other immune response genes, some of which function downstream from *Relish*. Interestingly, inflammatory markers such as cytokines and NF-κb are also up-regulated in sleep-deprived mammals and are thought to promote sleep (Majde and Krueger 2005).

Using the genetics available in *Drosophila*, we next asked whether NF-κb promotes sleep in flies (Williams et al. 2007). We found that sleep is not reduced in flies that lack *Relish*, but it is reduced in flies heterozygous for a mutation in *Relish*. The lack of a phenotype in the homozygote may reflect compensation of some sort, not unlike what occurs with other phenotypes associated with processes such as aging (Rogina et al. 2000). The sleep phenotype in *Relish* mutants was rescued by expressing a wild-type *Relish* transgene in the fat body (*Drosophila* equivalent of the liver and adipose tissue and a major site of immune signaling), suggesting that the activity of peripheral tissues can affect the sleep state. We also asked

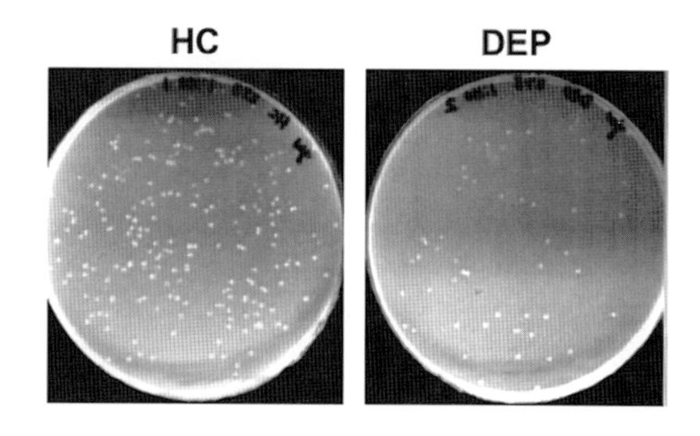

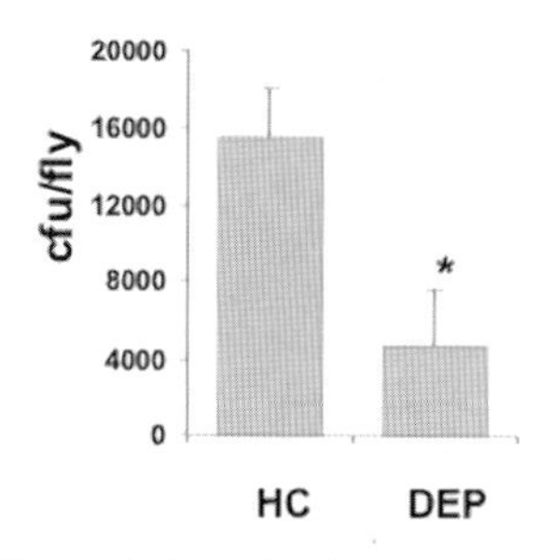

Figure 7. Effects of sleep deprivation on the immune assay. Flies were injected with bacteria, and 24 hours later, they were homogenized and plated onto culture plates containing LB medium and ampicillin. The number of colony-forming units was scored for flies sleep deprived for several hours (DEP) and handled controls (HC). Sleep-deprived flies formed fewer colonies, indicating increased resistance to infection.

the question of whether sleep deprivation affects the immune response. To our surprise, we found that resistance to infection was increased in sleep-deprived flies (Fig. 7). Although this seems to be counterintuitive, it is consistent with the up-regulated expression of immune genes under these conditions. We suggest that the enhancement of immune/inflammatory responses during acute sleep deprivation evolved as a mechanism to allow organisms to cope with intermittent periods of forced sleeplessness. We would predict that immune responses would be compromised with chronic sleep deprivation, but this has not yet been experimentally tested.

CONCLUSIONS

Using a *Drosophila* model, we have gained considerable insight into the mechanisms that constitute the clock and synchronize it to light. The importance of the transcription-based feedback loop is attested to by its conservation across species, but it is becoming increasingly clear that tightly controlled posttranslational modification of clock proteins is perhaps even more critical for timekeeping by the clock. The acute response of the *Drosophila* clock to light also involves changes at the posttranslational level, although these are probably followed rapidly by transcriptional effects. We are now also realizing that extrinsic factors other than light can have profound effects on the clock. With respect to the outputs controlled by the clock, although we still understand little about the mechanisms used by the clock to transmit time-of-day signals, studies have been initiated to directly investigate the multilevel regulation of one output, sleep. Since the first demonstration that *Drosophila* rest is a sleeplike state controlled by homeostatic mechanisms, in addition to the well-known circadian regulation, *Drosophila* has become popular as a model for sleep. Studies of homeostatic control are revealing important sleep-regulating sites in the fly brain and also identifying molecules that affect sleep. It is expected that future work will identify the cellular and molecular circuits that underlie circadian and homeostatic control of sleep and indicate the point at which these systems intersect.

ACKNOWLEDGMENTS

We thank Susan Kelchner for administrative assistance and other members of the laboratory for their effort on projects not reported here, but which impacted work described in this chapter. Work in the laboratory is supported by the National Institutes of Health and the Howard Hughes Medical Institute.

REFERENCES

Akten B., Jauch E., Genova G.K., Kim E.Y., Edery I., Raabe T., and Jackson F.R. 2003. A role for CK2 in the *Drosophila* circadian oscillator. *Nat. Neurosci.* **6:** 251.

Brennan C.A. and Anderson K.V. 2004. *Drosophila:* The genetics of innate immune recognition and response. *Annu. Rev. Immunol.* **22:** 457.

Cirelli C. and Tononi G. 1999. Differences in brain gene expression between sleep and waking as revealed by mRNA differential display and cDNA microarray technology. *J. Sleep Res.* (suppl. 1) **8:** 44.

Cyran S.A., Buchsbaum A.M., Reddy K.L., Lin M.C., Glossop N.R., Hardin P.E., Young M.W., Storti R.V., and Blau J. 2003. vrille, Pdp1, and dClock form a second feedback loop in the *Drosophila* circadian clock. *Cell* **112:** 329.

Damiola F., Le Minh N., Preitner N., Kornmann B., Fleury-Olela F., and Schibler U. 2000. Restricted feeding uncouples circadian oscillators in peripheral tissues from the central pacemaker in the suprachiasmatic nucleus. *Genes Dev.* **14:** 2950.

Emery P., Stanewsky R., Helfrich-Förster C., Emery-Le M., Hall J.C., and Rosbash M. 2000. *Drosophila* CRY is a deep brain circadian photoreceptor. *Neuron* **26:** 493.

Fang Y., Sathyanarayanan S., and Sehgal A. 2007. Post-translational regulation of the *Drosophila* circadian clock requires protein phosphatase 1 (PP1). *Genes Dev.* **21:** 1506.

Glossop N.R., Lyons L.C., and Hardin P.E. 1999. Interlocked feedback loops within the *Drosophila* circadian oscillator. *Science* **286:** 766.

Glossop N.R., Houl J.H., Zheng H., Ng F.S., Dudek S.M., and Hardin P.E. 2003. VRILLE feeds back to control circadian transcription of Clock in the *Drosophila* circadian oscillator. *Neuron* **37:** 249.

Hardin P.E. 2005. The circadian timekeeping system of *Drosophila*. *Curr. Biol.* **15:** R714.

Hendricks J.C., Sehgal A., and Pack A.I. 2000a. The need for a simple animal model to understand sleep. *Prog. Neurobiol.* **61:** 339.

Hendricks J.C., Finn S.M., Panckeri K.A., Chavkin J., Williams J.A., Sehgal A., and Pack A.I. 2000b. Rest in *Drosophila* is a sleep-like state. *Neuron* **25:** 129.

Hendricks J.C., Williams J.A., Panckeri K., Kirk D., Tello M., Yin J.C., and Sehgal A. 2001. A non-circadian role for cAMP signaling and CREB activity in *Drosophila* rest homeostasis. *Nat. Neurosci.* **4:** 1108.

Hunter-Ensor M., Ousley A., and Sehgal A. 1996. Regulation of the *Drosophila* protein timeless suggests a mechanism for resetting the circadian clock by light. *Cell* **84:** 677.

Joiner W.J., Crocker A., White B.H., and Sehgal A. 2006. Sleep in *Drosophila* is regulated by adult mushroom bodies. *Nature* **441:** 757.

Kaushik R., Nawathean P., Busza A., Murad A., Emery P., and Rosbash M. 2007. PER-TIM interactions with the photoreceptor cryptochrome mediate circadian temperature responses in *Drosophila*. *PLoS Biol.* **5:** e146.

Kloss B., Price J.L., Saez L., Blau J., Rothenfluh A., Wesley C.S., and Young M.W. 1998. The *Drosophila* clock gene double-time encodes a protein closely related to human casein kinase I epsilon. *Cell* **94:** 97.

Koh K., Zheng X., and Sehgal A. 2006a. JETLAG resets the *Drosophila* circadian clock by promoting light-induced degradation of TIMELESS. *Science* **312:** 1809.

Koh K., Evans J.M., Hendricks J.C., and Sehgal A. 2006b. A *Drosophila* model for age-associated changes in sleep:wake cycles. *Proc. Natl. Acad. Sci.* **103:** 13843.

Kops G.J., Dansen T.B., Polderman P.E., Saarloos I., Wirtz K.W., Coffer P.J., Huang T.T., Bos J.L., Medema R.H., and Burgering B.M. 2002. Forkhead transcription factor FOXO3a protects quiescent cells from oxidative stress. *Nature* **419:** 316.

Levine J.D., Funes P., Dowse H.B., and Hall J.C. 2002. Resetting the circadian clock by social experience in *Drosophila melanogaster*. *Science* **298:** 2010.

Lin F.J., Song W., Meyer-Bernstein E., Naidoo N., and Sehgal A. 2001. Photic signaling by cryptochrome in the *Drosophila* circadian system. *Mol. Cell. Biol.* **21:** 7287.

Lin J.M., Kilman V.L., Keegan K., Paddock B., Emery-Le M., Rosbash M., and Allada R. 2002. A role for casein kinase 2alpha in the *Drosophila* circadian clock. *Nature* **420:** 816.

Majde J.A. and Krueger J.M. 2005. Links between the innate immune system and sleep. *J. Allergy Clin. Immunol.* **116:** 1188.

Martinek S., Inonog S., Manoukian A.S., and Young M.W. 2001. A role for the segment polarity gene shaggy/GSK-3 in the *Drosophila* circadian clock. *Cell* **105:** 769.

Mayford M. and Kandel E.R. 1999. Genetic approaches to memory storage. *Trends Genet.* **15:** 463.

Myers M.P., Wager-Smith K., Rothenfluh-Hilfiker A., and Young M.W. 1996. Light-induced degradation of TIMELESS and entrainment of the *Drosophila* circadian clock. *Science* **271:** 1736.

Naidoo N., Song W., Hunter-Ensor M., and Sehgal A. 1999. A role for the proteasome in the light response of the timeless clock protein. *Science* **285:** 1737.

Neufeld T.P. 2003. Shrinkage control: Regulation of insulin-mediated growth by FOXO transcription factors. *J. Biol.* **2:** 18.

Nitz D.A., van Swinderen B., Tononi G., and Greenspan R.J. 2002. Electrophysiological correlates of rest and activity in *Drosophila melanogaster*. *Curr. Biol.* **12:** 1934.

Pandi-Perumal S.R., Seils L.K., Kayumov L., Ralph M.R., Lowec A.H., Moller H., and Swaab D.F. 2002. Senescence, sleep, and circadian rhythms. *Ageing Res. Rev.* **1:** 559.

Pappenheimer J.R., Miller T.B., and Goodrich C.A. 1967. Sleep-promoting effects of cerebrospinal fluid from sleep-deprived goats. *Proc. Natl. Acad. Sci.* **58:** 513.

Pappenheimer J.R., Koski G., Fencl V., Karnovsky M.L., and Krueger J. 1975. Extraction of sleep-promoting factor S from cerebrospinal fluid and from brains of sleep-deprived animals. *J. Neurophysiol.* **38:** 1299.

Price J.L., Dembinska M.E., Young M.W., and Rosbash M. 1995. Suppression of PERIOD protein abundance and circadian cycling by the *Drosophila* clock mutation timeless. *EMBO J.* **14:** 4044.

Rogina B., Reenan R.A., Nilsen S.P., and Helfand S.L. 2000. Extended life-span conferred by cotransporter gene mutations in *Drosophila*. *Science* **290:** 2137.

Sathyanarayanan S., Zheng X., Xiao R., and Sehgal A. 2004.

Post-translational modification of *Drosophila* PERIOD protein by protein phosphatase 2A. *Cell* **116:** 603.
Sehgal A., Ed. 2004. *Molecular biology of circadian rhythms*. Wiley-Liss, Hoboken, New Jersey.
Shaw P.J., Cirelli C., Greenspan R.J., and Tononi G. 2000. Correlates of sleep and waking in *Drosophila melanogaster*. *Science* **287:** 1834.
Stanewsky R., Kaneko M., Emery P., Beretta B., Wager-Smith K., Kay S.A., Rosbash M., and Hall J.C. 1998. The cryb mutation identifies cryptochrome as a circadian photoreceptor in *Drosophila*. *Cell* **95:** 681.
Stokkan K.A., Yamazaki S., Tei H., Sakaki Y., and Menaker M. 2001. Entrainment of the circadian clock in the liver by feeding. *Science* **291:** 490.
Toh K.L., Jones C.R., He Y., Eide E.J., Hinz W.A., Virshup D.M., Ptáček L.J., and Fu Y.H. 2001. An hPer2 phosphorylation site mutation in familial advanced sleep phase syndrome. *Science* **291:** 1040.
Williams J.A., Sathyanarayanan S., Hendricks J.C., and Sehgal A. 2007. Interaction between sleep and the immune response in *Drosophila:* A role for the NFkappaB relish. *Sleep* **30:** 389.
Xu Y., Padiath Q.S., Shapiro R.E., Jones C.R., Wu S.C., Saigoh N., Saigoh K., Ptáček L.J., and Fu Y.H. 2005. Functional consequences of a CKIdelta mutation causing familial advanced sleep phase syndrome. *Nature* **434:** 640.
Yang Z. and Sehgal A. 2001. Role of molecular oscillations in generating behavioral rhythms in *Drosophila*. *Neuron* **29:** 453.
Yuan Q., Joiner W.J., and Sehgal A. 2006. A sleep-promoting role for the *Drosophila* serotonin receptor 1A. *Curr. Biol.* **16:** 1051.
Zeng H., Qian Z., Myers M.P., and Rosbash M. 1996. A light-entrainment mechanism for the *Drosophila* circadian clock. *Nature* **380:** 129.
Zheng X., Yang Z., Yue Z., Alvarez, J.D., and Sehgal A. 2007. Foxo and the insulin signaling pathway regulate sensitivity of the circadian clock to oxidative stress. *Proc. Natl. Acad. Sci.* **104:** 15899.

Neurohormonal and Neuromodulatory Control of Sleep in *Drosophila*

K. Foltenyi,*† R. Andretic,† J.W. Newport,*‡ and R.J. Greenspan†
*Department of Biological Sciences, University of California, San Diego, La Jolla, California 92093; †The Neurosciences Institute, San Diego, California 92121

The fruit fly *Drosophila melanogaster* has emerged in recent years as a tractable system for studying sleep. The sleep-wake dichotomy represents one of the principal transitions in global brain state, and neurohormones and neuromodulators are well known for their ability to change global brain states. Here, we describe studies of two brain systems that regulate sleep in *Drosophila*, the neurohormonal epidermal growth factor receptor system and the neuromodulatory dopaminergic system, each of which acts through a discrete anatomical locus in the dorsal brain. Both control systems display considerable mechanistic similarity to those in mammals, suggesting possible functional homologies.

INTRODUCTION

Sleep is one of the most obvious aspects of physiology and behavior to be regulated by the circadian system. Originally assumed to include only mammals and birds, the circle of sleeping animals has been gradually expanded to include reptiles, amphibians, fish, and eventually invertebrates (Campbell and Tobler 1984; Siegel 1995). Despite many years of study, the fundamental function of sleep remains a mystery. In recent years, the fruit fly *Drosophila melanogaster* has come into its own as a model system for all factors human (with the possible exception of dermatology or dentistry), so it was only a matter of time before it too was tested and shown to exhibit bona fide sleep-like behavior (Hendricks et al. 2000; Shaw et al. 2000; Cirelli 2006).

A phenomenon such as sleep is played out on many levels in the nervous system and thus requires a multilevel analysis. The approach described here aims at manipulating the fly's signal transduction machinery and its neuromodulatory system in order to probe its sleep mechanisms. These studies have revealed requirements for the epidermal growth factor receptor (*Egf-r*)/EGFR-induced extracellular signal-regulated kinase (ERK) signal transduction pathway and for the dopaminergic system as essential elements of the sleep mechanism and have begun to place them in their anatomical context.

THE Egf-r PATHWAY

EGFR signal transduction is one of the classic pathways in development, where it mediates a wide array of cell fate and developmental polarity events in a broad range of organisms (Shilo 2005). The pathway has also been found to be involved in regulating some rhythmic behaviors in hamsters (Kramer et al. 2001), and ectopic overstimulation of the pathway could increase sleep levels in rabbits (Kushikata et al. 1998). Furthermore, ERK, which is activated by EGFR signaling among many other upstream signaling switches, has been shown to have a role in synaptic plasticity in mammals (Sweatt 2004) as well as in flies (Hoeffer et al. 2003). In the circadian system, ERK signaling is required for normal rhythmicity in *Drosophila* (Williams et al. 2001) and has been implicated in light entrainment in mammals (Coogan and Piggins 2004).

With these findings as background, and with the extensive armamentarium of molecular genetic tools for manipulating *Egf-r* signaling in *Drosophila*, we tested this system's role in sleep by activating and inhibiting upstream components of the pathway. Secretion of *Egf-r* ligands is regulated by the processing proteins Rhomboid and Star (Shilo 2005). Heat shock induction of these gene products was our tool for generating gain-of-function variants, and an RNA interference (RNAi) construct targeted at Rhomboid (Guichard et al. 2002) for generating loss-of-function variants, all under the control of the Gal4-UAS (upstream activating sequence) system (Brand and Perrimon 1993).

MANIPULATION OF THE Egf-r PATHWAY AND SLEEP

When the ligand-processing protein Rhomboid is transiently overexpressed, flies show a pronounced increase in total sleep once they have recovered from the heat shock treatment itself (Fig. 1, top/middle) (Foltenyi et al. 2007). When Star as well as Rhomboid is overexpressed, the effect is greater (Fig. 1, bottom), and when the dose of both is doubled (with two doses of the heat shock–Gal4 driver), the effect is further increased (Fig. 1, top/middle). Overexpression of an *Egf-r* ligand, a secreted form of the Spitz (s-Spitz) protein (Schweitzer et al. 1995), similarly increases sleep (Fig. 1, top/middle). An inactive mutant form of Rhomboid, rho^{H281Y} (Urban et al. 2002) had no effect other than a short-lived response to the heat shock (Fig. 1, top/middle). Blockade of *Egf-r* responsiveness, using the dominant-negative $Egf\text{-}r^{DN}$ (Freeman 1996),

‡Deceased 25 December 2005.

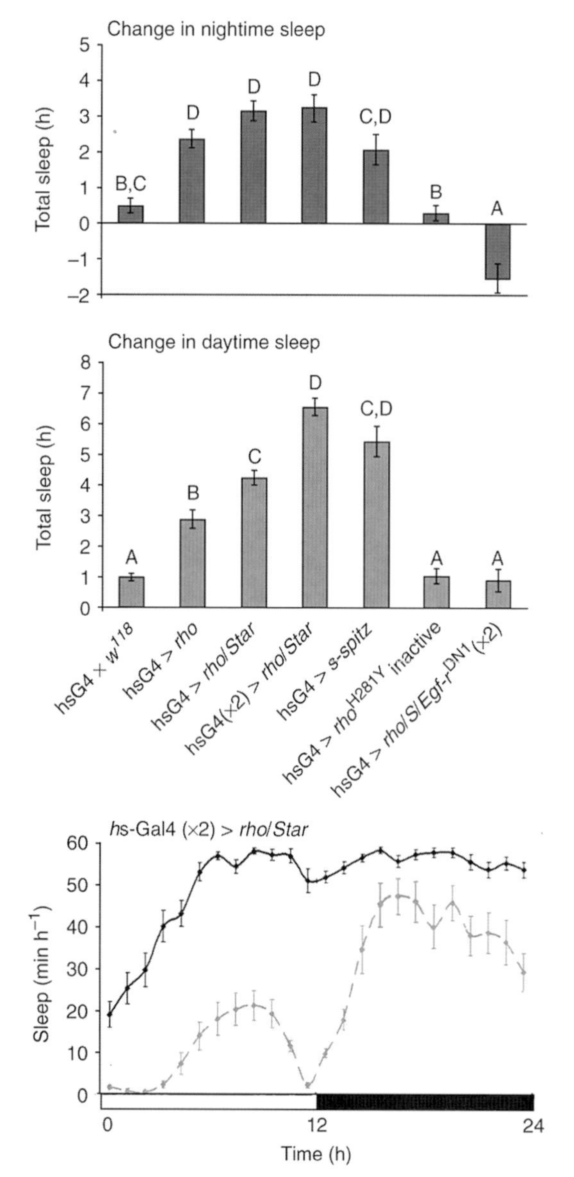

Figure 1. Dose-dependent increase in sleep relative to preheat shock baseline during the night (*top*) or the day (*middle*) after overexpression of Rho or Rho and Star, and blockade with *Egf-r^{DN1}*. Letters above bars represent statistically significant groups as determined by the Tukey–Kramer HSD test for normally distributed data ($P < 0.05$). (*Bottom*) Average activity traces of the effect of stimulating EGFR signaling with hs-Gal4 (2 doses) > *rho,Star*. (*Gray trace*) Preheat shock; (*black trace*) postheat shock. (Reprinted, with permission, from Foltenyi et al. 2007 [Nature Publishing Group].)

prevented and reversed the increase in sleep due to overexpression of Rhomboid and Star (Fig. 1, top). The increase in total sleep in the affected genotypes was due to an increase in both sleep bout number and bout duration, indicating that flies with up-regulated *Egf-r* activity initiated sleep more often and maintained the state longer than normal (Foltenyi et al. 2007). As an indicator of the general health of the flies, locomotor activity during waking bouts was within the range of control genotypes for all strains whose sleep was affected (Foltenyi et al. 2007).

When Rhomboid expression is attenuated throughout the nervous system, flies show a correspondingly pronounced decrease in total sleep (Fig. 2). This was accomplished by driving expression of an RNAi against *rhomboid*, *rho^{DN}*, that was driven by the *elav* panneural promoter (Robinow and White 1988). A screen of 48 driver strains with expression in the central nervous system narrowed down the focus of *rho^{DN}* action to a relatively restricted set of cells in the pars intercerebralis that project from the dorsal surface of the protocerebrum down past the esophagus into the tritocerebrum (Fig. 3, left shows one of these, c687). As in the case of overexpression, the number and duration of sleep bouts were affected with an increase in sleep bout number but a dramatic decrease in duration. Once again, their locomotor activity during waking bouts was within the normal range. Circadian rhythms, both period and phase, were also within the wild-type range for these strains (Foltenyi et al. 2007).

To confirm that the pertinent cells in the pars intercerebralis were expressing the *rhomboid* gene product, fly brains expressing a reporter *lacZ* gene under the control of another of the pars intercerebralis drivers, 50Y, were stained with an antibody to the *lacZ* gene product and also counterstained by in situ hybridization with *rho* antisense RNA. The merged image (Fig. 3, right) indicates the presence of the *rhomboid* gene product in the 50Y cells.

Having shown that restricted underexpression of Rhomboid decreases sleep, we wanted to confirm that the heat-shock-induced increase in sleep was at least in part acting through the same cells as the RNAi-induced decrease. To this end, a strain capable of temperature-sensitive activation of the restricted drivers was tested using $Gal80^{ts}$, a conditional Gal4 repressor, driven ubiquitously by a tubulin promoter (McGuire et al. 2004), and the *Egf-r* ligand s-Spitz (cf. Fig. 1). When Gal80 repression of s-Spitz expression in the pars intercerebralis is relieved in adults, the predicted increase in sleep is obtained (Foltenyi et al. 2007).

Finally, to address the question if *Egf-r* activation corresponds with the change in behavior, the time course of increased sleep was compared with the time course of molecular changes. When Rhomboid is overexpressed, the increase in its level far outlasts the duration of increased sleep (Fig. 4, top left). In contrast, the time course of activation of the principal intracellular target of *Egf-r*, ERK, matches that of the sleep increase very closely (Fig. 4, top right). This was measured using an antibody to the activated (phosphorylated) form of ERK (Foltenyi et al. 2007). Moreover, the principal site of ERK activation after heat shock is in the tritocerebrum (Fig. 4, bottom left), target region for the key cells in the pars intercerebralis (Fig. 3, left).

There is thus a convergence of phenotypes in which molecular effects on ERK activation map both temporally and spatially onto the functional effects on sleep of *Egf-r* manipulation. The upshot is that real-time physiological activation or inhibition of the neurohormonal *Egf-r* pathway profoundly affects the sleep state of *Drosophila*.

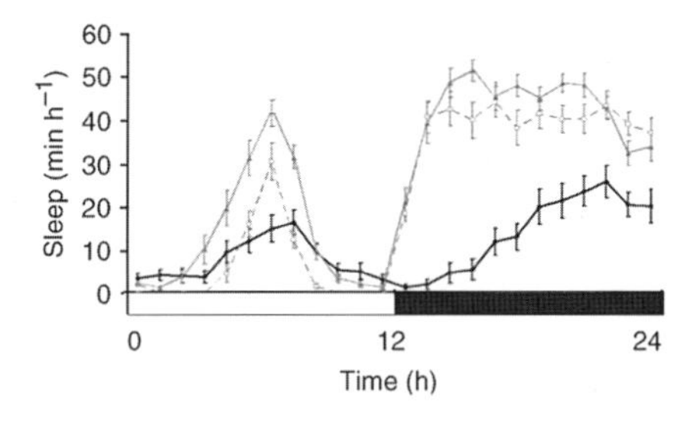

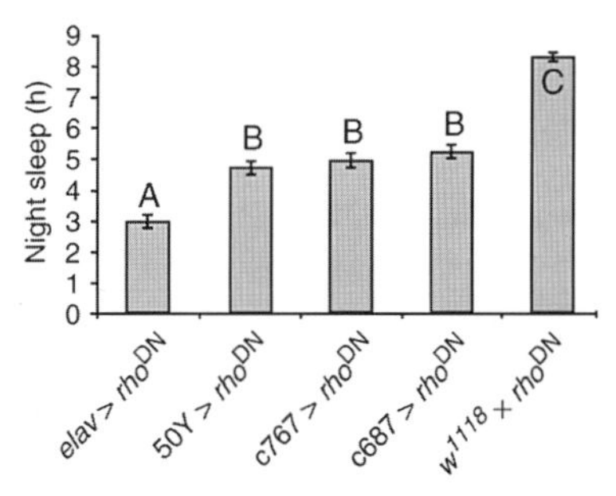

Figure 2. Reduced sleep after directed *rho*-RNAi expression. (*Left*) Representative traces of sleep levels for one 24-hour light:dark cycle for flies with UAS-rho^{DN} (*rho* RNAi) driven by the panneural *elav*-Gal4 driver (*black trace*) and its controls, *elav*-Gal4 (*dashed gray trace*), and UAS-rho^{DN} (*solid gray trace*). (*Right*) Histograms for the total nighttime sleep levels for *elav* > rho^{DN}, 50Y > rho^{DN}, c767 > rho^{DN}, and c687 > rho^{DN} (error bars represent ±S.E.M.) and their common control. Letters above bars represent statistically significant groups as determined by the Tukey–Kramer HSD test for normally distributed data ($P < 0.05$). (Reprinted, with permission, from Foltenyi et al. 2007 [Nature Publishing Group].)

THE DOPAMINERGIC SYSTEM IN SLEEP AND AROUSAL

Dopamine is one of several neurotransmitter systems that modulate arousal states in mammals (Robbins et al. 1998; Jones 2005; Monti and Monti 2007). In *Drosophila*, studies of diurnal and sleep-related gene expression have revealed a dopamine receptor and a dopamine catabolizing enzyme, arylalkylamine-*N*-acetyltransferase (Shaw et al. 2000; Cirelli et al. 2005; Zimmerman et al. 2006).

When flies are depleted of dopamine after feeding on 3-iodo-tyrosine (3IY) (Neckameyer 1998), the total amount of time spent sleeping in each 24-hour period increases nearly fourfold (Fig. 5A) (see Andretic et al. 2005). The general health of these flies is unimpaired as measured by the amount of locomotor activity they show during waking epochs. A pharmacological treatment that potentiates dopamine activity, although not as specifically as 3IY inhibits it, produces the opposite effect. Flies fed on methamphetamine sleep less than normal (Fig. 5B) and do not show a normal homeostatic response after sleep deprivation (Fig. 5C).

Genetic manipulations of dopamine levels have comparable effects. A fly mutant for the sleep-regulated enzyme arylalkylamine *N*-acetyltransferase takes an abnormally long time to recover from sleep deprivation, an effect that is dose-dependent (Shaw et al. 2000). A mutant in the fly dopamine transporter gene, *fumin* (Kume et al. 2005), sleeps less than normal and does not show a normal homeostatic response after sleep deprivation. Overexpression of the fly's vascular monoamine transporter protein (VMAT) in its monoaminergic cells produces a heightened level of motor activity (Chang et al. 2006).

Other measures of arousal are also affected by manipulations that modify dopaminergic cells. Methamphetamine substantially increases the courtship ardor of male flies (Andretic et al. 2005). Genetic manipulations of excitability and synaptic release in dopaminergic cells alter the fly's sensitivity to anesthesia: Hyperexcitation decreases sensitivity and blockade of dopamine release increases sensitivity (van Swinderen 2006). The fly's response to the salience of a stimulus is dependent on functional synaptic release from dopaminergic neurons (Andretic et al. 2005; Zhang et al. 2007), a property that is reflected in the defects in associative conditioning seen after blockade of

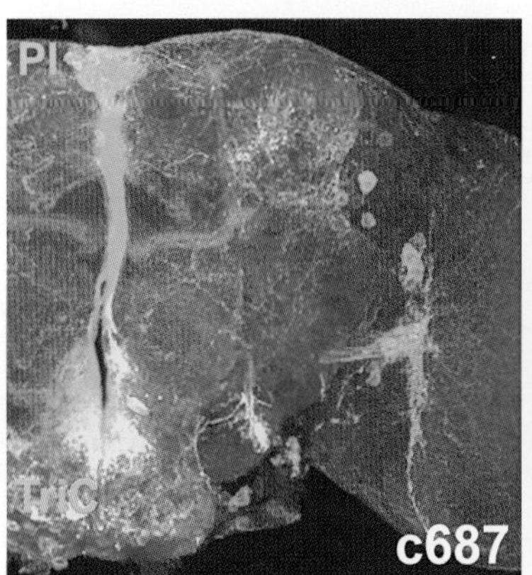

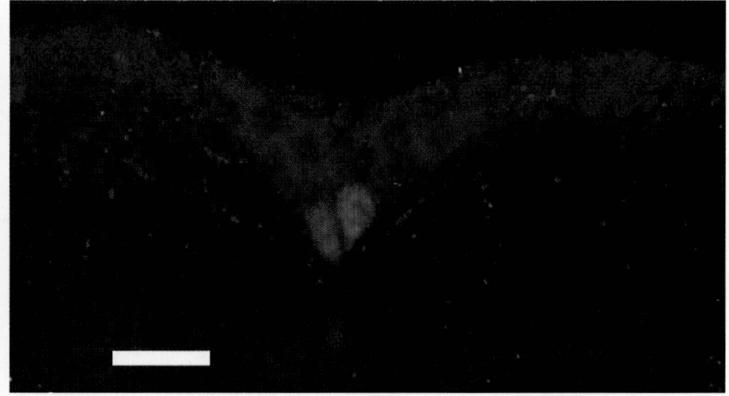

Figure 3. Confocal images of whole-mount brains, frontal view: (*Left*) the pars intercerebralis (PI) expression pattern of Gal4-c687 as revealed by a membrane-bound form of green fluorescent protein (GFP) (UAS-mCD8::GFP.L) (*green*), with the neuropil counterstained by antibody nc82 (*red*); (*right*) a close-up of the PI region of a brain expressing LacZ under 50Y control, costained for *rho* with antisense RNA (*red*), for an antibody to β-galactosidase (*green*), and with DAPI (*blue*) to show cell nuclei. Bar, 20 μm. (Reprinted, with permission, from Foltenyi et al. 2007 [Nature Publishing Group].)

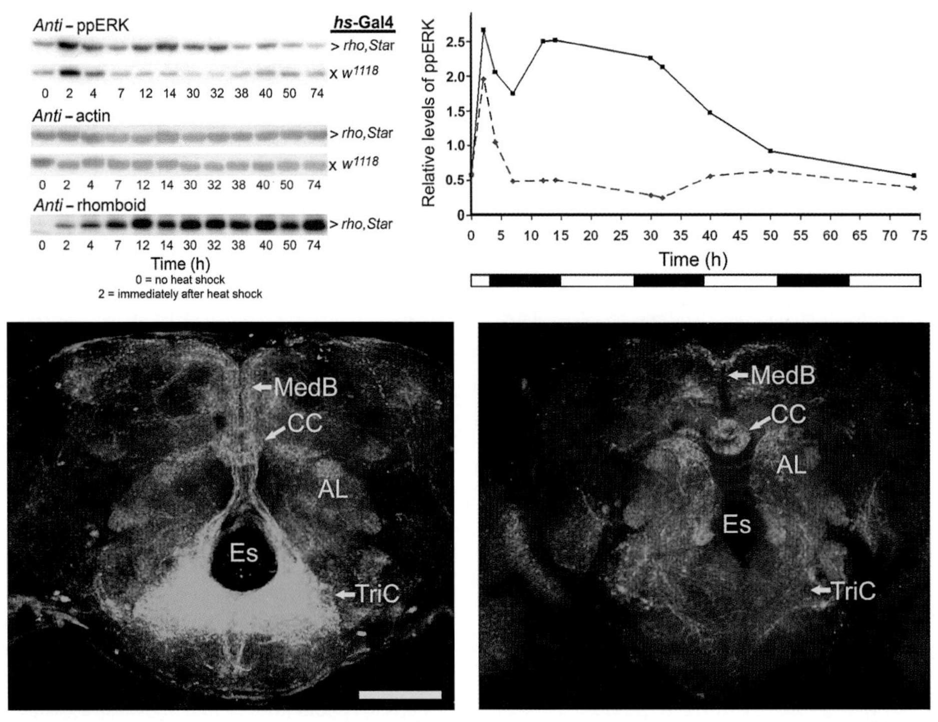

Figure 4. Increased sleep correlated with increased levels of ppERK. (*Top left*) Immunoblots for time course of heat-shock-induced expression in whole heads using antibodies against ppERK, actin, and Rho. (*Top right*) Time course of levels of ppERK, normalized to actin loading control in *top left*, prepared from fly heads. (*Bottom left*) A whole-mount hs-Gal4 > *rho,Star* brain after heat shock, stained for ppERK (*green*); (*bottom right*) its hs-Gal4 > w^{1118} control. (Reprinted, with permission, from Foltenyi et al. 2007 [Nature Publishing Group].)

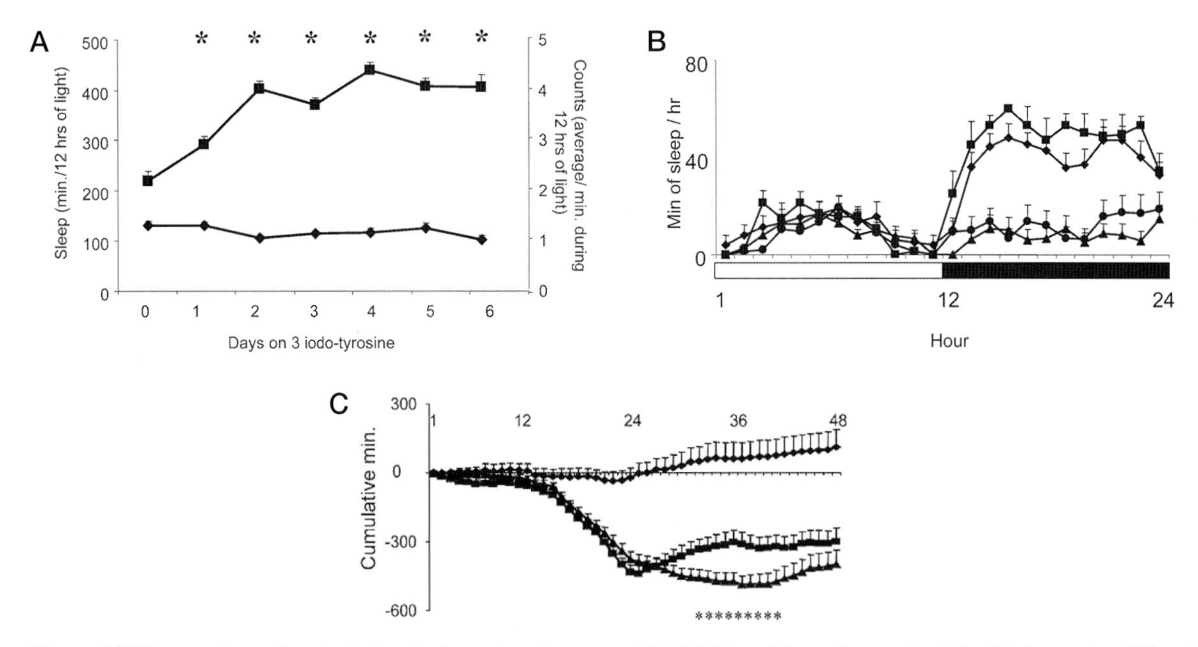

Figure 5. Effects on sleep of manipulating the dopaminergic system. (*A*) Inhibition of dopamine synthesis by 3-iodotyrosine. (*Closed square*) Total sleep during successive 24-hour periods; (*closed diamond*) counts per waking minute during each 24-hour period. *Asterisks* indicate statistically significant difference from baseline day 0 (*t*-test with Bonferroni correction, $P < 0.05$). (*B*) Potentiation of dopamine by methamphetamine. (*Closed square*) 0 mg/ml; (*closed diamond*) 0.32 mg/ml; (*closed circle*) 0.64 mg/ml; (*closed triangle*) 1.25 mg/ml mixed into the food. (*C*) Methamphetamine effect on rebound after sleep deprivation. Cumulative plot of sleep loss or gain for deprivation (day 1) and recovery (day 2) versus their respective baseline days. (*Closed diamond*) No treatment; (*closed triangle*) 9-hour sleep deprivation on day 1; (*closed square*) 9-hour sleep deprivation on day 1 + 1.25 mg/ml methamphetamine on day 2. *Asterisks* indicate statistically significant difference after sleep deprivation with (*closed square*) and without (*closed triangle*) methamphetamine (*t*-test with Bonferroni correction, $P < 0.05$). (Reprinted, with permission, from Andretic et al. 2005 [© Elsevier].)

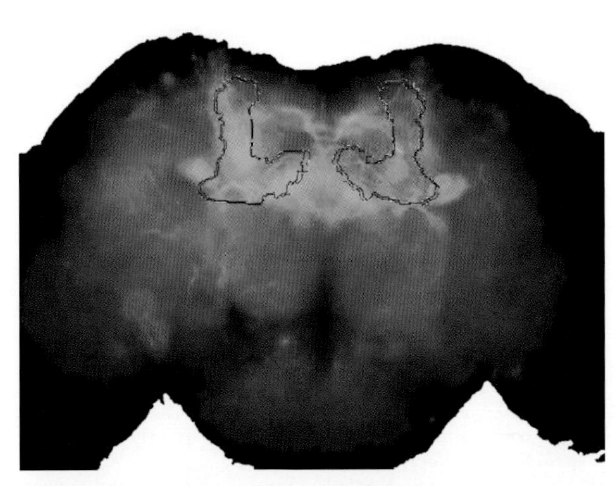

Figure 6. Dopaminergic system in the *Drosophila* brain. Whole-mount brain, frontal view, expressing EGFP in dopaminergic cells, driven by the promoter for tyrosine hydroxylase (*Th*-Gal4 > UAS-EGFP). Silhouettes of mushroom bodies are shown in black outline.

dopaminergic transmission (Schwaerzel et al. 2003) or dopamine receptor function (Kim et al. 2007).

The focus of all of these dopaminergic effects is likely to be the mushroom bodies, at least in part. Mushroom body output is necessary for the same responses that require dopamine: the salience response (van Swinderen and Greenspan 2003; Zhang et al. 2007) and associative conditioning (Connolly et al. 1996; Schwaerzel et al. 2003; Kim et al. 2007). Some of the most prominent dopaminergic projections in the fly brain terminate on the mushroom bodies (Fig. 6) (see Zhang et al. 2007), and mushroom bodies also have a role in modulating sleep (Joiner et al. 2006; Pitman et al. 2006).

INTEGRATING SLEEP SIGNALS IN THE FLY BRAIN

With the demonstration of two separate systems, *Egf-r* and dopamine, for regulating sleep, the question arises as to their relationship to each other. Are they sequential? Parallel? Independent? Linked? And what regulates them? The circadian system is a good place to start asking about their regulation, given the major circadian control of this behavior, as well as the effect of certain circadian mutants on it (Shaw et al. 2002; Hendricks et al. 2003b).

Anatomically, both systems are located in the dorsal medial brain and are thus well positioned to receive input from Pdf-containing cells that mediate locomotor rhythms in the brain (Fig. 7) (see Helfrich-Förster et al. 2007). Several of the Pdf-containing cells project to this area, and the suggestion has previously been made that their contacts with cells of the pars intercerebralis may regulate circadian neuropeptide secretion (Kaneko and Hall 2000). Moreover, the pars intercerebralis together with the corpus cardiacum have been suggested to be the developmental equivalent of the mammalian hypothalamic-pituitary axis (Veelaert et al. 1998; Chang et al. 2001; De Velasco et al. 2004; Hartenstein 2006). Because the hypothalamus is a major center in the mammalian brain for the regulation of arousal (Kilduff and Peyron 2000; Saper et al. 2005), and together with the pituitary gland secretes *Egf-r* ligands in mammals, it would appear that the pars intercerebralis shares some functional homology with the hypothalamus in its involvement in regulating arousal through neurohormones such as the *Egf-r* ligands.

The effects of manipulating dopaminergic transmission and the overt similarity between flies and mammals in the effects of drugs such as methamphetamine (see above) and modafinil (Hendricks et al. 2003a) that affect dopamine transporter activity suggest that there is also considerable

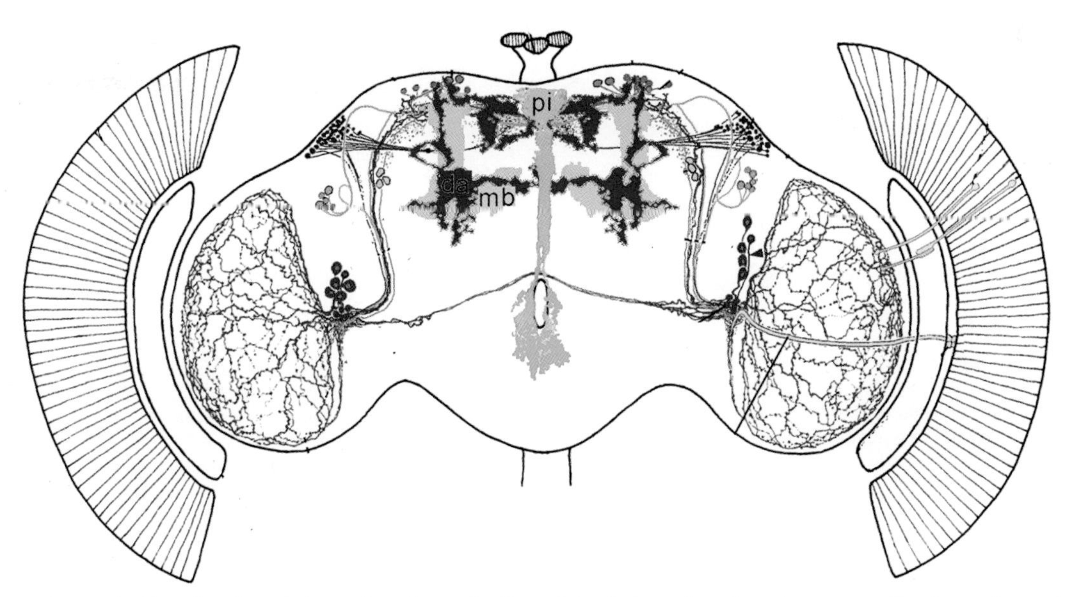

Figure 7. Anatomical location, frontal view, of mushroom bodies, mb (*tan*), dopaminergic system surrounding the mushroom bodies, da (*deep pink*), and pars intercerebralis neurons, pi (*green*) with respect to clock-expressing neurons: LN_d cells (*orange*); PDF-positive $l\text{-}LN_v$ and $s\text{-}LN_v$ cells (*red*); PDF-negative $s\text{-}LN_v$ cell (*violet*); DN_1, DN_2, and DN_3 cells (*blue*); PER/TIM-expressing LPNs (*olive green*). (Reprinted, with permission, from Helfrich-Förster et al. 2007 [©Wiley-Liss, Inc.].)

functional homology in this system. Dopamine is one of several neuromodulators making up the ascending systems of the mammalian brain (Jones 2005). Shared features of these systems include global effects on brain states, which can be long-lasting, and diffuse anatomical projections. The diffuseness of their projections has led to the suggestion that they function as a sprinkler system in the vertebrate brain, spraying their transmitters over relatively wide areas, rather than in point-to-point synaptic connections. In the fruit fly, these transmitter systems are also diffusely distributed (Monastirioti 1999), although they do not emanate from a centralized brain region and fan out into the brain as in vertebrates.

Whether these systems are closely linked to each other or are independent, as well as the targets of their activity, are open questions. Beyond these circuit considerations, there remains the issue of their relationship to the fundamental function of sleep.

ACKNOWLEDGMENTS

We thank Jenée Wagner and Esther J. Kim for technical assistance. This work was supported by a National Institutes of Health grant (J.W.N.) and a National Science Foundation grant 0432063 (R.J.G.). R.J.G. is the Dorothy and Lewis B. Cullman Fellow at the Neurosciences Institute, which is supported by the Neurosciences Research Foundation.

REFERENCES

Andretic R., van Swinderen B., and Greenspan R.J. 2005. Dopaminergic modulation of arousal in *Drosophila*. *Curr. Biol.* **15:** 1165.

Brand A.H. and Perrimon N. 1993. Targeted gene expression as a means of altering cell fates and generating dominant phenotypes. *Development* **118:** 401.

Campbell S.S. and Tobler I. 1984. Animal sleep: A review of sleep duration across phylogeny. *Neurosci. Biobehav. Rev.* **8:** 269.

Chang H.Y., Grygoruk A., Brooks E.S., Ackerson L.C., Maidment N.T., Bainton R.J., and Krantz D.E. 2006. Overexpression of the *Drosophila* vesicular monoamine transporter increases motor activity and courtship but decreases the behavioral response to cocaine. *Mol. Psychiatry* **11:** 99.

Chang T., Mazotta J., Dumstrei K., Dumitrescu A., and Hartenstein V. 2001. Dpp and Hh signaling in the *Drosophila* embryonic eye field. *Development* **128:** 4691.

Cirelli C. 2006. Sleep disruption, oxidative stress, and aging: New insights from fruit flies. *Proc. Natl. Acad. Sci.* **103:** 13901.

Cirelli C., LaVaute T.M., and Tononi G. 2005. Sleep and wakefulness modulate gene expression in *Drosophila*. *J. Neurochem.* **94:** 1411.

Connolly J.B., Roberts I.J., Armstrong J.D., Kaiser K. Forte M., Tully T., and O'Kane C.J. 1996. Associative learning disrupted by impaired Gs signaling in *Drosophila* mushroom bodies. *Science* **274:** 2104.

Coogan A.N. and Piggins H.D. 2004. MAP kinases in the mammalian circadian system—Key regulators of clock function. *J. Neurochem.* **90:** 769.

De Velasco B., Shen J., Go S., and Hartenstein V. 2004. Embryonic development of the *Drosophila* corpus cardiacum, a neuroendocrine gland with similarity to the vertebrate pituitary, is controlled by sine oculis and glass. *Dev. Biol.* **274:** 280.

Foltenyi K., Greenspan R.J., and Newport J.W. 2007. Activation of EGFR and ERK by rhomboid signaling regulates the consolidation and maintenance of sleep in *Drosophila*. *Nat. Neurosci.* **10:** 1160.

Freeman M. 1996. Reiterative use of the EGF receptor triggers differentiation of all cell types in the *Drosophila* eye. *Cell* **87:** 651.

Guichard A., Srinivasan S., Zimm G., and Bier E. 2002. A screen for dominant mutations applied to components in the *Drosophila* EGF-R pathway. *Proc. Natl. Acad. Sci.* **99:** 3752.

Hartenstein V. 2006. The neuroendocrine system of invertebrates: A developmental and evolutionary perspective. *J. Endocrinol.* **190:** 555.

Helfrich-Förster C., Shafer O.T., Wulbeck C., Grieshaber E., Rieger D., and Taghert P. 2007. Development and morphology of the clock-gene-expressing lateral neurons of *Drosophila melanogaster*. *J. Comp. Neurol.* **500:** 47.

Hendricks J.C., Kirk D., Panckeri K., Miller M.S., and Pack A.I. 2003a. Modafinil maintains waking in the fruit fly *Drosophila melanogaster*. *Sleep* **26:** 139.

Hendricks J.C., Lu S., Kume K., Yin J.C., Yang Z. and, Sehgal A. 2003b. Gender dimorphism in the role of cycle (BMAL1) in rest, rest regulation, and longevity in *Drosophila melanogaster*. *J. Biol. Rhythms* **18:** 12.

Hendricks J.C., Finn S.M., Panckeri K.A., Chavkin J., Williams J.A., Sehgal A., and Pack A.I. 2000. Rest in *Drosophila* is a sleep-like state. *Neuron* **25:** 129.

Hoeffer C.A., Sanyal S., and Ramaswami M. 2003. Acute induction of conserved synaptic signaling pathways in *Drosophila melanogaster*. *J. Neurosci.* **23:** 6362.

Joiner W.J., Crocker A., White B.H., and Sehgal A. 2006. Sleep in *Drosophila* is regulated by adult mushroom bodies. *Nature* **441:** 757.

Jones B.E. 2005. From waking to sleeping: Neuronal and chemical substrates. *Trends Pharmacol. Sci.* **26:** 578.

Kaneko M. and Hall J.C. 2000. Neuroanatomy of cells expressing clock genes in *Drosophila:* Transgenic manipulation of the period and timeless genes to mark the perikarya of circadian pacemaker neurons and their projections. *J. Comp. Neurol.* **422:** 66.

Kilduff T.S. and Peyron C. 2000. The hypocretin/orexin ligand-receptor system: Implications for sleep and sleep disorders. *Trends Neurosci.* **23:** 359.

Kim Y.C., Lee H.G., and Han K.A. 2007. D1 dopamine receptor dDA1 is required in the mushroom body neurons for aversive and appetitive learning in *Drosophila*. *J. Neurosci.* **27:** 7640.

Kramer A., Yang F.C., Snodgrass P., Li X., Scammell T.E., Davis F.C., and Weitz C.J. 2001. Regulation of daily locomotor activity and sleep by hypothalamic EGF receptor signaling. *Science* **294:** 2511.

Kume K., Kume S., Park S.K., Hirsh J., and Jackson F.R. 2005. Dopamine is a regulator of arousal in the fruit fly. *J. Neurosci.* **25:** 7377.

Kushikata T., Fang J., Chen Z., Wang Y., and Krueger J.M. 1998. Epidermal growth factor enhances spontaneous sleep in rabbits. *Am. J. Physiol.* **275:** R509.

McGuire S.E., Mao Z., and Davis R.L. 2004. Spatiotemporal gene expression targeting with the TARGET and gene-switch systems in *Drosophila*. *Sci. STKE* **2004:** l6.

Monastirioti M. 1999. Biogenic amine systems in the fruit fly *Drosophila melanogaster*. *Microsc. Res. Tech.* **45:** 106.

Monti J. and Monti D. 2007. The involvement of dopamine in the modulation of sleep and waking. *Sleep Med. Rev.* **11:** 113.

Neckameyer W.S. 1998. Dopamine modulates female sexual receptivity in *Drosophila melanogaster*. *J. Neurogenet.* **12:** 101.

Pitman J.L., McGill J.J., Keegan K.P., and Allada R. 2006. A dynamic role for the mushroom bodies in promoting sleep in *Drosophila*. *Nature* **441:** 753.

Robbins T.W., Granon S., Muir J.L., Durantou F., Harrison A., and Everitt B.J. 1998. Neural systems underlying arousal and attention. Implications for drug abuse. *Ann. N.Y. Acad. Sci.* **846:** 222.

Robinow S. and White K. 1988. The locus elav of *Drosophila melanogaster* is expressed in neurons at all developmental stages. *Dev. Biol.* **126:** 294.

Saper C.B. Scammell T.E., and Lu J. 2005. Hypothalamic regulation of sleep and circadian rhythms. *Nature* **437:** 1257.

Schwaerzel M., Monastirioti M., Scholz H., Friggi-Grelin F., Birman S., and Heisenberg M. 2003. Dopamine and octopamine differentiate between aversive and appetitive olfactory memories in *Drosophila*. *J. Neurosci.* **23:** 10495.

Schweitzer R., Shaharabany M., Seger R., and Shilo B.Z. 1995. Secreted Spitz triggers the DER signaling pathway and is a limiting component in embryonic ventral ectoderm determination. *Genes Dev.* **9:** 1518.

Shaw P.J., Cirelli C., Greenspan R.J., and Tononi G. 2000. Correlates of sleep and waking in *Drosophila melanogaster*. *Science* **287:** 1834.

Shaw P.J., Tononi G., Greenspan R.J., and Robinson D.F. 2002. Stress response genes protect against lethal effects of sleep deprivation in *Drosophila*. *Nature* **417:** 287.

Shilo B.Z. 2005. Regulating the dynamics of EGF receptor signaling in space and time. *Development* **132:** 4017.

Siegel J.M. 1995. Phylogeny and the function of REM sleep. *Behav. Brain Res.* **69:** 29.

Sweatt J.D. 2004. Mitogen-activated protein kinases in synaptic plasticity and memory. *Curr. Opin. Neurobiol.* **14:** 311.

Urban S., Lee J.R., and Freeman M. 2002. A family of Rhomboid intramembrane proteases activates all *Drosophila* membrane-tethered EGF ligands. *EMBO J.* **21:** 4277.

van Swinderen B. 2006. A succession of anesthetic endpoints in the *Drosophila* brain. *J. Neurobiol.* **66:** 1195.

van Swinderen B. and Greenspan R.J. 2003. Salience modulates 20-30 Hz brain activity in *Drosophila*. *Nat. Neurosci.* **6:** 579.

Veelaert D., Schoofs L., and De Loof A. 1998. Peptidergic control of the corpus cardiacum-corpora allata complex of locusts. *Int. Rev. Cytol.* **182:** 249.

Williams J.A., Su H.S., Bernards A., Field J., and Sehgal A. 2001. A circadian output in *Drosophila* mediated by neurofibromatosis-1 and Ras/MAPK. *Science* **293:** 2251.

Zhang K., Guo J..Z, Peng Y., Xi W., and Guo A. 2007. Dopamine-mushroom body circuit regulates saliency-based decision-making in *Drosophila*. *Science* **316:** 1901.

Zimmerman J.E., Rizzo W., Shockley K.R., Raizen D.M., Naidoo N., Mackiewicz M., Churchill G.A., and Pack A.I. 2006. Multiple mechanisms limit the duration of wakefulness in *Drosophila* brain. *Physiol. Genomics* **27:** 337.

Molecular Analysis of Sleep

M. Tafti and P. Franken
Center for Integrative Genomics, University of Lausanne, 1015 Lausanne, Switzerland

Rest or sleep in all animal species constitutes a period of quiescence necessary for recovery from activity. Whether rest and activity observed in all organisms share a similar fundamental molecular basis with sleep and wakefulness in mammals has not yet been established. In addition and in contrast to the circadian system, strong evidence that sleep is regulated at the transcriptional level is lacking. Nevertheless, several studies indicate that single genes may regulate some specific aspects of sleep. Efforts to better understand or confirm the role of known neurotransmission pathways in sleep-wake regulation using transgenic approaches resulted so far in only limited new insights. Recent gene expression profiling efforts in rats, mice, and fruit flies are promising and suggest that only a few gene categories are differentially regulated by behavioral state. How molecular analysis can help us to understand sleep is the focus of this chapter.

INTRODUCTION

Wakefulness, nonrapid eye movement (NREM), and rapid-eye movement (REM) sleep constitute the three states of vigilance in mammals and birds as defined by changes in brain electrical activity measured by the electroencephalogram (EEG), in muscle tone by the electromyogram (EMG), and in eye movements by the electro-oculogram (EOG). The EEG measures mainly cortical activity, and in species without a cortex or an underdeveloped cortex, changes in the EEG do not necessarily correlate with behavior. In addition, eye movements are absent in some species and muscle tone may not change in others, making these basic electrophysiological measures applicable only in mammals and birds. Other behavioral correlates of sleep must thus be used in most other species, including immobility, a stereotypic posture, reduced reactivity to external stimuli, reversibility to the waking state, and, of major importance, homeostatic regulation. Homeostatic regulation implies that the amount (or intensity) of rest should be correlated with that of the preceding duration of activity and therefore rest deprivation should result in a subsequent increase or rebound in rest (Campbell and Tobler 1984). These criteria can be applied to rest in many species, suggesting that sleep in mammals and birds shares some basic features with rest in species where sleep cannot be defined using the EEG and EMG. Sleep, as other behaviors, is also under the control of a circadian process that regulates its appropriate timing and structure. Lesions of the suprachiasmatic nucleus (SCN), the master clock in mammals, result in arrhythmic sleep and wakefulness, although the homeostatic response to sleep deprivation remains unchanged (Trachsel et al. 1992). However, recent evidence indicates a close relationship at the molecular level between the circadian and homeostatic processes regulating sleep (see Franken et al. 2007).

As early as the 1950s, twin studies in humans suggested that the amount and the structure of sleep are controlled by genetic factors (for review, see Dauvilliers et al. 2005; Tafti et al. 2005). More recent studies in inbred mice confirmed that not only the amount and the organization of sleep, but also specific features of the sleep EEG are under strong genetic control (Tafti et al. 1999; Tafti and Franken 2002). More specifically, single genes have been identified that affect EEG θ frequency (5–8 Hz) characteristic of REM sleep and δ (1–4 Hz) EEG activity characteristic of NREM sleep (Tafti et al. 2003; Maret et al. 2005). Molecular genetics of sleep disorders also made substantial progress recently with the identification of the genetic bases of fatal familial insomnia, narcolepsy, familial advanced sleep phase syndrome (FAPS), and restless legs and periodic limb movements in sleep (Stefansson et al. 2007; Tafti et al. 2007; Winkelmann et al. 2007). Nevertheless, the molecular bases of normal sleep and its regulation remain poorly understood. As summarized below, the endeavor is rendered difficult by the highly complex mechanisms underlying sleep regulation.

MOLECULAR NEUROCHEMISTRY OF SLEEP

Lesion and pharmacological studies show that sleep and wakefulness are controlled by multiple neuronal systems using different chemical neurotransmitters such as glutamate, acetylcholine, noradrenalin, dopamine, serotonin, histamine, adenosine, γ-amino-*n*-butyric acid (GABA), and orexin (Saper et al. 2005). Many of the neurons that are part of these control systems have widespread, direct, or diffuse projections to the cortex, subcortical relays, and brain stem or spinal cord, thus enabling them to influence the general state of arousal. Here, we give a brief overview of the results obtained in transgenic mouse models that were used to further elucidate the role of these signaling pathways in sleep.

The first neurotransmitter system that has been involved in sleep is serotonin. Serotonin neurons are active during wakefulness, diminish their activity during NREM, and are inactive during REM sleep. Sleep analyses in different serotonergic loss-of-function models indicated that the deletion of 5-HT1B and 5-HT1A receptors or the 5-HT transporter results in an increased amount of REM sleep (Boutrel et al. 1999, 2002; Wisor et al. 2003) without major effects on other aspects of sleep, confirming a selective role of the serotonergic system in the control of REM sleep.

The noradrenergic neurons of the locus coeruleus change activity very similar to that of serotonergic neurons and are critically involved in initiating and maintaining wakefulness, attention, and many other waking-related higher brain functions. Although pharmacological data have long implicated the noradrenergic system in the control of REM sleep, conflicting results have been reported with noradrenalin-deficient transgenic mice in which the dopamine β-hydroxylase gene was deleted (Hunsley and Palmiter 2003; Ouyang et al. 2004). The first study reported no major changes in sleep (Hunsley and Palmiter 2003), whereas the second reported a 2-hour reduction in time spent awake and a twofold reduction in REM sleep duration in mutant mice as compared to wild-type controls (Ouyang et al. 2004).

The role of dopamine is less well understood, but the deletion of the dopamine transporter decreased NREM sleep and increased wakefulness in mice (Wisor et al. 2001). Also in the fruit fly, the lack of the dopamine transporter resulted in decreased sleep time and increased wakefulness (Kume et al. 2005), strongly suggesting that dopamine is critically involved in the promotion and maintenance of wakefulness.

The cholinergic system is involved in wakefulness and REM sleep, but little molecular data are available to demonstrate its role in vigilance states. Mice lacking the β_2 subunit gene of the nicotinic acetycholine receptors show no change in any aspects of sleep, except for a slightly more consolidated REM sleep (Lena et al. 2004). The implication of the cholinergic system in REM sleep regulation is pharmacologically mediated through muscarinic receptors, and loss of function of many muscarinic receptors leads to hyperactivity, most probably due to increased dopaminergic tone (Wess 2004), but whether or not REM sleep is modified in these mice is unknown.

The role of histaminergic and adenosinergic systems in sleep has become increasingly recognized (McCarley 2007; Parmentier et al. 2007). Histaminergic neurons are selectively active during wakefulness and silent during sleep. Mice lacking the histidine decarboxylase gene that cannot synthesize histamine have increased REM sleep and seem to be hypersomnolent (Parmentier et al. 2002). Histamine H1 receptor knockout mice have consolidated sleep and do not react to the arousing effects of H3 receptor agonists (Huang et al. 2006), confirming the general interest in antihistaminergic drugs to promote sleep and H3 receptor agonists as stimulants (Parmentier et al. 2007). Adenosine is thought to induce and increase sleep in specific brain regions such as the basal forebrain and the ventral lateral preoptic area. However, whether adenosine acts at specific brain regions and through different receptors (A1 or A2a) or globally is a matter of debate (Blanco-Centurion et al. 2006; Noor Alam et al. 2006). Nevertheless, mice with A2a receptor loss of function have reduced sleep and reduced response to sleep deprivation and caffeine (Urade et al. 2003; Huang et al. 2005). In addition, in humans, genetic variations in adenosine deaminase and A2a receptors affect sleep duration and intensity (Retey et al. 2005) and sensitivity to caffeine (Retey et al. 2007), corroborating the findings in mouse models.

The neurotransmitters orexin and melanin-concentrating hormone (MCH) have been recently involved in sleep regulation (Modirrousta et al. 2005). Deletion of the orexin gene or orexin neurons results in a phenotype similar to human and canine narcolepsy with cataplexy (i.e., loss of muscle tone triggered by strong emotions) and abrupt transition from wakefulness to REM sleep (Chemelli et al. 1999; Hara et al. 2001). In addition, orexin receptor-2 mutations in dogs (Lin et al. 1999) and KO mice have narcolepsy-like symptoms, although receptor-2-deficient mice are less severely affected than ligand-deficient mice (Willie et al. 2003).

Overall, it has now been well established in several species that orexin deficiency results in an inability to maintain consolidated bouts of wakefulness and in inappropriate wake-to-REM sleep transitions. Although the MCH system has been involved in the regulation of REM sleep (Verret et al. 2003), the sleep phenotype of MCH and MCH receptor deletion has not yet been established.

A final neurotransmitter system important in sleep concerns GABA. GABA is the major inhibitory neurotransmitter acting on GABA-A and GABA-B receptors and is involved in both naturally occurring and benzodiazepine-induced sleep. Again, little genetic evidence is available, although point mutations in different GABA-A receptor subunits produce differential responses to sleep-inducing benzodiazepines (Tobler et al. 2001; Kopp et al. 2004). Overall, ligand or receptor knock out of the major neurotransmitter systems (if not lethal) have limited effects on sleep probably because of multiple neurotransmitter contributions and the compensatory effects of constitutional loss of function. Site- and time-specific gene deletion or knock down in these neurotransmitter pathways are necessary to delineate their respective contributions.

MOLECULAR CORRELATES OF SLEEP NEED

Sleep need builds during wakefulness and decreases during sleep. Therefore, genes for which the transcripts accumulate during wakefulness (and decrease during sleep) might be implicated in sleep regulation. Thus, as for circadian rhythms, sleep-wake regulation might also rely on an interplay of transcriptional regulators. Several aspects of sleep (e.g., the NREM-REM cycle) do not seem to be compatible with changes in gene expression because the time frame is too short. Changes in second messengers and posttranslational protein modifications could be considered instead (Schibler and Tafti 1999). Nevertheless, at least for sleep need, sleep-related genes may be induced and accumulate in a predictive manner. Different techniques have been used to isolate sleep- or wakefulness-specific transcripts (Tafti and Franken 2002). Here, we focus on the latest and most powerful approach, cDNA chip technology.

As opposed to the regulation of circadian rhythms for which a single brain structure seems to be responsible (i.e., the SCN), many brain regions have been implicated in the regulation of sleep, and for several, their role in sleep has not yet been well established. Some of these regions were briefly mentioned in the preceding paragraph, but in general, evidence for a specific role in sleep homeostasis is lacking with the exception, perhaps, of the locus coeruleus (Cirelli et al. 2005b) and the basal forebrain (Basheer et al. 2000). The EEG slow oscillations, the amplitude and prevalence of which are widely used as an EEG correlate of homeostatic sleep need (Borbely 1982), are generated by the interplay between the thalamus and the cortex. There is evidence that the hypothalamus, a highly specialized and complex brain region, is also involved in the homeostatic regulation of sleep (Saper et al. 2005).

Gene expression profiling has been attempted for several of these brain regions after spontaneous or enforced wakefulness in mice and rats. The first study (Terao et al. 2003) screened the expression on cDNA arrays of 1176 known genes in the cerebral cortex of C57BL/6J mice after 6 hours of sleep deprivation and identified two gene categories for which mRNA expression was up-regulated. These include immediate-early genes such as *Fos* and several heat shock proteins such as *Pdia4* (*Erp-72*) and *Hspa5* (*Grp78*). Cirelli et al. (2004) screened 15,459 of 24,000 potential transcripts on rat RGU34A,B,C chips for differentially regulated transcripts by sleep and waking as well as by time of day (circadian) in the cortex and the cerebellum and found that approximately a similar number of transcripts (5%) were affected by behavioral states and by time of day (day vs. night). In addition, they not only found that 490 transcripts were up-regulated by 8 hours of spontaneous or enforced wakefulness, but also that 261 transcripts were up-regulated after 8 hours of spontaneous sleep (Cirelli et al. 2004). The study confirmed the involvement of immediate-early and heat shock genes, and it also identified genes involved in synaptic plasticity, energy metabolism, and synaptic excitatory transmission being up-regulated during waking and genes involved in protein synthesis, lipid metabolism, and membrane trafficking being up-regulated during sleep (Cirelli et al. 2004).

Rest in the fruit fly *Drosophila melanogaster* shares all the important characteristics of sleep in mammals and birds, although they obviously lack the specific electrophysiological correlates that characterize sleep in mammals (Hendricks et al. 2000; Shaw et al. 2000). Of interest is the conservation of the homeostatic regulation of fly sleep as in other species. Therefore, it can be speculated that if sleep need is regulated at the transcriptional level, the same essential genes might also show changes in expression as a function of time spent in rest or activity. Accordingly, as in the rodent brain, rest deprivation and spontaneous wakefulness in the fly head induce up-regulation of genes belonging to specific categories. Wakefulness induces transcription factors, stress and immune response, and glutamatergic genes and genes involved in metabolism, whereas rest induces genes involved in lipid metabolism (Cirelli et al. 2005a), strongly suggesting a cross-species conservation of several molecular pathways involved in behavioral state control.

Another important question related to sleep function is the consequence of long-term sleep deprivation, which can result in death, for example, in rats after 2–3 weeks (Rechtschaffen and Bergmann 2002). Cirelli et al. (2006) addressed this question by sleep-depriving rats for either 8 hours or 1 week and profiling the expression of more than 24,000 transcripts in the cerebral cortex. Long-term sleep loss induced overexpression of genes involved in many inflammatory and stress response genes mainly in glial cells, suggesting a trend toward generalized brain pathology. In comparison, several synaptic plasticity genes were up-regulated only after short-term sleep deprivation (Cirelli et al. 2006).

The effects of a 6-hour sleep deprivation and 6-hour sleep deprivation followed by a 2-hour recovery sleep on gene expression were also investigated in the cortex, basal forebrain, and hypothalamus of rats and for 12 overexpressed genes replicated in the mouse (Terao et al. 2006). Although differences in the number of genes up- and down-regulated by sleep deprivation and recovery were found between the three brain regions, again, the most significant changes concerned immediate-early and heat shock protein genes in both species. Although different transcripts were found between the hypothalamus and cortex and basal forebrain, many genes changed expression in at least two of the three regions. In addition, several transcripts that were found to be up-regulated after sleep deprivation remained up-regulated following a 2-hour recovery sleep, suggesting a time course of dissipation not paralleling that of sleep need based on the dynamics of the amplitude and prevalence of EEG slow waves.

As opposed to the above-mentioned studies, two recent reports in flies and mice suggest that sleep deprivation results in a decrease in gene expression instead of an increase in gene expression (Zimmerman et al. 2006; Mackiewicz et al. 2007). In both studies, the expression data have been analyzed at different time points (with or without sleep deprivation) with a reference time point taken at the beginning of the consolidated rest or sleep period. In this analysis, circadian and sleep-wake–dependent effects on gene expression are difficult to separate. Interestingly, *Hspa5/Bip/Grp78*, which has been shown to invariably and most significantly increase expression after sleep deprivation in several species, was not detected by Zimmerman et al. in the flies (Zimmerman et al. 2006; Mackiewicz et al. 2007), although the same authors later claimed a major role for this heat shock protein in *Drosophila* sleep homeostasis (Naidoo et al. 2007).

The results summarized above raise several questions. Although brain structures thought to be implicated in sleep (cortex, hypothalamus) or not (cerebellum) have been investigated, it is not clear whether transcriptional changes occur in specific regions or in the whole brain (i.e., local vs. global). How specific transcriptional changes are to the brain has also not been determined because comparative analysis in other tissues is lacking. Even though time course analyses have been performed, systematic studies aiming at separating circadian and sleep-wake–dependent effects on gene transcription are also lacking. To overcome these limitations, we have performed a series of gene-profiling experiments in mice. We had already shown that sleep need varies with genetic background in mice (Franken et al. 2001); we therefore sleep-deprived three different mouse strains with differential recovery response to 6 hours of sleep deprivation. We also sampled the liver in the same animals as a reference peripheral organ for comparative transcriptome analysis with the whole brain. In both brain and liver, half of the differentially expressed transcripts were affected by genotype (S. Maret et al., in prep.). Transcriptional changes were much larger in the liver than in the brain, although, overall, few genes showed significant changes (less than 50 in the brain and 200 in the liver) after sleep deprivation. Most transcripts were up-regulated after sleep deprivation and belonged mainly to immediate-early, heat shock, and synaptic plasticity gene categories. However, by taking into account genetic background and tissue, very few genes showed consistent and brain-specific changes. Among these, the short splice variant of the *Homer1* gene (*Homer1a*) showed the most significant increase in all three strains after sleep loss. We also performed another whole-brain transcriptome analysis by sleep-depriving the same three mouse strains for 6 hours at four time points around the clock. This experiment revealed that under baseline conditions (without sleep deprivation) approximately 8% (2032) of all microarray probe sets detected in the brain showed a significant time-of-day pattern of expression. However, under sleep deprivation conditions, only 1.5% (390 probe sets) remained cyclic, strongly suggesting that a large majority of cycling transcripts is modulated by behavioral states rather than being directly driven by the circadian clock. Among all rhythmic transcripts under baseline conditions, *Homer1a* clearly showed the largest amplitude of variation. *Homer1a* is an activity-induced gene involved in metabotropic glutamatergic neurotransmission and its up-regulation results in the buffering of intracellular calcium. These two experiments strongly suggested that *Homer1*a represents a major molecular index of sleep need in the whole brain. Interestingly, *Homer1a* maps in the middle of a region on mouse chromosome 13 that we had already identified by linkage analysis to be associated with sleep need (Franken et al. 2001). To further assess brain transcriptional changes induced by sleep deprivation, we generated a transgenic mouse model in which a Flag-tagged poly(A)-binding protein (PABP) is expressed under the control of the *Homer1* promoter. Because PABP binds poly(A) tails of mRNA for processing, immunoprecipitation of PABP-mRNA complexes by an antiepitope antibody results in enrichment and purification of mature mRNAs expressed in a cell-specific manner (Roy et al. 2002). Transcriptome analysis of *Homer1*-expressing neurons after sleep deprivation revealed that in addition to *Homer1a*, very few other genes are overexpressed by sleep loss. These include the activity-induced genes Fos-like 2 (*Fosl2*), Jonctaphilin 3 (*Jph3*), prostaglandin-endoperoxide synthase 2 (*Ptgs2*), and neuronal pentraxin 2 (*Nptx2*). *Jph3* and *Nptx2* are activity-induced through either ryanodine receptor-mediated intracellular calcium mobilization (*Jph3*) or activity-induced AMPA (α-amino-3-hydroxy-5-methyl-4-isoazole) receptor synaptic clustering (*Nptx2*). Among the very few down-regulated transcripts, another activity-induced gene, 4-nitrophenylphosphatase domain and non-neuronal SNAP25-like protein homolog 1 (*Nipsnap1*), was identified, suggesting that plasticity genes can be up- or down-regulated by sleep deprivation. On the basis of these findings, we have proposed that *Homer1a* induction suggests a novel role for sleep in protecting and recovering from glutamatergic overstimulation imposed by wakefulness through transcriptional changes that ultimately regulate the intracellular calcium homeostasis.

CONCLUSIONS

As summarized here, the molecular basis of sleep remains poorly understood. Even at the very basic neurobiological level, no strong molecular evidence supports a specific role for any one of the major neurotransmission pathways. Whether sleep may be regulated at the transcriptional level remains unclear. Most studies indicate that immediate-early, heat shock, and synaptic plasticity genes are up-regulated by sleep loss. However, the specificity of these genes, except for *Homer1a*, has not yet been established. In addition, induction of gene expression, even if specific to sleep loss, does not constitute proof of a functional role of this gene in sleep-wake regulation. Genes of all three categories are induced by spontaneous and induced neuronal activity, as well as in reaction to pathological conditions such as stroke, inflammation, or seizure. In accordance with the concept of the homeostatic regulation of sleep, the results of several gene-profiling experiments found increased transcription of a few functionally relevant genes during extended wakefulness, thus reflecting the molecular correlate of accumulated sleep need. Zimmerman et al. (2006) proposed an alternative hypothesis: A decrease rather than an increase in gene expression would result in reduction of the cellular processes that promote wakefulness. That experiments aimed at answering the same basic question and using the same techniques can arrive at such diverse results and interpretations is disquieting and calls for standardized designs and analyses.

As discussed above, the structural complexity of the brain and the multitude of signaling pathways involved in sleep, together with the lack of a "sleep homeostasis center" render the identification of molecular substrates of sleep highly complicated. In addition, at the mRNA level, a high complexity in mammals makes any detection of differentially expressed mRNA a tedious task, even by using microarray technology. Alternative possibilities are either to use the less complex model species such as the fruit fly or to use the more specific mRNA techniques such as the mRNA tagging that we have introduced. Serial analysis of gene expression (SAGE) and highly specific substractive hybridization techniques such as the selective amplification via biotin and restriction-mediated enrichment (SABRE) might overcome the limitations to detect low-abundance mRNAs. Finally, the analyses discussed here make inferences about the involvement of genes in sleep regulation based solely on the temporal and/or state-dependent changes in gene transcription while the dynamics of translation and posttranslational modification of gene products could importantly differ. As for the genome-wide expression profiling progress, differential proteomic approaches will depend on technological advances. Proteomic sleep studies have just begun to be performed (O'Hara et al. 2007). Finally, all gene expression experiments are correlative and gain- or loss-of-function studies of identified genes are necessary to causally implicate any of the sleep target genes.

ACKNOWLEDGMENTS

This work is supported by the Swiss National Science Foundation and the State of Vaud, Switzerland.

REFERENCES

Basheer R., Porkka-Heiskanen T., Strecker R.E., Thakkar M.M., and McCarley R.W. 2000. Adenosine as a biological signal mediating sleepiness following prolonged wakefulness. *Biol. Signals Recept.* **9:** 319.

Blanco-Centurion C., Xu M., Murillo-Rodriguez E., Gerashchenko D., Shiromani A.M., Salin-Pascual R.J., Hof P.R., and Shiromani P.J. 2006. Adenosine and sleep homeostasis in the basal forebrain. *J. Neurosci.* **26:** 8092.

Borbely A.A. 1982. A two process model of sleep regulation. *Hum. Neurobiol.* **1:** 195.

Boutrel B., Franc B., Hen R., Hamon M., and Adrien J. 1999. Key role of 5-HT1B receptors in the regulation of paradoxical sleep as evidenced in 5-HT1B knock-out mice. *J. Neurosci.* **19:** 3204.

Boutrel B., Monaca C., Hen R., Hamon M., and Adrien J. 2002. Involvement of 5-HT1A receptors in homeostatic and stress-induced adaptive regulations of paradoxical sleep: Studies in 5-HT1A knock-out mice. *J. Neurosci.* **22:** 4686.

Campbell S.S. and Tobler I. 1984. Animal sleep: A review of sleep duration across phylogeny. *Neurosci. Biobehav. Rev.* **8:** 269.

Chemelli R.M., Willie J.T., Sinton C.M., Elmquist J.K., Scammell T., Lee C., Richardson J.A., Williams S.C., Xiong Y., Kisanuki Y., Fitch T.E., Nakazato M., Hammer R.E., Saper C.B., and Yanagisawa M. 1999. Narcolepsy in orexin knockout mice: Molecular genetics of sleep regulation. *Cell* **98:** 437.

Cirelli C., Faraguna U., and Tononi G. 2006. Changes in brain gene expression after long-term sleep deprivation. *J. Neurochem.* **98:** 1632.

Cirelli C., Gutierrez C.M., and Tononi G. 2004. Extensive and divergent effects of sleep and wakefulness on brain gene expression. *Neuron* **41:** 35.

Cirelli C., LaVaute T.M., and Tononi G. 2005a. Sleep and wakefulness modulate gene expression in *Drosophila*. *J. Neurochem.* **94**: 1411.

Cirelli C., Huber R., Gopalakrishnan A., Southard T.L., and Tononi G. 2005b. Locus ceruleus control of slow-wave homeostasis. *J. Neurosci.* **25:** 4503.

Dauvilliers Y., Maret S., and Tafti M. 2005. Genetics of normal and pathological sleep in humans. *Sleep Med. Rev.* **9:** 91.

Franken P., Chollet D., and Tafti M. 2001. The homeostatic regulation of sleep need is under genetic control. *J. Neurosci.* **21:** 2610.

Franken P., Thomason R., Heller H.C., and O'Hara B.F. 2007. A non-circadian role for clock-genes in sleep homeostasis: A strain comparison. *BMC Neurosci.* **8:** 87.

Hara J., Beuckmann C.T., Nambu T., Willie J.T., Chemelli R.M., Sinton C.M., Sugiyama F., Yagami K., Goto K., Yanagisawa M., and Sakurai T. 2001. Genetic ablation of orexin neurons in mice results in narcolepsy, hypophagia, and obesity. *Neuron* **30:** 345.

Hendricks J.C., Finn S.M., Panckeri K.A., Chavkin J., Williams J.A., Sehgal A., and Pack A.I. 2000. Rest in *Drosophila* is a sleep-like state. *Neuron* **25:** 129.

Huang Z.L., Mochizuki T., Qu W.M., Hong Z.Y., Watanabe T., Urade Y., and Hayaishi O. 2006. Altered sleep-wake characteristics and lack of arousal response to H3 receptor antagonist in histamine H1 receptor knockout mice. *Proc. Natl. Acad. Sci.* **103:** 4687.

Huang Z.L., Qu W.M., Eguchi N., Chen J.F., Schwarzschild M.A., Fredholm B.B., Urade Y., and Hayaishi O. 2005. Adenosine A2A, but not A1, receptors mediate the arousal effect of caffeine. *Nat. Neurosci.* **8:** 858.

Hunsley M.S. and Palmiter R.D. 2003. Norepinephrine-deficient mice exhibit normal sleep-wake states but have shorter sleep latency after mild stress and low doses of amphetamine. *Sleep* **26:** 521.

Kopp C., Rudolph U., Low K., and Tobler I. 2004. Modulation of rhythmic brain activity by diazepam: GABA(A) receptor subtype and state specificity. *Proc. Natl. Acad. Sci.* **101:** 3674.

Kume K., Kume S., Park S.K., Hirsh J., and Jackson F.R. 2005. Dopamine is a regulator of arousal in the fruit fly. *J. Neurosci.* **25:** 7377.

Lena C., Popa D., Grailhe R., Escourrou P., Changeux J.P., and Adrien J. 2004. Beta2-containing nicotinic receptors contribute to the organization of sleep and regulate putative micro-arousals in mice. *J. Neurosci.* **24:** 5711.

Lin L., Faraco J., Li R., Kadotani H., Rogers W., Lin X., Qiu X., de Jong P.J., Nishino S., and Mignot E. 1999. The sleep disorder canine narcolepsy is caused by a mutation in the hypocretin (orexin) receptor 2 gene. *Cell* **98:** 365.

Mackiewicz M., Shockley K.R., Romer M.A., Galante R.J., Zimmerman J.E., Naidoo N., Baldwin D.A., Jensen S.T., Churchill G.A., and Pack A.I. 2007. Macromolecule biosynthesis—A key function of sleep. *Physiol. Genomics* (in press).

Maret S., Franken P., Dauvilliers Y., Ghyselinck N.B., Chambon P., and Tafti M. 2005. Retinoic acid signaling affects cortical synchrony during sleep. *Science* **310:** 111.

McCarley R.W. 2007. Neurobiology of REM and NREM sleep. *Sleep Med.* **8:** 302.

Modirrousta M., Mainville L., and Jones B.E. 2005. Orexin and MCH neurons express c-Fos differently after sleep deprivation vs. recovery and bear different adrenergic receptors. *Eur. J. Neurosci.* **21:** 2807.

Naidoo N., Casiano V., Cater J., Zimmerman J., and Pack A.I. 2007. A role for the molecular chaperone protein BiP/GRP78 in *Drosophila* sleep homeostasis. *Sleep* **30:** 557.

Noor Alam M.D., Szymusiak R., and McGinty D. 2006. Adenosinergic regulation of sleep: Multiple sites of action in the brain. *Sleep* **29:** 1384.

O'Hara B.F., Ding J., Bernat R.L., and Franken P. 2007. Genomic and proteomic approaches towards an understanding of sleep. *CNS Neurol. Disord. Drug Targets* **6:** 71.

Ouyang M., Hellman K., Abel T., and Thomas S.A. 2004. Adrenergic signaling plays a critical role in the maintenance of waking and in the regulation of REM sleep. *J. Neurophysiol.* **92:** 2071.

Parmentier R., Ohtsu H., Djebbara-Hannas Z., Valatx J.L., Watanabe T., and Lin J.S. 2002. Anatomical, physiological, and pharmacological characteristics of histidine decarboxylase knock-out mice: Evidence for the role of brain histamine in behavioral and sleep-wake control. *J. Neurosci.* **22:** 7695.

Parmentier R., Anaclet C., Guhennec C., Brousseau E., Bricout D., Giboulot T., Bozyczko-Coyne D., Spiegel K., Ohtsu H., Williams M., and Lin J.S. 2007. The brain H3-receptor as a novel therapeutic target for vigilance and sleep-wake disorders. *Biochem. Pharmacol.* **73:** 1157.

Rechtschaffen A. and Bergmann B.M. 2002. Sleep deprivation in the rat: An update of the 1989 paper. *Sleep* **25:** 18.

Retey J.V., Adam M., Khatami R., Luhmann U.F., Jung H.H., Berger W., and Landolt H.P. 2007. A genetic variation in the adenosine A2A receptor gene (ADORA2A) contributes to individual sensitivity to caffeine effects on sleep. *Clin. Pharmacol. Ther.* **81:** 692.

Retey J.V., Adam M., Honegger E., Khatami R., Luhmann U.F., Jung H.H., Berger W., and Landolt H.P. 2005. A functional genetic variation of adenosine deaminase affects the duration and intensity of deep sleep in humans. *Proc. Natl. Acad. Sci.* **102:** 15676.

Roy P.J., Stuart J.M., Lund J., and Kim S.K. 2002. Chromosomal clustering of muscle-expressed genes in *Caenorhabditis elegans*. *Nature* **418:** 975.

Saper C.B., Scammell T.E., and Lu J. 2005. Hypothalamic regulation of sleep and circadian rhythms. *Nature* **437**: 1257.

Schibler U. and Tafti M. 1999. Molecular approaches towards the isolation of sleep-related genes. *J. Sleep. Res.* (suppl. 1) **8:** 1.

Shaw P.J., Cirelli C., Greenspan R.J., and Tononi G. 2000. Correlates of sleep and waking in *Drosophila melanogaster*. *Science* **287:** 1834.

Stefansson H., Rye D.B., Hicks A., Petursson H., Ingason A., Thorgeirsson T.E., Palsson S., Sigmundsson T., Sigurdsson A.P., Eiriksdottir I., Soebech E., Bliwise D., Beck J.M., Rosen A., Waddy S., Trotti L.M., Iranzo A., Thambisetty M., Hardarson G.A., Kristjansson K., Gudmundsson L.J., Thorsteinsdottir U., Kong A., Gulcher J.R., Gudbjartsson D., and Stefansson K. 2007. A genetic risk factor for periodic limb movements in sleep. *N. Engl. J. Med.* **357:** 639.

Tafti M. and Franken P. 2002. Invited review: Genetic dissection of sleep. *J. Appl. Physiol.* **92:** 1339.

Tafti M., Dauvilliers Y., and Overeem S. 2007. Narcolepsy and familial advanced sleep-phase syndrome: Molecular genetics of sleep disorders. *Curr. Opin. Genet. Dev.* **17:** 222.

Tafti M., Maret S., and Dauvilliers Y. 2005. Genes for normal sleep and sleep disorders. *Ann. Med.* **37:** 580.

Tafti M., Chollet D., Valatx J.L., and Franken P. 1999. Quantitative trait loci approach to the genetics of sleep in recombinant inbred mice. *J. Sleep Res.* (suppl. 1) **8:** 37.

Tafti M., Petit B., Chollet D., Neidhart E., de Bilbao F., Kiss J.Z., Wood P.A., and Franken P. 2003. Deficiency in short-chain fatty acid beta-oxidation affects theta oscillations during sleep. *Nat. Genet.* **34:** 320.

Terao A., Wisor J.P., Peyron C., Apte-Deshpande A., Wurts S.W., Edgar D.M., and Kilduff T.S. 2006. Gene expression in the rat brain during sleep deprivation and recovery sleep: An Affymetrix GeneChip study. *Neuroscience* **137:** 593.

Terao A., Steininger T.L., Hyder K., Apte-Deshpande A., Ding J., Rishipathak D., Davis R.W., Heller H.C., and Kilduff T.S. 2003. Differential increase in the expression of heat shock protein family members during sleep deprivation and during sleep. *Neuroscience* **116:** 187.

Tobler I., Kopp C., Deboer T., and Rudolph U. 2001. Diazepam-induced changes in sleep: Role of the alpha 1 GABA(A) receptor subtype. *Proc. Natl. Acad. Sci.* **98:** 6464.

Trachsel L., Edgar D.M., Seidel W.F., Heller H.C., and Dement W.C. 1992. Sleep homeostasis in suprachiasmatic nuclei-lesioned rats: Effects of sleep deprivation and triazolam administration. *Brain Res.* **589:** 253.

Urade Y., Eguchi N., Qu W.M., Sakata M., Huang Z.L., Chen J.F., Schwarzschild M.A., Fink J.S., and Hayaishi O. 2003. Sleep regulation in adenosine A2A receptor-deficient mice. *Neurology* **61:** S94.

Verret L., Goutagny R., Fort P., Cagnon L., Salvert D., Leger L., Boissard R., Salin P., Peyron C., and Luppi P.H. 2003. A role of melanin-concentrating hormone producing neurons in the central regulation of paradoxical sleep. *BMC Neurosci.* **4:** 19.

Wess J. 2004. Muscarinic acetylcholine receptor knockout mice: Novel phenotypes and clinical implications. *Annu. Rev. Pharmacol. Toxicol.* **44:** 423.

Willie J.T., Chemelli R.M., Sinton C.M., Tokita S., Williams S.C., Kisanuki Y.Y., Marcus J.N., Lee C., Elmquist J.K., Kohlmeier K.A., Leonard C.S., Richardson J.A., Hammer R.E., and Yanagisawa M. 2003. Distinct narcolepsy syndromes in Orexin receptor-2 and Orexin null mice: Molecular genetic dissection of Non-REM and REM sleep regulatory processes. *Neuron* **38:** 715.

Winkelmann J., Schormair B., Lichtner P., Ripke S., Xiong L., Jalilzadeh S., Fulda S., Putz B., Eckstein G., Hauk S., Trenkwalder C., Zimprich A., Stiasny-Kolster K., Oertel W., Bachmann C.G., Paulus W., Peglau I., Eisensehr I., Montplaisir J., Turecki G., Rouleau G., Gieger C., Illig T., Wichmann H.E., Holsboer F., Muller-Myhsok B., and Meitinger T. 2007. Genome-wide association study of restless legs syndrome identifies common variants in three genomic regions. *Nat. Genet.* **39:** 1000.

Wisor J.P., Nishino S., Sora I., Uhl G.H., Mignot E., and Edgar D.M. 2001. Dopaminergic role in stimulant-induced wakefulness. *J. Neurosci.* **21:** 1787.

Wisor J.P., Wurts S.W., Hall F.S., Lesch K.P., Murphy D.L., Uhl G.R., and Edgar D.M. 2003. Altered rapid eye movement sleep timing in serotonin transporter knockout mice. *Neuroreport* **14:** 233.

Zimmerman J.E., Rizzo W., Shockley K.R., Raizen D.M., Naidoo N., Mackiewicz M., Churchill G.A., and Pack A.I. 2006. Multiple mechanisms limit the duration of wakefulness in *Drosophila* brain. *Physiol. Genomics* **27:** 337.

Sleep and Circadian Rhythms in Humans

C.A. Czeisler and J.J. Gooley
Division of Sleep Medicine, Harvard Medical School and Division of Sleep Medicine, Department of Medicine, Brigham and Women's Hospital, Boston, Massachusetts 02115

During the past 50 years, converging evidence reveals that the fundamental properties of the human circadian system are shared in common with those of other organisms. Concurrent data from multiple physiological rhythms in humans revealed that under some conditions, rhythms oscillated at different periods within the same individuals and led to the conclusion 30 years ago that the human circadian system was composed of multiple oscillators organized hierarchically; this inference has recently been confirmed using molecular techniques in species ranging from unicellular marine organisms to mammals. Although humans were once thought to be insensitive to the resetting effects of light, light is now recognized as the principal circadian synchronizer in humans, capable of eliciting weak (Type 1) or strong (Type 0) resetting, depending on stimulus strength and timing. Realization that circadian photoreception could be maintained in the absence of sight was first recognized in blind humans, as was the property of adaptation of the sensitivity of circadian photoreception to prior light history. In sighted humans, the intrinsic circadian period is very tightly distributed around approximately 24.2 hours and exhibits aftereffects of prior entrainment. Phase angle of entrainment is dependent on circadian period, at least in young adults. Circadian pacemakers in humans drive daily variations in many physiologic and behavioral variables, including circadian rhythms in alertness and sleep propensity. Under entrained conditions, these rhythms interact with homeostatic regulation of the sleep/wake cycle to determine the ability to sustain vigilance during the day and to sleep at night. Quantitative understanding of the fundamental properties of the multioscillator circadian system in humans and their interaction with sleep/wake homeostasis has many applications to health and disease, including the development of treatments for circadian rhythm and sleep disorders.

INTRODUCTION

Seventy years ago, Kleitman (1963) was the first to study human circadian rhythms in human subjects shielded from periodic environmental cues. In 1938, he studied two subjects living on non-24-hour sleep/wake, light/dark, and meal schedules while living deep within Kentucky's Mammoth Cave, shielded from the influence of the Earth's 24-hour day. Measurement of the daily rhythm of body temperature in one of the subjects revealed the circadian temperature rhythm to be endogenously generated, persisting for a month with a near-24-hour period despite imposition of a 28-hour rest/activity schedule. That first underground cave study of human circadian rhythms, in the longest known cave on Earth, was far ahead of its time. The fact that a physiological rhythm could oscillate *not only* in the absence of periodic changes in the environment, *but also* at a period different from that of behavioral cyclicity established the endogenous and physiologic nature of human circadian rhythms for the first time. More than two decades later, in a paper on human circadian rhythms that was presented at the first CSHL symposium on circadian rhythmicity, Lobban reported on a series of field studies in which subjects lived on non-24-hour schedules during the continuous light of summer within the Arctic circle (Lewis and Lobban 1957a,b; Lobban 1961). At the time, controversy still persisted as to whether the circadian rhythm of body temperature in humans could be shifted by an inversion of the daily routine, such as is required by night-shift workers—a question that had been hotly debated since the beginning of the 20th century (Benedict 1904; Gibson 1905). The 1960 CSHL symposium on circadian clocks brought together scientists working on circadian rhythms in many different organisms, and a number of common properties of circadian clocks began to emerge.

Few at the 1960 CSHL symposium on circadian clocks, however, could have imagined that the then-recent discovery by Hastings and Sweeney (1958) of the circadian phase-dependent sensitivity to photic resetting of the circadian pacemaker in the unicellular marine dinoflagellate *Gonyaulax polyedra*, and its codependence on the intensity (Hastings and Sweeney 1958) and wavelength of light (Sweeney et al. 1959; Hastings and Sweeney 1960), would be found to apply to circadian clocks in a remarkably wide array of organisms, from cyanobacteria (Kondo et al. 1993) to humans (Czeisler et al. 1989; Jewett et al. 1992, 2000; Boivin et al. 1996; Zeitzer et al. 2000; Khalsa et al. 2003; Lockley et al. 2003). The purpose of this chapter is to review progress that has been made in understanding the properties of the human circadian pacemaker(s) and the relationship between circadian rhythms and the timing of the sleep/wake cycle. We also review the critical role of sleep and circadian rhythms in clinical and occupational medicine.

THE HUMAN CIRCADIAN SYSTEM

In humans, many aspects of human physiology and behavior vary with circadian phase (Czeisler and Jewett 1990; Johnson et al. 1992; Allan and Czeisler 1994; Waldstreicher et al. 1996; El-Hajj Fuleihan et al. 1997; Cajochen et al. 1999; Czeisler et al. 2000). Thirty-five years ago, the central neuroanatomic structures responsible for both the generation of endogenous circadian rhythms and their synchronization with the 24-hour day were identified (Moore 1972; Moore and Eichler 1972; Moore and Lenn 1972; Stephan and Zucker 1972). Deep within the brain, two bilaterally paired clusters of hypothalamic neurons comprising the suprachiasmatic nuclei (SCN) act as the central neural pacemaker for the generation and/or synchronization of circadian rhythms in mammals (Ralph et al. 1990; Klein et al. 1991; Edgar et al. 1993; Moore 1994; Welsh et al. 1995; Mumford et al. 1996; Weaver 1998). Evaluation of the impact and the formal properties of the circadian system in humans is aided by the number of variables that can be measured simultaneously but hampered by constraints on the interventions that can ethically be used to study the system, rendering it difficult to characterize the basic properties of the circadian system in humans.

Circadian rhythms are self-sustained and persist in the absence of environmental time cues with remarkable precision. As such, a circadian rhythm represents a cyclic process that can be described by the period (?), phase (?), and amplitude of the oscillation, together with its resetting sensitivity to various circadian synchronizers. These formal properties of the circadian oscillatory system are thought to be determined genetically by a core set of clock genes, which are critical for the generation, maintenance, and synchronization of circadian pacemaker output. Normally, circadian rhythms are entrained to the solar day, ensuring that behavioral, physiologic, and genetic rhythms are timed appropriately with daily changes in the environment. Circadian entrainment is achieved through daily resetting of the circadian pacemaker, such that $T = ? - \varnothing?$, whereby T is the *imposed period* of the environmental synchronizer (e.g., the 24-hour solar day), ? is the *intrinsic period* of the circadian oscillator, and ∅? is the daily phase shift required for stable entrainment. The relationship between the phase of an endogenous circadian rhythm (e.g., the peak of the melatonin rhythm) and the phase of the imposed environmental synchronizer (e.g., the light/dark cycle), termed the phase angle of entrainment (?), is a function of the difference between T and ? and the resetting sensitivity of the circadian system. Defining the period and resetting properties of the human circadian system is therefore critical for understanding how the timing of diverse behavioral and physiologic processes is established.

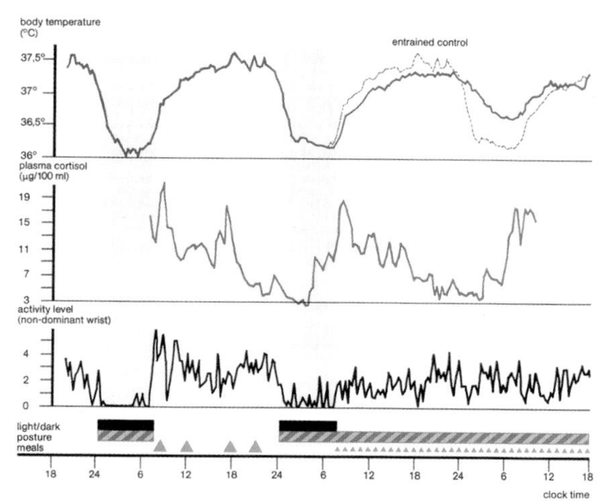

Figure 1. The constant routine procedure is used to assess circadian phase and amplitude. During normal entrainment to a 24-hour day, body temperature, cortisol, and activity exhibit a high-amplitude diurnal rhythm. The constant routine procedure allows for assessment of circadian phase and amplitude by reducing the effects of exogenous time cues and rest/activity cycles on the underlying circadian rhythm. During the constant routine procedure, subjects are kept awake continuously in bed in a constant posture, with continuous exposure to dim ambient light, and small meals are spread evenly throughout the protocol. The circadian rhythm of core body temperature during the constant routine procedure shows a decrease in measured amplitude, as compared to the observed amplitude of the rhythm assessed during baseline sleep/wake (*red trace*). In contrast, the cortisol rhythm shows a similar waveform during entrained conditions and the constant routine procedure (*orange trace*). The diurnal rhythm of rest/activity, as determined by wrist-actigraphy monitoring, is markedly reduced during the constant routine procedure (*black trace*). (*Gray shading*) Sleep episodes in darkness; (*blue hatched bars*) constant posture; (*green triangles*) timing of meals. (Reprinted, with permission, from Czeisler 1986 [© Boehringer Ingelheim].)

Circadian Phase

The phase of a circadian rhythm is defined with respect to an easily identifiable reference point of the endogenous circadian oscillation, such as the trough of the body temperature rhythm or the onset of the melatonin rhythm. Thus, a circadian phase shift can be determined by measuring the change in timing of the chosen phase marker from one cycle to the next. During ambulatory conditions, changes in environmental stimuli and behavior (e.g., light/dark, rest/activity, and temperature) often obscure the endogenous component of the underlying circadian oscillations that is being measured. Therefore, endogenous circadian phase is best assessed under environmental conditions that minimize exposure to stimuli that evoke response(s) in the physiologic variables being monitored for a minimum of one circadian cycle (Fig. 1). During a constant routine procedure, ambient light is continuously dim, metabolic intake is distributed evenly throughout day and night, and constant posture and wakefulness are maintained. The phase and amplitude of endogenous circadian components of daily rhythms in sleep propensity and in thermoregulatory, endocrine, cardiac, renal, respiratory, neurobehavioral, and gastrointestinal functions have been characterized under such constant routine conditions and have been found to be distinct from the profiles of those variables recorded in the presence of periodic stimuli that evoke responses in these functions.

Circadian Amplitude

The measured amplitude of a circadian rhythm refers to the half-distance from the maximum to the minimum of the observed rhythm. As first demonstrated by the late Arthur Winfree (1969, 1974, 1980, 1987), both oscillator phase and oscillator amplitude are required to describe adequately the resetting properties of the circadian system (see below). The absolute value of the measured amplitude of an observed rhythm does not necessarily equate with the endogenous amplitude of the circadian oscillator. Whereas the absolute value of the amplitude of the melatonin rhythm varies by more than tenfold among individuals, the absolute value of the amplitude of the core body temperature rhythm varies by less than twofold in those same individuals. Suppression of circadian amplitude is associated with a comparable reduction in the amplitude of both parameters, *relative to the initial measured amplitude* of each variable (Shanahan et al. 1997).

Entrainment to Light

Despite an initial report suggesting that light was an effective circadian synchronizer in humans (Aschoff et al. 1969), based on a subsequent study by that same group (Wever 1970), for many years, it was thought that the human circadian system was insensitive to light and that social interaction was the predominant synchronizer mediating circadian entrainment to the solar day (Wever 1970, 1974, 1979; Aschoff 1976; Aschoff and Wever 1981). This observation led to the belief that humans had evolved beyond the need to rely on periodic exposure to a physical stimulus such as light for circadian entrainment. However appealing this notion may appear, there is now overwhelming evidence that the human circadian pacemaker is in fact exquisitely sensitive to light as a circadian synchronizer—with a phase-response curve and sensitivity to light that are similar to those described in lower organisms such as adult *Drosophila pseudoobscura* and the cockroach *Leucophaea maderae*—and that light is the primary circadian synchronizer in humans (Czeisler 1978, 1995; Czeisler et al. 1981; Shanahan et al. 1997; Czeisler and Wright 1999; Shanahan and Czeisler 2000; Zeitzer et al. 2000; Wright et al. 2001; Gronfier et al. 2004, 2007; Lockley 2007). In mammals, light information is transmitted to the circadian pacemaker in the SCN via the retinohypothalamic tract, a monosynaptic pathway that originates from a small subset of retinal ganglion cells (Moore 1972; Moore and Lenn 1972). Resetting the SCN, in turn, shifts the timing of diverse behavioral and physiologic functions. To entrain stably to the light/dark cycle, the phase of the circadian pacemaker must be reset daily such that a phase shift is equivalent to the difference between the $T = 24$ hour cycle and the intrinsic circadian period ($\varnothing ? = ? - T$). For individuals with a circadian period longer than 24 hours, a daily phase advance of the circadian system is required for stable entrainment to the light/dark cycle. Conversely, individuals with a circadian period shorter than 24 hours require a daily phase delay in order to synchronize to the 24-hour environmental cycle. As discussed below, the resetting effects of light depend on several factors, including the circadian phase at which the light is administered; the pattern, duration, irradiance, and wavelength of the exposure to light; and recent photic history.

Circadian Phase-dependent Resetting to Light

The most important functional property of circadian oscillators is phase-dependent resetting; i.e., the magnitude and direction of phase resetting are dependent on the circadian phase at which a synchronizing stimulus occurs. This fundamental property of circadian clocks is summarized by the phase-response curve, a plot of the resetting response versus the circadian phase of the perturbation. The first phase-response curve was constructed by Hastings and Sweeney (1958) in the single-celled eukaryote *Gonyaulax polyhedra*. Following this pioneering study, phase-dependent resetting of circadian clocks in response to light and other synchronizers has been demonstrated in a diverse array of species, ranging from prokaryotes to humans. A conserved feature in all light-sensitive circadian oscillatory systems is that exposure to light during the early subjective night induces phase-delay shifts, whereas exposure to light in the late subjective night elicits phase-advance shifts.

On the basis of the topology of circadian resetting responses to light, Winfree classified phase resetting as being either Type 1 or Type 0. Weak Type-1 resetting is characterized by small phase shifts of only a few hours and little or no reduction in endogenous circadian pacemaker amplitude. In contrast, strong Type-0 resetting is characterized by large phase shifts of up to 12 hours and occurs via prior reduction of circadian pacemaker amplitude. On the basis of his phase-amplitude resetting model of circadian rhythms in *Drosophila*, Winfree predicted that a critical stimulus of intermediate strength, when administered at a critical circadian phase during the subjective night (the phase at which phase delays transition to phase advances), would drive the circadian oscillatory system to its singularity, characterized by a marked reduction in amplitude such that phase cannot be determined (Winfree 1969, 1980, 1987). Winfree demonstrated the property of critical resetting in *Drosophila*, in which the circadian amplitude of the eclosion rhythm approached zero in response to a carefully titrated light pulse. In summary, Winfree showed that (1) circadian resetting to light cannot be explained by a simple phase-only model and (2) organisms that display Type-0 resetting can also display Type-1 resetting to a stimulus of reduced strength.

Consistent with Winfree's predictions, Type-1 resetting, critical resetting, and Type-0 resetting have been demonstrated in humans (Czeisler et al. 1989; Jewett et al. 1991, 1992; Khalsa et al. 2003). The first human phase-response curve (PRC) to light was constructed in response to a three-cycle bright-light stimulus administered during 3 days (Fig. 2, left). Phase shifts of up to 12 hours were observed, and the Type-0 resetting contour closely matched that described in other organisms, such as the mosquito (Peterson 1980). Consistent with Winfree's phase-amplitude model of resetting, reducing the stimulus to a critical strength administered at a critical phase resulted in critical resetting in which circadian rhythm amplitude approached zero (Jewett et al. 1991). Reducing the stimulus strength further to a single-cycle exposure to bright light (6.5 hours, ~10,000 lux) resulted in Type-1 resetting (Fig. 2, right) (Khalsa et al. 2003). Consistent with Type-1 PRCs in other mammals, maximum phase delays of about −3.5 hours were observed in response to light administered during the early subjective night (before the body temperature rhythm nadir), and maximum phase advances of about +3.0 hours were observed following exposure to light during the late subjective night (after the body temperature minimum). Interestingly, the human PRC does not exhibit a dead zone of sensitivity during the subjective daytime, indicating that the human circadian

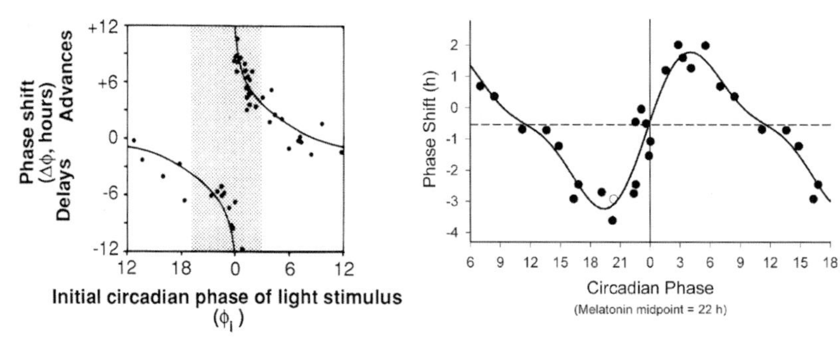

Figure 2. Circadian phase-dependent resetting of the human circadian system. (*Left*) The human circadian system exhibits Type-0 resetting in response to a three-cycle stimulus of bright light (5 hours, ~10,000 lux), characterized by large phase shifts of up to 12 hours. Phase shifts of the body temperature rhythm are plotted with respect to the initial circadian phase at which the light stimulus was given. (*Right*) Type-1 resetting of the melatonin rhythm is observed in response to a single cycle stimulus of bright light (6.5 hours, ~10,000 lux). The circadian phase of the light intervention is plotted relative to the nadir of the body temperature rhythm, defined as initial phase zero. (*Left panel,* Reprinted, with permission, from Czeisler et al. 1989 [© AAAS]; *right panel,* reprinted, with permission, from Khalsa et al. 2003 [© Physiological Society].)

system is sensitive to light across all circadian phases (Jewett et al. 1997; Khalsa et al. 2003).

Dose-dependent Resetting to Light

Resetting responses to light can be enhanced by increasing the duration or intensity of the stimulus. In humans, it has been demonstrated that phase resetting of the circadian system exhibits a saturating nonlinear dose-response curve to different levels of illuminance (Fig. 3a) (Boivin et al. 1996; Zeitzer et al. 2000). In the early subjective night, the dose response to 6.5 hours of white light saturates at about 500 lux, when administered on the background of constant dim light. Remarkably, exposure to ordinary indoor room light (~100 lux) elicits a half-maximal phase-shifting response (–1.5 hours) as compared to light that is 100 times brighter, indicating that the human circadian pacemaker is quite sensitive to light encountered in everyday life. These resetting responses have served as the basis of a mathematical model of the circadian resetting effect of light in humans (Kronauer 1990; Kronauer and Czeisler 1993; Jewett et al. 1999b; Kronauer et al. 1999). Moreover, like circadian rhythms in the mosquito and the tropical *Kalanchoe* plant, exposure to light of critical timing and intensity can drive the oscillator to its region of singularity, effectively "stopping" the circadian clock in humans (Jewett et al. 1991).

Most human subjects are able to maintain stable entrainment to a 24-hour cycle in which ambient room light was about 1.5 lux, suggesting that even candlelight can induce small shifts of the human circadian system (Fig. 3b) (Wright et al. 2001). To date, a systematic evaluation of the duration dependence of circadian resetting responses to light has not been conducted. However, in our preliminary analyses of the human PRC, maximum phase shifts to 1 hour of bright white light (~10,000 lux) were about 40% as effective as phase shifts measured in response to 6.5 hours of white light (~10,000 lux), despite

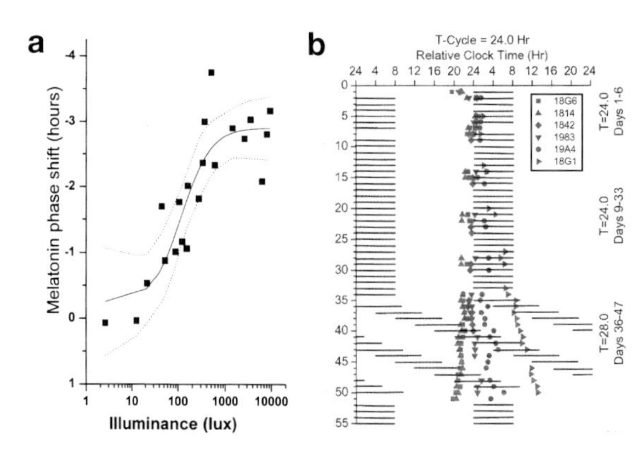

Figure 3. The human circadian system is exquisitely sensitive to light. (*a*) The dose response for phase resetting of the melatonin rhythm to white light (6.5 hours) is nonlinear, with a saturating phase-shift response at ~500 lux. Half-maximal phase resetting (–1.5 hours) is observed in response to ~100 lux, indicating that ordinary room light is highly effective at resetting the human circadian system. The light exposure was administered during the early biological night, and circadian phase of the melatonin rhythm was assessed during a constant routine procedure, before and after the light intervention. (*b*) Subjects are able to entrain to a 24-hour *T* cycle consisting of 16 hours of dim light (~1.5 lux) and 8 hours of sleep in darkness (Days 9–33), provided their intrinsic period is sufficiently close to 24 hours, as determined by forced desynchrony (*T* = 28 hours, Days 36–47). Symbols indicate the onset of the melatonin rhythm in individual subjects, and the sleep/wake schedule is double-plotted. (*a,* Reprinted, with permission, from Zeitzer et al. 2000 [© Physiological Society]; *b,* reprinted, with permission, from Wright et al. 2001 [© National Academy of Sciences.]

representing only 15% of the stimulus strength (1 hour/6.5 hours) (Khalsa et al. 2003; Lockley et al. 2006a). Hence, for the 6.5-hour light stimulus, the early part of a light stimulus is more effective at resetting the circadian system than the later part, as predicted by Kronauer's dynamic model of the resetting effect of light on the human circadian pacemaker (Kronauer et al. 1999).

Wavelength Sensitivity of Circadian Responses to Light

The wavelength sensitivity of a circadian system is dependent on the underlying photoreceptors that provide input to the circadian pacemaker. The human visual image-forming system consists of rods and cones that mediate night vision and color vision, respectively. Some blind individuals, with complete loss of conscious visual perception, show intact photic circadian entrainment and melatonin suppression in response to light (Czeisler et al. 1995; Klerman et al. 2002), suggesting that the classical visual photopigments are not required for circadian photoreception (Fig. 4). These results are consistent with findings in lower mammals demonstrating that rods and cones are dispensable for resetting of the circadian pacemaker (Provencio et al. 1994; Yoshimura and Ebihara 1996; Foster et al. 1998; Foster and Hankins 2002). Recently, it was shown that retinal ganglion cells that project to the SCN contain the blue-light-sensitive photopigment melanopsin (Gooley et al. 2001; Berson et al. 2002; Hannibal et al. 2002, 2004; Hattar et al. 2002). In the absence of rod and cone function, the melanopsin-containing cells mediate circadian entrainment (Yoshimura and Ebihara 1996; Freedman et al. 1999). In animals with intact retinae, however, the classical visual photopigments and melanopsin contribute to phase resetting of the circadian pacemaker (Hattar et al. 2003; Panda et al. 2003). In blind humans who show intact circadian photoreception, it is likely that the melanopsin cells reset the circadian clock, as evidenced by the short wavelength sensitivity of the circadian system (Czeisler et al. 1995; Zaidi et al. 2007). In normally sighted individuals, circadian phase resetting and melatonin suppression in response to bright monochromatic light is most sensitive to short-wavelength (blue) light (Fig. 4), indicating that melanopsin has a primary role in human circadian photoreception (Brainard et al. 2001; Thapan et al. 2001; Lockley et al. 2003). However, long-wavelength light is also effective at resetting human circadian rhythms, especially in response to lower irradiances, suggesting that the cones can function as

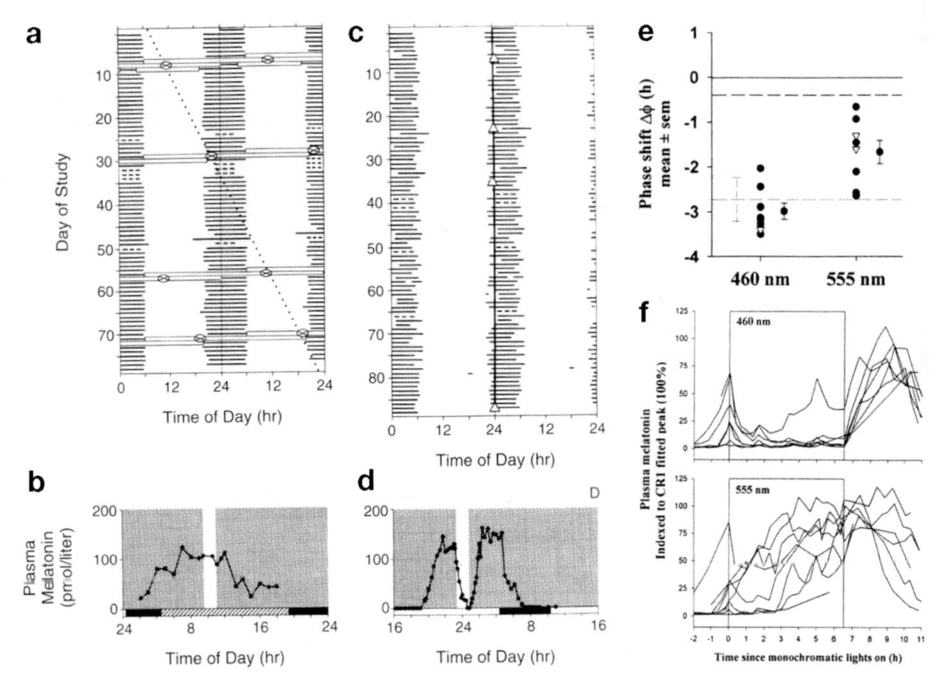

Figure 4. Visual photoreceptors are not required for circadian responses to light. (*a*) In a blind individual who kept a regular sleep/wake pattern for 78 days, the minimum of the body temperature rhythm (crosses) exhibited a free-running period of 24.5 hours (*dotted line*), indicating that the circadian system did not entrain to the solar day. (*b*) In the same individual, exposure to bright white light (~10,000 lux, *white vertical bar*) during the biological night (*hatched horizontal bar*) did not suppress the rhythm of melatonin. (*c*) In a different blind individual, the sleep/wake pattern and the peak of the melatonin rhythm (*triangles*) were synchronized for 88 days, indicating entrainment to the 24-hour day. (*d*) In this same subject, exposure to bright light during the night inhibited melatonin synthesis, indicating that nonvisual responses to light remained intact in some blind individuals. (*Black bars*) Sleep in darkness. (*e*) In normal-sighted subjects, circadian phase resetting of the melatonin rhythm is short-wavelength-sensitive, suggesting that the blue-light-sensitive melanopsin cells are the primary circadian photoreceptors in humans. A 6.5-hour monochromatic light stimulus (460 nm vs. 555 nm) was given during the early biological night. (*Upper dashed line*) Average drift in phase due to circadian period; (*lower dashed line*) average phase-shifting response to 6.7 hours of polychromatic white light (~10,000 lux) given at the same circadian phase. (*Closed circles*) Plasma melatonin; (*open triangles*) salivary melatonin. (*f*) In the same set of individuals, the 460-nm light stimulus elicited strong suppression of the melatonin rhythm across the 6.5-hour light intervention, whereas the 555-nm light stimulus elicited weak and transient suppression of melatonin. The *boxes* enclose the light intervention. (*a–d*, Reprinted, with permission, from Czeisler et al. 1995 [© Massachusetts Medical Society]; *e,f,* reprinted, with permission, from Lockley et al. 2003 [© Endocrine Society].)

circadian photoreceptors in humans (Zeitzer et al. 1997; J.J. Gooley et al., unpubl.). To determine the relative contributions of the three-cone photopic visual system and the melanopsin cells to circadian phase resetting, it will be important to characterize fully the spectral sensitivity of the human circadian system.

Resetting Responses to Intermittent Light

Exposure to intermittent light is highly effective at resetting the human circadian system. The phase-resetting effects of 6.5 hours of continuous bright white light (~10,000 lux) is comparable to a 6.5-hour intermittent exposure consisting of six cycles of 15 minutes of bright light (~10,000 lux) and 60 minutes of dim light (<3 lux) (Rimmer et al. 2000). Despite representing only 23% of continuous bright-light exposure conditions, the intermittent-light regimen elicited comparable phase shifts (Fig. 5a). Thus, a single sequence of intermittent bright-light pulses has a greater resetting efficacy on a per-minute basis than does continuous light exposure. In a subsequent study, exposure to two 45-minute pulses of bright light in the early subjective evening entrained the circadian system to a non-24-hour day (? + 1), indicating that intermittent pulses are highly efficient at resetting human circadian rhythms (Fig. 5b) (Gronfier et al. 2007).

Effects of Photic History on Resetting Responses to Light

Not surprisingly, most studies that have examined the resetting capacity of the human circadian pacemaker have focused on the light stimulus (i.e., duration, intensity, and wavelength) and the circadian phase at which the stimulus was administered. However, the effects of background lighting and photic history on the light resetting of the circadian pacemaker are largely unknown. Prior exposure to 3 days of indoor room light (~200 lux) attenuates suppression of the melatonin rhythm in response to 200 lux of light, as compared to prior exposure to 3 days of dim light (<3 lux) (Fig. 6) (Smith et al. 2004). It is currently being investigated whether preexposure to room light desensitizes circadian phase resetting. In future studies, it will be important to determine the time course and dose dependence of desensitization of circadian responses to light.

Period of the Human Circadian System

As noted above, Kleitman demonstrated in 1938 that one of his subjects exhibited a near-24-hour rhythm of body temperature, despite being scheduled on an imposed sleep/wake schedule of 28 hours (a forced desynchrony protocol), revealing that circadian rhythms could be separated from the influence of the timing of sleep/wake and light/dark schedules. Subsequently, Jürgen Aschoff and Rütger Wever attempted to determine the average period of the human circadian system by conducting a series of month-long studies of human subjects living in underground bunkers in Germany, beginning in 1960 and continuing for more than 25 years. In contrast to the finding from Kleitman's first cave study, they reported that under such conditions, human subjects exhibited rhythms of body temperature, urine volume, and sleep/wake with an average period of about 25 hours. Their studies of humans exposed to a self-selected, periodic light/dark cycle sug-

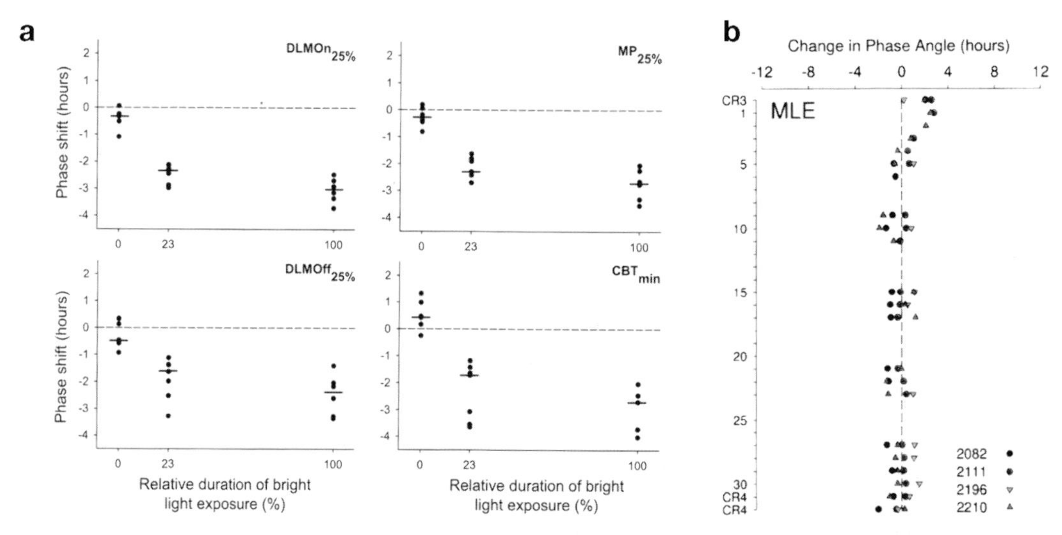

Figure 5. The human circadian system is highly sensitive to intermittent light. (*a*) Exposure to intermittent bright light (~9500 lux) is nearly as effective as continuous exposure to 6.7 hours of bright light at resetting the circadian rhythms of melatonin and body temperature, despite representing only 23% of the duration of the continuous-light exposure condition. Light intervention was administered during the early biological night. The intermittent bright-light stimulus consisted of six 15-minute pulses of light separated by 60 minutes of dim light (<1 lux). (CBT) Core body temperature minimum; ($DLMOn_{25\%}$) dim-light melatonin onset; ($DLMOff_{25\%}$) dim-light melatonin offset; ($MP_{25\%}$) melatonin midpoint. (*b*) Intermittent light is highly efficient at entraining the circadian system to a non-24-hour day (? + 1 hour). Subjects were exposed to two 45-minute pulses of bright light in the early subjective evening. The melatonin offset during the ? + 1-hour *T* cycle is plotted with respect to the phase measured on Day 3 of the protocol. Symbols correspond to the melatonin offset measured in different subjects. (*a*, Reprinted, with permission, from Gronfier et al. 2004 [© American Physiological Society]; *b*, reprinted, with permission, from Gronfier et al. 2007 [(c) National Academy of Sciences].)

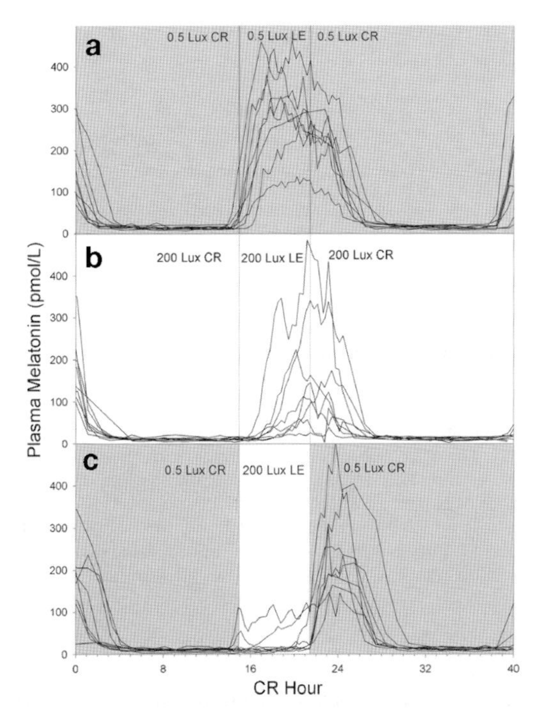

Figure 6. Adaptation of melatonin suppression in response to prior exposure to room light. (*a*) The baseline rhythm of melatonin is shown in eight subjects, as assessed during a constant routine procedure in dim ambient light. (*b*) Following exposure to room light (~200 lux) for 3 consecutive days during scheduled wake, the melatonin rhythm showed only minor suppression and a delayed onset in response to exposure to room light during the biological night. (*c*) In contrast, following exposure to dim light (<1 lux) for 3 consecutive days during scheduled wake, an acute exposure to room light (6.5 hours) resulted in strong suppression of the melatonin rhythm. (Reprinted, with permission, from Smith et al. 2004 [© Endocrine Society].)

gested that—unlike that of other mammals—the period of the activity rhythm in humans was highly variable, ranging from 13 to 65 hours (median 25.2 hours) and that the circadian period of the body temperature cycle averaged nearly 25 hours. These findings from Aschoff and Wever were consistent with similar reports that subsequently emerged from cave studies in France (Siffre 1964; Chouvet et al. 1974; Jouvet et al. 1974), England (Mills 1964), the United States (Siffre 1975), and from similar laboratory studies of humans shielded from external time cues conducted in Baltimore (Findley 1966), Gainesville, Florida (Webb and Agnew 1974a, b), and New York (Weitzman et al. 1981). However, in all of these studies, human subjects were given free access to artificial lighting and were therefore allowed to self-select their exposure to the light/dark cycle. Under these experimental conditions, human subjects self-selecting their bedtimes and wake times most commonly choose to retire near the nadir of the endogenous circadian temperature cycle and awaken on the rising slope of the temperature rhythm (Czeisler 1978; Czeisler et al. 1980a,b). Thus, they do not expose themselves to light equally across the circadian cycle (Klerman et al. 1996). Unrestricted exposure to ordinary indoor room light was permitted in those studies because, at that time, it had been incorrectly concluded that the human circadian system was not sensitive to the resetting effects of ordinary indoor room light intensity (Wever 1970, 1974; Aschoff 1976). Once it was demonstrated that human circadian pacemakers were exquisitely sensitive to the resetting effects of light, we realized that it was necessary to reassess circadian period in humans under conditions in which exposure to light was controlled (Czeisler et al. 1999). Subjects in prior studies designed to assess circadian period in humans had self-selected their exposure to light, the most powerful stimulus known to reset the circadian pacemaker. Moreover, analysis of those data revealed that subjects in those experiments had preferentially selected exposure to light during the phase-delay portion of the daily light sensitivity rhythm (Klerman et al. 1996; Khalsa et al. 2003). Mathematical modeling revealed that this would result in a net phase delay of the circadian system each day, leading the observed circadian period measured under such conditions to be longer than the actual intrinsic circadian period of the individuals at that time (Fig. 7a) (Klerman et al. 1996).

To assess the period of the human circadian pacemaker more precisely, Kleitman's forced desynchrony protocol was used to distribute circadian resetting stimuli more uniformly across the circadian cycle, and the strength of those synchronizers was minimized (Czeisler et al. 1990a, 1999). In the forced desynchrony method for assessment of circadian period, the imposed sleep/wake cycle is scheduled outside of the range of entrainment of the human circadian system (Fig. 7b). Importantly, subjects are only exposed to dim light during each scheduled wake episode, thereby minimizing the phase-resetting effects of light and ensuring that stable entrainment does not occur (i.e., T = ? – ∅? is not satisfied). As a result, the output of the endogenous circadian pacemaker becomes desynchronized from the imposed sleep/wake schedule. Because the forced desynchrony protocol is conducted over many circadian cycles, the timing of the sleep/wake and light/dark cycles is distributed much more uniformly across circadian phases, allowing for assessment of intrinsic circadian period.

Reevaluation of the circadian period under these controlled conditions has revealed that the intrinsic circadian period in sighted humans averages between 24.1 and 24.2 hours, with low interindividual variability as seen in other mammals, and that the intrinsic period remains stable with age in healthy adults. About 25% of human subjects exhibit a circadian period of less than 24 hours. The percent coefficient of circadian period variation in human subjects is about 0.55%, rather than 30% as previously estimated for free-running activity patterns (Czeisler et al. 1999). This tighter distribution is consistent with circadian period variability observed in other animals, such as the hamster, mouse, and gila monster (Czeisler et al. 1999). It has been argued that the circadian period estimated in forced desynchrony experiments may not be a better measure of the intrinsic period of the circadian oscillator, reflecting merely differences in experimental conditions. To test directly whether the average period of the human circadian pacemaker is closer to 24 hours (as measured on the forced desynchrony protocol) or to the classic value of 25 hours previously estimated in humans (as measured in

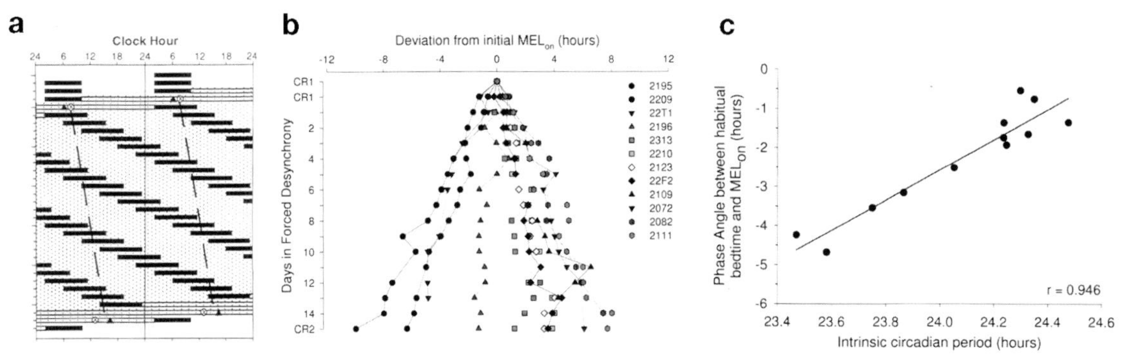

Figure 7. Assessment of circadian period by forced desynchrony. (*a*) The intrinsic period of a human subject was determined by exposure to a $T = 28$-hour cycle (18.67 hours wake, 9.33 hours sleep), which was outside the range of entrainment. The circadian period of this individual was 24.28 hours, as determined by the drift in phase of the melatonin peak (*closed triangles*) and the body temperature rhythm (*crosses*), assessed before and after the imposed forced desynchrony. During wake episodes, the light intensity was ~15 lux. (*Black bars*) Scheduled sleep episodes; data are double-plotted. (*b*) Daily melatonin phase estimates are shown for a group of subjects exposed to a $T = 28$-hour forced desynchrony protocol for 2 weeks. A quarter of subjects exhibited a circadian period of less than 24 hours. Data are plotted with respect to the timing of the dim-light melatonin onset at the beginning of the imposed forced desynchrony. (*c*) Phase angle of entrainment correlates strongly with intrinsic circadian period. The phase angle was determined as the difference in time between habitual bedtime (lights off) and the daily onset of melatonin secretion. Circadian period was determined by forced desynchrony. (*a*, Reprinted, with permission, from Czeisler et al. 1999 [© AAAS]; *b,c*, reprinted, with permission, from Gronfier et al. 2007 [© National Academy of Sciences].)

a self-selected light/dark cycle in the absence of external synchronizers), the range of entrainment in response to a weak resetting stimulus (~1.5 lux) was examined. It was found that in a dim light/dark cycle, most subjects were able to stably entrain to a 24.0-hour T cycle, whereas none were able to entrain to a 24.6-hour T cycle (Wright et al. 2001). These results indicate that the average circadian period in humans is much closer to 24 hours than to 25 hours, as predicted from the forced desynchrony studies. Moreover, small interindividual variations in intrinsic circadian period as measured in the forced desynchrony protocol are associated with remarkable differences in the phase angle of entrainment to the 24-hour day, as illustrated in Figure 7c (Gronfier et al. 2007). Thus, the conclusion that circadian period is very close to 24 hours in humans is both supported by and accounts for the ability of a very weak photic synchronizer (i.e., candlelight) to entrain circadian rhythms in most humans to a 24-hour day but not to a 24.65- or 23.5-hour day. Recently, it was discovered that the intrinsic period of the human circadian pacemaker was significantly longer in sighted subjects entrained to a 24.65-hour light/dark cycle than it was in those same subjects after entrainment to a 23.5-hour light/dark cycle (Scheer et al. 2007). Such aftereffects of prior entrainment, which have been observed in other species (Pittendrigh and Daan 1976), reveal the plasticity of the period of the human circadian system.

In blind individuals, whose circadian rhythms are not synchronized to the 24-hour day, the average circadian period is closer to 24.5 hours (Lockley et al. 1997), in both field and laboratory studies (Sack et al. 1992; Dijk and Lockley 2002; J.T. Hull et al., unpubl.). The somewhat shorter intrinsic period observed in sighted subjects may reflect aftereffects of prior entrainment in the sighted subjects (Scheer et al. 2007) and/or inclusion of only those blind subjects with longer than average circadian periods, who are therefore unable to maintain entrainment via weaker non-photic synchronizers (Czeisler et al. 1999). Collectively, these findings indicate that the period of the human circadian pacemaker, as measured immediately upon release from long-standing entrainment to the 24-hour day, is very close to 24 hours, with a tight distribution in the general population ranging from about 23.5 hours to 24.7 hours. Thus, in most individuals, daily phase shifts of less than 1 hour are required for stable entrainment to the 24-hour day.

REGULATION OF THE SLEEP/WAKE RHYTHM

The sleep/wake cycle is perhaps the most overt manifestation of circadian rhythms in humans. Because the rest/activity cycle is commonly used as an output variable in studies of circadian rhythms in vertebrates (Ralph et al. 1990), the timing of the sleep/wake cycle in humans is often presumed to be a simple reflection of the output of the circadian pacemaker. In actuality, many other factors affect sleep propensity in humans. This is because, like nutrition, sleep is an independent biological need in all animals. Without sleep, animals are unable to sustain life (Rechtschaffen et al. 1989). Therefore, sleep propensity is dependent not only circadian rhythmicity, but also on sleep satiety—also known as the level of homeostatic sleep drive (Dijk and Czeisler 1995). Acute sleep loss, sleep interruptions, sleep disorders, and chronic under-sleeping—i.e., sleeping less each day than the amount needed—increase homeostatic sleep drive. Increased homeostatic sleep drive due to any of these factors increases sleep propensity and impairs neurobehavioral performance (Jewett 1997). There is an interaction between the circadian and homeostatic systems, such that the amplitude of circadian oscillations in neurobehavioral performance is increased disproportionately when homeostatic sleep pressure is elevated (Jewett 1997; Jewett et al. 1999c). In particular, reaction time slows and the frequency of attentional failures increases when sleep propensity is elevated, regardless of whether that elevation is a result of acute sleep deprivation, chronic sleep loss, or misalignment of circadian phase (Cockley et al. 2004). Some exogenous agents, such as hypnotics, increase sleep propensity, whereas others, such as stimulants, decrease sleep propensity.

Homeostatic and Circadian Interaction

Just 25 years ago, Kronauer revealed that the interaction of two oscillatory processes with very different properties best explained the timing of human sleep/wake behavior in human subjects living for many months in the absence of time cues (Kronauer et al. 1982). Remarkably, that same year, Borbély (1982) hypothesized that each and every day in humans, the timing of sleep and wakefulness is governed by a subtle interplay between two such cyclic processes—one reflecting homeostatic sleep drive and the other the output of the circadian pacemaker. Homeostatic sleep drive accumulates with each waking hour and is only dissipated by sleep itself. This appetitive oscillatory process has properties very different from those of the circadian oscillator, which opposes the increasing homeostatic drive for sleep that builds near the end of our habitual waking day, leading Edgar et al. (1993) to propose an opponent process model of sleep regulation in primates.

Daily variations in alertness and neurobehavioral performance reflect the output of the circadian pacemaker(s) in humans (Czeisler et al. 1990b, 1994; Dijk et al. 1992; Johnson et al. 1992; Cajochen et al. 1999; Wyatt et al. 1999; Durmer and Dinges 2005). Sleepiness, vigilance, short-term memory, and attention or ability to concentrate are most impaired just after the body temperature nadir, near our regular wake time (Fig. 8) (Cajochen et al. 1999). Ironically, the circadian drive for wakefulness peaks just before habitual bedtime in humans entrained to the 24-hour day. This paradoxical phase relationship between the timing of the circadian sleep propensity rhythm and the timing of sleep and wakefulness during entrainment to the 24-hour day is postulated to facilitate consolidation of sleep and wakefulness in humans (Fig. 9) (Dijk and Czeisler 1994). During entrainment to the 24-hour day, the circadian pacemaker opposes this increasing drive for

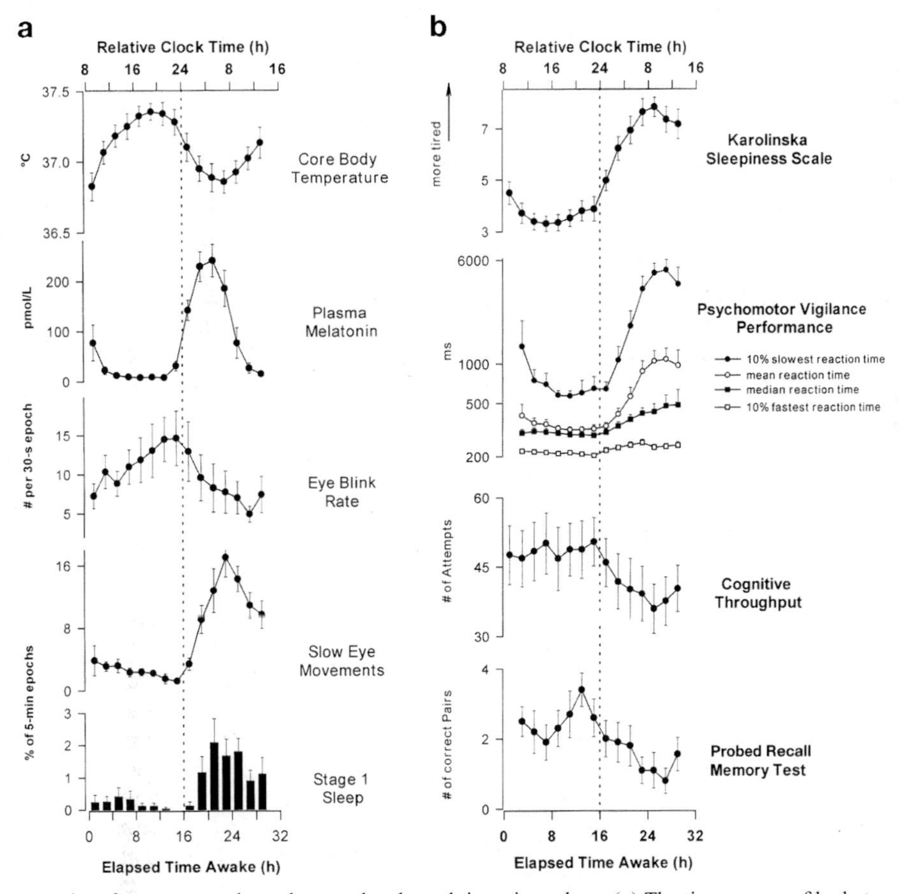

Figure 8. Alertness and performance are dependent on the elapsed time since sleep. (*a*) The time course of body temperature, melatonin, eye-blink rate, slow eye movements, and stage-1 sleep are shown for ten subjects during a constant routine procedure. Eye movements and sleep were assessed during the Karolinska drowsiness test, which was administered hourly. (*b*) In the same group of subjects, the time course is shown for sleepiness as assessed by the Karolinska sleepiness scale (KSS; highest possible score = 9, lowest possible score = 1), psychomotor vigilance (a 10-minute reaction time test), cognitive throughput (number of attempts in a 4-minute addition task), and memory performance (number of correct word pairs in a probed recall memory task). The mean ± S.E.M. is shown for each measure. All data were binned in 2-hour intervals and expressed with respect to elapsed time since scheduled wake time. (*Vertical dashed line* [16 hours]) Time at which each subject would normally go to bed. (Reprinted, with permission, from Cajochen et al. 1999 [©American Physiological Society].)

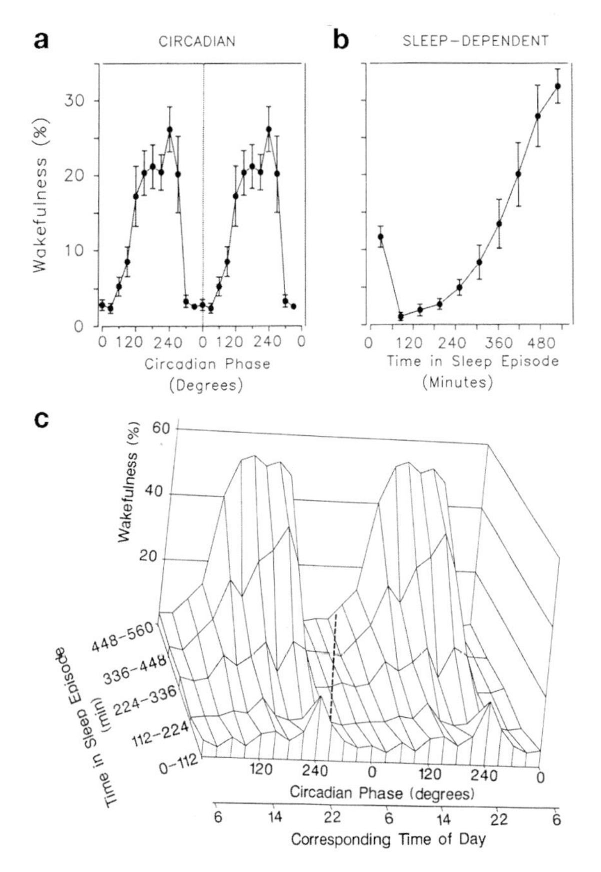

Figure 9. Percent wakefulness during scheduled sleep is determined by the interaction of circadian and homeostatic processes. The percentage of wakefulness during scheduled sleep was assessed in eight subjects during a forced desynchrony protocol (T = 28-hour cycle; 18.67 hours wake, 9.33 hours sleep). (*a*) Percent wakefulness shows a high-amplitude circadian rhythm that peaks just before normal bedtime. (*b*) After the first hour of scheduled sleep, percent wakefulness shows a nonlinear increase with respect to the elapsed time of the sleep episode. (*c*) Percent wakefulness during scheduled sleep is determined by interaction of the circadian rhythm in the drive for waking and the homeostatic drive for waking, which is dependent on the time elapsed since the start of the sleep episode. Circadian phase zero is defined as the phase of the body temperature minimum. For reference, clock times are shown that would correspond to the circadian phase of the body temperature rhythm under entrained conditions. (*Dashed line*) Trajectory of circadian phase and time elapsed since the start of scheduled sleep, corresponding to a nocturnal sleep episode under entrained conditions. (Reprinted, with permission, from Dijk and Czeisler 1994 [© Elsevier Science].)

sleep in the latter half of the usual waking day by sending out an increasingly stronger drive for waking. A couple of hours before bedtime, the pineal gland releases the sleep-promoting hormone melatonin into the bloodstream. Melatonin receptors on the SCN then suppress the firing of SCN neurons (Liu et al. 1997). This action of melatonin, which should not interfere with the ability of SCN to oscillate with a near-24-hour period (Schwartz et al. 1987), may serve to quiet the wake-promoting signal emanating from the SCN, thereby facilitating sleep just after the peak of the circadian drive for wakefulness (Barinaga 1997). The SCN, in turn, promotes sleep most strongly just before the habitual wake time, after many hours of sleep have dissipated homeostatic sleep pressure. The peak in the circadian rhythm of REM (rapid eye movement) sleep propensity is just before wake time, concurrent with the peak in the circadian rhythm of sleep propensity (Czeisler et al. 1980b).

Melatonin, Sleep, and Alertness

Forced desynchrony studies have revealed that sleep efficiency is greatest when subjects sleep at the circadian phase during which endogenous melatonin is released (Dijk et al. 1997). In contrast, wakefulness during the sleep episode is greatest when the sleep episode is scheduled to occur at a circadian phase during which endogenous melatonin secretion is absent (Dijk et al. 1997; Wyatt et al. 1999). This observation led us to hypothesize that exogenous administration of melatonin might improve sleep efficiency at such times. In a double-blind, placebo-controlled trial, it was found that melatonin administration 30 minutes before the scheduled sleep episode enabled volunteers to obtain about 30 minutes more sleep when they slept at a circadian phase at which they did not release endogenous melatonin (Fig. 10) (Wyatt et al. 2006). It was found that both 0.3 mg of melatonin, which raised plasma melatonin levels two to three times higher than endogenous secretion, and 5.0 mg of melatonin, which raised plasma melatonin

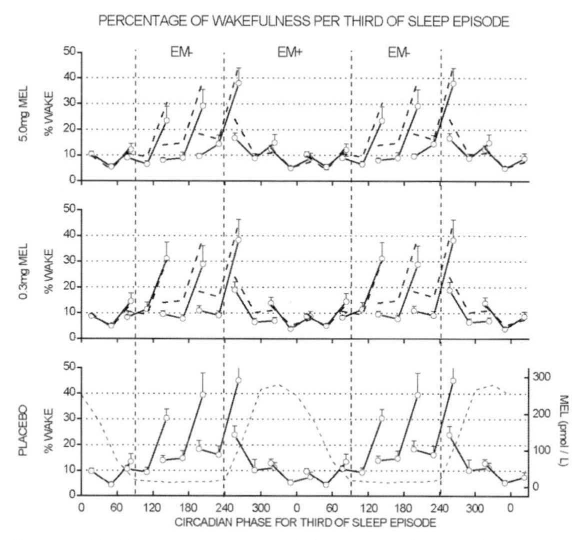

Figure 10. Exogenous melatonin improves sleep quality during the biological daytime but not during the biological night. The percentage of wakefulness during scheduled sleep was assessed in eight subjects during a 27-day forced desynchrony protocol (T = 20-hour cycle; 13.33 hours wake, 6.67 hours sleep). Melatonin (5.0 mg or 0.3 mg) or placebo was given 30 minutes before each scheduled sleep episode. The percentage of wakefulness during scheduled sleep was reduced during the biological daytime in subjects who were given melatonin before the sleep episode. In contrast, during the biological night, when melatonin levels are normally high, exogenous melatonin did not reduce percent wakefulness during scheduled sleep, as compared to subjects who were given the placebo. For reference, the percentage of wakefulness under the placebo condition (*lower panel*) is replotted in the *upper two panels* with a *dashed line*. (*Dashed trace*) Mean level of endogenous melatonin is plotted for the placebo group. Circadian phase zero is defined as the phase of the body temperature minimum. (Reprinted, with permission, from Wyatt et al. 2006 [© American Academy of Sleep Medicine].)

levels much higher, to be effective. In these healthy young adult subjects, no difference in efficacy between the two doses was found, although the study was not powered to detect such a difference. There was no effect of melatonin administration on sleep efficiency when the melatonin was administered at the circadian phase of endogenous melatonin production (Wyatt et al. 2006). Thus, the efficacy of melatonin as a hypnotic is dependent on circadian phase.

Similarly, exogenous melatonin acts as a soporific agent when it is administered during the daytime, when melatonin is usually absent from the bloodstream (Dollins et al. 1994; Cajochen et al. 1996, 1997). Photic suppression of endogenous melatonin levels at night results in an immediate improvement in alertness and performance (Campbell and Dawson 1990; Cajochen et al. 2005; Lockley et al. 2006c). Melatonin suppression appears to be critical to this immediate alerting effect of light, because retinal exposure to bright monochromatic green light (~555 nm) at the peak of the photopic sensitivity function is less effective in eliciting both melatonin suppression and improvements in alertness and performance than is exposure to shorter-wavelength blue light (~460 nm) near the peak of sensitivity of the novel photopigment melanopsin (Cajochen et al. 2005; Lockley et al. 2006c).

APPLICATIONS TO CLINICAL AND OCCUPATIONAL MEDICINE

Unlike most other diurnal animals, humans frequently attempt to forego sleep during nighttime hours for work or pleasure. The difficulty humans have in remaining alert while working at night is chronicled in ancient literature. There are many reasons why it is so difficult to do so. First of all, working at night requires individuals to perform when their circadian sleep propensity rhythm is misaligned with respect to the timing of wakefulness. Alertness and sleep propensity vary markedly with circadian phase (Czeisler et al. 1980a,b; Dijk and Czeisler 1994, 1995; Dijk et al. 1997, 1999; Wyatt et al. 1999; Czeisler and Khalsa 2000; Czeisler and Dijk 2001). Night work requires individuals to remain awake and alert at the peak of the circadian sleep propensity rhythm. In the absence of sleep, in the latter half of the night near the habitual wake time, elevated homeostatic drive for sleep interacts with the circadian peak of sleep propensity to create a critical zone of vulnerability. On the basis of laboratory studies that have quantified the contribution of the circadian pacemaker and the sleep homeostat to sleep duration and consolidation, subjective alertness, neurocognitive performance, and mood (Czeisler et al. 1980a,b; Dijk and Czeisler 1994, 1995; Dijk et al. 1997, 1999; Wyatt et al. 1999), the interaction of these two processes in determining alertness and performance has been successfully modeled mathematically (Jewett 1997; Jewett et al. 1999a,c; Jewett and Kronauer 1999). During extended durations of wakefulness, sleepiness generally increases and neurobehavioral performance deteriorates; at the same time, there is a circadian rhythm in these parameters (Dijk et al. 1992, 1999, 2000; Johnson et al. 1992; Klein et al. 1993; Czeisler et al. 1994; Dijk and Czeisler 1994, 1995; Boivin et al. 1997; Jewett 1997; Jewett et al. 1999a,d; Wyatt et al. 1999; Czeisler and Khalsa 2000; Czeisler and Dijk 2001; Cajochen et al. 2002; Wright et al. 2002). During sustained wakefulness coupled with circadian phase misalignment, 24 hours of sleep deprivation has been shown to impair neurobehavioral performance to an extent that is comparable to a level of 0.10% blood alcohol content (Dawson and Reid 1997; Lamond and Dawson 1999; Williamson and Feyer 2000; Powell et al. 2001; Falleti et al. 2003). Positron emission tomography (PET) imaging has revealed that such acute sleep deprivation is associated with decreased metabolism in the thalamus, prefrontal cortex, and parietal cortex (Thomas et al. 2000). In fact, the amount of time it takes to react to a visual stimulus (simple reaction time) averages three times larger after 24 hours of wakefulness at an adverse circadian phase than before an individual has stayed up all night (see Fig. 8) (Cajochen et al. 1999). Extended durations of wakefulness also increase the risk of attentional failures—in which the eyes begin rolling around in their sockets—heralding an involuntary transition from wakefulness to sleep, despite efforts to remain awake (Cajochen et al. 1999; Lockley et al. 2004).

Impact of Marathon Shifts on Safety

Despite advances in understanding the impact of sleep deprivation on performance, there are very few industries in the United States for which work hours are limited by law. Thus, individuals in many safety-sensitive industries routinely work marathon shifts, often as a consequence of direct or indirect monetary incentives. Firefighters in Los Angeles, for example, are routinely scheduled to work shifts that last 96 consecutive hours. The 100,000+ physicians-in-training in the United States are routinely required to work shifts of ≥30 consecutive hours, often twice per week (Czeisler 2006). Yet, physicians scheduled to work more than 24-hour shifts during their training experience twice as many attentional failures working at night (Lockley et al. 2004), and they make 36% more serious medical errors, including more than five times as many serious diagnostic mistakes while caring for patients in intensive care units, as those same interns when scheduled to work no more than 16 consecutive hours (Landrigan et al. 2004, 2007; Czeisler 2006; Lockley et al. 2006b, 2007). Despite averaging 2.6 hours of sleep during >24-hour marathon shifts, such physicians also have more than twice the risk of a motor vehicle crash driving home from such >24-hour shifts than from shifts averaging 12 hours in duration (Barger et al. 2005). When performing procedures during the daytime, these young physicians have a 73% increased risk of a percutaneous injury after a night on duty than after a night of sleep (Ayas et al. 2006). During the course of 1 year, one in five of these physicians reported making a fatigue-related mistake that seriously injured a patient and one in 20 of these physicians reported making a fatigue-related mistake that resulted in the death of a patient (Barger et al. 2006). These mistakes were, respectively, 600% and 300% more likely to occur during months in which those interns were scheduled to work extended duration (>24 hours) shifts more than once per week (Barger et al. 2006).

First Night Shift

The impairments associated with 24 consecutive hours of wakefulness are not limited to those scheduled to work for 24 consecutive hours. In fact, many employees scheduled to work a standard 8-hour night shift remain awake for 24 consecutive hours when they make the transition from the day shift to the night shift, because they are often awake all day (i.e., for 16 hours) before they even begin their first night shift. Thus, vigilance performance and attention are at their worst on the first night of a shift rotation sequence (Santhi et al. 2007).

Chronic Sleep Restriction

Ironically, after having struggled to stay awake throughout the night, those same night workers have difficulty sleeping during the day—again due to circadian misalignment, this time between the timing of the sleep opportunity and the timing of the sleep propensity rhythm. The resulting chronic sleep loss can itself have an adverse impact on performance. A number of consecutive nights of inadequate sleep have been shown to have detrimental effects on alertness, vigilance, psychomotor skills, and mood (Belenky et al. 2003; Czeisler 2003; Van Dongen and Dinges 2003; Van Dongen et al. 2003; Durmer and Dinges 2005). Objective measures of performance, including reaction time and memory, worsen. Within a week, loss of as little as 2 hours of sleep per day can impair performance—as measured by the ability to sustain vigilance on a reaction time task—by an amount equivalent to 24 consecutive hours of wakefulness (Belenky et al. 2003; Van Dongen et al. 2003). Such chronic sleep deprivation leads to an increased probability of experiencing lapses of attention, episodes of automatic behavior, and/or falling asleep while attempting to remain awake. Chronic sleep loss adversely affects neurobehavioral performance, even in individuals sleeping at night and attempting to perform during the daytime. In a condition of chronic sleep deprivation, even when wakefulness is scheduled during an appropriate circadian phase, the probability of a sleep-related attentional failure or neurocognitive performance failure while waking is markedly increased (Belenky et al. 2003; Van Dongen et al. 2003; Durmer and Dinges 2005). Six hours of time in bed per night for a week or two brings the average young adult to the same level of impairment as 24 hours of wakefulness, whereas 4 hours of time in bed per night rapidly induces a level of impairment comparable with 48 hours of wakefulness (i.e., two consecutive days and nights without sleep). Metabolic studies have demonstrated that such sleep curtailment has adverse effects on the metabolic and immune systems as well (Spiegel et al. 2000, 2001, 2002, 2004a,b, 2005; Mander et al. 2001; Van Cauter et al. 2007). Moreover, sleep loss interferes with memory consolidation and learning (Stickgold et al. 2000; Walker et al. 2002; Huber et al. 2004; Walker and Stickgold 2004; Stickgold 2005). As with alcohol intoxication, chronically sleep-deprived individuals tend to underestimate the extent to which their performance is impaired, despite increasing impairment evident in objective recordings of the rate of lapses of attention (Van Dongen et al. 2003). Importantly, the effects of recurrent nights of sleep restriction are not overcome with a single night of sleep (Belenky et al. 2003).

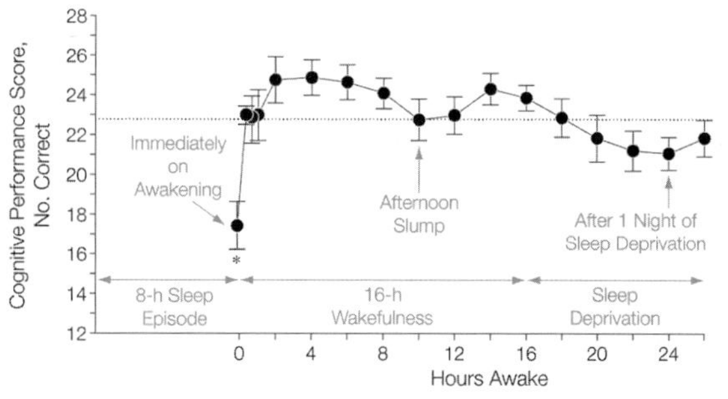

Figure 11. Cognitive performance is severely impaired by sleep inertia. Performance on a 2-minute addition task (2-digit numbers) was assessed in nine subjects during a 26-hour wake episode following 8 hours of sleep. Immediately upon awakening, cognitive performance was much worse than that observed following a night of sleep deprivation. (*Dotted line*) Group mean across the 26-hour period. Error bars indicate the S.E.M. and the asterisk indicates a significant difference from all subsequent time points (P ″ 0.01). (Reprinted, with permission, from Wertz et al. 2006 [© American Medical Association].)

Sleep Inertia

Performance is markedly degraded during the transition from wakefulness to sleep (Langdon and Hartman 1961; Hartman and Langdon 1965; Hartman et al. 1965; Koulack and Schultz 1974; Dinges et al. 1985; Balkin and Badia 1988; Dinges 1993; Naitoh et al. 1993; Achermann et al. 1995; Bruck and Pisani 1997; Ferrara and De Gennaro 2000; Balkin et al. 2002; Wertz et al. 2006). The extent to which this phenomenon, called sleep inertia, interferes with neurobehavioral performance is related to the depth of the prior sleep episode (Dinges 1993). Thus, agents that interfere with sleep, such as caffeine, can mute the effect of sleep inertia (Van Dongen et al. 2001). The adverse impact of sleep inertia on performance can exceed the impact of total sleep deprivation (Fig. 11) (Wertz et al. 2006). Individuals who are subjected to acute total sleep deprivation or chronic sleep restriction often experience very deep sleep, which will increase the effects of sleep inertia (Dinges 1993).

Sleep Disorders

Some medical conditions and medications increase homeostatic sleep drive indirectly by disrupting sleep; others increase sleep tendency directly. Either can increase the risk of attentional failures, sleep-related errors, and accidents (Carter et al. 2003; Colten and Altevogt 2006). These include primary sleep disorders, such as narcolepsy and sleep apnea. Patients with obstructive sleep apnea have a 6- to 13-fold increased risk of motor vehicle crashes (Teran-Santos et al. 1999). Repeated interruptions of sleep, such as is experienced by physicians when they are on call, degrade the restorative quality of sleep, compared to an equal amount of consolidated sleep. This is thought to be a primary basis for the excessive daytime sleepiness associated with sleep-disordered breathing, which induces many brief arousals during the night. Interestingly, just being on call itself disturbs sleep, even when the individual is not called (Torsvall and Åkerstedt 1988; Richardson et al. 1996). Age decreases the risk of sleep-related lapses of attention at night; in fact, young people are at the greatest risk of the hazards of sleep loss (Åkerstedt et al. 1994). However, as individuals age, it becomes increasingly more difficult to obtain the recovery sleep that is needed following sleep deprivation. Even when sleep deprived, older people have a great deal of difficulty sleeping at an adverse circadian phase (Dijk and Duffy 1999; Dijk et al. 1999, 2000).

Drowsy Driving

The United States is a drowsy nation. Individuals struggling to stay awake in the face of elevated sleep pressure—whether due to acute total sleep deprivation, chronic sleep restriction, or repeated interruption of sleep (due to external interruptions or the presence of a sleep disorder)—are not always able to do so. This is reflected by our performance on the single most safety-sensitive task that is shared in common by most American adults: driving motor vehicles on the nation's highways. This routine highly over-learned task performed in a moving vehicle, the motion of which stimulates the neurovestibular input to the ventrolateral preoptic (VLPO) area (Fuller et al. 2002), provides a setting that can unmask elevated homeostatic sleep drive. Sleep loss and misalignment of circadian phase increase the likelihood that the VLPO area of the hypothalamus will initiate an involuntary transition from wakefulness to sleep (Saper et al. 2005). The nonlinear interaction of the SCN and the sleep homeostat results in two times of day at which such sleep attacks are most probable: in the wee hours of the morning near the habitual wake time and in the mid afternoon. Once a sleep attack occurs, driving performance of an unresponsive drowsy driver is of course even worse than that of a drunk driver. Sometimes drowsy individuals linger in the transitional state between sleep and wakefulness, in which part of the brain is asleep while part of the brain remains awake. This transitional state of *automatic behavior* or *sleep drunkenness* is characterized by the ability to continue performing routine highly overlearned tasks such as driving—even providing semiautomatic responses to stimuli—without appropriate situational awareness or judgment (Guilleminault et al. 1975a,b).

The scope of the problem of drowsy driving in the United States is staggering. Data collected by the National Highway Transportation Safety Administration (NHTSA) indicate that at least 15 million drivers nationwide have nodded off or fallen asleep while driving in the past 6 months (Royal 2003). Data collected by NHTSA reveal that 250,000 drivers in the nation fall asleep at the wheel every day, or about three drivers every second throughout day and night, endangering themselves, their families, and their fellow citizens (Royal 2003). The outcome of these fall-asleep episodes is sobering. More than half of these drowsy drivers wandered into another lane, drifted onto the shoulder, or drove across the centerline during the incident. In another 10% of these incidents, the driver ran off the road. In fact, an estimated 1,350,000 drivers nationwide were involved in a drowsy-driving-related crash in the past 5 years—that is 30 drowsy driver crashes per hour or one every 2 minutes. NHTSA also sponsored a 100-car study in which drivers were video-monitored while driving in their cars for a year. Analysis of the data revealed that 22% of actual and near-miss crashes were caused by drowsiness—equal to the fraction of actual and near-miss crashes caused by all other sources of driver distraction combined (Klauer et al. 2006). Drowsy driving accounts for an estimated 20% of all motor vehicle crashes and injuries (Colten and Altevogt 2006). That means that there are more than 200,000 motor vehicle injuries every year due to drowsy driving crashes, 60,000 of which are debilitating injuries and more than 8000 motor vehicle fatalities every year.

FUTURE INITIATIVES

Twenty-five years ago, it was demonstrated that simple changes to the scheduling of night shift work designed to facilitate circadian adaptation could significantly improve both work schedule satisfaction and productivity (Czeisler et al. 1982). However, it has been recognized since the turn of the 20th century that the circadian rhythms of night shift workers often fail to adjust to the inversion of their sleep/wake schedules, even after years of permanent night shift work (Benedict 1904). Therefore, after the discovery that light could rapidly reset the human circadian pacemaker, the efficacy of its use in night shift workers was studied. Exposure to bright light and darkness was then found to be effective in resetting the circadian rhythms of night shift workers (Fig. 12) (Czeisler et al. 1990b), such that the sleep-promoting hormone melatonin was released during their scheduled daytime sleep episode, rather than during their scheduled nighttime shift (Czeisler et al. 1991). This circadian rhythm realignment improved performance during night shift hours and increased the dura-

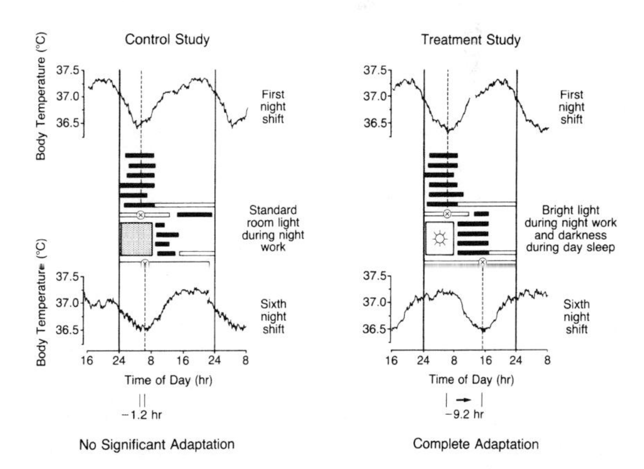

Figure 12. Adaptation to shift work with exposure to bright light and darkness. Exposure to room light (<150 lux; *stippled box*) during 5 nights of night work resulted in small shifts of the circadian rhythm of the body temperature rhythm, as assessed on the first and sixth nights of shift work. In contrast, exposure to bright light (7,000–12,000 lux; *solar symbol* in the open box) resulted in strong resetting and adaptation to the night shift schedule. The protocol is shown in the middle panels: (*Black bars*) Sleep; (*crosses*) body temperature minimum (also shown in the *vertical dashed lines*). Shift work was scheduled from midnight to 8:00 a.m. The temperature data are double-plotted. (Reprinted, with permission, from Czeisler et al. 1990b [© Massachusetts Medical Society].)

tion of sleep during the daytime hours by nearly 2 hours per day (Czeisler et al. 1990b).

This lighting technology was first applied to facilitate circadian adaptation of NASA astronauts to night launches of the space shuttle (Czeisler et al. 1991). NASA has installed bright-light facilities in its astronaut crew quarters at both the Johnson Space Center and the Kennedy Space Center. Since its introduction nearly 20 years ago, NASA has used bright-light shifting of crew members during the prelaunch quarantine period for all space shuttle flights requiring a shift of three or more hours in the timing of sleep. Bright-light shifting of circadian rhythms has also been used successfully to facilitate adaptation to the night shift on off-shore oil platforms in the North Sea and in other specific locations (Bjorvatn et al. 1999). High-fidelity simulations of the transition from day shift work to night shift work have revealed the extent to which exposure to bright light during night work, exposure to darkness during day sleep, or regularity in the timing of sleep contributed to the overall improvement that was observed when all three were optimized. Exposure to bright light and a fixed daytime sleep episode in darkness each contributed approximately equally to the resulting circadian adaptation (Horowitz et al. 2001).

Given recent discoveries that exposure to shorter-wavelength light is more effective than other wavelengths in resetting circadian rhythms (Lockley et al. 2003), future countermeasures for shift workers will likely involve timed exposure to light of specific wavelengths and intensities in order to facilitate most effectively adaptation to night shift work. In addition, it is critical that corporations develop appropriate policies regarding the maximum work episode duration, minimum time off between shifts, and maximum weekly work hours, along with policies regarding travel across time zones (Czeisler and Fryer 2006).

It is also very important for our society to establish work-hour limits to protect workers in our 24/7 culture. The European Union has led the way on this front with its adoption of the European Working Time Directive, which limits—in all occupations—the number of consecutive hours to which employees can be scheduled. In principle, all EU employees are limited to 13 consecutive hours of work, with a minimum of 11 hours of rest between work shifts, although there are a number of exceptions and opt-out policies that impact these rules. Given the toll that sleep deprivation is taking on the health, productivity, and safety of American workers, it is time that U.S. policy makers evaluate the evidence and implement comprehensive legislation on this issue in the United States. However, as the Institute of Medicine highlighted in its recent report on sleep deprivation and sleep disorders (Colten and Altevogt 2006), education is the most pressing hurdle that must be overcome to address this under-recognized public health problem.

ACKNOWLEDGMENTS

The research reviewed has been supported in part by grants from the National Institutes of Health (National Center for Research Resources; National Heart, Lung and Blood Institute; National Institute on Aging; National Institute of Mental Health; National Institute of Neurological Diseases and Stroke), the National Aeronautics and Space Administration; the National Space Biomedical Research Institute; and the Air Force Office of Scientific Research. The authors thank Ms. Lorna Preston for her editorial assistance. We are indebted to Drs. Diane B. Boivin, Emery N. Brown, Christian Cajochen, Derk-Jan Dijk, Jeanne F. Duffy, Claude Gronfier, Megan E. Jewett, Sat Bir S. Khalsa, Elizabeth B. Klerman, Richard E. Kronauer, Steven W. Lockley, Nayantara Santhi, Frank A. Scheer, Theresa L. Shanahan, Kenneth P. Wright, Jr., James K. Wyatt, and Jamie M. Zeitzer for their many contributions to this body of knowledge.

REFERENCES

Achermann P., Werth E., Dijk D.J., and Borbély A.A. 1995. Time course of sleep inertia after nighttime and daytime sleep episodes. *Arch. Ital. Biol.* **134:** 109.

Åkerstedt T., Czeisler C.A., Dinges D.F., and Horne J.A. 1994. Accidents and sleepiness: A consensus statement from the International Conference on Work Hours, Sleep and Accidents (Stockholm, September 8–10, 1994). *J. Sleep Res.* **3:** 195.

Allan J.S. and Czeisler C.A. 1994. Persistence of the circadian thyrotropin rhythm under constant conditions and after light-induced shifts of circadian phase. *J. Clin. Endocrinol. Metab.* **79:** 508.

Aschoff J. 1976. Circadian systems in man and their implications. *Hosp. Pract.* **11:** 51.

Aschoff J. and Wever R. 1981. The circadian system of man. In *Biological rhythms: Handbook of behavioral neurobiology* (ed. J. Aschoff), p. 311. Plenum Press, New York.

Aschoff J., Pöppel E., and Wever R. 1969. Circadiane perodik des menschen unter dem einfluss von licht-dunkel-wechseln unterschiedlicher periode. *Pflügers Arch.* **306:** 58.

Ayas N.T., Barger L.K., Cade B.E., Hashimoto D.M., Rosner B., Cronin J.W., Speizer F.E., and Czeisler C.A. 2006. Extended work duration and the risk of self-reported percutaneous injuries in interns. *J. Am. Med. Assoc.* **296:** 1055.

Balkin T.J. and Badia P. 1988. Relationship between sleep inertia and sleepiness: Cumulative effects of four nights of sleep disruption/restriction on performance following abrupt nocturnal awakenings. *Biol. Psychol.* **27:** 245.

Balkin T.J., Braun A.R., Wesensten N.J., Jeffries K., Varga M., Baldwin P., Belenky G., and Herscovitch P. 2002. The process of awakening: A PET study of regional brain activity patterns mediating the re-establishment of alertness and consciousness. *Brain* **125:** 2308.

Barger L.K., Ayas N.T., Cade B.E., Cronin J.W., Rosner B., Speizer F.E., and Czeisler C.A. 2006. Impact of extended-duration shifts on medical errors, adverse events, and attentional failures. *PLoS Med.* **3:** e487.

Barger L.K., Cade B.E., Ayas N.T., Cronin J.W., Rosner B., Speizer F.E., and Czeisler C.A. 2005. Extended work shifts and the risk of motor vehicle crashes among interns. *N. Engl. J. Med.* **352:** 125.

Barinaga M. 1997. How jet-lag hormone does double duty in the brain. *Science* **277:** 480.

Belenky G., Wesensten N.J., Thorne D.R., Thomas M.L., Sing H.C., Redmond D.P., Russo M.B., and Balkin T.J. 2003. Patterns of performance degradation and restoration during sleep restriction and subsequent recovery: A sleep dose-response study. *J. Sleep Res.* **12:** 1.

Benedict F.G. 1904. Studies in body-temperature. I. Influence of the inversion of the daily routine; the temperature of night-workers. *Am. J. Physiol.* **11:** 145.

Berson D.M., Dunn F.A., and Takao M. 2002. Phototransduction by retinal ganglion cells that set the circadian clock. *Science* **295:** 1070.

Bjorvatn B., Kecklund G., and Åkerstedt T. 1999. Bright light treatment used for adaptation to night work and re-adaptation back to day life. A field study at an oil platform in the North Sea. *J. Sleep. Res.* **8:** 105.

Boivin D.B., Duffy J.F., Kronauer R.E., and Czeisler C.A. 1996. Dose-response relationships for resetting of human circadian clock by light. *Nature* **379:** 540.

Boivin D.B., Czeisler C.A., Dijk D.J., Duffy J.F., Folkard S., Minors D.S., Totterdell P., and Waterhouse J.M. 1997. Complex interaction of the sleep-wake cycle and circadian phase modulates mood in healthy subjects. *Arch. Gen. Psychiatry* **54:** 145.

Borbély A.A. 1982. A two process model of sleep regulation. *Hum. Neurobiol.* **1:** 195.

Brainard G.C., Hanifin J.P., Greeson J.M., Byrne B., Glickman G., Gerner E., and Rollag M.D. 2001. Action spectrum for melatonin regulation in humans: Evidence for a novel circadian photoreceptor. *J. Neurosci.* **21:** 6405.

Bruck D. and Pisani D. 1997. The effects of sleep inertia on decision-making performance. *J. Sleep Res.* **8:** 95.

Cajochen C., Kräuchi K., and Wirz-Justice A. 1997. The acute soporific action of daytime melatonin administration: Effects on the EEG during wakefulness and subjective alertness. *J. Biol. Rhythms* **12:** 636.

Cajochen C., Wyatt J.K., Czeisler C.A., and Dijk D.J. 2002. Separation of circadian and wake duration-dependent modulation of EEG activation during wakefulness. *Neuroscience* **114:** 1047.

Cajochen C., Khalsa S.B.S., Wyatt J.K., Czeisler C.A., and Dijk D.J. 1999. EEG and ocular correlates of circadian melatonin phase and human performance decrements during sleep loss. *Am. J. Physiol.* **277:** R640.

Cajochen C., Kräuchi K., von Arx M.-A., Möri D., Graw P., and Wirz-Justice A. 1996. Daytime melatonin administration enhances sleepiness and ?/? activity in the waking EEG. *Neurosci. Lett.* **207:** 209.

Cajochen C., Munch M., Kobialka S., Kräuchi K., Steiner R., Oelhafen P., Orgul S., and Wirz-Justice A. 2005. High sensitivity of human melatonin, alertness, thermoregulation, and heart rate to short wavelength light. *J. Clin. Endocrinol. Metab.* **90:** 1311.

Campbell S.S. and Dawson D. 1990. Enhancement of nighttime alertness and performance with bright ambient light. *Physiol. Behav.* **48:** 317.

Carter N., Ulfberg J., Nystrom B., and Edling C. 2003. Sleep debt, sleepiness and accidents among males in the general population and male professional drivers. *Accid. Anal. Prev.* **35:** 613.

Chouvet G., Mouret J., Coindet J., Siffre M., and Jouvet M. 1974. Periodicite bicircadienne du cycle veille-sommeil dans des conditions hors du temps. Etude polygraphique. *Electroencephalogr. Clin. Neurophysiol.* **37:** 367.

Colten H.R. and Altevogt B.M., eds. 2006. *Sleep disorders and sleep deprivation: An unmet public health problem.* National Academies Press, Washington, D.C.

Czeisler C.A. 1978. "Human circadian physiology: Internal organization of temperature, sleep-wake, and neuroendocrine rhythms monitored in an environment free of time cues." Ph.D. thesis, Stanford University, Stanford, California.

———. 1986. Circadian rhythmicity and its disorders. In *Sleep and wakefulness: Pharmacology and pathology* (ed. A.N. Nicholson and I.B. Welbers), p. 1. Boehringer Ingelheim International, Ingelheim-am-Rhein, Germany.

———. 1995. The effect of light on the human circadian pacemaker. *Ciba Found. Symp.* **183:** 254.

———. 2003. Quantifying consequences of chronic sleep restriction. *Sleep* **26:** 247.

———. 2006. The Gordon Wilson Lecture: Work hours, sleep and patient safety in residency training. *Trans. Am. Clin. Climatol. Assoc.* **117:** 159.

Czeisler C.A. and Dijk D.J. 2001. Human circadian physiology and sleep-wake regulation. In *Handbook of behavioral neurobiology: Circadian clocks* (ed. J.S. Takahashi et al.), p. 531. Plenum Publishing, New York.

Czeisler C.A. and Fryer B. 2006. Sleep deficit: The performance killer. *Harv. Bus. Rev.* **84:** 53.

Czeisler C.A. and Jewett M.E. 1990. Human circadian physiology: Interaction of the behavioral rest—Activity cycle with the output of the endogenous circadian pacemaker. In *Handbook of sleep disorders* (ed. M.J. Thorpy), p. 117. Marcel Dekker, New York.

Czeisler C.A. and Khalsa S.B.S. 2000. The human circadian timing system and sleep-wake regulation. In *Principles and practice of sleep medicine* (ed. M.H. Kryger et al.), p. 353. W.B. Saunders, Philadelphia, Pennsylvania.

Czeisler C.A. and Wright K.P., Jr. 1999. Influence of light on circadian rhythmicity in humans. In *Neurobiology of sleep and circadian rhythms* (ed. F.W. Turek and P.C. Zee), p. 149. Marcel Dekker, New York.

Czeisler C.A., Allan J.S., and Kronauer R.E. 1990a. A method for assaying the effects of therapeutic agents on the period of the endogenous circadian pacemaker in man. In *Sleep and biological rhythms: Basic mechanisms and applications to psychiatry* (ed. J. Montplaisir and R. Godbout), p. 87. Oxford University Press, New York.

Czeisler C.A, Chiasera A.J., and Duffy J.F. 1991. Research on sleep, circadian rhythms and aging: Applications to manned spaceflight. *Exp. Gerontol.* **26:** 217.

Czeisler C.A., Dijk D.J., and Duffy J.F. 1994. Entrained phase of the circadian pacemaker serves to stabilize alertness and performance throughout the habitual waking day. In *Sleep onset: Normal and abnormal processes* (ed. R.D. Ogilvie and J.R. Harsh), p. 89. American Psychological Association, Washington, D.C.

Czeisler C.A., Moore-Ede M.C., and Coleman R.M. 1982. Rotating shift work schedules that disrupt sleep are improved by applying circadian principles. *Science* **217:** 460.

Czeisler C.A., Winkelman J.W., and Richardson G.S. 2000. Disorders of sleep and circadian rhythms. In *Harrison's principles of internal medicine* (ed. E. Braunwald et al.), p. 1. McGraw-Hill, New York.

Czeisler C.A., Richardson G.S., Zimmerman J.C., Moore-Ede M.C., and Weitzman E.D. 1981. Entrainment of human circadian rhythms by light/dark cycles: A reassessment. *Photochem. Photobiol.* **34:** 239.

Czeisler C.A., Weitzman E.D., Moore-Ede M.C., Zimmerman J.C., and Knauer R.S. 1980a. Human sleep: Its duration and organization depend on its circadian phase. *Science* **210:** 1264.

Czeisler C.A., Zimmerman J.C., Ronda J.M., Moore-Ede M.C., and Weitzman E.D. 1980b. Timing of REM sleep is coupled to the circadian rhythm of body temperature in man. *Sleep* **2:** 329.

Czeisler C.A., Johnson M.P., Duffy J.F., Brown E.N., Ronda J.M., and Kronauer R.E. 1990b. Exposure to bright light and darkness to treat physiologic maladaptation to night work. *N. Engl. J. Med.* **322:** 1253.

Czeisler C.A., Kronauer R.E., Allan J.S., Duffy J.F., Jewett M.E., Brown E.N., and Ronda J.M. 1989. Bright light induction of strong (type 0) resetting of the human circadian pacemaker. *Science* **244:** 1328.

Czeisler C.A., Shanahan T.L., Klerman E.B., Martens H., Brotman D.J., Emens J.S., Klein T., and Rizzo J.F., III. 1995. Suppression of melatonin secretion in some blind patients by exposure to bright light. *N. Engl. J. Med.* **332:** 6.

Czeisler C.A., Duffy J.F., Shanahan T.L., Brown E.N., Mitchell J.F., Rimmer D.W., Ronda J.M., Silva E.J., Allan J.S., Emens J.S., Dijk D.J., and Kronauer R.E. 1999. Stability, precision, and near-24-hour period of the human circadian pacemaker. *Science* **284:** 2177

Dawson D. and Reid K. 1997. Fatigue, alcohol and performance impairment. *Nature* **388:** 235.

Dijk D.J. and Czeisler C.A. 1994. Paradoxical timing of the circadian rhythm of sleep propensity serves to consolidate sleep and wakefulness in humans. *Neurosci. Lett.* **166:** 63.

———. 1995. Contribution of the circadian pacemaker and the sleep homeostat to sleep propensity, sleep structure, electroencephalographic slow waves, and sleep spindle activity in humans. *J. Neurosci.* **15:** 3526.

Dijk D.J. and Duffy J.F. 1999. Circadian regulation of human sleep and age-related changes in its timing, consolidation and EEG characteristics. *Ann. Med.* **31:** 130.

Dijk D.J. and Lockley S.W. 2002. Integration of human sleep-wake regulation and circadian rhythmicity. *J. Appl. Physiol.* **92:** 852.

Dijk D.J., Duffy J.F., and Czeisler C.A. 1992. Circadian and sleep/wake dependent aspects of subjective alertness and cognitive performance. *J. Sleep Res.* **1:** 112.

———. 2000. Contribution of circadian physiology and sleep homeostasis to age-related changes in human sleep. *Chronobiol. Int.* **17:** 285.

Dijk D.J., Duffy J.F., Riel E, Shanahan T.L., and Czeisler C.A. 1999. Ageing and the circadian and homeostatic regulation of human sleep during forced desynchrony of rest, melatonin and temperature rhythms. *J. Physiol.* **516:** 611.

Dijk D.J., Shanahan T.L., Duffy J.F., Ronda J.M., and Czeisler C.A. 1997. Variation of electroencephalographic activity during non-rapid eye movement and rapid eye movement sleep with phase of circadian melatonin rhythm in humans. *J. Physiol.* **505:** 851.

Dinges D.F. 1993. Sleep inertia. In *Encyclopedia of sleep and dreaming* (ed. M.A. Carskadon), p. 553. Macmillan, New York.

Dinges D.F., Orne M.T., and Orne E.C. 1985. Assessing performance upon abrupt awakening from naps during quasi-continuous operations. *Behav. Res. Methods Instrum. Comput.* **17:** 37.

Dollins A.B., Zhdanova I.V., Wurtman R.J., Lynch H.J., and Deng M.H. 1994. Effect of inducing nocturnal serum melatonin concentrations in daytime on sleep, mood, body temperature, and performance. *Proc. Natl. Acad. Sci.* **91:** 1824.

Durmer J.S. and Dinges D.F. 2005. Neurocognitive consequences of sleep deprivation. *Semin. Neurol.* **25:** 117.

Edgar D.M., Dement W.C., and Fuller C.A. 1993. Effect of SCN lesions on sleep in squirrel monkeys: Evidence for opponent processes in sleep-wake regulation. *J. Neurosci.* **13:** 1065.

El-Hajj Fuleihan G., Klerman E.B., Brown E.N., Choe Y., Brown E.M., and Czeisler C.A. 1997. The parathyroid hormone circadian rhythm is truly endogenous—A general clinical research center study. *J. Clin. Endocrinol. Metab.* **82:** 281.

Falleti M.G., Maruff P., Collie A., Darby D.G., and McStephen M. 2003. Qualitative similarities in cognitive impairment associated with 24 h of sustained wakefulness and a blood alcohol concentration of 0.05%. *J. Sleep Res.* **12:** 265.

Ferrara M. and De Gennaro L. 2000. The sleep inertia phenomenon during the sleep-wake transition: Theoretical and operational issues. *Aviat. Space Environ. Med.* **71:** 843.

Findley J.D. 1966. Programmed environments for the experimental analysis of human behavior. In *Operant behavior: Areas of research and application* (ed. W.K. Hong), p. 827. Appleton-Century-Crofts, New York.

Foster R.G. and Hankins M.W. 2002. Non-rod, non-cone photoreception in the vertebrates. *Prog. Retinal Eye Res.* **21:** 507.

Foster R.G., David-Gray Z.K., Freedman M.S., Lucas R.J., Lupi D., von schantz M., and Soni B.G. 1998. Photic regulation of the mammalian biological clock: The use of mutants and genetically engineered mice. In *Circadian clocks and entrainment: Proceedings of the 7th Sapporo Symposium on Biological Rhythms* (ed. K. Honma and S. Honma), p. 3. Hokkaido University Press, Sapporo, Japan.

Freedman M.S., Lucas R.J., Soni B., von schantz M., Muñoz M., David-Gray Z., and Foster R. 1999. Regulation of mammalian circadian behavior by non-rod, non-cone, ocular photoreceptors. *Science* **284:** 502.

Fuller P.M., Jones T.A., Jones S.M., and Fuller C.A. 2002. Neurovestibular modulation of circadian and homeostatic regulation: Vestibulohypothalamic connection. *Proc. Natl. Acad. Sci.* **99:** 15723.

Gibson R.B. 1905. The effects of transposition of the daily routine on the rhythm of temperature variation. *Am. J. Med. Sci.* **129:** 1048.

Gooley J.J., Lu J., Chou T.C., Scammell T.E., and Saper C.B. 2001. Melanopsin in cells of origin of the retinohypothalamic tract. *Nat. Neurosci.* **4:** 1165.

Gronfier C., Wright K.P., Jr., Kronauer R.E., and Czeisler C.A. 2007. Entrainment of the human circadian pacemaker to longer-than-24h days. *Proc. Natl. Acad. Sci.* **104:** 9081.

Gronfier C., Wright K.P., Jr., Kronauer R.E., Jewett M.E., and Czeisler C.A. 2004. Efficacy of a single sequence of intermittent bright light pulses for delaying circadian phase in humans. *Am. J. Physiol. Endocrinol. Metab.* **287:** E174.

Guilleminault C., Phillips R., and Dement W.C. 1975a. A syndrome of hypersomnia with automatic behavior. *Electroencephalogr. Clin. Neurophysiol.* **38:** 403.

Guilleminault C., Billiard M., Montplaisir J., and Dement W.C. 1975b. Altered states of consciousness in disorders of daytime sleepiness. *J. Neurol. Sci.* **26:** 377.

Hannibal J., Hindersson P., Knudson S.M., Georg B., and Fahrenkrug J. 2002. The photopigment melanopsin is exclusively present in pituitary adenylate cyclase-activiting polypeptide-containing retinal ganglion cells of the retinohypothalamic tract. *J. Neurosci.* **22:** 1.

Hannibal J., Hindersson P., Ostergaard J., Georg B., Heegaard S., Larsen P.J., and Fahrenkrug J. 2004. Melanopsin is expressed in PACAP-containing retinal ganglion cells of the human retinohypothalamic tract. *Invest. Ophthalmol. Vis. Sci.* **45:** 4202.

Hartman B.O. and Langdon D.E. 1965. A second study on performance upon sudden awakening. TR-65-61, 1–10. Brooks Air Force Base, Brooks, Texas.

Hartman B.O., Langdon D.E., and Mc Kenzie R.E. 1965. A third study on performance upon sudden awakening. TR-65-63, 1–4. Brooks Air Force Base, Brooks, Texas.

Hastings J.W. and Sweeney B.M. 1958. A persistent diurnal rhythm of luminescence in *Gonyaulax polyedra. Biol. Bull.* **115:** 440.

———. 1960. The action spectrum for shifting the phase of the rhythm of luminescence in *Gonyaulax polyedra. J. Gen. Physiol.* **43:** 697.

Hattar S., Liao H.-W., Takao M., Berson D.M., and Yau K.-W. 2002. Melanopsin-containing retinal ganglion cells: Architecture, projections, and intrinsic photosensitivity. *Science* **295:** 1065.

Hattar S., Lucas R.J., Mrosovsky N., Thompson S., Douglas R.H., Hankins M.W., Lem J., Biel M., Hofmann F., Foster R.G., and Yau K.-W. 2003. Melanopsin and rod-cone photoreceptive systems account for all major accessory visual functions in mice. *Nature* **424:** 75.

Horowitz T.S., Cade B.E., Wolfe J.M., and Czeisler C.A. 2001. Efficacy of bright light and sleep/darkness scheduling in alleviating circadian maladaptation to night work. *Am. J. Physiol. Endocrinol. Metab.* **281:** E384.

Huber R., Ghilardi M.F., Massimini M., and Tononi G. 2004. Local sleep and learning. *Nature* **430:** 78.

Jewett M.E. 1997. "Models of circadian and homeostatic regulation of human performance and alertness." Ph.D. thesis, Harvard University, Cambridge, Massachusetts.

Jewett M.E. and Kronauer R.E. 1999. Interactive mathematical models of subjective alertness and cognitive throughput in humans. *J. Biol. Rhythms* **14:** 588.

Jewett M.E., Borbély A.A., and Czeisler C.A. 1999a. Proceedings of the workshop on biomathematical models of circadian rhythmicity, sleep regulations, and neurobehavioral function in humans. *J. Biol. Rhythms* **14:** 429.

Jewett M.E., Forger D.B., and Kronauer R.E. 1999b. Revised limit cycle oscillator model of human circadian pacemaker. *J. Biol. Rhythms* **14:** 493.

Jewett M.E., Kronauer R.E., and Czeisler C.A. 1991. Light-induced suppression of endogenous circadian amplitude in humans. *Nature* **350:** 59.

Jewett M.E., Dijk D.J., Kronauer R.E., and Dinges D.F. 1999c. Dose-response relationship between sleep duration and human psychomotor vigilance and subjective alertness. *Sleep* **22:** 171.

Jewett M.E., Kronauer R.E., Klerman E.B., and Czeisler C.A. 1992. Phase/amplitude resetting maps: A comparison of model simulations and laboratory trials. *Soc. Res. Biol. Rhythms* **3:** 77.

Jewett M.E., Rimmer D.W., Duffy J.F., Klerman E.B., Kronauer R.E., and Czeisler C.A. 1997. Human circadian pacemaker is sensitive to light throughout subjective day without evidence of transients. *Am. J. Physiol.* **273:** R1800.

Jewett M.E., Wyatt J.K., Ritz-De Cecco A., Khalsa S.B., Dijk D.J., and Czeisler C.A. 1999d. Time course of sleep inertia dissipation in human performance and alertness. *J. Sleep Res.* **8:** 1.

Jewett M.E., Khalsa S.B.S., Klerman E.B., Duffy J.F., Rimmer D.W., Kronauer R.E., and Czeisler C.A. 2000. 3-cycle bright light stimulus induces type 0 resetting in human melatonin rhythm. *Soc. Res. Biol. Rhythms* **7:** 134.

Johnson M.P., Duffy J.F., Dijk D.J., Ronda J.M., Dyal C.M., and Czeisler C.A. 1992. Short-term memory, alertness and performance: A reappraisal of their relationship to body temperature. *J. Sleep Res.* **1:** 24.

Jouvet M., Mouret J., Chouvet G., and Siffre M. 1974. Toward a 48-hour day: Experimental bicardian rhythm in man. In *The neurosciences: Third study program* (ed. F.O. Schmitt and F. Worden), p. 491. MIT Press, Cambridge, Massachusetts.

Khalsa S.B.S., Jewett M.E., Cajochen C., and Czeisler C.A. 2003. A phase response curve to single bright light pulses in human subjects. *J. Physiol.* **549:** 945.

Klauer S.G., Dingus T.A., Neale V.L., Sudweeks J.D., and Ramsey D.J. 2006. *The impact of driver inattention on near-crash/crash risk: An analysis using the 100-car naturalistic driving study data* (HS810594, 1–192). National Highway Traffic Safety Administration, Washington, D.C.

Klein D.C., Moore R.Y., and Reppert S.M., eds. 1991. *Suprachiasmatic nucleus: The mind's clock.* Oxford University Press, New York.

Klein T., Martens H., Dijk D.J., Kronauer R.E., Seely E.W., and Czeisler C.A. 1993. Chronic non-24-hour circadian rhythm sleep disorder in a blind man with a regular 24-hour sleep-wake schedule. *Sleep* **16:** 333.

Kleitman N. 1963. *Sleep and wakefulness*. University of Chicago Press, Chicago, Illinois.

Klerman E.B., Dijk D.J., Kronauer R.E., and Czeisler C.A. 1996. Simulations of light effects on the human circadian pacemaker: Implications for assessment of intrinsic period. *Am. J. Physiol.* **270:** R271.

Klerman E.B., Shanahan T.L., Brotman D.J., Rimmer D.W., Emens J.S., Rizzo J.F., III, and Czeisler C.A. 2002. Photic resetting of the human circadian pacemaker in the absence of conscious vision. *J. Biol. Rhythms* **17:** 548.

Kondo T., Strayer C.A., Kulkarni R.D., Taylor W., Ishiura M., Golden S.S., and Johnson C.H. 1993. Circadian rhythms in prokaryotes: Luciferase as a reporter of circadian gene expression in cyanobacteria. *Proc. Natl. Acad. Sci.* **90:** 5672.

Koulack D. and Schultz K.J. 1974. Task performance after awakenings from different stages of sleep. *Percept. Mot. Skills* **39:** 792.

Kronauer R.E. 1990. A quantitative model for the effects of light on the amplitude and phase of the deep circadian pacemaker, based on human data. In *Sleep '90: Proceedings of the 10th European Congress on Sleep Research,* Dusseldorf (ed. J. Horne), p. 306. Pontenagel Press, Bochum, Germany.

Kronauer R.E. and Czeisler C.A. 1993. Understanding the use of light to control the circadian pacemaker in humans. In *Light and biological rhythms in man* (ed. L. Wetterberg), p. 217. Pergamon Press, Oxford, United Kingdom.

Kronauer R.E., Forger D.B., and Jewett M.E. 1999. Quantifying human circadian pacemaker response to brief, extended, and repeated light stimuli over the phototopic range. *J. Biol. Rhythms* **14:** 500.

Kronauer R.E., Czeisler C.A., Pilato S.F., Moore-Ede M.C., and Weitzman E.D. 1982. Mathematical model of the human circadian system with two interacting oscillators. *Am. J. Physiol.* **242:** R3.

Lamond N. and Dawson D. 1999. Quantifying the performance impairment associated with fatigue. *J. Sleep Res.* **8:** 255.

Landrigan C.P., Czeisler C.A., Barger L.K., Ayas N.T., Rothschild J.M., and Lockley S.W. 2007. Effective implementation of work-hour limits and systemic improvements. *Jt. Comm. J. Qual. Patient Saf.* **33:** 19.

Landrigan C.P., Rothschild J.M., Cronin J.W., Kaushal R., Burdick E., Katz J.T., Lilly C.M., Stone P.H., Lockley S.W., Bates D.W., and Czeisler C.A. 2004. Effect of reducing interns' work hours on serious medical errors in intensive care units. *N. Engl. J. Med.* **351:** 1838.

Langdon D.E. and Hartman B. 1961. Performance upon sudden awakening. *School of Aerospace Medicine* 62-17, 1–8. Brooks Air Force Base, Brooks, Texas.

Lewis P.R. and Lobban M.C. 1957a. Dissociation of diurnal rhythms in human subjects living on abnormal time routines. *Q. J. Exp. Physiol.* **42:** 371.

———. 1957b. The effects of prolonged periods of life on abnormal time routines upon excretory rhythms in human subjects. *Q. J. Exp. Physiol.* **42:** 356.

Liu C., Weaver D.R., Jin X., Shearman L.P., Pieschi R.L., Gribkoff V.K., and Reppert S.M. 1997. Molecular dissection of two distinct actions of melatonin on the suprachiasmatic circadian clock. *Neuron* **19:** 91.

Lobban M.C. 1961. The entrainment of circadian rhythms in man. *Cold Spring Harbor Symp. Quant. Biol.* **25:** 325.

Lockley S.W. 2008. Human circadian rhythms: Influence of light on circadian rhythmicity in humans. In *New encyclopedia of neuroscience* (ed. L. Squire). Elsevier, Oxford, United Kingdom. (In press.)

Lockley S.W., Brainard G.C., and Czeisler C.A. 2003. High sensitivity of the human circadian melatonin rhythm to resetting by short wavelength light. *J. Clin. Endocrinol. Metab.* **88:** 4502.

Lockley S.W., Gooley J.J., Kronauer R.E., and Czeisler C.A. 2006a. Phase response curve to single one-hour pulses of 10,000 lux bright white light in humans. In *Abstracts from the Proceedings of the 10th Annual Meeting of the Society for Research on Biological Rhythms,* Sandestin, Florida, p. 132. Society for Research on Biological Rhythms, University of Illinois, Urbana-Champaign.

Lockley S.W., Landrigan C.P., Barger L.K., and Czeisler C.A. 2006b. When policy meets physiology: The challenge of reducing resident work hours. *Clin. Orthop. Relat. Res.* **449:** 116.

Lockley S.W., Barger L.K., Ayas N.T., Rothschild J.M., Czeisler C.A., and Landrigan C.P. 2007. Effects of health care provider work hours and sleep deprivation on safety and performance. *Jt. Comm. J. Qual. Patient Saf.* **33:** 7.

Lockley S.W., Evans E.E., Scheer F.A., Brainard G.C., Czeisler C.A., and Aeschbach D. 2006c. Short-wavelength sensitivity for the direct effects of light on alertness, vigilance, and the waking electroencephalogram in humans. *Sleep* **29:** 161.

Lockley S.W., Skene D.J., Arendt J., Tabandeh H., Bird A.C., and Defrance R. 1997. Relationship between melatonin rhythms and visual loss in the blind. *J. Clin. Endocrinol. Metab.* **82:** 3763.

Lockley S.W., Cronin J.W., Evans E.E., Cade B.E., Lee C.J., Landrigan C.P., Rothschild J.M., Katz J.T., Lilly C.M., Stone P.H., Aeschbach D., and Czeisler C.A. 2004. Effect of reducing interns' weekly work hours on sleep and attentional failures. *N. Engl. J. Med.* **351:** 1829.

Mander B.A., Colecchia E.F., Spiegel K., and Van Cauter E. 2001. Short sleep: A risk factor for insulin resistance and obesity. *Sleep* (suppl.) **24:** A74.

Mills J.N. 1964. Circadian rhythms during and after three months in solitude underground. *J. Physiol.* **174:** 217.

Moore R.Y. 1972. Visual pathways and the central neural control of diurnal rhythms. In *The neurosciences: Third study program* (ed. F.O. Schmitt and F.G. Worden), p. 537. MIT Press, Cambridge, Massachusetts.

———. 1995. Organization of the mammalian circadian system. In *Circadian clocks and their adjustment* (ed. J.M. Waterhouse), p. 88. John Wiley and Sons, Inc., Chichester (Ciba Foundation Symposium 183).

Moore R.Y. and Eichler V.B. 1972. Loss of a circadian adrenal corticosterone rhythm following suprachiasmatic lesions in

the rat. *Brain Res.* **42:** 201.
Moore R.Y. and Lenn N.J. 1972. A retinohypothalamic projection in the rat. *J. Comp. Neurol.* **146:** 1.
Mumford G.K., Benowitz N.L., Evans S.M., Kaminski B.J., Preston K.L., Sannerud C.A., Silverman K., and Griffiths R.R. 1996. Absorption rate of methylxanthines following capsules, cola and chocolate. *Eur. J. Clin. Pharmacol.* **51:** 319.
Naitoh P., Kelly T., and Babkoff H. 1993. Sleep inertia: Best time not to wake up. *Chronobiol. Int.* **10:** 109.
Panda S., Provencio I., Tu D.C., Pires S.S., Rollag M.D., Castrucci A.M., Pletcher M.T., Sato T.K., Wiltshire T., Andahazy M., Kay S.A., Van Gelder R.N., and Hogenesch J.B. 2003. Melanopsin is required for non-image-forming photic responses in blind mice. *Science* **301:** 525.
Peterson E.L. 1980. Phase resetting a mosquito circadian oscillator. I. Phase resetting surface. *J. Comp. Physiol.* **138:** 201.
Pittendrigh C.S. and Daan S. 1976. A functional analysis of circadian pacemakers in nocturnal rodents. I. The stability and lability of spontaneous frequency. *J. Comp. Physiol. A* **106:** 223.
Powell N.B., Schechtman K.B., Riley R.W., Li K., Troell R., and Guilleminault C. 2001. The road to danger: The comparative risks of driving while sleepy. *Laryngoscope* **111:** 887.
Provencio I., Wong S., Lederman A.B., Argamaso S.M., and Foster R.G. 1994. Visual and circadian responses to light in aged retinally degenerate (*rd*) mice. *Vision Res.* **34:** 1799.
Ralph M.R., Foster R.G., Davis F.C., and Menaker M. 1990. Transplanted suprachiasmatic nucleus determines circadian period. *Science* **247:** 975.
Rechtschaffen A., Bergmann B.M., Everson C.A., Kushida C.A., and Gilliland M.A. 1989. Sleep deprivation in the rat: X. Integration and discussion of the findings. *Sleep* **12:** 68.
Richardson G.S., Wyatt J.K., Sullivan J.P., Orav E., Ward A., Wolf M.A., and Czeisler C.A. 1996. Objective assessment of sleep and alertness in medical house-staff and the impact of protected time for sleep. *Sleep* **19:** 718.
Rimmer D.W., Boivin D.B., Shanahan T.L., Kronauer R.E., Duffy J.F., and Czeisler C.A. 2000. Dynamic resetting of the human circadian pacemaker by intermittent bright light. *Am. J. Physiol. Regul. Integr. Comp. Physiol.* **279:** R1574.
Royal D. 2003. *National survey of distracted and drowsy driving attitudes and behavior: 2002. I. Findings* (DOT HS 809 566, 1–61). National Highway Traffic Safety Administration, Washington, D.C.
Sack R.L., Lewy A.J., Blood M.L., Keith L.D., and Nakagawa H. 1992. Circadian rhythm abnormalities in totally blind people: Incidence and clinical significance. *J. Clin. Endocrinol. Metab.* **75:** 127.
Santhi N., Horowitz T.S., Duffy J.F., and Czeisler C.A. 2007. Acute sleep deprivation and circadian misalignment associated with transition onto the first night of work impairs visual selective attention. *PLoS ONE* **2:** e1233.
Saper C.B., Scammell T.E., and Lu J. 2005. Hypothalamic regulation of sleep and circadian rhythms. *Nature* **437:** 1257.
Scheer F.A., Wright K.P., Jr., Kronauer R.E., and Czeisler C.A. 2007. Plasticity of the intrinsic period of the human circadian timing system. *PLoS ONE* **2:** e721.
Schwartz W.J., Gross R.A., and Morton M.T. 1987. The suprachiasmatic nuclei contain a tetrodotoxin-resistant circadian pacemaker. *Proc. Natl. Acad. Sci.* **84:** 1694.
Shanahan T.L. and Czeisler C.A. 2000. Physiological effects of light on the human circadian pacemaker. *Semin. Perinatol.* **24:** 299.
Shanahan T.L., Zeitzer J.M., and Czeisler C.A. 1997. Resetting the melatonin rhythm with light in humans. *J. Biol. Rhythms* **12:** 556.
Siffre M. 1964. *Beyond time*. McGraw Hill, New York.
———. 1975. Six months alone in a cave. *Natl. Geogr. Mag.* **147:** 426.
Smith K.A., Schoen M.W., and Czeisler C.A. 2004. Adaptation of human pineal melatonin suppression by recent photic history. *J. Clin. Endocrinol. Metab.* **89:** 3610.
Spiegel K., Sheridan J.F., and Van Cauter E. 2002. Effect of sleep deprivation on response to immunization. *J. Am. Med. Assoc.* **288:** 1471.
Spiegel K., Leproult R., Copinschi G., and Van Cauter E. 2001. Impact of sleep length on the 24-h leptin profile. *Sleep* (suppl.) **24:** A74.
Spiegel K., Tasali E., Penev P., and Van Cauter E. 2004a. Brief communication: Sleep curtailment in healthy young men is associated with decreased leptin levels, elevated ghrelin levels, and increased hunger and appetite. *Ann. Intern. Med.* **141:** 846.
Spiegel K., Knutson K., Leproult R., Tasali E., and Van Cauter E. 2005. Sleep loss: A novel risk factor for insulin resistance and Type 2 diabetes. *J. Appl. Physiol.* **99:** 2008.
Spiegel K., Leproult R., L'Hermite-Balériaux M., Copinschi G., Penev P.D., and Van Cauter E. 2004b. Leptin levels are dependent on sleep duration: Relationships with sympathovagal balance, carbohydrate regulation, cortisol, and thyrotropin. *J. Clin. Endocrinol. Metab.* **89:** 5762.
Spiegel K., Leproult R., Colecchia E.F., L'Hermite-Balériaux M., Nie Z., Copinschi G., and Van Cauter E. 2000. Adaptation of the 24-h growth hormone profile to a state of sleep debt. *Am. J. Physiol.* **279:** R874.
Stephan K. and Zucker I. 1972. Circadian rhythms in drinking behavior and locomotor activity of rats are eliminated by hypothalamic lesions. *Proc. Natl. Acad. Sci.* **69:** 1583.
Stickgold R. 2005. Sleep-dependent memory consolidation. *Nature* **437:** 1272.
Stickgold R., James L., and Hobson J.A. 2000. Visual discrimination learning requires sleep after training. *Nat. Neurosci.* **3:** 1237.
Sweeney B.M., Haxo F.T., and Hastings J.W. 1959. Action spectrum for two effects of light on luminescence in *Gonyaulax polyedra*. *J. Gen. Physiol.* **43:** 285.
Teran-Santos J., Jimenez-Gomez A., and Cordero-Guevara J. 1999. The association between sleep apnea and the risk of traffic accidents (Cooperative Group Burgos-Santander). *N. Engl. J. Med.* **340:** 847.
Thapan K., Arendt J., and Skene D.J. 2001. An action spectrum for melatonin suppression: Evidence for a novel non-rod, non-cone photoreceptor system in humans. *J. Physiol.* **535:** 261.
Thomas M., Sing H., Belenky G., Holcomb H., Mayberg H., Dannals R., Wagner H., Thorne D., Popp K., Rowland L., Welsh A., Balwinski S., and Redmond D. 2000. Neural basis of alertness and cognitive performance impairments during sleepiness. I. Effects of 24 h of sleep deprivation on waking human regional brain activity. *J. Sleep Res.* **9:** 335.
Torsvall L. and Åkerstedt T. 1988. Disturbed sleep while being on call: An EEG study of ships' engineers. *Sleep* **11:** 35.
Van Cauter E., Holmback U., Knutson K., Leproult R., Miller A., Nedeltcheva A., Pannain S., Penev P., Tasali E., and Spiegel K. 2007. Impact of sleep and sleep loss on neuroendocrine and metabolic function. *Horm. Res.* (suppl. 1) **67:** 2.
Van Dongen H.P.A. and Dinges D.F. 2003. Sleep debt and cumulative excess wakefulness. *Sleep* **26**: 249.
Van Dongen H.P.A., Maislin G., Mullington J.M., and Dinges D.F. 2003. The cumulative cost of additional wakefulness: Dose-response effects on neurobehavioral functions and sleep physiology from chronic sleep restriction and total sleep deprivation. *Sleep* **26:** 117.
Van Dongen H.P.A., Price N.J., Mullington J.M., Szuba M.P., Kapoor S.C., and Dinges D.F. 2001. Caffeine eliminates psychomotor vigilance deficits from sleep inertia. *Sleep* **24:** 813.
Waldstreicher J., Duffy J.F., Brown E.N., Rogacz S., Allan J.S., and Czeisler C.A. 1996. Gender differences in the temporal organization of prolactin (PRL) secretion: Evidence for a sleep-independent circadian rhythm of circulating PRL levels—A clinical research center study. *J. Clin. Endocrinol. Metab.* **81:** 1483.
Walker M.P. and Stickgold R. 2004. Sleep-dependent learning and memory consolidation. *Neuron* **44:** 121.
Walker M.P., Brakefield T., Morgan A., Hobson J.A., and Stickgold R. 2002. Practice with sleep makes perfect: Sleep-dependent motor skill learning. *Neuron* **35:** 205.
Weaver D.R. 1998. The suprachiasmatic nucleus: A 25-year retrospective. *J. Biol. Rhythms.* **13:** 100.
Webb W.B. and Agnew H.W., Jr. 1974a. Regularity in the control of the free-running sleep-wakefulness rhythm. *Aerosp. Med.* **45:** 701.

———. 1974b. Sleep and waking in a time-free environment. *Aerosp. Med.* **45:** 617.

Weitzman E.D., Czeisler C.A., and Moore-Ede M.C. 1981. Sleep-wake, endocrine and temperature rhythms in man during temporal isolation. In *The twenty-four hour workday: Proceedings of a NIOSH Symposium of Variations in Work-Sleep Schedules*, Cincinnati (ed. L.C. Johnson et al.), p. 105. U.S. Department of Health and Human Services, Washington, D.C.

Welsh D.K., Logothetis D.E., Meister M., and Reppert S.M. 1995. Individual neurons dissociated from rat suprachiasmatic nucleus express independently phased circadian firing rhythms. *Neuron* **14:** 697.

Wertz A.T., Ronda J.M., Czeisler C.A., and Wright K.P., Jr. 2006. Effects of sleep inertia on cognition. *J. Am. Med. Assoc.* **295:** 163.

Wever R. 1970. Zur zeitgeber-stärke eines licht-dunkel-wechsels für die circadiane periodik des menschen. *Eur. J. Physiol.* **321:** 133.

———. 1974. The influence of light on human circadian rhythms. *Nord. Council Arct. Med. Res. Rep.* **10:** 33.

———. 1979. *The circadian system of man: Results of experiments under temporal isolation*. Springer-Verlag, New York.

Williamson A.M. and Feyer A.M. 2000. Moderate sleep deprivation produces impairments in cognitive and motor performance equivalent to legally prescribed levels of alcohol intoxication. *Occup. Environ. Med.* **57:** 649.

Winfree A.T. 1969. Corkscrews and singularities in fruitflies: Resetting behavior of circadian eclosion rhythm. In *Biochronometry* (ed. M. Menaker), p. 81. National Academy of Sciences, Washington, D.C.

———. 1974. Suppressing *Drosophila* circadian rhythm with dim light. *Science* **183:** 970.

———. 1980. *The geometry of biological time*. Springer-Verlag, New York.

———. 1987. *The timing of biological clocks*. Scientific American Books, New York.

Wright K.P., Jr., Hull J.T., and Czeisler C.A. 2002. Relationship between alertness, performance, and body temperature in humans. *Am. J. Physiol. Regul. Integr. Comp. Physiol.* **283:** R1370.

Wright K.P., Jr., Hughes R.J., Kronauer R.E., Dijk D.J., and Czeisler C.A. 2001. Intrinsic near-24-h pacemaker period determines limits of circadian entrainment to a weak synchronizer in humans. *Proc. Natl. Acad. Sci.* **98:** 14027.

Wyatt J.K., Ritz-De Cecco A., Czeisler C.A., and Dijk D.J. 1999. Circadian temperature and melatonin rhythms, sleep, and neurobehavioral function in humans living on a 20-h day. *Am. J. Physiol.* **277:** R1152.

Wyatt J.K., Dijk D.J., Ritz-De Cecco A., Ronda J.M., and Czeisler C.A. 2006. Sleep facilitating effect of exogenous melatonin in healthy young men and women is circadian-phase dependent. *Sleep* **29:** 609.

Yoshimura T. and Ebihara S. 1996. Spectral sensitivity of photoreceptors mediating phase-shifts of circadian rhythms in retinally degenerate CBA/J (*rd/rd*) and normal CBA/N (+/+) mice. *J. Comp. Physiol. A* **178:** 797.

Zaidi F.H., Hull J.T., Peirson S.N., Wulff K., Aeschbach D., Gooley J.J., Brainard G.C., Gregory-Evans K., Rizzo J.F., III, Czeisler C.A., Foster R.G., Moseley M.J., and Lockley S.W. 2007. Short-wavelength light sensitivity of circadian, pupillary, and visual awareness in humans lacking an outer retina. *Curr. Biol.* **17:** 2122.

Zeitzer J.M., Kronauer R.E., and Czeisler C.A. 1997. Photopic transduction implicated in human circadian entrainment. *Neurosci. Lett.* **232:** 135.

Zeitzer J.M., Dijk D.J., Kronauer R.E., Brown E.N., and Czeisler C.A. 2000. Sensitivity of the human circadian pacemaker to nocturnal light: Melatonin phase resetting and suppression. *J. Physiol.* **526:** 695.

Thermosensitive Splicing of a Clock Gene and Seasonal Adaptation

W.-F. Chen, K.H. Low, C. Lim, and I. Edery
Department of Molecular Biology and Biochemistry, Rutgers University, Center for Advanced Biotechnology and Medicine, Piscataway, New Jersey 08854

Similar to many diurnal animals, the daily distribution of activity in *Drosophila* exhibits a bimodal pattern with clock-controlled morning and evening peaks separated by a midday "siesta." In prior work, we showed that the thermosensitive splicing of a 3′-terminal intron in the RNA product from the *Drosophila period* (*per*) gene (*dper*) is critical for temperature-induced adjustments in the timing of evening activity. Cold temperatures enhance the splicing efficiency of this intron (termed dmpi8, *Drosophila melanogaster per i*ntron *8*), an event that stimulates the daily accumulation of *dper* RNA and protein, leading to earlier evening activity. Conversely, warm temperatures attenuate dmpi8 splicing efficiency contributing to delayed evening activity, likely ensuring that flies avoid activity during the hot midday sun when they are at increased risk of desiccation. Here, we discuss the underlying molecular mechanisms governing the thermosensitive splicing of dmpi8 and how it contributes to seasonal changes in the daily activity patterns of *Drosophila*. On a broader perspective, RNA–RNA interactions likely have fundamental roles in the thermal adaptation of life forms to the daily and seasonal changes in temperature.

INTRODUCTION

Organisms on this planet have adapted to the daily rotation of the earth on its axis. By means of endogenous circadian (≅24 hours) "clocks" or pacemakers that can be synchronized to the daily and seasonal changes in external time cues, most notably, visible light and ambient temperature, life forms anticipate environmental transitions, perform activities at biologically advantageous times during the day, and undergo characteristic seasonal responses (for review, see Hastings et al. 1991; Edery 2000). Light is almost certainly the predominant entraining agent in nature. Under natural conditions the light/dark (LD) cycle aligns the phases of clocks and evokes daily adjustments in the approximately 24-hour endogenous periods of these oscillators such that they precisely match the 24-hour solar day. The duration of day length (photoperiod) can modify the temporal alignment between a circadian rhythm and local time. A physiologically relevant advantage of this inherent flexibility of clocks is that the daily distributions of physiological and behavioral rhythms are not rigidly locked to local time but can be adjusted for seasonal changes in day length. Despite the obvious importance of photoperiod, ambient temperature is also a key environmental modality regulating the timing of circadian rhythms (Sweeney and Hastings 1961; Rensing and Ruoff 2002). This makes intuitive sense because in temperate latitudes, seasonal changes in day length are also accompanied by predictable changes in average daily temperatures. The work presented here is mainly focused on using *Drosophila* as a model organism to address how a circadian system contributes to temperature-dependent changes in the distribution of daily activity.

Before discussing this work, however, it is important to recognize that temperature has multiple effects on circadian-regulated phenomena (Rensing and Ruoff 2002). Perhaps the best known but least understood response of circadian rhythms to temperature is that period length is relatively constant over a wide range of physiologically relevant temperatures, an active mechanism termed "temperature compensation." The ability of circadian timing systems to maintain stable periods despite changes in temperature is important because whether it is summer or winter, the solar day does not change in duration. Although period length is rather insensitive to temperature fluctuations, the phases and amplitudes of circadian rhythms can exhibit clear temperature-induced changes. Most notably, circadian rhythms can be phase-shifted by changes in temperature (pulses and step-up or -down) and entrained by daily temperature cycles (see, e.g., Sweeney and Hastings 1961; Wheeler et al. 1993; Liu et al. 1997, 1998; Sidote et al. 1998; and references therein). Other work indicates that the amplitude of a circadian oscillator changes as a function of temperature (usually perceived as increasing with rising temperature), suggested by some as a solution to temperature compensation of period length (Lakin-Thomas et al. 1991; Pittendrigh et al. 1991). In addition, clocks only function within a restricted temperature range, and outside these limits, the oscillatory mechanism appears to be held constant or "stops" (Liu et al. 1997; Revel et al. 2007; and references therein), which might be an extreme example of amplitude reduction.

A less-studied effect of temperature on clock function is based on the observation that the steady-state phases of behavioral rhythms can vary as a function of temperature, even during entrainment by daily light/dark cycles (Sweeney and Hastings 1961). Many animals display a bimodal distribution of activity, with peaks centered around the morning and evening hours that are separated by a midday "siesta" time of varying lengths. A classic example of how temperature affects daily activity patterns is that of garter snakes which are mainly nocturnal at warm temperatures and diurnal at cold temperatures and display a bimodal distribution of activity at intermediate temperatures (Heckrotte 1961). In general, diurnal animals respond to colder temperatures by displaying a greater proportion of their activity during daytime hours, whereas nighttime activity predominates at warmer temperatures. This directional response has a clear adaptive value, ensuring that the activity of an organism is maximal at a time of day when the temperature would be expected to be optimal for activity (Sweeney and Hastings 1961).

If there is such clear adaptive value, why is this effect of temperature on circadian rhythms less studied? There are several reasons, which also highlight some of the challenges in trying to understand the underlying mechanism and the choices of model systems. First, from a historical perspective, the entrée was based on observing how temperature affects the daily distribution of activity (i.e., movement such as locomotion), which naturally focused the work on animals. Second, it was thought that much of the thermal effect was likely due to the "direct" (or masking) effects of temperature on activity (see, e.g., Matsumoto et al. 1998), minimizing the notion of potential clock involvement. Finally, when it comes to trying to understand "seasonal" adaptation of clocks, much of the focus has been placed on the impact of changing day length or photoperiod. This is quite natural as light is the predominant daily synchronizer in nature, and it is almost certain that circadian rhythms in all organisms, even homeotherms, exhibit phase adjustments in response to changes in day length. Indeed, the dual oscillator model based on coupled morning (M) and evening (E) pacemakers was largely developed over the years as a framework to explain how the M and E peaks track sunrise and sunset, respectively, despite seasonal changes in day length (Pittendrigh and Daan 1976). This photocentric view of seasonal adaptation has for the most part not incorporated the effects of temperature on modulating clock dynamics.

Below, we briefly describe some of our work using *Drosophila* to explore the role of a circadian clock in the adaptation of daily activity patterns to seasonal changes in temperatures. Thermal-sensitive splicing of a clock mRNA has a key role in this adaptation (Majercak et al. 1999). Remarkably, a similar mechanism has also been shown to underlie some temperature responses in the *Neurospora* clock (Colot et al. 2005; Diernfellner et al. 2005; Brunner and Diernfellner 2006). On a broader perspective, the results suggest that naturally selected variations in the strengths of RNA–RNA interactions offer a simple basis for designing an adaptive molecular mechanism displaying highly calibrated thermal responses that can subsequently impact a wide array of physiological and behavioral programs.

DROSOPHILA AS A MODEL SYSTEM TO STUDY THE ROLE OF CIRCADIAN CLOCKS IN REGULATING THE DAILY DISTRIBUTION OF ACTIVITY AS A FUNCTION OF SEASONAL CHANGES IN TEMPERATURE

Several years ago, we used *D. melanogaster* as a model system to understand the role of a clock in regulating the pattern of daily activity as a function of changes in temperature and day length (Majercak et al. 1999). Under standard daily cycles of 12-hour light followed by 12-hour dark (12:12 L:D) at 25°C, *D. melanogaster* exhibit a bimodal activity pattern with clock-controlled peaks centered around the lights-on ("morning" peak) and lights-off ("evening" peak) transitions that are separated by a midday dip or "siesta" (Fig. 1) (see, e.g., Hamblen-Coyle et al. 1992). For the purposes of this discussion, it is important to note at the outset that despite the clock-controlled bimodal activity pattern, the evening activity component is largely assayed as a bona fide readout of the circadian system in *Drosophila*. This is mainly because the morning peak usually coincides with a direct stimulatory effect of light following dawn, and the evening peak most visibly persists in constant darkness (Wheeler et al. 1993). The photoperiod and ambient temperature have effects on the distribution of activity throughout a daily cycle. Over a wide range of photoperiods, the evening peak occurs around the light-to-dark transition (Qiu and Hardin 1996; Majercak et al. 1999; Rieger et al. 2003; Shafer et al. 2004). As a result, lengthening the photoperiod delays the evening activity relative to the most recent sunrise. Temperature appears to be the main environmental cue determining whether the evening activity will be mostly diurnal or nocturnal (Majercak et al. 1999). With increasing temperature, the evening activity becomes progressively more nocturnal and midday inactivity more pronounced (see Figs. 1 and 3) (Matsumoto et al. 1998; Majercak et al. 1999; Yoshii et al. 2002). This directional response is observed in many diurnal animals, displaying a greater proportion of their activity during the cooler nighttime hours on hot days and conversely the warmer daytime hours during cold days (Sweeney and Hastings 1961).

A Role for Thermal-sensitive Splicing of a Clock Gene in Seasonal Adaptation

We showed that splicing of the 3′-terminal intron in *dper* RNA (termed dmpi8) (Cheng et al. 1998) is an important aspect of how the *Drosophila* circadian clock adapts to seasonal changes in temperature (Majercak et al. 1999). At cold temperatures, the proportion of dmpi8 spliced *dper* mRNA (termed type B′) compared to the unspliced variant (termed type A) is enhanced, leading to more rapid daily increases in *dper* transcript levels and earlier evening activity. Splicing per se, as opposed to retention or removal of the dmpi8 intron, stimulates increases in *dper* mRNA levels. This led to the hypothesis that assembly of active spliceosomes at the 3′-terminal intron somehow produces more mature transcripts, possibly by facilitating 3′-end formation (Fig. 1). Transgenic flies bearing variant *dper* transgenes where splicing of dmpi8 was abrogated manifested nocturnal evening activity even on cold days (Majercak et al. 1999). Long photoperiods partially counteract cold-induced phase advances in the accumulation of *dper* mRNA and protein by delaying the upswing in TIMELESS (TIM) levels. The functional interrelationships between dPER and TIM ensure that information concerning ambient temperature and light is properly integrated, resulting in activity rhythms that are optimally aligned with the prevailing environmental conditions.

In more recent findings, we and other investigators

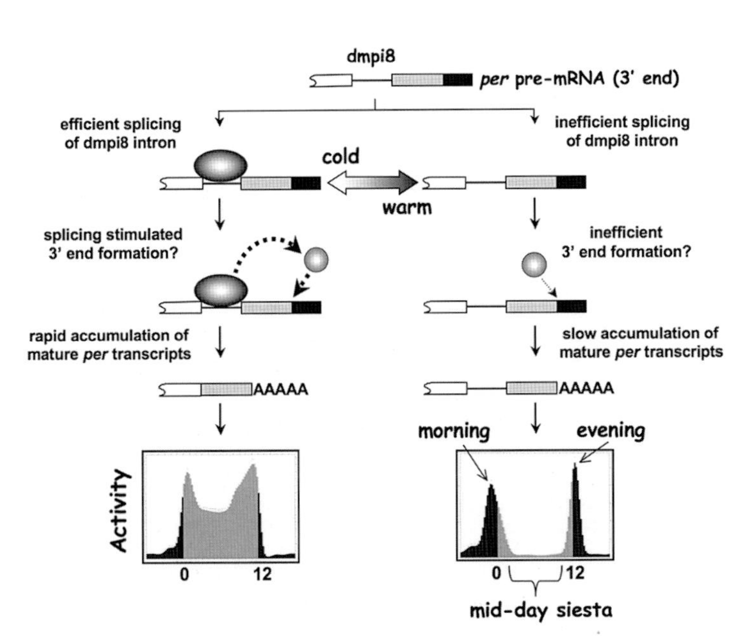

Figure 1. Model for how thermosensitive splicing of the *dper* 3′-terminal intron (dmpi8) regulates seasonal changes in the daily distribution of activity in *Drosophila melanogaster*. Shown at the top is the 3′ end of the *dper* precursor mRNA (pre-mRNA); (*white box*) Sequences from translation stop codon to just upstream of 5′ splice site of dmpi8; (*horizontal line*) dmpi8 intron; (*gray box*) sequences following 3′ splice site of dmpi8 until 3′ cleavage and polyadenylation site; (*black box*) transcribed sequences downstream from 3′ cleavage and polyadenylation site. Cold temperatures enhance binding of spliceosome (*large oval*), which stimulates binding of 3′-end formation factors (*small circle*). More rapid accumulation of *dper* transcripts leads to advanced evening activity and less prominent midday inactivity. Conversely, on warm days, the inefficient splicing of dmpi8 leads to an extended midday siesta and preferential nocturnal evening activity, events that enhance the ability of *Drosophila* to avoid the deleterious effects associated with the hot midday sun. For *Drosophila* activity profiles: (*black bars*) dark; (*gray bars*) light. Locomotor activity was collected in 15 min bins and relative values throughout a daily cycle (example shown, 12 hr light:12 hr dark) shown in arbitrary units. (*Left panel*) 18°C; (*right panel*) 29°C (also shown are morning and evening activity components, including midday siesta). (Adapted from Majercak et al. 1999.)

showed that daily fluctuations in the splicing efficiency of dmpi8 are regulated by the clock in a manner that depends on the photoperiod (day length) and temperature (Collins et al. 2004; Majercak et al. 2004). Shortening the photoperiod enhances dmpi8 splicing and advances its cycle, whereas the amplitude of the clock-regulated daytime decline in splicing increases as temperatures rise. This suggests that at elevated temperatures, the clock has a more pronounced role in maintaining low splicing during the day, an adaptive response that for small insects likely minimizes the risks of desiccation during the hot midday hours. Indeed, on entering direct sunlight, a 10-mg insect can heat up by about 10°C in less than 10 seconds (Sayeed and Benzer 1996), a potentially life-threatening event for *Drosophila* that only weigh in the vicinity of 1 mg. In summary, the splicing efficiency of dmpi8 is regulated by the clock, temperature, and day length, ultimately modulating the trajectory of daytime increases in *dper* mRNA levels and hence clock phase. We propose that the multimodal regulation in the splicing efficiency of dmpi8 acts as a "seasonal sensor" conveying *calendar* information to the animal. We and others also identified a novel nonphotic role for NORPA (*no receptor potential A*) in the regulation of dmpi8 splicing (Collins et al. 2004; Majercak et al. 2004). In this chapter, we do not discuss the role of *norpA* further, but we focus on the basis for temperature-dependent changes in dmpi8 splicing efficiency.

Weak 5′ and 3′ Splice Sites as a Basis for Thermosensitive Splicing

As an initial attempt to understand the basis for the temperature-dependent splicing of dmpi8, we sought to develop a simplified cell-culture-based assay that recapitulates this thermosensitivity phenotype. *D. melanogaster* (Oregon-R strain [OR]) *dper* genomic sequences encompassing the entire 3′-untranslated region (3′UTR) followed by 90 bp of 3′-flanking nontranscribed region were fused downstream from a *luciferase* (*luc*) reporter gene. Expression of the *luc-dper*/3′UTR hybrid gene was placed under the control of the constitutive actin 5C promoter (pAct) (Fig. 2A) (Angelichio et al. 1991). To enable the simple introduction of different intron and flanking exon sequences, this parent wild-type vector (termed dmpi8) also includes two engineered XhoI and KpnI restriction sites placed 10 bp upstream and downstream from the dmpi8 5′ and 3′ splice sites (ss), respectively. Following transfection of the popular *Drosophila* Schneider S2 cells, at least two independent lines of stable transformants were isolated for each construct. Subsequently, separate aliquots of cells were incubated for 24 hours at different temperatures in the dark. Total RNA was extracted, and the relative levels of spliced and nonspliced products were determined as previously described using reverse transcriptase–polymerase chain reaction (RT-PCR) whereby values are normalized to an internal RNA standard (Majercak et al. 2004). The dmpi8 control RNA exhibited approximately 2.5–3.2-fold increases ($n > 10$; $P < 0.001$ using Student's t-test for values obtained at 12°C and 24°C) in the proportion of spliced-to-unspliced RNA at 12°C compared to 24°C (Fig. 2D). This is similar to the 2-3-fold increase in the ratio of spliced type B′ to unspliced type-A *dper* RNA variants observed in wild-type flies at 18°C compared to 29°C (Majercak et al. 1999). Thus, the *dper* 3′UTR is sufficient to manifest thermosensitive splicing of dmpi8 in a

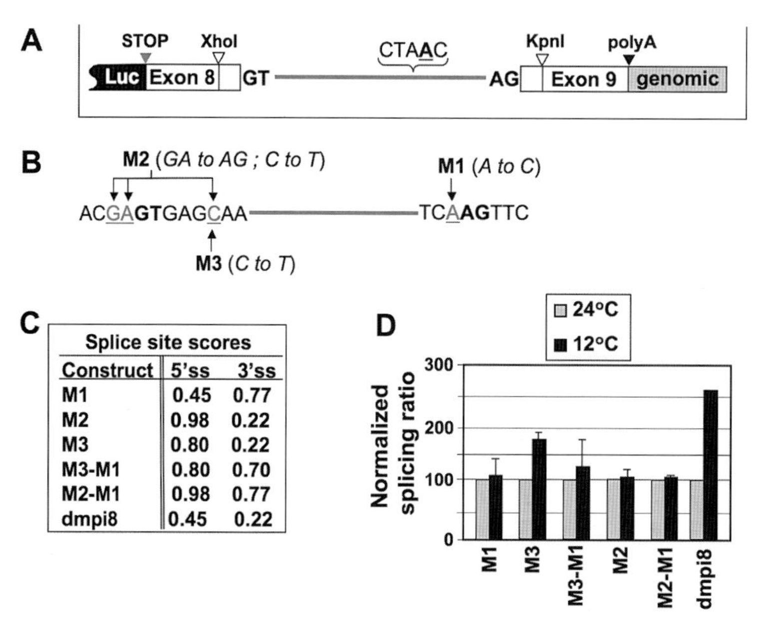

Splice site scores

Construct	5'ss	3'ss
M1	0.45	0.77
M2	0.98	0.22
M3	0.80	0.22
M3-M1	0.80	0.70
M2-M1	0.98	0.77
dmpi8	0.45	0.22

Figure 2. Multiple suboptimal splice sites underlie temperature-dependent splicing of the dmpi8 intron. (*A*) A luciferase (*luc*) reporter gene fused to 3′ *dper* sequences was used to transfect S2 cells. XhoI and KpnI sites were engineered into the fusion construct such that they flanked the 5′ and 3′ splice sites of dmpi8 to facilitate swapping intronic sequences. (*B*) Schematic of mutations made to increase the predicted strengths of the 5′ and/or 3′ splice sites of dmpi8. (*C*) Predicted scores for the 5′ and 3′ splice sites of the different constructs (range, 0–1, where 1 is the highest). (*D*) Cells were grown at either 12°C or 24°C, and the relative splicing efficiency was determined (values for 24°C were set to 1 and the rest of the values normalized). Note that whereas dmpi8 exhibits approximately 2.5-fold more efficient splicing at 12°C compared to 24°C, the other transcripts exhibit little (e.g., M3) to no (e.g., M2) temperature-sensitive splicing.

simplified cell culture system, providing a powerful approach to investigate mechanistic issues.

A feature of the *dper* 3′-terminal intron in *D. melanogaster* that caught our attention was that both the 5′ and 3′ splice sites were predicted to be weak. For example, using a program that predicts splice sites in *D. melanogaster* (www.fruitfly.org/seq_tools/splice.html), dmpi8 has the lowest scores for both the 5′ (score = 0.45) and 3′ (score = 0.21) splice sites of all the introns in *per* (the other *per* small introns had scores of 0.88–0.97 for the 5′ and 0.42–0.95 for the 3′ splice sites, with 1 being highly likely) (Fig. 2C). The presumptive branchpoint (BP) lies in a sequence that is very similar but not identical to the consensus sequence for *D. melanogaster* introns (CUAAU) (Fig. 2A). Multiple suboptimal splices sites for the dmpi8 intron was eerily reminiscent of prior work from Murphy and co-workers studying the basis for the thermosensitive splicing phenotype of the ts*110* mutant of Moloney murine sarcoma virus (Mo-MSVts110). A series of elegant studies showed that multiple suboptimal splicing signals were the basis for the temperature-sensitive splicing of Mo-MSVts110 (Cizdziel et al. 1988; Touchman et al. 1995; Ainsworth et al. 1996). For example, introducing a consensus branchpoint signal increased splicing efficiency and abrogated the thermodependency (Touchman et al. 1995). A similar result was observed by strengthening the 5′splice site (Ainsworth et al. 1996). Intriguingly, other random mutagenesis strategies to generate temperature-sensitive phenotypes were eventually shown to be a result of introducing mutations that weaken key splicing signals on precursor mRNAs (pre-mRNAs).

Indeed, increasing the predicted strength of either the 5′ splice site (e.g., M2) or the 3′ splice site (e.g., M1) in dmpi8 eliminated thermosensitivity, whereas in other cases, the thermal response was diminished (e.g., M3) (Fig. 2B–D). Thus, there is a graded response that correlates with the predicted strengths of the 5′ and 3′ signals. As with Mo-MSVts110, our findings suggest that temperature regulation of dmpi8 splicing is not driven by discrete thermosensitive elements but involves an intricate balance of multiple suboptimal splice sites, raising the possibility of intricate coevolution. Studies using transgenic flies whereby we replaced the wild-type *D. melanogaster* 3′UTR (from the standard Oregon-R strain) with the M2 version also exhibited near complete splicing at all test temperatures (data not shown). Moreover, in both S2 cells and flies, versions of the *dper* 3′-terminal intron with higher splicing efficiency also exhibited higher RNA levels (data not shown). Together, these results are highly consistent with our earlier model that the splicing efficiency of the *dper* 3′-terminal intron stimulates the production of mature transcripts.

A Similar Mechanism in the *Neurospora* Clock

Along with *Drosophila*, work in *Neurospora* laid the early foundation for mechanistic insights underlying clock mechanisms (Dunlap 1996). The initial biochemical characterization of the FREQUENCY (FRQ) protein led to a remarkable realization that has grown over the years and can be equally applied to other clock systems. Both dPER and FRQ exhibit dramatic progressive increases in phosphorylation throughout a daily cycle that are intimately linked to abundance, whereby the most hyperphosphorylated isoforms are associated with sharp decreases in levels (Edery et al. 1994b; Garceau et al. 1997). Today, we know that despite the lack of sequence conservation, there are uncanny similarities between dPER and FRQ proteins with respect to clock function. Importantly, both have been shown to be bona fide "state variables" in their respective clockworks (Aronson et al. 1994; Edery et al. 1994a), indicating that the daily fluctuations in the abundance/activity of these proteins, as opposed to simply their presence, are inextricably linked to the phase of the clock. dPER and FRQ proteins share similar modes of action, whereby they participate in transcriptional autoinhibitory loops crucial to clock function, most likely by acting as conduits for the timely delivery or regulation of *trans*-acting factors that more directly inhibit the key transcription factors which drive their own expression (Schafmeier et al. 2005; He et al. 2006; Yu et al. 2006; Kim et al. 2007). Although we have gotten used to the idea by now, it is still rather intriguing that virtually identical kinases, phosphatases, and F-box proteins regulate the daily life cycles of PER and FRQ proteins (Gallego and Virshup 2007).

Perhaps then, it should not be too unexpected that an aspect of *frq* expression regulated by temperature involves mRNA splicing. Earlier work from Dunlap and co-workers showed that temperature regulates the relative levels of two isoforms of FRQ protein, a short (s-FRQ) and long (l-FRQ) version that arise from alternative use of translation initiation sites (Liu et al. 1997). More recent work from the Brunner and Dunlap labs demonstrated that the ratio of l-FRQ versus s-FRQ is regulated by thermosensitive splicing of an intron that when excised removes the translation initiation site of l-FRQ (Colot et al. 2005; Diernfellner et al. 2005; Brunner and Diernfellner 2006). Moreover, inefficient recognition of nonconsensus splice sites at elevated temperatures was shown to underlie the thermosensitivity (Diernfellner et al. 2005; Brunner and Diernfellner 2006). Additional steps including temperature-dependent effects on the rate of ribosome scanning at the different translation initiation recognition sequences allow for adjustments in FRQ levels that are appropriate for the local ambient temperature (Diernfellner et al. 2005; Brunner and Diernfellner 2006). Thus, in two widely different species, the clockworks adapt to changes in temperature by thermal adjustments in the levels of key state variables via a mechanism involving an initial thermosensitive splicing event that has ramifications for other more downstream aspects of mRNA metabolism or utilization, such as the efficiency of 3′-end formation in the case of *dper* or translation for *frq*.

HOW DO WEAK SPLICING SIGNALS YIELD THERMOSENSITIVE SPLICING?

From the aforementioned examples, such as *dper*, *frq*, and Mo-MSVts110, it is clear that the presence of multiple suboptimal splicing signals can render the splicing efficiency of an intron temperature-dependent. In very general terms, pre-mRNA splicing in eukaryotes occurs via two major steps that differ in the need for ATP consumption (for review, see Hastings and Krainer 2001). The first step is an ATP-hydrolysis-independent assembly of the core splicing machinery or spliceosome to an intronic sequence. This involves the recognition of *cis*-acting signals on the mRNA such as the 5′ and 3′ splice sites and 3′ polypyrimidine tracts. Subsequently, in the second phase, which requires ATP hydrolysis, the pre-mRNA and splicing machinery undergo structural rearrangements, culminating in cleavage at the 5′ intronic site, lariat formation, and finally, removal of the intron with concomitant ligation of the exonic sequences. The initial binding of the spliceosome and the subsequent catalytic events are heavily dependent on dynamic interactions between the substrate transcript and the small nuclear RNAs (snRNAs) that are core constituents of the splicing machinery.

Given the ability of RNA to form intramolecular and intermolecular base-pairing interactions of varying strengths, splicing offers a rich opportunity for generating a molecular mechanism that can elicit calibrated thermal responses. On the basis of the important roles for the 5′ and 3′ splice sites in spliceosome docking, it is likely that for introns with multiple suboptimal splicing signals, low temperatures stabilize RNA–RNA and/or RNA–protein interactions between the splicing machinery and the RNA substrate, leading to more efficient splicing. For example, low temperatures could enhance the binding of the U1 snRNA to nonconsensus 5′ splice sites, modulating a major rate-limiting step in overall splicing efficiency. It is also possible that temperature has effects during the catalytic phase of splicing. For example, the U4 and U6 snRNAs undergo structural rearrangements between them that can be modulated by temperature (Shannon and Guthrie 1991). However, this type of mechanism lacks substrate specificity and would have global effects on splicing. Other possibilities include temperature-dependent changes in (1) RNA secondary structure that modulates the ability of an intron to be recognized and excised and (2) binding of *trans*-acting regulatory splicing factors, such as SR proteins that can enhance or suppress splicing (Bourgeois et al. 2004).

LIGHT-INDUCTION OF *TIM* EXPRESSION AT COLD TEMPERATURES: AN INTEGRATED MECHANISM

Other work from our lab also showed that light stimulates expression of *tim* at low but not warm temperatures (Chen et al. 2006). Intriguingly, although numerous lines of evidence indicate that *dper* and *tim* expression are activated by the same mechanism, light has no measurable acute effect on *dper* mRNA abundance (Chen et al. 2006). The acute effects of light on *tim* expression are temporally gated, essentially restricted to the daily rising phase in *tim* mRNA levels. Because the start of the daily upswing in *tim* expression begins several hours after dawn in long photoperiods (day length), this gating mechanism likely ensures that sunrise does not prematurely stimulate *tim* expression during unseasonally cold spring/summer days. We suggest that photic stimulation of *tim* expression at low temperatures is part of a seasonal adaptive response that helps advance the phase of the clock on cold days, enabling flies to exhibit preferential daytime activity

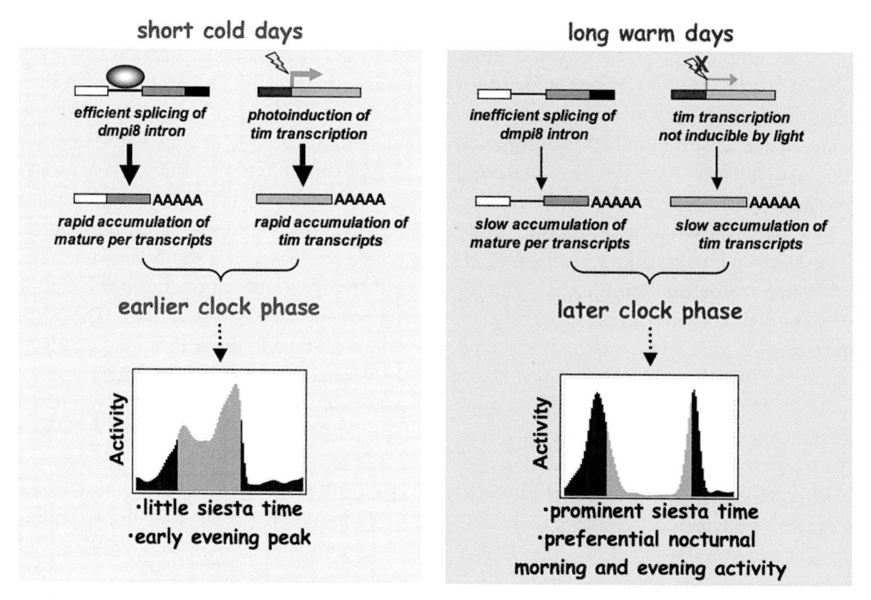

Figure 3. Integrated effects of photoperiod and temperature on *dper* and *tim* expression. Cold short days stimulate the rapid accumulation of both *dper* and *tim* transcripts, leading to advanced evening activity and little to no midday siesta. Long photoperiods and warm temperatures delay the daily upswings in *dper* and *tim* transcripts, contributing to preferential nocturnal evening activity and pronounced midday inactivity. Thus, temperature and day length have parallel effects on the daily accumulation profiles in *dper* and *tim* transcripts.

despite the (usually) earlier onset of dusk. Thus, we envisage an integrated mechanism whereby during autumn/winter seasons, splicing of dmpi8 is enhanced, leading to a more rapid rise in *dper* RNA levels, and light stimulates *tim* induction, responses that function in a coordinate manner to advance the timing of when de-novo-synthesized PER and TIM proteins begin to interact, leading to an earlier clock phase (Fig. 3). Otherwise stated, the ability of temperature and photoperiod to adjust trajectories in the rising phases of one or more clock RNAs constitutes a major mechanism contributing to seasonal adaptation of clock function in *Drosophila*.

This raises the role of rhythmic clock gene expression in light of recent findings in cyanobacteria which clearly show that a biochemical oscillator with circadian properties can be reconstituted in vitro (Nakajima et al. 2005). In this case, mixing purified recombinant versions of the key clock proteins KaiA, KaiB, and KaiC in the presence of ATP as an energy source resulted in daily cycles in the phosphorylated state of KaiC. This dramatic result convincingly shows that circadian molecular cycles can occur in the absence of cycling gene expression. KaiC could be considered the equivalent of dPER and FRQ in that these proteins undergo daily cycles in phosphorylation that are likely to be critical signals in period determination and defining the state of the clock. Although it is possible that reconstituted rhythms in dPER or FRQ phosphorylation might be attained, rhythmic clock gene expression is likely to be an important element in optimizing the dynamics of clock systems.

CONCLUSIONS

The work described herein focused on the use of *D. melanogaster* to study the role of a circadian system in modulating daily activity patterns in response to seasonal changes in temperature. This species has also contributed to other effects of temperature on circadian systems. For example, the elegant work from Kyriacou and colleagues showed that a stretch of Thr-Gly (TG) repeats centrally located in the dPER-coding region modulates temperature compensation of period length (Sawyer et al. 1997). Importantly, these TG repeats are subject to natural selection and exhibit a latitudinal cline in natural populations of *D. melanogaster* (Sawyer et al. 1997). Thus, dPER activity appears to be a focal point for thermal responses in the *Drosophila* clock. Ongoing studies are aimed at determining whether natural selection is also operating at the level of dmpi8 splicing efficiency and whether there is an association between the thermosensitivty of 3′-terminal splicing and the type of TG repeat present on *dper* (i.e., diurnal phase adjustments can also be attained by changes in period length). More recent work has also identified a factor termed *nocte* that specifically modulates circadian entrainment to daily changes in temperature but not light/dark cycles (Glaser and Stanewsky 2005). Intriguingly, *nocte* appears to be unrelated to dmpi8 splicing, indicating that multiple pathways convey thermal information to the circadian system in flies.

Among the most exciting recent findings in the study of circadian rhythms was the identification of long-sought "morning" and "evening" pacemakers in *D. melanogaster*. The M cells preferentially regulate the morning peak of activity, whereas the E cells drive the evening component (Grima et al. 2004; Stoleru et al. 2004). Storelu, Rosbash, and coworkers showed that day length can modulate the relative impact of either M and E cells on the overall circadian system (Stoleru et al. 2007). For example, during long nights, the M cells dominate, whereas during long days, the E cells are predominant. A current challenge is to integrate the photoperiodic effects on M-E pacemakers with temperature. In this regard, a curious feature is that although the evening activity is delayed during warm days, the morning peak is advanced (e.g., see Figs. 1 and 3). Although this is almost certainly linked to the "desire" of minimizing daytime activity on hot days, how does the clock delay evening activity while advancing morning activity? Recent findings suggest some possibilities, such as differential phase adjustments in M and E clocks (Bachleitner et al. 2007). Moreover, the possibility that some pacemaker cells in the *Drosophila* brain might be preferentially responsive to temperature compared to light (Yoshii et al. 2005; Miyasako et al. 2007) suggests that a complex web of neural networks is involved in integrating photoperiodic and thermal cues to yield seasonably appropriate daily activity patterns. Finally, genome-wide expression studies indicate that information concerning ambient temperature and photoperiod has widespread and coordinate effects on circadian regulation of gene expression in *Drosophila* (Boothroyd et al. 2007). This work also indicated that an intron in *tim* is regulated by temperature, suggesting a broader role for thermosensitive splicing in clock responses to temperature. In addition, the recent demonstration that two different isoforms of *tim* have differential effects on photoperiod-induced ovarian diapause (Sandrelli et al. 2007; Tauber et al. 2007) further indicates that TIM is also a significant clock component in contributing to the seasonal adaptation of *Drosophila* physiology and behavior. Thus, the partnership between dPER and TIM is a key step for the circadian integration of thermal and photic information.

Of course, temperature and day length are highly associated in temperate climates, making it unsurprising that these two modalities are closely intertwined at the level of clock function. Nonetheless, as Pittendrigh noted "although photoperiod is a more reliable indicator of season and organisms have evolved sophisticated methods to measure its length, the physiological value of such a timing device is likely to be mainly directed at anticipating and thus preparing for changes in daily temperatures" (Pittendrigh and Takamura 1987). Recent work in *Drosophila* and *Neurospora* indicates that the thermosensitive splicing of central clock components is a common feature of how circadian clocks respond to temperature. Although other possibilities are likely, weak or nonconsensus splice signals that are more efficiently recognized at lower temperatures appear to be a common theme in generating temperature-dependent splicing of precursor mRNAs. On a broader perspective, RNA–RNA interactions with the ability to form a multitude of short- and long-range base-pair interactions varying in strength appear to be ideal targets for constructing intracellular molecular mechanisms that can be targeted by natural selection to generate highly calibrated thermal responses.

ACKNOWLEDGMENTS

We thank all the members in the Edery lab, past and present, for their many contributions to studies aimed at understanding the role of the clock in seasonal adaptation, especially John Majercak who initiated the work.

REFERENCES

Ainsworth J.R., Rossi L.M., and Murphy E.C., Jr. 1996. The Moloney murine sarcoma virus ts110 5′ splice site signal contributes to the regulation of splicing efficiency and thermosensitivity. *J. Virol.* **70:** 6474.

Angelichio M.L., Beck J.A., Johansen H., and Ivey-Hoyle M. 1991. Comparison of several promoters and polyadenylation signals for use in heterologous gene expression in cultured *Drosophila* cells. *Nucleic Acids Res.* **19:** 5037.

Aronson B.D., Johnson K.A., Loros J.J., and Dunlap J.C. 1994. Negative feedback defining a circadian clock: Autoregulation of the clock gene *frequency*. *Science* **263:** 1578.

Bachleitner W., Kempinger L., Wulbeck C., Rieger D., and Helfrich-Förster C. 2007. Moonlight shifts the endogenous clock of *Drosophila melanogaster*. *Proc. Natl. Acad. Sci.* **104:** 3538.

Boothroyd C.E., Wijnen H., Naef F., Saez L., and Young M.W. 2007. Integration of light and temperature in the regulation of circadian gene expression in *Drosophila*. *PLoS Genet.* **3:** e54.

Bourgeois C.F., Lejeune F., and Stevenin J. 2004. Broad specificity of SR (serine/arginine) proteins in the regulation of alternative splicing of pre-messenger RNA. *Prog. Nucleic Acid Res. Mol. Biol.* **78:** 37.

Brunner M. and Diernfellner A. 2006. How temperature affects the circadian clock of *Neurospora crassa*. *Chronobiol. Int.* **23:** 81.

Chen W.F., Majercak J., and Edery I. 2006. Clock-gated photic stimulation of *timeless* expression at cold temperatures and seasonal adaptation in *Drosophila*. *J. Biol. Rhythms* **21:** 256.

Cheng Y., Gvakharia B., and Hardin P.E. 1998. Two alternatively spliced transcripts from the *Drosophila period* gene rescue rhythms having different molecular and behavioral characteristics. *Mol. Cell. Biol.* **18:** 6505.

Cizdziel P.E., de Mars M., and Murphy E.C., Jr. 1988. Exploitation of a thermosensitive splicing event to study pre-mRNA splicing in vivo. *Mol. Cell. Biol.* **8:** 1558.

Collins B.H., Rosato E., and Kyriacou C.P. 2004. Seasonal behavior in *Drosophila melanogaster* requires the photoreceptors, the circadian clock, and phospholipase C. *Proc. Natl. Acad. Sci.* **101:** 1945.

Colot H.V., Loros J.J., and Dunlap J.C. 2005. Temperature-modulated alternative splicing and promoter use in the Circadian clock gene *frequency*. *Mol. Biol. Cell* **16:** 5563.

Diernfellner A.C., Schafmeier T., Merrow M.W., and Brunner M. 2005. Molecular mechanism of temperature sensing by the circadian clock of *Neurospora crassa*. *Genes Dev.* **19:** 1968.

Dunlap J.C. 1996. Genetics and molecular analysis of circadian rhythms. *Annu. Rev. Genet.* **30:** 579.

Edery I. 2000. Circadian rhythms in a nutshell. *Physiol. Genomics* **3:** 59.

Edery I., Rutila J.E., and Rosbash M. 1994a. Phase shifting of the circadian clock by induction of the *Drosophila period* protein. *Science* **263:** 237.

Edery I., Zwiebel L.J., Dembinska M.E., and Rosbash M. 1994b. Temporal phosphorylation of the *Drosophila period* protein. *Proc. Natl. Acad. Sci.* **91:** 2260.

Gallego M. and Virshup D.M. 2007. Post-translational modifications regulate the ticking of the circadian clock. *Nat. Rev. Mol. Cell Biol.* **8:** 139.

Garceau N.Y., Liu Y., Loros J.J., and Dunlap J.C. 1997. Alternative initiation of translation and time-specific phosphorylation yield multiple forms of the essential clock protein FREQUENCY. *Cell* **89:** 469.

Glaser F.T. and Stanewsky R. 2005. Temperature synchronization of the *Drosophila* circadian clock. *Curr. Biol.* **15:** 1352.

Grima B., Chelot E., Xia R., and Rouyer F. 2004. Morning and evening peaks of activity rely on different clock neurons of the *Drosophila* brain. *Nature* **431:** 869.

Hamblen-Coyle M.J., Wheeler D.A., Rutila J.E., Rosbash M., and Hall J.C. 1992. Behavior of period-altered circadian rhythm mutants of *Drosophila* in light:dark cycles (Diptera: Drosophilidae). *J. Insect Behav.* **5:** 417.

Hastings J.W., Rusak B., and Boulos Z. 1991. Circadian rhythms: The physiology of biological timing. In *Neural and integrative animal physiology* (ed. C.L. Prosser), p. 435. Wiley-Liss, New York.

Hastings M.L. and Krainer A.R. 2001. Pre-mRNA splicing in the new millennium. *Curr. Opin. Cell Biol.* **13:** 302.

He Q., Cha J., He Q., Lee H.C., Yang Y., and Liu Y. 2006. CKI and CKII mediate the FREQUENCY-dependent phosphorylation of the WHITE COLLAR complex to close the *Neurospora* circadian negative feedback loop. *Genes Dev.* **20:** 2552.

Heckrotte C. 1961. The effect of the environmental factors in the locomotor activity of the plains garter snake (*Thamnophis radix* radix). *Anim. Behav.* **10:** 193.

Kim E.Y., Ko H.W., Yu W., Hardin P.E., and Edery I. 2007. A DOUBLETIME kinase binding domain on the *Drosophila* PERIOD protein is essential for its hyperphosphorylation, transcriptional repression, and circadian clock function. *Mol. Cell. Biol.* **27:** 5014.

Lakin-Thomas P.L., Brody S., and Cote G.G. 1991. Amplitude model for the effects of mutations and temperature on period and phase resetting of the *Neurospora* circadian oscillator. *J. Biol. Rhythms* **6:** 281.

Liu Y., Garceau N.Y., Loros J.J., and Dunlap J.C. 1997. Thermally regulated translational control of FRQ mediates aspects of temperature responses in the *Neurospora* circadian clock. *Cell* **89:** 1.

Liu Y., Merrow M., Loros J.J., and Dunlap J.C. 1998. How temperature changes reset a circadian oscillator. *Science* **281:** 825.

Majercak J., Chen W.F., and Edery I. 2004. Splicing of the *period* gene 3′-terminal intron is regulated by light, circadian clock factors, and phospholipase C. *Mol. Cell. Biol.* **24:** 3359.

Majercak J., Sidote D., Hardin P.E., and Edery I. 1999. How a circadian clock adapts to seasonal decreases in temperature and day length. *Neuron* **24:** 219.

Matsumoto A., Matsumoto N., Harui Y., Sakamoto M., and Tomioka K. 1998. Light and temperature cooperate to regulate the circadian locomotor rhythm of wild type and *period* mutants of *Drosophila melanogaster*. *J. Insect Physiol.* **44:** 587.

Miyasako Y., Umezaki Y., and Tomioka K. 2007. Separate sets of cerebral clock neurons are responsible for light and temperature entrainment of *Drosophila* circadian locomotor rhythms. *J. Biol. Rhythms* **22:** 115.

Nakajima M., Imai K., Ito H., Nishiwaki T., Murayama Y., Iwasaki H., Oyama T., and Kondo T. 2005. Reconstitution of circadian oscillation of cyanobacterial KaiC phosphorylation in vitro. *Science* **308:** 414.

Pittendrigh C.S. and Daan S. 1976. A functional analysis of circadian pacemakers in nocturnal rodents. V. Pacemaker structure: A clock for all seasons. *J. Comp. Physiol. A* **106:** 333.

Pittendrigh C.S. and Takamura T. 1987. Temperature dependence and evolutionary adjustment of critical night length in insect photoperiodism. *Proc. Natl. Acad. Sci.* **84:** 7169.

Pittendrigh C.S., Kyner W.T., and Takamura T. 1991. The amplitude of circadian oscillations: Temperature dependence, latitudinal clines, and the photoperiodic time measurement. *J. Biol. Rhythms* **6:** 299.

Qiu J. and Hardin P.E. 1996. *per* mRNA cycling is locked to lights-off under photoperiodic conditions that support circadian feedback loop function. *Mol. Cell. Biol.* **16:** 4182.

Rensing L. and Ruoff P. 2002. Temperature effect on entrainment, phase shifting, and amplitude of circadian clocks and its molecular bases. *Chronobiol. Int.* **19:** 807.

Revel F.G., Herwig A., Garidou M.L., Dardente H., Menet J.S., Masson-Pevet M., Simonneaux V., Saboureau M., and Pevet P. 2007. The circadian clock stops ticking during deep hibernation in the European hamster. *Proc. Natl. Acad. Sci.* **104:** 13816.

Rieger D., Stanewsky R., and Helfrich-Förster C. 2003. Cryptochrome, compound eyes, Hofbauer-Buchner eyelets, and ocelli play different roles in the entrainment and masking pathway of the locomotor activity rhythm in the fruit fly *Drosophila melanogaster*. *J. Biol. Rhythms* **18:** 377.

Sandrelli F., Tauber E., Pegoraro M., Mazzotta G., Cisotto P., Landskron J., Stanewsky R., Piccin A., Rosato E., Zordan M., Costa R., and Kyriacou C.P. 2007. A molecular basis for natural selection at the *timeless* locus in *Drosophila melanogaster*. *Science* **316:** 1898.

Sawyer L.A., Hennessy J.M., Peixoto A.A., Rosato E., Parkinson H., Costa R., and Kyriacou C.P. 1997. Natural variation in a *Drosophila* clock gene and temperature compensation. *Science* **278:** 2117.

Sayeed O. and Benzer S. 1996. Behavioral genetics of thermosensation and hygrosensation in *Drosophila*. *Proc. Natl. Acad. Sci.* **93:** 6079.

Schafmeier T., Haase A., Kaldi K., Scholz J., Fuchs M., and Brunner M. 2005. Transcriptional feedback of *Neurospora* circadian clock gene by phosphorylation-dependent inactivation of its transcription factor. *Cell* **122:** 235.

Shafer O.T., Levine J.D., Truman J.W., and Hall J.C. 2004. Flies by night: Effects of changing day length on *Drosophila's* circadian clock. *Curr. Biol.* **14:** 424.

Shannon K.W. and Guthrie C. 1991. Suppressors of a U4 snRNA mutation define a novel U6 snRNP protein with RNA-binding motifs. *Genes Dev.* **5:** 773.

Sidote D., Majercak J., Parikh V., and Edery I. 1998. Differential effects of light and heat on the *Drosophila* circadian clock proteins PER and TIM. *Mol. Cell. Biol.* **18:** 2004.

Stoleru D., Peng Y., Agosto J., and Rosbash M. 2004. Coupled oscillators control morning and evening locomotor behaviour of *Drosophila*. *Nature* **431:** 862.

Stoleru D., Nawathean P., de la Paz Fernández M., Menet J.S., Ceriani M.F., and Rosbash M. 2007. The *Drosophila* circadian network is a seasonal timer. *Cell* **129:** 207.

Sweeney B.M. and Hastings J.W. 1961. Effects of temperature upon diurnal rhythms. *Cold Spring Harbor Symp. Quant. Biol.* **25:** 87.

Tauber E., Zordan M., Sandrelli F., Pegoraro M., Osterwalder N., Breda C., Daga A., Selmin A., Monger K., Benna C., Rosato E., Kyriacou C.P., and Costa R. 2007. Natural selection favors a newly derived *timeless* allele in *Drosophila melanogaster*. *Science* **316:** 1895.

Touchman J.W., D'Souza I., Heckman C.A., Zhou R., Biggart N.W., and Murphy E.C., Jr. 1995. Branchpoint and polypyrimidine tract mutations mediating the loss and partial recovery of the Moloney murine sarcoma virus MuSVts110 thermosensitive splicing phenotype. *J. Virol.* **69:** 7724.

Wheeler D.A., Hamblen-Coyle M.J., Dushay M.S., and Hall J.C. 1993. Behavior in light-dark cycles of *Drosophila* mutants that are arrhythmic, blind, or both. *J. Biol. Rhythms* **8:** 67.

Yoshii T., Sakamoto M., and Tomioka K. 2002. A temperature-dependent timing mechanism is involved in the circadian system that drives locomotor rhythms in the fruit fly *Drosophila melanogaster*. *Zool. Sci.* **19:** 841.

Yoshii T., Heshiki Y., Ibuki-Ishibashi T., Matsumoto A., Tanimura T., and Tomioka K. 2005. Temperature cycles drive *Drosophila* circadian oscillation in constant light that otherwise induces behavioural arrhythmicity. *Eur. J. Neurosci.* **22:** 1176.

Yu W., Zheng H., Houl J.H., Dauwalder B., and Hardin P.E. 2006. PER-dependent rhythms in CLK phosphorylation and E-box binding regulate circadian transcription. *Genes Dev.* **20:** 723.

Endogenous Circannual Clock and HP Complex in a Hibernation Control System

N. Kondo
Mitsubishi Kagaku Institute of Life Sciences, Machida, Tokyo 194-8511, Japan

Hibernation in mammals is a mysterious biological phenomenon that appears on a seasonal basis for surviving a potentially lethal low body temperature (Tb) near 0°C and protecting organisms from various diseases and harmful events during hibernation. The exact mechanism by which such a unique ability is seasonally developed is still unknown. On the basis of our previous finding that the source of calcium ions for excitation-contraction coupling in myocardium of chipmunks, a rodent hibernator, is seasonally modulated for hibernation, the liver-derived hibernation-specific protein (HP) complex was discovered. Recently, the HP complex was identified as a promising candidate hormone that carries a hibernation signal to the brain independently of Tb and environmental changes for developing a capacity for tolerating low Tb. This finding will promote new approaches to understanding biological hibernation systems, including a circannual clock and its signaling pathway between the brain and the periphery. A new definition of hibernation and a possible model of a hibernation control system are proposed.

INTRODUCTION

Mammalian hibernation has long been of interest as a mysterious biological phenomenon. During hibernation, Tb decreases to about 0°C, and then metabolic responses are remarkably depressed. Such a profoundly depressed metabolism generally leads to the dysfunction of various cells and tissues in nonhibernators, resulting in lethal damage and eventual death. However, hibernating animals survive such a severely depressed physiological state without any damage. At a decreased Tb during hibernation, the heart rate dramatically decreases with greatly reduced blood flow, yet the animals can tolerate these hypothermic and ischemic stresses. Thus, hibernators possess the ability to develop a tolerance to a low-temperature disturbance of metabolism during hibernation in order to prevent cold injuries.

Hibernation has been further suggested to increase tolerances to hypoxia and hypoglycemia in the brain, bacterial infections, tumorigenesis, radiation, disuse muscle atrophy (see Kondo et al. 2006), and osteoporosis (Donahue et al. 2006). In immune functions in a hibernating state, the classical pathway of the serum complement system is active (Maniero 2002) and the splenic macrophages bind more bacterial lipopolysaccharide (Maniero 2005). Furthermore, hibernators live much longer even under warm conditions where a decreased Tb is prevented (Zivadinovic and Andjus 1996; Kondo and Narita 1998; Kondo 1999). Thus, hibernation is proposed to be a seasonally regulated physiological adaptation that is developed specifically for maintaining health and life by protecting organisms from severe physiological states. If better understood, hibernation would greatly contribute to a wide variety of medical applications as well as maintenance of health in humans. Here, our attempts to understand seasonally controlled hibernation as a physiological system for maintaining health and life are reviewed.

SEASONALLY REGULATED TOLERANCE OF HEARTS TO LOW Tb

As regulated heart function is the most important factor for maintaining life, there has been great interest in the physiological characteristics of hearts in a hibernating state that bestow tolerance to a remarkably decreased Tb (Wang 1988). The first evidence that the source of calcium ions (Ca^{2+}) for cardiac excitation-contraction coupling is modulated during hibernation was demonstrated by an electrophysiological and pharmacological study of the hearts of chipmunks (*Tamias sibiricus*), a rodent hibernator (Kondo and Shibata 1984). In cardiac muscles from hibernating chipmunks, the plateau phase of a membrane action potential, which is generated by Ca^{2+} currents carried by *trans*-sarcolemmal Ca^{2+} influx through Ca^{2+} channels, is markedly inhibited. This inhibited plateau phase is attributed to a remarkably decreased activation of voltage-dependent Ca^{2+} channels (Kondo 1986a) by a rapid repolarization below their activation threshold by enhanced transient K^+ outward currents that are sensitive to 4-aminopyridine (Kondo 1986b). However, despite much less Ca^{2+} influx through Ca^{2+} channels, the contraction was substantially unchanged between active and hibernating animal myocardium. The discrepancy between decreased *trans*-sarcolemmal Ca^{2+} inflow and unchanged contraction was solved by experiments using ryanodine and caffeine, which are inhibitors of Ca^{2+} release and uptake, respectively, by the sarcoplasmic reticulum (SR), which are internal Ca^{2+} store sites; the ability of the SR to take up and release Ca^{2+} is greatly augmented in myocardium from hibernating animals (Kondo 1988). It is believed that this augmentation compensates for decreased Ca^{2+} inflow through Ca^{2+} channels for maintaining homeostasis of cardiac function. Such an augmented ability of the SR is supported by a study on cardiac SR vesicles from hibernating ground squirrels which found that both the rate of Ca^{2+} uptake and level of Ca^{2+} accumulation are increased (Belke et al. 1991), as well as a recent study that reported the gene expression level of the SR Ca^{2+}-ATPase but not Ca^{2+}-binding protein calsequestrin is increased in hibernating woodchuck (Yatani et al. 2004).

The results of these studies suggest that in hibernating animals, intracellular Ca^{2+} homeostasis for cardiac contraction and relaxation is maintained primarily by internal Ca^{2+} release and reuptake by the modulated SR (Fig. 1, top) (Kondo 1997). This modulation would maintain regulated functions of the heart at low Tb and simultaneously greatly economize on energy that is consumed to remove cytosolic Ca^{2+} for relaxation because an excess amount of Ca^{2+} is not loaded in cells due to the supply of a moderate amount of Ca^{2+} from the SR that stores the limited Ca^{2+}. Thus, the modulation of Ca^{2+} sources in myocardium from hibernating animals establishes an ideal intracellular Ca^{2+} recycling system for the prevention of cellular Ca^{2+} overload by enhancing the SR ability and avoiding excess Ca^{2+} influx induced by an extensively delayed inactivation process of Ca^{2+} channels at low Tb. The switching to this recycling system would increase the capacity of hearts for tolerating low Tb. A similar cardiac modulation has been observed in two other hibernators, ground squirrels and woodchucks (Wang et al. 2002; Yatani et al. 2004), suggesting a common mechanism for the prevention of cold damage in mammalian hibernators.

The modulation of cardiac Ca^{2+} regulation found in hibernating chipmunks is observed even in normothermic chipmunks in which a decreased Tb is prevented throughout a year by keeping them under conditions of constant warmth (23°C) and a 12-hour light/dark cycle (Kondo 1987; Kondo and Kondo 1992b). The cardiac electromechanical properties are switched from extracellular Ca^{2+}-dependent properties to the SR Ca^{2+}-dependent properties only in the winter season under constant laboratory conditions (Fig. 1, bottom). This result clearly indicates that switching to unique properties seen in hibernating animals seasonally occurs even in normothermic animals, suggesting that the hibernation-like modulation is seasonally caused independently of Tb and environmental changes and completed before the onset of hibernation for tolerating low Tb. From these findings, it can be surmised that a seasonal timing mechanism functions endogenously in organisms and systemically controls the ability to adapt to severe physiological states during the hibernation season. To understand the mechanism by which hibernation is controlled, a seasonally regulated molecular probe specific for hibernation, which should be correlated with the seasonal modulation in hearts, has been explored.

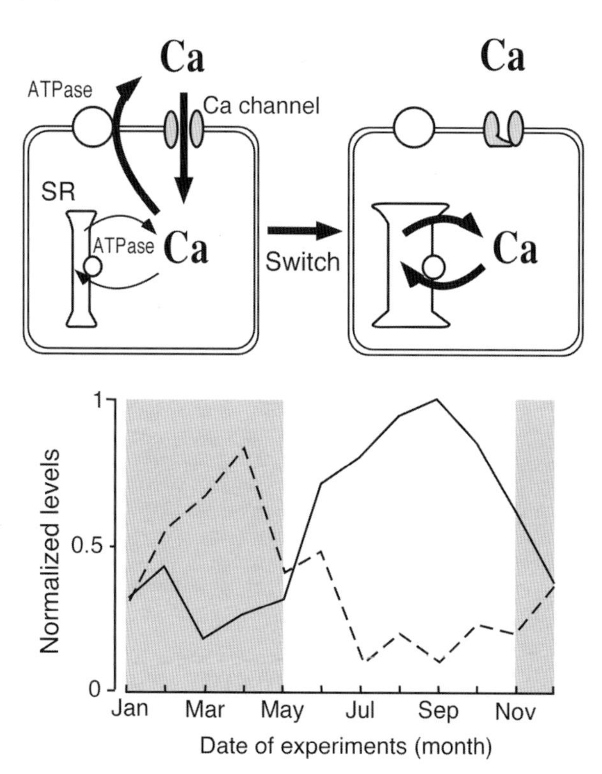

Figure 1. Modulation of Ca^{2+} source for cardiac excitation-contraction coupling during hibernation. (*Top*) Modulation is circannually switched from active (*left*) to hibernating (*right*) state. (*Bottom*) Circannual changes in electromechanical properties dependent on *trans*-sarcolemmal Ca^{2+} influx (*solid line*) and the SR Ca^{2+} release (*dashed line*) occur in animals kept under conditions of constant warmth and a 12-hour light/dark cycle. (*Gray area*) Period of hibernation-like changes; (*vertical axis*) normalized levels of dependence with respect to differences in the degree of dependence between active and hibernating animals myocardium.

HIBERNATION MARKER AND CIRCANNUAL HIBERNATION RHYTHM

In the blood of chipmunks, liver-derived protein complex has been isolated as a factor that is correlated with seasonally changed electromechanical properties in cardiac muscles (Kondo and Kondo 1992b). This complex is composed of four novel proteins termed hibernation-specific proteins (HP) (Kondo and Kondo 1992a). Three of four HPs (HP20, HP25, and HP27) have homologous structures to a collagen-like domain and form a heterotrimer complex (HP20c) through a triple-helix structure in amino-terminal regions (Fig. 2). HP20c further associates with the fourth protein, HP55, which is a pro-

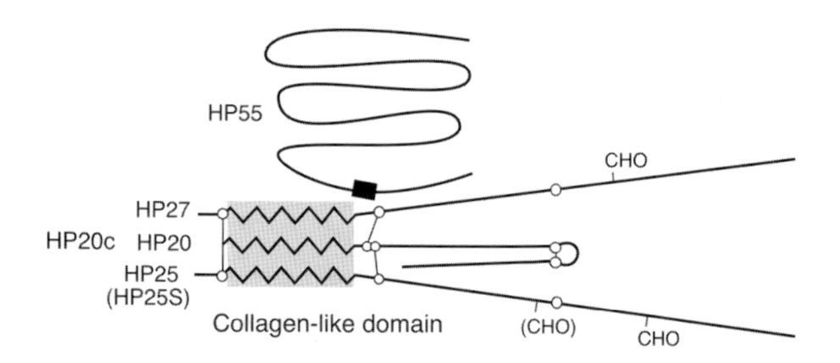

Figure 2. Characteristics of the HP complex structure. The lengths of the lines correspond to the number of amino acids. *Gray area* in HP20c indicates a collagenous domain forming a triple helix. *Open circles* and CHOs indicate cysteine residues and glycosylation sites, respectively. Lines between *open circles* show disulfide bonds. HP25S is a glycosylated HP25. *Closed box* in HP55 shows an active center and it is assumed it interacts with HP20c at this position.

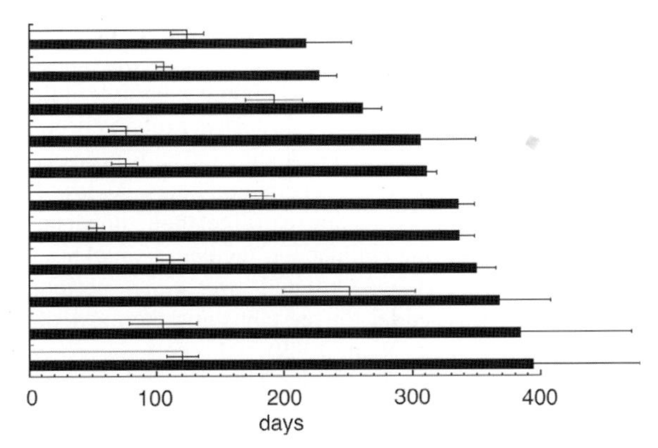

Figure 3. Circannual hibernation rhythms. Cycle periods (*closed bars*) and durations (*open bars*) of free-running hibernation in 11 animals maintained under conditions of constant cold and darkness. Lines attached to bars show mean ±S.E.M. (n = between 5 and 9).

tease inhibitor in the serpin superfamily (Kondo and Kondo 1996), in the blood by a hydrophobic bond (Kondo and Kondo 1992a). HP complex (complex of HP20c with HP55) in the blood is down-regulated in correlation with hibernation by depressed expression of genes in the liver (Kondo and Kondo 1992a; Takamatsu et al. 1993). On the basis of the structural characteristics of the HP complex, it seems likely that HP20c is inactivated by binding to HP55 in the blood (Kondo and Kondo 1992a).

To clarify the correlation between the HP complex and hibernation, the rhythmicity of hibernation in chipmunks has been examined throughout their lives under constant laboratory conditions. Under conditions of constant cold (5°C) and 24-hour darkness, the animals exhibited a free-running hibernation rhythm with each individual circannual timing (Fig. 3). The circannual cycle and the duration of hibernation range between 5 and 13 months and between 2 and 8 months, respectively, and are relatively constant in individuals. There is no correlation between the cycle period and the duration. Surprisingly, the maximum life span of these animals with a circannual hibernation rhythm was approximately 11 years, which is four times longer than rats and mice (Kondo et al. 2006). However, a few animals that were unable to hibernate throughout their life had only short life spans (up to 3 years). Thus, the timing of hibernation is controlled by a system generating endogenous circannual rhythm characterized in each individual. Animals that undergo hibernation have an exceptionally long life span.

The down-regulation of the HP complex is closely correlated with a circannual hibernation rhythm (Kondo et al. 2006); levels of HP complex in the blood are decreased prior to the onset of hibernation and kept low during hibernation, following which the decreased levels return to active-state levels in association with the termination of hibernation (Fig. 4). Such a down-regulation of the HP complex in the blood was due to depressed expression of HP genes in the liver. The fact that down-regulation of the HP complex is turned on and off before the onset and termination of hibernation, respectively, implies that a circannual timing mechanism for hibernation controls expression of the HP complex. Interestingly, in a few animals that are unable to hibernate under conditions of constant cold and darkness, the HP complex in the blood is never down-regulated, suggesting an important role for the HP complex in hibernation. This is supported by the species-specific expression of HP genes in hibernators; HP genes are expressed in hibernators, such as ground squirrels and woodchucks (Kondo and Kondo 1993), whereas in tree squirrels, which are nonhibernators of the same squirrel family as chipmunks, HP genes are not expressed due to a base substitution in the promoter region (Kojima et al. 2001).

More significant evidence for a circannual regulation of the HP complex and its role in hibernation has been shown in animals in which a decreased Tb was prevented by maintaining them under conditions of constant warmth (23°C) and a 12-hour light/dark cycle (Kondo et al. 2006). Even in such normothermic animals, the HP complex in the blood is circannually regulated, and the timing and duration of down-regulation of the HP complex are similar to those in circannually hibernating animals kept under conditions of constant cold and darkness, demonstrating that a circannual timing mechanism functions independently of Tb and environment changes. Only when the level of the HP complex in the blood is down-regulated do animals develop decreased Tb and begin hibernation due to exposure to cold. This indicates that a circannually regulated decrease in the HP complex in the blood is necessary for hibernation; during this decrease, the ability to tolerate low Tb may be developed in order to survive the hypothermic state. The similarities in the timing and duration of circannual HP down-regulation between animals kept in constant warm and cold environments imply that an endogenous circannual clock system that generates a circannual rhythm is insensitive to Tb and environmental changes.

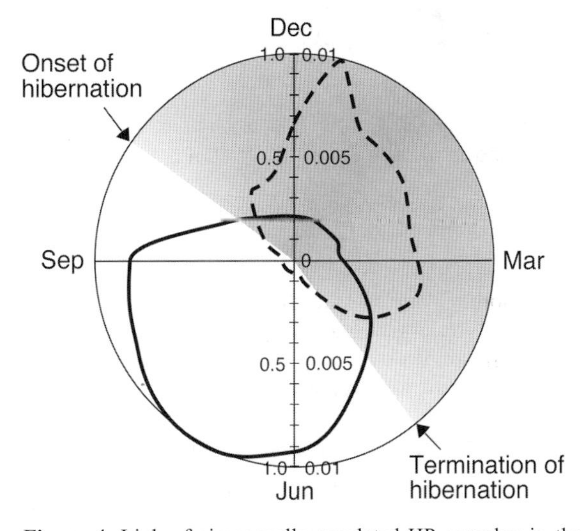

Figure 4. Link of circannually regulated HP complex in the blood and CSF to hibernation. A cycle of the HP complex and hibernation rhythm in animals kept under conditions of constant cold and darkness was fitted to a year. (*Solid lines*) HP levels in the blood; (*dashed lines*) HP levels in the CSF. (*Numbers on vertical axis*) HP levels in the blood (*left*) and CSF (*right*) normalized to their maximum content in the blood.

INCREASED HP COMPLEX IN THE BRAIN DURING HIBERNATION

By immunochemical analysis of cerebrospinal fluid (CSF) using antibodies against components of the HP complex, it has been demonstrated that the HP complex is markedly increased in CSF during hibernation despite its decrease in the blood (Fig. 4) (Kondo et al. 2006). An increase in the HP complex in CSF during hibernation is due to circannually up-regulated transcytosis of the HP complex through the choroidal epithelium (the blood-CSF barrier) but not to the HP gene expression in the brain. As levels of the HP complex in the blood are higher than those in the CSF even during hibernation, a sufficient amount of the HP complex exists in the blood for supplying to the CSF. The up-regulation of the HP complex in CSF is coincident with a circannual down-regulation of that in the blood (Fig. 4), and similar regulation in CSF and blood is observed in normothermic animals in which a decreased Tb was prevented by keeping them under conditions of constant warmth and a 12-hour light/dark cycle, indicating that a circannual timing mechanism simultaneously and conversely regulates the HP complex in the blood and CSF independently of Tb and environmental changes. This suggests that a circannual clock outputs a signal simultaneously to the liver for decreasing HP production and to the choroidal epithelium for increasing HP transport. Thus, the HP complex is increased in the brain under control of an endogenous circannual rhythm and then hibernation begins.

Circannually facilitated HP transport through the blood-CSF barrier may have a critical role in the functioning of the HP complex in the brain because the HP complex dissociates to HP20c and HP55 in CSF (Kondo et al. 2006). Because the HP complex exists in the blood as a complex consisting of an interaction between HP20c and HP55, which is probably an inactive form (Kondo and Kondo 1992a), HP20c that has dissociated from HP55 is likely to be an active form. Such an activation of the HP complex through the blood-CSF barrier may be controlled by a signal outputted by a circannual clock. This implies that the brain is a target of HP20c for control of hibernation.

THE HP COMPLEX ESSENTIAL FOR HIBERNATION

Studies on the effects of administration of antibody against HP20c into lateral ventricles on hibernation have demonstrated the essential role of circannually increased brain HP20c in hibernation (Kondo et al. 2006). The administered antibody decreases the amount of time spent in low Tb (hibernation time) in a dose-dependent manner. In some animals, hibernation was interrupted during antibody administration, whereas in other animals, antibody administered in the late stage of hibernation accelerated the termination of hibernation. Thus, blocking HP20c activity in the brain with antibody dramatically decreases hibernation time.

In this study, the concept of "hibernation time" may be primarily used for the quantitative evaluation of hibernation (Kondo et al. 2006). During hibernation, animals periodically arouse for several hours after spending several days at a continuous low Tb. In chipmunks, periods of this periodic arousal and continuous low Tb are relatively constant in each individual, and periodic arousal is necessary for surviving low Tb during hibernation bouts. As low Tb is potentially lethal in cells due to the metabolic imbalance ascribed to perturbing cellular ion regulation and energy metabolism that is accumulated over time, periodic arousal may be essential for reestablishing the metabolic balance for surviving (Wang 1989). If this periodic arousal is disturbed, the animal would die. Therefore, time spent in continuous low Tb during hibernation bouts means the limit of time survived at a decreased Tb. This limit of time may correspond to a capacity for tolerating low Tb. Even nonhibernators, including humans, can tolerate low Tb, although the length of time survived at low Tb is much less than that in hibernators. For example, humans can survive less than 1 hour at Tb near 20°C, whereas chipmunks survive approximately 7 days (168 hours) below 10°C during the hibernation season. Such a prolonged time may depend on the capacity for tolerating low Tb developed by a circannually regulated ability to hibernate. The fact that anti-HP20c antibody administered into the brain dramatically decreases hibernation time or interrupts hibernation suggests that brain HP20c is essential for developing a capacity for protecting organisms not only from low Tb, but also from other lethal factors and events during hibernation.

PROPOSED ROLE OF THE HP COMPLEX

Although the exact function of the HP complex in the brain and peripheral organs remains to be elucidated, some significant roles for HP20c have been proposed based on the results of studies on members of HP20c family proteins, which have structural and evolutional similarities to HP20c, such as complement protein C1q, precerebellin, ACRP30/adiponectin, and tumor necrosis factor (TNF) family (Kishore et al. 2004). The liver-derived C1q is involved in activation of the classical complement pathway responsible for antimicrobial defense and immunological processes such as immune tolerance, phagocytosis of bacteria, neutralization of retroviruses, and modulation of dendritic cells, B cells, and fibroblasts. Precerebellin is expressed in the cerebellum and may have a role in the development and stability of Purkinje cell synapses. Adiponectin derived from adipocytes is known to regulate lipid and carbohydrate catabolism for maintaining energy homeostasis and resolve inflammation. The TNF family members are involved in adaptive immunity, apoptosis selectively in tumor cells, inflammation, cell survival and proliferation, and bone and energy homeostasis. Thus, proteins that are structurally homologous to HP20c have been shown to contribute to energy homeostasis, immunity, tumor cell apoptosis, and cell survival. Because similar biological effects for HP20c have been proposed, an increase in brain HP20c during hibernation may protect organisms from depressed metabolism and lethal diseases as observed during hibernation in hibernators. An essential role for brain HP20c in tolerating low Tb (Kondo et al. 2006) is compatible with this possibility.

CIRCANNUAL HP RHYTHM FOR MAINTAINING LIFE

Studies of the hearts of hibernators have suggested that circannually modulated cardiac Ca^{2+} regulation independent of Tb and environmental changes develops cardiac tolerance to lethal low Tb by preventing cellular Ca^{2+} overload. This may be expanded into the protection of principal organs and tissues from low Tb because a Ca^{2+} overloaded state causes lethal damage in various cells by activation of Ca^{2+}-dependent necrotic processes such as proteolysis, lipid peroxidation, and mitochondrial swelling. Therefore, a circannual timing mechanism may have a central role in systemic modulation of cellular functions for protecting organisms from lowering of Tb via a signaling pathway. The HP complex is believed to be the first promising candidate hormone that carries a circannual signal into the brain for protecting organisms, most likely by maintaining cellular ion balance, energy homeostasis, and immune functions.

Our preliminary study of the longevity of chipmunks (Kondo and Narita 1998; Kondo 1999) led to the surprising involvement of circannually controlled HP complex in maintaining health and life. Animals with a circannual HP rhythm that undergo lowering of Tb during hibernation seasons by being kept under conditions of constant cold or those that maintain a normothermic state throughout by being kept under conditions of constant warmth have a much longer life span; many of these animals live close to 10 years and the maximum life span is over 10 years, which is approximately four or five times longer than rats and mice. On the other hand, animals without circannual HP rhythm have a short life span of only a few years, similar to rodent nonhibernators. This result suggests that animals in which brain HP20c is circannually increased live much longer and in good health, probably due to the prevention of lethal diseases and slowing of aging. Because such a long life span is not due to depressed metabolism by low Tb during hibernation, it is further suggested that hibernation should be redefined as the ability to govern modulation of cellular functions for maintaining a long life through hormonal signaling, rather than the ability to decrease Tb. Namely, mammalian hibernation may be controlled by a physiological system consisting of a circannual oscillator, its signaling pathway, and target sites. The HP complex may be an essential modulator for governing cellular tolerance to metabolic depression, diseases, and aging.

CIRCANNUAL RHYTHM AND CIRCADIAN RHYTHM

A circannual rhythm has a central role in the survival of homeothermic animals in nature; this rhythm is responsible for controlling the timing of reproduction in mammals and migration in birds, which are critical for survival. However, there are serious problems with respect to studying circannual timing mechanisms because of the long cycle of a circannual rhythm, influences of environmental changes such as light and temperature, and its expression at the organism level. By identifying the HP complex as a unique marker of a circannual rhythm independent of Tb and environmental changes, a circannual timing mechanism has been characterized without any influence by environmental and Tb changes.

Because HP rhythm is correlated with a circannual hibernation rhythm, the influences of Tb and environmental changes on a circannual rhythm can be explored by comparing animals subjected to a decreased Tb under conditions of constant cold (5°C) and darkness and normothermic animals under conditions of constant warmth (23°C) and a 12-hour light/dark cycle (Kondo et al. 2006). A circannual HP rhythm is generated independently of Tb change, indicating that lowering of Tb does not trigger the timing of the rhythm. Furthermore, cycle periods and duration of hibernation in animals maintained under cold conditions are similar to those of HP down-regulation in normothermic animals kept under constant warm conditions, indicating that the generation of circannual rhythms is not affected by differences in Tb, environmental temperature, and light between animals kept under cold and warm conditions. Thus, a circannual clock mechanism maintains homeostasis even at Tb near 0°C and is little affected by light.

Circadian rhythms are known to possess temperature compensation properties that are similar to those of circannual HP rhythms, and it has been proposed that seasonal rhythms depend on the circadian system. As circadian rhythms are generated by negative feedback loops that are formed by core circadian clock genes and their products, this or similar mechanisms are assumed to function in the generation of a circannual rhythm. However, although free-running circadian periods are longer than 24 hours and can be reset by light, cycle periods of free-running circannual rhythms are shorter than 1 year and unaffected by the difference between darkness and a 12-hour light/dark cycle. A circadian rhythm can be generated in a variety of organs and tissues in organisms and even in monocytes, whereas a circannual rhythm would be generated only at the organism level. A circannual rhythm in chipmunks is insensitive to aging (Kondo et al. 2006), whereas a circadian rhythm in rodents is affected by aging (Bentivoglio et al. 2006). The suggested direct link between circadian cycles and metabolic cycles (Rutter et al. 2002) may be incompatible with the result that a free-running circannual cycle of the HP complex is not affected by a decreased Tb during hibernation that dramatically depressed metabolic processes. These characteristics suggest a lesser degree of similarity between circannual and circadian timing mechanisms. Therefore, a circannual rhythm may not be generated by estimating time on the basis of circadian rhythms but could be modulated by circadian rhythms for adapting periods of free-running endogenous circannual rhythms to a seasonal cycle in nature. Systemic rather than cellular feedback loops may be involved in circannual rhythm generation because of the insensitivity of this rhythm to Tb changes and its systemic expression at the organism level. Our proposed cross-talk model for a hibernation control system (Kondo et al. 2006) implies the involvement of a systemic negative feedback loop formed by the HP complex and a signaling pathway between the brain and periphery in the generation of a circannual rhythm (Fig. 5).

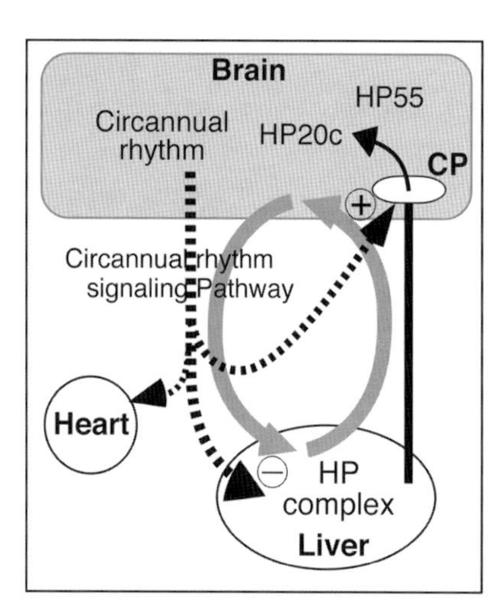

Figure 5. Proposed system for systemic circannual rhythm generation. Inverse regulation of HP complex in the brain (+: up-regulation) and liver (−: down-regulation) may systemically form a negative feedback loop between the brain and periphery (*gray arrow*). (*Dashed arrow*) Predicted circannual rhythm signaling pathway from the brain to periphery. (CP) Choroid plexus (blood-CSF barrier).

HIBERNATION AND SLEEP

Although similarities between hibernation and sleep have been suggested by a decreased Tb during either sleep or hibernation and reports that entry into hibernation is an extension of slow-wave sleep (Berger 1984; Krilowicz et al. 1988), there may be serious differences between them as shown below. The amplitude of decreased Tb during sleep is less than 1°C, whereas that during hibernation is more than 30°C, and sleep does not develop a capacity for tolerating low Tb, although hibernation does. Hibernation cycles are controlled by a circannual rhythm, and circadian rhythms may have a role in sleep homeostasis. Although the amount of time spent in sleep is in proportion to that spent in wakefulness, there is no correlation between periods of hibernation and active state. Furthermore, an analysis of electroencephalograms suggested that slow-wave sleep as restoration from wakefulness is disturbed during hibernation (Daan et al. 1991). Thus, the mechanism(s) of hibernation may not be based on sleep.

Our recent study on the role of the HP complex in hibernation (Kondo et al. 2006) demonstrated that a sleep-like inactive state induced by a decreased Tb is not necessary for the ability to hibernate because the ability to survive a state of hibernation is circannually developed independently of decreased Tb. Furthermore, the fact that chipmunks exhibiting circannual HP rhythm have a surprisingly long life span with or without Tb lowering reveals that an inactive state due to a low Tb during hibernation is not essential, although sleep is, for the maintenance of life.

The above discussion indicates that an inactive state during hibernation is not of primary importance for life processes, although sleep is critical and inevitable for these processes, and that in a hibernating state but not sleep, organisms maintain regulated function at the molecular, cellular, and organ levels at a potentially lethal low Tb. This implies that there is a minimal degree of similarity between the mechanisms of hibernation and sleep.

CONCLUSIONS

A circannual rhythm responsible for hibernation controls the modulation of Ca^{2+} sources for cardiac excitation-contraction coupling independently of Tb and environmental changes, and this rhythm establishes an intracellular Ca^{2+} recycling system for the prevention of cellular Ca^{2+} overload at low Tb. This modulation may lead to the development of a capacity for hearts to tolerate low Tb during the hibernation season. The liver-derived HP complex was identified as being a novel hormone essential for hibernation in the brain. Under control of an endogenous circannual rhythm, the HP complex is down-regulated in the blood and concurrently up-regulated in CSF through active transport via the blood-CSF barrier. The HP complex is activated in the brain for developing a capacity for tolerating a severely depressed metabolic state by low Tb and would help the animal live longer. These results suggest that a circannually regulated physiological system that controls hibernation via the HP complex maintains health and life and allow us to discuss circannual rhythms versus circadian rhythms and hibernation versus sleep. Simultaneous and inverse regulation of the HP complex in the blood and the brain led us to propose that a circannual rhythm is generated by a systemic negative feedback loop formed between the brain and periphery.

ACKNOWLEDGMENTS

This work was supported by the Kanagawa Academy of Science and Technology and the National Space Development Agency of Japan. I thank all of the researchers and research assistants for their valuable collaboration during this project.

REFERENCES

Belke D.D., Milner R.E., and Wang L.C.H. 1991. Seasonal variations in the rate and capacity of cardiac SR calcium accumulation in a hibernating species. *Cryobiology.* **28:** 354.

Bentivoglio M., Deng X.H., Nygard M., Sadki A., and Kristensson K. 2006. The aging suprachiasmatic nucleus and cytokines: Functional, molecular, and cellular changes in rodents. *Chronobiol. Int.* **23:** 437.

Berger R.J. 1984. Slow wave sleep, shallow torpor and hibernation: Homologous states of diminished metabolism and body temperature. *Biol. Psychol.* **19:** 305.

Daan S., Barnes B.M., and Strijkstra A.M. 1991. Warming up for sleep?: Ground squirrels sleep during arousals from hibernation. *Neurosci. Lett.* **128:** 265.

Donahue S.W., Galley S.A., Vaughan M.R., Patterson-Buckendahl P., Demers L.M., Vance J.L., and McGee M.E. 2006. Parathyroid hormone may maintain bone formation in hibernating black bears (*Ursus americanus*) to prevent disuse osteoporosis. *J. Exp. Biol.* **209:** 1630.

Kishore U., Gaboriaud C., Waters P., Shrive A.K., Greenhough T.J., Reid K.B.M., Sim R.B., and Arlaud G.J. 2004. C1q and tumor necrosis factor superfamily: Modularity and versatility. *Trends Immunol.* **25:** 551.

Kojima M., Shiba T., Kondo N., and Takamatsu N. 2001. The tree squirrel HP-25 gene is a pseudogene. *Eur. J. Biochem.* **268:** 5997.

Kondo J. and Kondo N. 1996. Structural aspects of complex of hibernation-specific proteins. In *Adaptation to the cold: Tenth International Hibernation Symposium* (ed. F. Geiser et al.), p. 351. University of New England Press, Armidale, Australia.

Kondo N. 1986a. Excitation-contraction coupling in myocardium of nonhibernating and hibernating chipmunks: Effects of isoprenaline, a high calcium medium and ryanodine. *Circ. Res.* **59:** 221.

———. 1986b. Excitation-contraction coupling in the myocardium of hibernating chipmunks. *Experientia* **42:** 1220.

———. 1987. Identification of a pre-hibernating state in myocardium from nonhibernating chipmunks. *Experientia* **43:** 873.

———. 1988. Comparison between effects of caffeine and ryanodine on electromechanical coupling in myocardium of hibernating chipmunks: Role of internal Ca stores. *Br. J. Pharmacol.* **95:** 1287.

———. 1997. Physiological and biochemical studies on hibernation control mechanism in mammalian hibernation. In *Sleep and sleep disorders: From molecule to behavior* (ed. O. Hayaishi and S. Inoue), p. 129. Academic Press, Tokyo, Japan.

———. 1999. Hibernation control mechanism and possible applications to humans. *J. Br. Interplanet. Soc.* **58:** 343.

Kondo N. and Kondo J. 1992a. Identification of novel blood proteins specific for mammalian hibernation. *J. Biol. Chem.* **267:** 473.

———. 1992b. Identification of novel types of protein specific for mammalian hibernation with circannual rhythm. In *Circadian clocks from cell to human* (ed. T. Hiroshige and K. Honma), p. 89. Hokkaido University Press, Sapporo, Japan.

———. 1993. Identification and characterization of novel types of plasma protein specific for hibernation in rodents. In *Life in the cold: Ecological, physiological and molecular mechanisms* (ed. C. Carey et al.), p. 467. Westview Press, Boulder, Colorado.

Kondo N. and Narita A. 1998. Does hibernation increase survival rate in chipmunks? *Zoolog. Sci.* (suppl.) **15:** 106.

Kondo N. and Shibata S. 1984. Calcium source for excitation-contraction coupling in myocardium of nonhibernating and hibernating chipmunks. *Science* **225:** 641.

Kondo N., Sekijima T., Kondo J., Takamatsu N., Tohya K., and Ohtsu T. 2006. Circannual control of hibernation by HP complex in the brain. *Cell* **125:** 161.

Krilowicz B.L., Glotzbach S.F., and Heller H.C. 1988. Neuronal activity during complete bouts of hibernation. *Am. J. Physiol.* **255:** R1008.

Maniero G.D. 2002. Classical pathway serum complement activity throughout various stages of the annual cycle of a mammalian hibernator, the golden-mantled ground squirrel, *Spermophilus lateralis*. *Dev. Comp. Immunol.* **26:** 563.

———. 2005. Ground squirrel splenic macrophages bind lipopolysaccharide over a wide range of temperatures at all phases of their annual hibernation cycle. *Comp. Immunol. Microbiol. Infect. Dis.* **28:** 297.

Rutter J., Reick M., and McKnight S.L. 2002. Metabolism and the control of circadian rhythms. *Annu. Rev. Biochem.* **71:** 307.

Takamatsu N., Ohba K.-I., Kondo J., Kondo N., and Shiba T. 1993. Hibernation-associated gene regulation of plasma proteins with a collagen-like domain in mammalian hibernators. *Mol. Cell. Biol.* **13:** 1516.

Wang L.C.H. 1988. Mammalian hibernation: An escape from the cold. In *Advances in comparative and environmental physiology* (ed. R. Gilles), vol. 2, p. 1. Springer-Verlag, Berlin.

———. 1989. Ecological, physiological, and biochemical aspects of torpor in mammals and birds. In *Advances in comparative and environmental physiology* (ed. R. Gilles), vol. 4, p. 361. Springer-Verlag, Berlin.

Wang S.Q., Lakatta E.G., Cheng H., and Zhou Z.Q. 2002. Adaptive mechanisms of intracellular calcium homeostasis in mammalian hibernators. *J. Exp. Biol.* **205:** 2957.

Yatani A., Kim S.J., Kudej R.K., Wang Q., Depre C., Irie K., Kranias E.G., Vatner S.F., and Vatner D.E. 2004. Insights into cardioprotection obtained from study of cellular Ca^{2+} handling in myocardium of true hibernating mammals. *Am. J. Physiol. Heart Circ. Physiol.* **286:** H2219.

Zivadinovic D. and Andjus R.K. 1996. Life span of the European ground squirrel *Spermophilus citellus* under free-running conditions and entrainment. In *Adaptations to the cold: Tenth International Hibernation Symposium* (ed. F. Geiser et al.), p. 103. University of New England Press, Armidale, Australia.

On the Chronobiology of Cohabitation

M.J. Paul and W.J. Schwartz
Department of Neurology, University of Massachusetts Medical School, Worcester, Massachusetts 01655

Social regulation of animal circadian rhythms may enable individuals in a population to temporally synchronize or segregate their activities within the community. Relatively little is known about the mechanisms for such interindividual temporal adaptations or how the circadian system might be involved. The literature suggests that actual prolonged cohabitation might lead to robust effects on the rhythmicity of cohoused individuals but that these effects are not easily reproduced by indirect or pulsatile social contacts. We have begun to study the conditions under which such cohabitation effects might be revealed in the laboratory, and we present and discuss initial data that cohousing pairs of golden hamsters can result in a persistent change in the free-running circadian period of one of the two hamsters of the pair. We believe that analyzing the societal level of temporal organization, and ultimately dissecting its underlying mechanisms, will enrich our understanding of the circadian clock and its role in establishing ecological communities.

INTRODUCTION

Our knowledge of the circadian system of animals at the molecular, cellular, tissue, and organismal levels is remarkable, and we are beginning to understand how each of these levels contributes to the emergent properties and increased complexity of the system as a whole. For the most part, analysis has been performed using singly housed animals in plastic cages with temperature, humidity, and access to food rigidly controlled. Of course, many species ordinarily would not live out their lives in such seclusion. Some live in colonies with highly developed social structures and a clear division of labor, requiring modifications to daily rhythms; in bees, for example, foragers periodically leave the hive and express robust circadian activity rhythms, whereas nurses care for the brood continuously and are active during both the day and the night (Shemesh et al. 2007). Social mediation of honeybee (*Apis mellifera*) temporal organization becomes apparent when the colony experiences a shortage of nurses. Foragers return to nursing and adopt an arrhythmic, persistently active state (Bloch and Robinson 2001). In the wild, animals living together might synchronize their behaviors to achieve common goals or, alternatively, actively avoid each other to lessen competition for limited resources. At this *supra*-organismal level of organization, how are such interindividual daily adaptations achieved in order to form real ecological communities? Is the circadian system involved, and what mechanisms are responsible?

SYNCHRONIZATION AND SEGREGATION OF ANIMALS IN THE FIELD AND THE LABORATORY

Two recent reviews (Davidson and Menaker 2003; Mistlberger and Skene 2004) summarize our knowledge of social cues and the expression of circadian rhythmicity, highlighting the conceptual, experimental, and species-specific complications of this research. In the field, temporal synchronization among animals has been inferred from studies of Canadian beavers (*Castor canadensis*) living together as a family unit, as they exhibit a single coherent free-running rhythm in daily noise while they overwinter in constant darkness (DD) under thick ice (Bovet and Oertli 1974; Potvin and Bovet 1975). The period (τ) of the free-running rhythm does not correspond to any environmental factor, suggesting that social interactions alone are responsible for the synchronous group activity. In the laboratory, several studies have demonstrated that cohoused or group-housed animals—including killifish (*Fundulus heteroclitus*; Kavaliers 1980), deer mice (*Peromyscus maniculatus*; Crowley and Bovet 1980), palm squirrels (*Funambulus pennanti*; Rajaratnam and Redman 1999), and fruit flies (*Drosophila melanogaster*; Levine et al. 2002)—may display coherent or converging activity rhythms, perhaps of altered amplitude or waveform, under constant lighting conditions (for reviews, see Regal and Connolly 1980; Davidson and Menaker 2003; Mistlberger and Skene 2004). Although these studies have established that cohabitation can alter the daily activity pattern in several species, their reliance on group activity data has prohibited any further chronobiological or mechanistic analyses. The expression of a single free-running rhythm in group-housed individuals could be due to social entrainment (mutual or unidirectional), but it could also reflect similar phases and τ values before cohabitation or a masking effect of one, possibly socially dominant, individual. To distinguish between these alternatives, the activity of each individual must be identified before, during, and after cohabitation.

Various studies have used different strategies to distinguish individual activity under social conditions. First, investigators have housed animals in close proximity to, but not in direct contact with, each other. Under these housing conditions, social effects on the circadian system depend on the species but appear to be quite modest. When animals are housed in the same cage but separated by a barrier (e.g., wire mesh, clear perforated plastic, or a restraining cage), studies have reported alterations in activity levels, reentrainment rates to a photically induced phase-shift, or τ values in DD (e.g., in *Octodon degus*; Goel and Lee 1995, 1996, 1997), but clear examples of temporal synchronization have not been obtained. One study has demonstrated synchronous activity due to social cues in individuals separated by a barrier. Caged microchiropteran bats (*Hipposideros speoris*) housed in a cave without access to light synchronize their activity rhythms to local time, in-phase with their free-living conspecifics that leave the nest daily around sunset; a control bat caged in a cave devoid of other bats instead exhibited a free-running rhythm (Marimuthu et al. 1981).

Some studies have shown circadian effects in some individuals upon presentation of a putative social cue alone. Supplying female degus with an odor cue from entrained degus has been reported to "temporarily" entrain some of the animals and to accelerate reentrainment to a phase-advanced light/dark cycle (Governale and Lee 2001; Jechura et al. 2006). Some bird species can entrain to playbacks of conspecific birdsong presented every 24 hours (Gwinner 1966; Menaker and Eskin 1966), but house sparrows (*Passer domesticus*) can also entrain to periodic noise (Reebs 1989), so it is not clear if entrainment to birdsong playback is truly a social phenomenon or the result of nonspecific arousal.

Some animals may segregate, rather than synchronize, their activity times, perhaps to avoid competition for a particular temporal niche. For example, two species of closely related spiny mice (*Acomys cahirinus* and *Acomys russatus*) are able to cohabit in the same rocky terrain of the Dead Sea Valley by assuming different activity phases: *A. cahirinus* is nocturnal and *A. russatus* is diurnal (Shkolnick 1971). When *A. cahirinus* is removed from the field site, *A. russatus* becomes nocturnal, indicating that the presence of *A. cahirinus* prevents *A. russatus* from being active at its preferred phase of the daily cycle. When *A. cahirinus* is released back into the field site, both species reestablish their temporal segregation. Importantly, light does not regulate this phenomenon. Similar examples of competitively displaced temporal activity patterns have been described between and across other species under seminatural conditions (Glass and Slade 1980; Ziv et al. 1993). In the laboratory, *A. cahirinus* and *A. russatus* are strictly and primarily nocturnal, respectively (Weber and Hohn 2005; Cohen and Kronfeld-Schor 2006), and chemical signals from *A. cahirinus* have been shown to alter the phase angle of entrainment of *A. russatus*, suggesting that these cues contribute to their temporal partitioning reported in the field (Haim and Rozenfeld 1993; Fluxman and Haim 1993). There is also a report of laboratory group housing of long-tailed field mice (*Apodemus sylvaticus*), in which three subordinate mice were observed to avoid the time of peak out-of-burrow activity of a dominant mouse (Bovet 1972).

SEARCHING FOR TEMPORAL COUPLING BETWEEN GOLDEN HAMSTERS

Golden hamsters (*Mesocricetus auratus*) would seem to be well-suited as a laboratory species for investigating the circadian effects of cohabitation. They are widely available, exhibit precise locomotor activity (wheel-running) rhythms with sharp activity onsets, respond predictably to a range of nonphotic phase-shifting cues (Mrosovsky 1996), and have been experimental subjects for dissecting the neurobiological and hormonal substrates that underlie social interactions, especially agonistic (conflict) behaviors (Albers et al. 2002; Huhman 2006).

There is an inconsistent literature thus far on social cues and circadian rhythm resetting in hamsters. Closely housed or cohoused hamsters not in physical contact do not affect each other's rhythms (Davis et al. 1987; Gattermann and Weinandy 1997), although more rapid reentrainment to a phase-advanced light/dark cycle has been reported when males are paired with estrous females (Honrado and Mrosovsky 1989). Other studies have restricted the interval of social contact to a few hours per day, ensuring that the activity during the remaining hours is due solely to the experimental animal. This approach has led to conflicting results (Aschoff and von Goetz 1988; Mrosovsky 1988; Refinetti et al. 1992; Mistlberger et al. 2003), perhaps due to differences in experimental design (blind vs. intact animals, DD vs. constant light, interaction via neutral cage vs. resident-intruder paradigm). Refinetti et al. (1992) cohoused two hamsters in physical contact with each other but with markedly different τ values: One was entrained to a short 23.3-hour T cycle, whereas the other was blinded and free-ran with a τ value of about 24 hours. There was no synchronization or relative coordination of the activity of the blind hamster to that of the entrained one. The conclusion from all these studies has been that social factors provide only weak inputs to the hamster circadian clock (Davidson and Menaker 2003; Mistlberger and Skene 2004).

Even though social interactions may be weak, their effects on behavior might be strong under certain circumstances. If we consider hamsters as individual oscillators, and their social interactions as a form of weak coupling, then one would predict that strong effects might be precipitated when cohabitants are in direct physical contact for a relatively long period of time and their respective τ values are close to one another (Winfree 2001). To our knowledge, no reported experiment in hamsters has incorporated such a design, which should be more sensitive for inducing and detecting the possible effects of social interactions than the studies previously described. We have now performed such an experiment and report our initial observations here.

COHOUSING HAMSTERS LEADS TO A PERSISTENT τ CHANGE IN ONE HAMSTER OF A PAIR

Baseline wheel-running activity was recorded in adult male golden hamsters singly housed in DD for 3–7 weeks. Sixteen hamsters with dissimilar τ values were selected to yield eight cohabiting pairs. The mean ($\pm$S.E.M.) τ for the population as a whole was 24.12 ± 0.03 hours (range 23.95–24.32 hours), and the mean difference in τ between the two hamsters of each pair before cohousing was 0.18 ± 0.04 hours. To decrease the incidence of fighting at the time of pairing, hamsters were cohoused in a fresh cage with nesting material available. A cage lid equipped with a single running wheel was positioned on top of the cage, and behavioral observations were recorded for 30 minutes under dim red light at the time of pairing, with specific attention given to aggressive behaviors and postures. During cohabitation, hamsters were checked regularly for behavioral interactions and signs of physical discomfort (initially once a day during the first week, then about 3 times per week thereafter).

The number of wheel revolutions for each hamster or hamster pair was binned every 15 minutes and plotted at 24-hour intervals with successive days stacked vertically to form single-plotted actograms. τ was determined by drawing an eye-fitted regression line over 7–10 successive activity onsets immediately before cohabitation and after separation using the τ cursor function in the Actiview program (Mini-Mitter Co., Inc., Oregon). Cycle-to-cycle variability was determined by first computing the periods of ten successive cycles using the same τ cursor function; onsets occurring after cage changing were excluded from analyses. The standard deviation of these ten cycle lengths was then calculated to provide a measure of τ variability for each hamster. Mean daily activity was calculated by averaging the number of wheel revolutions occurring in a 24-hour interval over 10 successive days.

Wheel-running activities before pairing, during cohabitation, and after separation for two pairs of hamsters are illustrated in Figure 1. In both cases, an obvious and sudden change in τ of one of the cohabitants was observed. Hamster activities appeared to temporally "collide" when the locomotor offsets of nos. 7 and 34 intersected the locomotor onsets of nos. 8 and 20, respectively. The effect was a clear shortening of τ in nos. 7 and 34 (–0.21 and –0.19 hour, respectively), resulting in a temporal separation of the hamsters' wheel running in their cages in DD; the τ change persisted after the hamsters were separated and singly housed.

Figure 2 shows the τ values for hamsters of all eight pairs before cohabitation and after separation. The τ change associated with cohabitation was generally asymmetric, with one hamster of the pair exhibiting a larger change in τ than that of its cohabitant; and this new τ continued to be expressed by the affected hamster after the pair was separated (and stably maintained in the cases we followed for weeks to months). The mean τ change in relatively "affected" versus "unaffected" hamsters of the pairs was 0.19 ± 0.02 hour versus 0.05 ± 0.01 hour, respectively, and the difference between these two populations was significant (Student's *t*-test, $p < 0.05$). Social rank, as assessed by dominance status at pairing and again just before separation, could not be confidently determined for three of the pairs; for the remaining five pairs, the affected hamster was dominant in three and subordinate in two.

It is interesting to note that in seven of our eight pairs, the affected hamster's τ was shortened by cohabitation; in these cases, the affected hamster was the animal that had exhibited the longer τ of the pair before cohabitation. There was a trend toward increased cycle-to-cycle variability (i.e., decreased precision of the locomotor rhythm) in the precohabitation activity of the hamster of each pair whose τ would later be shortened by cohabitation (τ variability before cohabitation, 0.45 ± 0.09 hour in hamsters whose τ was subsequently "affected" by cohabitation vs. 0.24 ± 0.05 hour in hamsters whose τ was subsequently "unaffected" by cohabitation; $p = 0.051$, Student's *t*-test), and a positive correlation between the degree of this imprecision and the magnitude of τ shortening ($r^2 = 0.29$; $p = 0.03$). There was no difference in precision after separation (τ variability after cohabitation, 0.24 ± 0.05 hour in hamsters whose τ was previously "affected" by cohab-

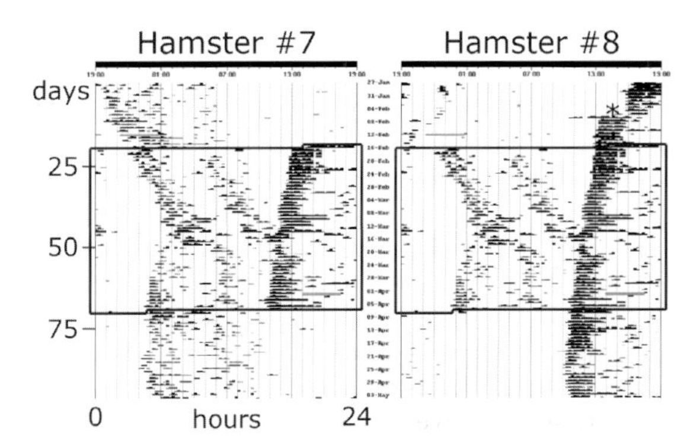

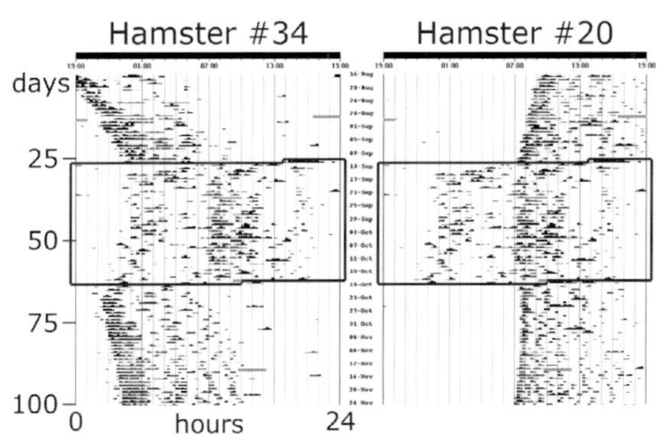

Figure 1. Wheel-running activity in DD of pairs of hamsters (*top*, no. 7 with no. 8; *bottom*, no. 34 with no. 20) before, during, and after cohabitation. Activity is illustrated as single-plotted actograms with consecutive 24-hour intervals drawn on successive lines. Activity enclosed within the boxes represents wheel-running behavior during cohabitation; note that this activity is identical for the hamsters within each pair and that individual activity cannot be conclusively determined. Asterisk (*) marks a large phase-shift associated with an anesthetization procedure on hamster no. 8 before cohabitation. (*Gray bars*) Intervals of lost computer function.

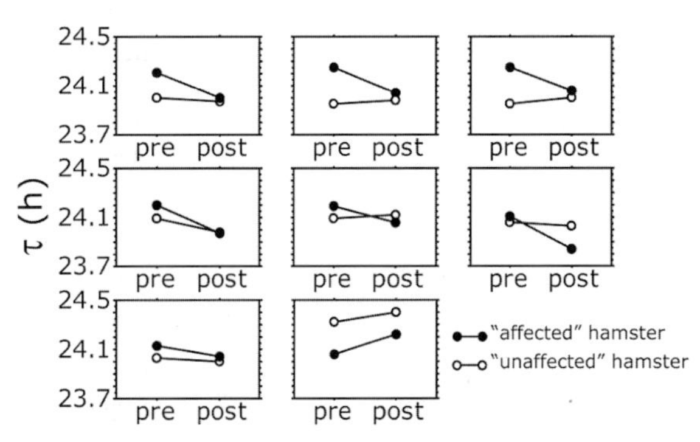

Figure 2. The value of τ before (pre) and after (post) cohabitation is shown in each box for the two hamsters of a pair, for a total of eight pairs.

itation vs. 0.25 ± 0.05 hour in hamsters whose τ was previously "unaffected" by cohabitation). Mean daily activity counts did not differ between affected and unaffected hamsters either before cohabitation or after separation (4513 ± 823 vs. 5059 ± 978 before and 3817 ± 507 vs. 3738 ± 613 after cohabitation, respectively, affected vs. unaffected hamsters; $p > 0.05$ for each comparison, Student's *t*-test).

We also have examined a few additional cases in which hamsters with similar τ values were paired at the same phase; under these circumstances, it is obvious that individual activity rhythms could not be discriminated during cohabitation (Fig. 3). For one such pair (Fig. 3, top, note that both cohabitants in this instance were females), no circadian effects were apparent, whereas for another pair (Fig. 3, bottom) the result was a gross, persistent shortening of τ in hamster no. 65. The hamsters of this particular pair were observed to be fighting 50 days after pairing, but they likely began fighting earlier than that because these hamsters, which initially shared the same nest, were seen on several occasions using separate nests beginning 28 days after pairing. Affected hamster no. 65 was the behaviorally dominant animal of this pair.

One other case of an initial pairing of hamsters with similar τ values at the same phase is shown in Figure 4. The τ of hamster no. 16 gradually shortened soon after pairing, after which the combined locomotor activities of hamsters nos. 16 and 5 were recorded for nearly 9 months. Visual inspection of the actogram shows that hamster no. 16 continued to be the affected hamster of the pair; the record hints that collisions of locomotor onsets and offsets might be a critical factor for inducing the change in τ, although in this case its direction was unpredictable—first shortened, then lengthened, and finally with no change at all (arrows, Fig. 4).

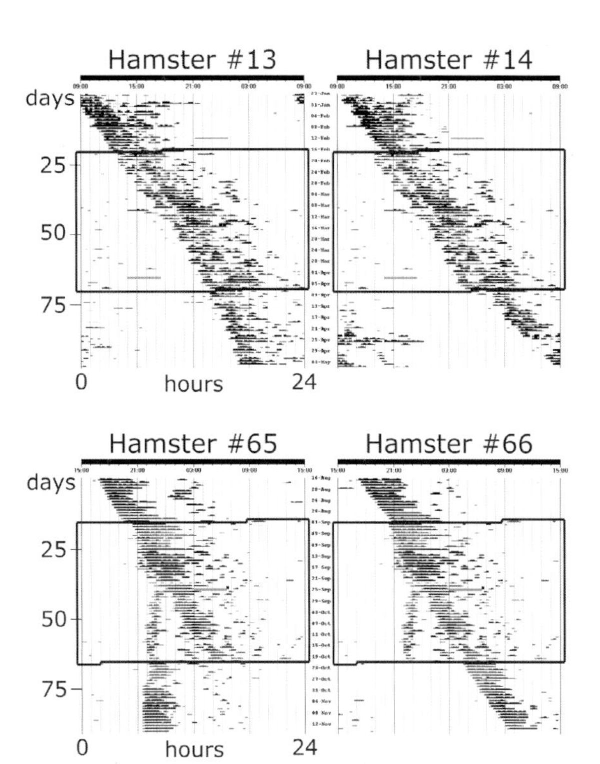

Figure 3. Wheel-running activity in DD of pairs of hamsters (*top*, no. 13 with no. 14; *bottom*, no. 65 with no. 66) before, during, and after cohabitation, illustrated as single-plotted actograms as in Fig. 1.

TOWARD A CHRONOBIOLOGY OF COHABITATION

Circadian rhythms are believed to be adaptive responses to the challenges of living on a rotating planet. The

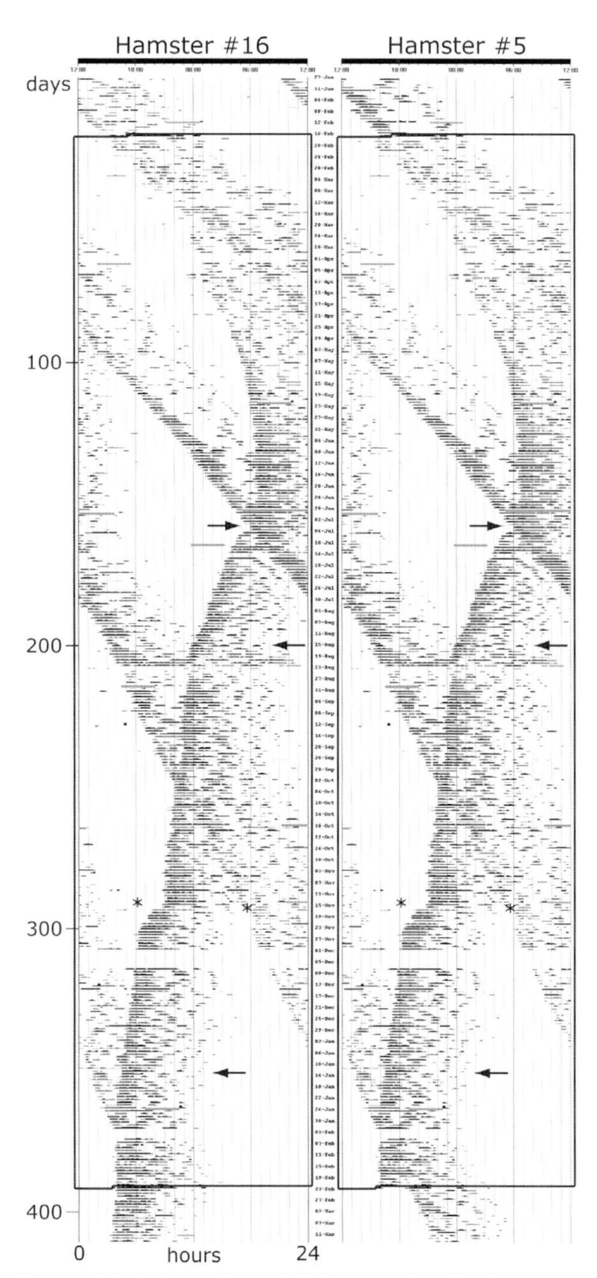

Figure 4. Wheel-running activity in DD of a pair of hamsters (no. 16 with no. 5) before, during, and after cohabitation, illustrated as single-plotted actograms as in Fig. 1. Arrows mark collisions of locomotor onsets and offsets between the hamsters. Asterisks (*) mark a phase-shift associated with food competition tests for dominance status.

generation of rhythmicity and its entrainment to the light/dark cycle have been the focus of much research, as represented by the reports in this volume, as well as by the work of this laboratory (W.J.S.) for the past few decades, from molecular to organismal levels. Relatively less attention has been directed to investigating the generation and entrainment of rhythmicity at the community level. The importance of social cues may be as a mechanism for adjusting the phase of individuals within a population to a shared photic environment. We have observed a persistent, cohabitation-associated, statistically significant shortening of τ in DD that might be functionally significant. If such a change were stable and affected individuals differentially, it could translate to an altered phase angle of entrainment to a light/dark cycle, with temporal partitioning of activity between cohabiting animals as a result.

One potential confound of our experiment is that the change in τ might reflect a seasonal change of behavior or physiology that accompanies the transfer of hamsters to short-day or DD conditions, rather than to cohabitation per se. For example, the effect might be secondary to the decreased testicular hormone secretion characteristic of the short-day condition. We believe that this is unlikely, however; castration lengthens, rather than shortens, τ in mice (Daan et al. 1975) and in fact does not seem to affect the value of τ in golden hamsters (Morin and Cummings 1981). Elliott and Tamarkin (1994) have shown no change in τ of hamsters maintained in DD for varying lengths of time (5 days, 7 weeks, 12 weeks, and 27 weeks).

It is noteworthy that the τ change predominantly affects one hamster of a pair and is sustained (for months in the few hamsters that we have followed long-term). Such a persistent behavioral change in response to a social encounter reminds us of the phenomenon of "conditioned defeat" in golden hamsters (for review, see Huhman 2006). After a hamster has been defeated in a single encounter with a more aggressive male, it remains submissive and defensive when tested by further agonistic encounters with smaller, nonaggressive opponents ("loser effect"), an outcome that can last for at least 1 month. We emphasize that in our experiments, the level of aggression generally appears to be modest, not leading to physical injury, and cage mates often huddle together in the same nest. It seems unlikely that our cohabitation effects are due to continuous or extreme arousal, and indeed, social defeat stress is not believed to perturb the period, phase, or amplitude of the central circadian pacemaker that regulates behavioral rhythms (Meerlo et al. 2002).

We do not understand why cohousing is associated with a preferential change in one of the two cohabitants' τ, although the affected individual's longer τ and less precise rhythm before cohabitation may be a clue. Our next step must include the capability to distinguish the activity of individual animals housed together, and implantable miniature temperature sensors are already commercially available for this purpose. We have also custom designed a "micrologger" that measures general locomotor activity and is small enough to implant in hamsters (manufactured by Sigma Delta Technologies, Perth, Australia). The logger senses movement using an accelerometer, collecting at two samples/second and integrating over 5 minutes using a microprocessor equipped with a 4-MHz clock. The 5-minute integration is then stored in a memory chip. The logger is powered by an attached battery and thus functions as a self-contained activity acquisition and storage unit enabling measurement in individual animals for up to 6 months, without interference from transmissions generated by nearby devices. The board is 1.5 x 1.7 cm and weighs approximately 3 grams; a Teflon tape wrap and custom casing fabricated from acrylonitrile butadiene styrene protect the device from damaging water vapor. The implanted device does not inhibit hamster wheel running (Fig. 5), encouraging us to begin to ask questions about individual hamster activities during cohabitation (e.g., is the change in τ dependent on the phase relationship of the cohabitants' rhythms, especially at the apparent temporal "collisions" seen in Figs. 1 and 4?). We also should be able to analyze the influence of the running wheel itself; we already know that several nonphotic zeitgebers are thought to rely on at least some level of locomotor activity (Mistlberger et al. 2003), and the effects of nonphotic zeitgebers are often correlated with the amount of wheel-running behavior these manipulations elicit (for review, see Mrosovsky 1996).

Ultimately, we would hope to implicate specific neurobiological circuits as mediators of our effects and at least one candidate system immediately comes to mind. In rodents, aggression, affiliative behavior, and social recognition are regulated, in part, by the peptide neurotransmitter vasopressin (for review, see Albers et al. 2002), and vasopressin V_{1a} receptor mRNA and radioligand binding have been mapped in the golden hamster and are abundantly present in the suprachiasmatic nucleus (Young et al. 2000), site of the central circadian pacemaker regulating locomotor rhythmicity.

CONCLUSIONS

As we begin this line of research, we have many more questions than conclusions. We do know that our expertise in circadian biology needs to be supplemented by

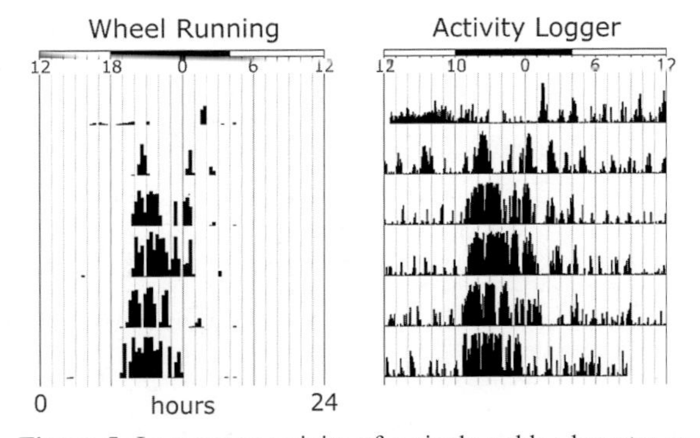

Figure 5. Locomotor activity of a single golden hamster as recorded by a running wheel (*left*) or subcutaneously implanted micrologger (*right*). Activity is illustrated as a single-plotted actogram. The hamster was maintained in a 14–10-hour light/dark cycle as indicated by the *white* and *black* bars at the top of the actogram. As expected, wheel running accounts for only a part of the animal's total 24-hour activity.

more knowledge of behavioral ecology and social biology. We also appreciate that our τ findings in hamsters may reflect an unnatural laboratory situation (e.g., being trapped in a relatively small enclosure with a cage mate) and might not be relevant for this species in its natural habitat. Right now, we are at the beginning, gathering data to rekindle interest in this problem, to encourage future experiments under more naturalistic conditions, and eventually to test hypotheses in field studies. We submit that analyzing this interorganismal and supraorganismal dimension of temporal organization, and ultimately dissecting its underlying mechanisms, will enrich our understanding of the circadian clock and the chronobiology of societies, perhaps providing insights into natural function, but certainly providing data on the social interactions of laboratory rodents that are routinely cohoused every day in our own biomedical facilities.

ACKNOWLEDGMENTS

We thank Dr. David Paydarfar for helpful discussions. This work was supported by NINDS F32 NS058135 (M.J.P.), and R01 NS46605 (W.J.S.). The contents of this report are solely the responsibility of the authors and do not necessarily represent the official views of the National Institutes of Health.

REFERENCES

Albers H.E., Huhman K.L., and Meisl R.L. 2002. Hormonal basis of social conflict and communication. In *Hormones, brain and behavior* (ed. D.W. Pfaff et al.), vol. 1, p. 393. Academic Press, San Diego, California.

Aschoff J. and von Goetz C. 1988. Masking of circadian activity rhythms in male golden hamsters by the presence of females. *Behav. Ecol. Sociobiol.* **22:** 409.

Bloch G. and Robinson G.E. 2001. Reversal of honeybee behavioural rhythms. *Nature* **410:** 1040.

Bovet J. 1972. On the social behavior in a stable group of long-tailed field mice (*Apodemus sylvaticus*). II. Its relations with distribution of daily activity. *Behaviour* **41:** 55.

Bovet J. and Oertli E.F. 1974. Free-running circadian activity rhythms in free living beaver (*Castor canadensis*). *J. Comp. Physiol.* **92:** 1.

Cohen R. and Kronfeld-Schor N. 2006. Individual variability and photic entrainment of circadian rhythms in golden spiny mice. *Physiol. Behav.* **87:** 563.

Crowley M. and Bovet J. 1980. Social synchronization of circadian rhythms in deer mice (*Peromyscus maniculatus*). *Behav. Ecol. Sociobiol.* **7**: 99.

Daan S., Damassa D., Pittendrigh C.S., and Smith E.R. 1975. An effect of castration and testosterone replacement on a circadian pacemaker in mice (*Mus musculus*). *Proc. Natl. Acad. Sci.* **72:** 3744.

Davidson A.J. and Menaker M. 2003. Birds of a feather clock together—Sometimes: Social synchronization of circadian rhythms. *Curr. Opin. Neurobiol.* **13:** 765.

Davis F.C., Stice S., and Menaker M. 1987. Activity and reproductive state in the hamster: Independent control by social stimuli and a circadian pacemaker. *Physiol. Behav.* **40:** 583.

Elliott J.A. and Tamarkin L. 1994. Complex circadian regulation of pineal melatonin and wheel-running in Syrian hamsters. *J. Comp. Physiol. A* **174:** 469.

Fluxman S. and Haim A. 1993. Daily rhythms of body temperature in *Acomys russatus:* The response to chemical signals released by *Acomys cahirinus*. *Chronobiol. Int.* **10:** 159.

Gattermann R. and Weinandy R. 1997. Lack of social entrainment of circadian activity rhythms in the solitary golden hamster and in the highly social Mongolian gerbil. *Biol. Rhythm Res.* **28:** 85.

Glass G.E. and Slade N.A. 1980. The effect of *Sigmodon hispidus* on spatial and temporal activity of *Microtus ochrogaster:* Evidence for competition. *Ecology* **61:** 358.

Goel N. and Lee T.M. 1995. Sex differences and effects of social cues on daily rhythms following phase advances in *Octodon degus*. *Physiol. Behav.* **58:** 205.

———. 1996. Relationship of circadian activity and social behaviors to reentrainment rates in diurnal *Octodon degus* (Rodentia). *Physiol. Behav.* **59:** 817.

———. 1997. Social cues modulate free-running circadian activity rhythms in the diurnal rodent, *Octodon degus*. *Am. J. Physiol.* **273:** R797.

Governale M.M. and Lee T.M. 2001. Olfactory cues accelerate reentrainment following phase shifts and entrain free-running rhythms in female *Octodon degus* (Rodentia). *J. Biol. Rhythms* **16:** 489.

Gwinner E. 1966. Periodicity of a circadian rhythm in birds by species-specific song cycles (Aves, Fringillidae: *Carduelis spinus, Serinus serinus*). *Experientia* **22:** 765.

Haim A. and Rozenfeld F.M. 1993. Temporal segregation in coexisting *Acomys* species: The role of odour. *Physiol. Behav.* **54:** 1159.

Honrado G.I. and Mrosovsky N. 1989. Arousal by sexual stimuli accelerates the re-entrainment of hamsters to phase advanced light-dark cycles. *Behav. Ecol. Sociobiol.* **25:** 57.

Huhman K.L. 2006. Social conflict models: Can they inform us about human psychopathology? *Horm. Behav.* **50:** 640.

Jechura T.J., Mahoney M.M., Stimpson C.D., and Lee T.M. 2006. Odor-specific effects on reentrainment following phase advances in the diurnal rodent, *Octodon degus*. *Am. J. Physiol.* **291:** R1808.

Kavaliers M. 1980. Social groupings and circadian activity of the killifish, *Fundulus heteroclitus*. *Biol. Bull.* **158:** 66.

Levine J.D., Funes P., Dowse H.B., and Hall J.C. 2002. Resetting the circadian clock by social experience in *Drosophila melanogaster*. *Science* **298:** 2010.

Marimuthu G., Rajan S., and Chandrashekaran M.K. 1981. Social entrainment of the circadian rhythm in the flight activity of the microchiropteran bat *Hipposideros speoris*. *Behav. Ecol. Sociobiol.* **8:** 147.

Meerlo P., Sgoifo A., and Turek F.W. 2002. The effects of social defeat and other stressors on the expression of circadian rhythms. *Stress* **5:** 15.

Menaker M. and Eskin A. 1966. Entrainment of circadian rhythms by sound in *Passer domesticus*. *Science* **154:** 1579.

Mistlberger R.E. and Skene D.J. 2004. Social influences on mammalian circadian rhythms: Animal and human studies. *Biol. Rev. Camb. Philos. Soc.* **79:** 533.

Mistlberger R.E., Antle M.C., Webb I.C., Jones M., Weinberg J., and Pollock M.S. 2003. Circadian clock resetting by arousal in Syrian hamsters: The role of stress and activity. *Am. J. Physiol.* **285:** R917.

Morin L.P. and Cummings L.A. 1981. Effect of surgical or photoperiodic castration, testosterone replacement or pinealectomy on male hamster running rhythmicity. *Physiol. Behav.* **26:** 825.

Mrosovsky N. 1988. Phase response curves for social entrainment. *J. Comp. Physiol. A* **162:** 35.

———. 1996. Locomotor activity and non-photic influences on circadian clocks. *Biol. Rev. Camb. Philos. Soc.* **71:** 343.

Potvin C.L. and Bovet J. 1975. Annual cycle of patterns of activity rhythms in beaver colonies (*Castor canadensis*). *J. Comp. Physiol.* **98:** 243.

Rajaratnam S.M. and Redman J.R. 1999. Social contact synchronizes free-running activity rhythms of diurnal palm squirrels. *Physiol. Behav.* **66:** 21.

Reebs S.G. 1989. Acoustical entrainment of circadian activity rhythms in house sparrows: Constant light is not necessary. *Ethology* **80:** 172.

Refinetti R., Nelson D.E., and Menaker M. 1992. Social stimuli fail to act as entraining agents of circadian rhythms in the golden hamster. *J. Comp. Physiol. A* **170:** 181.

Regal P.J. and Connolly M.S. 1980. Social influences on biological rhythms. *Behaviour* **72:** 171.
Shemesh Y., Cohen M., and Bloch G. 2007. Natural plasticity in circadian rhythms is mediated by reorganization in the molecular clockwork in honeybees. *FASEB J.* **21:** 2304.
Shkolnik A. 1971. Diurnal activity in a small desert rodent. *Int. J. Biometeorol.* **15:** 115.
Weber E.T. and Hohn V.M. 2005. Circadian activity rhythms in the spiny mouse, *Acomys cahirinus*. *Physiol. Behav.* **86:** 427.
Winfree A.T. 2001. *The Geometry of biological time,* 2nd edition. Springer-Verlag, New York.
Young L.J., Wang Z., Cooper T.T., and Albers H.E. 2000. Vasopressin (V_{1a}) receptor binding, mRNA expression and transcriptional regulation by androgen in the Syrian hamster brain. *J. Neuroendocrinol.* **12:** 1179.
Ziv Y., Abramsky Z., Kotler B.P., and Subach A. 1993. Interference competition and temporal and habitat partitioning in two gerbil species. *Oikos* **66:** 237.

Melatonin and Human Chronobiology

A.J. LEWY

Department of Psychiatry, Oregon Health & Science University, Portland, Oregon 97239-3098

With the development of accurate and sensitive assays for measuring melatonin in plasma and saliva, it has been possible to advance our understanding of human chronobiology. In particular, the dim light melatonin onset (DLMO) is expected to have an increasingly important role in the diagnosis of circadian phase disorders and their treatment with appropriately timed bright light exposure and/or low-dose melatonin administration. The phase angle difference (PAD) between DLMO and mid-sleep can be used as a marker for internal circadian alignment and may also be used to differentiate individuals who are phase advanced from those who are phase delayed (a long interval indicates the former and a short interval indicates the latter). To provide a corrective phase delay, light exposure should be scheduled in the evening and melatonin should be administered in the morning. To provide a corrective phase advance, light exposure should be scheduled in the morning and melatonin should be administered in the afternoon/evening. The study of patients with seasonal affective disorder (SAD), as well as individuals who are totally blind, has resulted in several findings of interest to basic scientists, as well as psychiatrists and sleep specialists.

INTRODUCTION: THE GCMS MELATONIN ASSAY

The longer duration of pineal melatonin production during winter nights compared to the summer is a time-of-year signal essential for the regulation of seasonal rhythms, which are ubiquitous, particularly in mammals, for example, seasonal reproductive cycles. Both long- and short-day breeders rely on melatonin to signal the "biological night." This function was elucidated by measuring melatonin production in extracts from whole pineal glands. Human studies, however, had to wait until plasma levels of melatonin could be reliably measured: The gas chromatographic–negative chemical ionization mass spectrometric (GCMS) assay using a deuterated internal standard achieved the requisite accuracy, sensitivity, and precision for measuring melatonin in humans (Lewy and Markey 1978). This assay was instrumental in settling the controversial issue of reports of extrapineal sources contributing to melatonin levels in the circulation (Ozaki and Lynch 1976): Exclusive derivation of melatonin from the pineal validated the use of circulating melatonin levels (Lewy et al. 1980a; Neuwelt and Lewy 1983) as an indicator of pineal melatonin production, of the timing of the endogenous circadian pacemaker (located in the suprachiasmatic nucleus of the hypothalamus), and of the effects of the light/dark cycle (mediated by the retinohypothalamic tract). This assay also directly and indirectly enabled the eventual development of widely used RIAs (radioimmunoassay) with sufficient specificity for measuring melatonin in human plasma and saliva.

DISCUSSION

Suppression of Nighttime Melatonin Production by Light

The GCMS assay was also used to show that nighttime human melatonin production could be suppressed by light, provided it was sufficiently intense; light of intermediate intensity produced an intermediate amount of suppression (i.e., the response was not binary) (Lewy et al. 1980b). Ordinary intensity light is usually insufficient in humans (Arendt 1978; Wetterberg 1978), although all other animals suppress with very little light (Illnerová et al. 1978; Illnerová 1979). We speculated that exposure to sunlight renders humans less sensitive to light (Lewy et al. 1980b). In fact, squirrels tested immediately after being caught in the wild require bright light for melatonin suppression (Reiter et al. 1981). Subsequent human studies have documented the relationship between history of prior exposure and sensitivity to light (Hebert et al. 2002; Smith et al. 2004). However, restricting ambient exposure to dim light for 1–2 days in the constant routine protocol can exaggerate melatonin suppression and circadian phase-shifting responses (Boivin et al. 1996; Zeitzer et al. 2000). Although all intensities should ideally be taken into account when relating circadian phase to the ambient light/dark cycle, intensities of 2,000–10,000 lux are standard for light treatment.

The Melatonin Suppression Test

The melatonin suppression test (MST) was first applied to demonstrate supersensitivity in patients who were actively manic or depressed (Lewy et al. 1981). Subsequently, 500 lux was shown to cause 50% suppression in healthy control subjects, whereas euthymic manic-depressive patients suppressed almost completely at this intensity (Lewy et al. 1985d), an intensity that is ideal for identifying supersensitivity. A version of the MST has been used in blind people with no conscious light perception: 10,000 lux causing 33% suppression has been proposed by some researchers as a way to inform ophthalmologists about the advisability of therapeutic enucleation (Czeisler et al. 1995), in that they interpret a positive test (i.e., suppression) to mean—even in cases of total and irreversible loss of vision and conscious light perception—that bilateral enucleation should not be done because the eyes should still be able to functionally mediate entrainment. However, several objections can be raised about this use of the MST. The arbitrary choices of the 10,000-lux light intensity and 33% suppression threshold do not take into account the fact that blind people are often exposed to sunlight as bright as 100,000 lux. Furthermore, because of the possibility of photoreceptors becoming up-regulated following blindness, people with no conscious light perception may still be capable of entraining to even low-intensity light. Another objection relates to safety: A blind person staring at a bright light fixture emitting 10,000 lux runs the risk of possibly damaging their few remaining photoreceptors, particularly if they have become up-regulated. Another objection is that some blind people with a positive MST nevertheless free-run (indeed, some sighted people free-run) (McArthur et al. 1996), whereas some blind people with a negative MST are naturally entrained to the 24-hour day (even some who are bilaterally enucleated; see below). A different type of MST that satisfies these objections must be configured before its use in blind people can be recommended.

Plasma Melatonin Profiles in Totally Blind People

The study of 24-hour melatonin rhythms in totally blind people and the exploration of what might be the consequences of light deprivation also resulted from the finding that light can suppress melatonin production in sighted humans. Previously, human chronobiology researchers had concluded that the light/dark cycle was relatively unimportant compared to social cues (Wever 1979). However, we described abnormally phased 24-hour plasma melatonin profiles in six of ten blind subjects studied in December, 1979 and January, 1980 (Lewy 1981). Fifteen months later, two of them were studied weekly on four occasions: The subject who was bilaterally enucleated appeared to be entrained but at a very abnormal phase, whereas the other subject was found to have a free-running melatonin rhythm with a circadian period (tau) of 24.7 hours (Lewy and Newsome 1983). Accordingly, we proposed that blind people can be categorized into three types: Those who are entrained at the normal phase, those who are entrained at an abnormal phase, and those who are free-running. Smith et al. (1981) reported that serum melatonin values in four blind people were greater at 1400 (86–142 pg/ml) than at 2300 (64–72 pg/ml). However, these data should be interpreted with care, because high daytime levels were routinely found with the nonspecific RIAs in use at the time (Smith et al. 1977). It should also be noted that Lynch et al. (1975) were the first to describe abnormal (urinary) melatonin levels in blind people (because of space constraints, historical notes will be limited in this volume).

Circadian Phase-shifting Effects of Light

Because sunlight is usually brighter than indoor light, two other implications of the melatonin suppression finding were that sighted people might have biological rhythms cued to the natural light/dark cycle (relatively unperturbed by ordinary-intensity room light) and that bright artificial light could be used to experimentally, and perhaps therapeutically, manipulate these rhythms. The first rhythms to be tested, however, were not thought to be circadian. Just after Kripke et al. (1983) began to treat major depressive disorder with morning light, based on his "critical interval" theory, we treated a patient with winter depression (Lewy et al. 1982), a disorder previously unknown to us and subsequently described as seasonal affective disorder, or SAD (Rosenthal et al. 1984). He responded after several days of receiving 2000 lux scheduled at 6–9 a.m. and at 4–7 p.m., so as to lengthen his perceived photoperiod. Subsequently, a group of patients were successfully treated under more controlled conditions (Rosenthal et al. 1984). SAD is discussed in detail later in this chapter.

Regarding circadian effects, Wever reported in 1983 that bright light scheduled throughout the photoperiod could increase the range of entrainment to a gradually lengthening T cycle (Wever et al. 1983). In the same year (Lewy et al. 1983), we proposed that bright light could be used according to our hypothesized human phase-response curve (PRC) to treat circadian phase disorders, such as delayed sleep phase syndrome (DSPS). At the time, DSPS was treated by scheduling sleep 3 hours later each day (termed "chronotherapy"; Czeisler et al. 1981), which was based on a two-pacemaker model (Kronauer et al. 1982). Their thinking was that there were two endogenous circadian pacemakers, one for the sleep/wake cycle (located in the suprachiasmatic nucleus) that could directly entrain a separate pacemaker for temperature (thought to be located elsewhere; Kronauer et al. 1982). This model explained internal desynchronization observed in temporal isolation, but so did the one-pacemaker model proposed by Eastman (1982). In any event, we began to treat patients with DSPS with morning bright light exposure (Lewy et al. 1983), an intervention that continues to be preferred over chronotherapy (Wright et al. 2006). Our hypothesized bright light PRC was also the basis for scheduling a sunlight exposure for two subjects who had flown across nine time zones (Daan and Lewy 1984). One subject avoided sunlight for the first 3 hours in the morning, obtaining it after 10 a.m., which according to our hypothesized light PRC would be the beginning of the advance zone before it began to adjust to the new time zone: The temperature rhythm quickly advanced nine hours. However, the subject who obtained sunlight exposure beginning at 7 a.m., thus stimulating the delay zone, shifted in that direction and did not adapt to the new time zone even after 2 weeks.

The Clock-Gate Model, the Light PRC, and the DLMO

In 1984 and 1985 (Lewy et al. 1984, 1985a), we published a test of our "clock-gate" model (Lewy 1983; Lewy et al. 1985b) for determining how light regulates the melatonin circadian rhythm in humans. Holding the sleep/wake cycle constant in four healthy control subjects also enabled the test of our proposed PRC to light (Lewy et al. 1983) and the first use of the DLMO. The DLMO is now acknowledged as the best marker for circadian phase position in humans, with a standard deviation that is less than half of the sampling interval (Klerman et al. 2002a).

Some researchers believe that the DLMO marks the phase of only one oscillator for melatonin production (Wehr et al. 1993; Parry et al. 1997; Benloucif et al. 2005), based on the elegant two-oscillator model proposed by Illnerová and Vanecek (1982). This model consists of an evening oscillator cued to dusk that controls the onset of melatonin production and a morning oscillator cued to dawn that controls the offset of melatonin production, which explains the finding in rats that a very short pulse of light in the middle of the night flattens melatonin levels for a few days. However, humans do not respond to light in this way (Vondrasova-Jelinkova et al. 1999).

The clock-gate model is the most parsimonious explanation for changes in melatonin duration measured under naturalistic ambient light conditions. Only the two-oscillator model can explain changes resulting from a light/dark cycle preceding measurement under dim light conditions. However, these changes (best assessed using the melatonin synthesis offset, or SynOff, rather than the DLMOff to mark the putative second oscillator) (Lewy et al. 1999) are small and can only be demonstrated using exotic light/dark cycles (Wehr et al. 1993). Therefore, if there are two oscillators for human melatonin production, they are tightly coupled under most circumstances.

The interventions required for minimizing masking effects when measuring phase markers can alter phase. This is one reason why we do not reduce light intensity below 10–30 lux, why we do not begin dim light before 5 p.m. (about 1 hour before the earliest expected DLMO), and why we obtain DLMOs no more frequently than once per week (Lewy et al. 2006a). Under these circumstances, if there is a same-day phase advance in the DLMO due to the dim light intervention, it is thought to be small. The phase-advancing effects of the constant routine have not been studied since an earlier report (Czeisler et al. 1985). The effects of several days of dim light, as well as other interventions that are part of the forced desynchrony (FD) protocol (Czeisler et al. 1999), could be even more profound in sighted individuals (see below).

The DLMO was used to compare the phase-shifting effects of morning versus evening light (Lewy et al. 1984, 1985a, 1987a). In our studies, light was scheduled no earlier than wake time and no later than bedtime, so as not to interfere with sleep. Subsequently, four complete PRCs testing exposure at other times of the day and night were published (Honma and Honma 1988; Czeisler et al. 1989; Wever 1989; Minors et al. 1991). The most detailed was also the most controversial (Czeisler et al. 1989), in that claims were made that two daily light pulses could suppress the amplitude of the endogenous circadian pacemaker. Suppression of circadian amplitude is at present not an area of much research or clinical interest, which is one reason why measurement of melatonin has replaced temperature as a research tool. The DLMO provides the most important information for assessing the circadian system.

Although the DLMO was originally measured in plasma, it is now more often measured in saliva

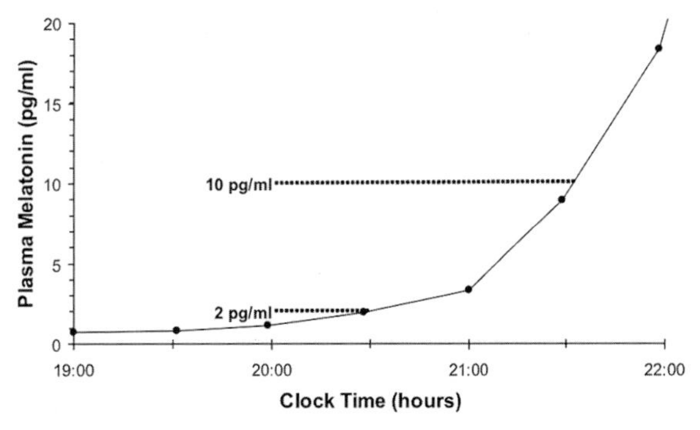

Figure 1. The dim light melatonin onset (DLMO) is measured by collecting plasma or saliva under dim light conditions usually every 30 minutes between 6 p.m. and bedtime. The 10 pg/ml plasma $DLMO_{10}$ is equivalent to the 3 pg/ml saliva $DLMO_3$ and indicates circadian time CT 14; the 2 pg/ml plasma $DLMO_2$, equivalent to the 0.7 pg/ml saliva $DLMO_{0.7}$, is on average about 1 hour earlier and indicates CT 13. Determining DLMO clock time and conversion to CT are important in optimizing treatment with phase-resetting agents (bright light and low-dose melatonin) (Fig. 2). $DLMO_{10}$ occurs on average about 6 hours before the time of mid sleep (Fig. 5); $DLMO_3$ occurs on average about 7 hours before the timing of mid sleep. (Reprinted, with permission, from Lewy et al. 2007a [©Les Laboratoires Servier].)

(Voultsios et al. 1997). Saliva levels are routinely about one-third those of plasma, when they are low, such as for the DLMO. The most common operational definition of the DLMO is the interpolated time when levels continuously rise above 10 pg/ml in plasma (3 pg/ml in saliva). In low secretors, we use the lower thresholds of 2 pg/ml plasma (0.7 pg/ml saliva), provided minimal detectable concentrations and basal levels of melatonin are low. The lower thresholds occur on average about 1 hour before the higher DLMO thresholds (Fig. 1). The 1-hour conversion is convenient when comparing DLMOs that have been calculated using these two thresholds. Different thresholds are sometimes used for the same person, for example, if assay parameters vary; however, they are more often of use when comparing the DLMO of one person to another. The above is important when calculating the time interval between the DLMO and mid-sleep, as well as for using the DLMO (or MO, in blind people) to mark circadian time (CT) for assessing the phase of the light and melatonin PRCs.

No matter what its clock time, the $DLMO_{10}$ is designated CT 14. This is because the $DLMO_{10}$ occurs on average about 14 hours after wake time in entrained sighted people. PRCs are often plotted with the abscissa given in CT. When one of the light PRCs (Czeisler et al. 1989) was published according to established conventions (Johnson 1990) (such as using the beginning of the light pulse as its phase reference point), the crossover times were at CT 6 and CT 18. In Figure 2, the crossover times are taken from this light PRC and the times for optimally and conveniently scheduling bright to cause phase advances and delays are taken from our work (Lewy et al. 1987a). Conversion of CT to clock times is also provided, based on an average wake time of 6 a.m. and an average $DLMO_{10}$ of 8 p.m. Optimal times to administer melatonin based on our melatonin PRC in

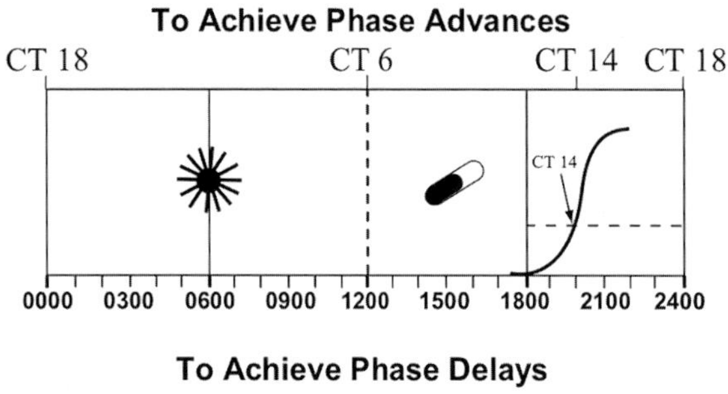

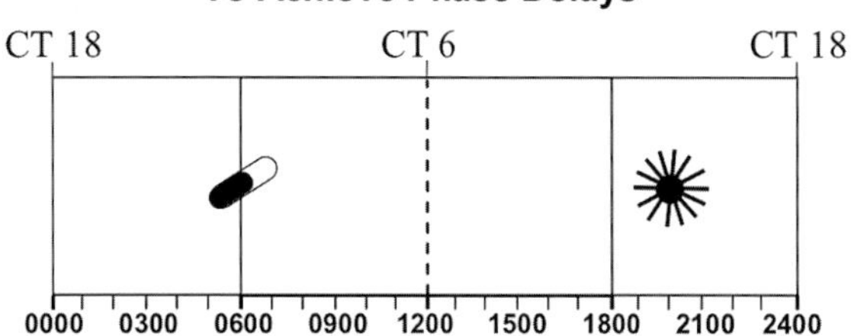

Figure 2. The phase-response curves (PRCs) for bright light and low-dose melatonin (see text) provide the best times to schedule these phase-resetting agents, according to circadian time (CT), which is optimally provided by the DLMO. The $DLMO_{10}$ indicates CT 14. Thus, the crossover times are 8 hours before and 4 hours after the $DLMO_{10}$. Also indicated are clock times typical for individuals who awaken at 6 a.m. (0600). More recent studies suggest that the crossover times between advance and delay zones for the light PRC might be occurring up to a few hours later (Khalsa et al. 2003; Revell and Eastman 2005). (Corrected and adapted, with permission, from Lewy et al. 2007a [©Les Laboratoires Servier].)

sighted people (discussed below) (Lewy et al. 1992, 1998a) are also provided in this figure. Wake time (CT 0) is a convenient way to estimate the phase of these PRCs in sighted, entrained people, although the $DLMO_{10}$ (CT 14) provides a more accurate estimate. Wake time is not used in blind people; we use the plasma MO_{10} (saliva MO_3) to indicate CT 14 in them. Wake time is also not useful if a (sighted) individual has had a very recent change in their sleep/wake cycle, such as occurs in shift workers. In people who have traveled across time zones, the pre-travel wake time can be used to estimate the phase of the PRCs on arrival at destination, which are expected to shift at least 1 hour per day until adaptation is complete (Lewy et al. 1995). With appropriately scheduled sunlight exposure and/or melatonin treatment, the rate of adjustment should be about 2–3 hours per day.

Fortunately, obtaining a DLMO does not interfere with normal sleep times, because it almost always occurs before habitual sleep onset, particularly the $DLMO_2$. Given the above, it is not surprising that the DLMO has become the circadian phase marker of choice in research studies. The only masking influence that needs to be controlled is light. Posture, which once was thought to be a potential confound by some investigators (Deacon and Arendt 1994), does not seem to have much effect on the DLMO (Cajochen et al. 2003).

Given its usefulness, reliability, simplicity, and convenience, there is little doubt that the DLMO will become a standard medical test, particularly after more researchers and clinicians gain experience with home saliva collections. Kept cool, samples can be express-mailed back to the laboratory; after centrifuging the absorbent material, the saliva is immediately frozen. In blind people, the MO can occur during the hours of sleep; however, sampling every 1–2 hours between wake time and bedtime for a "wake time circadian phase assessment" will result in an MO and/or an MOff, because the duration of active melatonin production is shorter than the duration of wakefulness. We can estimate the MO from the MOff, either because on another occasion we have determined either the MO/MOff or MOff/MO interval for that individual or by assuming these intervals to be about 10 or 14 hours, respectively, using the MO_3. Blind people have an easier time with collections, because the greatest challenge for sighted individuals is maintaining strict dim light conditions. However, amber goggles that filter out blue light may prove to be helpful. Incidentally, the advantages of blue or blue-enhanced white light for causing phase shifts versus the risk of the "blue light hazard" will not be discussed further here. No doubt these topics, as well as that of circadian photoreception, will be considered in detail elsewhere in this volume.

Entrainment of Blind People to Melatonin and the Melatonin Dose-response Curve

Before the landmark study of Redman, Armstrong, and Ng (1983), in which daily melatonin injections were shown to be able to entrain free-running rats, there was not much interest in the circadian effects of melatonin in mammals, despite the fact that in other animals (birds and reptiles), melatonin has profound circadian effects (Underwood 1986; Cassone 1990). In mammals, melatonin was thought to mediate primarily, if not exclusively, biological rhythms of the seasonal type (for review, see Arendt 1995). Humans, as well as some strains of rats, may be exceptional, in that melatonin may have more of a circadian effect than a seasonal one. In fact, it may be disadvantageous for a mammal to use melatonin for both circadian and seasonal timekeeping. Indeed, hamsters are more sensitive to the circadian phase-shifting effects of melatonin injections when they are perinatal and sexually incompetent (Davis and Mannion 1988).

The landmark study inspired Robert Sack and I to first investigate the effects of melatonin in BFRs (blind free-runners) (Sack et al. 1987). The same study may have also influenced the first two investigations of the phase-shifting effects of melatonin in sighted people (Arendt et al. 1985; Mallo et al. 1988; for review, see Lewy and Sack 1997). Although we published a case report showing entrainment of a blind person to melatonin in the early 1990s (Sack et al. 1990, 1991), it was not until the end of that decade that unequivocal entrainment (to 10 mg) could be demonstrated in a group of BFRs (Sack et al. 2000). In another study of the same number of subjects (Lockley et al. 2000), barely half entrained to 5 mg, leading the authors to conclude that starting the first bedtime dose as it happened to coincide with the advance zone of the melatonin PRC was critical for entrainment. However, extensive animal literature indicates that entrainment is as likely to (eventually) occur if the phase-resetting agent is started on the delay zone, which is what we have found in

BFRs (Emens et al. 2003; Lewy et al. 2004a). When the dose was switched from 5 to 0.5 mg, the advance versus delay zone difference was less (Hack et al. 2003) and the authors continued to recommend against initiating treatment on the delay zone. Recently, however, they have indicated agreement with our thinking about this issue, at least with respect to low doses (Lockley et al. 2007).

Too high a dose may make entrainment less likely if there is too much spillover of melatonin levels onto the wrong zone of the melatonin PRC. When given on the advance zone, an advance will always occur, but its magnitude will be reduced due to spillover onto the delay zone. Therefore, some low doses are capable of entraining a BFR than some higher doses, particularly if the BFR has a long tau and needs a large daily phase advance in order to entrain (Lewy et al. 2002). Melatonin administration was originally given at bedtime (Lockley et al. 2000; Sack et al. 2000). However, this results in an entrained MO at a delayed phase (Lewy et al. 2001). Although entrained, they will complain of difficulty falling asleep and getting up in the morning, even though the high doses used (5–10 mg) would be expected to have soporific side effects. Hence, we now give low doses of melatonin to most BFRs around 6 p.m., which results in an entrained MO at 8 or 9 p.m. (or about 2–3 hours before desired sleep onset). Low doses (≤0.3–0.5 mg) will not cause unpleasant soporific effects between 6 p.m. and bedtime.

In any event, treatment of both blind and sighted people is based on the melatonin PRC. When melatonin is given between CT 6 and CT 18, it causes phase advances; when it is given between CT 18 and CT 6, it causes phase delays. There was some skepticism about the delay zone in our first report (Lewy et al. 1992), despite independent confirmation (Zaidan et al. 1994; Middleton et al. 1997), leading some researchers to suggest that melatonin could only cause phase advances (Wirz-Justice et al. 2002) and that melatonin's main circadian effect in humans was to shorten tau (Arendt et al. 1997; Czeisler 1997). In a second study, we gave melatonin at more times during the night, leaving no doubt about the existence of a robust delay zone (Lewy et al. 1998a). Nevertheless, the tau-shortening effect of melatonin on human circadian rhythms remains an idea with some currency (Arendt 2006).

Once entrainment occurs, the time interval between the melatonin dose and the MO at steady state is the phase angle of entrainment (PAE), which correlates with tau: The longer the tau, the greater the PAE (Fig. 3) (Lewy et al. 2001). Given the relationship between tau and PAE, we have recommended that PAE be used to estimate tau (Lewy et al. 2003). However, in BFRs, measurement of tau in the field is relatively easy to do and is preferred. In sighted people, some investigators use the DLMO/sleep onset interval (melatonin/sleep interval, or MSI) to estimate the PAE in sighted people (Wright et al. 2005). We prefer to use the sleep offset/DLMO interval, which we have termed the ZT (zeitgeber time) of the DLMO. However, although DLMO ZT adheres to the traditional way of calculating PAE by using "lights on" for marking ZT 0, humans are exposed to a complex light-intensity contour that varies throughout the day. We will continue to use wake time as the phase reference for the light/dark

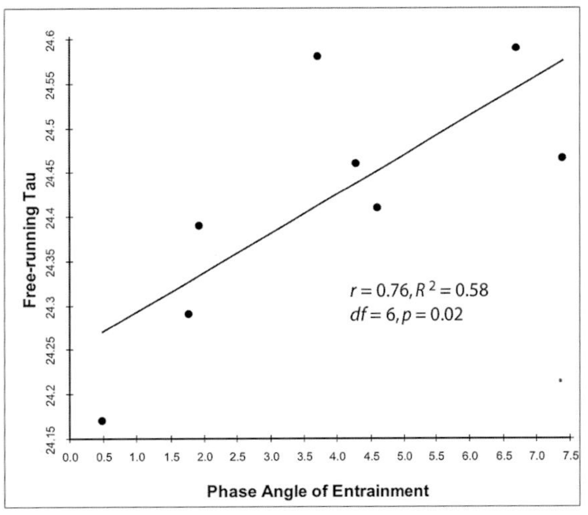

Figure 3. The phase-angle difference between the time of the entraining 10 mg of melatonin dose at bedtime and the time of the entrained melatonin onset (MO) occurring 0.5–7.4 hours later results in phase angles of entrainment (PAEs) depending on the blind free-runner's (BFR's) pretreatment circadian period (tau). We have proposed (Lewy et al. 2003) that PAE can used to estimate tau in BFRs. Once certain issues are resolved (see text), such as aftereffects in sighted people entrained to the 24-hour light/dark cycle (Scheer et al. 2007), PAE may be useful in estimating tau in sighted people. (Data, used with permission, from Lewy et al. 2001 [©Elsevier].)

zeitgeber, until a better way of calculating PAE in sighted people is found.

As mentioned above, for BFRs with taus greater than 24 hours, taking melatonin at about 6 p.m. results in an entrained MO at about 8–9 p.m. If a BFR with a tau less than 24 hours takes melatonin at 6 p.m., they will likely have an entrained MO at about 8–9 a.m.; taking melatonin as late as bedtime results in an entrained MO occurring no later than the afternoon. The correct clock time for administering melatonin to these individuals is wake time, which will result in an entrained MO at about 8–9 p.m. (Lewy et al. 2004b; Emens et al. 2006).

The above examples illustrate the importance of the melatonin PRC in the treatment of BFRs. The melatonin PRC is also important for treating the sighted: Entrained people should take melatonin about 5–6 hours before their $DLMO_{10}$ (saliva $DLMO_3$) for optimally causing a phase advance. On average, in people who awaken at 6 a.m., this would be an administration clock time of about 2–3 p.m. However, administration later in the afternoon will also cause robust phase advances.

To achieve phase delays, melatonin is not given before wake time, so as to not interfere with sleep. However, an individual who awakens spontaneously in the middle of the night can take melatonin to cause a phase delay if it is after about CT 19 (1 a.m.). A sophisticated understanding of the melatonin PRC will be important when crafting melatonin treatment regimens and formulations.

In addition to the human melatonin PRC, treatment recommendations are expected to be guided by the human melatonin dose-response curve (Fig. 4) (Lewy et al.

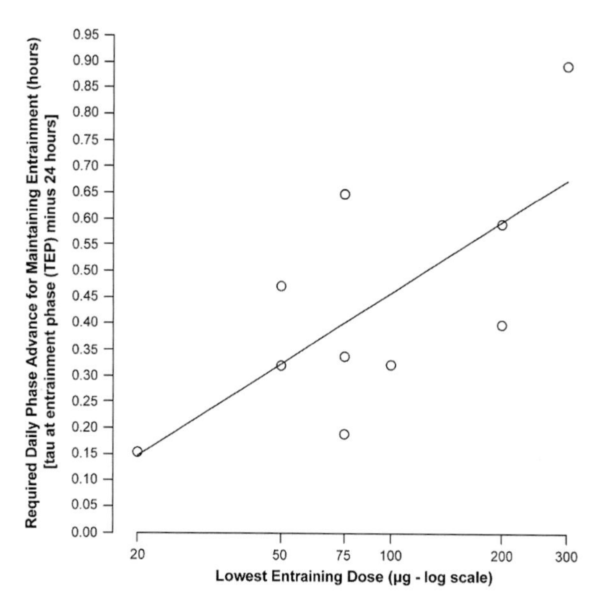

Figure 4. Phase-advance dose-response curve for melatonin in humans. The lowest dose found to entrain ten blind free-runners (BFRs) is plotted on the abscissa. The daily phase advance required for entrainment is plotted on the ordinate and is calculated for each BFR by subtracting 24 hours from tau at entrainment phase (TEP) (see text). In four BFRs, TEP was less than 24.40 hours. Their average taus were, in increasing order of TEP, 24.20, 24.30, 24.34, and 24.33 hours. All BFRs were given melatonin in the early evening and entrained at the normal phase, which is further evidence that all BFRs had taus greater than 24.00 hours (see text). (Reprinted, with permission, from Lewy et al. 2005 [©Taylor & Francis].)

2005). We constructed this curve by determining the lowest entraining dose in a group of BFRs with different taus. Circadian status (entrained vs. free-running) is dichotomous and thus enables calculation of very small phase shifts: A BFR with a tau of 24.1 hours will entrain to a melatonin dose that causes a daily phase advance of 6 minutes. The curve is log-linear, at least within the range of 20 and 300 µg (which produced a 1-hour phase shift). In this study, care was taken to avoid all possible confounds, in particular, the potential influence of nonphotic or weak zeitgebers. Thus, melatonin was administered at a time when these were minimal (about 6 p.m.), and the tau at entrainment phase (TEP, see below) was used instead of average tau to estimate the daily phase-advance necessary for entrainment (TEP-24 hours).

For the melatonin dose-response curve study, we assumed that there would not be much melatonin receptor desensitization secondary to exposure to melatonin, which has been reported in rats (Gerdin et al. 2004). This assumption was based on the fact that we could step the dose of melatonin down to at least 0.5 mg without loss of entrainment in BFRs who were initially entrained to 10 mg (Sack et al. 2000). We also assumed that what we originally thought (Sack et al. 2000) was evidence for aftereffects (a somewhat persistent shortening of tau following release into a free-run) is actually evidence for relative coordination (RC) to weak (probably nonphotic) environmental zeitgebers (see below).

Relative Coordination to Weak Zeitgebers, the Tau Response Curve, and Tau at Entrainment Phase

In our first five BFRs assessed in a high-resolution protocol (MOs obtained as frequently as every 2 weeks throughout a complete circadian beat cycle [CBC], in which the MO traverses 24 hours), we observed a pattern typical of RC when the MOs were plotted longitudinally against clock time (Emens et al. 2005). Similar to free-running animals exposed to a daily zeitgeber too weak to effect entrainment, tau is longer than average in one half of the CBC and shorter than average in the other half. To precisely describe RC, we calculate the slope between each overlapping pair of successive MOs throughout one or more complete CBCs. The "two-point tau" is calculated as the slope plus 24 hours. The tau response curve (tauRC) is a plot of two-point taus against the midpoint between the clock times for the pair of MOs. For example, if the MO is at 8 a.m. on day 1 and at 10 a.m. on day 5, the tau between days 1 and 5 would be 24.5 hours. As the MO drifts later each day between about 8 a.m. and about 8 p.m., the two-point taus are longer than the average tau for each individual (the long-tau zone of the tauRC); as the MO drifts between 8 p.m. and 8 a.m., it is shorter than average (the short-tau zone of the tauRC), i.e., the crossover times of the tauRC are on average at about 8 a.m. and 8 p.m. Because the latter coincides with the average time of the DLMO in entrained sighted people, it appears that the weak zeitgebers are working to effect entrainment at the normal phase or a few hours later. BFRs with taus less than 24 hours have the same tauRCs; for them, the weak zeitgebers are perversely working to effect entrainment at an undesirable phase. Fortunately, there are relatively few such individuals (and they can be successfully treated with low-dose melatonin taken at wake time). Astronauts adapting to the Martian day of 24.65 hours are another group of individuals that might benefit from a small daily dose of melatonin taken at wake time (Lewy et al. 2004b).

In the dose-response study described above, a tauRC was calculated for each BFR, all of whom had taus greater than 24 hours and were administered melatonin about 6 p.m. in order to entrain them at the normal phase. The tauRC was used to calculate TEP. TEP is the two-point tau of an individual at the clock time of their now-entrained MO. This takes into account advances or delays from the weak zeitgebers that would otherwise confound attributing the daily advance necessary for entrainment exclusively to the melatonin dose.

Average Tau and Human Tau Phenotypes

TEP, not average tau, was used to calculate the daily phase advance required for entrainment in the dose-response study, although when the MO is about 8 p.m., it is not much different from average tau. The standard method of calculating average tau is by linear regression through a longitudinal plot of the MOs. There is another way to calculate tau in BFRs, first reported by the Czeisler group, although there are errors in their algebraic equations (Klein et al. 1993). In any event, average tau can also be calculated by dividing 24 hours by the number of days of a complete CBC and then adding (for BFRs with taus greater than 24 hours) 24 hours. For example, if the MO is occurring at 2 p.m. on day 1 and then next occurs at 2 p.m. on day 48, average tau is 24.5 hours. Interestingly, these two ways of calculating average tau produce nearly identical results and may suggest that RC does not affect the average tau of an individual, another reason why the average tau in a blind person may be the most accurate estimate of intrinsic tau. The CBC method is most accurate when the beat cycle begins and ends (at the same time) in the long-tau zone. However, if any two-point taus overlap 24 hours (i.e., for most BFRs, become <24 hours), a confound will be introduced when calculating average tau (using any method). In most blind people, this "transient entrainment" would shorten tau, as would the effects (if any) of a 24-hour forcing period. Thus, the average tau in blind people of about 24.5 hours may be slightly less than the average of their "true" intrinsic taus, another reason why their taus compared to the shorter taus of sighted people (studied under FD) is such an important scientific issue (see below). Therefore, average tau should be corrected in the few BFRs who have transient entrainment (which can be detected with frequent phase assessments in the short-tau zone of the tauRC).

Aftereffects in sighted people resulting from entrainment to the 24-hour light/dark cycle (Scheer et al. 2007) have been offered as a possible explanation for their shorter average tau (most recently calculated to be 24.07 hours in 12 subjects studied in an FD T cycle of 28 hours under 1.5 lux during wakefulness; Gronfier et al. 2007) compared to blind people. However, aftereffects cannot explain why the range in taus in the sighted now appears to be the same as in the blind (about 1 hour) and why the proportion of people with taus less than 24 hours is now one-third (Gronfier et al. 2007), which is much lower in the blind (less than about 10%), in whom average tau is about 24.5 hours (Kendall and Sack 2000; Lockley et al. 1997; Sack et al. 1992b); however, definite statements require more sophisticated models about how exactly aftereffects are generated. Other explanations for the differences in tau between blind and sighted people have been sampling bias in the blind (excluding taus that are too close to 24 hours to be designated as free-running) or exposure of blind people to nonphotic time cues in field studies; however, sampling bias is probably not a sufficient explanation and the other explanation is unlikely (Kendall and Sack 2000; Emens et al. 2005).

The differences in average tau and the percentage of taus less than 24 hours between the blind and the sighted can be rectified by another explanation: Either ambient light of 1.5 lux FD shortens tau by up to 0.4 hour on average and/or becoming blind lengthens tau by up to 0.4 hour on average. In a thought experiment, we subtracted 0.4 hour from TEP (and from average tau) that we had calculated in our dose-response study (see Fig. 4) (Lewy et al. 2005). Even subtracting 0.2 hour produces a result contrary to what we found: Determining the baseline taus of these BFRs to be greater than 24 hours must have been correct, or they would not have entrained their MOs at the normal phase when taking exogenous melatonin in the early evening (BFRs with taus less than 24 hours entrain to an evening dose of melatonin with their MOs occurring in the afternoon). Therefore, it appears to be prudent to conclude that FD in dim light confounds the accurate measure of tau with an average shortening of 0.4 hour. This confound is probably consistent, given that these taus are apparently useful in correlational analyses (Wright et al. 2005; Gronfier et al. 2007); however, consistency remains to be documented. In any event, the study of tau in sighted people is problematic, given the recent reports of aftereffects (Scheer et al. 2007), which curiously appear to be different from those found in animals, in that the latter exhibit a gradual return to baseline tau which does not appear to be the case over several weeks of follow-up in humans. Studies (Wever 1979) using temporal isolation have been criticized (Czeisler et al. 1999) for introducing an artifact of lengthening tau of at least 0.4 hour. Now it appears that FD in dim light may be introducing a similar artifact in the other direction, perhaps the lower the light intensity, the greater the artifact (possibly a parametric effect of light). However, as light intensity increases above a certain threshold, it will tend to alter tau in the direction of the T cycle (i.e., $T = 28$ hours lengthens tau if light intensity is above a certain threshold).

The above discussion may help develop more accurate tau phenotyping in sighted people, which will be necessary for identifying the human clock genes that may be responsible for circadian phase disorders (for review, see Lamont et al. 2007). Similar to melatonin entrainment of blind people (see Fig. 3) (Lewy et al. 2001), PAE in sighted people entrained to the light/dark cycle was subsequently reported to correlate with tau under FD (Wright et al. 2005). When the issues discussed above, including those of aftereffects due to entrainment to the 24-hour light/dark cycle (as well as the best way to calculate PAE in sighted people), are settled, tau phenotyping can proceed apace and help in the identification of human clock genes and their disorders. Until then, molecular biologists can focus on those people with extremely short and extremely long taus, in whom the "true" intrinsic tau is less important. In the meantime, the measurement of tau in totally blind humans is closest to the established convention in which animals are studied under conditions of constant darkness. Perhaps the dichotomous difference between the very few BFRs with taus less than 24 hours and the vast majority who have taus greater than 24 hours will be particularly important in identifying the human clock genes.

It has also been suggested that altered innervation of the SCN from the retinohypothalamic tract might lengthen tau (Klerman 2001). However, animal studies do not support this (Bobbert and Riethoven 1991; Mistlberger 1991; Yamazaki et al. 2002). Furthermore, a change in tau correlating with the duration of blindness has not been observed cross-sectionally in either our data or those of others (Lockley et al. 1997). Whatever are the reason(s) for the discrepancy in average taus between sighted people studied under FD and blind people, their elucidation may lead to profoundly important advances in our understanding of human chronobiology. In any event, knowledge of a BFR's tau—at least whether it is less than or greater than 24 hours (perhaps its most intrinsic characteristic)—is critical when treating BFRs: For entrainment at the normal phase, melatonin should be administered in the early evening to most blind people; however, if a blind person's tau is less than 24 hours, even just slightly less (i.e., 23.94 hours), tau must be administered in the morning for normal entrainment, which most conveniently would be wake time.

Women May Be More Sensitive to Social Cues than Men

Average tau is not needed for calculating the range of oscillation, or RC amplitude, which we define as the shortest two-point tau subtracted from the longest two-point tau in a BFR's tauRC. RC amplitude varies greatly among blind people. Studies of this variability may help identify the as yet unknown weak zeitgebers responsible for the daily phase advances that appear to shorten observed two-point taus and daily phase delays that appear to lengthen them (we do not believe that intrinsic tau is actually changing). In fact, RC amplitude may be a "bioassay" for sensitivity to these weak zeitgebers, particularly if they have consistent strength and timing.

Women appear to be more sensitive to the weak zeitgebers than men: Free-running blind women have twice the RC amplitude than men; comparing circadian status, using more rigorous definitions of entrainment and free-running than was customary (Sack et al. 1992b; Lockley et al. 1997; Klerman et al. 2002b; Lewy et al. 2003) before our report of RC in BFRs (Emens et al. 2005), we found no totally blind men who are naturally entrained, compared to 25% of women (Lewy et al. 2007c). The difference in circadian status proportion occurs after puberty; in fact, the proportion of naturally entrained prepubertal boys appears to be greater than that of girls. Circadian status may turn out to be another bioassay (in addition to RC amplitude) for social sensitivity, given that putative candidates for these zeitgebers are probably dependent in some way on social cues. These include sleep, activity, exercise, meal times, temperature, acoustic signals, and pheromones (Goel et al. 1998). Because we believe that the overwhelming majority of humans have taus greater than 24 hours, we hypothesized that sighted women would entrain at an earlier PAE. A search of the literature with this hypothesis in mind was fruitful (Campbell et al. 1989; Mongrain et al. 2004), as well as analysis of one of our own data sets (Emens et al. 2007). Interestingly, RC amplitude for our group of blind people (up to 56 minutes) appears to be greater than what has been reported (Scheer et al. 2007) for the magnitude of experimentally induced aftereffects on tau in sighted people (up to 10 minutes).

Circadian Disorders in Sighted People

Knowledge of a sighted person's tau is not needed for treating circadian disorders, which are mostly, if not exclusively, disorders of circadian phase position. In some cases, such as in ASPS and DSPS, the timing of sleep is sufficient for diagnosis and treatment. The first patient with DSPS treated with morning light was reported in 1983 (Lewy et al. 1983), and the first patient with ASPS treated with evening light was anecdotally reported in 1985 (Lewy et al. 1985b). Melatonin was first used to treat DSPS in 1990 (Dahlitz et al. 1991), although not at the optimal time to cause phase advances, according to the melatonin PRC (Lewy et al. 1998a). Aaron Lerner may be the first person reported to have been successfully treated for ASPS with the neurohormone he discovered (0.3 mg of melatonin at wake time) (Lewy 2007). Other circadian phase disorders amenable to melatonin (and light) treatment include jet lag (Arendt et al. 1987; Lewy et al. 1995; Burgess et al. 2003) and maladaptation to shift work (Sack et al. 1992a). Treatment has not progressed rapidly in the area of shift work, due to many factors, such as rapidly changing schedules and the fact that DLMO time cannot even be grossly estimated from sleep times in these individuals.

The Phase-shift Hypothesis

Perhaps the most complex circadian disorders are psychiatric. The precise mechanism for the circadian component of psychiatric disorders, as well as of poorly maintained and nonrestorative sleep, is likely based on internal circadian misalignment between the sleep/wake cycle (and rhythms related to it) and those circadian rhythms that are more tightly coupled to the endogenous pacemaker (Kripke et al. 1978; Wehr et al. 1979). Although melatonin is probably not mediating pathology, its circadian rhythm is the most useful for marking the phase of the endogenous circadian pacemaker that regulates circadian rhythms which may indeed be pathogenic if misaligned with the sleep/wake cycle. SAD, because of the predictable winter depressions and the antidepressant response to bright light, is an ideal model for understanding the role of circadian rhythms in psychiatric disorders, as are totally blind people for studying another unfortunate situation (indeed, experiment of nature) involving light deprivation.

According to our PSH, most patients with SAD become depressed in the winter at least in part because of a phase delay with respect to the sleep/wake cycle (Lewy et al. 1985c, 1987a,b, 1989) and that bright light should optimally be scheduled in the morning in order to provide a corrective phase advance. Following our publication that morning light is more antidepressant than evening light (Lewy et al. 1987a), a vigorous debate ensued with some (Avery et al. 1990) but not all (Wirz-Justice et al. 1993) investigators supporting the PSH. Superiority of morning light was finally resolved in 1998 (Eastman et al. 1998; Lewy et al. 1998b; Terman et al. 1998). This still did not establish the PSH, because it could be argued that there is an overall increase in light sensitivity in the morning. Even correlational studies between circadian phase and treatment response (for review, see Terman et al. 2001; Lewy et al. 2006b), although important, did not establish causality. This was done recently using melatonin, rather than light, to treat SAD patients (Lewy et al. 2006b). Our study appears to be the first report of psychiatric symptom severity correlating with a biological marker before and in the course of treatment in the same patients.

Unlike light, low-dose (≤0.3 mg) melatonin cannot be detected by research subjects and has less of a placebo component. Furthermore, it would not be expected to be antidepressant in SAD unless used at the opposite time of the day (afternoon/evening) to cause phase advances similar to those caused by morning light. An early corollary of the PSH is that a smaller subgroup of SAD patients phase advance in the winter and require bright light in the evening that would provide a corrective phase delay; therefore, the correct time of administration is in the afternoon/evening for the prototypical phase-delayed patient and in the morning for the phase-advanced types. The phase angle difference (PAD) between the DLMO and mid-sleep assesses internal circadian alignment (Fig. 5). Average PAD in healthy subjects, using the $DLMO_{10}$, is 6 hours (Lewy et al. 1998b). Accordingly, we hypothesized that depression ratings would be lowest in patients with PAD 6, which turned out to be the vertex of parabolic curves that significantly correlated depressive symptom scores plotted against PAD before and after treatment (subjects were randomly assigned to placebo, morning melatonin treatment or afternoon/melatonin treatment). Subjects with PAD ≤6 at baseline were designated as phase-delayed types, i.e., their DLMOs were delayed with respect to the sleep/wake cycle. Subjects with PAD >6 were designated as phase-advanced types. More than two decades after proposing that subjects be phase typed before treating them with a phase-resetting agent (Lewy et al. 1985c), PAD now provides a heuristically useful and operationally precise way of phase typing.

Despite the fact that all subjects would be expected to improve slightly as the days lengthened during January and February, treatment at the wrong time made some patients worse. Phase-delayed subjects treated with afternoon/evening melatonin who shifted across the parabolic minimum did not do as well as those who shifted closer to PAD 6 and phase-advanced subjects did even less well. Among the prototypical (phase-delayed) subjects given the treatment of choice (afternoon/evening melatonin to cause a phase advance), 65% of the variance in depression ratings was explained by PAD. Furthermore, pretreatment versus posttreatment change toward PAD 6 also correlated with improvement. Thus, the "sweet spot" of PAD 6 represents optimal internal circadian alignment.

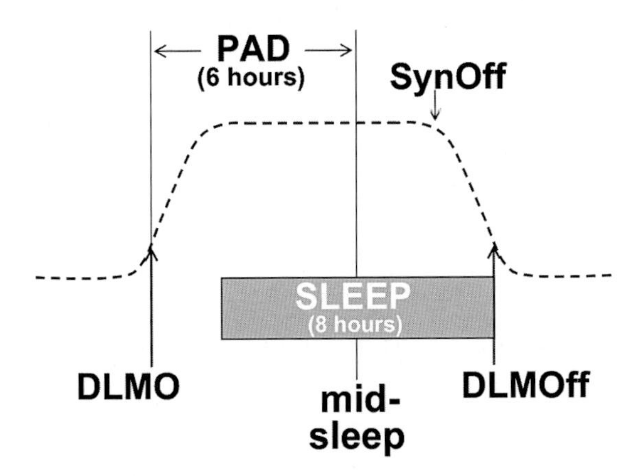

Figure 5. The phase-angle difference (PAD) between the plasma $DLMO_{10}$ and the time of mid sleep is on average about 6 hours in healthy controls. PAD 6 can be used to phase type individuals and to assess internal circadian misalignment. (Adapted, with permission, from Lewy et al. 2006b [©National Academy of Sciences].)

Circadian misalignment is a necessary, but not sufficient, cause for SAD. Patients do not have a mean PAD that is much different from that of healthy controls. The range and distribution of PADs also appear to be similar for both groups. A biological marker need not be different in patients compared to healthy controls; in fact, a marker that correlates with symptom severity may actually be more useful, particularly when the diagnosis can be made by interview. Analysis of an extant baseline data set in another group of SAD patients (Lewy et al. 1998b) replicated the parabolic findings precisely, including the 2:1 ratio of phase-delayed to -advanced patients (Lewy et al. 2007b).

Terman et al. (2001) have proposed a variant of the PSH for SAD that assumes all patients are of the phase-delayed type and should preferentially respond to morning light in order to cause a phase advance, the greater the phase advance, the better. They may consider PAD to be an epiphenomenon. Indeed, they recommend using mid-sleep (or, alternatively, Horne–Ostberg morningness/eveningness ratings) to estimate the DLMO. They then recommend, even if it involves interfering with sleep, scheduling exposure to begin 8.5 hours after the estimated DLMO in order to induce what we (Lewy and Sack 1989) and they (Terman et al. 2001) have found to be about a 1.5-hour phase advance. Although earlier light and earlier sleep times would cause greater phase advances in the DLMO, we would advise not requiring SAD subjects to wake up earlier because this would tend to shorten, not lengthen, the DLMO/mid-sleep PAD (the Terman group only assessed the DLMO/sleep onset interval (MSI) as their marker for PAD). Another difference between the original PSH and the Terman variant is that the latter leads to a recommendation of the earliest morning light in whom they would consider to be their least phase-delayed patients, whereas we would consider at least some of these patients to be of the phase-advanced type and would recommend that light be scheduled in the evening in order to provide a corrective phase delay.

We consider the clock time of the DLMO, not PAD, to be an epiphenomenon: If sleep times are invariant, a change in PAD will result in a commensurate change in DLMO clock time, but it is the change in PAD that is clinically important. In addition to representing internal circadian alignment, PAD is less likely than real (or estimated) DLMO clock time to be confounded by changes in sleep times (either because of masking effects or because of an altered light/dark cycle) or to be accurately guessed by research subjects or their raters. Regarding the latter, if a person is phase typed as delayed because of a DLMO that is delayed with respect to sleep, then sleep will be advanced with respect to the DLMO in this individual, thus making it difficult to guess phase type correctly and thereby improving the double-blind integrity of the study.

Goodness of fit to PAD can also be used to validate clinical assessment and improve how best to refine them, as well as help to identify a clinical endophenotype. In an item-by-item analysis, we found that three items in the SIGH-SAD instrument could account for all of the statistically significant findings originally reported using the entire 29-item scale (Lewy et al. 2006b). These three items rated subjective depression severity, subjective anxiety severity, and amount of observed agitation at the interview (Lewy et al. 2007b). Therefore, it would not be surprising if mixed anxiety and depressive states had a circadian component that can be treated, probably adjunctively, with a phase-resetting agent. Even if the circadian component explains only 10% of the variance, melatonin—and to a lesser extent bright light—can be added to just about any treatment regimen with impunity. We are just beginning to appreciate the significance of correlating PAD with symptom severity in sleep, psychiatric and perhaps other medical disorders, as well as interindividual and intraindividual changes in cognition, mood, and attention in healthy controls.

CONCLUSION: THE FUNCTION OF ENDOGENOUS MELATONIN PRODUCTION IN HUMANS

Before concluding, we might speculate about the function of melatonin in humans. When we first described the melatonin PRC, we hypothesized that humans have retained the acute suppressant effect of light in order to facilitate the circadian phase-resetting effects of melatonin and to therefore augment entrainment of the endogenous circadian pacemaker by the phase-shifting effects of the light/dark cycle (Lewy et al. 1992). Our work with BFRs, however, now provides the basis for possibly a more important role to all sighted babies (Fig. 6). Exogenous melatonin entrains BFRs in part because the administration time is provided by the researcher or clinician. Similarly, the sighted mother sends her melatonin signal to the third trimester fetus (Reppert 1979) and to the suckling newborn (Illnerová et al. 1993). Once a few months old, "circadian blindness" ends (Kleitman and Engelmann 1953; Tomioka and Tomioka 1991; McGraw et al. 1999), and the 24-hour light/dark cycle via the retinohypothalamic tract can entrain the endogenous circadian pacemaker.

Melatonin receptors in the SCN are functional by the third trimester (Reppert et al. 1988). Melatonin freely crosses the placenta (Reppert 1979) and is metabolized by the mother, so that both mother and fetus receive the same chemical signal for nighttime darkness. Whether or not melatonin levels in breast milk (which are about the same or are at least two-thirds those of plasma; Illnerová et al. 1993) are sufficient for the entraining infant is not presently known. Some investigators have calculated the levels to be too low, as least for rat pups (Rowe and Kennaway 2002). However, with substantial activity of the catabolic enzyme converting melatonin to 6-hydroxymelatonin (Skene 2001) not occurring until after age 2–3 months (Sonnier and Cresteil 1998), suckling infants until this age may have a melatonin bioavailability closer to 100% than the adult's 3–15%, as well as a smaller volume of distribution, resulting in a barely sufficient level for this function. Exposure of the mother to ambient light could reduce melatonin levels further. Thus, studies of BFRs may be applicable for synchronizing the sleep/wake cycles of perinates to their mother's. Systems theory would suggest that, even if it makes no difference to the baby to be entrained or free-running, a less sleep-deprived mother will deliver better care. It may not be premature to

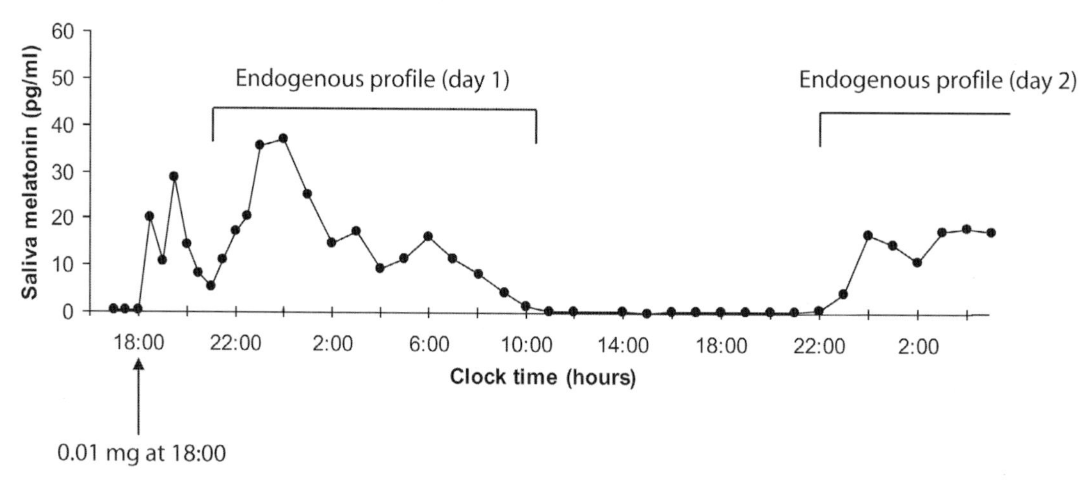

Figure 6. Salivary pharmacokinetic (PK) profile and endogenous melatonin levels in a BFR entrained to a daily dose of 0.01 mg of exogenous melatonin. The dose is given at 18:00 (1 hour earlier than usual in order to better discern its PK profile), and the endogenous melatonin onset (MO) occurs a few hours later. In the steady-state situation, there is no melatonin-free interval between the exogenous and endogenous profiles, maximizing the phase-shifting effects of this very low dose, which must produce a daily phase advance of least 0.1 hour to entrain this subject. Although saliva levels are less reliable than plasma levels (at least for higher concentrations, when the threefold difference between them becomes greater), there is no doubt that this dose is producing no more than a physiologic level for this subject. This finding suggests that very low doses may be therapeutic and that there may be a circadian function for endogenous melatonin production.

recommend that mothers avoid light exposure at night and to label pumped milk as to whether it was collected during the day or the night, so that the infant receives it at the appropriate time.

In conclusion, the DMLO can be used to phase type circadian disorders. The DLMO also marks the phase of the light and melatonin PRCs, informing treatment. Measurement of the DLMO is expected to become increasingly important, particularly relating its PAD to the sleep/wake cycle. Assessing the DLMO in the course of treatment provides an objective measure of treatment efficacy and can indicate if too much of a phase shift has occurred. Thus, the DLMO provides information about whether or not a circadian phase disorder is present, its phase type, the appropriate choice and timing of treatment, and objective monitoring of treatment efficacy.

NOTE ADDED IN PROOF

A.J.L. is coinventor on several melatonin use-patents owned by Oregon Health & Science University currently not licensed to any company.

ACKNOWLEDGMENTS

We thank the research subjects and the nursing staff of the Oregon Health & Science University (OHSU) Clinical and Translational Research Center, Robert Sack, Jonathan Emens, Kyle Johnson, Mark Kinzie, Jeannie Songer, Krista Yuhas, Rebecca Bernert, Jennifer Rough, Neelam Sims, Anne Rutherford, Allie Buti, Diana Arntz, Cameron Brick, Cara Bussell, and Kitt Woods. This work was supported by Public Health Service grants R01 MH55703, R01 MH56874, R01 AG21826, and R01 HD42125 (to A.J.L.) and 5 M01 RR000334 (to the Clinical and Translational Research Center of OHSU). A.J.L. was supported by the National Alliance for Research on Schizophrenia and Depression 2000 Distinguished Investigator Award. J.S.E. was supported by Public Health Service grant K23 RR017636-01.

REFERENCES

Arendt J. 1978. Melatonin assays in body fluids. *J. Neural Transm. Suppl.* **13:** 265.

———. 1995. *Melatonin and the mammalian pineal gland.* Chapman and Hall, London, United Kingdom.

———. 2006. Melatonin and human rhythms. *Chronobiol. Int.* **23:** 21.

Arendt J., Skene D.J., Middleton B., Lockley S.W., and Deacon S. 1997. Efficacy of melatonin treatment in jet lag, shift work, and blindness. *J. Biol. Rhythms* **12:** 604.

Arendt J., Aldhous M., English J., Marks V., Arendt J. H., Marks M., and Folkard S. 1987. Some effects of jet-lag and their alleviation by melatonin. *Ergonomics* **30:** 1379.

Arendt J., Bojkowski C., Folkard S., Franey C., Marks V., Minors D., Waterhouse J.

Wever R.A., Wildgruber C., and Wright J. 1985. Some effects of melatonin and the control of its secretion in humans. *Ciba Found. Symp.* **117:** 266.

Avery D.H., Khan A., Dager S.R., Cox G.B., and Dunner D.L. 1990. Bright light treatment of winter depression: Morning versus evening light. *Acta Psychiatr. Scand.* **82:** 335.

Benloucif S., Guico M.J., Reid K.J., Wolfe L.F., L'Hermite-Balériaux M., and Zee P.C. 2005. Stability of melatonin and temperature as circadian phase markers and their relation to sleep times in humans. *J. Biol. Rhythms* **20:** 178.

Bobbert A.C. and Riethoven J.J. 1991. Feedback in the rabbit's central circadian system, revealed by the changes in its free-running food intake pattern induced by blinding, cervical sympathectomy, pinealectomy, and melatonin administration. *J. Biol. Rhythms* **6:** 263.

Boivin D.B., Duffey J.F., Kronauer R.E., and Czeisler C.A. 1996. Dose-response relationships for resetting of human circadian clock by light. *Nature* **379:** 540.

Burgess H.J., Crowley S.J., Gazda C.J., Fogg L.F., and Eastman C.I. 2003. Preflight adjustment to eastward travel: 3 days of advancing sleep with and without morning bright light. *J. Biol. Rhythms* **18:** 318.

Cajochen C., Jewett M.E., and Dijk D.J. 2003. Human circadian melatonin rhythm phase delay during a fixed sleep-wake schedule interspersed with nights of sleep deprivation. *J. Pineal Res.* **35:** 149.

Campbell S.S., Gillin J.C., Kripke D.F., Erikson P., and Clopton P. 1989. Gender differences in the circadian temperature rhythms of healthy elderly subjects: Relationships to sleep quality. *Sleep* **12:** 529.

Cassone V.M. 1990. Effects of melatonin on vertebrate circadian systems. *Trends Neurosci.* **13:** 457.

Czeisler C.A. 1997. Commentary: Evidence for melatonin as a circadian phase-shifting agent. *J. Biol. Rhythms* **12:** 618.

Czeisler C.A., Kronauer R.E., Ronda J.M., and Rios C.D. 1985. Sleep deprivation in constant light phase advance shifts and shortens the free-running period of the human circadian timing system. *Sleep Res.* **14:** 252.

Czeisler C.A., Kronauer R.E., Allan J.S., Duffy J.F., Jewett M.E., Brown E.N., and Ronda J.M. 1989. Bright light induction of strong (type O) resetting of the human circadian pacemaker. *Science* **244:** 1328.

Czeisler C.A., Richardson G.S., Coleman R.M., Zimmerman J.C., Moore-Ede M.C., Dement W.C., and Weitzman E.D. 1981. Chronotherapy: Resetting the circadian clocks of patients with delayed sleep phase insomnia. *Sleep* **4:** 1.

Czeisler C.A., Shanahan T.L., Klerman E.B., Martens H., Brotman D.J., Emens J.S., Klein T., and Rizzo J.F., III. 1995. Suppression of melatonin secretion in some blind patients by exposure to bright light. *N. Engl. J. Med.* **332:** 6.

Czeisler C.A., Duffy J.F., Shanahan T.L., Brown E.N., Mitchell J.F., Rimmer D.W., Ronda J.M., Silva E.J., Allan J.S., Emens J.S., Dijk D.-J., and Kronauer R.E. 1999. Stability, precision and near-24-hour period of the human circadian pacemaker. *Science* **284:** 2177.

Daan S. and Lewy A.J. 1984. Scheduled exposure to daylight: A potential strategy to reduce "jet lag" following transmeridian flight. *Psychopharmacol. Bull.* **20:** 566.

Dahlitz M., Alvarez B., Vignau J., English J., Arendt J., and Parkes J.D. 1991. Delayed sleep phase syndrome response to melatonin. *Lancet* **337:** 1121.

Davis F.C. and Mannion J. 1988. Entrainment of hamster pup circadian rhythms by prenatal melatonin injections to the mother. *Am. J. Physiol.* **255:** R439.

Deacon S. and Arendt J. 1994. Posture influences melatonin concentrations in plasma and saliva in humans. *Neurosci. Lett.* **167:** 191.

Eastman C.I. 1982. The phase-shift model of spontaneous internal desychronization in humans. In *Vertebrate circadian systems: Structure and physiology* (ed. J. Aschoff et al.), p. 262. Springer-Verlag, Berlin.

Eastman C.I., Young M.A., Fogg L.F., Liu L., and Meaden P.M. 1998. Bright light treatment of winter depression: A placebo-controlled trial. *Arch. Gen. Psychiatry* **55:** 883.

Emens J., Lewy A., and Yuhas K. 2007. Women have a longer phase angle of entrainment than men. *Sleep* (suppl.) **30:** A63. (Abstr.)

Emens J.S., Lewy A.J., Bernert R.A., and Lefler B.J. 2003. Entrainment of the human circadian pacemaker by melatonin is independent of the circadian phase of treatment initiation. *Chronobiol. Int.* **20:** 1189.

Emens J.S., Lewy A.J., Lefler B.J., and Sack R.L. 2005. Relative coordination to unknown "weak zeitgebers" in free-running blind individuals. *J. Biol. Rhythms* **20:** 159.

Emens J., Lewy A.J., Yuhas K., Jackman A.R., and Johnson K.P. 2006. Melatonin entrains free-running blind individuals with circadian periods less than 24 hours. *Sleep* (suppl.) **29:** A62. (Abstr.)

Gerdin M.J., Masana M.I., Rivera-Bermúdez M.A., Hudson R.L., Earnest D.J., Gillette M.U., and Dubocovich M.L. 2004. Melatonin desensitizes endogenous MT2 melatonin receptors in the rat suprachiasmatic nucleus: Relevance for defining the periods of sensitivity of the mammalian circadian clock to melatonin. *FASEB J.* **18:** 1646.

Goel N., Lee T.M., and Pieper D.R. 1998. Removal of the olfactory bulbs delays photic reentrainment of circadian activity rhythms and modifies the reproductive axis in male *Octodon degus*. *Brain Res.* **792:** 229.

Gronfier C., Wright K.P., Kronauer R.E., and Czeisler C.A. 2007. Entrainment of the human circadian pacemaker to longer-than-24-h days. *Proc. Natl. Acad. Sci.* **104:** 9081.

Hack L.M., Lockley S.W., Arendt J., and Skene D.J. 2003. The effects of low-dose 0.5-mg melatonin on the free-running circadian rhythms of blind subjects. *J. Biol. Rhythms* **18:** 420.

Hebert M., Martin S.K., Lee C., and Eastman C.I. 2002. The effects of prior light history on the suppression of melatonin by light in humans. *J. Pineal Res.* **33:** 198.

Honma K. and Honma S. 1988. A human phase response curve for bright light pulses. *Jpn. J. Psychiatry Neurol.* **42:** 167.

Illnerová H. 1979. Effect of one minute exposure to light at night on rat pineal serotonin N-acetyltransferase and melatonin. *J. Neurochem.* **32:** 673.

Illnerová H. and Vanecek J. 1982. Two-oscillator structure of the pacemaker controlling the circadian rhythm of *N*-acetyltransferase in the rat pineal gland. *J. Comp. Physiol. A* **145:** 539.

Illnerová H., Buresová M., and Presl J. 1993. Melatonin rhythm in human milk. *J. Clin. Endocrinol. Metab.* **77:** 838.

Illnerová H., Wetterberg L., and Saav J. 1978. Effect of light on nocturnal melatonin concentration in the rat pineal gland and serum. *Physiol. Bohemoslov.* **27:** 544.

Johnson C.H. 1990. *An atlas of phase response curves for circadian and circatidal rhythms.* Vanderbilt University, Nashville, Tennessee.

Kendall A.R. and Sack R.L. 2000. The effects of age on human circadian period. *Sleep* **23:** A184.

Khalsa S.B., Jewett M.E., Cajochen C., and Czeisler C.A. 2003. A phase-response curve to single bright light pulses in human subjects. *J. Physiol.* **549:** 945.

Klein T., Martens H., Dijk D.J., Kronauer R.E., Seely E.W., and Czeisler C.A. 1993. Circadian sleep regulation in the absence of light perception: Chronic non-24-hour circadian rhythm sleep disorder in a blind man with a regular 24-hour sleep-wake schedule. *Sleep* **16:** 333.

Kleitman N. and Engelmann T.G. 1953. Sleep characteristics of infants. *J. Appl. Physiol.* **6:** 269.

Klerman E.B. 2001. Non-photic effects on the circadian system: Results from experiments in blind and sighted individuals. In *Zeitgebers, entrainment and masking of the circadian system* (ed. K. Honma and S. Honma), p. 155. Hokkaido University Press, Sapporo, Japan.

Klerman E.B., Gershengorn H.B., Duffy J.F., and Kronauer R.E. 2002a. Comparisons of the variability of three markers of the human circadian pacemaker. *J. Biol. Rhythms* **17:** 181.

Klerman E.B., Shanahan T.L., Brotman D.J., Rimmer D.W., Emens J.S., Rizzo J.F., III, and Czeisler C.A. 2002b. Photic resetting of the human circadian pacemaker in the absence of conscious vision. *J. Biol. Rhythms* **17:** 548.

Kripke D.F., Risch S.C., and Janowsky D. 1983. Bright white light alleviates depression. *Psychiatry Res.* **10:** 105.

Kripke D.F., Mullaney D.J., Atkinson M., and Wolf S. 1978. Circadian rhythm disorders in manic-depressives. *Biol. Psychiatry*, **13:** 335.

Kronauer R.E., Czeisler C.A., Pilato S.F., Moore-Ede M.C., and Weitzman E.D. 1982. Mathematical model of the human circadian system with two interacting oscillators. *Am. J. Physiol.* **242:** R3.

Lamont E.W., Legault-Coutu D., Cermakian N., and Boivin D.B. 2007. The role of circadian clock genes in mental disorders. *Dialogues Clin. Neurosci.* **9:** 333.

Lewy A.J. 1981. Human plasma melatonin studies: Effects of light and implications for biological rhythm research. *Adv. Biosci.* **29:** 397.

———. 1983. Biochemistry and regulation of mammalian melatonin production. In *The pineal gland* (ed. R.M. Relkin), p. 77. Elsevier, North-Holland, New York.

———. 2007. Current understanding and future implications of the circadian uses of melatonin, a neurohormone discovered by Aaron B. Lerner. *J. Invest. Dermatol.* **127:** 2082.

Lewy A.J. and Markey S.P. 1978. Analysis of melatonin in human plasma by gas chromatography negative chemical ionization mass spectrometry. *Science* **201:** 741.

Lewy A.J. and Newsome D.A. 1983. Different types of melatonin circadian secretory rhythms in some blind subjects. *J. Clin. Endocrinol. Metab.* **56:** 1103.

Lewy A.J. and Sack R.L. 1989. The dim light melatonin onset (DLMO) as a marker for circadian phase position. *Chronobiol. Int.* **6:** 93.

———. 1997. Exogenous melatonin's phase shifting effects on the endogenous melatonin profile in sighted humans: A brief review and critique of the literature. *J. Biol. Rhythms* **12:** 595.

Lewy A.J., Cutler N.L., and Sack R. L. 1999. The endogenous melatonin profile as a marker for circadian phase position. *J. Biol. Rhythms* **14:** 227.

Lewy A.J., Sack R.L., and Singer C.M. 1984. Assessment and treatment of chronobiologic disorders using plasma melatonin levels and bright light exposure: The clock-gate model and the phase response curve. *Psychopharmacol. Bull.* **20:** 561.

———. 1985a. Immediate and delayed effects of bright light on human melatonin production: Shifting "dawn" and "dusk" shifts the dim light melatonin onset (DLMO). *Ann. N.Y. Acad. Sci.* **453:** 253.

———. 1985b. Melatonin, light and chronobiological disorders. *Ciba Found Symp.* **117:** 231.

———. 1985c. Treating phase typed chronobiologic sleep and mood disorders using appropriately timed bright artificial light. *Psychopharmacol. Bull.* **21:** 368.

Lewy A.J., Ahmed S., Jackson J.M.L., and Sack R.L. 1992. Melatonin shifts circadian rhythms according to a phase-response curve. *Chronobiol. Int.* **9:** 380.

Lewy A.J., Emens J.S., Bernert R.A., and Lefler B. J. 2004b. Eventual entrainment of the human circadian pacemaker by melatonin is independent of the circadian phase of treatment initiation: Clinical implications. *J. Biol. Rhythms* **19:** 68.

Lewy A.J., Emens J., Jackman A., and Yuhas K. 2006a. Circadian uses of melatonin in humans. *Chronobiol. Int.* **23:** 403.

Lewy A.J., Hasler B.P., Emens J.S., and Sack R.L. 2001. Pretreatment circadian period in free-running blind people may predict the phase angle of entrainment to melatonin. *Neurosci. Lett.* **313:** 158.

Lewy A.J., Kern H.A., Rosenthal N.E., and Wehr T.A. 1982. Bright artificial light treatment of a manic-depressive patient with a seasonal mood cycle. *Am. J. Psychiatry* **139:** 1496.

Lewy A.J., Lefler B.J., Emens J.S., and Bauer V.K. 2006b. The circadian basis of winter depression. *Proc. Natl. Acad. Sci.* **103:** 7414.

Lewy A.J., Sack R.L., Miller S., and Hoban T.M. 1987a. Antidepressant and circadian phase-shifting effects of light. *Science* **235:** 352.

Lewy A.J., Sack R.L., Singer C.M., and White D.M. 1987b. The phase shift hypothesis for bright light's therapeutic mechanism of action: Theoretical considerations and experimental evidence. *Psychopharmacol. Bull.* **23:** 349.

Lewy A.J., Emens J.S., Lefler B.J., Yuhas K., and Jackman A.R. 2005. Melatonin entrains free-running blind people according to a physiological dose-response curve. *Chronobiol. Int.* **22:** 1093.

Lewy A.J., Emens J.S., Sack R.L., Hasler B.P., and Bernert R.A. 2002. Low, but not high, doses of melatonin entrained a free-running blind person with a long circadian period.

Chronobiol. Int. **19:** 649.

———. 2003. Zeitgeber hierarchy in humans: Resetting the circadian phase positions of blind people using melatonin. *Chronobiol. Int.* **20:** 837.

Lewy A.J., Sack R.L., Singer C.M., White D.M., and Hoban T.M. 1989. Winter depression: The phase angle between sleep and other circadian rhythms may be critical. In *Seasonal affective disorder* (ed. C. Thompson and T. Silverstone), p. 205. CNS Publishers, London.

Lewy A.J., Tetsuo M., Markey S.P., Goodwin F.K., and Kopin I.J. 1980a. Pinealectomy abolishes plasma melatonin in the rat. *J. Clin. Endocrinol. Metab.* **50:** 204.

Lewy A.J., Wehr T.A., Goodwin F.K., Newsome D.A., and Markey S.P. 1980b. Light suppresses melatonin secretion in humans. *Science* **210:** 1267.

Lewy A.J., Wehr T.A., Goodwin F.K., Newsome D.A., and Rosenthal N.E. 1981. Manic-depressive patients may be supersensitive to light. *Lancet* **I:** 383.

Lewy A.J., Rough J., Songer J., Mishra N., Yuhas K., and Emens J. 2007a. The phase shift hypothesis for the circadian component of winter depression. *Dialogues Clin. Neurosci.* **9:** 291.

Lewy A.J., Sack R.L., Blood M.L., Bauer V.K., Cutler N.L., and Thomas K.H. 1995. Melatonin marks circadian phase position and resets the endogenous circadian pacemaker in humans. *Ciba Found. Symp.* **183:** 303.

Lewy A.J., Sack R.L., Fredrickson R.H., Reaves M., Denney D., and Zielske D.R. 1983. The use of bright light in the treatment of chronobiologic sleep and mood disorders: The phase-response curve. *Psychopharmacol. Bull.* **19:** 523.

Lewy A.J., Woods K., Kinzie J., Emens J., Songer J., and Yuhas K. 2007b. DLMO/Mid-sleep interval of six hours phase types SAD patients and parabolically correlates with symptom severity. *Sleep* (suppl.) **30:** A63. (Abstr.)

Lewy A.J., Emens J., Lefler B.J., Koenig A.R., Yuhas K., Johnson K.P., and Giger P.T. 2004b. Melatonin-induced phase delays of the human circadian pacemaker. *Sleep* (suppl.) **27:** A79. (Abstr.)

Lewy A.J., Nurnberger J.I., Wehr T.A., Pack D., Becker L.E., Powell R., and Newsome D.A. 1985d. Supersensitivity to light: Possible trait marker for manic-depressive illness. *Am. J. Psychiatry* **142:** 725.

Lewy A.J., Bauer V.K., Ahmed S., Thomas K.H., Cutler N.L., Singer C.M., Moffit M.T., and Sack R.L.1998a. The human phase response curve (PRC) to melatonin is about 12 hours out of phase with the PRC to light. *Chronobiol. Int.* **15:** 71.

Lewy A.J., Bauer V.K., Cutler N.L., Sack R.L., Ahmed S., Thomas K.H., Blood M.L., and Latham J.M. 1998b. Morning versus evening light treatment of patients with winter depression. *Arch. Gen. Psychiatry* **55:** 890.

Lewy A.J., Yuhas K., Emens J., Woods K., Arntz D., Songer J., Johnson K., Rough J., Brick C., and Bussell C. 2007c. Are the circadian rhythms of blind adult males less sensitive to social cues than females? *Sleep* (suppl.) **30:** A63. (Abstr.)

Lockley S., Arendt J., and Skene D. 2007. Visual impairment and circadian rhythm disorders. *Dialogues Clin. Neurosci.* **9:** 301.

Lockley S.W., Skene D.J., Arendt J., Tabandeh H., Bird A.C., and Defrance R. 1997. Relationship between melatonin rhythms and visual loss in the blind. *J. Clin. Endocrinol. Metab.* **82:** 3763.

Lockley S.W., Skene D.J., James K., Thapan K., Wright J., and Arendt J. 2000. Melatonin administration can entrain the free-running circadian system of blind subjects. *J. Endocrinol.* **164:** R1.

Lynch H.J., Ozaki Y., Shakal D., and Wurtman R.J. 1975. Melatonin excretion of man and rats: Effect of time of day, sleep, pinealectomy and food consumption. *Int. J. Biometeorol.* **19:** 267.

Mallo C., Zaidan R., Faure A., Brun J., Chazot G., and Claustrat B. 1988. Effects of a four-day nocturnal melatonin treatment on the 24 h plasma melatonin, cortisol and prolactin profiles in humans. *Acta Endocrinol.* **119:** 474.

McArthur A.J., Lewy A.J., and Sack R.L. 1996. Non-24-hour sleep-wake syndrome in a sighted man: Circadian rhythm studies and efficacy of melatonin treatment. *Sleep* **19:** 544.

McGraw K., Hoffmann R., Harker C., and Herman J.H. 1999. The development of circadian rhythms in a human infant. *Sleep* **22:** 303.

Middleton B., Arendt J., and Stone B.M. 1997. Complex effects of melatonin on human circadian rhythms in constant dim light. *J. Biol. Rhythms* **12:** 467.

Minors D.S., Waterhouse J.M., and Wirz-Justice A. 1991. A human phase-response curve to light. *Neurosci. Lett.* **133:** 36.

Mistlberger R. 1991. Effects of daily schedules of forced activity on free-running rhythms in the rat. *J. Biol. Rhythms* **6:** 71.

Mongrain V., Lavoie S., Selmaoui B., Paquet J., and Dumont M. 2004. Phase relationships between sleep-wake cycle and underlying circadian rhythms in morningness-eveningness. *J. Biol. Rhythms* **19:** 248.

Neuwelt E.A. and Lewy A.J. 1983. Disappearance of plasma melatonin after removal of a neoplastic pineal gland. *N. Engl. J. Med.* **308:** 1132.

Ozaki Y. and Lynch H.J. 1976. Presence of melatonin in plasma and urine of pinealectomized rats. *Endocrinology* **99:** 641.

Parry B.L., Berga S.L., Mostofi N., Klauber M.R., and Resnick A. 1997. Plasma melatonin circadian rhythms during the menstrual cycle and after light therapy in premenstrual dysphoric disorder and normal control subjects. *J. Biol. Rhythms* **12:** 47.

Redman J., Armstrong S., and Ng K.T. 1983. Free-running activity rhythms in the rat: Entrainment by melatonin. *Science* **219:** 1089.

Reiter R., Richardson B., and Hurlbut E. 1981. Pineal, retinal, and harderian gland melatonin in a diurnal species, the Richardson's ground squirrel. *Neurosci. Lett.* **22:** 285.

Reppert S. 1979. Maternal-fetal transfer of melatonin in the non-human primate. *Pediatr. Res.* **13:** 788.

Reppert S.M., Weaver D.R., Rivkees S.A., and Stopa E.G. 1988. Putative melatonin receptors are located in a human biological clock. *Science* **242:** 78.

Revell V.L. and Eastman C.I. 2005. How to trick mother nature into letting you fly around or stay up all night. *J. Biol. Rhythms* **20:** 353.

Rosenthal N.E., Sack D.A., Gillin J.C., Lewy A.J., Goodwin F.K., Davenport Y., Mueller P.S., Newsome D.A., and Wehr T.A. 1984. Seasonal affective disorder: A description of the syndrome and preliminary findings with light therapy. *Arch. Gen. Psychiatry* **41:** 72.

Rowe S.A. and Kennaway D.J. 2002. Melatonin in rat milk and the likelihood of its role in postnatal maternal entrainment of rhythms. *Am. J. Physiol. Regul. Integr. Comp. Physiol.* **282:** R797.

Sack R.L., Blood M.L., and Lewy A.J. 1992a. Melatonin rhythms in night shift workers. *Sleep* **15:** 434.

Sack R.L., Lewy A.J., and Hoban T.M. 1987. Free-running melatonin rhythms in blind people: Phase shifts with melatonin and triazolam administration. In *Temporal disorder in human oscillatory systems* (ed. L. Rensing et al.), p. 219. Springer-Verlag, Heidelberg, Germany.

Sack R.L., Stevenson J., and Lewy A.J. 1990. Entrainment of a previously free-running blind human with melatonin administration. *Sleep Res.* **19:** 404.

Sack R.L., Brandes R.W., Kendall A.R., and Lewy A.J. 2000. Entrainment of free-running circadian rhythms by melatonin in blind people. *N. Engl. J. Med.* **343:** 1070.

Sack R.L., Lewy A.J., Blood M.L., Keith L.D., and Nakagawa H. 1992b. Circadian rhythm abnormalities in totally blind people: Incidence and clinical significance. *J. Clin. Endocrinol. Metab.* **75:** 127.

Sack R.L., Lewy A.J., Blood M.L., Stevenson J., and Keith L.D. 1991. Melatonin administration to blind people: Phase advances and entrainment. *J. Biol. Rhythms* **6:** 249.

Scheer F., Wright K.P., Kronauer R.E., and Czeisler C.A. 2007. Plasticity of the intrinsic period of the human circadian timing system. *PLos ONE* **2:** E721.

Skene D.J. 2001. Contribution of CYP1A2 in the hepatic metabolism of melatonin: Studies with isolated microsomal preparations and liver slices. *J. Pineal Res.* **31:** 333.

Smith J.A., O'Hara J., and Schiff A.A. 1981. Altered diurnal

serum melatonin rhythm in blind men. *Lancet* **II:** 933.
Smith J.A., Padwick D., Mee T.J.X., Minneman K.P., and Bird E.D. 1977. Synchronous nyctohemeral rhythms in human blood melatonin and in human post-mortem pineal enzyme. *Clin. Endocrinol.* **6:** 219.
Smith K.A., Schoen M.W., and Czeisler C.A. 2004. Adaptation of human pineal melatonin suppression by recent photic history. *J. Clin. Endocrinol. Metab.* **89:** 3610.
Sonnier M. and Cresteil T. 1998. Delayed ontogenesis of CYP1A2 in human liver. *Eur. J. Biochem.* **251:** 893..
Terman J.S., Terman M., Lo E.-S., and Cooper T.B. 2001. Circadian time of morning light administration and therapeutic response in winter depression. *Arch. Gen. Psychiatry* **58:** 69.
Terman M., Terman J.S., and Ross D.C. 1998. A controlled trial of timed bright light and negative air ionization for treatment of winter depression. *Arch. Gen. Psychiatry* **55:** 875.
Tomioka K. and Tomioka F. 1991. Development of circadian sleep-wakefulness rhythmicity of three infants. *J. Interdiscip. Cycle Res.* **22:** 71.
Underwood H. 1986. Circadian rhythms in lizards: Phase response curve for melatonin. *J. Pineal Res.* **3:** 187.
Vondrasova-Jelinkova D., Hajek I., and Illnerová H. 1999. Adjustment of the human melatonin and cortisol rhythms to shortening of the natural summer photoperiod. *Brain Res.* **816:** 249.
Voultsios A., Kennaway D.J., and Dawson D. 1997. Salivary melatonin as a circadian phase marker: Validation and comparison to plasma melatonin. *J. Biol. Rhythms* **12:** 457.
Wehr T.A., Wirz-Justice A., Goodwin F.K., Duncan W., and Gillin J.C. 1979. Phase advance of the circadian sleep-wake cycle as an antidepressant. *Science* **206:** 710.
Wehr T.A., Moul D.E., Barbato G., Giesen H.A., Seidel J.A., Barker C., and Bender C. 1993. Conservation of photoperiod-responsive mechanisms in humans. *Am. J. Physiol.* **265:** R846.
Wetterberg L. 1978. Melatonin in humans: Physiological and clinical studies (review). *J. Neural Transm. Suppl.* **13:** 289.
Wever R.A. 1979. *The circadian system of man. Results of experiments under temporal isolation.* Springer-Verlag, New York.
———. 1989. Light effects on human circadian rhythms. A review of recent Andech's experiments. *J. Biol. Rhythms* **4:** 161.
Wever R.A., Polasek J., and Wildgruber C.M. 1983. Bright light affects human circadian rhythms. *Eur. J. Physiol.* **396:** 85.
Wirz-Justice A., Werth E., Renz C., Muller S., and Krauchi K. 2002. No evidence for a phase delay in human circadian rhythms after a single morning melatonin administration. *J. Pineal Res.* **32:** 1.
Wirz-Justice A., Graw P., Krauchi K., Gisin B., Jochum A., Arendt J., Fisch H., Buddeberg C., and Poldinger W. 1993. Light therapy in seasonal affective disorder is independent of time of day or circadian phase. *Arch. Gen. Psychiatry* **50:** 929.
Wright H., Lack L., and Bootzin R. 2006. Relationship between dim light melatonin onset and the timing of sleep in sleep onset insomniacs. *Sleep Biol. Rhythms* **4:** 78.
Wright K.P., Jr., Gronfier C., Duffy J.F., and Czeisler C.A. 2005. Intrinsic period and light intensity determine the phase relationship between melatonin and sleep in humans. *J. Biol. Rhythms* **20:** 178.
Yamazaki S., Alones V., and Menaker M. 2002. Interaction of the retina with suprachiasmatic pacemakers in the control of circadian behavior. *J. Biol. Rhythms* **17:** 315.
Zaidan R., Geoffriau M., Brun J., Taillard J., Bureau C., Chazot G., and Claustrat B. 1994. Melatonin is able to influence its secretion in humans: Description of a phase-response curve. *Neuroendocrinology* **60:** 105.
Zeitzer J.M., Dijk D.J., Kronauer R.E., Brown E.N., and Czeisler C.A. 2000. Sensitivity of the human circadian pacemaker to nocturnal light: Melatonin phase resetting and suppression. *J. Physiol.* **526:** 695.

Role for the *Clock* Gene in Bipolar Disorder

C.A. McClung
Department of Psychiatry and Center for Basic Neuroscience, University of Texas Southwestern Medical Center, Dallas, Texas 75390-9070

Nearly all patients with bipolar disorder have severely disrupted circadian rhythms. Treatment with mood stabilizers can restore these daily rhythms, and this is correlated with patient recovery. However, it is still uncertain whether clock abnormalities are the cause of bipolar disorder or if these rhythm disruptions are secondary to alterations in other circuits. Furthermore, the mechanism by which the circadian clock might influence mood is still unclear. With cloning and characterization of the circadian genes and recent advances in molecular biology, we are starting to understand this strong association between circadian rhythms and bipolar disorder. Recent human genetic and mouse behavioral studies indicate that the *Clock* gene is particularly relevant in the mood disruptions associated with this disorder. Furthermore, it appears that *Clock* expression outside of the central pacemaker of the suprachiasmatic nucleus (SCN) is involved in mood regulation. In this chapter, the evidence linking circadian rhythms, the *Clock* gene, and bipolar disorder is discussed, along with the possible biology that underlies this connection.

INTRODUCTION TO RHYTHMS AND MOOD

It has been known for some time that disruptions in biological rhythms are strongly associated with nearly all mood disorders. In fact, some of the most common symptoms of diseases such as major depressive disorder and bipolar disorder are abnormal sleep/wake, appetite, and social rhythms (Boivin 2000; Bunney and Bunney 2000; Lenox et al. 2002; Grandin et al. 2006; McClung 2007). Depressed individuals often experience the most severe symptoms in the early morning with gradual improvement throughout the day (Rusting and Larsen 1998). Furthermore, depression is more prevalent in areas of the world that receive little sunlight for extended periods of time (Booker et al. 1991). In fact, one of the most common mood disorders, affecting approximately 2–5% of the population in temperate climates, is seasonal affective disorder (SAD), a syndrome where depressive symptoms occur only in the winter months when there are shorter days and a later dawn (Lam and Levitan 2000; Magnusson and Boivin 2003).

Mood disorders may be caused by an abnormally shifted or arrhythmic clock. It is known that blunted or abnormal circadian rhythms in a variety of bodily functions including body temperature, plasma cortisol, norepinephrine, thyroid stimulating hormone, blood pressure, pulse, and melatonin are common in depressed and bipolar patients (Atkinson et al. 1975; Kripke et al. 1978; Souetre et al. 1989). Interestingly, these rhythms usually return to normal with antidepressant or mood stabilizer treatment and patient recovery. Furthermore, genetic sleep disorders such as familial advanced phase sleep syndrome (FASPS) or delayed sleep phase syndrome (DSPS) are both highly comorbid with depression and anxiety (Shirayama et al. 2003; Xu et al. 2005; Hamet and Tremblay 2006;). Even individuals that are genetically predisposed toward "eveningness" (a preference for the evening) versus "morningness" (a preference for the morning) are more likely to develop mood disorders (Drennan et al. 1991; Chelminski et al. 1999).

Given the cyclic nature of bipolar disorder, many researchers have speculated that circadian abnormalities underlie its development (Kripke et al. 1978; Mitterauer 2000; Mansour et al. 2005). Bipolar patients have severely disrupted rhythms in nearly all measures, and this is often used as part of the disease diagnosis. In addition, it is very common for symptoms to be seasonal, in that patients are more likely to have depressive symptoms in the winter and manic symptoms during the summer (Mitterauer 2000). In many bipolar patients, manic or depressive episodes are stimulated by disruptions in their sleep/wake cycle (Frank et al. 2000). For these patients, shift work or jobs with erratic work schedules can be severely detrimental. These individuals may have a molecular clock that is unable to properly adapt to changes in the environment, and this underlies the precipitation of these episodes (Grandin et al. 2006). Often, dramatic mood stabilizing effects in bipolar patients can be obtained through strict regulation of the sleep/wake cycle (Wirz-Justice et al. 2005).

Manic or depressive episodes can also be brought on by periods of stress and this may be related to the effect of stress on the clock. The Social Zeitgeber Theory proposes that in vulnerable individuals, life stress affects sleep/wake and social rhythms, leading to circadian clock disruption and subsequent depressive or manic episodes (Grandin et al. 2006). Interpersonal and Social Rhythm Therapy has been used successfully to prevent the recurrence of these episodes in bipolar patients (Frank et al. 2000). Therefore, these disruptions in rhythms in bipolar patients could underlie their extreme mood fluctuations brought on by periods of stress.

TREATMENTS FOR MOOD DISORDERS INVOLVE THE CLOCK

Virtually all of the successful treatments for mood disorders alter circadian rhythms, and it appears that these rhythm changes are important for therapeutic efficacy.

Total sleep deprivation (TSD) is a rapid and effective short-term treatment for depression that is often used in hospitals. Similar to treatment with antidepressant drugs, it improves depressive symptoms in about 40–60% of patients (Wirz-Justice and Van den Hoofdakker 1999; Giedke and Schwarzler 2002). However, in bipolar patients, TSD can not only reverse the depression, but it can lead to a manic episode. It is thought that TSD acts by resetting the circadian clock, and several circadian phase-setting and sleep-phase hypotheses have been put forth to explain its therapeutic action (Wirz-Justice and Van den Hoofdakker 1999). Unfortunately, the effects of TSD are short, and patients often relapse after a few days. The effects can be extended by drug treatment or by a phase-advance in rhythms brought about by light therapy and regulation of the sleep/wake cycle.

Bright-light therapy on its own has been used to successfully treat seasonal depression for more than 20 years. Several studies also indicate that light therapy can be equally effective in treating nonseasonal depression, as well as other mood disorders (Terman and Terman 2005). Initially, SAD patients were treated with light therapy both in the morning and in the evening to mimic the longer days of spring and summer; however, it later was found that light therapy given exclusively in the morning (producing a phase-advance in rhythms) had a much greater therapeutic effect in most patients (Magnusson and Boivin 2003). Interestingly, a recent study by Lewy et al. (2006) found that most SAD patients are naturally phase-delayed; however, a small subgroup is phase-advanced. Specific phase-shifting treatments (either light or melatonin) given to these individuals at certain times of day lead to an optimal circadian period and alleviated their depressive symptoms. These data strongly suggest that specific circadian phase-shifting is crucial for therapeutic success in the treatment of depression.

In bipolar patients, the mood stabilizers lithium and valproate are commonly used for treatment. Both of these drugs have been repeatedly shown to alter the circadian period, leading to a long period in *Drosophila*, nonhuman primates, rodents, and humans (Johnsson et al. 1983; Welsh and Moore-Ede 1990; Klemfuss 1992; Hafen and Wollnik 1994; Dokucu et al. 2005). Lithium is also able to slow the abnormally fast circadian rhythms found in most bipolar patients (Atkinson et al. 1975; Kripke et al. 1978). Furthermore, patients that have an abnormally fast clock respond positively to lithium treatment, whereas the few bipolar patients that begin with an abnormally slow clock do not respond favorably to lithium treatment (Atkinson et al. 1975; Kripke et al. 1978). These actions of lithium on the free-running period appear to be suprachiasmatic nucleus (SCN)-dependent (LeSauter and Silver 1993).

Similar to morning bright-light therapy, the antidepressant, fluoxetine, can affect circadian output by producing a phase-advance in the firing of SCN neurons as shown in rat slice cultures (Sprouse et al. 2006). Other studies have shown that serotonin neurons from the midbrain raphe nuclei innervate the SCN, and local applications of 5-hydroxytryptamine (5-HT) or 5-HT1A and 5-HT7 receptor agonists to the SCN will also produce a phase-advance in circadian activity (Dudley et al. 1999; Ehlen et al. 2001). Thus, it is possible that antidepressants in the selective serotonin reuptake inhibitor (SSRI) class may also exert some of their effects on depression through modulation of the circadian clock. Interestingly, SSRIs and mood stabilizers can have opposing therapeutic actions in bipolar patients (Thase 2005). This could be linked to their opposing actions on rhythms because SSRIs cause a phase-advance in rhythms, whereas lithium can cause a phase-delay (Campbell et al. 1989; Sprouse et al. 2006).

Recently, agomelatine, a potent agonist of the melatonin receptors, has proven to be highly effective in animal models of depression and in several ongoing clinical trials involving patients with major depression and bipolar depression (den Boer et al. 2006; Hamon and Bourgoin 2006; Zupancic and Guilleminault 2006). As expected by its pharmacologic profile, agomelatine resynchronizes the circadian rhythms in body temperature, cortisol, and other hormones in animal models and in humans (Leproult et al. 2005). Curiously, agomelatine is much more effective as an antidepressant than melatonin itself (Srinivasan et al. 2006). Agomelatine could have different binding or kinetic properties at the melatonin receptors, which makes the response different from that seen with melatonin. Furthermore, agomelatine seems to have no effect on central serotonin transmission or the density and function of 5-HT1A receptors as seen with SSRI administration (Hanoun et al. 2004; Millan et al. 2005). However, it does have 5-HT2C antagonistic properties, and it enhances mesolimbic dopaminergic and noradrenergic transmission (Millan et al. 2003; Serretti et al. 2004). Moreover, chronic, but not acute, treatment with agomelatine induces neurogenesis in the hippocampus similar to other antidepressants (Banasr et al. 2006). Therefore, it is uncertain exactly how agomelatine alleviates depression, and if the circadian rhythm changes produced by this drug are involved.

EVIDENCE SUGGESTING A SPECIFIC ROLE FOR CLOCK AND BMAL1 IN BIPOLAR DISORDER

Several human genetic studies have implicated specific genes that make up the molecular clock in the manifestation of mood disorders. For example, genetic variants in *Npas2*, *Per2*, and *Bmal1* have been found to associate with the development of SAD (Johansson et al. 2003; Partonen et al. 2007). Bipolar disorder has been most strongly linked to variations in *Clock* and *Bmal1*. A single-nucleotide polymorphism (SNP) in the 3′-flanking region of the *Clock* gene (3111 T to C) associates with a higher recurrence rate of bipolar episodes (Benedetti et al. 2003). This SNP is also associated with greater insomnia and decreased need for sleep in bipolar patients (Serretti et al. 2003, 2005). Other genetic studies have identified haplotypes and SNPs in *Bmal1* that significantly associate with bipolar disorder in general (Mansour et al. 2006; Nievergelt et al. 2006). These genetic studies suggest an association between the CLOCK/BMAL1 complex and certain aspects of bipolar disorder.

To determine more specifically how the *Clock* gene is involved in mood regulation we tested mice that have a point mutation in this gene which creates a dominant-neg-

Table 1. Comparison of Behaviors in Humans and Mice

Symptoms of mania	*Clock* mutant mice
Disrupted circadian rhythms	disrupted circadian rhythms
Hyperactivity	hyperactivity
Decreased sleep	decreased sleep
Feelings of extreme euphoria	hyperhedonia/less helplessness
Increased risk-taking	reduced anxiety
Propensity toward drug abuse	increased preference for cocaine

Reprinted, with permission, from Roybal et al. (2007 [© National Academy of Sciences]).

ative protein (King et al. 1997). Interestingly, we found that these mice display a complete behavioral profile that is remarkably similar to that of human mania (Roybal et al. 2007). A comparison between human subjects and the *Clock* mutant mice is given in Table 1. These behaviors include general hyperactivity; a decrease in anxiety when measured in multiple tests; an increase in the reward value for sucrose, cocaine, and brain stimulation; and a decrease in measures of depression-like behavior (Roybal et al. 2007). Other groups have reported a decrease in sleep in these mice and an increase in exploratory behavior, adding to their manic-like profile (Naylor et al. 2000; Easton et al. 2003). Importantly, these mice display an increase in goal-directed activities with the potential for adverse consequences as shown by intracranial self-stimulation measures. This is a key feature in the diagnosis of the manic component of human bipolar disorder. Their increase in preference for cocaine also correlates with the high comorbidity for substance abuse in bipolar patients. Some of the largest rates of addiction occur in bipolar patients, with the greatest risk occurring in patients with frequent manic episodes (Regier et al. 1990; Kessler et al. 1996; Maremmani et al. 2006). The abuse of nearly all mood-altering drugs is significantly higher in bipolar patients than in the general population (Regier et al. 1990; Maremmani et al. 2006). However, the use of drugs such as alcohol and heroin seems to associate most strongly with the depression phase of the disorder, and patients in the manic state seem strongly drawn to psychostimulants such as amphetamines and cocaine (Estroff et al. 1985; Regier et al. 1990).

Lithium is still the most commonly prescribed mood stabilizer, and it is particularly effective in treating mania (Shastry 1997). To determine if the manic-like behavior of the *Clock* mutants can be rescued by lithium treatment, we gave LiCl in the drinking water at 600 mg/liter for 10 days as described previously (Dehpour et al. 2002). We found that this treatment produces a stable, serum Li^+ concentration of 0.41 ± 0.06 mmole/liter which is at the low end of the therapeutic range for human patients (Gelenberg et al. 1989). Chronic lithium treatment restored the levels of behavioral despair of the *Clock* mutant mice to near wild-type levels. In addition, their responses in measures of anxiety also returned to normal levels after lithium treatment (Fig. 1).

Interestingly, behavior of the wild-type mice was not significantly affected by this concentration of LiCl in most measures. However, there was a significant decrease in the levels of anxiety in the elevated plus maze. Previous studies that looked at the effect of lithium on various behavioral measures in wild-type mice have reported mixed results. A recent study by Bersudsky et al. (2007) found that lithium treatments that produced high serum levels of 1.3–1.4 mmole/liter decreased the immobility time of wild-type mice in the forced swim test; however, levels of 0.8 mmole/liter or lower had no effect. Our mice have lithium serum levels that are much lower (~0.4 mmole/liter); thus, it is consistent that the wild-type mice may not have significant differences in behavioral measures of mood. Similar to the situation in wild-type mice, lithium treatment at therapeutic doses does not have a significant mood-altering effect on healthy human volunteers (Calil et al. 1990). Therefore, the *Clock* mutant mice provide us with an excellent opportunity to study the development of mania, as well as the molecular mechanisms that underlie the efficacy of lithium as a treatment in bipolar patients.

DOES CLOCK IN THE SCN INFLUENCE MOOD?

It is unclear as to what extent the manic-like phenotypes of the *Clock* mutant mice are due to the loss of *Clock* function in the central circadian pacemaker of the SCN or to the loss of *Clock* in other brain regions. A study by Tataroglu et al. (2004) found that bilateral SCN lesions in rats have an antidepressant-like effect in the forced swim

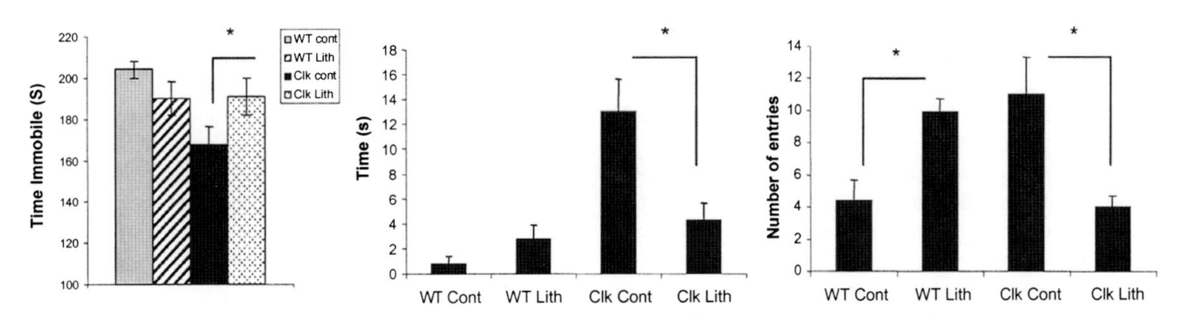

Figure 1. Effect of lithium treatment on the *Clock* mutants in behavioral measures (*left*) LiCl (600 mg/liter, 10 days) leads to an increase in the time immobile of the *Clock* mutants in the FST. LiCl treatment decreases the time spent by the *Clock* mutants in the center of an open field (*middle*) and in the open arms of the EPM (*right*) (In all tests, $*P < 0.05$, Student's *t*-test, $n = 8$.) (Reprinted, with permission, from Roybal et al. 2007 [© National Academy of Sciences].)

test. Furthermore, SCN lesions prevent the anxiolytic effects of agomelatine following repeated bouts of social defeat (Tuma et al. 2005). Therefore, the SCN might be involved in the regulation of mood. However, these lesions had no effect on general activity or baseline measures of anxiety, suggesting that other brain regions are also involved in the full behavioral spectrum associated with mania. One intriguing possible connection between mania and central clock function comes from recent reports showing phosphorylation of multiple circadian genes by glycogen synthase kinase 3β (GSK3β). Lithium is known to inhibit the actions of GSK3β, and there are indications that this action is important in mood stabilization (Gould and Manji 2005). Indeed, transgenic mice overexpressing *Gsk3*β are hyperactive, have reduced immobility in the forced swim test, and an increased startle response, reminiscent of human mania (Prickaerts et al. 2006). The *Drosophila* ortholog of GSK3β, SHAGGY, promotes the nuclear translocation of PER/TIM by phosphorylating TIM (Harms et al. 2003). In mammals, GSK3β is expressed in the SCN, and there is a robust circadian rhythm in its phosphorylation and activity in this region (Iitaka et al. 2005). Furthermore, GSK3β can phosphorylate PER2, CRY2, and Rev-erbα, leading to the proper regulation of circadian rhythms (Iitaka et al. 2005; Kurabayashi et al. 2006; Yin et al. 2006). Lithium inhibits this activity and promotes a long circadian period (Padiath et al. 2004). Therefore, it is possible that some of the therapeutic effects of lithium on mood stabilization are derived from this inhibition of GSK3β in the SCN. Intriguingly, the *Clock* mutant mice also have a long period in circadian activity rhythms when mice are kept in constant darkness (Vitaterna et al. 1994; King et al. 1997). It will be interesting to determine how lithium treatment affects the rhythms of these mice.

IMPORTANT RHYTHMS IN OTHER CIRCUITS

Some of the major neurotransmitters that have been implicated in mood regulation, including serotonin, norepinephrine, and dopamine, have a circadian rhythm in their levels, release, and synthesis-related enzymes (Weiner et al. 1992; Shieh et al. 1997; Aston-Jones et al. 2001; Barassin et al. 2002; Khaldy et al. 2002; Castaneda et al. 2004; Weber et al. 2004; Malek et al. 2005). There are also circadian rhythms in the expression and activity of several of the receptors that bind these neurotransmitters, suggesting that these entire circuits are under circadian control (Kafka et al. 1983; Wesemann and Weiner 1990; Witte and Lemmer 1991; Coon et al. 1997; Akhisaroglu et al. 2005).

The hippocampus, and in particular, the neurogenesis that occurs in this region, is thought to be important in the development and treatment of mood disorders because chronic stress inhibits neurogenesis, whereas antidepressant treatment enhances it (Dranovsky and Hen 2006). Neurogenesis and cell proliferation are under circadian control, and recent studies have found that the circadian gene, *Per2*, is involved in this process (Kochman et al. 2006). In addition, two genes highly expressed in the hippocampus, brain-derived neurotrophic factor (*Bdnf*) and its receptor, tyrosine receptor kinase B (*TrkB*), have been strongly implicated in the regulation of mood and neurogenesis, and both have a robust circadian rhythm in their expression in this region (Liang et al. 1998; Berchtold et al. 1999; Schaaf et al. 2000; Dolci et al. 2003; Kuipers and Bramham 2006). It is possible that disruptions of the rhythms in the hippocampus or in any of these other systems could lead to mood destabilization.

REGULATION OF THESE RHYTHMS BY THE SCN

The modulation of circadian neurotransmission in other brain regions may occur through indirect projections from the SCN. For example, an indirect projection from the SCN to the locus coeruleus (LC) regulates the circadian rhythm in noradrenergic neuronal activity (Aston-Jones et al. 2001). A potential indirect pathway from the SCN to the ventral tegmental area (VTA) and dorsal raphe nucleus via the dorsomedial hypothalamic nucleus has also been described (Deurveilher and Semba 2005). There is also a dependence on the SCN for the circadian regulation of certain genes in the dopaminergic neurons of the VTA (Sleipness et al. 2007). The SCN could also influence rhythms in other brain regions through its control of circulating hormones and peptides. Melatonin receptors are widely expressed, and certain types of antidepressant treatments alter MT1 and MT2 receptor levels in the hippocampus and striatum (Hirsch-Rodriguez et al. 2007). The stress hormone, corticosterone, also has a strong circulating rhythm controlled through an indirect SCN projection to the adrenal cortex, and alterations in its rhythm could influence long-term anxiety and mood (Engeland and Arnhold 2005). Furthermore, SCN's regulation of hormones and hypothalamic peptides involved in metabolism such as insulin, orexin, leptin, and ghrelin could also modulate neurotransmission in brain regions associated with mood regulation. Recent studies have found that leptin-mediated signaling in the mesolimbic dopaminergic pathway is involved in modulating neuronal activity, reward, and motivational behavior (Fulton et al. 2006; Hommel et al. 2006). There is also evidence to suggest that the orexin system regulates not only arousal states, but also dopaminergic activity, drug, and food reward (Harris and Aston-Jones 2006; Narita et al. 2006). Interestingly the *Clock* mutant mice have a metabolic phenotype leading to obesity (Turek et al. 2005). Obesity is very common in bipolar patients, and this is often worsened by lithium treatment (Newcomer 2006). In the *Clock* mutant mice, the levels and circadian rhythms of multiple hypothalamic peptides are severely attenuated (Turek et al. 2005). In addition, levels of serum leptin are significantly higher during the light phase and levels of corticosterone are lower throughout the light/dark cycle (Turek et al. 2005). It is possible that these altered hormone and peptide rhythms are involved in the manic-like state we see in these mice.

THE CIRCADIAN GENES IN OTHER BRAIN REGIONS

Circadian gene expression outside of the SCN, in key "mood-related" brain regions, may be important. These genes can form peripheral clocks that respond to SCN signals or function independently in response to certain stimuli. For example, circadian activity rhythms in rodents can be entrained to daytime methamphetamine injections, even in SCN lesioned animals (Iijima et al. 2002). This treatment shifts the expression of the *period* genes in striatal regions in a manner that matches the shift in activity rhythms (Iijima et al. 2002). This same shift in *period* gene expression does not occur in the SCN with methamphetamine treatment; thus, there is a disconnect between the molecular rhythms in the SCN and striatum.

Interestingly, a microarray study by Ogden et al. (2004) found that the mood stabilizer, valproate, decreased the expression of the circadian genes *CK1δ* and *Cry2* in the amygdala, a region of the brain associated with emotional behavior and fear. These changes were prevented by cotreatment with methamphetamine, which was given to induce manic-like symptoms, suggesting that these changes may be involved in the treatment of mania (Ogden et al. 2004). Therefore, mood stabilizer treatment may involve a change in rhythms in the amygdala.

A recent study by Uz et al. (2005) found that treatment with the antidepressant fluoxetine altered the expression of *Clock* and *Bmal1*, along with *Npas2* in the mouse hippocampus. These same changes did not occur in striatal regions, indicating that they may be brain-region-specific. Furthermore, they were all induced by chronic and not acute fluoxetine, suggesting that these changes may be therapeutically relevant because fluoxetine must be administered for days to weeks to see significant antidepressant effects in humans (Uz et al. 2005). Therefore, this complex may be involved in rhythm and mood regulation through expression in the hippocampus. In addition, our recent studies suggest that *Clock* functions in the dopamine cells of the VTA to regulate the expression of other circadian genes and several genes involved in dopaminergic transmission including the rate-limiting enzyme in dopamine synthesis tyrosine hydroxylase (TH) (McClung et al. 2005; Roybal et al. 2007). Interestingly, TH and a number of other genes involved in dopaminergic transmission such as cholecystokinin, the dopamine transporter, and various dopamine receptors, all have a robust circadian rhythm in expression, suggesting that they might be regulated by circadian genes (Weber et al. 2004; Sleipness et al. 2007). We have found that the *Clock* mutant mice have an increase in dopaminergic activity in the VTA that correlates with their manic-like behavior (Fig. 2) (McClung et al. 2005). Furthermore, when we express a functional CLOCK protein specifically in the VTA using virus-mediated gene transfer, we are able to rescue at least a portion of their manic behavior (Fig. 3) (Roybal et al. 2007). The effects of CLOCK restoration in the VTA in other behaviors associated with mania have yet to be tested. These studies suggest that proper CLOCK expression in the VTA is important in the regulation of dopaminergic activity and at least some of

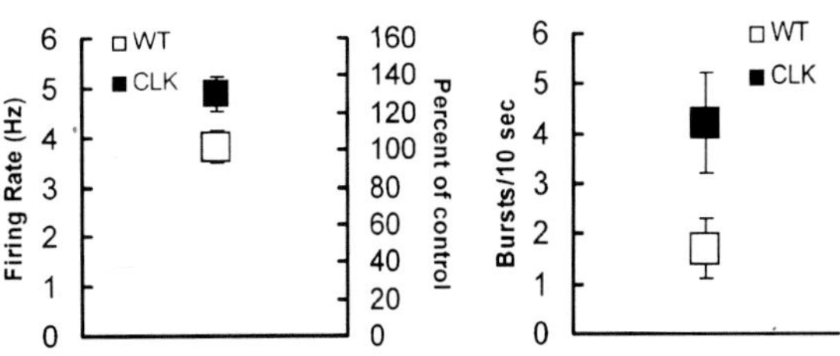

Figure 2. Dopamine cell-firing rates and bursting are increased in *Clock* mutant mice. (*Top left*) Representative recording of a mouse dopamine neuron. (*Right*) Example of an averaged dopamine triphasic waveform. (*Bottom left*) *Clock* mutant mice (*Clk*) ($n = 24$) exhibited higher basal dopamine firing rates compared with wild type (WT) ($n = 26$) mice ($P < 0.03$). Squares represent the mean ±S.E.M. of each group. (*Bottom right*) *Clock* mutant mice exhibited more burst events per 10 seconds compared with wild type controls ($P < 0.01$). Each vertical bar represents the mean ±S.E.M. of each group. The percentage of bursting activity, i.e., spikes emitted in bursts, was greater compared with wild-type mice (Clk, 23.5 ± 4.3 vs. 10.6 ± 3.6 in WT; $P < 0.03$). *Clock* mutant mice had larger burst sizes (number of spikes/burst) compared with controls (2.8 ± 0.1 vs. 2.0 ± 0.1; $P < 0.008$). (Reprinted, with permission, from McClung et al. 2005 [© National Academy of Sciences].)

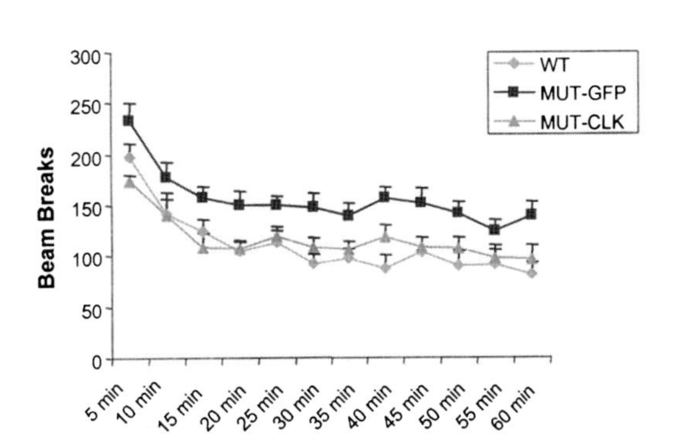

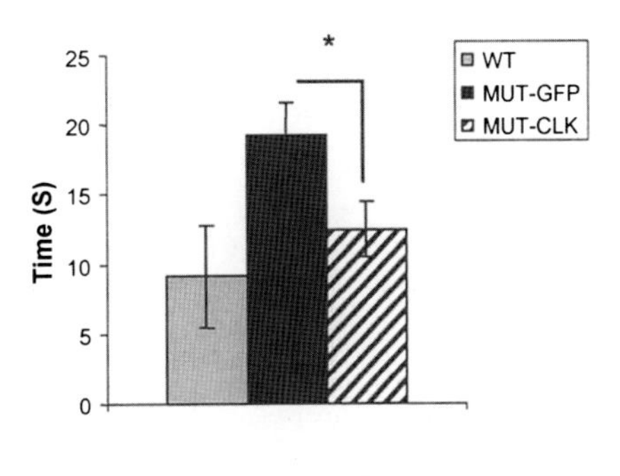

Figure 3. Effect of CLOCK viral expression in the VTA of *Clock* mutant mice. (*Left*) Locomotor hyperactivity in response to novelty and time spent in the center of an open field (*right*) is reduced following two weeks of functional CLOCK overexpression in the VTA of *Clock* mutant mice (For (*left*) total locomotor activity, levels are significantly different, $*P < 0.05$, Student's *t*-test, $n = 9–12$). (Reprinted, with permission, from Roybal et al. 2007 [© National Academy of Sciences].)

the behavior associated with mania. Future studies will determine to what extent other circadian genes and other brain regions are involved in these responses.

CONCLUSIONS

There is considerable evidence suggesting a connection between circadian rhythms and bipolar disorder. Studies are starting to unravel this complex association between disruptions in rhythms and mood regulation. Some of this regulation might be SCN-dependent, whereas other components are not. The *Clock* gene appears to have an important role in this disorder, however, it is still unclear how a disruption in CLOCK function leads to a manic-like state. Our data suggest that at least a subset of these behavioral abnormalities are due to a loss of CLOCK function in the dopamine cells of the VTA. Because CLOCK is a transcription factor, it is likely that its regulation of gene expression is centrally involved. CLOCK is widely expressed and regulates the expression of many different genes in many different tissues including regions of the brain implicated in mood-associated behavior (Miller et al. 2007). The large task ahead is to determine the target genes of CLOCK in specific brain regions that are the most relevant in mood regulation.

ACKNOWLEDGMENTS

I thank Joseph Peevey and Edgardo Falcon for assistance with the manuscript. We also thank the National Institute on Drug Abuse (NIDA), National Institute of Mental Health (NIMH), and National Alliance for Research on Schizophrenia and Depression (NARSAD) for financial support.

REFERENCES

Akhisaroglu M., Kurtuncu M., Manev H., and Uz T. 2005. Diurnal rhythms in quinpirole-induced locomotor behaviors and striatal D2/D3 receptor levels in mice. *Pharmacol. Biochem. Behav.* **80:** 371.

Aston-Jones G., Chen S., Zhu Y., and Oshinsky M.L. 2001. A neural circuit for circadian regulation of arousal. *Nat. Neurosci.* **4:** 732.

Atkinson M., Kripke D.F., and Wolf S.R. 1975. Autorhythmometry in manic-depressives. *Chronobiologia* **2:** 325.

Banasr M., Soumier A., Hery M., Mocaer E., and Daszuta A. 2006. Agomelatine, a new antidepressant, induces regional changes in hippocampal neurogenesis. *Biol. Psychiatry* **59 :** 1087.

Barassin S., Raison S., Saboureau M., Bienvenu C., Maitre M., Malan A., and Pevet P. 2002. Circadian tryptophan hydroxylase levels and serotonin release in the suprachiasmatic nucleus of the rat. *Eur. J. Neurosci.* **15:** 833.

Benedetti F., Serretti A., Colombo C., Barbini B., Lorenzi C., Campori E., and Smeraldi E. 2003. Influence of CLOCK gene polymorphism on circadian mood fluctuation and illness recurrence in bipolar depression. *Am. J. Med. Genet. B Neuropsychiatr. Genet.* **123:** 23.

Berchtold N.C., Oliff H.S., Isackson P., and Cotman C.W. 1999. Hippocampal BDNF mRNA shows a diurnal regulation, primarily in the exon III transcript. *Brain Res. Mol. Brain Res.* **71:** 11.

Bersudsky Y., Shaldubina A., and Belmaker R.H. 2007. Lithium's effect in forced-swim test is blood level dependent but not dependent on weight loss. *Behav. Pharmacol.* **18:** 77.

Boivin D.B. 2000. Influence of sleep-wake and circadian rhythm disturbances in psychiatric disorders. *J Psychiatry Neurosci.* **25:** 446.

Booker J.M., Hellekson C.J., Putilov A.A., and Danilenko K.V. 1991. Seasonal depression and sleep disturbances in Alaska and Siberia: A pilot study. *Arctic Med. Res.* **Suppl.:** 281.

Bunney W.E. and Bunney B.G. 2000. Molecular clock genes in man and lower animals: Possible implications for circadian abnormalities in depression. *Neuropsychopharmacology* **22:** 335.

Calil H.M., Zwicker A.P., and Klepacz S. 1990. The effects of lithium carbonate on healthy volunteers: Mood stabilization? *Biol. Psychiatry* **27:** 711.

Campbell S.S., Gillin J.C., Kripke D.F., Janowsky D.S., and Risch S.C. 1989. Lithium delays circadian phase of temperature and REM sleep in a bipolar depressive: A case report. *Psychiatry Res.* **27:** 23.

Castaneda T.R., de Prado B.M., Prieto D., and Mora F. 2004. Circadian rhythms of dopamine, glutamate and GABA in the striatum and nucleus accumbens of the awake rat: Modulation by light. *J. Pineal Res.* **36:** 177.

Chelminski I., Ferraro F.R., Petros T.V., and Plaud J.J. 1999. An analysis of the "eveningness-morningness" dimension in "depressive" college students. *J. Affect. Disord.* **52:** 19.

Coon S.L., McCune S.K., Sugden D., and Klein D.C. 1997. Regulation of pineal alpha1B-adrenergic receptor mRNA: Day/night rhythm and beta-adrenergic receptor/cyclic AMP control. *Mol. Pharmacol.* **51:** 551.

Dehpour A.R., Sadr S.S., Azizi M.R., Namiranian K., Farahani M., and Javidan A.N. 2002. Lithium inhibits the development of physical dependence to clonidine in mice. *Pharmacol. Toxicol.* **90:** 89.

den Boer J.A., Bosker F.J., and Meesters Y. 2006. Clinical efficacy of agomelatine in depression: The evidence. *Int. Clin. Psychopharmacol.* (suppl. 1) **21:** S21.

Deurveilher S. and Semba K. 2005. Indirect projections from the suprachiasmatic nucleus to major arousal-promoting cell groups in rat: Implications for the circadian control of behavioural state. *Neuroscience* **130:** 165.

Dokucu M.E., Yu L., and Taghert P.H. 2005. Lithium- and valproate-induced alterations in circadian locomotor behavior in *Drosophila*. *Neuropsychopharmacology* **30:** 2216.

Dolci C., Montaruli A., Roveda E., Barajon I., Vizzotto L., Grassi Zucconi G., and Carandente F. 2003. Circadian variations in expression of the trkB receptor in adult rat hippocampus. *Brain Res.* **994:** 67.

Dranovsky A. and Hen R. 2006. Hippocampal neurogenesis: Regulation by stress and antidepressants. *Biol. Psychiatry* **59:** 1136.

Drennan M.D., Klauber M.R., Kripke D.F., and Goyette L.M. 1991. The effects of depression and age on the Horne-Ostberg morningness-eveningness score. *J. Affect. Disord.* **23:** 93.

Dudley T.E., Dinardo L.A., and Glass J.D. 1999. In vivo assessment of the midbrain raphe nuclear regulation of serotonin release in the hamster suprachiasmatic nucleus. *J. Neurophysiol.* **81:** 1469.

Easton A., Arbuzova J., and Turek F.W. 2003. The circadian Clock mutation increases exploratory activity and escape-seeking behavior. *Genes Brain Behav.* **2:** 11.

Ehlen J.C., Grossman G.H., and Glass J.D. 2001. In vivo resetting of the hamster circadian clock by 5-HT7 receptors in the suprachiasmatic nucleus. *J. Neurosci.* **21:** 5351.

Engeland W.C. and Arnhold M.M. 2005. Neural circuitry in the regulation of adrenal corticosterone rhythmicity. *Endocrine* **28:** 325.

Estroff T.W., Dackis C.A., Gold M.S., and Pottash A.L. 1985. Drug abuse and bipolar disorders. *Int. J. Psychiatry Med.* **15:** 37.

Frank E., Swartz H.A., and Kupfer D.J. 2000. Interpersonal and social rhythm therapy: Managing the chaos of bipolar disorder. *Biol. Psychiatry* **48:** 593.

Fulton S., Pissios P., Manchon R.P., Stiles L., Frank L., Pothos E.N., Maratos-Flier E., and Flier J.S. 2006. Leptin regulation of the mesoaccumbens dopamine pathway. *Neuron* **51:** 811.

Gelenberg A.J., Kane J.M., Keller M.B., Lavori P., Rosenbaum

J.F., Cole K., and Lavelle J. 1989. Comparison of standard and low serum levels of lithium for maintenance treatment of bipolar disorder. *N. Engl. J. Med.* **321:** 1489.

Giedke H. and Schwarzler F. 2002. Therapeutic use of sleep deprivation in depression. *Sleep Med. Rev.* **6:** 361.

Gould T.D. and Manji H.K. 2005. Glycogen synthase kinase-3: A putative molecular target for lithium mimetic drugs. *Neuropsychopharmacology* **30:** 1223.

Grandin L.D., Alloy L.B., and Abramson L.Y. 2006. The social zeitgeber theory, circadian rhythms, and mood disorders: Review and evaluation. *Clin. Psychol. Rev.* **26:** 679.

Hafen T. and Wollnik F. 1994. Effect of lithium carbonate on activity level and circadian period in different strains of rats. *Pharmacol. Biochem. Behav.* **49:** 975.

Hamet P. and Tremblay J. 2006. Genetics of the sleep-wake cycle and its disorders. *Metabolism* **55:** S7.

Hamon M. and Bourgoin S. 2006. Pharmacological profile of antidepressants: A likely basis for their efficacy and side effects? *Eur. Neuropsychopharmacol.* (suppl. 5) **16:** S625.

Hanoun N., Mocaer E., Boyer P.A., Hamon M., and Lanfumey L. 2004. Differential effects of the novel antidepressant agomelatine (S 20098) versus fluoxetine on 5-HT1A receptors in the rat brain. *Neuropharmacology* **47:** 515.

Harms E., Young M.W., and Saez L. 2003. CK1 and GSK3 in the *Drosophila* and mammalian circadian clock. *Novartis Found. Symp.* **253:** 267.

Harris G.C. and Aston-Jones G. 2006. Arousal and reward: A dichotomy in orexin function. *Trends Neurosci.* **29:** 571.

Hirsch-Rodriguez E., Imbesi M., Manev R., Uz T., and Manev H. 2007. The pattern of melatonin receptor expression in the brain may influence antidepressant treatment. *Med. Hypotheses* **69:** 120.

Hommel J.D., Trinko R., Sears R.M., Georgescu D., Liu Z.W., Gao X.B., Thurmon J.J., Marinelli M., and DiLeone R.J. 2006. Leptin receptor signaling in midbrain dopamine neurons regulates feeding. *Neuron* **51:** 801.

Iijima M., Nikaido T., Akiyama M., Moriya T., and Shibata S. 2002. Methamphetamine-induced, suprachiasmatic nucleus-independent circadian rhythms of activity and mPer gene expression in the striatum of the mouse. *Eur. J. Neurosci.* **16:** 921.

Iitaka C., Miyazaki K., Akaike T., and Ishida N. 2005. A role for glycogen synthase kinase-3beta in the mammalian circadian clock. *J. Biol. Chem.* **280:** 29397.

Johansson C., Willeit M., Smedh C., Ekholm J., Paunio T., Kieseppa T., Lichtermann D., Praschak-Rieder N., Neumeister A., Nilsson L.G., Kasper S., Peltonen L., Adolfsson R., Schalling M., and Partonen T. 2003. Circadian clock-related polymorphisms in seasonal affective disorder and their relevance to diurnal preference. *Neuropsychopharmacology* **28:** 734.

Johnsson A., Engelmann W., Pflug B., and Klemke W. 1983. Period lengthening of human circadian rhythms by lithium carbonate, a prophylactic for depressive disorders. *Int. J. Chronobiol.* **8:** 129.

Kafka M.S., Wirz-Justice A., Naber D., Moore R.Y., and Benedito M.A. 1983. Circadian rhythms in rat brain neurotransmitter receptors. *Fed. Proc.* **42:** 2796.

Kessler R.C., Nelson C.B., McGonagle K.A., Edlund M.J., Frank R.G., and Leaf P.J. 1996. The epidemiology of co-occurring addictive and mental disorders: Implications for prevention and service utilization. *Am. J. Orthopsychiatry* **66:** 17.

Khaldy H., Leon J., Escames G., Bikjdaouene L., Garcia J.J., and Acuna-Castroviejo D. 2002. Circadian rhythms of dopamine and dihydroxyphenyl acetic acid in the mouse striatum: Effects of pinealectomy and of melatonin treatment. *Neuroendocrinology* **75:** 201.

King D.P., Zhao Y., Sangoram A.M., Wilsbacher L.D., Tanaka M., Antoch M.P., Steeves T.D., Vitaterna M.H., Kornhauser J.M., Lowrey P.L., Turek F.W., and Takahashi J.S. 1997. Positional cloning of the mouse circadian clock gene. *Cell* **89:** 641.

Klemfuss H. 1992. Rhythms and the pharmacology of lithium. *Pharmacol. Ther.* **56:** 53.

Kochman L.J., Weber E.T., Fornal C.A., and Jacobs B.L. 2006. Circadian variation in mouse hippocampal cell proliferation. *Neurosci. Lett.* **406:** 256.

Kripke D.F., Mullaney D.J., Atkinson M., and Wolf S. 1978. Circadian rhythm disorders in manic-depressives. *Biol. Psychiatry* **13:** 335.

Kuipers S.D. and Bramham C.R. 2006. Brain-derived neurotrophic factor mechanisms and function in adult synaptic plasticity: New insights and implications for therapy. *Curr. Opin. Drug Discov. Dev.* **9:** 580.

Kurabayashi N., Hirota T., Harada Y., Sakai M., and Fukada Y. 2006. Phosphorylation of mCRY2 at Ser557 in the hypothalamic suprachiasmatic nucleus of the mouse. *Chronobiol. Int.* **23:** 129.

Lam R.W. and Levitan R.D. 2000. Pathophysiology of seasonal affective disorder: A review. *J. Psychiatry Neurosci.* **25:** 469.

Lenox R.H., Gould T.D., and Manji H.K. 2002. Endophenotypes in bipolar disorder. *Am. J. Med. Genet.* **114:** 391.

Leproult R., Van Onderbergen A., L'Hermite-Baleriaux M., Van Cauter E., and Copinschi G. 2005. Phase-shifts of 24-h rhythms of hormonal release and body temperature following early evening administration of the melatonin agonist agomelatine in healthy older men. *Clin. Endocrinol.* **63:** 298.

LeSauter J. and Silver R. 1993. Lithium lengthens the period of circadian rhythms in lesioned hamsters bearing SCN grafts. *Biol. Psychiatry* **34:** 75.

Lewy A.J., Lefler B.J., Emens J.S., and Bauer V.K. 2006. The circadian basis of winter depression. *Proc. Natl. Acad. Sci.* **103:** 7414.

Liang F.Q., Walline R., and Earnest D.J. 1998. Circadian rhythm of brain-derived neurotrophic factor in the rat suprachiasmatic nucleus. *Neurosci. Lett.* **242:** 89.

Magnusson A. and Boivin D. 2003. Seasonal affective disorder: An overview. *Chronobiol. Int.* **20:** 189.

Malek Z.S., Dardente H., Pevet P., and Raison S. 2005. Tissue-specific expression of tryptophan hydroxylase mRNAs in the rat midbrain: Anatomical evidence and daily profiles. *Eur. J. Neurosci.* **22:** 895.

Mansour H.A., Monk T.H., and Nimgaonkar V.L. 2005. Circadian genes and bipolar disorder. *Ann. Med.* **37:** 196.

Mansour H.A., Wood J., Logue T., Chowdari K.V., Dayal M., Kupfer D.J., Monk T.H., Devlin B., and Nimgaonkar V.L. 2006. Association study of eight circadian genes with bipolar I disorder, schizoaffective disorder and schizophrenia. *Genes Brain Behav.* **5:** 150.

Maremmani I., Perugi G., Pacini M., and Akiskal H. S. 2006. Toward a unitary perspective on the bipolar spectrum and substance abuse: opiate addiction as a paradigm. *J. Affect. Disord.* **93:** 1.

McClung C.A. 2007. Circadian genes, rhythms and the biology of mood disorders. *Pharmacol. Ther.* **114:** 222.

McClung C.A., Sidiropoulou K., Vitaterna M., Takahashi J.S., White F.J., Cooper D.C., and Nestler E.J. 2005. Regulation of dopaminergic transmission and cocaine reward by the Clock gene. *Proc. Natl. Acad. Sci.* **102:** 9377.

Millan M.J., Brocco M., Gobert A., and Dekeyne A. 2005. Anxiolytic properties of agomelatine, an antidepressant with melatoninergic and serotonergic properties: Role of 5-HT2C receptor blockade. *Psychopharmacology* **177:** 448.

Millan M.J., Gobert A., Lejeune F., Dekeyne A., Newman-Tancredi A., Pasteau V., Rivet J.M., and Cussac D. 2003. The novel melatonin agonist agomelatine (S20098) is an antagonist at 5-hydroxytryptamine2C receptors, blockade of which enhances the activity of frontocortical dopaminergic and adrenergic pathways. *J. Pharmacol. Exp. Ther.* **306:** 954.

Miller B.H., McDearmon E.L., Panda S., Hayes K.R., Zhang J., Andrews J.L., Antoch M.P., Walker J.R., Esser K.A., Hogenesch J.B., and Takahashi J. S. 2007. Circadian and CLOCK-controlled regulation of the mouse transcriptome and cell proliferation. *Proc. Natl. Acad. Sci.* **104:** 3342.

Mitterauer B. 2000. Clock genes, feedback loops and their possible role in the etiology of bipolar disorders: An integrative model. *Med. Hypotheses* **55:** 155.

Narita M., Nagumo Y., Hashimoto S., Khotib J., Miyatake M.,

Sakurai T., Yanagisawa M., Nakamachi T., Shioda S., and Suzuki T. 2006. Direct involvement of orexinergic systems in the activation of the mesolimbic dopamine pathway and related behaviors induced by morphine. *J. Neurosci.* **26:** 398.

Naylor E., Bergmann B.M., Krauski K., Zee P.C., Takahashi J.S., Vitaterna M.H., and Turek F.W. 2000. The circadian clock mutation alters sleep homeostasis in the mouse. *J. Neurosci.* **20:** 8138.

Newcomer J.W. 2006. Medical risk in patients with bipolar disorder and schizophrenia. *J. Clin. Psychiatry* **67:** e16.

Nievergelt C.M., Kripke D.F., Barrett T.B., Burg E., Remick R.A., Sadovnick A.D., McElroy S.L., Keck P.E., Jr., Schork N.J., and Kelsoe J.R. 2006. Suggestive evidence for association of the circadian genes PERIOD3 and ARNTL with bipolar disorder. *Am. J. Med. Genet. B Neuropsychiatr. Genet.* **141:** 234.

Ogden C.A., Rich M.E., Schork N.J., Paulus M.P., Geyer M.A., Lohr J.B., Kuczenski R., and Niculescu A.B. 2004. Candidate genes, pathways and mechanisms for bipolar (manic-depressive) and related disorders: An expanded convergent functional genomics approach. *Mol. Psychiatry* **9:** 1007.

Padiath Q.S., Paranjpe D., Jain S., and Sharma V.K. 2004. Glycogen synthase kinase 3beta as a likely target for the action of lithium on circadian clocks. *Chronobiol. Int.* **21:** 43.

Partonen T., Treutlein J., Alpman A., Frank J., Johansson C., Depner M., Aron L., Rietschel M., Wellek S., Soronen P., Paunio T., Koch A., Chen P., Lathrop M., Adolfsson R., Persson M.L., Kasper S., Schalling M., Peltonen L., and Schumann G. 2007. Three circadian clock genes Per2, Arntl, and Npas2 contribute to winter depression. *Ann. Med.* **39:** 229.

Prickaerts J., Moechars D., Cryns K., Lenaerts I., van Craenendonck H., Goris I., Daneels G., Bouwknecht J.A., and Steckler T. 2006. Transgenic mice overexpressing glycogen synthase kinase 3beta: A putative model of hyperactivity and mania. *J. Neurosci.* **26:** 9022.

Regier D.A., Farmer M.E., Rae D.S., Locke B.Z., Keith S.J., Judd L.L., and Goodwin F.K. 1990. Comorbidity of mental disorders with alcohol and other drug abuse. Results from the Epidemiologic Catchment Area (ECA) Study. *J. Am. Med. Assoc.* **264:** 2511.

Roybal K., Theobold D., Graham A., Dinieri J.A., Russo S.J., Krishnan V., Chakravarty S., Peevey J., Oehrlein N., Birnbaum S., Vitaterna M.H., Orsulak P., Takahashi J.S., Nestler E.J., Carlezon W.A., Jr., and McClung C.A. 2007. Mania-like behavior induced by disruption of CLOCK (from the cover). *Proc. Natl. Acad. Sci.* **104:** 6406.

Rusting C.L. and Larsen R.J. 1998. Diurnal patterns of unpleasant mood: Associations with neuroticism, depression, and anxiety. *J. Pers.* **66:** 85.

Schaaf M.J., Duurland R., de Kloet E.R., and Vreugdenhil E. 2000. Circadian variation in BDNF mRNA expression in the rat hippocampus. *Brain Res. Mol. Brain Res.* **75:** 342.

Serretti A., Artioli P., and De Ronchi D. 2004. The 5-HT2C receptor as a target for mood disorders. *Expert Opin. Ther. Targets* **8:** 15.

Serretti A., Benedetti F., Mandelli L., Lorenzi C., Pirovano A., Colombo C., and Smeraldi E. 2003. Genetic dissection of psychopathological symptoms: Insomnia in mood disorders and CLOCK gene polymorphism. *Am. J. Med. Genet. B Neuropsychiatr. Genet.* **121:** 35.

Serretti A., Cusin C., Benedetti F., Mandelli L., Pirovano A., Zanardi R., Colombo C., and Smeraldi E. 2005. Insomnia improvement during antidepressant treatment and CLOCK gene polymorphism. *Am. J. Med. Genet. B Neuropsychiatr. Genet.* **137:** 36.

Shastry B.S. 1997. On the functions of lithium: The mood stabilizer. *Bioessays* **19:** 199.

Shieh K.R., Chu Y.S., and Pan J.T. 1997. Circadian change of dopaminergic neuron activity: Effects of constant light and melatonin. *Neuroreport* **8:** 2283.

Shirayama M., Shirayama Y., Iida H., Kato M., Kajimura N., Watanabe T., Sekimoto M., Shirakawa S., Okawa M., and Takahashi K. 2003. The psychological aspects of patients with delayed sleep phase syndrome (DSPS). *Sleep Med.* **4:** 427.

Sleipness E.P., Sorg B.A., and Jansen H.T. 2007. Diurnal differences in dopamine transporter and tyrosine hydroxylase levels in rat brain: Dependence on the suprachiasmatic nucleus. *Brain Res.* **1129:** 34.

Souetre E., Salvati E., Belugou J.L., Pringuey D., Candito M., Krebs B., Ardisson J.L., and Darcourt G. 1989. Circadian rhythms in depression and recovery: Evidence for blunted amplitude as the main chronobiological abnormality. *Psychiatry Res.* **28:** 263.

Sprouse J., Braselton J., and Reynolds L. 2006. Fluoxetine modulates the circadian biological clock via phase advances of suprachiasmatic nucleus neuronal firing. *Biol. Psychiatry* **60:** 896.

Srinivasan V., Smits M., Spence W., Lowe A.D., Kayumov L., Pandi-Perumal S.R., Parry B., and Cardinali D.P. 2006. Melatonin in mood disorders. *World J. Biol. Psychiatry* **7:** 138.

Tataroglu O., Aksoy A., Yilmaz A., and Canbeyli R. 2004. Effect of lesioning the suprachiasmatic nuclei on behavioral despair in rats. *Brain Res.* **1001:** 118.

Terman M. and Terman J.S. 2005. Light therapy for seasonal and nonseasonal depression: Efficacy, protocol, safety, and side effects. *CNS Spectr.* **10:** 647.

Thase, M. E. 2005. Bipolar depression: Issues in diagnosis and treatment. *Harv. Rev. Psychiatry* **13:** 257.

Tuma J., Strubbe J.H., Mocaer E., and Koolhaas J.M. 2005. Anxiolytic-like action of the antidepressant agomelatine (S 20098) after a social defeat requires the integrity of the SCN. *Eur. Neuropsychopharmacol.* **15:** 545.

Turek F.W., Joshu C., Kohsaka A., Lin E., Ivanova G., McDearmon E., Laposky A., Losee-Olson S., Easton A., Jensen D.R., Eckel R.H., Takahashi J.S., and Bass J. 2005. Obesity and metabolic syndrome in circadian Clock mutant mice. *Science* **308:** 1043.

Uz T., Ahmed R., Akhisaroglu M., Kurtuncu M., Imbesi M., Dirim Arslan A., and Manev H. 2005. Effect of fluoxetine and cocaine on the expression of clock genes in the mouse hippocampus and striatum. *Neuroscience* **134:** 1309.

Vitaterna M.H., King D.P., Chang A.M., Kornhauser J.M., Lowrey P.L., McDonald J.D., Dove W.F., Pinto L.H., Turek F.W., and Takahashi J.S. 1994. Mutagenesis and mapping of a mouse gene, Clock, essential for circadian behavior. *Science* **264:** 719.

Weber M., Lauterburg T., Tobler I., and Burgunder J.M. 2004. Circadian patterns of neurotransmitter related gene expression in motor regions of the rat brain. *Neurosci. Lett.* **358:** 17.

Weiner N., Clement H.W., Gemsa D., and Wesemann W. 1992. Circadian and seasonal rhythms of 5-HT receptor subtypes, membrane anisotropy and 5-HT release in hippocampus and cortex of the rat. *Neurochem. Int.* **21:** 7.

Welsh D.K. and Moore-Ede M.C. 1990. Lithium lengthens circadian period in a diurnal primate, *Saimiri sciureus*. *Biol. Psychiatry* **28:** 117.

Wesemann W. and Weiner N. 1990. Circadian rhythm of serotonin binding in rat brain. *Prog. Neurobiol.* **35:** 405.

Wirz-Justice A. and Van den Hoofdakker R.H. 1999. Sleep deprivation in depression: What do we know, where do we go? *Biol. Psychiatry* **46:** 445.

Wirz-Justice A., Benedetti F., Berger M., Lam R.W., Martiny K., Terman M., and Wu J.C. 2005. Chronotherapeutics (light and wake therapy) in affective disorders. *Psychol. Med.* **35:** 939.

Witte K. and Lemmer B. 1991. Rhythms in second messenger mechanisms. *Pharmacol. Ther.* **51:** 231.

Xu Y., Padiath Q.S., Shapiro R.E., Jones C.R., Wu S.C., Saigoh N., Saigoh K., Ptacek L.J., and Fu Y.H. 2005. Functional consequences of a CKIdelta mutation causing familial advanced sleep phase syndrome. *Nature* **434:** 640.

Yin L., Wang J., Klein P.S., and Lazar M. A. 2006. Nuclear receptor Rev-erbα is a critical lithium-sensitive component of the circadian clock. *Science* **311:** 1002.

Zupancic M. and Guilleminault C. 2006. Agomelatine: A preliminary review of a new antidepressant. *CNS Drugs* **20:** 981.

The Possible Interplay of Synaptic and Clock Genes in Autism Spectrum Disorders

T. Bourgeron
Human Genetics and Cognitive Functions Unit, Department of Neuroscience, Institut Pasteur, Paris, France, and Université Denis Diderot Paris 7, Paris, France

Autism spectrum disorders (ASD) are complex neurodevelopmental conditions characterized by deficits in social communication, absence or delay in language, and stereotyped and repetitive behaviors. Results from genetic studies reveal one pathway associated with susceptibility to ASD, which includes the synaptic cell adhesion molecules NLGN3, NLGN4, and NRXN1 and a postsynaptic scaffolding protein SHANK3. This protein complex is crucial for the maintenance of functional synapses as well as the adequate balance between neuronal excitation and inhibition. Among the factors that could modulate this pathway are the genes controlling circadian rhythms. Indeed, sleep disorders and low melatonin levels are frequently observed in ASD. In this context, an alteration of both this synaptic pathway and the setting of the clock would greatly increase the risk of ASD. In this chapter, I report genetic and neurobiological findings that highlight the major role of synaptic and clock genes in the susceptibility to ASD. On the basis of these lines of evidence, I propose that future studies of ASD should investigate the circadian modulation of synaptic function as a focus for functional analyses and the development of new therapeutic strategies.

INTRODUCTION

As a result of a highly complex phylogenetic and ontogenic process, humans have acquired the ability to communicate with language and highly specialized skills to recognize social cues (eye gaze, joint attention, theory of mind, empathy). In some individuals, this ability to communicate is hampered by the occurrence of genetic/epigenetic variations and/or environmental insults. After exclusion of known biological diseases (e.g., deafness) and known environmental causes (e.g., social and teaching), approximately 8–10% of school age children suffer from language and/or communication difficulties (Shaywitz et al. 1990; Fombonne 2005). One of the most severe syndromes associated with an alteration of language and social communication is autism.

Autism was first described by the psychiatrist Leo Kanner in 1943 and is diagnosed on the basis of three behaviorally altered domains, namely, social deficits, impaired language and communication, and stereotyped and repetitive behaviors (Kanner 1943). Beyond this unifying definition lies an extreme degree of clinical heterogeneity, ranging from debilitating impairments to mild personality traits. Hence, autism is not a single disease entity, but rather a complex phenotype encompassing either multiple "autistic disorders" or a continuum of autistic-like traits and behaviors. To take into account this heterogeneity, the term autism spectrum disorders (ASD) is now used and includes autistic syndrome, pervasive development disorder not otherwise specified (PDD-NOS), Asperger syndrome, childhood disintegrative disorder (CDD), and Rett syndrome (APA 1994). Whereas CDD and Rett syndrome are severe neurological disorders, Asperger syndrome refers to the portion of the ASD continuum characterized by higher cognitive abilities and by more normal language function.

The behavioral singularities that occur in ASD are related to a wide spectrum of cognitive functions such as language, memory, and visual and auditive attentions. Two of these cognitive deficits are rather characteristic of ASD: a weak "central coherence" and the lack of a "theory of mind." "Central coherence" defines our ability to understand context. Individuals with ASD are sometimes better than age-matched controls in detecting details in a picture, but they have great difficulty in seeing "the bigger picture" and in understanding the context of the situation (Frith 1998). The term "theory of mind" describes an individual's understanding of the motives, knowledge, and beliefs of others. Individuals with ASD have an absence of theory of mind or a delay in the acquisition of it. This deficit could be a major cause of their difficulties in social interactions (Baron-Cohen et al. 1985).

Epidemiologic studies report a dramatic rise of ASD during the last two decades (from 2–5 to 60/10,000 children). However, this recent increase is most likely explained by the use of broader diagnostic criteria and the increased attention by the medical community (Fombonne 2005). For still unknown reasons, males are more frequently affected than females. The male-to-female ratio is 4:1, but it increases to 23:1 in individuals without identified morphological or brain abnormalities.

GENETIC CAUSES OF AUTISM SPECTRUM DISORDERS

Since the original reports by Kanner and Asperger, many studies have advocated a genetic etiology for autism (Freitag 2007). Familial cases are substantially more frequent than expected by chance. Hence, the recurrence risk of autism in sibships is approximately 45 times greater than in the general population. Furthermore, twin studies have documented a higher concordance rate in monozygotic (60–91%) than in dizygotic twins (0–6%) (Bailey et al. 1995). However, due to the heterogeneity of the syndrome and the absence of apparent mendelian segregation, the mode of inheritance of ASD is still a matter of debate.

The long-time postulated polygenic model was recently challenged by the identification of rare cases of an apparently monogenic form of ASD caused by a single-gene mutation or de novo copy-number variants (CNVs) (Jamain et al. 2003; Jacquemont et al. 2006; Durand et al. 2007; Sebat et al. 2007, Szatmari et al. 2007). Although currently restricted to a limited number of patients, these apparent monogenic forms of ASD may be more frequent than originally expected. In addition, epigenetic anomalies are strongly suspected, but their actual impact on autism remains largely unknown.

In approximately 30% of the cases, ASD are associated with a known genetic syndrome or with chromosomal rearrangements (Fig. 1) (Freitag 2007). The association with known genetic disorders indicates that anomalies in distinct physiological processes such as chromatin remodeling (Rett syndrome; MECP2), synaptic gene regulation (Fragile-X syndrome, FMRP), protein synthesis and actin cytoskeleton dynamics (tuberous sclerosis, TSC1/TSC2; neurofibromatosis NF1), cell growth (Cowden syndrome, PTEN), and calcium signaling (Timothy syndrome, CACNA1C) can increase the risk for an individual to have ASD. However, if the causative genes are numerous and diverse, they might all interfere with a more restricted number of downstream pathways at the origin of ASD. Consistent with this hypothesis, several synaptic genes were found to be strongly associated with nonsyndromic ASD, providing a better view of the complex pathways, which alter properties of the neuronal networks and likely contribute to the disorders (Belmonte and Bourgeron 2006).

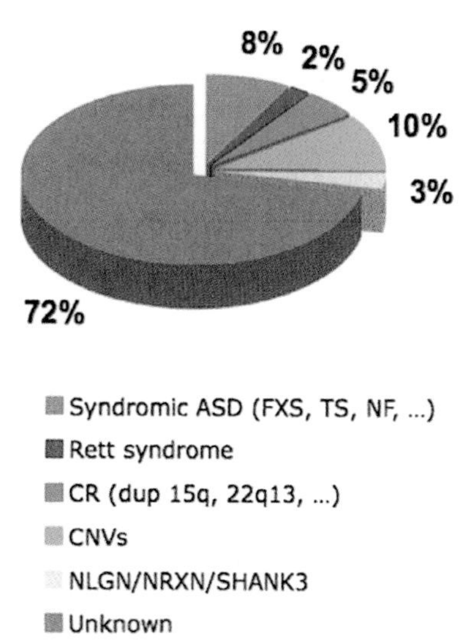

Figure 1. Broad estimation of the heterogeneous causes of ASD. ASD includes about 8% of known genetic syndromes (e.g., Fragile-X syndrome [FXS], tuberous sclerosis [TS], neurofibromatosis [NF]), about 2% of Rett syndrome, about 5% of chromosomal rearrangements (CR), about 10% of copy-number variants (CNVs), about 3% of mutations in the NLGN/NRXN/SHANK3 pathway, and about 72% of unknown causes. These numbers are only a broad estimation because epidemiological data concerning the causes of ASD are missing. In addition, the percentage may vary for sporadic or familial cases and if the affected individual has dysmorphic features.

SYNAPSES AND AUTISM SPECTRUM DISORDERS

Synaptic Genes and Autism Spectrum Disorders

The first synaptic genes associated with autism and Asperger syndrome to be discovered were the X-linked neuroligins *NLGN3* and *NLGN4*. Neuroligins are cell adhesion molecules with an esterase domain that have a crucial role in the formation of functional synapses. They are located at the postsynaptic side of the synapse and bind to other cell adhesion molecules called neurexins located on the presynaptic side of the synapse (Fig. 2). There are five neuroligin genes, *NLGN1*, *NLGN2*, *NLGN3*, *NLGN4*, and *NLGN4Y*, in the human genome. Searching for mutations in X-linked genes, Jamain et al. (2003) identified a frame-shift mutation in the *NLGN4* gene in two brothers, one with autism and the other with Asperger syndrome (Jamain et al. 2003). This mutation

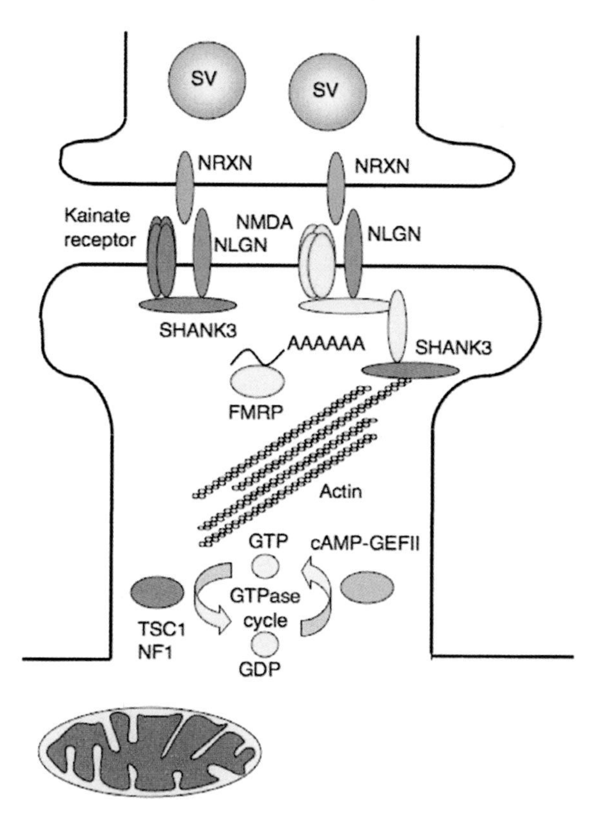

Figure 2. Synaptic genes associated with ASD. Synaptic vesicles (SV) and neurexins (NRXN) are present at the presynaptic side of a glutamatergic synapse. At the postsynaptic side, the NLGN and the glutamate receptors bind to scaffolding proteins of the postsynaptic density (PSD) such as SHANK3. FMRP controls the translation of several synaptic proteins. TSC1 and NF1 are regulating the actin dynamics and the morphology of the neuron. MECP2 (not shown here) regulates gene expression by modifying chromatin structure.

(D396X) was located in the esterase domain, leading to premature termination of the protein before the transmembrane domain. In the same study, a nonsynonymous mutation (R451C) of the X-linked *NLGN3*, which affects a highly conserved amino acid from the esterase domain, was identified in a second family with two brothers, one with autism and one with Asperger syndrome. These mutations (*NLGN4* D396X and *NLGN3* R451C) were intensively studied at the functional level and were found to cause abnormal synaptogenesis in cultured neuronal cells (Chih et al. 2004; Comoletti et al. 2004). Although mutations in *NLGN3*- and *NLGN4*-coding sequences are rare in individuals with ASD (<1% of the individuals) (Vincent et al. 2004; Blasi et al. 2006), other groups have identified independent *NLGN4* mutations in individuals with ASD and/or mental retardation (Laumonnier et al. 2004; Yan et al. 2005). In addition, abnormal *NLGN3* and *NLGN4* spliced isoforms were detected in blood cells from individuals with ASD (Talebizadeh et al. 2006). If these abnormal transcripts are actually present in the brain of the affected individuals, this finding may represent a new type of *NLGN* alteration in ASD.

The second gene identified within this pathway was *SHANK3* located on chromosome 22, a region deleted in several individuals with ASD (Manning et al. 2004). SHANK3 is a scaffolding protein of the postsynaptic density (PSD), which binds to the NLGN, and known to regulate the structural organization of dendritic spines (Boeckers et al. 2002; Meyer et al. 2004). Durand et al. (2007) sequenced the coding sequence of *SHANK3* in 227 individuals with ASD and showed that mutations, or loss of one copy of the gene, are associated with autism, whereas the presence of an extra copy might be associated with Asperger syndrome. Among the variations identified, a de novo frame-shift mutation that originated in a mother with germinal mosaicism was present in two brothers with autism. Expression in cultured neurons of the rat *Shank3* cDNA carrying the frame-shift mutation indicated that the truncated protein, in contrast to the full-length protein, is absent in the dendritic spines. These results provided further genetic and functional evidence that the synaptic pathway, which includes *NLGN*, *NRXN*, and *SHANK3*, is associated with ASD.

Finally, the third gene identified within this pathway is *NRXN1* on chromosome 2p. Using a whole-genome approach, an international collaborative effort—the Autism Genome Project Consortium—investigated 1168 multiplex families for the presence of linkage and CNVs (Szatmari et al. 2007). This analysis detected a new locus for autism on chromosome 11 and the presence of a de novo deletion of the *NRXN1* gene located on chromosome 2p16 in two sisters with ASD. The *NRXN1* gene encodes neurexin, the presynaptic partner of the *NLGN*, confirming that a defect in this synaptic pathway could cause ASD.

Other synaptic genes may be associated with ASD (Persico and Bourgeron 2006). Three of these candidate genes encode glutamate receptor subunits, the *glutamate receptor ionotropic kainate 2* gene (*GRIK2* or *GluR6*) (Jamain et al. 2002), the *N*-methyl-D-aspartate (NMDA) receptor-2 subunit (*GRIN2A*) (Barnby et al. 2005), and glutamate receptor ionotropic AMPA 3 (*GRIA3*) (Jacquemont et al. 2006). These proteins are direct (or close) binding partners of the NLGN/NRXN1/SHANK3 protein complex, but the functional consequence of the "risk alleles" remains largely unknown.

Atypical Synapses in Autism Spectrum Disorders?

Taken together, these results strongly suggest that the NRXN/NLGN/SHANK protein complex has an important function in ASD. Although little data exists on the specific role of this pathway in the human brain, studies on neuronal cell culture and animal models provided crucial information on its function.

First, neuroligins and neurexins enhance synapse formation in vitro (Scheiffele et al. 2000) but, surprisingly, are not required for the generation of synapses in vivo (Varoqueaux et al. 2006). Indeed, results from knockout (KO) mice demonstrate that neither NLGNs nor neurexins are required for the initial formation of synapses, but both are essential for synaptic function and mouse survival (Missler 2003; Varoqueaux et al. 2006). Therefore, neuroligins may not establish but may specify and validate synapses via an activity-dependent mechanism, with different neuroligins acting on distinct types of synapses (Chubykin et al. 2007). This model, proposed by Chubykin et al. (2007), reconciles the overexpression and knockout phenotypes and suggests that neuroligins contribute to the activity-dependent formation of neural circuits.

Second, neuroligins and neurexins are also emerging as central organizing molecules for excitatory glutamatergic and inhibitory GABAergic synapses in mammalian brain (Graf et al. 2004; Prange et al. 2004). NLGN1, NLGN3 and NLGN4 are specific to glutamatergic synapses, whereas NLGN2 is restricted to GABAergic synapses (Varoqueaux et al. 2004). A selectivity for glutamatergic versus GABAergic synapse is also conferred by alternative splicing of both partners (Craig and Kang 2007). An insertion of an alternatively spliced exon in β-neurexins selectively promotes GABAergic synaptic function, whereas an insertion of an alternative spliced exon in neuroligin 1 selectively promotes glutamatergic synaptic function. This role in synaptic specificity is highly relevant to ASD because imbalance between excitation and inhibition could lead to epilepsy, a disease observed in almost 25% of individuals with ASD (Freitag 2007).

Finally, at least in humans, this pathway appears to be highly sensitive to gene dosage. Indeed, a deletion of a single copy of *SHANK3* or *NRXN1* seems to be sufficient to cause ASD (Durand et al. 2007; Szatmari et al. 2007). Although the change in the number and/or the quality of the synaptic contacts may be subtle, the consequences at the cognitive level are obvious because patients sometimes present with complete absence of speech. This gene-dosage sensitivity of the complex is an important feature that should be taken into account when reconsidering the mode of inheritance of ASD and, more generally, for understanding the evolution of higher brain functions.

Taken together, these results indicate that the NLGN/NRXN/SHANK3 complex is actually associated with ASD. Nevertheless, it was shown that the severity of the syndrome associated with mutations within this pathway could greatly differ from one individual to another (even if they carry identical or similar mutations) (Jamain et al. 2003; Laumonnier et al. 2004). This relative heterogeneity in the phenotype indicates that this pathway could be modulated by other genetics and/or environmental factors. Among these factors, I propose that abnormal circadian rhythms may increase the risk as well as the severity of the disorder.

Circadian Rhythms and Autism Spectrum Disorders

Despite the fact that sleep is one of the major concerns for families having a child with ASD, this problem was often considered as an epiphenomenon and therefore did not catch the attention of the scientific community. However, recent results showing abnormal melatonin synthesis, as well as an efficacy of melatonin therapy for sleep problems observed in ASD, may change this initial disregard from a possible key role of the clock and circadian regulations in ASD.

Clock Genes and Autism Spectrum Disorders

The involvement of clock genes in ASD was first proposed by Wimpory et al. (2002). To test this hypothesis, Nicholas et al. (2007) screened single-nucleotide polymorphisms (SNPs) in 11 clock/clock-related genes in 110 individuals with ASD and their parents. A significant allelic association was detected for *PER1* and *NPAS2*. Haplotype analysis within *PER1* gave a single significant result ($P = 0.03$), whereas for *NPAS2,* 40 of the 136 possible two-marker combinations were significant at the $P < 0.05$ level, with the best result between markers rs1811399 and rs2117714 ($P = 0.001$). This first study of clock genes in ASD is promising, but the relatively small sample studied and the absence of significance after correction for multiple testing warrant the extension of these studies to larger cohorts.

The most consistent results reporting abnormal circadian rhythms in ASD concern the melatonin synthesis pathway. Melatonin is considered to be the hormonal message for darkness because it is released during the night in all vertebrates examined, independent of whether the animal is diurnally or nocturnally active. It is produced mainly in the pineal gland by the conversion of serotonin to *N*-acetylserotonin (NAS) by the rate-limiting enzyme AA-NAT (arylalkylamine *N*-acetyltransferase), followed by the conversion of NAS to melatonin by HIOMT (hydroxyindole *O*-methyltransferase) (Fig. 3, top) (Axelrod 1974). At least five independent groups detected abnormal melatonin levels in ASD (Rivto et al. 1993; Nir et al. 1995; Kulman et al. 2000; Tordjman et al. 2005; Melke et al. 2007). With the exception of Ritvo et al. (1993), who reported increased daytime urinary melatonin levels and similar nocturnal values, all of the remaining studies found an abnormal decrease of melatonin concentration in individuals with ASD.

Nir et al. (1995) observed an abnormal circadian pattern of melatonin in a group of 10 young adult males with ASD. Although not out of phase, the serum melatonin lev-

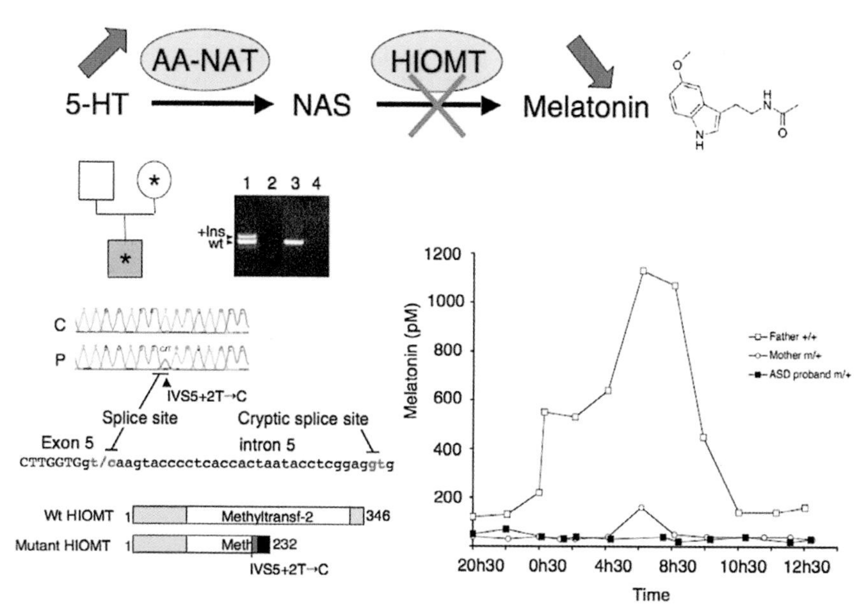

Figure 3. Abnormal melatonin synthesis in ASD. (*Top*) Melatonin is produced by the conversion of serotonin to *N*-acetylserotonin by the rate-limiting enzyme AA-NAT, followed by the conversion of *N*-acetylserotonin (NAS) to melatonin by HIOMT. A single deficiency of HIOMT may be responsible for the low melatonin level and the accumulation of serotonin observed in children with ASD. (*Bottom left*) Pedigree structure of a family carrying the splice site mutation IVS5 + 2T > C. (*Orange*) Child with high-functioning autism. Reverse transcriptase–polymerase chain reaction (RT-PCR)-amplifying exons 4 to 6 of the *ASMT* cDNA from BLCL of the ASD1 proband carry the splice site mutation IVS5 + 2T > C (lane *1* +RT; lane *2* –RT) and a control (lane *3* +RT; lane *4* –RT). The insertion (+Ins) of 31 bp in the *ASMT* transcript should lead to premature truncation of the protein, lacking the methyltransferase domain. (Wt) Wild type. (*Bottom right*) Nocturnal melatonin profile of the proband (male, 24 years old), his mother (53 years old), and his father (55 years old). The proband (P) and the mother (M) are heterozygous (m/+) for the splice site mutation IVS5 + 2T > C. The father does not carry the *ASMT* mutation (+/+). (Adapted from Melke et al. 2007.)

els differed from normal in amplitude and concentration. Marginal changes in diurnal rhythms of serum thyroid-stimulating hormone (TSH) and possibly prolactin were also recorded. Subjects with seizures tended to have an abnormal pattern of melatonin that correlated with electroencephalogram (EEG) changes.

Kulman et al. (2000) investigated 14 children with ASD and 20 age-matched controls for their whole 24-hour circadian rhythms by collecting venous blood samples at 4-hour intervals. In their cohort, no autistic children showed a normal circadian rhythm. In more detail, 10 of 14 showed no daily variation in melatonin blood levels, whereas the remaining 4 patients had higher levels of melatonin during the day rather than during the night. The children with inverted circadian rhythms of melatonin may have had a disorder similar to Smith–Magenis syndrome (SMS). Indeed, patients with SMS present with inverted melatonin circadian rhythms and severe behavior problems similar to autism (De Leersnyder 2006). In this study, melatonin (5 mg/day orally in the evening) was found to improve the sleep disturbance in two of three patients with a more pronounced disorder.

Tordjman et al. (2005) measured the overnight urinary output of the predominant melatonin metabolite, 6-sulphatoxymelatonin (6-SM) in the urine of 49 individuals with ASD and 88 controls. The nocturnal urinary excretion of 6-SM was significantly reduced in patients with ASD compared to controls (mean ±S.E.M., 0.75 ± 0.11 vs. 1.8 ± 0.17 g/hr, $P = 0.0001$). A large proportion of the individuals with autistic disorder (31 of 49, 63%) had 6-SM excretion values that were less than half of the mean excretion rate observed in the control group. Furthermore, the low level of melatonin was correlated with the severity of autistic impairments in verbal communication and play ($p < 0.05$).

Taken together, these studies performed on different cohorts and using different methodologies indicate that an abnormal melatonin level is a frequent trait in ASD. Nevertheless, both the underlying cause of this anomaly and its relationship with ASD (cause or consequence?) were still unexplained. This dilemma was partly resolved by Melke et al. (2007), who used a combination of genetics and biochemical approaches to address these issues. The original aim of this study was to identify susceptibility genes for ASD on the pseudo-autosomal region 1 (PAR1). PAR1 is a short region of 2.7 Mb located on both the X and Y chromosomes and deleted in several individuals with ASD. Among the 12 PAR1 genes, *ASMT* was considered to be an excellent candidate because it encodes HIOMT, the last enzyme in the melatonin biosynthesis pathway (Simonneaux and Ribelayga 2003). All *ASMT* exons and promoters were sequenced in 250 individuals with ASD. Variations affecting the protein sequence of HIOMT (N17K, K81E, G306A, and L326F) were enriched in the ASD group, and a splicing mutation (IVS5 + 2T > C) was present in two families with ASD, but not in controls (Fig. 3, bottom left). In addition, two polymorphisms (rs4446909 and rs5989681) located in the promoter were more frequent in ASD compared to controls ($P = 0.0006$) and were associated with a decrease in *ASMT* transcripts in B-lymphoblastoid cell lines ($P = 2 \times 10^{-10}$). Biochemical analyses performed on blood platelets of 43 individuals with ASD and 48 controls revealed a highly significant decrease in HIOMT activity ($P = 2 \times 10^{-12}$) and melatonin level ($P = 3 \times 10^{-11}$) in ASD. The HIOMT deficit was also detected in cultured cells of the patients, ruling out inhibitory effects of environmental factors or regulation acting at a higher physiological level.

All together, these results confirm that a large percentage of patients (65%) have less than half the mean of the melatonin control values, a proportion very similar (63%) to that previously reported (Tordjman et al. 2005). Furthermore, they demonstrate that this deficit was not the consequence of ASD but a primary trait caused by a deficiency in HIOMT activity. Consistent with this primary deficit, parents were found to have lower melatonin than controls. Hence, in one family, the mother and the affected son carried an *ASMT* splice site mutation, and both have virtually no circadian variation of melatonin level (Fig. 3, bottom right). This anomaly observed in some parents indicated that the melatonin deficit by itself is not sufficient to cause ASD.

Interestingly, one of the most consistent biochemical findings in ASD is the observation of high serotonin levels in affected individuals and their relatives (Cook and Leventhal 1996). Variations within the serotonin transporter *SLC6A4* have been suspected to have a role in this anomaly, but this has never been formally proven. Therefore, one attractive hypothesis to explain the high serotonin levels in ASD is that a primary deficit of melatonin synthesis could indirectly lead to an accumulation of the upstream substrates NAS and serotonin (Fig. 3, top). In this context, a single defect within the melatonin synthesis pathway could lead to both low melatonin and high serotonin levels in ASD. This hypothesis reconciles both results, but it raises the critical question of the real nature of the pathological process. Is it high serotonin or low melatonin or the combination of both that increases the risk of having ASD? Moreover, the pathological role of other compounds produced by the HIOMT and/or intermediate metabolites such as NAS cannot be excluded. The putative roles of serotonin on the susceptibility to ASD have been discussed extensively elsewhere (Cook and Leventhal 1996). The next part of this chapter concentrates on the potential roles of melatonin in the susceptibility to ASD.

ATYPICAL SLEEP AND CIRCADIAN RHYTHMS IN AUTISM SPECTRUM DISORDERS

Melatonin is one of the factors that sets the internal clock to a 24-hour cycle and therefore is crucial for appropriate regulation of the sleep/wake (S/W) cycle. In humans, the developmental course of the S/W cycle consists of three epochs, i.e., development of the circadian S/W cycle by 4 months (Fig. 4), decrease of the daytime sleep by 1.5 years, and establishment of the biphasic S/W cycle by 5 years (Segawa 2006). Normal S/W rhythms seem to be crucial for appropriate cognitive development in children. Moreover, sleep is associated with intense

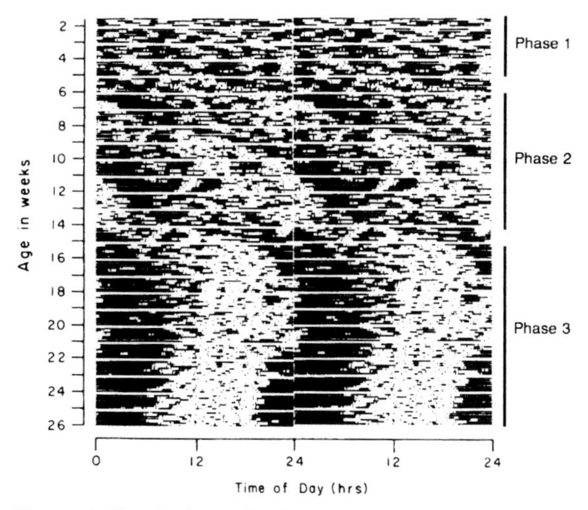

Figure 4. Sleep/wake cycle of a typically developed child during the first 6 months of life. The numbers plotted at the top are 24-hour clock times. Each bar represents 2 days, with the time spent in sleep (*black*) or awake (*white*). During phase 1 (from birth until 2 months), the S/W cycle of the child shows no circadian rhythm. In phase 2 (from 2 to 4 months), a free-running pattern is observed. The S/W cycle reflects the setting of the internal clock and is out of phase with night and day. In phase 3, the child has a normal circadian S/W cycle in phase with night and day. (Adapted from Kleitman and Engelmann 1953.)

neuronal function and has an important role on how our memory is formed and ultimately shaped (Stickgold 2005).

Interviews with the parents of patients with ASD revealed that more than 70% of patients had delayed development of the circadian S/W cycle by at least 5 months (Segawa 2006). This problem persists through childhood because 50–80% of children with ASD show highly significant increased sleep latency and nocturnal awakenings (Richdale and Prior 1995; Taira et al. 1998; Takase et al. 1998; Hering et al. 1999; Elia et al. 2000; Hayashi 2000; Schreck and Mulick 2000; Gail Williams et al. 2004; Wiggs and Stores 2004; Limoges et al. 2005; Oyane and Bjorvatn 2005; Polimeni et al. 2005; Allik et al. 2006; Hare et al. 2006; Liu et al. 2006; Malow et al. 2006). An extensive survey of the sleep problems observed in ASD was recently reported by Liu et al. (2006). In their cohort of 167 children (age 8.8 ± 4.2, 86% males), 86% had at least one sleep problem almost everyday, including 54% with bedtime resistance, 56% with insomnia, 53% with parasomnias, 25% with sleep disordered breathing, 45% with morning rise problems, and 31% with daytime sleepiness.

Although sleep problems are repeatedly observed in ASD (predominantly in initiating and maintaining sleep), the occurrence of abnormal sleep patterns, such as fewer rapid eye movement (REM), is still a matter of debate. Some studies could not detect a difference in the sleep architecture of ASD (Tanguay et al. 1976; Elia et al. 2000). Other studies reported increased duration of stage-1 sleep, decreased non-REM sleep and slow-wave sleep, fewer stage-2 EEG sleep spindles, and a lower number of rapid eye movements during REM sleep than controls (Diomedi et al. 1999; Daoust et al. 2004; Limoges et al. 2005). This absence of consensus on the sleep architecture in ASD could be due to the difference in the severity of the disorder and in the age of the patients.

Interestingly, when a sleep diary was completed, some individuals with ASD showed a free-running pattern of the S/W rhythm (Fig. 5) (Hayashi 2000; Segawa 2006). Such a pattern is observed in some blind individuals, who cannot perceive the light information of night and day transmitted from the retina to the suprachiasmatic nucleus (SCN) (Skene et al. 1999). Therefore, free-running patterns are highly suggestive of a problem in the circadian setting of the clock. On the basis of this observation, Segawa (2006) postulated that the abnormality in ASD occurs as the child is entering into the day/night cycle. This assumption, made solely on the basis of the sleep pattern, is highly relevant to the melatonin deficit described previously. Therefore, if the anomalies of the S/W cycle are caused by a melatonin deficiency, a melatonin treatment should imrove the sleep of the patients.

MELATONIN TREATMENT IN AUTISM SPECTRUM DISORDERS

Only two small-scale studies have tested the efficacy of melatonin in ASD (Garstang and Wallis 2006; Giannotti et al. 2006). The first study evaluated openly the long-term effectiveness of controlled-release melatonin in 25 children, aged 2.6 to 9.6 years with ASD without other coexistent pathologies (Giannotti et al. 2006). During treatment, sleep patterns of all children improved. After discontinuation, 16 children returned to the pretreatment score; however, readministration of melatonin was again effective. Treatment gains were maintained at 12- and 24-month follow-ups. No adverse side effects were reported. In the second study, a randomized, placebo-controlled double-blind crossover trial of melatonin was undertaken in 11 children with ASD (Garstang and Wallis 2006). All sleep parameters (sleep latency, waking per night, total sleep duration) were improved after treatment with melatonin. Hence, both studies provide evidence that melatonin is an effective and well-tolerated treatment for children with ASD and sleep difficulties. However, although sleep was improved, none of these studies could show a significant improvement in cognitive functions and social communication after melatonin treatment. This absence of recovery of cognitive function most likely reflects the presence of other anomalies in the brains of the children with autism. Furthermore, the regulation of the S/W cycle in the first years of age may be crucial for this critical phase of development, and therefore, developmental problems may not be reversed by melatonin treatment occurring afterward. For that reason, the early detection of a melatonin deficit in children may greatly improve the efficacy of melatonin treatment in ASD.

WHO CONTROLS WHOM? THE CLOCK OR THE SYNAPSE ?

The results presented above indicate that at least two pathways are altered in children with autism. One is a synaptic pathway sensitive to gene dosage and has a key

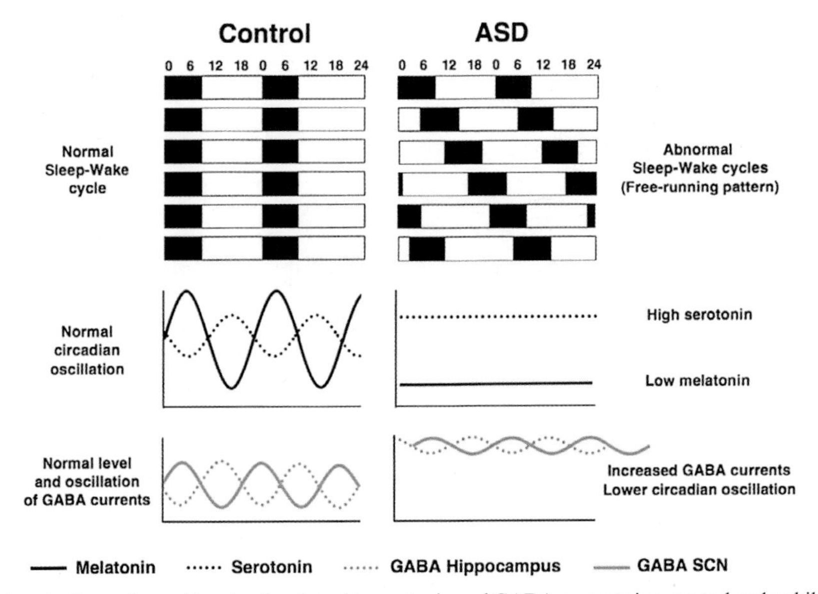

Figure 5. Model of sleep/wake cycles and levels of melatonin, serotonin and GABA currents in a control and a child with ASD. (*Upper panel*) Schematic diary of a control and an individual with ASD. The numbers plotted at the top are 24-hour clock times. Each bar represents 2 days, with the time spent in sleep (*black*) or awake (*white*). Each line is two successive days. The melatonin and serotonin levels are shown below. (*Lower panel*) Possible circadian oscillation of the GABA currents within the hippocampus and in the SCN. A defect in the NLGN/NRXN/SHANK3 pathway may increase the GABAergic currents and the defect in melatonin synthesis may reduce their oscillations.

role in the establishment of the neuronal network. The second pathway is related to the setting of the clock and is crucial for circadian rhythms of the S/W cycle. Currently, there is no experimental evidence that these pathways interact with each other. However, their critical biological functions strongly suggest that an anomaly in one of these pathways may perturb the other.

Several lines of evidence suggest that melatonin could modulate neuronal networks by influencing both the strength and the circadian oscillation of neuronal transmission (Liu et al. 1997; Jin et al. 2003; Weil et al. 2006). First, melatonin was shown to modulate the phosphorylation of CREB (CRE-binding protein) by decreasing the activity of the cAMP-dependent protein kinase A (PKA) through its binding to melatonin receptor MT1 (von Gall et al. 2000; Jin et al. 2003). Considering the key role of CREB in memory consolidation (Mizuno et al. 2002), the absence of melatonin may alter this process. Second, melatonin significantly inhibits synaptic transmission and long-term potentiation (LTP) in the CA1 region of the hippocampus (Ozcan et al. 2006) through a mechanism involving MT2-receptor-mediated regulation of the PKA pathway (Wang et al. 2005). This inhibition of LTP by melatonin is not mediated by blockade of NMDA receptors (Collins and Davies 1997) or the cholinergic system (Feng et al. 2002) but through the modulation of GABAergic system (Feng et al. 2002).

This emerging role of melatonin in the modulation the GABAergic system (Golombek et al. 1996) may be the most interesting characteristic linking melatonin with the susceptibility to ASD. Indeed, melatonin seems to influence the day/night variations in glutamate and GABA neurotransmitter levels (Marquez de Prado et al. 2000), as well as to modulate GABA-induced currents (Wan et al. 1999; Wu et al. 1999; Prada et al. 2005). Interestingly, melatonin can exert opposite effects: It increases $GABA_A$-receptor-mediated current in the rat SCN via the MT1 subtype, but it inhibits this current in the hippocampus via MT2. In this line, abnormal melatonin levels could reveal a subtle defect in the NRXN/NLGN/SHANK3 pathway that would not have been perceived if GABA oscillations were normal (Figs. 5 and 6).

An alternative possibility is that the NLGN/NRXN/SHANK3 synaptic pathway could alter the clock and the circadian rhythms in individuals with ASD. Unfortunately, to my knowledge, there is no information on the sleep pattern and circadian rhythms of the patients with mutations within the NLGN/NRXN/SHANK3 synaptic pathway. The same is true for the mice carrying mutations with this synaptic pathway. Nevertheless, mutations in the *FMR1* and *MECP2* genes were shown to alter the circadian rhythms and S/W cycles of patients with Fragile-X or Rett syndrome, respectively. Moreover, *Drosophila* carrying a mutation in the orthologous gene for *FMR1* show strong circadian anomalies (Dockendorff et al. 2002). Therefore, the possibility that a defect in the NLGN/ NRXN/SHANK3 synaptic pathway could alter the circadian rhythms in patients with ASD is plausible and further research is warranted to address this issue.

CONCLUSIONS

Even if genetic and functional data are still sparse, a picture is starting to emerge that at least in some cases, ASD may be due to a problem in the development of specific pathways controlling the wiring of neuronal networks. My hypothesis is that a second anomaly in the setting of the circadian rhythms may reveal, or increase, this synaptic defect in ASD. Many questions remain

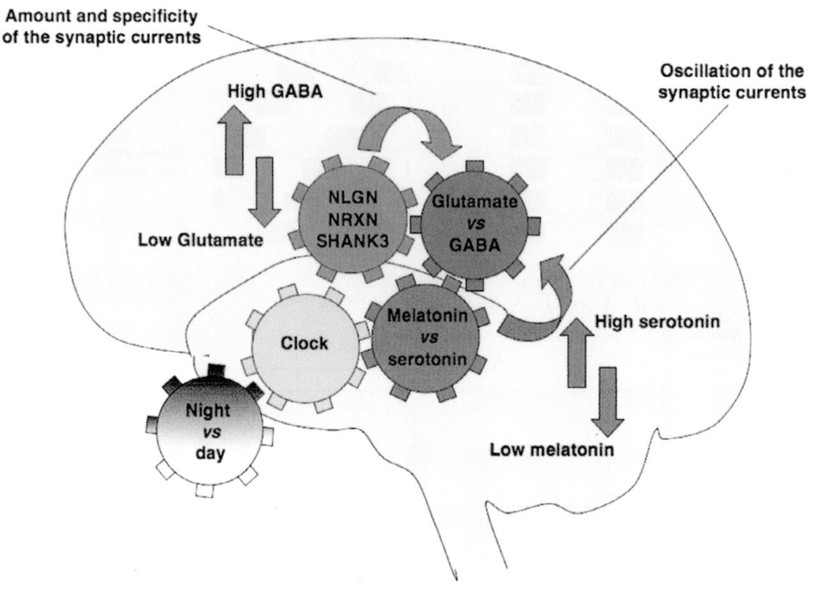

Figure 6. Interplay between the NLGN/NRXN/SHANK3 pathway and melatonin circadian oscillation in ASD. (*Cog wheels*) Possible interactions between the different pathways. A defect in the NLGN/NRXN/SHANK3 may alter the amount of synaptic contacts as well as the balance between glutamate and GABA. The deficit in melatonin synthesis may alter the oscillation as well as the balance between different synaptic currents (serotonin and GABA). The cumulative effect of both anomalies may greatly increase the risk to have ASD.

unanswered such as the nature of the affected neuronal networks and the potential of reversing the phenotype induced by the mutations. In this line, animal models carrying mutations in the NLGN/NRXN/SHANK3 pathway should provide key information. Recent studies on the mutant mice for *Fmr1* (Fragile-X syndrome) and *Mecp2* (Rett syndrome) are encouraging because the neurological phenotype could be reversed in mice both by pharmacological treatment (Hayashi et al. 2007) or reactivation of the gene (Guy et al. 2007). On the basis of these lines of evidence, the interactive role between circadian modulation and synaptic function might be one of the most fascinating areas for future functional analyses and for the development of new therapeutic strategies for ASD.

ACKNOWLEDGMENTS

I thank Moshe Yaniv, Bernard Lakowski, and Ken McElreavey for helpful discussions and comments on the manuscript. This work was supported by the Pasteur Institute, INSERM, Assistance Publique-Hôpitaux de Paris, Fondation France Télécom, Cure Autism Now, Fondation de France, Fondation Biomédicale de la Mairie de Paris, Fondation pour la Recherche Médicale, EUSynapse European Commission FP6, AUTISM MOLGEN European Commission FP6, and ENI-NET European Commission FP6.

REFERENCES

Allik H., Larsson J.O., and Smedje H. 2006. Sleep patterns of school-age children with Asperger syndrome or high-functioning autism. *J. Autism Dev. Disord.* **36:** 585.

American Psychiatric Association (APA). 1994. *Diagnostic and statistical manual of mental disorders,* 4th edition. American Psychiatric Association, Washington, D.C.

Axelrod J. 1974. The pineal gland: A neurochemical transducer. *Science* **184:** 1341.

Bailey A., Le Couteur A., Gottesman I., Bolton P., Simonoff E., Yuzda E., and Rutter M. 1995. Autism as a strongly genetic disorder: Evidence from a British twin study. *Psychol. Med.* **25:** 63.

Barnby G., Abbott A., Sykes N., Morris A., Weeks D.E., Mott R., Lamb J., Bailey A.J., and Monaco A.P. 2005. Candidate-gene screening and association analysis at the autism-susceptibility locus on chromosome 16p: Evidence of association at GRIN2A and ABAT. *Am. J. Hum. Genet.* **76:** 950.

Baron-Cohen S., Leslie A.M., and Frith U. 1985. Does the autistic child have a "theory of mind"? *Cognition* **21:** 37.

Belmonte M.K. and Bourgeron T. 2006. Fragile X syndrome and autism at the intersection of genetic and neural networks. *Nat. Neurosci.* **9:** 1221.

Blasi F., Bacchelli E., Pesaresi G., Carone S., Bailey A.J., and Maestrini E. 2006. Absence of coding mutations in the X-linked genes neuroligin 3 and neuroligin 4 in individuals with autism from the IMGSAC collection. *Am. J. Med. Genet. B Neuropsychiatr. Genet.* **141:** 220.

Boeckers T.M., Bockmann J., Kreutz M.R., and Gundelfinger E.D. 2002. ProSAP/Shank proteins—A family of higher order organizing molecules of the postsynaptic density with an emerging role in human neurological disease. *J. Neurochem.* **81:** 903.

Chih B., Afridi S.K., Clark L., and Scheiffele P. 2004. Disorder-associated mutations lead to functional inactivation of neuroligins. *Hum. Mol. Genet.* **13:** 1471.

Chubykin A.A., Atasoy D., Etherton M.R., Brose N., Kavalali E.T., Gibson J.R., and Sudhof T.C. 2007. Activity-dependent validation of excitatory versus inhibitory synapses by neuroligin-1 versus neuroligin-2. *Neuron* **54:** 919.

Collins D.R. and Davies S.N. 1997. Melatonin blocks the induction of long-term potentiation in an N-methyl-D-aspartate independent manner. *Brain Res.* **767:** 162.

Comoletti D., De Jaco A., Jennings L.L., Flynn R.E., Gaietta G., Tsigelny I., Ellisman M.H., and Taylor P. 2004. The Arg451Cys-neuroligin-3 mutation associated with autism reveals a defect in protein processing. *J. Neurosci.* **24:** 4889.

Cook E.H. and Leventhal B.L. 1996. The serotonin system in autism. *Curr. Opin. Pediatr.* **8:** 348.

Craig A.M. and Kang Y. 2007. Neurexin-neuroligin signaling in synapse development. *Curr. Opin. Neurobiol.* **17:** 43.

Daoust A.M., Limoges E., Bolduc C., Mottron L., and Godbout R. 2004. EEG spectral analysis of wakefulness and REM sleep in high functioning autistic spectrum disorders. *Clin. Neurophysiol.* **115:** 1368.

De Leersnyder H. 2006. Inverted rhythm of melatonin secretion in Smith-Magenis syndrome: From symptoms to treatment. *Trends Endocrinol. Metab.* **17:** 291.

Diomedi M., Curatolo P., Scalise A., Placidi F., Caretto F., and Gigli G.L. 1999. Sleep abnormalities in mentally retarded autistic subjects: Down's syndrome with mental retardation and normal subjects. *Brain Dev.* **21:** 548.

Dockendorff T.C., Su H.S., McBride S.M., Yang Z., Choi C.H., Siwicki K.K., Sehgal A., and Jongens T.A. 2002. *Drosophila* lacking dfmr1 activity show defects in circadian output and fail to maintain courtship interest. *Neuron* **34:** 973.

Durand C.M., Betancur C., Boeckers T.M., Bockmann J., Chaste P., Fauchereau F., Nygren G., Rastam M., Gillberg I.C., Anckarsater H., Sponheim E., Goubran-Botros H., Delorme R., Chabane N., Mouren-Simeoni M.C., de Mas P., Bieth E., Roge B., Heron D., Burglen L., Gillberg C., Leboyer M., and Bourgeron T. 2007. Mutations in the gene encoding the synaptic scaffolding protein SHANK3 are associated with autism spectrum disorders. *Nat. Genet.* **39:** 25.

Elia M., Ferri R., Musumeci S.A., Del Gracco S., Bottitta M., Scuderi C., Miano G., Panerai S., Bertrand T., and Grubar J.C. 2000. Sleep in subjects with autistic disorder: A neurophysiological and psychological study. *Brain Dev.* **22:** 88.

Feng Y., Zhang L.X., and Chao D.M. 2002. Role of melatonin in spatial learning and memory in rats and its mechanism (translation). *Sheng Li Xue Bao* **54:** 65.

Fombonne E. 2005. Epidemiology of autistic disorder and other pervasive developmental disorders. *J. Clin. Psychiatry* (suppl. 10) **66:** 3.

Freitag CM. 2007. The genetics of autistic disorders and its clinical relevance: A review of the literature. *Mol. Psychiatry* **12:** 2.

Frith U. 1998. Cognitive deficits in developmental disorders. *Scand. J. Psychol.* **39:** 191.

Gail Williams P., Sears L.L., and Allard A. 2004. Sleep problems in children with autism. *J. Sleep Res.* **13:** 265.

Garstang J. and Wallis M. 2006. Randomized controlled trial of melatonin for children with autistic spectrum disorders and sleep problems. *Child Care Health Dev.* **32:** 585.

Giannotti F., Cortesi F., Cerquiglini A., and Bernabei P. 2006. An open-label study of controlled-release melatonin in treatment of sleep disorders in children with autism. *J. Autism Dev. Disord.* **36:** 741.

Golombek D.A., Pevet P., and Cardinali D.P. 1996. Melatonin effects on behavior: Possible mediation by the central GABAergic system. *Neurosci. Biobehav. Rev.* **20:** 403.

Graf E.R., Zhang X., Jin S.X., Linhoff M.W., and Craig A.M. 2004. Neurexins induce differentiation of GABA and glutamate postsynaptic specializations via neuroligins. *Cell* **119:** 1013.

Guy J., Gan J., Selfridge J., Cobb S., and Bird A. 2007. Reversal of neurological defects in a mouse model of Rett syndrome. *Science* **315:** 1143.

Hare D.J., Jones S., and Evershed K. 2006. Objective investigation of the sleep-wake cycle in adults with intellectual disabilities and autistic spectrum disorders. *J. Intellect. Disabil. Res.* **50:** 701.

Hayashi E. 2000. Effect of melatonin on sleep-wake rhythm: The sleep diary of an autistic male. *Psychiatry Clin. Neurosci.* **54:** 383.

Hayashi M.L., Rao B.S., Seo J.S., Choi H.S., Dolan B.M., Choi S.Y., Chattarji S., and Tonegawa S. 2007. Inhibition of p21-activated kinase rescues symptoms of fragile X syndrome in mice. *Proc. Natl. Acad. Sci.* **104:** 11489.

Hering E., Epstein R., Elroy S., Iancu D.R., and Zelnik N. 1999. Sleep patterns in autistic children. *J. Autism Dev. Disord.* **29:** 143.

Jacquemont M.L., Sanlaville D., Redon R., Raoul O., Cormier-Daire V., Lyonnet S., Amiel J., Le Merrer M., Heron D., de Blois M.C., Prieur M., Vekemans M., Carter N.P., Munnich A., Colleaux L., and Philippe A. 2006. Array-based comparative genomic hybridisation identifies high frequency of cryptic chromosomal rearrangements in patients with syndromic autism spectrum disorders. *J. Med. Genet.* **43:** 843.

Jamain S., Betancur C., Quach H., Philippe A., Fellous M., Giros B., Gillberg C., Leboyer M., Bourgeron T., and the Paris Autism Research International Sibpair (PARIS) Study. 2002. Linkage and association of the glutamate receptor 6 gene with autism. *Mol. Psychiatry* **7:** 302.

Jamain S., Quach H., Betancur C., Rastam M., Colineaux C., Gillberg I.C., Soderstrom H., Giros B., Leboyer M., Gillberg C., Bourgeron T., and the Paris Autism Research International Sibpair Study. 2003. Mutations of the X-linked genes encoding neuroligins NLGN3 and NLGN4 are associated with autism. *Nat. Genet.* **34:** 27.

Jin X., von Gall C., Pieschl R.L., Gribkoff V.K., Stehle J.H., Reppert S.M., and Weaver D.R. 2003. Targeted disruption of the mouse Mel(1b) melatonin receptor. *Mol. Cell. Biol.* **23:** 1054.

Kanner L. 1943 Autistic disturbances of affective contact. *Nerv. Child* **2:** 217.

Kleitman N. and Engelmann T.G. 1953. Sleep characteristics of infants. *J. Appl. Physiol.* **6:** 269.

Kulman G., Lissoni P., Rovelli F., Roselli M.G., Brivio F., and Sequeri P. 2000. Evidence of pineal endocrine hypofunction in autistic children. *Neuroendocrinol. Lett.* **21:** 31.

Laumonnier F., Bonnet-Brilhault F., Gomot M., Blanc R., David A., Moizard M.P., Raynaud M., Ronce N., Lemonnier E., Calvas P., Laudier B., Chelly J., Fryns J.P., Ropers H.H., Hamel B.C., Andres C., Barthelemy C., Moraine C., and Briault S. 2004. X-linked mental retardation and autism are associated with a mutation in the NLGN4 gene, a member of the neuroligin family. *Am. J. Hum. Genet.* **74:** 552.

Limoges E., Mottron L., Bolduc C., Berthiaume C., and Godbout R. 2005. Atypical sleep architecture and the autism phenotype. *Brain* **128:** 1049.

Liu C., Weaver D.R., Jin X., Shearman L.P., Pieschl R.L., Gribkoff V.K., and Reppert S.M. 1997. Molecular dissection of two distinct actions of melatonin on the suprachiasmatic circadian clock. *Neuron* **19:** 91.

Liu X., Hubbard J.A., Fabes R.A., and Adam J.B. 2006. Sleep disturbances and correlates of children with autism spectrum disorders. *Child Psychiatry Hum. Dev.* **37:** 179.

Malow B.A., Marzec M.L., McGrew S.G., Wang L., Henderson L.M., and Stone W.L. 2006. Characterizing sleep in children with autism spectrum disorders: A multidimensional approach. *Sleep* **29:** 1563.

Manning M.A., Cassidy S.B., Clericuzio C., Cherry A.M., Schwartz S., Hudgins L., Enns G.M., and Hoyme H.E. 2004. Terminal 22q deletion syndrome: A newly recognized cause of speech and language disability in the autism spectrum. *Pediatrics* **114:** 451.

Marquez de Prado B., Castaneda T.R., Galindo A., del Arco A., Segovia G., Reiter R.J., and Mora F. 2000. Melatonin disrupts circadian rhythms of glutamate and GABA in the neostriatum of the aware rat: A microdialysis study. *J. Pineal Res.* **29:** 209.

Melke J., Goubran-Botros H., Chaste P., Betancur C., Nygren G., Anckarsäter H., Rastam M., Ståhlberg O., Gillberg I.C., Delorme R., Chabane N., Mouren-Simeoni M.C., Fauchereau F., Durand C.M., Chevalier F., Drouot X., Collet C., Launay J.M., Leboyer M., Gillberg C., and Bourgeron T. 2007. Abnormal melatonin synthesis in autism spectrum disorders. *Mol. Psychiatry* (in press).

Meyer G., Varoqueaux F., Neeb A., Oschlies M., and Brose N. 2004. The complexity of PDZ domain-mediated interactions at glutamatergic synapses: A case study on neuroligin. *Neuropharmacology* **47:** 724.

Missler M. 2003. Synaptic cell adhesion goes functional. *Trends Neurosci.* **26:** 176.

Mizuno M., Yamada K., Maekawa N., Saito K., Seishima M., and Nabeshima T. 2002. CREB phosphorylation as a molecular marker of memory processing in the hippocampus for spatial learning. *Behav. Brain Res.* **133:** 135.

Nicholas B., Rudrasingham V., Nash S., Kirov G., Owen M.J., and Wimpory D.C. 2007. Association of Per1 and Npas2 with autistic disorder: Support for the clock genes/social timing hypothesis. *Mol. Psychiatry* **12:** 581.

Nir I., Meir D., Zilber N., Knobler H., Hadjez J., and Lerner Y. 1995. Brief report: Circadian melatonin, thyroid-stimulating hormone, prolactin, and cortisol levels in serum of young adults with autism. *J. Autism Dev. Disord.* **25:** 641.

Oyane N.M. and Bjorvatn B. 2005. Sleep disturbances in adolescents and young adults with autism and Asperger syndrome. *Autism* **9:** 83.

Ozcan M., Yilmaz B., and Carpenter D.O. 2006. Effects of melatonin on synaptic transmission and long-term potentiation in two areas of mouse hippocampus. *Brain Res.* **1111:** 90.

Persico A.M. and Bourgeron T. 2006. Searching for ways out of the autism maze: Genetic, epigenetic and environmental clues. *Trends Neurosci.* **29:** 349.

Polimeni M.A., Richdale A.L., and Francis A.J. 2005. A survey of sleep problems in autism, Asperger's disorder and typically developing children. *J. Intellect. Disabil. Res.* **49:** 260.

Prada C., Udin S.B., Wiechmann A.F., and Zhdanova I.V. 2005. Stimulation of melatonin receptors decreases calcium levels in *Xenopus* tectal cells by activating GABA(C) receptors. *J. Neurophysiol.* **94:** 968.

Prange O., Wong T.P., Gerrow K., Wang Y.T., and El-Husseini A. 2004. A balance between excitatory and inhibitory synapses is controlled by PSD-95 and neuroligin. *Proc. Natl. Acad. Sci.* **101:** 13915.

Richdale A.L. and Prior M.R. 1995. The sleep/wake rhythm in children with autism. *Eur. Child Adolesc. Psychiatry* **4:** 175.

Rivto E.R., Ritvo R., Yuwiler A., Brothers A., Freeman B.J., and Plotkin S. 1993. Elevated daytime melatonin in autism: A pilot study. *Eur. Child Adolesc. Psychiatry* **2:** 75.

Scheiffele P., Fan J., Choih J., Fetter R., and Serafini T. 2000. Neuroligin expressed in nonneuronal cells triggers presynaptic development in contacting axons. *Cell* **101:** 657.

Schreck K.A. and Mulick J.A. 2000. Parental report of sleep problems in children with autism. *J. Autism Dev. Disord.* **30:** 127.

Sebat J., Lakshmi B., Malhotra D., Troge J., Lese-Martin C., Walsh T., Yamrom B., Yoon S., Krasnitz A., Kendall J., Leotta A., Pai D., Zhang R., Lee Y.H., Hicks J., Spence S.J., Lee A.T., Puura K., Lehtimaki T., Ledbetter D., Gregersen P.K., Bregman J., Sutcliffe J.S., Jobanputra V., and Chung W., et al. 2007. Strong association of de novo copy number mutations with autism. *Science* **316:** 445.

Segawa M. 2006. Epochs of development of the sleep-wake cycle reflect the modulation of the higher cortical function particular for each epoch. *Sleep Biol. Rhythms* **4:** 4.

Shaywitz S.E., Shaywitz B.A., Fletcher J.M., and Escobar M.D. 1990. Prevalence of reading disability in boys and girls. Results of the Connecticut Longitudinal Study. *J. Am. Med. Assoc.* **264:** 998.

Simonneaux V. and Ribelayga C. 2003. Generation of the melatonin endocrine message in mammals: A review of the complex regulation of melatonin synthesis by norepinephrine, peptides, and other pineal transmitters. *Pharmacol. Rev.* **55:** 325.

Skene D.J., Lockley S.W., and Arendt J. 1999. Melatonin in circadian sleep disorders in the blind. *Biol. Signals Recept.* **8:** 90.

Stickgold R. 2005. Sleep-dependent memory consolidation. *Nature* **437:** 1272.

Szatmari P., Paterson A.D., Zwaigenbaum L., Roberts W., Brian J., Liu X.Q., Vincent J.B., Skaug J.L., Thompson A.P., Senman L., Feuk L., Qian C., Bryson S.E., Jones M.B., Marshall C.R., Scherer S.W., Vieland V.J., Bartlett C., Mangin L.V., Goedken R., Segre A., Pericak-Vance M.A., Cuccaro M.L., Gilbert J.R., and Wright H.H., et al. (Autism Genome Project Consortium). 2007. Mapping autism risk loci using genetic linkage and chromosomal rearrangements. *Nat. Genet.* **39:** 319.

Taira M., Takase M., and Sasaki H. 1998. Sleep disorder in children with autism. *Psychiatry Clin. Neurosci.* **52:** 182.

Takase M., Taira M., and Sasaki H. 1998. Sleep-wake rhythm of autistic children. *Psychiatry Clin. Neurosci.* **52:** 181.

Talebizadeh Z., Lam D.Y., Theodoro M.F., Bittel D.C., Lushington G.H., and Butler M.G. 2006. Novel splice isoforms for NLGN3 and NLGN4 with possible implications in autism. *J. Med. Genet.* **43:** e21.

Tanguay P.E., Ornitz E.M., Forsythe A.B., and Ritvo E.R. 1976. Rapid eye movement (REM) activity in normal and autistic children during REM sleep. *J. Autism Child. Schizophr.* **6:** 275.

Tordjman S., Anderson G.M., Pichard N., Charbuy H., and Touitou Y. 2005. Nocturnal excretion of 6-sulphatoxymelatonin in children and adolescents with autistic disorder. *Biol. Psychiatry* **57:** 134.

Varoqueaux F., Jamain S., and Brose N. 2004. Neuroligin 2 is exclusively localized to inhibitory synapses. *Eur. J. Cell Biol.* **83:** 449.

Varoqueaux F., Aramuni G., Rawson R.L., Mohrmann R., Missler M., Gottmann K., Zhang W., Sudhof T.C., and Brose N. 2006. Neuroligins determine synapse maturation and function. *Neuron* **51:** 741.

Vincent J.B., Kolozsvari D., Roberts W.S., Bolton P.F., Gurling H.M., and Scherer S.W. 2004. Mutation screening of X-chromosomal neuroligin genes: No mutations in 196 autism probands. *Am. J. Med. Genet. B Neuropsychiatr. Genet.* **129:** 82.

von Gall C., Weaver D.R., Kock M., Korf H.W., and Stehle J.H. 2000. Melatonin limits transcriptional impact of phosphoCREB in the mouse SCN via the Mel1a receptor. *Neuroreport* **11:** 1803.

Wan Q., Man H.Y., Liu F., Braunton J., Niznik H.B., Pang S.F., Brown G.M., and Wang Y.T. 1999. Differential modulation of GABAA receptor function by Mel1a and Mel1b receptors. *Nat. Neurosci.* **2:** 401.

Wang L.M., Suthana N.A., Chaudhury D., Weaver D.R., and Colwell C.S. 2005. Melatonin inhibits hippocampal long-term potentiation. *Eur. J. Neurosci.* **22:** 2231.

Weil Z.M., Hotchkiss A.K., Gatien M.L., Pieke-Dahl S., and Nelson R.J. 2006. Melatonin receptor (MT1) knockout mice display depression-like behaviors and deficits in sensorimotor gating. *Brain Res. Bull.* **68:** 425.

Wiggs L. and Stores G. 2004. Sleep patterns and sleep disorders in children with autistic spectrum disorders: Insights using parent report and actigraphy. *Dev. Med. Child Neurol.* **46:** 372.

Wimpory D., Nicholas B., and Nash S. 2002. Social timing, clock genes and autism: A new hypothesis. *J. Intellect. Disabil. Res.* **46:** 352.

Wu F.S., Yang Y.C., and Tsai J.J. 1999. Melatonin potentiates the GABA(A) receptor-mediated current in cultured chick spinal cord neurons. *Neurosci. Lett.* **260:** 177.

Yan J., Oliveira G., Coutinho A., Yang C., Feng J., Katz C., Sram J., Bockholt A., Jones I.R., Craddock N., Cook E.H., Vicente A., and Sommer S.S. 2005. Analysis of the neuroligin 3 and 4 genes in autism and other neuropsychiatric patients. *Mol. Psychiatry* **10:** 329.

Circadian Clocks: 50 Years On

M. Menaker
Department of Biology and Center for Biological Timing, University of Virginia, Charlottesville, Virginia 22904-4328

Since the first Cold Spring Harbor meeting on "Biological Clocks" in 1960, the field has progressed from the study of a fascinating but esoteric set of phenomena of interest primarily to a relatively small group of prescient biologists to become recognized as defining a centrally important aspect of biological organization. This change is the consequence of a profound increase in understanding of the mechanisms that generate and control circadian rhythmicity, coupled with the realization that circadian temporal organization is an important component of much of what most organisms do. As such, it impinges on human health, agriculture, and biological conservation, as well as on many more basic aspects of biology at every level. Many of the seminal discoveries of the last 47 years were presented and discussed at this exciting meeting.

Well, almost 50 years have gone by—the 25th Cold Spring Harbor Symposium on Quantitative Biology "Biological Clocks" was held in June of 1960. The 72nd, only the second devoted to this subject, followed it by 47 years. During those years, our understanding of circadian phenomena has grown exponentially, in parallel with much of the rest of biology. Although we now know a great deal more than we did in 1960, the seeds of our current understanding are present in volume 25, and it is instructive both scientifically and historically to ask what caused them to germinate, to grow, and, in some cases, to flower.

Most importantly, they were very good seeds. Although what we knew then was almost exclusively phenomenological, the phenomena were extraordinarily interesting. Circadian timekeeping was precise, in some cases almost unbelievably so; Pat DeCoursey's flying squirrels began their nightly activity with an accuracy of a few minutes, about 1 part in 30,000, in the complete absence of external time referents. Other overt rhythms were almost as good. Such precision demands a highly evolved control system that, almost by definition, can be unraveled when appropriate techniques are available. Although not available in 1960, such techniques were already on the way.

Circadian rhythmicity—with very similar formal properties—was found in nearly all organisms: protists, fungi, plants, and a variety of animal species. Once you leave the realm of cells and subcellular organization, that level of generality is rare in biology and can only mean that the phenomenon is of fundamental importance. Several different protists *(Euglena, Gonyaulax)* had circadian rhythms, demonstrating that the requisite machinery could be packed into a single cell.

The fact that the circadian period was close to a day, could be synchronized by environmental cues to exactly 24 hours, and, importantly, to a determinate phase relationship with those cues, suggested that the mechanism could function as a clock. This suggestion was supported by some dramatic examples, in particular, sun compass orientation, which could be manipulated by manipulating circadian timing. The idea of a biological clock made of cells, and ultimately, of molecules, caught people's imaginations and led Pittendrigh to search for and then clearly demonstrate the unusual property of temperature-compensated period.

Finally, it was clear that a great deal of biochemistry, physiology, and behavior was rhythmic with circadian periodicity, confirming the fundamental importance of the phenomenon, underlining its potential adaptive significance, and suggesting its possible involvement in various pathologies. The importance of these possibilities was brought home by the demonstration that human beings had circadian rhythms that were indistinguishable from those of other mammals.

In 1960, the study of biological clocks was at the same point in its logical development as was the study of genetics 60 years earlier, before the chromosome theory of heredity. Clocks, like "heredity factors," were locked in a black box that could be studied only by manipulating its outputs. One could study the results of crosses in the one case and light pulses in the other, deriving in both cases information that would become a vital foundation for subsequent analysis, but what was inside the black box was completely unknown. Not only its contents, but even its location was a mystery and in neither case could the box be unlocked until it was found. Because of the demonstration that hereditary factors were located on chromosomes, the "chromosome theory" quickly generated a large body of new and exciting work; for circadian rhythms, the path to the core oscillator was longer and more convoluted.

Tremendous progress has been made since 1960. Most obviously, there has been an explosive increase in our understanding of circadian mechanisms at several levels of organization. This began with the identification of circadian pacemakers in multicellular organisms: silk moths, cockroaches, *Aplysia*, birds, and mammals. It is still ongoing in *Drosophila*, where painstaking neuroanatomy combined with genetics and behavior is revealing the circadian function of specific neurons in the brain

All authors cited here without dates refer to papers in this volume.

(Helfrich-Förster et al.), and in mammals, where inputs to and outputs from the suprachiasmatic nucleus (SCN) are being mapped and their functions pursued (Güler et al.; Doyle and Menaker; Yan et al.; Saper and Fuller). There is still much of importance to learn at this level of organization, especially in mammals; the location of circadian oscillators that respond to food and drugs such as methamphetamine is unknown, and output pathways that connect central pacemakers with peripheral organs and rhythmic behaviors such as sleep are under intense investigation. Work at this level is likely to yield medically important insights as the relationship of circadian rhythmicity to a large range of normal physiological processes and to many specific pathologies becomes more widely appreciated.

The initial identification of pacemaking structures preceded the genetic approaches pioneered in *Drosophila* by Konopka and Benzer (1971) and in *Neurospora* by Dunlap and Feldman (1988). For a while, genetics and functional anatomy proceeded along parallel paths, but when genetic insights began to produce molecular tools, the paths merged. This merger has recently produced a cornucopia of new data that has enabled a rapid increase in our understanding of fundamental circadian mechanisms. Summaries of that increase form a large portion of the content of this volume.

Outlines of the central molecular loops that generate circadian oscillations have been worked out during the past several years for flies, mice, *Neurospora*, *Arabidopsis,* and the cyanobacterium *Synechococcus elongatus.* The outline is most complete for *Synechococcus*, least complete for *Arabidopsis.* At the 1960 meeting, there was much speculation about the possible existence of circadian rhythms in bacteria and discussion of the great analytical advantages of a bacterial clock system if one could be found. It took quite a while, but when circadian rhythmicity was discovered in *Synechococcus*, the predicted analytical progress came rapidly. Current understanding of the mechanism of the circadian oscillator in *Synechococcus* has become the gold standard to which students committed to the reductive analysis of other circadian systems aspire. The circadian cycle can be generated in vitro by incubating three bacterial proteins with ATP. The "artificial" rhythm is temperature-compensated and its period matches that of wild-type mutant strains of the bacteria from which the proteins are derived (Kondo; Johnson). Since the Cold Spring Harbor Laboratory meeting took place, a new paper has appeared proposing a detailed quantitative model of the phosphorylation events that drive the oscillation (Rust et al. 2007).

These results have settled several 1960 questions. Obviously, bacteria, although probably a very limited subset, can have circadian rhythms with all their essential properties. Circadian rhythmicity is likely to be very very old because what we know about cyanobacteria suggests that they have not changed much in the last 2–3 billion years. Because there are no mechanistic homologies between clocks in cyanobacteria and those in other groups of organisms, they have almost certainly evolved independently. Indeed, what we know about clock mechanisms in general suggests at least three independent origins in cyanobacteria, plants, and animals and possibly four, depending on how one feels about the fungi. If that is true, convergence at the formal level, i.e., temperature-compensation, response to light, range of period, has been remarkable.

Although timekeeping in cyanobacteria does not appear to require a transcription-translation feedback loop, virtually all transcription in this organism is regulated in a circadian manner. Fascinating questions remain concerning the ways in which the protein clock regulates gene expression and other cellular activities (Golden). Circadian timekeeping in other organisms certainly involves transcription-translation mechanisms, but are they absolutely essential? Experiments in flies suggest that they probably are (Rosbash et al.), but the existence of protein clocks in complex organisms cannot be excluded.

A group of papers addresses our still incomplete understanding of the circadian-rhythm-generating mechanisms in eukaryotes. Clearly, we have not yet identified many of the genes that regulate circadian rhythmicity in mammals, as there are as many as 20 or more mutant phenotypes with unknown genetic bases (Siepka et al.). In *Drosophila*, studies of the properties of gene networks suggest that many genes that do not show up in mutant screens have important effects on circadian phenotype and that genes involved in the clock mechanism can be greatly influenced by many other genes (Foltenyi et al.; Hall et al.). Among these may be clock-controlled genes that regulate processes that evade particular screens but may feed back on core oscillators and modify their properties. Gene-chip analysis of clock neurons has been used to identify candidate genes of this kind (Blau et al.).

Unraveling the mechanisms that extend the time required to complete a circadian loop remains a high priority. Not surprisingly, as in cyanobacteria, phosphorylation is centrally important, probably in all such systems. It is an important regulator of circadian period and is responsible for some of the built-in delays that produce near 24-hour cycles from biochemical oscillations that would otherwise run much faster (Virshup et al.; Querfurth et al.; Vanselow and Kramer; Maywood et al.). It may also be involved in the still mysterious mechanism of temperature-compensation of period length (Dunlap et al.). Other period-extending mechanisms under study include the incorporation of fixed interval timers within the circadian cycle (Saez et al.) and chromatin remodeling by modification of histones (Grimaldi et al.).

Posttranscriptional mechanisms act both within the core circadian loop and on target mRNAs to regulate rhythmic expression patterns and thus clock outputs. A variety of such mechanisms have major effects on the period of the circadian cycle (Vanselow and Kramer) and on its maintenance (Somers et al.). Posttranscriptional mechanisms also have a significant role in shaping the circadian profiles of clock-controlled genes (Garbarino-Pico and Green; Keene) and are thus important and, until recently, underappreciated components of the output pathways that link the central circadian loop to many critical aspects of physiology and behavior (Foltenyi et al.; Chen et al.; Loros et al.; de Paula et al.). Degradation of cryptochromes (and probably other clock proteins) is involved in regulating both period and amplitude of circadian rhythms in mice (M. Pagano, unpubl.; Siepka et al.; Maywood et al.).

During the past 10 years, it has become clear that clocks are widely distributed *within* multicellular organisms. This was anticipated in 1960 because of the many different rhythms that could be observed in an individual and the presence of clocks in single-celled protists. However, the unequivocal demonstration that there were self-sustained oscillators with a full range of circadian properties in the cells, tissues, and organs of eukaryotes awaited the advent of dynamic molecular reporters, chiefly luciferase. These, when combined with transgenic and transient transfection technology, made it possible to satisfy the basic condition for efficient circadian experimentation: long-term automatic recording of rhythmicity. Cells and tissues from transgenic animals with luciferase reporting the circadian transcription of clock genes were cultured and rhythms of light output measured with photomultipliers. Most displayed rhythmicity in vitro with varying degrees of robustness. These results brought home to the field, in ways that no amount of inference could, that we were dealing not with a single molecular or neural oscillator, not with a single measured behavior, but with a complex system. As students of biological organization, we have to understand its system properties as well as the properties of its individual components.

Complete analysis of any biological system involves at least five steps: identifying its components, discovering their individual properties, understanding the links among them and their interactions with each other, learning how the system responds to the environment, and, finally, how it functions adaptively in nature. This is clearly a daunting task and one which, for "the" circadian system, has barely begun. It is important to recognize that despite the significant molecular homologies among the cell-autonomous circadian oscillators of eukaryotes, many important system-level details will vary widely among species. For mammals, the currently available tools dictate an emphasis on mice and, to a lesser extent, on rats, but while acknowledging their advantages, we should also be aware of their limitations, of which two are major. First, these laboratory rodents are no longer real animals and so it is almost meaningless to ask how their circadian systems function adaptively in nature. The second limitation is particularly important in light of the use of these animals as models of human disease. It is clear that that aspect of circadian research promises new and important insights into a wide variety of pathologies (see more below). For many aspects of mammalian physiology, rats and mice are reasonable first approximations of humans, but for studies aimed at identifying circadian influences, an important distinction between these rodent models and humans must be kept in mind: We are diurnal, and the rodents are strongly nocturnal. Because circadian systems evolve under strong selective pressure to maximize the adaptive significance of phase control, this distinction is likely to have important consequences. These may crop up not only in circadian studies, but also as aspects of classical homeostatic physiology are examined at greater depth. We cannot abandon these models, but we do need to make comparisons with the admittedly less convenient diurnal models where possible.

Several contributions to the present volume report real progress in exploring the links between central and peripheral clocks. Such links may well involve tissue-specific nuclear receptors which are shown to oscillate with circadian periods and could function as part of a network coupling circadian clock outputs to metabolism, to reproduction, and, secondarily, to hormonally regulated behavior (Yang et al.). Circadian oscillators in peripheral organs regulate the expression of large numbers of so-called clock-controlled genes (CCGs). Most of the genes so regulated are tissue-specific; for example, of the more than 300 CCGs in heart and liver, only about 10% are common to both organs (Storch et al. 2002). This strongly suggests that each organ has its own functionally significant circadian gene expression profile, although it has been difficult to connect the details of such gene expression with organ-specific functions. Experiments with transgenic mice engineered to conditionally ablate clock function in the liver alone or in the entire animal with the exception of the liver have led to the interesting conclusion that cyclic expression of most (90%) liver CCGs depends on a functional clock in the liver itself (Kornmann et al.). These results leave open the question of whether the liver clock directly drives circadian gene expression or interprets signals from the SCN or elsewhere. Without a liver clock mice fail to regulate glucose levels normally. When clock function is limited to the liver alone, its cells can still be synchronized to daily cycles of food availability (Storch et al.). These results suggest that the liver operates with more independence from central oscillators than might be expected in a highly integrated system. That in itself might be an advantage in a world in which food availability may be quasirhythmic and not always phase-locked to the day/night cycle. The ability of a liver clock to respond flexibly to changing rhythms of food availability could be particularly useful to "weed" species like rats and mice. It will be interesting to see if similar patterns of control operate in other tissues and in the livers of organisms that have rigidly timed feeding opportunities in nature.

If, as seems almost certain, circadian clocks pervasively regulate important aspects of cell and organ function, it would not be surprising to discover that they are involved in a wide range of pathologies. Hints of such involvement come from several sources. The deleterious effects of time shifts, be they the result of jet lag or shift work, are well known to most people from personal experience, although in most of the scientifically controlled studies, it has not been possible to separate effects on the integrity of the circadian system or on rhythmicity of specific functions from the effects of fatigue produced by sleep disruption.

Rhythmic sleep may be simply an output of the circadian system, such as rhythmic body temperature, or its interaction with central circadian oscillators may be more intimate (M. Yanagisawa, unpubl.; Saper and Fuller; Tafti and Franken). Sleep deprivation is such a drastic treatment for most mammals that it is difficult to untangle its many effects. One way to study the interaction of sleep and circadian rhythms is to use model organisms that normally have consolidated sleep but seem able to do without it. Flies (Sehgal et al.) and zebra fish may fall into this category, but with a few exceptions, the field has neglected a promising model in Passerine birds that appear perfectly healthy after months without consolidated sleep (Gaston and Menaker 1968; Rattenborg et al. 2004).

Mutations in genes that are part of the circadian rhythm-generating loop have major effects not only on molecular events in the loop, but also on a variety of other processes. Some of these are clearly related to the circadian function of the gene involved. Perhaps the best examples are the mutations in humans that cause familial advanced sleep phase syndrome (FASPS) (Ptâĉek et al.) and their orthologs in model organisms (Loudon et al.). Such mutations are especially useful because they open the connections between molecular and physiological processes to further analysis. Other more general effects of circadian mutations are more difficult to interpret because it is usually unclear whether the effect of the mutation is only or even primarily on the circadian system or on some noncircadian function of the gene. In this category are circadian mutations that produce phenotypes that may mimic psychiatric disorders of humans (McClung) as well as reproductive disorders, bone and muscle defects, and cancer (Gery and Koeffler). Although these phenotypes are difficult to ascribe to particular clock-related mechanisms, their existence underlines the wide influence that this system, taken in its broadest sense, exerts on normal function.

A different health-related aspect of circadian organization derives from the fact, clear already in 1960, that there are robust circadian rhythms of sensitivity to a variety of environmental insults (Kondratov and Antoch). Efforts to take advantage of such rhythms for therapeutic purposes, e.g., chronotherapy for cancer treatment, are showing promising results (Lévi et al.) and should improve as we learn more about the ways in which the circadian system interacts with many aspects of basic physiology and, in particular, with the cell cycle.

Modeling at some level is implicit in every scientific undertaking, and explicit modeling of circadian organization has been an important aspect of the field since its inception. Many of the early models rested on analogies with physical oscillators and were particularly useful in suggesting experiments designed to explore how far those analogies could be pushed, e.g., phase-response curves, limits of entrainment, aftereffects, and frequency demultiplication. Now that so much more is known about detailed circadian mechanisms, modelers are faced with the task of incorporating into their models what is known about the interactions among multiple negative feedback loops involving many genes and proteins. Such models have been developed for plant (Millar et al.) and mammalian circadian systems (Ueda). They are useful for organizing large bodies of data and inferring logical structure, but the challenge is to use them to predict unanticipated system properties or components. Some success has already been achieved. Another approach to understanding the basic logic of oscillating networks is to compare several and attempt to extract common features that are essential to their function. This comparative approach has been a staple of biological analysis for hundreds of years, but it is now possible to apply it at the fundamental molecular level. It has the potential to provide important insights into the structure of both circadian (Mockler et al.) and higher-frequency metabolic cycles (Tu and McKnight; Hughes et al.).

From a comparative point of view, new biological models are always welcome. Work with lepidopteran species and other insects suggests that *Drosophila* may be atypical in having only one *Cry* gene. The differences between the molecular aspects of circadian organization in *Drosophila* and the monarch butterfly (Reppert) suggest that the *Drosophila* pattern may be highly derived and point to the dangers of inferring evolutionary history from a small number of model species chosen for convenience.

Implied comparison on yet another level is exemplified by a group of papers dealing with noncircadian aspects of temporal organization. These deal with the "clock" that underlies the development of body segmentation (Kageyama et al.; Pourquié) and with aging (C. Kenyon, unpubl.; Guarente; Ruvkun et al.). Although there is as yet no evidence of a direct relationship between these processes and circadian rhythmicity, it is not out of the question that some of the same genes may be involved in their control. There is at least one good example of that kind of pleiotropy in the regulation by the *Drosophila Per* gene of the period length of both the circadian rhythm and the much higher-frequency rhythm of wing vibration used by courting male flies (Konopka et al. 1996). Other circadian genes are expressed during development of *Drosophila* in both oscillator cell precursors and nonoscillator cells; however, their function in these latter cells is unknown (Benito et al.).

The importance of timing to events in development was beautifully underlined by Martin Raff in the Reginald B. Harris Lecture. He described in vitro experiments with oligodendrocyte precursor cells which contain an internal timer that schedules cessation of cell division and initiation of differentiation (Raff). The timing in vitro parallels the timing in vivo and depends in part on the levels of two proteins. Control is at both the transcriptional and post-transcriptional levels.

The circadian rhythms of human beings are very much the same as those of other mammals. Their responses to the physical environment (chiefly photic) are similar to those of other diurnal mammals. However, the interaction of circadian mechanisms with the social environment produces unique behavioral and physiological responses. Furthermore, the social environment produces major modifications of the photic environment, in particular, extension of the photoperiod by artificial light and concomitant reduction in overall light intensity as a consequence of indoor living. The obvious disadvantages of humans as experimental subjects are at least partially offset by some unique advantages. Humans are more cooperative than mice. They sit still, answer questions, fill out questionnaires, spit in tubes, and urinate in cups on command, and they have lots of blood. Even though breeding experiments are out, there are many natural experiments going on all the time; the breeding population is very large and variable, it is found in geographically diverse locations, and its genome is known in great detail. Some of these advantages have been exploited in circadian studies. The distribution of phases of sleep and activity rhythms in a very large sample has been determined by using a simple questionnaire (Roenneberg and Merrow). The large sample size enables a useful descriptive analysis of human chronotype by age, sex, occupation, etc. At the extremes of an almost normal distribution are "larks" and "owls" waiting for genetic/molecular analysis which has already been successfully initiated in the study of FASPS (Ptáĉek et al.).

Compared to its central role in the circadian systems of many nonmammalian vertebrates (Menaker and Tosini 1996) and in the control of reproduction in photoperiodically regulated, seasonally breeding mammals (Goldman 2001), the role of melatonin in the physiology of mammalian species who do not breed seasonally is disappointingly minor. Melatonin has small effects on the entrainment of mammalian behavioral rhythms which are nonetheless useful clinically in helping blind humans to synchronize to their environment, but its most significant use is as a reliable marker of circadian phase (Lewy). It may be involved at some level in psychiatric disorders such as depression and autism (Bourgeron), perhaps in memory formation (Rawashdeh et al. 2007), and in some cancers, but it is hard to escape the feeling that despite a great deal of work, we have still not identified its basic function in mammals.

The importance of understanding the detailed interaction of the human circadian system with the social and physical environment that we have created for ourselves was made dramatically clear in the Dorcas Cumming Lecture presented by Charles Czeisler. Using hard data collected primarily from doctors at stages in their careers at which they were required to work long noncircadian schedules, he described their involvement in driving mishaps and potential for medical mistakes (Czeisler et al.). As fatigue increases, judgment declines progressively, so that severely fatigued individuals, like people who have had too much to drink, do not realize that they are impaired. Understanding of the social costs of fatigue is one of the most important practical benefits that could be derived from our current knowledge of the circadian regulation of sleep. It is frustrating to see it ignored by those who design work schedules for pilots, truck drivers, shift workers, and medical residents.

Circadian rhythmicity is one of the most obvious and easily studied aspects of the much broader problem of understanding the temporal organization of living systems. The temporal program that underlies biological clocks is particularly amenable to analysis, and a mere 50 years of work has revealed a great many of its secrets—at an unprecedented array of organization levels from behavior to molecular structure. This may prove to be a model for future work on the temporal structure of other biological systems; at the least, it cannot help but draw attention to the importance of time in biology.

The field of biological clocks has always been exceptionally broad both in terms of the model systems studied and in the endpoints measured. It has often seemed on the verge of subdividing along either organism or process lines, but it has been repeatedly rescued by appreciation of the deep formal similarities among its subjects. Its breadth has made it a unique meeting ground for scientists with very different backgrounds and goals. The tendency to draw people in from other fields with new approaches and fresh ideas has contributed in a major way to its rapid growth. That tendency is likely to accelerate as the multiple dynamic connections between circadian temporal programs and other aspects of biological organization become more widely recognized. This will have important practical consequences for medicine, for agriculture, and for species conservation. It would be a shame if the field continued its neglect of its defining but admittedly most difficult question: How do animals, plants, fungi, and bacteria make adaptive use of their biological clocks in the worlds in which they live?

REFERENCES

Dunlap J.C. and Feldman J.F. 1988. On the involvement of protein synthesis in the circadian clock in *Neurospora*. *Proc. Natl. Acad. Sci.* **85:** 1096.

Gaston S. and Menaker M. 1968. Pineal function: The biological clock in the sparrow? *Science* **160:** 1125.

Goldman B.D. 2001. Mammalian photoperiodic system: Formal properties and neuroendocrine mechanisms of photoperiodic time measurement. *J. Biol. Rhythms* **16:** 283.

Konopka R.J. and Benzer S. 1971. Clock mutants of *Drosophila melanogaster*. *Proc. Natl. Acad. Sci.* **68:** 2112.

Konopka R.J., Kyriacou C.P., and Hall J.C. 1996. Mosaic analysis in the *Drosophila* CNS of circadian and courtship-song rhythms affected by a period clock mutation. *J. Neurogenet.* **11:** 117.

Menaker M. and Tosini G. 1996. Evolution of vertebrate circadian systems. In *Circadian organization and oscillatory coupling: 6th Sapporo Symposium on Biological Rhythms* (ed. K. Honma and S. Honma), p.39. Hokkaido University Press, Sapporo, Japan.

Rattenborg N.C., Mandt B.H., Obermeyer W.H., Winsauer P.J., Huber R., Wikelski M., and Benca R.M. 2004. Migratory sleeplessness in the white-crowned sparrow (*Zonotrichia leucophrys gambelii*). *PLoS Biol.* **2:** E212.

Rawashdeh O., Hernandez de Boursetti N., Roman G., and Cahill G.M. 2007. Melatonin suppresses nighttime memory formation in zebrafish. *Science* **318:** 1144.

Rust M.J., Markson J.S., Lane W.S., Fisher D.S., and O'Shea E.K. 2007. Ordered phosphorylation governs oscillation of a three-protein circadian clock. *Science* **318:** 809.

Storch K.-F., Lipan O., Leykin I., Viswanathan N., Davis F.C., Wong W.H., and Weitz C.J. 2002. Extensive and divergent circdian gene expression in liver and heart. *Nature* **417:** 78.

Author Index

Subject Index

D

E

F

G

H

I

J

K

L